PEARSON

Biology
Bio Sci 93

Fourth Custom Edition for University of California, Irvine

Taken from:
Biology, Ninth Edition
By Neil A. Campbell, Jane B. Reece, Lisa A. Urry, Michael L. Cain, Steven A Wasserman, Peter V. Minorsky, and Robert B. Jackson

Cover Art: Courtesy of Lenart Nilsson/Albert Bonniers Forlag/Boehringer Ingelheim/Scanpix.

Excerpts taken from:

Biology, Ninth Edition
By Neil A. Campbell, Jane B. Reece, Lisa A. Urry, Michael L. Cain, Steven A Wasserman, Peter V. Minorsky, and Robert B. Jackson

Published by Benjamin Cummings
San Francisco, CA 94111

This special edition published in cooperation with Pearson Learning Solutions.

Pearson Learning Solutions, 501 Boylston Street, Suite 900, Boston, MA 02116
A Pearson Education Company
www.pearsoned.com

Printed in the United States of America

2 3 4 5 6 7 8 9 10 V363 16 15 14 13 12

000200010270753515

RR/JG

PEARSON

ISBN 10: 1-256-13135-0
ISBN 13: 978-1-256-13135-9

Brief Contents

About the Authors

The Ninth Edition author team's contributions reflect their biological expertise as researchers and teaching sensibilities gained from years of experience as instructors at diverse institutions. The team's highly collaborative style continues to be evident in the cohesiveness and consistency of the Ninth Edition.

Jane B. Reece

The head of the Ninth Edition author team, Jane Reece was Neil Campbell's longtime collaborator. She has participated in every edition of *BIOLOGY*. Earlier, Jane taught biology at Middlesex County College and Queensborough Community College. She holds an A.B. in Biology from Harvard University, an M.S. in Microbiology from Rutgers University, and a Ph.D. in Bacteriology from the University of California, Berkeley. Jane's research as a doctoral student and postdoctoral fellow focused on genetic recombination in bacteria. Besides her work on *CAMPBELL BIOLOGY*, she has been a coauthor of *Biology: Concepts & Connections*, *Essential Biology*, and *The World of the Cell*.

Lisa A. Urry

Lisa Urry (Chapter 1 and Units 1–3) is a professor and developmental biologist and recent Chair of the Biology Department at Mills College. After graduating from Tufts University with a double major in Biology and French, Lisa completed her Ph.D. in molecular and developmental biology at Massachusetts Institute of Technology (MIT). She has published a number of research papers, most of them focused on gene expression during embryonic and larval development in sea urchins. Lisa is also deeply committed to promoting opportunities for women in science education and research.

Michael L. Cain

Michael Cain (Units 4 and 5) is an ecologist and evolutionary biologist who is now writing full-time. Michael earned a joint degree in Biology and Math at Bowdoin College, an M.Sc. from Brown University, and a Ph.D. in Ecology and Evolutionary Biology from Cornell University. As a faculty member at New Mexico State University and Rose-Hulman Institute of Technology, he taught a wide range of courses, including introductory biology, ecology, evolution, botany, and conservation biology. Michael is the author of dozens of scientific papers on topics that include foraging behavior in insects and plants, long-distance seed dispersal, and speciation in crickets. In addition to his work on *CAMPBELL BIOLOGY*, Michael is also the lead author of an ecology textbook.

Steven A. Wasserman

Steve Wasserman (Unit 7) is a professor at the University of California, San Diego (UCSD). He earned his A.B. in Biology from Harvard University and his Ph.D. in Biological Sciences from MIT. Through his research on regulatory pathway mechanisms in the fruit fly *Drosophila*, Steve has contributed to the fields of developmental biology, reproduction, and immunity. As a faculty member at the University of Texas Southwestern Medical Center and UCSD, he has taught genetics, development, and physiology to undergraduate, graduate, and medical students. He has also served as the research mentor for more than a dozen doctoral students and more than 50 aspiring scientists at the undergraduate and high school levels. Steve has been the recipient of distinguished scholar awards from both the Markey Charitable Trust and the David and Lucille Packard Foundation. In 2007, he received UCSD's Distinguished Teaching Award for undergraduate teaching.

Peter V. Minorsky

Peter Minorsky (Unit 6) is a professor at Mercy College in New York, where he teaches evolution, ecology, botany, and introductory biology. He received his B.A. in Biology from Vassar College and his Ph.D. in Plant Physiology from Cornell University. He is also the science writer for the journal *Plant Physiology*. After a postdoctoral fellowship at the University of Wisconsin at Madison, Peter taught at Kenyon College, Union College, Western Connecticut State University, and Vassar College. He is an electrophysiologist who studies responses of plants to stress. Peter received the 2008 Award for Teaching Excellence at Mercy College.

Robert B. Jackson

Rob Jackson (Unit 8) is a professor of biology and Nicholas Chair of Environmental Sciences at Duke University. Rob holds a B.S. in Chemical Engineering from Rice University, as well as M.S. degrees in Ecology and Statistics and a Ph.D. in Ecology from Utah State University. Rob directed Duke's Program in Ecology for many years and just finished a term as the Vice President of Science for the Ecological Society of America. Rob has received numerous awards, including a Presidential Early Career Award in Science and Engineering from the National Science Foundation. He also enjoys popular writing, having published a trade book about the environment, *The Earth Remains Forever*, and two books of poetry for children, *Animal Mischief* and *Weekend Mischief*.

Neil A. Campbell

Neil Campbell combined the investigative nature of a research scientist with the soul of an experienced and caring teacher. He earned his M.A. in Zoology from UCLA and his Ph.D. in Plant Biology from the University of California, Riverside, where he received the Distinguished Alumnus Award in 2001. Neil published numerous research articles on desert and coastal plants and how the sensitive plant (*Mimosa*) and other legumes move their leaves. His 30 years of teaching in diverse environments included general biology courses at Cornell University, Pomona College, and San Bernardino Valley College, where he received the college's first Outstanding Professor Award in 1986. Neil was a visiting scholar in the Department of Botany and Plant Sciences at the University of California, Riverside. In addition to his authorship of this book, he coauthored *Biology: Concepts & Connections* and *Essential Biology* with Jane Reece. For the Ninth Edition of this book, we honor Neil's contributions to biology education by adopting the title *CAMPBELL BIOLOGY*.

Preface

Biology is an enormous subject, one that can seem overwhelming to students and scientists alike. Moreover, discoveries are being made at an unprecedented pace—from new kinds of small RNA molecules to the Neanderthal genome, from new biofuels to communities of organisms thriving beneath vast glaciers, from emerging infectious diseases to cancer vaccines. As a result, a general biology course faces a daunting challenge: to keep students from suffocating under an avalanche of information. *CAMPBELL BIOLOGY* addresses this challenge by providing a strong foundation for understanding both current knowledge and new developments in the context of underlying biological concepts.

Key Concepts and Unifying Themes

In each chapter of this textbook, a framework of three to six carefully chosen **Key Concepts** provide context for supporting details, helping students distinguish the "forest" from the "trees." The numbered Key Concepts are presented at the beginning of the chapter and then serve as headings for each chapter section. *Concept Check Questions* at the end of each section provide a hierarchical framework for self-assessment that builds students' confidence and then challenges them to push the limits of their understanding with several types of critical thinking questions. The *Summary of Key Concepts* at the end of the chapter refocuses students on the main points. *CAMPBELL BIOLOGY* also helps students organize and make sense of what they learn on a grander scale by emphasizing **evolution and other unifying themes** that pervade biology. These themes are introduced in Chapter 1 and integrated throughout the book.

New to This Edition: An Emphasis on Making Connections

In addition to Key Concepts and themes, we've created new features for the Ninth Edition that help students see the big picture by *making connections*. These include the following:

New Make Connections Questions: Making connections across chapters

New **Make Connections Questions** help students see how different areas of biology tie together, helping them overcome the tendency to compartmentalize information. Each question challenges students to move beyond memorization and gain a deeper understanding of biological principles by asking them to relate chapter content to material they learned earlier in the course. For example, we ask students to connect

- DNA replication (Chapter 16, see p. 319) to the cell cycle (Chapter 12);
- Soil formation (Chapter 37, see p. 789) to the properties of water (Chapter 3); and
- Aquatic biomes (Chapter 52, see p. 1163) to osmoregulation (Chapter 44).

At least three Make Connections Questions appear in each chapter. In addition, online *Make Connections Tutorials* in MasteringBiology® (see p. xi) connect content from two different chapters using figures from the book.

Expanded Evolution Coverage: Making connections to evolution in every chapter

Evolution is the core theme of biology, and in this edition it is more evident than ever. At least one **Evolution section in every chapter** focuses on evolutionary aspects of the chapter material, highlighted by a new Evolution banner. See, for example, the new discussions of enzyme evolution (p. 157), coevolution of flowers and pollinators (p. 806), and evolution of hormone function in animals (pp. 988–989).

New Impact Figures: Making connections between scientific advances and the real world

Our new **Impact Figures** motivate students by highlighting the dramatic impact of recent discoveries in biology. These figures feature high-interest topics such as induced pluripotent stem cells and regenerative medicine (Chapter 20, p. 417), the discovery of *Tiktaalik* (Chapter 34, p. 710), and the use of forensic ecology to track elephant poaching (Chapter 56, p. 1243). The *Why It Matters* section of each figure explains the relevance of the research to students' lives, global problems, or the field of biology itself. Each Impact Figure ends with a suggestion for *Further Reading* and a *What If?* or *Make Connections Question* to develop critical thinking skills.

New Visual Organizers and 3-D Art: Making connections visually

The new **Visual Organizer** format highlights the main parts of a figure, helping students see the key categories at a glance. See, for instance, Figure 17.24 on types of small-scale mutations (p. 345) or Figure 27.3, Gram staining (p. 557). Throughout the book, selected figures have been rendered in a more **3-D art style** while keeping an appropriate balance between realism and teaching effectiveness. Figure 52.3, Exploring Global Climate Patterns (p. 1146), is one example.

Restructured Chapter Reviews: Making connections at a higher level

In the chapter summaries, each concept section now concludes with a new **Summary of Key Concepts Question** that is tied to a major learning goal. Also, this edition increases student awareness of different levels of thinking by organizing the end-of-chapter questions into three levels based on **Bloom's taxonomy**, which classifies types of thinking that are important in learning. Our levels are (1) Knowledge/Comprehension, (2) Application/Analysis, and (3) Synthesis/Evaluation. (These same levels are used in the Campbell Test Bank.) The range of question types helps students develop critical thinking skills and prepare for the kinds of questions they'll encounter on exams. New **Write About a Theme Questions** give students practice writing short, coherent essays that connect the chapter's content to one of the book's themes. (A suggested grading rubric can be found on p. xv, and sample answers are provided for instructors in the MasteringBiology Instructor Resources area.) A new **MasteringBiology preview section** at the end of each chapter lists *Assignments*—tutorials, activities, and questions that instructors can assign. This section also directs students to the *eText* and *Study Area* for online self study.

New Content: Making connections to advances in science

As in each new edition, the Ninth Edition incorporates **new scientific content** and **organizational improvements**. These are summarized on pp. viii–ix, following this Preface.

MasteringBiology®: Making connections outside of class

MasteringBiology, the most widely used online assessment and tutorial program for biology, provides an extensive library of homework assignments that are graded automatically. In addition to BioFlix® Tutorials, other Tutorials, Activities, Reading Quiz Questions in every chapter, and 4,500 Test Bank Questions, MasteringBiology for the Ninth Edition features an **improved user interface** and the following **new Tutorials and Questions**: *Make Connections Tutorials, Student Misconceptions Questions* for every chapter, *Data Analysis Tutorials, Experimental Inquiry Tutorials, Video Tutor Sessions,* and *MasteringBiology: Virtual Labs*. For more information, see pp. xvi–xix and www.masteringbiology.com.

Our Hallmark Features

Besides our Key Concepts and unifying themes, several other features have contributed to the success of *CAMPBELL BIOLOGY*. Because text and illustrations are equally important for learning biology, **integration of figures and text** has been a hallmark of this book since its inception. Our popular *Exploring Figures* on selected topics epitomize this approach: Each is a learning unit of core content that brings together related illustrations and text. Another example is our *Guided Tour Figures*, which use descriptions in blue type to walk students through complex figures like an instructor would, pointing out key structures, functions, and steps of processes.

To encourage **active learning**, recent editions have incorporated new types of questions: *What If? Questions, Figure Legend Questions*, and *Draw It Questions* that ask students to sketch a structure, annotate a figure, or graph data. In the Ninth Edition, these questions are augmented by the new *Make Connections Questions*. Online, the highly interactive *MasteringBiology tutorials* are sophisticated active-learning tools.

Finally, *CAMPBELL BIOLOGY* features **scientific inquiry**, an essential component of any biology course. Complementing stories of scientific discovery in the text narrative and the unit-opening interviews, *Inquiry Figures* help students understand "how we know what we know" and provide a model of how to think like a scientist. Each one begins with a research question and then describes how researchers designed an experiment, interpreted their results, and drew conclusions. The source article is referenced, and a What If? Question asks students to consider an alternative scenario. Selected Inquiry Figures invite students to read and analyze the original research article in the supplement *Inquiry in Action: Interpreting Scientific Papers* (see p. xxi). At the end of each chapter, *Scientific Inquiry Questions* give students additional opportunities to practice critical thinking by developing hypotheses, designing experiments, and analyzing real research data. Beyond the book, activities involving scientific inquiry are featured in MasteringBiology and other supplements, both print and electronic (see pp. xviii–xxi).

Our Partnership with Instructors

A core value underlying all our work as authors is our belief in the importance of our partnership with instructors. Our primary way of serving instructors, of course, is providing a textbook, supplements, and media resources that serve their students well. In addition, Benjamin Cummings makes available a rich variety of instructor resources, in both print and electronic form (see p. xx). In our continuing efforts to improve the book and its supplements, we benefit tremendously from instructor feedback, not only in formal reviews from hundreds of scientists, but also via informal communication, often by e-mail.

The real test of any textbook is how well it helps instructors teach and students learn. We welcome comments from the students and instructors who use *CAMPBELL BIOLOGY*. Please address your suggestions to any of us:

Jane Reece at janereece@cal.berkeley.edu
Lisa Urry (Chapter 1 and Units 1–3) at lurry@mills.edu
Michael Cain (Units 4 and 5) at mcain@bowdoin.edu
Peter Minorsky (Unit 6) at pminorsky@mercy.edu
Steve Wasserman (Unit 7) at stevenw@ucsd.edu
Rob Jackson (Unit 8) at jackson@duke.edu

New Content

This section provides just a few highlights of new content and organizational improvements in *CAMPBELL BIOLOGY*, Ninth Edition.

CHAPTER 1 Introduction: Themes in the Study of Life

We have added a separate new theme on energy flow while retaining a theme on environmental interactions. Concept 1.3, on the scientific method, has been reframed to more accurately reflect the scientific process, with a focus on observations and hypotheses. A new Concept 1.4 discusses the value of technology to society while emphasizing the cooperative nature of science and the value of diversity among scientists.

UNIT ONE The Chemistry of Life

For this edition, the basic chemistry is enlivened by new content connecting it to evolution, ecology, and other areas of biology. Examples of new material include omega-3 fatty acids, the isomeric forms of methamphetamine, arsenic contamination of groundwater, and the basis of mad cow disease. The burgeoning importance of nucleic acids throughout biology has prompted us to expand our coverage of DNA and RNA structures in this first unit. In fact, a general aim for the first two units is to infuse the chapters with more detail about nucleic acids, genes, and related topics. Another enhancement, in this and the next two units, is the inclusion of more computer models of important proteins in contexts where they support students' understanding of molecular function.

UNIT TWO The Cell

For Chapter 6, we developed an Exploring Figure on microscopy, which includes new types of microscopy, and we added micrographs of various cell types to the Exploring Figure on eukaryotic cells. We also expanded our description of chromosome composition, with the goal of preempting some common student misconceptions about chromosomes and DNA. New connections to evolution include an introduction to the endosymbiont theory in Chapter 6 and some interesting evolutionary adaptations of cell membranes in Chapter 7. We've added a new section to Chapter 8 on the evolution of enzymes with new functions, which not only strengthens enzyme coverage but also provides an early introduction to the concept that mutations contribute to molecular evolution. In Chapter 9, we simplified the glycolysis figure and emphasized pyruvate oxidation as a separate step to help students focus on the main ideas. In keeping with our increased focus on global issues in the Ninth Edition, Chapter 10 has an Impact Figure on biofuels and a discussion of the possible effect of climate change on the distribution of C_3 and C_4 plants. In Chapter 11, we have added an Impact Figure to highlight the importance and medical relevance of G protein-coupled receptors.

UNIT THREE Genetics

In Chapters 13–17, we have added material to stimulate student interest—for example, a new Impact Figure on genetic testing for disease-associated mutations. As done throughout the Ninth Edition, we ask students to make connections between chapters so that they avoid the trap of compartmentalizing the information in each chapter. For instance, Chapter 15 discusses the Philadelphia chromosome associated with chronic myelogenous leukemia and asks students to connect this information to what they learned about signaling in the cell cycle in Chapter 12. Also, we encourage students to connect what they learn about DNA replication and chromosome structure in Chapter 16 to the material on chromosome behavior during the cell cycle in Chapter 12. Chapter 16 has a new figure showing a current 3-D model of the DNA replication complex, with the lagging strand looping back through it.

Chapters 18–21 are extensively updated, with the changes dominated by new genomic sequence data and discoveries about the regulation of gene expression. (The introduction to genes, genomes, and gene expression in Units One and Two should help prepare students for these revisions.) Chapter 18 includes a new section on nuclear architecture, which describes the organization of chromatin in the nucleus in relation to gene expression. The roles of various types of RNA molecules in regulation also receive special attention. In the section on cancer, we describe how technical advances can contribute to personalized cancer treatments based on the molecular characteristics of an individual's tumor. Chapter 19 discusses the 2009 H1N1 flu pandemic. Chapter 20 includes advances in techniques for DNA sequencing and for obtaining induced pluripotent stem (iPS) cells. Finally, the heavily revised Chapter 21 describes what has been learned from the sequencing of many genomes, including those of a number of human individuals.

UNIT SEVEN Animal Form and Function

In revising this unit, we strove to introduce physiological systems through a comparative approach that underscores how adaptations are linked to shared physiological challenges. In particular, we have highlighted the interrelationship of the endocrine and nervous systems at multiple points in the unit, helping students appreciate how these two forms of communication link tissues, organs, and individuals. Other revisions aim to keep students focused on fundamental concepts amid the details of complex systems. For example, many figures have been reconceived to emphasize key information, including new figures comparing single and double circulation (Chapter 42) and examining the function of antigen receptors (Chapter 43), as well as new Exploring Figures on the vertebrate kidney (Chapter 44) and the structure and function of the eye (Chapter 50). Chapter 43 has been significantly revised to support students' conceptual understanding of basic immunological responses and the key cellular players. Throughout the unit, new state-of-the-art images and material on current and compelling topics—such as circadian rhythms (Chapter 40), novel strains of influenza (Chapter 43), the effects of climate change on animal reproductive cycles (Chapter 46), and advances in understanding brain plasticity and function (Chapter 49)—will help engage students and encourage them to make connections beyond the text.

To the Student: How to Use This Book

Focus on the Key Concepts.

Each chapter is organized around a framework of 3 to 6 **Key Concepts** that will help you stay focused on the big picture and give you a context for the supporting details.

52

An Introduction to Ecology and the Biosphere

▲ **Figure 52.1 What threatens this amphibian's survival?**

Before you begin reading the chapter, get oriented by reading the **list of Key Concepts,** which introduces the big ideas covered in the chapter.

KEY CONCEPTS

52.1 Earth's climate varies by latitude and season and is changing rapidly
52.2 The structure and distribution of terrestrial biomes are controlled by climate and disturbance
52.3 Aquatic biomes are diverse and dynamic systems that cover most of Earth
52.4 Interactions between organisms and the environment limit the distribution of species

Each **Key Concept** serves as the heading for a major section of the chapter.

OVERVIEW

Discovering Ecology

When University of Delaware undergraduate Justin Yeager spent his summer abroad in Costa Rica, all he wanted was to see the tropical rain forest and to practice his Spanish. Instead, he rediscovered the variable harlequin toad (*Atelopus varius*), a species thought to be extinct in the mountain slopes of Costa Rica and Panama where it once lived (**Figure 52.1**). During the 1980s and 1990s, roughly two-thirds of the 82 known species of harlequin toads vanished. Scientists think that a disease-causing chytrid fungus, *Batrachochytrium dendrobatidis* (see Figure 31.26), contributed to many of these extinctions. Why was the fungus suddenly thriving in the rain forest? Cloudier days and warmer nights associated with global warming appear to have created an environment ideal for its success. As of 2009, the species that Yeager found was surviving as a single known population of fewer than 100 individuals.

What environmental factors limit the geographic distribution of harlequin toads? How do variations in their food supply or interactions with other species, such as pathogens, affect the size of their population? Questions like these are the subject of **ecology** (from the Greek *oikos*, home, and *logos*, study), the scientific study of the interactions between organisms and the environment. Ecological interactions occur at a hierarchy of scales that ecologists study, from single organisms to the globe (**Figure 52.2**).

Ecology's roots are in our basic human interest in observing other organisms. Naturalists, including Aristotle and Darwin, have long studied the living world and systematically recorded their observations. However, modern ecology involves more than observation. It is a rigorous experimental science that requires a breadth of biological knowledge. Ecologists generate hypotheses, manipulate environmental variables, and observe the outcome. In this unit, you will encounter many examples of ecological experiments, whose complex challenges have made ecologists innovators in experimental design and statistical inference.

In addition to providing a conceptual framework for understanding the field of ecology, Figure 52.2 provides the organizational framework for our final unit. In this chapter, we first describe Earth's climate and the importance of climate and other physical factors in determining the location of major life zones on land and in the oceans. We then examine how ecologists determine what controls the distribution and abundance of individual species. The next three chapters investigate population, community, and ecosystem ecology in detail, including approaches for restoring degraded ecosystems. The final chapter explores conservation biology and global ecology as we consider how ecologists apply biological knowledge to predict the global consequences of human activities and to conserve Earth's biodiversity.

CONCEPT 52.1

Earth's climate varies by latitude and season and is changing rapidly

The most significant influence on the distribution of organisms on land and in the oceans is **climate**, the long-term, prevailing weather conditions in a given area. Four physical

After reading a concept section, check your understanding using the **Concept Check Questions** at the end of the section. Work through these questions on your own or in a study group—they're good practice for the kinds of questions you might be asked on an exam. ▶

What If? Questions ask you to apply what you've learned. New **Make Connections Questions** ask you to relate content in the chapter to a concept you learned earlier in the course.

If you can answer these questions (see Appendix A to check your work), you're ready to move on. ▶

CONCEPT CHECK 52.1

1. Explain how the sun's unequal heating of Earth's surface leads to the development of deserts around 30° north and south of the equator.
2. What are some of the differences in microclimate between an unplanted agricultural field and a nearby stream corridor with trees?
3. WHAT IF? Changes in Earth's climate at the end of the last ice age happened gradually, taking centuries to thousands of years. If the current global warming happens very quickly, as predicted, how may this rapid climate change affect the ability of long-lived trees to evolve, compared with annual plants, which have much shorter generation times?
4. MAKE CONNECTIONS In Concept 10.4 (pp. 199–201), you learned about the important differences between C_3 and C_4 plants. Focusing just on the effects of temperature, would you expect the global distribution of C_4 plants to expand or contract as Earth becomes warmer? Why?

For suggested answers, see Appendix A.

Make connections across biology.

By relating the content of a chapter to material you learned earlier in the course, new **Make Connections Questions** help you develop a deeper understanding of biological principles.

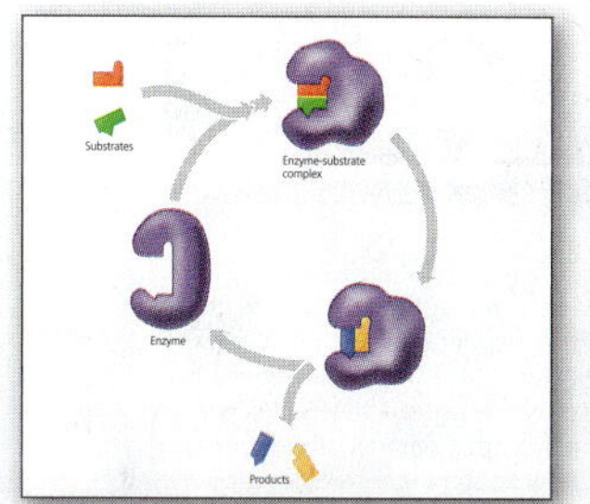

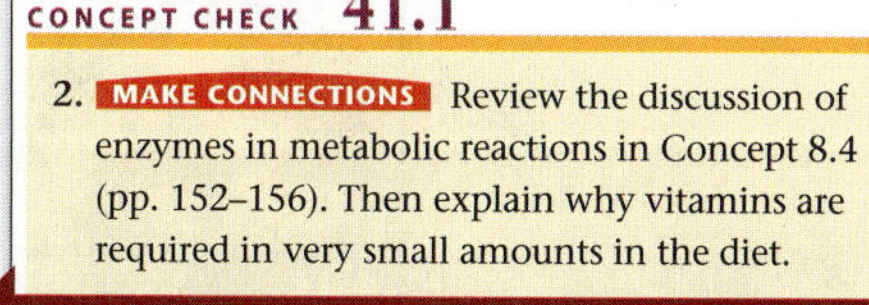

CONCEPT CHECK 41.1

2. **MAKE CONNECTIONS** Review the discussion of enzymes in metabolic reactions in Concept 8.4 (pp. 152–156). Then explain why vitamins are required in very small amounts in the diet.

Enzymes (Chapter 8) ⟷ **Animal nutrition** (Chapter 41)

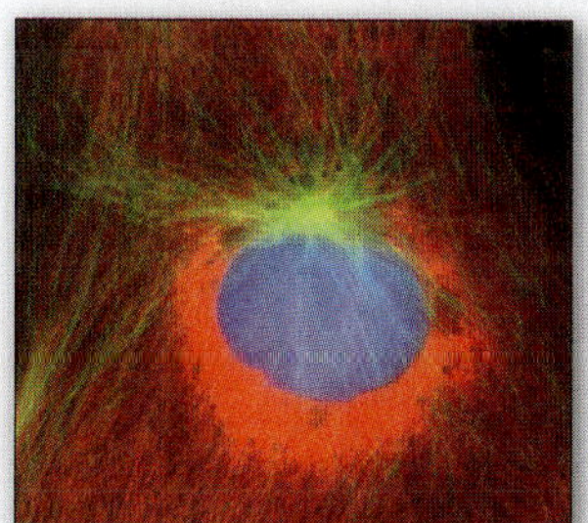

CONCEPT CHECK 16.2

3. **MAKE CONNECTIONS** What is the relationship between DNA replication and the S phase of the cell cycle? See Figure 12.6, page 231.

Cell cycle (Chapter 12) ⟷ **DNA replication** (Chapter 16)

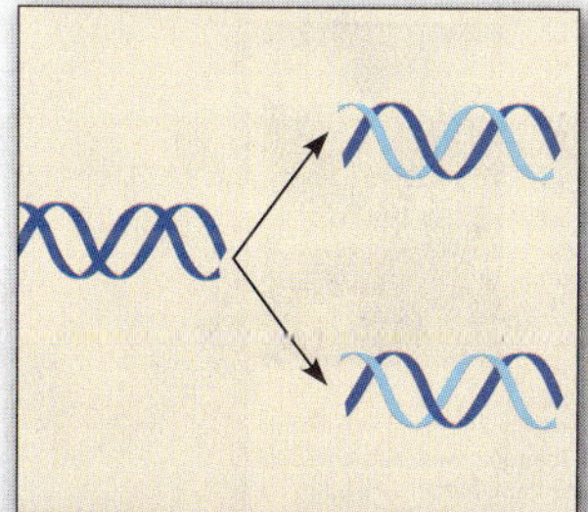

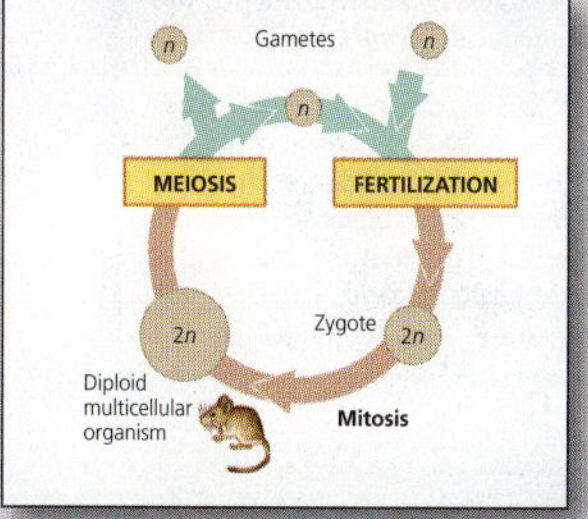

CONCEPT CHECK 31.2

1. **MAKE CONNECTIONS** Compare Figure 31.5 with Figure 13.6 (p. 252). In terms of haploidy versus diploidy, how do the life cycles of fungi and animals differ?

Meiosis (Chapter 13) ⟷ **Fungi** (Chapter 31)

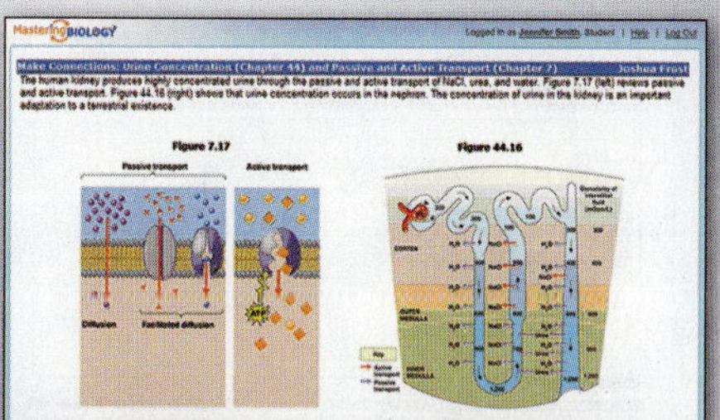

New **Make Connections Tutorials** help you connect biological concepts across chapters in an interactive way.

Make connections to **evolution,** the fundamental theme of biology.

Look for new **Evolution banners** highlighting sections in each chapter that focus on evolutionary aspects of the topic.

The Evolutionary Origins of Mitochondria and Chloroplasts

EVOLUTION Mitochondria and chloroplasts display similarities with bacteria that led to the **endosymbiont theory**, illustrated in **Figure 6.16**. This theory states that an early ancestor of eukaryotic cells engulfed an oxygen-using nonphotosynthetic prokaryotic cell. Eventually, the engulfed cell formed a relationship with the host cell in which it was enclosed, becoming an *endosymbiont* (a cell living within another cell). Indeed, over the course of evolution, the host cell and its endosymbiont merged into a single organism, a eukaryotic cell with a mitochondrion. At least one of these cells may have then taken up a photosynthetic prokaryote, becoming the ancestor of eukaryotic cells that contain chloroplasts.

This is a widely accepted theory, which we will discuss in more detail in Chapter 25. The model it proposes is consistent ... structural features of mitochondria and chloro... rather than being bounded by a single membrane ... of the endomembrane system, mitochondria ... chloroplasts have two membranes surrounding ... chloroplasts also have an internal system of membra... There is evidence that the ancestral engulfed ... had two outer membranes, which became the ... of mitochondria and chloroplasts. Second, ... mitochondria and chloroplasts contain ribo... well as circular DNA molecules attached to their ... The DNA in these organelles programs the ... of some of their own proteins, which are made on ... inside the organelles. Third, also consistent ... probable evolutionary origins as cells, mitochondria ... are autonomous (somewhat independent) or... that grow and reproduce within the cell.

Practice thinking like a scientist.

New **Impact Figures** demonstrate the dramatic impact of recent discoveries in biology and show that biology is constantly changing as new discoveries add to our understanding. ▶

▼ Figure 56.9

IMPACT

Forensic Ecology and Elephant Poaching

This array of severed tusks is part of an illegal shipment of 6,000 kg of ivory intercepted on its way from Africa to Singapore in 2002. Investigators wondered whether the elephants slaughtered for the ivory—perhaps as many as 6,500—were killed in the area where the shipment originated, primarily Zambia, or instead were killed across Africa, indicating a broader smuggling ring. Samuel Wasser, of the University of Washington, and colleagues amplified specific segments of DNA from the tusks using the polymerase chain reaction (PCR). These segments included stretches of DNA containing short tandem repeats (STRs; see Concept 20.4, pp. 420–421), the number of which varies among different elephant populations. The researchers then compared alleles at seven or more loci with a reference DNA database they had generated for elephants of known geographic origin. Their results showed conclusively that the elephants came from a narrow east-west band centered on Zambia rather than from across Africa.

Why It Matters explains the relevance of the research. ▶

WHY IT MATTERS The DNA analyses suggested that poaching rates were 30 times higher in Zambia than previously estimated. This news led to improved antipoaching efforts by the Zambian government. Techniques like those used in this study are being employed by conservation biologists to track the harvesting of many endangered species, including whales, sharks, and orchids.

Further Reading directs you to articles to explore. ▶

FURTHER READING S. K. Wasser et al., Forensic tools battle ivory poachers, *Scientific American* 399:68–76 (2009); S. K. Wasser et al., Using DNA to track the origin of the largest ivory seizure since the 1989 trade ban, *Proceedings of the National Academy of Sciences USA* 104:4228–4233 (2007).

A **Make Connections** or **What If? Question** encourages critical thinking. ▶

MAKE CONNECTIONS Figure 26.6 (p. 539) describes another example in which conservation biologists used DNA analyses to compare harvested samples with a reference DNA database. How are these examples similar, and how are they different? What limitations might there be to using such forensic methods in other suspected cases of poaching?

Inquiry Figures reveal "how we know what we know" by highlighting how researchers designed an experiment, interpreted their results, and drew conclusions. ▼

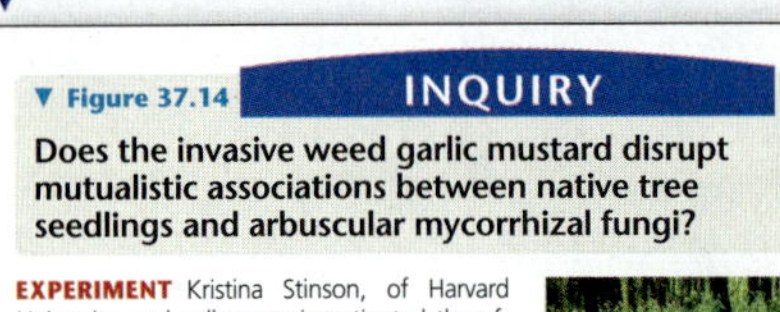

▼ Figure 37.14 **INQUIRY**

Does the invasive weed garlic mustard disrupt mutualistic associations between native tree seedlings and arbuscular mycorrhizal fungi?

EXPERIMENT Kristina Stinson, of Harvard University, and colleagues investigated the effect of invasive garlic mustard on the growth of native tree seedlings and associated mycorrhizal fungi. In one experiment, they grew seedlings of three North American trees—sugar maple, red maple, and white ash—in four different soils. Two of the soil samples were collected from a location where garlic mustard was growing, and one of these samples was sterilized. The other two soil samples were collected from a location devoid of garlic mustard, and one was then sterilized. After four months of growth, the researchers harvested the shoots and roots and determined the dried biomass. The roots were also analyzed for percent colonization by arbuscular mycorrhizal fungi.

RESULTS Native tree seedlings grew more slowly and were less able to form mycorrhizal associations when grown either in sterilized soil or in unsterilized soil collected from a location that had been invaded by garlic mustard.

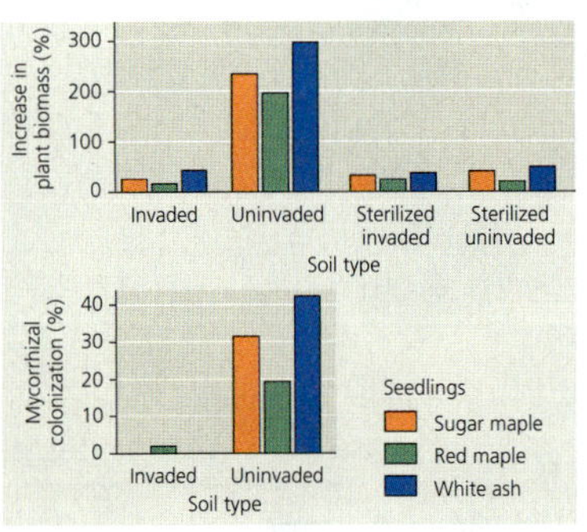

CONCLUSION The data support the hypothesis that garlic mustard suppresses growth of native trees by affecting the soil in a way that disrupts mutualistic associations between the trees and arbuscular mycorrhizal fungi.

SOURCE K. A. Stinson et al., Invasive plant suppresses the growth of native tree seedlings by disrupting belowground mutualisms, *PLoS Biol* (*Public Library of Science: Biology*) 4(5): e140 (2006).

INQUIRY IN ACTION Read and analyze the original paper in *Inquiry in Action: Interpreting Scientific Papers*.

WHAT IF? What effect would applying inorganic phosphate to soil invaded by garlic mustard have on the plant's ability to outcompete native species?

◀ Some **Inquiry Figures** invite you to read and analyze the original research paper in its complete form. You can find the journal article, along with a worksheet guiding you through it, in the separate book ***Inquiry in Action: Interpreting Scientific Papers.***

▲ After exploring the featured experiment, test your analytical skills by answering the **What If? Question**. Suggested answers are provided in Appendix A to help you gauge your understanding.

MasteringBIOLOGY®
www.masteringbiology.com

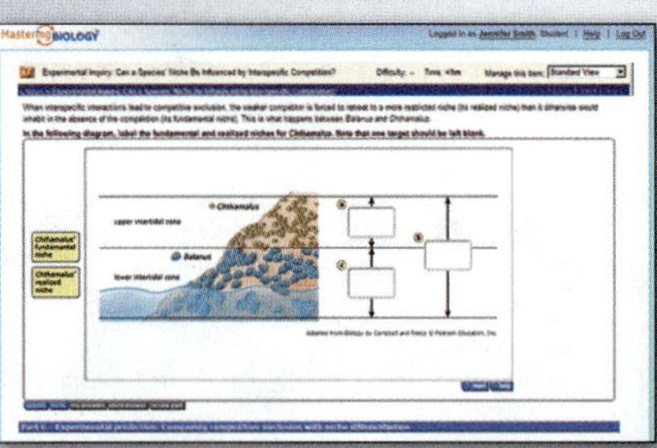

New **Experimental Inquiry Tutorials** give you practice analyzing experimental design and data and drawing conclusions.

Study the figures as you read the text.

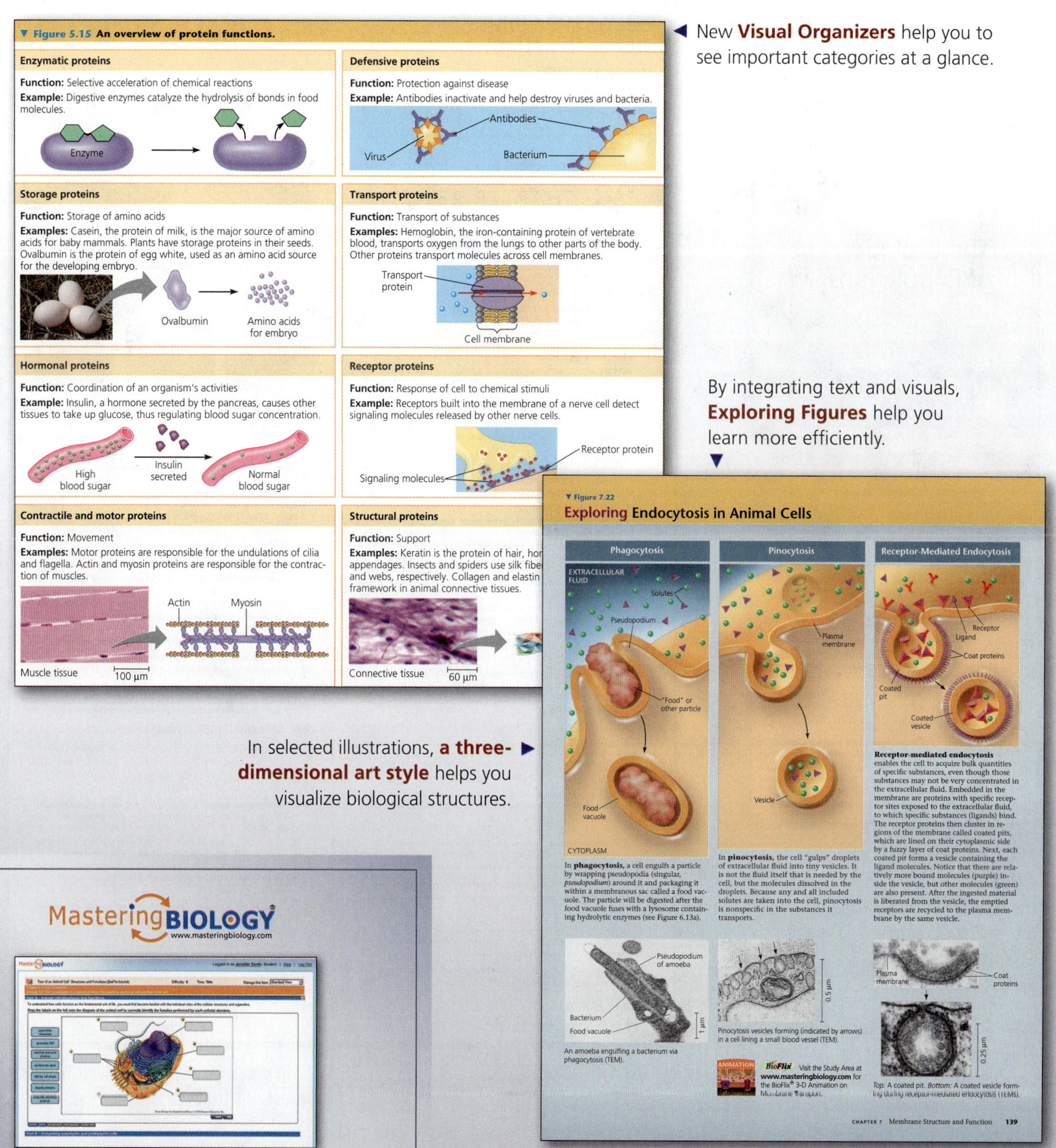

◀ New **Visual Organizers** help you to see important categories at a glance.

By integrating text and visuals, **Exploring Figures** help you learn more efficiently. ▼

In selected illustrations, **a three-dimensional art style** helps you visualize biological structures. ▶

Many **Tutorials** and **Activities** integrate art from the textbook, providing a unified learning experience.

▲ **BioFlix®** icons direct you to high-impact 3-D animations in the Study Area at www.masteringbiology.com.

Review what you've learned.

Chapter Reviews help you efficiently master the chapter content by focusing on the main points of the chapter and offering opportunities to practice for exams.

23

The Evolution of Populations

▲ **Figure 23.1** Is this finch evolving?

EVOLUTION

KEY CONCEPTS

23.1 Genetic variation makes evolution possible
23.2 The Hardy-Weinberg equation can be used to test whether a population is evolving
23.3 Natural selection, genetic drift, and gene flow can alter allele frequencies in a population
23.4 Natural selection is the only mechanism that consistently causes adaptive evolution

OVERVIEW

The Smallest Unit of Evolution

One common misconception about evolution is that individual organisms evolve. It is true that natural selection acts on individuals: Each organism's traits affect its survival and reproductive success compared with other individuals. But the evolutionary impact of natural selection is only apparent in the changes in a *population* of organisms over time.

Key Concepts, which were introduced in the beginning of the chapter and developed in the text, are summarized in the Chapter Review.

New **Summary of Key Concepts Questions** appear at the end of each concept summary. Check your answers using Appendix A.

Summary Figures present key information in a visual way.

23 CHAPTER REVIEW

SUMMARY OF KEY CONCEPTS

CONCEPT 23.1

Genetic variation makes evolution possible (pp. 469–473)

- **Genetic variation** refers to genetic differences among individuals within a population.
- The nucleotide differences that provide the basis of genetic variation arise by mutation and other processes that produce new alleles and new genes.
- New genetic variants are produced rapidly in organisms with short generation times. In sexually reproducing organisms, most of the genetic differences among individuals result from crossing over, the independent assortment of chromosomes, and fertilization.

? *Why do biologists estimate gene variability and nucleotide variability, and what do these estimates represent?*

CONCEPT 23.2

The Hardy-Weinberg equation can be used to test whether a population is evolving (pp. 473–476)

- A **population**, a localized group of organisms belonging to one species, is united by its **gene pool**, the aggregate of all the alleles in the population.
- The **Hardy-Weinberg principle** states that the allele and genotype frequencies of a population will remain constant if the population is large, mating is random, mutation is negligible, there is no gene flow, and there is no natural selection. For such a population, if p and q represent the frequencies of the only two possible alleles at a particular locus, then p^2 is the frequency of one kind of homozygote, q^2 is the frequency of the other kind of homozygote, and $2pq$ is the frequency of the heterozygous genotype.

? *Is it circular reasoning to calculate* p *and* q *from observed genotype frequencies and then use those values of* p *and* q *to test if the population is in Hardy-Weinberg equilibrium? Explain your answer. (Hint: Consider a specific case, such as a population with 195 individuals of genotype* AA, *10 of genotype* Aa, *and 195 of genotype* aa.)

CONCEPT 23.3

Natural selection, genetic drift, and gene flow can alter allele frequencies in a population (pp. 476–480)

- In natural selection, individuals that have certain inherited traits tend to survive and reproduce at higher rates than other individuals *because of* those traits.
- In **genetic drift**, chance fluctuations in allele frequencies over generations tend to reduce genetic variation.
- **Gene flow**, the transfer of alleles between populations, tends to reduce genetic differences between populations over time.

? *Would two small, geographically isolated populations in very different environments be likely to evolve in similar ways? Explain.*

CONCEPT 23.4

Natural selection is the only mechanism that consistently causes adaptive evolution (pp. 480–485)

- One organism has greater **relative fitness** than a second organism if it leaves more fertile descendants than the second organism. The modes of natural selection differ in how selection acts on phenotype (the white arrows in the summary diagram below represent selective pressure on a population).

Original population

Evolved population

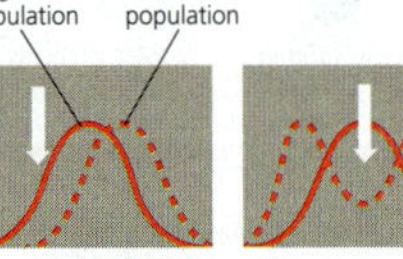

Directional selection

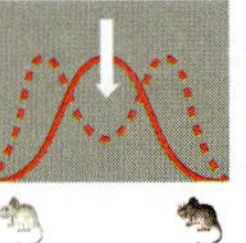

Disruptive selection

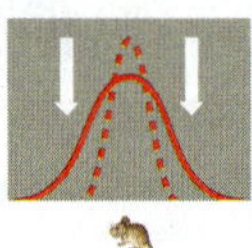

Stabilizing selection

- Unlike genetic drift and gene flow, natural selection consistently increases the frequencies of alleles that enhance survival and reproduction, thus improving the match between organisms and their environment.
- **Sexual selection** influences evolutionary change in secondary sex characteristics that can give individuals advantages in mating.
- Despite the winnowing effects of selection, populations have considerable genetic variation. Some of this variation represents **neutral variation**; additional variation can be maintained by diploidy and balancing selection.
- There are constraints to evolution: Natural selection can act only on available variation; structures result from modified ancestral anatomy; adaptations are often compromises; and chance, natural selection, and the environment interact.

? *How might secondary sex characteristics differ between males and females in a species in which females compete for mates?*

TEST YOUR UNDERSTANDING

LEVEL 1: KNOWLEDGE/COMPREHENSION

1. Natural selection changes allele frequencies because some ______ survive and reproduce more successfully than others.
 a. alleles c. gene pools e. individuals
 b. loci d. species
2. No two people are genetically identical, except for identical twins. The main source of genetic variation among human individuals is
 a. new mutations that occurred in the preceding generation.
 b. genetic drift due to the small size of the population.
 c. the reshuffling of alleles in sexual reproduction.
 d. geographic variation within the population.
 e. environmental effects.
3. Sparrows with average-sized wings survive severe storms better than those with longer or shorter wings, illustrating
 a. the bottleneck effect.
 b. disruptive selection.
 c. frequency-dependent selection.
 d. neutral variation.
 e. stabilizing selection.

486 UNIT FOUR Mechanisms of Evolution

To help you prepare for the various kinds of questions that may appear on a test, the end-of-chapter questions are now organized into three levels based on Bloom's Taxonomy:

Level 1: Knowledge/Comprehension
Level 2: Application/Analysis
Level 3: Synthesis/Evaluation

Evolution Connection Questions in the Chapter Review ask you to think critically about how an aspect of the chapter relates to evolution. ▶

Scientific Inquiry Questions ▶ at the end of each chapter give you opportunities to practice scientific thinking by developing hypotheses, designing experiments, and analyzing real research data.

Draw It Exercises in each chapter ask you to put pencil to paper and draw a structure, annotate a figure, or graph experimental data.

LEVEL 2: APPLICATION/ANALYSIS

4. If the nucleotide variability of a locus equals 0%, what is the gene variability and number of alleles at that locus?
   a. gene variability = 0%; number of alleles = 0
   b. gene variability = 0%; number of alleles = 1
   c. gene variability = 0%; number of alleles = 2
   d. gene variability > 0%; number of alleles = 2
   e. Without more information, gene variability and number of alleles cannot be determined.

5. There are 40 individuals in population 1, all with genotype *A1A1*, and there are 25 individuals in population 2, all with genotype *A2A2*. Assume that these populations are located far from each other and that their environmental conditions are very similar. Based on the information given here, the observed genetic variation is most likely an example of
   a. genetic drift.
   b. gene flow.
   c. disruptive selection.
   d. discrete variation.
   e. directional selection.

6. A fruit fly population has a gene with two alleles, *A1* and *A2*. Tests show that 70% of the gametes produced in the population contain the *A1* allele. If the population is in Hardy-Weinberg equilibrium, what proportion of the flies carry both *A1* and *A2*?
   a. 0.7 b. 0.49 c. 0.21 d. 0.42 e. 0.09

LEVEL 3: SYNTHESIS/EVALUATION

7. **EVOLUTION CONNECTION**
   How is the process of evolution revealed by the imperfections of living organisms?

8. **SCIENTIFIC INQUIRY**
   DRAW IT Richard Koehn, of the State University of New York, Stony Brook, and Thomas Hilbish, of the University of South Carolina, studied genetic variation in the marine mussel *Mytilus edulis* around Long Island, New York. They measured the frequency of a particular allele (lap^{94}) for an enzyme involved in regulating the mussel's internal saltwater balance. The researchers presented their data as a series of pie charts linked to sampling sites within Long Island Sound, where the salinity is highly variable, and along the coast of the open ocean, where salinity is constant:

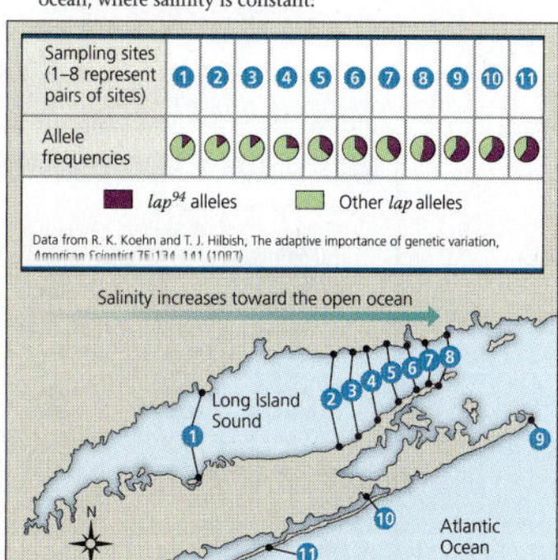

(Question 8, continued)
Create a data table for the 11 sampling sites by estimating the frequency of lap^{94} from the pie charts. (*Hint*: Think of each pie chart as a clock face to help you estimate the proportion of the shaded area.) Then graph the frequencies for sites 1–8 to show how the frequency of this allele changes with increasing salinity in Long Island Sound (from southwest to northeast). How do the data from sites 9–11 compare with the data from the sites within the Sound?

Construct a hypothesis that explains the patterns you observe in the data and that accounts for the following observations: (1) the lap^{94} allele helps mussels maintain osmotic balance in water with a high salt concentration but is costly to use in less salty water; and (2) mussels produce larvae that can disperse long distances before they settle on rocks and grow into adults.

9. **WRITE ABOUT A THEME**
   Emergent Properties Heterozygotes at the sickle-cell locus produce both normal and abnormal (sickle-cell) hemoglobin (see Concept 14.4). When hemoglobin molecules are packed into a heterozygote's red blood cells, some cells receive relatively large quantities of abnormal hemoglobin, making these cells prone to sickling. In a short essay (approximately 100–150 words), explain how these molecular and cellular events lead to emergent properties at the individual and population levels of biological organization.

For selected answers, see Appendix A.

MasteringBIOLOGY www.masteringbiology.com

1. MasteringBiology® Assignments

Make Connections Tutorial Hardy-Weinberg Principle (Chapter 23) and Inheritance of Alleles (Chapter 14)

Experimental Inquiry Tutorial Did Natural Selection of Ground Finches Occur When the Environment Changed?

BioFlix **Tutorial** Mechanisms of Evolution

Tutorial Hardy-Weinberg Principle

Activities Genetic Variation from Sexual Recombination • The Hardy-Weinberg Principle • Causes of Evolutionary Change • Three Modes of Natural Selection

Questions Student Misconceptions • Reading Quiz • Multiple Choice • End-of-Chapter

2. eText

Read your book online, search, take notes, highlight text, and more.

3. The Study Area

Practice Tests • Cumulative Test • *BioFlix* 3-D Animations • MP3 Tutor Sessions • Videos • Activities • Investigations • Lab Media • Audio Glossary • Word Study Tools • Art

CHAPTER 23 The Evolution of Populations 487

◀ New **Write About a Theme Questions** give you practice writing a short essay that connects the chapter's content to one of the bookwide themes introduced in Chapter 1.

◀ Each chapter now ends with a preview of the **MasteringBiology®** resources that can help you succeed in the course.

Suggested Grading Rubric for "Write About a Theme" Short-Answer Essays

	Understanding of Theme and Relationship to Topic	Use of Supporting Examples or Details	Appropriate Use of Terminology	Quality of Writing
4	Evidence of full and complete understanding	Examples well chosen, details accurate and applied to theme	Accurate scientific terminology enhances the essay	Excellent organization, sentence structure, and grammar
3	Evidence of good understanding	Examples or details are generally well applied to theme	Terminology is correctly used	Good sentence flow, sentence structure, and grammar
2	Evidence of a basic understanding	Supporting examples and details are adequate	Terminology used is not totally accurate or appropriate	Some organizational and grammatical problems
1	Evidence of limited understanding	Examples and details are minimal	Appropriate terminology is not present	Poorly organized. Grammatical and spelling errors detract from essay
0	Essay shows no understanding of theme	Examples lacking or incorrect	Terminology lacking or incorrect	Essay is very poorly written

◀ This **Writing Rubric** explains criteria on which your writing may be graded. The rubric and tips for writing good short-answer essays can be found in the Study Area at www.masteringbiology.com.

To the Student: How to Effectively

The Mastering system empowers you to take charge of your learning—at your convenience, 24/7.

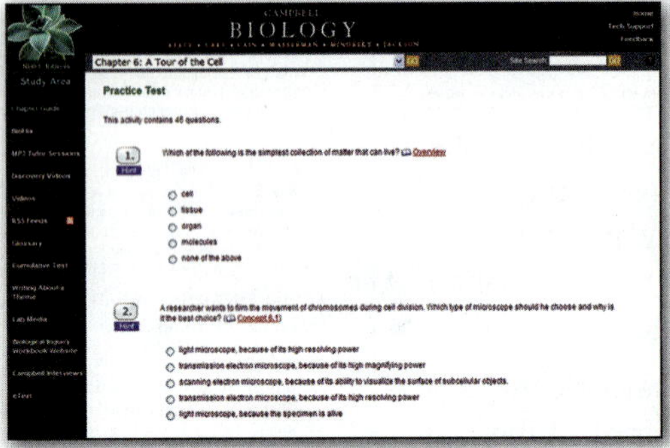

Practice Tests help you assess your understanding of each chapter, providing feedback for right and wrong answers.

Use the Study Area on your own or in a study group.

BioFlix

BioFlix® 3-D animations explore the most difficult biology topics, reinforced with tutorials, quizzes, and more.

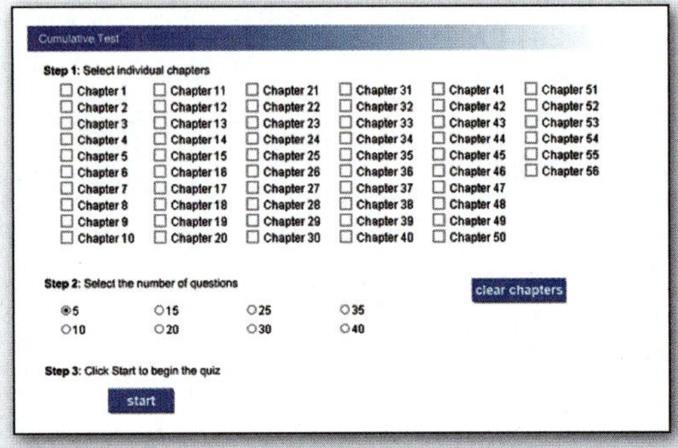

The **Cumulative Test** allows you to build a practice test with questions from multiple chapters.

Access your book online.

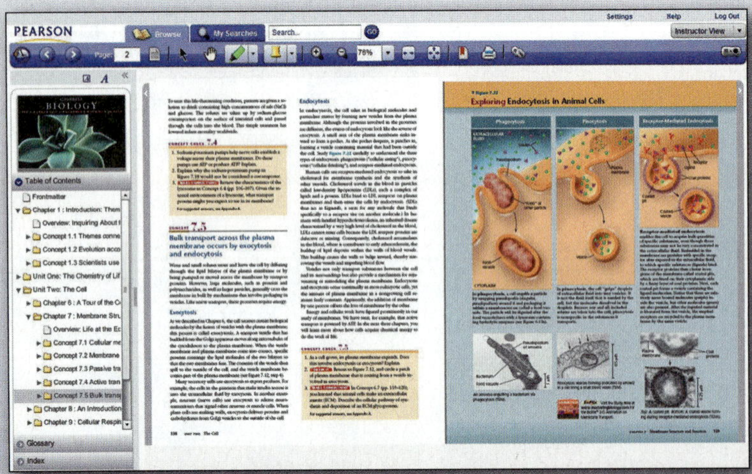

The **Pearson eText** gives you access to the text whenever and wherever you can access the Internet. The eText includes powerful interactive and customization functions:

- write notes
- highlight text
- bookmark pages
- zoom
- click hyperlinked words to view definitions
- search
- link to media activities and quizzes

Your professor can also write notes for the class and highlight important material using a new tool that works like an electronic pen on a whiteboard.

Use MasteringBiology®

Get personalized coaching & feedback.

Your instructor may assign self-paced **MasteringBiology® tutorials** that provide individualized coaching with specific hints and feedback on the toughest topics in the course.

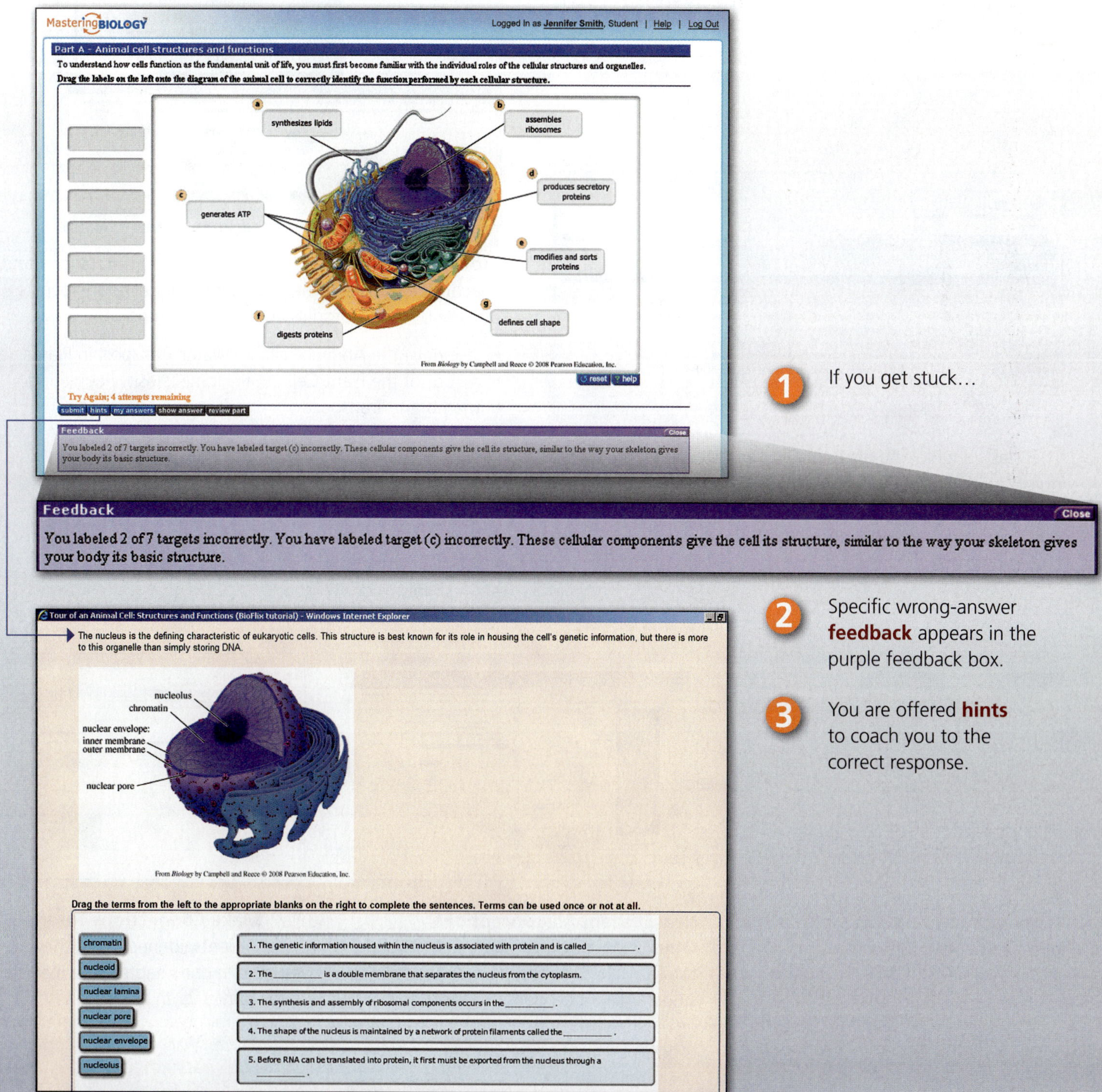

1. If you get stuck…
2. Specific wrong-answer **feedback** appears in the purple feedback box.
3. You are offered **hints** to coach you to the correct response.

To the Instructor: New Content in

MasteringBiology is the most effective and widely used online science tutorial, homework, and assessment system available.

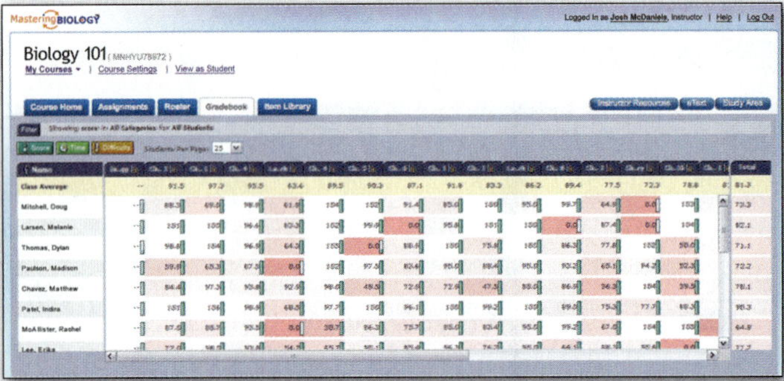

The MasteringBiology **gradebook** provides you with quick results and easy-to-interpret insights into student performance. Every assignment is **automatically graded** and shades of red highlight vulnerable students and challenging assignments.

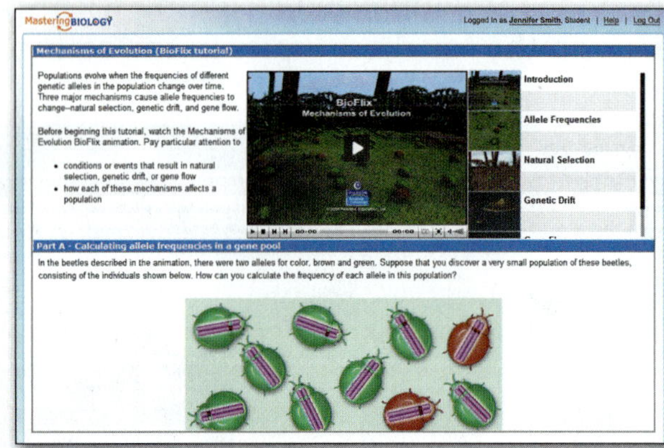

BioFlix® Tutorials use 3-D, movie-quality animations and coaching exercises to help students master tough topics outside of class. Tutorials and animations include:

- A Tour of the Animal Cell
- A Tour of the Plant Cell
- Membrane Transport
- Cellular Respiration
- Photosynthesis
- Mitosis
- Meiosis
- DNA Replication
- Protein Synthesis
- Mechanisms of Evolution
- Water Transport in Plants
- Homeostasis: Regulating Blood Sugar
- Gas Exchange
- How Neurons Work
- How Synapses Work
- Muscle Contraction
- Population Ecology
- The Carbon Cycle

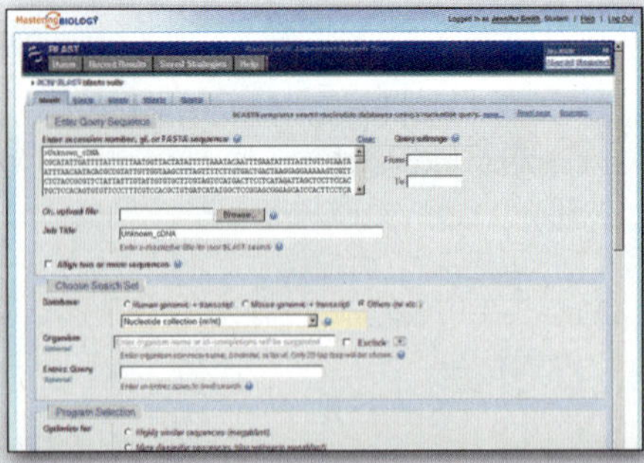

New **Data Analysis Tutorials** allow students to analyze real data from online databases.

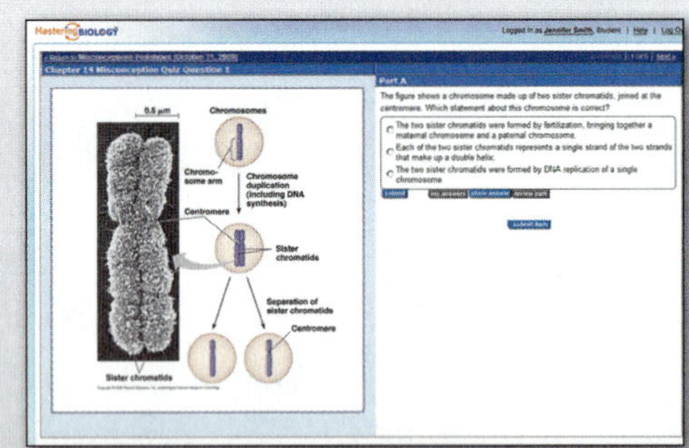

New **Student Misconceptions Questions** provide assignable quizzes for each chapter based on common student misconceptions. Students are provided with feedback, and the instructor is provided with in-class strategies for overcoming these misconceptions.

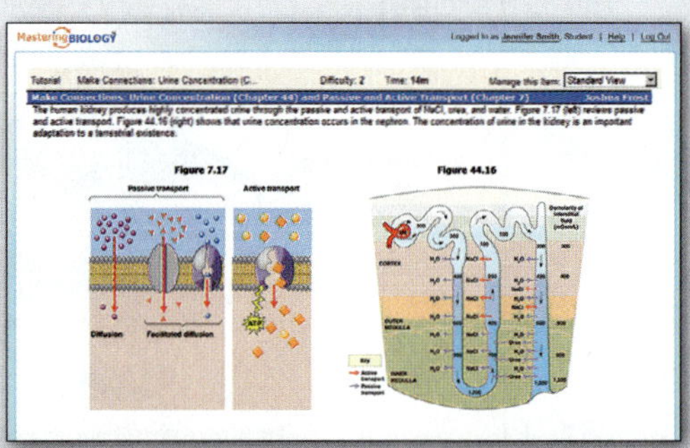

New **Make Connections Tutorials** help students connect what they are learning in one chapter with material they learned in an earlier chapter.

MasteringBiology®

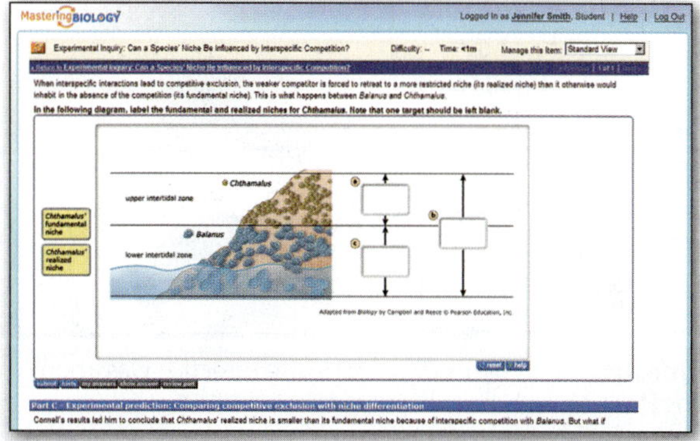

◀ New **Experimental Inquiry Tutorials,** based on some of biology's most influential experiments, give students practice analyzing experimental design and data, and help students understand how to reach conclusions based on collected data. Topics include:

- What Can You Learn About the Process of Science from Investigating a Cricket's Chirp?
- Which Wavelengths of Light Drive Photosynthesis?
- What Is the Inheritance Pattern of Sex-Linked Traits?
- Does DNA Replication Follow the Conservative, Semiconservative, or Dispersive Model?
- Did Natural Selection of Ground Finches Occur When the Environment Changed?
- What Effect Does Auxin Have on Coleoptile Growth?
- What Role Do Genes Play in Appetite Regulation?
- How Do Calcium Ions Help to Prevent Polyspermy During Egg Fertilization?
- Can a Species' Niche Be Influenced by Interspecific Competition?
- What Factors Influence the Loss of Nutrients from a Forest Ecosystem?

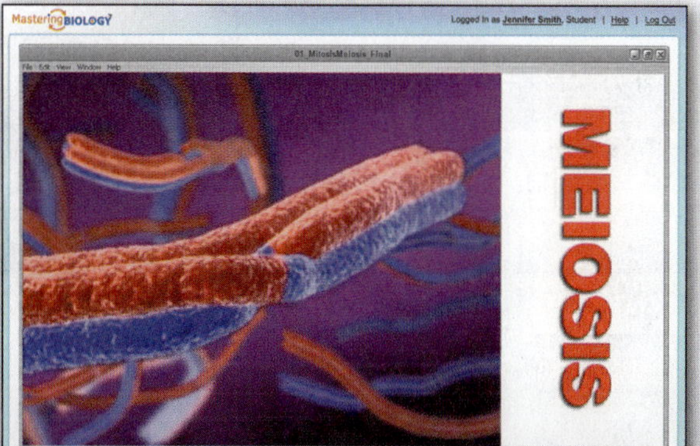

◀ The New **Video Tutor Sessions** walk students through tough topics with clearly explained visuals and demonstrations. Topics include:

- Mitosis and Meiosis
- Sex-Linked Pedigrees
- DNA Structure
- DNA Profiling Techniques
- Biodiversity
- Phylogenetic Trees

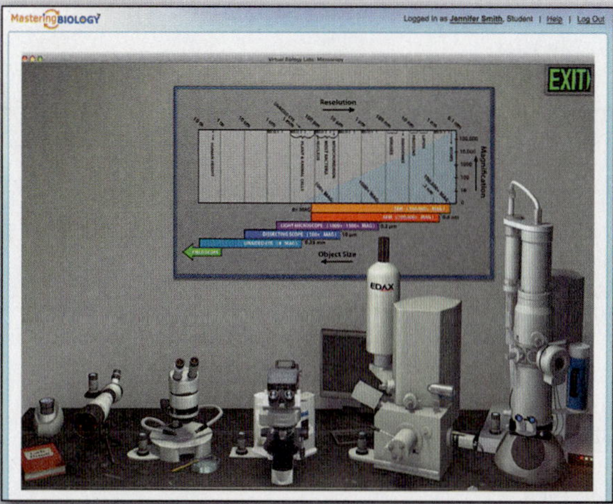

◀ The new **MasteringBiology: Virtual Biology Labs** online environment promotes critical thinking skills using virtual experiments and explorations that may be difficult to perform in a wet lab environment due to time, cost, or safety concerns.

- MasteringBiology: Virtual Biology Labs offer unique learning experiences in microscopy, molecular biology, genetics, ecology, and systematics.
- Choose from 20–30 automatically graded, "pre-set" lab activities that are ready to assign to students, or create your own from scratch.
- Each "pre-set" lab provides an assignable lab report with questions that are automatically graded and recorded in the MasteringBiology gradebook.
- Student subscriptions are available standalone or packaged with the *Campbell Biology* textbook.

Supplements

For Instructors

Instructor Resource DVD

978-0-321-67786-0 • 0-321-67786-2

Assets for each chapter include:

- All figures, photos, and tables in JPEG and PowerPoint®
- Prepared PowerPoint Presentations for each chapter, with lecture notes, editable figures from the text, and links to animations and videos
- Clicker Questions in PowerPoint
- 250+ Instructor Animations, including 3-D BioFlix®
- Discovery Channel™ Videos
- Test Bank questions in TestGen® software and Microsoft® Word

BioFlix® animations invigorate classroom lectures with 3-minute "movie quality" 3-D graphics (see list on p. xviii).

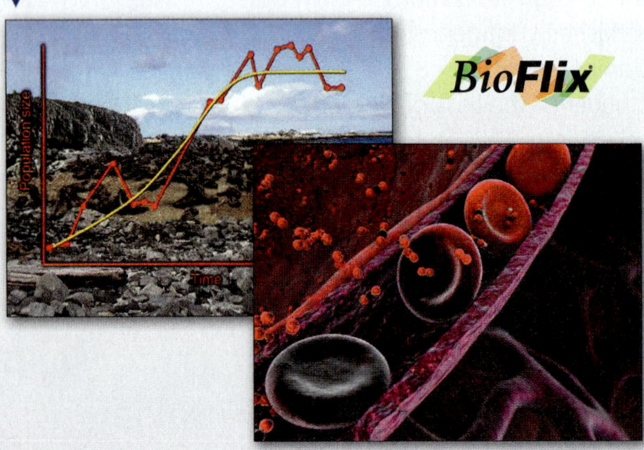

Customizable PowerPoints provide a jumpstart for each lecture.

Chloroplasts: The Sites of Photosynthesis in Plants

- Leaves are the major locations of photosynthesis
- Their green color is from **chlorophyll**, the green pigment within chloroplasts
- Light energy absorbed by chlorophyll drives the synthesis of organic molecules in the chloroplast
- CO_2 enters and O_2 exits the leaf through microscopic pores called **stomata**

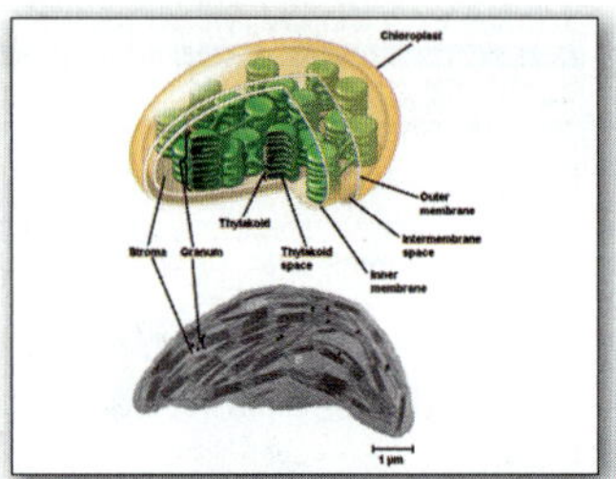

Clicker Questions can be used to stimulate effective classroom discussions (for use with or without clickers).

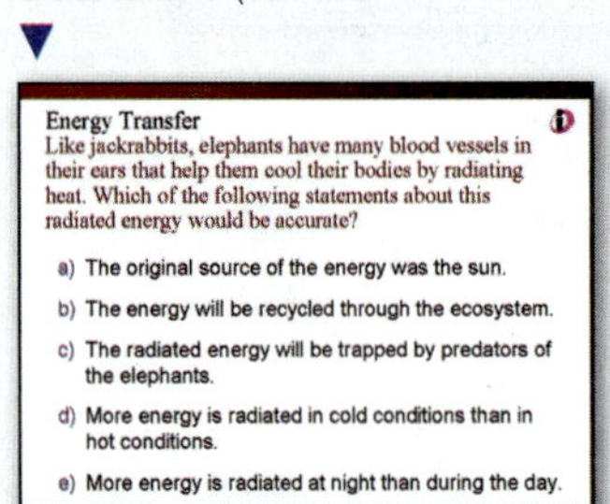

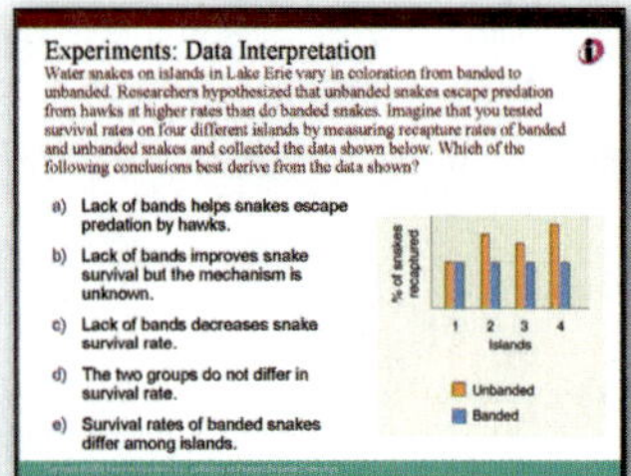

All of the art and photos from the book are provided with customizable labels. More than 1,600 photos from the text and other sources are included.

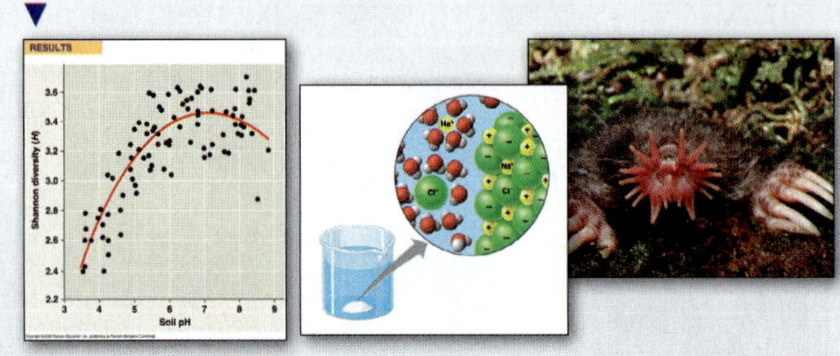

Printed Test Bank

by Louise Paquin, *McDaniel College*, Michael Dini, *Texas Tech University*, John Lepri, *University of North Carolina, Greensboro*, Jung Choi, *Georgia Institute of Technology*, John Zarnetske, *Hoosick Falls Central School*, and Ronald Balsamo, *Villanova University*

978-0-321-69729-5 • 0-321-69729-4

For the Ninth Edition, the Test Bank authors have increased the number of high-level application and synthesis questions.

Transparency Acetates

978-0-321-69708-0 • 0-321-69708-1

Course Management Systems

Test Bank questions, quizzes, and selected content from the Study Area of MasteringBiology are available in these popular course management systems:

CourseCompass™ (www.pearsonhighered.com)

Blackboard (www.pearsonhighered.com)

Instructor Resources Area in MasteringBiology

This area includes:

- PowerPoint® Lectures
- Clicker Questions
- JPEG Images
- Animations
- Videos
- Lecture Outlines
- Learning Objectives
- Strategies for Overcoming Common Student Misconceptions
- Instructor Guides for Supplements
- Write About a Theme: Tips for Grading Short-Answer Essays
- Suggested Answers for Essay Questions
- Test Bank Files
- Lab Media

For Students

***Student Study Guide*, Ninth Edition**
by Martha R. Taylor
978-0-321-62992-0 • 0-321-62992-2

This popular study guide helps students extract key ideas from the textbook and organize their knowledge of biology. Exercises include concept maps for each chapter, chapter summaries, word roots, chapter tests, and a variety of interactive questions in various formats.

Inquiry in Action: Interpreting Scientific Papers*, Second Edition
by Ruth Buskirk, *University of Texas at Austin*, and Christopher M. Gillen, *Kenyon College*
978-0-321-68336-6 • 0-321-68336-6

Selected Inquiry Figures in the Ninth Edition direct students to read and analyze the complete original research paper. In this supplement, those articles are reprinted and accompanied by questions that help students analyze the article. The Inquiry Figures from the book are reprinted in the supplement.

Practicing Biology: A Student Workbook*, Fourth Edition
by Jean Heitz and Cynthia Giffen, *University of Wisconsin, Madison*
978-0-321-68328-1 • 0-321-68328-5

This workbook offers a variety of activities to suit different learning styles. Activities such as modeling and mapping allow students to visualize and understand biological processes. Other activities focus on basic skills, such as reading and drawing graphs.

Biological Inquiry: A Workbook of Investigative Cases*, Third Edition
by Margaret Waterman, *Southeast Missouri State University,* and Ethel Stanley, *BioQUEST Curriculum Consortium and Beloit College*
978-0-321-68320-5 • 0-321-68320-X

This workbook offers ten investigative cases. A student website is in the Study Area at **www.masteringbiology.com.**

***Study Card*, Ninth Edition**
978-0-321-68322-9 • 0-321-68322-6

This quick-reference card provides an overview of the entire field of biology and helps students see the connections between topics.

***Spanish Glossary*, Ninth Edition**
by Laura P. Zanello, *University of California, Riverside*
978-0-321-68321-2 • 0-321-68321-8

This resource provides definitions in Spanish for all the glossary terms.

Into the Jungle: Great Adventures in the Search for Evolution
by Sean B. Carroll, *University of Wisconsin, Madison*
978-0-321-55671-4 • 0-321-55671-2

These nine short tales vividly depict key discoveries in evolutionary biology and the excitement of the scientific process. Online resources available at www.aw-bc.com/carroll.

Get Ready for Biology
978-0-321-50057-1 • 0-321-50057-1

This engaging workbook helps students brush up on important math and study skills and get up to speed on biological terminology and the basics of chemistry and cell biology.

***A Short Guide to Writing About Biology*, Seventh Edition**
by Jan A. Pechenik, *Tufts University*
978-0-321-66838-7 • 0-321-66838-3

This best-selling writing guide teaches students to think as biologists and to express ideas clearly and concisely through their writing.

***An Introduction to Chemistry for Biology Students*, Ninth Edition**
by George I. Sackheim, *University of Illinois, Chicago*
978-0- 8053-9571-6 • 0-8053-9571-7

This text/workbook helps students review and master all the basic facts, concepts, and terminology of chemistry that they need for their life science course.

***The Chemistry of Life CD-ROM, Biology Version*, Second Edition**
by Robert M. Thornton, *University of California, Davis*
978-0-8053-3063-2 • 0-8053-3063-1

This CD-ROM uses animations, interactive simulations, and quizzes with feedback to help students learn or review the chemistry needed to succeed in introductory biology.

For Lab

***Investigating Biology Laboratory Manual*, Seventh Edition**
by Judith G. Morgan, *Emory University,* and M. Eloise Brown Carter, *Oxford College of Emory University*
978-0-321-66821-9 • 0-321-66821-9

The Seventh Edition emphasizes connections to recurring themes in biology, including structure and function, unity and diversity, and the overarching theme of evolution.

***Annotated Instructor Edition for Investigating Biology Laboratory Manual*, Seventh Edition**
by Judith G. Morgan, *Emory University,* and M. Eloise Brown Carter, *Oxford College of Emory University*
978-0-321-67668-9 • 0-321-67668-8

***Preparation Guide for Investigating Biology Laboratory Manual*, Seventh Edition**
by Judith G. Morgan, *Emory University*, and M. Eloise Brown Carter, *Oxford College of Emory University*
978-0-321-67669-6 • 0-321-67669-6

Symbiosis: The Benjamin Cummings Custom Laboratory Program for the Biological Sciences

www.pearsoncustom.com/database/symbiosis/bc.html

MasteringBiology®: Virtual Biology Labs

www.masteringbiology.com
This online environment promotes critical thinking skills using virtual experiments and explorations that may be difficult to perform in a wet lab environment due to time, cost, or safety concerns. Designed to supplement or substitute for existing wet labs, this product offers students unique learning experiences and critical thinking exercises in the areas of microscopy, molecular biology, genetics, ecology, and systematics.

*** An Instructor Guide is available for download in the Instructor Resources Area at www.masteringbiology.com.**

Featured Figures

Impact Figures

Exploring Figures

Inquiry Figures

* The Inquiry Figure, original research paper, and a worksheet to guide you through the paper are provided in *Inquiry in Action: Interpreting Scientific Papers*, Second Edition.
† See the related Experimental Inquiry Tutorial in MasteringBiology® (www.masteringbiology.com).

Research Method Figures

Interviews

UNIT 

UNIT 

UNIT 

UNIT 

UNIT 

UNIT 

UNIT 

UNIT 

Acknowledgments

The authors wish to express their gratitude to the global community of instructors, researchers, students, and publishing professionals who have contributed to this edition.

As authors of this text, we are mindful of the daunting challenge of keeping up to date in all areas of our rapidly expanding subject. We are grateful to the many scientists who helped shape this edition by discussing their research fields with us, answering specific questions in their areas of expertise, and sharing their ideas about biology education. We are especially grateful to the following, listed alphabetically: John Archibald, John Armour, Kristian Axelsen, Scott Bowling, Barbara Bowman, Andy Cameron, Scott Carroll, Amy Cheng-Vollmer, Michele Clamp, David DeRosier, Doug DeSimone, Binh An Diep, David Ehrhardt, Robert Fowler, Peter Fraser, Matt Friedman, Tom Gingeras, Anita Gondor, Ken Halanych, Jeff Hardin, Catherine Hurlbut, Adam Johnson, Dale Kaiser, Patrick Keeling, Emir Khatipov, Chris Killian, Andrew Knoll, Nikos Kyrpides, Teri Liegler, Zhe-Xi Luo, Kent MacDonald, Nick Matzke, Melissa Michael, Nadia Naffakh, Rolf Ohlsson, Aharon Oren, Tom Owens, Kevin Padian, Nathalie Pardigon, Bruce Pavlik, Kevin J. Peterson, Michael Pollock, Rebekah Rasooly, Andrew Roger, Ole Seehausen, Alastair Simpson, Betty Smocovitis, Frank Solomon, Pam Soltis, Hans Thewissen, Mark Uhen, Vance Vredenburg, Elisabeth Wade, Phil Zamore, and Christine Zardecki. In addition, a total of 168 biologists, listed on pages xxvi–xxvii, provided detailed reviews of chapters for this edition, helping us ensure the book's scientific accuracy and improve its pedagogical effectiveness. And finally, we thank Marty Taylor, author of all nine editions of the *Student Study Guide*, for her many contributions to the accuracy, clarity, and consistency of the book.

Thanks also to the other professors and students, from all over the world, who offered suggestions directly to the authors. We alone bear the responsibility for any errors that remain in the text, but the dedication of our consultants, reviewers, and other correspondents makes us confident in the accuracy and effectiveness of this edition.

Interviews with prominent scientists have been a hallmark of *CAMPBELL BIOLOGY* since its inception, and conducting these interviews was again one of the great pleasures of revising the book. To open the eight units of this edition, we are proud to include interviews with Susan Solomon, Bonnie Bassler, Joan Steitz, Geerat Vermeij, Ford Doolittle, Luis Herrera-Estrella, Toto Olivera, and Camille Parmesan.

The value of *CAMPBELL BIOLOGY* as a learning tool is greatly enhanced by the supplementary materials that have been created for instructors and students. We recognize that the dedicated authors of these materials are essentially writing mini (and not so mini) books. We much appreciate the hard work and creativity of all the authors listed, with their creations, on pages xx–xxi. In addition, we are grateful to Joan Sharp (*Lecture Outlines*, *Learning Objectives*, and *Student Misconceptions*) and Erin Barley and Kathleen Fitzpatrick (*PowerPoint® Lectures*).

The electronic media for this text are invaluable teaching and learning aids. We thank the hardworking and creative authors of the new material for this edition: Tom Owens, Joan Sharp, and Jennifer Yeh (*MasteringBiology®*); Richard Cowlishaw, Tod Duncan, and Stephanie Pandolfi (*Study Area Practice Tests*); and Eric Simon (*VideoTutors*). And we thank Brian White for his work. Also, we are grateful to the many other people—biology instructors, editors, artists, production experts, and narrators—who are listed in the credits for these and other elements of the electronic media that accompany the book. And we thank the reviewers and class testers of BioFlix® and MasteringBiology.

CAMPBELL BIOLOGY, Ninth Edition, results from an unusually strong synergy between a team of scientists and a team of publishing professionals. Our editorial team at Benjamin Cummings again demonstrated unmatched talents, commitment, and pedagogical insights. Our new Acquisitions Editor, Josh Frost, brought publishing savvy, intelligence, and a much appreciated level head to leading the whole team. The clarity and effectiveness of every page owe much to our extraordinary Supervising Editors Pat Burner and Beth Winickoff, who headed the top-notch development team: Developmental Editors John Burner, Matt Lee, and Mary Catherine Hager; and Developmental Artists Hilair Chism, Carla Simmons, Andrew Recher, and Jay McElroy. Our unsurpassed Senior Editorial Manager Ginnie Simione Jutson, Executive Director of Development Deborah Gale, Assistant Editor Logan Triglia, and Editor-in-Chief Beth Wilbur were indispensable in moving the project in the right direction. We also want to thank Robin Heyden for organizing the annual Biology Leadership Conferences and keeping us in touch with the world of AP Biology.

You would not have this beautiful book in your hands today if not for the work of the book production team: Executive Managing Editor Erin Gregg; Managing Editor Michael Early; Senior Production Project Manager Shannon Tozier; Senior Photo Editor Donna Kalal; Photo Researcher Maureen Spuhler; Copy Editor Janet Greenblatt; Art Editor Laura Murray; Proofreaders Joanna Dinsmore and Pete Shanks; Permissions Editors Sue Ewing and Beth Keister; Senior Project Editor Emily Bush, S4Carlisle; Composition Manager Holly Paige, S4Carlisle; Art Production Manager Kristina Seymour, Precision Graphics; Design Director Mark Ong; Designer Gary Hespenheide; and Manufacturing Buyer Michael Penne. We also thank those who worked on the book's supplements: Susan Berge, Nina Lewallen Hufford, Brady Golden, Jane Brundage, James Bruce, and John Hammett. And for creating the wonderful package of electronic media that accompanies the book, we are grateful to Tania Mlawer, Director of Editorial Content for MasteringBiology, and Jonathan Ballard, Deb Greco, Sarah Jensen, Mary Catherine Hager, Alice Fugate, Juliana Tringali, Josh Gentry, Steve Wright, Kristen Sutton, Katie Foley, Karen Sheh, and David Kokorowski, as well as Director of Media Development Lauren Fogel and Director of Media Strategy Stacy Treco.

For their important roles in marketing the book, we thank Christy Lawrence, Lauren Harp, Scott Dustan, Lillian Carr, Jane Campbell, Jessica Perry, Nora Massuda, and Jessica Tree.

We are grateful to Linda Davis, President of Pearson Math and Science, who has shared our commitment to excellence and provided strong support for five editions now. Moreover, we thank Paul Corey, President of Pearson Benjamin Cummings, and Editorial Director Frank Ruggirello for their enthusiasm, encouragement, and support.

The Pearson sales team, which represents *CAMPBELL BIOLOGY* on campus, is an essential link to the users of the text. They tell us what you like and don't like about the book, communicate the features of the book, and provide prompt service. We thank them for their hard work and professionalism. For representing our book to our international audience, we thank our sales and marketing partners throughout the world. They are all strong allies in biology education.

Finally, we wish to thank our families and friends for their encouragement and patience throughout this long project. Our special thanks to Paul, Dan, Maria, Armelle, and Sean (J.B.R.); Lily, Ross, Lily-too, and Alex (L.A.U.); Debra and Hannah (M.L.C.); Harry, Elga, Aaron, Sophie, Noah, and Gabriele (S.A.W.); Natalie (P.V.M.); and Sally, Robert, David, and Will (R.B.J.). And, as always, thanks to Rochelle, Allison, Jason, and McKay.

Jane Reece, Lisa Urry, Michael Cain,
Steve Wasserman, Peter Minorsky, and Rob Jackson

Reviewers

Ninth Edition Reviewers

Ann Aguanno, *Marymount Manhattan College*
Marc Albrecht, *University of Nebraska*
John Alcock, *Arizona State University*
Eric Alcorn, *Acadia University*
Terry Austin, *Temple College*
Brian Bagatto, *University of Akron*
Virginia Baker, *Chipola College*
Bonnie Baxter, *Westminster College*
Marilee Benore, *University of Michigan, Dearborn*
Catherine Black, *Idaho State University*
William Blaker, *Furman University*
Edward Blumenthal, *Marquette University*
David Bos, *Purdue University*
Scott Bowling, *Auburn University*
Beth Burch, *Huntington University*
Ragan Callaway, *The University of Montana*
Kenneth M. Cameron, *University of Wisconsin, Madison*
Patrick Canary, *Northland Pioneer College*
Cheryl Keller Capone, *Pennsylvania State University*
Karen I. Champ, *Central Florida Community College*
David Champlin, *University of Southern Maine*
Brad Chandler, *Palo Alto College*
Wei-Jen Chang, *Hamilton College*
Jung Choi, *Georgia Institute of Technology*
Steve Christensen, *Brigham Young University, Idaho*
James T. Colbert, *Iowa State University*
William Cushwa, *Clark College*
Shannon Datwyler, *California State University, Sacramento*
Eugene Delay, *University of Vermont*
Daniel DerVartanian, *University of Georgia*
Janet De Souza-Hart, *Massachusetts College of Pharmacy & Health Sciences*
Kathryn A. Durham, *Lorain Community College*
Curt Elderkin, *College of New Jersey*
Mary Ellard-Ivey, *Pacific Lutheran University*
George Ellmore, *Tufts University*
Robert C. Evans, *Rutgers University, Camden*
Sam Fan, *Bradley University*
Paul Farnsworth, *University of New Mexico*
Myriam Alhadeff Feldman, *Cascadia Community College*
Teresa Fischer, *Indian River Community College*
David Fitch, *New York University*
T. Fleming, *Bradley University*
Robert Fowler, *San Jose State University*
Robert Franklin, *College of Charleston*
Art Fredeen, *University of Northern British Columbia*
Matt Friedman, *University of Chicago*
Cynthia M. Galloway, *Texas A&M University, Kingsville*
Simon Gilroy, *University of Wisconsin, Madison*
Jim Goetze, *Laredo Community College*
Lynda Goff, *University of California, Santa Cruz*
Roy Golsteyn, *University of Lethbridge*
Barbara E. Goodman, *University of South Dakota*
David Grise, *Texas A&M University, Corpus Christi*
Devney Hamilton, *Stanford University (student)*
Matthew B. Hamilton, *Georgetown University*
Jeanne M. Harris, *University of Vermont*
Stephanie Harvey, *Georgia Southwestern State University*
Bernard Hauser, *University of Florida*
Andreas Hejnol, *Sars International Centre for Marine Molecular Biology*
Jason Hodin, *Stanford University*
Sara Huang, *Los Angeles Valley College*
Catherine Hurlbut, *Florida State College, Jacksonville*
Diane Husic, *Moravian College*
Thomas Jacobs, *University of Illinois*
Mark Jaffe, *Nova Southeastern University*
Douglas Jensen, *Converse College*
Lance Johnson, *Midland Lutheran College*
Cheryl Jorcyk, *Boise State University*
Caroline Kane, *University of California, Berkeley*
Jennifer Katcher, *Pima Community College*
Eric G. Keeling, *Cary Institute of Ecosystem Studies*
Chris Kennedy, *Simon Fraser University*
Hillar Klandorf, *West Virginia University*
Mark Knauss, *Georgia Highlands College*
Roger Koeppe, *University of Arkansas*
Peter Kourtev, *Central Michigan University*
Eliot Krause, *Seton Hall University*
Steven Kristoff, *Ivy Tech Community College*
William Kroll, *Loyola University*
Rukmani Kuppuswami, *Laredo Community College*
Lee Kurtz, *Georgia Gwinnett College*
Michael P. Labare, *United States Military Academy, West Point*
Ellen Lamb, *University of North Carolina, Greensboro*
William Lamberts, *College of St Benedict and St John's University*
Tali D. Lee, *University of Wisconsin, Eau Claire*
Hugh Lefcort, *Gonzaga University*
Alcinda Lewis, *University of Colorado, Boulder*
Graeme Lindbeck, *Valenica Community College*
Hannah Lui, *University of California, Irvine*
Cindy Malone, *California State University, Northridge*
Julia Marrs, *Barnard College (student)*
Kathleen Marrs, *Indiana University-Purdue University, Indianapolis*
Mike Mayfield, *Ball State University*
Kamau Mbuthia, *Bowling Green State University*
Tanya McGhee, *Craven Community College*
Darcy Medica, *Pennsylvania State University*
Susan Meiers, *Western Illinois University*
Alex Mills, *University of Windsor*
Eli Minkoff, *Bates College*
Subhash Minocha, *University of New Hampshire*
Ivona Mladenovic, *Simon Fraser University*
Courtney Murren, *College of Charleston*
Kimberlyn Nelson, *Pennsylvania State University*
Jacalyn Newman, *University of Pittsburgh*
Kathleen Nolta, *University of Michigan*
Aharon Oren, *The Hebrew University*
Henry R. Owen, *Eastern Illinois University*
Stephanie Pandolfi, *Michigan State University*
Nathalie Pardigon, *Institut Pasteur*
Cindy Paszkowski, *University of Alberta*
Andrew Pease, *Stevenson University*
Nancy Pelaez, *Purdue University*
Irene Perry, *University of Texas of the Permian Basin*
Roger Persell, *Hunter College*
Mark Pilgrim, *College of Coastal Georgia*
Vera M. Piper, *Shenadoah University*
Crima Pogge, *City College of San Francisco*
Michael Pollock, *Mount Royal University*
Roberta Pollock, *Occidental College*
Therese M. Poole, *Georgia State University*
Angela R. Porta, *Kean University*
Robert Powell, *Avila University*
Elena Pravosudova, *University of Nevada, Reno*
Terrell Pritts, *University of Arkansas, Little Rock*
Monica Ranes-Goldberg, *University of California, Berkeley*
Robert S. Rawding, *Gannon University*
Sarah Richart, *Azusa Pacific University*
Kenneth Robinson, *Purdue University*
Heather Roffey, *Marianopolis College*
Patricia Rugaber, *College of Coastal Georgia*
Scott Russell, *University of Oklahoma*
Louis Santiago, *University of California, Riverside*
Tom Sawicki, *Spartanburg Community College*
Thomas W. Schoener, *University of California, Davis*
Patricia Schulte, *University of British Colombia*
Brenda Schumpert, *Valencia Community College*
David Schwartz, *Houston Community College*
Brent Selinger, *University of Lethbridge*
Alison M. Shakarian, *Salve Regina University*
Robin L. Sherman, *Nova Southeastern University*
Sedonia Sipes, *Southern Illinois University, Carbondale*
Joel Stafstrom, *Northern Illinois University*
Alam Stam, *Capital University*
Judy Stone, *Colby College*
Cynthia Surmacz, *Bloomsburg University*
David Tam, *University of North Texas*
Yves Tan, *Cabrillo College*
Emily Taylor, *California Polytechnic State University*
Franklyn Tan Te, *Miami Dade College*
Kent Thomas, *Wichita State University*

Saba Valadkhan, *Center for RNA Molecular Biology*
Sarah VanVickle-Chavez, *Washington University, St. Louis*
William Velhagen, *New York University*
Janice Voltzow, *University of Scranton*
Margaret Voss, *Penn State Erie*
Charles Wade, *C.S. Mott Community College*
Claire Walczak, *Indiana University*
Jerry Waldvogel, *Clemson University*
Robert Lee Wallace, *Ripon College*
Fred Wasserman, *Boston University*
John Weishampel, *University of Central Florida*
Susan Whittemore, *Keene State College*
Janet Wolkenstein, *Hudson Valley Community College*
Grace Wyngaard, *James Madison University*
Paul Yancey, *Whitman College*
Anne D. Yoder, *Duke University*
Nina Zanetti, *Siena College*
Sam Zeveloff, *Weber State University*
Theresa Zucchero, *Methodist University*

Reviewers of Previous Editions

Kenneth Able *(State University of New York, Albany)*
Thomas Adams *(Michigan State University)*
Martin Adamson *(University of British Columbia)*
Dominique Adriaens *(Ghent University)*
Shylaja Akkaraju *(Bronx Community College of CUNY)*
John Alcock *(Arizona State University)*
George R. Aliaga *(Tarrant County College)*
Richard Almon *(State University of New York, Buffalo)*
Bonnie Amos *(Angelo State University)*
Katherine Anderson *(University of California, Berkeley)*
Richard J. Andren *(Montgomery County Community College)*
Estry Ang *(University of Pittsburgh, Greensburg)*
Jeff Appling *(Clemson University)*
J. David Archibald *(San Diego State University)*
David Armstrong *(University of Colorado, Boulder)*
Howard J. Arnott *(University of Texas, Arlington)*
Mary Ashley *(University of Illinois, Chicago)*
Angela S. Aspbury *(Texas State University)*
Robert Atherton *(University of Wyoming)*
Karl Aufderheide *(Texas A&M University)*
Leigh Auleb *(San Francisco State University)*
P. Stephen Baenziger *(University of Nebraska)*
Ellen Baker *(Santa Monica College)*
Katherine Baker *(Millersville University)*
William Barklow *(Framingham State College)*
Susan Barman *(Michigan State University)*
Steven Barnhart *(Santa Rosa Junior College)*
Andrew Barton *(University of Maine Farmington)*
Rebecca A. Bartow *(Western Kentucky University)*
Ron Basmajian *(Merced College)*
David Bass *(University of Central Oklahoma)*
Bonnie Baxter *(Hobart & William Smith)*
Tim Beagley *(Salt Lake Community College)*
Margaret E. Beard *(College of the Holy Cross)*
Tom Beatty *(University of British Columbia)*
Chris Beck *(Emory University)*
Wayne Becker *(University of Wisconsin, Madison)*
Patricia Bedinger *(Colorado State University)*
Jane Beiswenger *(University of Wyoming)*
Anne Bekoff *(University of Colorado, Boulder)*
Marc Bekoff *(University of Colorado, Boulder)*
Tania Beliz *(College of San Mateo)*
Adrianne Bendich *(Hoffman-La Roche, Inc.)*
Barbara Bentley *(State University of New York, Stony Brook)*
Darwin Berg *(University of California, San Diego)*
Werner Bergen *(Michigan State University)*
Gerald Bergstrom *(University of Wisconsin, Milwaukee)*
Anna W. Berkovitz *(Purdue University)*
Dorothy Berner *(Temple University)*
Annalisa Berta *(San Diego State University)*
Paulette Bierzychudek *(Pomona College)*
Charles Biggers *(Memphis State University)*
Kenneth Birnbaum *(New York University)*
Michael W. Black *(California Polytechnic State University, San Luis Obispo)*
Robert Blanchard *(University of New Hampshire)*
Andrew R. Blaustein *(Oregon State University)*
Judy Bluemer *(Morton College)*
Edward Blumenthal *(Marquette University)*
Robert Blystone *(Trinity University)*
Robert Boley *(University of Texas, Arlington)*
Jason E. Bond *(East Carolina University)*
Eric Bonde *(University of Colorado, Boulder)*
Cornelius Bondzi *(Hampton University)*
Richard Boohar *(University of Nebraska, Omaha)*
Carey L. Booth *(Reed College)*
Allan Bornstein *(Southeast Missouri State University)*
Oliver Bossdorf *(State University of New York, Stony Book)*
James L. Botsford *(New Mexico State University)*
Lisa Boucher *(University of Nebraska, Omaha)*
J. Michael Bowes *(Humboldt State University)*
Richard Bowker *(Alma College)*
Robert Bowker *(Glendale Community College, Arizona)*
Barbara Bowman *(Mills College)*
Barry Bowman *(University of California, Santa Cruz)*
Deric Bownds *(University of Wisconsin, Madison)*
Robert Boyd *(Auburn University)*
Sunny Boyd *(University of Notre Dame)*
Jerry Brand *(University of Texas, Austin)*
Edward Braun *(Iowa State University)*
Theodore A. Bremner *(Howard University)*
James Brenneman *(University of Evansville)*
Charles H. Brenner *(Berkeley, California)*
Lawrence Brewer *(University of Kentucky)*
Donald P. Briskin *(University of Illinois, Urbana)*
Paul Broady *(University of Canterbury)*
Chad Brommer *(Emory University)*
Judith L. Bronstein *(University of Arizona)*
Danny Brower *(University of Arizona)*
Carole Browne *(Wake Forest University)*
Mark Browning *(Purdue University)*
David Bruck *(San Jose State University)*
Robb T. Brumfield *(Louisiana State University)*
Herbert Bruneau *(Oklahoma State University)*
Gary Brusca *(Humboldt State University)*
Richard C. Brusca *(University of Arizona, Arizona-Sonora Desert Museum)*
Alan H. Brush *(University of Connecticut, Storrs)*
Howard Buhse *(University of Illinois, Chicago)*
Arthur Buikema *(Virginia Tech)*
Al Burchsted *(College of Staten Island)*
Meg Burke *(University of North Dakota)*
Edwin Burling *(De Anza College)*
William Busa *(Johns Hopkins University)*
Jorge Busciglio *(University of California, Irvine)*
John Bushnell *(University of Colorado)*
Linda Butler *(University of Texas, Austin)*
David Byres *(Florida Community College, Jacksonville)*
Guy A. Caldwell *(University of Alabama)*
Jane Caldwell *(West Virginia University)*
Kim A. Caldwell *(University of Alabama)*
R. Andrew Cameron *(California Institute of Technology)*
Alison Campbell *(University of Waikato)*
Iain Campbell *(University of Pittsburgh)*
W. Zacheus Cande *(University of California, Berkeley)*
Robert E. Cannon *(University of North Carolina, Greensboro)*
Deborah Canington *(University of California, Davis)*
Frank Cantelmo *(St. John's University)*
John Capeheart *(University of Houston, Downtown)*
Gregory Capelli *(College of William and Mary)*
Richard Cardullo *(University of California, Riverside)*
Nina Caris *(Texas A&M University)*
Jeffrey Carmichael *(University of North Dakota)*
Robert Carroll *(East Carolina University)*
Laura L. Carruth *(Georgia State University)*
J. Aaron Cassill *(University of Texas, San Antonio)*
David Champlin *(University of Southern Maine)*
Bruce Chase *(University of Nebraska, Omaha)*
P. Bryant Chase *(Florida State University)*
Doug Cheeseman *(De Anza College)*
Shepley Chen *(University of Illinois, Chicago)*
Giovina Chinchar *(Tougaloo College)*
Joseph P. Chinnici *(Virginia Commonwealth University)*
Jung H. Choi *(Georgia Institute of Technology)*
Geoffrey Church *(Fairfield University)*
Henry Claman *(University of Colorado Health Science Center)*
Anne Clark *(Binghamton University)*
Greg Clark *(University of Texas)*
Patricia J. Clark *(Indiana University-Purdue University, Indianapolis)*
Ross C. Clark *(Eastern Kentucky University)*
Lynwood Clemens *(Michigan State University)*
Janice J. Clymer *(San Diego Mesa College)*
William P. Coffman *(University of Pittsburgh)*
Austin Randy Cohen *(California State University, Northridge)*
J. John Cohen *(University of Colorado Health Science Center)*
Jim Colbert *(Iowa State University)*
Jan Colpaert *(Hasselt University)*

Robert Colvin *(Ohio University)*
Jay Comeaux *(McNeese State University)*
David Cone *(Saint Mary's University)*
Elizabeth Connor *(University of Massachusetts)*
Joanne Conover *(University of Connecticut)*
Gregory Copenhaver *(University of North Carolina, Chapel Hill)*
John Corliss *(University of Maryland)*
James T. Costa *(Western Carolina University)*
Stuart J. Coward *(University of Georgia)*
Charles Creutz *(University of Toledo)*
Bruce Criley *(Illinois Wesleyan University)*
Norma Criley *(Illinois Wesleyan University)*
Joe W. Crim *(University of Georgia)*
Greg Crowther *(University of Washington)*
Karen Curto *(University of Pittsburgh)*
Anne Cusic *(University of Alabama, Birmingham)*
Richard Cyr *(Pennsylvania State University)*
Marymegan Daly *(The Ohio State University)*
W. Marshall Darley *(University of Georgia)*
Cynthia Dassler *(The Ohio State University)*
Marianne Dauwalder *(University of Texas, Austin)*
Larry Davenport *(Samford University)*
Bonnie J. Davis *(San Francisco State University)*
Jerry Davis *(University of Wisconsin, La Crosse)*
Michael A. Davis *(Central Connecticut State University)*
Thomas Davis *(University of New Hampshire)*
John Dearn *(University of Canberra)*
Maria E. de Bellard *(California State University, Northridge)*
Teresa DeGolier *(Bethel College)*
James Dekloe *(University of California, Santa Cruz)*
Patricia A. DeLeon *(University of Delaware)*
Veronique Delesalle *(Gettysburg College)*
T. Delevoryas *(University of Texas, Austin)*
Roger Del Moral *(University of Washington)*
Charles F. Delwiche *(University of Maryland)*
Diane C. DeNagel *(Northwestern University)*
William L. Dentler *(University of Kansas)*
Daniel Dervartanian *(University of Georgia)*
Jean DeSaix *(University of North Carolina, Chapel Hill)*
Biao Ding *(Ohio State University)*
Michael Dini *(Texas Tech University)*
Andrew Dobson *(Princeton University)*
Stanley Dodson *(University of Wisconsin, Madison)*
Mark Drapeau *(University of California, Irvine)*
John Drees *(Temple University School of Medicine)*
Charles Drewes *(Iowa State University)*
Marvin Druger *(Syracuse University)*
Gary Dudley *(University of Georgia)*
Susan Dunford *(University of Cincinnati)*
Betsey Dyer *(Wheaton College)*
Robert Eaton *(University of Colorado)*
Robert S. Edgar *(University of California, Santa Cruz)*
Douglas Eernisse *(California State University, Fullerton)*
Douglas J. Eernisse *(California State University, Fullerton)*
Betty J. Eidemiller *(Lamar University)*
Brad Elder *(Doane College)*
William D. Eldred *(Boston University)*
Michelle Elekonich *(University of Nevada, Las Vegas)*
Mary Ellard-Ivey *(Pacific Lutheran University)*
Norman Ellstrand *(University of California, Riverside)*
Johnny El-Rady *(University of South Florida)*
Dennis Emery *(Iowa State University)*
John Endler *(University of California, Santa Barbara)*
Margaret T. Erskine *(Lansing Community College)*
Gerald Esch *(Wake Forest University)*
Frederick B. Essig *(University of South Florida)*
Mary Eubanks *(Duke University)*
David Evans *(University of Florida)*
Robert C. Evans *(Rutgers University, Camden)*
Sharon Eversman *(Montana State University)*
Olukemi Fadayomi *(Ferris State University)*
Lincoln Fairchild *(Ohio State University)*
Peter Fajer *(Florida State University)*
Bruce Fall *(University of Minnesota)*
Lynn Fancher *(College of DuPage)*
Ellen H. Fanning *(Vanderbilt University)*
Paul Farnsworth *(University of Texas, San Antonio)*
Larry Farrell *(Idaho State University)*
Jerry F. Feldman *(University of California, Santa Cruz)*
Lewis Feldman *(University of California, Berkeley)*
Eugene Fenster *(Longview Community College)*
Russell Fernald *(University of Oregon)*
Rebecca Ferrell *(Metropolitan State College of Denver)*
Kim Finer *(Kent State University)*
Milton Fingerman *(Tulane University)*
Barbara Finney *(Regis College)*
Frank Fish *(West Chester University)*
David Fisher *(University of Hawaii, Manoa)*
Jonathan S. Fisher *(St. Louis University)*
Steven Fisher *(University of California, Santa Barbara)*
Kirk Fitzhugh *(Natural History Museum of Los Angeles County)*
Lloyd Fitzpatrick *(University of North Texas)*
William Fixsen *(Harvard University)*
Abraham Flexer *(Manuscript Consultant, Boulder, Colorado)*
Kerry Foresman *(University of Montana)*
Norma Fowler *(University of Texas, Austin)*
Robert G. Fowler *(San Jose State University)*
David Fox *(University of Tennessee, Knoxville)*
Carl Frankel *(Pennsylvania State University, Hazleton)*
James Franzen *(University of Pittsburgh)*
Bill Freedman *(Dalhousie University)*
Otto Friesen *(University of Virginia)*
Frank Frisch *(Chapman University)*
Virginia Fry *(Monterey Peninsula College)*
Bernard Frye *(University of Texas, Arlington)*
Jed Fuhrman *(University of Southern California)*
Alice Fulton *(University of Iowa)*
Chandler Fulton *(Brandeis University)*
Sara Fultz *(Stanford University)*
Berdell Funke *(North Dakota State University)*
Anne Funkhouser *(University of the Pacific)*
Zofia E. Gagnon *(Marist College)*
Michael Gaines *(University of Miami)*
Arthur W. Galston *(Yale University)*
Stephen Gammie *(University of Wisconsin, Madison)*
Carl Gans *(University of Michigan)*
John Gapter *(University of Northern Colorado)*
Andrea Gargas *(University of Wisconsin, Madison)*
Lauren Garner *(California Polytechnic State University, San Luis Obispo)*
Reginald Garrett *(University of Virginia)*
Patricia Gensel *(University of North Carolina)*
Chris George *(California Polytechnic State University, San Luis Obispo)*
Robert George *(University of Wyoming)*
J. Whitfield Gibbons *(University of Georgia)*
J. Phil Gibson *(Agnes Scott College)*
Frank Gilliam *(Marshall University)*
Simon Gilroy *(Pennsylvania State University)*
Alan D. Gishlick *(Gustavus Adolphus College)*
Todd Gleeson *(University of Colorado)*
Jessica Gleffe *(University of California, Irvine)*
John Glendinning *(Barnard College)*
David Glenn-Lewin *(Wichita State University)*
William Glider *(University of Nebraska)*
Tricia Glidewell *(Marist School)*
Elizabeth A. Godrick *(Boston University)*
Lynda Goff *(University of California, Santa Cruz)*
Elliott Goldstein *(Arizona State University)*
Paul Goldstein *(University of Texas, El Paso)*
Sandra Gollnick *(State University of New York, Buffalo)*
Anne Good *(University of California, Berkeley)*
Judith Goodenough *(University of Massachusetts, Amherst)*
Wayne Goodey *(University of British Columbia)*
Robert Goodman *(University of Wisconsin, Madison)*
Ester Goudsmit *(Oakland University)*
Linda Graham *(University of Wisconsin, Madison)*
Robert Grammer *(Belmont University)*
Joseph Graves *(Arizona State University)*
Phyllis Griffard *(University of Houston, Downtown)*
A. J. F. Griffiths *(University of British Columbia)*
William Grimes *(University of Arizona)*
Mark Gromko *(Bowling Green State University)*
Serine Gropper *(Auburn University)*
Katherine L. Gross *(Ohio State University)*
Gary Gussin *(University of Iowa)*
Mark Guyer *(National Human Genome Research Institute)*
Ruth Levy Guyer *(Bethesda, Maryland)*
R. Wayne Habermehl *(Montgomery County Community College)*
Mac Hadley *(University of Arizona)*
Joel Hagen *(Radford University)*
Jack P. Hailman *(University of Wisconsin)*
Leah Haimo *(University of California, Riverside)*
Ken Halanych *(Auburn University)*
Jody Hall *(Brown University)*
Douglas Hallett, *(Northern Arizona University)*
Rebecca Halyard *(Clayton State College)*
E. William Hamilton *(Washington and Lee University)*
Sam Hammer *(Boston University)*
Penny Hanchey-Bauer *(Colorado State University)*

William F. Hanna *(Massasoit Community College)*
Laszlo Hanzely *(Northern Illinois University)*
Jeff Hardin *(University of Wisconsin, Madison)*
Lisa Harper *(University of California, Berkeley)*
Richard Harrison *(Cornell University)*
Carla Hass *(Pennsylvania State University)*
Chris Haufler *(University of Kansas)*
Bernard A. Hauser *(University of Florida)*
Chris Haynes *(Shelton State Community College)*
Evan B. Hazard *(Bemidji State University, Emeritus)*
H. D. Heath *(California State University, Hayward)*
George Hechtel *(State University of New York, Stony Brook)*
S. Blair Hedges *(Pennsylvania State University)*
Brian Hedlund *(University of Nevada, Las Vegas)*
David Heins *(Tulane University)*
Jean Heitz *(University of Wisconsin, Madison)*
John D. Helmann *(Cornell University)*
Colin Henderson *(University of Montana)*
Susan Hengeveld *(Indiana University)*
Michelle Henricks *(University of California, Los Angeles)*
Caroll Henry *(Chicago State University)*
Frank Heppner *(University of Rhode Island)*
Albert Herrera *(University of Southern California)*
Scott Herrick *(Missouri Western State College)*
Ira Herskowitz *(University of California, San Francisco)*
Paul E. Hertz *(Barnard College)*
David Hibbett *(Clark University)*
R. James Hickey *(Miami University)*
William Hillenius *(College of Charleston)*
Kenneth Hillers *(California Polytechnic State University, San Luis Obispo)*
Ralph Hinegardner *(University of California, Santa Cruz)*
William Hines *(Foothill College)*
Robert Hinrichsen *(Indiana University of Pennsylvania)*
Helmut Hirsch *(State University of New York, Albany)*
Tuan-hua David Ho *(Washington University)*
Carl Hoagstrom *(Ohio Northern University)*
James Hoffman *(University of Vermont)*
A. Scott Holaday *(Texas Tech University)*
N. Michele Holbrook *(Harvard University)*
James Holland *(Indiana State University, Bloomington)*
Charles Holliday *(Lafayette College)*
Lubbock Karl Holte *(Idaho State University)*
Alan R. Holyoak *(Brigham Young University, Idaho)*
Laura Hoopes *(Occidental College)*
Nancy Hopkins *(Massachusetts Institute of Technology)*
Sandra Horikami *(Daytona Beach Community College)*
Kathy Hornberger *(Widener University)*
Pius F. Horner *(San Bernardino Valley College)*
Becky Houck *(University of Portland)*
Margaret Houk *(Ripon College)*
Daniel J. Howard *(New Mexico State University)*
Ronald R. Hoy *(Cornell University)*
Cristin Hulslander *(University of Oregon)*
Donald Humphrey *(Emory University School of Medicine)*
Robert J. Huskey *(University of Virginia)*
Steven Hutcheson *(University of Maryland, College Park)*
Sandra Hsu *(Skyline College)*
Linda L. Hyde *(Gordon College)*
Bradley Hyman *(University of California, Riverside)*
Mark Iked *(San Bernardino Valley College)*
Jeffrey Ihara *(Mira Costa College)*
Cheryl Ingram-Smith *(Clemson University)*
Alice Jacklet *(State University of New York, Albany)*
John Jackson *(North Hennepin Community College)*
John C. Jahoda *(Bridgewater State College)*
Dan Johnson *(East Tennessee State University)*
Lee Johnson *(The Ohio State University)*
Randall Johnson *(University of California, San Diego)*
Stephen Johnson *(William Penn University)*
Wayne Johnson *(Ohio State University)*
Kenneth C. Jones *(California State University, Northridge)*
Russell Jones *(University of California, Berkeley)*
Chad Jordan *(North Carolina State University)*
Alan Journet *(Southeast Missouri State University)*
Walter Judd *(University of Florida)*
Thomas W. Jurik *(Iowa State University)*
Caroline M. Kane *(University of California, Berkeley)*
Thomas C. Kane *(University of Cincinnati)*
Tamos Kapros *(University of Missouri)*
E. L. Karlstrom *(University of Puget Sound)*
Jennifer Katcher *(Pima Community College)*
Laura A. Katz *(Smith College)*
Maureen Kearney *(Field Museum of Natural History)*
Patrick Keeling *(University of British Columbia)*
Elizabeth A. Kellogg *(University of Missouri, St. Louis)*
Norm Kenkel *(University of Manitoba)*
Chris Kennedy *(Simon Fraser University)*
George Khoury *(National Cancer Institute)*
Rebecca T. Kimball *(University of Florida)*
Mark Kirk *(University of Missouri, Columbia)*
Robert Kitchin *(University of Wyoming)*
Attila O. Klein *(Brandeis University)*
Daniel Klionsky *(University of Michigan)*
Jennifer Knight *(University of Colorado)*
Ned Knight *(Linfield College)*
David Kohl *(University of California, Santa Barbara)*
Greg Kopf *(University of Pennsylvania School of Medicine)*
Thomas Koppenheffer *(Trinity University)*
Margareta Krabbe *(Uppsala University)*
Anselm Kratochwil *(Universität Osnabrück)*
Deborah M. Kristan *(California State University, San Marcos)*
William Kroll *(Loyola University, Chicago)*
Janis Kuby *(San Francisco State University)*
Justin P. Kumar *(Indiana University)*
David Kurijaka *(Ohio University)*
Marc-André Lachance *(University of Western Ontario)*
J. A. Lackey *(State University of New York, Oswego)*
Elaine Lai *(Brandeis University)*
Mohamed Lakrim *(Kingsborough Community College)*
Lynn Lamoreux *(Texas A&M University)*
William L'Amoreaux *(College of Staten Island)*
Carmine A. Lanciani *(University of Florida)*
Kenneth Lang *(Humboldt State University)*
Dominic Lannutti *(El Paso Community College)*
Allan Larson *(Washington University)*
John Latto *(University of California, Santa Barbara)*
Diane K. Lavett *(State University of New York, Cortland, and Emory University)*
Charles Leavell *(Fullerton College)*
C. S. Lee *(University of Texas)*
Daewoo Lee *(Ohio University)*
Robert Leonard *(University of California, Riverside)*
Michael R. Leonardo *(Coe College)*
John Lepri *(University of North Carolina, Greensboro)*
Donald Levin *(University of Texas, Austin)*
Mike Levine *(University of California, Berkeley)*
Joseph Levine *(Boston College)*
Bill Lewis *(Shoreline Community College)*
John Lewis *(Loma Linda University)*
Lorraine Lica *(California State University, Hayward)*
Harvey Liftin *(Broward Community College)*
Harvey Lillywhite *(University of Florida, Gainesville)*
Graeme Lindbeck *(Valencia Community College)*
Clark Lindgren *(Grinnell College)*
Diana Lipscomb *(George Washington University)*
Christopher Little *(The University of Texas, Pan American)*
Kevin D. Livingstone *(Trinity University)*
Andrea Lloyd *(Middlebury College)*
Sam Loker *(University of New Mexico)*
Christopher A. Loretz *(State University of New York, Buffalo)*
Jane Lubchenco *(Oregon State University)*
Douglas B. Luckie *(Michigan State University)*
Margaret A. Lynch *(Tufts University)*
Steven Lynch *(Louisiana State University, Shreveport)*
Richard Machemer Jr. *(St. John Fisher College)*
Elizabeth Machunis-Masuoka *(University of Virginia)*
James MacMahon *(Utah State University)*
Christine R. Maher *(University of Southern Maine)*
Linda Maier *(University of Alabama, Huntsville)*
Jose Maldonado *(El Paso Community College)*
Richard Malkin *(University of California, Berkeley)*
Charles Mallery *(University of Miami)*
Keith Malmos *(Valencia Community College, East Campus)*
Cindy Malone *(California State University, Northridge)*
Carol Mapes *(Kutztown University of Pennsylvania)*
William Margolin *(University of Texas Medical School)*
Lynn Margulis *(Boston University)*
Kathleen A. Marrs *(Indiana University-Purdue University, Indianapolis)*
Edith Marsh *(Angelo State University)*
Diane L. Marshall *(University of New Mexico)*
Linda Martin Morris *(University of Washington)*
Karl Mattox *(Miami University of Ohio)*
Joyce Maxwell *(California State University, Northridge)*
Jeffrey D. May *(Marshall University)*
Lee McClenaghan *(San Diego State University)*
Richard McCracken *(Purdue University)*
Andrew McCubbin *(Washington State University)*
Kerry McDonald *(University of Missouri, Columbia)*
Jacqueline McLaughlin *(Pennsylvania State University, Lehigh Valley)*

Neal McReynolds *(Texas A&M International)*
Lisa Marie Meffert *(Rice University)*
Michael Meighan *(University of California, Berkeley)*
Scott Meissner *(Cornell University)*
Paul Melchior *(North Hennepin Community College)*
Phillip Meneely *(Haverford College)*
John Merrill *(Michigan State University)*
Brian Metscher *(University of California, Irvine)*
Ralph Meyer *(University of Cincinnati)*
James Mickle *(North Carolina State University)*
Roger Milkman *(University of Iowa)*
Helen Miller *(Oklahoma State University)*
John Miller *(University of California, Berkeley)*
Kenneth R. Miller *(Brown University)*
John E. Minnich *(University of Wisconsin, Milwaukee)*
Michael J. Misamore *(Texas Christian University)*
Kenneth Mitchell *(Tulane University School of Medicine)*
Alan Molumby *(University of Illinois, Chicago)*
Nicholas Money *(Miami University)*
Russell Monson *(University of Colorado, Boulder)*
Joseph P. Montoya *(Georgia Institute of Technology)*
Frank Moore *(Oregon State University)*
Janice Moore *(Colorado State University)*
Randy Moore *(Wright State University)*
William Moore *(Wayne State University)*
Carl Moos *(Veterans Administration Hospital, Albany, New York)*
Michael Mote *(Temple University)*
Alex Motten *(Duke University)*
Jeanette Mowery *(Madison Area Technical College)*
Deborah Mowshowitz *(Columbia University)*
Rita Moyes *(Texas A&M College Station)*
Darrel L. Murray *(University of Illinois, Chicago)*
John Mutchmor *(Iowa State University)*
Elliot Myerowitz *(California Institute of Technology)*
Gavin Naylor *(Iowa State University)*
John Neess *(University of Wisconsin, Madison)*
Tom Neils *(Grand Rapids Community College)*
Raymond Neubauer *(University of Texas, Austin)*
Todd Newbury *(University of California, Santa Cruz)*
James Newcomb *(New England College)*
Harvey Nichols *(University of Colorado, Boulder)*
Deborah Nickerson *(University of South Florida)*
Bette Nicotri *(University of Washington)*
Caroline Niederman *(Tomball College)*
Maria Nieto *(California State University, Hayward)*
Anders Nilsson *(University of Umeå)*
Greg Nishiyama *(College of the Canyons)*
Charles R. Noback *(College of Physicians and Surgeons, Columbia University)*
Jane Noble-Harvey *(Delaware University)*
Mary C. Nolan *(Irvine Valley College)*
Peter Nonacs *(University of California, Los Angeles)*
Mohamed A. F. Noor *(Duke University)*
Shawn Nordell *(St. Louis University)*
Richard S. Norman *(University of Michigan, Dearborn, Emeritus)*
David O. Norris *(University of Colorado, Boulder)*
Steven Norris *(California State, Channel Islands)*
Gretchen North *(Occidental College)*
Cynthia Norton *(University of Maine, Augusta)*
Steve Norton *(East Carolina University)*
Steve Nowicki *(Duke University)*
Bette H. Nybakken *(Hartnell College)*
Brian O'Conner *(University of Massachusetts, Amherst)*
Gerard O'Donovan *(University of North Texas)*
Eugene Odum *(University of Georgia)*
Mark P. Oemke *(Alma College)*
Linda Ogren *(University of California, Santa Cruz)*
Patricia O'Hern *(Emory University)*
Nathan O. Okia *(Auburn University, Montgomery)*
Jeanette Oliver *(St. Louis Community College, Florissant Valley)*
Gary P. Olivetti *(University of Vermont)*
John Olsen *(Rhodes College)*
Laura J. Olsen *(University of Michigan)*
Sharman O'Neill *(University of California, Davis)*
Wan Ooi *(Houston Community College)*
John Oross *(University of California (Riverside)*
Gay Ostarello *(Diablo Valley College)*
Catherine Ortega *(Fort Lewis College)*
Charissa Osborne *(Butler University)*
Thomas G. Owens *(Cornell University)*
Penny Padgett *(University of North Carolina, Chapel Hill)*
Kevin Padian *(University of California, Berkeley)*
Dianna Padilla *(State University of New York, Stony Brook)*
Anthony T. Paganini *(Michigan State University)*
Barry Palevitz *(University of Georgia)*
Michael A. Palladino *(Monmouth University)*
Daniel Papaj *(University of Arizona)*
Peter Pappas *(County College of Morris)*
Bulah Parker *(North Carolina State University)*
Stanton Parmeter *(Chemeketa Community College)*
Robert Patterson *(San Francisco State University)*
Ronald Patterson *(Michigan State University)*
Crellin Pauling *(San Francisco State University)*
Kay Pauling *(Foothill Community College)*
Daniel Pavuk *(Bowling Green State University)*
Debra Pearce *(Northern Kentucky University)*
Patricia Pearson *(Western Kentucky University)*
Shelley Penrod *(North Harris College)*
Imara Y. Perera *(North Carolina State University)*
Beverly Perry *(Houston Community College)*
David Pfennig *(University of North Carolina, Chapel Hill)*
David S. Pilliod *(California Polytechnic State University, San Luis Obispo)*
J. Chris Pires *(University of Missouri, Columbia)*
Bob Pittman *(Michigan State University)*
James Platt *(University of Denver)*
Martin Poenie *(University of Texas, Austin)*
Scott Poethig *(University of Pennsylvania)*
Jeffrey Pommerville *(Texas A&M University)*
Angela R. Porta *(Kean University)*
Warren Porter *(University of Wisconsin)*
Daniel Potter *(University of California, Davis)*
Donald Potts *(University of California, Santa Cruz)*
Andy Pratt *(University of Canterbury)*
David Pratt *(University of California, Davis)*
Halina Presley *(University of Illinois, Chicago)*
Mary V. Price *(University of California, Riverside)*
Mitch Price *(Pennsylvania State University)*
Rong Sun Pu *(Kean University)*
Rebecca Pyles *(East Tennessee State University)*
Scott Quackenbush *(Florida International University)*
Ralph Quatrano *(Oregon State University)*
Peter Quinby *(University of Pittsburgh)*
Val Raghavan *(Ohio State University)*
Deanna Raineri *(University of Illinois, Champaign-Urbana)*
Talitha Rajah *(Indiana University Southeast)*
Charles Ralph *(Colorado State University)*
Thomas Rand *(Saint Mary's University)*
Robert H. Reaves *(Glendale Community College)*
Kurt Redborg *(Coe College)*
Ahnya Redman *(Pennsylvania State)*
Brian Reeder *(Morehead State University)*
Bruce Reid *(Kean University)*
David Reid *(Blackburn College)*
C. Gary Reiness *(Lewis & Clark College)*
Charles Remington *(Yale University)*
Erin Rempala *(San Diego Mesa College)*
David Reznick *(University of California, Riverside)*
Douglas Rhoads *(University of Arkansas)*
Fred Rhoades *(Western Washington State University)*
Eric Ribbens *(Western Illinois University)*
Christina Richards *(New York University)*
Christopher Riegle *(Irvine Valley College)*
Loren Rieseberg *(University of British Columbia)*
Bruce B. Riley *(Texas A&M University)*
Donna Ritch *(Pennsylvania State University)*
Carol Rivin *(Oregon State University East)*
Laurel Roberts *(University of Pittsburgh)*
Thomas Rodella *(Merced College)*
Rodney Rogers *(Drake University)*
William Roosenburg *(Ohio University)*
Mike Rosenzweig *(Virginia Polytechnic Institute and State University)*
Wayne Rosing *(Middle Tennessee State University)*
Thomas Rost *(University of California, Davis)*
Stephen I. Rothstein *(University of California, Santa Barbara)*
John Ruben *(Oregon State University)*
Albert Ruesink *(Indiana University)*
Neil Sabine *(Indiana University)*
Tyson Sacco *(Cornell University)*
Rowan F. Sage *(University of Toronto)*
Tammy Lynn Sage *(University of Toronto)*
Don Sakaguchi *(Iowa State University)*
Walter Sakai *(Santa Monica College)*
Mark F. Sanders *(University of California, Davis)*
Ted Sargent *(University of Massachusetts, Amherst)*
K. Sathasivan *(University of Texas, Austin)*
Gary Saunders *(University of New Brunswick)*
Thomas R. Sawicki *(Spartanburg Community College)*
Inder Saxena *(University of Texas, Austin)*
Carl Schaefer *(University of Connecticut)*

Maynard H. Schaus *(Virginia Wesleyan College)*
Renate Scheibe *(University of Osnabrück)*
David Schimpf *(University of Minnesota, Duluth)*
William H. Schlesinger *(Duke University)*
Mark Schlissel *(University of California, Berkeley)*
Christopher J. Schneider *(Boston University)*
Thomas W. Schoener *(University of California, Davis)*
Robert Schorr *(Colorado State University)*
Patricia M. Schulte *(University of British Columbia)*
Karen S. Schumaker *(University of Arizona)*
David J. Schwartz *(Houston Community College)*
Christa Schwintzer *(University of Maine)*
Erik P. Scully *(Towson State University)*
Robert W. Seagull *(Hofstra University)*
Edna Seaman *(Northeastern University)*
Duane Sears *(University of California, Santa Barbara)*
Orono Shukdeb Sen *(Bethune-Cookman College)*
Wendy Sera *(Seton Hill University)*
Timothy E. Shannon *(Francis Marion University)*
Joan Sharp *(Simon Fraser University)*
Victoria C. Sharpe *(Blinn College)*
Elaine Shea *(Loyola College, Maryland)*
Stephen Sheckler *(Virginia Polytechnic Institute and State University)*
Richard Sherwin *(University of Pittsburgh)*
Lisa Shimeld *(Crafton Hills College)*
James Shinkle *(Trinity University)*
Barbara Shipes *(Hampton University)*
Richard M. Showman *(University of South Carolina)*
Peter Shugarman *(University of Southern California)*
Alice Shuttey *(DeKalb Community College)*
James Sidie *(Ursinus College)*
Daniel Simberloff *(Florida State University)*
Rebecca Simmons *(University of North Dakota)*
Anne Simon *(University of Maryland, College Park)*
Robert Simons *(University of California (Los Angeles)*
Alastair Simpson *(Dalhousie University)*
Susan Singer *(Carleton College)*
Roger Sloboda *(Dartmouth University)*
John Smarrelli *(Le Moyne College)*
Andrew T. Smith *(Arizona State University)*
Kelly Smith *(University of North Florida)*
Nancy Smith-Huerta *(Miami Ohio University)*
John Smol *(Queen's University)*
Andrew J. Snope *(Essex Community College)*
Julio G. Soto *(San Jose State University)*
Mitchell Sogin *(Woods Hole Marine Biological Laboratory)*
Susan Sovonick-Dunford *(University of Cincinnati)*
Frederick W. Spiegel *(University of Arkansas)*
John Stachowicz *(University of California, Davis)*
Amanda Starnes *(Emory University)*
Karen Steudel *(University of Wisconsin)*
Barbara Stewart *(Swarthmore College)*
Gail A. Stewart *(Camden County College)*
Cecil Still *(Rutgers University, New Brunswick)*
Margery Stinson *(Southwestern College)*
James Stockand *(University of Texas Health Science Center, San Antonio)*
John Stolz *(California Institute of Technology)*
Richard D. Storey *(Colorado College)*
Stephen Strand *(University of California, Los Angeles)*
Eric Strauss *(University of Massachusetts, Boston)*
Antony Stretton *(University of Wisconsin, Madison)*
Russell Stullken *(Augusta College)*
Mark Sturtevant *(University of Michigan, Flint)*
John Sullivan *(Southern Oregon State University)*
Gerald Summers *(University of Missouri)*
Judith Sumner *(Assumption College)*
Marshall D. Sundberg *(Emporia State University)*
Lucinda Swatzell *(Southeast Missouri State University)*
Daryl Sweeney *(University of Illinois, Champaign-Urbana)*
Samuel S. Sweet *(University of California, Santa Barbara)*
Janice Swenson *(University of North Florida)*
Michael A. Sypes *(Pennsylvania State University)*
Lincoln Taiz *(University of California, Santa Cruz)*
Samuel Tarsitano *(Southwest Texas State University)*
David Tauck *(Santa Clara University)*
Emily Taylor *(California Polytechnic State University, San Luis Obispo)*
James Taylor *(University of New Hampshire)*
John W. Taylor *(University of California, Berkeley)*
Martha R. Taylor *(Cornell University)*
Thomas Terry *(University of Connecticut)*
Roger Thibault *(Bowling Green State University)*
William Thomas *(Colby-Sawyer College)*
Cyril Thong *(Simon Fraser University)*
John Thornton *(Oklahoma State University)*
Robert Thornton *(University of California, Davis)*
William Thwaites *(Tillamook Bay Community College)*
Stephen Timme *(Pittsburg State University)*
Eric Toolson *(University of New Mexico)*
Leslie Towill *(Arizona State University)*
James Traniello *(Boston University)*
Paul Q. Trombley *(Florida State University)*
Nancy J. Trun *(Duquesne University)*
Constantine Tsoukas *(San Diego State University)*
Marsha Turell *(Houston Community College)*
Robert Tuveson *(University of Illinois, Urbana)*
Maura G. Tyrrell *(Stonehill College)*
Catherine Uekert *(Northern Arizona University)*
Claudia Uhde-Stone *(California State University, East Bay)*
Gordon Uno *(University of Oklahoma)*
Lisa A. Urry *(Mills College)*
Saba Valadkhan *(Case Western Reserve University School of Medicine)*
James W. Valentine *(University of California, Santa Barbara)*
Joseph Vanable *(Purdue University)*
Theodore Van Bruggen *(University of South Dakota)*
Kathryn VandenBosch *(Texas A&M University)*
Gerald Van Dyke *(North Carolina State University)*
Brandi Van Roo *(Framingham State College)*
Moira Van Staaden *(Bowling Green State)*
Steven D. Verhey *(Central Washington University)*
Kathleen Verville *(Washington College)*
Sara Via *(University of Maryland)*
Frank Visco *(Orange Coast College)*
Laurie Vitt *(University of California, Los Angeles)*
Neal Voelz *(St. Cloud State University)*
Thomas J. Volk *(University of Wisconsin, La Crosse)*
Leif Asbjørn Vøllestad *(University of Oslo)*
Susan D. Waaland *(University of Washington)*
William Wade *(Dartmouth Medical College)*
D. Alexander Wait *(Southwest Missouri State University)*
John Waggoner *(Loyola Marymount University)*
Jyoti Wagle *(Houston Community College)*
Edward Wagner *(University of California, Irvine)*
Dan Walker *(San Jose State University)*
Robert L. Wallace *(Ripon College)*
Jeffrey Walters *(North Carolina State University)*
Linda Walters *(University of Central Florida)*
Nickolas M. Waser *(University of California, Riverside)*
Margaret Waterman *(University of Pittsburgh)*
Charles Webber *(Loyola University of Chicago)*
Peter Webster *(University of Massachusetts, Amherst)*
Terry Webster *(University of Connecticut, Storrs)*
Beth Wee *(Tulane University)*
Andrea Weeks *(George Mason University)*
Peter Wejksnora *(University of Wisconsin, Milwaukee)*
Kentwood Wells *(University of Connecticut)*
David J. Westenberg, *(University of Missouri, Rolla)*
Richard Wetts *(University of California, Irvine)*
Matt White *(Ohio University)*
Ernest H. Williams *(Hamilton College)*
Kathy Williams *(San Diego State University)*
Stephen Williams *(Glendale Community College)*
Elizabeth Willott *(University of Arizona)*
Christopher Wills *(University of California, San Diego)*
Paul Wilson *(California State University, Northridge)*
Fred Wilt *(University of California, Berkeley)*
Peter Wimberger *(University of Puget Sound)*
Robert Winning *(Eastern Michigan University)*
E. William Wischusen *(Louisiana State University)*
Susan Whittemore *(Keene State College)*
Clarence Wolfe *(Northern Virginia Community College)*
Vickie L. Wolfe *(Marshall University)*
Robert T. Woodland *(University of Massachusetts Medical School)*
Joseph Woodring *(Louisiana State University)*
Denise Woodward *(Pennsylvania State University)*
Patrick Woolley *(East Central College)*
Sarah E. Wyatt *(Ohio University)*
Ramin Yadegari *(University of Arizona)*
Paul Yancey *(Whitman College)*
Philip Yant *(University of Michigan)*
Linda Yasui *(Northern Illinois University)*
Hideo Yonenaka *(San Francisco State University)*
Gina M. Zainelli *(Loyola University, Chicago)*
Edward Zalisko *(Blackburn College)*
Zai Ming Zhao *(University of Texas, Austin)*
John Zimmerman *(Kansas State University)*
Miriam Zolan *(Indiana University)*
Uko Zylstra *(Calvin College)*

Detailed Contents

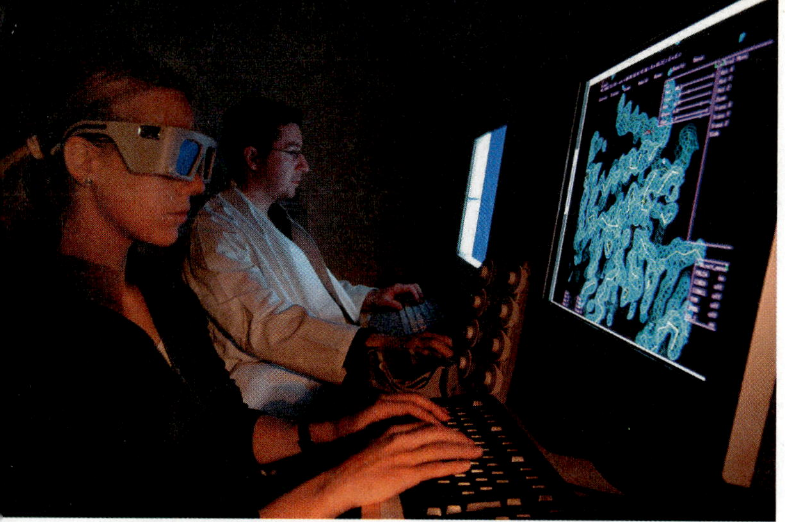

5 The Structure and Function of Large Biological Molecules 68

UNIT

2 The Cell 92

6 A Tour of the Cell 94

7 Membrane Structure and Function 125

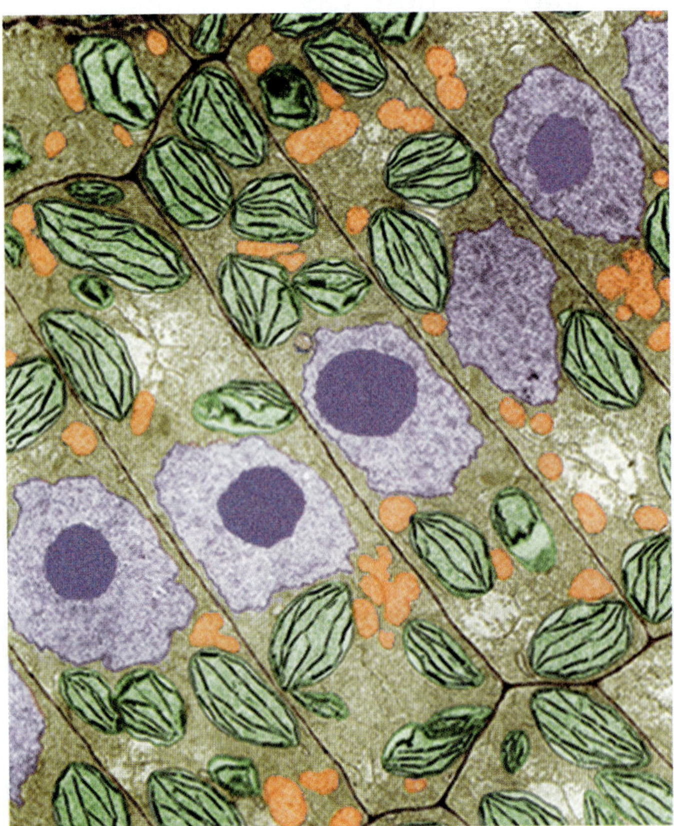

11 Cell Communication 206

12 The Cell Cycle 228

UNIT 3

Genetics 246

13 Meiosis and Sexual Life Cycles 248

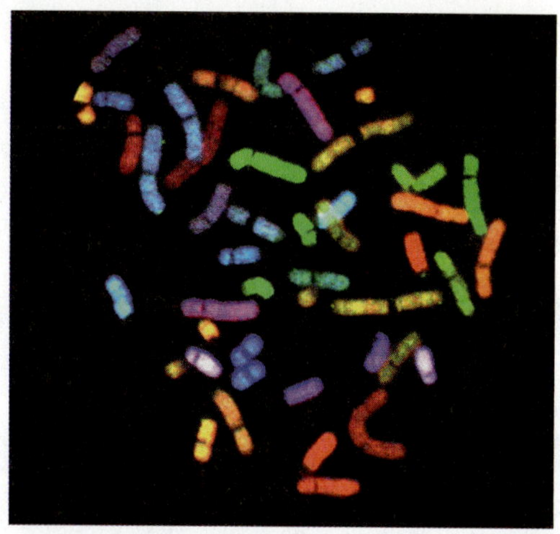

14 Mendel and the Gene Idea 262

1

Introduction: Themes in the Study of Life

▲ **Figure 1.1 How is the mother-of-pearl plant adapted to its environment?**

KEY CONCEPTS

1.1 The themes of this book make connections across different areas of biology

1.2 The Core Theme: Evolution accounts for the unity and diversity of life

1.3 In studying nature, scientists make observations and then form and test hypotheses

1.4 Science benefits from a cooperative approach and diverse viewpoints

OVERVIEW

Inquiring About Life

The mother-of-pearl plant, or ghost plant (**Figure 1.1** and cover), is native to a single mountain in northeastern Mexico. Its fleshy, succulent leaves and other features allow this plant to store and conserve water. Even when rain falls, the plant's access to water is limited because it grows in crevices of vertical rock walls, where little soil is present to hold rainwater (**Figure 1.2**). The plant's water-conserving characteristics help it survive and thrive in these nooks and crannies. Similar features are found in many plants that live in dry environments, allowing them to eke out a living where rain is unpredictable.

An organism's adaptations to its environment, such as adaptations for conserving water, are the result of **evolution**, the process of change that has transformed life on Earth from its earliest beginnings to the diversity of organisms living today. Evolution is the fundamental organizing principle of biology and the core theme of this book.

Although biologists know a great deal about life on Earth, many mysteries remain. For instance, what exactly led to the origin of flowering among plants such as the one pictured here? Posing questions about the living world and seeking science-based answers—scientific inquiry—are the central activities of **biology**, the scientific study of life. Biologists' questions can be ambitious. They may ask how a single tiny cell becomes a tree or a dog, how the human mind works, or how the different forms of life in a forest interact. Most people wonder about the organisms living around them, and many interesting questions probably occur to you when you are out-of-doors, surrounded by the natural world. When they do, you are already thinking like a biologist. More than anything else, biology is a quest, an ongoing inquiry about the nature of life.

What is life? Even a small child realizes that a dog or a plant is alive, while a rock or a lawn mower is not. Yet the phenomenon we call life defies a simple, one-sentence definition. We recognize life by what living things do. **Figure 1.3**, on the next page, highlights some of the properties and processes we associate with life.

While limited to a handful of images, Figure 1.3 reminds us that the living world is wondrously varied. How do biologists

▲ **Figure 1.2 The mother-of-pearl plant (*Graptopetalum paraguayense*).** This plant's thick leaves hold water, enabling it to live where soil is scarce. The leaves vary in color, as seen here.

▼ **Order.** This close-up of a sunflower illustrates the highly ordered structure that characterizes life.

▲ **Evolutionary adaptation.** The appearance of this pygmy sea horse camouflages the animal in its environment. Such adaptations evolve over many generations by the reproductive success of those individuals with heritable traits that are best suited to their environments.

▲ **Response to the environment.** This Venus flytrap closed its trap rapidly in response to the environmental stimulus of a damselfly landing on the open trap.

► **Reproduction.** Organisms (living things) reproduce their own kind. Here, a baby giraffe stands close to its mother.

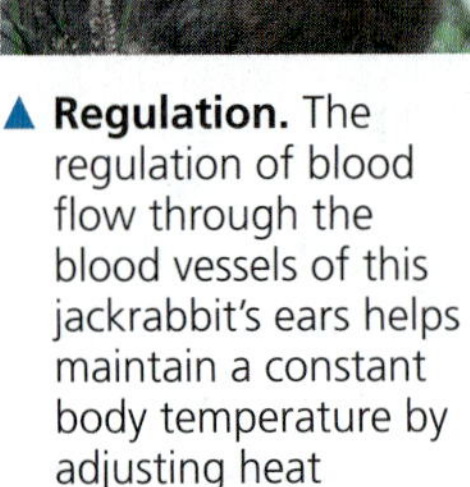

▲ **Regulation.** The regulation of blood flow through the blood vessels of this jackrabbit's ears helps maintain a constant body temperature by adjusting heat exchange with the surrounding air.

▲ **Energy processing.** This hummingbird obtains fuel in the form of nectar from flowers. The hummingbird will use chemical energy stored in its food to power flight and other work.

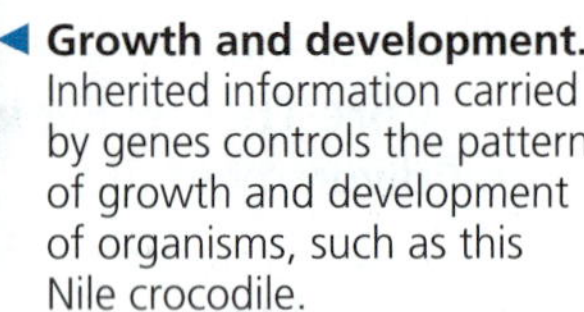

◄ **Growth and development.** Inherited information carried by genes controls the pattern of growth and development of organisms, such as this Nile crocodile.

▲ **Figure 1.3 Some properties of life.**

make sense of this diversity and complexity? This opening chapter sets up a framework for answering this question. The first part of the chapter provides a panoramic view of the biological "landscape," organized around some unifying themes. We then focus on biology's core theme, evolution, with an introduction to the reasoning that led Charles Darwin to his explanatory theory. Next, we look at scientific inquiry—how scientists raise and attempt to answer questions about the natural world. Finally, we address the culture of science and its effects on society.

CONCEPT 1.1

The themes of this book make connections across different areas of biology

Biology is a subject of enormous scope, and news reports reveal exciting new biological discoveries being made every day. Simply memorizing the factual details of this huge subject is most likely not the best way to develop a coherent view of life. A better approach is to take a more active role by connecting the many things you learn to a set of themes that pervade all of biology. Focusing on a few big ideas—ways of thinking about life that will still hold true decades from now—will help you organize and make sense of all the information you'll encounter as you study biology. To help you, we have selected eight unifying themes to serve as touchstones as you proceed through this book.

Theme: New Properties Emerge at Each Level in the Biological Hierarchy

The study of life extends from the microscopic scale of the molecules and cells that make up organisms to the global scale of the entire living planet. We can divide this enormous range into different levels of biological organization.

Imagine zooming in from space to take a closer and closer look at life on Earth. It is spring in Ontario, Canada, and our destination is a local forest, where we will eventually explore a maple leaf right down to the molecular level. **Figure 1.4**, on the next two pages, narrates this journey into life, with the numbers leading you through the levels of biological organization illustrated by the photographs.

Emergent Properties

If we now zoom back out from the molecular level in Figure 1.4, we can see that novel properties emerge at each step, properties that are not present at the preceding level. These **emergent properties** are due to the arrangement and interactions of parts as complexity increases. For example, although photosynthesis occurs in an intact chloroplast, it will not take place in a disorganized test-tube mixture of chlorophyll and other chloroplast molecules. Photosynthesis requires a specific organization of these molecules in the chloroplast. To take another example, if a blow to the head disrupts the intricate architecture of a human brain, the mind may cease to function properly even though all of the brain tissues are still present. Our thoughts and memories are emergent properties of a complex network of nerve cells. At a much higher level of biological organization—at the ecosystem level—the recycling of chemical elements essential to life, such as carbon, depends on a network of diverse organisms interacting with each other and with the soil, water, and air.

Emergent properties are not unique to life. A box of bicycle parts won't take you anywhere, but if they are arranged in a certain way, you can pedal to your chosen destination. And while the graphite in a pencil "lead" and the diamond in a wedding ring are both pure carbon, they have very different appearances and properties due to the different arrangements of their carbon atoms. Both of these examples point out the importance of arrangement. Compared to such nonliving examples, however, the unrivaled complexity of biological systems makes the emergent properties of life especially challenging to study.

The Power and Limitations of Reductionism

Because the properties of life emerge from complex organization, scientists seeking to understand biological systems confront a dilemma. On the one hand, we cannot fully explain a higher level of order by breaking it down into its parts. A dissected animal no longer functions; a cell reduced to its chemical ingredients is no longer a cell. Disrupting a living system interferes with its functioning. On the other hand, something as complex as an organism or a cell cannot be analyzed without taking it apart.

Reductionism—the approach of reducing complex systems to simpler components that are more manageable to study—is a powerful strategy in biology. For example, by studying the molecular structure of DNA that had been extracted from cells, James Watson and Francis Crick inferred, in 1953, how this molecule could serve as the chemical basis of inheritance. The central role of DNA in cells and organisms became better understood, however, when scientists were able to study the interactions of DNA with other molecules. Biologists must balance the reductionist strategy with the larger-scale, holistic objective of understanding emergent properties—how the parts of cells, organisms, and higher levels of order, such as ecosystems, work together. This is the goal of an approach developed over the last 50 years called systems biology.

Systems Biology

A system is simply a combination of components that function together. A biologist can study a system at any level of organization. A single leaf cell can be considered a system, as can a frog, an ant colony, or a desert ecosytem. To understand how such systems work, it is not enough to have a "parts list," even a complete one. Realizing this, many researchers are now complementing the reductionist approach with new strategies for studying whole systems. This change in perspective is analogous to moving from ground level on a street corner, where you can observe local traffic, to a helicopter high above a city, from which you can see how variables such as time of day, construction projects, accidents, and traffic-signal malfunctions affect traffic throughout the city.

Systems biology is an approach that attempts to model the dynamic behavior of whole biological systems based on a study of the interactions among the system's parts. Successful models enable biologists to predict how a change in one or more variables will affect other components and the whole system. Thus, the systems approach enables us to pose new kinds of questions. How might a drug that lowers blood pressure affect the functioning of organs throughout the human body? How might increasing a crop's water supply affect processes in the plants, such as the storage of molecules essential for human nutrition? How might a gradual increase in atmospheric carbon dioxide alter ecosystems and the entire biosphere? The ultimate aim of systems biology is to answer large-scale questions like the last one.

▼ Figure 1.4

Exploring Levels of Biological Organization

◀ 1 The Biosphere

As soon as we are near enough to Earth to make out its continents and oceans, we begin to see signs of life—in the green mosaic of the planet's forests, for example. This is our first view of the biosphere, which consists of all life on Earth and all the places where life exists—most regions of land, most bodies of water, the atmosphere to an altitude of several kilometers, and even sediments far below the ocean floor and rocks many kilometers below Earth's surface.

◀ 2 Ecosystems

As we approach Earth's surface for an imaginary landing in Ontario, we can begin to make out a forest with an abundance of trees that lose their leaves in one season and grow new ones in another (deciduous trees). Such a deciduous forest is an example of an ecosystem. Grasslands, deserts, and the ocean's coral reefs are other types of ecosystems. An ecosystem consists of all the living things in a particular area, along with all the nonliving components of the environment with which life interacts, such as soil, water, atmospheric gases, and light. All of Earth's ecosystems combined make up the biosphere.

▶ 3 Communities

The entire array of organisms inhabiting a particular ecosystem is called a biological community. The community in our forest ecosystem includes many kinds of trees and other plants, a diversity of animals, various mushrooms and other fungi, and enormous numbers of diverse microorganisms, which are living forms, such as bacteria, that are too small to see without a microscope. Each of these forms of life is called a *species*.

▶ 4 Populations

A population consists of all the individuals of a species living within the bounds of a specified area. For example, our Ontario forest includes a population of sugar maple trees and a population of white-tailed deer. We can now refine our definition of a community as the set of populations that inhabit a particular area.

▲ 5 Organisms

Individual living things are called organisms. Each of the maple trees and other plants in the forest is an organism, and so is each forest animal—whether deer, squirrel, frog, or beetle. The soil teems with microorganisms such as bacteria.

▼6 Organs and Organ Systems

The structural hierarchy of life continues to unfold as we explore the architecture of the more complex organisms. A maple leaf is an example of an organ, a body part that carries out a particular function in the body. Stems and roots are the other major organs of plants. Examples of human organs are the brain, heart, and kidney. The organs of humans, other complex animals, and plants are organized into organ systems, each a team of organs that cooperate in a larger function. For example, the human digestive system includes such organs as the tongue, stomach, and intestines. Organs consist of multiple tissues.

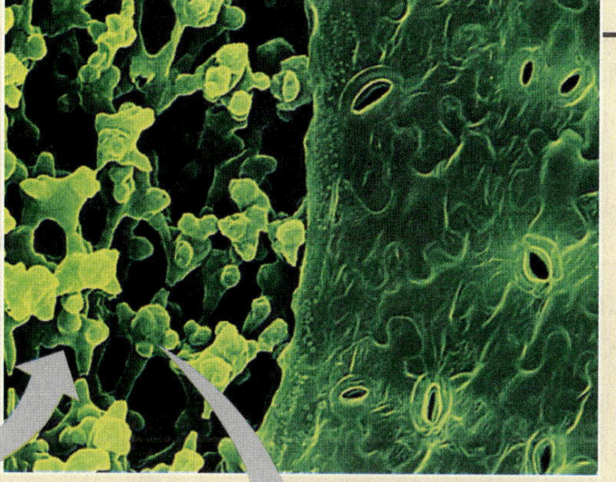

◄7 Tissues

Our next scale change—to see the tissues of a leaf—requires a microscope. Each tissue is made up of a group of cells that work together, performing a specialized function. The leaf shown here has been cut on an angle. The honeycombed tissue in the interior of the leaf (left portion of photo) is the main location of photosynthesis, the process that converts light energy to the chemical energy of sugar and other food. We are viewing the sliced leaf from a perspective that also enables us to see the jigsaw puzzle–like "skin" on the surface of the leaf, a tissue called epidermis (right part of photo). The pores through the epidermis allow the gas carbon dioxide, a raw material for sugar production, to reach the photosynthetic tissue inside the leaf. At this scale, we can also see that each tissue has a distinct cellular structure.

Cell

10 µm

◄8 Cells

The cell is life's fundamental unit of structure and function. Some organisms, such as amoebas and most bacteria, are single cells. Other organisms, including plants and animals, are multicellular. Instead of a single cell performing all the functions of life, a multicellular organism has a division of labor among specialized cells. A human body consists of trillions of microscopic cells of many different kinds, such as muscle cells and nerve cells, which are organized into the various specialized tissues. For example, muscle tissue consists of bundles of muscle cells. In the photo at the upper left, we see a more highly magnified view of some cells in a leaf tissue. One cell is only about 40 micrometers (µm) across. It would take about 500 of these cells to reach across a small coin. As tiny as these cells are, you can see that each contains numerous green structures called chloroplasts, which are responsible for photosynthesis.

►9 Organelles

Chloroplasts are examples of organelles, the various functional components present in cells. In this image, a very powerful tool called an electron microscope brings a single chloroplast into sharp focus.

Chloroplast

1 µm

►10 Molecules

Our last scale change drops us into a chloroplast for a view of life at the molecular level. A molecule is a chemical structure consisting of two or more small chemical units called atoms, which are represented as balls in this computer graphic of a chlorophyll molecule. Chlorophyll is the pigment molecule that makes a maple leaf green. One of the most important molecules on Earth, chlorophyll absorbs sunlight during the first step of photosynthesis. Within each chloroplast, millions of chlorophyll molecules, together with accessory molecules, are organized into the equipment that converts light energy to the chemical energy of food.

Atoms

Chlorophyll molecule

Systems biology is relevant to the study of life at all levels. During the early years of the 20th century, biologists studying how animal bodies function (animal physiology) began integrating data on how multiple organs coordinate processes such as the regulation of sugar concentration in the blood. And in the 1960s, scientists investigating ecosystems pioneered a more mathematically sophisticated systems approach with elaborate models diagramming the network of interactions between organisms and nonliving components of ecosystems, such as salt marshes. More recently, with the sequencing of DNA from many species, systems biology has taken hold at the cellular and molecular levels, as we'll describe later when we discuss DNA.

Theme: Organisms Interact with Other Organisms and the Physical Environment

Turn back again to Figure 1.4, this time focusing on the forest. In an ecosystem, each organism interacts continuously with its environment, which includes both other organisms and physical factors. The leaves of a tree, for example, absorb light from the sun, take in carbon dioxide from the air, and release oxygen to the air **(Figure 1.5)**. Both the organism and the environment are affected by the interactions between them. For example, a plant takes up water and minerals from the soil through its roots, and its roots help form soil by breaking up rocks. On a global scale, plants and other photosynthetic organisms have generated all the oxygen in the air.

▲ **Figure 1.5 Interactions of an African acacia tree with other organisms and the physical environment.**

A tree also interacts with other organisms, such as soil microorganisms associated with its roots, insects that live in the tree, and animals that eat its leaves and fruit. Interactions between organisms ultimately result in the cycling of nutrients in ecosystems. For example, minerals acquired by a tree will eventually be returned to the soil by other organisms that decompose leaf litter, dead roots, and other organic debris. The minerals are then available to be taken up by plants again.

Like all organisms, we humans interact with our environment. Unfortunately, our interactions sometimes have drastic consequences. For example, since the Industrial Revolution in the 1800s, the burning of fossil fuels (coal, oil, and gas) has been increasing at an ever-accelerating pace. This practice releases gaseous compounds into the atmosphere, including prodigious amounts of carbon dioxide (CO_2). About half the human-generated CO_2 stays in the atmosphere, acting like a layer of glass around the planet that admits radiation that warms the Earth but prevents heat from radiating into outer space. Scientists estimate that the average temperature of the planet has risen 1°C since 1900 due to this "greenhouse effect," and they project an additional rise in average global temperature of at least 3°C over the course of the 21st century.

This global warming, a major aspect of **global climate change**, has already had dire effects on life-forms and their habitats all over planet Earth. Polar bears have lost a significant portion of the ice platform from which they hunt, and there are examples of small rodents and plant species that have shifted their ranges to higher altitudes, as well as bird populations that have altered their migration schedules. Only time will reveal the consequences of these changes. Scientists predict that even if we stopped burning fossil fuels today, it would take several centuries to return to preindustrial CO_2 levels. That scenario is highly improbable, so it is imperative that we learn all we can about the effects of global climate change on Earth and its populations. Acting as the stewards of our planet, we must strive to find ways to address this problem.

Theme: Life Requires Energy Transfer and Transformation

As you saw in Figure 1.5, a tree's leaves absorb sunlight. The input of energy from the sun makes life possible: A fundamental characteristic of living organisms is their use of energy to carry out life's activities. Moving, growing, reproducing, and the other activities of life are work, and work requires energy. In the business of living, organisms often

(a) Energy flow from sunlight to producers to consumers **(b) Using energy to do work**

▲ **Figure 1.6 Energy flow in an ecosystem.** This endangered Red Colobus monkey lives in Tanzania.

transform one form of energy to another. Chlorophyll molecules within the tree's leaves harness the energy of sunlight and use it to drive photosynthesis, converting carbon dioxide and water to sugar and oxygen. The chemical energy in sugar is then passed along by plants and other photosynthetic organisms (producers) to consumers. Consumers are organisms, such as animals, that feed on producers and other consumers **(Figure 1.6a)**.

An animal's muscle cells use sugar as fuel to power movements, converting chemical energy to kinetic energy, the energy of motion **(Figure 1.6b)**. The cells in a leaf use sugar to drive the process of cell proliferation during leaf growth, transforming stored chemical energy into cellular work. In both cases, some of the energy is converted to thermal energy, which dissipates to the surroundings as heat. In contrast to chemical nutrients, which recycle within an ecosystem, energy flows through an ecosystem, usually entering as light and exiting as heat.

Theme: Structure and Function Are Correlated at All Levels of Biological Organization

Another theme evident in Figure 1.4 is the idea that form fits function, which you'll recognize from everyday life. For example, a screwdriver is suited to tighten or loosen screws, a hammer to pound nails. How a device works is correlated with its structure. Applied to biology, this theme is a guide to the anatomy of life at all its structural levels. An example from Figure 1.4 is seen in the leaf: Its thin, flat shape maximizes the amount of sunlight that can be captured by its chloroplasts. Analyzing a biological structure gives us clues about what it does and how it works. Conversely, knowing the function of something provides insight into its construction. An example from the animal kingdom, the wing of a bird, provides additional instances of the structure-function theme **(Figure 1.7)**. In exploring life on its different structural levels, we discover functional beauty at every turn.

(a) A bird's wings have an aerodynamically efficient shape.

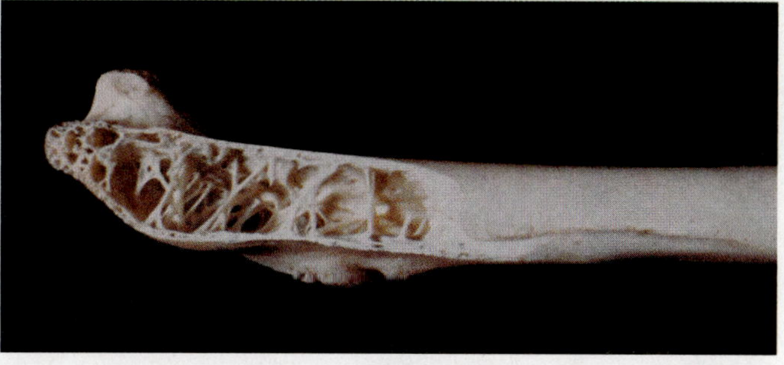

(b) Wing bones have a honeycombed internal structure that is strong but lightweight.

▲ **Figure 1.7 Form fits function in a gull's wing. (a)** The shape of a bird's wings and **(b)** the structure of its bones make flight possible.

? *How does form fit function in a human hand?*

Theme: The Cell Is an Organism's Basic Unit of Structure and Function

In life's structural hierarchy, the cell has a special place as the lowest level of organization that can perform all activities required for life. Moreover, the activities of organisms are all based on the activities of cells. For instance, the movement of your eyes as you read this line is based on activities of muscle and nerve cells. Even a global process such as the recycling of carbon is the cumulative product of cellular activities, including the photosynthesis that occurs in the chloroplasts of leaf cells. Understanding how cells work is a major focus of biological research.

All cells share certain characteristics. For example, every cell is enclosed by a membrane that regulates the passage of materials between the cell and its surroundings. And every cell uses DNA as its genetic information. However, we can distinguish between two main forms of cells: prokaryotic cells and eukaryotic cells. The cells of two groups of microorganisms, called bacteria (singular, *bacterium*) and archaea (singular, *archaean*), are prokaryotic. All other forms of life, including plants and animals, are composed of eukaryotic cells.

A **eukaryotic cell** is subdivided by internal membranes into various membrane-enclosed organelles **(Figure 1.8)**. In most eukaryotic cells, the largest organelle is the nucleus, which contains the cell's DNA. The other organelles are located in the cytoplasm, the entire region between the nucleus and outer membrane of the cell. The chloroplast you saw in Figure 1.4 is an organelle found in eukaryotic cells that carry out photosynthesis. Prokaryotic cells are much simpler and generally smaller than eukaryotic cells, as seen clearly in Figure 1.8. In a **prokaryotic cell**, the DNA is not separated from the rest of the cell by enclosure in a membrane-bounded nucleus. Prokaryotic cells also lack the other kinds of membrane-enclosed organelles that characterize eukaryotic cells. The properties of all organisms, whether prokaryotic or eukaryotic, are based in the structure and function of cells.

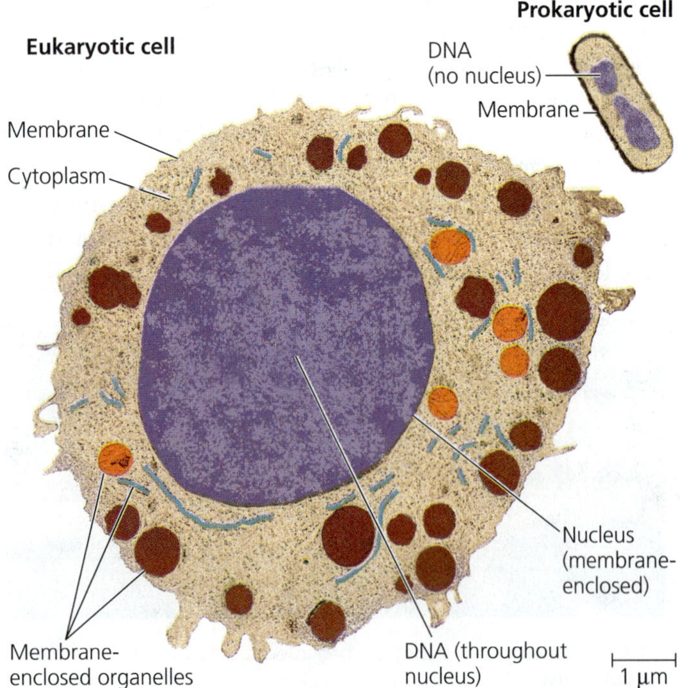

▲ **Figure 1.8 Contrasting eukaryotic and prokaryotic cells in size and complexity.**

Theme: The Continuity of Life Is Based on Heritable Information in the Form of DNA

The division of cells to form new cells is the foundation for all reproduction and for the growth and repair of multicellular organisms. Inside the dividing cell in **Figure 1.9**, you can see structures called chromosomes, which are stained with a blue-glowing dye. The chromosomes have almost all of the cell's genetic material, its **DNA** (short for deoxyribonucleic acid). DNA is the substance of **genes**, the units of inheritance that transmit information from parents to offspring. Your blood group (A, B, AB, or O), for example, is the result of certain genes that you inherited from your parents.

DNA Structure and Function

Each chromosome contains one very long DNA molecule, with hundreds or thousands of genes arranged along its length. The genes encode the information necessary to build other molecules in the cell, most notably proteins. Proteins play structural roles and are also responsible for carrying out cellular work. They thus establish a cell's identity.

The DNA of chromosomes replicates as a cell prepares to divide, and each of the two cellular offspring inherits a complete set of genes, identical to that of the parent cell. Each of us began life as a single cell stocked with DNA inherited from our parents. Replication of that DNA with each round of cell division transmitted copies of the DNA to our trillions of cells. The DNA controls the development and maintenance of the entire organism and, indirectly, everything the organism does **(Figure 1.10)**. The DNA serves as a central database.

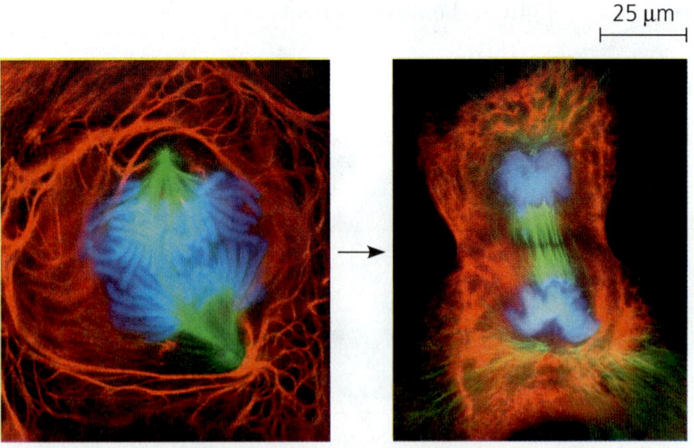

▲ **Figure 1.9 A lung cell from a newt divides into two smaller cells that will grow and divide again.**

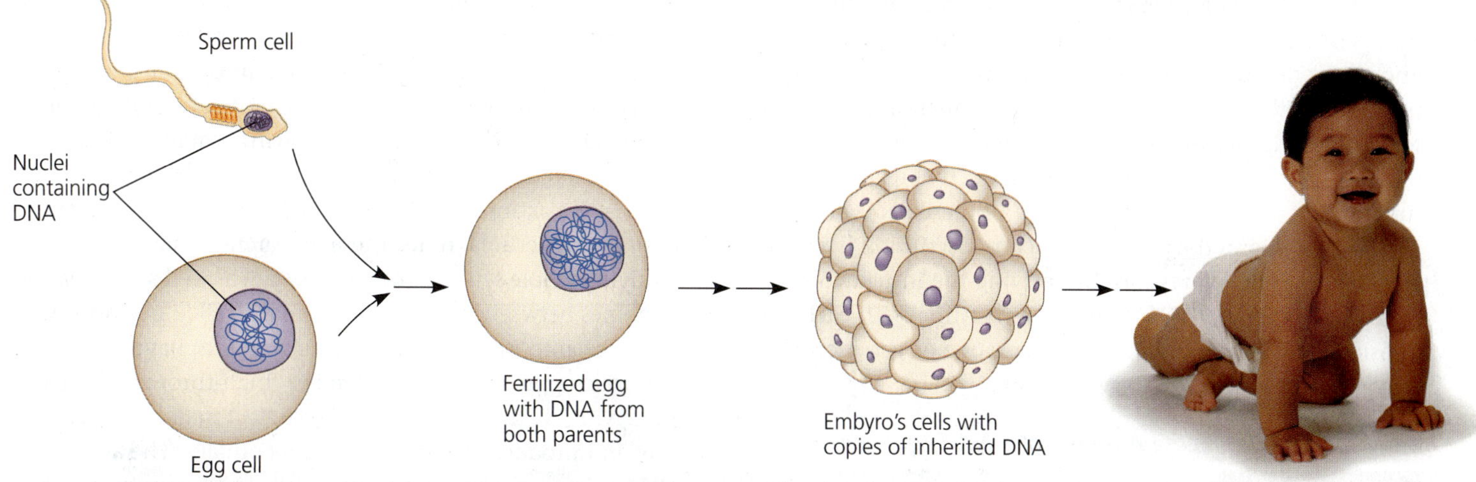

▲ **Figure 1.10 Inherited DNA directs development of an organism.**

The molecular structure of DNA accounts for its ability to store information. Each DNA molecule is made up of two long chains, called strands, arranged in a double helix. Each chain is made up of four kinds of chemical building blocks called nucleotides, abbreviated A, T, C, and G **(Figure 1.11)**. The way DNA encodes information is analogous to how we arrange the letters of the alphabet into precise sequences with specific meanings. The word *rat*, for example, evokes a rodent; the words *tar* and *art*, which contain the same letters, mean very different things. We can think of nucleotides as a four-letter alphabet of inheritance. Specific sequential arrangements of these four nucleotide letters encode the information in genes, which are typically hundreds or thousands of nucleotides long.

DNA provides the blueprints for making proteins, and proteins are the main players in building and maintaining the cell and carrying out its activities. For instance, the information carried in a bacterial gene may specify a certain protein in a bacterial cell membrane, while the information in a human gene may denote a protein hormone that stimulates growth. Other human proteins include proteins in a muscle cell that drive contraction and the defensive proteins called antibodies. Enzymes, which catalyze (speed up) specific chemical reactions, are mostly proteins and are crucial to all cells.

The DNA of genes controls protein production indirectly, using a related kind of molecule called RNA as an intermediary. The sequence of nucleotides along a gene is transcribed into RNA, which is then translated into a specific protein with a unique shape and function. This entire process, by which the information in a gene directs the production of a cellular product, is called **gene expression**. In translating genes into proteins, all forms of life employ essentially the same genetic code. A particular sequence of nucleotides says the same thing in one organism as it does in another. Differences between organisms reflect differences between their nucleotide sequences rather than between their genetic codes.

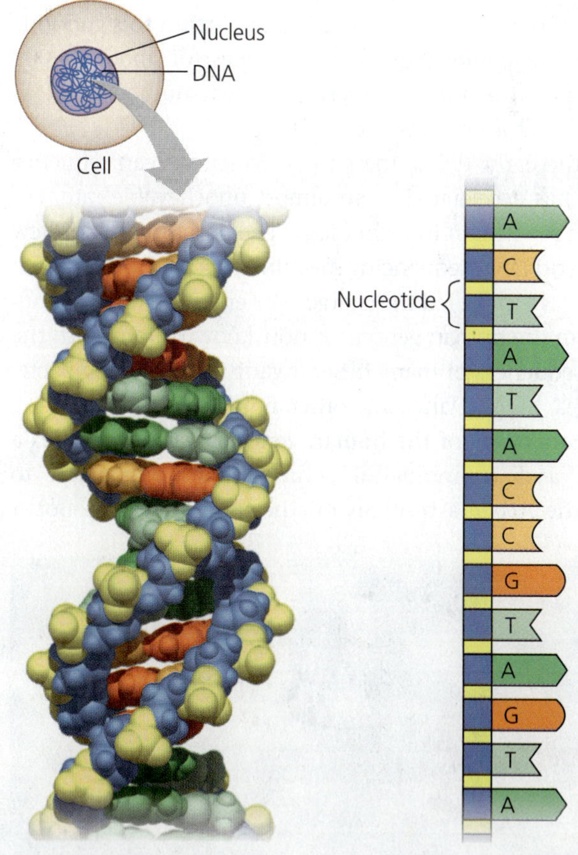

(a) DNA double helix. This model shows each atom in a segment of DNA. Made up of two long chains of building blocks called nucleotides, a DNA molecule takes the three-dimensional form of a double helix.

(b) Single strand of DNA. These geometric shapes and letters are simple symbols for the nucleotides in a small section of one chain of a DNA molecule. Genetic information is encoded in specific sequences of the four types of nucleotides. (Their names are abbreviated A, T, C, and G.)

▲ **Figure 1.11 DNA: The genetic material.**

Not all RNA molecules in the cell are translated into protein; some RNAs carry out other important tasks. We have known for decades that some types of RNA are actually components of the cellular machinery that manufactures proteins. Recently, scientists have discovered whole new classes of RNA that play other roles in the cell, such as regulating the functioning of protein-coding genes. All these RNAs are specified by genes, and the process of their transcription is also referred to as gene expression. By carrying the instructions for making proteins and RNAs and by replicating with each cell division, DNA ensures faithful inheritance of genetic information from generation to generation.

Genomics: Large-Scale Analysis of DNA Sequences

The entire "library" of genetic instructions that an organism inherits is called its **genome**. A typical human cell has two similar sets of chromosomes, and each set has DNA totaling about 3 billion nucleotide pairs. If the one-letter abbreviations for the nucleotides of one strand were written in letters the size of those you are now reading, the genetic text would fill about 600 books the size of this one. Within this genomic library of nucleotide sequences are genes for about 75,000 kinds of proteins and an as yet unknown number of RNA molecules that do not code for proteins.

Since the early 1990s, the pace at which we can sequence genomes has accelerated at an almost unbelievable rate, enabled by a revolution in technology. The development of new methods and DNA-sequencing machines, such as those shown in **Figure 1.12**, have led the charge. The entire sequence of nucleotides in the human genome is now known, along with the genome sequences of many other organisms, including bacteria, archaea, fungi, plants, and other animals.

The sequencing of the human genome was heralded as a scientific and technological achievement comparable to landing the *Apollo* astronauts on the moon in 1969. But it was only the beginning of an even bigger research endeavor, an effort to learn how the activities of the myriad proteins encoded by the DNA are coordinated in cells and whole organisms. To make sense of the deluge of data from genome-sequencing projects and the growing catalog of known protein functions, scientists are applying a systems approach at the cellular and molecular levels. Rather than investigating a single gene at a time, these researchers have shifted to studying whole sets of genes of a species as well as comparing genomes between species—an approach called **genomics**.

Three important research developments have made the genomic approach possible. One is "high-throughput" technology, tools that can analyze biological materials very rapidly and produce enormous amounts of data. The automatic DNA-sequencing machines that made the sequencing of the human genome possible are examples of high-throughput devices (see Figure 1.12). The second major development is **bioinformatics**, the use of computational tools to store, organize, and analyze the huge volume of data that result from high-throughput methods. The third key development is the formation of interdisciplinary research teams—melting pots of diverse specialists that may include computer scientists, mathematicians, engineers, chemists, physicists, and, of course, biologists from a variety of fields.

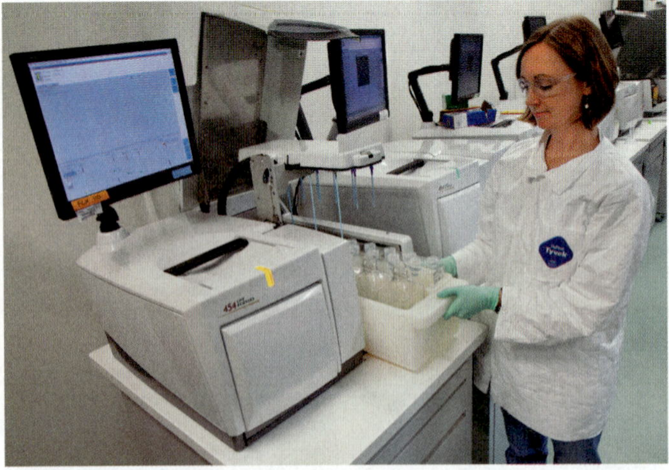

▲ **Figure 1.12 Biology as an information science.** Automatic DNA-sequencing machines and abundant computing power make the sequencing of genomes possible. This facility in Walnut Creek, California, is part of the Joint Genome Institute.

Theme: Feedback Mechanisms Regulate Biological Systems

Just as a coordinated control of traffic flow is necessary for a city to function smoothly, regulation of biological processes is crucial to the operation of living systems. Consider your muscles, for instance. When your muscle cells require more energy during exercise, they increase their consumption of the sugar molecules that serve as fuel. In contrast, when you rest, a different set of chemical reactions converts surplus sugar to storage molecules.

Like most of the cell's chemical processes, those that either decompose or store sugar are accelerated, or catalyzed, by proteins called enzymes. Each type of enzyme catalyzes a specific chemical reaction. In many cases, these reactions are linked into chemical pathways, each step with its own enzyme. How does the cell coordinate its various chemical pathways? In our example of sugar management, how does the cell match fuel supply to demand, regulating its opposing pathways of sugar consumption and storage? The key is the ability of many biological processes to self-regulate by a mechanism called feedback.

In feedback regulation, the output, or product, of a process regulates that very process. The most common form of regulation in living systems is **negative feedback**, in which accumulation of an end product of a process slows that process. For example, the cell's breakdown of sugar generates chemical energy in the form of a substance called ATP. When a cell makes more ATP than it can use, the excess ATP "feeds back"

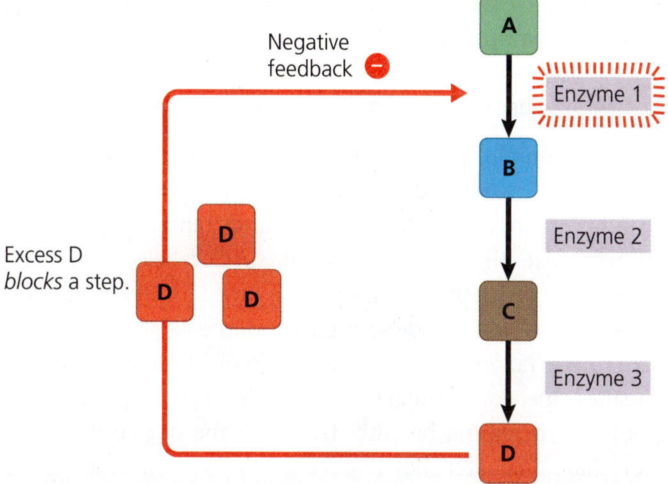

(a) Negative feedback. This three-step chemical pathway converts substance A to substance D. A specific enzyme catalyzes each chemical reaction. Accumulation of the final product (D) inhibits the first enzyme in the sequence, thus slowing down production of more D.

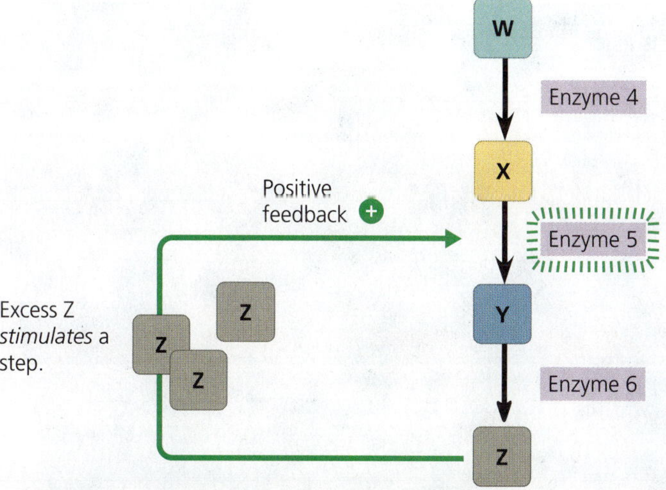

(b) Positive feedback. In a biochemical pathway regulated by positive feedback, a product stimulates an enzyme in the reaction sequence, increasing the rate of production of the product.

▲ **Figure 1.13 Regulation by feedback mechanisms.**

? *What would happen to the feedback system if enzyme 2 were missing?*

and inhibits an enzyme near the beginning of the pathway **(Figure 1.13a)**.

Though less common than processes regulated by negative feedback, there are also many biological processes regulated by **positive feedback**, in which an end product *speeds up* its own production **(Figure 1.13b)**. The clotting of your blood in response to injury is an example. When a blood vessel is damaged, structures in the blood called platelets begin to aggregate at the site. Positive feedback occurs as chemicals released by the platelets attract *more* platelets. The platelet pileup then initiates a complex process that seals the wound with a clot.

Feedback is a regulatory motif common to life at all levels, from the molecular level to ecosystems and the biosphere. Such regulation is an example of the integration that makes living systems much greater than the sum of their parts.

Evolution, the Overarching Theme of Biology

Having considered all the other themes that run through this book, let's now turn to biology's core theme—evolution. Evolution is the one idea that makes sense of everything we know about living organisms. Life has been evolving on Earth for billions of years, resulting in a vast diversity of past and present organisms. But along with the diversity we find many shared features. For example, while the sea horse, jackrabbit, hummingbird, crocodile, and giraffes in Figure 1.3 look very different, their skeletons are basically similar. The scientific explanation for this unity and diversity—and for the suitability of organisms for their environments—is evolution: the idea that the organisms living on Earth today are the modified descendants of common ancestors. In other words, we can explain traits shared by two organisms with the idea that they have descended from a common ancestor, and we can account for differences with the idea that heritable changes have occurred along the way. Many kinds of evidence support the occurrence of evolution and the theory that describes how it takes place. In the next section, we'll consider the fundamental concept of evolution in greater detail.

CONCEPT CHECK 1.1

1. For each biological level in Figure 1.4, write a sentence that includes the next "lower" level. Example: "A community consists of *populations* of the various species inhabiting a specific area."
2. What theme or themes are exemplified by (a) the sharp spines of a porcupine, (b) the cloning of a plant from a single cell, and (c) a hummingbird using sugar to power its flight?
3. **WHAT IF?** For each theme discussed in this section, give an example not mentioned in the book.

For suggested answers, see Appendix A.

CONCEPT **1.2**

The Core Theme: Evolution accounts for the unity and diversity of life

EVOLUTION The list of biological themes discussed in Concept 1.1 is not absolute; some people might find a shorter or longer list more useful. There is consensus among biologists, however, as to the core theme of biology: It is evolution. To quote one of the founders of modern evolutionary theory, Theodosius Dobzhansky, "Nothing in biology makes sense except in the light of evolution."

In addition to encompassing a hierarchy of size scales from molecules to the biosphere, biology extends across the great diversity of species that have ever lived on Earth. To understand Dobzhansky's statement, we need to discuss how biologists think about this vast diversity.

Classifying the Diversity of Life

Diversity is a hallmark of life. Biologists have so far identified and named about 1.8 million species. To date, this diversity of life is known to include at least 100,000 species of fungi, 290,000 plant species, 52,000 vertebrate species (animals with backbones), and 1 million insect species (more than half of all known forms of life)—not to mention the myriad types of single-celled organisms. Researchers identify thousands of additional species each year. Estimates of the total number of species range from about 10 million to over 100 million. Whatever the actual number, the enormous variety of life gives biology a very broad scope. Biologists face a major challenge in attempting to make sense of this variety.

Grouping Species: The Basic Idea

There is a human tendency to group diverse items according to their similarities and their relationships to each other. For instance, we may speak of "squirrels" and "butterflies," though we recognize that many different species belong to each group. We may even sort groups into broader categories, such as rodents (which include squirrels) and insects (which include butterflies). Taxonomy, the branch of biology that names and classifies species, formalizes this ordering of species into groups of increasing breadth, based on the degree to which they share characteristics **(Figure 1.14)**. You will learn more about the details of this taxonomic scheme in Chapter 26. For

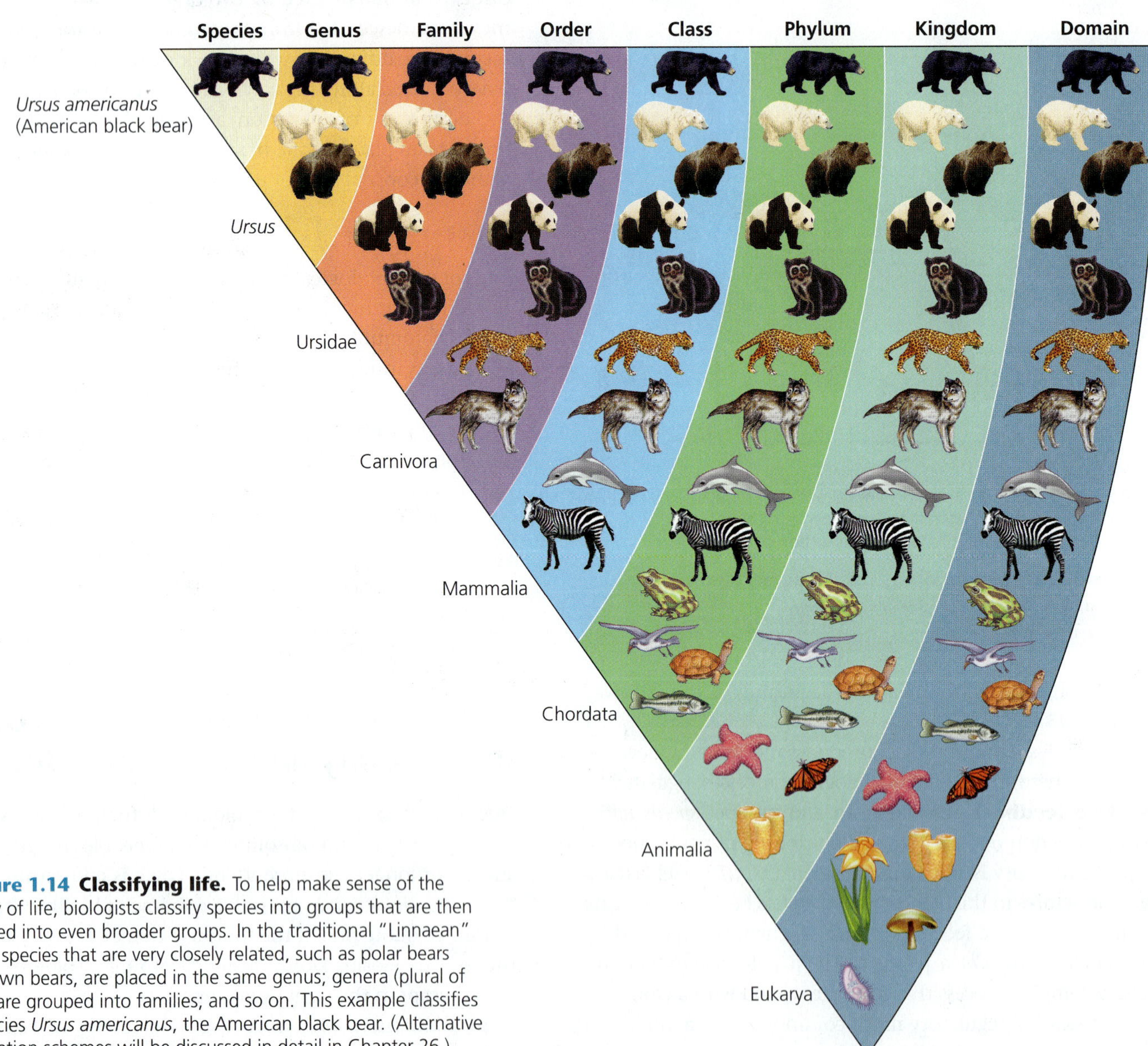

▲ **Figure 1.14 Classifying life.** To help make sense of the diversity of life, biologists classify species into groups that are then combined into even broader groups. In the traditional "Linnaean" system, species that are very closely related, such as polar bears and brown bears, are placed in the same genus; genera (plural of genus) are grouped into families; and so on. This example classifies the species *Ursus americanus*, the American black bear. (Alternative classification schemes will be discussed in detail in Chapter 26.)

now, we will focus on the big picture by considering the broadest units of classification, kingdoms and domains.

The Three Domains of Life

Historically, scientists have classified the diversity of lifeforms into kingdoms and finer groupings by careful comparisons of structure, function, and other obvious features. In the last few decades, new methods of assessing species relationships, such as comparisons of DNA sequences, have led to an ongoing reevaluation of the number and boundaries of kingdoms. Researchers have proposed anywhere from six kingdoms to dozens of kingdoms. While debate continues at the kingdom level, there is consensus among biologists that the kingdoms of life can be grouped into three even higher levels of classification called domains. The three domains are named Bacteria, Archaea, and Eukarya **(Figure 1.15)**.

The organisms making up two of the three domains—domain **Bacteria** and domain **Archaea**—are all prokaryotic. Most prokaryotes are single-celled and microscopic. Previously, bacteria and archaea were combined in a single kingdom because they shared the prokaryotic form of cell structure. But much evidence now supports the view that bacteria and archaea represent two very distinct branches of prokaryotic life, different in key ways that you'll learn about in Chapter 27. There is also evidence that archaea are at least as closely related to eukaryotic organisms as they are to bacteria.

All the eukaryotes (organisms with eukaryotic cells) are now grouped in domain **Eukarya**. This domain includes three kingdoms of multicellular eukaryotes: kingdoms Plantae,

▼ Figure 1.15 The three domains of life.

(a) Domain Bacteria

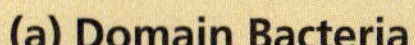

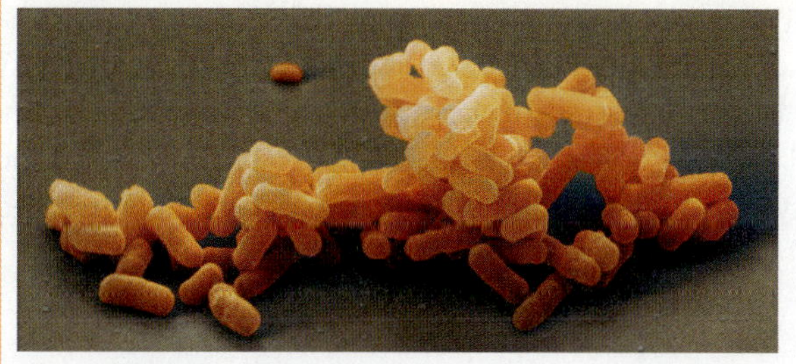

Bacteria are the most diverse and widespread prokaryotes and are now classified into multiple kingdoms. Each rod-shaped structure in this photo is a bacterial cell.

(b) Domain Archaea

Many of the prokaryotes known as **archaea** live in Earth's extreme environments, such as salty lakes and boiling hot springs. Domain Archaea includes multiple kingdoms. Each round structure in this photo is an archaeal cell.

(c) Domain Eukarya

▲ **Kingdom Plantae** consists of terrestrial multicellular eukaryotes (land plants) that carry out photosynthesis, the conversion of light energy to the chemical energy in food.

► **Kingdom Fungi** is defined in part by the nutritional mode of its members (such as this mushroom), which absorb nutrients from outside their bodies.

◄ **Kingdom Animalia** consists of multicellular eukaryotes that ingest other organisms.

100 µm

► **Protists** are mostly unicellular eukaryotes and some relatively simple multicellular relatives. Pictured here is an assortment of protists inhabiting pond water. Scientists are currently debating how to classify protists in a way that accurately reflects their evolutionary relationships.

Fungi, and Animalia. These three kingdoms are distinguished partly by their modes of nutrition. Plants produce their own sugars and other food molecules by photosynthesis. Fungi absorb dissolved nutrients from their surroundings; many decompose dead organisms and organic wastes (such as leaf litter and animal feces) and absorb nutrients from these sources. Animals obtain food by ingestion, which is the eating and digesting of other organisms. Animalia is, of course, the kingdom to which we belong. But neither animals, plants, nor fungi are as numerous or diverse as the single-celled eukaryotes we call protists. Although protists were once placed in a single kingdom, biologists now realize that they do not form a single natural group of species. And recent evidence shows that some protist groups are more closely related to multicellular eukaryotes such as animals and fungi than they are to each other. Thus, the recent taxonomic trend has been to split the protists into several groups.

Unity in the Diversity of Life

As diverse as life is, it also displays remarkable unity. Earlier we mentioned both the similar skeletons of different vertebrate animals and the universal genetic language of DNA (the genetic code). In fact, similarities between organisms are evident at all levels of the biological hierarchy. For example, unity is obvious in many features of cell structure **(Figure 1.16)**.

How can we account for life's dual nature of unity and diversity? The process of evolution, explained next, illuminates both the similarities and differences in the world of life and introduces another dimension of biology: historical time.

Charles Darwin and the Theory of Natural Selection

The history of life, as documented by fossils and other evidence, is the saga of a changing Earth billions of years old, inhabited by an evolving cast of living forms **(Figure 1.17)**. This evolutionary view of life came into sharp focus in November 1859, when Charles Robert Darwin published one of the most important and influential books ever written. Entitled *On the Origin of Species by Means of Natural Selection*, Darwin's book was an immediate bestseller and soon made "Darwinism," as it was dubbed at the time, almost synonymous with the concept of evolution **(Figure 1.18)**.

The Origin of Species articulated two main points. The first point was that contemporary species arose from a succession of ancestors, an idea that Darwin supported with a large amount of evidence. (We will discuss the evidence for evolution in detail in Chapter 22.) Darwin called this evolutionary history of species "descent with modification." It was an insightful phrase, as it captured the duality of life's unity and diversity—unity in the kinship among species that descended

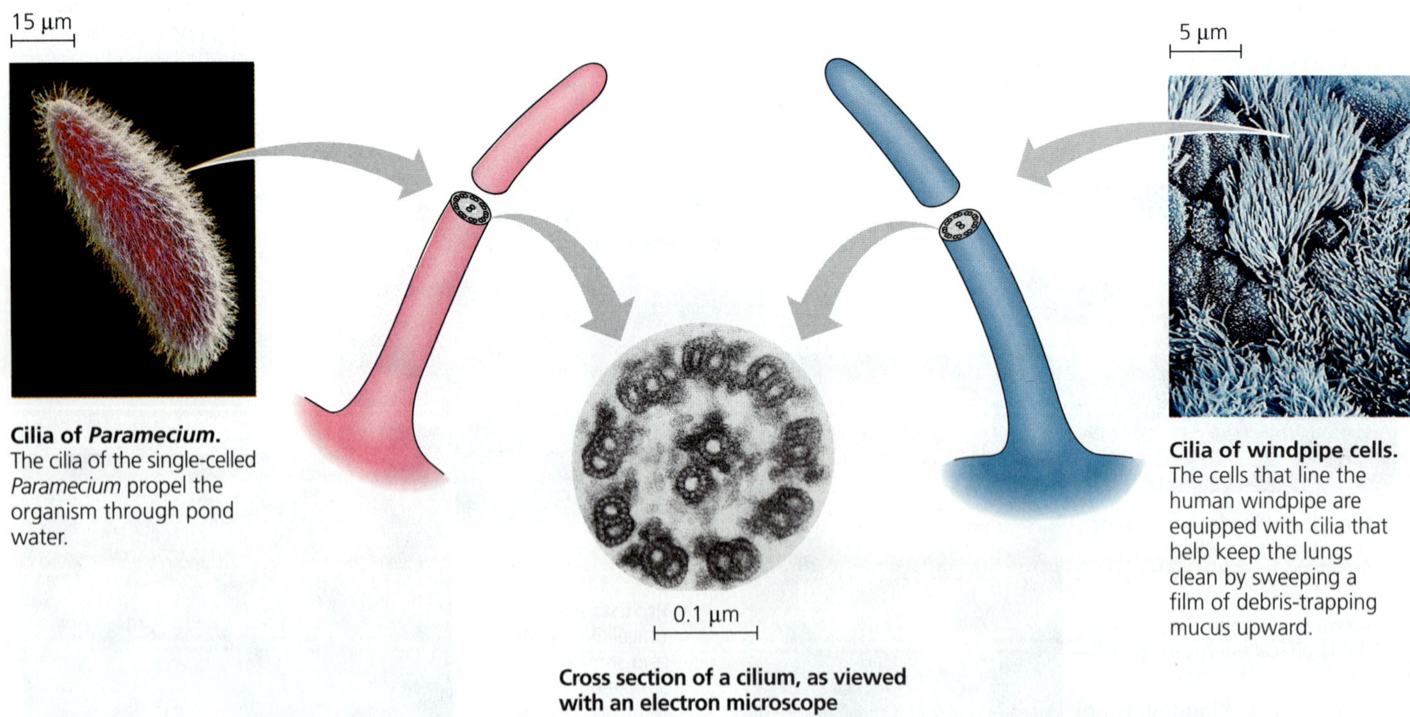

▲ **Figure 1.16 An example of unity underlying the diversity of life: the architecture of cilia in eukaryotes.** Cilia (singular, *cilium*) are extensions of cells that function in locomotion. They occur in eukaryotes as diverse as *Paramecium* and humans. Even organisms so different share a common architecture for their cilia, which have an elaborate system of tubules that is striking in cross-sectional views.

▲ **Figure 1.17 Digging into the past.** Paleontologists carefully excavate the hind leg of a long-necked dinosaur (*Rapetosaurus krausei*) from rocks in Madagascar.

from common ancestors, diversity in the modifications that evolved as species branched from their common ancestors **(Figure 1.19)**. Darwin's second main point was a proposed mechanism for descent with modification. He called this evolutionary mechanism "natural selection."

Darwin synthesized his theory of natural selection from observations that by themselves were neither new nor profound. Others had the pieces of the puzzle, but Darwin saw how they fit together. He started with the following three observations from nature: First, individuals in a population vary in their traits, many of which seem to be heritable (passed on from parents to offspring). Second, a population can produce far more offspring than can survive to produce offspring of their own. With more individuals than the environment is able to support, competition is inevitable. Third, species generally suit their environments—in other words, they are adapted to their environments. For instance, a common adaptation among birds with tough seeds as their major food source is that they have especially strong beaks.

▲ **Figure 1.18 Charles Darwin as a young man.**

Darwin made inferences from these observations to arrive at his theory of evolution. He reasoned that individuals with inherited traits that are best suited to the local environment are more likely to survive and reproduce than less suited individuals. Over many generations, a higher and higher proportion of individuals in a population will have the advantageous traits. Evolution occurs as the unequal reproductive success of individuals ultimately leads to adaptation to their environment, as long as the environment remains the same.

Darwin called this mechanism of evolutionary adaptation **natural selection** because the natural environment "selects" for the propagation of certain traits among naturally occurring variant traits in the population. The example

◄ **Figure 1.19 Unity and diversity in the orchid family.** These three orchids are variations on a common floral theme. For example, each of these flowers has a liplike petal that helps attract pollinating insects and provides a landing platform for the pollinators.

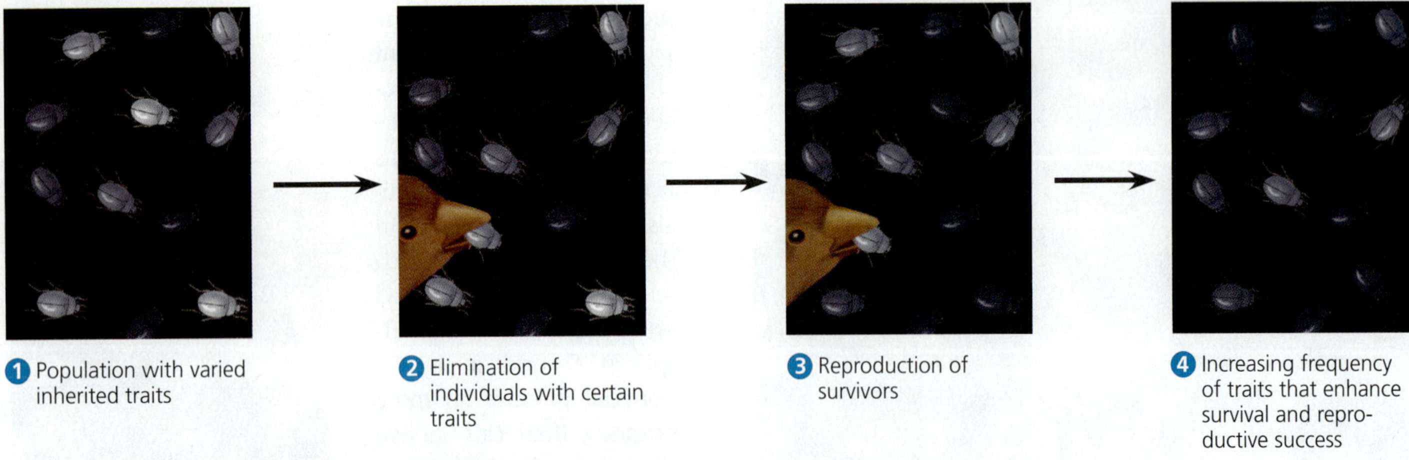

▲ **Figure 1.20 Natural selection.** This imaginary beetle population has colonized a locale where the soil has been blackened by a recent brush fire. Initially, the population varies extensively in the inherited coloration of the individuals, from very light gray to charcoal. For hungry birds that prey on the beetles, it is easiest to spot the beetles that are lightest in color.

in **Figure 1.20** illustrates the ability of natural selection to "edit" a population's heritable variations in color. We see the products of natural selection in the exquisite adaptations of various organisms to the special circumstances of their way of life and their environment. The wings of the bat shown in **Figure 1.21** are an excellent example of adaptation.

The Tree of Life

Take another look at the skeletal architecture of the bat's wings in Figure 1.21. These forelimbs, though adapted for flight, actually have all the same bones, joints, nerves, and blood vessels found in other limbs as diverse as the human arm, the horse's foreleg, and the whale's flipper. Indeed, all mammalian forelimbs are anatomical variations of a common architecture, much as the flowers in Figure 1.19 are variations on an underlying "orchid" theme. Such examples of kinship connect life's unity in diversity to the Darwinian concept of descent with modification. In this view, the unity of mammalian limb anatomy reflects inheritance of that structure from a common ancestor—the "prototype" mammal from which all other mammals descended. The diversity of mammalian forelimbs results from modification by natural selection operating over millions of generations in different environmental contexts. Fossils and other evidence corroborate anatomical unity in supporting this view of mammalian descent from a common ancestor.

▲ **Figure 1.21 Evolutionary adaptation.** Bats, the only mammals capable of active flight, have wings with webbing between extended "fingers." In the Darwinian view of life, such adaptations are refined over time by natural selection.

Darwin proposed that natural selection, by its cumulative effects over long periods of time, could cause an ancestral species to give rise to two or more descendant species. This could occur, for example, if one population fragmented into several subpopulations isolated in different environments. In these separate arenas of natural selection, one species could gradually radiate into multiple species as the geographically isolated populations adapted over many generations to different sets of environmental factors.

The "family tree" of 14 finches in **Figure 1.22** illustrates a famous example of adaptive radiation of new species from a common ancestor. Darwin collected specimens of these birds during his 1835 visit to the remote Galápagos Islands, 900 kilometers (km) off the Pacific coast of South America. These relatively young, volcanic islands are home to many species of plants and animals found nowhere else in the world, though most Galápagos organisms are clearly related to species on the South American mainland. After volcanism built the Galápagos several million years ago, finches probably diversified on the various islands from an ancestral finch species that by chance reached the archipelago from elsewhere. (Once thought to have originated on the mainland of South America like many Galápagos organisms, the ancestral finches are now thought to have come from the West Indies—islands of the Caribbean that were once much closer to the Galápagos than they are now.)

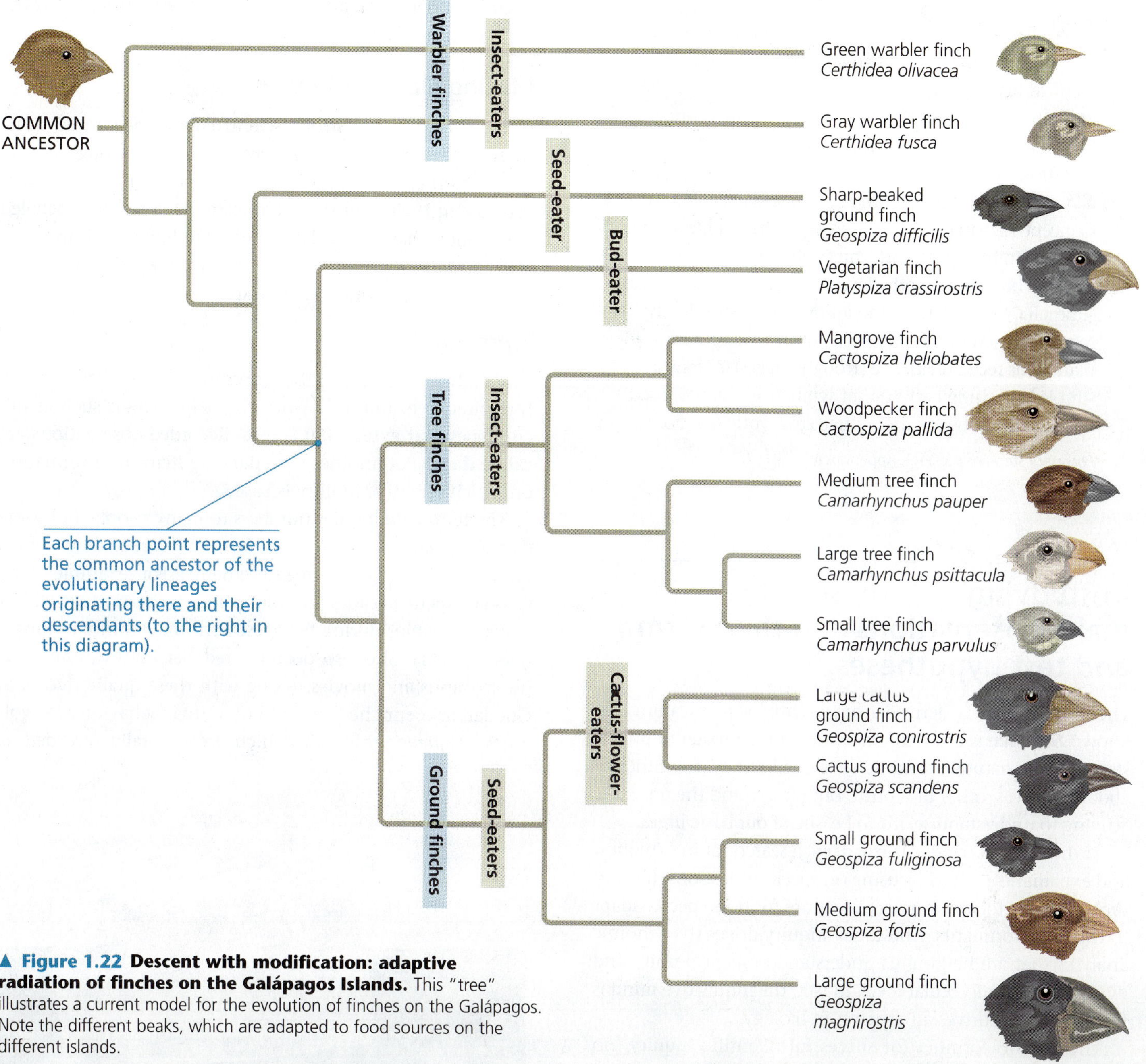

▲ **Figure 1.22 Descent with modification: adaptive radiation of finches on the Galápagos Islands.** This "tree" illustrates a current model for the evolution of finches on the Galápagos. Note the different beaks, which are adapted to food sources on the different islands.

Years after Darwin's collection of Galápagos finches, researchers began to sort out the relationships among the finch species, first from anatomical and geographic data and more recently with the help of DNA sequence comparisons.

Biologists' diagrams of evolutionary relationships generally take treelike forms, though today biologists usually turn the trees sideways as in Figure 1.22. Tree diagrams make sense: Just as an individual has a genealogy that can be diagrammed as a family tree, each species is one twig of a branching tree of life extending back in time through ancestral species more and more remote. Species that are very similar, such as the Galápagos finches, share a common ancestor at a relatively recent branch point on the tree of life. But through an ancestor that lived much farther back in time, finches are related to sparrows, hawks, penguins, and all other birds. And birds, mammals, and all other vertebrates share a common ancestor even more ancient. We find evidence of still broader relationships in such similarities as the identical construction of all eukaryotic cilia (see Figure 1.16). Trace life back far enough, and there are only fossils of the primeval prokaryotes that inhabited Earth over 3.5 billion years ago. We can recognize their vestiges in our own cells—in the universal genetic code, for example. All of life is connected through its long evolutionary history.

CONCEPT CHECK 1.2

1. How is a mailing address analogous to biology's hierarchical taxonomic system?
2. Explain why "editing" is an appropriate metaphor for how natural selection acts on a population's heritable variation.
3. **WHAT IF?** The three domains you learned about in Concept 1.2 can be represented in the tree of life as the three main branches, with three subbranches on the eukaryotic branch being the kingdoms Plantae, Fungi, and Animalia. What if fungi and animals are more closely related to each other than either of these kingdoms is to plants—as recent evidence strongly suggests? Draw a simple branching pattern that symbolizes the proposed relationship between these three eukaryotic kingdoms.

For suggested answers, see Appendix A.

CONCEPT 1.3

In studying nature, scientists make observations and then form and test hypotheses

The word *science* is derived from a Latin verb meaning "to know." **Science** is a way of knowing—an approach to understanding the natural world. It developed out of our curiosity about ourselves, other life-forms, our planet, and the universe. Striving to understand seems to be one of our basic urges.

At the heart of science is **inquiry**, a search for information and explanation, often focusing on specific questions. Inquiry drove Darwin to seek answers in nature for how species adapt to their environments. And today inquiry drives the genomic analyses that are helping us understand biological unity and diversity at the molecular level. In fact, the inquisitive mind is the engine that drives all progress in biology.

There is no formula for successful scientific inquiry, no single scientific method with a rule book that researchers must rigidly follow. As in all quests, science includes elements of challenge, adventure, and luck, along with careful planning, reasoning, creativity, cooperation, competition, patience, and the persistence to overcome setbacks. Such diverse elements of inquiry make science far less structured than most people realize. That said, it is possible to distill certain characteristics that help to distinguish science from other ways of describing and explaining nature.

Scientists attempt to understand how natural phenomena work using a process of inquiry that includes making observations, forming logical hypotheses, and testing them. The process is necessarily repetitive: In testing a hypothesis, more observations may force formation of a new hypothesis or revision of the original one, and further testing. In this way, scientists circle closer and closer to their best estimation of the laws governing nature.

Making Observations

In the course of their work, scientists describe natural structures and processes as accurately as possible through careful observation and analysis of data. The observations are often valuable in their own right. For example, a series of detailed observations have shaped our understanding of cell structure, and another set of observations are currently expanding our databases of genomes of diverse species.

Types of Data

Observation is the use of the senses to gather information, either directly or indirectly with the help of tools such as microscopes that extend our senses. Recorded observations are called **data**. Put another way, data are items of information on which scientific inquiry is based.

The term *data* implies numbers to many people. But some data are *qualitative*, often in the form of recorded descriptions rather than numerical measurements. For example, Jane Goodall spent decades recording her observations of chimpanzee behavior during field research in a Tanzanian jungle **(Figure 1.23)**. She also documented her observations with photographs and movies. Along with these qualitative data, Goodall also enriched the field of animal behavior with volumes of *quantitative* data, which are generally recorded as

▲ **Figure 1.23 Jane Goodall collecting qualitative data on chimpanzee behavior.** Goodall recorded her observations in field notebooks, often with sketches of the animals' behavior.

measurements. Skim through any of the scientific journals in your college library, and you'll see many examples of quantitative data organized into tables and graphs.

Inductive Reasoning

Collecting and analyzing observations can lead to important conclusions based on a type of logic called **inductive reasoning**. Through induction, we derive generalizations from a large number of specific observations. "The sun always rises in the east" is an example. And so is "All organisms are made of cells." The latter generalization, part of the so-called cell theory, was based on two centuries of microscopic observations by biologists of cells in diverse biological specimens. Careful observations and data analyses, along with the generalizations reached by induction, are fundamental to our understanding of nature.

Forming and Testing Hypotheses

Observations and inductive reasoning stimulate us to seek natural causes and explanations for those observations. What *caused* the diversification of finches on the Galápagos Islands? What *causes* the roots of a plant seedling to grow downward and the leaf-bearing shoot to grow upward? What *explains* the generalization that the sun always rises in the east? In science, such inquiry usually involves the proposing and testing of hypothetical explanations—that is, hypotheses.

The Role of Hypotheses in Inquiry

In science, a **hypothesis** is a tentative answer to a well-framed question—an explanation on trial. It is usually a rational accounting for a set of observations, based on the available data and guided by inductive reasoning. A scientific hypothesis leads to predictions that can be tested by making additional observations or by performing experiments.

We all use hypotheses in solving everyday problems. Let's say, for example, that your flashlight fails during a camp-out. That's an observation. The question is obvious: Why doesn't the flashlight work? Two reasonable hypotheses based on your experience are that (1) the batteries in the flashlight are dead or (2) the bulb is burnt out. Each of these alternative hypotheses leads to predictions you can test with experiments. For example, the dead-battery hypothesis predicts that replacing the batteries will fix the problem. **Figure 1.24** diagrams this campground inquiry. Of course, we rarely dissect our thought processes this way when we are solving a problem using hypotheses, predictions, and experiments. But the hypothesis-based nature of science clearly has its origins in the human tendency to figure things out by trial and error.

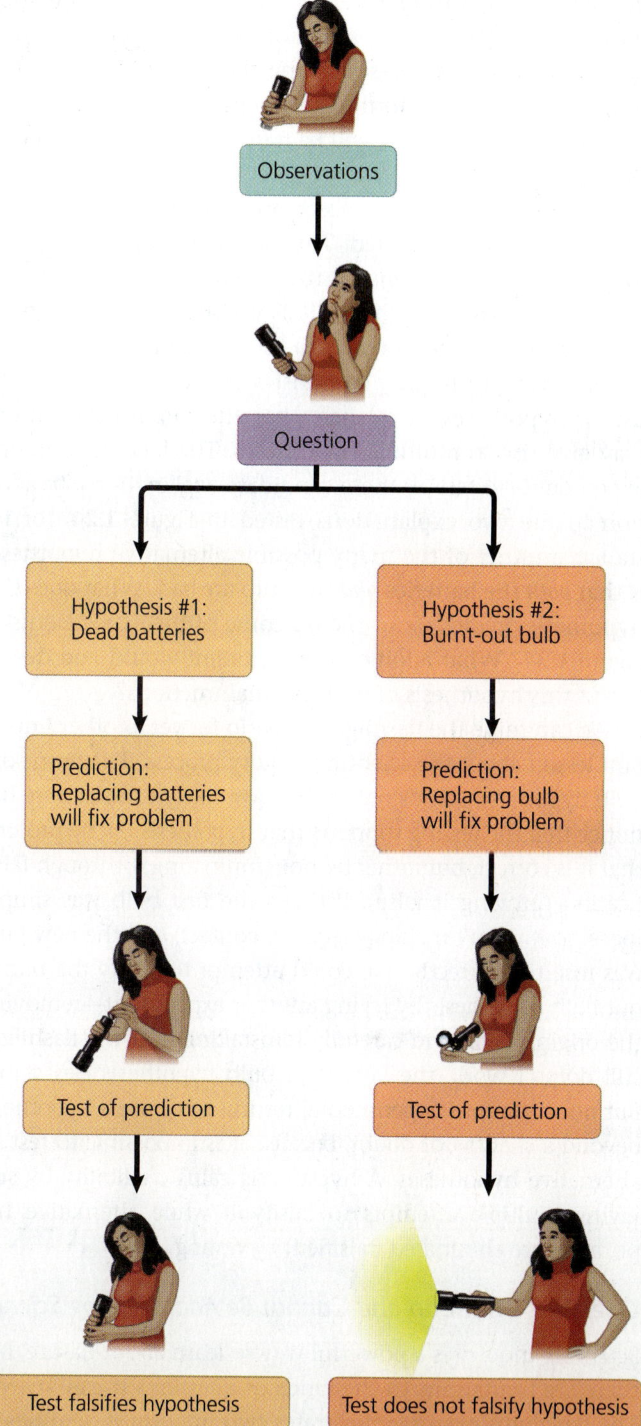

▲ **Figure 1.24 A campground example of hypothesis-based inquiry.**

Deductive Reasoning and Hypothesis Testing

A type of logic called deduction is built into the use of hypotheses in science. Deduction contrasts with induction, which, remember, is reasoning from a set of specific observations to reach a general conclusion—a process that feeds into hypothesis formation. **Deductive reasoning** is generally used after the hypothesis has been developed and involves logic that flows in the opposite direction, from the general to the specific. From general premises, we extrapolate to the specific results we should expect if the premises are true. If all organisms are made of cells (premise 1), and humans are organisms (premise 2), then humans are composed of cells (deductive prediction about a specific case).

When using hypotheses in the scientific process, deductions usually take the form of predictions of experimental or observational results that will be found if a particular hypothesis (premise) is correct. We then test the hypothesis by carrying out the experiments or observations to see whether or not the results are as predicted. This deductive testing takes the form of "*If . . . then*" logic. In the case of the flashlight example: *If* the dead-battery hypothesis is correct and you replace the batteries with new ones, *then* the flashlight should work.

The flashlight inquiry demonstrates a key point about the use of hypotheses in science: that the initial observations may give rise to multiple hypotheses. The ideal is to design experiments to test all these candidate explanations. In addition to the two explanations tested in Figure 1.24, for instance, another of the many possible alternative hypotheses is that *both* the batteries *and* the bulb are bad. What does this hypothesis predict about the outcome of the experiments in Figure 1.24? What additional experiment would you design to test this hypothesis of multiple malfunctions?

We can mine the flashlight scenario for yet another important lesson about the scientific inquiry process. The burnt-out bulb hypothesis stands out as the most likely explanation, but notice that the testing supports that hypothesis *not* by proving that it is correct, but rather by not eliminating it through falsification (proving it false). Perhaps the first bulb was simply loose, so it wasn't making electrical contact, and the new bulb was inserted correctly. We could attempt to falsify the burnt-out bulb hypothesis by trying another experiment—removing the original bulb and carefully reinstalling it. If the flashlight still doesn't work, the burnt-out bulb hypothesis can stand. But no amount of experimental testing can *prove* a hypothesis beyond a shadow of doubt, because it is impossible to test *all* alternative hypotheses. A hypothesis gains credibility by surviving multiple attempts to falsify it while alternative hypotheses are eliminated (falsified) by testing.

Questions That Can and Cannot Be Addressed by Science

Scientific inquiry is a powerful way to learn about nature, but there are limitations to the kinds of questions it can answer. The flashlight example illustrates two important qualities of scientific hypotheses. First, a hypothesis must be *testable*; there must be some way to check the validity of the idea. Second, a hypothesis must be *falsifiable*; there must be some observation or experiment that could reveal if such an idea is actually *not* true. The hypothesis that dead batteries are the sole cause of the broken flashlight could be falsified by replacing the old batteries with new ones and finding that the flashlight still doesn't work.

Not all hypotheses meet the criteria of science: You wouldn't be able to devise a test to falsify the hypothesis that invisible campground ghosts are fooling with your flashlight! Because science requires natural explanations for natural phenomena, it can neither support nor falsify hypotheses that angels, ghosts, or spirits, whether benevolent or evil, cause storms, rainbows, illnesses, and cures. Such supernatural explanations are simply outside the bounds of science, as are religious matters, which are issues of personal faith.

The Flexibility of the Scientific Method

The flashlight example of Figure 1.24 traces an idealized process of inquiry called *the scientific method*. We can recognize the elements of this process in most of the research articles published by scientists, but rarely in such structured form. Very few scientific inquiries adhere rigidly to the sequence of steps prescribed by the "textbook" scientific method. For example, a scientist may start to design an experiment, but then backtrack upon realizing that more preliminary observations are necessary. In other cases, puzzling observations simply don't prompt well-defined questions until other research places those observations in a new context. For example, Darwin collected specimens of the Galápagos finches, but it wasn't until years later, as the idea of natural selection began to gel, that biologists began asking key questions about the history of those birds.

Moreover, scientists sometimes redirect their research when they realize they have been asking the wrong question. For example, in the early 20th century, much research on schizophrenia and manic-depressive disorder (now called bipolar disorder) got sidetracked by focusing too much on the question of how life experiences might cause these serious maladies. Research on the causes and potential treatments became more productive when it was refocused on questions of how certain chemical imbalances in the brain contribute to mental illness. To be fair, we acknowledge that such twists and turns in scientific inquiry become more evident with the advantage of historical perspective.

It is important for you to get some experience with the power of the scientific method—by using it for some of the laboratory inquiries in your biology course, for example. But it is also important to avoid stereotyping science as a lock-step adherence to this method.

A Case Study in Scientific Inquiry: Investigating Mimicry in Snake Populations

Now that we have highlighted the key features of scientific inquiry—making observations and forming and testing hypotheses—you should be able to recognize these features in a case study of actual scientific research.

The story begins with a set of observations and inductive generalizations. Many poisonous animals are brightly colored, often with distinctive patterns that stand out against the background. This is called *warning coloration* because it apparently signals "dangerous species" to potential predators. But there are also mimics. These imposters look like poisonous species but are actually harmless. A question that follows from these observations is: What is the function of such mimicry? A reasonable hypothesis is that the "deception" is an evolutionary adaptation that reduces the harmless animal's risk of being eaten because predators mistake it for the poisonous species. This hypothesis was first formulated by British scientist Henry Bates in 1862.

As obvious as this hypothesis may seem, it has been relatively difficult to test, especially with field experiments. But in 2001, biologists David and Karin Pfennig, of the University of North Carolina, along with William Harcombe, an undergraduate, designed a simple but elegant set of field experiments to test Bates's mimicry hypothesis.

The team investigated a case of mimicry among snakes that live in North and South Carolina **(Figure 1.25)**. A venomous snake called the eastern coral snake has warning coloration: bold, alternating rings of red, yellow (or white), and black. (The word *venomous* is used when a poisonous species delivers their poison actively, by stinging, stabbing, or biting.) Predators rarely attack these coral snakes. It is unlikely that the predators learn this avoidance behavior by trial and error, as a first encounter with a coral snake is usually deadly. In areas where coral snakes live, natural selection has apparently increased the frequency of predators that have inherited an instinctive avoidance of the coral snake's coloration. A nonvenomous snake named the scarlet kingsnake mimics the ringed coloration of the coral snake.

Both types of snakes live in the Carolinas, but the kingsnakes' geographic range also extends into regions where no coral snakes are found (see Figure 1.25). The geographic distribution of the snakes made it possible to test the key prediction of the mimicry hypothesis. Avoiding snakes with warning coloration is an adaptation we expect to be present only in predator populations that evolved in areas where the venomous coral snakes are present. Therefore, mimicry should help protect kingsnakes from predators *only in regions where coral snakes also live*. The mimicry hypothesis predicts that predators adapted to the warning coloration of coral snakes will attack kingsnakes less frequently than will predators in areas where coral snakes are absent.

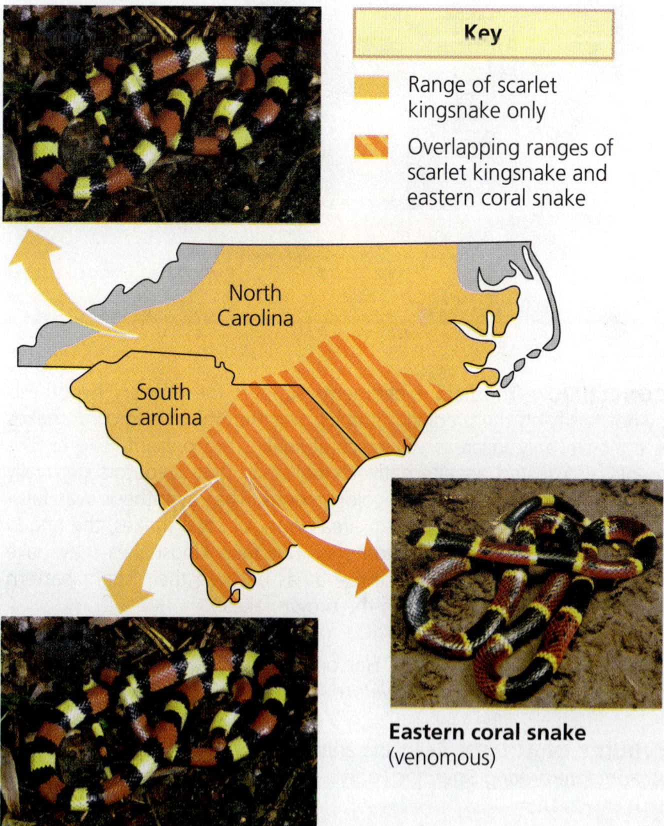

▲ **Figure 1.25 The geographic ranges of a venomous snake and its mimic.** The scarlet kingsnake (*Lampropeltis triangulum*) mimics the warning coloration of the venomous eastern coral snake (*Micrurus fulvius*).

Field Experiments with Artificial Snakes

To test the prediction, Harcombe made hundreds of artificial snakes out of wire covered with plasticine. He fashioned two versions of fake snakes: an *experimental group* with the red, black, and white ring pattern of kingsnakes and a *control group* of plain brown artificial snakes as a basis of comparison **(Figure 1.26)**.

The researchers placed equal numbers of the two types of artificial snakes in field sites throughout North and South

(a) Artificial kingsnake

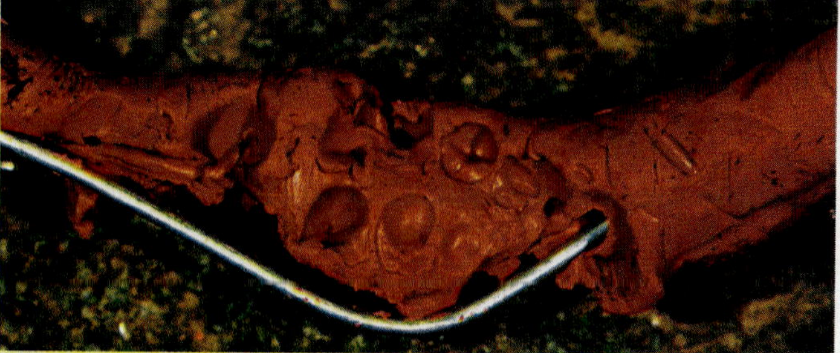

(b) Brown artificial snake that has been attacked

▲ **Figure 1.26 Artificial snakes used in field experiments to test the mimicry hypothesis.** A bear has chewed on the brown artificial snake in **(b)**.

Carolina, including the region where coral snakes are absent. After four weeks, the scientists retrieved the fake snakes and recorded how many had been attacked by looking for bite or claw marks. The most common predators were foxes, coyotes, and raccoons, but black bears also attacked some of the artificial snakes (see Figure 1.26b).

The data fit the key prediction of the mimicry hypothesis. Compared to the brown artificial snakes, the ringed artificial snakes were attacked by predators less frequently only in field sites within the geographic range of the venomous coral snakes. **Figure 1.27** summarizes the field experiments that the researchers carried out. This figure also introduces a format we will use throughout the book for other examples of biological inquiry.

Experimental Controls and Repeatability

The snake mimicry experiment is an example of a **controlled experiment**, one that is designed to compare an experimental group (the artificial kingsnakes, in this case) with a control group (the brown artificial snakes). Ideally, the experimental and control groups differ only in the one factor the experiment is designed to test—in our example, the effect of the snakes' coloration on the behavior of predators. Without the control group, the researchers would not have been able to rule out other factors as causes of the more frequent attacks on the artificial kingsnakes—such as different numbers of predators or different temperatures in the different test areas. The clever experimental design left coloration as the only factor that could account for the low predation rate on the artificial kingsnakes placed within the range of coral snakes. It was not the absolute number of attacks on the artificial kingsnakes that counted, but the difference between that number and the number of attacks on the brown snakes.

A common misconception is that the term *controlled experiment* means that scientists control the experimental environment to keep everything constant except the one variable being tested. But that's impossible in field research and not realistic even in highly regulated laboratory environments. Researchers usually "control" unwanted variables not by *eliminating* them through environmental regulation, but by *canceling out* their effects by using control groups.

Another hallmark of science is that the observations and experimental results must be repeatable. Observations that can't be verified may be interesting or even entertaining, but they cannot count as evidence in scientific inquiry. The headlines of supermarket tabloids would have you believe that humans are occasionally born with the head of a dog and that some of your classmates are extraterrestrials. The unconfirmed eyewitness accounts and the computer-rigged photos are amusing but unconvincing. In science, evidence from observations and experiments is only convincing if it stands up to the criterion of repeatability. The scientists who investigated snake mimicry in the Carolinas obtained similar data when they

▼ Figure 1.27 INQUIRY

Does the presence of venomous coral snakes affect predation rates on their mimics, kingsnakes?

EXPERIMENT David Pfennig and his colleagues made artificial snakes to test a prediction of the mimicry hypothesis: that kingsnakes benefit from mimicking the warning coloration of venomous coral snakes only in regions where coral snakes are present. The researchers placed equal numbers of artificial kingsnakes (experimental group) and brown artificial snakes (control group) at 14 field sites, half in the area the two snakes cohabit and half in the area where coral snakes are absent. The researchers recovered the artificial snakes after four weeks and tabulated predation data based on teeth and claw marks on the snakes.

RESULTS In field sites where coral snakes are absent, most attacks were on artificial kingsnakes. Where coral snakes were present, most attacks were on brown artificial snakes.

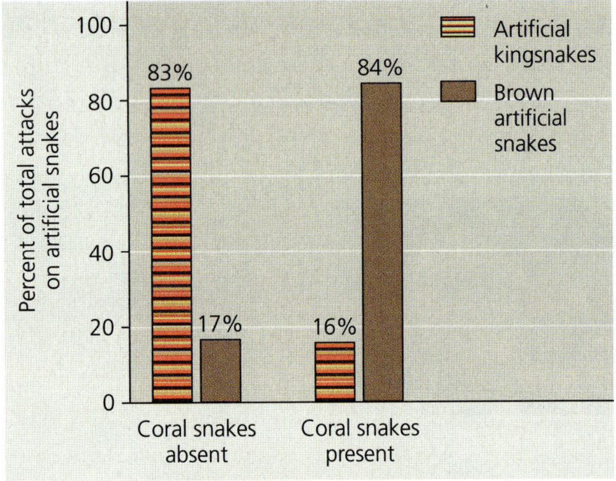

CONCLUSION The field experiments support the mimicry hypothesis by not falsifying the prediction, which was that mimicking coral snakes is effective only in areas where coral snakes are present. The experiments also tested an alternative hypothesis: that predators generally avoid all snakes with brightly colored rings. That hypothesis was falsified by the data showing that in areas without coral snakes, the ringed coloration failed to repel predators. (The fake kingsnakes may have been attacked more often in those areas because their bright pattern made them easier to spot than the brown fakes.)

SOURCE D. W. Pfennig, W. R. Harcombe, and K. S. Pfennig, Frequency-dependent Batesian mimicry, *Nature* 410:323 (2001).

INQUIRY IN ACTION Read and analyze the original paper in *Inquiry in Action: Interpreting Scientific Papers.*

MB See the related Experimental Inquiry Tutorial in MasteringBiology.

WHAT IF? What experimental results would you predict if predators throughout the Carolinas avoided all snakes with brightly colored ring patterns?

repeated their experiments with different species of coral snakes and kingsnakes in Arizona. And *you* should be able to obtain similar results if you were to repeat the snake experiments.

Theories in Science

"It's just a theory!" Our everyday use of the term *theory* often implies an untested speculation. But the term *theory* has a different meaning in science. What is a scientific theory, and how is it different from a hypothesis or from mere speculation?

First, a scientific **theory** is much broader in scope than a hypothesis. *This* is a hypothesis: "Mimicking the coloration of venomous snakes is an adaptation that protects nonvenomous snakes from predators." But *this* is a theory: "Evolutionary adaptations arise by natural selection." Darwin's theory of natural selection accounts for an enormous diversity of adaptations, including mimicry.

Second, a theory is general enough to spin off many new, specific hypotheses that can be tested. For example, two researchers at Princeton University, Peter and Rosemary Grant, were motivated by the theory of natural selection to test the specific hypothesis that the beaks of Galápagos finches evolve in response to changes in the types of available food. (Their results supported their hypothesis; see p. 469.)

And third, compared to any one hypothesis, a theory is generally supported by a much greater body of evidence. Those theories that become widely adopted in science (such as the theory of natural selection) explain a great diversity of observations and are supported by a vast accumulation of evidence. In fact, scrutiny of theories continues through testing of the specific, falsifiable hypotheses they spawn.

In spite of the body of evidence supporting a widely accepted theory, scientists must sometimes modify or even reject theories when new research methods produce results that don't fit. For example, the theory of biological diversity that lumped bacteria and archaea together as a kingdom of prokaryotes began to erode when new methods for comparing cells and molecules made it possible to test some of the hypothetical relationships between organisms that were based on the theory. If there is "truth" in science, it is conditional, based on the preponderance of available evidence.

CONCEPT CHECK 1.3

1. Contrast inductive reasoning with deductive reasoning.
2. In the snake mimicry experiment, what is the variable?
3. Why is natural selection called a theory?
4. **WHAT IF?** Suppose you extended the snake mimicry experiment to an area of Virginia where neither type of snake is known to live. What results would you predict at your field site?

For suggested answers, see Appendix A.

CONCEPT 1.4

Science benefits from a cooperative approach and diverse viewpoints

Movies and cartoons sometimes portray scientists as loners working in isolated labs. In reality, science is an intensely social activity. Most scientists work in teams, which often include both graduate and undergraduate students **(Figure 1.28)**. And to succeed in science, it helps to be a good communicator. Research results have no impact until shared with a community of peers through seminars, publications, and websites.

Building on the Work of Others

The great scientist Sir Isaac Newton once said: "To explain all nature is too difficult a task for any one man or even for any one age. 'Tis much better to do a little with certainty, and leave the rest for others that come after you. . . ." Anyone who becomes a scientist, driven by curiosity about how nature works, is sure to benefit greatly from the rich storehouse of discoveries by others who have come before.

Scientists working in the same research field often check one another's claims by attempting to confirm observations or repeat experiments. If experimental results cannot be repeated by scientific colleagues, this failure may reflect some underlying weakness in the original claim, which will then have to be revised. In this sense, science polices itself. Integrity and adherence to high professional standards in reporting results are central to the scientific endeavor. After all, the validity of experimental data is key to designing further lines of inquiry.

It is not unusual for several scientists to converge on the same research question. Some scientists enjoy the challenge of being first with an important discovery or key experiment, while others derive more satisfaction from cooperating with fellow scientists working on the same problem.

▲ **Figure 1.28 Science as a social process.** In laboratory meetings, lab members help each other interpret data, troubleshoot experiments, and plan future lines of inquiry.

Cooperation is facilitated when scientists use the same organism. Often it is a widely used **model organism**—a species that is easy to grow in the lab and lends itself particularly well to the questions being investigated. Because all organisms are evolutionarily related, lessons learned from a model organism are often widely applicable. For example, genetic studies of the fruit fly *Drosophila melanogaster* have taught us a lot about how genes work in other species, including humans. Some other popular model organisms are the mustard plant *Arabidopsis thaliana*, the soil worm *Caenorhabditis elegans*, the zebrafish *Danio rerio*, the mouse *Mus musculus*, and the bacterium *Escherichia coli*. As you read through this book, note the many contributions that these and other model organisms have made to the study of life.

Biologists may come at interesting questions from different angles. Some biologists focus on ecosystems, while others study natural phenomena at the level of organisms or cells. This book is divided into units that look at biology from different levels. Yet any given problem can be addressed from many perspectives, which in fact complement each other.

As a beginning biology student, you can benefit from making connections between the different levels of biology. You can begin to develop this skill by noticing when certain topics crop up again and again in different units. One such topic is sickle-cell disease, a well-understood genetic condition that is prevalent among native inhabitants of Africa and other warm regions and their descendants. Another topic viewed at different levels in this book is global climate change, mentioned earlier in this chapter. Sickle-cell disease and global climate change will appear in several units of the book, each time addressed at a new level. We hope these recurring topics will help you integrate the material you're learning and enhance your enjoyment of biology by helping you keep the "big picture" in mind.

Science, Technology, and Society

The biology community is part of society at large, embedded in the cultural milieu of the times. Some philosophers of science argue that scientists are so influenced by cultural and political values that science is no more objective than other ways of understanding nature. At the other extreme are people who speak of scientific theories as though they were natural laws instead of human interpretations of nature. The reality of science is probably somewhere in between—rarely perfectly objective, but continuously vetted through the expectation that observations and experiments be repeatable and hypotheses be testable and falsifiable.

The relationship of science to society becomes clearer when we add technology to the picture. Though science and technology sometimes employ similar inquiry patterns, their basic goals differ. The goal of science is to understand natural phenomena. In contrast, **technology** generally *applies* scientific knowledge for some specific purpose. Biologists and other scientists usually speak of "discoveries," while engineers and other technologists more usually speak of "inventions." And the beneficiaries of those inventions include scientists, who put new technology to work in their research. Thus, science and technology are interdependent.

▲ **Figure 1.29 DNA technology and crime scene investigation.** In 2008, forensic analysis of DNA samples from a crime scene led to the release of Charles Chatman from prison after he had served nearly 27 years for a rape he didn't commit. The photo shows Judge John Creuzot hugging Mr. Chatman after his conviction was overturned. The details of forensic analysis of DNA will be described in Chapter 20.

The potent combination of science and technology can have dramatic effects on society. Sometimes, the applications of basic research that turn out to be the most beneficial come out of the blue, from completely unanticipated observations in the course of scientific exploration. For example, discovery of the structure of DNA by Watson and Crick 60 years ago and subsequent achievements in DNA science led to the technologies of DNA manipulation that are transforming applied fields such as medicine, agriculture, and forensics **(Figure 1.29)**. Perhaps Watson and Crick envisioned that their discovery would someday lead to important applications, but it is unlikely that they could have predicted exactly what all those applications would be.

The directions that technology takes depend less on the curiosity that drives basic science than on the current needs and wants of people and on the social environment of the times. Debates about technology center more on "*should* we do it" than "*can* we do it." With advances in technology come difficult choices. For example, under what circumstances is it acceptable to use DNA technology to find out if particular people have genes for hereditary diseases? Should such tests always be voluntary, or are there circumstances when genetic testing should be mandatory? Should insurance companies or employers have access to the information, as they do for many other types of personal health data? These questions are becoming much more urgent as the sequencing of individual genomes becomes quicker and cheaper.

Such ethical issues have as much to do with politics, economics, and cultural values as with science and technology. All citizens—not only professional scientists—have a responsibility to be informed about how science works and about the potential benefits and risks of technology. The relationship between science, technology, and society increases the significance and value of any biology course.

The Value of Diverse Viewpoints in Science

Many of the technological innovations with the most profound impact on human society originated in settlements along trade routes, where a rich mix of different cultures ignited new ideas. For example, the printing press, which helped spread knowledge to all social classes and ultimately led to the book in your hands, was invented by the German Johannes Gutenberg around 1440. This invention relied on several innovations from China, including paper and ink. Paper traveled along trade routes from China to Baghdad, where technology was developed for its mass production. This technology then migrated to Europe, as did water-based ink from China, which was modified by Gutenberg to become oil-based ink. We have the cross-fertilization of diverse cultures to thank for the printing press, and the same can be said for other important inventions.

Along similar lines, science stands to gain much from embracing a diversity of backgrounds and viewpoints among its practitioners. But just how diverse a population are scientists in relation to gender, race, ethnicity, and other attributes?

The scientific community reflects the cultural standards and behaviors of society at large. It is therefore not surprising that until recently, women and certain minorities have faced huge obstacles in their pursuit to become professional scientists in many countries around the world. Over the past 50 years, changing attitudes about career choices have increased the proportion of women in biology and some other sciences, so that now women constitute roughly half of undergraduate biology majors and biology Ph.D. students. The pace has been slow at higher levels in the profession, however, and women and many racial and ethnic groups are still significantly underrepresented in many branches of science. This lack of diversity hampers the progress of science. The more voices that are heard at the table, the more robust, valuable, and productive the scientific interchange will be. The authors of this textbook welcome all students to the community of biologists, wishing you the joys and satisfactions of this very exciting and satisfying field of science—biology.

CONCEPT CHECK 1.4

1. How does science differ from technology?
2. **WHAT IF?** The gene that causes sickle-cell disease is present in a higher percentage of residents of sub-Saharan Africa than it is among those of African descent living in the United States. The presence of this gene provides some protection from malaria, a serious disease that is widespread in sub-Saharan Africa. Discuss an evolutionary process that could account for the different percentages among residents of the two regions.

For suggested answers, see Appendix A.

1 CHAPTER REVIEW

SUMMARY OF KEY CONCEPTS

CONCEPT 1.1

The themes of this book make connections across different areas of biology (pp. 2–11)

- **Theme: New properties emerge at each level in the biological hierarchy** The hierarchy of life unfolds as follows: biosphere > ecosystem > community > population > organism > organ system > organ > tissue > cell > organelle > molecule > atom. With each step upward from atoms, new properties emerge as a result of interactions among components at the lower levels. In an approach called reductionism, complex systems are broken down to simpler components that are more manageable to study. In **systems biology**, scientists attempt to model the dynamic behavior of whole biological systems based on a study of the interactions among the system's parts.

- **Theme: Organisms interact with other organisms and the physical environment** Plants take up nutrients from the soil and chemicals from the air and use energy from the sun. Interactions between plants and other organisms result in cycling of chemical nutrients within an ecosystem. One harmful outcome of human interactions with the environment has been global climate change, caused by burning of fossil fuels and increasing atmospheric CO_2.

- **Theme: Life requires energy transfer and transformation** Energy flows through an ecosystem. All organisms must perform work, which requires energy. Energy from sunlight is converted to chemical energy by producers, which is then passed on to consumers.

- **Theme: Structure and function are correlated at all levels of biological organization**
The form of a biological structure suits its function and vice versa.

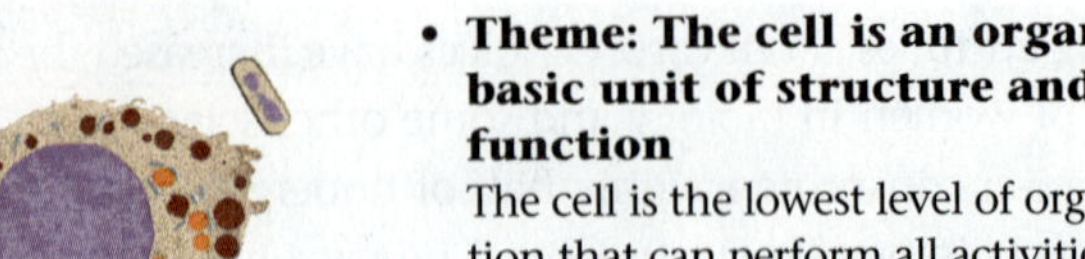

- **Theme: The cell is an organism's basic unit of structure and function**
The cell is the lowest level of organization that can perform all activities required for life. Cells are either prokaryotic or eukaryotic. **Eukaryotic cells** contain membrane-enclosed organelles, including a DNA-containing nucleus. **Prokaryotic cells** lack such organelles.

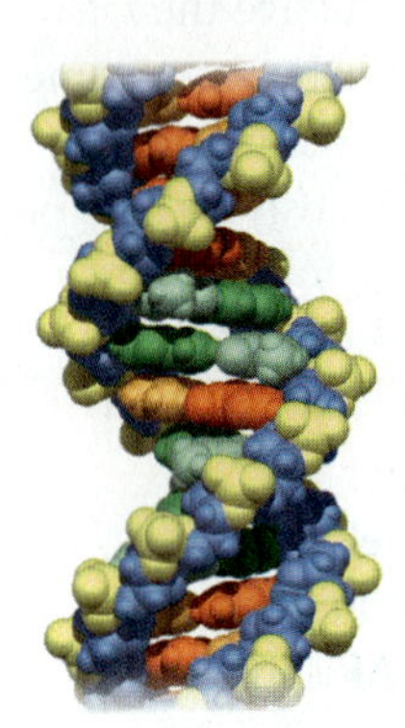

- **Theme: The continuity of life is based on heritable information in the form of DNA**
Genetic information is encoded in the nucleotide sequences of **DNA**. It is DNA that transmits heritable information from parents to offspring. DNA sequences program a cell's protein production by being transcribed into RNA and then translated into specific proteins, a process called **gene expression**. Gene expression also results in RNAs that are not translated into protein but serve other important functions. **Genomics** is the large-scale analysis of the DNA sequences within a species as well as the comparison of sequences between species.

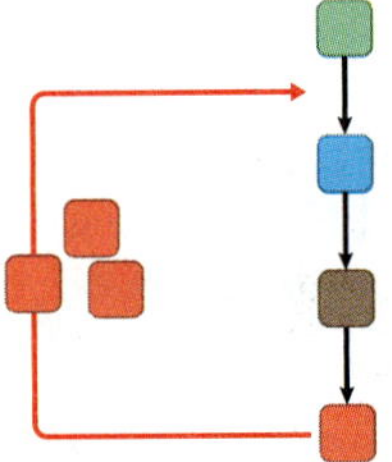

- **Theme: Feedback mechanisms regulate biological systems**
In **negative feedback**, accumulation of an end product slows the process that makes that product. In **positive feedback**, the end product stimulates the production of more product. Feedback is a type of regulation common to life at all levels, from molecules to ecosystems.

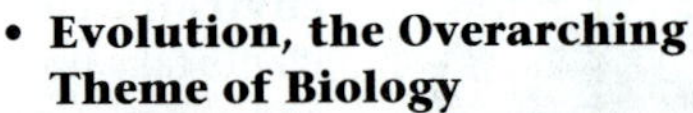

- **Evolution, the Overarching Theme of Biology**
Evolution accounts for the unity and diversity of life and also for the match of organisms to their environments.

? *Why is evolution considered the core theme of biology?*

CONCEPT 1.2

The Core Theme: Evolution accounts for the unity and diversity of life (pp. 11–18)

- Biologists classify species according to a system of broader and broader groups. Domain **Bacteria** and domain **Archaea** consist of prokaryotes. Domain **Eukarya**, the eukaryotes, includes various groups of protists and the kingdoms Plantae, Fungi, and Animalia. As diverse as life is, there is also evidence of remarkable unity, which is revealed in the similarities between different kinds of organisms.
- Darwin proposed **natural selection** as the mechanism for evolutionary adaptation of populations to their environments.

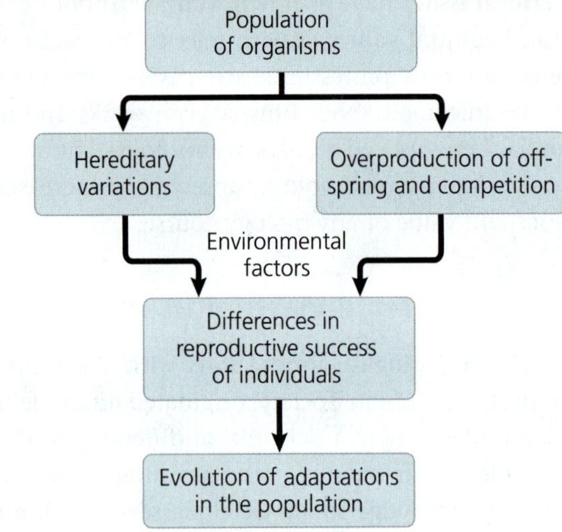

- Each species is one twig of a branching tree of life extending back in time through ancestral species more and more remote. All of life is connected through its long evolutionary history.

? *How could natural selection have led to the evolution of adaptations such as the thick, water-conserving leaves of the mother-of-pearl plant on the cover of this book?*

CONCEPT 1.3

In studying nature, scientists make observations and then form and test hypotheses (pp. 18–23)

- In scientific **inquiry**, scientists make observations (collect **data**) and use **inductive reasoning** to draw a general conclusion, which can be developed into a testable **hypothesis**. **Deductive reasoning** makes predictions that can be used to test hypotheses: If a hypothesis is correct, and we test it, then we can expect the predictions to come true. Hypotheses must be testable and falsifiable; science can address neither the possibility of supernatural phenomena nor the validity of religious beliefs.
- **Controlled experiments**, such as the study investigating mimicry in snake populations, are designed to demonstrate the effect of one variable by testing control groups and experimental groups that differ in only that one variable.
- A scientific **theory** is broad in scope, generates new hypotheses, and is supported by a large body of evidence.

? *What are the roles of inductive and deductive reasoning in the process of scientific inquiry?*

CONCEPT 1.4

Science benefits from a cooperative approach and diverse viewpoints (pp. 23–25)

- Science is a social activity. The work of each scientist builds on the work of others that have come before. Scientists must be able to repeat each other's results, so integrity is key. Biologists approach questions at different levels; their approaches complement each other.
- **Technology** is a method or device that applies scientific knowledge for some specific purpose that affects society. The ultimate impact of basic research is not always immediately obvious.
- Diversity among scientists promotes progress in science.

? *Explain why different approaches and diverse backgrounds among scientists are important.*

TEST YOUR UNDERSTANDING

LEVEL 1: KNOWLEDGE/COMPREHENSION

1. All the organisms on your campus make up
   a. an ecosystem.
   b. a community.
   c. a population.
   d. an experimental group.
   e. a taxonomic domain.

2. Which of the following is a correct sequence of levels in life's hierarchy, proceeding downward from an individual animal?
   a. brain, organ system, nerve cell, nervous tissue
   b. organ system, nervous tissue, brain
   c. organism, organ system, tissue, cell, organ
   d. nervous system, brain, nervous tissue, nerve cell
   e. organ system, tissue, molecule, cell

3. Which of the following is *not* an observation or inference on which Darwin's theory of natural selection is based?
   a. Poorly adapted individuals never produce offspring.
   b. There is heritable variation among individuals.
   c. Because of overproduction of offspring, there is competition for limited resources.
   d. Individuals whose inherited characteristics best fit them to the environment will generally produce more offspring.
   e. A population can become adapted to its environment over time.

4. Systems biology is mainly an attempt to
   a. analyze genomes from different species.
   b. simplify complex problems by reducing the system into smaller, less complex units.
   c. understand the behavior of entire biological systems.
   d. build high-throughput machines for the rapid acquisition of biological data.
   e. speed up the technological application of scientific knowledge.

5. Protists and bacteria are grouped into different domains because
   a. protists eat bacteria.
   b. bacteria are not made of cells.
   c. protists have a membrane-bounded nucleus, which bacterial cells lack.
   d. bacteria decompose protists.
   e. protists are photosynthetic.

6. Which of the following best demonstrates the unity among all organisms?
   a. matching DNA nucleotide sequences
   b. descent with modification
   c. the structure and function of DNA
   d. natural selection
   e. emergent properties

7. A controlled experiment is one that
   a. proceeds slowly enough that a scientist can make careful records of the results.
   b. tests experimental and control groups in parallel.
   c. is repeated many times to make sure the results are accurate.
   d. keeps all variables constant.
   e. is supervised by an experienced scientist.

8. Which of the following statements best distinguishes hypotheses from theories in science?
   a. Theories are hypotheses that have been proved.
   b. Hypotheses are guesses; theories are correct answers.
   c. Hypotheses usually are relatively narrow in scope; theories have broad explanatory power.
   d. Hypotheses and theories are essentially the same thing.
   e. Theories are proved true; hypotheses are often falsified.

LEVEL 2: APPLICATION/ANALYSIS

9. Which of the following is an example of qualitative data?
   a. The temperature decreased from 20°C to 15°C.
   b. The plant's height is 25 centimeters (cm).
   c. The fish swam in a zigzag motion.
   d. The six pairs of robins hatched an average of three chicks.
   e. The contents of the stomach are mixed every 20 seconds.

10. Which of the following best describes the logic of scientific inquiry?
    a. If I generate a testable hypothesis, tests and observations will support it.
    b. If my prediction is correct, it will lead to a testable hypothesis.
    c. If my observations are accurate, they will support my hypothesis.
    d. If my hypothesis is correct, I can expect certain test results.
    e. If my experiments are set up right, they will lead to a testable hypothesis.

11. **DRAW IT** With rough sketches, draw a biological hierarchy similar to the one in Figure 1.4 but using a coral reef as the ecosystem, a fish as the organism, its stomach as the organ, and DNA as the molecule. Include all levels in the hierarchy.

LEVEL 3: SYNTHESIS/EVALUATION

12. **EVOLUTION CONNECTION**
    A typical prokaryotic cell has about 3,000 genes in its DNA, while a human cell has about 20,500 genes. About 1,000 of these genes are present in both types of cells. Based on your understanding of evolution, explain how such different organisms could have this same subset of genes. What sorts of functions might these shared genes have?

13. **SCIENTIFIC INQUIRY**
    Based on the results of the snake mimicry case study, suggest another hypothesis researchers might use to extend the investigation.

14. **WRITE ABOUT A THEME**
    Evolution In a short essay (100–150 words), discuss Darwin's view of how natural selection resulted in both unity and diversity of life on Earth. Include in your discussion some of his evidence. (See p. xv for a suggested grading rubric. The rubric and tips for writing good essays can also be found in the Study Area of MasteringBiology.)

For selected answers, see Appendix A.

MasteringBIOLOGY www.masteringbiology.com

1. MasteringBiology® Assignments

Experimental Inquiry Tutorial What Can You Learn About the Process of Science from Investigating a Cricket's Chirp?
Tutorial The Scientific Method
Activities The Levels of Life Card Game • Form Fits Function: Cells • Heritable Information: DNA • Introduction to Experimental Design • GraphIt!: An Introduction to Graphing
Questions Student Misconceptions • Reading Quiz • Multiple Choice • End-of-Chapter

2. eText

Read your book online, search, take notes, highlight text, and more.

3. The Study Area

Practice Tests • Cumulative Test •  3-D Animations • MP3 Tutor Sessions • Videos • Activities • Investigations • Lab Media • Audio Glossary • Word Study Tools • Art

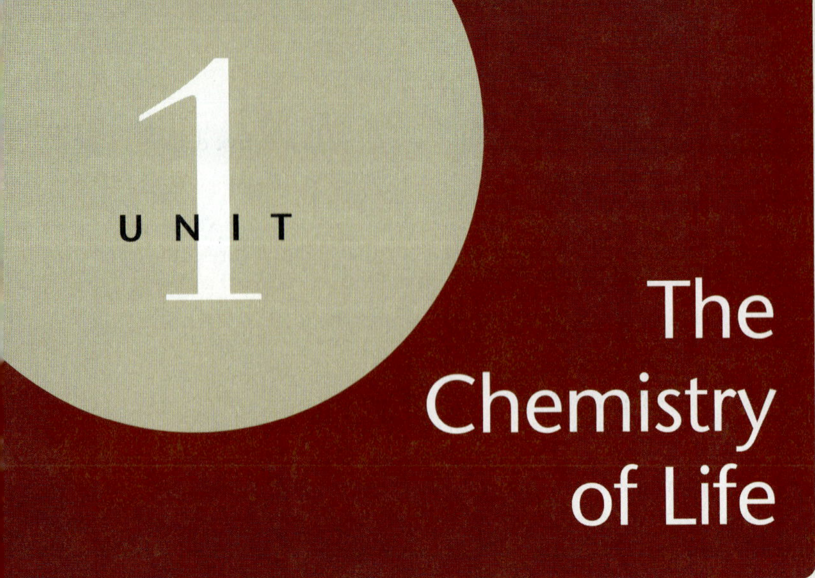

The Chemistry of Life

An Interview with

Susan Solomon

Although Susan Solomon is not a biologist, her research as an atmospheric chemist has profound implications for life on Earth. Since earning degrees from the Illinois Institute of Technology and the University of California, Berkeley, Dr. Solomon has been a leader in determining the cause of the Antarctic ozone hole and in producing the 2007 report of the United Nations Intergovernmental Panel on Climate Change (IPCC), which concluded that warming of Earth's climate is unequivocal. These activities have given her a public role in communicating science to policymakers and society at large. In recognition of her scientific accomplishments, she has been awarded the U.S. National Medal of Science, the Blue Planet Prize, and, with Al Gore and the other IPCC members, the Nobel Peace Prize. A member of the U.S. National Academy of Sciences, the European Academy of Sciences, the Academy of Sciences of France, and the Royal Society of the United Kingdom, she works for the National Oceanic and Atmospheric Administration in Boulder, Colorado.

How is Earth's atmosphere important to life?

Life on Earth today could not have evolved without an atmosphere. We all know that we and many other organisms require oxygen (O_2) from the atmosphere, and plants use carbon dioxide (CO_2) to grow. The atmosphere also contains a form of oxygen called ozone that has three oxygen atoms (O_3) instead of two. Organisms would never have been able to leave the ocean and survive on land without the development of an ozone layer in the upper atmosphere. Ozone has the important property of absorbing ultraviolet (UV) light, which would otherwise cause DNA damage. Damage from UV can lead to skin cancer and cataracts; it can also harm many crops and even phytoplankton [small photosynthetic aquatic organisms].

Early in your career, you led an expedition to make measurements of the atmosphere in Antarctica. Tell us about that.

In 1985, the British Antarctic Survey reported a surprising discovery: that the springtime ozone at their station in Antarctica had fallen by 30–50% since the late 1970s, resulting in an "ozone hole"! People had begun to be worried about whether the ozone layer might be vulnerable to changes caused by human activity, but only very minor changes had been expected. In 1986, I had the chance to lead a new Antarctic expedition to help confirm the British data and to study the problem further. We didn't just measure ozone; we measured about a dozen other atmospheric molecules that allowed us to tell *why* the ozone was being destroyed.

What did you find out?

It turns out that the ozone chemistry in Antarctica is extremely different from what it is anywhere else. That's because Antarctica is very cold—it really is the coldest place on Earth. It's so cold that clouds form in its upper stratosphere, about 10–30 kilometers above sea level, and those clouds help convert chemicals from chlorofluorocarbons (CFCs) to ozone-damaging substances.

CFCs are synthetic compounds, made only by humans. They were used back then for a variety of purposes—for example, in refrigeration, as solvents, and as propellants for sprays. Many tons of CFCs were emitted every year. I came up with the idea that the reason an ozone hole developed in Antarctica had to do with chemical reactions that happen between a gas and a surface and that the surface in this case was the small particles that make up those stratospheric clouds. Our data supported this hypothesis. The reactions on those particles make the CFCs hundreds of times more damaging than they would be otherwise. The absence of such clouds in most other parts of the world is why we don't have ozone holes elsewhere, although stratospheric clouds form occasionally in the Arctic and there is significant ozone loss there. Scientists had been concerned since the mid-1970s that human use of CFCs might cause some ozone depletion, but they had expected a loss of only about 3–5% in 100 years.

How do CFCs destroy ozone?

When CFCs arrive at the stratosphere, which typically takes a few years, high-energy radiation up there can break them down, releasing chlorine atoms. The chlorine atoms destroy the ozone catalytically, which means that the atoms don't get used up in the process. So even if only a small amount of CFC is broken down, the tiny bit of chlorine produced can destroy an enormous amount of ozone.

In the U.N. Montreal Protocol of 1987, the nations of the world agreed to stop producing CFCs. However, the CFCs in the atmosphere disappear only very slowly; typically they hang around for 50 to 100 years. What that means is that the CFCs we've already put in the atmosphere will continue to produce an ozone hole for many decades to come, even though we're not using these substances any more. Global emissions are very near zero now, and we're beginning to see the ozone hole slowly diminish. But it will probably not go away completely until around 2060.

While the ozone hole remains, it continues to cause damage. For example, there is evidence that the phytoplankton in the Antarctic Ocean are being affected by increased UV, and the phytoplankton are the base of the main Antarctic food chain: They feed the krill, which feed the fish, which feed the penguins, seals, and whales.

Let's talk about an effect that other atmospheric changes are having—climate change.

There's no question that the planet is getting warmer. We know that, on average, our planet is about 1.4°F (0.8°C) warmer than it was 100 years ago, and this past decade has been the warmest decade in at least the last 100 years. We also know that glaciers worldwide are retreating and that sea level is rising. There's a breadth of scientific data, acquired by different techniques, that tells us that global warming really is unequivocal.

The warming has to do with the greenhouse effect, right?

We're lucky that this planet has a greenhouse effect, because if it didn't, we would be very cold indeed! Our planet is heated by the sun, and much of the infrared radiation (heat) that would otherwise be released back into space is trapped by "greenhouse gases" in the atmosphere. This makes the planet about 30°C hotter than it would be otherwise. But of course anything can be bad if you have too much of it, and what we're doing now is increasing the greenhouse effect of our atmosphere beyond its natural state. If we keep emitting the greenhouse gases that are causing the warming, then we will see some very significant warming in the coming century.

The main greenhouse gas we're adding to the atmosphere is CO_2, from burning fossil fuel and to a lesser extent deforestation. We have increased the atmospheric concentration of CO_2 by about 30% compared to any value that has been found for the last 800,000 years. This has been determined by digging up ice cores in Antarctica and measuring the gases in the air bubbles trapped in the ice. So we know that we have perturbed the atmosphere in a way that the planet hasn't seen in at least 800,000 years.

The CFCs we discussed earlier are actually the third most important greenhouse gas at present, after CO_2 and methane. Pound for pound, CFCs are much more potent as greenhouse gases than CO_2. The phase-out of CFCs since the signing of the Montreal Protocol has not only avoided a lot of ozone destruction that would otherwise have happened, but has also reduced our input of gases that cause climate change.

How is life on Earth being affected by climate change?

There are some things that we can already begin to see and talk about, but there's an enormous amount that we still don't know. We do know that the oceans are getting more acidic because CO_2 is taken up by the ocean and converted to carbonic acid, which can affect the ability of shellfish to make their shells. Other ocean life is also likely to be harmed by the increased acidity, such as the organisms of coral reefs. But there's also emerging evidence that some other marine organisms may do better—lobsters, perhaps.

As a westerner I'm extremely concerned about the greatly increased population of pine beetles in the western United States. These beetles are killing pine trees in unprecedented numbers. There's good evidence that a contributing factor to this explosion of pine beetles is global warming. I think we're going to see more of this kind of thing. Also, it is clear that bird migration is already being affected by global warming. Whether global warming will lead to extinction of some animals is an important question. The signature extinction issue is the polar bear; as the sea ice of the Arctic decreases, the polar bear could become extinct. We don't really know yet how much biological adaptation is possible in the time available. We'll probably find out that there are some winner species out there and some loser species. In agriculture, many crops are sensitive to increasing temperatures. One of the relevant findings about corn is that for every degree of warming, about 10% of crop production is lost—a big change.

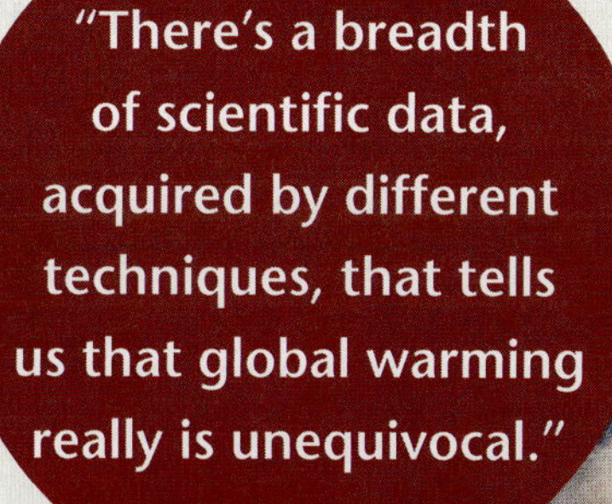

Does less precipitation always go along with higher temperatures?

In some places there will probably be less precipitation and in other places more. There's a band of subtropical and tropical regions where we are pretty confident that it will get drier—for example, Mexico, the Mediterranean region, parts of Australia. In the higher latitudes, places like Canada and Norway will likely get wetter. In between, it's harder to predict.

Tell us about the IPCC and your work on it.

The IPCC is fundamentally a mechanism for the communication of information about climate change from the science community to the policy community. It was set up in 1988 when people were beginning to recognize that climate change was a real possibility. Policymakers decided that they needed to get reliable scientific information so they could begin to talk about what to do, if anything. Every six or seven years, scientists are asked by their governments to get together and assess what we know and don't know on the basis of the published scientific literature.

I have been involved in the IPCC since 1992, and in 2001 I was elected by the panel, representing over 100 governments, to co-chair the scientific assessment team. In a process lasting several years, we generated a detailed report summarizing the state of climate science. Our report was then reviewed by dozens of governments and more than 600 scientists. The report itself and every one of their 30,000 comments are available on the Internet. We refined and refined the draft in consideration of those comments and finalized the document in 2007.

What were the main conclusions of your 2007 report?

The first conclusion, based on many independent lines of evidence, was that the Earth is warming. There's no doubt we are now living on a planet that is warmer than a century ago. The second main conclusion was that most—more than half—of the warming is very likely due to increases in greenhouse gases, primarily CO_2. We did a careful analysis of the uncertainties: When we say "very likely," we mean that there's a 90% chance or greater that most of the warming is due to emissions of greenhouse gases by human activity.

What have you learned about working at the interface of science and policy?

It's one of the most difficult things a scientist can do. Science normally takes us into a laboratory or out into the field or into scholarly discussions with colleagues. Getting involved with policy is quite different: It takes us out of the lab and makes us much more aware of the strong emotions around many issues. In that sense, it's a bit daunting. But it's uplifting to see how valuable science can be in helping society make more informed choices. Scientists can help make sure that whatever it is we choose to do as a society we're doing knowingly, not in ignorance. I appreciate all the reasons why people ask tough questions about the science. How much do we know? What really are the uncertainties? Yes, there's a lot at stake here, all the more reason why there has to be really good science going into it.

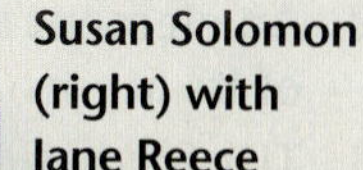
Susan Solomon (right) with Jane Reece

2

The Chemical Context of Life

▲ **Figure 2.1 Who tends this "garden"?**

KEY CONCEPTS

2.1 Matter consists of chemical elements in pure form and in combinations called compounds

2.2 An element's properties depend on the structure of its atoms

2.3 The formation and function of molecules depend on chemical bonding between atoms

2.4 Chemical reactions make and break chemical bonds

OVERVIEW

A Chemical Connection to Biology

The Amazon rain forest in South America is a showcase for the diversity of life on Earth. Colorful birds, insects, and other animals live in a densely-packed environment of trees, shrubs, vines, and wildflowers, and an excursion along a waterway or a forest path typically reveals a lush variety of plant life. Visitors traveling near the Amazon's headwaters in Peru are therefore surprised to come across tracts of forest like that seen in the foreground of the photo in **Figure 2.1**. This patch is almost completely dominated by a single plant species—a small flowering tree called *Duroia hirsuta*. Travelers may wonder if the plot of land is planted and maintained by local people, but the indigenous people are as mystified as the visitors. They call these stands of *Duroia* trees "devil's gardens," from a legend attributing them to an evil forest spirit.

Seeking a scientific explanation, a research team at Stanford University recently solved the "devil's garden" mystery. **Figure 2.2** describes their main experiment. The researchers showed that the "farmers" who create and maintain these gardens are actually ants that live in the hollow stems of the *Duroia* trees. The ants do not plant the *Duroia* trees, but they prevent other plant species from growing in the garden by injecting intruders with a poisonous chemical. In this way, the ants create space for the growth of the *Duroia* trees that serve as their home. With the ability to maintain and expand its habitat, a single colony of devil's garden ants can live for hundreds of years.

The chemical used by the ants to weed their garden turns out to be formic acid. This substance is produced by many species of ants and in fact got its name from the Latin word for ant, *formica*. For many ant species, the formic acid probably serves as a disinfectant that protects the ants against microbial parasites. The devil's garden ant is the first ant species found to use formic acid as an herbicide, an important addition to the list of functions mediated by chemicals in the insect world. Scientists have long known that chemicals play a major role in insect communication, attraction of mates, and defense against predators.

Research on devil's gardens is only one example of the relevance of chemistry to the study of life. Unlike a list of college courses, nature is not neatly packaged into the individual natural sciences—biology, chemistry, physics, and so forth. Biologists specialize in the study of life, but organisms and their environments are natural systems to which the concepts of chemistry and physics apply. Biology is a multidisciplinary science.

This unit of chapters introduces some basic concepts of chemistry that apply to the study of life. We will make many connections to the themes introduced in Chapter 1. One of these themes is the organization of life into a hierarchy of structural levels, with additional properties emerging at each successive level. In this unit, we will see how emergent properties are apparent at the lowest levels of biological organization—such as the ordering of atoms into molecules and the interactions of those molecules within cells. Somewhere in the transition from molecules to cells, we will cross the blurry boundary between nonlife and life. This chapter focuses on the chemical components that make up all matter.

▼ **Figure 2.2** INQUIRY

What creates "devil's gardens" in the rain forest?

EXPERIMENT Working under Deborah Gordon and with Michael Greene, graduate student Megan Frederickson sought the cause of "devil's gardens," stands of a single species of tree, *Duroia hirsuta*. One hypothesis was that ants living in these trees, *Myrmelachista schumanni*, produce a poisonous chemical that kills trees of other species; another was that the *Duroia* trees themselves kill competing trees, perhaps by means of a chemical.

To test these hypotheses, Frederickson did field experiments in Peru. Two saplings of a local nonhost tree species, *Cedrela odorata*, were planted inside each of ten devil's gardens. At the base of one sapling, a sticky insect barrier was applied; the other was unprotected. Two more *Cedrela* saplings, with and without barriers, were planted about 50 meters outside each garden.

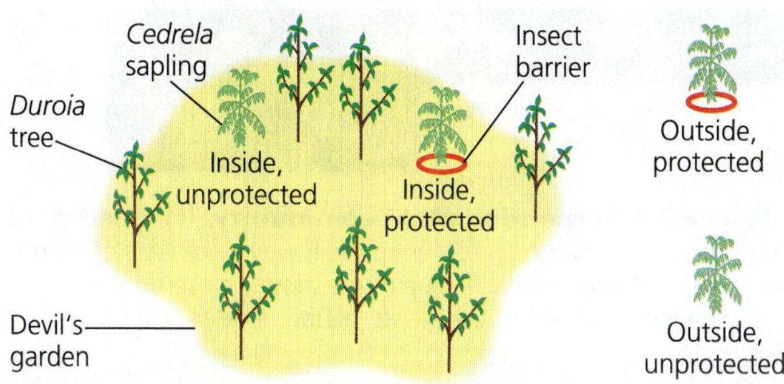

The researchers observed ant activity on the *Cedrela* leaves and measured areas of dead leaf tissue after one day. They also chemically analyzed contents of the ants' poison glands.

RESULTS The ants made injections from the tips of their abdomens into leaves of unprotected saplings in their gardens (see photo). Within one day, these leaves developed dead areas (see graph). The protected saplings were uninjured, as were the saplings planted outside the gardens. Formic acid was the only chemical detected in the poison glands of the ants.

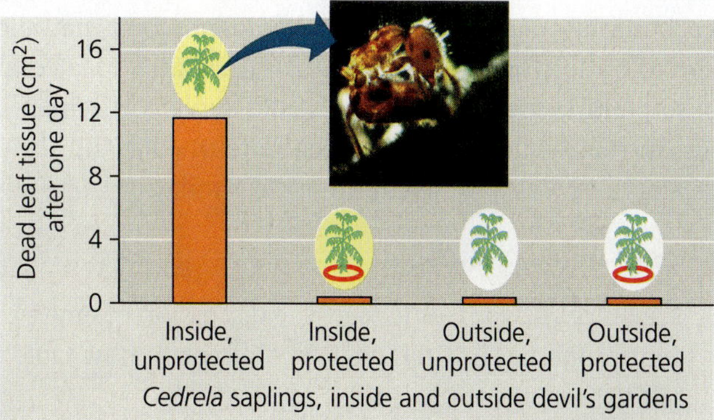

CONCLUSION Ants of the species *Myrmelachista schumanni* kill nonhost trees by injecting the leaves with formic acid, thus creating hospitable habitats (devil's gardens) for the ant colony.

SOURCE M. E. Frederickson, M. J. Greene, and D. M. Gordon, "Devil's gardens" bedevilled by ants, *Nature* 437:495–496 (2005).

INQUIRY IN ACTION Read and analyze the original paper in *Inquiry in Action: Interpreting Scientific Papers*.

WHAT IF? What would be the results if the unprotected saplings' inability to grow in the devil's gardens was caused by a chemical released by the *Duroia* trees rather than by the ants?

CONCEPT **2.1**

Matter consists of chemical elements in pure form and in combinations called compounds

Organisms are composed of **matter**, which is defined as anything that takes up space and has mass.* Matter exists in many diverse forms. Rocks, metals, oils, gases, and humans are just a few examples of what seems an endless assortment of matter.

Elements and Compounds

Matter is made up of elements. An **element** is a substance that cannot be broken down to other substances by chemical reactions. Today, chemists recognize 92 elements occurring in nature; gold, copper, carbon, and oxygen are examples. Each element has a symbol, usually the first letter or two of its name. Some symbols are derived from Latin or German; for instance, the symbol for sodium is Na, from the Latin word *natrium*.

A **compound** is a substance consisting of two or more different elements combined in a fixed ratio. Table salt, for example, is sodium chloride (NaCl), a compound composed of the elements sodium (Na) and chlorine (Cl) in a 1:1 ratio. Pure sodium is a metal, and pure chlorine is a poisonous gas. When chemically combined, however, sodium and chlorine form an edible compound. Water (H_2O), another compound, consists of the elements hydrogen (H) and oxygen (O) in a 2:1 ratio. These are simple examples of organized matter having emergent properties: A compound has characteristics different from those of its elements **(Figure 2.3)**.

▲ **Figure 2.3 The emergent properties of a compound.** The metal sodium combines with the poisonous gas chlorine, forming the edible compound sodium chloride, or table salt.

*Sometimes we substitute the term weight for mass, although the two are not identical. Mass is the amount of matter in an object, whereas the weight of an object is how strongly that mass is pulled by gravity. The weight of an astronaut walking on the moon is approximately 1/6 the astronaut's weight on Earth, but his or her mass is the same. However, as long as we are earthbound, the weight of an object is a measure of its mass; in everyday language, therefore, we tend to use the terms interchangeably.

The Elements of Life

Of the 92 natural elements, about 20–25% are **essential elements** that an organism needs to live a healthy life and reproduce. The essential elements are similar among organisms, but there is some variation—for example, humans need 25 elements, but plants need only 17.

Just four elements—oxygen (O), carbon (C), hydrogen (H), and nitrogen (N)—make up 96% of living matter. Calcium (Ca), phosphorus (P), potassium (K), sulfur (S), and a few other elements account for most of the remaining 4% of an organism's mass. **Trace elements** are required by an organism in only minute quantities. Some trace elements, such as iron (Fe), are needed by all forms of life; others are required only by certain species. For example, in vertebrates (animals with backbones), the element iodine (I) is an essential ingredient of a hormone produced by the thyroid gland. A daily intake of only 0.15 milligram (mg) of iodine is adequate for normal activity of the human thyroid. An iodine deficiency in the diet causes the thyroid gland to grow to abnormal size, a condition called goiter. Where it is available, eating seafood or iodized salt reduces the incidence of goiter. All the elements needed by the human body are listed in **Table 2.1**.

Some naturally occurring elements are toxic to organisms. In humans, for instance, the element arsenic has been linked to numerous diseases and can be lethal. In some areas of the world, arsenic occurs naturally and can make its way into the groundwater. As a result of using water from drilled wells in southern Asia, millions of people have been inadvertently exposed to arsenic-laden water. Efforts are under way to reduce arsenic levels in their water supply.

Table 2.1 Elements in the Human Body

Element	Symbol	Percentage of Body Mass (including water)	
Oxygen	O	65.0%	96.3%
Carbon	C	18.5%	
Hydrogen	H	9.5%	
Nitrogen	N	3.3%	
Calcium	Ca	1.5%	3.7%
Phosphorus	P	1.0%	
Potassium	K	0.4%	
Sulfur	S	0.3%	
Sodium	Na	0.2%	
Chlorine	Cl	0.2%	
Magnesium	Mg	0.1%	

Trace elements (less than 0.01% of mass): Boron (B), chromium (Cr), cobalt (Co), copper (Cu), fluorine (F), iodine (I), iron (Fe), manganese (Mn), molybdenum (Mo), selenium (Se), silicon (Si), tin (Sn), vanadium (V), zinc (Zn)

▲ **Figure 2.4 Serpentine plant community.** The plants in the large photo are growing on serpentine soil, which contains elements that are usually toxic to plants. The insets show a close-up of serpentine rock and one of the plants, a Tiburon Mariposa lily.

Case Study: Evolution of Tolerance to Toxic Elements

EVOLUTION Some species have become adapted to environments containing elements that are usually toxic. A compelling example is found in serpentine plant communities. Serpentine is a jade-like mineral that contains toxic elements such as chromium, nickel, and cobalt. Although most plants cannot survive in soil that forms from serpentine rock, a small number of plant species have adaptations that allow them to do so **(Figure 2.4)**. Presumably, variants of ancestral, nonserpentine species arose that could survive in serpentine soils, and subsequent natural selection resulted in the distinctive array of species we see in these areas today.

CONCEPT CHECK 2.1

1. **MAKE CONNECTIONS** Review the discussion of emergent properties in Chapter 1 (p. 3). Explain how table salt has emergent properties.
2. Is a trace element an essential element? Explain.
3. In humans, iron is a trace element required for the proper functioning of hemoglobin, the molecule that carries oxygen in red blood cells. What might be the effects of an iron deficiency?
4. **MAKE CONNECTIONS** Review the discussion of natural selection in Chapter 1 (pp. 14–16) and explain how natural selection might have played a role in the evolution of species that are tolerant of serpentine soils.

For suggested answers, see Appendix A.

CONCEPT 2.2

An element's properties depend on the structure of its atoms

Each element consists of a certain type of atom that is different from the atoms of any other element. An **atom** is the smallest unit of matter that still retains the properties of an element. Atoms are so small that it would take about a million of them to stretch across the period printed at the end of this sentence. We symbolize atoms with the same abbreviation used for the element that is made up of those atoms. For example, the symbol C stands for both the element carbon and a single carbon atom.

Subatomic Particles

Although the atom is the smallest unit having the properties of an element, these tiny bits of matter are composed of even smaller parts, called *subatomic particles*. Physicists have split the atom into more than a hundred types of particles, but only three kinds of particles are relevant here: **neutrons**, **protons**, and **electrons**. Protons and electrons are electrically charged. Each proton has one unit of positive charge, and each electron has one unit of negative charge. A neutron, as its name implies, is electrically neutral.

Protons and neutrons are packed together tightly in a dense core, or **atomic nucleus**, at the center of an atom; protons give the nucleus a positive charge. The electrons form a sort of cloud of negative charge around the nucleus, and it is the attraction between opposite charges that keeps the electrons in the vicinity of the nucleus. **Figure 2.5** shows two commonly used models of the structure of the helium atom as an example.

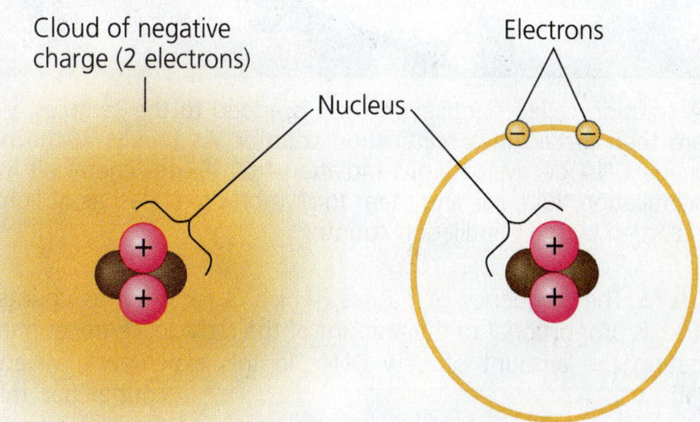

(a) This model represents the two electrons as a cloud of negative charge.

(b) In this more simplified model, the electrons are shown as two small yellow spheres on a circle around the nucleus.

▲ **Figure 2.5 Simplified models of a helium (He) atom.** The helium nucleus consists of 2 neutrons (brown) and 2 protons (pink). Two electrons (yellow) exist outside the nucleus. These models are not to scale; they greatly overestimate the size of the nucleus in relation to the electron cloud.

The neutron and proton are almost identical in mass, each about 1.7×10^{-24} gram (g). Grams and other conventional units are not very useful for describing the mass of objects so minuscule. Thus, for atoms and subatomic particles (and for molecules, too), we use a unit of measurement called the **dalton**, in honor of John Dalton, the British scientist who helped develop atomic theory around 1800. (The dalton is the same as the *atomic mass unit*, or *amu*, a unit you may have encountered elsewhere.) Neutrons and protons have masses close to 1 dalton. Because the mass of an electron is only about 1/2,000 that of a neutron or proton, we can ignore electrons when computing the total mass of an atom.

Atomic Number and Atomic Mass

Atoms of the various elements differ in their number of subatomic particles. All atoms of a particular element have the same number of protons in their nuclei. This number of protons, which is unique to that element, is called the **atomic number** and is written as a subscript to the left of the symbol for the element. The abbreviation $_2$He, for example, tells us that an atom of the element helium has 2 protons in its nucleus. Unless otherwise indicated, an atom is neutral in electrical charge, which means that its protons must be balanced by an equal number of electrons. Therefore, the atomic number tells us the number of protons and also the number of electrons in an electrically neutral atom.

We can deduce the number of neutrons from a second quantity, the **mass number**, which is the sum of protons plus neutrons in the nucleus of an atom. The mass number is written as a superscript to the left of an element's symbol. For example, we can use this shorthand to write an atom of helium as $^{4}_{2}$He. Because the atomic number indicates how many protons there are, we can determine the number of neutrons by subtracting the atomic number from the mass number: The helium atom, $^{4}_{2}$He, has 2 neutrons. For sodium (Na):

Mass number = number of protons + neutrons
= 23 for sodium

$^{23}_{11}$Na

Atomic number = number of protons
= number of electrons in a neutral atom
= 11 for sodium

Number of neutrons = mass number − atomic number
= 23 − 11 = 12 for sodium

The simplest atom is hydrogen, $^{1}_{1}$H, which has no neutrons; it consists of a single proton with a single electron.

As mentioned earlier, the contribution of electrons to mass is negligible. Therefore, almost all of an atom's mass is concentrated in its nucleus. Because neutrons and protons each have a mass very close to 1 dalton, the mass number is an approximation of the total mass of an atom, called its **atomic mass**. So we might say that the atomic mass of sodium ($^{23}_{11}$Na) is 23 daltons, although more precisely it is 22.9898 daltons.

Isotopes

All atoms of a given element have the same number of protons, but some atoms have more neutrons than other atoms of the same element and therefore have greater mass. These different atomic forms of the same element are called **isotopes** of the element. In nature, an element occurs as a mixture of its isotopes. For example, consider the three isotopes of the element carbon, which has the atomic number 6. The most common isotope is carbon-12, $^{12}_{6}C$, which accounts for about 99% of the carbon in nature. The isotope $^{12}_{6}C$ has 6 neutrons. Most of the remaining 1% of carbon consists of atoms of the isotope $^{13}_{6}C$, with 7 neutrons. A third, even rarer isotope, $^{14}_{6}C$, has 8 neutrons. Notice that all three isotopes of carbon have 6 protons; otherwise, they would not be carbon. Although the isotopes of an element have slightly different masses, they behave identically in chemical reactions. (The number usually given as the atomic mass of an element, such as 22.9898 daltons for sodium, is actually an average of the atomic masses of all the element's naturally occurring isotopes.)

Both ^{12}C and ^{13}C are stable isotopes, meaning that their nuclei do not have a tendency to lose particles. The isotope ^{14}C, however, is unstable, or radioactive. A **radioactive isotope** is one in which the nucleus decays spontaneously, giving off particles and energy. When the decay leads to a change in the number of protons, it transforms the atom to an atom of a different element. For example, when a radioactive carbon atom decays, it becomes an atom of nitrogen.

Radioactive isotopes have many useful applications in biology. In Chapter 25, you will learn how researchers use measurements of radioactivity in fossils to date these relics of past life. As shown in **Figure 2.6**, radioactive isotopes are also useful as tracers to follow atoms through metabolism, the chemical processes of an organism. Cells use the radioactive atoms as they would use nonradioactive isotopes of the same element, but the radioactive tracers can be readily detected.

Radioactive tracers are important diagnostic tools in medicine. For example, certain kidney disorders can be diagnosed by injecting small doses of substances containing radioactive isotopes into the blood and then measuring the amount of tracer excreted in the urine. Radioactive tracers are also used in combination with sophisticated imaging instruments. PET scanners, for instance, can monitor chemical processes, such as those involved in cancerous growth, as they actually occur in the body **(Figure 2.7)**.

Although radioactive isotopes are very useful in biological research and medicine, radiation from decaying isotopes also poses a hazard to life by damaging cellular molecules. The severity of this damage depends on the type and amount of radiation an organism absorbs. One of the most serious environmental threats is radioactive fallout from nuclear accidents. The doses of most isotopes used in medical diagnosis, however, are relatively safe.

▼ Figure 2.6 RESEARCH METHOD

Radioactive Tracers

APPLICATION Scientists use radioactive isotopes to label certain chemical compounds, creating tracers that allow them to follow a metabolic process or locate the compound within an organism. In this example, radioactive tracers are utilized to determine the effect of temperature on the rate at which cells make copies of their DNA.

TECHNIQUE

1 Compounds used by cells to make DNA are added to human cells. One ingredient is labeled with ^{3}H, a radioactive isotope of hydrogen. Nine dishes of cells are incubated at different temperatures. The cells make new DNA, incorporating the radioactive tracer.

2 Cells from each incubator are placed in tubes; their DNA is isolated; and unused labeled compounds are removed.

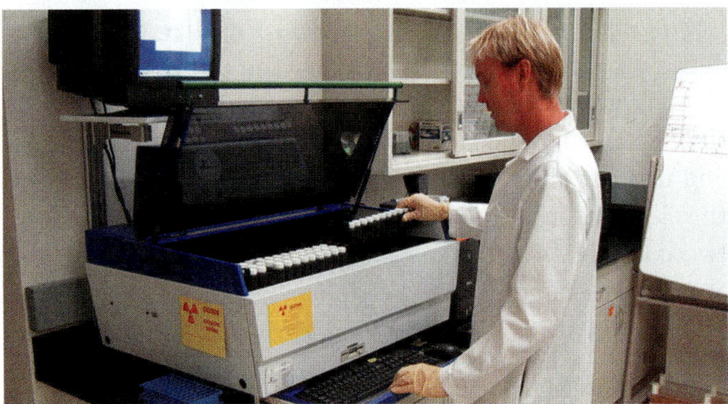

3 A solution called scintillation fluid is added to the samples, which are then placed in a scintillation counter. As the ^{3}H in the newly made DNA decays, it emits radiation that excites chemicals in the scintillation fluid, causing them to give off light. Flashes of light are recorded by the scintillation counter.

RESULTS The frequency of flashes, which is recorded as counts per minute, is proportional to the amount of the radioactive tracer present, indicating the amount of new DNA. In this experiment, when the counts per minute are plotted against temperature, it is clear that temperature affects the rate of DNA synthesis; the most DNA was made at 35°C.

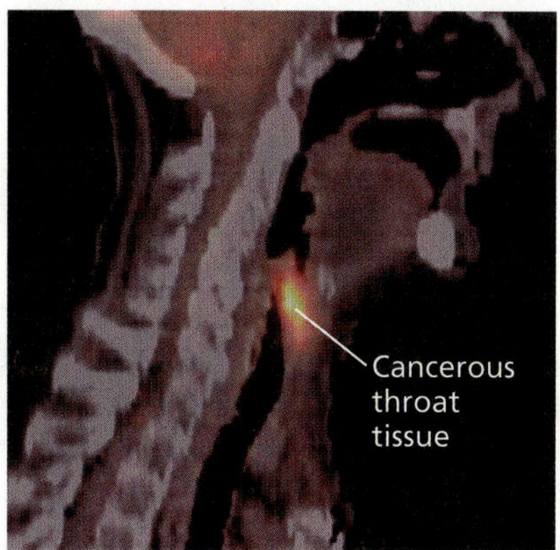

◀ **Figure 2.7 A PET scan, a medical use for radioactive isotopes.** PET, an acronym for positron-emission tomography, detects locations of intense chemical activity in the body. The bright yellow spot marks an area with an elevated level of radioactively labeled glucose, which in turn indicates high metabolic activity, a hallmark of cancerous tissue.

(a) A ball bouncing down a flight of stairs provides an analogy for energy levels of electrons, because the ball can come to rest only on each step, not between steps.

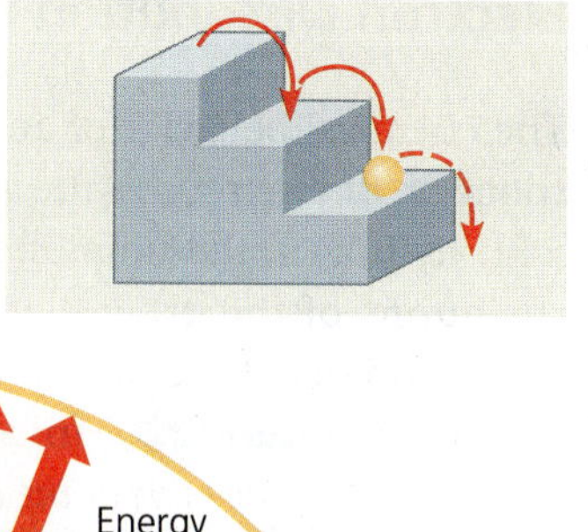

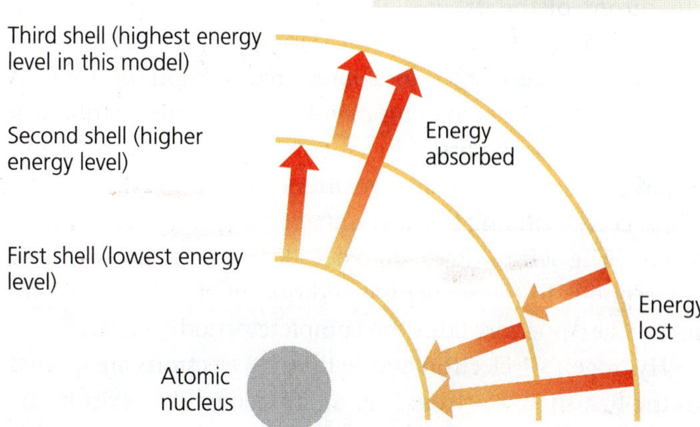

(b) An electron can move from one shell to another only if the energy it gains or loses is exactly equal to the difference in energy between the energy levels of the two shells. Arrows in this model indicate some of the stepwise changes in potential energy that are possible.

▲ **Figure 2.8 Energy levels of an atom's electrons.** Electrons exist only at fixed levels of potential energy called electron shells.

The Energy Levels of Electrons

The simplified models of the atom in Figure 2.5 greatly exaggerate the size of the nucleus relative to the volume of the whole atom. If an atom of helium were the size of a typical football stadium, the nucleus would be the size of a pencil eraser in the center of the field. Moreover, the electrons would be like two tiny gnats buzzing around the stadium. Atoms are mostly empty space.

When two atoms approach each other during a chemical reaction, their nuclei do not come close enough to interact. Of the three kinds of subatomic particles we have discussed, only electrons are directly involved in the chemical reactions between atoms.

An atom's electrons vary in the amount of energy they possess. **Energy** is defined as the capacity to cause change—for instance, by doing work. **Potential energy** is the energy that matter possesses because of its location or structure. For example, water in a reservoir on a hill has potential energy because of its altitude. When the gates of the reservoir's dam are opened and the water runs downhill, the energy can be used to do work, such as turning generators. Because energy has been expended, the water has less energy at the bottom of the hill than it did in the reservoir. Matter has a natural tendency to move to the lowest possible state of potential energy; in this example, the water runs downhill. To restore the potential energy of a reservoir, work must be done to elevate the water against gravity.

The electrons of an atom have potential energy because of how they are arranged in relation to the nucleus. The negatively charged electrons are attracted to the positively charged nucleus. It takes work to move a given electron farther away from the nucleus, so the more distant an electron is from the nucleus, the greater its potential energy. Unlike the continuous flow of water downhill, changes in the potential energy of electrons can occur only in steps of fixed amounts. An electron having a certain amount of energy is something like a ball on a staircase **(Figure 2.8a)**. The ball can have different amounts of potential energy, depending on which step it is on, but it cannot spend much time between the steps. Similarly, an electron's potential energy is determined by its energy level. An electron cannot exist between energy levels.

An electron's energy level is correlated with its average distance from the nucleus. Electrons are found in different **electron shells**, each with a characteristic average distance and energy level. In diagrams, shells can be represented by concentric circles **(Figure 2.8b)**. The first shell is closest to the nucleus, and electrons in this shell have the lowest potential energy. Electrons in the second shell have more energy, and electrons in the third shell even more energy. An electron can change the shell it occupies, but only by absorbing or losing an amount of energy equal to the difference in potential energy between its position in the old shell and that in the new shell. When an electron absorbs energy, it moves to a shell farther out from the nucleus. For example, light energy can excite an electron to a higher energy level. (Indeed, this is the first step taken when plants harness the energy of sunlight for photosynthesis, the process that produces food from carbon dioxide and water.) When an electron loses energy, it "falls back" to a shell closer to the nucleus, and the lost energy is usually released to the environment as heat. For example, sunlight excites electrons in the surface of a car to higher energy levels. When the electrons fall back to their original levels, the car's surface heats up. This thermal energy can be transferred to the air or to your hand if you touch the car.

Electron Distribution and Chemical Properties

The chemical behavior of an atom is determined by the distribution of electrons in the atom's electron shells. Beginning with hydrogen, the simplest atom, we can imagine building the atoms of the other elements by adding 1 proton and 1 electron at a time (along with an appropriate number of neutrons). **Figure 2.9**, an abbreviated version of what is called the *periodic table of the elements*, shows this distribution of electrons for the first 18 elements, from hydrogen ($_1$H) to argon ($_{18}$Ar). The elements are arranged in three rows, or periods, corresponding to the number of electron shells in their atoms. The left-to-right sequence of elements in each row corresponds to the sequential addition of electrons and protons. (See Appendix B for the complete periodic table.)

Hydrogen's 1 electron and helium's 2 electrons are located in the first shell. Electrons, like all matter, tend to exist in the lowest available state of potential energy. In an atom, this state is in the first shell. However, the first shell can hold no more than 2 electrons; thus, hydrogen and helium are the only elements in the first row of the table. An atom with more than 2 electrons must use higher shells because the first shell is full. The next element, lithium, has 3 electrons. Two of these electrons fill the first shell, while the third electron occupies the second shell. The second shell holds a maximum of 8 electrons. Neon, at the end of the second row, has 8 electrons in the second shell, giving it a total of 10 electrons.

The chemical behavior of an atom depends mostly on the number of electrons in its *outermost* shell. We call those outer electrons **valence electrons** and the outermost electron shell the **valence shell**. In the case of lithium, there is only 1 valence electron, and the second shell is the valence shell. Atoms with the same number of electrons in their valence shells exhibit similar chemical behavior. For example, fluorine (F) and chlorine (Cl) both have 7 valence electrons, and both form compounds when combined with the element sodium (see Figure 2.3). An atom with a completed valence shell is unreactive; that is, it will not interact readily with other atoms. At the far right of the periodic table are helium, neon, and argon, the only three elements shown in Figure 2.9 that have full valence shells. These elements are said to be *inert*, meaning chemically unreactive. All the other atoms in Figure 2.9 are chemically reactive because they have incomplete valence shells.

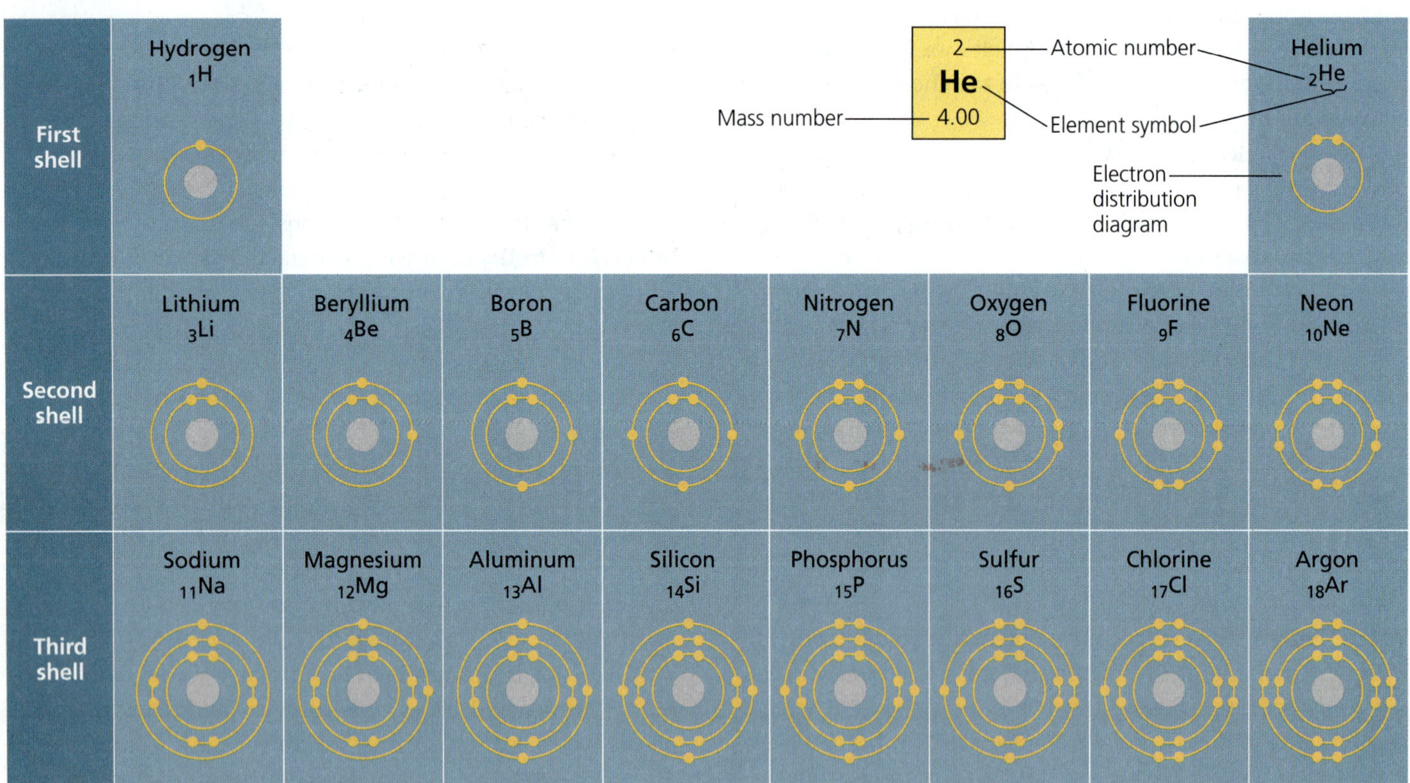

▲ **Figure 2.9 Electron distribution diagrams for the first 18 elements in the periodic table.** In a standard periodic table (see Appendix B), information for each element is presented as shown for helium in the inset. In the diagrams in this table, electrons are represented as yellow dots and electron shells as concentric circles. These diagrams are a convenient way to picture the distribution of an atom's electrons among its electron shells, but these simplified models do not accurately represent the shape of the atom or the location of its electrons. The elements are arranged in rows, each representing the filling of an electron shell. As electrons are added, they occupy the lowest available shell.

? *What is the atomic number of magnesium? How many protons and electrons does it have? How many electron shells? How many valence electrons?*

Electron Orbitals

In the early 1900s, the electron shells of an atom were visualized as concentric paths of electrons orbiting the nucleus, somewhat like planets orbiting the sun. It is still convenient to use two-dimensional concentric-circle diagrams, as in Figure 2.9, to symbolize three-dimensional electron shells. However, you need to remember that each concentric circle represents only the *average* distance between an electron in that shell and the nucleus. Accordingly, the concentric-circle diagrams do not give a real picture of an atom. In reality, we can never know the exact location of an electron. What we can do instead is describe the space in which an electron spends most of its time. The three-dimensional space where an electron is found 90% of the time is called an **orbital**.

Each electron shell contains electrons at a particular energy level, distributed among a specific number of orbitals of distinctive shapes and orientations. **Figure 2.10** shows the orbitals of neon as an example, with its electron distribution diagram for reference. You can think of an orbital as a component of an electron shell. The first electron shell has only one spherical *s* orbital (called 1*s*), but the second shell has four orbitals: one large spherical *s* orbital (called 2*s*) and three dumbbell-shaped *p* orbitals (called 2*p* orbitals). (The third shell and other higher electron shells also have *s* and *p* orbitals, as well as orbitals of more complex shapes.)

No more than 2 electrons can occupy a single orbital. The first electron shell can therefore accommodate up to 2 electrons in its *s* orbital. The lone electron of a hydrogen atom occupies the 1*s* orbital, as do the 2 electrons of a helium atom. The four orbitals of the second electron shell can hold up to 8 electrons, 2 in each orbital. Electrons in each of the four orbitals have nearly the same energy, but they move in different volumes of space.

The reactivity of atoms arises from the presence of unpaired electrons in one or more orbitals of their valence shells. As you will see in the next section, atoms interact in a way that completes their valence shells. When they do so, it is the *unpaired* electrons that are involved.

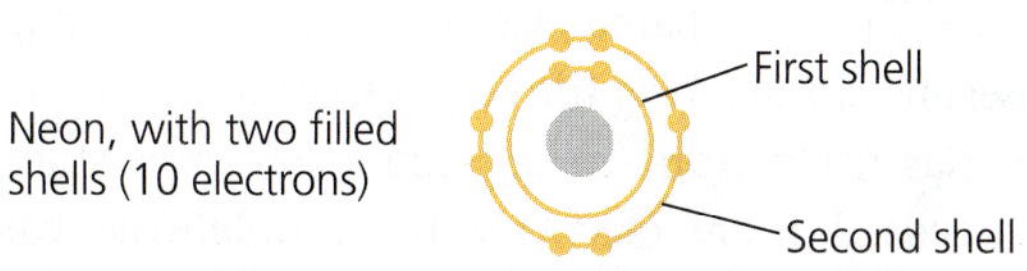

(a) Electron distribution diagram. An electron distribution diagram is shown here for a neon atom, which has a total of 10 electrons. Each concentric circle represents an electron shell, which can be subdivided into electron orbitals.

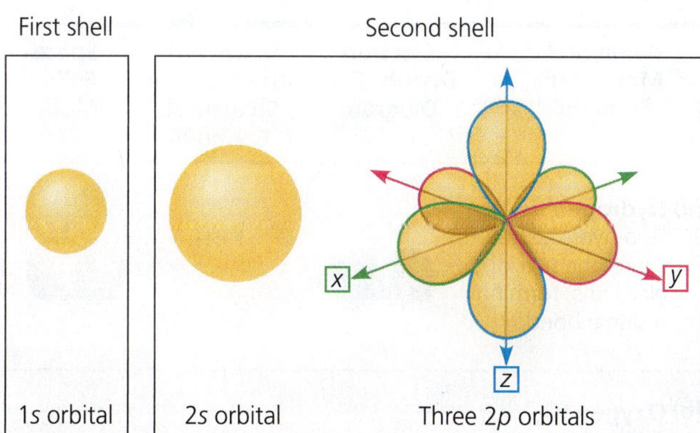

(b) Separate electron orbitals. The three-dimensional shapes represent electron orbitals—the volumes of space where the electrons of an atom are most likely to be found. Each orbital holds a maximum of 2 electrons. The first electron shell, on the left, has one spherical (*s*) orbital, designated 1*s*. The second shell, on the right, has one larger *s* orbital (designated 2*s* for the second shell) plus three dumbbell-shaped orbitals called *p* orbitals (2*p* for the second shell). The three 2*p* orbitals lie at right angles to one another along imaginary *x*-, *y*-, and *z*-axes of the atom. Each 2*p* orbital is outlined here in a different color.

(c) Superimposed electron orbitals. To reveal the complete picture of the electron orbitals of neon, we superimpose the 1*s* orbital of the first shell and the 2*s* and three 2*p* orbitals of the second shell.

▲ **Figure 2.10 Electron orbitals.**

CONCEPT CHECK 2.2

1. A lithium atom has 3 protons and 4 neutrons. What is its atomic mass in daltons?
2. A nitrogen atom has 7 protons, and the most common isotope of nitrogen has 7 neutrons. A radioactive isotope of nitrogen has 8 neutrons. Write the atomic number and mass number of this radioactive nitrogen as a chemical symbol with a subscript and superscript.
3. How many electrons does fluorine have? How many electron shells? Name the orbitals that are occupied. How many electrons are needed to fill the valence shell?
4. **WHAT IF?** In Figure 2.9, if two or more elements are in the same row, what do they have in common? If two or more elements are in the same column, what do they have in common?

For suggested answers, see Appendix A.

CONCEPT 2.3

The formation and function of molecules depend on chemical bonding between atoms

Now that we have looked at the structure of atoms, we can move up the hierarchy of organization and see how atoms combine to form molecules and ionic compounds. Atoms with incomplete valence shells can interact with certain other atoms in such a way that each partner completes its valence shell: The atoms either share or transfer valence electrons. These interactions usually result in atoms staying close together, held by attractions called **chemical bonds**. The strongest kinds of chemical bonds are covalent bonds and ionic bonds.

Covalent Bonds

A **covalent bond** is the sharing of a pair of valence electrons by two atoms. For example, let's consider what happens when two hydrogen atoms approach each other. Recall that hydrogen has 1 valence electron in the first shell, but the shell's capacity is 2 electrons. When the two hydrogen atoms come close enough for their 1*s* orbitals to overlap, they can share their electrons **(Figure 2.11)**. Each hydrogen atom is now associated with 2 electrons in what amounts to a completed valence shell. Two or more atoms held together by covalent bonds constitute a **molecule**, in this case a hydrogen molecule.

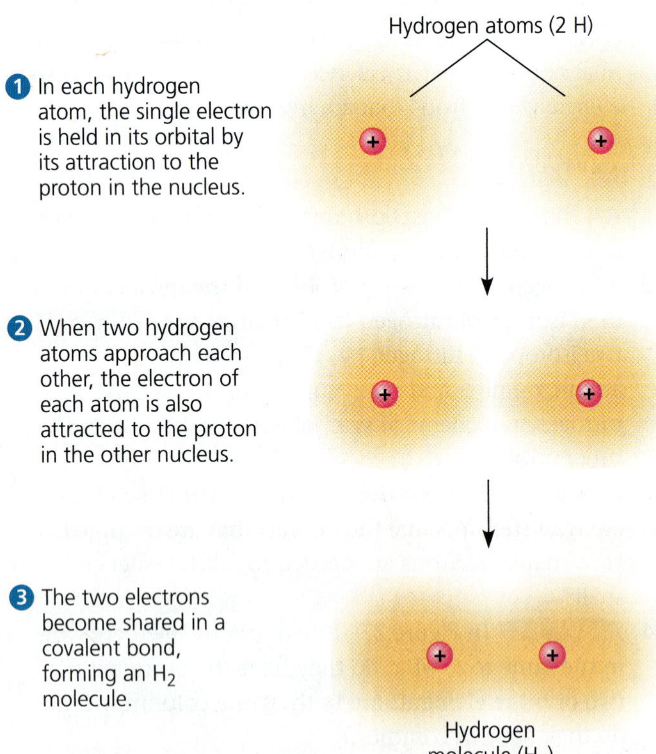

▲ **Figure 2.11 Formation of a covalent bond.**

Figure 2.12a shows several ways of representing a hydrogen molecule. Its *molecular formula*, H_2, simply indicates that the molecule consists of two atoms of hydrogen. Electron sharing can be depicted by an electron distribution diagram or by a *Lewis dot structure*, in which element symbols are surrounded by dots that represent the valence electrons (H:H). We can also use a *structural formula*, H—H, where the line represents a **single bond**, a pair of shared electrons. A space-filling model comes closest to representing the actual shape of the molecule.

Oxygen has 6 electrons in its second electron shell and therefore needs 2 more electrons to complete its valence shell. Two oxygen atoms form a molecule by sharing *two* pairs of valence electrons **(Figure 2.12b)**. The atoms are thus joined by a **double bond** (O═O).

Name and Molecular Formula	Electron Distribution Diagram	Lewis Dot Structure and Structural Formula	Space-Filling Model
(a) Hydrogen (H_2). Two hydrogen atoms share one pair of electrons, forming a single bond.	H H	H:H H—H	
(b) Oxygen (O_2). Two oxygen atoms share two pairs of electrons, forming a double bond.	O O	Ö::Ö O═O	
(c) Water (H_2O). Two hydrogen atoms and one oxygen atom are joined by single bonds, forming a molecule of water.	O H H	:Ö:H H O—H | H	
(d) Methane (CH_4). Four hydrogen atoms can satisfy the valence of one carbon atom, forming methane.	H H C H H	H H:C:H H H | H—C—H | H	

▲ **Figure 2.12 Covalent bonding in four molecules.** The number of electrons required to complete an atom's valence shell generally determines how many covalent bonds that atom will form. This figure shows several ways of indicating covalent bonds.

Each atom that can share valence electrons has a bonding capacity corresponding to the number of covalent bonds the atom can form. When the bonds form, they give the atom a full complement of electrons in the valence shell. The bonding capacity of oxygen, for example, is 2. This bonding capacity is called the atom's **valence** and usually equals the number of unpaired electrons required to complete the atom's outermost (valence) shell. See if you can determine the valences of hydrogen, oxygen, nitrogen, and carbon by studying the electron distribution diagrams in Figure 2.9. You can see that the valence of hydrogen is 1; oxygen, 2; nitrogen, 3; and carbon, 4. However, the situation is more complicated for elements in the third row of the periodic table. Phosphorus, for example, can have a valence of 3, as we would predict from the presence of 3 unpaired electrons in its valence shell. In some molecules that are biologically important, however, phosphorus can form three single bonds and one double bond. Therefore, it can also have a valence of 5.

The molecules H_2 and O_2 are pure elements rather than compounds because a compound is a combination of two or more *different* elements. Water, with the molecular formula H_2O, is a compound. Two atoms of hydrogen are needed to satisfy the valence of one oxygen atom. **Figure 2.12c** shows the structure of a water molecule. Water is so important to life that Chapter 3 is devoted entirely to its structure and behavior.

Methane, the main component of natural gas, is a compound with the molecular formula CH_4. It takes four hydrogen atoms, each with a valence of 1, to complement one atom of carbon, with its valence of 4 **(Figure 2.12d)**. We will look at many other compounds of carbon in Chapter 4.

Atoms in a molecule attract shared electrons to varying degrees, depending on the element. The attraction of a particular atom for the electrons of a covalent bond is called its **electronegativity**. The more electronegative an atom is, the more strongly it pulls shared electrons toward itself. In a covalent bond between two atoms of the same element, the electrons are shared equally because the two atoms have the same electronegativity—the tug-of-war is at a standoff. Such a bond is called a **nonpolar covalent bond**. For example, the single bond of H_2 is nonpolar, as is the double bond of O_2. However, when one atom is bonded to a more electronegative atom, the electrons of the bond are not shared equally. This type of bond is called a **polar covalent bond**. Such bonds vary in their polarity, depending on the relative electronegativity of the two atoms. For example, the bonds between the oxygen and hydrogen atoms of a water molecule are quite polar **(Figure 2.13)**.

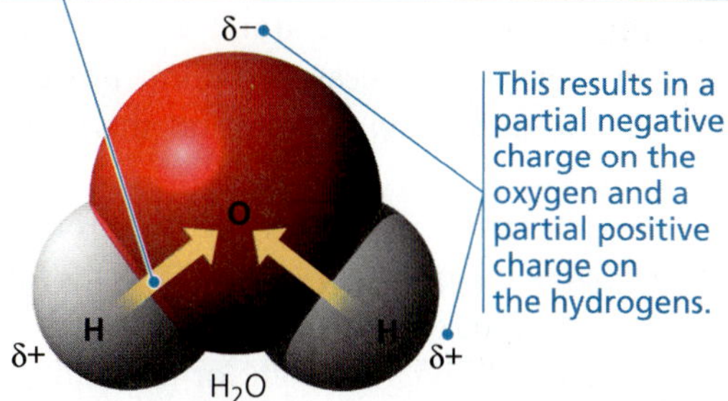

▲ **Figure 2.13 Polar covalent bonds in a water molecule.**

Oxygen is one of the most electronegative of all the elements, attracting shared electrons much more strongly than hydrogen does. In a covalent bond between oxygen and hydrogen, the electrons spend more time near the oxygen nucleus than they do near the hydrogen nucleus. Because electrons have a negative charge and are pulled toward oxygen in a water molecule, the oxygen atom has a partial negative charge (indicated by the Greek letter δ with a minus sign, δ–, or "delta minus"), and each hydrogen atom has a partial positive charge (δ+, or "delta plus"). In contrast, the individual bonds of methane (CH_4) are much less polar because the electronegativities of carbon and hydrogen are similar.

Ionic Bonds

In some cases, two atoms are so unequal in their attraction for valence electrons that the more electronegative atom strips an electron completely away from its partner. This is what happens when an atom of sodium ($_{11}Na$) encounters an atom of chlorine ($_{17}Cl$) **(Figure 2.14)**. A sodium atom has a total of 11 electrons, with its single valence electron in the third electron shell. A chlorine atom has a total of 17 electrons,

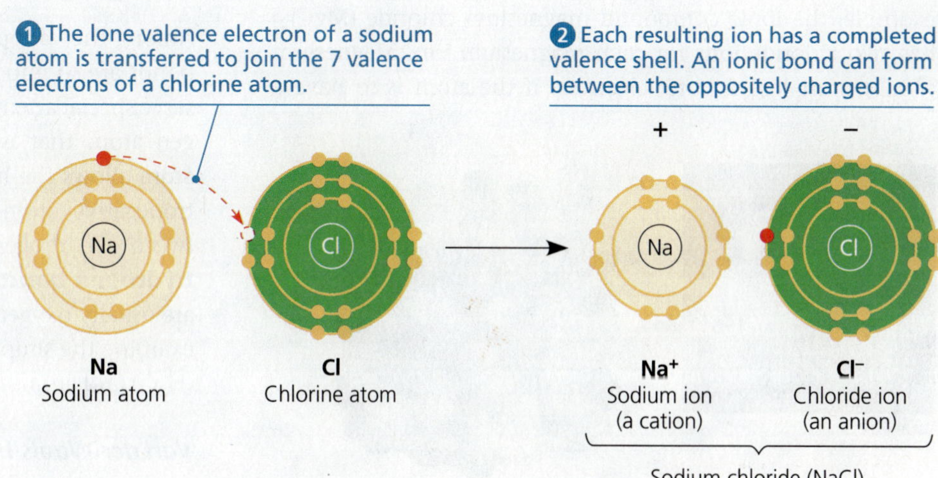

▲ **Figure 2.14 Electron transfer and ionic bonding.** The attraction between oppositely charged atoms, or ions, is an ionic bond. An ionic bond can form between any two oppositely charged ions, even if they have not been formed by transfer of an electron from one to the other.

with 7 electrons in its valence shell. When these two atoms meet, the lone valence electron of sodium is transferred to the chlorine atom, and both atoms end up with their valence shells complete. (Because sodium no longer has an electron in the third shell, the second shell is now the valence shell.)

The electron transfer between the two atoms moves one unit of negative charge from sodium to chlorine. Sodium, now with 11 protons but only 10 electrons, has a net electrical charge of 1+. A charged atom (or molecule) is called an **ion**. When the charge is positive, the ion is specifically called a **cation**; the sodium atom has become a cation. Conversely, the chlorine atom, having gained an extra electron, now has 17 protons and 18 electrons, giving it a net electrical charge of 1–. It has become a chloride ion—an **anion**, or negatively charged ion. Because of their opposite charges, cations and anions attract each other; this attraction is called an **ionic bond**. The transfer of an electron is not the formation of a bond; rather, it allows a bond to form because it results in two ions of opposite charge. Any two ions of opposite charge can form an ionic bond. The ions do not need to have acquired their charge by an electron transfer with each other.

Compounds formed by ionic bonds are called **ionic compounds**, or **salts**. We know the ionic compound sodium chloride (NaCl) as table salt **(Figure 2.15)**. Salts are often found in nature as crystals of various sizes and shapes. Each salt crystal is an aggregate of vast numbers of cations and anions bonded by their electrical attraction and arranged in a three-dimensional lattice. Unlike a covalent compound, which consists of molecules having a definite size and number of atoms, an ionic compound does not consist of molecules. The formula for an ionic compound, such as NaCl, indicates only the ratio of elements in a crystal of the salt. "NaCl" by itself is not a molecule.

Not all salts have equal numbers of cations and anions. For example, the ionic compound magnesium chloride ($MgCl_2$) has two chloride ions for each magnesium ion. Magnesium ($_{12}Mg$) must lose 2 outer electrons if the atom is to have a complete valence shell, so it tends to become a cation with a net charge of 2+ (Mg^{2+}). One magnesium cation can therefore form ionic bonds with two chloride anions.

▲ **Figure 2.15 A sodium chloride (NaCl) crystal.** The sodium ions (Na^+) and chloride ions (Cl^-) are held together by ionic bonds. The formula NaCl tells us that the ratio of Na^+ to Cl^- is 1:1.

The term *ion* also applies to entire molecules that are electrically charged. In the salt ammonium chloride (NH_4Cl), for instance, the anion is a single chloride ion (Cl^-), but the cation is ammonium (NH_4^+), a nitrogen atom with four covalently bonded hydrogen atoms. The whole ammonium ion has an electrical charge of 1+ because it is 1 electron short.

Environment affects the strength of ionic bonds. In a dry salt crystal, the bonds are so strong that it takes a hammer and chisel to break enough of them to crack the crystal in two. If the same salt crystal is dissolved in water, however, the ionic bonds are much weaker because each ion is partially shielded by its interactions with water molecules. Most drugs are manufactured as salts because they are quite stable when dry but can dissociate (come apart) easily in water. In the next chapter, you will learn how water dissolves salts.

Weak Chemical Bonds

In organisms, most of the strongest chemical bonds are covalent bonds, which link atoms to form a cell's molecules. But weaker bonding within and between molecules is also indispensable in the cell, contributing greatly to the emergent properties of life. Many large biological molecules are held in their functional form by weak bonds. In addition, when two molecules in the cell make contact, they may adhere temporarily by weak bonds. The reversibility of weak bonding can be an advantage: Two molecules can come together, respond to one another in some way, and then separate.

Several types of weak chemical bonds are important in organisms. One is the ionic bond as it exists between ions dissociated in water, which we just discussed. Hydrogen bonds and van der Waals interactions are also crucial to life.

Hydrogen Bonds

Among the various kinds of weak chemical bonds, hydrogen bonds are so important in the chemistry of life that they deserve special attention. The partial positive charge on a hydrogen atom that is covalently bonded to an electronegative atom allows the hydrogen to be attracted to a different electronegative atom nearby. This noncovalent attraction between a hydrogen and an electronegative atom is called a **hydrogen bond**. In living cells, the electronegative partners are usually oxygen or nitrogen atoms. Refer to **Figure 2.16** to examine the simple case of hydrogen bonding between water (H_2O) and ammonia (NH_3).

Van der Waals Interactions

Even a molecule with nonpolar covalent bonds may have positively and negatively charged regions. Electrons are not always symmetrically distributed in such a molecule; at any

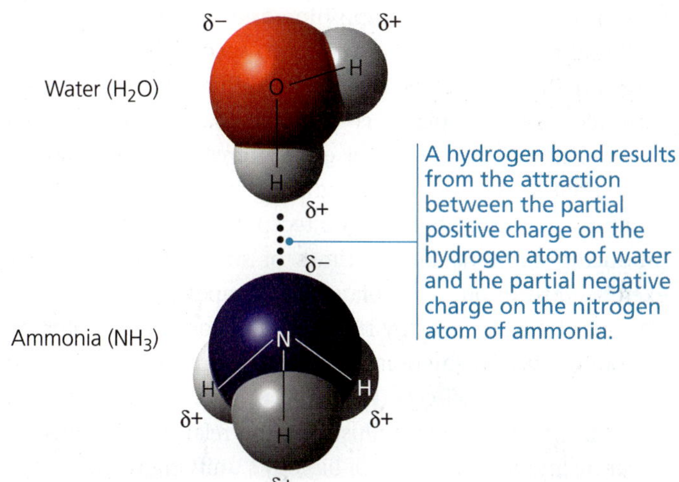

▲ **Figure 2.16 A hydrogen bond.**

DRAW IT *Draw five water molecules using structural formulas and indicating partial charges, and show how they can make hydrogen bonds with each other.*

instant, they may accumulate by chance in one part of the molecule or another. The results are ever-changing regions of positive and negative charge that enable all atoms and molecules to stick to one another. These **van der Waals interactions** are individually weak and occur only when atoms and molecules are very close together. When many such interactions occur simultaneously, however, they can be powerful: Van der Waals interactions are the reason a gecko lizard (right) can walk straight up a wall! Each gecko toe has hundreds of thousands of tiny hairs, with multiple projections at each hair's tip that increase surface area. Apparently, the van der Waals interactions between the hair tip molecules and the molecules of the wall's surface are so numerous that despite their individual weakness, together they can support the gecko's body weight.

Van der Waals interactions, hydrogen bonds, ionic bonds in water, and other weak bonds may form not only between molecules but also between parts of a large molecule, such as a protein. The cumulative effect of weak bonds is to reinforce the three-dimensional shape of the molecule. You will learn more about the very important biological roles of weak bonds in Chapter 5.

Molecular Shape and Function

A molecule has a characteristic size and shape. The precise shape of a molecule is usually very important to its function in the living cell.

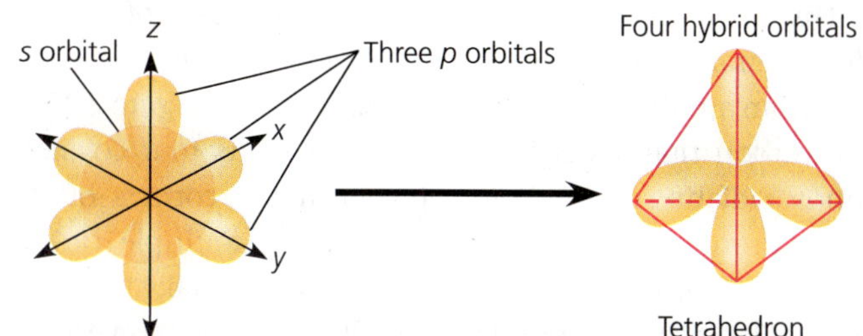

(a) Hybridization of orbitals. The single *s* and three *p* orbitals of a valence shell involved in covalent bonding combine to form four teardrop-shaped hybrid orbitals. These orbitals extend to the four corners of an imaginary tetrahedron (outlined in pink).

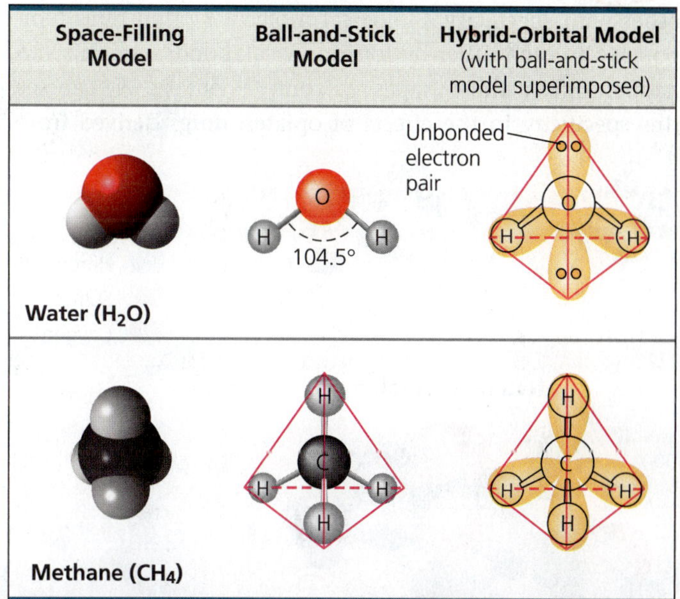

(b) Molecular-shape models. Three models representing molecular shape are shown for water and methane. The positions of the hybrid orbitals determine the shapes of the molecules.

▲ **Figure 2.17 Molecular shapes due to hybrid orbitals.**

A molecule consisting of two atoms, such as H_2 or O_2, is always linear, but most molecules with more than two atoms have more complicated shapes. These shapes are determined by the positions of the atoms' orbitals. When an atom forms covalent bonds, the orbitals in its valence shell undergo rearrangement. For atoms with valence electrons in both *s* and *p* orbitals (review Figure 2.10), the single *s* and three *p* orbitals form four new hybrid orbitals shaped like identical teardrops extending from the region of the atomic nucleus **(Figure 2.17a)**. If we connect the larger ends of the teardrops with lines, we have the outline of a geometric shape called a tetrahedron, a pyramid with a triangular base.

For the water molecule (H_2O), two of the hybrid orbitals in the oxygen atom's valence shell are shared with hydrogen atoms **(Figure 2.17b)**. The result is a molecule shaped roughly like a V, with its two covalent bonds spread apart at an angle of 104.5°.

The methane molecule (CH_4) has the shape of a completed tetrahedron because all four hybrid orbitals of the carbon atom are shared with hydrogen atoms (see Figure 2.17b). The carbon nucleus is at the center, with its four covalent bonds radiating to hydrogen nuclei at the corners of the tetrahedron. Larger molecules containing multiple carbon atoms, including many of the molecules that make up living matter, have more complex overall shapes. However, the tetrahedral shape of a carbon atom bonded to four other atoms is often a repeating motif within such molecules.

Molecular shape is crucial in biology because it determines how biological molecules recognize and respond to one another with specificity. Biological molecules often bind temporarily to each other by forming weak bonds, but this can happen only if their shapes are complementary. We can see this specificity in the effects of opiates, drugs derived from opium. Opiates, such as morphine and heroin, relieve pain and alter mood by weakly binding to specific receptor molecules on the surfaces of brain cells. Why would brain cells carry receptors for opiates, compounds that are not made by our bodies? The discovery of endorphins in 1975 answered this question. Endorphins are signaling molecules made by the pituitary gland that bind to the receptors, relieving pain and producing euphoria during times of stress, such as intense exercise. It turns out that opiates have shapes similar to endorphins and mimic them by binding to endorphin receptors in the brain. That is why opiates (such as morphine) and endorphins have similar effects **(Figure 2.18)**. The role of molecular shape in brain chemistry illustrates the relationship between structure and function, one of biology's unifying themes.

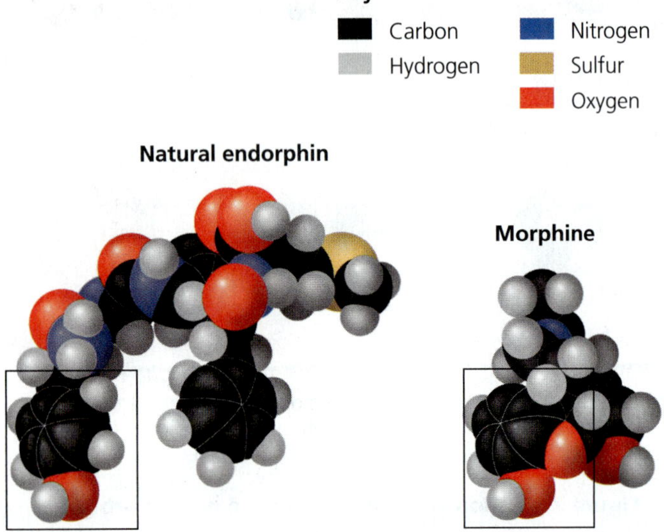

(a) Structures of endorphin and morphine. The boxed portion of the endorphin molecule (left) binds to receptor molecules on target cells in the brain. The boxed portion of the morphine molecule (right) is a close match.

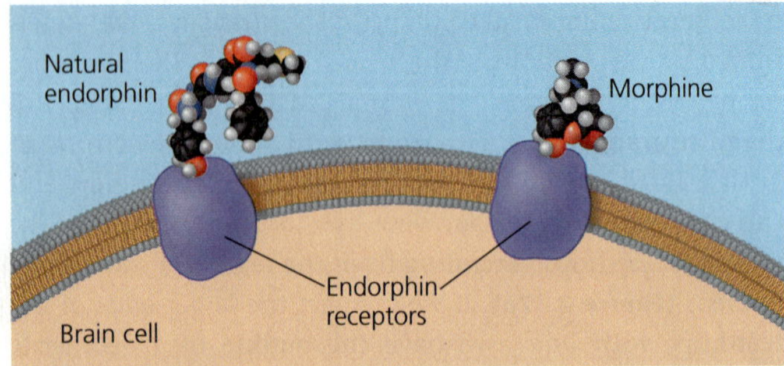

(b) Binding to endorphin receptors. Both endorphin and morphine can bind to endorphin receptors on the surface of a brain cell.

▲ **Figure 2.18 A molecular mimic.** Morphine affects pain perception and emotional state by mimicking the brain's natural endorphins.

CONCEPT CHECK 2.3

1. Why does the structure H—C=C—H fail to make sense chemically?
2. What holds the atoms together in a crystal of magnesium chloride ($MgCl_2$)?
3. **WHAT IF?** If you were a pharmaceutical researcher, why would you want to learn the three-dimensional shapes of naturally occurring signaling molecules?

For suggested answers, see Appendix A.

CONCEPT 2.4

Chemical reactions make and break chemical bonds

The making and breaking of chemical bonds, leading to changes in the composition of matter, are called **chemical reactions**. An example is the reaction between hydrogen and oxygen molecules that forms water:

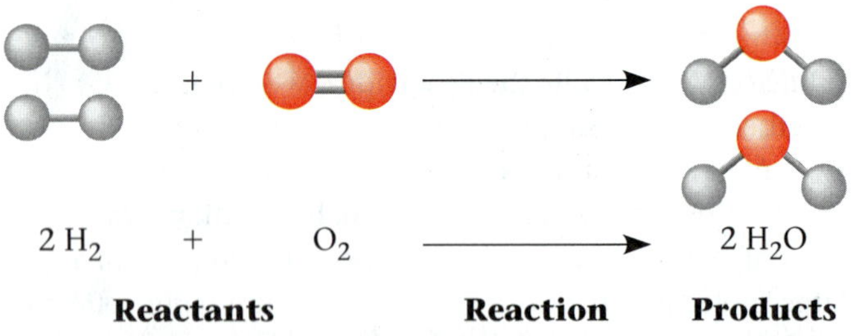

This reaction breaks the covalent bonds of H_2 and O_2 and forms the new bonds of H_2O. When we write a chemical reaction, we use an arrow to indicate the conversion of the starting materials, called the **reactants**, to the **products**. The coefficients indicate the number of molecules involved; for example, the coefficient 2 in front of the H_2 means that

▲ **Figure 2.19 Photosynthesis: a solar-powered rearrangement of matter.** *Elodea*, a freshwater plant, produces sugar by rearranging the atoms of carbon dioxide and water in the chemical process known as photosynthesis, which is powered by sunlight. Much of the sugar is then converted to other food molecules. Oxygen gas (O_2) is a by-product of photosynthesis; notice the bubbles of oxygen escaping from the leaves in the photo.

? *Explain how this photo relates to the reactants and products in the equation for photosynthesis given in the text. (You will learn more about photosynthesis in Chapter 10.)*

the reaction starts with two molecules of hydrogen. Notice that all atoms of the reactants must be accounted for in the products. Matter is conserved in a chemical reaction: Reactions cannot create or destroy matter but can only rearrange it.

Photosynthesis, which takes place within the cells of green plant tissues, is a particularly important example of how chemical reactions rearrange matter. Humans and other animals ultimately depend on photosynthesis for food and oxygen, and this process is at the foundation of almost all ecosystems. The following chemical shorthand summarizes the process of photosynthesis:

$$6\ CO_2 + 6\ H_2O \rightarrow C_6H_{12}O_6 + 6\ O_2$$

The raw materials of photosynthesis are carbon dioxide (CO_2), which is taken from the air, and water (H_2O), which is absorbed from the soil. Within the plant cells, sunlight powers the conversion of these ingredients to a sugar called glucose ($C_6H_{12}O_6$) and oxygen molecules (O_2), a by-product that the plant releases into the surroundings **(Figure 2.19)**. Although photosynthesis is actually a sequence of many chemical reactions, we still end up with the same number and types of atoms that we had when we started. Matter has simply been rearranged, with an input of energy provided by sunlight.

All chemical reactions are reversible, with the products of the forward reaction becoming the reactants for the reverse reaction. For example, hydrogen and nitrogen molecules can combine to form ammonia, but ammonia can also decompose to regenerate hydrogen and nitrogen:

$$3\ H_2 + N_2 \rightleftharpoons 2\ NH_3$$

The two opposite-headed arrows indicate that the reaction is reversible.

One of the factors affecting the rate of a reaction is the concentration of reactants. The greater the concentration of reactant molecules, the more frequently they collide with one another and have an opportunity to react and form products. The same holds true for products. As products accumulate, collisions resulting in the reverse reaction become more frequent. Eventually, the forward and reverse reactions occur at the same rate, and the relative concentrations of products and reactants stop changing. The point at which the reactions offset one another exactly is called **chemical equilibrium**. This is a dynamic equilibrium; reactions are still going on, but with no net effect on the concentrations of reactants and products. Equilibrium does *not* mean that the reactants and products are equal in concentration, but only that their concentrations have stabilized at a particular ratio. The reaction involving ammonia reaches equilibrium when ammonia decomposes as rapidly as it forms. In some chemical reactions, the equilibrium point may lie so far to the right that these reactions go essentially to completion; that is, virtually all the reactants are converted to products.

We will return to the subject of chemical reactions after more detailed study of the various types of molecules that are important to life. In the next chapter, we focus on water, the substance in which all the chemical processes of organisms occur.

CONCEPT CHECK 2.4

1. **MAKE CONNECTIONS** Consider the reaction between hydrogen and oxygen that forms water, shown with ball-and-stick models on page 42. Study Figure 2.12 and draw the Lewis dot structures representing this reaction.
2. Which type of chemical reaction occurs faster at equilibrium, the formation of products from reactants or reactants from products?
3. **WHAT IF?** Write an equation that uses the products of photosynthesis as reactants and the reactants of photosynthesis as products. Add energy as another product. This new equation describes a process that occurs in your cells. Describe this equation in words. How does this equation relate to breathing?

For suggested answers, see Appendix A.

2 CHAPTER REVIEW

SUMMARY OF KEY CONCEPTS

CONCEPT 2.1

Matter consists of chemical elements in pure form and in combinations called compounds (pp. 31–32)

- **Elements** cannot be broken down chemically to other substances. A **compound** contains two or more different elements in a fixed ratio. Oxygen, carbon, hydrogen, and nitrogen make up approximately 96% of living matter.

? *In what way does the need for iodine or iron in your diet differ from your need for calcium or phosphorus?*

CONCEPT 2.2

An element's properties depend on the structure of its atoms (pp. 33–37)

- An **atom**, the smallest unit of an element, has the following components:

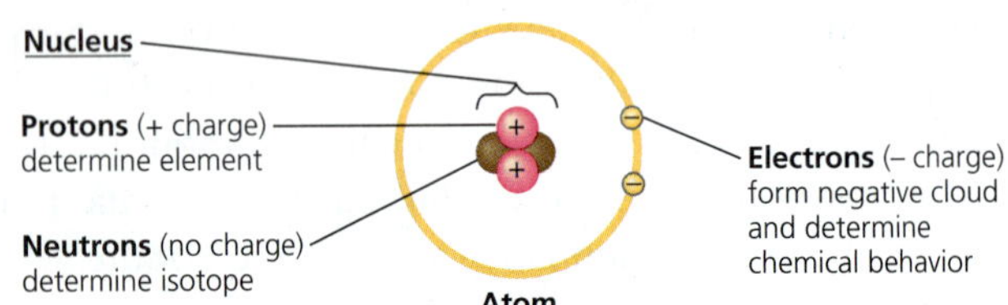

- An electrically neutral atom has equal numbers of electrons and protons; the number of protons determines the **atomic number**. The **atomic mass** is measured in **daltons** and is roughly equal to the sum of protons plus neutrons. **Isotopes** of an element differ from each other in neutron number and therefore mass. Unstable isotopes give off particles and energy as radioactivity.
- In an atom, electrons occupy specific **electron shells**; the electrons in a shell have a characteristic energy level. Electron distribution in shells determines the chemical behavior of an atom. An atom that has an incomplete outer shell, the **valence shell**, is reactive.
- Electrons exist in **orbitals**, three-dimensional spaces with specific shapes that are components of electron shells.

Electron orbitals

DRAW IT *Draw the electron distribution diagrams for neon ($_{10}$Ne) and argon ($_{18}$Ar). Use these diagrams to explain why these elements are chemically unreactive.*

CONCEPT 2.3

The formation and function of molecules depend on chemical bonding between atoms (pp. 38–42)

- **Chemical bonds** form when atoms interact and complete their valence shells. **Covalent bonds** form when pairs of electrons are shared.

H· + H· ⟶ H:H
Single covalent bond

:Ö· + ·Ö: ⟶ Ö::Ö
Double covalent bond

- **Molecules** consist of two or more covalently bonded atoms. The attraction of an atom for the electrons of a covalent bond is its **electronegativity**. If both atoms are the same, they have the same electronegativity and share a **nonpolar covalent bond**. Electrons of a **polar covalent bond** are pulled closer to the more electronegative atom.
- An **ion** forms when an atom or molecule gains or loses an electron and becomes charged. An **ionic bond** is the attraction between two oppositely charged ions.

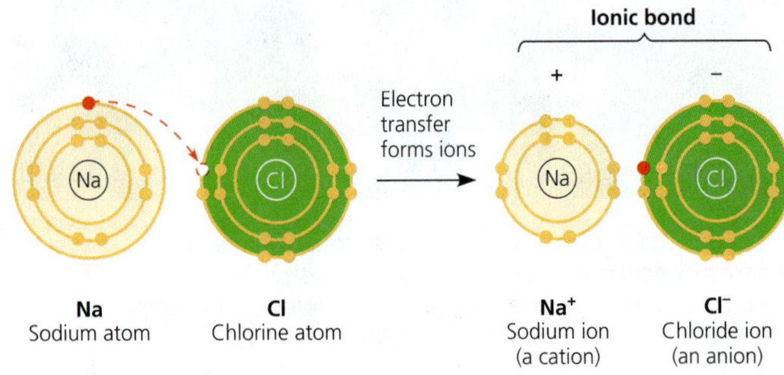

- Weak bonds reinforce the shapes of large molecules and help molecules adhere to each other. A **hydrogen bond** is an attraction between a hydrogen atom carrying a partial positive charge ($\delta+$) and an electronegative atom ($\delta-$). **Van der Waals interactions** occur between transiently positive and negative regions of molecules.
- A molecule's shape is determined by the positions of its atoms' valence orbitals. Covalent bonds result in hybrid orbitals, which are responsible for the shapes of H_2O, CH_4, and many more complex biological molecules. Shape is usually the basis for the recognition of one biological molecule by another.

? *In terms of electron sharing between atoms, compare nonpolar covalent bonds, polar covalent bonds, and the formation of ions.*

CONCEPT 2.4

Chemical reactions make and break chemical bonds (pp. 42–43)

- **Chemical reactions** change **reactants** into **products** while conserving matter. All chemical reactions are theoretically reversible. **Chemical equilibrium** is reached when the forward and reverse reaction rates are equal.

? *What would happen to the concentration of products if more reactants were added to a reaction that was in chemical equilibrium? How would this addition affect the equilibrium?*

TEST YOUR UNDERSTANDING

LEVEL 1: KNOWLEDGE/COMPREHENSION

1. In the term *trace element*, the adjective *trace* means that
   a. the element is required in very small amounts.
   b. the element can be used as a label to trace atoms through an organism's metabolism.
   c. the element is very rare on Earth.
   d. the element enhances health but is not essential for the organism's long-term survival.
   e. the element passes rapidly through the organism.

2. Compared with ^{31}P, the radioactive isotope ^{32}P has
   a. a different atomic number.
   b. a different charge.
   c. one more proton.
   d. one more electron.
   e. one more neutron.

3. The reactivity of an atom arises from
   a. the average distance of the outermost electron shell from the nucleus.
   b. the existence of unpaired electrons in the valence shell.
   c. the sum of the potential energies of all the electron shells.
   d. the potential energy of the valence shell.
   e. the energy difference between the *s* and *p* orbitals.

4. Which statement is true of all atoms that are anions?
   a. The atom has more electrons than protons.
   b. The atom has more protons than electrons.
   c. The atom has fewer protons than does a neutral atom of the same element.
   d. The atom has more neutrons than protons.
   e. The net charge is 1−.

5. Which of the following statements correctly describes any chemical reaction that has reached equilibrium?
   a. The concentrations of products and reactants are equal.
   b. The reaction is now irreversible.
   c. Both forward and reverse reactions have halted.
   d. The rates of the forward and reverse reactions are equal.
   e. No reactants remain.

LEVEL 2: APPLICATION/ANALYSIS

6. We can represent atoms by listing the number of protons, neutrons, and electrons—for example, $2p^+, 2n^0, 2e^-$ for helium. Which of the following represents the ^{18}O isotope of oxygen?
   a. $6p^+, 8n^0, 6e^-$
   b. $8p^+, 10n^0, 8e^-$
   c. $9p^+, 9n^0, 9e^-$
   d. $7p^+, 2n^0, 9e^-$
   e. $10p^+, 8n^0, 9e^-$

7. The atomic number of sulfur is 16. Sulfur combines with hydrogen by covalent bonding to form a compound, hydrogen sulfide. Based on the number of valence electrons in a sulfur atom, predict the molecular formula of the compound.
   a. HS b. HS_2 c. H_2S d. H_3S_2 e. H_4S

8. What coefficients must be placed in the following blanks so that all atoms are accounted for in the products?

   $C_6H_{12}O_6 \rightarrow$ ____ C_2H_6O + ____ CO_2

   a. 1; 2 b. 3; 1 c. 1; 3 d. 1; 1 e. 2; 2

9. **DRAW IT** Draw Lewis dot structures for each hypothetical molecule shown below, using the correct number of valence electrons for each atom. Determine which molecule makes sense because each atom has a complete valence shell and each bond has the correct number of electrons. Explain what makes the other molecules nonsensical, considering the number of bonds each type of atom can make.

```
(a)  O=C—H

        H   H
        |   |
     H—O—C—C=O
        |
(b)     H

        H       H
        |       |
(c)  H—C—H—C=O
        |
        H

          O
          |
(d)  H—N=H
```

LEVEL 3: SYNTHESIS/EVALUATION

10. **EVOLUTION CONNECTION**

    The percentages of naturally occurring elements making up the human body (see Table 2.1) are similar to the percentages of these elements found in other organisms. How could you account for this similarity among organisms?

11. **SCIENTIFIC INQUIRY**

    Female silkworm moths (*Bombyx mori*) attract males by emitting chemical signals that spread through the air. A male hundreds of meters away can detect these molecules and fly toward their source. The sensory organs responsible for this behavior are the comblike antennae visible in the photograph shown here. Each filament of an antenna is equipped with thousands of receptor cells that detect the sex attractant. Based on what you learned in this chapter, propose a hypothesis to account for the ability of the male moth to detect a specific molecule in the presence of many other molecules in the air. What predictions does your hypothesis make? Design an experiment to test one of these predictions.

12. **WRITE ABOUT A THEME**

    Emergent Properties While waiting at an airport, Neil Campbell once overheard this claim: "It's paranoid and ignorant to worry about industry or agriculture contaminating the environment with their chemical wastes. After all, this stuff is just made of the same atoms that were already present in our environment." Drawing on your knowledge of electron distribution, bonding, and the theme of emergent properties (pp. 3–5), write a short essay (100–150 words) countering this argument.

For selected answers, see Appendix A.

MasteringBIOLOGY www.masteringbiology.com

1. MasteringBiology® Assignments

Tutorials The Anatomy of Atoms • Atomic Number and Mass Number

Activities Structure of the Atomic Nucleus • Electron Arrangement • Covalent Bonds • Nonpolar and Polar Molecules • Ionic Bonds • Hydrogen Bonds

Questions Student Misconceptions • Reading Quiz • Multiple Choice • End-of-Chapter

2. eText

Read your book online, search, take notes, highlight text, and more.

3. The Study Area

Practice Tests • Cumulative Test • *BioFlix* 3-D Animations • MP3 Tutor Sessions • Videos • Activities • Investigations • Lab Media • Audio Glossary • Word Study Tools • Art

3

Water and Life

▲ **Figure 3.1 How does the habitat of a polar bear depend on the chemistry of water?**

KEY CONCEPTS

3.1 **Polar covalent bonds in water molecules result in hydrogen bonding**

3.2 **Four emergent properties of water contribute to Earth's suitability for life**

3.3 **Acidic and basic conditions affect living organisms**

OVERVIEW

The Molecule That Supports All of Life

As astronomers study newly discovered planets orbiting distant stars, they hope to find evidence of water on these far-off celestial bodies, for water is the substance that makes possible life as we know it here on Earth. All organisms familiar to us are made mostly of water and live in an environment dominated by water. Water is the biological medium here on Earth, and possibly on other planets as well.

Three-quarters of Earth's surface is covered by water. Although most of this water is in liquid form, water is also present on Earth as a solid (ice) and a gas (water vapor). Water is the only common substance to exist in the natural environment in all three physical states of matter. Furthermore, the solid state of water floats on the liquid, a rare property emerging from the chemistry of the water molecule. Ice can thus provide a hunting platform for the polar bear in **Figure 3.1**.

The abundance of water is a major reason Earth is habitable. In a classic book called *The Fitness of the Environment*, ecologist Lawrence Henderson highlighted the importance of water to life. While acknowledging that life adapts to its environment through natural selection, Henderson emphasized that for life to exist at all, the environment must first be suitable.

Life on Earth began in water and evolved there for 3 billion years before spreading onto land. Modern life, even terrestrial (land-dwelling) life, remains tied to water. All living organisms require water more than any other substance. Human beings, for example, can survive for quite a few weeks without food, but only a week or so without water. Molecules of water participate in many chemical reactions necessary to sustain life. Most cells are surrounded by water, and cells themselves are about 70–95% water.

What properties of the simple water molecule make it so indispensable to life on Earth? In this chapter, you will learn how the structure of a water molecule allows it to interact with other molecules, including other water molecules. This ability leads to water's unique emergent properties that help make Earth suitable for life.

CONCEPT **3.1**

Polar covalent bonds in water molecules result in hydrogen bonding

Water is so common that it is easy to overlook the fact that it is an exceptional substance with many extraordinary qualities. Following the theme of emergent properties, we can trace water's unique behavior to the structure and interactions of its molecules.

Studied on its own, the water molecule is deceptively simple. It is shaped like a wide V, with its two hydrogen atoms joined to the oxygen atom by single covalent bonds. Oxygen is more electronegative than hydrogen, so the electrons of the covalent bonds spend more time closer to oxygen than to hydrogen; these are **polar covalent bonds** (see Figure 2.13). This unequal sharing of electrons and water's V-like shape make it a **polar molecule**, meaning that its overall charge is unevenly distributed: The oxygen region of the molecule has a partial negative charge ($\delta-$), and each hydrogen has a partial positive charge ($\delta+$).

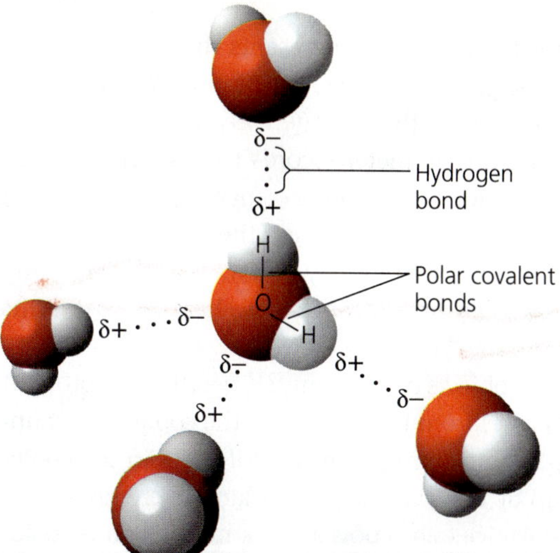

▲ **Figure 3.2 Hydrogen bonds between water molecules.** The charged regions in a water molecule are due to its polar covalent bonds. Oppositely charged regions of neighboring water molecules are attracted to each other, forming hydrogen bonds. Each molecule can hydrogen-bond to multiple partners, and these associations are constantly changing.

DRAW IT *Draw partial charges on all the atoms of the water molecule on the far left above, and draw two more water molecules hydrogen-bonded to it.*

The properties of water arise from attractions between oppositely charged atoms of different water molecules: The slightly positive hydrogen of one molecule is attracted to the slightly negative oxygen of a nearby molecule. The two molecules are thus held together by a hydrogen bond (**Figure 3.2**). When water is in its liquid form, its hydrogen bonds are very fragile, each about 1/20 as strong as a covalent bond. The hydrogen bonds form, break, and re-form with great frequency. Each lasts only a few trillionths of a second, but the molecules are constantly forming new hydrogen bonds with a succession of partners. Therefore, at any instant, a substantial percentage of all the water molecules are hydrogen-bonded to their neighbors. The extraordinary qualities of water are emergent properties resulting in large part from the hydrogen bonding that organizes water molecules into a higher level of structural order.

CONCEPT CHECK 3.1

1. **MAKE CONNECTIONS** What is electronegativity, and how does it affect interactions between water molecules? Review p. 39 and Figure 2.13.
2. Why is it unlikely that two neighboring water molecules would be arranged like this?

$$\begin{matrix} & H\ H & \\ O & & O \\ & H\ H & \end{matrix}$$

3. **WHAT IF?** What would be the effect on the properties of the water molecule if oxygen and hydrogen had equal electronegativity?

For suggested answers, see Appendix A.

CONCEPT **3.2**

Four emergent properties of water contribute to Earth's suitability for life

We will examine four emergent properties of water that contribute to Earth's suitability as an environment for life: cohesive behavior, ability to moderate temperature, expansion upon freezing, and versatility as a solvent.

Cohesion of Water Molecules

Water molecules stay close to each other as a result of hydrogen bonding. Although the arrangement of molecules in a sample of liquid water is constantly changing, at any given moment many of the molecules are linked by multiple hydrogen bonds. These linkages make water more structured than most other liquids. Collectively, the hydrogen bonds hold the substance together, a phenomenon called **cohesion**.

Cohesion due to hydrogen bonding contributes to the transport of water and dissolved nutrients against gravity in plants (**Figure 3.3**). Water from the roots reaches the leaves through a network of water-conducting cells. As water evaporates from a

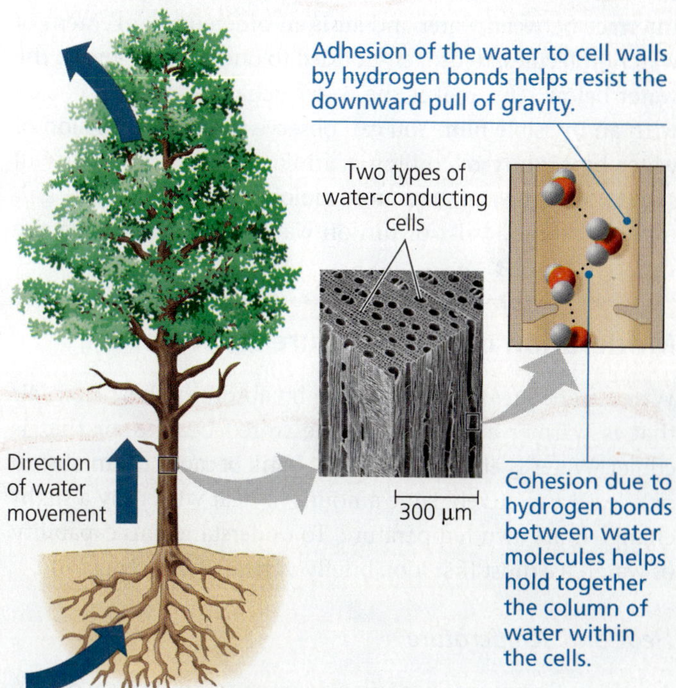

▲ **Figure 3.3 Water transport in plants.** Evaporation from leaves pulls water upward from the roots through water-conducting cells. Because of the properties of cohesion and adhesion, the tallest trees can transport water more than 100 m upward—approximately one-quarter the height of the Empire State Building in New York City.

BioFlix Visit the Study Area at **www.masteringbiology.com** for the BioFlix® 3-D Animation on Water Transport in Plants.

▲ **Figure 3.4 Walking on water.** The high surface tension of water, resulting from the collective strength of its hydrogen bonds, allows this raft spider to walk on the surface of a pond.

leaf, hydrogen bonds cause water molecules leaving the veins to tug on molecules farther down, and the upward pull is transmitted through the water-conducting cells all the way to the roots. **Adhesion**, the clinging of one substance to another, also plays a role. Adhesion of water to cell walls by hydrogen bonds helps counter the downward pull of gravity (see Figure 3.3).

Related to cohesion is **surface tension**, a measure of how difficult it is to stretch or break the surface of a liquid. Water has a greater surface tension than most other liquids. At the interface between water and air is an ordered arrangement of water molecules, hydrogen-bonded to one another and to the water below. This makes the water behave as though coated with an invisible film. You can observe the surface tension of water by slightly overfilling a drinking glass; the water will stand above the rim. In a more biological example, some animals can stand, walk, or run on water without breaking the surface (**Figure 3.4**).

Moderation of Temperature by Water

Water moderates air temperature by absorbing heat from air that is warmer and releasing the stored heat to air that is cooler. Water is effective as a heat bank because it can absorb or release a relatively large amount of heat with only a slight change in its own temperature. To understand this capability of water, we must first look briefly at heat and temperature.

Heat and Temperature

Anything that moves has **kinetic energy**, the energy of motion. Atoms and molecules have kinetic energy because they are always moving, although not necessarily in any particular direction. The faster a molecule moves, the greater its kinetic energy. **Heat** is a form of energy. For a given body of matter, the amount of heat is a measure of the matter's *total* kinetic energy due to motion of its molecules; thus, heat depends in part on the matter's volume. Although heat is related to temperature, they are not the same thing. **Temperature** is a measure of heat intensity that represents the *average* kinetic energy of the molecules, regardless of volume. When water is heated in a coffeemaker, the average speed of the molecules increases, and the thermometer records this as a rise in temperature of the liquid. The amount of heat also increases in this case. Note, however, that although the pot of coffee has a much higher temperature than, say, the water in a swimming pool, the swimming pool contains more heat because of its much greater volume.

Whenever two objects of different temperature are brought together, heat passes from the warmer to the cooler object until the two are the same temperature. Molecules in the cooler object speed up at the expense of the kinetic energy of the warmer object. An ice cube cools a drink not by adding coldness to the liquid, but by absorbing heat from the liquid as the ice itself melts.

In general, we will use the **Celsius scale** to indicate temperature. (Celsius degrees are abbreviated °C; Appendix C shows how to convert between Celsius and Fahrenheit.) At sea level, water freezes at 0°C and boils at 100°C. The temperature of the human body averages 37°C, and comfortable room temperature is about 20–25°C.

One convenient unit of heat used in this book is the **calorie (cal)**. A calorie is the amount of heat it takes to raise the temperature of 1 g of water by 1°C. Conversely, a calorie is also the amount of heat that 1 g of water releases when it cools by 1°C. A **kilocalorie (kcal)**, 1,000 cal, is the quantity of heat required to raise the temperature of 1 kilogram (kg) of water by 1°C. (The "calories" on food packages are actually kilocalories.) Another energy unit used in this book is the **joule (J)**. One joule equals 0.239 cal; one calorie equals 4.184 J.

Water's High Specific Heat

The ability of water to stabilize temperature stems from its relatively high specific heat. The **specific heat** of a substance is defined as the amount of heat that must be absorbed or lost for 1 g of that substance to change its temperature by 1°C. We already know water's specific heat because we have defined a calorie as the amount of heat that causes 1 g of water to change its temperature by 1°C. Therefore, the specific heat of water is 1 calorie per gram and per degree Celsius, abbreviated as 1 cal/g·°C. Compared with most other substances, water has an unusually high specific heat. For example, ethyl alcohol, the type of alcohol in alcoholic beverages, has a specific heat of 0.6 cal/g·°C; that is, only 0.6 cal is required to raise the temperature of 1 g of ethyl alcohol by 1°C.

Because of the high specific heat of water relative to other materials, water will change its temperature less when it absorbs or loses a given amount of heat. The reason you can burn your fingers by touching the side of an iron pot on the stove when the water in the pot is still lukewarm is that the specific heat of water is ten times greater than that of iron.

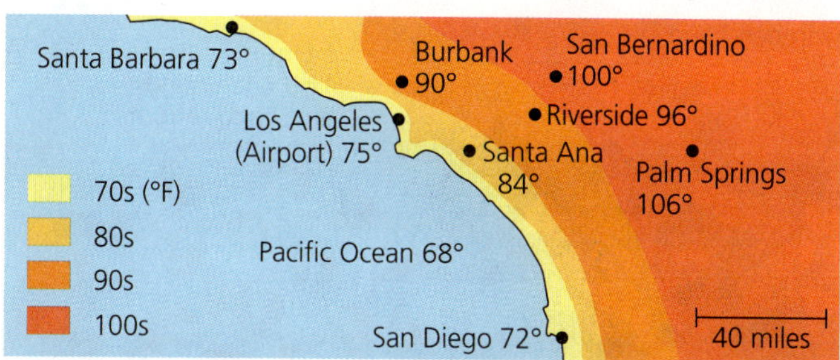

▲ **Figure 3.5 Effect of a large body of water on climate.** By absorbing or releasing heat, oceans moderate coastal climates. In this example from an August day in Southern California, the relatively cool ocean reduces coastal air temperatures by absorbing heat.

In other words, the same amount of heat will raise the temperature of 1 g of the iron much faster than it will raise the temperature of 1 g of the water. Specific heat can be thought of as a measure of how well a substance resists changing its temperature when it absorbs or releases heat. Water resists changing its temperature; when it does change its temperature, it absorbs or loses a relatively large quantity of heat for each degree of change.

We can trace water's high specific heat, like many of its other properties, to hydrogen bonding. Heat must be absorbed in order to break hydrogen bonds; by the same token, heat is released when hydrogen bonds form. A calorie of heat causes a relatively small change in the temperature of water because much of the heat is used to disrupt hydrogen bonds before the water molecules can begin moving faster. And when the temperature of water drops slightly, many additional hydrogen bonds form, releasing a considerable amount of energy in the form of heat.

What is the relevance of water's high specific heat to life on Earth? A large body of water can absorb and store a huge amount of heat from the sun in the daytime and during summer while warming up only a few degrees. At night and during winter, the gradually cooling water can warm the air. This is the reason coastal areas generally have milder climates than inland regions **(Figure 3.5)**. The high specific heat of water also tends to stabilize ocean temperatures, creating a favorable environment for marine life. Thus, because of its high specific heat, the water that covers most of Earth keeps temperature fluctuations on land and in water within limits that permit life. Also, because organisms are made primarily of water, they are better able to resist changes in their own temperature than if they were made of a liquid with a lower specific heat.

Evaporative Cooling

Molecules of any liquid stay close together because they are attracted to one another. Molecules moving fast enough to overcome these attractions can depart the liquid and enter the air as a gas. This transformation from a liquid to a gas is called vaporization, or *evaporation*. Recall that the speed of molecular movement varies and that temperature is the *average* kinetic energy of molecules. Even at low temperatures, the speediest molecules can escape into the air. Some evaporation occurs at any temperature; a glass of water at room temperature, for example, will eventually evaporate completely. If a liquid is heated, the average kinetic energy of molecules increases and the liquid evaporates more rapidly.

Heat of vaporization is the quantity of heat a liquid must absorb for 1 g of it to be converted from the liquid to the gaseous state. For the same reason that water has a high specific heat, it also has a high heat of vaporization relative to most other liquids. To evaporate 1 g of water at 25°C, about 580 cal of heat is needed—nearly double the amount needed to vaporize a gram of alcohol or ammonia. Water's high heat of vaporization is another emergent property resulting from the strength of its hydrogen bonds, which must be broken before the molecules can make their exodus from the liquid.

The high amount of energy required to vaporize water has a wide range of effects. On a global scale, for example, it helps moderate Earth's climate. A considerable amount of solar heat absorbed by tropical seas is consumed during the evaporation of surface water. Then, as moist tropical air circulates poleward, it releases heat as it condenses and forms rain. On an organismal level, water's high heat of vaporization accounts for the severity of steam burns. These burns are caused by the heat energy released when steam condenses into liquid on the skin.

As a liquid evaporates, the surface of the liquid that remains behind cools down. This **evaporative cooling** occurs because the "hottest" molecules, those with the greatest kinetic energy, are the most likely to leave as gas. It is as if the hundred fastest runners at a college transferred to another school; the average speed of the remaining students would decline.

Evaporative cooling of water contributes to the stability of temperature in lakes and ponds and also provides a mechanism that prevents terrestrial organisms from overheating. For example, evaporation of water from the leaves of a plant helps keep the tissues in the leaves from becoming too warm in the sunlight. Evaporation of sweat from human skin dissipates body heat and helps prevent overheating on a hot day or when excess heat is generated by strenuous activity. High humidity on a hot day increases discomfort because the high concentration of water vapor in the air inhibits the evaporation of sweat from the body.

Floating of Ice on Liquid Water

Water is one of the few substances that are less dense as a solid than as a liquid. In other words, ice floats on liquid water. While other materials contract and become denser when they solidify, water expands. The cause of this exotic behavior is, once again, hydrogen bonding. At temperatures above

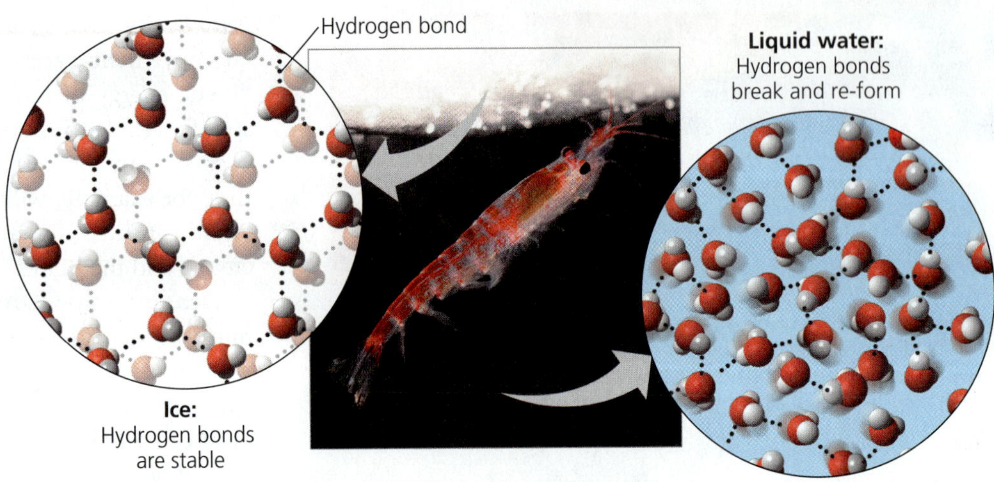

▶ **Figure 3.6 Ice: crystalline structure and floating barrier.** In ice, each molecule is hydrogen-bonded to four neighbors in a three-dimensional crystal. Because the crystal is spacious, ice has fewer molecules than an equal volume of liquid water. In other words, ice is less dense than liquid water. Floating ice becomes a barrier that protects the liquid water below from the colder air. The marine organism shown here is a type of shrimp called krill; it was photographed beneath floating ice in the Southern Ocean near Antarctica.

WHAT IF? *If water did not form hydrogen bonds, what would happen to the shrimp's environment?*

4°C, water behaves like other liquids, expanding as it warms and contracting as it cools. As the temperature falls from 4°C to 0°C, water begins to freeze because more and more of its molecules are moving too slowly to break hydrogen bonds. At 0°C, the molecules become locked into a crystalline lattice, each water molecule hydrogen-bonded to four partners **(Figure 3.6)**. The hydrogen bonds keep the molecules at "arm's length," far enough apart to make ice about 10% less dense (10% fewer molecules for the same volume) than liquid water at 4°C. When ice absorbs enough heat for its temperature to rise above 0°C, hydrogen bonds between molecules are disrupted. As the crystal collapses, the ice melts, and molecules are free to slip closer together. Water reaches its greatest density at 4°C and then begins to expand as the molecules move faster. Even in liquid water, many of the molecules are connected by hydrogen bonds, though only transiently: The hydrogen bonds are constantly breaking and re-forming.

The ability of ice to float due to its lower density is an important factor in the suitability of the environment for life. If ice sank, then eventually all ponds, lakes, and even oceans would freeze solid, making life as we know it impossible on Earth. During summer, only the upper few inches of the ocean would thaw. Instead, when a deep body of water cools, the floating ice insulates the liquid water below, preventing it from freezing and allowing life to exist under the frozen surface, as shown in the photo in Figure 3.6. Besides insulating the water below, ice also provides solid habitat for some animals, such as polar bears and seals (see Figure 3.1).

Along with many other scientists, Susan Solomon, the interviewee for this unit (see pp. 28-29), is worried that these bodies of ice are at risk of disappearing. Global warming, which is caused by carbon dioxide and other "greenhouse" gases in the atmosphere, is having a profound effect on icy environments around the globe. In the Arctic, the average air temperature has risen 1.4°C just since 1961. This temperature increase has affected the seasonal balance between Arctic sea ice and liquid water, causing ice to form later in the year, to melt earlier, and to cover a smaller area. The alarming rate at which glaciers and Arctic sea ice are disappearing is posing an extreme challenge to animals that depend on ice for their survival.

Water: The Solvent of Life

A sugar cube placed in a glass of water will dissolve. The glass will then contain a uniform mixture of sugar and water; the concentration of dissolved sugar will be the same everywhere in the mixture. A liquid that is a completely homogeneous mixture of two or more substances is called a **solution**. The dissolving agent of a solution is the **solvent**, and the substance that is dissolved is the **solute**. In this case, water is the solvent and sugar is the solute. An **aqueous solution** is one in which water is the solvent.

The medieval alchemists tried to find a universal solvent, one that would dissolve anything. They learned that nothing works better than water. Yet, water is not a universal solvent; if it were, it would dissolve any container in which it was stored, including our cells. Water is a very versatile solvent, however, a quality we can trace to the polarity of the water molecule.

Suppose, for example, that a spoonful of table salt, the ionic compound sodium chloride (NaCl), is placed in water **(Figure 3.7)**. At the surface of each grain, or crystal, of salt, the sodium and chloride ions are exposed to the solvent. These ions and the water molecules have a mutual affinity owing to the attraction between opposite charges. The oxygen regions of the water molecules are negatively charged and are attracted to sodium cations. The hydrogen regions are positively charged and are attracted to chloride anions. As a result, water molecules surround the individual sodium and chloride ions, separating and shielding them from one another. The sphere of water molecules around each dissolved ion is called a **hydration shell**. Working inward from the surface of each salt crystal, water eventually dissolves all the ions. The result is a solution of two solutes, sodium cations and chloride anions, homogeneously mixed with water, the solvent. Other ionic compounds also dissolve in water. Seawater, for instance, contains a great variety of dissolved ions, as do living cells.

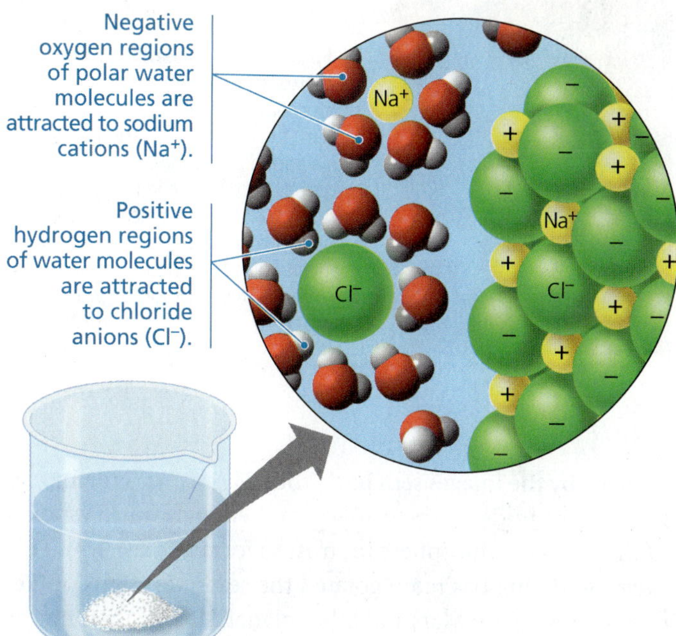

▲ **Figure 3.7 Table salt dissolving in water.** A sphere of water molecules, called a hydration shell, surrounds each solute ion.

WHAT IF? *What would happen if you heated this solution for a long time?*

A compound does not need to be ionic to dissolve in water; many compounds made up of nonionic polar molecules, such as sugars, are also water-soluble. Such compounds dissolve when water molecules surround each of the solute molecules, forming hydrogen bonds with them. Even molecules as large as proteins can dissolve in water if they have ionic and polar regions on their surface **(Figure 3.8)**. Many different kinds of polar compounds are dissolved (along with ions) in the water of such biological fluids as blood, the sap of plants, and the liquid within all cells. Water is the solvent of life.

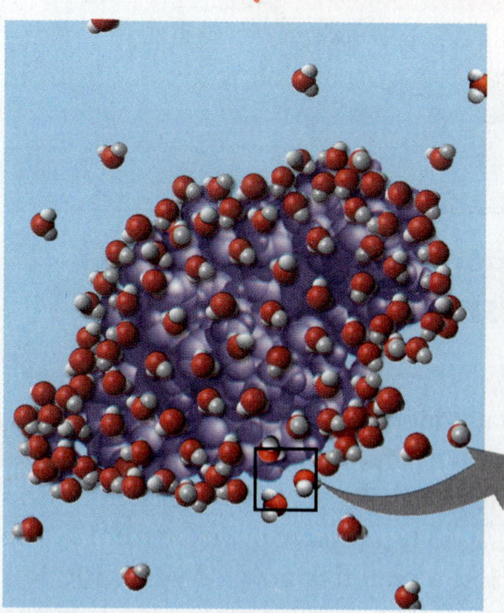

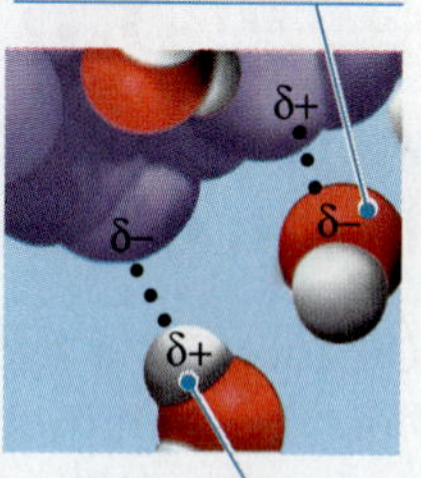

▲ **Figure 3.8 A water-soluble protein.** Human lysozyme is a protein found in tears and saliva that has antibacterial action. This model shows the lysozyme molecule (purple) in an aqueous environment. Ionic and polar regions on the protein's surface attract water molecules.

Hydrophilic and Hydrophobic Substances

Any substance that has an affinity for water is said to be **hydrophilic** (from the Greek *hydro*, water, and *philios*, loving). In some cases, substances can be hydrophilic without actually dissolving. For example, some molecules in cells are so large that they do not dissolve. Instead, they remain suspended in the aqueous liquid of the cell. Such a mixture is an example of a **colloid**, a stable suspension of fine particles in a liquid. Another example of a hydrophilic substance that does not dissolve is cotton, a plant product. Cotton consists of giant molecules of cellulose, a compound with numerous regions of partial positive and partial negative charges that can form hydrogen bonds with water. Water adheres to the cellulose fibers. Thus, a cotton towel does a great job of drying the body, yet it does not dissolve in the washing machine. Cellulose is also present in the walls of water-conducting cells in a plant; you read earlier how the adhesion of water to these hydrophilic walls allows water transport to occur.

There are, of course, substances that do not have an affinity for water. Substances that are nonionic and nonpolar (or otherwise cannot form hydrogen bonds) actually seem to repel water; these substances are said to be **hydrophobic** (from the Greek *phobos*, fearing). An example from the kitchen is vegetable oil, which, as you know, does not mix stably with water-based substances such as vinegar. The hydrophobic behavior of the oil molecules results from a prevalence of relatively nonpolar covalent bonds, in this case bonds between carbon and hydrogen, which share electrons almost equally. Hydrophobic molecules related to oils are major ingredients of cell membranes. (Imagine what would happen to a cell if its membrane dissolved!)

Solute Concentration in Aqueous Solutions

Biological chemistry is "wet" chemistry. Most of the chemical reactions in organisms involve solutes dissolved in water. To understand such reactions, we must know how many atoms and molecules are involved and be able to calculate the concentration of solutes in an aqueous solution (the number of solute molecules in a volume of solution).

When carrying out experiments, we use mass to calculate the number of molecules. We know the mass of each atom in a given molecule, so we can calculate the **molecular mass**, which is simply the sum of the masses of all the atoms in a molecule. As an example, let's calculate the molecular mass of table sugar (sucrose), which has the molecular formula $C_{12}H_{22}O_{11}$. In round numbers of daltons, the mass of a carbon atom is 12, the mass of a hydrogen atom is 1, and the mass of an oxygen atom is 16. Thus, sucrose has a molecular mass of $(12 \times 12) + (22 \times 1) + (11 \times 16) = 342$ daltons. Of course, weighing out small numbers of molecules is not practical. For this reason, we usually measure substances in units called moles. Just as a dozen always means 12 objects, a **mole (mol)** represents an exact number of objects: 6.02×10^{23}, which is called Avogadro's number. Because of the way in which Avogadro's number and the unit *dalton* were originally defined, there are 6.02×10^{23} daltons in 1 g. This is significant because once we determine the molecular mass of a molecule such as sucrose, we can use the same number (342), but with the unit *gram*, to represent the mass of 6.02×10^{23} molecules of sucrose, or 1 mol of sucrose (this is sometimes called the *molar mass*). To obtain 1 mol of sucrose in the lab, therefore, we weigh out 342 g.

The practical advantage of measuring a quantity of chemicals in moles is that a mole of one substance has exactly the same number of molecules as a mole of any other substance. If the molecular mass of substance A is 342 daltons and that of substance B is 10 daltons, then 342 g of A will have the same number of molecules as 10 g of B. A mole of ethyl alcohol (C_2H_6O) also contains 6.02×10^{23} molecules, but its mass is only 46 g because the mass of a molecule of ethyl alcohol is less than that of a molecule of sucrose. Measuring in moles makes it convenient for scientists working in the laboratory to combine substances in fixed ratios of molecules.

How would we make a liter (L) of solution consisting of 1 mol of sucrose dissolved in water? We would measure out 342 g of sucrose and then gradually add water, while stirring, until the sugar was completely dissolved. We would then add enough water to bring the total volume of the solution up to 1 L. At that point, we would have a 1-molar (1 *M*) solution of sucrose. **Molarity**—the number of moles of solute per liter of solution—is the unit of concentration most often used by biologists for aqueous solutions.

Water's capacity as a versatile solvent complements the other properties discussed in this chapter. Since these remarkable properties allow water to support life on Earth so well, scientists who seek life elsewhere in the universe look for water as a sign that a planet might sustain life.

Possible Evolution of Life on Other Planets with Water

EVOLUTION Humans have probably always gazed skyward, wondering whether other living beings exist beyond Earth. And if life has arisen on other planets, into what form or forms has it evolved? Biologists who look for life elsewhere in the universe (known as *astrobiologists*) have concentrated their search on planets that might have water. To date, more than 200 planets have been found outside our solar system, and there is evidence for the presence of water vapor on one or two of them. In our own solar system, Mars has been most compelling to astrobiologists as a focus of study.

Like Earth, Mars has an ice cap at both poles. And in the decades since the age of space exploration began, scientists have found intriguing signs that water may exist elsewhere on Mars. Finally, in 2008, the robotic spacecraft *Phoenix* landed on Mars and began to sample its surface. Years of debate were resolved by the images sent back from *Phoenix*: Ice is definitely present just under Mars's surface, and enough water vapor is in the Martian atmosphere for frost to form **(Figure 3.9)**. This exciting finding has reinvigorated the search for signs of life, past or present, on Mars and other planets. If any life-forms or fossils are found, their study will shed light on the process of evolution from an entirely new perspective.

◀ **Figure 3.9 Subsurface ice and morning frost on Mars.** This photograph was taken by the Mars lander *Phoenix* in 2008. The trench was scraped by a robotic arm, uncovering ice (white in rectangle near bottom) below the surface material. Frost also appears as a white coating in several places in the upper half of the image. This photograph was colorized by NASA to highlight the ice.

CONCEPT CHECK 3.2

1. Describe how properties of water contribute to the upward movement of water in a tree.
2. Explain the saying "It's not the heat; it's the humidity."
3. How can the freezing of water crack boulders?
4. The concentration of the appetite-regulating hormone ghrelin is about $1.3 \times 10^{-10} M$ in a fasting person. How many molecules of ghrelin are in 1 L of blood?
5. **WHAT IF?** A water strider (which can walk on water) has legs that are coated with a hydrophobic substance. What might be the benefit? What would happen if the substance were hydrophilic?

For suggested answers, see Appendix A.

CONCEPT 3.3

Acidic and basic conditions affect living organisms

Occasionally, a hydrogen atom participating in a hydrogen bond between two water molecules shifts from one molecule to the other. When this happens, the hydrogen atom leaves its electron behind, and what is actually transferred is a **hydrogen ion** (H^+), a single proton with a charge of 1+. The water molecule that lost a proton is now a **hydroxide ion** (OH^-), which has a charge of 1−. The proton binds to the other water molecule, making that molecule a **hydronium ion** (H_3O^+). We can picture the chemical reaction as shown at the top of the next page.

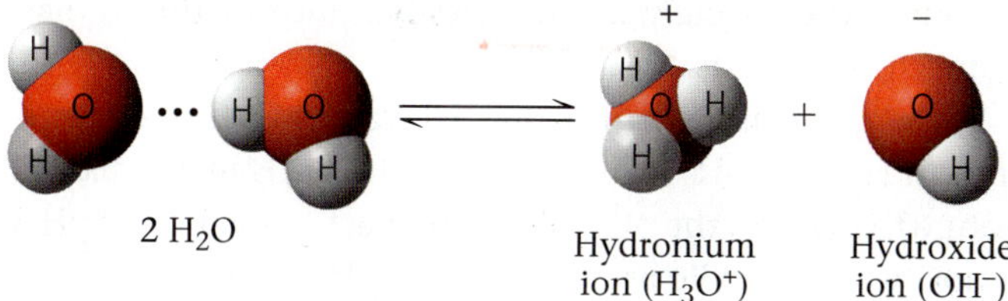

By convention, H^+ (the hydrogen ion) is used to represent H_3O^+ (the hydronium ion), and we follow that practice here. Keep in mind, though, that H^+ does not exist on its own in an aqueous solution. It is always associated with another water molecule in the form of H_3O^+.

As indicated by the double arrows, this is a reversible reaction that reaches a state of dynamic equilibrium when water molecules dissociate at the same rate that they are being re-formed from H^+ and OH^-. At this equilibrium point, the concentration of water molecules greatly exceeds the concentrations of H^+ and OH^-. In pure water, only one water molecule in every 554 million is dissociated; the concentration of each ion in pure water is 10^{-7} *M* (at 25°C). This means there is only one ten-millionth of a mole of hydrogen ions per liter of pure water and an equal number of hydroxide ions.

Although the dissociation of water is reversible and statistically rare, it is exceedingly important in the chemistry of life. H^+ and OH^- are very reactive. Changes in their concentrations can drastically affect a cell's proteins and other complex molecules. As we have seen, the concentrations of H^+ and OH^- are equal in pure water, but adding certain kinds of solutes, called acids and bases, disrupts this balance. Biologists use something called the pH scale to describe how acidic or basic (the opposite of acidic) a solution is. In the remainder of this chapter, you will learn about acids, bases, and pH and why changes in pH can adversely affect organisms.

Acids and Bases

What would cause an aqueous solution to have an imbalance in H^+ and OH^- concentrations? When acids dissolve in water, they donate additional H^+ to the solution. An **acid** is a substance that increases the hydrogen ion concentration of a solution. For example, when hydrochloric acid (HCl) is added to water, hydrogen ions dissociate from chloride ions:

$$HCl \rightarrow H^+ + Cl^-$$

This source of H^+ (dissociation of water is the other source) results in an acidic solution—one having more H^+ than OH^-.

A substance that reduces the hydrogen ion concentration of a solution is called a **base**. Some bases reduce the H^+ concentration directly by accepting hydrogen ions. Ammonia (NH_3), for instance, acts as a base when the unshared electron pair in nitrogen's valence shell attracts a hydrogen ion from the solution, resulting in an ammonium ion (NH_4^+):

$$NH_3 + H^+ \rightleftharpoons NH_4^+$$

Other bases reduce the H^+ concentration indirectly by dissociating to form hydroxide ions, which combine with hydrogen ions and form water. One such base is sodium hydroxide (NaOH), which in water dissociates into its ions:

$$NaOH \rightarrow Na^+ + OH^-$$

In either case, the base reduces the H^+ concentration. Solutions with a higher concentration of OH^- than H^+ are known as basic solutions. A solution in which the H^+ and OH^- concentrations are equal is said to be neutral.

Notice that single arrows were used in the reactions for HCl and NaOH. These compounds dissociate completely when mixed with water, so hydrochloric acid is called a strong acid and sodium hydroxide a strong base. In contrast, ammonia is a relatively weak base. The double arrows in the reaction for ammonia indicate that the binding and release of hydrogen ions are reversible reactions, although at equilibrium there will be a fixed ratio of NH_4^+ to NH_3.

There are also weak acids, which reversibly release and accept back hydrogen ions. An example is carbonic acid:

$$\underset{\text{Carbonic acid}}{H_2CO_3} \rightleftharpoons \underset{\text{Bicarbonate ion}}{HCO_3^-} + \underset{\text{Hydrogen ion}}{H^+}$$

Here the equilibrium so favors the reaction in the left direction that when carbonic acid is added to pure water, only 1% of the molecules are dissociated at any particular time. Still, that is enough to shift the balance of H^+ and OH^- from neutrality.

The pH Scale

In any aqueous solution at 25°C, the *product* of the H^+ and OH^- concentrations is constant at 10^{-14}. This can be written

$$[H^+][OH^-] = 10^{-14}$$

In such an equation, brackets indicate molar concentration. In a neutral solution at room temperature (25°C), $[H^+] = 10^{-7}$ and $[OH^-] = 10^{-7}$, so in this case, 10^{-14} is the product of $10^{-7} \times 10^{-7}$. If enough acid is added to a solution to increase $[H^+]$ to 10^{-5} *M*, then $[OH^-]$ will decline by an equivalent amount to 10^{-9} *M* (note that $10^{-5} \times 10^{-9} = 10^{-14}$). This constant relationship expresses the behavior of acids and bases in an aqueous solution. An acid not only adds hydrogen ions to a solution, but also removes hydroxide ions because of the tendency for H^+ to combine with OH^-, forming water. A base has the opposite effect, increasing OH^- concentration but also reducing H^+ concentration by the formation of water. If enough of a base is added to raise the OH^- concentration to 10^{-4} *M*, it will cause the H^+ concentration to drop to 10^{-10} *M*. Whenever we know the concentration of either H^+ or OH^- in an aqueous solution, we can deduce the concentration of the other ion.

Because the H^+ and OH^- concentrations of solutions can vary by a factor of 100 trillion or more, scientists have

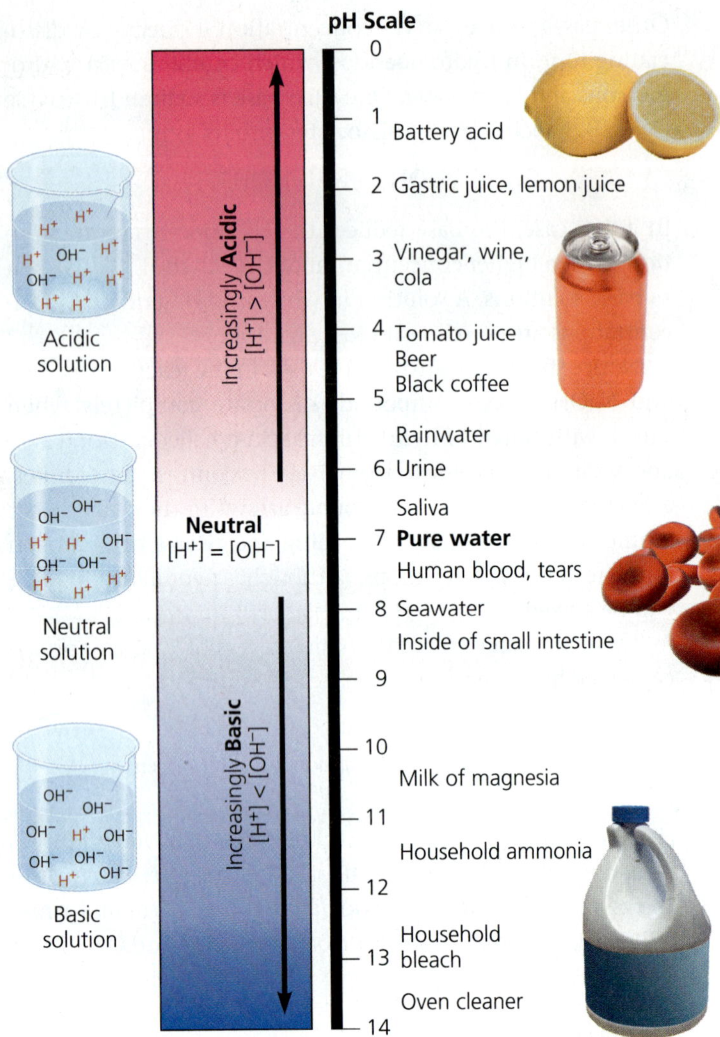

▲ **Figure 3.10 The pH scale and pH values of some aqueous solutions.**

developed a way to express this variation more conveniently than in moles per liter. The pH scale **(Figure 3.10)** compresses the range of H^+ and OH^- concentrations by employing logarithms. The **pH** of a solution is defined as the negative logarithm (base 10) of the hydrogen ion concentration:

$$pH = -\log [H^+]$$

For a neutral aqueous solution, $[H^+]$ is 10^{-7} *M*, giving us

$$-\log 10^{-7} = -(-7) = 7$$

Notice that pH *declines* as H^+ concentration *increases*. Notice, too, that although the pH scale is based on H^+ concentration, it also implies OH^- concentration. A solution of pH 10 has a hydrogen ion concentration of 10^{-10} *M* and a hydroxide ion concentration of 10^{-4} *M*.

The pH of a neutral aqueous solution at 25°C is 7, the midpoint of the pH scale. A pH value less than 7 denotes an acidic solution; the lower the number, the more acidic the solution. The pH for basic solutions is above 7. Most biological fluids are within the range pH 6–8. There are a few exceptions, however, including the strongly acidic digestive juice of the human stomach, which has a pH of about 2.

Remember that each pH unit represents a tenfold difference in H^+ and OH^- concentrations. It is this mathematical feature that makes the pH scale so compact. A solution of pH 3 is not twice as acidic as a solution of pH 6, but a thousand times ($10 \times 10 \times 10$) more acidic. When the pH of a solution changes slightly, the actual concentrations of H^+ and OH^- in the solution change substantially.

Buffers

The internal pH of most living cells is close to 7. Even a slight change in pH can be harmful, because the chemical processes of the cell are very sensitive to the concentrations of hydrogen and hydroxide ions. The pH of human blood is very close to 7.4, or slightly basic. A person cannot survive for more than a few minutes if the blood pH drops to 7 or rises to 7.8, and a chemical system exists in the blood that maintains a stable pH. If you add 0.01 mol of a strong acid to a liter of pure water, the pH drops from 7.0 to 2.0. If the same amount of acid is added to a liter of blood, however, the pH decrease is only from 7.4 to 7.3. Why does the addition of acid have so much less of an effect on the pH of blood than it does on the pH of water?

The presence of substances called buffers allows biological fluids to maintain a relatively constant pH despite the addition of acids or bases. A **buffer** is a substance that minimizes changes in the concentrations of H^+ and OH^- in a solution. It does so by accepting hydrogen ions from the solution when they are in excess and donating hydrogen ions to the solution when they have been depleted. Most buffer solutions contain a weak acid and its corresponding base, which combine reversibly with hydrogen ions.

There are several buffers that contribute to pH stability in human blood and many other biological solutions. One of these is carbonic acid (H_2CO_3), formed when CO_2 reacts with water in blood plasma. As mentioned earlier, carbonic acid dissociates to yield a bicarbonate ion (HCO_3^-) and a hydrogen ion (H^+):

$$\underset{\substack{H^+ \text{ donor} \\ \text{(acid)}}}{H_2CO_3} \underset{\text{Response to a drop in pH}}{\overset{\text{Response to a rise in pH}}{\rightleftharpoons}} \underset{\substack{H^+ \text{ acceptor} \\ \text{(base)}}}{HCO_3^-} + \underset{\substack{\text{Hydrogen} \\ \text{ion}}}{H^+}$$

The chemical equilibrium between carbonic acid and bicarbonate acts as a pH regulator, the reaction shifting left or right as other processes in the solution add or remove hydrogen ions. If the H^+ concentration in blood begins to fall (that is, if pH rises), the reaction proceeds to the right and more carbonic acid dissociates, replenishing hydrogen ions. But when H^+ concentration in blood begins to rise (when pH drops), the reaction proceeds to the left, with HCO_3^- (the base) removing the hydrogen ions from the solution and forming H_2CO_3. Thus, the carbonic acid–bicarbonate buffering system consists of an acid and a base in equilibrium with each other. Most other buffers are also acid-base pairs.

Acidification: A Threat to Water Quality

Among the many threats to water quality posed by human activities is the burning of fossil fuels, which releases gaseous compounds into the atmosphere. When certain of these compounds react with water, the water becomes more acidic, altering the delicate balance of conditions for life on Earth.

Carbon dioxide is the main product of fossil fuel combustion. About 25% of human-generated CO_2 is absorbed by the oceans. In spite of the huge volume of water in the oceans, scientists worry that the absorption of so much CO_2 will harm marine ecosystems.

Recent data have shown that such fears are well founded. When CO_2 dissolves in seawater, it reacts with water to form carbonic acid, which lowers ocean pH, a process known as **ocean acidification**. Based on measurements of CO_2 levels in air bubbles trapped in ice over thousands of years, scientists calculate that the pH of the oceans is 0.1 pH unit lower now than at any time in the past 420,000 years. Recent studies predict that it will drop another 0.3–0.5 pH unit by the end of this century.

As seawater acidifies, the extra hydrogen ions combine with carbonate ions (CO_3^{2-}) to form bicarbonate ions (HCO_3^-), thereby reducing the carbonate concentration **(Figure 3.11)**.

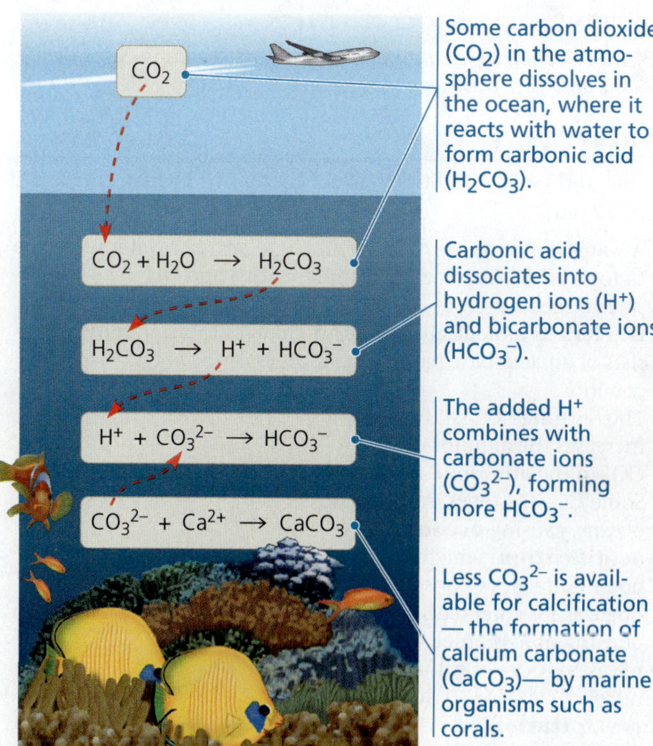

▲ **Figure 3.11 Atmospheric CO_2 from human activities and its fate in the ocean.**

▼ **Figure 3.12**

IMPACT

The Threat of Ocean Acidification to Coral Reef Ecosystems

Recently, scientists have sounded the alarm about the effects of ocean acidification, the process in which oceans become more acidic due to increased atmospheric carbon dioxide levels (see Figure 3.11). They predict that the resulting decrease in the concentration of carbonate ion (CO_3^{2-}) will take a serious toll on coral reef calcification. Taking many studies into account, and including the effects of ocean warming as well, one group of scientists defined three scenarios for coral reefs during this century, depending on whether the concentration of atmospheric CO_2 (a) stays at today's level, (b) increases at the current rate, or (c) increases more rapidly. The photographs below show coral reefs resembling those predicted under each scenario.

(a)

(b)

(c)

The healthy coral reef in (a) supports a highly diverse group of species and bears little resemblance to the damaged coral reef in (c).

WHY IT MATTERS The disappearance of coral reef ecosystems would be a tragic loss of biological diversity. In addition, coral reefs provide shoreline protection, a feeding ground for many commercial fishery species, and a popular tourist draw, so coastal human communities would suffer from greater wave damage, collapsed fisheries, and reduced tourism.

FURTHER READING O. Hoegh-Guldberg et al., Coral reefs under rapid climate change and ocean acidification, *Science* 318:1737–1742 (2007). S. C. Doney, The dangers of ocean acidification, *Scientific American*, March 2006, 58–65.

WHAT IF? Would lowering the ocean's carbonate concentration have any effect, even indirectly, on organisms that don't form $CaCO_3$? Explain.

Scientists predict that ocean acidification will cause the carbonate concentration to decrease by 40% by the year 2100. This is of great concern because carbonate is required for calcification, the production of calcium carbonate ($CaCO_3$) by many marine organisms, including reef-building corals and animals that build shells. Coral reefs are sensitive ecosystems that act as havens for a great diversity of marine life **(Figure 3.12)**.

The burning of fossil fuels is also a major source of sulfur oxides and nitrogen oxides. These compounds react with water in the air to form strong acids, which fall to Earth with rain or snow. **Acid precipitation** refers to rain, snow, or fog with a pH lower (more acidic) than 5.2. (Uncontaminated rain has a pH of about 5.6, which is slightly acidic due to the formation of carbonic acid from CO_2 and water.) Acid precipitation can damage life in lakes and streams, and it adversely affects plants on land by changing soil chemistry. To address this problem, the U.S. Congress amended the Clean Air Act in 1990, and the mandated improvements in industrial technologies have been largely responsible for improving the health of most North American lakes and forests.

If there is any reason for optimism about the future quality of water resources on our planet, it is that we have made progress in learning about the delicate chemical balances in oceans, lakes, and rivers. Continued progress can come only from the actions of informed individuals, like yourselves, who are concerned about environmental quality. This requires understanding the crucial role that water plays in the suitability of the environment for continued life on Earth.

CONCEPT CHECK 3.3

1. Compared with a basic solution at pH 9, the same volume of an acidic solution at pH 4 has ____ times as many hydrogen ions (H^+).
2. HCl is a strong acid that dissociates in water: $HCl \rightarrow H^+ + Cl^-$. What is the pH of 0.01 *M* HCl?
3. Acetic acid (CH_3COOH) can be a buffer, similar to carbonic acid. Write the dissociation reaction, identifying the acid, base, H^+ acceptor, and H^+ donor.
4. **WHAT IF?** Given a liter of pure water and a liter solution of acetic acid, what would happen to the pH if you added 0.01 mol of a strong acid to each? Use the reaction equation from question 3 to explain the result.

For suggested answers, see Appendix A.

3 CHAPTER REVIEW

SUMMARY OF KEY CONCEPTS

CONCEPT 3.1

Polar covalent bonds in water molecules result in hydrogen bonding (pp. 46–47)

- A hydrogen bond forms when the slightly negatively charged oxygen of one water molecule is attracted to the slightly positively charged hydrogen of a nearby water molecule. Hydrogen bonding between water molecules is the basis for water's properties.

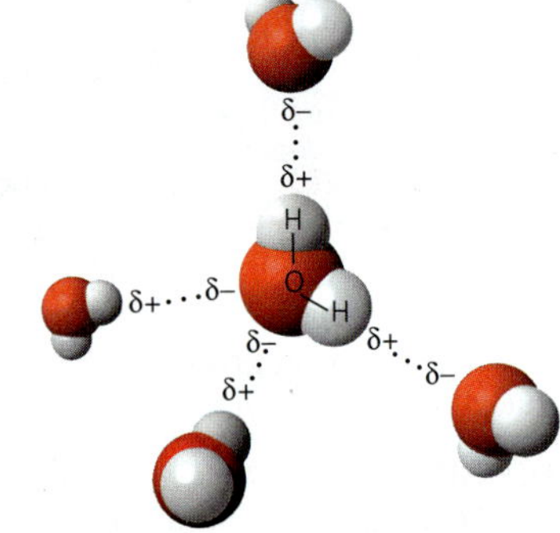

DRAW IT *Label a hydrogen bond and a polar covalent bond in this figure. How many hydrogen bonds can each water molecule make?*

CONCEPT 3.2

Four emergent properties of water contribute to Earth's suitability for life (pp. 47–52)

- Hydrogen bonding keeps water molecules close to each other, and this **cohesion** helps pull water upward in the microscopic water-conducting cells of plants. Hydrogen bonding is also responsible for water's **surface tension**.
- Water has a high **specific heat**: Heat is absorbed when hydrogen bonds break and is released when hydrogen bonds form. This helps keep temperatures relatively steady, within limits that permit life. **Evaporative cooling** is based on water's high **heat of vaporization**. The evaporative loss of the most energetic water molecules cools a surface.
- Ice floats because it is less dense than liquid water. This allows life to exist under the frozen surfaces of lakes and polar seas.

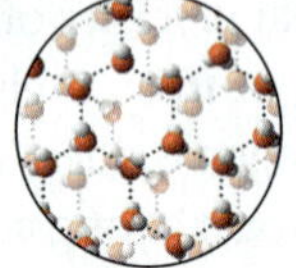

Ice: stable hydrogen bonds

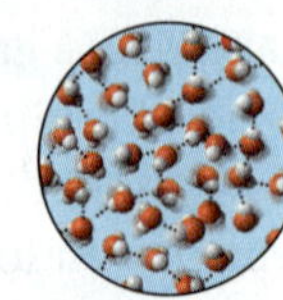

Liquid water: transient hydrogen bonds

- Water is an unusually versatile **solvent** because its polar molecules are attracted to charged and polar substances capable of forming hydrogen bonds. **Hydrophilic** substances have an affinity for water; **hydrophobic** substances do not. **Molarity**, the number of moles of **solute** per liter of **solution**, is used as a measure of solute concentration in solutions. A **mole** is a certain number of molecules of a substance. The mass of a mole of a substance in grams is the same as the **molecular mass** in daltons.
- The emergent properties of water support life on Earth and may contribute to the potential for life to have evolved on other planets.

? *Describe how different types of solutes dissolve in water. Explain the difference between a solution and a colloid.*

CONCEPT 3.3

Acidic and basic conditions affect living organisms (pp. 52–56)

- A water molecule can transfer an H^+ to another water molecule to form H_3O^+ (represented simply by H^+) and OH^-.
- The concentration of H^+ is expressed as **pH**; $pH = -\log [H^+]$. **Buffers** in biological fluids resist changes in pH. A buffer consists of an acid-base pair that combines reversibly with hydrogen ions.
- The burning of fossil fuels increases the amount of CO_2 in the atmosphere. Some CO_2 dissolves in the oceans, causing **ocean acidification**, which has potentially grave consequences for coral reefs. The burning of fossil fuels also releases oxides of sulfur and nitrogen, leading to **acid precipitation**.

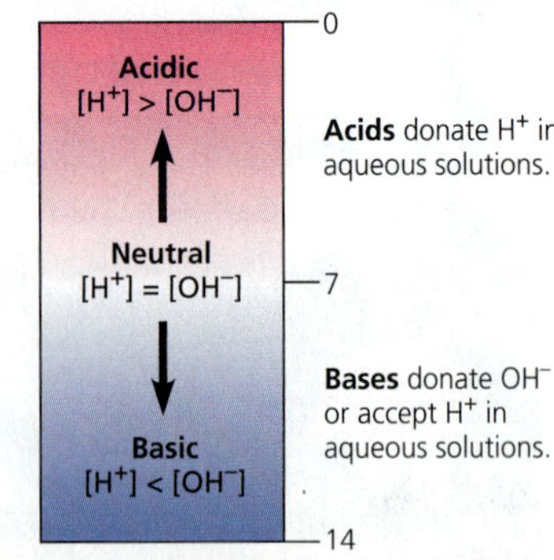

? *Explain how increasing amounts of CO_2 dissolving in the ocean leads to ocean acidification. How does this change in pH affect carbonate ion concentration and the rate of calcification?*

TEST YOUR UNDERSTANDING

LEVEL 1: KNOWLEDGE/COMPREHENSION

1. Many mammals control their body temperature by sweating. Which property of water is most directly responsible for the ability of sweat to lower body temperature?
   a. water's change in density when it condenses
   b. water's ability to dissolve molecules in the air
   c. the release of heat by the formation of hydrogen bonds
   d. the absorption of heat by the breaking of hydrogen bonds
   e. water's high surface tension

2. The bonds that are broken when water vaporizes are
   a. ionic bonds.
   b. hydrogen bonds between water molecules.
   c. covalent bonds between atoms within water molecules.
   d. polar covalent bonds.
   e. nonpolar covalent bonds.

3. Which of the following is a hydrophobic material?
   a. paper d. sugar
   b. table salt e. pasta
   c. wax

4. We can be sure that a mole of table sugar and a mole of vitamin C are equal in their
   a. mass in daltons. d. number of atoms.
   b. mass in grams. e. number of molecules.
   c. volume.

5. Measurements show that the pH of a particular lake is 4.0. What is the hydrogen ion concentration of the lake?
   a. 4.0 *M* b. 10^{-10} *M* c. 10^{-4} *M* d. 10^{4} *M* e. 4%

6. What is the *hydroxide* ion concentration of the lake described in question 5?
   a. 10^{-10} *M* b. 10^{-4} *M* c. 10^{-7} *M* d. 10^{-14}*M* e. 10 *M*

LEVEL 2: APPLICATION/ANALYSIS

7. A slice of pizza has 500 kcal. If we could burn the pizza and use all the heat to warm a 50-L container of cold water, what would be the approximate increase in the temperature of the water? (*Note*: A liter of cold water weighs about 1 kg.)
   a. 50°C b. 5°C c. 1°C d. 100°C e. 10°C

8. How many grams of acetic acid ($C_2H_4O_2$) would you use to make 10 L of a 0.1 *M* aqueous solution of acetic acid? (*Note*: The atomic masses, in daltons, are approximately 12 for carbon, 1 for hydrogen, and 16 for oxygen.)
   a. 10 g b. 0.1 g c. 6.0 g d. 60 g e. 0.6 g

9. **DRAW IT** Draw the hydration shells that form around a potassium ion and a chloride ion when potassium chloride (KCl) dissolves in water. Label the positive, negative, and partial charges on the atoms.

10. **MAKE CONNECTIONS** What do global warming (see Chapter 1, p. 6) and ocean acidification have in common?

LEVEL 3: SYNTHESIS/EVALUATION

11. In agricultural areas, farmers pay close attention to the weather forecast. Right before a predicted overnight freeze, farmers spray water on crops to protect the plants. Use the properties of water to explain how this method works. Be sure to mention why hydrogen bonds are responsible for this phenomenon.

12. **EVOLUTION CONNECTION**
    This chapter explains how the emergent properties of water contribute to the suitability of the environment for life. Until fairly recently, scientists assumed that other physical requirements for life included a moderate range of temperature, pH, atmospheric pressure, and salinity, as well as low levels of toxic chemicals. That view has changed with the discovery of organisms known as extremophiles, which have been found flourishing in hot, acidic sulfur springs, around hydrothermal vents deep in the ocean, and in soils with high levels of toxic metals. Why would astrobiologists be interested in studying extremophiles? What does the existence of life in such extreme environments say about the possibility of life on other planets?

13. **SCIENTIFIC INQUIRY**
    Design a controlled experiment to test the hypothesis that acid precipitation inhibits the growth of *Elodea,* a common freshwater plant (see Figure 2.19, p. 43).

14. **SCIENTIFIC INQUIRY**
    In a study reported in 2000, C. Langdon and colleagues used an artificial coral reef system to test the effect of carbonate concentration on the rate of calcification by reef organisms. The graph on the right presents one set of their results. Describe what these data show. How do these results relate to the ocean acidification that is associated with increasing atmospheric CO_2 levels?

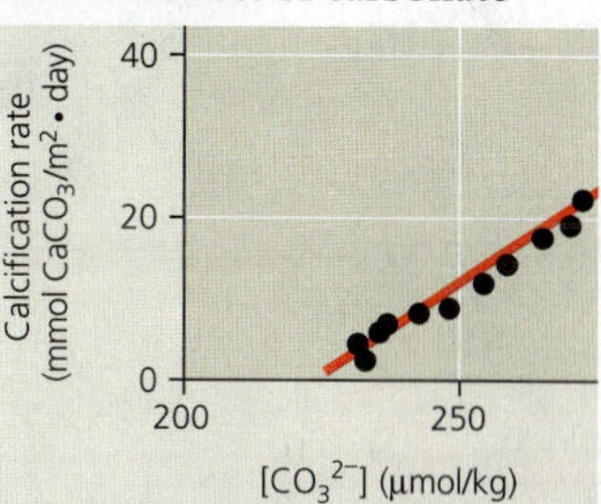

15. **SCIENCE, TECHNOLOGY, AND SOCIETY**
    Agriculture, industry, and the growing populations of cities all compete, through political influence, for water. If you were in charge of water resources in an arid region, what would your priorities be for allocating the limited water supply for various uses? How would you try to build consensus among the different special-interest groups?

16. **WRITE ABOUT A THEME**
    Emergent Properties Several emergent properties of water contribute to the suitability of the environment for life. In a short essay (100–150 words), describe how the ability of water to function as a versatile solvent arises from the structure of water molecules.

For selected answers, see Appendix A.

MasteringBIOLOGY www.masteringbiology.com

1. MasteringBiology® Assignments
Tutorials Hydrogen Bonding and Water • The pH Scale
Activities The Polarity of Water • Cohesion of Water • Dissociation of Water Molecules • Acids, Bases, and pH
Questions Student Misconceptions • Reading Quiz • Multiple Choice • End-of-Chapter

2. eText
Read your book online, search, take notes, highlight text, and more.

3. The Study Area
Practice Tests • Cumulative Test • ***BioFlix*** 3-D Animations • MP3 Tutor Sessions • Videos • Activities • Investigations • Lab Media • Audio Glossary • Word Study Tools • Art

4

Carbon and the Molecular Diversity of Life

▲ **Figure 4.1 What properties make carbon the basis of all life?**

KEY CONCEPTS

4.1 Organic chemistry is the study of carbon compounds

4.2 Carbon atoms can form diverse molecules by bonding to four other atoms

4.3 A few chemical groups are key to the functioning of biological molecules

OVERVIEW

Carbon: The Backbone of Life

Water is the universal medium for life on Earth, but living organisms, such as the plants and Roosevelt elk in **Figure 4.1**, are made up of chemicals based mostly on the element carbon. Carbon enters the biosphere through the action of plants. Plants use solar energy to transform atmospheric CO_2 into the molecules of life, which are then taken in by plant-eating animals.

Of all chemical elements, carbon is unparalleled in its ability to form molecules that are large, complex, and varied, making possible the diversity of organisms that have evolved on Earth. Proteins, DNA, carbohydrates, and other molecules that distinguish living matter from inanimate material are all composed of carbon atoms bonded to one another and to atoms of other elements. Hydrogen (H), oxygen (O), nitrogen (N), sulfur (S), and phosphorus (P) are other common ingredients of these compounds, but it is the element carbon (C) that accounts for the enormous variety of biological molecules.

Large biological molecules, such as proteins, are the main focus of Chapter 5. In this chapter, we investigate the properties of smaller molecules. We will use these small molecules to illustrate concepts of molecular architecture that will help explain why carbon is so important to life, at the same time highlighting the theme that emergent properties arise from the organization of matter in living organisms.

CONCEPT 4.1

Organic chemistry is the study of carbon compounds

For historical reasons, compounds containing carbon are said to be organic, and the branch of chemistry that specializes in the study of carbon compounds is called **organic chemistry**. Organic compounds range from simple molecules, such as methane (CH_4), to colossal ones, such as proteins, with thousands of atoms. Most organic compounds contain hydrogen atoms in addition to carbon atoms.

The overall percentages of the major elements of life—C, H, O, N, S, and P—are quite uniform from one organism to another. Because of carbon's versatility, however, this limited assortment of atomic building blocks can be used to build an inexhaustible variety of organic molecules. Different species of organisms, and different individuals within a species, are distinguished by variations in their organic molecules.

Since the dawn of human history, people have used other organisms as sources of valued substances—from foods and medicines to fabrics. The science of organic chemistry originated in attempts to purify and improve the yield of such products. By the early 1800s, chemists had learned to make many simple compounds in the laboratory by combining elements under the right conditions. Artificial synthesis of the complex molecules extracted from living matter seemed impossible, however. At that time, the Swedish chemist Jöns Jakob Berzelius made the distinction between organic compounds, those thought to arise only in living organisms, and inorganic compounds, those found only in the nonliving world. *Vitalism*, the belief in a life force outside the jurisdiction of physical and chemical laws, provided the foundation for the new discipline of organic chemistry.

Chemists began to chip away at the support for vitalism when they finally learned to synthesize organic compounds in the laboratory. In 1828, Friedrich Wöhler, a German chemist who had studied with Berzelius, tried to make an "inorganic" salt, ammonium cyanate, by mixing solutions of ammonium ions (NH_4^+) and cyanate ions (CNO^-). Wöhler was astonished to find that instead he had made urea, an organic compound present in the urine of animals. Wöhler challenged the vitalists when he wrote, "I must tell you that I can prepare urea without requiring a kidney or an animal, either man or dog." However, one of the ingredients used in the synthesis, the cyanate, had been extracted from animal blood, and the vitalists were not swayed by Wöhler's discovery. A few years later, however, Hermann Kolbe, a student of Wöhler's, made the organic compound acetic acid from inorganic substances that could be prepared directly from pure elements. Vitalism crumbled completely after several decades of laboratory synthesis of increasingly complex organic compounds.

Organic Molecules and the Origin of Life on Earth

EVOLUTION In 1953, Stanley Miller, a graduate student of Harold Urey's at the University of Chicago, helped bring the abiotic (nonliving) synthesis of organic compounds into the context of evolution. Study **Figure 4.2** to learn about his classic experiment. From his results, Miller concluded that complex organic molecules could arise spontaneously under conditions thought to have existed on the early Earth. Miller also performed experiments designed to mimic volcanic conditions, with roughly similar results. In 2008, a former graduate student of Miller's discovered some samples from these experiments. Reanalyzing them using modern equipment, he identified additional organic compounds that had not been found by Miller. Although the jury is still out, these experiments support the idea that abiotic synthesis of organic compounds, perhaps near volcanoes, could have been an early stage in the origin of life (see Chapter 25).

The pioneers of organic chemistry helped shift the mainstream of biological thought from vitalism to *mechanism*, the view that physical and chemical laws govern all natural phenomena, including the processes of life. Organic chemistry was redefined as the study of carbon compounds, regardless of origin. Organisms produce most of the naturally occurring organic compounds, and these molecules represent a diversity and range of complexity unrivaled by inorganic compounds. However, the rules of chemistry apply to all molecules. The foundation of organic chemistry is not some intangible life force, but the unique chemical versatility of the element carbon.

CONCEPT CHECK 4.1

1. Why was Wöhler astonished to find he had made urea?
2. **WHAT IF?** When Miller tried his experiment without the electrical discharge, no organic compounds were found. What might explain this result?

For suggested answers, see Appendix A.

▼ Figure 4.2 **INQUIRY**

Can organic molecules form under conditions believed to simulate those on the early Earth?

EXPERIMENT In 1953, Stanley Miller set up a closed system to mimic conditions thought to have existed on the early Earth. A flask of water simulated the primeval sea. The water was heated so that some vaporized and moved into a second, higher flask containing the "atmosphere"—a mixture of gases. Sparks were discharged in the synthetic atmosphere to mimic lightning.

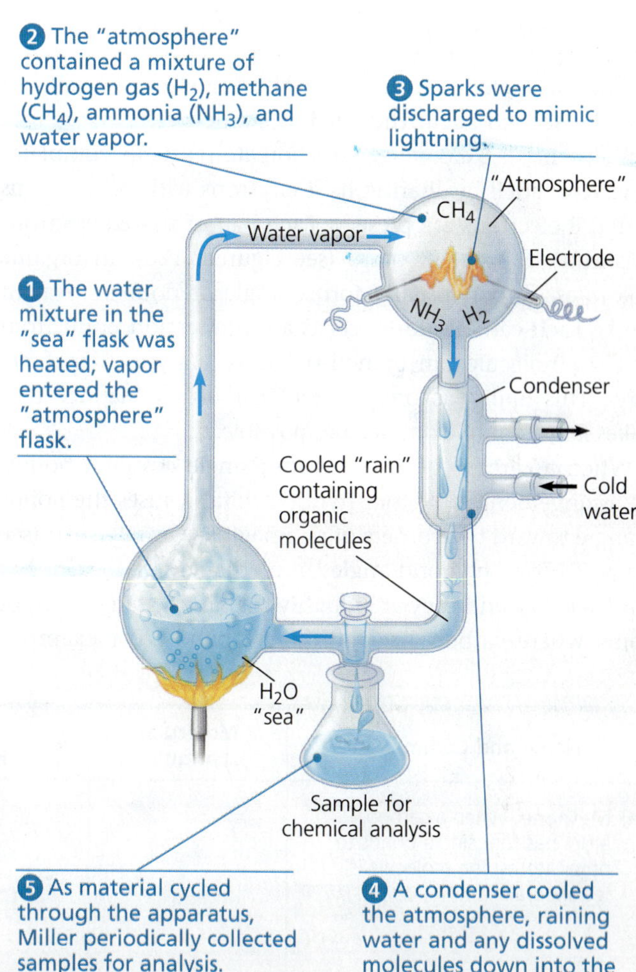

RESULTS Miller identified a variety of organic molecules that are common in organisms. These included simple compounds, such as formaldehyde (CH_2O) and hydrogen cyanide (HCN), and more complex molecules, such as amino acids and long chains of carbon and hydrogen known as hydrocarbons.

CONCLUSION Organic molecules, a first step in the origin of life, may have been synthesized abiotically on the early Earth. (We will explore this hypothesis in more detail in Chapter 25.)

SOURCE S. L. Miller, A production of amino acids under possible primitive Earth conditions, *Science* 117:528–529 (1953).

WHAT IF? If Miller had increased the concentration of NH_3 in his experiment, how might the relative amounts of the products HCN and CH_2O have differed?

CONCEPT 4.2

Carbon atoms can form diverse molecules by bonding to four other atoms

The key to an atom's chemical characteristics is its electron configuration. This configuration determines the kinds and number of bonds an atom will form with other atoms.

The Formation of Bonds with Carbon

Carbon has 6 electrons, with 2 in the first electron shell and 4 in the second shell; thus, it has 4 valence electrons in a shell that holds 8 electrons. A carbon atom usually completes its valence shell by sharing its 4 electrons with other atoms so that 8 electrons are present. Each pair of shared electrons constitutes a covalent bond (see Figure 2.12d). In organic molecules, carbon usually forms single or double covalent bonds. Each carbon atom acts as an intersection point from which a molecule can branch off in as many as four directions. This ability is one facet of carbon's versatility that makes large, complex molecules possible.

When a carbon atom forms four single covalent bonds, the arrangement of its four hybrid orbitals causes the bonds to angle toward the corners of an imaginary tetrahedron (see Figure 2.17b). The bond angles in methane (CH_4) are 109.5° **(Figure 4.3a)**, and they are roughly the same in any group of atoms where carbon has four single bonds. For example, ethane (C_2H_6) is shaped like two overlapping tetrahedrons **(Figure 4.3b)**. In molecules with more carbons, every grouping of a carbon bonded to four other atoms has a tetrahedral shape. But when two carbon atoms are joined by a double bond, as in ethene (C_2H_4), the atoms joined to those carbons are in the same plane as the carbons **(Figure 4.3c)**. We find it convenient to write molecules as structural formulas, as if the molecules being represented are two-dimensional, but keep in mind that molecules are three-dimensional and that the shape of a molecule often determines its function.

The electron configuration of carbon gives it covalent compatibility with many different elements. **Figure 4.4** shows the valences of carbon and its most frequent partners—hydrogen, oxygen, and nitrogen. These are the four major atomic components of organic molecules. These valences are the basis for the rules of covalent bonding in organic chemistry—the building code for the architecture of organic molecules.

Let's consider how the rules of covalent bonding apply to carbon atoms with partners other than hydrogen. We'll look at two examples, the simple molecules carbon dioxide and urea.

In the carbon dioxide molecule (CO_2), a single carbon atom is joined to two atoms of oxygen by double covalent bonds. The structural formula for CO_2 is shown here:

$$O=C=O$$

Each line in a structural formula represents a pair of shared electrons. Thus, the two double bonds in CO_2 have the same number of shared electrons as four single bonds. The arrangement completes the valence shells of all atoms in the molecule.

Name and Comment	Molecular Formula	Structural Formula	Ball-and-Stick Model (molecular shape in pink)	Space-Filling Model
(a) Methane. When a carbon atom has four single bonds to other atoms, the molecule is tetrahedral.	CH_4	H—C(—H)(—H)—H		
(b) Ethane. A molecule may have more than one tetrahedral group of single-bonded atoms. (Ethane consists of two such groups.)	C_2H_6	H—C(H)(H)—C(H)(H)—H		
(c) Ethene (ethylene). When two carbon atoms are joined by a double bond, all atoms attached to those carbons are in the same plane; the molecule is flat.	C_2H_4	$H_2C=CH_2$		

▲ **Figure 4.3 The shapes of three simple organic molecules.**

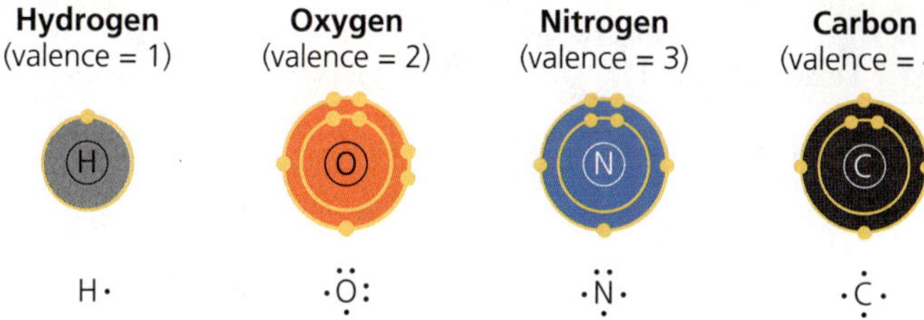

▲ **Figure 4.4 Valences of the major elements of organic molecules.** Valence is the number of covalent bonds an atom can form. It is generally equal to the number of electrons required to complete the valence (outermost) shell (see Figure 2.9). All the electrons are shown for each atom in the electron distribution diagrams (top). Only the valence shell electrons are shown in the Lewis dot structures (bottom). Note that carbon can form four bonds.

MAKE CONNECTIONS *Refer to Figure 2.9 (p. 36) and draw the Lewis dot structures for sodium, phosphorus, sulfur, and chlorine.*

Because CO_2 is a very simple molecule and lacks hydrogen, it is often considered inorganic, even though it contains carbon. Whether we call CO_2 organic or inorganic, however, it is clearly important to the living world as the source of carbon for all organic molecules in organisms.

Urea, $CO(NH_2)_2$, is the organic compound found in urine that Wöhler synthesized in the early 1800s. Again, each atom has the required number of covalent bonds. In this case, one carbon atom participates in both single and double bonds.

Urea and carbon dioxide are molecules with only one carbon atom. But as Figure 4.3 shows, a carbon atom can also use one or more valence electrons to form covalent bonds to other carbon atoms, linking the atoms into chains of seemingly infinite variety.

Molecular Diversity Arising from Carbon Skeleton Variation

Carbon chains form the skeletons of most organic molecules. The skeletons vary in length and may be straight, branched, or arranged in closed rings (**Figure 4.5**). Some carbon skeletons have double bonds, which vary in number and location. Such variation in carbon skeletons is one important source of the molecular complexity and diversity that characterize living matter. In addition, atoms of other elements can be bonded to the skeletons at available sites.

Hydrocarbons

All of the molecules shown in Figures 4.3 and 4.5 are **hydrocarbons**, organic molecules consisting of only carbon and hydrogen. Atoms of hydrogen are attached to the carbon skeleton wherever electrons are available for covalent bonding. Hydrocarbons are the major components of petroleum, which is called a fossil fuel because it consists of the partially decomposed remains of organisms that lived millions of years ago.

▼ **Figure 4.5 Four ways that carbon skeletons can vary.**

(a) Length

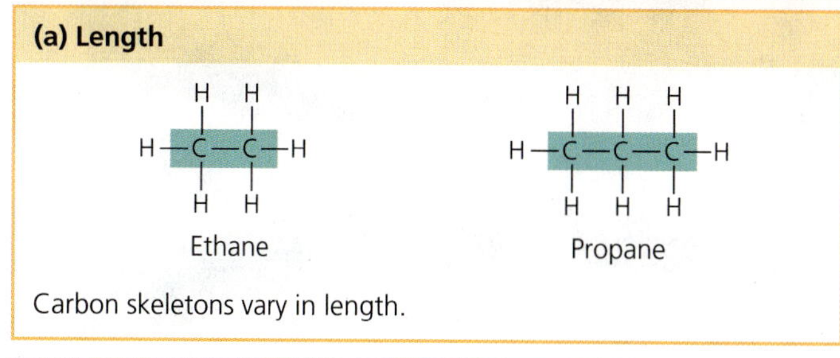

Carbon skeletons vary in length.

(b) Branching

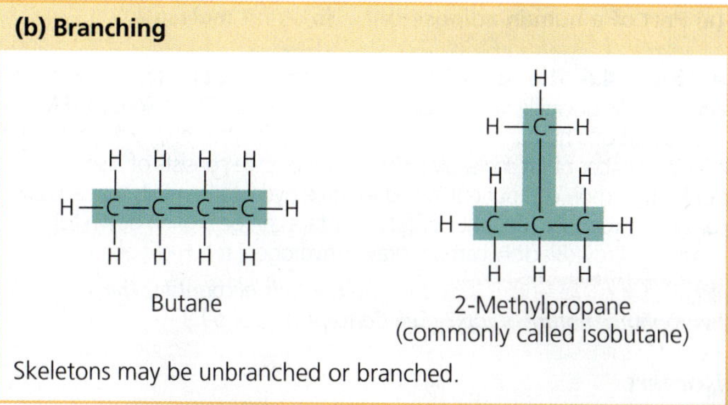

Skeletons may be unbranched or branched.

(c) Double bond position

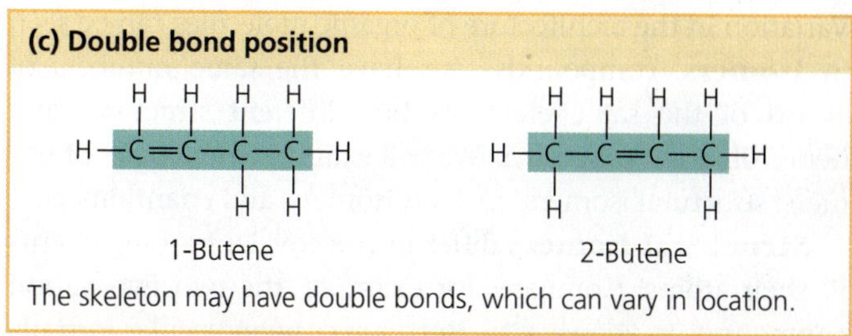

The skeleton may have double bonds, which can vary in location.

(d) Presence of rings

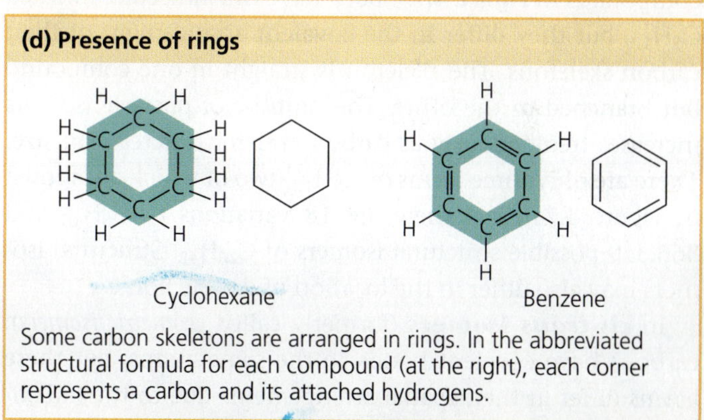

Some carbon skeletons are arranged in rings. In the abbreviated structural formula for each compound (at the right), each corner represents a carbon and its attached hydrogens.

Although hydrocarbons are not prevalent in most living organisms, many of a cell's organic molecules have regions consisting of only carbon and hydrogen. For example, the molecules known as fats have long hydrocarbon tails attached to a nonhydrocarbon component (**Figure 4.6**, on the next page). Neither petroleum nor fat dissolves in water; both are hydrophobic compounds because the great majority of their bonds are relatively nonpolar carbon-to-hydrogen linkages. Another characteristic of hydrocarbons is that they can undergo reactions that release a relatively large amount of energy. The gasoline that fuels a car consists of hydrocarbons, and the hydrocarbon tails of fats serve as stored fuel for animals.

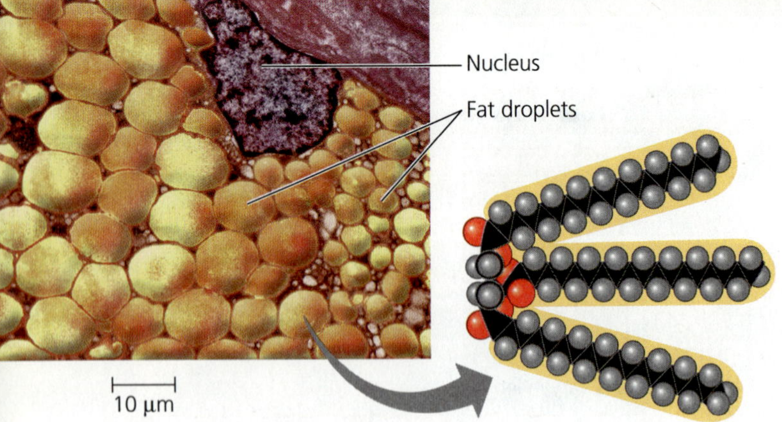

(a) Part of a human adipose cell **(b) A fat molecule**

▲ **Figure 4.6 The role of hydrocarbons in fats.** **(a)** Mammalian adipose cells stockpile fat molecules as a fuel reserve. This colorized TEM shows part of a human adipose cell with many fat droplets, each containing a large number of fat molecules. **(b)** A fat molecule consists of a small, nonhydrocarbon component joined to three hydrocarbon tails that account for the hydrophobic behavior of fats. The tails can be broken down to provide energy. (Black = carbon; gray = hydrogen; red = oxygen.)

MAKE CONNECTIONS *How do the tails account for the hydrophobic nature of fats? (See Concept 3.2, p. 51.)*

▼ **Figure 4.7 Three types of isomers, compounds with the same molecular formula but different structures.**

(a) Structural isomers

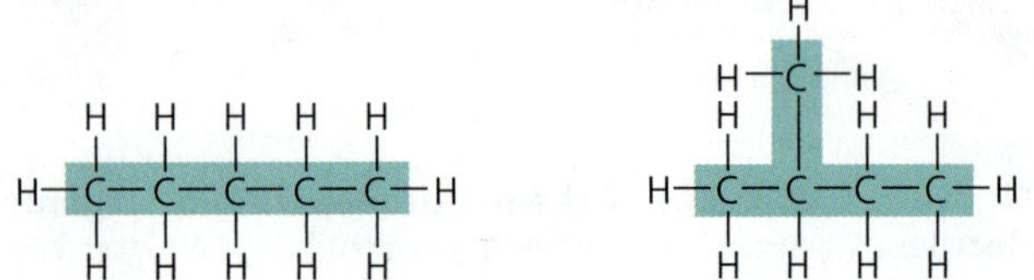

Structural isomers differ in covalent partners, as shown in this example of two isomers of C_5H_{12}: pentane (left) and 2-methyl butane (right).

(b) *Cis-trans* isomers

cis isomer: The two Xs are on the same side.

trans isomer: The two Xs are on opposite sides.

Cis-trans isomers differ in arrangement about a double bond. In these diagrams, X represents an atom or group of atoms attached to a double-bonded carbon.

(c) Enantiomers

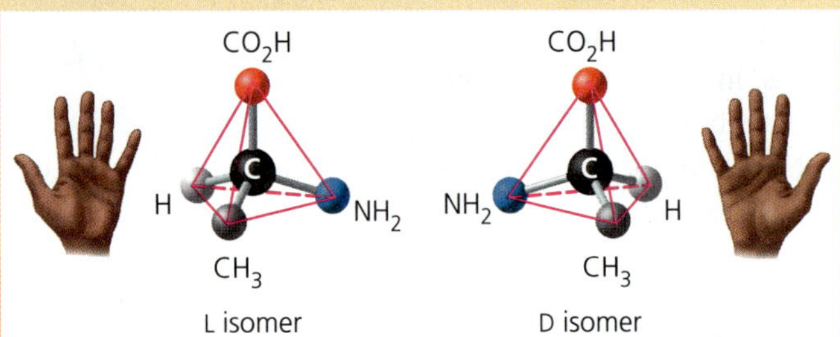

Enantiomers differ in spatial arrangement around an asymmetric carbon, resulting in molecules that are mirror images, like left and right hands. The two isomers are designated the L and D isomers from the Latin for "left" and "right" (*levo* and *dextro*). Enantiomers cannot be superimposed on each other.

DRAW IT *There are three structural isomers of C_5H_{12}; draw the one not shown in (a).*

Isomers

Variation in the architecture of organic molecules can be seen in **isomers**, compounds that have the same numbers of atoms of the same elements but different structures and hence different properties. We will examine three types of isomers: structural isomers, *cis-trans* isomers, and enantiomers.

Structural isomers differ in the covalent arrangements of their atoms. Compare, for example, the two five-carbon compounds in **Figure 4.7a**. Both have the molecular formula C_5H_{12}, but they differ in the covalent arrangement of their carbon skeletons. The skeleton is straight in one compound but branched in the other. The number of possible isomers increases tremendously as carbon skeletons increase in size. There are only three forms of C_5H_{12} (two of which are shown in Figure 4.7a), but there are 18 variations of C_8H_{18} and 366,319 possible structural isomers of $C_{20}H_{42}$. Structural isomers may also differ in the location of double bonds.

In ***cis-trans*** **isomers** (formerly called *geometric isomers*), carbons have covalent bonds to the same atoms, but these atoms differ in their spatial arrangements due to the inflexibility of double bonds. Single bonds allow the atoms they join to rotate freely about the bond axis without changing the compound. In contrast, double bonds do not permit such rotation. If a double bond joins two carbon atoms, and each C also has two different atoms (or groups of atoms) attached to it, then two distinct *cis-trans* isomers are possible. Consider a simple molecule with two double-bonded carbons, each of which has an H and an X attached to it **(Figure 4.7b)**. The arrangement with both Xs on the same side of the double bond is called a *cis* isomer, and that with the Xs on opposite sides is called a *trans* isomer. The subtle difference in shape between such isomers can dramatically affect the biological activities of organic molecules. For example, the biochemistry of vision involves a light-induced change of rhodopsin, a chemical compound in the eye, from the *cis* isomer to the *trans* isomer (see Figure 50.17). Another example involves *trans* fats, which are discussed in Chapter 5.

Enantiomers are isomers that are mirror images of each other and that differ in shape due to the presence of an *asymmetric carbon*, one that is attached to four different atoms or groups of atoms. (See the middle carbon in the ball-and-stick models shown in **Figure 4.7c**.) The four groups can be arranged in space around the asymmetric carbon in two different ways that are mirror images. Enantiomers are, in a way, left-handed and right-handed versions of the molecule. Just as your right hand won't fit into a left-handed glove, a "right-handed" molecule won't fit into the same space as the "left-handed" version.

Drug	Condition	Effective Enantiomer	Ineffective Enantiomer
Ibuprofen	Pain; inflammation	*S*-Ibuprofen	*R*-Ibuprofen
Albuterol	Asthma	*R*-Albuterol	*S*-Albuterol

▲ **Figure 4.8 The pharmacological importance of enantiomers.** Ibuprofen and albuterol are examples of drugs whose enantiomers have different effects. (*S* and *R* are letters used in one system to distinguish between enantiomers.) Ibuprofen reduces inflammation and pain. It is commonly sold as a mixture of the two enantiomers. The *S* enantiomer is 100 times more effective than the other. Albuterol is used to relax bronchial muscles, improving airflow in asthma patients. Only *R*-albuterol is synthesized and sold as a drug; the *S* form counteracts the active *R* form.

Usually, only one isomer is biologically active because only that form can bind to specific molecules in an organism.

The concept of enantiomers is important in the pharmaceutical industry because the two enantiomers of a drug may not be equally effective, as is the case for both ibuprofen and the asthma medication albuterol **(Figure 4.8)**. Methamphetamine also occurs in two enantiomers that have very different effects. One enantiomer is the highly addictive stimulant drug known as "crank," sold illegally in the street drug trade. The other has a much weaker effect and is even found as an ingredient in an over-the-counter vapor inhaler for treatment of nasal congestion! The differing effects of enantiomers in the body demonstrate that organisms are sensitive to even the most subtle variations in molecular architecture. Once again, we see that molecules have emergent properties that depend on the specific arrangement of their atoms.

CONCEPT CHECK 4.2

1. **DRAW IT** Draw a structural formula for C_2H_4.
2. Which molecules in Figure 4.5 are isomers? For each pair, identify the type of isomer.
3. How are gasoline and fat chemically similar?
4. **WHAT IF?** Can propane (C_3H_8) form isomers?

For suggested answers, see Appendix A.

CONCEPT 4.3

A few chemical groups are key to the functioning of biological molecules

The distinctive properties of an organic molecule depend not only on the arrangement of its carbon skeleton but also on the chemical groups attached to that skeleton. We can think of hydrocarbons, the simplest organic molecules, as the underlying framework for more complex organic molecules. A number of chemical groups can replace one or more of the hydrogens bonded to the carbon skeleton of the hydrocarbon. (Some groups include atoms of the carbon skeleton, as we will see.) These groups may participate in chemical reactions or may contribute to function indirectly by their effects on molecular shape. The number and arrangement of the groups help give each molecule its unique properties.

The Chemical Groups Most Important in the Processes of Life

Consider the differences between estradiol (a type of estrogen) and testosterone. These compounds are female and male sex hormones, respectively, in humans and other vertebrates. Both are steroids, organic molecules with a common carbon skeleton in the form of four fused rings. These sex hormones differ only in the chemical groups attached to the rings (shown here in abbreviated form); the distinctions in molecular architecture are shaded in blue:

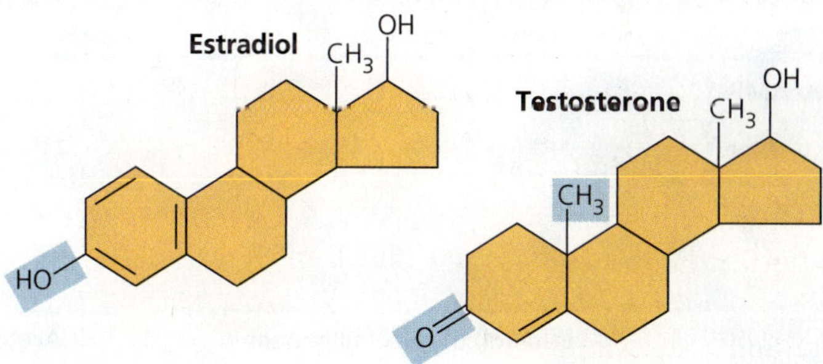

The different actions of these two molecules on many targets throughout the body help produce the contrasting anatomical and physiological features of male and female vertebrates. Thus, even our sexuality has its biological basis in variations of molecular architecture.

In the example of sex hormones, different chemical groups contribute to function by affecting the molecule's shape. In other cases, the chemical groups affect molecular function by being directly involved in chemical reactions; these important chemical groups are known as **functional groups**. Each functional group participates in chemical reactions in a characteristic way from one organic molecule to another.

The seven chemical groups most important in biological processes are the hydroxyl, carbonyl, carboxyl, amino, sulfhydryl, phosphate, and methyl groups. The first six groups can act as functional groups; they are also hydrophilic and thus increase the solubility of organic compounds in water. The methyl group is not reactive, but instead often serves as a recognizable tag on biological molecules. Before reading further, study **Figure 4.9** on the next two pages to familiarize yourself with these biologically important chemical groups.

▼ **Figure 4.9**

Exploring Some Biologically Important Chemical Groups

CHEMICAL GROUP	Hydroxyl	Carbonyl	Carboxyl
STRUCTURE	—OH (may be written HO—) In a **hydroxyl group** (—OH), a hydrogen atom is bonded to an oxygen atom, which in turn is bonded to the carbon skeleton of the organic molecule. (Do not confuse this functional group with the hydroxide ion, OH^-.)	$-\overset{\displaystyle O}{\overset{\Vert}{C}}-$ The **carbonyl group** (>CO) consists of a carbon atom joined to an oxygen atom by a double bond.	$-\overset{\displaystyle O}{\overset{\Vert}{C}}-OH$ When an oxygen atom is double-bonded to a carbon atom that is also bonded to an —OH group, the entire assembly of atoms is called a **carboxyl group** (—COOH).
NAME OF COMPOUND	**Alcohols** (Their specific names usually end in *-ol.*)	**Ketones** if the carbonyl group is within a carbon skeleton **Aldehydes** if the carbonyl group is at the end of the carbon skeleton	**Carboxylic acids**, or organic acids
EXAMPLE	CH_3CH_2OH **Ethanol**, the alcohol present in alcoholic beverages	CH_3COCH_3 **Acetone**, the simplest ketone CH_3CH_2CHO **Propanal**, an aldehyde	CH_3COOH **Acetic acid**, which gives vinegar its sour taste
FUNCTIONAL PROPERTIES	• Is polar as a result of the electrons spending more time near the electronegative oxygen atom. • Can form hydrogen bonds with water molecules, helping dissolve organic compounds such as sugars. (Sugars are shown in Figure 5.3.)	• A ketone and an aldehyde may be structural isomers with different properties, as is the case for acetone and propanal. • Ketone and aldehyde groups are also found in sugars, giving rise to two major groups of sugars: ketoses (containing ketone groups) and aldoses (containing aldehyde groups).	• Acts as an acid; can donate an H^+ because the covalent bond between oxygen and hydrogen is so polar: $-COOH \rightleftharpoons -COO^- + H^+$ Nonionized Ionized • Found in cells in the ionized form with a charge of 1− and called a carboxylate ion.

Amino

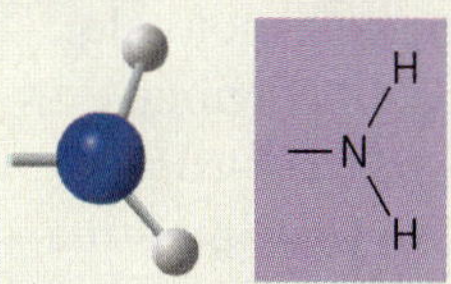

The **amino group** (—NH_2) consists of a nitrogen atom bonded to two hydrogen atoms and to the carbon skeleton.

Amines

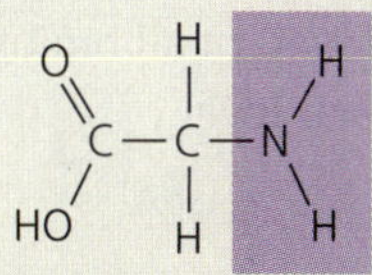

Glycine, a compound that is both an amine and a carboxylic acid because it has both an amino group and a carboxyl group; compounds with both groups are called **amino acids**

- Acts as a base; can pick up an H^+ from the surrounding solution (water, in living organisms):

$$H^+ + —NH_2 \rightleftharpoons —{}^{+}NH_3$$

Nonionized Ionized

- Found in cells in the ionized form with a charge of 1+.

Sulfhydryl

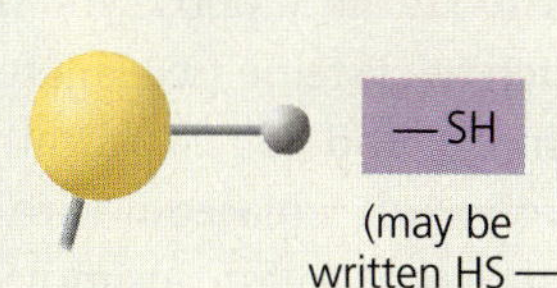

(may be written HS—)

The **sulfhydryl group** (—SH) consists of a sulfur atom bonded to an atom of hydrogen; it resembles a hydroxyl group in shape.

Thiols

Cysteine, an important sulfur-containing amino acid

- Two sulfhydryl groups can react, forming a covalent bond. This "cross-linking" helps stabilize protein structure (see Figure 5.20, Tertiary Structure).
- Cross-linking of cysteines in hair proteins maintains the curliness or straightness of hair. Straight hair can be "permanently" curled by shaping it around curlers and then breaking and re-forming the cross-linking bonds.

Phosphate

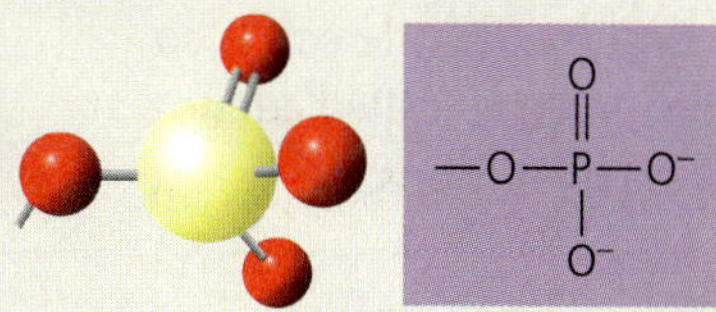

In the **phosphate group** shown here, a phosphorus atom is bonded to four oxygen atoms; one oxygen is bonded to the carbon skeleton; two oxygens carry negative charges (—OPO_3^{2-}). In this text, Ⓟ represents an attached phosphate group.

Organic phosphates

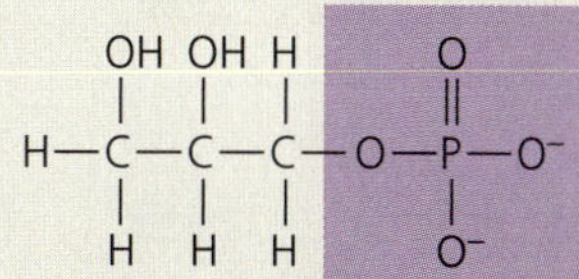

Glycerol phosphate, which takes part in many important chemical reactions in cells; glycerol phosphate also provides the backbone for phospholipids, the most prevalent molecules in cell membranes

- Contributes negative charge to the molecule of which it is a part (2– when at the end of a molecule, as above; 1– when located internally in a chain of phosphates).
- Molecules containing phosphate groups have the potential to react with water, releasing energy.

Methyl

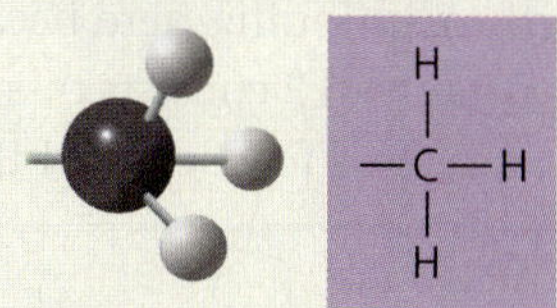

A **methyl group** (—CH_3) consists of a carbon bonded to three hydrogen atoms. The carbon of a methyl group may be attached to a carbon or to a different atom.

Methylated compounds

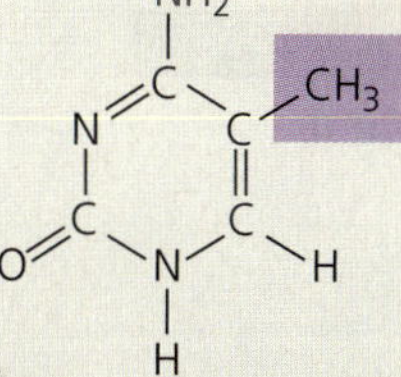

5-Methyl cytidine, a component of DNA that has been modified by addition of a methyl group

- Addition of a methyl group to DNA, or to molecules bound to DNA, affects the expression of genes.
- Arrangement of methyl groups in male and female sex hormones affects their shape and function (see p. 63).

MAKE CONNECTIONS *Given the information in this figure and what you know about the electronegativity of oxygen (see Concept 2.3, p. 39), predict which of the following molecules would be the stronger acid (see Concept 3.3, p. 53). Explain your answer.*

a. CH_3CH_2COOH b. $CH_3COCOOH$

ATP: An Important Source of Energy for Cellular Processes

The "Phosphate" column in Figure 4.9 shows a simple example of an organic phosphate molecule. A more complicated organic phosphate, **adenosine triphosphate**, or **ATP**, is worth mentioning here because its function in the cell is so important. ATP consists of an organic molecule called adenosine attached to a string of three phosphate groups:

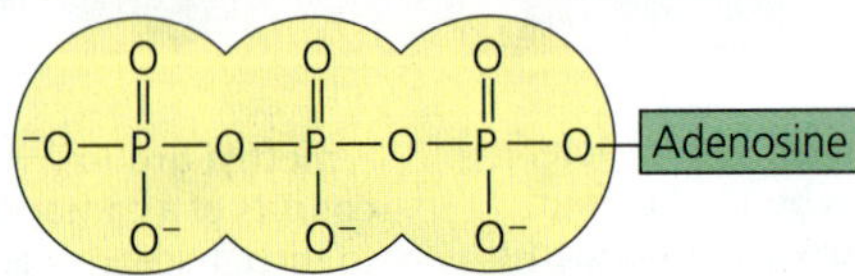

Where three phosphates are present in series, as in ATP, one phosphate may be split off as a result of a reaction with water. This inorganic phosphate ion, $HOPO_3^{2-}$, is often abbreviated Ⓟ$_i$ in this book. Having lost one phosphate, ATP becomes adenosine *di*phosphate, or ADP. Although ATP is sometimes said to store energy, it is more accurate to think of it as storing the potential to react with water. This reaction releases energy that can be used by the cell. You will learn about this in more detail in Chapter 8.

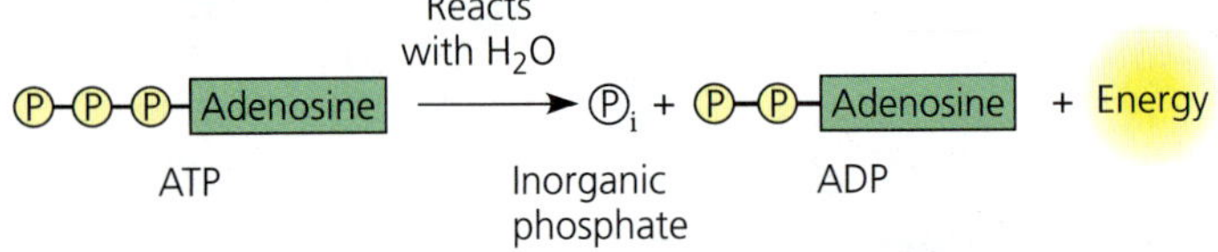

CONCEPT CHECK 4.3

1. What does the term *amino acid* signify about the structure of such a molecule?
2. What chemical change occurs to ATP when it reacts with water and releases energy?
3. **WHAT IF?** Suppose you had an organic molecule such as cysteine (see Figure 4.9, sulfhydryl group example), and you chemically removed the —NH_2 group and replaced it with —COOH. Draw the structural formula for this molecule and speculate about its chemical properties. Is the central carbon asymmetric before the change? After?

For suggested answers, see Appendix A.

The Chemical Elements of Life: *A Review*

Living matter, as you have learned, consists mainly of carbon, oxygen, hydrogen, and nitrogen, with smaller amounts of sulfur and phosphorus. These elements all form strong covalent bonds, an essential characteristic in the architecture of complex organic molecules. Of all these elements, carbon is the virtuoso of the covalent bond. The versatility of carbon makes possible the great diversity of organic molecules, each with particular properties that emerge from the unique arrangement of its carbon skeleton and the chemical groups appended to that skeleton. At the foundation of all biological diversity lies this variation at the molecular level.

4 CHAPTER REVIEW

SUMMARY OF KEY CONCEPTS

CONCEPT 4.1

Organic chemistry is the study of carbon compounds (pp. 58–59)

- Living matter is made mostly of carbon, oxygen, hydrogen, and nitrogen, with some sulfur and phosphorus. Biological diversity has its molecular basis in carbon's ability to form a huge number of molecules with particular shapes and chemical properties.
- Organic compounds were once thought to arise only within living organisms, but this idea (vitalism) was disproved when chemists were able to synthesize organic compounds in the laboratory.

? *How did Stanley Miller's experiments extend the idea of mechanism to the origin of life?*

CONCEPT 4.2

Carbon atoms can form diverse molecules by bonding to four other atoms (pp. 60–63)

- Carbon, with a valence of 4, can bond to various other atoms, including O, H, and N. Carbon can also bond to other carbon atoms, forming the carbon skeletons of organic compounds. These skeletons vary in length and shape and have bonding sites for atoms of other elements. **Hydrocarbons** consist only of carbon and hydrogen.
- **Isomers** are compounds with the same molecular formula but different structures and properties. Three types of isomers are **structural isomers**, ***cis-trans*** **isomers**, and **enantiomers**.

? *Refer back to Figure 4.9. What type of isomers are acetone and propanal? How many asymmetric carbons are present in acetic acid, glycine, and glycerol phosphate? Can these three molecules exist as forms that are enantiomers?*

CONCEPT 4.3

A few chemical groups are key to the functioning of biological molecules (pp. 63–66)

- Chemical groups attached to the carbon skeletons of organic molecules participate in chemical reactions (**functional groups**) or contribute to function by affecting molecular shape (see Figure 4.9).
- **ATP (adenosine triphosphate)** consists of adenosine attached to three phosphate groups. ATP can react with water, forming inorganic phosphate and ADP (adenosine diphosphate). This reaction releases energy that can be used by the cell (see the equation at the top of the next page).

Reacts with H_2O

Ⓟ-Ⓟ-Ⓟ-Adenosine ⟶ $Ⓟ_i$ + Ⓟ-Ⓟ-Adenosine + Energy

ATP Inorganic phosphate ADP

? *In what ways does a methyl group differ chemically from the other six important chemical groups shown in Figure 4.9?*

TEST YOUR UNDERSTANDING

LEVEL 1: KNOWLEDGE/COMPREHENSION

1. Organic chemistry is currently defined as
   a. the study of compounds made only by living cells.
   b. the study of carbon compounds.
   c. the study of vital forces.
   d. the study of natural (as opposed to synthetic) compounds.
   e. the study of hydrocarbons.

2. Which functional group is *not* present in this molecule?
   a. carboxyl
   b. sulfhydryl
   c. hydroxyl
   d. amino

   (Structure: HO–C(=O)–C(H)(NH₂)–C(H)(H)–OH)

3. **MAKE CONNECTIONS** Which chemical group is most likely to be responsible for an organic molecule behaving as a base (see Concept 3.3, p. 53)?
   a. hydroxyl c. carboxyl e. phosphate
   b. carbonyl d. amino

LEVEL 2: APPLICATION/ANALYSIS

4. Which of the following hydrocarbons has a double bond in its carbon skeleton?
   a. C_3H_8 b. C_2H_6 c. CH_4 d. C_2H_4 e. C_2H_2

5. Choose the term that correctly describes the relationship between these two sugar molecules:
   a. structural isomers
   b. *cis-trans* isomers
   c. enantiomers
   d. isotopes

   (Structures: H–C(H)(OH)–C(=O)–C(H)(OH)–H and H–C(=O)–C(H)(OH)–C(H)(OH)–H)

6. Identify the asymmetric carbon in this molecule:

   (Structure: O=C(H)–C(OH)(H)–C(H)(H)–C(H)(H)–C(H)(H)–H, with carbons labeled a, b, c, d, e)

7. Which action could produce a carbonyl group?
   a. the replacement of the —OH of a carboxyl group with hydrogen
   b. the addition of a thiol to a hydroxyl
   c. the addition of a hydroxyl to a phosphate
   d. the replacement of the nitrogen of an amine with oxygen
   e. the addition of a sulfhydryl to a carboxyl

8. Which of the molecules shown in question 5 has an asymmetric carbon? Which carbon is asymmetric?

LEVEL 3: SYNTHESIS/EVALUATION

9. **EVOLUTION CONNECTION**

   DRAW IT Some scientists think that life elsewhere in the universe might be based on the element silicon, rather than on carbon, as on Earth. Look at the electron distribution diagram for silicon in Figure 2.9 and draw the Lewis dot structure for silicon. What properties does silicon share with carbon that would make silicon-based life more likely than, say, neon-based life or aluminum-based life?

10. **SCIENTIFIC INQUIRY**

    Thalidomide achieved notoriety 50 years ago because of a wave of birth defects among children born to women who took this drug during pregnancy as a treatment for morning sickness. Thalidomide is a mixture of two enantiomers; one reduces morning sickness, but the other causes severe birth defects. (Although the beneficial enantiomer can be synthesized and given to patients, it is converted in the body to the harmful enantiomer.) The U.S. Food and Drug Administration (FDA) withheld approval of thalidomide in 1960. Since then, however, the FDA has approved this drug for the treatment of conditions associated with Hansen's disease (leprosy) and newly diagnosed multiple myeloma, a blood and bone marrow cancer. In clinical trials, thalidomide also shows promise as a treatment for AIDS, tuberculosis, inflammatory diseases, and some other types of cancer. Assuming that molecules related to thalidomide could be synthesized in the laboratory, describe in a broad way the type of experiments you would do to improve the benefits of this drug and minimize its harmful effects.

11. **WRITE ABOUT A THEME**

    Structure and Function In 1918, an epidemic of sleeping sickness caused an unusual rigid paralysis in some survivors, similar to symptoms of advanced Parkinson's disease. Years later, L-dopa (below, left), a chemical used to treat Parkinson's disease, was given to some of these patients, as dramatized in the movie *Awakenings*, starring Robin Williams. L-dopa was remarkably effective at eliminating the paralysis, at least temporarily. However, its enantiomer, D-dopa (right), was subsequently shown to have no effect at all, as is the case for Parkinson's disease. In a short essay (100–150 words), discuss how the effectiveness of one enantiomer and not the other illustrates the theme of structure and function.

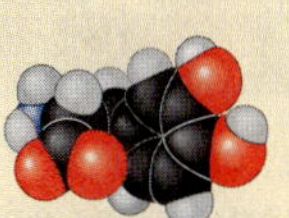

L-dopa

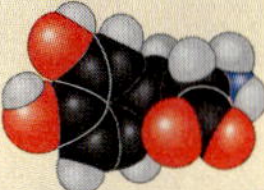

D-dopa

For selected answers, see Appendix A.

5

The Structure and Function of Large Biological Molecules

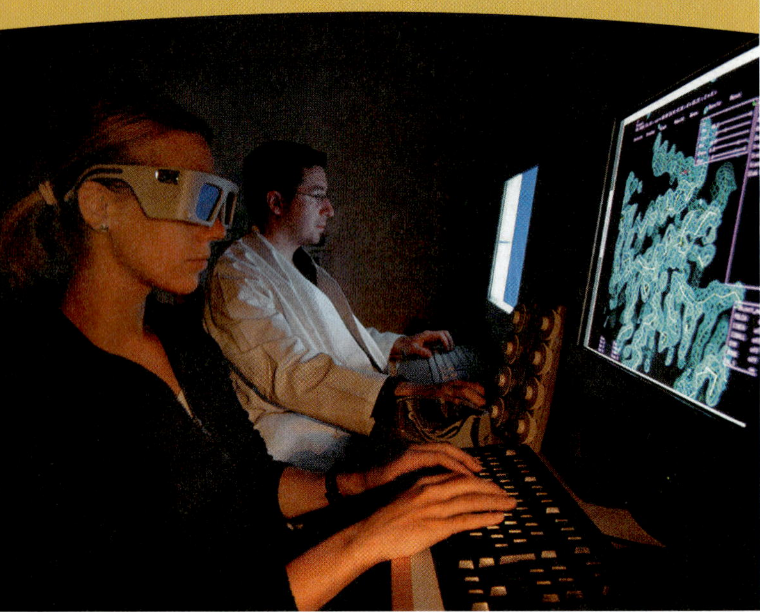

▲ **Figure 5.1 Why do scientists study the structures of macromolecules?**

KEY CONCEPTS

5.1 Macromolecules are polymers, built from monomers
5.2 Carbohydrates serve as fuel and building material
5.3 Lipids are a diverse group of hydrophobic molecules
5.4 Proteins include a diversity of structures, resulting in a wide range of functions
5.5 Nucleic acids store, transmit, and help express hereditary information

OVERVIEW

The Molecules of Life

Given the rich complexity of life on Earth, we might expect organisms to have an enormous diversity of molecules. Remarkably, however, the critically important large molecules of all living things—from bacteria to elephants—fall into just four main classes: carbohydrates, lipids, proteins, and nucleic acids. On the molecular scale, members of three of these classes—carbohydrates, proteins, and nucleic acids—are huge and are therefore called **macromolecules**. For example, a protein may consist of thousands of atoms that form a molecular colossus with a mass well over 100,000 daltons. Considering the size and complexity of macromolecules, it is noteworthy that biochemists have determined the detailed structure of so many of them. The scientist in the foreground of **Figure 5.1** is using 3-D glasses to help her visualize the structure of the protein displayed on her screen.

The architecture of a large biological molecule helps explain how that molecule works. Like water and simple organic molecules, large biological molecules exhibit unique emergent properties arising from the orderly arrangement of their atoms. In this chapter, we'll first consider how macromolecules are built. Then we'll examine the structure and function of all four classes of large biological molecules: carbohydrates, lipids, proteins, and nucleic acids.

CONCEPT 5.1

Macromolecules are polymers, built from monomers

The macromolecules in three of the four classes of life's organic compounds—carbohydrates, proteins, and nucleic acids—are chain-like molecules called polymers (from the Greek *polys*, many, and *meros*, part). A **polymer** is a long molecule consisting of many similar or identical building blocks linked by covalent bonds, much as a train consists of a chain of cars. The repeating units that serve as the building blocks of a polymer are smaller molecules called **monomers** (from the Greek *monos*, single). Some of the molecules that serve as monomers also have other functions of their own.

The Synthesis and Breakdown of Polymers

Although each class of polymer is made up of a different type of monomer, the chemical mechanisms by which cells make and break down polymers are basically the same in all cases. In cells, these processes are facilitated by **enzymes**, specialized macromolecules that speed up chemical reactions. Monomers are connected by a reaction in which two molecules are covalently bonded to each other, with the loss of a water molecule; this is known as a **dehydration reaction** (**Figure 5.2a**). When a bond forms between two monomers, each monomer contributes part of the water molecule that is released during the reaction: One monomer provides a hydroxyl group (—OH), while the other provides a hydrogen (—H). This reaction is repeated as monomers are added to the chain one by one, making a polymer.

Polymers are disassembled to monomers by **hydrolysis**, a process that is essentially the reverse of the dehydration reac-

▼ **Figure 5.2 The synthesis and breakdown of polymers.**

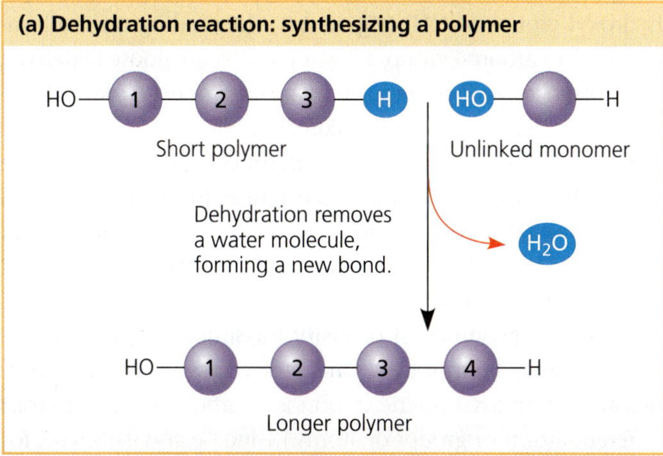

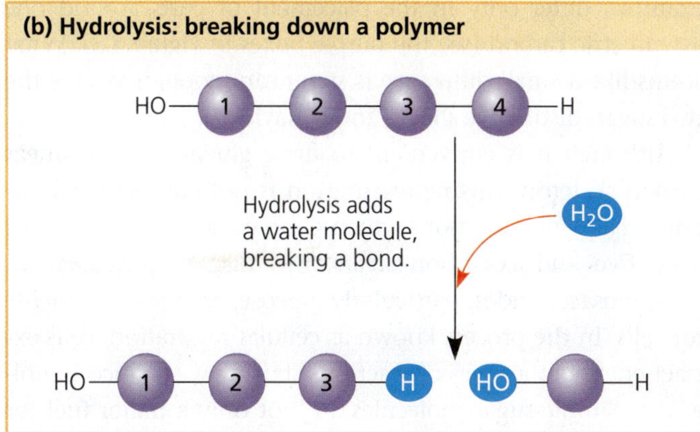

tion (**Figure 5.2b**). Hydrolysis means to break using water (from the Greek *hydro*, water, and *lysis*, break). The bond between the monomers is broken by the addition of a water molecule, with the hydrogen from the water attaching to one monomer and the hydroxyl group attaching to the adjacent monomer. An example of hydrolysis working within our bodies is the process of digestion. The bulk of the organic material in our food is in the form of polymers that are much too large to enter our cells. Within the digestive tract, various enzymes attack the polymers, speeding up hydrolysis. The released monomers are then absorbed into the bloodstream for distribution to all body cells. Those cells can then use dehydration reactions to assemble the monomers into new, different polymers that can perform specific functions required by the cell.

The Diversity of Polymers

Each cell has thousands of different macromolecules; the collection varies from one type of cell to another even in the same organism. The inherent differences between human siblings reflect small variations in polymers, particularly DNA and proteins. Molecular differences between unrelated individuals are more extensive and those between species greater still. The diversity of macromolecules in the living world is vast, and the possible variety is effectively limitless.

What is the basis for such diversity in life's polymers? These molecules are constructed from only 40 to 50 common monomers and some others that occur rarely. Building a huge variety of polymers from such a limited number of monomers is analogous to constructing hundreds of thousands of words from only 26 letters of the alphabet. The key is arrangement—the particular linear sequence that the units follow. However, this analogy falls far short of describing the great diversity of macromolecules because most biological polymers have many more monomers than the number of letters in the longest word. Proteins, for example, are built from 20 kinds of amino acids arranged in chains that are typically hundreds of amino acids long. The molecular logic of life is simple but elegant: Small molecules common to all organisms are ordered into unique macromolecules.

Despite this immense diversity, molecular structure and function can still be grouped roughly by class. Let's examine each of the four major classes of large biological molecules. For each class, the large molecules have emergent properties not found in their individual building blocks.

CONCEPT CHECK 5.1

1. What are the four main classes of large biological molecules? Which class does not consist of polymers?
2. How many molecules of water are needed to completely hydrolyze a polymer that is ten monomers long?
3. **WHAT IF?** Suppose you eat a serving of fish. What reactions must occur for the amino acid monomers in the protein of the fish to be converted to new proteins in your body?

For suggested answers, see Appendix A.

CONCEPT 5.2

Carbohydrates serve as fuel and building material

Carbohydrates include both sugars and polymers of sugars. The simplest carbohydrates are the monosaccharides, or simple sugars; these are the monomers from which more complex carbohydrates are constructed. Disaccharides are double sugars, consisting of two monosaccharides joined by a covalent bond. Carbohydrates also include macromolecules called polysaccharides, polymers composed of many sugar building blocks.

Sugars

Monosaccharides (from the Greek *monos*, single, and *sacchar*, sugar) generally have molecular formulas that are some multiple of the unit CH_2O. Glucose ($C_6H_{12}O_6$), the most common monosaccharide, is of central importance in the chemistry

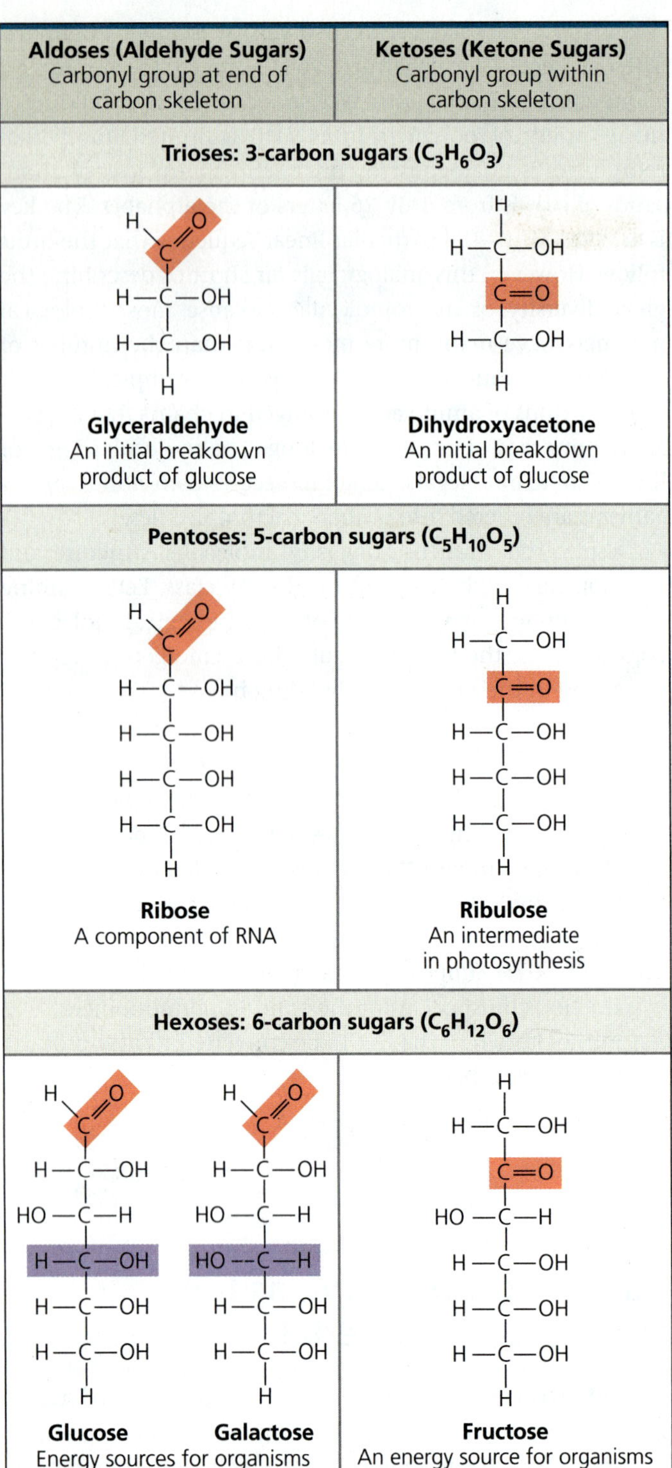

▲ **Figure 5.3 The structure and classification of some monosaccharides.** Sugars vary in the location of their carbonyl groups (orange), the length of their carbon skeletons, and the spatial arrangement around asymmetric carbons (compare, for example, the purple portions of glucose and galactose).

MAKE CONNECTIONS *In the 1970s, a process was developed that converts the glucose in corn syrup to its sweeter isomer, fructose. High-fructose corn syrup, a common ingredient in soft drinks and processed food, is a mixture of glucose and fructose. What type of isomers are glucose and fructose? See Figure 4.7, p. 62.*

of life. In the structure of glucose, we can see the trademarks of a sugar: The molecule has a carbonyl group (C=O) and multiple hydroxyl groups (—OH) **(Figure 5.3)**. Depending on the location of the carbonyl group, a sugar is either an aldose (aldehyde sugar) or a ketose (ketone sugar). Glucose, for example, is an aldose; fructose, an isomer of glucose, is a ketose. (Most names for sugars end in *-ose*.) Another criterion for classifying sugars is the size of the carbon skeleton, which ranges from three to seven carbons long. Glucose, fructose, and other sugars that have six carbons are called hexoses. Trioses (three-carbon sugars) and pentoses (five-carbon sugars) are also common.

Still another source of diversity for simple sugars is in the spatial arrangement of their parts around asymmetric carbons. (Recall that an asymmetric carbon is a carbon attached to four different atoms or groups of atoms.) Glucose and galactose, for example, differ only in the placement of parts around one asymmetric carbon (see the purple boxes in Figure 5.3). What seems like a small difference is significant enough to give the two sugars distinctive shapes and behaviors.

Although it is convenient to draw glucose with a linear carbon skeleton, this representation is not completely accurate. In aqueous solutions, glucose molecules, as well as most other five- and six-carbon sugars, form rings **(Figure 5.4)**.

Monosaccharides, particularly glucose, are major nutrients for cells. In the process known as cellular respiration, cells extract energy in a series of reactions starting with glucose molecules. Simple-sugar molecules are not only a major fuel for cellular work, but their carbon skeletons also serve as raw material for the synthesis of other types of small organic molecules, such as amino acids and fatty acids. Sugar molecules that are not immediately used in these ways are generally incorporated as monomers into disaccharides or polysaccharides.

A **disaccharide** consists of two monosaccharides joined by a **glycosidic linkage**, a covalent bond formed between two monosaccharides by a dehydration reaction. For example, maltose is a disaccharide formed by the linking of two molecules of glucose **(Figure 5.5a)**. Also known as malt sugar, maltose is an ingredient used in brewing beer. The most prevalent disaccharide is sucrose, which is table sugar. Its two monomers are glucose and fructose **(Figure 5.5b)**. Plants generally transport carbohydrates from leaves to roots and other nonphotosynthetic organs in the form of sucrose. Lactose, the sugar present in milk, is another disaccharide, in this case a glucose molecule joined to a galactose molecule.

Polysaccharides

Polysaccharides are macromolecules, polymers with a few hundred to a few thousand monosaccharides joined by glycosidic linkages. Some polysaccharides serve as storage material, hydrolyzed as needed to provide sugar for cells. Other polysaccharides serve as building material for structures that

(a) Linear and ring forms. Chemical equilibrium between the linear and ring structures greatly favors the formation of rings. The carbons of the sugar are numbered 1 to 6, as shown. To form the glucose ring, carbon 1 bonds to the oxygen attached to carbon 5.

(b) Abbreviated ring structure. Each corner represents a carbon. The ring's thicker edge indicates that you are looking at the ring edge-on; the components attached to the ring lie above or below the plane of the ring.

▲ **Figure 5.4 Linear and ring forms of glucose.**

DRAW IT *Start with the linear form of fructose (see Figure 5.3) and draw the formation of the fructose ring in two steps. First, number the carbons starting at the top of the linear structure. Then attach carbon 5 via its oxygen to carbon 2. Compare the number of carbons in the fructose and glucose rings.*

(a) Dehydration reaction in the synthesis of maltose. The bonding of two glucose units forms maltose. The glycosidic linkage joins the number 1 carbon of one glucose to the number 4 carbon of the second glucose. Joining the glucose monomers in a different way would result in a different disaccharide.

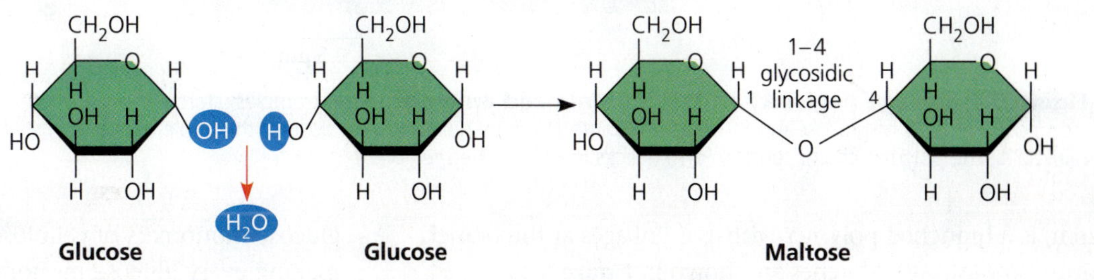

(b) Dehydration reaction in the synthesis of sucrose. Sucrose is a disaccharide formed from glucose and fructose. Notice that fructose, though a hexose like glucose, forms a five-sided ring.

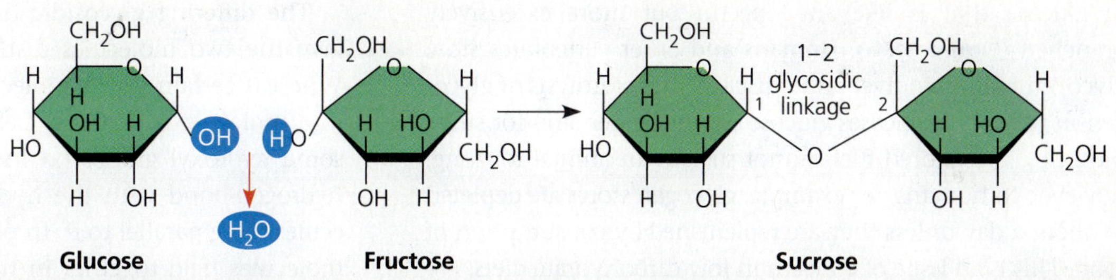

▲ **Figure 5.5 Examples of disaccharide synthesis.**

DRAW IT *Referring to Figure 5.4, number the carbons in each sugar in this figure. Show how the numbering is consistent with the name of the glycosidic linkage in each disaccharide.*

protect the cell or the whole organism. The architecture and function of a polysaccharide are determined by its sugar monomers and by the positions of its glycosidic linkages.

Storage Polysaccharides

Both plants and animals store sugars for later use in the form of storage polysaccharides. Plants store **starch**, a polymer of glucose monomers, as granules within cellular structures known as plastids, which include chloroplasts. Synthesizing starch enables the plant to stockpile surplus glucose. Because glucose is a major cellular fuel, starch represents stored energy. The sugar can later be withdrawn from this carbohydrate "bank" by hydrolysis, which breaks the bonds between the glucose monomers. Most animals, including humans, also have enzymes that can hydrolyze plant starch, making glucose available as a nutrient for cells. Potato tubers and grains—the fruits of wheat, maize (corn), rice, and other grasses—are the major sources of starch in the human diet.

Most of the glucose monomers in starch are joined by 1–4 linkages (number 1 carbon to number 4 carbon), like the glucose units in maltose (see Figure 5.5a). The simplest form of starch, amylose, is unbranched. Amylopectin, a more complex

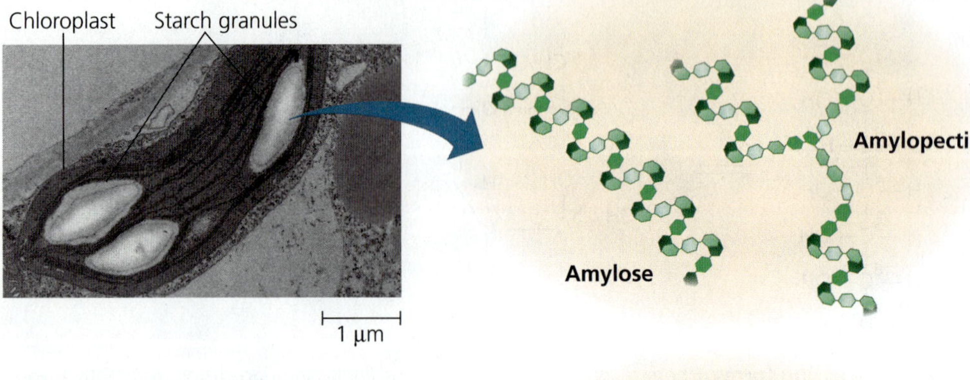

(a) Starch: a plant polysaccharide. This micrograph shows part of a plant cell with a chloroplast, the cellular organelle where glucose is made and then stored as starch granules. Amylose (unbranched) and amylopectin (branched) are two forms of starch.

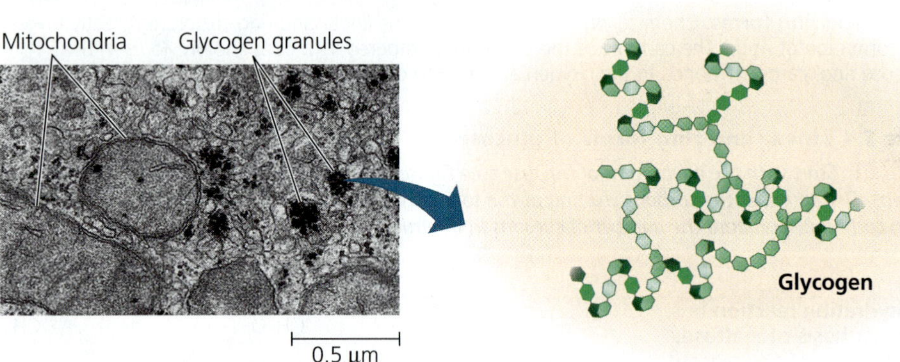

(b) Glycogen: an animal polysaccharide. Animal cells stockpile glycogen as dense clusters of granules within liver and muscle cells, as shown in this micrograph of part of a liver cell. Mitochondria are cellular organelles that help break down glucose released from glycogen. Note that glycogen is more branched than amylopectin starch.

▲ **Figure 5.6 Storage polysaccharides of plants and animals.** These examples, starch and glycogen, are composed entirely of glucose monomers, represented here by hexagons. Because of the angle of the 1–4 linkages, the polymer chains tend to form helices in unbranched regions.

starch, is a branched polymer with 1–6 linkages at the branch points. Both of these starches are shown in **Figure 5.6a**.

Animals store a polysaccharide called **glycogen**, a polymer of glucose that is like amylopectin but more extensively branched **(Figure 5.6b)**. Humans and other vertebrates store glycogen mainly in liver and muscle cells. Hydrolysis of glycogen in these cells releases glucose when the demand for sugar increases. This stored fuel cannot sustain an animal for long, however. In humans, for example, glycogen stores are depleted in about a day unless they are replenished by consumption of food. This is an issue of concern in low-carbohydrate diets.

Structural Polysaccharides

Organisms build strong materials from structural polysaccharides. For example, the polysaccharide called **cellulose** is a major component of the tough walls that enclose plant cells. On a global scale, plants produce almost 10^{14} kg (100 billion tons) of cellulose per year; it is the most abundant organic compound on Earth. Like starch, cellulose is a polymer of glucose, but the glycosidic linkages in these two polymers differ. The difference is based on the fact that there are actually two slightly different ring structures for glucose **(Figure 5.7a)**. When glucose forms a ring, the hydroxyl group attached to the number 1 carbon is positioned either below or above the plane of the ring. These two ring forms for glucose are called alpha (α) and beta (β), respectively. In starch, all the glucose monomers are in the α configuration **(Figure 5.7b)**, the arrangement we saw in Figures 5.4 and 5.5. In contrast, the glucose monomers of cellulose are all in the β configuration, making every glucose monomer "upside down" with respect to its neighbors **(Figure 5.7c)**.

The differing glycosidic linkages in starch and cellulose give the two molecules distinct three-dimensional shapes. Whereas certain starch molecules are largely helical, a cellulose molecule is straight. Cellulose is never branched, and some hydroxyl groups on its glucose monomers are free to hydrogen-bond with the hydroxyls of other cellulose molecules lying parallel to it. In plant cell walls, parallel cellulose molecules held together in this way are grouped into units called microfibrils **(Figure 5.8)**. These cable-like microfibrils are a strong building material for plants and an important substance for humans because cellulose is the major constituent of paper and the only component of cotton.

Enzymes that digest starch by hydrolyzing its α linkages are unable to hydrolyze the β linkages of cellulose because of the distinctly different shapes of these two molecules. In fact, few organisms possess enzymes that can digest cellulose. Animals, including humans, do not; the cellulose in our food passes through the digestive tract and is eliminated with the feces. Along the way, the cellulose abrades the wall of the digestive tract and stimulates the lining to secrete mucus, which aids in the smooth passage of food through the tract. Thus, although cellulose is not a nutrient for humans, it is an important part of a healthful diet. Most fresh fruits, vegetables, and whole grains are rich in cellulose. On food packages, "insoluble fiber" refers mainly to cellulose.

(a) **α and β glucose ring structures.** These two interconvertible forms of glucose differ in the placement of the hydroxyl group (highlighted in blue) attached to the number 1 carbon.

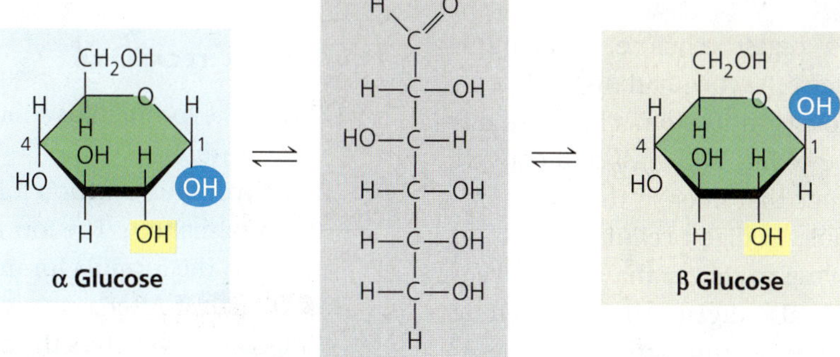

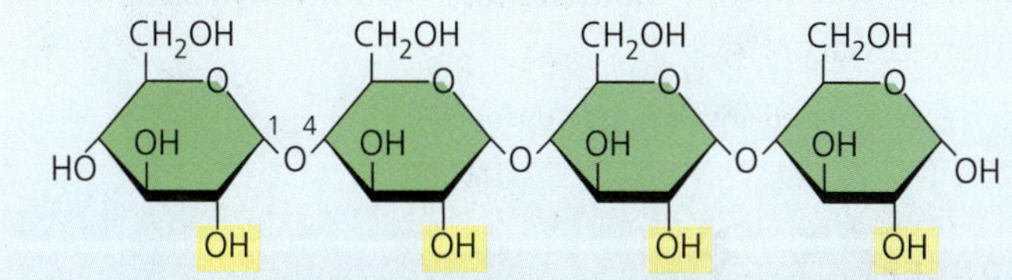

(b) **Starch: 1–4 linkage of α glucose monomers.** All monomers are in the same orientation. Compare the positions of the —OH groups highlighted in yellow with those in cellulose (c).

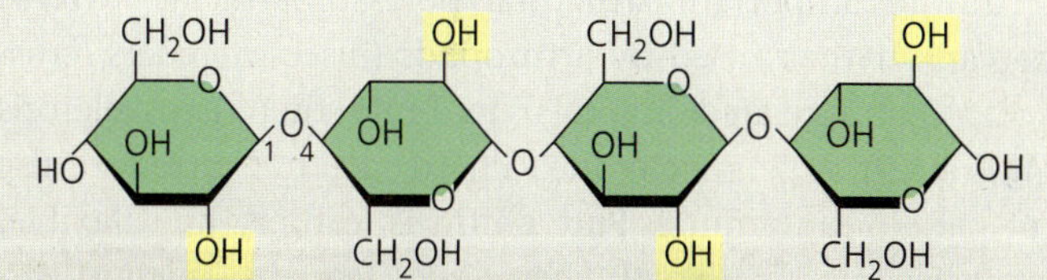

(c) **Cellulose: 1–4 linkage of β glucose monomers.** In cellulose, every β glucose monomer is upside down with respect to its neighbors.

▲ **Figure 5.7 Starch and cellulose structures.**

▲ **Figure 5.8 The arrangement of cellulose in plant cell walls.**

Some microorganisms can digest cellulose, breaking it down into glucose monomers. A cow harbors cellulose-digesting prokaryotes and protists in its stomach. These microbes hydrolyze the cellulose of hay and grass and convert the glucose to other compounds that nourish the cow. Similarly, a termite, which is unable to digest cellulose by itself, has prokaryotes or protists living in its gut that can make a meal of wood. Some fungi can also digest cellulose, thereby helping recycle chemical elements within Earth's ecosystems.

Another important structural polysaccharide is **chitin**, the carbohydrate used by arthropods (insects, spiders, crustaceans, and related animals) to build their exoskeletons **(Figure 5.9)**. An exoskeleton is a hard case that surrounds the soft parts of an animal. Pure chitin is leathery and flexible, but it becomes hardened when encrusted with calcium carbonate, a salt. Chitin is also found in many fungi, which use this polysaccharide rather than cellulose as the building material for their cell walls. Chitin is similar to cellulose, with β linkages, except that the glucose monomer of chitin has a nitrogen-containing appendage (see Figure 5.9, top right).

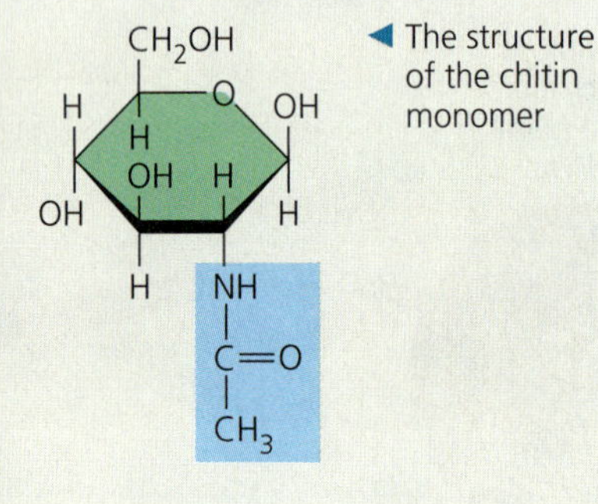

◀ The structure of the chitin monomer

◀ Chitin forms the exoskeleton of arthropods. This cicada is molting, shedding its old exoskeleton and emerging in adult form.

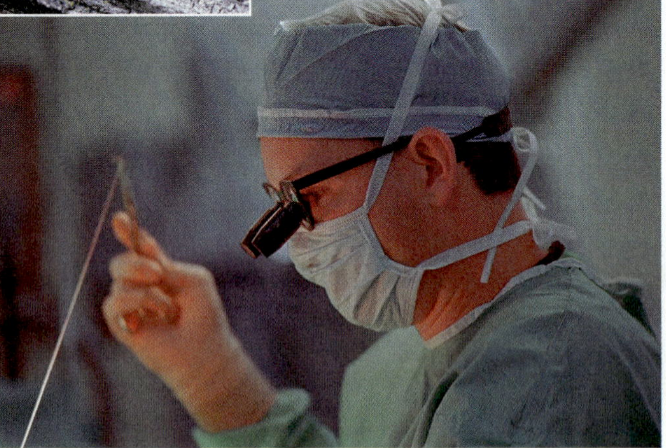

▲ Chitin is used to make a strong and flexible surgical thread that decomposes after the wound or incision heals.

▲ **Figure 5.9 Chitin, a structural polysaccharide.**

CONCEPT CHECK 5.2

1. Write the formula for a monosaccharide that has three carbons.
2. A dehydration reaction joins two glucose molecules to form maltose. The formula for glucose is $C_6H_{12}O_6$. What is the formula for maltose?
3. **WHAT IF?** After a cow is given antibiotics to treat an infection, a vet gives the animal a drink of "gut culture" containing various prokaryotes. Why is this necessary?

For suggested answers, see Appendix A.

CONCEPT **5.3**

Lipids are a diverse group of hydrophobic molecules

Lipids are the one class of large biological molecules that does not include true polymers, and they are generally not big enough to be considered macromolecules. The compounds called **lipids** are grouped together because they share one important trait: They mix poorly, if at all, with water. The hydrophobic behavior of lipids is based on their molecular structure. Although they may have some polar bonds associated with oxygen, lipids consist mostly of hydrocarbon regions. Lipids are varied in form and function. They include waxes and certain pigments, but we will focus on the most biologically important types of lipids: fats, phospholipids, and steroids.

Fats

Although fats are not polymers, they are large molecules assembled from smaller molecules by dehydration reactions. A **fat** is constructed from two kinds of smaller molecules: glycerol and fatty acids **(Figure 5.10a)**. Glycerol is an alcohol; each of its three carbons bears a hydroxyl group. A **fatty acid** has a long carbon skeleton, usually 16 or 18 carbon atoms in length. The carbon at one end of the skeleton is part of a carboxyl group, the functional group that gives these molecules the name fatty *acid.* The rest of the skeleton consists of a hydrocarbon chain. The relatively nonpolar C—H bonds in the hydrocarbon chains of fatty acids are the reason fats are hydrophobic. Fats separate from water because the water molecules hydrogen-bond to one another and exclude the fats. This is the reason that vegetable oil (a liquid fat) separates from the aqueous vinegar solution in a bottle of salad dressing.

In making a fat, three fatty acid molecules are each joined to glycerol by an ester linkage, a bond between a hydroxyl group and a carboxyl group. The resulting fat, also called a **triacylglycerol**, thus consists of three fatty acids linked to one glycerol molecule. (Still another name for a fat is

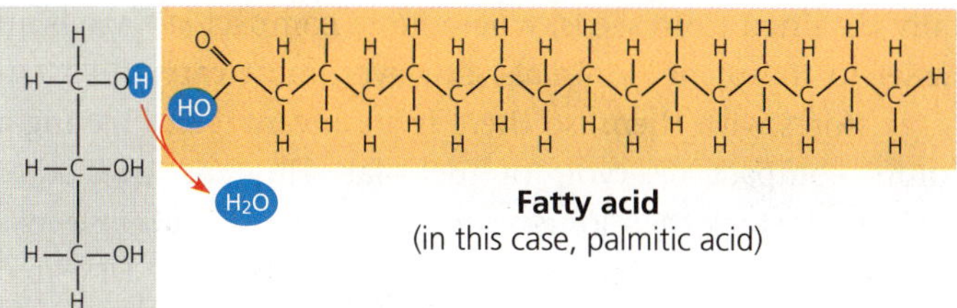

(a) One of three dehydration reactions in the synthesis of a fat

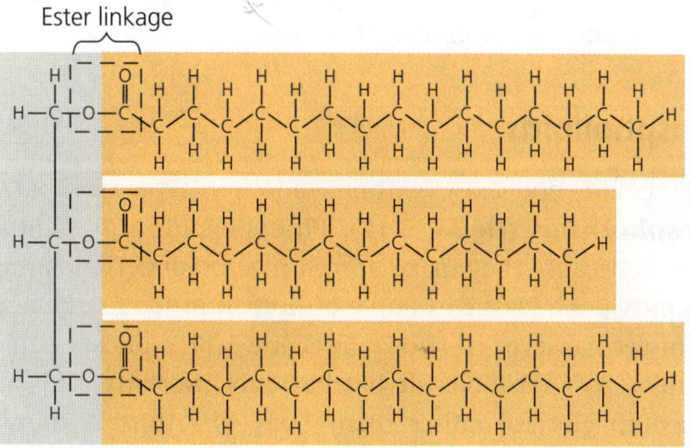

(b) Fat molecule (triacylglycerol)

▲ **Figure 5.10 The synthesis and structure of a fat, or triacylglycerol.** The molecular building blocks of a fat are one molecule of glycerol and three molecules of fatty acids. **(a)** One water molecule is removed for each fatty acid joined to the glycerol. **(b)** A fat molecule with three fatty acid units, two of them identical. The carbons of the fatty acids are arranged zigzag to suggest the actual orientations of the four single bonds extending from each carbon (see Figure 4.3a).

triglyceride, a word often found in the list of ingredients on packaged foods.) The fatty acids in a fat can be the same, or they can be of two or three different kinds, as in **Figure 5.10b**.

The terms *saturated fats* and *unsaturated fats* are commonly used in the context of nutrition **(Figure 5.11)**. These terms refer to the structure of the hydrocarbon chains of the fatty acids. If there are no double bonds between carbon atoms composing a chain, then as many hydrogen atoms as possible are bonded to the carbon skeleton. Such a structure is said to be *saturated* with hydrogen, and the resulting fatty acid therefore called a **saturated fatty acid (Figure 5.11a)**. An **unsaturated fatty acid** has one or more double bonds, with one fewer hydrogen atom on each double-bonded carbon. Nearly all double bonds in naturally occurring fatty acids are *cis* double bonds, which cause a kink in the hydrocarbon chain wherever they occur **(Figure 5.11b)**. (See Figure 4.7 to remind yourself about *cis* and *trans* double bonds.)

A fat made from saturated fatty acids is called a saturated fat. Most animal fats are saturated: The hydrocarbon chains of their fatty acids—the "tails" of the fat molecules—lack double bonds, and their flexibility allows the fat molecules to pack together tightly. Saturated animal fats—such as lard and butter—are solid at room temperature. In contrast, the fats of plants

▼ **Figure 5.11 Saturated and unsaturated fats and fatty acids.**

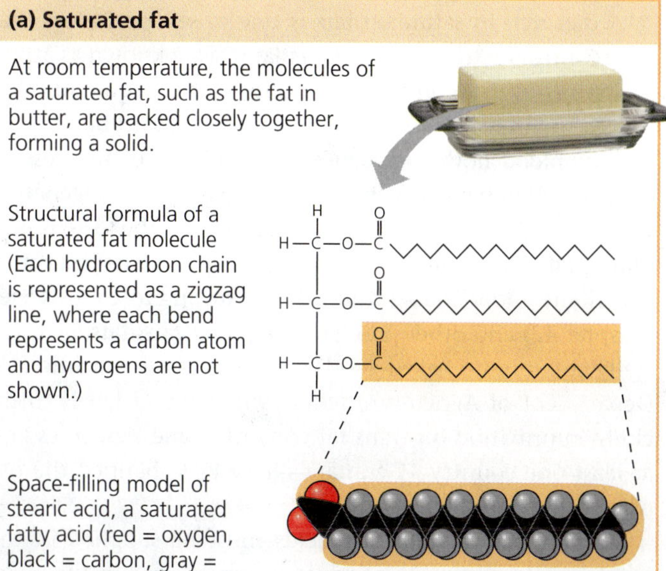

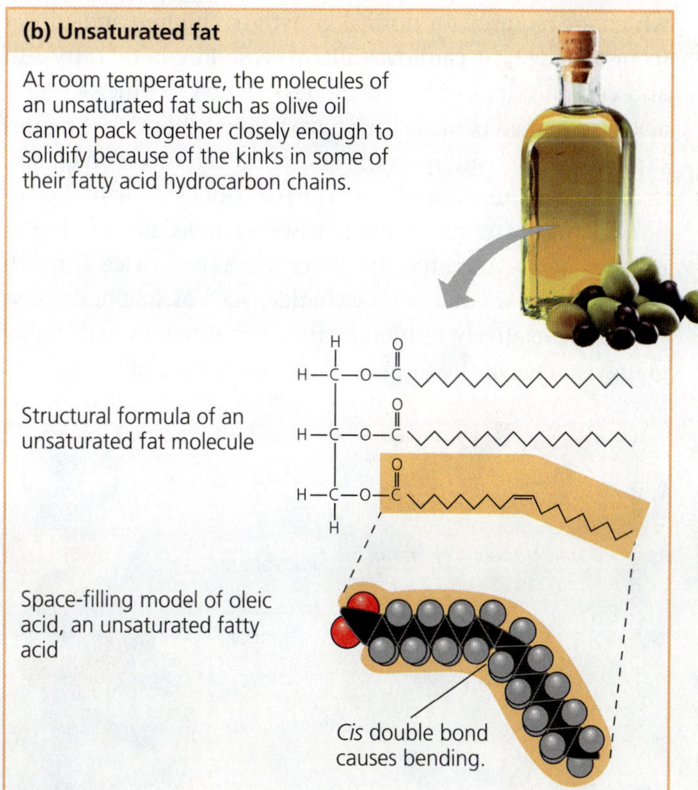

and fishes are generally unsaturated, meaning that they are built of one or more types of unsaturated fatty acids. Usually liquid at room temperature, plant and fish fats are referred to as oils—olive oil and cod liver oil are examples. The kinks where the *cis* double bonds are located prevent the molecules from packing together closely enough to solidify at room temperature. The phrase "hydrogenated vegetable oils" on food labels means that unsaturated fats have been synthetically converted to saturated fats by adding hydrogen. Peanut butter, margarine, and many other products are hydrogenated to prevent lipids from separating out in liquid (oil) form.

A diet rich in saturated fats is one of several factors that may contribute to the cardiovascular disease known as atherosclerosis. In this condition, deposits called plaques develop within the walls of blood vessels, causing inward bulges that impede blood flow and reduce the resilience of the vessels. Recent studies have shown that the process of hydrogenating vegetable oils produces not only saturated fats but also unsaturated fats with *trans* double bonds. These **trans fats** may contribute more than saturated fats to atherosclerosis (see Chapter 42) and other problems. Because trans fats are especially common in baked goods and processed foods, the U.S. Department of Agriculture requires nutritional labels to include information on trans fat content. Some U.S. cities and at least one country—Denmark—have even banned the use of trans fats in restaurants.

Certain unsaturated fatty acids must be supplied in the human diet because they cannot be synthesized in the body. These essential fatty acids include the omega-3 fatty acids, which are required for normal growth in children and appear to protect against cardiovascular disease in adults. Fatty fish and certain nuts and vegetable oils are rich in omega-3 fatty acids (so named because they have a double bond at the third carbon-carbon bond from the end of the hydrocarbon chain).

The major function of fats is energy storage. The hydrocarbon chains of fats are similar to gasoline molecules and just as rich in energy. A gram of fat stores more than twice as much energy as a gram of a polysaccharide, such as starch. Because plants are relatively immobile, they can function with bulky energy storage in the form of starch. (Vegetable oils are generally obtained from seeds, where more compact storage is an asset to the plant.) Animals, however, must carry their energy stores with them, so there is an advantage to having a more compact reservoir of fuel—fat. Humans and other mammals stock their long-term food reserves in adipose cells (see Figure 4.6a), which swell and shrink as fat is deposited and withdrawn from storage. In addition to storing energy, adipose tissue also cushions such vital organs as the kidneys, and a layer of fat beneath the skin insulates the body. This subcutaneous layer is especially thick in whales, seals, and most other marine mammals, protecting them from cold ocean water.

Phospholipids

Cells could not exist without another type of lipid—**phospholipids** **(Figure 5.12)**. Phospholipids are essential for cells because they make up cell membranes. Their structure provides a classic example of how form fits function at the molecular level. As shown in Figure 5.12, a phospholipid is similar to a fat molecule but has only two fatty acids attached to glycerol rather than three. The third hydroxyl group of glycerol is joined to a phosphate group, which has a negative electrical charge in the cell. Additional small molecules, which are usually charged or polar, can be linked to the phosphate group to form a variety of phospholipids.

The two ends of phospholipids show different behavior toward water. The hydrocarbon tails are hydrophobic and are excluded from water. However, the phosphate group and its attachments form a hydrophilic head that has an affinity for water. When phospholipids are added to water, they self-assemble into double-layered structures called "bilayers," shielding their hydrophobic portions from water **(Figure 5.13)**.

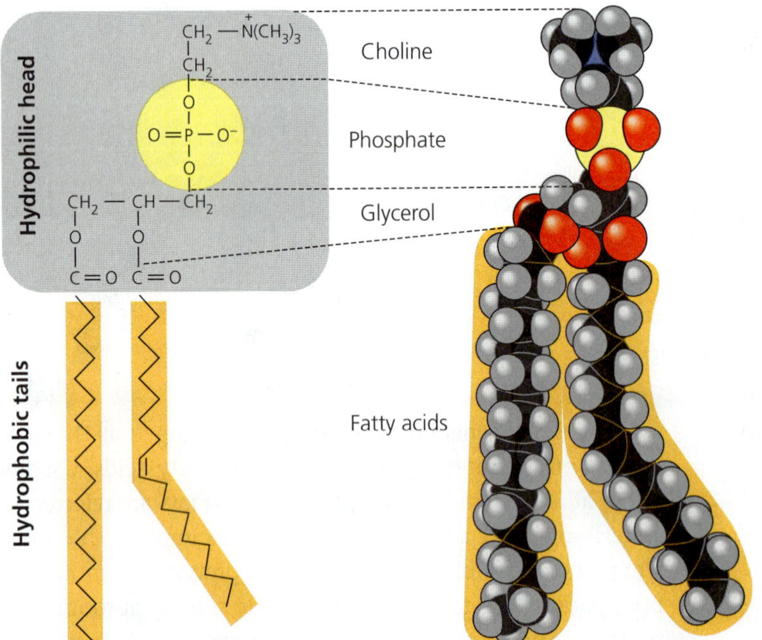

(a) Structural formula **(b) Space-filling model**

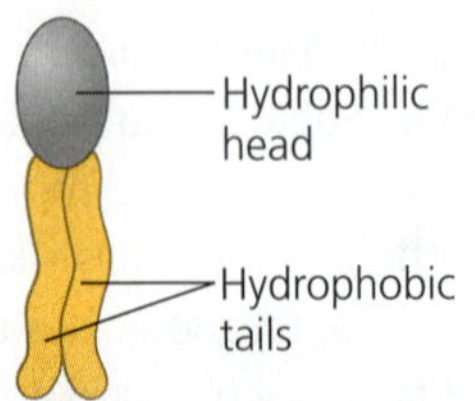

(c) Phospholipid symbol

◄ **Figure 5.12 The structure of a phospholipid.** A phospholipid has a hydrophilic (polar) head and two hydrophobic (nonpolar) tails. Phospholipid diversity is based on differences in the two fatty acids and in the groups attached to the phosphate group of the head. This particular phospholipid, called a phosphatidylcholine, has an attached choline group. The kink in one of its tails is due to a *cis* double bond. Shown here are **(a)** the structural formula, **(b)** the space-filling model (yellow = phosphorus, blue = nitrogen), and **(c)** the symbol for a phospholipid that will appear throughout this book. (In most figures, this symbol will be used to represent a phospholipid with either saturated or unsaturated tails.)

DRAW IT *Draw an oval around the hydrophilic head of the space-filling model.*

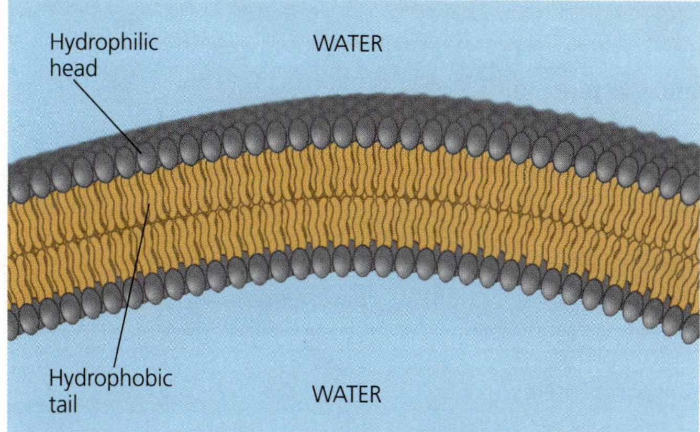

▲ **Figure 5.13 Bilayer structure formed by self-assembly of phospholipids in an aqueous environment.** The phospholipid bilayer shown here is the main fabric of biological membranes. Note that the hydrophilic heads of the phospholipids are in contact with water in this structure, whereas the hydrophobic tails are in contact with each other and remote from water.

At the surface of a cell, phospholipids are arranged in a similar bilayer. The hydrophilic heads of the molecules are on the outside of the bilayer, in contact with the aqueous solutions inside and outside of the cell. The hydrophobic tails point toward the interior of the bilayer, away from the water. The phospholipid bilayer forms a boundary between the cell and its external environment; in fact, cells could not exist without phospholipids.

Steroids

Steroids are lipids characterized by a carbon skeleton consisting of four fused rings. Different steroids, such as cholesterol and the vertebrate sex hormones, are distinguished by the particular chemical groups attached to this ensemble of rings (**Figure 5.14**). **Cholesterol** is a crucial molecule in animals. It is a common component of animal cell membranes and is also the precursor from which other steroids are synthesized. In vertebrates, cholesterol is synthesized in the liver and obtained from the diet. A high level of cholesterol in the blood may contribute to atherosclerosis. In fact, both saturated fats and trans fats exert their negative impact on health by affecting cholesterol levels.

▲ **Figure 5.14 Cholesterol, a steroid.** Cholesterol is the molecule from which other steroids, including the sex hormones, are synthesized. Steroids vary in the chemical groups attached to their four interconnected rings (shown in gold).

MAKE CONNECTIONS *Compare cholesterol with the sex hormones shown in Concept 4.3 on p. 63. Circle the chemical groups that cholesterol has in common with estradiol; put a square around the chemical groups that cholesterol has in common with testosterone.*

CONCEPT CHECK 5.3

1. Compare the structure of a fat (triglyceride) with that of a phospholipid.
2. Why are human sex hormones considered lipids?
3. **WHAT IF?** Suppose a membrane surrounded an oil droplet, as it does in the cells of plant seeds. Describe and explain the form it might take.

For suggested answers, see Appendix A.

CONCEPT 5.4

Proteins include a diversity of structures, resulting in a wide range of functions

Nearly every dynamic function of a living being depends on proteins. In fact, the importance of proteins is underscored by their name, which comes from the Greek word *proteios*, meaning "first," or "primary." Proteins account for more than 50% of the dry mass of most cells, and they are instrumental in almost everything organisms do. Some proteins speed up chemical reactions, while others play a role in defense, storage, transport, cellular communication, movement, or structural support. **Figure 5.15**, on the next page, shows examples of proteins with these functions, which you'll learn more about in later chapters.

Life would not be possible without enzymes, most of which are proteins. Enzymatic proteins regulate metabolism by acting as **catalysts**, chemical agents that selectively speed up chemical reactions without being consumed by the reaction. Because an enzyme can perform its function over and over again, these molecules can be thought of as workhorses that keep cells running by carrying out the processes of life.

A human has tens of thousands of different proteins, each with a specific structure and function; proteins, in fact, are the most structurally sophisticated molecules known. Consistent with their diverse functions, they vary extensively in structure, each type of protein having a unique three-dimensional shape.

Polypeptides

Diverse as proteins are, they are all unbranched polymers constructed from the same set of 20 amino acids. Polymers of amino acids are called **polypeptides**. A **protein** is a biologically functional molecule that consists of one or more polypeptides, each folded and coiled into a specific three-dimensional structure.

▼ Figure 5.15 An overview of protein functions.

Enzymatic proteins

Function: Selective acceleration of chemical reactions

Example: Digestive enzymes catalyze the hydrolysis of bonds in food molecules.

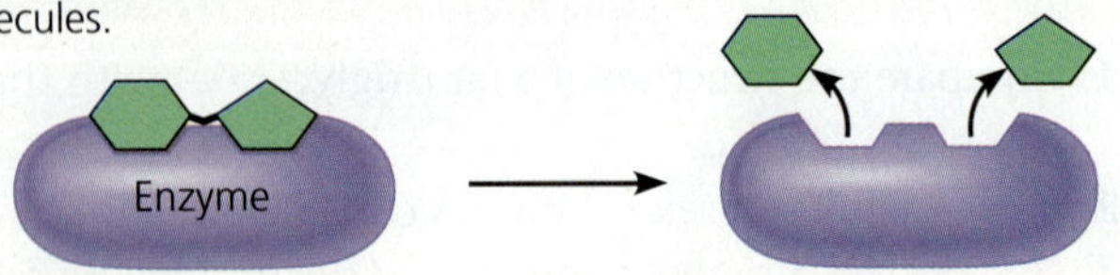

Defensive proteins

Function: Protection against disease

Example: Antibodies inactivate and help destroy viruses and bacteria.

Antibodies
Virus
Bacterium

Storage proteins

Function: Storage of amino acids

Examples: Casein, the protein of milk, is the major source of amino acids for baby mammals. Plants have storage proteins in their seeds. Ovalbumin is the protein of egg white, used as an amino acid source for the developing embryo.

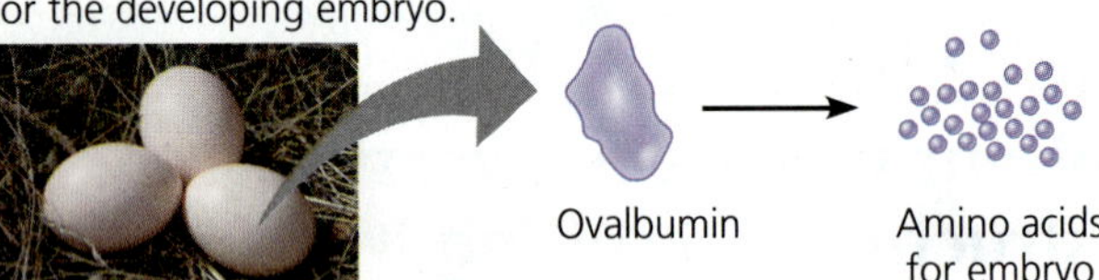

Transport proteins

Function: Transport of substances

Examples: Hemoglobin, the iron-containing protein of vertebrate blood, transports oxygen from the lungs to other parts of the body. Other proteins transport molecules across cell membranes.

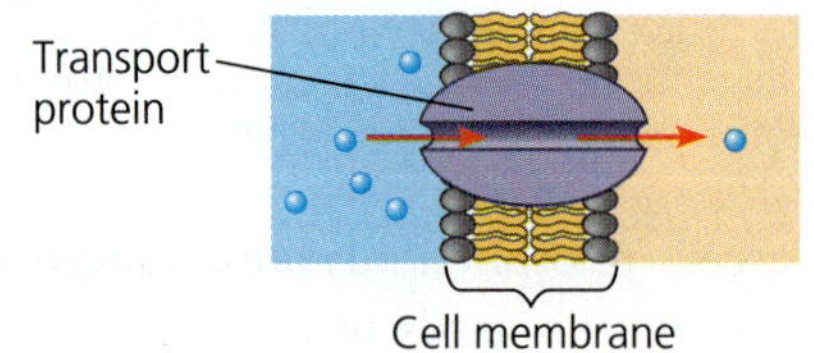

Hormonal proteins

Function: Coordination of an organism's activities

Example: Insulin, a hormone secreted by the pancreas, causes other tissues to take up glucose, thus regulating blood sugar concentration.

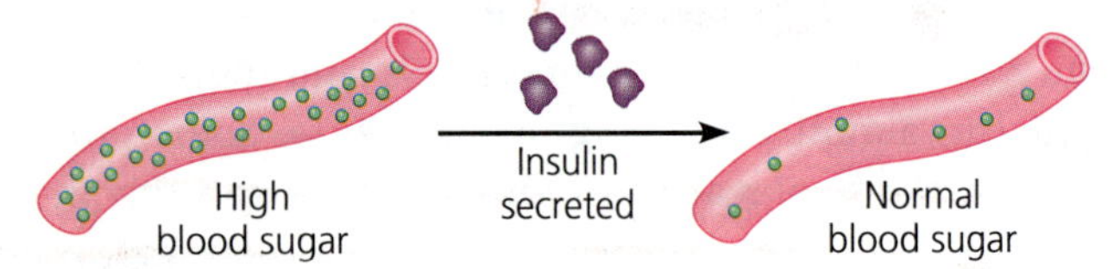

Receptor proteins

Function: Response of cell to chemical stimuli

Example: Receptors built into the membrane of a nerve cell detect signaling molecules released by other nerve cells.

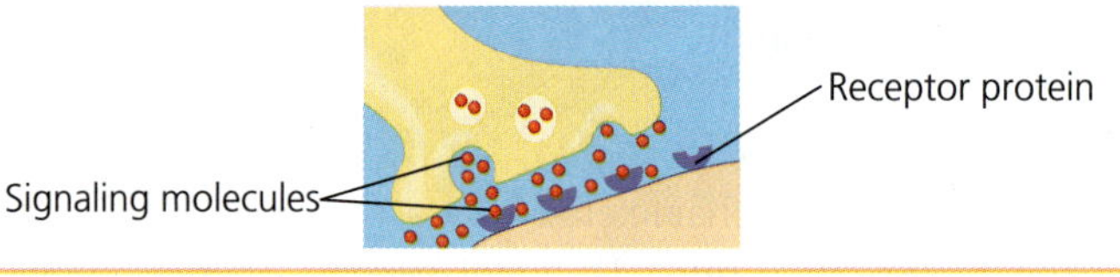

Contractile and motor proteins

Function: Movement

Examples: Motor proteins are responsible for the undulations of cilia and flagella. Actin and myosin proteins are responsible for the contraction of muscles.

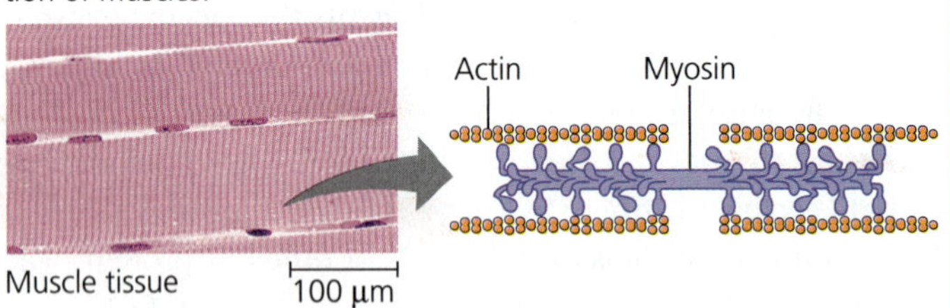

Structural proteins

Function: Support

Examples: Keratin is the protein of hair, horns, feathers, and other skin appendages. Insects and spiders use silk fibers to make their cocoons and webs, respectively. Collagen and elastin proteins provide a fibrous framework in animal connective tissues.

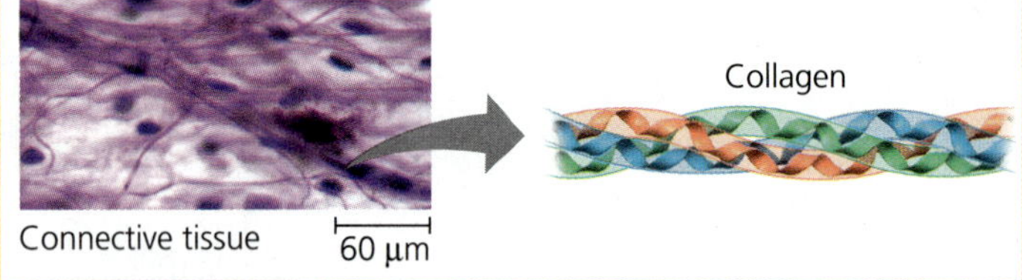

Amino Acid Monomers

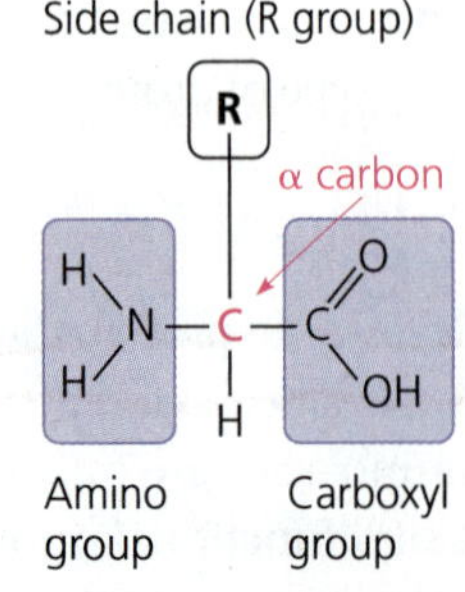

All amino acids share a common structure. An **amino acid** is an organic molecule possessing both an amino group and a carboxyl group (see Figure 4.9). The illustration at the right shows the general formula for an amino acid. At the center of the amino acid is an asymmetric carbon atom called the *alpha (α) carbon.* Its four different partners are an amino group, a carboxyl group, a hydrogen atom, and a variable group symbolized by R. The R group, also called the side chain, differs with each amino acid.

Figure 5.16 shows the 20 amino acids that cells use to build their thousands of proteins. Here the amino groups and carboxyl groups are all depicted in ionized form, the way they usually exist at the pH found in a cell. The side chain (R group) may be as simple as a hydrogen atom, as in the amino acid glycine, or it may be a carbon skeleton with various functional groups attached, as in glutamine.

▼ Figure 5.16 The 20 amino acids of proteins. The amino acids are grouped here according to the properties of their side chains (R groups) and shown in their prevailing ionic forms at pH 7.2, the pH within a cell. The three-letter and one-letter abbreviations for the amino acids are in parentheses. All amino acids used in proteins are L enantiomers, the form shown here (see Figure 4.7).

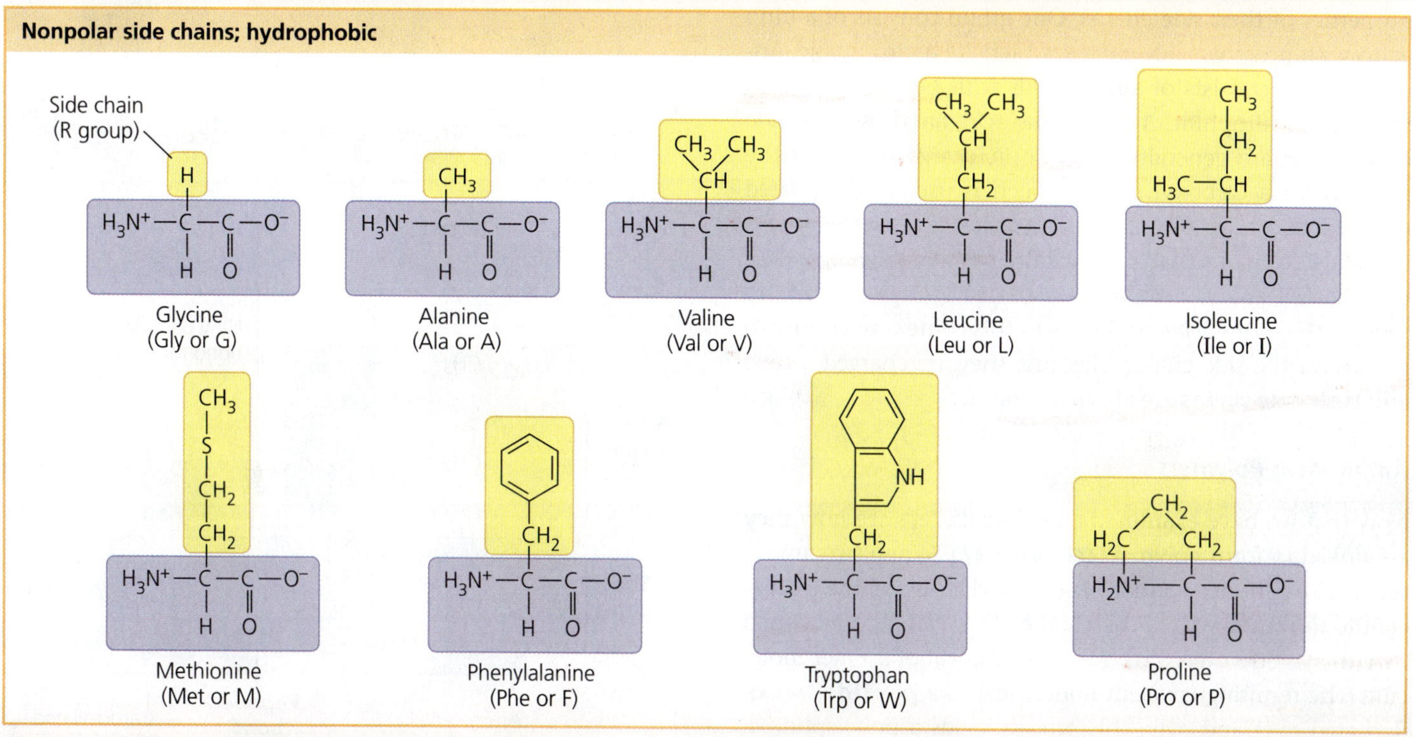

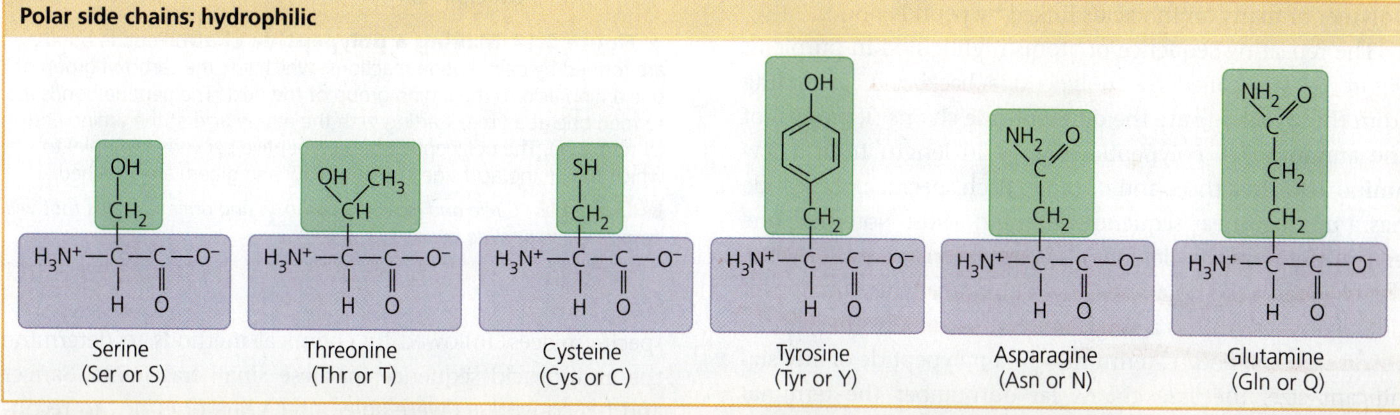

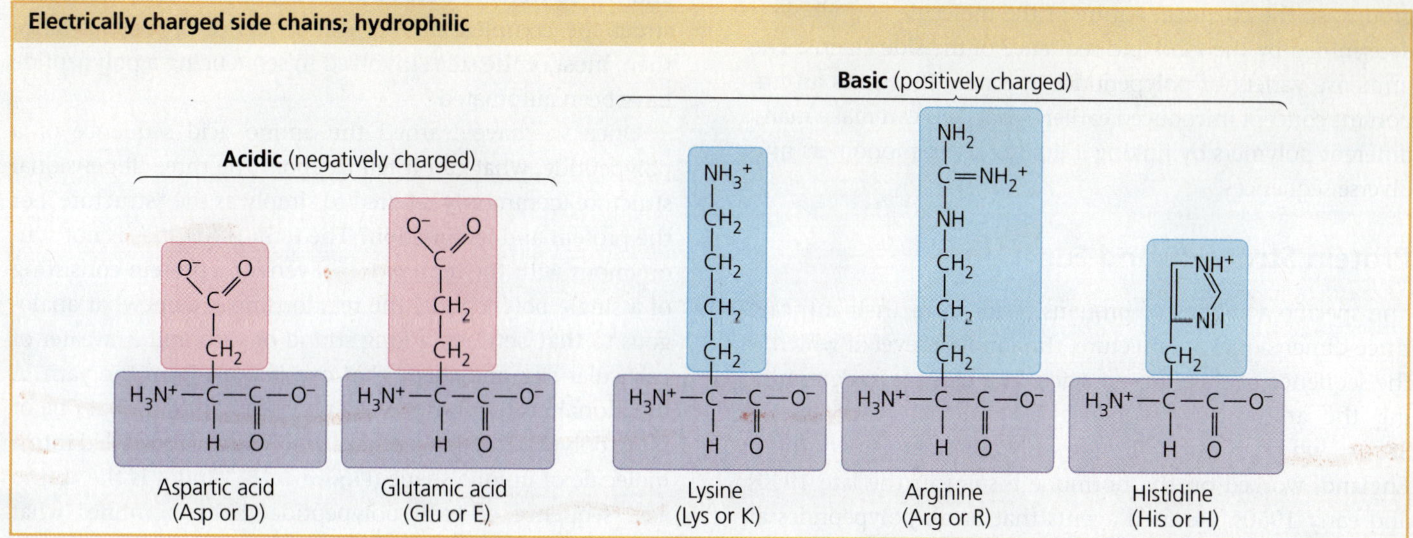

The physical and chemical properties of the side chain determine the unique characteristics of a particular amino acid, thus affecting its functional role in a polypeptide. In Figure 5.16, the amino acids are grouped according to the properties of their side chains. One group consists of amino acids with nonpolar side chains, which are hydrophobic. Another group consists of amino acids with polar side chains, which are hydrophilic. Acidic amino acids are those with side chains that are generally negative in charge owing to the presence of a carboxyl group, which is usually dissociated (ionized) at cellular pH. Basic amino acids have amino groups in their side chains that are generally positive in charge. (Notice that *all* amino acids have carboxyl groups and amino groups; the terms *acidic* and *basic* in this context refer only to groups on the side chains.) Because they are charged, acidic and basic side chains are also hydrophilic.

Amino Acid Polymers

Now that we have examined amino acids, let's see how they are linked to form polymers **(Figure 5.17)**. When two amino acids are positioned so that the carboxyl group of one is adjacent to the amino group of the other, they can become joined by a dehydration reaction, with the removal of a water molecule. The resulting covalent bond is called a **peptide bond**. Repeated over and over, this process yields a polypeptide, a polymer of many amino acids linked by peptide bonds.

The repeating sequence of atoms highlighted in purple in Figure 5.17 is called the polypeptide backbone. Extending from this backbone are the different side chains (R groups) of the amino acids. Polypeptides range in length from a few amino acids to a thousand or more. Each specific polypeptide has a unique linear sequence of amino acids. Note that one end of the polypeptide chain has a free amino group, while the opposite end has a free carboxyl group. Thus, a polypeptide of any length has a single amino end (N-terminus) and a single carboxyl end (C-terminus). In a polypeptide of any significant size, the side chains far outnumber the terminal groups, so the chemical nature of the molecule as a whole is determined by the kind and sequence of the side chains. The immense variety of polypeptides in nature illustrates an important concept introduced earlier—that cells can make many different polymers by linking a limited set of monomers into diverse sequences.

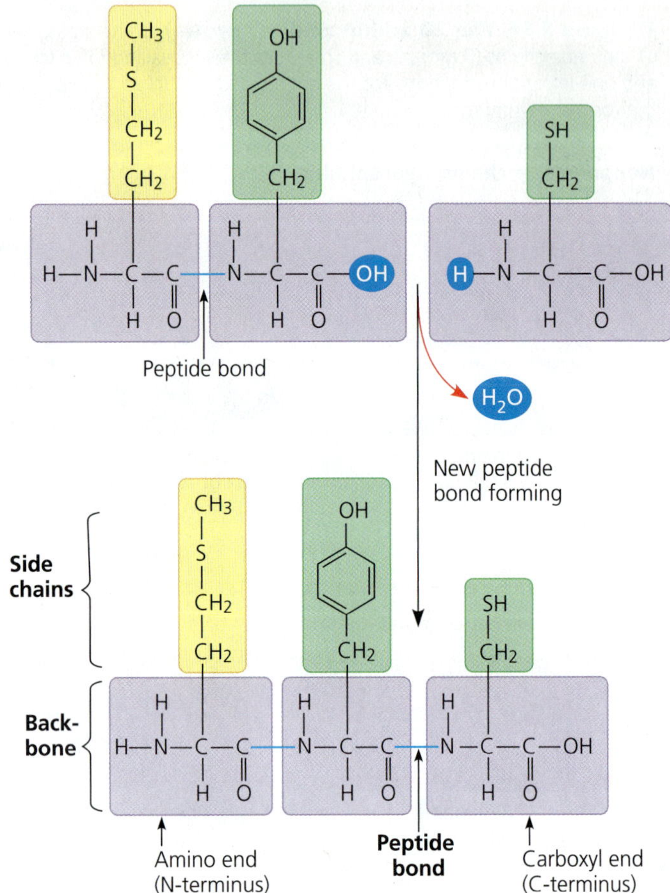

▲ **Figure 5.17 Making a polypeptide chain.** Peptide bonds are formed by dehydration reactions, which link the carboxyl group of one amino acid to the amino group of the next. The peptide bonds are formed one at a time, starting with the amino acid at the amino end (N-terminus). The polypeptide has a repetitive backbone (purple) to which the amino acid side chains (yellow and green) are attached.

DRAW IT *Circle and label the carboxyl and amino groups that will form the new peptide bond.*

Protein Structure and Function

The specific activities of proteins result from their intricate three-dimensional architecture, the simplest level of which is the sequence of their amino acids. The pioneer in determining the amino acid sequence of proteins was Frederick Sanger, who, with his colleagues at Cambridge University in England, worked on the hormone insulin in the late 1940s and early 1950s. He used agents that break polypeptides at specific places, followed by chemical methods to determine the amino acid sequence in these small fragments. Sanger and his co-workers were able, after years of effort, to reconstruct the complete amino acid sequence of insulin. Since then, most of the steps involved in sequencing a polypeptide have been automated.

Once we have learned the amino acid sequence of a polypeptide, what can it tell us about the three-dimensional structure (commonly referred to simply as the "structure") of the protein and its function? The term *polypeptide* is not synonymous with the term *protein*. Even for a protein consisting of a single polypeptide, the relationship is somewhat analogous to that between a long strand of yarn and a sweater of particular size and shape that can be knit from the yarn. A functional protein is not *just* a polypeptide chain, but one or more polypeptides precisely twisted, folded, and coiled into a molecule of unique shape **(Figure 5.18)**. And it is the amino acid sequence of each polypeptide that determines what

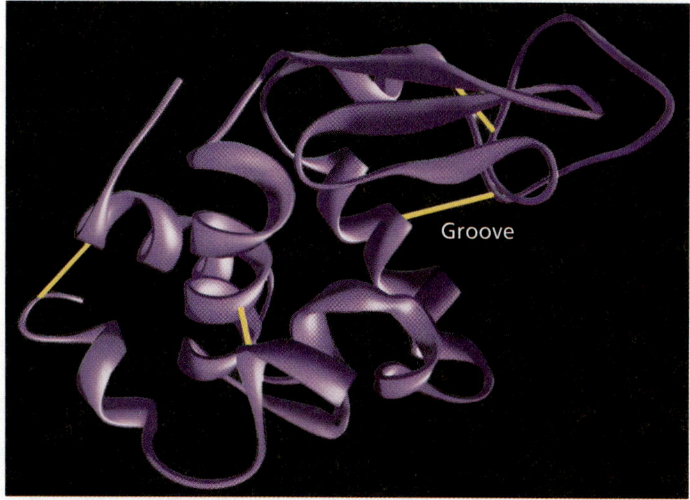

(a) A **ribbon model** shows how the single polypeptide chain folds and coils to form the functional protein. (The yellow lines represent disulfide bridges that stabilize the protein's shape; see Figure 5.20.)

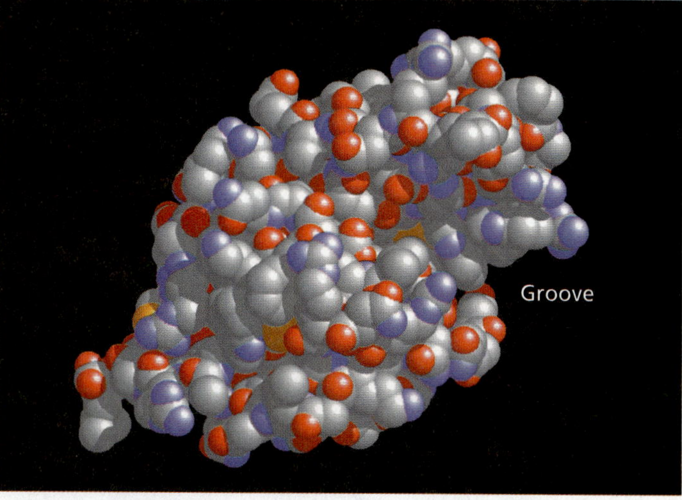

(b) A **space-filling model** shows more clearly the globular shape seen in many proteins, as well as the specific three-dimensional structure unique to lysozyme.

▲ **Figure 5.18 Structure of a protein, the enzyme lysozyme.** Present in our sweat, tears, and saliva, lysozyme is an enzyme that helps prevent infection by binding to and destroying specific molecules on the surface of many kinds of bacteria. The groove is the part of the protein that recognizes and binds to the target molecules on bacterial walls.

three-dimensional structure the protein will have under normal cellular conditions.

When a cell synthesizes a polypeptide, the chain generally folds spontaneously, assuming the functional structure for that protein. This folding is driven and reinforced by the formation of a variety of bonds between parts of the chain, which in turn depends on the sequence of amino acids. Many proteins are roughly spherical (*globular proteins*), while others are shaped like long fibers (*fibrous proteins*). Even within these broad categories, countless variations exist.

A protein's specific structure determines how it works. In almost every case, the function of a protein depends on its ability to recognize and bind to some other molecule. In an especially striking example of the marriage of form and function, **Figure 5.19** shows the exact match of shape between an antibody (a protein in the body) and the particular foreign substance on a flu virus that the antibody binds to and marks for destruction. In Chapter 43, you'll learn more about how the immune system generates antibodies that match the shapes of specific foreign molecules so well. Also, you may recall from Chapter 2 that natural signaling molecules called endorphins bind to specific receptor proteins on the surface of brain cells in humans, producing euphoria and relieving pain. Morphine, heroin, and other opiate drugs are able to mimic endorphins because they all share a similar shape with endorphins and can thus fit into and bind to endorphin receptors in the brain. This fit is very specific, something like a lock and key (see Figure 2.18). Thus, the function of a protein—for instance, the ability of a receptor protein to bind to a particular pain-relieving signaling molecule—is an emergent property resulting from exquisite molecular order.

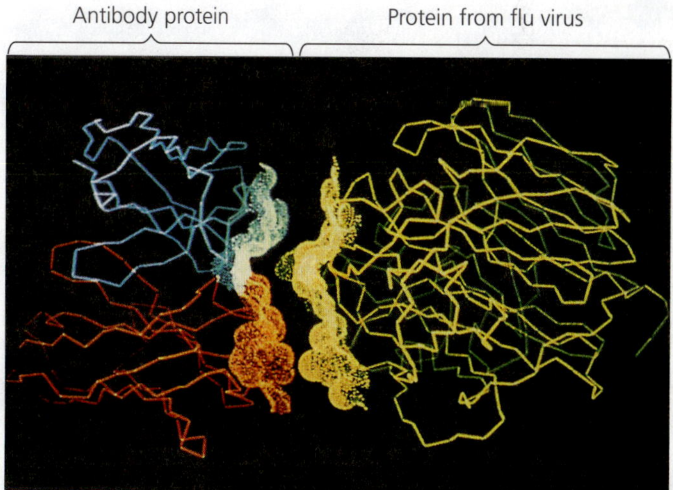

▲ **Figure 5.19 An antibody binding to a protein from a flu virus.** A technique called X-ray crystallography was used to generate a computer model of an antibody protein (blue and orange, left) bound to a flu virus protein (green and yellow, right). Computer software was then used to back the images away from each other, revealing the exact complementarity of shape between the two protein surfaces.

Four Levels of Protein Structure

With the goal of understanding the function of a protein, learning about its structure is often productive. In spite of their great diversity, all proteins share three superimposed levels of structure, known as primary, secondary, and tertiary structure. A fourth level, quaternary structure, arises when a protein consists of two or more polypeptide chains. **Figure 5.20**, on the following two pages, describes these four levels of protein structure. Be sure to study this figure thoroughly before going on to the next section.

▼ **Figure 5.20**

Exploring Levels of Protein Structure

Primary Structure

Linear chain of amino acids

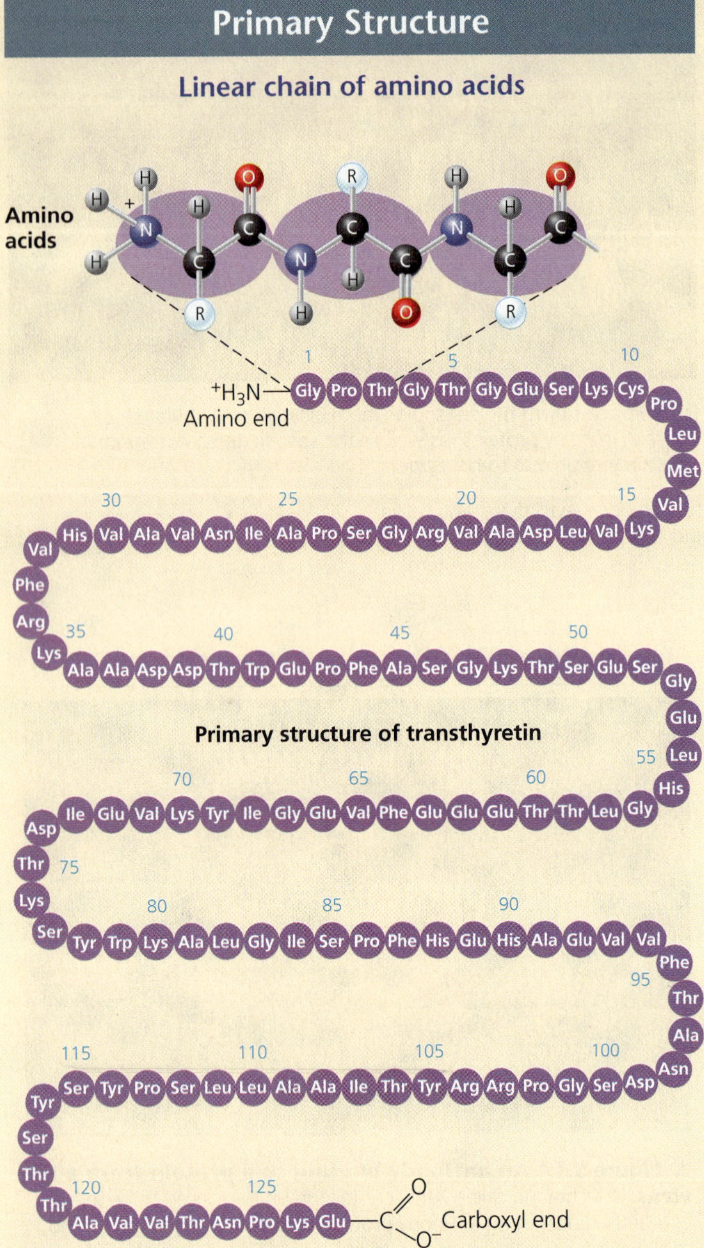

The **primary structure** of a protein is a linked series of amino acids with a unique sequence. As an example, let's consider transthyretin, a globular blood protein that transports vitamin A and one of the thyroid hormones throughout the body. Transthyretin is made up of four identical polypeptide chains, each composed of 127 amino acids. Shown here is one of these chains unraveled for a closer look at its primary structure. Each of the 127 positions along the chain is occupied by one of the 20 amino acids, indicated here by its three-letter abbreviation.

The primary structure is like the order of letters in a very long word. If left to chance, there would be 20^{127} different ways of making a polypeptide chain 127 amino acids long. However, the precise primary structure of a protein is determined not by the random linking of amino acids, but by inherited genetic information. The primary structure in turn dictates secondary and tertiary structure, due to the chemical nature of the backbone and the side chains (R groups) of the amino acids positioned along the chain.

Secondary Structure

Regions stabilized by hydrogen bonds between atoms of the polypeptide backbone

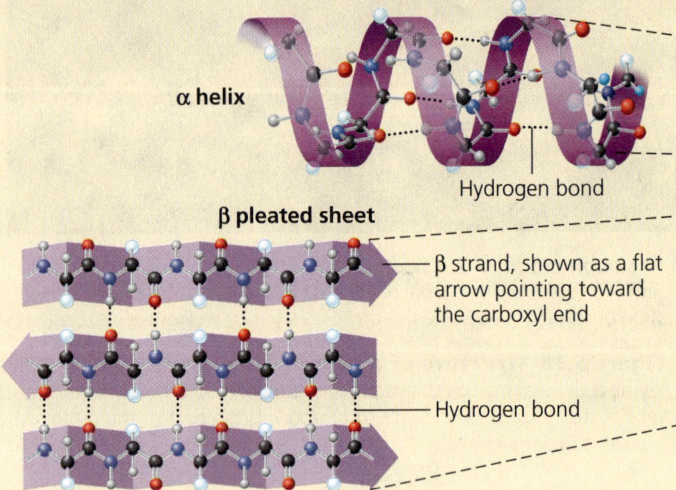

Most proteins have segments of their polypeptide chains repeatedly coiled or folded in patterns that contribute to the protein's overall shape. These coils and folds, collectively referred to as **secondary structure**, are the result of hydrogen bonds between the repeating constituents of the polypeptide backbone (not the amino acid side chains). Within the backbone, the oxygen atoms have a partial negative charge, and the hydrogen atoms attached to the nitrogens have a partial positive charge (see Figure 2.16); therefore, hydrogen bonds can form between these atoms. Individually, these hydrogen bonds are weak, but because they are repeated many times over a relatively long region of the polypeptide chain, they can support a particular shape for that part of the protein.

One such secondary structure is the **α helix**, a delicate coil held together by hydrogen bonding between every fourth amino acid, shown above. Although each transthyretin polypeptide has only one α helix region (see tertiary structure on the next page), other globular proteins have multiple stretches of α helix separated by nonhelical regions (see hemoglobin on the next page). Some fibrous proteins, such as α-keratin, the structural protein of hair, have the α helix formation over most of their length.

The other main type of secondary structure is the **β pleated sheet**. As shown above, in this structure two or more strands of the polypeptide chain lying side by side (called β strands) are connected by hydrogen bonds between parts of the two parallel polypeptide backbones. β pleated sheets make up the core of many globular proteins, as is the case for transthyretin (see tertiary structure on the next page), and dominate some fibrous proteins, including the silk protein of a spider's web. The teamwork of so many hydrogen bonds makes each spider silk fiber stronger than a steel strand of the same weight.

▼ Spiders secrete silk fibers made of a structural protein containing β pleated sheets, which allow the spider web to stretch and recoil.

Tertiary Structure

Three-dimensional shape stabilized by interactions between side chains

Transthyretin polypeptide

Superimposed on the patterns of secondary structure is a protein's tertiary structure, shown above in a ribbon model of the transthyretin polypeptide. While secondary structure involves interactions between backbone constituents, **tertiary structure** is the overall shape of a polypeptide resulting from interactions between the side chains (R groups) of the various amino acids. One type of interaction that contributes to tertiary structure is—somewhat misleadingly—called a **hydrophobic interaction**. As a polypeptide folds into its functional shape, amino acids with hydrophobic (nonpolar) side chains usually end up in clusters at the core of the protein, out of contact with water. Thus, a "hydrophobic interaction" is actually caused by the exclusion of nonpolar substances by water molecules. Once nonpolar amino acid side chains are close together, van der Waals interactions help hold them together. Meanwhile, hydrogen bonds between polar side chains and ionic bonds between positively and negatively charged side chains also help stabilize tertiary structure. These are all weak interactions in the aqueous cellular environment, but their cumulative effect helps give the protein a unique shape.

Covalent bonds called **disulfide bridges** may further reinforce the shape of a protein. Disulfide bridges form where two cysteine monomers, which have sulfhydryl groups (—SH) on their side chains (see Figure 4.9), are brought close together by the folding of the protein. The sulfur of one cysteine bonds to the sulfur of the second, and the disulfide bridge (—S—S—) rivets parts of the protein together (see yellow lines in Figure 5.18a). All of these different kinds of interactions can contribute to the tertiary structure of a protein, as shown here in a small part of a hypothetical protein:

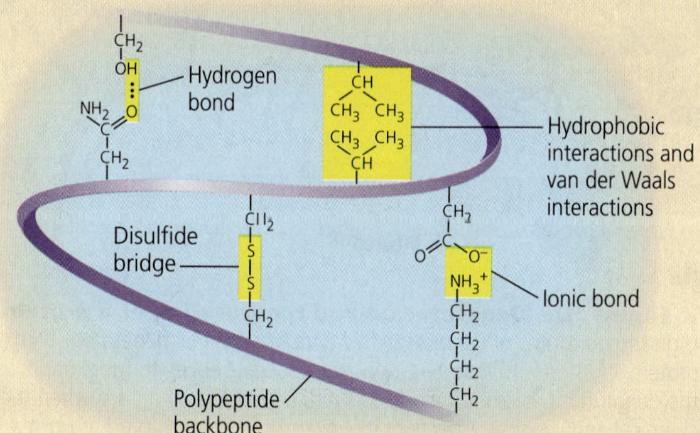

Quaternary Structure

Association of multiple polypeptides, forming a functional protein

Transthyretin protein (four identical polypeptides)

Some proteins consist of two or more polypeptide chains aggregated into one functional macromolecule. **Quaternary structure** is the overall protein structure that results from the aggregation of these polypeptide subunits. For example, shown above is the complete globular transthyretin protein, made up of its four polypeptides.

Another example is collagen, shown below, which is a fibrous protein that has three identical helical polypeptides intertwined into a larger triple helix, giving the long fibers great strength. This suits collagen fibers to their function as the girders of connective tissue in skin, bone, tendons, ligaments, and other body parts. (Collagen accounts for 40% of the protein in a human body.)

Collagen

Hemoglobin, the oxygen-binding protein of red blood cells shown below, is another example of a globular protein with quaternary structure. It consists of four polypeptide subunits, two of one kind (α) and two of another kind (β). Both α and β subunits consist primarily of α-helical secondary structure. Each subunit has a nonpolypeptide component, called heme, with an iron atom that binds oxygen.

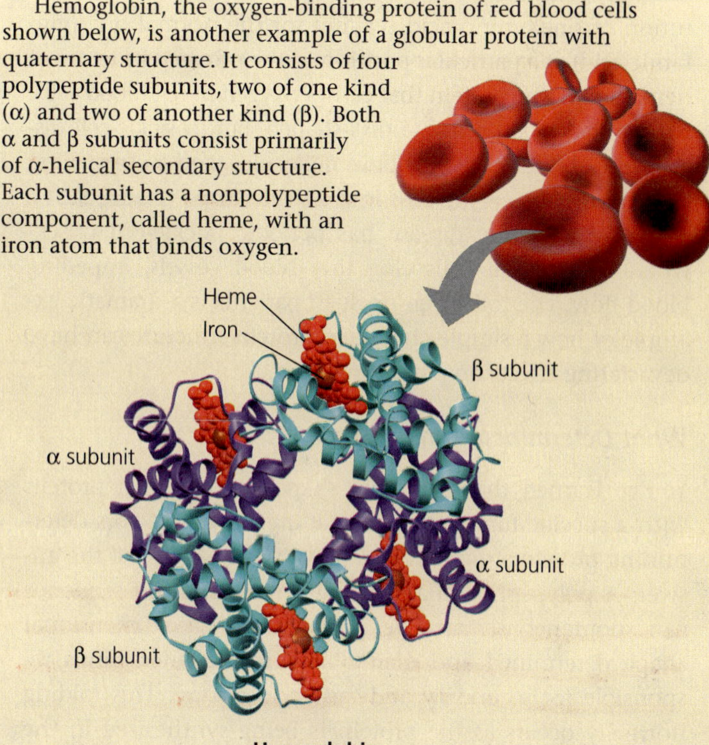

	Primary Structure	Secondary and Tertiary Structures	Quaternary Structure	Function	Red Blood Cell Shape
Normal hemoglobin	1 Val 2 His 3 Leu 4 Thr 5 Pro 6 Glu 7 Glu	β subunit	Normal hemoglobin α β β α	Molecules do not associate with one another; each carries oxygen.	Normal red blood cells are full of individual hemoglobin molecules, each carrying oxygen. 10 µm
Sickle-cell hemoglobin	1 Val 2 His 3 Leu 4 Thr 5 Pro 6 Val 7 Glu	Exposed hydrophobic region β subunit	Sickle-cell hemoglobin α β β α	Molecules interact with one another and crystallize into a fiber; capacity to carry oxygen is greatly reduced.	Fibers of abnormal hemoglobin deform red blood cell into sickle shape. 10 µm

▲ **Figure 5.21 A single amino acid substitution in a protein causes sickle-cell disease.**

MAKE CONNECTIONS *Considering the chemical characteristics of the amino acids valine and glutamic acid (see Figure 5.16), propose a possible explanation for the dramatic effect on protein function that occurs when valine is substituted for glutamic acid.*

Sickle-Cell Disease: A Change in Primary Structure

Even a slight change in primary structure can affect a protein's shape and ability to function. For instance, **sickle-cell disease**, an inherited blood disorder, is caused by the substitution of one amino acid (valine) for the normal one (glutamic acid) at a particular position in the primary structure of hemoglobin, the protein that carries oxygen in red blood cells. Normal red blood cells are disk-shaped, but in sickle-cell disease, the abnormal hemoglobin molecules tend to crystallize, deforming some of the cells into a sickle shape **(Figure 5.21)**. A person with the disease has periodic "sickle-cell crises" when the angular cells clog tiny blood vessels, impeding blood flow. The toll taken on such patients is a dramatic example of how a simple change in protein structure can have devastating effects on protein function.

What Determines Protein Structure?

You've learned that a unique shape endows each protein with a specific function. But what are the key factors determining protein structure? You already know most of the answer: A polypeptide chain of a given amino acid sequence can spontaneously arrange itself into a three-dimensional shape determined and maintained by the interactions responsible for secondary and tertiary structure. This folding normally occurs as the protein is being synthesized in the crowded environment within a cell, aided by other proteins. However, protein structure also depends on the physical and chemical conditions of the protein's environment. If the pH, salt concentration, temperature, or other aspects of its environment are altered, the weak chemical bonds and interactions within a protein may be destroyed, causing the protein to unravel and lose its native shape, a change called **denaturation (Figure 5.22)**. Because it is misshapen, the denatured protein is biologically inactive.

Most proteins become denatured if they are transferred from an aqueous environment to a nonpolar solvent, such as

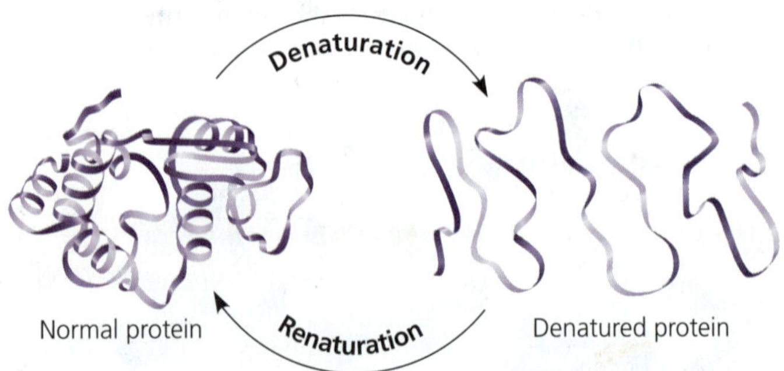

▲ **Figure 5.22 Denaturation and renaturation of a protein.** High temperatures or various chemical treatments will denature a protein, causing it to lose its shape and hence its ability to function. If the denatured protein remains dissolved, it can often renature when the chemical and physical aspects of its environment are restored to normal.

ether or chloroform; the polypeptide chain refolds so that its hydrophobic regions face outward toward the solvent. Other denaturation agents include chemicals that disrupt the hydrogen bonds, ionic bonds, and disulfide bridges that maintain a protein's shape. Denaturation can also result from excessive heat, which agitates the polypeptide chain enough to overpower the weak interactions that stabilize the structure. The white of an egg becomes opaque during cooking because the denatured proteins are insoluble and solidify. This also explains why excessively high fevers can be fatal: Proteins in the blood can denature at very high body temperatures.

When a protein in a test-tube solution has been denatured by heat or chemicals, it can sometimes return to its functional shape when the denaturing agent is removed. We can conclude that the information for building specific shape is intrinsic to the protein's primary structure. The sequence of amino acids determines the protein's shape—where an α helix can form, where β pleated sheets can exist, where disulfide bridges are located, where ionic bonds can form, and so on. But how does protein folding occur in the cell?

Protein Folding in the Cell

Biochemists now know the amino acid sequence for more than 10 million proteins and the three-dimensional shape for more than 20,000. Researchers have tried to correlate the primary structure of many proteins with their three-dimensional structure to discover the rules of protein folding. Unfortunately, however, the protein-folding process is not that simple. Most proteins probably go through several intermediate structures on their way to a stable shape, and looking at the mature structure does not reveal the stages of folding required to achieve that form. However, biochemists have developed methods for tracking a protein through such stages.

Crucial to the folding process are **chaperonins** (also called chaperone proteins), protein molecules that assist in the proper folding of other proteins (**Figure 5.23**). Chaperonins do not specify the final structure of a polypeptide. Instead, they keep the new polypeptide segregated from "bad influences" in the cytoplasmic environment while it folds spontaneously. The chaperonin shown in Figure 5.23, from the bacterium *E. coli*, is a giant multiprotein complex shaped like a hollow cylinder. The cavity provides a shelter for folding polypeptides. In the past decade, researchers have discovered molecular systems that interact with chaperonins and check whether proper folding has occurred. Such systems either refold the misfolded proteins correctly or mark them for destruction.

Misfolding of polypeptides is a serious problem in cells. Many diseases, such as Alzheimer's, Parkinson's, and mad cow disease, are associated with an accumulation of misfolded proteins. In fact, misfolded versions of the transthyretin protein featured in Figure 5.20 have been implicated in several diseases, including one form of senile dementia.

Even when scientists have a correctly folded protein in hand, determining its exact three-dimensional structure is not simple, for a single protein molecule has thousands of atoms. The first 3-D structures were worked out in 1959 for hemoglobin and a related protein. The method that made these feats possible was **X-ray crystallography**, which has since been used to determine the 3-D structure of many other proteins. In a recent example, Roger Kornberg and his colleagues at Stanford University used this method to elucidate the structure of RNA polymerase, an enzyme that plays a crucial role in the expression of genes (**Figure 5.24**, on the next page). Another method for analyzing protein structure is nuclear magnetic resonance (NMR) spectroscopy, which does not require protein crystallization. A still newer approach employs bioinformatics (see Chapter 1) to predict the 3-D structure of polypeptides from their amino acid sequence. X-ray crystallography, NMR spectroscopy, and bioinformatics are complementary approaches to understanding protein structure and function.

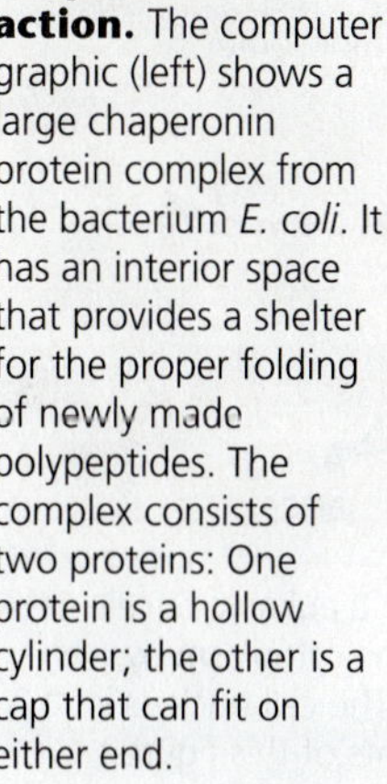

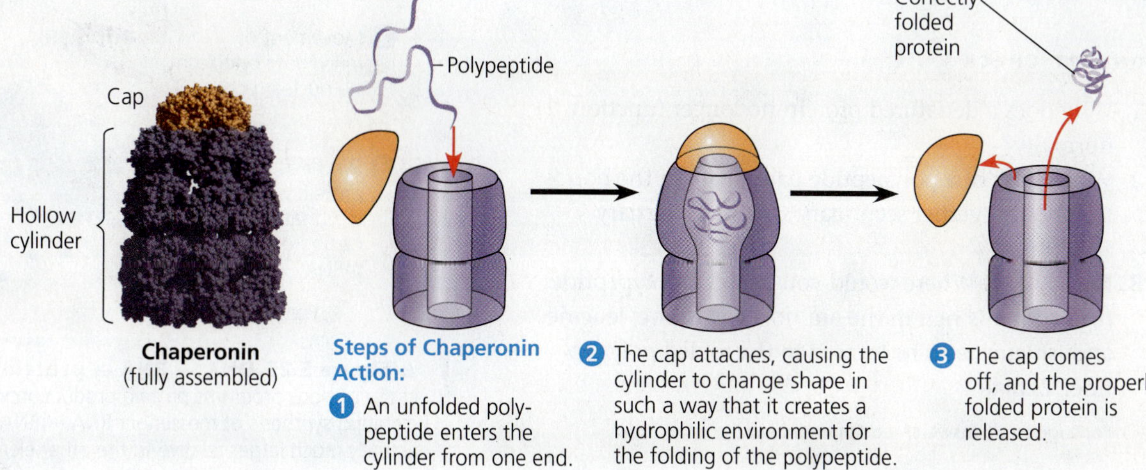

▶ Figure 5.23 A chaperonin in action. The computer graphic (left) shows a large chaperonin protein complex from the bacterium *E. coli*. It has an interior space that provides a shelter for the proper folding of newly made polypeptides. The complex consists of two proteins: One protein is a hollow cylinder; the other is a cap that can fit on either end.

▼ **Figure 5.24** INQUIRY

What can the 3-D shape of the enzyme RNA polymerase II tell us about its function?

EXPERIMENT In 2006, Roger Kornberg was awarded the Nobel Prize in Chemistry for using X-ray crystallography to determine the 3-D shape of RNA polymerase II, which binds to the DNA double helix and synthesizes RNA. After crystallizing a complex of all three components, Kornberg and his colleagues aimed an X-ray beam through the crystal. The atoms of the crystal diffracted (bent) the X-rays into an orderly array that a digital detector recorded as a pattern of spots called an X-ray diffraction pattern.

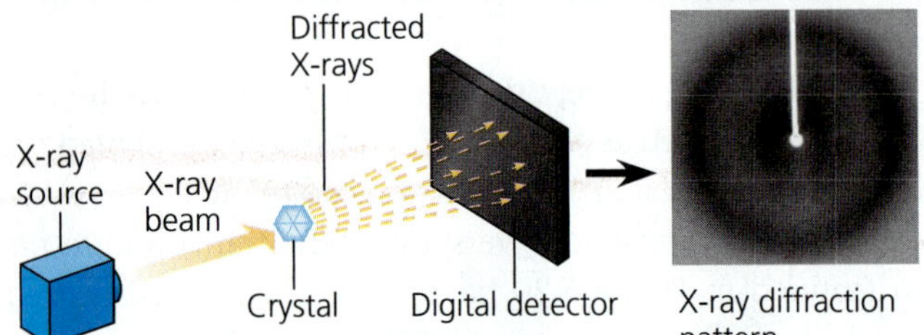

RESULTS Using data from X-ray diffraction patterns, as well as the amino acid sequence determined by chemical methods, Kornberg and colleagues built a 3-D model of the complex with the help of computer software.

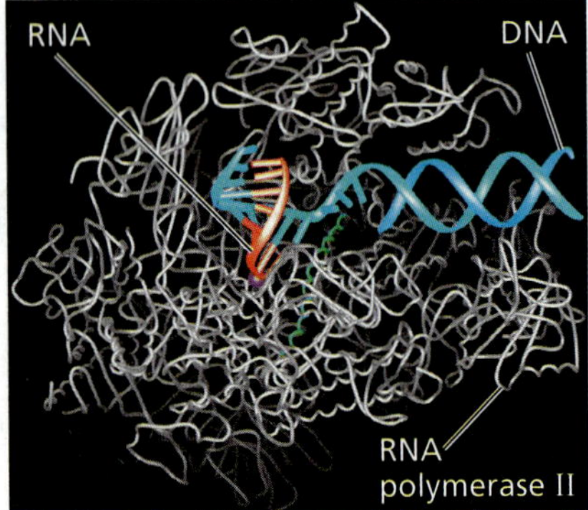

CONCLUSION By analyzing their model, the researchers developed a hypothesis about the functions of different regions of RNA polymerase II. For example, the region above the DNA may act as a clamp that holds the nucleic acids in place. (You'll learn more about this enzyme in Chapter 17.)

SOURCE A. L. Gnatt et al., Structural basis of transcription: an RNA polymerase II elongation complex at 3.3Å, *Science* 292:1876–1882 (2001).

WHAT IF? If you were an author of the paper and were describing the model, what type of protein structure would you call the small polypeptide spirals in RNA polymerase II?

CONCEPT CHECK 5.4

1. Why does a denatured protein no longer function normally?
2. What parts of a polypeptide participate in the bonds that hold together secondary structure? Tertiary structure?
3. **WHAT IF?** Where would you expect a polypeptide region that is rich in the amino acids valine, leucine, and isoleucine to be located in the folded polypeptide? Explain.

For suggested answers, see Appendix A.

CONCEPT 5.5

Nucleic acids store, transmit, and help express hereditary information

If the primary structure of polypeptides determines a protein's shape, what determines primary structure? The amino acid sequence of a polypeptide is programmed by a discrete unit of inheritance known as a **gene**. Genes consist of DNA, which belongs to the class of compounds called nucleic acids. **Nucleic acids** are polymers made of monomers called *nucleotides*.

The Roles of Nucleic Acids

The two types of nucleic acids, **deoxyribonucleic acid (DNA)** and **ribonucleic acid (RNA)**, enable living organisms to reproduce their complex components from one generation to the next. Unique among molecules, DNA provides directions for its own replication. DNA also directs RNA synthesis and, through RNA, controls protein synthesis (**Figure 5.25**).

DNA is the genetic material that organisms inherit from their parents. Each chromosome contains one long DNA molecule, usually carrying several hundred or more genes. When a cell reproduces itself by dividing, its DNA molecules are copied and passed along from one generation of cells to the next. Encoded in the structure of DNA is the information that

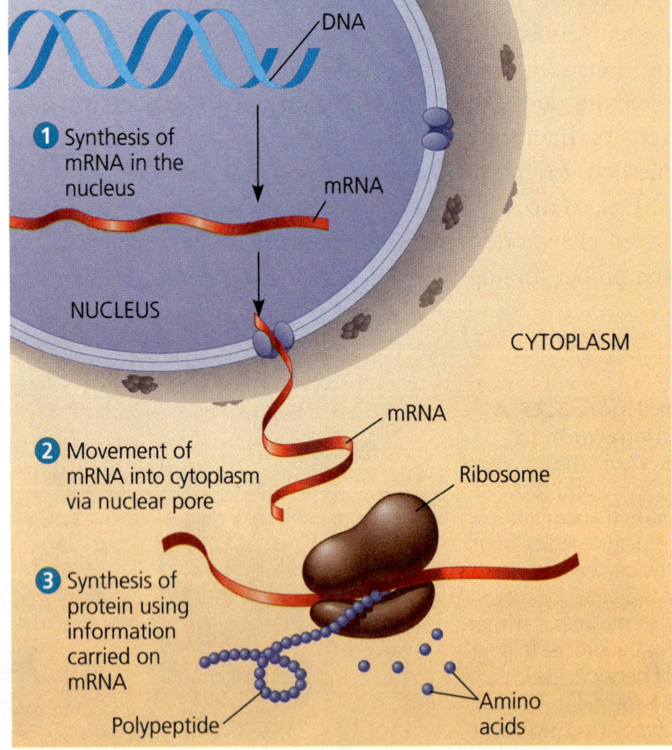

▲ **Figure 5.25 DNA → RNA → protein.** In a eukaryotic cell, DNA in the nucleus programs protein production in the cytoplasm by dictating synthesis of messenger RNA (mRNA). (The cell nucleus is actually much larger relative to the other elements of this figure.)

programs all the cell's activities. The DNA, however, is not directly involved in running the operations of the cell, any more than computer software by itself can print a bank statement or read the bar code on a box of cereal. Just as a printer is needed to print out a statement and a scanner is needed to read a bar code, proteins are required to implement genetic programs. The molecular hardware of the cell—the tools for biological functions—consists mostly of proteins. For example, the oxygen carrier in red blood cells is the protein hemoglobin, not the DNA that specifies its structure.

How does RNA, the other type of nucleic acid, fit into gene expression, the flow of genetic information from DNA to proteins? Each gene along a DNA molecule directs synthesis of a type of RNA called *messenger RNA (mRNA)*. The mRNA molecule interacts with the cell's protein-synthesizing machinery to direct production of a polypeptide, which folds into all or part of a protein. We can summarize the flow of genetic information as DNA → RNA → protein (see Figure 5.25). The sites of protein synthesis are tiny structures called ribosomes. In a eukaryotic cell, ribosomes are in the cytoplasm, but DNA resides in the nucleus. Messenger RNA conveys genetic instructions for building proteins from the nucleus to the cytoplasm. Prokaryotic cells lack nuclei but still use mRNA to convey a message from the DNA to ribosomes and other cellular equipment that translate the coded information into amino acid sequences. In recent years, the spotlight has been turned on other, previously unknown types of RNA that play many other roles in the cell. As is so often true in biology, the story is still being written! You'll hear more about the newly discovered functions of RNA molecules in Chapter 18.

The Components of Nucleic Acids

Nucleic acids are macromolecules that exist as polymers called **polynucleotides** (**Figure 5.26a**). As indicated by the name, each polynucleotide consists of monomers called **nucleotides**. A nucleotide, in general, is composed of three parts: a nitrogen-containing (nitrogenous) base, a five-carbon sugar (a pentose), and one or more phosphate groups (**Figure 5.26b**). In a polynucleotide, each monomer has only one phosphate group. The portion of a nucleotide without any phosphate groups is called a *nucleoside*.

To build a nucleotide, let's first consider the nitrogenous bases (**Figure 5.26c**). Each nitrogenous base has one or two rings that include nitrogen atoms. (They are called nitrogenous *bases* because the nitrogen atoms tend to take up H^+

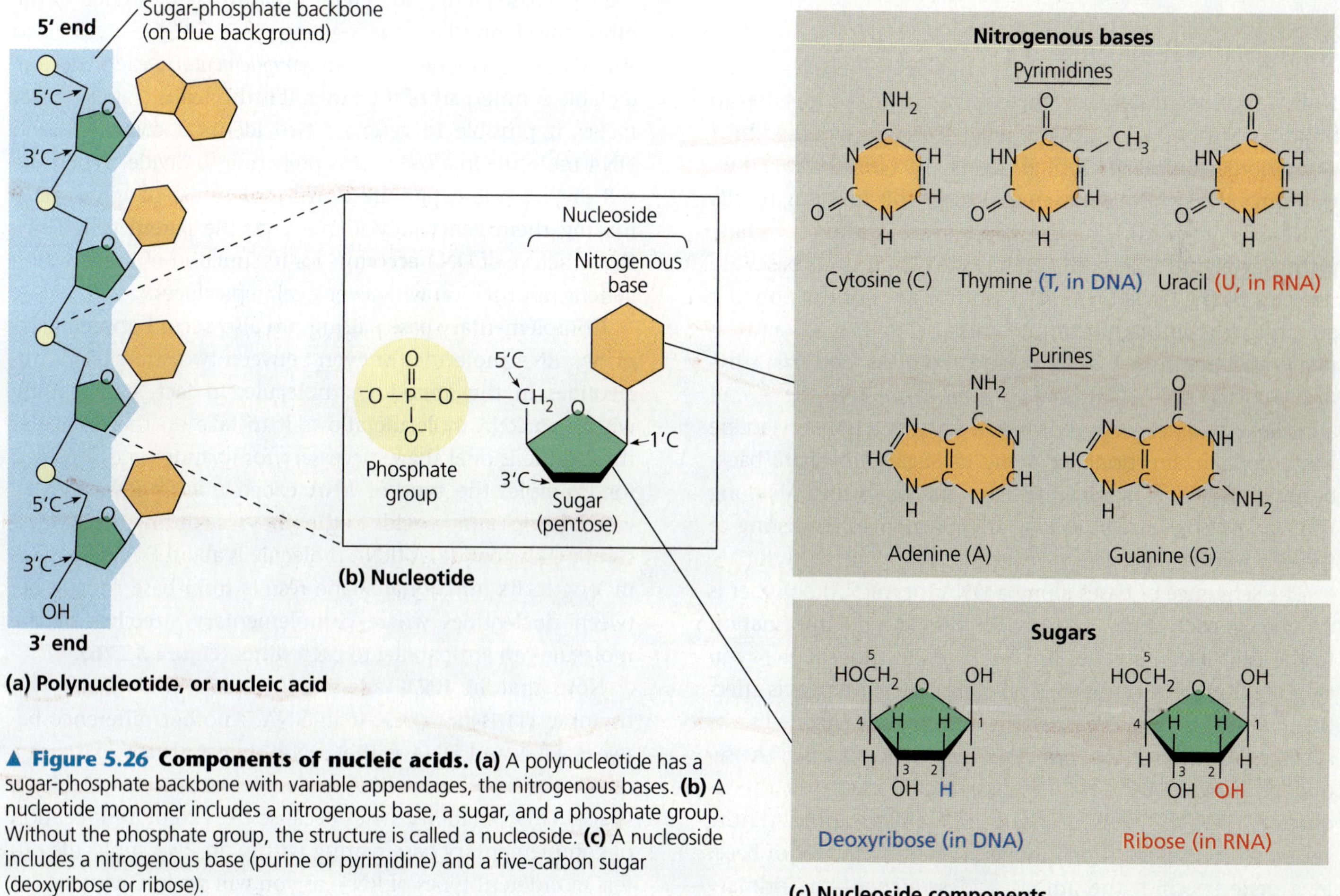

▲ **Figure 5.26 Components of nucleic acids. (a)** A polynucleotide has a sugar-phosphate backbone with variable appendages, the nitrogenous bases. **(b)** A nucleotide monomer includes a nitrogenous base, a sugar, and a phosphate group. Without the phosphate group, the structure is called a nucleoside. **(c)** A nucleoside includes a nitrogenous base (purine or pyrimidine) and a five-carbon sugar (deoxyribose or ribose).

from solution, thus acting as bases.) There are two families of nitrogenous bases: pyrimidines and purines. A **pyrimidine** has one six-membered ring of carbon and nitrogen atoms. The members of the pyrimidine family are cytosine (C), thymine (T), and uracil (U). **Purines** are larger, with a six-membered ring fused to a five-membered ring. The purines are adenine (A) and guanine (G). The specific pyrimidines and purines differ in the chemical groups attached to the rings. Adenine, guanine, and cytosine are found in both DNA and RNA; thymine is found only in DNA and uracil only in RNA.

Now let's add a sugar to the nitrogenous base. In DNA the sugar is **deoxyribose**; in RNA it is **ribose** (see Figure 5.26c). The only difference between these two sugars is that deoxyribose lacks an oxygen atom on the second carbon in the ring; hence the name *deoxy*ribose. To distinguish the numbers of the sugar carbons from those used for the ring atoms of the attached nitrogenous base, the sugar carbon numbers of a nucleoside or nucleotide have a prime (′) after them. Thus, the second carbon in the sugar ring is the 2′ ("2 prime") carbon, and the carbon that sticks up from the ring is called the 5′ carbon.

So far, we have built a nucleoside (nitrogenous base plus sugar). To complete the construction of a nucleotide, we attach a phosphate group to the 5′ carbon of the sugar (see Figure 5.26b). The molecule is now a nucleoside monophosphate, better known as a nucleotide.

Nucleotide Polymers

Now we can see how these nucleotides are linked together to build a polynucleotide. Adjacent nucleotides are joined by a phosphodiester linkage, which consists of a phosphate group that links the sugars of two nucleotides. This bonding results in a backbone with a repeating pattern of sugar-phosphate units (see Figure 5.26a). (Note that the nitrogenous bases are not part of the backbone.) The two free ends of the polymer are distinctly different from each other. One end has a phosphate attached to a 5′ carbon, and the other end has a hydroxyl group on a 3′ carbon; we refer to these as the 5′ end and the 3′ end, respectively. We can say that a polynucleotide has a built-in directionality along its sugar-phosphate backbone, from 5′ to 3′, somewhat like a one-way street. All along this sugar-phosphate backbone are appendages consisting of the nitrogenous bases.

The sequence of bases along a DNA (or mRNA) polymer is unique for each gene and provides very specific information to the cell. Because genes are hundreds to thousands of nucleotides long, the number of possible base sequences is effectively limitless. A gene's meaning to the cell is encoded in its specific sequence of the four DNA bases. For example, the sequence 5′-AGGTAACTT-3′ means one thing, whereas the sequence 5′-CGCTTTAAC-3′ has a different meaning. (Entire genes, of course, are much longer.) The linear order of bases in a gene specifies the amino acid sequence—the primary structure—of a protein, which in turn specifies that protein's three-dimensional structure and its function in the cell.

The Structures of DNA and RNA Molecules

RNA molecules usually exist as single polynucleotide chains like the one shown in Figure 5.26a. In contrast, DNA molecules have two polynucleotides, or "strands," that spiral around an imaginary axis, forming a **double helix** **(Figure 5.27a)**. The two sugar-phosphate backbones run in opposite 5′ → 3′ directions from each other; this arrangement is referred to as **antiparallel**, somewhat like a divided highway. The sugar-phosphate backbones are on the outside of the helix, and the nitrogenous bases are paired in the interior of the helix. The two strands are held together by hydrogen bonds between the paired bases (see Figure 5.27a). Most DNA molecules are very long, with thousands or even millions of base pairs. One long DNA double helix includes many genes, each one a particular segment of the molecule.

Only certain bases in the double helix are compatible with each other. Adenine (A) always pairs with thymine (T), and guanine (G) always pairs with cytosine (C). If we were to read the sequence of bases along one strand of the double helix, we would know the sequence of bases along the other strand. If a stretch of one strand has the base sequence 5′-AGGTCCG-3′, then the base-pairing rules tell us that the same stretch of the other strand must have the sequence 3′-TCCAGGC-5′. The two strands of the double helix are *complementary*, each the predictable counterpart of the other. It is this feature of DNA that makes it possible to generate two identical copies of each DNA molecule in a cell that is preparing to divide. When the cell divides, the copies are distributed to the daughter cells, making them genetically identical to the parent cell. Thus, the structure of DNA accounts for its function of transmitting genetic information whenever a cell reproduces.

Complementary base pairing can also occur between parts of two RNA molecules or even between two stretches of nucleotides in the *same* RNA molecule. In fact, base pairing within an RNA molecule allows it to take on the particular three-dimensional shape necessary for its function. Consider, for example, the type of RNA called *transfer RNA* (*tRNA*), which brings amino acids to the ribosome during the synthesis of a polypeptide. A tRNA molecule is about 80 nucleotides in length. Its functional shape results from base pairing between nucleotides where complementary stretches of the molecule run antiparallel to each other **(Figure 5.27b)**.

Note that in RNA, adenine (A) pairs with uracil (U); thymine (T) is not present in RNA. Another difference between RNA and DNA is that DNA almost always exists as a double helix, whereas RNA molecules are more variable in shape. This variability arises because the extent and location of complementary base pairing within an RNA molecule differs in different types of RNA, as you will see in Chapter 17.

▶ **Figure 5.27 The structures of DNA and tRNA molecules. (a)** The DNA molecule is usually a double helix, with the sugar-phosphate backbones of the antiparallel polynucleotide strands (symbolized here by blue ribbons) on the outside of the helix. Holding the two strands together are pairs of nitrogenous bases attached to each other by hydrogen bonds. As illustrated here with symbolic shapes for the bases, adenine (A) can pair only with thymine (T), and guanine (G) can pair only with cytosine (C). Each DNA strand in this figure is the structural equivalent of the polynucleotide diagrammed in Figure 5.26a. **(b)** A tRNA molecule has a roughly L-shaped structure, with complementary base pairing of antiparallel stretches of RNA. In RNA, A pairs with U.

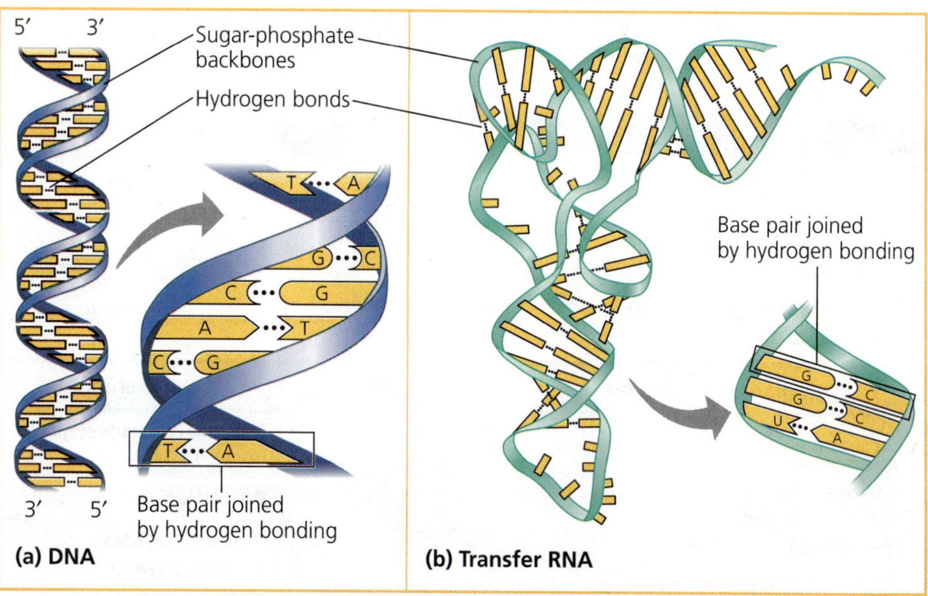

DNA and Proteins as Tape Measures of Evolution

EVOLUTION We are accustomed to thinking of shared traits, such as hair and milk production in mammals, as evidence of shared ancestors. Because we now understand that DNA carries heritable information in the form of genes, we can see that genes and their products (proteins) document the hereditary background of an organism. The linear sequences of nucleotides in DNA molecules are passed from parents to offspring; these sequences determine the amino acid sequences of proteins. Siblings have greater similarity in their DNA and proteins than do unrelated individuals of the same species. If the evolutionary view of life is valid, we should be able to extend this concept of "molecular genealogy" to relationships between species: We should expect two species that appear to be closely related based on fossil and anatomical evidence to also share a greater proportion of their DNA and protein sequences than do more distantly related species. In fact, that is the case. An example is the comparison of the β polypeptide chain of human hemoglobin with the corresponding hemoglobin polypeptide in other vertebrates. In this chain of 146 amino acids, humans and gorillas differ in just 1 amino acid, while humans and frogs differ in 67 amino acids. Molecular biology has added a new tape measure to the toolkit biologists use to assess evolutionary kinship.

The Theme of Emergent Properties in the Chemistry of Life: *A Review*

Recall that life is organized along a hierarchy of structural levels (see Figure 1.4). With each increasing level of order, new properties emerge. In Chapters 2–5, we have dissected the chemistry of life. But we have also begun to develop a more integrated view of life, exploring how properties emerge with increasing order.

We have seen that water's behavior results from the interactions of its molecules, each an ordered arrangement of hydrogen and oxygen atoms. We reduced the complexity and diversity of organic compounds to carbon skeletons and appended chemical groups. We saw that macromolecules are assembled from small organic molecules, taking on new properties. By completing our overview with an introduction to macromolecules and lipids, we have built a bridge to Unit Two, where we will study cell structure and function. We will keep a balance between the need to reduce life to simpler processes and the ultimate satisfaction of viewing those processes in their integrated context.

CONCEPT CHECK 5.5

1. **DRAW IT** Go to Figure 5.26a and, for the top three nucleotides, number all the carbons in the sugars, circle the nitrogenous bases, and star the phosphates.
2. **DRAW IT** In a DNA double helix, a region along one DNA strand has this sequence of nitrogenous bases: 5′-TAGGCCT-3′. Copy this sequence, and write down its complementary strand, clearly indicating the 5′ and 3′ ends of the complementary strand.
3. **WHAT IF?** (a) Suppose a substitution occurred in one DNA strand of the double helix in question 2, resulting in

   5′-TAAGCCT-3′
   3′-ATCCGGA-5′

   Copy these two strands, and circle and label the mismatched bases. (b) If the modified top strand is used by the cell to construct a complementary strand, what would that matching strand be?

For suggested answers, see Appendix A.

5 CHAPTER REVIEW

SUMMARY OF KEY CONCEPTS

CONCEPT 5.1

Macromolecules are polymers, built from monomers (pp. 68–69)

- Carbohydrates, proteins, and nucleic acids are **polymers**, chains of **monomers**. The components of lipids vary.

Monomers form larger molecules by **dehydration reactions**, in which water molecules are released. Polymers can disassemble by the reverse process, **hydrolysis**. An immense variety of polymers can be built from a small set of monomers.

? *What is the fundamental basis for the differences between carbohydrates, proteins, and nucleic acids?*

Large Biological Molecules	Components	Examples	Functions
CONCEPT 5.2 Carbohydrates serve as fuel and building material (pp. 69–74) **?** *Compare the composition, structure, and function of starch and cellulose. What role do starch and cellulose play in the human body?*	CH_2OH, H, O, H, H, OH, H, HO, OH, H, OH Monosaccharide monomer	**Monosaccharides:** glucose, fructose	Fuel; carbon sources that can be converted to other molecules or combined into polymers
		Disaccharides: lactose, sucrose	
		Polysaccharides: • Cellulose (plants) • Starch (plants) • Glycogen (animals) • Chitin (animals and fungi)	• Strengthens plant cell walls • Stores glucose for energy • Stores glucose for energy • Strengthens exoskeletons and fungal cell walls
CONCEPT 5.3 Lipids are a diverse group of hydrophobic molecules (pp. 74–77) **?** *Why are lipids not considered to be macromolecules or polymers?*	Glycerol 3 fatty acids	**Triacylglycerols** (fats or oils): glycerol + 3 fatty acids	Important energy source
	Head with (P) 2 fatty acids	**Phospholipids:** phosphate group + 2 fatty acids	Lipid bilayers of membranes Hydrophobic tails Hydrophilic heads
	Steroid backbone	**Steroids:** four fused rings with attached chemical groups	• Component of cell membranes (cholesterol) • Signaling molecules that travel through the body (hormones)
CONCEPT 5.4 Proteins include a diversity of structures, resulting in a wide range of functions (pp. 77–86) **?** *Proteins are the most structurally and functionally diverse class of biological molecules. Explain the basis for this diversity.*	R, H, N, C, C, O, H, H, OH Amino acid monomer (20 types)	• Enzymes • Structural proteins • Storage proteins • Transport proteins • Hormones • Receptor proteins • Motor proteins • Defensive proteins	• Catalyze chemical reactions • Provide structural support • Store amino acids • Transport substances • Coordinate organismal responses • Receive signals from outside cell • Function in cell movement • Protect against disease
CONCEPT 5.5 Nucleic acids store, transmit, and help express hereditary information (pp. 86–89) **?** *What role does complementary base pairing play in the functions of nucleic acids?*	Nitrogenous base Phosphate group (P) CH_2 O Sugar	**DNA:** • Sugar = deoxyribose • Nitrogenous bases = C, G, A, T • Usually double-stranded	Stores hereditary information
		RNA: • Sugar = ribose • Nitrogenous bases = C, G, A, U • Usually single-stranded	Various functions during gene expression, including carrying instructions from DNA to ribosomes

TEST YOUR UNDERSTANDING

LEVEL 1: KNOWLEDGE/COMPREHENSION

1. Which of the following categories includes all others in the list?
   a. monosaccharide
   b. disaccharide
   c. starch
   d. carbohydrate
   e. polysaccharide

2. The enzyme amylase can break glycosidic linkages between glucose monomers only if the monomers are in the α form. Which of the following could amylase break down?
   a. glycogen, starch, and amylopectin
   b. glycogen and cellulose
   c. cellulose and chitin
   d. starch and chitin
   e. starch, amylopectin, and cellulose

3. Which of the following statements concerning *unsaturated* fats is true?
   a. They are more common in animals than in plants.
   b. They have double bonds in the carbon chains of their fatty acids.
   c. They generally solidify at room temperature.
   d. They contain more hydrogen than do saturated fats having the same number of carbon atoms.
   e. They have fewer fatty acid molecules per fat molecule.

4. The structural level of a protein *least* affected by a disruption in hydrogen bonding is the
   a. primary level.
   b. secondary level.
   c. tertiary level
   d. quaternary level.
   e. All structural levels are equally affected.

5. Enzymes that break down DNA catalyze the hydrolysis of the covalent bonds that join nucleotides together. What would happen to DNA molecules treated with these enzymes?
   a. The two strands of the double helix would separate.
   b. The phosphodiester linkages of the polynucleotide backbone would be broken.
   c. The purines would be separated from the deoxyribose sugars.
   d. The pyrimidines would be separated from the deoxyribose sugars.
   e. All bases would be separated from the deoxyribose sugars.

LEVEL 2: APPLICATION/ANALYSIS

6. The molecular formula for glucose is $C_6H_{12}O_6$. What would be the molecular formula for a polymer made by linking ten glucose molecules together by dehydration reactions?
   a. $C_{60}H_{120}O_{60}$
   b. $C_6H_{12}O_6$
   c. $C_{60}H_{102}O_{51}$
   d. $C_{60}H_{100}O_{50}$
   e. $C_{60}H_{111}O_{51}$

7. Which of the following pairs of base sequences could form a short stretch of a normal double helix of DNA?
   a. 5′-purine-pyrimidine-purine-pyrimidine-3′ with 3′-purine-pyrimidine-purine-pyrimidine-5′
   b. 5′-AGCT-3′ with 5′-TCGA-3′
   c. 5′-GCGC-3′ with 5′-TATA-3′
   d. 5′-ATGC-3′ with 5′-GCAT-3′
   e. All of these pairs are correct.

8. Construct a table that organizes the following terms, and label the columns and rows.

phosphodiester linkages	polypeptides	monosaccharides
peptide bonds	triacylglycerols	nucleotides
glycosidic linkages	polynucleotides	amino acids
ester linkages	polysaccharides	fatty acids

9. **DRAW IT** Copy the polynucleotide strand in Figure 5.26a and label the bases G, T, C, and T, starting from the 5′ end. Assuming this is a DNA polynucleotide, now draw the complementary strand, using the same symbols for phosphates (circles), sugars (pentagons), and bases. Label the bases. Draw arrows showing the 5′ → 3′ direction of each strand. Use the arrows to make sure the second strand is antiparallel to the first. *Hint*: After you draw the first strand vertically, turn the paper upside down; it is easier to draw the second strand from the 5′ toward the 3′ direction as you go from top to bottom.

LEVEL 3: SYNTHESIS/EVALUATION

10. **EVOLUTION CONNECTION**
    Comparisons of amino acid sequences can shed light on the evolutionary divergence of related species. If you were comparing two living species, would you expect all proteins to show the same degree of divergence? Why or why not?

11. **SCIENTIFIC INQUIRY**
    Suppose you are a research assistant in a lab studying DNA-binding proteins. You have been given the amino acid sequences of all the proteins encoded by the genome of a certain species and have been asked to find candidate proteins that could bind DNA. What type of amino acids would you expect to see in such proteins? Why?

12. **SCIENCE, TECHNOLOGY, AND SOCIETY**
    Some amateur and professional athletes take anabolic steroids to help them "bulk up" or build strength. The health risks of this practice are extensively documented. Apart from health considerations, how do you feel about the use of chemicals to enhance athletic performance? Is an athlete who takes anabolic steroids cheating, or is such use part of the preparation required to succeed in competition? Explain.

13. **WRITE ABOUT A THEME**
    Structure and Function Proteins, which have diverse functions in a cell, are all polymers of the same subunits—amino acids. Write a short essay (100–150 words) that discusses how the structure of amino acids allows this one type of polymer to perform so many functions.

For selected answers, see Appendix A.

UNIT 2 The Cell

An Interview with

Bonnie L. Bassler

Bonnie Bassler loves her life as a biologist scrutinizing the secret lives of bacteria. For the past 20 years or so, Bonnie and her lab (her "gang," as she calls them) have made momentous discoveries about how bacterial cells use chemicals to communicate with each other in a process called quorum sensing. Dr. Bassler has a B.S. in Biochemistry from the University of California at Davis and a Ph.D. in Biochemistry from The Johns Hopkins University. Among her many awards and honors, she has received a MacArthur Foundation Fellowship and is a member of the National Academy of Sciences. She is the 2010–2011 President of the American Society for Microbiology, the largest specialized life science organization in the world. At Princeton University since 1994, she is currently the Squibb Professor in Molecular Biology and an Investigator of the Howard Hughes Medical Institute.

How did you get started in science?

I've always been interested in nature and animals, and in puzzles and mystery books—I really like figuring things out. As an undergraduate at UC Davis, I worked in a lab on a bacterial project while taking courses in both biochemistry and genetics. Then, as a graduate student at Johns Hopkins, I learned a lot of biochemistry while studying marine bacteria. The bacteria belong to the genus *Vibrio*, and I was working on chemotaxis, movement by cells toward food or away from noxious chemicals in the environment.

What are the advantages of using bacteria for research in cell biology?

Bacteria have been the foundation of molecular biology for the last 100 years because they're accessible. They grow fast, they form clones of identical cells, and they're amenable to biochemical and genetic analyses. Most of what we initially learned about molecular biology—about genes and proteins and other biomolecules—came from work done on bacteria. Because of evolutionary history, the most basic and ancient life processes that happen in bacteria also happen in humans and other higher organisms. Humans have more bells and whistles, more proteins, more sophistication, more complexity. But if you want to understand the basic components of a process, very often you can use bacteria to do that. Also, working with bacteria fits my personality. I prefer having 10 billion bacterial offspring the day after an experiment to having to wait weeks or months for a small number of baby mice. Every morning there's a surprise waiting for me in the incubator!

What is quorum sensing, and how did you first hear about it?

When I was finishing my graduate work, I heard a talk by Mike Silverman, a scientist with the Agouron Institute in San Diego, about how bacteria "talk" to each other, "count" their own numbers, and coordinate their behavior. Mike had been working on a light-producing (bioluminescent) marine bacterium called *Vibrio fischeri* that lives symbiotically inside a variety of marine animals. The animal provides nutrients for the bacteria, which live in an enclosed space within the animal's body. In return, the bacteria provide light that benefits the animal—by scaring away predators or attracting prey or a mate. But if only a small number of bacteria are present, they do not make light—producing light would waste energy because the light wouldn't be visible. The word "quorum" means "the number needed to do something," and bacteria can sense whether there is a quorum or not and act accordingly.

The way quorum sensing works is that bacteria release certain signaling chemicals into the environment. As the bacterial cells increase in number, the molecules reach a concentration at which many of them bind to receptor proteins on the surface of or inside the bacteria. The signaling molecule fits together with the receptor like a key in a lock. In the case of the surface receptors, each receptor molecule has a part on the outside of the cell and a part on the inside. The signaling chemical binds to the outer part of the receptor, "tickling" the protein so that it makes something happen inside the cell. For instance, in *Vibrio fischeri*, binding of signaling molecules ultimately turns on genes that code for enzymes that make light. Mike had worked out this mechanism of how cells of *Vibrio fischeri* turn on light in synchrony.

It's important to understand that back then, we just didn't think about bacteria like that—we thought bacteria ignored each other and did their own thing as solitary cells. I was totally fascinated. I thought, "He's either crazy or he's brilliant—but I just have to work on that." I went up to the podium after his talk and begged him to let me be his postdoc. Finally he said yes, even though he was a geneticist and I was a biochemist! He took a chance on me.

How does a genetic approach differ from a biochemical one?

Geneticists make lots of mutant organisms, then think up clever strategies to find the ones with mutations in the genes they're interested in. In the case of quorum sensing in bioluminescent bacteria, you look for cells that remain dark. If you have mutated genes involved in quorum sensing, you would expect the bacteria not to make light because light emission depends on the cells communicating with each other. Eventually, you would hope to identify the components that function in normal light-emitting bacteria but not in the mutants. Biochemists, on the other hand, start by isolating molecules and studying their properties directly. Genetics and biochemistry are complementary approaches. I'm glad I know both because the combination is more effective than either approach by itself.

What did you learn about quorum sensing as a postdoc?

In Mike's lab, I worked on another species of bioluminescent *Vibrio* called *Vibrio harveyi*. Because these bacteria are free-living in the ocean, we thought their quorum-sensing molecular circuitry might be more complicated than that of *Vibrio fischeri*. What I found was that *Vibrio harveyi* has two parallel systems for quorum sensing, one that senses cells of the same species and one that counts bacteria of other species. Fast-forwarding a decade or so, this second system seems to be present in *many* bacteria, and the second signaling molecule appears to be universal. So, apparently bacteria can measure the ratios of these two signals, and they're saying, "How many of us and how many of you are there?" Then they do different things, depending on who is in the majority. And this isn't just restricted to bioluminescence. Other bacterial behaviors are also controlled by quorum sensing, such as forming an organized thin layer (called a biofilm) on your teeth or coordinating a virulent infection.

Tell us more about biofilms.

We used to think that most bacteria lived as individual cells suspended in liquid environments. But we now understand that in the wild, they live attached to surfaces in biofilms, and they secrete carbohydrates and other molecules that form a protective slime on the biofilm surface. Most of us have noticed the biofilm coating our teeth every morning. Believe it or not, there are about 600 bacterial species in that biofilm just trying to make a living, getting nutrients from us, but the side effect is that we get cavities. And when someone has a lung infection or an implant or heart valve that harbors an infection, the bacteria are growing as a biofilm in the lungs or on the introduced device. So we now understand why these infections are so hard to treat: It's because the slime on the biofilm is providing a protective shield that antibiotics can't penetrate.

What questions are you and your lab asking now?

My group is interested in how information outside an organism gets inside so that the organism does the right thing at the right time. We work on bacteria because they're simple, but we hope that we will have insights for people working on higher organisms. And we're curious about how collective behaviors first evolved on Earth. How did multicellularity come about? We know that the first organisms were bacteria, but how did they begin to do things together? How did groups of cells in your body come to act like a liver or a heart? We're very interested in how the flow of information through networks facilitates multicellularity.

Are there applications for the basic research you do?

When you're asking fundamental questions, you hope that surprises, things you never thought of, will come out of it. Now that we know that bacteria talk to each other and perform group activities, the question is whether we could interfere with these conversations for therapeutic purposes. Could we make molecules that keep bacteria from "talking" or "hearing"? Maybe these would be new antibiotics. Biofilms are a terrible problem in medicine and dental health, and now that we are starting to know the molecular basis for their formation, maybe we can learn how to prevent them from forming.

Bacteria get a lot of bad press for the negative things they do. On the other hand, bacteria also do many miraculous things that keep us alive; they are working for us every instant of our lives. You are covered with a bacterial biofilm that acts as invisible body armor—these good bacteria occupy all the spaces on your skin, preventing invading bacteria from attaching. Throughout your gut you have a huge mass of bacteria, and they're making vitamins for you and helping you digest your food. So all biofilms aren't bad—and for the good biofilms, what if we could find a molecule to make quorum sensing better? Rather than an antibiotic, this would be a probiotic.

"The sky's the limit for biologists because biology is the science of the 21st century and it touches every part of our lives."

Where do you think this field is going?

I think we'll be turning our attention to the possibility of communication between organisms from different kingdoms and different domains. Bacteria have been around for over 4 billion years and have probably been living with multicellular eukaryotic hosts for hundreds of millions of years. So why wouldn't these hosts have evolved strategies to listen in, say, to the conversation being carried out by a group of pathogenic bacteria? Does our immune system "hear" bacterial signaling molecules? Do hosts actively prevent quorum sensing among pathogenic bacteria? Do they tune in and help the good bacteria? I think this is going to be a dialogue, not a monologue.

What do you enjoy most about your life as a scientist?

I love what I work on. I figured out as a postdoc how much fun this life in science is—that it is not about me against other scientists or who is going to discover something first. Instead, it's me against this bacterium, and we are in it head-to-head for the rest of our lives, in a contest of wills between bacteria trying to keep their secrets and me trying to discover them. Also, the basic question of how groups work together fascinates me. I work with a fantastic group of students and we share everything—everybody gives everybody everything, and then we all get more. That's quorum sensing! Both my molecular and nonmolecular lives involve getting the group to do more than the individual. I love that parallelism! My gang of students show me their data, and it's my job to help them figure out the science and get on with their careers. I'm so lucky—having 24 hands and 12 brains is so much better than two hands and one brain. The science is always changing, and trying to keep up with these young and tireless people is hugely challenging and rewarding.

What is your advice to an undergraduate who is considering a career in biology?

For undergraduates who are considering a life in science, my advice is to work on something that you are passionate about. Don't be limited by thinking that bench science is the only thing a scientist can do. There are so many potential careers for a biologist. You could work on Capitol Hill as a scientific advisor or policymaker. You could teach. You could be a lawyer. You could be a writer who helps the public understand science. You could work on science education at the kindergarten level. Figure out your particular combination of personality traits and what you really love doing as a scientist; then make that niche for yourself and bring science to that career. The sky's the limit for biologists because biology is the science of the 21st century and it touches every part of our lives.

Bonnie Bassler (left) with Lisa Urry (center) and Jane Reece

6

A Tour of the Cell

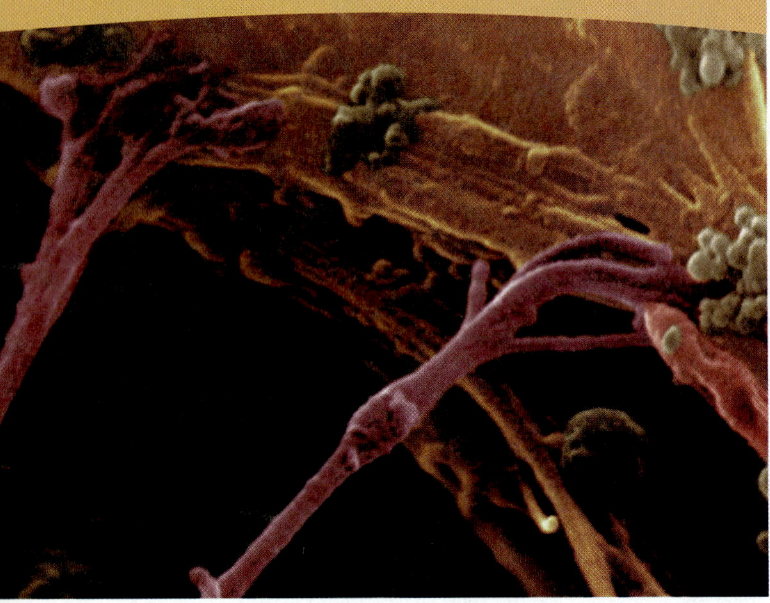

▲ **Figure 6.1 How do your brain cells help you learn about biology?**

KEY CONCEPTS

- **6.1** Biologists use microscopes and the tools of biochemistry to study cells
- **6.2** Eukaryotic cells have internal membranes that compartmentalize their functions
- **6.3** The eukaryotic cell's genetic instructions are housed in the nucleus and carried out by the ribosomes
- **6.4** The endomembrane system regulates protein traffic and performs metabolic functions in the cell
- **6.5** Mitochondria and chloroplasts change energy from one form to another
- **6.6** The cytoskeleton is a network of fibers that organizes structures and activities in the cell
- **6.7** Extracellular components and connections between cells help coordinate cellular activities

OVERVIEW

The Fundamental Units of Life

Given the scope of biology, you may wonder sometimes how you will ever learn all the material in this course! The answer involves cells, which are as fundamental to the living systems of biology as the atom is to chemistry. The contraction of muscle cells moves your eyes as you read this sentence. The words on the page are translated into signals that nerve cells carry to your brain. **Figure 6.1** shows extensions from one nerve cell (purple) making contact with another nerve cell (orange) in the brain. As you study, your goal is to make connections like these that solidify memories and permit learning to occur.

All organisms are made of cells. In the hierarchy of biological organization, the cell is the simplest collection of matter that can be alive. Indeed, many forms of life exist as single-celled organisms. More complex organisms, including plants and animals, are multicellular; their bodies are cooperatives of many kinds of specialized cells that could not survive for long on their own. Even when cells are arranged into higher levels of organization, such as tissues and organs, the cell remains the organism's basic unit of structure and function.

All cells are related by their descent from earlier cells. However, they have been modified in many different ways during the long evolutionary history of life on Earth. But although cells can differ substantially from one another, they share common features. In this chapter, we'll first examine the tools and techniques that allow us to understand cells, then tour the cell and become acquainted with its components.

CONCEPT 6.1

Biologists use microscopes and the tools of biochemistry to study cells

How can cell biologists investigate the inner workings of a cell, usually too small to be seen by the unaided eye? Before we tour the cell, it will be helpful to learn how cells are studied.

Microscopy

The development of instruments that extend the human senses has gone hand in hand with the advance of science. The discovery and early study of cells progressed with the invention of microscopes in 1590 and their refinement during the 1600s. Cell walls were first seen by Robert Hooke in 1665 as he looked through a microscope at dead cells from the bark of an oak tree. But it took the wonderfully crafted lenses of Antoni van Leeuwenhoek to visualize living cells. Imagine Hooke's awe when he visited van Leeuwenhoek in 1674 and the world of microorganisms—what his host called "very little animalcules"—was revealed to him.

The microscopes first used by Renaissance scientists, as well as the microscopes you are likely to use in the laboratory, are all light microscopes. In a **light microscope (LM)**, visible light is passed through the specimen and then through glass lenses. The lenses refract (bend) the light in such a way that the image of the specimen is magnified as it is projected into the eye or into a camera (see Appendix D).

Three important parameters in microscopy are magnification, resolution, and contrast. *Magnification* is the ratio of an object's image size to its real size. Light microscopes can magnify effectively to about 1,000 times the actual size of the specimen; at greater magnifications, additional details cannot be seen clearly. *Resolution* is a measure of the clarity of the image; it is the minimum distance two points can be separated and still be distinguished as two points. For example, what appears to the unaided eye as one star in the sky may be resolved as twin stars with a telescope, which has a higher resolving ability than the eye. Similarly, using standard techniques, the light microscope cannot resolve detail finer than about 0.2 micrometer (μm), or 200 nanometers (nm), regardless of the magnification **(Figure 6.2)**. The third parameter, *contrast*, accentuates differences in parts of the sample. Improvements in light microscopy have included new methods for enhancing contrast, such as staining or labeling cell components to stand out visually. **Figure 6.3**, on the next page, shows different types of microscopy; study this figure as you read the rest of this section.

Until recently, the resolution barrier prevented cell biologists from using standard light microscopy to study **organelles**, the membrane-enclosed structures within eukaryotic cells. To see these structures in any detail required the development of a new instrument. In the 1950s, the electron microscope was introduced to biology. Rather than light, the **electron microscope (EM)** focuses a beam of electrons through the specimen or onto its surface (see Appendix D). Resolution is inversely related to the wavelength of the radiation a microscope uses for imaging, and electron beams have much shorter wavelengths than visible light. Modern electron microscopes can theoretically achieve a resolution of about 0.002 nm, though in practice they usually cannot resolve structures smaller than about 2 nm across. Still, this is a hundredfold improvement over the standard light microscope.

The **scanning electron microscope (SEM)** is especially useful for detailed study of the topography of a specimen (see Figure 6.3). The electron beam scans the surface of the sample, usually coated with a thin film of gold. The beam excites electrons on the surface, and these secondary electrons are detected by a device that translates the pattern of electrons into an electronic signal to a video screen. The result is an image of the specimen's surface that appears three-dimensional.

The **transmission electron microscope (TEM)** is used to study the internal structure of cells (see Figure 6.3). The TEM aims an electron beam through a very thin section of the specimen, similar to the way a light microscope transmits light through a slide. The specimen has been stained with atoms of heavy metals, which attach to certain cellular structures, thus enhancing the electron density of some parts of the cell more than others. The electrons passing through the specimen are scattered more in the denser regions, so fewer are transmitted. The image displays the pattern of transmitted electrons. Instead of using glass lenses, the TEM uses electromagnets as lenses to bend the paths of the electrons, ultimately focusing the image onto a monitor for viewing.

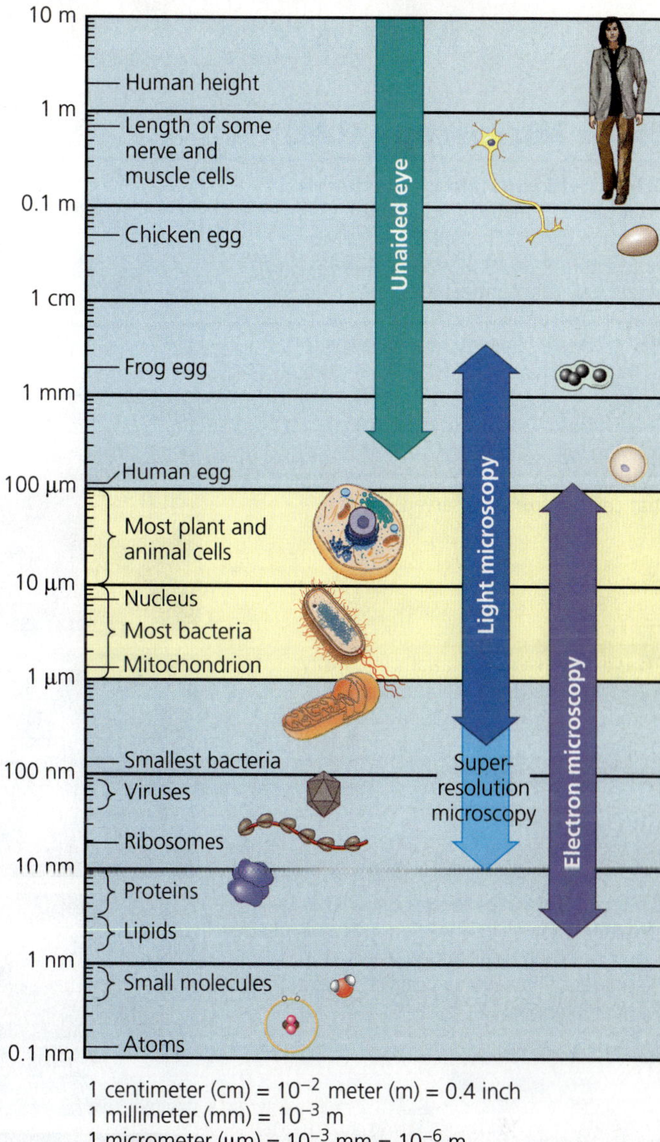

▲ **Figure 6.2 The size range of cells.** Most cells are between 1 and 100 μm in diameter (yellow region of chart) and are therefore visible only under a microscope. Notice that the scale along the left side is logarithmic to accommodate the range of sizes shown. Starting at the top of the scale with 10 m and going down, each reference measurement marks a tenfold decrease in diameter or length. For a complete table of the metric system, see Appendix C.

Electron microscopes have revealed many organelles and other subcellular structures that were impossible to resolve with the light microscope. But the light microscope offers advantages, especially in studying living cells. A disadvantage of

▼ **Figure 6.3**

Exploring Microscopy

Light Microscopy (LM)

Brightfield (unstained specimen). Light passes directly through the specimen. Unless the cell is naturally pigmented or artificially stained, the image has little contrast. (The first four light micrographs show human cheek epithelial cells; the scale bar pertains to all four micrographs.)

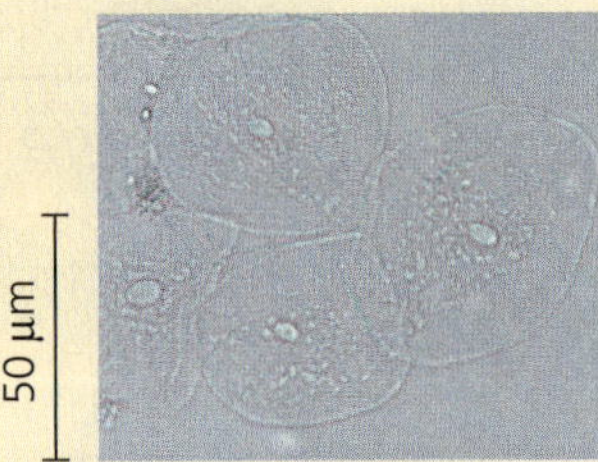

Brightfield (stained specimen). Staining with various dyes enhances contrast. Most staining procedures require that cells be fixed (preserved).

Phase-contrast. Variations in density within the specimen are amplified to enhance contrast in unstained cells, which is especially useful for examining living, unpigmented cells.

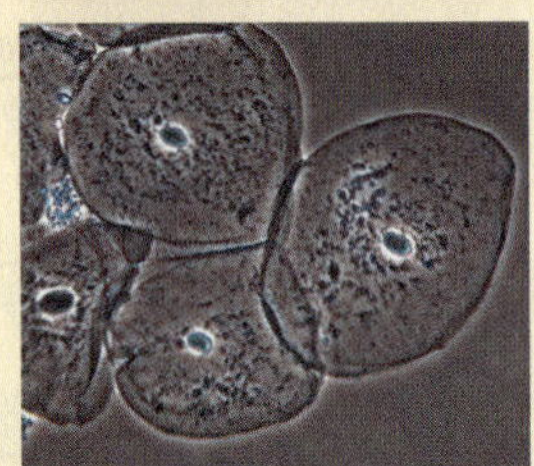

Differential-interference-contrast (Nomarski). As in phase-contrast microscopy, optical modifications are used to exaggerate differences in density, making the image appear almost 3-D.

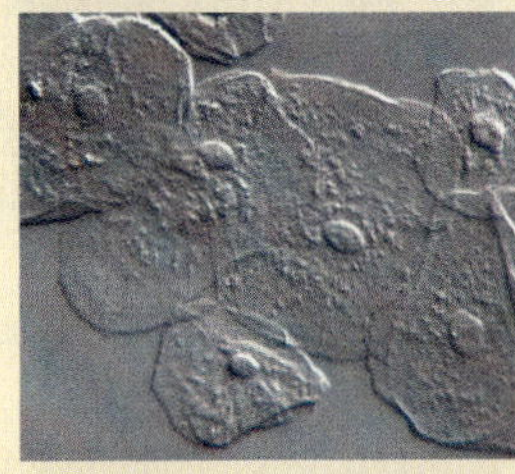

Fluorescence. The locations of specific molecules in the cell can be revealed by labeling the molecules with fluorescent dyes or antibodies; some cells have molecules that fluoresce on their own. Fluorescent substances absorb ultraviolet radiation and emit visible light. In this fluorescently labeled uterine cell, nuclear material is blue, organelles called mitochondria are orange, and the cell's "skeleton" is green.

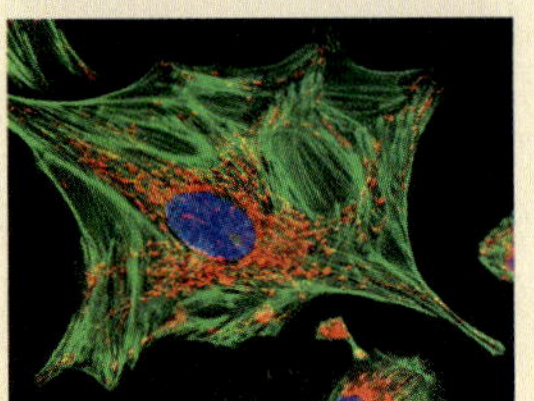

Confocal. The top image is a standard fluorescence micrograph of fluorescently labeled nervous tissue (nerve cells are green, support cells are orange, and regions of overlap are yellow); below it is a confocal image of the same tissue. Using a laser, this "optical sectioning" technique eliminates out-of-focus light from a thick sample, creating a single plane of fluorescence in the image. By capturing sharp images at many different planes, a 3-D reconstruction can be created. The standard image is blurry because out-of-focus light is not excluded.

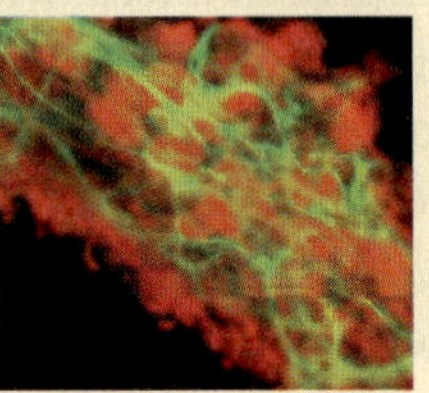

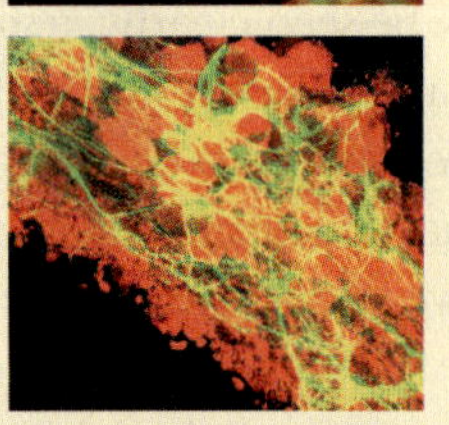

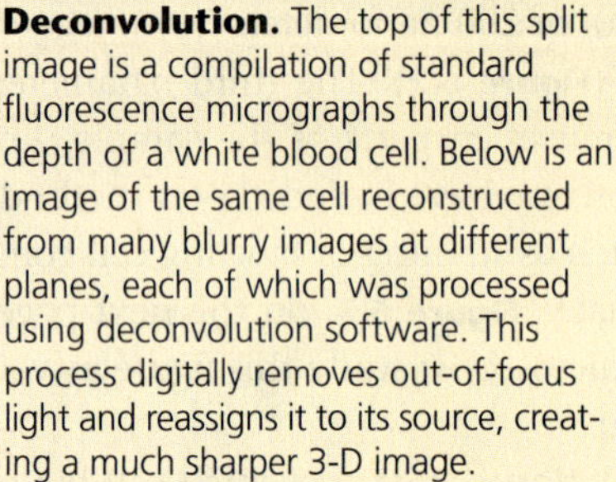

Deconvolution. The top of this split image is a compilation of standard fluorescence micrographs through the depth of a white blood cell. Below is an image of the same cell reconstructed from many blurry images at different planes, each of which was processed using deconvolution software. This process digitally removes out-of-focus light and reassigns it to its source, creating a much sharper 3-D image.

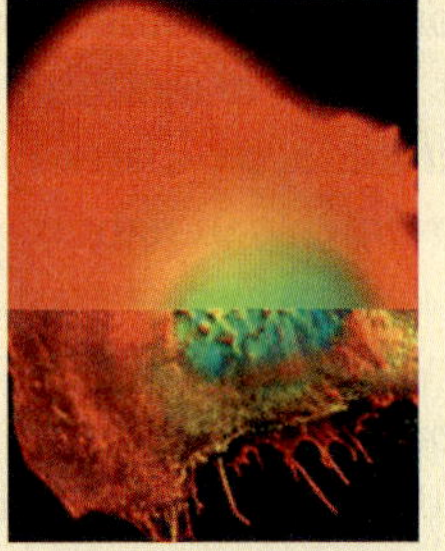

Super-resolution. On the top is a confocal image of part of a nerve cell, using a fluorescent label that binds to a molecule clustered in small sacs in the cell (vesicles) that are 40 nm in diameter. The greenish-yellow spots are blurry because 40 nm is below the 200-nm limit of resolution for standard light microscopy. Below is an image of the same part of the cell, seen using a new "super-resolution" technique. Sophisticated equipment is used to light up individual fluorescent molecules and record their position. Combining information from many molecules in different places "breaks" the limit of resolution, resulting in the sharp greenish-yellow dots seen here. (Each dot is a 40-nm vesicle.)

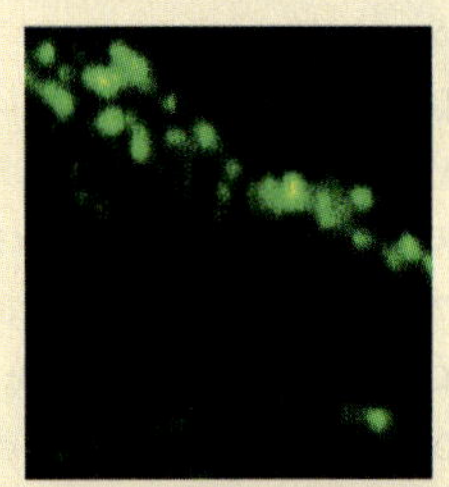

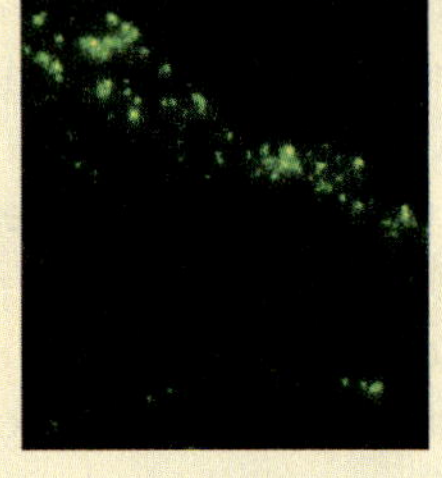

Electron Microscopy (EM)

Scanning electron microscopy (SEM). Micrographs taken with a scanning electron microscope show a 3-D image of the surface of a specimen. This SEM shows the surface of a cell from a trachea (windpipe) covered with cilia. Beating of the cilia helps move inhaled debris upward toward the throat. The SEM and TEM shown here have been artificially colorized. (Electron micrographs are black and white, but are often artificially colorized to highlight particular structures.)

Abbreviations used in this book:
LM = Light Micrograph
SEM = Scanning Electron Micrograph
TEM = Transmission Electron Micrograph

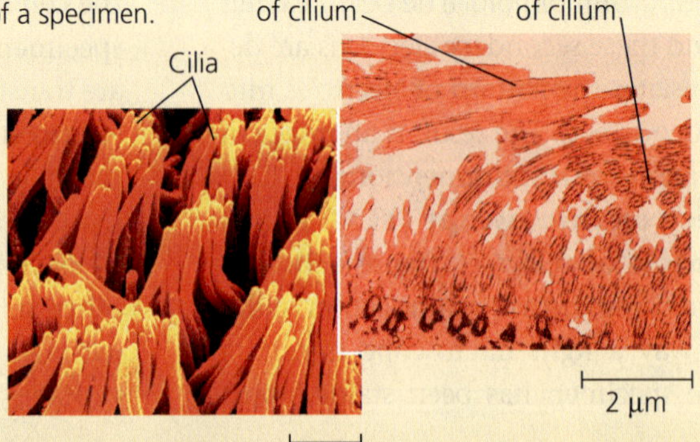

Transmission electron microscopy (TEM). A transmission electron microscope profiles a thin section of a specimen. Here we see a section through a tracheal cell, revealing its internal structure. In preparing the TEM, some cilia were cut along their lengths, creating longitudinal sections, while other cilia were cut straight across, creating cross sections.

electron microscopy is that the methods used to prepare the specimen kill the cells. For all microscopy techniques, in fact, specimen preparation can introduce artifacts, structural features seen in micrographs that do not exist in the living cell.

In the past several decades, light microscopy has been revitalized by major technical advances (see Figure 6.3). Labeling individual cellular molecules or structures with fluorescent markers has made it possible to see such structures with increasing detail. In addition, both confocal and deconvolution microscopy have sharpened images of three-dimensional tissues and cells. Finally, over the past ten years, a group of new techniques and labeling molecules have allowed researchers to "break" the resolution barrier and distinguish subcellular structures as small as 10–20 nm across. As this "super-resolution microscopy" becomes more widespread, the images we'll see of living cells may well be as awe-inspiring to us as van Leeuwenhoek's were to Robert Hooke 350 years ago.

Microscopes are the most important tools of *cytology*, the study of cell structure. To understand the function of each structure, however, required the integration of cytology and *biochemistry*, the study of the chemical processes (metabolism) of cells.

Cell Fractionation

A useful technique for studying cell structure and function is **cell fractionation**, which takes cells apart and separates major organelles and other subcellular structures from one another **(Figure 6.4)**. The instrument used is the centrifuge, which spins test tubes holding mixtures of disrupted cells at a series of increasing speeds. At each speed, the resulting force causes a fraction of the cell components to settle to the bottom of the tube, forming a pellet. At lower speeds, the pellet consists of larger components, and higher speeds yield a pellet with smaller components.

Cell fractionation enables researchers to prepare specific cell components in bulk and identify their functions, a task not usually possible with intact cells. For example, on one of the cell fractions, biochemical tests showed the presence of enzymes involved in cellular respiration, while electron microscopy revealed large numbers of the organelles called mitochondria. Together, these data helped biologists determine that mitochondria are the sites of cellular respiration. Biochemistry and cytology thus complement each other in correlating cell function with structure.

CONCEPT CHECK 6.1

1. How do stains used for light microscopy compare with those used for electron microscopy?
2. **WHAT IF?** Which type of microscope would you use to study (a) the changes in shape of a living white blood cell and (b) the details of surface texture of a hair?

For suggested answers, see Appendix A.

▼ Figure 6.4 RESEARCH METHOD

Cell Fractionation

APPLICATION Cell fractionation is used to isolate (fractionate) cell components based on size and density.

TECHNIQUE Cells are homogenized in a blender to break them up. The resulting mixture (homogenate) is centrifuged. The supernatant (liquid) is poured into another tube and centrifuged at a higher speed for a longer time. This process is repeated several times. This "differential centrifugation" results in a series of pellets, each containing different cell components.

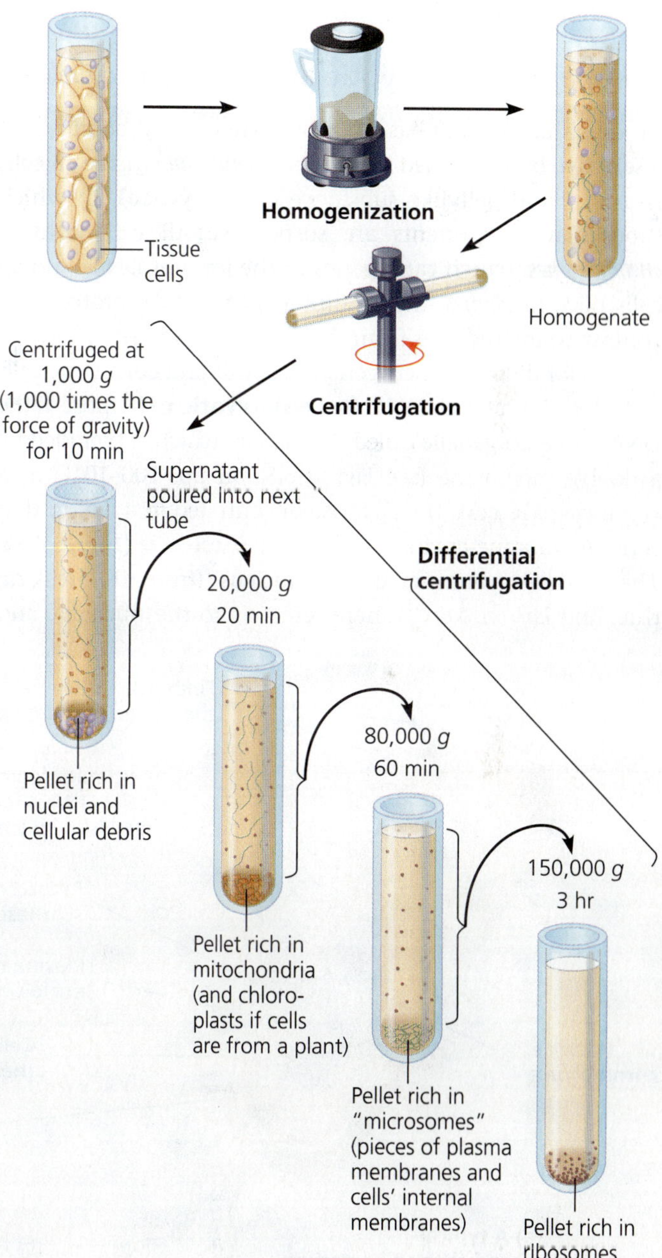

RESULTS In early experiments, researchers used microscopy to identify the organelles in each pellet and biochemical methods to determine their metabolic functions. These identifications established a baseline for this method, enabling today's researchers to know which cell fraction they should collect in order to isolate and study particular organelles.

CONCEPT 6.2

Eukaryotic cells have internal membranes that compartmentalize their functions

Cells—the basic structural and functional units of every organism—are of two distinct types: prokaryotic and eukaryotic. Organisms of the domains Bacteria and Archaea consist of prokaryotic cells. Protists, fungi, animals, and plants all consist of eukaryotic cells.

Comparing Prokaryotic and Eukaryotic Cells

All cells share certain basic features: They are all bounded by a selective barrier, called the *plasma membrane*. Inside all cells is a semifluid, jellylike substance called **cytosol**, in which subcellular components are suspended. All cells contain *chromosomes*, which carry genes in the form of DNA. And all cells have *ribosomes*, tiny complexes that make proteins according to instructions from the genes.

A major difference between prokaryotic and eukaryotic cells is the location of their DNA. In a **eukaryotic cell**, most of the DNA is in an organelle called the *nucleus*, which is bounded by a double membrane (see Figure 6.8, on pp. 100–101). In a **prokaryotic cell**, the DNA is concentrated in a region that is not membrane-enclosed, called the **nucleoid** **(Figure 6.5)**. The word *eukaryotic* means "true nucleus" (from the Greek *eu*, true, and *karyon*, kernel, here referring to the nucleus), and the word *prokaryotic* means "before nucleus" (from the Greek *pro*, before), reflecting the fact that prokaryotic cells evolved before eukaryotic cells.

The interior of either type of cell is called the **cytoplasm**; in eukaryotic cells, this term refers only to the region between the nucleus and the plasma membrane. Within the cytoplasm of a eukaryotic cell, suspended in cytosol, are a variety of organelles of specialized form and function. These membrane-bounded structures are absent in prokaryotic cells. Thus, the presence or absence of a true nucleus is just one aspect of the disparity in structural complexity between the two types of cells.

Eukaryotic cells are generally much larger than prokaryotic cells (see Figure 6.2). Size is a general feature of cell structure that relates to function. The logistics of carrying out cellular metabolism sets limits on cell size. At the lower limit, the smallest cells known are bacteria called mycoplasmas, which have diameters between 0.1 and 1.0 μm. These are perhaps the smallest packages with enough DNA to program metabolism and enough enzymes and other cellular equipment to carry out the activities necessary for a cell to sustain itself and reproduce. Typical bacteria are 1–5 μm in diameter, about ten times the size of mycoplasmas. Eukaryotic cells are typically 10–100 μm in diameter.

Metabolic requirements also impose theoretical upper limits on the size that is practical for a single cell. At the boundary of every cell, the **plasma membrane** functions as a selective barrier that allows passage of enough oxygen, nutrients, and wastes to service the entire cell **(Figure 6.6)**. For each square micrometer of membrane, only a limited amount of a particular

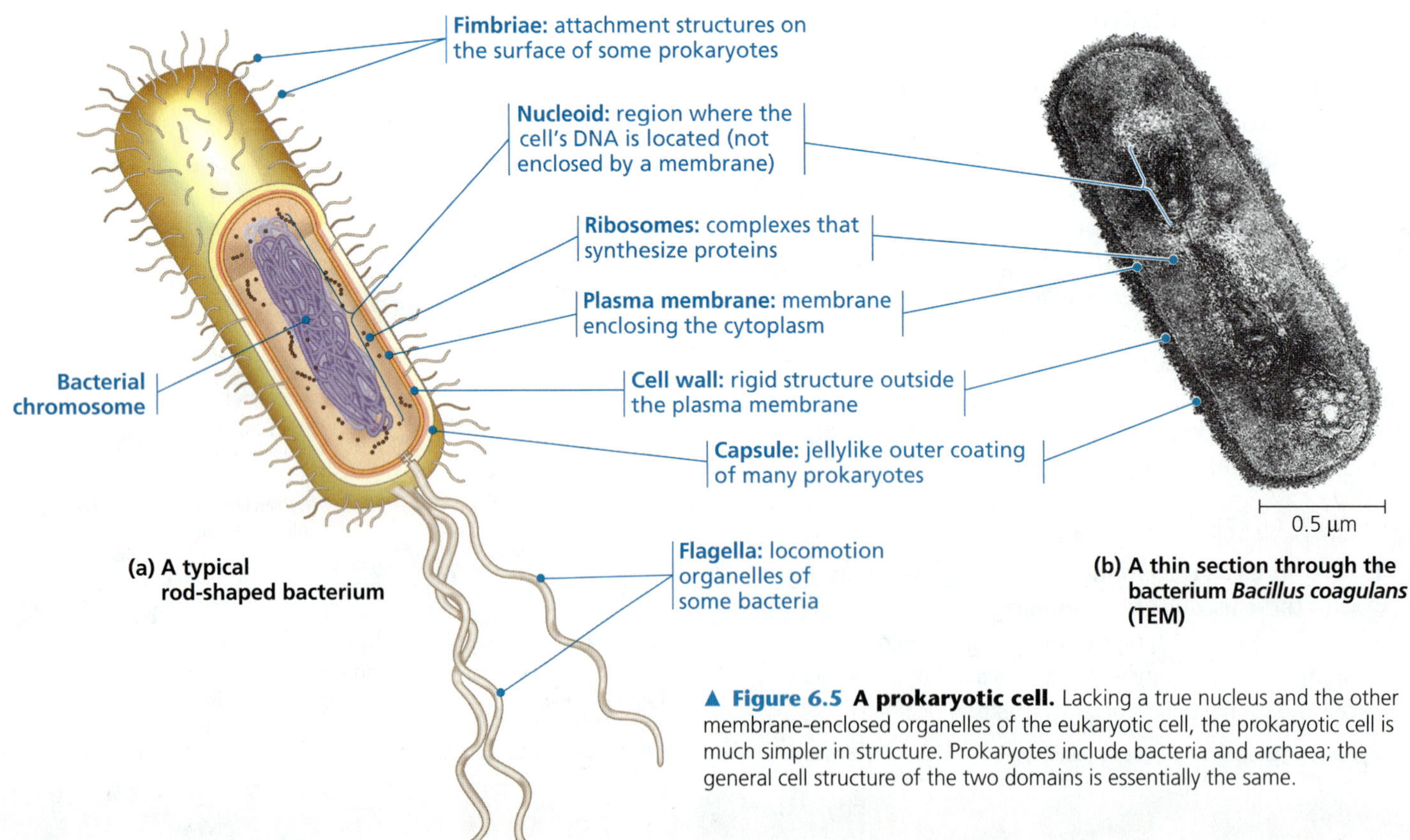

▲ **Figure 6.5 A prokaryotic cell.** Lacking a true nucleus and the other membrane-enclosed organelles of the eukaryotic cell, the prokaryotic cell is much simpler in structure. Prokaryotes include bacteria and archaea; the general cell structure of the two domains is essentially the same.

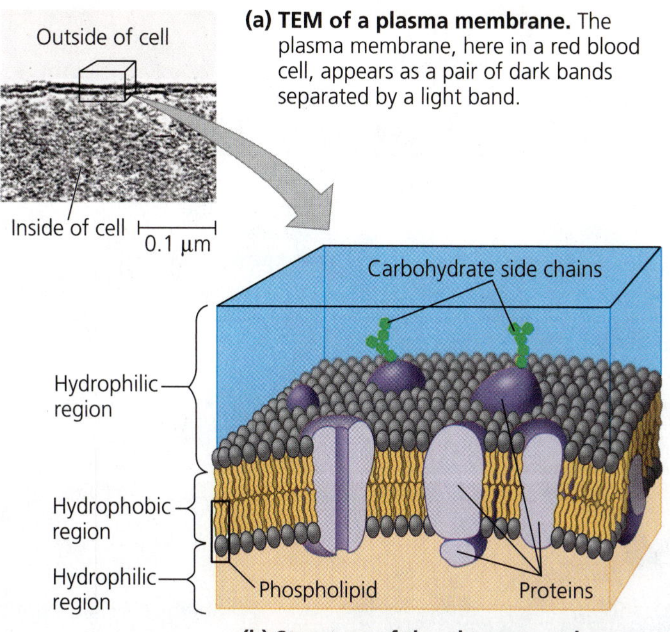

(a) TEM of a plasma membrane. The plasma membrane, here in a red blood cell, appears as a pair of dark bands separated by a light band.

(b) Structure of the plasma membrane

▲ **Figure 6.6 The plasma membrane.** The plasma membrane and the membranes of organelles consist of a double layer (bilayer) of phospholipids with various proteins attached to or embedded in it. The hydrophobic parts, including phospholipid tails and interior portions of membrane proteins, are found in the interior of the membrane. The hydrophilic parts, including phospholipid heads, exterior portions of proteins, and channels of proteins, are in contact with the aqueous solution. Carbohydrate side chains may be attached to proteins or lipids on the outer surface of the plasma membrane.

MAKE CONNECTIONS *Review Figure 5.12 (p. 76) and describe the characteristics of a phospholipid that allow it to function as the major component in the plasma membrane.*

Surface area increases while total volume remains constant

	1	5	1
Total surface area [sum of the surface areas (height × width) of all box sides × number of boxes]	6	150	750
Total volume [height × width × length × number of boxes]	1	125	125
Surface-to-volume (S-to-V) ratio [surface area ÷ volume]	6	1.2	6

▲ **Figure 6.7 Geometric relationships between surface area and volume.** In this diagram, cells are represented as boxes. Using arbitrary units of length, we can calculate the cell's surface area (in square units, or units^2), volume (in cubic units, or units^3), and ratio of surface area to volume. A high surface-to-volume ratio facilitates the exchange of materials between a cell and its environment.

substance can cross per second, so the ratio of surface area to volume is critical. As a cell (or any other object) increases in size, its volume grows proportionately more than its surface area. (Area is proportional to a linear dimension squared, whereas volume is proportional to the linear dimension cubed.) Thus, a smaller object has a greater ratio of surface area to volume **(Figure 6.7)**.

The need for a surface area sufficiently large to accommodate the volume helps explain the microscopic size of most cells and the narrow, elongated shapes of others, such as nerve cells. Larger organisms do not generally have *larger* cells than smaller organisms—they simply have *more* cells (see Figure 6.7). A sufficiently high ratio of surface area to volume is especially important in cells that exchange a lot of material with their surroundings, such as intestinal cells. Such cells may have many long, thin projections from their surface called *microvilli*, which increase surface area without an appreciable increase in volume.

The evolutionary relationships between prokaryotic and eukaryotic cells will be discussed later in this chapter, and prokaryotic cells will be described in detail in Chapter 27. Most of the discussion of cell structure that follows in this chapter applies to eukaryotic cells.

A Panoramic View of the Eukaryotic Cell

In addition to the plasma membrane at its outer surface, a eukaryotic cell has extensive and elaborately arranged internal membranes that divide the cell into compartments—the organelles mentioned earlier. The cell's compartments provide different local environments that facilitate specific metabolic functions, so incompatible processes can go on simultaneously inside a single cell. The plasma membrane and organelle membranes also participate directly in the cell's metabolism, because many enzymes are built right into the membranes.

Because membranes are so fundamental to the organization of the cell, Chapter 7 will discuss them in detail. The basic fabric of most biological membranes is a double layer of phospholipids and other lipids. Embedded in this lipid bilayer or attached to its surfaces are diverse proteins (see Figure 6.6). However, each type of membrane has a unique composition of lipids and proteins suited to that membrane's specific functions. For example, enzymes embedded in the membranes of the organelles called mitochondria function in cellular respiration.

Before continuing with this chapter, examine the eukaryotic cells in **Figure 6.8**, on the next two pages. The generalized diagrams of an animal cell and a plant cell introduce the various organelles and highlight the key differences between animal and plant cells. The micrographs at the bottom of the figure give you a glimpse of cells from different types of eukaryotic organisms.

▼ Figure 6.8

Exploring Eukaryotic Cells

Animal Cell (cutaway view of generalized cell)

ENDOPLASMIC RETICULUM (ER): network of membranous sacs and tubes; active in membrane synthesis and other synthetic and metabolic processes; has rough (ribosome-studded) and smooth regions

Rough ER

Smooth ER

Nuclear envelope: double membrane enclosing the nucleus; perforated by pores; continuous with ER

Nucleolus: nonmembranous structure involved in production of ribosomes; a nucleus has one or more nucleoli

Chromatin: material consisting of DNA and proteins; visible in a dividing cell as individual condensed chromosomes

NUCLEUS

Flagellum: motility structure present in some animal cells, composed of a cluster of microtubules within an extension of the plasma membrane

Centrosome: region where the cell's microtubules are initiated; contains a pair of centrioles

CYTOSKELETON: reinforces cell's shape; functions in cell movement; components are made of protein. Includes:

Microfilaments

Intermediate filaments

Microtubules

Plasma membrane: membrane enclosing the cell

Ribosomes (small brown dots)**:** complexes that make proteins; free in cytosol or bound to rough ER or nuclear envelope

Microvilli: projections that increase the cell's surface area

Golgi apparatus: organelle active in synthesis, modification, sorting, and secretion of cell products

Peroxisome: organelle with various specialized metabolic functions; produces hydrogen peroxide as a by-product, then converts it to water

Mitochondrion: organelle where cellular respiration occurs and most ATP is generated

Lysosome: digestive organelle where macromolecules are hydrolyzed

In animal cells but not plant cells:
Lysosomes
Centrosomes, with centrioles
Flagella (but present in some plant sperm)

Animal Cells

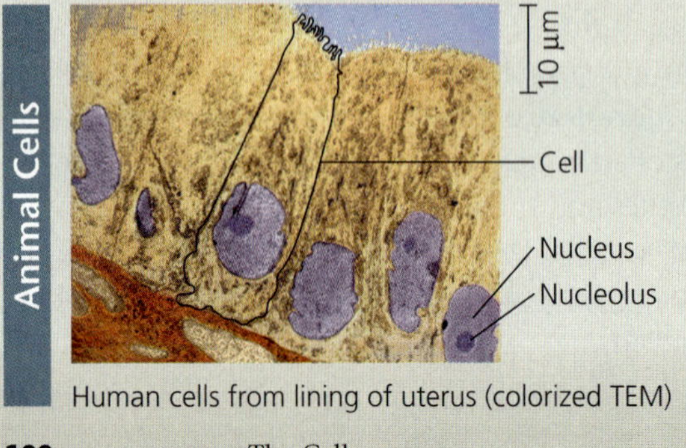

Human cells from lining of uterus (colorized TEM)

Fungal Cells

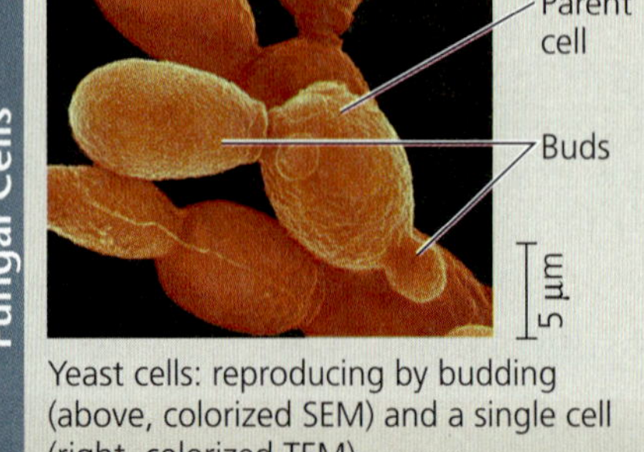

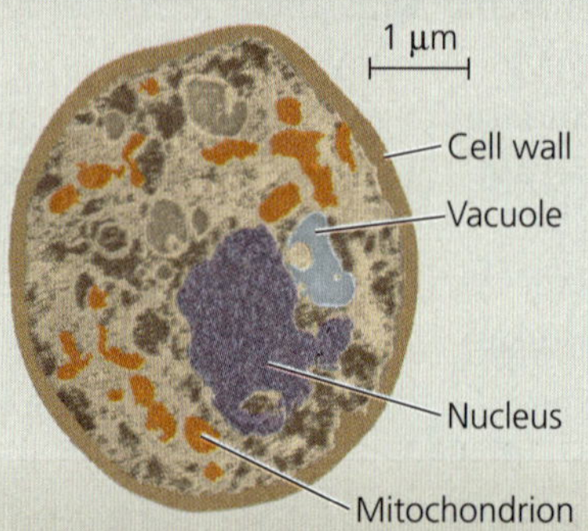

Yeast cells: reproducing by budding (above, colorized SEM) and a single cell (right, colorized TEM)

Plant Cell (cutaway view of generalized cell)

ANIMATION **BioFlix** Visit the Study Area at **www.masteringbiology.com** for the BioFlix® 3-D Animations Tour of an Animal Cell and Tour of a Plant Cell.

NUCLEUS
- **Nuclear envelope**
- **Nucleolus**
- **Chromatin**

Rough endoplasmic reticulum

Smooth endoplasmic reticulum

Ribosomes (small brown dots)

Central vacuole: prominent organelle in older plant cells; functions include storage, breakdown of waste products, hydrolysis of macromolecules; enlargement of vacuole is a major mechanism of plant growth

Golgi apparatus

CYTOSKELETON
- **Microfilaments**
- **Intermediate filaments**
- **Microtubules**

Mitochondrion

Peroxisome

Plasma membrane

Cell wall: outer layer that maintains cell's shape and protects cell from mechanical damage; made of cellulose, other polysaccharides, and protein

Wall of adjacent cell

Chloroplast: photosynthetic organelle; converts energy of sunlight to chemical energy stored in sugar molecules

Plasmodesmata: cytoplasmic channels through cell walls that connect the cytoplasms of adjacent cells

In plant cells but not animal cells:
Chloroplasts
Central vacuole
Cell wall
Plasmodesmata

Plant Cells

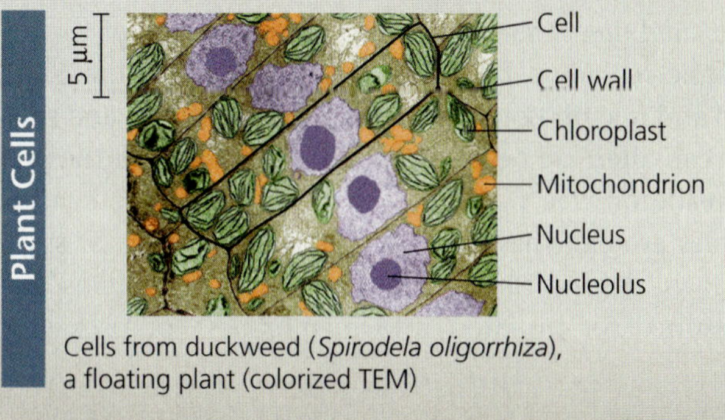

Cells from duckweed (*Spirodela oligorrhiza*), a floating plant (colorized TEM)

Protistan Cells

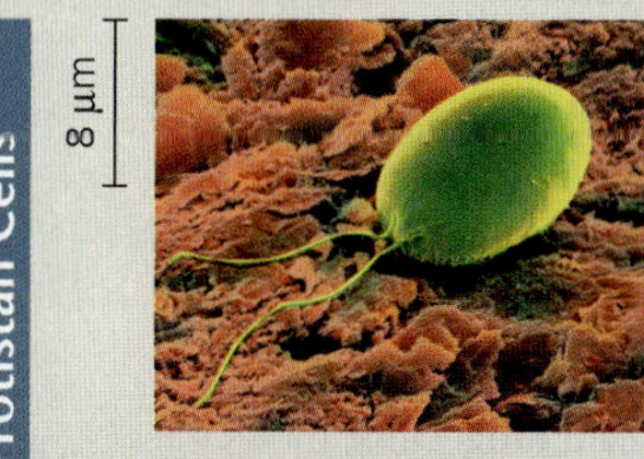

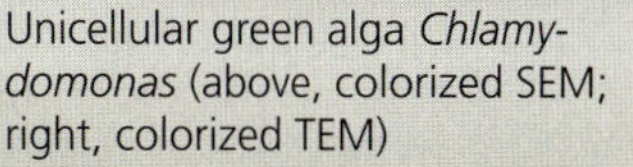

Unicellular green alga *Chlamydomonas* (above, colorized SEM; right, colorized TEM)

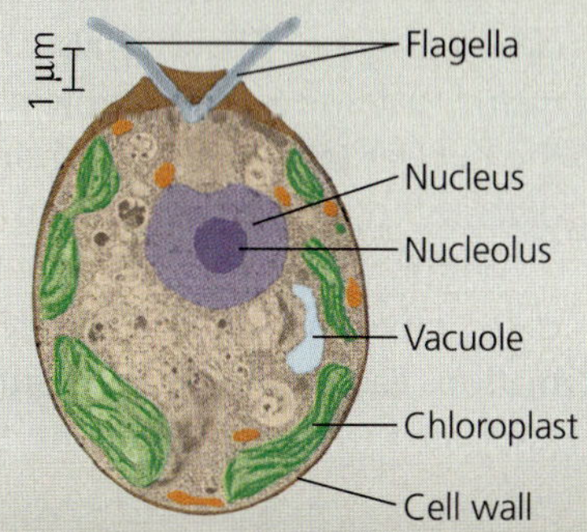

CONCEPT CHECK 6.2

1. After carefully reviewing Figure 6.8, briefly describe the structure and function of the nucleus, the mitochondrion, the chloroplast, and the endoplasmic reticulum.
2. **WHAT IF?** Imagine an elongated cell (such as a nerve cell) that measures $125 \times 1 \times 1$ arbitrary units. Predict how its surface-to-volume ratio would compare with those in Figure 6.7. Then calculate the ratio and check your prediction.

For suggested answers, see Appendix A.

CONCEPT 6.3

The eukaryotic cell's genetic instructions are housed in the nucleus and carried out by the ribosomes

On the first stop of our detailed tour of the cell, let's look at two cellular components involved in the genetic control of the cell: the nucleus, which houses most of the cell's DNA, and the ribosomes, which use information from the DNA to make proteins.

The Nucleus: Information Central

The **nucleus** contains most of the genes in the eukaryotic cell. (Some genes are located in mitochondria and chloroplasts.) It is generally the most conspicuous organelle in a eukaryotic cell, averaging about 5 μm in diameter. The **nuclear envelope** encloses the nucleus **(Figure 6.9)**, separating its contents from the cytoplasm.

The nuclear envelope is a *double* membrane. The two membranes, each a lipid bilayer with associated proteins, are separated by a space of 20–40 nm. The envelope is perforated by pore structures that are about 100 nm in diameter. At the lip of each pore, the inner and outer membranes of the nuclear envelope are continuous. An intricate protein structure called a *pore complex* lines each pore and plays an important role in the cell by regulating the entry and exit of proteins and RNAs, as well as large complexes of macromolecules. Except at the pores, the nuclear side of the envelope is lined by the **nuclear lamina**, a netlike array of protein filaments that maintains the shape of the nucleus by mechanically supporting the nuclear envelope. There is also much evidence for a *nuclear matrix*, a framework of protein fibers extending throughout the nuclear interior. The nuclear lamina and matrix may help organize the genetic material so it functions efficiently.

Within the nucleus, the DNA is organized into discrete units called **chromosomes**, structures that carry the genetic information. Each chromosome contains one long DNA molecule associated with many proteins. Some of the proteins help coil the DNA molecule of each chromosome, reducing its length and allowing it to fit into the nucleus. The complex of DNA and proteins making up chromosomes is called **chromatin**. When a cell is not dividing, stained chromatin appears as a diffuse mass in micrographs, and the chromosomes cannot be distinguished from one another, even though discrete chromosomes are present. As a cell prepares to divide, however, the chromosomes coil (condense) further, becoming thick enough to be distinguished as separate structures. Each eukaryotic species has a characteristic number of chromosomes. For example, a typical human cell has 46 chromosomes in its nucleus; the exceptions are the sex cells (eggs and sperm), which have only 23 chromosomes in humans. A fruit fly cell has 8 chromosomes in most cells and 4 in the sex cells.

A prominent structure within the nondividing nucleus is the **nucleolus** (plural, *nucleoli*), which appears through the electron microscope as a mass of densely stained granules and fibers adjoining part of the chromatin. Here a type of RNA called *ribosomal RNA* (rRNA) is synthesized from instructions in the DNA. Also in the nucleolus, proteins imported from the cytoplasm are assembled with rRNA into large and small subunits of ribosomes. These subunits then exit the nucleus through the nuclear pores to the cytoplasm, where a large and a small subunit can assemble into a ribosome. Sometimes there are two or more nucleoli; the number depends on the species and the stage in the cell's reproductive cycle.

As we saw in Figure 5.25, the nucleus directs protein synthesis by synthesizing messenger RNA (mRNA) according to instructions provided by the DNA. The mRNA is then transported to the cytoplasm via the nuclear pores. Once an mRNA molecule reaches the cytoplasm, ribosomes translate the mRNA's genetic message into the primary structure of a specific polypeptide. This process of transcribing and translating genetic information is described in detail in Chapter 17.

Ribosomes: Protein Factories

Ribosomes, which are complexes made of ribosomal RNA and protein, are the cellular components that carry out protein synthesis **(Figure 6.10)**. Cells that have high rates of protein synthesis have particularly large numbers of ribosomes. For example, a human pancreas cell has a few million ribosomes. Not surprisingly, cells active in protein synthesis also have prominent nucleoli.

Ribosomes build proteins in two cytoplasmic locales. At any given time, *free ribosomes* are suspended in the cytosol, while *bound ribosomes* are attached to the outside of the endoplasmic reticulum or nuclear envelope (see Figure 6.10). Bound and free ribosomes are structurally identical, and ribosomes can alternate between the two roles. Most of the proteins made on free ribosomes function within the cytosol; examples are enzymes that catalyze the first steps of sugar breakdown. Bound ribosomes generally make proteins that are destined for insertion into membranes, for packaging

1 µm

Nucleus

Nucleus

Nucleolus

Chromatin

Nuclear envelope:

Inner membrane

Outer membrane

Nuclear pore

Rough ER

▲ **Surface of nuclear envelope.** TEM of a specimen prepared by a technique known as freeze-fracture.

Pore complex

Ribosome

◀ **Close-up of nuclear envelope**

0.25 µm

▲ **Pore complexes (TEM).** Each pore is ringed by protein particles.

1 µm

◀ **Nuclear lamina (TEM).** The netlike lamina lines the inner surface of the nuclear envelope.

▲ **Chromatin.** Part of a chromosome from a non-dividing cell shows two states of coiling of the DNA (blue) and protein (purple) complex. The thicker form is sometimes also organized into long loops.

▲ **Figure 6.9 The nucleus and its envelope.** Within the nucleus are the chromosomes, which appear as a mass of chromatin (DNA and associated proteins), and one or more nucleoli (singular, *nucleolus*), which function in ribosome synthesis. The nuclear envelope, which consists of two membranes separated by a narrow space, is perforated with pores and lined by the nuclear lamina.

MAKE CONNECTIONS *Since the chromosomes contain the genetic material and reside in the nucleus, how does the rest of the cell get access to the information they carry? See Figure 5.25, page 86.*

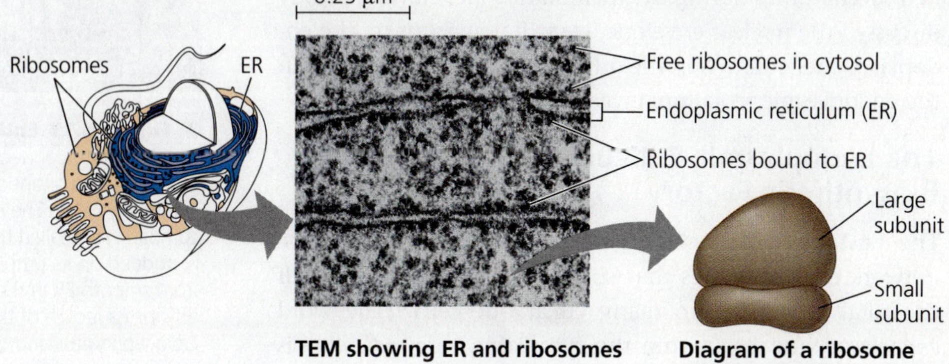

▶ **Figure 6.10 Ribosomes.** This electron micrograph of part of a pancreas cell shows many ribosomes, both free (in the cytosol) and bound (to the endoplasmic reticulum). The simplified diagram of a ribosome shows its two subunits.

DRAW IT *After you have read the section on ribosomes, circle a ribosome in the micrograph that might be making a protein that will be secreted.*

within certain organelles such as lysosomes (see Figure 6.8), or for export from the cell (secretion). Cells that specialize in protein secretion—for instance, the cells of the pancreas that secrete digestive enzymes—frequently have a high proportion of bound ribosomes. You will learn more about ribosome structure and function in Chapter 17.

CONCEPT CHECK 6.3

1. What role do ribosomes play in carrying out genetic instructions?
2. Describe the molecular composition of nucleoli and explain their function.
3. **WHAT IF?** As a cell begins the process of dividing, its chromatin becomes more and more condensed. Does the number of chromosomes change during this process? Explain.

For suggested answers, see Appendix A.

CONCEPT 6.4

The endomembrane system regulates protein traffic and performs metabolic functions in the cell

Many of the different membranes of the eukaryotic cell are part of the **endomembrane system**, which includes the nuclear envelope, the endoplasmic reticulum, the Golgi apparatus, lysosomes, various kinds of vesicles and vacuoles, and the plasma membrane. This system carries out a variety of tasks in the cell, including synthesis of proteins, transport of proteins into membranes and organelles or out of the cell, metabolism and movement of lipids, and detoxification of poisons. The membranes of this system are related either through direct physical continuity or by the transfer of membrane segments as tiny **vesicles** (sacs made of membrane). Despite these relationships, the various membranes are not identical in structure and function. Moreover, the thickness, molecular composition, and types of chemical reactions carried out in a given membrane are not fixed, but may be modified several times during the membrane's life. Having already discussed the nuclear envelope, we will now focus on the endoplasmic reticulum and the other endomembranes to which the endoplasmic reticulum gives rise.

The Endoplasmic Reticulum: Biosynthetic Factory

The **endoplasmic reticulum (ER)** is such an extensive network of membranes that it accounts for more than half the total membrane in many eukaryotic cells. (The word *endoplasmic* means "within the cytoplasm," and *reticulum* is Latin for "little net.") The ER consists of a network of membranous tubules and sacs called cisternae (from the Latin *cisterna*, a reservoir for a liquid). The ER membrane separates the internal compartment of the ER, called the ER lumen (cavity) or cisternal space, from the cytosol. And because the ER membrane is continuous with the nuclear envelope, the space between the two membranes of the envelope is continuous with the lumen of the ER **(Figure 6.11)**.

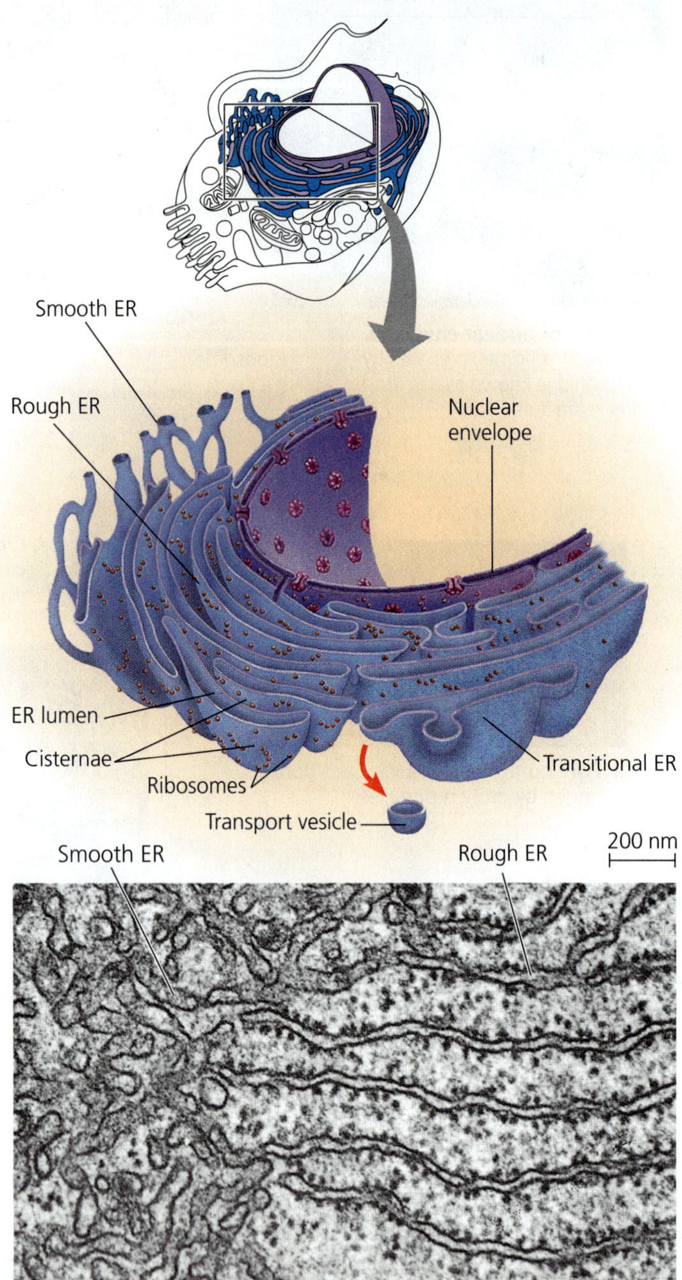

▲ **Figure 6.11 Endoplasmic reticulum (ER).** A membranous system of interconnected tubules and flattened sacs called cisternae, the ER is also continuous with the nuclear envelope. (The drawing is a cutaway view.) The membrane of the ER encloses a continuous compartment called the ER lumen (or cisternal space). Rough ER, which is studded on its outer surface with ribosomes, can be distinguished from smooth ER in the electron micrograph (TEM). Transport vesicles bud off from a region of the rough ER called transitional ER and travel to the Golgi apparatus and other destinations.

There are two distinct, though connected, regions of the ER that differ in structure and function: smooth ER and rough ER. **Smooth ER** is so named because its outer surface lacks ribosomes. **Rough ER** is studded with ribosomes on the outer surface of the membrane and thus appears rough through the electron microscope. As already mentioned, ribosomes are also attached to the cytoplasmic side of the nuclear envelope's outer membrane, which is continuous with rough ER.

Functions of Smooth ER

The smooth ER functions in diverse metabolic processes, which vary with cell type. These processes include synthesis of lipids, metabolism of carbohydrates, detoxification of drugs and poisons, and storage of calcium ions.

Enzymes of the smooth ER are important in the synthesis of lipids, including oils, phospholipids, and steroids. Among the steroids produced by the smooth ER in animal cells are the sex hormones of vertebrates and the various steroid hormones secreted by the adrenal glands. The cells that synthesize and secrete these hormones—in the testes and ovaries, for example—are rich in smooth ER, a structural feature that fits the function of these cells.

Other enzymes of the smooth ER help detoxify drugs and poisons, especially in liver cells. Detoxification usually involves adding hydroxyl groups to drug molecules, making them more soluble and easier to flush from the body. The sedative phenobarbital and other barbiturates are examples of drugs metabolized in this manner by smooth ER in liver cells. In fact, barbiturates, alcohol, and many other drugs induce the proliferation of smooth ER and its associated detoxification enzymes, thus increasing the rate of detoxification. This, in turn, increases tolerance to the drugs, meaning that higher doses are required to achieve a particular effect, such as sedation. Also, because some of the detoxification enzymes have relatively broad action, the proliferation of smooth ER in response to one drug can increase tolerance to other drugs as well. Barbiturate abuse, for example, can decrease the effectiveness of certain antibiotics and other useful drugs.

The smooth ER also stores calcium ions. In muscle cells, for example, the smooth ER membrane pumps calcium ions from the cytosol into the ER lumen. When a muscle cell is stimulated by a nerve impulse, calcium ions rush back across the ER membrane into the cytosol and trigger contraction of the muscle cell. In other cell types, calcium ion release from the smooth ER triggers different responses, such as secretion of vesicles carrying newly synthesized proteins.

Functions of Rough ER

Many types of cells secrete proteins produced by ribosomes attached to rough ER. For example, certain pancreatic cells synthesize the protein insulin in the ER and secrete this hormone into the bloodstream. As a polypeptide chain grows from a bound ribosome, the chain is threaded into the ER lumen through a pore formed by a protein complex in the ER membrane. As the new polypeptide enters the ER lumen, it folds into its native shape. Most secretory proteins are **glycoproteins**, proteins that have carbohydrates covalently bonded to them. The carbohydrates are attached to the proteins in the ER by enzymes built into the ER membrane.

After secretory proteins are formed, the ER membrane keeps them separate from proteins that are produced by free ribosomes and that will remain in the cytosol. Secretory proteins depart from the ER wrapped in the membranes of vesicles that bud like bubbles from a specialized region called transitional ER (see Figure 6.11). Vesicles in transit from one part of the cell to another are called **transport vesicles**; we will discuss their fate shortly.

In addition to making secretory proteins, rough ER is a membrane factory for the cell; it grows in place by adding membrane proteins and phospholipids to its own membrane. As polypeptides destined to be membrane proteins grow from the ribosomes, they are inserted into the ER membrane itself and anchored there by their hydrophobic portions. Like the smooth ER, the rough ER also makes membrane phospholipids; enzymes built into the ER membrane assemble phospholipids from precursors in the cytosol. The ER membrane expands and portions of it are transferred in the form of transport vesicles to other components of the endomembrane system.

The Golgi Apparatus: Shipping and Receiving Center

After leaving the ER, many transport vesicles travel to the **Golgi apparatus**. We can think of the Golgi as a warehouse for receiving, sorting, shipping, and even some manufacturing. Here, products of the ER, such as proteins, are modified and stored and then sent to other destinations. Not surprisingly, the Golgi apparatus is especially extensive in cells specialized for secretion.

The Golgi apparatus consists of flattened membranous sacs—cisternae—looking like a stack of pita bread (**Figure 6.12**, on the next page). A cell may have many, even hundreds, of these stacks. The membrane of each cisterna in a stack separates its internal space from the cytosol. Vesicles concentrated in the vicinity of the Golgi apparatus are engaged in the transfer of material between parts of the Golgi and other structures.

A Golgi stack has a distinct structural directionality, with the membranes of cisternae on opposite sides of the stack differing in thickness and molecular composition. The two sides of a Golgi stack are referred to as the *cis* face and the *trans* face; these act, respectively, as the receiving and shipping departments of the Golgi apparatus. The *cis* face is usually located near the ER. Transport vesicles move material from the ER to the Golgi apparatus. A vesicle that buds from the ER can add its membrane and the contents of its lumen to the *cis* face by fusing with a Golgi membrane. The *trans* face gives rise to vesicles that pinch off and travel to other sites.

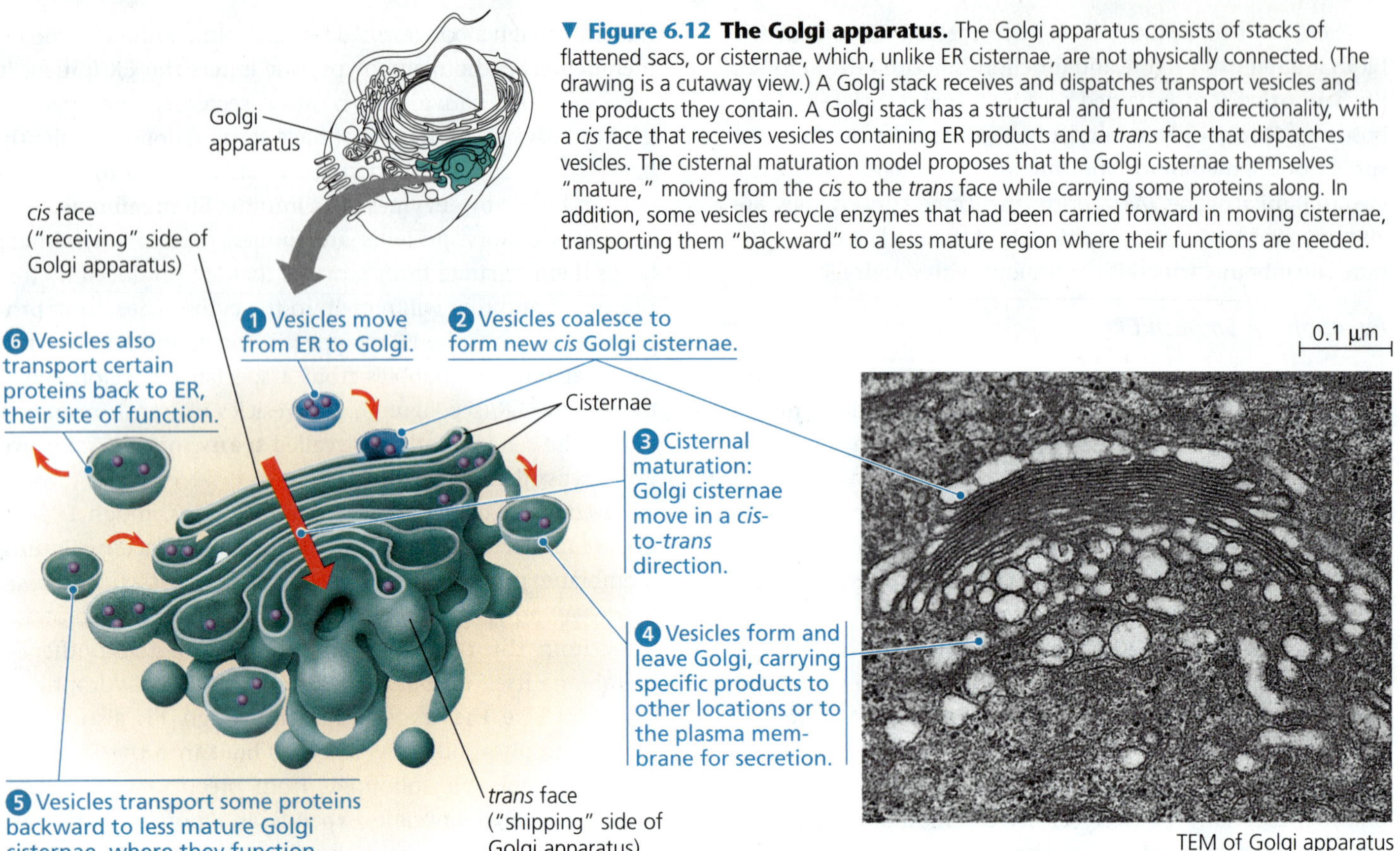

▼ **Figure 6.12 The Golgi apparatus.** The Golgi apparatus consists of stacks of flattened sacs, or cisternae, which, unlike ER cisternae, are not physically connected. (The drawing is a cutaway view.) A Golgi stack receives and dispatches transport vesicles and the products they contain. A Golgi stack has a structural and functional directionality, with a *cis* face that receives vesicles containing ER products and a *trans* face that dispatches vesicles. The cisternal maturation model proposes that the Golgi cisternae themselves "mature," moving from the *cis* to the *trans* face while carrying some proteins along. In addition, some vesicles recycle enzymes that had been carried forward in moving cisternae, transporting them "backward" to a less mature region where their functions are needed.

Products of the endoplasmic reticulum are usually modified during their transit from the *cis* region to the *trans* region of the Golgi apparatus. For example, glycoproteins formed in the ER have their carbohydrates modified, first in the ER itself, then as they pass through the Golgi. The Golgi removes some sugar monomers and substitutes others, producing a large variety of carbohydrates. Membrane phospholipids may also be altered in the Golgi.

In addition to its finishing work, the Golgi apparatus also manufactures some macromolecules. Many polysaccharides secreted by cells are Golgi products. For example, pectins and certain other noncellulose polysaccharides are made in the Golgi of plant cells and then incorporated along with cellulose into their cell walls. Like secretory proteins, nonprotein Golgi products that will be secreted depart from the *trans* face of the Golgi inside transport vesicles that eventually fuse with the plasma membrane.

The Golgi manufactures and refines its products in stages, with different cisternae containing unique teams of enzymes. Until recently, biologists viewed the Golgi as a static structure, with products in various stages of processing transferred from one cisterna to the next by vesicles. While this may occur, recent research has given rise to a new model of the Golgi as a more dynamic structure. According to the *cisternal maturation model*, the cisternae of the Golgi actually progress forward from the *cis* to the *trans* face, carrying and modifying their cargo as they move. Figure 6.12 shows the details of this model.

Before a Golgi stack dispatches its products by budding vesicles from the *trans* face, it sorts these products and targets them for various parts of the cell. Molecular identification tags, such as phosphate groups added to the Golgi products, aid in sorting by acting like ZIP codes on mailing labels. Finally, transport vesicles budded from the Golgi may have external molecules on their membranes that recognize "docking sites" on the surface of specific organelles or on the plasma membrane, thus targeting the vesicles appropriately.

Lysosomes: Digestive Compartments

A **lysosome** is a membranous sac of hydrolytic enzymes that an animal cell uses to digest (hydrolyze) macromolecules. Lysosomal enzymes work best in the acidic environment found in lysosomes. If a lysosome breaks open or leaks its contents, the released enzymes are not very active because the cytosol has a neutral pH. However, excessive leakage from a large number of lysosomes can destroy a cell by self-digestion.

Hydrolytic enzymes and lysosomal membrane are made by rough ER and then transferred to the Golgi apparatus for further processing. At least some lysosomes probably arise by budding from the *trans* face of the Golgi apparatus (see Figure 6.12). How are the proteins of the inner surface of the lysosomal membrane and the digestive enzymes themselves spared from destruction? Apparently, the three-dimensional shapes of these proteins protect vulnerable bonds from enzymatic attack.

Lysosomes carry out intracellular digestion in a variety of circumstances. Amoebas and many other protists eat by engulfing smaller organisms or food particles, a process called **phagocytosis** (from the Greek *phagein*, to eat, and *kytos*, vessel, referring here to the cell). The *food vacuole* formed in this way then fuses with a lysosome, whose enzymes digest the food (**Figure 6.13a**, bottom). Digestion products, including simple sugars, amino acids, and other monomers, pass into the cytosol and become nutrients for the cell. Some human cells also carry out phagocytosis. Among them are macrophages, a type of white blood cell that helps defend the body by engulfing and destroying bacteria and other invaders (see Figure 6.13a, top, and Figure 6.33).

Lysosomes also use their hydrolytic enzymes to recycle the cell's own organic material, a process called *autophagy*. During autophagy, a damaged organelle or small amount of cytosol becomes surrounded by a double membrane (of unknown origin), and a lysosome fuses with the outer membrane of this vesicle **(Figure 6.13b)**. The lysosomal enzymes dismantle the enclosed material, and the organic monomers are returned to the cytosol for reuse. With the help of lysosomes, the cell continually renews itself. A human liver cell, for example, recycles half of its macromolecules each week.

The cells of people with inherited lysosomal storage diseases lack a functioning hydrolytic enzyme normally present in lysosomes. The lysosomes become engorged with indigestible substrates, which begin to interfere with other cellular activities. In Tay-Sachs disease, for example, a lipid-digesting enzyme is missing or inactive, and the brain becomes impaired by an accumulation of lipids in the cells. Fortunately, lysosomal storage diseases are rare in the general population.

Vacuoles: Diverse Maintenance Compartments

Vacuoles are large vesicles derived from the endoplasmic reticulum and Golgi apparatus. Thus, vacuoles are an integral part of a cell's endomembrane system. Like all cellular membranes, the vacuolar membrane is selective in transporting solutes; as a result, the solution inside a vacuole differs in composition from the cytosol.

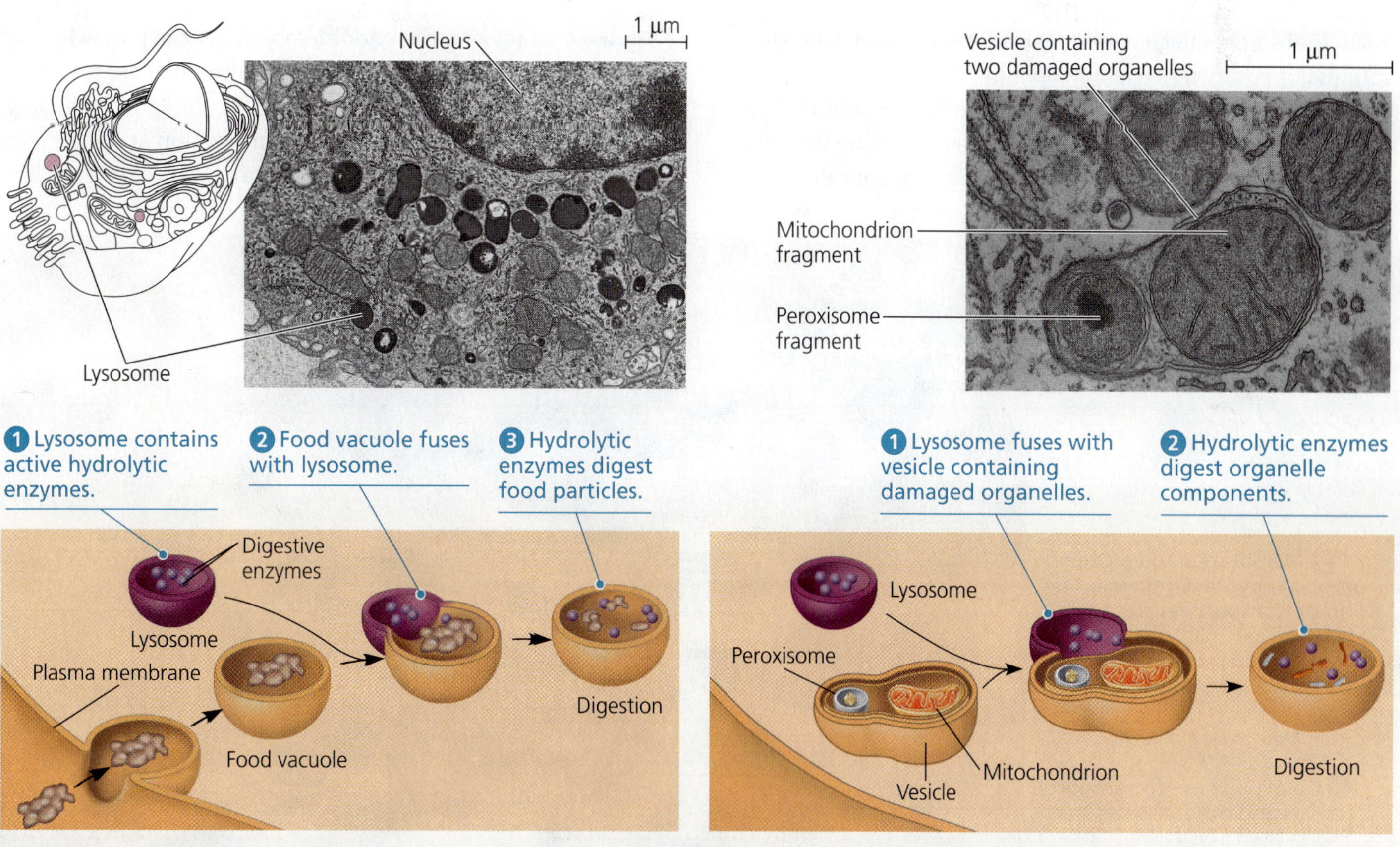

▲ **Figure 6.13 Lysosomes.** Lysosomes digest (hydrolyze) materials taken into the cell and recycle intracellular materials. **(a)** *Top*: In this macrophage (a type of white blood cell) from a rat, the lysosomes are very dark because of a stain that reacts with one of the products of digestion within the lysosome (TEM). Macrophages ingest bacteria and viruses and destroy them using lysosomes. *Bottom*: This diagram shows one lysosome fusing with a food vacuole during the process of phagocytosis by a protist. **(b)** *Top*: In the cytoplasm of this rat liver cell is a vesicle containing two disabled organelles; the vesicle will fuse with a lysosome in the process of autophagy (TEM). *Bottom*: This diagram shows fusion of such a vesicle with a lysosome. This type of vesicle has a double membrane of unknown origin. The outer membrane fuses with the lysosome, and the inner membrane is degraded along with the damaged organelles.

Vacuoles perform a variety of functions in different kinds of cells. **Food vacuoles**, formed by phagocytosis, have already been mentioned (see Figure 6.13a). Many freshwater protists have **contractile vacuoles** that pump excess water out of the cell, thereby maintaining a suitable concentration of ions and molecules inside the cell (see Figure 7.16). In plants and fungi, certain vacuoles carry out enzymatic hydrolysis, a function shared by lysosomes in animal cells. (In fact, some biologists consider these hydrolytic vacuoles to be a type of lysosome.) In plants, smaller vacuoles can hold reserves of important organic compounds, such as the proteins stockpiled in the storage cells in seeds. Vacuoles may also help protect the plant against herbivores by storing compounds that are poisonous or unpalatable to animals. Some plant vacuoles contain pigments, such as the red and blue pigments of petals that help attract pollinating insects to flowers.

Mature plant cells generally contain a large **central vacuole** **(Figure 6.14)**, which develops by the coalescence of smaller vacuoles. The solution inside the central vacuole, called cell sap, is the plant cell's main repository of inorganic ions, including potassium and chloride. The central vacuole plays a major role in the growth of plant cells, which enlarge as the vacuole absorbs water, enabling the cell to become larger with a minimal investment in new cytoplasm. The cytosol often occupies only a thin layer between the central vacuole and the plasma membrane, so the ratio of plasma membrane surface to cytosolic volume is sufficient, even for a large plant cell.

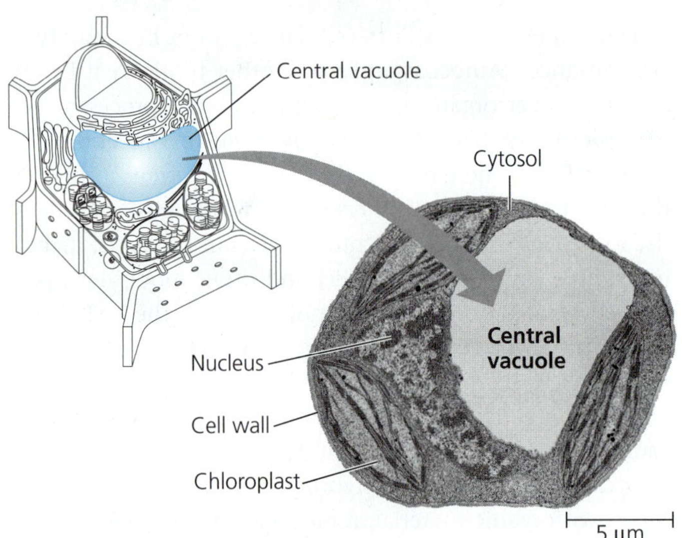

▲ **Figure 6.14 The plant cell vacuole.** The central vacuole is usually the largest compartment in a plant cell; the rest of the cytoplasm is often confined to a narrow zone between the vacuolar membrane and the plasma membrane (TEM).

The Endomembrane System: *A Review*

Figure 6.15 reviews the endomembrane system, showing the flow of membrane lipids and proteins through the various organelles. As the membrane moves from the ER to the Golgi and then elsewhere, its molecular composition and metabolic functions are modified, along with those of its contents. The

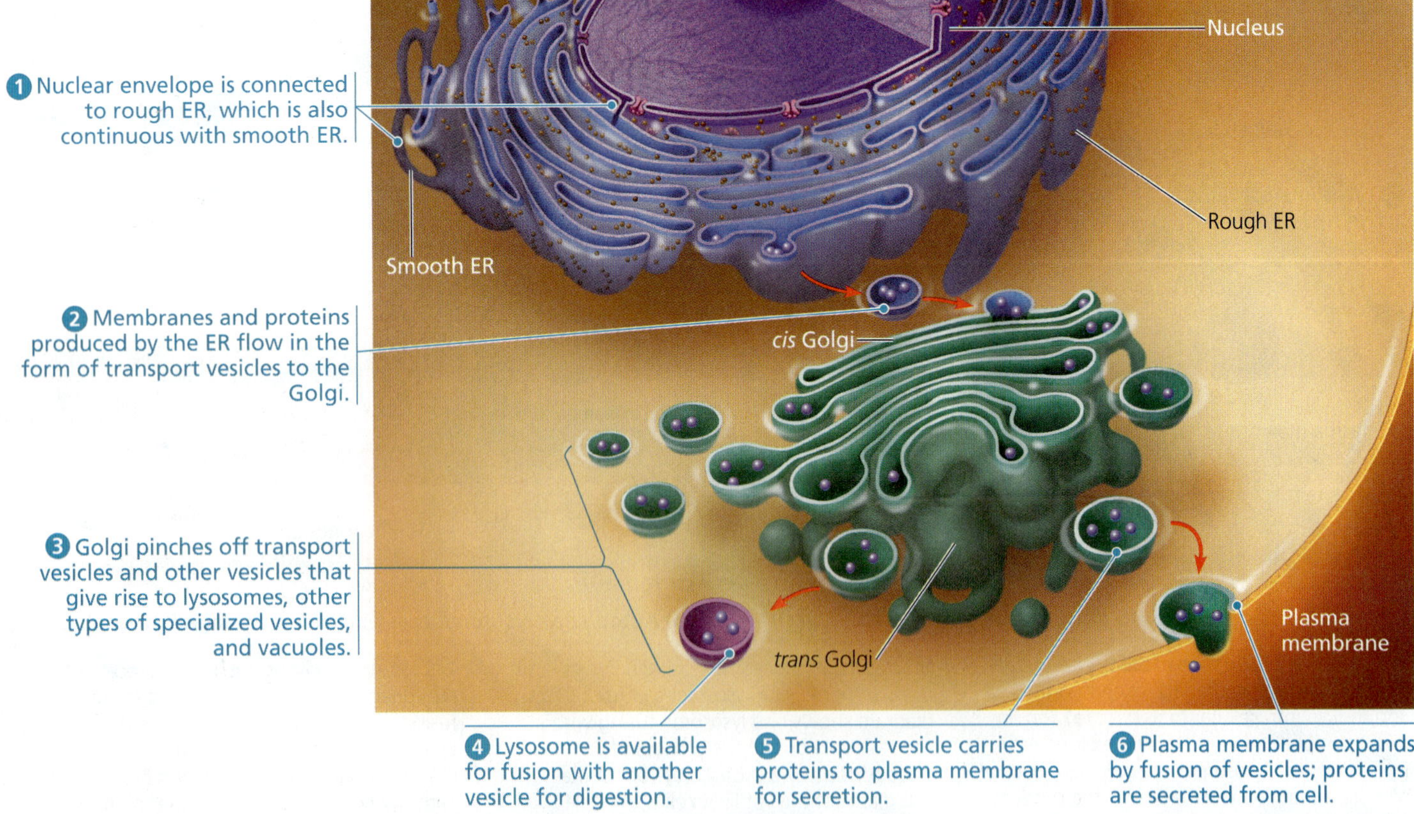

▲ **Figure 6.15 Review: relationships among organelles of the endomembrane system.** The red arrows show some of the migration pathways for membranes and the materials they enclose.

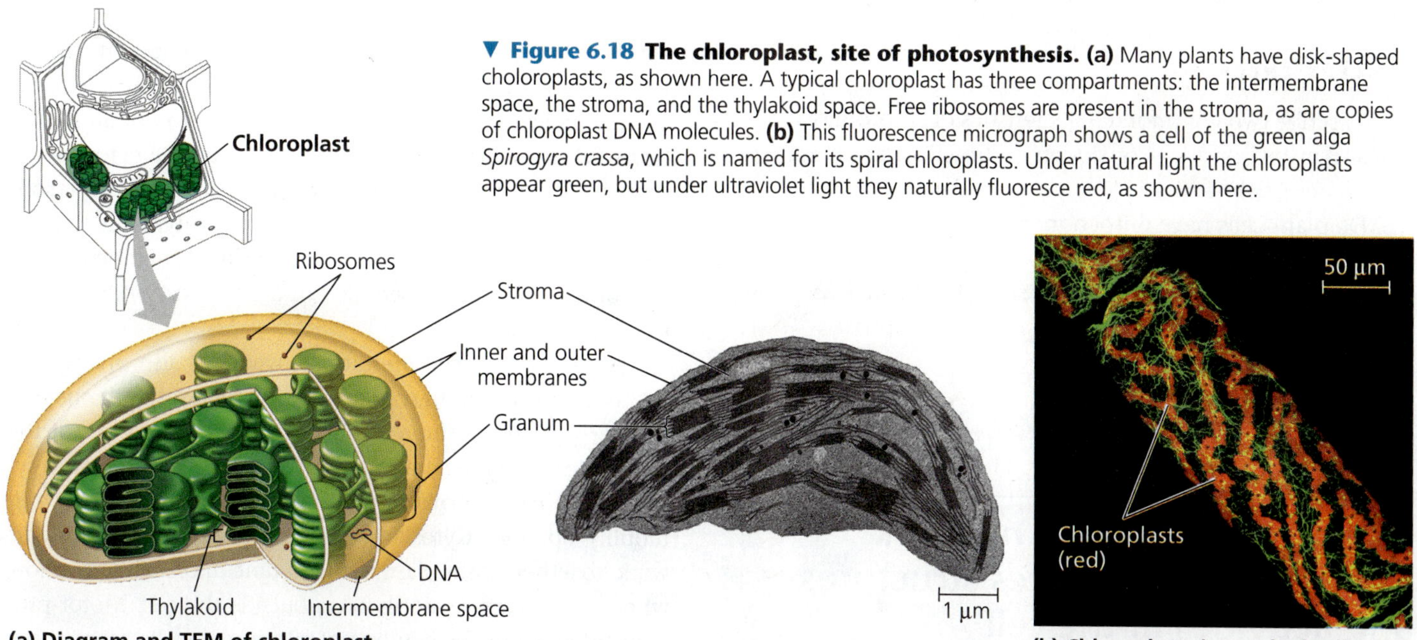

▼ **Figure 6.18 The chloroplast, site of photosynthesis. (a)** Many plants have disk-shaped choloroplasts, as shown here. A typical chloroplast has three compartments: the intermembrane space, the stroma, and the thylakoid space. Free ribosomes are present in the stroma, as are copies of chloroplast DNA molecules. **(b)** This fluorescence micrograph shows a cell of the green alga *Spirogyra crassa*, which is named for its spiral chloroplasts. Under natural light the chloroplasts appear green, but under ultraviolet light they naturally fluoresce red, as shown here.

(a) Diagram and TEM of chloroplast

(b) Chloroplasts in an algal cell

to their dynamic behavior in the living cell. Their shape is changeable, and they grow and occasionally pinch in two, reproducing themselves. They are mobile and, with mitochondria and other organelles, move around the cell along tracks of the cytoskeleton, a structural network we will consider later in this chapter.

The chloroplast is a specialized member of a family of closely related plant organelles called **plastids**. One type of plastid, the *amyloplast*, is a colorless organelle that stores starch (amylose), particularly in roots and tubers. Another is the *chromoplast*, which has pigments that give fruits and flowers their orange and yellow hues.

Peroxisomes: Oxidation

The **peroxisome** is a specialized metabolic compartment bounded by a single membrane **(Figure 6.19)**. Peroxisomes contain enzymes that remove hydrogen atoms from various substrates and transfer them to oxygen (O_2), thus producing hydrogen peroxide (H_2O_2) as a by-product (from which the organelle derives its name). These reactions have many different functions. Some peroxisomes use oxygen to break fatty acids down into smaller molecules that are transported to mitochondria and used as fuel for cellular respiration. Peroxisomes in the liver detoxify alcohol and other harmful compounds by transferring hydrogen from the poisons to oxygen. The H_2O_2 formed by peroxisomes is itself toxic, but the organelle also contains an enzyme that converts H_2O_2 to water. This is an excellent example of how the cell's compartmental structure is crucial to its functions: The enzymes that produce hydrogen peroxide and those that dispose of this toxic compound are sequestered away from other cellular components that could be damaged.

Specialized peroxisomes called *glyoxysomes* are found in the fat-storing tissues of plant seeds. These organelles contain enzymes that initiate the conversion of fatty acids to sugar, which the emerging seedling uses as a source of energy and carbon until it can produce its own sugar by photosynthesis.

How peroxisomes are related to other organelles is still an open question. They grow larger by incorporating proteins made in the cytosol and ER, as well as lipids made in the ER and within the peroxisome itself. Peroxisomes may increase in number by splitting in two when they reach a certain size, sparking the suggestion of an endosymbiotic evolutionary origin, but others argue against this scenario. The debate continues.

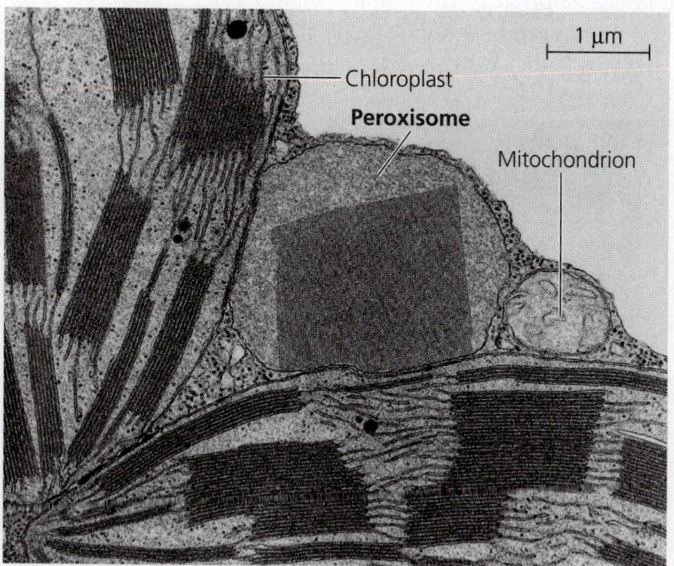

▲ **Figure 6.19 A peroxisome.** Peroxisomes are roughly spherical and often have a granular or crystalline core that is thought to be a dense collection of enzyme molecules. This peroxisome is in a leaf cell (TEM). Notice its proximity to two chloroplasts and a mitochondrion. These organelles cooperate with peroxisomes in certain metabolic functions.

CONCEPT CHECK 6.5

1. Describe two common characteristics of chloroplasts and mitochondria. Consider both function and membrane structure.
2. Do plant cells have mitochondria? Explain.
3. **WHAT IF?** A classmate proposes that mitochondria and chloroplasts should be classified in the endomembrane system. Argue against the proposal.

For suggested answers, see Appendix A.

CONCEPT 6.6

The cytoskeleton is a network of fibers that organizes structures and activities in the cell

In the early days of electron microscopy, biologists thought that the organelles of a eukaryotic cell floated freely in the cytosol. But improvements in both light microscopy and electron microscopy have revealed the **cytoskeleton**, a network of fibers extending throughout the cytoplasm (**Figure 6.20**). The cytoskeleton, which plays a major role in organizing the structures and activities of the cell, is composed of three types of molecular structures: microtubules, microfilaments, and intermediate filaments.

Roles of the Cytoskeleton: Support and Motility

The most obvious function of the cytoskeleton is to give mechanical support to the cell and maintain its shape. This is especially important for animal cells, which lack walls. The remarkable strength and resilience of the cytoskeleton as a whole is based on its architecture. Like a dome tent, the cytoskeleton is stabilized by a balance between opposing forces exerted by its elements. And just as the skeleton of an animal helps fix the positions of other body parts, the cytoskeleton provides anchorage for many organelles and even cytosolic enzyme molecules. The cytoskeleton is more dynamic than an animal skeleton, however. It can be quickly dismantled in one part of the cell and reassembled in a new location, changing the shape of the cell.

Several types of cell motility (movement) also involve the cytoskeleton. The term *cell motility* encompasses both changes in cell location and more limited movements of parts of the cell. Cell motility generally requires the interaction of the cytoskeleton with **motor proteins**. Examples of such cell motility abound. Cytoskeletal elements and motor proteins work together with plasma membrane molecules to allow whole cells to move along fibers outside the cell. Motor proteins bring about the bending of cilia and flagella by gripping microtubules within those organelles and sliding them against each other. A similar mechanism involving microfilaments causes muscle cells to contract. Inside the cell, vesicles and other organelles often use motor protein "feet" to "walk" to their destinations along a track provided by the cytoskeleton. For example, this is how vesicles containing neurotransmitter molecules migrate to the tips of axons, the long extensions of nerve cells that release these molecules as chemical signals to adjacent nerve cells (**Figure 6.21**). The vesicles that bud off

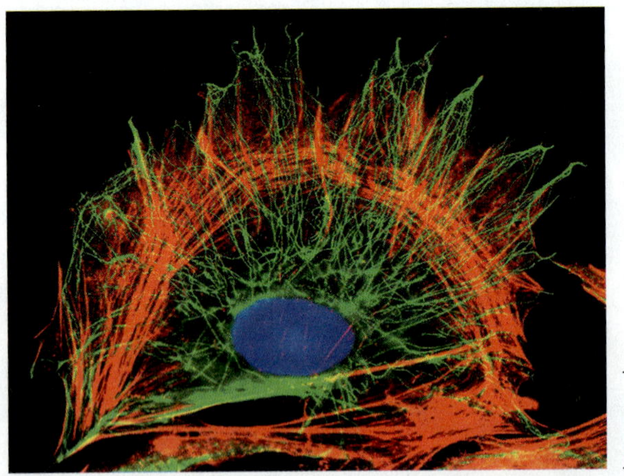

▲ **Figure 6.20 The cytoskeleton.** As shown in this fluorescence micrograph, the cytoskeleton extends throughout the cell. The cytoskeletal elements have been tagged with different fluorescent molecules: green for microtubules and red for microfilaments. A third component of the cytoskeleton, intermediate filaments, is not evident here. (The DNA in the nucleus is blue.)

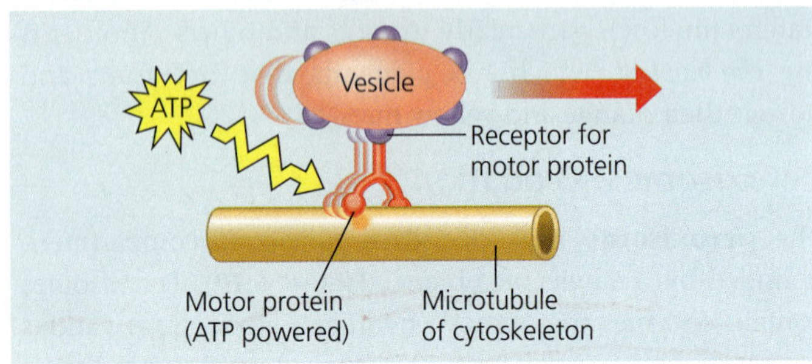

(a) Motor proteins that attach to receptors on vesicles can "walk" the vesicles along microtubules or, in some cases, microfilaments.

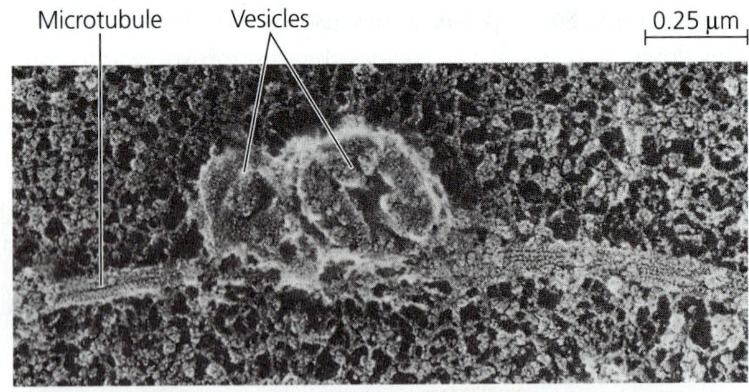

(b) In this SEM of a squid giant axon (a nerve cell extension), two vesicles containing neurotransmitters migrate toward the tip of the axon via the mechanism shown in (a).

▲ **Figure 6.21 Motor proteins and the cytoskeleton.**

Table 6.1 The Structure and Function of the Cytoskeleton

Property	Microtubules (Tubulin Polymers)	Microfilaments (Actin Filaments)	Intermediate Filaments
Structure	Hollow tubes; wall consists of 13 columns of tubulin molecules	Two intertwined strands of actin, each a polymer of actin subunits	Fibrous proteins supercoiled into thicker cables
Diameter	25 nm with 15-nm lumen	7 nm	8–12 nm
Protein subunits	Tubulin, a dimer consisting of α-tubulin and β-tubulin	Actin	One of several different proteins (such as keratins), depending on cell type
Main functions	Maintenance of cell shape (compression-resisting "girders") Cell motility (as in cilia or flagella) Chromosome movements in cell division Organelle movements	Maintenance of cell shape (tension-bearing elements) Changes in cell shape Muscle contraction Cytoplasmic streaming Cell motility (as in pseudopodia) Cell division (cleavage furrow formation)	Maintenance of cell shape (tension-bearing elements) Anchorage of nucleus and certain other organelles Formation of nuclear lamina

Fluorescence micrographs of fibroblasts, a favorite cell type for cell biology studies. In each, the structure of interest has been tagged with fluorescent molecules. In the first and third micrographs, the DNA in the nucleus has also been tagged (blue or orange).

from the ER travel to the Golgi along cytoskeletal tracks. The cytoskeleton also manipulates the plasma membrane, making it bend inward to form food vacuoles or other phagocytic vesicles. And the streaming of cytoplasm that circulates materials within many large plant cells is yet another kind of cellular movement brought about by the cytoskeleton.

Components of the Cytoskeleton

Now let's look more closely at the three main types of fibers that make up the cytoskeleton: *Microtubules* are the thickest of the three types; *microfilaments* (also called actin filaments) are the thinnest; and *intermediate filaments* are fibers with diameters in a middle range (**Table 6.1**).

Microtubules

All eukaryotic cells have **microtubules**, hollow rods measuring about 25 nm in diameter and from 200 nm to 25 µm in length. The wall of the hollow tube is constructed from a globular protein called tubulin. Each tubulin protein is a *dimer*, a molecule made up of two subunits. A tubulin dimer consists of two slightly different polypeptides, α-tubulin and β-tubulin. Microtubules grow in length by adding tubulin dimers; they can also be disassembled and their tubulin used to build microtubules elsewhere in the cell. Because of the orientation of tubulin dimers, the two ends of a microtubule are slightly different. One end can accumulate or release tubulin dimers at a much higher rate than the other, thus growing and shrinking significantly during cellular activities. (This is called the "plus end," not because it can only add tubulin proteins but because it's the end where both "on" and "off" rates are much higher.)

Microtubules shape and support the cell and also serve as tracks along which organelles equipped with motor proteins can move. In addition to the example in Figure 6.21, microtubules guide secretory vesicles from the Golgi apparatus to the plasma membrane. Microtubules are also involved in the separation of chromosomes during cell division, which will be discussed in Chapter 12.

Centrosomes and Centrioles In animal cells, microtubules grow out from a **centrosome**, a region that is often located near the nucleus and is considered a "microtubule-organizing center." These microtubules function as compression-resisting girders of the cytoskeleton. Within the centrosome is a pair of **centrioles**, each composed of nine sets of triplet microtubules arranged in a ring **(Figure 6.22)**. Before an animal cell divides, the centrioles replicate. Although centrosomes with centrioles may help organize microtubule assembly in animal cells, they are not essential for this function in all eukaryotes; fungi and almost all plant cells lack centrosomes with centrioles but have well-organized microtubules. Apparently, other microtubule-organizing centers play the role of centrosomes in these cells.

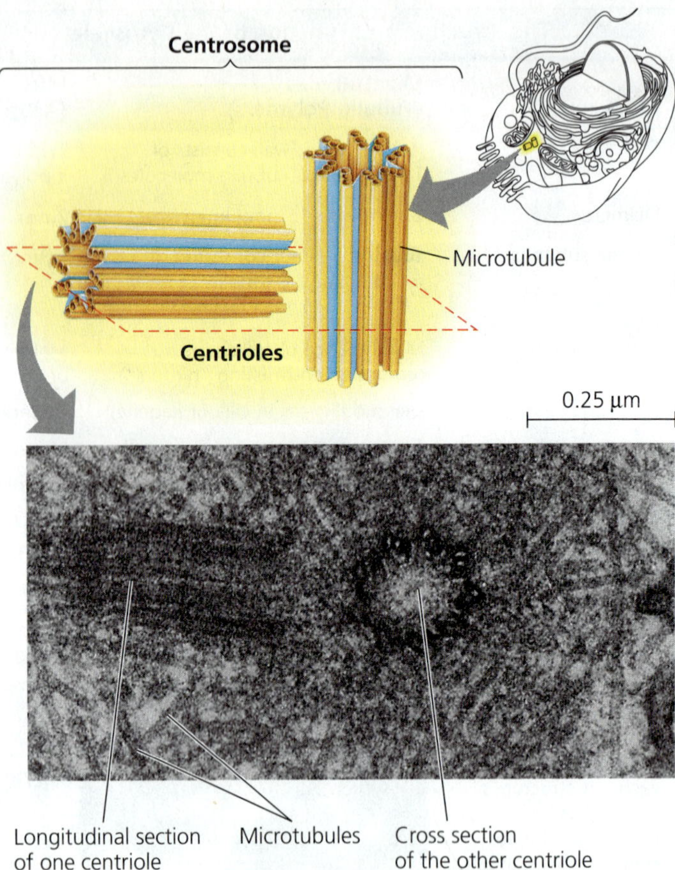

▲ **Figure 6.22 Centrosome containing a pair of centrioles.** Most animal cells have a centrosome, a region near the nucleus where the cell's microtubules are initiated. Within the centrosome is a pair of centrioles, each about 250 nm (0.25 µm) in diameter. The two centrioles are at right angles to each other, and each is made up of nine sets of three microtubules. The blue portions of the drawing represent nontubulin proteins that connect the microtubule triplets (TEM).

? *How many microtubules are in a centrosome? In the drawing, circle and label one microtubule and describe its structure. Circle and label a triplet.*

Cilia and Flagella In eukaryotes, a specialized arrangement of microtubules is responsible for the beating of **flagella** (singular, *flagellum*) and **cilia** (singular, *cilium*), microtubule-containing extensions that project from some cells. (The bacterial flagellum, shown in Figure 6.5, has a completely different structure.) Many unicellular eukaryotes are propelled through water by cilia or flagella that act as locomotor appendages, and the sperm of animals, algae, and some plants have flagella. When cilia or flagella extend from cells that are held in place as part of a tissue layer, they can move fluid over the surface of the tissue. For example, the ciliated lining of the trachea (windpipe) sweeps mucus containing trapped debris out of the lungs (see the EMs in Figure 6.3). In a woman's reproductive tract, the cilia lining the oviducts help move an egg toward the uterus.

Motile cilia usually occur in large numbers on the cell surface. They are about 0.25 µm in diameter and about 2–20 µm long. Flagella are the same diameter but longer, 10–200 µm. Also, flagella are usually limited to just one or a few per cell.

Flagella and cilia differ in their beating patterns **(Figure 6.23)**. A flagellum has an undulating motion that generates force in the same direction as the flagellum's axis, like the tail of a fish. In contrast, cilia work more like oars, with alternating power and recovery strokes generating force in a direction perpendicular to the cilium's axis, much as the oars of a racing crew boat extend outward at a right angle to the boat's forward movement.

A cilium may also act as a signal-receiving "antenna" for the cell. Cilia that have this function are generally nonmotile, and there is only one per cell. (In fact, in vertebrate animals, it appears that almost all cells have such a cilium, which is called a *primary cilium*.) Membrane proteins on this kind of cilium transmit molecular signals from the cell's environment to its interior, triggering signaling pathways that may lead to changes in the cell's activities. Cilium-based signaling appears to be crucial to brain function and to embryonic development.

Though different in length, number per cell, and beating pattern, motile cilia and flagella share a common structure. Each motile cilium and flagellum has a group of microtubules sheathed in an extension of the plasma membrane **(Figure 6.24)**. Nine doublets of microtubules are arranged in a ring; in the center of the ring are two single microtubules.

▶ **Figure 6.23 A comparison of the beating of flagella and motile cilia.**

(a) Motion of flagella. A flagellum usually undulates, its snakelike motion driving a cell in the same direction as the axis of the flagellum. Propulsion of a human sperm cell is an example of flagellate locomotion (LM).

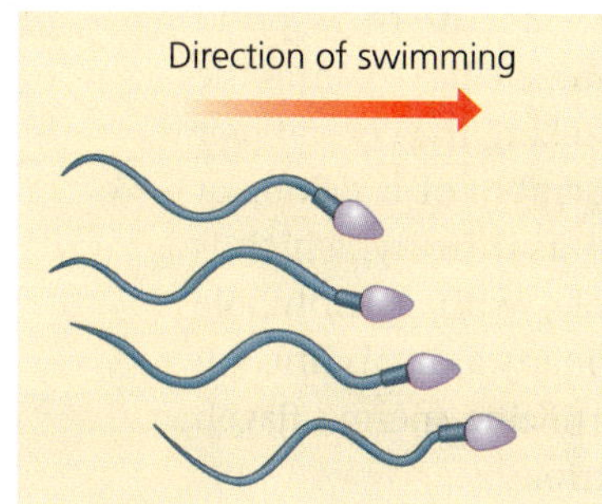

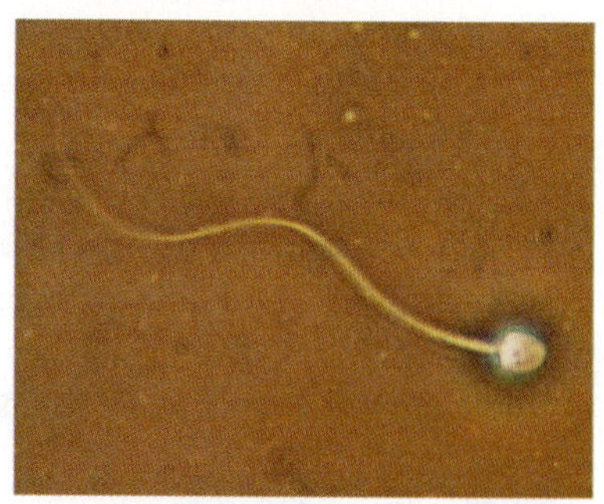

5 µm

(b) Motion of cilia. Cilia have a back-and-forth motion. The rapid power stroke moves the cell in a direction perpendicular to the axis of the cilium. Then, during the slower recovery stroke, the cilium bends and sweeps sideways, closer to the cell surface. A dense nap of cilia, beating at a rate of about 40 to 60 strokes a second, covers this *Colpidium*, a freshwater protist (colorized SEM).

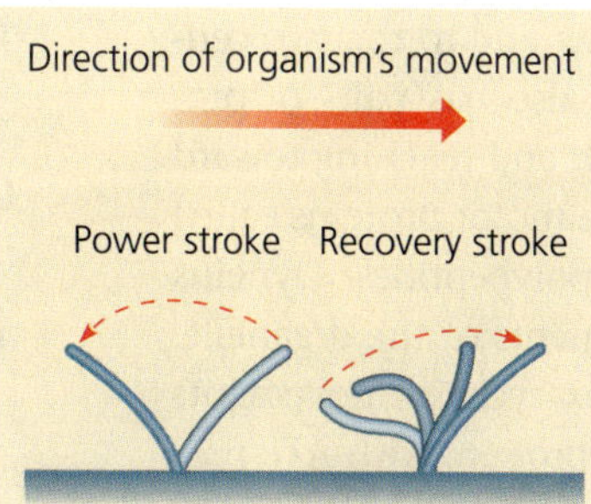

15 µm

Microtubules

Plasma membrane

Basal body

0.5 µm

(a) A longitudinal section of a motile cilium shows microtubules running the length of the structure (TEM).

0.1 µm

Outer microtubule doublet

Dynein proteins

Central microtubule

Radial spoke

Cross-linking proteins between outer doublets

Plasma membrane

(b) A cross section through a motile cilium shows the "9 + 2" arrangement of microtubules (TEM). The outer microtubule doublets and the two central microtubules are held together by flexible cross-linking proteins (blue in art), including the radial spokes. The doublets also have attached motor proteins called dyneins (red in art).

0.1 µm

Triplet

Cross section of basal body

(c) Basal body: The nine outer doublets of a cilium or flagellum extend into the basal body, where each doublet joins another microtubule to form a ring of nine triplets. Each triplet is connected to the next by nontubulin proteins (thinner blue lines in diagram). This is a "9 + 0" arrangement: the two central microtubules are not present because they terminate above the basal body (TEM).

▲ **Figure 6.24 Structure of a flagellum or motile cilium.**

DRAW IT *In (a), circle the central pair of microtubules. Show where they terminate, and explain why they aren't seen in the cross section of the basal body in (c).*

This arrangement, referred to as the "9 + 2" pattern, is found in nearly all eukaryotic flagella and motile cilia. (Nonmotile primary cilia have a "9 + 0" pattern, lacking the central pair of microtubules.) The microtubule assembly of a cilium or flagellum is anchored in the cell by a **basal body**, which is structurally very similar to a centriole, with microtubule triplets in a "9 + 0" pattern. In fact, in many animals (including humans), the basal body of the fertilizing sperm's flagellum enters the egg and becomes a centriole.

In flagella and motile cilia, flexible cross-linking proteins, evenly spaced along the length of the cilium or flagellum, connect the outer doublets to each other and to the two central microtubules. Each outer doublet also has pairs of protruding proteins spaced along its length and reaching toward the neighboring doublet; these are large motor proteins called **dyneins**, each composed of several polypeptides. Dyneins are responsible for the bending movements of the organelle. A dynein molecule performs a complex cycle of movements caused by changes in the shape of the protein, with ATP providing the energy for these changes (**Figure 6.25**).

The mechanics of dynein-based bending involve a process that resembles walking. A typical dynein protein has two "feet" that "walk" along the microtubule of the adjacent doublet, one foot maintaining contact while the other releases and reattaches one step farther along the microtubule. Without any restraints on the movement of the microtubule doublets, one doublet would continue to "walk" along and slide past the surface of the other, elongating the cilium or flagellum rather than bending it (see Figure 6.25a). For lateral movement of a cilium or flagellum, the dynein "walking" must have something to pull against, as when the muscles in your leg pull against your bones to move your knee. In cilia and flagella, the microtubule doublets seem to be held in place by the cross-linking proteins just inside the outer doublets and by the radial spokes and other structural elements. Thus, neighboring doublets cannot slide past each other very far. Instead, the forces exerted by dynein "walking" cause the doublets to curve, bending the cilium or flagellum (see Figure 6.25b and c).

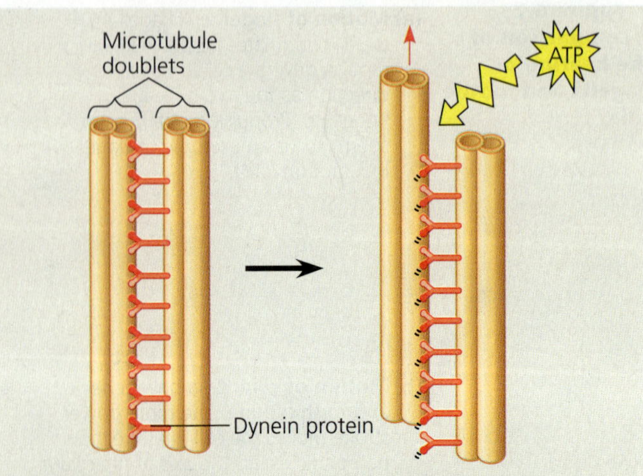

(a) Effect of unrestrained dynein movement. If a cilium or flagellum had no cross-linking proteins, the two feet of each dynein along one doublet (powered by ATP) would alternately grip and release the adjacent doublet. This "walking" motion would push the adjacent doublet up. Instead of bending, the doublets would slide past each other.

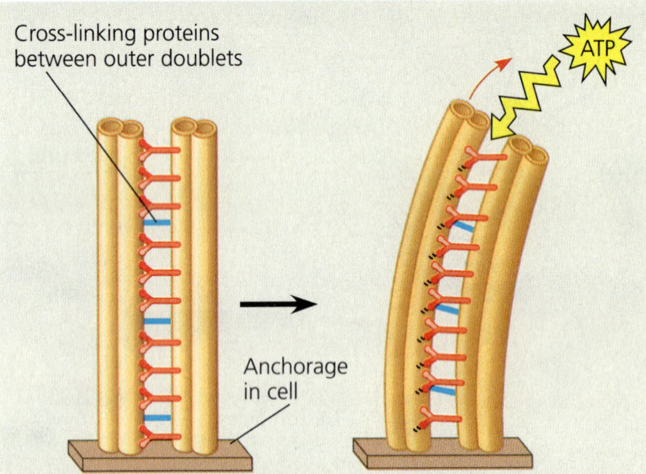

(b) Effect of cross-linking proteins. In a cilium or flagellum, two adjacent doublets cannot slide far because they are physically restrained by proteins, so they bend. (Only two of the nine outer doublets in Figure 6.24b are shown here.)

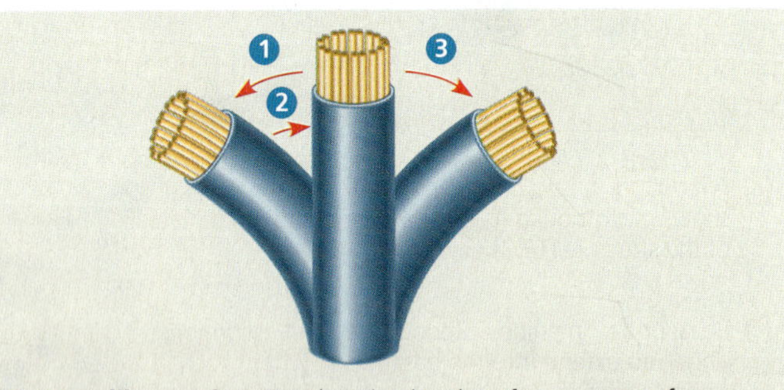

(c) Wavelike motion. Synchronized cycles of movement of many dyneins probably cause a bend to begin at the base of the cilium or flagellum and move outward toward the tip. Many successive bends, such as the ones shown here to the left and right, result in a wavelike motion. In this diagram, the two central microtubules and the cross-linking proteins are not shown.

▲ **Figure 6.25 How dynein "walking" moves flagella and cilia.**

Microfilaments (Actin Filaments)

Microfilaments are solid rods about 7 nm in diameter. They are also called actin filaments because they are built from molecules of **actin**, a globular protein. A microfilament is a twisted double chain of actin subunits (see Table 6.1). Besides occurring as linear filaments, microfilaments can form structural networks when certain proteins bind along the side of an actin filament and allow a new filament to extend as a branch. Like microtubules, microfilaments seem to be present in all eukaryotic cells.

In contrast to the compression-resisting role of microtubules, the structural role of microfilaments in the cytoskeleton is to bear tension (pulling forces). A three-dimensional network formed by microfilaments just inside the plasma

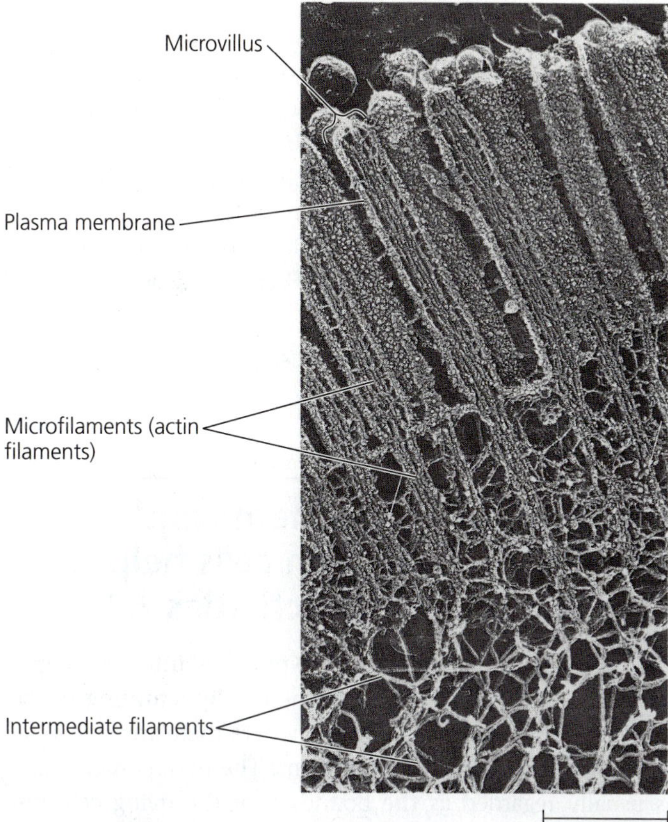

▲ **Figure 6.26 A structural role of microfilaments.** The surface area of this nutrient-absorbing intestinal cell is increased by its many microvilli (singular, *microvillus*), cellular extensions reinforced by bundles of microfilaments. These actin filaments are anchored to a network of intermediate filaments (TEM).

membrane (*cortical microfilaments*) helps support the cell's shape (see Figure 6.8). This network gives the outer cytoplasmic layer of a cell, called the **cortex**, the semisolid consistency of a gel, in contrast with the more fluid (*sol*) state of the interior cytoplasm. In animal cells specialized for transporting materials across the plasma membrane, such as intestinal cells, bundles of microfilaments make up the core of microvilli, delicate projections that increase the cell's surface area **(Figure 6.26)**.

Microfilaments are well known for their role in cell motility, particularly as part of the contractile apparatus of muscle cells. Thousands of actin filaments are arranged parallel to one another along the length of a muscle cell, interdigitated with thicker filaments made of a protein called **myosin** **(Figure 6.27a)**. Like dynein when it interacts with microtubules, myosin acts as a motor protein by means of projections that "walk" along the actin filaments. Contraction of the muscle cell results from the actin and myosin filaments sliding past one another in this way, shortening the cell. In other kinds of cells, actin filaments are associated with myosin in miniature and less elaborate versions of the arrangement in muscle cells. These actin-myosin aggregates are responsible for

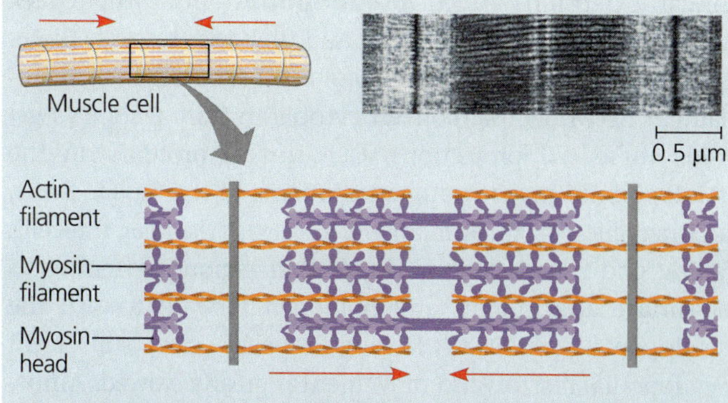

(a) Myosin motors in muscle cell contraction. The "walking" of myosin projections (the so-called heads) drives the parallel myosin and actin filaments past each other so that the actin filaments approach each other in the middle (red arrows). This shortens the muscle cell. Muscle contraction involves the shortening of many muscle cells at the same time (TEM).

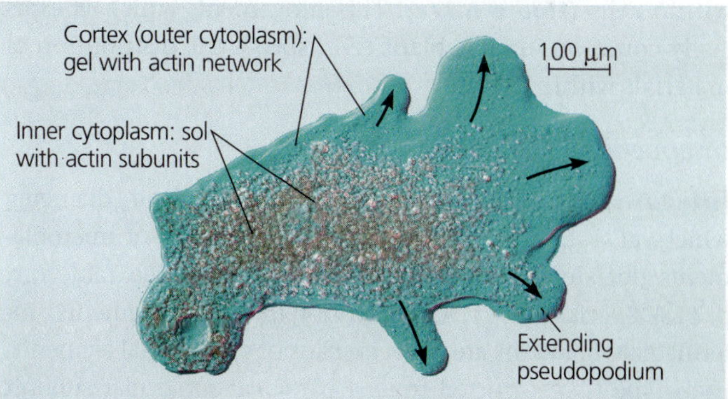

(b) Amoeboid movement. Interaction of actin filaments with myosin causes contraction of the cell, pulling the cell's trailing end (at left) forward (to the right) (LM).

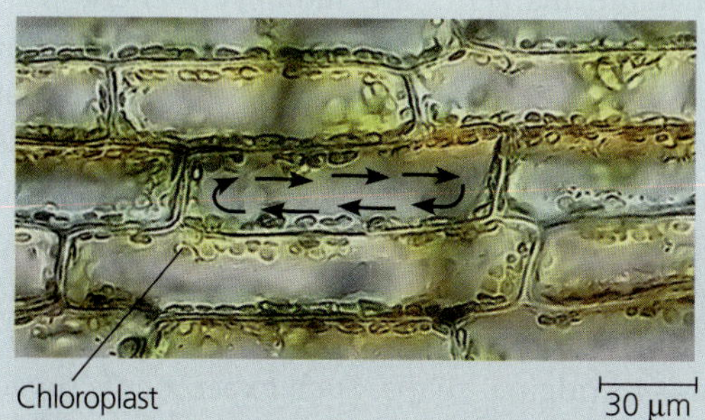

(c) Cytoplasmic streaming in plant cells. A layer of cytoplasm cycles around the cell, moving over a carpet of parallel actin filaments. Myosin motors attached to organelles in the fluid cytosol may drive the streaming by interacting with the actin (LM).

▲ **Figure 6.27 Microfilaments and motility.** In these three examples, interactions between actin filaments and motor proteins bring about cell movement.

localized contractions of cells. For example, a contracting belt of microfilaments forms a cleavage furrow that pinches a dividing animal cell into two daughter cells.

Localized contraction brought about by actin and myosin also plays a role in amoeboid movement **(Figure 6.27b)**. A cell such as an amoeba crawls along a surface by extending cellular extensions called **pseudopodia** (from the Greek *pseudes*, false, and *pod*, foot), and moving toward them. Pseudopodia extend by assembly of actin subunits into microfilament networks that convert cytoplasm from a sol to a gel inside these cell projections. Cell surface proteins on the pseudopodium make strong attachments to the "road." Next, the interaction of microfilaments with myosin near the cell's trailing end causes contraction of that region, loosening its cell-surface attachments and pulling it forward toward the pseudopodia. Amoebae lacking myosin can still form pseudopodia, but forward movement is greatly slowed. Amoebas are not the only cells that move by crawling; so do many cells in the animal body, including some white blood cells.

In plant cells, both actin-myosin interactions and sol-gel transformations brought about by actin may be involved in **cytoplasmic streaming**, a circular flow of cytoplasm within cells **(Figure 6.27c)**. This movement, which is especially common in large plant cells, speeds the distribution of materials within the cell.

Intermediate Filaments

Intermediate filaments are named for their diameter, which, at 8–12 nm, is larger than the diameter of microfilaments but smaller than that of microtubules (see Table 6.1, p. 113). Specialized for bearing tension (like microfilaments), intermediate filaments are a diverse class of cytoskeletal elements. Each type is constructed from a particular molecular subunit belonging to a family of proteins whose members include the keratins. Microtubules and microfilaments, in contrast, are consistent in diameter and composition in all eukaryotic cells.

Intermediate filaments are more permanent fixtures of cells than are microfilaments and microtubules, which are often disassembled and reassembled in various parts of a cell. Even after cells die, intermediate filament networks often persist; for example, the outer layer of our skin consists of dead skin cells full of keratin proteins. Chemical treatments that remove microfilaments and microtubules from the cytoplasm of living cells leave a web of intermediate filaments that retains its original shape. Such experiments suggest that intermediate filaments are especially sturdy and that they play an important role in reinforcing the shape of a cell and fixing the position of certain organelles. For instance, the nucleus typically sits within a cage made of intermediate filaments, fixed in location by branches of the filaments that extend into the cytoplasm. Other intermediate filaments make up the nuclear lamina, which lines the interior of the nuclear envelope (see Figure 6.9). By supporting a cell's shape, intermediate filaments help the cell carry out its specific function. For example, the long extensions (axons) of nerve cells that transmit impulses are strengthened by intermediate filaments. Thus, the various kinds of intermediate filaments may function together as the permanent framework of the entire cell.

CONCEPT CHECK 6.6

1. Describe shared features of microtubule-based motion of flagella and microfilament-based muscle contraction.
2. How do cilia and flagella bend?
3. **WHAT IF?** Males afflicted with Kartagener's syndrome are sterile because of immotile sperm, and they tend to suffer from lung infections. This disorder has a genetic basis. Suggest what the underlying defect might be.

For suggested answers, see Appendix A.

CONCEPT 6.7

Extracellular components and connections between cells help coordinate cellular activities

Having crisscrossed the cell to explore its interior components, we complete our tour of the cell by returning to the surface of this microscopic world, where there are additional structures with important functions. The plasma membrane is usually regarded as the boundary of the living cell, but most cells synthesize and secrete materials that are extracellular, or external to the plasma membrane. Although these materials and the structures they form are outside the cell, their study is important to cell biology because they are involved in a great many cellular functions.

Cell Walls of Plants

The **cell wall** is an extracellular structure of plant cells that distinguishes them from animal cells (see Figure 6.8). The wall protects the plant cell, maintains its shape, and prevents excessive uptake of water. On the level of the whole plant, the strong walls of specialized cells hold the plant up against the force of gravity. Prokaryotes, fungi, and some protists also have cell walls, as you saw in Figures 6.5 and 6.8, but we will postpone discussion of them until Unit Five.

Plant cell walls are much thicker than the plasma membrane, ranging from 0.1 μm to several micrometers. The exact chemical composition of the wall varies from species to species and even from one cell type to another in the same plant, but the basic design of the wall is consistent. Microfibrils made of the polysaccharide cellulose (see Figure 5.8) are synthesized by an enzyme called cellulose synthase and secreted to the extracellular space, where they become embedded in a matrix of other polysaccharides and proteins. This combination of materials, strong fibers in a "ground substance" (matrix), is the same basic architectural design found in steel-reinforced concrete and in fiberglass.

A young plant cell first secretes a relatively thin and flexible wall called the **primary cell wall (Figure 6.28)**. In actively

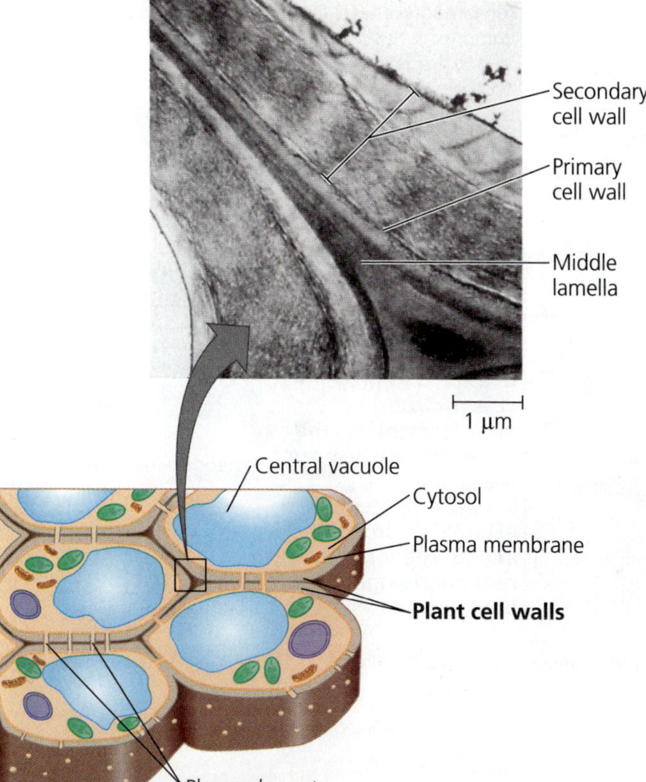

▲ **Figure 6.28 Plant cell walls.** The drawing shows several cells, each with a large vacuole, a nucleus, and several chloroplasts and mitochondria. The transmission electron micrograph shows the cell walls where two cells come together. The multilayered partition between plant cells consists of adjoining walls individually secreted by the cells.

growing cells, the cellulose fibrils are oriented at right angles to the direction of cell expansion. Researchers investigated the role of microtubules in orienting these cellulose fibrils **(Figure 6.29)**. Their observations strongly support the idea that microtubules in the cell cortex guide cellulose synthase as it synthesizes and deposits cellulose fibrils. By orienting cellulose deposition, microtubules thus affect the growth pattern of the cells.

Between primary walls of adjacent cells is the **middle lamella**, a thin layer rich in sticky polysaccharides called pectins. The middle lamella glues adjacent cells together (see Figure 6.28). (Pectin is used as a thickening agent in jams and jellies.) When the cell matures and stops growing, it strengthens its wall. Some plant cells do this simply by secreting hardening substances into the primary wall. Other cells add a **secondary cell wall** between the plasma membrane and the primary wall. The secondary wall, often deposited in several laminated layers, has a strong and durable matrix that affords the cell protection and support. Wood, for example, consists mainly of secondary walls. Plant cell walls are usually perforated by channels between adjacent cells called plasmodesmata (see Figure 6.28), which will be discussed shortly.

▼ **Figure 6.29** INQUIRY

What role do microtubules play in orienting deposition of cellulose in cell walls?

EXPERIMENT Previous experiments on preserved plant tissues had shown alignment of microtubules in the cell cortex with cellulose fibrils in the cell wall. Also, drugs that disrupted microtubules were observed to cause disoriented cellulose fibrils. To further investigate the possible role of cortical microtubules in guiding cellulose fibril deposition, David Ehrhardt and colleagues at Stanford University used a type of confocal microscopy to study cell wall deposition in living cells. In these cells, they labeled both cellulose synthase and microtubules with fluorescent markers and observed them over time.

RESULTS Each fluorescence image below represents a combination of 30 images taken over a 5-minute period to detect the movement of cellulose synthase and microtubules. These two coincided highly over time. The labeling molecules caused cellulose synthase to fluoresce green and the microtubules to fluoresce red. The arrowheads indicate prominent areas where the two are seen to align.

10 μm

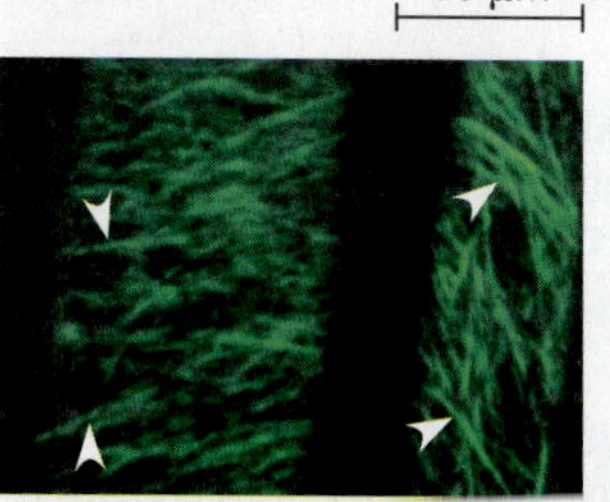

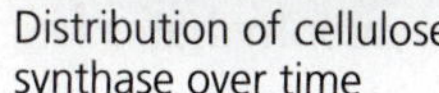

Distribution of cellulose synthase over time

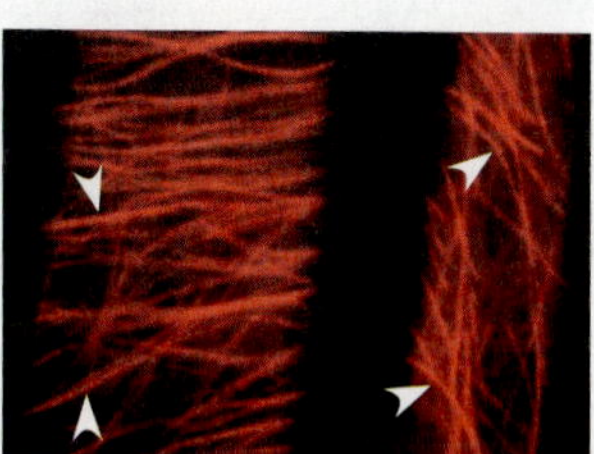

Distribution of microtubules over time

CONCLUSION The organization of microtubules appears to directly guide the path of cellulose synthase as it lays down cellulose, thus determining the orientation of cellulose fibrils.

SOURCE A. R. Paradez et al., Visualization of cellulose synthase demonstrates functional association with microtubules, *Science* 312:1491–1495 (2006).

WHAT IF? In a second experiment, the researchers exposed the plant cells to blue light, previously shown to cause reorientation of microtubules. What events would you predict would follow blue light exposure?

The Extracellular Matrix (ECM) of Animal Cells

Although animal cells lack walls akin to those of plant cells, they do have an elaborate **extracellular matrix (ECM)**. The main ingredients of the ECM are glycoproteins and other carbohydrate-containing molecules secreted by the cells. (Recall that glycoproteins are proteins with covalently bonded carbohydrate, usually short chains of sugars.) The most abundant glycoprotein in the ECM of most animal cells is **collagen**, which forms strong fibers outside the cells (see Figure 5.20). In fact, collagen accounts for about 40% of the total protein in the human body. The collagen fibers are embedded in a network woven out of **proteoglycans** secreted

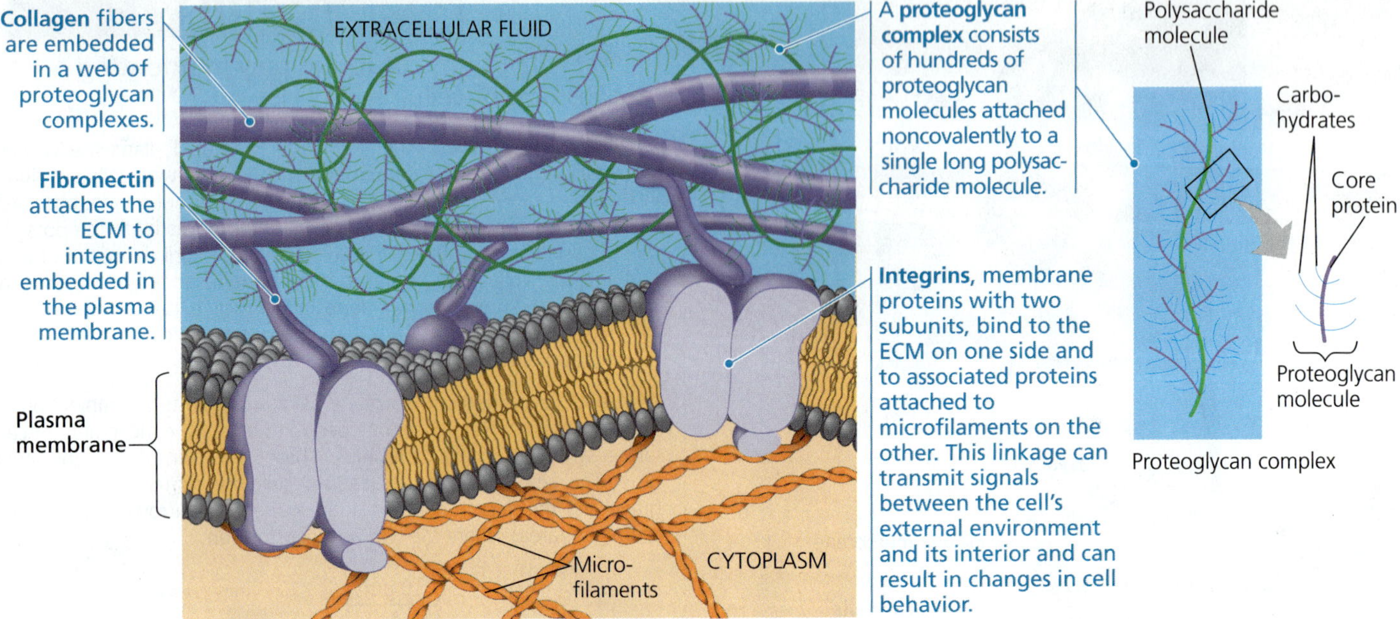

▲ **Figure 6.30 Extracellular matrix (ECM) of an animal cell.** The molecular composition and structure of the ECM vary from one cell type to another. In this example, three different types of glycoproteins are present: proteoglycans, collagen, and fibronectin.

by cells **(Figure 6.30)**. A proteoglycan molecule consists of a small core protein with many carbohydrate chains covalently attached, so that it may be up to 95% carbohydrate. Large proteoglycan complexes can form when hundreds of proteoglycan molecules become noncovalently attached to a single long polysaccharide molecule, as shown in Figure 6.30. Some cells are attached to the ECM by still other ECM glycoproteins, such as **fibronectin**. Fibronectin and other ECM proteins bind to cell-surface receptor proteins called **integrins** that are built into the plasma membrane. Integrins span the membrane and bind on their cytoplasmic side to associated proteins attached to microfilaments of the cytoskeleton. The name *integrin* is based on the word *integrate*: Integrins are in a position to transmit signals between the ECM and the cytoskeleton and thus to integrate changes occurring outside and inside the cell.

Current research on fibronectin, other ECM molecules, and integrins is revealing the influential role of the extracellular matrix in the lives of cells. By communicating with a cell through integrins, the ECM can regulate a cell's behavior. For example, some cells in a developing embryo migrate along specific pathways by matching the orientation of their microfilaments to the "grain" of fibers in the extracellular matrix. Researchers have also learned that the extracellular matrix around a cell can influence the activity of genes in the nucleus. Information about the ECM probably reaches the nucleus by a combination of mechanical and chemical signaling pathways. Mechanical signaling involves fibronectin, integrins, and microfilaments of the cytoskeleton. Changes in the cytoskeleton may in turn trigger chemical signaling pathways inside the cell, leading to changes in the set of proteins being made by the cell and therefore changes in the cell's function. In this way, the extracellular matrix of a particular tissue may help coordinate the behavior of all the cells of that tissue. Direct connections between cells also function in this coordination, as we discuss next.

Cell Junctions

Cells in an animal or plant are organized into tissues, organs, and organ systems. Neighboring cells often adhere, interact, and communicate via sites of direct physical contact.

Plasmodesmata in Plant Cells

It might seem that the nonliving cell walls of plants would isolate plant cells from one another. But in fact, as shown in **Figure 6.31**, cell walls are perforated with **plasmodesmata** (singular, *plasmodesma*; from the Greek *desmos*, to bind),

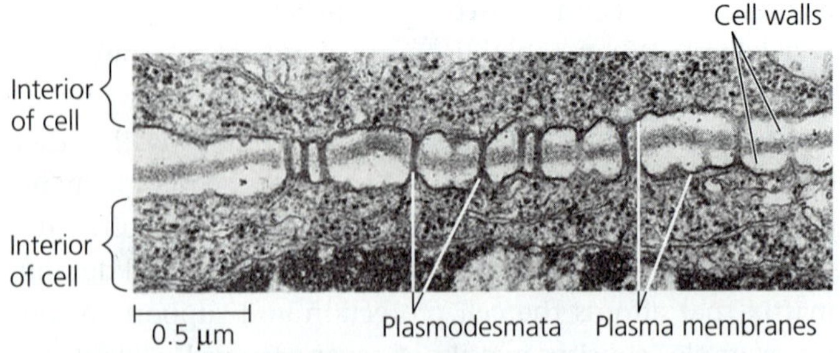

▲ **Figure 6.31 Plasmodesmata between plant cells.** The cytoplasm of one plant cell is continuous with the cytoplasm of its neighbors via plasmodesmata, cytoplasmic channels through the cell walls (TEM).

membrane-lined channels filled with cytoplasm. Cytosol passes through the plasmodesmata and joins the internal chemical environments of adjacent cells. These connections unify most of the plant into one living continuum. The plasma membranes of adjacent cells line the channel of each plasmodesma and thus are continuous. Water and small solutes can pass freely from cell to cell, and recent experiments have shown that in some circumstances, certain proteins and RNA molecules can also do this (see Concept 36.6). The macromolecules transported to neighboring cells appear to reach the plasmodesmata by moving along fibers of the cytoskeleton.

Tight Junctions, Desmosomes, and Gap Junctions in Animal Cells

In animals, there are three main types of cell junctions: *tight junctions*, *desmosomes*, and *gap junctions*. (Gap junctions are most like the plasmodesmata of plants, although gap junction pores are not lined with membrane.) All three types of cell junctions are especially common in epithelial tissue, which lines the external and internal surfaces of the body. **Figure 6.32** uses epithelial cells of the intestinal lining to illustrate these junctions.

▼ Figure 6.32

Exploring Cell Junctions in Animal Tissues

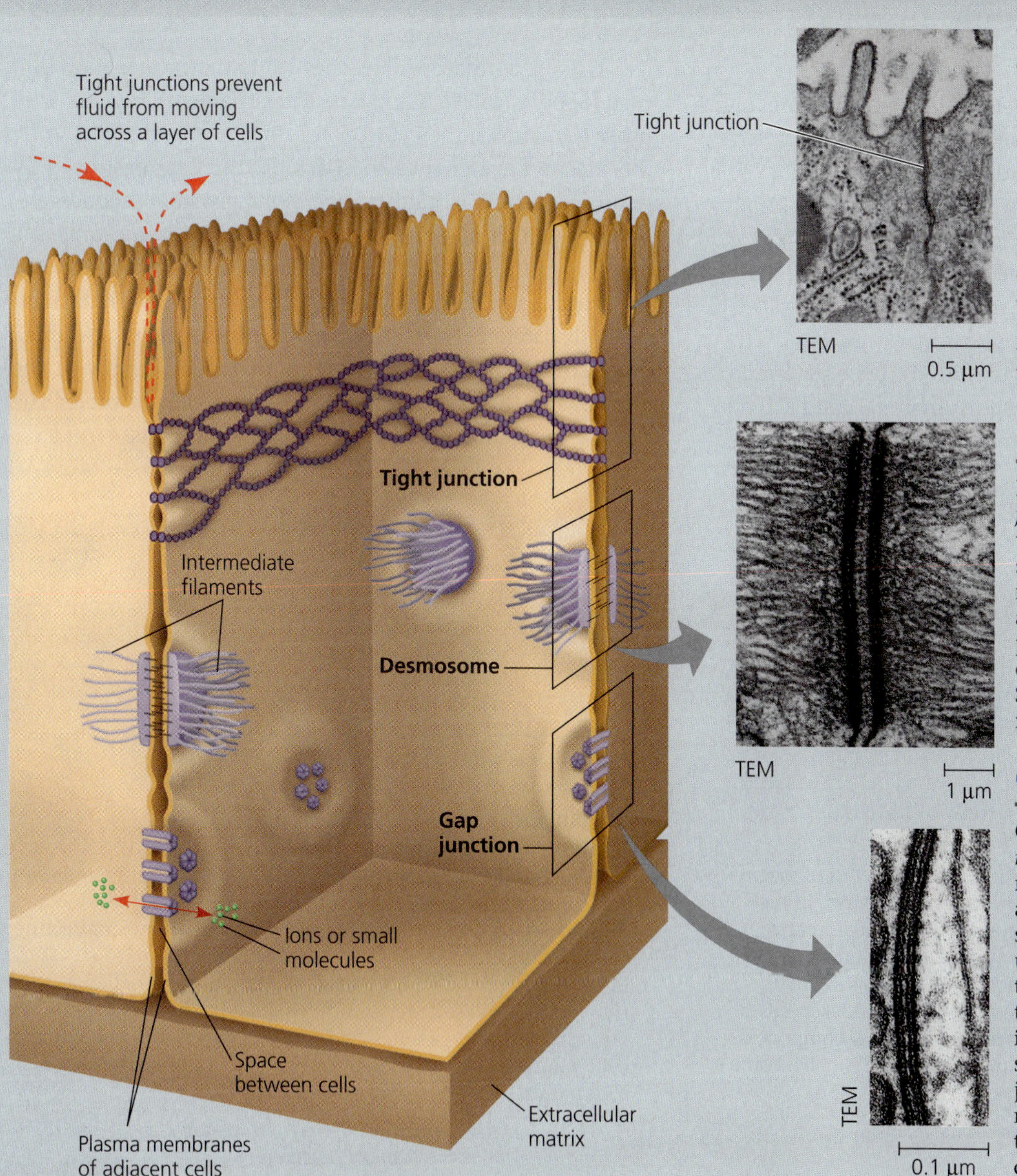

Tight Junctions

At **tight junctions**, the plasma membranes of neighboring cells are very tightly pressed against each other, bound together by specific proteins (purple). Forming continuous seals around the cells, tight junctions prevent leakage of extracellular fluid across a layer of epithelial cells. For example, tight junctions between skin cells make us watertight by preventing leakage between cells in our sweat glands.

Desmosomes

Desmosomes (also called *anchoring junctions*) function like rivets, fastening cells together into strong sheets. Intermediate filaments made of sturdy keratin proteins anchor desmosomes in the cytoplasm. Desmosomes attach muscle cells to each other in a muscle. Some "muscle tears" involve the rupture of desmosomes.

Gap Junctions

Gap junctions (also called *communicating junctions*) provide cytoplasmic channels from one cell to an adjacent cell and in this way are similar in their function to the plasmodesmata in plants. Gap junctions consist of membrane proteins that surround a pore through which ions, sugars, amino acids, and other small molecules may pass. Gap junctions are necessary for communication between cells in many types of tissues, such as heart muscle, and in animal embryos.

CONCEPT CHECK 6.7

1. In what way are the cells of plants and animals structurally different from single-celled eukaryotes?
2. **WHAT IF?** If the plant cell wall or the animal extracellular matrix were impermeable, what effect would this have on cell function?
3. **MAKE CONNECTIONS** The polypeptide chain that makes up a tight junction weaves back and forth through the membrane four times, with two extracellular loops, and one loop plus short C-terminal and N-terminal tails in the cytoplasm. Looking at Figure 5.16 (p. 79), what would you predict about the amino acid sequence of the tight-junction protein?

For suggested answers, see Appendix A.

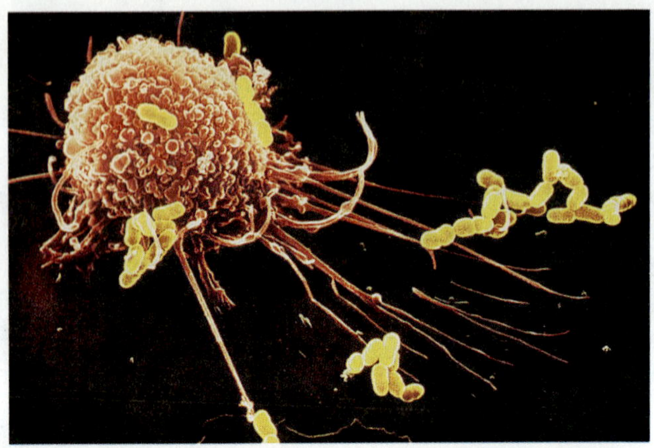

▲ **Figure 6.33 The emergence of cellular functions.** The ability of this macrophage (brown) to recognize, apprehend, and destroy bacteria (yellow) is a coordinated activity of the whole cell. Its cytoskeleton, lysosomes, and plasma membrane are among the components that function in phagocytosis (colorized SEM).

The Cell: A Living Unit Greater Than the Sum of Its Parts

From our panoramic view of the cell's compartmental organization to our close-up inspection of each organelle's architecture, this tour of the cell has provided many opportunities to correlate structure with function. (This would be a good time to review cell structure by returning to Figure 6.8, on pp. 100 and 101.) But even as we dissect the cell, remember that none of its components works alone. As an example of cellular integration, consider the microscopic scene in **Figure 6.33**. The large cell is a macrophage (see Figure 6.13a). It helps defend the mammalian body against infections by ingesting bacteria (the smaller cells) into phagocytic vesicles. The macrophage crawls along a surface and reaches out to the bacteria with thin pseudopodia (called filopodia). Actin filaments interact with other elements of the cytoskeleton in these movements. After the macrophage engulfs the bacteria, they are destroyed by lysosomes. The elaborate endomembrane system produces the lysosomes. The digestive enzymes of the lysosomes and the proteins of the cytoskeleton are all made on ribosomes. And the synthesis of these proteins is programmed by genetic messages dispatched from the DNA in the nucleus. All these processes require energy, which mitochondria supply in the form of ATP. Cellular functions arise from cellular order: The cell is a living unit greater than the sum of its parts.

6 CHAPTER REVIEW

SUMMARY OF KEY CONCEPTS

CONCEPT 6.1

Biologists use microscopes and the tools of biochemistry to study cells (pp. 94–97)

- Improvements in microscopy that affect the parameters of magnification, resolution, and contrast have catalyzed progress in the study of cell structure. **Light microscopy** (LM) and **electron microscopy** (EM), as well as other types, remain important tools.
- Cell biologists can obtain pellets enriched in particular cellular components by centrifuging disrupted cells at sequential speeds, a process known as **cell fractionation**. Larger cellular components are in the pellet after lower-speed centrifugation, and smaller components are in the pellet after higher-speed centrifugation.

? *How do microscopy and biochemistry complement each other to reveal cell structure and function?*

CONCEPT 6.2

Eukaryotic cells have internal membranes that compartmentalize their functions (pp. 98–102)

- All cells are bounded by a **plasma membrane**.
- **Prokaryotic cells** lack nuclei and other membrane-enclosed **organelles**, while **eukaryotic cells** have internal membranes that compartmentalize cellular functions.
- The surface-to-volume ratio is an important parameter affecting cell size and shape.
- Plant and animal cells have most of the same organelles: a nucleus, endoplasmic reticulum, Golgi apparatus, and mitochondria. Some organelles are found only in plant or in animal cells. Chloroplasts are present only in cells of photosynthetic eukaryotes.

? *Explain how the compartmental organization of a eukaryotic cell contributes to its biochemical functioning.*

	Cell Component	Structure	Function
CONCEPT 6.3 **The eukaryotic cell's genetic instructions are housed in the nucleus and carried out by the ribosomes (pp. 102–104)** ? *Describe the relationship between the nucleus and ribosomes.*	Nucleus (ER)	Surrounded by nuclear envelope (double membrane) perforated by nuclear pores; nuclear envelope continuous with endoplasmic reticulum (ER)	Houses chromosomes, which are made of chromatin (DNA and proteins); contains nucleoli, where ribosomal subunits are made; pores regulate entry and exit of materials
	Ribosome	Two subunits made of ribosomal RNA and proteins; can be free in cytosol or bound to ER	Protein synthesis
CONCEPT 6.4 **The endomembrane system regulates protein traffic and performs metabolic functions in the cell (pp. 104–109)** ? *Describe the key role played by transport vesicles in the endomembrane system.*	Endoplasmic reticulum (Nuclear envelope)	Extensive network of membrane-bounded tubules and sacs; membrane separates lumen from cytosol; continuous with nuclear envelope	Smooth ER: synthesis of lipids, metabolism of carbohydrates, Ca^{2+} storage, detoxification of drugs and poisons Rough ER: aids in synthesis of secretory and other proteins from bound ribosomes; adds carbohydrates to proteins to make glycoproteins; produces new membrane
	Golgi apparatus	Stacks of flattened membranous sacs; has polarity (*cis* and *trans* faces)	Modification of proteins, carbohydrates on proteins, and phospholipids; synthesis of many polysaccharides; sorting of Golgi products, which are then released in vesicles
	Lysosome	Membranous sac of hydrolytic enzymes (in animal cells)	Breakdown of ingested substances, cell macromolecules, and damaged organelles for recycling
	Vacuole	Large membrane-bounded vesicle	Digestion, storage, waste disposal, water balance, cell growth, and protection
CONCEPT 6.5 **Mitochondria and chloroplasts change energy from one form to another (pp. 109–112)** ? *What is the endosymbiont theory?*	Mitochondrion	Bounded by double membrane; inner membrane has infoldings (cristae)	Cellular respiration
	Chloroplast	Typically two membranes around fluid stroma, which contains thylakoids stacked into grana (in cells of photosynthetic eukaryotes, including plants)	Photosynthesis
	Peroxisome	Specialized metabolic compartment bounded by a single membrane	Contains enzymes that transfer hydrogen atoms from substrates to oxygen, producing hydrogen peroxide (H_2O_2) as a by-product; H_2O_2 is converted to water by another enzyme

CONCEPT 6.6

The cytoskeleton is a network of fibers that organizes structures and activities in the cell (pp. 112–118)

- The **cytoskeleton** functions in structural support for the cell and in motility and signal transmission.
- **Microtubules** shape the cell, guide organelle movement, and separate chromosomes in dividing cells. **Cilia** and **flagella** are motile appendages containing microtubules. Primary cilia also play sensory and signaling roles. **Microfilaments** are thin rods functioning in muscle contraction, amoeboid movement, **cytoplasmic streaming**, and microvillus support. **Intermediate filaments** support cell shape and fix organelles in place.

? *Describe the role of motor proteins inside the eukaryotic cell and in whole-cell movement.*

CONCEPT 6.7

Extracellular components and connections between cells help coordinate cellular activities (pp. 118–122)

- Plant **cell walls** are made of cellulose fibers embedded in other polysaccharides and proteins. Cellulose deposition is oriented along microtubules.
- Animal cells secrete glycoproteins that form the **extracellular matrix (ECM)**, which functions in support, adhesion, movement, and regulation.
- Cell junctions connect neighboring cells in plants and animals. Plants have **plasmodesmata** that pass through adjoining cell walls. Animal cells have **tight junctions**, **desmosomes**, and **gap junctions**.

? *Compare the composition and functions of a plant cell wall and the extracellular matrix of an animal cell.*

TEST YOUR UNDERSTANDING

LEVEL 1: KNOWLEDGE/COMPREHENSION

1. Which structure is *not* part of the endomembrane system?
   a. nuclear envelope
   b. chloroplast
   c. Golgi apparatus
   d. plasma membrane
   e. ER

2. Which structure is common to plant *and* animal cells?
   a. chloroplast
   b. wall made of cellulose
   c. central vacuole
   d. mitochondrion
   e. centriole

3. Which of the following is present in a prokaryotic cell?
   a. mitochondrion
   b. ribosome
   c. nuclear envelope
   d. chloroplast
   e. ER

4. Which structure-function pair is *mismatched*?
   a. nucleolus; production of ribosomal subunits
   b. lysosome; intracellular digestion
   c. ribosome; protein synthesis
   d. Golgi; protein trafficking
   e. microtubule; muscle contraction

LEVEL 2: APPLICATION/ANALYSIS

5. Cyanide binds to at least one molecule involved in producing ATP. If a cell is exposed to cyanide, most of the cyanide will be found within the
   a. mitochondria.
   b. ribosomes.
   c. peroxisomes.
   d. lysosomes.
   e. endoplasmic reticulum.

6. What is the most likely pathway taken by a newly synthesized protein that will be secreted by a cell?
   a. ER → Golgi → nucleus
   b. Golgi → ER → lysosome
   c. nucleus → ER → Golgi
   d. ER → Golgi → vesicles that fuse with plasma membrane
   e. ER → lysosomes → vesicles that fuse with plasma membrane

7. Which cell would be best for studying lysosomes?
   a. muscle cell
   b. nerve cell
   c. phagocytic white blood cell
   d. leaf cell of a plant
   e. bacterial cell

8. **DRAW IT** From memory, draw two eukaryotic cells, labeling the structures listed here and showing any physical connections between the internal structures of each cell: nucleus, rough ER, smooth ER, mitochondrion, centrosome, chloroplast, vacuole, lysosome, microtubule, cell wall, ECM, microfilament, Golgi apparatus, intermediate filament, plasma membrane, peroxisome, ribosome, nucleolus, nuclear pore, vesicle, flagellum, microvilli, plasmodesma.

LEVEL 3: SYNTHESIS/EVALUATION

9. **EVOLUTION CONNECTION**
   Which aspects of cell structure best reveal evolutionary unity? What are some examples of specialized modifications?

10. **SCIENTIFIC INQUIRY**
    Imagine protein X, destined to span the plasma membrane. Assume that the mRNA carrying the genetic message for protein X has already been translated by ribosomes in a cell culture. If you fractionate the cells (see Figure 6.4), in which fraction would you find protein X? Explain by describing its transit through the cell.

11. **WRITE ABOUT A THEME**
    Emergent Properties Considering some of the characteristics that define life and drawing on your new knowledge of cellular structures and functions, write a short essay (100–150 words) that discusses this statement: Life is an emergent property that appears at the level of the cell. (Review pp. 3–5 in Chapter 1.)

For selected answers, see Appendix A.

www.masteringbiology.com

1. MasteringBiology® Assignments

BioFlix Tutorials Tour of an Animal Cell: The Endomembrane System • Structures and Functions; Tour of a Plant Cell: Structures and Functions

Tutorials Connections Between Cells

Activities Metric System Review • Prokaryotic Cell Structure and Function • Discovery Channel Video: Cells • Role of the Nucleus and Ribosomes in Protein Synthesis • Transport into the Nucleus • A Pulse Chase Experiment • The Endomembrane System • Cilia and Flagella • Cell Junctions • Review: Animal Cell Structure and Function

Questions Student Misconceptions • Reading Quiz • Multiple Choice • End-of-Chapter

2. eText

Read your book online, search, take notes, highlight text, and more.

3. The Study Area

Practice Tests • Cumulative Test • BioFlix 3-D Animations • MP3 Tutor Sessions • Videos • Activities • Investigations • Lab Media • Audio Glossary • Word Study Tools • Art

7

Membrane Structure and Function

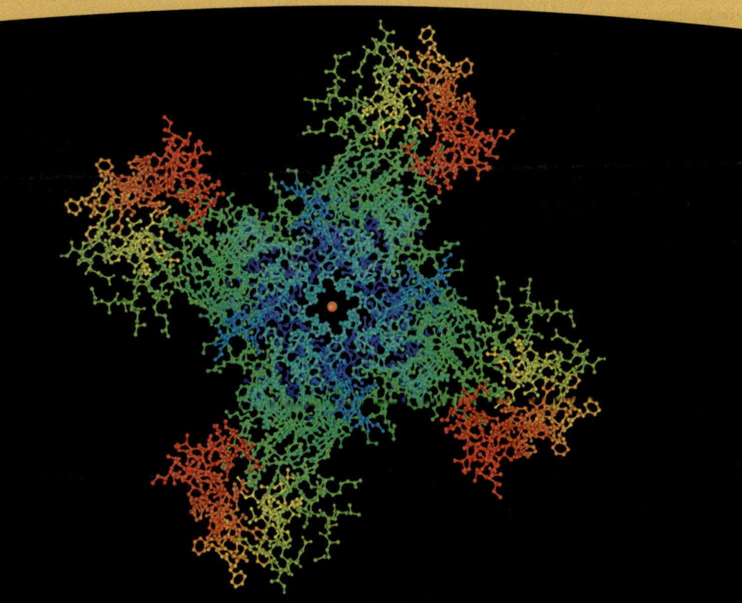

▲ **Figure 7.1 How do cell membrane proteins help regulate chemical traffic?**

KEY CONCEPTS

7.1 Cellular membranes are fluid mosaics of lipids and proteins

7.2 Membrane structure results in selective permeability

7.3 Passive transport is diffusion of a substance across a membrane with no energy investment

7.4 Active transport uses energy to move solutes against their gradients

7.5 Bulk transport across the plasma membrane occurs by exocytosis and endocytosis

OVERVIEW

Life at the Edge

The plasma membrane is the edge of life, the boundary that separates the living cell from its surroundings. A remarkable film only about 8 nm thick—it would take over 8,000 plasma membranes to equal the thickness of this page—the plasma membrane controls traffic into and out of the cell it surrounds. Like all biological membranes, the plasma membrane exhibits **selective permeability**; that is, it allows some substances to cross it more easily than others. One of the earliest episodes in the evolution of life may have been the formation of a membrane that enclosed a solution different from the surrounding solution while still permitting the uptake of nutrients and elimination of waste products. The ability of the cell to discriminate in its chemical exchanges with its environment is fundamental to life, and it is the plasma membrane and its component molecules that make this selectivity possible.

In this chapter, you will learn how cellular membranes control the passage of substances. The image in **Figure 7.1** shows the elegant structure of a eukaryotic plasma membrane protein that plays a crucial role in nerve cell signaling. This protein provides a channel for a stream of potassium ions (K^+) to exit a nerve cell at a precise moment after nerve stimulation, restoring the cell's ability to fire again. (The orange ball in the center represents one potassium ion moving through the channel.) In this way, the plasma membrane and its proteins not only act as an outer boundary but also enable the cell to carry out its functions. The same applies to the many varieties of internal membranes that partition the eukaryotic cell: The molecular makeup of each membrane allows compartmentalized specialization in cells. To understand how membranes work, we'll begin by examining their architecture.

CONCEPT 7.1

Cellular membranes are fluid mosaics of lipids and proteins

Lipids and proteins are the staple ingredients of membranes, although carbohydrates are also important. The most abundant lipids in most membranes are phospholipids. The ability of phospholipids to form membranes is inherent in their molecular structure. A phospholipid is an **amphipathic** molecule, meaning it has both a hydrophilic region and a hydrophobic region (see Figure 5.12). Other types of membrane lipids are also amphipathic. Furthermore, most of the proteins within membranes have both hydrophobic and hydrophilic regions.

How are phospholipids and proteins arranged in the membranes of cells? In the **fluid mosaic model**, the membrane is a fluid structure with a "mosaic" of various proteins embedded in or attached to a double layer (bilayer) of phospholipids. Scientists propose models as hypotheses, ways of organizing and explaining existing information. Let's explore how the fluid mosaic model was developed.

Membrane Models: *Scientific Inquiry*

Scientists began building molecular models of the membrane decades before membranes were first seen with the electron microscope (in the 1950s). In 1915, membranes isolated from red blood cells were chemically analyzed and found to be composed of lipids and proteins. Ten years later, two Dutch scientists reasoned that cell membranes must be phospholipid bilayers. Such a double layer of molecules could exist as a stable boundary between two aqueous compartments because the molecular arrangement shelters the hydrophobic tails of the phospholipids from water while exposing the hydrophilic heads to water **(Figure 7.2)**.

If a phospholipid bilayer was the main fabric of a membrane, where were the proteins located? Although the heads of phospholipids are hydrophilic, the surface of a pure phospholipid bilayer adheres less strongly to water than does the surface of a biological membrane. Given this difference, Hugh Davson and James Danielli suggested in 1935 that the membrane might be coated on both sides with hydrophilic proteins. They proposed a sandwich model: a phospholipid bilayer between two layers of proteins.

When researchers first used electron microscopes to study cells in the 1950s, the pictures seemed to support the Davson-Danielli model. By the late 1960s, however, many cell biologists recognized two problems with the model. First, inspection of a variety of membranes revealed that membranes with different functions differ in structure and chemical composition. A second, more serious problem became apparent once membrane proteins were better characterized. Unlike proteins dissolved in the cytosol, membrane proteins are not very soluble in water because they are amphipathic. If such proteins were layered on the surface of the membrane, their hydrophobic parts would be in aqueous surroundings.

Taking these observations into account, S. J. Singer and G. Nicolson proposed in 1972 that membrane proteins reside in the phospholipid bilayer with their hydrophilic regions protruding **(Figure 7.3)**. This molecular arrangement would maximize contact of hydrophilic regions of proteins and phospholipids with water in the cytosol and extracellular fluid, while providing their hydrophobic parts with a nonaqueous environment. In this fluid mosaic model, the membrane is a mosaic of protein molecules bobbing in a fluid bilayer of phospholipids.

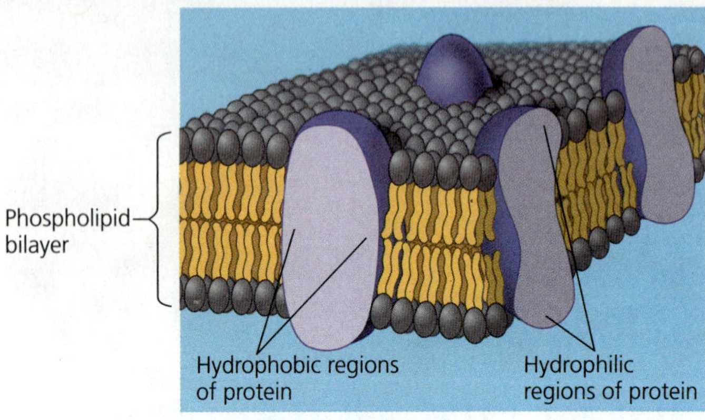

▲ **Figure 7.3 The original fluid mosaic model for membranes.**

A method of preparing cells for electron microscopy called freeze-fracture has demonstrated visually that proteins are indeed embedded in the phospholipid bilayer of the membrane **(Figure 7.4)**. Freeze-fracture splits a membrane along the middle of the bilayer, somewhat like pulling apart a chunky peanut butter sandwich. When the membrane layers are viewed in the electron microscope, the interior of the

▼ **Figure 7.4 Phospholipid bilayer (cross section).**

▼ **Figure 7.2 Phospholipid bilayer (cross section).**

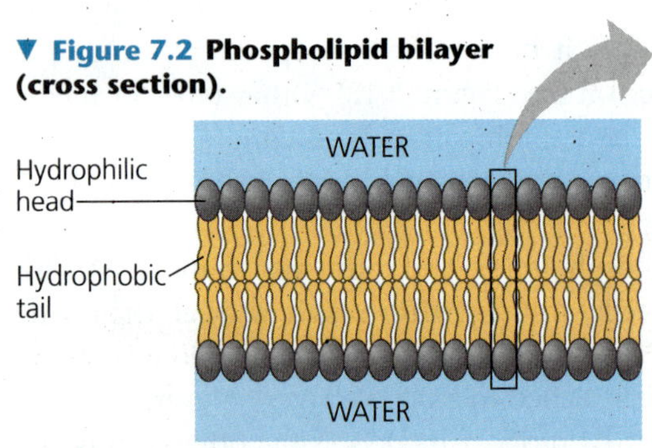

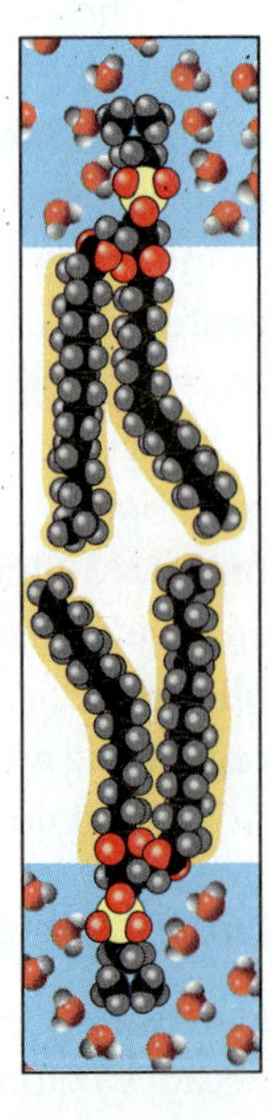

MAKE CONNECTIONS *Consulting Figure 5.12 (p. 76), circle the hydrophilic and hydrophobic portions of the enlarged phospholipids on the right. Explain what each portion contacts when the phospholipids are in the plasma membrane.*

▼ **Figure 7.4** RESEARCH METHOD

Freeze-fracture

APPLICATION A cell membrane can be split into its two layers, revealing the structure of the membrane's interior.

TECHNIQUE A cell is frozen and fractured with a knife. The fracture plane often follows the hydrophobic interior of a membrane, splitting the phospholipid bilayer into two separated layers. Each membrane protein goes wholly with one of the layers.

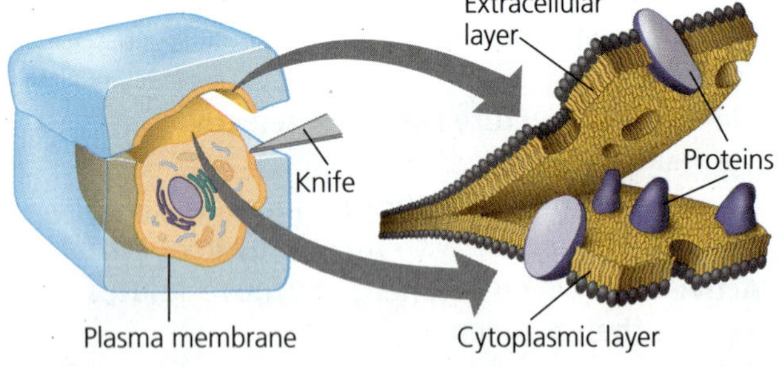

RESULTS These SEMs show membrane proteins (the "bumps") in the two layers, demonstrating that proteins are embedded in the phospholipid bilayer.

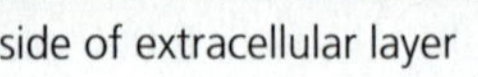

Inside of extracellular layer

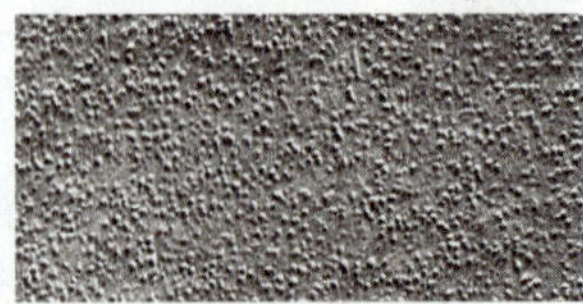

Inside of cytoplasmic layer

▲ **Figure 7.5 Updated model of an animal cell's plasma membrane (cutaway view).**

bilayer appears cobblestoned, with protein particles interspersed in a smooth matrix, in agreement with the fluid mosaic model. Some proteins remain attached to one layer or the other, like the peanut chunks in the sandwich.

Because models are hypotheses, replacing one model of membrane structure with another does not imply that the original model was worthless. The acceptance or rejection of a model depends on how well it fits observations and explains experimental results. New findings may make a model obsolete; even then, it may not be totally scrapped, but revised to incorporate the new observations. The fluid mosaic model is continually being refined. For example, groups of proteins are often found associated in long-lasting, specialized patches, where they carry out common functions. The lipids themselves appear to form defined regions as well. Also, the membrane may be much more packed with proteins than imagined in the classic fluid mosaic model—compare the updated model in **Figure 7.5** with the original model in Figure 7.3. Let's now take a closer look at membrane structure.

The Fluidity of Membranes

Membranes are not static sheets of molecules locked rigidly in place. A membrane is held together primarily by hydrophobic interactions, which are much weaker than covalent bonds (see Figure 5.20). Most of the lipids and some of the proteins can shift about laterally—that is, in the plane of the membrane, like partygoers elbowing their way through a crowded room **(Figure 7.6)**. It is quite rare, however, for a molecule to flip-flop transversely across the membrane, switching from one phospholipid layer to the other; to do so, the hydrophilic part of the molecule must cross the hydrophobic interior of the membrane.

The lateral movement of phospholipids within the membrane is rapid. Adjacent phospholipids switch positions about 10^7 times per second, which means that a phospholipid can travel about 2 μm—the length of many bacterial cells—in 1 second. Proteins are much larger than lipids and move more slowly, but some membrane proteins do drift, as shown in a classic experiment described in **Figure 7.7**, on the next page.

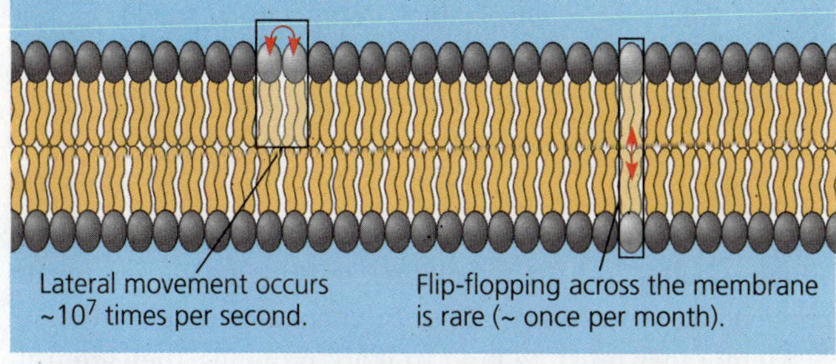

▲ **Figure 7.6 The movement of phospholipids.**

▼ Figure 7.7 **INQUIRY**

Do membrane proteins move?

EXPERIMENT Larry Frye and Michael Edidin, at Johns Hopkins University, labeled the plasma membrane proteins of a mouse cell and a human cell with two different markers and fused the cells. Using a microscope, they observed the markers on the hybrid cell.

RESULTS

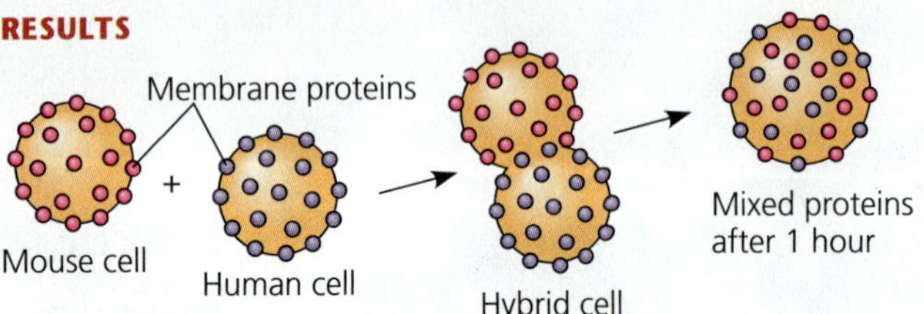

CONCLUSION The mixing of the mouse and human membrane proteins indicates that at least some membrane proteins move sideways within the plane of the plasma membrane.

SOURCE L. D. Frye and M. Edidin, The rapid intermixing of cell surface antigens after formation of mouse-human heterokaryons, *Journal of Cell Science* 7:319 (1970).

WHAT IF? Suppose the proteins did not mix in the hybrid cell, even many hours after fusion. Would you be able to conclude that proteins don't move within the membrane? What other explanation could there be?

And some membrane proteins seem to move in a highly directed manner, perhaps driven along cytoskeletal fibers by motor proteins connected to the membrane proteins' cytoplasmic regions. However, many other membrane proteins seem to be held immobile by their attachment to the cytoskeleton or to the extracellular matrix (see Figure 7.5).

A membrane remains fluid as temperature decreases until finally the phospholipids settle into a closely packed arrangement and the membrane solidifies, much as bacon grease forms lard when it cools. The temperature at which a membrane solidifies depends on the types of lipids it is made of. The membrane remains fluid to a lower temperature if it is rich in phospholipids with unsaturated hydrocarbon tails (see Figures 5.11 and 5.12). Because of kinks in the tails where double bonds are located, unsaturated hydrocarbon tails cannot pack together as closely as saturated hydrocarbon tails, and this makes the membrane more fluid **(Figure 7.8a)**.

The steroid cholesterol, which is wedged between phospholipid molecules in the plasma membranes of animal cells, has different effects on membrane fluidity at different temperatures **(Figure 7.8b)**. At relatively high temperatures—at 37°C, the body temperature of humans, for example—cholesterol makes the membrane less fluid by restraining phospholipid movement. However, because cholesterol also hinders the close packing of phospholipids, it lowers the temperature required for the membrane to solidify. Thus, cholesterol can be thought of as a "fluidity buffer" for the membrane, resisting changes in membrane fluidity that can be caused by changes in temperature.

Membranes must be fluid to work properly; they are usually about as fluid as salad oil. When a membrane solidifies, its permeability changes, and enzymatic proteins in the membrane may become inactive if their activity requires them to be able to move within the membrane. However, membranes that are too fluid cannot support protein function either. Therefore, extreme environments pose a challenge for life, resulting in evolutionary adaptations that include differences in membrane lipid composition.

Evolution of Differences in Membrane Lipid Composition

EVOLUTION Variations in the cell membrane lipid compositions of many species appear to be evolutionary adaptations that maintain the appropriate membrane fluidity under specific environmental conditions. For instance, fishes that live in extreme cold have membranes with a high proportion of unsaturated hydrocarbon tails, enabling their membranes to remain fluid (see Figure 7.8a). At the other extreme, some bacteria and archaea thrive at temperatures greater than 90°C (194°F) in thermal hot springs and geysers. Their membranes include unusual lipids that may prevent excessive fluidity at such high temperatures.

The ability to change the lipid composition of cell membranes in response to changing temperatures has evolved in organisms that live where temperatures vary. In many plants that tolerate extreme cold, such as winter wheat, the percentage of unsaturated phospholipids increases in autumn, an adjustment that keeps the membranes from solidifying during winter. Certain bacteria and archaea can also change the proportion of unsaturated phospholipids in their cell membranes, depending on the temperature at which they are growing. Overall, natural selection has apparently favored organisms whose mix of membrane lipids ensures an appropriate level of membrane fluidity for their environment.

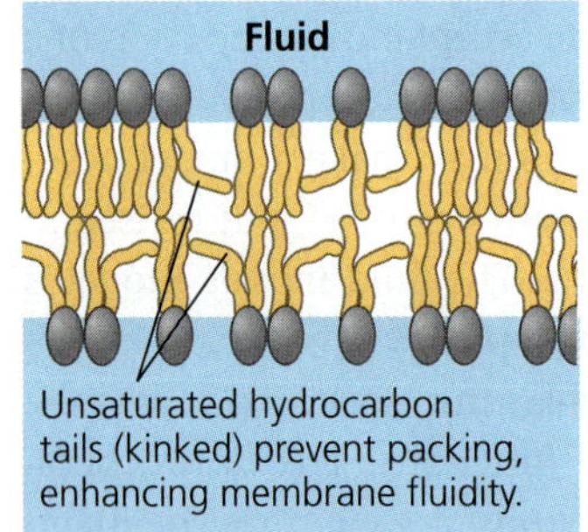

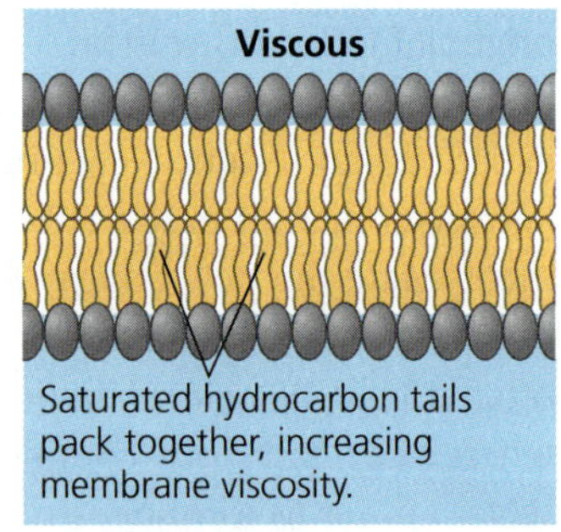

(a) Unsaturated versus saturated hydrocarbon tails.

(b) Cholesterol within the animal cell membrane. Cholesterol reduces membrane fluidity at moderate temperatures by reducing phospholipid movement, but at low temperatures it hinders solidification by disrupting the regular packing of phospholipids.

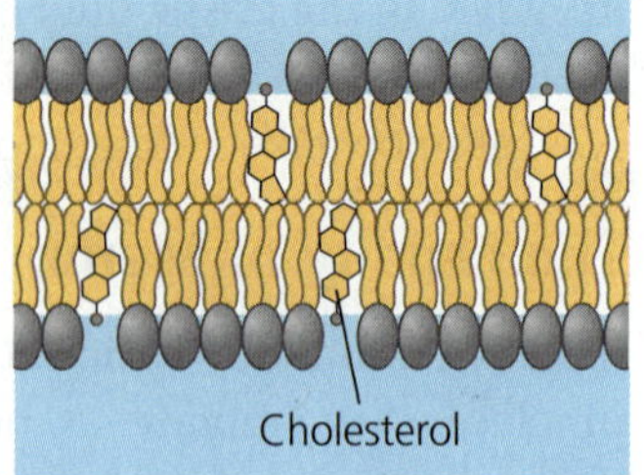

▲ **Figure 7.8 Factors that affect membrane fluidity.**

Membrane Proteins and Their Functions

Now we come to the *mosaic* aspect of the fluid mosaic model. Somewhat like a tile mosaic, a membrane is a collage of different proteins, often clustered together in groups, embedded in the fluid matrix of the lipid bilayer (see Figure 7.5). More than 50 kinds of proteins have been found so far in the plasma membrane of red blood cells, for example. Phospholipids form the main fabric of the membrane, but proteins determine most of the membrane's functions. Different types of cells contain different sets of membrane proteins, and the various membranes within a cell each have a unique collection of proteins.

Notice in Figure 7.5 that there are two major populations of membrane proteins: integral proteins and peripheral proteins. **Integral proteins** penetrate the hydrophobic interior of the lipid bilayer. The majority are *transmembrane proteins*, which span the membrane; other integral proteins extend only partway into the hydrophobic interior. The hydrophobic regions of an integral protein consist of one or more stretches of nonpolar amino acids (see Figure 5.16), usually coiled into α helices **(Figure 7.9)**. The hydrophilic parts of the molecule are exposed to the aqueous solutions on either side of the membrane. Some proteins also have a hydrophilic channel through their center that allows passage of hydrophilic substances (see Figure 7.1). **Peripheral proteins** are not embedded in the lipid bilayer at all; they are appendages loosely bound to the surface of the membrane, often to exposed parts of integral proteins (see Figure 7.5).

On the cytoplasmic side of the plasma membrane, some membrane proteins are held in place by attachment to the cytoskeleton. And on the extracellular side, certain membrane proteins are attached to fibers of the extracellular matrix (see Figure 6.30; *integrins* are one type of integral protein). These attachments combine to give animal cells a stronger framework than the plasma membrane alone could provide.

Figure 7.10 gives an overview of six major functions performed by proteins of the plasma membrane. A single cell

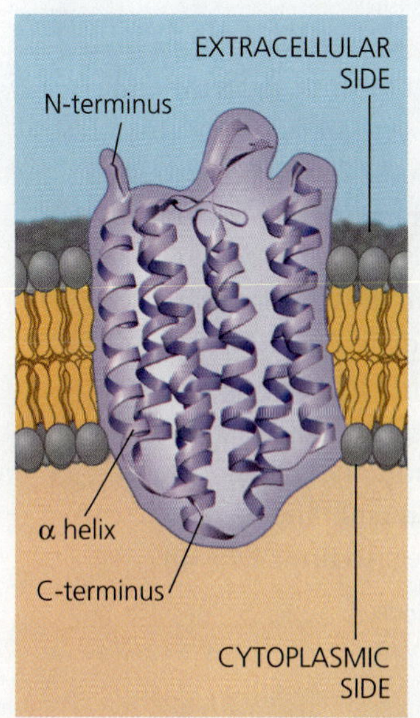

◀ **Figure 7.9 The structure of a transmembrane protein.** Bacteriorhodopsin (a bacterial transport protein) has a distinct orientation in the membrane, with its N-terminus outside the cell and its C-terminus inside. This ribbon model highlights the α-helical secondary structure of the hydrophobic parts, which lie mostly within the hydrophobic interior of the membrane. The protein includes seven transmembrane helices. The nonhelical hydrophilic segments are in contact with the aqueous solutions on the extracellular and cytoplasmic sides of the membrane.

(a) Transport. *Left:* A protein that spans the membrane may provide a hydrophilic channel across the membrane that is selective for a particular solute. *Right:* Other transport proteins shuttle a substance from one side to the other by changing shape (see Figure 7.17). Some of these proteins hydrolyze ATP as an energy source to actively pump substances across the membrane.

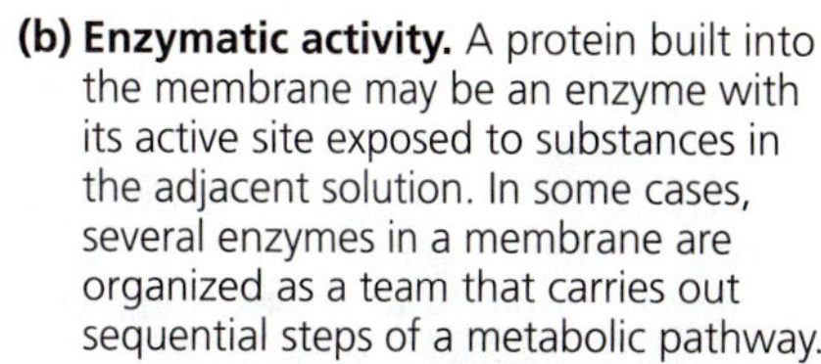

(b) Enzymatic activity. A protein built into the membrane may be an enzyme with its active site exposed to substances in the adjacent solution. In some cases, several enzymes in a membrane are organized as a team that carries out sequential steps of a metabolic pathway.

(c) Signal transduction. A membrane protein (receptor) may have a binding site with a specific shape that fits the shape of a chemical messenger, such as a hormone. The external messenger (signaling molecule) may cause the protein to change shape, allowing it to relay the message to the inside of the cell, usually by binding to a cytoplasmic protein (see Figure 11.6).

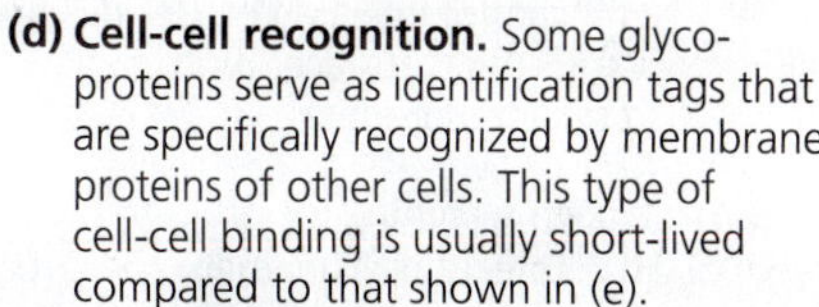

(d) Cell-cell recognition. Some glycoproteins serve as identification tags that are specifically recognized by membrane proteins of other cells. This type of cell-cell binding is usually short-lived compared to that shown in (e).

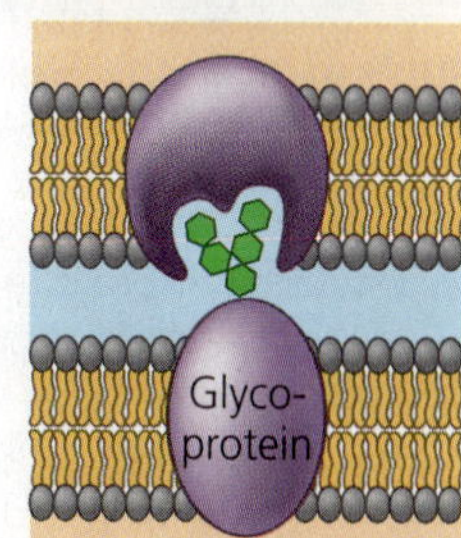

(e) Intercellular joining. Membrane proteins of adjacent cells may hook together in various kinds of junctions, such as gap junctions or tight junctions (see Figure 6.32). This type of binding is more long-lasting than that shown in (d).

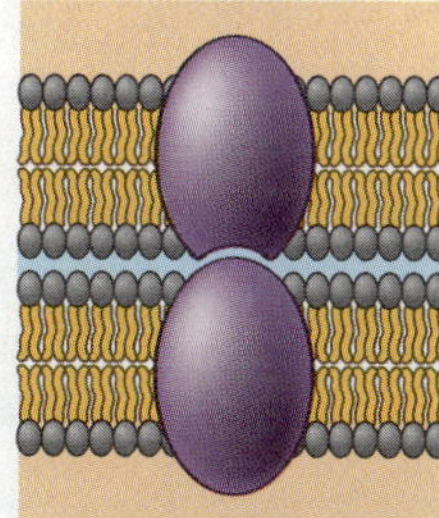

(f) Attachment to the cytoskeleton and extracellular matrix (ECM). Microfilaments or other elements of the cytoskeleton may be noncovalently bound to membrane proteins, a function that helps maintain cell shape and stabilizes the location of certain membrane proteins. Proteins that can bind to ECM molecules can coordinate extracellular and intracellular changes (see Figure 6.30).

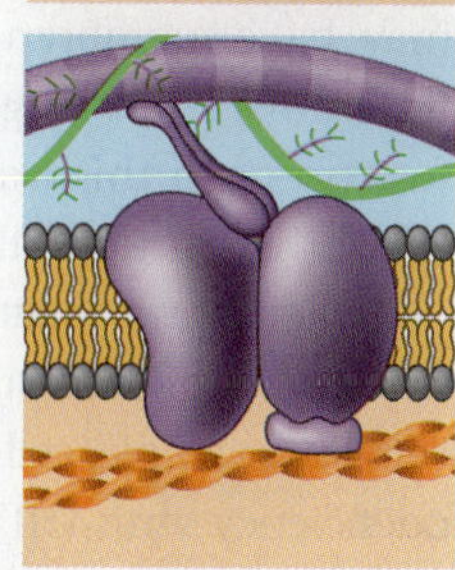

▲ **Figure 7.10 Some functions of membrane proteins.** In many cases, a single protein performs multiple tasks.

? *Some transmembrane proteins can bind to a particular ECM molecule and, when bound, transmit a signal into the cell. Use the proteins shown here to explain how this might occur.*

▼ Figure 7.11
IMPACT

Blocking HIV Entry into Cells as a Treatment for HIV Infections

Despite multiple exposures to HIV, a small number of people do not develop AIDS and show no evidence of HIV-infected cells. Comparing their genes with the genes of infected individuals, researchers discovered that resistant individuals have an unusual form of a gene that codes for an immune cell-surface protein called CCR5. Further work showed that HIV binds to a main protein receptor (CD4) on an immune cell, but most types of HIV also need to bind to CCR5 as a "co-receptor" to actually infect the cell (below, left). An absence of CCR5 on the cells of resistant individuals, due to the gene alteration, prevents the virus from entering the cells (below, right).

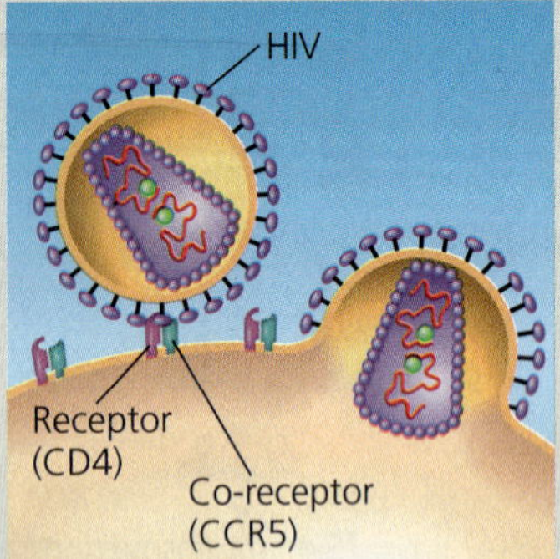

HIV can infect a cell that has CCR5 on its surface, as in most people.

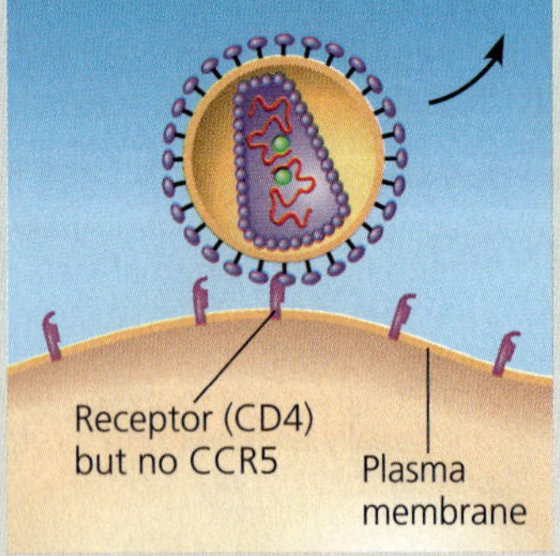

HIV cannot infect a cell lacking CCR5 on its surface, as in resistant individuals.

WHY IT MATTERS Researchers have been searching for drugs to block cell-surface receptors involved in HIV infection. The main receptor protein, CD4, performs many important functions for cells, so interfering with it could cause dangerous side effects. Discovery of the CCR5 co-receptor provided a safer target for development of drugs that mask CCR5 and block HIV entry. One such drug, maraviroc (brand name Selzentry), was approved for treatment of HIV infection in 2007.

FURTHER READING T. Kenakin, New bull's-eyes for drugs, *Scientific American* 293(4):50–57 (2005).

MAKE CONNECTIONS Study Figures 2.18 (p. 42) and 5.19 (p. 81), both of which show pairs of molecules binding to each other. What would you predict about CCR5 that would allow HIV to bind to it? How could a drug molecule interfere with this binding?

may have membrane proteins carrying out several of these functions, and a single membrane protein may have multiple functions. In this way, the membrane is a functional mosaic as well as a structural one.

Proteins on the surface of a cell are important in the medical field because some proteins can help outside agents invade the cell. For example, cell-surface proteins help the human immunodeficiency virus (HIV) infect immune system cells, leading to acquired immune deficiency syndrome (AIDS). (You'll read more about HIV in Chapter 19.) Learning about the proteins that HIV binds to on immune cells has been central to developing a treatment for HIV infection **(Figure 7.11)**.

The Role of Membrane Carbohydrates in Cell-Cell Recognition

Cell-cell recognition, a cell's ability to distinguish one type of neighboring cell from another, is crucial to the functioning of an organism. It is important, for example, in the sorting of cells into tissues and organs in an animal embryo. It is also the basis for the rejection of foreign cells by the immune system, an important line of defense in vertebrate animals (see Chapter 43). Cells recognize other cells by binding to molecules, often containing carbohydrates, on the extracellular surface of the plasma membrane (see Figure 7.10d).

Membrane carbohydrates are usually short, branched chains of fewer than 15 sugar units. Some are covalently bonded to lipids, forming molecules called **glycolipids**. (Recall that *glyco* refers to the presence of carbohydrate.) However, most are covalently bonded to proteins, which are thereby **glycoproteins** (see Figure 7.5).

The carbohydrates on the extracellular side of the plasma membrane vary from species to species, among individuals of the same species, and even from one cell type to another in a single individual. The diversity of the molecules and their location on the cell's surface enable membrane carbohydrates to function as markers that distinguish one cell from another. For example, the four human blood types designated A, B, AB, and O reflect variation in the carbohydrate part of glycoproteins on the surface of red blood cells.

Synthesis and Sidedness of Membranes

Membranes have distinct inside and outside faces. The two lipid layers may differ in specific lipid composition, and each protein has directional orientation in the membrane (see Figure 7.9). **Figure 7.12** shows how membrane sidedness arises: The asymmetrical arrangement of proteins, lipids, and their associated carbohydrates in the plasma membrane is determined as the membrane is being built by the endoplasmic reticulum (ER) and Golgi apparatus.

CONCEPT CHECK 7.1

1. The carbohydrates attached to some proteins and lipids of the plasma membrane are added as the membrane is made and refined in the ER and Golgi apparatus. The new membrane then forms transport vesicles that travel to the cell surface. On which side of the vesicle membrane are the carbohydrates?
2. **WHAT IF?** The soil immediately around hot springs is much warmer than that in neighboring regions. Two closely related species of native grasses are found, one in the warmer region and one in the cooler region. If you analyzed their membrane lipid compositions, what would you expect to find? Explain.

For suggested answers, see Appendix A.

▼ **Figure 7.12 Synthesis of membrane components and their orientation in the membrane.** The cytoplasmic (orange) face of the plasma membrane differs from the extracellular (aqua) face. The latter arises from the inside face of ER, Golgi, and vesicle membranes.

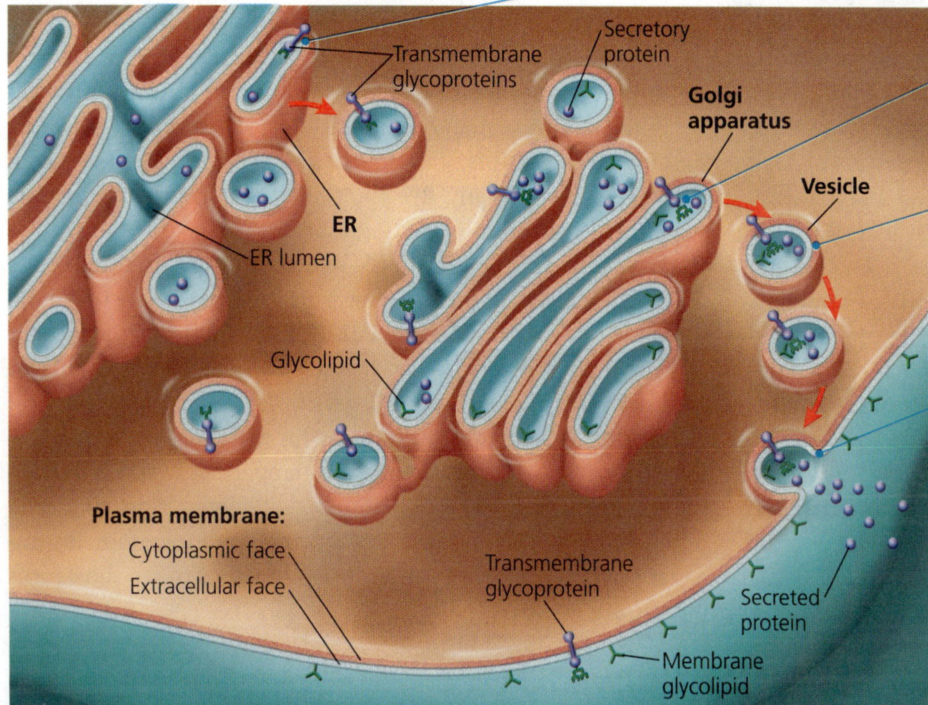

❶ Membrane proteins and lipids are synthesized in the endoplasmic reticulum (ER). Carbohydrates (green) are added to the transmembrane proteins (purple dumbbells), making them glycoproteins. The carbohydrate portions may then be modified.

❷ Inside the Golgi apparatus, the glycoproteins undergo further carbohydrate modification, and lipids acquire carbohydrates, becoming glycolipids.

❸ The glycoproteins, glycolipids, and secretory proteins (purple spheres) are transported in vesicles to the plasma membrane.

❹ As vesicles fuse with the plasma membrane, the outside face of the vesicle becomes continuous with the inside (cytoplasmic) face of the plasma membrane. This releases the secretory proteins from the cell, a process called *exocytosis*, and positions the carbohydrates of membrane glycoproteins and glycolipids on the outside (extracellular) face of the plasma membrane.

DRAW IT *Draw an integral membrane protein extending from partway through the ER membrane into the ER lumen. Next, draw the protein where it would be located in a series of numbered steps ending at the plasma membrane. Would the protein contact the cytoplasm or the extracellular fluid?*

CONCEPT **7.2**

Membrane structure results in selective permeability

The biological membrane is an exquisite example of a supramolecular structure—many molecules ordered into a higher level of organization—with emergent properties beyond those of the individual molecules. The remainder of this chapter focuses on one of the most important of those properties: the ability to regulate transport across cellular boundaries, a function essential to the cell's existence. We will see once again that form fits function: The fluid mosaic model helps explain how membranes regulate the cell's molecular traffic.

A steady traffic of small molecules and ions moves across the plasma membrane in both directions. Consider the chemical exchanges between a muscle cell and the extracellular fluid that bathes it. Sugars, amino acids, and other nutrients enter the cell, and metabolic waste products leave it. The cell takes in O_2 for use in cellular respiration and expels CO_2. Also, the cell regulates its concentrations of inorganic ions, such as Na^+, K^+, Ca^{2+}, and Cl^-, by shuttling them one way or the other across the plasma membrane. In spite of heavy traffic through them, cell membranes are selectively permeable, and substances do not cross the barrier indiscriminately. The cell is able to take up some small molecules and ions and exclude others. Also, substances that move through the membrane do so at different rates.

The Permeability of the Lipid Bilayer

Nonpolar molecules, such as hydrocarbons, carbon dioxide, and oxygen, are hydrophobic and can therefore dissolve in the lipid bilayer of the membrane and cross it easily, without the aid of membrane proteins. However, the hydrophobic interior of the membrane impedes the direct passage of ions and polar molecules, which are hydrophilic, through the membrane. Polar molecules such as glucose and other sugars pass only slowly through a lipid bilayer, and even water, an extremely small polar molecule, does not cross very rapidly. A charged atom or molecule and its surrounding shell of water (see Figure 3.7) find the hydrophobic interior of the membrane even more difficult to penetrate. Furthermore, the lipid bilayer is only one aspect of the gatekeeper system responsible for the selective permeability of a cell. Proteins built into the membrane play key roles in regulating transport.

Transport Proteins

Cell membranes *are* permeable to specific ions and a variety of polar molecules. These hydrophilic substances can avoid contact with the lipid bilayer by passing through **transport proteins** that span the membrane.

Some transport proteins, called *channel proteins*, function by having a hydrophilic channel that certain molecules or atomic ions use as a tunnel through the membrane (see Figure 7.10a, left). For example, the passage of water molecules through the membrane in certain cells is greatly facilitated by channel proteins known as **aquaporins**. Each aquaporin allows entry of up to *3 billion* (3×10^9) water molecules per second, passing single file through its central channel, which fits ten at a time. Without aquaporins, only a tiny fraction of these water molecules would pass through the same area of the cell membrane in a second, so the channel protein brings about a tremendous increase in rate. Other transport proteins, called *carrier proteins*, hold onto their passengers and change shape in a way that shuttles them across the membrane (see Figure 7.10a, right). A transport protein is specific for the substance it translocates (moves), allowing only a certain substance (or a small group of related substances) to cross the membrane. For example, a specific carrier protein in the plasma membrane of red blood cells transports glucose across the membrane 50,000 times faster than glucose can pass through on its own. This "glucose transporter" is so selective that it even rejects fructose, a structural isomer of glucose.

Thus, the selective permeability of a membrane depends on both the discriminating barrier of the lipid bilayer and the specific transport proteins built into the membrane. But what establishes the *direction* of traffic across a membrane? At a given time, what determines whether a particular substance will enter the cell or leave the cell? And what mechanisms actually drive molecules across membranes? We will address these questions next as we explore two modes of membrane traffic: passive transport and active transport.

CONCEPT CHECK 7.2

1. Two molecules that can cross a lipid bilayer without help from membrane proteins are O_2 and CO_2. What property allows this to occur?
2. Why is a transport protein needed to move water molecules rapidly and in large quantities across a membrane?
3. **MAKE CONNECTIONS** Aquaporins exclude passage of hydronium ions (H_3O^+; see pp. 52–53). Recent research on fat metabolism has shown that some aquaporins allow passage of glycerol, a three-carbon alcohol (see Figure 5.10, p. 75), as well as H_2O. Since H_3O^+ is much closer in size to water than is glycerol, what do you suppose is the basis of this selectivity?

For suggested answers, see Appendix A.

CONCEPT 7.3

Passive transport is diffusion of a substance across a membrane with no energy investment

Molecules have a type of energy called thermal energy (heat), due to their constant motion. One result of this motion is **diffusion**, the movement of molecules of any substance so that they spread out evenly into the available space. Each molecule moves randomly, yet diffusion of a *population* of molecules may be directional. To understand this process, let's imagine a synthetic membrane separating pure water from a solution of a dye in water. Study **Figure 7.13a** carefully to appreciate how diffusion would result in both solutions having equal concentrations of the dye molecules. Once that point is reached, there will be a dynamic equilibrium, with as many dye molecules crossing the membrane each second in one direction as in the other.

We can now state a simple rule of diffusion: In the absence of other forces, a substance will diffuse from where it is more concentrated to where it is less concentrated. Put another way, any substance will diffuse down its **concentration gradient**, the region along which the density of a chemical substance increases or decreases (in this case, decreases). No work must be done to make this happen; diffusion is a spontaneous process, needing no input of energy. Note that each substance diffuses down its *own* concentration gradient, unaffected by the concentration gradients of other substances **(Figure 7.13b)**.

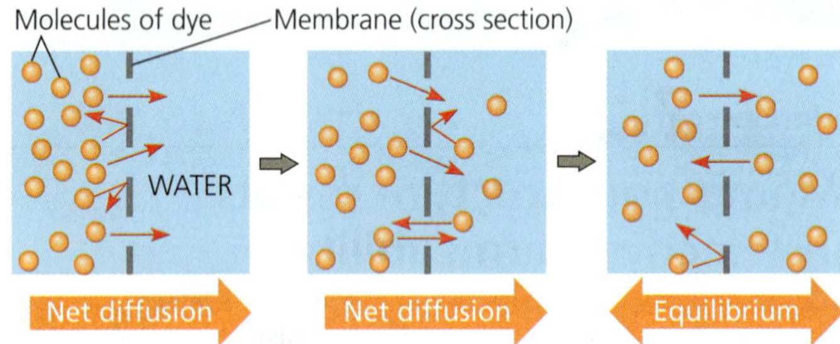

(a) Diffusion of one solute. The membrane has pores large enough for molecules of dye to pass through. Random movement of dye molecules will cause some to pass through the pores; this will happen more often on the side with more dye molecules. The dye diffuses from where it is more concentrated to where it is less concentrated (called diffusing down a concentration gradient). This leads to a dynamic equilibrium: The solute molecules continue to cross the membrane, but at equal rates in both directions.

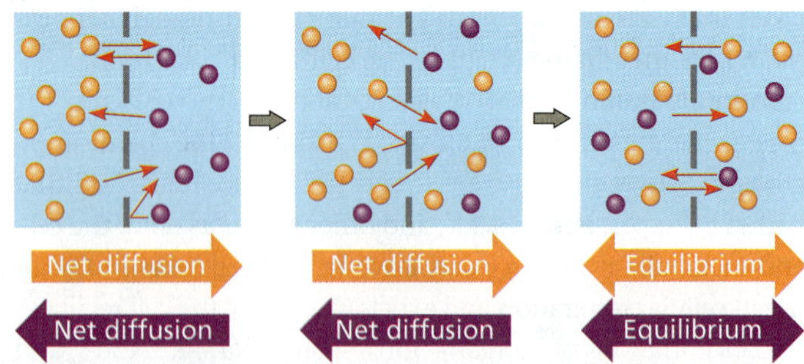

(b) Diffusion of two solutes. Solutions of two different dyes are separated by a membrane that is permeable to both. Each dye diffuses down its own concentration gradient. There will be a net diffusion of the purple dye toward the left, even though the *total* solute concentration was initially greater on the left side.

▲ **Figure 7.13 The diffusion of solutes across a synthetic membrane.** Each of the large arrows under the diagrams shows the net diffusion of the dye molecules of that color.

Much of the traffic across cell membranes occurs by diffusion. When a substance is more concentrated on one side of a membrane than on the other, there is a tendency for the substance to diffuse across the membrane down its concentration gradient (assuming that the membrane is permeable to that substance). One important example is the uptake of oxygen by a cell performing cellular respiration. Dissolved oxygen diffuses into the cell across the plasma membrane. As long as cellular respiration consumes the O_2 as it enters, diffusion into the cell will continue because the concentration gradient favors movement in that direction.

The diffusion of a substance across a biological membrane is called **passive transport** because the cell does not have to expend energy to make it happen. The concentration gradient itself represents potential energy (see Chapter 2, p. 35) and drives diffusion. Remember, however, that membranes are selectively permeable and therefore have different effects on the rates of diffusion of various molecules. In the case of water, aquaporins allow water to diffuse very rapidly across the membranes of certain cells. As we'll see next, the movement of water across the plasma membrane has important consequences for cells.

Effects of Osmosis on Water Balance

To see how two solutions with different solute concentrations interact, picture a U-shaped glass tube with a selectively permeable artificial membrane separating two sugar solutions **(Figure 7.14)**. Pores in this synthetic membrane are too small for sugar molecules to pass through but large enough for water molecules. How does this affect the *water* concentration? It seems logical that the solution with the higher concentration of solute would have the lower concentration of water and that water would diffuse into it from the other side for that reason. However, for a dilute solution like most biological fluids, solutes do not affect the water concentration significantly. Instead, tight clustering of water molecules around the hydrophilic solute molecules makes some of the water unavailable to cross the membrane. It is the difference in *free* water concentration that is important. In the end, the effect is the same: Water diffuses across the membrane from the region of lower solute concentration (higher free water concentration) to that of higher solute concentration (lower free water concentration) until the solute concentrations on both sides of the membrane are equal. The diffusion of free water across a selectively permeable membrane, whether artificial or cellular, is called **osmosis**. The movement of water across cell membranes and the balance of water between the cell and its environment are crucial to organisms. Let's now apply to living cells what we have learned about osmosis in artificial systems.

Water Balance of Cells Without Walls

To explain the behavior of a cell in a solution, we must consider both solute concentration and membrane permeability.

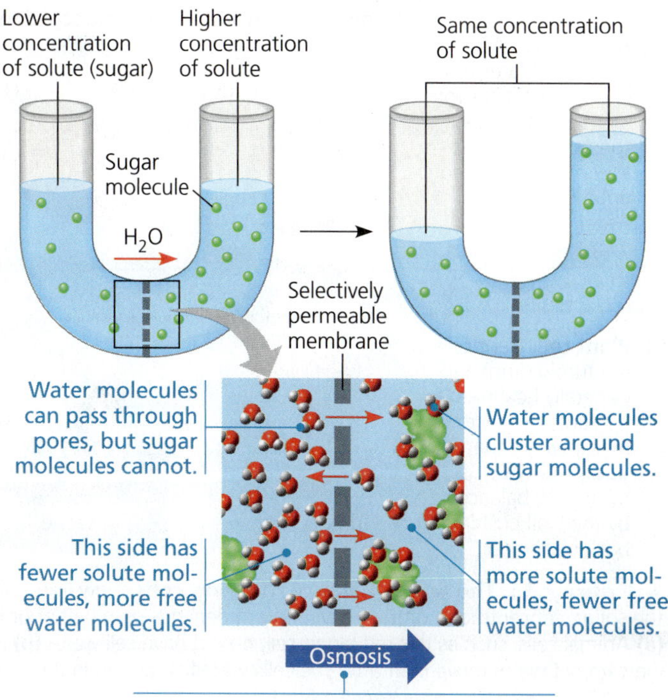

▲ **Figure 7.14 Osmosis.** Two sugar solutions of different concentrations are separated by a membrane that the solvent (water) can pass through but the solute (sugar) cannot. Water molecules move randomly and may cross in either direction, but overall, water diffuses from the solution with less concentrated solute to that with more concentrated solute. This diffusion of water, or osmosis, equalizes the sugar concentrations on both sides.

WHAT IF? *If an orange dye capable of passing through the membrane was added to the left side of the tube above, how would it be distributed at the end of the experiment? (See Figure 7.13.) Would the final solution levels in the tube be affected?*

Both factors are taken into account in the concept of **tonicity**, the ability of a surrounding solution to cause a cell to gain or lose water. The tonicity of a solution depends in part on its concentration of solutes that cannot cross the membrane (nonpenetrating solutes) relative to that inside the cell. If there is a higher concentration of nonpenetrating solutes in the surrounding solution, water will tend to leave the cell, and vice versa.

If a cell without a wall, such as an animal cell, is immersed in an environment that is **isotonic** to the cell (*iso* means "same"), there will be no *net* movement of water across the plasma membrane. Water diffuses across the membrane, but at the same rate in both directions. In an isotonic environment, the volume of an animal cell is stable (**Figure 7.15a**, on the next page).

Now let's transfer the cell to a solution that is **hypertonic** to the cell (*hyper* means "more," in this case referring to nonpenetrating solutes). The cell will lose water, shrivel, and probably die. This is one way an increase in the salinity (saltiness) of a lake can kill animals there; if the lake water becomes hypertonic to the animals' cells, the cells might shrivel and

(a) **Animal cell.** An animal cell fares best in an isotonic environment unless it has special adaptations that offset the osmotic uptake or loss of water.

(b) **Plant cell.** Plant cells are turgid (firm) and generally healthiest in a hypotonic environment, where the uptake of water is eventually balanced by the wall pushing back on the cell.

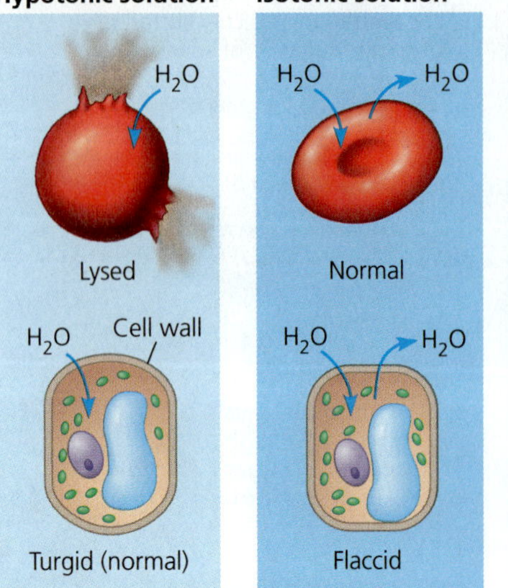

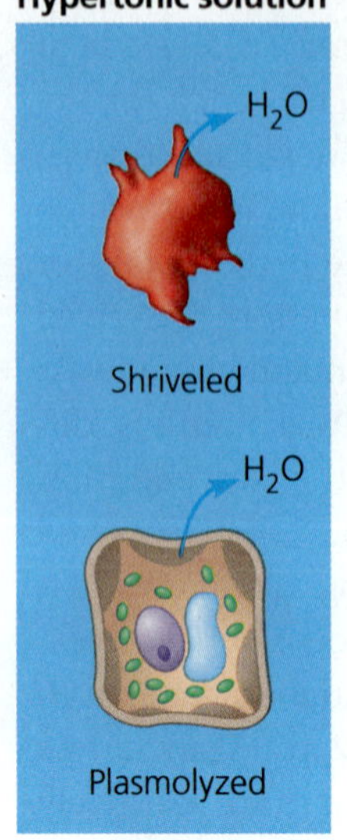

▲ **Figure 7.15 The water balance of living cells.** How living cells react to changes in the solute concentration of their environment depends on whether or not they have cell walls. **(a)** Animal cells, such as this red blood cell, do not have cell walls. **(b)** Plant cells do. (Arrows indicate net water movement after the cells were first placed in these solutions.)

die. However, taking up too much water can be just as hazardous to an animal cell as losing water. If we place the cell in a solution that is **hypotonic** to the cell (*hypo* means "less"), water will enter the cell faster than it leaves, and the cell will swell and lyse (burst) like an overfilled water balloon.

A cell without rigid walls can tolerate neither excessive uptake nor excessive loss of water. This problem of water balance is automatically solved if such a cell lives in isotonic surroundings. Seawater is isotonic to many marine invertebrates. The cells of most terrestrial (land-dwelling) animals are bathed in an extracellular fluid that is isotonic to the cells. In hypertonic or hypotonic environments, however, organisms that lack rigid cell walls must have other adaptations for **osmoregulation**, the control of solute concentrations and water balance. For example, the unicellular protist *Paramecium caudatum* lives in pond water, which is hypotonic to the cell. *P. caudatum* has a plasma membrane that is much less permeable to water than the membranes of most other cells, but this only slows the uptake of water, which continually enters the cell. The *P. caudatum* cell doesn't burst because it is also equipped with a contractile vacuole, an organelle that functions as a bilge pump to force water out of the cell as fast as it enters by osmosis **(Figure 7.16)**. We will examine other evolutionary adaptations for osmoregulation in Chapter 44.

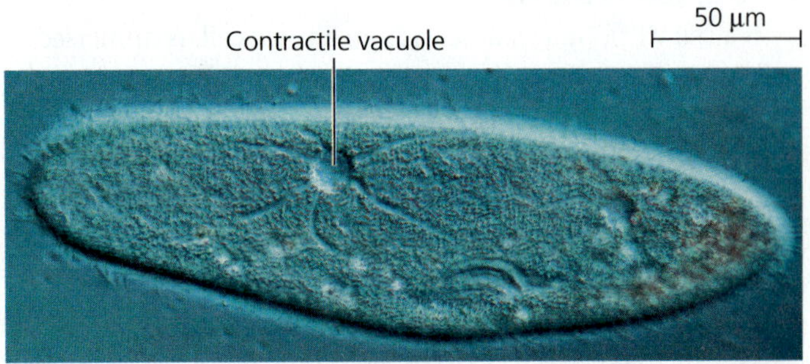

▲ **Figure 7.16 The contractile vacuole of *Paramecium caudatum*.** The vacuole collects fluid from a system of canals in the cytoplasm. When full, the vacuole and canals contract, expelling fluid from the cell (LM).

Water Balance of Cells with Walls

The cells of plants, prokaryotes, fungi, and some protists are surrounded by walls (see Figure 6.28). When such a cell is immersed in a hypotonic solution—bathed in rainwater, for example—the wall helps maintain the cell's water balance. Consider a plant cell. Like an animal cell, the plant cell swells as water enters by osmosis **(Figure 7.15b)**. However, the relatively inelastic wall will expand only so much before it exerts a back pressure on the cell, called *turgor pressure*, that opposes further water uptake. At this point, the cell is **turgid** (very firm), which is the healthy state for most plant cells. Plants that are not woody, such as most houseplants, depend for mechanical support on cells kept turgid by a surrounding hypotonic solution. If a plant's cells and their surroundings are isotonic, there is no net tendency for water to enter, and the cells become **flaccid** (limp).

However, a wall is of no advantage if the cell is immersed in a hypertonic environment. In this case, a plant cell, like an animal cell, will lose water to its surroundings and shrink. As the plant cell shrivels, its plasma membrane pulls away from the wall. This phenomenon, called **plasmolysis**, causes the plant to wilt and can lead to plant death. The walled cells of bacteria and fungi also plasmolyze in hypertonic environments.

Facilitated Diffusion: Passive Transport Aided by Proteins

Let's look more closely at how water and certain hydrophilic solutes cross a membrane. As mentioned earlier, many polar molecules and ions impeded by the lipid bilayer of the membrane diffuse passively with the help of transport proteins that span the membrane. This phenomenon is called **facilitated diffusion**. Cell biologists are still trying to learn exactly how various transport proteins facilitate diffusion. Most transport proteins are very specific: They transport some substances but not others.

As described earlier, the two types of transport proteins are channel proteins and carrier proteins. Channel proteins simply provide corridors that allow specific molecules or ions to cross the membrane **(Figure 7.17a)**. The hydrophilic passageways

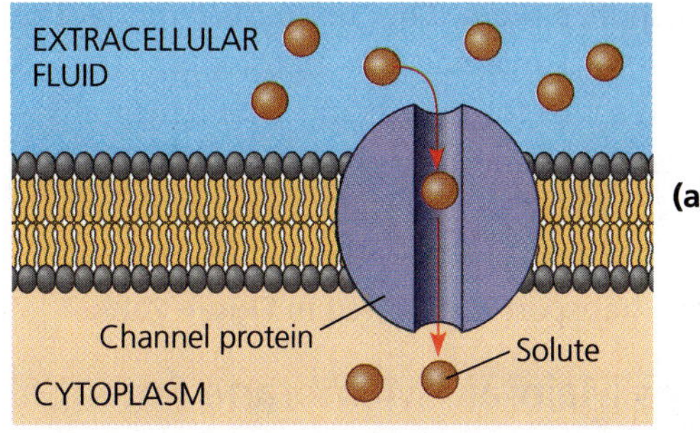

(a) A channel protein (purple) has a channel through which water molecules or a specific solute can pass.

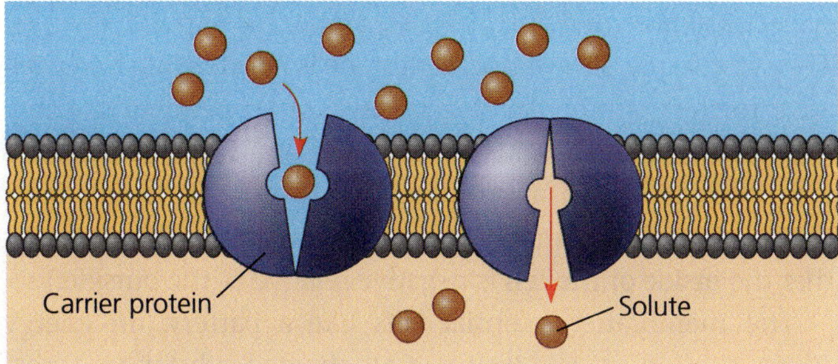

(b) A carrier protein alternates between two shapes, moving a solute across the membrane during the shape change.

▲ **Figure 7.17 Two types of transport proteins that carry out facilitated diffusion.** In both cases, the protein can transport the solute in either direction, but the net movement is down the concentration gradient of the solute.

provided by these proteins can allow water molecules or small ions to diffuse very quickly from one side of the membrane to the other. Aquaporins, the water channel proteins, facilitate the massive amounts of diffusion that occur in plant cells and in animal cells such as red blood cells (see Figure 7.15). Certain kidney cells also have a high number of aquaporins, allowing them to reclaim water from urine before it is excreted. If the kidneys did not perform this function, you would excrete about 180 L of urine per day—and have to drink an equal volume of water!

Channel proteins that transport ions are called **ion channels**. Many ion channels function as **gated channels**, which open or close in response to a stimulus. For some gated channels, the stimulus is electrical. The ion channel shown in Figure 7.1, for example, opens in response to an electrical stimulus, allowing potassium ions to leave the cell. Other gated channels open or close when a specific substance other than the one to be transported binds to the channel. Both types of gated channels are important in the functioning of the nervous system, as you'll learn in Chapter 48.

Carrier proteins, such as the glucose transporter mentioned earlier, seem to undergo a subtle change in shape that somehow translocates the solute-binding site across the membrane **(Figure 7.17b)**. Such a change in shape may be triggered by the binding and release of the transported molecule. Like ion channels, carrier proteins involved in facilitated diffusion result in the net movement of a substance down its concentration gradient. No energy input is thus required: This is passive transport.

In certain inherited diseases, specific transport systems are either defective or missing altogether. An example is cystinuria, a human disease characterized by the absence of a carrier protein that transports cysteine and some other amino acids across the membranes of kidney cells. Kidney cells normally reabsorb these amino acids from the urine and return them to the blood, but an individual afflicted with cystinuria develops painful stones from amino acids that accumulate and crystallize in the kidneys.

CONCEPT CHECK 7.3

1. How do you think a cell performing cellular respiration rids itself of the resulting CO_2?
2. In the supermarket, produce is often sprayed with water. Explain why this makes vegetables look crisp.
3. **WHAT IF?** If a *Paramecium caudatum* swims from a hypotonic to an isotonic environment, will its contractile vacuole become more active or less? Why?

For suggested answers, see Appendix A.

CONCEPT **7.4**

Active transport uses energy to move solutes against their gradients

Despite the help of transport proteins, facilitated diffusion is considered passive transport because the solute is moving down its concentration gradient, a process that requires no energy. Facilitated diffusion speeds transport of a solute by providing efficient passage through the membrane, but it does not alter the direction of transport. Some transport proteins, however, can move solutes against their concentration gradients, across the plasma membrane from the side where they are less concentrated (whether inside or outside) to the side where they are more concentrated.

The Need for Energy in Active Transport

To pump a solute across a membrane against its gradient requires work; the cell must expend energy. Therefore, this type of membrane traffic is called **active transport**. The transport proteins that move solutes against their concentration gradients are all carrier proteins rather than channel proteins. This makes sense because when channel proteins are open, they merely allow solutes to diffuse down their concentration gradients rather than picking them up and transporting them against their gradients.

Active transport enables a cell to maintain internal concentrations of small solutes that differ from concentrations in its environment. For example, compared with its surroundings,

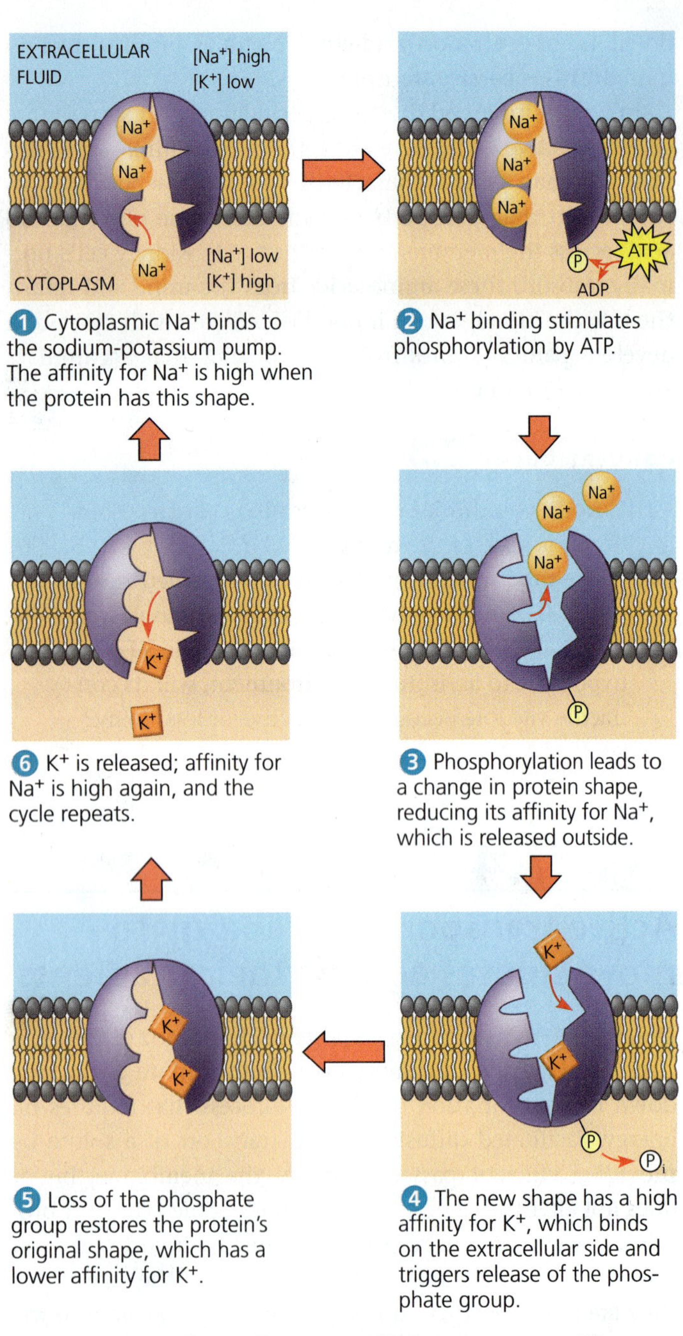

▲ **Figure 7.18 The sodium-potassium pump: a specific case of active transport.** This transport system pumps ions against steep concentration gradients: Sodium ion concentration ([Na^+]) is high outside the cell and low inside, while potassium ion concentration ([K^+]) is low outside the cell and high inside. The pump oscillates between two shapes in a cycle that moves 3 Na^+ out of the cell for every 2 K^+ pumped into the cell. The two shapes have different affinities for Na^+ and K^+. ATP powers the shape change by transferring a phosphate group to the transport protein (phosphorylating the protein).

an animal cell has a much higher concentration of potassium ions (K^+) and a much lower concentration of sodium ions (Na^+). The plasma membrane helps maintain these steep gradients by pumping Na^+ out of the cell and K^+ into the cell.

As in other types of cellular work, ATP supplies the energy for most active transport. One way ATP can power active transport is by transferring its terminal phosphate group directly to the transport protein. This can induce the protein to change its shape in a manner that translocates a solute bound to the protein across the membrane. One transport system that works this way is the **sodium-potassium pump**, which exchanges Na^+ for K^+ across the plasma membrane of animal cells **(Figure 7.18)**. The distinction between passive transport and active transport is reviewed in **Figure 7.19**.

How Ion Pumps Maintain Membrane Potential

All cells have voltages across their plasma membranes. Voltage is electrical potential energy—a separation of opposite charges. The cytoplasmic side of the membrane is negative in charge relative to the extracellular side because of an unequal distribution of anions and cations on the two sides. The voltage across a membrane, called a **membrane potential**, ranges from about -50 to -200 millivolts (mV). (The minus sign indicates that the inside of the cell is negative relative to the outside.)

The membrane potential acts like a battery, an energy source that affects the traffic of all charged substances across the membrane. Because the inside of the cell is negative compared with the outside, the membrane potential favors the passive transport of cations into the cell and anions out of the cell. Thus, *two* forces drive the diffusion of ions across a membrane: a chemical force (the ion's concentration gradient) and an electrical force (the effect of the membrane potential on

▼ **Figure 7.19 Review: passive and active transport.**

Passive transport. Substances diffuse spontaneously down their concentration gradients, crossing a membrane with no expenditure of energy by the cell. The rate of diffusion can be greatly increased by transport proteins in the membrane.

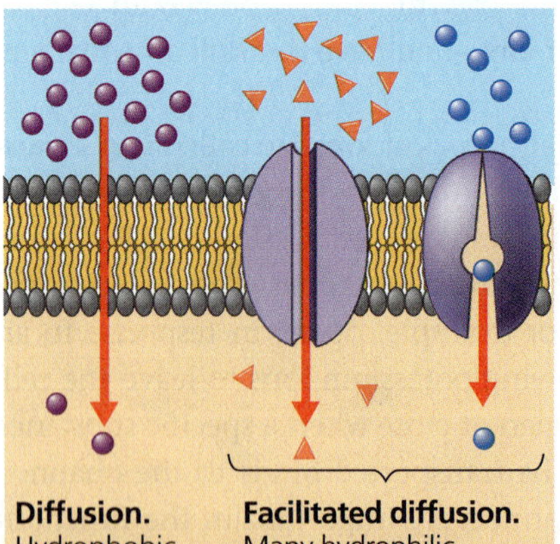

Diffusion. Hydrophobic molecules and (at a slow rate) very small uncharged polar molecules can diffuse through the lipid bilayer.

Facilitated diffusion. Many hydrophilic substances diffuse through membranes with the assistance of transport proteins, either channel proteins (left) or carrier proteins (right).

Active transport. Some transport proteins act as pumps, moving substances across a membrane against their concentration (or electrochemical) gradients. Energy for this work is usually supplied by ATP.

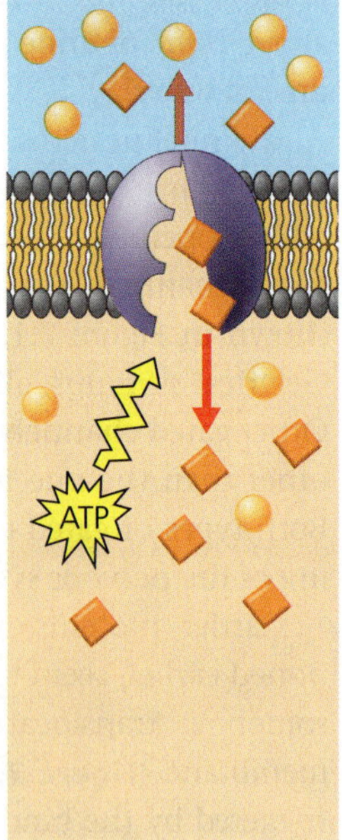

? *For each solute in the right panel, describe its direction of movement, and state whether it is going with or against its concentration gradient.*

the ion's movement). This combination of forces acting on an ion is called the **electrochemical gradient.**

In the case of ions, then, we must refine our concept of passive transport: An ion diffuses not simply down its *concentration* gradient but, more exactly, down its *electrochemical* gradient. For example, the concentration of Na^+ inside a resting nerve cell is much lower than outside it. When the cell is stimulated, gated channels open that facilitate Na^+ diffusion. Sodium ions then "fall" down their electrochemical gradient, driven by the concentration gradient of Na^+ and by the attraction of these cations to the negative side (inside) of the membrane. In this example, both electrical and chemical contributions to the electrochemical gradient act in the same direction across the membrane, but this is not always so. In cases where electrical forces due to the membrane potential oppose the simple diffusion of an ion down its concentration gradient, active transport may be necessary. In Chapter 48, you'll learn about the importance of electrochemical gradients and membrane potentials in the transmission of nerve impulses.

Some membrane proteins that actively transport ions contribute to the membrane potential. An example is the sodium-potassium pump. Notice in Figure 7.18 that the pump does not translocate Na^+ and K^+ one for one, but pumps three sodium ions out of the cell for every two potassium ions it pumps into the cell. With each "crank" of the pump, there is a net transfer of one positive charge from the cytoplasm to the extracellular fluid, a process that stores energy as voltage. A transport protein that generates voltage across a membrane is called an **electrogenic pump**. The sodium-potassium pump appears to be the major electrogenic pump of animal cells. The main electrogenic pump of plants, fungi, and bacteria is a **proton pump**, which actively transports protons (hydrogen ions, H^+) out of the cell. The pumping of H^+ transfers positive charge from the cytoplasm to the extracellular solution **(Figure 7.20)**. By generating voltage across membranes, electrogenic pumps help store energy that can be tapped for cellular work. One important use of proton gradients in the cell is for ATP synthesis during cellular respiration, as you will see in Chapter 9. Another is a type of membrane traffic called cotransport.

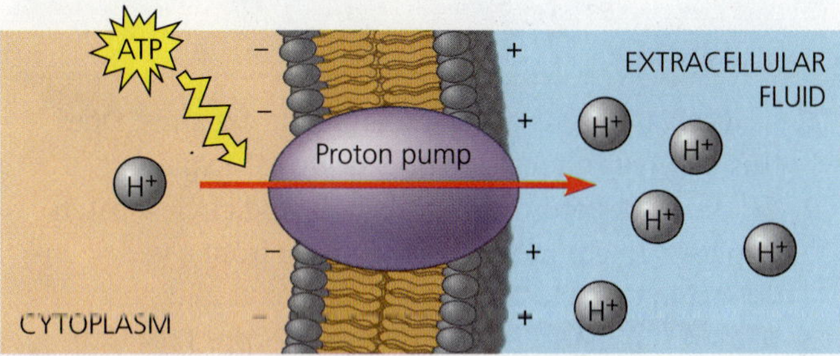

▲ **Figure 7.20 A proton pump.** Proton pumps are electrogenic pumps that store energy by generating voltage (charge separation) across membranes. A proton pump translocates positive charge in the form of hydrogen ions. The voltage and H^+ concentration gradient represent a dual energy source that can drive other processes, such as the uptake of nutrients. Most proton pumps are powered by ATP.

Cotransport: Coupled Transport by a Membrane Protein

A single ATP-powered pump that transports a specific solute can indirectly drive the active transport of several other solutes in a mechanism called **cotransport**. A substance that has been pumped across a membrane can do work as it moves back across the membrane by diffusion, analogous to water that has been pumped uphill and performs work as it flows back down. Another transport protein, a cotransporter separate from the pump, can couple the "downhill" diffusion of this substance to the "uphill" transport of a second substance against its own concentration (or electrochemical) gradient. For example, a plant cell uses the gradient of H^+ generated by its proton pumps to drive the active transport of amino acids, sugars, and several other nutrients into the cell. One transport protein couples the return of H^+ to the transport of sucrose into the cell **(Figure 7.21)**. This protein can translocate sucrose into the cell against a concentration gradient, but only if the sucrose molecule travels in the company of a hydrogen ion. The hydrogen ion uses the transport protein as an avenue to diffuse down the electrochemical gradient maintained by the proton pump. Plants use sucrose-H^+ cotransport to load sucrose produced by photosynthesis into cells in the veins of leaves. The vascular tissue of the plant can then distribute the sugar to nonphotosynthetic organs, such as roots.

What we know about cotransport proteins in animal cells has helped us find more effective treatments for diarrhea, a serious problem in developing countries. Normally, sodium in waste is reabsorbed in the colon, maintaining constant levels in the body, but diarrhea expels waste so rapidly that reabsorption is not possible, and sodium levels fall precipitously.

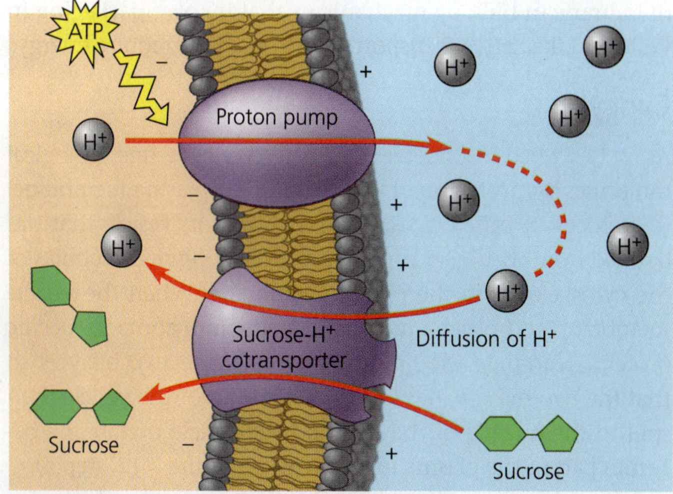

▲ **Figure 7.21 Cotransport: active transport driven by a concentration gradient.** A carrier protein, such as this sucrose-H^+ cotransporter in a plant cell, is able to use the diffusion of H^+ down its electrochemical gradient into the cell to drive the uptake of sucrose. The H^+ gradient is maintained by an ATP-driven proton pump that concentrates H^+ outside the cell, thus storing potential energy that can be used for active transport, in this case of sucrose. Thus, ATP indirectly provides the energy necessary for cotransport. (The cell wall is not shown.)

To treat this life-threatening condition, patients are given a solution to drink containing high concentrations of salt (NaCl) and glucose. The solutes are taken up by sodium-glucose cotransporters on the surface of intestinal cells and passed through the cells into the blood. This simple treatment has lowered infant mortality worldwide.

CONCEPT CHECK 7.4

1. Sodium-potassium pumps help nerve cells establish a voltage across their plasma membranes. Do these pumps use ATP or produce ATP? Explain.
2. Explain why the sodium-potassium pump in Figure 7.18 would not be considered a cotransporter.
3. **MAKE CONNECTIONS** Review the characteristics of the lysosome in Concept 6.4 (pp. 106–107). Given the internal environment of a lysosome, what transport protein might you expect to see in its membrane?

For suggested answers, see Appendix A.

CONCEPT 7.5

Bulk transport across the plasma membrane occurs by exocytosis and endocytosis

Water and small solutes enter and leave the cell by diffusing through the lipid bilayer of the plasma membrane or by being pumped or moved across the membrane by transport proteins. However, large molecules, such as proteins and polysaccharides, as well as larger particles, generally cross the membrane in bulk by mechanisms that involve packaging in vesicles. Like active transport, these processes require energy.

Exocytosis

As we described in Chapter 6, the cell secretes certain biological molecules by the fusion of vesicles with the plasma membrane; this process is called **exocytosis**. A transport vesicle that has budded from the Golgi apparatus moves along microtubules of the cytoskeleton to the plasma membrane. When the vesicle membrane and plasma membrane come into contact, specific proteins rearrange the lipid molecules of the two bilayers so that the two membranes fuse. The contents of the vesicle then spill to the outside of the cell, and the vesicle membrane becomes part of the plasma membrane (see Figure 7.12, step 4).

Many secretory cells use exocytosis to export products. For example, the cells in the pancreas that make insulin secrete it into the extracellular fluid by exocytosis. In another example, neurons (nerve cells) use exocytosis to release neurotransmitters that signal other neurons or muscle cells. When plant cells are making walls, exocytosis delivers proteins and carbohydrates from Golgi vesicles to the outside of the cell.

Endocytosis

In **endocytosis**, the cell takes in biological molecules and particulate matter by forming new vesicles from the plasma membrane. Although the proteins involved in the processes are different, the events of endocytosis look like the reverse of exocytosis. A small area of the plasma membrane sinks inward to form a pocket. As the pocket deepens, it pinches in, forming a vesicle containing material that had been outside the cell. Study **Figure 7.22** carefully to understand the three types of endocytosis: phagocytosis ("cellular eating"), pinocytosis ("cellular drinking"), and receptor-mediated endocytosis.

Human cells use receptor-mediated endocytosis to take in cholesterol for membrane synthesis and the synthesis of other steroids. Cholesterol travels in the blood in particles called low-density lipoproteins (LDLs), each a complex of lipids and a protein. LDLs bind to LDL receptors on plasma membranes and then enter the cells by endocytosis. (LDLs thus act as **ligands**, a term for any molecule that binds specifically to a receptor site on another molecule.) In humans with familial hypercholesterolemia, an inherited disease characterized by a very high level of cholesterol in the blood, LDLs cannot enter cells because the LDL receptor proteins are defective or missing. Consequently, cholesterol accumulates in the blood, where it contributes to early atherosclerosis, the buildup of lipid deposits within the walls of blood vessels. This buildup causes the walls to bulge inward, thereby narrowing the vessels and impeding blood flow.

Vesicles not only transport substances between the cell and its surroundings but also provide a mechanism for rejuvenating or remodeling the plasma membrane. Endocytosis and exocytosis occur continually in most eukaryotic cells, yet the amount of plasma membrane in a nongrowing cell remains fairly constant. Apparently, the addition of membrane by one process offsets the loss of membrane by the other.

Energy and cellular work have figured prominently in our study of membranes. We have seen, for example, that active transport is powered by ATP. In the next three chapters, you will learn more about how cells acquire chemical energy to do the work of life.

CONCEPT CHECK 7.5

1. As a cell grows, its plasma membrane expands. Does this involve endocytosis or exocytosis? Explain.
2. **DRAW IT** Return to Figure 7.12, and circle a patch of plasma membrane that is coming from a vesicle involved in exocytosis.
3. **MAKE CONNECTIONS** In Concept 6.7 (pp. 119–120), you learned that animal cells make an extracellular matrix (ECM). Describe the cellular pathway of synthesis and deposition of an ECM glycoprotein.

For suggested answers, see Appendix A.

▼ **Figure 7.22**

Exploring Endocytosis in Animal Cells

Phagocytosis

EXTRACELLULAR FLUID

Solutes

Pseudopodium

"Food" or other particle

Food vacuole

CYTOPLASM

In **phagocytosis,** a cell engulfs a particle by wrapping pseudopodia (singular, *pseudopodium*) around it and packaging it within a membranous sac called a food vacuole. The particle will be digested after the food vacuole fuses with a lysosome containing hydrolytic enzymes (see Figure 6.13a).

Pseudopodium of amoeba

Bacterium

Food vacuole

1 µm

An amoeba engulfing a bacterium via phagocytosis (TEM).

Pinocytosis

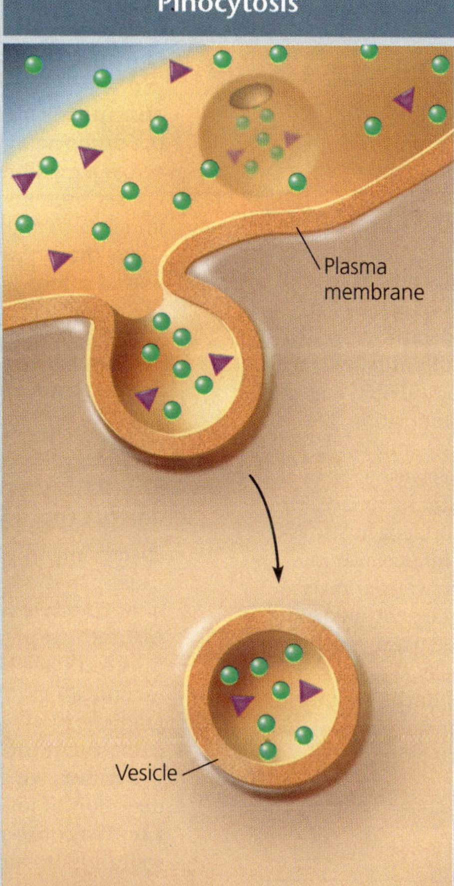

In **pinocytosis,** the cell "gulps" droplets of extracellular fluid into tiny vesicles. It is not the fluid itself that is needed by the cell, but the molecules dissolved in the droplets. Because any and all included solutes are taken into the cell, pinocytosis is nonspecific in the substances it transports.

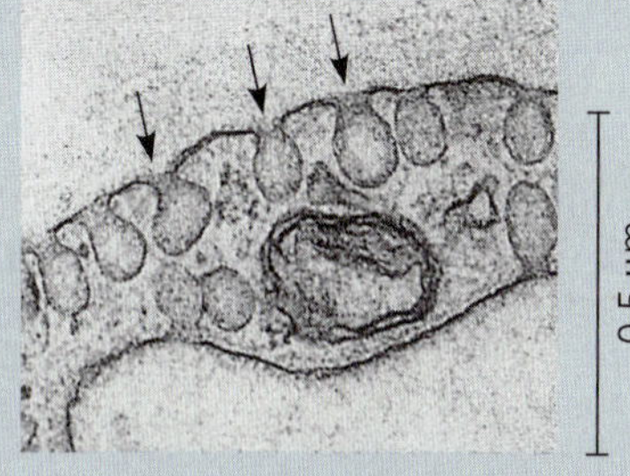

Pinocytosis vesicles forming (indicated by arrows) in a cell lining a small blood vessel (TEM).

BioFlix Visit the Study Area at **www.masteringbiology.com** for the BioFlix® 3-D Animation on Membrane Transport.

Receptor-Mediated Endocytosis

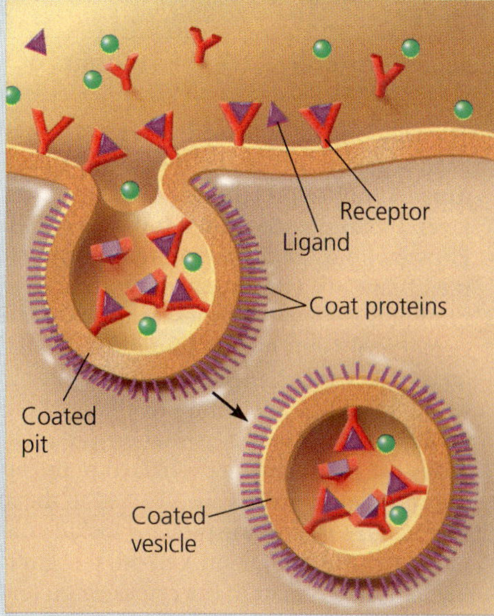

Receptor-mediated endocytosis enables the cell to acquire bulk quantities of specific substances, even though those substances may not be very concentrated in the extracellular fluid. Embedded in the membrane are proteins with specific receptor sites exposed to the extracellular fluid, to which specific substances (ligands) bind. The receptor proteins then cluster in regions of the membrane called coated pits, which are lined on their cytoplasmic side by a fuzzy layer of coat proteins. Next, each coated pit forms a vesicle containing the ligand molecules. Notice that there are relatively more bound molecules (purple) inside the vesicle, but other molecules (green) are also present. After the ingested material is liberated from the vesicle, the emptied receptors are recycled to the plasma membrane by the same vesicle.

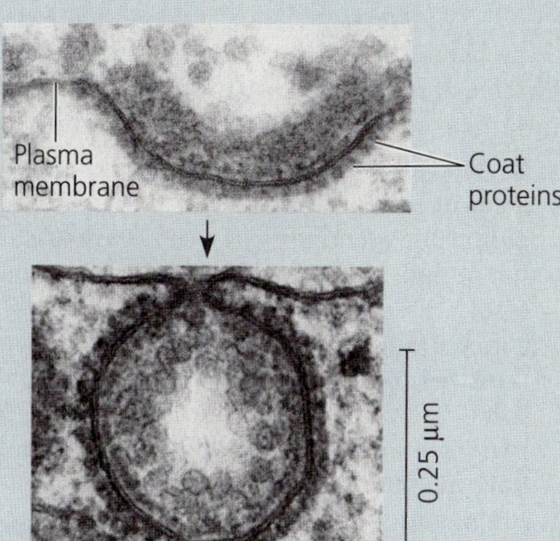

Top: A coated pit. *Bottom:* A coated vesicle forming during receptor-mediated endocytosis (TEMs).

7 CHAPTER REVIEW

SUMMARY OF KEY CONCEPTS

CONCEPT 7.1

Cellular membranes are fluid mosaics of lipids and proteins (pp. 125–131)

- The Davson-Danielli sandwich model of the membrane has been replaced by the **fluid mosaic model**, in which **amphipathic** proteins are embedded in the phospholipid bilayer. Proteins with related functions often cluster in patches.
- Phospholipids and some proteins move laterally within the membrane. The unsaturated hydrocarbon tails of some phospholipids keep membranes fluid at lower temperatures, while cholesterol helps membranes resist changes in fluidity caused by temperature changes. Differences in membrane lipid composition, as well as the ability to change lipid composition, are evolutionary adaptations that ensure membrane fluidity.
- **Integral proteins** are embedded in the lipid bilayer; **peripheral proteins** are attached to the membrane surface. The functions of membrane proteins include transport, enzymatic activity, signal transduction, cell-cell recognition, intercellular joining, and attachment to the cytoskeleton and extracellular matrix. Short chains of sugars linked to proteins (in **glycoproteins**) and lipids (in **glycolipids**) on the exterior side of the plasma membrane interact with surface molecules of other cells.
- Membrane proteins and lipids are synthesized in the ER and modified in the ER and Golgi apparatus. The inside and outside faces of membranes differ in molecular composition.

? *In what ways are membranes crucial to life?*

CONCEPT 7.2

Membrane structure results in selective permeability (pp. 131–132)

- A cell must exchange molecules and ions with its surroundings, a process controlled by the **selective permeability** of the plasma membrane. Hydrophobic substances are soluble in lipid and pass through membranes rapidly, whereas polar molecules and ions generally require specific **transport proteins** to cross the membrane.

? *How do **aquaporins** affect the permeability of a membrane?*

CONCEPT 7.3

Passive transport is diffusion of a substance across a membrane with no energy investment (pp. 132–135)

- **Diffusion** is the spontaneous movement of a substance down its **concentration gradient**. Water diffuses out through the permeable membrane of a cell (**osmosis**) if the solution outside has a higher solute concentration (**hypertonic**) than the cytosol; water enters the cell if the solution has a lower solute concentration (**hypotonic**). If the concentrations are equal (**isotonic**), no net osmosis occurs. Cell survival depends on balancing water uptake and loss. Cells lacking walls (as in animals and some protists) are isotonic with their environments or have adaptations for **osmoregulation**. Plants, prokaryotes, fungi, and some protists have relatively inelastic cell walls, so the cells don't burst in a hypotonic environment.
- In a type of **passive transport** called **facilitated diffusion**, a transport protein speeds the movement of water or a solute across a membrane down its concentration gradient. **Ion channels**, some of which are **gated channels**, facilitate the diffusion of ions across a membrane. Carrier proteins can undergo changes in shape that translocate bound solutes across the membrane.

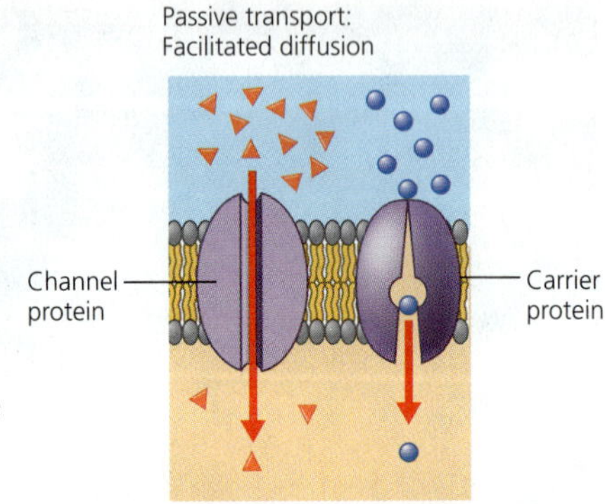

? *What happens to a cell placed in a hypertonic solution? Describe the free water concentration inside and out.*

CONCEPT 7.4

Active transport uses energy to move solutes against their gradients (pp. 135–138)

Active transport

- Specific membrane proteins use energy, usually in the form of ATP, to do the work of **active transport**. The **sodium-potassium pump** is an example.
- Ions can have both a concentration (chemical) gradient and an electrical gradient (voltage). These gradients combine in the **electrochemical gradient**, which determines the net direction of ionic diffusion. **Electrogenic pumps**, such as the sodium-potassium pump and **proton pumps**, are transport proteins that contribute to electrochemical gradients.
- **Cotransport** of two solutes occurs when a membrane protein enables the "downhill" diffusion of one solute to drive the "uphill" transport of the other.

? *ATP is not directly involved in the functioning of a cotransporter. Why, then, is cotransport considered active transport?*

CONCEPT 7.5

Bulk transport across the plasma membrane occurs by exocytosis and endocytosis (p. 138)

- In **exocytosis**, transport vesicles migrate to the plasma membrane, fuse with it, and release their contents. In **endocytosis**, molecules enter cells within vesicles that pinch inward from the plasma membrane. The three types of endocytosis are **phagocytosis**, **pinocytosis**, and **receptor-mediated endocytosis**.

? *Which type of endocytosis involves ligands? What does this type of transport enable a cell to do?*

TEST YOUR UNDERSTANDING

LEVEL 1: KNOWLEDGE/COMPREHENSION

1. In what way do the membranes of a eukaryotic cell vary?
   a. Phospholipids are found only in certain membranes.
   b. Certain proteins are unique to each membrane.

c. Only certain membranes of the cell are selectively permeable.
d. Only certain membranes are constructed from amphipathic molecules.
e. Some membranes have hydrophobic surfaces exposed to the cytoplasm, while others have hydrophilic surfaces facing the cytoplasm.

2. According to the fluid mosaic model of membrane structure, proteins of the membrane are mostly
   a. spread in a continuous layer over the inner and outer surfaces of the membrane.
   b. confined to the hydrophobic interior of the membrane.
   c. embedded in a lipid bilayer.
   d. randomly oriented in the membrane, with no fixed inside-outside polarity.
   e. free to depart from the fluid membrane and dissolve in the surrounding solution.

3. Which of the following factors would tend to increase membrane fluidity?
   a. a greater proportion of unsaturated phospholipids
   b. a greater proportion of saturated phospholipids
   c. a lower temperature
   d. a relatively high protein content in the membrane
   e. a greater proportion of relatively large glycolipids compared with lipids having smaller molecular masses

LEVEL 2: APPLICATION/ANALYSIS

4. Which of the following processes includes all others?
   a. osmosis
   b. diffusion of a solute across a membrane
   c. facilitated diffusion
   d. passive transport
   e. transport of an ion down its electrochemical gradient

5. Based on Figure 7.21, which of these experimental treatments would increase the rate of sucrose transport into the cell?
   a. decreasing extracellular sucrose concentration
   b. decreasing extracellular pH
   c. decreasing cytoplasmic pH
   d. adding an inhibitor that blocks the regeneration of ATP
   e. adding a substance that makes the membrane more permeable to hydrogen ions

6. **DRAW IT** An artificial "cell" consisting of an aqueous solution enclosed in a selectively permeable membrane is immersed in a beaker containing a different solution, the "environment," as shown below. The membrane is permeable to water and to the simple sugars glucose and fructose but impermeable to the disaccharide sucrose.
   a. Draw solid arrows to indicate the net movement of solutes into and/or out of the cell.
   b. Is the solution outside the cell isotonic, hypotonic, or hypertonic?

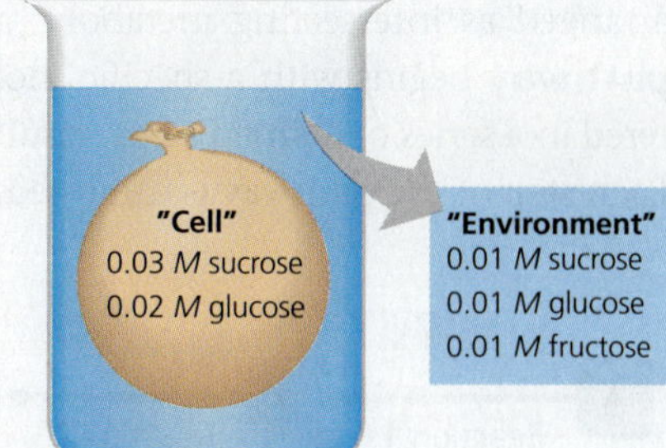

   c. Draw a dashed arrow to show the net osmosis, if any.
   d. Will the artificial cell become more flaccid, more turgid, or stay the same?
   e. Eventually, will the two solutions have the same or different solute concentrations?

LEVEL 3: SYNTHESIS/EVALUATION

7. **EVOLUTION CONNECTION**
   Paramecium and other protists that live in hypotonic environments have cell membranes that limit water uptake, while those living in isotonic environments have membranes that are more permeable to water. What water regulation adaptations might have evolved in protists in hypertonic habitats such as Great Salt Lake? In habitats with changing salt concentration?

8. **SCIENTIFIC INQUIRY**
   An experiment is designed to study the mechanism of sucrose uptake by plant cells. Cells are immersed in a sucrose solution, and the pH of the solution is monitored. Samples of the cells are taken at intervals, and their sucrose concentration is measured. After a decrease in the pH of the solution to a steady, slightly acidic level, sucrose uptake begins. Propose a hypothesis for these results. What do you think would happen if an inhibitor of ATP regeneration by the cell were added to the beaker once the pH is at a steady level? Explain.

9. **SCIENCE, TECHNOLOGY, AND SOCIETY**
   Extensive irrigation in arid regions causes salts to accumulate in the soil. (When water evaporates, salts that were dissolved in the water are left behind in the soil.) Based on what you learned about water balance in plant cells, explain why increased soil salinity (saltiness) might be harmful to crops. Suggest ways to minimize damage. What costs are attached to your solutions?

10. **WRITE ABOUT A THEME**
    Environmental Interactions A human pancreatic cell obtains O_2, fuel molecules such as glucose, and building materials such as amino acids and cholesterol from its environment, and it releases CO_2 as a waste product of cellular respiration. In response to hormonal signals, the cell secretes digestive enzymes. It also regulates its ion concentrations by exchange with its environment. Based on what you have just learned about the structure and function of cellular membranes, write a short essay (100–150 words) that describes how such a cell accomplishes these interactions with its environment.

For selected answers, see Appendix A.

8

An Introduction to Metabolism

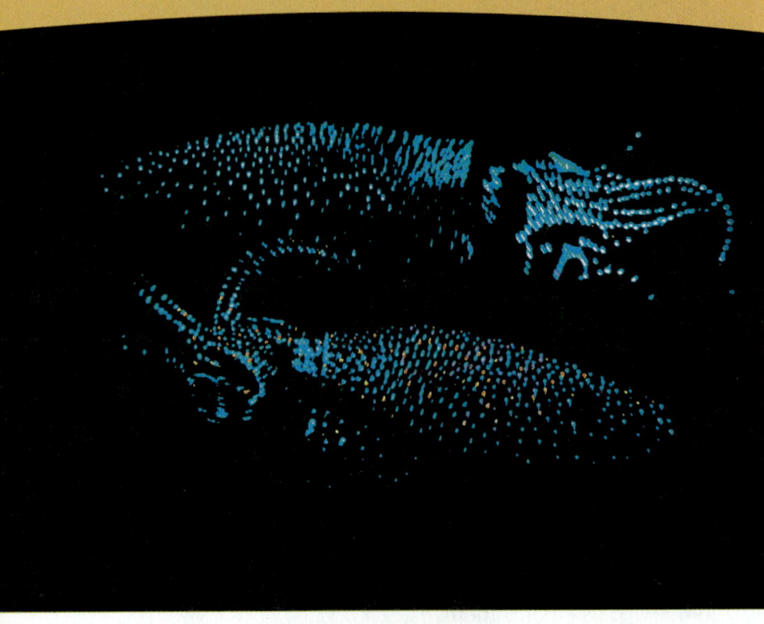

▲ **Figure 8.1 What causes these two squid to glow?**

KEY CONCEPTS

8.1 An organism's metabolism transforms matter and energy, subject to the laws of thermodynamics

8.2 The free-energy change of a reaction tells us whether or not the reaction occurs spontaneously

8.3 ATP powers cellular work by coupling exergonic reactions to endergonic reactions

8.4 Enzymes speed up metabolic reactions by lowering energy barriers

8.5 Regulation of enzyme activity helps control metabolism

OVERVIEW

The Energy of Life

The living cell is a chemical factory in miniature, where thousands of reactions occur within a microscopic space. Sugars can be converted to amino acids that are linked together into proteins when needed, and when food is digested, proteins are dismantled into amino acids that can be converted to sugars. Small molecules are assembled into polymers, which may be hydrolyzed later as the needs of the cell change. In multicellular organisms, many cells export chemical products that are used in other parts of the organism. The process called cellular respiration drives the cellular economy by extracting the energy stored in sugars and other fuels. Cells apply this energy to perform various types of work, such as the transport of solutes across the plasma membrane, which we discussed in Chapter 7. In a more exotic example, cells of the two firefly squid (*Watasenia scintillans*) shown mating in **Figure 8.1** convert the energy stored in certain organic molecules to light, a process called bioluminescence. (The light pattern aids in mate recognition and protection from predators lurking below.) Bioluminescence and other metabolic activities carried out by a cell are precisely coordinated and controlled. In its complexity, its efficiency, and its responsiveness to subtle changes, the cell is peerless as a chemical factory. The concepts of metabolism that you learn in this chapter will help you understand how matter and energy flow during life's processes and how that flow is regulated.

CONCEPT **8.1**

An organism's metabolism transforms matter and energy, subject to the laws of thermodynamics

The totality of an organism's chemical reactions is called **metabolism** (from the Greek *metabole*, change). Metabolism is an emergent property of life that arises from orderly interactions between molecules.

Organization of the Chemistry of Life into Metabolic Pathways

We can picture a cell's metabolism as an elaborate road map of the thousands of chemical reactions that occur in a cell, arranged as intersecting metabolic pathways. A **metabolic pathway** begins with a specific molecule, which is then altered in a series of defined steps, resulting in a certain product. Each step of the pathway is catalyzed by a specific enzyme:

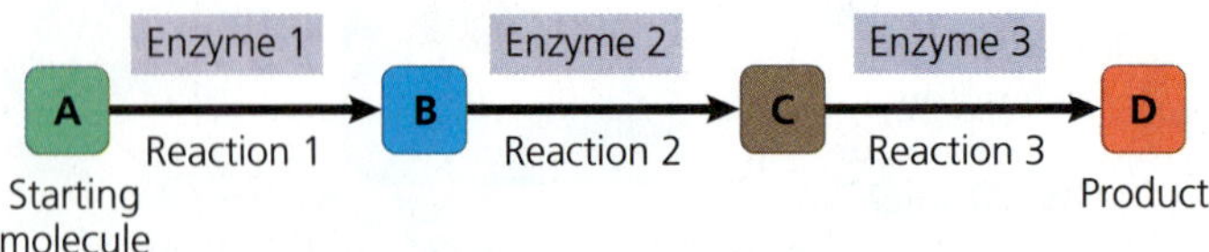

Analogous to the red, yellow, and green stoplights that control the flow of automobile traffic, mechanisms that regulate enzymes balance metabolic supply and demand.

Metabolism as a whole manages the material and energy resources of the cell. Some metabolic pathways release energy by breaking down complex molecules to simpler compounds. These degradative processes are called **catabolic pathways**, or breakdown pathways. A major pathway of catabolism is cellular respiration, in which the sugar glucose and other organic fuels are broken down in the presence of oxygen to carbon dioxide and water. (Pathways can have more than one starting molecule and/or product.) Energy that was stored in the organic molecules becomes available to do the work of the cell, such as ciliary beating or membrane transport. **Anabolic pathways**, in contrast, consume energy to build complicated molecules from simpler ones; they are sometimes called biosynthetic pathways. Examples of anabolism are the synthesis of an amino acid from simpler molecules and the synthesis of a protein from amino acids. Catabolic and anabolic pathways are the "downhill" and "uphill" avenues of the metabolic landscape. Energy released from the downhill reactions of catabolic pathways can be stored and then used to drive the uphill reactions of anabolic pathways.

In this chapter, we will focus on mechanisms common to metabolic pathways. Because energy is fundamental to all metabolic processes, a basic knowledge of energy is necessary to understand how the living cell works. Although we will use some nonliving examples to study energy, the concepts demonstrated by these examples also apply to **bioenergetics**, the study of how energy flows through living organisms.

Forms of Energy

Energy is the capacity to cause change. In everyday life, energy is important because some forms of energy can be used to do work—that is, to move matter against opposing forces, such as gravity and friction. Put another way, energy is the ability to rearrange a collection of matter. For example, you expend energy to turn the pages of this book, and your cells expend energy in transporting certain substances across membranes. Energy exists in various forms, and the work of life depends on the ability of cells to transform energy from one form to another.

Energy can be associated with the relative motion of objects; this energy is called **kinetic energy**. Moving objects can perform work by imparting motion to other matter: A pool player uses the motion of the cue stick to push the cue ball, which in turn moves the other balls; water gushing through a dam turns turbines; and the contraction of leg muscles pushes bicycle pedals. **Heat**, or **thermal energy**, is kinetic energy associated with the random movement of atoms or molecules. Light is also a type of energy that can be harnessed to perform work, such as powering photosynthesis in green plants.

An object not presently moving may still possess energy. Energy that is not kinetic is called **potential energy**; it is energy that matter possesses because of its location or structure. Water behind a dam, for instance, possesses energy because of its altitude above sea level. Molecules possess energy because of the arrangement of electrons in the bonds between their atoms. **Chemical energy** is a term used by biologists to refer to the potential energy available for release in a chemical reaction. Recall that catabolic pathways release energy by breaking down complex molecules. Biologists say that these complex molecules, such as glucose, are high in chemical energy. During a catabolic reaction, some bonds are broken and others formed, releasing energy and resulting in lower-energy breakdown products. This transformation also occurs, for example, in the engine of a car when the hydrocarbons of gasoline react explosively with oxygen, releasing the energy that pushes the pistons and producing exhaust. Although less explosive, a similar reaction of food molecules with oxygen provides chemical energy in biological systems, producing carbon dioxide and water as waste products. Biochemical pathways, carried out in the context of cellular structures, enable cells to release chemical energy from food molecules and use the energy to power life processes.

How is energy converted from one form to another? Consider the divers in **Figure 8.2**. The young woman climbing the ladder to the diving platform is releasing chemical energy from the food she ate for lunch and using some of that energy to perform the work of climbing. The kinetic energy of muscle movement is thus being transformed into potential energy due to her increasing height above the water. The young man diving is converting his potential energy to kinetic energy, which is then transferred to the water as he enters it. A small amount of energy is lost as heat due to friction.

▲ **Figure 8.2 Transformations between potential and kinetic energy.**

Now let's go back one step and consider the original source of the organic food molecules that provided the necessary chemical energy for the diver to climb the steps. This chemical energy was itself derived from light energy by plants during photosynthesis. Organisms are energy transformers.

The Laws of Energy Transformation

The study of the energy transformations that occur in a collection of matter is called **thermodynamics**. Scientists use the word *system* to denote the matter under study; they refer to the rest of the universe—everything outside the system—as the *surroundings*. An *isolated system*, such as that approximated by liquid in a thermos bottle, is unable to exchange either energy or matter with its surroundings. In an *open system*, energy and matter can be transferred between the system and its surroundings. Organisms are open systems. They absorb energy—for instance, light energy or chemical energy in the form of organic molecules—and release heat and metabolic waste products, such as carbon dioxide, to the surroundings. Two laws of thermodynamics govern energy transformations in organisms and all other collections of matter.

The First Law of Thermodynamics

According to the **first law of thermodynamics**, the energy of the universe is constant: *Energy can be transferred and transformed, but it cannot be created or destroyed.* The first law is also known as the *principle of conservation of energy*. The electric company does not make energy, but merely converts it to a form that is convenient for us to use. By converting sunlight to chemical energy, a plant acts as an energy transformer, not an energy producer.

The brown bear in **Figure 8.3a** will convert the chemical energy of the organic molecules in its food to kinetic and other forms of energy as it carries out biological processes. What happens to this energy after it has performed work? The second law of thermodynamics helps to answer this question.

The Second Law of Thermodynamics

If energy cannot be destroyed, why can't organisms simply recycle their energy over and over again? It turns out that during every energy transfer or transformation, some energy becomes unavailable to do work. In most energy transformations, more usable forms of energy are at least partly converted to heat, which is the energy associated with the random motion of atoms or molecules. Only a small fraction of the chemical energy from the food in Figure 8.3a is transformed into the motion of the brown bear shown in **Figure 8.3b**; most is lost as heat, which dissipates rapidly through the surroundings.

In the process of carrying out chemical reactions that perform various kinds of work, living cells unavoidably convert other forms of energy to heat. A system can put heat to work only when there is a temperature difference that results in the heat flowing from a warmer location to a cooler one. If temperature is uniform, as it is in a living cell, then the only use for heat energy generated during a chemical reaction is to warm a body of matter, such as the organism. (This can make a room crowded with people uncomfortably warm, as each person is carrying out a multitude of chemical reactions!)

A logical consequence of the loss of usable energy during energy transfer or transformation is that each such event makes the universe more disordered. Scientists use a quantity called **entropy** as a measure of disorder, or randomness.

(a) First law of thermodynamics: Energy can be transferred or transformed but neither created nor destroyed. For example, chemical reactions in this brown bear (*Ursus arctos*) will convert the chemical (potential) energy in the fish into the kinetic energy of running, shown in (b).

(b) Second law of thermodynamics: Every energy transfer or transformation increases the disorder (entropy) of the universe. For example, as it runs, disorder is increased around the bear by the release of heat and small molecules that are the by-products of metabolism. A brown bear can run at speeds up to 35 miles per hour (56 km/hr)—as fast as a racehorse.

▲ **Figure 8.3 The two laws of thermodynamics.**

The more randomly arranged a collection of matter is, the greater its entropy. We can now state the **second law of thermodynamics**: *Every energy transfer or transformation increases the entropy of the universe.* Although order can increase locally, there is an unstoppable trend toward randomization of the universe as a whole.

In many cases, increased entropy is evident in the physical disintegration of a system's organized structure. For example, you can observe increasing entropy in the gradual decay of an unmaintained building. Much of the increasing entropy of the universe is less apparent, however, because it appears as increasing amounts of heat and less ordered forms of matter. As the bear in Figure 8.3b converts chemical energy to kinetic energy, it is also increasing the disorder of its surroundings by producing heat and small molecules, such as the CO_2 it exhales, that are the breakdown products of food.

The concept of entropy helps us understand why certain processes occur without any input of energy. It turns out that for a process to occur on its own, without outside help, it must increase the entropy of the universe. A process that can occur without an input of energy is called a **spontaneous process**. Note that as we're using it here, the word *spontaneous* does not imply that such a process would occur quickly; rather, the word signifies that the process is energetically favorable. (In fact, it may be helpful for you to think of the phrase "energetically favorable" when you read the formal term "spontaneous.") Some spontaneous processes, such as an explosion, may be virtually instantaneous, while others, such as the rusting of an old car over time, are much slower. A process that cannot occur on its own is said to be nonspontaneous; it will happen only if energy is added to the system. We know from experience that certain events occur spontaneously and others do not. For instance, we know that water flows downhill spontaneously but moves uphill only with an input of energy, such as when a machine pumps the water against gravity. This understanding gives us another way to state the second law: *For a process to occur spontaneously, it must increase the entropy of the universe.*

Biological Order and Disorder

Living systems increase the entropy of their surroundings, as predicted by thermodynamic law. It is true that cells create ordered structures from less organized starting materials. For example, simpler molecules are ordered into the more complex structure of an amino acid, and amino acids are ordered into polypeptide chains. At the organismal level as well, complex and beautifully ordered structures result from biological processes that use simpler starting materials **(Figure 8.4)**. However, an organism also takes in organized forms of matter and energy from the surroundings and replaces them with less ordered forms. For example, an animal obtains starch, proteins, and other complex molecules from the food it eats. As catabolic pathways break these molecules down, the animal releases carbon dioxide and water—small molecules that possess less chemical energy than the food did. The depletion of chemical energy is accounted for by heat generated during metabolism. On a larger scale, energy flows into most ecosystems in the form of light and exits in the form of heat (see Figure 1.6).

▲ **Figure 8.4 Order as a characteristic of life.** Order is evident in the detailed structures of the sea urchin skeleton and the succulent plant shown here. As open systems, organisms can increase their order as long as the order of their surroundings decreases.

During the early history of life, complex organisms evolved from simpler ancestors. For example, we can trace the ancestry of the plant kingdom from much simpler organisms called green algae to more complex flowering plants. However, this increase in organization over time in no way violates the second law. The entropy of a particular system, such as an organism, may actually decrease as long as the total entropy of the *universe*—the system plus its surroundings—increases. Thus, organisms are islands of low entropy in an increasingly random universe. The evolution of biological order is perfectly consistent with the laws of thermodynamics.

CONCEPT CHECK 8.1

1. **MAKE CONNECTIONS** How does the second law of thermodynamics help explain the diffusion of a substance across a membrane? See Figure 7.13 on page 132.
2. Describe the forms of energy found in an apple as it grows on a tree, then falls, then is digested by someone who eats it.
3. **WHAT IF?** If you place a teaspoon of sugar in the bottom of a glass of water, it will dissolve completely over time. Left longer, eventually the water will disappear and the sugar crystals will reappear. Explain these observations in terms of entropy.

For suggested answers, see Appendix A.

CONCEPT 8.2

The free-energy change of a reaction tells us whether or not the reaction occurs spontaneously

The laws of thermodynamics that we've just discussed apply to the universe as a whole. As biologists, we want to understand the chemical reactions of life—for example, which reactions occur spontaneously and which ones require some input of energy from outside. But how can we know this without assessing the energy and entropy changes in the entire universe for each separate reaction?

Free-Energy Change, ΔG

Recall that the universe is really equivalent to "the system" plus "the surroundings." In 1878, J. Willard Gibbs, a professor at Yale, defined a very useful function called the Gibbs free energy of a system (without considering its surroundings), symbolized by the letter G. We'll refer to the Gibbs free energy simply as free energy. **Free energy** is the portion of a system's energy that can perform work when temperature and pressure are uniform throughout the system, as in a living cell. Let's consider how we determine the free-energy change that occurs when a system changes—for example, during a chemical reaction.

The change in free energy, ΔG, can be calculated for a chemical reaction by applying the following equation:

$$\Delta G = \Delta H - T\Delta S$$

This equation uses only properties of the system (the reaction) itself: ΔH symbolizes the change in the system's *enthalpy* (in biological systems, equivalent to total energy); ΔS is the change in the system's entropy; and T is the absolute temperature in Kelvin (K) units (K = °C + 273; see Appendix C).

Once we know the value of ΔG for a process, we can use it to predict whether the process will be spontaneous (that is, whether it is energetically favorable and will occur without an input of energy). More than a century of experiments has shown that only processes with a negative ΔG are spontaneous. For ΔG to be negative, either ΔH must be negative (the system gives up enthalpy and H decreases) or $T\Delta S$ must be positive (the system gives up order and S increases), or both: When ΔH and $T\Delta S$ are tallied, ΔG has a negative value ($\Delta G < 0$) for all spontaneous processes. In other words, every spontaneous process decreases the system's free energy, and processes that have a positive or zero ΔG are never spontaneous.

This information is immensely interesting to biologists, for it gives us the power to predict which kinds of change can happen without help. Such spontaneous changes can be harnessed to perform work. This principle is very important in the study of metabolism, where a major goal is to determine which reactions can supply energy for cellular work.

Free Energy, Stability, and Equilibrium

As we saw in the previous section, when a process occurs spontaneously in a system, we can be sure that ΔG is negative. Another way to think of ΔG is to realize that it represents the difference between the free energy of the final state and the free energy of the initial state:

$$\Delta G = G_{\text{final state}} - G_{\text{initial state}}$$

Thus, ΔG can be negative only when the process involves a loss of free energy during the change from initial state to final state. Because it has less free energy, the system in its final state is less likely to change and is therefore more stable than it was previously.

We can think of free energy as a measure of a system's instability—its tendency to change to a more stable state. Unstable systems (higher G) tend to change in such a way that they become more stable (lower G). For example, a diver on top of a platform is less stable (more likely to fall) than when floating in the water; a drop of concentrated dye is less stable (more likely to disperse) than when the dye is spread randomly through the liquid; and a glucose molecule is less stable (more likely to break down) than the simpler molecules into which it can be split **(Figure 8.5)**. Unless something prevents it, each of these systems will move toward greater stability: The diver falls, the solution becomes uniformly colored, and the glucose molecule is broken down.

Another term that describes a state of maximum stability is *equilibrium*, which you learned about in Chapter 2 in connection with chemical reactions. There is an important relationship between free energy and equilibrium, including chemical equilibrium. Recall that most chemical reactions are reversible and proceed to a point at which the forward and backward reactions occur at the same rate. The reaction is then said to be at chemical equilibrium, and there is no further net change in the relative concentration of products and reactants.

As a reaction proceeds toward equilibrium, the free energy of the mixture of reactants and products decreases. Free energy increases when a reaction is somehow pushed away from equilibrium, perhaps by removing some of the products (and thus changing their concentration relative to that of the reactants). For a system at equilibrium, G is at its lowest possible value in that system. We can think of the equilibrium state as a free-energy valley. Any change from the equilibrium position will have a positive ΔG and will not be spontaneous. For this reason, systems never spontaneously move away from equilibrium. Because a system at equilibrium cannot spontaneously change, it can do no work. *A process is spontaneous and can perform work only when it is moving toward equilibrium.*

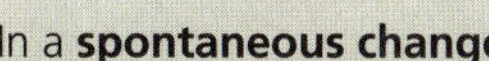

- More free energy (higher G)
- Less stable
- Greater work capacity

In a **spontaneous change**
- The free energy of the system decreases ($\Delta G < 0$)
- The system becomes more stable
- The released free energy can be harnessed to do work

- Less free energy (lower G)
- More stable
- Less work capacity

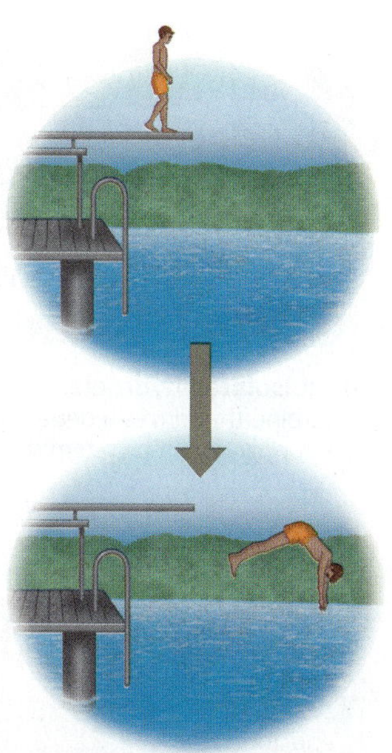

(a) Gravitational motion. Objects move spontaneously from a higher altitude to a lower one.

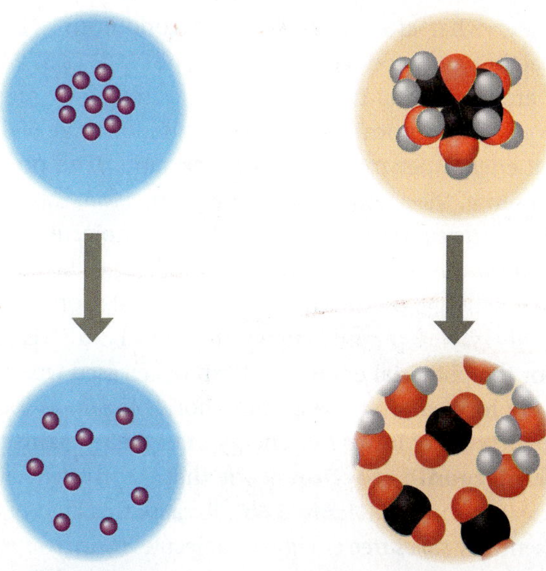

(b) Diffusion. Molecules in a drop of dye diffuse until they are randomly dispersed.

(c) Chemical reaction. In a cell, a glucose molecule is broken down into simpler molecules.

▲ **Figure 8.5 The relationship of free energy to stability, work capacity, and spontaneous change.** Unstable systems (top) are rich in free energy, G. They have a tendency to change spontaneously to a more stable state (bottom), and it is possible to harness this "downhill" change to perform work.

Free Energy and Metabolism

We can now apply the free-energy concept more specifically to the chemistry of life's processes.

Exergonic and Endergonic Reactions in Metabolism

Based on their free-energy changes, chemical reactions can be classified as either exergonic ("energy outward") or endergonic ("energy inward"). An **exergonic reaction** proceeds with a net release of free energy **(Figure 8.6a)**. Because the chemical mixture loses free energy (G decreases), ΔG is negative for an exergonic reaction. Using ΔG as a standard for spontaneity, exergonic reactions are those that occur spontaneously. (Remember, the word *spontaneous* implies that it is energetically favorable, not that it will occur rapidly.) The magnitude of ΔG for an exergonic reaction represents the maximum amount of work the reaction can perform.* The greater the decrease in free energy, the greater the amount of work that can be done.

We can use the overall reaction for cellular respiration as an example:

$$C_6H_{12}O_6 + 6\ O_2 \rightarrow 6\ CO_2 + 6\ H_2O$$
$$\Delta G = -686\ \text{kcal/mol}\ (-2{,}870\ \text{kJ/mol})$$

*The word *maximum* qualifies this statement, because some of the free energy is released as heat and cannot do work. Therefore, ΔG represents a theoretical upper limit of available energy.

▼ **Figure 8.6 Free energy changes (ΔG) in exergonic and endergonic reactions.**

(a) Exergonic reaction: energy released, spontaneous

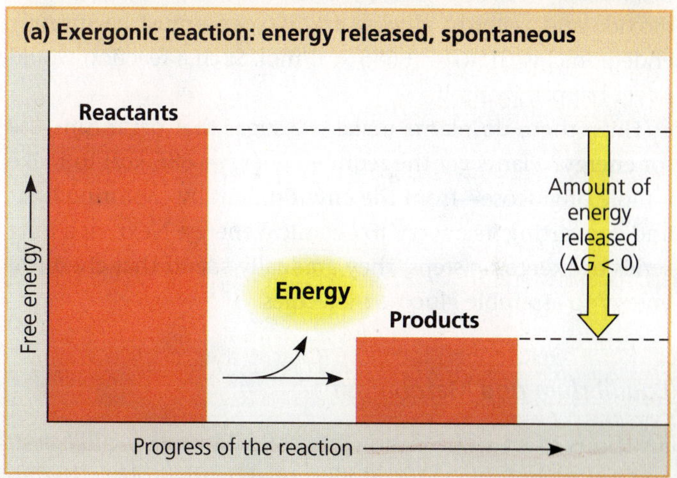

(b) Endergonic reaction: energy required, nonspontaneous

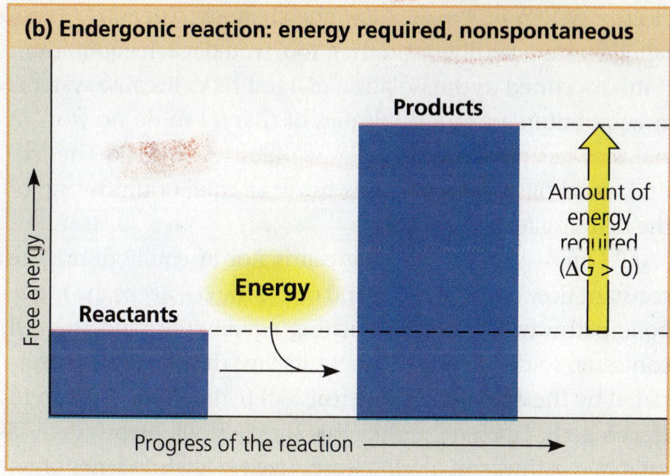

For each mole (180 g) of glucose broken down by respiration under what are called "standard conditions" (1 *M* of each reactant and product, 25°C, pH 7), 686 kcal (2,870 kJ) of energy are made available for work. Because energy must be conserved, the chemical products of respiration store 686 kcal less free energy per mole than the reactants. The products are, in a sense, the spent exhaust of a process that tapped the free energy stored in the bonds of the sugar molecules.

It is important to realize that the breaking of bonds does not release energy; on the contrary, as you will soon see, it requires energy. The phrase "energy stored in bonds" is shorthand for the potential energy that can be released when new bonds are formed after the original bonds break, as long as the products are of lower free energy than the reactants.

An **endergonic reaction** is one that absorbs free energy from its surroundings **(Figure 8.6b)**. Because this kind of reaction essentially *stores* free energy in molecules (*G* increases), ΔG is positive. Such reactions are nonspontaneous, and the magnitude of ΔG is the quantity of energy required to drive the reaction. If a chemical process is exergonic (downhill), releasing energy in one direction, then the reverse process must be endergonic (uphill), using energy. A reversible process cannot be downhill in both directions. If $\Delta G = -686$ kcal/mol for respiration, which converts glucose and oxygen to carbon dioxide and water, then the reverse process—the conversion of carbon dioxide and water to glucose and oxygen—must be strongly endergonic, with $\Delta G = +686$ kcal/mol. Such a reaction would never happen by itself.

How, then, do plants make the sugar that organisms use for energy? Plants get the required energy—686 kcal to make a mole of glucose—from the environment by capturing light and converting its energy to chemical energy. Next, in a long series of exergonic steps, they gradually spend that chemical energy to assemble glucose molecules.

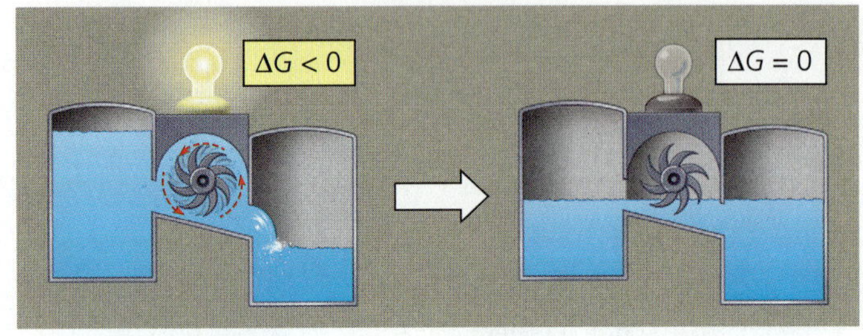

(a) An isolated hydroelectric system. Water flowing downhill turns a turbine that drives a generator providing electricity to a lightbulb, but only until the system reaches equilibrium.

(b) An open hydroelectric system. Flowing water keeps driving the generator because intake and outflow of water keep the system from reaching equilibrium.

(c) A multistep open hydroelectric system. Cellular respiration is analogous to this system: Glucose is broken down in a series of exergonic reactions that power the work of the cell. The product of each reaction becomes the reactant for the next, so no reaction reaches equilibrium.

▲ **Figure 8.7 Equilibrium and work in isolated and open systems.**

Equilibrium and Metabolism

Reactions in an isolated system eventually reach equilibrium and can then do no work, as illustrated by the isolated hydroelectric system in **Figure 8.7a**. The chemical reactions of metabolism are reversible, and they, too, would reach equilibrium if they occurred in the isolation of a test tube. Because systems at equilibrium are at a minimum of *G* and can do no work, a cell that has reached metabolic equilibrium is dead! The fact that metabolism as a whole is never at equilibrium is one of the defining features of life.

Like most systems, a living cell is not in equilibrium. The constant flow of materials in and out of the cell keeps the metabolic pathways from ever reaching equilibrium, and the cell continues to do work throughout its life. This principle is illustrated by the open (and more realistic) hydroelectric system in **Figure 8.7b**. However, unlike this simple single-step system, a catabolic pathway in a cell releases free energy in a series of reactions. An example is cellular respiration, illustrated by analogy in **Figure 8.7c**. Some of the reversible reactions of respiration are constantly "pulled" in one direction—that is, they are kept out of equilibrium. The key to maintaining this lack of equilibrium is that the product of a reaction does not accumulate but instead becomes a reactant in the next step; finally, waste products are expelled from the cell. The overall sequence of reactions is kept going by the huge free-energy difference between glucose and oxygen at the top of the energy "hill" and carbon dioxide and water at the "downhill" end. As long as our cells have a steady supply of glucose or other fuels and oxygen and are able to expel waste products to the surroundings, their metabolic pathways never reach equilibrium and can continue to do the work of life.

We see once again how important it is to think of organisms as open systems. Sunlight provides a daily source of free energy for an ecosystem's plants and other photosynthetic organisms. Animals and other nonphotosynthetic organisms in an ecosystem must have a source of free energy in the form of the organic products of photosynthesis. Now that we have applied the free-energy concept to metabolism, we are ready to see how a cell actually performs the work of life.

CONCEPT CHECK 8.2

1. Cellular respiration uses glucose and oxygen, which have high levels of free energy, and releases CO_2 and water, which have low levels of free energy. Is cellular respiration spontaneous or not? Is it exergonic or endergonic? What happens to the energy released from glucose?
2. **MAKE CONNECTIONS** As you saw in Figure 7.20 on page 137, a key process in metabolism is the transport of hydrogen ions (H^+) across a membrane to create a concentration gradient. Other processes can result in an equal concentration of H^+ on each side. Which situation allows the H^+ to perform work in this system? How is the answer consistent with what is shown in regard to energy in Figure 7.20?
3. **WHAT IF?** Some night-time partygoers wear glow-in-the-dark necklaces. The necklaces start glowing once they are "activated," which usually involves snapping the necklace in a way that allows two chemicals to react and emit light in the form of chemiluminescence. Is the chemical reaction exergonic or endergonic? Explain your answer.

For suggested answers, see Appendix A.

CONCEPT 8.3

ATP powers cellular work by coupling exergonic reactions to endergonic reactions

A cell does three main kinds of work:

- *Chemical work,* the pushing of endergonic reactions that would not occur spontaneously, such as the synthesis of polymers from monomers (chemical work will be discussed further here and in Chapters 9 and 10)
- *Transport work,* the pumping of substances across membranes against the direction of spontaneous movement (see Chapter 7)
- *Mechanical work,* such as the beating of cilia (see Chapter 6), the contraction of muscle cells, and the movement of chromosomes during cellular reproduction

A key feature in the way cells manage their energy resources to do this work is **energy coupling**, the use of an exergonic process to drive an endergonic one. ATP is responsible for mediating most energy coupling in cells, and in most cases it acts as the immediate source of energy that powers cellular work.

The Structure and Hydrolysis of ATP

ATP (adenosine triphosphate) was introduced in Chapter 4 when we discussed the phosphate group as a functional group. ATP contains the sugar ribose, with the nitrogenous base adenine and a chain of three phosphate groups bonded to it **(Figure 8.8a)**. In addition to its role in energy coupling, ATP is also one of the nucleoside triphosphates used to make RNA (see Figure 5.26).

The bonds between the phosphate groups of ATP can be broken by hydrolysis. When the terminal phosphate bond is broken by addition of a water molecule, a molecule of inorganic phosphate ($HOPO_3^{2-}$, abbreviated Ⓟ$_i$ throughout this book) leaves the ATP, which becomes adenosine diphosphate,

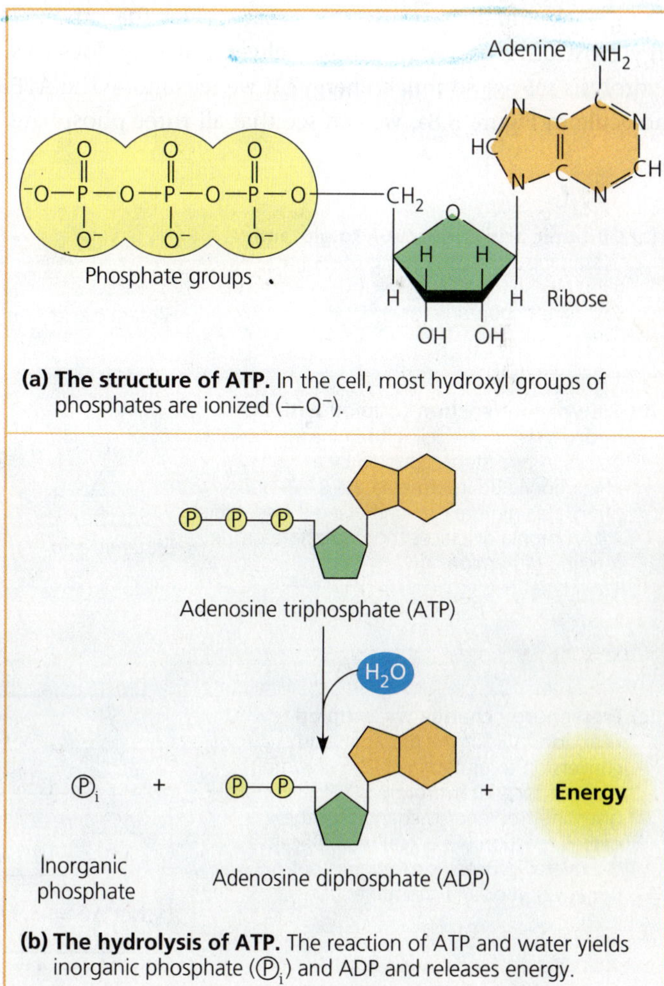

(a) The structure of ATP. In the cell, most hydroxyl groups of phosphates are ionized (—O⁻).

(b) The hydrolysis of ATP. The reaction of ATP and water yields inorganic phosphate (Ⓟ$_i$) and ADP and releases energy.

▲ **Figure 8.8 The structure and hydrolysis of adenosine triphosphate (ATP).**

or ADP (**Figure 8.8b**). The reaction is exergonic and releases 7.3 kcal of energy per mole of ATP hydrolyzed:

$$\text{ATP} + H_2O \rightarrow \text{ADP} + \text{Ⓟ}_i$$
$$\Delta G = -7.3 \text{ kcal/mol } (-30.5 \text{ kJ/mol})$$

This is the free-energy change measured under standard conditions. In the cell, conditions do not conform to standard conditions, primarily because reactant and product concentrations differ from 1 *M*. For example, when ATP hydrolysis occurs under cellular conditions, the actual ΔG is about −13 kcal/mol, 78% greater than the energy released by ATP hydrolysis under standard conditions.

Because their hydrolysis releases energy, the phosphate bonds of ATP are sometimes referred to as high-energy phosphate bonds, but the term is misleading. The phosphate bonds of ATP are not unusually strong bonds, as "high-energy" may imply; rather, the reactants (ATP and water) themselves have high energy relative to the energy of the products (ADP and Ⓟ$_i$). The release of energy during the hydrolysis of ATP comes from the chemical change to a state of lower free energy, not from the phosphate bonds themselves.

ATP is useful to the cell because the energy it releases on losing a phosphate group is somewhat greater than the energy most other molecules could deliver. But why does this hydrolysis release so much energy? If we reexamine the ATP molecule in Figure 8.8a, we can see that all three phosphate groups are negatively charged. These like charges are crowded together, and their mutual repulsion contributes to the instability of this region of the ATP molecule. The triphosphate tail of ATP is the chemical equivalent of a compressed spring.

How the Hydrolysis of ATP Performs Work

When ATP is hydrolyzed in a test tube, the release of free energy merely heats the surrounding water. In an organism, this same generation of heat can sometimes be beneficial. For instance, the process of shivering uses ATP hydrolysis during muscle contraction to generate heat and warm the body. In most cases in the cell, however, the generation of heat alone would be an inefficient (and potentially dangerous) use of a valuable energy resource. Instead, the cell's proteins harness the energy released during ATP hydrolysis in several ways to perform the three types of cellular work—chemical, transport, and mechanical.

For example, with the help of specific enzymes, the cell is able to use the energy released by ATP hydrolysis directly to drive chemical reactions that, by themselves, are endergonic. If the ΔG of an endergonic reaction is less than the amount of energy released by ATP hydrolysis, then the two reactions can be coupled so that, overall, the coupled reactions are exergonic (**Figure 8.9**). This usually involves the transfer of a phosphate

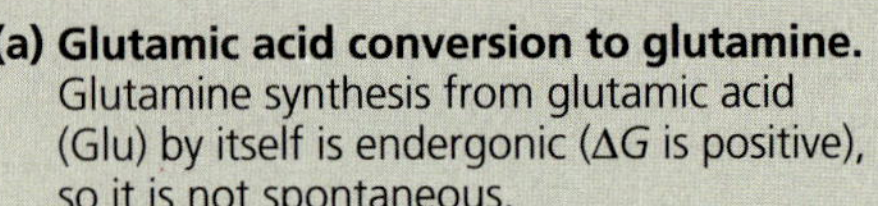

(a) Glutamic acid conversion to glutamine. Glutamine synthesis from glutamic acid (Glu) by itself is endergonic (ΔG is positive), so it is not spontaneous.

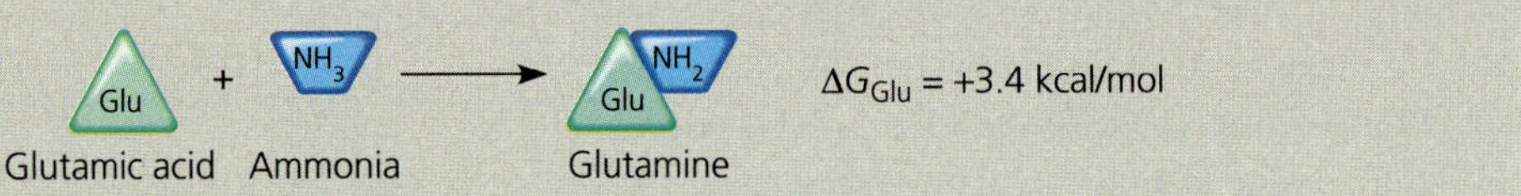

(b) Conversion reaction coupled with ATP hydrolysis. In the cell, glutamine synthesis occurs in two steps, coupled by a phosphorylated intermediate. ❶ ATP phosphorylates glutamic acid, making it less stable. ❷ Ammonia displaces the phosphate group, forming glutamine.

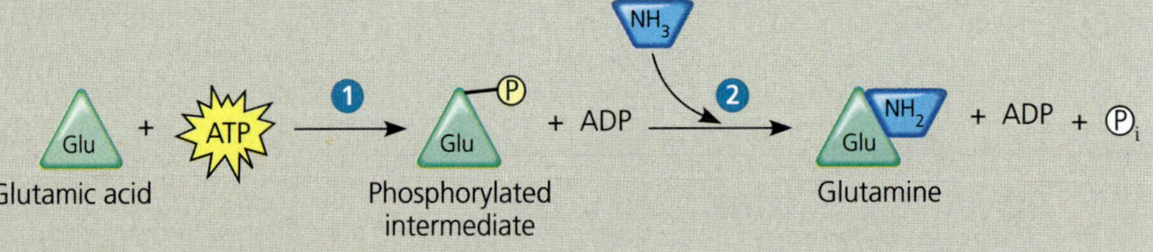

(c) Free-energy change for coupled reaction. ΔG for the glutamic acid conversion to glutamine (+3.4 kcal/mol) plus ΔG for ATP hydrolysis (–7.3 kcal/mol) gives the free-energy change for the overall reaction (–3.9 kcal/mol). Because the overall process is exergonic (net ΔG is negative), it occurs spontaneously.

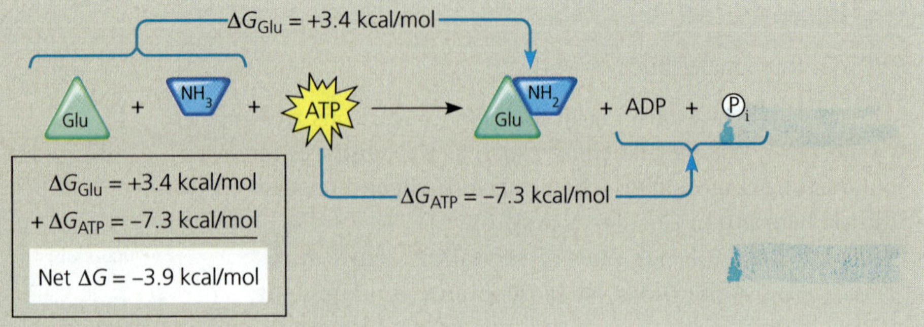

▲ **Figure 8.9 How ATP drives chemical work: Energy coupling using ATP hydrolysis.** In this example, the exergonic process of ATP hydrolysis is used to drive an endergonic process—the cellular synthesis of the amino acid glutamine from glutamic acid and ammonia.

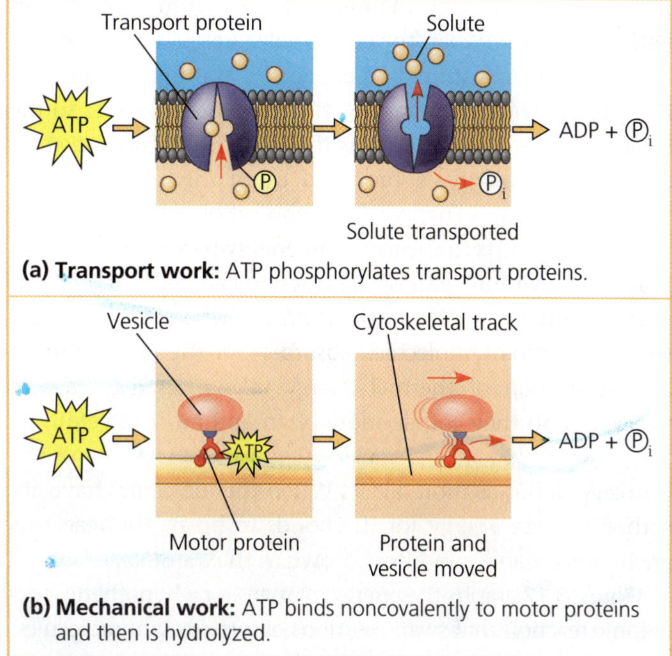

(a) Transport work: ATP phosphorylates transport proteins.

(b) Mechanical work: ATP binds noncovalently to motor proteins and then is hydrolyzed.

▲ **Figure 8.10 How ATP drives transport and mechanical work.** ATP hydrolysis causes changes in the shapes and binding affinities of proteins. This can occur either **(a)** directly, by phosphorylation, as shown for a membrane protein carrying out active transport of a solute (see also Figure 7.18), or **(b)** indirectly, via noncovalent binding of ATP and its hydrolytic products, as is the case for motor proteins that move vesicles (and other organelles) along cytoskeletal "tracks" in the cell (see also Figure 6.21).

group from ATP to some other molecule, such as the reactant. The recipient with the phosphate group covalently bonded to it is then called a **phosphorylated intermediate**. The key to coupling exergonic and endergonic reactions is the formation of this phosphorylated intermediate, which is more reactive (less stable) than the original unphosphorylated molecule.

Transport and mechanical work in the cell are also nearly always powered by the hydrolysis of ATP. In these cases, ATP hydrolysis leads to a change in a protein's shape and often its ability to bind another molecule. Sometimes this occurs via a phosphorylated intermediate, as seen for the transport protein in **Figure 8.10a**. In most instances of mechanical work involving motor proteins "walking" along cytoskeletal elements **(Figure 8.10b)**, a cycle occurs in which ATP is first bound noncovalently to the motor protein. Next, ATP is hydrolyzed, releasing ADP and Ⓟ$_i$. Another ATP molecule can then bind. At each stage, the motor protein changes its shape and ability to bind the cytoskeleton, resulting in movement of the protein along the cytoskeletal track.

The Regeneration of ATP

An organism at work uses ATP continuously, but ATP is a renewable resource that can be regenerated by the addition of phosphate to ADP **(Figure 8.11)**. The free energy required to

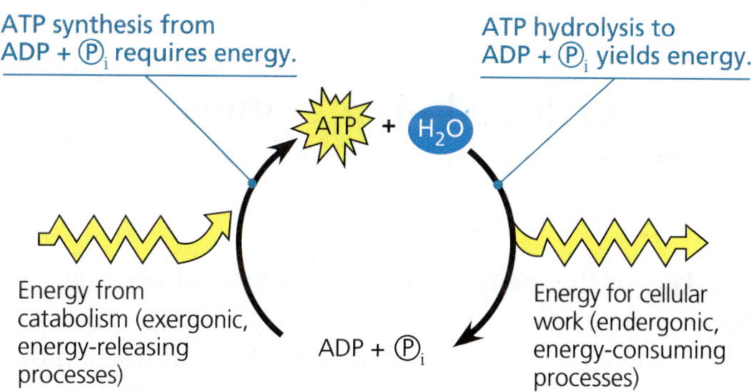

▲ **Figure 8.11 The ATP cycle.** Energy released by breakdown reactions (catabolism) in the cell is used to phosphorylate ADP, regenerating ATP. Chemical potential energy stored in ATP drives most cellular work.

phosphorylate ADP comes from exergonic breakdown reactions (catabolism) in the cell. This shuttling of inorganic phosphate and energy is called the ATP cycle, and it couples the cell's energy-yielding (exergonic) processes to the energy-consuming (endergonic) ones. The ATP cycle proceeds at an astonishing pace. For example, a working muscle cell recycles its entire pool of ATP in less than a minute. That turnover represents 10 million molecules of ATP consumed and regenerated per second per cell. If ATP could not be regenerated by the phosphorylation of ADP, humans would use up nearly their body weight in ATP each day.

Because both directions of a reversible process cannot be downhill, the regeneration of ATP from ADP and Ⓟ$_i$ is necessarily endergonic:

$$\text{ADP} + \text{Ⓟ}_i \rightarrow \text{ATP} + H_2O$$
$$\Delta G = +7.3 \text{ kcal/mol } (+30.5 \text{ kJ/mol}) \text{ (standard conditions)}$$

Since ATP formation from ADP and Ⓟ$_i$ is not spontaneous, free energy must be spent to make it occur. Catabolic (exergonic) pathways, especially cellular respiration, provide the energy for the endergonic process of making ATP. Plants also use light energy to produce ATP. Thus, the ATP cycle is a revolving door through which energy passes during its transfer from catabolic to anabolic pathways.

CONCEPT CHECK 8.3

1. How does ATP typically transfer energy from exergonic to endergonic reactions in the cell?
2. Which of the following combinations has more free energy: glutamic acid + ammonia + ATP, or glutamine + ADP + Ⓟ$_i$? Explain your answer.
3. **MAKE CONNECTIONS** Considering what you learned in Concepts 7.3 and 7.4 (pp. 134–136), does Figure 8.10a show passive or active transport? Explain.

For suggested answers, see Appendix A.

CONCEPT **8.4**

Enzymes speed up metabolic reactions by lowering energy barriers

The laws of thermodynamics tell us what will and will not happen under given conditions but say nothing about the rate of these processes. A spontaneous chemical reaction occurs without any requirement for outside energy, but it may occur so slowly that it is imperceptible. For example, even though the hydrolysis of sucrose (table sugar) to glucose and fructose is exergonic, occurring spontaneously with a release of free energy ($\Delta G = -7$ kcal/mol), a solution of sucrose dissolved in sterile water will sit for years at room temperature with no appreciable hydrolysis. However, if we add a small amount of the enzyme sucrase to the solution, then all the sucrose may be hydrolyzed within seconds, as shown below:

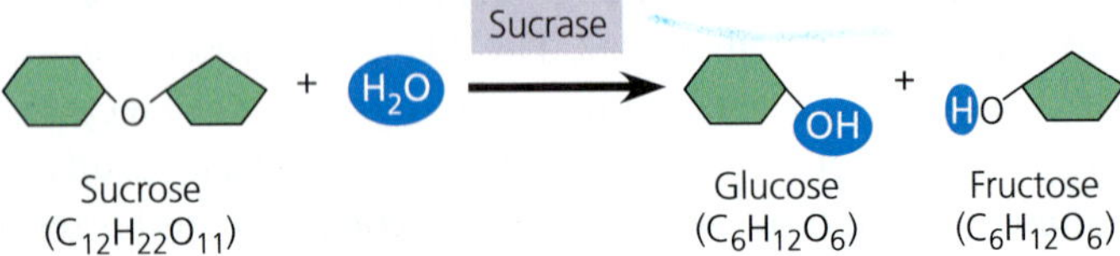

How does the enzyme do this?

An **enzyme** is a macromolecule that acts as a **catalyst**, a chemical agent that speeds up a reaction without being consumed by the reaction. (In this chapter, we are focusing on enzymes that are proteins. RNA enzymes, also called ribozymes, are discussed in Chapters 17 and 25.) Without regulation by enzymes, chemical traffic through the pathways of metabolism would become terribly congested because many chemical reactions would take such a long time. In the next two sections, we will see what prevents a spontaneous reaction from occurring faster and how an enzyme changes the situation.

The Activation Energy Barrier

Every chemical reaction between molecules involves both bond breaking and bond forming. For example, the hydrolysis of sucrose involves breaking the bond between glucose and fructose and one of the bonds of a water molecule and then forming two new bonds, as shown above. Changing one molecule into another generally involves contorting the starting molecule into a highly unstable state before the reaction can proceed. This contortion can be compared to the bending of a metal key ring when you pry it open to add a new key. The key ring is highly unstable in its opened form but returns to a stable state once the key is threaded all the way onto the ring. To reach the contorted state where bonds can change, reactant molecules must absorb energy from their surroundings. When the new bonds of the product molecules form, energy is released as heat, and the molecules return to stable shapes with lower energy than the contorted state.

The initial investment of energy for starting a reaction—the energy required to contort the reactant molecules so the bonds can break—is known as the *free energy of activation*, or **activation energy**, abbreviated E_A in this book. We can think of activation energy as the amount of energy needed to push the reactants to the top of an energy barrier, or uphill, so that the "downhill" part of the reaction can begin. Activation energy is often supplied in the form of thermal energy (heat) that the reactant molecules absorb from the surroundings. The absorption of thermal energy accelerates the reactant molecules, so they collide more often and more forcefully. It also agitates the atoms within the molecules, making the breakage of bonds more likely. When the molecules have absorbed enough energy for the bonds to break, the reactants are in an unstable condition known as the *transition state*.

Figure 8.12 graphs the energy changes for a hypothetical exergonic reaction that swaps portions of two reactant molecules:

$$\underset{\text{Reactants}}{AB + CD} \rightarrow \underset{\text{Products}}{AC + BD}$$

The activation of the reactants is represented by the uphill portion of the graph, in which the free-energy content of the

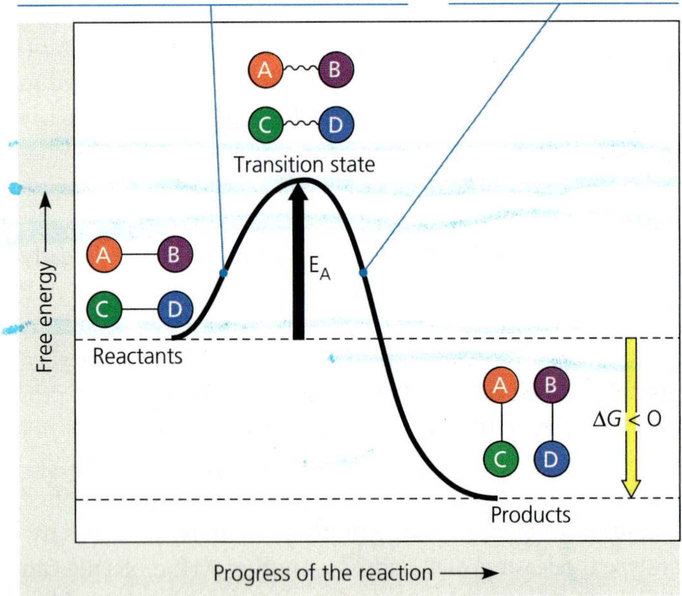

▲ **Figure 8.12 Energy profile of an exergonic reaction.** The "molecules" are hypothetical, with A, B, C, and D representing portions of the molecules. Thermodynamically, this is an exergonic reaction, with a negative ΔG, and the reaction occurs spontaneously. However, the activation energy (E_A) provides a barrier that determines the rate of the reaction.

DRAW IT *Graph the progress of an endergonic reaction in which EF and GH form products EG and FH, assuming that the reactants must pass through a transition state.*

reactant molecules is increasing. At the summit, when energy equivalent to E_A has been absorbed, the reactants are in the transition state: They are activated, and their bonds can be broken. As the atoms then settle into their new, more stable bonding arrangements, energy is released to the surroundings. This corresponds to the downhill part of the curve, which shows the loss of free energy by the molecules. The overall decrease in free energy means that E_A is repaid with interest, as the formation of new bonds releases more energy than was invested in the breaking of old bonds..

The reaction shown in Figure 8.12 is exergonic and occurs spontaneously ($\Delta G < 0$). However, the activation energy provides a barrier that determines the rate of the reaction. The reactants must absorb enough energy to reach the top of the activation energy barrier before the reaction can occur. For some reactions, E_A is modest enough that even at room temperature there is sufficient thermal energy for many of the reactant molecules to reach the transition state in a short time. In most cases, however, E_A is so high and the transition state is reached so rarely that the reaction will hardly proceed at all. In these cases, the reaction will occur at a noticeable rate only if the reactants are heated. For example, the reaction of gasoline and oxygen is exergonic and will occur spontaneously, but energy is required for the molecules to reach the transition state and react. Only when the spark plugs fire in an automobile engine can there be the explosive release of energy that pushes the pistons. Without a spark, a mixture of gasoline hydrocarbons and oxygen will not react because the E_A barrier is too high.

How Enzymes Lower the E_A Barrier

Proteins, DNA, and other complex molecules of the cell are rich in free energy and have the potential to decompose spontaneously; that is, the laws of thermodynamics favor their breakdown. These molecules persist only because at temperatures typical for cells, few molecules can make it over the hump of activation energy. However, the barriers for selected reactions must occasionally be surmounted for cells to carry out the processes needed for life. Heat speeds a reaction by allowing reactants to attain the transition state more often, but this solution would be inappropriate for biological systems. First, high temperature denatures proteins and kills cells. Second, heat would speed up *all* reactions, not just those that are needed. Instead of heat, organisms use catalysis to speed up reactions.

An enzyme catalyzes a reaction by lowering the E_A barrier (**Figure 8.13**), enabling the reactant molecules to absorb enough energy to reach the transition state even at moderate temperatures. An enzyme cannot change the ΔG for a reaction; it cannot make an endergonic reaction exergonic. Enzymes can only hasten reactions that would eventually occur anyway, but this function makes it possible for the cell to have a dynamic

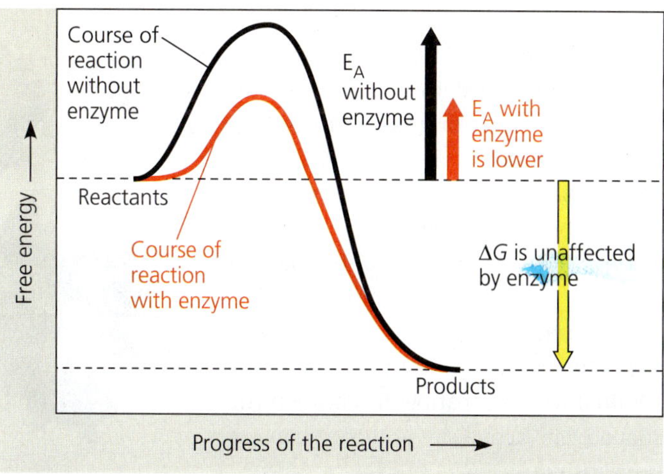

▲ **Figure 8.13 The effect of an enzyme on activation energy.** Without affecting the free-energy change (ΔG) for a reaction, an enzyme speeds the reaction by reducing its activation energy (E_A).

metabolism, routing chemicals smoothly through the cell's metabolic pathways. And because enzymes are very specific for the reactions they catalyze, they determine which chemical processes will be going on in the cell at any particular time.

Substrate Specificity of Enzymes

The reactant an enzyme acts on is referred to as the enzyme's **substrate**. The enzyme binds to its substrate (or substrates, when there are two or more reactants), forming an **enzyme-substrate complex**. While enzyme and substrate are joined, the catalytic action of the enzyme converts the substrate to the product (or products) of the reaction. The overall process can be summarized as follows:

Enzyme + Substrate(s)	$\rightleftharpoons$	Enzyme-substrate complex	$\rightleftharpoons$	Enzyme + Product(s)

For example, the enzyme sucrase (most enzyme names end in *-ase*) catalyzes the hydrolysis of the disaccharide sucrose into its two monosaccharides, glucose and fructose (see p. 152):

Sucrase + Sucrose + H_2O	$\rightleftharpoons$	Sucrase-sucrose-H_2O complex	$\rightleftharpoons$	Sucrase + Glucose + Fructose

The reaction catalyzed by each enzyme is very specific; an enzyme can recognize its specific substrate even among closely related compounds. For instance, sucrase will act only on sucrose and will not bind to other disaccharides, such as maltose. What accounts for this molecular recognition? Recall that most enzymes are proteins, and proteins are macromolecules with unique three-dimensional configurations. The specificity of an enzyme results from its shape, which is a consequence of its amino acid sequence.

Only a restricted region of the enzyme molecule actually binds to the substrate. This region, called the **active site**, is typically a pocket or groove on the surface of the enzyme where catalysis occurs **(Figure 8.14a)**. Usually, the active site is formed by only a few of the enzyme's amino acids, with the rest of the protein molecule providing a framework that determines the configuration of the active site. The specificity of an enzyme is attributed to a compatible fit between the shape of its active site and the shape of the substrate.

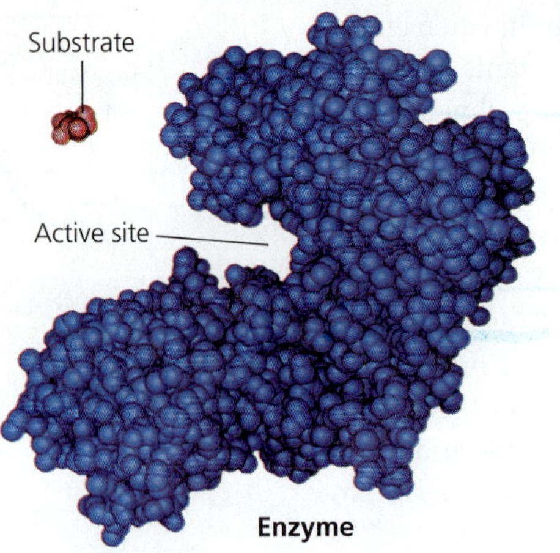

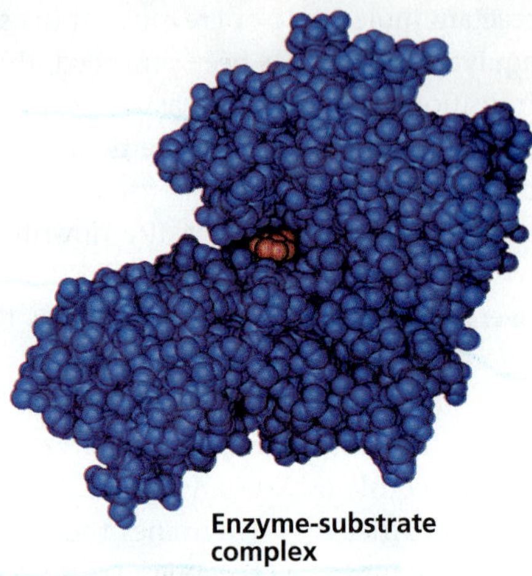

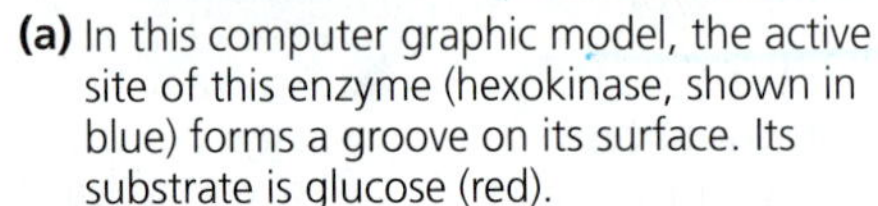

(a) In this computer graphic model, the active site of this enzyme (hexokinase, shown in blue) forms a groove on its surface. Its substrate is glucose (red).

(b) When the substrate enters the active site, it forms weak bonds with the enzyme, inducing a change in the shape of the protein. This change allows additional weak bonds to form, causing the active site to enfold the substrate and hold it in place.

▲ **Figure 8.14 Induced fit between an enzyme and its substrate.**

An enzyme is not a stiff structure locked into a given shape. In fact, recent work by biochemists has shown clearly that enzymes (and other proteins as well) seem to "dance" between subtly different shapes in a dynamic equilibrium, with slight differences in free energy for each "pose." The shape that best fits the substrate isn't necessarily the one with the lowest energy, but during the very short time the enzyme takes on this shape, its active site can bind to the substrate. It has been known for more than 50 years that the active site itself is also not a rigid receptacle for the substrate. As the substrate enters the active site, the enzyme changes shape slightly due to interactions between the substrate's chemical groups and chemical groups on the side chains of the amino acids that form the active site. This shape change makes the active site fit even more snugly around the substrate **(Figure 8.14b)**. This **induced fit** is like a clasping handshake. Induced fit brings chemical groups of the active site into positions that enhance their ability to catalyze the chemical reaction.

Catalysis in the Enzyme's Active Site

In most enzymatic reactions, the substrate is held in the active site by so-called weak interactions, such as hydrogen bonds and ionic bonds. R groups of a few of the amino acids that make up the active site catalyze the conversion of substrate to product, and the product departs from the active site. The enzyme is then free to take another substrate molecule into its active site. The entire cycle happens so fast that a single enzyme molecule typically acts on about a thousand substrate molecules per second. Some enzymes are much faster. Enzymes, like other catalysts, emerge from the reaction in their original form. Therefore, very small amounts of enzyme can have a huge metabolic impact by functioning over and over again in catalytic cycles. **Figure 8.15** shows a catalytic cycle involving two substrates and two products.

Most metabolic reactions are reversible, and an enzyme can catalyze either the forward or the reverse reaction, depending on which direction has a negative ΔG. This in turn depends mainly on the relative concentrations of reactants and products. The net effect is always in the direction of equilibrium.

Enzymes use a variety of mechanisms that lower activation energy and speed up a reaction (see Figure 8.15, step ❸). First, in reactions involving two or more reactants, the active site provides a template on which the substrates can come together in the proper orientation for a reaction to occur between them. Second, as the active site of an enzyme clutches the bound substrates, the enzyme may stretch the substrate molecules toward their transition-state form, stressing and bending critical chemical bonds that must be broken during the reaction. Because E_A is proportional to the difficulty of breaking the bonds, distorting the substrate helps it approach the transition state and thus reduces the amount of free energy that must be absorbed to achieve that state.

Third, the active site may also provide a microenvironment that is more conducive to a particular type of reaction than the solution itself would be without the enzyme. For example, if the active site has amino acids with acidic R groups, the active site may be a pocket of low pH in an otherwise neutral cell. In such cases, an acidic amino acid may facilitate H^+ transfer to the substrate as a key step in catalyzing the reaction.

A fourth mechanism of catalysis is the direct participation of the active site in the chemical reaction. Sometimes this process even involves brief covalent bonding between the substrate and the side chain of an amino acid of the enzyme. Subsequent steps of the reaction restore the side chains to their original states, so that the active site is the same after the reaction as it was before.

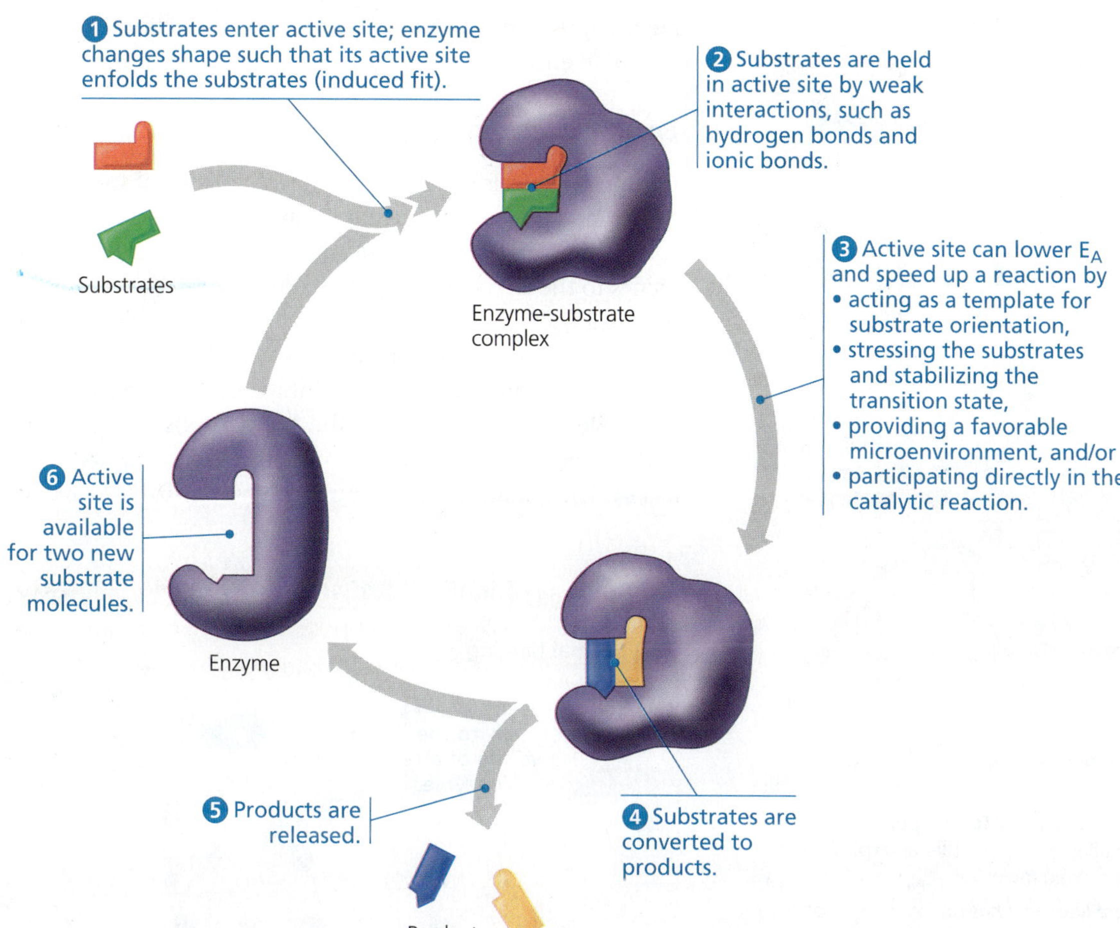

◀ **Figure 8.15 The active site and catalytic cycle of an enzyme.** An enzyme can convert one or more reactant molecules to one or more product molecules. The enzyme shown here converts two substrate molecules to two product molecules.

The rate at which a particular amount of enzyme converts substrate to product is partly a function of the initial concentration of the substrate: The more substrate molecules that are available, the more frequently they access the active sites of the enzyme molecules. However, there is a limit to how fast the reaction can be pushed by adding more substrate to a fixed concentration of enzyme. At some point, the concentration of substrate will be high enough that all enzyme molecules have their active sites engaged. As soon as the product exits an active site, another substrate molecule enters. At this substrate concentration, the enzyme is said to be *saturated*, and the rate of the reaction is determined by the speed at which the active site converts substrate to product. When an enzyme population is saturated, the only way to increase the rate of product formation is to add more enzyme. Cells often increase the rate of a reaction by producing more enzyme molecules.

Effects of Local Conditions on Enzyme Activity

The activity of an enzyme—how efficiently the enzyme functions—is affected by general environmental factors, such as temperature and pH. It can also be affected by chemicals that specifically influence that enzyme. In fact, researchers have learned much about enzyme function by employing such chemicals.

Effects of Temperature and pH

Recall from Chapter 5 that the three-dimensional structures of proteins are sensitive to their environment. As a consequence, each enzyme works better under some conditions than under other conditions, because these *optimal conditions* favor the most active shape for the enzyme molecule.

Temperature and pH are environmental factors important in the activity of an enzyme. Up to a point, the rate of an enzymatic reaction increases with increasing temperature, partly because substrates collide with active sites more frequently when the molecules move rapidly. Above that temperature, however, the speed of the enzymatic reaction drops sharply. The thermal agitation of the enzyme molecule disrupts the hydrogen bonds, ionic bonds, and other weak interactions that stabilize the active shape of the enzyme, and the protein molecule eventually denatures. Each enzyme has an optimal temperature at which its reaction rate is greatest. Without denaturing the enzyme, this temperature allows the greatest number of molecular collisions and the fastest conversion of the reactants to product molecules. Most human enzymes have optimal temperatures of about 35–40°C (close to human body temperature). The thermophilic bacteria that live in hot springs contain enzymes with optimal temperatures of 70°C or higher (**Figure 8.16a** on the next page).

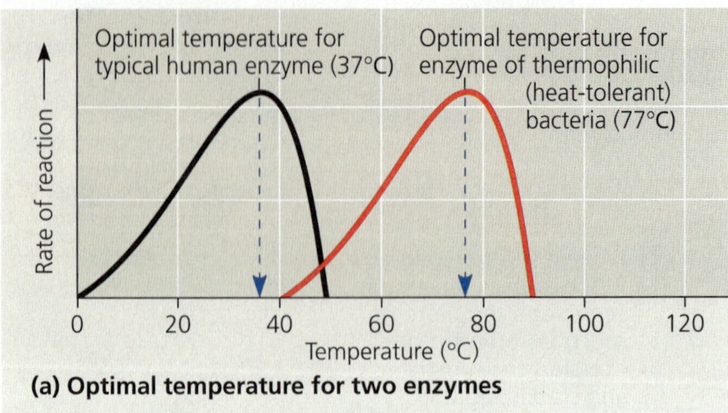

(a) Optimal temperature for two enzymes

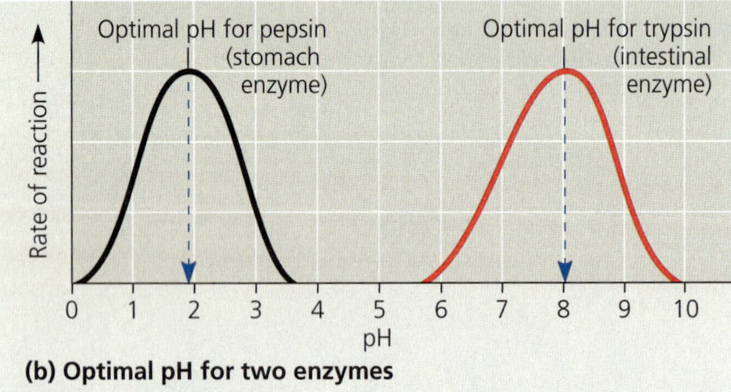

(b) Optimal pH for two enzymes

▲ **Figure 8.16 Environmental factors affecting enzyme activity.** Each enzyme has an optimal **(a)** temperature and **(b)** pH that favor the most active shape of the protein molecule.

DRAW IT *Given that a mature lysosome has an internal pH of around 4.5, draw a curve in (b) showing what you would predict for a lysosomal enzyme, labeling its optimal pH.*

Just as each enzyme has an optimal temperature, it also has a pH at which it is most active. The optimal pH values for most enzymes fall in the range of pH 6–8, but there are exceptions. For example, pepsin, a digestive enzyme in the human stomach, works best at pH 2. Such an acidic environment denatures most enzymes, but pepsin is adapted to maintain its functional three-dimensional structure in the acidic environment of the stomach. In contrast, trypsin, a digestive enzyme residing in the alkaline environment of the human intestine, has an optimal pH of 8 and would be denatured in the stomach **(Figure 8.16b)**.

Cofactors

Many enzymes require nonprotein helpers for catalytic activity. These adjuncts, called **cofactors**, may be bound tightly to the enzyme as permanent residents, or they may bind loosely and reversibly along with the substrate. The cofactors of some enzymes are inorganic, such as the metal atoms zinc, iron, and copper in ionic form. If the cofactor is an organic molecule, it is more specifically called a **coenzyme**. Most vitamins are important in nutrition because they act as coenzymes or raw materials from which coenzymes are made. Cofactors function in various ways, but in all cases where they are used, they perform a crucial chemical function in catalysis. You'll encounter examples of cofactors later in the book.

Enzyme Inhibitors

Certain chemicals selectively inhibit the action of specific enzymes, and we have learned a lot about enzyme function by studying the effects of these molecules. If the inhibitor attaches to the enzyme by covalent bonds, inhibition is usually irreversible.

Many enzyme inhibitors, however, bind to the enzyme by weak interactions, in which case inhibition is reversible. Some reversible inhibitors resemble the normal substrate molecule and compete for admission into the active site **(Figure 8.17a** and **b)**. These mimics, called **competitive inhibitors**, reduce

▼ **Figure 8.17 Inhibition of enzyme activity.**

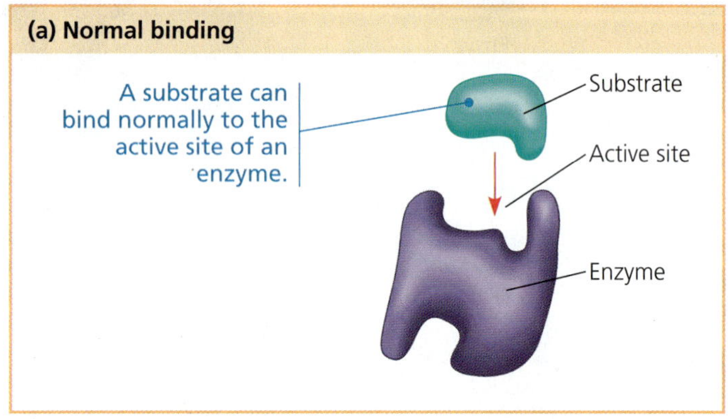

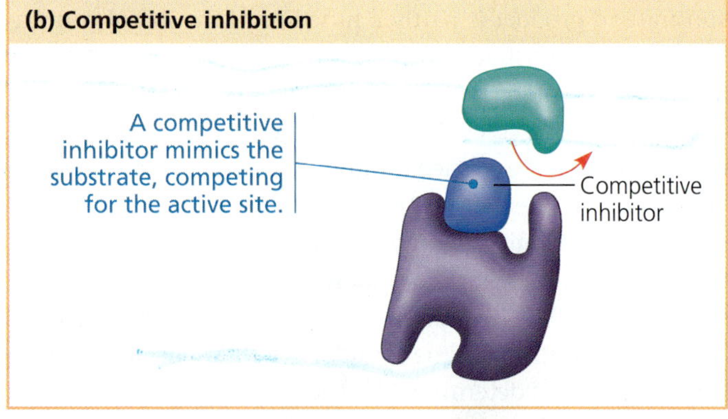

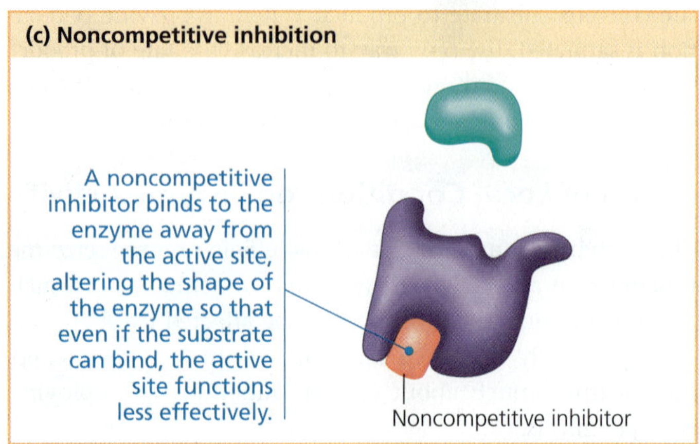

the productivity of enzymes by blocking substrates from entering active sites. This kind of inhibition can be overcome by increasing the concentration of substrate so that as active sites become available, more substrate molecules than inhibitor molecules are around to gain entry to the sites.

In contrast, **noncompetitive inhibitors** do not directly compete with the substrate to bind to the enzyme at the active site (**Figure 8.17c**). Instead, they impede enzymatic reactions by binding to another part of the enzyme. This interaction causes the enzyme molecule to change its shape in such a way that the active site becomes less effective at catalyzing the conversion of substrate to product.

Toxins and poisons are often irreversible enzyme inhibitors. An example is sarin, a nerve gas that caused the death of several people and injury to many others when it was released by terrorists in the Tokyo subway in 1995. This small molecule binds covalently to the R group on the amino acid serine, which is found in the active site of acetylcholinesterase, an enzyme important in the nervous system. Other examples include the pesticides DDT and parathion, inhibitors of key enzymes in the nervous system. Finally, many antibiotics are inhibitors of specific enzymes in bacteria. For instance, penicillin blocks the active site of an enzyme that many bacteria use to make their cell walls.

Citing enzyme inhibitors that are metabolic poisons may give the impression that enzyme inhibition is generally abnormal and harmful. In fact, molecules naturally present in the cell often regulate enzyme activity by acting as inhibitors. Such regulation—selective inhibition—is essential to the control of cellular metabolism, as we will discuss in Concept 8.5.

The Evolution of Enzymes

EVOLUTION Thus far, biochemists have discovered and named more than 4,000 different enzymes in various species, and this list probably represents the tip of the proverbial iceberg. How did this grand profusion of enzymes arise? Recall that most enzymes are proteins, and proteins are encoded by genes. A permanent change in a gene, known as a *mutation*, can result in a protein with one or more changed amino acids. In the case of an enzyme, if the changed amino acids are in the active site or some other crucial region, the altered enzyme might have a novel activity or might bind to a different substrate. Under environmental conditions where the new function benefits the organism, natural selection would tend to favor the mutated form of the gene, causing it to persist in the population. This simplified model is generally accepted as the main way in which the multitude of different enzymes arose over the past few billion years of life's history.

Data supporting this model have been collected by researchers using a lab procedure that mimics evolution in natural populations. One group tested whether the function of an enzyme called β-galactosidase could change over time in populations of the bacterium *Escherichia coli* (*E. coli*). β-galactosidase breaks down the disaccharide lactose into the simple sugars glucose and galactose. Using molecular techniques, the researchers introduced random mutations into *E. coli* genes and then tested the bacteria for their ability to break down a slightly different disaccharide (one that has the sugar fucose in place of galactose). They selected the mutant bacteria that could do this best and exposed them to another round of mutation and selection. After seven rounds, the "evolved" enzyme bound the new substrate several hundred times more strongly, and broke it down 10 to 20 times more quickly, than did the original enzyme.

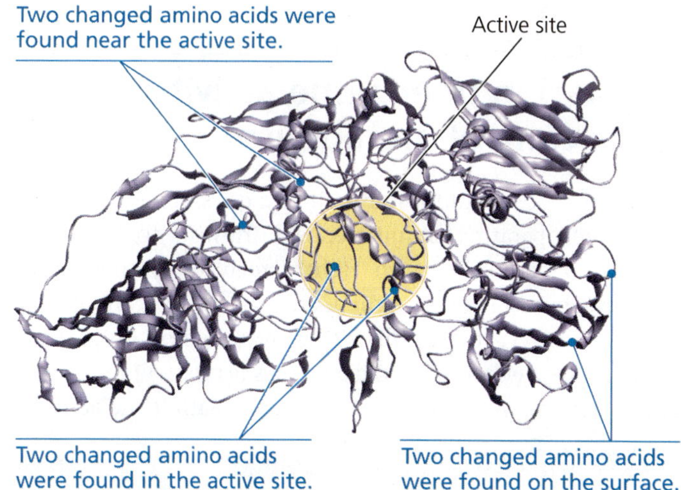

▲ **Figure 8.18 Mimicking evolution of an enzyme with a new function.** After seven rounds of mutation and selection in a lab, the enzyme β-galactosidase evolved into an enzyme specialized for breaking down a sugar different from lactose. This ribbon model shows one subunit of the altered enzyme; six amino acids were different.

The researchers found that six amino acids had changed in the enzyme altered in this experiment. Two of these changed amino acids were in the active site, two were nearby, and two were on the surface of the protein (**Figure 8.18**). This experiment and others like it strengthen the notion that a few changes can indeed alter enzyme function.

CONCEPT CHECK 8.4

1. Many spontaneous reactions occur very slowly. Why don't all spontaneous reactions occur instantly?
2. Why do enzymes act only on very specific substrates?
3. **WHAT IF?** Malonate is an inhibitor of the enzyme succinate dehydrogenase. How would you determine whether malonate is a competitive or noncompetitive inhibitor?
4. **MAKE CONNECTIONS** In nature, what conditions could lead to natural selection favoring bacteria with enzymes that could break down the fucose-containing disaccharide discussed above? See the discussion of natural selection in Concept 1.2, pages 14–16.

For suggested answers, see Appendix A.

CONCEPT 8.5

Regulation of enzyme activity helps control metabolism

Chemical chaos would result if all of a cell's metabolic pathways were operating simultaneously. Intrinsic to life's processes is a cell's ability to tightly regulate its metabolic pathways by controlling when and where its various enzymes are active. It does this either by switching on and off the genes that encode specific enzymes (as we will discuss in Unit Three) or, as we discuss here, by regulating the activity of enzymes once they are made.

Allosteric Regulation of Enzymes

In many cases, the molecules that naturally regulate enzyme activity in a cell behave something like reversible noncompetitive inhibitors (see Figure 8.17c): These regulatory molecules change an enzyme's shape and the functioning of its active site by binding to a site elsewhere on the molecule, via noncovalent interactions. **Allosteric regulation** is the term used to describe any case in which a protein's function at one site is affected by the binding of a regulatory molecule to a separate site. It may result in either inhibition or stimulation of an enzyme's activity.

Allosteric Activation and Inhibition

Most enzymes known to be allosterically regulated are constructed from two or more subunits, each composed of a polypeptide chain with its own active site. The entire complex oscillates between two different shapes, one catalytically active and the other inactive **(Figure 8.19a)**. In the simplest kind of allosteric regulation, an activating or inhibiting regulatory molecule binds to a regulatory site (sometimes called an allosteric site), often located where subunits join. The binding of an *activator* to a regulatory site stabilizes the shape that has functional active sites, whereas the binding of an *inhibitor* stabilizes the inactive form of the enzyme. The subunits of an allosteric enzyme fit together in such a way that a shape change in one subunit is transmitted to all others. Through this interaction of subunits, a single activator or inhibitor molecule that binds to one regulatory site will affect the active sites of all subunits.

▼ **Figure 8.19 Allosteric regulation of enzyme activity.**

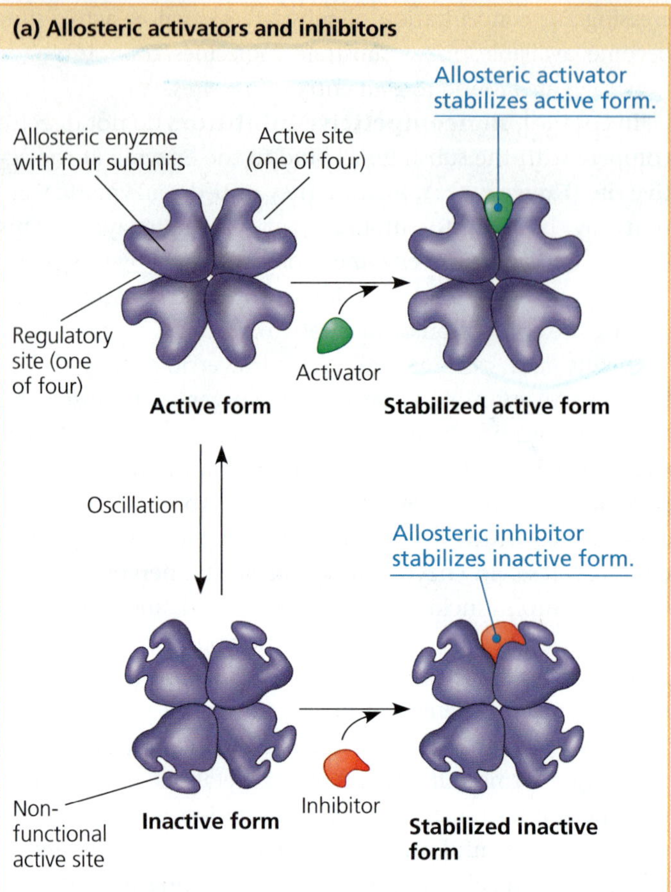

At low concentrations, activators and inhibitors dissociate from the enzyme. The enzyme can then oscillate again.

(b) Cooperativity: another type of allosteric activation

Binding of one substrate molecule to active site of one subunit locks all subunits in active conformation.

Substrate

Inactive form

Stabilized active form

The inactive form shown on the left oscillates with the active form when the active form is not stabilized by substrate.

Fluctuating concentrations of regulators can cause a sophisticated pattern of response in the activity of cellular enzymes. The products of ATP hydrolysis (ADP and Ⓟ$_i$), for example, play a complex role in balancing the flow of traffic between anabolic and catabolic pathways by their effects on key enzymes. ATP binds to several catabolic enzymes allosterically, lowering their affinity for substrate and thus inhibiting their activity. ADP, however, functions as an activator of the same enzymes. This is logical because catabolism functions in regenerating ATP. If ATP production lags behind its use, ADP accumulates and activates the enzymes that speed up catabolism, producing more ATP. If the supply of ATP exceeds demand, then catabolism slows down as ATP molecules accumulate and bind to the same enzymes, inhibiting them. (You'll see specific examples of this type of regulation when you learn about cellular respiration in the next chapter.) ATP, ADP, and other related molecules also affect key enzymes in anabolic pathways. In this way, allosteric enzymes control the rates of important reactions in both sorts of metabolic pathways.

In another kind of allosteric activation, a *substrate* molecule binding to one active site in a multisubunit enzyme triggers a shape change in all the subunits, thereby increasing catalytic activity at the other active sites **(Figure 8.19b)**. Called **cooperativity**, this mechanism amplifies the response of enzymes to substrates: One substrate molecule primes an enzyme to act on additional substrate molecules more readily. Cooperativity is considered "allosteric" regulation because binding of the substrate to one active site affects catalysis in another active site.

Although the vertebrate oxygen transport protein hemoglobin is not an enzyme, classic studies of cooperative binding in this protein have elucidated the principle of cooperativity. Hemoglobin is made up of four subunits, each of which has an oxygen-binding site (see Figure 5.20). The binding of an oxygen molecule to one binding site increases the affinity for oxygen of the remaining binding sites. Thus, where oxygen is at high levels, such as in the lungs or gills, hemoglobin's affinity for oxygen increases as more binding sites are filled. In oxygen-deprived tissues, however, the release of each oxygen molecule decreases the oxygen affinity of the other binding sites, resulting in the release of oxygen where it is most needed. Cooperativity works similarly in multisubunit enzymes that have been studied.

Identification of Allosteric Regulators

Although allosteric regulation is probably quite widespread, relatively few of the many known metabolic enzymes have been shown to be regulated in this way. Allosteric regulatory molecules are hard to characterize, in part because they tend to bind the enzyme at low affinity and are therefore hard to isolate. Recently, however, pharmaceutical companies have turned their attention to allosteric regulators. These molecules are attractive drug candidates for enzyme regulation because they exhibit higher specificity for particular enzymes than do inhibitors that bind to the active site. (An active site may be similar to the active site in another, related enzyme, whereas allosteric regulatory sites appear to be quite distinct between enzymes.)

Figure 8.20 describes a search for allosteric regulators, carried out as a collaboration between researchers at the University of California at San Francisco and a company called Sunesis Pharmaceuticals. The study was designed to find allosteric inhibitors of *caspases*, protein-digesting enzymes that play an active role in inflammation and cell death. (You'll learn more about caspases and cell death in Chapter 11.) By specifically regulating these enzymes, we may be able to better manage inappropriate inflammatory responses, such as those commonly seen in vascular and neurodegenerative diseases.

▼ Figure 8.20 INQUIRY

Are there allosteric inhibitors of caspase enzymes?

EXPERIMENT In an effort to identify allosteric inhibitors of caspases, Justin Scheer and co-workers screened close to 8,000 compounds for their ability to bind to a possible allosteric binding site in caspase 1 and inhibit the enzyme's activity. Each compound was designed to form a disulfide bond with a cysteine near the site in order to stabilize the low-affinity interaction that is expected of an allosteric inhibitor. As the caspases are known to exist in both active and inactive forms, the researchers hypothesized that this linkage might lock the enzyme in the inactive form.

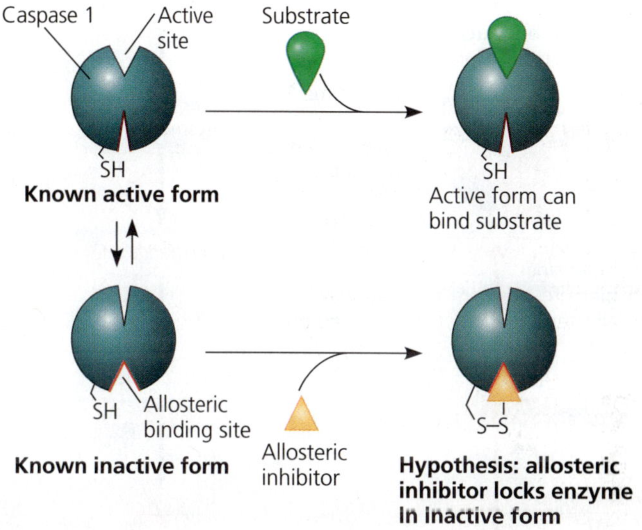

To test this model, X-ray diffraction analysis was used to determine the structure of caspase 1 when bound to one of the inhibitors and to compare it with the active and inactive structures.

RESULTS Fourteen compounds were identified that could bind to the proposed allosteric site (red) of caspase 1 and block enzymatic activity. The enzyme's shape when one such inhibitor was bound resembled the inactive caspase 1 more than the active form.

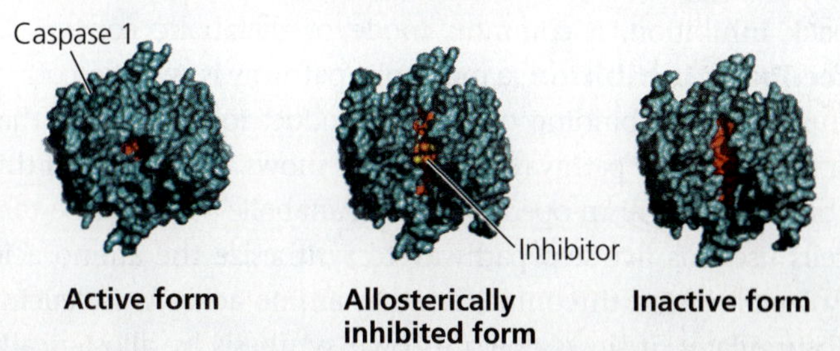

CONCLUSION That particular inhibitory compound apparently locks the enzyme in its inactive form, as expected for a true allosteric regulator. The data therefore support the existence of an allosteric inhibitory site on caspase 1 that can be used to control enzymatic activity.

SOURCE J. M. Scheer et al., A common allosteric site and mechanism in caspases, *Proceedings of the National Academy of Sciences* 103: 7595–7600 (2006).

WHAT IF? As a control, the researchers broke the disulfide linkage between one of the inhibitors and the caspase. Assuming that the experimental solution contains no other inhibitors, how would you expect the caspase 1 activity to be affected?

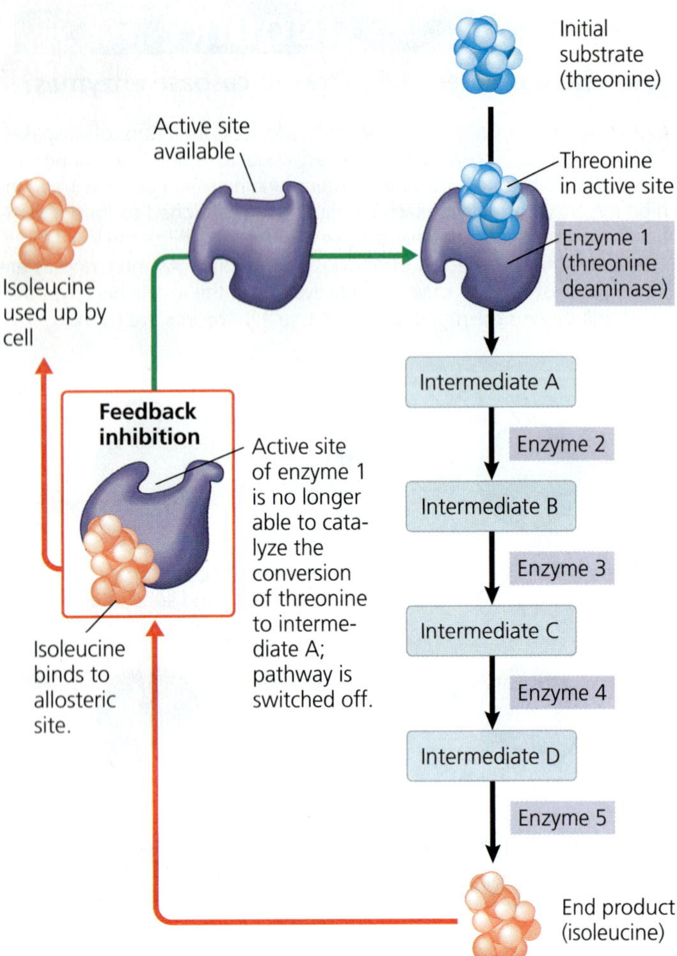

▲ **Figure 8.21 Feedback inhibition in isoleucine synthesis.**

Feedback Inhibition

When ATP allosterically inhibits an enzyme in an ATP-generating pathway, as we discussed earlier, the result is feedback inhibition, a common mode of metabolic control. In **feedback inhibition**, a metabolic pathway is switched off by the inhibitory binding of its end product to an enzyme that acts early in the pathway. **Figure 8.21** shows an example of this control mechanism operating on an anabolic pathway. Certain cells use this five-step pathway to synthesize the amino acid isoleucine from threonine, another amino acid. As isoleucine accumulates, it slows down its own synthesis by allosterically inhibiting the enzyme for the first step of the pathway. Feedback inhibition thereby prevents the cell from wasting chemical resources by making more isoleucine than is necessary.

Specific Localization of Enzymes Within the Cell

The cell is not just a bag of chemicals with thousands of different kinds of enzymes and substrates in a random mix. The cell is compartmentalized, and cellular structures help bring order to metabolic pathways. In some cases, a team of enzymes for several steps of a metabolic pathway are assembled into a multienzyme complex. The arrangement facilitates the sequence of reactions, with the product from the first enzyme becoming the substrate for an adjacent enzyme in the complex, and so on, until the end product is released. Some enzymes and enzyme complexes have fixed locations within the cell and act as structural components of particular membranes. Others are in solution within particular membrane-enclosed eukaryotic organelles, each with its own internal chemical environment. For example, in eukaryotic cells, the enzymes for cellular respiration reside in specific locations within mitochondria **(Figure 8.22)**.

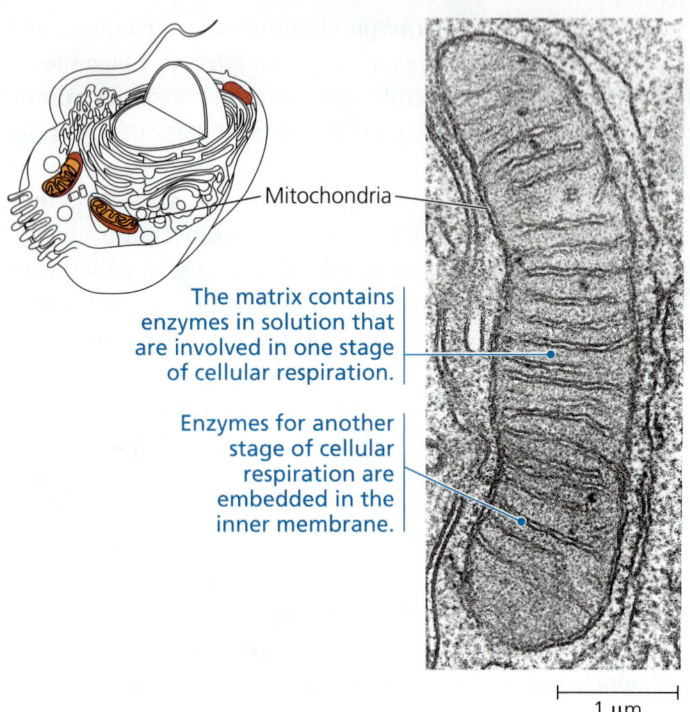

▲ **Figure 8.22 Organelles and structural order in metabolism.** Organelles such as the mitochondrion (TEM) contain enzymes that carry out specific functions, in this case cellular respiration.

In this chapter, you have learned that metabolism, the intersecting set of chemical pathways characteristic of life, is a choreographed interplay of thousands of different kinds of cellular molecules. In the next chapter, we explore cellular respiration, the major catabolic pathway that breaks down organic molecules, releasing energy for the crucial processes of life.

CONCEPT CHECK 8.5

1. How do an activator and an inhibitor have different effects on an allosterically regulated enzyme?
2. **WHAT IF?** Imagine you are a pharmacological researcher who wants to design a drug that inhibits a particular enzyme. Upon reading the scientific literature, you find that the enzyme's active site is similar to that of several other enzymes. What might be a good approach to developing your inhibitor drug?

For suggested answers, see Appendix A.

8 CHAPTER REVIEW

SUMMARY OF KEY CONCEPTS

CONCEPT 8.1

An organism's metabolism transforms matter and energy, subject to the laws of thermodynamics (pp. 142–145)

- **Metabolism** is the collection of chemical reactions that occur in an organism. **Enzymes** catalyze reactions in intersecting **metabolic pathways**, which may be **catabolic** (breaking down molecules, releasing energy) or **anabolic** (building molecules, consuming energy).
- **Energy** is the capacity to cause change; some forms of energy do work by moving matter. **Kinetic energy** is associated with motion and includes **thermal energy (heat)** associated with random motion of atoms or molecules. **Potential energy** is related to the location or structure of matter and includes **chemical energy** possessed by a molecule due to its structure.
- **The first law of thermodynamics**, conservation of energy, states that energy cannot be created or destroyed, only transferred or transformed. The **second law of thermodynamics** states that **spontaneous processes**, those requiring no outside input of energy, increase the **entropy** (disorder) of the universe.

? *Explain how the highly ordered structure of a cell does not conflict with the second law of thermodynamics.*

CONCEPT 8.2

The free-energy change of a reaction tells us whether or not the reaction occurs spontaneously (pp. 146–149)

- A living system's **free energy** is energy that can do work under cellular conditions. The change in free energy (ΔG) during a biological process is related directly to enthalpy change (ΔH) and to the change in entropy (ΔS): $\Delta G = \Delta H - T\Delta S$. Organisms live at the expense of free energy. During a spontaneous change, free energy decreases and the stability of a system increases. At maximum stability, the system is at equilibrium and can do no work.
- In an **exergonic** (spontaneous) chemical reaction, the products have less free energy than the reactants ($-\Delta G$). **Endergonic** (nonspontaneous) reactions require an input of energy ($+\Delta G$). The addition of starting materials and the removal of end products prevent metabolism from reaching equilibrium.

? *Explain the meaning of each component in the equation for the change in free energy of a spontaneous chemical reaction. Why are spontaneous reactions important in the metabolism of a cell?*

CONCEPT 8.3

ATP powers cellular work by coupling exergonic reactions to endergonic reactions (pp. 149–151)

- **ATP** is the cell's energy shuttle. Hydrolysis of its terminal phosphate yields ADP and phosphate and releases free energy.
- Through **energy coupling**, the exergonic process of ATP hydrolysis drives endergonic reactions by transfer of a phosphate group to specific reactants, forming a **phosphorylated intermediate** that is more reactive. ATP hydrolysis (sometimes with protein phosphorylation) also causes changes in the shape and binding affinities of transport and motor proteins.
- Catabolic pathways drive regeneration of ATP from ADP + Ⓟ$_i$.

? *Describe the ATP cycle: How is ATP used and regenerated in a cell?*

CONCEPT 8.4

Enzymes speed up metabolic reactions by lowering energy barriers (pp. 152–157)

- In a chemical reaction, the energy necessary to break the bonds of the reactants is the **activation energy**, E_A.
- **Enzymes** lower the E_A barrier:

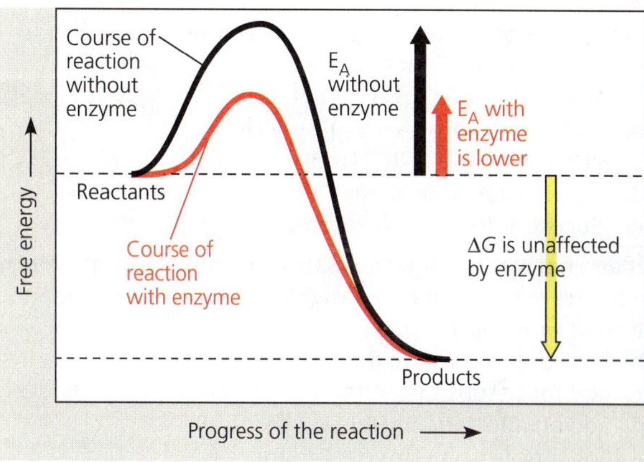

- Each type of enzyme has a unique **active site** that combines specifically with its **substrate(s)**, the reactant molecule(s) on which it acts. The enzyme changes shape slightly when it binds the substrate(s) (**induced fit**).
- The active site can lower an E_A barrier by orienting substrates correctly, straining their bonds, providing a favorable microenvironment, or even covalently bonding with the substrate.
- Each enzyme has an optimal temperature and pH. Inhibitors reduce enzyme function. A **competitive inhibitor** binds to the active site, whereas a **noncompetitive inhibitor** binds to a different site on the enzyme.
- Natural selection, acting on organisms with mutant genes encoding altered enzymes, is a major evolutionary force responsible for the diverse array of enzymes found in organisms.

? *How do both activation energy barriers and enzymes help maintain the structural and metabolic order of life?*

CONCEPT 8.5

Regulation of enzyme activity helps control metabolism (pp. 158–160)

- Many enzymes are subject to **allosteric regulation**: Regulatory molecules, either activators or inhibitors, bind to specific regulatory sites, affecting the shape and function of the enzyme. In **cooperativity**, binding of one substrate molecule can stimulate binding or activity at other active sites. In **feedback inhibition**, the end product of a metabolic pathway allosterically inhibits the enzyme for a previous step in the pathway.
- Some enzymes are grouped into complexes, some are incorporated into membranes, and some are contained inside organelles, increasing the efficiency of metabolic processes.

? *What roles do allosteric regulation and feedback inhibition play in the metabolism of a cell?*

TEST YOUR UNDERSTANDING

LEVEL 1: KNOWLEDGE/COMPREHENSION

1. Choose the pair of terms that correctly completes this sentence: Catabolism is to anabolism as _______ is to _______.
   a. exergonic; spontaneous
   b. exergonic; endergonic
   c. free energy; entropy
   d. work; energy
   e. entropy; enthalpy
2. Most cells cannot harness heat to perform work because
   a. heat is not a form of energy.
   b. cells do not have much heat; they are relatively cool.
   c. temperature is usually uniform throughout a cell.
   d. heat can never be used to do work.
   e. heat must remain constant during work.
3. Which of the following metabolic processes can occur without a net influx of energy from some other process?
   a. ADP + Ⓟ$_i$ $\rightarrow$ ATP + H_2O
   b. $C_6H_{12}O_6 + 6\ O_2 \rightarrow 6\ CO_2 + 6\ H_2O$
   c. $6\ CO_2 + 6\ H_2O \rightarrow C_6H_{12}O_6 + 6\ O_2$
   d. amino acids $\rightarrow$ protein
   e. glucose + fructose $\rightarrow$ sucrose
4. If an enzyme in solution is saturated with substrate, the most effective way to obtain a faster yield of products is to
   a. add more of the enzyme.
   b. heat the solution to 90°C.
   c. add more substrate.
   d. add an allosteric inhibitor.
   e. add a noncompetitive inhibitor.
5. Some bacteria are metabolically active in hot springs because
   a. they are able to maintain a lower internal temperature.
   b. high temperatures make catalysis unnecessary.
   c. their enzymes have high optimal temperatures.
   d. their enzymes are completely insensitive to temperature.
   e. they use molecules other than proteins or RNAs as their main catalysts.

LEVEL 2: APPLICATION/ANALYSIS

6. If an enzyme is added to a solution where its substrate and product are in equilibrium, what will occur?
   a. Additional product will be formed.
   b. Additional substrate will be formed.
   c. The reaction will change from endergonic to exergonic.
   d. The free energy of the system will change.
   e. Nothing; the reaction will stay at equilibrium.

LEVEL 3: SYNTHESIS/EVALUATION

7. **DRAW IT** Using a series of arrows, draw the branched metabolic reaction pathway described by the following statements, and then answer the question at the end. Use red arrows and minus signs to indicate inhibition.

   L can form either M or N.
   M can form O.
   O can form either P or R.
   P can form Q.
   R can form S.
   O inhibits the reaction of L to form M.
   Q inhibits the reaction of O to form P.
   S inhibits the reaction of O to form R.

   Which reaction would prevail if both Q and S were present in the cell in high concentrations?
   a. L $\rightarrow$ M
   b. M $\rightarrow$ O
   c. L $\rightarrow$ N
   d. O $\rightarrow$ P
   e. R $\rightarrow$ S

8. **EVOLUTION CONNECTION**
   A recent revival of the antievolutionary "intelligent design" argument holds that biochemical pathways are too complex to have evolved, because all intermediate steps in a given pathway must be present to produce the final product. Critique this argument. How could you use the diversity of metabolic pathways that produce the same or similar products to support your case?

9. **SCIENTIFIC INQUIRY**
   DRAW IT A researcher has developed an assay to measure the activity of an important enzyme present in liver cells growing in culture. She adds the enzyme's substrate to a dish of cells and then measures the appearance of reaction products. The results are graphed as the amount of product on the *y*-axis versus time on the *x*-axis. The researcher notes four sections of the graph. For a short period of time, no products appear (section A). Then (section B) the reaction rate is quite high (the slope of the line is steep). Next, the reaction gradually slows down (section C). Finally, the graph line becomes flat (section D). Draw and label the graph, and propose a model to explain the molecular events occurring at each stage of this reaction profile.

10. **SCIENCE, TECHNOLOGY, AND SOCIETY**
    Organophosphates (organic compounds containing phosphate groups) are commonly used as insecticides to improve crop yield. Organophosphates typically interfere with nerve signal transmission by inhibiting the enzymes that degrade transmitter molecules. They affect humans and other vertebrates as well as insects. Thus, the use of organophosphate pesticides poses some health risks. On the other hand, these molecules break down rapidly upon exposure to air and sunlight. As a consumer, what level of risk are you willing to accept in exchange for an abundant and affordable food supply?

11. **WRITE ABOUT A THEME**
    Energy Transfer Life requires energy. In a short essay (100–150 words), describe the basic principles of bioenergetics in an animal cell. How is the flow and transformation of energy different in a photosynthesizing cell? Include the role of ATP and enzymes in your discussion.

For selected answers, see Appendix A.

www.masteringbiology.com

1. MasteringBiology® Assignments

Tutorials ATP and Energy • How Enzymes Function • Enzyme and Substrate Concentrations • Factors That Affect Reaction Rate • Enzyme Inhibition • Regulating Enzyme Action
Activities Energy Transformations • The Structure of ATP • Chemical Reactions and ATP • How Enzymes Work
Questions Student Misconceptions • Reading Quiz • Multiple Choice • End-of-Chapter

2. eText

Read your book online, search, take notes, highlight text, and more.

3. The Study Area

Practice Tests • Cumulative Test • ***BioFlix*** 3-D Animations • MP3 Tutor Sessions • Videos • Activities • Investigations • Lab Media • Audio Glossary • Word Study Tools • Art

9

Cellular Respiration and Fermentation

▲ **Figure 9.1 How do these leaves power the work of life for this chimpanzee?**

KEY CONCEPTS

9.1 Catabolic pathways yield energy by oxidizing organic fuels

9.2 Glycolysis harvests chemical energy by oxidizing glucose to pyruvate

9.3 After pyruvate is oxidized, the citric acid cycle completes the energy-yielding oxidation of organic molecules

9.4 During oxidative phosphorylation, chemiosmosis couples electron transport to ATP synthesis

9.5 Fermentation and anaerobic respiration enable cells to produce ATP without the use of oxygen

9.6 Glycolysis and the citric acid cycle connect to many other metabolic pathways

OVERVIEW

Life Is Work

Living cells require transfusions of energy from outside sources to perform their many tasks—for example, assembling polymers, pumping substances across membranes, moving, and reproducing. The chimpanzee in **Figure 9.1** obtains energy for its cells by eating plants; some animals feed on other organisms that eat plants. The energy stored in the organic molecules of food ultimately comes from the sun. Energy flows into an ecosystem as sunlight and exits as heat; in contrast, the chemical elements essential to life are recycled **(Figure 9.2)**. Photosynthesis generates oxygen and organic molecules used by the mitochondria of eukaryotes (including plants and algae) as fuel for cellular respiration. Respiration breaks this fuel down, generating ATP. The waste products of this type of respiration, carbon dioxide and water, are the raw materials for photosynthesis. In this chapter, we consider how cells harvest the chemical energy stored in organic molecules and use it to generate ATP, the molecule that drives most cellular work. After presenting some basics about respiration, we will focus on three key pathways of respiration: glycolysis, the citric acid cycle, and oxidative phosphorylation. We'll also consider fermentation, a somewhat simpler pathway coupled to glycolysis that has deep evolutionary roots.

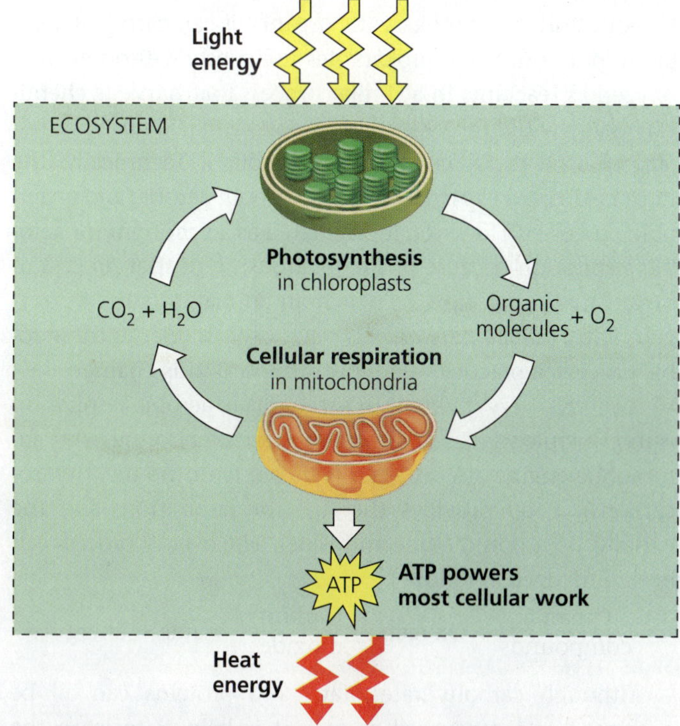

▲ **Figure 9.2 Energy flow and chemical recycling in ecosystems.** Energy flows into an ecosystem as sunlight and ultimately leaves as heat, while the chemical elements essential to life are recycled.

BioFlix Visit the Study Area at **www.masteringbiology.com** for the BioFlix® 3-D Animation on The Carbon Cycle.

CONCEPT 9.1

Catabolic pathways yield energy by oxidizing organic fuels

As you learned in Chapter 8, metabolic pathways that release stored energy by breaking down complex molecules are called catabolic pathways. Electron transfer plays a major role in these pathways. In this section, we consider these processes, which are central to cellular respiration.

Catabolic Pathways and Production of ATP

Organic compounds possess potential energy as a result of the arrangement of electrons in the bonds between their atoms. Compounds that can participate in exergonic reactions can act as fuels. With the help of enzymes, a cell systematically degrades complex organic molecules that are rich in potential energy to simpler waste products that have less energy. Some of the energy taken out of chemical storage can be used to do work; the rest is dissipated as heat.

One catabolic process, **fermentation**, is a partial degradation of sugars or other organic fuel that occurs without the use of oxygen. However, the most prevalent and efficient catabolic pathway is **aerobic respiration**, in which oxygen is consumed as a reactant along with the organic fuel (*aerobic* is from the Greek *aer*, air, and *bios*, life). The cells of most eukaryotic and many prokaryotic organisms can carry out aerobic respiration. Some prokaryotes use substances other than oxygen as reactants in a similar process that harvests chemical energy without oxygen; this process is called *anaerobic respiration* (the prefix *an-* means "without"). Technically, the term **cellular respiration** includes both aerobic and anaerobic processes. However, it originated as a synonym for aerobic respiration because of the relationship of that process to organismal respiration, in which an animal breathes in oxygen. Thus, *cellular respiration* is often used to refer to the aerobic process, a practice we follow in most of this chapter.

Although very different in mechanism, aerobic respiration is in principle similar to the combustion of gasoline in an automobile engine after oxygen is mixed with the fuel (hydrocarbons). Food provides the fuel for respiration, and the exhaust is carbon dioxide and water. The overall process can be summarized as follows:

$$\text{Organic compounds} + \text{Oxygen} \rightarrow \text{Carbon dioxide} + \text{Water} + \text{Energy}$$

Although carbohydrates, fats, and proteins can all be processed and consumed as fuel, it is helpful to learn the steps of cellular respiration by tracking the degradation of the sugar glucose ($C_6H_{12}O_6$):

$$C_6H_{12}O_6 + 6\ O_2 \rightarrow 6\ CO_2 + 6\ H_2O + \text{Energy (ATP + heat)}$$

Glucose is the fuel that cells most often use; we will discuss other organic molecules contained in foods later in the chapter.

This breakdown of glucose is exergonic, having a free-energy change of −686 kcal (2,870 kJ) per mole of glucose decomposed ($\Delta G = -686$ kcal/mol). Recall that a negative ΔG indicates that the products of the chemical process store less energy than the reactants and that the reaction can happen spontaneously—in other words, without an input of energy.

Catabolic pathways do not directly move flagella, pump solutes across membranes, polymerize monomers, or perform other cellular work. Catabolism is linked to work by a chemical drive shaft—ATP, which you learned about in Chapter 8. To keep working, the cell must regenerate its supply of ATP from ADP and Ⓟ$_i$ (see Figure 8.11). To understand how cellular respiration accomplishes this, let's examine the fundamental chemical processes known as oxidation and reduction.

Redox Reactions: Oxidation and Reduction

How do the catabolic pathways that decompose glucose and other organic fuels yield energy? The answer is based on the transfer of electrons during the chemical reactions. The relocation of electrons releases energy stored in organic molecules, and this energy ultimately is used to synthesize ATP.

The Principle of Redox

In many chemical reactions, there is a transfer of one or more electrons (e^-) from one reactant to another. These electron transfers are called oxidation-reduction reactions, or **redox reactions** for short. In a redox reaction, the loss of electrons from one substance is called **oxidation**, and the addition of electrons to another substance is known as **reduction**. (Note that *adding* electrons is called *reduction*; negatively charged electrons added to an atom *reduce* the amount of positive charge of that atom.) To take a simple, nonbiological example, consider the reaction between the elements sodium (Na) and chlorine (Cl) that forms table salt:

becomes oxidized (loses electron): Na → Na^+

$$Na + Cl \longrightarrow Na^+ + Cl^-$$

becomes reduced (gains electron): Cl → Cl^-

We could generalize a redox reaction this way:

becomes oxidized: Xe^- → X

$$Xe^- + Y \longrightarrow X + Ye^-$$

becomes reduced: Y → Ye^-

In the generalized reaction, substance Xe^-, the electron donor, is called the **reducing agent**; it reduces Y, which accepts the donated electron. Substance Y, the electron acceptor, is the **oxidizing agent**; it oxidizes Xe^- by removing its electron. Because an electron transfer requires both a donor and an acceptor, oxidation and reduction always go together.

Not all redox reactions involve the complete transfer of electrons from one substance to another; some change the degree of electron sharing in covalent bonds. The reaction

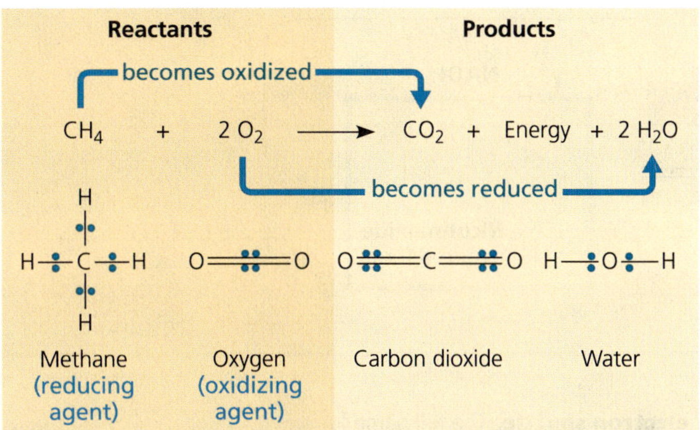

▲ **Figure 9.3 Methane combustion as an energy-yielding redox reaction.** The reaction releases energy to the surroundings because the electrons lose potential energy when they end up being shared unequally, spending more time near electronegative atoms such as oxygen.

between methane and oxygen, shown in **Figure 9.3**, is an example. As explained in Chapter 2, the covalent electrons in methane are shared nearly equally between the bonded atoms because carbon and hydrogen have about the same affinity for valence electrons; they are about equally electronegative. But when methane reacts with oxygen, forming carbon dioxide, electrons end up shared less equally between the carbon atom and its new covalent partners, the oxygen atoms, which are very electronegative. In effect, the carbon atom has partially "lost" its shared electrons; thus, methane has been oxidized.

Now let's examine the fate of the reactant O_2. The two atoms of the oxygen molecule (O_2) share their electrons equally. But when oxygen reacts with the hydrogen from methane, forming water, the electrons of the covalent bonds spend more time near the oxygen (see Figure 9.3). In effect, each oxygen atom has partially "gained" electrons, so the oxygen molecule has been reduced. Because oxygen is so electronegative, it is one of the most potent of all oxidizing agents.

Energy must be added to pull an electron away from an atom, just as energy is required to push a ball uphill. The more electronegative the atom (the stronger its pull on electrons), the more energy is required to take an electron away from it. An electron loses potential energy when it shifts from a less electronegative atom toward a more electronegative one, just as a ball loses potential energy when it rolls downhill. A redox reaction that moves electrons closer to oxygen, such as the burning (oxidation) of methane, therefore releases chemical energy that can be put to work.

Oxidation of Organic Fuel Molecules During Cellular Respiration

The oxidation of methane by oxygen is the main combustion reaction that occurs at the burner of a gas stove. The combustion of gasoline in an automobile engine is also a redox reaction; the energy released pushes the pistons. But the energy-yielding redox process of greatest interest to biologists is respiration: the oxidation of glucose and other molecules in food. Examine again the summary equation for cellular respiration, but this time think of it as a redox process:

$$\overbrace{C_6H_{12}O_6 \quad + \quad 6\,O_2 \longrightarrow 6\,CO_2}^{\text{becomes oxidized}} \quad + \quad \underbrace{6\,H_2O}_{\text{becomes reduced}} \quad + \quad \text{Energy}$$

As in the combustion of methane or gasoline, the fuel (glucose) is oxidized and oxygen is reduced. The electrons lose potential energy along the way, and energy is released.

In general, organic molecules that have an abundance of hydrogen are excellent fuels because their bonds are a source of "hilltop" electrons, whose energy may be released as these electrons "fall" down an energy gradient when they are transferred to oxygen. The summary equation for respiration indicates that hydrogen is transferred from glucose to oxygen. But the important point, not visible in the summary equation, is that the energy state of the electron changes as hydrogen (with its electron) is transferred to oxygen. In respiration, the oxidation of glucose transfers electrons to a lower energy state, liberating energy that becomes available for ATP synthesis.

The main energy-yielding foods, carbohydrates and fats, are reservoirs of electrons associated with hydrogen. Only the barrier of activation energy holds back the flood of electrons to a lower energy state (see Figure 8.12). Without this barrier, a food substance like glucose would combine almost instantaneously with O_2. If we supply the activation energy by igniting glucose, it burns in air, releasing 686 kcal (2,870 kJ) of heat per mole of glucose (about 180 g). Body temperature is not high enough to initiate burning, of course. Instead, if you swallow some glucose, enzymes in your cells will lower the barrier of activation energy, allowing the sugar to be oxidized in a series of steps.

Stepwise Energy Harvest via NAD^+ and the Electron Transport Chain

If energy is released from a fuel all at once, it cannot be harnessed efficiently for constructive work. For example, if a gasoline tank explodes, it cannot drive a car very far. Cellular respiration does not oxidize glucose in a single explosive step either. Rather, glucose and other organic fuels are broken down in a series of steps, each one catalyzed by an enzyme. At key steps, electrons are stripped from the glucose. As is often the case in oxidation reactions, each electron travels with a proton—thus, as a hydrogen atom. The hydrogen atoms are not transferred directly to oxygen, but instead are usually passed first to an electron carrier, a coenzyme called **NAD^+** (nicotinamide adenine dinucleotide, a derivative of the vitamin niacin). NAD^+ is well suited as an electron carrier because it can cycle easily between oxidized (NAD^+) and reduced (NADH) states. As an electron acceptor, NAD^+ functions as an oxidizing agent during respiration.

How does NAD^+ trap electrons from glucose and other organic molecules? Enzymes called dehydrogenases remove a

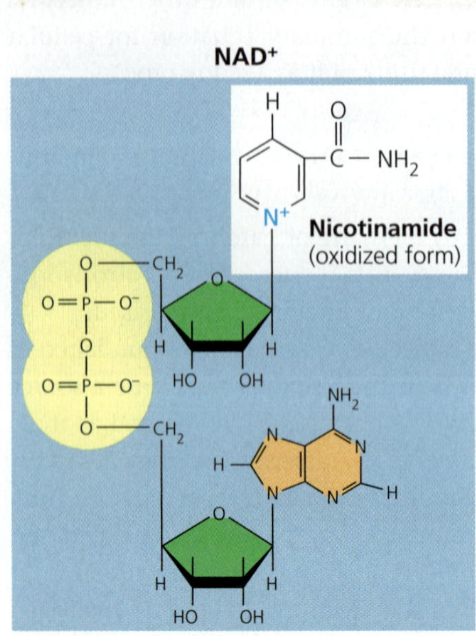

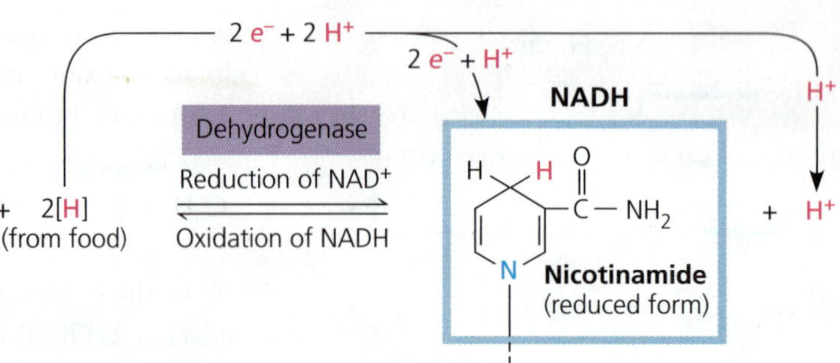

◀ **Figure 9.4 NAD$^+$ as an electron shuttle.** The full name for NAD$^+$, nicotinamide adenine dinucleotide, describes its structure: The molecule consists of two nucleotides joined together at their phosphate groups (shown in yellow). (Nicotinamide is a nitrogenous base, although not one that is present in DNA or RNA; see Figure 5.26.) The enzymatic transfer of 2 electrons and 1 proton (H$^+$) from an organic molecule in food to NAD$^+$ reduces the NAD$^+$ to NADH; the second proton (H$^+$) is released. Most of the electrons removed from food are transferred initially to NAD$^+$.

pair of hydrogen atoms (2 electrons and 2 protons) from the substrate (glucose, in this example), thereby oxidizing it. The enzyme delivers the 2 electrons along with 1 proton to its coenzyme, NAD$^+$ **(Figure 9.4)**. The other proton is released as a hydrogen ion (H$^+$) into the surrounding solution:

$$H{-}\overset{|}{\underset{|}{C}}{-}OH + NAD^+ \xrightarrow{\text{Dehydrogenase}} \overset{|}{\underset{|}{C}}{=}O + NADH + H^+$$

By receiving 2 negatively charged electrons but only 1 positively charged proton, NAD$^+$ has its charge neutralized when it is reduced to NADH. The name NADH shows the hydrogen that has been received in the reaction. NAD$^+$ is the most versatile electron acceptor in cellular respiration and functions in several of the redox steps during the breakdown of glucose.

Electrons lose very little of their potential energy when they are transferred from glucose to NAD$^+$. Each NADH molecule formed during respiration represents stored energy that can be tapped to make ATP when the electrons complete their "fall" down an energy gradient from NADH to oxygen.

How do electrons that are extracted from glucose and stored as potential energy in NADH finally reach oxygen? It will help to compare the redox chemistry of cellular respiration to a much simpler reaction: the reaction between hydrogen and oxygen to form water **(Figure 9.5a)**. Mix H_2 and O_2, provide a spark for activation energy, and the gases combine explosively. In fact, combustion of liquid H_2 and O_2 is harnessed to power the main engines of the space shuttle after it is launched, boosting it into orbit. The explosion represents a release of energy as the electrons of hydrogen "fall" closer to the electronegative oxygen atoms. Cellular respiration also brings hydrogen and oxygen together to form water, but there are two important differences. First, in cellular respiration, the hydrogen that reacts with oxygen is derived from organic molecules rather than H_2. Second, instead of occurring in one explosive reaction, respiration uses an

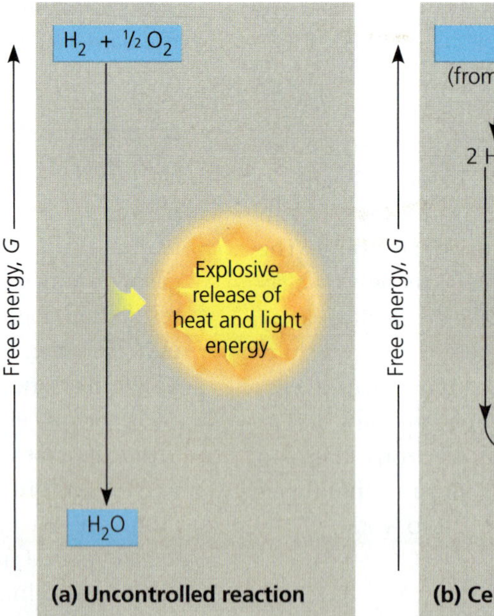

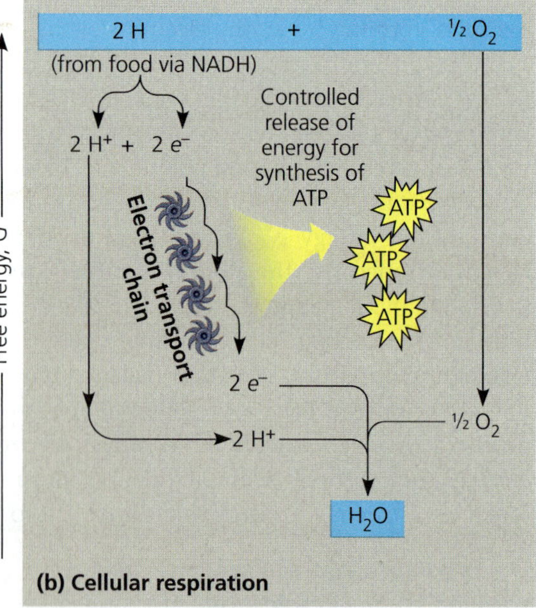

▲ **Figure 9.5 An introduction to electron transport chains. (a)** The one-step exergonic reaction of hydrogen with oxygen to form water releases a large amount of energy in the form of heat and light: an explosion. **(b)** In cellular respiration, the same reaction occurs in stages: An electron transport chain breaks the "fall" of electrons in this reaction into a series of smaller steps and stores some of the released energy in a form that can be used to make ATP. (The rest of the energy is released as heat.)

electron transport chain to break the fall of electrons to oxygen into several energy-releasing steps (**Figure 9.5b**). An **electron transport chain** consists of a number of molecules, mostly proteins, built into the inner membrane of the mitochondria of eukaryotic cells and the plasma membrane of aerobically respiring prokaryotes. Electrons removed from glucose are shuttled by NADH to the "top," higher-energy end of the chain. At the "bottom," lower-energy end, O_2 captures these electrons along with hydrogen nuclei (H^+), forming water.

Electron transfer from NADH to oxygen is an exergonic reaction with a free-energy change of −53 kcal/mol (−222 kJ/mol). Instead of this energy being released and wasted in a single explosive step, electrons cascade down the chain from one carrier molecule to the next in a series of redox reactions, losing a small amount of energy with each step until they finally reach oxygen, the terminal electron acceptor, which has a very great affinity for electrons. Each "downhill" carrier is more electronegative than, and thus capable of oxidizing, its "uphill" neighbor, with oxygen at the bottom of the chain. Therefore, the electrons removed from glucose by NAD^+ fall down an energy gradient in the electron transport chain to a far more stable location in the electronegative oxygen atom. Put another way, oxygen pulls electrons down the chain in an energy-yielding tumble analogous to gravity pulling objects downhill.

In summary, during cellular respiration, most electrons travel the following "downhill" route: glucose → NADH → electron transport chain → oxygen. Later in this chapter, you will learn more about how the cell uses the energy released from this exergonic electron fall to regenerate its supply of ATP. For now, having covered the basic redox mechanisms of cellular respiration, let's look at the entire process by which energy is harvested from organic fuels.

The Stages of Cellular Respiration: *A Preview*

The harvesting of energy from glucose by cellular respiration is a cumulative function of three metabolic stages:

1. Glycolysis (color-coded teal throughout the chapter)
2. Pyruvate oxidation and the citric acid cycle (color-coded salmon)
3. Oxidative phosphorylation: electron transport and chemiosmosis (color-coded violet)

Biochemists usually reserve the term *cellular respiration* for stages 2 and 3. We include glycolysis, however, because most respiring cells deriving energy from glucose use glycolysis to produce the starting material for the citric acid cycle.

As diagrammed in **Figure 9.6**, glycolysis and pyruvate oxidation followed by the citric acid cycle are the catabolic pathways that break down glucose and other organic fuels. **Glycolysis**, which occurs in the cytosol, begins the degradation process by breaking glucose into two molecules of a compound called pyruvate. In eukaryotes, pyruvate enters the mitochondrion and is oxidized to a compound called acetyl CoA, which enters the **citric acid cycle**. There, the breakdown of glucose to carbon dioxide is completed. (In prokaryotes, these processes take place in the cytosol.) Thus, the carbon dioxide produced by respiration represents fragments of oxidized organic molecules.

Some of the steps of glycolysis and the citric acid cycle are redox reactions in which dehydrogenases transfer electrons from substrates to NAD^+, forming NADH. In the third stage of respiration, the electron transport chain accepts electrons from the breakdown products of the first two stages (most often via NADH) and passes these electrons from one molecule to another. At the end of the chain, the electrons are combined with molecular oxygen and hydrogen ions (H^+), forming water (see

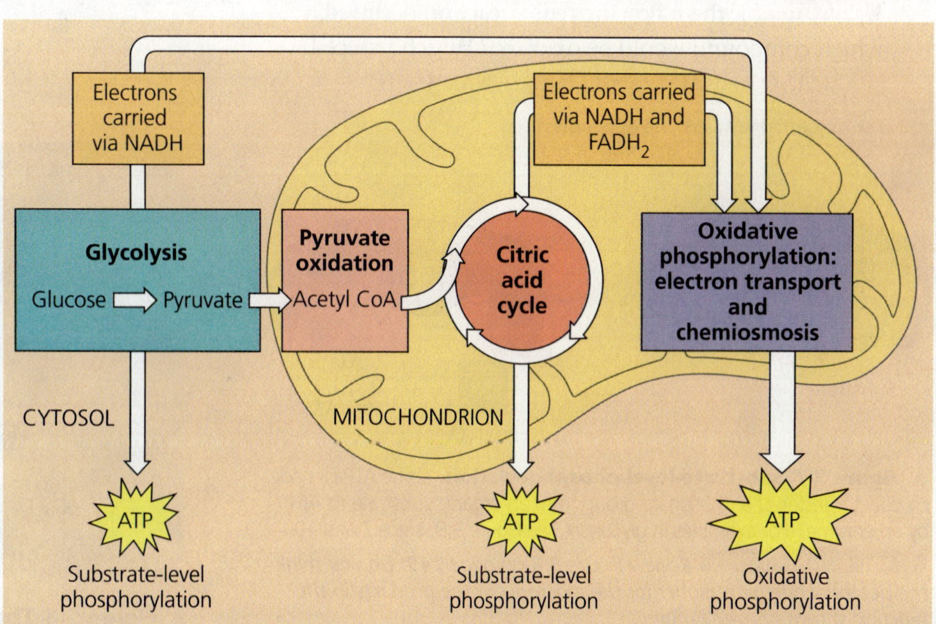

▶ **Figure 9.6 An overview of cellular respiration.** During glycolysis, each glucose molecule is broken down into two molecules of the compound pyruvate. In eukaryotic cells, as shown here, the pyruvate enters the mitochondrion. There it is oxidized to acetyl CoA, which is further oxidized to CO_2 in the citric acid cycle. NADH and a similar electron carrier, a coenzyme called $FADH_2$, transfer electrons derived from glucose to electron transport chains, which are built into the inner mitochondrial membrane. (In prokaryotes, the electron transport chains are located in the plasma membrane.) During oxidative phosphorylation, electron transport chains convert the chemical energy to a form used for ATP synthesis in the process called chemiosmosis.

BioFlix Visit the Study Area at **www.masteringbiology.com** for the BioFlix® 3-D Animation on Cellular Respiration.

Figure 9.5b). The energy released at each step of the chain is stored in a form the mitochondrion (or prokaryotic cell) can use to make ATP from ADP. This mode of ATP synthesis is called **oxidative phosphorylation** because it is powered by the redox reactions of the electron transport chain.

In eukaryotic cells, the inner membrane of the mitochondrion is the site of electron transport and chemiosmosis, the processes that together constitute oxidative phosphorylation. (In prokaryotes, these processes take place in the plasma membrane.) Oxidative phosphorylation accounts for almost 90% of the ATP generated by respiration. A smaller amount of ATP is formed directly in a few reactions of glycolysis and the citric acid cycle by a mechanism called **substrate-level phosphorylation** (**Figure 9.7**). This mode of ATP synthesis occurs when an enzyme transfers a phosphate group from a substrate molecule to ADP, rather than adding an inorganic phosphate to ADP as in oxidative phosphorylation. "Substrate molecule" here refers to an organic molecule generated as an intermediate during the catabolism of glucose.

For each molecule of glucose degraded to carbon dioxide and water by respiration, the cell makes up to about 32 molecules of ATP, each with 7.3 kcal/mol of free energy. Respiration cashes in the large denomination of energy banked in a single molecule of glucose (686 kcal/mol) for the small change of many molecules of ATP, which is more practical for the cell to spend on its work.

This preview has introduced you to how glycolysis, the citric acid cycle, and oxidative phosphorylation fit into the process of cellular respiration. We are now ready to take a closer look at each of these three stages of respiration.

CONCEPT CHECK 9.1

1. Compare and contrast aerobic and anaerobic respiration.
2. **WHAT IF?** If the following redox reaction occurred, which compound would be oxidized? Which reduced?

$$C_4H_6O_5 + NAD^+ \longrightarrow C_4H_4O_5 + NADH + H^+$$

For suggested answers, see Appendix A.

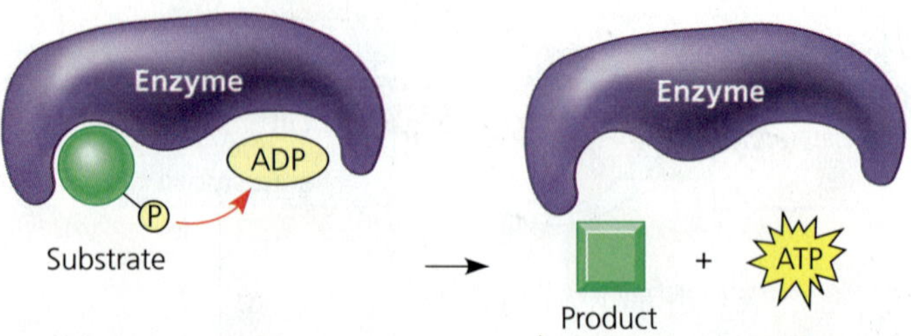

▲ **Figure 9.7 Substrate-level phosphorylation.** Some ATP is made by direct transfer of a phosphate group from an organic substrate to ADP by an enzyme. (For examples in glycolysis, see Figure 9.9, steps 7 and 10.)

MAKE CONNECTIONS *Review Figure 8.8 on page 149. Do you think the potential energy is higher for the reactants or the products in the reaction shown above? Explain.*

CONCEPT 9.2

Glycolysis harvests chemical energy by oxidizing glucose to pyruvate

The word *glycolysis* means "sugar splitting," and that is exactly what happens during this pathway. Glucose, a six-carbon sugar, is split into two three-carbon sugars. These smaller sugars are then oxidized and their remaining atoms rearranged to form two molecules of pyruvate. (Pyruvate is the ionized form of pyruvic acid.)

As summarized in **Figure 9.8**, glycolysis can be divided into two phases: energy investment and energy payoff. During the energy investment phase, the cell actually spends ATP. This investment is repaid with interest during the energy payoff phase, when ATP is produced by substrate-level phosphorylation and NAD^+ is reduced to NADH by electrons released from the oxidation of glucose. The net energy yield from glycolysis, per glucose molecule, is 2 ATP plus 2 NADH. The ten steps of the glycolytic pathway are shown in **Figure 9.9**.

All of the carbon originally present in glucose is accounted for in the two molecules of pyruvate; no carbon is released as CO_2 during glycolysis. Glycolysis occurs whether or not O_2 is present. However, if O_2 *is* present, the chemical energy stored in pyruvate and NADH can be extracted by pyruvate oxidation, the citric acid cycle, and oxidative phosphorylation.

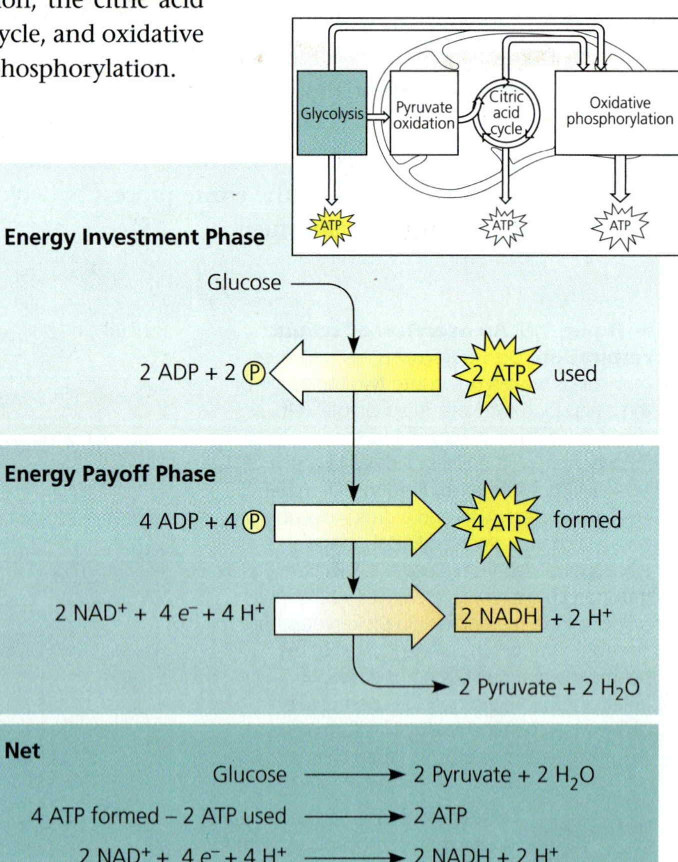

▲ **Figure 9.8 The energy input and output of glycolysis.**

▼ **Figure 9.9 A closer look at glycolysis.** The orientation diagram on the left relates glycolysis to the entire process of respiration. Note that glycolysis is a source of ATP and NADH.

WHAT IF? *What would happen if you removed the dihydroxyacetone phosphate generated in step 4 as fast as it was produced?*

Glycolysis | Pyruvate oxidation | Citric acid cycle | Oxidative phosphorylation

ATP | ATP | ATP

Glycolysis: Energy Investment Phase

Glucose → (ATP → ADP; Hexokinase) → Glucose 6-phosphate → (Phosphoglucoisomerase) → Fructose 6-phosphate → (ATP → ADP; Phosphofructokinase) → Fructose 1,6-bisphosphate → (Aldolase 4) → Dihydroxyacetone phosphate ⇄ (Isomerase 5) Glyceraldehyde 3-phosphate → To step 6

1 Hexokinase transfers a phosphate group from ATP to glucose, making it more chemically reactive. The charge on the phosphate also traps the sugar in the cell.

2 Glucose 6-phosphate is converted to its isomer, fructose 6-phosphate.

3 Phosphofructokinase transfers a phosphate group from ATP to the opposite end of the sugar, investing a second molecule of ATP. This is a key step for regulation of glycolysis.

4 Aldolase cleaves the sugar molecule into two different three-carbon sugars (isomers).

5 Isomerase catalyzes the reversible conversion between the two isomers. This reaction never reaches equilibrium: Glyceraldehyde 3-phosphate is used as the substrate of the next reaction (step 6) as fast as it forms.

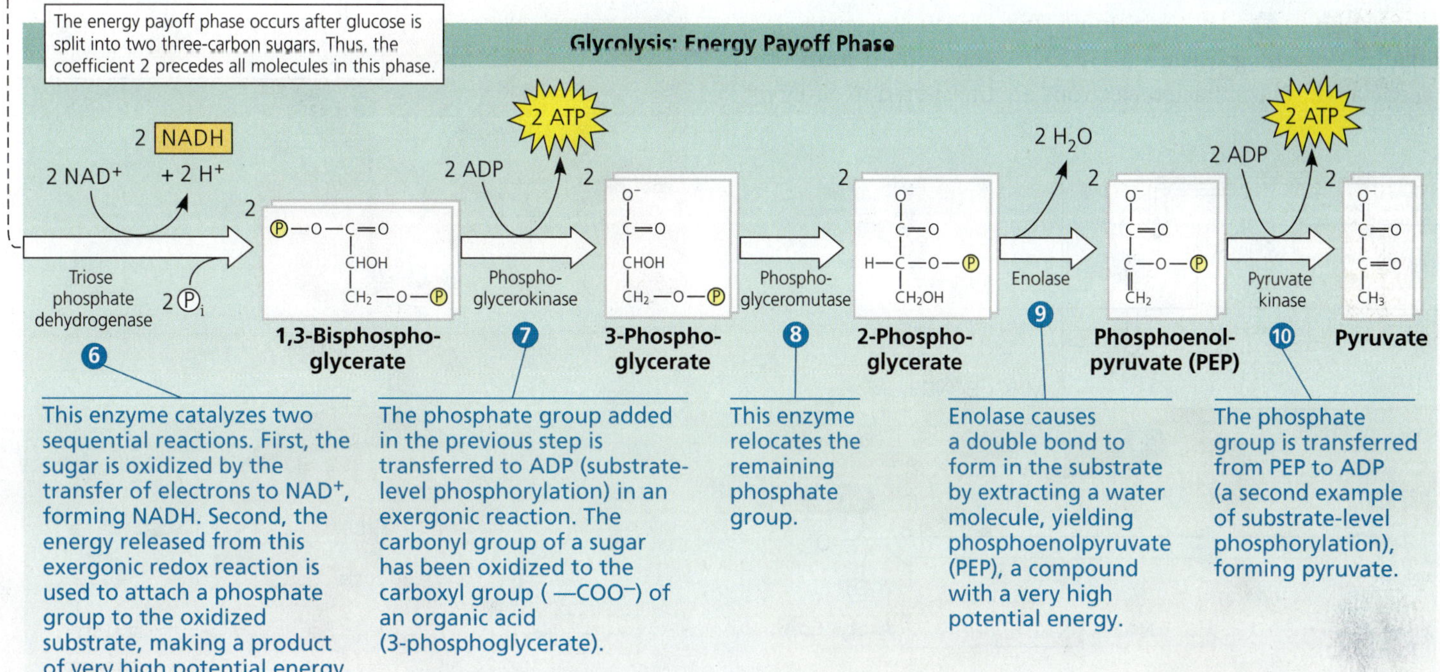

CONCEPT CHECK 9.2

1. During the redox reaction in glycolysis (step 6 in Figure 9.9), which molecule acts as the oxidizing agent? The reducing agent?
2. **MAKE CONNECTIONS** Step 3 in Figure 9.9 is a major point of regulation of glycolysis. The enzyme phosphofructokinase is allosterically regulated by ATP and related molecules (see Concept 8.5, p. 158). Considering the overall result of glycolysis, would you expect ATP to inhibit or stimulate activity of this enzyme? (*Hint*: Make sure you consider the role of ATP as an allosteric regulator, not as a substrate of the enzyme.)

For suggested answers, see Appendix A.

CONCEPT 9.3

After pyruvate is oxidized, the citric acid cycle completes the energy-yielding oxidation of organic molecules

Glycolysis releases less than a quarter of the chemical energy in glucose that can be released by cells; most of the energy remains stockpiled in the two molecules of pyruvate. If molecular oxygen is present, the pyruvate enters a mitochondrion (in eukaryotic cells), where the oxidation of glucose is completed. (In prokaryotic cells, this process occurs in the cytosol.)

Oxidation of Pyruvate to Acetyl CoA

Upon entering the mitochondrion via active transport, pyruvate is first converted to a compound called acetyl coenzyme A, or **acetyl CoA** (**Figure 9.10**). This step, linking glycolysis and the citric acid cycle, is carried out by a multienzyme complex that catalyzes three reactions: ❶ Pyruvate's carboxyl group (—COO⁻), which is already fully oxidized and thus has little chemical energy, is removed and given off as a molecule of CO_2. (This is the first step in which CO_2 is released during respiration.) ❷ The remaining two-carbon fragment is oxidized, forming acetate (CH_3COO^-, the ionized form of acetic acid). The extracted electrons are transferred to NAD^+, storing energy in the form of NADH. ❸ Finally, coenzyme A (CoA), a sulfur-containing compound derived from a B vitamin, is attached via its sulfur atom to the acetate, forming acetyl CoA, which has a high potential energy; in other words, the reaction of acetyl CoA to yield lower-energy products is highly exergonic. This molecule will now feed its acetyl group into the citric acid cycle for further oxidation.

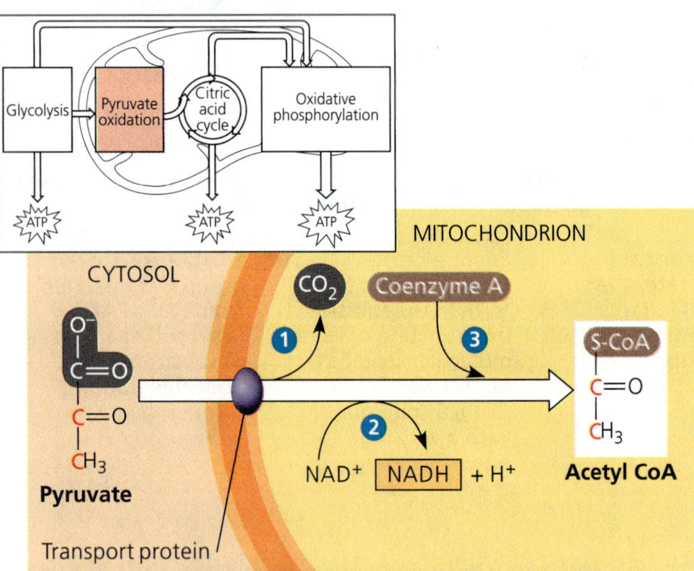

▲ **Figure 9.10 Oxidation of pyruvate to acetyl CoA, the step before the citric acid cycle.** Pyruvate is a charged molecule, so in eukaryotic cells it must enter the mitochondrion via active transport, with the help of a transport protein. Next, a complex of several enzymes (the pyruvate dehydrogenase complex) catalyzes the three numbered steps, which are described in the text. The acetyl group of acetyl CoA will enter the citric acid cycle. The CO_2 molecule will diffuse out of the cell. By convention, coenzyme A is abbreviated S-CoA when it is attached to a molecule, emphasizing the sulfur atom (S).

The Citric Acid Cycle

The citric acid cycle is also called the tricarboxylic acid cycle or the Krebs cycle, the latter honoring Hans Krebs, the German-British scientist who was largely responsible for working out the pathway in the 1930s. The cycle functions as a metabolic furnace that oxidizes organic fuel derived from pyruvate. **Figure 9.11** summarizes the inputs and outputs as pyruvate is broken down to three CO_2 molecules, including the molecule of CO_2 released during the conversion of pyruvate to acetyl CoA. The cycle generates 1 ATP per turn by

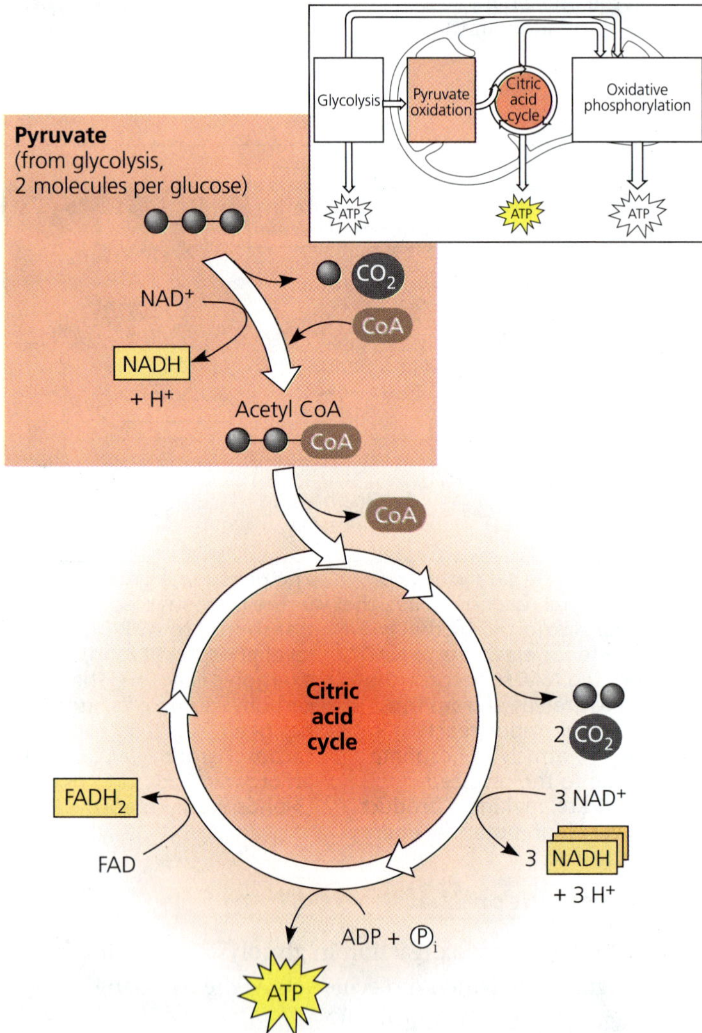

▲ **Figure 9.11 An overview of pyruvate oxidation and the citric acid cycle.** The inputs and outputs per pyruvate molecule are shown. To calculate on a per-glucose basis, multiply by 2, because each glucose molecule is split during glycolysis into two pyruvate molecules.

substrate-level phosphorylation, but most of the chemical energy is transferred to NAD$^+$ and a related electron carrier, the coenzyme FAD (flavin adenine dinucleotide, derived from riboflavin, a B vitamin), during the redox reactions. The reduced coenzymes, NADH and FADH$_2$, shuttle their cargo of high-energy electrons into the electron transport chain.

Now let's look at the citric acid cycle in more detail. The cycle has eight steps, each catalyzed by a specific enzyme. You can see in **Figure 9.12** that for each turn of the citric acid cycle, two carbons (red) enter in the relatively reduced form of an acetyl group (step 1), and two different carbons (blue) leave in the completely oxidized form of CO_2 molecules

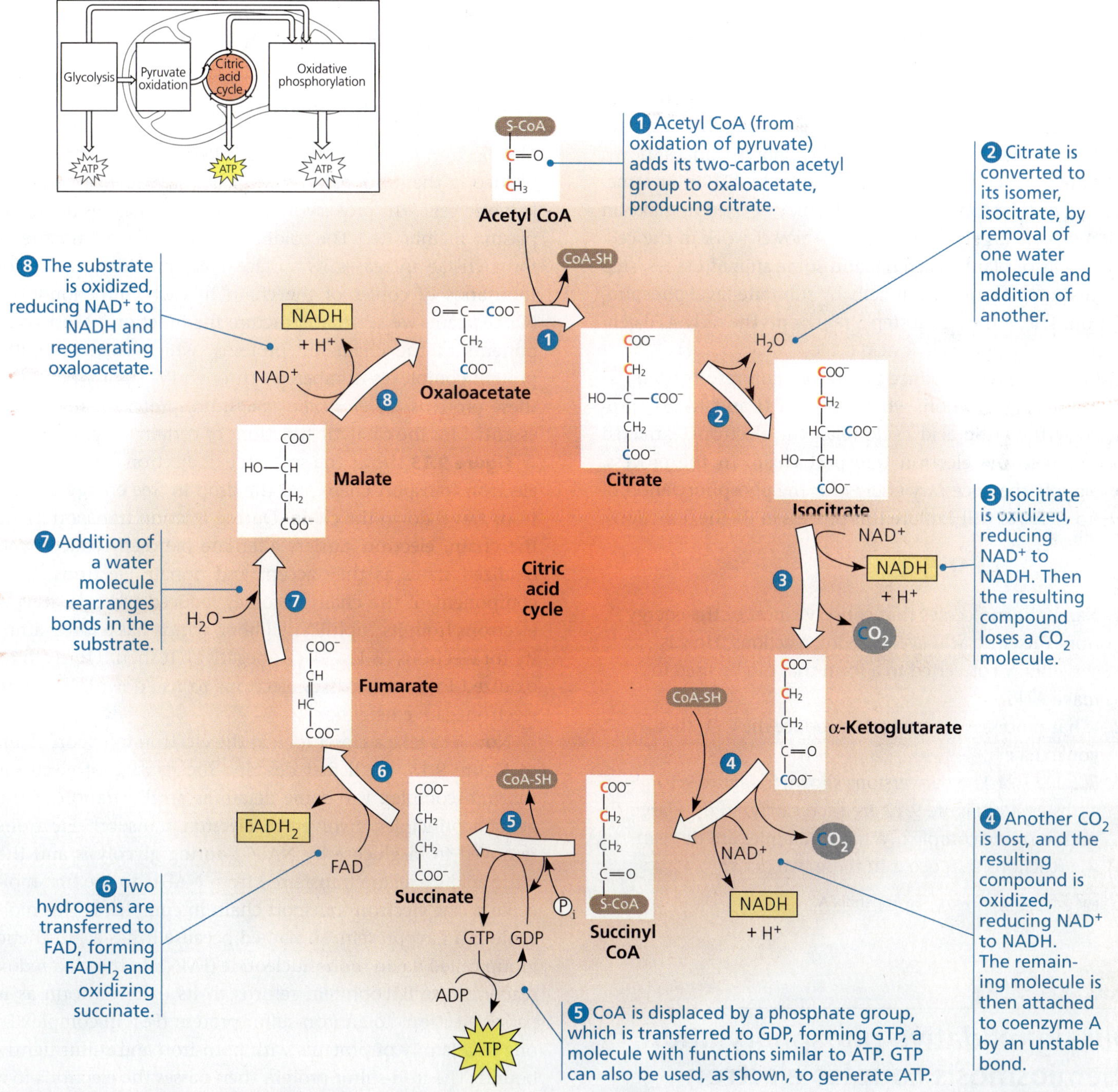

▲ **Figure 9.12 A closer look at the citric acid cycle.** In the chemical structures, red type traces the fate of the two carbon atoms that enter the cycle via acetyl CoA (step 1), and blue type indicates the two carbons that exit the cycle as CO_2 in steps 3 and 4. (The red labeling goes only through step 5 because the succinate molecule is symmetrical; the two ends cannot be distinguished from each other.) Notice that the carbon atoms that enter the cycle from acetyl CoA do not leave the cycle in the same turn. They remain in the cycle, occupying a different location in the molecules on their next turn, after another acetyl group is added. As a consequence, the oxaloacetate that is regenerated at step 8 is composed of different carbon atoms each time around. In eukaryotic cells, all the citric acid cycle enzymes are located in the mitochondrial matrix except for the enzyme that catalyzes step 6, which resides in the inner mitochondrial membrane. Carboxylic acids are represented in their ionized forms, as —COO$^-$, because the ionized forms prevail at the pH within the mitochondrion. For example, citrate is the ionized form of citric acid.

(steps 3 and 4). The acetyl group of acetyl CoA joins the cycle by combining with the compound oxaloacetate, forming citrate (step 1). (Citrate is the ionized form of citric acid, for which the cycle is named.) The next seven steps decompose the citrate back to oxaloacetate. It is this regeneration of oxaloacetate that makes this process a *cycle*.

Now let's tally the energy-rich molecules produced by the citric acid cycle. For each acetyl group entering the cycle, 3 NAD^+ are reduced to NADH (steps 3, 4, and 8). In step 6, electrons are transferred not to NAD^+, but to FAD, which accepts 2 electrons and 2 protons to become $FADH_2$. In many animal tissue cells, step 5 produces a guanosine triphosphate (GTP) molecule by substrate-level phosphorylation, as shown in Figure 9.12. GTP is a molecule similar to ATP in its structure and cellular function. This GTP may be used to make an ATP molecule (as shown) or directly power work in the cell. In the cells of plants, bacteria, and some animal tissues, step 5 forms an ATP molecule directly by substrate-level phosphorylation. The output from step 5 represents the only ATP generated directly by the citric acid cycle.

Most of the ATP produced by respiration results from oxidative phosphorylation, when the NADH and $FADH_2$ produced by the citric acid cycle relay the electrons extracted from food to the electron transport chain. In the process, they supply the necessary energy for the phosphorylation of ADP to ATP. We will explore this process in the next section.

CONCEPT CHECK 9.3

1. Name the molecules that conserve most of the energy from the citric acid cycle's redox reactions. How is this energy converted to a form that can be used to make ATP?
2. What processes in your cells produce the CO_2 that you exhale?
3. **WHAT IF?** The conversions shown in Figure 9.10 and step 4 of Figure 9.12 are each catalyzed by a large multienzyme complex. What similarities are there in the reactions that occur in these two cases?

For suggested answers, see Appendix A.

CONCEPT 9.4

During oxidative phosphorylation, chemiosmosis couples electron transport to ATP synthesis

Our main objective in this chapter is to learn how cells harvest the energy of glucose and other nutrients in food to make ATP. But the metabolic components of respiration we have dissected so far, glycolysis and the citric acid cycle, produce only 4 ATP molecules per glucose molecule, all by substrate-level phosphorylation: 2 net ATP from glycolysis and 2 ATP from the citric acid cycle. At this point, molecules of NADH (and $FADH_2$) account for most of the energy extracted from the glucose. These electron escorts link glycolysis and the citric acid cycle to the machinery of oxidative phosphorylation, which uses energy released by the electron transport chain to power ATP synthesis. In this section, you will learn first how the electron transport chain works and then how electron flow down the chain is coupled to ATP synthesis.

The Pathway of Electron Transport

The electron transport chain is a collection of molecules embedded in the inner membrane of the mitochondrion in eukaryotic cells. (In prokaryotes, these molecules reside in the plasma membrane.) The folding of the inner membrane to form cristae increases its surface area, providing space for thousands of copies of the chain in each mitochondrion. (Once again, we see that structure fits function.) Most components of the chain are proteins, which exist in multiprotein complexes numbered I through IV. Tightly bound to these proteins are *prosthetic groups*, nonprotein components essential for the catalytic functions of certain enzymes.

Figure 9.13 shows the sequence of electron carriers in the electron transport chain and the drop in free energy as electrons travel down the chain. During electron transport along the chain, electron carriers alternate between reduced and oxidized states as they accept and donate electrons. Each component of the chain becomes reduced when it accepts electrons from its "uphill" neighbor, which has a lower affinity for electrons (is less electronegative). It then returns to its oxidized form as it passes electrons to its "downhill," more electronegative neighbor.

Now let's take a closer look at the electron transport chain in Figure 9.13. We'll first describe the passage of electrons through complex I in some detail, as an illustration of the general principles involved in electron transport. Electrons removed from glucose by NAD^+, during glycolysis and the citric acid cycle, are transferred from NADH to the first molecule of the electron transport chain in complex I. This molecule is a flavoprotein, so named because it has a prosthetic group called flavin mononucleotide (FMN). In the next redox reaction, the flavoprotein returns to its oxidized form as it passes electrons to an iron-sulfur protein (Fe·S in complex I), one of a family of proteins with both iron and sulfur tightly bound. The iron-sulfur protein then passes the electrons to a compound called ubiquinone (Q in Figure 9.13). This electron carrier is a small hydrophobic molecule, the only member of the electron transport chain that is not a protein. Ubiquinone is individually mobile within the membrane rather than residing in a particular complex. (Another name for ubiquinone is coenzyme Q, or CoQ; you may have seen it sold as a nutritional supplement.)

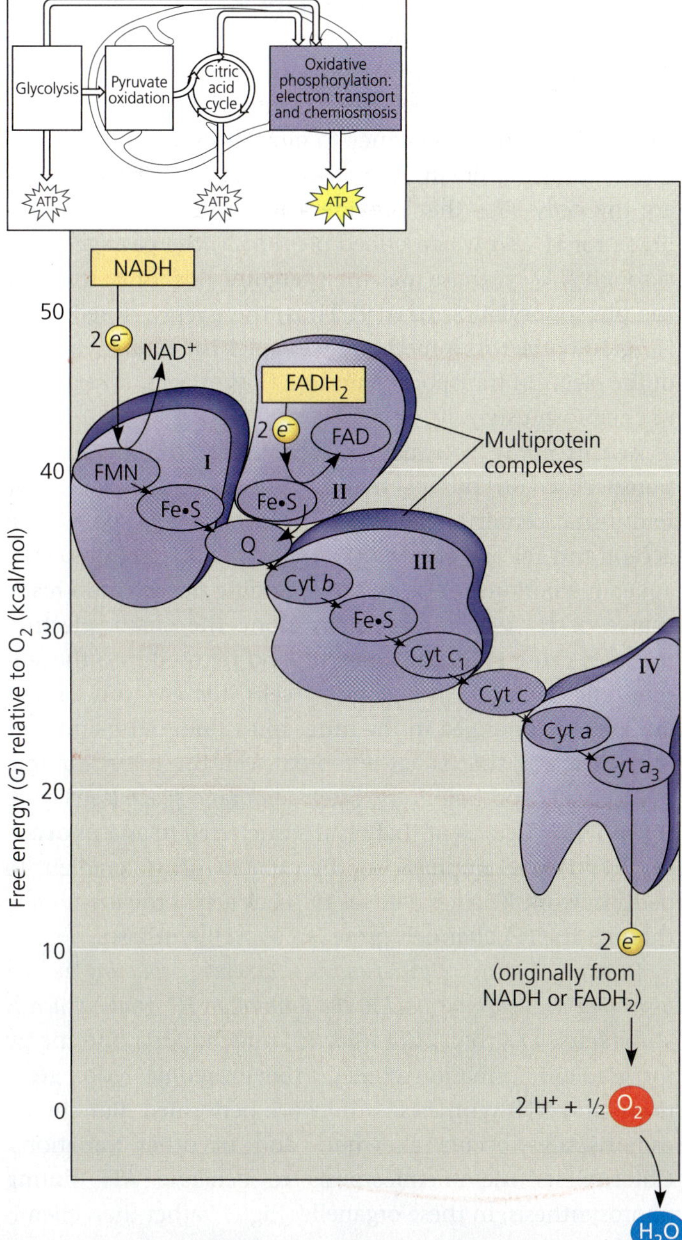

▲ **Figure 9.13 Free-energy change during electron transport.** The overall energy drop (ΔG) for electrons traveling from NADH to oxygen is 53 kcal/mol, but this "fall" is broken up into a series of smaller steps by the electron transport chain. (An oxygen atom is represented here as ½ O_2 to emphasize that the electron transport chain reduces molecular oxygen, O_2, not individual oxygen atoms.)

Most of the remaining electron carriers between ubiquinone and oxygen are proteins called **cytochromes**. Their prosthetic group, called a heme group, has an iron atom that accepts and donates electrons. (It is similar to the heme group in hemoglobin, the protein of red blood cells, except that the iron in hemoglobin carries oxygen, not electrons.) The electron transport chain has several types of cytochromes, each a different protein with a slightly different electron-carrying heme group. The last cytochrome of the chain, cyt a_3, passes its electrons to oxygen, which is *very* electronegative. Each oxygen atom also picks up a pair of hydrogen ions from the aqueous solution, forming water.

Another source of electrons for the transport chain is $FADH_2$, the other reduced product of the citric acid cycle. Notice in Figure 9.13 that $FADH_2$ adds its electrons to the electron transport chain from within complex II, at a lower energy level than NADH does. Consequently, although NADH and $FADH_2$ each donate an equivalent number of electrons (2) for oxygen reduction, the electron transport chain provides about one-third less energy for ATP synthesis when the electron donor is $FADH_2$ rather than NADH. We'll see why in the next section.

The electron transport chain makes no ATP directly. Instead, it eases the fall of electrons from food to oxygen, breaking a large free-energy drop into a series of smaller steps that release energy in manageable amounts. How does the mitochondrion (or the prokaryotic plasma membrane) couple this electron transport and energy release to ATP synthesis? The answer is a mechanism called chemiosmosis.

Chemiosmosis: The Energy-Coupling Mechanism

Populating the inner membrane of the mitochondrion or the prokaryotic plasma membrane are many copies of a protein complex called **ATP synthase**, the enzyme that actually makes ATP from ADP and inorganic phosphate. ATP synthase works like an ion pump running in reverse. Recall from Chapter 7 that ion pumps usually use ATP as an energy source to transport ions against their gradients. In fact, the proton pump shown in Figure 7.20 is an ATP synthase. As we mentioned in Chapter 8, enzymes can catalyze a reaction in either direction, depending on the ΔG for the reaction, which is affected by the local concentrations of reactants and products. Rather than hydrolyzing ATP to pump protons against their concentration gradient, under the conditions of cellular respiration ATP synthase uses the energy of an existing ion gradient to power ATP synthesis. The power source for the ATP synthase is a difference in the concentration of H^+ on opposite sides of the inner mitochondrial membrane. (We can also think of this gradient as a difference in pH, since pH is a measure of H^+ concentration.) This process, in which energy stored in the form of a hydrogen ion gradient across a membrane is used to drive cellular work such as the synthesis of ATP, is called **chemiosmosis** (from the Greek *osmos*, push). We have previously used the word *osmosis* in discussing water transport, but here it refers to the flow of H^+ across a membrane.

From studying the structure of ATP synthase, scientists have learned how the flow of H^+ through this large enzyme powers ATP generation. ATP synthase is a multisubunit complex with four main parts, each made up of multiple polypeptides. Protons move one by one into binding sites on one of the parts (the rotor), causing it to spin in a way that catalyzes ATP production from ADP and inorganic phosphate (**Figure 9.14**). The flow of protons thus behaves somewhat like a rushing stream that turns a waterwheel. ATP synthase is the smallest molecular rotary motor known in nature.

How does the inner mitochondrial membrane or the prokaryotic plasma membrane generate and maintain the H^+ gradient that drives ATP synthesis by the ATP synthase protein complex? Establishing the H^+ gradient is a major function of the electron transport chain, which is shown in its mitochondrial location in **Figure 9.15**. The chain is an energy converter that uses the exergonic flow of electrons from NADH and $FADH_2$ to pump H^+ across the membrane, from the mitochondrial matrix into the intermembrane space. The H^+ has a tendency to move back across the membrane, diffusing down its gradient. And the ATP synthases are the only sites that provide a route through the membrane for H^+. As we described previously, the passage of H^+ through ATP synthase uses the exergonic flow of H^+ to drive the phosphorylation of ADP. Thus, the energy stored in an H^+ gradient across a membrane couples the redox reactions of the electron transport chain to ATP synthesis, an example of chemiosmosis.

▲ **Figure 9.14 ATP synthase, a molecular mill.** The ATP synthase protein complex functions as a mill, powered by the flow of hydrogen ions. Multiple copies of this complex reside in mitochondrial and chloroplast membranes of eukaryotes and in the plasma membranes of prokaryotes. Each of the four parts of ATP synthase consists of a number of polypeptide subunits.

At this point, you may be wondering how the electron transport chain pumps hydrogen ions. Researchers have found that certain members of the electron transport chain accept and release protons (H^+) along with electrons. (The aqueous solutions inside and surrounding the cell are a ready source of H^+.) At certain steps along the chain, electron transfers cause H^+ to be taken up and released into the surrounding solution. In eukaryotic cells, the electron carriers are spatially arranged in the inner mitochondrial membrane in such a way that H^+ is accepted from the mitochondrial matrix and deposited in the intermembrane space (see Figure 9.15). The H^+ gradient that results is referred to as a **proton-motive force**, emphasizing the capacity of the gradient to perform work. The force drives H^+ back across the membrane through the H^+ channels provided by ATP synthases.

In general terms, *chemiosmosis is an energy-coupling mechanism that uses energy stored in the form of an H^+ gradient across a membrane to drive cellular work*. In mitochondria, the energy for gradient formation comes from exergonic redox reactions, and ATP synthesis is the work performed. But chemiosmosis also occurs elsewhere and in other variations. Chloroplasts use chemiosmosis to generate ATP during photosynthesis; in these organelles, light (rather than chemical energy) drives both electron flow down an electron transport chain and the resulting H^+ gradient formation. Prokaryotes, as already mentioned, generate H^+ gradients across their plasma membranes. They then tap the proton-motive force not only to make ATP inside the cell but also to rotate their flagella and to pump nutrients and waste products across the membrane. Because of its central importance to energy conversions in prokaryotes and eukaryotes, chemiosmosis has helped unify the study of bioenergetics. Peter Mitchell was awarded the Nobel Prize in 1978 for originally proposing the chemiosmotic model.

An Accounting of ATP Production by Cellular Respiration

In the last few sections, we have looked rather closely at the key processes of cellular respiration. Now let's take a step

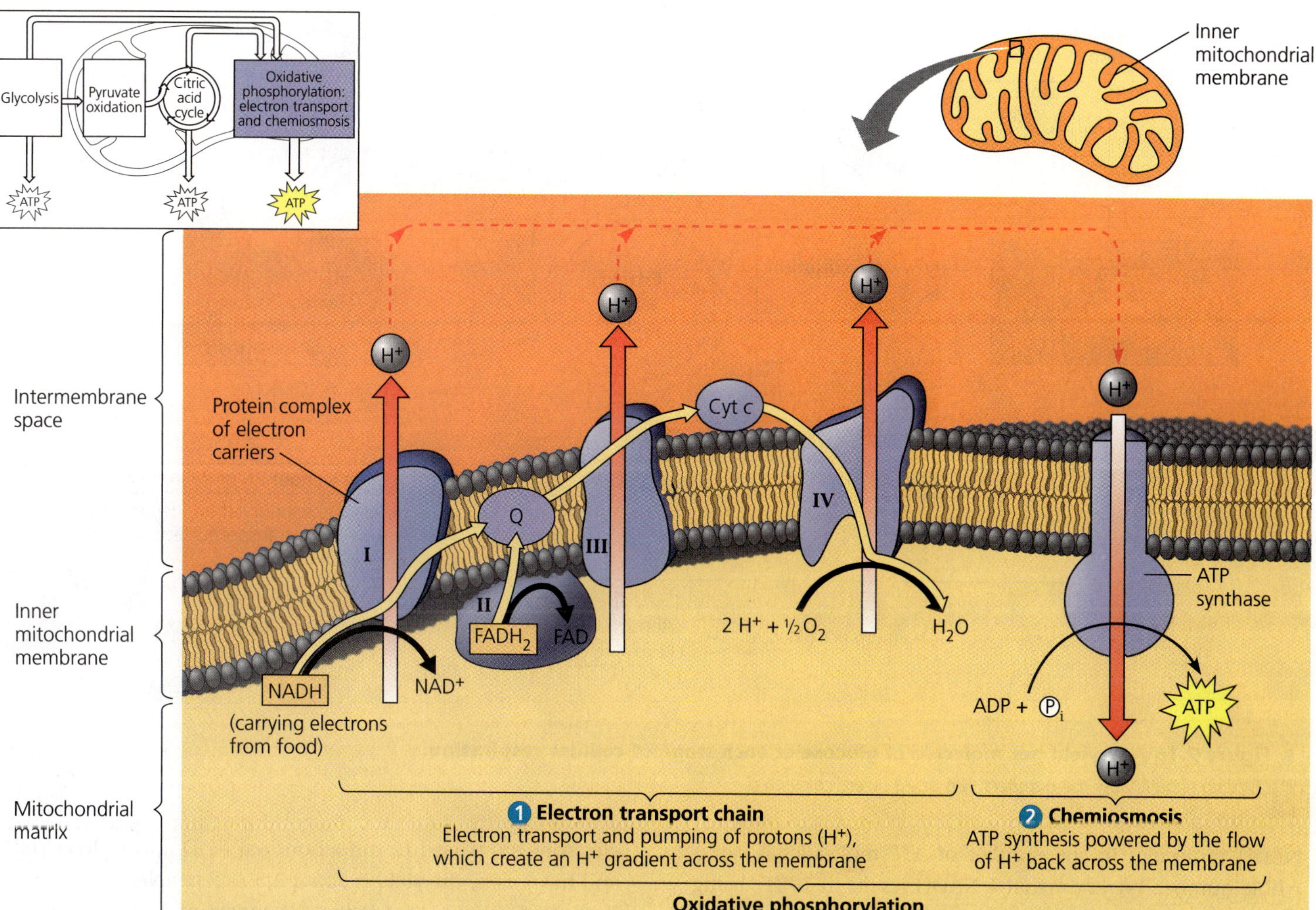

▲ **Figure 9.15 Chemiosmosis couples the electron transport chain to ATP synthesis.** ① NADH and $FADH_2$ shuttle high-energy electrons extracted from food during glycolysis and the citric acid cycle into an electron transport chain built into the inner mitochondrial membrane. The gold arrows trace the transport of electrons, which finally pass to oxygen at the "downhill" end of the chain, forming water. As Figure 9.13 showed, most of the electron carriers of the chain are grouped into four complexes. Two mobile carriers, ubiquinone (Q) and cytochrome *c* (Cyt *c*), move rapidly, ferrying electrons between the large complexes. As complexes I, III, and IV accept and then donate electrons, they pump protons from the mitochondrial matrix into the intermembrane space. (In prokaryotes, protons are pumped outside the plasma membrane.) Note that $FADH_2$ deposits its electrons via complex II and so results in fewer protons being pumped into the intermembrane space than occurs with NADH. Chemical energy originally harvested from food is transformed into a proton-motive force, a gradient of H^+ across the membrane. ② During chemiosmosis, the protons flow back down their gradient via ATP synthase, which is built into the membrane nearby. The ATP synthase harnesses the proton-motive force to phosphorylate ADP, forming ATP. Together, electron transport and chemiosmosis make up oxidative phosphorylation.

WHAT IF? *If complex IV were nonfunctional, could chemiosmosis produce any ATP, and if so, how would the rate of synthesis differ?*

back and remind ourselves of its overall function: harvesting the energy of glucose for ATP synthesis.

During respiration, most energy flows in this sequence: glucose → NADH → electron transport chain → proton-motive force → ATP. We can do some bookkeeping to calculate the ATP profit when cellular respiration oxidizes a molecule of glucose to six molecules of carbon dioxide. The three main departments of this metabolic enterprise are glycolysis, the citric acid cycle, and the electron transport chain, which drives oxidative phosphorylation. **Figure 9.16**, on the next page, gives a detailed accounting of the ATP yield per glucose molecule oxidized. The tally adds the 4 ATP produced directly by substrate-level phosphorylation during glycolysis and the citric acid cycle to the many more molecules of ATP generated by oxidative phosphorylation. Each NADH that transfers a pair of electrons from glucose to the electron transport chain contributes enough to the proton-motive force to generate a maximum of about 3 ATP.

Why are the numbers in Figure 9.16 inexact? There are three reasons we cannot state an exact number of ATP molecules generated by the breakdown of one molecule of glucose. First, phosphorylation and the redox reactions are not directly coupled to each other, so the ratio of the number of

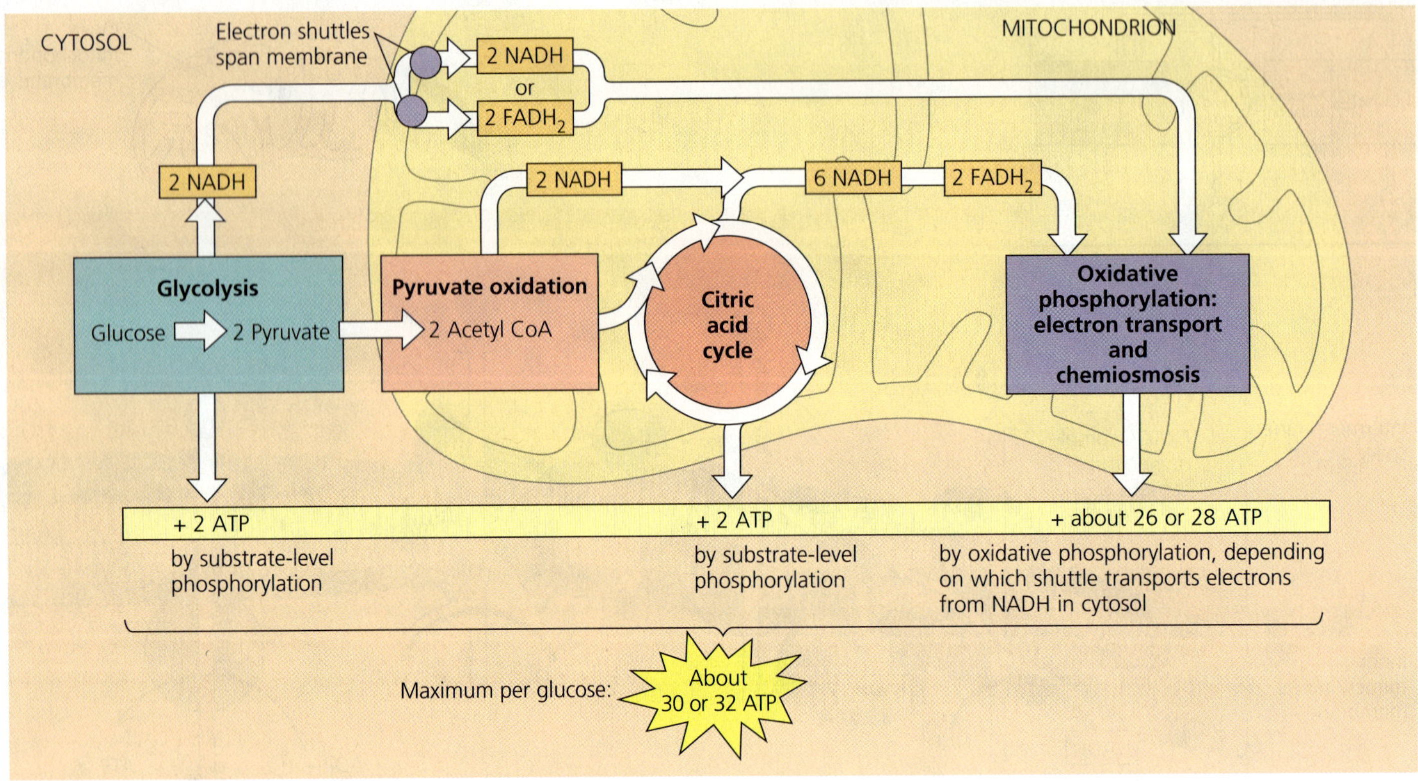

▲ **Figure 9.16 ATP yield per molecule of glucose at each stage of cellular respiration.**

? *Explain exactly how the numbers "26 or 28" were calculated.*

NADH molecules to the number of ATP molecules is not a whole number. We know that 1 NADH results in 10 H^+ being transported out across the inner mitochondrial membrane, but the exact number of H^+ that must reenter the mitochondrial matrix via ATP synthase to generate 1 ATP has long been debated. Based on experimental data, however, most biochemists now agree that the most accurate number is 4 H^+. Therefore, a single molecule of NADH generates enough proton-motive force for the synthesis of 2.5 ATP. The citric acid cycle also supplies electrons to the electron transport chain via $FADH_2$, but since its electrons enter later in the chain, each molecule of this electron carrier is responsible for transport of only enough H^+ for the synthesis of 1.5 ATP. These numbers also take into account the slight energetic cost of moving the ATP formed in the mitochondrion out into the cytosol, where it will be used.

Second, the ATP yield varies slightly depending on the type of shuttle used to transport electrons from the cytosol into the mitochondrion. The mitochondrial inner membrane is impermeable to NADH, so NADH in the cytosol is segregated from the machinery of oxidative phosphorylation. The 2 electrons of NADH captured in glycolysis must be conveyed into the mitochondrion by one of several electron shuttle systems. Depending on the kind of shuttle in a particular cell type, the electrons are passed either to NAD^+ or to FAD in the mitochondrial matrix (see Figure 9.16). If the electrons are passed to FAD, as in brain cells, only about 1.5 ATP can result from each NADH that was originally generated in the cytosol. If the electrons are passed to mitochondrial NAD^+, as in liver cells and heart cells, the yield is about 2.5 ATP per NADH.

A third variable that reduces the yield of ATP is the use of the proton-motive force generated by the redox reactions of respiration to drive other kinds of work. For example, the proton-motive force powers the mitochondrion's uptake of pyruvate from the cytosol. However, if *all* the proton-motive force generated by the electron transport chain were used to drive ATP synthesis, one glucose molecule could generate a maximum of 28 ATP produced by oxidative phosphorylation plus 4 ATP (net) from substrate-level phosphorylation to give a total yield of about 32 ATP (or only about 30 ATP if the less efficient shuttle were functioning).

We can now roughly estimate the efficiency of respiration—that is, the percentage of chemical energy in glucose that has been transferred to ATP. Recall that the complete oxidation of a mole of glucose releases 686 kcal of energy under standard conditions ($\Delta G = -686$ kcal/mol). Phosphorylation of ADP to form ATP stores at least 7.3 kcal per mole of ATP. Therefore, the efficiency of respiration is 7.3 kcal per mole of ATP times 32 moles of ATP per mole of glucose divided by 686 kcal per mole of glucose, which equals 0.34. Thus, about 34% of the potential chemical energy in glucose has been transferred to ATP; the actual percentage is probably higher because ΔG is lower under cellular conditions. Cellular respiration is remarkably efficient in its energy conversion. By comparison, the most efficient automobile converts only about 25% of the energy stored in gasoline to energy that moves the car.

The rest of the energy stored in glucose is lost as heat. We humans use some of this heat to maintain our relatively high body temperature (37°C), and we dissipate the rest through sweating and other cooling mechanisms.

Under certain conditions, it may be beneficial to reduce the efficiency of cellular respiration. A remarkable adaptation is shown by hibernating mammals, which overwinter in a state of inactivity and lowered metabolism. Although their internal body temperature is lower than normal, it still must be kept significantly higher than the external air temperature. One type of tissue, called brown fat, is made up of cells packed full of mitochondria. The inner mitochondrial membrane contains a channel protein called the uncoupling protein, which allows protons to flow back down their concentration gradient without generating ATP. Activation of these proteins in hibernating mammals results in ongoing oxidation of stored fuel stores (fats), generating heat without any ATP production. In the absence of such an adaptation, the ATP level would build up to a point that cellular respiration would be shut down due to regulatory mechanisms to be discussed later.

CONCEPT CHECK 9.4

1. What effect would an absence of O_2 have on the process shown in Figure 9.15?
2. **WHAT IF?** In the absence of O_2, as in question 1, what do you think would happen if you decreased the pH of the intermembrane space of the mitochondrion? Explain your answer.
3. **MAKE CONNECTIONS** In Concept 7.1 (pp. 127–128), you learned that membranes must be fluid to function properly. How does the operation of the electron transport chain support that assertion?

For suggested answers, see Appendix A.

CONCEPT 9.5

Fermentation and anaerobic respiration enable cells to produce ATP without the use of oxygen

Because most of the ATP generated by cellular respiration is due to the work of oxidative phosphorylation, our estimate of ATP yield from aerobic respiration is contingent on an adequate supply of oxygen to the cell. Without the electronegative oxygen to pull electrons down the transport chain, oxidative phosphorylation eventually ceases. However, there are two general mechanisms by which certain cells can oxidize organic fuel and generate ATP *without* the use of oxygen: anaerobic respiration and fermentation. The distinction between these two is that an electron transport chain is used in anaerobic respiration but not in fermentation. (The electron transport chain is also called the respiratory chain because of its role in both types of cellular respiration.)

We have already mentioned anaerobic respiration, which takes place in certain prokaryotic organisms that live in environments without oxygen. These organisms have an electron transport chain but do not use oxygen as a final electron acceptor at the end of the chain. Oxygen performs this function very well because it is extremely electronegative, but other, less electronegative substances can also serve as final electron acceptors. Some "sulfate-reducing" marine bacteria, for instance, use the sulfate ion (SO_4^{2-}) at the end of their respiratory chain. Operation of the chain builds up a proton-motive force used to produce ATP, but H_2S (hydrogen sulfide) is produced as a by-product rather than water. The rotten-egg odor you may have smelled while walking through a salt marsh or a mudflat signals the presence of sulfate-reducing bacteria.

Fermentation is a way of harvesting chemical energy without using either oxygen or any electron transport chain—in other words, without cellular respiration. How can food be oxidized without cellular respiration? Remember, oxidation simply refers to the loss of electrons to an electron acceptor, so it does not need to involve oxygen. Glycolysis oxidizes glucose to two molecules of pyruvate. The oxidizing agent of glycolysis is NAD^+, and neither oxygen nor any electron transfer chain is involved. Overall, glycolysis is exergonic, and some of the energy made available is used to produce 2 ATP (net) by substrate-level phosphorylation. If oxygen *is* present, then additional ATP is made by oxidative phosphorylation when NADH passes electrons removed from glucose to the electron transport chain. But glycolysis generates 2 ATP whether oxygen is present or not—that is, whether conditions are aerobic or anaerobic.

As an alternative to respiratory oxidation of organic nutrients, fermentation is an extension of glycolysis that allows continuous generation of ATP by the substrate-level phosphorylation of glycolysis. For this to occur, there must be a sufficient supply of NAD^+ to accept electrons during the oxidation step of glycolysis. Without some mechanism to recycle NAD^+ from NADH, glycolysis would soon deplete the cell's pool of NAD^+ by reducing it all to NADH and would shut itself down for lack of an oxidizing agent. Under aerobic conditions, NAD^+ is recycled from NADH by the transfer of electrons to the electron transport chain. An anaerobic alternative is to transfer electrons from NADH to pyruvate, the end product of glycolysis.

Types of Fermentation

Fermentation consists of glycolysis plus reactions that regenerate NAD^+ by transferring electrons from NADH to pyruvate or derivatives of pyruvate. The NAD^+ can then be reused to oxidize sugar by glycolysis, which nets two molecules of ATP by substrate-level phosphorylation. There are many types of fermentation, differing in the end products formed from pyruvate. Two common types are alcohol fermentation and lactic acid fermentation.

In **alcohol fermentation** **(Figure 9.17a)**, pyruvate is converted to ethanol (ethyl alcohol) in two steps. The first step releases carbon dioxide from the pyruvate, which is converted to the two-carbon compound acetaldehyde. In the second step, acetaldehyde is reduced by NADH to ethanol. This regenerates the supply of NAD^+ needed for the continuation of glycolysis. Many bacteria carry out alcohol fermentation under anaerobic conditions. Yeast (a fungus) also carries out alcohol fermentation. For thousands of years, humans have used yeast in brewing, winemaking, and baking. The CO_2 bubbles generated by baker's yeast during alcohol fermentation allow bread to rise.

During **lactic acid fermentation** **(Figure 9.17b)**, pyruvate is reduced directly by NADH to form lactate as an end product, with no release of CO_2. (Lactate is the ionized form of lactic acid.) Lactic acid fermentation by certain fungi and bacteria is used in the dairy industry to make cheese and yogurt.

Human muscle cells make ATP by lactic acid fermentation when oxygen is scarce. This occurs during the early stages of strenuous exercise, when sugar catabolism for ATP production outpaces the muscle's supply of oxygen from the blood. Under these conditions, the cells switch from aerobic respiration to fermentation. The lactate that accumulates was previously thought to cause muscle fatigue and pain, but recent research suggests instead that increased levels of potassium ions (K^+) may be to blame, while lactate appears to enhance muscle performance. In any case, the excess lactate is gradually carried away by the blood to the liver, where it is converted back to pyruvate by liver cells. Because oxygen is available, this pyruvate can then enter the mitochondria in liver cells and complete cellular respiration.

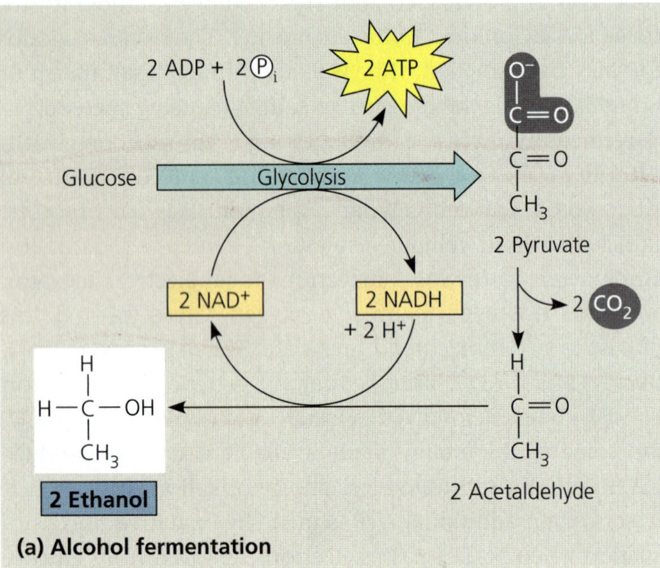

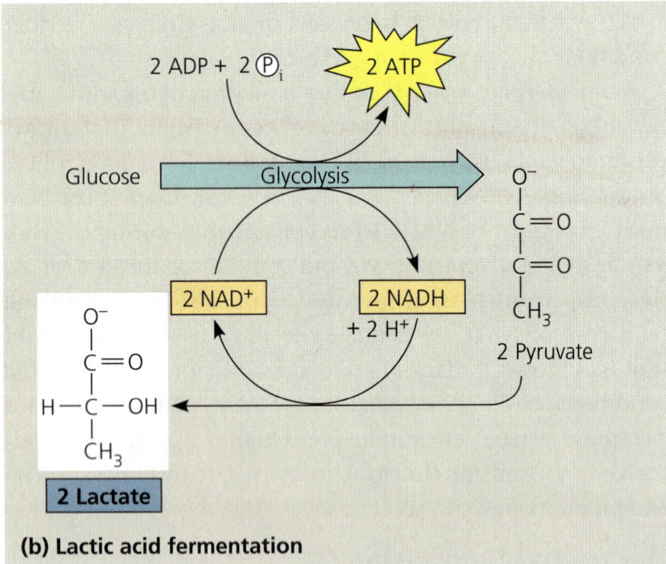

▲ **Figure 9.17 Fermentation.** In the absence of oxygen, many cells use fermentation to produce ATP by substrate-level phosphorylation. Pyruvate, the end product of glycolysis, serves as an electron acceptor for oxidizing NADH back to NAD^+, which can then be reused in glycolysis. Two of the common end products formed from fermentation are **(a)** ethanol and **(b)** lactate, the ionized form of lactic acid.

Comparing Fermentation with Anaerobic and Aerobic Respiration

Fermentation, anaerobic respiration, and aerobic respiration are three alternative cellular pathways for producing ATP by harvesting the chemical energy of food. All three use glycolysis to oxidize glucose and other organic fuels to pyruvate, with a net production of 2 ATP by substrate-level phosphorylation. And in all three pathways, NAD^+ is the oxidizing agent that accepts electrons from food during glycolysis.

A key difference among the three pathways is the contrasting mechanisms for oxidizing NADH back to NAD^+, which is required to sustain glycolysis. In fermentation, the final electron acceptor is an organic molecule such as pyruvate (lactic acid fermentation) or acetaldehyde (alcohol fermentation). In cellular respiration, by contrast, electrons carried by NADH are transferred to an electron transport chain, where they move stepwise down a series of redox reactions to a final electron acceptor. In aerobic respiration, the final electron acceptor is oxygen; in anaerobic respiration, the final acceptor is another molecule that is electronegative (although invariably less so than oxygen). Passage of electrons from NADH to the electron transport chain not only regenerates the NAD^+ required for glycolysis but pays an ATP bonus when the stepwise electron transport from this NADH to oxygen drives oxidative phosphorylation. An even bigger ATP payoff comes from the oxidation of pyruvate in the mitochondrion, which is unique to respiration. Without an electron transport chain, the energy still stored in pyruvate is unavailable to most cells. Thus, cellular respiration harvests much more energy from each sugar molecule than fermentation can. In fact, aerobic respiration yields up to 16 times as much ATP per glucose molecule as does fermentation—up to 32 molecules of ATP for respiration, compared with 2 molecules of ATP produced by substrate-level phosphorylation in fermentation.

Some organisms, called **obligate anaerobes**, carry out only fermentation or anaerobic respiration. In fact, these organisms cannot survive in the presence of oxygen, some forms of which can actually be toxic if protective systems are not present in the cell. A few cell types, such as cells of the vertebrate brain, can carry out only aerobic oxidation of pyruvate, not fermentation. Other organisms, including yeasts and many bacteria, can make enough ATP to survive using either fermentation or respiration. Such species are called **facultative anaerobes**. On the cellular level, our muscle cells behave as facultative anaerobes. In such cells, pyruvate is a fork in the metabolic road that leads to two alternative catabolic routes (**Figure 9.18**). Under aerobic conditions, pyruvate can be converted to acetyl CoA, and oxidation continues in the citric acid cycle via aerobic respiration. Under anaerobic conditions, lactic acid fermentation occurs: Pyruvate is diverted from the citric acid cycle, serving instead as an electron acceptor to recycle NAD^+. To make the same amount of ATP, a facultative anaerobe has to consume sugar at a much faster rate when fermenting than when respiring.

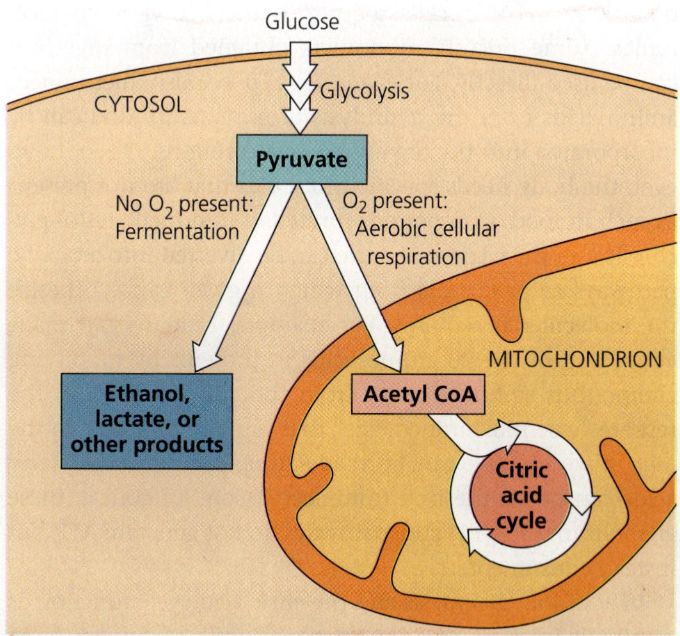

▲ **Figure 9.18 Pyruvate as a key juncture in catabolism.** Glycolysis is common to fermentation and cellular respiration. The end product of glycolysis, pyruvate, represents a fork in the catabolic pathways of glucose oxidation. In a facultative anaerobe or a muscle cell, which are capable of both aerobic cellular respiration and fermentation, pyruvate is committed to one of those two pathways, usually depending on whether or not oxygen is present.

The Evolutionary Significance of Glycolysis

EVOLUTION The role of glycolysis in both fermentation and respiration has an evolutionary basis. Ancient prokaryotes are thought to have used glycolysis to make ATP long before oxygen was present in Earth's atmosphere. The oldest known fossils of bacteria date back 3.5 billion years, but appreciable quantities of oxygen probably did not begin to accumulate in the atmosphere until about 2.7 billion years ago. Cyanobacteria produced this O_2 as a by-product of photosynthesis. Therefore, early prokaryotes may have generated ATP exclusively from glycolysis. The fact that glycolysis is today the most widespread metabolic pathway among Earth's organisms suggests that it evolved very early in the history of life. The cytosolic location of glycolysis also implies great antiquity; the pathway does not require any of the membrane-bounded organelles of the eukaryotic cell, which evolved approximately 1 billion years after the prokaryotic cell. Glycolysis is a metabolic heirloom from early cells that continues to function in fermentation and as the first stage in the breakdown of organic molecules by respiration.

CONCEPT CHECK 9.5

1. Consider the NADH formed during glycolysis. What is the final acceptor for its electrons during fermentation? What is the final acceptor for its electrons during aerobic respiration?
2. **WHAT IF?** A glucose-fed yeast cell is moved from an aerobic environment to an anaerobic one. How would its rate of glucose consumption change if ATP were to be generated at the same rate?

For suggested answers, see Appendix A.

CONCEPT 9.6

Glycolysis and the citric acid cycle connect to many other metabolic pathways

So far, we have treated the oxidative breakdown of glucose in isolation from the cell's overall metabolic economy. In this section, you will learn that glycolysis and the citric acid cycle are major intersections of the cell's catabolic and anabolic (biosynthetic) pathways.

The Versatility of Catabolism

Throughout this chapter, we have used glucose as the fuel for cellular respiration. But free glucose molecules are not common in the diets of humans and other animals. We obtain most of our calories in the form of fats, proteins, sucrose and other disaccharides, and starch, a polysaccharide. All these

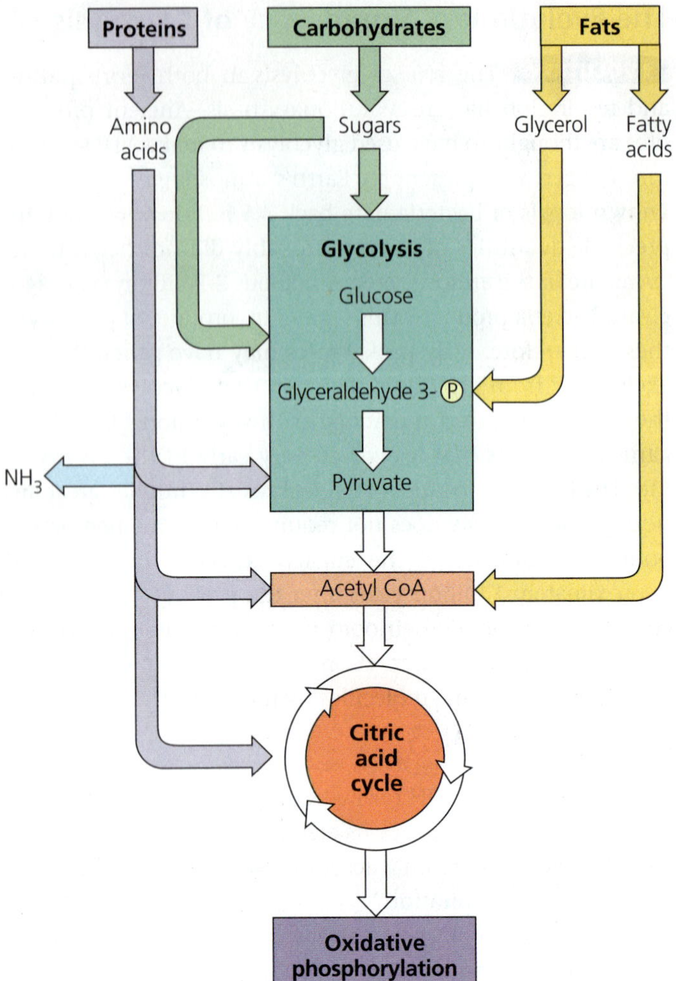

▲ **Figure 9.19 The catabolism of various molecules from food.** Carbohydrates, fats, and proteins can all be used as fuel for cellular respiration. Monomers of these molecules enter glycolysis or the citric acid cycle at various points. Glycolysis and the citric acid cycle are catabolic funnels through which electrons from all kinds of organic molecules flow on their exergonic fall to oxygen.

organic molecules in food can be used by cellular respiration to make ATP (**Figure 9.19**).

Glycolysis can accept a wide range of carbohydrates for catabolism. In the digestive tract, starch is hydrolyzed to glucose, which can then be broken down in the cells by glycolysis and the citric acid cycle. Similarly, glycogen, the polysaccharide that humans and many other animals store in their liver and muscle cells, can be hydrolyzed to glucose between meals as fuel for respiration. The digestion of disaccharides, including sucrose, provides glucose and other monosaccharides as fuel for respiration.

Proteins can also be used for fuel, but first they must be digested to their constituent amino acids. Many of the amino acids are used by the organism to build new proteins. Amino acids present in excess are converted by enzymes to intermediates of glycolysis and the citric acid cycle. Before amino acids can feed into glycolysis or the citric acid cycle, their amino groups must be removed, a process called *deamination*. The nitrogenous refuse is excreted from the animal in the form of ammonia (NH_3), urea, or other waste products.

Catabolism can also harvest energy stored in fats obtained either from food or from storage cells in the body. After fats are digested to glycerol and fatty acids, the glycerol is converted to glyceraldehyde 3-phosphate, an intermediate of glycolysis. Most of the energy of a fat is stored in the fatty acids. A metabolic sequence called **beta oxidation** breaks the fatty acids down to two-carbon fragments, which enter the citric acid cycle as acetyl CoA. NADH and $FADH_2$ are also generated during beta oxidation; they can enter the electron transport chain, leading to further ATP production. Fats make excellent fuel, in large part due to their chemical structure and the high energy level of their electrons (equally shared between carbon and hydrogen) compared to those of carbohydrates. A gram of fat oxidized by respiration produces more than twice as much ATP as a gram of carbohydrate. Unfortunately, this also means that a person trying to lose weight must work hard to use up fat stored in the body because so many calories are stockpiled in each gram of fat.

Biosynthesis (Anabolic Pathways)

Cells need substance as well as energy. Not all the organic molecules of food are destined to be oxidized as fuel to make ATP. In addition to calories, food must also provide the carbon skeletons that cells require to make their own molecules. Some organic monomers obtained from digestion can be used directly. For example, as previously mentioned, amino acids from the hydrolysis of proteins in food can be incorporated into the organism's own proteins. Often, however, the body needs specific molecules that are not present as such in food. Compounds formed as intermediates of glycolysis and the citric acid cycle can be diverted into anabolic pathways as precursors from which the cell can synthesize the molecules it requires. For example, humans can make about half of the 20 amino acids in proteins by modifying compounds siphoned away from the citric acid cycle; the rest are "essential amino acids" that must be obtained in the diet. Also, glucose can be made from pyruvate, and fatty acids can be synthesized from acetyl CoA. Of course, these anabolic, or biosynthetic, pathways do not generate ATP, but instead consume it.

In addition, glycolysis and the citric acid cycle function as metabolic interchanges that enable our cells to convert some kinds of molecules to others as we need them. For example, an intermediate compound generated during glycolysis, dihydroxyacetone phosphate (see Figure 9.9, step 5), can be converted to one of the major precursors of fats. If we eat more food than we need, we store fat even if our diet is fatfree. Metabolism is remarkably versatile and adaptable.

Regulation of Cellular Respiration via Feedback Mechanisms

Basic principles of supply and demand regulate the metabolic economy. The cell does not waste energy making more of a particular substance than it needs. If there is a glut of a certain amino acid, for example, the anabolic pathway that synthesizes that amino acid from an intermediate of the citric acid cycle is switched off. The most common mechanism for this control is feedback inhibition: The end product of the anabolic pathway inhibits the enzyme that catalyzes an early step of the pathway (see Figure 8.21). This prevents the needless diversion of key metabolic intermediates from uses that are more urgent.

The cell also controls its catabolism. If the cell is working hard and its ATP concentration begins to drop, respiration speeds up. When there is plenty of ATP to meet demand, respiration slows down, sparing valuable organic molecules for other functions. Again, control is based mainly on regulating the activity of enzymes at strategic points in the catabolic pathway. As shown in **Figure 9.20**, one important switch is phosphofructokinase, the enzyme that catalyzes step 3 of glycolysis (see Figure 9.9). That is the first step that commits the substrate irreversibly to the glycolytic pathway. By controlling the rate of this step, the cell can speed up or slow down the entire catabolic process. Phosphofructokinase can thus be considered the pacemaker of respiration.

Phosphofructokinase is an allosteric enzyme with receptor sites for specific inhibitors and activators. It is inhibited by ATP and stimulated by AMP (adenosine monophosphate), which the cell derives from ADP. As ATP accumulates, inhibition of the enzyme slows down glycolysis. The enzyme becomes active again as cellular work converts ATP to ADP (and AMP) faster than ATP is being regenerated. Phosphofructokinase is also sensitive to citrate, the first product of the citric acid cycle. If citrate accumulates in mitochondria, some of it passes into the cytosol and inhibits phosphofructokinase. This mechanism helps synchronize the rates of glycolysis and the citric acid cycle. As citrate accumulates, glycolysis slows down, and the supply of acetyl groups to the citric acid cycle decreases. If citrate consumption increases, either because of a demand for more ATP or because anabolic pathways are draining off intermediates of the citric acid cycle, glycolysis accelerates and meets the demand. Metabolic balance is augmented by the control of enzymes that catalyze other key steps of glycolysis and the citric acid cycle. Cells are thrifty, expedient, and responsive in their metabolism.

Cellular respiration and metabolic pathways play a role of central importance in organisms. Examine Figure 9.2 again to put cellular respiration into the broader context of energy flow and chemical cycling in ecosystems. The energy that keeps us alive is *released*, not *produced*, by cellular respiration. We are tapping energy that was stored in food by photosynthesis. In the next chapter, you will learn how photosynthesis captures light and converts it to chemical energy.

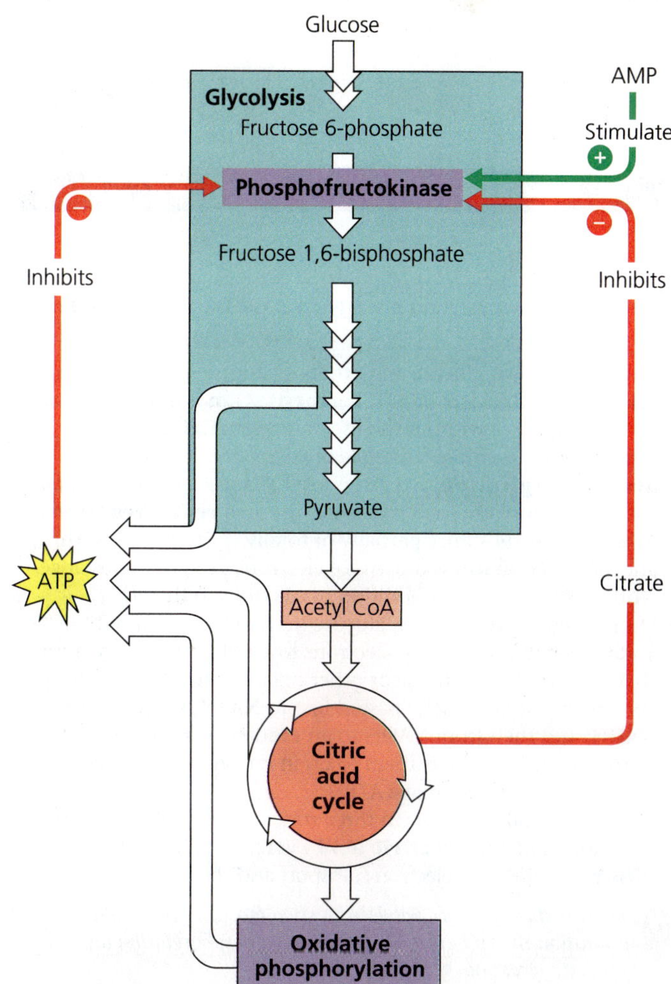

▲ **Figure 9.20 The control of cellular respiration.** Allosteric enzymes at certain points in the respiratory pathway respond to inhibitors and activators that help set the pace of glycolysis and the citric acid cycle. Phosphofructokinase, which catalyzes an early step in glycolysis (see Figure 9.9), is one such enzyme. It is stimulated by AMP (derived from ADP) but is inhibited by ATP and by citrate. This feedback regulation adjusts the rate of respiration as the cell's catabolic and anabolic demands change.

CONCEPT CHECK 9.6

1. **MAKE CONNECTIONS** Compare the structure of a fat (see Figure 5.10, p. 75) with that of a carbohydrate (see Figure 5.3, p. 70). What features of their structures make fat a much better fuel?
2. Under what circumstances might your body synthesize fat molecules?
3. **MAKE CONNECTIONS** Return to Figure 5.6b on page 72 and look at the arrangement of glycogen and mitochondria in the micrograph. What is the connection between glycogen and mitochondria?
4. **WHAT IF?** What will happen in a muscle cell that has used up its supply of oxygen and ATP? (Review Figures 9.18 and 9.20.)
5. **WHAT IF?** During intense exercise, can a muscle cell use fat as a concentrated source of chemical energy? Explain. (Review Figures 9.18 and 9.19.)

For suggested answers, see Appendix A.

9 CHAPTER REVIEW

SUMMARY OF KEY CONCEPTS

CONCEPT 9.1

Catabolic pathways yield energy by oxidizing organic fuels (pp. 164–168)

- Cells break down glucose and other organic fuels to yield chemical energy in the form of ATP. **Fermentation** is a partial degradation of glucose without the use of oxygen. **Cellular respiration** is a more complete breakdown of glucose; in **aerobic respiration**, oxygen is used as a reactant. The cell taps the energy stored in food molecules through **redox reactions**, in which one substance partially or totally shifts electrons to another. **Oxidation** is the loss of electrons from one substance, while **reduction** is the addition of electrons to the other.
- During aerobic respiration, glucose ($C_6H_{12}O_6$) is oxidized to CO_2, and O_2 is reduced to H_2O. Electrons lose potential energy during their transfer from glucose or other organic compounds to oxygen. Electrons are usually passed first to $\mathbf{NAD^+}$, reducing it to NADH, and then from NADH to an **electron transport chain**, which conducts them to O_2 in energy-releasing steps. The energy is used to make ATP.
- Aerobic respiration occurs in three stages: (1) **glycolysis**, (2) pyruvate oxidation and the **citric acid cycle**, and (3) **oxidative phosphorylation** (electron transport and chemiosmosis).

? *Describe the difference between the two processes in cellular respiration that produce ATP: oxidative phosphorylation and substrate-level phosphorylation.*

CONCEPT 9.2

Glycolysis harvests chemical energy by oxidizing glucose to pyruvate (pp. 168–169)

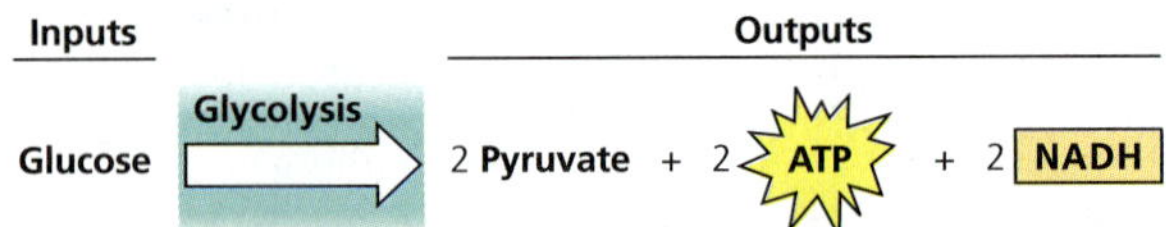

? *What is the source of energy for the formation of ATP and NADH in glycolysis?*

CONCEPT 9.3

After pyruvate is oxidized, the citric acid cycle completes the energy-yielding oxidation of organic molecules (pp. 170–172)

- In eukaryotic cells, pyruvate enters the mitochondrion and is oxidized to **acetyl CoA**, which is further oxidized in the citric acid cycle.

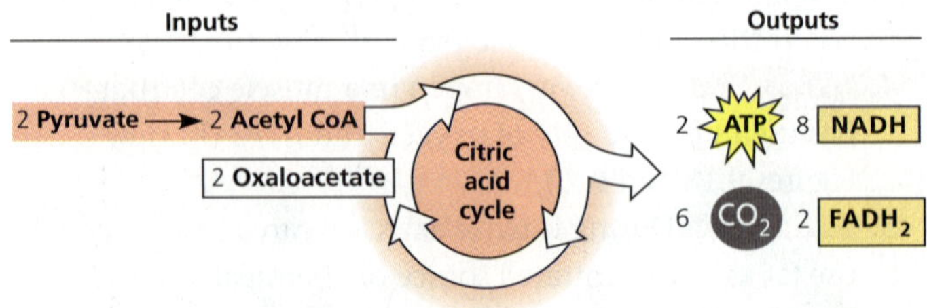

? *What molecular products indicate the complete oxidation of glucose during cellular respiration?*

CONCEPT 9.4

During oxidative phosphorylation, chemiosmosis couples electron transport to ATP synthesis (pp. 172–177)

- NADH and $FADH_2$ transfer electrons to the electron transport chain. Electrons move down the chain, losing energy in several energy-releasing steps. Finally, electrons are passed to O_2, reducing it to H_2O.

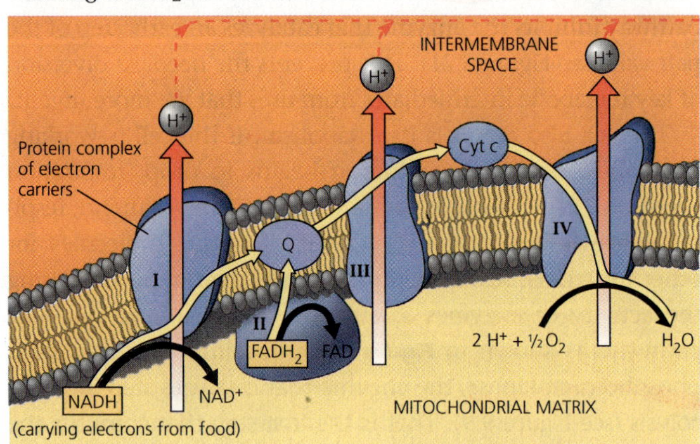

- At certain steps along the electron transport chain, electron transfer causes protein complexes to move H^+ from the mitochondrial matrix (in eukaryotes) to the intermembrane space, storing energy as a **proton-motive force** (H^+ gradient). As H^+ diffuses back into the matrix through **ATP synthase**, its passage drives the phosphorylation of ADP, a process called **chemiosmosis**.

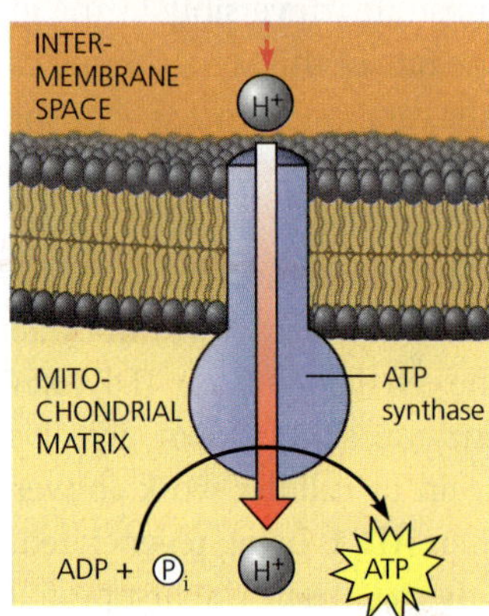

- About 34% of the energy stored in a glucose molecule is transferred to ATP during cellular respiration, producing a maximum of about 32 ATP.

? *Briefly explain the mechanism by which ATP synthase produces ATP. List three locations in which ATP synthases are found.*

CONCEPT 9.5

Fermentation and anaerobic respiration enable cells to produce ATP without the use of oxygen (pp. 177–179)

- Glycolysis nets 2 ATP by substrate-level phosphorylation, whether oxygen is present or not. Under anaerobic conditions, either anaerobic respiration or fermentation can take place. In anaerobic respiration, an electron transport chain is present with a final electron acceptor other than oxygen. In fermentation, the electrons from NADH are passed to pyruvate or a derivative of pyruvate, regenerating the NAD^+ required to oxidize more glucose. Two common types of fermentation are **alcohol fermentation** and **lactic acid fermentation**.
- Fermentation and anaerobic or aerobic respiration all use glycolysis to oxidize glucose, but they differ in their final electron acceptor and whether an electron transport chain is used (respiration) or not (fermentation). Respiration yields more ATP;

aerobic respiration, with O_2 as the final electron acceptor, yields about 16 times as much ATP as does fermentation.
- Glycolysis occurs in nearly all organisms and is thought to have evolved in ancient prokaryotes before there was O_2 in the atmosphere.

? *Which process yields more ATP, fermentation or anaerobic respiration? Explain.*

CONCEPT 9.6

Glycolysis and the citric acid cycle connect to many other metabolic pathways (pp. 179–181)

- Catabolic pathways funnel electrons from many kinds of organic molecules into cellular respiration. Many carbohydrates can enter glycolysis, most often after conversion to glucose. Amino acids of proteins must be deaminated before being oxidized. The fatty acids of fats undergo **beta oxidation** to two-carbon fragments and then enter the citric acid cycle as acetyl CoA. Anabolic pathways can use small molecules from food directly or build other substances using intermediates of glycolysis or the citric acid cycle.
- Cellular respiration is controlled by allosteric enzymes at key points in glycolysis and the citric acid cycle.

? *Describe how the catabolic pathways of glycolysis and the citric acid cycle intersect with anabolic pathways in the metabolism of a cell.*

TEST YOUR UNDERSTANDING

LEVEL 1: KNOWLEDGE/COMPREHENSION

1. The *immediate* energy source that drives ATP synthesis by ATP synthase during oxidative phosphorylation is the
   a. oxidation of glucose and other organic compounds.
   b. flow of electrons down the electron transport chain.
   c. affinity of oxygen for electrons.
   d. H^+ concentration across the membrane holding ATP synthase.
   e. transfer of phosphate to ADP.
2. Which metabolic pathway is common to both fermentation and cellular respiration of a glucose molecule?
   a. the citric acid cycle
   b. the electron transport chain
   c. glycolysis
   d. synthesis of acetyl CoA from pyruvate
   e. reduction of pyruvate to lactate
3. In mitochondria, exergonic redox reactions
   a. are the source of energy driving prokaryotic ATP synthesis.
   b. are directly coupled to substrate-level phosphorylation.
   c. provide the energy that establishes the proton gradient.
   d. reduce carbon atoms to carbon dioxide.
   e. are coupled via phosphorylated intermediates to endergonic processes.
4. The final electron acceptor of the electron transport chain that functions in aerobic oxidative phosphorylation is
   a. oxygen. b. water. c. NAD^+. d. pyruvate. e. ADP.

LEVEL 2: APPLICATION/ANALYSIS

5. What is the oxidizing agent in the following reaction?
   Pyruvate + NADH + H^+ $\rightarrow$ Lactate + NAD^+
   a. oxygen b. NADH c. NAD^+ d. lactate e. pyruvate
6. When electrons flow along the electron transport chains of mitochondria, which of the following changes occurs?
   a. The pH of the matrix increases.
   b. ATP synthase pumps protons by active transport.
   c. The electrons gain free energy.
   d. The cytochromes phosphorylate ADP to form ATP.
   e. NAD^+ is oxidized.
7. Most CO_2 from catabolism is released during
   a. glycolysis.
   b. the citric acid cycle.
   c. lactate fermentation.
   d. electron transport.
   e. oxidative phosphorylation.

LEVEL 3: SYNTHESIS/EVALUATION

8. **DRAW IT** The graph here shows the pH difference across the inner mitochondrial membrane over time in an actively respiring cell. At the time indicated by the vertical arrow, a metabolic poison is added that specifically and completely inhibits all function of mitochondrial ATP synthase. Draw what you would expect to see for the rest of the graphed line.

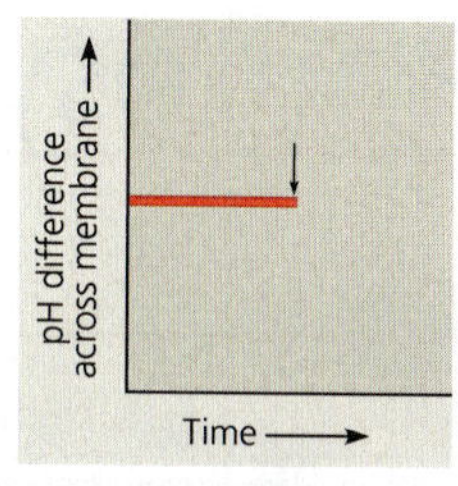

9. **EVOLUTION CONNECTION**
   ATP synthases are found in the prokaryotic plasma membrane and in mitochondria and chloroplasts. What does this suggest about the evolutionary relationship of these eukaryotic organelles to prokaryotes? How might the amino acid sequences of the ATP synthases from the different sources support or refute your hypothesis?
10. **SCIENTIFIC INQUIRY**
    In the 1930s, some physicians prescribed low doses of a compound called dinitrophenol (DNP) to help patients lose weight. This unsafe method was abandoned after some patients died. DNP uncouples the chemiosmotic machinery by making the lipid bilayer of the inner mitochondrial membrane leaky to H^+. Explain how this could cause weight loss and death.
11. **WRITE ABOUT A THEME**
    Emergent Properties In a short essay (100–150 words), explain how oxidative phosphorylation—the production of ATP using energy derived from the redox reactions of a spatially organized electron transport chain followed by chemiosmosis—is an example of how new properties emerge at each level of the biological hierarchy.

For selected answers, see Appendix A.

10

Photosynthesis

▲ **Figure 10.1 How can sunlight, seen here as a spectrum of colors in a rainbow, power the synthesis of organic substances?**

KEY CONCEPTS

10.1 Photosynthesis converts light energy to the chemical energy of food

10.2 The light reactions convert solar energy to the chemical energy of ATP and NADPH

10.3 The Calvin cycle uses the chemical energy of ATP and NADPH to reduce CO_2 to sugar

10.4 Alternative mechanisms of carbon fixation have evolved in hot, arid climates

OVERVIEW

The Process That Feeds the Biosphere

Life on Earth is solar powered. The chloroplasts of plants capture light energy that has traveled 150 million kilometers from the sun and convert it to chemical energy that is stored in sugar and other organic molecules. This conversion process is called **photosynthesis**. Let's begin by placing photosynthesis in its ecological context.

Photosynthesis nourishes almost the entire living world directly or indirectly. An organism acquires the organic compounds it uses for energy and carbon skeletons by one of two major modes: autotrophic nutrition or heterotrophic nutrition. **Autotrophs** are "self-feeders" (*auto-* means "self," and *trophos* means "feeder"); they sustain themselves without eating anything derived from other living beings. Autotrophs produce their organic molecules from CO_2 and other inorganic raw materials obtained from the environment. They are the ultimate sources of organic compounds for all nonautotrophic organisms, and for this reason, biologists refer to autotrophs as the *producers* of the biosphere.

Almost all plants are autotrophs; the only nutrients they require are water and minerals from the soil and carbon dioxide from the air. Specifically, plants are *photo*autotrophs, organisms that use light as a source of energy to synthesize organic substances (**Figure 10.1**). Photosynthesis also occurs in algae, certain other protists, and some prokaryotes (**Figure 10.2**). In this chapter, we will touch on these other groups in passing, but our emphasis will be on plants. Variations in autotrophic nutrition that occur in prokaryotes and algae will be described in Chapters 27 and 28.

Heterotrophs obtain their organic material by the second major mode of nutrition. Unable to make their own food, they live on compounds produced by other organisms (*hetero-* means "other"). Heterotrophs are the biosphere's *consumers*. The most obvious form of this "other-feeding" occurs when an animal eats plants or other animals. But heterotrophic nutrition may be more subtle. Some heterotrophs consume the remains of dead organisms by decomposing and feeding on organic litter such as carcasses, feces, and fallen leaves; they are known as decomposers. Most fungi and many types of prokaryotes get their nourishment this way. Almost all heterotrophs, including humans, are completely dependent, either directly or indirectly, on photoautotrophs for food—and also for oxygen, a by-product of photosynthesis.

The Earth's supply of fossil fuels was formed from remains of organisms that died hundreds of millions of years ago. In a sense, then, fossil fuels represent stores of the sun's energy from the distant past. Because these resources are being used at a much higher rate than they are replenished, researchers

▲ **Figure 10.2 Photoautotrophs.** These organisms use light energy to drive the synthesis of organic molecules from carbon dioxide and (in most cases) water. They feed themselves and the entire living world. **(a)** On land, plants are the predominant producers of food. In aquatic environments, photoautotrophs include unicellular and **(b)** multicellular algae, such as this kelp; **(c)** some non-algal unicellular protists, such as *Euglena*; **(d)** the prokaryotes called cyanobacteria; and **(e)** other photosynthetic prokaryotes, such as these purple sulfur bacteria, which produce sulfur (the yellow globules within the cells) (c–e, LMs).

▼ **Figure 10.3**

IMPACT

Alternative Fuels from Plants and Algae

Biofuels from crops such as corn, soybeans, and cassava have been proposed as a supplement or even replacement for fossil fuels. To produce "bioethanol," the starch made naturally by the plants is simply converted to glucose and then fermented to ethanol by microorganisms. Alternatively, a simple chemical process can yield "biodiesel" from plant oils. Either product can be mixed with gasoline or used alone to power vehicles. Some species of unicellular algae are especially prolific oil producers, and they can be easily cultured in containers such as the tubular plastic bags shown below.

WHY IT MATTERS The rate of fossil fuel use by humans far outpaces its formation in the earth: Fossil fuels are a nonrenewable source of energy. Tapping the power of sunlight by using products of photosynthesis to generate energy is a sustainable alternative if cost-effective techniques can be developed. It is generally agreed that using algae is preferable to growing crops for this purpose because this use of cropland diminishes the food supply and drives up food prices.

FURTHER READING A. L. Haag, Algae bloom again, *Nature* 447:520–521 (2007).

WHAT IF? The main product of fossil fuel combustion is CO_2, and this combustion is the source of the increase in atmospheric CO_2 concentration. Scientists have proposed strategically situating containers of these algae near industrial plants, as shown above, or near highly congested city streets. Why does this arrangement make sense?

are exploring methods of capitalizing on the photosynthetic process to provide alternative fuels **(Figure 10.3)**.

In this chapter, you will learn how photosynthesis works. After a discussion of the general principles of photosynthesis, we will consider the two stages of photosynthesis: the light reactions, in which solar energy is captured and transformed into chemical energy; and the Calvin cycle, in which the chemical energy is used to make organic molecules of food. Finally, we will consider a few aspects of photosynthesis from an evolutionary perspective.

CONCEPT **10.1**

Photosynthesis converts light energy to the chemical energy of food

The remarkable ability of an organism to harness light energy and use it to drive the synthesis of organic compounds emerges from structural organization in the cell: Photosynthetic enzymes and other molecules are grouped together in a biological membrane, enabling the necessary series of chemical reactions to be carried out efficiently. The process of photosynthesis most likely originated in a group of bacteria that had infolded regions of the plasma membrane containing clusters of such molecules. In existing photosynthetic bacteria, infolded photosynthetic membranes function similarly to the internal membranes of the chloroplast, a eukaryotic organelle. According to the endosymbiont theory, the original chloroplast was a photosynthetic prokaryote that lived inside an ancestor of eukaryotic cells. (You learned about this theory in Chapter 6 and it will be described more fully in Chapter 25.) Chloroplasts are present in a variety of photosynthesizing organisms (see Figure 10.2), but here we will focus on plants.

Chloroplasts: The Sites of Photosynthesis in Plants

All green parts of a plant, including green stems and unripened fruit, have chloroplasts, but the leaves are the major sites of photosynthesis in most plants **(Figure 10.4)**. There are about half a million chloroplasts in a chunk of leaf with a top surface area of 1 mm^2. Chloroplasts are found mainly in the cells of the **mesophyll**, the tissue in the interior of the leaf. Carbon dioxide enters the leaf, and oxygen exits, by way of microscopic pores called **stomata** (singular, *stoma*; from the Greek, meaning "mouth"). Water absorbed by the roots is delivered to the leaves in veins. Leaves also use veins to export sugar to roots and other nonphotosynthetic parts of the plant.

A typical mesophyll cell has about 30–40 chloroplasts, each organelle measuring about 2–4 μm by 4–7 μm. A chloroplast has an envelope of two membranes surrounding a dense fluid called the **stroma**. Suspended within the stroma is a third membrane system, made up of sacs called **thylakoids**, which segregates the stroma from the *thylakoid space* inside these sacs. In some places, thylakoid sacs are stacked in columns called *grana* (singular, *granum*). **Chlorophyll**, the green pigment that gives leaves their color, resides in the thylakoid membranes of the chloroplast. (The internal photosynthetic membranes of some prokaryotes are also called thylakoid membranes; see Figure 27.7b.) It is the light energy absorbed

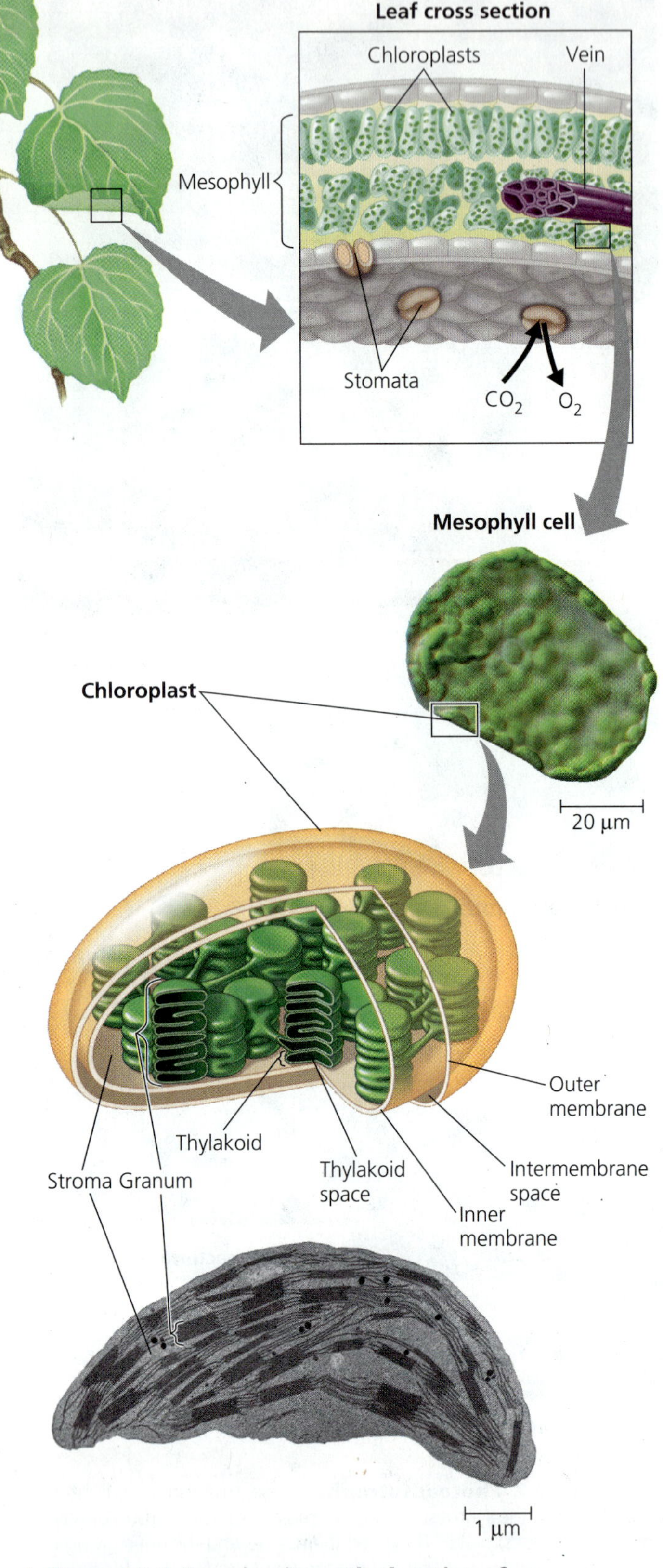

▲ **Figure 10.4 Zooming in on the location of photosynthesis in a plant.** Leaves are the major organs of photosynthesis in plants. These pictures take you into a leaf, then into a cell, and finally into a chloroplast, the organelle where photosynthesis occurs (middle, LM; bottom, TEM).

by chlorophyll that drives the synthesis of organic molecules in the chloroplast. Now that we have looked at the sites of photosynthesis in plants, we are ready to look more closely at the process of photosynthesis.

Tracking Atoms Through Photosynthesis: *Scientific Inquiry*

Scientists have tried for centuries to piece together the process by which plants make food. Although some of the steps are still not completely understood, the overall photosynthetic equation has been known since the 1800s: In the presence of light, the green parts of plants produce organic compounds and oxygen from carbon dioxide and water. Using molecular formulas, we can summarize the complex series of chemical reactions in photosynthesis with this chemical equation:

$$6\ CO_2 + 12\ H_2O + \text{Light energy} \rightarrow C_6H_{12}O_6 + 6\ O_2 + 6\ H_2O$$

We use glucose ($C_6H_{12}O_6$) here to simplify the relationship between photosynthesis and respiration, but the direct product of photosynthesis is actually a three-carbon sugar that can be used to make glucose. Water appears on both sides of the equation because 12 molecules are consumed and 6 molecules are newly formed during photosynthesis. We can simplify the equation by indicating only the net consumption of water:

$$6\ CO_2 + 6\ H_2O + \text{Light energy} \rightarrow C_6H_{12}O_6 + 6\ O_2$$

Writing the equation in this form, we can see that the overall chemical change during photosynthesis is the reverse of the one that occurs during cellular respiration. Both of these metabolic processes occur in plant cells. However, as you will soon learn, chloroplasts do not synthesize sugars by simply reversing the steps of respiration.

Now let's divide the photosynthetic equation by 6 to put it in its simplest possible form:

$$CO_2 + H_2O \rightarrow [CH_2O] + O_2$$

Here, the brackets indicate that CH_2O is not an actual sugar but represents the general formula for a carbohydrate. In other words, we are imagining the synthesis of a sugar molecule one carbon at a time. Six repetitions would theoretically produce a glucose molecule. Let's now use this simplified formula to see how researchers tracked the elements C, H, and O from the reactants of photosynthesis to the products.

The Splitting of Water

One of the first clues to the mechanism of photosynthesis came from the discovery that the O_2 given off by plants is derived from H_2O and not from CO_2. The chloroplast splits water into hydrogen and oxygen. Before this discovery, the prevailing hypothesis was that photosynthesis split carbon dioxide ($CO_2 \rightarrow C + O_2$) and then added water to the carbon ($C + H_2O \rightarrow [CH_2O]$). This hypothesis predicted that the O_2 released during photosynthesis came from CO_2. This idea was challenged in the 1930s by C. B. van Niel, of Stanford University. Van Niel was investigating photosynthesis in bacteria that make their carbohydrate from CO_2 but do not release O_2. He concluded that, at least in these bacteria, CO_2 is not split into carbon and oxygen. One group of bacteria used hydrogen sulfide (H_2S) rather than water for photosynthesis, forming yellow globules of sulfur as a waste product (these globules are visible in Figure 10.2e). Here is the chemical equation for photosynthesis in these sulfur bacteria:

$$CO_2 + 2\ H_2S \rightarrow [CH_2O] + H_2O + 2\ S$$

Van Niel reasoned that the bacteria split H_2S and used the hydrogen atoms to make sugar. He then generalized that idea, proposing that all photosynthetic organisms require a hydrogen source but that the source varies:

$$\begin{aligned} \text{Sulfur bacteria:}\quad & CO_2 + 2\ H_2S \rightarrow [CH_2O] + H_2O + 2\ S \\ \text{Plants:}\quad & CO_2 + 2\ H_2O \rightarrow [CH_2O] + H_2O + O_2 \\ \text{General:}\quad & CO_2 + 2\ H_2X \rightarrow [CH_2O] + H_2O + 2\ X \end{aligned}$$

Thus, van Niel hypothesized that plants split H_2O as a source of electrons from hydrogen atoms, releasing O_2 as a by-product.

Nearly 20 years later, scientists confirmed van Niel's hypothesis by using oxygen-18 (^{18}O), a heavy isotope, as a tracer to follow the fate of oxygen atoms during photosynthesis. The experiments showed that the O_2 from plants was labeled with ^{18}O *only* if water was the source of the tracer (experiment 1). If the ^{18}O was introduced to the plant in the form of CO_2, the label did not turn up in the released O_2 (experiment 2). In the following summary, red denotes labeled atoms of oxygen (^{18}O):

$$\begin{aligned} \text{Experiment 1:}\ & CO_2 + 2\ H_2O \rightarrow [CH_2O] + H_2O + O_2 \\ \text{Experiment 2:}\ & CO_2 + 2\ H_2O \rightarrow [CH_2O] + H_2O + O_2 \end{aligned}$$

A significant result of the shuffling of atoms during photosynthesis is the extraction of hydrogen from water and its incorporation into sugar. The waste product of photosynthesis, O_2, is released to the atmosphere. **Figure 10.5** shows the fates of all atoms in photosynthesis.

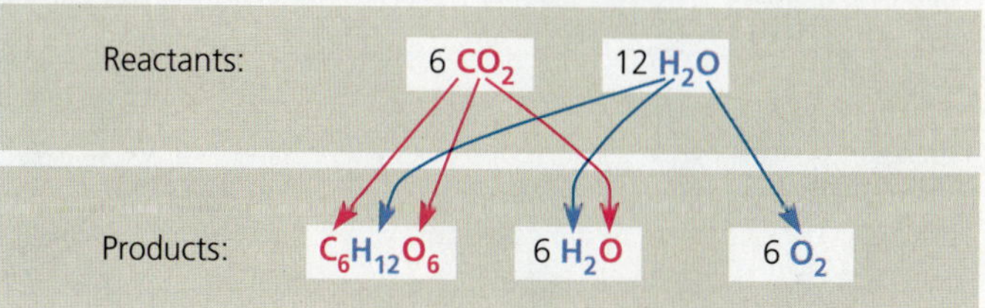

▲ **Figure 10.5 Tracking atoms through photosynthesis.** The atoms from CO_2 are shown in magenta, and the atoms from H_2O are shown in blue.

Photosynthesis as a Redox Process

Let's briefly compare photosynthesis with cellular respiration. Both processes involve redox reactions. During cellular respiration, energy is released from sugar when electrons associated with hydrogen are transported by carriers to oxygen, forming water as a by-product (see p. 164). The electrons lose potential energy as they "fall" down the electron transport chain toward electronegative oxygen, and the mitochondrion harnesses that energy to synthesize ATP (see Figure 9.15). Photosynthesis reverses the direction of electron flow. Water is split, and electrons are transferred along with hydrogen ions from the water to carbon dioxide, reducing it to sugar.

$$\text{Energy} + 6\,CO_2 + 6\,H_2O \longrightarrow C_6H_{12}O_6 + 6\,O_2$$

(CO_2 becomes reduced; H_2O becomes oxidized)

Because the electrons increase in potential energy as they move from water to sugar, this process requires energy—in other words is endergonic. This energy boost is provided by light.

The Two Stages of Photosynthesis: *A Preview*

The equation for photosynthesis is a deceptively simple summary of a very complex process. Actually, photosynthesis is not a single process, but two processes, each with multiple steps. These two stages of photosynthesis are known as the **light reactions** (the *photo* part of photosynthesis) and the **Calvin cycle** (the *synthesis* part) **(Figure 10.6)**.

The light reactions are the steps of photosynthesis that convert solar energy to chemical energy. Water is split, providing a source of electrons and protons (hydrogen ions, H^+) and giving off O_2 as a by-product. Light absorbed by chlorophyll drives a transfer of the electrons and hydrogen ions from water to an acceptor called **NADP**$^+$ (nicotinamide adenine dinucleotide phosphate), where they are temporarily stored. The electron acceptor NADP$^+$ is first cousin to NAD$^+$, which functions as an electron carrier in cellular respiration; the two molecules differ only by the presence of an extra phosphate group in the NADP$^+$ molecule. The light reactions use solar power to reduce NADP$^+$ to NADPH by adding a pair of electrons along with an H^+. The light reactions also generate ATP, using chemiosmosis to power the addition of a phosphate group to ADP, a process called **photophosphorylation**. Thus, light energy is initially converted to chemical energy in the form of two compounds: NADPH, a source of electrons as "reducing power" that can be passed along to an electron acceptor, reducing it, and ATP, the versatile energy currency of cells. Notice that the light reactions produce no sugar; that happens in the second stage of photosynthesis, the Calvin cycle.

The Calvin cycle is named for Melvin Calvin, who, along with his colleagues, began to elucidate its steps in the late 1940s. The cycle begins by incorporating CO_2 from the air

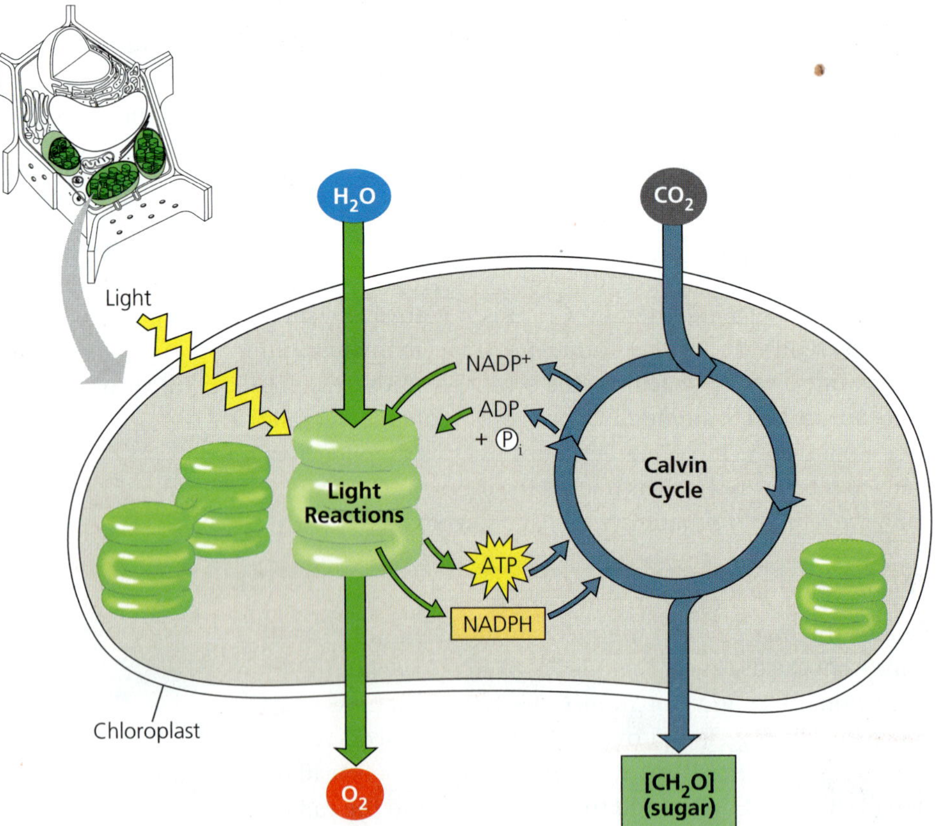

▶ **Figure 10.6 An overview of photosynthesis: cooperation of the light reactions and the Calvin cycle.** In the chloroplast, the thylakoid membranes are the sites of the light reactions, whereas the Calvin cycle occurs in the stroma. The light reactions use solar energy to make ATP and NADPH, which supply chemical energy and reducing power, respectively, to the Calvin cycle. The Calvin cycle incorporates CO_2 into organic molecules, which are converted to sugar. (Recall that most simple sugars have formulas that are some multiple of CH_2O.)

BioFlix Visit the Study Area at **www.masteringbiology.com** for the BioFlix® 3-D Animation on Photosynthesis.

into organic molecules already present in the chloroplast. This initial incorporation of carbon into organic compounds is known as **carbon fixation**. The Calvin cycle then reduces the fixed carbon to carbohydrate by the addition of electrons. The reducing power is provided by NADPH, which acquired its cargo of electrons in the light reactions. To convert CO_2 to carbohydrate, the Calvin cycle also requires chemical energy in the form of ATP, which is also generated by the light reactions. Thus, it is the Calvin cycle that makes sugar, but it can do so only with the help of the NADPH and ATP produced by the light reactions. The metabolic steps of the Calvin cycle are sometimes referred to as the dark reactions, or light-independent reactions, because none of the steps requires light *directly*. Nevertheless, the Calvin cycle in most plants occurs during daylight, for only then can the light reactions provide the NADPH and ATP that the Calvin cycle requires. In essence, the chloroplast uses light energy to make sugar by coordinating the two stages of photosynthesis.

As Figure 10.6 indicates, the thylakoids of the chloroplast are the sites of the light reactions, while the Calvin cycle occurs in the stroma. On the outside of the thylakoids, molecules of $NADP^+$ and ADP pick up electrons and phosphate, respectively, and NADPH and ATP are then released to the stroma, where they play crucial roles in the Calvin cycle. The two stages of photosynthesis are treated in this figure as metabolic modules that take in ingredients and crank out products. In the next two sections, we'll look more closely at how the two stages work, beginning with the light reactions.

CONCEPT CHECK 10.1

1. How do the reactant molecules of photosynthesis reach the chloroplasts in leaves?
2. How did the use of an oxygen isotope help elucidate the chemistry of photosynthesis?
3. **WHAT IF?** The Calvin cycle requires ATP and NADPH, products of the light reactions. If a classmate asserted that the light reactions don't depend on the Calvin cycle and, with continual light, could just keep on producing ATP and NADPH, how would you respond?

For suggested answers, see Appendix A.

CONCEPT **10.2**

The light reactions convert solar energy to the chemical energy of ATP and NADPH

Chloroplasts are chemical factories powered by the sun. Their thylakoids transform light energy into the chemical energy of ATP and NADPH. To understand this conversion better, we need to know about some important properties of light.

The Nature of Sunlight

Light is a form of energy known as electromagnetic energy, also called electromagnetic radiation. Electromagnetic energy travels in rhythmic waves analogous to those created by dropping a pebble into a pond. Electromagnetic waves, however, are disturbances of electric and magnetic fields rather than disturbances of a material medium such as water.

The distance between the crests of electromagnetic waves is called the **wavelength**. Wavelengths range from less than a nanometer (for gamma rays) to more than a kilometer (for radio waves). This entire range of radiation is known as the **electromagnetic spectrum** (**Figure 10.7**). The segment most important to life is the narrow band from about 380 nm to 750 nm in wavelength. This radiation is known as **visible light** because it can be detected as various colors by the human eye.

The model of light as waves explains many of light's properties, but in certain respects light behaves as though it consists of discrete particles, called **photons**. Photons are not tangible objects, but they act like objects in that each of them has a fixed quantity of energy. The amount of energy is inversely related to the wavelength of the light: the shorter the wavelength, the greater the energy of each photon of that light. Thus, a photon of violet light packs nearly twice as much energy as a photon of red light.

Although the sun radiates the full spectrum of electromagnetic energy, the atmosphere acts like a selective window, allowing visible light to pass through while screening out a substantial fraction of other radiation. The part of the spectrum we can see—visible light—is also the radiation that drives photosynthesis.

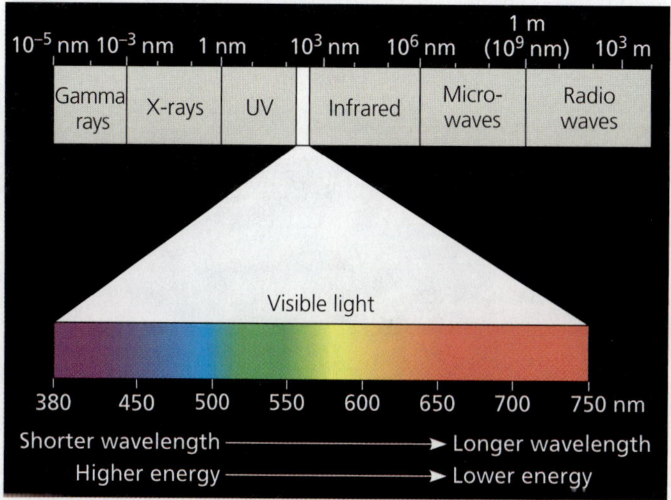

▲ **Figure 10.7 The electromagnetic spectrum.** White light is a mixture of all wavelengths of visible light. A prism can sort white light into its component colors by bending light of different wavelengths at different angles. (Droplets of water in the atmosphere can act as prisms, forming a rainbow; see Figure 10.1.) Visible light drives photosynthesis.

Photosynthetic Pigments: The Light Receptors

When light meets matter, it may be reflected, transmitted, or absorbed. Substances that absorb visible light are known as *pigments*. Different pigments absorb light of different wavelengths, and the wavelengths that are absorbed disappear. If a pigment is illuminated with white light, the color we see is the color most reflected or transmitted by the pigment. (If a pigment absorbs all wavelengths, it appears black.) We see green when we look at a leaf because chlorophyll absorbs violet-blue and red light while transmitting and reflecting green light (**Figure 10.8**). The ability of a pigment to absorb various wavelengths of light can be measured with an instrument called a **spectrophotometer**. This machine directs beams of light of different wavelengths through a solution of the pigment and measures the fraction of the light transmitted at each wavelength. A graph plotting a pigment's light absorption versus wavelength is called an **absorption spectrum** (**Figure 10.9**).

The absorption spectra of chloroplast pigments provide clues to the relative effectiveness of different wavelengths for driving photosynthesis, since light can perform work in chloroplasts only if it is absorbed. **Figure 10.10a** shows the absorption spectra of three types of pigments in chloroplasts: **chlorophyll *a***, which participates directly in the light reactions; the accessory pigment *chlorophyll b*; and a group of accessory pigments called carotenoids. The spectrum of chlorophyll *a* suggests that violet-blue and red light work best for photosynthesis, since they are absorbed, while green is the least effective

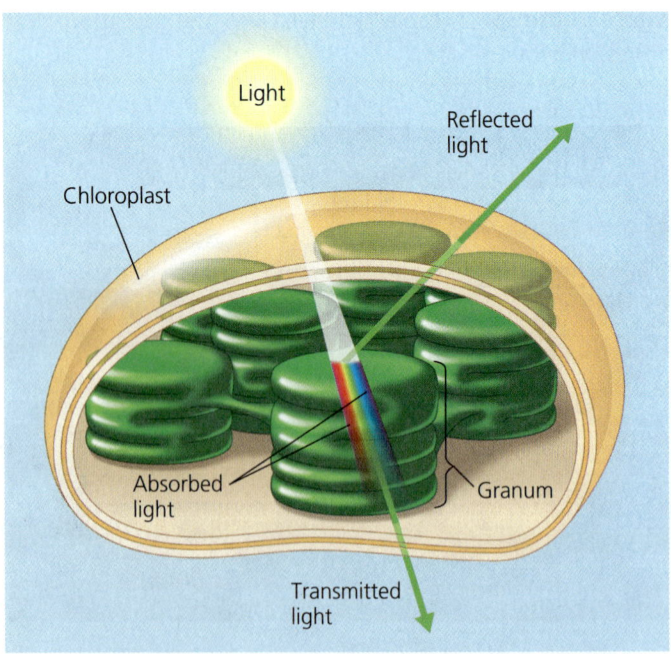

▲ **Figure 10.8 Why leaves are green: interaction of light with chloroplasts.** The chlorophyll molecules of chloroplasts absorb violet-blue and red light (the colors most effective in driving photosynthesis) and reflect or transmit green light. This is why leaves appear green.

▼ **Figure 10.9** **RESEARCH METHOD**

Determining an Absorption Spectrum

APPLICATION An absorption spectrum is a visual representation of how well a particular pigment absorbs different wavelengths of visible light. Absorption spectra of various chloroplast pigments help scientists decipher each pigment's role in a plant.

TECHNIQUE A spectrophotometer measures the relative amounts of light of different wavelengths absorbed and transmitted by a pigment solution.

1. White light is separated into colors (wavelengths) by a prism.
2. One by one, the different colors of light are passed through the sample (chlorophyll in this example). Green light and blue light are shown here.
3. The transmitted light strikes a photoelectric tube, which converts the light energy to electricity.
4. The electric current is measured by a galvanometer. The meter indicates the fraction of light transmitted through the sample, from which we can determine the amount of light absorbed.

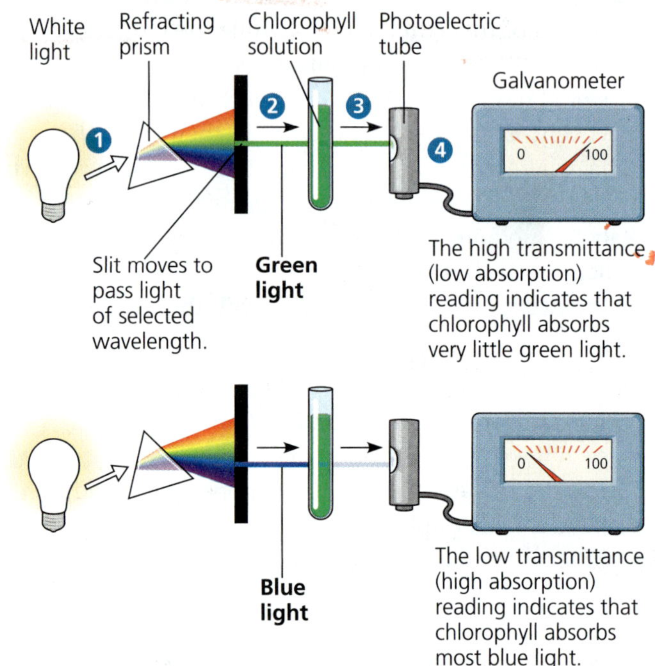

RESULTS See Figure 10.10a for absorption spectra of three types of chloroplast pigments.

color. This is confirmed by an **action spectrum** for photosynthesis (**Figure 10.10b**), which profiles the relative effectiveness of different wavelengths of radiation in driving the process. An action spectrum is prepared by illuminating chloroplasts with light of different colors and then plotting wavelength against some measure of photosynthetic rate, such as CO_2 consumption or O_2 release. The action spectrum for photosynthesis was first demonstrated by Theodor W. Engelmann, a German botanist, in 1883. Before equipment for measuring O_2 levels had even been invented, Engelmann performed a

▼ **Figure 10.10** INQUIRY

Which wavelengths of light are most effective in driving photosynthesis?

EXPERIMENT Absorption and action spectra, along with a classic experiment by Theodor W. Engelmann, reveal which wavelengths of light are photosynthetically important.

RESULTS

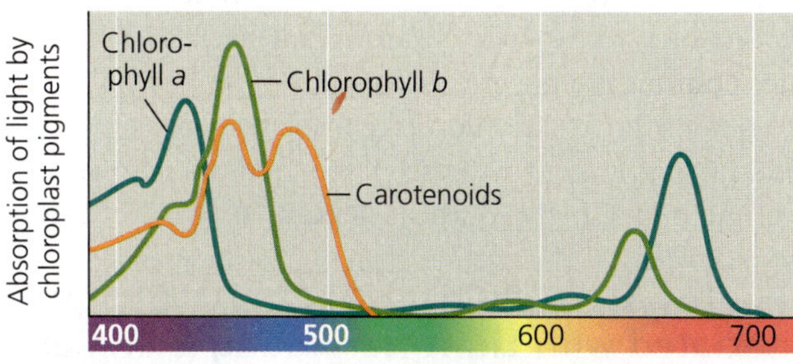

(a) Absorption spectra. The three curves show the wavelengths of light best absorbed by three types of chloroplast pigments.

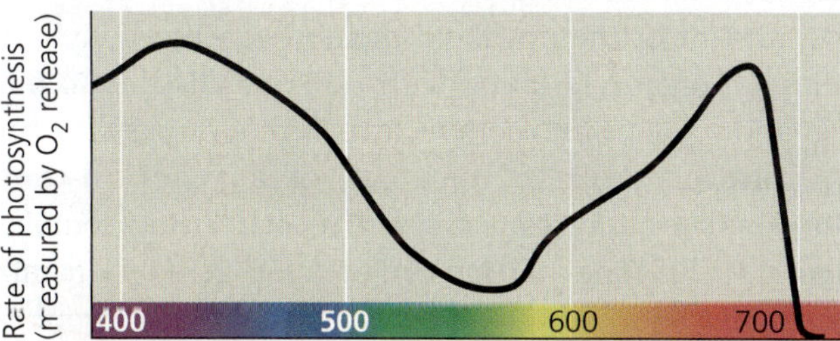

(b) Action spectrum. This graph plots the rate of photosynthesis versus wavelength. The resulting action spectrum resembles the absorption spectrum for chlorophyll *a* but does not match exactly (see part a). This is partly due to the absorption of light by accessory pigments such as chlorophyll *b* and carotenoids.

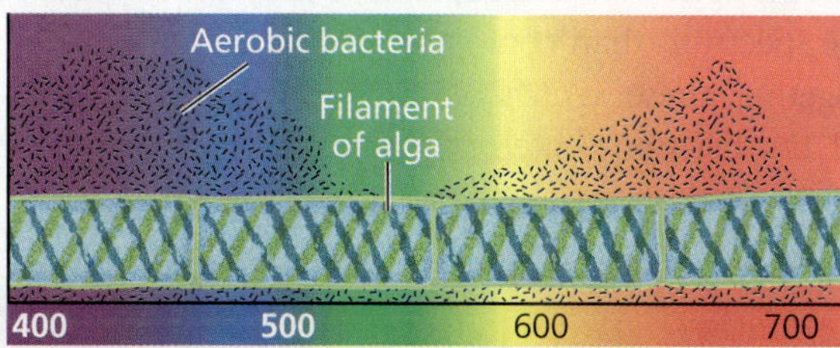

(c) Engelmann's experiment. In 1883, Theodor W. Engelmann illuminated a filamentous alga with light that had been passed through a prism, exposing different segments of the alga to different wavelengths. He used aerobic bacteria, which concentrate near an oxygen source, to determine which segments of the alga were releasing the most O_2 and thus photosynthesizing most. Bacteria congregated in greatest numbers around the parts of the alga illuminated with violet-blue or red light.

CONCLUSION Light in the violet-blue and red portions of the spectrum is most effective in driving photosynthesis.

SOURCE T. W. Engelmann, *Bacterium photometricum.* Ein Betraq zur vergleichenden Physiologie des Licht- und farbensinnes, *Archiv. für Physiologie* 30:95–124 (1883).

See the related Experimental Inquiry Tutorial in MasteringBiology.

WHAT IF? If Engelmann had used a filter that allowed only red light to pass through, how would the results have differed?

clever experiment in which he used bacteria to measure rates of photosynthesis in filamentous algae (**Figure 10.10c**). His results are a striking match to the modern action spectrum shown in Figure 10.10b.

Notice by comparing Figures 10.10a and 10.10b that the action spectrum for photosynthesis does not exactly match the absorption spectrum of chlorophyll *a*. The absorption spectrum of chlorophyll *a* alone underestimates the effectiveness of certain wavelengths in driving photosynthesis. This is partly because accessory pigments with different absorption spectra are also photosynthetically important in chloroplasts and broaden the spectrum of colors that can be used for photosynthesis. **Figure 10.11** shows the structure of chlorophyll *a* compared with that of ***chlorophyll b***. A slight structural difference between them is enough to cause the two pigments to absorb at slightly different wavelengths in the red and blue parts of the spectrum (see Figure 10.10a). As a result, chlorophyll *a* is blue green and chlorophyll *b* is olive green.

Other accessory pigments include **carotenoids**, hydrocarbons that are various shades of yellow and orange because they absorb violet and blue-green light (see Figure 10.10a). Carotenoids may broaden the spectrum of colors that can drive photosynthesis. However, a more important function of at least some carotenoids seems to be *photoprotection*: These

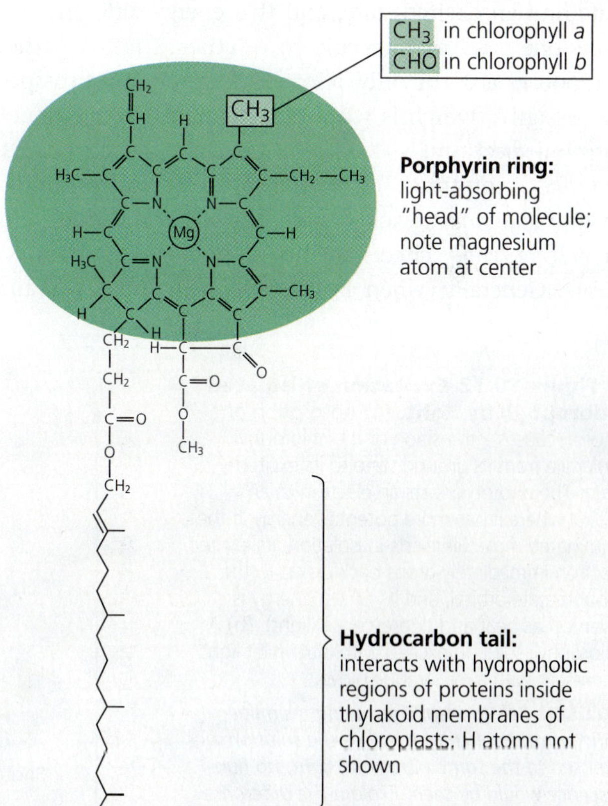

▲ **Figure 10.11 Structure of chlorophyll molecules in chloroplasts of plants.** Chlorophyll *a* and chlorophyll *b* differ only in one of the functional groups bonded to the porphyrin ring. (Also see the space-filling model of chlorophyll in Figure 1.4, p. 5.)

compounds absorb and dissipate excessive light energy that would otherwise damage chlorophyll or interact with oxygen, forming reactive oxidative molecules that are dangerous to the cell. Interestingly, carotenoids similar to the photoprotective ones in chloroplasts have a photoprotective role in the human eye. These and related molecules, often found in health food products, are valued as "phytochemicals" (from the Greek *phyton*, plant), compounds with antioxidant properties. Plants can synthesize all the antioxidants they require, but humans and other animals must obtain some of them from their diets.

Excitation of Chlorophyll by Light

What exactly happens when chlorophyll and other pigments absorb light? The colors corresponding to the absorbed wavelengths disappear from the spectrum of the transmitted and reflected light, but energy cannot disappear. When a molecule absorbs a photon of light, one of the molecule's electrons is elevated to an orbital where it has more potential energy. When the electron is in its normal orbital, the pigment molecule is said to be in its ground state. Absorption of a photon boosts an electron to an orbital of higher energy, and the pigment molecule is then said to be in an excited state. The only photons absorbed are those whose energy is exactly equal to the energy difference between the ground state and an excited state, and this energy difference varies from one kind of molecule to another. Thus, a particular compound absorbs only photons corresponding to specific wavelengths, which is why each pigment has a unique absorption spectrum.

Once absorption of a photon raises an electron from the ground state to an excited state, the electron cannot remain there long. The excited state, like all high-energy states, is unstable. Generally, when isolated pigment molecules absorb light, their excited electrons drop back down to the ground-state orbital in a billionth of a second, releasing their excess energy as heat. This conversion of light energy to heat is what makes the top of an automobile so hot on a sunny day. (White cars are coolest because their paint reflects all wavelengths of visible light, although it may absorb ultraviolet and other invisible radiation.) In isolation, some pigments, including chlorophyll, emit light as well as heat after absorbing photons. As excited electrons fall back to the ground state, photons are given off. This afterglow is called fluorescence. If a solution of chlorophyll isolated from chloroplasts is illuminated, it will fluoresce in the red-orange part of the spectrum and also give off heat **(Figure 10.12)**.

A Photosystem: A Reaction-Center Complex Associated with Light-Harvesting Complexes

Chlorophyll molecules excited by the absorption of light energy produce very different results in an intact chloroplast than they do in isolation (see Figure 10.12). In their native environment of the thylakoid membrane, chlorophyll molecules are organized along with other small organic molecules and proteins into complexes called photosystems.

A **photosystem** is composed of a **reaction-center complex** surrounded by several light-harvesting complexes **(Figure 10.13)**. The reaction-center complex is an organized association of proteins holding a special pair of chlorophyll *a* molecules. Each **light-harvesting complex** consists of various pigment molecules (which may include chlorophyll *a*, chlorophyll *b*, and carotenoids) bound to proteins. The number and variety of pigment molecules enable a photosystem to harvest light over a larger surface area and a larger portion of the spectrum than could any single pigment molecule alone. Together, these light-harvesting complexes act as an antenna for the reaction-center complex. When a pigment molecule

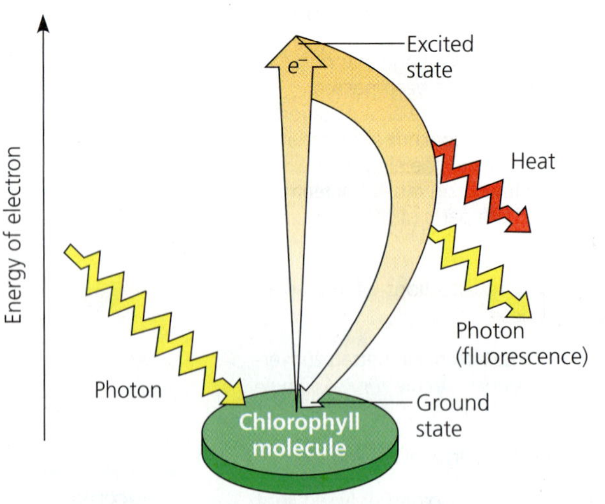

(a) Excitation of isolated chlorophyll molecule

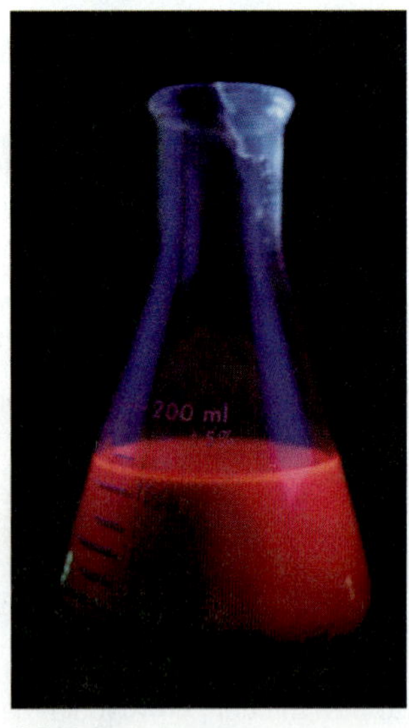

(b) Fluorescence

▶ **Figure 10.12 Excitation of isolated chlorophyll by light. (a)** Absorption of a photon causes a transition of the chlorophyll molecule from its ground state to its excited state. The photon boosts an electron to an orbital where it has more potential energy. If the illuminated molecule exists in isolation, its excited electron immediately drops back down to the ground-state orbital, and its excess energy is given off as heat and fluorescence (light). **(b)** A chlorophyll solution excited with ultraviolet light fluoresces with a red-orange glow.

WHAT IF? *If a leaf containing a similar concentration of chlorophyll as the solution was exposed to the same ultraviolet light, no fluorescence would be seen. Explain the difference in fluorescence emission between the solution and the leaf.*

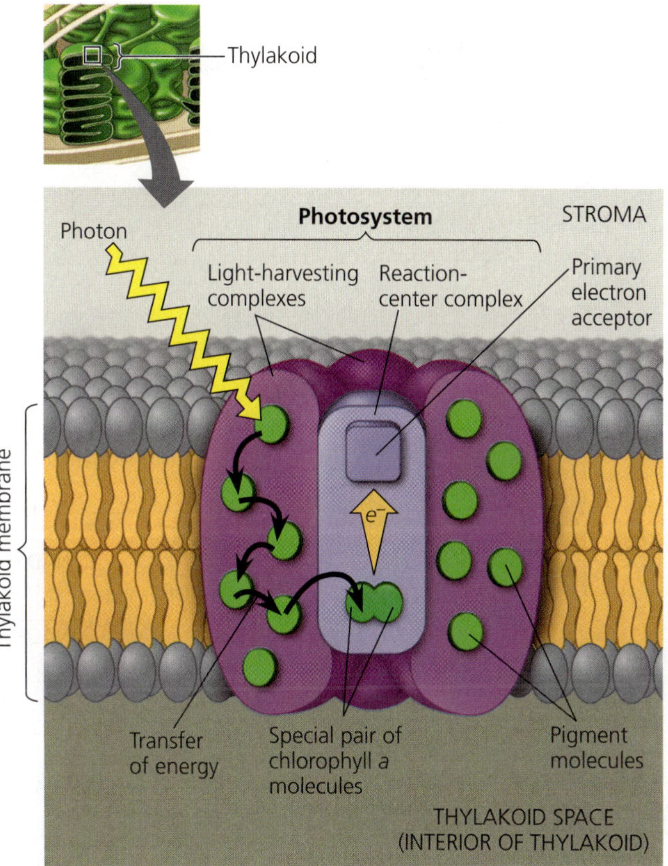

(a) How a photosystem harvests light. When a photon strikes a pigment molecule in a light-harvesting complex, the energy is passed from molecule to molecule until it reaches the reaction-center complex. Here, an excited electron from the special pair of chlorophyll *a* molecules is transferred to the primary electron acceptor.

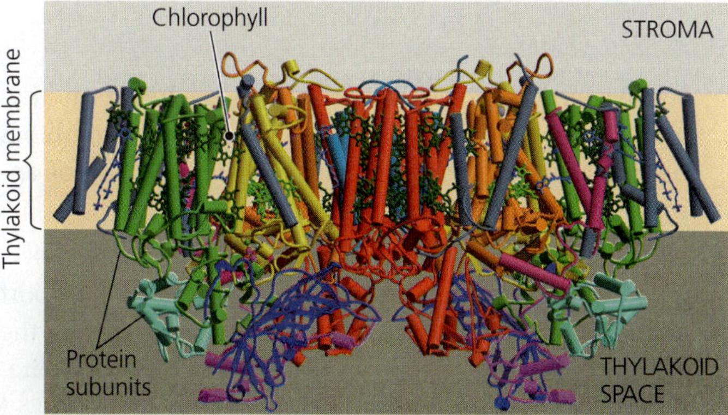

(b) Structure of photosystem II. This computer model of photosystem II, based on X-ray crystallography, shows two photosystem complexes side by side. Chlorophyll molecules (small green ball-and-stick models) are interspersed with protein subunits (cylinders and ribbons). For simplicity, photosystem II will be shown as a single complex in the rest of the chapter.

▲ **Figure 10.13 The structure and function of a photosystem.**

absorbs a photon, the energy is transferred from pigment molecule to pigment molecule within a light-harvesting complex, somewhat like a human "wave" at a sports arena, until it is passed into the reaction-center complex. The reaction-center complex also contains a molecule capable of accepting electrons and becoming reduced; this is called the **primary electron acceptor**. The pair of chlorophyll *a* molecules in the reaction-center complex are special because their molecular environment—their location and the other molecules with which they are associated—enables them to use the energy from light not only to boost one of their electrons to a higher energy level, but also to transfer it to a different molecule—the primary electron acceptor.

The solar-powered transfer of an electron from the reaction-center chlorophyll *a* pair to the primary electron acceptor is the first step of the light reactions. As soon as the chlorophyll electron is excited to a higher energy level, the primary electron acceptor captures it; this is a redox reaction. In the flask shown in Figure 10.12, isolated chlorophyll fluoresces because there is no electron acceptor, so electrons of photoexcited chlorophyll drop right back to the ground state. In the structured environment of a chloroplast, however, an electron acceptor is readily available, and the potential energy represented by the excited electron is not dissipated as light and heat. Thus, each photosystem—a reaction-center complex surrounded by light-harvesting complexes—functions in the chloroplast as a unit. It converts light energy to chemical energy, which will ultimately be used for the synthesis of sugar.

The thylakoid membrane is populated by two types of photosystems that cooperate in the light reactions of photosynthesis. They are called **photosystem II (PS II)** and **photosystem I (PS I)**. (They were named in order of their discovery, but photosystem II functions first in the light reactions.) Each has a characteristic reaction-center complex—a particular kind of primary electron acceptor next to a special pair of chlorophyll *a* molecules associated with specific proteins. The reaction-center chlorophyll *a* of photosystem II is known as P680 because this pigment is best at absorbing light having a wavelength of 680 nm (in the red part of the spectrum). The chlorophyll *a* at the reaction-center complex of photosystem I is called P700 because it most effectively absorbs light of wavelength 700 nm (in the far-red part of the spectrum). These two pigments, P680 and P700, are nearly identical chlorophyll *a* molecules. However, their association with different proteins in the thylakoid membrane affects the electron distribution in the two pigments and accounts for the slight differences in their light-absorbing properties. Now let's see how the two photosystems work together in using light energy to generate ATP and NADPH, the two main products of the light reactions.

Linear Electron Flow

Light drives the synthesis of ATP and NADPH by energizing the two photosystems embedded in the thylakoid membranes of chloroplasts. The key to this energy transformation is a flow of electrons through the photosystems and other molecular components built into the thylakoid membrane. This is called

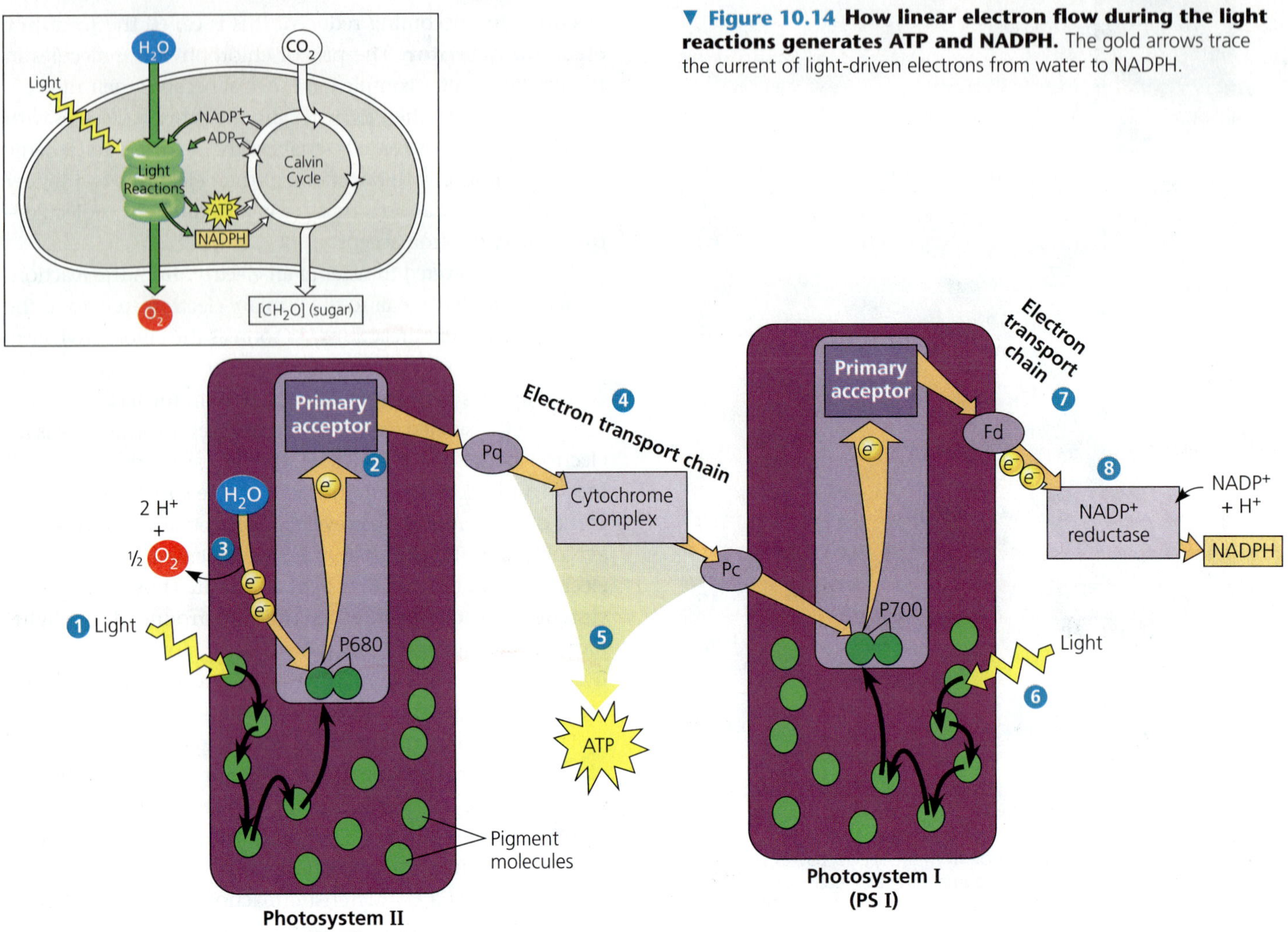

▼ **Figure 10.14 How linear electron flow during the light reactions generates ATP and NADPH.** The gold arrows trace the current of light-driven electrons from water to NADPH.

linear electron flow, and it occurs during the light reactions of photosynthesis, as shown in **Figure 10.14**. The following steps correspond to the numbered steps in the figure.

1. A photon of light strikes a pigment molecule in a light-harvesting complex of PS II, boosting one of its electrons to a higher energy level. As this electron falls back to its ground state, an electron in a nearby pigment molecule is simultaneously raised to an excited state. The process continues, with the energy being relayed to other pigment molecules until it reaches the P680 pair of chlorophyll *a* molecules in the PS II reaction-center complex. It excites an electron in this pair of chlorophylls to a higher energy state.
2. This electron is transferred from the excited P680 to the primary electron acceptor. We can refer to the resulting form of P680, missing an electron, as $P680^+$.
3. An enzyme catalyzes the splitting of a water molecule into two electrons, two hydrogen ions (H^+), and an oxygen atom. The electrons are supplied one by one to the $P680^+$ pair, each electron replacing one transferred to the primary electron acceptor. ($P680^+$ is the strongest biological oxidizing agent known; its electron "hole" must be filled. This greatly facilitates the transfer of electrons from the split water molecule.) The H^+ are released into the thylakoid lumen. The oxygen atom immediately combines with an oxygen atom generated by the splitting of another water molecule, forming O_2.
4. Each photoexcited electron passes from the primary electron acceptor of PS II to PS I via an electron transport chain, the components of which are similar to those of the electron transport chain that functions in cellular respiration. The electron transport chain between PS II and PS I is made up of the electron carrier plastoquinone (Pq), a cytochrome complex, and a protein called plastocyanin (Pc).
5. The exergonic "fall" of electrons to a lower energy level provides energy for the synthesis of ATP. As electrons pass through the cytochrome complex, H^+ are pumped into the thylakoid lumen, contributing to the proton gradient that is subsequently used in chemiosmosis.
6. Meanwhile, light energy has been transferred via light-harvesting complex pigments to the PS I reaction-center complex, exciting an electron of the P700 pair of chlorophyll *a* molecules located there. The photoexcited electron was then transferred to PS I's primary electron acceptor, creating an electron "hole" in the P700—which

we now can call P700$^+$. In other words, P700$^+$ can now act as an electron acceptor, accepting an electron that reaches the bottom of the electron transport chain from PS II.

7. Photoexcited electrons are passed in a series of redox reactions from the primary electron acceptor of PS I down a second electron transport chain through the protein ferredoxin (Fd). (This chain does not create a proton gradient and thus does not produce ATP.)

8. The enzyme NADP$^+$ reductase catalyzes the transfer of electrons from Fd to NADP$^+$. Two electrons are required for its reduction to NADPH. This molecule is at a higher energy level than water, and its electrons are more readily available for the reactions of the Calvin cycle than were those of water. This process also removes an H$^+$ from the stroma.

As complicated as the scheme shown in Figure 10.14 is, do not lose track of its functions. The light reactions use solar power to generate ATP and NADPH, which provide chemical energy and reducing power, respectively, to the carbohydrate-synthesizing reactions of the Calvin cycle. The energy changes of electrons during their linear flow through the light reactions are shown in a mechanical analogy in **Figure 10.15**.

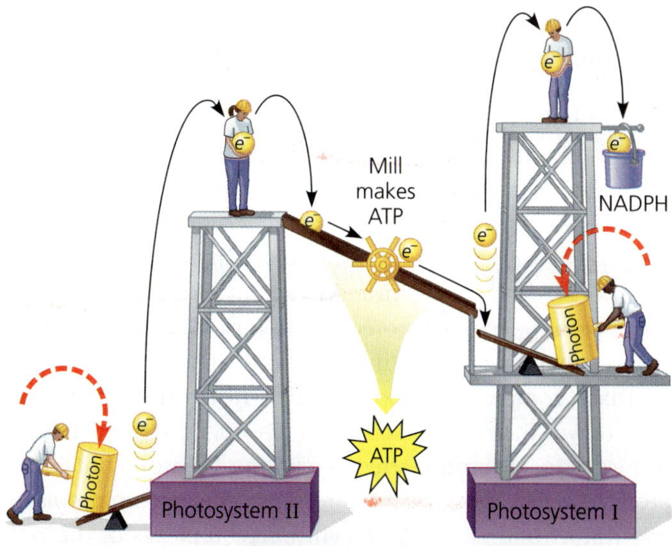

▲ **Figure 10.15 A mechanical analogy for linear electron flow during the light reactions.**

Cyclic Electron Flow

In certain cases, photoexcited electrons can take an alternative path called **cyclic electron flow**, which uses photosystem I but not photosystem II. You can see in **Figure 10.16** that cyclic flow is a short circuit: The electrons cycle back from ferredoxin (Fd) to the cytochrome complex and from there continue on to a P700 chlorophyll in the PS I reaction-center complex. There is no production of NADPH and no release of oxygen. Cyclic flow does, however, generate ATP.

Several of the currently existing groups of photosynthetic bacteria are known to have photosystem I but not photosystem II; for these species, which include the purple sulfur bacteria (see Figure 10.2e), cyclic electron flow is the sole means of generating ATP in photosynthesis. Evolutionary biologists hypothesize that these bacterial groups are descendants of the bacteria in which photosynthesis first evolved, in a form similar to cyclic electron flow.

Cyclic electron flow can also occur in photosynthetic species that possess both photosystems; this includes some prokaryotes, such as the cyanobacteria shown in Figure 10.2d, as well as the eukaryotic photosynthetic species that have been tested so far. Although the process is probably in part an "evolutionary leftover," it clearly plays at least one beneficial role for these organisms. Mutant plants that are not able to carry out cyclic electron flow are capable of growing well in low light, but do not grow well where light is intense. This is evidence for the idea that cyclic electron flow may be photoprotective. Later you'll learn more about cyclic electron flow as it relates to a particular adaptation of photosynthesis (C_4 plants; see Concept 10.4).

Whether ATP synthesis is driven by linear or cyclic electron flow, the actual mechanism is the same. Before we move on to consider the Calvin cycle, let's review chemiosmosis, the process that uses membranes to couple redox reactions to ATP production.

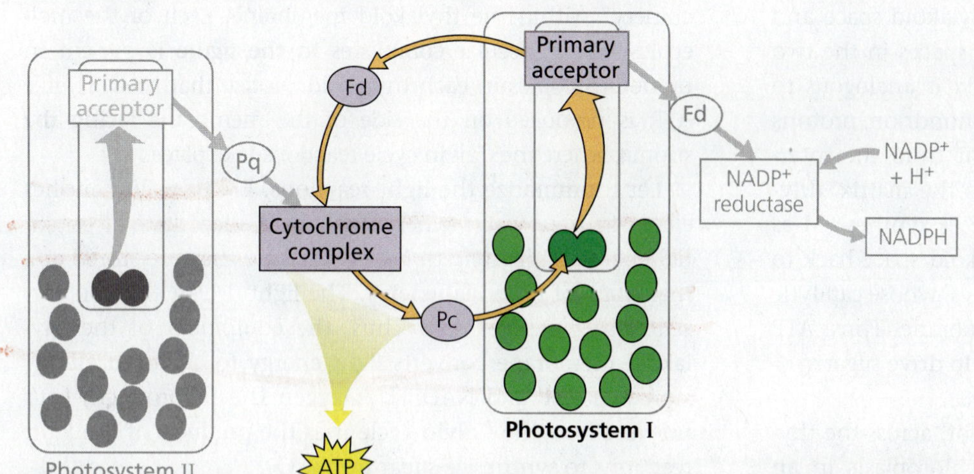

◀ **Figure 10.16 Cyclic electron flow.** Photoexcited electrons from PS I are occasionally shunted back from ferredoxin (Fd) to chlorophyll via the cytochrome complex and plastocyanin (Pc). This electron shunt supplements the supply of ATP (via chemiosmosis) but produces no NADPH. The "shadow" of linear electron flow is included in the diagram for comparison with the cyclic route. The two ferredoxin molecules shown in this diagram are actually one and the same—the final electron carrier in the electron transport chain of PS I.

? *Look at Figure 10.15, and explain how you would alter it to show a mechanical analogy for cyclic electron flow.*

A Comparison of Chemiosmosis in Chloroplasts and Mitochondria

Chloroplasts and mitochondria generate ATP by the same basic mechanism: chemiosmosis. An electron transport chain assembled in a membrane pumps protons across the membrane as electrons are passed through a series of carriers that are progressively more electronegative. In this way, electron transport chains transform redox energy to a proton-motive force, potential energy stored in the form of an H^+ gradient across a membrane. Built into the same membrane is an ATP synthase complex that couples the diffusion of hydrogen ions down their gradient to the phosphorylation of ADP. Some of the electron carriers, including the iron-containing proteins called cytochromes, are very similar in chloroplasts and mitochondria. The ATP synthase complexes of the two organelles are also very much alike. But there are noteworthy differences between oxidative phosphorylation in mitochondria and photophosphorylation in chloroplasts. In mitochondria, the high-energy electrons dropped down the transport chain are extracted from organic molecules (which are thus oxidized), while in chloroplasts, the source of electrons is water. Chloroplasts do not need molecules from food to make ATP; their photosystems capture light energy and use it to drive the electrons from water to the top of the transport chain. In other words, mitochondria use chemiosmosis to transfer chemical energy from food molecules to ATP, whereas chloroplasts transform light energy into chemical energy in ATP.

Although the spatial organization of chemiosmosis differs slightly between chloroplasts and mitochondria, it is easy to see similarities in the two **(Figure 10.17)**. The inner membrane of the mitochondrion pumps protons from the mitochondrial matrix out to the intermembrane space, which then serves as a reservoir of hydrogen ions. The thylakoid membrane of the chloroplast pumps protons from the stroma into the thylakoid space (interior of the thylakoid), which functions as the H^+ reservoir. If you imagine the cristae of mitochondria pinching off from the inner membrane, this may help you see how the thylakoid space and the intermembrane space are comparable spaces in the two organelles, while the mitochondrial matrix is analogous to the stroma of the chloroplast. In the mitochondrion, protons diffuse down their concentration gradient from the intermembrane space through ATP synthase to the matrix, driving ATP synthesis. In the chloroplast, ATP is synthesized as the hydrogen ions diffuse from the thylakoid space back to the stroma through ATP synthase complexes, whose catalytic knobs are on the stroma side of the membrane. Thus, ATP forms in the stroma, where it is used to help drive sugar synthesis during the Calvin cycle **(Figure 10.18)**.

The proton (H^+) gradient, or pH gradient, across the thylakoid membrane is substantial. When chloroplasts in an experimental setting are illuminated, the pH in the thylakoid space drops to about 5 (the H^+ concentration increases), and the pH in the stroma increases to about 8 (the H^+ concentration decreases). This gradient of three pH units corresponds to a thousandfold difference in H^+ concentration. If in the laboratory the lights are turned off, the pH gradient is abolished, but it can quickly be restored by turning the lights back on. Experiments such as this provided strong evidence in support of the chemiosmotic model.

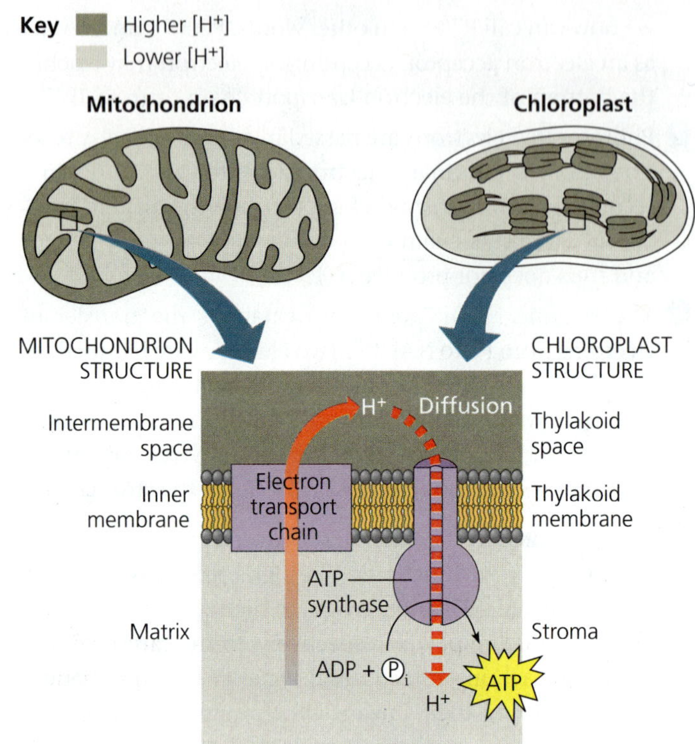

▲ **Figure 10.17 Comparison of chemiosmosis in mitochondria and chloroplasts.** In both kinds of organelles, electron transport chains pump protons (H^+) across a membrane from a region of low H^+ concentration (light gray in this diagram) to one of high H^+ concentration (dark gray). The protons then diffuse back across the membrane through ATP synthase, driving the synthesis of ATP.

Based on studies in several laboratories, Figure 10.18 shows a current model for the organization of the light-reaction "machinery" within the thylakoid membrane. Each of the molecules and molecular complexes in the figure is present in numerous copies in each thylakoid. Notice that NADPH, like ATP, is produced on the side of the membrane facing the stroma, where the Calvin cycle reactions take place.

Let's summarize the light reactions. Electron flow pushes electrons from water, where they are at a low state of potential energy, ultimately to NADPH, where they are stored at a high state of potential energy. The light-driven electron current also generates ATP. Thus, the equipment of the thylakoid membrane converts light energy to chemical energy stored in ATP and NADPH. (Oxygen is a by-product.) Let's now see how the Calvin cycle uses the products of the light reactions to synthesize sugar from CO_2.

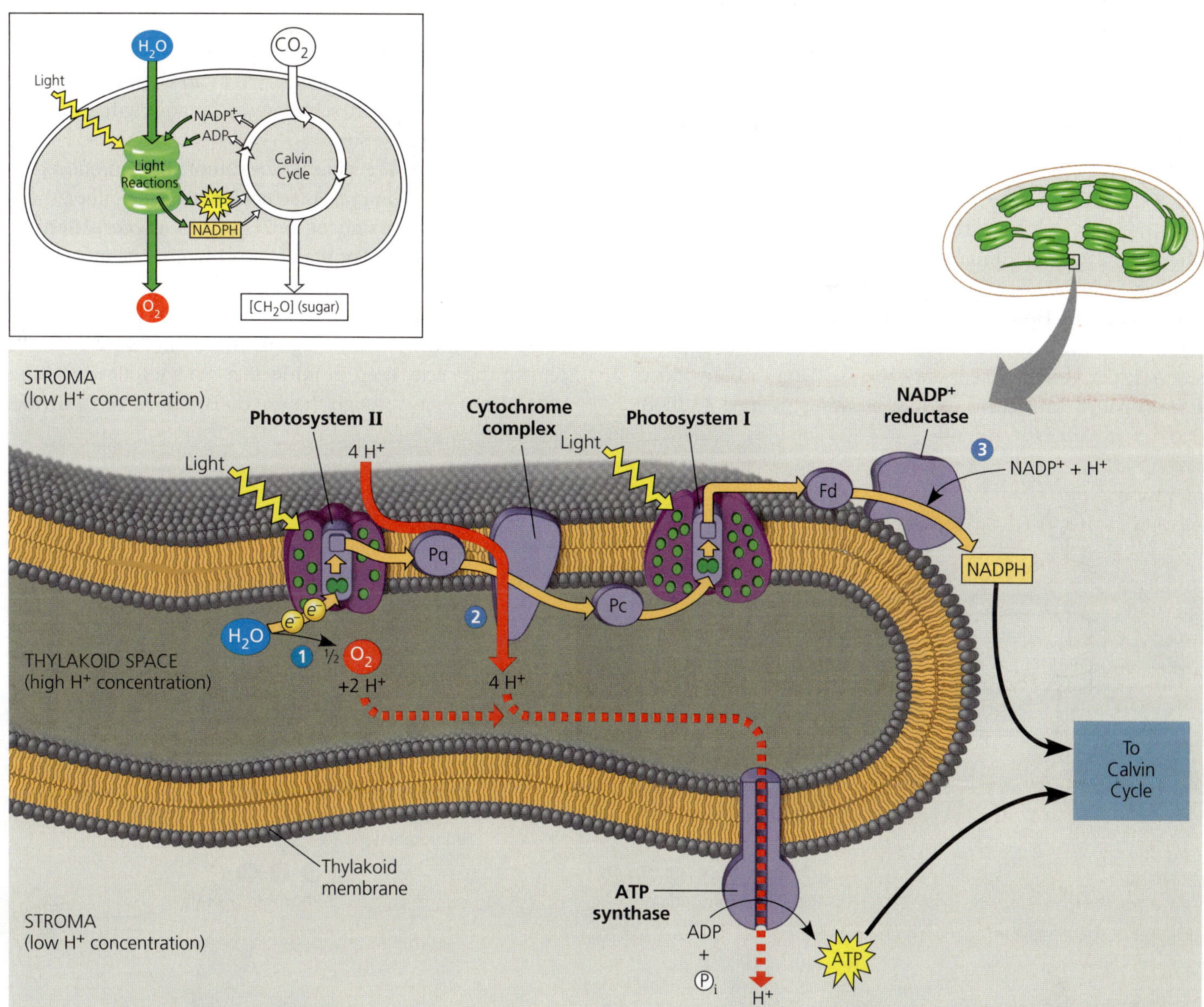

▲ **Figure 10.18 The light reactions and chemiosmosis: the organization of the thylakoid membrane.** This diagram shows a current model for the organization of the thylakoid membrane. The gold arrows track the linear electron flow outlined in Figure 10.14. As electrons pass from carrier to carrier in redox reactions, hydrogen ions removed from the stroma are deposited in the thylakoid space, storing energy as a proton-motive force (H^+ gradient). At least three steps in the light reactions contribute to the proton gradient: 1 Water is split by photosystem II on the side of the membrane facing the thylakoid space; 2 as plastoquinone (Pq), a mobile carrier, transfers electrons to the cytochrome complex, four protons are translocated across the membrane into the thylakoid space; and 3 a hydrogen ion is removed from the stroma when it is taken up by $NADP^+$. Notice that in step 2, hydrogen ions are being pumped from the stroma into the thylakoid space, as in Figure 10.17. The diffusion of H^+ from the thylakoid space back to the stroma (along the H^+ concentration gradient) powers the ATP synthase. These light-driven reactions store chemical energy in NADPH and ATP, which shuttle the energy to the carbohydrate-producing Calvin cycle.

CONCEPT CHECK 10.2

1. What color of light is *least* effective in driving photosynthesis? Explain.
2. Compared to a solution of isolated chlorophyll, why do intact chloroplasts release less heat and fluorescence when illuminated?
3. In the light reactions, what is the initial electron donor? Where do the electrons finally end up?
4. **WHAT IF?** In an experiment, isolated chloroplasts placed in an illuminated solution with the appropriate chemicals can carry out ATP synthesis. Predict what would happen to the rate of synthesis if a compound is added to the solution that makes membranes freely permeable to hydrogen ions.

For suggested answers, see Appendix A.

CONCEPT **10.3**

The Calvin cycle uses the chemical energy of ATP and NADPH to reduce CO_2 to sugar

The Calvin cycle is similar to the citric acid cycle in that a starting material is regenerated after molecules enter and leave the cycle. However, while the citric acid cycle is catabolic, oxidizing acetyl CoA and using the energy to synthesize ATP, the Calvin cycle is anabolic, building carbohydrates from smaller molecules and consuming energy. Carbon enters the Calvin cycle in the form of CO_2 and leaves in the form of sugar. The cycle spends ATP as an energy source and consumes NADPH as reducing power for adding high-energy electrons to make the sugar.

As we mentioned previously, the carbohydrate produced directly from the Calvin cycle is actually not glucose, but a three-carbon sugar; the name of this sugar is **glyceraldehyde 3-phosphate (G3P)**. For the net synthesis of one molecule of G3P, the cycle must take place three times, fixing three molecules of CO_2. (Recall that carbon fixation refers to the initial incorporation of CO_2 into organic material.) As we trace the steps of the cycle, keep in mind that we are following three molecules of CO_2 through the reactions. **Figure 10.19** divides

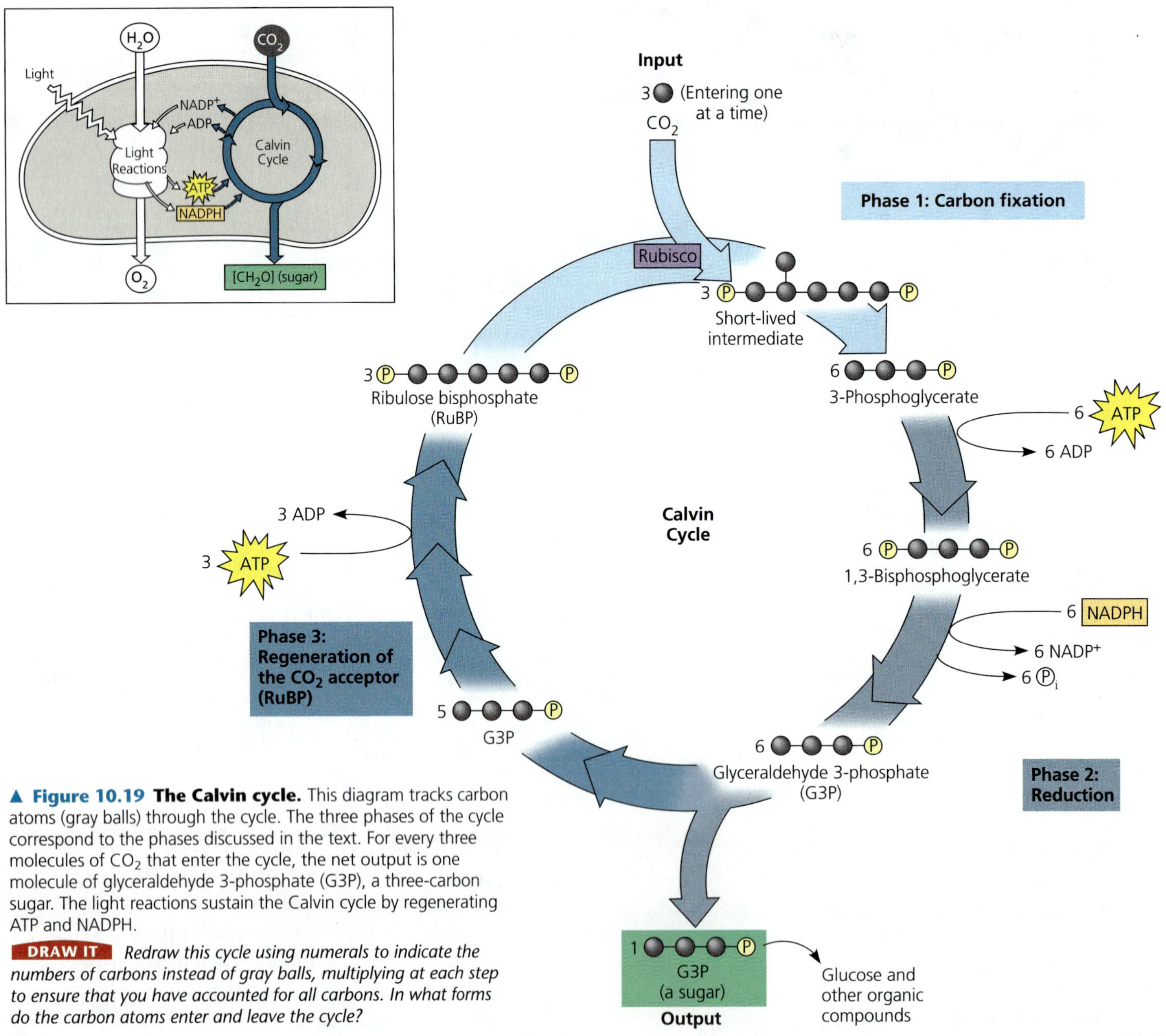

▲ **Figure 10.19 The Calvin cycle.** This diagram tracks carbon atoms (gray balls) through the cycle. The three phases of the cycle correspond to the phases discussed in the text. For every three molecules of CO_2 that enter the cycle, the net output is one molecule of glyceraldehyde 3-phosphate (G3P), a three-carbon sugar. The light reactions sustain the Calvin cycle by regenerating ATP and NADPH.

DRAW IT *Redraw this cycle using numerals to indicate the numbers of carbons instead of gray balls, multiplying at each step to ensure that you have accounted for all carbons. In what forms do the carbon atoms enter and leave the cycle?*

the Calvin cycle into three phases: carbon fixation, reduction, and regeneration of the CO_2 acceptor.

Phase 1: Carbon fixation. The Calvin cycle incorporates each CO_2 molecule, one at a time, by attaching it to a five-carbon sugar named ribulose bisphosphate (abbreviated RuBP). The enzyme that catalyzes this first step is RuBP carboxylase, or **rubisco**. (This is the most abundant protein in chloroplasts and is also thought to be the most abundant protein on Earth.) The product of the reaction is a six-carbon intermediate so unstable that it immediately splits in half, forming two molecules of 3-phosphoglycerate (for each CO_2 fixed).

Phase 2: Reduction. Each molecule of 3-phosphoglycerate receives an additional phosphate group from ATP, becoming 1,3-bisphosphoglycerate. Next, a pair of electrons donated from NADPH reduces 1,3-bisphosphoglycerate, which also loses a phosphate group, becoming G3P. Specifically, the electrons from NADPH reduce a carboyxl group on 1,3-bisphosphoglycerate to the aldehyde group of G3P, which stores more potential energy. G3P is a sugar—the same three-carbon sugar formed in glycolysis by the splitting of glucose (see Figure 9.9). Notice in Figure 10.19 that for every *three* molecules of CO_2 that enter the cycle, there are *six* molecules of G3P formed. But only one molecule of this three-carbon sugar can be counted as a net gain of carbohydrate. The cycle began with 15 carbons' worth of carbohydrate in the form of three molecules of the five-carbon sugar RuBP. Now there are 18 carbons' worth of carbohydrate in the form of six molecules of G3P. One molecule exits the cycle to be used by the plant cell, but the other five molecules must be recycled to regenerate the three molecules of RuBP.

Phase 3: Regeneration of the CO_2 acceptor (RuBP). In a complex series of reactions, the carbon skeletons of five molecules of G3P are rearranged by the last steps of the Calvin cycle into three molecules of RuBP. To accomplish this, the cycle spends three more molecules of ATP. The RuBP is now prepared to receive CO_2 again, and the cycle continues.

For the net synthesis of one G3P molecule, the Calvin cycle consumes a total of nine molecules of ATP and six molecules of NADPH. The light reactions regenerate the ATP and NADPH. The G3P spun off from the Calvin cycle becomes the starting material for metabolic pathways that synthesize other organic compounds, including glucose and other carbohydrates. Neither the light reactions nor the Calvin cycle alone can make sugar from CO_2. Photosynthesis is an emergent property of the intact chloroplast, which integrates the two stages of photosynthesis.

CONCEPT CHECK 10.3

1. To synthesize one glucose molecule, the Calvin cycle uses ________ molecules of CO_2, ________ molecules of ATP, and ________ molecules of NADPH.
2. Explain why the large numbers of ATP and NADPH molecules used during the Calvin cycle are consistent with the high value of glucose as an energy source.
3. **WHAT IF?** Explain why a poison that inhibits an enzyme of the Calvin cycle will also inhibit the light reactions.
4. **MAKE CONNECTIONS** Review Figures 9.9 (p. 169) and 10.19. Discuss the roles of intermediate and product played by glyceraldehyde 3-phosphate (G3P) in the two processes shown in these figures.

For suggested answers, see Appendix A.

CONCEPT 10.4

Alternative mechanisms of carbon fixation have evolved in hot, arid climates

EVOLUTION Ever since plants first moved onto land about 475 million years ago, they have been adapting to the problems of terrestrial life, particularly the problem of dehydration. In Chapters 29 and 36, we will consider anatomical adaptations that help plants conserve water, while in this chapter we are concerned with metabolic adaptations. The solutions often involve trade-offs. An important example is the compromise between photosynthesis and the prevention of excessive water loss from the plant. The CO_2 required for photosynthesis enters a leaf via stomata, the pores on the leaf surface (see Figure 10.4). However, stomata are also the main avenues of transpiration, the evaporative loss of water from leaves. On a hot, dry day, most plants close their stomata, a response that conserves water. This response also reduces photosynthetic yield by limiting access to CO_2. With stomata even partially closed, CO_2 concentrations begin to decrease in the air spaces within the leaf, and the concentration of O_2 released from the light reactions begins to increase. These conditions within the leaf favor an apparently wasteful process called photorespiration.

Photorespiration: An Evolutionary Relic?

In most plants, initial fixation of carbon occurs via rubisco, the Calvin cycle enzyme that adds CO_2 to ribulose bisphosphate. Such plants are called **C_3 plants** because the first organic product of carbon fixation is a three-carbon compound,

3-phosphoglycerate (see Figure 10.19). Rice, wheat, and soybeans are C_3 plants that are important in agriculture. When their stomata partially close on hot, dry days, C_3 plants produce less sugar because the declining level of CO_2 in the leaf starves the Calvin cycle. In addition, rubisco can bind O_2 in place of CO_2. As CO_2 becomes scarce within the air spaces of the leaf, rubisco adds O_2 to the Calvin cycle instead of CO_2. The product splits, and a two-carbon compound leaves the chloroplast. Peroxisomes and mitochondria rearrange and split this compound, releasing CO_2. The process is called **photorespiration** because it occurs in the light (*photo*) and consumes O_2 while producing CO_2 (*respiration*). However, unlike normal cellular respiration, photorespiration generates no ATP; in fact, photorespiration consumes ATP. And unlike photosynthesis, photorespiration produces no sugar. In fact, photorespiration *decreases* photosynthetic output by siphoning organic material from the Calvin cycle and releasing CO_2 that would otherwise be fixed.

How can we explain the existence of a metabolic process that seems to be counterproductive for the plant? According to one hypothesis, photorespiration is evolutionary baggage—a metabolic relic from a much earlier time when the atmosphere had less O_2 and more CO_2 than it does today. In the ancient atmosphere that prevailed when rubisco first evolved, the inability of the enzyme's active site to exclude O_2 would have made little difference. The hypothesis suggests that modern rubisco retains some of its chance affinity for O_2, which is now so concentrated in the atmosphere that a certain amount of photorespiration is inevitable.

We now know that, at least in some cases, photorespiration plays a protective role in plants. Plants that are impaired in their ability to carry out photorespiration (due to defective genes) are more susceptible to damage induced by excess light. Researchers consider this clear evidence that photorespiration acts to neutralize the otherwise damaging products of the light reactions, which build up when a low CO_2 concentration limits the progress of the Calvin cycle. Whether there are other benefits of photorespiration is still unknown. In many types of plants—including a significant number of crop plants—photorespiration drains away as much as 50% of the carbon fixed by the Calvin cycle. As heterotrophs that depend on carbon fixation in chloroplasts for our food, we naturally view photorespiration as wasteful. Indeed, if photorespiration could be reduced in certain plant species without otherwise affecting photosynthetic productivity, crop yields and food supplies might increase.

In some plant species, alternate modes of carbon fixation have evolved that minimize photorespiration and optimize the Calvin cycle—even in hot, arid climates. The two most important of these photosynthetic adaptations are C_4 photosynthesis and crassulacean acid metabolism (CAM).

C_4 Plants

The **C_4 plants** are so named because they preface the Calvin cycle with an alternate mode of carbon fixation that forms a four-carbon compound as its first product. Several thousand species in at least 19 plant families use the C_4 pathway. Among the C_4 plants important to agriculture are sugarcane and corn, members of the grass family.

A unique leaf anatomy is correlated with the mechanism of C_4 photosynthesis (**Figure 10.20**; compare with Figure 10.4). In C_4 plants, there are two distinct types of photosynthetic cells: bundle-sheath cells and mesophyll cells. **Bundle-sheath cells** are arranged into tightly packed sheaths around the veins of the leaf. Between the bundle sheath and the leaf surface are the more loosely arranged mesophyll cells. The Calvin cycle is confined to the chloroplasts of the bundle-sheath cells. However, the Calvin cycle is preceded by incorporation of CO_2 into organic compounds in the mesophyll cells. See the numbered steps in Figure 10.20, which are also described here:

1. The first step is carried out by an enzyme present only in mesophyll cells called **PEP carboxylase**. This enzyme adds CO_2 to phosphoenolpyruvate (PEP), forming the four-carbon product oxaloacetate. PEP carboxylase has a much higher affinity for CO_2 than does rubisco and no affinity for O_2. Therefore, PEP carboxylase can fix carbon efficiently when rubisco cannot—that is, when it is hot and dry and stomata are partially closed, causing CO_2 concentration in the leaf to fall and O_2 concentration to rise.
2. After the C_4 plant fixes carbon from CO_2, the mesophyll cells export their four-carbon products (malate in the example shown in Figure 10.20) to bundle-sheath cells through plasmodesmata (see Figure 6.31).
3. Within the bundle-sheath cells, the four-carbon compounds release CO_2, which is reassimilated into organic material by rubisco and the Calvin cycle. The same reaction regenerates pyruvate, which is transported to mesophyll cells. There, ATP is used to convert pyruvate to PEP, allowing the reaction cycle to continue; this ATP can be thought of as the "price" of concentrating CO_2 in the bundle-sheath cells. To generate this extra ATP, bundle-sheath cells carry out cyclic electron flow, the process described earlier in this chapter (see Figure 10.16). In fact, these cells contain PS I but no PS II, so cyclic electron flow is their only photosynthetic mode of generating ATP.

In effect, the mesophyll cells of a C_4 plant pump CO_2 into the bundle sheath, keeping the CO_2 concentration in the bundle-sheath cells high enough for rubisco to bind carbon

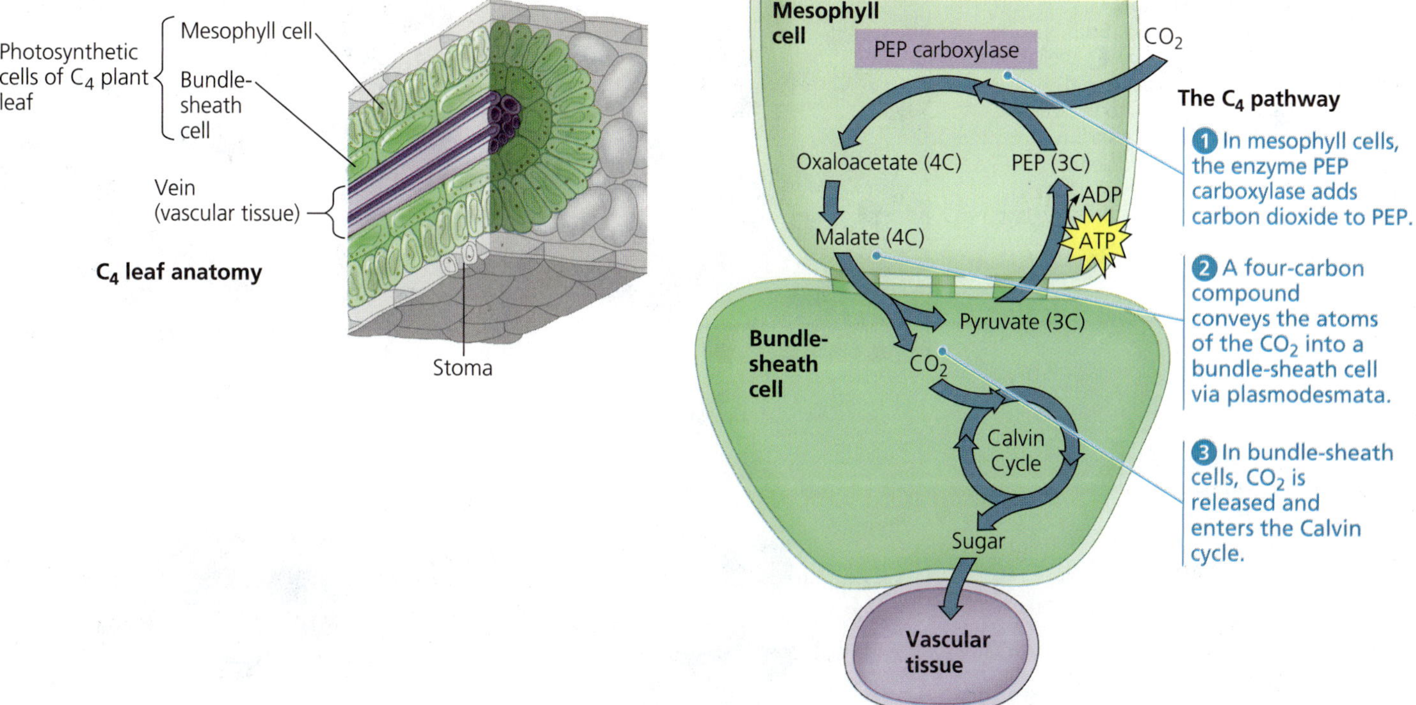

▲ **Figure 10.20 C_4 leaf anatomy and the C_4 pathway.** The structure and biochemical functions of the leaves of C_4 plants are an evolutionary adaptation to hot, dry climates. This adaptation maintains a CO_2 concentration in the bundle sheath that favors photosynthesis over photorespiration.

dioxide rather than oxygen. The cyclic series of reactions involving PEP carboxylase and the regeneration of PEP can be thought of as a CO_2-concentrating pump that is powered by ATP. In this way, C_4 photosynthesis minimizes photorespiration and enhances sugar production. This adaptation is especially advantageous in hot regions with intense sunlight, where stomata partially close during the day, and it is in such environments that C_4 plants evolved and thrive today.

Since the Industrial Revolution began in the 1800s, human activities such as the burning of fossil fuels have drastically increased the concentration of CO_2 in the atmosphere. The resulting global climate change, including an increase in average temperatures around the planet, may have far-reaching effects on plant species. Scientists are concerned that increasing CO_2 concentration and temperature may affect C_3 and C_4 plants differently, thus changing the relative abundance of these species in a given plant community.

Which type of plant would stand to gain more from increasing CO_2 levels? Recall that in C_3 plants, the binding of O_2 rather than CO_2 by rubisco leads to photorespiration, lowering the efficiency of photosynthesis. C_4 plants overcome this problem by concentrating CO_2 in the bundle-sheath cells at the cost of ATP. Rising CO_2 levels should benefit C_3 plants by lowering the amount of photorespiration that occurs. At the same time, rising temperatures have the opposite effect, increasing photorespiration. (Other factors such as water availability may also come into play.) In contrast, many C_4 plants could be largely unaffected by increasing CO_2 levels or temperature. In different regions, the particular combination of these two factors is likely to alter the balance of C_3 and C_4 plants in varying ways. The effects of such a widespread and variable change in community structure are unpredictable and thus a cause of legitimate concern.

CAM Plants

A second photosynthetic adaptation to arid conditions has evolved in many succulent (water-storing) plants, numerous cacti, pineapples, and representatives of several other plant families. These plants open their stomata during the night and close them during the day, just the reverse of how other plants behave. Closing stomata during the day helps desert plants conserve water, but it also prevents CO_2 from entering the leaves. During the night, when their stomata are open, these plants take up CO_2 and incorporate it into a variety

▶ **Figure 10.21 C_4 and CAM photosynthesis compared.** Both adaptations are characterized by ❶ preliminary incorporation of CO_2 into organic acids, followed by ❷ transfer of CO_2 to the Calvin cycle. The C_4 and CAM pathways are two evolutionary solutions to the problem of maintaining photosynthesis with stomata partially or completely closed on hot, dry days.

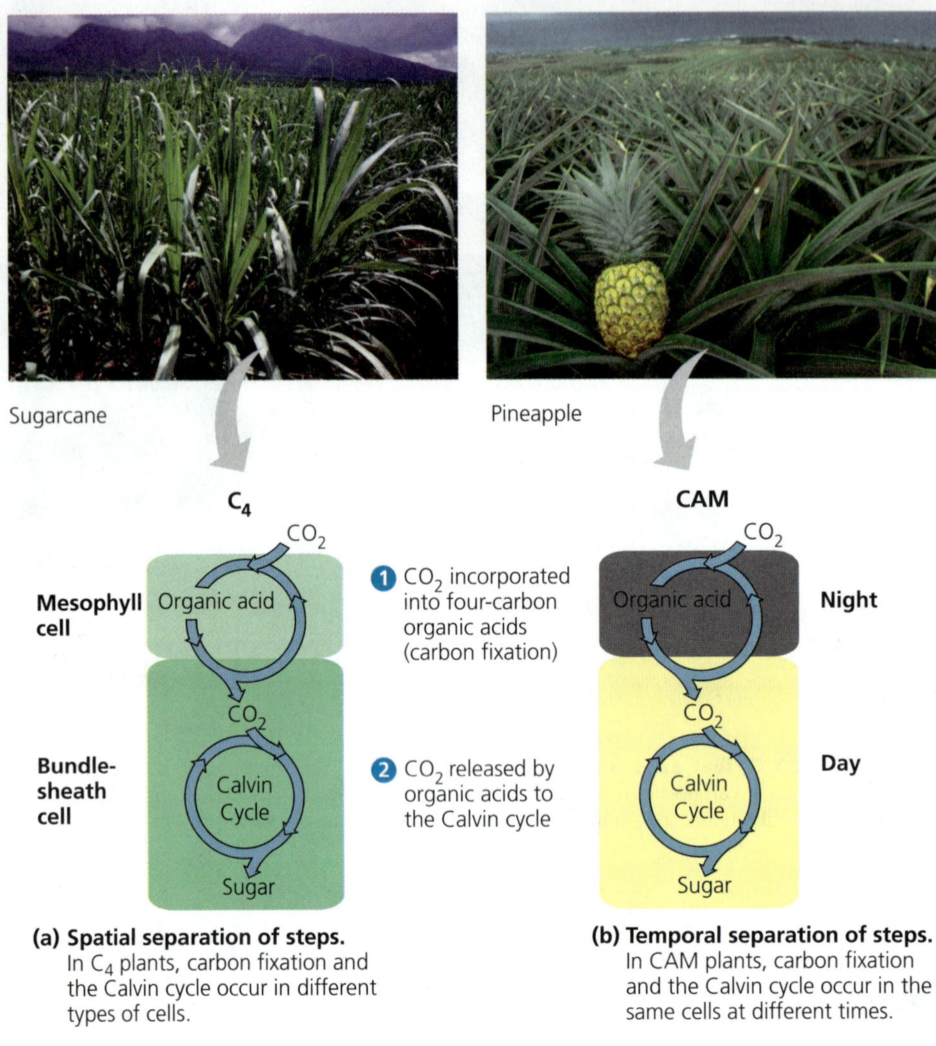

(a) Spatial separation of steps. In C_4 plants, carbon fixation and the Calvin cycle occur in different types of cells.

(b) Temporal separation of steps. In CAM plants, carbon fixation and the Calvin cycle occur in the same cells at different times.

of organic acids. This mode of carbon fixation is called **crassulacean acid metabolism**, or **CAM**, after the plant family Crassulaceae, the succulents in which the process was first discovered. The mesophyll cells of **CAM plants** store the organic acids they make during the night in their vacuoles until morning, when the stomata close. During the day, when the light reactions can supply ATP and NADPH for the Calvin cycle, CO_2 is released from the organic acids made the night before to become incorporated into sugar in the chloroplasts.

Notice in **Figure 10.21** that the CAM pathway is similar to the C_4 pathway in that carbon dioxide is first incorporated into organic intermediates before it enters the Calvin cycle. The difference is that in C_4 plants, the initial steps of carbon fixation are separated structurally from the Calvin cycle, whereas in CAM plants, the two steps occur at separate times but within the same cell. (Keep in mind that CAM, C_4, and C_3 plants all eventually use the Calvin cycle to make sugar from carbon dioxide.)

CONCEPT CHECK 10.4

1. Explain why photorespiration lowers photosynthetic output for plants.
2. The presence of only PS I, not PS II, in the bundle-sheath cells of C_4 plants has an effect on O_2 concentration. What is that effect, and how might that benefit the plant?
3. **MAKE CONNECTIONS** Refer to the discussion of ocean acidification in Concept 3.3 (p. 55). Ocean acidification and changes in the distribution of C_3 and C_4 plants may seem to be two very different problems, but what do they have in common? Explain.
4. **WHAT IF?** How would you expect the relative abundance of C_3 versus C_4 and CAM species to change in a geographic region whose climate becomes much hotter and drier, with no change in CO_2 concentration?

For suggested answers, see Appendix A.

The Importance of Photosynthesis: *A Review*

In this chapter, we have followed photosynthesis from photons to food. The light reactions capture solar energy and use it to make ATP and transfer electrons from water to $NADP^+$, forming NADPH. The Calvin cycle uses the ATP and NADPH to produce sugar from carbon dioxide. The energy that enters the chloroplasts as sunlight becomes stored as chemical energy in organic compounds. See **Figure 10.22** for a review of the entire process.

What are the fates of photosynthetic products? The sugar made in the chloroplasts supplies the entire plant with chemical energy and carbon skeletons for the synthesis of all the major organic molecules of plant cells. About 50% of the organic material made by photosynthesis is consumed as fuel for cellular respiration in the mitochondria of the plant cells. Sometimes there is a loss of photosynthetic products to photorespiration.

Technically, green cells are the only autotrophic parts of the plant. The rest of the plant depends on organic molecules exported from leaves via veins. In most plants, carbohydrate is transported out of the leaves in the form of sucrose, a disaccharide. After arriving at nonphotosynthetic cells, the sucrose provides raw material for cellular respiration and a multitude of anabolic pathways that synthesize proteins, lipids, and other products. A considerable amount of sugar in the form of glucose is linked together to make the polysaccharide cellulose, especially in plant cells that are still growing and maturing. Cellulose, the main ingredient of cell walls, is the most abundant organic molecule in the plant—and probably on the surface of the planet.

Most plants manage to make more organic material each day than they need to use as respiratory fuel and precursors for biosynthesis. They stockpile the extra sugar by synthesizing starch, storing some in the chloroplasts themselves and some in storage cells of roots, tubers, seeds, and fruits. In accounting for the consumption of the food molecules produced by photosynthesis, let's not forget that most plants lose leaves, roots, stems, fruits, and sometimes their entire bodies to heterotrophs, including humans.

On a global scale, photosynthesis is the process responsible for the presence of oxygen in our atmosphere. Furthermore, in terms of food production, the collective productivity of the minuscule chloroplasts is prodigious: Photosynthesis makes an estimated 160 billion metric tons of carbohydrate per year (a metric ton is 1,000 kg, about 1.1 tons). That's organic matter equivalent in mass to a stack of about 60 trillion copies of this textbook—17 stacks of books reaching from Earth to the sun! No other chemical process on the planet can match the output of photosynthesis. And as we mentioned earlier, researchers are seeking ways to capitalize on photosynthetic production to produce alternative fuels. No process is more important than photosynthesis to the welfare of life on Earth.

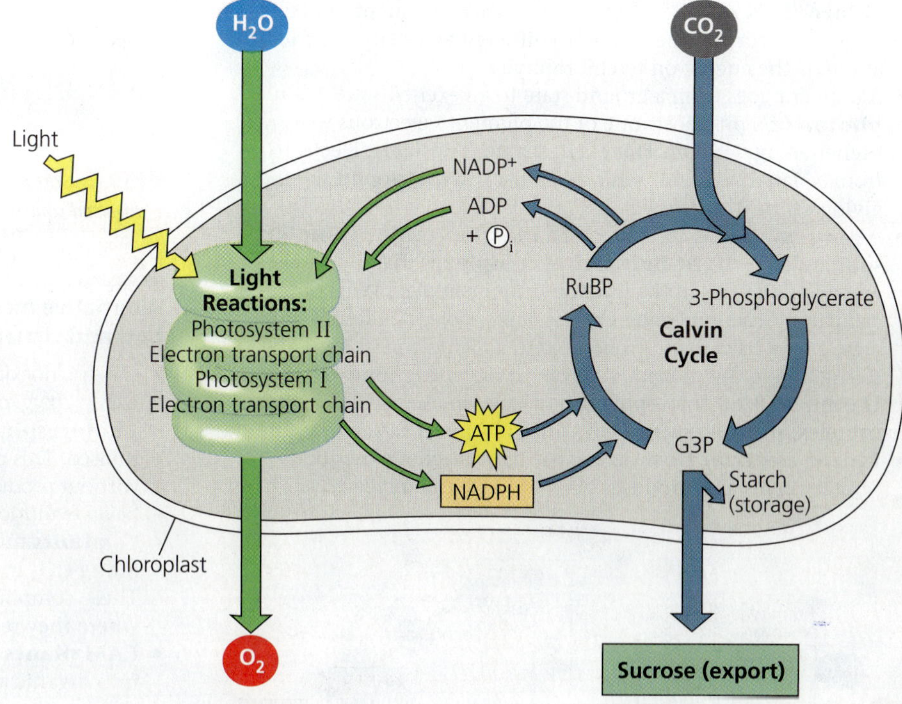

▶ **Figure 10.22 A review of photosynthesis.** This diagram outlines the main reactants and products of the light reactions and the Calvin cycle as they occur in the chloroplasts of plant cells. The entire ordered operation depends on the structural integrity of the chloroplast and its membranes. Enzymes in the chloroplast and cytosol convert glyceraldehyde 3-phosphate (G3P), the direct product of the Calvin cycle, to many other organic compounds.

MAKE CONNECTIONS *Return to the micrograph in Figure 5.6a, on page 72. Label and describe where the light reactions and the Calvin cycle take place. Also explain where the starch granules in the micrograph came from.*

10 CHAPTER REVIEW

SUMMARY OF KEY CONCEPTS

CONCEPT 10.1

Photosynthesis converts light energy to the chemical energy of food (pp. 186–189)

- In **autotrophic** eukaryotes, photosynthesis occurs in **chloroplasts**, organelles containing **thylakoids**. Stacks of thylakoids form grana. **Photosynthesis** is summarized as

$$6\ CO_2 + 12\ H_2O + \text{Light energy} \rightarrow C_6H_{12}O_6 + 6\ O_2 + 6\ H_2O.$$

Chloroplasts split water into hydrogen and oxygen, incorporating the electrons of hydrogen into sugar molecules. Photosynthesis is a redox process: H_2O is oxidized, and CO_2 is reduced. The **light reactions** in the thylakoid membranes split water, releasing O_2, producing ATP, and forming **NADPH**. The **Calvin cycle** in the **stroma** forms sugar from CO_2, using ATP for energy and NADPH for reducing power.

? *Compare and describe the roles of CO_2 and H_2O in respiration and photosynthesis.*

CONCEPT 10.2

The light reactions convert solar energy to the chemical energy of ATP and NADPH (pp. 189–197)

- Light is a form of electromagnetic energy. The colors we see as **visible light** include those **wavelengths** that drive photosynthesis. A pigment absorbs light of specific wavelengths; **chlorophyll *a*** is the main photosynthetic pigment in plants. Other accessory pigments absorb different wavelengths of light and pass the energy on to chlorophyll *a*.
- A pigment goes from a ground state to an excited state when a **photon** of light boosts one of the pigment's electrons to a higher-energy orbital. This excited state is unstable. Electrons from isolated pigments tend to fall back to the ground state, giving off heat and/or light.
- A **photosystem** is composed of a **reaction-center complex** surrounded by **light-harvesting complexes** that funnel the energy of photons to the reaction-center complex. When a special pair of reaction-center chlorophyll *a* molecules absorbs energy, one of its electrons is boosted to a higher energy level and transferred to the **primary electron acceptor**. **Photosystem II** contains P680 chlorophyll *a* molecules in the reaction-center complex; **photosystem I** contains P700 molecules.
- **Linear electron flow** during the light reactions uses both photosystems and produces NADPH, ATP, and oxygen:

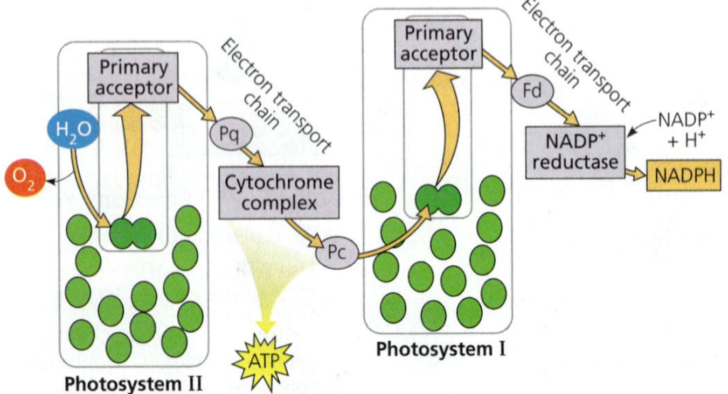

- **Cyclic electron flow** employs only photosystem I, producing ATP but no NADPH or O_2.
- During chemiosmosis in both mitochondria and chloroplasts, electron transport chains generate an H^+ gradient across a membrane. ATP synthase uses this proton-motive force to make ATP.

? *The absorption spectrum of chlorophyll* a *differs from the action spectrum of photosynthesis. Explain this observation.*

CONCEPT 10.3

The Calvin cycle uses the chemical energy of ATP and NADPH to reduce CO_2 to sugar (pp. 198–199)

- The Calvin cycle occurs in the stroma, using electrons from NADPH and energy from ATP. One molecule of **G3P** exits the cycle per three CO_2 molecules fixed and is converted to glucose and other organic molecules.

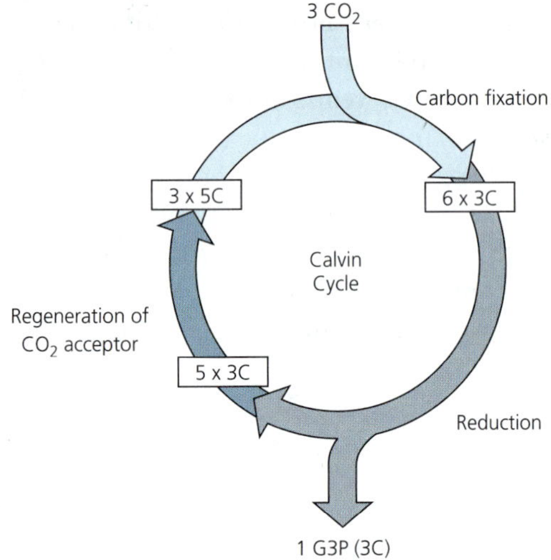

DRAW IT *On the diagram above, draw where ATP and NADPH are used and where rubisco functions. Describe these steps.*

CONCEPT 10.4

Alternative mechanisms of carbon fixation have evolved in hot, arid climates (pp. 199–202)

- On dry, hot days, **C_3 plants** close their stomata, conserving water. Oxygen from the light reactions builds up. In **photorespiration**, O_2 substitutes for CO_2 in the active site of rubisco. This process consumes organic fuel and releases CO_2 without producing ATP or carbohydrate. Photorespiration may be an evolutionary relic, and it may play a photoprotective role.
- **C_4 plants** minimize the cost of photorespiration by incorporating CO_2 into four-carbon compounds in mesophyll cells. These compounds are exported to **bundle-sheath cells**, where they release carbon dioxide for use in the Calvin cycle.
- **CAM plants** open their stomata at night, incorporating CO_2 into organic acids, which are stored in mesophyll cells. During the day, the stomata close, and the CO_2 is released from the organic acids for use in the Calvin cycle.
- Organic compounds produced by photosynthesis provide the energy and building material for ecosystems.

? *Why are C_4 and CAM photosynthesis more energetically expensive than C_3 photosynthesis? What climate conditions would favor C_4 and CAM plants?*

TEST YOUR UNDERSTANDING

LEVEL 1: KNOWLEDGE/COMPREHENSION

1. The light reactions of photosynthesis supply the Calvin cycle with
   a. light energy.
   b. CO_2 and ATP.
   c. H_2O and NADPH.
   d. ATP and NADPH.
   e. sugar and O_2.

2. Which of the following sequences correctly represents the flow of electrons during photosynthesis?
   a. NADPH $\rightarrow$ O_2 $\rightarrow$ CO_2
   b. H_2O $\rightarrow$ NADPH $\rightarrow$ Calvin cycle
   c. NADPH $\rightarrow$ chlorophyll $\rightarrow$ Calvin cycle
   d. H_2O $\rightarrow$ photosystem I $\rightarrow$ photosystem II
   e. NADPH $\rightarrow$ electron transport chain $\rightarrow$ O_2

3. How is photosynthesis similar in C_4 plants and CAM plants?
   a. In both cases, only photosystem I is used.
   b. Both types of plants make sugar without the Calvin cycle.
   c. In both cases, rubisco is not used to fix carbon initially.
   d. Both types of plants make most of their sugar in the dark.
   e. In both cases, thylakoids are not involved in photosynthesis.

4. Which of the following statements is a correct distinction between autotrophs and heterotrophs?
   a. Only heterotrophs require chemical compounds from the environment.
   b. Cellular respiration is unique to heterotrophs.
   c. Only heterotrophs have mitochondria.
   d. Autotrophs, but not heterotrophs, can nourish themselves beginning with CO_2 and other nutrients that are inorganic.
   e. Only heterotrophs require oxygen.

5. Which of the following does *not* occur during the Calvin cycle?
   a. carbon fixation
   b. oxidation of NADPH
   c. release of oxygen
   d. regeneration of the CO_2 acceptor
   e. consumption of ATP

LEVEL 2: APPLICATION/ANALYSIS

6. In mechanism, photophosphorylation is most similar to
   a. substrate-level phosphorylation in glycolysis.
   b. oxidative phosphorylation in cellular respiration.
   c. the Calvin cycle.
   d. carbon fixation.
   e. reduction of $NADP^+$.

7. Which process is most directly driven by light energy?
   a. creation of a pH gradient by pumping protons across the thylakoid membrane
   b. carbon fixation in the stroma
   c. reduction of $NADP^+$ molecules
   d. removal of electrons from chlorophyll molecules
   e. ATP synthesis

LEVEL 3: SYNTHESIS/EVALUATION

8. **EVOLUTION CONNECTION**
   Photorespiration can decrease soybeans' photosynthetic output by about 50%. Would you expect this figure to be higher or lower in wild relatives of soybeans? Why?

9. **SCIENTIFIC INQUIRY**
   MAKE CONNECTIONS **DRAW IT** The following diagram represents an experiment with isolated thylakoids. The thylakoids were first made acidic by soaking them in a solution at pH 4. After the thylakoid space reached pH 4, the thylakoids were transferred to a basic solution at pH 8. The thylakoids then made ATP in the dark. (See Concept 3.3, pp. 53–54, to review pH).

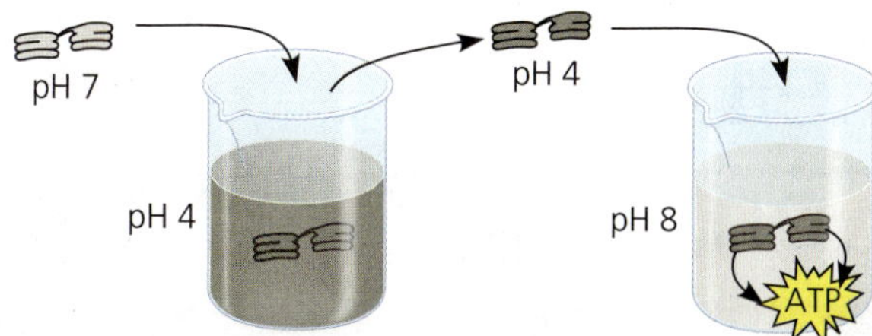

   Draw an enlargement of part of the thylakoid membrane in the beaker with the solution at pH 8. Draw ATP synthase. Label the areas of high H^+ concentration and low H^+ concentration. Show the direction protons flow through the enzyme, and show the reaction where ATP is synthesized. Would ATP end up in the thylakoid or outside of it? Explain why the thylakoids in the experiment were able to make ATP in the dark.

10. **SCIENCE, TECHNOLOGY, AND SOCIETY**
    Scientific evidence indicates that the CO_2 added to the air by the burning of wood and fossil fuels is contributing to global warming, a rise in global temperature. Tropical rain forests are estimated to be responsible for approximately 20% of global photosynthesis, yet the consumption of large amounts of CO_2 by living trees is thought to make little or no *net* contribution to reduction of global warming. Why might this be? (*Hint*: What processes in both living and dead trees produce CO_2?)

11. **WRITE ABOUT A THEME**
    Energy Transfer Life is solar powered. Almost all the producers of the biosphere depend on energy from the sun to produce the organic molecules that supply the energy and carbon skeletons needed for life. In a short essay (100–150 words), describe how the process of photosynthesis in the chloroplasts of plants transforms the energy of sunlight into the chemical energy of sugar molecules.

For selected answers, see Appendix A.

11

Cell Communication

▲ **Figure 11.1 How does cell signaling trigger the desperate flight of this gazelle?**

KEY CONCEPTS

11.1 External signals are converted to responses within the cell

11.2 Reception: A signaling molecule binds to a receptor protein, causing it to change shape

11.3 Transduction: Cascades of molecular interactions relay signals from receptors to target molecules in the cell

11.4 Response: Cell signaling leads to regulation of transcription or cytoplasmic activities

11.5 Apoptosis integrates multiple cell-signaling pathways

OVERVIEW

Cellular Messaging

The Thomson's gazelle in **Figure 11.1** is fleeing for its life, seeking to escape the predatory cheetah nipping at its heels. The gazelle's heart is racing, its breathing accelerated and its muscles performing at their highest level. These physiological functions are all part of the "fight-or-flight" response, driven by hormones released from the adrenal glands at times of stress—in this case, when the gazelle first sensed the cheetah. Hormonal signaling and the subsequent response by cells and tissues throughout the gazelle's body illustrate how cell-to-cell communication allows the trillions of cells in a multicellular organism to "talk" to each other, coordinating their activities. Communication between cells is essential not only for multicellular organisms such as gazelles and oak trees but for many unicellular organisms as well.

In studying how cells signal to each other and how they interpret the signals they receive, biologists have discovered some universal mechanisms of cellular regulation, additional evidence for the evolutionary relatedness of all life. The same small set of cell-signaling mechanisms shows up again and again in diverse species, in biological processes ranging from hormone action to embryonic development to cancer. The signals received by cells, whether originating from other cells or from changes in the physical environment, take various forms, including light and touch. However, cells most often communicate with each other by chemical signals. For instance, the fight-or-flight response is triggered by a signaling molecule called epinephrine. In this chapter, we focus on the main mechanisms by which cells receive, process, and respond to chemical signals sent from other cells. We will also take a look at *apoptosis*, a type of programmed cell death that integrates input from multiple signaling pathways.

CONCEPT **11.1**

External signals are converted to responses within the cell

What does a "talking" cell say to a "listening" cell, and how does the latter cell respond to the message? Let's approach these questions by first looking at communication among microorganisms, for microbes living today provide a glimpse into the role of cell signaling in the evolution of life on Earth.

Evolution of Cell Signaling

EVOLUTION One topic of cell "conversation" is sex—at least for the yeast *Saccharomyces cerevisiae*, which people have used for millennia to make bread, wine, and beer. Researchers have learned that cells of this yeast identify their mates by chemical signaling. There are two sexes, or mating types, called **a** and **α** (**Figure 11.2**). Cells of mating type **a** secrete a signaling

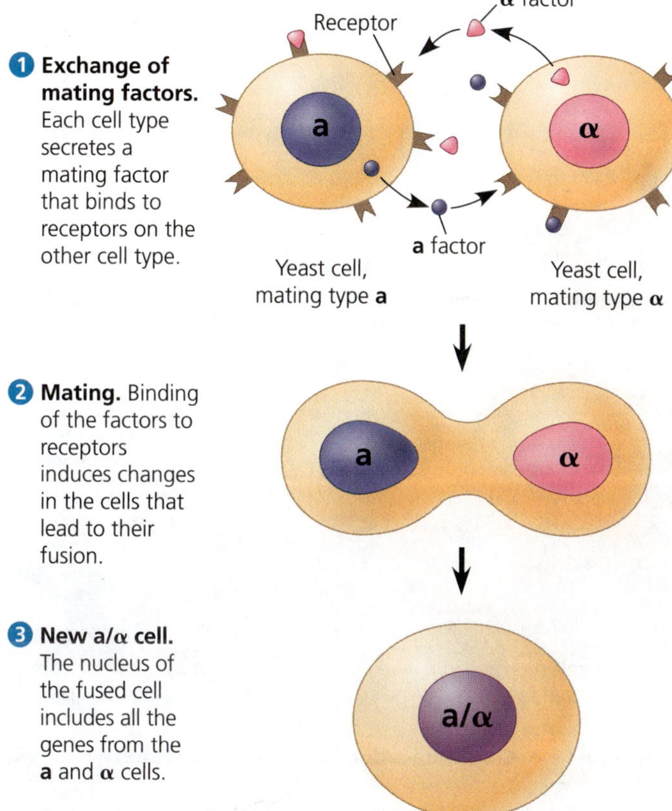

▲ **Figure 11.2 Communication between mating yeast cells.** *Saccharomyces cerevisiae* cells use chemical signaling to identify cells of opposite mating type and initiate the mating process. The two mating types and their corresponding chemical signaling molecules, or mating factors, are called **a** and **α**.

molecule called **a** factor, which can bind to specific receptor proteins on nearby α cells. At the same time, α cells secrete α factor, which binds to receptors on **a** cells. Without actually entering the cells, the two mating factors cause the cells to grow toward each other and also bring about other cellular changes. The result is the fusion, or mating, of two cells of opposite type. The new **a**/α cell contains all the genes of both original cells, a combination of genetic resources that provides advantages to the cell's descendants, which arise by subsequent cell divisions.

Once received at the yeast cell surface, how is the mating signal changed, or *transduced*, into a form that brings about the cellular response of mating? The received signal is converted to a specific cellular response in a series of steps called a **signal transduction pathway**. Many such pathways have been extensively studied in both yeast and animal cells. Amazingly, the molecular details of signal transduction in yeast and mammals are strikingly similar, even though the last common ancestor of these two groups of organisms lived over a billion years ago. These similarities—and others more recently uncovered between signaling systems in bacteria and plants—suggest that early versions of today's cell-signaling mechanisms evolved well before the first multicellular creatures appeared on Earth.

Scientists such as Bonnie Bassler, the interviewee for Unit 2 (see pp. 92–93), think that signaling mechanisms first evolved in ancient prokaryotes and single-celled eukaryotes, then were adopted for new uses by their multicellular descendants. Cell signaling is critical in the microbial world; a classic example in one bacterial species is shown in **Figure 11.3**. Bacterial cells secrete small molecules that can be detected by other bacterial cells. The concentration of such signaling molecules, sensed by the bacteria, allows them to monitor the local density of cells, a phenomenon called *quorum sensing*. Quorum sensing allows bacterial populations to coordinate their behaviors so they can carry out activities that are only productive when performed by a given number of cells in synchrony. One example is formation of a *biofilm*, an aggregation of bacterial cells adhered to a surface; the cells in the biofilm generally derive nutrition from the surface they are on. You have probably encountered biofilms many times, perhaps without realizing it. The slimy coating on a fallen log or on leaves lying on a forest path, or on your teeth each morning, are examples of bacterial biofilms. Biofilms are responsible for cavities—a good argument for daily tooth brushing and flossing to disrupt them!

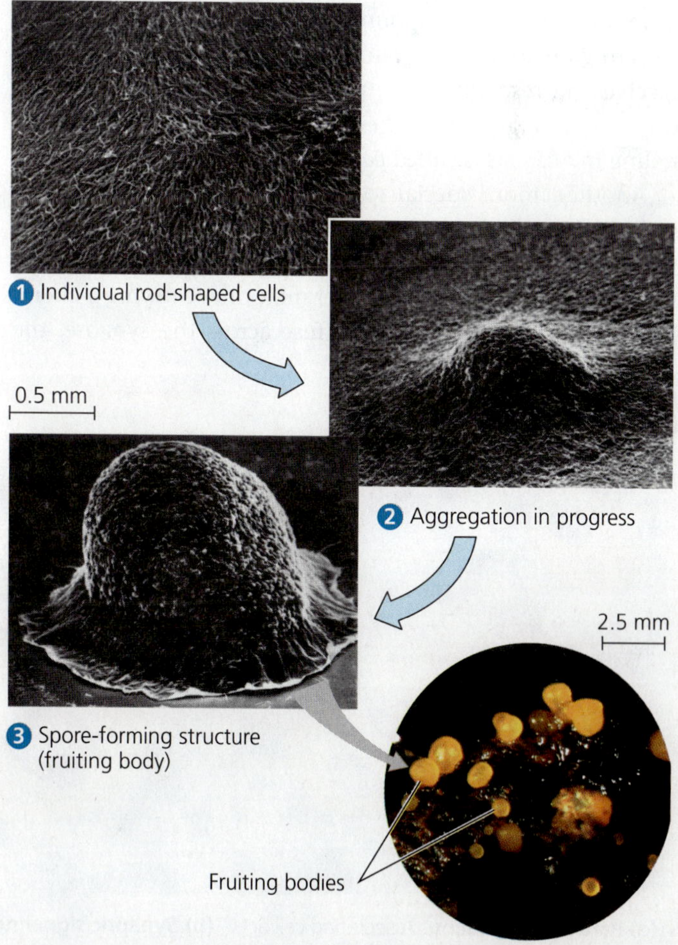

▲ **Figure 11.3 Communication among bacteria.** Soil-dwelling bacteria called myxobacteria ("slime bacteria") use chemical signals to share information about nutrient availability. When food is scarce, starving cells secrete a molecule that stimulates neighboring cells to aggregate. The cells form a structure, called a fruiting body, that produces thick-walled spores capable of surviving until the environment improves. The bacteria shown here are *Myxococcus xanthus* (steps 1–3, SEMs; lower photo, LM).

Local and Long-Distance Signaling

Like bacteria or yeast cells, cells in a multicellular organism usually communicate via chemical messengers targeted for cells that may or may not be immediately adjacent. As we saw in Chapters 6 and 7, eukaryotic cells may communicate by direct contact **(Figure 11.4)**, one type of local signaling. Both animals and plants have cell junctions that, where present, directly connect the cytoplasms of adjacent cells **(Figure 11.4a)**. In these cases, signaling substances dissolved in the cytosol can pass freely between adjacent cells. Moreover, animal cells may communicate via direct contact between membrane-bound cell-surface molecules in a process called cell-cell recognition **(Figure 11.4b)**. This sort of local signaling is important in embryonic development and the immune response.

In many other cases of local signaling, messenger molecules are secreted by the signaling cell. Some of these travel only short distances; such **local regulators** influence cells in the vicinity. One class of local regulators in animals, *growth factors*, consists of compounds that stimulate nearby target cells to grow and divide. Numerous cells can simultaneously receive and respond to the molecules of growth factor produced by a single cell in their vicinity. This type of local signaling in animals is called *paracrine signaling* **(Figure 11.5a)**.

Another, more specialized type of local signaling called *synaptic signaling* occurs in the animal nervous system **(Figure 11.5b)**. An electrical signal along a nerve cell triggers the secretion of neurotransmitter molecules carrying a chemical signal. These molecules diffuse across the synapse, the narrow space between the nerve cell and its target cell (often another nerve cell), triggering a response in the target cell.

Beyond communication through plasmodesmata (plant cell junctions), local signaling in plants is not as well understood. Because of their cell walls, plants use mechanisms somewhat different from those operating locally in animals.

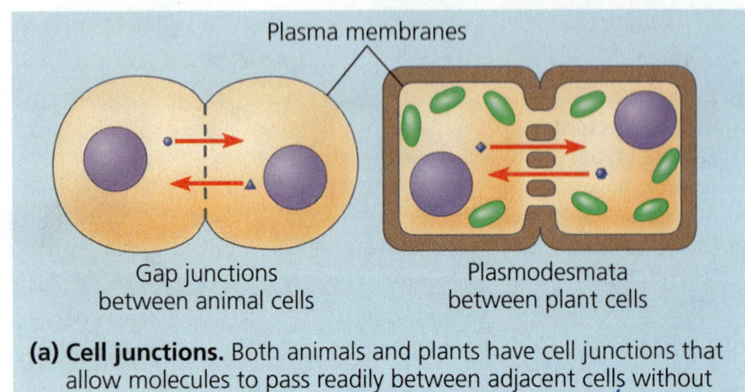

(a) Cell junctions. Both animals and plants have cell junctions that allow molecules to pass readily between adjacent cells without crossing plasma membranes.

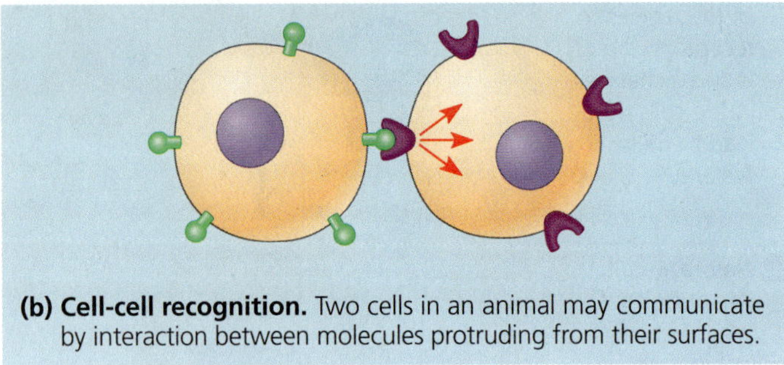

(b) Cell-cell recognition. Two cells in an animal may communicate by interaction between molecules protruding from their surfaces.

▲ **Figure 11.4 Communication by direct contact between cells.**

Local signaling

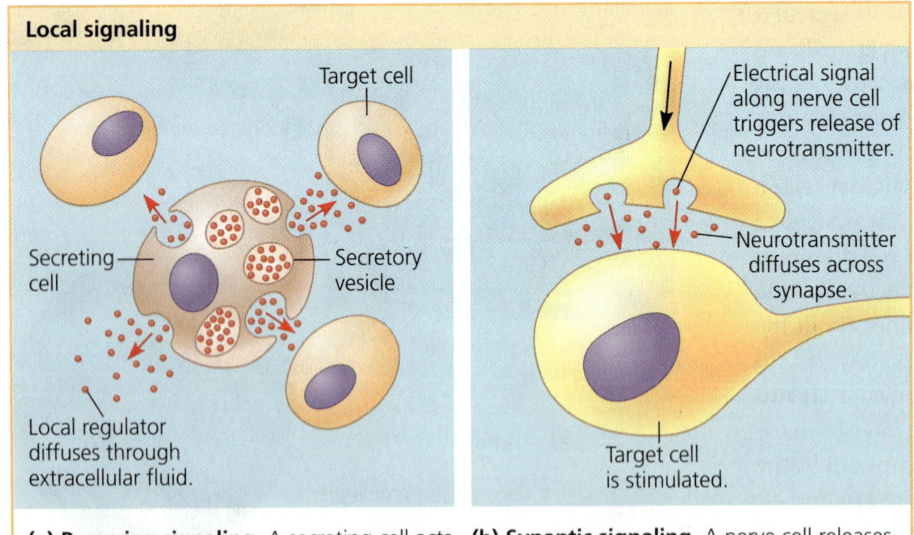

(a) Paracrine signaling. A secreting cell acts on nearby target cells by discharging molecules of a local regulator (a growth factor, for example) into the extracellular fluid.

(b) Synaptic signaling. A nerve cell releases neurotransmitter molecules into a synapse, stimulating the target cell.

Long-distance signaling

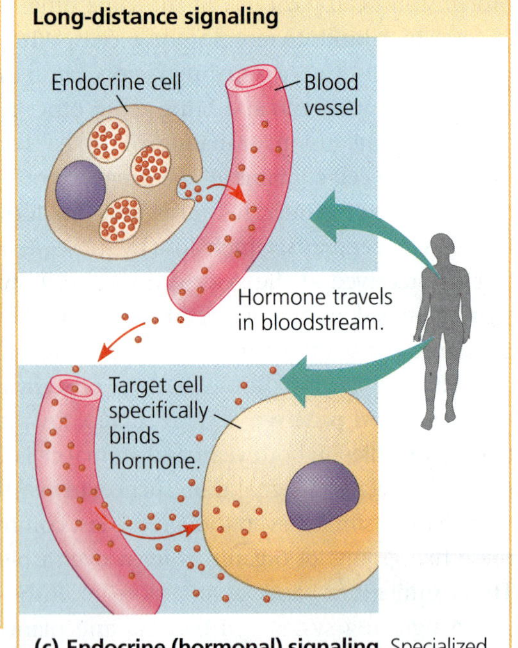

(c) Endocrine (hormonal) signaling. Specialized endocrine cells secrete hormones into body fluids, often blood. Hormones reach virtually all body cells, but are bound only by some cells.

▲ **Figure 11.5 Local and long-distance cell signaling by secreted molecules in animals.** In both local and long-distance signaling, only specific target cells that can recognize a given signaling molecule will respond to it.

Both animals and plants use chemicals called **hormones** for long-distance signaling. In hormonal signaling in animals, also known as *endocrine signaling*, specialized cells release hormone molecules, which travel via the circulatory system to other parts of the body, where they reach target cells that can recognize and respond to the hormones **(Figure 11.5c)**. Plant hormones (often called *plant growth regulators*) sometimes travel in vessels but more often reach their targets by moving through cells or by diffusing through the air as a gas (see Chapter 39). Hormones vary widely in molecular size and type, as do local regulators. For instance, the plant hormone ethylene, a gas that promotes fruit ripening and helps regulate growth, is a hydrocarbon of only six atoms (C_2H_4), small enough to pass through cell walls. In contrast, the mammalian hormone insulin, which regulates sugar levels in the blood, is a protein with thousands of atoms.

The transmission of a signal through the nervous system can also be considered an example of long-distance signaling. An electrical signal travels the length of a nerve cell and is then converted back to a chemical signal when a signaling molecule is released and crosses the synapse to another nerve cell. Here it is converted back to an electrical signal. In this way, a nerve signal can travel along a series of nerve cells. Because some nerve cells are quite long, the nerve signal can quickly travel great distances—from your brain to your big toe, for example. This type of long-distance signaling will be covered in detail in Chapter 48.

What happens when a cell encounters a secreted signaling molecule? The ability of a cell to respond is determined by whether it has a specific receptor molecule that can bind to the signaling molecule. The information conveyed by this binding, the signal, must then be changed into another form—transduced—inside the cell before the cell can respond. The remainder of the chapter discusses this process, primarily as it occurs in animal cells.

The Three Stages of Cell Signaling: *A Preview*

Our current understanding of how chemical messengers act via signal transduction pathways had its origins in the pioneering work of Earl W. Sutherland, whose research led to a Nobel Prize in 1971. Sutherland and his colleagues at Vanderbilt University were investigating how the animal hormone epinephrine (also called adrenaline) stimulates the breakdown of the storage polysaccharide glycogen within liver cells and skeletal muscle cells. Glycogen breakdown releases the sugar glucose 1-phosphate, which the cell converts to glucose 6-phosphate. The cell (a liver cell, for example) can then use this compound, an early intermediate in glycolysis, for energy production. Alternatively, the compound can be stripped of phosphate and released from the liver cell into the blood as glucose, which can fuel cells throughout the body. Thus, one effect of epinephrine is the mobilization of fuel reserves, which can be used by the animal to either defend itself (fight) or escape whatever elicited a scare (flight). (The gazelle in Figure 11.1 is clearly engaged in the latter.)

Sutherland's research team discovered that epinephrine stimulates glycogen breakdown by somehow activating a cytosolic enzyme, glycogen phosphorylase. However, when epinephrine was added to a test-tube mixture containing the enzyme and its substrate, glycogen, no breakdown occurred. Epinephrine could activate glycogen phosphorylase only when the hormone was added to a solution containing *intact* cells. This result told Sutherland two things. First, epinephrine does not interact directly with the enzyme responsible for glycogen breakdown; an intermediate step or series of steps must be occurring inside the cell. Second, the plasma membrane is somehow involved in transmitting the signal.

Sutherland's early work suggested that the process going on at the receiving end of a cellular conversation can be dissected into three stages: reception, transduction, and response **(Figure 11.6)**:

1. **Reception.** Reception is the target cell's detection of a signaling molecule coming from outside the cell. A chemical signal is "detected" when the signaling molecule binds to a receptor protein located at the cell's surface or inside the cell.

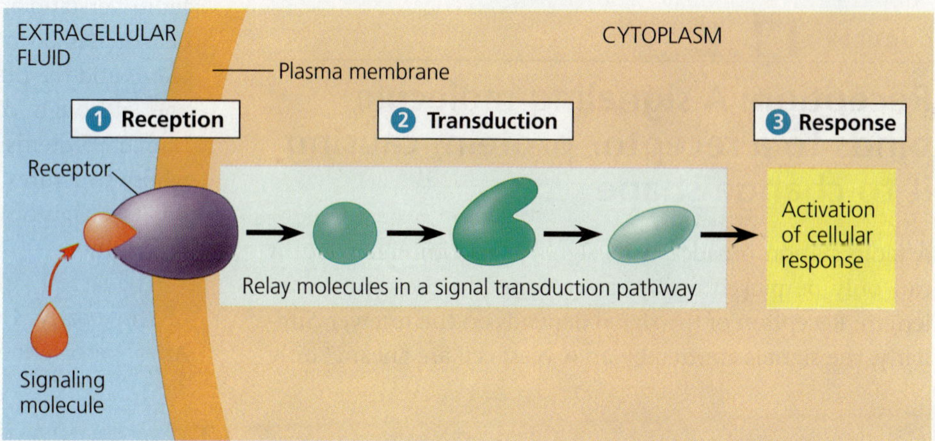

▶ **Figure 11.6 Overview of cell signaling.** From the perspective of the cell receiving the message, cell signaling can be divided into three stages: signal reception, signal transduction, and cellular response. When reception occurs at the plasma membrane, as shown here, the transduction stage is usually a pathway of several steps, with each relay molecule in the pathway bringing about a change in the next molecule. The final molecule in the pathway triggers the cell's response. The three stages are explained in more detail in the text.

? *How does the epinephrine in Sutherland's experiment fit into this diagram of cell signaling?*

2. **Transduction.** The binding of the signaling molecule changes the receptor protein in some way, initiating the process of transduction. The transduction stage converts the signal to a form that can bring about a specific cellular response. In Sutherland's system, the binding of epinephrine to a receptor protein in a liver cell's plasma membrane leads to activation of glycogen phosphorylase. Transduction sometimes occurs in a single step but more often requires a sequence of changes in a series of different molecules—a *signal transduction pathway*. The molecules in the pathway are often called relay molecules.

3. **Response.** In the third stage of cell signaling, the transduced signal finally triggers a specific cellular response. The response may be almost any imaginable cellular activity—such as catalysis by an enzyme (for example, glycogen phosphorylase), rearrangement of the cytoskeleton, or activation of specific genes in the nucleus. The cell-signaling process helps ensure that crucial activities like these occur in the right cells, at the right time, and in proper coordination with the activities of other cells of the organism. We'll now explore the mechanisms of cell signaling in more detail, including a discussion of fine-tuning and termination of the process.

CONCEPT CHECK 11.1

1. Explain how signaling is involved in ensuring that yeast cells fuse only with cells of the opposite mating type.
2. Explain how nerve cells provide examples of both local and long-distance signaling.
3. **WHAT IF?** When epinephrine is mixed with glycogen phosphorylase and glycogen in a test tube, is glucose 1-phosphate generated? Why or why not?
4. In liver cells, glycogen phosphorylase acts in which of the three stages of the signaling pathway associated with an epinephrine-initiated signal?

For suggested answers, see Appendix A.

CONCEPT 11.2

Reception: A signaling molecule binds to a receptor protein, causing it to change shape

A radio station broadcasts its signal indiscriminately, but it can only be picked up by radios tuned to the right wavelength: Reception of the signal depends on the receiver. Similarly, the signals emitted by an **a** yeast cell are "heard" only by its prospective mates, α cells. In the case of epinephrine, the hormone encounters many types of cells as it circulates in the blood, but only certain target cells detect and react to the hormone molecule. A receptor protein on or in the target cell allows the cell to "hear" the signal and respond to it. The signaling molecule is complementary in shape to a specific site on the receptor and attaches there, like a key in a lock or a substrate in the catalytic site of an enzyme. The signaling molecule behaves as a **ligand**, the term for a molecule that specifically binds to another molecule, often a larger one. Ligand binding generally causes a receptor protein to undergo a change in shape. For many receptors, this shape change directly activates the receptor, enabling it to interact with other cellular molecules. For other kinds of receptors, the immediate effect of ligand binding is to cause the aggregation of two or more receptor molecules, which leads to further molecular events inside the cell.

Most signal receptors are plasma membrane proteins. Their ligands are water-soluble and generally too large to pass freely through the plasma membrane. Other signal receptors, however, are located inside the cell. We discuss both of these types next.

Receptors in the Plasma Membrane

Most water-soluble signaling molecules bind to specific sites on receptor proteins that span the cell's plasma membrane. Such a transmembrane receptor transmits information from the extracellular environment to the inside of the cell by changing shape or aggregating when a specific ligand binds to it. We can see how cell-surface transmembrane receptors work by looking at three major types: G protein-coupled receptors, receptor tyrosine kinases, and ion channel receptors. These receptors are discussed and illustrated in **Figure 11.7**, on the next three pages; study this figure before going on.

Cell-surface receptor molecules play crucial roles in the biological systems of animals, and not surprisingly, their malfunctions are associated with many human diseases, including cancer, heart disease, and asthma. Working out the structure and function of these receptors will allow us to better understand and treat these conditions. Therefore, this endeavor has been a major focus of both university research teams and the pharmaceutical industry. In spite of this effort, and although cell-surface receptors make up 30% of all human proteins, they make up only 1% of the proteins whose structures have been determined by X-ray crystallography (see Figure 5.24): Their structures are very challenging to determine.

The largest family of human cell-surface receptors consists of the nearly 1,000 G protein-coupled receptors (GPCRs). After persistent efforts, researchers have made significant

▼ **Figure 11.7**

Exploring Cell-Surface Transmembrane Receptors

G Protein-Coupled Receptors

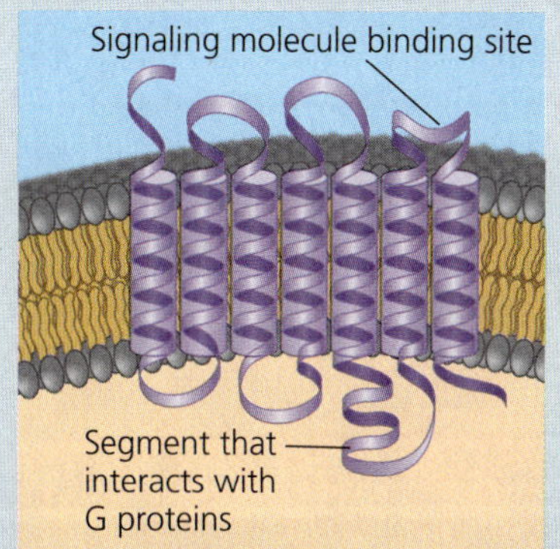

G protein-coupled receptor

A **G protein-coupled receptor** (GPCR) is a cell-surface transmembrane receptor that works with the help of a **G protein**, a protein that binds the energy-rich molecule GTP. Many different signaling molecules, including yeast mating factors, epinephrine and many other hormones, and neurotransmitters, use G protein-coupled receptors. These receptors vary in the binding sites for their signaling molecules (often referred to as their ligands) and also for different types of G proteins inside the cell. Nevertheless, G protein-coupled receptor proteins are all remarkably similar in structure.

In fact, they make up a large family of eukaryotic receptor proteins with a secondary structure in which the single polypeptide, represented here as a ribbon, has seven transmembrane α helices, outlined with cylinders and depicted in a row for clarity. Specific loops between the helices form binding sites for signaling and G protein molecules.

G protein-coupled receptor systems are extremely widespread and diverse in their functions, including roles in embryonic development and sensory reception. In humans, for example, vision, smell, and taste depend on such systems. Similarities in structure in G proteins and G protein-coupled receptors in diverse organisms suggest that G proteins and associated receptors evolved very early.

G protein systems are involved in many human diseases, including bacterial infections. The bacteria that cause cholera, pertussis (whooping cough), and botulism, among others, make their victims ill by producing toxins that interfere with G protein function. Pharmacologists now realize that up to 60% of all medicines used today exert their effects by influencing G protein pathways.

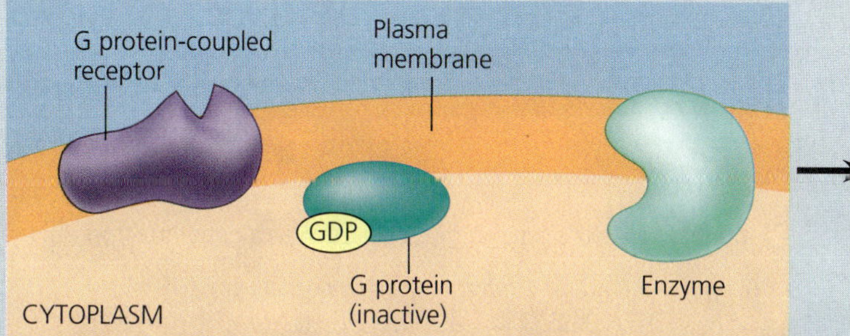

1 Loosely attached to the cytoplasmic side of the membrane, the G protein functions as a molecular switch that is either on or off, depending on which of two guanine nucleotides is attached, GDP or GTP—hence the term *G protein*. (GTP, or guanosine triphosphate, is similar to ATP.) When GDP is bound to the G protein, as shown above, the G protein is inactive. The receptor and G protein work together with another protein, usually an enzyme.

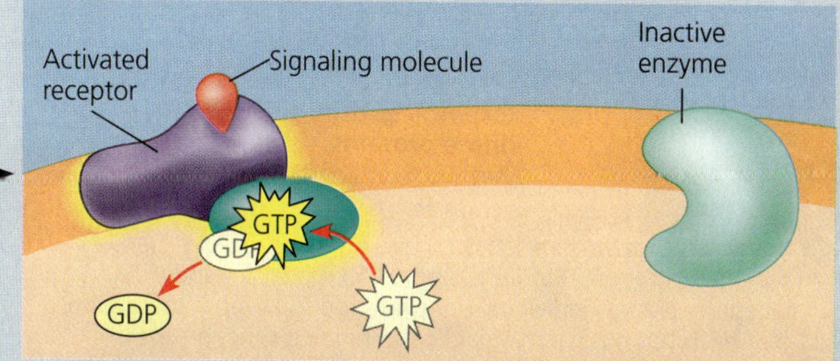

2 When the appropriate signaling molecule binds to the extracellular side of the receptor, the receptor is activated and changes shape. Its cytoplasmic side then binds an inactive G protein, causing a GTP to displace the GDP. This activates the G protein.

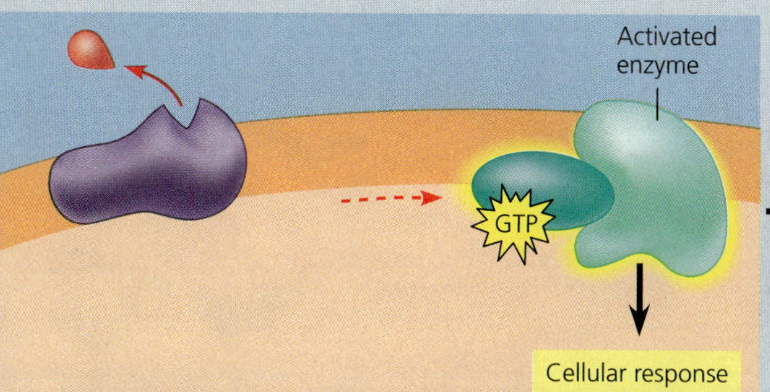

3 The activated G protein dissociates from the receptor, diffuses along the membrane, and then binds to an enzyme, altering the enzyme's shape and activity. Once activated, the enzyme can trigger the next step leading to a cellular response. (Binding of signaling molecules is reversible: Like other ligands, they bind and dissociate many times. The ligand concentration outside the cell determines how often a ligand is bound and causes signaling.)

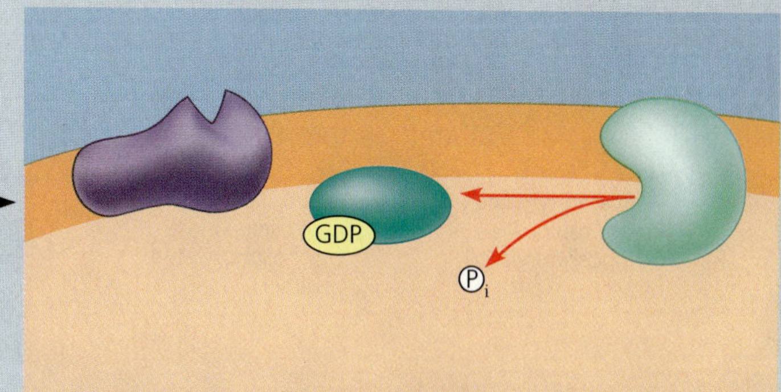

4 The changes in the enzyme and G protein are only temporary because the G protein also functions as a GTPase enzyme—in other words, it then hydrolyzes its bound GTP to GDP. Now inactive again, the G protein leaves the enzyme, which returns to its original state. The G protein is now available for reuse. The GTPase function of the G protein allows the pathway to shut down rapidly when the signaling molecule is no longer present.

Continued on next page

Exploring Cell-Surface Transmembrane Receptors

Receptor Tyrosine Kinases

Receptor tyrosine kinases (RTKs) belong to a major class of plasma membrane receptors characterized by having enzymatic activity. A *kinase* is an enzyme that catalyzes the transfer of phosphate groups. The part of the receptor protein extending into the cytoplasm functions as a tyrosine kinase, an enzyme that catalyzes the transfer of a phosphate group from ATP to the amino acid tyrosine on a substrate protein. Thus, receptor tyrosine kinases are membrane receptors that attach phosphates to tyrosines.

One receptor tyrosine kinase complex may activate ten or more different transduction pathways and cellular responses. Often, more than one signal transduction pathway can be triggered at once, helping the cell regulate and coordinate many aspects of cell growth and cell reproduction. The ability of a single ligand-binding event to trigger so many pathways is a key difference between receptor tyrosine kinases and G protein-coupled receptors. Abnormal receptor tyrosine kinases that function even in the absence of signaling molecules are associated with many kinds of cancer.

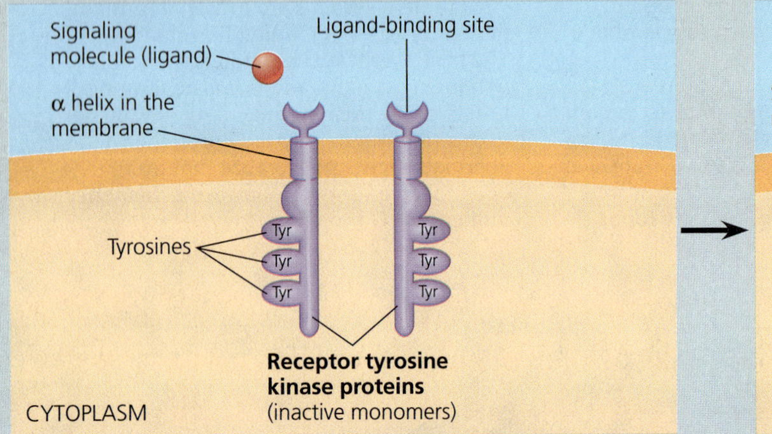

1 Many receptor tyrosine kinases have the structure depicted schematically here. Before the signaling molecule binds, the receptors exist as individual units referred to as monomers. Notice that each has an extracellular ligand-binding site, an α helix spanning the membrane, and an intracellular tail containing multiple tyrosines.

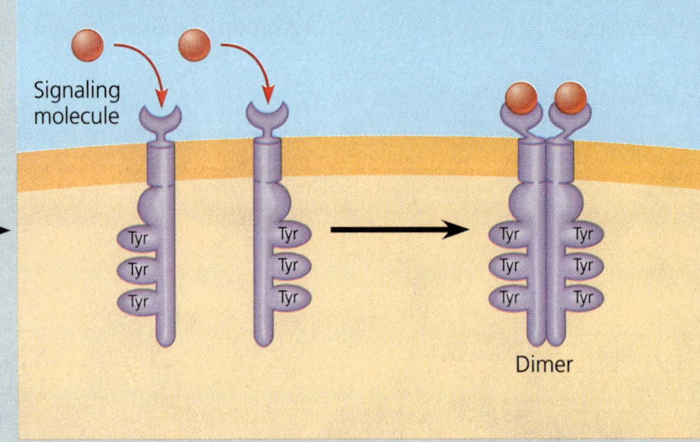

2 The binding of a signaling molecule (such as a growth factor) causes two receptor monomers to associate closely with each other, forming a complex known as a dimer (dimerization).

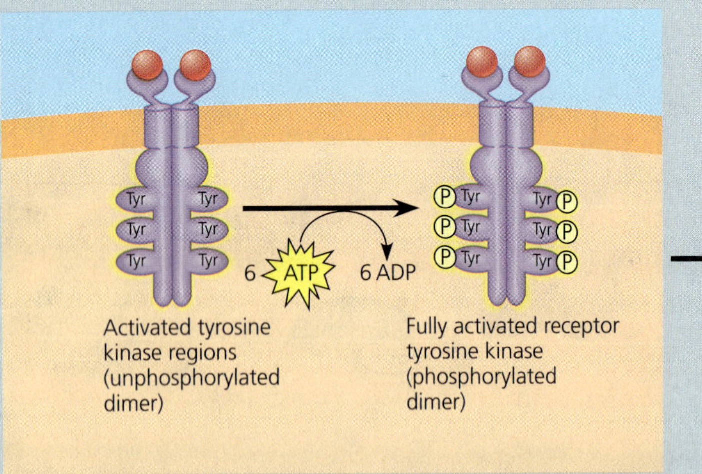

3 Dimerization activates the tyrosine kinase region of each monomer; each tyrosine kinase adds a phosphate from an ATP molecule to a tyrosine on the tail of the other monomer.

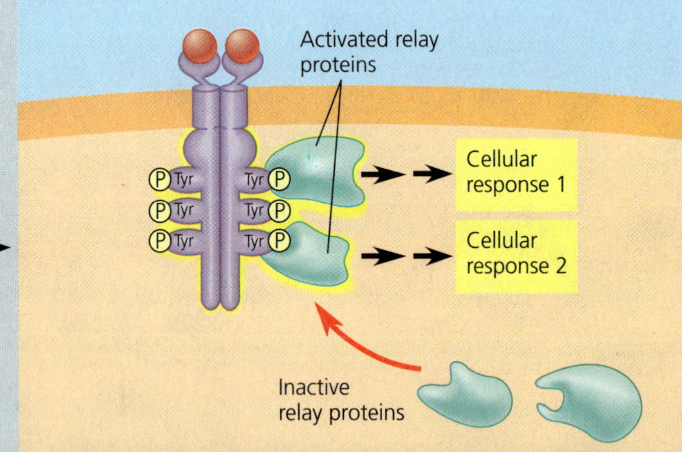

4 Now that the receptor is fully activated, it is recognized by specific relay proteins inside the cell. Each such protein binds to a specific phosphorylated tyrosine, undergoing a resulting structural change that activates the bound protein. Each activated protein triggers a transduction pathway, leading to a cellular response.

Ion Channel Receptors

A **ligand-gated ion channel** is a type of membrane receptor containing a region that can act as a "gate" when the receptor changes shape. When a signaling molecule binds as a ligand to the receptor protein, the gate opens or closes, allowing or blocking the flow of specific ions, such as Na^+ or Ca^{2+}, through a channel in the receptor. Like the other receptors we have discussed, these proteins bind the ligand at a specific site on their extracellular sides.

1 Here we show a ligand-gated ion channel receptor in which the gate remains closed until a ligand binds to the receptor.

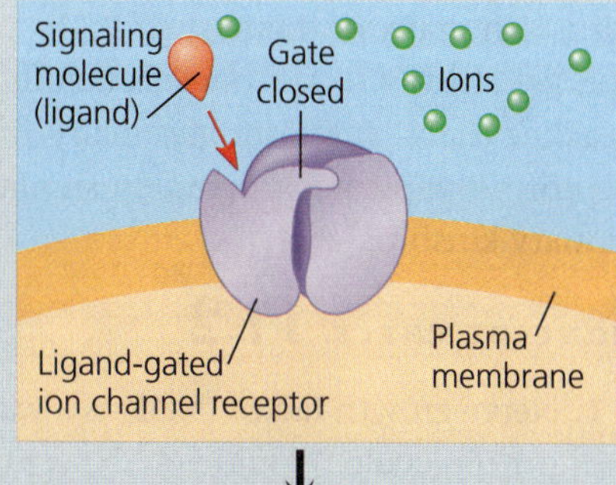

2 When the ligand binds to the receptor and the gate opens, specific ions can flow through the channel and rapidly change the concentration of that particular ion inside the cell. This change may directly affect the activity of the cell in some way.

Gate open
Cellular response

3 When the ligand dissociates from this receptor, the gate closes and ions no longer enter the cell.

Gate closed

Ligand-gated ion channels are very important in the nervous system. For example, the neurotransmitter molecules released at a synapse between two nerve cells (see Figure 11.5b) bind as ligands to ion channels on the receiving cell, causing the channels to open. Ions flow in (or, in some cases, out), triggering an electrical signal that propagates down the length of the receiving cell. Some gated ion channels are controlled by electrical signals instead of ligands; these *voltage gated ion channels* are also crucial to the functioning of the nervous system, as we will discuss in Chapter 48.

MAKE CONNECTIONS *Examine the ion channel protein in Figure 7.1 (p. 125) and the discussion of it on page 135. What type of stimulus opens that ion channel? According to the information given above, what type of ion channel is described?*

▼ **Figure 11.8**

IMPACT

Determining the Structure of a G Protein-Coupled Receptor (GPCR)

GPCRs are flexible and inherently unstable, so they have been difficult to crystallize, a required step in determining their structure by X-ray crystallography. Recently, however, researchers have crystallized the human β_2-adrenergic receptor in the presence of a ligand similar to the natural one (green in the model below) and cholesterol (orange), which stabilized the receptor enough for its structure to be determined. Two receptor molecules (blue) are shown here as ribbon models in a side view within the plasma membrane.

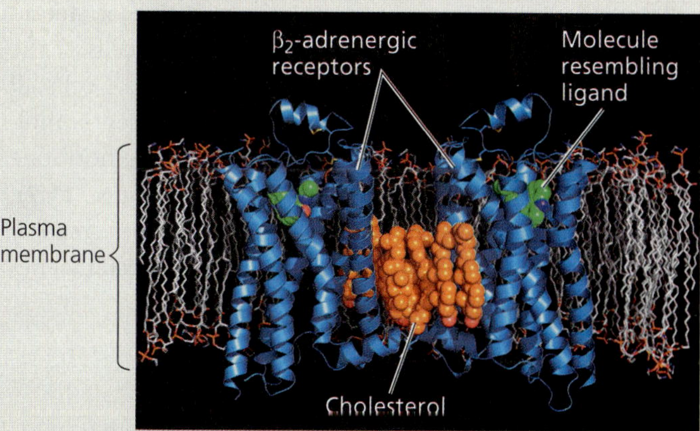

WHY IT MATTERS The β_2-adrenergic receptor is found on smooth muscle cells throughout the body, and abnormal forms of it are associated with diseases such as asthma, hypertension, and heart failure. Current drugs used for these conditions produce unwanted side effects, and further research may yield better drugs. Also, since GPCRs share structural similarities, this work on the β_2-adrenergic receptor will aid development of treatments for diseases associated with other GPCRs.

FURTHER READING R. Ranganathan, Signaling across the cell membrane, *Science* 318:1253–1254 (2007).

WHAT IF? The model shown above represents the receptor in an inactive state, not bound to a G protein. Can you suggest conditions for crystallizing the protein that would reveal the structure of the receptor while it is actively signaling to the inside of the cell?

breakthroughs in elucidating the structure of several G protein-coupled receptors over the past few years **(Figure 11.8)**.

Abnormal functioning of receptor tyrosine kinases (RTKs) is associated with many types of cancers. For example, patients with breast cancer cells that have excessive levels of a receptor tyrosine kinase called HER2 have a poor prognosis. Using molecular biological techniques, researchers have developed a protein called Herceptin that binds to HER2 on cells and inhibits their growth, thus thwarting further tumor development. In some clinical studies, treatment with Herceptin improved patient survival rates by more than one-third. One goal of ongoing research into these cell-surface receptors and other cell-signaling proteins is development of additional successful treatments.

Intracellular Receptors

Intracellular receptor proteins are found in either the cytoplasm or nucleus of target cells. To reach such a receptor, a chemical messenger passes through the target cell's plasma membrane. A number of important signaling molecules can do this because they are either hydrophobic enough or small enough to cross the hydrophobic interior of the membrane. Such hydrophobic chemical messengers include the steroid hormones and thyroid hormones of animals. Another chemical signaling molecule with an intracellular receptor is nitric oxide (NO), a gas; its very small molecules readily pass between the membrane phospholipids.

The behavior of testosterone is representative of steroid hormones. In males, the hormone is secreted by cells of the testes. It then travels through the blood and enters cells all over the body. However, only cells that contain receptor molecules for testosterone respond. In these cells, the hormone binds to the receptor protein, activating it **(Figure 11.9)**. With the hormone attached, the active form of the receptor protein then enters the nucleus and turns on specific genes that control male sex characteristics.

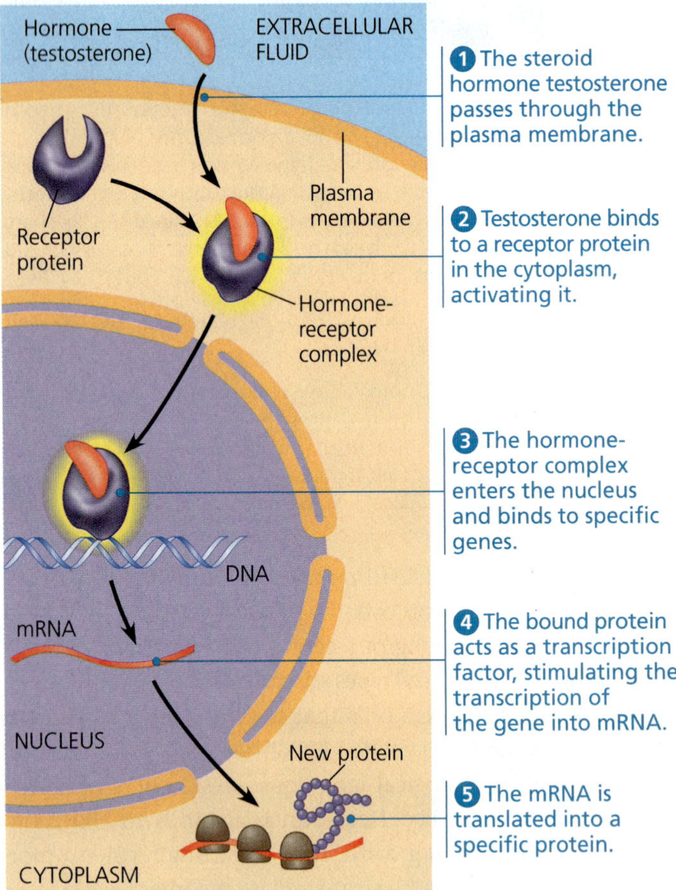

▲ **Figure 11.9 Steroid hormone interacting with an intracellular receptor.**

? *Why is a cell-surface receptor protein not required for this steroid hormone to enter the cell?*

How does the activated hormone-receptor complex turn on genes? Recall that the genes in a cell's DNA function by being transcribed and processed into messenger RNA (mRNA), which leaves the nucleus and is translated into a specific protein by ribosomes in the cytoplasm (see Figure 5.25). Special proteins called *transcription factors* control which genes are turned on—that is, which genes are transcribed into mRNA—in a particular cell at a particular time. The testosterone receptor, when activated, acts as a transcription factor that turns on specific genes.

By acting as a transcription factor, the testosterone receptor itself carries out the complete transduction of the signal. Most other intracellular receptors function in the same way, although many of them, such as the thyroid hormone receptor, are already in the nucleus before the signaling molecule reaches them. Interestingly, many of these intracellular receptor proteins are structurally similar, suggesting an evolutionary kinship.

CONCEPT CHECK 11.2

1. Nerve growth factor (NGF) is a water-soluble signaling molecule. Would you expect the receptor for NGF to be intracellular or in the plasma membrane? Why?
2. **WHAT IF?** What would the effect be if a cell made defective receptor tyrosine kinase proteins that were unable to dimerize?
3. **MAKE CONNECTIONS** How is ligand binding similar to the process of allosteric regulation of enzymes? See Figure 8.19 on page 158.

For suggested answers, see Appendix A.

CONCEPT 11.3

Transduction: Cascades of molecular interactions relay signals from receptors to target molecules in the cell

When receptors for signaling molecules are plasma membrane proteins, like most of those we have discussed, the transduction stage of cell signaling is usually a multistep pathway. Steps often include activation of proteins by addition or removal of phosphate groups or release of other small molecules or ions that act as messengers. One benefit of multiple steps is the possibility of greatly amplifying a signal. If some of the molecules in a pathway transmit the signal to numerous molecules at the next step in the series, the result can be a large number of activated molecules at the end of the pathway. Moreover, multistep pathways provide more opportunities for coordination and regulation than simpler systems do. This allows fine-tuning of the response, in both unicellular and multicellular organisms, as we'll discuss later in the chapter.

Signal Transduction Pathways

The binding of a specific signaling molecule to a receptor in the plasma membrane triggers the first step in the chain of molecular interactions—the signal transduction pathway—that leads to a particular response within the cell. Like falling dominoes, the signal-activated receptor activates another molecule, which activates yet another molecule, and so on, until the protein that produces the final cellular response is activated. The molecules that relay a signal from receptor to response, which we call relay molecules in this book, are often proteins. The interaction of proteins is a major theme of cell signaling. Indeed, protein interaction is a unifying theme of all regulation at the cellular level.

Keep in mind that the original signaling molecule is not physically passed along a signaling pathway; in most cases, it never even enters the cell. When we say that the signal is relayed along a pathway, we mean that certain information is passed on. At each step, the signal is transduced into a different form, commonly a shape change in a protein. Very often, the shape change is brought about by phosphorylation.

Protein Phosphorylation and Dephosphorylation

Previous chapters introduced the concept of activating a protein by adding one or more phosphate groups to it (see Figure 8.10a). In Figure 11.7, we have already seen how phosphorylation is involved in the activation of receptor tyrosine kinases. In fact, the phosphorylation and dephosphorylation of proteins is a widespread cellular mechanism for regulating protein activity. An enzyme that transfers phosphate groups from ATP to a protein is generally known as a **protein kinase**. Recall that a receptor tyrosine kinase phosphorylates tyrosines on the other receptor tyrosine kinase in a dimer. Most cytoplasmic protein kinases, however, act on proteins different from themselves. Another distinction is that most cytoplasmic protein kinases phosphorylate either of two other amino acids, serine or threonine, rather than tyrosine. Such serine/threonine kinases are widely involved in signaling pathways in animals, plants, and fungi.

Many of the relay molecules in signal transduction pathways are protein kinases, and they often act on other protein kinases in the pathway. **Figure 11.10** depicts a hypothetical

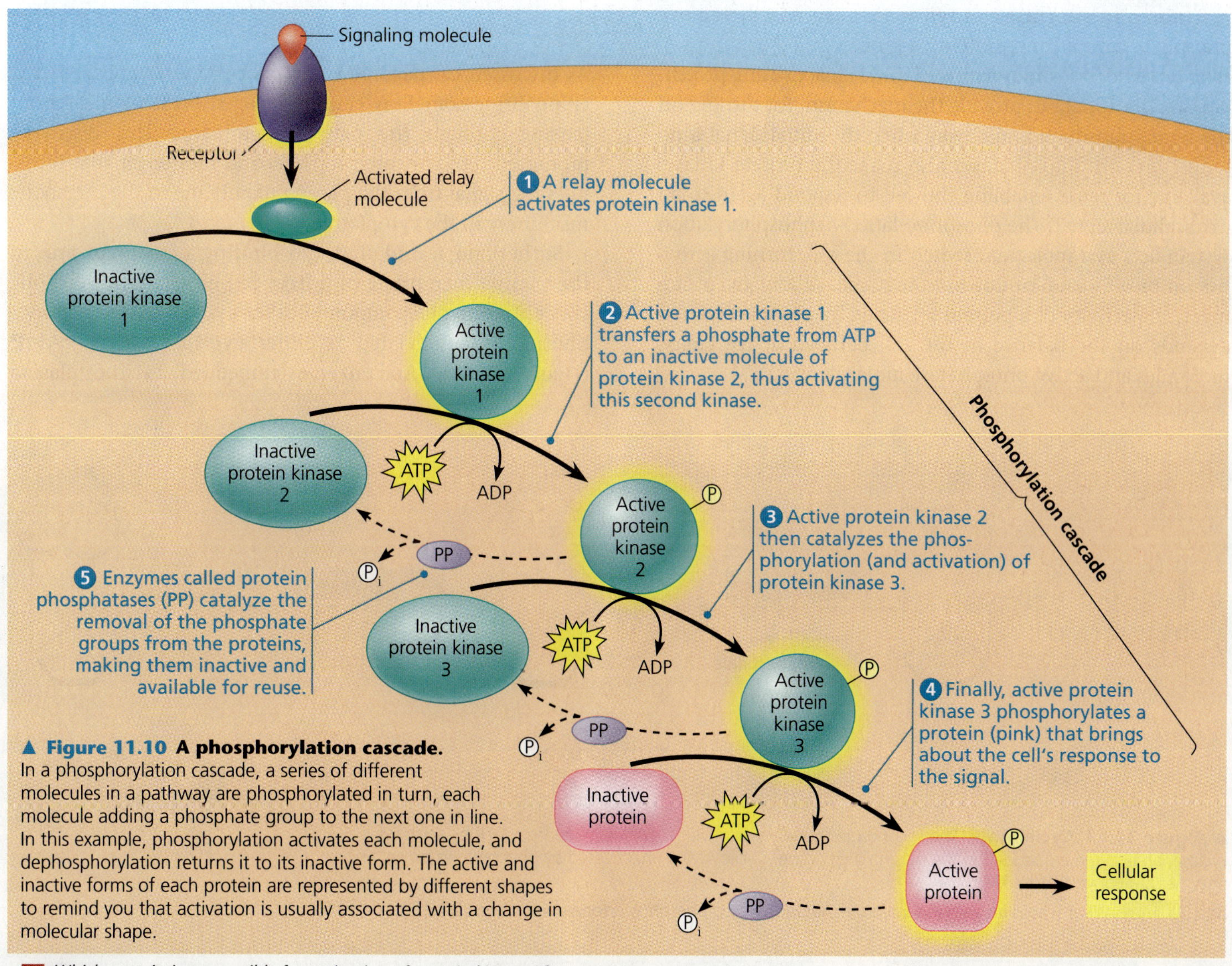

▲ **Figure 11.10 A phosphorylation cascade.** In a phosphorylation cascade, a series of different molecules in a pathway are phosphorylated in turn, each molecule adding a phosphate group to the next one in line. In this example, phosphorylation activates each molecule, and dephosphorylation returns it to its inactive form. The active and inactive forms of each protein are represented by different shapes to remind you that activation is usually associated with a change in molecular shape.

? *Which protein is responsible for activation of protein kinase 3?*

pathway containing three different protein kinases that create a "phosphorylation cascade." The sequence shown is similar to many known pathways, including those triggered in yeast by mating factors and in animal cells by many growth factors. The signal is transmitted by a cascade of protein phosphorylations, each bringing with it a shape change. Each such shape change results from the interaction of the newly added phosphate groups with charged or polar amino acids (see Figure 5.16). The addition of phosphate groups often changes a protein from an inactive form to an active form. In other cases, though, phosphorylation *decreases* the activity of the protein.

The importance of protein kinases can hardly be overstated. About 2% of our own genes are thought to code for protein kinases. A single cell may have hundreds of different kinds, each specific for a different substrate protein. Together, they probably regulate a large proportion of the thousands of proteins in a cell. Among these are most of the proteins that, in turn, regulate cell reproduction. Abnormal activity of such a kinase can cause abnormal cell growth and contribute to the development of cancer.

Equally important in the phosphorylation cascade are the **protein phosphatases**, enzymes that can rapidly remove phosphate groups from proteins, a process called dephosphorylation. By dephosphorylating and thus inactivating protein kinases, phosphatases provide the mechanism for turning off the signal transduction pathway when the initial signal is no longer present. Phosphatases also make the protein kinases available for reuse, enabling the cell to respond again to an extracellular signal. The phosphorylation-dephosphorylation system acts as a molecular switch in the cell, turning activities on or off, or up or down, as required. At any given moment, the activity of a protein regulated by phosphorylation depends on the balance in the cell between active kinase molecules and active phosphatase molecules.

Small Molecules and Ions as Second Messengers

Not all components of signal transduction pathways are proteins. Many signaling pathways also involve small, nonprotein, water-soluble molecules or ions called **second messengers**. (This term is used because the pathway's "first messenger" is considered to be the extracellular signaling molecule—the ligand—that binds to the membrane receptor.) Because second messengers are small and water-soluble, they can readily spread throughout the cell by diffusion. For example, as we'll see shortly, a second messenger called cyclic AMP carries the signal initiated by epinephrine from the plasma membrane of a liver or muscle cell into the cell's interior, where the signal eventually brings about glycogen breakdown. Second messengers participate in pathways that are initiated by both G protein-coupled receptors and receptor tyrosine kinases. The two most widely used second messengers are cyclic AMP and calcium ions, Ca^{2+}. A large variety of relay proteins are sensitive to the cytosolic concentration of one or the other of these second messengers.

Cyclic AMP

As discussed on page 209, Earl Sutherland established that epinephrine somehow causes glycogen breakdown without passing through the plasma membrane. This discovery prompted him to search for a second messenger that transmits the signal from the plasma membrane to the metabolic machinery in the cytoplasm.

Sutherland found that the binding of epinephrine to the plasma membrane of a liver cell elevates the cytosolic concentration of a compound called cyclic adenosine monophosphate, abbreviated as either **cyclic AMP** or **cAMP** (**Figure 11.11**). An enzyme embedded in the plasma

▲ **Figure 11.11 Cyclic AMP.** The second messenger cyclic AMP (cAMP) is made from ATP by adenylyl cyclase, an enzyme embedded in the plasma membrane. Cyclic AMP is inactivated by phosphodiesterase, an enzyme that converts it to AMP.

WHAT IF? *What would happen if a molecule that inactivated phosphodiesterase were introduced into the cell?*

membrane, **adenylyl cyclase**, converts ATP to cAMP in response to an extracellular signal—in this case, provided by epinephrine. But epinephrine doesn't stimulate adenylyl cyclase directly. When epinephrine outside the cell binds to a specific receptor protein, the protein activates adenylyl cyclase, which in turn can catalyze the synthesis of many molecules of cAMP. In this way, the normal cellular concentration of cAMP can be boosted 20-fold in a matter of seconds. The cAMP broadcasts the signal to the cytoplasm. It does not persist for long in the absence of the hormone because another enzyme, called phosphodiesterase, converts cAMP to AMP. Another surge of epinephrine is needed to boost the cytosolic concentration of cAMP again.

Subsequent research has revealed that epinephrine is only one of many hormones and other signaling molecules that trigger the formation of cAMP. It has also brought to light the other components of cAMP pathways, including G proteins, G protein-coupled receptors, and protein kinases **(Figure 11.12)**. The immediate effect of cAMP is usually the activation of a serine/threonine kinase called *protein kinase A*. The activated protein kinase A then phosphorylates various other proteins, depending on the cell type. (The complete pathway for epinephrine's stimulation of glycogen breakdown is shown later, in Figure 11.16.)

Further regulation of cell metabolism is provided by other G protein systems that *inhibit* adenylyl cyclase. In these systems, a different signaling molecule activates a different receptor, which in turn activates an *inhibitory* G protein.

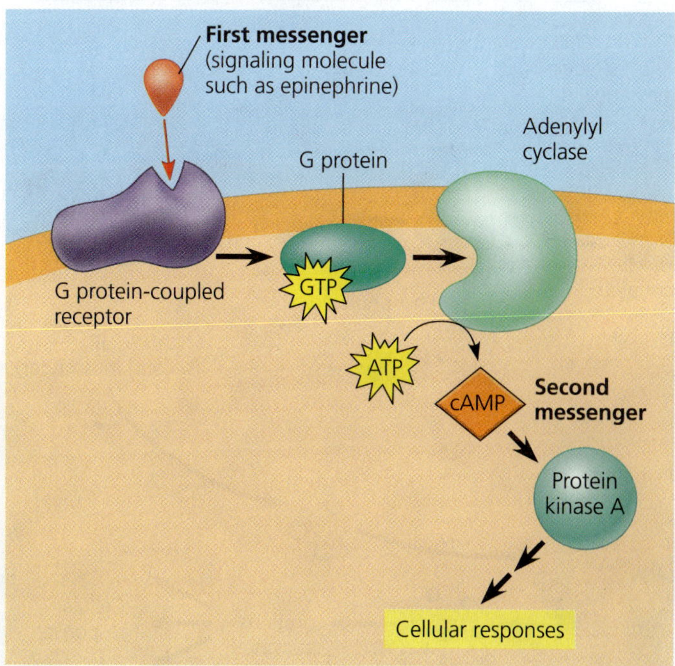

▲ **Figure 11.12 cAMP as a second messenger in a G protein signaling pathway.** The first messenger activates a G protein-coupled receptor, which activates a specific G protein. In turn, the G protein activates adenylyl cyclase, which catalyzes the conversion of ATP to cAMP. The cAMP then acts as a second messenger and activates another protein, usually protein kinase A, leading to cellular responses.

Now that we know about the role of cAMP in G protein signaling pathways, we can explain in molecular detail how certain microbes cause disease. Consider cholera, a disease that is frequently epidemic in places where the water supply is contaminated with human feces. People acquire the cholera bacterium, *Vibrio cholerae*, by drinking contaminated water. The bacteria form a biofilm on the lining of the small intestine and produce a toxin. The cholera toxin is an enzyme that chemically modifies a G protein involved in regulating salt and water secretion. Because the modified G protein is unable to hydrolyze GTP to GDP, it remains stuck in its active form, continuously stimulating adenylyl cyclase to make cAMP. The resulting high concentration of cAMP causes the intestinal cells to secrete large amounts of salts into the intestines, with water following by osmosis. An infected person quickly develops profuse diarrhea and if left untreated can soon die from the loss of water and salts.

Our understanding of signaling pathways involving cyclic AMP or related messengers has allowed us to develop treatments for certain conditions in humans. In one pathway, *cyclic GMP*, or *cGMP*, acts as a signaling molecule whose effects include relaxation of smooth muscle cells in artery walls. A compound that inhibits the hydrolysis of cGMP to GMP, thus prolonging the signal, was originally prescribed for chest pains because it increased blood flow to the heart muscle. Under the trade name Viagra, this compound is now widely used as a treatment for erectile dysfunction in human males. Because Viagra leads to dilation of blood vessels, it also allows increased blood flow to the penis, optimizing physiological conditions for penile erections.

Calcium Ions and Inositol Trisphosphate (IP_3)

Many signaling molecules in animals, including neurotransmitters, growth factors, and some hormones, induce responses in their target cells via signal transduction pathways that increase the cytosolic concentration of calcium ions (Ca^{2+}). Calcium is even more widely used than cAMP as a second messenger. Increasing the cytosolic concentration of Ca^{2+} causes many responses in animal cells, including muscle cell contraction, secretion of certain substances, and cell division. In plant cells, a wide range of hormonal and environmental stimuli can cause brief increases in cytosolic Ca^{2+} concentration, triggering various signaling pathways, such as the pathway for greening in response to light (see Figure 39.4). Cells use Ca^{2+} as a second messenger in both G protein and receptor tyrosine kinase pathways.

Although cells always contain some Ca^{2+}, this ion can function as a second messenger because its concentration in the cytosol is normally much lower than the concentration outside the cell **(Figure 11.13)**. In fact, the level of Ca^{2+} in the blood and extracellular fluid of an animal is often more than 10,000 times higher than that in the cytosol. Calcium ions are actively transported out of the cell and are actively imported from the cytosol into the endoplasmic reticulum (and, under some conditions, into mitochondria and chloroplasts) by various protein pumps. As a result, the calcium concentration in the ER is usually much higher than that in the cytosol. Because the cytosolic calcium level is low, a small change in absolute numbers of ions represents a relatively large percentage change in calcium concentration.

In response to a signal relayed by a signal transduction pathway, the cytosolic calcium level may rise, usually by a mechanism that releases Ca^{2+} from the cell's ER. The pathways leading to calcium release involve still other second messengers, **inositol trisphosphate (IP_3)** and **diacylglycerol (DAG)**. These two messengers are produced by cleavage of a certain kind of phospholipid in the plasma membrane. **Figure 11.14** shows how this occurs and how IP_3 stimulates the release of calcium from the ER. Because IP_3 acts before calcium in these pathways, calcium could be considered a "*third* messenger." However, scientists use the term *second messenger* for all small, nonprotein components of signal transduction pathways.

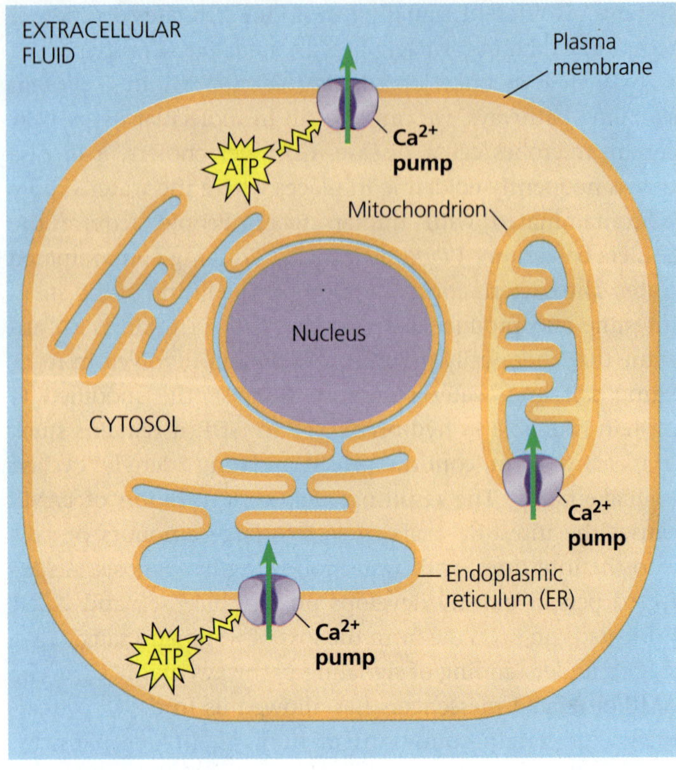

▲ **Figure 11.13 The maintenance of calcium ion concentrations in an animal cell.** The Ca^{2+} concentration in the cytosol is usually much lower (beige) than in the extracellular fluid and ER (blue). Protein pumps in the plasma membrane and the ER membrane, driven by ATP, move Ca^{2+} from the cytosol into the extracellular fluid and into the lumen of the ER. Mitochondrial pumps, driven by chemiosmosis (see Chapter 9), move Ca^{2+} into mitochondria when the calcium level in the cytosol rises significantly.

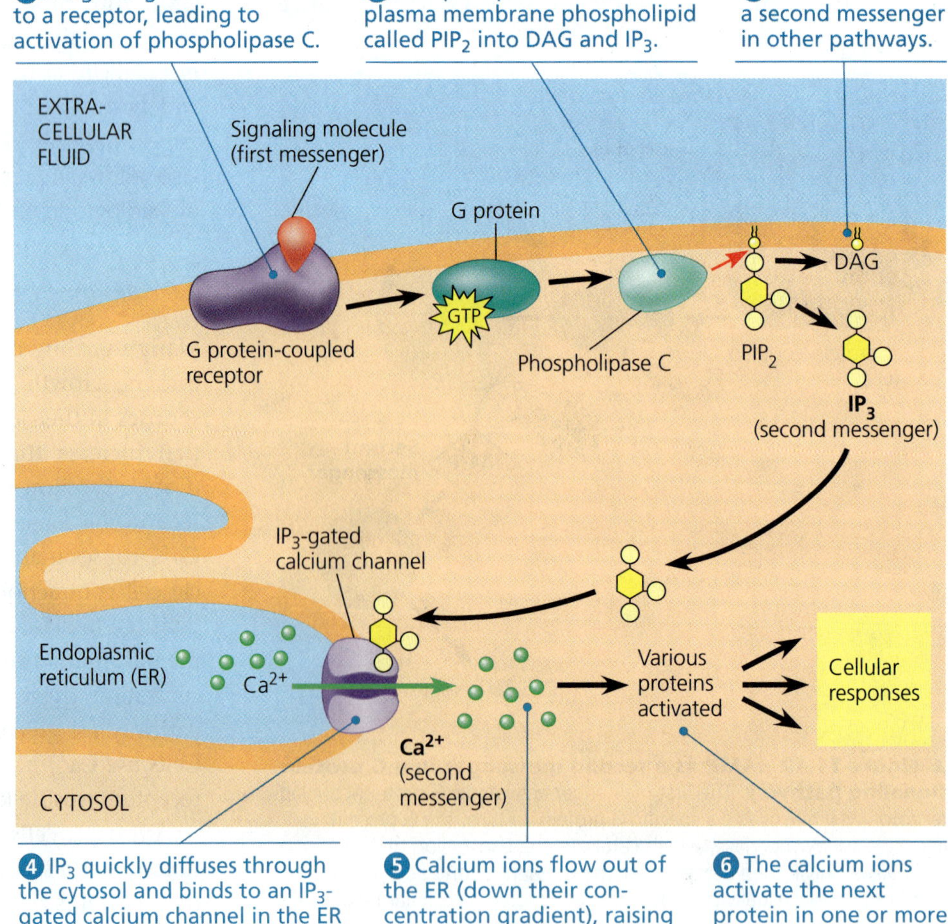

► **Figure 11.14 Calcium and IP_3 in signaling pathways.** Calcium ions (Ca^{2+}) and inositol trisphosphate (IP_3) function as second messengers in many signal transduction pathways. In this figure, the process is initiated by the binding of a signaling molecule to a G protein-coupled receptor. A receptor tyrosine kinase could also initiate this pathway by activating phospholipase C.

CONCEPT CHECK 11.3

1. What is a protein kinase, and what is its role in a signal transduction pathway?
2. When a signal transduction pathway involves a phosphorylation cascade, how does the cell's response get turned off?
3. What is the actual "signal" that is being transduced in any signal transduction pathway, such as those shown in Figures 11.6 and 11.10? In what way is this information being passed from the exterior to the interior of the cell?
4. **WHAT IF?** Upon activation of phospholipase C by the binding of a ligand to a receptor, what effect does the IP_3-gated calcium channel have on Ca^{2+} concentration in the cytosol?

For suggested answers, see Appendix A.

CONCEPT 11.4

Response: Cell signaling leads to regulation of transcription or cytoplasmic activities

We now take a closer look at the cell's subsequent response to an extracellular signal—what some researchers call the "output response." What is the nature of the final step in a signaling pathway?

Nuclear and Cytoplasmic Responses

Ultimately, a signal transduction pathway leads to the regulation of one or more cellular activities. The response at the end of the pathway may occur in the nucleus of the cell or in the cytoplasm.

Many signaling pathways ultimately regulate protein synthesis, usually by turning specific genes on or off in the nucleus. Like an activated steroid receptor (see Figure 11.9), the final activated molecule in a signaling pathway may function as a transcription factor. **Figure 11.15** shows an example in which a signaling pathway activates a transcription factor that turns a gene on: The response to the growth factor signal is transcription, the synthesis of mRNA, which will be translated in the cytoplasm into a specific protein. In other cases, the transcription factor might regulate a gene by turning it off. Often a transcription factor regulates several different genes.

Sometimes a signaling pathway may regulate the *activity* of proteins rather than their *synthesis*, directly affecting proteins that function outside the nucleus. For example, a signal may cause the opening or closing of an ion channel in the plasma membrane or a change in cell metabolism. As we have seen, the response of liver cells to the hormone epinephrine helps regulate cellular energy metabolism by affecting the activity of an enzyme. The final step in the signaling pathway that begins with epinephrine binding activates the enzyme that catalyzes the breakdown of glycogen. **Figure 11.16**, on the next page, shows the complete pathway leading to the release of glucose 1-phosphate molecules from glycogen. Notice that as each molecule is activated, the response is amplified, a subject we'll return to shortly.

In addition to controlling enzymes, signaling events may regulate other cellular attributes, even activities of the cell as a whole. An example of the latter can be found in the processes leading to the mating of yeast cells (see Figure 11.2). Yeast cells are not motile; their mating process depends on the growth of localized projections of one cell toward a cell of the

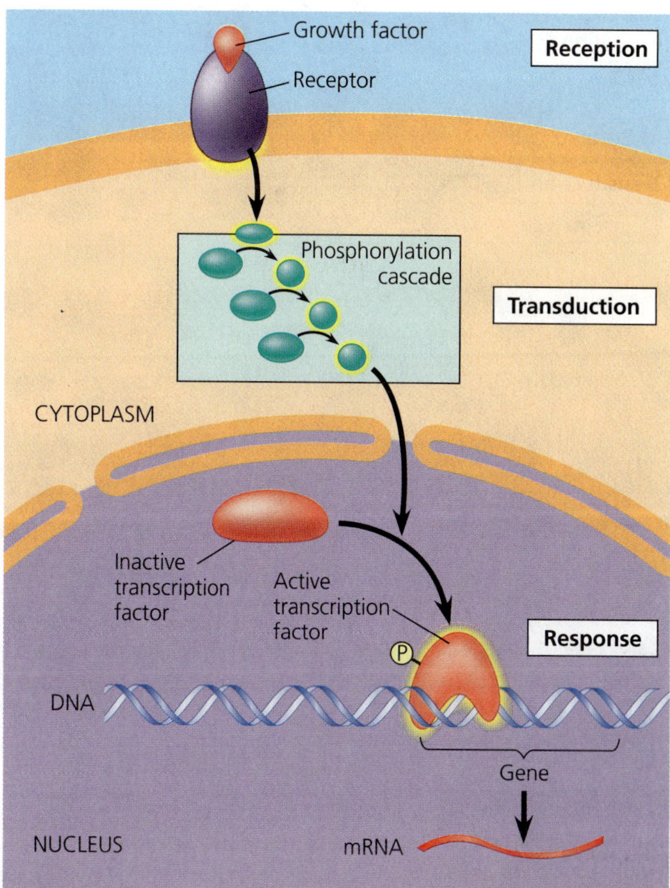

▲ **Figure 11.15 Nuclear responses to a signal: the activation of a specific gene by a growth factor.** This diagram is a simplified representation of a typical signaling pathway that leads to the regulation of gene activity in the cell nucleus. The initial signaling molecule, a local regulator called a growth factor, triggers a phosphorylation cascade, as in Figure 11.10. (The ATP molecules and phosphate groups are not shown.) Once phosphorylated, the last kinase in the sequence enters the nucleus and there activates a gene-regulating protein, a transcription factor. This protein stimulates transcription of a specific gene (or genes). The resulting mRNA then directs the synthesis of a particular protein in the cytoplasm.

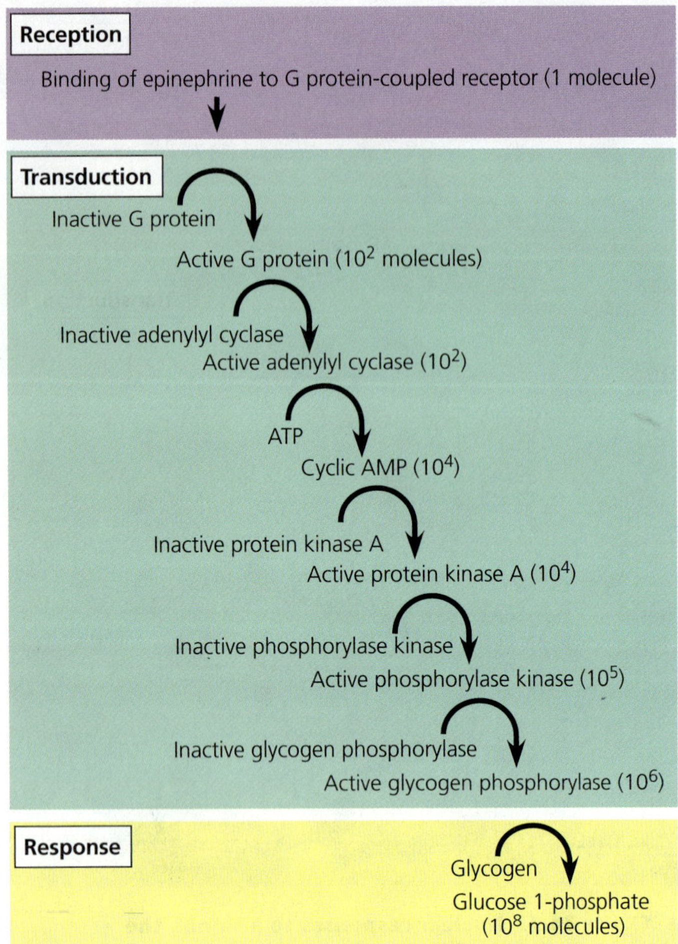

▲ **Figure 11.16 Cytoplasmic response to a signal: the stimulation of glycogen breakdown by epinephrine.** In this signaling system, the hormone epinephrine acts through a G protein-coupled receptor to activate a succession of relay molecules, including cAMP and two protein kinases (see also Figure 11.12). The final protein activated is the enzyme glycogen phosphorylase, which uses inorganic phosphate to release glucose monomers from glycogen in the form of glucose 1-phosphate molecules. This pathway amplifies the hormonal signal: One receptor protein can activate about 100 molecules of G protein, and each enzyme in the pathway, once activated, can act on many molecules of its substrate, the next molecule in the cascade. The number of activated molecules given for each step is approximate.

opposite mating type. As shown in **Figure 11.17**, binding of the mating factor causes this directional growth. When the mating factor binds, it activates signaling pathway kinases that affect the growth and orientation of cytoskeletal microfilaments. Because activation of signaling kinases is coupled in this way to cytoskeletal dynamics, cell projections emerge from regions of the plasma membrane exposed to the highest concentration of the mating factor. As a result, these projections are oriented toward the cell of the opposite mating type, which is the source of the signaling molecule.

The signal receptors, relay molecules, and second messengers introduced so far in this chapter participate in a variety of pathways, leading to both nuclear and cytoplasmic responses. Some of these pathways lead to cell division. The molecular messengers that initiate cell division pathways include growth factors and certain plant and animal hormones. Malfunctioning of growth factor pathways like the one in Figure 11.15 can contribute to the development of cancer, as we will see in Chapter 18.

Fine-Tuning of the Response

Regardless of whether the response occurs in the nucleus or in the cytoplasm, it is fine-tuned at multiple points rather than simply being turned "on" or "off." Here we'll consider four aspects of fine-tuning. First, as mentioned earlier, a signaling pathway with numerous steps between the initial signaling event at the cell surface and the cell's response results in amplification of the signal and thus the response. Second, such a multistep pathway has many different points at which the cell's response can be regulated, contributing to the specificity of the response and allowing coordination with other signaling pathways. Third, the overall efficiency of the response is enhanced by the presence of proteins known as scaffolding proteins. Finally, a crucial point in fine-tuning the response is the termination of the signal.

Signal Amplification

Elaborate enzyme cascades amplify the cell's response to a signal. At each catalytic step in the cascade, the number of activated products is much greater than in the preceding step. For example, in the epinephrine-triggered pathway in Figure 11.16, each adenylyl cyclase molecule catalyzes the formation of many cAMP molecules, each molecule of protein kinase A phosphorylates many molecules of the next kinase in the pathway, and so on. The amplification effect stems from the fact that these proteins persist in the active form long enough to process numerous molecules of substrate before they become inactive again. As a result of the signal's amplification, a small number of epinephrine molecules binding to receptors on the surface of a liver cell or muscle cell can lead to the release of hundreds of millions of glucose molecules from glycogen.

The Specificity of Cell Signaling and Coordination of the Response

Consider two different cells in your body—a liver cell and a heart muscle cell, for example. Both are in contact with your bloodstream and are therefore constantly exposed to many different hormone molecules, as well as to local regulators secreted by nearby cells. Yet the liver cell responds to some signals but ignores others, and the same is true for the heart cell. And some kinds of signals trigger responses in both cells—but different responses. For instance, epinephrine stimulates the liver cell to break down glycogen, but the main response of the heart cell to epinephrine is contraction, leading to a more rapid heartbeat. How do we account for this difference?

▼ **Figure 11.17** INQUIRY

How do signals induce directional cell growth during mating in yeast?

EXPERIMENT When a yeast cell binds mating factor molecules from a cell of the opposite mating type, a signaling pathway causes it to grow a projection toward the potential mate. The cell with the projection is called a "shmoo" because it resembles a 1950s cartoon character by that name. Dina Matheos and colleagues in Mark Rose's lab at Princeton University sought to determine how mating factor signaling is linked to this asymmetrical growth. Previous work had shown that activation of Fus3, one of the kinases in the signaling cascade, caused it to move to the membrane near where the factor bound. Preliminary experiments by these researchers identified formin, a protein that directs the construction of microfilaments, as a phosphorylation target of Fus3 kinase. To examine the role of Fus3 and formin in shmoo formation, the researchers generated two mutant yeast strains: one that no longer had the kinase (this strain is called ΔFus3) and one that lacked the formin (Δformin). To observe the effects of these mutations on cell growth induced by the mating factor, the cell walls of each strain were first stained with a green fluorescent dye. These green-stained cells were then exposed to mating factor and stained with a red fluorescent dye that labeled new cell wall growth. Images taken of the cells after the staining procedure were then compared with a similarly treated strain that expressed Fus3 and formin (the wild type).

RESULTS The cells of the wild-type strain showed shmoo projections, whose walls were stained red, while the rest of their cell walls were green, indicating asymmetrical growth. Cells of both the ΔFus3 and Δformin strains showed no shmoo formation, and their cell walls were stained almost uniformly yellow. This color resulted from merged green and red stains, indicating symmetrical growth, characteristic of cells not exposed to mating factor.

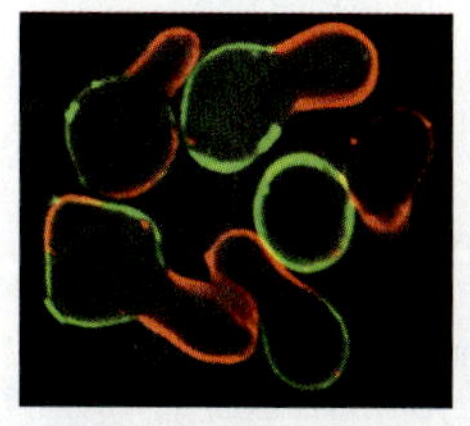
Wild type (with shmoos)

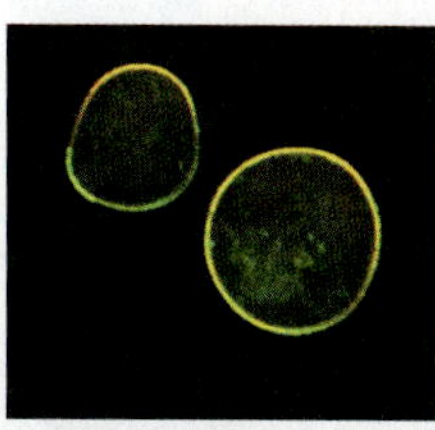
ΔFus3

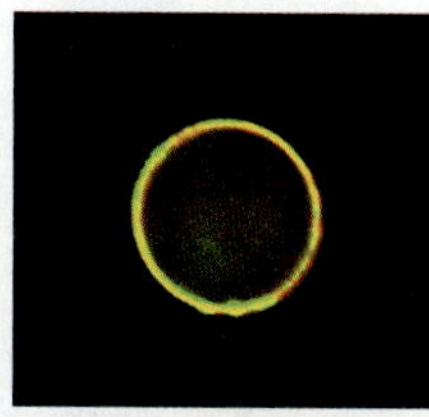
Δformin

CONCLUSION The similar defect (lack of ability to form shmoos) in strains lacking either Fus3 or formin suggests that both proteins are required for shmoo formation. These results led the investigators to propose the model shown here for the induction of asymmetrical growth in the receiving cell directed toward the cell of the opposite mating type.

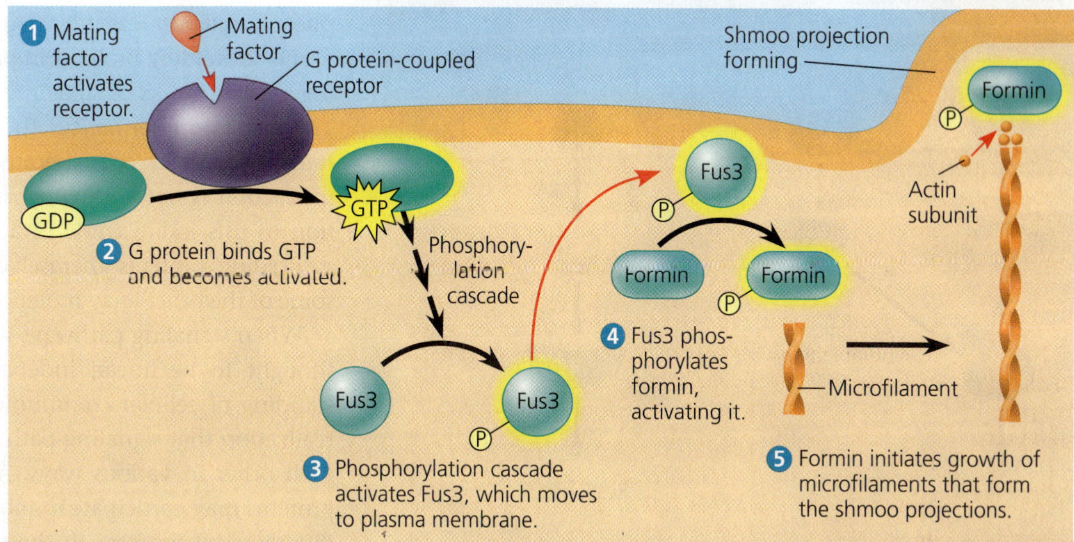

SOURCE D. Matheos et al., Pheromone-induced polarization is dependent on the Fus3p MAPK acting through the formin Bni1p, *Journal of Cell Biology* 165:99–109 (2004).

WHAT IF? Based on these results and the proposed model from this work, what would happen to a cell if its Fus3 kinase were not able to associate with the membrane upon activation?

The explanation for the specificity exhibited in cellular responses to signals is the same as the basic explanation for virtually all differences between cells: Because different kinds of cells turn on different sets of genes, *different kinds of cells have different collections of proteins* (**Figure 11.18**, on the next page). The response of a particular cell to a signal depends on its particular collection of signal receptor proteins, relay proteins, and proteins needed to carry out the response. A liver cell, for example, is poised to respond appropriately to epinephrine by having the proteins listed in Figure 11.16 as well as those needed to manufacture glycogen.

Thus, two cells that respond differently to the same signal differ in one or more of the proteins that handle and respond to the signal. Notice in Figure 11.18 that different pathways may have some molecules in common. For example, cells A, B, and C all use the same receptor protein for the red signaling molecule; differences in other proteins account for their differing responses. In cell D, a different receptor protein is used for the same signaling molecule, leading to yet another response. In cell B, a pathway that is triggered by a single kind of signal diverges to produce two responses; such branched pathways often involve receptor tyrosine kinases (which can activate multiple relay proteins) or second messengers (which can regulate numerous proteins). In cell C, two pathways triggered by separate signals converge to modulate a single response. Branching of pathways and "cross-talk" (interaction)

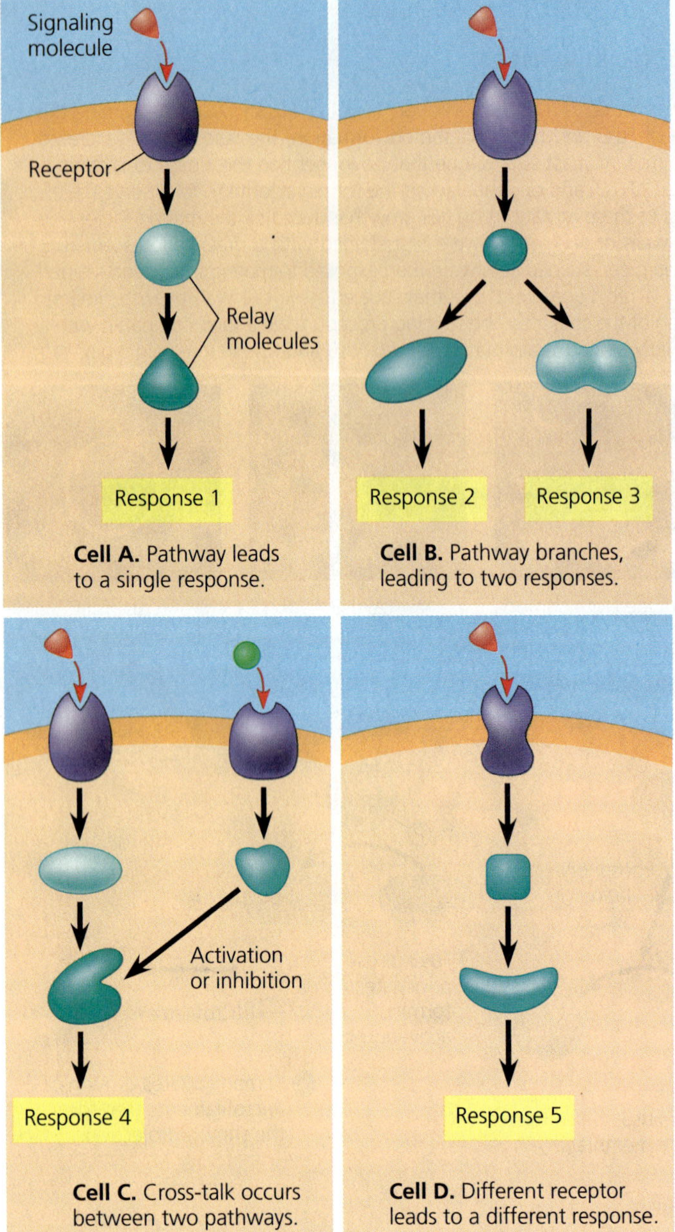

▲ **Figure 11.18 The specificity of cell signaling.** The particular proteins a cell possesses determine what signaling molecules it responds to and the nature of the response. The four cells in these diagrams respond to the same signaling molecule (red) in different ways because each has a different set of proteins (purple and teal). Note, however, that the same kinds of molecules can participate in more than one pathway.

MAKE CONNECTIONS *Study the signaling pathway shown in Figure 11.14 (p. 218), and explain how the situation pictured for cell B above could apply to that pathway.*

between pathways are important in regulating and coordinating a cell's responses to information coming in from different sources in the body. (You'll learn more about this coordination in Concept 11.5.) Moreover, the use of some of the same proteins in more than one pathway allows the cell to economize on the number of different proteins it must make.

Signaling Efficiency: Scaffolding Proteins and Signaling Complexes

The illustrations of signaling pathways in Figure 11.18 (as well as diagrams of other pathways in this chapter) are greatly simplified. The diagrams show only a few relay molecules and, for clarity's sake, display these molecules spread out in the cytosol. If this were true in the cell, signaling pathways would operate very inefficiently because most relay molecules are proteins, and proteins are too large to diffuse quickly through the viscous cytosol. How does a particular protein kinase, for instance, find its substrate?

In many cases, the efficiency of signal transduction is apparently increased by the presence of **scaffolding proteins**, large relay proteins to which several other relay proteins are simultaneously attached. For example, one scaffolding protein isolated from mouse brain cells holds three protein kinases and carries these kinases with it when it binds to an appropriately activated membrane receptor; it thus facilitates a specific phosphorylation cascade **(Figure 11.19)**. Researchers have found scaffolding proteins in brain cells that *permanently* hold together networks of signaling pathway proteins at synapses. This hardwiring enhances the speed and accuracy of signal transfer between cells, because the rate of protein-protein interaction is not limited by diffusion. Furthermore, in addition to this indirect role in activation of relay proteins, the scaffolding proteins themselves may more directly activate some of the other relay proteins.

When signaling pathways were first discovered, they were thought to be linear, independent pathways. Our understanding of cellular communication has benefited from the realization that signaling-pathway components interact with each other in various ways. As seen in Figure 11.18, some proteins may participate in more than one pathway, either in different cell types or in the same cell at different times or under different conditions. These observations underscore

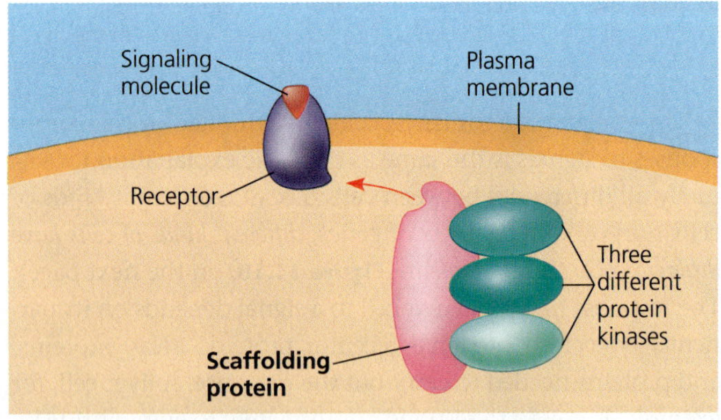

▲ **Figure 11.19 A scaffolding protein.** The scaffolding protein shown here (pink) simultaneously binds to a specific activated membrane receptor and three different protein kinases. This physical arrangement facilitates signal transduction by these molecules and may directly activate relay molecules in some cases.

the importance of transient—or, in some cases, permanent—protein complexes in the process of cell signaling.

The importance of the relay proteins that serve as points of branching or intersection in signaling pathways is highlighted by the problems arising when these proteins are defective or missing. For instance, in an inherited disorder called Wiskott-Aldrich syndrome (WAS), the absence of a single relay protein leads to such diverse effects as abnormal bleeding, eczema, and a predisposition to infections and leukemia. These symptoms are thought to arise primarily from the absence of the protein in cells of the immune system. By studying normal cells, scientists found that the WAS protein is located just beneath the cell surface. The protein interacts both with microfilaments of the cytoskeleton and with several different components of signaling pathways that relay information from the cell surface, including pathways regulating immune cell proliferation. This multifunctional relay protein is thus both a branch point and an important intersection point in a complex signal transduction network that controls immune cell behavior. When the WAS protein is absent, the cytoskeleton is not properly organized and signaling pathways are disrupted, leading to the WAS symptoms.

Termination of the Signal

To keep Figure 11.18 simple, we did not indicate the *inactivation* mechanisms that are an essential aspect of cell signaling. For a cell of a multicellular organism to remain capable of responding to incoming signals, each molecular change in its signaling pathways must last only a short time. As we saw in the cholera example, if a signaling pathway component becomes locked into one state, whether active or inactive, consequences for the organism can be dire.

The ability of a cell to receive new signals depends on reversibility of the changes produced by prior signals. The binding of signaling molecules to receptors is reversible. As the external concentration of signaling molecules falls, fewer receptors are bound at any given moment, and the unbound receptors revert to their inactive form. The cellular response occurs only when the concentration of receptors with bound signaling molecules is above a certain threshold. When the number of active receptors falls below that threshold, the cellular response ceases. Then, by a variety of means, the relay molecules return to their inactive forms: The GTPase activity intrinsic to a G protein hydrolyzes its bound GTP; the enzyme phosphodiesterase converts cAMP to AMP; protein phosphatases inactivate phosphorylated kinases and other proteins; and so forth. As a result, the cell is soon ready to respond to a fresh signal.

In this section, we explored the complexity of signaling initiation and termination in a single pathway, and we saw the potential for pathways to intersect with each other. In the next section, we'll consider one especially important network of interacting pathways in the cell.

CONCEPT CHECK 11.4

1. How can a target cell's response to a single hormone molecule result in a response that affects a million other molecules?
2. **WHAT IF?** If two cells have different scaffolding proteins, explain how they might behave differently in response to the same signaling molecule.
3. **MAKE CONNECTIONS** Review the discussion of protein phosphatases on page 216, and see Figure 11.10 on page 215. Some human diseases are associated with malfunctioning protein phosphatases. How would such proteins affect signaling pathways?

For suggested answers, see Appendix A.

CONCEPT 11.5

Apoptosis integrates multiple cell-signaling pathways

To be or not to be? One of the most elaborate networks of signaling pathways in the cell seems to ask and answer this question posed by Hamlet. Cells that are infected, damaged, or have reached the end of their functional life span often undergo "programmed cell death." The best-understood type of this controlled cell suicide is **apoptosis** (from the Greek, meaning "falling off," and used in a classic Greek poem to refer to leaves falling from a tree). During this process, cellular agents chop up the DNA and fragment the organelles and other cytoplasmic components. The cell shrinks and becomes lobed (a change called "blebbing"; **Figure 11.20**), and the cell's parts are packaged up in vesicles that are engulfed and digested by specialized scavenger cells, leaving no trace. Apoptosis protects neighboring cells from damage that they would otherwise suffer if a dying cell merely leaked out all its contents, including its many digestive enzymes.

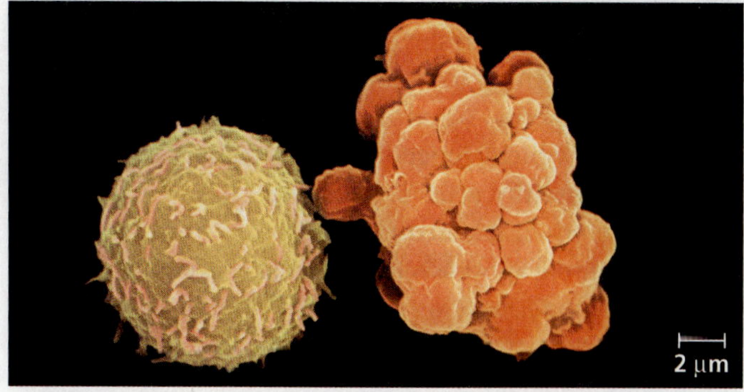

▲ **Figure 11.20 Apoptosis of a human white blood cell.** We can compare a normal white blood cell (left) with a white blood cell undergoing apoptosis (right). The apoptotic cell is shrinking and forming lobes ("blebs"), which eventually are shed as membrane-bounded cell fragments (colorized SEMs).

Apoptosis in the Soil Worm *Caenorhabditis elegans*

Embryonic development is a period during which apoptosis is widespread and plays a crucial role. The molecular mechanisms underlying apoptosis were worked out in detail by researchers studying embryonic development of a small soil worm, a nematode called *Caenorhabditis elegans*. Because the adult worm has only about a thousand cells, the researchers were able to work out the entire ancestry of each cell. The timely suicide of cells occurs exactly 131 times during normal development of *C. elegans*, at precisely the same points in the cell lineage of each worm. In worms and other species, apoptosis is triggered by signals that activate a cascade of "suicide" proteins in the cells destined to die.

Genetic research on *C. elegans* has revealed two key apoptosis genes, called *ced-3* and *ced-4* (*ced* stands for "cell death"), which encode proteins essential for apoptosis. The proteins are called Ced-3 and Ced-4, respectively. These and most other proteins involved in apoptosis are continually present in cells, but in inactive form; thus, regulation occurs at the level of protein activity rather than through gene activity and protein synthesis. In *C. elegans*, a protein in the outer mitochondrial membrane, called Ced-9 (the product of the *ced-9* gene), serves as a master regulator of apoptosis, acting as a brake in the absence of a signal promoting apoptosis **(Figure 11.21)**. When a death signal is received by the cell, it overrides the brake, and the apoptotic pathway activates proteases and nucleases, enzymes that cut up the proteins and DNA of the cell. The main proteases of apoptosis are called *caspases*; in the nematode, the chief caspase is Ced-3.

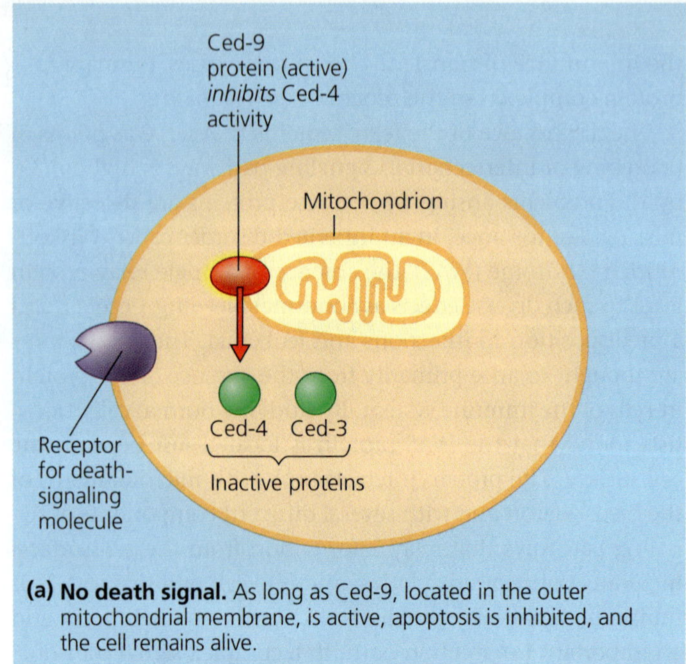

(a) No death signal. As long as Ced-9, located in the outer mitochondrial membrane, is active, apoptosis is inhibited, and the cell remains alive.

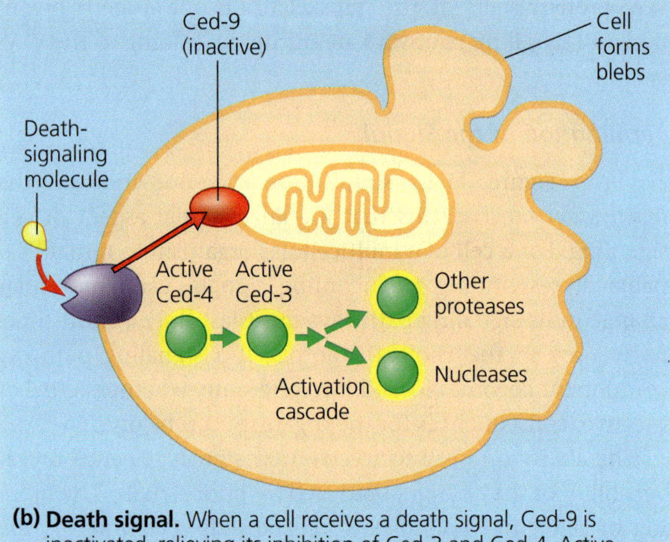

(b) Death signal. When a cell receives a death signal, Ced-9 is inactivated, relieving its inhibition of Ced-3 and Ced-4. Active Ced-3, a protease, triggers a cascade of reactions leading to activation of nucleases and other proteases. The action of these enzymes causes the changes seen in apoptotic cells and eventual cell death.

▲ **Figure 11.21 Molecular basis of apoptosis in *C. elegans*.** Three proteins, Ced-3, Ced-4, and Ced-9, are critical to apoptosis and its regulation in the nematode. Apoptosis is more complicated in mammals but involves proteins similar to those in the nematode.

Apoptotic Pathways and the Signals That Trigger Them

In humans and other mammals, several different pathways, involving about 15 different caspases, can carry out apoptosis. The pathway that is used depends on the type of cell and on the particular signal that initiates apoptosis. One major pathway involves certain mitochondrial proteins that are triggered to form molecular pores in the mitochondrial outer membrane, causing it to leak and release other proteins that promote apoptosis. Surprisingly, these latter include cytochrome *c*, which functions in mitochondrial electron transport in healthy cells (see Figure 9.15) but acts as a cell death factor when released from mitochondria. The process of mitochondrial apoptosis in mammals uses proteins similar to the nematode proteins Ced-3, Ced-4, and Ced-9. These can be thought of as relay proteins capable of transducing the apoptotic signal.

At key gateways into the apoptotic program, relay proteins integrate signals from several different sources and can send a cell down an apoptotic pathway. Often, the signal originates outside the cell, like the death-signaling molecule depicted in Figure 11.21b, which presumably was released by a neighboring cell. When a death-signaling ligand occupies a cell-surface receptor, this binding leads to activation of caspases and other enzymes that carry out apoptosis, without involving the mitochondrial pathway. This process of signal reception, transduction, and response is similar to what we discussed earlier in this chapter. In a twist on the classic scenario, two other types of alarm signals that can lead to apoptosis originate from *inside* the cell rather than from a cell-surface receptor.

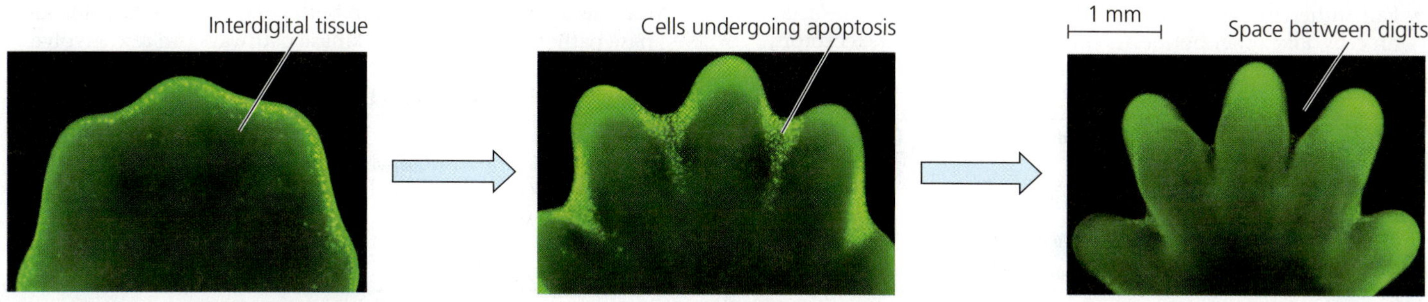

▲ **Figure 11.22 Effect of apoptosis during paw development in the mouse.** In mice, humans, other mammals, and land birds, the embryonic region that develops into feet or hands initially has a solid, platelike structure. Apoptosis eliminates the cells in the interdigital regions, thus forming the digits. The embryonic mouse paws shown in these fluorescence light micrographs are stained so that cells undergoing apoptosis appear a bright yellowish green. Apoptosis of cells begins at the margin of each interdigital region (left), peaks as the tissue in these regions is reduced (middle), and is no longer visible when the interdigital tissue has been eliminated (right).

One signal comes from the nucleus, generated when the DNA has suffered irreparable damage, and a second comes from the endoplasmic reticulum when excessive protein misfolding occurs. Mammalian cells make life-or-death "decisions" by somehow integrating the death signals and life signals they receive from these external and internal sources.

A built-in cell suicide mechanism is essential to development and maintenance in all animals. The similarities between apoptosis genes in nematodes and mammals, as well as the observation that apoptosis occurs in multicellular fungi and even in single-celled yeasts, indicate that the basic mechanism evolved early in the evolution of eukaryotes. In vertebrates, apoptosis is essential for normal development of the nervous system, for normal operation of the immune system, and for normal morphogenesis of hands and feet in humans and paws in other mammals **(Figure 11.22)**. The level of apoptosis between the developing digits is lower in the webbed feet of ducks and other water birds than in the nonwebbed feet of land birds, such as chickens. In the case of humans, the failure of appropriate apoptosis can result in webbed fingers and toes.

Significant evidence points to the involvement of apoptosis in certain degenerative diseases of the nervous system, such as Parkinson's disease and Alzheimer's disease. Also, cancer can result from a failure of cell suicide; some cases of human melanoma, for example, have been linked to faulty forms of the human version of the *C. elegans* Ced-4 protein. It is not surprising, therefore, that the signaling pathways feeding into apoptosis are quite elaborate. After all, the life-or-death question is the most fundamental one imaginable for a cell.

This chapter has introduced you to many of the general mechanisms of cell communication, such as ligand binding, protein-protein interactions and shape changes, cascades of interactions, and protein phosphorylation. As you continue through the text, you will encounter numerous examples of cell signaling.

CONCEPT CHECK 11.5

1. Give an example of apoptosis during embryonic development, and explain its function in the developing embryo.
2. **WHAT IF?** What types of protein defects could result in apoptosis occurring when it should not? What types could result in apoptosis not occurring when it should?

For suggested answers, see Appendix A.

11 CHAPTER REVIEW

SUMMARY OF KEY CONCEPTS

CONCEPT 11.1

External signals are converted to responses within the cell (pp. 206–210)

- **Signal transduction pathways** are crucial for many processes, including the mating of yeast cells. In fact, signaling in microbes has much in common with processes in multicellular organisms, suggesting an early evolutionary origin of signaling mechanisms. Bacterial cells can sense the local density of bacterial cells (quorum sensing) by binding molecules secreted by other cells. In some cases, such signals lead to aggregation of these cells into biofilms.
- In local signaling, animal cells may communicate by direct contact or by secreting **local regulators**, such as growth factors or neurotransmitters. For long-distance signaling, both animals and plants use **hormones**; animals also pass signals electrically.

- Earl Sutherland discovered how the hormone epinephrine acts on cells. Like other hormones that bind to membrane receptors, it triggers a three-stage cell-signaling pathway:

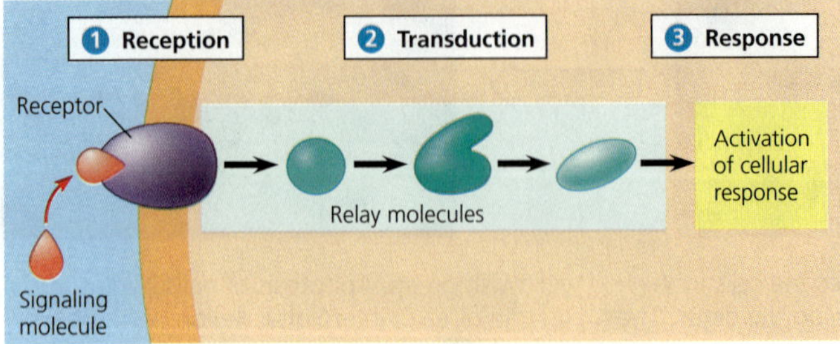

? *What determines* whether *a cell responds to a hormone such as epinephrine? What determines* how *a cell responds to such a hormone?*

CONCEPT 11.2

Reception: A signaling molecule binds to a receptor protein, causing it to change shape (pp. 210–214)

- The binding between signaling molecule (**ligand**) and receptor is highly specific. A specific shape change in a receptor is often the initial transduction of the signal.
- There are three major types of cell-surface transmembrane receptors: (1) **G protein-coupled receptors (GPCRs)** work with the help of cytoplasmic **G proteins**. Ligand binding activates the receptor, which then activates a specific G protein, which activates yet another protein, thus propagating the signal along a signal transduction pathway. (2) **Receptor tyrosine kinases (RTKs)** react to the binding of signaling molecules by forming dimers and then adding phosphate groups to tyrosines on the cytoplasmic part of the other monomer making up the dimer. Relay proteins in the cell can then be activated by binding to different phosphorylated tyrosines, allowing this receptor to trigger several pathways at once. (3) **Ligand-gated ion channels** open or close in response to binding by specific signaling molecules, regulating the flow of specific ions across the membrane.
- The activity of all three types of receptors is crucial to proper cell functioning, and abnormal GPCRs and RTKs are associated with many human diseases.
- Intracellular receptors are cytoplasmic or nuclear proteins. Signaling molecules that are hydrophobic or small enough to cross the plasma membrane bind to these receptors inside the cell.

? *How are the structures of a G protein-coupled receptor and a receptor tyrosine kinase similar? In what key way does the triggering of signal transduction pathways differ for these two types of receptors?*

CONCEPT 11.3

Transduction: Cascades of molecular interactions relay signals from receptors to target molecules in the cell (pp. 214–219)

- At each step in a signal transduction pathway, the signal is transduced into a different form, which commonly involves a shape change in a protein. Many signal transduction pathways include phosphorylation cascades, in which a series of **protein kinases** each add a phosphate group to the next one in line, activating it. Enzymes called **protein phosphatases** remove the phosphate groups. The balance between phosphorylation and dephosphorylation regulates the activity of proteins involved in the sequential steps of a signal transduction pathway.
- **Second messengers**, such as the small molecule **cyclic AMP (cAMP)** and the ion Ca^{2+}, diffuse readily through the cytosol and thus help broadcast signals quickly. Many G proteins activate **adenylyl cyclase**, which makes cAMP from ATP. Cells use Ca^{2+} as a second messenger in both G protein and tyrosine kinase pathways. The tyrosine kinase pathways can also involve two other second messengers, **diacylglycerol (DAG)** and **inositol trisphosphate (IP_3)**. IP_3 can trigger a subsequent increase in Ca^{2+} levels.

? *What is the difference between a protein kinase and a second messenger? Can both types of molecules operate in the same signal transduction pathway?*

CONCEPT 11.4

Response: Cell signaling leads to regulation of transcription or cytoplasmic activities (pp. 219–223)

- Some pathways lead to a nuclear response: Specific genes are turned on or off by activation of proteins called transcription factors. In other pathways, the response involves cytoplasmic regulation, including cytoskeletal rearrangement (which can lead to cell shape changes) or changes in enzyme activity.
- Cellular responses are not simply on or off; they are fine-tuned at many steps in the process. Each catalytic protein in a signaling pathway amplifies the signal by activating multiple copies of the next component of the pathway; for long pathways, the total amplification may be a millionfold or more. The particular combination of proteins in a cell gives the cell great specificity in both the signals it detects and the responses it carries out. **Scaffolding proteins** can increase signal transduction efficiency. Pathway branching and cross-talk further help the cell coordinate incoming signals and responses. Signal response is terminated quickly by the reversal of ligand binding.

? *What mechanisms in the cell terminate its response to a signal and maintain its ability to respond to new signals?*

CONCEPT 11.5

Apoptosis integrates multiple cell-signaling pathways (pp. 223–225)

- **Apoptosis** is a type of programmed cell death in which cell components are disposed of in an orderly fashion, without damage to neighboring cells. Studies of the soil worm *Caenorhabditis elegans* showed that apoptosis occurs at defined times during embryonic development and clarified molecular details of the signaling pathway involved in the process. A protein (Ced-9) in the mitochondrial membrane acts as a brake; when released by a death signal, it allows activation of caspases, the main proteases that carry out apoptosis, and nucleases.
- Several apoptotic signaling pathways exist in the cells of humans and other mammals, and these pathways may be triggered in several ways. A major pathway involves pore formation in the outer mitochondrial membrane, which leads to release of factors that activate caspases. Signals eliciting this response can originate from outside or inside the cell.

? *What is an explanation for the similarities between genes in yeasts, nematodes, and mammals that control apoptosis?*

TEST YOUR UNDERSTANDING

LEVEL 1: KNOWLEDGE/COMPREHENSION

1. Phosphorylation cascades involving a series of protein kinases are useful for cellular signal transduction because
   a. they are species specific.
   b. they always lead to the same cellular response.
   c. they amplify the original signal manyfold.
   d. they counter the harmful effects of phosphatases.
   e. the number of molecules used is small and fixed.

2. Binding of a signaling molecule to which type of receptor leads directly to a change in the distribution of ions on opposite sides of the membrane?
   a. receptor tyrosine kinase
   b. G protein-coupled receptor
   c. phosphorylated receptor tyrosine kinase dimer
   d. ligand-gated ion channel
   e. intracellular receptor

3. The activation of receptor tyrosine kinases is characterized by
   a. dimerization and phosphorylation.
   b. dimerization and IP_3 binding.
   c. a phosphorylation cascade.
   d. GTP hydrolysis.
   e. channel protein shape change.

4. Lipid-soluble signaling molecules, such as testosterone, cross the membranes of all cells but affect only target cells because
   a. only target cells retain the appropriate DNA segments.
   b. intracellular receptors are present only in target cells.
   c. most cells lack the Y chromosome required.
   d. only target cells possess the cytosolic enzymes that transduce the testosterone.
   e. only in target cells is testosterone able to initiate the phosphorylation cascade leading to activated transcription factor.

5. Consider this pathway: epinephrine → G protein-coupled receptor → G protein → adenylyl cyclase → cAMP. Identify the second messenger.
   a. cAMP
   b. G protein
   c. GTP
   d. adenylyl cyclase
   e. G protein-coupled receptor

6. Apoptosis involves all but which of the following?
   a. fragmentation of the DNA
   b. cell-signaling pathways
   c. activation of cellular enzymes
   d. lysis of the cell
   e. digestion of cellular contents by scavenger cells

LEVEL 2: APPLICATION/ANALYSIS

7. Which observation suggested to Sutherland the involvement of a second messenger in epinephrine's effect on liver cells?
   a. Enzymatic activity was proportional to the amount of calcium added to a cell-free extract.
   b. Receptor studies indicated that epinephrine was a ligand.
   c. Glycogen breakdown was observed only when epinephrine was administered to intact cells.
   d. Glycogen breakdown was observed when epinephrine and glycogen phosphorylase were combined.
   e. Epinephrine was known to have different effects on different types of cells.

8. Protein phosphorylation is commonly involved with all of the following *except*
   a. regulation of transcription by extracellular signaling molecules.
   b. enzyme activation.
   c. activation of G protein-coupled receptors.
   d. activation of receptor tyrosine kinases.
   e. activation of protein kinase molecules.

LEVEL 3: SYNTHESIS/EVALUATION

9. **DRAW IT** Draw the following apoptotic pathway, which operates in human immune cells. A death signal is received when a molecule called Fas binds its cell-surface receptor. The binding of many Fas molecules to receptors causes receptor clustering. The intracellular regions of the receptors, when together, bind proteins called adaptor proteins. These in turn bind to inactive molecules of caspase-8, which become activated and then activate caspase-3. Once activated, caspase-3 initiates apoptosis.

10. **EVOLUTION CONNECTION**
    What evolutionary mechanisms might account for the origin and persistence of cell-to-cell signaling systems in unicellular prokaryotes?

11. **SCIENTIFIC INQUIRY**
    Epinephrine initiates a signal transduction pathway that involves production of cyclic AMP (cAMP) and leads to the breakdown of glycogen to glucose, a major energy source for cells. But glycogen breakdown is actually only part of the fight-or-flight response that epinephrine brings about; the overall effect on the body includes increased heart rate and alertness, as well as a burst of energy. Given that caffeine blocks the activity of cAMP phosphodiesterase, propose a mechanism by which caffeine ingestion leads to heightened alertness and sleeplessness.

12. **SCIENCE, TECHNOLOGY, AND SOCIETY**
    The aging process is thought to be initiated at the cellular level. Among the changes that can occur after a certain number of cell divisions is the loss of a cell's ability to respond to growth factors and other chemical signals. Much research into aging is aimed at understanding such losses, with the ultimate goal of significantly extending the human life span. Not everyone, however, agrees that this is a desirable goal. If life expectancy were greatly increased, what might be the social and ecological consequences?

13. **WRITE ABOUT A THEME**
    Emergent Properties The property of life emerges at the biological level of the cell. The highly regulated process of apoptosis is not simply the destruction of a cell; it is also an emergent property. Write a short essay (about 100–150 words) that briefly explains the role of apoptosis in the development and proper functioning of an animal and then describes how this form of programmed cell death is a process that emerges from the orderly integration of signaling pathways.

For selected answers, see Appendix A.

12

The Cell Cycle

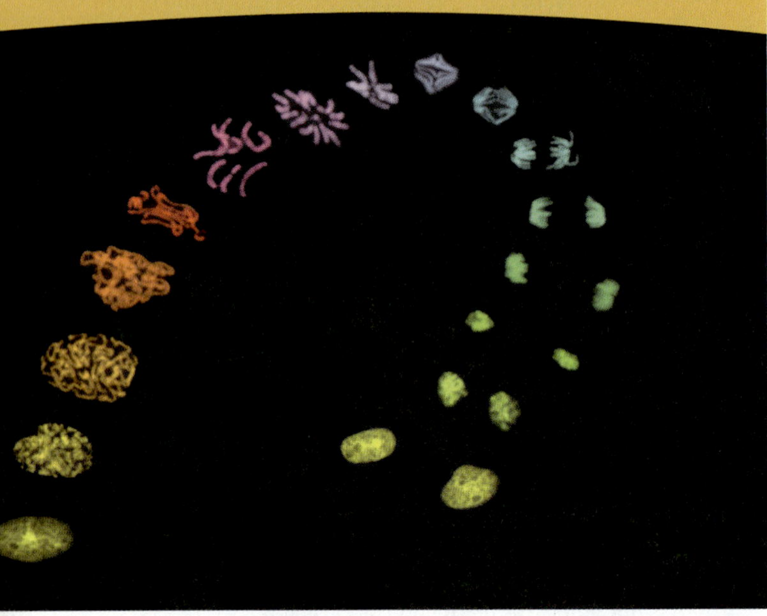

▲ **Figure 12.1 How do a cell's chromosomes change during cell division?**

KEY CONCEPTS

12.1 Most cell division results in genetically identical daughter cells

12.2 The mitotic phase alternates with interphase in the cell cycle

12.3 The eukaryotic cell cycle is regulated by a molecular control system

OVERVIEW

The Key Roles of Cell Division

The ability of organisms to produce more of their own kind is the one characteristic that best distinguishes living things from nonliving matter. This unique capacity to procreate, like all biological functions, has a cellular basis. Rudolf Virchow, a German physician, put it this way in 1855: "Where a cell exists, there must have been a preexisting cell, just as the animal arises only from an animal and the plant only from a plant." He summarized this concept with the Latin axiom "*Omnis cellula e cellula*," meaning "Every cell from a cell." The continuity of life is based on the reproduction of cells, or **cell division**. The series of fluorescence micrographs in **Figure 12.1** follows an animal cell's chromosomes, from lower left to lower right, as one cell divides into two.

Cell division plays several important roles in life. The division of one prokaryotic cell reproduces an entire organism. The same is true of a unicellular eukaryote **(Figure 12.2a)**. Cell division also enables multicellular eukaryotes to develop from a single cell, like the fertilized egg that gave rise to the two-celled embryo in **Figure 12.2b**. And after such an organism is fully grown, cell division continues to function in renewal and repair, replacing cells that die from normal wear and tear or accidents. For example, dividing cells in your bone marrow continuously make new blood cells **(Figure 12.2c)**.

The cell division process is an integral part of the **cell cycle**, the life of a cell from the time it is first formed from a dividing parent cell until its own division into two daughter cells. (Our use of the words *daughter* or *sister* in relation to cells is not meant to imply gender.) Passing identical genetic material to cellular offspring is a crucial function of cell division. In this chapter, you will learn how this process occurs. After studying the cellular mechanics of cell division in eukaryotes and bacteria, you will learn about the molecular control system that regulates progress through the eukaryotic cell cycle and what happens when the control system malfunctions. Because a breakdown in cell cycle control plays a major role in cancer development, this aspect of cell biology is an active area of research.

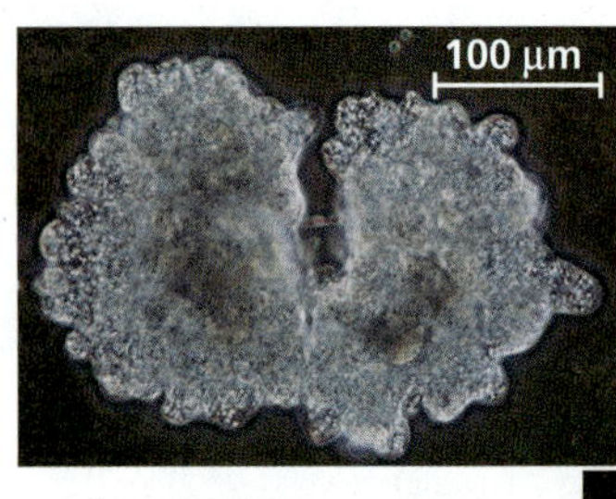

◀ **(a) Reproduction.** An amoeba, a single-celled eukaryote, is dividing into two cells. Each new cell will be an individual organism (LM).

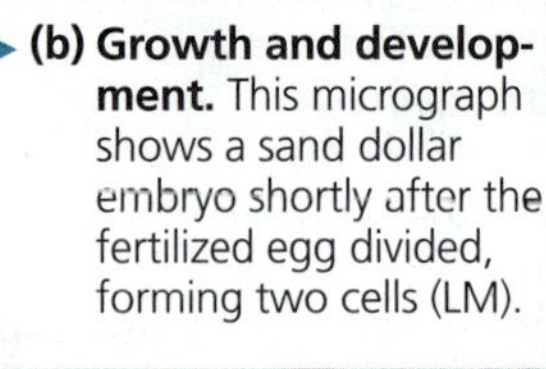

▶ **(b) Growth and development.** This micrograph shows a sand dollar embryo shortly after the fertilized egg divided, forming two cells (LM).

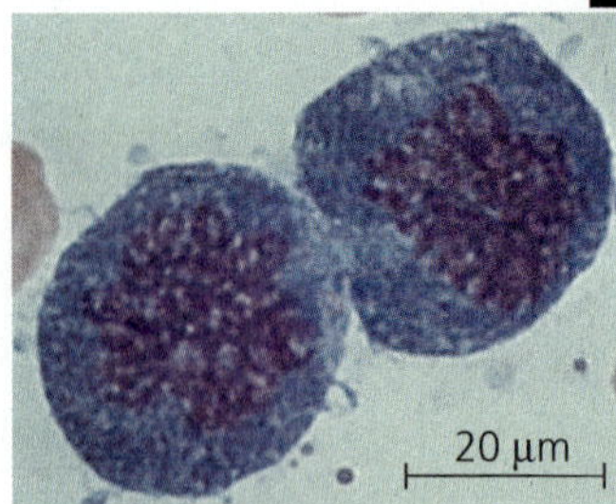

◀ **(c) Tissue renewal.** These dividing bone marrow cells will give rise to new blood cells (LM).

▲ **Figure 12.2 The functions of cell division.**

CONCEPT **12.1**

Most cell division results in genetically identical daughter cells

The reproduction of an assembly as complex as a cell cannot occur by a mere pinching in half; a cell is not like a soap bubble that simply enlarges and splits in two. In both prokaryotes and eukaryotes, most cell division involves the distribution of identical genetic material—DNA—to two daughter cells. (The exception is meiosis, the special type of eukaryotic cell division that can produce sperm and eggs.) What is most remarkable about cell division is the fidelity with which the DNA is passed along from one generation of cells to the next. A dividing cell duplicates its DNA, allocates the two copies to opposite ends of the cell, and only then splits into daughter cells. After we describe the distribution of DNA during cell division in animal and plant cells, we'll consider the process in other eukaryotes as well as in bacteria.

Cellular Organization of the Genetic Material

A cell's endowment of DNA, its genetic information, is called its **genome**. Although a prokaryotic genome is often a single DNA molecule, eukaryotic genomes usually consist of a number of DNA molecules. The overall length of DNA in a eukaryotic cell is enormous. A typical human cell, for example, has about 2 m of DNA—a length about 250,000 times greater than the cell's diameter. Yet before the cell can divide to form genetically identical daughter cells, all of this DNA must be copied, or replicated, and then the two copies must be separated so that each daughter cell ends up with a complete genome.

The replication and distribution of so much DNA is manageable because the DNA molecules are packaged into structures called **chromosomes**, so named because they take up certain dyes used in microscopy (from the Greek *chroma*, color, and *soma*, body) **(Figure 12.3)**. Each eukaryotic chromosome consists of one very long, linear DNA molecule associated with many proteins (see Figure 6.9). The DNA molecule carries several hundred to a few thousand genes, the units of information that specify an organism's inherited traits. The associated proteins maintain the structure of the chromosome and help control the activity of the genes. Together, the entire complex of DNA and proteins that is the building material of chromosomes is referred to as **chromatin**. As you will soon see, the chromatin of a chromosome varies in its degree of condensation during the process of cell division.

Every eukaryotic species has a characteristic number of chromosomes in each cell nucleus. For example, the nuclei of human **somatic cells** (all body cells except the reproductive cells) each contain 46 chromosomes, made up of two sets of 23, one set inherited from each parent. Reproductive cells, or **gametes**—sperm and eggs—have half as many chromosomes as somatic cells, or one set of 23 chromosomes in humans. The

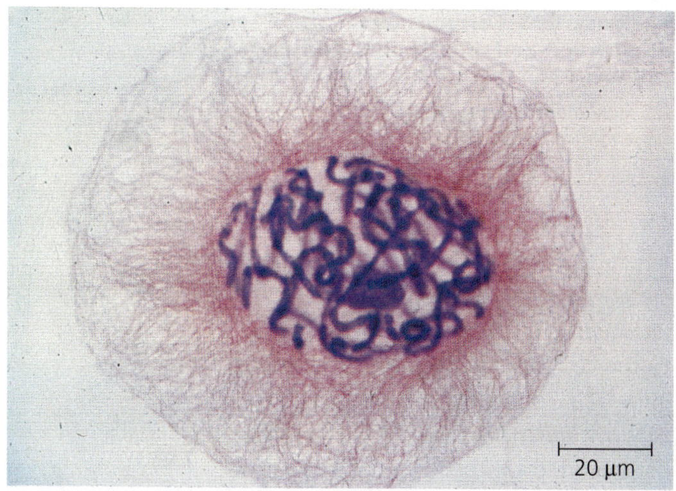

▲ **Figure 12.3 Eukaryotic chromosomes.** Chromosomes (stained purple) are visible within the nucleus of this cell from an African blood lily. The thinner red threads in the surrounding cytoplasm are the cytoskeleton. The cell is preparing to divide (LM).

number of chromosomes in somatic cells varies widely among species: 18 in cabbage plants, 48 in chimpanzees, 56 in elephants, 90 in hedgehogs, and 148 in one species of alga. We'll now consider how these chromosomes behave during cell division.

Distribution of Chromosomes During Eukaryotic Cell Division

When a cell is not dividing, and even as it replicates its DNA in preparation for cell division, each chromosome is in the form of a long, thin chromatin fiber. After DNA replication, however, the chromosomes condense as a part of cell division: Each chromatin fiber becomes densely coiled and folded, making the chromosomes much shorter and so thick that we can see them with a light microscope.

Each duplicated chromosome has two **sister chromatids**, which are joined copies of the original chromosome **(Figure 12.4)**. The two chromatids, each containing an identical DNA molecule, are initially attached all along their lengths by protein complexes called *cohesins*; this attachment is known as *sister chromatid cohesion*. Each sister chromatid has a **centromere**, a region containing specific DNA sequences

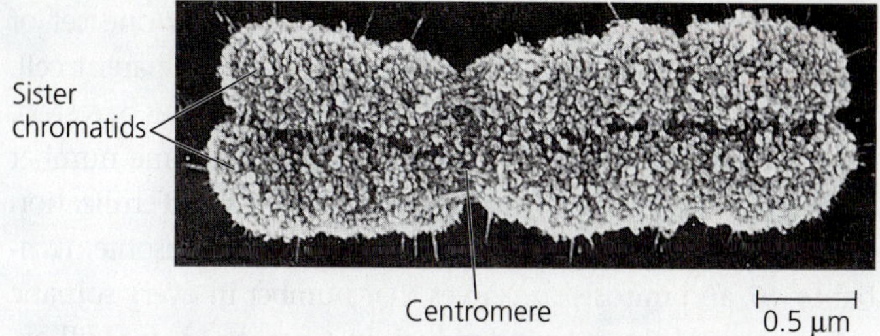

▲ **Figure 12.4 A highly condensed, duplicated human chromosome (SEM).**

DRAW IT *Circle one sister chromatid of the chromosome in this micrograph.*

▶ **Figure 12.5 Chromosome duplication and distribution during cell division.**

? *How many chromatid arms does the chromosome in ❷ have?*

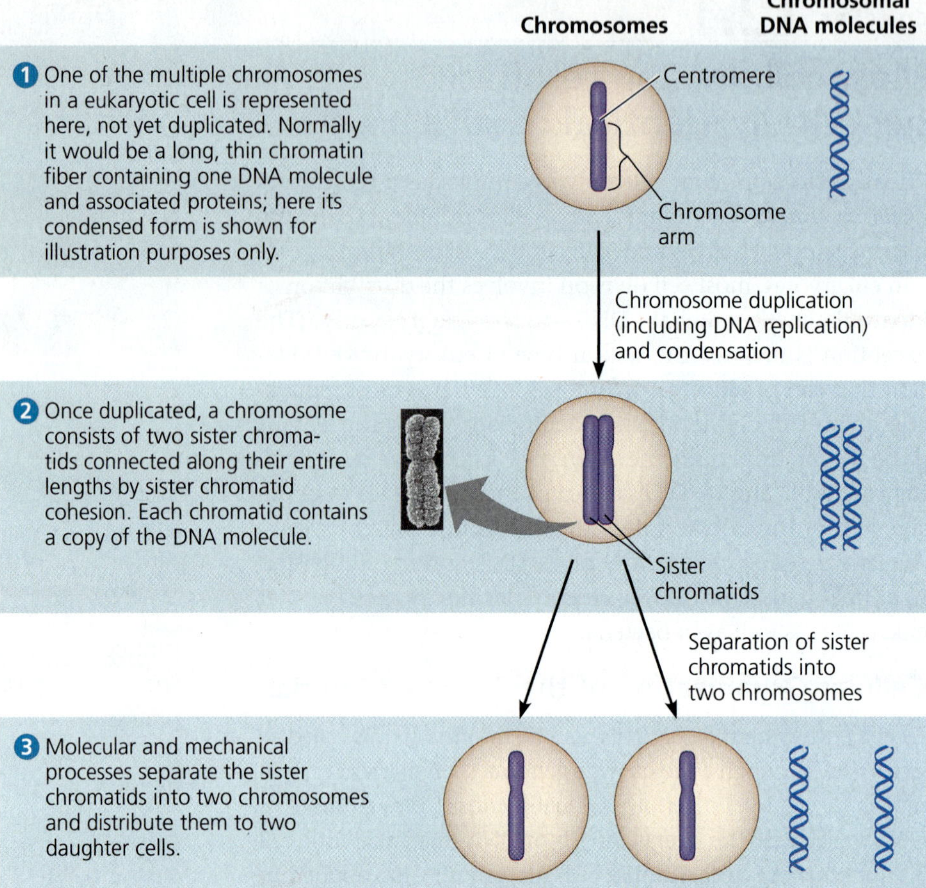

where the chromatid is attached most closely to its sister chromatid. This attachment is mediated by proteins bound to the centromeric DNA sequences and gives the condensed, duplicated chromosome a narrow "waist." The part of a chromatid on either side of the centromere is referred to as an *arm* of the chromatid. (An uncondensed, unduplicated chromosome has a single centromere and two arms.)

Later in the cell division process, the two sister chromatids of each duplicated chromosome separate and move into two new nuclei, one forming at each end of the cell. Once the sister chromatids separate, they are no longer called sister chromatids but are considered individual chromosomes. Thus, each new nucleus receives a collection of chromosomes identical to that of the parent cell **(Figure 12.5)**. **Mitosis**, the division of the genetic material in the nucleus, is usually followed immediately by **cytokinesis**, the division of the cytoplasm. One cell has become two, each the genetic equivalent of the parent cell.

What happens to the chromosome number as we follow the human life cycle through the generations? You inherited 46 chromosomes, one set of 23 from each parent. They were combined in the nucleus of a single cell when a sperm from your father united with an egg from your mother, forming a fertilized egg, or zygote. Mitosis and cytokinesis produced the 200 trillion somatic cells that now make up your body, and the same processes continue to generate new cells to replace dead and damaged ones. In contrast, you produce gametes—eggs or sperm—by a variation of cell division called *meiosis*, which yields nonidentical daughter cells that have only one set of chromosomes, half as many chromosomes as the parent cell. Meiosis in humans occurs only in the gonads (ovaries or testes). In each generation, meiosis reduces the chromosome number from 46 (two sets of chromosomes) to 23 (one set). Fertilization fuses two gametes together and returns the chromosome number to 46, and mitosis conserves that number in every somatic cell nucleus of the new individual. In Chapter 13, we will examine the role of meiosis in reproduction and inheritance in more detail. In the remainder of this chapter, we focus on mitosis and the rest of the cell cycle in eukaryotes.

CONCEPT CHECK 12.1

1. How many chromatids are in a duplicated chromosome?
2. **WHAT IF?** A chicken has 78 chromosomes in its somatic cells. How many chromosomes did the chicken inherit from each parent? How many chromosomes are in each of the chicken's gametes? How many chromosomes will be in each somatic cell of the chicken's offspring?

For suggested answers, see Appendix A.

CONCEPT **12.2**

The mitotic phase alternates with interphase in the cell cycle

In 1882, a German anatomist named Walther Flemming developed dyes that allowed him to observe, for the first time, the behavior of chromosomes during mitosis and cytokinesis. (In fact, Flemming coined the terms *mitosis* and *chromatin*.) During the period between one cell division and the next, it appeared to Flemming that the cell was simply growing larger. But we now know that many critical events occur during this stage in the life of a cell.

Phases of the Cell Cycle

Mitosis is just one part of the cell cycle **(Figure 12.6)**. In fact, the **mitotic (M) phase**, which includes both mitosis and cytokinesis, is usually the shortest part of the cell cycle. Mitotic cell division alternates with a much longer stage called **interphase**, which often accounts for about 90% of the cycle. During interphase, a cell that is about to divide grows and copies its chromosomes in preparation for cell division. Interphase can be divided into subphases: the **G_1 phase** ("first gap"), the **S phase** ("synthesis"), and the **G_2 phase** ("second gap"). During all three subphases, a cell that will eventually divide grows by producing proteins and cytoplasmic organelles such as mitochondria and endoplasmic reticulum. However, chromosomes are duplicated only during the S phase. (We will discuss synthesis of DNA in Chapter 16.) Thus, a cell grows (G_1), continues to grow as it copies its chromosomes (S), grows more as it completes preparations for cell division (G_2), and divides (M). The daughter cells may then repeat the cycle.

A particular human cell might undergo one division in 24 hours. Of this time, the M phase would occupy less than 1 hour, while the S phase might occupy about 10–12 hours, or about half the cycle. The rest of the time would be apportioned between the G_1 and G_2 phases. The G_2 phase usually takes 4–6 hours; in our example, G_1 would occupy about 5–6 hours. G_1 is the most variable in length in different types of cells. Some cells in a multicellular organism divide very infrequently or not at all. These cells spend their time in G_1 (or a related phase called G_0) doing their job in the organism—a nerve cell carries impulses, for example.

Mitosis is conventionally broken down into five stages: **prophase**, **prometaphase**, **metaphase**, **anaphase**, and **telophase**. Overlapping with the latter stages of mitosis, cytokinesis completes the mitotic phase. **Figure 12.7**, on the next two pages, describes these stages in an animal cell. Study this figure thoroughly before progressing to the next two sections, which examine mitosis and cytokinesis more closely.

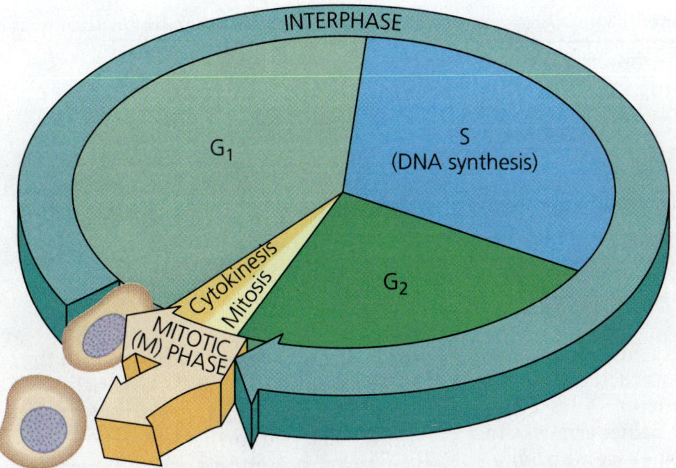

▲ **Figure 12.6 The cell cycle.** In a dividing cell, the mitotic (M) phase alternates with interphase, a growth period. The first part of interphase (G_1) is followed by the S phase, when the chromosomes duplicate; G_2 is the last part of interphase. In the M phase, mitosis distributes the daughter chromosomes to daughter nuclei, and cytokinesis divides the cytoplasm, producing two daughter cells. The relative durations of G_1, S, and G_2 may vary.

The Mitotic Spindle: *A Closer Look*

Many of the events of mitosis depend on the **mitotic spindle**, which begins to form in the cytoplasm during prophase. This structure consists of fibers made of microtubules and associated proteins. While the mitotic spindle assembles, the other microtubules of the cytoskeleton partially disassemble, providing the material used to construct the spindle. The spindle microtubules elongate (polymerize) by incorporating more subunits of the protein tubulin (see Table 6.1) and shorten (depolymerize) by losing subunits.

In animal cells, the assembly of spindle microtubules starts at the **centrosome**, a subcellular region containing material that functions throughout the cell cycle to organize the cell's microtubules. (It is also called the *microtubule-organizing center*.) A pair of centrioles is located at the center of the centrosome, but they are not essential for cell division: If the centrioles are destroyed with a laser microbeam, a spindle nevertheless forms during mitosis. In fact, centrioles are not even present in plant cells, which do form mitotic spindles.

During interphase in animal cells, the single centrosome duplicates, forming two centrosomes, which remain together near the nucleus. The two centrosomes move apart during prophase and prometaphase of mitosis as spindle microtubules grow out from them. By the end of prometaphase, the two centrosomes, one at each pole of the spindle, are at opposite ends of the cell. An **aster**, a radial array of short microtubules, extends from each centrosome. The spindle includes the centrosomes, the spindle microtubules, and the asters.

Each of the two sister chromatids of a duplicated chromosome has a **kinetochore**, a structure of proteins associated with specific sections of chromosomal DNA at each centromere. The chromosome's two kinetochores face in opposite directions. During prometaphase, some of the spindle microtubules attach to the kinetochores; these are called kinetochore microtubules. (The number of microtubules attached to a kinetochore varies among species, from one microtubule in yeast cells to 40 or so in some mammalian cells.) When one of a chromosome's kinetochores is "captured" by microtubules, the chromosome begins to move toward the pole from which those microtubules extend. However, this movement is checked as soon as microtubules from the opposite pole attach to the other kinetochore. What happens next is like a tug-of-war that ends in a draw. The chromosome moves first in one direction, then the other, back and forth, finally settling midway between the two ends of the cell. At metaphase, the centromeres of all the duplicated chromosomes are on a plane midway between the spindle's

▼ Figure 12.7

Exploring Mitosis in an Animal Cell

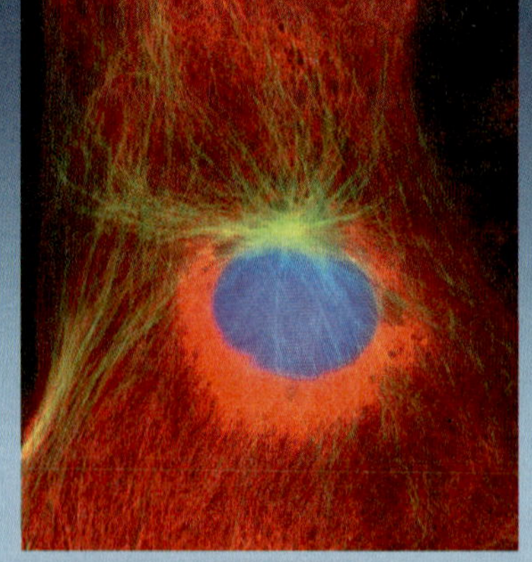

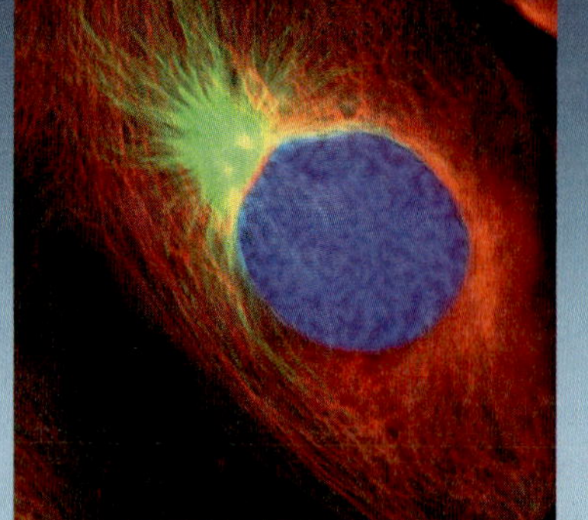

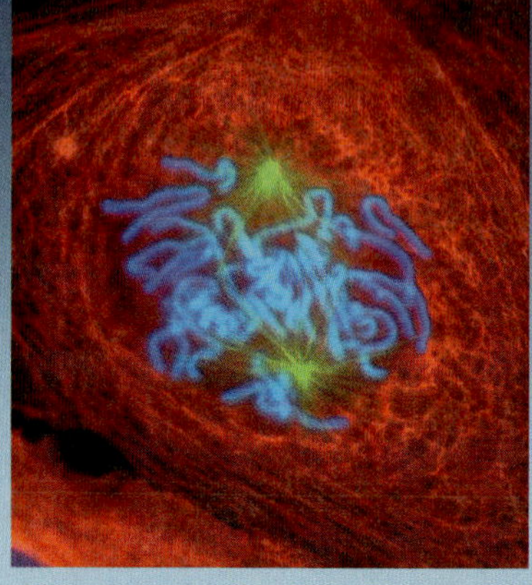

G_2 of Interphase

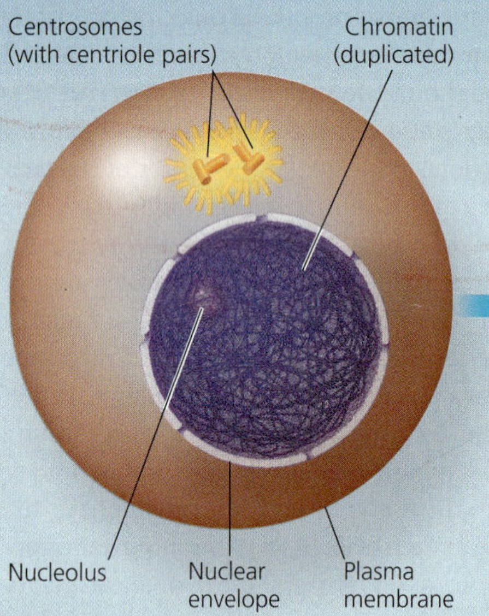

Prophase

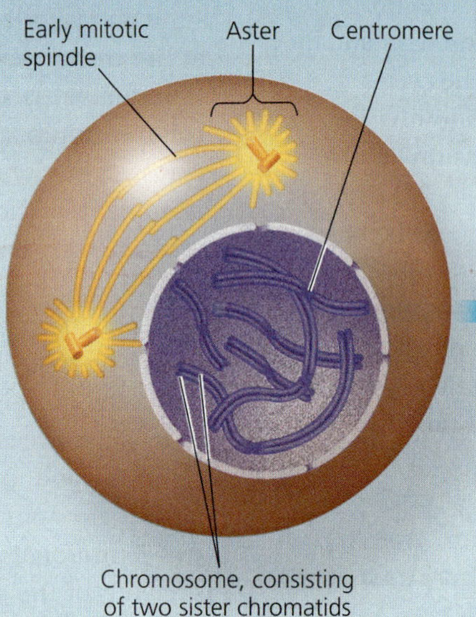

Prometaphase

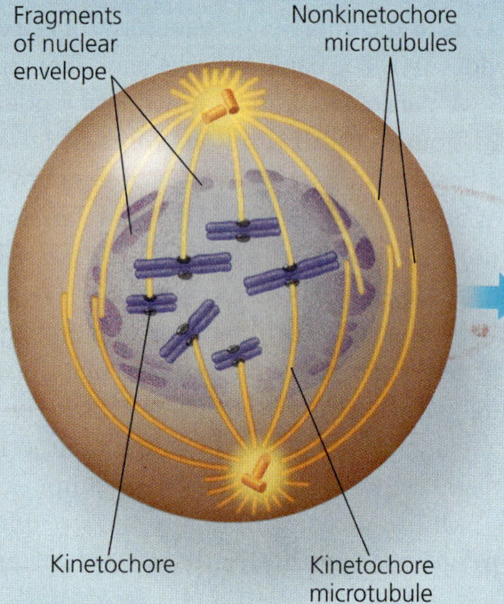

G_2 of Interphase

- A nuclear envelope encloses the nucleus.
- The nucleus contains one or more nucleoli (singular, *nucleolus*).
- Two centrosomes have formed by duplication of a single centrosome. Centrosomes are regions in animal cells that organize the microtubules of the spindle. Each centrosome contains two centrioles.
- Chromosomes, duplicated during S phase, cannot be seen individually because they have not yet condensed.

The light micrographs show dividing lung cells from a newt, which has 22 chromosomes in its somatic cells. Chromosomes appear blue, microtubules green, and intermediate filaments red. For simplicity, the drawings show only 6 chromosomes.

Prophase

- The chromatin fibers become more tightly coiled, condensing into discrete chromosomes observable with a light microscope.
- The nucleoli disappear.
- Each duplicated chromosome appears as two identical sister chromatids joined at their centromeres and, in some species, all along their arms by cohesins (sister chromatid cohesion).
- The mitotic spindle (named for its shape) begins to form. It is composed of the centrosomes and the microtubules that extend from them. The radial arrays of shorter microtubules that extend from the centrosomes are called asters ("stars").
- The centrosomes move away from each other, propelled partly by the lengthening microtubules between them.

Prometaphase

- The nuclear envelope fragments.
- The microtubules extending from each centrosome can now invade the nuclear area.
- The chromosomes have become even more condensed.
- Each of the two chromatids of each chromosome now has a kinetochore, a specialized protein structure at the centromere.
- Some of the microtubules attach to the kinetochores, becoming "kinetochore microtubules," which jerk the chromosomes back and forth.
- Nonkinetochore microtubules interact with those from the opposite pole of the spindle.

? *How many molecules of DNA are in the prometaphase drawing? How many molecules per chromosome? How many double helices are there per chromosome? Per chromatid?*

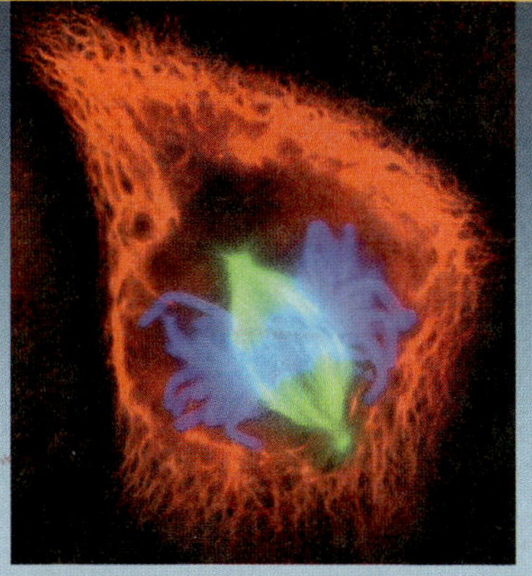

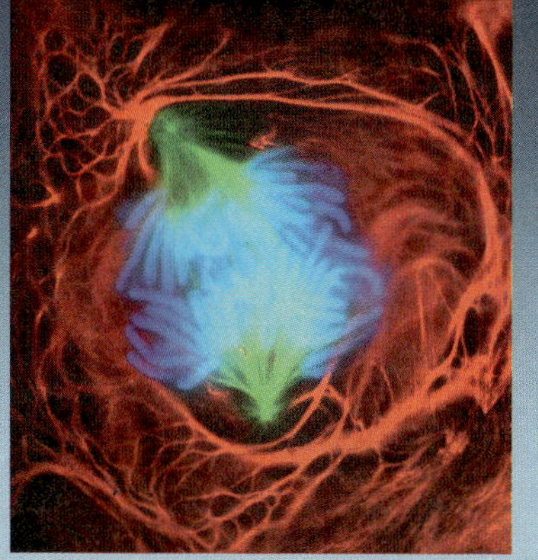

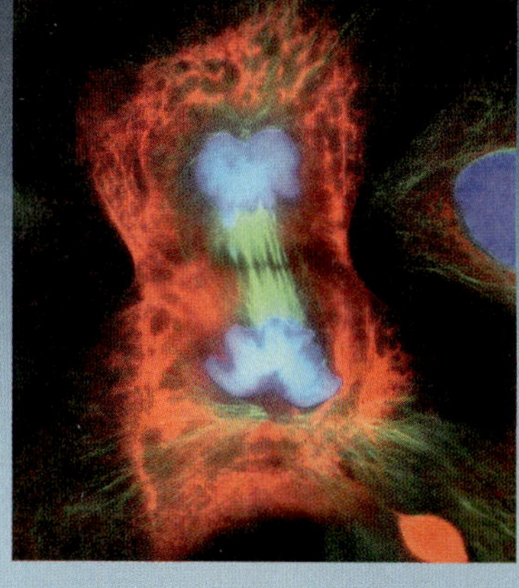

Metaphase

Anaphase

Telophase and Cytokinesis

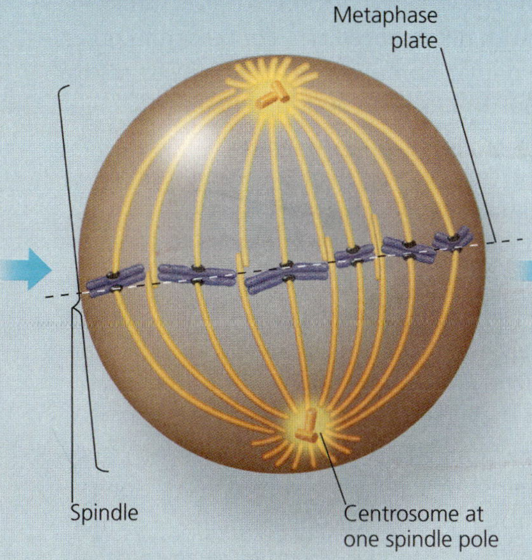

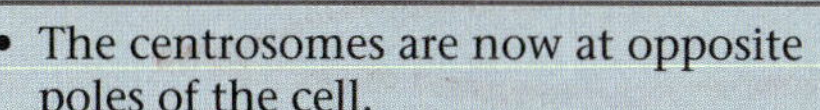

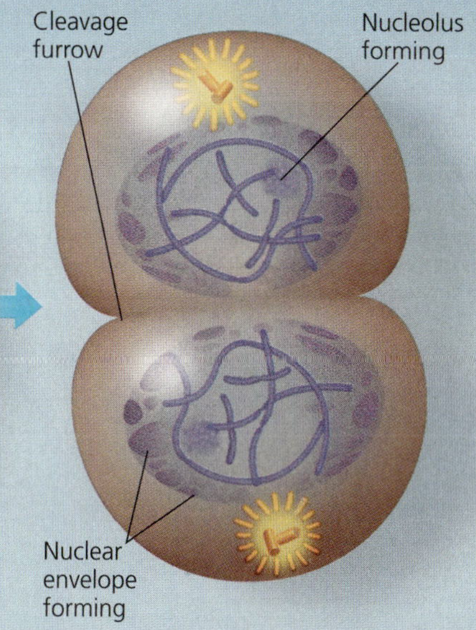

Metaphase

- The centrosomes are now at opposite poles of the cell.
- The chromosomes convene at the *metaphase plate*, a plane that is equidistant between the spindle's two poles. The chromosomes' centromeres lie at the metaphase plate.
- For each chromosome, the kinetochores of the sister chromatids are attached to kinetochore microtubules coming from opposite poles.

Anaphase

- Anaphase is the shortest stage of mitosis, often lasting only a few minutes.
- Anaphase begins when the cohesin proteins are cleaved. This allows the two sister chromatids of each pair to part suddenly. Each chromatid thus becomes a full-fledged chromosome.
- The two liberated daughter chromosomes begin moving toward opposite ends of the cell as their kinetochore microtubules shorten. Because these microtubules are attached at the centromere region, the chromosomes move centromere first (at about 1 μm/min).
- The cell elongates as the nonkinetochore microtubules lengthen.
- By the end of anaphase, the two ends of the cell have equivalent—and complete—collections of chromosomes.

Telophase

- Two daughter nuclei form in the cell. Nuclear envelopes arise from the fragments of the parent cell's nuclear envelope and other portions of the endomembrane system.
- Nucleoli reappear.
- The chromosomes become less condensed.
- Any remaining spindle microtubules are depolymerized.
- Mitosis, the division of one nucleus into two genetically identical nuclei, is now complete.

Cytokinesis

- The division of the cytoplasm is usually well under way by late telophase, so the two daughter cells appear shortly after the end of mitosis.
- In animal cells, cytokinesis involves the formation of a cleavage furrow, which pinches the cell in two.

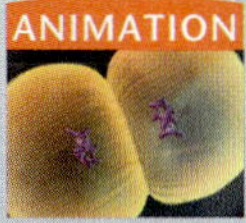

BioFlix Visit the Study Area at **www.masteringbiology.com** for the BioFlix® 3-D Animation on Mitosis.

two poles. This plane is called the **metaphase plate**, which is an imaginary rather than an actual cellular structure **(Figure 12.8)**. Meanwhile, microtubules that do not attach to kinetochores have been elongating, and by metaphase they overlap and interact with other nonkinetochore microtubules from the opposite pole of the spindle. (These are sometimes called "polar" microtubules.) By metaphase, the microtubules of the asters have also grown and are in contact with the plasma membrane. The spindle is now complete.

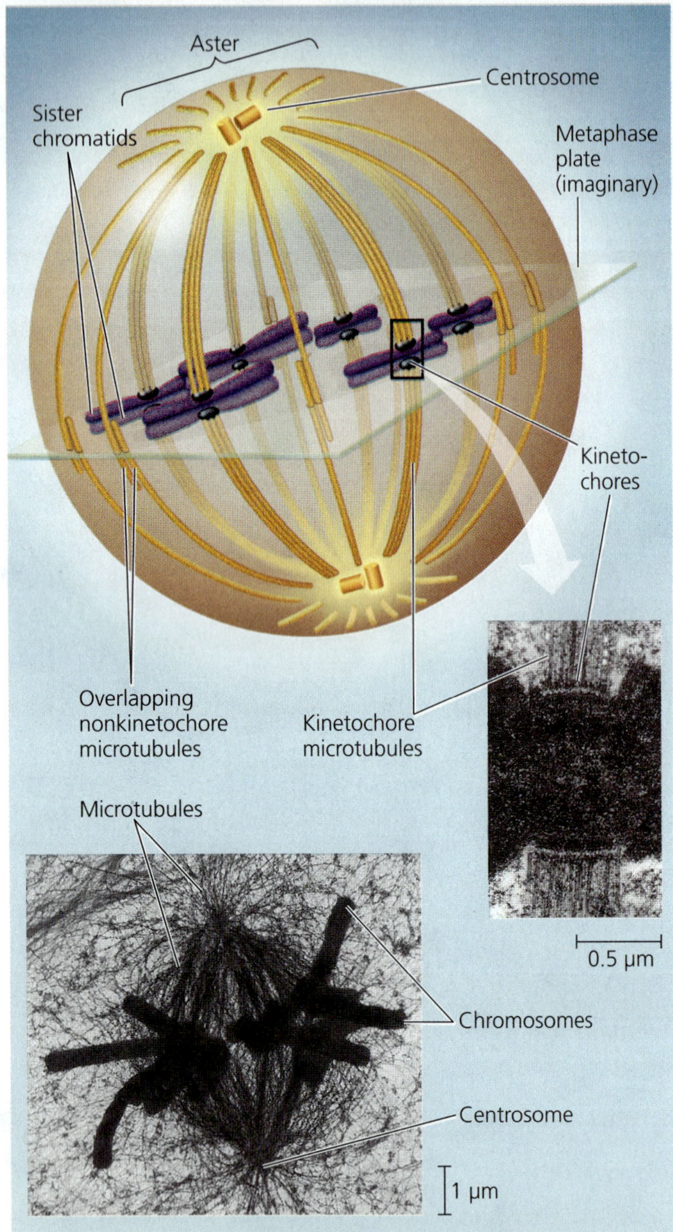

▲ **Figure 12.8 The mitotic spindle at metaphase.** The kinetochores of each chromosome's two sister chromatids face in opposite directions. Here, each kinetochore is attached to a *cluster* of kinetochore microtubules extending from the nearest centrosome. Nonkinetochore microtubules overlap at the metaphase plate (TEMs).

DRAW IT *On the lower micrograph, draw a line indicating the position of the metaphase plate. Circle an aster. Draw arrows indicating the directions of chromosome movement once anaphase begins.*

The structure of the completed spindle correlates well with its function during anaphase. Anaphase commences suddenly when the cohesins holding together the sister chromatids of each chromosome are cleaved by an enzyme called *separase*. Once the chromatids become separate, full-fledged chromosomes, they move toward opposite ends of the cell.

How do the kinetochore microtubules function in this poleward movement of chromosomes? Apparently, two mechanisms are in play, both involving motor proteins. (To review how motor proteins move an object along a microtubule, see Figure 6.21.) A clever experiment carried out in 1987 suggested that motor proteins on the kinetochores "walk" the chromosomes along the microtubules, which depolymerize at their kinetochore ends after the motor proteins have passed **(Figure 12.9)**. (This is referred to as the "Pacman" mechanism because of its resemblance to the arcade game character that moves by eating all the dots in its path.) However, other researchers, working with different cell types or cells from other species, have shown that chromosomes are "reeled in" by motor proteins at the spindle poles and that the microtubules depolymerize after they pass by these motor proteins. The general consensus now is that both mechanisms are used and that their relative contributions vary among cell types.

In a dividing animal cell, the nonkinetochore microtubules are responsible for elongating the whole cell during anaphase. Nonkinetochore microtubules from opposite poles overlap each other extensively during metaphase (see Figure 12.8). During anaphase, the region of overlap is reduced as motor proteins attached to the microtubules walk them away from one another, using energy from ATP. As the microtubules push apart from each other, their spindle poles are pushed apart, elongating the cell. At the same time, the microtubules lengthen somewhat by the addition of tubulin subunits to their overlapping ends. As a result, the microtubules continue to overlap.

At the end of anaphase, duplicate groups of chromosomes have arrived at opposite ends of the elongated parent cell. Nuclei re-form during telophase. Cytokinesis generally begins during anaphase or telophase, and the spindle eventually disassembles by depolymerization of microtubules.

Cytokinesis: *A Closer Look*

In animal cells, cytokinesis occurs by a process known as **cleavage**. The first sign of cleavage is the appearance of a **cleavage furrow**, a shallow groove in the cell surface near the old metaphase plate **(Figure 12.10a)**. On the cytoplasmic side of the furrow is a contractile ring of actin microfilaments associated with molecules of the protein myosin. The actin microfilaments interact with the myosin molecules, causing the ring to contract. The contraction of the dividing cell's ring of microfilaments is like the pulling of a drawstring. The cleavage furrow deepens until the parent cell is pinched in two, producing two completely separated cells, each with its own nucleus and share of cytosol, organelles, and other subcellular structures.

▼ **Figure 12.9** INQUIRY

At which end do kinetochore microtubules shorten during anaphase?

EXPERIMENT Gary Borisy and colleagues at the University of Wisconsin wanted to determine whether kinetochore microtubules depolymerize at the kinetochore end or the pole end as chromosomes move toward the poles during mitosis. First they labeled the microtubules of a pig kidney cell in early anaphase with a yellow fluorescent dye.

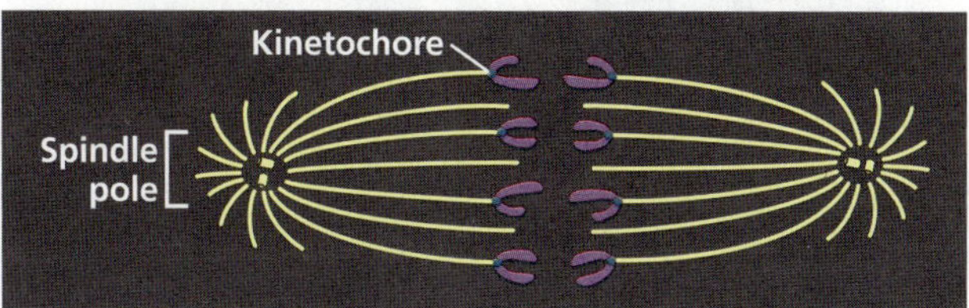

Then they marked a region of the kinetochore microtubules between one spindle pole and the chromosomes by using a laser to eliminate the fluorescence from that region, while leaving the microtubules intact (see below). As anaphase proceeded, they monitored the changes in microtubule length on either side of the mark.

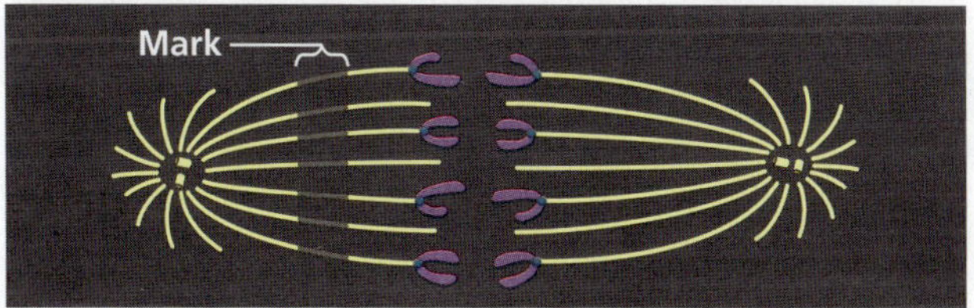

RESULTS As the chromosomes moved poleward, the microtubule segments on the kinetochore side of the mark shortened, while those on the spindle pole side stayed the same length.

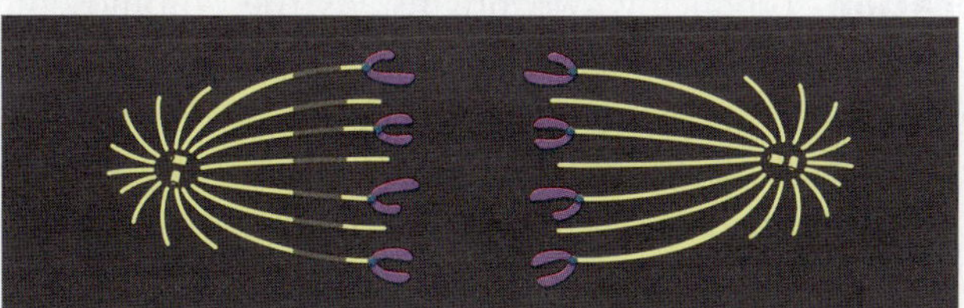

CONCLUSION During anaphase in this cell type, chromosome movement is correlated with kinetochore microtubules shortening at their kinetochore ends and not at their spindle pole ends. This experiment supports the hypothesis that during anaphase, a chromosome is walked along a microtubule as the microtubule depolymerizes at its kinetochore end, releasing tubulin subunits.

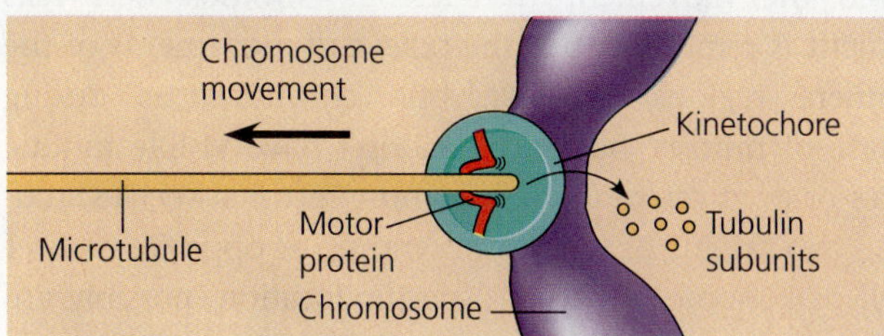

SOURCE G. J. Gorbsky, P. J. Sammak, and G. G. Borisy, Chromosomes move poleward in anaphase along stationary microtubules that coordinately disassemble from their kinetochore ends, *Journal of Cell Biology* 104:9–18 (1987).

WHAT IF? If this experiment had been done on a cell type in which "reeling in" at the poles was the main cause of chromosome movement, how would the mark have moved relative to the poles? How would the microtubule lengths have changed?

▼ **Figure 12.10 Cytokinesis in animal and plant cells.**

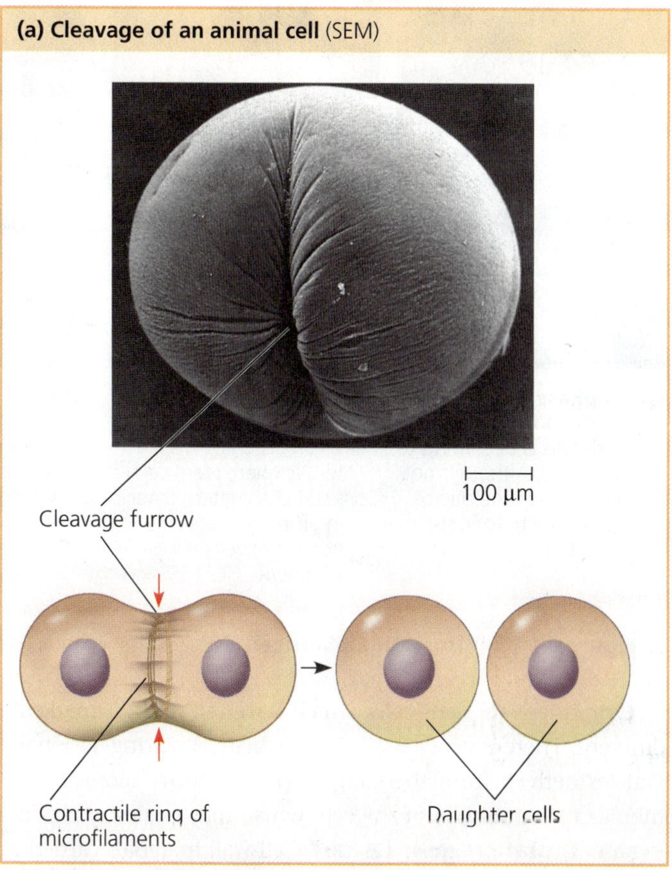

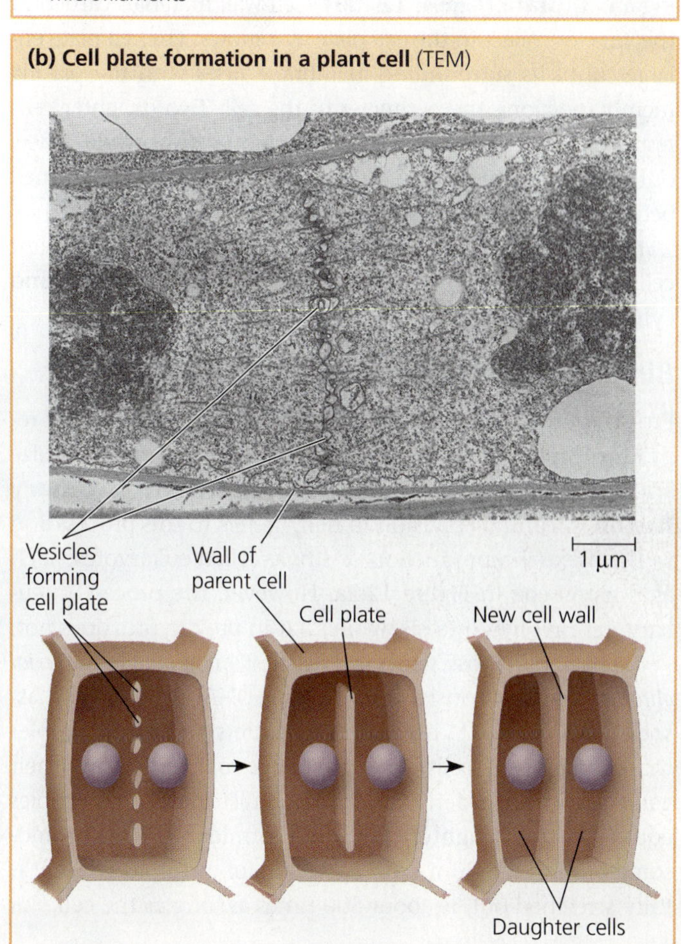

Nucleus
Nucleolus
Chromatin condensing
Chromosomes
Cell plate
10 μm

1 **Prophase.** The chromatin is condensing and the nucleolus is beginning to disappear. Although not yet visible in the micrograph, the mitotic spindle is starting to form.

2 **Prometaphase.** Discrete chromosomes are now visible; each consists of two aligned, identical sister chromatids. Later in prometaphase, the nuclear envelope will fragment.

3 **Metaphase.** The spindle is complete, and the chromosomes, attached to microtubules at their kinetochores, are all at the metaphase plate.

4 **Anaphase.** The chromatids of each chromosome have separated, and the daughter chromosomes are moving to the ends of the cell as their kinetochore microtubules shorten.

5 **Telophase.** Daughter nuclei are forming. Meanwhile, cytokinesis has started: The cell plate, which will divide the cytoplasm in two, is growing toward the perimeter of the parent cell.

▲ **Figure 12.11 Mitosis in a plant cell.** These light micrographs show mitosis in cells of an onion root.

Cytokinesis in plant cells, which have cell walls, is markedly different. There is no cleavage furrow. Instead, during telophase, vesicles derived from the Golgi apparatus move along microtubules to the middle of the cell, where they coalesce, producing a **cell plate** **(Figure 12.10b)**. Cell wall materials carried in the vesicles collect in the cell plate as it grows. The cell plate enlarges until its surrounding membrane fuses with the plasma membrane along the perimeter of the cell. Two daughter cells result, each with its own plasma membrane. Meanwhile, a new cell wall arising from the contents of the cell plate has formed between the daughter cells.

Figure 12.11 is a series of micrographs of a dividing plant cell. Examining this figure will help you review mitosis and cytokinesis.

Binary Fission in Bacteria

Prokaryotes (bacteria and archaea) can undergo a type of reproduction in which the cell grows to roughly double its size and then divides to form two cells. The term **binary fission**, meaning "division in half," refers to this process and to the asexual reproduction of single-celled eukaryotes, such as the amoeba in Figure 12.2a. However, the process in eukaryotes involves mitosis, while that in prokaryotes does not.

In bacteria, most genes are carried on a single *bacterial chromosome* that consists of a circular DNA molecule and associated proteins. Although bacteria are smaller and simpler than eukaryotic cells, the challenge of replicating their genomes in an orderly fashion and distributing the copies equally to two daughter cells is still formidable. The chromosome of the bacterium *Escherichia coli*, for example, when it is fully stretched out, is about 500 times as long as the cell. For such a long chromosome to fit within the cell requires that it be highly coiled and folded.

In *E. coli*, the process of cell division is initiated when the DNA of the bacterial chromosome begins to replicate at a specific place on the chromosome called the **origin of replication**, producing two origins. As the chromosome continues to replicate, one origin moves rapidly toward the opposite end of the cell **(Figure 12.12)**. While the chromosome is replicating, the cell elongates. When replication is complete and the bacterium has reached about twice its initial size, its plasma membrane pinches inward, dividing the parent *E. coli* cell into two daughter cells. In this way, each cell inherits a complete genome.

Using the techniques of modern DNA technology to tag the origins of replication with molecules that glow green in fluorescence microscopy (see Figure 6.3), researchers have directly observed the movement of bacterial chromosomes. This movement is reminiscent of the poleward movements of the centromere regions of eukaryotic chromosomes during anaphase of mitosis, but bacteria don't have visible mitotic spindles or even microtubules. In most bacterial species studied, the two origins of replication end up at opposite ends of the cell or in some other very specific location, possibly anchored there by one or more proteins. How bacterial chromosomes move and how their specific location is established and maintained are still not fully understood. However, several proteins have been identified that play important roles: One resembling eukaryotic actin apparently functions in bacterial chromosome movement during cell division, and another that is related to tubulin seems to help pinch the plasma membrane inward, separating the two bacterial daughter cells.

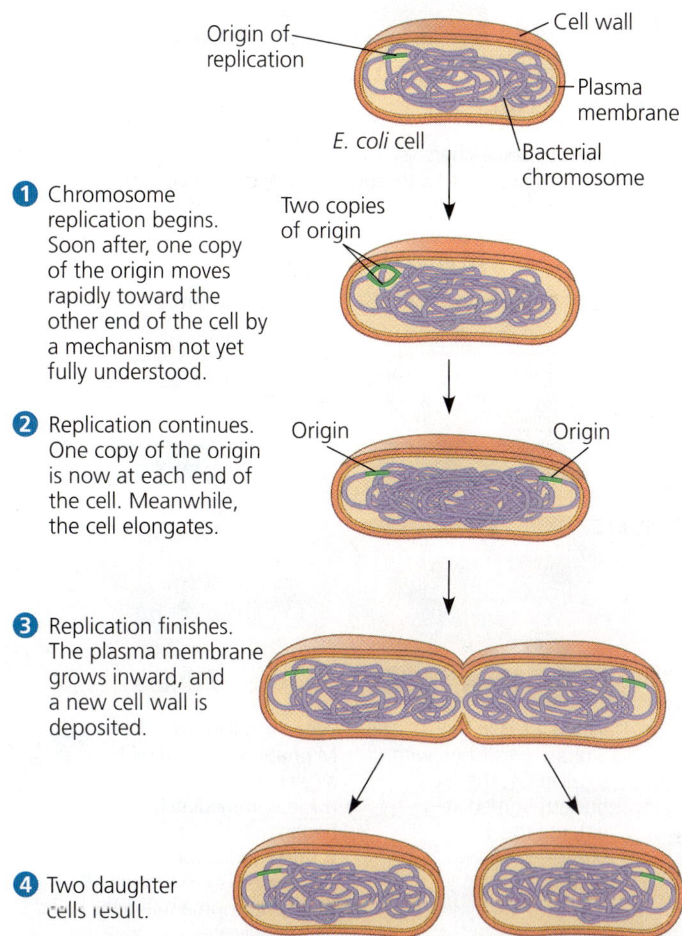

▲ **Figure 12.12 Bacterial cell division by binary fission.** The bacterium *E. coli*, shown here, has a single, circular chromosome.

The Evolution of Mitosis

EVOLUTION Given that prokaryotes preceded eukaryotes on Earth by more than a billion years, we might hypothesize that mitosis evolved from simpler prokaryotic mechanisms of cell reproduction. The fact that some of the proteins involved in bacterial binary fission are related to eukaryotic proteins that function in mitosis supports that hypothesis.

As eukaryotes evolved, along with their larger genomes and nuclear envelopes, the ancestral process of binary fission, seen today in bacteria, somehow gave rise to mitosis. **Figure 12.13** shows some variations on cell division in different groups of organisms. These processes may be similar to mechanisms used by ancestral species and thus may resemble steps in the evolution of mitosis from a binary fission-like process presumably carried out by very early bacteria. Possible intermediate stages are suggested by two unusual types of nuclear division found today in certain unicellular eukaryotes—dinoflagellates, diatoms, and some yeasts. These two modes of nuclear division are thought to be cases where ancestral mechanisms have remained relatively unchanged over evolutionary time. In both types, the nuclear envelope remains intact, in contrast to what happens in most eukaryotic cells.

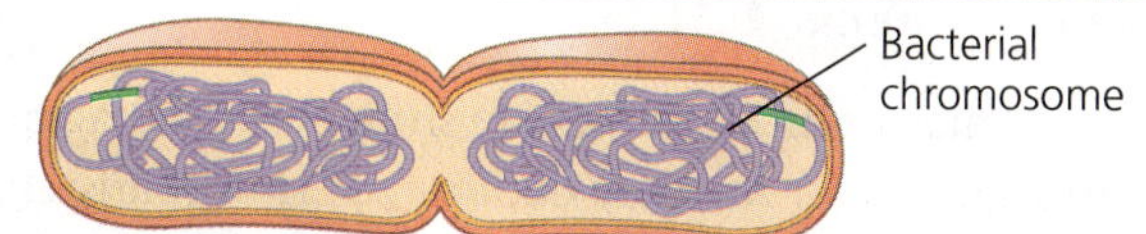

(a) Bacteria. During binary fission in bacteria, the origins of the daughter chromosomes move to opposite ends of the cell. The mechanism is not fully understood, but proteins may anchor the daughter chromosomes to specific sites on the plasma membrane.

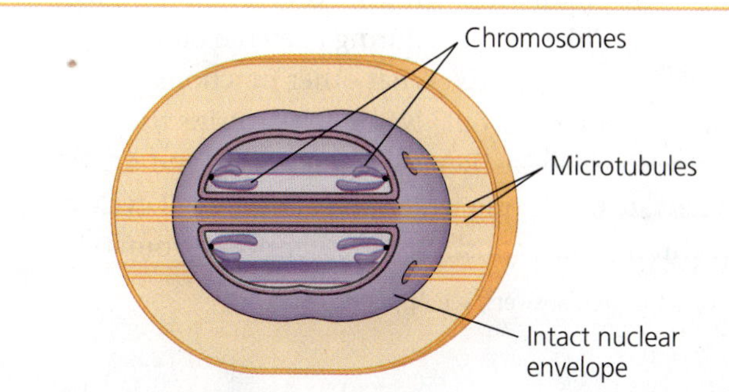

(b) Dinoflagellates. In unicellular protists called dinoflagellates, the chromosomes attach to the nuclear envelope, which remains intact during cell division. Microtubules pass through the nucleus inside cytoplasmic tunnels, reinforcing the spatial orientation of the nucleus, which then divides in a process reminiscent of bacterial binary fission.

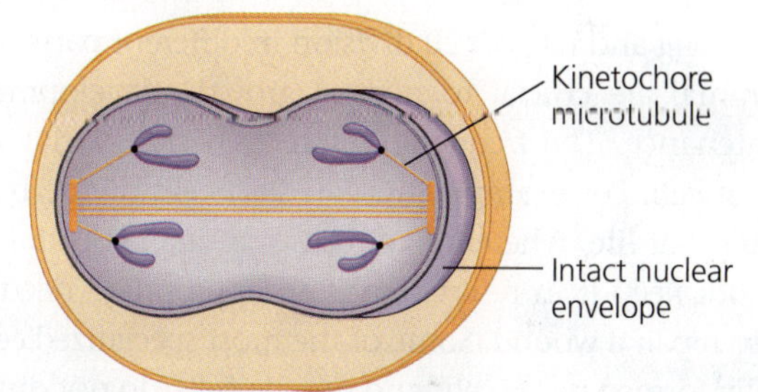

(c) Diatoms and some yeasts. In two other groups of unicellular protists, diatoms and some yeasts, the nuclear envelope also remains intact during cell division. In these organisms, the microtubules form a spindle *within* the nucleus. Microtubules separate the chromosomes, and the nucleus splits into two daughter nuclei.

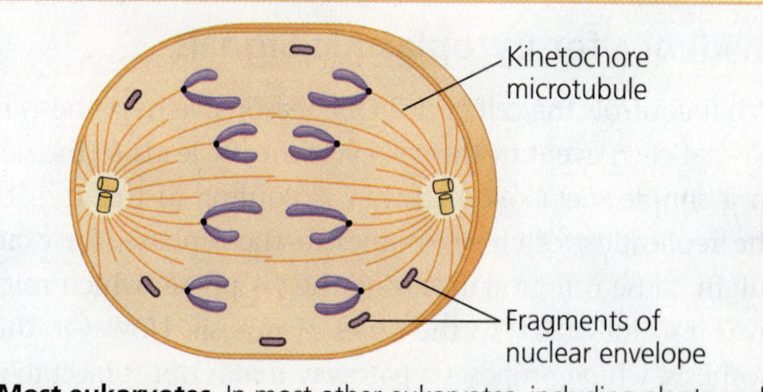

(d) Most eukaryotes. In most other eukaryotes, including plants and animals, the spindle forms outside the nucleus, and the nuclear envelope breaks down during mitosis. Microtubules separate the chromosomes, and two nuclear envelopes then form.

▲ **Figure 12.13 Mechanisms of cell division in several groups of organisms.** Some unicellular eukaryotes existing today have mechanisms of cell division that may resemble intermediate steps in the evolution of mitosis. Except for (a), these schematic diagrams do not show cell walls.

CONCEPT CHECK 12.2

1. How many chromosomes are shown in the diagram in Figure 12.8? Are they duplicated? How many chromatids are shown?
2. Compare cytokinesis in animal cells and plant cells.
3. What is the function of nonkinetochore microtubules?
4. Compare the roles of tubulin and actin during eukaryotic cell division with the roles of tubulin-like and actin-like proteins during bacterial binary fission.
5. **MAKE CONNECTIONS** What other functions do actin and tubulin carry out? Name the proteins they interact with to do so. (Review Figures 6.21a and 6.27a.)
6. **WHAT IF?** During which stages of the cell cycle does a chromosome consist of two identical chromatids?

For suggested answers, see Appendix A.

CONCEPT 12.3

The eukaryotic cell cycle is regulated by a molecular control system

The timing and rate of cell division in different parts of a plant or animal are crucial to normal growth, development, and maintenance. The frequency of cell division varies with the type of cell. For example, human skin cells divide frequently throughout life, whereas liver cells maintain the ability to divide but keep it in reserve until an appropriate need arises—say, to repair a wound. Some of the most specialized cells, such as fully formed nerve cells and muscle cells, do not divide at all in a mature human. These cell cycle differences result from regulation at the molecular level. The mechanisms of this regulation are of intense interest, not only for understanding the life cycles of normal cells but also for understanding how cancer cells manage to escape the usual controls.

▼ Figure 12.14 INQUIRY

Do molecular signals in the cytoplasm regulate the cell cycle?

EXPERIMENT Researchers at the University of Colorado wondered whether a cell's progression through the cell cycle is controlled by cytoplasmic molecules. To investigate this, they selected cultured mammalian cells that were at different phases of the cell cycle and induced them to fuse. Two such experiments are shown here.

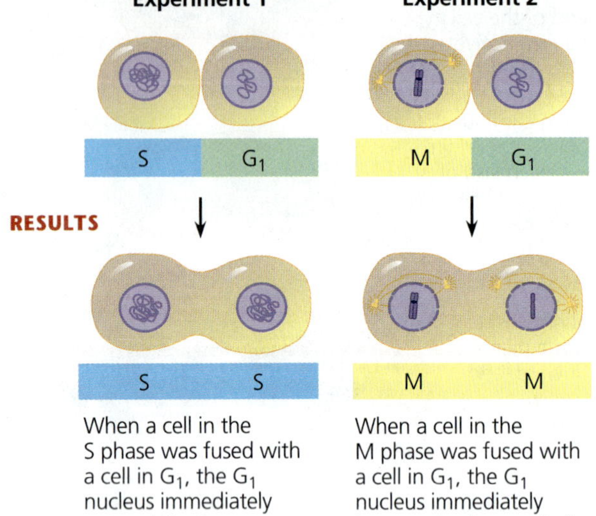

When a cell in the S phase was fused with a cell in G_1, the G_1 nucleus immediately entered the S phase—DNA was synthesized.

When a cell in the M phase was fused with a cell in G_1, the G_1 nucleus immediately began mitosis—a spindle formed and chromatin condensed, even though the chromosome had not been duplicated.

CONCLUSION The results of fusing a G_1 cell with a cell in the S or M phase of the cell cycle suggest that molecules present in the cytoplasm during the S or M phase control the progression to those phases.

SOURCE R. T. Johnson and P. N. Rao, Mammalian cell fusion: Induction of premature chromosome condensation in interphase nuclei, *Nature* 226:717–722 (1970).

WHAT IF? If the progression of phases did not depend on cytoplasmic molecules and each phase began when the previous one was complete, how would the results have differed?

Evidence for Cytoplasmic Signals

What controls the cell cycle? One reasonable hypothesis might be that each event in the cell cycle merely leads to the next, as in a simple metabolic pathway. According to this hypothesis, the replication of chromosomes in the S phase, for example, might cause cell growth during the G_2 phase, which might in turn lead inevitably to the onset of mitosis. However, this hypothesis, which proposes a pathway that is not subject to either internal or external regulation, turns out to be incorrect.

In the early 1970s, a variety of experiments led to an alternative hypothesis: that the cell cycle is driven by specific signaling molecules present in the cytoplasm. Some of the first strong evidence for this hypothesis came from experiments with mammalian cells grown in culture. In these experiments, two cells in different phases of the cell cycle were fused to form a single cell with two nuclei. If one of the original cells was in the S phase and the other was in G_1, the G_1 nucleus immediately entered the S phase, as though stimulated by signaling molecules present in the cytoplasm of the first cell. Similarly, if a cell undergoing mitosis (M phase) was fused with another cell in any stage of its cell cycle, even G_1, the second nucleus immediately entered mitosis, with condensation of the chromatin and formation of a mitotic spindle **(Figure 12.14)**.

The Cell Cycle Control System

The experiment shown in Figure 12.14 and other experiments on animal cells and yeasts demonstrated that the sequential events of the cell cycle are directed by a distinct **cell cycle control system**, a cyclically operating set of molecules in the cell that both triggers and coordinates key events

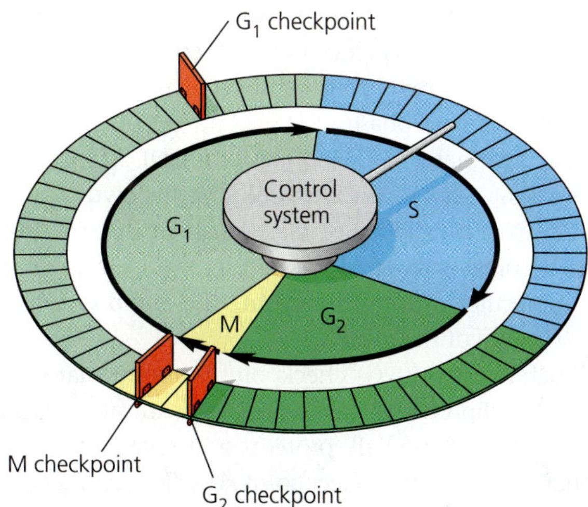

▲ **Figure 12.15 Mechanical analogy for the cell cycle control system.** In this diagram of the cell cycle, the flat "stepping stones" around the perimeter represent sequential events. Like the control device of an automatic washer, the cell cycle control system proceeds on its own, driven by a built-in clock. However, the system is subject to internal and external regulation at various checkpoints, of which three are shown (red).

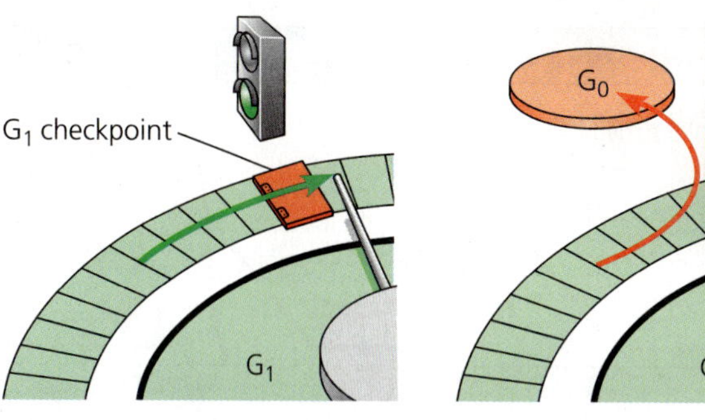

(a) If a cell receives a go-ahead signal at the G_1 checkpoint, the cell continues on in the cell cycle.

(b) If a cell does not receive a go-ahead signal at the G_1 checkpoint, the cell exits the cell cycle and goes into G_0, a nondividing state.

▲ **Figure 12.16 The G_1 checkpoint.**

WHAT IF? *What might be the result if the cell ignored the checkpoint and progressed through the cell cycle?*

in the cell cycle. The cell cycle control system has been compared to the control device of an automatic washing machine (**Figure 12.15**). Like the washer's timing device, the cell cycle control system proceeds on its own, according to a built-in clock. However, just as a washer's cycle is subject to both internal control (such as the sensor that detects when the tub is filled with water) and external adjustment (such as activation of the start mechanism), the cell cycle is regulated at certain checkpoints by both internal and external signals.

A **checkpoint** in the cell cycle is a control point where stop and go-ahead signals can regulate the cycle. (The signals are transmitted within the cell by the kinds of signal transduction pathways discussed in Chapter 11.) Animal cells generally have built-in stop signals that halt the cell cycle at checkpoints until overridden by go-ahead signals. Many signals registered at checkpoints come from cellular surveillance mechanisms inside the cell. These signals report whether crucial cellular processes that should have occurred by that point have in fact been completed correctly and thus whether or not the cell cycle should proceed. Checkpoints also register signals from outside the cell, as we will discuss later. Three major checkpoints are found in the G_1, G_2, and M phases (see Figure 12.15).

For many cells, the G_1 checkpoint—dubbed the "restriction point" in mammalian cells—seems to be the most important. If a cell receives a go-ahead signal at the G_1 checkpoint, it will usually complete the G_1, S, G_2, and M phases and divide. If it does not receive a go-ahead signal at that point, it will exit the cycle, switching into a nondividing state called the **G_0 phase** (**Figure 12.16**). Most cells of the human body are actually in the G_0 phase. As mentioned earlier, mature nerve cells and muscle cells never divide. Other cells, such as liver cells, can be "called back" from the G_0 phase to the cell cycle by external cues, such as growth factors released during injury.

To understand how cell cycle checkpoints work, we first need to see what kinds of molecules make up the cell cycle control system (the molecular basis for the cell cycle clock) and how a cell progresses through the cycle. Then we will consider the internal and external checkpoint signals that can make the clock pause or continue.

The Cell Cycle Clock: Cyclins and Cyclin-Dependent Kinases

Rhythmic fluctuations in the abundance and activity of cell cycle control molecules pace the sequential events of the cell cycle. These regulatory molecules are mainly proteins of two types: protein kinases and cyclins. Protein kinases are enzymes that activate or inactivate other proteins by phosphorylating them (see Chapter 11). Particular protein kinases give the go-ahead signals at the G_1 and G_2 checkpoints.

Many of the kinases that drive the cell cycle are actually present at a constant concentration in the growing cell, but much of the time they are in an inactive form. To be active, such a kinase must be attached to a **cyclin**, a protein that gets its name from its cyclically fluctuating concentration in the cell. Because of this requirement, these kinases are called **cyclin-dependent kinases**, or **Cdks**. The activity of a Cdk rises and falls with changes in the concentration of its cyclin partner. **Figure 12.17a**, on the next page, shows the fluctuating activity of **MPF**, the cyclin-Cdk complex that was discovered first (in frog eggs). Note that the peaks of MPF activity correspond to the peaks of cyclin concentration. The cyclin level rises during the S and G_2 phases and then falls abruptly during M phase.

The initials MPF stand for "maturation-promoting factor," but we can think of MPF as "M-phase-promoting factor" because it triggers the cell's passage past the G_2 checkpoint into

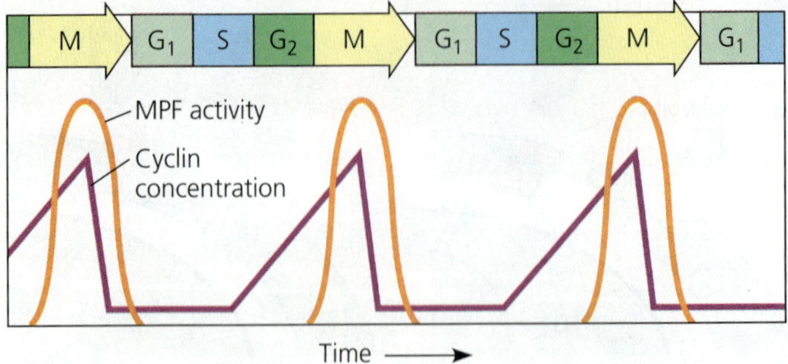

(a) Fluctuation of MPF activity and cyclin concentration during the cell cycle

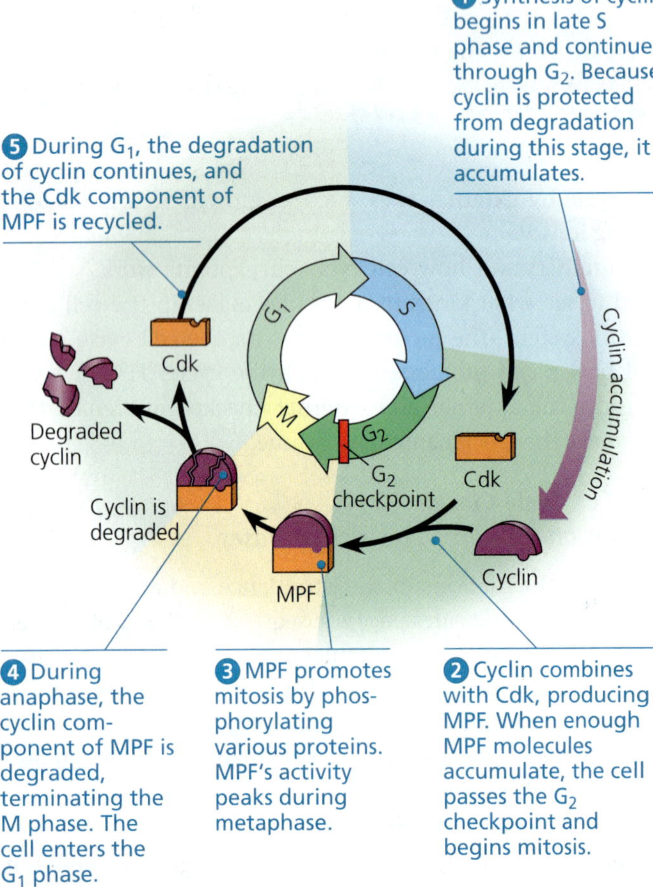

(b) Molecular mechanisms that help regulate the cell cycle

▲ **Figure 12.17 Molecular control of the cell cycle at the G_2 checkpoint.** The steps of the cell cycle are timed by rhythmic fluctuations in the activity of cyclin-dependent kinases (Cdks). Here we focus on a cyclin-Cdk complex in animal cells called MPF, which acts at the G_2 checkpoint as a go-ahead signal, triggering the events of mitosis.

? *Explain how the events in the diagram in (b) are related to the "Time" axis of the graph in (a).*

M phase **(Figure 12.17b)**. When cyclins that accumulate during G_2 associate with Cdk molecules, the resulting MPF complex phosphorylates a variety of proteins, initiating mitosis. MPF acts both directly as a kinase and indirectly by activating other kinases. For example, MPF causes phosphorylation of various proteins of the nuclear lamina (see Figure 6.9), which promotes fragmentation of the nuclear envelope during prometaphase of mitosis. There is also evidence that MPF contributes to molecular events required for chromosome condensation and spindle formation during prophase.

During anaphase, MPF helps switch itself off by initiating a process that leads to the destruction of its own cyclin. The noncyclin part of MPF, the Cdk, persists in the cell, inactive until it becomes part of MPF again by associating with new cyclin molecules synthesized during the S and G_2 phases of the next round of the cycle.

Cell behavior at the G_1 checkpoint is also regulated by the activity of cyclin-Cdk protein complexes. Animal cells appear to have at least three Cdk proteins and several different cyclins that operate at this checkpoint. The fluctuating activities of different cyclin-Cdk complexes are of major importance in controlling all the stages of the cell cycle.

Stop and Go Signs: Internal and External Signals at the Checkpoints

Research scientists are currently working out the pathways that link signals originating inside and outside the cell with the responses by cyclin-dependent kinases and other proteins. An example of an internal signal occurs at the third important checkpoint, the M phase checkpoint. Anaphase, the separation of sister chromatids, does not begin until all the chromosomes are properly attached to the spindle at the metaphase plate. Researchers have learned that as long as some kinetochores are unattached to spindle microtubules, the sister chromatids remain together, delaying anaphase. Only when the kinetochores of all the chromosomes are properly attached to the spindle does the appropriate regulatory protein complex become activated. (In this case, the regulatory molecule is not a cyclin-Cdk complex but, instead, a different complex made up of several proteins.) Once activated, the complex sets off a chain of molecular events that activates the enzyme separase, which cleaves the cohesins, allowing the sister chromatids to separate. This mechanism ensures that daughter cells do not end up with missing or extra chromosomes.

Studies using animal cells in culture have led to the identification of many external factors, both chemical and physical, that can influence cell division. For example, cells fail to divide if an essential nutrient is lacking in the culture medium. (This is analogous to trying to run an automatic washing machine without the water supply hooked up; an internal sensor won't allow the machine to continue past the point where water is needed.) And even if all other conditions are favorable, most types of mammalian cells divide in culture only if the growth medium includes specific growth factors. As mentioned in Chapter 11, a **growth factor** is a protein released by certain cells that stimulates other cells to divide. Researchers have discovered more than 50 growth factors. Different cell types respond specifically to different growth factors or combinations of growth factors.

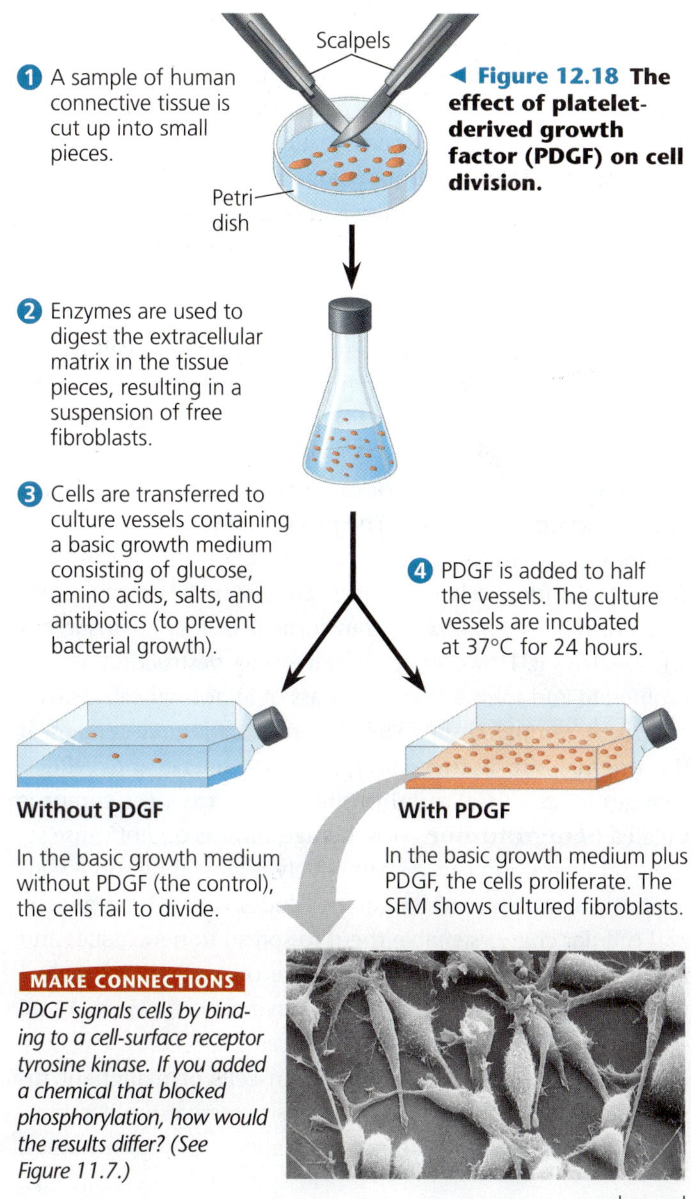

◄ **Figure 12.18 The effect of platelet-derived growth factor (PDGF) on cell division.**

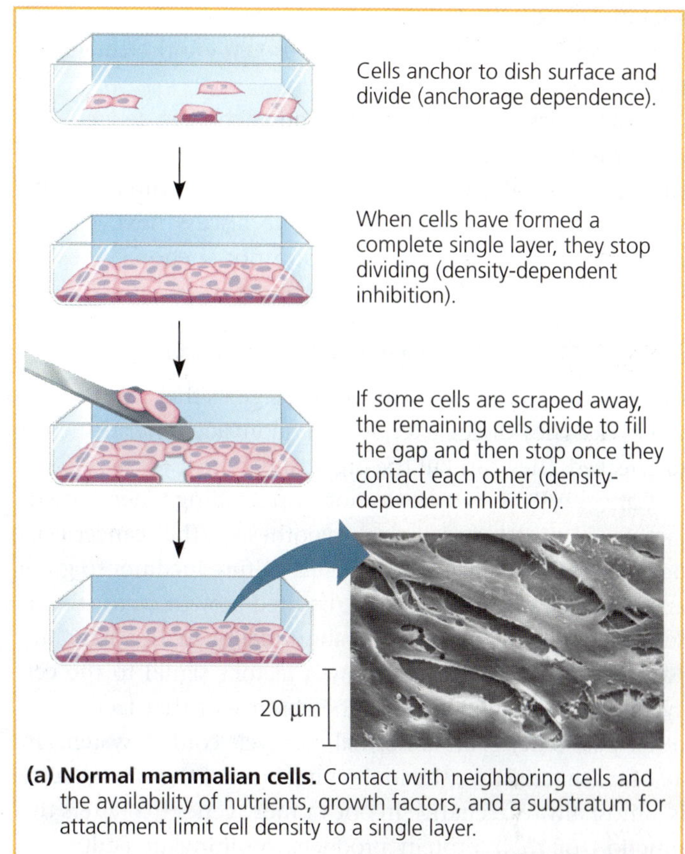

(a) Normal mammalian cells. Contact with neighboring cells and the availability of nutrients, growth factors, and a substratum for attachment limit cell density to a single layer.

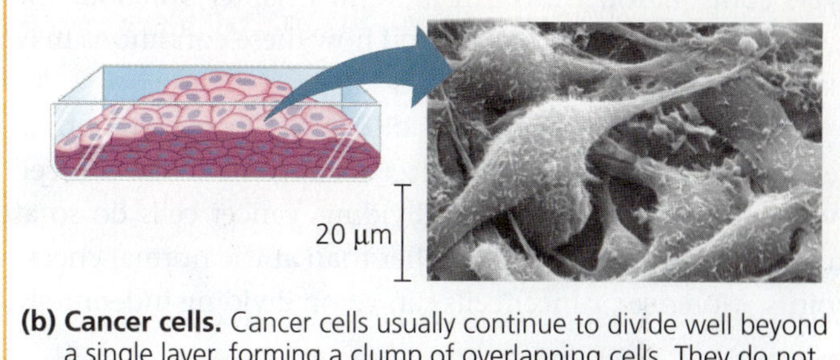

(b) Cancer cells. Cancer cells usually continue to divide well beyond a single layer, forming a clump of overlapping cells. They do not exhibit anchorage dependence or density-dependent inhibition.

▲ **Figure 12.19 Density-dependent inhibition and anchorage dependence of cell division.** Individual cells are shown disproportionately large in the drawings.

Consider, for example, *platelet-derived growth factor (PDGF)*, which is made by blood cell fragments called platelets. The experiment illustrated in **Figure 12.18** demonstrates that PDGF is required for the division of cultured fibroblasts, a type of connective tissue cell. Fibroblasts have PDGF receptors on their plasma membranes. The binding of PDGF molecules to these receptors (which are receptor tyrosine kinases; see Chapter 11) triggers a signal transduction pathway that allows the cells to pass the G_1 checkpoint and divide. PDGF stimulates fibroblast division not only in the artificial conditions of cell culture, but also in an animal's body. When an injury occurs, platelets release PDGF in the vicinity. The resulting proliferation of fibroblasts helps heal the wound.

The effect of an external physical factor on cell division is clearly seen in **density-dependent inhibition**, a phenomenon in which crowded cells stop dividing **(Figure 12.19a)**. As first observed many years ago, cultured cells normally divide until they form a single layer of cells on the inner surface of the culture container, at which point the cells stop dividing. If some cells are removed, those bordering the open space begin dividing again and continue until the vacancy is filled. Follow-up studies revealed that the binding of a cell-surface protein to its counterpart on an adjoining cell sends a growth-inhibiting signal to both cells, preventing them from moving forward in the cell cycle, even in the presence of growth factors.

Most animal cells also exhibit **anchorage dependence** (see Figure 12.19a). To divide, they must be attached to a substratum, such as the inside of a culture jar or the extracellular matrix of a tissue. Experiments suggest that like cell density, anchorage is signaled to the cell cycle control system via pathways involving plasma membrane proteins and elements of the cytoskeleton linked to them.

Density-dependent inhibition and anchorage dependence appear to function in the body's tissues as well as in cell culture, checking the growth of cells at some optimal density and location. Cancer cells, which we discuss next, exhibit neither density-dependent inhibition nor anchorage dependence **(Figure 12.19b)**.

Loss of Cell Cycle Controls in Cancer Cells

Cancer cells do not heed the normal signals that regulate the cell cycle. They divide excessively and invade other tissues. If unchecked, they can kill the organism.

Cancer cells in culture do not stop dividing when growth factors are depleted. A logical hypothesis is that cancer cells do not need growth factors in their culture medium to grow and divide. They may make a required growth factor themselves, or they may have an abnormality in the signaling pathway that conveys the growth factor's signal to the cell cycle control system even in the absence of that factor. Another possibility is an abnormal cell cycle control system. In all of these scenarios, the underlying basis of the abnormality is almost always a change in one or more genes that alters the function of their protein products, resulting in faulty cell cycle control. You will learn more in Chapter 18 about the genetic bases of these changes and how these conditions may lead to cancer.

There are other important differences between normal cells and cancer cells that reflect derangements of the cell cycle. If and when they stop dividing, cancer cells do so at random points in the cycle, rather than at the normal checkpoints. Moreover, cancer cells can go on dividing indefinitely in culture if they are given a continual supply of nutrients; in essence, they are "immortal." A striking example is a cell line that has been reproducing in culture since 1951. Cells of this line are called HeLa cells because their original source was a tumor removed from a woman named Henrietta Lacks. By contrast, nearly all normal mammalian cells growing in culture divide only about 20 to 50 times before they stop dividing, age, and die. (We'll see a possible reason for this phenomenon when we discuss DNA replication in Chapter 16.) Finally, cancer cells evade the normal controls that trigger a cell to undergo apoptosis when something is wrong—for example, when an irreparable mistake has occurred during DNA replication preceding mitosis.

The abnormal behavior of cancer cells can be catastrophic when it occurs in the body. The problem begins when a single cell in a tissue undergoes **transformation**, the process that converts a normal cell to a cancer cell. The body's immune system normally recognizes a transformed cell as an insurgent and destroys it. However, if the cell evades destruction, it may proliferate and form a tumor, a mass of abnormal cells within otherwise normal tissue. The abnormal cells may remain at the original site if they have too few genetic and cellular changes to survive at another site. In that case, the tumor is called a **benign tumor**. Most benign tumors do not cause serious problems and can be completely removed by surgery. In contrast, a **malignant tumor** includes cells whose genetic and cellular changes enable them to spread to new tissues and impair the functions of one or more organs. An individual with a malignant tumor is said to have cancer; **Figure 12.20** shows the development of breast cancer.

The changes that have occurred in cells of malignant tumors show up in many ways besides excessive proliferation. These cells may have unusual numbers of chromosomes,

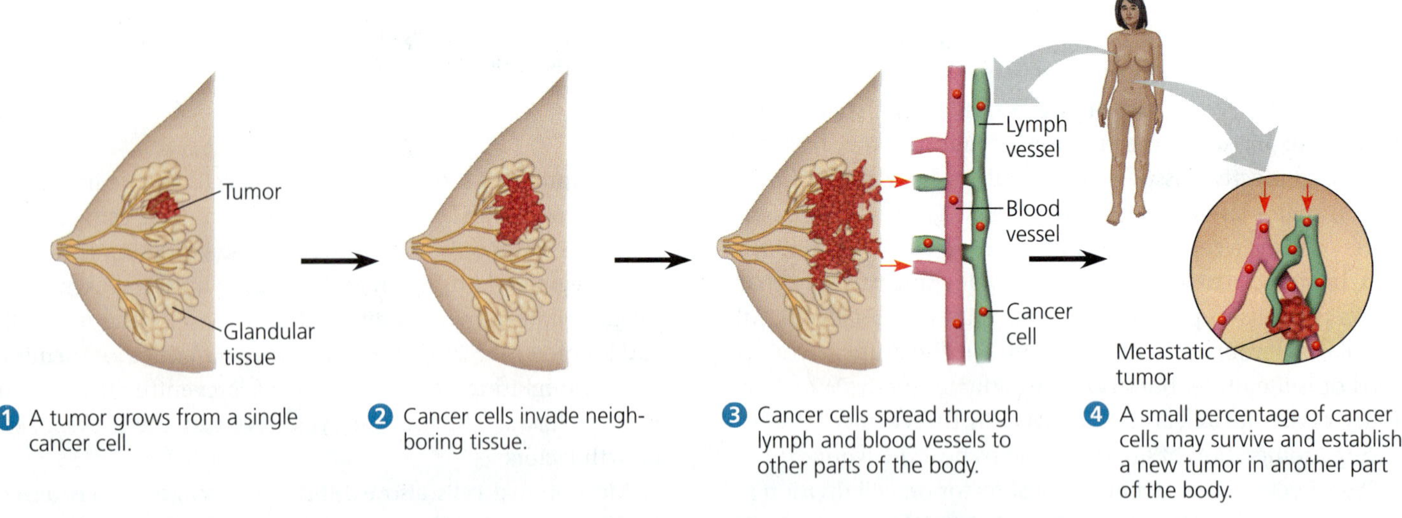

▲ **Figure 12.20 The growth and metastasis of a malignant breast tumor.** The cells of malignant (cancerous) tumors grow in an uncontrolled way and can spread to neighboring tissues and, via lymph and blood vessels, to other parts of the body. The spread of cancer cells beyond their original site is called metastasis.

though whether this is a cause or an effect of transformation is a current topic of debate. Their metabolism may be disabled, and they may cease to function in any constructive way. Abnormal changes on the cell surface cause cancer cells to lose attachments to neighboring cells and the extracellular matrix, allowing them to spread into nearby tissues. Cancer cells may also secrete signaling molecules that cause blood vessels to grow toward the tumor. A few tumor cells may separate from the original tumor, enter blood vessels and lymph vessels, and travel to other parts of the body. There, they may proliferate and form a new tumor. This spread of cancer cells to locations distant from their original site is called **metastasis** (see Figure 12.20).

A tumor that appears to be localized may be treated with high-energy radiation, which damages DNA in cancer cells much more than it does in normal cells, apparently because the majority of cancer cells have lost the ability to repair such damage. To treat known or suspected metastatic tumors, chemotherapy is used, in which drugs that are toxic to actively dividing cells are administered through the circulatory system. As you might expect, chemotherapeutic drugs interfere with specific steps in the cell cycle. For example, the drug Taxol freezes the mitotic spindle by preventing microtubule depolymerization, which stops actively dividing cells from proceeding past metaphase. The side effects of chemotherapy are due to the drugs' effects on normal cells that divide often. For example, nausea results from chemotherapy's effects on intestinal cells, hair loss from effects on hair follicle cells, and susceptibility to infection from effects on immune system cells.

Over the past several decades, researchers have produced a flood of valuable information about cell-signaling pathways and how their malfunction contributes to the development of cancer through effects on the cell cycle. Coupled with new molecular techniques, such as the ability to rapidly sequence the DNA of cells in a particular tumor, medical treatments for cancer are beginning to become more "personalized" to a particular patient's tumor. Breast cancer provides a good example. Basic research on the processes described in Chapters 11 and 12 has augmented our understanding of the molecular events underlying development of breast cancer. Proteins functioning in cell signaling pathways that affect the cell cycle are often found to be altered in breast cancer cells. Analyzing the level and sequences of such proteins has allowed physicians to better tailor the treatment to the cancers of some individuals, as shown in **Figure 12.21**.

One of the big lessons we've learned about the development of cancer, though, is how very complex the process is. There are many areas that remain to be explored. Perhaps the reason we have so many unanswered questions about cancer cells is that there is still so much to learn about how normal cells function. The cell, life's basic unit of structure and function, holds enough secrets to engage researchers well into the future.

▼ **Figure 12.21**

IMPACT

Advances in Treatment of Breast Cancer

Cancer cells, such as the breast cancer cell shown below, are analyzed by DNA sequencing and other molecular techniques to look for alterations in the level or sequence of specific proteins associated with cancer. For example, the cells of roughly 20–25% of breast cancer tumors show abnormally high amounts of a cell-surface receptor tyrosine kinase called HER2, and many show an increase in the number of estrogen receptor (ER) molecules, intracellular receptors that can trigger cell division. Based on lab findings, a physician can prescribe chemotherapy with a molecule that blocks the function of the specific protein (Herceptin for HER2 and tamoxifen for ERs). Treatment using these agents, when appropriate, has led to increased survival rates and fewer cancer recurrences.

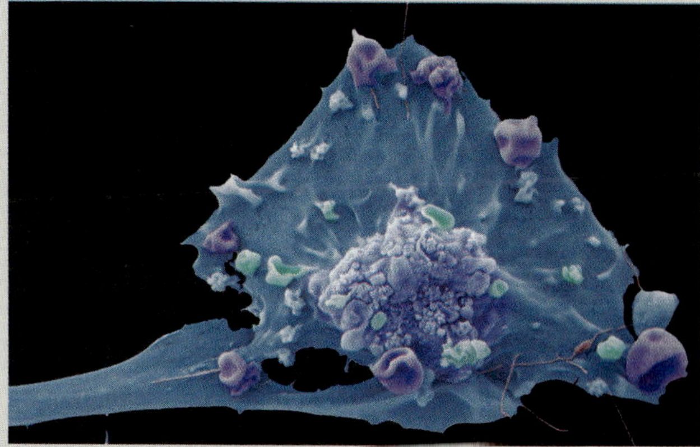

WHY IT MATTERS Approximately one out of every eight women will develop breast cancer, the most common cancer among women. Worldwide, the incidence of breast cancer has been increasing annually. However, the mortality rate from this disease is falling in the United States and elsewhere, probably a result of earlier detection and improved treatment. Furthermore, what we are learning from the study of breast cancer also enhances our understanding of the development and treatment of other types of cancer.

FURTHER READING F. J. Esteva and G. N. Hortobagyi, Gaining ground on breast cancer, *Scientific American* 298:58–65 (2008).

MAKE CONNECTIONS Review the material in Chapter 11 on receptor tyrosine kinases and intracellular receptors (Figures 11.7 and 11.9 on pp. 212–214). Explain in general how these receptors might function in triggering cell division.

CONCEPT CHECK 12.3

1. In Figure 12.14, why do the nuclei resulting from experiment 2 contain different amounts of DNA?
2. How does MPF allow a cell to pass the G_2 phase checkpoint and enter mitosis? (See Figure 12.17.)
3. What phase are most of your body cells in?
4. Compare and contrast a benign tumor and a malignant tumor.
5. **WHAT IF?** What would happen if you performed the experiment in Figure 12.18 with cancer cells?

For suggested answers, see Appendix A.

12 CHAPTER REVIEW

SUMMARY OF KEY CONCEPTS

- Unicellular organisms reproduce by **cell division**; multicellular organisms depend on cell division for their development from a fertilized egg and for growth and repair. Cell division is part of the **cell cycle**, an ordered sequence of events in the life of a cell from its origin until it divides into daughter cells.

CONCEPT 12.1

Most cell division results in genetically identical daughter cells (pp. 229–230)

- The genetic material (DNA) of a cell—its **genome**—is partitioned among **chromosomes**. Each eukaryotic chromosome consists of one DNA molecule associated with many proteins that maintain chromosome structure and help control the activity of genes. Together, the complex of DNA and associated proteins is called **chromatin**. The chromatin of a chromosome exists in different states of condensation at different times. In animals, **gametes** have one set of chromosomes and **somatic cells** have two sets.
- Cells replicate their genetic material before they divide, ensuring that each daughter cell can receive a copy of the DNA. In preparation for cell division, chromosomes are duplicated, each one then consisting of two identical **sister chromatids** joined along their lengths by sister chromatid cohesion and held most tightly together at a constricted region at the **centromeres** of the chromatids. When this cohesion is broken, the chromatids separate during cell division, becoming the chromosomes of the new daughter cells. Eukaryotic cell division consists of **mitosis** (division of the nucleus) and **cytokinesis** (division of the cytoplasm).

? *Differentiate between these terms: chromosome, chromatin, and chromatid.*

CONCEPT 12.2

The mitotic phase alternates with interphase in the cell cycle (pp. 230–238)

- Between divisions, a cell is in **interphase**: the $\mathbf{G_1}$, **S**, and $\mathbf{G_2}$ phases. The cell grows throughout interphase, but DNA is replicated only during the synthesis (S) phase. Mitosis and cytokinesis make up the **mitotic (M) phase** of the cell cycle.

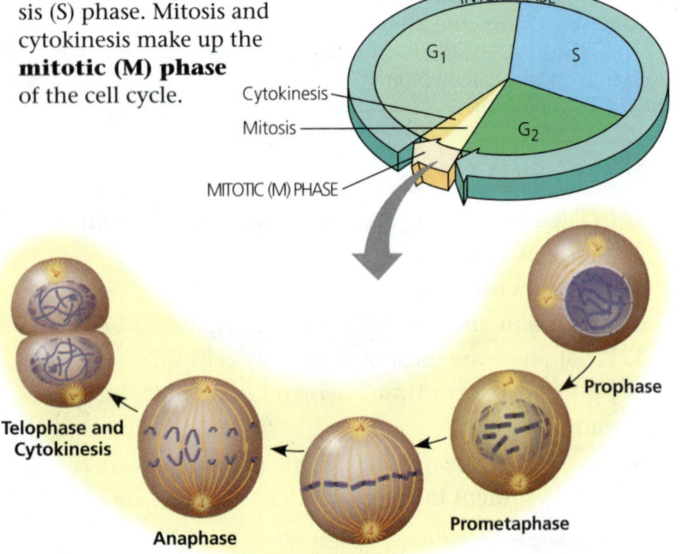

- The **mitotic spindle** is an apparatus of microtubules that controls chromosome movement during mitosis. In animal cells, the spindle arises from the **centrosomes** and includes spindle microtubules and **asters**. Some spindle microtubules attach to the **kinetochores** of chromosomes and move the chromosomes to the **metaphase plate**. In anaphase, sister chromatids separate, and motor proteins move them along the kinetochore microtubules toward opposite ends of the cell. Meanwhile, motor proteins push nonkinetochore microtubules from opposite poles away from each other, elongating the cell. In telophase, genetically identical daughter nuclei form at opposite ends of the cell.
- Mitosis is usually followed by cytokinesis. Animal cells carry out cytokinesis by **cleavage**, and plant cells form a **cell plate**.
- During **binary fission** in bacteria, the chromosome replicates and the two daughter chromosomes actively move apart. Some of the proteins involved in bacterial binary fission are related to eukaryotic actin and tubulin.
- Since prokaryotes preceded eukaryotes by more than a billion years, it is likely that mitosis evolved from prokaryotic cell division. Certain unicellular eukaryotes exhibit mechanisms of cell division that may be similar to those of ancestors of existing eukaryotes. Such mechanisms might have been intermediate steps in the evolution of mitosis from bacterial binary fission.

? *In which of the three subphases of interphase and the stages of mitosis do chromosomes exist as single DNA molecules?*

CONCEPT 12.3

The eukaryotic cell cycle is regulated by a molecular control system (pp. 238–243)

- Signaling molecules present in the cytoplasm regulate progress through the cell cycle.
- The **cell cycle control system** is molecularly based. Cyclic changes in regulatory proteins work as a cell cycle clock. The key molecules are **cyclins** and **cyclin-dependent kinases (Cdks)**. The clock has specific **checkpoints** where the cell cycle stops until a go-ahead signal is received. Cell culture has enabled researchers to study the molecular details of cell division. Both internal signals and external signals control the cell cycle checkpoints via signal transduction pathways. Most cells exhibit **density-dependent inhibition** of cell division as well as **anchorage dependence**.
- Cancer cells elude normal cell cycle regulation and divide out of control, forming tumors. **Malignant tumors** invade surrounding tissues and can undergo **metastasis**, exporting cancer cells to other parts of the body, where they may form secondary tumors. Recent advances in understanding the cell cycle and cell signaling, as well as techniques for sequencing DNA, have allowed improvements in cancer treatment.

? *Explain the significance of the G_1, G_2, and M checkpoints and the go-ahead signals involved in the cell cycle control system.*

TEST YOUR UNDERSTANDING

LEVEL 1: KNOWLEDGE/COMPREHENSION

1. Through a microscope, you can see a cell plate beginning to develop across the middle of a cell and nuclei forming on either side of the cell plate. This cell is most likely
   a. an animal cell in the process of cytokinesis.
   b. a plant cell in the process of cytokinesis.

c. an animal cell in the S phase of the cell cycle.
d. a bacterial cell dividing.
e. a plant cell in metaphase.

2. Vinblastine is a standard chemotherapeutic drug used to treat cancer. Because it interferes with the assembly of microtubules, its effectiveness must be related to
a. disruption of mitotic spindle formation.
b. inhibition of regulatory protein phosphorylation.
c. suppression of cyclin production.
d. myosin denaturation and inhibition of cleavage furrow formation.
e. inhibition of DNA synthesis.

3. One difference between cancer cells and normal cells is that cancer cells
a. are unable to synthesize DNA.
b. are arrested at the S phase of the cell cycle.
c. continue to divide even when they are tightly packed together.
d. cannot function properly because they are affected by density-dependent inhibition.
e. are always in the M phase of the cell cycle.

4. The decline of MPF activity at the end of mitosis is due to
a. the destruction of the protein kinase Cdk.
b. decreased synthesis of Cdk.
c. the degradation of cyclin.
d. the accumulation of cyclin.
e. synthesis of DNA.

5. In the cells of some organisms, mitosis occurs without cytokinesis. This will result in
a. cells with more than one nucleus.
b. cells that are unusually small.
c. cells lacking nuclei.
d. destruction of chromosomes.
e. cell cycles lacking an S phase.

6. Which of the following does *not* occur during mitosis?
a. condensation of the chromosomes
b. replication of the DNA
c. separation of sister chromatids
d. spindle formation
e. separation of the spindle poles

LEVEL 2: APPLICATION/ANALYSIS

7. In the light micrograph below of dividing cells near the tip of an onion root, identify a cell in each of the following stages: prophase, prometaphase, metaphase, anaphase, and telophase. Describe the major events occurring at each stage.

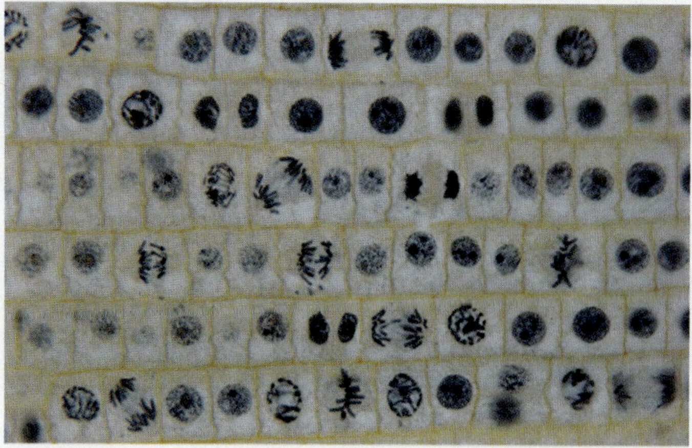

8. A particular cell has half as much DNA as some other cells in a mitotically active tissue. The cell in question is most likely in
a. G_1.
b. G_2.
c. prophase.
d. metaphase.
e. anaphase.

9. The drug cytochalasin B blocks the function of actin. Which of the following aspects of the animal cell cycle would be most disrupted by cytochalasin B?
a. spindle formation
b. spindle attachment to kinetochores
c. DNA synthesis
d. cell elongation during anaphase
e. cleavage furrow formation and cytokinesis

10. **DRAW IT** Draw one eukaryotic chromosome as it would appear during interphase, during each of the stages of mitosis, and during cytokinesis. Also draw and label the nuclear envelope and any microtubules attached to the chromosome(s).

LEVEL 3: SYNTHESIS/EVALUATION

11. **EVOLUTION CONNECTION**
The result of mitosis is that the daughter cells end up with the same number of chromosomes that the parent cell had. Another way to maintain the number of chromosomes would be to carry out cell division first and then duplicate the chromosomes in each daughter cell. Do you think this would be an equally good way of organizing the cell cycle? Why do you suppose that evolution has not led to this alternative?

12. **SCIENTIFIC INQUIRY**
Although both ends of a microtubule can gain or lose subunits, one end (called the plus end) polymerizes and depolymerizes at a higher rate than the other end (the minus end). For spindle microtubules, the plus ends are in the center of the spindle, and the minus ends are at the poles. Motor proteins that move along microtubules specialize in walking either toward the plus end or toward the minus end; the two types are called plus end–directed and minus end–directed motor proteins, respectively. Given what you know about chromosome movement and spindle changes during anaphase, predict which type of motor proteins would be present on (a) kinetochore microtubules and (b) nonkinetochore microtubules.

13. **WRITE ABOUT A THEME**
The Genetic Basis of Life The continuity of life is based on heritable information in the form of DNA. In a short essay (100–150 words), explain how the process of mitosis faithfully parcels out exact copies of this heritable information in the production of genetically identical daughter cells.

For selected answers, see Appendix A.

www.masteringbiology.com

1. MasteringBiology® Assignments
BioFlix **Tutorials** Mitosis: Mitosis and the Cell Cycle • Mechanism of Mitosis • Comparing Cell Division in Animals, Plants, and Bacteria
Activities The Cell Cycle • Mitosis and Cytokinesis Animation • The Phases of Mitosis • Four Phases of the Cell Cycle • Causes of Cancer
Questions Student Misconceptions • Reading Quiz • Multiple Choice • End-of-Chapter

2. eText
Read your book online, search, take notes, highlight text, and more.

3. The Study Area
Practice Tests • Cumulative Test • ***BioFlix*** 3-D Animations • MP3 Tutor Sessions • Videos · Activities • Investigations • Lab Media • Audio Glossary • Word Study Tools • Art

An Interview with

Joan A. Steitz

RNA is Joan Steitz's favorite molecule, and her research into its structures and functions has made contributions of enormous importance to our understanding of genetics at the molecular level. Raised in Minnesota, Dr. Steitz has a B.S. in Chemistry from Antioch College and a Ph.D. in Biochemistry and Molecular Biology from Harvard, where she worked in the laboratory of James D. Watson. Among her many awards and honors are the National Medal of Science, the Gairdner International Award, and 12 honorary doctorates. She is a member of the National Academy of Sciences and the Institute of Medicine. A teacher and researcher at Yale University since 1970, she is now Sterling Professor of Molecular Biophysics and Biochemistry and an Investigator of the Howard Hughes Medical Institute.

How did you get started in molecular genetics?

I first learned about the structure of DNA in my third year of college, during a co-op job at MIT. I was enthralled with the idea that DNA might be the molecular basis for all of the genetics—red hair, wrinkled peas, and so forth—that I had learned about in high school. After that, I worked in a molecular biology lab in Germany as a student abroad. Nevertheless, I decided to go to medical school.

I didn't apply to a Ph.D. program because I'd never seen a woman heading up a research lab, and it didn't enter my mind that I could do that. But I did know some women physicians, so I applied to medical school and was admitted to Harvard. However, the summer before I was supposed to enter, I ended up working in the lab of cell biologist Joe Gall, then at the University of Minnesota. For the first time, I had my own project, and I loved it. By August 1st, I decided that I didn't care if I would never be the head of a lab; I just wanted to do research. Luckily, I was able to switch from the medical school to a graduate program at Harvard.

How did you end up as a graduate student of Jim Watson?

I was interested in the question of whether all cellular organelles have DNA, like mitochondria do. So I first approached a cell biologist, a famous microscopist who nevertheless reserved a bench in the corner of his lab for biochemistry. He conceded that his lab might be suitable, then gave me an unencouraging look and said, "But you're a woman. What are you going to do when you get married and have kids?" I barely made it out of his office before bursting into tears. Then I went to my second-choice thesis advisor, Jim Watson. I had done very well in his course, and he accepted me into his lab. So I became his first female graduate student, something I didn't discover until months later.

What was it like being in Watson's lab?

The Watson lab was a very exciting place at that time, in 1964. We knew that genes in DNA were transcribed into complementary RNA (a process called transcription) and that RNA called messenger RNA (mRNA) was translated into protein by ribosomes (translation). Besides mRNA, the only kinds of cellular RNA that were known were transfer RNA (tRNA) and ribosomal RNA (rRNA), although it was also known that some viruses had RNA instead of DNA as their genomes. But when I started grad school, we didn't yet know the genetic code—how the nucleotide sequence in mRNA corresponds to the amino acid sequence in protein—or much of anything about how transcription or translation occurred.

Jim would go off to meetings, and when he came back, everybody would crowd around him in the hall to find out what was new. Imagine the excitement when we heard, at an international biochemistry congress I actually attended, that the genetic code had been figured out! Or when someone in our lab discovered that a special kind of tRNA initiated protein synthesis. Things were happening very, very rapidly! The atmosphere was fiercely competitive but paradoxically collegial—the three or four labs that were working on the mechanisms of transcription and translation were all in contact with each other.

What was your research as a graduate student?

I worked on a newly discovered virus, R17, that infects the bacterium *E. coli*. Like other simple viruses, R17 is just a small amount of nucleic acid inside a protein coat. Throughout that era, molecular biologists fervently believed that unless you worked on something really simple, you would never figure out the molecular basis of life. So a virus that had only three genes (later found to be four) was the perfect thing to study.

The nucleic acid of R17, its genome, is RNA. This RNA gets into bacterial cells, and about an hour later out come 10,000 copies of the virus. So lots of things are happening in those cells. I studied a viral protein called the A-protein. For my thesis, I characterized the A-protein and what happened if there were mutations in its gene: You got virus particles that looked normal in the electron microscope but couldn't infect a bacterium. It turned out that the A-protein was needed for the virus to attach to the cell.

What did you do after graduate school?

I was married by then, and my husband had arranged to do a postdoc at the Medical Research Council (MRC) at Cambridge University, a mecca for structural and molecular biology. Jim Watson had written to Francis Crick asking him to find a place for me, but when I arrived at Cambridge, Francis suggested I do library research. Eventually, however, I found a bit of bench space for a lab project.

Fred Sanger's lab was nearby, and he was just working out his method for sequencing RNA. There was a lot of interchange with the people in Fred's lab, and they were very interested in the sequence of the R17 genome. Since it was very small, it was a really good molecule to work on. Previously, a paper had been published describing a method for isolating the particular stretches of mRNA bound to a functioning ribosome: You treated the mRNA-ribosome complex with ribonuclease, an enzyme that breaks down unprotected RNA, and you ended up with the part of the mRNA that had been bound and therefore protected by the ribosome, about 30 nucleotides long. The project I took on was to make ribosomes bind to R17 RNA (which functions as mRNA in normal virus infection) under conditions where they start but do not elongate proteins, and then isolate the ribosome-bound RNA segments. I would then determine the sequence of the parts of this RNA where translation started. Other people had considered and rejected this project. They were all male postdocs with wives and children who knew that in two years they would have to interview for tenure-track jobs, and this project had little chance of quick success. But since I thought I couldn't aim higher than a research position in somebody else's lab, I felt free to take on a risky project. (So, being a woman determined the two most important decisions of my early scientific career: ending up in Watson's lab and choosing my project at Cambridge.)

I determined the RNA nucleotide sequences at the beginning of the three R17 genes known at the time. These sequences included AUG, already known to be the "start codon" in mRNA (the first nucleotide triplet translated). And the sequences that followed AUG fit what was already known about the protein sequences, according to the genetic code. We also established that there were spaces between genes in the viral genome. And we figured out that sometimes the virus RNA folded into secondary structures that were important in regulating how many ribosomes would get on at a particular start site. This work at Cambridge—and better academic opportunities for women in the United States—led to my faculty position at Yale.

When you arrived at Yale, what was your first big discovery?

I found out how ribosomes locate the regions on mRNA where they attach and start translation. At Cambridge I had worked out the three 30-nucleotide sequences where ribosomes bind to R17 RNA, but it still wasn't clear how ribosomes honed in on these sequences out of the virus's 3,500 nucleotides. One idea was that a stretch of mRNA rich in purines, just upstream of where translation actually starts, would base-pair with the 3′ end of the rRNA molecule in the small ribosomal subunit of bacteria. So I went to work testing that hypothesis. I soon had direct evidence that there actually is a physical interaction between the end of the "16S" rRNA molecule and the regions of mRNA that are bound by ribosomes. So this RNA-RNA base pairing, along with RNA-RNA base pairing between tRNA and mRNA, is the basis of polypeptide initiation.

You then turned to eukaryotic mRNA. What is different about mRNA production in eukaryotic cells, compared with bacteria?

The main difference comes from the fact that the genes of humans and other eukaryotes have interruptions in them, stretches of nucleic acid that are not translated. These interruptions, called *introns*, have to be removed from the RNA transcript before it is translated. But we didn't know this when I got interested in the subject. At that time, all we knew was that only 5–10% of the RNA transcribed from eukaryotic genes got out of the nucleus as mRNA. I was intrigued by this mystery and decided to switch from prokaryotes to eukaryotes to try to study it. Then, when introns were discovered, the reason for the loss of RNA became clear—though not how the extra RNA was removed. To make mRNA, somehow the introns have to be precisely removed and the coding bits have to be glued back together—a process called RNA splicing.

What have you learned since then about RNA splicing?

The most important molecular players are small RNA molecules that base-pair with sequences at the ends of RNA introns. This base pairing initiates the assembly of a ribosome-sized machine called a *spliceosome* made of RNA-protein subunits called snRNPs (pronounced "snurps") and other proteins. A spliceosome removes introns and joins together the protein-coding pieces. So RNA-RNA base pairing is the basis of the whole splicing process, just like it's the basis of the initiation of translation. Now there is more and more evidence that the RNAs are the catalytic components of the spliceosome, with the proteins playing supporting roles.

Does your research have any medical relevance?

We learned early on that people with lupus, an autoimmune disease, make antibodies to snRNPs, the RNA-protein subunits of spliceosomes. This discovery has been useful for the diagnosis of a number of autoimmune diseases and even for the prognosis of individual patients—although it hasn't led to cures. What we do in my lab, however, is very basic research. Somebody's got to figure out the basics in order for somebody else to figure out how to apply it.

What's going on now in the RNA field?

Lots of new classes of small RNA molecules have been discovered that, like rRNA, tRNA, and the RNAs in snRNPs, do not themselves code for protein. All these RNAs are important in getting information out of the DNA and into the functioning proteins of the cell. For instance, tiny RNAs called microRNAs, which associate with particular proteins, are involved in regulating translation. Again, it's RNA-RNA base pairing that determines the specificity. The theme of my research over my entire career has been finding out how RNAs interact with other RNAs to provide specificity along the pathway of gene expression. Proteins play important auxiliary roles, but it's basically been one RNA interacting with another RNA. I started working on RNA while I was a student, and it has continued to be my favorite molecule! There's enough to learn to last for many more lifetimes.

What do the discoveries about RNA suggest about the early stages of life on Earth?

Most biologists think that RNA was the first and most important genetic material, probably serving the first cells as both genome and the means by which the information in the genome directed cellular functions. Over time, cells have replaced the RNA genome with DNA, and many of the other RNA molecules with proteins. But the crucial processes of gene expression and its regulation are still dependent on various RNAs—4 billion years after life first arose!

"I started working on RNA while I was a student, and it has continued to be my favorite molecule!"

Joan Steitz (center) with Lisa Urry (right) and Jane Reece

13

Meiosis and Sexual Life Cycles

▲ **Figure 13.1 What accounts for family resemblance?**

KEY CONCEPTS

- **13.1** Offspring acquire genes from parents by inheriting chromosomes
- **13.2** Fertilization and meiosis alternate in sexual life cycles
- **13.3** Meiosis reduces the number of chromosome sets from diploid to haploid
- **13.4** Genetic variation produced in sexual life cycles contributes to evolution

OVERVIEW

Variations on a Theme

Most people who send out birth announcements mention the sex of the baby, but they don't feel the need to specify that their offspring is a human being! One of the characteristics of life is the ability of organisms to reproduce their own kind—elephants produce little elephants, and oak trees generate oak saplings. Exceptions to this rule show up only as sensational but highly suspect stories in tabloid newspapers.

Another rule often taken for granted is that offspring resemble their parents more than they do unrelated individuals. If you examine the family members shown in **Figure 13.1**—actress Sissy Spacek and her husband Jack Fisk with daughters Madison and Schuyler Fisk—you can pick out some similar features among them. The transmission of traits from one generation to the next is called inheritance, or **heredity** (from the Latin *heres*, heir). However, sons and daughters are not identical copies of either parent or of their siblings. Along with inherited similarity, there is also **variation**. Farmers have exploited the principles of heredity and variation for thousands of years, breeding plants and animals for desired traits. But what are the biological mechanisms leading to the hereditary similarity and variation that we call a "family resemblance"? The answer to this question eluded biologists until the advance of genetics in the 20th century.

Genetics is the scientific study of heredity and hereditary variation. In this unit, you will learn about genetics at multiple levels, from organisms to cells to molecules. On the practical side, you will see how genetics continues to revolutionize medicine and agriculture, and you will be asked to consider some social and ethical questions raised by our ability to manipulate DNA, the genetic material. At the end of the unit, you will be able to stand back and consider the whole genome, an organism's entire complement of DNA. Rapid acquisition and analysis of the genome sequences of many species, including our own, have taught us a great deal about evolution on the molecular level—in other words, evolution of the genome itself. In fact, genetic methods and discoveries are catalyzing progress in all areas of biology, from cell biology to physiology, developmental biology, behavior, and even ecology.

We begin our study of genetics in this chapter by examining how chromosomes pass from parents to offspring in sexually reproducing organisms. The processes of meiosis (a special type of cell division) and fertilization (the fusion of sperm and egg) maintain a species' chromosome count during the sexual life cycle. We will describe the cellular mechanics of meiosis and explain how this process differs from mitosis. Finally, we will consider how both meiosis and fertilization contribute to genetic variation, such as the variation obvious in the family shown in Figure 13.1.

CONCEPT **13.1**

Offspring acquire genes from parents by inheriting chromosomes

Family friends may tell you that you have your mother's freckles or your father's eyes. Of course, parents do not, in any literal sense, give their children freckles, eyes, hair, or any other traits. What, then, *is* actually inherited?

Inheritance of Genes

Parents endow their offspring with coded information in the form of hereditary units called **genes**. The genes we inherit from our mothers and fathers are our genetic link to our parents, and they account for family resemblances such as shared eye color or freckles. Our genes program the specific traits that emerge as we develop from fertilized eggs into adults.

The genetic program is written in the language of DNA, the polymer of four different nucleotides you learned about in Chapters 1 and 5. Inherited information is passed on in the form of each gene's specific sequence of DNA nucleotides, much as printed information is communicated in the form of meaningful sequences of letters. In both cases, the language is symbolic. Just as your brain translates the word *apple* into a mental image of the fruit, cells translate genes into freckles and other features. Most genes program cells to synthesize specific enzymes and other proteins, whose cumulative action produces an organism's inherited traits. The programming of these traits in the form of DNA is one of the unifying themes of biology.

The transmission of hereditary traits has its molecular basis in the precise replication of DNA, which produces copies of genes that can be passed from parents to offspring. In animals and plants, reproductive cells called **gametes** are the vehicles that transmit genes from one generation to the next. During fertilization, male and female gametes (sperm and eggs) unite, thereby passing on genes of both parents to their offspring.

Except for small amounts of DNA in mitochondria and chloroplasts, the DNA of a eukaryotic cell is packaged into chromosomes within the nucleus. Every species has a characteristic number of chromosomes. For example, humans have 46 chromosomes in their **somatic cells**—all cells of the body except the gametes and their precursors. Each chromosome consists of a single long DNA molecule elaborately coiled in association with various proteins. One chromosome includes several hundred to a few thousand genes, each of which is a specific sequence of nucleotides within the DNA molecule. A gene's specific location along the length of a chromosome is called the gene's **locus** (plural, *loci*; from the Latin, meaning "place"). Our genetic endowment consists of the genes that are part of the chromosomes we inherited from our parents.

Comparison of Asexual and Sexual Reproduction

Only organisms that reproduce asexually have offspring that are exact genetic copies of themselves. In **asexual reproduction**, a single individual is the sole parent and passes copies of all its genes to its offspring without the fusion of gametes. For example, single-celled eukaryotic organisms can reproduce asexually by mitotic cell division, in which DNA is copied and allocated equally to two daughter cells. The genomes of the offspring are virtually exact copies of the parent's genome. Some multicellular organisms are also capable of reproducing asexually (**Figure 13.2**). Because the cells of the offspring are derived by mitosis in the parent, the "chip off the old block" is usually genetically identical to its parent. An individual that reproduces asexually gives rise to a **clone**, a group of genetically identical individuals. Genetic differences occasionally arise in asexually reproducing organisms as a result of changes in the DNA called mutations, which we will discuss in Chapter 17.

(a) Hydra

(b) Redwoods

▲ **Figure 13.2 Asexual reproduction in two multicellular organisms. (a)** This relatively simple animal, a hydra, reproduces by budding. The bud, a localized mass of mitotically dividing cells, develops into a small hydra, which detaches from the parent (LM). **(b)** All the trees in this circle of redwoods arose asexually from a single parent tree, whose stump is in the center of the circle.

In **sexual reproduction**, two parents give rise to offspring that have unique combinations of genes inherited from the two parents. In contrast to a clone, offspring of sexual reproduction vary genetically from their siblings and both parents: They are variations on a common theme of family resemblance, not exact replicas. Genetic variation like that shown in Figure 13.1 is an important consequence of sexual reproduction. What mechanisms generate this genetic variation? The key is the behavior of chromosomes during the sexual life cycle.

CONCEPT CHECK 13.1

1. Explain what causes the traits of parents (such as hair color) to show up in their offspring.
2. How do asexually reproducing organisms produce offspring that are genetically identical to each other and to their parents?
3. **WHAT IF?** A horticulturalist breeds orchids, trying to obtain a plant with a unique combination of desirable traits. After many years, she finally succeeds. To produce more plants like this one, should she cross-breed it with another plant or clone it? Why?

For suggested answers, see Appendix A.

CONCEPT 13.2

Fertilization and meiosis alternate in sexual life cycles

A **life cycle** is the generation-to-generation sequence of stages in the reproductive history of an organism, from conception to production of its own offspring. In this section, we use humans as an example to track the behavior of chromosomes through the sexual life cycle. We begin by considering the chromosome count in human somatic cells and gametes. We will then explore how the behavior of chromosomes relates to the human life cycle and other types of sexual life cycles.

Sets of Chromosomes in Human Cells

In humans, each somatic cell has 46 chromosomes. During mitosis, the chromosomes become condensed enough to be visible under a light microscope. At this point, they can be distinguished from one another by their size, the positions of their centromeres, and the pattern of colored bands produced by certain stains.

Careful examination of a micrograph of the 46 human chromosomes from a single cell in mitosis reveals that there are two chromosomes of each of 23 types. This becomes clear when images of the chromosomes are arranged in pairs, starting with the longest chromosomes. The resulting ordered display is called a **karyotype** (**Figure 13.3**). The two chromosomes composing a pair have the same length, centromere position, and staining pattern: These are called **homologous chromosomes**, or homologs. Both chromosomes of each pair carry genes controlling the same inherited characters. For example, if a gene for eye color is situated at a particular locus on a certain chromosome, then the homolog of that chromosome will also have a version of the same gene specifying eye color at the equivalent locus.

The two distinct chromosomes referred to as X and Y are an important exception to the general pattern of homologous chromosomes in human somatic cells. Human females have a homologous pair of X chromosomes (XX), but males have one X and one Y chromosome (XY). Only small parts of the X and Y are homologous. Most of the genes carried on the X chromosome do not have counterparts on the tiny Y, and the Y chromosome has genes lacking on the X. Because they determine an individual's sex, the X and Y chromosomes are called **sex chromosomes**. The other chromosomes are called **autosomes**.

The occurrence of pairs of homologous chromosomes in each human somatic cell is a consequence of our sexual origins. We inherit one chromosome of each pair from each parent. Thus, the 46 chromosomes in our somatic cells are actually two sets of 23 chromosomes—a maternal set (from our mother) and a paternal set (from our father). The number of chromosomes in

▼ **Figure 13.3** RESEARCH METHOD

Preparing a Karyotype

APPLICATION A karyotype is a display of condensed chromosomes arranged in pairs. Karyotyping can be used to screen for defective chromosomes or abnormal numbers of chromosomes associated with certain congenital disorders, such as Down syndrome.

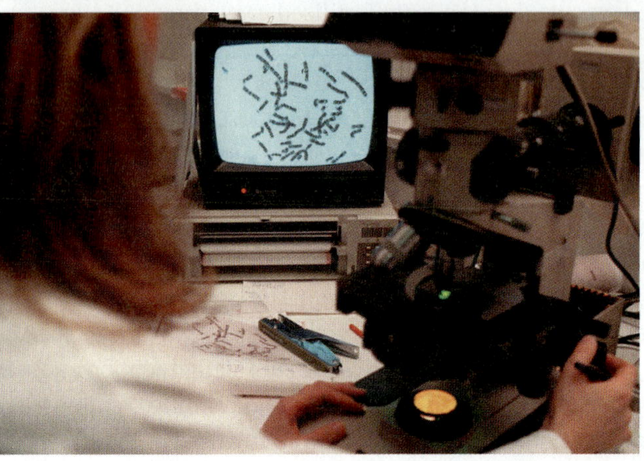

TECHNIQUE Karyotypes are prepared from isolated somatic cells, which are treated with a drug to stimulate mitosis and then grown in culture for several days. Cells arrested in metaphase, when chromosomes are most highly condensed, are stained and then viewed with a microscope equipped with a digital camera. A photograph of the chromosomes is displayed on a computer monitor, and the images of the chromosomes are arranged into pairs according to their appearance.

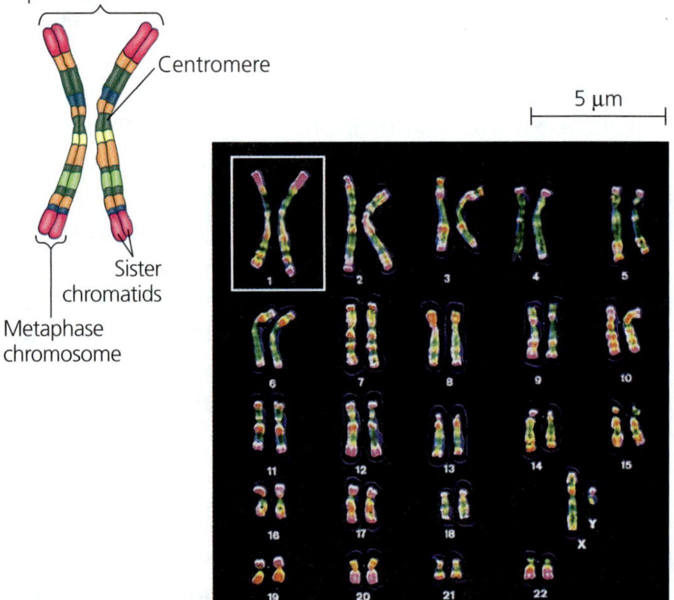

RESULTS This karyotype shows the chromosomes from a normal human male. The size of the chromosome, position of the centromere, and pattern of stained bands help identify specific chromosomes. Although difficult to discern in the karyotype, each metaphase chromosome consists of two closely attached sister chromatids (see the diagram of a pair of homologous duplicated chromosomes).

a single set is represented by *n*. Any cell with two chromosome sets is called a **diploid cell** and has a diploid number of chromosomes, abbreviated $2n$. For humans, the diploid number is 46 ($2n = 46$), the number of chromosomes in our somatic cells. In a cell in which DNA synthesis has occurred, all the chromosomes are duplicated, and therefore each consists of two identical sister chromatids, associated closely at the centromere and along the arms. **Figure 13.4** helps clarify the various terms that we use to describe duplicated chromosomes in a diploid cell. Study this figure so that you understand the differences between homologous chromosomes, sister chromatids, nonsister chromatids, and chromosome sets.

Unlike somatic cells, gametes contain a single set of chromosomes. Such cells are called **haploid cells**, and each has a haploid number of chromosomes (*n*). For humans, the haploid number is 23 ($n = 23$). The set of 23 consists of the 22 autosomes plus a single sex chromosome. An unfertilized egg contains an X chromosome, but a sperm may contain an X or a Y chromosome.

Note that each sexually reproducing species has a characteristic diploid number and haploid number. For example, the fruit fly, *Drosophila melanogaster*, has a diploid number ($2n$) of 8 and a haploid number (*n*) of 4, while dogs have a diploid number of 78 and a haploid number of 39.

Now that you have learned the concepts of diploid and haploid numbers of chromosomes, let's consider chromosome behavior during sexual life cycles. We'll use the human life cycle as an example.

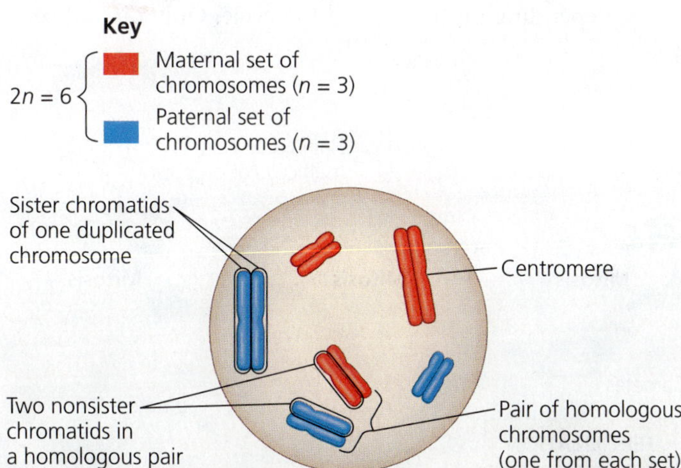

▲ **Figure 13.4 Describing chromosomes.** A cell from an organism with a diploid number of 6 ($2n = 6$) is depicted here following chromosome duplication and condensation. Each of the six duplicated chromosomes consists of two sister chromatids associated closely along their lengths. Each homologous pair is composed of one chromosome from the maternal set (red) and one from the paternal set (blue). Each set is made up of three chromosomes in this example. Nonsister chromatids are any two chromatids in a pair of homologous chromosomes that are not sister chromatids—in other words, one maternal and one paternal chromatid.

? *What is the haploid number of this cell? Is a "set" of chromosomes haploid or diploid?*

Behavior of Chromosome Sets in the Human Life Cycle

The human life cycle begins when a haploid sperm from the father fuses with a haploid egg from the mother. This union of gametes, culminating in fusion of their nuclei, is called **fertilization**. The resulting fertilized egg, or **zygote**, is diploid because it contains two haploid sets of chromosomes bearing genes representing the maternal and paternal family lines. As a human develops into a sexually mature adult, mitosis of the zygote and its descendant cells generates all the somatic cells of the body. Both chromosome sets in the zygote and all the genes they carry are passed with precision to the somatic cells.

The only cells of the human body not produced by mitosis are the gametes, which develop from specialized cells called *germ cells* in the gonads—ovaries in females and testes in males (**Figure 13.5**). Imagine what would happen if human gametes were made by mitosis: They would be diploid like the somatic cells. At the next round of fertilization, when two gametes fused, the normal chromosome number of 46 would

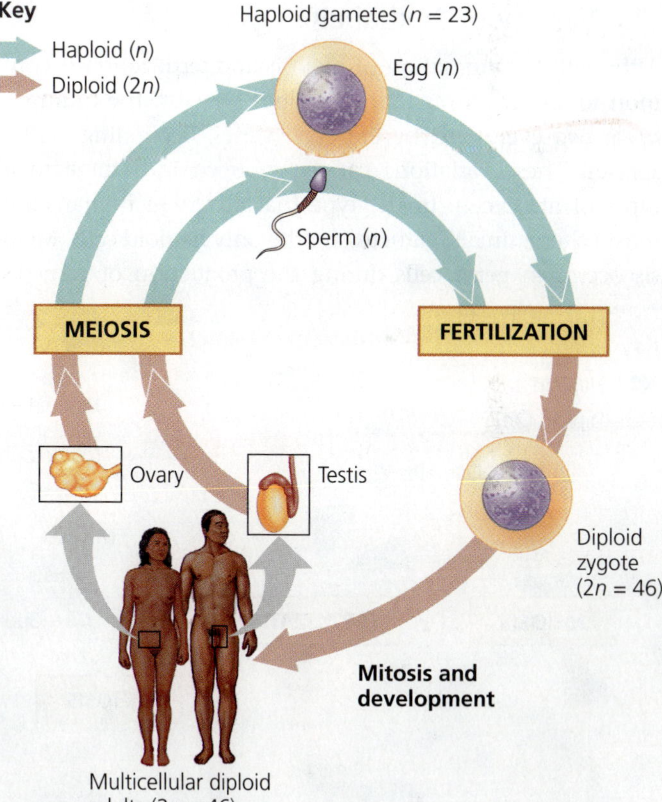

▲ **Figure 13.5 The human life cycle.** In each generation, the number of chromosome sets doubles at fertilization but is halved during meiosis. For humans, the number of chromosomes in a haploid cell is 23, consisting of one set ($n = 23$); the number of chromosomes in the diploid zygote and all somatic cells arising from it is 46, consisting of two sets ($2n = 46$).

This figure introduces a color code that will be used for other life cycles later in this book. The aqua arrows identify haploid stages of a life cycle, and the tan arrows identify diploid stages.

double to 92, and each subsequent generation would double the number of chromosomes yet again. This does not happen, however, because in sexually reproducing organisms, gamete formation involves a type of cell division called **meiosis**. This type of cell division reduces the number of sets of chromosomes from two to one in the gametes, counterbalancing the doubling that occurs at fertilization. In animals, meiosis occurs only in germ cells, which are in the ovaries or testes. As a result of meiosis, each human sperm and egg is haploid ($n = 23$). Fertilization restores the diploid condition by combining two haploid sets of chromosomes, and the human life cycle is repeated, generation after generation (see Figure 13.5). You will learn more about the production of sperm and eggs in Chapter 46.

In general, the steps of the human life cycle are typical of many sexually reproducing animals. Indeed, the processes of fertilization and meiosis are the hallmarks of sexual reproduction in plants, fungi, and protists as well as in animals. Fertilization and meiosis alternate in sexual life cycles, maintaining a constant number of chromosomes in each species from one generation to the next.

The Variety of Sexual Life Cycles

Although the alternation of meiosis and fertilization is common to all organisms that reproduce sexually, the timing of these two events in the life cycle varies, depending on the species. These variations can be grouped into three main types of life cycles. In the type that occurs in humans and most other animals, gametes are the only haploid cells. Meiosis occurs in germ cells during the production of gametes, which undergo no further cell division prior to fertilization. After fertilization, the diploid zygote divides by mitosis, producing a multicellular organism that is diploid **(Figure 13.6a)**.

Plants and some species of algae exhibit a second type of life cycle called **alternation of generations**. This type includes both diploid and haploid stages that are multicellular. The multicellular diploid stage is called the *sporophyte*. Meiosis in the sporophyte produces haploid cells called *spores*. Unlike a gamete, a haploid spore doesn't fuse with another cell but divides mitotically, generating a multicellular haploid stage called the *gametophyte*. Cells of the gametophyte give rise to gametes by mitosis. Fusion of two haploid gametes at fertilization results in a diploid zygote, which develops into the next sporophyte generation. Therefore, in this type of life cycle, the sporophyte generation produces a gametophyte as its offspring, and the gametophyte generation produces the next sporophyte generation **(Figure 13.6b)**. Clearly, the term *alternation of generations* is a fitting name for this type of life cycle.

A third type of life cycle occurs in most fungi and some protists, including some algae. After gametes fuse and form a diploid zygote, meiosis occurs without a multicellular diploid offspring developing. Meiosis produces not gametes but haploid cells that then divide by mitosis and give rise to either unicellular descendants or a haploid multicellular adult organism. Subsequently, the haploid organism carries out further mitoses, producing the cells that develop into gametes. The only diploid stage found in these species is the single-celled zygote **(Figure 13.6c)**.

Note that *either* haploid or diploid cells can divide by mitosis, depending on the type of life cycle. Only diploid cells,

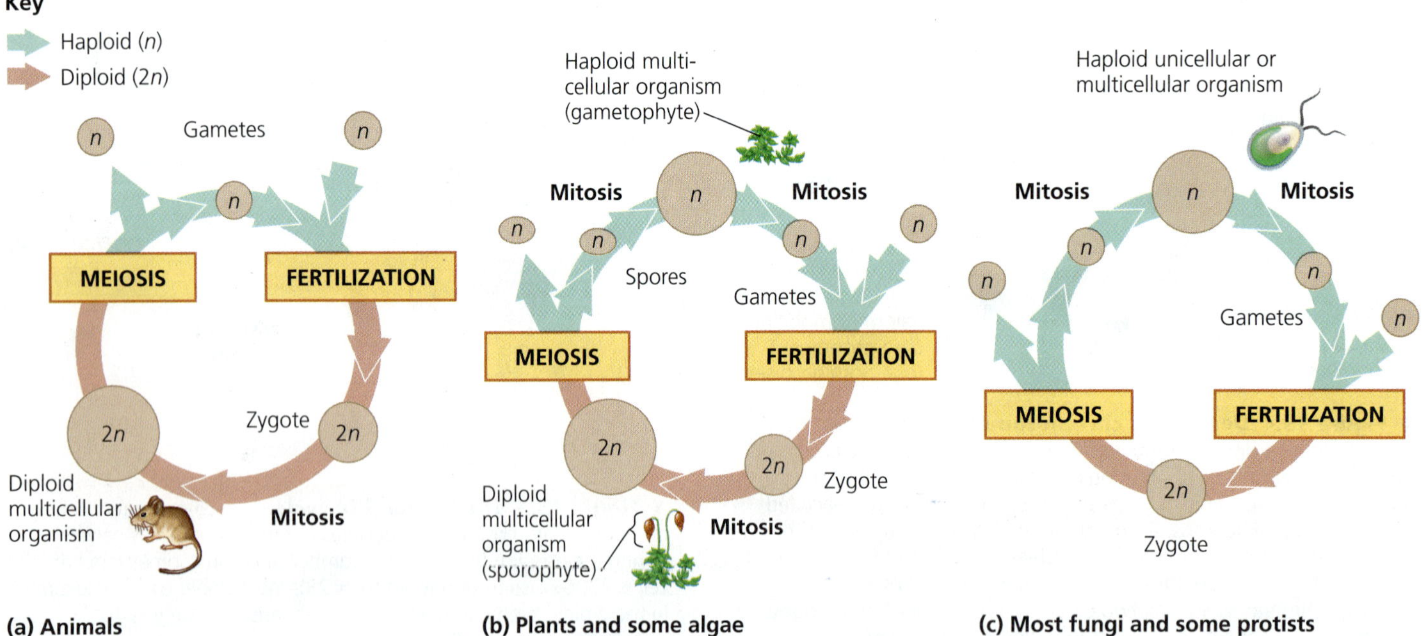

▲ **Figure 13.6 Three types of sexual life cycles.** The common feature of all three cycles is the alternation of meiosis and fertilization, key events that contribute to genetic variation among offspring. The cycles differ in the timing of these two key events.

however, can undergo meiosis because haploid cells have a single set of chromosomes that cannot be further reduced. Though the three types of sexual life cycles differ in the timing of meiosis and fertilization, they share a fundamental result: genetic variation among offspring. A closer look at meiosis will reveal the sources of this variation.

CONCEPT CHECK 13.2

1. **MAKE CONNECTIONS** In Figure 13.4, how many DNA molecules (double helices) are present (see Figure 12.5)?
2. How does the alternation of meiosis and fertilization in the life cycles of sexually reproducing organisms maintain the normal chromosome count for each species?
3. Each sperm of a pea plant contains seven chromosomes. What are the haploid and diploid numbers for this species?
4. **WHAT IF?** A certain eukaryote lives as a unicellular organism, but during environmental stress, it produces gametes. The gametes fuse, and the resulting zygote undergoes meiosis, generating new single cells. What type of organism could this be?

For suggested answers, see Appendix A.

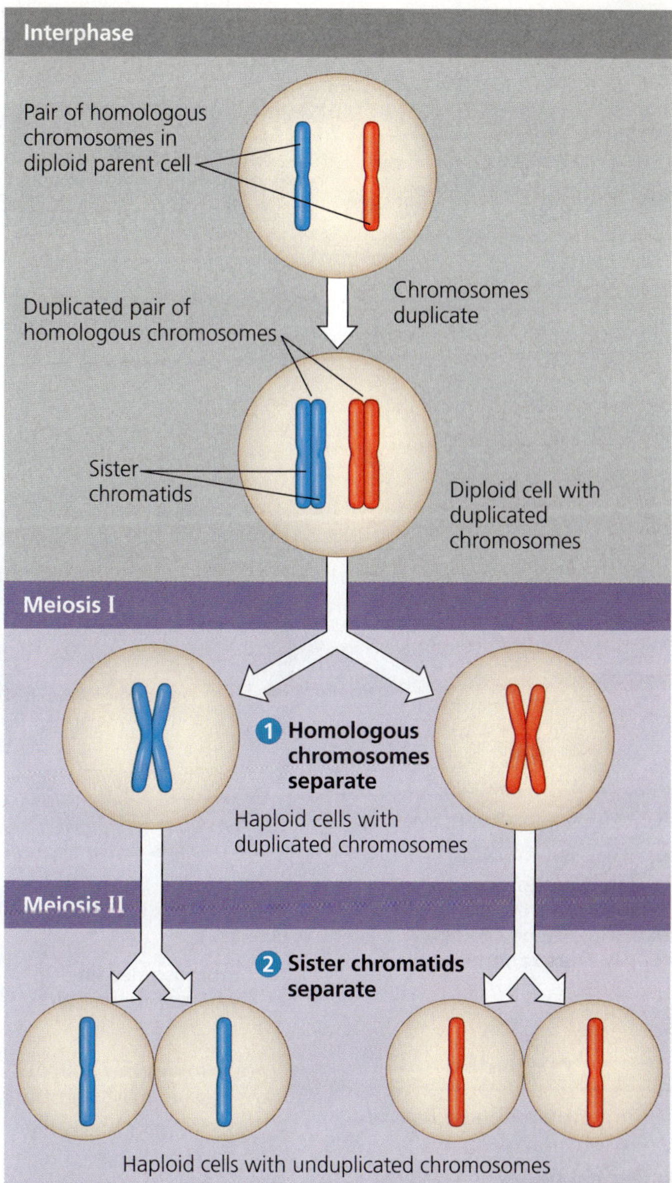

▲ **Figure 13.7 Overview of meiosis: how meiosis reduces chromosome number.** After the chromosomes duplicate in interphase, the diploid cell divides *twice*, yielding four haploid daughter cells. This overview tracks just one pair of homologous chromosomes, which for the sake of simplicity are drawn in the condensed state throughout. (They would not normally be condensed during interphase.) The red chromosome was inherited from the female parent, the blue chromosome from the male parent.

DRAW IT *Redraw the cells in this figure using a simple double helix to represent each DNA molecule.*

CONCEPT 13.3

Meiosis reduces the number of chromosome sets from diploid to haploid

Many of the steps of meiosis closely resemble corresponding steps in mitosis. Meiosis, like mitosis, is preceded by the duplication of chromosomes. However, this single duplication is followed by not one but two consecutive cell divisions, called **meiosis I** and **meiosis II**. These two divisions result in four daughter cells (rather than the two daughter cells of mitosis), each with only half as many chromosomes as the parent cell.

The Stages of Meiosis

The overview of meiosis in **Figure 13.7** shows, for a single pair of homologous chromosomes in a diploid cell, that both members of the pair are duplicated and the copies sorted into four haploid daughter cells. Recall that sister chromatids are two copies of *one* chromosome, closely associated all along their lengths; this association is called *sister chromatid cohesion*. Together, the sister chromatids make up one duplicated chromosome (see Figure 13.4). In contrast, the two chromosomes of a homologous pair are individual chromosomes that were inherited from different parents. Homologs appear alike in the microscope, but they may have different versions of genes, each called an *allele*, at corresponding loci (for example, an allele for freckles on one chromosome and an allele for the absence of freckles at the same locus on the homolog). Homologs are not associated with each other in any obvious way except during meiosis, as you will soon see.

Figure 13.8, on the next two pages, describes in detail the stages of the two divisions of meiosis for an animal cell whose diploid number is 6. Meiosis halves the total number of chromosomes in a very specific way, reducing the number of sets from two to one, with each daughter cell receiving one set of chromosomes. Study Figure 13.8 thoroughly before going on.

▼ Figure 13.8

Exploring Meiosis in an Animal Cell

MEIOSIS I: Separates homologous chromosomes

Prophase I | Metaphase I | Anaphase I | Telophase I and Cytokinesis

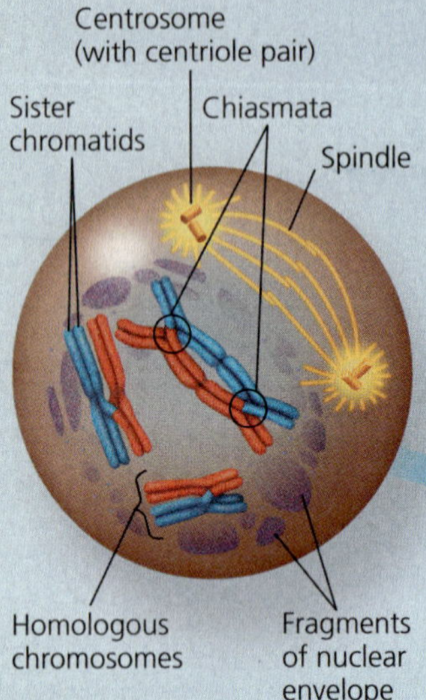

Duplicated homologous chromosomes (red and blue) pair and exchange segments; $2n = 6$ in this example.

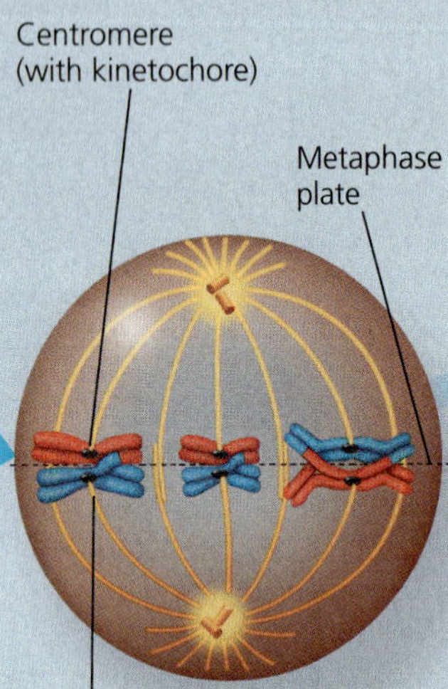

Chromosomes line up by homologous pairs.

Each pair of homologous chromosomes separates.

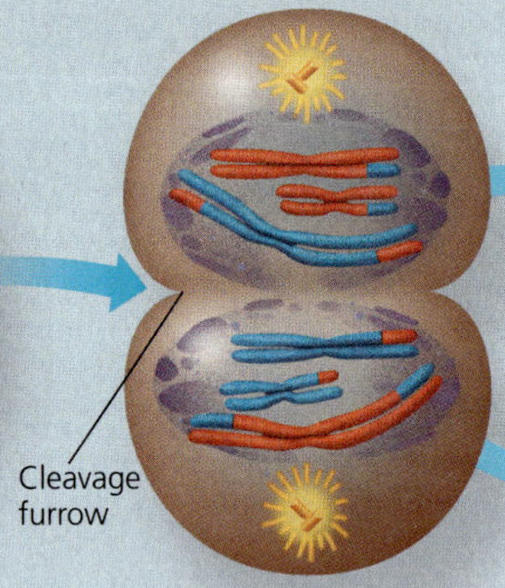

Two haploid cells form; each chromosome still consists of two sister chromatids.

Prophase I

During early prophase I, before the stage shown above:

- Chromosomes begin to condense, and homologs loosely pair along their lengths, aligned gene by gene.
- Paired homologs become physically connected to each other along their lengths by a zipper-like protein structure, the *synaptonemal complex;* this state is called **synapsis.**
- **Crossing over**, a genetic rearrangement between nonsister chromatids involving the exchange of corresponding segments of DNA molecules, begins during pairing and synaptonemal complex formation, and is completed while homologs are in synapsis.

At the stage shown above:

- Synapsis has ended with the disassembly of the synaptonemal complex in mid-prophase, and the chromosomes in each pair have moved apart slightly.
- Each homologous pair has one or more X-shaped regions called **chiasmata** (singular, *chiasma*). A chiasma exists at the point where a crossover has occurred. It appears as a cross because sister chromatid cohesion still holds the two original sister chromatids together, even in regions beyond the crossover point, where one chromatid is now part of the other homolog.
- Centrosome movement, spindle formation, and nuclear envelope breakdown occur as in mitosis.

Later in prophase I, after the stage shown above:

- Microtubules from one pole or the other attach to the two kinetochores, protein structures at the centromeres of the two homologs. The homologous pairs then move toward the metaphase plate.

Metaphase I

- Pairs of homologous chromosomes are now arranged at the metaphase plate, with one chromosome in each pair facing each pole.
- Both chromatids of one homolog are attached to kinetochore microtubules from one pole; those of the other homolog are attached to microtubules from the opposite pole.

Anaphase I

- Breakdown of proteins responsible for sister chromatid cohesion along chromatid arms allows homologs to separate.
- The homologs move toward opposite poles, guided by the spindle apparatus.
- Sister chromatid cohesion persists at the centromere, causing chromatids to move as a unit toward the same pole.

Telophase I and Cytokinesis

- At the beginning of telophase I, each half of the cell has a complete haploid set of duplicated chromosomes. Each chromosome is composed of two sister chromatids; one or both chromatids include regions of nonsister chromatid DNA.
- Cytokinesis (division of the cytoplasm) usually occurs simultaneously with telophase I, forming two haploid daughter cells.
- In animal cells like these, a cleavage furrow forms. (In plant cells, a cell plate forms.)
- In some species, chromosomes decondense and nuclear envelopes form.
- No chromosome duplication occurs between meiosis I and meiosis II.

MEIOSIS II: Separates sister chromatids

Prophase II	Metaphase II	Anaphase II	Telophase II and Cytokinesis

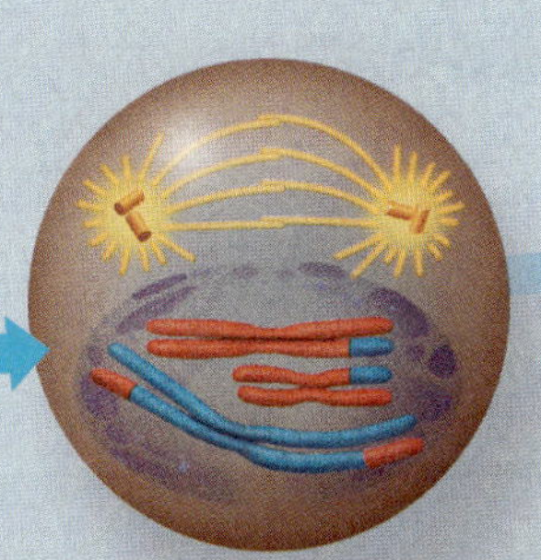
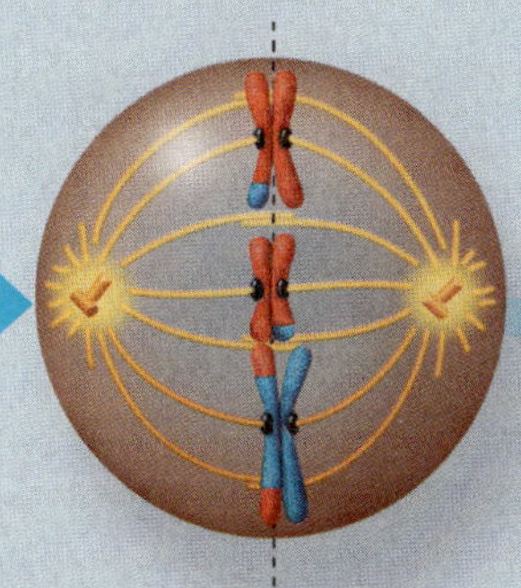
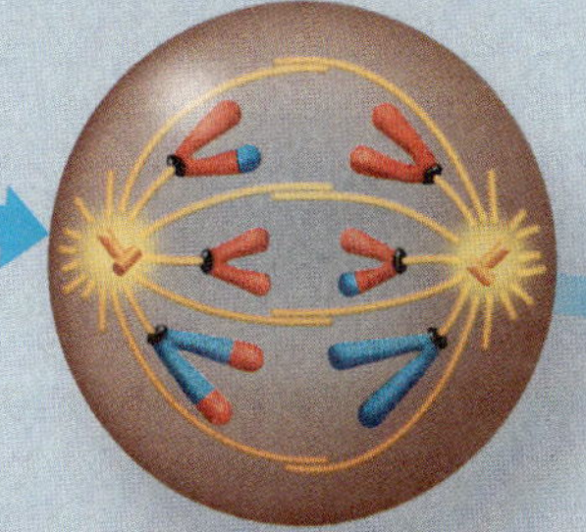
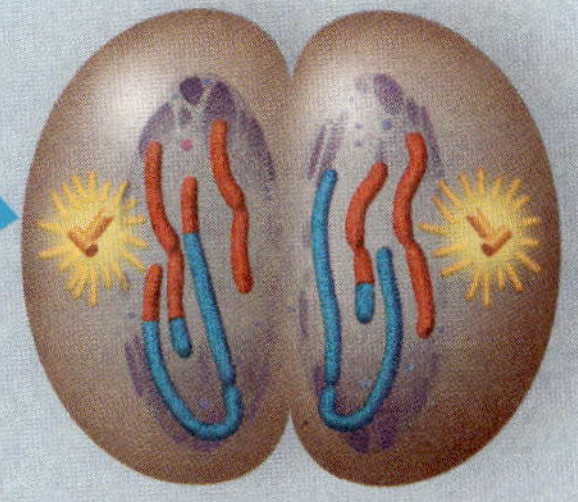

During another round of cell division, the sister chromatids finally separate; four haploid daughter cells result, containing unduplicated chromosomes.

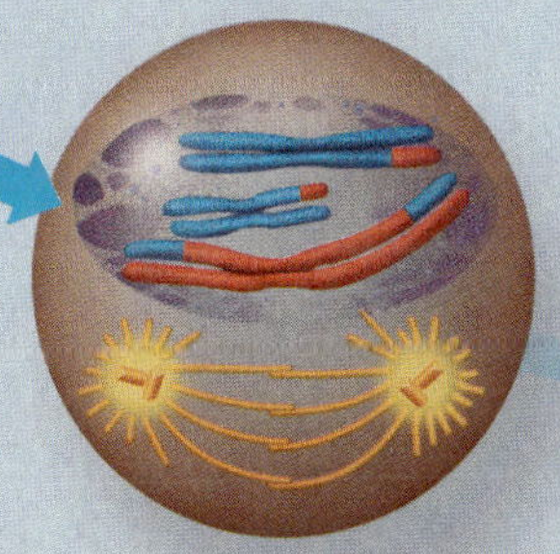
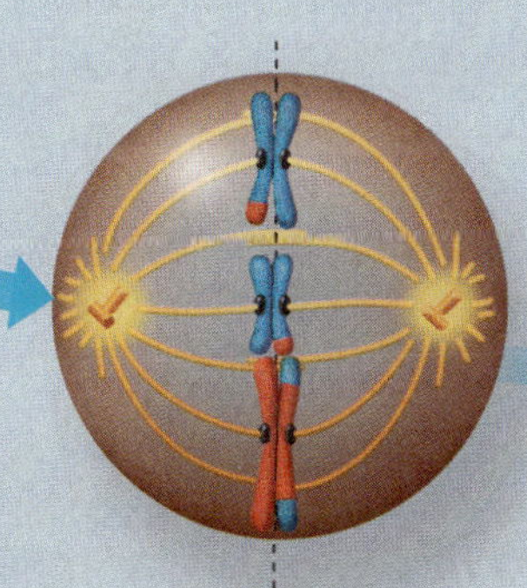
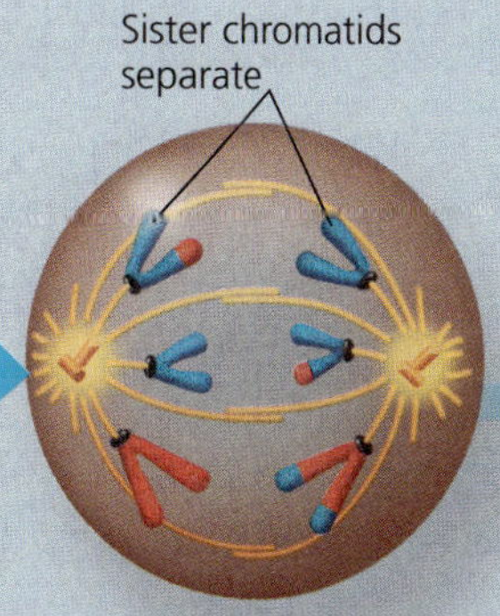

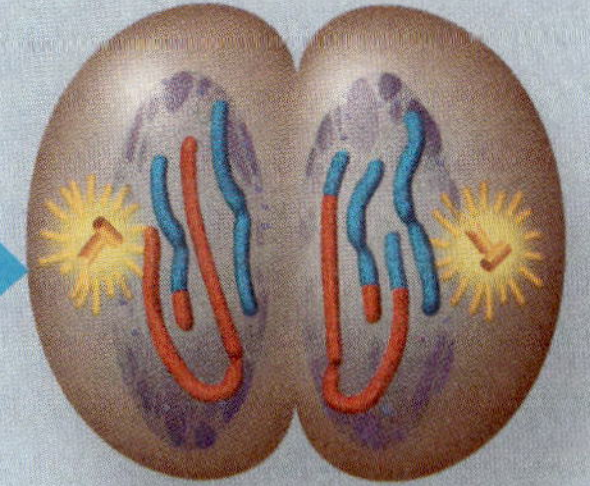

Prophase II

- A spindle apparatus forms.
- In late prophase II (not shown here), chromosomes, each still composed of two chromatids associated at the centromere, move toward the metaphase II plate.

Metaphase II

- The chromosomes are positioned at the metaphase plate as in mitosis.
- Because of crossing over in meiosis I, the two sister chromatids of each chromosome are *not* genetically identical.
- The kinetochores of sister chromatids are attached to microtubules extending from opposite poles.

Anaphase II

- Breakdown of proteins holding the sister chromatids together at the centromere allows the chromatids to separate. The chromatids move toward opposite poles as individual chromosomes.

Telophase II and Cytokinesis

- Nuclei form, the chromosomes begin decondensing, and cytokinesis occurs.
- The meiotic division of one parent cell produces four daughter cells, each with a haploid set of (unduplicated) chromosomes.
- The four daughter cells are genetically distinct from one another and from the parent cell.

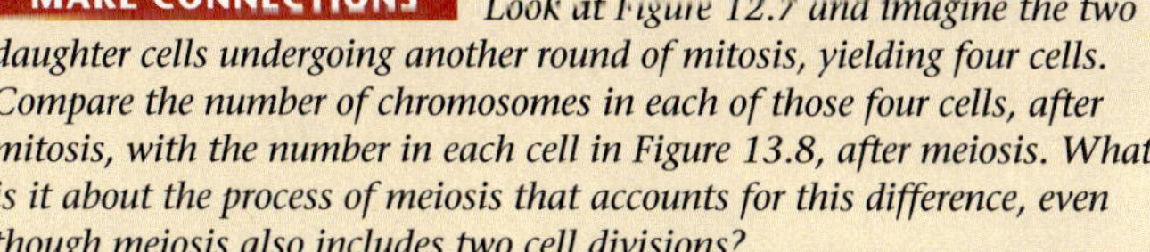

MAKE CONNECTIONS *Look at Figure 12.7 and imagine the two daughter cells undergoing another round of mitosis, yielding four cells. Compare the number of chromosomes in each of those four cells, after mitosis, with the number in each cell in Figure 13.8, after meiosis. What is it about the process of meiosis that accounts for this difference, even though meiosis also includes two cell divisions?*

BioFlix Visit the Study Area at **www.masteringbiology.com** for the BioFlix® 3-D Animation on Meiosis.

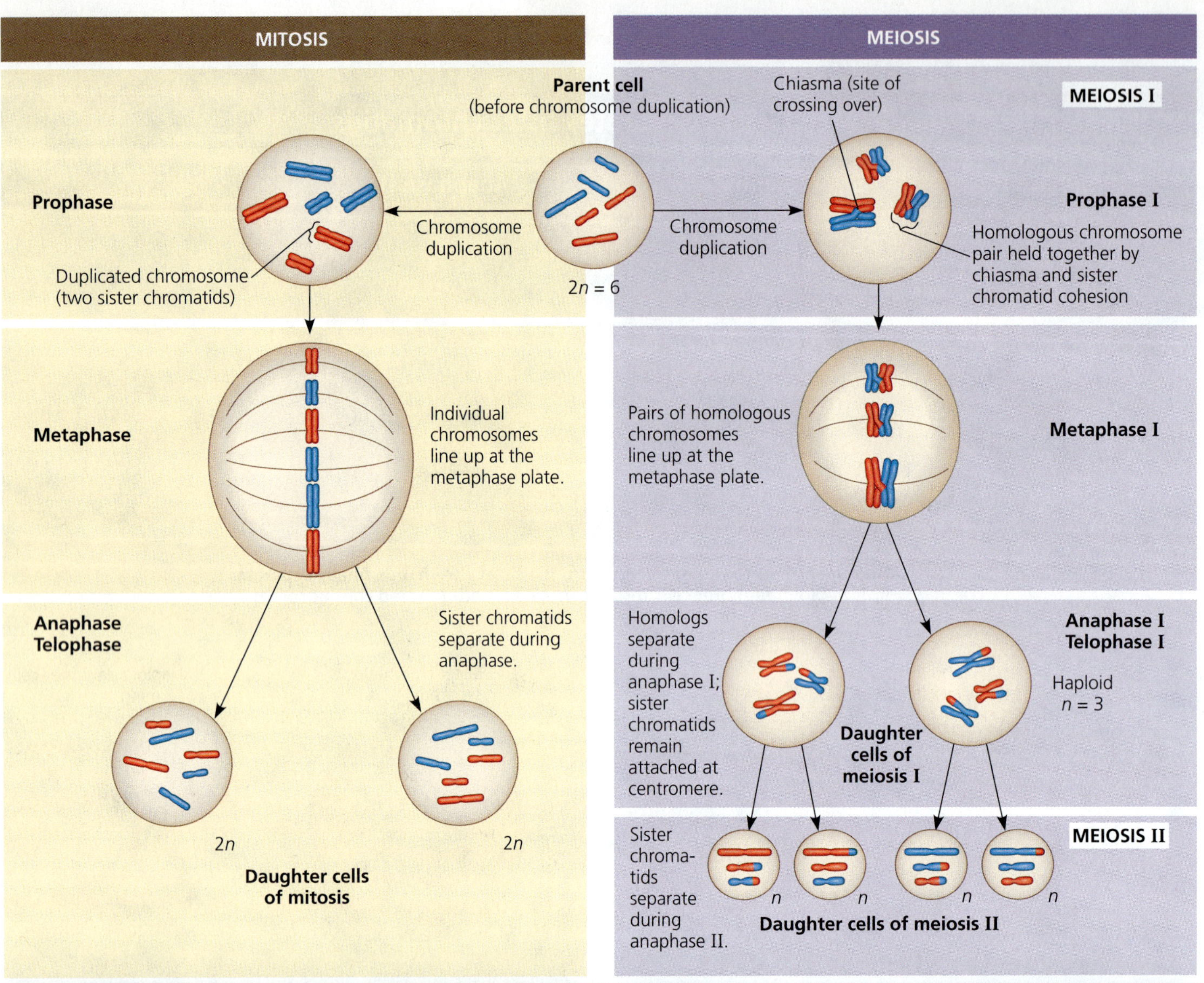

SUMMARY

Property	Mitosis	Meiosis
DNA replication	Occurs during interphase before mitosis begins	Occurs during interphase before meiosis I begins
Number of divisions	One, including prophase, metaphase, anaphase, and telophase	Two, each including prophase, metaphase, anaphase, and telophase
Synapsis of homologous chromosomes	Does not occur	Occurs during prophase I along with crossing over between nonsister chromatids; resulting chiasmata hold pairs together due to sister chromatid cohesion
Number of daughter cells and genetic composition	Two, each diploid ($2n$) and genetically identical to the parent cell	Four, each haploid (n), containing half as many chromosomes as the parent cell; genetically different from the parent cell and from each other
Role in the animal body	Enables multicellular adult to arise from zygote; produces cells for growth, repair, and, in some species, asexual reproduction	Produces gametes; reduces number of chromosomes by half and introduces genetic variability among the gametes

▲ **Figure 13.9 A comparison of mitosis and meiosis in diploid cells.**

DRAW IT *Could any other combinations of chromosomes be generated during meiosis II from the specific cells shown in telophase I? Explain. (Hint: Draw the cells as they would appear in metaphase II.)*

A Comparison of Mitosis and Meiosis

Figure 13.9 summarizes the key differences between meiosis and mitosis in diploid cells. Basically, meiosis reduces the number of chromosome sets from two (diploid) to one (haploid), whereas mitosis conserves the number of chromosome sets. Therefore, meiosis produces cells that differ genetically from their parent cell and from each other, whereas mitosis produces daughter cells that are genetically identical to their parent cell and to each other.

Three events unique to meiosis occur during meiosis I:

1. **Synapsis and crossing over.** During prophase I, duplicated homologs pair up, and the formation of the synaptonemal complex between them holds them in synapsis. Crossing over also occurs during prophase I. Synapsis and crossing over normally do not occur during prophase of mitosis.
2. **Homologous pairs at the metaphase plate.** At metaphase I of meiosis, chromosomes are positioned at the metaphase plate as pairs of homologs, rather than individual chromosomes, as in metaphase of mitosis.
3. **Separation of homologs.** At anaphase I of meiosis, the duplicated chromosomes of each homologous pair move toward opposite poles, but the sister chromatids of each duplicated chromosome remain attached. In anaphase of mitosis, by contrast, sister chromatids separate.

How do sister chromatids stay together through meiosis I but separate from each other in meiosis II and mitosis? Sister chromatids are attached along their lengths by protein complexes called *cohesins*. In mitosis, this attachment lasts until the end of metaphase, when enzymes cleave the cohesins, freeing the sister chromatids to move to opposite poles of the cell. In meiosis, sister chromatid cohesion is released in two steps, one at the start of anaphase I and one at anaphase II. In metaphase I, homologs are held together by cohesion between sister chromatid arms in regions beyond points of crossing over, where stretches of sister chromatids now belong to different chromosomes. As shown in Figure 13.8, the combination of crossing over and sister chromatid cohesion along the arms results in the formation of a chiasma. Chiasmata hold homologs together as the spindle forms for the first meiotic division. At the onset of anaphase I, the release of cohesion along sister chromatid arms allows homologs to separate. At anaphase II, the release of sister chromatid cohesion at the centromeres allows the sister chromatids to separate. Thus, sister chromatid cohesion and crossing over, acting together, play an essential role in the lining up of chromosomes by homologous pairs at metaphase I.

Meiosis I is called the *reductional division* because it halves the number of chromosome sets per cell—a reduction from two sets (the diploid state) to one set (the haploid state). During the second meiotic division, meiosis II (sometimes called the *equational division*), the sister chromatids separate, producing haploid daughter cells. The mechanism for separating sister chromatids is virtually identical in meiosis II and mitosis. The molecular basis of chromosome behavior during meiosis continues to be a focus of intense research.

CONCEPT CHECK 13.3

1. **MAKE CONNECTIONS** How are the chromosomes in a cell at metaphase of mitosis similar to and different from the chromosomes in a cell at metaphase of meiosis II? (Compare Figures 12.7 and 13.8.)
2. **WHAT IF?** Given that the synaptonemal complex disappears by the end of prophase, how would the two homologs be associated if crossing over did not occur? What effect might this ultimately have on gamete formation?

For suggested answers, see Appendix A.

CONCEPT 13.4

Genetic variation produced in sexual life cycles contributes to evolution

How do we account for the genetic variation illustrated in Figure 13.1? As you will learn in more detail in later chapters, mutations are the original source of genetic diversity. These changes in an organism's DNA create the different versions of genes known as *alleles*. Once these differences arise, reshuffling of the alleles during sexual reproduction produces the variation that results in each member of a sexually reproducing population having a unique combination of traits.

Origins of Genetic Variation Among Offspring

In species that reproduce sexually, the behavior of chromosomes during meiosis and fertilization is responsible for most of the variation that arises in each generation. Let's examine three mechanisms that contribute to the genetic variation arising from sexual reproduction: independent assortment of chromosomes, crossing over, and random fertilization.

Independent Assortment of Chromosomes

One aspect of sexual reproduction that generates genetic variation is the random orientation of pairs of homologous chromosomes at metaphase of meiosis I. At metaphase I, the homologous pairs, each consisting of one maternal and one paternal chromosome, are situated at the metaphase plate. (Note that the terms *maternal* and *paternal* refer, respectively, to the mother and father of the individual whose cells are undergoing meiosis.) Each pair may orient with either its maternal or paternal homolog closer to a given pole—its orientation is as random as the flip of a coin. Thus, there is a 50% chance that a particular daughter cell of meiosis I will get the maternal chromosome of a certain homologous pair and a 50% chance that it will get the paternal chromosome.

Because each pair of homologous chromosomes is positioned independently of the other pairs at metaphase I, the first meiotic division results in each pair sorting its maternal and paternal homologs into daughter cells independently of every other pair. This is called *independent assortment*. Each daughter cell represents one outcome of all possible combinations of maternal and paternal chromosomes. As shown in **Figure 13.10**, the number of combinations possible for daughter cells formed by meiosis of a diploid cell with $n = 2$ (two pairs of homologous chromosomes) is four: two possible arrangements for the first pair times two possible arrangements for the second pair. Note that only two of the four combinations of daughter cells shown in the figure would result from meiosis of a *single* diploid cell, because a single parent cell would have one or the other possible chromosomal arrangement at metaphase I, but not both. However, the population of daughter cells resulting from meiosis of a large number of diploid cells contains all four types in approximately equal numbers. In the case of $n = 3$, eight combinations of chromosomes are possible for daughter cells. More generally, the number of possible combinations when chromosomes sort independently during meiosis is 2^n, where n is the haploid number of the organism.

In the case of humans ($n = 23$), the number of possible combinations of maternal and paternal chromosomes in the resulting gametes is 2^{23}, or about 8.4 million. Each gamete that you produce in your lifetime contains one of roughly 8.4 million possible combinations of chromosomes.

Crossing Over

As a consequence of the independent assortment of chromosomes during meiosis, each of us produces a collection of gametes differing greatly in their combinations of the chromosomes we inherited from our two parents. Figure 13.10 suggests that each chromosome in a gamete is exclusively maternal or paternal in origin. In fact, this is *not* the case, because crossing over produces **recombinant chromosomes**, individual chromosomes that carry genes (DNA) derived from two different parents **(Figure 13.11)**. In meiosis in humans, an average of one to three crossover events occur per chromosome pair, depending on the size of the chromosomes and the position of their centromeres.

Crossing over begins very early in prophase I as homologous chromosomes pair loosely along their lengths. Each gene on one homolog is aligned precisely with the corresponding gene on the other homolog. In a single crossover event, the DNA of two *nonsister* chromatids—one maternal and one paternal chromatid of a homologous pair—is broken by specific proteins at precisely corresponding points, and the two segments beyond the crossover point are each joined to the other chromatid. Thus, a paternal chromatid is joined to a piece of maternal chromatid beyond the crossover point, and vice versa. In this way, crossing over produces chromosomes with new combinations of maternal and paternal alleles (see Figure 13.11).

At metaphase II, chromosomes that contain one or more recombinant chromatids can be oriented in two alternative, nonequivalent ways with respect to other chromosomes, because their sister chromatids are no longer identical. The different possible arrangements of nonidentical sister chromatids during meiosis II further increase the number of genetic types of daughter cells that can result from meiosis.

You will learn more about crossing over in Chapter 15. The important point for now is that crossing over, by combining DNA inherited from two parents into a single chromosome, is an important source of genetic variation in sexual life cycles.

Random Fertilization

The random nature of fertilization adds to the genetic variation arising from meiosis. In humans, each male and female gamete represents one of about 8.4 million (2^{23}) possible chromosome combinations due to independent assortment. The fusion of a male gamete with a female gamete during

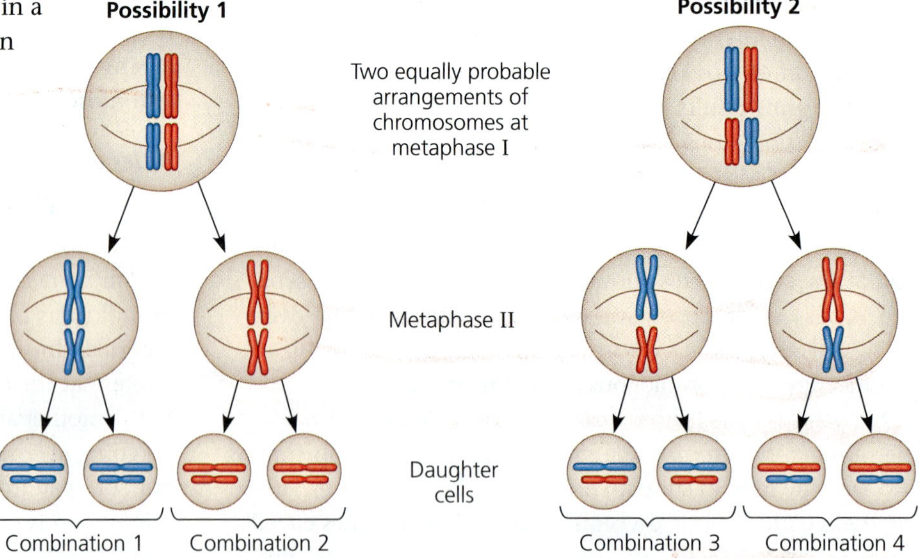

▶ **Figure 13.10 The independent assortment of homologous chromosomes in meiosis.**

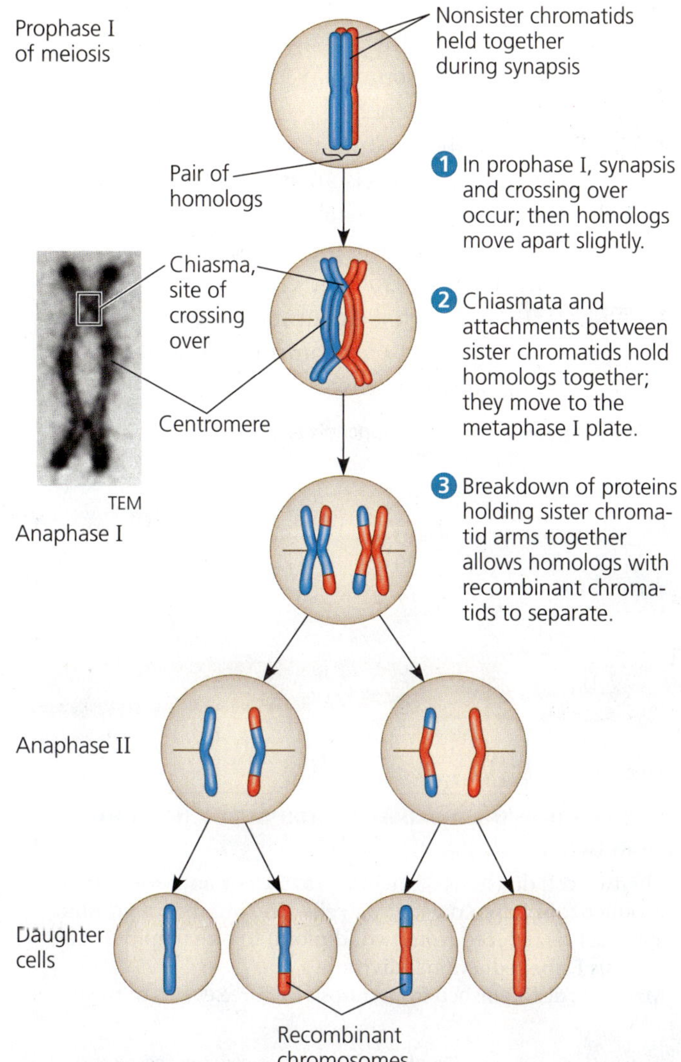

▲ **Figure 13.11 The results of crossing over during meiosis.**

fertilization will produce a zygote with any of about 70 trillion ($2^{23} \times 2^{23}$) diploid combinations. If we factor in the variation brought about by crossing over, the number of possibilities is truly astronomical. It may sound trite, but you really *are* unique.

The Evolutionary Significance of Genetic Variation Within Populations

EVOLUTION Now that you've learned how new combinations of genes arise among offspring in a sexually reproducing population, let's see how the genetic variation in a population relates to evolution. Darwin recognized that a population evolves through the differential reproductive success of its variant members. On average, those individuals best suited to the local environment leave the most offspring, thereby transmitting their genes. Thus, natural selection results in the accumulation of genetic variations favored by the environment. As the environment changes, the population may survive if, in each generation, at least some of its members can cope effectively with the new conditions. Mutations are the original source of different alleles, which are then mixed and matched during meiosis. New and different combinations of alleles may work better than those that previously prevailed. The ability of sexual reproduction to generate genetic diversity is one of the most commonly proposed explanations for the evolutionary persistence of this process.

On the other hand, in a stable environment, asexual reproduction would seem to be more advantageous, because it ensures perpetuation of successful combinations of alleles. Furthermore, asexual reproduction is less expensive; its energy costs to the organism are lower than those of sexual reproduction, for reasons that will be discussed in Chapter 46.

In spite of these apparent disadvantages, sexual reproduction is almost universal among animals as far as we know. While a few species are capable of reproducing asexually under unusual circumstances, animals that always reproduce asexually are quite rare. The best-established example, to date, is a group of microscopic animals called bdelloid rotifers (the "b" in "bdelloid" is silent), shown in **Figure 13.12**. This group includes about 400 species that live in a great variety of environments around the world. They inhabit streams, lake bottoms, puddles, lichens, tree bark, and masses of decaying vegetation. Recent studies have provided convincing evidence that these animals reproduce only asexually and probably haven't engaged in sex in the 40 million years since their evolutionary origins!

Does the discovery of the evolutionarily successful, asexually reproducing bdelloid rotifer cast doubt on the advantage of genetic variation arising from sexual reproduction? On the contrary, this group may be considered an exception that proves the rule. In studies of bdelloid rotifers, biologists have found mechanisms other than sexual reproduction that increase genetic diversity in these organisms. For example, they live in environments that can dry up for long periods of time, during which they can enter a state of suspended animation. In this state, their cell membranes may crack in places, allowing entry of DNA from other rotifers and even other species. Evidence suggests that this DNA can become incorporated into the genome of the rotifer, leading to increased genetic diversity. (You'll learn more about this process, called *horizontal gene transfer*, in Chapter 26.) Taken as a whole, these studies support the idea that genetic variation is evolutionarily advantageous and that a

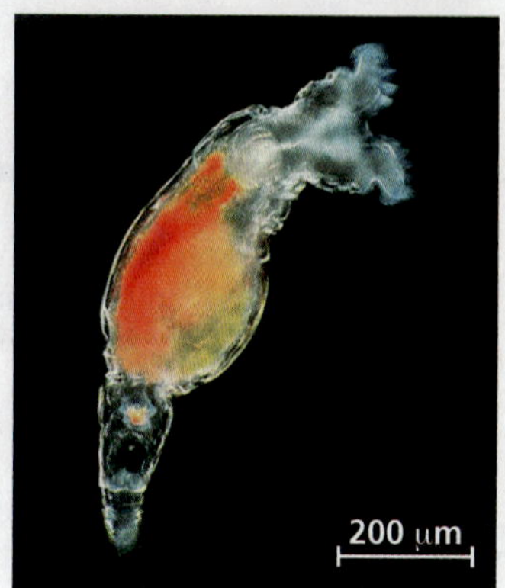

▲ **Figure 13.12 A bdelloid rotifer, an animal that reproduces only asexually.**

different mechanism to generate genetic variation has evolved in bdelloid rotifers.

In this chapter, we have seen how sexual reproduction greatly increases the genetic variation present in a population. Although Darwin realized that heritable variation is what makes evolution possible, he could not explain why offspring resemble—but are not identical to—their parents. Ironically, Gregor Mendel, a contemporary of Darwin, published a theory of inheritance that helps explain genetic variation, but his discoveries had no impact on biologists until 1900, more than 15 years after Darwin (1809–1882) and Mendel (1822–1884) had died. In the next chapter, you will learn how Mendel discovered the basic rules governing the inheritance of specific traits.

CONCEPT CHECK 13.4

1. What is the original source of variation among the different alleles of a gene?
2. The diploid number for fruit flies is 8, and the diploid number for grasshoppers is 46. If no crossing over took place, would the genetic variation among offspring from a given pair of parents be greater in fruit flies or grasshoppers? Explain.
3. **WHAT IF?** Under what circumstances would crossing over during meiosis *not* contribute to genetic variation among daughter cells?

For suggested answers, see Appendix A.

13 CHAPTER REVIEW

SUMMARY OF KEY CONCEPTS

CONCEPT 13.1

Offspring acquire genes from parents by inheriting chromosomes (pp. 248–249)

- Each **gene** in an organism's DNA exists at a specific **locus** on a certain chromosome. We inherit one set of chromosomes from our mother and one set from our father.
- In **asexual reproduction**, a single parent produces genetically identical offspring by mitosis. **Sexual reproduction** combines sets of genes from two different parents, leading to genetically diverse offspring.

? *Explain why human offspring resemble their parents but are not identical to them.*

CONCEPT 13.2

Fertilization and meiosis alternate in sexual life cycles (pp. 250–253)

- As seen in a **karyotype**, normal human **somatic cells** are **diploid**. They have 46 chromosomes made up of two sets of 23—one set from each parent. In human diploid cells, there are 22 **homologous** pairs of **autosomes**, each with a maternal and a paternal homolog. The 23rd pair, the **sex chromosomes**, determines whether the person is female (XX) or male (XY).
- At sexual maturity in the human **life cycle**, ovaries and testes (the gonads) produce **haploid gametes** by **meiosis**, each gamete containing a single set of 23 chromosomes ($n = 23$). During **fertilization**, an egg and sperm unite, forming a diploid ($2n = 46$) single-celled **zygote**, which develops into a multicellular organism by mitosis.
- Sexual life cycles differ in the timing of meiosis relative to fertilization and in the point(s) of the cycle at which a multicellular organism is produced by mitosis.

? *Compare the life cycles of animals and plants, mentioning their similarities and differences.*

CONCEPT 13.3

Meiosis reduces the number of chromosome sets from diploid to haploid (pp. 253–257)

- The two cell divisions of meiosis, **meiosis I** and **meiosis II**, produce four haploid daughter cells. The number of chromosome sets is reduced from two (diploid) to one (haploid) during meiosis I, the reductional division.
- Meiosis is distinguished from mitosis by three events of meiosis I:

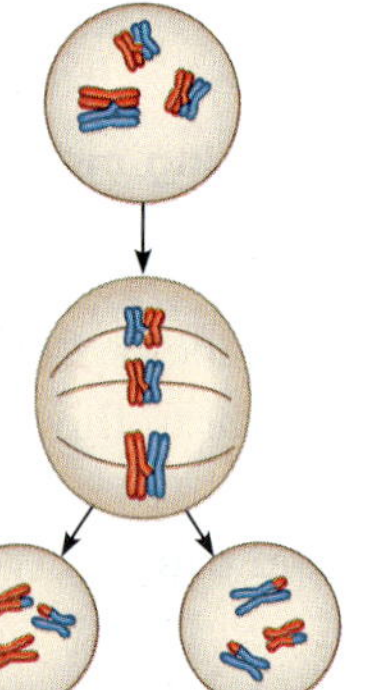

Prophase I: Each homologous pair undergoes **synapsis** and **crossing over** between nonsister chromatids with the subsequent appearance of **chiasmata**.

Metaphase I: Chromosomes line up as homologous pairs on the metaphase plate.

Anaphase I: Homologs separate from each other; sister chromatids remain joined at the centromere.

Meiosis II separates the sister chromatids.

- The combination of sister chromatid cohesion and crossing over leads to chiasmata, which hold homologs together until anaphase I. Cohesins are cleaved along the chromatid arms at anaphase I, allowing the homologs to separate, and at the centromeres in anaphase II, allowing sister chromatids to separate.

? *During prophase I, homologous chromosomes pair up and undergo synapsis and crossing over. Explain why this cannot also occur during prophase II.*

CONCEPT 13.4

Genetic variation produced in sexual life cycles contributes to evolution (pp. 257–260)

- Three events in sexual reproduction contribute to genetic variation in a population: independent assortment of chromosomes during meiosis, crossing over during meiosis I, and random fertilization of egg cells by sperm. Crossing over involves breakage and rejoining of the DNA of nonsister chromatids in a homologous pair, resulting in recombinant chromatids that will become **recombinant chromosomes**.

- Genetic variation is the raw material for evolution by natural selection. Mutations are the original source of this variation; the production of new combinations of variant genes in sexual reproduction generates additional genetic diversity. Animals that reproduce only asexually are quite rare, underscoring the apparently great advantage of genetic diversity.

? *Explain how three processes unique to meiosis generate a great deal of genetic variation.*

TEST YOUR UNDERSTANDING

LEVEL 1: KNOWLEDGE/COMPREHENSION

1. A human cell containing 22 autosomes and a Y chromosome is
   a. a sperm.
   b. an egg.
   c. a zygote.
   d. a somatic cell of a male.
   e. a somatic cell of a female.

2. Which life cycle stage is found in plants but not animals?
   a. gamete
   b. zygote
   c. multicellular diploid
   d. multicellular haploid
   e. unicellular diploid

3. Homologous chromosomes move toward opposite poles of a dividing cell during
   a. mitosis.
   b. meiosis I.
   c. meiosis II.
   d. fertilization.
   e. binary fission.

LEVEL 2: APPLICATION/ANALYSIS

4. Meiosis II is similar to mitosis in that
   a. sister chromatids separate during anaphase.
   b. DNA replicates before the division.
   c. the daughter cells are diploid.
   d. homologous chromosomes synapse.
   e. the chromosome number is reduced.

5. If the DNA content of a diploid cell in the G_1 phase of the cell cycle is x, then the DNA content of the same cell at metaphase of meiosis I would be
   a. $0.25x$.
   b. $0.5x$.
   c. x.
   d. $2x$.
   e. $4x$.

6. If we continued to follow the cell lineage from question 5, then the DNA content of a single cell at metaphase of meiosis II would be
   a. $0.25x$.
   b. $0.5x$.
   c. x.
   d. $2x$.
   e. $4x$.

7. How many different combinations of maternal and paternal chromosomes can be packaged in gametes made by an organism with a diploid number of 8 ($2n = 8$)?
   a. 2
   b. 4
   c. 8
   d. 16
   e. 32

8. **DRAW IT** The diagram at right shows a cell in meiosis.

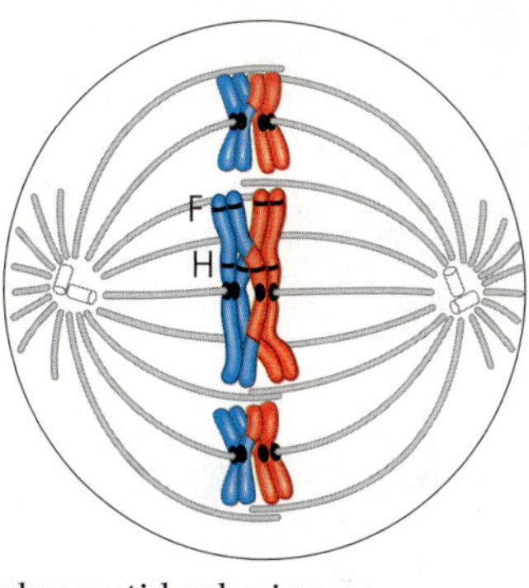

   (a) Copy the drawing to a separate sheet of paper and label appropriate structures with these terms, drawing lines or brackets as needed: chromosome (label as duplicated or unduplicated), centromere, kinetochore, sister chromatids, nonsister chromatids, homologous pair, homologs, chiasma, sister chromatid cohesion.
   (b) Describe the makeup of a haploid set and a diploid set.
   (c) Identify the stage of meiosis shown.

LEVEL 3: SYNTHESIS/EVALUATION

9. How can you tell the cell in question 8 is undergoing meiosis, not mitosis?

10. **EVOLUTION CONNECTION**
    Many species can reproduce either asexually or sexually. What might be the evolutionary significance of the switch from asexual to sexual reproduction that occurs in some organisms when the environment becomes unfavorable?

11. **SCIENTIFIC INQUIRY**
    The diagram above represents a meiotic cell in a certain individual. A previous study has shown that the freckles gene is located at the locus marked F, and the hair-color gene is located at the locus marked H, both on the long chromosome. The individual from whom this cell was taken has inherited different alleles for each gene ("freckles" and "black hair" from one parent, and "no freckles" and "blond hair" from the other). Predict allele combinations in the gametes resulting from this meiotic event. (It will help if you draw out the rest of meiosis, labeling alleles by name.) List other possible combinations of these alleles in this individual's gametes.

12. **WRITE ABOUT A THEME**
    The Genetic Basis of Life The continuity of life is based on heritable information in the form of DNA. In a short essay (100–150 words), explain how chromosome behavior during sexual reproduction in animals ensures perpetuation of parental traits in offspring and, at the same time, genetic variation among offspring.

For selected answers, see Appendix A.

MasteringBIOLOGY www.masteringbiology.com

1. MasteringBiology® Assignments

Video Tutor Session Mitosis and Meiosis

BioFlix Tutorials Meiosis: Genes, Chromosomes, and Sexual Reproduction • The Mechanism • Determinants of Heredity and Genetic Variation

Activities Asexual and Sexual Life Cycles • Meiosis • Meiosis Animation • Origins of Genetic Variation

Questions Student Misconceptions • Reading Quiz • Multiple Choice • End-of-Chapter

2. eText

Read your book online, search, take notes, highlight text, and more.

3. The Study Area

Practice Tests • Cumulative Test • BioFlix 3-D Animations • MP3 Tutor Sessions • Videos • Activities • Investigations • Lab Media • Audio Glossary • Word Study Tools • Art

14

Mendel and the Gene Idea

▲ **Figure 14.1 What principles of inheritance did Gregor Mendel discover by breeding garden pea plants?**

KEY CONCEPTS

14.1 Mendel used the scientific approach to identify two laws of inheritance

14.2 The laws of probability govern Mendelian inheritance

14.3 Inheritance patterns are often more complex than predicted by simple Mendelian genetics

14.4 Many human traits follow Mendelian patterns of inheritance

OVERVIEW

Drawing from the Deck of Genes

If you spotted a woman with bright purple hair walking down the street, you would probably deduce that she hadn't inherited her striking hair color from either parent. Consciously or not, you have transformed a lifetime of observations of hair color and other features into a list of possible variations that occur naturally among people. Brown, blue, green, or gray eyes; black, brown, blond, or red hair—these are just a few examples of heritable variations that we may observe among individuals in a population. What are the genetic principles that account for the transmission of such traits from parents to offspring in humans and other organisms?

The explanation of heredity most widely in favor during the 1800s was the "blending" hypothesis, the idea that genetic material contributed by the two parents mixes in a manner analogous to the way blue and yellow paints blend to make green. This hypothesis predicts that over many generations, a freely mating population will give rise to a uniform population of individuals. However, our everyday observations and the results of breeding experiments with animals and plants contradict that prediction. The blending hypothesis also fails to explain other phenomena of inheritance, such as traits reappearing after skipping a generation.

An alternative to the blending model is a "particulate" hypothesis of inheritance: the gene idea. According to this model, parents pass on discrete heritable units—genes—that retain their separate identities in offspring. An organism's collection of genes is more like a deck of cards than a pail of paint. Like playing cards, genes can be shuffled and passed along, generation after generation, in undiluted form.

Modern genetics had its genesis in an abbey garden, where a monk named Gregor Mendel documented a particulate mechanism for inheritance. **Figure 14.1** shows Mendel (back row, holding a sprig of fuchsia) with his fellow monks. Mendel developed his theory of inheritance several decades before chromosomes were observed under the microscope and the significance of their behavior was understood. In this chapter, we will step into Mendel's garden to re-create his experiments and explain how he arrived at his theory of inheritance. We will also explore inheritance patterns more complex than those observed by Mendel in garden peas. Finally, we will see how the Mendelian model applies to the inheritance of human variations, including hereditary disorders such as sickle-cell disease.

CONCEPT **14.1**

Mendel used the scientific approach to identify two laws of inheritance

Mendel discovered the basic principles of heredity by breeding garden peas in carefully planned experiments. As we retrace his work, you will recognize the key elements of the scientific process that were introduced in Chapter 1.

Mendel's Experimental, Quantitative Approach

Mendel grew up on his parents' small farm in a region of Austria that is now part of the Czech Republic. In this agricultural area, Mendel and the other children received agricultural training in school along with their basic education. As an adolescent, Mendel overcame financial hardship and illness to excel in high school and, later, at the Olmutz Philosophical Institute.

In 1843, at the age of 21, Mendel entered an Augustinian monastery, a reasonable choice at that time for someone who valued the life of the mind. He considered becoming a teacher but failed the necessary examination. In 1851, he left the monastery to pursue two years of study in physics and chemistry at the University of Vienna. These were very important years for Mendel's development as a scientist, in large part due to the strong influence of two professors. One was the physicist Christian Doppler, who encouraged his students to learn science through experimentation and trained Mendel to use mathematics to help explain natural phenomena. The other was a botanist named Franz Unger, who aroused Mendel's interest in the causes of variation in plants. The instruction Mendel received from these two mentors later played a critical role in his experiments with garden peas.

After attending the university, Mendel returned to the monastery and was assigned to teach at a local school, where several other instructors were enthusiastic about scientific research. In addition, his fellow monks shared a long-standing fascination with the breeding of plants. The monastery therefore provided fertile soil in more ways than one for Mendel's scientific endeavors. Around 1857, Mendel began breeding garden peas in the abbey garden to study inheritance. Although the question of heredity had long been a focus of curiosity at the monastery, Mendel's fresh approach allowed him to deduce principles that had remained elusive to others.

One reason Mendel probably chose to work with peas is that they are available in many varieties. For example, one variety has purple flowers, while another variety has white flowers. A heritable feature that varies among individuals, such as flower color, is called a **character**. Each variant for a character, such as purple or white color for flowers, is called a **trait**.

Other advantages of using peas are their short generation time and the large number of offspring from each mating. Furthermore, Mendel could strictly control mating between plants. The reproductive organs of a pea plant are in its flowers, and each pea flower has both pollen-producing organs (stamens) and an egg-bearing organ (carpel).* In nature, pea plants usually self-fertilize: Pollen grains from the stamens land on the carpel of the same flower, and sperm released from the pollen grains fertilize eggs present in the carpel. To achieve cross-pollination (fertilization between different plants), Mendel removed the immature stamens of a plant before they produced pollen and then dusted pollen from another plant onto the altered flowers (**Figure 14.2**). Each resulting zygote then developed into a plant embryo encased in a seed (pea). Mendel could thus always be sure of the parentage of new seeds.

Mendel chose to track only those characters that occurred in two distinct, alternative forms. For example, his plants had either purple flowers or white flowers; there were no colors intermediate between these two varieties. Had Mendel focused instead on characters that varied in a continuum among individuals—seed weight, for example—he would not have discovered the particulate nature of inheritance. (You'll learn why later.)

*As you learned in Figure 13.6b, meiosis in plants produces spores, not gametes. In flowering plants like the pea, each spore develops into a microscopic haploid gametophyte that contains only a few cells and is located on the parent plant. The gametophyte produces sperm, in pollen grains, and eggs, in the carpel. For simplicity, we will not include the gametophyte stage in our discussion of fertilization in plants.

▼ Figure 14.2 RESEARCH METHOD

Crossing Pea Plants

APPLICATION By crossing (mating) two true-breeding varieties of an organism, scientists can study patterns of inheritance. In this example, Mendel crossed pea plants that varied in flower color.

TECHNIQUE

RESULTS When pollen from a white flower was transferred to a purple flower, the first-generation hybrids all had purple flowers. The result was the same for the reciprocal cross, which involved the transfer of pollen from purple flowers to white flowers.

Mendel also made sure that he started his experiments with varieties that, over many generations of self-pollination, had produced only the same variety as the parent plant. Such plants are said to be **true-breeding**. For example, a plant with purple flowers is true-breeding if the seeds produced by self-pollination in successive generations all give rise to plants that also have purple flowers.

In a typical breeding experiment, Mendel cross-pollinated two contrasting, true-breeding pea varieties—for example, purple-flowered plants and white-flowered plants (see Figure 14.2). This mating, or *crossing*, of two true-breeding varieties is called **hybridization**. The true-breeding parents are referred to as the **P generation** (parental generation), and their hybrid offspring are the **F_1 generation** (first filial generation, the word *filial* from the Latin word for "son"). Allowing these F_1 hybrids to self-pollinate (or to cross-pollinate with other F_1 hybrids) produces an **F_2 generation** (second filial generation). Mendel usually followed traits for at least the P, F_1, and F_2 generations. Had Mendel stopped his experiments with the F_1 generation, the basic patterns of inheritance would have escaped him. Mendel's quantitative analysis of the F_2 plants from thousands of genetic crosses like these allowed him to deduce two fundamental principles of heredity, which have come to be called the law of segregation and the law of independent assortment.

The Law of Segregation

If the blending model of inheritance were correct, the F_1 hybrids from a cross between purple-flowered and white-flowered pea plants would have pale purple flowers, a trait intermediate between those of the P generation. Notice in Figure 14.2 that the experiment produced a very different result: All the F_1 offspring had flowers just as purple as the purple-flowered parents. What happened to the white-flowered plants' genetic contribution to the hybrids? If it were lost, then the F_1 plants could produce only purple-flowered offspring in the F_2 generation. But when Mendel allowed the F_1 plants to self-pollinate and planted their seeds, the white-flower trait reappeared in the F_2 generation.

Mendel used very large sample sizes and kept accurate records of his results: 705 of the F_2 plants had purple flowers, and 224 had white flowers. These data fit a ratio of approximately three purple to one white (**Figure 14.3**). Mendel reasoned that the heritable factor for white flowers did not disappear in the F_1 plants, but was somehow hidden, or masked, when the purple-flower factor was present. In Mendel's terminology, purple flower color is a *dominant* trait, and white flower color is a *recessive* trait. The reappearance of white-flowered plants in the F_2 generation was evidence that the heritable factor causing white flowers had not been diluted or destroyed by coexisting with the purple-flower factor in the F_1 hybrids.

Mendel observed the same pattern of inheritance in six other characters, each represented by two distinctly different traits (**Table 14.1**). For example, when Mendel crossed a true-breeding variety that produced smooth, round pea seeds with one that produced wrinkled seeds, all the F_1 hybrids produced round seeds; this is the dominant trait for seed shape. In the F_2 generation, approximately 75% of the seeds were round and 25% were wrinkled—a 3:1 ratio, as in Figure 14.3. Now let's see how Mendel deduced the law of

▼ **Figure 14.3** **INQUIRY**

When F_1 hybrid pea plants self- or cross-pollinate, which traits appear in the F_2 generation?

EXPERIMENT Around 1860, in a monastery garden in Brünn, Austria, Gregor Mendel used the character of flower color in pea plants to follow traits through two generations. He crossed true-breeding purple-flowered plants and white-flowered plants (crosses are symbolized by ×). The resulting F_1 hybrids were allowed to self-pollinate or were cross-pollinated with other F_1 hybrids. The F_2 generation plants were then observed for flower color.

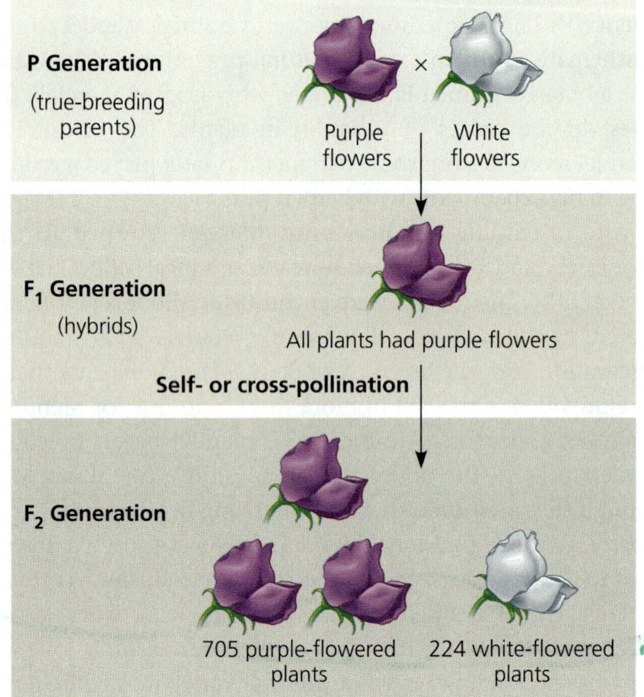

RESULTS Both purple-flowered and white-flowered plants appeared in the F_2 generation, in a ratio of approximately 3:1.

CONCLUSION The "heritable factor" for the recessive trait (white flowers) had not been destroyed, deleted, or "blended" in the F_1 generation but was merely masked by the presence of the factor for purple flowers, which is the dominant trait.

SOURCE G. Mendel, Experiments in plant hybridization, *Proceedings of the Natural History Society of Brünn* 4:3–47 (1866).

WHAT IF? If you mated two purple-flowered plants from the P generation, what ratio of traits would you expect to observe in the offspring? Explain.

Table 14.1 The Results of Mendel's F_1 Crosses for Seven Characters in Pea Plants

Character	Dominant Trait	×	Recessive Trait	F_2 Generation Dominant: Recessive	Ratio
Flower color	Purple	×	White	705:224	3.15:1
Flower position	Axial	×	Terminal	651:207	3.14:1
Seed color	Yellow	×	Green	6,022:2,001	3.01:1
Seed shape	Round	×	Wrinkled	5,474:1,850	2.96:1
Pod shape	Inflated	×	Constricted	882:299	2.95:1
Pod color	Green	×	Yellow	428:152	2.82:1
Stem length	Tall	×	Dwarf	787:277	2.84:1

segregation from his experimental results. In the discussion that follows, we will use modern terms instead of some of the terms used by Mendel. (For example, we'll use "gene" instead of Mendel's "heritable factor.")

Mendel's Model

Mendel developed a model to explain the 3:1 inheritance pattern that he consistently observed among the F_2 offspring in his pea experiments. We describe four related concepts making up this model, the fourth of which is the law of segregation.

First, *alternative versions of genes account for variations in inherited characters.* The gene for flower color in pea plants, for example, exists in two versions, one for purple flowers and the other for white flowers. These alternative versions of a gene are

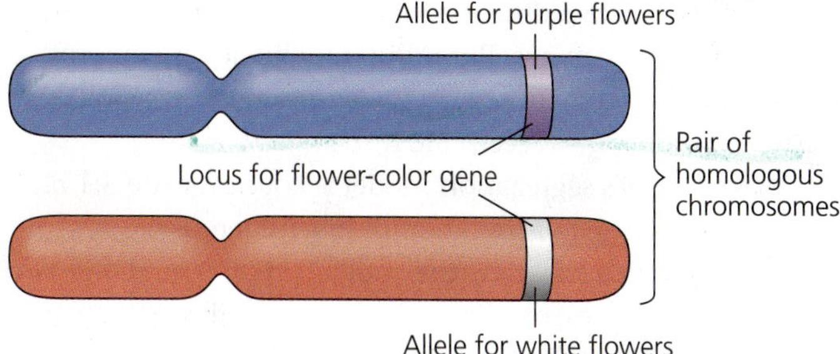

▲ **Figure 14.4 Alleles, alternative versions of a gene.** A somatic cell has two copies of each chromosome (forming a homologous pair) and thus two copies of each gene; the alleles may be identical or different. This figure depicts a pair of homologous chromosomes in an F_1 hybrid pea plant. The paternally inherited chromosome (blue), which was present in the sperm within a pollen grain, has an allele for purple flowers, and the maternally inherited chromosome (red), which was present in an egg within a carpel, has an allele for white flowers.

called **alleles** (**Figure 14.4**). Today, we can relate this concept to chromosomes and DNA. As noted in Chapter 13, each gene is a sequence of nucleotides at a specific place, or locus, along a particular chromosome. The DNA at that locus, however, can vary slightly in its nucleotide sequence and hence in its information content. The purple-flower allele and the white-flower allele are two DNA sequence variations possible at the flower-color locus on one of a pea plant's chromosomes.

Second, *for each character, an organism inherits two copies of a gene, one from each parent.* (These are also called alleles of that gene.) Remarkably, Mendel made this deduction without knowing about the role, or even the existence, of chromosomes. Recall from Chapter 13 that each somatic cell in a diploid organism has two sets of chromosomes, one set inherited from each parent. Thus, a genetic locus is actually represented twice in a diploid cell, once on each homolog of a specific pair of chromosomes. The two alleles at a particular locus may be identical, as in the true-breeding plants of Mendel's P generation. Or the alleles may differ, as in the F_1 hybrids (see Figure 14.4).

Third, *if the two alleles at a locus differ, then one, the* ***dominant allele****, determines the organism's appearance; the other, the* ***recessive allele****, has no noticeable effect on the organism's appearance.* Accordingly, Mendel's F_1 plants had purple flowers because the allele for that trait is dominant and the allele for white flowers is recessive.

The fourth and final part of Mendel's model, the **law of segregation**, states that *the two alleles for a heritable character segregate (separate from each other) during gamete formation and end up in different gametes.* Thus, an egg or a sperm gets only one of the two alleles that are present in the somatic cells of the organism making the gamete. In terms of chromosomes, this segregation corresponds to the distribution of the two members of a pair of homologous chromosomes to different gametes in meiosis (see Figure 13.7). Note that if an organism has identical alleles for a particular character—that is, the organism is true-breeding for that character—then that allele is present in all gametes. But if different alleles are present, as in the F_1 hybrids, then 50% of the gametes receive the dominant allele and 50% receive the recessive allele.

Does Mendel's segregation model account for the 3:1 ratio he observed in the F_2 generation of his numerous crosses? For the flower-color character, the model predicts that the two different alleles present in an F_1 individual will segregate into gametes such that half the gametes will have the purple-flower allele and half will have the white-flower allele. During self-pollination, gametes of each class unite randomly. An egg with a purple-flower allele has an equal chance of being fertilized by a sperm with a purple-flower allele or one with a white-flower allele. Since the same is true for an egg with a white-flower allele, there are four equally likely combinations of sperm and egg. **Figure 14.5** illustrates these combinations using a **Punnett square**, a handy diagrammatic device for predicting the allele composition of offspring from a cross between individuals of known genetic makeup. Notice that we use a capital letter to symbolize a dominant allele and a lowercase letter for a recessive allele. In our example, *P* is the purple-flower allele, and *p* is the white-flower allele; the gene itself is sometimes referred to as the *P*/*p* gene.

In the F_2 offspring, what color will the flowers be? One-fourth of the plants have inherited two purple-flower alleles; clearly, these plants will have purple flowers. One-half of the F_2 offspring have inherited one purple-flower allele and one white-flower allele; these plants will also have purple flowers, the dominant trait. Finally, one-fourth of the F_2 plants have inherited two white-flower alleles and will express the recessive trait. Thus, Mendel's model accounts for the 3:1 ratio of traits that he observed in the F_2 generation.

Useful Genetic Vocabulary

An organism that has a pair of identical alleles for a character is said to be **homozygous** for the gene controlling that character. In the parental generation in Figure 14.5, the purple pea plant is homozygous for the dominant allele (*PP*), while the white plant is homozygous for the recessive allele (*pp*). Homozygous plants "breed true" because all of their gametes contain the same allele—either *P* or *p* in this example. If we cross dominant homozygotes with recessive homozygotes, every offspring will have two different alleles—*Pp* in the case of the F_1 hybrids of our flower-color experiment (see Figure 14.5). An organism that has two different alleles for a gene is said to be **heterozygous** for that gene. Unlike homozygotes, heterozygotes produce gametes with different alleles, so they are not true-breeding. For example, *P*- and *p*-containing gametes are both produced by our F_1 hybrids. Self-pollination of the F_1 hybrids thus produces both purple-flowered and white-flowered offspring.

Because of the different effects of dominant and recessive alleles, an organism's traits do not always reveal its genetic composition. Therefore, we distinguish between an organism's appearance or observable traits, called its **phenotype**, and its genetic makeup, its **genotype**. In the case of flower color in pea plants, *PP* and *Pp* plants have the same phenotype (purple) but different genotypes. **Figure 14.6** reviews these terms. Note that "phenotype" refers to physiological traits as well as traits that relate directly to appearance.

Each true-breeding plant of the parental generation has identical alleles, *PP* or *pp*.

Gametes (circles) each contain only one allele for the flower-color gene. In this case, every gamete produced by one parent has the same allele.

P Generation

Appearance: Purple flowers × White flowers
Genetic makeup: *PP* *pp*
Gametes: *P* *p*

Union of parental gametes produces F_1 hybrids having a *Pp* combination. Because the purple-flower allele is dominant, all these hybrids have purple flowers.

When the hybrid plants produce gametes, the two alleles segregate. Half of the gametes receive the *P* allele and the other half the *p* allele.

F_1 Generation

Appearance: Purple flowers
Genetic makeup: *Pp*
Gametes: 1/2 *P* 1/2 *p*

This box, a Punnett square, shows all possible combinations of alleles in offspring that result from an $F_1 \times F_1$ (*Pp* × *Pp*) cross. Each square represents an equally probable product of fertilization. For example, the bottom left box shows the genetic combination resulting from a *p* egg fertilized by a *P* sperm.

Random combination of the gametes results in the 3:1 ratio that Mendel observed in the F_2 generation.

F_2 Generation

Sperm from F_1 (*Pp*) plant

Eggs from F_1 (*Pp*) plant

	P	*p*
P	*PP*	*Pp*
p	*Pp*	*pp*

3 : 1

▲ **Figure 14.5 Mendel's law of segregation.** This diagram shows the genetic makeup of the generations in Figure 14.3. It illustrates Mendel's model for inheritance of the alleles of a single gene. Each plant has two alleles for the gene controlling flower color, one allele inherited from each of the plant's parents. To construct a Punnett square that predicts the F_2 generation offspring, we list all the possible gametes from one parent (here, the F_1 female) along the left side of the square and all the possible gametes from the other parent (here, the F_1 male) along the top. The boxes represent the offspring resulting from all the possible unions of male and female gametes.

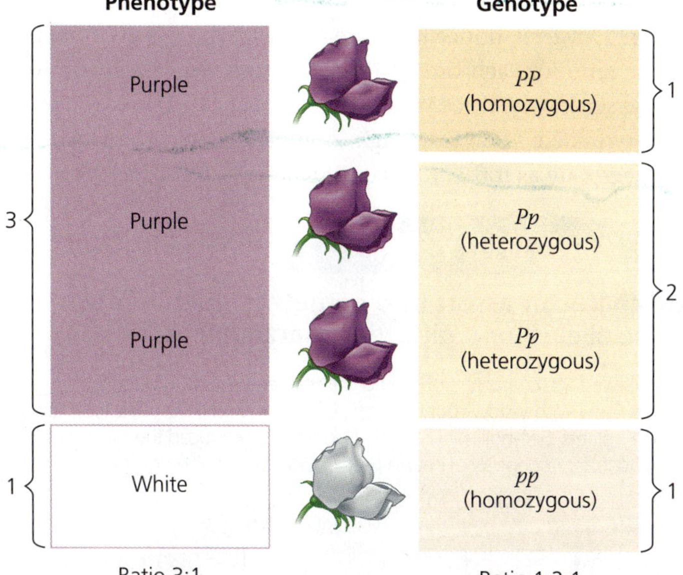

▲ **Figure 14.6 Phenotype versus genotype.** Grouping F_2 offspring from a cross for flower color according to phenotype results in the typical 3:1 phenotypic ratio. In terms of genotype, however, there are actually two categories of purple-flowered plants, *PP* (homozygous) and *Pp* (heterozygous), giving a 1:2:1 genotypic ratio.

For example, there is a pea variety that lacks the normal ability to self-pollinate. This physiological variation (non-self-pollination) is a phenotypic trait.

The Testcross

Suppose we have a "mystery" pea plant that has purple flowers. We cannot tell from its flower color if this plant is homozygous (*PP*) or heterozygous (*Pp*) because both genotypes result in the same purple phenotype. To determine the genotype, we can cross this plant with a white-flowered plant (*pp*), which will make only gametes with the recessive allele (*p*). The allele in the gamete contributed by the mystery plant will therefore determine the appearance of the offspring **(Figure 14.7)**. If all the offspring of the cross have purple flowers, then the purple-flowered mystery plant must be homozygous for the dominant allele, because a *PP* × *pp* cross produces all *Pp* offspring. But if both the purple and the white phenotypes appear among the offspring, then the purple-flowered parent must be heterozygous. The offspring of a *Pp* × *pp* cross will be expected to have a 1:1 phenotypic ratio. Breeding an organism of unknown genotype with a recessive homozygote is called a **testcross** because it can reveal the genotype of that organism. The testcross was devised by Mendel and continues to be an important tool of geneticists.

▼ **Figure 14.7 RESEARCH METHOD**

The Testcross

APPLICATION An organism that exhibits a dominant trait, such as purple flowers in pea plants, can be either homozygous for the dominant allele or heterozygous. To determine the organism's genotype, geneticists can perform a testcross.

TECHNIQUE In a testcross, the individual with the unknown genotype is crossed with a homozygous individual expressing the recessive trait (white flowers in this example), and Punnett squares are used to predict the possible outcomes.

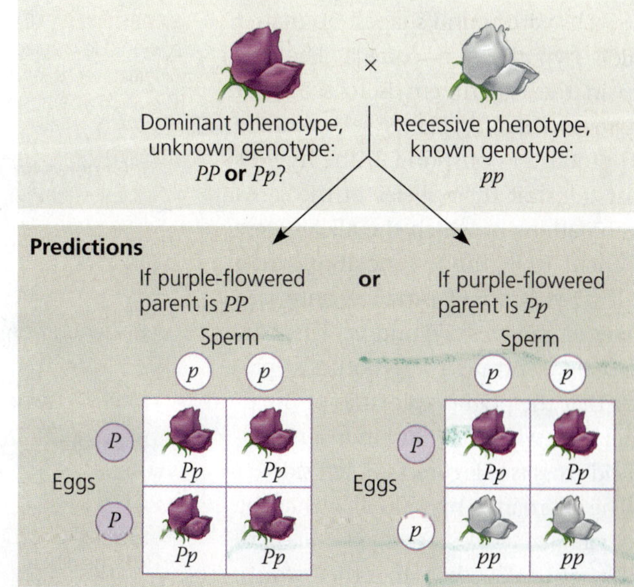

RESULTS Matching the results to either prediction identifies the unknown parental genotype (either *PP* or *Pp* in this example). In this testcross, we transferred pollen from a white-flowered plant to the carpels of a purple-flowered plant; the opposite (reciprocal) cross would have led to the same results.

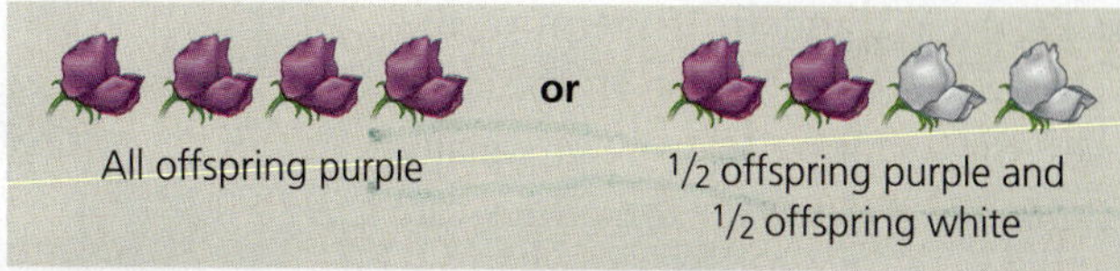

The Law of Independent Assortment

Mendel derived the law of segregation from experiments in which he followed only a *single* character, such as flower color. All the F_1 progeny produced in his crosses of true-breeding parents were **monohybrids**, meaning that they were heterozygous for the one particular character being followed in the cross. We refer to a cross between such heterozygotes as a **monohybrid cross**.

Mendel identified his second law of inheritance by following *two* characters at the same time, such as seed color and seed shape. Seeds (peas) may be either yellow or green. They also may be either round (smooth) or wrinkled. From single-character crosses, Mendel knew that the allele for yellow seeds is dominant (*Y*), and the allele for green seeds is recessive (*y*). For the seed-shape character, the allele for round is dominant (*R*), and the allele for wrinkled is recessive (*r*).

Imagine crossing two true-breeding pea varieties that differ in *both* of these characters—a cross between a plant with yellow-round seeds (*YYRR*) and a plant with green-wrinkled seeds (*yyrr*).

The F_1 plants will be **dihybrids**, individuals heterozygous for the two characters being followed in the cross (*YyRr*). But are these two characters transmitted from parents to offspring as a package? That is, will the *Y* and *R* alleles always stay together, generation after generation? Or are seed color and seed shape inherited independently? **Figure 14.8** shows how a **dihybrid cross**, a cross between F_1 dihybrids, can determine which of these two hypotheses is correct.

The F_1 plants, of genotype *YyRr*, exhibit both dominant phenotypes, yellow seeds with round shapes, no matter which hypothesis is correct. The key step in the experiment is to see what happens when F_1 plants self-pollinate and produce F_2 offspring. If the hybrids must transmit their alleles in the same combinations in which the alleles were inherited from the P generation, then the F_1 hybrids will produce only two classes of gametes: *YR* and *yr*. This "dependent assortment" hypothesis predicts that the phenotypic ratio of the F_2 generation will be 3:1, just as in a monohybrid cross (see Figure 14.8, left side).

The alternative hypothesis is that the two pairs of alleles segregate independently of each other. In other words, genes are packaged into gametes in all possible allelic combinations, as long as each gamete has one allele for each gene. In our example, an F_1 plant will produce four classes of gametes in equal quantities: *YR*, *Yr*, *yR*, and *yr*. If sperm of the four classes fertilize eggs of the four classes, there will be 16 (4 × 4) equally probable ways in which the alleles can combine in the F_2 generation, as shown in Figure 14.8, right side. These combinations result in four phenotypic categories with a ratio of 9:3:3:1 (nine yellow-round to three green-round to three yellow-wrinkled to one green-wrinkled). When Mendel did the experiment and classified the F_2 offspring, his results were close to the predicted 9:3:3:1 phenotypic ratio, supporting the hypothesis that the alleles for one gene—controlling seed color or seed shape, in this example—are sorted into gametes independently of the alleles of other genes.

Mendel tested his seven pea characters in various dihybrid combinations and always observed a 9:3:3:1 phenotypic ratio in the F_2 generation. However, notice in Figure 14.8 that there is a 3:1 phenotypic ratio for each one of the two characters if you consider them separately: three yellow to one green, and three round to one wrinkled. As far as a single character is concerned, the alleles segregate as if this were a monohybrid cross. The results of

▼ **Figure 14.8** INQUIRY

Do the alleles for one character assort into gametes dependently or independently of the alleles for a different character?

EXPERIMENT Gregor Mendel followed the characters of seed color and seed shape through the F_2 generation. He crossed a true-breeding plant with yellow-round seeds with a true-breeding plant with green-wrinkled seeds, producing dihybrid F_1 plants. Self-pollination of the F_1 dihybrids produced the F_2 generation. The two hypotheses (dependent and independent assortment) predict different phenotypic ratios.

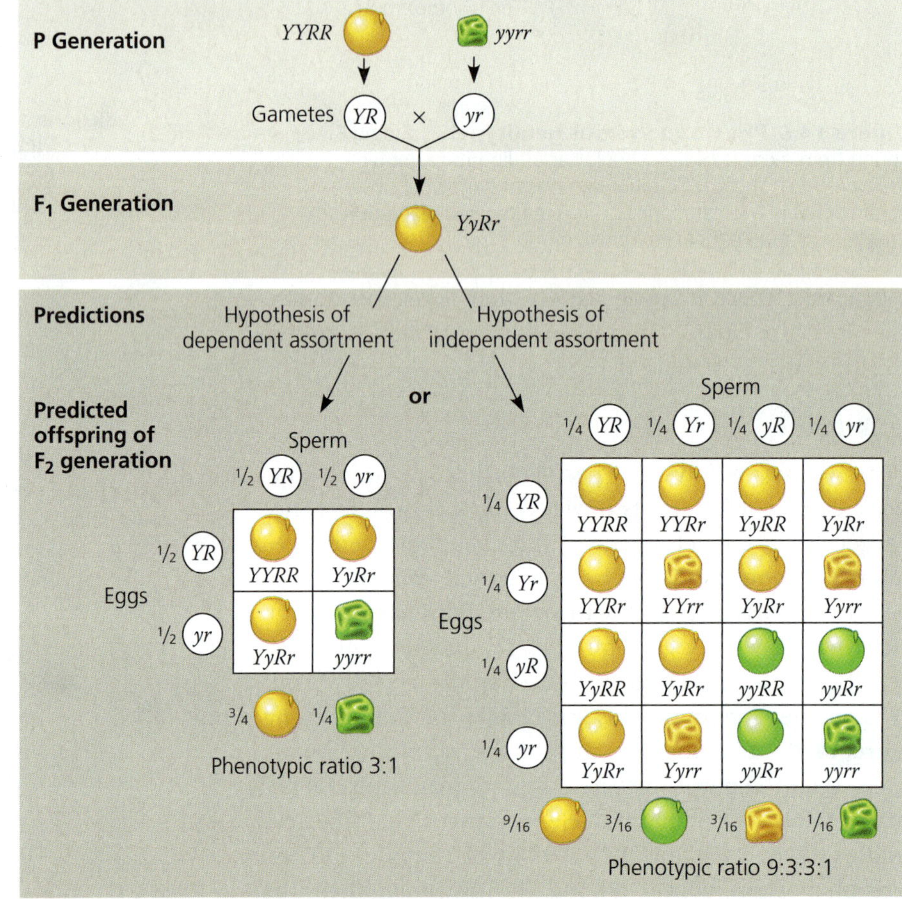

RESULTS

315 108 101 32 Phenotypic ratio approximately 9:3:3:1

CONCLUSION Only the hypothesis of independent assortment predicts the appearance of two of the observed phenotypes: green-round seeds and yellow-wrinkled seeds (see the right-hand Punnett square). The alleles for seed color and seed shape sort into gametes independently of each other.

SOURCE G. Mendel, Experiments in plant hybridization, *Proceedings of the Natural History Society of Brünn* 4:3–47 (1866).

WHAT IF? Suppose Mendel had transferred pollen from an F_1 plant to the carpel of a plant that was homozygous recessive for both genes. Set up the cross and draw Punnett squares that predict the offspring for both hypotheses. Would this cross have supported the hypothesis of independent assortment equally well?

Mendel's dihybrid experiments are the basis for what we now call the **law of independent assortment**, which states that *each pair of alleles segregates independently of each other pair of alleles during gamete formation.*

This law applies only to genes (allele pairs) located on different chromosomes—that is, on chromosomes that are not homologous—or very far apart on the same chromosome. (The latter case will be explained in Chapter 15, along with the more complex inheritance patterns of genes located near each other, which tend to be inherited together.) All the pea characters Mendel chose for analysis were controlled by genes on different chromosomes (or far apart on one chromosome); this situation greatly simplified interpretation of his multicharacter pea crosses. All the examples we consider in the rest of this chapter involve genes located on different chromosomes.

CONCEPT CHECK 14.1

1. **DRAW IT** Pea plants heterozygous for flower position and stem length (*AaTt*) are allowed to self-pollinate, and 400 of the resulting seeds are planted. Draw a Punnett square for this cross. How many offspring would be predicted to have terminal flowers and be dwarf? (See Table 14.1.)
2. **WHAT IF?** List all gametes that could be made by a pea plant heterozygous for seed color, seed shape, and pod shape (*YyRrIi*; see Table 14.1). How large a Punnett square would you need to draw to predict the offspring of a self-pollination of this "trihybrid"?
3. **MAKE CONNECTIONS** In some pea plant crosses, the plants are self-pollinated. Refer back to Concept 13.1 (pp. 248–249) and explain whether self-pollination is considered asexual or sexual reproduction.

For suggested answers, see Appendix A.

CONCEPT **14.2**

The laws of probability govern Mendelian inheritance

Mendel's laws of segregation and independent assortment reflect the same rules of probability that apply to tossing coins, rolling dice, and drawing cards from a deck. The probability scale ranges from 0 to 1. An event that is certain to occur has a probability of 1, while an event that is certain *not* to occur has a probability of 0. With a coin that has heads on both sides, the probability of tossing heads is 1, and the probability of tossing tails is 0. With a normal coin, the chance of tossing heads is $\frac{1}{2}$, and the chance of tossing tails is $\frac{1}{2}$. The probability of drawing the ace of spades from a 52-card deck is $\frac{1}{52}$. The probabilities of all possible outcomes for an event must add up to 1. With a deck of cards, the chance of picking a card other than the ace of spades is $\frac{51}{52}$.

Tossing a coin illustrates an important lesson about probability. For every toss, the probability of heads is $\frac{1}{2}$. The outcome of any particular toss is unaffected by what has happened on previous trials. We refer to phenomena such as coin tosses as independent events. Each toss of a coin, whether done sequentially with one coin or simultaneously with many, is independent of every other toss. And like two separate coin tosses, the alleles of one gene segregate into gametes independently of another gene's alleles (the law of independent assortment). Two basic rules of probability can help us predict the outcome of the fusion of such gametes in simple monohybrid crosses and more complicated crosses.

The Multiplication and Addition Rules Applied to Monohybrid Crosses

How do we determine the probability that two or more independent events will occur together in some specific combination? For example, what is the chance that two coins tossed simultaneously will both land heads up? The **multiplication rule** states that to determine this probability, we multiply the probability of one event (one coin coming up heads) by the probability of the other event (the other coin coming up heads). By the multiplication rule, then, the probability that both coins will land heads up is $\frac{1}{2} \times \frac{1}{2} = \frac{1}{4}$.

We can apply the same reasoning to an F_1 monohybrid cross. With seed shape in pea plants as the heritable character, the genotype of F_1 plants is *Rr*. Segregation in a heterozygous plant is like flipping a coin in terms of calculating the probability of each outcome: Each egg produced has a $\frac{1}{2}$ chance of carrying the dominant allele (*R*) and a $\frac{1}{2}$ chance of carrying the recessive allele (*r*). The same odds apply to each sperm cell produced. For a particular F_2 plant to have wrinkled seeds, the recessive trait, both the egg and the sperm that come together must carry the *r* allele. The probability that an *r* allele will be present in both gametes at fertilization is found by multiplying $\frac{1}{2}$ (the probability that the egg will have an *r*) $\times$ $\frac{1}{2}$ (the probability that the sperm will have an *r*). Thus, the multiplication rule tells us that the probability of an F_2 plant having wrinkled seeds (*rr*) is $\frac{1}{4}$ (**Figure 14.9**, on the next page). Likewise, the probability of an F_2 plant carrying both dominant alleles for seed shape (*RR*) is $\frac{1}{4}$.

To figure out the probability that an F_2 plant from a monohybrid cross will be heterozygous rather than homozygous, we need to invoke a second rule. Notice in Figure 14.9 that the dominant allele can come from the egg and the recessive allele from the sperm, or vice versa. That is, F_1 gametes can combine to produce *Rr* offspring in two *mutually exclusive* ways: For any particular heterozygous F_2 plant, the dominant allele can come from the egg *or* the sperm, but not from both. According to the **addition rule**, the probability that any one of two or more mutually exclusive events will occur is calculated by adding their individual probabilities. As we have just seen, the multiplication rule gives us the individual

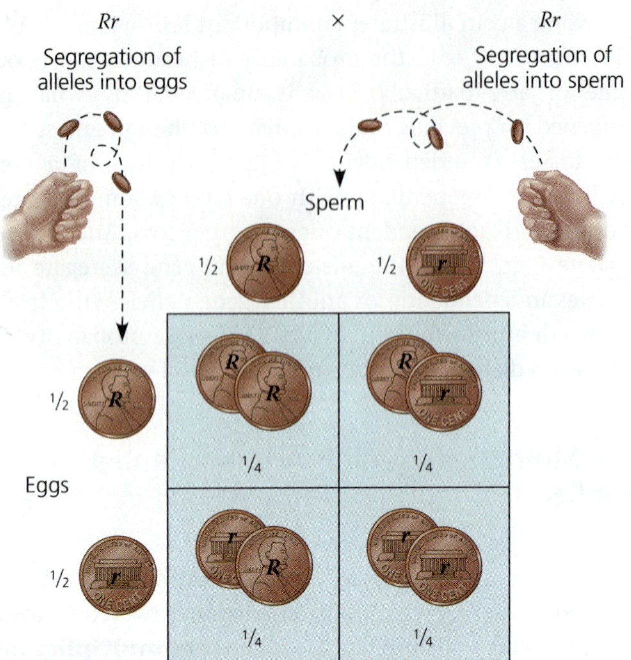

▲ **Figure 14.9 Segregation of alleles and fertilization as chance events.** When a heterozygote (*Rr*) forms gametes, whether a particular gamete ends up with an *R* or an *r* is like the toss of a coin. We can determine the probability for any genotype among the offspring of two heterozygotes by multiplying together the individual probabilities of an egg and sperm having a particular allele (*R* or *r* in this example).

probabilities that we will now add together. The probability for one possible way of obtaining an F_2 heterozygote—the dominant allele from the egg and the recessive allele from the sperm—is ¼. The probability for the other possible way—the recessive allele from the egg and the dominant allele from the sperm—is also ¼ (see Figure 14.9). Using the rule of addition, then, we can calculate the probability of an F_2 heterozygote as ¼ + ¼ = ½.

Solving Complex Genetics Problems with the Rules of Probability

We can also apply the rules of probability to predict the outcome of crosses involving multiple characters. Recall that each allelic pair segregates independently during gamete formation (the law of independent assortment). Thus, a dihybrid or other multicharacter cross is equivalent to two or more independent monohybrid crosses occurring simultaneously. By applying what we have learned about monohybrid crosses, we can determine the probability of specific genotypes occurring in the F_2 generation without having to construct unwieldy Punnett squares.

Consider the dihybrid cross between *YyRr* heterozygotes shown in Figure 14.8. We will focus first on the seed-color character. For a monohybrid cross of *Yy* plants, we can use a simple Punnett square to determine that the probabilities of the offspring genotypes are ¼ for *YY*, ½ for *Yy*, and ¼ for *yy*. We can draw a second Punnett square to determine that the same probabilities apply to the offspring genotypes for seed shape: ¼ *RR*, ½ *Rr*, and ¼ *rr*. Knowing these probabilities, we can simply use the multiplication rule to determine the probability of each of the genotypes in the F_2 generation. To give two examples, the calculations for finding the probabilities of two of the possible F_2 genotypes (*YYRR* and *YyRR*) are shown below:

Probability of *YYRR* = 1/4 (probability of *YY*) × 1/4 (*RR*) = 1/16

Probability of *YyRR* = 1/2 (*Yy*) × 1/4 (*RR*) = 1/8

The *YYRR* genotype corresponds to the upper left box in the larger Punnett square in Figure 14.8 (one box = 1/16). Looking closely at the larger Punnett square in Figure 14.8, you will see that 2 of the 16 boxes (⅛) correspond to the *YyRR* genotype.

Now let's see how we can combine the multiplication and addition rules to solve even more complex problems in Mendelian genetics. Imagine a cross of two pea varieties in which we track the inheritance of three characters. Let's cross a trihybrid with purple flowers and yellow, round seeds (heterozygous for all three genes) with a plant with purple flowers and green, wrinkled seeds (heterozygous for flower color but homozygous recessive for the other two characters). Using Mendelian symbols, our cross is *PpYyRr* × *Ppyyrr*. What fraction of offspring from this cross are predicted to exhibit the recessive phenotypes for *at least two* of the three characters?

To answer this question, we can start by listing all genotypes that fulfill this condition: *ppyyRr*, *ppYyrr*, *Ppyyrr*, *PPyyrr*, and *ppyyrr*. (Because the condition is *at least two* recessive traits, it includes the last genotype, which shows all three recessive traits.) Next, we calculate the probability for each of these genotypes resulting from our *PpYyRr* × *Ppyyrr* cross by multiplying together the individual probabilities for the allele pairs, just as we did in our dihybrid example. Note that in a cross involving heterozygous and homozygous allele pairs (for example, *Yy* × *yy*), the probability of heterozygous offspring is ½ and the probability of homozygous offspring is ½. Finally, we use the addition rule to add the probabilities for all the different genotypes that fulfill the condition of at least two recessive traits, as shown below:

ppyyRr	1/4 (probability of *pp*) × 1/2 (*yy*) × 1/2 (*Rr*)	= 1/16
ppYyrr	1/4 × 1/2 × 1/2	= 1/16
Ppyyrr	1/2 × 1/2 × 1/2	= 2/16
PPyyrr	1/4 × 1/2 × 1/2	= 1/16
ppyyrr	1/4 × 1/2 × 1/2	= 1/16
Chance of *at least two* recessive traits		= 6/16 or 3/8

In time, you'll be able to solve genetics problems faster by using the rules of probability than by filling in Punnett squares.

We cannot predict with certainty the exact numbers of progeny of different genotypes resulting from a genetic cross. But the rules of probability give us the *chance* of various outcomes. Usually, the larger the sample size, the closer the results will conform to our predictions. The reason Mendel counted so many offspring from his crosses is that he understood this statistical feature of inheritance and had a keen sense of the rules of chance.

CONCEPT CHECK 14.2

1. For any gene with a dominant allele *A* and recessive allele *a*, what proportions of the offspring from an *AA* × *Aa* cross are expected to be homozygous dominant, homozygous recessive, and heterozygous?
2. Two organisms, with genotypes *BbDD* and *BBDd*, are mated. Assuming independent assortment of the *B/b* and *D/d* genes, write the genotypes of all possible offspring from this cross and use the rules of probability to calculate the chance of each genotype occurring.
3. **WHAT IF?** Three characters (flower color, seed color, and pod shape) are considered in a cross between two pea plants (*PpYyIi* × *ppYyii*). What fraction of offspring are predicted to be homozygous recessive for at least two of the three characters?

For suggested answers, see Appendix A.

CONCEPT 14.3

Inheritance patterns are often more complex than predicted by simple Mendelian genetics

In the 20th century, geneticists extended Mendelian principles not only to diverse organisms, but also to patterns of inheritance more complex than those described by Mendel. For the work that led to his two laws of inheritance, Mendel chose pea plant characters that turn out to have a relatively simple genetic basis: Each character is determined by one gene, for which there are only two alleles, one completely dominant and the other completely recessive. (There is one exception: Mendel's pod-shape character is actually determined by two genes.) Not all heritable characters are determined so simply, and the relationship between genotype and phenotype is rarely so straightforward. Mendel himself realized that he could not explain the more complicated patterns he observed in crosses involving other pea characters or other plant species. This does not diminish the utility of Mendelian genetics (also called Mendelism), however, because the basic principles of segregation and independent assortment apply even to more complex patterns of inheritance. In this section, we will extend Mendelian genetics to hereditary patterns that were not reported by Mendel.

Extending Mendelian Genetics for a Single Gene

The inheritance of characters determined by a single gene deviates from simple Mendelian patterns when alleles are not completely dominant or recessive, when a particular gene has more than two alleles, or when a single gene produces multiple phenotypes. We will describe examples of each of these situations in this section.

Degrees of Dominance

Alleles can show different degrees of dominance and recessiveness in relation to each other. In Mendel's classic pea crosses, the F_1 offspring always looked like one of the two parental varieties because one allele in a pair showed **complete dominance** over the other. In such situations, the phenotypes of the heterozygote and the dominant homozygote are indistinguishable.

For some genes, however, neither allele is completely dominant, and the F_1 hybrids have a phenotype somewhere between those of the two parental varieties. This phenomenon, called **incomplete dominance**, is seen when red snapdragons are crossed with white snapdragons: All the F_1 hybrids have pink flowers (**Figure 14.10**). This third, intermediate phenotype results from flowers of the heterozygotes having

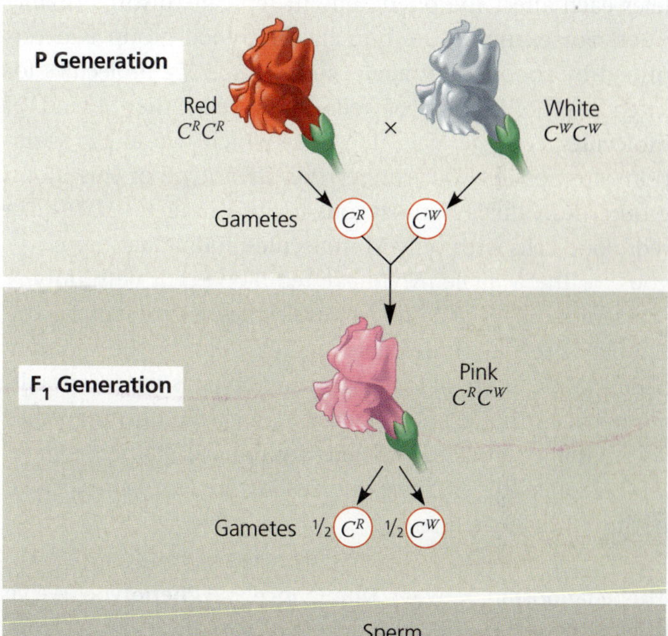

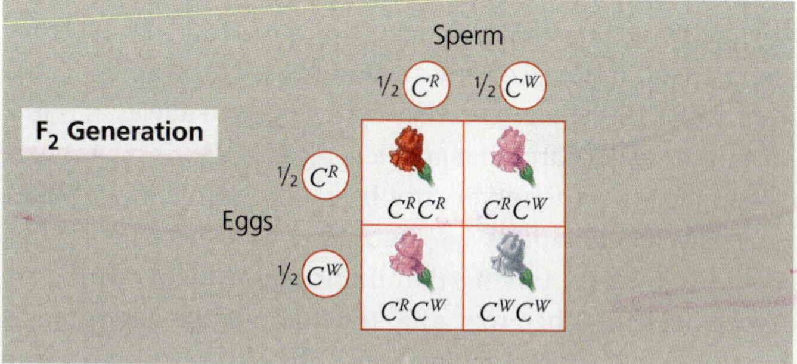

▲ **Figure 14.10 Incomplete dominance in snapdragon color.** When red snapdragons are crossed with white ones, the F_1 hybrids have pink flowers. Segregation of alleles into gametes of the F_1 plants results in an F_2 generation with a 1:2:1 ratio for both genotype and phenotype. Neither allele is dominant, so rather than using upper- and lowercase letters, we use the letter *C* with a superscript to indicate an allele for flower color: C^R for red and C^W for white.

? *Suppose a classmate argues that this figure supports the blending hypothesis for inheritance. What might your classmate say, and how would you respond?*

less red pigment than the red homozygotes. (This is unlike the case of Mendel's pea plants, where the *Pp* heterozygotes make enough pigment for the flowers to be purple, indistinguishable from those of *PP* plants.)

At first glance, incomplete dominance of either allele seems to provide evidence for the blending hypothesis of inheritance, which would predict that the red or white trait could never be retrieved from the pink hybrids. In fact, interbreeding F_1 hybrids produces F_2 offspring with a phenotypic ratio of one red to two pink to one white. (Because heterozygotes have a separate phenotype, the genotypic and phenotypic ratios for the F_2 generation are the same, 1:2:1.) The segregation of the red-flower and white-flower alleles in the gametes produced by the pink-flowered plants confirms that the alleles for flower color are heritable factors that maintain their identity in the hybrids; that is, inheritance is particulate.

Another variation on dominance relationships between alleles is called **codominance**; in this variation, the two alleles each affect the phenotype in separate, distinguishable ways. For example, the human MN blood group is determined by codominant alleles for two specific molecules located on the surface of red blood cells, the M and N molecules. A single gene locus, at which two allelic variations are possible, determines the phenotype of this blood group. Individuals homozygous for the *M* allele (*MM*) have red blood cells with only M molecules; individuals homozygous for the *N* allele (*NN*) have red blood cells with only N molecules. But *both* M and N molecules are present on the red blood cells of individuals heterozygous for the *M* and *N* alleles (*MN*). Note that the MN phenotype is *not* intermediate between the M and N phenotypes, which distinguishes codominance from incomplete dominance. Rather, *both* M and N phenotypes are exhibited by heterozygotes, since both molecules are present.

The Relationship Between Dominance and Phenotype We've now seen that the relative effects of two alleles range from complete dominance of one allele, through incomplete dominance of either allele, to codominance of both alleles. It is important to understand that an allele is called *dominant* because it is seen in the phenotype, not because it somehow subdues a recessive allele. Alleles are simply variations in a gene's nucleotide sequence. When a dominant allele coexists with a recessive allele in a heterozygote, they do not actually interact at all. It is in the pathway from genotype to phenotype that dominance and recessiveness come into play.

To illustrate the relationship between dominance and phenotype, we can use one of the characters Mendel studied—round versus wrinkled pea seed shape. The dominant allele (round) codes for an enzyme that helps convert an unbranched form of starch to a branched form in the seed. The recessive allele (wrinkled) codes for a defective form of this enzyme, leading to an accumulation of unbranched starch, which causes excess water to enter the seed by osmosis. Later, when the seed dries, it wrinkles. If a dominant allele is present, no excess water enters the seed and it does not wrinkle when it dries. One dominant allele results in enough of the enzyme to synthesize adequate amounts of branched starch, which means that dominant homozygotes and heterozygotes have the same phenotype: round seeds.

A closer look at the relationship between dominance and phenotype reveals an intriguing fact: For any character, the observed dominant/recessive relationship of alleles depends on the level at which we examine phenotype. **Tay-Sachs disease**, an inherited disorder in humans, provides an example. The brain cells of a child with Tay-Sachs disease cannot metabolize certain lipids because a crucial enzyme does not work properly. As these lipids accumulate in brain cells, the child begins to suffer seizures, blindness, and degeneration of motor and mental performance and dies within a few years.

Only children who inherit two copies of the Tay-Sachs allele (homozygotes) have the disease. Thus, at the *organismal* level, the Tay-Sachs allele qualifies as recessive. However, the activity level of the lipid-metabolizing enzyme in heterozygotes is intermediate between that in individuals homozygous for the normal allele and that in individuals with Tay-Sachs disease. The intermediate phenotype observed at the *biochemical* level is characteristic of incomplete dominance of either allele. Fortunately, the heterozygote condition does not lead to disease symptoms, apparently because half the normal enzyme activity is sufficient to prevent lipid accumulation in the brain. Extending our analysis to yet another level, we find that heterozygous individuals produce equal numbers of normal and dysfunctional enzyme molecules. Thus, at the *molecular* level, the normal allele and the Tay-Sachs allele are codominant. As you can see, whether alleles appear to be completely dominant, incompletely dominant, or codominant depends on the level at which the phenotype is analyzed.

Frequency of Dominant Alleles Although you might assume that the dominant allele for a particular character would be more common in a population than the recessive allele, this is not a given. For example, about one baby out of 400 in the United States is born with extra fingers or toes, a condition known as polydactyly. Some cases are caused by the presence of a dominant allele. The low frequency of polydactyly indicates that the recessive allele, which results in five digits per appendage, is far more prevalent than the dominant allele in the population. In Chapter 23, you will learn how relative frequencies of alleles in a population are affected by natural selection.

Multiple Alleles

Only two alleles exist for the pea characters that Mendel studied, but most genes exist in more than two allelic forms. The ABO blood groups in humans, for instance, are determined by three alleles of a single gene: I^A, I^B, and *i*. A person's blood

(a) The three alleles for the ABO blood groups and their carbohydrates. Each allele codes for an enzyme that may add a specific carbohydrate (designated by the superscript on the allele and shown as a triangle or circle) to red blood cells.

Allele	I^A	I^B	i
Carbohydrate	A △	B ○	none

(b) Blood group genotypes and phenotypes. There are six possible genotypes, resulting in four different phenotypes.

Genotype	I^AI^A or I^Ai	I^BI^B or I^Bi	I^AI^B	ii
Red blood cell appearance				
Phenotype (blood group)	A	B	AB	O

▲ **Figure 14.11 Multiple alleles for the ABO blood groups.** The four blood groups result from different combinations of three alleles.

? *Based on the surface carbohydrate phenotype in (b), what are the dominance relationships among the alleles?*

group (phenotype) may be one of four types: A, B, AB, or O. These letters refer to two carbohydrates—A and B—that may be found on the surface of red blood cells. A person's blood cells may have carbohydrate A (type A blood), carbohydrate B (type B), both (type AB), or neither (type O), as shown schematically in **Figure 14.11**. Matching compatible blood groups is critical for safe blood transfusions (see Chapter 43).

Pleiotropy

So far, we have treated Mendelian inheritance as though each gene affects only one phenotypic character. Most genes, however, have multiple phenotypic effects, a property called **pleiotropy** (from the Greek *pleion*, more). In humans, for example, pleiotropic alleles are responsible for the multiple symptoms associated with certain hereditary diseases, such as cystic fibrosis and sickle-cell disease, discussed later in this chapter. In the garden pea, the gene that determines flower color also affects the color of the coating on the outer surface of the seed, which can be gray or white. Given the intricate molecular and cellular interactions responsible for an organism's development and physiology, it isn't surprising that a single gene can affect a number of characteristics in an organism.

Extending Mendelian Genetics for Two or More Genes

Dominance relationships, multiple alleles, and pleiotropy all have to do with the effects of the alleles of a single gene. We now consider two situations in which two or more genes are involved in determining a particular phenotype.

Epistasis

In **epistasis** (from the Greek for "standing upon"), the phenotypic expression of a gene at one locus alters that of a gene at a second locus. An example will help clarify this concept. In Labrador retrievers (commonly called "Labs"), black coat color is dominant to brown. Let's designate *B* and *b* as the two alleles for this character. For a Lab to have brown fur, its genotype must be *bb*; these dogs are called chocolate Labs. But there is more to the story. A second gene determines whether or not pigment will be deposited in the hair. The dominant allele, symbolized by *E*, results in the deposition of either black or brown pigment, depending on the genotype at the first locus. But if the Lab is homozygous recessive for the second locus (*ee*), then the coat is yellow, regardless of the genotype at the black/brown locus. In this case, the gene for pigment deposition (*E/e*) is said to be epistatic to the gene that codes for black or brown pigment (*B/b*).

What happens if we mate black Labs that are heterozygous for both genes (*BbEe*)? Although the two genes affect the same phenotypic character (coat color), they follow the law of independent assortment. Thus, our breeding experiment represents an F_1 dihybrid cross, like those that produced a 9:3:3:1 ratio in Mendel's experiments. We can use a Punnett square to represent the genotypes of the F_2 offspring (**Figure 14.12**). As a result

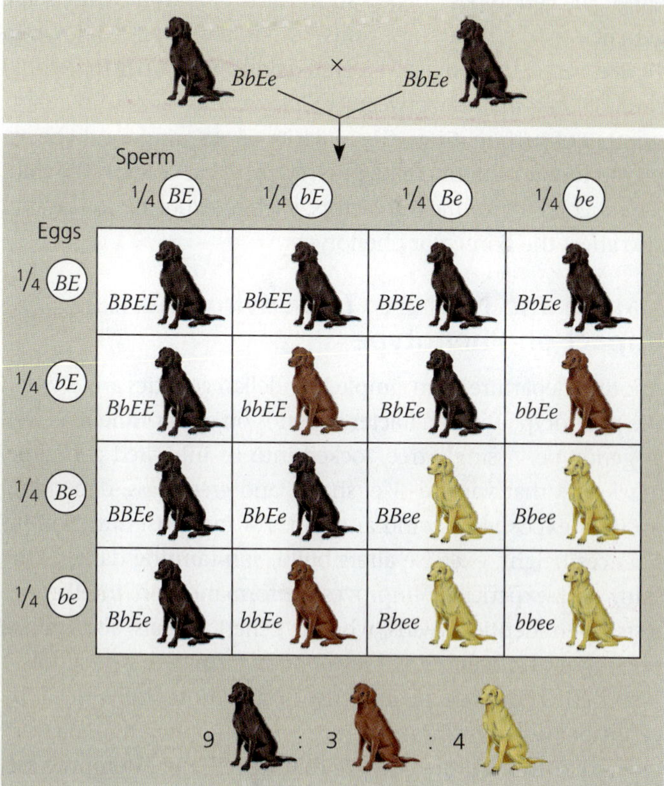

▲ **Figure 14.12 An example of epistasis.** This Punnett square illustrates the genotypes and phenotypes predicted for offspring of matings between two black Labrador retrievers of genotype *BbEe*. The *E/e* gene, which is epistatic to the *B/b* gene coding for hair pigment, controls whether or not pigment of any color will be deposited in the hair.

of epistasis, the phenotypic ratio among the F_2 offspring is nine black to three chocolate (brown) to four yellow. Other types of epistatic interactions produce different ratios, but all are modified versions of 9:3:3:1.

Polygenic Inheritance

Mendel studied characters that could be classified on an either-or basis, such as purple versus white flower color. But for many characters, such as human skin color and height, an either-or classification is impossible because the characters vary in the population in gradations along a continuum. These are called **quantitative characters**. Quantitative variation usually indicates **polygenic inheritance**, an additive effect of two or more genes on a single phenotypic character (the converse of pleiotropy, where a single gene affects several phenotypic characters).

There is evidence, for instance, that skin pigmentation in humans is controlled by at least three separately inherited genes (probably more, but we will simplify). Let's consider three genes, with a dark-skin allele for each gene (*A*, *B*, or *C*) contributing one "unit" of darkness (also a simplification) to the phenotype and being incompletely dominant to the other allele (*a*, *b*, or *c*). An *AABBCC* person would be very dark, while an *aabbcc* individual would be very light. An *AaBbCc* person would have skin of an intermediate shade. Because the alleles have a cumulative effect, the genotypes *AaBbCc* and *AABbcc* would make the same genetic contribution (three units) to skin darkness. As shown in **Figure 14.13**, there are seven skin-color phenotypes that could result from a mating between *AaBbCc* heterozygotes. In a large number of such matings, the majority of offspring would be expected to have intermediate phenotypes (skin color in the middle range). Environmental factors, such as exposure to the sun, also affect the skin-color phenotype.

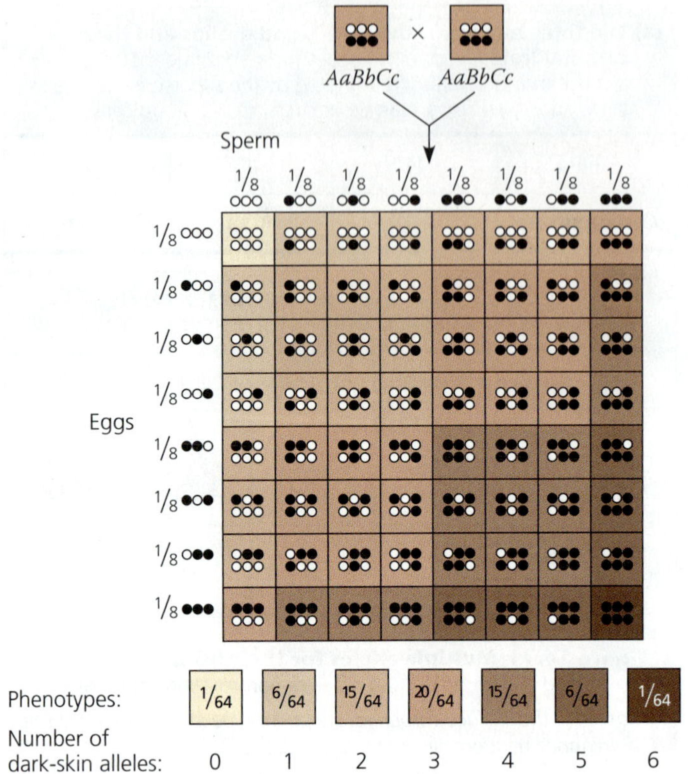

▲ **Figure 14.13 A simplified model for polygenic inheritance of skin color.** According to this model, three separately inherited genes affect the darkness of skin. The heterozygous individuals (*AaBbCc*) represented by the two rectangles at the top of this figure each carry three dark-skin alleles (black circles, which represent *A*, *B*, or *C*) and three light-skin alleles (white circles, which represent *a*, *b*, or *c*). The Punnett square shows all the possible genetic combinations in gametes and in offspring of a large number of hypothetical matings between these heterozygotes. The results are summarized by the phenotypic ratios under the Punnett square.

DRAW IT *Make a bar graph of the results, with skin color (number of dark-skin alleles) along the x-axis and fraction of offspring along the y-axis. Draw a rough curve corresponding to the results and discuss what it shows about the relative proportions of different phenotypes among the offspring.*

Nature and Nurture: The Environmental Impact on Phenotype

Another departure from simple Mendelian genetics arises when the phenotype for a character depends on environment as well as genotype. A single tree, locked into its inherited genotype, has leaves that vary in size, shape, and greenness, depending on their exposure to wind and sun. For humans, nutrition influences height, exercise alters build, sun-tanning darkens the skin, and experience improves performance on intelligence tests. Even identical twins, who are genetic equals, accumulate phenotypic differences as a result of their unique experiences.

Whether human characteristics are more influenced by genes or the environment—nature versus nurture—is a very old and hotly contested debate that we will not attempt to settle here. We can say, however, that a genotype generally is not associated with a rigidly defined phenotype, but rather with a range of phenotypic possibilities due to environmental influences. This phenotypic range is called the **norm of reaction** for a genotype **(Figure 14.14)**. For some characters, such as the

▲ **Figure 14.14 The effect of environment on phenotype.** The outcome of a genotype lies within its norm of reaction, a phenotypic range that depends on the environment in which the genotype is expressed. For example, hydrangea flowers of the same genetic variety range in color from blue-violet to pink, with the shade and intensity of color depending on the acidity and aluminum content of the soil.

ABO blood group system, the norm of reaction has no breadth whatsoever; that is, a given genotype mandates a very specific phenotype. Other characteristics, such as a person's blood count of red and white cells, vary quite a bit, depending on such factors as the altitude, the customary level of physical activity, and the presence of infectious agents.

Generally, norms of reaction are broadest for polygenic characters. Environment contributes to the quantitative nature of these characters, as we have seen in the continuous variation of skin color. Geneticists refer to such characters as **multifactorial**, meaning that many factors, both genetic and environmental, collectively influence phenotype.

Integrating a Mendelian View of Heredity and Variation

We have now broadened our view of Mendelian inheritance by exploring degrees of dominance as well as multiple alleles, pleiotropy, epistasis, polygenic inheritance, and the phenotypic impact of the environment. How can we integrate these refinements into a comprehensive theory of Mendelian genetics? The key is to make the transition from the reductionist emphasis on single genes and phenotypic characters to the emergent properties of the organism as a whole, one of the themes of this book.

The term *phenotype* can refer not only to specific characters, such as flower color and blood group, but also to an organism in its entirety—*all* aspects of its physical appearance, internal anatomy, physiology, and behavior. Similarly, the term *genotype* can refer to an organism's entire genetic makeup, not just its alleles for a single genetic locus. In most cases, a gene's impact on phenotype is affected by other genes and by the environment. In this integrated view of heredity and variation, an organism's phenotype reflects its overall genotype and unique environmental history.

Considering all that can occur in the pathway from genotype to phenotype, it is indeed impressive that Mendel could uncover the fundamental principles governing the transmission of individual genes from parents to offspring. Mendel's two laws, segregation and independent assortment, explain heritable variations in terms of alternative forms of genes (hereditary "particles," now known as the alleles of genes) that are passed along, generation after generation, according to simple rules of probability. This theory of inheritance is equally valid for peas, flies, fishes, birds, and human beings—indeed, for any organism with a sexual life cycle. Furthermore, by extending the principles of segregation and independent assortment to help explain such hereditary patterns as epistasis and quantitative characters, we begin to see how broadly Mendelism applies. From Mendel's abbey garden came a theory of particulate inheritance that anchors modern genetics. In the last section of this chapter, we will apply Mendelian genetics to human inheritance, with emphasis on the transmission of hereditary diseases.

CONCEPT CHECK 14.3

1. *Incomplete dominance* and *epistasis* are both terms that define genetic relationships. What is the most basic distinction between these terms?
2. If a man with type AB blood marries a woman with type O, what blood types would you expect in their children? What fraction would you expect of each type?
3. **WHAT IF?** A rooster with gray feathers and a hen of the same phenotype produce 15 gray, 6 black, and 8 white chicks. What is the simplest explanation for the inheritance of these colors in chickens? What phenotypes would you expect in the offspring of a cross between a gray rooster and a black hen?

For suggested answers, see Appendix A.

CONCEPT 14.4

Many human traits follow Mendelian patterns of inheritance

Peas are convenient subjects for genetic research, but humans are not. The human generation span is long—about 20 years—and human parents produce many fewer offspring than peas and most other species. Even more important, it wouldn't be ethical to ask pairs of humans to breed so that the phenotypes of their offspring could be analyzed! In spite of these constraints, the study of human genetics continues, spurred on by our desire to understand our own inheritance. New molecular biological techniques have led to many breakthrough discoveries, as we will see in Chapter 20, but basic Mendelism endures as the foundation of human genetics.

Pedigree Analysis

Unable to manipulate the mating patterns of people, geneticists must analyze the results of matings that have already occurred. They do so by collecting information about a family's history for a particular trait and assembling this information into a family tree describing the traits of parents and children across the generations—the family **pedigree**.

Figure 14.15a, on the next page, shows a three-generation pedigree that traces the occurrence of a pointed contour of the hairline on the forehead. This trait, called a widow's peak, is due to a dominant allele, *W*. Because the widow's-peak allele is dominant, all individuals who lack a widow's peak must be homozygous recessive (*ww*). The two grandparents with widow's peaks must have the *Ww* genotype, since some of their offspring are homozygous recessive. The offspring in the second generation who *do* have widow's peaks must also be heterozygous, because they are the products of *Ww* × *ww* matings. The third generation in this pedigree consists of two sisters. The one

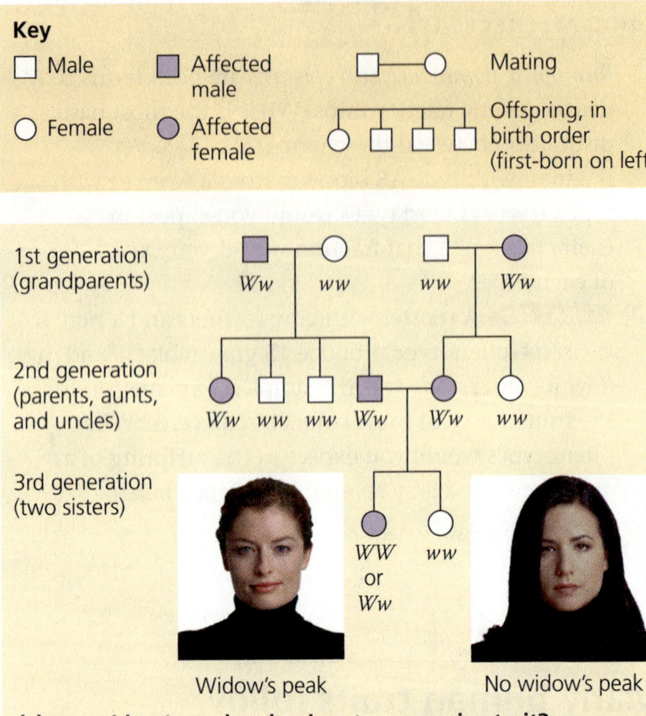

(a) Is a widow's peak a dominant or recessive trait?

Tips for pedigree analysis: Notice in the third generation that the second-born daughter lacks a widow's peak, although both of her parents had the trait. Such a pattern of inheritance supports the hypothesis that the trait is due to a dominant allele. If the trait were due to a *recessive* allele, and both parents had the recessive phenotype, then *all* of their offspring would also have the recessive phenotype.

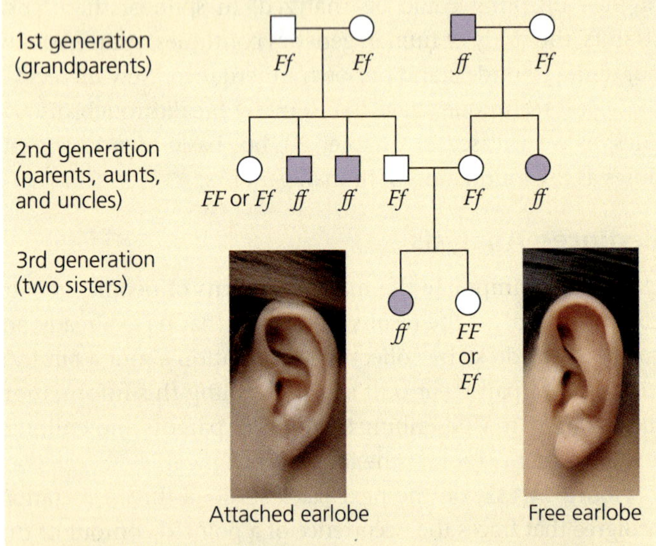

(b) Is an attached earlobe a dominant or recessive trait?

Tips for pedigree analysis: Notice that the first-born daughter in the third generation has attached earlobes, although both of her parents lack that trait (they have free earlobes). Such a pattern is easily explained if the attached-lobe phenotype is due to a recessive allele. If it were due to a *dominant* allele, then at least one parent would also have had the trait.

▲ **Figure 14.15 Pedigree analysis.** Each of these pedigrees traces a trait through three generations of the same family. The two traits have different inheritance patterns, as seen by analysis of the pedigrees.

who has a widow's peak could be either homozygous (*WW*) or heterozygous (*Ww*), given what we know about the genotypes of her parents (both *Ww*).

Figure 14.15b is a pedigree of the same family, but this time we focus on a recessive trait, attached earlobes. We'll use *f* for the recessive allele and *F* for the dominant allele, which results in free earlobes. As you work your way through the pedigree, notice once again that you can apply what you have learned about Mendelian inheritance to understand the genotypes shown for the family members.

An important application of a pedigree is to help us calculate the probability that a future child will have a particular genotype and phenotype. Suppose that the couple represented in the second generation of Figure 14.15 decides to have one more child. What is the probability that the child will have a widow's peak? This is equivalent to a Mendelian F_1 monohybrid cross (*Ww* × *Ww*), and thus the probability that a child will inherit a dominant allele and have a widow's peak is $\frac{3}{4}$ ($\frac{1}{4}$ *WW* + $\frac{1}{2}$ *Ww*). What is the probability that the child will have attached earlobes? Again, we can treat this as a monohybrid cross (*Ff* × *Ff*), but this time we want to know the chance that the offspring will be homozygous recessive (*ff*). That probability is $\frac{1}{4}$. Finally, what is the chance that the child will have a widow's peak *and* attached earlobes? Assuming that the genes for these two characters are on different chromosomes, the two pairs of alleles will assort independently in this dihybrid cross (*WwFf* × *WwFf*). Thus, we can use the multiplication rule: $\frac{3}{4}$ (chance of widow's peak) × $\frac{1}{4}$ (chance of attached earlobes) = $\frac{3}{16}$ (chance of widow's peak and attached earlobes).

Pedigrees are a more serious matter when the alleles in question cause disabling or deadly diseases instead of innocuous human variations such as hairline or earlobe configuration. However, for disorders inherited as simple Mendelian traits, the same techniques of pedigree analysis apply.

Recessively Inherited Disorders

Thousands of genetic disorders are known to be inherited as simple recessive traits. These disorders range in severity from relatively mild, such as albinism (lack of pigmentation, which results in susceptibility to skin cancers and vision problems), to life-threatening, such as cystic fibrosis.

The Behavior of Recessive Alleles

How can we account for the behavior of alleles that cause recessively inherited disorders? Recall that genes code for proteins of specific function. An allele that causes a genetic disorder (let's call it allele *a*) codes for either a malfunctioning protein or no protein at all. In the case of disorders classified as recessive, heterozygotes (*Aa*) are typically normal in phenotype because one copy of the normal allele (*A*) produces a sufficient amount of the specific protein. Thus, a recessively inherited disorder shows up only in the homozygous individuals (*aa*)

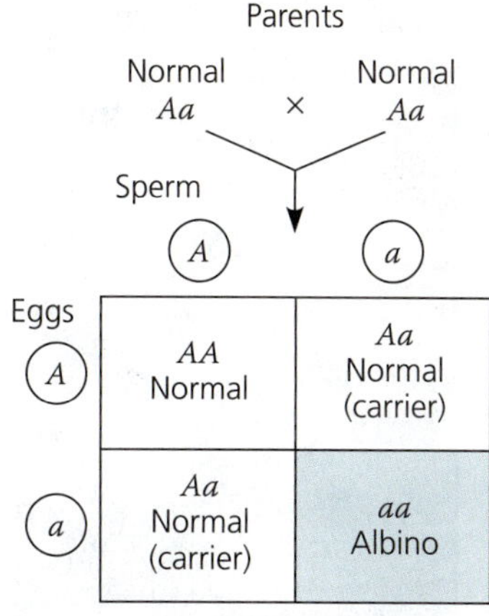

▲ **Figure 14.16 Albinism: a recessive trait.** One of the two sisters shown here has normal coloration; the other is albino. Most recessive homozygotes are born to parents who are carriers of the disorder but themselves have a normal phenotype, the case shown in the Punnett square.

? *What is the probability that the sister with normal coloration is a carrier of the albinism allele?*

who inherit one recessive allele from each parent. Although phenotypically normal with regard to the disorder, heterozygotes may transmit the recessive allele to their offspring and thus are called **carriers**. **Figure 14.16** illustrates these ideas using albinism as an example.

Most people who have recessive disorders are born to parents who are carriers of the disorder but have a normal phenotype, as is the case shown in the Punnett square in Figure 14.16. A mating between two carriers corresponds to a Mendelian F_1 monohybrid cross, so the predicted genotypic ratio for the offspring is 1 *AA* : 2 *Aa* : 1 *aa*. Thus, each child has a ¼ chance of inheriting a double dose of the recessive allele; in the case of albinism, such a child will be albino. From the genotypic ratio, we also can see that out of three offspring with the *normal* phenotype (one *AA* plus two *Aa*), two are predicted to be heterozygous carriers, a ⅔ chance. Recessive homozygotes could also result from *Aa* × *aa* and *aa* × *aa* matings, but if the disorder is lethal before reproductive age or results in sterility (neither of which is true for albinism), no *aa* individuals will reproduce. Even if recessive homozygotes are able to reproduce, such individuals will still account for a much smaller percentage of the population than heterozygous carriers (for reasons we will examine in Chapter 23).

In general, genetic disorders are not evenly distributed among all groups of people. For example, the incidence of Tay-Sachs disease, which we described earlier in this chapter, is disproportionately high among Ashkenazic Jews, Jewish people whose ancestors lived in central Europe. In that population, Tay-Sachs disease occurs in one out of 3,600 births, an incidence about 100 times greater than that among non-Jews or Mediterranean (Sephardic) Jews. This uneven distribution results from the different genetic histories of the world's peoples during less technological times, when populations were more geographically (and hence genetically) isolated.

When a disease-causing recessive allele is rare, it is relatively unlikely that two carriers of the same harmful allele will meet and mate. However, if the man and woman are close relatives (for example, siblings or first cousins), the probability of passing on recessive traits increases greatly. These are called consanguineous ("same blood") matings, and they are indicated in pedigrees by double lines. Because people with recent common ancestors are more likely to carry the same recessive alleles than are unrelated people, it is more likely that a mating of close relatives will produce offspring homozygous for recessive traits—including harmful ones. Such effects can be observed in many types of domesticated and zoo animals that have become inbred.

There is debate among geneticists about the extent to which human consanguinity increases the risk of inherited diseases. Many deleterious alleles have such severe effects that a homozygous embryo spontaneously aborts long before birth. Still, most societies and cultures have laws or taboos forbidding marriages between close relatives. These rules may have evolved out of empirical observation that in most populations, stillbirths and birth defects are more common when parents are closely related. Social and economic factors have also influenced the development of customs and laws against consanguineous marriages.

Cystic Fibrosis

The most common lethal genetic disease in the United States is **cystic fibrosis**, which strikes one out of every 2,500 people of European descent but is much rarer in other groups. Among people of European descent, one out of 25 (4%) are carriers of the cystic fibrosis allele. The normal allele for this gene codes for a membrane protein that functions in the transport of chloride ions between certain cells and the extracellular fluid. These chloride transport channels are defective or absent in the plasma membranes of children who inherit two recessive alleles for cystic fibrosis. The result is an abnormally high concentration of extracellular chloride, which causes the mucus that coats certain cells to become thicker and stickier than normal. The mucus builds up in the pancreas, lungs, digestive tract, and other organs, leading to multiple (pleiotropic) effects, including poor absorption of nutrients from the intestines, chronic bronchitis, and recurrent bacterial infections.

If untreated, most children with cystic fibrosis die before their 5th birthday. But daily doses of antibiotics to prevent infection, gentle pounding on the chest to clear mucus from clogged airways, and other preventive treatments can prolong life. In the United States, more than half of those with cystic fibrosis now survive into their late 20s or even 30s and beyond.

Sickle-Cell Disease: A Genetic Disorder with Evolutionary Implications

EVOLUTION The most common inherited disorder among people of African descent is **sickle-cell disease**, which affects one out of 400 African-Americans. Sickle-cell disease is caused by the substitution of a single amino acid in the hemoglobin protein of red blood cells; in homozygous individuals, all hemoglobin is of the sickle-cell (abnormal) variety. When the oxygen content of an affected individual's blood is low (at high altitudes or under physical stress, for instance), the sickle-cell hemoglobin molecules aggregate into long rods that deform the red cells into a sickle shape (see Figure 5.21). Sickled cells may clump and clog small blood vessels, often leading to other symptoms throughout the body, including physical weakness, pain, organ damage, and even paralysis. Regular blood transfusions can ward off brain damage in children with sickle-cell disease, and new drugs can help prevent or treat other problems, but there is no cure.

Although two sickle-cell alleles are necessary for an individual to manifest full-blown sickle-cell disease, the presence of one sickle-cell allele can affect the phenotype. Thus, at the organismal level, the normal allele is incompletely dominant to the sickle-cell allele. Heterozygotes (carriers), said to have *sickle-cell trait*, are usually healthy, but they may suffer some sickle-cell symptoms during prolonged periods of reduced blood oxygen. At the molecular level, the two alleles are codominant; both normal and abnormal (sickle-cell) hemoglobins are made in heterozygotes.

About one out of ten African-Americans have sickle-cell trait, an unusually high frequency of heterozygotes for an allele with severe detrimental effects in homozygotes. Why haven't evolutionary processes resulted in the disappearance of this allele among this population? One explanation is that having a single copy of the sickle-cell allele reduces the frequency and severity of malaria attacks, especially among young children. The malaria parasite spends part of its life cycle in red blood cells (see Figure 28.10), and the presence of even heterozygous amounts of sickle-cell hemoglobin results in lower parasite densities and hence reduced malaria symptoms. Thus, in tropical Africa, where infection with the malaria parasite is common, the sickle-cell allele confers an advantage to heterozygotes even though it is harmful in the homozygous state. (The balance between these two effects will be discussed in Chapter 23, p. 484.) The relatively high frequency of African-Americans with sickle-cell trait is a vestige of their African roots.

Dominantly Inherited Disorders

Although many harmful alleles are recessive, a number of human disorders are due to dominant alleles. One example is *achondroplasia*, a form of dwarfism that occurs in one of every 25,000 people. Heterozygous individuals have the dwarf phenotype **(Figure 14.17)**. Therefore, all people who are not achondroplastic dwarfs—99.99% of the population—are homozygous for the recessive allele. Like the presence of extra fingers or toes mentioned earlier, achondroplasia is a trait for which the recessive allele is much more prevalent than the corresponding dominant allele.

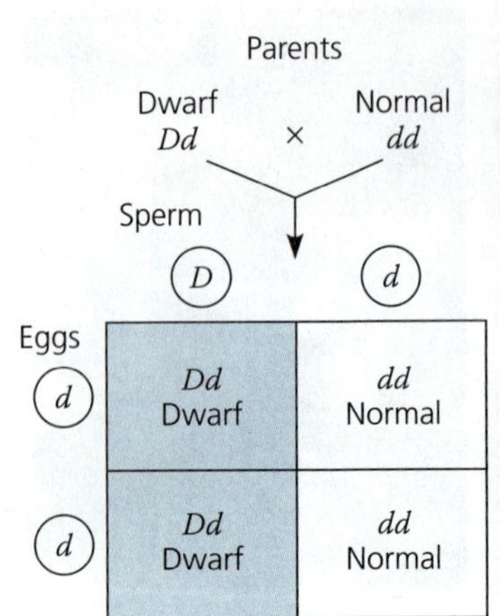

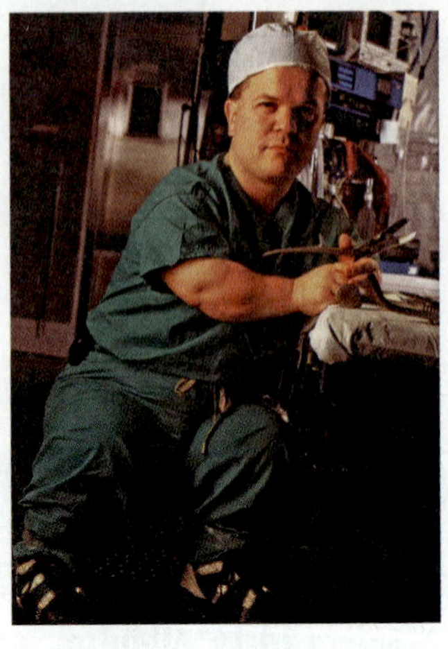

▲ **Figure 14.17 Achondroplasia: a dominant trait.** Dr. Michael C. Ain has achondroplasia, a form of dwarfism caused by a dominant allele. This has inspired his work: He is a specialist in the repair of bone defects caused by achondroplasia and other disorders. The dominant allele (*D*) might have arisen as a mutation in the egg or sperm of a parent or could have been inherited from an affected parent, as shown for an affected father in the Punnett square.

Dominant alleles that cause a lethal disease are much less common than recessive alleles that have lethal effects. All lethal alleles arise by mutations (changes to the DNA) in cells that produce sperm or eggs; presumably, such mutations are equally likely to be recessive or dominant. A lethal recessive allele can be passed from one generation to the next by heterozygous carriers because the carriers themselves have normal phenotypes. A lethal dominant allele, however, often causes the death of afflicted individuals before they can mature and reproduce, so the allele is not passed on to future generations.

Huntington's Disease: A Late-Onset Lethal Disease

The timing of onset of a disease significantly affects its inheritance. A lethal dominant allele is able to be passed on if it causes death at a relatively advanced age. By the time symptoms are evident, the individual with the allele may have already transmitted it to his or her children. For example, **Huntington's disease**, a degenerative disease of the nervous system, is caused by a lethal dominant allele that has no obvious phenotypic effect until the individual is about 35 to 45 years old. Once the deterioration of the nervous system begins, it is irreversible and inevitably fatal. As with other dominant traits, a child born to a parent with the Huntington's disease allele has a 50% chance of inheriting the allele and the disorder (see the Punnett square in Figure 14.17). In the United States, this devastating disease afflicts about one in 10,000 people.

At one time, the onset of symptoms was the only way to know if a person had inherited the Huntington's allele, but this is no longer the case. By analyzing DNA samples from a large family with a high incidence of the disorder, geneticists tracked the Huntington's allele to a locus near the tip of chromosome 4, and the gene was sequenced in 1993. This information led to the development of a test that could detect the presence of the Huntington's allele in an individual's genome. (The methods that make such tests possible are discussed in Chapter 20.) The availability of this test poses an agonizing dilemma for those with a family history of Huntington's disease. Some individuals may want to be tested for this disease before planning a family, whereas others may decide it would be too stressful to find out. Clearly, this is a highly personal decision.

Multifactorial Disorders

The hereditary diseases we have discussed so far are sometimes described as simple Mendelian disorders because they result from abnormality of one or both alleles at a single genetic locus. Many more people are susceptible to diseases that have a multifactorial basis—a genetic component plus a significant environmental influence. Heart disease, diabetes, cancer, alcoholism, certain mental illnesses such as schizophrenia and bipolar disorder, and many other diseases are multifactorial. In many cases, the hereditary component is polygenic. For example, many genes affect cardiovascular health, making some of us more prone than others to heart attacks and strokes. No matter what our genotype, however, our lifestyle has a tremendous effect on phenotype for cardiovascular health and other multifactorial characters. Exercise, a healthful diet, abstinence from smoking, and an ability to handle stressful situations all reduce our risk of heart disease and some types of cancer.

At present, so little is understood about the genetic contributions to most multifactorial diseases that the best public health strategy is to educate people about the importance of environmental factors and to promote healthful behavior.

Genetic Testing and Counseling

Avoiding simple Mendelian disorders is possible when the risk of a particular genetic disorder can be assessed before a child is conceived or during the early stages of the pregnancy. Many hospitals have genetic counselors who can provide information to prospective parents concerned about a family history for a specific disease.

Counseling Based on Mendelian Genetics and Probability Rules

Consider the case of a hypothetical couple, John and Carol. Each had a brother who died from the same recessively inherited lethal disease. Before conceiving their first child, John and Carol seek genetic counseling to determine the risk of having a child with the disease. From the information about their brothers, we know that both parents of John and both parents of Carol must have been carriers of the recessive allele. Thus, John and Carol are both products of *Aa* × *Aa* crosses, where *a* symbolizes the allele that causes this particular disease. We also know that John and Carol are not homozygous recessive (*aa*), because they do not have the disease. Therefore, their genotypes are either *AA* or *Aa*.

Given a genotypic ratio of 1 *AA* : 2 *Aa* : 1 *aa* for offspring of an *Aa* × *Aa* cross, John and Carol each have a ⅔ chance of being carriers (*Aa*). According to the rule of multiplication, the overall probability of their firstborn having the disorder is ⅔ (the chance that John is a carrier) times ⅔ (the chance that Carol is a carrier) times ¼ (the chance of two carriers having a child with the disease), which equals ⅑. Suppose that Carol and John decide to have a child—after all, there is an ⅞ chance that their baby will not have the disorder. If, despite these odds, their child is born with the disease, then we would know that *both* John and Carol are, in fact, carriers (*Aa* genotype). If both John and Carol are carriers, there is a ¼ chance that any subsequent child this couple has will have the disease. The probability is higher for subsequent children because the diagnosis of the disease in the first child established that both parents are carriers, not because the genotype of the first child affects in any way that of future children.

When we use Mendel's laws to predict possible outcomes of matings, it is important to remember that each child represents an independent event in the sense that its genotype is unaffected by the genotypes of older siblings. Suppose that John and Carol have three more children, and *all three* have the hypothetical hereditary disease. There is only one chance in 64 (¼ × ¼ × ¼) that such an outcome will occur. Despite this run of misfortune, the chance that still another child of this couple will have the disease remains ¼.

Tests for Identifying Carriers

Most children with recessive disorders are born to parents with normal phenotypes. The key to accurately assessing the genetic risk for a particular disease is therefore to find out whether the prospective parents are heterozygous carriers of the recessive allele. For an increasing number of heritable disorders, tests are available that can distinguish individuals of normal phenotype who are dominant homozygotes from those who are heterozygous carriers (**Figure 14.18**, on the next page). There are now tests that can identify carriers of the alleles for Tay-Sachs disease, sickle-cell disease, and the most common form of cystic fibrosis.

These tests for identifying carriers enable people with family histories of genetic disorders to make informed decisions about having children, but raise other issues. Could carriers be denied health or life insurance or lose the jobs providing those benefits, even though they themselves are healthy? The Genetic Information Nondiscrimination Act, signed into law in the United States in 2008, allays these concerns by prohibiting discrimination in employment or insurance coverage based on genetic test results. A question that remains is whether sufficient genetic counseling is available to help large numbers of individuals understand their genetic test results. Even when test results are clearly understood, affected individuals may still face difficult decisions. Advances in biotechnology offer

▼ **Figure 14.18**

IMPACT

Genetic Testing

Since the sequencing of the human genome was completed in 2003, there has been a virtual explosion in the number and kinds of DNA-based genetic tests. As of 2010, genetic testing for over 2,000 different disease-causing alleles is available.

WHY IT MATTERS For prospective parents with a family history of a recessive or late-onset dominant disorder, deciding whether to have children can be a difficult decision. Genetic testing can eliminate some of the uncertainty and allow better predictions of the probabilities and risks involved.

FURTHER READING Designing rules for designer babies, *Scientific American* 300:29 (2009).

WHAT IF? If one parent tests positive and the other tests negative for a recessive allele associated with a disorder, what is the probability that their first child will have the disorder? That their first child will be a carrier? That, if their first child is a carrier, the second will also be a carrier?

the potential to reduce human suffering, but along with them come ethical issues that require conscientious deliberation.

Fetal Testing

Suppose a couple expecting a child learns that they are both carriers of the Tay-Sachs allele. In the 14th–16th week of pregnancy, tests performed along with a technique called **amniocentesis** can determine whether the developing fetus has Tay-Sachs disease **(Figure 14.19a)**. In this procedure, a physician inserts a needle into the uterus and extracts about 10 mL of amniotic fluid, the liquid that bathes the fetus. Some genetic disorders can be detected from the presence of certain molecules in the amniotic fluid itself. Tests for other disorders, including Tay-Sachs disease, are performed on the DNA of cells cultured in the laboratory, descendants of fetal cells sloughed off into the amniotic fluid. A karyotype of these cultured cells can also identify certain chromosomal defects (see Figure 13.3).

In an alternative technique called **chorionic villus sampling (CVS)**, a physician inserts a narrow tube through the cervix into the uterus and suctions out a tiny sample of tissue from the placenta, the organ that transmits nutrients and fetal wastes between the fetus and the mother **(Figure 14.19b)**. The cells of the chorionic villi of the placenta, the portion sampled, are derived from the fetus and have the same genotype and DNA sequence as the new individual. These cells are proliferating rapidly enough to allow karyotyping to be carried out immediately. This rapid analysis represents an advantage over amniocentesis, in which the cells must be cultured for several weeks before karyotyping. Another advantage of CVS is that it can be performed as early as the 8th–10th week of pregnancy.

Recently, medical scientists have developed methods for isolating fetal cells, or even fetal DNA, that have escaped into the mother's blood. Although very few are present, the cells can be cultured and tested, and the fetal DNA can also be analyzed.

Imaging techniques allow a physician to examine a fetus directly for major anatomical abnormalities that might not show up in genetic tests. In the *ultrasound* technique, reflected sound waves are used to produce an image of the fetus by a simple noninvasive procedure. In *fetoscopy*, a needle-thin tube containing a viewing scope and fiber optics (to transmit light) is inserted into the uterus.

Ultrasound and isolation of fetal cells or DNA from maternal blood pose no known risk to either mother or fetus, while the other procedures can cause complications in a small percentage of cases. Amniocentesis or CVS for diagnostic testing is generally offered to women over age 35, due to their increased risk of bearing a child with Down syndrome, and may also be offered to younger women if there are known concerns. If the fetal tests reveal a serious disorder, the parents face the difficult choice of either terminating the pregnancy or preparing to care for a child with a genetic disorder.

Newborn Screening

Some genetic disorders can be detected at birth by simple biochemical tests that are now routinely performed in most hospitals in the United States. One common screening program is for phenylketonuria (PKU), a recessively inherited disorder that occurs in about one out of every 10,000–15,000 births in the United States. Children with this disease cannot properly metabolize the amino acid phenylalanine. This compound and its by-product, phenylpyruvate, can accumulate to toxic levels in the blood, causing severe intellectual disability (mental retardation). However, if PKU is detected in the newborn, a special diet low in phenylalanine will usually allow normal development. (Among many other substances, this diet excludes the artificial sweetener aspartame, which contains phenylalanine.) Unfortunately, few other genetic disorders are treatable at present.

Fetal and newborn screening for serious inherited diseases, tests for identifying carriers, and genetic counseling all rely on the Mendelian model of inheritance. We owe the "gene idea"—the concept of heritable factors transmitted according

▼ **Figure 14.19 Testing a fetus for genetic disorders.** Biochemical tests may detect substances associated with particular disorders, and genetic testing can detect many genetic abnormalities. Karyotyping shows whether the chromosomes of the fetus are normal in number and appearance.

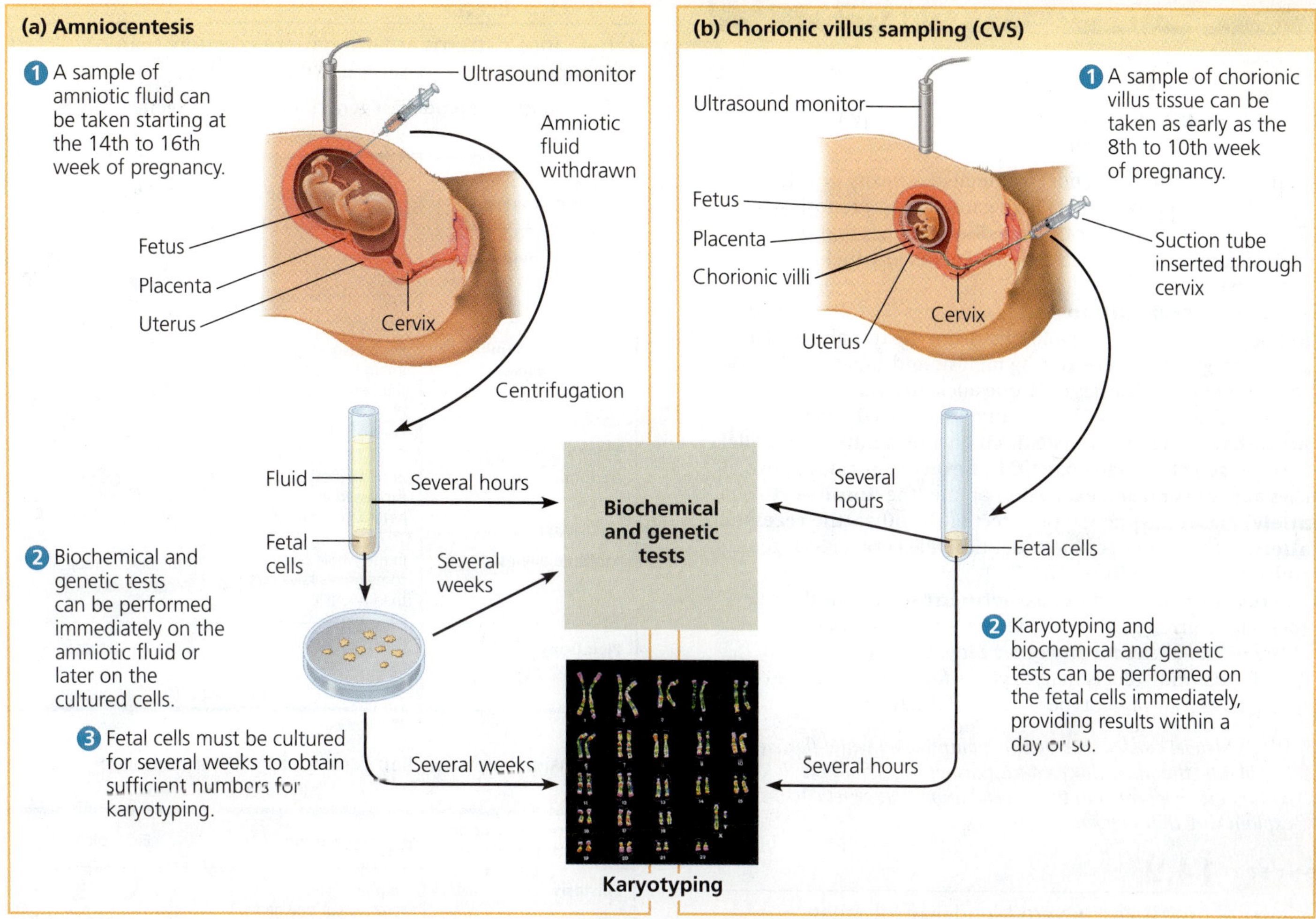

to simple rules of chance—to the elegant quantitative experiments of Gregor Mendel. The importance of his discoveries was overlooked by most biologists until early in the 20th century, decades after he reported his findings. In the next chapter, you will learn how Mendel's laws have their physical basis in the behavior of chromosomes during sexual life cycles and how the synthesis of Mendelism and a chromosome theory of inheritance catalyzed progress in genetics.

CONCEPT CHECK 14.4

1. Beth and Tom each have a sibling with cystic fibrosis, but neither Beth nor Tom nor any of their parents have the disease. Calculate the probability that if this couple has a child, the child will have cystic fibrosis. What would be the probability if a test revealed that Tom is a carrier but Beth is not? Explain your answers.
2. **MAKE CONNECTIONS** Review Figures 5.16, 5.20, and 5.21 (pp. 79 and 82–84). Explain how the change of a single amino acid in hemoglobin leads to the aggregation of hemoglobin into long rods.
3. Joan was born with six toes on each foot, a dominant trait called polydactyly. Two of her five siblings and her mother, but not her father, also have extra digits. What is Joan's genotype for the number-of-digits character? Explain your answer. Use *D* and *d* to symbolize the alleles for this character.
4. **MAKE CONNECTIONS** In Table 14.1 (p. 265), note the phenotypic ratio of the dominant to recessive trait in the F_2 generation for the monohybrid cross involving flower color. Then determine the phenotypic ratio for the offspring of the second-generation couple in Figure 14.15b. What accounts for the difference in the two ratios?

For suggested answers, see Appendix A.

14 CHAPTER REVIEW

SUMMARY OF KEY CONCEPTS

CONCEPT 14.1

Mendel used the scientific approach to identify two laws of inheritance (pp. 262–269)

- In the 1860s, Gregor Mendel formulated a theory of inheritance based on experiments with garden peas, proposing that parents pass on to their offspring discrete genes that retain their identity through generations. This theory includes two "laws."
- The **law of segregation** states that genes have alternative forms, or **alleles**. In a diploid organism, the two alleles of a gene segregate (separate) during meiosis and gamete formation; each sperm or egg carries only one allele of each pair. This law explains the 3:1 ratio of F_2 phenotypes observed when **monohybrids** self-pollinate. Each organism inherits one allele for each gene from each parent. In **heterozygotes**, the two alleles are different, and expression of one (the **dominant allele**) masks the phenotypic effect of the other (the **recessive allele**). **Homozygotes** have identical alleles of a given gene and are **true-breeding**.
- The **law of independent assortment** states that the pair of alleles for a given gene segregates into gametes independently of the pair of alleles for any other gene. In a cross between **dihybrids** (individuals heterozygous for two genes), the offspring have four phenotypes in a 9:3:3:1 ratio.

? *When Mendel crossed true-breeding purple- and white-flowered pea plants, the white-flowered trait disappeared from the F_1 generation but reappeared in the F_2 generation. Use genetic terms to explain why that happened.*

CONCEPT 14.2

The laws of probability govern Mendelian inheritance (pp. 269–271)

- The **multiplication rule** states that the probability of two or more events occurring together is equal to the product of the individual probabilities of the independent single events. The **addition rule** states that the probability of an event that can occur in two or more independent, mutually exclusive ways is the sum of the individual probabilities.
- The rules of probability can be used to solve complex genetics problems. A dihybrid or other multicharacter cross is equivalent to two or more independent monohybrid crosses occurring simultaneously. In calculating the chances of the various offspring genotypes from such crosses, each character is first considered separately and then the individual probabilities are multiplied.

DRAW IT *Redraw the Punnett square on the right side of Figure 14.8 as two smaller monohybrid Punnett squares, one for each gene. Below each square, list the fractions of each phenotype produced. Use the rule of multiplication to compute the overall fraction of each possible dihybrid phenotype. What is the phenotypic ratio?*

CONCEPT 14.3

Inheritance patterns are often more complex than predicted by simple Mendelian genetics (pp. 271–275)

- Extensions of Mendelian genetics for a single gene:

Relationship among alleles of a single gene	Description	Example
Complete dominance of one allele	Heterozygous phenotype same as that of homozygous dominant	*PP* *Pp*
Incomplete dominance of either allele	Heterozygous phenotype intermediate between the two homozygous phenotypes	C^RC^R C^RC^W C^WC^W
Codominance	Both phenotypes expressed in heterozygotes	I^AI^B
Multiple alleles	In the whole population, some genes have more than two alleles	ABO blood group alleles I^A, I^B, i
Pleiotropy	One gene is able to affect multiple phenotypic characters	Sickle-cell disease

- Extensions of Mendelian genetics for two or more genes:

Relationship among two or more genes	Description	Example
Epistasis	The phenotypic expression of one gene affects that of another	*BbEe* × *BbEe*; *BE*, *bE*, *Be*, *be*; 9 : 3 : 4
Polygenic inheritance	A single phenotypic character is affected by two or more genes	*AaBbCc* × *AaBbCc*

- The expression of a genotype can be affected by environmental influences, the "nurture" in nature versus nurture. The phenotypic range of a particular genotype is called its **norm of reaction**.

Polygenic characters that are also influenced by the environment are called **multifactorial** characters.

- An organism's overall phenotype, including its physical appearance, internal anatomy, physiology, and behavior, reflects its overall genotype and unique environmental history. Even in more complex inheritance patterns, Mendel's fundamental laws of segregation and independent assortment still apply.

? *Which of the following are demonstrated by the inheritance patterns of the ABO blood group alleles: complete dominance, incomplete dominance, codominance, multiple alleles, pleiotropy, epistasis, and/or polygenic inheritance? Explain how, for each of your answers.*

CONCEPT 14.4

Many human traits follow Mendelian patterns of inheritance (pp. 275–281)

- Analysis of family **pedigrees** can be used to deduce the possible genotypes of individuals and make predictions about future offspring. Predictions are statistical probabilities rather than certainties.
- Many genetic disorders are inherited as simple recessive traits. Most affected (homozygous recessive) individuals are children of phenotypically normal, heterozygous **carriers**.
- Lethal dominant alleles are eliminated from the population if affected people die before reproducing. Nonlethal dominant alleles and lethal ones that strike relatively late in life can be inherited in a Mendelian way.
- Many human diseases are multifactorial—that is, they have both genetic and environmental components and do not follow simple Mendelian inheritance patterns.
- Using family histories, genetic counselors help couples determine the probability that their children will have genetic disorders. Genetic testing of prospective parents to reveal whether they are carriers of recessive alleles associated with specific disorders has become widely available. **Amniocentesis** and **chorionic villus sampling** can indicate whether a suspected genetic disorder is present in a fetus. Other genetic tests can be performed after birth.

? *Both members of a couple know that they are carriers of the cystic fibrosis allele. None of their three children has cystic fibrosis, but any one of them might be a carrier. They would like to have a fourth child but are worried that it would very likely have the disease, since the first three do not. What would you tell the couple? Would it remove some more uncertainty in their prediction if they could find out from genetic tests whether the three children are carriers?*

TIPS FOR GENETICS PROBLEMS

1. Write down symbols for the alleles. (These may be given in the problem.) When represented by single letters, the dominant allele is uppercase and the recessive is lowercase.
2. Write down the possible genotypes, as determined by the phenotype.
   a. If the phenotype is that of the dominant trait (for example, purple flowers), then the genotype is either homozygous dominant or heterozygous (*PP* or *Pp*, in this example).
   b. If the phenotype is that of the recessive trait, the genotype must be homozygous recessive (for example, *pp*).
   c. If the problem says "true-breeding," the genotype is homozygous.
3. Determine what the problem is asking for. If asked to do a cross, write it out in the form [Genotype] × [Genotype], using the alleles you've decided on.
4. To figure out the outcome of a cross, set up a Punnett square.
   a. Put the gametes of one parent at the top and those of the other on the left. To determine the allele(s) in each gamete for a given genotype, set up a systematic way to list all the possibilities. (Remember, each gamete has one allele of each gene.) Note that there are 2^n possible types of gametes, where n is the number of gene loci that are heterozygous. For example, an individual with genotype *AaBbCc* would produce $2^3 = 8$ types of gametes. Write the genotypes of the gametes in circles above the columns and to the left of the rows.
   b. Fill in the Punnett square as if each possible sperm were fertilizing each possible egg, making all of the possible offspring. In a cross of *AaBbCc* × *AaBbCc*, for example, the Punnett square would have 8 columns and 8 rows, so there are 64 different offspring; you would know the genotype of each and thus the phenotype. Count genotypes and phenotypes to obtain the genotypic and phenotypic ratios. Because the Punnett square is so large, this method is not the most efficient. Instead, see tip 5.
5. You can use the rules of probability if the Punnett square would be too big. (For example, see the question at the end of the summary for Concept 14.2 and question 7 on the next page.) You can consider each gene separately (see pp. 270–271).
6. If, instead, the problem gives you the phenotypic ratios of offspring, but not the genotypes of the parents in a given cross, the phenotypes can help you deduce the parents' unknown genotypes.
   a. For example, if ½ of the offspring have the recessive phenotype and ½ the dominant, you know that the cross was between a heterozygote and a homozygous recessive.
   b. If the ratio is 3:1, the cross was between two heterozygotes.
   c. If two genes are involved and you see a 9:3:3:1 ratio in the offspring, you know that each parent is heterozygous for both genes. Caution: Don't assume that the reported numbers will exactly equal the predicted ratios. For example, if there are 13 offspring with the dominant trait and 11 with the recessive, assume that the ratio is one dominant to one recessive.
7. For pedigree problems, use the tips in Figure 14.15 and below to determine what kind of trait is involved.
   a. If parents without the trait have offspring with the trait, the trait must be recessive and the parents both carriers.
   b. If the trait is seen in every generation, it is most likely dominant (see the next possibility, though).
   c. If both parents have the trait, then in order for it to be recessive, all offspring must show the trait.
   d. To determine the likely genotype of a certain individual in a pedigree, first label the genotypes of all the family members you can. Even if some of the genotypes are incomplete, label what you do know. For example, if an individual has the dominant phenotype, the genotype must be *AA* or *Aa*; you can write this as *A_*. Try different possibilities to see which fits the results. Use the rules of probability to calculate the probability of each possible genotype being the correct one.

TEST YOUR UNDERSTANDING

LEVEL 1: KNOWLEDGE/COMPREHENSION

1. Match each term on the left with a statement on the right.

Term	Statement
__ Gene	a. Has no effect on phenotype in a heterozygote
__ Allele	b. A variant for a character
__ Character	c. Having two identical alleles for a gene
__ Trait	d. A cross between individuals heterozygous for a single character
__ Dominant allele	e. An alternative version of a gene
__ Recessive allele	f. Having two different alleles for a gene
__ Genotype	g. A heritable feature that varies among individuals
__ Phenotype	h. An organism's appearance or observable traits
__ Homozygous	i. A cross between an individual with an unknown genotype and a homozygous recessive individual
__ Heterozygous	j. Determines phenotype in a heterozygote
__ Testcross	k. The genetic makeup of an individual
__ Monohybrid cross	l. A heritable unit that determines a character; can exist in different forms

2. **DRAW IT** Two pea plants heterozygous for the characters of pod color and pod shape are crossed. Draw a Punnett square to determine the phenotypic ratios of the offspring.

3. In some plants, a true-breeding, red-flowered strain gives all pink flowers when crossed with a white-flowered strain: C^RC^R (red) $\times$ C^WC^W (white) $\rightarrow$ C^RC^W (pink). If flower position (axial or terminal) is inherited as it is in peas (see Table 14.1), what will be the ratios of genotypes and phenotypes of the F_1 generation resulting from the following cross: axial-red (true-breeding) $\times$ terminal-white? What will be the ratios in the F_2 generation?

4. A man with type A blood marries a woman with type B blood. Their child has type O blood. What are the genotypes of these three individuals? What genotypes, and in what frequencies, would you expect in future offspring from this marriage?

5. A man has six fingers on each hand and six toes on each foot. His wife and their daughter have the normal number of digits. Remember that extra digits is a dominant trait. What fraction of this couple's children would be expected to have extra digits?

6. **DRAW IT** A pea plant heterozygous for inflated pods (*Ii*) is crossed with a plant homozygous for constricted pods (*ii*). Draw a Punnett square for this cross. Assume that pollen comes from the *ii* plant.

LEVEL 2: APPLICATION/ANALYSIS

7. Flower position, stem length, and seed shape are three characters that Mendel studied. Each is controlled by an independently assorting gene and has dominant and recessive expression as follows:

Character	*Dominant*	*Recessive*
Flower position	Axial (*A*)	Terminal (*a*)
Stem length	Tall (*T*)	Dwarf (*t*)
Seed shape	Round (*R*)	Wrinkled (*r*)

If a plant that is heterozygous for all three characters is allowed to self-fertilize, what proportion of the offspring would you expect to be as follows? (*Note*: Use the rules of probability instead of a huge Punnett square.)

(a) homozygous for the three dominant traits
(b) homozygous for the three recessive traits
(c) heterozygous for all three characters
(d) homozygous for axial and tall, heterozygous for seed shape

8. A black guinea pig crossed with an albino guinea pig produces 12 black offspring. When the albino is crossed with a second black one, 7 blacks and 5 albinos are obtained. What is the best explanation for this genetic outcome? Write genotypes for the parents, gametes, and offspring.

9. In sesame plants, the one-pod condition (*P*) is dominant to the three-pod condition (*p*), and normal leaf (*L*) is dominant to wrinkled leaf (*l*). Pod type and leaf type are inherited independently. Determine the genotypes for the two parents for all possible matings producing the following offspring:

(a) 318 one-pod, normal leaf and 98 one-pod, wrinkled leaf
(b) 323 three-pod, normal leaf and 106 three-pod, wrinkled leaf
(c) 401 one-pod, normal leaf
(d) 150 one-pod, normal leaf, 147 one-pod, wrinkled leaf, 51 three-pod, normal leaf, and 48 three-pod, wrinkled leaf
(e) 223 one-pod, normal leaf, 72 one-pod, wrinkled leaf, 76 three-pod, normal leaf, and 27 three-pod, wrinkled leaf

10. Phenylketonuria (PKU) is an inherited disease caused by a recessive allele. If a woman and her husband, who are both carriers, have three children, what is the probability of each of the following?

(a) All three children are of normal phenotype.
(b) One or more of the three children have the disease.
(c) All three children have the disease.
(d) At least one child is phenotypically normal.
(*Note*: It will help to remember that the probabilities of all possible outcomes always add up to 1.)

11. The genotype of F_1 individuals in a tetrahybrid cross is *AaBbCcDd*. Assuming independent assortment of these four genes, what are the probabilities that F_2 offspring will have the following genotypes?

(a) *aabbccdd*
(b) *AaBbCcDd*
(c) *AABBCCDD*
(d) *AaBBccDd*
(e) *AaBBCCdd*

12. What is the probability that each of the following pairs of parents will produce the indicated offspring? (Assume independent assortment of all gene pairs.)

(a) $AABBCC \times aabbcc \rightarrow AaBbCc$
(b) $AABbCc \times AaBbCc \rightarrow AAbbCC$
(c) $AaBbCc \times AaBbCc \rightarrow AaBbCc$
(d) $aaBbCC \times AABbcc \rightarrow AaBbCc$

13. Karen and Steve each have a sibling with sickle-cell disease. Neither Karen nor Steve nor any of their parents have the disease, and none of them have been tested to see if they have the sickle-cell trait. Based on this incomplete information, calculate the probability that if this couple has a child, the child will have sickle-cell disease.

14. In 1981, a stray black cat with unusual rounded, curled-back ears was adopted by a family in California. Hundreds of descendants of the cat have since been born, and cat fanciers hope to develop the curl cat into a show breed. Suppose you

owned the first curl cat and wanted to develop a true-breeding variety. How would you determine whether the curl allele is dominant or recessive? How would you obtain true-breeding curl cats? How could you be sure they are true-breeding?

15. Imagine that a newly discovered, recessively inherited disease is expressed only in individuals with type O blood, although the disease and blood group are independently inherited. A normal man with type A blood and a normal woman with type B blood have already had one child with the disease. The woman is now pregnant for a second time. What is the probability that the second child will also have the disease? Assume that both parents are heterozygous for the gene that causes the disease.

16. In tigers, a recessive allele causes an absence of fur pigmentation (a white tiger) and a cross-eyed condition. If two phenotypically normal tigers that are heterozygous at this locus are mated, what percentage of their offspring will be cross-eyed? What percentage of cross-eyed tigers will be white?

17. In maize (corn) plants, a dominant allele *I* inhibits kernel color, while the recessive allele *i* permits color when homozygous. At a different locus, the dominant allele *P* causes purple kernel color, while the homozygous recessive genotype *pp* causes red kernels. If plants heterozygous at both loci are crossed, what will be the phenotypic ratio of the offspring?

18. The pedigree below traces the inheritance of alkaptonuria, a biochemical disorder. Affected individuals, indicated here by the colored circles and squares, are unable to metabolize a substance called alkapton, which colors the urine and stains body tissues. Does alkaptonuria appear to be caused by a dominant allele or by a recessive allele? Fill in the genotypes of the individuals whose genotypes can be deduced. What genotypes are possible for each of the other individuals?

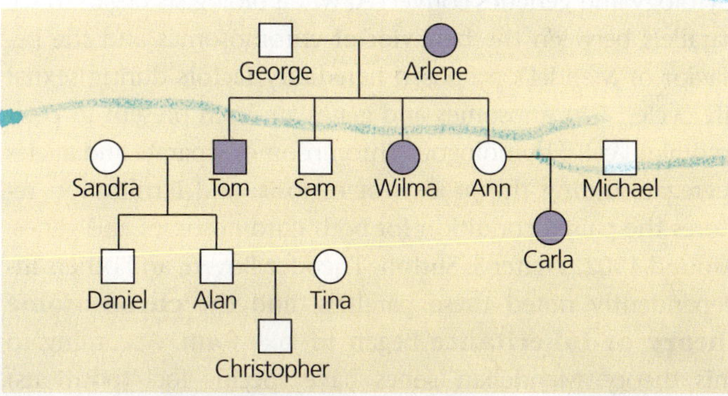

19. Imagine that you are a genetic counselor, and a couple planning to start a family comes to you for information. Charles was married once before, and he and his first wife had a child with cystic fibrosis. The brother of his current wife, Elaine, died of cystic fibrosis. What is the probability that Charles and Elaine will have a baby with cystic fibrosis? (Neither Charles, Elaine, nor their parents have cystic fibrosis.)

20. In mice, black fur (*B*) is dominant to white (*b*). At a different locus, a dominant allele (*A*) produces a band of yellow just below the tip of each hair in mice with black fur. This gives a frosted appearance known as agouti. Expression of the recessive allele (*a*) results in a solid coat color. If mice that are heterozygous at both loci are crossed, what is the expected phenotypic ratio of their offspring?

LEVEL 3: SYNTHESIS/EVALUATION

21. **EVOLUTION CONNECTION**
Over the past half century, there has been a trend in the United States and other developed countries for people to marry and start families later in life than did their parents and grandparents. What effects might this trend have on the incidence (frequency) of late-acting dominant lethal alleles in the population?

22. **SCIENTIFIC INQUIRY**
You are handed a mystery pea plant with tall stems and axial flowers and asked to determine its genotype as quickly as possible. You know that the allele for tall stems (*T*) is dominant to that for dwarf stems (*t*) and that the allele for axial flowers (*A*) is dominant to that for terminal flowers (*a*).
(a) What are *all* the possible genotypes for your mystery plant?
(b) Describe the *one* cross you would do, out in your garden, to determine the exact genotype of your mystery plant.
(c) While waiting for the results of your cross, you predict the results for each possible genotype listed in part a. How do you do this? Why is this not called "performing a cross"?
(d) Explain how the results of your cross and your predictions will help you learn the genotype of your mystery plant.

23. **SCIENCE, TECHNOLOGY, AND SOCIETY**
Imagine that one of your parents has Huntington's disease. What is the probability that you, too, will someday manifest the disease? There is no cure for Huntington's. Would you want to be tested for the Huntington's allele? Why or why not?

24. **WRITE ABOUT A THEME**
The Genetic Basis of Life The continuity of life is based on heritable information in the form of DNA. In a short essay (100–150 words), explain how the passage of genes from parents to offspring, in the form of particular alleles, ensures perpetuation of parental traits in offspring and, at the same time, genetic variation among offspring. Use genetic terms in your explanation.

For selected answers, see Appendix A.

MasteringBIOLOGY www.masteringbiology.com

1. MasteringBiology® Assignments
Tutorials Determining Genotype: Pea Pod Color • Inheritance of Fur Color in Mice • Pedigree Analysis: Galactosemia
Activities Monohybrid Cross • Dihybrid Cross • Mendel's Experiments • The Principle of Independent Assortment • Gregor's Garden • Incomplete Dominance
Questions Student Misconceptions • Reading Quiz • Multiple Choice • End-of-Chapter

2. eText
Read your book online, search, take notes, highlight text, and more.

3. The Study Area
Practice Tests • Cumulative Test • *BioFlix* 3-D Animations • MP3 Tutor Sessions • Videos • Activities • Investigations • Lab Media • Audio Glossary • Word Study Tools • Art

15

The Chromosomal Basis of Inheritance

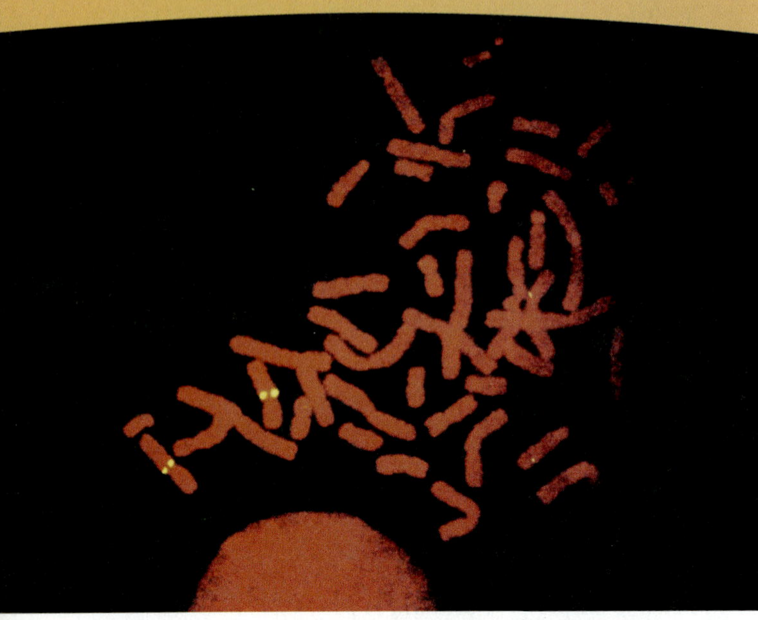

▲ **Figure 15.1 Where are Mendel's hereditary factors located in the cell?**

KEY CONCEPTS

15.1 Mendelian inheritance has its physical basis in the behavior of chromosomes

15.2 Sex-linked genes exhibit unique patterns of inheritance

15.3 Linked genes tend to be inherited together because they are located near each other on the same chromosome

15.4 Alterations of chromosome number or structure cause some genetic disorders

15.5 Some inheritance patterns are exceptions to standard Mendelian inheritance

OVERVIEW

Locating Genes Along Chromosomes

Gregor Mendel's "hereditary factors" were purely an abstract concept when he proposed their existence in 1860. At that time, no cellular structures were known that could house these imaginary units. Even after chromosomes were first observed, many biologists remained skeptical about Mendel's laws of segregation and independent assortment until there was sufficient evidence that these principles of heredity had a physical basis in chromosomal behavior.

Today, we know that genes—Mendel's "factors"—are located along chromosomes. We can see the location of a particular gene by tagging chromosomes with a fluorescent dye that highlights that gene. For example, the four yellow dots in **Figure 15.1** mark the locus of a specific gene on the sister chromatids of a homologous pair of replicated human chromosomes. This chapter extends what you learned in the past two chapters: We describe the chromosomal basis for the transmission of genes from parents to offspring, along with some important exceptions to the standard mode of inheritance.

CONCEPT 15.1

Mendelian inheritance has its physical basis in the behavior of chromosomes

Using improved techniques of microscopy, cytologists worked out the process of mitosis in 1875 and meiosis in the 1890s. Cytology and genetics converged when biologists began to see parallels between the behavior of chromosomes and the behavior of Mendel's proposed hereditary factors during sexual life cycles: Chromosomes and genes are both present in pairs in diploid cells; homologous chromosomes separate and alleles segregate during the process of meiosis; and fertilization restores the paired condition for both chromosomes and genes. Around 1902, Walter S. Sutton, Theodor Boveri, and others independently noted these parallels, and the **chromosome theory of inheritance** began to take form. According to this theory, Mendelian genes have specific loci (positions) along chromosomes, and it is the chromosomes that undergo segregation and independent assortment.

Figure 15.2 shows that the behavior of homologous chromosomes during meiosis can account for the segregation of the alleles at each genetic locus to different gametes. The figure also shows that the behavior of nonhomologous chromosomes can account for the independent assortment of the alleles for two or more genes located on different chromosomes. By carefully studying this figure, which traces the same dihybrid pea cross you learned about in Figure 14.8, you can see how the behavior of chromosomes during meiosis in the F_1 generation and subsequent random fertilization give rise to the F_2 phenotypic ratio observed by Mendel.

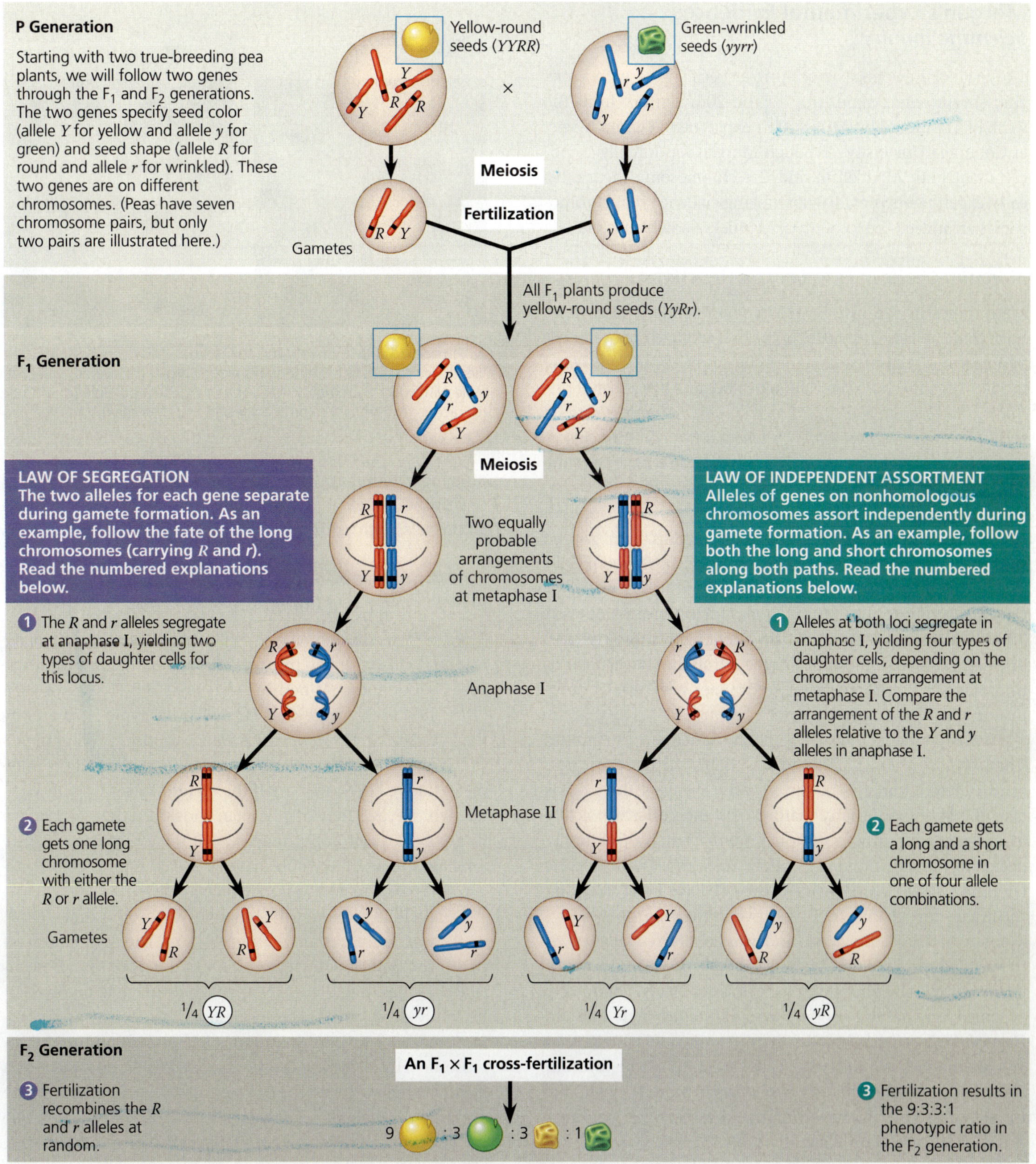

▲ **Figure 15.2 The chromosomal basis of Mendel's laws.** Here we correlate the results of one of Mendel's dihybrid crosses (see Figure 14.8) with the behavior of chromosomes during meiosis (see Figure 13.8). The arrangement of chromosomes at metaphase I of meiosis and their movement during anaphase I account for the segregation and independent assortment of the alleles for seed color and shape. Each cell that undergoes meiosis in an F_1 plant produces two kinds of gametes. If we count the results for all cells, however, each F_1 plant produces equal numbers of all four kinds of gametes because the alternative chromosome arrangements at metaphase I are equally likely.

? *If you crossed an F_1 plant with a plant that was homozygous recessive for both genes (yyrr), how would the phenotypic ratio of the offspring compare with the 9:3:3:1 ratio seen here?*

Morgan's Experimental Evidence: *Scientific Inquiry*

The first solid evidence associating a specific gene with a specific chromosome came early in the 20th century from the work of Thomas Hunt Morgan, an experimental embryologist at Columbia University. Although Morgan was initially skeptical about both Mendelism and the chromosome theory, his early experiments provided convincing evidence that chromosomes are indeed the location of Mendel's heritable factors.

Morgan's Choice of Experimental Organism

Many times in the history of biology, important discoveries have come to those insightful or lucky enough to choose an experimental organism suitable for the research problem being tackled. Mendel chose the garden pea because a number of distinct varieties were available. For his work, Morgan selected a species of fruit fly, *Drosophila melanogaster*, a common insect that feeds on the fungi growing on fruit. Fruit flies are prolific breeders; a single mating will produce hundreds of offspring, and a new generation can be bred every two weeks. Morgan's laboratory began using this convenient organism for genetic studies in 1907 and soon became known as "the fly room."

Another advantage of the fruit fly is that it has only four pairs of chromosomes, which are easily distinguishable with a light microscope. There are three pairs of autosomes and one pair of sex chromosomes. Female fruit flies have a pair of homologous X chromosomes, and males have one X chromosome and one Y chromosome.

While Mendel could readily obtain different pea varieties from seed suppliers, Morgan was probably the first person to want different varieties of the fruit fly. He faced the tedious task of carrying out many matings and then microscopically inspecting large numbers of offspring in search of naturally occurring variant individuals. After many months of this, he lamented, "Two years' work wasted. I have been breeding those flies for all that time and I've got nothing out of it." Morgan persisted, however, and was finally rewarded with the discovery of a single male fly with white eyes instead of the usual red. The phenotype for a character most commonly observed in natural populations, such as red eyes in *Drosophila*, is called the **wild type** (**Figure 15.3**). Traits that are alternatives to the wild type, such as white eyes in *Drosophila*, are called *mutant phenotypes* because they are due to alleles assumed to have originated as changes, or mutations, in the wild-type allele.

Morgan and his students invented a notation for symbolizing alleles in *Drosophila* that is still widely used for fruit flies. For a given character in flies, the gene takes its symbol from the first mutant (non–wild type) discovered. Thus, the allele for white eyes in *Drosophila* is symbolized by w. A superscript + identifies the allele for the wild-type trait—w^+ for the allele for red eyes, for example. Over the years, a variety

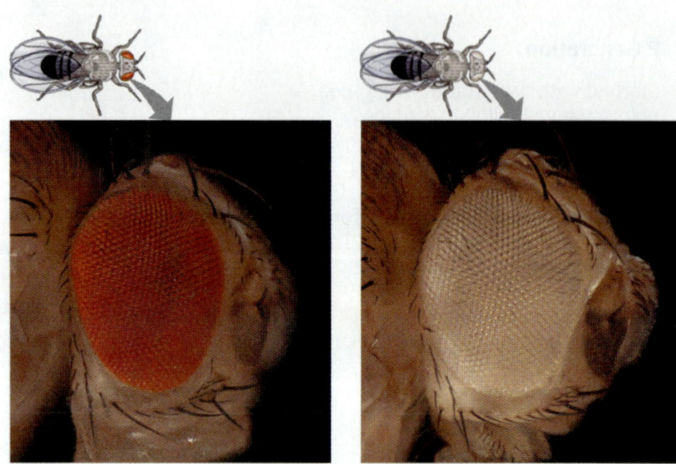

▲ **Figure 15.3 Morgan's first mutant.** Wild-type *Drosophila* flies have red eyes (left). Among his flies, Morgan discovered a mutant male with white eyes (right). This variation made it possible for Morgan to trace a gene for eye color to a specific chromosome (LMs).

of gene notation systems have been developed for different organisms. For example, human genes are usually written in all capitals, such as *HD* for the allele for Huntington's disease.

Correlating Behavior of a Gene's Alleles with Behavior of a Chromosome Pair

Morgan mated his white-eyed male fly with a red-eyed female. All the F_1 offspring had red eyes, suggesting that the wild-type allele is dominant. When Morgan bred the F_1 flies to each other, he observed the classical 3:1 phenotypic ratio among the F_2 offspring. However, there was a surprising additional result: The white-eye trait showed up only in males. All the F_2 females had red eyes, while half the males had red eyes and half had white eyes. Therefore, Morgan concluded that somehow a fly's eye color was linked to its sex. (If the eye-color gene were unrelated to sex, he would have expected half of the white-eyed flies to be male and half female.)

Recall that a female fly has two X chromosomes (XX), while a male fly has an X and a Y (XY). The correlation between the trait of white eye color and the male sex of the affected F_2 flies suggested to Morgan that the gene involved in his white-eyed mutant was located exclusively on the X chromosome, with no corresponding allele present on the Y chromosome. His reasoning can be followed in **Figure 15.4**. For a male, a single copy of the mutant allele would confer white eyes; since a male has only one X chromosome, there can be no wild-type allele (w^+) present to mask the recessive allele. On the other hand, a female could have white eyes only if both her X chromosomes carried the recessive mutant allele (w). This was impossible for the F_2 females in Morgan's experiment because all the F_1 fathers had red eyes.

Morgan's finding of the correlation between a particular trait and an individual's sex provided support for the chromosome theory of inheritance: namely, that a specific gene is

▼ Figure 15.4 **INQUIRY**

In a cross between a wild-type female fruit fly and a mutant white-eyed male, what color eyes will the F_1 and F_2 offspring have?

EXPERIMENT Thomas Hunt Morgan wanted to analyze the behavior of two alleles of a fruit fly eye-color gene. In crosses similar to those done by Mendel with pea plants, Morgan and his colleagues mated a wild-type (red-eyed) female with a mutant white-eyed male.

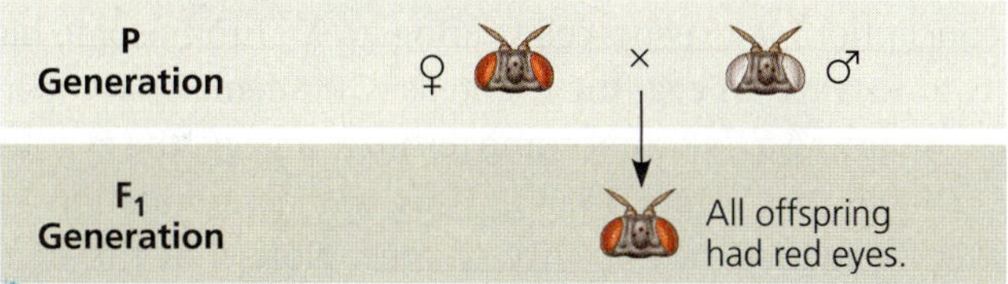

Morgan then bred an F_1 red-eyed female to an F_1 red-eyed male to produce the F_2 generation.

RESULTS The F_2 generation showed a typical Mendelian ratio of 3 red-eyed flies : 1 white-eyed fly. However, no females displayed the white-eye trait; all white-eyed flies were males.

CONCLUSION All F_1 offspring had red eyes, so the mutant white-eye trait (w) must be recessive to the wild-type red-eye trait (w^+). Since the recessive trait—white eyes—was expressed only in males in the F_2 generation, Morgan deduced that this eye-color gene is located on the X chromosome and that there is no corresponding locus on the Y chromosome.

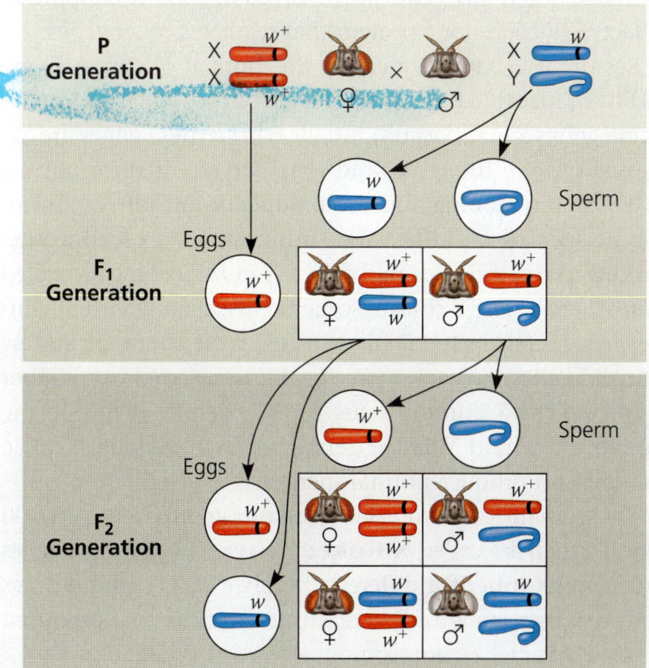

SOURCE: T. H. Morgan, Sex-limited inheritance in *Drosophila, Science* 32:120–122 (1910).

(MB) See the related Experimental Inquiry Tutorial in MasteringBiology.

WHAT IF? Suppose this eye-color gene were located on an autosome. Predict the phenotypes (including gender) of the F_2 flies in this hypothetical cross. (*Hint*: Draw a Punnett square.)

carried on a specific chromosome (in this case, an eye-color gene on the X chromosome). In addition, Morgan's work indicated that genes located on a sex chromosome exhibit unique inheritance patterns, which we will discuss in the next section. Recognizing the importance of Morgan's early work, many bright students were attracted to his fly room.

CONCEPT CHECK 15.1

1. Which one of Mendel's laws relates to the inheritance of alleles for a single character? Which law relates to the inheritance of alleles for two characters in a dihybrid cross?
2. **MAKE CONNECTIONS** Review the description of meiosis in Figure 13.8 (pp. 254–255) and Mendel's two laws in Concept 14.1 (pp. 264–269). What is the physical basis for each of Mendel's laws?
3. **WHAT IF?** Propose a possible reason that the first naturally occurring mutant fruit fly Morgan saw involved a gene on a sex chromosome.

For suggested answers, see Appendix A.

CONCEPT **15.2**

Sex-linked genes exhibit unique patterns of inheritance

As you just learned, Morgan's discovery of a trait (white eyes) that correlated with the sex of flies was a key episode in the development of the chromosome theory of inheritance. Because the identity of the sex chromosomes in an individual could be inferred by observing the sex of the fly, the behavior of the two members of the pair of sex chromosomes could be correlated with the behavior of the two alleles of the eye-color gene. In this section, we consider the role of sex chromosomes in inheritance in more detail. We begin by reviewing the chromosomal basis of sex determination in humans and some other animals.

The Chromosomal Basis of Sex

Whether we are male or female is one of our more obvious phenotypic characters. Although the anatomical and physiological differences between women and men are numerous, the chromosomal basis for determining sex is rather simple. In humans and other mammals, there are two varieties of sex chromosomes, designated X and Y. The Y chromosome is much smaller than the X chromosome **(Figure 15.5)**. A person who inherits two

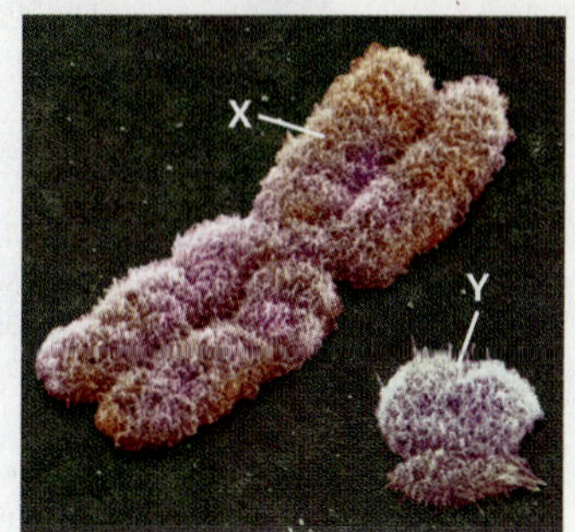

▲ Figure 15.5 Human sex chromosomes.

X chromosomes, one from each parent, usually develops as a female. A male develops from a zygote containing one X chromosome and one Y chromosome (**Figure 15.6a**). Short segments at either end of the Y chromosome are the only regions that are homologous with corresponding regions of the X. These homologous regions allow the X and Y chromosomes in males to pair and behave like homologous chromosomes during meiosis in the testes.

In mammalian testes and ovaries, the two sex chromosomes segregate during meiosis, and each gamete receives one. Each egg contains one X chromosome. In contrast, sperm fall into two categories: Half the sperm cells a male produces contain an X chromosome, and half contain a Y chromosome. We can trace the sex of each offspring to the events of conception: If a sperm cell bearing an X chromosome happens to fertilize an egg, the zygote is XX, a female; if a sperm cell containing a Y chromosome fertilizes an egg, the zygote is XY, a male (see Figure 15.6a). Thus, sex determination is a matter of chance—a fifty-fifty chance. Note that the mammalian X-Y system isn't the only chromosomal system for determining sex. **Figure 15.6b–d** illustrates three other systems.

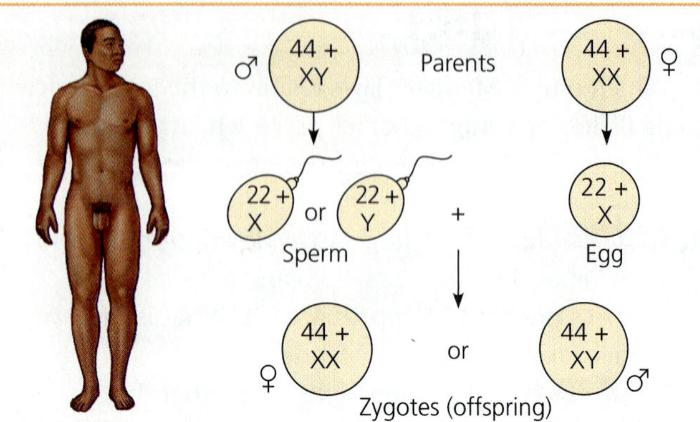

(a) The X-Y system. In mammals, the sex of an offspring depends on whether the sperm cell contains an X chromosome or a Y.

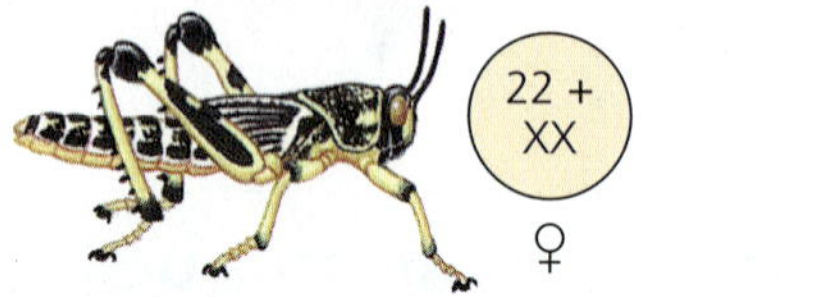
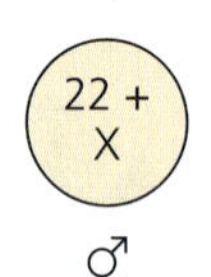

(b) The X-0 system. In grasshoppers, cockroaches, and some other insects, there is only one type of sex chromosome, the X. Females are XX; males have only one sex chromosome (X0). Sex of the offspring is determined by whether the sperm cell contains an X chromosome or no sex chromosome.

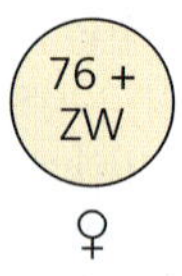
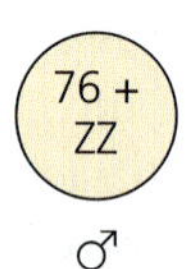

(c) The Z-W system. In birds, some fishes, and some insects, the sex chromosomes present in the egg (not the sperm) determine the sex of offspring. The sex chromosomes are designated Z and W. Females are ZW and males are ZZ.

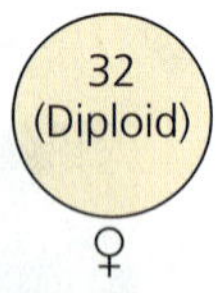
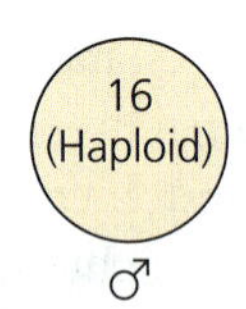

(d) The haplo-diploid system. There are no sex chromosomes in most species of bees and ants. Females develop from fertilized eggs and are thus diploid. Males develop from unfertilized eggs and are haploid; they have no fathers.

▲ **Figure 15.6 Some chromosomal systems of sex determination.** Numerals indicate the number of autosomes in the species pictured. In *Drosophila*, males are XY, but sex depends on the ratio between the number of X chromosomes and the number of autosome sets, not simply on the presence of a Y chromosome.

In humans, the anatomical signs of sex begin to emerge when the embryo is about 2 months old. Before then, the rudiments of the gonads are generic—they can develop into either testes or ovaries, depending on whether or not a Y chromosome is present. In 1990, a British research team identified a gene on the Y chromosome required for the development of testes. They named the gene *SRY*, for sex-determining region of Y. In the absence of *SRY*, the gonads develop into ovaries. The biochemical, physiological, and anatomical features that distinguish males and females are complex, and many genes are involved in their development. In fact, *SRY* codes for a protein that regulates other genes.

Researchers have sequenced the human Y chromosome and have identified 78 genes that code for about 25 proteins (some genes are duplicates). About half of these genes are expressed only in the testis, and some are required for normal testicular functioning and the production of normal sperm. A gene located on either sex chromosome is called a **sex-linked gene**; those located on the Y chromosome are called *Y-linked genes*. The Y chromosome is passed along virtually intact from a father to all his sons. Because there are so few Y-linked genes, very few disorders are transferred from father to son on the Y chromosome. A rare example is that in the absence of certain Y-linked genes, an XY individual is male but does not produce normal sperm.

The human X chromosome contains approximately 1,100 genes, which are called **X-linked genes**. The fact that males and females inherit a different number of X chromosomes leads to a pattern of inheritance different from that produced by genes located on autosomes.

Inheritance of X-Linked Genes

While most Y-linked genes help determine sex, the X chromosomes have genes for many characters unrelated to sex. X-linked genes in humans follow the same pattern of inheritance that Morgan observed for the eye-color locus he studied in *Drosophila* (see Figure 15.4). Fathers pass X-linked

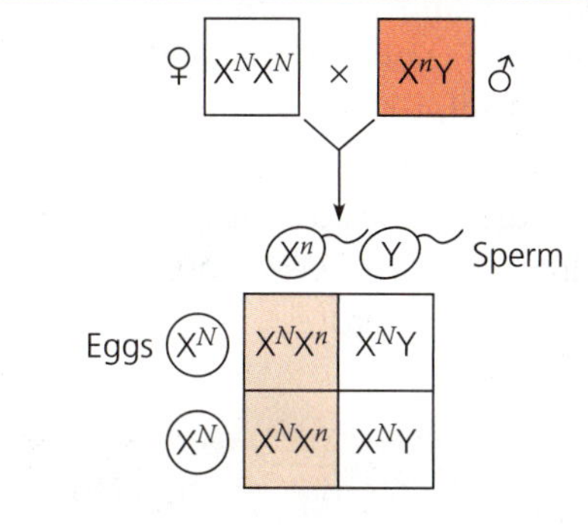

(a) A color-blind father will transmit the mutant allele to all daughters but to no sons. When the mother is a dominant homozygote, the daughters will have the normal phenotype but will be carriers of the mutation.

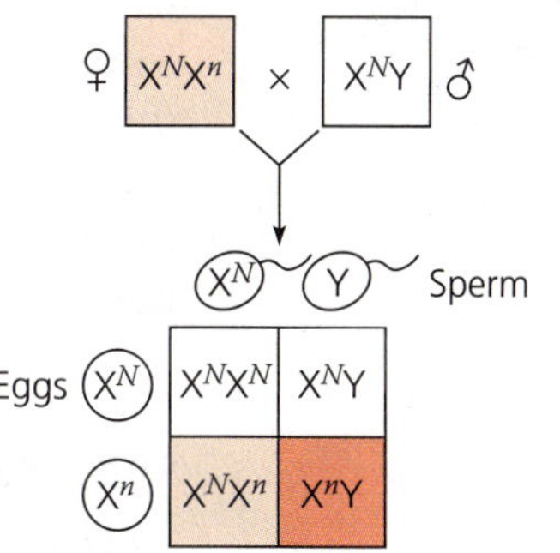

(b) If a carrier mates with a male who has normal color vision, there is a 50% chance that each daughter will be a carrier like her mother and a 50% chance that each son will have the disorder.

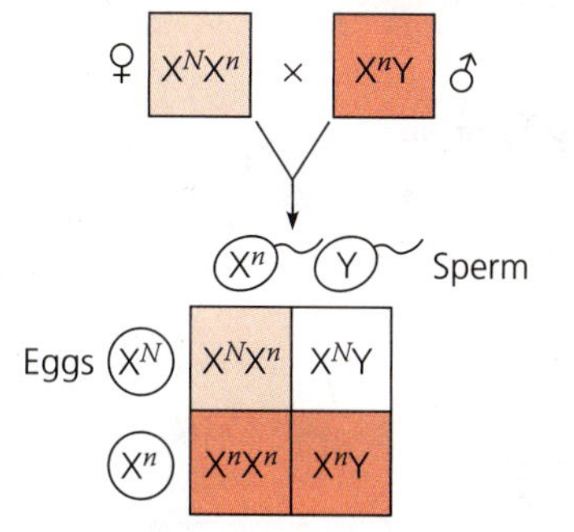

(c) If a carrier mates with a color-blind male, there is a 50% chance that each child born to them will have the disorder, regardless of sex. Daughters who have normal color vision will be carriers, whereas males who have normal color vision will be free of the recessive allele.

▲ **Figure 15.7 The transmission of X-linked recessive traits.** In this diagram, color blindness is used as an example. The superscript *N* represents the dominant allele for normal color vision carried on the X chromosome, and the superscript *n* represents the recessive allele, which has a mutation causing color blindness. White boxes indicate unaffected individuals, light orange boxes indicate carriers, and dark orange boxes indicate color-blind individuals.

? *If a color-blind woman married a man who had normal color vision, what would be the probable phenotypes of their children?*

alleles to all of their daughters but to none of their sons. In contrast, mothers can pass X-linked alleles to both sons and daughters, as shown in **Figure 15.7**.

If an X-linked trait is due to a recessive allele, a female will express the phenotype only if she is homozygous for that allele. Because males have only one locus, the terms *homozygous* and *heterozygous* lack meaning for describing their X-linked genes; the term *hemizygous* is used in such cases. Any male receiving the recessive allele from his mother will express the trait. For this reason, far more males than females have X-linked recessive disorders. However, even though the chance of a female inheriting a double dose of the mutant allele is much less than the probability of a male inheriting a single dose, there *are* females with X-linked disorders. For instance, color blindness is a mild disorder almost always inherited as an X-linked trait. A color-blind daughter may be born to a color-blind father whose mate is a carrier (see Figure 15.7c). Because the X-linked allele for color blindness is relatively rare, though, the probability that such a man and woman will mate is low.

A number of human X-linked disorders are much more serious than color blindness. An example is **Duchenne muscular dystrophy**, which affects about one out of every 3,500 males born in the United States. The disease is characterized by a progressive weakening of the muscles and loss of coordination. Affected individuals rarely live past their early 20s. Researchers have traced the disorder to the absence of a key muscle protein called dystrophin and have mapped the gene for this protein to a specific locus on the X chromosome.

Hemophilia is an X-linked recessive disorder defined by the absence of one or more of the proteins required for blood clotting. When a person with hemophilia is injured, bleeding is prolonged because a firm clot is slow to form. Small cuts in the skin are usually not a problem, but bleeding in the muscles or joints can be painful and can lead to serious damage. In the 1800s, hemophilia was widespread among the royal families of Europe. Queen Victoria of England is known to have passed the allele to several of her descendants. Subsequent intermarriage with royal family members of other nations, such as Spain and Russia, further spread this X-linked trait, and its incidence is well documented in royal pedigrees. Today, people with hemophilia are treated as needed with intravenous injections of the protein that is missing.

X Inactivation in Female Mammals

Female mammals, including humans, inherit two X chromosomes—twice the number inherited by males—so you may wonder whether females make twice as much as males of the proteins encoded by X-linked genes. In fact, most of one X chromosome in each cell in female mammals becomes inactivated during early embryonic development. As a result, the cells of females and males have the same effective dose (one copy) of most X-linked genes. The inactive X in each cell of a female condenses into a compact object called a **Barr body** (discovered by Canadian anatomist Murray Barr), which lies along the inside of the nuclear envelope. Most of the genes of the X chromosome that forms the Barr body are not expressed. In the ovaries, Barr-body chromosomes are reactivated in the cells that give rise to eggs, so every female gamete has an active X.

British geneticist Mary Lyon demonstrated that the selection of which X chromosome will form the Barr body occurs randomly and independently in each embryonic cell present at the time of X inactivation. As a consequence, females consist

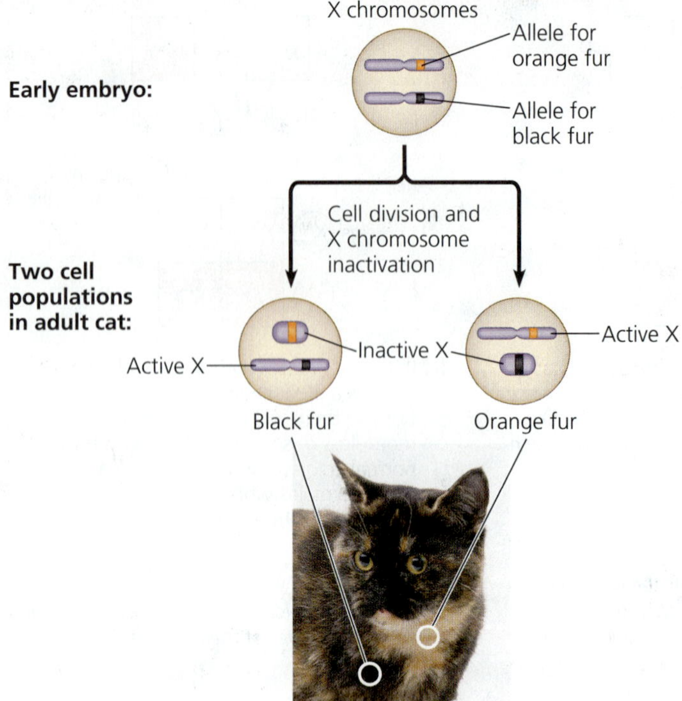

▲ **Figure 15.8 X inactivation and the tortoiseshell cat.** The tortoiseshell gene is on the X chromosome, and the tortoiseshell phenotype requires the presence of two different alleles, one for orange fur and one for black fur. Normally, only females can have both alleles, because only they have two X chromosomes. If a female cat is heterozygous for the tortoiseshell gene, she is tortoiseshell. Orange patches are formed by populations of cells in which the X chromosome with the orange allele is active; black patches have cells in which the X chromosome with the black allele is active. ("Calico" cats also have white areas, which are determined by yet another gene.)

of a *mosaic* of two types of cells: those with the active X derived from the father and those with the active X derived from the mother. After an X chromosome is inactivated in a particular cell, all mitotic descendants of that cell have the same inactive X. Thus, if a female is heterozygous for a sex-linked trait, about half her cells will express one allele, while the others will express the alternate allele. **Figure 15.8** shows how this mosaicism results in the mottled coloration of a tortoiseshell cat. In humans, mosaicism can be observed in a recessive X-linked mutation that prevents the development of sweat glands. A woman who is heterozygous for this trait has patches of normal skin and patches of skin lacking sweat glands.

Inactivation of an X chromosome involves modification of the DNA and the histone proteins bound to it, including attachment of methyl groups (—CH_3) to one of the nitrogenous bases of DNA nucleotides. (The regulatory role of DNA methylation is discussed further in Chapter 18.) A particular region of each X chromosome contains several genes involved in the inactivation process. The two regions, one on each X chromosome, associate briefly with each other in each cell at an early stage of embryonic development. Then one of the genes, called *XIST* (for X-inactive specific transcript) becomes active *only* on the chromosome that will become the Barr body. Multiple copies of the RNA product of this gene apparently attach to the X chromosome on which they are made, eventually almost covering it. Interaction of this RNA with the chromosome seems to initiate X inactivation, and the RNA products of other genes nearby on the X chromosome help to regulate the process.

CONCEPT CHECK 15.2

1. A white-eyed female *Drosophila* is mated with a red-eyed (wild-type) male, the reciprocal cross of the one shown in Figure 15.4. What phenotypes and genotypes do you predict for the offspring?
2. Neither Tim nor Rhoda has Duchenne muscular dystrophy, but their firstborn son does have it. What is the probability that a second child of this couple will have the disease? What is the probability if the second child is a boy? A girl?
3. **MAKE CONNECTIONS** Consider what you learned about dominant and recessive alleles in Concept 14.1 (p. 265). If a disorder were caused by a dominant X-linked allele, how would the inheritance pattern differ from what we see for recessive X-linked disorders?

For suggested answers, see Appendix A.

CONCEPT 15.3

Linked genes tend to be inherited together because they are located near each other on the same chromosome

The number of genes in a cell is far greater than the number of chromosomes; in fact, each chromosome has hundreds or thousands of genes. (The Y chromosome is an exception.) Genes located near each other on the same chromosome tend to be inherited together in genetic crosses; such genes are said to be genetically linked and are called **linked genes**. (Note the distinction between the terms *sex-linked gene*, referring to a single gene on a sex chromosome, and *linked genes*, referring to two or more genes on the same chromosome that tend to be inherited together.) When geneticists follow linked genes in breeding experiments, the results deviate from those expected from Mendel's law of independent assortment.

How Linkage Affects Inheritance

To see how linkage between genes affects the inheritance of two different characters, let's examine another of Morgan's *Drosophila* experiments. In this case, the characters are body color and wing size, each with two different phenotypes.

Wild-type flies have gray bodies and normal-sized wings. In addition to these flies, Morgan had managed to obtain, through breeding, doubly mutant flies with black bodies and wings much smaller than normal, called vestigial wings. The mutant alleles are recessive to the wild-type alleles, and neither gene is on a sex chromosome. In his investigation of these two genes, Morgan carried out the crosses shown in **Figure 15.9**. The first was a P generation cross to generate F_1 dihybrid flies, and the second was a testcross.

The resulting flies had a much higher proportion of the combinations of traits seen in the P generation flies (called parental phenotypes) than would be expected if the two genes assorted independently. Morgan thus concluded that body color and wing size are usually inherited together in

▼ Figure 15.9 **INQUIRY**

How does linkage between two genes affect inheritance of characters?

EXPERIMENT Morgan wanted to know whether the genes for body color and wing size were genetically linked, and if so, how this affected their inheritance. The alleles for body color are b^+ (gray) and b (black), and those for wing size are vg^+ (normal) and vg (vestigial).

Morgan mated true-breeding P (parental) generation flies—wild-type flies with black, vestigial-winged flies—to produce heterozygous F_1 dihybrids ($b^+ b\ vg^+ vg$), all of which are wild-type in appearance.

He then mated wild-type F_1 dihybrid females with black, vestigial-winged males. This testcross will reveal the genotype of the eggs made by the dihybrid female.

The male's sperm contributes only recessive alleles, so the phenotype of the offspring reflects the genotype of the female's eggs.

Note: Although only females (with pointed abdomens) are shown, half the offspring in each class would be males (with rounded abdomens).

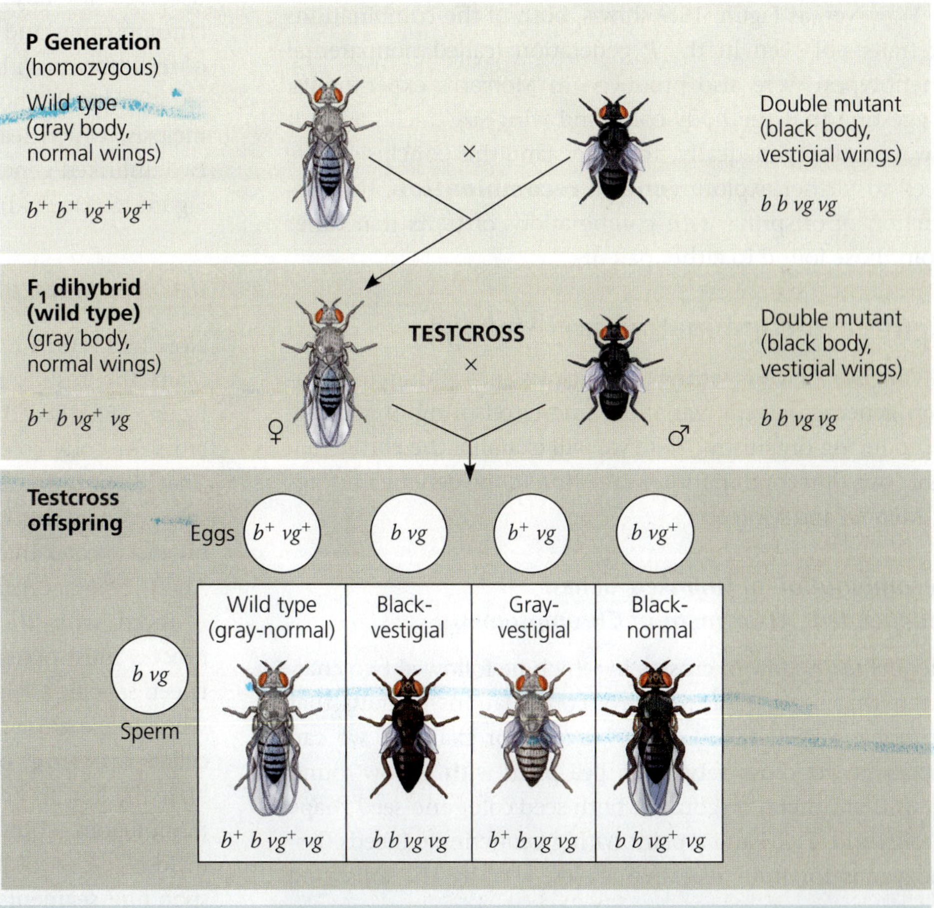

PREDICTED RATIOS

	Wild type		Black-vestigial		Gray-vestigial		Black-normal
If genes are located on different chromosomes:	1	:	1	:	1	:	1
If genes are located on the same chromosome *and* parental alleles are always inherited together:	1	:	1	:	0	:	0
RESULTS	965	:	944	:	206	:	185

CONCLUSION Since most offspring had a parental (P generation) phenotype, Morgan concluded that the genes for body color and wing size are genetically linked on the same chromosome. However, the production of a relatively small number of offspring with nonparental phenotypes indicated that some mechanism occasionally breaks the linkage between specific alleles of genes on the same chromosome.

SOURCE: T. H. Morgan and C. J. Lynch, The linkage of two factors in *Drosophila* that are not sex-linked, *Biological Bulletin* 23:174–182 (1912).

WHAT IF? If the parental (P generation) flies had been true-breeding for gray body with vestigial wings and black body with normal wings, which phenotypic class(es) would be largest among the testcross offspring?

specific (parental) combinations because the genes for these characters are near each other on the same chromosome:

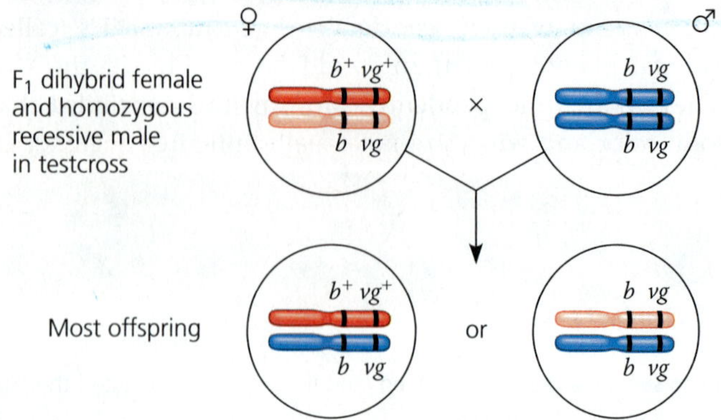

However, as Figure 15.9 shows, both of the combinations of traits not seen in the P generation (called nonparental phenotypes) were also produced in Morgan's experiments, suggesting that the body-color and wing-size alleles are not always linked genetically. To understand this conclusion, we need to further explore **genetic recombination**, the production of offspring with combinations of traits that differ from those found in either parent.

Genetic Recombination and Linkage

In Chapter 13, you learned that meiosis and random fertilization generate genetic variation among offspring of sexually reproducing organisms. Here we will examine the chromosomal basis of recombination in relation to the genetic findings of Mendel and Morgan.

Recombination of Unlinked Genes: Independent Assortment of Chromosomes

Mendel learned from crosses in which he followed two characters that some offspring have combinations of traits that do not match those of either parent. For example, we can represent the cross between a pea plant with yellow-round seeds that is heterozygous for both seed color and seed shape (a dihybrid, *YyRr*) and a plant with green-wrinkled seeds (homozygous for both recessive alleles, *yyrr*) by the following Punnett square:

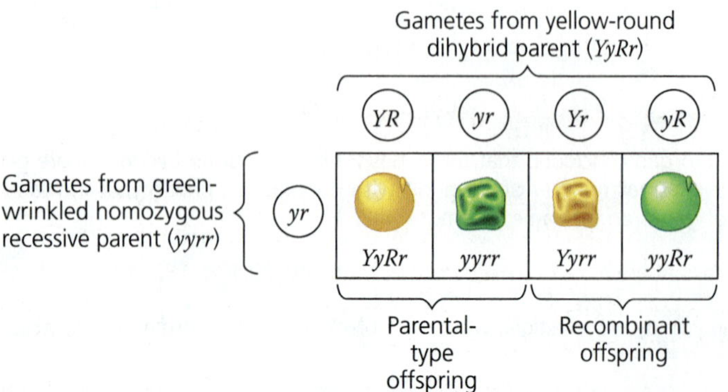

Notice in this Punnett square that one-half of the offspring are expected to inherit a phenotype that matches either of the parental (P generation) phenotypes. These offspring are called **parental types**. But two nonparental phenotypes are also found among the offspring. Because these offspring have new combinations of seed shape and color, they are called **recombinant types**, or **recombinants** for short. When 50% of all offspring are recombinants, as in this example, geneticists say that there is a 50% frequency of recombination. The predicted phenotypic ratios among the offspring are similar to what Mendel actually found in *YyRr* × *yyrr* crosses (a type of testcross because it reveals the genotype of the gametes made by the dihybrid *YyRr* plant).

A 50% frequency of recombination in such testcrosses is observed for any two genes that are located on different chromosomes and thus cannot be linked. The physical basis of recombination between unlinked genes is the random orientation of homologous chromosomes at metaphase I of meiosis, which leads to the independent assortment of the two unlinked genes (see Figure 13.10 and the question in the Figure 15.2 legend).

Recombination of Linked Genes: Crossing Over

Now let's return to Morgan's fly room to see how we can explain the results of the *Drosophila* testcross illustrated in Figure 15.9. Recall that most of the offspring from the testcross for body color and wing size had parental phenotypes. That suggested that the two genes were on the same chromosome, since the occurrence of parental types with a frequency greater than 50% indicates that the genes are linked. About 17% of offspring, however, were recombinants.

Faced with these results, Morgan proposed that some process must occasionally break the physical connection between specific alleles of genes on the same chromosome. Subsequent experiments demonstrated that this process, now called **crossing over**, accounts for the recombination of linked genes. In crossing over, which occurs while replicated homologous chromosomes are paired during prophase of meiosis I, a set of proteins orchestrates an exchange of corresponding segments of one maternal and one paternal chromatid (see Figure 13.11). In effect, end portions of two nonsister chromatids trade places each time a crossover occurs.

Figure 15.10 shows how crossing over in a dihybrid female fly resulted in recombinant eggs and ultimately recombinant offspring in Morgan's testcross. Most of the eggs had a chromosome with either the b^+ vg^+ or *b vg* parental genotype for body color and wing size, but some eggs had a recombinant chromosome (b^+ *vg* or *b* vg^+). Fertilization of these various classes of eggs by homozygous recessive sperm (*b vg*) produced an offspring population in which 17% exhibited a nonparental, recombinant phenotype, reflecting combinations of alleles not seen before in either P generation parent.

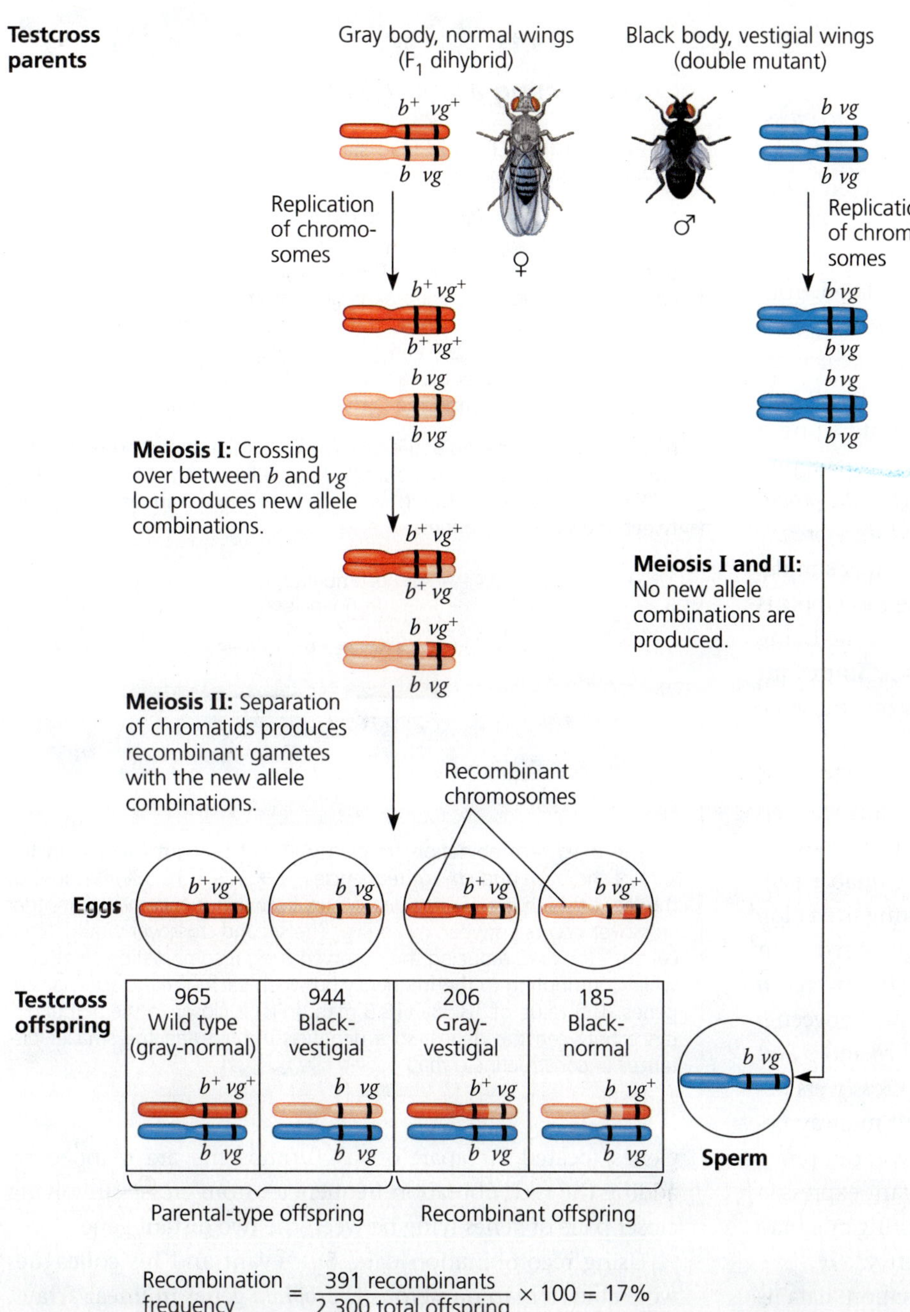

$$\text{Recombination frequency} = \frac{391 \text{ recombinants}}{2{,}300 \text{ total offspring}} \times 100 = 17\%$$

◀ **Figure 15.10 Chromosomal basis for recombination of linked genes.** In these diagrams re-creating the testcross in Figure 15.9, we track chromosomes as well as genes. The maternal chromosomes are color-coded red and pink to distinguish one homolog from the other before any meiotic crossing over has taken place. Because crossing over between the *b* and *vg* loci occurs in some, but not all, egg-producing cells, more eggs with parental-type chromosomes than with recombinant ones are produced in the mating females. Fertilization of the eggs by sperm of genotype *b vg* gives rise to some recombinant offspring. The recombination frequency is the percentage of recombinant flies in the total pool of offspring.

DRAW IT *Suppose, as in the question at the bottom of Figure 15.9, the parental (P generation) flies were true-breeding for gray body with vestigial wings and black body with normal wings. Draw the chromosomes in each of the four possible kinds of eggs from an F_1 female, and label each chromosome as "parental" or "recombinant."*

New Combinations of Alleles: Variation for Natural Selection

EVOLUTION In Chapter 13, you learned how the physical behavior of chromosomes during meiosis contributes to the generation of variation in offspring. Each pair of homologous chromosomes lines up independently of other pairs during metaphase I, and crossing over prior to that, during prophase I, can mix and match parts of maternal and paternal homologs. Chapter 14 described Mendel's elegant experiments showing that the behavior of the abstract entities known as genes (or, more concretely, alleles of genes) also leads to variation in offspring. Now, putting these different ideas together, you can see that the recombinant chromosomes resulting from crossing over may bring alleles together in new combinations, and the subsequent events of meiosis distribute to gametes the recombinant chromosomes in a multitude of combinations, such as the new variants seen in Figures 15.9 and 15.10. Random fertilization then increases even further the number of variant allele combinations that can be created.

This abundance of genetic variation provides the raw material on which natural selection works. If the traits conferred by particular combinations of alleles are better suited for a given environment, organisms possessing those genotypes will be expected to thrive and leave more offspring, ensuring the continuation of their genetic complement. In the next generation, of course, the alleles will be shuffled anew. Ultimately, the interplay between environment and genotype will determine which genetic combinations persist over time.

Mapping the Distance Between Genes Using Recombination Data: *Scientific Inquiry*

The discovery of linked genes and recombination due to crossing over led one of Morgan's students, Alfred H. Sturtevant, to a method for constructing a **genetic map**, an ordered list of the genetic loci along a particular chromosome.

Sturtevant hypothesized that the percentage of recombinant offspring, the *recombination frequency*, calculated from experiments like the one in Figures 15.9 and 15.10, depends on the distance between genes on a chromosome. He assumed that crossing over is a random event, with the chance of crossing over approximately equal at all points along a chromosome. Based on these assumptions, Sturtevant predicted that *the farther apart two genes are, the higher the probability that a crossover will occur between them and therefore the higher the recombination frequency*. His reasoning was simple: The greater the distance between two genes, the more points there are between them where crossing over can occur. Using recombination data from various fruit fly crosses, Sturtevant proceeded to assign relative positions to genes on the same chromosomes—that is, to *map* genes.

A genetic map based on recombination frequencies is called a **linkage map**. **Figure 15.11** shows Sturtevant's linkage map of three genes: the body-color (*b*) and wing-size (*vg*) genes depicted in Figure 15.10 and a third gene, called cinnabar (*cn*). Cinnabar is one of many *Drosophila* genes affecting eye color. Cinnabar eyes, a mutant phenotype, are a brighter red than the wild-type color. The recombination frequency between *cn* and *b* is 9%; that between *cn* and *vg*, 9.5%; and that between *b* and *vg*, 17%. In other words, crossovers between *cn* and *b* and between *cn* and *vg* are about half as frequent as crossovers between *b* and *vg*. Only a map that locates *cn* about midway between *b* and *vg* is consistent with these data, as you can prove to yourself by drawing alternative maps. Sturtevant expressed the distances between genes in **map units**, defining one map unit as equivalent to a 1% recombination frequency.

In practice, the interpretation of recombination data is more complicated than this example suggests. Some genes on a chromosome are so far from each other that a crossover between them is virtually certain. The observed frequency of recombination in crosses involving two such genes can have a maximum value of 50%, a result indistinguishable from that for genes on different chromosomes. In this case, the physical connection between genes on the same chromosome is not reflected in the results of genetic crosses. Despite being on the same chromosome and thus being *physically connected*, the genes are *genetically unlinked*; alleles of such genes assort independently, as if they were on different chromosomes. In fact, at least two of the genes for pea characters that Mendel studied are now known to be on the same chromosome, but the distance between them is so great that linkage is not observed in genetic crosses. Consequently, the two genes behaved as if they were on different chromosomes in Mendel's experiments.

▼ **Figure 15.11** RESEARCH METHOD

Constructing a Linkage Map

APPLICATION A linkage map shows the relative locations of genes along a chromosome.

TECHNIQUE A linkage map is based on the assumption that the probability of a crossover between two genetic loci is proportional to the distance separating the loci. The recombination frequencies used to construct a linkage map for a particular chromosome are obtained from experimental crosses, such as the cross depicted in Figures 15.9 and 15.10. The distances between genes are expressed as map units, with one map unit equivalent to a 1% recombination frequency. Genes are arranged on the chromosome in the order that best fits the data.

RESULTS In this example, the observed recombination frequencies between three *Drosophila* gene pairs (*b–cn* 9%, *cn–vg* 9.5%, and *b–vg* 17%) best fit a linear order in which *cn* is positioned about halfway between the other two genes:

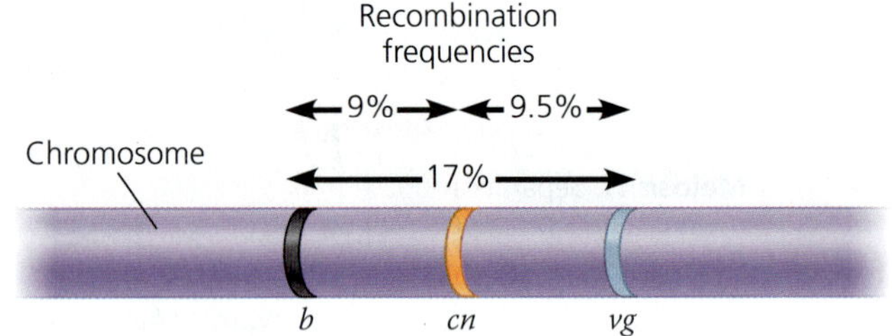

The *b–vg* recombination frequency (17%) is slightly less than the sum of the *b–cn* and *cn–vg* frequencies (9 + 9.5 = 18.5%) because of the few times that one crossover occurs between *b* and *cn* and another crossover occurs between *cn* and *vg*. The second crossover would "cancel out" the first, reducing the observed *b–vg* recombination frequency while contributing to the frequency between each of the closer pairs of genes. The value of 18.5% (18.5 map units) is closer to the actual distance between the genes, so a geneticist would add the smaller distances in constructing a map.

Genes located far apart on a chromosome are mapped by adding the recombination frequencies from crosses involving closer pairs of genes lying between the two distant genes.

Using recombination data, Sturtevant and his colleagues were able to map numerous *Drosophila* genes in linear arrays. They found that the genes clustered into four groups of linked genes (*linkage groups*). Light microscopy had revealed four pairs of chromosomes in *Drosophila*, so the linkage map provided additional evidence that genes are located on chromosomes. Each chromosome has a linear array of specific genes, each gene with its own locus **(Figure 15.12)**.

Because a linkage map is based strictly on recombination frequencies, it gives only an approximate picture of a chromosome. The frequency of crossing over is not actually uniform over the length of a chromosome, as Sturtevant assumed, and therefore map units do not correspond to actual physical distances (in nanometers, for instance). A linkage map does portray the order of genes along a chromosome, but it does not accurately portray the precise locations of those genes. Other methods enable geneticists to construct **cytogenetic maps** of chromosomes, which locate genes with respect to chromosomal

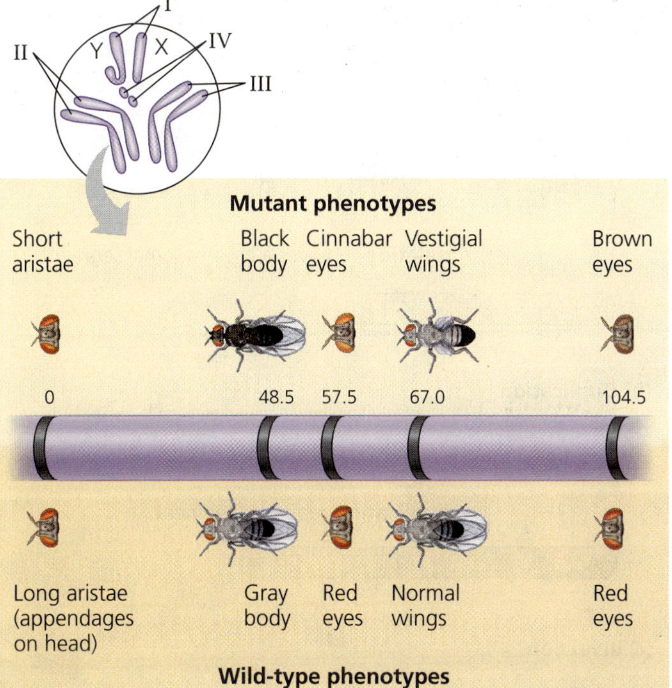

▲ **Figure 15.12 A partial genetic (linkage) map of a *Drosophila* chromosome.** This simplified map shows just a few of the genes that have been mapped on *Drosophila* chromosome II. The number at each gene locus indicates the number of map units between that locus and the locus for arista length (left). Notice that more than one gene can affect a given phenotypic characteristic, such as eye color. Also, note that in contrast to the homologous autosomes (II–IV), the X and Y sex chromosomes (I) have distinct shapes.

features, such as stained bands, that can be seen in the microscope. The ultimate maps, which we will discuss in Chapter 21, display the physical distances between gene loci in DNA nucleotides. Comparing a linkage map with such a physical map or with a cytogenetic map of the same chromosome, we find that the linear order of genes is identical in all the maps, but the spacing between genes is not.

CONCEPT CHECK 15.3

1. When two genes are located on the same chromosome, what is the physical basis for the production of recombinant offspring in a testcross between a dihybrid parent and a double-mutant (recessive) parent?
2. For each type of offspring of the testcross in Figure 15.9, explain the relationship between its phenotype and the alleles contributed by the female parent.
3. **WHAT IF?** Genes *A*, *B*, and *C* are located on the same chromosome. Testcrosses show that the recombination frequency between *A* and *B* is 28% and between *A* and *C* is 12%. Can you determine the linear order of these genes? Explain.

For suggested answers, see Appendix A.

CONCEPT **15.4**

Alterations of chromosome number or structure cause some genetic disorders

As you have learned so far in this chapter, the phenotype of an organism can be affected by small-scale changes involving individual genes. Random mutations are the source of all new alleles, which can lead to new phenotypic traits.

Large-scale chromosomal changes can also affect an organism's phenotype. Physical and chemical disturbances, as well as errors during meiosis, can damage chromosomes in major ways or alter their number in a cell. Large-scale chromosomal alterations in humans and other mammals often lead to spontaneous abortion (miscarriage) of a fetus, and individuals born with these types of genetic defects commonly exhibit various developmental disorders. Plants may tolerate such genetic defects better than animals do.

Abnormal Chromosome Number

Ideally, the meiotic spindle distributes chromosomes to daughter cells without error. But there is an occasional mishap, called a **nondisjunction**, in which the members of a pair of homologous chromosomes do not move apart properly during meiosis I or sister chromatids fail to separate during meiosis II **(Figure 15.13)**. In these cases, one gamete

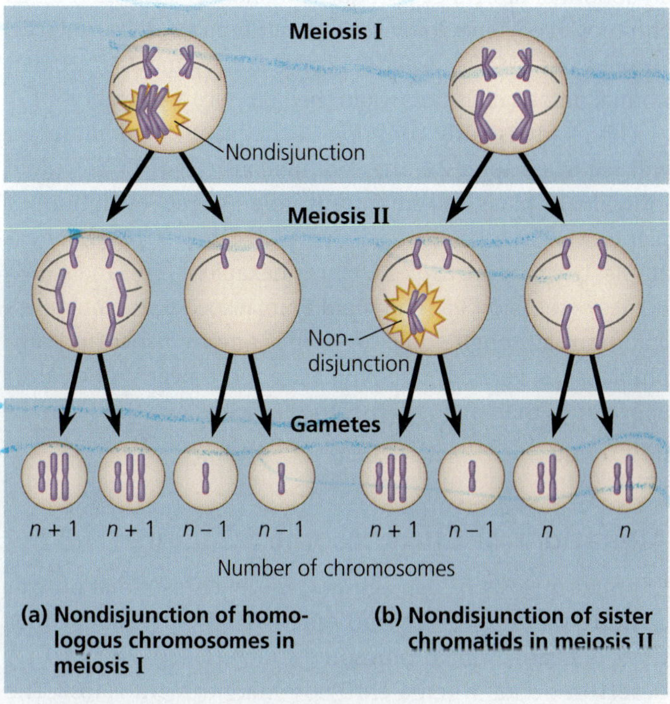

▲ **Figure 15.13 Meiotic nondisjunction.** Gametes with an abnormal chromosome number can arise by nondisjunction in either meiosis I or meiosis II. For simplicity, the figure does not show the spores formed by meiosis in plants. Ultimately, spores form gametes that have the defects shown. (See Figure 13.6.)

receives two of the same type of chromosome and another gamete receives no copy. The other chromosomes are usually distributed normally.

If either of the aberrant gametes unites with a normal one at fertilization, the zygote will also have an abnormal number of a particular chromosome, a condition known as **aneuploidy**. (Aneuploidy may involve more than one chromosome.) Fertilization involving a gamete that has no copy of a particular chromosome will lead to a missing chromosome in the zygote (so that the cell has $2n - 1$ chromosomes); the aneuploid zygote is said to be **monosomic** for that chromosome. If a chromosome is present in triplicate in the zygote (so that the cell has $2n + 1$ chromosomes), the aneuploid cell is **trisomic** for that chromosome. Mitosis will subsequently transmit the anomaly to all embryonic cells. If the organism survives, it usually has a set of traits caused by the abnormal dose of the genes associated with the extra or missing chromosome. Down syndrome is an example of trisomy in humans that will be discussed later. Nondisjunction can also occur during mitosis. If such an error takes place early in embryonic development, then the aneuploid condition is passed along by mitosis to a large number of cells and is likely to have a substantial effect on the organism.

Some organisms have more than two complete chromosome sets in all somatic cells. The general term for this chromosomal alteration is **polyploidy**; the specific terms *triploidy* ($3n$) and *tetraploidy* ($4n$) indicate three or four chromosomal sets, respectively. One way a triploid cell may arise is by the fertilization of an abnormal diploid egg produced by nondisjunction of all its chromosomes. Tetraploidy could result from the failure of a $2n$ zygote to divide after replicating its chromosomes. Subsequent normal mitotic divisions would then produce a $4n$ embryo.

Polyploidy is fairly common in the plant kingdom. As we will see in Chapter 24, the spontaneous origin of polyploid individuals plays an important role in the evolution of plants. Many of the plant species we eat are polyploid; for example, bananas are triploid, wheat hexaploid ($6n$), and strawberries octoploid ($8n$). Polyploid animal species are much less common, although some are found among fishes and amphibians. In general, polyploids are more nearly normal in appearance than aneuploids. One extra (or missing) chromosome apparently disrupts genetic balance more than does an entire extra set of chromosomes.

Alterations of Chromosome Structure

Errors in meiosis or damaging agents such as radiation can cause breakage of a chromosome, which can lead to four types of changes in chromosome structure (**Figure 15.14**). A **deletion** occurs when a chromosomal fragment is lost. The affected chromosome is then missing certain genes. (If the centromere is deleted, the entire chromosome will be lost.) The "deleted" fragment may become attached as an extra segment to a sister chromatid, producing a **duplication**.

▼ Figure 15.14 Alterations of chromosome structure. Red arrows indicate breakage points. Dark purple highlights the chromosomal parts affected by the rearrangements.

(a) Deletion

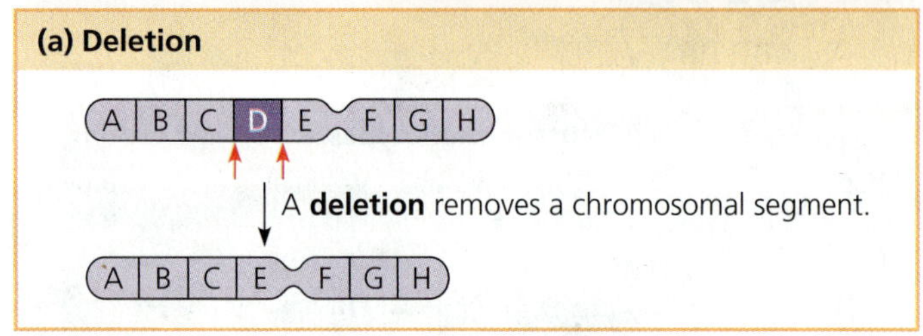

(b) Duplication

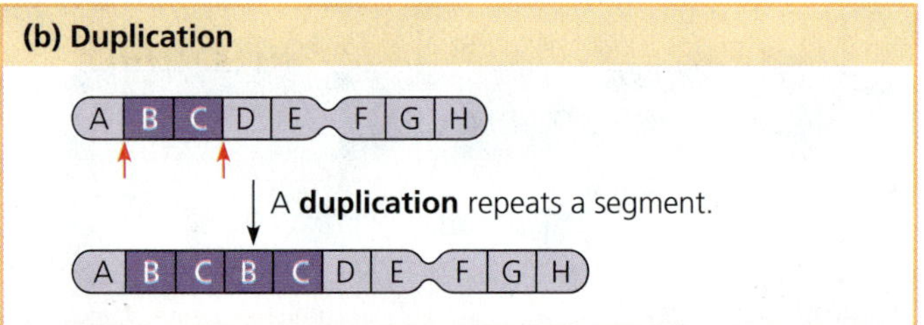

(c) Inversion

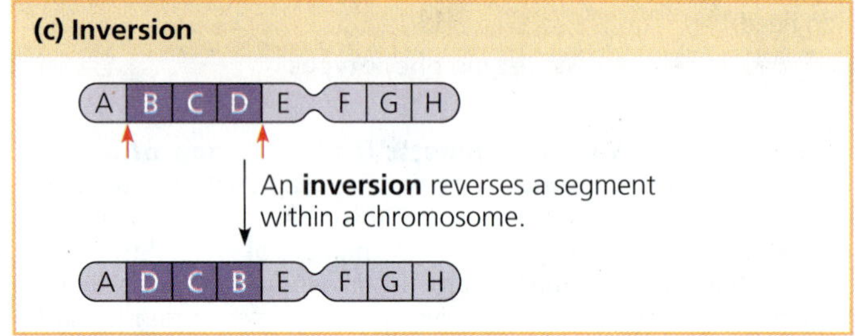

(d) Translocation

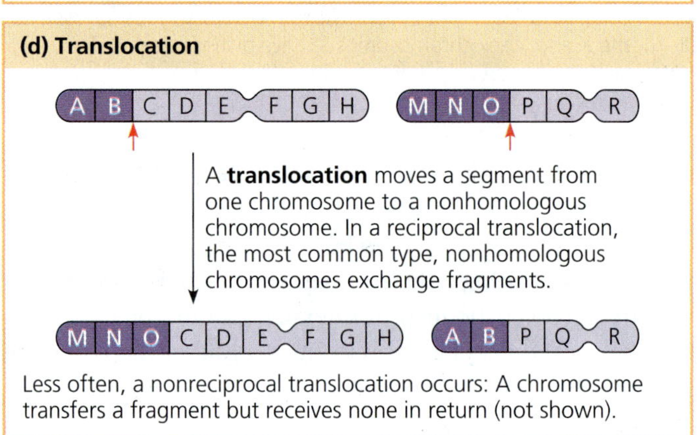

Less often, a nonreciprocal translocation occurs: A chromosome transfers a fragment but receives none in return (not shown).

Alternatively, a detached fragment could attach to a nonsister chromatid of a homologous chromosome. In that case, though, the "duplicated" segments might not be identical because the homologs could carry different alleles of certain genes. A chromosomal fragment may also reattach to the original chromosome but in the reverse orientation, producing an **inversion**. A fourth possible result of chromosomal breakage is for the fragment to join a nonhomologous chromosome, a rearrangement called a **translocation**.

Deletions and duplications are especially likely to occur during meiosis. In crossing over, nonsister chromatids sometimes exchange unequal-sized segments of DNA, so that one partner gives up more genes than it receives. The products of such an unequal crossover are one chromosome with a deletion and one chromosome with a duplication.

A diploid embryo that is homozygous for a large deletion (or has a single X chromosome with a large deletion, in a male) is usually missing a number of essential genes, a condition that is ordinarily lethal. Duplications and translocations also tend to be harmful. In reciprocal translocations, in which segments are exchanged between nonhomologous chromosomes, and in inversions, the balance of genes is not abnormal—all genes are present in their normal doses. Nevertheless, translocations and inversions can alter phenotype because a gene's expression can be influenced by its location among neighboring genes; such events sometimes have devastating effects.

Human Disorders Due to Chromosomal Alterations

Alterations of chromosome number and structure are associated with a number of serious human disorders. As described earlier, nondisjunction in meiosis results in aneuploidy in gametes and any resulting zygotes. Although the frequency of aneuploid zygotes may be quite high in humans, most of these chromosomal alterations are so disastrous to development that the affected embryos are spontaneously aborted long before birth. However, some types of aneuploidy appear to upset the genetic balance less than others, with the result that individuals with certain aneuploid conditions can survive to birth and beyond. These individuals have a set of traits—a *syndrome*—characteristic of the type of aneuploidy. Genetic disorders caused by aneuploidy can be diagnosed before birth by fetal testing (see Figure 14.19).

Down Syndrome (Trisomy 21)

One aneuploid condition, **Down syndrome**, affects approximately one out of every 700 children born in the United States **(Figure 15.15)**. Down syndrome is usually the result of an extra chromosome 21, so that each body cell has a total of 47 chromosomes. Because the cells are trisomic for chromosome 21, Down syndrome is often called *trisomy 21*. Down syndrome includes characteristic facial features, short stature, correctable heart defects, and developmental delays. Individuals with Down syndrome have an increased chance of developing leukemia and Alzheimer's disease but have a lower rate of high blood pressure, atherosclerosis (hardening of the arteries), stroke, and many types of solid tumors. Although people with Down syndrome, on average, have a life span shorter than normal, most, with proper medical treatment, live to middle age and beyond. Many live independently or at home with their families, are employed, and are valuable contributors to their communities. Almost all males and about half of females with Down syndrome are sexually underdeveloped and sterile.

The frequency of Down syndrome increases with the age of the mother. While the disorder occurs in just 0.04% of children born to women under age 30, the risk climbs to 0.92% for mothers at age 40 and is even higher for older mothers. The correlation of Down syndrome with maternal age has not yet been explained. Most cases result from nondisjunction during meiosis I, and some research points to an age-dependent abnormality in a meiosis checkpoint that normally delays anaphase until all the kinetochores are attached to the spindle (like the M phase checkpoint of the mitotic cell cycle; see Chapter 12). Trisomies of some other chromosomes also increase in incidence with maternal age, although infants with other autosomal trisomies rarely survive for long. Due to its low risk and its potential for providing useful information, prenatal screening for trisomies in the embryo is now offered to all pregnant women. In 2008, the Prenatally and Postnatally Diagnosed Conditions Awareness Act was signed into law in the United States. This law stipulates that medical practitioners give accurate, up-to-date information about any prenatal or postnatal diagnosis received by parents and that they connect parents with appropriate support services.

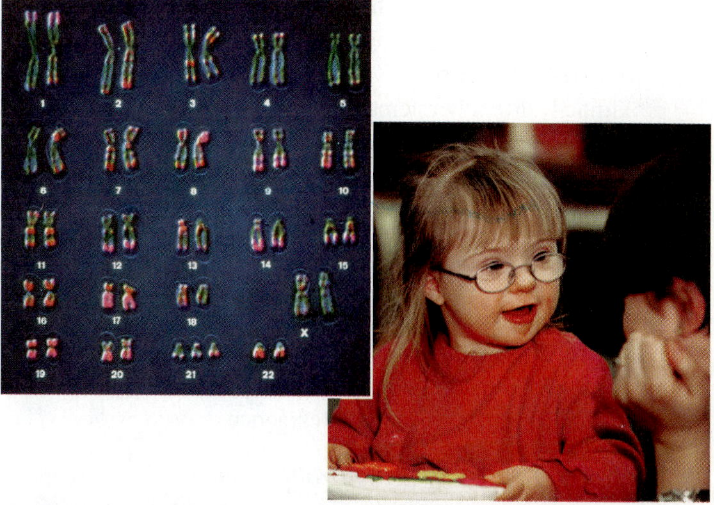

▲ **Figure 15.15 Down syndrome.** The karyotype shows trisomy 21, the most common cause of Down syndrome. The child exhibits the facial features characteristic of this disorder.

Aneuploidy of Sex Chromosomes

Nondisjunction of sex chromosomes produces a variety of aneuploid conditions. Most of these conditions appear to upset the genetic balance less than aneuploid conditions involving autosomes. This may be because the Y chromosome carries relatively few genes and because extra copies of the X chromosome become inactivated as Barr bodies in somatic cells.

An extra X chromosome in a male, producing XXY, occurs approximately once in every 500 to 1,000 live male births. People with this disorder, called *Klinefelter syndrome*, have male sex organs, but the testes are abnormally small and the man is sterile. Even though the extra X is inactivated, some breast enlargement and other female body characteristics are common. Affected individuals may have subnormal intelligence. About 1 of every 1,000 males is born with an extra Y chromosome (XYY). These males undergo normal sexual development and do not exhibit any well-defined syndrome, but they tend to be somewhat taller than average.

Females with trisomy X (XXX), which occurs once in approximately 1,000 live female births, are healthy and have no unusual physical features other than being slightly taller than average. Triple-X females are at risk for learning disabilities but are fertile. Monosomy X, called *Turner syndrome*, occurs about once in every 2,500 female births and is the only known viable monosomy in humans. Although these X0 individuals are phenotypically female, they are sterile because their sex organs do not mature. When provided with estrogen replacement therapy, girls with Turner syndrome do develop secondary sex characteristics. Most have normal intelligence.

Disorders Caused by Structurally Altered Chromosomes

Many deletions in human chromosomes, even in a heterozygous state, cause severe problems. One such syndrome, known as *cri du chat* ("cry of the cat"), results from a specific deletion in chromosome 5. A child born with this deletion is severely intellectually disabled, has a small head with unusual facial features, and has a cry that sounds like the mewing of a distressed cat. Such individuals usually die in infancy or early childhood.

Chromosomal translocations have been implicated in certain cancers, including *chronic myelogenous leukemia* (CML). This disease occurs when a reciprocal translocation happens during mitosis of cells that will become white blood cells. In these cells, the exchange of a large portion of chromosome 22 with a small fragment from a tip of chromosome 9 produces a much shortened, easily recognized chromosome 22, called the *Philadelphia chromosome* (**Figure 15.16**). Such an exchange causes cancer by activating a gene that leads to uncontrolled cell cycle progression. The mechanism of gene activation will be discussed in Chapter 18.

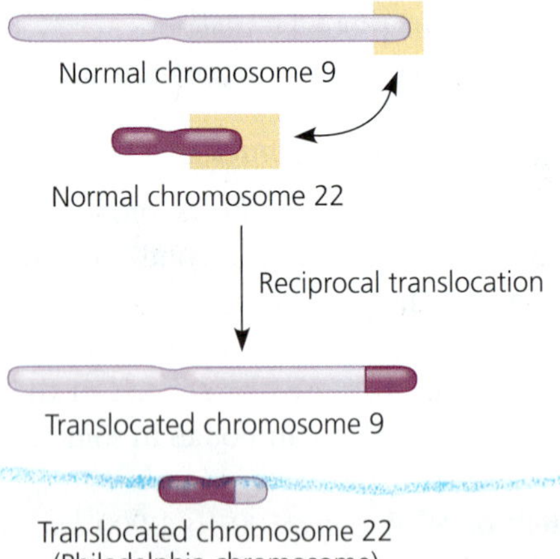

▲ **Figure 15.16 Translocation associated with chronic myelogenous leukemia (CML).** The cancerous cells in nearly all CML patients contain an abnormally short chromosome 22, the so-called Philadelphia chromosome, and an abnormally long chromosome 9. These altered chromosomes result from the reciprocal translocation shown here, which presumably occurred in a single white blood cell precursor undergoing mitosis and was then passed along to all descendant cells.

CONCEPT CHECK 15.4

1. About 5% of individuals with Down syndrome have a chromosomal translocation in which a third copy of chromosome 21 is attached to chromosome 14. If this translocation occurred in a parent's gonad, how could it lead to Down syndrome in a child?
2. **WHAT IF?** The ABO blood type locus has been mapped on chromosome 9. A father who has type AB blood and a mother who has type O blood have a child with trisomy 9 and type A blood. Using this information, can you tell in which parent the nondisjunction occurred? Explain your answer.
3. **MAKE CONNECTIONS** The gene that is activated on the Philadelphia chromosome codes for an intracellular tyrosine kinase. Review the discussion of cell cycle control and cancer in Concept 12.3 (pp. 242–243), and explain how the activation of this gene could contribute to the development of cancer.

For suggested answers, see Appendix A.

CONCEPT 15.5

Some inheritance patterns are exceptions to standard Mendelian inheritance

In the previous section, you learned about deviations from the usual patterns of chromosomal inheritance due to abnormal events in meiosis and mitosis. We conclude this chapter by describing two normally occurring exceptions to Mendelian genetics, one involving genes located in the nucleus and the other involving genes located outside the nucleus. In both cases, the sex of the parent contributing an allele is a factor in the pattern of inheritance.

Genomic Imprinting

Throughout our discussions of Mendelian genetics and the chromosomal basis of inheritance, we have assumed that a given allele will have the same effect whether it was inherited from the mother or the father. This is probably a safe assumption most of the time. For example, when Mendel crossed purple-flowered pea plants with white-flowered pea plants, he observed the same results regardless of whether the purple-flowered parent supplied the eggs or the sperm. In recent years, however, geneticists have identified two to three dozen traits in mammals that depend on which parent passed along the alleles for those traits. Such variation in phenotype depending on whether an allele is inherited from the male or female parent is called **genomic imprinting**. (Note that unlike sex-linked genes, most imprinted genes are on autosomes.)

Genomic imprinting occurs during gamete formation and results in the silencing of a particular allele of certain genes. Because these genes are imprinted differently in sperm and eggs, a zygote expresses only one allele of an imprinted gene, that inherited from either the female or the male parent. The imprints are then transmitted to all body cells during development. In each generation, the old imprints are "erased" in gamete-producing cells, and the chromosomes of the developing gametes are newly imprinted according to the sex of the individual forming the gametes. In a given species, the imprinted genes are always imprinted in the same way. For instance, a gene imprinted for maternal allele expression is always imprinted this way, generation after generation.

Consider, for example, the mouse gene for insulin-like growth factor 2 (*Igf2*), one of the first imprinted genes to be identified. Although this growth factor is required for normal prenatal growth, only the paternal allele is expressed **(Figure 15.17a)**. Evidence that the *Igf2* gene is imprinted came initially from crosses between normal-sized (wild-type) mice and dwarf (mutant) mice homozygous for a recessive mutation in the *Igf2* gene. The phenotypes of heterozygous offspring (with one normal allele and one mutant) differed, depending on whether the mutant allele came from the mother or the father **(Figure 15.17b)**.

What exactly is a genomic imprint? In many cases, it seems to consist of methyl (—CH_3) groups that are added to cytosine nucleotides of one of the alleles. Such methylation may silence the allele, an effect consistent with evidence that heavily methylated genes are usually inactive (see Chapter 18). However, for a few genes, methylation has been shown to *activate* expression of the allele. This is the case for the *Igf2* gene: Methylation of certain cytosines on the paternal chromosome leads to expression of the paternal *Igf2* allele. The apparent inconsistency as to whether methylation activates or silences alleles was resolved in part when researchers found that DNA methylation operates indirectly by recruiting enzymes that modify DNA-associated proteins (histones), leading to condensation of the local DNA. Depending on the original function of the condensed DNA in regulating allele expression, the result is either silencing or activation of a given allele.

Genomic imprinting is thought to affect only a small fraction of the genes in mammalian genomes, but most of the known imprinted genes are critical for embryonic development. In experiments with mice, for example, embryos engineered to inherit both copies of certain chromosomes from the same parent usually die before birth, whether that parent is male or female. A few years ago, however, scientists in Japan combined the genetic material from two eggs in a zygote while allowing expression of the *Igf2* gene from only one of the egg nuclei. The zygote developed into an apparently healthy mouse. Normal development seems to require that embryonic cells have exactly one active copy—not zero, not two—of certain genes. The association of aberrant imprinting

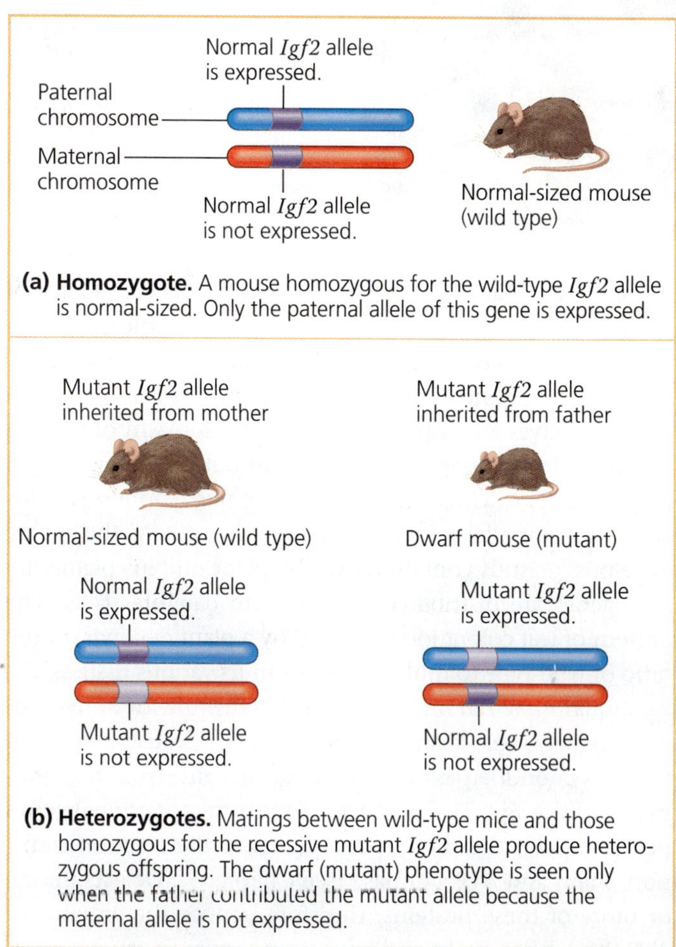

▲ **Figure 15.17 Genomic imprinting of the mouse *Igf2* gene.**

with abnormal development and certain cancers has stimulated numerous studies of how different genes are imprinted.

Inheritance of Organelle Genes

Although our focus in this chapter has been on the chromosomal basis of inheritance, we end with an important amendment: Not all of a eukaryotic cell's genes are located on nuclear chromosomes, or even in the nucleus; some genes are located in organelles in the cytoplasm. Because they are outside the nucleus, these genes are sometimes called *extranuclear genes* or *cytoplasmic genes*. Mitochondria, as well as chloroplasts and other plastids in plants, contain small circular DNA molecules that carry a number of genes. These organelles reproduce themselves and transmit their genes to daughter organelles. Organelle genes are not distributed to offspring according to the same rules that direct the distribution of nuclear chromosomes during meiosis, so they do not display Mendelian inheritance.

The first hint that extranuclear genes exist came from studies by the German scientist Karl Correns on the inheritance of yellow or white patches on the leaves of an otherwise green plant. In 1909, he observed that the coloration of the offspring was determined only by the maternal parent (the

◀ **Figure 15.18 Variegated leaves from English holly (*Ilex aquifolium*).** Variegated (striped or spotted) leaves result from mutations in pigment genes located in plastids, which generally are inherited from the maternal parent.

source of eggs) and not by the paternal parent (the source of sperm). Subsequent research showed that such coloration patterns, or variegation, are due to mutations in plastid genes that control pigmentation **(Figure 15.18)**. In most plants, a zygote receives all its plastids from the cytoplasm of the egg and none from the sperm, which contributes little more than a haploid set of chromosomes. An egg may contain plastids with different alleles for a pigmentation gene. As the zygote develops, plastids containing wild-type or mutant pigmentation genes are distributed randomly to daughter cells. The pattern of leaf coloration exhibited by a plant depends on the ratio of wild-type to mutant plastids in its various tissues.

Similar maternal inheritance is also the rule for mitochondrial genes in most animals and plants, because almost all the mitochondria passed on to a zygote come from the cytoplasm of the egg. The products of most mitochondrial genes help make up the protein complexes of the electron transport chain and ATP synthase (see Chapter 9). Defects in one or more of these proteins, therefore, reduce the amount of ATP the cell can make and have been shown to cause a number of rare human disorders. Because the parts of the body most susceptible to energy deprivation are the nervous system and the muscles, most mitochondrial diseases primarily affect these systems. For example, *mitochondrial myopathy* causes weakness, intolerance of exercise, and muscle deterioration. Another mitochondrial disorder is *Leber's hereditary optic neuropathy*, which can produce sudden blindness in people as young as their 20s or 30s. The four mutations found thus far to cause this disorder affect oxidative phosphorylation during cellular respiration, a crucial function for the cell.

In addition to the rare diseases clearly caused by defects in mitochondrial DNA, mitochondrial mutations inherited from a person's mother may contribute to at least some cases of diabetes and heart disease, as well as to other disorders that commonly debilitate the elderly, such as Alzheimer's disease. In the course of a lifetime, new mutations gradually accumulate in our mitochondrial DNA, and some researchers think that these mutations play a role in the normal aging process.

Wherever genes are located in the cell—in the nucleus or in cytoplasmic organelles—their inheritance depends on the precise replication of DNA, the genetic material. In the next chapter, you will learn how this molecular reproduction occurs.

CONCEPT CHECK 15.5

1. Gene dosage, the number of active copies of a gene, is important to proper development. Identify and describe two processes that establish the proper dosage of certain genes.
2. Reciprocal crosses between two primrose varieties, A and B, produced the following results: A female × B male → offspring with all green (nonvariegated) leaves; B female × A male → offspring with spotted (variegated) leaves. Explain these results.
3. **WHAT IF?** Mitochondrial genes are critical to the energy metabolism of cells, but mitochondrial disorders caused by mutations in these genes are generally not lethal. Why not?

For suggested answers, see Appendix A.

15 CHAPTER REVIEW

SUMMARY OF KEY CONCEPTS

CONCEPT 15.1

Mendelian inheritance has its physical basis in the behavior of chromosomes (pp. 286–289)

- The **chromosome theory of inheritance** states that genes are located on chromosomes and that the behavior of chromosomes during meiosis accounts for Mendel's laws of segregation and independent assortment.
- Morgan's discovery that transmission of the X chromosome in *Drosophila* correlates with inheritance of an eye-color trait was the first solid evidence indicating that a specific gene is associated with a specific chromosome.

? *What characteristic of the sex chromosomes allowed Morgan to correlate their behavior with that of the alleles of the eye-color gene?*

CONCEPT 15.2

Sex-linked genes exhibit unique patterns of inheritance (pp. 289–292)

- Sex is an inherited phenotypic character usually determined by which sex chromosomes are present. Humans and other mammals have an X-Y system in which sex is determined by whether a Y chromosome is present. Other systems are found in birds, fishes, and insects.
- The sex chromosomes carry **sex-linked genes** for some traits that are unrelated to sex characteristics. For instance, recessive alleles causing color blindness are **X-linked** (carried on the X chromosome). Fathers transmit this and other X-linked alleles to all daughters but to no sons. Any male who inherits such an allele from his mother will express the trait.
- In mammalian females, one of the two X chromosomes in each cell is randomly inactivated during early embryonic development,

becoming highly condensed into a **Barr body**. The descendant cells inherit the same inactivated X chromosome. If a female is heterozygous for a particular gene located on the X chromosome, she will be mosaic for that character, with about half her cells expressing the maternal allele and about half expressing the paternal allele.

? *Why are males affected much more often than females by X-linked disorders?*

CONCEPT 15.3

Linked genes tend to be inherited together because they are located near each other on the same chromosome (pp. 292–297)

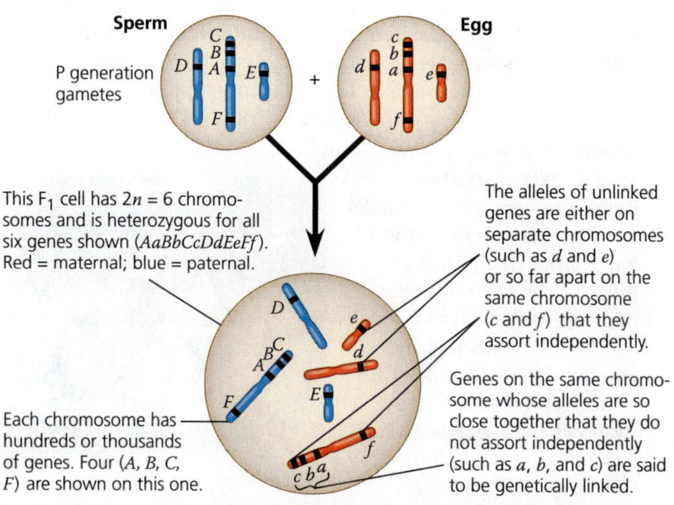

- Among offspring from an F_1 testcross, **parental types** have the same combination of traits as those in the P generation parents. **Recombinant types (recombinants)** exhibit new combinations of traits not seen in either P generation parent. Because of the independent assortment of chromosomes, unlinked genes exhibit a 50% frequency of recombination in the gametes. For genetically **linked genes**, **crossing over** between nonsister chromatids during meiosis I accounts for the observed recombinants, always less than 50% of the total.
- The order of genes on a chromosome and the relative distances between them can be deduced from recombination frequencies observed in genetic crosses. These data allow construction of a **linkage map** (a type of **genetic map**). The farther apart genes are, the more likely their allele combinations will be recombined during crossing over.

? *Why are specific alleles of two genes that are farther apart more likely to show recombination than those of two closer genes?*

CONCEPT 15.4

Alterations of chromosome number or structure cause some genetic disorders (pp. 297–300)

- **Aneuploidy**, an abnormal chromosome number, can result from **nondisjunction** during meiosis. When a normal gamete unites with one containing two copies or no copies of a particular chromosome, the resulting zygote and its descendant cells either have one extra copy of that chromosome (**trisomy**, $2n + 1$) or are missing a copy (**monosomy**, $2n - 1$). **Polyploidy** (more than two complete sets of chromosomes) can result from complete nondisjunction during gamete formation.
- Chromosome breakage can result in alterations of chromosome structure: **deletions**, **duplications**, **inversions**, and **translocations**. Translocations can be reciprocal or nonreciprocal.
- Changes in the number of chromosomes per cell or in the structure of individual chromosomes can affect the phenotype and, in some cases, lead to human disorders. Such alterations cause **Down syndrome** (usually due to trisomy of chromosome 21), certain cancers associated with chromosomal translocations, and various other human disorders.

? *Why are inversions and reciprocal translocations less likely to be lethal than are aneuploidy, duplications, deletions, and nonreciprocal translocations?*

CONCEPT 15.5

Some inheritance patterns are exceptions to standard Mendelian inheritance (pp. 300–302)

- In mammals, the phenotypic effects of a small number of particular genes depend on which allele is inherited from each parent, a phenomenon called **genomic imprinting**. Imprints are formed during gamete production, with the result that one allele (either maternal or paternal) is not expressed in offspring.
- The inheritance of traits controlled by the genes present in mitochondria and plastids depends solely on the maternal parent because the zygote's cytoplasm containing these organelles comes from the egg. Some diseases affecting the nervous and muscular systems are caused by defects in mitochondrial genes that prevent cells from making enough ATP.

? *Explain how genomic imprinting and inheritance of mitochondrial and chloroplast DNA are exceptions to standard Mendelian inheritance.*

TEST YOUR UNDERSTANDING

LEVEL 1: KNOWLEDGE/COMPREHENSION

1. A man with hemophilia (a recessive, sex-linked condition) has a daughter of normal phenotype. She marries a man who is normal for the trait. What is the probability that a daughter of this mating will be a hemophiliac? That a son will be a hemophiliac? If the couple has four sons, what is the probability that all four will be born with hemophilia?
2. Pseudohypertrophic muscular dystrophy is an inherited disorder that causes gradual deterioration of the muscles. It is seen almost exclusively in boys born to apparently normal parents and usually results in death in the early teens. Is this disorder caused by a dominant or a recessive allele? Is its inheritance sex-linked or autosomal? How do you know? Explain why this disorder is almost never seen in girls.
3. A wild-type fruit fly (heterozygous for gray body color and normal wings) is mated with a black fly with vestigial wings. The offspring have the following phenotypic distribution: wild-type, 778; black-vestigial, 785; black-normal, 158; gray-vestigial, 162. What is the recombination frequency between these genes for body color and wing size?
4. What pattern of inheritance would lead a geneticist to suspect that an inherited disorder of cell metabolism is due to a defective mitochondrial gene?
5. A space probe discovers a planet inhabited by creatures that reproduce with the same hereditary patterns seen in humans. Three phenotypic characters are height (T = tall, t = dwarf), head appendages (A = antennae, a = no antennae), and nose morphology (S = upturned snout, s = downturned snout). Since the creatures are not "intelligent," Earth scientists are able to do some controlled breeding experiments using various heterozygotes in testcrosses. For tall heterozygotes with antennae, the offspring are

tall-antennae, 46; dwarf-antennae, 7; dwarf-no antennae, 42; tall-no antennae, 5. For heterozygotes with antennae and an upturned snout, the offspring are antennae-upturned snout, 47; antennae-downturned snout, 2; no antennae-downturned snout, 48; no antennae-upturned snout, 3. Calculate the recombination frequencies for both experiments.

LEVEL 2: APPLICATION/ANALYSIS

6. Using the information from problem 5, scientists do a further testcross using a heterozygote for height and nose morphology. The offspring are: tall-upturned snout, 40; dwarf-upturned snout, 9; dwarf-downturned snout, 42; tall-downturned snout, 9. Calculate the recombination frequency from these data; then use your answer from problem 5 to determine the correct sequence of the three linked genes.
7. Red-green color blindness is caused by a sex-linked recessive allele. A color-blind man marries a woman with normal vision whose father was color-blind. What is the probability that they will have a color-blind daughter? What is the probability that their first son will be color-blind? (Note the different wording in the two questions.)
8. A wild-type fruit fly (heterozygous for gray body color and red eyes) is mated with a black fruit fly with purple eyes. The offspring are wild-type, 721; black-purple, 751; gray-purple, 49; black-red, 45. What is the recombination frequency between these genes for body color and eye color? Using information from problem 3, what fruit flies (genotypes and phenotypes) would you mate to determine the sequence of the body-color, wing-size, and eye-color genes on the chromosome?
9. **DRAW IT** A fruit fly that is true-breeding for gray body with vestigial wings ($b^+ \, b^+ \, vg \, vg$) is mated with one that is true-breeding for black body with normal wings ($b \, b \, vg^+ \, vg^+$).
   (a) Draw the chromosomes for the P generation flies, using red for the gray fly and pink for the black one. Show the position of each allele.
   (b) Draw the chromosomes and label the alleles of an F_1 fly.
   (c) Suppose an F_1 female is testcrossed. Draw the chromosomes of the resulting offspring in a Punnett square.
   (d) Knowing that the distance between these two genes is 17 map units, predict the phenotypic ratios of these offspring.
10. Women born with an extra X chromosome (XXX) are generally healthy and indistinguishable in appearance from normal XX women. What is a likely explanation for this finding? How could you test this explanation?
11. Determine the sequence of genes along a chromosome based on the following recombination frequencies: *A–B*, 8 map units; *A–C*, 28 map units; *A–D*, 25 map units; *B–C*, 20 map units; *B–D*, 33 map units.
12. Assume that genes *A* and *B* are on the same chromosome and are 50 map units apart. An animal heterozygous at both loci is crossed with one that is homozygous recessive at both loci. What percentage of the offspring will show recombinant phenotypes resulting from crossovers? Without knowing these genes are on the same chromosome, how would you interpret the results of this cross?
13. Two genes of a flower, one controlling blue (*B*) versus white (*b*) petals and the other controlling round (*R*) versus oval (*r*) stamens, are linked and are 10 map units apart. You cross a homozygous blue-oval plant with a homozygous white-round plant. The resulting F_1 progeny are crossed with homozygous white-oval plants, and 1,000 F_2 progeny are obtained. How many F_2 plants of each of the four phenotypes do you expect?
14. You design *Drosophila* crosses to provide recombination data for gene *a*, which is located on the chromosome shown in Figure 15.12. Gene *a* has recombination frequencies of 14% with the vestigial-wing locus and 26% with the brown-eye locus. Approximately where is *a* located along the chromosome?

LEVEL 3: SYNTHESIS/EVALUATION

15. Banana plants, which are triploid, are seedless and therefore sterile. Propose a possible explanation.
16. **EVOLUTION CONNECTION**
    You have seen that crossing over, or recombination, is thought to be evolutionarily advantageous because it continually shuffles genetic alleles into novel combinations, allowing evolutionary processes to occur. Until recently, it was thought that the genes on the Y chromosome might degenerate because they lack homologous genes on the X chromosome with which to recombine. However, when the Y chromosome was sequenced, eight large regions were found to be internally homologous to each other, and quite a few of the 78 genes represent duplicates. (Y chromosome researcher David Page has called it a "hall of mirrors.") What might be a benefit of these regions?
17. **SCIENTIFIC INQUIRY**

    Butterflies have an X-Y sex determination system that is different from that of flies or humans. Female butterflies may be either XY or XO, while butterflies with two or more X chromosomes are males. This photograph shows a tiger swallowtail *gynandromorph*, an individual that is half male (left side) and half female (right side). Given that the first division of the zygote divides the embryo into the future right and left halves of the butterfly, propose a hypothesis that explains how nondisjunction during the first mitosis might have produced this unusual-looking butterfly.
18. **WRITE ABOUT A THEME**
    The Genetic Basis of Life The continuity of life is based on heritable information in the form of DNA. In a short essay (100–150 words), relate the structure and behavior of chromosomes to inheritance in both asexually and sexually reproducing species.

For selected answers, see Appendix A.

16

The Molecular Basis of Inheritance

▲ **Figure 16.1 How was the structure of DNA determined?**

KEY CONCEPTS

16.1 DNA is the genetic material

16.2 Many proteins work together in DNA replication and repair

16.3 A chromosome consists of a DNA molecule packed together with proteins

OVERVIEW

Life's Operating Instructions

In April 1953, James Watson and Francis Crick shook the scientific world with an elegant double-helical model for the structure of deoxyribonucleic acid, or DNA. **Figure 16.1** shows Watson (left) and Crick admiring their DNA model, which they built from tin and wire. Over the past 60 years or so, their model has evolved from a novel proposition to an icon of modern biology. Mendel's heritable factors and Morgan's genes on chromosomes are, in fact, composed of DNA. Chemically speaking, your genetic endowment is the DNA you inherited from your parents. DNA, the substance of inheritance, is the most celebrated molecule of our time.

Of all nature's molecules, nucleic acids are unique in their ability to direct their own replication from monomers. Indeed, the resemblance of offspring to their parents has its basis in the precise replication of DNA and its transmission from one generation to the next. Hereditary information is encoded in the chemical language of DNA and reproduced in all the cells of your body. It is this DNA program that directs the development of your biochemical, anatomical, physiological, and, to some extent, behavioral traits. In this chapter, you will discover how biologists deduced that DNA is the genetic material and how Watson and Crick worked out its structure. You will also learn about **DNA replication**, the process by which a DNA molecule is copied, and how cells repair their DNA. Finally, you will explore how a molecule of DNA is packaged together with proteins in a chromosome.

CONCEPT **16.1**

DNA is the genetic material

Today, even schoolchildren have heard of DNA, and scientists routinely manipulate DNA in the laboratory, often to change the heritable traits of cells in their experiments. Early in the 20th century, however, identifying the molecules of inheritance loomed as a major challenge to biologists.

The Search for the Genetic Material: *Scientific Inquiry*

Once T. H. Morgan's group showed that genes exist as parts of chromosomes (described in Chapter 15), the two chemical components of chromosomes—DNA and protein—became the candidates for the genetic material. Until the 1940s, the case for proteins seemed stronger, especially since biochemists had identified them as a class of macromolecules with great heterogeneity and specificity of function, essential requirements for the hereditary material. Moreover, little was known about nucleic acids, whose physical and chemical properties seemed far too uniform to account for the multitude of specific inherited traits exhibited by every organism. This view gradually changed as experiments with microorganisms yielded unexpected results. As with the work of Mendel and Morgan, a key factor in determining the identity of the genetic material was the choice of appropriate experimental organisms. The role of DNA in heredity was first worked out while studying bacteria and the viruses that infect them, which are far simpler than pea plants, fruit flies, or humans. In this section, we will trace the search for the genetic material in some detail as a case study in scientific inquiry.

Evidence That DNA Can Transform Bacteria

The discovery of the genetic role of DNA dates back to 1928. While attempting to develop a vaccine against pneumonia, a British medical officer named Frederick Griffith was studying *Streptococcus pneumoniae*, a bacterium that causes pneumonia in mammals. Griffith had two strains (varieties) of the bacterium, one pathogenic (disease-causing) and one nonpathogenic (harmless). He was surprised to find that when he killed the pathogenic bacteria with heat and then mixed the cell remains with living bacteria of the nonpathogenic strain, some of the living cells became pathogenic (**Figure 16.2**). Furthermore, this newly acquired trait of pathogenicity was inherited by all the descendants of the transformed bacteria. Clearly, some chemical component of the dead pathogenic cells caused this heritable change, although the identity of the substance was not known. Griffith called the phenomenon **transformation**, now defined as a change in genotype and phenotype due to the assimilation of external DNA by a cell. (This use of the word *transformation* should not be confused with the conversion of a normal animal cell to a cancerous one, discussed near the end of Concept 12.3)

▼ Figure 16.2 INQUIRY

Can a genetic trait be transferred between different bacterial strains?

EXPERIMENT Frederick Griffith studied two strains of the bacterium *Streptococcus pneumoniae*. Bacteria of the S (smooth) strain can cause pneumonia in mice; they are pathogenic because they have an outer capsule that protects them from an animal's defense system. Bacteria of the R (rough) strain lack a capsule and are nonpathogenic. To test for the trait of pathogenicity, Griffith injected mice with the two strains:

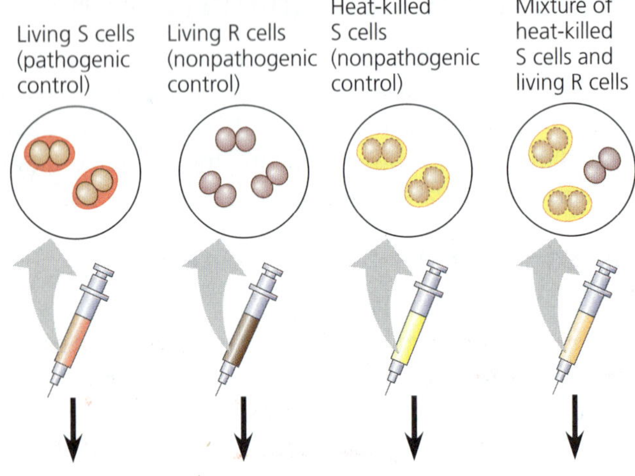

RESULTS

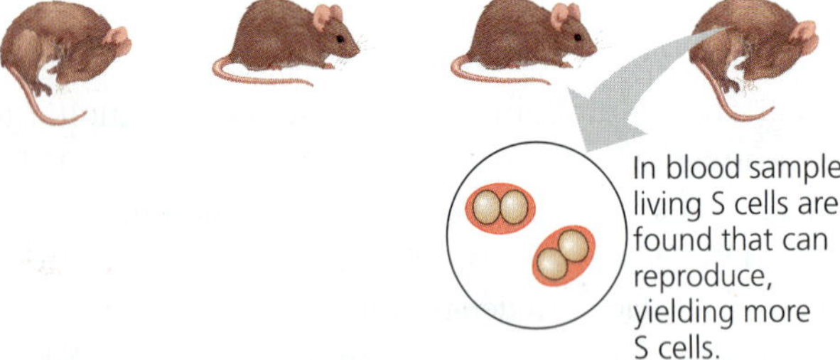

CONCLUSION Griffith concluded that the living R bacteria had been transformed into pathogenic S bacteria by an unknown, heritable substance from the dead S cells that allowed the R cells to make capsules.

SOURCE F. Griffith, The significance of pneumococcal types, *Journal of Hygiene* 27:113–159 (1928).

WHAT IF? How did this experiment rule out the possibility that the R cells could have simply used the capsules of the dead S cells to become pathogenic?

Griffith's work set the stage for a 14-year effort by American bacteriologist Oswald Avery to identify the transforming substance. Avery focused on three main candidates: DNA, RNA (the other nucleic acid in cells), and protein. Avery broke open the heat-killed pathogenic bacteria and extracted the cellular contents. He treated each of three samples with an agent that inactivated one type of molecule, then tested the sample for its ability to transform live nonpathogenic bacteria. Only when DNA was allowed to remain active did transformation occur. In 1944, Avery and his colleagues Maclyn McCarty and Colin MacLeod announced that the transforming agent was DNA. Their discovery was greeted with interest but considerable skepticism, in part because of the lingering belief that proteins were better candidates for the genetic material. Moreover, many biologists were not convinced that the genes of bacteria would be similar in composition and function to those of more complex organisms. But the major reason for the continued doubt was that so little was known about DNA.

Evidence That Viral DNA Can Program Cells

Additional evidence for DNA as the genetic material came from studies of viruses that infect bacteria (**Figure 16.3**). These viruses are called **bacteriophages** (meaning "bacteria-eaters"), or **phages** for short. Viruses are much simpler than

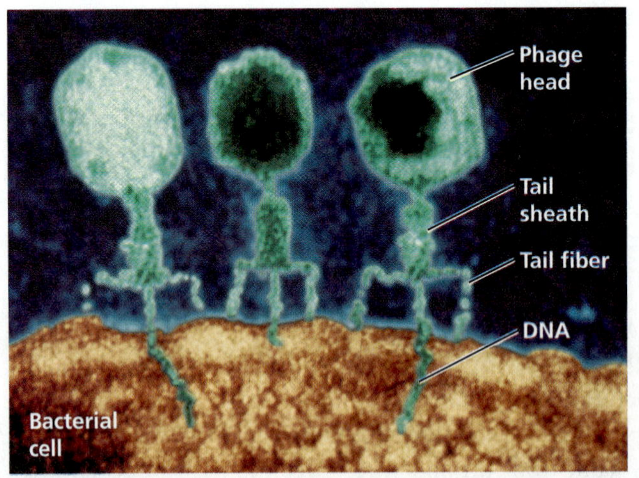

▲ **Figure 16.3 Viruses infecting a bacterial cell.** Phages called T2 attach to the host cell and inject their genetic material through the plasma membrane while the head and tail parts remain on the outer bacterial surface (colorized TEM).

cells. A **virus** is little more than DNA (or sometimes RNA) enclosed by a protective coat, which is often simply protein. To produce more viruses, a virus must infect a cell and take over the cell's metabolic machinery.

Phages have been widely used as tools by researchers in molecular genetics. In 1952, Alfred Hershey and Martha Chase performed experiments showing that DNA is the genetic material of a phage known as T2. This is one of many phages that infect *Escherichia coli* (*E. coli*), a bacterium that normally lives in the intestines of mammals and is a model organism for molecular biologists. At that time, biologists already knew that T2, like many other phages, was composed almost entirely of DNA and protein. They also knew that the T2 phage could quickly turn an *E. coli* cell into a T2-producing factory that released many copies when the cell ruptured. Somehow, T2 could reprogram its host cell to produce viruses. But which viral component—protein or DNA—was responsible?

Hershey and Chase answered this question by devising an experiment showing that only one of the two components of T2 actually enters the *E. coli* cell during infection (**Figure 16.4**). In their experiment, they used a radioactive isotope of sulfur to tag protein in one batch of T2 and a radioactive isotope of

▼ Figure 16.4 INQUIRY

Is protein or DNA the genetic material of phage T2?

EXPERIMENT Alfred Hershey and Martha Chase used radioactive sulfur and phosphorus to trace the fates of protein and DNA, respectively, of T2 phages that infected bacterial cells. They wanted to see which of these molecules entered the cells and could reprogram them to make more phages.

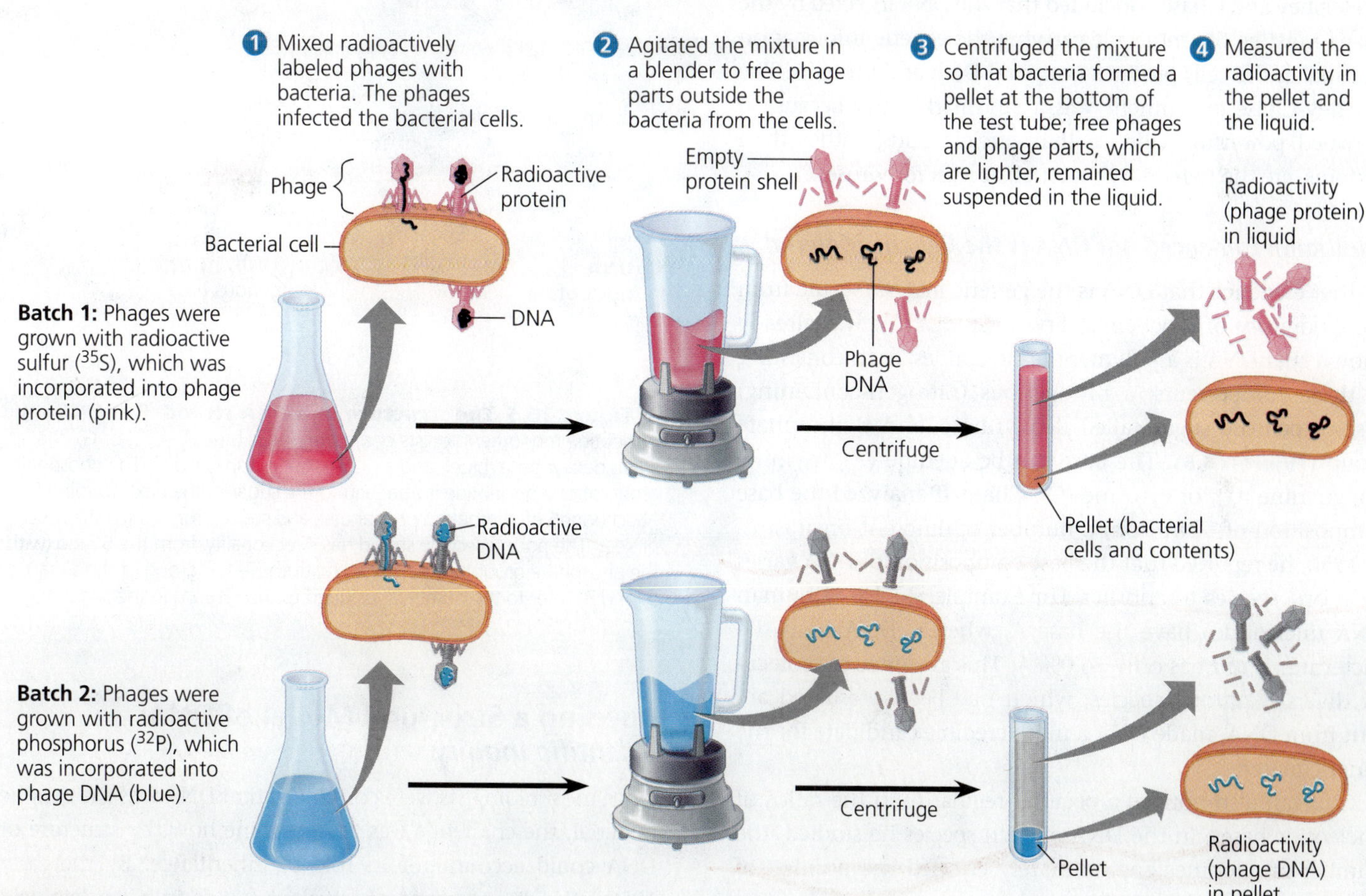

RESULTS When proteins were labeled (batch 1), radioactivity remained outside the cells; but when DNA was labeled (batch 2), radioactivity was found inside the cells. Bacterial cells with radioactive phage DNA released new phages with some radioactive phosphorus.

CONCLUSION Phage DNA entered bacterial cells, but phage proteins did not. Hershey and Chase concluded that DNA, not protein, functions as the genetic material of phage T2.

SOURCE A. D. Hershey and M. Chase, Independent functions of viral protein and nucleic acid in growth of bacteriophage, *Journal of General Physiology* 36:39–56 (1952).

WHAT IF? How would the results have differed if proteins carried the genetic information?

phosphorus to tag DNA in a second batch. Because protein, but not DNA, contains sulfur, radioactive sulfur atoms were incorporated only into the protein of the phage. In a similar way, the atoms of radioactive phosphorus labeled only the DNA, not the protein, because nearly all the phage's phosphorus is in its DNA. In the experiment, separate samples of nonradioactive *E. coli* cells were allowed to be infected by the protein-labeled and DNA-labeled batches of T2. The researchers then tested the two samples shortly after the onset of infection to see which type of molecule—protein or DNA—had entered the bacterial cells and would therefore be capable of reprogramming them.

Hershey and Chase found that the phage DNA entered the host cells but the phage protein did not. Moreover, when these bacteria were returned to a culture medium, the infection ran its course, and the *E. coli* released phages that contained some radioactive phosphorus, further showing that the DNA inside the cell played an ongoing role during the infection process.

Hershey and Chase concluded that the DNA injected by the phage must be the molecule carrying the genetic information that makes the cells produce new viral DNA and proteins. The Hershey-Chase experiment was a landmark study because it provided powerful evidence that nucleic acids, rather than proteins, are the hereditary material, at least for viruses.

Additional Evidence That DNA Is the Genetic Material

Further evidence that DNA is the genetic material came from the laboratory of biochemist Erwin Chargaff. It was already known that DNA is a polymer of nucleotides, each consisting of three components: a nitrogenous (nitrogen-containing) base, a pentose sugar called deoxyribose, and a phosphate group (**Figure 16.5**). The base can be adenine (A), thymine (T), guanine (G), or cytosine (C). Chargaff analyzed the base composition of DNA from a number of different organisms. In 1950, he reported that the base composition of DNA varies from one species to another. For example, 30.3% of human DNA nucleotides have the base A, whereas DNA from the bacterium *E. coli* has only 26.0% A. This evidence of molecular diversity among species, which had been presumed absent from DNA, made DNA a more credible candidate for the genetic material.

Chargaff also noticed a peculiar regularity in the ratios of nucleotide bases. In the DNA of each species he studied, the number of adenines approximately equaled the number of thymines, and the number of guanines approximately equaled the number of cytosines. In human DNA, for example, the four bases are present in these percentages: A = 30.3% and T = 30.3%; G = 19.5% and C = 19.9%.

These two findings became known as *Chargaff's rules*: (1) the base composition varies between species, and (2) within a species, the number of A and T bases are equal and the number of G and C bases are equal. The basis for these rules remained unexplained until the discovery of the double helix.

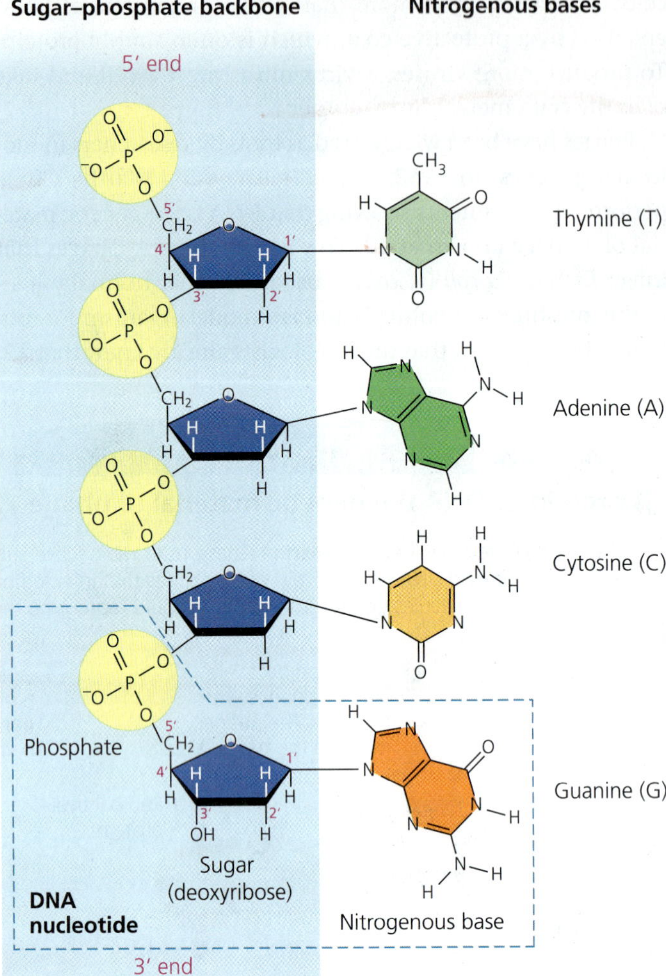

▲ **Figure 16.5 The structure of a DNA strand.** Each DNA nucleotide monomer consists of a nitrogenous base (T, A, C, or G), the sugar deoxyribose (blue), and a phosphate group (yellow). The phosphate group of one nucleotide is attached to the sugar of the next, forming a "backbone" of alternating phosphates and sugars from which the bases project. The polynucleotide strand has directionality, from the 5′ end (with the phosphate group) to the 3′ end (with the —OH group of the sugar). 5′ and 3′ refer to the numbers assigned to the carbons in the sugar ring.

Building a Structural Model of DNA: *Scientific Inquiry*

Once most biologists were convinced that DNA was the genetic material, the challenge was to determine how the structure of DNA could account for its role in inheritance. By the early 1950s, the arrangement of covalent bonds in a nucleic acid polymer was well established (see Figure 16.5), and researchers focused on discovering the three-dimensional structure of DNA. Among the scientists working on the problem were Linus Pauling, at the California Institute of Technology, and Maurice Wilkins and Rosalind Franklin, at King's College in London. First to come up with the correct answer, however, were two scientists who were relatively unknown at the time—the American James Watson and the Englishman Francis Crick.

The brief but celebrated partnership that solved the puzzle of DNA structure began soon after Watson journeyed to Cambridge University, where Crick was studying protein structure with a technique called X-ray crystallography (see Figure 5.24). While visiting the laboratory of Maurice Wilkins, Watson saw an X-ray diffraction image of DNA produced by Wilkins's accomplished colleague Rosalind Franklin **(Figure 16.6a)**. Images produced by X-ray crystallography are not actually pictures of molecules. The spots and smudges in **Figure 16.6b** were produced by X-rays that were diffracted (deflected) as they passed through aligned fibers of purified DNA. Watson was familiar with the type of X-ray diffraction pattern that helical molecules produce, and an examination of the photo that Wilkins showed him confirmed that DNA was helical in shape. It also augmented earlier data obtained by Franklin and others suggesting the width of the helix and the spacing of the nitrogenous bases along it. The pattern in this photo implied that the helix was made up of two strands, contrary to a three-stranded model that Linus Pauling had proposed a short time earlier. The presence of two strands accounts for the now-familiar term **double helix** **(Figure 16.7)**.

Watson and Crick began building models of a double helix that would conform to the X-ray measurements and what was then known about the chemistry of DNA, including Chargaff's rule of base equivalences. Having also read an unpublished annual report summarizing Franklin's work, they knew she had concluded that the sugar-phosphate backbones were on the outside of the DNA molecule, contrary to their working model. Franklin's arrangement was appealing because it put the

(a) Rosalind Franklin

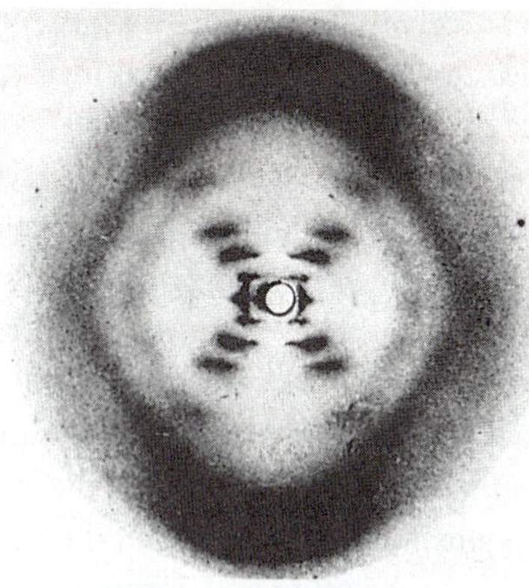

(b) Franklin's X-ray diffraction photograph of DNA

▲ **Figure 16.6 Rosalind Franklin and her X-ray diffraction photo of DNA.** Franklin, a very accomplished X-ray crystallographer, conducted critical experiments resulting in the photograph that allowed Watson and Crick to deduce the double-helical structure of DNA.

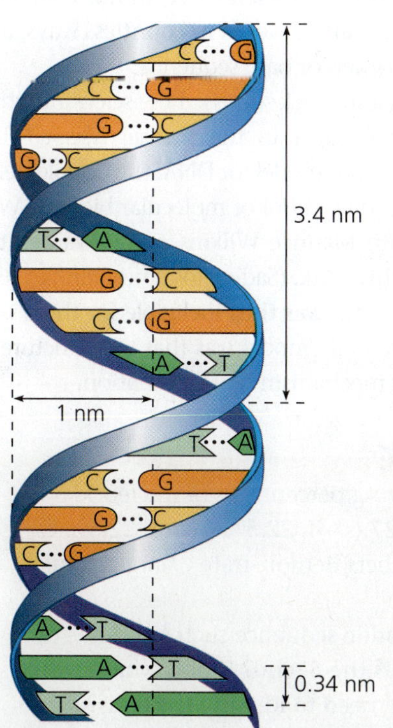

(a) Key features of DNA structure. The "ribbons" in this diagram represent the sugar-phosphate backbones of the two DNA strands. The helix is "right-handed," curving up to the right. The two strands are held together by hydrogen bonds (dotted lines) between the nitrogenous bases, which are paired in the interior of the double helix.

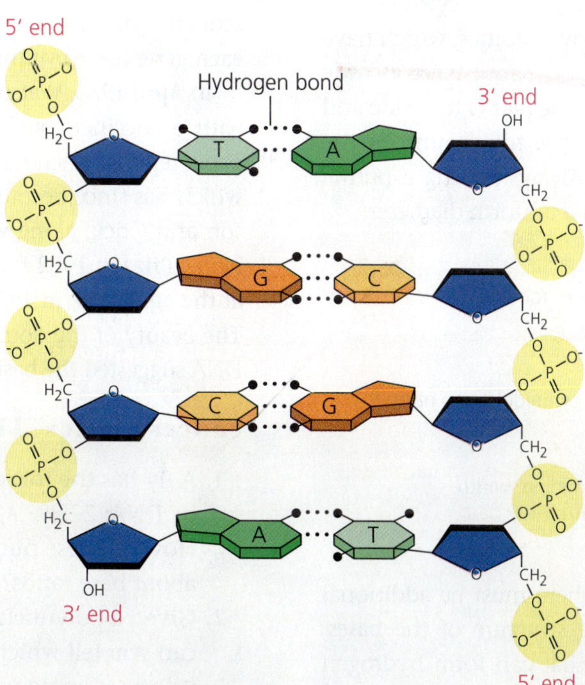

(b) Partial chemical structure. For clarity, the two DNA strands are shown untwisted in this partial chemical structure. Strong covalent bonds link the units of each strand, while weaker hydrogen bonds hold one strand to the other. Notice that the strands are antiparallel, meaning that they are oriented in opposite directions.

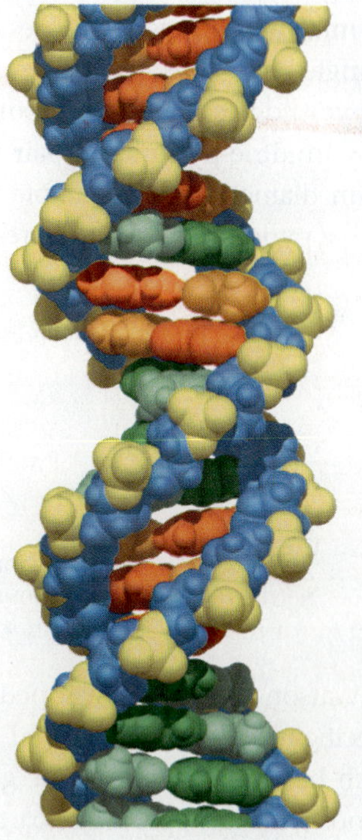

(c) Space-filling model. The tight stacking of the base pairs is clear in this computer model. Van der Waals interactions between the stacked pairs play a major role in holding the molecule together (see Chapter 2).

▲ **Figure 16.7 The double helix.**

relatively hydrophobic nitrogenous bases in the molecule's interior, away from the surrounding aqueous solution, and the negatively charged phosphate groups wouldn't be forced together in the interior. Watson constructed a model with the nitrogenous bases facing the interior of the double helix. In this model, the two sugar-phosphate backbones are **antiparallel**—that is, their subunits run in opposite directions (see Figure 16.7). You can imagine the overall arrangement as a rope ladder with rigid rungs. The side ropes represent the sugar-phosphate backbones, and the rungs represent pairs of nitrogenous bases. Now imagine holding one end of the ladder and twisting the other end, forming a spiral. Franklin's X-ray data indicated that the helix makes one full turn every 3.4 nm along its length. With the bases stacked just 0.34 nm apart, there are ten layers of base pairs, or rungs of the ladder, in each full turn of the helix.

The nitrogenous bases of the double helix are paired in specific combinations: adenine (A) with thymine (T), and guanine (G) with cytosine (C). It was mainly by trial and error that Watson and Crick arrived at this key feature of DNA. At first, Watson imagined that the bases paired like with like—for example, A with A and C with C. But this model did not fit the X-ray data, which suggested that the double helix had a uniform diameter. Why is this requirement inconsistent with like-with-like pairing of bases? Adenine and guanine are purines, nitrogenous bases with two organic rings, while cytosine and thymine are nitrogenous bases called pyrimidines, which have a single ring. Thus, purines (A and G) are about twice as wide as pyrimidines (C and T). A purine-purine pair is too wide and a pyrimidine-pyrimidine pair too narrow to account for the 2-nm diameter of the double helix. Always pairing a purine with a pyrimidine, however, results in a uniform diameter:

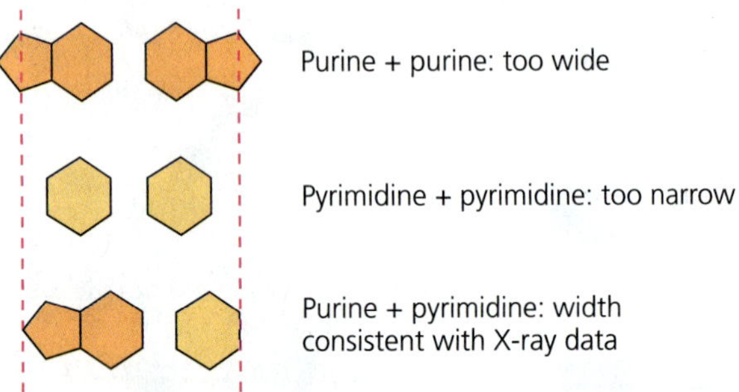

Watson and Crick reasoned that there must be additional specificity of pairing dictated by the structure of the bases. Each base has chemical side groups that can form hydrogen bonds with its appropriate partner: Adenine can form two hydrogen bonds with thymine and only thymine; guanine forms three hydrogen bonds with cytosine and only cytosine. In shorthand, A pairs with T, and G pairs with C (**Figure 16.8**).

The Watson-Crick model took into account Chargaff's ratios and ultimately explained them. Wherever one strand of a DNA molecule has an A, the partner strand has a T. And a G in one strand is always paired with a C in the complementary strand. Therefore, in the DNA of any organism, the amount of adenine equals the amount of thymine, and the amount of guanine equals the amount of cytosine. Although the base-pairing rules dictate the combinations of nitrogenous bases that form the "rungs" of the double helix, they do not restrict the sequence of nucleotides *along* each DNA strand. The linear sequence of the four bases can be varied in countless ways, and each gene has a unique order, or base sequence.

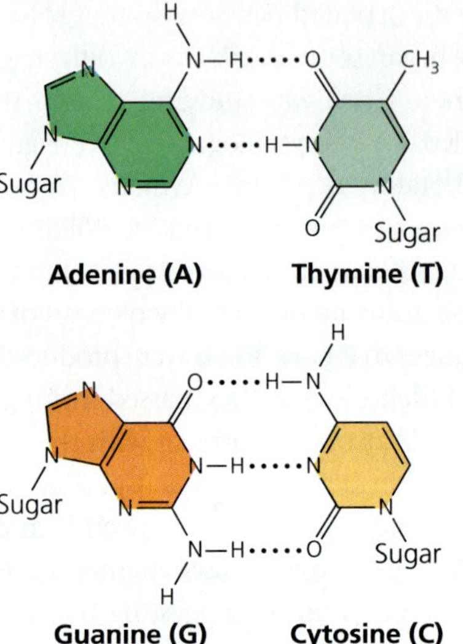

▲ **Figure 16.8 Base pairing in DNA.** The pairs of nitrogenous bases in a DNA double helix are held together by hydrogen bonds, shown here as black dotted lines.

In April 1953, Watson and Crick surprised the scientific world with a succinct, one-page paper in the journal *Nature*.* The paper reported their molecular model for DNA: the double helix, which has since become the symbol of molecular biology. Watson and Crick, along with Maurice Wilkins, were awarded the Nobel Prize in 1962 for this work. (Sadly, Rosalind Franklin died at the age of 38, in 1958, and was thus ineligible for the prize.) The beauty of the double helix model was that the structure of DNA suggested the basic mechanism of its replication.

CONCEPT CHECK 16.1

1. A fly has the following percentages of nucleotides in its DNA: 27.3% A, 27.6% T, 22.5% G, and 22.5% C. How do these numbers demonstrate Chargaff's rule about base ratios?
2. Given a polynucleotide sequence such as GAATTC, can you tell which is the 5′ end? If not, what further information do you need to identify the ends? (See Figure 16.5.)
3. **WHAT IF?** Griffith did not expect transformation to occur in his experiment. What results was he expecting? Explain.

For suggested answers, see Appendix A.

*J. D. Watson and F. H. C. Crick, Molecular structure of nucleic acids: a structure for deoxyribose nucleic acids, *Nature* 171:737–738 (1953).

CONCEPT **16.2**

Many proteins work together in DNA replication and repair

The relationship between structure and function is manifest in the double helix. The idea that there is specific pairing of nitrogenous bases in DNA was the flash of inspiration that led Watson and Crick to the double helix. At the same time, they saw the functional significance of the base-pairing rules. They ended their classic paper with this wry statement: "It has not escaped our notice that the specific pairing we have postulated immediately suggests a possible copying mechanism for the genetic material." In this section, you will learn about the basic principle of DNA replication, as well as some important details of the process.

The Basic Principle: Base Pairing to a Template Strand

In a second paper, Watson and Crick stated their hypothesis for how DNA replicates:

> Now our model for deoxyribonucleic acid is, in effect, a pair of templates, each of which is complementary to the other. We imagine that prior to duplication the hydrogen bonds are broken, and the two chains unwind and separate. Each chain then acts as a template for the formation onto itself of a new companion chain, so that eventually we shall have two pairs of chains, where we only had one before. Moreover, the sequence of the pairs of bases will have been duplicated exactly.*

Figure 16.9 illustrates Watson and Crick's basic idea. To make it easier to follow, we show only a short section of double helix in untwisted form. Notice that if you cover one of the two DNA strands of Figure 16.9a, you can still determine its linear sequence of nucleotides by referring to the uncovered strand and applying the base-pairing rules. The two strands are complementary; each stores the information necessary to reconstruct the other. When a cell copies a DNA molecule, each strand serves as a template for ordering nucleotides into a new, complementary strand. Nucleotides line up along the template strand according to the base-pairing rules and are linked to form the new strands. Where there was one double-stranded DNA molecule at the beginning of the process, there are soon two, each an exact replica of the "parental" molecule. The copying mechanism is analogous to using a photographic negative to make a positive image, which can in turn be used to make another negative, and so on.

This model of DNA replication remained untested for several years following publication of the DNA structure. The requisite experiments were simple in concept but difficult to perform. Watson and Crick's model predicts that when a double helix replicates, each of the two daughter molecules will have one old strand, from the parental molecule, and one newly made strand. This **semiconservative model** can be distinguished from a conservative model of replication, in which the two parental strands somehow come back together after the process (that is, the parental molecule is conserved). In yet a third model, called the dispersive model, all four strands of DNA following replication have a mixture of old and new DNA. These three models are shown in

*F. H. C. Crick and J. D. Watson, The complementary structure of deoxyribonucleic acid, *Proceedings of the Royal Society of London A* 223:80 (1954).

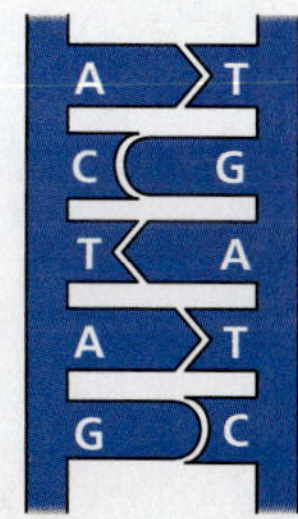

(a) The parental molecule has two complementary strands of DNA. Each base is paired by hydrogen bonding with its specific partner, A with T and G with C.

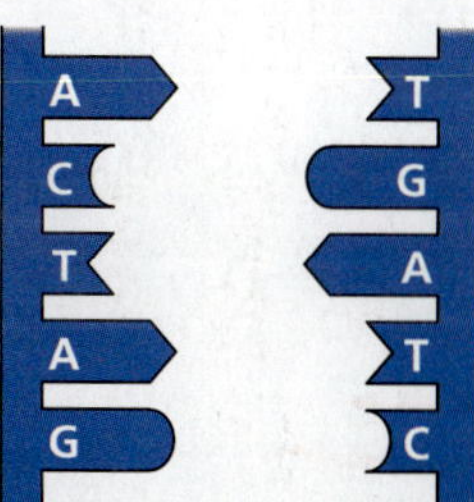

(b) The first step in replication is separation of the two DNA strands. Each parental strand can now serve as a template that determines the order of nucleotides along a new, complementary strand.

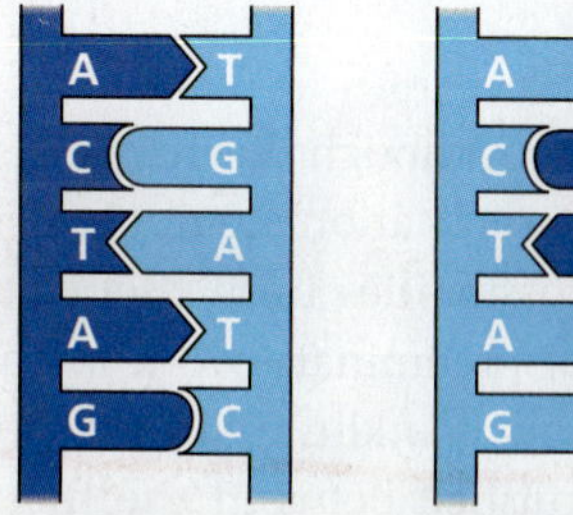

(c) The complementary nucleotides line up and are connected to form the sugar-phosphate backbones of the new strands. Each "daughter" DNA molecule consists of one parental strand (dark blue) and one new strand (light blue).

▲ **Figure 16.9 A model for DNA replication: the basic concept.** In this simplified illustration, a short segment of DNA has been untwisted into a structure that resembles a ladder. The side rails of the ladder are the sugar-phosphate backbones of the two DNA strands; the rungs are the pairs of nitrogenous bases. Simple shapes symbolize the four kinds of bases. Dark blue represents DNA strands present in the parental molecule; light blue represents newly synthesized DNA.

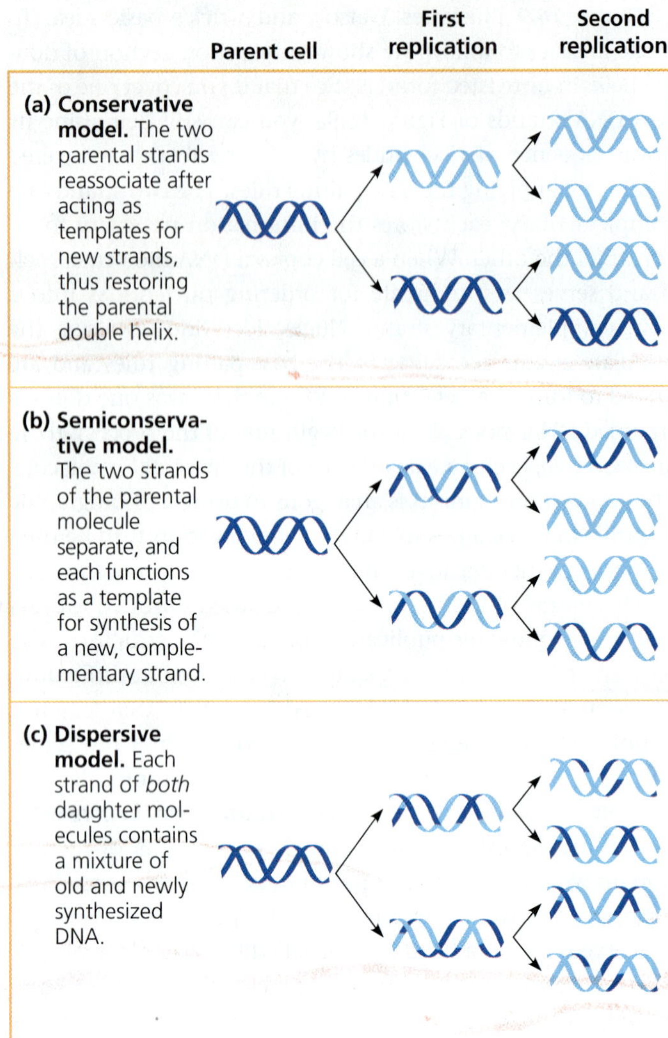

▲ **Figure 16.10 Three alternative models of DNA replication.** Each short segment of double helix symbolizes the DNA within a cell. Beginning with a parent cell, we follow the DNA for two more generations of cells—two rounds of DNA replication. Newly made DNA is light blue.

Figure 16.10. Although mechanisms for conservative or dispersive DNA replication are not easy to devise, these models remained possibilities until they could be ruled out. After two years of preliminary work in the late 1950s, Matthew Meselson and Franklin Stahl devised a clever experiment that distinguished between the three models, described in detail in **Figure 16.11**. Their experiment supported the semiconservative model of DNA replication, as predicted by Watson and Crick, and is widely acknowledged among biologists to be a classic example of elegant experimental design.

The basic principle of DNA replication is conceptually simple. However, the actual process involves some complicated biochemical gymnastics, as we will now see.

DNA Replication: *A Closer Look*

The bacterium *E. coli* has a single chromosome of about 4.6 million nucleotide pairs. In a favorable environment, an *E. coli* cell

▼ **Figure 16.11** INQUIRY

Does DNA replication follow the conservative, semiconservative, or dispersive model?

EXPERIMENT At the California Institute of Technology, Matthew Meselson and Franklin Stahl cultured *E. coli* for several generations in a medium containing nucleotide precursors labeled with a heavy isotope of nitrogen, ^{15}N. They then transferred the bacteria to a medium with only ^{14}N, a lighter isotope. A sample was taken after DNA replicated once; another sample was taken after DNA replicated again. They extracted DNA from the bacteria in the samples and then centrifuged each DNA sample to separate DNA of different densities.

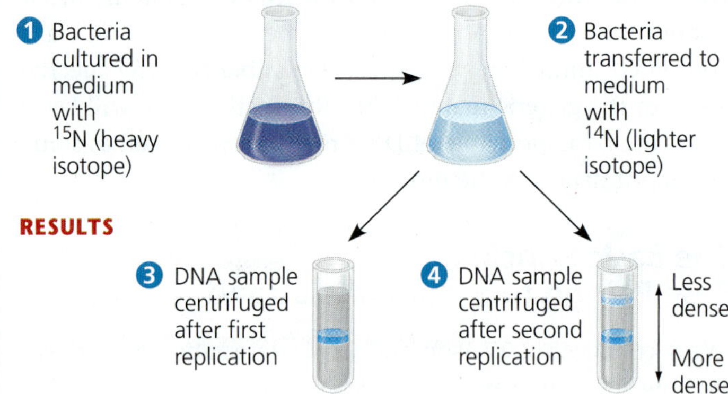

CONCLUSION Meselson and Stahl compared their results to those predicted by each of the three models in Figure 16.10, as shown below. The first replication in the ^{14}N medium produced a band of hybrid (^{15}N-^{14}N) DNA. This result eliminated the conservative model. The second replication produced both light and hybrid DNA, a result that refuted the dispersive model and supported the semiconservative model. They therefore concluded that DNA replication is semiconservative.

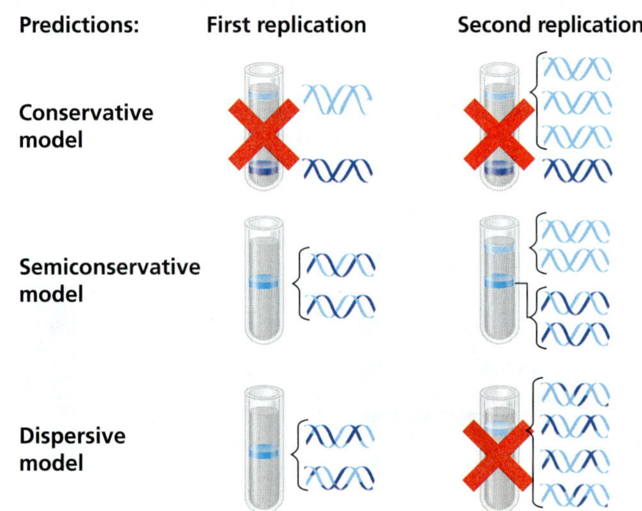

SOURCE M. Meselson and F. W. Stahl, The replication of DNA in *Escherichia coli, Proceedings of the National Academy of Sciences USA* 44:671–682 (1958).

INQUIRY IN ACTION Read and analyze the original paper in *Inquiry in Action: Interpreting Scientific Papers.*

MB See the related Experimental Inquiry Tutorial in MasteringBiology.

WHAT IF? If Meselson and Stahl had first grown the cells in ^{14}N-containing medium and then moved them into ^{15}N-containing medium before taking samples, what would have been the result?

can copy all this DNA and divide to form two genetically identical daughter cells in less than an hour. Each of *your* cells has 46 DNA molecules in its nucleus, one long double-helical molecule per chromosome. In all, that represents about 6 billion nucleotide pairs, or over a thousand times more DNA than is found in a bacterial cell. If we were to print the one-letter symbols for these bases (A, G, C, and T) the size of the type you are now reading, the 6 billion nucleotide pairs of information in a diploid human cell would fill about 1,200 books as thick as this text. Yet it takes one of your cells just a few hours to copy all of this DNA. This replication of an enormous amount of genetic information is achieved with very few errors—only about one per 10 billion nucleotides. The copying of DNA is remarkable in its speed and accuracy.

More than a dozen enzymes and other proteins participate in DNA replication. Much more is known about how this "replication machine" works in bacteria (such as *E. coli*) than in eukaryotes, and we will describe the basic steps of the process for *E. coli*, except where otherwise noted. What scientists have learned about eukaryotic DNA replication suggests, however, that most of the process is fundamentally similar for prokaryotes and eukaryotes.

Getting Started

The replication of a DNA molecule begins at particular sites called **origins of replication**, short stretches of DNA having a specific sequence of nucleotides. The *E. coli* chromosome, like many other bacterial chromosomes, is circular and has a single origin. Proteins that initiate DNA replication recognize this sequence and attach to the DNA, separating the two strands and opening up a replication "bubble." Replication of DNA then proceeds in both directions until the entire molecule is copied **(Figure 16.12a)**. In contrast to a bacterial chromosome, a eukaryotic chromosome may have hundreds

▼ Figure 16.12 Origins of replication in *E. coli* and eukaryotes. The red arrows indicate the movement of the replication forks and thus the overall directions of DNA replication within each bubble.

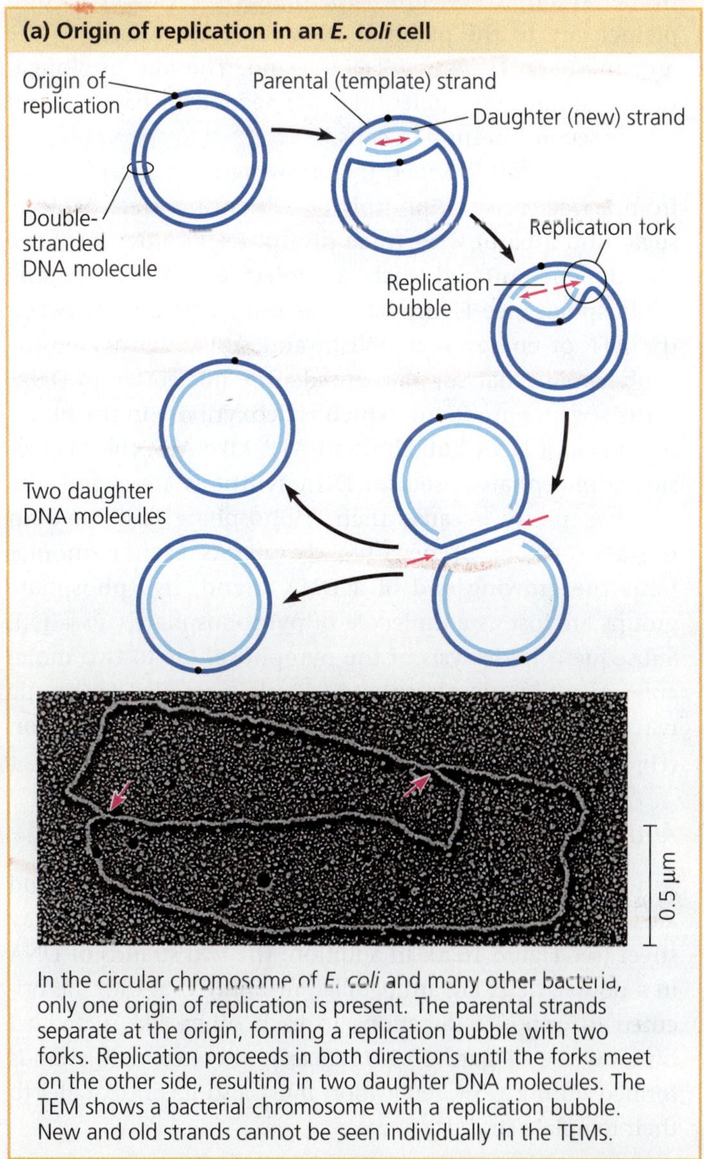

In the circular chromosome of *E. coli* and many other bacteria, only one origin of replication is present. The parental strands separate at the origin, forming a replication bubble with two forks. Replication proceeds in both directions until the forks meet on the other side, resulting in two daughter DNA molecules. The TEM shows a bacterial chromosome with a replication bubble. New and old strands cannot be seen individually in the TEMs.

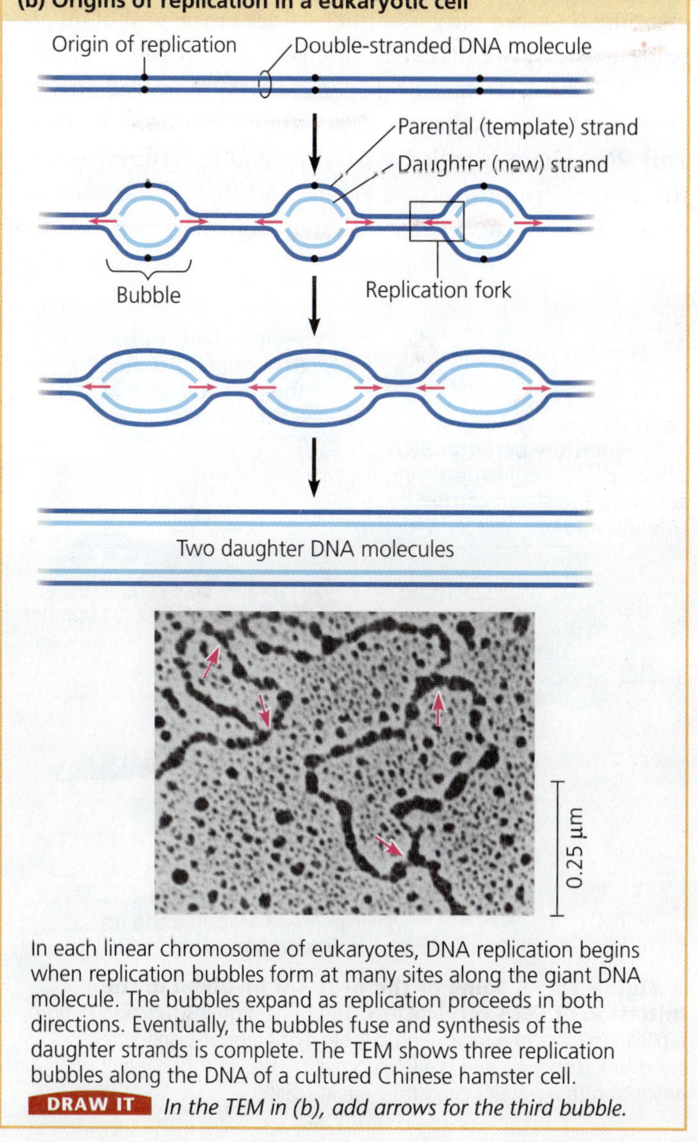

In each linear chromosome of eukaryotes, DNA replication begins when replication bubbles form at many sites along the giant DNA molecule. The bubbles expand as replication proceeds in both directions. Eventually, the bubbles fuse and synthesis of the daughter strands is complete. The TEM shows three replication bubbles along the DNA of a cultured Chinese hamster cell.

DRAW IT *In the TEM in (b), add arrows for the third bubble.*

or even a few thousand replication origins. Multiple replication bubbles form and eventually fuse, thus speeding up the copying of the very long DNA molecules (**Figure 16.12b**). As in bacteria, eukaryotic DNA replication proceeds in both directions from each origin.

At each end of a replication bubble is a **replication fork**, a Y-shaped region where the parental strands of DNA are being unwound. Several kinds of proteins participate in the unwinding (**Figure 16.13**). **Helicases** are enzymes that untwist the double helix at the replication forks, separating the two parental strands and making them available as template strands. After the parental strands separate, **single-strand binding proteins** bind to the unpaired DNA strands, keeping them from re-pairing. The untwisting of the double helix causes tighter twisting and strain ahead of the replication fork. **Topoisomerase** helps relieve this strain by breaking, swiveling, and rejoining DNA strands.

The unwound sections of parental DNA strands are now available to serve as templates for the synthesis of new complementary DNA strands. However, the enzymes that synthesize DNA cannot *initiate* the synthesis of a polynucleotide; they can only add nucleotides to the end of an already existing chain that is base-paired with the template strand. The initial nucleotide chain that is produced during DNA synthesis is actually a short stretch of RNA, not DNA. This RNA chain is called a **primer** and is synthesized by the enzyme **primase** (see Figure 16.13). Primase starts a complementary RNA chain from a single RNA nucleotide, adding RNA nucleotides one at a time, using the parental DNA strand as a template. The completed primer, generally 5–10 nucleotides long, is thus base-paired to the template strand. The new DNA strand will start from the 3′ end of the RNA primer.

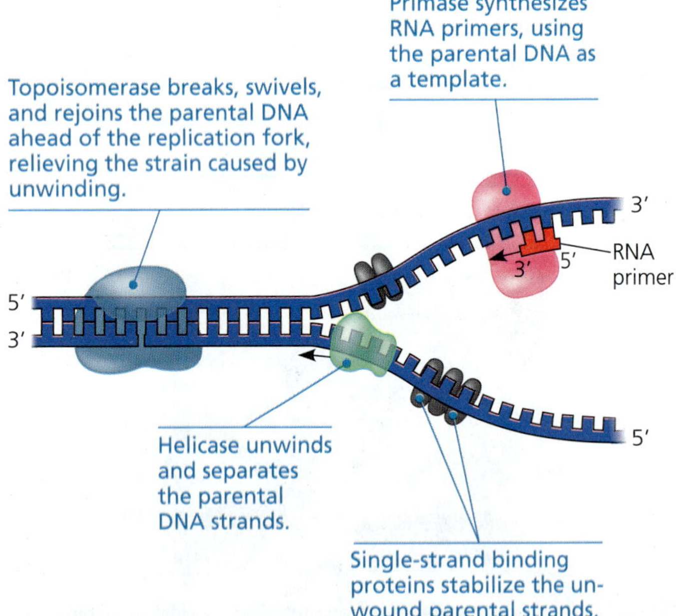

▲ **Figure 16.13 Some of the proteins involved in the initiation of DNA replication.** The same proteins function at both replication forks in a replication bubble. For simplicity, only the left-hand fork is shown, and the DNA bases are drawn much larger in relation to the proteins than they are in reality.

Synthesizing a New DNA Strand

Enzymes called **DNA polymerases** catalyze the synthesis of new DNA by adding nucleotides to a preexisting chain. In *E. coli*, there are several different DNA polymerases, but two appear to play the major roles in DNA replication: DNA polymerase III and DNA polymerase I. The situation in eukaryotes is more complicated, with at least 11 different DNA polymerases discovered so far; however, the general principles are the same.

Most DNA polymerases require a primer and a DNA template strand, along which complementary DNA nucleotides line up. In *E. coli*, DNA polymerase III (abbreviated DNA pol III) adds a DNA nucleotide to the RNA primer and then continues adding DNA nucleotides, complementary to the parental DNA template strand, to the growing end of the new DNA strand. The rate of elongation is about 500 nucleotides per second in bacteria and 50 per second in human cells.

Each nucleotide added to a growing DNA strand comes from a nucleoside triphosphate, which is a nucleoside (a sugar and a base) with three phosphate groups. You have already encountered such a molecule—ATP (adenosine triphosphate; see Figure 8.8). The only difference between the ATP of energy metabolism and dATP, the nucleoside triphosphate that supplies an adenine nucleotide to DNA, is the sugar component, which is deoxyribose in the building block of DNA but ribose in ATP. Like ATP, the nucleoside triphosphates used for DNA synthesis are chemically reactive, partly because their triphosphate tails have an unstable cluster of negative charge. As each monomer joins the growing end of a DNA strand, two phosphate groups are lost as a molecule of pyrophosphate (Ⓟ—Ⓟ$_i$). Subsequent hydrolysis of the pyrophosphate to two molecules of inorganic phosphate (Ⓟ$_i$) is a coupled exergonic reaction that helps drive the polymerization reaction (**Figure 16.14**).

Antiparallel Elongation

As we have noted previously, the two ends of a DNA strand are different, giving each strand directionality, like a one-way street (see Figure 16.5). In addition, the two strands of DNA in a double helix are antiparallel, meaning that they are oriented in opposite directions to each other, like a divided highway (see Figure 16.14). Clearly, the two new strands formed during DNA replication must also be antiparallel to their template strands.

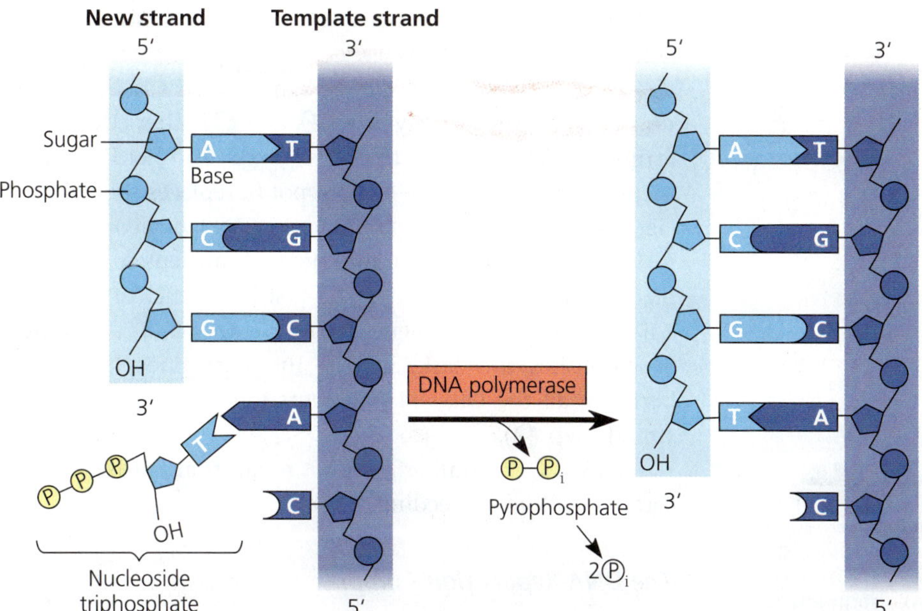

◀ **Figure 16.14 Incorporation of a nucleotide into a DNA strand.** DNA polymerase catalyzes the addition of a nucleoside triphosphate to the 3′ end of a growing DNA strand, with the release of two phosphates.

? *Use this diagram to explain what we mean when we say that each DNA strand has directionality.*

How does the antiparallel arrangement of the double helix affect replication? Because of their structure, DNA polymerases can add nucleotides only to the free 3′ end of a primer or growing DNA strand, never to the 5′ end (see Figure 16.14). Thus, a new DNA strand can elongate only in the 5′ → 3′ direction. With this in mind, let's examine one of the two replication forks in a bubble **(Figure 16.15)**. Along one template strand, DNA polymerase III can synthesize a complementary strand continuously by elongating the new DNA in the mandatory 5′ → 3′ direction. DNA pol III remains in the replication fork on that template strand and continuously adds nucleotides to the new complementary strand as the fork progresses. The DNA strand made by this mechanism is called the **leading strand**. Only one primer is required for DNA pol III to synthesize the leading strand (see Figure 16.15).

To elongate the other new strand of DNA in the mandatory 5′ → 3′ direction, DNA pol III must work along the other template strand in the direction *away from* the replication fork. The DNA strand elongating in this direction is called the **lagging strand**.* In contrast to the leading strand, which elongates continuously, the lagging strand is synthesized discontinuously, as a series of segments. These segments of the lagging strand are called **Okazaki fragments**, after the Japanese scientist who discovered them. The fragments are about 1,000–2,000 nucleotides long in *E. coli* and 100–200 nucleotides long in eukaryotes.

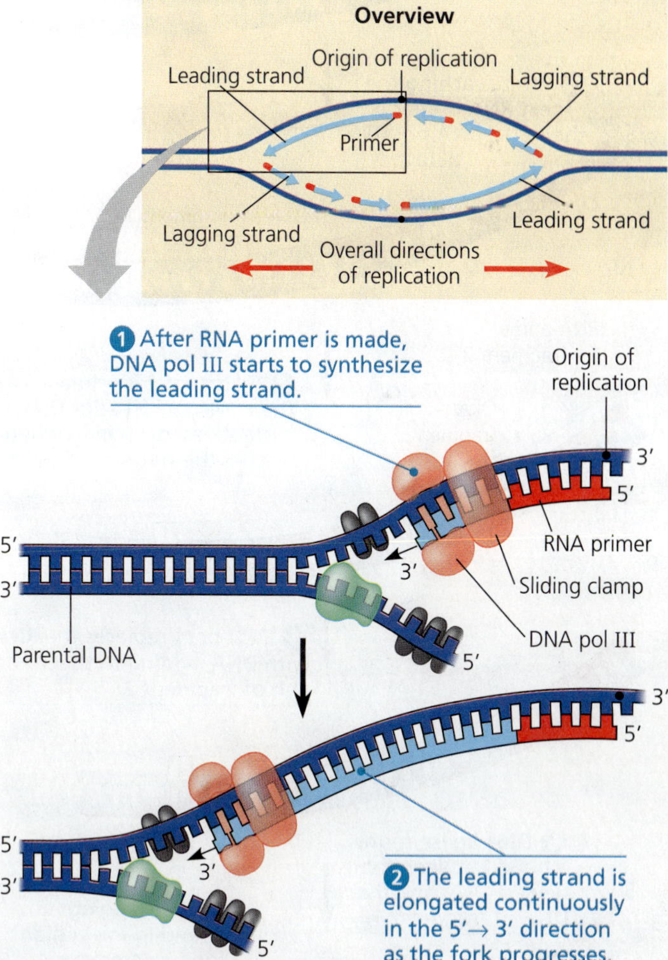

▲ **Figure 16.15 Synthesis of the leading strand during DNA replication.** This diagram focuses on the left replication fork shown in the overview box. DNA polymerase III (DNA pol III), shaped like a cupped hand, is shown closely associated with a protein called the "sliding clamp" that encircles the newly synthesized double helix like a doughnut. The sliding clamp moves DNA pol III along the DNA template strand.

*Synthesis of the leading strand and synthesis of the lagging strand occur concurrently and at the same rate. The lagging strand is so named because its synthesis is delayed slightly relative to synthesis of the leading strand; each new fragment of the lagging strand cannot be started until enough template has been exposed at the replication fork.

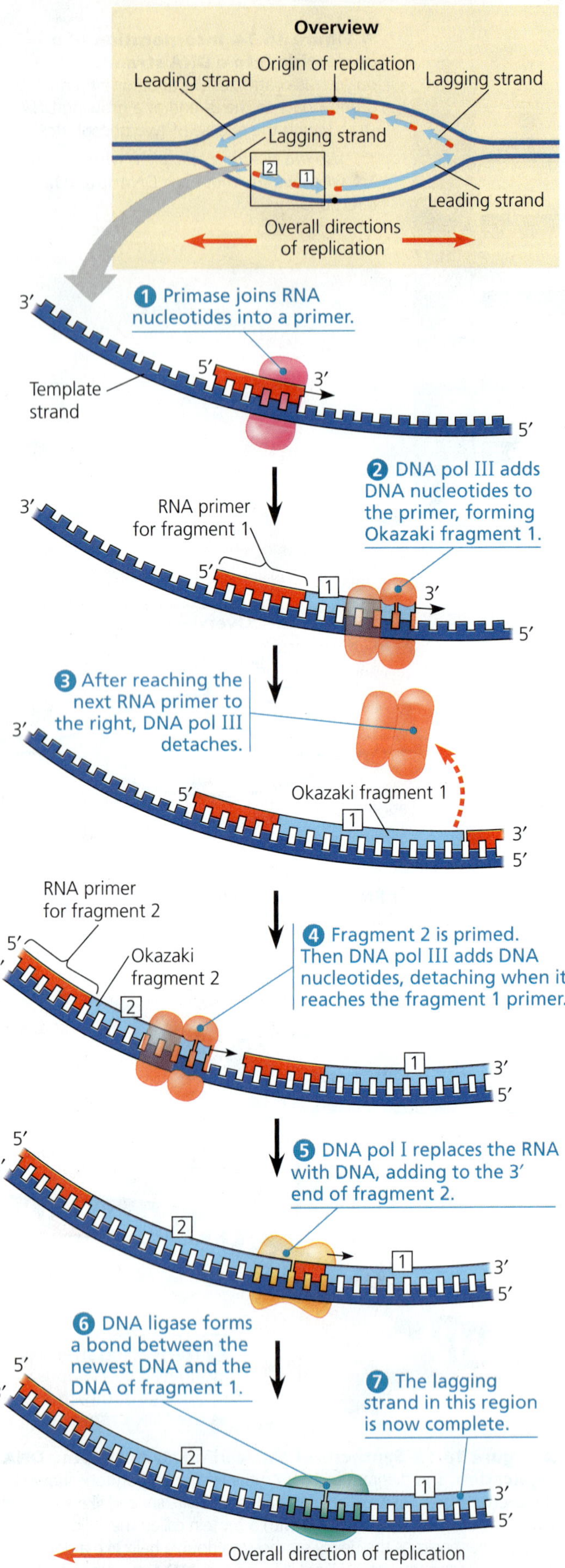

▲ **Figure 16.16 Synthesis of the lagging strand.**

Figure 16.16 illustrates the steps in the synthesis of the lagging strand at one fork. Whereas only one primer is required on the leading strand, each Okazaki fragment on the lagging strand must be primed separately (steps 1 and 4). After DNA pol III forms an Okazaki fragment (steps 2–4), another DNA polymerase, DNA polymerase I (DNA pol I), replaces the RNA nucleotides of the adjacent primer with DNA nucleotides (step 5). But DNA pol I cannot join the final nucleotide of this replacement DNA segment to the first DNA nucleotide of the adjacent Okazaki fragment. Another enzyme, **DNA ligase**, accomplishes this task, joining the sugar-phosphate backbones of all the Okazaki fragments into a continuous DNA strand (step 6).

Figure 16.17 summarizes DNA replication. Please study it carefully before proceeding.

The DNA Replication Complex

It is traditional—and convenient—to represent DNA polymerase molecules as locomotives moving along a DNA "railroad track," but such a model is inaccurate in two important ways. First, the various proteins that participate in DNA replication actually form a single large complex, a "DNA replication machine." Many protein-protein interactions facilitate the efficiency of this complex. For example, by interacting with other proteins at the fork, primase apparently acts as a molecular brake, slowing progress of the replication fork and coordinating the placement of primers and the rates of replication on the leading and lagging strands. Second, the DNA replication complex may not move along the DNA; rather, the DNA may move through the complex during the replication process. In eukaryotic cells, multiple copies of the complex, perhaps grouped into "factories," may be anchored to the nuclear matrix, a framework of fibers extending through the interior of the nucleus. Recent studies support a model in which two DNA polymerase molecules, one on each template strand, "reel in" the parental DNA and extrude newly made daughter DNA molecules. Additional evidence suggests that the lagging strand is looped back through the complex **(Figure 16.18).**

Proofreading and Repairing DNA

We cannot attribute the accuracy of DNA replication solely to the specificity of base pairing. Although errors in the completed DNA molecule amount to only one in 10^{10} (10 billion) nucleotides, initial pairing errors between incoming nucleotides and those in the template strand are 100,000 times more common—an error rate of one in 10^5 nucleotides. During DNA replication, DNA polymerases proofread each nucleotide against its template as soon as it is added to the growing strand. Upon finding an incorrectly paired nucleotide, the polymerase removes the nucleotide and then resumes synthesis. (This action is similar to fixing a typing error by deleting the wrong letter and then entering the correct letter.)

Overview

Origin of replication

Leading strand

Lagging strand

Lagging strand

Leading strand

Overall directions of replication

❶ Helicase unwinds the parental double helix.

❷ Molecules of single-strand binding protein stabilize the unwound template strands.

❸ The leading strand is synthesized continuously in the 5′ → 3′ direction by DNA pol III.

❹ Primase begins synthesis of the RNA primer for the fifth Okazaki fragment.

❺ DNA pol III is completing synthesis of the fourth fragment. When it reaches the RNA primer on the third fragment, it will detach and begin adding DNA nucleotides to the 3′ end of the fifth fragment primer in the replication fork.

❻ DNA pol I removes the primer from the 5′ end of the second fragment, replacing it with DNA nucleotides that it adds one by one to the 3′ end of the third fragment. The replacement of the last RNA nucleotide with DNA leaves the sugar-phosphate backbone with a free 3′ end.

❼ DNA ligase joins the 3′ end of the second fragment to the 5′ end of the first fragment.

Leading strand

Parental DNA

DNA pol III

Primer

Primase

DNA pol III

Lagging strand

DNA pol I

DNA ligase

▲ **Figure 16.17 A summary of bacterial DNA replication.** The detailed diagram shows the left-hand replication fork of the replication bubble in the overview (upper right). Viewing each daughter strand in its entirety in the overview, you can see that half of it is made continuously as the leading strand, while the other half (on the other side of the origin) is synthesized in fragments as the lagging strand.

DRAW IT *Draw a similar diagram showing the right-hand fork of this bubble, numbering the Okazaki fragments appropriately.*

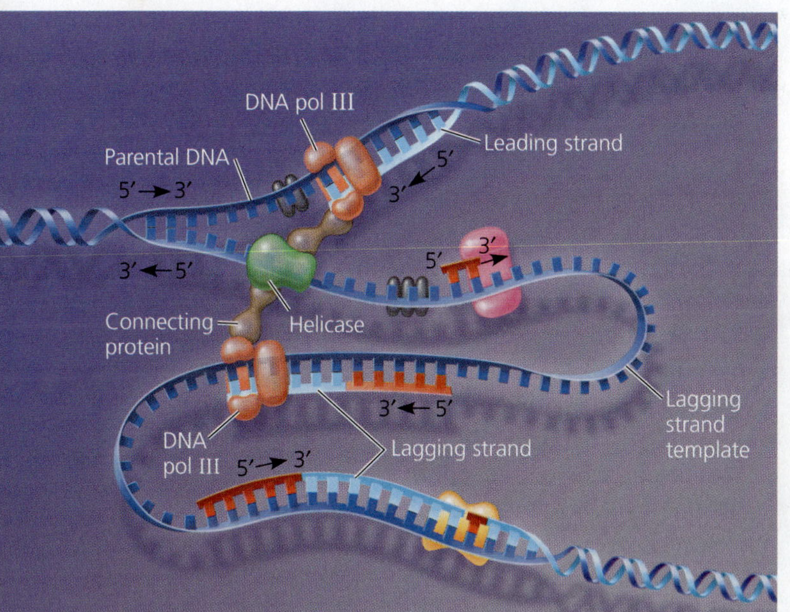

▲ **Figure 16.18 A current model of the DNA replication complex.** Two DNA polymerase III molecules work together in a complex, one on each template strand. The lagging strand template DNA loops through the complex.

BioFlix Visit the Study Area at **www.masteringbiology.com** for the BioFlix® 3-D Animation on DNA Replication.

Mismatched nucleotides sometimes evade proofreading by a DNA polymerase. In **mismatch repair**, other enzymes remove and replace incorrectly paired nucleotides that have resulted from replication errors. Researchers spotlighted the importance of such repair enzymes when they found that a hereditary defect in one of them is associated with a form of colon cancer. Apparently, this defect allows cancer-causing errors to accumulate in the DNA faster than normal.

Incorrectly paired or altered nucleotides can also arise after replication. In fact, maintenance of the genetic information encoded in DNA requires frequent repair of various kinds of damage to existing DNA. DNA molecules are constantly subjected to potentially harmful chemical and physical agents, such as cigarette smoke and X-rays, as we'll discuss in Chapter 17. In addition, DNA bases often undergo spontaneous chemical changes under normal cellular conditions. However, these changes in DNA are usually corrected before they become permanent changes—*mutations*—perpetuated through successive replications. Each cell continuously monitors and repairs its genetic material. Because repair of damaged DNA is so important to the survival of an organism, it is no surprise that many different DNA repair enzymes have evolved. Almost 100 are known in *E. coli*, and about 130 have been identified so far in humans.

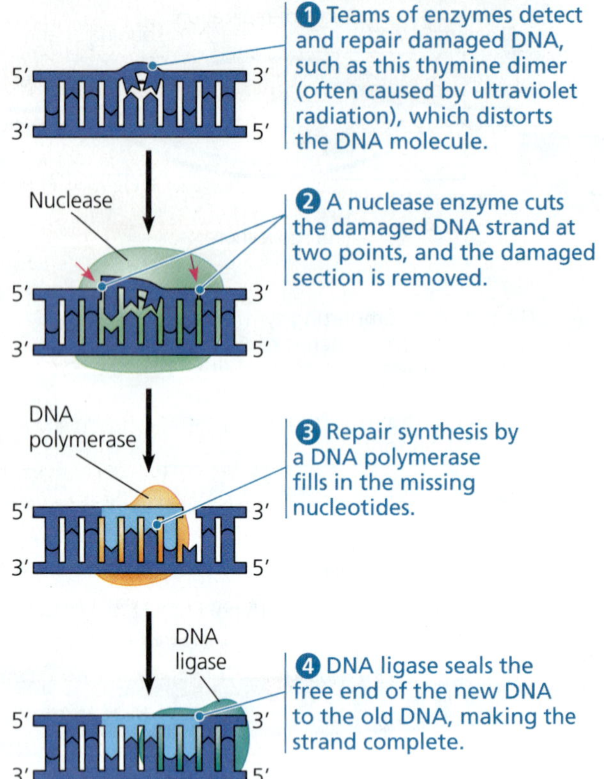

▲ **Figure 16.19 Nucleotide excision repair of DNA damage.**

Most cellular systems for repairing incorrectly paired nucleotides, whether they are due to DNA damage or to replication errors, use a mechanism that takes advantage of the base-paired structure of DNA. In many cases, a segment of the strand containing the damage is cut out (excised) by a DNA-cutting enzyme—a **nuclease**—and the resulting gap is then filled in with nucleotides, using the undamaged strand as a template. The enzymes involved in filling the gap are a DNA polymerase and DNA ligase. One such DNA repair system is called **nucleotide excision repair (Figure 16.19)**.

An important function of the DNA repair enzymes in our skin cells is to repair genetic damage caused by the ultraviolet rays of sunlight. One type of damage, shown in Figure 16.19, is the covalent linking of thymine bases that are adjacent on a DNA strand. Such *thymine dimers* cause the DNA to buckle and interfere with DNA replication. The importance of repairing this kind of damage is underscored by the disorder xeroderma pigmentosum, which in most cases is caused by an inherited defect in a nucleotide excision repair enzyme. Individuals with this disorder are hypersensitive to sunlight; mutations in their skin cells caused by ultraviolet light are left uncorrected, resulting in skin cancer.

Evolutionary Significance of Altered DNA Nucleotides

EVOLUTION Faithful replication of the genome and repair of DNA damage are important for the functioning of the organism and for passing on a complete, accurate genome to the next generation. The error rate after proofreading and repair is extremely low, but rare mistakes do slip through. Once a mismatched nucleotide pair is replicated, the sequence change is permanent in the daughter molecule that has the incorrect nucleotide as well as in any subsequent copies. As you know, a permanent change in the DNA sequence is called a mutation.

As you'll learn in Chapter 17, mutations can change the phenotype of an organism. And if they occur in germ cells (which give rise to gametes), mutations can be passed on from generation to generation. The vast majority of such changes are harmful, but a very small percentage can be beneficial. In either case, mutations are the source of the variation on which natural selection operates during evolution and are ultimately responsible for the appearance of new species. (You'll learn more about this process in Unit Four.) The balance between complete fidelity of DNA replication or repair and a low mutation rate has, over long periods of time, allowed the evolution of the rich diversity of species we see on Earth today.

Replicating the Ends of DNA Molecules

In spite of the impressive capabilities of DNA polymerases, there is a small portion of the cell's DNA that DNA polymerases can neither replicate nor repair. For linear DNA, such as the DNA of eukaryotic chromosomes, the fact that a DNA polymerase can add nucleotides only to the 3′ end of a preexisting polynucleotide leads to what might appear to be a problem. The usual replication machinery provides no way to complete the 5′ ends of daughter DNA strands. Even if an Okazaki fragment can be started with an RNA primer bound to the very end of the template strand, once that primer is removed, it cannot be replaced with DNA because there is no 3′ end available for nucleotide addition **(Figure 16.20)**. As a result, repeated rounds of replication produce shorter and shorter DNA molecules with uneven ("staggered") ends.

Most prokaryotes have a circular chromosome, with no ends, so the shortening of DNA does not occur. But what protects the genes of linear eukaryotic chromosomes from being eroded away during successive rounds of DNA replication? It turns out that eukaryotic chromosomal DNA molecules have special nucleotide sequences called **telomeres** at their ends **(Figure 16.21)**. Telomeres do not contain genes; instead, the DNA typically consists of multiple repetitions of one short nucleotide sequence. In each human telomere, for example, the six-nucleotide sequence TTAGGG is repeated between 100 and 1,000 times. Telomeric DNA acts as a kind of buffer zone that protects the organism's genes. In addition, specific proteins associated with telomeric DNA prevent the staggered ends of the daughter molecule from activating the cell's systems for monitoring DNA damage. (Staggered ends of a DNA molecule, which often result from double-strand breaks, can trigger signal transduction pathways leading to cell cycle arrest or cell death.)

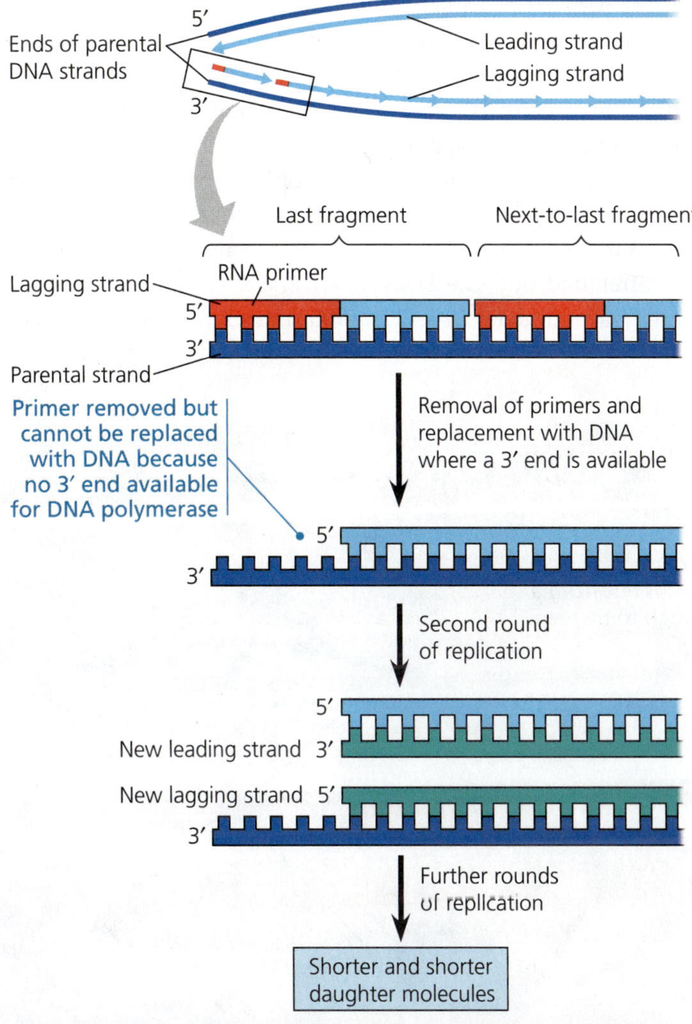

▲ **Figure 16.20 Shortening of the ends of linear DNA molecules.** Here we follow the end of one strand of a DNA molecule through two rounds of replication. After the first round, the new lagging strand is shorter than its template. After a second round, both the leading and lagging strands have become shorter than the original parental DNA. Although not shown here, the other ends of these DNA molecules also become shorter.

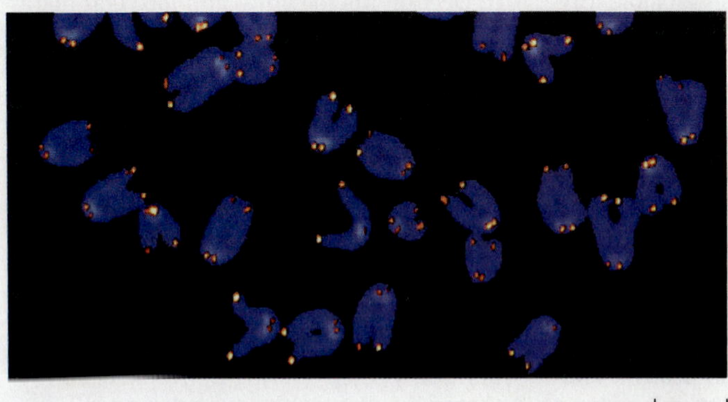

▲ **Figure 16.21 Telomeres.** Eukaryotes have repetitive, noncoding sequences called telomeres at the ends of their DNA. Telomeres are stained orange in these mouse chromosomes (LM).

Telomeres provide their protective function by postponing the erosion of genes located near the ends of DNA molecules. As shown in Figure 16.20, telomeres become shorter during every round of replication. As we would expect, telomeric DNA tends to be shorter in dividing somatic cells of older individuals and in cultured cells that have divided many times. It has been proposed that shortening of telomeres is somehow connected to the aging process of certain tissues and even to aging of the organism as a whole.

But what about cells whose genome must persist virtually unchanged from an organism to its offspring over many generations? If the chromosomes of germ cells became shorter in every cell cycle, essential genes would eventually be missing from the gametes they produce. However, this does not occur: An enzyme called **telomerase** catalyzes the lengthening of telomeres in eukaryotic germ cells, thus restoring their original length and compensating for the shortening that occurs during DNA replication. Telomerase is not active in most human somatic cells, but its activity in germ cells results in telomeres of maximum length in the zygote.

Normal shortening of telomeres may protect organisms from cancer by limiting the number of divisions that somatic cells can undergo. Cells from large tumors often have unusually short telomeres, as we would expect for cells that have undergone many cell divisions. Further shortening would presumably lead to self-destruction of the tumor cells. Intriguingly, researchers have found telomerase activity in cancerous somatic cells, suggesting that its ability to stabilize telomere length may allow these cancer cells to persist. Many cancer cells do seem capable of unlimited cell division, as do immortal strains of cultured cells (see Chapter 12). If telomerase is indeed an important factor in many cancers, it may provide a useful target for both cancer diagnosis and chemotherapy.

Thus far in this chapter, you have learned about the structure and replication of a DNA molecule. In the next section, we'll take a step back and examine how DNA is packaged into chromosomes, the structures that carry the genetic information.

CONCEPT CHECK 16.2

1. What role does complementary base pairing play in the replication of DNA?
2. Make a table listing the functions of seven proteins involved in DNA replication in *E. coli*.
3. **MAKE CONNECTIONS** What is the relationship between DNA replication and the S phase of the cell cycle? See Figure 12.6, page 231.
4. **WHAT IF?** If the DNA pol I in a given cell were nonfunctional, how would that affect the synthesis of a *leading* strand? In the overview box in Figure 16.17, point out where DNA pol I would normally function on the top leading strand.

For suggested answers, see Appendix A.

CONCEPT 16.3

A chromosome consists of a DNA molecule packed together with proteins

The main component of the genome in most bacteria is one double-stranded, circular DNA molecule that is associated with a small amount of protein. Although we refer to this structure as the *bacterial chromosome*, it is very different from a eukaryotic chromosome, which consists of one linear DNA molecule associated with a large amount of protein. In *E. coli*, the chromosomal DNA consists of about 4.6 million nucleotide pairs, representing about 4,400 genes. This is 100 times more DNA than is found in a typical virus, but only about one-thousandth as much DNA as in a human somatic cell. Still, that is a lot of DNA to be packaged in such a small container.

Stretched out, the DNA of an *E. coli* cell would measure about a millimeter in length, 500 times longer than the cell.

▼ **Figure 16.22**

Exploring Chromatin Packing in a Eukaryotic Chromosome

This series of diagrams and transmission electron micrographs depicts a current model for the progressive levels of DNA coiling and folding. The illustration zooms out from a single molecule of DNA to a metaphase chromosome, which is large enough to be seen with a light microscope.

DNA, the double helix

Shown here is a ribbon model of DNA, with each ribbon representing one of the sugar-phosphate backbones. As you will recall from Figure 16.7, the phosphate groups along the backbone contribute a negative charge along the outside of each strand. The TEM shows a molecule of naked DNA; the double helix alone is 2 nm across.

Histones

Proteins called **histones** are responsible for the first level of DNA packing in chromatin. Although each histone is small—containing only about 100 amino acids—the total mass of histone in chromatin approximately equals the mass of DNA. More than a fifth of a histone's amino acids are positively charged (lysine or arginine) and therefore bind tightly to the negatively charged DNA.

Four types of histones are most common in chromatin: H2A, H2B, H3, and H4. The histones are very similar among eukaryotes; for example, all but two of the amino acids in cow H4 are identical to those in pea H4. The apparent conservation of histone genes during evolution probably reflects the important role of histones in organizing DNA within cells.

The four main types of histones are critical to the next level of DNA packing. (A fifth type of histone, called H1, is involved in a further stage of packing.)

Nucleosomes, or "beads on a string" (10-nm fiber)

In electron micrographs, unfolded chromatin is 10 nm in diameter (the *10-nm fiber*). Such chromatin resembles beads on a string (see the TEM). Each "bead" is a **nucleosome**, the basic unit of DNA packing; the "string" between beads is called *linker DNA*.

A nucleosome consists of DNA wound twice around a protein core composed of two molecules each of the four main histone types. The amino end (N-terminus) of each histone (the *histone tail*) extends outward from the nucleosome.

In the cell cycle, the histones leave the DNA only briefly during DNA replication. Generally, they do the same during transcription, another process that requires access to the DNA by the cell's molecular machinery. Chapter 18 will discuss some recent findings about the role of histone tails and nucleosomes in the regulation of gene expression.

Within a bacterium, however, certain proteins cause the chromosome to coil and "supercoil," densely packing it so that it fills only part of the cell. Unlike the nucleus of a eukaryotic cell, this dense region of DNA in a bacterium, called the **nucleoid**, is not bounded by membrane (see Figure 6.5).

Eukaryotic chromosomes each contain a single linear DNA double helix that, in humans, averages about 1.5×10^8 nucleotide pairs. This is an enormous amount of DNA relative to a chromosome's condensed length. If completely stretched out, such a DNA molecule would be about 4 cm long, thousands of times the diameter of a cell nucleus—and that's not even considering the DNA of the other 45 human chromosomes!

In the cell, eukaryotic DNA is precisely combined with a large amount of protein. Together, this complex of DNA and protein, called **chromatin**, fits into the nucleus through an elaborate, multilevel system of packing. Our current view of the successive levels of DNA packing in a chromosome is outlined in **Figure 16.22**. Study this figure carefully before reading further.

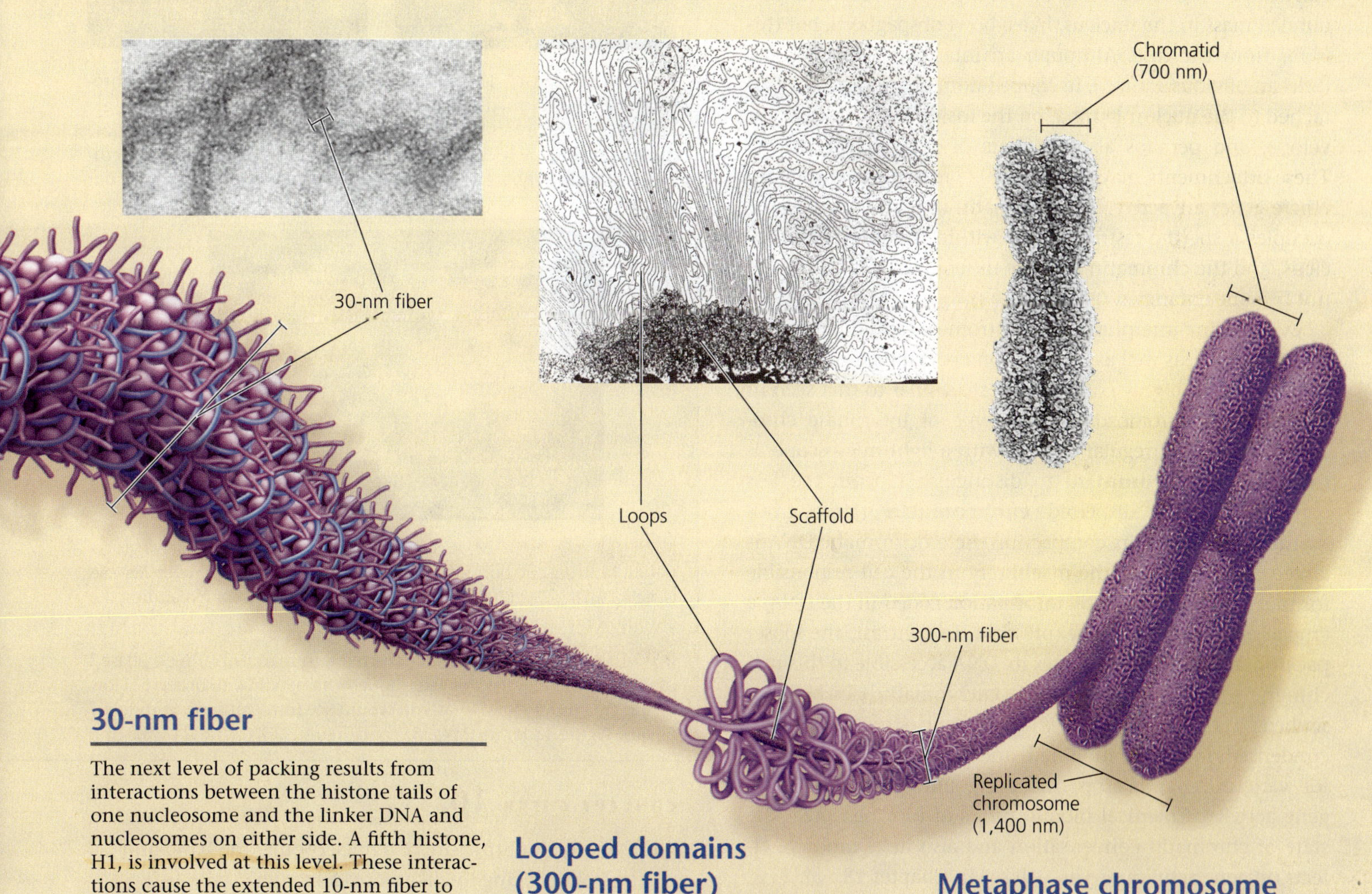

30-nm fiber

The next level of packing results from interactions between the histone tails of one nucleosome and the linker DNA and nucleosomes on either side. A fifth histone, H1, is involved at this level. These interactions cause the extended 10-nm fiber to coil or fold, forming a chromatin fiber roughly 30 nm in thickness, the *30-nm fiber*. Although the 30-nm fiber is quite prevalent in the interphase nucleus, the packing arrangement of nucleosomes in this form of chromatin is still a matter of some debate.

Looped domains (300-nm fiber)

The 30-nm fiber, in turn, forms loops called *looped domains* attached to a chromosome scaffold made of proteins, thus making up a *300-nm fiber*. The scaffold is rich in one type of topoisomerase, and H1 molecules also appear to be present.

Metaphase chromosome

In a mitotic chromosome, the looped domains themselves coil and fold in a manner not yet fully understood, further compacting all the chromatin to produce the characteristic metaphase chromosome shown in the micrograph above. The width of one chromatid is 700 nm. Particular genes always end up located at the same places in metaphase chromosomes, indicating that the packing steps are highly specific and precise.

Chromatin undergoes striking changes in its degree of packing during the course of the cell cycle (see Figure 12.7). In interphase cells stained for light microscopy, the chromatin usually appears as a diffuse mass within the nucleus, suggesting that the chromatin is highly extended. As a cell prepares for mitosis, its chromatin coils and folds up (condenses), eventually forming a characteristic number of short, thick metaphase chromosomes that are distinguishable from each other with the light microscope.

Though interphase chromatin is generally much less condensed than the chromatin of mitotic chromosomes, it shows several of the same levels of higher-order packing. Some of the chromatin comprising a chromosome seems to be present as a 10-nm fiber, but much is compacted into a 30-nm fiber, which in some regions is further folded into looped domains. Early on, biologists assumed that interphase chromatin was a tangled mass in the nucleus, like a bowl of spaghetti, but this is far from the case. Although an interphase chromosome lacks an obvious scaffold, its looped domains appear to be attached to the nuclear lamina, on the inside of the nuclear envelope, and perhaps also to fibers of the nuclear matrix. These attachments may help organize regions of chromatin where genes are active. The chromatin of each chromosome occupies a specific restricted area within the interphase nucleus, and the chromatin fibers of different chromosomes do not become entangled **(Figure 16.23)**.

Even during interphase, the centromeres and telomeres of chromosomes, as well as other chromosomal regions in some cells, exist in a highly condensed state similar to that seen in a metaphase chromosome. This type of interphase chromatin, visible as irregular clumps with a light microscope, is called **heterochromatin**, to distinguish it from the less compacted, more dispersed **euchromatin** ("true chromatin"). Because of its compaction, heterochromatic DNA is largely inaccessible to the machinery in the cell responsible for transcribing the genetic information coded in the DNA, a crucial early step in gene expression. In contrast, the looser packing of euchromatin makes its DNA accessible to this machinery, so the genes present in euchromatin can be transcribed. The chromosome is a dynamic structure that is condensed, loosened, modified, and remodeled as necessary for various cell processes, including mitosis, meiosis, and gene activity. Chemical modifications of histones affect the state of chromatin condensation and also have multiple effects on gene activity, as you will see in Chapter 18.

In this chapter, you have learned how DNA molecules are arranged in chromosomes and how DNA replication provides the copies of genes that parents pass to offspring. However, it is not enough that genes be copied and transmitted; the information they carry must be used by the cell. In other words, genes must also be "expressed." In the next chapter, we will examine how the cell expresses the genetic information encoded in DNA.

▼ **Figure 16.23**

IMPACT

Painting Chromosomes

Using techniques you'll learn about in Chapter 20, researchers have been able to treat human chromosomes with special molecular tags, such that each chromosome pair can be seen as a different color. Below on the left is a spread of chromosomes treated in this way; on the right they are organized into a karyotype.

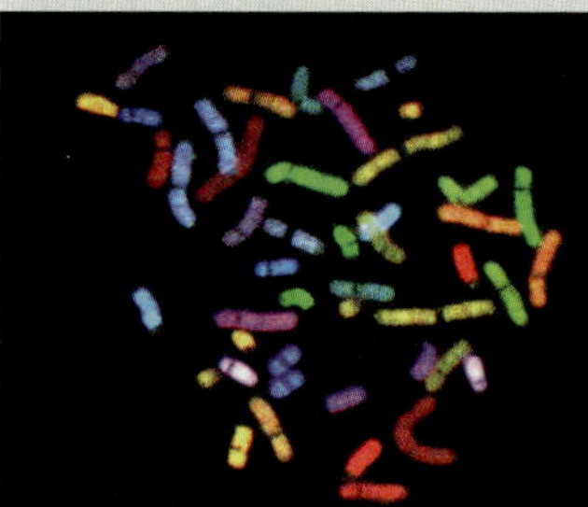

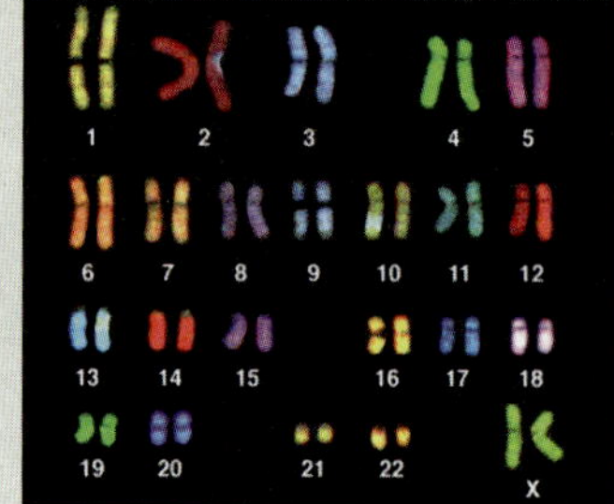

WHY IT MATTERS The ability to visually distinguish among chromosomes has allowed researchers to see how the chromosomes are arranged in the interphase nucleus. As you can see in the interphase nucleus below, each chromosome appears to occupy a specific territory during interphase. In general, the two homologs of a pair are not located together.

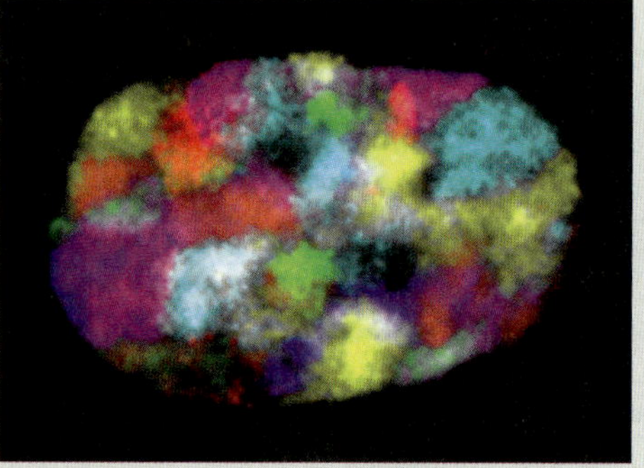

FURTHER READING M. R. Speicher and N. P. Carter, The new cytogenetics: blurring the boundaries with molecular biology, *Nature Reviews Genetics* 6:782–792 (2005); J. L. Marx, New methods for expanding the chromosomal paint kit, *Science* 273:430 (1996).

MAKE CONNECTIONS If you arrested a human cell in metaphase I of meiosis and applied this technique, what would you observe? How would this differ from what you would see in metaphase of mitosis? Review Figure 13.8 (pp. 254–255) and Figure 12.7 (pp. 232–233).

CONCEPT CHECK 16.3

1. Describe the structure of a nucleosome, the basic unit of DNA packing in eukaryotic cells.
2. What two properties, one structural and one functional, distinguish heterochromatin from euchromatin?
3. **MAKE CONNECTIONS** Interphase chromosomes appear to be attached to the nuclear lamina and perhaps also the nuclear matrix. Describe these two structures. See page 102 and Figure 6.9 on page 103.

For suggested answers, see Appendix A.

16 CHAPTER REVIEW

SUMMARY OF KEY CONCEPTS

CONCEPT 16.1

DNA is the genetic material (pp. 305–310)

- Experiments with bacteria and with **phages** provided the first strong evidence that the genetic material is DNA.
- Watson and Crick deduced that DNA is a **double helix** and built a structural model. Two **antiparallel** sugar-phosphate chains wind around the outside of the molecule; the nitrogenous bases project into the interior, where they hydrogen-bond in specific pairs, A with T, G with C.

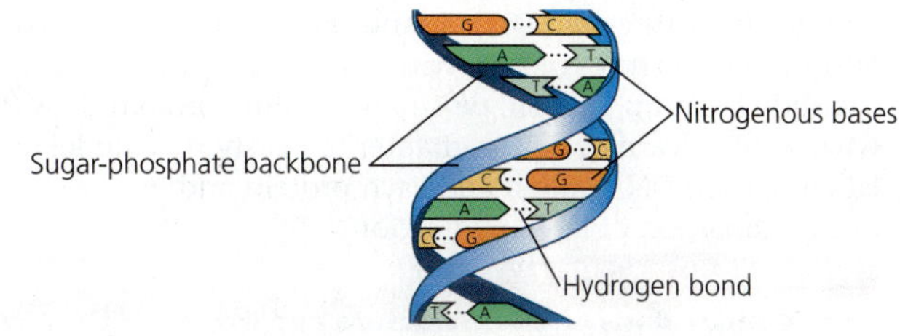

? *What does it mean when we say that the two DNA strands in the double helix are antiparallel? What would an end of the double helix look like if the strands were parallel?*

CONCEPT 16.2

Many proteins work together in DNA replication and repair (pp. 311–319)

- The Meselson-Stahl experiment showed that **DNA replication** is **semiconservative**: The parental molecule unwinds, and each strand then serves as a template for the synthesis of a new strand according to base-pairing rules.
- DNA replication at one **replication fork** is summarized here:

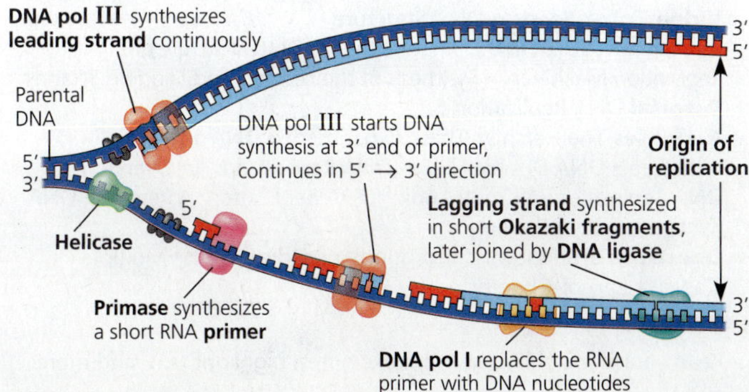

- DNA polymerases proofread new DNA, replacing incorrect nucleotides. In **mismatch repair**, enzymes correct errors that persist. **Nucleotide excision repair** is a general process by which **nucleases** cut out and replace damaged stretches of DNA.
- The ends of eukaryotic chromosomal DNA get shorter with each round of replication. The presence of **telomeres**, repetitive sequences at the ends of linear DNA molecules, postpones the erosion of genes. **Telomerase** catalyzes the lengthening of telomeres in germ cells.

? *Compare DNA replication on the leading and lagging strands, including both similarities and differences.*

CONCEPT 16.3

A chromosome consists of a DNA molecule packed together with proteins (pp. 320–322)

- The bacterial chromosome is usually a circular DNA molecule with some associated proteins, making up the **nucleoid** of the cell. Eukaryotic **chromatin** making up a chromosome is composed of DNA, **histones**, and other proteins. The histones bind to each other and to the DNA to form **nucleosomes**, the most basic units of DNA packing. Histone tails extend outward from each bead-like nucleosome core. Additional coiling and folding leads ultimately to the highly condensed chromatin of the metaphase chromosome. In interphase cells, most chromatin is less compacted (**euchromatin**), but some remains highly condensed (**heterochromatin**). Euchromatin, but not heterochromatin, is generally accessible for transcription of genes.

? *Describe the levels of chromatin packing you'd expect to see in an interphase nucleus.*

TEST YOUR UNDERSTANDING

LEVEL 1: KNOWLEDGE/COMPREHENSION

1. In his work with pneumonia-causing bacteria and mice, Griffith found that
   a. the protein coat from pathogenic cells was able to transform nonpathogenic cells.
   b. heat-killed pathogenic cells caused pneumonia.
   c. some substance from pathogenic cells was transferred to nonpathogenic cells, making them pathogenic.
   d. the polysaccharide coat of bacteria caused pneumonia.
   e. bacteriophages injected DNA into bacteria.
2. What is the basis for the difference in how the leading and lagging strands of DNA molecules are synthesized?
   a. The origins of replication occur only at the 5′ end.
   b. Helicases and single-strand binding proteins work at the 5′ end.
   c. DNA polymerase can join new nucleotides only to the 3′ end of a growing strand.
   d. DNA ligase works only in the 3′ → 5′ direction.
   e. Polymerase can work on only one strand at a time.
3. In analyzing the number of different bases in a DNA sample, which result would be consistent with the base-pairing rules?
   a. A = G
   b. A + G = C + T
   c. A + T = G + T
   d. A = C
   e. G = T
4. The elongation of the leading strand during DNA synthesis
   a. progresses away from the replication fork.
   b. occurs in the 3′ → 5′ direction.
   c. produces Okazaki fragments.
   d. depends on the action of DNA polymerase.
   e. does not require a template strand.
5. In a nucleosome, the DNA is wrapped around
   a. polymerase molecules.
   b. ribosomes.
   c. histones.
   d. a thymine dimer.
   e. satellite DNA.

LEVEL 2: APPLICATION/ANALYSIS

6. *E. coli* cells grown on ^{15}N medium are transferred to ^{14}N medium and allowed to grow for two more generations (two rounds of DNA replication). DNA extracted from these cells is centrifuged. What density distribution of DNA would you expect in this experiment?
   a. one high-density and one low-density band
   b. one intermediate-density band
   c. one high-density and one intermediate-density band
   d. one low-density and one intermediate-density band
   e. one low-density band

7. A biochemist isolates, purifies, and combines in a test tube a variety of molecules needed for DNA replication. When she adds some DNA to the mixture, replication occurs, but each DNA molecule consists of a normal strand paired with numerous segments of DNA a few hundred nucleotides long. What has she probably left out of the mixture?
   a. DNA polymerase
   b. DNA ligase
   c. nucleotides
   d. Okazaki fragments
   e. primase

8. The spontaneous loss of amino groups from adenine in DNA results in hypoxanthine, an uncommon base, opposite thymine. What combination of proteins could repair such damage?
   a. nuclease, DNA polymerase, DNA ligase
   b. telomerase, primase, DNA polymerase
   c. telomerase, helicase, single-strand binding protein
   d. DNA ligase, replication fork proteins, adenylyl cyclase
   e. nuclease, telomerase, primase

9. **MAKE CONNECTIONS** Although the proteins that cause the *E. coli* chromosome to coil are not histones, what property would you expect them to share with histones, given their ability to bind to DNA (see Figure 5.16, p. 79)?

LEVEL 3: SYNTHESIS/EVALUATION

10. The table below shows the base composition of DNA in several species. Explain how these data demonstrate Chargaff's rules.

Source	Adenine	Guanine	Cytosine	Thymine
E. coli	24.7%	26.0%	25.7%	23.6%
Wheat	28.1	21.8	22.7	27.4
Sea urchin	32.8	17.7	17.3	32.1
Salmon	29.7	20.8	20.4	29.1
Human	30.4	19.6	19.9	30.1
Ox	29.0	21.2	21.2	28.7

11. **EVOLUTION CONNECTION**
Some bacteria may be able to respond to environmental stress by increasing the rate at which mutations occur during cell division. How might this be accomplished? Might there be an evolutionary advantage of this ability? Explain.

12. **SCIENTIFIC INQUIRY**

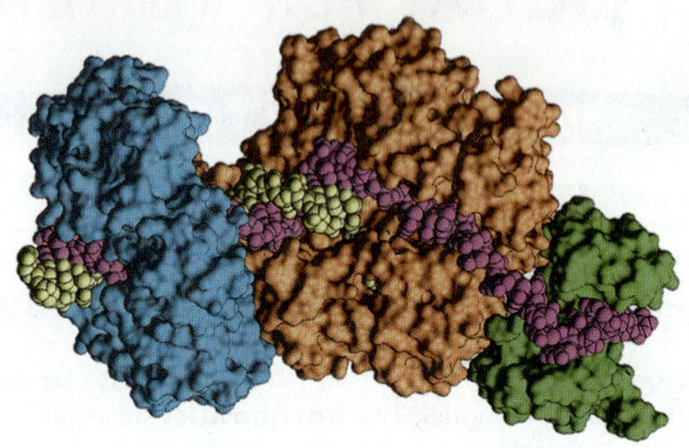

DRAW IT Model building can be an important part of the scientific process. The illustration shown above is a computer-generated model of a DNA replication complex. The parental and newly synthesized DNA strands are color-coded differently, as are each of the following three proteins: DNA pol III, the sliding clamp, and single-strand binding protein. Use what you've learned in this chapter to clarify this model by labeling each DNA strand and each protein and indicating the overall direction of DNA replication.

13. **WRITE ABOUT A THEME**
The Genetic Basis of Life; Structure and Function The continuity of life is based on heritable information in the form of DNA, and structure and function are correlated at all levels of biological organization. In a short essay (100–150 words), describe how the structure of DNA is correlated with its role as the molecular basis of inheritance.

For selected answers, see Appendix A.

www.masteringbiology.com

1. MasteringBiology® Assignments

Experimental Inquiry Tutorial Does DNA Replication Follow the Conservative, Semiconservative, or Dispersive Model?
Video Tutor Session DNA Structure
BioFlix **Tutorials** DNA Replication: DNA Structure and Replication Machinery • Synthesis of the Leading and Lagging Strands
Tutorial DNA Replication
Activities The Hershey-Chase Experiment • DNA and RNA Structure • DNA Double Helix • DNA Replication: An Overview • DNA Replication: A Closer Look • DNA Replication: A Review • DNA Synthesis • DNA Packing
Questions Student Misconceptions • Reading Quiz • Multiple Choice • End-of-Chapter

2. eText

Read your book online, search, take notes, highlight text, and more.

3. The Study Area

Practice Tests • Cumulative Test • ***BioFlix*** 3-D Animations • MP3 Tutor Sessions • Videos • Activities • Investigations • Lab Media • Audio Glossary • Word Study Tools • Art

17

From Gene to Protein

▲ **Figure 17.1 How does a single faulty gene result in the dramatic appearance of an albino deer?**

KEY CONCEPTS

- **17.1** Genes specify proteins via transcription and translation
- **17.2** Transcription is the DNA-directed synthesis of RNA: *a closer look*
- **17.3** Eukaryotic cells modify RNA after transcription
- **17.4** Translation is the RNA-directed synthesis of a polypeptide: *a closer look*
- **17.5** Mutations of one or a few nucleotides can affect protein structure and function
- **17.6** While gene expression differs among the domains of life, the concept of a gene is universal

OVERVIEW

The Flow of Genetic Information

In 2006, a young albino deer seen frolicking with several brown deer in the mountains of eastern Germany elicited a public outcry **(Figure 17.1)**. A local hunting organization announced that the albino deer suffered from a "genetic disorder" and should be shot. Some argued that the deer should merely be prevented from mating with other deer to safeguard the population's gene pool. Others favored relocating the albino deer to a nature reserve because they worried that it might be more noticeable to predators if left in the wild. A German rock star even held a benefit concert to raise funds for the relocation. What led to the striking phenotype of this deer, the cause of this lively debate?

You learned in Chapter 14 that inherited traits are determined by genes and that the trait of albinism is caused by a recessive allele of a pigmentation gene. The information content of genes is in the form of specific sequences of nucleotides along strands of DNA, the genetic material. But how does this information determine an organism's traits? Put another way, what does a gene actually say? And how is its message translated by cells into a specific trait, such as brown hair, type A blood, or, in the case of an albino deer, a total lack of pigment? The albino deer has a faulty version of a key protein, an enzyme required for pigment synthesis, and this protein is faulty because the gene that codes for it contains incorrect information.

This example illustrates the main point of this chapter: The DNA inherited by an organism leads to specific traits by dictating the synthesis of proteins and of RNA molecules involved in protein synthesis. In other words, proteins are the link between genotype and phenotype. **Gene expression** is the process by which DNA directs the synthesis of proteins (or, in some cases, just RNAs). The expression of genes that code for proteins includes two stages: transcription and translation. This chapter describes the flow of information from gene to protein in detail and explains how genetic mutations affect organisms through their proteins. Understanding the processes of gene expression, which are similar in all three domains of life, will allow us to revisit the concept of the gene in more detail at the end of the chapter.

CONCEPT 17.1

Genes specify proteins via transcription and translation

Before going into the details of how genes direct protein synthesis, let's step back and examine how the fundamental relationship between genes and proteins was discovered.

Evidence from the Study of Metabolic Defects

In 1902, British physician Archibald Garrod was the first to suggest that genes dictate phenotypes through enzymes that catalyze specific chemical reactions in the cell. Garrod postulated that the symptoms of an inherited disease reflect a person's inability to make a particular enzyme. He later referred to such diseases as "inborn errors of metabolism." Garrod gave as one example the hereditary condition called alkaptonuria. In this disorder, the urine is black because it contains the chemical alkapton, which darkens upon exposure to air. Garrod reasoned that most people have an enzyme that metabolizes alkapton, whereas people with alkaptonuria have inherited an inability to make that enzyme.

Garrod may have been the first to recognize that Mendel's principles of heredity apply to humans as well as peas. Garrod's realization was ahead of its time, but research several decades later supported his hypothesis that a gene dictates the production of a specific enzyme. Biochemists accumulated much evidence that cells synthesize and degrade most organic molecules via metabolic pathways, in which each chemical reaction in a sequence is catalyzed by a specific enzyme (see p. 142). Such metabolic pathways lead, for instance, to the synthesis of the pigments that give the brown deer in Figure 17.1 their fur color or fruit flies (*Drosophila*) their eye color (see Figure 15.3). In the 1930s, the American biochemist and geneticist George Beadle and his French colleague Boris Ephrussi speculated that in *Drosophila*, each of the various mutations affecting eye color blocks pigment synthesis at a specific step by preventing production of the enzyme that catalyzes that step. But neither the chemical reactions nor the enzymes that catalyze them were known at the time.

Nutritional Mutants in Neurospora: Scientific Inquiry

A breakthrough in demonstrating the relationship between genes and enzymes came a few years later at Stanford University, where Beadle and Edward Tatum began working with a bread mold, *Neurospora crassa*. They bombarded *Neurospora* with X-rays, shown in the 1920s to cause genetic changes, and then looked among the survivors for mutants that differed in their nutritional needs from the wild-type bread mold. Wild-type *Neurospora* has modest food requirements. It can grow in the laboratory on a simple solution of inorganic salts, glucose, and the vitamin biotin, incorporated into agar, a support medium. From this *minimal medium*, the mold cells use their metabolic pathways to produce all the other molecules they need. Beadle and Tatum identified mutants that could not survive on minimal medium, apparently because they were unable to synthesize certain essential molecules from the minimal ingredients. To ensure survival of these nutritional mutants, Beadle and Tatum allowed them to grow on a *complete growth medium*, which consisted of minimal medium supplemented with all 20 amino acids and a few other nutrients. The complete growth medium could support any mutant that couldn't synthesize one of the supplements.

To characterize the metabolic defect in each nutritional mutant, Beadle and Tatum took samples from the mutant growing on complete medium and distributed them to a number of different vials. Each vial contained minimal medium plus a single additional nutrient. The particular supplement that allowed growth indicated the metabolic defect. For example, if the only supplemented vial that supported growth of the mutant was the one fortified with the amino acid arginine, the researchers could conclude that the mutant was defective in the biochemical pathway that wild-type cells use to synthesize arginine.

In fact, such arginine-requiring mutants were obtained and studied by two colleagues of Beadle and Tatum, Adrian Srb and Norman Horowitz, who wanted to investigate the biochemical pathway for arginine synthesis in *Neurospora* (**Figure 17.2**). Srb and Horowitz pinned down each mutant's defect more specifically, using additional tests to distinguish among three classes of arginine-requiring mutants. Mutants in each class required a different set of compounds along the arginine-synthesizing pathway, which has three steps. These results, and those of many similar experiments done by Beadle and Tatum, suggested that each class was blocked at a different step in this pathway because mutants in that class lacked the enzyme that catalyzes the blocked step.

Because each mutant was defective in a single gene, Beadle and Tatum saw that, taken together, the collected results provided strong support for a working hypothesis they had proposed earlier. The *one gene–one enzyme hypothesis*, as they dubbed it, states that the function of a gene is to dictate the production of a specific enzyme. Further support for this hypothesis came from experiments that identified the specific enzymes lacking in the mutants. Beadle and Tatum shared a Nobel Prize in 1958 for "their discovery that genes act by regulating definite chemical events" (in the words of the Nobel committee).

The Products of Gene Expression: A Developing Story

As researchers learned more about proteins, they made revisions to the one gene–one enzyme hypothesis. First of all, not all proteins are enzymes. Keratin, the structural protein of animal hair, and the hormone insulin are two examples of nonenzyme proteins. Because proteins that are not enzymes are nevertheless gene products, molecular biologists began to think in terms of one gene–one protein. However, many proteins are constructed from two or more different polypeptide chains, and each polypeptide is specified by its own gene. For example, hemoglobin, the oxygen-transporting protein of vertebrate red blood cells, contains two kinds of polypeptides, and thus two genes code for this protein (see Figure 5.20). Beadle and Tatum's idea was therefore restated as the *one gene–one polypeptide hypothesis*. Even this description is not entirely accurate, though. First, many eukaryotic genes can each code for a set of closely related polypeptides via a process called alternative splicing, which you will learn about later in this chapter. Second, quite a few genes code for RNA molecules that have important functions in cells

▼ Figure 17.2 INQUIRY

Do individual genes specify the enzymes that function in a biochemical pathway?

EXPERIMENT Working with the mold *Neurospora crassa*, Adrian Srb and Norman Horowitz, then at Stanford University, used Beadle and Tatum's experimental approach to isolate mutants that required arginine in their growth medium. The researchers showed that these mutants fell into three classes, each defective in a different gene. From other considerations, they suspected that the metabolic pathway of arginine biosynthesis involved a precursor nutrient and the intermediate molecules ornithine and citrulline. Their most famous experiment, shown here, tested both the one gene–one enzyme hypothesis and their postulated arginine-synthesizing pathway. In this experiment, they grew their three classes of mutants under the four different conditions shown in the Results section below. They included minimal medium (MM) as a control because they knew that wild-type cells could grow on MM but mutant cells could not. (See test tubes on the right.)

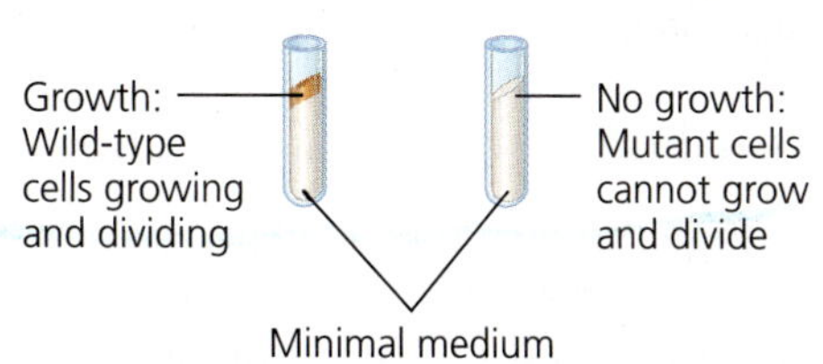

RESULTS The wild-type strain was capable of growth under all experimental conditions, requiring only the minimal medium. The three classes of mutants each had a specific set of growth requirements. For example, class II mutants could not grow when ornithine alone was added but could grow when either citrulline or arginine was added.

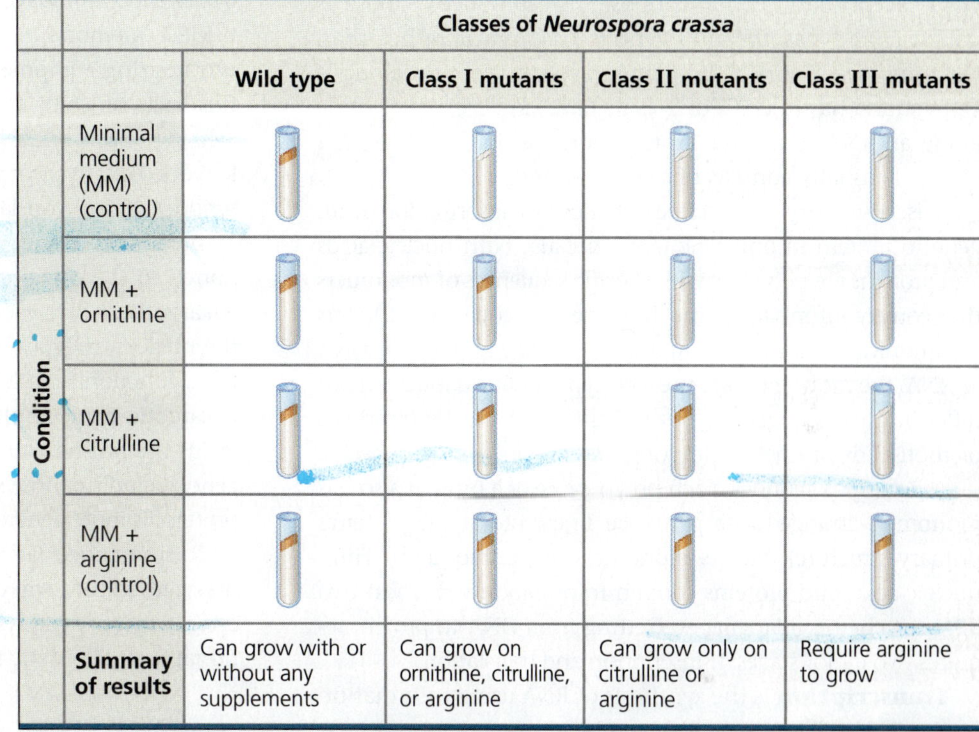

Condition	Wild type	Class I mutants	Class II mutants	Class III mutants
	Classes of *Neurospora crassa*			
Minimal medium (MM) (control)				
MM + ornithine				
MM + citrulline				
MM + arginine (control)				
Summary of results	Can grow with or without any supplements	Can grow on ornithine, citrulline, or arginine	Can grow only on citrulline or arginine	Require arginine to grow

CONCLUSION From the growth requirements of the mutants, Srb and Horowitz deduced that each class of mutant was unable to carry out one step in the pathway for synthesizing arginine, presumably because it lacked the necessary enzyme. Because each of their mutants was mutated in a single gene, they concluded that each mutated gene must normally dictate the production of one enzyme. Their results supported the one gene–one enzyme hypothesis proposed by Beadle and Tatum and also confirmed that the arginine pathway described in the mammalian liver also operates in *Neurospora*. (Notice in the Results that a mutant can grow only if supplied with a compound made *after* the defective step because this bypasses the defect.)

Gene (codes for enzyme)	**Wild type**	**Class I mutants** (mutation in gene *A*)	**Class II mutants** (mutation in gene *B*)	**Class III mutants** (mutation in gene *C*)
	Precursor	Precursor	Precursor	Precursor
Gene *A* →	Enzyme A	Enzyme A (✗)	Enzyme A	Enzyme A
	Ornithine	Ornithine	Ornithine	Ornithine
Gene *B* →	Enzyme B	Enzyme B	Enzyme B (✗)	Enzyme B
	Citrulline	Citrulline	Citrulline	Citrulline
Gene *C* →	Enzyme C	Enzyme C	Enzyme C	Enzyme C (✗)
	Arginine	Arginine	Arginine	Arginine

SOURCE A. M. Srb and N. H. Horowitz, The ornithine cycle in *Neurospora* and its genetic control, *Journal of Biological Chemistry* 154:129–139 (1944).

WHAT IF? Suppose the experiment had shown that class I mutants could grow only in MM supplemented by ornithine or arginine and that class II mutants could grow in MM supplemented by citrulline, ornithine, or arginine. What conclusions would the researchers have drawn from those results regarding the biochemical pathway and the defect in class I and class II mutants?

even though they are never translated into protein. For now, we will focus on genes that do code for polypeptides. (Note that it is common to refer to these gene products as proteins—a practice you will encounter in this book—rather than more precisely as polypeptides.)

Basic Principles of Transcription and Translation

Genes provide the instructions for making specific proteins. But a gene does not build a protein directly. The bridge between DNA and protein synthesis is the nucleic acid RNA. You learned in Chapter 5 that RNA is chemically similar to DNA except that it contains ribose instead of deoxyribose as its sugar and has the nitrogenous base uracil rather than thymine (see Figure 5.26). Thus, each nucleotide along a DNA strand has A, G, C, or T as its base, and each nucleotide along an RNA strand has A, G, C, or U as its base. An RNA molecule usually consists of a single strand.

It is customary to describe the flow of information from gene to protein in linguistic terms because both nucleic acids and proteins are polymers with specific sequences of monomers that convey information, much as specific sequences of letters communicate information in a language like English. In DNA or RNA, the monomers are the four types of nucleotides, which differ in their nitrogenous bases. Genes are typically hundreds or thousands of nucleotides long, each gene having a specific sequence of nucleotides. Each polypeptide of a protein also has monomers arranged in a particular linear order (the protein's primary structure), but its monomers are amino acids. Thus, nucleic acids and proteins contain information written in two different chemical languages. Getting from DNA to protein requires two major stages: transcription and translation.

Transcription is the synthesis of RNA using information in the DNA. The two nucleic acids are written in different forms of the same language, and the information is simply transcribed, or "rewritten," from DNA to RNA. Just as a DNA strand provides a template for making a new complementary strand during DNA replication, it also can serve as a template for assembling a complementary sequence of RNA nucleotides. For a protein-coding gene, the resulting RNA molecule is a faithful transcript of the gene's protein-building instructions. This type of RNA molecule is called **messenger RNA (mRNA)** because it carries a genetic message from the DNA to the protein-synthesizing machinery of the cell. (Transcription is the general term for the synthesis of *any* kind of RNA on a DNA template. Later, you will learn about some other types of RNA produced by transcription.)

Translation is the synthesis of a polypeptide using the information in the mRNA. During this stage, there is a change in language: The cell must translate the nucleotide sequence of an mRNA molecule into the amino acid sequence of a polypeptide. The sites of translation are **ribosomes**, complex particles that facilitate the orderly linking of amino acids into polypeptide chains.

Transcription and translation occur in all organisms, both those that lack a membrane-bounded nucleus (bacteria and archaea) and those that have one (eukaryotes). Because most studies of transcription and translation have used bacteria and eukaryotic cells, these are our main focus in this chapter. Our understanding of transcription and translation in archaea lags behind, but in the last section of the chapter we will discuss a few aspects of archaeal gene expression.

The basic mechanics of transcription and translation are similar for bacteria and eukaryotes, but there is an important difference in the flow of genetic information within the cells. Because bacteria do not have nuclei, their DNA is not separated by nuclear membranes from ribosomes and the other protein-synthesizing equipment **(Figure 17.3a)**. As you will see later, this lack of compartmentalization allows translation of an mRNA to begin while its transcription is still in progress. In a eukaryotic cell, by contrast, the nuclear envelope separates transcription from translation in space and time **(Figure 17.3b)**. Transcription occurs in the nucleus, and mRNA is then transported to the cytoplasm, where translation occurs. But before eukaryotic RNA transcripts from protein-coding genes can leave the nucleus, they are modified in various ways to produce the final, functional mRNA. The transcription of a protein-coding eukaryotic gene results in *pre-mRNA*, and further processing yields the finished mRNA. The initial RNA transcript from any gene, including those specifying RNA that is not translated into protein, is more generally called a **primary transcript**.

To summarize: Genes program protein synthesis via genetic messages in the form of messenger RNA. Put another way, cells are governed by a molecular chain of command with a directional flow of genetic information, shown here by arrows:

This concept was dubbed the *central dogma* by Francis Crick in 1956. How has the concept held up over time? In the 1970s, scientists were surprised to discover that some RNA molecules can act as templates for DNA synthesis, a process you'll read about in Chapter 19. However, these exceptions do not invalidate the idea that, in general, genetic information flows from DNA to RNA to protein. In the next section, we discuss how the instructions for assembling amino acids into a specific order are encoded in nucleic acids.

The Genetic Code

When biologists began to suspect that the instructions for protein synthesis were encoded in DNA, they recognized a problem: There are only four nucleotide bases to specify 20 amino acids. Thus, the genetic code cannot be a language like Chinese, where each written symbol corresponds to a word. How many nucleotides, then, correspond to an amino acid?

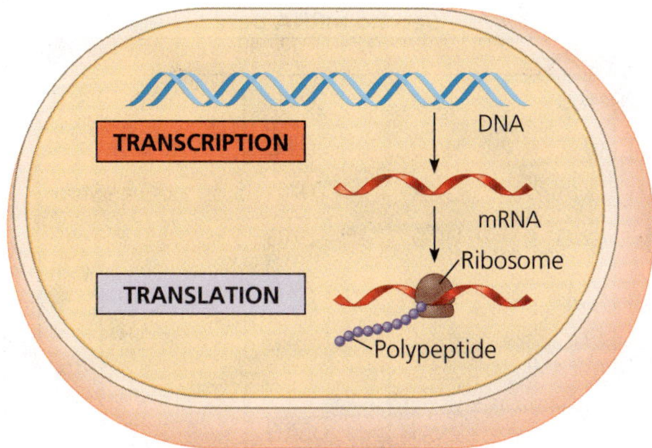

(a) Bacterial cell. In a bacterial cell, which lacks a nucleus, mRNA produced by transcription is immediately translated without additional processing.

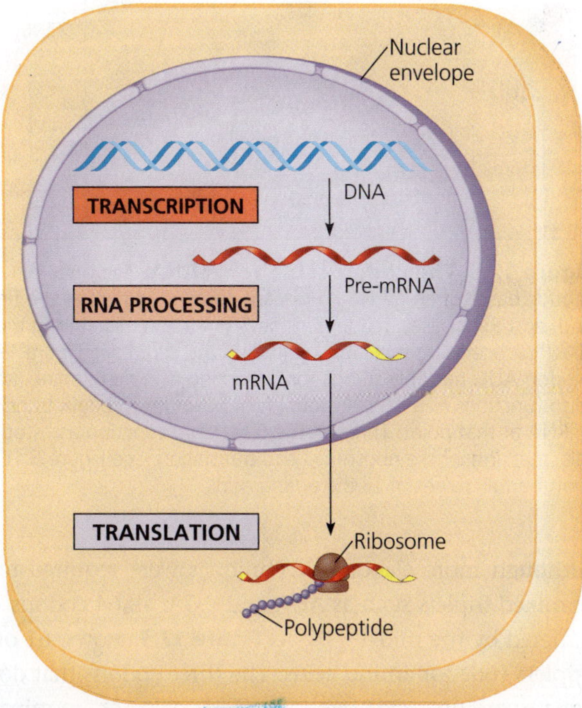

(b) Eukaryotic cell. The nucleus provides a separate compartment for transcription. The original RNA transcript, called pre-mRNA, is processed in various ways before leaving the nucleus as mRNA.

▲ **Figure 17.3 Overview: the roles of transcription and translation in the flow of genetic information.** In a cell, inherited information flows from DNA to RNA to protein. The two main stages of information flow are transcription and translation. A miniature version of part (a) or (b) accompanies several figures later in the chapter as an orientation diagram to help you see where a particular figure fits into the overall scheme.

Codons: Triplets of Nucleotides

If each kind of nucleotide base were translated into an amino acid, only 4 of the 20 amino acids could be specified. Would a language of two-letter code words suffice? The two-nucleotide sequence AG, for example, could specify one amino acid, and GT could specify another. Since there are four possible nucleotide bases in each position, this would give us 16 (that is, 4^2) possible arrangements—still not enough to code for all 20 amino acids.

Triplets of nucleotide bases are the smallest units of uniform length that can code for all the amino acids. If each arrangement of three consecutive nucleotide bases specifies an amino acid, there can be 64 (that is, 4^3) possible code words—more than enough to specify all the amino acids. Experiments have verified that the flow of information from gene to protein is based on a **triplet code**: The genetic instructions for a polypeptide chain are written in the DNA as a series of nonoverlapping, three-nucleotide words. The series of words in a gene is transcribed into a complementary series of nonoverlapping, three-nucleotide words in mRNA, which is then translated into a chain of amino acids **(Figure 17.4)**.

During transcription, the gene determines the sequence of nucleotide bases along the length of the RNA molecule that is being synthesized. For each gene, only one of the two DNA strands is transcribed. This strand is called the **template strand** because it provides the pattern, or template, for the sequence of nucleotides in an RNA transcript. For any given gene, the same strand is used as the template every time the gene is transcribed. For other genes on the same DNA molecule, however, the opposite strand may be the one that always functions as the template.

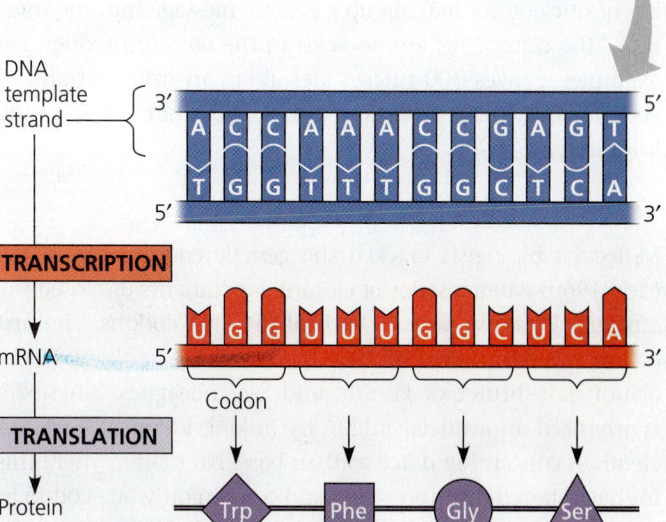

▲ **Figure 17.4 The triplet code.** For each gene, one DNA strand functions as a template for transcription of RNAs, such as mRNA. The base-pairing rules for DNA synthesis also guide transcription, except that uracil (U) takes the place of thymine (T) in RNA. During translation, the mRNA is read as a sequence of nucleotide triplets, called codons. Each codon specifies an amino acid to be added to the growing polypeptide chain. The mRNA is read in the 5′ → 3′ direction.

? *Compare the sequence of the mRNA to that of the nontemplate DNA strand, in both cases reading from 5′ → 3′.*

An mRNA molecule is complementary rather than identical to its DNA template because RNA nucleotides are assembled on the template according to base-pairing rules (see Figure 17.4). The pairs are similar to those that form during DNA replication, except that U, the RNA substitute for T, pairs with A and the mRNA nucleotides contain ribose instead of deoxyribose. Like a new strand of DNA, the RNA molecule is synthesized in an antiparallel direction to the template strand of DNA. (To review what is meant by "antiparallel" and the 5′ and 3′ ends of a nucleic acid chain, see Figure 16.7.) In the example in Figure 17.4, the nucleotide triplet ACC along the DNA (written as 3′-ACC-5′) provides a template for 5′-UGG-3′ in the mRNA molecule. The mRNA nucleotide triplets are called **codons**, and they are customarily written in the 5′ → 3′ direction. In our example, UGG is the codon for the amino acid tryptophan (abbreviated Trp). The term *codon* is also used for the DNA nucleotide triplets along the *nontemplate* strand. These codons are complementary to the template strand and thus identical in sequence to the mRNA, except that they have T instead of U. (For this reason, the nontemplate DNA strand is sometimes called the "coding strand.")

During translation, the sequence of codons along an mRNA molecule is decoded, or translated, into a sequence of amino acids making up a polypeptide chain. The codons are read by the translation machinery in the 5′ → 3′ direction along the mRNA. Each codon specifies which one of the 20 amino acids will be incorporated at the corresponding position along a polypeptide. Because codons are nucleotide triplets, the number of nucleotides making up a genetic message must be three times the number of amino acids in the protein product. For example, it takes 300 nucleotides along an mRNA strand to code for the amino acids in a polypeptide that is 100 amino acids long.

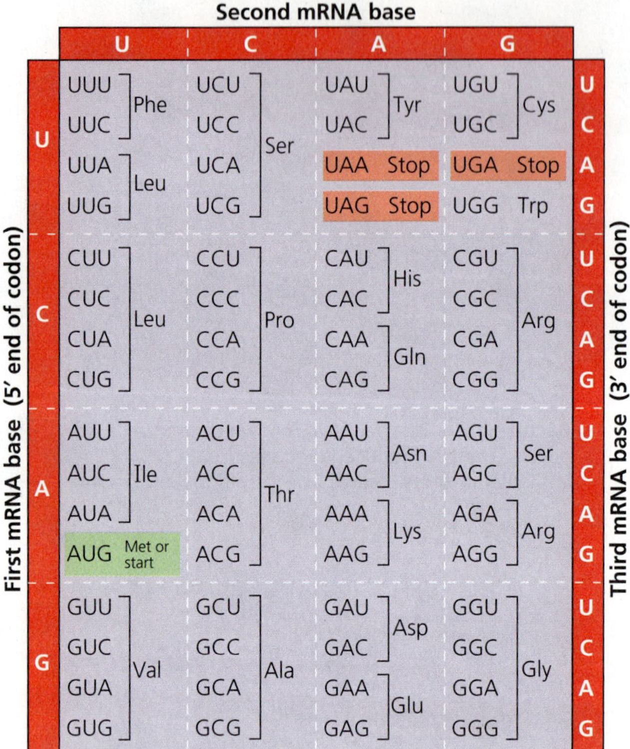

First mRNA base (5′ end of codon)	Second mRNA base: U	C	A	G	Third mRNA base (3′ end of codon)
U	UUU Phe	UCU Ser	UAU Tyr	UGU Cys	U
	UUC	UCC	UAC	UGC	C
	UUA Leu	UCA	UAA Stop	UGA Stop	A
	UUG	UCG	UAG Stop	UGG Trp	G
C	CUU Leu	CCU Pro	CAU His	CGU Arg	U
	CUC	CCC	CAC	CGC	C
	CUA	CCA	CAA Gln	CGA	A
	CUG	CCG	CAG	CGG	G
A	AUU Ile	ACU Thr	AAU Asn	AGU Ser	U
	AUC	ACC	AAC	AGC	C
	AUA	ACA	AAA Lys	AGA Arg	A
	AUG Met or start	ACG	AAG	AGG	G
G	GUU Val	GCU Ala	GAU Asp	GGU Gly	U
	GUC	GCC	GAC	GGC	C
	GUA	GCA	GAA Glu	GGA	A
	GUG	GCG	GAG	GGG	G

▲ **Figure 17.5 The codon table for mRNA.** The three nucleotide bases of an mRNA codon are designated here as the first, second, and third bases, reading in the 5′ → 3′ direction along the mRNA. (Practice using this table by finding the codons in Figure 17.4.) The codon AUG not only stands for the amino acid methionine (Met) but also functions as a "start" signal for ribosomes to begin translating the mRNA at that point. Three of the 64 codons function as "stop" signals, marking where ribosomes end translation. See Figure 5.16 for a list of the full names of all the amino acids.

Cracking the Code

Molecular biologists cracked the genetic code of life in the early 1960s when a series of elegant experiments disclosed the amino acid translations of each of the RNA codons. The first codon was deciphered in 1961 by Marshall Nirenberg, of the National Institutes of Health, and his colleagues. Nirenberg synthesized an artificial mRNA by linking identical RNA nucleotides containing uracil as their base. No matter where this message started or stopped, it could contain only one codon in repetition: UUU. Nirenberg added this "poly-U" to a test-tube mixture containing amino acids, ribosomes, and the other components required for protein synthesis. His artificial system translated the poly-U into a polypeptide containing many units of the amino acid phenylalanine (Phe), strung together as a long polyphenylalanine chain. Thus, Nirenberg determined that the mRNA codon UUU specifies the amino acid phenylalanine. Soon, the amino acids specified by the codons AAA, GGG, and CCC were also determined.

Although more elaborate techniques were required to decode mixed triplets such as AUA and CGA, all 64 codons were deciphered by the mid-1960s. As **Figure 17.5** shows, 61 of the 64 triplets code for amino acids. The three codons that do not designate amino acids are "stop" signals, or termination codons, marking the end of translation. Notice that the codon AUG has a dual function: It codes for the amino acid methionine (Met) and also functions as a "start" signal, or initiation codon. Genetic messages usually begin with the mRNA codon AUG, which signals the protein-synthesizing machinery to begin translating the mRNA at that location. (Because AUG also stands for methionine, polypeptide chains begin with methionine when they are synthesized. However, an enzyme may subsequently remove this starter amino acid from the chain.)

Notice in Figure 17.5 that there is redundancy in the genetic code, but no ambiguity. For example, although codons GAA and GAG both specify glutamic acid (redundancy), neither of them ever specifies any other amino acid (no ambiguity). The redundancy in the code is not altogether random. In many cases, codons that are synonyms for a particular amino acid differ only in the third nucleotide base of the triplet. We will consider a possible benefit of this redundancy later in the chapter.

Our ability to extract the intended message from a written language depends on reading the symbols in the correct groupings—that is, in the correct **reading frame**. Consider this statement: "The red dog ate the bug." Group the letters incorrectly by starting at the wrong point, and the result will probably be gibberish: for example, "her edd oga tet heb ug." The reading frame is also important in the molecular language of cells. The short stretch of polypeptide shown in Figure 17.4, for instance, will be made correctly only if the mRNA nucleotides are read from left to right (5′ → 3′) in the groups of three shown in the figure: UGG UUU GGC UCA. Although a genetic message is written with no spaces between the codons, the cell's protein-synthesizing machinery reads the message as a series of nonoverlapping three-letter words. The message is *not* read as a series of overlapping words—UGGUUU, and so on—which would convey a very different message.

Evolution of the Genetic Code

EVOLUTION The genetic code is nearly universal, shared by organisms from the simplest bacteria to the most complex plants and animals. The RNA codon CCG, for instance, is translated as the amino acid proline in all organisms whose genetic code has been examined. In laboratory experiments, genes can be transcribed and translated after being transplanted from one species to another, sometimes with quite striking results, as shown in **Figure 17.6**! Bacteria can be programmed by the insertion of human genes to synthesize certain human proteins for medical use, such as insulin. Such applications have produced many exciting developments in the area of biotechnology (see Chapter 20).

Exceptions to the universality of the genetic code include translation systems in which a few codons differ from the standard ones. Slight variations in the genetic code exist in certain unicellular eukaryotes and in the organelle genes of some species. Despite these exceptions, the evolutionary significance of the code's *near* universality is clear. A language shared by all living things must have been operating very early in the history of life—early enough to be present in the common ancestor of all present-day organisms. A shared genetic vocabulary is a reminder of the kinship that bonds all life on Earth.

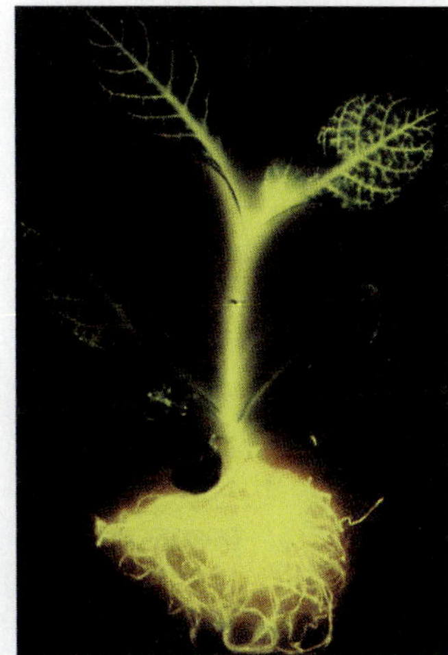

(a) Tobacco plant expressing a firefly gene. The yellow glow is produced by a chemical reaction catalyzed by the protein product of the firefly gene.

(b) Pig expressing a jellyfish gene. Researchers injected the gene for a fluorescent protein into fertilized pig eggs. One of the eggs developed into this fluorescent pig.

▲ **Figure 17.6 Expression of genes from different species.** Because diverse forms of life share a common genetic code, one species can be programmed to produce proteins characteristic of a second species by introducing DNA from the second species into the first.

CONCEPT CHECK 17.1

1. **MAKE CONNECTIONS** In a research article about alkaptonuria published in 1902, Garrod suggested that humans inherit two "characters" (alleles) for a particular enzyme and that both parents must contribute a faulty version for the offspring to have the disorder. Today, would this disorder be called dominant or recessive? See Concept 14.4, pages 276–278.
2. What polypeptide product would you expect from a poly-G mRNA that is 30 nucleotides long?
3. **DRAW IT** The template strand of a gene contains the sequence 3′-TTCAGTCGT-5′. Draw the nontemplate sequence and the mRNA sequence, indicating 5′ and 3′ ends of each. Compare the two sequences.
4. **WHAT IF?** **DRAW IT** Imagine that the nontemplate sequence in question 3 was transcribed instead of the template sequence. Draw the mRNA sequence and translate it using Figure 17.5. (Be sure to pay attention to the 5′ and 3′ ends.) Predict how well the protein synthesized from the nontemplate strand would function, if at all.

For suggested answers, see Appendix A.

CONCEPT **17.2**

Transcription is the DNA-directed synthesis of RNA: *a closer look*

Now that we have considered the linguistic logic and evolutionary significance of the genetic code, we are ready to reexamine transcription, the first stage of gene expression, in more detail.

Molecular Components of Transcription

Messenger RNA, the carrier of information from DNA to the cell's protein-synthesizing machinery, is transcribed from the template strand of a gene. An enzyme called an **RNA polymerase** pries the two strands of DNA apart and joins together RNA nucleotides complementary to the DNA template strand, thus elongating the RNA polynucleotide (**Figure 17.7**). Like the DNA polymerases that function in DNA replication, RNA polymerases can assemble a polynucleotide only in its 5′ → 3′ direction. Unlike DNA polymerases, however, RNA polymerases are able to start a chain from scratch; they don't need a primer.

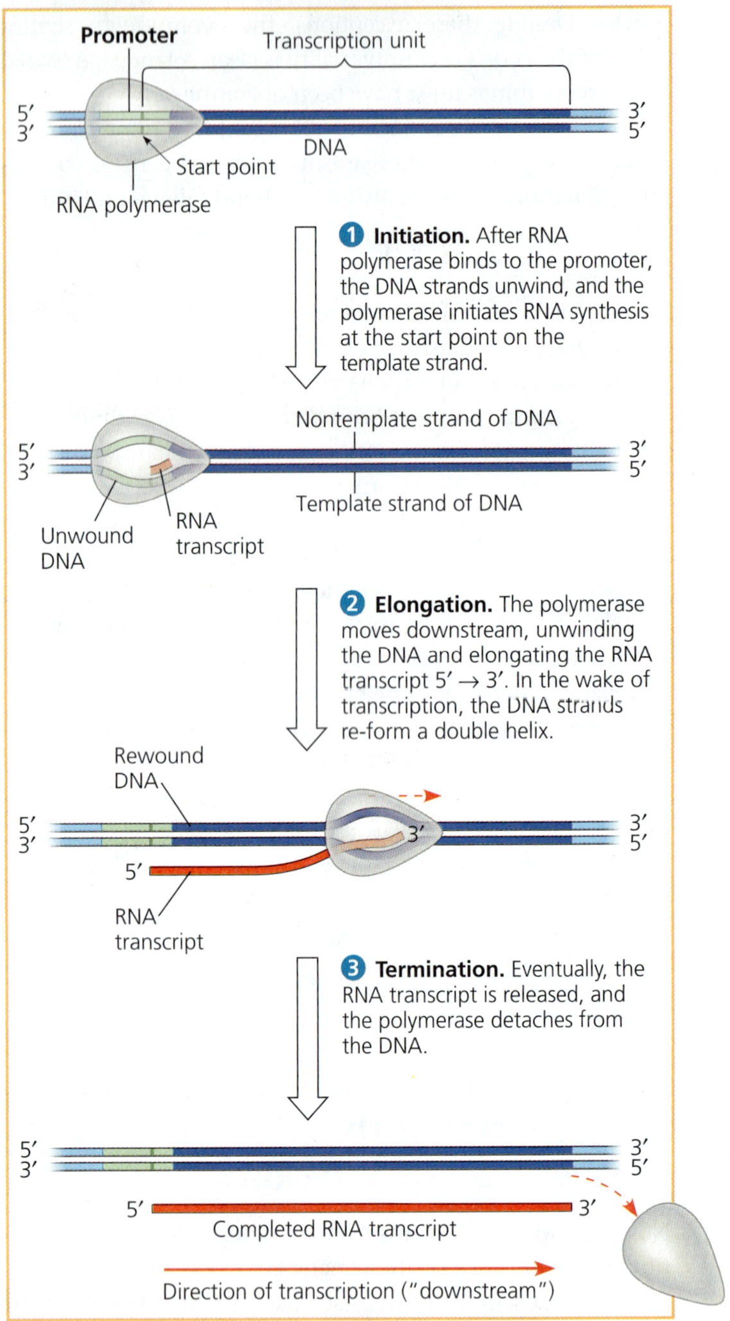

▲ **Figure 17.7 The stages of transcription: initiation, elongation, and termination.** This general depiction of transcription applies to both bacteria and eukaryotes, but the details of termination differ, as described in the text. Also, in a bacterium, the RNA transcript is immediately usable as mRNA; in a eukaryote, the RNA transcript must first undergo processing.

MAKE CONNECTIONS *Compare the use of a template strand during transcription and replication. See Figure 16.17, page 317.*

Specific sequences of nucleotides along the DNA mark where transcription of a gene begins and ends. The DNA sequence where RNA polymerase attaches and initiates transcription is known as the **promoter**; in bacteria, the sequence that signals the end of transcription is called the **terminator**. (The termination mechanism is different in eukaryotes; we'll describe it later.) Molecular biologists refer to the direction of transcription as "downstream" and the other direction as "upstream." These terms are also used to describe the positions of nucleotide sequences within the DNA or RNA. Thus, the promoter sequence in DNA is said to be upstream from the terminator. The stretch of DNA that is transcribed into an RNA molecule is called a **transcription unit**.

Bacteria have a single type of RNA polymerase that synthesizes not only mRNA but also other types of RNA that function in protein synthesis, such as ribosomal RNA. In contrast, eukaryotes have at least three types of RNA polymerase in their nuclei. The one used for mRNA synthesis is called RNA polymerase II. The other RNA polymerases transcribe RNA molecules that are not translated into protein. In the discussion of transcription that follows, we start with the features of mRNA synthesis common to both bacteria and eukaryotes and then describe some key differences.

Synthesis of an RNA Transcript

The three stages of transcription, as shown in Figure 17.7 and described next, are initiation, elongation, and termination of the RNA chain. Study Figure 17.7 to familiarize yourself with the stages and the terms used to describe them.

RNA Polymerase Binding and Initiation of Transcription

The promoter of a gene includes within it the transcription **start point** (the nucleotide where RNA synthesis actually begins) and typically extends several dozen or more nucleotide pairs upstream from the start point. RNA polymerase binds in a precise location and orientation on the promoter, therefore determining where transcription starts and which of the two strands of the DNA helix is used as the template.

Certain sections of a promoter are especially important for binding RNA polymerase. In bacteria, the RNA polymerase itself specifically recognizes and binds to the promoter. In eukaryotes, a collection of proteins called **transcription factors** mediate the binding of RNA polymerase and the initiation of transcription. Only after transcription factors are attached to the promoter does RNA polymerase II bind to it. The whole complex of transcription factors and RNA polymerase II bound to the promoter is called a **transcription initiation complex**. **Figure 17.8** shows the role of transcription factors and a crucial promoter DNA sequence called a **TATA box** in forming the initiation complex at a eukaryotic promoter.

The interaction between eukaryotic RNA polymerase II and transcription factors is an example of the importance of protein-protein interactions in controlling eukaryotic

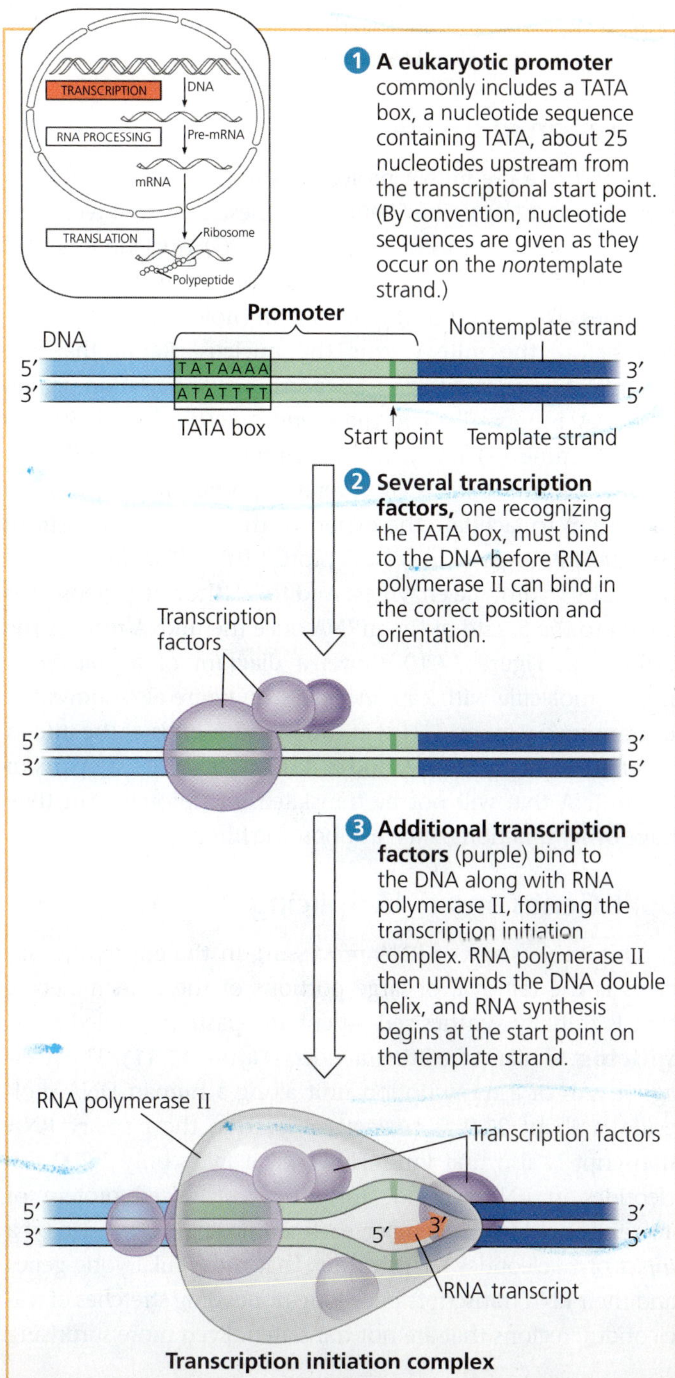

▲ **Figure 17.8 The initiation of transcription at a eukaryotic promoter.** In eukaryotic cells, proteins called transcription factors mediate the initiation of transcription by RNA polymerase II.

? *Explain how the interaction of RNA polymerase with the promoter would differ if the figure showed transcription initiation for bacteria.*

transcription. (And as you learned in Figure 16.22, the DNA of a eukaryotic chromosome is complexed with histones and other proteins in the form of chromatin. The roles of these proteins in making the DNA accessible to transcription factors will be discussed in Chapter 18). Once the appropriate transcription factors are firmly attached to the promoter DNA and the polymerase is bound in the correct orientation, the enzyme unwinds the two DNA strands and starts transcribing the template strand.

Elongation of the RNA Strand

As RNA polymerase moves along the DNA, it continues to untwist the double helix, exposing about 10–20 DNA nucleotides at a time for pairing with RNA nucleotides **(Figure 17.9)**. The enzyme adds nucleotides to the 3′ end of the growing RNA molecule as it continues along the double helix. In the wake of this advancing wave of RNA synthesis, the new RNA molecule peels away from its DNA template, and the DNA double helix re-forms. Transcription progresses at a rate of about 40 nucleotides per second in eukaryotes.

A single gene can be transcribed simultaneously by several molecules of RNA polymerase following each other like trucks in a convoy. A growing strand of RNA trails off from each polymerase, with the length of each new strand reflecting how far along the template the enzyme has traveled from the start point (see the mRNA molecules in Figure 17.25). The congregation of many polymerase molecules simultaneously transcribing a single gene increases the amount of mRNA transcribed from it, which helps the cell make the encoded protein in large amounts.

Termination of Transcription

The mechanism of termination differs between bacteria and eukaryotes. In bacteria, transcription proceeds through a terminator sequence in the DNA. The transcribed terminator (an RNA sequence) functions as the termination signal,

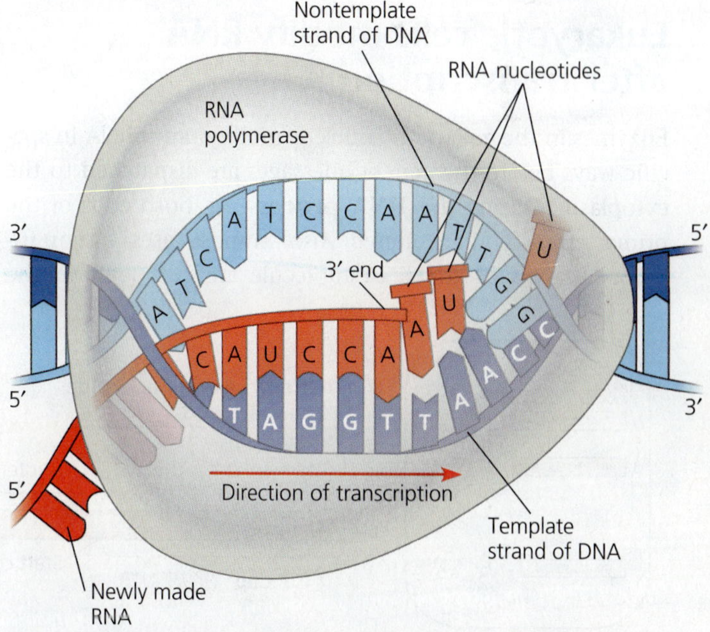

▲ **Figure 17.9 Transcription elongation.** RNA polymerase moves along the DNA template strand, joining complementary RNA nucleotides to the 3′ end of the growing RNA transcript. Behind the polymerase, the new RNA peels away from the template strand, which re-forms a double helix with the nontemplate strand.

causing the polymerase to detach from the DNA and release the transcript, which requires no further modification before translation. In eukaryotes, RNA polymerase II transcribes a sequence on the DNA called the polyadenylation signal sequence, which codes for a polyadenylation signal (AAUAAA) in the pre-mRNA. Then, at a point about 10–35 nucleotides downstream from the AAUAAA signal, proteins associated with the growing RNA transcript cut it free from the polymerase, releasing the pre-mRNA. The pre-mRNA then undergoes processing, the topic of the next section.

CONCEPT CHECK 17.2

1. **MAKE CONNECTIONS** Compare DNA polymerase and RNA polymerase in terms of how they function, the requirement for a template and primer, the direction of synthesis, and the type of nucleotides used. See Figure 16.17, page 317.
2. What is a promoter, and is it located at the upstream or downstream end of a transcription unit?
3. What enables RNA polymerase to start transcribing a gene at the right place on the DNA in a bacterial cell? In a eukaryotic cell?
4. **WHAT IF?** Suppose X-rays caused a sequence change in the TATA box of a particular gene's promoter. How would that affect transcription of the gene? (See Figure 17.8.)

For suggested answers, see Appendix A.

CONCEPT 17.3

Eukaryotic cells modify RNA after transcription

Enzymes in the eukaryotic nucleus modify pre-mRNA in specific ways before the genetic messages are dispatched to the cytoplasm. During this **RNA processing**, both ends of the primary transcript are altered. Also, in most cases, certain interior sections of the RNA molecule are cut out and the remaining parts spliced together. These modifications produce an mRNA molecule ready for translation.

Alteration of mRNA Ends

Each end of a pre-mRNA molecule is modified in a particular way **(Figure 17.10)**. The 5′ end is synthesized first; it receives a **5′ cap**, a modified form of a guanine (G) nucleotide added onto the 5′ end after transcription of the first 20–40 nucleotides. The 3′ end of the pre-mRNA molecule is also modified before the mRNA exits the nucleus. Recall that the pre-mRNA is released soon after the polyadenylation signal, AAUAAA, is transcribed. At the 3′ end, an enzyme adds 50–250 more adenine (A) nucleotides, forming a **poly-A tail**. The 5′ cap and poly-A tail share several important functions. First, they seem to facilitate the export of the mature mRNA from the nucleus. Second, they help protect the mRNA from degradation by hydrolytic enzymes. And third, they help ribosomes attach to the 5′ end of the mRNA once the mRNA reaches the cytoplasm. Figure 17.10 shows a diagram of a eukaryotic mRNA molecule with cap and tail. The figure also shows the untranslated regions (UTRs) at the 5′ and 3′ ends of the mRNA (referred to as the 5′ UTR and 3′ UTR). The UTRs are parts of the mRNA that will not be translated into protein, but they have other functions, such as ribosome binding.

Split Genes and RNA Splicing

A remarkable stage of RNA processing in the eukaryotic nucleus is the removal of large portions of the RNA molecule that is initially synthesized—a cut-and-paste job called **RNA splicing**, similar to editing a video **(Figure 17.11)**. The average length of a transcription unit along a human DNA molecule is about 27,000 nucleotide pairs, so the primary RNA transcript is also that long. However, it takes only 1,200 nucleotides in RNA to code for the average-sized protein of 400 amino acids. (Remember, each amino acid is encoded by a *triplet* of nucleotides.) This means that most eukaryotic genes and their RNA transcripts have long noncoding stretches of nucleotides, regions that are not translated. Even more surprising

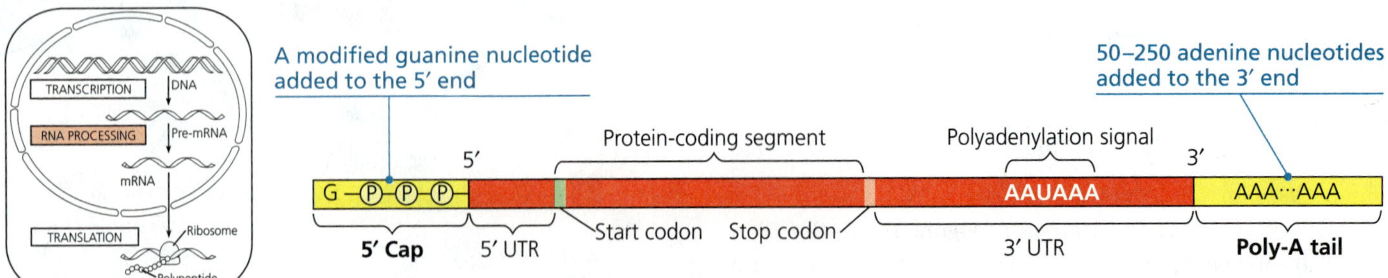

▲ **Figure 17.10 RNA processing: Addition of the 5′ cap and poly-A tail.** Enzymes modify the two ends of a eukaryotic pre-mRNA molecule. The modified ends may promote the export of mRNA from the nucleus, and they help protect the mRNA from degradation. When the mRNA reaches the cytoplasm, the modified ends, in conjunction with certain cytoplasmic proteins, facilitate ribosome attachment. The 5′ cap and poly-A tail are not translated into protein, nor are the regions called the 5′ untranslated region (5′ UTR) and 3′ untranslated region (3′ UTR).

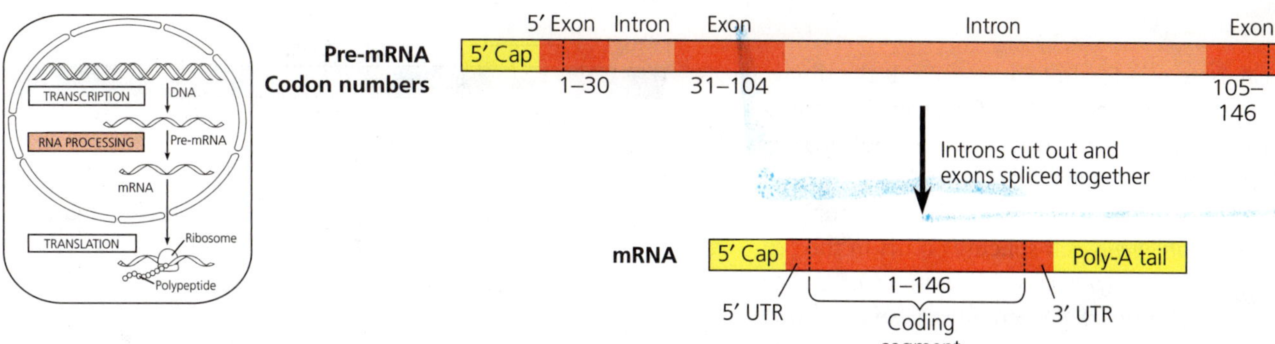

▲ **Figure 17.11 RNA processing: RNA splicing.** The RNA molecule shown here codes for β-globin, one of the polypeptides of hemoglobin. The numbers under the RNA refer to codons; β-globin is 146 amino acids long. The β-globin gene and its pre-mRNA transcript have three exons, corresponding to sequences that will leave the nucleus as mRNA. (The 5′ UTR and 3′ UTR are parts of exons because they are included in the mRNA; however, they do not code for protein.) During RNA processing, the introns are cut out and the exons spliced together. In many genes, the introns are much larger than the exons.

is that most of these noncoding sequences are interspersed between coding segments of the gene and thus between coding segments of the pre-mRNA. In other words, the sequence of DNA nucleotides that codes for a eukaryotic polypeptide is usually not continuous; it is split into segments. The noncoding segments of nucleic acid that lie between coding regions are called *int*ervening sequences, or **introns**. The other regions are called **exons**, because they are eventually *ex*pressed, usually by being translated into amino acid sequences. (Exceptions include the UTRs of the exons at the ends of the RNA, which make up part of the mRNA but are not translated into protein. Because of these exceptions, you may find it helpful to think of exons as sequences of RNA that *exit* the nucleus.) The terms *intron* and *exon* are used for both RNA sequences and the DNA sequences that encode them.

In making a primary transcript from a gene, RNA polymerase II transcribes both introns and exons from the DNA, but the mRNA molecule that enters the cytoplasm is an abridged version. The introns are cut out from the molecule and the exons joined together, forming an mRNA molecule with a continuous coding sequence. This is the process of RNA splicing.

How is pre-mRNA splicing carried out? Researchers have learned that the signal for RNA splicing is a short nucleotide sequence at each end of an intron. Joan Steitz, our interviewee for this unit (see pp. 246–247), discovered in 1979 that particles called *small nuclear ribonucleoproteins*, abbreviated *snRNPs* (pronounced "snurps"), recognize these splice sites. As the full name implies, snRNPs are located in the cell nucleus and are composed of RNA and protein molecules. The RNA in a snRNP particle is called a *small nuclear RNA (snRNA)*; each snRNA molecule is about 150 nucleotides long. Several different snRNPs join with additional proteins to form an even larger assembly called a **spliceosome**, which is almost as big as a ribosome. The spliceosome interacts with certain sites along an intron, releasing the intron, which is rapidly degraded, and joining together the two exons that flanked the intron (**Figure 17.12**). It turns out that snRNAs catalyze these processes, as well as participating in spliceosome assembly and splice site recognition.

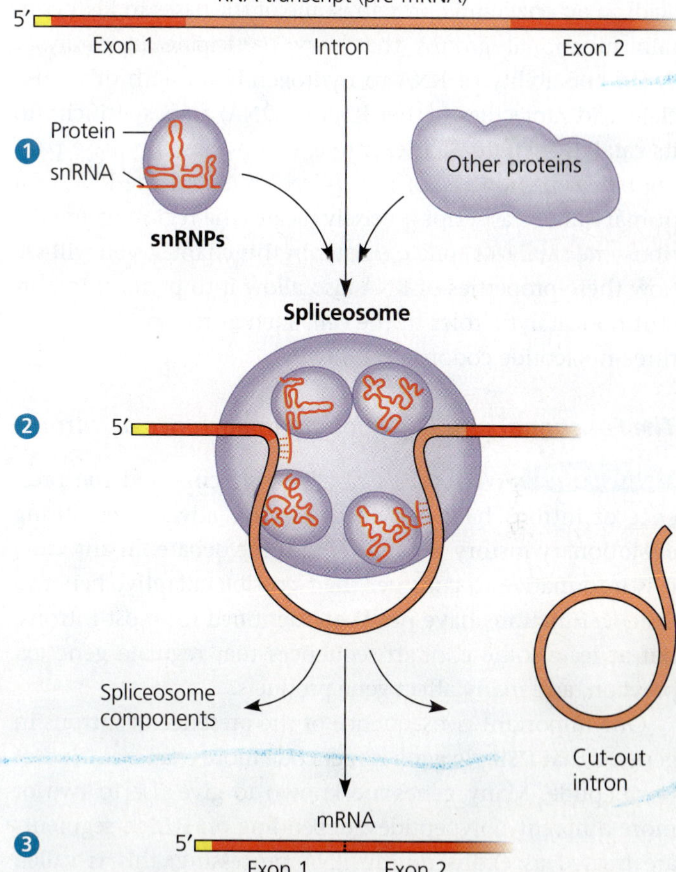

▲ **Figure 17.12 The roles of snRNPs and spliceosomes in pre-mRNA splicing.** The diagram shows only a portion of the pre-mRNA transcript; additional introns and exons lie downstream from the ones pictured here. **1** Small nuclear ribonucleoproteins (snRNPs) and other proteins form a molecular complex called a spliceosome on a pre-mRNA molecule containing exons and introns. **2** Within the spliceosome, snRNA base-pairs with nucleotides at specific sites along the intron. **3** The spliceosome cuts the pre-mRNA, releasing the intron for rapid degradation, and at the same time splices the exons together. The spliceosome then comes apart, releasing mRNA, which now contains only exons.

Ribozymes

The idea of a catalytic role for snRNA arose from the discovery of **ribozymes**, RNA molecules that function as enzymes. In some organisms, RNA splicing can occur without proteins or even additional RNA molecules: The intron RNA functions as a ribozyme and catalyzes its own excision! For example, in the ciliate protist *Tetrahymena*, self-splicing occurs in the production of ribosomal RNA (rRNA), a component of the organism's ribosomes. The pre-rRNA actually removes its own introns. The discovery of ribozymes rendered obsolete the idea that all biological catalysts are proteins.

Three properties of RNA enable some RNA molecules to function as enzymes. First, because RNA is single-stranded, a region of an RNA molecule may base-pair with a complementary region elsewhere in the same molecule, which gives the molecule a particular three-dimensional structure. A specific structure is essential to the catalytic function of ribozymes, just as it is for enzymatic proteins. Second, like certain amino acids in an enzymatic protein, some of the bases in RNA contain functional groups that may participate in catalysis. Third, the ability of RNA to hydrogen-bond with other nucleic acid molecules (either RNA or DNA) adds specificity to its catalytic activity. For example, complementary base pairing between the RNA of the spliceosome and the RNA of a primary RNA transcript precisely locates the region where the ribozyme catalyzes splicing. Later in this chapter, you will see how these properties of RNA also allow it to perform important noncatalytic roles in the cell, such as recognition of the three-nucleotide codons on mRNA.

The Functional and Evolutionary Importance of Introns

EVOLUTION Whether or not RNA splicing and the presence of introns have provided selective advantages during evolutionary history is a matter of some debate. In any case, it is informative to consider their possible adaptive benefits. Specific functions have not been identified for most introns, but at least some contain sequences that regulate gene expression, and many affect gene products.

One important consequence of the presence of introns in genes is that a single gene can encode more than one kind of polypeptide. Many genes are known to give rise to two or more different polypeptides, depending on which segments are treated as exons during RNA processing; this is called **alternative RNA splicing** (see Figure 18.13). For example, sex differences in fruit flies are largely due to differences in how males and females splice the RNA transcribed from certain genes. Results from the Human Genome Project (discussed in Chapter 21) suggest that alternative RNA splicing is one reason humans can get along with about the same number of genes as a nematode (roundworm). Because of alternative splicing, the number of different protein products an organism produces can be much greater than its number of genes.

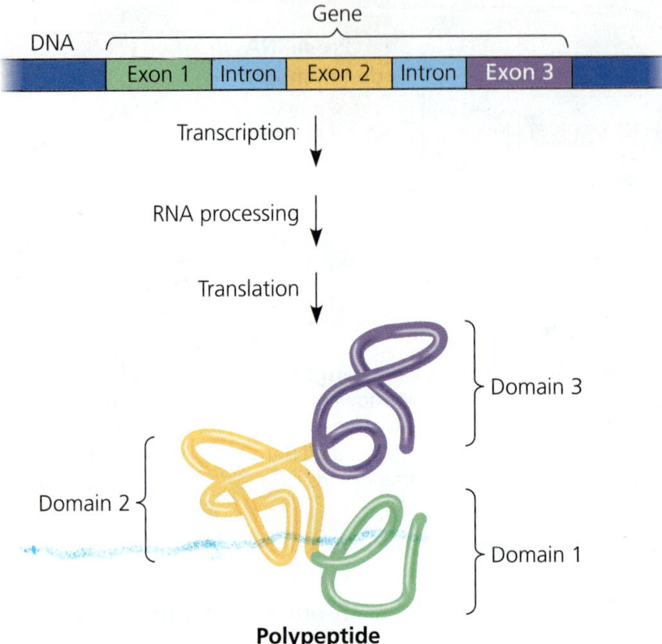

▲ **Figure 17.13 Correspondence between exons and protein domains.**

Proteins often have a modular architecture consisting of discrete structural and functional regions called **domains**. One domain of an enzyme, for example, might include the active site, while another might allow the enzyme to bind to a cellular membrane. In quite a few cases, different exons code for the different domains of a protein **(Figure 17.13)**.

The presence of introns in a gene may facilitate the evolution of new and potentially beneficial proteins as a result of a process known as *exon shuffling*. Introns increase the probability of crossing over between the exons of alleles of a gene—simply by providing more terrain for crossovers without interrupting coding sequences. This might result in new combinations of exons and proteins with altered structure and function. We can also imagine the occasional mixing and matching of exons between completely different (nonallelic) genes. Exon shuffling of either sort could lead to new proteins with novel combinations of functions. While most of the shuffling would result in nonbeneficial changes, occasionally a beneficial variant might arise.

CONCEPT CHECK 17.3

1. How can human cells make 75,000–100,000 different proteins, given that there are about 20,000 human genes?
2. How is RNA splicing similar to editing a video? What would introns correspond to in this analogy?
3. **WHAT IF?** What would be the effect of treating cells with an agent that removed the cap from mRNAs?

For suggested answers, see Appendix A.

CONCEPT 17.4

Translation is the RNA-directed synthesis of a polypeptide: *a closer look*

We will now examine in greater detail how genetic information flows from mRNA to protein—the process of translation. As we did for transcription, we'll concentrate on the basic steps of translation that occur in both bacteria and eukaryotes, while pointing out key differences.

Molecular Components of Translation

In the process of translation, a cell "reads" a genetic message and builds a polypeptide accordingly. The message is a series of codons along an mRNA molecule, and the translator is called **transfer RNA (tRNA)**. The function of tRNA is to transfer amino acids from the cytoplasmic pool of amino acids to a growing polypeptide in a ribosome. A cell keeps its cytoplasm stocked with all 20 amino acids, either by synthesizing them from other compounds or by taking them up from the surrounding solution. The ribosome, a structure made of proteins and RNAs, adds each amino acid brought to it by tRNA to the growing end of a polypeptide chain (**Figure 17.14**).

Translation is simple in principle but complex in its biochemistry and mechanics, especially in the eukaryotic cell. In dissecting translation, we'll concentrate on the slightly less complicated version of the process that occurs in bacteria. We'll begin by looking at the major players in this cellular process and then see how they act together in making a polypeptide.

The Structure and Function of Transfer RNA

The key to translating a genetic message into a specific amino acid sequence is the fact that molecules of tRNA are not all identical, and each type of tRNA molecule translates a particular mRNA codon into a particular amino acid. A tRNA molecule arrives at a ribosome bearing a specific amino acid at one end. At the other end of the tRNA is a nucleotide triplet called an **anticodon**, which base-pairs with a complementary codon on mRNA. For example, consider the mRNA codon GGC, which is translated as the amino acid glycine. The tRNA that base-pairs with this codon by hydrogen bonding has CCG as its anticodon and carries glycine at its other end (see the incoming tRNA approaching the ribosome in Figure 17.14). As an mRNA molecule is moved through a ribosome, glycine will be added to the polypeptide chain whenever the codon GGC is presented for translation. Codon by codon, the genetic message is translated as tRNAs deposit amino acids in the order prescribed, and the ribosome joins the amino acids into a chain. The tRNA molecule is a translator in the sense that it can read a nucleic acid word (the mRNA codon) and interpret it as a protein word (the amino acid).

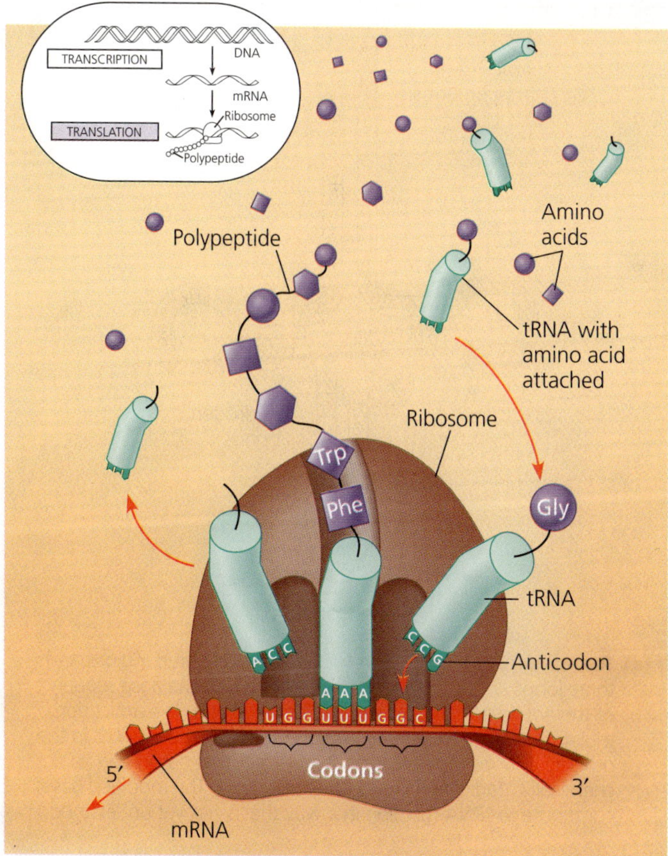

▲ **Figure 17.14 Translation: the basic concept.** As a molecule of mRNA is moved through a ribosome, codons are translated into amino acids, one by one. The interpreters are tRNA molecules, each type with a specific anticodon at one end and a corresponding amino acid at the other end. A tRNA adds its amino acid cargo to a growing polypeptide chain when the anticodon hydrogen-bonds to a complementary codon on the mRNA. The figures that follow show some of the details of translation in a bacterial cell.

BioFlix Visit the Study Area at **www.masteringbiology.com** for the BioFlix® 3-D Animation on Protein Synthesis.

Like mRNA and other types of cellular RNA, transfer RNA molecules are transcribed from DNA templates. In a eukaryotic cell, tRNA, like mRNA, is made in the nucleus and then travels from the nucleus to the cytoplasm, where translation occurs. In both bacterial and eukaryotic cells, each tRNA molecule is used repeatedly, picking up its designated amino acid in the cytosol, depositing this cargo onto a polypeptide chain at the ribosome, and then leaving the ribosome, ready to pick up another amino acid.

A tRNA molecule consists of a single RNA strand that is only about 80 nucleotides long (compared to hundreds of nucleotides for most mRNA molecules). Because of the presence of complementary stretches of nucleotide bases that can hydrogen-bond to each other, this single strand can fold back upon itself

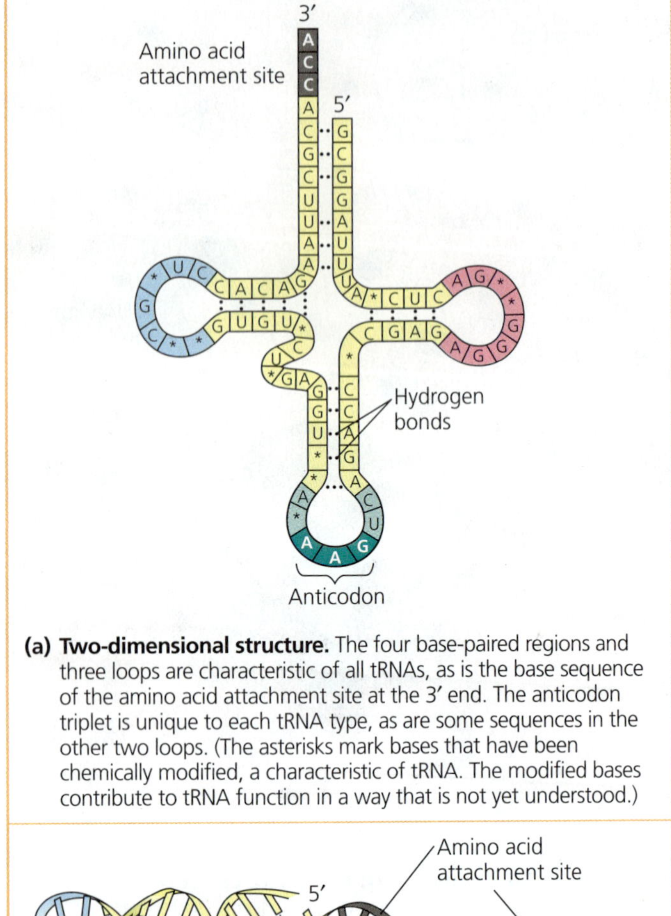

(a) Two-dimensional structure. The four base-paired regions and three loops are characteristic of all tRNAs, as is the base sequence of the amino acid attachment site at the 3′ end. The anticodon triplet is unique to each tRNA type, as are some sequences in the other two loops. (The asterisks mark bases that have been chemically modified, a characteristic of tRNA. The modified bases contribute to tRNA function in a way that is not yet understood.)

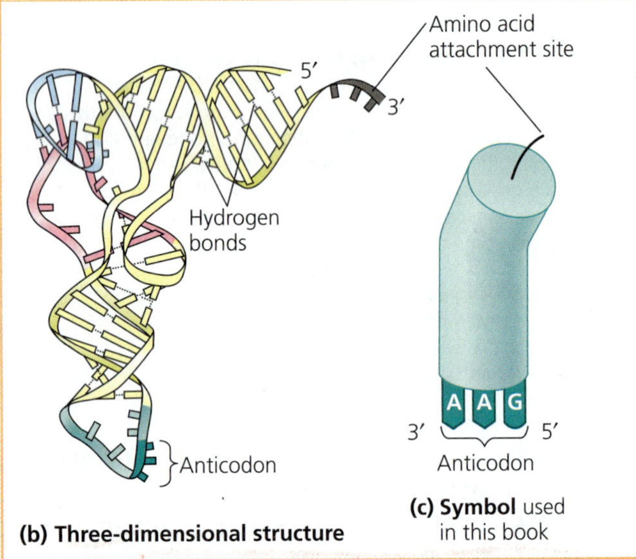

(b) Three-dimensional structure

(c) Symbol used in this book

▲ **Figure 17.15 The structure of transfer RNA (tRNA).** Anticodons are conventionally written 3′ → 5′ to align properly with codons written 5′ → 3′ (see Figure 17.14). For base pairing, RNA strands must be antiparallel, like DNA. For example, anticodon 3′-AAG-5′ pairs with mRNA codon 5′-UUC-3′.

and form a molecule with a three-dimensional structure. Flattened into one plane to clarify this base pairing, a tRNA molecule looks like a cloverleaf **(Figure 17.15a)**. The tRNA actually twists and folds into a compact three-dimensional structure that is roughly L-shaped **(Figure 17.15b)**. The loop extending from one end of the L includes the anticodon, the particular nucleotide triplet that base-pairs to a specific mRNA codon.

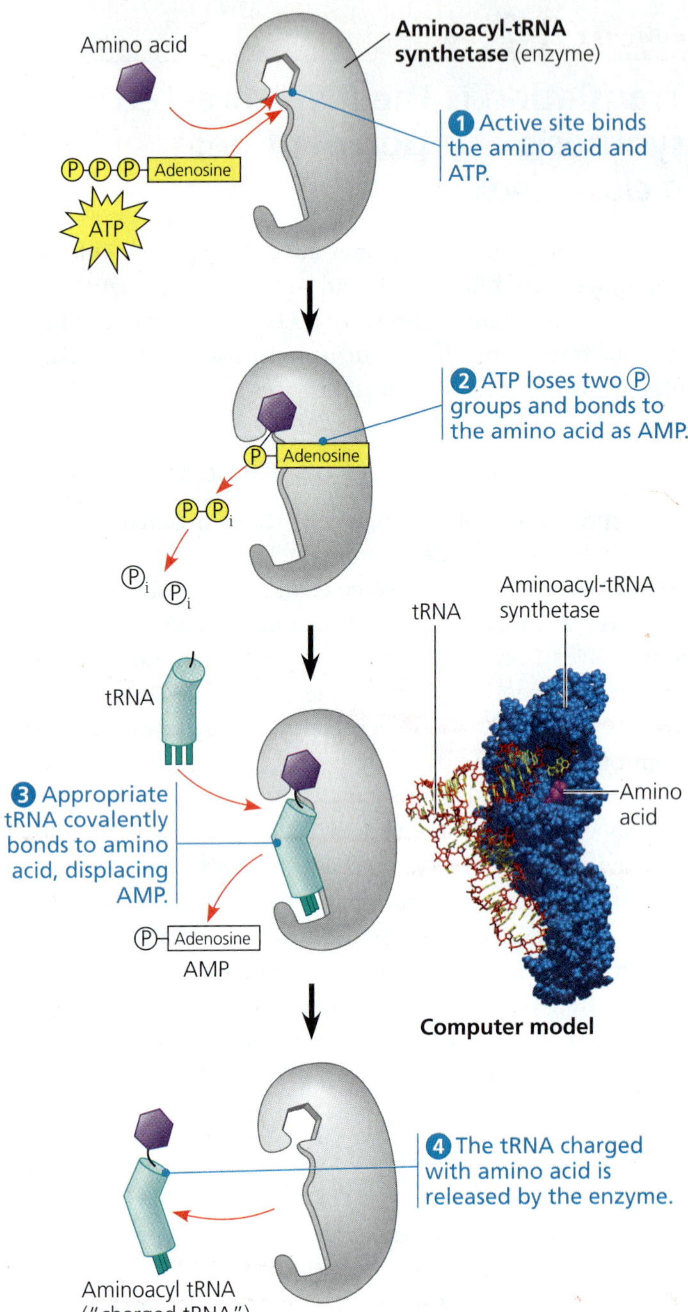

▲ **Figure 17.16 An aminoacyl-tRNA synthetase joining a specific amino acid to a tRNA.** Linkage of the tRNA and amino acid is an endergonic process that occurs at the expense of ATP. The ATP loses two phosphate groups, becoming AMP (adenosine monophosphate).

From the other end of the L-shaped tRNA molecule protrudes its 3′ end, which is the attachment site for an amino acid. Thus, the structure of a tRNA molecule fits its function.

The accurate translation of a genetic message requires two instances of molecular recognition. First, a tRNA that binds to an mRNA codon specifying a particular amino acid must carry that amino acid, and no other, to the ribosome. The correct matching up of tRNA and amino acid is carried out by a family of related enzymes called **aminoacyl-tRNA synthetases** **(Figure 17.16)**. The active site of each type of aminoacyl-tRNA synthetase fits only a specific combination of amino acid and tRNA. (Regions of both the amino acid attachment end and the anticodon end of the tRNA are instrumental in ensuring the specific fit.) There are 20 different synthetases, one for each amino acid; each synthetase is able to bind all the different tRNAs that code for its particular amino acid. The synthetase catalyzes the covalent attachment of the amino acid to its tRNA in a process driven by the hydrolysis of ATP. The resulting aminoacyl tRNA, also called a charged tRNA, is released from the enzyme and is then available to deliver its amino acid to a growing polypeptide chain on a ribosome.

The second instance of molecular recognition is the pairing of the tRNA anticodon with the appropriate mRNA codon. If one tRNA variety existed for each mRNA codon specifying an amino acid, there would be 61 tRNAs (see Figure 17.5). In fact, there are only about 45, signifying that some tRNAs must be able to bind to more than one codon. Such versatility is possible because the rules for base pairing between the third nucleotide base of a codon and the corresponding base of a tRNA anticodon are relaxed compared to those at other codon positions. For example, the nucleotide base U at the 5′ end of a tRNA anticodon can pair with either A or G in the third position (at the 3′ end) of an mRNA codon. The flexible base pairing at this codon position is called **wobble**. Wobble explains why the synonymous codons for a given amino acid most often differ in their third nucleotide base, but not in the other bases. For example, a tRNA with the anticodon 3′-UCU-5′ can base-pair with either the mRNA codon 5′-AGA-3′ or 5′-AGG-3′, both of which code for arginine (see Figure 17.5).

Ribosomes

Ribosomes facilitate the specific coupling of tRNA anticodons with mRNA codons during protein synthesis. A ribosome consists of a large subunit and a small subunit, each made up of proteins and one or more **ribosomal RNAs (rRNAs)** **(Figure 17.17)**. In eukaryotes, the subunits are made in the nucleolus. Ribosomal RNA genes are transcribed, and the RNA is processed and assembled with proteins imported from the cytoplasm. The resulting ribosomal subunits are then exported via nuclear pores to the cytoplasm. In both bacteria and eukaryotes, large and small subunits join to form a functional ribosome only when they attach to an mRNA molecule. About one-third of the mass of a ribosome is made up of proteins; the rest consists of rRNAs, either three molecules (in bacteria) or four (in eukaryotes). Because most cells contain thousands of ribosomes, rRNA is the most abundant type of cellular RNA.

Although the ribosomes of bacteria and eukaryotes are very similar in structure and function, those of eukaryotes are slightly larger and differ somewhat from bacterial ribosomes in their molecular composition. The differences are medically significant. Certain antibiotic drugs can inactivate bacterial ribosomes without inhibiting the ability of eukaryotic ribosomes to make proteins. These drugs, including tetracycline and streptomycin, are used to combat bacterial infections.

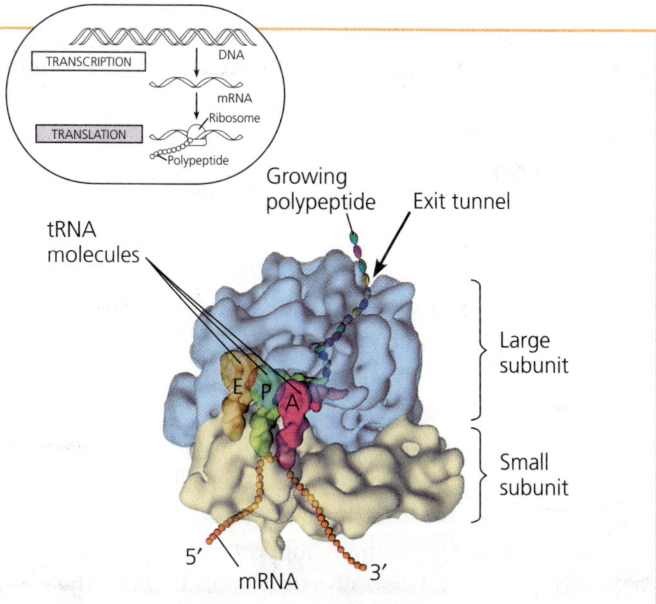

(a) Computer model of functioning ribosome. This is a model of a bacterial ribosome, showing its overall shape. The eukaryotic ribosome is roughly similar. A ribosomal subunit is a complex of ribosomal RNA molecules and proteins.

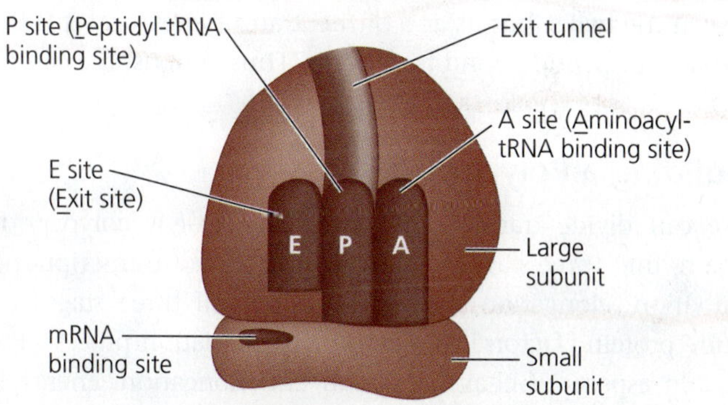

(b) Schematic model showing binding sites. A ribosome has an mRNA binding site and three tRNA binding sites, known as the A, P, and E sites. This schematic ribosome will appear in later diagrams.

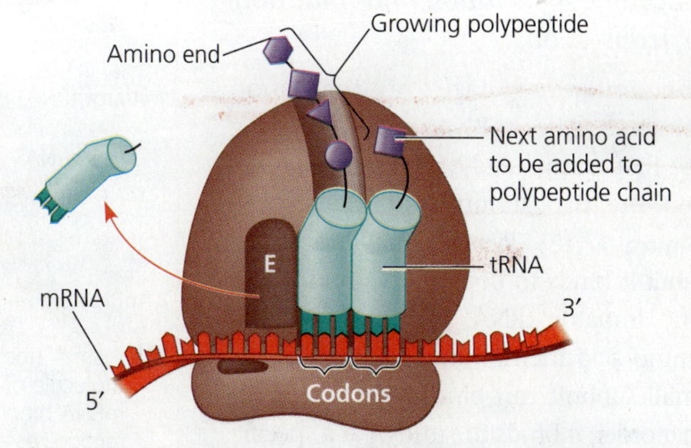

(c) Schematic model with mRNA and tRNA. A tRNA fits into a binding site when its anticodon base-pairs with an mRNA codon. The P site holds the tRNA attached to the growing polypeptide. The A site holds the tRNA carrying the next amino acid to be added to the polypeptide chain. Discharged tRNA leaves from the E site.

▲ **Figure 17.17 The anatomy of a functioning ribosome.**

The structure of a ribosome reflects its function of bringing mRNA together with tRNAs carrying amino acids. In addition to a binding site for mRNA, each ribosome has three binding sites for tRNA, as described in Figure 17.17. The **P site** (**p**eptidyl-tRNA binding site) holds the tRNA carrying the growing polypeptide chain, while the **A site** (**a**minoacyl-tRNA binding site) holds the tRNA carrying the next amino acid to be added to the chain. Discharged tRNAs leave the ribosome from the **E site** (**e**xit site). The ribosome holds the tRNA and mRNA in close proximity and positions the new amino acid for addition to the carboxyl end of the growing polypeptide. It then catalyzes the formation of the peptide bond. As the polypeptide becomes longer, it passes through an *exit tunnel* in the ribosome's large subunit. When the polypeptide is complete, it is released through the exit tunnel.

A lot of evidence strongly supports the hypothesis that rRNA, not protein, is primarily responsible for both the structure and the function of the ribosome. The proteins, which are largely on the exterior, support the shape changes of the rRNA molecules as they carry out catalysis during translation. Ribosomal RNA is the main constituent of the interface between the two subunits and of the A and P sites, and it is the catalyst of peptide bond formation. Thus, a ribosome can be regarded as one colossal ribozyme!

Building a Polypeptide

We can divide translation, the synthesis of a polypeptide chain, into three stages (analogous to those of transcription): initiation, elongation, and termination. All three stages require protein "factors" that aid in the translation process. For certain aspects of chain initiation and elongation, energy is also required. It is provided by the hydrolysis of guanosine triphosphate (GTP), a molecule closely related to ATP.

Ribosome Association and Initiation of Translation

The initiation stage of translation brings together mRNA, a tRNA bearing the first amino acid of the polypeptide, and the two subunits of a ribosome (**Figure 17.18**). First, a small ribosomal subunit binds to both mRNA and a specific initiator tRNA, which carries the amino acid methionine. In bacteria, the small subunit can bind these two in either order; it binds the mRNA at a specific RNA sequence, just upstream of the start codon, AUG. (Joan Steitz, our Unit Three interviewee, discovered the binding site on the mRNA and showed that complementary base pairing between this site and a ribosomal RNA was involved.) In eukaryotes, the small subunit, with the initiator tRNA already bound, binds to the 5′ cap of the mRNA and then moves, or *scans*, downstream along the mRNA until it reaches the start codon; the initiator tRNA then hydrogen-bonds to the AUG start codon. In either case, the start codon signals the start of translation; this is important because it establishes the codon reading frame for the mRNA.

The union of mRNA, initiator tRNA, and a small ribosomal subunit is followed by the attachment of a large ribosomal subunit, completing the *translation initiation complex*. Proteins called *initiation factors* are required to bring all these components together. The cell also expends energy obtained by hydrolysis of a GTP molecule to form the initiation complex. At the completion of the initiation process, the initiator tRNA sits in the P site of the ribosome, and the vacant A site is ready for the next aminoacyl tRNA. Note that a polypeptide is always synthesized in one direction, from the initial methionine at the amino end, also called the N-terminus, toward the final amino acid at the carboxyl end, also called the C-terminus (see Figure 5.17).

Elongation of the Polypeptide Chain

In the elongation stage of translation, amino acids are added one by one to the previous amino acid at the C-terminus of the growing chain. Each addition involves the participation of several proteins called *elongation factors* and occurs in a three-step cycle described in **Figure 17.19**. Energy expenditure occurs in the first and third steps. Codon recognition requires hydrolysis of one molecule of GTP, which increases the accuracy and efficiency of this step. One more GTP is hydrolyzed to provide energy for the translocation step.

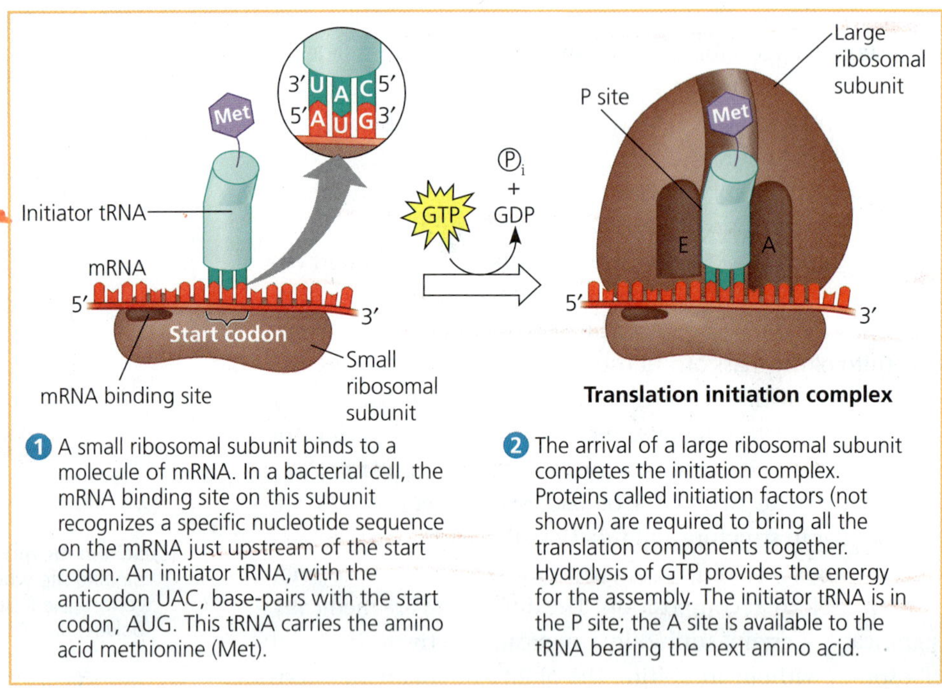

▲ **Figure 17.18 The initiation of translation.**

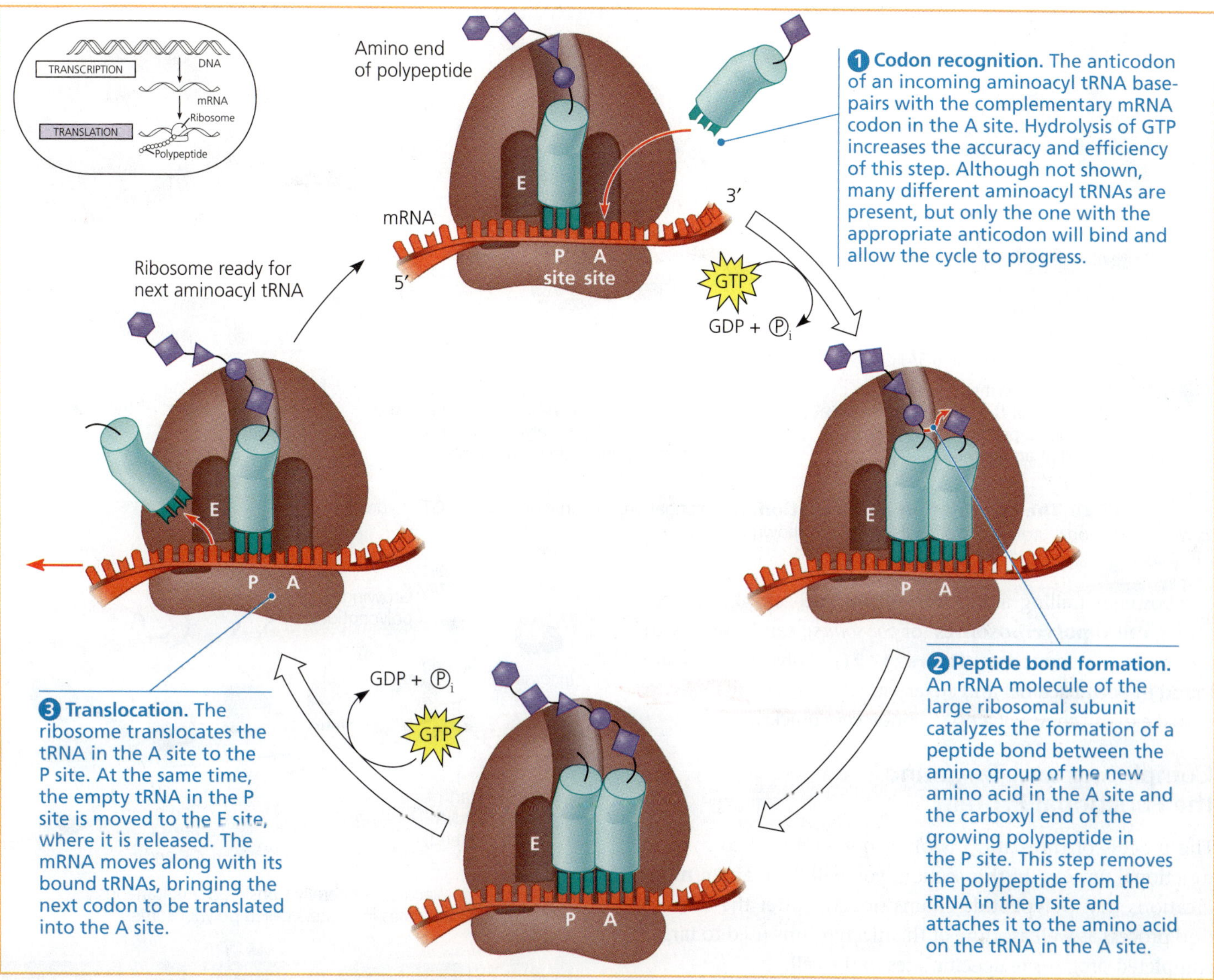

▲ **Figure 17.19 The elongation cycle of translation.** The hydrolysis of GTP plays an important role in the elongation process. Not shown are the proteins called elongation factors.

The mRNA is moved through the ribosome in one direction only, 5′ end first; this is equivalent to the ribosome moving 5′ → 3′ on the mRNA. The important point is that the ribosome and the mRNA move relative to each other, unidirectionally, codon by codon. The elongation cycle takes less than a tenth of a second in bacteria and is repeated as each amino acid is added to the chain until the polypeptide is completed.

Termination of Translation

The final stage of translation is termination (**Figure 17.20**, on the next page). Elongation continues until a stop codon in the mRNA reaches the A site of the ribosome. The nucleotide base triplets UAG, UAA, and UGA do not code for amino acids but instead act as signals to stop translation. A *release factor*, a protein shaped like an aminoacyl tRNA, binds directly to the stop codon in the A site. The release factor causes the addition of a water molecule instead of an amino acid to the polypeptide chain. (There are plenty of water molecules available in the aqueous cellular environment.) This reaction breaks (hydrolyzes) the bond between the completed polypeptide and the tRNA in the P site, releasing the polypeptide through the exit tunnel of the ribosome's large subunit. The remainder of the translation assembly then comes apart in a multistep process, aided by other protein factors. Breakdown of the translation assembly requires the hydrolysis of two more GTP molecules.

Polyribosomes

A single ribosome can make an average-sized polypeptide in less than a minute. Typically, however, multiple ribosomes translate an mRNA at the same time; that is, a single mRNA is used to make many copies of a polypeptide simultaneously. Once a ribosome is far enough past the start codon, a second ribosome can attach to the mRNA, eventually resulting in a number

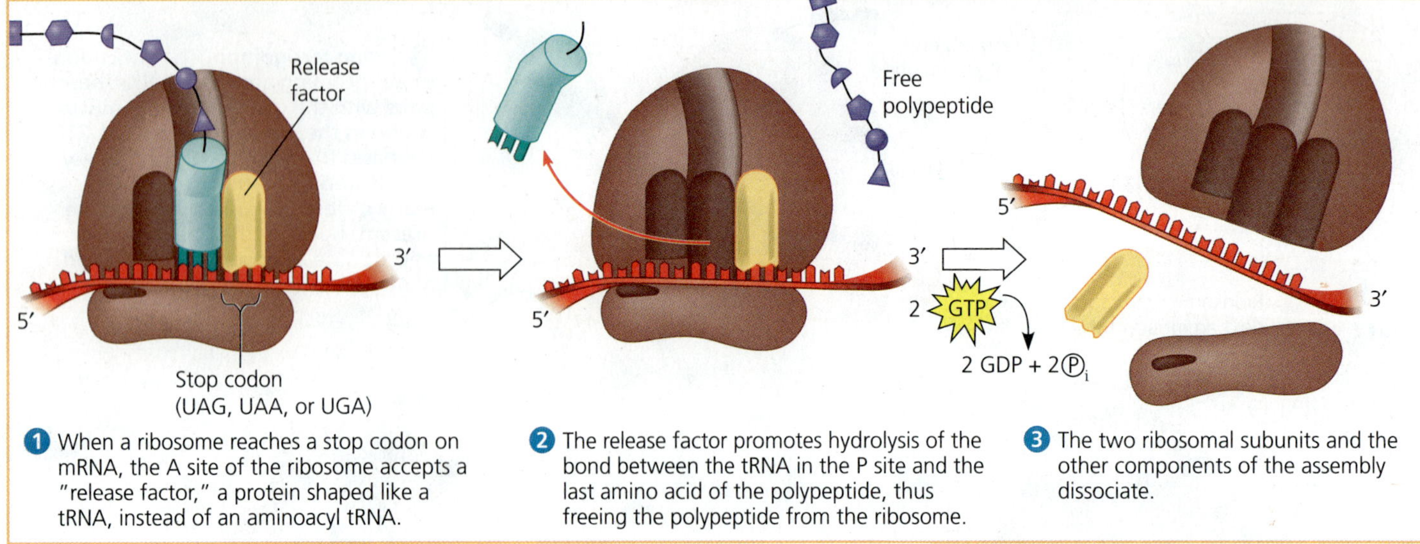

▲ **Figure 17.20 The termination of translation.** Like elongation, termination requires GTP hydrolysis as well as additional protein factors, which are not shown here.

of ribosomes trailing along the mRNA. Such strings of ribosomes, called **polyribosomes** (or *polysomes*), can be seen with an electron microscope **(Figure 17.21)**. Polyribosomes are found in both bacterial and eukaryotic cells. They enable a cell to make many copies of a polypeptide very quickly.

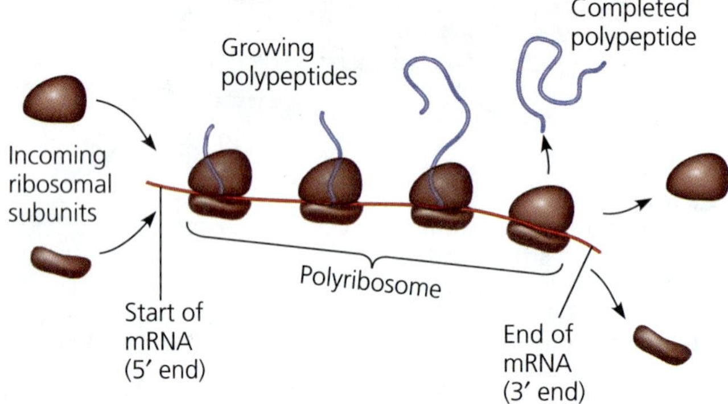

(a) An mRNA molecule is generally translated simultaneously by several ribosomes in clusters called polyribosomes.

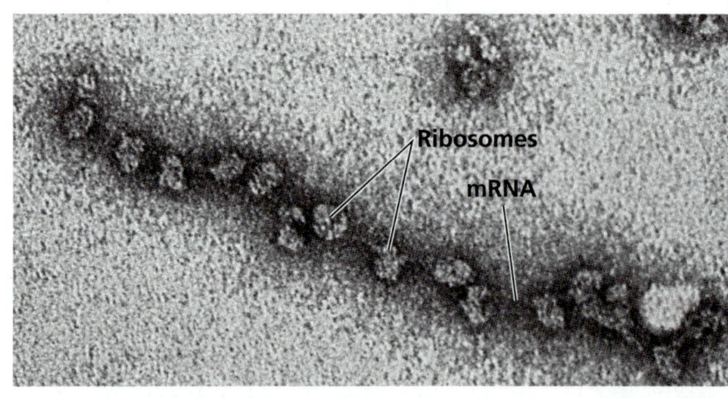

(b) This micrograph shows a large polyribosome in a bacterial cell. Growing polypeptides are not visible here (TEM).

▲ **Figure 17.21 Polyribosomes.**

Completing and Targeting the Functional Protein

The process of translation is often not sufficient to make a functional protein. In this section, you will learn about modifications that polypeptide chains undergo after the translation process as well as some of the mechanisms used to target completed proteins to specific sites in the cell.

Protein Folding and Post-Translational Modifications

During its synthesis, a polypeptide chain begins to coil and fold spontaneously as a consequence of its amino acid sequence (primary structure), forming a protein with a specific shape: a three-dimensional molecule with secondary and tertiary structure (see Figure 5.20). Thus, a gene determines primary structure, and primary structure in turn determines shape. In many cases, a chaperone protein (chaperonin) helps the polypeptide fold correctly (see Figure 5.23).

Additional steps—*post-translational modifications*—may be required before the protein can begin doing its particular job in the cell. Certain amino acids may be chemically modified by the attachment of sugars, lipids, phosphate groups, or other additions. Enzymes may remove one or more amino acids from the leading (amino) end of the polypeptide chain. In some cases, a polypeptide chain may be enzymatically cleaved into two or more pieces. For example, the protein insulin is first synthesized as a single polypeptide chain but becomes active only after an enzyme cuts out a central part of the chain, leaving a protein made up of two polypeptide chains connected by disulfide bridges. In other cases, two or more polypeptides that are synthesized separately may come together, becoming the subunits of a protein that has quaternary structure. A familiar example is hemoglobin (see Figure 5.20).

Targeting Polypeptides to Specific Locations

In electron micrographs of eukaryotic cells active in protein synthesis, two populations of ribosomes (and polyribosomes) are evident: free and bound (see Figure 6.10). Free ribosomes are suspended in the cytosol and mostly synthesize proteins that stay in the cytosol and function there. In contrast, bound ribosomes are attached to the cytosolic side of the endoplasmic reticulum (ER) or to the nuclear envelope. Bound ribosomes make proteins of the endomembrane system (the nuclear envelope, ER, Golgi apparatus, lysosomes, vacuoles, and plasma membrane) as well as proteins secreted from the cell, such as insulin. It is important to note that the ribosomes themselves are identical and can switch their status from free to bound.

What determines whether a ribosome is free in the cytosol or bound to rough ER? Polypeptide synthesis always begins in the cytosol as a free ribosome starts to translate an mRNA molecule. There the process continues to completion—*unless* the growing polypeptide itself cues the ribosome to attach to the ER. The polypeptides of proteins destined for the endomembrane system or for secretion are marked by a **signal peptide**, which targets the protein to the ER **(Figure 17.22)**. The signal peptide, a sequence of about 20 amino acids at or near the leading end (N-terminus) of the polypeptide, is recognized as it emerges from the ribosome by a protein-RNA complex called a **signal-recognition particle (SRP)**. This particle functions as an escort that brings the ribosome to a receptor protein built into the ER membrane. The receptor is part of a multiprotein translocation complex. Polypeptide synthesis continues there, and the growing polypeptide snakes across the membrane into the ER lumen via a protein pore. The signal peptide is usually removed by an enzyme. The rest of the completed polypeptide, if it is to be secreted from the cell, is released into solution within the ER lumen (as in Figure 17.22). Alternatively, if the polypeptide is to be a membrane protein, it remains partially embedded in the ER membrane.

Other kinds of signal peptides are used to target polypeptides to mitochondria, chloroplasts, the interior of the nucleus, and other organelles that are not part of the endomembrane system. The critical difference in these cases is that translation is completed in the cytosol before the polypeptide is imported into the organelle. The mechanisms of translocation also vary, but in all cases studied to date, the "postal zip codes" that address proteins for secretion or to cellular locations are signal peptides of some sort. Bacteria also employ signal peptides to target proteins to the plasma membrane for secretion.

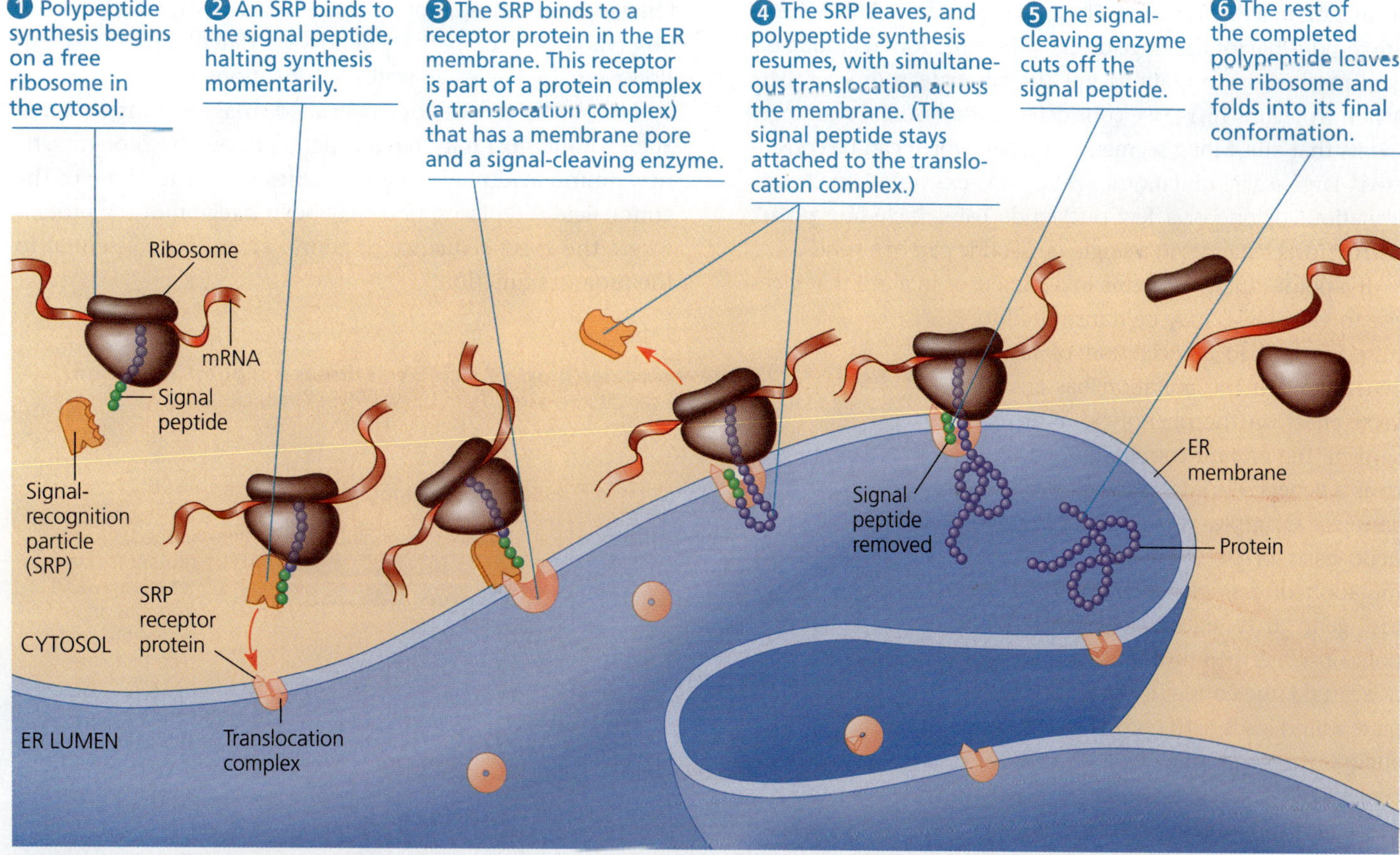

▲ **Figure 17.22 The signal mechanism for targeting proteins to the ER.** A polypeptide destined for the endomembrane system or for secretion from the cell begins with a signal peptide, a series of amino acids that targets it for the ER. This figure shows the synthesis of a secretory protein and its simultaneous import into the ER. In the ER and then in the Golgi, the protein will be processed further. Finally, a transport vesicle will convey it to the plasma membrane for release from the cell (see Figure 7.12).

CONCEPT CHECK 17.4

1. What two processes ensure that the correct amino acid is added to a growing polypeptide chain?
2. Discuss the ways in which rRNA structure likely contributes to ribosomal function.
3. Describe how a polypeptide to be secreted is transported to the endomembrane system.
4. **WHAT IF?** **DRAW IT** If a tRNA has the anticodon 3′-CGU-5′, what two different codons could it bind to? Draw each on an mRNA, labeling all 5′ and 3′ ends. Add the amino acid carried by this tRNA.

For suggested answers, see Appendix A.

CONCEPT 17.5

Mutations of one or a few nucleotides can affect protein structure and function

Now that you have explored the process of gene expression, you are ready to understand the effects of changes to the genetic information of a cell (or virus). These changes, called **mutations**, are responsible for the huge diversity of genes found among organisms because mutations are the ultimate source of new genes. In Figure 15.14, we considered chromosomal rearrangements that affect long segments of DNA, which can be considered large-scale mutations. Here we examine small-scale mutations of one or a few nucleotide pairs, including **point mutations**, changes in a single nucleotide pair of a gene.

If a point mutation occurs in a gamete or in a cell that gives rise to gametes, it may be transmitted to offspring and to a succession of future generations. If the mutation has an adverse effect on the phenotype of an organism, the mutant condition is referred to as a genetic disorder or hereditary disease. For example, we can trace the genetic basis of sickle-cell disease to the mutation of a single nucleotide pair in the gene that encodes the β-globin polypeptide of hemoglobin. The change of a single nucleotide in the DNA's template strand leads to the production of an abnormal protein (**Figure 17.23**; also see Figure 5.21). In individuals who are homozygous for the mutant allele, the sickling of red blood cells caused by the altered hemoglobin produces the multiple symptoms associated with sickle-cell disease (see Chapter 14). Another disorder caused by a point mutation is a heart condition, familial cardiomyopathy, that is responsible for some incidents of sudden death in young athletes. Point mutations in several genes have been identified, any of which can lead to this disorder.

Types of Small-Scale Mutations

Let's now consider how small-scale mutations affect proteins. Point mutations within a gene can be divided into two general categories: (1) single nucleotide-pair substitutions and (2) nucleotide-pair insertions or deletions. Insertions and deletions can involve one or more nucleotide pairs.

Substitutions

A **nucleotide-pair substitution** is the replacement of one nucleotide and its partner with another pair of nucleotides (**Figure 17.24a**). Some substitutions have no effect on the encoded protein, owing to the redundancy of the genetic code. For example, if 3′-CCG-5′ on the template strand mutated to 3′-CCA-5′, the mRNA codon that used to be GGC would become GGU, but a glycine would still be inserted at the proper location in the protein (see Figure 17.5). In other words, a change in a nucleotide pair may transform one codon into another that is translated into the same amino acid. Such a change is an example of a **silent mutation**, which has no observable effect on the phenotype. (Silent mutations can occur outside genes as well.) Substitutions that change one amino acid to another one are called **missense mutations**. Such a mutation may have little effect on the protein: The new amino acid may have properties similar to those of the amino acid it replaces, or it may be in a region of the protein where the exact sequence of amino acids is not essential to the protein's function.

▼ Figure 17.23 The molecular basis of sickle-cell disease: a point mutation. The allele that causes sickle-cell disease differs from the wild-type (normal) allele by a single DNA nucleotide pair.

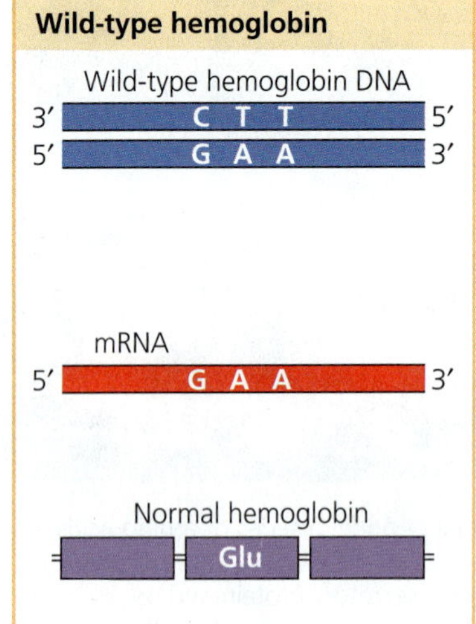

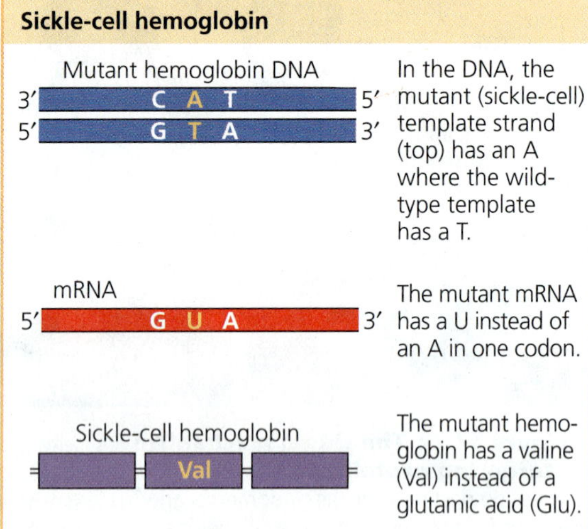

▼ Figure 17.24 Types of small-scale mutations that affect mRNA sequence. All but one of the types shown here also affect the amino acid sequence of the encoded polypeptide.

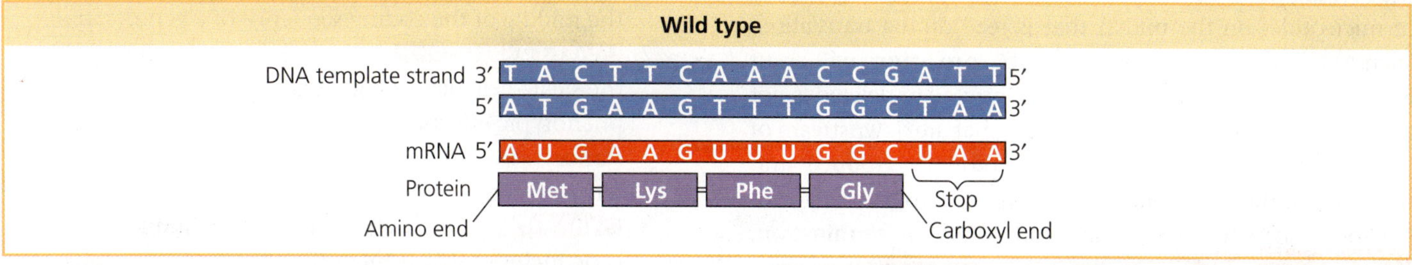

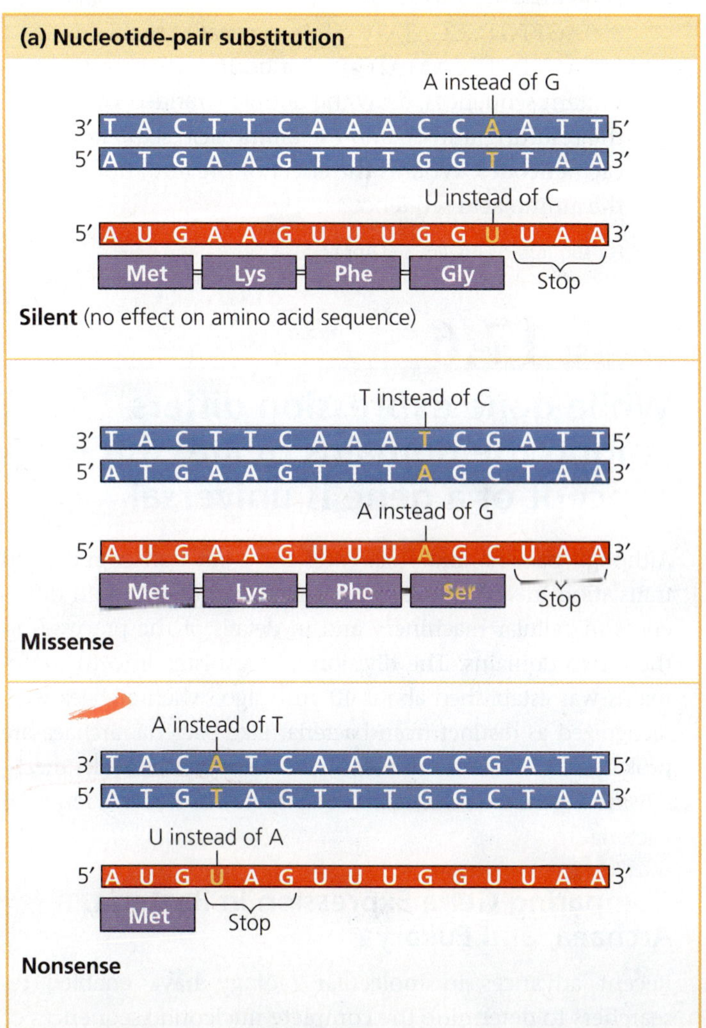

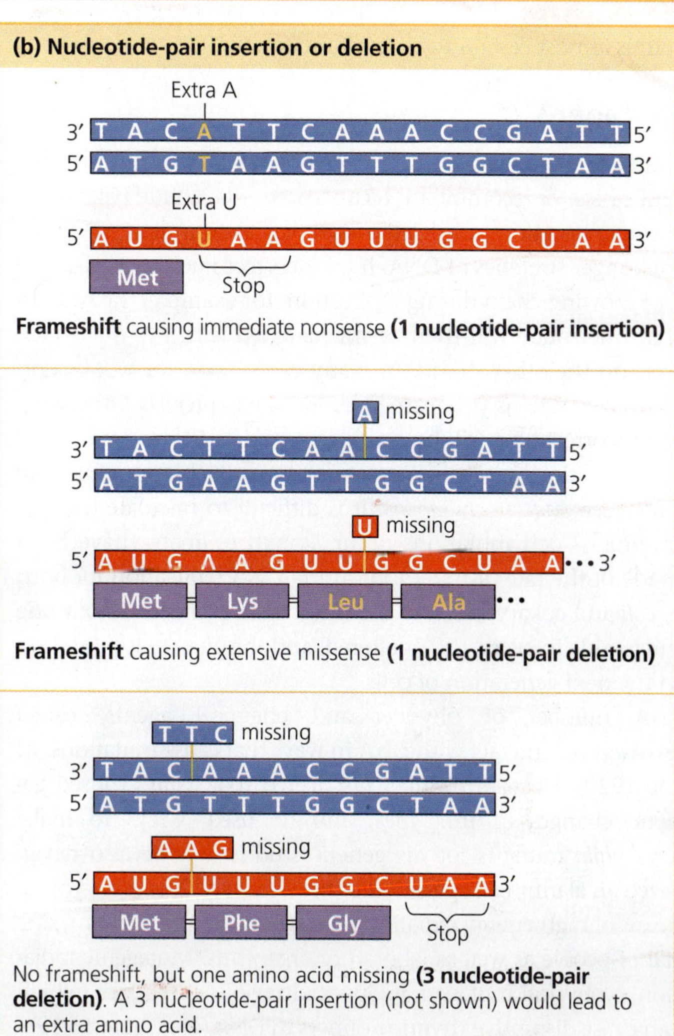

However, the nucleotide-pair substitutions of greatest interest are those that cause a major change in a protein. The alteration of a single amino acid in a crucial area of a protein—such as in the part of hemoglobin shown in Figure 17.23 or in the active site of an enzyme as shown in Figure 8.18—will significantly alter protein activity. Occasionally, such a mutation leads to an improved protein or one with novel capabilities, but much more often such mutations are detrimental, leading to a useless or less active protein that impairs cellular function.

Substitution mutations are usually missense mutations; that is, the altered codon still codes for an amino acid and thus makes sense, although not necessarily the *right* sense. But a point mutation can also change a codon for an amino acid into a stop codon. This is called a **nonsense mutation**, and it causes translation to be terminated prematurely; the resulting polypeptide will be shorter than the polypeptide encoded by the normal gene. Nearly all nonsense mutations lead to nonfunctional proteins.

Insertions and Deletions

Insertions and **deletions** are additions or losses of nucleotide pairs in a gene **(Figure 17.24b)**. These mutations have a disastrous effect on the resulting protein more often than substitutions do. Insertion or deletion of nucleotides may alter the reading frame of the genetic message, the triplet grouping of nucleotides on the mRNA that is read during translation. Such a mutation, called a **frameshift mutation**, will occur whenever the number of nucleotides inserted or deleted is not a multiple of three. All the nucleotides that are downstream of the deletion or insertion will be improperly grouped into codons, and the result will be extensive missense, usually ending sooner or later in nonsense and premature termination. Unless the frameshift is very near the end of the gene, the protein is almost certain to be nonfunctional.

Mutagens

Mutations can arise in a number of ways. Errors during DNA replication or recombination can lead to nucleotide-pair substitutions, insertions, or deletions, as well as to mutations affecting longer stretches of DNA. If an incorrect nucleotide is added to a growing chain during replication, for example, the base on that nucleotide will then be mismatched with the nucleotide base on the other strand. In many cases, the error will be corrected by systems you learned about in Chapter 16. Otherwise, the incorrect base will be used as a template in the next round of replication, resulting in a mutation. Such mutations are called *spontaneous mutations*. It is difficult to calculate the rate at which such mutations occur. Rough estimates have been made of the rate of mutation during DNA replication for both *E. coli* and eukaryotes, and the numbers are similar: About one nucleotide in every 10^{10} is altered, and the change is passed on to the next generation of cells.

A number of physical and chemical agents, called **mutagens**, interact with DNA in ways that cause mutations. In the 1920s, Hermann Muller discovered that X-rays caused genetic changes in fruit flies, and he used X-rays to make *Drosophila* mutants for his genetic studies. But he also recognized an alarming implication of his discovery: X-rays and other forms of high-energy radiation pose hazards to the genetic material of people as well as laboratory organisms. Mutagenic radiation, a physical mutagen, includes ultraviolet (UV) light, which can cause disruptive thymine dimers in DNA (see Figure 16.19).

Chemical mutagens fall into several categories. Nucleotide analogs are chemicals that are similar to normal DNA nucleotides but that pair incorrectly during DNA replication. Some other chemical mutagens interfere with correct DNA replication by inserting themselves into the DNA and distorting the double helix. Still other mutagens cause chemical changes in bases that change their pairing properties.

Researchers have developed a variety of methods to test the mutagenic activity of chemicals. A major application of these tests is the preliminary screening of chemicals to identify those that may cause cancer. This approach makes sense because most carcinogens (cancer-causing chemicals) are mutagenic, and conversely, most mutagens are carcinogenic.

CONCEPT CHECK 17.5

1. What happens when one nucleotide pair is lost from the middle of the coding sequence of a gene?
2. **MAKE CONNECTIONS** Individuals heterozygous for the sickle-cell allele are generally healthy but show phenotypic effects of the allele under some circumstances; see Concept 14.4, pages 277–278. Explain in terms of gene expression.
3. **WHAT IF?** **DRAW IT** The template strand of a gene includes this sequence: 3′-TACTTGTCCGATATC-5′. It is mutated to 3′-TACTTGTCCAATATC-5′. For both normal and mutant sequences, draw the double-stranded DNA, the resulting mRNA, and the amino acid sequence each encodes. What is the effect of the mutation on the amino acid sequence?

For suggested answers, see Appendix A.

CONCEPT 17.6

While gene expression differs among the domains of life, the concept of a gene is universal

Although bacteria and eukaryotes carry out transcription and translation in very similar ways, we have noted certain differences in cellular machinery and in details of the processes in these two domains. The division of organisms into three domains was established about 40 years ago, when archaea were recognized as distinct from bacteria. Like bacteria, archaea are prokaryotes. However, archaea share many aspects of the mechanisms of gene expression with eukaryotes, as well as a few with bacteria.

Comparing Gene Expression in Bacteria, Archaea, and Eukarya

Recent advances in molecular biology have enabled researchers to determine the complete nucleotide sequences of hundreds of genomes, including many genomes from each domain. This wealth of data allows us to compare gene and protein sequences across domains. Foremost among genes of interest are those that encode components of such fundamental biological processes as transcription and translation.

Bacterial and eukaryotic RNA polymerases differ significantly from each other, while the single archaeal RNA polymerase resembles the three eukaryotic ones. Archaea and eukaryotes use a complex set of transcription factors, unlike the smaller set of accessory proteins in bacteria. Transcription is terminated differently in bacteria and eukaryotes. The little that is known about archaeal transcription termination suggests that it is similar to the eukaryotic process.

As far as translation is concerned, archaeal ribosomes are the same size as bacterial ribosomes, but their sensitivities to chemical inhibitors more closely match those of eukaryotic ribosomes. We mentioned earlier that initiation of translation is slightly different in bacteria and eukaryotes. In this respect, the archaeal process is more like that of bacteria.

The most important differences between bacteria and eukaryotes with regard to gene expression arise from the bacterial cell's lack of compartmental organization. Like a one-room workshop, a bacterial cell ensures a streamlined operation. In the absence of a nucleus, it can simultaneously transcribe and translate the same gene (**Figure 17.25**), and the newly made protein can quickly diffuse to its site of function. Most researchers suspect that transcription and translation are coupled like this in archaeal cells as well, since archaea lack a nuclear envelope. In contrast, the eukaryotic cell's nuclear envelope segregates transcription from translation and provides a compartment for extensive RNA processing. This processing stage includes additional steps whose regulation can help coordinate the eukaryotic cell's elaborate activities (see Chapter 18).

Learning more about the proteins and RNAs involved in archaeal transcription and translation will tell us much about the evolution of these processes in all three domains. In spite of the differences in gene expression cataloged here, however, the idea of the gene itself is a unifying concept among all forms of life.

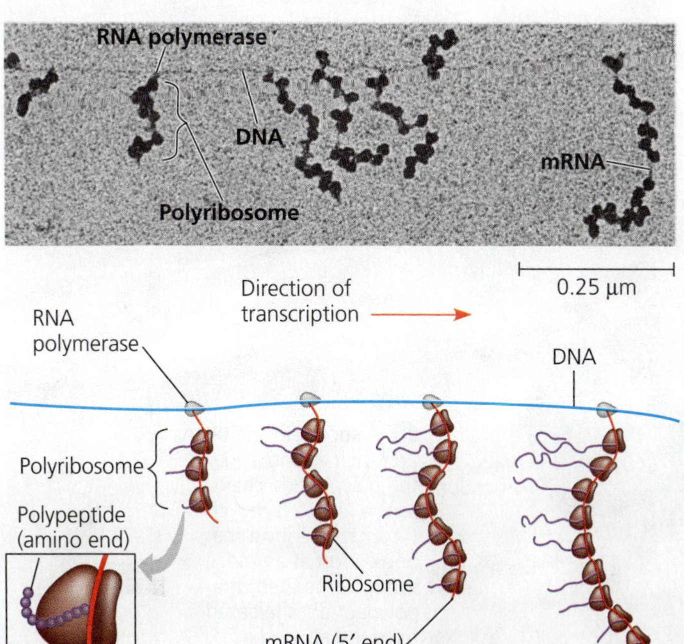

▲ **Figure 17.25 Coupled transcription and translation in bacteria.** In bacterial cells, the translation of mRNA can begin as soon as the leading (5′) end of the mRNA molecule peels away from the DNA template. The micrograph (TEM) shows a strand of *E. coli* DNA being transcribed by RNA polymerase molecules. Attached to each RNA polymerase molecule is a growing strand of mRNA, which is already being translated by ribosomes. The newly synthesized polypeptides are not visible in the micrograph but are shown in the diagram.

? *Which one of the mRNA molecules started transcription first? On that mRNA, which ribosome started translating first?*

What Is a Gene? *Revisiting the Question*

Our definition of a gene has evolved over the past few chapters, as it has through the history of genetics. We began with the Mendelian concept of a gene as a discrete unit of inheritance that affects a phenotypic character (Chapter 14). We saw that Morgan and his colleagues assigned such genes to specific loci on chromosomes (Chapter 15). We went on to view a gene as a region of specific nucleotide sequence along the length of the DNA molecule of a chromosome (Chapter 16). Finally, in this chapter, we have considered a functional definition of a gene as a DNA sequence that codes for a specific polypeptide chain. (**Figure 17.26**, on the next page, summarizes the path from gene to polypeptide in a eukaryotic cell.) All these definitions are useful, depending on the context in which genes are being studied.

Clearly, the statement that a gene codes for a polypeptide is too simple. Most eukaryotic genes contain noncoding segments (such as introns), so large portions of these genes have no corresponding segments in polypeptides. Molecular biologists also often include promoters and certain other regulatory regions of DNA within the boundaries of a gene. These DNA sequences are not transcribed, but they can be considered part of the functional gene because they must be present for transcription to occur. Our definition of a gene must also be broad enough to include the DNA that is transcribed into rRNA, tRNA, and other RNAs that are not translated. These genes have no polypeptide products but play crucial roles in the cell. Thus, we arrive at the following definition: *A gene is a region of DNA that can be expressed to produce a final functional product that is either a polypeptide or an RNA molecule.*

When considering phenotypes, however, it is often useful to start by focusing on genes that code for polypeptides. In this chapter, you have learned in molecular terms how a typical gene is expressed—by transcription into RNA and then translation into a polypeptide that forms a protein of specific structure and function. Proteins, in turn, bring about an organism's observable phenotype.

A given type of cell expresses only a subset of its genes. This is an essential feature in multicellular organisms: You'd be in trouble if the lens cells in your eyes started expressing the genes for hair proteins, which are normally expressed only in hair follicle cells! Gene expression is precisely regulated. We'll explore gene regulation in the next chapter, beginning with the simpler case of bacteria and continuing with eukaryotes.

CONCEPT CHECK 17.6

1. Would the coupling of processes shown in Figure 17.25 be found in a eukaryotic cell? Explain.
2. **WHAT IF?** In eukaryotic cells, mRNAs have been found to have a circular arrangement in which proteins hold the poly-A tail near the 5′ cap. How might this increase translation efficiency?

For suggested answers, see Appendix A.

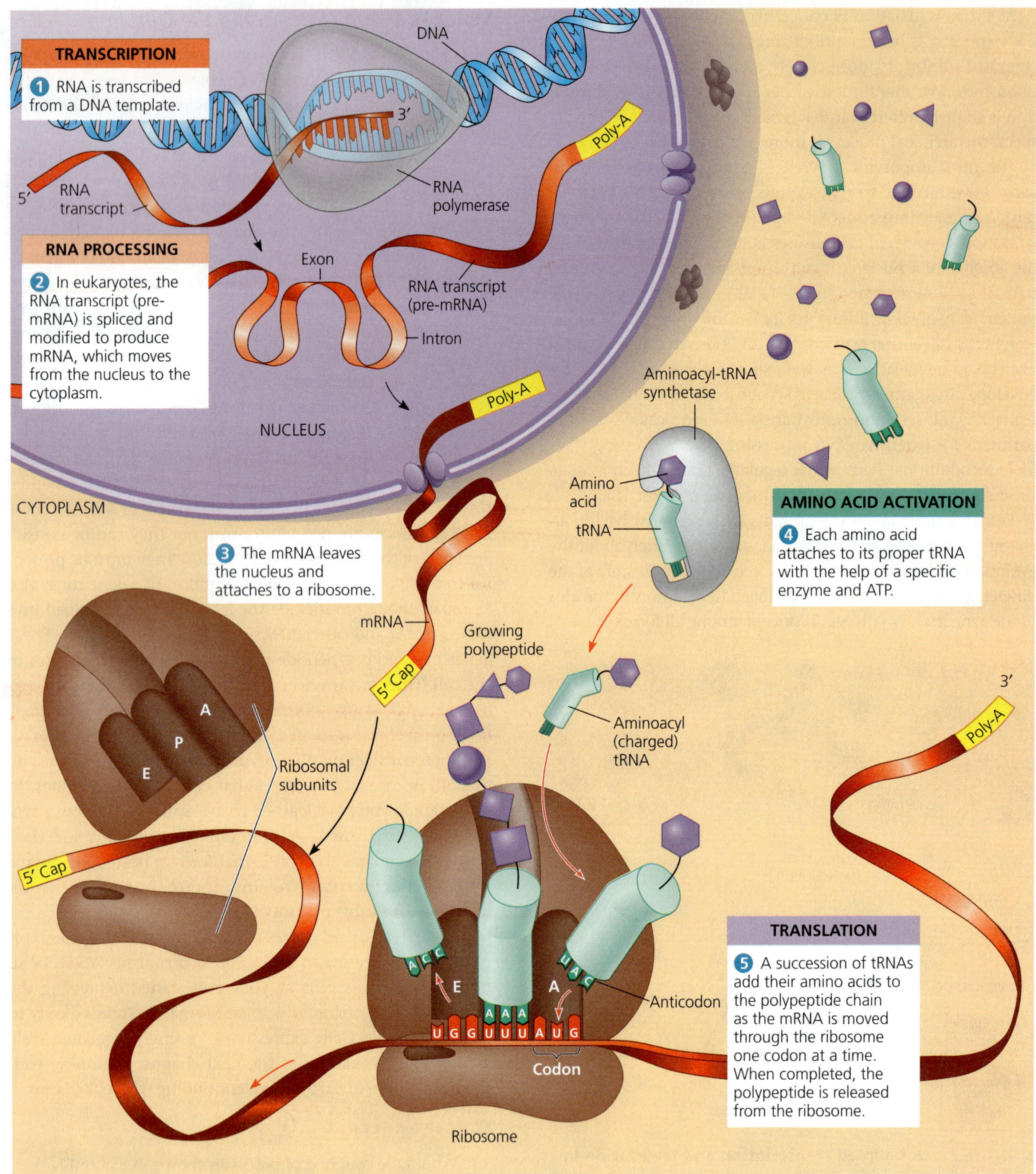

▲ **Figure 17.26 A summary of transcription and translation in a eukaryotic cell.** This diagram shows the path from one gene to one polypeptide. Keep in mind that each gene in the DNA can be transcribed repeatedly into many identical RNA molecules and that each mRNA can be translated repeatedly to yield many identical polypeptide molecules. (Also, remember that the final products of some genes are not polypeptides but RNA molecules, including tRNA and rRNA.) In general, the steps of transcription and translation are similar in bacterial, archaeal, and eukaryotic cells. The major difference is the occurrence of RNA processing in the eukaryotic nucleus. Other significant differences are found in the initiation stages of both transcription and translation and in the termination of transcription.

17 CHAPTER REVIEW

SUMMARY OF KEY CONCEPTS

CONCEPT 17.1

Genes specify proteins via transcription and translation (pp. 325–331)

- DNA controls metabolism by directing cells to make specific enzymes and other proteins, via the process of **gene expression**. Beadle and Tatum's studies of mutant strains of *Neurospora* led to the one gene–one polypeptide hypothesis. Genes code for polypeptide chains or specify RNA molecules.
- **Transcription** is the synthesis of RNA complementary to a **template strand** of DNA, providing a nucleotide-to-nucleotide transfer of information. **Translation** is the synthesis of a polypeptide whose amino acid sequence is specified by the nucleotide sequence in **mRNA**; this informational transfer thus involves a change of language, from that of nucleotides to that of amino acids.
- Genetic information is encoded as a sequence of nonoverlapping nucleotide triplets, or **codons**. A codon in messenger RNA (mRNA) either is translated into an amino acid (61 of the 64 codons) or serves as a stop signal (3 codons). Codons must be read in the correct **reading frame**.

? *Describe the process of gene expression, by which a gene affects the phenotype of an organism.*

CONCEPT 17.2

Transcription is the DNA-directed synthesis of RNA: *a closer look* (pp. 331–334)

- RNA synthesis is catalyzed by **RNA polymerase**, which links together RNA nucleotides complementary to a DNA template strand. This process follows the same base-pairing rules as DNA replication, except that in RNA, uracil substitutes for thymine.

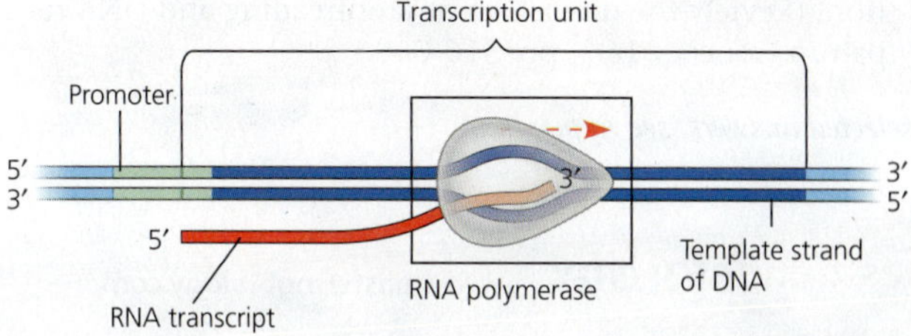

- The three stages of transcription are initiation, elongation, and termination. A **promoter**, often including a **TATA box** in eukaryotes, establishes where RNA synthesis is initiated. **Transcription factors** help eukaryotic RNA polymerase recognize promoter sequences, forming a **transcription initiation complex**. The mechanisms of termination are different in bacteria and eukaryotes.

? *What are the similarities and differences in the initiation of gene transcription in bacteria and eukaryotes?*

CONCEPT 17.3

Eukaryotic cells modify RNA after transcription (pp. 334–336)

- Before leaving the nucleus, eukaryotic mRNA molecules undergo **RNA processing**, which includes RNA splicing, the addition of a modified nucleotide **5′ cap** to the 5′ end, and the addition of a **poly-A tail** to the 3′ end.

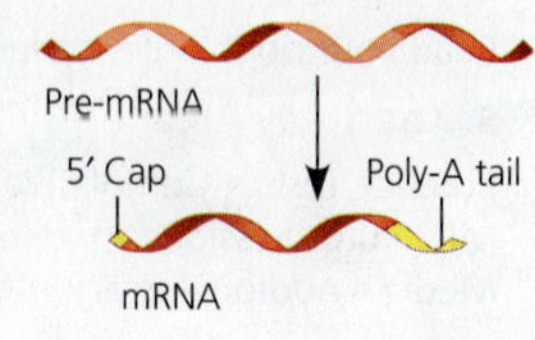

- Most eukaryotic genes are split into segments: They have **introns** interspersed among the **exons** (the regions included in the mRNA). In **RNA splicing**, introns are removed and exons joined. RNA splicing is typically carried out by **spliceosomes**, but in some cases, RNA alone catalyzes its own splicing. The catalytic ability of some RNA molecules, called **ribozymes**, derives from the inherent properties of RNA. The presence of introns allows for **alternative RNA splicing**.

? *What function do the 5′ cap and the poly-A tail serve on a eukaryotic mRNA?*

CONCEPT 17.4

Translation is the RNA-directed synthesis of a polypeptide: *a closer look* (pp. 337–344)

- A cell translates an mRNA message into protein using **transfer RNAs (tRNAs)**. After being bound to a specific amino acid by an **aminoacyl-tRNA synthetase**, a tRNA lines up via its **anticodon** at the complementary codon on mRNA. A **ribosome**, made up of **ribosomal RNAs (rRNAs)** and proteins, facilitates this coupling with binding sites for mRNA and tRNA.
- Ribosomes coordinate the three stages of translation: initiation, elongation, and termination. The formation of peptide bonds between amino acids is catalyzed by rRNA as tRNAs move through the **A** and **P sites** and exit through the **E site**.

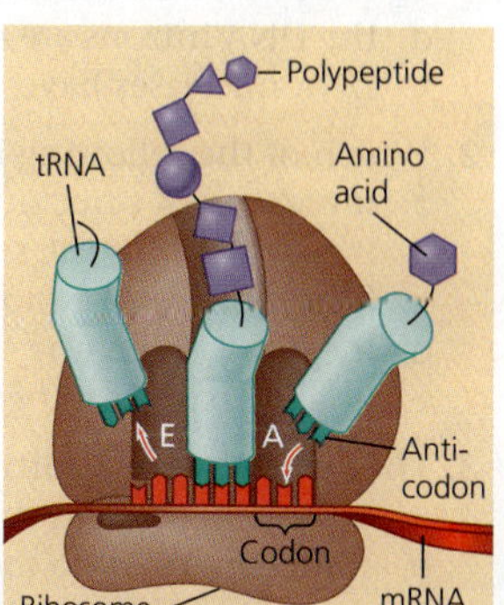

- A single mRNA molecule can be translated simultaneously by a number of ribosomes, forming a **polyribosome**.
- After translation, modifications to proteins can affect their three-dimensional shape. Free ribosomes in the cytosol initiate synthesis of all proteins, but proteins destined for the endomembrane system or for secretion are transported into the ER. Such proteins have a **signal peptide** to which a **signal-recognition particle (SRP)** binds, enabling the translating ribosome to bind to the ER.

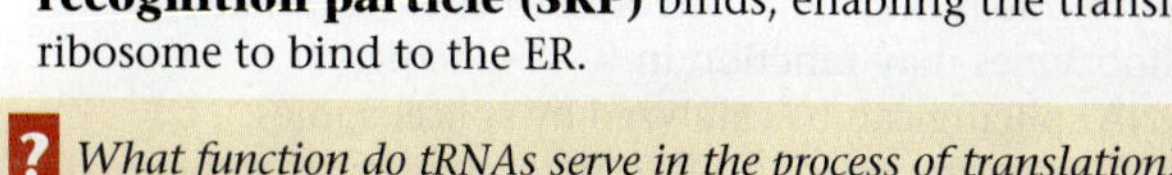

? *What function do tRNAs serve in the process of translation?*

CONCEPT 17.5

Mutations of one or a few nucleotides can affect protein structure and function (pp. 344–346)

- Small-scale **mutations** include **point mutations**, changes in one DNA nucleotide pair, which may lead to production of nonfunctional proteins. **Nucleotide-pair substitutions** can cause **missense** or **nonsense mutations**. Nucleotide-pair **insertions** or **deletions** may produce **frameshift mutations**.
- Spontaneous mutations can occur during DNA replication, recombination, or repair. Chemical and physical **mutagens** cause DNA damage that can alter genes.

? *What will be the results of chemically modifying one nucleotide base of a gene? What role is played by DNA repair systems in the cell?*

CONCEPT 17.6

While gene expression differs among the domains of life, the concept of a gene is universal (pp. 346–348)

- There are some differences in gene expression among bacteria, archaea, and eukaryotes. Because bacterial cells lack a nuclear envelope, translation can begin while transcription is still in progress. Archaeal cells show similarities to both eukaryotic and bacterial cells in their processes of gene expression. In a eukaryotic cell, the nuclear envelope separates transcription from translation, and extensive RNA processing occurs in the nucleus.
- A gene is a region of DNA whose final functional product is either a polypeptide or an RNA molecule.

? *How does the presence of a nuclear envelope affect gene expression in eukaryotes?*

TEST YOUR UNDERSTANDING

LEVEL 1: KNOWLEDGE/COMPREHENSION

1. In eukaryotic cells, transcription cannot begin until
   a. the two DNA strands have completely separated and exposed the promoter.
   b. several transcription factors have bound to the promoter.
   c. the 5′ caps are removed from the mRNA.
   d. the DNA introns are removed from the template.
   e. DNA nucleases have isolated the transcription unit.

2. Which of the following is *not* true of a codon?
   a. It consists of three nucleotides.
   b. It may code for the same amino acid as another codon.
   c. It never codes for more than one amino acid.
   d. It extends from one end of a tRNA molecule.
   e. It is the basic unit of the genetic code.

3. The anticodon of a particular tRNA molecule is
   a. complementary to the corresponding mRNA codon.
   b. complementary to the corresponding triplet in rRNA.
   c. the part of tRNA that bonds to a specific amino acid.
   d. changeable, depending on the amino acid that attaches to the tRNA.
   e. catalytic, making the tRNA a ribozyme.

4. Which of the following is *not* true of RNA processing?
   a. Exons are cut out before mRNA leaves the nucleus.
   b. Nucleotides may be added at both ends of the RNA.
   c. Ribozymes may function in RNA splicing.
   d. RNA splicing can be catalyzed by spliceosomes.
   e. A primary transcript is often much longer than the final RNA molecule that leaves the nucleus.

5. Which component is *not* directly involved in translation?
   a. mRNA b. DNA c. tRNA d. ribosomes e. GTP

LEVEL 2: APPLICATION/ANALYSIS

6. Using Figure 17.5, identify a 5′ → 3′ sequence of nucleotides in the DNA template strand for an mRNA coding for the polypeptide sequence Phe-Pro-Lys.
   a. 5′-UUUGGGAAA-3′
   b. 5′-GAACCCCTT-3′
   c. 5′-AAAACCTTT-3′
   d. 5′-CTTCGGGAA-3′
   e. 5′-AAACCCUUU-3′

7. Which of the following mutations would be *most* likely to have a harmful effect on an organism?
   a. a nucleotide-pair substitution
   b. a deletion of three nucleotides near the middle of a gene
   c. a single nucleotide deletion in the middle of an intron
   d. a single nucleotide deletion near the end of the coding sequence
   e. a single nucleotide insertion downstream of, and close to, the start of the coding sequence

8. **DRAW IT** Fill in the following table:

Type of RNA	Functions
Messenger RNA (mRNA)	
Transfer RNA (tRNA)	
	Plays catalytic (ribozyme) roles and structural roles in ribosomes
Primary transcript	
Small nuclear RNA (snRNA)	

LEVEL 3: SYNTHESIS/EVALUATION

9. **EVOLUTION CONNECTION** Most amino acids are coded for by a set of similar codons (see Figure 17.5). What evolutionary explanations can you give for this pattern? (*Hint*: There is one explanation relating to ancestry, and some less obvious ones of a "form-fits-function" type.)

10. **SCIENTIFIC INQUIRY**
    Knowing that the genetic code is almost universal, a scientist uses molecular biological methods to insert the human β-globin gene (shown in Figure 17.11) into bacterial cells, hoping the cells will express it and synthesize functional β-globin protein. Instead, the protein produced is nonfunctional and is found to contain many fewer amino acids than does β-globin made by a eukaryotic cell. Explain why.

11. **WRITE ABOUT A THEME**
    Evolution and The Genetic Basis of Life Evolution accounts for the unity and diversity of life, and the continuity of life is based on heritable information in the form of DNA. In a short essay (100–150 words), discuss how the fidelity with which DNA is inherited is related to the processes of evolution. (Review the discussion of proofreading and DNA repair in Concept 16.2, pp. 316–318.)

For selected answers, see Appendix A.

www.masteringbiology.com

1. MasteringBiology® Assignments

Make Connections Tutorial Point Mutations (Chapter 17) and Protein Structure (Chapter 5)

BioFlix **Tutorials** Protein Synthesis: Overview • Transcription and RNA Processing • Translation and Protein Targeting Pathways

Tutorials The Genetic Code • Following the Instructions in DNA • Types of RNA • Point Mutations

Activities Overview of Protein Synthesis • RNA Synthesis • Transcription • RNA Processing • Synthesizing Proteins • Translation • The Triplet Nature of the Genetic Code

Questions Student Misconceptions • Reading Quiz • Multiple Choice • End-of-Chapter

2. eText

Read your book online, search, take notes, highlight text, and more.

3. The Study Area

Practice Tests • Cumulative Test • ***BioFlix*** 3-D Animations • MP3 Tutor Sessions • Videos • Activities • Investigations • Lab Media • Audio Glossary • Word Study Tools • Art

18

Regulation of Gene Expression

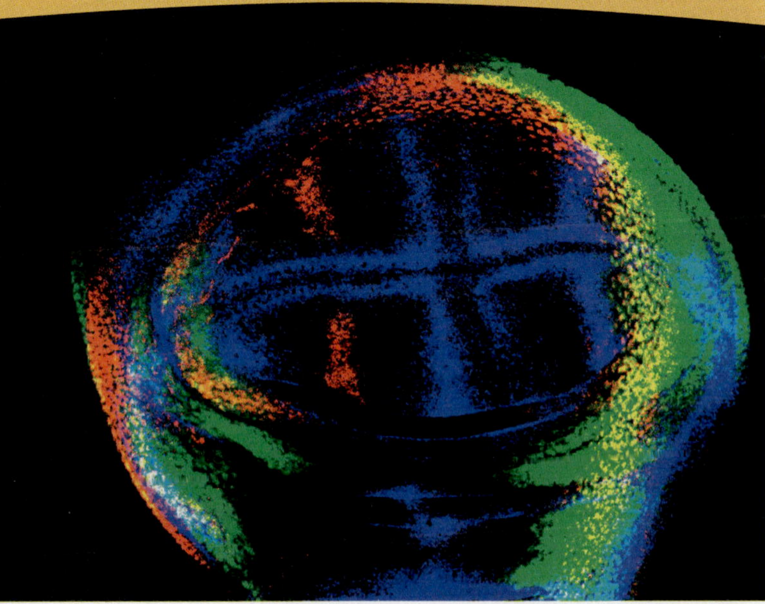

▲ **Figure 18.1 What regulates the precise pattern of gene expression in the developing wing of a fly embryo?**

KEY CONCEPTS

18.1 Bacteria often respond to environmental change by regulating transcription

18.2 Eukaryotic gene expression is regulated at many stages

18.3 Noncoding RNAs play multiple roles in controlling gene expression

18.4 A program of differential gene expression leads to the different cell types in a multicellular organism

18.5 Cancer results from genetic changes that affect cell cycle control

OVERVIEW

Conducting the Genetic Orchestra

It's almost concert time! Dissonance reigns as the orchestra members individually tune their instruments. Then, after a brief hush, the conductor's baton rises, pauses, and begins a series of elaborate movements, directing specific instruments to join in and others to raise or lower their volume at defined moments. Properly balanced and timed, discordant sounds are thus transformed into a beautiful symphony that enraptures the audience.

In a similar way, cells intricately and precisely regulate their gene expression. Both prokaryotes and eukaryotes must alter their patterns of gene expression in response to changes in environmental conditions. Multicellular eukaryotes must also develop and maintain multiple cell types. Each cell type contains the same genome but expresses a different subset of genes, a significant challenge in gene regulation.

An adult fruit fly, for example, develops from a single fertilized egg, passing through a wormlike stage called a larva. At every stage, gene expression is carefully regulated, ensuring that the right genes are expressed only at the correct time and place. In the larva, the adult wing forms in a disk-shaped pocket of several thousand cells, shown in **Figure 18.1**. The tissue in this image has been treated to reveal the mRNA for three genes—labeled red, blue, and green—using techniques covered in Chapter 20. (Red and green together appear yellow.) The intricate pattern of expression for each gene is the same from larva to larva at this stage, and it provides a graphic display of the precision of gene regulation. But what is the molecular basis for this pattern? Why is one particular gene expressed only in the few hundred cells that appear blue in this image and not in the other cells?

In this chapter, we first explore how bacteria regulate expression of their genes in response to different environmental conditions. We then examine how eukaryotes regulate gene expression to maintain different cell types. Gene expression in eukaryotes, as in bacteria, is often regulated at the stage of transcription, but control at other stages is also important. In recent years, researchers have been surprised to discover the many roles played by RNA molecules in regulating eukaryotic gene expression, a topic we cover next. We then consider what happens when a complex program of gene regulation works properly during embryonic development: A single cell—the fertilized egg—becomes a fully functioning organism made up of many different cell types. Finally, we investigate how cancer can result when gene regulation goes awry. Orchestrating proper gene expression by all cells is crucial to the functions of life.

CONCEPT **18.1**

Bacteria often respond to environmental change by regulating transcription

Bacterial cells that can conserve resources and energy have a selective advantage over cells that are unable to do so. Thus, natural selection has favored bacteria that express only the genes whose products are needed by the cell.

Consider, for instance, an individual *E. coli* cell living in the erratic environment of a human colon, dependent for its nutrients on the whimsical eating habits of its host. If the environment is lacking in the amino acid tryptophan, which the bacterium needs to survive, the cell responds by activating a metabolic pathway that makes tryptophan from another compound. Later, if the human host eats a tryptophan-rich meal, the bacterial cell stops producing tryptophan, thus saving itself from squandering its resources to produce a substance that is available from the surrounding solution in prefabricated form. This is just one example of how bacteria tune their metabolism to changing environments.

Metabolic control occurs on two levels, as shown for the synthesis of tryptophan in **Figure 18.2**. First, cells can adjust the activity of enzymes already present. This is a fairly fast response, which relies on the sensitivity of many enzymes to chemical cues that increase or decrease their catalytic activity (see Chapter 8). The activity of the first enzyme in the tryptophan synthesis pathway is inhibited by the pathway's end product **(Figure 18.2a)**. Thus, if tryptophan accumulates in a cell, it shuts down the synthesis of more tryptophan by inhibiting enzyme activity. Such *feedback inhibition*, typical of anabolic (biosynthetic) pathways, allows a cell to adapt to short-term fluctuations in the supply of a substance it needs.

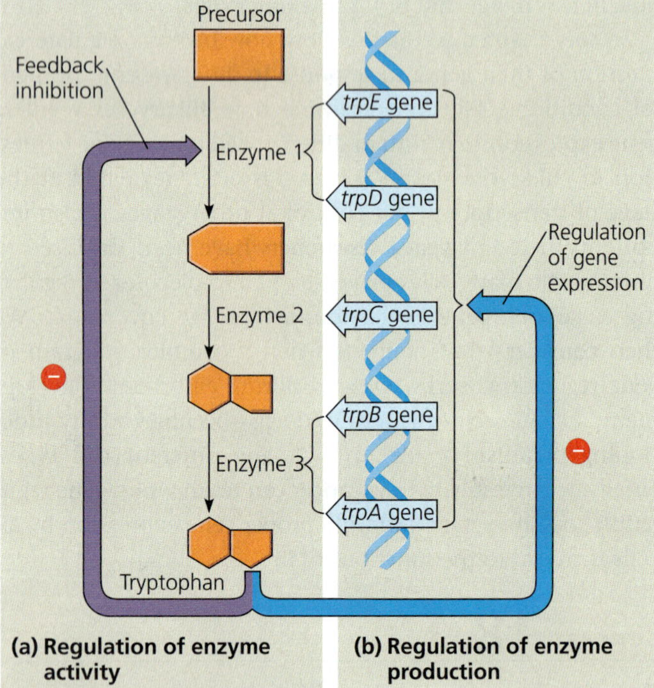

▲ **Figure 18.2 Regulation of a metabolic pathway.** In the pathway for tryptophan synthesis, an abundance of tryptophan can both **(a)** inhibit the activity of the first enzyme in the pathway (feedback inhibition), a rapid response, and **(b)** repress expression of the genes encoding all subunits of the enzymes in the pathway, a longer-term response. Genes *trpE* and *trpD* encode the two subunits of enzyme 1, and genes *trpB* and *trpA* encode the two subunits of enzyme 3. (The genes were named before the order in which they functioned in the pathway was determined.) The ⊖ symbol stands for inhibition.

Second, cells can adjust the production level of certain enzymes; that is, they can regulate the expression of the genes encoding the enzymes. If, in our example, the environment provides all the tryptophan the cell needs, the cell stops making the enzymes that catalyze the synthesis of tryptophan **(Figure 18.2b)**. In this case, the control of enzyme production occurs at the level of transcription, the synthesis of messenger RNA coding for these enzymes. More generally, many genes of the bacterial genome are switched on or off by changes in the metabolic status of the cell. One basic mechanism for this control of gene expression in bacteria, described as the *operon model*, was discovered in 1961 by François Jacob and Jacques Monod at the Pasteur Institute in Paris. Let's see what an operon is and how it works, using the control of tryptophan synthesis as our first example.

Operons: The Basic Concept

E. coli synthesizes the amino acid tryptophan from a precursor molecule in the multistep pathway shown in Figure 18.2. Each reaction in the pathway is catalyzed by a specific enzyme, and the five genes that code for the subunits of these enzymes are clustered together on the bacterial chromosome. A single promoter serves all five genes, which together constitute a transcription unit. (Recall from Chapter 17 that a promoter is a site where RNA polymerase can bind to DNA and begin transcription.) Thus, transcription gives rise to one long mRNA molecule that codes for the five polypeptides making up the enzymes in the tryptophan pathway. The cell can translate this one mRNA into five separate polypeptides because the mRNA is punctuated with start and stop codons that signal where the coding sequence for each polypeptide begins and ends.

A key advantage of grouping genes of related function into one transcription unit is that a single "on-off switch" can control the whole cluster of functionally related genes; in other words, these genes are *coordinately controlled*. When an *E. coli* cell must make tryptophan for itself because the nutrient medium lacks this amino acid, all the enzymes for the metabolic pathway are synthesized at one time. The switch is a segment of DNA called an **operator**. Both its location and name suit its function: Positioned within the promoter or, in some cases, between the promoter and the enzyme-coding genes, the operator controls the access of RNA polymerase to the genes. All together, the operator, the promoter, and the genes they control—the entire stretch of DNA required for enzyme production for the tryptophan pathway—constitute an **operon**. The *trp* operon (*trp* for tryptophan) is one of many operons in the *E. coli* genome **(Figure 18.3)**.

If the operator is the operon's switch for controlling transcription, how does this switch work? By itself, the *trp* operon is turned on; that is, RNA polymerase can bind to the promoter and transcribe the genes of the operon. The operon

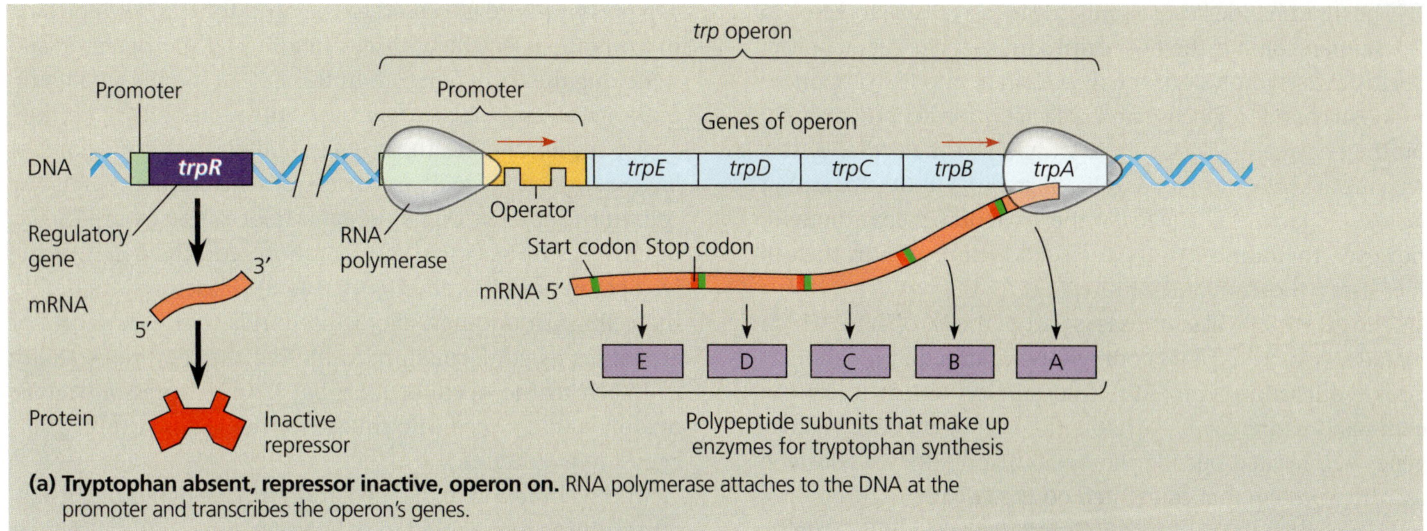

(a) Tryptophan absent, repressor inactive, operon on. RNA polymerase attaches to the DNA at the promoter and transcribes the operon's genes.

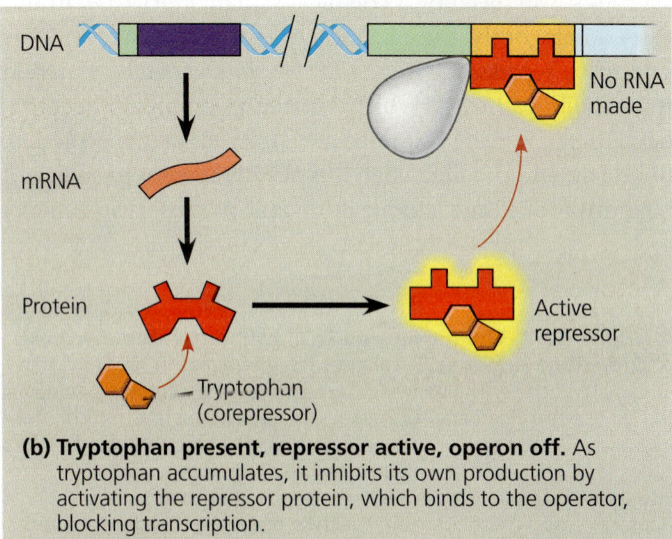

(b) Tryptophan present, repressor active, operon off. As tryptophan accumulates, it inhibits its own production by activating the repressor protein, which binds to the operator, blocking transcription.

▲ **Figure 18.3 The *trp* operon in *E. coli*: regulated synthesis of repressible enzymes.** Tryptophan is an amino acid produced by an anabolic pathway catalyzed by repressible enzymes. **(a)** The five genes encoding the polypeptide subunits of the enzymes in this pathway (see Figure 18.2) are grouped, along with a promoter, into the *trp* operon. The *trp* operator (the repressor binding site) is located within the *trp* promoter (the RNA polymerase binding site). **(b)** Accumulation of tryptophan, the end product of the pathway, represses transcription of the *trp* operon, thus blocking synthesis of all the enzymes in the pathway and shutting down tryptophan production.

? *Describe what happens to the* trp *operon as the cell uses up its store of tryptophan.*

can be switched off by a protein called the *trp* **repressor**. The repressor binds to the operator and blocks attachment of RNA polymerase to the promoter, preventing transcription of the genes. A repressor protein is specific for the operator of a particular operon. For example, the repressor that switches off the *trp* operon by binding to the *trp* operator has no effect on other operons in the *E. coli* genome.

The *trp* repressor is the protein product of a **regulatory gene** called *trpR*, which is located some distance from the *trp* operon and has its own promoter. Regulatory genes are expressed continuously, although at a low rate, and a few *trp* repressor molecules are always present in *E. coli* cells. Why, then, is the *trp* operon not switched off permanently? First, the binding of repressors to operators is reversible. An operator vacillates between two states: one without the repressor bound and one with the repressor bound. The relative duration of each state depends on the number of active repressor molecules around. Second, the *trp* repressor, like most regulatory proteins, is an allosteric protein, with two alternative shapes, active and inactive (see Figure 8.20). The *trp* repressor is synthesized in an inactive form with little affinity for the *trp* operator. Only if tryptophan binds to the *trp* repressor at an allosteric site does the repressor protein change to the active form that can attach to the operator, turning the operon off.

Tryptophan functions in this system as a **corepressor**, a small molecule that cooperates with a repressor protein to switch an operon off. As tryptophan accumulates, more tryptophan molecules associate with *trp* repressor molecules, which can then bind to the *trp* operator and shut down production of the tryptophan pathway enzymes. If the cell's tryptophan level drops, transcription of the operon's genes resumes. The *trp* operon is one example of how gene expression can respond to changes in the cell's internal and external environment.

Repressible and Inducible Operons: Two Types of Negative Gene Regulation

The *trp* operon is said to be a *repressible operon* because its transcription is usually on but can be inhibited (repressed) when a specific small molecule (in this case, tryptophan) binds allosterically to a regulatory protein. In contrast, an *inducible operon* is usually off but can be stimulated (induced) when a specific small molecule interacts with a regulatory protein. The classic example of an inducible operon is the *lac* operon (*lac* for lactose), which was the subject of Jacob and Monod's pioneering research.

The disaccharide lactose (milk sugar) is available to *E. coli* in the human colon if the host drinks milk. Lactose metabolism begins with hydrolysis of the disaccharide into its component monosaccharides, glucose and galactose, a reaction catalyzed by the enzyme β-galactosidase. Only a few molecules of this enzyme are present in an *E. coli* cell growing in the absence of lactose. If lactose is added to the bacterium's environment, however, the number of β-galactosidase molecules in the cell increases a thousandfold within about 15 minutes.

The gene for β-galactosidase is part of the *lac* operon, which includes two other genes coding for enzymes that function in lactose utilization. The entire transcription unit is under the command of one main operator and promoter. The regulatory gene, *lacI*, located outside the operon, codes for an allosteric repressor protein that can switch off the *lac* operon by binding to the operator. So far, this sounds just like regulation of the *trp* operon, but there is one important difference. Recall that the *trp* repressor is inactive by itself and requires tryptophan as a corepressor in order to bind to the operator. The *lac* repressor, in contrast, is active by itself, binding to the operator and switching the *lac* operon off. In this case, a specific small molecule, called an **inducer**, *inactivates* the repressor.

For the *lac* operon, the inducer is allolactose, an isomer of lactose formed in small amounts from lactose that enters the cell. In the absence of lactose (and hence allolactose), the *lac* repressor is in its active configuration, and the genes of the *lac* operon are silenced **(Figure 18.4a)**. If lactose is added to the cell's surroundings, allolactose binds to the *lac* repressor and alters its conformation, nullifying the repressor's ability to attach to the operator. Without bound repressor, the *lac* operon is transcribed into mRNA for the lactose-utilizing enzymes **(Figure 18.4b)**.

In the context of gene regulation, the enzymes of the lactose pathway are referred to as *inducible enzymes* because their synthesis is induced by a chemical signal (allolactose, in this case). Analogously, the enzymes for tryptophan synthesis are said to be repressible. *Repressible enzymes* generally function in anabolic pathways, which synthesize essential end products from raw materials (precursors). By suspending production of an end product when it is already present in sufficient quantity, the cell can allocate its organic precursors and energy

(a) Lactose absent, repressor active, operon off. The *lac* repressor is innately active, and in the absence of lactose it switches off the operon by binding to the operator.

▼ **Figure 18.4 The *lac* operon in *E. coli*: regulated synthesis of inducible enzymes.** *E. coli* uses three enzymes to take up and metabolize lactose. The genes for these three enzymes are clustered in the *lac* operon. One gene, *lacZ*, codes for β-galactosidase, which hydrolyzes lactose to glucose and galactose. The second gene, *lacY*, codes for a permease, the membrane protein that transports lactose into the cell. The third gene, *lacA*, codes for an enzyme called transacetylase, whose function in lactose metabolism is still unclear. The gene for the *lac* repressor, *lacI*, happens to be adjacent to the *lac* operon, an unusual situation. The function of the teal region at the upstream end of the promoter (the left end in these diagrams) will be revealed in Figure 18.5.

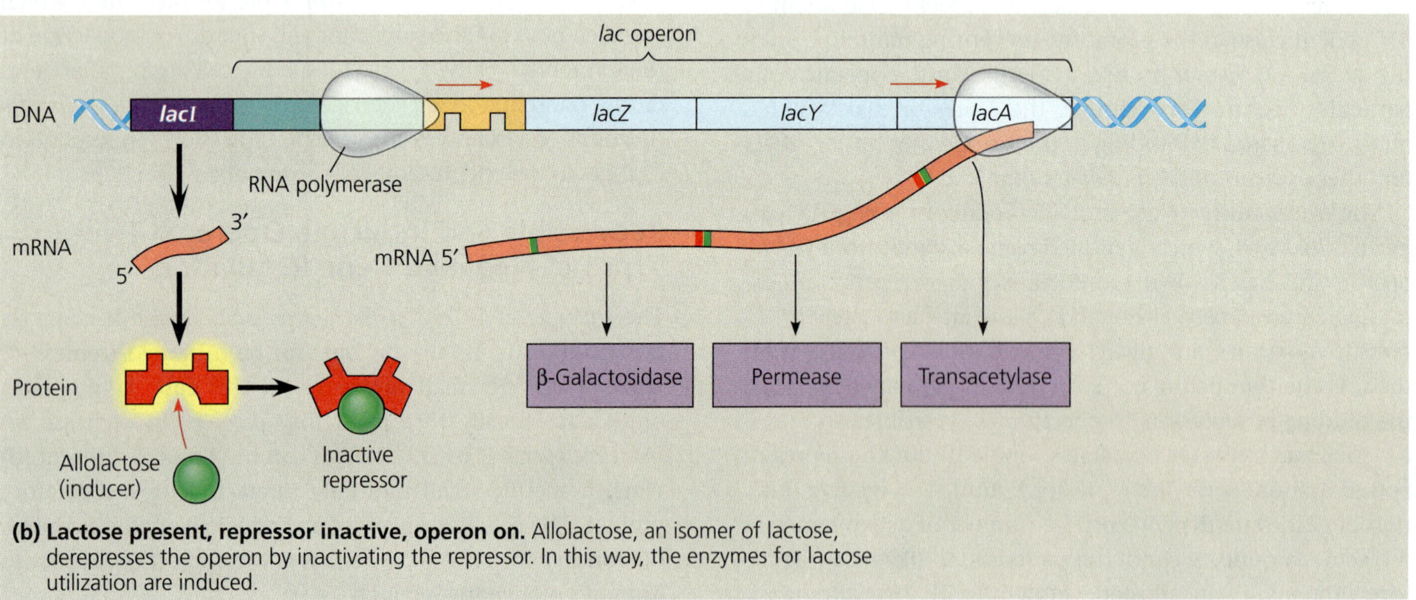

(b) Lactose present, repressor inactive, operon on. Allolactose, an isomer of lactose, derepresses the operon by inactivating the repressor. In this way, the enzymes for lactose utilization are induced.

for other uses. In contrast, inducible enzymes usually function in catabolic pathways, which break down a nutrient to simpler molecules. By producing the appropriate enzymes only when the nutrient is available, the cell avoids wasting energy and precursors making proteins that are not needed.

Regulation of both the *trp* and *lac* operons involves the *negative* control of genes, because the operons are switched off by the active form of the repressor protein. It may be easier to see this for the *trp* operon, but it is also true for the *lac* operon. Allolactose induces enzyme synthesis not by acting directly on the genome, but by freeing the *lac* operon from the negative effect of the repressor. Gene regulation is said to be *positive* only when a regulatory protein interacts directly with the genome to switch transcription on. Let's look at an example of the positive control of genes, again involving the *lac* operon.

Positive Gene Regulation

When glucose and lactose are both present in its environment, *E. coli* preferentially uses glucose. The enzymes for glucose breakdown in glycolysis (see Figure 9.9) are continually present. Only when lactose is present *and* glucose is in short supply does *E. coli* use lactose as an energy source, and only then does it synthesize appreciable quantities of the enzymes for lactose breakdown.

How does the *E. coli* cell sense the glucose concentration and relay this information to the genome? Again, the mechanism depends on the interaction of an allosteric regulatory protein with a small organic molecule, in this case **cyclic AMP (cAMP)**, which accumulates when glucose is scarce (see Figure 11.11 for the structure of cAMP). The regulatory protein, called *catabolite activator protein (CAP)*, is an **activator**, a protein that binds to DNA and stimulates transcription of a gene. When cAMP binds to this regulatory protein, CAP assumes its active shape and can attach to a specific site at the upstream end of the *lac* promoter **(Figure 18.5a)**. This attachment increases the affinity of RNA polymerase for the promoter, which is actually rather low even when no repressor is bound to the operator. By facilitating the binding of RNA polymerase to the promoter and thereby increasing the rate of transcription, the attachment of CAP to the promoter directly stimulates gene expression. Therefore, this mechanism qualifies as positive regulation.

If the amount of glucose in the cell increases, the cAMP concentration falls, and without cAMP, CAP detaches from the operon. Because CAP is inactive, RNA polymerase binds less efficiently to the promoter, and transcription of the *lac* operon proceeds at only a low level, even in the presence of lactose **(Figure 18.5b)**. Thus, the *lac* operon is under dual control: negative control by the *lac* repressor and positive control by CAP. The state of the *lac* repressor (with or without bound allolactose) determines whether or not transcription of the *lac* operon's genes occurs at all; the state of CAP (with or without bound cAMP) controls the *rate* of transcription if the operon is repressor-free. It is as though the operon has both an on-off switch and a volume control.

In addition to regulating the *lac* operon, CAP helps regulate other operons that encode enzymes used in catabolic pathways. All told, it may affect the expression of more than 100 genes in *E. coli*. When glucose is plentiful and CAP is inactive, the synthesis of enzymes that catabolize compounds other than glucose generally slows down. The ability to catabolize other compounds, such as lactose, enables a cell deprived of glucose to survive. The compounds present in the cell at the moment determine which operons are switched on—the result of simple interactions of activator and repressor proteins with the promoters of the genes in question.

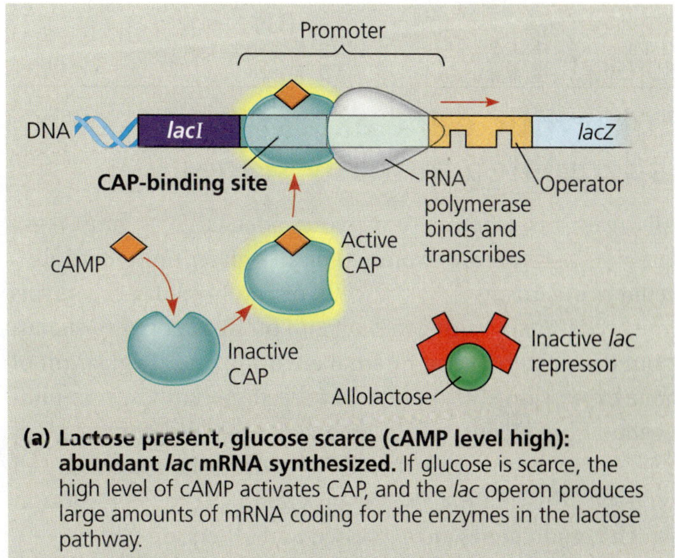

(a) Lactose present, glucose scarce (cAMP level high): abundant *lac* mRNA synthesized. If glucose is scarce, the high level of cAMP activates CAP, and the *lac* operon produces large amounts of mRNA coding for the enzymes in the lactose pathway.

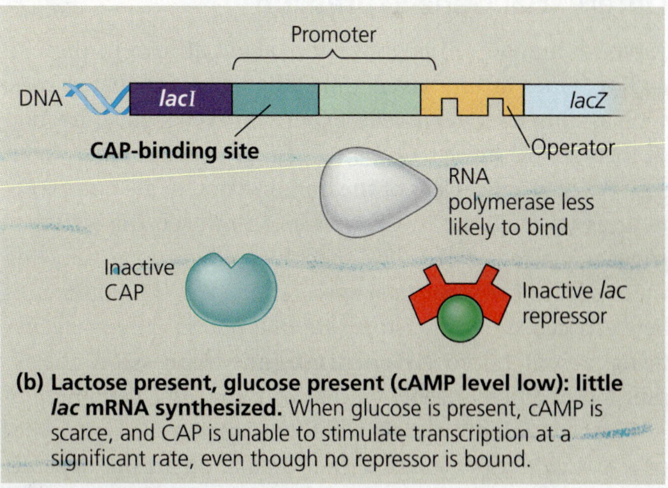

(b) Lactose present, glucose present (cAMP level low): little *lac* mRNA synthesized. When glucose is present, cAMP is scarce, and CAP is unable to stimulate transcription at a significant rate, even though no repressor is bound.

▲ **Figure 18.5 Positive control of the *lac* operon by catabolite activator protein (CAP).** RNA polymerase has high affinity for the *lac* promoter only when catabolite activator protein (CAP) is bound to a DNA site at the upstream end of the promoter. CAP attaches to its DNA site only when associated with cyclic AMP (cAMP), whose concentration in the cell rises when the glucose concentration falls. Thus, when glucose is present, even if lactose also is available, the cell preferentially catabolizes glucose and makes very little of the lactose-utilizing enzymes.

CONCEPT CHECK 18.1

1. How does binding of the *trp* corepressor and the *lac* inducer to their respective repressor proteins alter repressor function and transcription in each case?
2. Describe the binding of RNA polymerase, repressors, and activators to the *lac* operon when both lactose and glucose are scarce. What is the effect of these scarcities on transcription of the *lac* operon?
3. **WHAT IF?** A certain mutation in *E. coli* changes the *lac* operator so that the active repressor cannot bind. How would this affect the cell's production of β-galactosidase?

For suggested answers, see Appendix A.

CONCEPT 18.2

Eukaryotic gene expression is regulated at many stages

All organisms, whether prokaryotes or eukaryotes, must regulate which genes are expressed at any given time. Both unicellular organisms and the cells of multicellular organisms must continually turn genes on and off in response to signals from their external and internal environments. Regulation of gene expression is also essential for cell specialization in multicellular organisms, which are made up of different types of cells, each with a distinct role. To perform its role, each cell type must maintain a specific program of gene expression in which certain genes are expressed and others are not.

Differential Gene Expression

A typical human cell might express about 20% of its protein-coding genes at any given time. Highly differentiated cells, such as muscle or nerve cells, express an even smaller fraction of their genes. Almost all the cells in an organism contain an identical genome. (Cells of the immune system are one exception, as you will see in Chapter 43.) However, the subset of genes expressed in the cells of each type is unique, allowing these cells to carry out their specific function. The differences between cell types, therefore, are due not to different genes being present, but to **differential gene expression**, the expression of different genes by cells with the same genome.

The function of any cell, whether a single-celled eukaryote or a particular cell type in a multicellular organism, depends on the appropriate set of genes being expressed. The transcription factors of a cell must locate the right genes at the right time, a task on a par with finding a needle in a haystack. When gene expression proceeds abnormally, serious imbalances and diseases, including cancer, can arise.

Figure 18.6 summarizes the entire process of gene expression in a eukaryotic cell, highlighting key stages in the expression of a protein-coding gene. Each stage depicted in

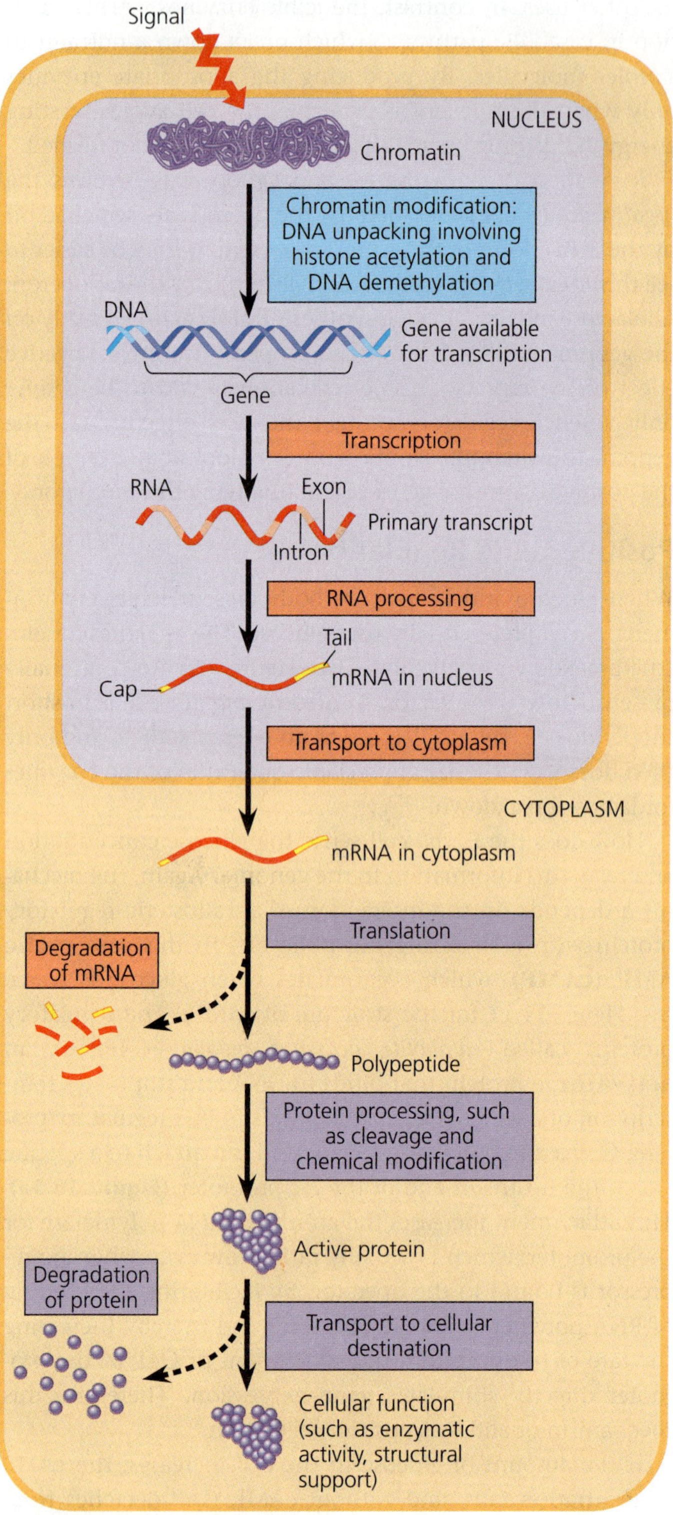

▲ **Figure 18.6 Stages in gene expression that can be regulated in eukaryotic cells.** In this diagram, the colored boxes indicate the processes most often regulated; each color indicates the type of molecule that is affected (blue = DNA, orange = RNA, purple = protein). The nuclear envelope separating transcription from translation in eukaryotic cells offers an opportunity for post-transcriptional control in the form of RNA processing that is absent in prokaryotes. In addition, eukaryotes have a greater variety of control mechanisms operating before transcription and after translation. The expression of any given gene, however, does not necessarily involve every stage shown; for example, not every polypeptide is cleaved.

Figure 18.6 is a potential control point at which gene expression can be turned on or off, accelerated, or slowed down.

Only 50 years ago, an understanding of the mechanisms that control gene expression in eukaryotes seemed almost hopelessly out of reach. Since then, new research methods, notably advances in DNA technology (see Chapter 20), have enabled molecular biologists to uncover many of the details of eukaryotic gene regulation. In all organisms, a common control point for gene expression is at transcription; regulation at this stage often occurs in response to signals coming from outside the cell, such as hormones or other signaling molecules. For this reason, the term *gene expression* is often equated with transcription for both bacteria and eukaryotes. While this is most often the case for bacteria, the greater complexity of eukaryotic cell structure and function provides opportunities for regulating gene expression at many additional stages (see Figure 18.6). In the remainder of this section, we'll examine some of the important control points of eukaryotic gene expression more closely.

Regulation of Chromatin Structure

Recall that the DNA of eukaryotic cells is packaged with proteins in an elaborate complex known as chromatin, the basic unit of which is the nucleosome (see Figure 16.22). The structural organization of chromatin not only packs a cell's DNA into a compact form that fits inside the nucleus, but also helps regulate gene expression in several ways. The location of a gene's promoter relative to nucleosomes and to the sites where the DNA attaches to the chromosome scaffold or nuclear lamina can affect whether the gene is transcribed. In addition, genes within heterochromatin, which is highly condensed, are usually not expressed. Lastly, certain chemical modifications to the histone proteins and to the DNA of chromatin can influence both chromatin structure and gene expression. Here we examine the effects of these modifications, which are catalyzed by specific enzymes.

Histone Modifications

There is abundant evidence that chemical modifications to histones, the proteins around which the DNA is wrapped in nucleosomes, play a direct role in the regulation of gene transcription. The N-terminus of each histone molecule in a nucleosome protrudes outward from the nucleosome **(Figure 18.7a)**. These histone tails are accessible to various modifying enzymes that catalyze the addition or removal of specific chemical groups.

In **histone acetylation,** acetyl groups (—$COCH_3$) are attached to lysines in histone tails; deacetylation is the removal of acetyl groups. When the lysines are acetylated, their positive charges are neutralized and the histone tails no longer bind to neighboring nucleosomes **(Figure 18.7b)**. Such binding promotes the folding of chromatin into a more compact structure; when this binding does not occur, chromatin has a looser structure. As a result, transcription proteins have easier

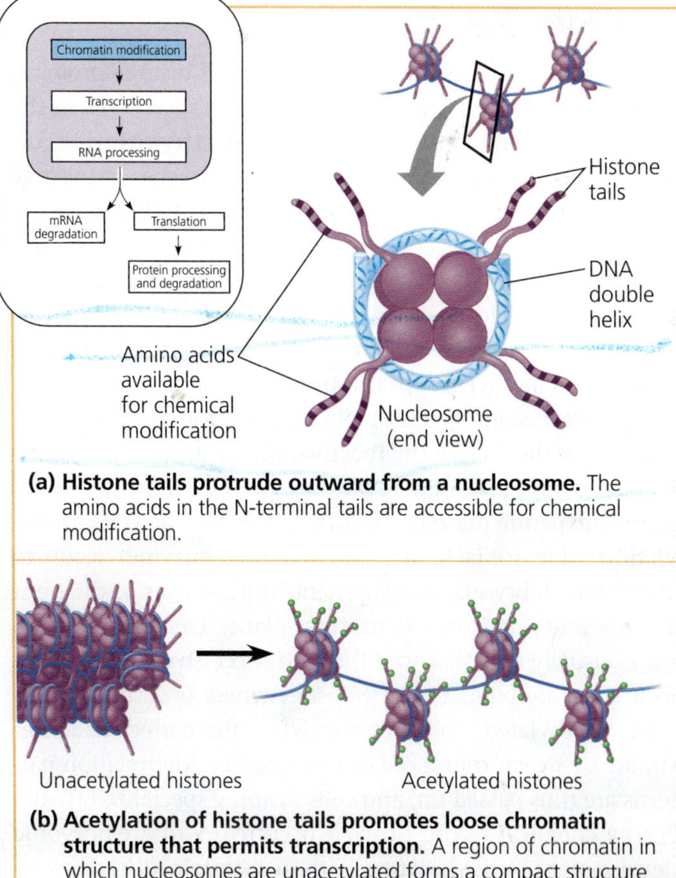

(a) Histone tails protrude outward from a nucleosome. The amino acids in the N-terminal tails are accessible for chemical modification.

(b) Acetylation of histone tails promotes loose chromatin structure that permits transcription. A region of chromatin in which nucleosomes are unacetylated forms a compact structure (left) in which the DNA is not transcribed. When nucleosomes are highly acetylated (right), the chromatin becomes less compact, and the DNA is accessible for transcription.

▲ **Figure 18.7 A simple model of histone tails and the effect of histone acetylation.** In addition to acetylation, histones can undergo several other types of modifications that also help determine the chromatin configuration in a region.

access to genes in an acetylated region. Researchers have shown that some enzymes that acetylate or deacetylate histones are closely associated with or even components of the transcription factors that bind to promoters (see Figure 17.8). These observations suggest that histone acetylation enzymes may promote the initiation of transcription not only by remodeling chromatin structure, but also by binding to and thus "recruiting" components of the transcription machinery.

Other chemical groups, such as methyl and phosphate groups, can be reversibly attached to amino acids in histone tails. Addition of methyl groups (—CH_3) to histone tails (histone methylation) can promote condensation of the chromatin, while addition of a phosphate group (phosphorylation) to an amino acid next to a methylated amino acid can have the opposite effect. The recent discovery that modifications to histone tails can affect chromatin structure and gene expression has led to the *histone code hypothesis*. This hypothesis proposes that specific combinations of modifications, as well as the order in which they have occurred, help determine the chromatin configuration, which in turn influences transcription.

DNA Methylation

While some enzymes methylate the tails of histone proteins, a different set of enzymes can methylate certain bases in the DNA itself, usually cytosine. Such **DNA methylation** occurs in most plants, animals, and fungi. Long stretches of inactive DNA, such as that of inactivated mammalian X chromosomes (see Figure 15.8), are generally more methylated than regions of actively transcribed DNA, although there are exceptions. On a smaller scale, individual genes are usually more heavily methylated in cells in which they are not expressed. Removal of the extra methyl groups can turn on some of these genes.

At least in some species, DNA methylation seems to be essential for the long-term inactivation of genes that occurs during normal cell differentiation in the embryo. For instance, experiments have shown that deficient DNA methylation (due to lack of a methylating enzyme) leads to abnormal embryonic development in organisms as different as mice and *Arabidopsis* (a mustard plant). Once methylated, genes usually stay that way through successive cell divisions in a given individual. At DNA sites where one strand is already methylated, enzymes methylate the correct daughter strand after each round of DNA replication. Methylation patterns are thus passed on, and cells forming specialized tissues keep a chemical record of what occurred during embryonic development. A methylation pattern maintained in this way also accounts for *genomic imprinting* in mammals, where methylation permanently regulates expression of either the maternal or paternal allele of particular genes at the start of development (see Figure 15.17).

Epigenetic Inheritance

The chromatin modifications that we have just discussed do not entail a change in the DNA sequence, yet they may be passed along to future generations of cells. Inheritance of traits transmitted by mechanisms not directly involving the nucleotide sequence is called **epigenetic inheritance**. Whereas mutations in the DNA are permanent changes, modifications to the chromatin can be reversed, by processes that are not yet fully understood. The molecular systems for chromatin modification may well interact with each other in a regulated way. In *Drosophila*, for example, experiments have suggested that a particular histone-modifying enzyme recruits a DNA methylation enzyme to one region and that the two enzymes collaborate to silence a particular set of genes. Working in the opposite order, proteins have also been found that first bind to methylated DNA and then recruit histone deacetylation enzymes. Thus, a dual mechanism, involving both DNA methylation and histone deacetylation, can repress transcription.

Researchers are amassing more and more evidence for the importance of epigenetic information in the regulation of gene expression. Epigenetic variations might help explain why one identical twin acquires a genetically based disease, such as schizophrenia, but the other does not, despite their identical genomes. Alterations in normal patterns of DNA methylation are seen in some cancers, where they are associated with inappropriate gene expression. Evidently, enzymes that modify chromatin structure are integral parts of the eukaryotic cell's machinery for regulating transcription.

Regulation of Transcription Initiation

Chromatin-modifying enzymes provide initial control of gene expression by making a region of DNA either more or less able to bind the transcription machinery. Once the chromatin of a gene is optimally modified for expression, the initiation of transcription is the next major step at which gene expression is regulated. As in bacteria, the regulation of transcription initiation in eukaryotes involves proteins that bind to DNA and either facilitate or inhibit binding of RNA polymerase. The process is more complicated in eukaryotes, however. Before looking at how eukaryotic cells control their transcription, let's review the structure of a typical eukaryotic gene and its transcript.

Organization of a Typical Eukaryotic Gene

A eukaryotic gene and the DNA elements (segments) that control it are typically organized as shown in **Figure 18.8**, which extends what you learned about eukaryotic genes in Chapter 17. Recall that a cluster of proteins called a *transcription initiation complex* assembles on the promoter sequence at the "upstream" end of the gene. One of these proteins, RNA polymerase II, then proceeds to transcribe the gene, synthesizing a primary RNA transcript (pre-mRNA). RNA processing includes enzymatic addition of a 5′ cap and a poly-A tail, as well as splicing out of introns, to yield a mature mRNA. Associated with most eukaryotic genes are multiple **control elements**, segments of noncoding DNA that serve as binding sites for the proteins called transcription factors, which in turn regulate transcription. Control elements and the transcription factors they bind are critical to the precise regulation of gene expression seen in different cell types.

The Roles of Transcription Factors

To initiate transcription, eukaryotic RNA polymerase requires the assistance of transcription factors. Some transcription factors, such as those illustrated in Figure 17.8, are essential for the transcription of *all* protein-coding genes; therefore, they are often called *general transcription factors*. Only a few general transcription factors independently bind a DNA sequence, such as the TATA box within the promoter; the others primarily bind proteins, including each other and RNA polymerase II. Protein-protein interactions are crucial to the initiation of eukaryotic transcription. Only when the complete initiation complex has assembled can the polymerase begin to move along the DNA template strand, producing a complementary strand of RNA.

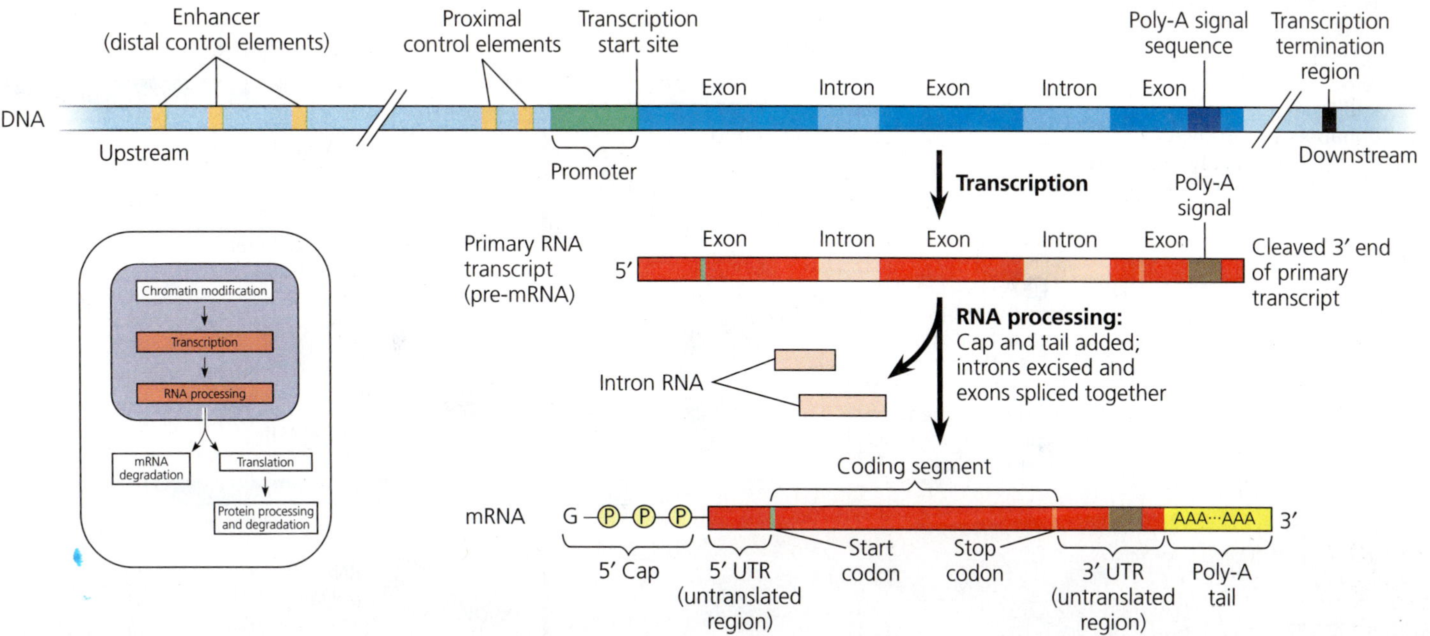

▲ **Figure 18.8 A eukaryotic gene and its transcript.** Each eukaryotic gene has a promoter, a DNA sequence where RNA polymerase binds and starts transcription, proceeding "downstream." A number of control elements (gold) are involved in regulating the initiation of transcription; these are DNA sequences located near (proximal to) or far from (distal to) the promoter. Distal control elements can be grouped together as enhancers, one of which is shown for this gene. A polyadenylation (poly-A) signal sequence in the last exon of the gene is transcribed into an RNA sequence that signals where the transcript is cleaved and the poly-A tail added. Transcription may continue for hundreds of nucleotides beyond the poly-A signal before terminating. RNA processing of the primary transcript into a functional mRNA involves three steps: addition of the 5′ cap, addition of the poly-A tail, and splicing. In the cell, the 5′ cap is added soon after transcription is initiated; splicing and poly-A tail addition may also occur while transcription is still under way (see Figure 17.10).

The interaction of general transcription factors and RNA polymerase II with a promoter usually leads to only a low rate of initiation and production of few RNA transcripts. In eukaryotes, high levels of transcription of particular genes at the appropriate time and place depend on the interaction of control elements with another set of proteins, which can be thought of as *specific transcription factors*.

Enhancers and Specific Transcription Factors As you can see in Figure 18.8, some control elements, named *proximal control elements*, are located close to the promoter. (Although some biologists consider proximal control elements part of the promoter, in this book we do not.) The more distant *distal control elements*, groupings of which are called **enhancers**, may be thousands of nucleotides upstream or downstream of a gene or even within an intron. A given gene may have multiple enhancers, each active at a different time or in a different cell type or location in the organism. Each enhancer, however, is generally associated with only that gene and no other.

In eukaryotes, the rate of gene expression can be strongly increased or decreased by the binding of specific transcription factors, either activators or repressors, to the control elements of enhancers. Hundreds of transcription activators have been discovered in eukaryotes; the structure of one example is shown in **Figure 18.9**. Researchers have identified two common structural elements in a large number of activator proteins: a DNA-binding domain—a part of the protein's three-dimensional structure that binds to DNA—and one or more activation domains. Activation domains bind other regulatory proteins or components of the transcription machinery, facilitating a series of protein-protein interactions that result in transcription of a given gene.

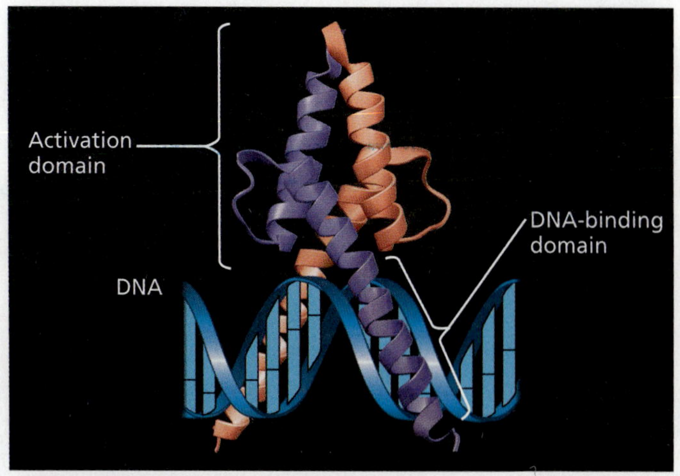

▲ **Figure 18.9 The structure of MyoD, a specific transcription factor that acts as an activator.** The MyoD protein is made up of two subunits (purple and salmon) with extensive regions of α helix. Each subunit has a DNA-binding domain and an activation domain (indicated by brackets for the purple subunit). The activation domain includes binding sites for the other subunit as well as other proteins. MyoD is involved in muscle development in vertebrate embryos and will be discussed further in Concept 18.4.

1 Activator proteins bind to distal control elements grouped as an enhancer in the DNA. This enhancer has three binding sites, each called a distal control element.

2 A DNA-bending protein brings the bound activators closer to the promoter. General transcription factors, mediator proteins, and RNA polymerase II are nearby.

3 The activators bind to certain mediator proteins and general transcription factors, helping them form an active transcription initiation complex on the promoter.

DNA
Activators
Enhancer
Distal control element
Promoter
Gene
TATA box
General transcription factors
DNA-bending protein
Group of mediator proteins
RNA polymerase II
RNA polymerase II
RNA synthesis
Transcription initiation complex

Chromatin modification
Transcription
RNA processing
mRNA degradation
Translation
Protein processing and degradation

▲ **Figure 18.10 A model for the action of enhancers and transcription activators.** Bending of the DNA by a protein enables enhancers to influence a promoter hundreds or even thousands of nucleotides away. Specific transcription factors called activators bind to the enhancer DNA sequences and then to a group of mediator proteins, which in turn bind to general transcription factors, assembling the transcription initiation complex. These protein-protein interactions facilitate the correct positioning of the complex on the promoter and the initiation of RNA synthesis. Only one enhancer (with three orange control elements) is shown here, but a gene may have several enhancers that act at different times or in different cell types.

Figure 18.10 shows a current model for how binding of activators to an enhancer located far from the promoter can influence transcription. Protein-mediated bending of the DNA is thought to bring the bound activators into contact with a group of *mediator proteins*, which in turn interact with proteins at the promoter. These multiple protein-protein interactions help assemble and position the initiation complex on the promoter. Support for this model includes a study showing that the proteins regulating a mouse globin gene contact both the gene's promoter and an enhancer located about 50,000 nucleotides upstream. Evidently, these two regions in the DNA must come together in a very specific fashion for this interaction to occur.

Specific transcription factors that function as repressors can inhibit gene expression in several different ways. Some repressors bind directly to control element DNA (in enhancers or elsewhere), blocking activator binding or, in some cases, turning off transcription even when activators are bound. Other repressors block the binding of activators to proteins that allow the activators to bind to DNA.

In addition to influencing transcription directly, some activators and repressors act indirectly by affecting chromatin structure. Studies using yeast and mammalian cells show that some activators recruit proteins that acetylate histones near the promoters of specific genes, thus promoting transcription (see Figure 18.7). Similarly, some repressors recruit proteins that deacetylate histones, leading to reduced transcription, a phenomenon referred to as *silencing*. Indeed, recruitment of chromatin-modifying proteins seems to be the most common mechanism of repression in eukaryotes.

Combinatorial Control of Gene Activation In eukaryotes, the precise control of transcription depends largely on the binding of activators to DNA control elements. Considering the great number of genes that must be regulated in a typical animal or plant cell, the number of completely different nucleotide sequences found in control elements is surprisingly small. A dozen or so short nucleotide sequences appear again and again in the control elements for different genes. On average, each enhancer is composed of about ten control elements, each of which can bind only one or two specific transcription factors. It is the particular *combination* of control elements in an enhancer associated with a gene, rather than the presence of a single unique control element, that is important in regulating transcription of the gene.

Even with only a dozen control element sequences available, a very large number of combinations are possible. A particular combination of control elements will be able to activate transcription only when the appropriate activator proteins are present, which may occur at a precise time during development or in a particular cell type. **Figure 18.11** illustrates how the use of different combinations of just a few control elements can allow differential regulation of transcription in two cell types. This can occur because each cell type contains a different group of activator proteins. How these groups came to differ will be explored in Concept 18.4.

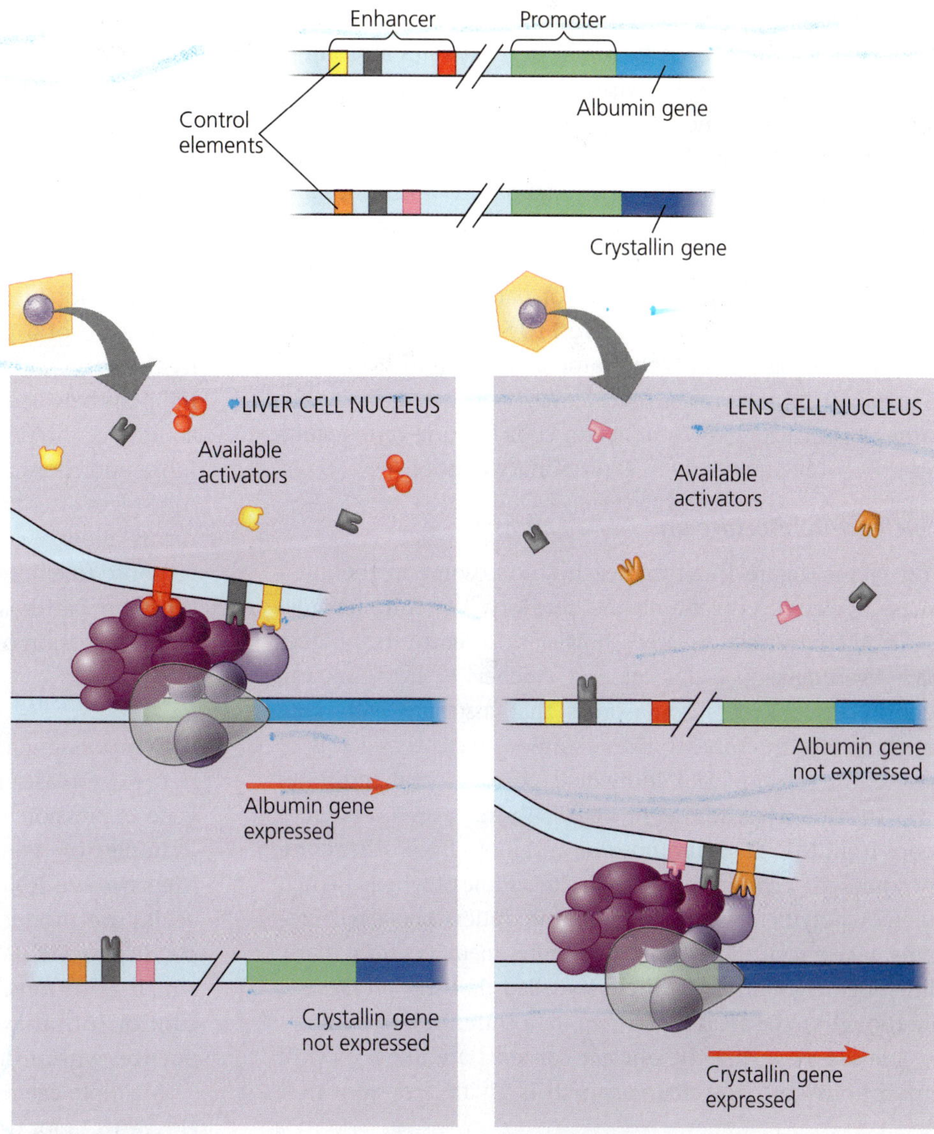

(a) Liver cell. The albumin gene is expressed, and the crystallin gene is not.

(b) Lens cell. The crystallin gene is expressed, and the albumin gene is not.

▲ **Figure 18.11 Cell type–specific transcription.** Both liver cells and lens cells have the genes for making the proteins albumin and crystallin, but only liver cells make albumin (a blood protein) and only lens cells make crystallin (the main protein of the lens of the eye). The specific transcription factors made in a cell determine which genes are expressed. In this example, the genes for albumin and crystallin are shown at the top, each with an enhancer made up of three different control elements. Although the enhancers for the two genes share one control element (gray), each enhancer has a unique combination of elements. All the activators required for high-level expression of the albumin gene are present only in liver cells (a), whereas the activators needed for expression of the crystallin gene are present only in lens cells (b). For simplicity, we consider only the role of activators here, although the presence or absence of repressors may also influence transcription in certain cell types.

? *Describe the enhancer for the albumin gene in each cell. How would the nucleotide sequence of this enhancer in the liver cell compare with that in the lens cell?*

Coordinately Controlled Genes in Eukaryotes

How does the eukaryotic cell deal with genes of related function that need to be turned on or off at the same time? Earlier in this chapter, you learned that in bacteria, such coordinately controlled genes are often clustered into an operon, which is regulated by a single promoter and transcribed into a single mRNA molecule. Thus, the genes are expressed together, and the encoded proteins are produced concurrently. With a few minor exceptions, operons that work in this way have *not* been found in eukaryotic cells.

Co-expressed eukaryotic genes, such as genes coding for the enzymes of a metabolic pathway, are typically scattered over different chromosomes. In these cases, coordinate gene expression depends on the association of a specific combination of control elements with every gene of a dispersed group. The presence of these elements can be compared to the raised flags on a few mailboxes out of many, signaling to the mail carrier to check those boxes. Copies of the activators that recognize the control elements bind to them, promoting simultaneous transcription of the genes, no matter where they are in the genome.

Coordinate control of dispersed genes in a eukaryotic cell often occurs in response to chemical signals from outside the cell. A steroid hormone, for example, enters a cell and binds to a specific intracellular receptor protein, forming a hormone-receptor complex that serves as a transcription activator (see Figure 11.9). Every gene whose transcription is stimulated by a particular steroid hormone, regardless of its chromosomal location, has a control element recognized by that hormone-receptor complex. This is how estrogen activates a group of genes that stimulate cell division in uterine cells, preparing the uterus for pregnancy.

Many signaling molecules, such as nonsteroid hormones and growth factors, bind to receptors on a cell's surface and never actually enter the cell. Such molecules can control gene expression indirectly by triggering signal transduction pathways that lead to activation of particular transcription activators or repressors (see Figure 11.15). Coordinate regulation in such pathways is the same as for steroid hormones: Genes with the same control elements are activated by the same chemical signals. Systems for coordinating gene regulation probably arose early in evolutionary history.

Nuclear Architecture and Gene Expression

You saw in Figure 16.23 that each chromosome in the interphase nucleus occupies a distinct territory. The chromosomes are not completely isolated, however. Recently, techniques have been developed that allow researchers to cross-link and identify regions of chromosomes that associate with each other during interphase. These studies reveal that loops of chromatin extend from individual chromosomal territories into specific sites in the nucleus (**Figure 18.12**). Different loops from the same chromosome and loops from other chromosomes may congregate in such sites, some of which are rich in RNA polymerases and other transcription-associated proteins. Like a recreation center that draws members from many different neighborhoods, these so-called *transcription factories* are thought to be areas specialized for a common function.

The old view that the nuclear contents are like a bowl of amorphous chromosomal spaghetti is giving way to a new model of a nucleus with a defined architecture and regulated movements of chromatin. Relocation of particular genes from their chromosomal territories to transcription factories may be part of the process of readying genes for transcription. This is an exciting area of current research that raises many fascinating questions for future study.

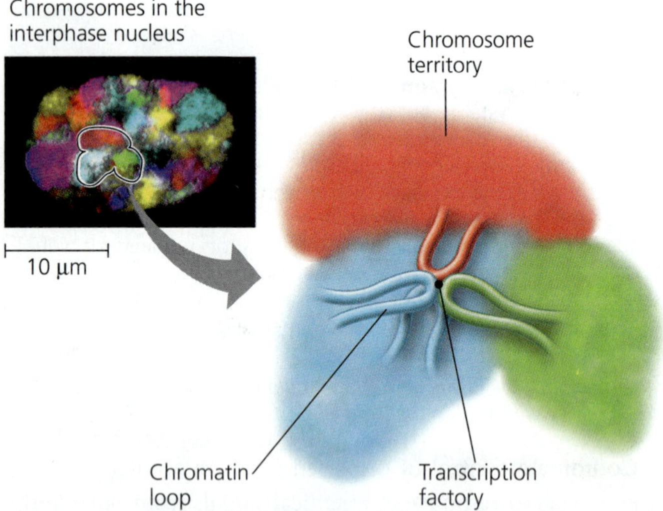

▲ **Figure 18.12 Chromosomal interactions in the interphase nucleus.** Although each chromosome has its own territory (see Figure 16.23), loops of chromatin may extend into other sites in the nucleus. Some of these sites are transcription factories that are occupied by multiple chromatin loops from the same chromosome (blue loops) or other chromosomes (red and green loops).

Mechanisms of Post-Transcriptional Regulation

Transcription alone does not constitute gene expression. The expression of a protein-coding gene is ultimately measured by the amount of functional protein a cell makes, and much happens between the synthesis of the RNA transcript and the activity of the protein in the cell. Researchers are discovering more and more regulatory mechanisms that operate at various stages after transcription (see Figure 18.6). These mechanisms allow a cell to fine-tune gene expression rapidly in response to environmental changes without altering its transcription patterns. Here we discuss how cells can regulate gene expression once a gene has been transcribed.

RNA Processing

RNA processing in the nucleus and the export of mature RNA to the cytoplasm provide several opportunities for regulating gene expression that are not available in prokaryotes. One example of regulation at the RNA-processing level is **alternative RNA splicing**, in which different mRNA molecules are produced from the same primary transcript, depending on which RNA segments are treated as exons and which as introns. Regulatory proteins specific to a cell type control intron-exon choices by binding to regulatory sequences within the primary transcript.

A simple example of alternative RNA splicing is shown in **Figure 18.13** for the troponin T gene, which encodes two different (though related) proteins. Other genes offer possibilities for far greater numbers of products. For instance, researchers have found a *Drosophila* gene with enough alternatively spliced exons to generate about 19,000 membrane proteins that have different extracellular domains. At least 17,500 (94%) of the alternative mRNAs are actually synthesized. Each developing nerve cell in the fly appears to synthesize a unique form of the protein, which acts as an identification badge on the cell surface.

It is clear that alternative RNA splicing can significantly expand the repertoire of a eukaryotic genome. In fact, alternative splicing was proposed as one explanation for the surprisingly low number of human genes counted when the human genome was sequenced about ten years ago. The number of human genes was found to be similar to that of a soil worm (nematode), a mustard plant, or a sea anemone. This discovery prompted questions about what, if not the number of genes, accounts for the more complex morphology (external form) of humans. It turns out that 75–100% of human genes that have multiple exons probably undergo alternative splicing. Thus, the extent of alternative splicing

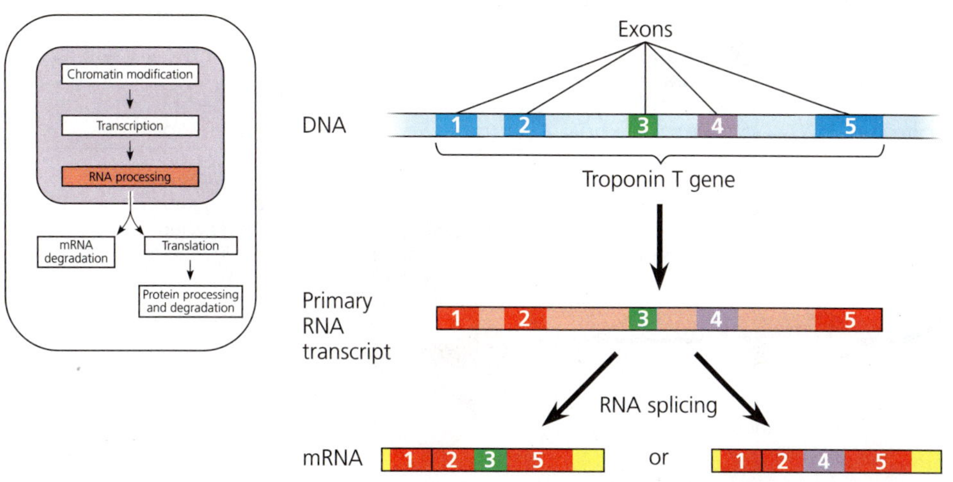

◄ **Figure 18.13 Alternative RNA splicing of the troponin T gene.** The primary transcript of this gene can be spliced in more than one way, generating different mRNA molecules. Notice that one mRNA molecule has ended up with exon 3 (green) and the other with exon 4 (purple). These two mRNAs are translated into different but related muscle proteins.

greatly multiplies the number of possible human proteins, which may be better correlated with complexity of form.

mRNA Degradation

The life span of mRNA molecules in the cytoplasm is important in determining the pattern of protein synthesis in a cell. Bacterial mRNA molecules typically are degraded by enzymes within a few minutes of their synthesis. This short life span of mRNAs is one reason bacteria can change their patterns of protein synthesis so quickly in response to environmental changes. In contrast, mRNAs in multicellular eukaryotes typically survive for hours, days, or even weeks. For instance, the mRNAs for the hemoglobin polypeptides (α-globin and β-globin) in developing red blood cells are unusually stable, and these long-lived mRNAs are translated repeatedly in these cells.

Nucleotide sequences that affect how long an mRNA remains intact are often found in the untranslated region (UTR) at the 3′ end of the molecule (see Figure 18.8). In one experiment, researchers transferred such a sequence from the short-lived mRNA for a growth factor to the 3′ end of a normally stable globin mRNA. The globin mRNA was quickly degraded.

During the past few years, other mechanisms that degrade or block expression of mRNA molecules have come to light. These mechanisms involve an important group of newly discovered RNA molecules that regulate gene expression at several levels, and we will discuss them later in this chapter.

Initiation of Translation

Translation presents another opportunity for regulating gene expression; such regulation occurs most commonly at the initiation stage (see Figure 17.18). For some mRNAs, the initiation of translation can be blocked by regulatory proteins that bind to specific sequences or structures within the untranslated region at the 5′ or 3′ end (5′ or 3′ UTR), preventing the attachment of ribosomes. (Recall from Chapter 17 that both the 5′ cap and the poly-A tail of an mRNA molecule are important for ribosome binding.) A different mechanism for blocking translation is seen in a variety of mRNAs present in the eggs of many organisms: Initially, these stored mRNAs lack poly-A tails of sufficient length to allow translation initiation. At the appropriate time during embryonic development, however, a cytoplasmic enzyme adds more adenine (A) nucleotides, prompting translation to begin.

Alternatively, translation of *all* the mRNAs in a cell may be regulated simultaneously. In a eukaryotic cell, such "global" control usually involves the activation or inactivation of one or more of the protein factors required to initiate translation. This mechanism plays a role in starting translation of mRNAs that are stored in eggs. Just after fertilization, translation is triggered by the sudden activation of translation initiation factors. The response is a burst of synthesis of the proteins encoded by the stored mRNAs. Some plants and algae store mRNAs during periods of darkness; light then triggers the reactivation of the translational apparatus.

Protein Processing and Degradation

The final opportunities for controlling gene expression occur after translation. Often, eukaryotic polypeptides must be processed to yield functional protein molecules. For instance, cleavage of the initial insulin polypeptide (pro-insulin) forms the active hormone. In addition, many proteins undergo chemical modifications that make them functional. Regulatory proteins are commonly activated or inactivated by the reversible addition of phosphate groups, and proteins destined for the surface of animal cells acquire sugars. Cell-surface proteins and many others must also be transported to target destinations in the cell in order to function. Regulation might occur at any of the steps involved in modifying or transporting a protein.

Finally, the length of time each protein functions in the cell is strictly regulated by means of selective degradation. Many proteins, such as the cyclins involved in regulating the cell cycle, must be relatively short-lived if the cell is to function appropriately (see Figure 12.17). To mark a particular protein for destruction, the cell commonly attaches molecules of a small protein called ubiquitin to the protein. Giant protein complexes called **proteasomes** then recognize the

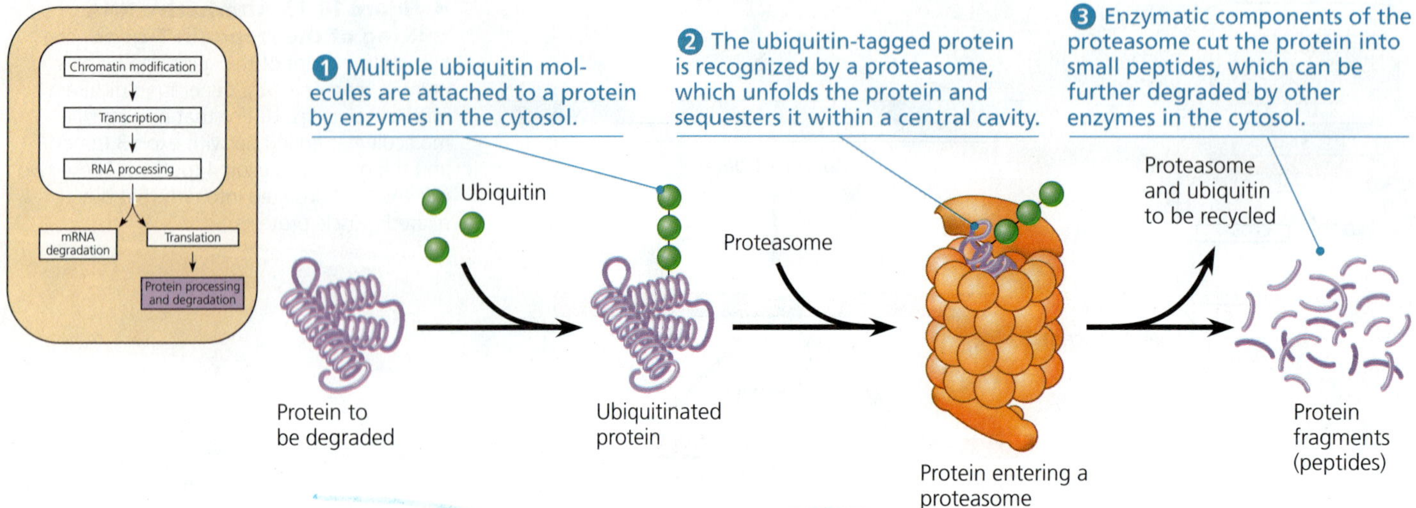

▲ **Figure 18.14 Degradation of a protein by a proteasome.** A proteasome, an enormous protein complex shaped like a trash can, chops up unneeded proteins in the cell. In most cases, the proteins attacked by a proteasome have been tagged with short chains of ubiquitin, a small protein. Steps 1 and 3 require ATP. Eukaryotic proteasomes are as massive as ribosomal subunits and are distributed throughout the cell. Their shape somewhat resembles that of chaperone proteins, which protect protein structure rather than destroy it (see Figure 5.23).

ubiquitin-tagged proteins and degrade them **(Figure 18.14)**. The importance of proteasomes is underscored by the finding that mutations making specific cell cycle proteins impervious to proteasome degradation can lead to cancer. The 2004 Nobel Prize in Chemistry was awarded to three scientists—two from Israel and one from the United States—who worked out the regulated process of protein degradation.

CONCEPT CHECK 18.2

1. In general, what is the effect of histone acetylation and DNA methylation on gene expression?
2. Compare the roles of general and specific transcription factors in regulating gene expression.
3. Suppose you compared the nucleotide sequences of the distal control elements in the enhancers of three genes that are expressed only in muscle tissue. What would you expect to find? Why?
4. Once mRNA encoding a particular protein reaches the cytoplasm, what are four mechanisms that can regulate the amount of the protein that is active in the cell?
5. **WHAT IF?** Examine Figure 18.11 and suggest a mechanism by which the yellow activator protein comes to be present in the liver cell but not in the lens cell.

For suggested answers, see Appendix A.

CONCEPT 18.3

Noncoding RNAs play multiple roles in controlling gene expression

Genome sequencing has revealed that protein-coding DNA accounts for only 1.5% of the human genome and a similarly small percentage of the genomes of many other multicellular eukaryotes. A very small fraction of the non-protein-coding DNA consists of genes for RNAs such as ribosomal RNA and transfer RNA. Until recently, most of the remaining DNA was assumed to be untranscribed. The idea was that since it didn't specify proteins or the few known types of RNA, such DNA didn't contain meaningful genetic information. However, a flood of recent data has contradicted this idea. For example, an in-depth study of a region comprising 1% of the human genome showed that more than 90% of that region was transcribed. Introns accounted for only a fraction of this transcribed, nontranslated RNA. These and other results suggest that a significant amount of the genome may be transcribed into non-protein-coding RNAs (also called *noncoding RNAs*, or *ncRNAs*), including a variety of small RNAs. While many questions about the functions of these RNAs remain unanswered, researchers are uncovering more evidence of their biological roles every day.

Biologists are excited about these recent discoveries, which hint at a large, diverse population of RNA molecules in the cell that play crucial roles in regulating gene expression—and have gone largely unnoticed until now. Clearly, we must revise our long-standing view that because mRNAs code for proteins, they are the most important RNAs functioning in the cell. This represents a major shift in the thinking of biologists, one that you are witnessing as students entering this field of study. It's as if our exclusive focus on a famous rock star has blinded us to the many backup musicians and songwriters working behind the scenes.

Regulation by both small and large ncRNAs is known to occur at several points in the pathway of gene expression, including mRNA translation and chromatin modification. We will focus mainly on two types of small ncRNAs that have been extensively studied in the past few years; the importance of these RNAs was acknowledged when they were the focus of the 2006 Nobel Prize in Physiology or Medicine.

Effects on mRNAs by MicroRNAs and Small Interfering RNAs

Since 1993, a number of research studies have uncovered small single-stranded RNA molecules, called **microRNAs (miRNAs)**, that are capable of binding to complementary sequences in mRNA molecules. The miRNAs are made from longer RNA precursors that fold back on themselves, forming one or more short double-stranded hairpin structures, each held together by hydrogen bonds **(Figure 18.15)**. After each hairpin is cut away from the precursor, it is trimmed by an enzyme (fittingly called Dicer) into a short double-stranded fragment of about 22 nucleotide pairs. One of the two strands is degraded, while the other strand, which is the miRNA, forms a complex with one or more proteins; the miRNA allows the complex to bind to any mRNA molecule with 7–8 nucleotides of complementary sequence. The miRNA-protein complex then either degrades the target mRNA or blocks its translation. It has been estimated that expression of at least one-half of all human genes may be regulated by miRNAs, a remarkable figure given that the existence of miRNAs was unknown a mere two decades ago.

A growing understanding of the miRNA pathway provided an explanation for a perplexing observation: Researchers had found that injecting double-stranded RNA molecules into a cell somehow turned off expression of a gene with the same sequence as the RNA. They called this experimental phenomenon **RNA interference (RNAi)**. It was later shown to be due to **small interfering RNAs (siRNAs)**, which are similar in size and function to miRNAs. In fact, subsequent research showed that the same cellular machinery generates miRNAs and siRNAs and that both can associate with the same proteins, producing similar results. The distinction between miRNAs and siRNAs is based on the nature of the precursor molecule for each. While an miRNA is usually formed from a single hairpin in a precursor RNA (see Figure 18.15), multiple siRNAs are formed from a much longer, linear, double-stranded RNA molecule.

We mentioned that laboratory investigators had injected double-stranded RNAs into cells, and you may wonder whether such molecules are ever found naturally. As you will learn in Chapter 19, some viruses have double-stranded RNA genomes. Because the cellular RNAi pathway can lead to the destruction of RNAs with sequences complementary to those found in double-stranded RNAs, this pathway may have evolved as a natural defense against infection by such viruses. However, the fact that RNAi can also affect the expression of nonviral cellular genes may reflect a different evolutionary origin for the RNAi pathway. Moreover, many species, including mammals, apparently produce their own long, double-stranded RNA precursors to small RNAs such as siRNAs. Once produced, these RNAs can interfere with gene expression at stages other than translation, as we'll discuss next.

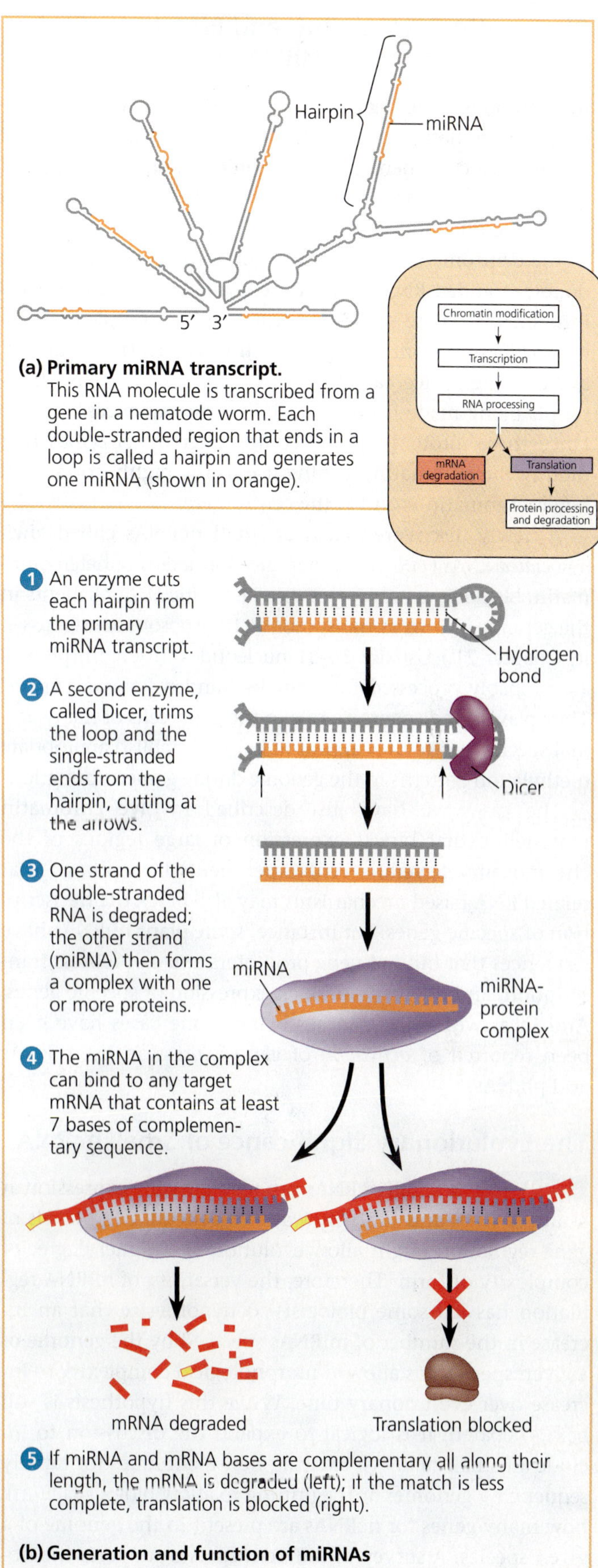

▲ **Figure 18.15 Regulation of gene expression by miRNAs.**

Chromatin Remodeling and Effects on Transcription by ncRNAs

In addition to affecting mRNAs, small RNAs can cause remodeling of chromatin structure. In some yeasts, siRNAs produced by the yeast cells themselves are required for the formation of heterochromatin at the centromeres of chromosomes. According to one model, an RNA transcript produced from DNA in the centromeric region of the chromosome is copied into double-stranded RNA by a yeast enzyme and then processed into siRNAs. These siRNAs associate with a complex of proteins (different from the one shown in Figure 18.15) and act as a homing device, targeting the complex back to RNA transcripts being made from the centromeric sequences of DNA. Once there, proteins in the complex recruit enzymes that modify the chromatin, turning it into the highly condensed heterochromatin found at the centromere.

A newly discovered class of small ncRNAs called *piwi-associated RNAs* (*piRNAs*) also induce formation of heterochromatin, blocking expression of some parasitic DNA elements in the genome known as transposons. (Transposons are discussed in Chapter 21.) Usually 24–31 nucleotides in length, piRNAs are probably processed from single-stranded RNA precursors. They play an indispensable role in the germ cells of many animal species, where they appear to help re-establish appropriate methylation patterns in the genome during gamete formation.

The cases we have just described involve chromatin remodeling that blocks expression of large regions of the chromosome. Several recent experiments have shown that related RNA-based mechanisms may also block the transcription of specific genes. For instance, some plant miRNAs have sequences that bind to gene promoters and can repress transcription, and piRNAs can block expression of specific genes. And in a twist on the same theme, some cases have even been reported of *activation* of gene expression by miRNAs and piRNAs.

The Evolutionary Significance of Small ncRNAs

EVOLUTION Small ncRNAs can regulate gene expression at multiple steps and in many ways. In general, extra levels of gene regulation might allow evolution of a higher degree of complexity of form. Therefore, the versatility of miRNA regulation has led some biologists to hypothesize that an increase in the number of miRNAs specified by the genome of a given species has allowed morphological complexity to increase over evolutionary time. While this hypothesis is still being debated, it is logical to expand the discussion to include all small ncRNAs. Exciting new techniques for rapidly sequencing genomes are beginning to allow biologists to ask how many genes for ncRNAs are present in the genome of a given species. A survey of different species supports the notion that siRNAs evolved first, followed by miRNAs and later piRNAs, which are found only in animals. And while there are hundreds of types of miRNAs, there appear to be many thousands of types of piRNAs, allowing the potential for very sophisticated gene regulation by piRNAs.

Given the extensive functions of ncRNAs, it is not surprising that many of the ncRNAs characterized thus far play important roles in embryonic development—the topic we turn to in the next section. Embryonic development is perhaps the ultimate example of precisely regulated gene expression.

CONCEPT CHECK 18.3

1. Compare and contrast miRNAs and siRNAs.
2. **WHAT IF?** If the mRNA being degraded in Figure 18.15 coded for a protein that promotes cell division in a multicellular organism, what would happen if a mutation disabled the gene encoding the miRNA that triggers this degradation?
3. **MAKE CONNECTIONS** In Concept 15.2 (pp. 291–292), you learned about inactivation of one of the X chromosomes in female mammals. Reread those pages, and suggest a model for how the *XIST* noncoding RNA functions to cause Barr body formation.

For suggested answers, see Appendix A.

CONCEPT 18.4

A program of differential gene expression leads to the different cell types in a multicellular organism

In the embryonic development of multicellular organisms, a fertilized egg (a zygote) gives rise to cells of many different types, each with a different structure and corresponding function. Typically, cells are organized into tissues, tissues into organs, organs into organ systems, and organ systems into the whole organism. Thus, any developmental program must produce cells of different types that form higher-level structures arranged in a particular way in three dimensions. The processes that occur during development in plants and animals are detailed in Chapters 35 and 47, respectively. In this chapter, we focus instead on the program of regulation of gene expression that orchestrates development, using a few animal species as examples.

A Genetic Program for Embryonic Development

The photos in **Figure 18.16** illustrate the dramatic difference between a zygote and the organism it becomes. This remarkable transformation results from three interrelated processes: cell division, cell differentiation, and morphogenesis. Through a succession of mitotic cell divisions, the zygote gives rise to a

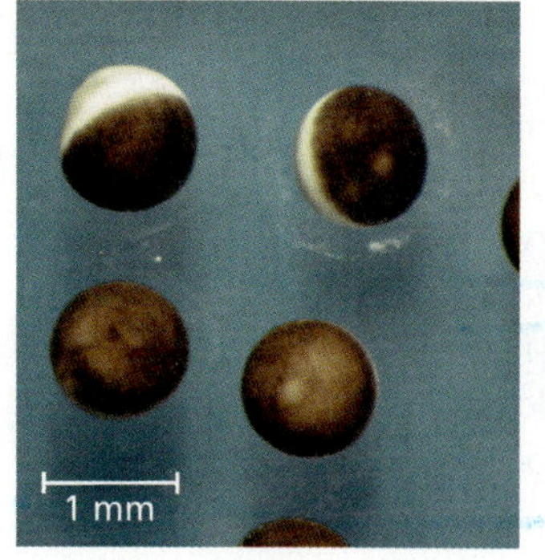

(a) Fertilized eggs of a frog **(b) Newly hatched tadpole**

▲ **Figure 18.16 From fertilized egg to animal: What a difference four days makes.** It takes just four days for cell division, differentiation, and morphogenesis to transform each of the fertilized frog eggs shown in (a) into a tadpole like the one in (b).

large number of cells. Cell division alone, however, would merely produce a great ball of identical cells, nothing like a tadpole. During embryonic development, cells not only increase in number, but also undergo cell **differentiation**, the process by which cells become specialized in structure and function. Moreover, the different kinds of cells are not randomly distributed but are organized into tissues and organs in a particular three-dimensional arrangement. The physical processes that give an organism its shape constitute **morphogenesis**, meaning "creation of form."

All three processes have their basis in cellular behavior. Even morphogenesis, the shaping of the organism, can be traced back to changes in the shape, motility, and other characteristics of the cells that make up various regions of the embryo. As you have seen, the activities of a cell depend on the genes it expresses and the proteins it produces. Almost all cells in an organism have the same genome; therefore, differential gene expression results from the genes being regulated differently in each cell type.

In Figure 18.11, you saw a simplified view of how differential gene expression occurs in two cell types, a liver cell and a lens cell. Each of these fully differentiated cells has a particular mix of specific activators that turn on the collection of genes whose products are required in the cell. The fact that both cells arose through a series of mitoses from a common fertilized egg inevitably leads to a question: How do different sets of activators come to be present in the two cells?

It turns out that materials placed into the egg by the mother set up a sequential program of gene regulation that is carried out as cells divide, and this program makes the cells become different from each other in a coordinated fashion. To understand how this works, we will consider two basic developmental processes: First, we'll explore how cells that arise from early embryonic mitoses develop the differences that start each cell along its own differentiation pathway. Second, we'll see how cellular differentiation leads to one particular cell type, using muscle development as an example.

Cytoplasmic Determinants and Inductive Signals

What generates the first differences among cells in an early embryo? And what controls the differentiation of all the various cell types as development proceeds? By this point in the chapter, you can probably deduce the answer: The specific genes expressed in any particular cell of a developing organism determine its path. Two sources of information, used to varying extents in different species, "tell" a cell which genes to express at any given time during embryonic development.

One important source of information early in development is the egg's cytoplasm, which contains both RNA and proteins encoded by the mother's DNA. The cytoplasm of an unfertilized egg is not homogeneous. Messenger RNA, proteins, other substances, and organelles are distributed unevenly in the unfertilized egg, and this unevenness has a profound impact on the development of the future embryo in many species. Maternal substances in the egg that influence the course of early development are called **cytoplasmic determinants** (**Figure 18.17a**, on the next page). After fertilization, early mitotic divisions distribute the zygote's cytoplasm into separate cells. The nuclei of these cells may thus be exposed to different cytoplasmic determinants, depending on which portions of the zygotic cytoplasm a cell received. The combination of cytoplasmic determinants in a cell helps determine its developmental fate by regulating expression of the cell's genes during the course of cell differentiation.

The other major source of developmental information, which becomes increasingly important as the number of embryonic cells increases, is the environment around a particular cell. Most influential are the signals impinging on an embryonic cell from other embryonic cells in the vicinity, including contact with cell-surface molecules on neighboring cells and the binding of growth factors secreted by neighboring cells. Such signals cause changes in the target cells, a process called **induction** (**Figure 18.17b**). The molecules conveying these signals within the target cell are cell-surface receptors and other proteins expressed by the embryo's own genes. In general, the signaling molecules send a cell down a specific developmental path by causing changes in its gene expression that eventually result in observable cellular changes. Thus, interactions between embryonic cells help induce differentiation of the many specialized cell types making up a new organism.

Sequential Regulation of Gene Expression During Cellular Differentiation

As the tissues and organs of an embryo develop and their cells differentiate, the cells become noticeably different in structure and function. These observable changes are actually the outcome of a cell's developmental history beginning at the first mitotic division of the zygote, as we have just seen. The

▼ Figure 18.17 Sources of developmental information for the early embryo.

(a) Cytoplasmic determinants in the egg

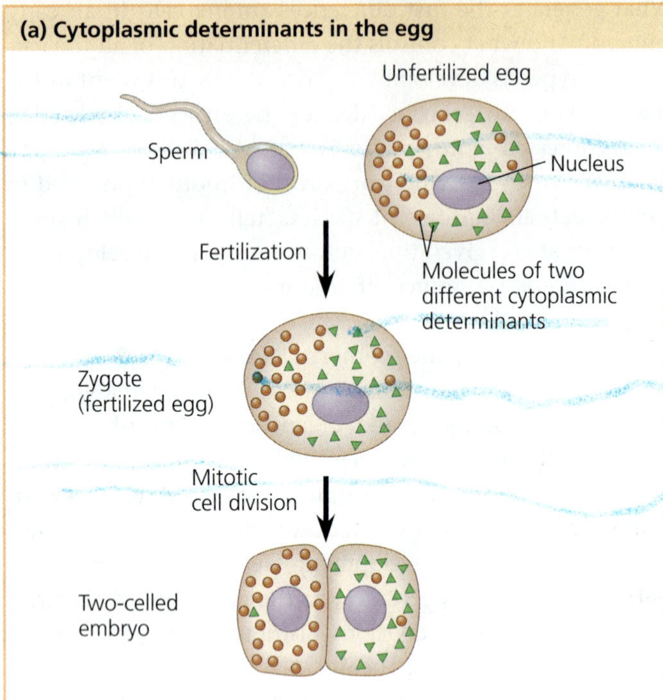

The unfertilized egg has molecules in its cytoplasm, encoded by the mother's genes, that influence development. Many of these cytoplasmic determinants, like the two shown here, are unevenly distributed in the egg. After fertilization and mitotic division, the cell nuclei of the embryo are exposed to different sets of cytoplasmic determinants and, as a result, express different genes.

(b) Induction by nearby cells

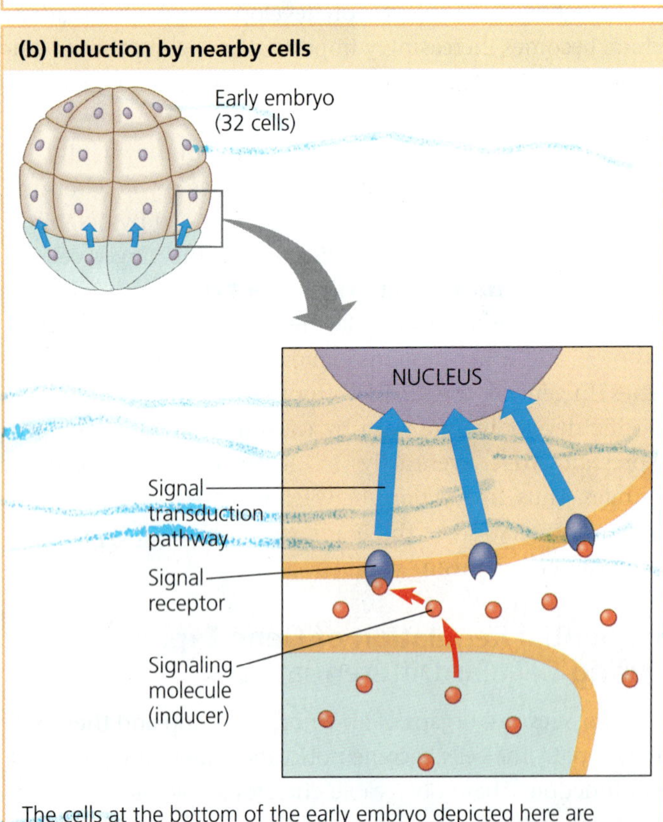

The cells at the bottom of the early embryo depicted here are releasing chemicals that signal nearby cells to change their gene expression.

earliest changes that set a cell on a path to specialization are subtle ones, showing up only at the molecular level. Before biologists knew much about the molecular changes occurring in embryos, they coined the term **determination** to refer to the events that lead to the observable differentiation of a cell. Once it has undergone determination, an embryonic cell is irreversibly committed to its final fate. If a committed cell is experimentally placed in another location in the embryo, it will still differentiate into the cell type that is its normal fate.

Today we understand determination in terms of molecular changes. The outcome of determination, observable cell differentiation, is marked by the expression of genes for *tissue-specific proteins*. These proteins are found only in a specific cell type and give the cell its characteristic structure and function. The first evidence of differentiation is the appearance of mRNAs for these proteins. Eventually, differentiation is observable with a microscope as changes in cellular structure. On the molecular level, different sets of genes are sequentially expressed in a regulated manner as new cells arise from division of their precursors. A number of the steps in gene expression may be regulated during differentiation, with transcription among the most important. In the fully differentiated cell, transcription remains the principal regulatory point for maintaining appropriate gene expression.

Differentiated cells are specialists at making tissue-specific proteins. For example, as a result of transcriptional regulation, liver cells specialize in making albumin, and lens cells specialize in making crystallin (see Figure 18.11). Skeletal muscle cells in vertebrates are another instructive example. Each of these cells is a long fiber containing many nuclei within a single plasma membrane. Skeletal muscle cells have high concentrations of muscle-specific versions of the contractile proteins myosin and actin, as well as membrane receptor proteins that detect signals from nerve cells.

Muscle cells develop from embryonic precursor cells that have the potential to develop into a number of cell types, including cartilage cells and fat cells, but particular conditions commit them to becoming muscle cells. Although the committed cells appear unchanged under the microscope, determination has occurred, and they are now *myoblasts*. Eventually, myoblasts start to churn out large amounts of muscle-specific proteins and fuse to form mature, elongated, multinucleate skeletal muscle cells **(Figure 18.18)**.

Researchers have worked out what happens at the molecular level during muscle cell determination by growing myoblasts in culture and analyzing them using molecular biological techniques you will learn about in Chapter 20. In a series of experiments, they isolated different genes, caused each to be expressed in a separate embryonic precursor cell, and then looked for differentiation into myoblasts and muscle cells. In this way, they identified several so-called "master regulatory genes" whose protein products commit the cells to becoming skeletal muscle. Thus, in the case of muscle cells,

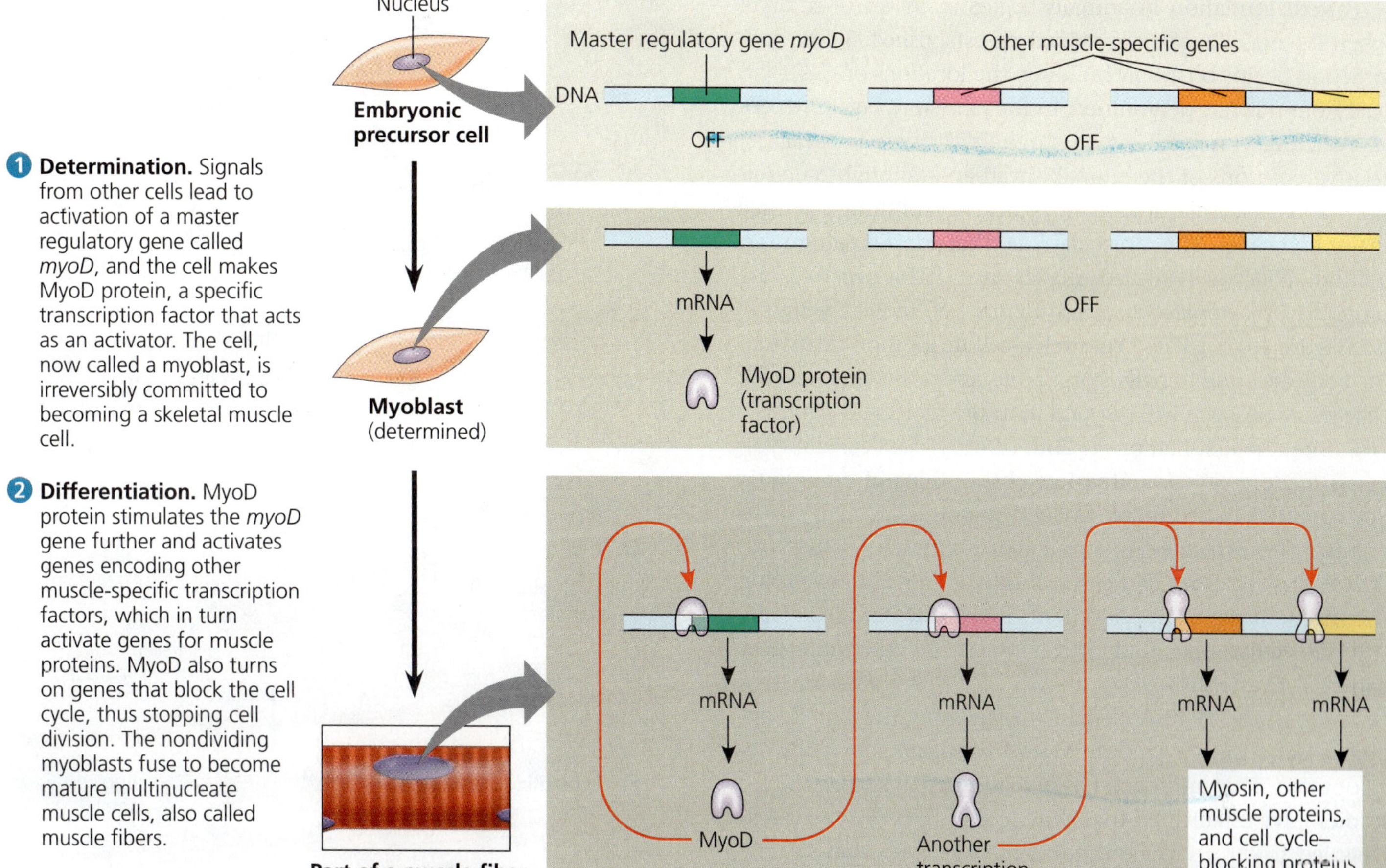

▲ **Figure 18.18 Determination and differentiation of muscle cells.** Skeletal muscle cells arise from embryonic cells as a result of changes in gene expression. (In this depiction, the process of gene activation is greatly simplified.)

WHAT IF? *What would happen if a mutation in the* myoD *gene resulted in a MyoD protein that could not activate the* myoD *gene?*

the molecular basis of determination is the expression of one or more of these master regulatory genes.

To understand more about how commitment occurs in muscle cell differentiation, let's focus on the master regulatory gene called *myoD* (see Figure 18.18). This gene encodes MyoD protein, a transcription factor that binds to specific control elements in the enhancers of various target genes and stimulates their expression (see Figure 18.9). Some target genes for MyoD encode still other muscle-specific transcription factors. MyoD also stimulates expression of the *myoD* gene itself, thus perpetuating its effect in maintaining the cell's differentiated state. Presumably, all the genes activated by MyoD have enhancer control elements recognized by MyoD and are thus coordinately controlled. Finally, the secondary transcription factors activate the genes for proteins such as myosin and actin that confer the unique properties of skeletal muscle cells.

The MyoD protein deserves its designation as a master regulatory gene. Researchers have shown that it is even capable of changing some kinds of fully differentiated nonmuscle cells, such as fat cells and liver cells, into muscle cells. Why doesn't it work on *all* kinds of cells? One likely explanation is that activation of the muscle-specific genes is not solely dependent on MyoD but requires a particular *combination* of regulatory proteins, some of which are lacking in cells that do not respond to MyoD. The determination and differentiation of other kinds of tissues may play out in a similar fashion.

We have now seen how different programs of gene expression that are activated in the fertilized egg can result in differentiated cells and tissues. But for the tissues to function effectively in the organism as a whole, the organism's *body plan*—its overall three-dimensional arrangement—must be established and superimposed on the differentiation process. Next we'll investigate the molecular basis for the establishment of the body plan, using the well-studied *Drosophila* as an example.

Pattern Formation: Setting Up the Body Plan

Cytoplasmic determinants and inductive signals both contribute to the development of a spatial organization in which the tissues and organs of an organism are all in their characteristic places. This process is called **pattern formation**.

Pattern formation in animals begins in the early embryo, when the major axes of an animal are established. Before construction begins on a new building, the locations of the front, back, and sides are determined. In the same way, before the tissues and organs of a bilaterally symmetrical animal appear, the relative positions of the animal's head and tail, right and left sides, and back and front are set up, thus establishing the three major body axes. The molecular cues that control pattern formation, collectively called **positional information**, are provided by cytoplasmic determinants and inductive signals (see Figure 18.17). These cues tell a cell its location relative to the body axes and to neighboring cells and determine how the cell and its progeny will respond to future molecular signals.

During the first half of the 20th century, classical embryologists made detailed anatomical observations of embryonic development in a number of species and performed experiments in which they manipulated embryonic tissues. Although this research laid the groundwork for understanding the mechanisms of development, it did not reveal the specific molecules that guide development or determine how patterns are established.

Then, in the 1940s, scientists began using the genetic approach—the study of mutants—to investigate *Drosophila* development. That approach has had spectacular success. These studies have established that genes control development and have led to an understanding of the key roles that specific molecules play in defining position and directing differentiation. By combining anatomical, genetic, and biochemical approaches to the study of *Drosophila* development, researchers have discovered developmental principles common to many other species, including humans.

The Life Cycle of Drosophila

Fruit flies and other arthropods have a modular construction, an ordered series of segments. These segments make up the body's three major parts: the head, the thorax (the midbody, from which the wings and legs extend), and the abdomen (**Figure 18.19a**). Like other bilaterally symmetrical animals, *Drosophila* has an anterior-posterior (head-to-tail) axis, a dorsal-ventral (back-to-belly) axis, and a right-left axis. In *Drosophila*, cytoplasmic determinants that are localized in the unfertilized egg provide positional information for the placement of anterior-posterior and dorsal-ventral axes even before fertilization. We'll focus here on the molecules involved in establishing the anterior-posterior axis.

The *Drosophila* egg develops in the female's ovary, surrounded by ovarian cells called nurse cells and follicle cells (**Figure 18.19b**, top). These support cells supply the egg with nutrients, mRNAs, and other substances needed for development and make the egg shell. After fertilization and laying of the egg, embryonic development results in the formation of a segmented larva, which goes through three larval stages. Then, in a process much like that by which a caterpillar

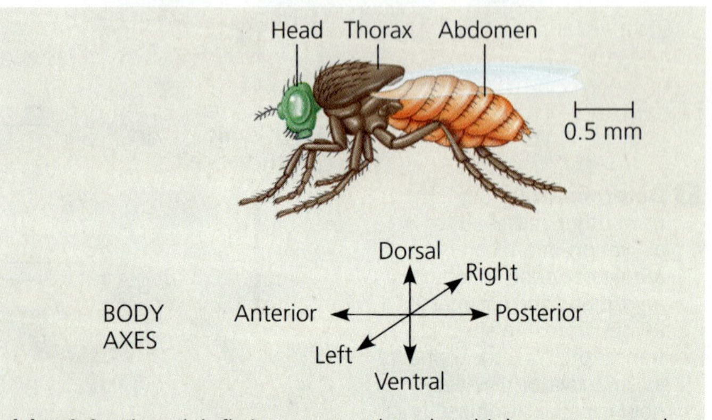

(a) Adult. The adult fly is segmented, and multiple segments make up each of the three main body parts—head, thorax, and abdomen. The body axes are shown by arrows.

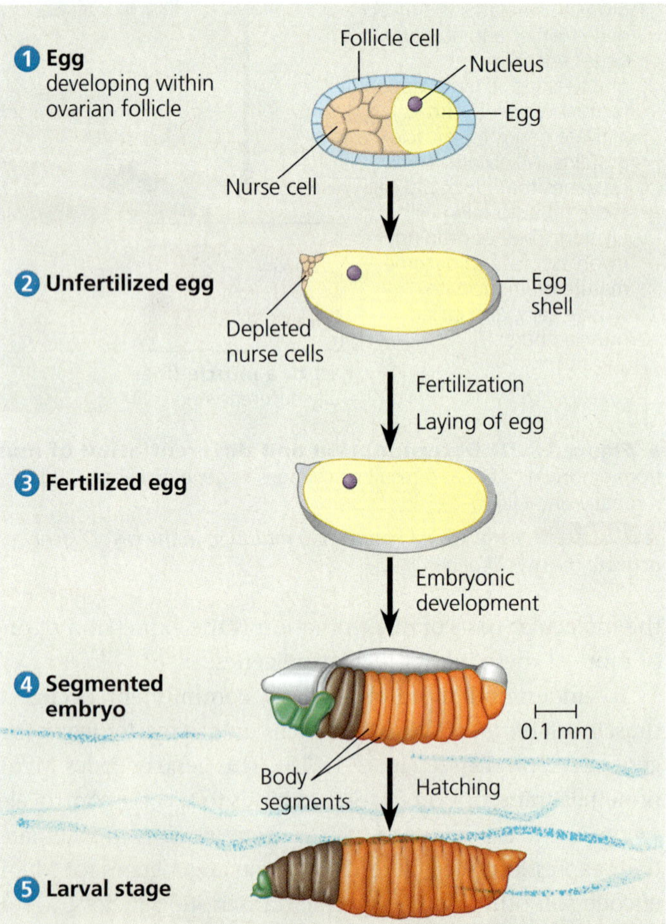

(b) Development from egg to larva. 1 The yellow egg is surrounded by other cells that form a structure called the follicle within one of the mother's ovaries. 2 The nurse cells shrink as they supply nutrients and mRNAs to the developing egg, which grows larger. Eventually, the mature egg fills the egg shell that is secreted by the follicle cells. 3 The egg is fertilized within the mother and then laid. 4 Embryonic development forms 5 a larva, which goes through three stages. The third stage forms a cocoon (not shown), within which the larva metamorphoses into the adult shown in (a).

▲ **Figure 18.19 Key developmental events in the life cycle of *Drosophila*.**

becomes a butterfly, the fly larva forms a cocoon in which it metamorphoses into the adult fly pictured in Figure 18.19a.

Genetic Analysis of Early Development: Scientific Inquiry

Edward B. Lewis was a visionary American biologist who, in the 1940s, first showed the value of the genetic approach to studying embryonic development in *Drosophila*. Lewis studied bizarre mutant flies with developmental defects that led to extra wings or legs in the wrong place **(Figure 18.20)**. He located the mutations on the fly's genetic map, thus connecting the developmental abnormalities to specific genes. This research supplied the first concrete evidence that genes somehow direct the developmental processes studied by embryologists. The genes Lewis discovered, called **homeotic genes**, control pattern formation in the late embryo, larva, and adult.

Insight into pattern formation during early embryonic development did not come for another 30 years, when two researchers in Germany, Christiane Nüsslein-Volhard and Eric Wieschaus, set out to identify *all* the genes that affect segment formation in *Drosophila*. The project was daunting for three reasons. The first was the sheer number of *Drosophila* genes, now known to total about 13,700. The genes affecting segmentation might be just a few needles in a haystack or might be so numerous and varied that the scientists would be unable to make sense of them. Second, mutations affecting a process as fundamental as segmentation would surely be **embryonic lethals**, mutations with phenotypes causing death at the embryonic or larval stage. Because organisms with embryonic lethal mutations never reproduce, they cannot be bred for study. The researchers dealt with this problem by looking for recessive mutations, which can be propagated in heterozygous flies that act as genetic carriers. Third, cytoplasmic determinants in the egg were known to play a role in axis formation, so the researchers knew they would have to study the mother's genes as well as those of the embryo. It is the mother's genes that we will discuss further as we focus on how the anterior-posterior body axis is set up in the developing egg.

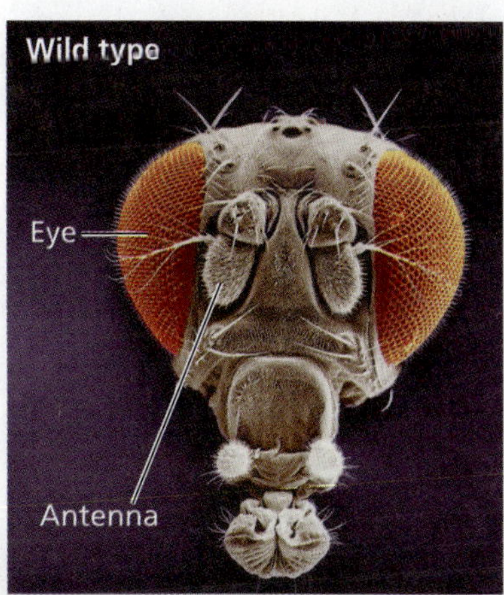

◀ **Figure 18.20 Abnormal pattern formation in *Drosophila*.** Mutations in certain regulatory genes, called homeotic genes, cause a misplacement of structures in an animal. These scanning electron micrographs contrast the head of a wild-type fly, bearing a pair of small antennae, with that of a homeotic mutant (a fly with a mutation in a single gene), bearing a pair of legs in place of antennae.

Nüsslein-Volhard and Wieschaus began their search for segmentation genes by exposing flies to a mutagenic chemical that affected the flies' gametes. They mated the mutagenized flies and then scanned their descendants for dead embryos or larvae with abnormal segmentation or other defects. For example, to find genes that might set up the anterior-posterior axis, they looked for embryos or larvae with abnormal ends, such as two heads or two tails, predicting that such abnormalities would arise from mutations in maternal genes required for correctly setting up the offspring's head or tail end.

Using this approach, Nüsslein-Volhard and Wieschaus eventually identified about 1,200 genes essential for pattern formation during embryonic development. Of these, about 120 were essential for normal segmentation. Over several years, the researchers were able to group these segmentation genes by general function, to map them, and to clone many of them for further study in the lab. The result was a detailed molecular understanding of the early steps in pattern formation in *Drosophila*.

When the results of Nüsslein-Volhard and Wieschaus were combined with Lewis's earlier work, a coherent picture of *Drosophila* development emerged. In recognition of their discoveries, the three researchers were awarded a Nobel Prize in 1995.

Let's consider further the genes that Nüsslein-Volhard, Wieschaus, and co-workers found for cytoplasmic determinants deposited in the egg by the mother. These genes set up the initial pattern of the embryo by regulating gene expression in broad regions of the early embryo.

Axis Establishment

As we mentioned earlier, cytoplasmic determinants in the egg are the substances that initially establish the axes of the *Drosophila* body. These substances are encoded by genes of the mother, fittingly called maternal effect genes. A **maternal effect gene** is a gene that, when mutant in the mother, results in a mutant phenotype in the offspring, regardless of the offspring's own genotype. In fruit fly development, the mRNA or protein products of maternal effect genes are placed in the egg while it is still in the mother's ovary. When the mother has a mutation in such a gene, she makes a defective gene product (or none at all), and her eggs are defective; when these eggs are fertilized, they fail to develop properly.

Because they control the orientation (polarity) of the egg and consequently of the fly, maternal effect genes are also called **egg-polarity genes**. One group of these genes sets up the anterior-posterior axis of the embryo, while a second group establishes the dorsal-ventral axis. Like mutations in segmentation genes, mutations in maternal effect genes are generally embryonic lethals.

Bicoid: A Morphogen Determining Head Structures To see how maternal effect genes determine the body axes of the offspring, we will focus on one such gene, called ***bicoid***, a term meaning "two-tailed." An embryo whose mother has two mutant alleles of the *bicoid* gene lacks the front half of its body and has posterior structures at both ends (**Figure 18.21**). This phenotype suggested to Nüsslein-Volhard and her colleagues that the product of the mother's *bicoid* gene is essential for setting up the anterior end of the fly and might be concentrated at the future anterior end of the embryo. This hypothesis is an example of the *morphogen gradient hypothesis* first proposed by embryologists a century ago; in this hypothesis, gradients of substances called **morphogens** establish an embryo's axes and other features of its form.

DNA technology and other modern biochemical methods enabled the researchers to test whether the *bicoid* product, a protein called Bicoid, is in fact a morphogen that determines the anterior end of the fly. The first question they asked was whether the mRNA and protein products of these genes are located in the egg in a position consistent with the hypothesis.

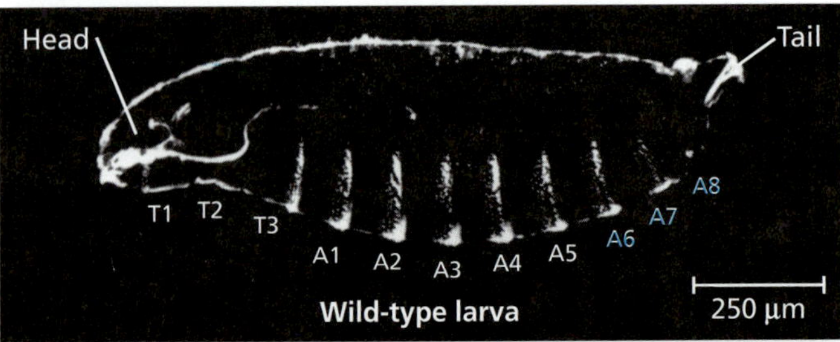

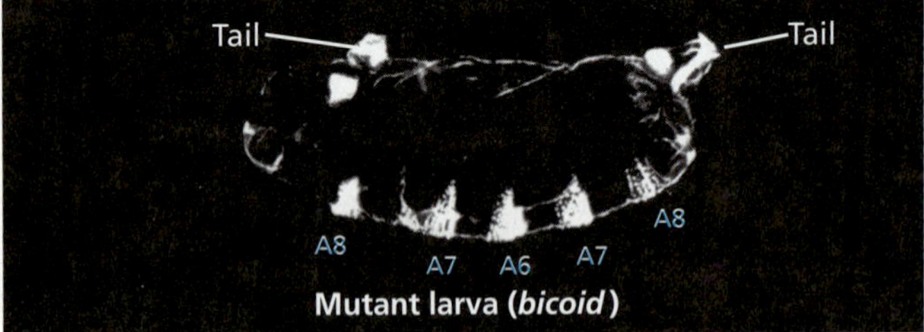

▲ **Figure 18.21 Effect of the *bicoid* gene on *Drosophila* development.** A wild-type fruit fly larva has a head, three thoracic (T) segments, eight abdominal (A) segments, and a tail. A larva whose mother has two mutant alleles of the *bicoid* gene has two tails and lacks all anterior structures (LMs).

They found that *bicoid* mRNA is highly concentrated at the extreme anterior end of the mature egg, as predicted by the hypothesis (**Figure 18.22**). After the egg is fertilized, the mRNA is translated into protein. The Bicoid protein then diffuses from the anterior end toward the posterior, resulting in a gradient of protein within the early embryo, with the highest concentration at the anterior end. These results are consistent with the hypothesis that Bicoid protein specifies the fly's anterior end. To test the hypothesis more specifically, scientists injected pure *bicoid* mRNA into various regions of early embryos. The protein that resulted from its translation caused anterior structures to form at the injection sites.

▼ **Figure 18.22** **INQUIRY**

Is Bicoid a morphogen that determines the anterior end of a fruit fly?

EXPERIMENT Using a genetic approach to study *Drosophila* development, Christiane Nüsslein-Volhard and colleagues at the European Molecular Biology Laboratory in Heidelberg, Germany, analyzed expression of the *bicoid* gene. The researchers hypothesized that *bicoid* normally codes for a morphogen that specifies the head (anterior) end of the embryo. To test this hypothesis, they used molecular techniques to determine where the mRNA and protein encoded by this gene were found in the fertilized egg and early embryo of wild-type flies.

RESULTS *Bicoid* mRNA (dark blue) was confined to the anterior end of the unfertilized egg. Later in development, Bicoid protein (dark orange) was seen to be concentrated in cells at the anterior end of the embryo.

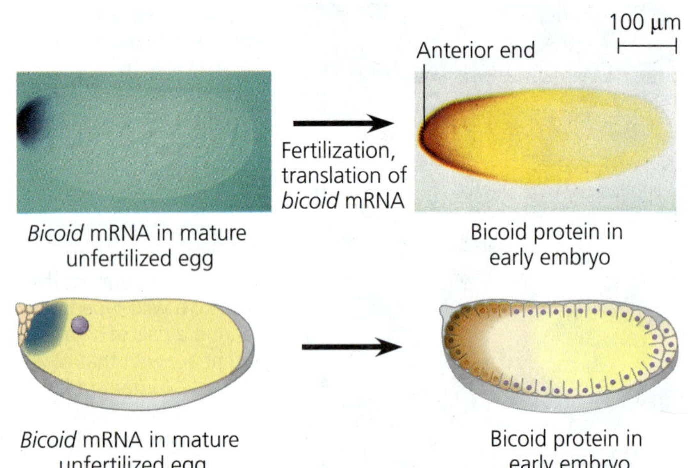

CONCLUSION The location of *bicoid* mRNA and the diffuse gradient of Bicoid protein seen later support the hypothesis that Bicoid protein is a morphogen specifying formation of head-specific structures.

SOURCE: C. Nüsslein-Volhard et al., Determination of anteroposterior polarity in *Drosophila*, *Science* 238:1675–1681 (1987); W. Driever and C. Nüsslein-Volhard, A gradient of *bicoid* protein in *Drosophila* embryos, *Cell* 54:83–93 (1988); T. Berleth et al., The role of localization of *bicoid* RNA in organizing the anterior pattern of the *Drosophila* embryo, *EMBO Journal* 7:1749–1756 (1988).

WHAT IF? If the hypothesis is correct, predict what would happen if you injected *bicoid* mRNA into the anterior end of an egg from a female with a mutation disabling the *bicoid* gene.

The *bicoid* research was groundbreaking for several reasons. First, it led to the identification of a specific protein required for some of the earliest steps in pattern formation. It thus helped us understand how different regions of the egg can give rise to cells that go down different developmental pathways. Second, it increased our understanding of the mother's critical role in the initial phases of embryonic development. Finally, the principle that a gradient of morphogens can determine polarity and position has proved to be a key developmental concept for a number of species, just as early embryologists had thought.

Maternal mRNAs are crucial during development of many species. In *Drosophila*, gradients of specific proteins encoded by maternal mRNAs determine the posterior and anterior ends and establish the dorsal-ventral axis. As the fly embryo grows, it reaches a point when the embryonic program of gene expression takes over, and the maternal mRNAs must be destroyed. (This process involves miRNAs in *Drosophila* and other species.) Later, positional information encoded by the embryo's genes, operating on an ever finer scale, establishes a specific number of correctly oriented segments and triggers the formation of each segment's characteristic structures. When the genes operating in this final step are abnormal, the pattern of the adult is abnormal, as you saw in Figure 18.20.

In this section, we have seen how a carefully orchestrated program of sequential gene regulation controls the transformation of a fertilized egg into a multicellular organism. The program is carefully balanced between turning on the genes for differentiation in the right place and turning off other genes. Even when an organism is fully developed, gene expression is regulated in a similarly fine-tuned manner. In the final section of the chapter, we'll consider how fine this tuning is by looking at how specific changes in expression of one or a few genes can lead to the development of cancer.

CONCEPT CHECK 18.4

1. As you learned in Chapter 12, mitosis gives rise to two daughter cells that are genetically identical to the parent cell. Yet you, the product of many mitotic divisions, are not composed of identical cells. Why?
2. MAKE CONNECTIONS Explain how the signaling molecules released by an embryonic cell can induce changes in a neighboring cell without entering the cell. (See Figures 11.15 and 11.16, pp. 219 and 220.)
3. Why are fruit fly maternal effect genes also called egg-polarity genes?
4. WHAT IF? In the blowup box in Figure 18.17b, the lower cell is synthesizing signaling molecules, whereas the upper cell is expressing receptors for these molecules. In terms of gene regulation, explain how these cells came to synthesize different molecules.

For suggested answers, see Appendix A.

CONCEPT 18.5

Cancer results from genetic changes that affect cell cycle control

In Chapter 12, we considered cancer as a set of diseases in which cells escape from the control mechanisms that normally limit their growth. Now that we have discussed the molecular basis of gene expression and its regulation, we are ready to look at cancer more closely. The gene regulation systems that go wrong during cancer turn out to be the very same systems that play important roles in embryonic development, the immune response, and many other biological processes. Thus, research into the molecular basis of cancer has both benefited from and informed many other fields of biology.

Types of Genes Associated with Cancer

The genes that normally regulate cell growth and division during the cell cycle include genes for growth factors, their receptors, and the intracellular molecules of signaling pathways. (To review the cell cycle, see Chapter 12.) Mutations that alter any of these genes in somatic cells can lead to cancer. The agent of such change can be random spontaneous mutation. However, it is likely that many cancer-causing mutations result from environmental influences, such as chemical carcinogens, X-rays and other high-energy radiation, and some viruses.

Cancer research led to the discovery of cancer-causing genes called **oncogenes** (from the Greek *onco*, tumor) in certain types of viruses (see Chapter 19). Subsequently, close counterparts of viral oncogenes were found in the genomes of humans and other animals. The normal versions of the cellular genes, called **proto-oncogenes**, code for proteins that stimulate normal cell growth and division.

How might a proto-oncogene—a gene that has an essential function in normal cells—become an oncogene, a cancer-causing gene? In general, an oncogene arises from a genetic change that leads to an increase either in the amount of the proto-oncogene's protein product or in the intrinsic activity of each protein molecule. The genetic changes that convert proto-oncogenes to oncogenes fall into three main categories: movement of DNA within the genome, amplification of a proto-oncogene, and point mutations in a control element or in the proto-oncogene itself (**Figure 18.23**, on the next page).

Cancer cells are frequently found to contain chromosomes that have broken and rejoined incorrectly, translocating fragments from one chromosome to another (see Figure 15.14). Now that you have learned how gene expression is regulated, you can understand the possible consequences of such translocations. If a translocated proto-oncogene ends up near an especially active promoter (or other control element), its transcription may increase, making it an oncogene. The second main type of genetic change, amplification, increases the

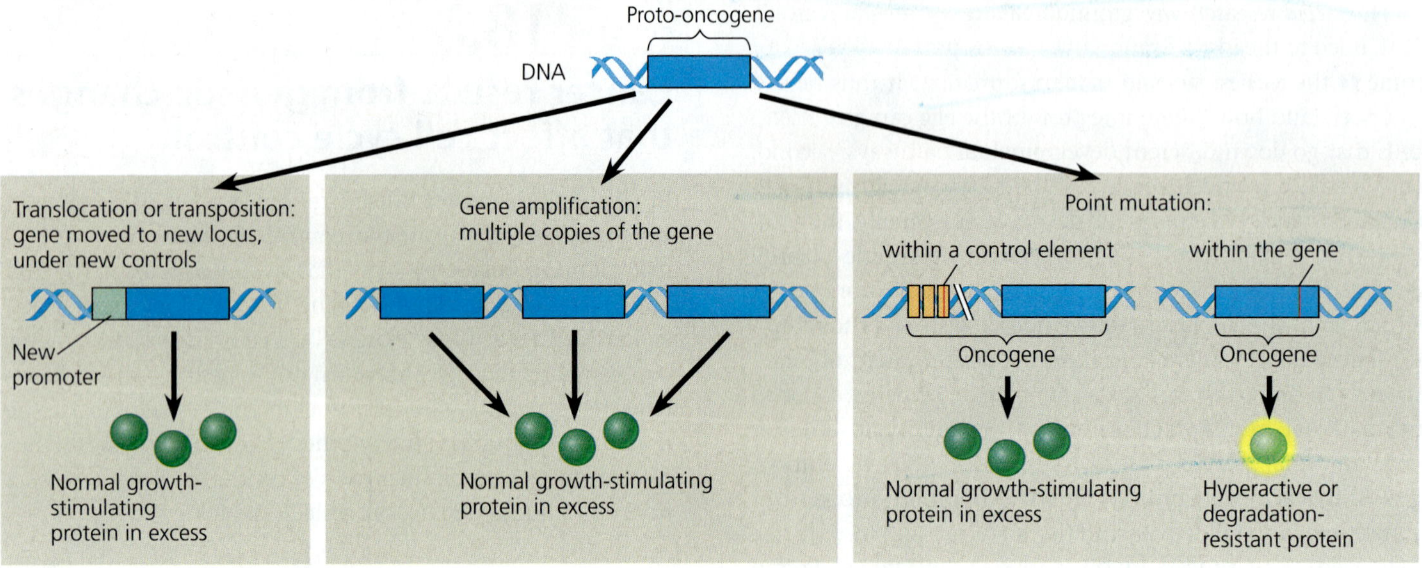

▲ **Figure 18.23 Genetic changes that can turn proto-oncogenes into oncogenes.**

number of copies of the proto-oncogene in the cell through repeated gene duplication (discussed in Chapter 21). The third possibility is a point mutation either (1) in the promoter or an enhancer that controls a proto-oncogene, causing an increase in its expression, or (2) in the coding sequence, changing the gene's product to a protein that is more active or more resistant to degradation than the normal protein. All these mechanisms can lead to abnormal stimulation of the cell cycle and put the cell on the path to malignancy.

Tumor-Suppressor Genes

In addition to genes whose products normally promote cell division, cells contain genes whose normal products *inhibit* cell division. Such genes are called **tumor-suppressor genes** because the proteins they encode help prevent uncontrolled cell growth. Any mutation that decreases the normal activity of a tumor-suppressor protein may contribute to the onset of cancer, in effect stimulating growth through the absence of suppression.

The protein products of tumor-suppressor genes have various functions. Some tumor-suppressor proteins repair damaged DNA, a function that prevents the cell from accumulating cancer-causing mutations. Other tumor-suppressor proteins control the adhesion of cells to each other or to the extracellular matrix; proper cell anchorage is crucial in normal tissues—and is often absent in cancers. Still other tumor-suppressor proteins are components of cell-signaling pathways that inhibit the cell cycle.

Interference with Normal Cell-Signaling Pathways

The proteins encoded by many proto-oncogenes and tumor-suppressor genes are components of cell-signaling pathways. Let's take a closer look at how such proteins function in normal cells and what goes wrong with their function in cancer cells. We will focus on the products of two key genes, the *ras* proto-oncogene and the *p53* tumor-suppressor gene. Mutations in *ras* occur in about 30% of human cancers, and mutations in *p53* in more than 50%.

The Ras protein, encoded by the ***ras* gene** (named for rat sarcoma, a connective tissue cancer), is a G protein that relays a signal from a growth factor receptor on the plasma membrane to a cascade of protein kinases (see Figure 11.7). The cellular response at the end of the pathway is the synthesis of a protein that stimulates the cell cycle **(Figure 18.24a)**. Normally, such a pathway will not operate unless triggered by the appropriate growth factor. But certain mutations in the *ras* gene can lead to production of a hyperactive Ras protein that triggers the kinase cascade even in the absence of growth factor, resulting in increased cell division. In fact, hyperactive versions or excess amounts of any of the pathway's components can have the same outcome: excessive cell division.

Figure 18.24b shows a pathway in which a signal leads to the synthesis of a protein that suppresses the cell cycle. In this case, the signal is damage to the cell's DNA, perhaps as the result of exposure to ultraviolet light. Operation of this signaling pathway blocks the cell cycle until the damage has been repaired. Otherwise, the damage might contribute to tumor formation by causing mutations or chromosomal abnormalities. Thus, the genes for the components of the pathway act as tumor-suppressor genes. The ***p53* gene**, named for the 53,000-dalton molecular weight of its protein product, is a tumor-suppressor gene. The protein it encodes is a specific transcription factor that promotes the synthesis of cell cycle–inhibiting proteins. That is why a mutation that knocks out the *p53* gene, like a mutation that leads to a hyperactive Ras protein, can lead to excessive cell growth and cancer **(Figure 18.24c)**.

(a) Cell cycle–stimulating pathway. This pathway is triggered by ① a growth factor that binds to ② its receptor in the plasma membrane. The signal is relayed to ③ a G protein called Ras. Like all G proteins, Ras is active when GTP is bound to it. Ras passes the signal to ④ a series of protein kinases. The last kinase activates ⑤ a transcription activator that turns on one or more genes for proteins that stimulate the cell cycle. If a mutation makes Ras or any other pathway component abnormally active, excessive cell division and cancer may result.

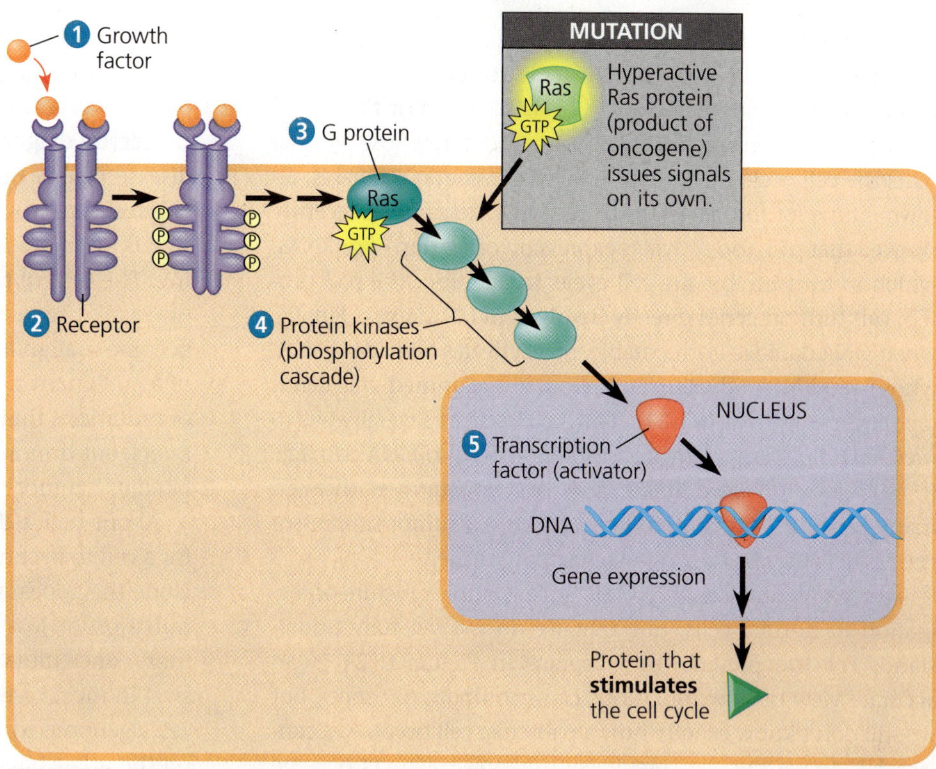

(b) Cell cycle–inhibiting pathway. In this pathway, ① DNA damage is an intracellular signal that is passed via ② protein kinases and leads to activation of ③ p53. Activated p53 promotes transcription of the gene for a protein that inhibits the cell cycle. The resulting suppression of cell division ensures that the damaged DNA is not replicated. If the DNA damage is irreparable, the p53 signal leads to programmed cell death (apoptosis). Mutations causing deficiencies in any pathway component can contribute to the development of cancer.

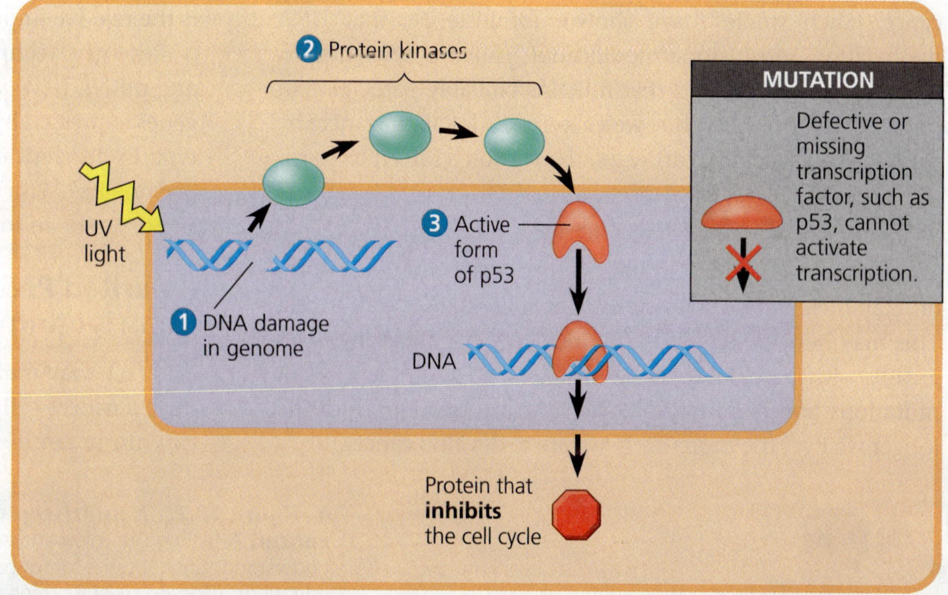

(c) Effects of mutations. Increased cell division, possibly leading to cancer, can result if the cell cycle is overstimulated, as in (a), or not inhibited when it normally would be, as in (b).

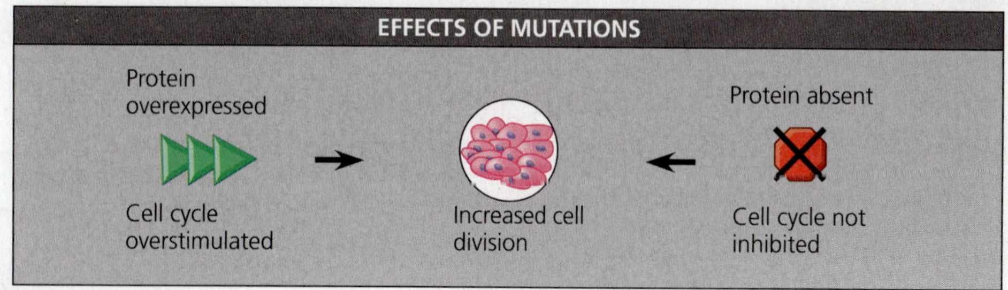

▲ **Figure 18.24 Signaling pathways that regulate cell division.** Both stimulatory and inhibitory pathways regulate the cell cycle, commonly by influencing transcription. Cancer can result from aberrations in such pathways, which may be caused by mutations, either spontaneous or environmentally triggered.

? *Looking at the pathway in (b), explain whether a cancer-causing mutation in a tumor-suppressor gene, such as* p53, *is more likely to be a recessive or a dominant mutation.*

The *p53* gene has been called the "guardian angel of the genome." Once the gene is activated—for example, by DNA damage—the p53 protein functions as an activator for several other genes. Often it activates a gene called *p21*, whose product halts the cell cycle by binding to cyclin-dependent kinases, allowing time for the cell to repair the DNA. Researchers recently showed that p53 also activates expression of a group of miRNAs, which in turn inhibit the cell cycle. In addition, the p53 protein can turn on genes directly involved in DNA repair. Finally, when DNA damage is irreparable, p53 activates "suicide" genes, whose protein products bring about programmed cell death (apoptosis; see Figure 11.21). Thus, p53 acts in several ways to prevent a cell from passing on mutations due to DNA damage. If mutations do accumulate and the cell survives through many divisions—as is more likely if the *p53* tumor-suppressor gene is defective or missing—cancer may ensue.

The many functions of p53 suggest a complex picture of regulation in normal cells, one that we do not yet fully understand. For the present, the diagram in Figure 18.24 is an accurate view of how mutations can contribute to cancer, but we still don't know exactly how a particular cell becomes a cancer cell. As we discover previously unknown aspects of gene regulation, it is informative to study their role in the onset of cancer. Such studies have shown, for instance, that DNA methylation and histone modification patterns differ in normal and cancer cells and that miRNAs probably participate in cancer development. While we've learned a lot about cancer by studying cell-signaling pathways, there is still a lot left to learn.

The Multistep Model of Cancer Development

More than one somatic mutation is generally needed to produce all the changes characteristic of a full-fledged cancer cell. This may help explain why the incidence of cancer increases greatly with age. If cancer results from an accumulation of mutations and if mutations occur throughout life, then the longer we live, the more likely we are to develop cancer.

The model of a multistep path to cancer is well supported by studies of one of the best-understood types of human cancer, colorectal cancer. About 135,000 new cases of colorectal cancer are diagnosed each year in the United States, and the disease causes 60,000 deaths each year. Like most cancers, colorectal cancer develops gradually **(Figure 18.25)**. The first sign is often a polyp, a small, benign growth in the colon lining. The cells of the polyp look normal, although they divide unusually frequently. The tumor grows and may eventually become malignant, invading other tissues. The development of a malignant tumor is paralleled by a gradual accumulation of mutations that convert proto-oncogenes to oncogenes and knock out tumor-suppressor genes. A *ras* oncogene and a mutated *p53* tumor-suppressor gene are often involved.

About half a dozen changes must occur at the DNA level for a cell to become fully cancerous. These changes usually include the appearance of at least one active oncogene and the mutation or loss of several tumor-suppressor genes. Furthermore, since mutant tumor-suppressor alleles are usually recessive, in most cases mutations must knock out *both* alleles in a cell's genome to block tumor suppression. (Most oncogenes, on the other hand, behave as dominant alleles.) The order in which these changes must occur is still under investigation, as is the relative importance of different mutations.

Recently, technical advances in the sequencing of DNA and mRNA have allowed medical researchers to compare the genes expressed by different types of tumors and by the same type in different individuals. These comparisons have led to personalized cancer treatments based on the molecular characteristics of an individual's tumor (see Figure 12.21.)

Inherited Predisposition and Other Factors Contributing to Cancer

The fact that multiple genetic changes are required to produce a cancer cell helps explain the observation that cancers can run in families. An individual inheriting an oncogene or

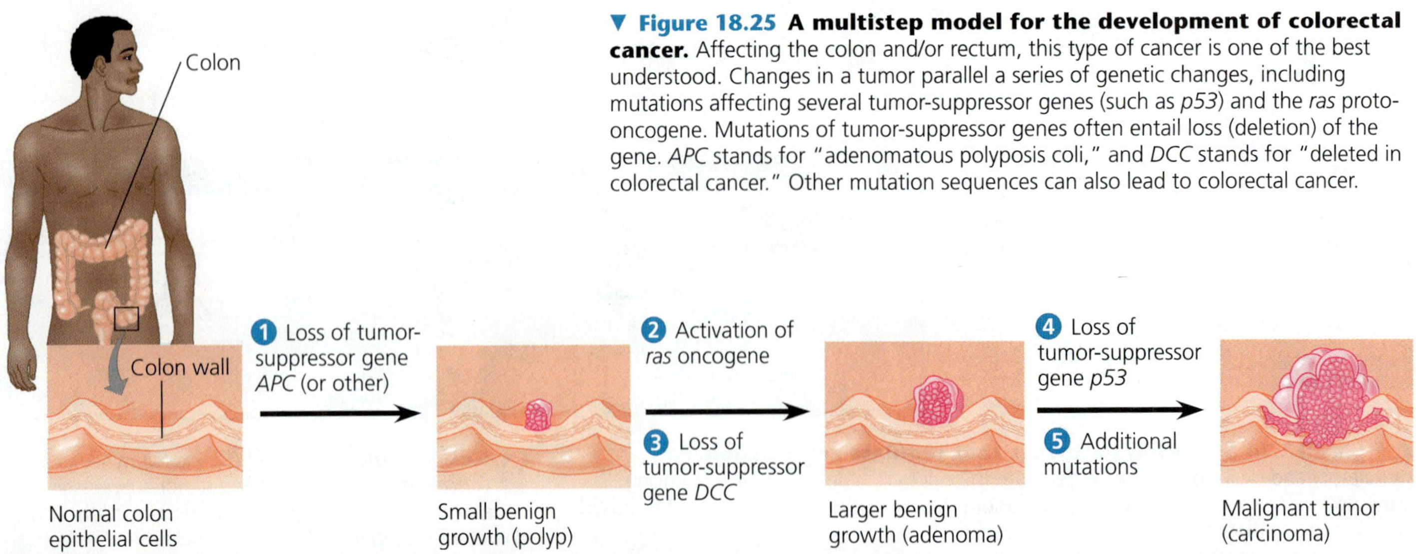

▼ **Figure 18.25 A multistep model for the development of colorectal cancer.** Affecting the colon and/or rectum, this type of cancer is one of the best understood. Changes in a tumor parallel a series of genetic changes, including mutations affecting several tumor-suppressor genes (such as *p53*) and the *ras* proto-oncogene. Mutations of tumor-suppressor genes often entail loss (deletion) of the gene. *APC* stands for "adenomatous polyposis coli," and *DCC* stands for "deleted in colorectal cancer." Other mutation sequences can also lead to colorectal cancer.

a mutant allele of a tumor-suppressor gene is one step closer to accumulating the necessary mutations for cancer to develop than is an individual without any such mutations.

Geneticists are devoting much effort to identifying inherited cancer alleles so that predisposition to certain cancers can be detected early in life. About 15% of colorectal cancers, for example, involve inherited mutations. Many of these affect the tumor-suppressor gene called *adenomatous polyposis coli*, or *APC* (see Figure 18.25). This gene has multiple functions in the cell, including regulation of cell migration and adhesion. Even in patients with no family history of the disease, the *APC* gene is mutated in 60% of colorectal cancers. In these individuals, new mutations must occur in both *APC* alleles before the gene's function is lost. Since only 15% of colorectal cancers are associated with known inherited mutations, researchers continue in their efforts to identify "markers" that could predict the risk of developing this type of cancer.

There is evidence of a strong inherited predisposition in 5–10% of patients with breast cancer. This is the second most common type of cancer in the United States, striking over 180,000 women (and some men) annually and killing 40,000 each year. In 1990, after 16 years of research, geneticist Mary-Claire King convincingly demonstrated that mutations in one gene—*BRCA1*—were associated with increased susceptibility to breast cancer, a finding that flew in the face of medical opinion at the time. (*BRCA* stands for *br*east *ca*ncer.) Mutations in that gene or the related *BRCA2* gene are found in at least half of inherited breast cancers, and tests using DNA sequencing can detect these mutations **(Figure 18.26)**. A woman who inherits one mutant *BRCA1* allele has a 60% probability of developing breast cancer before the age of 50, compared with only a 2% probability for an individual homozygous for the normal allele. Both *BRCA1* and *BRCA2* are considered tumor-suppressor genes because their wild-type alleles protect against breast cancer and their mutant alleles are recessive. Apparently, the BRCA1 and BRCA2 proteins both function in the cell's DNA damage repair pathway. More is known about BRCA2, which, in association with another protein, helps repair breaks that occur in both strands of DNA; it is crucial for maintaining undamaged DNA in a cell's nucleus.

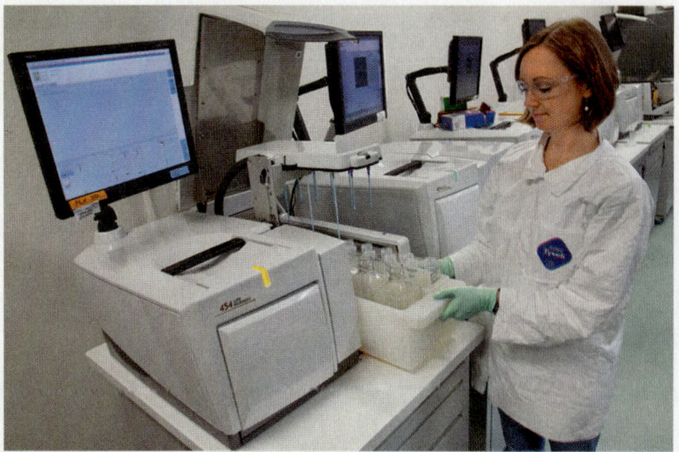

▲ **Figure 18.26 Testing for mutations in *BRCA1* and *BRCA2*.** Genetic testing for mutations that increase the risk of breast cancer is available for individuals with a family history of breast cancer. New "high-throughput" sequencing techniques can sequence many DNA samples at once, as shown here.

Because DNA breakage can contribute to cancer, it makes sense that the risk of cancer can be lowered by minimizing exposure to DNA-damaging agents, such as the ultraviolet radiation in sunlight and chemicals found in cigarette smoke. Novel methods for early diagnosis and treatment of specific cancers are being developed that rely on new techniques for analyzing, and perhaps interfering with, gene expression in tumors. Ultimately, such approaches may lower the death rate from cancer.

The study of genes associated with cancer, inherited or not, increases our basic understanding of how disruption of normal gene regulation can result in this disease. In addition to the mutations and other genetic alterations described in this section, a number of *tumor viruses* can cause cancer in various animals, including humans. In fact, one of the earliest breakthroughs in understanding cancer came in 1911, when Peyton Rous, an American pathologist, discovered a virus that causes cancer in chickens. The Epstein-Barr virus, which causes infectious mononucleosis, has been linked to several types of cancer in humans, notably Burkitt's lymphoma. Papillomaviruses are associated with cancer of the cervix, and a virus called HTLV-1 causes a type of adult leukemia. Worldwide, viruses seem to play a role in about 15% of the cases of human cancer.

Viruses may at first seem very different from mutations as a cause of cancer. However, we now know that viruses can interfere with gene regulation in several ways if they integrate their genetic material into the DNA of a cell. Viral integration may donate an oncogene to the cell, disrupt a tumor-suppressor gene, or convert a proto-oncogene to an oncogene. In addition, some viruses produce proteins that inactivate p53 and other tumor-suppressor proteins, making the cell more prone to becoming cancerous. Viruses are powerful biological agents, and you'll learn more about their function in Chapter 19.

CONCEPT CHECK 18.5

1. **MAKE CONNECTIONS** The p53 protein can activate genes involved in apoptosis, or programmed cell death. Review Concept 11.5 (pp. 223–225) and discuss how mutations in genes coding for proteins that function in apoptosis could contribute to cancer.
2. Under what circumstances is cancer considered to have a hereditary component?
3. **WHAT IF?** Explain how the types of mutations that lead to cancer are different for a proto-oncogene and a tumor-suppressor gene, in terms of the effect of the mutation on the activity of the gene product.

For suggested answers, see Appendix A.

18 CHAPTER REVIEW

SUMMARY OF KEY CONCEPTS

CONCEPT 18.1

Bacteria often respond to environmental change by regulating transcription (pp. 351–356)

- Cells control metabolism by regulating enzyme activity or the expression of genes coding for enzymes. In bacteria, genes are often clustered into **operons,** with one promoter serving several adjacent genes. An **operator** site on the DNA switches the operon on or off, resulting in coordinate regulation of the genes.

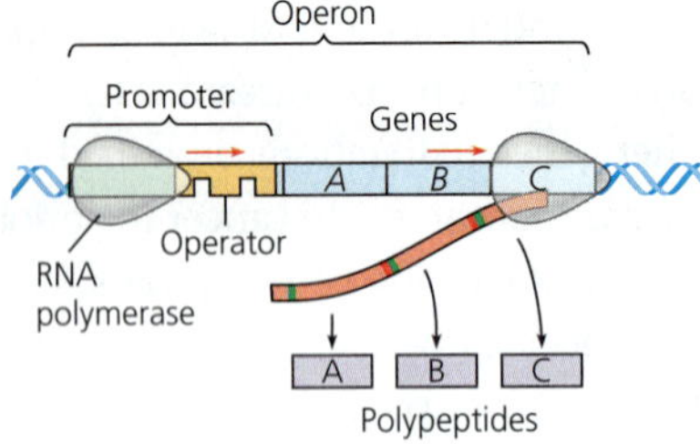

- Both repressible and inducible operons are examples of negative gene regulation. In either type of operon, binding of a specific **repressor** protein to the operator shuts off transcription. (The repressor is encoded by a separate **regulatory gene.**) In a repressible operon, the repressor is active when bound to a **corepressor**, usually the end product of an anabolic pathway.

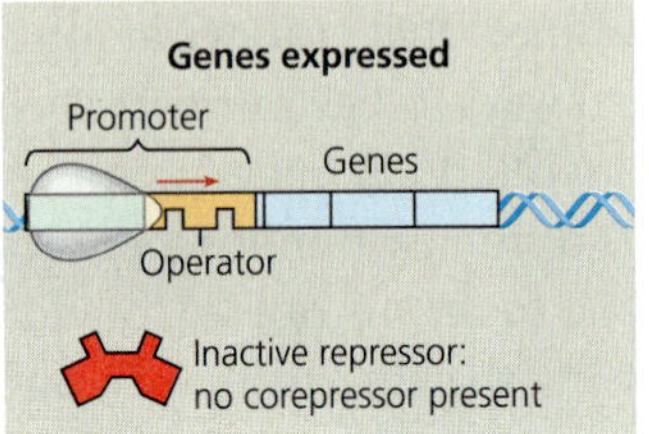

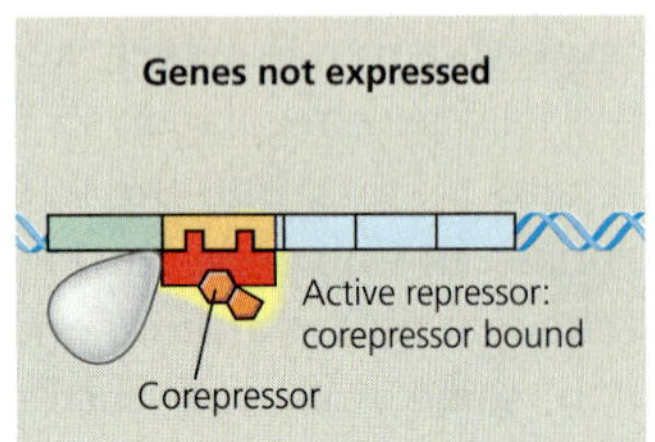

In an inducible operon, binding of an **inducer** to an innately active repressor inactivates the repressor and turns on transcription. Inducible enzymes usually function in catabolic pathways.

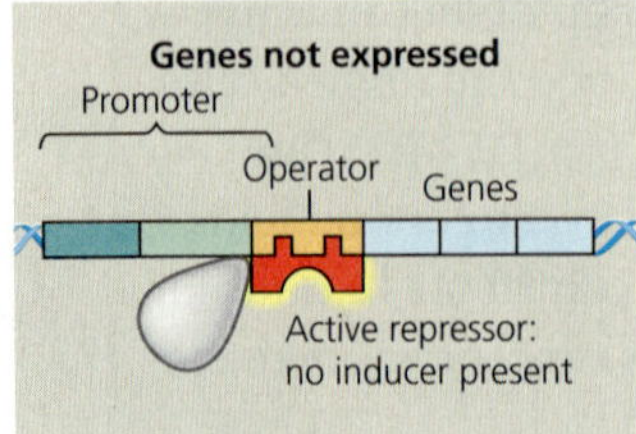

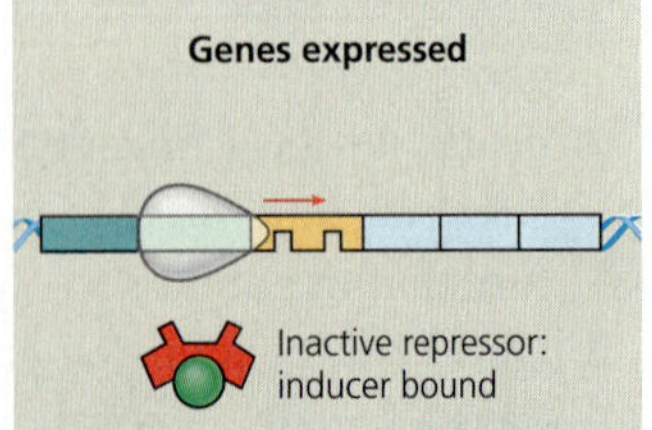

- Some operons are also subject to positive gene regulation via a stimulatory **activator** protein, such as catabolite activator protein (CAP), which, when activated by **cyclic AMP**, binds to a site within the promoter and stimulates transcription.

? *Compare and contrast the roles of the corepressor and the inducer in negative regulation of an operon.*

CONCEPT 18.2

Eukaryotic gene expression is regulated at many stages (pp. 356–364)

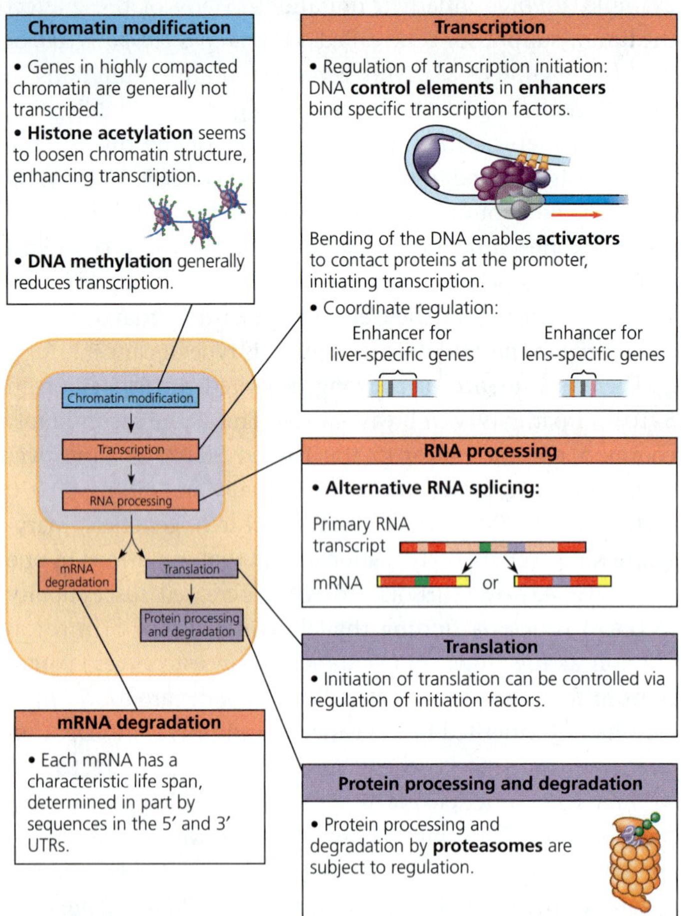

? *Describe what must happen for a cell-type-specific gene to be transcribed in a cell of that type.*

CONCEPT 18.3

Noncoding RNAs play multiple roles in controlling gene expression (pp. 364–366)

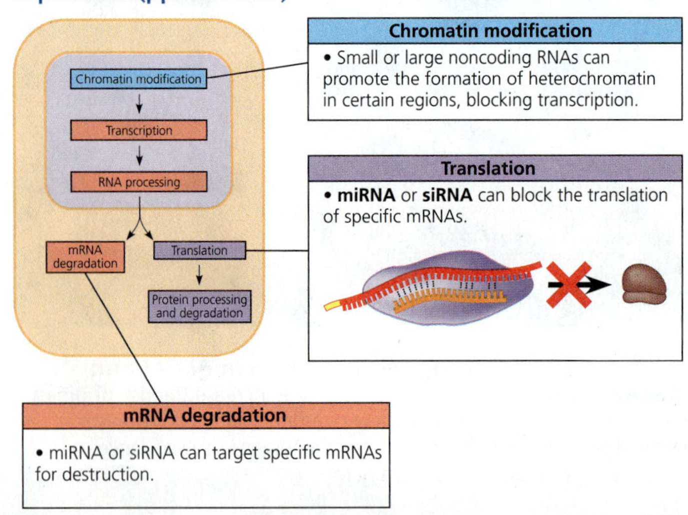

? *Why are miRNAs called noncoding RNAs? Explain how they participate in gene regulation.*

CONCEPT 18.4

A program of differential gene expression leads to the different cell types in a multicellular organism (pp. 366–373)

- Embryonic cells undergo **differentiation**, becoming specialized in structure and function. **Morphogenesis** encompasses the processes that give shape to the organism and its various parts. Cells differ in structure and function not because they contain different genes but because they express different portions of a common genome.
- **Cytoplasmic determinants** in the unfertilized egg regulate the expression of genes in the zygote that affect the developmental fate of embryonic cells. In the process called **induction**, signaling molecules from embryonic cells cause transcriptional changes in nearby target cells.
- Differentiation is heralded by the appearance of tissue-specific proteins, which enable differentiated cells to carry out their specialized roles.
- In animals, **pattern formation**, the development of a spatial organization of tissues and organs, begins in the early embryo. **Positional information**, the molecular cues that control pattern formation, tells a cell its location relative to the body's axes and to other cells. In *Drosophila*, gradients of **morphogens** encoded by **maternal effect genes** determine the body axes. For example, the gradient of **Bicoid** protein determines the anterior-posterior axis.

? *Describe the two main processes that cause embryonic cells to head down different pathways to their final fates.*

CONCEPT 18.5

Cancer results from genetic changes that affect cell cycle control (pp. 373–377)

- The products of **proto-oncogenes** and **tumor-suppressor genes** control cell division. A DNA change that makes a proto-oncogene excessively active converts it to an **oncogene**, which may promote excessive cell division and cancer. A tumor-suppressor gene encodes a protein that inhibits abnormal cell division. A mutation in such a gene that reduces the activity of its protein product may also lead to excessive cell division and possibly to cancer.
- Many proto-oncogenes and tumor-suppressor genes encode components of growth-stimulating and growth-inhibiting signaling pathways, respectively, and mutations in these genes can interfere with normal cell-signaling pathways. A hyperactive version of a protein in a stimulatory pathway, such as **Ras** (a G protein), functions as an oncogene protein. A defective version of a protein in an inhibitory pathway, such as **p53** (a transcription activator), fails to function as a tumor suppressor.
- In the multistep model of cancer development, normal cells are converted to cancer cells by the accumulation of mutations affecting proto-oncogenes and tumor-suppressor genes. Technical advances in DNA and mRNA sequencing are enabling cancer treatments that are more individually based.
- Individuals who inherit a mutant oncogene or tumor-suppressor allele have a predisposition to develop a particular cancer. Certain viruses promote cancer by integration of viral DNA into a cell's genome.

? *Compare the usual functions of proteins encoded by proto-oncogenes with the functions of proteins encoded by tumor-suppressor genes.*

TEST YOUR UNDERSTANDING

LEVEL 1: KNOWLEDGE/COMPREHENSION

1. If a particular operon encodes enzymes for making an essential amino acid and is regulated like the *trp* operon, then
   a. the amino acid inactivates the repressor.
   b. the enzymes produced are called inducible enzymes.
   c. the repressor is active in the absence of the amino acid.
   d. the amino acid acts as a corepressor.
   e. the amino acid turns on transcription of the operon.

2. Muscle cells differ from nerve cells mainly because they
   a. express different genes.
   b. contain different genes.
   c. use different genetic codes.
   d. have unique ribosomes.
   e. have different chromosomes.

3. The functioning of enhancers is an example of
   a. transcriptional control of gene expression.
   b. a post-transcriptional mechanism to regulate mRNA.
   c. the stimulation of translation by initiation factors.
   d. post-translational control that activates certain proteins.
   e. a eukaryotic equivalent of prokaryotic promoter functioning.

4. Cell differentiation always involves
   a. the production of tissue-specific proteins, such as muscle actin.
   b. the movement of cells.
   c. the transcription of the *myoD* gene.
   d. the selective loss of certain genes from the genome.
   e. the cell's sensitivity to environmental cues, such as light or heat.

5. Which of the following is an example of post-transcriptional control of gene expression?
   a. the addition of methyl groups to cytosine bases of DNA
   b. the binding of transcription factors to a promoter
   c. the removal of introns and alternative splicing of exons
   d. gene amplification contributing to cancer
   e. the folding of DNA to form heterochromatin

LEVEL 2: APPLICATION/ANALYSIS

6. What would occur if the repressor of an inducible operon were mutated so it could not bind the operator?
   a. irreversible binding of the repressor to the promoter
   b. reduced transcription of the operon's genes
   c. buildup of a substrate for the pathway controlled by the operon
   d. continuous transcription of the operon's genes
   e. overproduction of catabolite activator protein (CAP)

7. Absence of *bicoid* mRNA from a *Drosophila* egg leads to the absence of anterior larval body parts and mirror-image duplication of posterior parts. This is evidence that the product of the *bicoid* gene
   a. is transcribed in the early embryo.
   b. normally leads to formation of tail structures.
   c. normally leads to formation of head structures.
   d. is a protein present in all head structures.
   e. leads to programmed cell death.

8. Which of the following statements about the DNA in one of your brain cells is true?
   a. Most of the DNA codes for protein.
   b. The majority of genes are likely to be transcribed.
   c. Each gene lies immediately adjacent to an enhancer.
   d. Many genes are grouped into operon-like clusters.
   e. It is the same as the DNA in one of your heart cells.

9. Within a cell, the amount of protein made using a given mRNA molecule depends partly on
   a. the degree of DNA methylation.
   b. the rate at which the mRNA is degraded.
   c. the presence of certain transcription factors.
   d. the number of introns present in the mRNA.
   e. the types of ribosomes present in the cytoplasm.

10. Proto-oncogenes can change into oncogenes that cause cancer. Which of the following best explains the presence of these potential time bombs in eukaryotic cells?
   a. Proto-oncogenes first arose from viral infections.
   b. Proto-oncogenes normally help regulate cell division.
   c. Proto-oncogenes are genetic "junk."
   d. Proto-oncogenes are mutant versions of normal genes.
   e. Cells produce proto-oncogenes as they age.

LEVEL 3: SYNTHESIS/EVALUATION

11. **DRAW IT** The diagram below shows five genes, including their enhancers, from the genome of a certain species. Imagine that orange, blue, green, black, red, and purple activator proteins exist that can bind to the appropriately color-coded control elements in the enhancers of these genes.

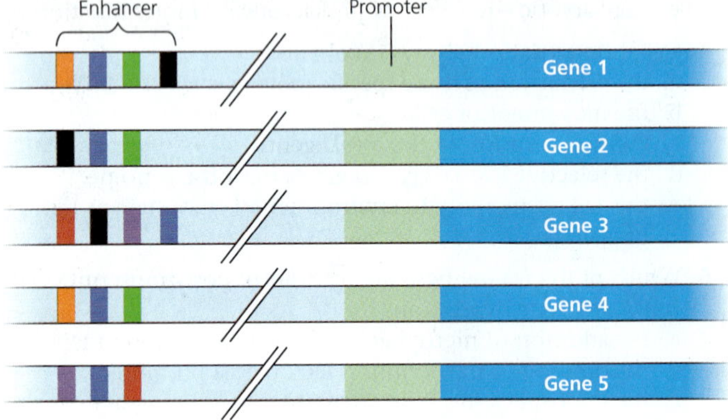

   a. Draw an X above enhancer elements (of all the genes) that would have activators bound in a cell in which only gene 5 is transcribed. Which colored activators would be present?
   b. Draw a dot above all enhancer elements that would have activators bound in a cell in which the green, blue, and orange activators are present. Which gene(s) would be transcribed?
   c. Imagine that genes 1, 2, and 4 code for nerve-specific proteins, and genes 3 and 5 are skin specific. Which activators would have to be present in each cell type to ensure transcription of the appropriate genes?

12. **EVOLUTION CONNECTION**
   DNA sequences can act as "tape measures of evolution" (see Chapter 5). Scientists analyzing the human genome sequence were surprised to find that some of the regions of the human genome that are most highly conserved (similar to comparable regions in other species) don't code for proteins. Propose a possible explanation for this observation.

13. **SCIENTIFIC INQUIRY**
   Prostate cells usually require testosterone and other androgens to survive. But some prostate cancer cells thrive despite treatments that eliminate androgens. One hypothesis is that estrogen, often considered a female hormone, may be activating genes normally controlled by an androgen in these cancer cells. Describe one or more experiments to test this hypothesis. (See Figure 11.9, p. 214, to review the action of these steroid hormones.)

14. **SCIENCE, TECHNOLOGY, AND SOCIETY**
   Trace amounts of dioxin were present in Agent Orange, a defoliant sprayed on vegetation during the Vietnam War. Animal tests suggest that dioxin can cause birth defects, cancer, liver and thymus damage, and immune system suppression, sometimes leading to death. But the animal tests are equivocal; a hamster is not affected by a dose that can kill a guinea pig. Dioxin acts somewhat like a steroid hormone, entering a cell and binding to a receptor protein that then attaches to the cell's DNA. How might this mechanism help explain the variety of dioxin's effects on different body systems and in different animals? How might you determine whether a type of illness is related to dioxin exposure? How might you determine whether a particular individual became ill as a result of exposure to dioxin? Which would be more difficult to demonstrate? Why?

15. **WRITE ABOUT A THEME**
   Feedback Regulation In a short essay (100–150 words), discuss how the processes shown in Figure 18.24a and b are examples of feedback mechanisms regulating biological systems.

For selected answers, see Appendix A.

19

Viruses

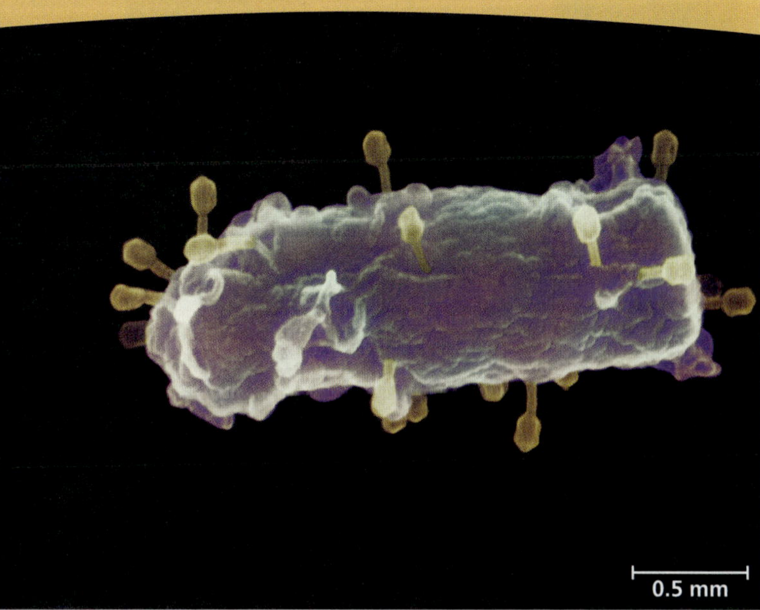

▲ **Figure 19.1 Are the tiny viruses infecting this *E. coli* cell alive?**

KEY CONCEPTS

19.1 A virus consists of a nucleic acid surrounded by a protein coat
19.2 Viruses replicate only in host cells
19.3 Viruses, viroids, and prions are formidable pathogens in animals and plants

OVERVIEW

A Borrowed Life

The photo in **Figure 19.1** shows a remarkable event: the attack of a bacterial cell by numerous structures that resemble miniature lollipops. These structures, a type of virus called T4 bacteriophage, are seen infecting the bacterium *Escherichia coli* in this colorized SEM. By injecting its DNA into the cell, the virus sets in motion a genetic takeover of the bacterium, recruiting cellular machinery to mass-produce many new viruses.

Recall that bacteria and other prokaryotes are cells much smaller and more simply organized than the cells of eukaryotes, such as plants and animals. Viruses are smaller and simpler still. Lacking the structures and metabolic machinery found in a cell, a **virus** is an infectious particle consisting of little more than genes packaged in a protein coat.

Are viruses living or nonliving? Early on, they were considered biological chemicals; in fact, the Latin root for the word *virus* means "poison." Because viruses are capable of causing a wide variety of diseases and can be spread between organisms, researchers in the late 1800s saw a parallel with bacteria and proposed that viruses were the simplest of living forms. However, viruses cannot reproduce or carry out metabolic activities outside of a host cell. Most biologists studying viruses today would probably agree that they are not alive but exist in a shady area between life-forms and chemicals. The simple phrase used recently by two researchers describes them aptly enough: Viruses lead "a kind of borrowed life."

To a large extent, molecular biology was born in the laboratories of biologists studying viruses that infect bacteria. Experiments with viruses provided important evidence that genes are made of nucleic acids, and they were critical in working out the molecular mechanisms of the fundamental processes of DNA replication, transcription, and translation.

Beyond their value as experimental systems, viruses have unique genetic mechanisms that are interesting in their own right and that also help us understand how viruses cause disease. In addition, the study of viruses has led to the development of techniques that enable scientists to manipulate genes and transfer them from one organism to another. These techniques play an important role in basic research, biotechnology, and medical applications. For instance, viruses are used as agents of gene transfer in gene therapy (see Chapter 20).

In this chapter, we will explore the biology of viruses. We will begin with the structure of these simplest of all genetic systems and then describe the cycles by which they replicate. Next, we will discuss the role of viruses as disease-causing agents, or pathogens, and conclude by considering some even simpler infectious agents, viroids and prions.

CONCEPT 19.1

A virus consists of a nucleic acid surrounded by a protein coat

Scientists were able to detect viruses indirectly long before they were actually able to see them. The story of how viruses were discovered begins near the end of the 19th century.

The Discovery of Viruses: *Scientific Inquiry*

Tobacco mosaic disease stunts the growth of tobacco plants and gives their leaves a mottled, or mosaic, coloration. In 1883, Adolf Mayer, a German scientist, discovered that he could transmit the disease from plant to plant by rubbing sap extracted from diseased leaves onto healthy plants. After an unsuccessful search for an infectious microbe in the sap, Mayer suggested that the disease was caused by unusually small bacteria that were invisible under a microscope. This hypothesis was tested a decade later by Dimitri Ivanowsky, a Russian biologist who passed sap from infected tobacco leaves through a filter designed to remove bacteria. After filtration, the sap still produced mosaic disease.

But Ivanowsky clung to the hypothesis that bacteria caused tobacco mosaic disease. Perhaps, he reasoned, the bacteria were small enough to pass through the filter or made a toxin that could do so. The second possibility was ruled out when the Dutch botanist Martinus Beijerinck carried out a classic series of experiments that showed that the infectious agent in the filtered sap could replicate **(Figure 19.2)**.

In fact, the pathogen replicated only within the host it infected. In further experiments, Beijerinck showed that unlike bacteria used in the lab at that time, the mysterious agent of mosaic disease could not be cultivated on nutrient media in test tubes or petri dishes. Beijerinck imagined a replicating particle much smaller and simpler than a bacterium, and he is generally credited with being the first scientist to voice the concept of a virus. His suspicions were confirmed in 1935 when the American scientist Wendell Stanley crystallized the infectious particle, now known as tobacco mosaic virus (TMV). Subsequently, TMV and many other viruses were actually seen with the help of the electron microscope.

▼ **Figure 19.2** INQUIRY

What causes tobacco mosaic disease?

EXPERIMENT In the late 1800s, Martinus Beijerinck, of the Technical School in Delft, the Netherlands, investigated the properties of the agent that causes tobacco mosaic disease (then called spot disease).

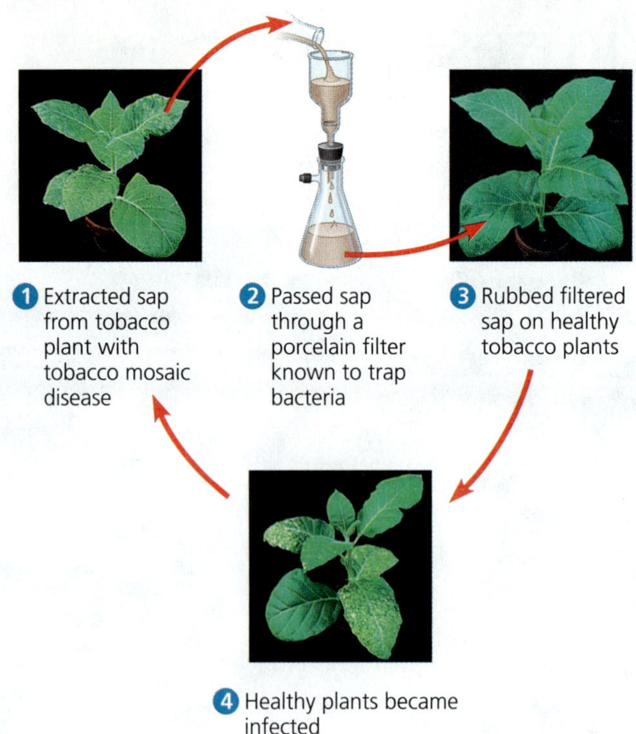

RESULTS When the filtered sap was rubbed on healthy plants, they became infected. Their sap, when extracted and filtered, could then act as the source of infection for another group of plants. Each successive group of plants developed the disease to the same extent as earlier groups.

CONCLUSION The infectious agent was apparently not a bacterium because it could pass through a bacterium-trapping filter. The pathogen must have been replicating in the plants because its ability to cause disease was undiluted after several transfers from plant to plant.

SOURCE M. J. Beijerinck, Concerning a *contagium vivum fluidum* as cause of the spot disease of tobacco leaves, *Verhandelingen der Koninkyke akademie Wettenschappen te Amsterdam* 65:3–21 (1898). Translation published in English as Phytopathological Classics Number 7 (1942), American Phytopathological Society Press, St. Paul, MN.

WHAT IF? If Beijerinck had observed that the infection of each group was weaker than that of the previous group and that ultimately the sap could no longer cause disease, what might he have concluded?

Structure of Viruses

The tiniest viruses are only 20 nm in diameter—smaller than a ribosome. Millions could easily fit on a pinhead. Even the largest known virus, which has a diameter of several hundred nanometers, is barely visible under the light microscope. Stanley's discovery that some viruses could be crystallized was exciting and puzzling news. Not even the simplest of cells can aggregate into regular crystals. But if viruses are not cells, then what are they? Examining the structure of a virus more closely reveals that it is an infectious particle consisting of nucleic acid enclosed in a protein coat and, for some viruses, surrounded by a membranous envelope.

Viral Genomes

We usually think of genes as being made of double-stranded DNA—the conventional double helix—but many viruses defy this convention. Their genomes may consist of double-stranded DNA, single-stranded DNA, double-stranded RNA, or single-stranded RNA, depending on the type of virus. A virus is called a DNA virus or an RNA virus, based on the kind of nucleic acid that makes up its genome. In either case, the genome is usually organized as a single linear or circular molecule of nucleic acid, although the genomes of some viruses consist of multiple molecules of nucleic acid. The smallest viruses known have only four genes in their genome, while the largest have several hundred to a thousand. For comparison, bacterial genomes contain about 200 to a few thousand genes.

Capsids and Envelopes

The protein shell enclosing the viral genome is called a **capsid**. Depending on the type of virus, the capsid may be rod-shaped, polyhedral, or more complex in shape (like T4). Capsids are built from a large number of protein subunits called *capsomeres*, but the number of different *kinds* of proteins in a capsid is usually small. Tobacco mosaic virus has a rigid, rod-shaped capsid made from over a thousand molecules of a single type of protein arranged in a helix; rod-shaped viruses are commonly called *helical viruses* for this reason **(Figure 19.3a)**. Adenoviruses, which infect the respiratory tracts of animals, have 252 identical protein molecules arranged in a polyhedral capsid with 20 triangular facets—an icosahedron; thus, these and other similarly shaped viruses are referred to as *icosahedral viruses* **(Figure 19.3b)**.

Some viruses have accessory structures that help them infect their hosts. For instance, a membranous envelope surrounds the capsids of influenza viruses and many other viruses found in animals **(Figure 19.3c)**. These **viral envelopes**, which are derived from the membranes of the host cell, contain host cell phospholipids and membrane proteins. They also contain proteins and glycoproteins of viral origin. (Glycoproteins are proteins with carbohydrates covalently attached.) Some viruses carry a few viral enzyme molecules within their capsids.

Many of the most complex capsids are found among the viruses that infect bacteria, called **bacteriophages**, or simply **phages**. The first phages studied included seven that infect *E. coli*. These seven phages were named type 1 (T1), type 2 (T2), and so forth, in the order of their discovery. The three T-even phages (T2, T4, and T6) turned out to be very similar in structure. Their capsids have elongated icosahedral heads

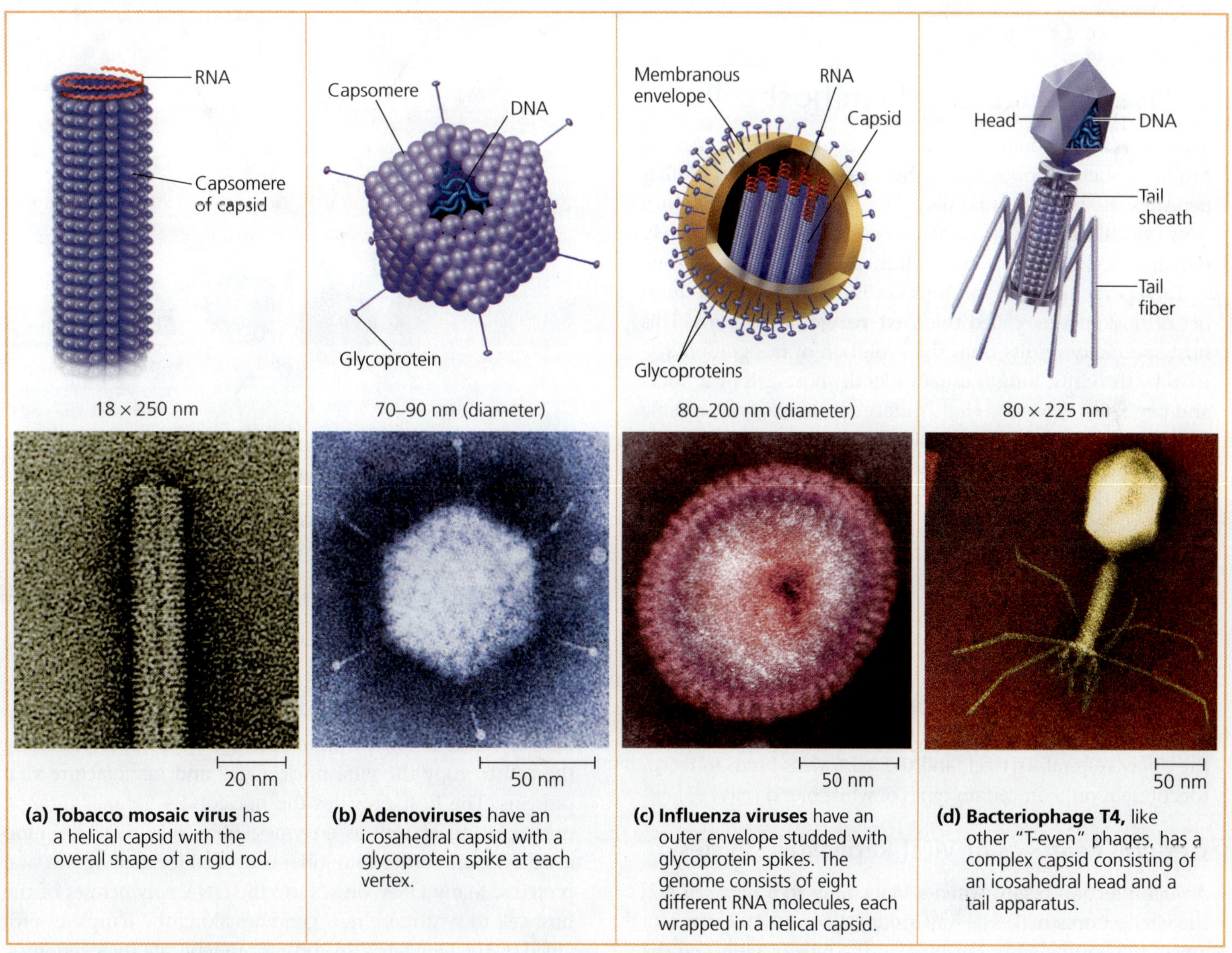

(a) Tobacco mosaic virus has a helical capsid with the overall shape of a rigid rod.

(b) Adenoviruses have an icosahedral capsid with a glycoprotein spike at each vertex.

(c) Influenza viruses have an outer envelope studded with glycoprotein spikes. The genome consists of eight different RNA molecules, each wrapped in a helical capsid.

(d) Bacteriophage T4, like other "T-even" phages, has a complex capsid consisting of an icosahedral head and a tail apparatus.

▲ **Figure 19.3 Viral structure.** Viruses are made up of nucleic acid (DNA or RNA) enclosed in a protein coat (the capsid) and sometimes further wrapped in a membranous envelope. The individual protein subunits making up the capsid are called capsomeres. Although diverse in size and shape, viruses have many common structural features. (All micrographs are colorized TEMs.)

enclosing their DNA. Attached to the head is a protein tail piece with fibers by which the phages attach to a bacterium **(Figure 19.3d)**. In the next section, we'll examine how these few viral parts function together with cellular components to produce large numbers of viral progeny.

CONCEPT CHECK 19.1

1. Compare the structures of tobacco mosaic virus (TMV) and influenza virus (see Figure 19.3).
2. **MAKE CONNECTIONS** In Figure 16.4 (p. 307), you learned how bacteriophages were used to provide evidence that DNA carries genetic information. Briefly describe the experiment carried out by Hershey and Chase, including in your description why the researchers chose to use phages.

For suggested answers, see Appendix A.

CONCEPT **19.2**

Viruses replicate only in host cells

Viruses lack metabolic enzymes and equipment for making proteins, such as ribosomes. They are obligate intracellular parasites; in other words, they can replicate only within a host cell. It is fair to say that viruses in isolation are merely packaged sets of genes in transit from one host cell to another.

Each particular virus can infect cells of only a limited number of host species, called the **host range** of the virus. This host specificity results from the evolution of recognition systems by the virus. Viruses usually identify host cells by a "lock-and-key" fit between viral surface proteins and specific receptor molecules on the outside of cells. (According to one model, such receptor molecules originally carried out functions that benefited the host cell but were co-opted later by viruses as portals of entry.) Some viruses have broad host ranges. For example, West Nile virus and equine encephalitis virus are distinctly different viruses that can each infect mosquitoes, birds, horses, and humans. Other viruses have host ranges so narrow that they infect only a single species. Measles virus, for instance, can infect only humans. Furthermore, viral infection of multicellular eukaryotes is usually limited to particular tissues. Human cold viruses infect only the cells lining the upper respiratory tract, and the AIDS virus binds to receptors present only on certain types of white blood cells.

General Features of Viral Replicative Cycles

A viral infection begins when a virus binds to a host cell and the viral genome makes its way inside **(Figure 19.4)**. The mechanism of genome entry depends on the type of virus and the type of host cell. For example, T-even phages use their elaborate tail apparatus to inject DNA into a bacterium (see Figure 19.3d). Other viruses are taken up by endocytosis or, in the case of

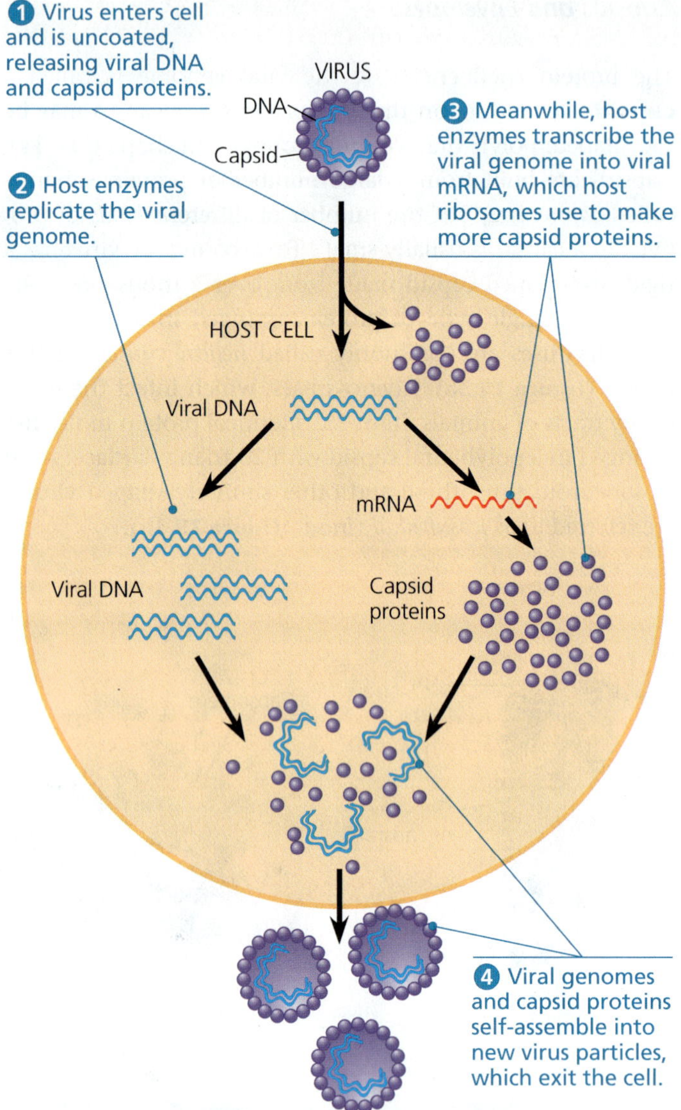

▲ **Figure 19.4 A simplified viral replicative cycle.** A virus is an obligate intracellular parasite that uses the equipment and small molecules of its host cell to replicate. In this simplest of viral cycles, the parasite is a DNA virus with a capsid consisting of a single type of protein.

MAKE CONNECTIONS *Label each of the straight black arrows with one word representing the name of the process that is occurring. Review Figure 17.26 on page 348.*

enveloped viruses, by fusion of the viral envelope with the plasma membrane. Once the viral genome is inside, the proteins it encodes can commandeer the host, reprogramming the cell to copy the viral nucleic acid and manufacture viral proteins. The host provides the nucleotides for making viral nucleic acids, as well as enzymes, ribosomes, tRNAs, amino acids, ATP, and other components needed for making the viral proteins. Many DNA viruses use the DNA polymerases of the host cell to synthesize new genomes along the templates provided by the viral DNA. In contrast, to replicate their genomes, RNA viruses use virally encoded RNA polymerases that can use RNA as a template. (Uninfected cells generally make no enzymes for carrying out this process.)

After the viral nucleic acid molecules and capsomeres are produced, they spontaneously self-assemble into new viruses. In fact, researchers can separate the RNA and capsomeres of TMV and then reassemble complete viruses simply by mixing the components together under the right conditions. The simplest type of viral replicative cycle ends with the exit of hundreds or thousands of viruses from the infected host cell, a process that often damages or destroys the cell. Such cellular damage and death, as well as the body's responses to this destruction, cause many of the symptoms associated with viral infections. The viral progeny that exit a cell have the potential to infect additional cells, spreading the viral infection.

There are many variations on the simplified viral replicative cycle we have just described. We will now take a look at some of these variations in bacterial viruses (phages) and animal viruses; later in the chapter, we will consider plant viruses.

Replicative Cycles of Phages

Phages are the best understood of all viruses, although some of them are also among the most complex. Research on phages led to the discovery that some double-stranded DNA viruses can replicate by two alternative mechanisms: the lytic cycle and the lysogenic cycle.

The Lytic Cycle

A phage replicative cycle that culminates in death of the host cell is known as a **lytic cycle**. The term refers to the last stage of infection, during which the bacterium lyses (breaks open) and releases the phages that were produced within the cell. Each of these phages can then infect a healthy cell, and a few successive lytic cycles can destroy an entire bacterial population in just a few hours. A phage that replicates only by a lytic cycle is a **virulent phage**. **Figure 19.5** illustrates the major steps in the lytic cycle of T4, a typical virulent phage. Study this figure before proceeding.

After reading about the lytic cycle, you may wonder why phages haven't exterminated all bacteria. In fact, phage treatments have been used medically in some countries to help control bacterial infections in humans. Bacteria are not defenseless, however. First, natural selection favors bacterial mutants with receptors that are no longer recognized by a particular type of phage. Second, when phage DNA successfully enters a bacterium, the DNA often is identified as foreign and cut up by cellular enzymes called **restriction enzymes**, which are so named because their activity *restricts* the ability of the phage to infect the bacterium. The bacterial cell's own DNA is methylated in a way that prevents attack

▶ **Figure 19.5 The lytic cycle of phage T4, a virulent phage.** Phage T4 has almost 300 genes, which are transcribed and translated using the host cell's machinery. One of the first phage genes translated after the viral DNA enters the host cell codes for an enzyme that degrades the host cell's DNA (step 2); the phage DNA is protected from breakdown because it contains a modified form of cytosine that is not recognized by the enzyme. The entire lytic cycle, from the phage's first contact with the cell surface to cell lysis, takes only 20–30 minutes at 37°C.

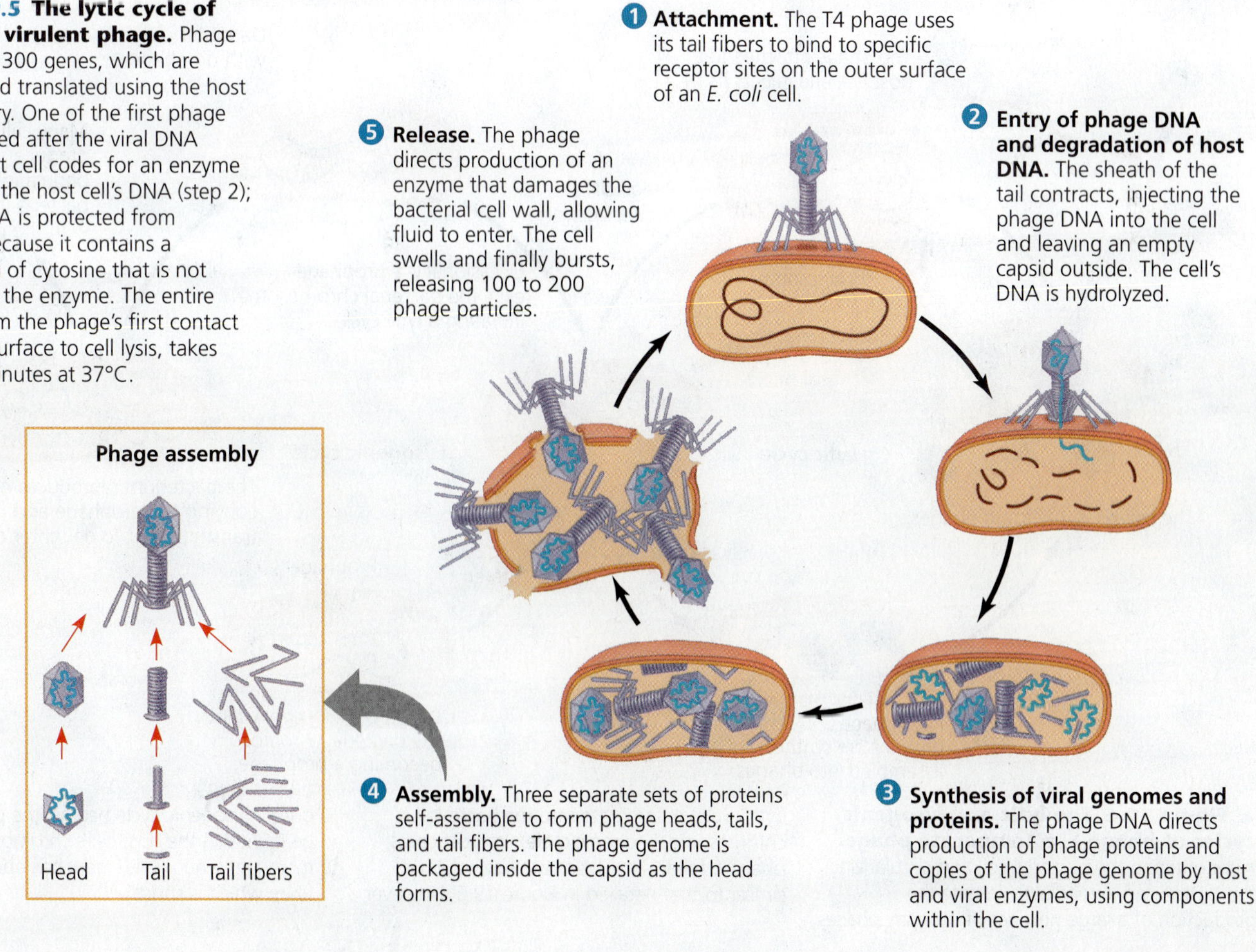

by its own restriction enzymes. But just as natural selection favors bacteria with mutant receptors or effective restriction enzymes, it also favors phage mutants that can bind the altered receptors or are resistant to particular restriction enzymes. Thus, the parasite-host relationship is in constant evolutionary flux.

There is yet a third important reason bacteria have been spared from extinction as a result of phage activity. Instead of lysing their host cells, many phages coexist with them in a state called lysogeny, which we'll now discuss.

The Lysogenic Cycle

In contrast to the lytic cycle, which kills the host cell, the **lysogenic cycle** allows replication of the phage genome without destroying the host. Phages capable of using both modes of replicating within a bacterium are called **temperate phages**. A temperate phage called lambda, written with the Greek letter λ, is widely used in biological research. Phage λ resembles T4, but its tail has only one short tail fiber.

Infection of an *E. coli* cell by phage λ begins when the phage binds to the surface of the cell and injects its linear DNA genome **(Figure 19.6)**. Within the host, the λ DNA molecule forms a circle. What happens next depends on the replicative mode: lytic cycle or lysogenic cycle. During a lytic cycle, the viral genes immediately turn the host cell into a λ-producing factory, and the cell soon lyses and releases its viral products. During a lysogenic cycle, however, the λ DNA molecule is incorporated into a specific site on the *E. coli* chromosome by viral proteins that break both circular DNA molecules and join them to each other. When integrated into the bacterial chromosome in this way, the viral DNA is known as a **prophage**. One prophage gene codes for a protein that prevents transcription of most of the other prophage genes. Thus, the phage genome is mostly silent within the bacterium. Every time the *E. coli* cell prepares to divide, it replicates the phage DNA along with its own and passes the copies on to daughter cells. A single infected cell can quickly give rise to a large population of bacteria carrying the virus in prophage form. This mechanism enables viruses to propagate without killing the host cells on which they depend.

The term *lysogenic* implies that prophages are capable of generating active phages that lyse their host cells. This occurs when the λ genome is induced to exit the bacterial chromosome and initiate a lytic cycle. An environmental signal, such as a certain chemical or high-energy radiation, usually triggers the switchover from the lysogenic to the lytic mode.

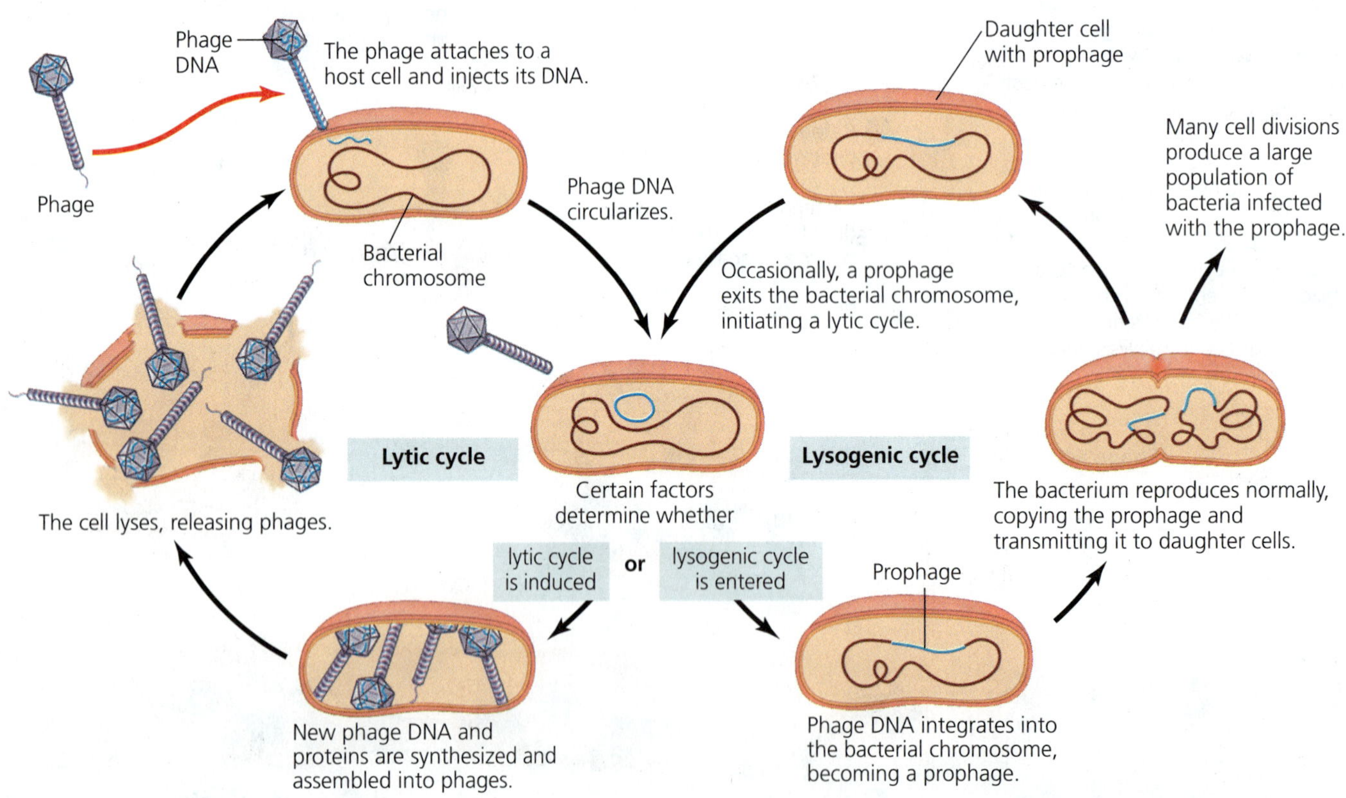

▲ **Figure 19.6 The lytic and lysogenic cycles of phage λ, a temperate phage.** After entering the bacterial cell and circularizing, the λ DNA can immediately initiate the production of a large number of progeny phages (lytic cycle) or integrate into the bacterial chromosome (lysogenic cycle). In most cases, phage λ follows the lytic pathway, which is similar to that detailed in Figure 19.5. However, once a lysogenic cycle begins, the prophage may be carried in the host cell's chromosome for many generations. Phage λ has one main tail fiber, which is short.

Table 19.1 Classes of Animal Viruses		
Class/Family	**Envelope**	**Examples That Cause Human Diseases**
I. Double-Stranded DNA (dsDNA)		
Adenovirus (see Figure 19.3b)	No	Respiratory viruses; tumor-causing viruses
Papovavirus	No	Papillomavirus (warts, cervical cancer); polyomavirus (tumors)
Herpesvirus	Yes	Herpes simplex I and II (cold sores, genital sores); varicella zoster (shingles, chicken pox); Epstein-Barr virus (mononucleosis, Burkitt's lymphoma)
Poxvirus	Yes	Smallpox virus; cowpox virus
II. Single-Stranded DNA (ssDNA)		
Parvovirus	No	B19 parvovirus (mild rash)
III. Double-Stranded RNA (dsRNA)		
Reovirus	No	Rotavirus (diarrhea); Colorado tick fever virus
IV. Single-Stranded RNA (ssRNA); Serves as mRNA		
Picornavirus	No	Rhinovirus (common cold); poliovirus; hepatitis A virus; other enteric (intestinal) viruses
Coronavirus	Yes	Severe acute respiratory syndrome (SARS)
Flavivirus	Yes	Yellow fever virus; West Nile virus; hepatitis C virus
Togavirus	Yes	Rubella virus; equine encephalitis viruses
V. ssRNA; Template for mRNA Synthesis		
Filovirus	Yes	Ebola virus (hemorrhagic fever)
Orthomyxovirus (see Figures 19.3c and 19.9a)	Yes	Influenza virus
Paramyxovirus	Yes	Measles virus; mumps virus
Rhabdovirus	Yes	Rabies virus
VI. ssRNA; Template for DNA Synthesis		
Retrovirus (see Figure 19.8)	Yes	Human immunodeficiency virus (HIV/AIDS); RNA tumor viruses (leukemia)

In addition to the gene for the transcription-preventing protein, a few other prophage genes may be expressed during lysogeny. Expression of these genes may alter the host's phenotype, a phenomenon that can have important medical significance. For example, the three species of bacteria that cause the human diseases diphtheria, botulism, and scarlet fever would not be so harmful to humans without certain prophage genes that cause the host bacteria to make toxins. And the difference between the *E. coli* strain that resides in our intestines and the O157:H7 strain that has caused several deaths by food poisoning appears to be the presence of prophages in the O157:H7 strain.

Replicative Cycles of Animal Viruses

Everyone has suffered from viral infections, whether cold sores, influenza, or the common cold. Like all viruses, those that cause illness in humans and other animals can replicate only inside host cells. Many variations on the basic scheme of viral infection and replication are represented among the animal viruses. One key variable is the nature of the viral genome: Is it composed of DNA or RNA? Is it double-stranded or single-stranded? The nature of the genome is the basis for the common classification of viruses shown in **Table 19.1**. Single-stranded RNA viruses are further classified into three classes (IV–VI) according to how the RNA genome functions in a host cell.

Whereas few bacteriophages have an envelope or RNA genome, many animal viruses have both. In fact, nearly all animal viruses with RNA genomes have an envelope, as do some with DNA genomes (see Table 19.1). Rather than consider all the mechanisms of viral infection and replication, we will focus on the roles of viral envelopes and on the functioning of RNA as the genetic material of many animal viruses.

Viral Envelopes

An animal virus equipped with an envelope—that is, an outer membrane—uses it to enter the host cell. Protruding from the outer surface of this envelope are viral glycoproteins that bind to specific receptor molecules on the surface of a host cell. **Figure 19.7**, on the next page, outlines the events in the replicative cycle of an enveloped virus with an RNA genome. Ribosomes bound to the endoplasmic reticulum (ER) of the host cell make the protein parts of the envelope glycoproteins; cellular enzymes in the ER and Golgi apparatus then add the sugars. The resulting viral glycoproteins, embedded in host cell–derived membrane, are transported to the cell surface. In a process much like exocytosis, new viral capsids are wrapped in membrane as they bud from the cell. In other words, the viral envelope is derived from the host cell's plasma membrane, although some of the molecules of this membrane are specified by viral genes. The enveloped viruses are now free to infect other cells. This replicative cycle does not necessarily kill the host cell, in contrast to the lytic cycles of phages.

Some viruses have envelopes that are not derived from plasma membrane. Herpesviruses, for example, are temporarily cloaked in membrane derived from the nuclear envelope of the host; they then shed this membrane in the cytoplasm and

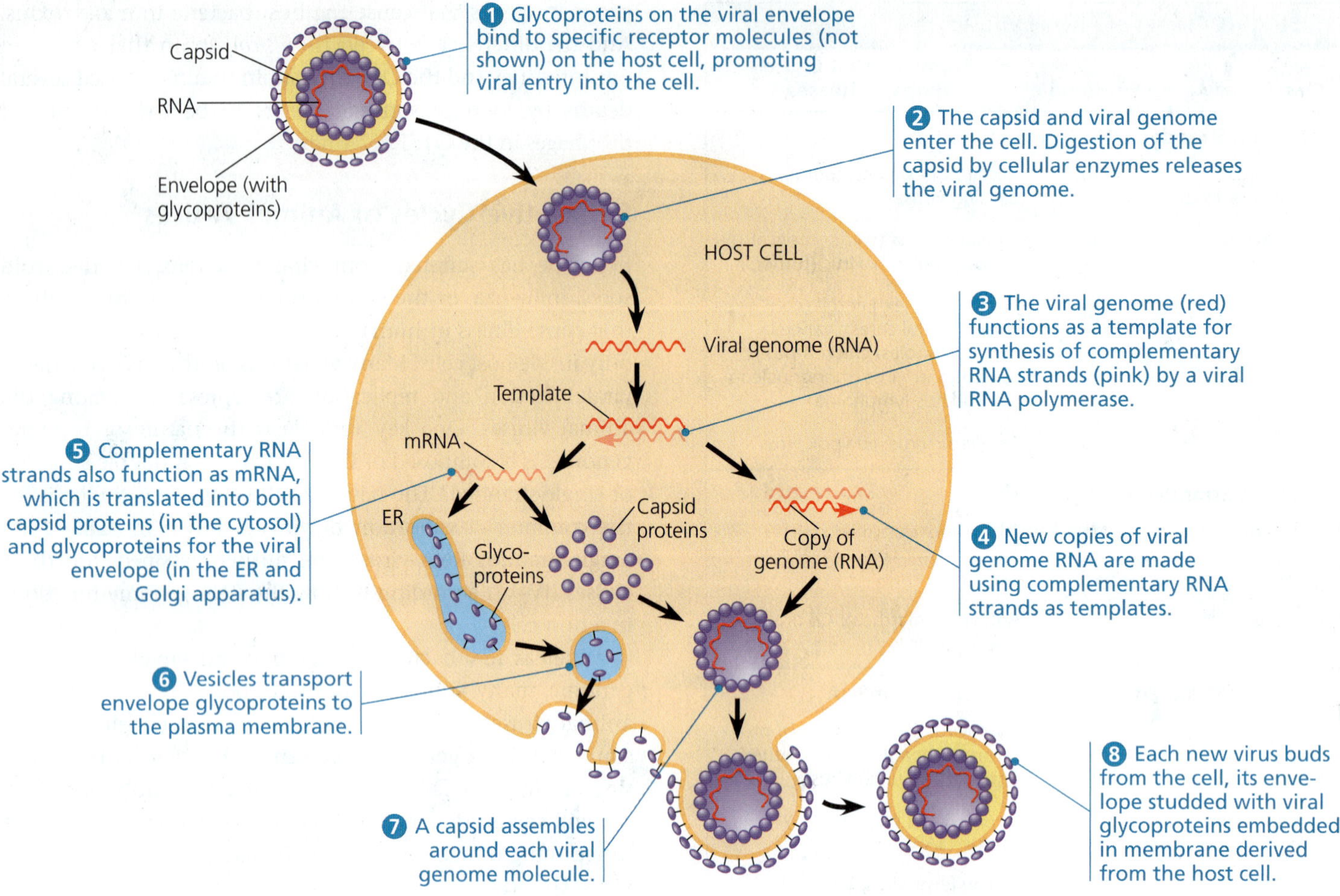

▲ **Figure 19.7 The replicative cycle of an enveloped RNA virus.** Shown here is a virus with a single-stranded RNA genome that functions as a template for synthesis of mRNA. Some enveloped viruses enter the host cell by fusion of the envelope with the cell's plasma membrane; others enter by endocytosis. For all enveloped RNA viruses, the formation of new envelopes for progeny viruses occurs by the mechanism depicted in this figure.

? *Name a virus that has infected you and has a replicative cycle matching this one. (Hint: See Table 19.1.)*

acquire a new envelope made from membrane of the Golgi apparatus. These viruses have a double-stranded DNA genome and replicate within the host cell nucleus, using a combination of viral and cellular enzymes to replicate and transcribe their DNA. In the case of herpesviruses, copies of the viral DNA can remain behind as mini-chromosomes in the nuclei of certain nerve cells. There they remain latent until some sort of physical or emotional stress triggers a new round of active virus production. The infection of other cells by these new viruses causes the blisters characteristic of herpes, such as cold sores or genital sores. Once someone acquires a herpesvirus infection, flare-ups may recur throughout the person's life.

RNA as Viral Genetic Material

Although some phages and most plant viruses are RNA viruses, the broadest variety of RNA genomes is found among the viruses that infect animals. Among the three types of single-stranded RNA genomes found in animal viruses, the genome of class IV viruses can directly serve as mRNA and thus can be translated into viral protein immediately after infection. Figure 19.7 shows a virus of class V, in which the RNA genome serves as a *template* for mRNA synthesis. The RNA genome is transcribed into complementary RNA strands, which function both as mRNA and as templates for the synthesis of additional copies of genomic RNA. All viruses that require RNA → RNA synthesis to make mRNA use a viral enzyme capable of carrying out this process; there are no such enzymes in most cells. The viral enzyme is packaged with the genome inside the viral capsid.

The RNA animal viruses with the most complicated replicative cycles are the **retroviruses** (class VI). These viruses are equipped with an enzyme called **reverse transcriptase**, which transcribes an RNA template into DNA, providing an RNA → DNA information flow, the opposite of the usual direction. This unusual phenomenon is the source of the name retroviruses (*retro* means "backward"). Of particular medical importance is **HIV (human immunodeficiency virus)**, the retrovirus that causes **AIDS (acquired immunodeficiency syndrome)**. HIV and other retroviruses are enveloped viruses that contain two identical molecules of single-stranded RNA and two molecules of reverse transcriptase.

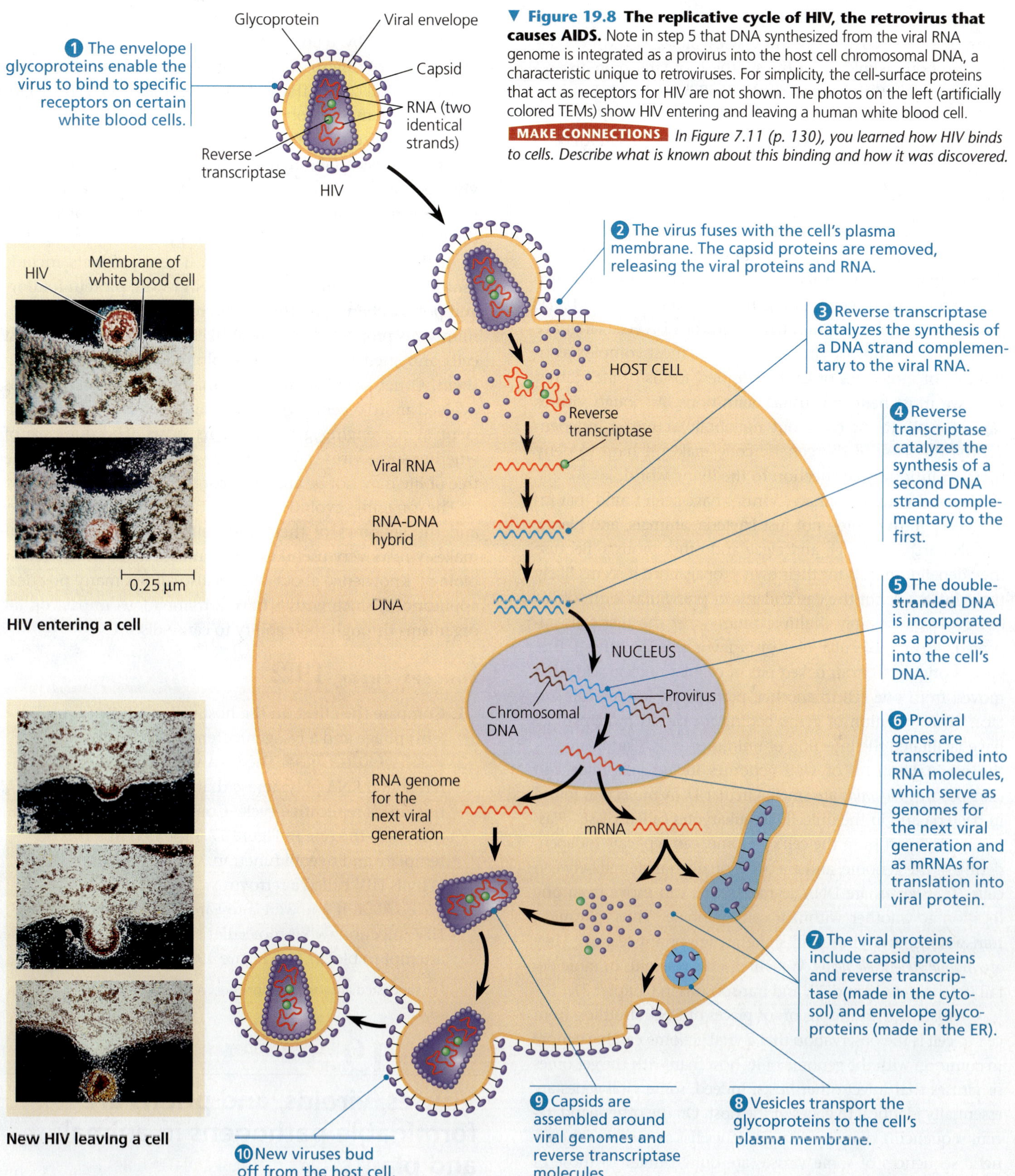

HIV entering a cell

New HIV leaving a cell

▼ **Figure 19.8 The replicative cycle of HIV, the retrovirus that causes AIDS.** Note in step 5 that DNA synthesized from the viral RNA genome is integrated as a provirus into the host cell chromosomal DNA, a characteristic unique to retroviruses. For simplicity, the cell-surface proteins that act as receptors for HIV are not shown. The photos on the left (artificially colored TEMs) show HIV entering and leaving a human white blood cell.

MAKE CONNECTIONS *In Figure 7.11 (p. 130), you learned how HIV binds to cells. Describe what is known about this binding and how it was discovered.*

Figure 19.8 traces the HIV replicative cycle, which is typical of a retrovirus. After HIV enters a host cell, its reverse transcriptase molecules are released into the cytoplasm, where they catalyze synthesis of viral DNA. The newly made viral DNA then enters the cell's nucleus and integrates into the DNA of a chromosome. The integrated viral DNA, called a **provirus**, never leaves the host's genome, remaining a permanent resident of the cell. (Recall that a prophage, in contrast, leaves the host's genome at the start of a lytic cycle.) The host's RNA polymerase transcribes the proviral DNA into RNA molecules, which can function both as mRNA for the synthesis of viral proteins and as genomes for the new viruses that will be assembled and released from the cell. In Chapter 43, we describe how HIV causes the deterioration of the immune system that occurs in AIDS.

Evolution of Viruses

EVOLUTION We began this chapter by asking whether or not viruses are alive. Viruses do not really fit our definition of living organisms. An isolated virus is biologically inert, unable to replicate its genes or regenerate its own supply of ATP. Yet it has a genetic program written in the universal language of life. Do we think of viruses as nature's most complex associations of molecules or as the simplest forms of life? Either way, we must bend our usual definitions. Although viruses cannot replicate or carry out metabolic activities independently, their use of the genetic code makes it hard to deny their evolutionary connection to the living world.

How did viruses originate? Viruses have been found that infect every form of life—not just bacteria, animals, and plants, but also archaea, fungi, and algae and other protists. Because they depend on cells for their own propagation, it seems likely that viruses are not the descendants of precellular forms of life but evolved—possibly multiple times—*after* the first cells appeared. Most molecular biologists favor the hypothesis that viruses originated from naked bits of cellular nucleic acids that moved from one cell to another, perhaps via injured cell surfaces. The evolution of genes coding for capsid proteins may have facilitated the infection of uninjured cells. Candidates for the original sources of viral genomes include plasmids and transposons. *Plasmids* are small, circular DNA molecules found in bacteria and in the unicellular eukaryotes called yeasts. Plasmids exist apart from the cell's genome, can replicate independently of the genome, and are occasionally transferred between cells. *Transposons* are DNA segments that can move from one location to another within a cell's genome. Thus, plasmids, transposons, and viruses all share an important feature: They are *mobile genetic elements*. We will discuss plasmids in more detail in Chapters 20 and 27, and transposons in Chapter 21.

Consistent with this vision of pieces of DNA shuttling from cell to cell is the observation that a viral genome can have more in common with the genome of its host than with the genomes of viruses that infect other hosts. Indeed, some viral genes are essentially identical to genes of the host. On the other hand, recent sequencing of many viral genomes has shown that the genetic sequences of some viruses are quite similar to those of seemingly distantly related viruses; for example, some animal viruses share similar sequences with plant viruses. This genetic similarity may reflect the persistence of groups of viral genes that were favored by natural selection during the early evolution of viruses and the eukaryotic cells that served as their hosts.

The debate about the origin of viruses has been reinvigorated recently by reports of mimivirus, the largest virus yet discovered. Mimivirus is a double-stranded DNA virus with an icosahedral capsid that is 400 nm in diameter. (The beginning of its name is short for *mi*micking *mi*crobe because the virus is the size of a small bacterium.) Its genome contains 1.2 million bases (about 100 times as many as the influenza virus genome) and an estimated 1,000 genes. Perhaps the most surprising aspect of mimivirus, however, is that some of the genes appear to code for products previously thought to be hallmarks of cellular genomes. These products include proteins involved in translation, DNA repair, protein folding, and polysaccharide synthesis. The researchers who described mimivirus propose that it most likely evolved *before* the first cells and then developed an exploitative relationship with them. Other scientists disagree, maintaining that the virus evolved more recently than cells and has simply been efficient at scavenging genes from its hosts. The question of whether some viruses deserve their own early branch on the tree of life may not be answered for some time.

The ongoing evolutionary relationship between viruses and the genomes of their host cells is an association that makes viruses very useful experimental systems in molecular biology. Knowledge about viruses also allows many practical applications, since viruses have a tremendous impact on all organisms through their ability to cause disease.

CONCEPT CHECK 19.2

1. Compare the effect on the host cell of a lytic (virulent) phage and a lysogenic (temperate) phage.
2. **MAKE CONNECTIONS** The RNA virus in Figure 19.7 has a viral RNA polymerase that functions in step 3 of the virus's replicative cycle. Compare this RNA polymerase to the one in Figure 17.9 (p. 333) in terms of template and overall function.
3. Why is HIV called a retrovirus?
4. **WHAT IF?** If you were a researcher trying to combat HIV infection, what molecular processes could you attempt to block? (See Figure 19.8.)

For suggested answers, see Appendix A.

CONCEPT 19.3

Viruses, viroids, and prions are formidable pathogens in animals and plants

Diseases caused by viral infections afflict humans, agricultural crops, and livestock worldwide. Other smaller, less complex entities known as viroids and prions also cause disease in plants and animals, respectively.

Viral Diseases in Animals

A viral infection can produce symptoms by a number of different routes. Viruses may damage or kill cells by causing the release of hydrolytic enzymes from lysosomes. Some viruses cause infected cells to produce toxins that lead to disease symptoms, and some have molecular components that are toxic, such as envelope proteins. How much damage a virus causes depends partly on the ability of the infected tissue to regenerate by cell division. People usually recover completely from colds because the epithelium of the respiratory tract, which the viruses infect, can efficiently repair itself. In contrast, damage inflicted by poliovirus to mature nerve cells is permanent because these cells do not divide and usually cannot be replaced. Many of the temporary symptoms associated with viral infections, such as fever and aches, actually result from the body's own efforts at defending itself against infection rather than from cell death caused by the virus.

The immune system is a complex and critical part of the body's natural defenses (see Chapter 43). It is also the basis for the major medical tool for preventing viral infections—vaccines. A **vaccine** is a harmless variant or derivative of a pathogen that stimulates the immune system to mount defenses against the harmful pathogen. Smallpox, a viral disease that was at one time a devastating scourge in many parts of the world, was eradicated by a vaccination program carried out by the World Health Organization (WHO). The very narrow host range of the smallpox virus—it infects only humans—was a critical factor in the success of this program. Similar worldwide vaccination campaigns are currently under way to eradicate polio and measles. Effective vaccines are also available to protect against rubella, mumps, hepatitis B, and a number of other viral diseases.

Although vaccines can prevent certain viral illnesses, medical technology can do little, at present, to cure most viral infections once they occur. The antibiotics that help us recover from bacterial infections are powerless against viruses. Antibiotics kill bacteria by inhibiting enzymes specific to bacteria but have no effect on eukaryotic or virally encoded enzymes. However, the few enzymes that are encoded by viruses have provided targets for other drugs. Most antiviral drugs resemble nucleosides and as a result interfere with viral nucleic acid synthesis. One such drug is acyclovir, which impedes herpesvirus replication by inhibiting the viral polymerase that synthesizes viral DNA. Similarly, azidothymidine (AZT) curbs HIV replication by interfering with the synthesis of DNA by reverse transcriptase. In the past two decades, much effort has gone into developing drugs against HIV. Currently, multidrug treatments, sometimes called "cocktails," have been found to be most effective. Such treatments commonly include a combination of two nucleoside mimics and a protease inhibitor, which interferes with an enzyme required for assembly of the viruses.

Emerging Viruses

Viruses that suddenly become apparent are often referred to as *emerging viruses*. HIV, the AIDS virus, is a classic example: This virus appeared in San Francisco in the early 1980s, seemingly out of nowhere, although later studies uncovered a case in the Belgian Congo in 1959. The deadly Ebola virus, recognized initially in 1976 in central Africa, is one of several emerging viruses that cause *hemorrhagic fever*, an often fatal syndrome (set of symptoms) characterized by fever, vomiting, massive bleeding, and circulatory system collapse. A number of other dangerous emerging viruses cause encephalitis, inflammation of the brain. One example is the West Nile virus, which appeared in North America for the first time in 1999 and has spread to all 48 contiguous states in the United States.

In April 2009, a general outbreak, or **epidemic**, of a flu-like illness appeared in Mexico and the United States. The infectious agent was quickly identifed as an influenza virus related to viruses that cause the seasonal flu **(Figure 19.9a)**. This particular virus was named H1N1 for reasons that will be

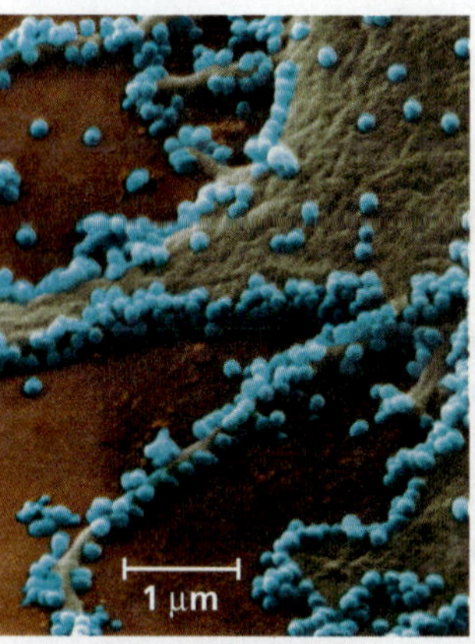

(a) 2009 pandemic H1N1 influenza A virus. Viruses (blue) are seen on an infected cell (green) in this colorized SEM.

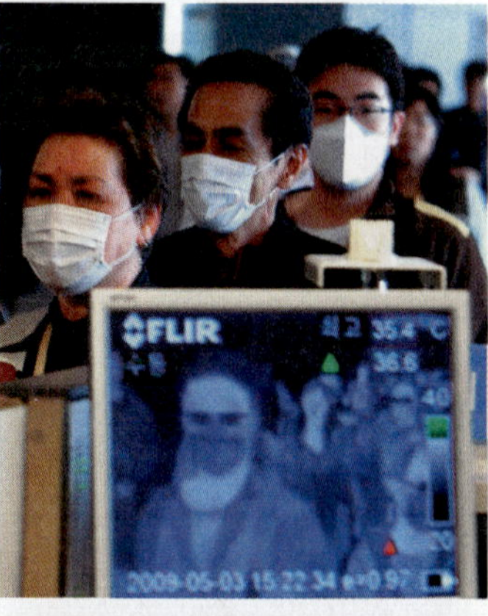

(b) 2009 pandemic screening. At a South Korean airport, thermal scans were used to detect passengers with a fever who might have the H1N1 flu.

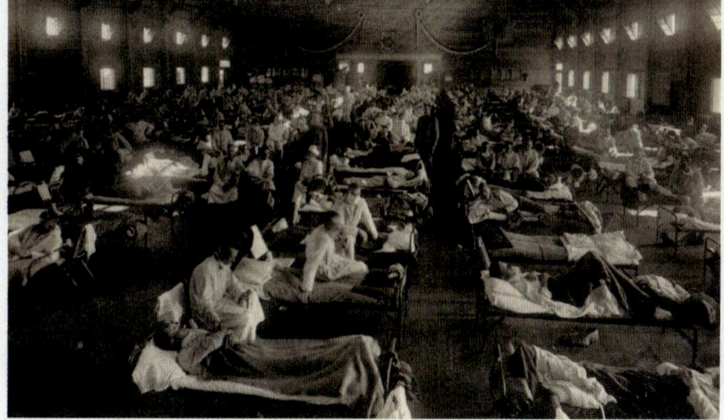

(c) 1918 flu pandemic. Many of those infected during the worst flu epidemic in the last 100 years were treated in large makeshift hospitals, such as this one.

▲ **Figure 19.9 Influenza in humans.**

explained shortly. The viral disease spread rapidly, prompting WHO to declare a global epidemic, or **pandemic**, in June 2009. By November, the disease had reached 207 countries, infecting over 600,000 people and killing almost 8,000. Public health agencies responded rapidly with guidelines for shutting down schools and other public places, and vaccine development and screening efforts were accelerated **(Figure 19.9b)**.

How do such viruses burst on the human scene, giving rise to harmful diseases that were previously rare or even unknown? Three processes contribute to the emergence of viral diseases. The first, and perhaps most important, is the mutation of existing viruses. RNA viruses tend to have an unusually high rate of mutation because errors in replicating their RNA genomes are not corrected by proofreading. Some mutations change existing viruses into new genetic varieties (strains) that can cause disease, even in individuals who are immune to the ancestral virus. For instance, seasonal flu epidemics are caused by new strains of influenza virus genetically different enough from earlier strains that people have little immunity to them.

A second process that can lead to the emergence of viral diseases is the dissemination of a viral disease from a small, isolated human population. For instance, AIDS went unnamed and virtually unnoticed for decades before it began to spread around the world. In this case, technological and social factors, including affordable international travel, blood transfusions, sexual promiscuity, and the abuse of intravenous drugs, allowed a previously rare human disease to become a global scourge.

A third source of new viral diseases in humans is the spread of existing viruses from other animals. Scientists estimate that about three-quarters of new human diseases originate in this way. Animals that harbor and can transmit a particular virus but are generally unaffected by it are said to act as a natural reservoir for that virus. For example, the 2009 flu pandemic mentioned earlier was likely passed to humans from pigs; for this reason, it was originally called "swine flu."

In general, flu epidemics provide an instructive example of the effects of viruses moving between species. There are three types of influenza virus: types B and C, which infect only humans and have never caused an epidemic, and type A, which infects a wide range of animals, including birds, pigs, horses, and humans. Influenza A strains have caused four major flu epidemics among humans in the last 100 years. The worst was the first one, the "Spanish flu" pandemic of 1918–1919, which killed about 40 million people, including many World War I soldiers **(Figure 19.9c)**.

Different strains of influenza A are given standardized names; for example, both the strain that caused the 1918 flu and the one that caused the 2009 pandemic flu are called H1N1. The name identifies which forms of two viral surface proteins are present: hemagglutinin (H) and neuraminidase (N). There are 16 different types of hemagglutinin, a protein that helps the flu virus attach to host cells, and 9 types of neuraminidase, an enzyme that helps release new virus particles from infected cells. Waterbirds have been found that carry viruses with all possible combinations of H and N.

A likely scenario for the 1918 pandemic and others is that the virus mutated as it passed from one host species to another. When an animal like a pig or a bird is infected with more than one strain of flu virus, the different strains can undergo genetic recombination if the RNA molecules making up their genomes mix and match during viral assembly. Pigs are thought to have been the breeding ground for the 2009 flu virus, which contains sequences from bird, pig, and human flu viruses. Coupled with mutation, these reassortments can lead to the emergence of a viral strain that is capable of infecting human cells. Humans who have never been exposed to that particular strain before will lack immunity, and the recombinant virus has the potential to be highly pathogenic. If such a flu virus recombines with viruses that circulate widely among humans, it may acquire the ability to spread easily from person to person, dramatically increasing the potential for a major human outbreak.

Although the 2009 H1N1 flu was declared a pandemic, its toll in lives was significantly lower than that of the 1918 flu. Significantly, however, 79% of the confirmed H1N1 cases in 2009 occurred in people under 30 years of age, and the highest mortality rates occurred in people under 64, opposite to patterns seen for seasonal flu. Some scientists hypothesize that the 1918 flu virus was the ancestor of most subsequent H1N1 epidemic-causing viruses, including that responsible for the 2009 pandemic. Older people are likely to have been exposed to earlier H1N1 viruses and have probably built up immunity to them. This could explain why contracting the 2009 H1N1 virus was more deadly for younger people, who were less likely to have been exposed to H1N1 viruses and to have built up immune defenses.

Perhaps a greater long-term threat is the avian flu caused by an H5N1 virus carried by wild and domestic birds. The first documented transmission to humans was in 1997, when 18 people in Hong Kong were infected and 6 subsequently died. While the 2009 H1N1 flu virus spread easily from human to human, reports of human-to-human transmission of the H5N1 avian flu are quite rare. More alarming, however, is the overall mortality rate of the H5N1 virus, which is greater than 50%. Furthermore, the host range of H5N1 is expanding, which provides increasing opportunities for different strains of the virus to reassort their genetic material and for new strains to emerge. If the H5N1 avian flu virus evolves so that it can spread easily from person to person, it could represent a major global health threat akin to that of the 1918 pandemic.

As we have seen, emerging viruses are generally not new; rather, they are existing viruses that mutate, disseminate more widely in the current host species, or spread to new host species. Changes in host behavior or environmental changes can increase the viral traffic responsible for emerging diseases. For example, new roads built through remote areas can allow viruses to spread between previously isolated human populations. Also, the destruction of forests to expand cropland can bring humans into contact with other animals that may host viruses capable of infecting humans.

Viral Diseases in Plants

More than 2,000 types of viral diseases of plants are known, and together they account for an estimated annual loss of $15 billion worldwide due to their destruction of agricultural and horticultural crops. Common signs of viral infection include bleached or brown spots on leaves and fruits, stunted growth, and damaged flowers or roots, all tending to diminish the yield and quality of crops **(Figure 19.10)**.

Plant viruses have the same basic structure and mode of replication as animal viruses. Most plant viruses discovered thus far, including tobacco mosaic virus (TMV), have an RNA genome. Many have a helical capsid, like TMV, while others have an icosahedral capsid (see Figure 19.3).

Viral diseases of plants spread by two major routes. In the first route, called *horizontal transmission*, a plant is infected from an external source of the virus. Because the invading virus must get past the plant's outer protective layer of cells (the epidermis), a plant becomes more susceptible to viral infections if it has been damaged by wind, injury, or herbivores. Herbivores, especially insects, pose a double threat because they can also act as carriers of viruses, transmitting disease from plant to plant. Moreover, farmers and gardeners may transmit plant viruses inadvertently on pruning shears and other tools. The other route of viral infection is *vertical transmission*, in which a plant inherits a viral infection from a parent. Vertical transmission can occur in asexual propagation (for example, through cuttings) or in sexual reproduction via infected seeds.

Once a virus enters a plant cell and begins replicating, viral genomes and associated proteins can spread throughout the plant by means of plasmodesmata, the cytoplasmic connections that penetrate the walls between adjacent plant cells (see Figure 36.20). The passage of viral macromolecules from cell to cell is facilitated by virally encoded proteins that cause enlargement of plasmodesmata. Scientists have not yet devised cures for most viral plant diseases. Consequently, research efforts are focused largely on reducing the transmission of such diseases and on breeding resistant varieties of crop plants.

▶ **Figure 19.10 Viral infection of plants.** Infection with particular viruses causes irregular brown patches on tomatoes (left), black blotching on squash (center), and streaking in tulips due to redistribution of pigment granules (right).

Viroids and Prions: The Simplest Infectious Agents

As small and simple as viruses are, they dwarf another class of pathogens: **viroids**. These are circular RNA molecules, only a few hundred nucleotides long, that infect plants. Viroids do not encode proteins but can replicate in host plant cells, apparently using host cell enzymes. These small RNA molecules seem to cause errors in the regulatory systems that control plant growth; the typical signs of viroid diseases are abnormal development and stunted growth. One viroid disease, called cadang-cadang, has killed more than 10 million coconut palms in the Philippines.

An important lesson from viroids is that a single molecule can be an infectious agent that spreads a disease. But viroids are nucleic acids, whose ability to be replicated is well known. Even more surprising is the evidence for infectious *proteins*, called **prions**, which appear to cause a number of degenerative brain diseases in various animal species. These diseases include scrapie in sheep; mad cow disease, which has plagued the European beef industry in recent years; and Creutzfeldt-Jakob disease in humans, which has caused the death of some 150 people in Great Britain over the past decade. Prions are most likely transmitted in food, as may occur when people eat prion-laden beef from cattle with mad cow disease. Kuru, another human disease caused by prions, was identified in the early 1900s among the South Fore natives of New Guinea. A kuru epidemic peaked there in the 1960s, puzzling scientists, who at first thought the disease had a genetic basis. Eventually, however, anthropological investigations ferreted out how the disease was spread: ritual cannibalism, a widespread practice among South Fore natives at that time.

Two characteristics of prions are especially alarming. First, prions act very

▶ **Figure 19.11 Model for how prions propagate.** Prions are misfolded versions of normal brain proteins. When a prion contacts a normally folded version of the same protein, it may induce the normal protein to assume the abnormal shape. The resulting chain reaction may continue until high levels of prion aggregation cause cellular malfunction and eventual degeneration of the brain.

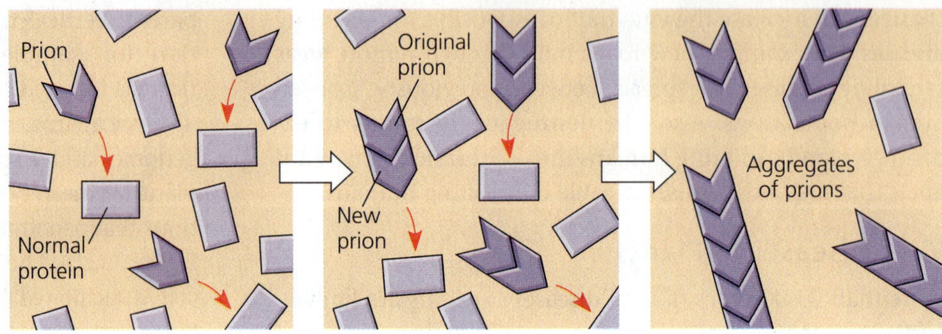

slowly, with an incubation period of at least ten years before symptoms develop. The lengthy incubation period prevents sources of infection from being identified until long after the first cases appear, allowing many more infections to occur. Second, prions are virtually indestructible; they are not destroyed or deactivated by heating to normal cooking temperatures. To date, there is no known cure for prion diseases, and the only hope for developing effective treatments lies in understanding the process of infection.

How can a protein, which cannot replicate itself, be a transmissible pathogen? According to the leading model, a prion is a misfolded form of a protein normally present in brain cells. When the prion gets into a cell containing the normal form of the protein, the prion somehow converts normal protein molecules to the misfolded prion versions. Several prions then aggregate into a complex that can convert other normal proteins to prions, which join the chain **(Figure 19.11)**. Prion aggregation interferes with normal cellular functions and causes disease symptoms. This model was greeted with much skepticism when it was first proposed by Stanley Prusiner in the early 1980s, but it is now widely accepted. Prusiner was awarded the Nobel Prize in 1997 for his work on prions.

CONCEPT CHECK 19.3

1. Describe two ways a preexisting virus can become an emerging virus.
2. Contrast horizontal and vertical transmission of viruses in plants.
3. **WHAT IF?** TMV has been isolated from virtually all commercial tobacco products. Why, then, is TMV infection not an additional hazard for smokers?

For suggested answers, see Appendix A.

19 CHAPTER REVIEW

SUMMARY OF KEY CONCEPTS

CONCEPT 19.1

A virus consists of a nucleic acid surrounded by a protein coat (pp. 381–384)

- Researchers discovered viruses in the late 1800s by studying a plant disease, tobacco mosaic disease.
- A **virus** is a small nucleic acid genome enclosed in a protein **capsid** and sometimes a membranous **viral envelope** containing viral proteins that help viruses enter cells. The genome may be single- or double-stranded DNA or RNA.

? *Are viruses generally considered living or nonliving? Explain.*

CONCEPT 19.2

Viruses replicate only in host cells (pp. 384–390)

- Viruses use enzymes, ribosomes, and small molecules of host cells to synthesize progeny viruses during replication. Each type of virus has a characteristic **host range**.
- **Phages** (viruses that infect bacteria) can replicate by two alternative mechanisms: the **lytic cycle** and the **lysogenic cycle**.

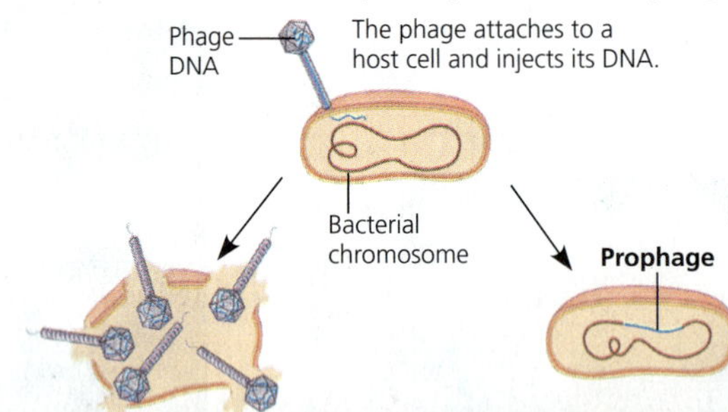

Lytic cycle
- **Virulent** or **temperate phage**
- Destruction of host DNA
- Production of new phages
- Lysis of host cell causes release of progeny phages

Lysogenic cycle
- **Temperate phage** only
- Genome integrates into bacterial chromosome as **prophage**, which (1) is replicated and passed on to daughter cells and (2) can be induced to leave the chromosome and initiate a lytic cycle

- Many animal viruses have an envelope. **Retroviruses** (such as **HIV**) use the enzyme **reverse transcriptase** to copy their RNA genome into DNA, which can be integrated into the host genome as a **provirus**.

- Since viruses can replicate only within cells, they probably evolved after the first cells appeared, perhaps as packaged fragments of cellular nucleic acid. The origin of viruses is still being debated.

? *Describe enzymes that are not found in most cells but are necessary for the replication of viruses of certain types.*

CONCEPT 19.3

Viruses, viroids, and prions are formidable pathogens in animals and plants (pp. 390–394)

- Symptoms of viral diseases may be caused by direct viral harm to cells or by the body's immune response. **Vaccines** stimulate the immune system to defend the host against specific viruses.
- Outbreaks of "new" viral diseases in humans are usually caused by existing viruses that expand their host territory. The H1N1 2009 flu virus was a new combination of pig, human, and avian viral genes that caused a pandemic. The H5N1 avian flu virus has the potential to cause a high-mortality flu pandemic.
- Viruses enter plant cells through damaged cell walls (horizontal transmission) or are inherited from a parent (vertical transmission).
- **Viroids** are naked RNA molecules that infect plants and disrupt their growth. **Prions** are slow-acting, virtually indestructible infectious proteins that cause brain diseases in mammals.

? *What aspect of an RNA virus makes it more likely than a DNA virus to become an emerging virus?*

TEST YOUR UNDERSTANDING

LEVEL 1: KNOWLEDGE/COMPREHENSION

1. Which of the following characteristics, structures, or processes is common to both bacteria and viruses?
   a. metabolism
   b. ribosomes
   c. genetic material composed of nucleic acid
   d. cell division
   e. independent existence

2. Emerging viruses arise by
   a. mutation of existing viruses.
   b. the spread of existing viruses to new host species.
   c. the spread of existing viruses more widely within their host species.
   d. all of the above
   e. none of the above

3. To cause a human pandemic, the H5N1 avian flu virus would have to
   a. spread to primates such as chimpanzees.
   b. develop into a virus with a different host range.
   c. become capable of human-to-human transmission.
   d. arise independently in chickens in North and South America.
   e. become much more pathogenic.

LEVEL 2: APPLICATION/ANALYSIS

4. A bacterium is infected with an experimentally constructed bacteriophage composed of the T2 phage protein coat and T4 phage DNA. The new phages produced would have
   a. T2 protein and T4 DNA.
   b. T2 protein and T2 DNA.
   c. a mixture of the DNA and proteins of both phages.
   d. T4 protein and T4 DNA.
   e. T4 protein and T2 DNA.

5. RNA viruses require their own supply of certain enzymes because
   a. host cells rapidly destroy the viruses.
   b. host cells lack enzymes that can replicate the viral genome.
   c. these enzymes translate viral mRNA into proteins.
   d. these enzymes penetrate host cell membranes.
   e. these enzymes cannot be made in host cells.

6. **DRAW IT** Redraw Figure 19.7 to show the replicative cycle of a virus with a single-stranded genome that can function as mRNA (a class IV virus).

LEVEL 3: SYNTHESIS/EVALUATION

7. **EVOLUTION CONNECTION**
   The success of some viruses lies in their ability to evolve rapidly within the host. Such a virus evades the host's defenses by mutating and producing many altered progeny viruses before the body can mount an attack. Thus, the viruses present late in infection differ from those that initially infected the body. Discuss this as an example of evolution in microcosm. Which viral lineages tend to predominate?

8. **SCIENTIFIC INQUIRY**
   When bacteria infect an animal, the number of bacteria in the body increases in an exponential fashion (graph A). After infection by a virulent animal virus with a lytic replicative cycle, there is no evidence of infection for a while. Then the number of viruses rises suddenly and subsequently increases in a series of steps (graph B). Explain the difference in the curves.

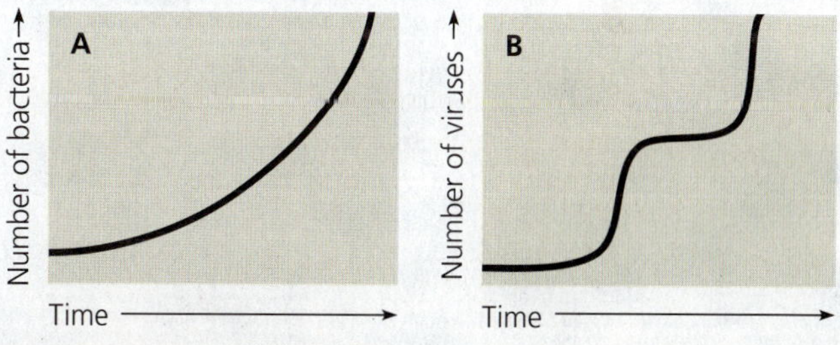

9. **WRITE ABOUT A THEME**
   Structure and Function While viruses are considered by most scientists to be nonliving, they do show some characteristics of life, including the correlation of structure and function. In a short essay (100–150 words), discuss how the structure of a virus correlates with its function.

For selected answers, see Appendix A.

20

Biotechnology

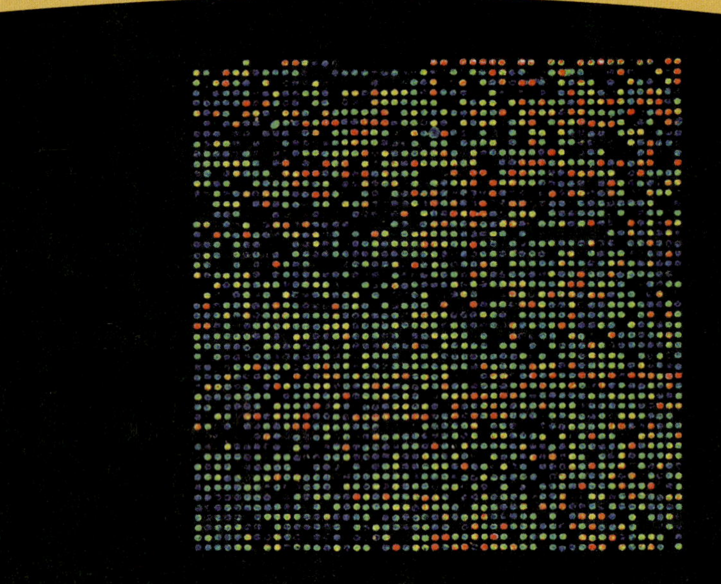

▲ **Figure 20.1 How can this array of spots be used to compare normal and cancerous tissues?**

KEY CONCEPTS

20.1 DNA cloning yields multiple copies of a gene or other DNA segment

20.2 DNA technology allows us to study the sequence, expression, and function of a gene

20.3 Cloning organisms may lead to production of stem cells for research and other applications

20.4 The practical applications of DNA technology affect our lives in many ways

OVERVIEW

The DNA Toolbox

In 2001, a major scientific milestone was announced: Researchers had completed a "first draft" sequence of all 3 billion base pairs of the human genome—only the fourth eukaryotic genome to be sequenced. This news electrified the scientific community. Few among them would have dared to dream that a mere nine years later, genome sequencing would be under way for more than 7,000 species. By 2010, researchers had completed sequencing more than 1,000 bacterial, 80 archaeal, and 100 eukaryotic genomes, with many more in progress.

Ultimately, these achievements are attributable to advances in DNA technology—methods of working with and manipulating DNA—that had their roots in the 1970s. A key accomplishment was the invention of techniques for making **recombinant DNA**, DNA molecules formed when segments of DNA from two different sources—often different species—are combined *in vitro* (in a test tube). This advance gave rise to the development of powerful techniques for analyzing genes and gene expression. How scientists prepare recombinant DNA and use DNA technology to answer fundamental biological questions are one focus of this chapter. In the next chapter (Chapter 21), we'll see how these techniques have allowed the sequencing of whole genomes, and we'll consider what we've learned from these sequences about the evolution of species and of the genome itself.

Another focus of this chapter is how our lives are affected by **biotechnology**, the manipulation of organisms or their components to make useful products. Biotechnology includes such early practices as selective breeding of farm animals and using microorganisms to make wine and cheese. Today, biotechnology also encompasses **genetic engineering**, the direct manipulation of genes for practical purposes. Genetic engineering has launched a revolution in biotechnology, greatly expanding the scope of its potential applications. Tools from the DNA toolbox are now applied in ways that affect everything from agriculture to criminal law to medical research. For instance, on the DNA microarray in **Figure 20.1**, the colored spots represent the relative level of expression of 2,400 human genes in normal and cancerous tissue. Using microarray analysis, researchers can quickly compare gene expression in different samples, such as those tested here. The knowledge gained from such gene expression studies is making a significant contribution to the study of cancer and other diseases.

In this chapter, we'll first describe the main techniques for manipulating DNA and analyzing gene expression and function. Next, we'll explore advances in cloning organisms and producing stem cells, techniques that have both expanded our basic understanding of biology and enhanced our ability to apply this understanding to global problems. Finally, we'll survey the practical applications of biotechnology and consider some of the social and ethical issues that arise as biotechnology becomes more pervasive in our lives.

CONCEPT 20.1

DNA cloning yields multiple copies of a gene or other DNA segment

The molecular biologist studying a particular gene faces a challenge. Naturally occurring DNA molecules are very long, and a single molecule usually carries many genes. Moreover, in many eukaryotic genomes, genes occupy only a small proportion of the chromosomal DNA, the rest being noncoding nucleotide sequences. A single human gene, for example, might constitute only 1/100,000 of a chromosomal DNA molecule. As a further complication, the distinctions between a gene and the surrounding DNA are subtle, consisting only of differences in nucleotide sequence. To work directly with specific genes, scientists have developed methods for preparing well-defined segments of DNA in multiple identical copies, a process called *DNA cloning*.

DNA Cloning and Its Applications: *A Preview*

Most methods for cloning pieces of DNA in the laboratory share certain general features. One common approach uses bacteria, most often *Escherichia coli*. Recall from Figure 16.12 that the *E. coli* chromosome is a large circular molecule of DNA. In addition, *E. coli* and many other bacteria have **plasmids**, small circular DNA molecules that replicate separately from the bacterial chromosome. A plasmid has only a small number of genes; these genes may be useful when the bacterium is in a particular environment but may not be required for survival or reproduction under most conditions.

To clone pieces of DNA in the laboratory, researchers first obtain a plasmid (originally isolated from a bacterial cell and genetically engineered for efficient cloning) and insert DNA from another source ("foreign" DNA) into it **(Figure 20.2)**. The resulting plasmid is now a recombinant DNA molecule. The plasmid is then returned to a bacterial cell, producing a *recombinant bacterium*. This single cell reproduces through repeated cell divisions to form a clone of cells, a population of genetically identical cells. Because the dividing bacteria replicate the recombinant plasmid and pass it on to their descendants, the foreign DNA and any genes it carries are cloned at the same time. The production of multiple copies of a single gene is called **gene cloning**.

Gene cloning is useful for two basic purposes: to make many copies of, or *amplify*, a particular gene and to produce a protein product. Researchers can isolate copies of a cloned gene from bacteria for use in basic research or to endow an organism with a new metabolic capability, such as pest resistance. For example, a resistance gene present in one crop species might be cloned and transferred into plants of another species. Alternatively, a protein with medical uses, such as human growth hormone, can be harvested in large quantities from cultures of bacteria carrying the cloned gene for the protein.

A single gene is usually a very small part of the total DNA in a cell. For example, a typical gene makes up only about one-millionth of the DNA in a human cell. The ability to

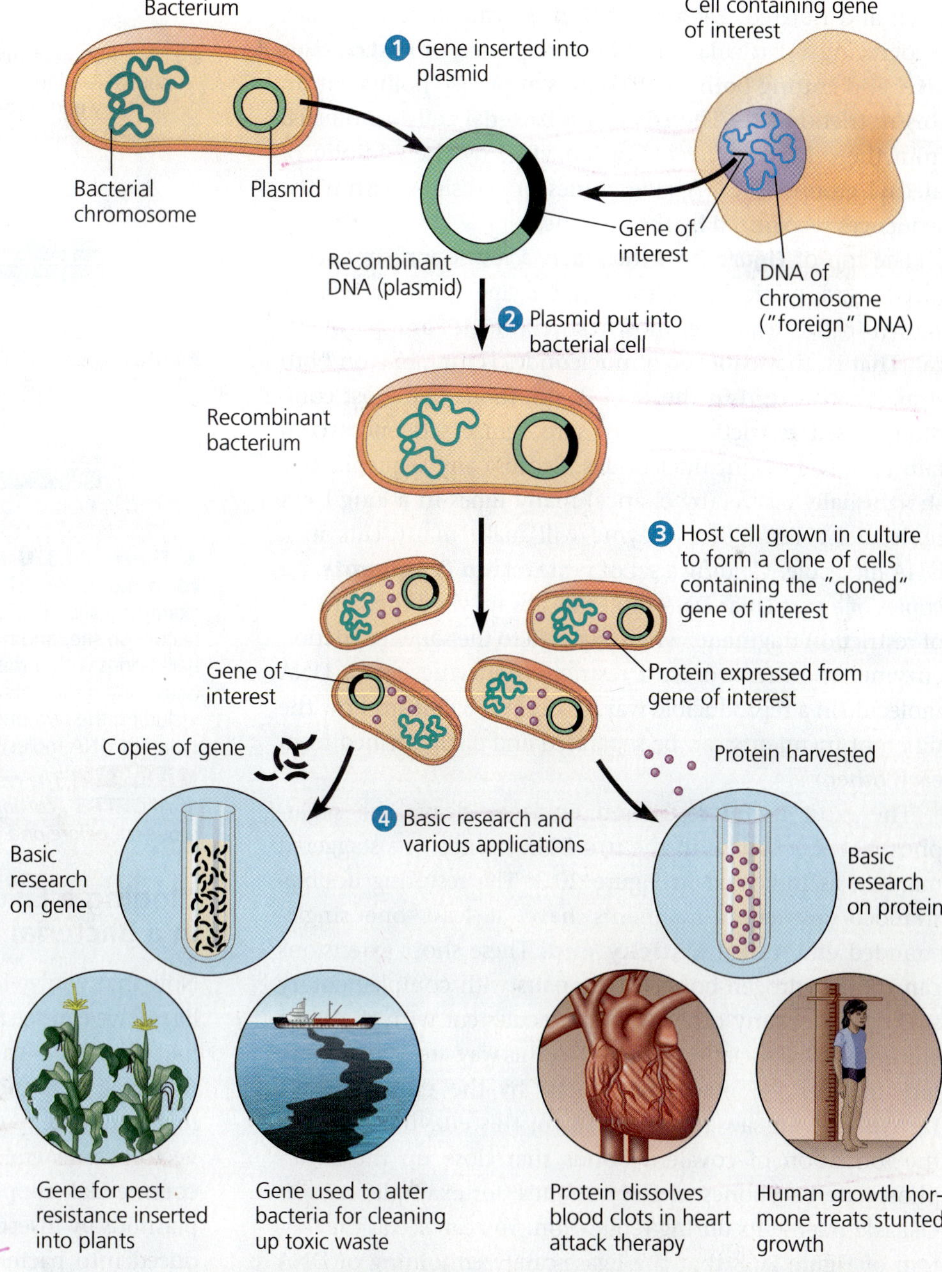

▲ **Figure 20.2 A preview of gene cloning and some uses of cloned genes.** In this simplified diagram of gene cloning, we start with a plasmid (originally isolated from a bacterial cell) and a gene of interest from another organism. Only one plasmid and one copy of the gene of interest are shown at the top of the figure, but the starting materials would include many of each.

amplify such rare DNA fragments is therefore crucial for any application involving a single gene.

Using Restriction Enzymes to Make Recombinant DNA

Gene cloning and genetic engineering rely on the use of enzymes that cut DNA molecules at a limited number of specific locations. These enzymes, called restriction endonucleases, or **restriction enzymes**, were discovered in the late 1960s by biologists doing basic research on bacteria. Restriction enzymes protect the bacterial cell by cutting up foreign DNA from other organisms or phages (see Chapter 19).

Hundreds of different restriction enzymes have been identified and isolated. Each restriction enzyme is very specific, recognizing a particular short DNA sequence, or **restriction site**, and cutting both DNA strands at precise points within this restriction site. The DNA of a bacterial cell is protected from the cell's own restriction enzymes by the addition of methyl groups (—CH_3) to adenines or cytosines within the sequences recognized by the enzymes.

The top of **Figure 20.3** illustrates a restriction site recognized by a particular restriction enzyme from *E. coli*. As shown in this example, most restriction sites are symmetrical. That is, the sequence of nucleotides is the same on both strands when read in the 5′ → 3′ direction. The most commonly used restriction enzymes recognize sequences containing four to eight nucleotides. Because any sequence this short usually occurs (by chance) many times in a long DNA molecule, a restriction enzyme will make many cuts in a DNA molecule, yielding a set of **restriction fragments**. All copies of a particular DNA molecule always yield the same set of restriction fragments when exposed to the same restriction enzyme. In other words, a restriction enzyme cuts a DNA molecule in a reproducible way. (Later you will learn how the different fragments can be separated and distinguished from each other.)

The most useful restriction enzymes cleave the sugar-phosphate backbones in the two DNA strands in a staggered manner, as indicated in Figure 20.3. The resulting double-stranded restriction fragments have at least one single-stranded end, called a **sticky end**. These short extensions can form hydrogen-bonded base pairs with complementary sticky ends on any other DNA molecules cut with the same enzyme. The associations formed in this way are only temporary but can be made permanent by the enzyme **DNA ligase**. As you saw in Figure 16.16, this enzyme catalyzes the formation of covalent bonds that close up the sugar-phosphate backbones of DNA strands; for example, it joins Okazaki fragments during replication. You can see at the bottom of Figure 20.3 that the ligase-catalyzed joining of DNA from two different sources produces a stable recombinant DNA molecule.

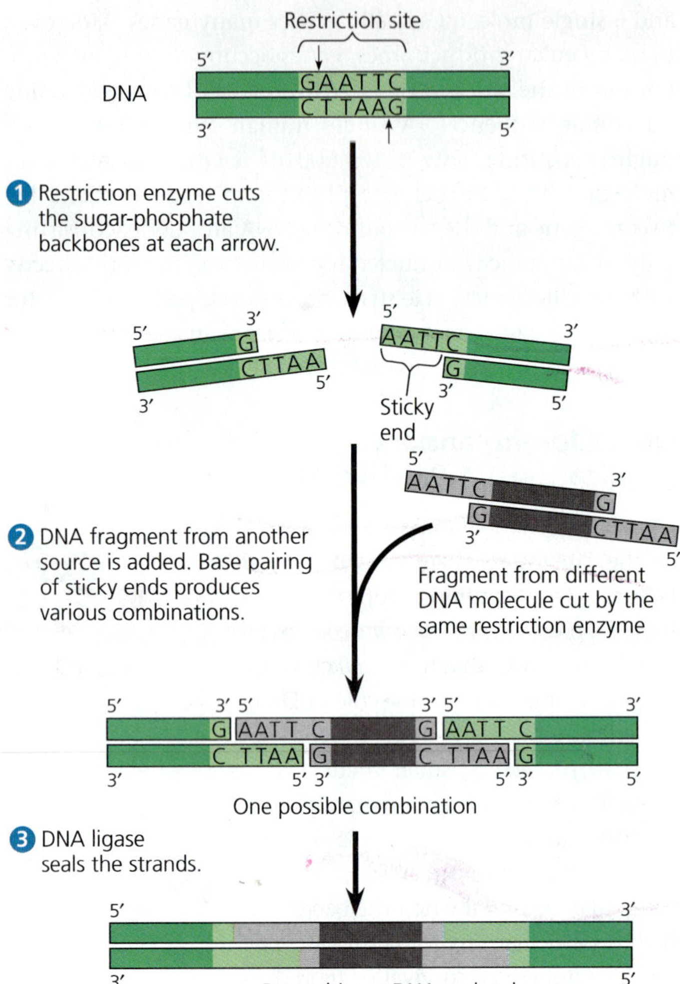

▲ **Figure 20.3 Using a restriction enzyme and DNA ligase to make recombinant DNA.** The restriction enzyme in this example (called *Eco*RI) recognizes a specific six-base-pair sequence, the restriction site, and makes staggered cuts in the sugar-phosphate backbones within this sequence, producing fragments with sticky ends. Any fragments with complementary sticky ends can base-pair, including the two original fragments. If the fragments come from different DNA molecules, the ligated product is recombinant DNA.

DRAW IT *The restriction enzyme* Hind*III recognizes the sequence 5′-AAGCTT-3′, cutting between the two 'A's. Draw the double-stranded sequence before and after the enzyme cuts.*

Cloning a Eukaryotic Gene in a Bacterial Plasmid

Now that you've learned about restriction enzymes and DNA ligase, we can see how genes are cloned in plasmids. The original plasmid is called a **cloning vector**, defined as a DNA molecule that can carry foreign DNA into a host cell and replicate there. Bacterial plasmids are widely used as cloning vectors for several reasons: They can be readily obtained from commercial suppliers, manipulated to form recombinant plasmids by insertion of foreign DNA *in vitro*, and then introduced into bacterial cells. Moreover, recombinant bacterial plasmids (and the foreign DNA they carry) multiply rapidly owing to the high reproductive rate of their host cells.

Producing Clones of Cells Carrying Recombinant Plasmids

Let's say we are researchers interested in studying the β-globin gene in a particular species of hummingbird. We start by cloning all the hummingbird genes; later we'll isolate the β-globin gene from all the others, a task very much like finding a needle in a haystack. **Figure 20.4** details one method for cloning hummingbird genes using a bacterial plasmid as the cloning vector.

❶ First, we isolate hummingbird genomic DNA from hummingbird cells. We also obtain our chosen vector, a particular bacterial plasmid from *E. coli* cells. The plasmid has been engineered to carry two genes that will later prove useful: amp^R, which makes *E. coli* cells resistant to the antibiotic ampicillin, and *lacZ*, which encodes the enzyme β-galactosidase, which hydrolyzes lactose (see p. 354). This enzyme can also hydrolyze a similar synthetic molecule called X-gal to form a blue product. The plasmid contains only one copy of the restriction site recognized by the restriction enzyme used in the next step, and that site is within the *lacZ* gene.

❷ Both the plasmid and the hummingbird DNA are cut with the same restriction enzyme, and then ❸ the fragments are mixed together, allowing base pairing between their complementary sticky ends. We then add DNA ligase, which covalently bonds the sugar-phosphate backbones of the fragments whose sticky ends have base-paired. Many of the resulting recombinant plasmids contain single hummingbird DNA fragments (three are shown in Figure 20.4), and at least one of them is expected to carry all or part of the β-globin gene. This step will also generate other products, such as plasmids containing multiple hummingbird DNA fragments, a combination of two plasmids, or a rejoined, nonrecombinant version of the original plasmid.

❹ The DNA mixture is then added to bacteria that have a mutation in the *lacZ* gene on their own chromosome, making them unable to hydrolyze

▼ Figure 20.4 RESEARCH METHOD

Cloning Genes in Bacterial Plasmids

APPLICATION Gene cloning is a process that produces many copies of a gene of interest. These copies can be used in sequencing the gene, in producing its encoded protein, or in basic research or other applications.

TECHNIQUE In this example, hummingbird genes are inserted into plasmids from *E. coli*. Only three plasmids and three hummingbird DNA fragments are shown, but millions of copies of the plasmid and a mixture of millions of different hummingbird DNA fragments would be present in the samples.

❶ Obtain engineered plasmid DNA and DNA from hummingbird cells. The hummingbird DNA contains the gene of interest.

❷ Cut both DNA samples with the same restriction enzyme, one that makes a single cut within the *lacZ* gene and many cuts within the hummingbird DNA.

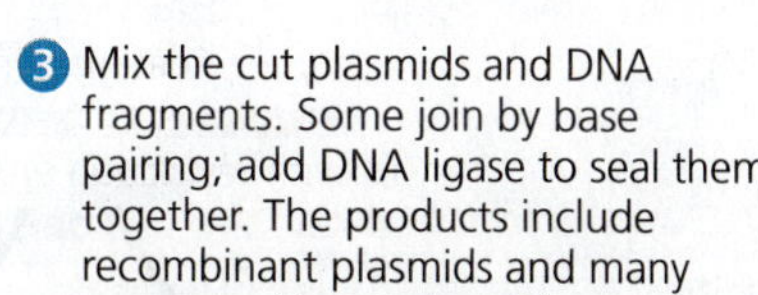

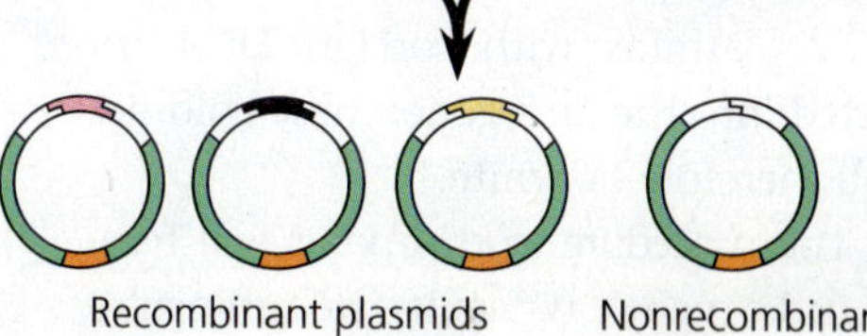

❸ Mix the cut plasmids and DNA fragments. Some join by base pairing; add DNA ligase to seal them together. The products include recombinant plasmids and many nonrecombinant plasmids.

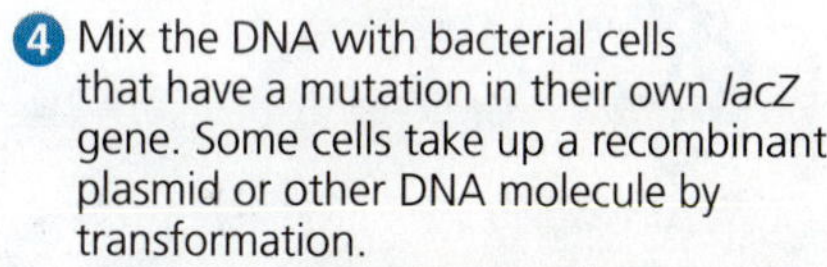

❹ Mix the DNA with bacterial cells that have a mutation in their own *lacZ* gene. Some cells take up a recombinant plasmid or other DNA molecule by transformation.

❺ Plate the bacteria on nutrient-containing agar medium supplemented with ampicillin and X-gal, a molecule resembling lactose. Incubate until colonies grow.

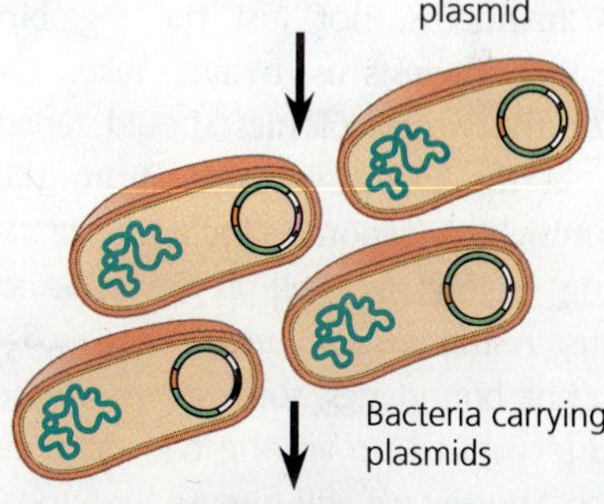

RESULTS Only a cell that took up a plasmid, which has the amp^R gene, will reproduce and form a colony. Colonies with nonrecombinant plasmids will be blue because they can hydrolyze X-gal, forming a blue product. Colonies with recombinant plasmids, in which *lacZ* is disrupted, will be white because they cannot hydrolyze X-gal.

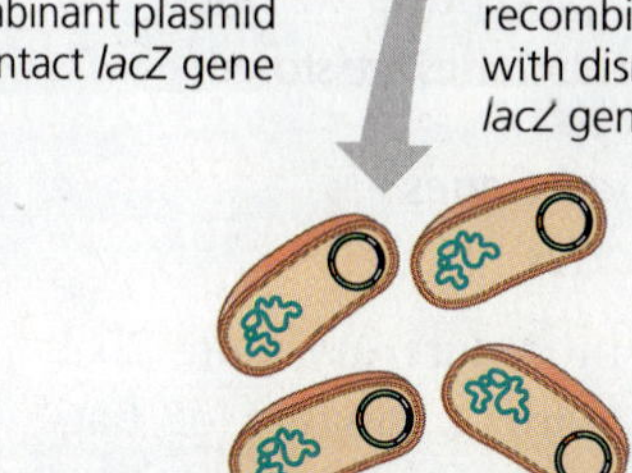

WHAT IF? If the medium used in step 5 did not contain ampicillin, what other colonies would grow? What color would they be?

lactose or X-gal. Under suitable experimental conditions, the cells take up foreign DNA by transformation (see p. 306). Some cells acquire a recombinant plasmid carrying a gene, while others may take up a nonrecombinant plasmid, a fragment of noncoding hummingbird DNA, or nothing at all. The amp^R and *lacZ* genes on the plasmid can help us sort out these possibilities.

5 First, plating out all the bacteria on solid nutrient medium containing ampicillin allows us to distinguish the cells that have taken up plasmids, whether recombinant or not, from the other cells. Under these conditions, only cells with a plasmid will reproduce because only they have the amp^R gene, conferring resistance to the ampicillin in the medium. Each reproducing bacterium forms a clone of cells. Once the clone contains between 10^5 and 10^8 cells, it is visible as a mass, or *colony*, on the agar. As cells reproduce, any foreign genes carried by recombinant plasmids are also copied (cloned).

Second, the presence of X-gal in the medium allows us to distinguish colonies with recombinant plasmids from those with nonrecombinant plasmids. Colonies containing nonrecombinant plasmids have the *lacZ* gene intact and will produce functional β-galactosidase. These colonies will be blue because the enzyme hydrolyzes the X-gal in the medium, forming a blue product. In contrast, no functional β-galactosidase is produced in colonies containing recombinant plasmids with foreign DNA inserted into the *lacZ* gene; these colonies will therefore be white.

The procedure to this point will have cloned many different hummingbird DNA fragments, not just the β-globin gene that interests us. In fact, taken together, the white colonies should represent all the DNA sequences from the hummingbird genome, including noncoding regions as well as genes. And because restriction enzymes do not recognize gene boundaries, some genes will be cut and divided up among two or more clones. Shortly, we will discuss the procedure we use to find the colony (cell clone) or colonies carrying the β-globin gene sequences among the many clones carrying other pieces of hummingbird DNA. To understand that procedure, we must first consider how the clones are stored.

Storing Cloned Genes in DNA Libraries

The cloning procedure in Figure 20.4, which starts with a mixture of fragments from the entire genome of an organism, is called a "shotgun" approach; no single gene is targeted for cloning. Numerous different recombinant plasmids are produced in step 3, and a clone of cells carrying each type of plasmid ends up as a white colony in step 5. The complete set of plasmid-containing cell clones, each carrying copies of a particular segment from the initial genome, is referred to as a **genomic library** (**Figure 20.5a**). Each "plasmid clone" in the library is like a book containing specific information. Today, scientists often obtain such libraries (or even particular cloned genes) from another researcher, a commercial source, or a sequencing center.

Historically, certain bacteriophages have also been used as cloning vectors for making genomic libraries. Fragments of foreign DNA can be spliced into a trimmed-down version of a phage genome, as into a plasmid, by using a restriction enzyme and DNA ligase. The normal infection process allows production of many new phage particles, each carrying the foreign DNA insert. Today, phages are generally used for making genomic libraries only in special cases.

Another type of vector widely used in library construction is a **bacterial artificial chromosome (BAC)**. In spite of the name, these are simply large plasmids, trimmed down so they contain just the genes necessary to ensure replication. An advantage of using BACs as vectors is that while a standard plasmid can carry a DNA insert no larger than 10,000 base pairs (10 kb), a BAC can carry an insert of 100–300 kb

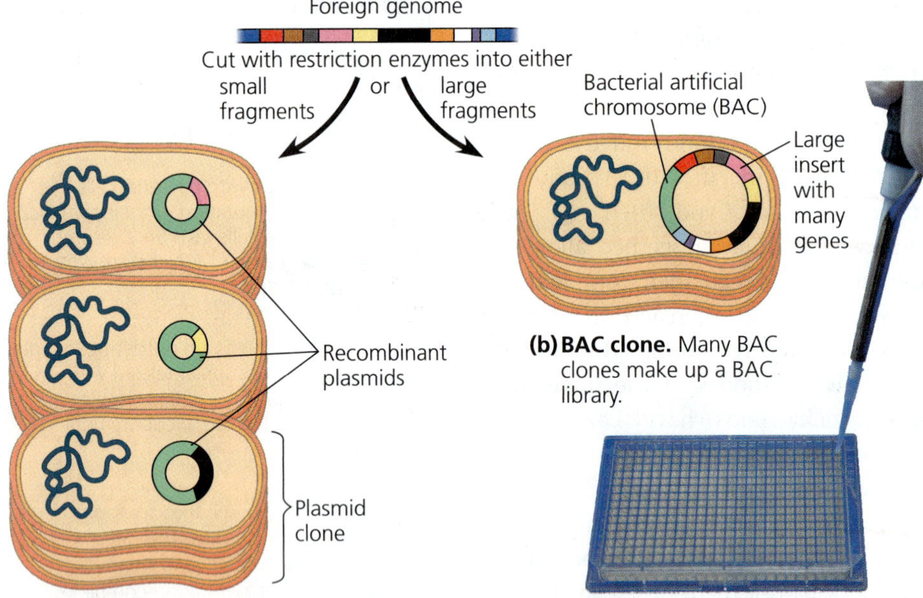

(a) Plasmid library. Shown are three of the thousands of "books" in a plasmid library. Each "book" is a clone of bacterial cells, which contain copies of a particular foreign genome fragment (pink, yellow, black segments) in their recombinant plasmids.

(b) BAC clone. Many BAC clones make up a BAC library.

(c) Storing genome libraries. Both plasmid and BAC genomic libraries are usually stored in a "multiwell" plastic plate; a 384-well plate is shown here. Each clone occupies one well. (The library of an entire genome would require many such plates.)

▲ **Figure 20.5 Genomic libraries.** A genomic library is a collection of many clones. Each clone carries copies of a particular DNA segment from a foreign genome, integrated into an appropriate DNA vector, such as a plasmid or a bacterial artificial chromosome (BAC). In a complete genomic library, the foreign DNA segments cover the entire genome of an organism. Note that the bacterial chromosomes are not drawn to scale; they are actually about 1,000 times larger than the vectors.

(Figure 20.5b). The very large insert size minimizes the number of clones needed to make up the genomic library, but it also makes them more challenging to work with in the lab, so the insert may later be cut up into smaller pieces that are "subcloned" into plasmid vectors.

Clones are usually stored in multiwelled plastic plates, with one clone per well **(Figure 20.5c)**. This orderly storage of clones, identified by their location in the plate, makes screening for the gene of interest very efficient, as you will see.

In a genomic library, the cloned β-globin gene would include not just exons containing the coding sequence, but also the promoter, untranslated regions, and any introns. Some biologists might be interested in the β-globin protein itself—they might wonder, for instance, if this oxygen-carrying protein is different from its counterpart in other, less metabolically active species. Such researchers can make another kind of DNA library by starting with fully processed mRNA extracted from cells where the gene is expressed **(Figure 20.6)**. The enzyme reverse transcriptase (obtained from retroviruses) is used *in vitro* to make a single-stranded DNA *reverse transcript* of each mRNA molecule. Recall that the 3′ end of the mRNA has a stretch of adenine (A) ribonucleotides called a poly-A tail. This feature allows use of a short strand of thymine deoxyribonucleotides (dT's) as a primer for the reverse transcriptase. Following enzymatic degradation of the mRNA, a second DNA strand, complementary to the first, is synthesized by DNA polymerase. The resulting double-stranded DNA is called **complementary DNA (cDNA)**. To create a library, the researchers must now modify the cDNA by adding restriction enzyme recognition sequences at each end. Then the cDNA is inserted into vector DNA in a manner similar to the insertion of genomic DNA fragments. The extracted mRNA is a mixture of all the mRNA molecules in the original cells, transcribed from many different genes. Therefore, the cDNAs that are cloned make up a **cDNA library** containing a collection of genes. However, a cDNA library represents only part of the genome—only the subset of genes that were transcribed in the cells from which the mRNA was isolated.

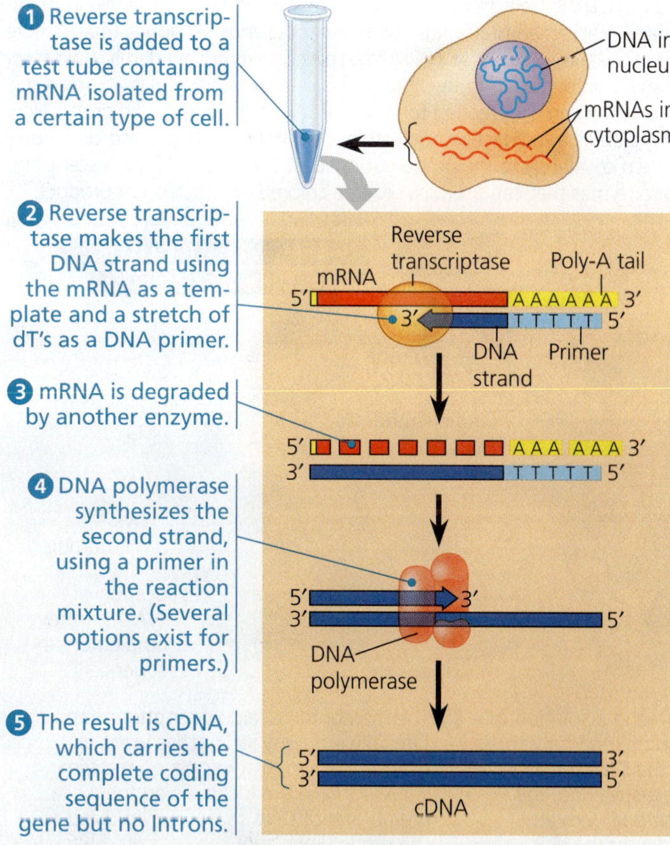

▲ **Figure 20.6 Making complementary DNA (cDNA) from eukaryotic genes.** Complementary DNA is DNA made *in vitro* using mRNA as a template for the first strand. Because the mRNA contains only exons, the resulting double-stranded cDNA carries the complete coding sequence of the gene but no introns. Although only one mRNA is shown here, the final collection of cDNAs would reflect all the mRNAs that were present in the cell.

Genomic and cDNA libraries each have advantages, depending on what is being studied. If you want to clone a gene but don't know what cell type expresses it or cannot obtain enough cells of the appropriate type, a genomic library is almost certain to contain the gene. Also, if you are interested in the regulatory sequences or introns associated with a gene, a genomic library is necessary, since these sequences are absent from mRNAs used in making a cDNA library. On the other hand, to study a specific protein (like β-globin), a cDNA library made from cells expressing the gene (like red blood cells) is ideal. A cDNA library can also be used to study sets of genes expressed in particular cell types, such as brain or liver cells. Finally, by making cDNA from cells of the same type at different times in the life of an organism, researchers can trace changes in patterns of gene expression during development.

Screening a Library for Clones Carrying a Gene of Interest

Now, returning to the results in Figure 20.4, we're ready to screen all the colonies with recombinant plasmids (the white colonies) for a clone of cells containing the hummingbird β-globin gene. We can detect this gene's DNA by its ability to base-pair with a complementary sequence on another nucleic acid molecule, using **nucleic acid hybridization**. The complementary molecule, a short, single-stranded nucleic acid that can be either RNA or DNA, is called a **nucleic acid probe**. If we know at least part of the nucleotide sequence of the gene of interest (perhaps from knowing the amino acid sequence of the protein it encodes or, as in our case, the gene's nucleotide sequence in a closely related species), we can synthesize a probe complementary to it. For example, if part of the sequence on one strand of the desired gene were

5′ ···CTCATCACCGGC··· 3′

then we would synthesize this probe:

3′ GAGTAGTGGCCG 5′

Each probe molecule, which will hydrogen-bond specifically to a complementary sequence in the desired gene, is labeled with a radioactive isotope, a fluorescent tag, or another molecule so we can track it.

Recall that the clones in our hummingbird genomic library have been stored in a multiwell plate (see Figure 20.5c). If we transfer a few cells from each well to a defined location on a membrane made of nylon or nitrocellulose, we can screen a large number of clones simultaneously for the presence of DNA complementary to our DNA probe **(Figure 20.7)**.

After we've identified the location of a clone carrying the β-globin gene, we can grow some cells from that colony in liquid culture in a large tank and then easily isolate many copies of the gene for our studies. We can also use the cloned gene as a probe to identify similar or identical genes in DNA from other sources, such as other species of birds.

Expressing Cloned Eukaryotic Genes

Once a particular gene has been cloned in host cells, its protein product can be produced in large amounts for research purposes or valuable practical applications, which we'll explore in Concept 20.4. Cloned genes can be expressed as protein in either bacterial or eukaryotic cells; each option has advantages and disadvantages.

Bacterial Expression Systems

Getting a cloned eukaryotic gene to function in bacterial host cells can be difficult because certain aspects of gene expression are different in eukaryotes and bacteria. To overcome differences in promoters and other DNA control sequences, scientists usually employ an **expression vector**, a cloning vector that contains a highly active bacterial promoter just upstream of a restriction site where the eukaryotic gene can be inserted in the correct reading frame. The bacterial host cell will recognize the promoter and proceed to express the foreign gene now linked to that promoter. Such expression vectors allow the synthesis of many eukaryotic proteins in bacterial cells.

▼ **Figure 20.7** RESEARCH METHOD

Detecting a Specific DNA Sequence by Hybridization with a Nucleic Acid Probe

APPLICATION Hybridization with a complementary nucleic acid probe detects a specific DNA sequence within a mixture of DNA molecules. In this example, a collection of bacterial clones from a hummingbird genomic library is screened to identify clones that carry a recombinant plasmid bearing the gene of interest. The library is stored in many multiwell plates, with one clone per well (see Figure 20.5c).

TECHNIQUE Cells from each clone are applied to a special nylon membrane. Each membrane has room for thousands of clones (many more than are shown here), so only a few membranes are needed to hold samples of all the clones in the library. This set of membranes is an *arrayed library* that can be screened for a specific gene using a labeled probe. Here the label is a radioactive nucleotide, but other labels are also commonly linked covalently to the probe nucleotides. These include fluorescent tags or enzymes that can produce either a colored or luminescent product.

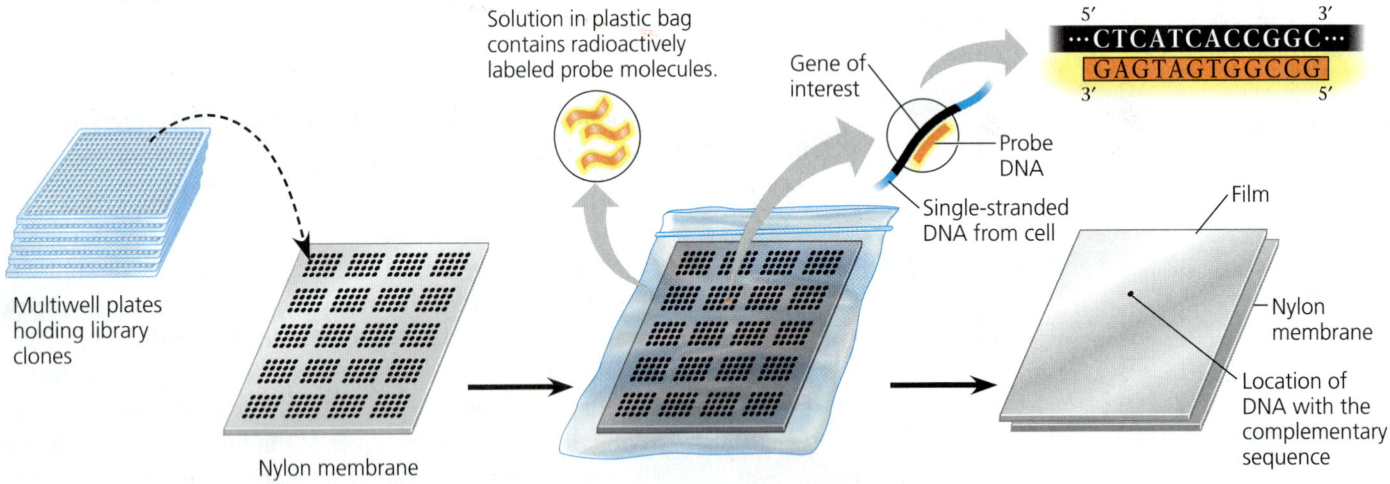

❶ Plate by plate, cells from each well, representing one clone, are transferred to a defined spot on a special nylon membrane. The nylon membrane is treated to break open the cells and denature their DNA; the resulting single-stranded DNA molecules stick to the membrane.

❷ The membrane is then incubated in a solution of radioactive probe molecules complementary to the gene of interest. Because the DNA immobilized on the membrane is single-stranded, the single-stranded probe can base-pair with any complementary DNA on the membrane. Excess DNA is then rinsed off. (One spot with radioactive probe–DNA hybrids is shown here in orange but would not be distinguishable yet.)

❸ The membrane is laid under photographic film, allowing any radioactive areas to expose the film. Black spots on the film correspond to the locations on the membrane of DNA that has hybridized to the probe. Each spot can be traced back to the original well containing the bacterial clone that holds the gene of interest.

RESULTS For a radioactive probe, the location of the black spot on a piece of photographic film identifies the clone containing the gene of interest. (Probes labeled in other ways use other detection systems.) By using probes with different nucleotide sequences in different experiments, researchers can screen the collection of bacterial clones for different genes.

Another problem with expressing cloned eukaryotic genes in bacteria is the presence of noncoding regions (introns) in most eukaryotic genes. Introns can make a eukaryotic gene very long and unwieldy, and they prevent correct expression of the gene by bacterial cells, which do not have RNA-splicing machinery. This problem can be surmounted by using a cDNA form of the gene, which includes only the exons.

Eukaryotic Cloning and Expression Systems

Molecular biologists can avoid eukaryotic-bacterial incompatibility by using eukaryotic cells such as yeasts, rather than bacteria, as hosts for cloning and/or expressing eukaryotic genes of interest. Yeasts, single-celled fungi, offer two advantages: They are as easy to grow as bacteria, and they have plasmids, a rarity among eukaryotes. Scientists have even constructed recombinant plasmids that combine yeast and bacterial DNA and can replicate in either type of cell.

Another reason to use eukaryotic host cells for expressing a cloned eukaryotic gene is that many eukaryotic proteins will not function unless they are modified after translation, for example, by the addition of carbohydrate (glycosylation) or lipid groups. Bacterial cells cannot carry out these modifications, and if the gene product requiring such processing is from a mammal, even yeast cells may not be able to modify the protein correctly. Several cultured cell types have proved successful as host cells for this purpose, including some mammalian cell lines and an insect cell line that can be infected by a virus (called baculovirus) carrying recombinant DNA.

Besides using vectors, scientists have developed a variety of other methods for introducing recombinant DNA into eukaryotic cells. In **electroporation**, a brief electrical pulse applied to a solution containing cells creates temporary holes in their plasma membranes, through which DNA can enter. (This technique is now commonly used for bacteria as well.) Alternatively, scientists can inject DNA directly into single eukaryotic cells using microscopically thin needles. To get DNA into plant cells, the soil bacterium *Agrobacterium* can be used, as well as other methods you will learn about later. If the introduced DNA is incorporated into a cell's genome by genetic recombination, then it may be expressed by the cell.

Cross-Species Gene Expression and Evolutionary Ancestry

EVOLUTION The ability to express eukaryotic proteins in bacteria (even if the proteins aren't glycosylated properly) is quite remarkable when we consider how different eukaryotic and bacterial cells are. Examples abound of genes that are taken from one species and function perfectly well when transferred into another very different species. These observations underscore the shared evolutionary ancestry of species living today.

One example involves a gene called *Pax-6*, which has been found in animals as diverse as vertebrates and fruit flies. The vertebrate *Pax-6* gene product (the PAX-6 protein) triggers a complex program of gene expression resulting in formation of the vertebrate eye, which has a single lens. The fly *Pax-6* gene leads to formation of the compound fly eye, which is quite different from the vertebrate eye. When scientists cloned the mouse *Pax-6* gene and introduced it into a fly embryo, they were surprised to see that it led to formation of a compound fly eye (see Figure 50.16). Conversely, when the fly *Pax-6* gene was transferred into a vertebrate embryo—a frog, in this case—a frog eye formed. Although the genetic programs triggered in vertebrates and flies generate very different eyes, both versions of the *Pax-6* gene can substitute for each other, evidence of their evolution from a gene in a common ancestor.

Simpler examples are seen in Figure 17.6, where a firefly gene is expressed in a tobacco plant, and a jellyfish gene product is seen in a pig. The basic mechanisms of gene expression have ancient evolutionary roots, which is the basis of many recombinant DNA techniques described in this chapter.

Amplifying DNA *in Vitro*: The Polymerase Chain Reaction (PCR)

DNA cloning in cells remains the best method for preparing large quantities of a particular gene or other DNA sequence. However, when the source of DNA is scanty or impure, the **polymerase chain reaction**, or **PCR**, is quicker and more selective. In this technique, any specific target segment within one or many DNA molecules can be quickly amplified in a test tube. With automation, PCR can make billions of copies of a target segment of DNA in a few hours, significantly faster than the days it would take to obtain the same number of copies by screening a DNA library for a clone with the desired gene and letting it replicate within host cells. In fact, PCR is being used increasingly to make enough of a specific DNA fragment to insert it directly into a vector, entirely skipping the steps of making and screening a library. To continue our literary analogy, PCR is like photocopying a few pages rather than checking out a book from the library.

In the PCR procedure, a three-step cycle brings about a chain reaction that produces an exponentially growing population of identical DNA molecules. During each cycle, the reaction mixture is heated to denature (separate) the DNA strands and then cooled to allow annealing (hydrogen bonding) of short, single-stranded DNA primers complementary to sequences on opposite strands at each end of the target sequence; finally, a heat-stable DNA polymerase extends the primers in the $5' \rightarrow 3'$ direction. If a standard DNA polymerase were used, the protein would be denatured along with the DNA during the first heating step and would have to be replaced after each cycle. The key to automating PCR was the discovery of an unusual heat-stable DNA polymerase called Taq polymerase, named after the bacterial species from which it was first isolated. This bacterial species, *Thermus aquaticus*, lives in hot springs, so natural selection has resulted in a heat-stable DNA polymerase that can withstand the heat at the start of each cycle. **Figure 20.8** illustrates the steps in PCR.

Just as impressive as the speed of PCR is its specificity. Only minute amounts of DNA need be present in the starting material, and this DNA can be in a partially degraded state, as long as a few molecules contain the complete target sequence. The key to this high specificity is the primers, which hydrogen-bond *only* to sequences at opposite ends of the target segment. (For high specificity, the primers must be at least 15 or so nucleotides long.) By the end of the third cycle, one-fourth of the molecules are identical to the target segment, with both strands the appropriate length. With each successive cycle, the number of target segment molecules of the correct length doubles, so the number of molecules equals 2^n, where n is the number of cycles. After 30 more cycles, about a billion copies of the target sequence are present!

Despite its speed and specificity, PCR amplification cannot substitute for gene cloning in cells when large amounts of a gene are desired. Occasional errors during PCR replication impose limits on the number of good copies that can be made by this method. When PCR is used to provide the specific DNA fragment for cloning, the resulting clones are sequenced to select clones with error-free inserts. PCR errors also impose limits on the length of DNA fragments that can be copied.

Devised in 1985, PCR has had a major impact on biological research and biotechnology. PCR has been used to amplify DNA from a wide variety of sources: fragments of ancient DNA from a 40,000-year-old frozen woolly mammoth; DNA from fingerprints or from tiny amounts of blood, tissue, or semen found at crime scenes; DNA from single embryonic cells for rapid prenatal diagnosis of genetic disorders; and DNA of viral genes from cells infected with viruses that are difficult to detect, such as HIV. We'll return to applications of PCR later in the chapter.

▼ Figure 20.8 RESEARCH METHOD

The Polymerase Chain Reaction (PCR)

APPLICATION With PCR, any specific segment—the target sequence—within a DNA sample can be copied many times (amplified), completely *in vitro*.

TECHNIQUE PCR requires double-stranded DNA containing the target sequence, a heat-resistant DNA polymerase, all four nucleotides, and two 15- to 20-nucleotide DNA strands that serve as primers. One primer is complementary to one end of the target sequence on one strand; the second primer is complementary to the other end of the sequence on the other strand.

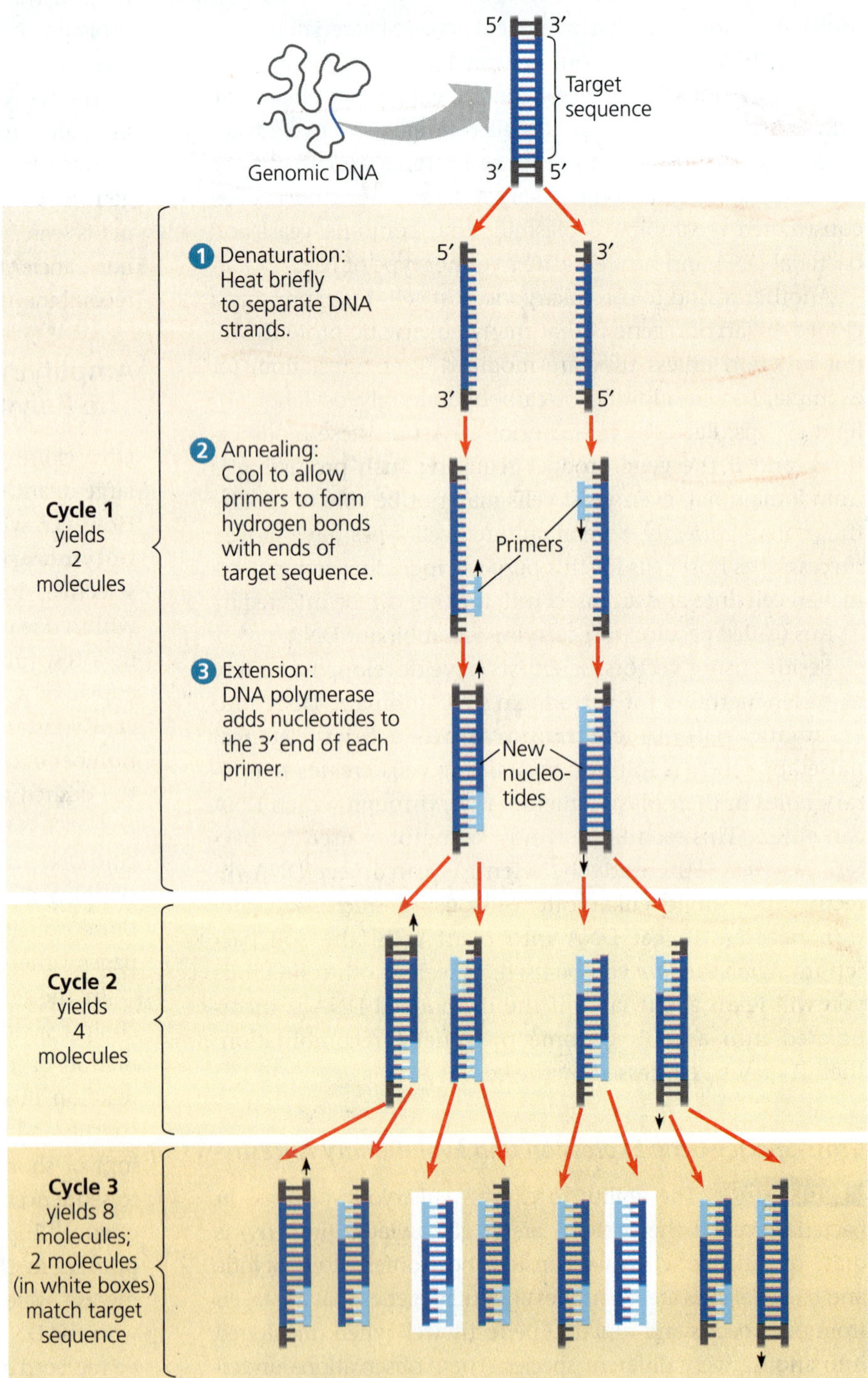

RESULTS After 3 cycles, two molecules match the target sequence exactly. After 30 more cycles, over 1 billion (10^9) molecules match the target sequence.

CONCEPT CHECK 20.1

1. The restriction site for an enzyme called *PvuI* is the following sequence:
   5′-C G A T C G-3′
   3′-G C T A G C-5′
   Staggered cuts are made between the T and C on each strand. What type of bonds are being cleaved?
2. **DRAW IT** One strand of a DNA molecule has the following sequence: 5′-CCTTGACGATCGTTACCG-3′. Draw the other strand. Will *PvuI* cut this molecule? If so, draw the products.
3. What are some potential difficulties in using plasmid vectors and bacterial host cells to produce large quantities of proteins from cloned eukaryotic genes?
4. **MAKE CONNECTIONS** Compare Figure 20.8 with Figure 16.20 (p. 319). How does replication of DNA ends during PCR proceed without shortening the fragments each time?

For suggested answers, see Appendix A.

CONCEPT 20.2

DNA technology allows us to study the sequence, expression, and function of a gene

Once DNA cloning has provided us with large quantities of specific DNA segments, we can tackle some interesting questions about a particular gene and its function. For example, does the sequence of the hummingbird β-globin gene suggest a protein structure that can carry oxygen more efficiently than its counterpart in less metabolically active species? Does a particular human gene differ from person to person, and are certain alleles of that gene associated with a hereditary disorder? Where in the body and when is a given gene expressed? And, ultimately, what role does a certain gene play in an organism?

Before we can begin to address such compelling questions, we must consider a few standard laboratory techniques that are used to analyze the DNA of genes.

Gel Electrophoresis and Southern Blotting

Many approaches for studying DNA molecules involve **gel electrophoresis**. This technique uses a gel made of a polymer, such as the polysaccharide *agarose*. The gel acts as a molecular sieve to separate nucleic acids or proteins on the basis of size, electrical charge, and other physical properties **(Figure 20.9)**. Because nucleic acid molecules carry negative charges on their phosphate groups, they all travel toward the positive pole in an

▼ **Figure 20.9** **RESEARCH METHOD**

Gel Electrophoresis

APPLICATION Gel electrophoresis is used for separating nucleic acids or proteins that differ in size, electrical charge, or other physical properties. DNA molecules are separated by gel electrophoresis in restriction fragment analysis of both cloned genes (see Figure 20.10) and genomic DNA (see Figure 20.11).

TECHNIQUE Gel electrophoresis separates macromolecules on the basis of their rate of movement through an agarose gel in an electric field: The distance a DNA molecule travels is inversely proportional to its length. A mixture of DNA molecules, usually fragments produced by restriction enzyme digestion (cutting) or PCR amplification, is separated into bands. Each band contains thousands of molecules of the same length.

1. Each sample, a mixture of DNA molecules, is placed in a separate well near one end of a thin slab of agarose gel. The gel is set into a small plastic support and immersed in an aqueous, buffered solution in a tray with electrodes at each end.

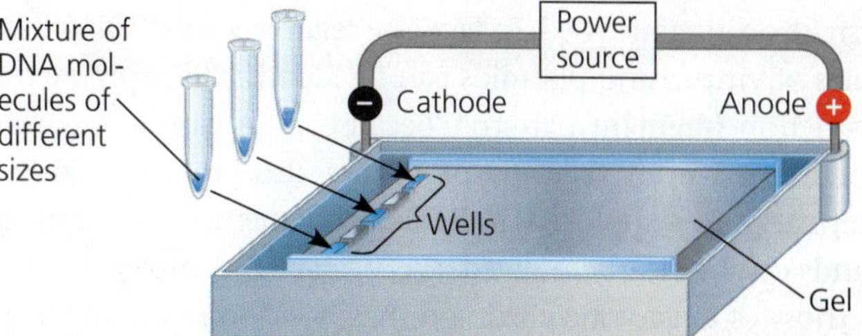

2. When the current is turned on, the negatively charged DNA molecules move toward the positive electrode, with shorter molecules moving faster than longer ones. Bands are shown here in blue, but in an actual gel, the bands would not be visible at this time.

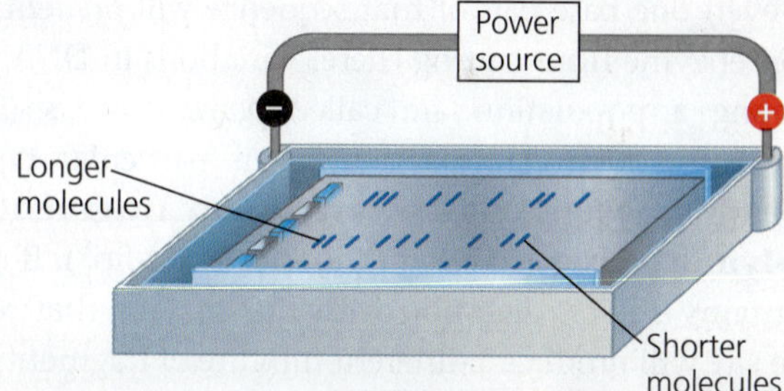

RESULTS After the current is turned off, a DNA-binding dye (ethidium bromide) is added. This dye fluoresces pink in ultraviolet light, revealing the separated bands to which it binds. In the gel below, the pink bands correspond to DNA fragments of different lengths separated by electrophoresis. If all the samples were initially cut with the same restriction enzyme, then the different band patterns indicate that they came from different sources.

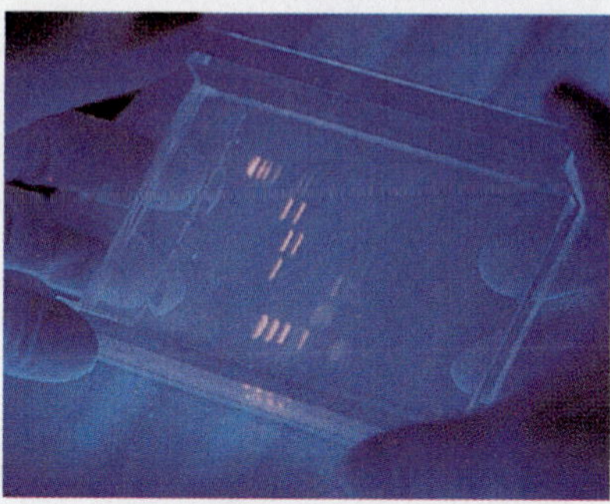

electric field. As they move, the thicket of agarose fibers impedes longer molecules more than it does shorter ones, separating them by length. Thus, gel electrophoresis separates a mixture of linear DNA molecules into bands, each band consisting of many thousands of DNA molecules of the same length.

One historically useful application of this technique has been *restriction fragment analysis*, which rapidly provides information about DNA sequences. With advances in sequencing technology, the approach taken in labs today is often simply to sequence the DNA sample in question. However, restriction fragment analysis is still done in some cases, and understanding how it is done will give you a better grasp of recombinant DNA technology. In this type of analysis, the DNA fragments produced by restriction enzyme digestion (cutting) of a DNA molecule are separated by gel electrophoresis. When the mixture of restriction fragments undergoes electrophoresis, it yields a band pattern characteristic of the starting molecule and the restriction enzyme used. In fact, the relatively small DNA molecules of viruses and plasmids can be identified simply by their restriction fragment patterns. Because DNA can be recovered undamaged from gels, the procedure also provides a way to prepare pure samples of individual fragments—assuming the bands can be clearly resolved. (Very large DNA molecules, such as those of eukaryotic chromosomes, yield so many fragments that they appear as a smear instead of distinct bands.)

Restriction fragment analysis can be used to compare two different DNA molecules—for example, two alleles of a gene—if the nucleotide difference affects a restriction site. A change in even one base pair of that sequence will prevent a restriction enzyme from cutting there. Variations in DNA sequence among a population are called *polymorphisms* (from the Greek for "many forms"), and this particular type of sequence change is called a **restriction fragment length polymorphism** (**RFLP**, pronounced "Rif-lip"). If one allele contains a RFLP, digestion with the enzyme that recognizes the site will produce a different mixture of fragments for each of the two alleles. Each mixture will give its own band pattern in gel electrophoresis.

For example, sickle-cell disease is caused by mutation of a single nucleotide located within a restriction sequence (a RFLP) in the human β-globin gene (see pp. 277–278 and Figure 17.23). Consequently, while other assays are preferred today, restriction fragment analysis was used for many years to distinguish the normal and sickle-cell alleles of the β-globin gene, as shown in **Figure 20.10**.

The starting materials in Figure 20.10 are samples of the cloned and purified β-globin alleles. But how could this test be done if we didn't have purified alleles to start with? If we wanted to determine whether a person is a heterozygous carrier of the mutant allele for sickle-cell disease, we would directly compare the genomic DNA from that person with DNA from both a person who has sickle-cell disease (and is homozygous for the mutant allele) and a person who is

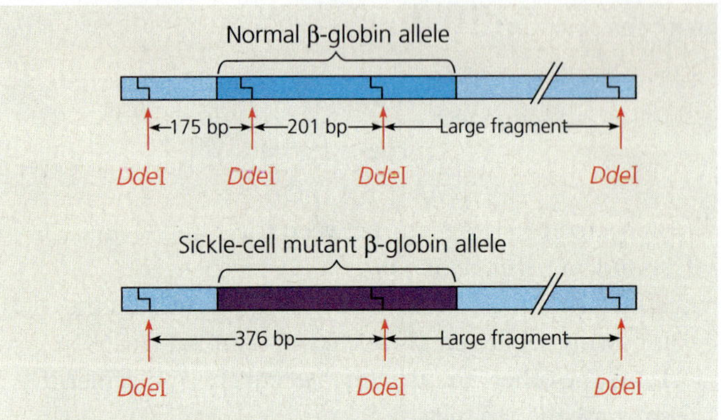

(a) *Dde*I restriction sites in normal and sickle-cell alleles of the β-globin gene. Shown here are the cloned alleles, separated from the vector DNA but including some DNA next to the coding sequence. The normal allele contains two sites within the coding sequence recognized by the *Dde*I restriction enzyme. The sickle-cell allele lacks one of these sites.

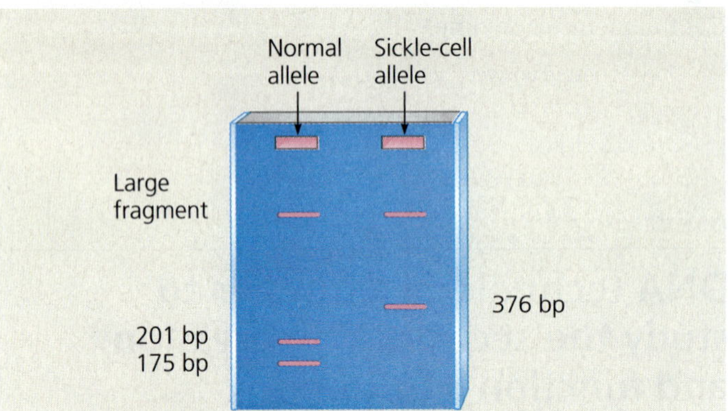

(b) Electrophoresis of restriction fragments from normal and sickle-cell alleles. Samples of each purified allele were cut with the *Dde*I enzyme and then subjected to gel electrophoresis, resulting in three bands for the normal allele and two bands for the sickle-cell allele. (The tiny fragments on the ends of both initial DNA molecules are identical and are not seen here.)

▲ **Figure 20.10 Using restriction fragment analysis to distinguish the normal and sickle-cell alleles of the human β-globin gene. (a)** The sickle-cell mutation destroys one of the *Dde*I restriction sites within the gene. **(b)** As a result, digestion with the *Dde*I enzyme generates different sets of fragments from the normal and sickle-cell alleles.

WHAT IF? *Given bacterial clones with recombinant plasmids carrying each of these alleles, how would you isolate the pure DNA samples run on the gel in (b)? (Hint: Study Figures 20.4 and 20.9.)*

homozygous for the normal allele. As we mentioned already, electrophoresis of genomic DNA digested with a restriction enzyme and stained with a DNA-binding dye yields too many bands to distinguish them individually. However, a classic method called **Southern blotting** (developed by British biochemist Edwin Southern), which combines gel electrophoresis and nucleic acid hybridization, allows us to detect just those bands that include parts of the β-globin gene. The principle is the same as in nucleic acid hybridization for screening bacterial clones (see Figure 20.7). In Southern blotting, the probe is usually a radioactively or otherwise labeled single-stranded DNA molecule that is complementary to the gene of interest. **Figure 20.11** outlines the entire procedure and demonstrates how it can differentiate a heterozygote (in this case, for the sickle-cell allele) from an individual homozygous for the normal allele.

The identification of carriers of mutant alleles associated with genetic diseases is only one of the ways Southern blotting has been used. In fact, this technique was a laboratory workhorse for many years. Recently, however, it has been supplanted by more rapid methods, often involving PCR amplification of the specific parts of genomes that may differ.

DNA Sequencing

Once a gene is cloned, its complete nucleotide sequence can be determined. Today, sequencing is automated, carried out by sequencing machines (see Figure 1.12). The first automated procedure was based on a technique called the *dideoxyribonucleotide*

▼ **Figure 20.11** RESEARCH METHOD

Southern Blotting of DNA Fragments

APPLICATION Researchers can detect specific nucleotide sequences within a complex DNA sample with this method. In particular, Southern blotting can be used to compare the restriction fragments produced from different samples of genomic DNA.

TECHNIQUE In this example, we compare genomic DNA samples from three individuals: a homozygote for the normal β-globin allele (I), a homozygote for the mutant sickle-cell allele (II), and a heterozygote (III). As in Figure 20.7, we show a radioactively labeled probe, but other methods of probe labeling and detection are also used.

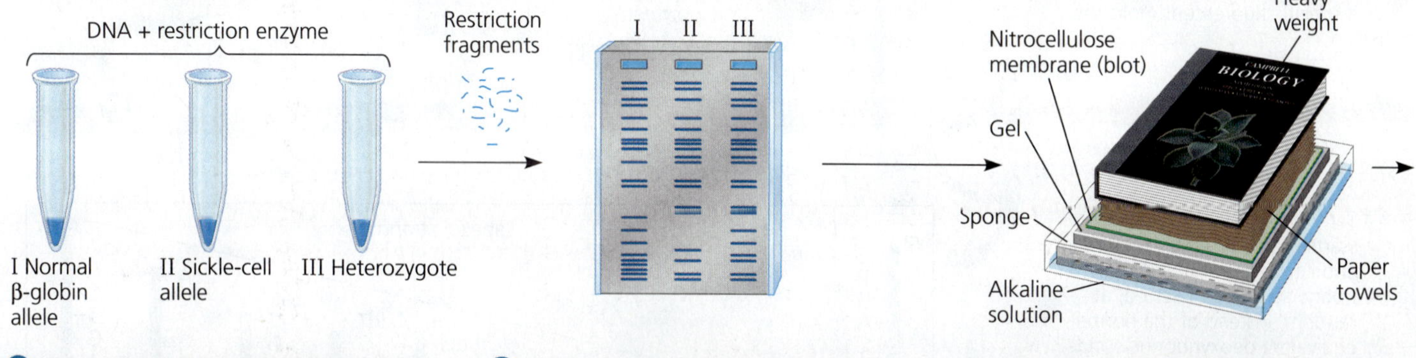

1 Preparation of restriction fragments. Each DNA sample is mixed with the same restriction enzyme, in this case *Dde*I. Digestion of each sample yields a mixture of thousands of restriction fragments.

2 Gel electrophoresis. The restriction fragments in each sample are separated by electrophoresis, forming a characteristic pattern of bands. (In reality, there would be many more bands than shown here, and they would be invisible unless stained.)

3 DNA transfer (blotting). With the gel arranged as shown above, capillary action pulls the alkaline solution upward through the gel, denaturing and transferring the DNA to a nitrocellulose membrane. This produces a blot with a pattern of DNA bands exactly like that of the gel.

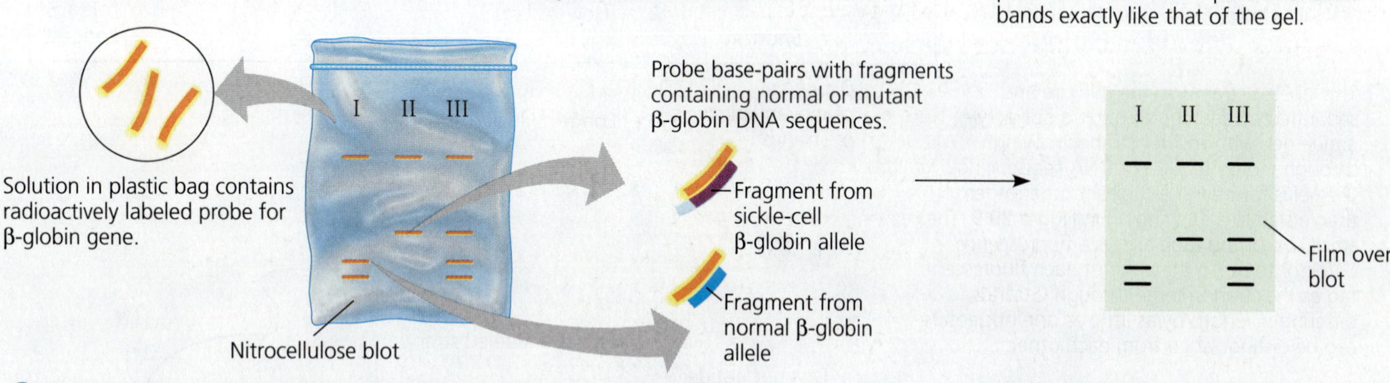

4 Hybridization with labeled probe. The nitrocellulose blot is exposed to a solution containing a probe labeled in some way. In this example, the probe is radioactively labeled, single-stranded DNA complementary to the β-globin gene. Probe molecules attach by base-pairing to any restriction fragments containing a part of the β-globin gene. (The bands would not be visible yet.)

5 Probe detection. A sheet of photographic film is laid over the blot. The radioactivity in the bound probe exposes the film to form an image corresponding to those bands containing DNA that base-paired with the probe.

RESULTS The band patterns for the three samples are clearly different, so this method can be used to identify heterozygous carriers of the sickle-cell allele (III), as well as those with the disease, who have two mutant alleles (II), and unaffected individuals, who have two normal alleles (I). Band patterns for samples I and II resemble those seen for the purified normal and mutant alleles, respectively, seen in Figure 20.10b. The band pattern for the sample from the heterozygote (III) is a combination of the patterns for the two homozygotes (I and II).

▼ **Figure 20.12** RESEARCH METHOD

Dideoxy Chain Termination Method for Sequencing DNA

APPLICATION The sequence of nucleotides in any cloned DNA fragment of up to 800–1,000 base pairs in length can be determined rapidly with machines that carry out sequencing reactions and separate the labeled reaction products by length.

TECHNIQUE This method synthesizes a set of DNA strands complementary to the original DNA fragment. Each strand starts with the same primer and ends with a dideoxyribonucleotide (ddNTP), a modified nucleotide. Incorporation of a ddNTP terminates a growing DNA strand because it lacks a 3′ —OH group, the site for attachment of the next nucleotide (see Figure 16.14). In the set of strands synthesized, each nucleotide position along the original sequence is represented by strands ending at that point with the complementary ddNTP. Because each type of ddNTP is tagged with a distinct fluorescent label, the identity of the ending nucleotides of the new strands, and ultimately the entire original sequence, can be determined.

1. The fragment of DNA to be sequenced is denatured into single strands and incubated in a test tube with the necessary ingredients for DNA synthesis: a primer designed to base-pair with the known 3′ end of the template strand, DNA polymerase, the four deoxyribonucleotides, and the four dideoxyribonucleotides, each tagged with a specific fluorescent molecule.

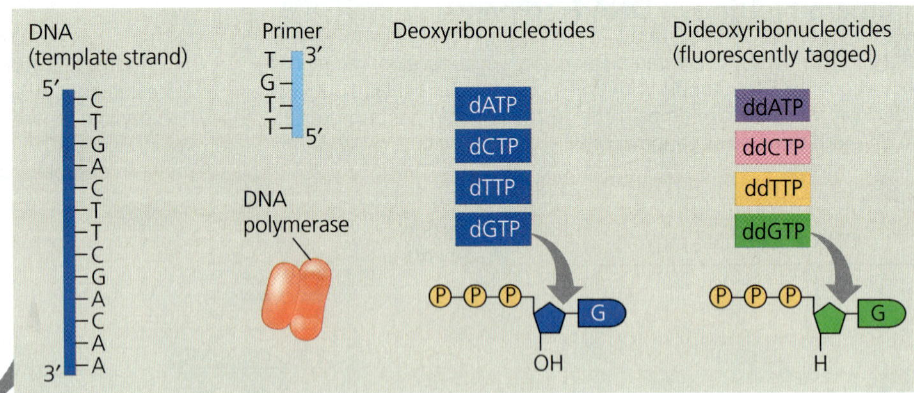

2. Synthesis of each new strand starts at the 3′ end of the primer and continues until a dideoxyribonucleotide is inserted, at random, instead of the normal equivalent deoxyribonucleotide. This prevents further elongation of the strand. Eventually, a set of labeled strands of various lengths is generated, with the color of the tag representing the last nucleotide in the sequence.

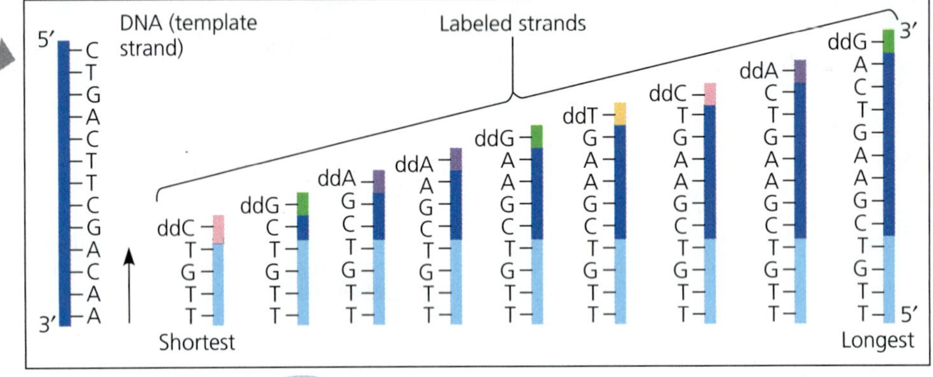

3. The labeled strands in the mixture are separated by passage through a polyacrylamide gel, with shorter strands moving through more quickly. For DNA sequencing, the gel is formed in a capillary tube rather than a slab like that shown in Figure 20.9. The small size of the tube allows a fluorescence detector to sense the color of each fluorescent tag as the strands come through. Strands differing in length by as little as one nucleotide can be distinguished from each other.

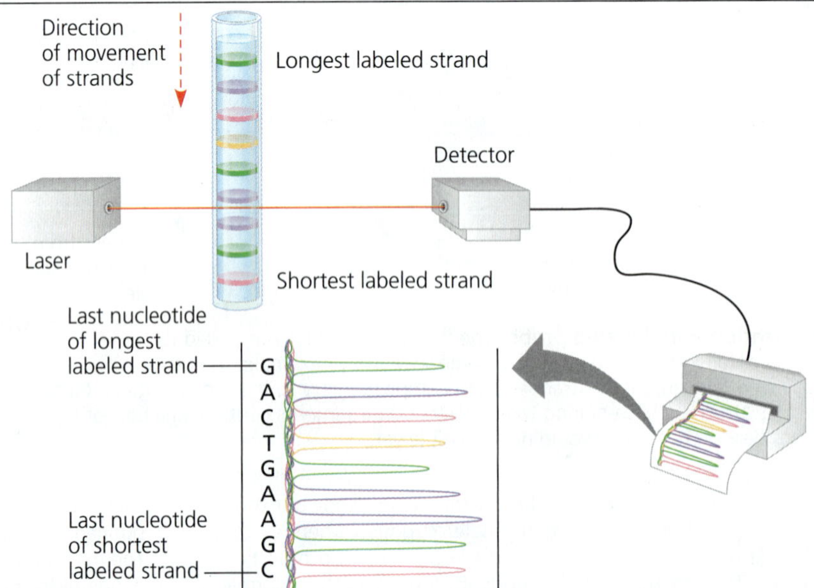

RESULTS The color of the fluorescent tag on each strand indicates the identity of the nucleotide at its end. The results can be printed out as a spectrogram, and the sequence, which is complementary to the template strand, can then be read from bottom (shortest strand) to top (longest strand). (Notice that the sequence here begins after the primer.)

(or *dideoxy*) *chain termination method*, for reasons you can see in **Figure 20.12**. This method was developed by British biochemist Frederick Sanger, who received the Nobel Prize in 1980 for this accomplishment. (One of only four people to win two Nobel Prizes, Sanger also won one in 1975 for determining the amino acid sequence of insulin.)

In the last ten years, "next-generation sequencing" techniques have been developed that do not rely on chain termination. Instead, a single template strand is immobilized, and reagents are added that allow so-called *sequencing by synthesis* of a complementary strand, one nucleotide at a time. A chemical trick enables electronic monitors to identify which of the four nucleotides is added, allowing determination of the sequence. Further technical changes have given rise to "third-generation sequencing," with each new technique being faster and less expensive than the previous. In Chapter 21, you'll learn more about how this rapid acceleration of sequencing technology has enhanced our study of genes and whole genomes.

Knowing the sequence of a gene allows researchers to compare it directly with genes in other species, where the function of the gene product may be known. If two genes from different species are quite similar in sequence, it is reasonable to suppose that their gene products perform similar functions. In this way, sequence comparisons provide clues to a gene's function, a topic we'll return to shortly. Another set of clues is provided by experimental approaches that analyze when and where a gene is expressed.

Analyzing Gene Expression

Having cloned a given gene, researchers can make labeled nucleic acid probes that will hybridize with mRNAs transcribed from the gene. The probes can provide information about when or where in the organism the gene is transcribed. Transcription levels are commonly used as a measure of gene expression.

Studying the Expression of Single Genes

Suppose we want to find out how the expression of the β-globin gene changes during the embryonic development of the hummingbird. There are at least two ways to do this.

The first is called **Northern blotting** (a play on words based on this method's close similarity to Southern blotting). In this method, we carry out gel electrophoresis on samples of mRNA from hummingbird embryos at different stages of development, transfer the samples to a nitrocellulose membrane, and then allow the mRNAs on the membrane to hybridize with a labeled probe recognizing β-globin mRNA. If we expose a film to the membrane, the resulting image will look similar to the Southern blot in Figure 20.11, with one band of a given size showing up in each sample. If the mRNA band is seen at a particular stage, we can hypothesize that the protein functions during events taking place at that stage. Like Southern blotting, Northern blotting has been a mainstay over the years, but it is being supplanted in many labs by other techniques.

A method that is quicker and more sensitive than Northern blotting (because it requires less mRNA) and therefore becoming more widely used is the **reverse transcriptase–polymerase chain reaction**, or **RT-PCR** (**Figure 20.13**). Analysis of hummingbird β-globin gene expression with RT-PCR begins similarly to Northern blotting, with the isolation of mRNAs from different developmental stages of hummingbird embryos. Reverse transcriptase is added next to make cDNA, which then serves as a template for PCR amplification using primers from the β-globin gene. When the products are run on a gel, copies of the amplified region will be observed as bands only in samples that originally contained the β-globin mRNA. In the case of hummingbird β-globin, for instance, we might expect to see a band appear at the stage when red blood cells begin forming, with all subsequent stages

▼ **Figure 20.13** RESEARCH METHOD

RT-PCR Analysis of the Expression of Single Genes

APPLICATION RT-PCR uses the enzyme reverse transcriptase (RT) in combination with PCR and gel electrophoresis. RT-PCR can be used to compare gene expression between samples—for instance, in different embryonic stages, in different tissues, or in the same type of cell under different conditions.

TECHNIQUE In this example, samples containing mRNAs from six embryonic stages of hummingbird were processed as shown below. (The mRNA from only one stage is shown.)

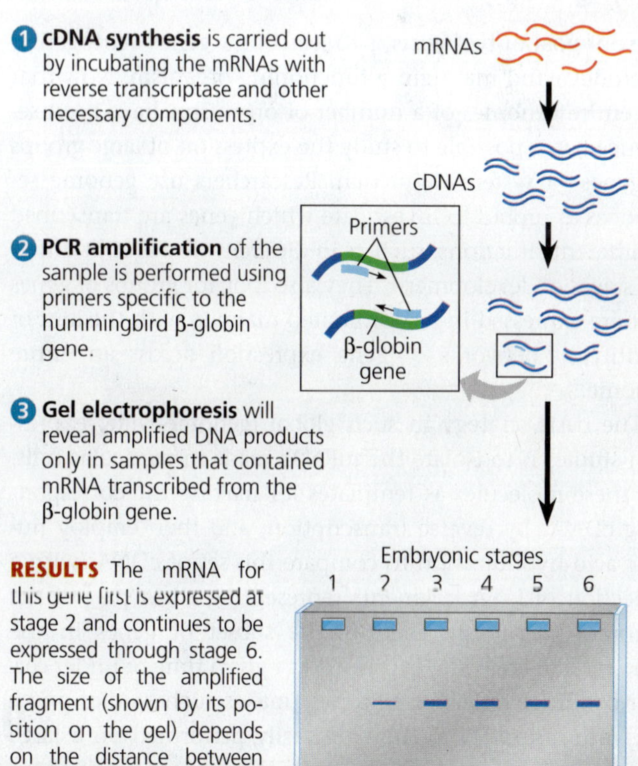

RESULTS The mRNA for this gene first is expressed at stage 2 and continues to be expressed through stage 6. The size of the amplified fragment (shown by its position on the gel) depends on the distance between the primers that were used.

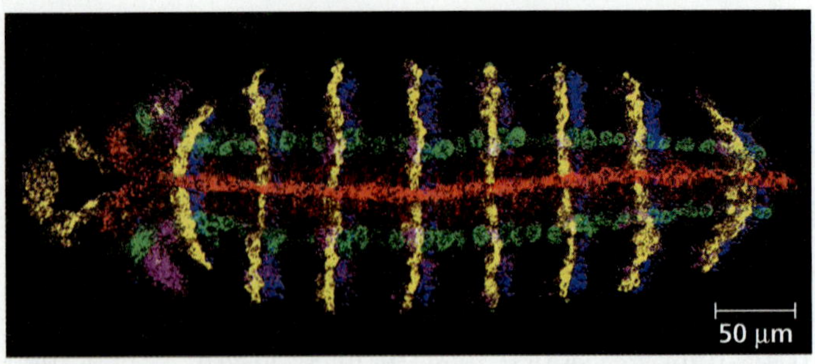

▲ **Figure 20.14 Determining where genes are expressed by *in situ* hybridization analysis.** This *Drosophila* embryo was incubated in a solution containing probes for five different mRNAs, each probe labeled with a different fluorescently colored tag. The embryo was then viewed using fluorescence microscopy. Each color marks where a specific gene is expressed as mRNA.

showing the same band. RT-PCR can also be carried out with mRNAs collected from different tissues at one time to discover which tissue is producing a specific mRNA.

An alternative way to determine which tissues or cells are expressing certain genes is to track down the location of specific mRNAs using labeled probes in place, or *in situ*, in the intact organism. This technique, called ***in situ* hybridization**, is most often carried out with probes labeled by attachment of fluorescent dyes (see Chapter 6). Different probes can be labeled with different dyes, sometimes with strikingly beautiful results **(Figure 20.14)**.

Studying the Expression of Interacting Groups of Genes

A major goal of biologists is to learn how genes act together to produce and maintain a functioning organism. Now that the entire genomes of a number of organisms have been sequenced, it is possible to study the expression of large groups of genes—a systems approach. Researchers use genome sequences as probes to investigate which genes are transcribed in different situations, such as in different tissues or at different stages of development. They also look for groups of genes that are expressed in a coordinated manner, with the aim of identifying networks of gene expression across an entire genome.

The basic strategy in such global (genome-wide) expression studies is to isolate the mRNAs made in particular cells, use these molecules as templates for making the corresponding cDNAs by reverse transcription, and then employ nucleic acid hybridization to compare this set of cDNAs with a collection of DNA fragments representing all or part of the genome. The results identify the subset of genes in the genome that are being expressed at a given time or under certain conditions. DNA technology makes such studies possible; with automation, they are easily performed on a large scale. Scientists can now measure the expression of thousands of genes at one time.

Genome-wide expression studies are made possible by **DNA microarray assays**. A DNA microarray consists of tiny amounts of a large number of single-stranded DNA fragments representing different genes fixed to a glass slide in a tightly spaced array, or grid (see Figure 20.1). (The microarray is also called a *DNA chip* by analogy to a computer chip.) Ideally, these fragments represent all the genes of an organism. **Figure 20.15** outlines how the DNA fragments on a microarray are tested for hybridization with cDNA molecules that have been prepared from the mRNAs in particular cells of interest and labeled with fluorescent dyes.

Using this technique, researchers have performed DNA microarray assays on more than 90% of the genes of the nematode *Caenorhabditis elegans* during every stage of its life cycle. The results show that expression of nearly 60% of *C. elegans* genes changes dramatically during development and that many genes are expressed in a sex-specific pattern. This study supports the model held by most developmental biologists that embryonic development involves a complex and elaborate program of gene expression, rather than simply the expression of a small number of important genes. This example illustrates the ability of DNA microarrays to reveal general profiles of gene expression over the lifetime of an organism.

In addition to uncovering gene interactions and providing clues to gene function, DNA microarray assays may contribute to a better understanding of diseases and suggest new diagnostic techniques or therapies. For instance, comparing patterns of gene expression in breast cancer tumors and noncancerous breast tissue has already resulted in more informed and effective treatment protocols. Ultimately, information from DNA microarray assays should provide a grander view of how ensembles of genes interact to form an organism and maintain its vital systems.

Determining Gene Function

How do scientists determine the function of a gene identified by the techniques described thus far in the chapter? Perhaps the most common approach is to disable the gene and then observe the consequences in the cell or organism. In one application of this approach, called ***in vitro* mutagenesis**, specific mutations are introduced into a cloned gene, and then the mutated gene is returned to a cell in such a way that it disables ("knocks out") the normal cellular copies of the same gene. If the introduced mutations alter or destroy the function of the gene product, the phenotype of the mutant cell may help reveal the function of the missing normal protein. Using molecular and genetic techniques worked out in the 1980s, researchers can even generate mice with any given gene disabled, in order to study the role of that gene in development and in the adult. Mario Capecchi, Martin Evans, and Oliver Smithies received the Nobel Prize in 2007 for first accomplishing this feat.

▼ Figure 20.15 RESEARCH METHOD

DNA Microarray Assay of Gene Expression Levels

APPLICATION With this method, researchers can test thousands of genes simultaneously to determine which ones are expressed in a particular tissue, under different environmental conditions, in various disease states, or at different developmental stages. They can also look for coordinated gene expression.

TECHNIQUE

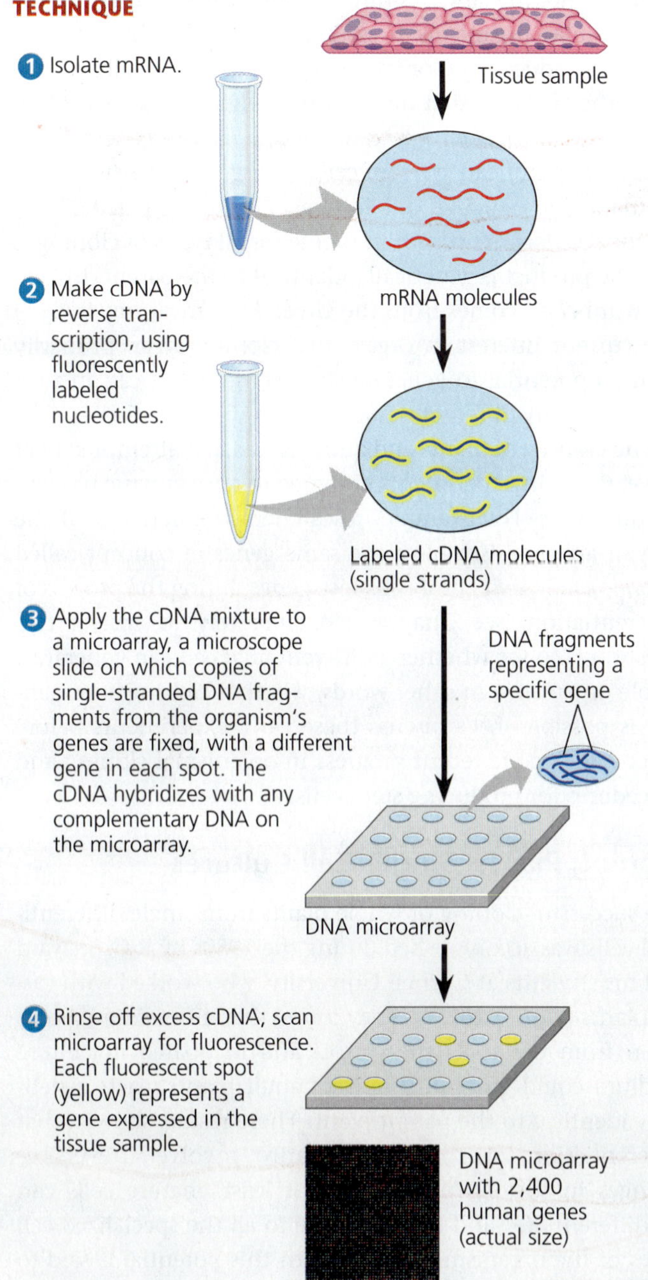

RESULTS The intensity of fluorescence at each spot is a measure of the expression in the tissue sample of the gene represented by that spot. Most often, as in the actual microarray above, two different samples are tested together by labeling the cDNAs prepared from each sample with labels of different colors, often green and red. The resulting color at a spot reveals the relative levels of expression of a particular gene in the two samples: Green indicates expression in one sample, red in the other, yellow in both, and black in neither. (See Figure 20.1 for a larger view.)

A newer method for silencing expression of selected genes exploits the phenomenon of **RNA interference (RNAi)**, described in Chapter 18. This experimental approach uses synthetic double-stranded RNA molecules matching the sequence of a particular gene to trigger breakdown of the gene's messenger RNA or to block its translation. In organisms such as the nematode and the fruit fly, RNAi has already proved valuable for analyzing the functions of genes on a large scale. In one study, RNAi was used to prevent expression of 86% of the genes in early nematode embryos, one gene at a time. Analysis of the phenotypes of the worms that developed from these embryos allowed the researchers to classify most of the genes into a small number of groups by function. This type of analysis, in which the functions of multiple genes are considered in a single study, is sure to become more common as research focuses on the importance of interactions between genes in the system as a whole. This is the basis of systems biology (see Chapter 21).

In humans, ethical considerations prohibit knocking out genes to determine their functions. An alternative approach is to analyze the genomes of large numbers of people with a certain phenotypic condition or disease, such as heart disease or diabetes, to try to find differences they all share compared with people without that condition. These large-scale analyses, called **genome-wide association studies**, do not require complete sequencing of all the genomes in the two groups. Instead, researchers test for *genetic markers*, DNA sequences that vary in the population. In a gene, such sequence variation is the basis of different alleles, as we saw earlier for sickle-cell disease. And just like coding sequences, noncoding DNA at a specific locus on a chromosome may exhibit small nucleotide differences (polymorphisms) among individuals.

Among the most useful of these genetic markers are single base-pair variations in the genomes of the human population. A single base-pair site where variation is found in at least 1% of the population is called a **single nucleotide polymorphism** (**SNP**, pronounced "snip"). A few million SNPs occur in the human genome, about once in 100–300 base pairs of both coding and noncoding DNA sequences. (Roughly 98.5% of our genome doesn't code for protein, as you will learn in Chapter 21.) It isn't necessary to sequence the DNA of multiple individuals to find SNPs; today they can be detected by very sensitive microarray analysis or by PCR.

Once a region is found that has a SNP shared by affected but not unaffected people, researchers focus on that region and sequence it. In the vast majority of cases, the SNP itself does not contribute to the disease, and most SNPs are in noncoding regions. Instead, if the SNP and a disease-causing allele are close enough, scientists can take advantage of the fact that crossing over between the marker and the gene is very unlikely during gamete formation. Therefore, the marker and gene will almost always be inherited together, even though the marker is not

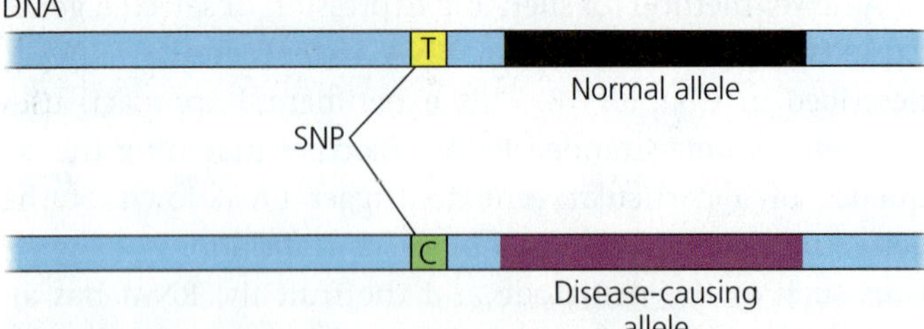

▲ **Figure 20.16 Single nucleotide polymorphisms (SNPs) as genetic markers for disease-causing alleles.** This diagram depicts homologous segments of DNA from two groups of individuals, those in one group having a particular disease or condition with a genetic basis. Unaffected people have a T at a particular SNP locus, while affected people have a C at that locus. A SNP that varies in this way is likely to be closely linked to one or more alleles of genes that contribute to the disease in question. (Here, only a single strand is shown for each DNA molecule.)

MAKE CONNECTIONS *What does it mean for a SNP to be "closely linked" to a disease-causing allele, and how does this allow the SNP to be used as a genetic marker? (See Concept 15.3, p. 296.)*

part of the gene (**Figure 20.16**). SNPs have been found that correlate with diabetes, heart disease, and several types of cancer, and the search is on for genes that might be involved.

The techniques and experimental approaches you have learned about thus far have already taught us a great deal about genes and the functions of their products. This research is now being augmented by the development of powerful techniques for cloning whole multicellular organisms. An aim of this work is to obtain special types of cells, called stem cells, that give rise to all the different kinds of tissues. On a basic level, stem cells would allow scientists to use the DNA-based methods previously discussed to study the process of cell differentiation. On a more applied level, recombinant DNA techniques could be used to alter stem cells for the treatment of disease. Methods involving the cloning of organisms and production of stem cells are the subject of the next section.

CONCEPT CHECK 20.2

1. If you isolated DNA from human cells, treated it with a restriction enzyme, and analyzed the sample by gel electrophoresis, what would you see? Explain.
2. Describe the role of complementary base pairing during Southern blotting, DNA sequencing, Northern blotting, RT-PCR, and microarray analysis.
3. Distinguish between a SNP and a RFLP.
4. **WHAT IF?** Consider the microarray in Figure 20.1, a larger image of the one in Figure 20.15. If a sample from normal tissue is labeled with a green fluorescent dye, and a sample from cancerous tissue is labeled red, what can you conclude about a spot that is green? Red? Yellow? Black? Which genes would you be interested in examining further if you were studying cancer? Explain.

For suggested answers, see Appendix A.

CONCEPT 20.3

Cloning organisms may lead to production of stem cells for research and other applications

In parallel with advances in DNA technology, scientists have been developing and refining methods for cloning whole multicellular organisms from single cells. In this context, cloning produces one or more organisms genetically identical to the "parent" that donated the single cell. This is often called *organismal cloning* to differentiate it from gene cloning and, more significantly, from cell cloning—the division of an asexually reproducing cell into a collection of genetically identical cells. (The common theme for all types of cloning is that the product is genetically identical to the parent. In fact, the word *clone* comes from the Greek *klon*, meaning "twig.") The current interest in organismal cloning arises primarily from its potential to generate stem cells, which can in turn generate many different tissues.

The cloning of plants and animals was first attempted over 50 years ago in experiments designed to answer basic biological questions. For example, researchers wondered if all the cells of an organism have the same genes (a concept called *genomic equivalence*) or if cells lose genes during the process of differentiation (see Chapter 18). One way to answer this question is to see whether a differentiated cell can generate a whole organism—in other words, whether cloning an organism is possible. Let's discuss these early experiments before we consider more recent progress in organismal cloning and procedures for producing stem cells.

Cloning Plants: Single-Cell Cultures

The successful cloning of whole plants from single differentiated cells was accomplished during the 1950s by F. C. Steward and his students at Cornell University, who worked with carrot plants (**Figure 20.17**). They found that differentiated cells taken from the root (the carrot) and incubated in culture medium could grow into normal adult plants, each genetically identical to the parent plant. These results showed that differentiation does not necessarily involve irreversible changes in the DNA. In plants, at least, mature cells can "dedifferentiate" and then give rise to all the specialized cell types of the organism. Any cell with this potential is said to be **totipotent**.

Plant cloning is now used extensively in agriculture. For some plants, such as orchids, cloning is the only commercially practical means of reproducing plants. In other cases, cloning has been used to reproduce a plant with valuable characteristics, such as the ability to resist a plant pathogen. In fact, you yourself may be a plant cloner: If you have ever grown a new plant from a cutting, you have practiced cloning!

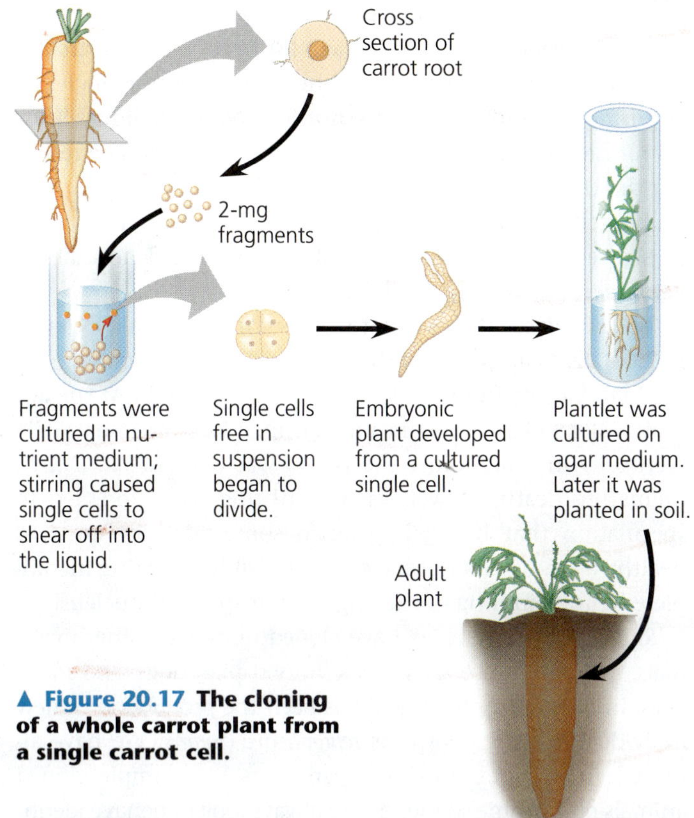

▲ **Figure 20.17 The cloning of a whole carrot plant from a single carrot cell.**

Cloning Animals: Nuclear Transplantation

Differentiated cells from animals generally do not divide in culture, much less develop into the multiple cell types of a new organism. Therefore, early researchers had to use a different approach to the question of whether differentiated animal cells can be totipotent. Their approach was to remove the nucleus of an unfertilized or fertilized egg and replace it with the nucleus of a differentiated cell, a procedure called *nuclear transplantation*. If the nucleus from the differentiated donor cell retains its full genetic capability, then it should be able to direct development of the recipient cell into all the tissues and organs of an organism.

Such experiments were conducted on one species of frog (*Rana pipiens*) by Robert Briggs and Thomas King in the 1950s and on another (*Xenopus laevis*) by John Gurdon in the 1970s. These researchers transplanted a nucleus from an embryonic or tadpole cell into an enucleated (nucleus-lacking) egg of the same species. In Gurdon's experiments, the transplanted nucleus was often able to support normal development of the egg into a tadpole **(Figure 20.18)**. However, he found that the potential of a transplanted nucleus to direct normal development was inversely related to the age of the donor: the older the donor nucleus, the lower the percentage of normally developing tadpoles.

From these results, Gurdon concluded that something in the nucleus *does* change as animal cells differentiate. In frogs and most other animals, nuclear potential tends to be restricted more and more as embryonic development and cell differentiation progress.

▼ **Figure 20.18** **INQUIRY**

Can the nucleus from a differentiated animal cell direct development of an organism?

EXPERIMENT John Gurdon and colleagues at Oxford University, in England, destroyed the nuclei of frog (*Xenopus laevis*) eggs by exposing the eggs to ultraviolet light. They then transplanted nuclei from cells of frog embryos and tadpoles into the enucleated eggs.

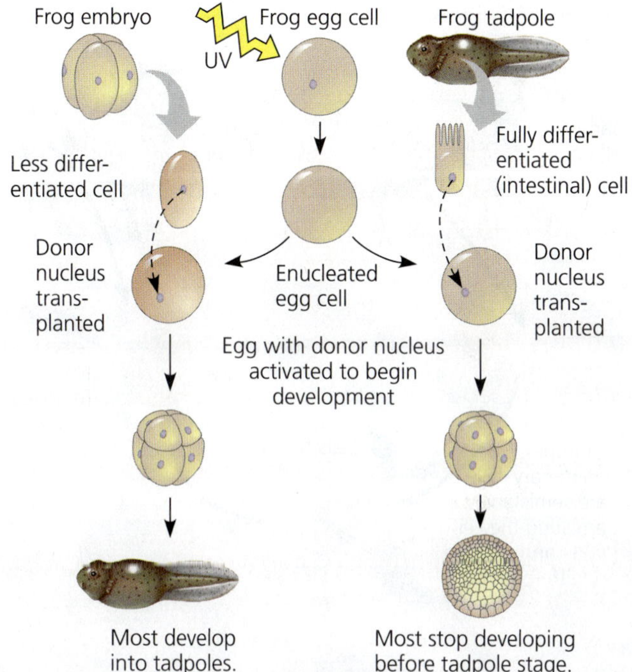

RESULTS When the transplanted nuclei came from an early embryo, whose cells are relatively undifferentiated, most of the recipient eggs developed into tadpoles. But when the nuclei came from the fully differentiated intestinal cells of a tadpole, fewer than 2% of the eggs developed into normal tadpoles, and most of the embryos stopped developing at a much earlier stage.

CONCLUSION The nucleus from a differentiated frog cell can direct development of a tadpole. However, its ability to do so decreases as the donor cell becomes more differentiated, presumably because of changes in the nucleus.

SOURCE J. B. Gurdon et al., The developmental capacity of nuclei transplanted from keratinized cells of adult frogs, *Journal of Embryology and Experimental Morphology* 34:93–112 (1975).

WHAT IF? If each cell in a four-cell embryo was already so specialized that it was not totipotent, what results would you predict for the experiment on the left side of the figure?

Reproductive Cloning of Mammals

In addition to cloning frogs, researchers have long been able to clone mammals by transplanting nuclei or cells from a variety of early embryos. But it was not known whether a nucleus from a fully differentiated cell could be reprogrammed to succeed in acting as a donor nucleus. In 1997, however, researchers at the Roslin Institute in Scotland captured newspaper headlines when they announced the birth of Dolly, a

▼ **Figure 20.19** RESEARCH METHOD

Reproductive Cloning of a Mammal by Nuclear Transplantation

APPLICATION This method is used to produce cloned animals whose nuclear genes are identical to those of the animal supplying the nucleus.

TECHNIQUE Shown here is the procedure used to produce Dolly, the first reported case of a mammal cloned using the nucleus of a differentiated cell.

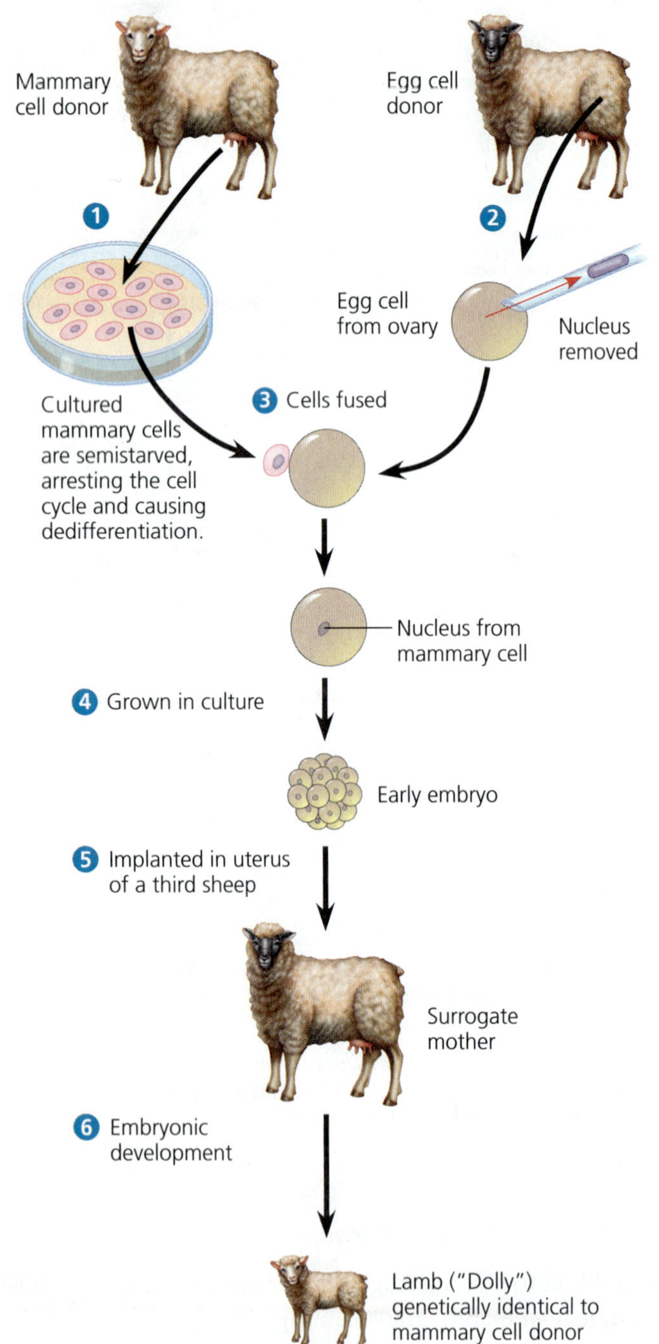

RESULTS The genetic makeup of the cloned animal is identical to that of the animal supplying the nucleus but differs from that of the egg donor and surrogate mother. (The latter two are "Scottish blackface" sheep, with dark faces.)

lamb cloned from an adult sheep by nuclear transplantation from a differentiated cell **(Figure 20.19)**. These researchers achieved the necessary dedifferentiation of donor nuclei by culturing mammary cells in nutrient-poor medium. They then fused these cells with enucleated sheep eggs. The resulting diploid cells divided to form early embryos, which were implanted into surrogate mothers. Out of several hundred implanted embryos, one successfully completed normal development, and Dolly was born.

Later analyses showed that Dolly's chromosomal DNA was indeed identical to that of the nucleus donor. (Her mitochondrial DNA came from the egg donor, as expected.) At the age of 6, Dolly suffered complications from a lung disease usually seen only in much older sheep and was euthanized. Dolly's premature death, as well as her arthritic condition, led to speculation that her cells were in some way not quite as healthy as those of a normal sheep, possibly reflecting incomplete reprogramming of the original transplanted nucleus.

Since 1997, researchers have cloned numerous other mammals, including mice, cats, cows, horses, pigs, dogs, and monkeys. In most cases, their goal has been the production of new individuals; this is known as *reproductive cloning*. We have already learned a lot from such experiments. For example, cloned animals of the same species do *not* always look or behave identically. In a herd of cows cloned from the same line of cultured cells, certain cows are dominant in behavior and others are more submissive. Another example of nonidentity in clones is the first cloned cat, named CC for Carbon Copy **(Figure 20.20)**. She has a calico coat, like her single female parent, but the color and pattern are different because of random X chromosome inactivation, which is a normal occurrence during embryonic development (see Figure 15.8). And identical human twins, which

▲ **Figure 20.20 CC, the first cloned cat, and her single parent.** Rainbow (left) donated the nucleus in a cloning procedure that resulted in CC (right). However, the two cats are not identical: Rainbow is a classic calico cat with orange patches on her fur and has a "reserved personality," while CC has a gray and white coat and is more playful.

are naturally occurring "clones," are always slightly different. Clearly, environmental influences and random phenomena can play a significant role during development.

The successful cloning of so many mammals has heightened speculation about the cloning of humans. Scientists in several labs around the world have tackled the first steps of human cloning. In the most common approach, nuclei from differentiated human cells are transplanted into unfertilized enucleated eggs, and the eggs are stimulated to divide. In 2001, a research group at a biotechnology company in Massachusetts observed a few early cell divisions in such an experiment. A few years later, researchers at Seoul National University, in South Korea, reported cloning embryos to an early stage called the blastocyst stage, but the scientists were later found guilty of research misconduct and data fabrication. This episode sent shock waves through the scientific community. In 2007, the first primate (macaque) embryos were cloned by researchers at the Oregon National Primate Research Center; these clones reached the blastocyst stage. This achievement has moved the field one step closer to human cloning, the prospect of which raises unprecedented ethical issues.

Problems Associated with Animal Cloning

In most nuclear transplantation studies thus far, only a small percentage of cloned embryos develop normally to birth. And like Dolly, many cloned animals exhibit defects. Cloned mice, for instance, are prone to obesity, pneumonia, liver failure, and premature death. Scientists assert that even cloned animals that appear normal are likely to have subtle defects.

In recent years, we have begun to uncover some reasons for the low efficiency of cloning and the high incidence of abnormalities. In the nuclei of fully differentiated cells, a small subset of genes is turned on and expression of the rest is repressed. This regulation often is the result of epigenetic changes in chromatin, such as acetylation of histones or methylation of DNA (see Figure 18.7). During the nuclear transfer procedure, many of these changes must be reversed in the later-stage nucleus from a donor animal for genes to be expressed or repressed appropriately in early stages of development. Researchers have found that the DNA in cells from cloned embryos, like that of differentiated cells, often has more methyl groups than does the DNA in equivalent cells from normal embryos of the same species. This finding suggests that the reprogramming of donor nuclei requires chromatin restructuring, which occurs incompletely during cloning procedures. Because DNA methylation helps regulate gene expression, misplaced methyl groups in the DNA of donor nuclei may interfere with the pattern of gene expression necessary for normal embryonic development. In fact, the success of a cloning attempt may depend in large part on whether or not the chromatin in the donor nucleus can be artificially modified to resemble that of a newly fertilized egg.

Stem Cells of Animals

The major goal of cloning human embryos is not reproduction, but the production of stem cells for treating human diseases. A **stem cell** is a relatively unspecialized cell that can both reproduce itself indefinitely and, under appropriate conditions, differentiate into specialized cells of one or more types. Thus, stem cells are able both to replenish their own population and to generate cells that travel down specific differentiation pathways.

Many early animal embryos contain stem cells capable of giving rise to differentiated embryonic cells of any type. Stem cells can be isolated from early embryos at a stage called the blastula stage or its human equivalent, the blastocyst stage (**Figure 20.21**). In culture, these *embryonic stem (ES) cells*

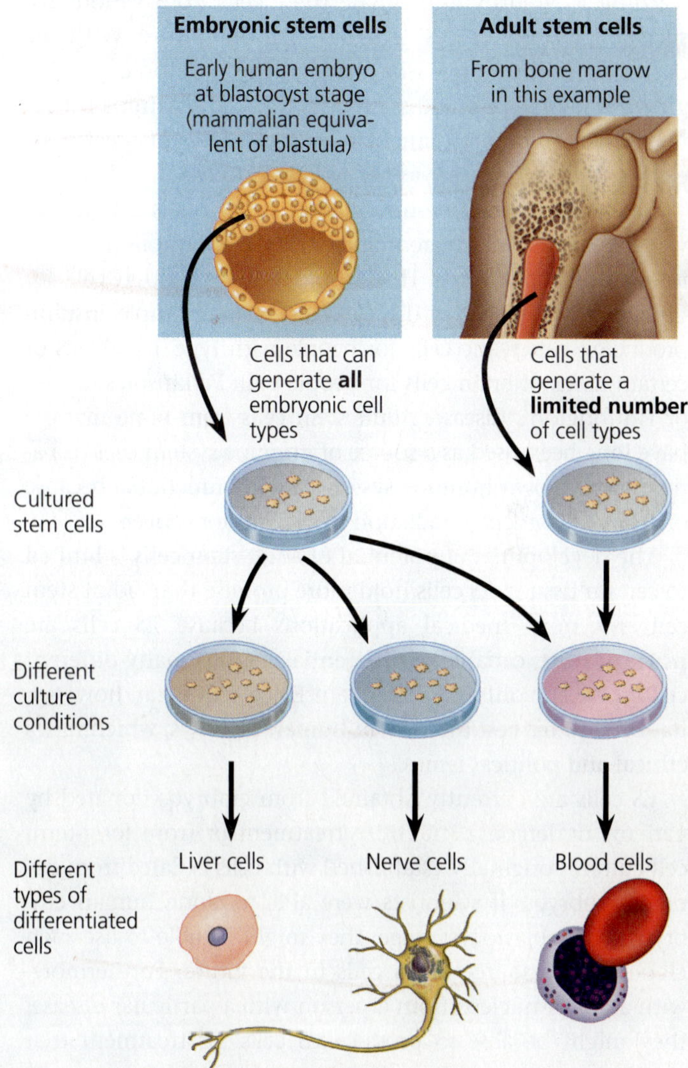

▲ **Figure 20.21 Working with stem cells.** Animal stem cells, which can be isolated from early embryos or adult tissues and grown in culture, are self-perpetuating, relatively undifferentiated cells. Embryonic stem cells are easier to grow than adult stem cells and can theoretically give rise to *all* types of cells in an organism. The range of cell types that can arise from adult stem cells is not yet fully understood.

reproduce indefinitely; and depending on culture conditions, they can be made to differentiate into a wide variety of specialized cells, including even eggs and sperm.

The adult body also has stem cells, which serve to replace nonreproducing specialized cells as needed. In contrast to ES cells, *adult stem cells* are not able to give rise to all cell types in the organism, though they can generate multiple types. For example, one of the several types of stem cells in bone marrow can generate all the different kinds of blood cells (see Figure 20.21), and another can differentiate into bone, cartilage, fat, muscle, and the linings of blood vessels. To the surprise of many, the adult brain has been found to contain stem cells that continue to produce certain kinds of nerve cells there. And recently, researchers have reported finding stem cells in skin, hair, eyes, and dental pulp. Although adult animals have only tiny numbers of stem cells, scientists are learning to identify and isolate these cells from various tissues and, in some cases, to grow them in culture. With the right culture conditions (for instance, the addition of specific growth factors), cultured stem cells from adult animals have been made to differentiate into multiple types of specialized cells, although none are as versatile as ES cells.

Research with embryonic or adult stem cells is a source of valuable data about differentiation and has enormous potential for medical applications. The ultimate aim is to supply cells for the repair of damaged or diseased organs: for example, insulin-producing pancreatic cells for people with type 1 diabetes or certain kinds of brain cells for people with Parkinson's disease or Huntington's disease. Adult stem cells from bone marrow have long been used as a source of immune system cells in patients whose own immune systems are nonfunctional because of genetic disorders or radiation treatments for cancer.

The developmental potential of adult stem cells is limited to certain tissues. ES cells hold more promise than adult stem cells for most medical applications because ES cells are **pluripotent**, capable of differentiating into many different cell types. The only way to obtain ES cells thus far, however, has been to harvest them from human embryos, which raises ethical and political issues.

ES cells are currently obtained from embryos donated by patients undergoing infertility treatment or from long-term cell cultures originally established with cells isolated from donated embryos. If scientists were able to clone human embryos to the blastocyst stage, they might be able to use such clones as the source of ES cells in the future. Furthermore, with a donor nucleus from a person with a particular disease, they might be able to produce ES cells for treatment that match the patient and are thus not rejected by his or her immune system. When the main aim of cloning is to produce ES cells to treat disease, the process is called *therapeutic cloning*. Although most people believe that reproductive cloning of humans is unethical, opinions vary about the morality of therapeutic cloning.

Resolving the debate now seems less imperative because researchers have been able to turn back the clock in fully differentiated cells, reprogramming them to act like ES cells. The accomplishment of this feat, which posed formidable obstacles, was announced in 2007, first by labs using mouse skin cells and then by additional groups using cells from human skin and other organs or tissues. In all these cases, researchers transformed the differentiated cells into ES cells by using retroviruses to introduce extra cloned copies of four "stem cell" master regulatory genes. All the tests that were carried out at the time indicated that the transformed cells, known as *induced pluripotent stem (iPS) cells*, could do everything ES cells can do. More recently, however, several research groups have uncovered differences between iPS and ES cells in gene expression and other cellular functions, such as cell division. At least until these differences are fully understood, the study of ES cells will continue to make important contributions to the development of stem cell therapies. (In fact, ES cells will likely always be a focus of basic research as well.) In the meantime, work is proceeding using the iPS cells in hand.

There are two major potential uses for human iPS cells. First, cells from patients suffering from diseases can be reprogrammed to become iPS cells, which can act as model cells for studying the disease and potential treatments. Human iPS cell lines have already been developed from individuals with type 1 diabetes, Parkinson's disease, and at least a dozen other diseases. Second, in the field of regenerative medicine, a patient's own cells could be reprogrammed into iPS cells and then used to replace nonfunctional tissues (**Figure 20.22**). Developing techniques that direct iPS cells to become specific cell types for this purpose is an area of intense research, one that has already seen some success. The iPS cells created in this way could eventually provide tailor-made "replacement" cells for patients without using any human eggs or embryos, thus circumventing most ethical objections.

CONCEPT CHECK 20.3

1. Based on current knowledge, how would you explain the difference in the percentage of tadpoles that developed from the two kinds of donor nuclei in Figure 20.18?
2. If you were to clone a carrot using the technique shown in Figure 20.17, would all the progeny plants ("clones") look identical? Why or why not?
3. **WHAT IF?** If you were a doctor who wanted to use iPS cells to treat a patient with severe type 1 diabetes, what new technique would have to be developed?
4. **MAKE CONNECTIONS** Compare an individual carrot cell in Figure 20.17 with the fully differentiated muscle cell in Figure 18.18 (p. 369) in terms of their potential to develop into different cell types.

For suggested answers, see Appendix A.

▼ Figure 20.22
IMPACT

The Impact of Induced Pluripotent Stem (iPS) Cells on Regenerative Medicine

While embryonic stem (ES) cells can generate every cell in an organism, the use of human embryos as their source is controversial. Several research groups have developed similar procedures for reprogramming fully differentiated cells to become induced pluripotent stem (iPS) cells, which act like ES cells. The technique is based on introducing transcription factors that are characteristic of stem cells into differentiated cells, such as skin cells.

WHY IT MATTERS Patients with diseases such as heart disease, diabetes, or Alzheimer's could have their own skin cells reprogrammed to become iPS cells. Once procedures have been developed for converting iPS cells into heart, pancreatic, or nervous system cells, the patients' own iPS cells might be used to treat their disease. This technique has already been used successfully to treat sickle-cell disease in a mouse that had been genetically engineered to have the disease. Shown below is how this therapy could work in humans, once researchers learn how iPS cells can be triggered to differentiate as desired (step 3).

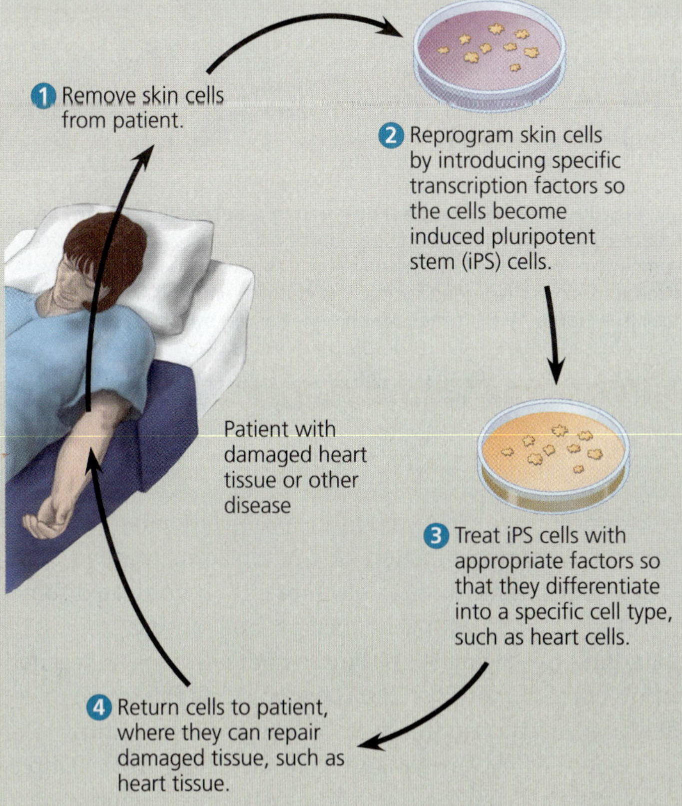

FURTHER READING G. Vogel and C. Holden, Field leaps forward with new stem cell advances, *Science* 318:1224–1225 (2007); K. Hochedlinger, Your inner healers, *Scientific American* 302:46–53 (2010).

WHAT IF? When organs are transplanted from a donor to a diseased recipient, the recipient's immune system may reject the transplant, a condition with serious and often fatal consequences. Would using converted iPS cells be expected to carry the same risk? Why or why not?

CONCEPT 20.4

The practical applications of DNA technology affect our lives in many ways

DNA technology is in the news almost every day. Most often, the topic is a new and promising application in medicine, but this is just one of numerous fields benefiting from DNA technology and genetic engineering.

Medical Applications

One important use of DNA technology is the identification of human genes whose mutation plays a role in genetic diseases. These discoveries may lead to ways of diagnosing, treating, and even preventing such conditions. DNA technology is also contributing to our understanding of "nongenetic" diseases, from arthritis to AIDS, since a person's genes influence susceptibility to these diseases. Furthermore, diseases of all sorts involve changes in gene expression within the affected cells and often within the patient's immune system. By using DNA microarray assays or other techniques to compare gene expression in healthy and diseased tissues, as seen in Figure 20.1, researchers hope to find many of the genes that are turned on or off in particular diseases. These genes and their products are potential targets for prevention or therapy.

Diagnosis and Treatment of Diseases

A new chapter in the diagnosis of infectious diseases has been opened by DNA technology, in particular the use of PCR and labeled nucleic acid probes to track down pathogens. For example, because the sequence of the RNA genome of HIV is known, RT-PCR can be used to amplify, and thus detect, HIV RNA in blood or tissue samples (see Figure 20.13). RT-PCR is often the best way to detect an otherwise elusive infective agent.

Medical scientists can now diagnose hundreds of human genetic disorders by using PCR with primers that target the genes associated with these disorders. The amplified DNA product is then sequenced to reveal the presence or absence of the disease-causing mutation. Among the genes for human diseases that have been identified are those for sickle-cell disease, hemophilia, cystic fibrosis, Huntington's disease, and Duchenne muscular dystrophy. Individuals afflicted with such diseases can often be identified before the onset of symptoms, even before birth. PCR can also be used to identify symptomless carriers of potentially harmful recessive alleles, essentially replacing Southern blotting for this purpose.

As you learned earlier, genome-wide association studies have pinpointed SNPs (single nucleotide polymorphisms) that are linked to disease-causing alleles. Individuals can be tested for the presence of a SNP that indicates the presence of the abnormal allele. The presence of particular SNPs is correlated with increased risk for conditions such as heart disease, Alzheimer's, and some types of cancer. Companies that offer individual genetic testing for risk factors like these are looking for the presence of previously identified, linked SNPs. It may be helpful for an individual to learn about their health risks, with the understanding that such genetic tests merely reflect correlations and do not make predictions.

The techniques described in this chapter have also prompted improvements in disease treatments. By analyzing the expression of many genes in breast cancer patients, researchers carrying out one genome-wide association study were able to identify 70 genes whose expression pattern could be correlated with the likelihood that the cancer would recur. Given that low-risk patients have a 96% survival rate over a ten-year period with no treatment, gene expression analysis allows doctors and patients access to valuable information when they are considering treatment options.

Many envision a future of "personalized medicine" where each person's genetic health profile can inform them about diseases or conditions for which they are especially at risk and help them make treatment choices. A genetic profile is currently taken to mean a set of genetic markers such as SNPs, but ultimately it could mean the complete DNA sequence of each individual—once sequencing becomes inexpensive enough.

Human Gene Therapy

Gene therapy—introducing genes into an afflicted individual for therapeutic purposes—holds great potential for treating the relatively small number of disorders traceable to a single defective gene. In theory, a normal allele of the defective gene could be inserted into the somatic cells of the tissue affected by the disorder.

For gene therapy of somatic cells to be permanent, the cells that receive the normal allele must be ones that multiply throughout the patient's life. Bone marrow cells, which include the stem cells that give rise to all the cells of the blood and immune system, are prime candidates. **Figure 20.23** outlines one possible procedure for gene therapy of an individual whose bone marrow cells do not produce a vital enzyme because of a single defective gene. One type of severe combined immunodeficiency (SCID) is caused by just this kind of defect. If the treatment is successful, the patient's bone marrow cells will begin producing the missing protein, and the patient may be cured.

The procedure shown in Figure 20.23 has been used in gene therapy trials for SCID. In a trial begun in France in 2000, ten young children with SCID were treated by the same procedure. Nine of these patients showed significant, definitive improvement after two years, the first indisputable success of gene therapy. However, three of the patients subsequently developed leukemia, a type of blood cell cancer, and one of them died. Two factors may have contributed to the development of leukemia: the insertion of the retroviral

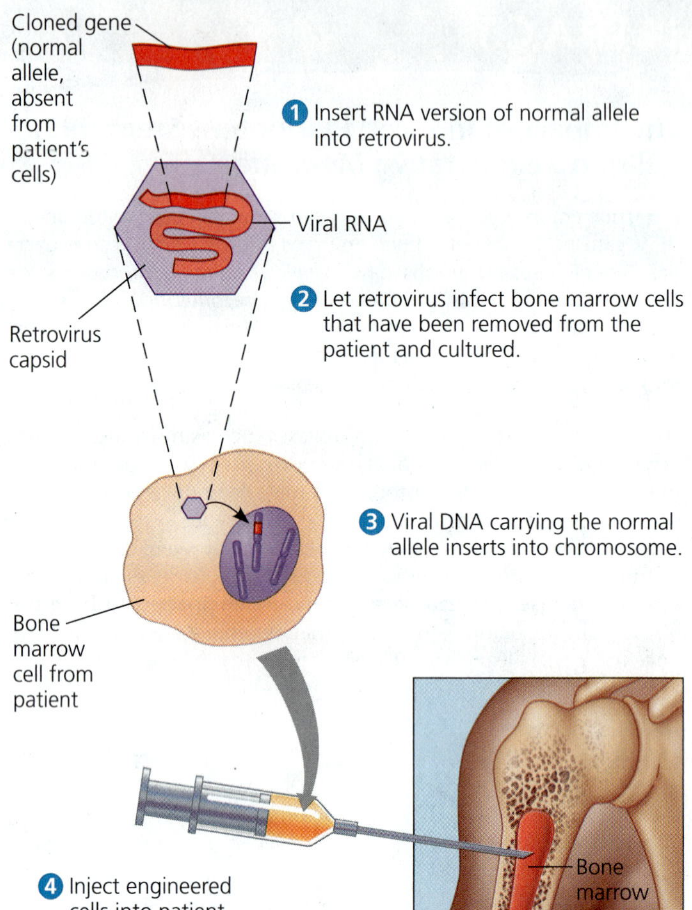

▲ **Figure 20.23 Gene therapy using a retroviral vector.** A retrovirus that has been rendered harmless is used as a vector in this procedure, which exploits the ability of a retrovirus to insert a DNA transcript of its RNA genome into the chromosomal DNA of its host cell (see Figure 19.8). If the foreign gene carried by the retroviral vector is expressed, the cell and its descendants will possess the gene product. Cells that reproduce throughout life, such as bone marrow cells, are ideal candidates for gene therapy.

vector near a gene involved in the proliferation of blood cells and an unknown function of the replacement gene itself. Two other genetic diseases have recently been treated somewhat successfully with gene therapy: one causing progressive blindness (see Figure 50.21) and the other leading to degeneration of the nervous system. The successful trials involve very few patients, but are still cause for cautious optimism.

Gene therapy raises many technical issues. For example, how can the activity of the transferred gene be controlled so that cells make appropriate amounts of the gene product at the right time and in the right place? How can we be sure that the insertion of the therapeutic gene does not harm some other necessary cell function? As more is learned about DNA control elements and gene interactions, researchers may be able to answer such questions.

In addition to technical challenges, gene therapy provokes ethical questions. Some critics believe that tampering with human genes in any way is immoral. Other observers see no fundamental difference between the transplantation of genes into somatic cells and the transplantation of organs. You might wonder whether scientists are considering engineering human germ-line cells in the hope of correcting a defect in future generations. At present, no one in the mainstream scientific community is pursuing this goal—it is considered much too risky. Such genetic engineering is routinely done in laboratory mice, though, and the technical problems relating to similar genetic engineering in humans will eventually be solved. Under what circumstances, if any, should we alter the genomes of human germ lines? Would this inevitably lead to the practice of eugenics, a deliberate effort to control the genetic makeup of human populations? While we may not have to resolve these questions right now, considering them is worthwhile because they will probably arise at some point in the future.

Pharmaceutical Products

The pharmaceutical industry derives significant benefit from advances in DNA technology and genetic research, applying them to the development of useful drugs to treat diseases. Pharmaceutical products are synthesized using methods of either organic chemistry or biotechnology, depending on the nature of the product.

Synthesis of Small Molecules for Use as Drugs Determining the sequence and structure of proteins crucial for tumor cell survival has led to the identification of small molecules that combat certain cancers by blocking the function of these proteins. One drug, imatinib (trade name Gleevec), is a small molecule that inhibits a specific receptor tyrosine kinase (see Figure 11.7). The overexpression of this receptor, resulting from a chromosomal translocation, is instrumental in causing chronic myelogenous leukemia (CML; see Figure 15.16). Patients in the early stages of CML who are treated with imatinib have exhibited nearly complete, sustained remission from the cancer. Drugs that work like this have also been used with success to treat a few types of lung and breast cancers. This approach is feasible only for cancers for which the molecular basis is fairly well understood.

Pharmaceutical products that are proteins can be synthesized on a large scale, using cells or whole organisms. Cell cultures are more widely used at present.

Protein Production in Cell Cultures You learned earlier in the chapter about DNA cloning and gene expression systems for producing large quantities of proteins that are present naturally in only minute amounts. The host cells used in such expression systems can even be engineered to secrete a protein as it is made, thereby simplifying the task of purifying it by traditional biochemical methods.

Among the first pharmaceutical products "manufactured" in this way were human insulin and human growth hormone (HGH). Some 2 million people with diabetes in the United States depend on insulin treatment to control their disease. Human growth hormone has been a boon to children born with a form of dwarfism caused by inadequate amounts of HGH. Another important pharmaceutical product produced by genetic engineering is tissue plasminogen activator (TPA). If administered shortly after a heart attack, TPA helps dissolve blood clots and reduces the risk of subsequent heart attacks.

Protein Production by "Pharm" Animals In some cases, instead of using cell systems to produce large quantities of protein products, pharmaceutical scientists can use whole animals. They can introduce a gene from an animal of one genotype into the genome of another individual, often of a different species. This individual is then called a **transgenic** animal. To do this, they first remove eggs from a female of the recipient species and fertilize them *in vitro*. Meanwhile, they have cloned the desired gene from the donor organism. They then inject the cloned DNA directly into the nuclei of the fertilized eggs. Some of the cells integrate the foreign DNA, the *transgene*, into their genomes and are able to express the foreign gene. The engineered embryos are then surgically implanted in a surrogate mother. If an embryo develops successfully, the result is a transgenic animal that expresses its new, "foreign" gene.

Assuming that the introduced gene encodes a protein desired in large quantities, these transgenic animals can act as pharmaceutical "factories." For example, a transgene for a human blood protein such as antithrombin can be inserted into the genome of a goat in such a way that the transgene's product is secreted in the animal's milk **(Figure 20.24)**. The protein is then purified from the milk (which is easier than purification from a cell culture). Researchers have also engineered transgenic chickens that express large amounts of the transgene's product in eggs. Biotechnology companies consider the characteristics of candidate animals in deciding which to use for engineering. For

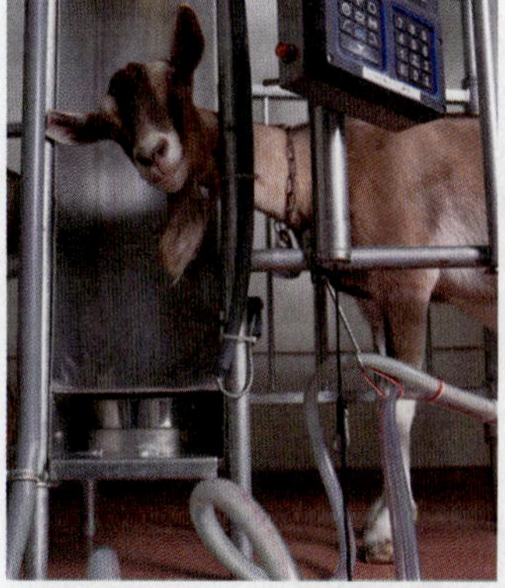

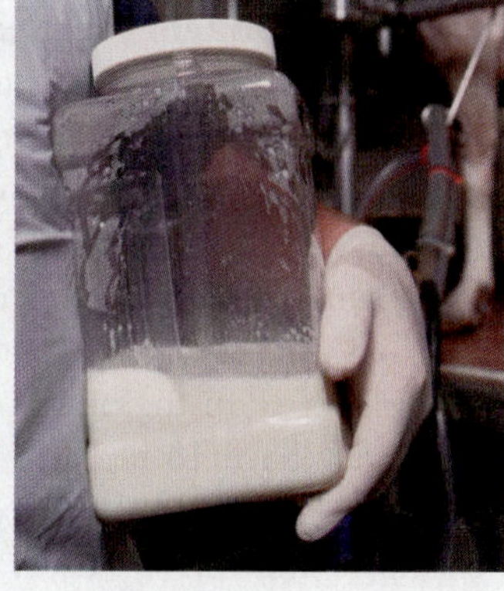

▲ **Figure 20.24 Goats as "pharm" animals.** This transgenic goat carries a gene for a human blood protein, antithrombin, which she secretes in her milk. Patients with a rare hereditary disorder in which this protein is lacking suffer from formation of blood clots in their blood vessels. Easily purified from the goat's milk, the protein has been approved in the United States and Europe for treating these patients.

example, goats reproduce faster than cows, and it is possible to harvest more protein from goat milk than from the milk of other rapidly reproducing mammals, such as rabbits.

Human proteins produced in transgenic farm animals for use in humans may differ in some ways from the naturally produced human proteins, possibly because of subtle differences in protein modification. Therefore, such proteins must be tested very carefully to ensure that they (or contaminants from the farm animals) will not cause allergic reactions or other adverse effects in patients who receive them.

Forensic Evidence and Genetic Profiles

In violent crimes, body fluids or small pieces of tissue may be left at the scene or on the clothes or other possessions of the victim or assailant. If enough blood, semen, or tissue is available, forensic laboratories can determine the blood type or tissue type by using antibodies to detect specific cell-surface proteins. However, such tests require fairly fresh samples in relatively large amounts. Also, because many people have the same blood or tissue type, this approach can only exclude a suspect; it cannot provide strong evidence of guilt.

DNA testing, on the other hand, can identify the guilty individual with a high degree of certainty, because the DNA sequence of every person is unique (except for identical twins). Genetic markers that vary in the population can be analyzed for a given person to determine that individual's unique set of genetic markers, or **genetic profile**. (This term is preferred over "DNA fingerprint" by forensic scientists, who want to emphasize the heritable aspect of these markers rather than the fact that they produce a pattern on a gel that, like a fingerprint, is visually recognizable.) The FBI started applying DNA technology to forensics in 1988, using RFLP analysis by Southern blotting to detect similarities and differences in DNA samples. This method required much smaller samples of blood or tissue than earlier methods—only about 1,000 cells.

Today, forensic scientists use an even more sensitive method that takes advantage of variations in length of genetic markers called **short tandem repeats (STRs)**. These are tandemly repeated units of two- to five-base sequences in specific regions of the genome. The number of repeats present in these regions is highly variable from person to person (polymorphic), and for one individual, the two alleles of an STR may even differ from each other. For example, one individual may have the sequence ACAT repeated 30 times at one genome locus and 15 times at the same locus on the other homolog, whereas another individual may have 18 repeats at this locus on each homolog. (These two genotypes can be expressed by the two repeat numbers: 30,15 and 18,18.) PCR is used to amplify particular STRs, using sets of primers that are labeled with different-colored fluorescent tags; the length of the region, and thus the number of repeats, can then be determined by electrophoresis. Because Southern blotting is not required, this method is quicker than RFLP analysis. And the PCR step allows use of the method even when the DNA is in poor condition or available only in minute quantities. A tissue sample containing as few as 20 cells can be sufficient for PCR amplification.

In a murder case, for example, this method can be used to compare DNA samples from the suspect, the victim, and a small amount of blood found at the crime scene. The forensic scientist tests only a few selected portions of the DNA—usually 13 STR markers. However, even this small set of markers can provide a forensically useful genetic profile because the probability that two people (who are not identical twins) would have exactly the same set of STR markers is vanishingly small. The Innocence Project, a nonprofit organization dedicated to overturning wrongful convictions, uses STR analysis of archived samples from crime scenes to revisit old cases. As of 2010, more than 250 innocent people had been released from prison as a result of forensic and legal work by this group **(Figure 20.25)**.

Genetic profiles can also be useful for other purposes. A comparison of the DNA of a mother, her child, and the purported father can conclusively settle a question of paternity. Sometimes paternity is of historical interest: Genetic profiles

(a) In 1984, Earl Washington was convicted and sentenced to death for the 1982 rape and murder of Rebecca Williams. His sentence was commuted to life in prison in 1993 due to new doubts about the evidence. In 2000, STR analysis by forensic scientists associated with The Innocence Project showed conclusively that he was innocent. This photo shows Washington just before his release in 2001, after 17 years in prison.

Source of sample	STR marker 1	STR marker 2	STR marker 3
Semen on victim	17,19	13,16	12,12
Earl Washington	16,18	14,15	11,12
Kenneth Tinsley	17,19	13,16	12,12

(b) In STR analysis, selected STR markers in a DNA sample are amplified by PCR, and the PCR products are separated by electrophoresis. The procedure reveals how many repeats are present for each STR locus in the sample. An individual has two alleles per STR locus, each with a certain number of repeats. This table shows the number of repeats for three STR markers in three samples: from semen found on the victim, from Washington, and from another man named Kenneth Tinsley, who was in prison because of an unrelated conviction. These and other STR data (not shown) exonerated Washington and led Tinsley to plead guilty to the murder.

▲ **Figure 20.25 STR analysis used to release an innocent man from prison.**

provided strong evidence that Thomas Jefferson or one of his close male relatives fathered at least one of the children of his slave Sally Hemings. Genetic profiles can also identify victims of mass casualties. The largest such effort occurred after the attack on the World Trade Center in 2001; more than 10,000 samples of victims' remains were compared with DNA samples from personal items, such as toothbrushes, provided by families. Ultimately, forensic scientists succeeded in identifying almost 3,000 victims using these methods.

Just how reliable is a genetic profile? The greater the number of markers examined in a DNA sample, the more likely it is that the profile is unique to one individual. In forensic cases using STR analysis with 13 markers, the probability of two people having identical DNA profiles is somewhere between one chance in 10 billion and one in several trillion. (For comparison, the world's population in 2009 was about 6.8 billion.) The exact probability depends on the frequency of those markers in the general population. Information on how common various markers are in different ethnic groups is critical because these marker frequencies may vary considerably among ethnic groups and between a particular ethnic group and the population as a whole. With the increasing availability of frequency data, forensic scientists can make extremely accurate statistical calculations. Thus, despite problems that can still arise from insufficient data, human error, or flawed evidence, genetic profiles are now accepted as compelling evidence by legal experts and scientists alike.

Environmental Cleanup

Increasingly, the remarkable ability of certain microorganisms to transform chemicals is being exploited for environmental cleanup. If the growth needs of such microbes make them unsuitable for direct use, scientists can now transfer the genes for their valuable metabolic capabilities into other microorganisms, which can then be used to treat environmental problems. For example, many bacteria can extract heavy metals, such as copper, lead, and nickel, from their environments and incorporate the metals into compounds such as copper sulfate or lead sulfate, which are readily recoverable. Genetically engineered microbes may become important in both mining minerals (especially as ore reserves are depleted) and cleaning up highly toxic mining wastes. Biotechnologists are also trying to engineer microbes that can degrade chlorinated hydrocarbons and other harmful compounds. These microbes could be used in wastewater treatment plants or by manufacturers before the compounds are ever released into the environment.

Agricultural Applications

Scientists are working to learn more about the genomes of agriculturally important plants and animals. For a number of years, they have been using DNA technology in an effort to improve agricultural productivity. The selective breeding of both livestock (animal husbandry) and crops has exploited naturally occurring mutations and genetic recombination for thousands of years.

As we described earlier, DNA technology enables scientists to produce transgenic animals, which speeds up the selective breeding process. The goals of creating a transgenic animal are often the same as the goals of traditional breeding—for instance, to make a sheep with better quality wool, a pig with leaner meat, or a cow that will mature in a shorter time. Scientists might, for example, identify and clone a gene that causes the development of larger muscles (muscles make up most of the meat we eat) in one breed of cattle and transfer it to other cattle or even to sheep. However, problems such as low fertility or increased susceptibility to disease are not uncommon among farm animals carrying genes from other species. Animal health and welfare are important issues to consider when developing transgenic animals.

Agricultural scientists have already endowed a number of crop plants with genes for desirable traits, such as delayed ripening and resistance to spoilage and disease. In one striking way, plants are easier to genetically engineer than most animals. For many plant species, a single tissue cell grown in culture can give rise to an adult plant (see Figure 20.17). Thus, genetic manipulations can be performed on an ordinary somatic cell and the cell then used to generate an organism with new traits.

The most commonly used vector for introducing new genes into plant cells is a plasmid, called the **Ti plasmid**, from the soil bacterium *Agrobacterium tumefaciens*. This plasmid integrates a segment of its DNA, known as T DNA, into the chromosomal DNA of its host plant cells. For vector purposes, researchers work with versions of the plasmid that do not cause disease (unlike the wild-type version) and that have been engineered to carry genes of interest within the borders of the T DNA. **Figure 20.26** (on the next page) outlines one method for using the Ti plasmid to produce transgenic plants.

Genetic engineering is rapidly replacing traditional plant-breeding programs, especially for useful traits, such as herbicide or pest resistance, determined by one or a few genes. Crops engineered with a bacterial gene making the plants resistant to herbicides can grow while weeds are destroyed, and genetically engineered crops that can resist destructive insects reduce the need for chemical insecticides. In India, the insertion of a salinity resistance gene from a coastal mangrove plant into the genomes of several rice varieties has resulted in rice plants that can grow in water three times as salty as seawater. The research foundation that carried out this feat of genetic engineering estimates that one-third of all irrigated land has high salinity owing to overirrigation and intensive use of chemical fertilizers, representing a serious threat to the food supply. Thus, salinity-resistant crop plants would be enormously valuable worldwide.

▼ **Figure 20.26** RESEARCH METHOD

Using the Ti Plasmid to Produce Transgenic Plants

APPLICATION Genes conferring useful traits, such as pest resistance, herbicide resistance, delayed ripening, and increased nutritional value, can be transferred from one plant variety or species to another using the Ti plasmid as a vector.

TECHNIQUE

1. The Ti plasmid is isolated from the bacterium *Agrobacterium tumefaciens*. The segment of the plasmid that integrates into the genome of host cells is called T DNA.

2. The foreign gene of interest is inserted into the middle of the T DNA using methods shown in Figure 20.4.

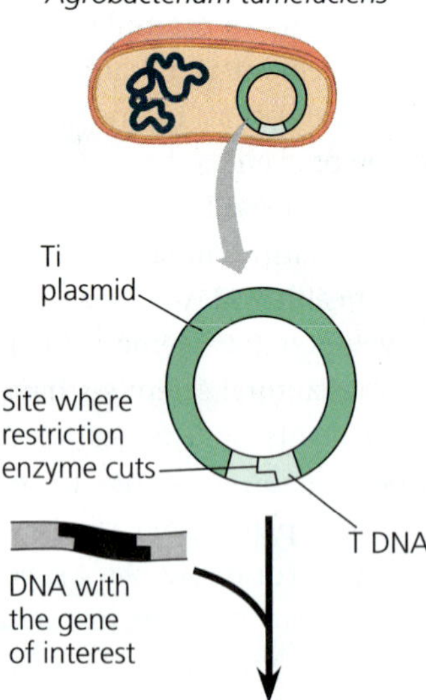

3. Recombinant plasmids can be introduced into cultured plant cells by electroporation. Or plasmids can be returned to *Agrobacterium*, which is then applied as a liquid suspension to the leaves of susceptible plants, infecting them. Once a plasmid is taken into a plant cell, its T DNA integrates into the cell's chromosomal DNA.

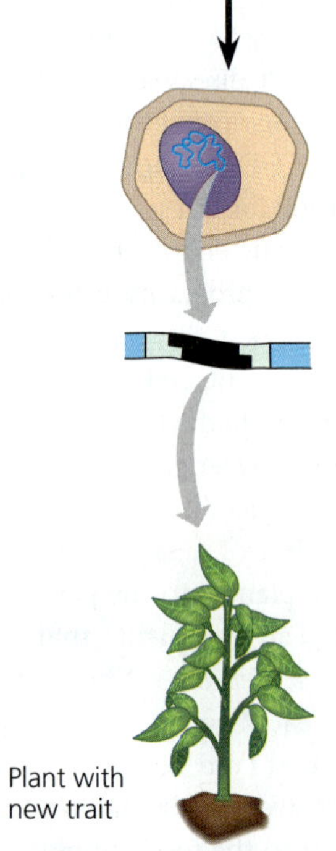

RESULTS Transformed cells carrying the transgene of interest can regenerate complete plants that exhibit the new trait conferred by the transgene.

Safety and Ethical Questions Raised by DNA Technology

Early concerns about potential dangers associated with recombinant DNA technology focused on the possibility that hazardous new pathogens might be created. What might happen, for instance, if cancer cell genes were transferred into bacteria or viruses? To guard against such rogue microbes, scientists developed a set of guidelines that were adopted as formal government regulations in the United States and some other countries. One safety measure is a set of strict laboratory procedures designed to protect researchers from infection by engineered microbes and to prevent the microbes from accidentally leaving the laboratory. In addition, strains of microorganisms to be used in recombinant DNA experiments are genetically crippled to ensure that they cannot survive outside the laboratory. Finally, certain obviously dangerous experiments have been banned.

Today, most public concern about possible hazards centers not on recombinant microbes but on **genetically modified (GM) organisms** used as food. A GM organism is one that has acquired by artificial means one or more genes from another species or even from another variety of the same species. Some salmon, for example, have been genetically modified by addition of a more active salmon growth hormone gene. However, the majority of the GM organisms that contribute to our food supply are not animals, but crop plants.

GM crops are widespread in the United States, Argentina, and Brazil; together these countries account for over 80% of the world's acreage devoted to such crops. In the United States, most corn, soybean, and canola crops are genetically modified, and GM products are not required to be labeled. However, the same foods are an ongoing subject of controversy in Europe, where the GM revolution has met with strong opposition. Many Europeans are concerned about the safety of GM foods and the possible environmental consequences of growing GM plants. Early in 2000, negotiators from 130 countries agreed on a Biosafety Protocol that requires exporters to identify GM organisms present in bulk food shipments and allows importing countries to decide whether the products pose environmental or health risks. (Although the United States declined to sign the agreement, it went into effect anyway because the majority of countries were in favor of it.) Since then, European countries have, on occasion, refused crops from the United States and other countries, leading to trade disputes. Although a small number of GM crops have been grown on European soil, these products have generally failed in local markets, and the future of GM crops in Europe is uncertain.

Advocates of a cautious approach toward GM crops fear that transgenic plants might pass their new genes to close relatives in nearby wild areas. We know that lawn and crop grasses, for example, commonly exchange genes with wild relatives via pollen transfer. If crop plants carrying genes for resistance to herbicides, diseases, or insect pests pollinated wild ones, the offspring might become "super weeds" that are very difficult to control. Another worry concerns possible risks to human health from GM foods. Some people fear that the protein products of transgenes might lead to allergic reactions. Although there is some evidence that this could happen, advocates claim that these proteins could be tested in advance to avoid producing ones that cause allergic reactions.

Today, governments and regulatory agencies throughout the world are grappling with how to facilitate the use of biotechnology in agriculture, industry, and medicine while ensuring that new products and procedures are safe. In the United States, such applications of biotechnology are evaluated for potential risks by various regulatory agencies, including the Food and Drug Administration, the Environmental Protection Agency, the National Institutes of Health, and the Department of Agriculture. Meanwhile, these same agencies and the public must consider the ethical implications of biotechnology.

Advances in biotechnology have allowed us to obtain complete genome sequences for humans and many other species, providing a vast treasure trove of information about genes. We can ask how certain genes differ from species to species, as well as how genes and, ultimately, entire genomes have evolved. (These are the subjects of Chapter 21.) At the same time, the increasing speed and falling cost of sequencing the genomes of individuals are raising significant ethical questions. Who should have the right to examine someone else's genetic information? How should that information be used? Should a person's genome be a factor in determining eligibility for a job or insurance? Ethical considerations, as well as concerns about potential environmental and health hazards, will likely slow some applications of biotechnology. There is always a danger that too much regulation will stifle basic research and its potential benefits. However, the power of DNA technology and genetic engineering—our ability to profoundly and rapidly alter species that have been evolving for millennia—demands that we proceed with humility and caution.

CONCEPT CHECK 20.4

1. What is the advantage of using stem cells for gene therapy?
2. List at least three different properties that have been acquired by crop plants via genetic engineering.
3. **WHAT IF?** As a physician, you have a patient with symptoms that suggest a hepatitis A infection, but you have not been able to detect viral proteins in the blood. Knowing that hepatitis A is an RNA virus, what lab tests could you perform to support your diagnosis? Explain what the results would mean.

For suggested answers, see Appendix A.

20 CHAPTER REVIEW

SUMMARY OF KEY CONCEPTS

CONCEPT 20.1

DNA cloning yields multiple copies of a gene or other DNA segment (pp. 396–405)

- **Gene cloning** and other techniques, collectively termed DNA technology, can be used to manipulate and analyze DNA and to produce useful new products and organisms.
- In **genetic engineering**, bacterial **restriction enzymes** are used to cut DNA molecules within short, specific nucleotide sequences (**restriction sites**), yielding a set of double-stranded DNA fragments with single-stranded **sticky ends**.

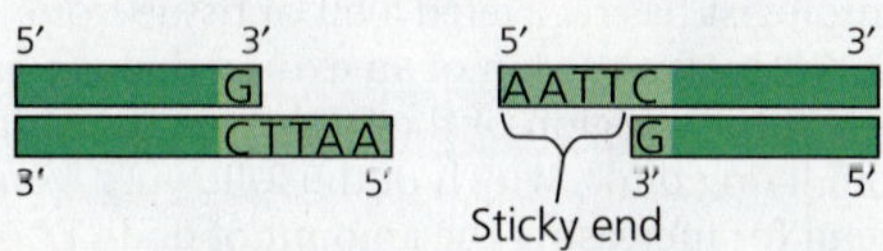

- The sticky ends on **restriction fragments** from one DNA source can base-pair with complementary sticky ends on fragments from other DNA molecules; sealing the base-paired fragments with **DNA ligase** produces **recombinant DNA** molecules.
- Cloning a eukaryotic gene in a bacterial plasmid:

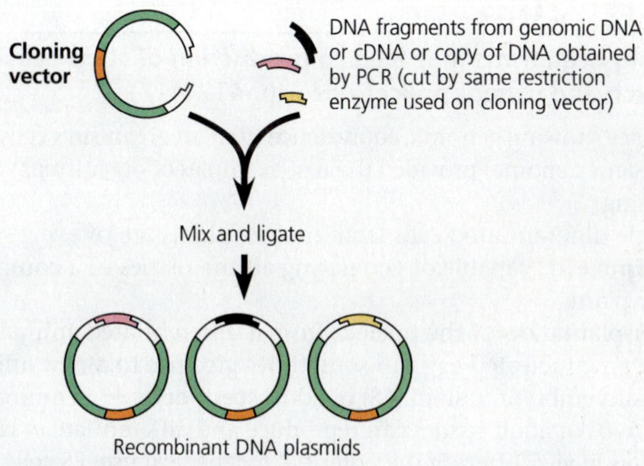

Cloning vectors include **plasmids** and **bacterial artificial chromosomes (BACs)**. Recombinant plasmids are returned to host cells, each of which divides to form a clone of cells. Collections of clones are stored as **genomic** or **complementary DNA (cDNA) libraries**. Libraries can be screened for a gene of interest using **nucleic acid hybridization** with a **nucleic acid probe**.
- Several technical difficulties hinder the expression of cloned eukaryotic genes in bacterial host cells. The use of cultured

eukaryotic cells (such as yeasts, insect cells, or cultured mammalian cells) as host cells, coupled with appropriate **expression vectors**, helps avoid these problems.
- The **polymerase chain reaction (PCR)** can produce many copies of (amplify) a specific target segment of DNA *in vitro*, using primers that bracket the desired sequence and a heat-resistant DNA polymerase.

? *Describe how the process of gene cloning results in a cell clone containing a recombinant plasmid.*

CONCEPT 20.2

DNA technology allows us to study the sequence, expression, and function of a gene (pp. 405–412)

- DNA restriction fragments of different lengths can be separated by **gel electrophoresis**. Specific fragments can be identified by **Southern blotting**, using labeled probes that hybridize to the DNA immobilized on a "blot" of the gel. Historically, **restriction fragment length polymorphisms (RFLPs)** were used to screen for some disease-causing alleles, such as the sickle-cell allele.
- Relatively short DNA fragments can be sequenced by the dideoxy chain termination method, which can be performed in automated sequencing machines. The rapid development of faster and cheaper methods is ongoing.
- Expression of a gene can be investigated using hybridization with labeled probes to look for specific mRNAs, either on a gel (**Northern blotting**) or in a whole organism (***in situ* hybridization**). Also, RNA can be transcribed into cDNA by reverse transcriptase and the cDNA amplified by PCR with specific primers (**RT-PCR**). **DNA microarrays** allow researchers to compare the expression of many genes at once in different tissues, at different times, or under different conditions.
- For a gene of unknown function, experimental inactivation of the gene and observation of the resulting phenotypic effects can provide clues to its function. In humans, **genome-wide association studies** use **single nucleotide polymorphisms (SNPs)** as genetic markers for alleles that are associated with particular conditions.

? *Complementary base pairing is the basis of most procedures used to analyze gene expression. Explain.*

CONCEPT 20.3

Cloning organisms may lead to production of stem cells for research and other applications (pp. 412–417)

- Studies showing genomic equivalence (that an organism's cells have the same genome) provided the first examples of organismal cloning.
- Single differentiated cells from mature plants are often **totipotent**: capable of generating all the tissues of a complete new plant.
- Transplantation of the nucleus from a differentiated animal cell into an enucleated egg can sometimes give rise to a new animal.
- Certain embryonic stem (ES) or adult **stem cells** from animal embryos or adult tissues can reproduce and differentiate *in vitro* as well as *in vivo*, offering the potential for medical use. ES cells are **pluripotent** but difficult to acquire. Induced pluripotent stem (iPS) cells resemble ES cells in their capacity to differentiate; they can be generated by reprogramming differentiated cells. iPS cells hold promise for medical research and regenerative medicine.

? *Describe how a researcher could carry out organismal cloning, production of ES cells, and generation of iPS cells, focusing on how the cells are reprogrammed and using mice as an example. (The procedures are basically the same in humans and mice.)*

CONCEPT 20.4

The practical applications of DNA technology affect our lives in many ways (pp. 417–423)

- DNA technology, including the analysis of genetic markers such as SNPs, is increasingly being used in the diagnosis of genetic and other diseases and offers potential for better treatment of genetic disorders (or even permanent cures through **gene therapy**), as well as more informed cancer therapies. Large-scale production of protein hormones and other proteins with therapeutic uses is possible with DNA technology. Some therapeutic proteins are being produced in **transgenic** "pharm" animals.
- Analysis of genetic markers such as **short tandem repeats (STRs)** in DNA isolated from tissue or body fluids found at crime scenes leads to a **genetic profile** that can provide definitive evidence that a suspect is innocent or strong evidence of guilt. Such analysis is also useful in parenthood disputes and in identifying the remains of crime victims.
- Genetically engineered microorganisms can be used to extract minerals from the environment or degrade various types of toxic waste materials.
- The aims of developing transgenic plants and animals are to improve agricultural productivity and food quality.
- The potential benefits of genetic engineering must be carefully weighed against the potential for harm to humans or the environment.

? *What factors affect whether a given genetic disease would be a good candidate for successful gene therapy?*

TEST YOUR UNDERSTANDING

LEVEL 1: KNOWLEDGE/COMPREHENSION

1. Which of the following tools of recombinant DNA technology is *incorrectly* paired with its use?
   a. restriction enzyme—analysis of RFLPs
   b. DNA ligase—cutting DNA, creating sticky ends of restriction fragments
   c. DNA polymerase—polymerase chain reaction to amplify sections of DNA
   d. reverse transcriptase—production of cDNA from mRNA
   e. electrophoresis—separation of DNA fragments
2. Plants are more readily manipulated by genetic engineering than are animals because
   a. plant genes do not contain introns.
   b. more vectors are available for transferring recombinant DNA into plant cells.
   c. a somatic plant cell can often give rise to a complete plant.
   d. genes can be inserted into plant cells by microinjection.
   e. plant cells have larger nuclei.
3. A paleontologist has recovered a bit of tissue from the 400-year-old preserved skin of an extinct dodo (a bird). To compare a specific region of the DNA from the sample with DNA from living birds, which of the following would be most useful for increasing the amount of dodo DNA available for testing?
   a. RFLP analysis
   b. polymerase chain reaction (PCR)
   c. electroporation
   d. gel electrophoresis
   e. Southern blotting

4. DNA technology has many medical applications. Which of the following is *not* done routinely at present?
   a. production of hormones for treating diabetes and dwarfism
   b. production of microbes that can metabolize toxins
   c. introduction of genetically engineered genes into human gametes
   d. prenatal identification of genetic disease alleles
   e. genetic testing for carriers of harmful alleles

5. In recombinant DNA methods, the term *vector* can refer to
   a. the enzyme that cuts DNA into restriction fragments.
   b. the sticky end of a DNA fragment.
   c. a SNP marker.
   d. a plasmid used to transfer DNA into a living cell.
   e. a DNA probe used to identify a particular gene.

LEVEL 2: APPLICATION/ANALYSIS

6. Which of the following would *not* be true of cDNA produced using human brain tissue as the starting material?
   a. It could be amplified by the polymerase chain reaction.
   b. It could be used to create a complete genomic library.
   c. It was produced from mRNA using reverse transcriptase.
   d. It could be used as a probe to detect genes expressed in the brain.
   e. It lacks the introns of the human genes.

7. Expression of a cloned eukaryotic gene in a bacterial cell involves many challenges. The use of mRNA and reverse transcriptase is part of a strategy to solve the problem of
   a. post-transcriptional processing.
   b. electroporation.
   c. post-translational processing.
   d. nucleic acid hybridization.
   e. restriction fragment ligation.

8. Which of the following sequences in double-stranded DNA is most likely to be recognized as a cutting site for a restriction enzyme?

a. AAGG	b. AGTC	c. GGCC	d. ACCA	e. AAAA
TTCC	TCAG	CCGG	TGGT	TTTT

9. **DRAW IT** You are making a genomic library for the aardvark, using a bacterial plasmid as a vector. The green diagram below shows the plasmid, which contains the restriction site for the enzyme used in Figure 20.3. Above the plasmid is a segment of linear aardvark DNA. Diagram your cloning procedure, showing what would happen to these two molecules during each step. Use one color for the aardvark DNA and its bases and another color for those of the plasmid. Label each step and all 5′ and 3′ ends.

5′ TCCATGAATTCTAAAGCGCTTATGAATTCACGGC 3′
3′ AGGTACTTAAGATTTCGCGAATACTTAAGTGCCG 5′

Aardvark DNA

Plasmid

LEVEL 3: SYNTHESIS/EVALUATION

10. **WHAT IF?** Imagine you want to study one of the human crystallins, proteins present in the lens of the eye. To obtain a sufficient amount of the protein of interest, you decide to clone the gene that codes for it. Would you construct a genomic library or a cDNA library? What material would you use as a source of DNA or RNA?

11. **EVOLUTION CONNECTION**
Ethical considerations aside, if DNA-based technologies became widely used, how might they change the way evolution proceeds, as compared with the natural evolutionary mechanisms of the past 4 billion years?

12. **SCIENTIFIC INQUIRY**
You hope to study a gene that codes for a neurotransmitter protein produced in human brain cells. You know the amino acid sequence of the protein. Explain how you might (a) identify what genes are expressed in a specific type of brain cell, (b) identify (isolate) the neurotransmitter gene, (c) produce multiple copies of the gene for study, and (d) produce large quantities of the neurotransmitter for evaluation as a potential medication.

13. **SCIENCE, TECHNOLOGY, AND SOCIETY**
Is there danger of discrimination based on testing for "harmful" genes? What policies can you suggest that would prevent such abuses?

14. **SCIENCE, TECHNOLOGY, AND SOCIETY**
Government funding of embryonic stem cell research has been a contentious political issue. Why has this debate been so heated? Summarize the arguments for and against embryonic stem cell research, and explain your own position on the issue.

15. **WRITE ABOUT A THEME**
The Genetic Basis of Life In a short essay (100–150 words), discuss how the genetic basis of life plays a central role in biotechnology.

For selected answers, see Appendix A.

21

Genomes and Their Evolution

▲ **Figure 21.1 What genomic information distinguishes a human from a chimpanzee?**

KEY CONCEPTS

21.1 New approaches have accelerated the pace of genome sequencing

21.2 Scientists use bioinformatics to analyze genomes and their functions

21.3 Genomes vary in size, number of genes, and gene density

21.4 Multicellular eukaryotes have much noncoding DNA and many multigene families

21.5 Duplication, rearrangement, and mutation of DNA contribute to genome evolution

21.6 Comparing genome sequences provides clues to evolution and development

OVERVIEW

Reading the Leaves from the Tree of Life

The chimpanzee (*Pan troglodytes*) is our closest living relative on the evolutionary tree of life. The boy in **Figure 21.1** and his chimpanzee companion are intently studying the same leaf, but only one of them is able to talk about it. What accounts for this difference between two primates that share so much of their evolutionary history? With the advent of recent techniques for rapidly sequencing complete genomes, we can now start to address the genetic basis of intriguing questions like this.

The chimpanzee genome was sequenced in 2005, two years after sequencing of the human genome was largely completed. Now that we can compare our genome with that of the chimpanzee base by base, we can tackle the more general issue of what differences in the genetic information account for the distinct characteristics of these two species of primates.

In addition to determining the sequences of the human and chimpanzee genomes, researchers have obtained complete genome sequences for *E. coli* and numerous other prokaryotes, as well as many eukaryotes, including *Zea mays* (corn), *Drosophila melanogaster* (fruit fly), *Mus musculus* (house mouse), and *Macaca mulatta* (rhesus macaque). In 2010, a draft sequence was announced for the genome of *Homo neanderthalensis*, an extinct species closely related to present-day humans. These whole and partial genomes are of great interest in their own right and are also providing important insights into evolution and other biological processes. Broadening the human-chimpanzee comparison to the genomes of other primates and more distantly related animals should reveal the sets of genes that control group-defining characteristics. Beyond that, comparisons with the genomes of bacteria, archaea, fungi, protists, and plants should enlighten us about the long evolutionary history of shared ancient genes and their products.

With the genomes of many species fully sequenced, scientists can study whole sets of genes and their interactions, an approach called **genomics**. The sequencing efforts that feed this approach have generated, and continue to generate, enormous volumes of data. The need to deal with this ever-increasing flood of information has spawned the field of **bioinformatics**, the application of computational methods to the storage and analysis of biological data.

We will begin this chapter by discussing two approaches to genome sequencing and some of the advances in bioinformatics and its applications. We will then summarize what has been learned from the genomes that have been sequenced thus far. Next, we will describe the composition of the human genome as a representative genome of a complex multicellular eukaryote. Finally, we will explore current ideas about how genomes evolve and about how the evolution of developmental mechanisms could have generated the great diversity of life on Earth today.

CONCEPT 21.1

New approaches have accelerated the pace of genome sequencing

Sequencing of the human genome, an ambitious undertaking, officially began as the **Human Genome Project** in 1990. Organized by an international, publicly funded consortium of scientists at universities and research institutes, the project involved 20 large sequencing centers in six countries plus a host of other labs working on small projects.

After sequencing of the human genome was largely completed in 2003, the sequence of each chromosome was carefully analyzed and described in a series of papers, the last of which covered chromosome 1 and was published in 2006. With this refinement, researchers termed the sequencing "virtually complete." To reach these milestones, the project proceeded through three stages that provided progressively more detailed views of the human genome: linkage mapping, physical mapping, and DNA sequencing.

Three-Stage Approach to Genome Sequencing

Even before the Human Genome Project began, earlier research had sketched a rough picture of the organization of the genomes of many organisms. For instance, the karyotyping of many species had revealed their chromosome numbers and banding patterns (see Figure 13.3). And some human genes had already been located on a particular region of a chromosome by fluorescence *in situ* hybridization (FISH), a method in which fluorescently labeled nucleic acid probes are allowed to hybridize to an immobilized array of whole chromosomes (see Figure 15.1). Cytogenetic maps based on this type of information provided the starting point for more detailed mapping of the human genome.

With these cytogenetic maps of the chromosomes in hand, the initial stage in sequencing the human genome was to construct a **linkage map** (a type of genetic map; see Figure 15.11) of several thousand genetic markers spaced throughout the chromosomes (**Figure 21.2**, stage 1). The order of the markers and the relative distances between them on such a map are based on recombination frequencies. The markers can be genes or any other identifiable sequences in the DNA, such as RFLPs or short tandem repeats (STRs), both discussed in Chapter 20. By 1992, researchers had compiled a human linkage map with some 5,000 markers. Such a map enabled them to locate other markers, including genes, by testing for genetic linkage to the known markers. It was also valuable as a framework for organizing more detailed maps of particular regions. Remember from Chapter 15, however, that absolute distances between genes cannot be determined using this approach.

The next stage was the physical mapping of the human genome. In a **physical map**, the distances between markers are expressed by some physical measure, usually the number of base pairs along the DNA. For whole-genome mapping, a physical map is made by cutting the DNA of each chromosome into a number of restriction fragments and then determining the original order of the fragments in the chromosomal DNA. The key is to make fragments that overlap and then use probes or automated nucleotide sequencing of the ends to find the overlaps (see Figure 21.2, stage 2). In this way, fragments can be assigned to a sequential order that corresponds to their order in a chromosome.

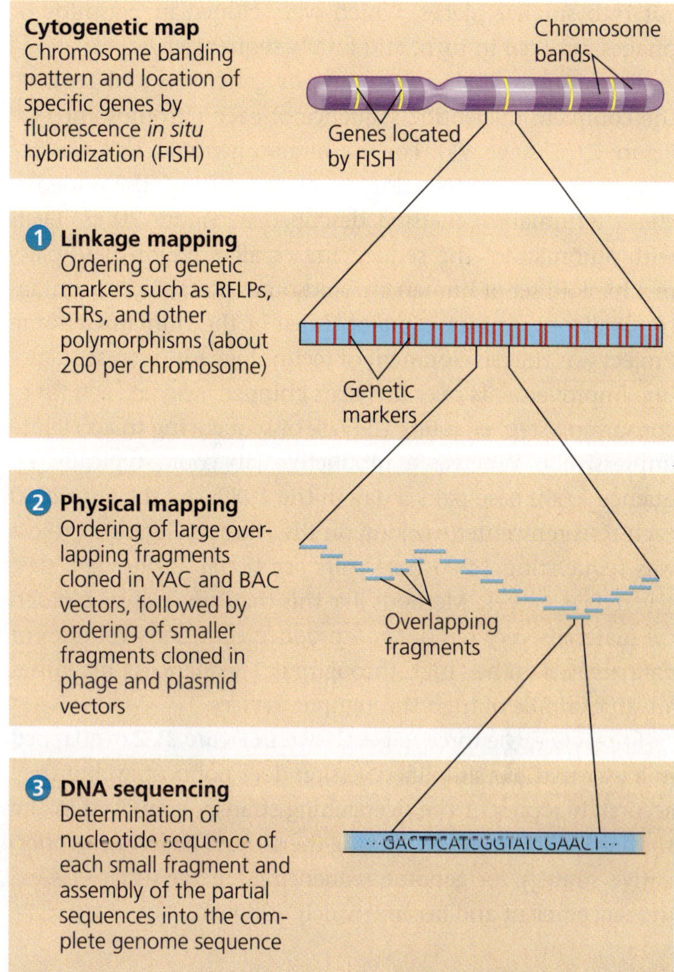

▲ **Figure 21.2 Three-stage approach to sequencing an entire genome.** Starting with a cytogenetic map of each chromosome, researchers with the Human Genome Project proceeded through three stages to reach the ultimate goal, the virtually complete nucleotide sequence of every chromosome.

The DNA fragments used for physical mapping were prepared by DNA cloning. With such a large genome, researchers had to carry out several rounds of DNA cutting, cloning, and physical mapping. In this approach, the first cloning vector was often a yeast artificial chromosome (YAC), which can carry inserted fragments a million base pairs long, or a bacterial artificial chromosome (BAC), which typically carries inserts of 100,000–300,000 base pairs. After such long fragments were put in order, each fragment was cut into smaller pieces, which were cloned in plasmids or phages, ordered in turn, and finally sequenced.

The ultimate goal in mapping any genome is to determine the complete nucleotide sequence of each chromosome (see Figure 21.2, stage 3). For the human genome, this was accomplished by sequencing machines, using the dideoxy chain termination method described in Figure 20.12. Even with automation, the sequencing of all 3 billion base pairs in a haploid set of human chromosomes presented a formidable challenge. In fact, a major thrust of the Human Genome Project was the development of technology for faster sequencing. Improvements over the years chipped away at each time-consuming step, enabling the rate of sequencing to accelerate impressively: Whereas a productive lab could typically sequence 1,000 base pairs a day in the 1980s, by the year 2000 each research center working on the Human Genome Project was sequencing 1,000 base pairs *per second*, 24 hours a day, seven days a week. Methods like this that can analyze biological materials very rapidly and produce enormous volumes of data are said to be "high-throughput." Sequencing machines are an example of high-throughput devices.

In practice, the three stages shown in Figure 21.2 overlapped in a way that our simplified version does not portray, but they accurately represent the overarching strategy employed in the Human Genome Project. During the project, however, an alternative strategy for genome sequencing emerged that was extremely efficient and became widely adopted.

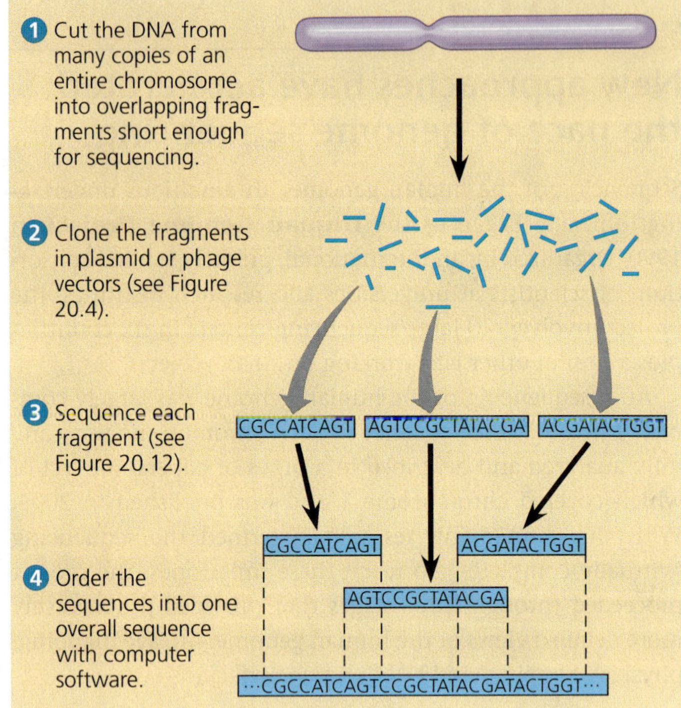

▲ **Figure 21.3 Whole-genome shotgun approach to sequencing.** In this approach, developed by Craig Venter and colleagues at the company he founded, Celera Genomics, random DNA fragments are sequenced and then ordered relative to each other. Compare this approach with the hierarchical, three-stage approach shown in Figure 21.2.

? *The fragments in stage 2 of this figure are depicted as scattered, whereas those in stage 2 of Figure 21.2 are drawn in a much more orderly fashion. How do these depictions reflect the two approaches?*

Whole-Genome Shotgun Approach to Genome Sequencing

In 1992, emboldened by advances in sequencing and computer technology, molecular biologist J. Craig Venter devised an alternative approach to the sequencing of whole genomes. Called the *whole-genome shotgun approach*, it essentially skips the linkage mapping and physical mapping stages and starts directly with the sequencing of DNA fragments from randomly cut DNA. Powerful computer programs then assemble the resulting very large number of overlapping short sequences into a single continuous sequence **(Figure 21.3)**. In 1998, despite the skepticism of many scientists, Venter set up a company (Celera Genomics) and declared his intention to sequence the entire human genome. Five years later, and 13 years after the Human Genome Project began, Celera Genomics and the public consortium jointly announced that sequencing of the human genome was largely complete.

Representatives of the public consortium point out that Celera's accomplishment relied heavily on the consortium's maps and sequence data and that the infrastructure established by their approach was a tremendous aid to Celera's efforts. Venter, on the other hand, has argued for the efficiency and economy of Celera's methods, and indeed, the public consortium made some use of them as well. Evidently, both approaches made valuable contributions.

Today, the whole-genome shotgun approach is widely used. Also, the development of newer sequencing techniques, generally called *sequencing by synthesis* (see Chapter 20), has resulted in massive increases in speed and decreases in the cost of sequencing entire genomes. In these new techniques, many very small fragments (fewer than 100 base pairs) are sequenced at the same time, and computer software rapidly assembles the complete sequence. Because of the sensitivity of these techniques, the fragments can be sequenced directly; the cloning step (stage 2 in Figure 21.3) is unnecessary. Whereas sequencing the first human genome took 13 years and cost $100 million, James Watson's genome was sequenced during four months in 2007 for about $1 million, and a group of researchers reported in 2010 that they had rapidly sequenced three human genomes for approximately $4,400 each!

These technological advances have also facilitated an approach called **metagenomics** (from the Greek *meta*, beyond), in which DNA from a group of species (a *metagenome*) is collected from an environmental sample and sequenced. Again, computer software accomplishes the task of sorting out the partial sequences and assembling them into specific genomes. So far, this approach has been applied to microbial communities found in environments as diverse as the Sargasso Sea and the human intestine. The ability to sequence the DNA of mixed populations eliminates the need to culture each species separately in the lab, a difficulty that has limited the study of many microbial species.

At first glance, genome sequences of humans and other organisms are simply dry lists of nucleotide bases—millions of A's, T's, C's, and G's in mind-numbing succession. Crucial to making sense of this massive amount of data have been new analytical approaches, which we discuss next.

CONCEPT CHECK 21.1

1. What is the major difference between a linkage map and a physical map of a chromosome?
2. In general, how does the approach to genome mapping used in the Human Genome Project differ from the whole-genome shotgun approach?

For suggested answers, see Appendix A.

CONCEPT 21.2

Scientists use bioinformatics to analyze genomes and their functions

Each of the 20 or so sequencing centers around the world working on the Human Genome Project churned out voluminous amounts of DNA sequence day after day. As the data began to accumulate, the need to coordinate efforts to keep track of all the sequences became clear. Thanks to the foresight of research scientists and government officials involved in the Human Genome Project, its goals included the establishment of banks of data, or databases, and the refining of analytical software. These databases and software programs would then be centralized and made readily accessible on the Internet. Accomplishing this aim has accelerated progress in DNA sequence analysis by making bioinformatics resources available to researchers worldwide and by speeding up the dissemination of information.

Centralized Resources for Analyzing Genome Sequences

Government-funded agencies carried out their mandate to establish databases and provide software with which scientists could analyze the sequence data. For example, in the United States, a joint endeavor between the National Library of Medicine and the National Institutes of Health (NIH) created the National Center for Biotechnology Information (NCBI), which maintains a website (www.ncbi.nlm.nih.gov) with extensive bioinformatics resources. On this site are links to databases, software, and a wealth of information about genomics and related topics. Similar websites have also been established by the European Molecular Biology Laboratory, the DNA Data Bank of Japan, and BGI (formerly known as the Beijing Genome Institute) in Shenzhen, China, three genome centers with which the NCBI collaborates. These large, comprehensive websites are complemented by others maintained by individual or small groups of laboratories. Smaller websites often provide databases and software designed for a narrower purpose, such as studying genetic and genomic changes in one particular type of cancer.

The NCBI database of sequences is called GenBank. As of May 2010, it included the sequences of 119 million fragments of genomic DNA, totaling 114 billion base pairs! GenBank is constantly updated, and the amount of data it contains is estimated to double approximately every 18 months. Any sequence in the database can be retrieved and analyzed using software from the NCBI website or elsewhere.

One software program available on the NCBI website, called BLAST, allows the visitor to compare a DNA sequence with every sequence in GenBank, base by base, to look for similar regions. Another program allows comparison of predicted protein sequences. Yet a third can search any protein sequence for common stretches of amino acids (domains) for which a function is known or suspected, and it can show a three-dimensional model of the domain alongside other relevant information (**Figure 21.4**, on the next page). There is even a software program that can compare a collection of sequences, either nucleic acids or polypeptides, and diagram them in the form of an evolutionary tree based on the sequence relationships. (One such diagram is shown in Figure 21.16.)

Two research institutions, Rutgers University and the University of California, San Diego, also maintain a worldwide Protein Data Bank, a database of all three-dimensional protein structures that have been determined. (The database is accessible at www.wwpdb.org.) These structures can be rotated by the viewer to show all sides of the protein.

There is a vast array of resources available for researchers anywhere in the world to use. Let us now consider the types of questions scientists can address using these resources.

Identifying Protein-Coding Genes and Understanding Their Functions

Using available DNA sequences, geneticists can study genes directly, without having to infer genotype from phenotype as in classical genetics. But this approach, called *reverse genetics*, poses a new challenge: determining the phenotype from the genotype. Given a long DNA sequence from a database such as GenBank, the aim of scientists is to identify all protein-coding genes in the sequence and ultimately their functions. This process is called **gene annotation**.

In the past, gene annotation was carried out laboriously by individual scientists interested in particular genes, but the process has now been largely automated. The usual approach is to use software to scan the stored sequences for transcriptional and translational start and stop signals, for RNA-splicing sites, and for other telltale signs of protein-coding genes. The software also looks for certain short sequences that specify known mRNAs. Thousands of such sequences, called

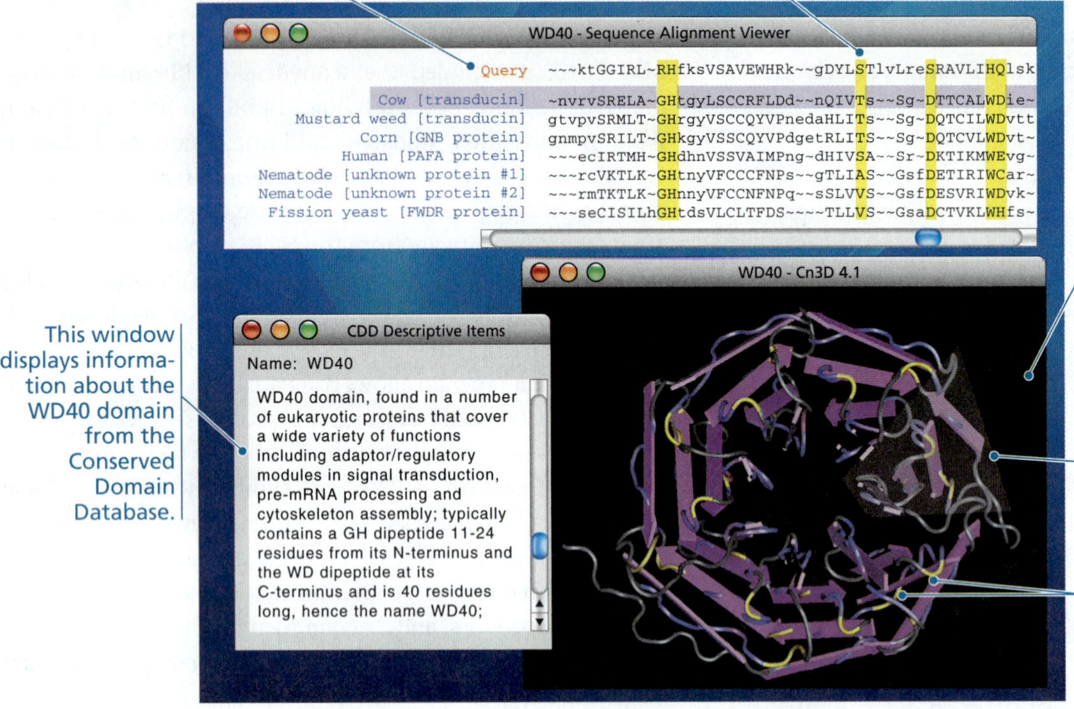

▲ **Figure 21.4 Bioinformatics tools available on the Internet.** A website maintained by the National Center for Biotechnology Information allows scientists and the public to access DNA and protein sequences and other stored data. The site includes a link to a protein structure database (Conserved Domain Database, CDD) that can find and describe similar domains in related proteins, as well as software (Cn3D, "See in 3D") that displays three-dimensional models of domains for which the structure has been determined. Some results are shown from a search for regions of proteins similar to an amino acid sequence in a muskmelon protein.

expressed sequence tags, or *ESTs*, have been collected from cDNA sequences and are cataloged in computer databases. This type of analysis identifies sequences that may be previously unknown protein-coding genes.

The identities of about half of the human genes were known before the Human Genome Project began. But what about the others, the previously unknown genes revealed by analysis of DNA sequences? Clues about their identities and functions come from comparing sequences that might be genes with known genes from other organisms, using the software described previously. Due to redundancy in the genetic code, the DNA sequence itself may vary more than the protein sequence does. Thus, scientists interested in proteins often compare the predicted amino acid sequence of a protein to that of other proteins.

Sometimes a newly identified sequence will match, at least partially, the sequence of a gene or protein whose function is well known. For example, part of a new gene may match a known gene that encodes an important signaling pathway protein such as a protein kinase (see Chapter 11), suggesting that the new gene does, too. Alternatively, the new gene sequence may be similar to a previously encountered sequence whose function is still unknown. Another possibility is that the sequence is entirely unlike anything ever seen before. This was true for about a third of the genes of *E. coli* when its genome was sequenced. In the last case, protein function is usually deduced through a combination of biochemical and functional studies. The biochemical approach aims to determine the three-dimensional structure of the protein as well as other attributes, such as potential binding sites for other molecules. Functional studies usually involve blocking or disabling the gene to see how the phenotype is affected. RNAi, described in Chapter 20, is an example of an experimental technique used to block gene function.

Understanding Genes and Gene Expression at the Systems Level

The impressive computational power provided by the tools of bioinformatics allows the study of whole sets of genes and their interactions, as well as the comparison of genomes from different species. Genomics is a rich source of new insights into fundamental questions about genome organization, regulation of gene expression, growth and development, and evolution.

One informative approach has been taken by a research project called ENCODE (Encyclopedia of DNA Elements), which began in 2003. First, researchers focused intensively on 1% of the human genome and attempted to learn all they could about the functionally important elements in that sequence. They looked for protein-coding genes and genes for noncoding RNAs as well as sequences that regulate DNA replication, gene expression (such as enhancers and promoters), and chromatin modifications. The pilot project was completed in 2007, yielding a wealth of information. One big surprise, discussed in Concept 18.3, was that over 90% of the region was transcribed into RNA, even though less than 2% codes for proteins. The success of this approach has led to two follow-up studies, one extending the analysis to the entire human genome and the other analyzing in a similar fashion the genomes of two model organisms, the soil nematode *Caenorhabditis elegans* and the fruit fly *Drosophila melanogaster*. Because genetic and molecular biological experiments can be performed on these species, testing the activities of potentially functional DNA elements in their genomes will reveal much about how the human genome works.

The success in sequencing genomes and studying entire sets of genes has encouraged scientists to attempt similar systematic study of the full protein sets (*proteomes*) encoded by genomes, an approach called **proteomics**. Proteins, not the genes that encode them, actually carry out most of the activities of the cell. Therefore, we must study when and where proteins are produced in an organism, as well as how they interact in networks, if we are to understand the functioning of cells and organisms.

How Systems Are Studied: An Example

Genomics and proteomics are enabling molecular biologists to approach the study of life from an increasingly global perspective. Using the tools we have described, biologists have begun to compile catalogs of genes and proteins—listings of all the "parts" that contribute to the operation of cells, tissues, and organisms. With such catalogs in hand, researchers have shifted their attention from the individual parts to their functional integration in biological systems. As you may recall, in Chapter 1 we discussed this systems biology approach, which aims to model the dynamic behavior of whole biological systems.

One important use of the systems biology approach is to define gene circuits and protein interaction networks. To map the protein interaction network in the yeast *Saccharomyces cerevisiae*, for instance, researchers used sophisticated techniques to knock out (disable) pairs of genes, one pair at a time, creating doubly mutant cells. They then compared the fitness of each double mutant (based in part on the size of the cell colony it formed) to that predicted from the fitnesses of the two single mutants. The researchers reasoned that if the observed fitness matched the prediction, then the products of the two genes didn't interact with each other, but if the observed fitness was greater or less than predicted, then the gene products interacted in the cell. Computer software then mapped genes based on the similarity of their interactions; a network-like "functional map" of these genetic interactions is displayed in **Figure 21.5**. To process the vast number of protein-protein interactions generated by this experiment and integrate them

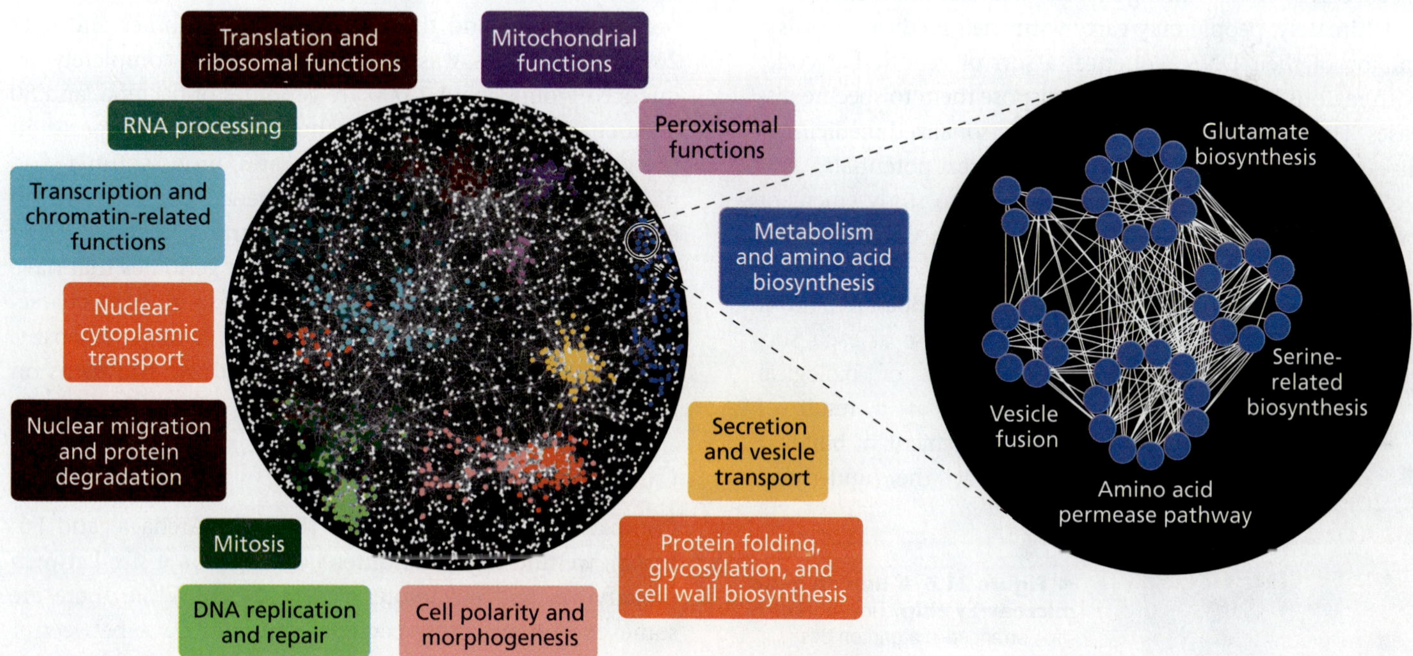

▲ **Figure 21.5 The systems biology approach to protein interactions.** This global protein interaction map shows the likely interactions (lines) among about 4,500 gene products (circles) in the yeast *Saccharomyces cerevisiae*. Circles of the same color represent gene products involved in one of the 13 cellular functions listed around the map. The blowup shows additional details of one map region where the gene products (blue circles) carry out amino acid biosynthesis, uptake, and related functions.

into the completed map required powerful computers, mathematical tools, and newly developed software. Thus, the systems biology approach has really been made possible by advances in computer technology and bioinformatics.

Application of Systems Biology to Medicine

The Cancer Genome Atlas is another example of systems biology in which a large group of interacting genes and gene products are analyzed together. This project, under the joint leadership of the National Cancer Institute and the NIH, aims to determine how changes in biological systems lead to cancer. A three-year pilot project beginning in 2007 set out to find all the common mutations in three types of cancer—lung cancer, ovarian cancer, and glioblastoma of the brain—by comparing gene sequences and patterns of gene expression in cancer cells with those in normal cells. Work on glioblastoma has confirmed the role of several suspected genes and identified a few unknown ones, suggesting possible new targets for therapies. The approach has proved so fruitful for these three types of cancer that it has been extended to ten other types, chosen because they are common and often lethal in humans.

Systems biology has tremendous potential in human medicine that is just starting to be explored. Silicon and glass "chips" have been developed that hold a microarray of most of the known human genes (**Figure 21.6**). Such chips are being used to analyze gene expression patterns in patients suffering from various cancers and other diseases, with the eventual aim of tailoring their treatment to their unique genetic makeup and the specifics of their cancers. This approach has had modest success in characterizing subsets of several cancers.

Ultimately, people may carry with their medical records a catalog of their DNA sequence, a sort of genetic bar code, with regions highlighted that predispose them to specific diseases. The use of such sequences for personalized medicine—disease prevention and treatment—has great potential.

Systems biology is a very efficient way to study emergent properties at the molecular level. Recall from Chapter 1 that according to the theme of emergent properties, novel properties arise at each successive level of biological complexity as a result of the arrangement of building blocks at the underlying level. The more we can learn about the arrangement and interactions of the components of genetic systems, the deeper will be our understanding of whole organisms. The rest of this chapter will survey what we've learned from genomic studies thus far.

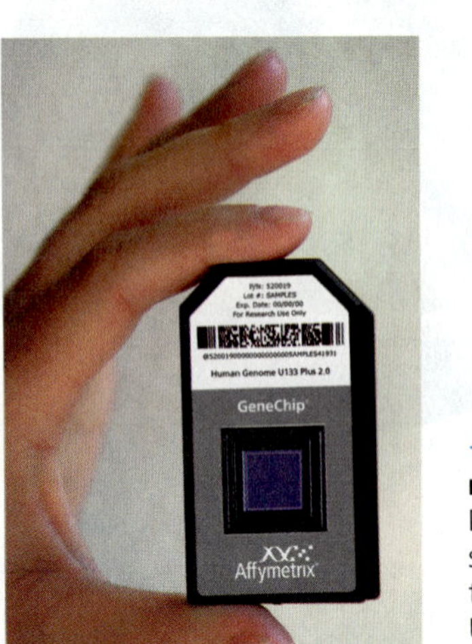

◄ **Figure 21.6 A human gene microarray chip.** Tiny spots of DNA arranged in a grid on this silicon wafer represent almost all of the genes in the human genome. Using this chip, researchers can analyze expression patterns for all these genes at the same time.

CONCEPT CHECK 21.2

1. What role does the Internet play in current genomics and proteomics research?
2. Explain the advantage of the systems biology approach to studying cancer versus the approach of studying a single gene at a time.
3. **MAKE CONNECTIONS** The ENCODE pilot project found that more than 90% of the genomic region being studied was transcribed into RNAs, far more than could be accounted for by protein-coding genes. Review Concept 18.3 (pp. 364–366) and suggest some roles that these RNAs might play.
4. **MAKE CONNECTIONS** In Concept 20.2 (p. 411), you learned about genome-wide association studies. Explain how these studies use the systems biology approach.

For suggested answers, see Appendix A.

CONCEPT **21.3**

Genomes vary in size, number of genes, and gene density

By early 2010, the sequencing of about 1,200 genomes had been completed and that of over 5,500 genomes and over 200 metagenomes was in progress. In the completely sequenced group, about 1,000 are genomes of bacteria, and 80 are archaeal genomes. Among the 124 eukaryotic species in the group are vertebrates, invertebrates, protists, fungi, and plants. The accumulated genome sequences contain a wealth of information that we are now beginning to mine. What have we learned so far by comparing the genomes that have been sequenced? In this section, we will examine the characteristics of genome size, number of genes, and gene density. Because these characteristics are so broad, we will focus on general trends, for which there are often exceptions.

Genome Size

Comparing the three domains (Bacteria, Archaea, and Eukarya), we find a general difference in genome size between prokaryotes and eukaryotes (**Table 21.1**). While there are some exceptions, most bacterial genomes have between 1 and 6 million base pairs (Mb); the genome of *E. coli*, for instance, has 4.6 Mb. Genomes of archaea are, for the most part, within the size range of bacterial genomes. (Keep in mind, however, that many fewer archaeal genomes have

Table 21.1 Genome Sizes and Estimated Numbers of Genes*

Organism	Haploid Genome Size (Mb)	Number of Genes	Genes per Mb
Bacteria			
Haemophilus influenzae	1.8	1,700	940
Escherichia coli	4.6	4,400	950
Archaea			
Archaeoglobus fulgidus	2.2	2,500	1,130
Methanosarcina barkeri	4.8	3,600	750
Eukaryotes			
Saccharomyces cerevisiae (yeast, a fungus)	12	6,300	525
Caenorhabditis elegans (nematode)	100	20,100	200
Arabidopsis thaliana (mustard family plant)	120	27,000	225
Drosophila melanogaster (fruit fly)	165	13,700	83
Oryza sativa (rice)	430	42,000	98
Zea mays (corn)	2,300	32,000	14
Mus musculus (house mouse)	2,600	22,000	11
Ailuropoda melanoleuca (giant panda)	2,400	21,000	9
Homo sapiens (human)	3,000	<21,000	7
Fritillaria assyriaca (lily family plant)	124,000	ND	ND

*Some values given here are likely to be revised as genome analysis continues. Mb = million base pairs. ND = not determined.

been completely sequenced, so this picture may change.) Eukaryotic genomes tend to be larger: The genome of the single-celled yeast *Saccharomyces cerevisiae* (a fungus) has about 12 Mb, while most animals and plants, which are multicellular, have genomes of at least 100 Mb. There are 165 Mb in the fruit fly genome, while humans have 3,000 Mb, about 500 to 3,000 times as many as a typical bacterium.

Aside from this general difference between prokaryotes and eukaryotes, a comparison of genome sizes among eukaryotes fails to reveal any systematic relationship between genome size and the organism's phenotype. For instance, the genome of *Fritillaria assyriaca*, a flowering plant in the lily family, contains 124 billion base pairs (124,000 Mb), about 40 times the size of the human genome. Even more striking, there is a single-celled amoeba, *Polychaos dubia*, whose genome size has been estimated at 670,000 Mb. (This genome has not yet been sequenced.) On a finer scale, comparing two insect species, the cricket (*Anabrus simplex*) genome turns out to have 11 times as many base pairs as the *Drosophila melanogaster* genome. There is a wide range of genome sizes within the groups of protists, insects, amphibians, and plants and less of a range within mammals and reptiles.

Number of Genes

The number of genes also varies between prokaryotes and eukaryotes: Bacteria and archaea, in general, have fewer genes than eukaryotes. Free-living bacteria and archaea have from 1,500 to 7,500 genes, while the number of genes in eukaryotes ranges from about 5,000 for unicellular fungi to at least 40,000 for some multicellular eukaryotes (see Table 21.1).

Within the eukaryotes, the number of genes in a species is often lower than expected from simply considering the size of its genome. Looking at Table 21.1, you can see that the genome of the nematode *C. elegans* is 100 Mb in size and contains roughly 20,000 genes. The *Drosophila* genome, in comparison, is much bigger (165 Mb) but has about two-thirds the number of genes—only 13,700 genes.

Considering an example closer to home, we noted that the human genome contains 3,000 Mb, well over ten times the size of either the *Drosophila* or *C. elegans* genome. At the outset of the Human Genome Project, biologists expected somewhere between 50,000 and 100,000 genes to be identified in the completed sequence, based on the number of known human proteins. As the project progressed, the estimate was revised downward several times, and in 2010, the most reliable count placed the number at fewer than 21,000. This relatively low number, similar to the number of genes in the nematode *C. elegans*, has surprised biologists, who had clearly expected many more human genes.

What genetic attributes allow humans (and other vertebrates) to get by with no more genes than nematodes? An important factor is that vertebrate genomes "get more bang for the buck" from their coding sequences because of extensive alternative splicing of RNA transcripts. Recall that this process generates more than one functional protein from a single gene (see Figure 18.13). A typical human gene contains about ten exons, and an estimated 93% or so of these multi-exon genes are spliced in at least two different ways. Some genes are expressed in hundreds of alternatively spliced forms, others in just two. It is not yet possible to catalog all of the different forms, but it is clear that the number of different proteins encoded in the human genome far exceeds the proposed number of genes.

Additional polypeptide diversity could result from post-translational modifications such as cleavage or the addition of carbohydrate groups in different cell types or at different developmental stages. Finally, the discovery of miRNAs and other small RNAs that play regulatory roles have added a new variable to the mix (see Concept 18.3). Some scientists think that this added level of regulation, when present, may contribute to greater organismal complexity for a given number of genes.

Gene Density and Noncoding DNA

In addition to genome size and number of genes, we can compare gene density in different species—in other words, how many genes there are in a given length of DNA. When we compare the genomes of bacteria, archaea, and eukaryotes, we see that eukaryotes generally have larger genomes but fewer genes in a given number of base pairs. Humans have hundreds or thousands of times as many base pairs in their genome as most bacteria, as we already noted, but only 5 to 15 times as many genes; thus, gene density is lower in humans (see Table 21.1). Even unicellular eukaryotes, such as yeasts, have fewer genes per million base pairs than bacteria and archaea. Among the genomes that have been sequenced completely thus far, humans and other mammals have the lowest gene density.

In all bacterial genomes studied so far, most of the DNA consists of genes for protein, tRNA, or rRNA; the small amount remaining consists mainly of nontranscribed regulatory sequences, such as promoters. The sequence of nucleotides along a bacterial protein-coding gene proceeds from start to finish without interruption by noncoding sequences (introns). In eukaryotic genomes, by contrast, most of the DNA neither encodes protein nor is transcribed into RNA molecules of known function, and the DNA includes more complex regulatory sequences. In fact, humans have 10,000 times as much noncoding DNA as bacteria. Some of this DNA in multicellular eukaryotes is present as introns within genes. Indeed, introns account for most of the difference in average length between human genes (27,000 base pairs) and bacterial genes (1,000 base pairs).

In addition to introns, multicellular eukaryotes have a vast amount of non-protein-coding DNA between genes. In the next section, we will describe the composition and arrangement of these great stretches of DNA in the human genome.

CONCEPT CHECK 21.3

1. According to the best current estimate, the human genome contains fewer than 21,000 genes. However, there is evidence that human cells produce many more than 21,000 different polypeptides. What processes might account for this discrepancy?
2. The number of sequenced genomes is constantly being updated. Go to www.genomesonline.org to find the current number of completed genomes for each domain as well as the number of genomes whose sequencing is in progress. (*Hint*: Click on "Enter GOLD," and then click on "Published Complete Genomes" for extra information.)
3. **WHAT IF?** What evolutionary processes might account for prokaryotes having smaller genomes than eukaryotes?

For suggested answers, see Appendix A.

CONCEPT 21.4

Multicellular eukaryotes have much noncoding DNA and many multigene families

We have spent most of this chapter, and indeed this unit, focusing on genes that code for proteins. Yet the coding regions of these genes and the genes for RNA products such as rRNA, tRNA, and miRNA make up only a small portion of the genomes of most multicellular eukaryotes. The bulk of many eukaryotic genomes consists of DNA sequences that neither code for proteins nor are transcribed to produce RNAs with known functions; this noncoding DNA was often described in the past as "junk DNA." However, much evidence is accumulating that this DNA plays important roles in the cell, an idea supported by its persistence in diverse genomes over many hundreds of generations. For example, comparison of the genomes of humans, rats, and mice has revealed the presence of almost 500 regions of noncoding DNA that are identical in sequence in all three species. This is a higher level of sequence conservation than is seen for protein-coding regions in these species, strongly suggesting that the noncoding regions have important functions. In this section, we examine how genes and noncoding DNA sequences are organized within genomes of multicellular eukaryotes, using the human genome as our main example. Genome organization tells us much about how genomes have evolved and continue to evolve, the next subject we'll consider.

Once the sequencing of the human genome was completed, it became clear that only a tiny part—1.5%—codes for proteins or is transcribed into rRNAs or tRNAs. **Figure 21.7** shows what is known about the makeup of the remaining 98.5%. Gene-related regulatory sequences and introns account, respectively, for 5% and about 20% of the human genome. The rest, located between functional genes, includes some unique noncoding DNA, such as gene fragments and **pseudogenes**, former genes that have accumulated mutations over a long time and no longer produce functional proteins. (The genes that produce small noncoding RNAs are a tiny percentage of the genome, distributed between the 20% introns and the 15% unique noncoding DNA.) Most intergenic DNA, however, is **repetitive DNA**, which consists of sequences that are present in multiple copies in the genome. Somewhat surprisingly, about 75% of this repetitive DNA (44% of the entire human genome) is made up of units called transposable elements and sequences related to them.

Transposable Elements and Related Sequences

Both prokaryotes and eukaryotes have stretches of DNA that can move from one location to another within the genome. These stretches are known as *transposable genetic elements*, or

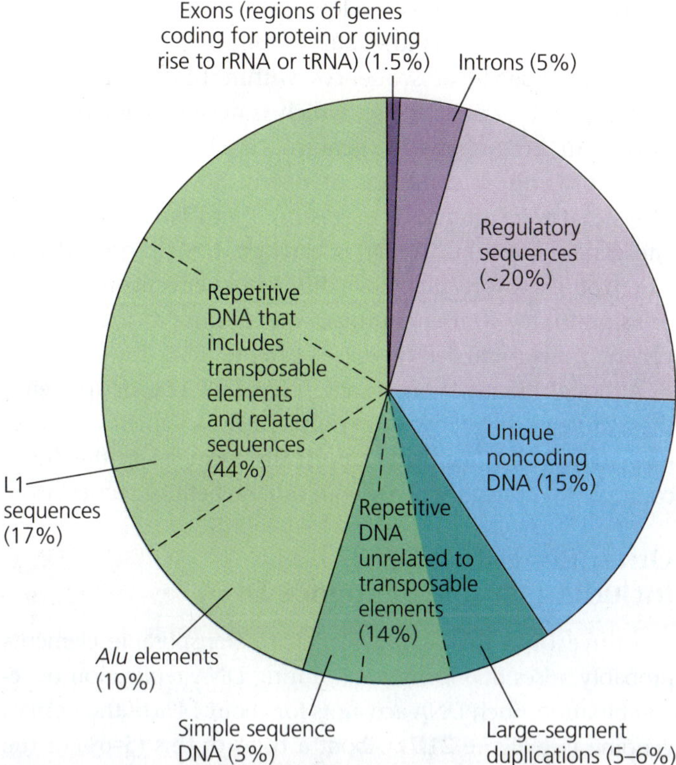

▲ **Figure 21.7 Types of DNA sequences in the human genome.** The gene sequences that code for proteins or are transcribed into rRNA or tRNA molecules make up only about 1.5% of the human genome (dark purple in the pie chart), while introns and regulatory sequences associated with genes (light purple) make up about a quarter. The vast majority of the human genome does not code for proteins or give rise to known RNAs, and much of it is repetitive DNA (dark and light green). Because repetitive DNA is the most difficult to sequence and analyze, classification of some portions is tentative, and the percentages given here may shift slightly as genome analysis proceeds. The genes that are transcribed into small noncoding RNAs such as miRNAs, which were recently discovered, are found among unique noncoding DNA sequences and within introns and thus would be included in two segments of this chart.

simply **transposable elements**. During the process called *transposition*, a transposable element moves from one site in a cell's DNA to a different target site by a type of recombination process. Transposable elements are sometimes called "jumping genes," but it should be kept in mind that they never completely detach from the cell's DNA. Instead, the original and new DNA sites are brought together by enzymes and other proteins that bend the DNA.

The first evidence for wandering DNA segments came from American geneticist Barbara McClintock's breeding experiments with Indian corn (maize) in the 1940s and 1950s **(Figure 21.8)**. As she tracked corn plants through multiple generations, McClintock identified changes in the color of corn kernels that made sense only if she postulated the existence of genetic elements capable of moving from other locations in the genome into the genes for kernel color, disrupting the genes so that the kernel color was changed. McClintock's discovery was met with great skepticism and virtually discounted at the time. Her

▲ **Figure 21.8 The effect of transposable elements on corn kernel color.** Barbara McClintock first proposed the idea of mobile genetic elements after observing variegations in corn kernel color (right).

careful work and insightful ideas were finally validated many years later when transposable elements were found in bacteria. In 1983, at the age of 81, McClintock received the Nobel Prize for her pioneering research.

Movement of Transposons and Retrotransposons

Eukaryotic transposable elements are of two types. The first type are **transposons**, which move within a genome by means of a DNA intermediate. Transposons can move by a "cut-and-paste" mechanism, which removes the element from the original site, or by a "copy-and-paste" mechanism, which leaves a copy behind **(Figure 21.9)**. Both mechanisms require an enzyme called *transposase*, which is generally encoded by the transposon.

Most transposable elements in eukaryotic genomes are of the second type, **retrotransposons**, which move by means of an RNA intermediate that is a transcript of the retrotransposon DNA. Retrotransposons always leave a copy at the original site during transposition, since they are initially transcribed

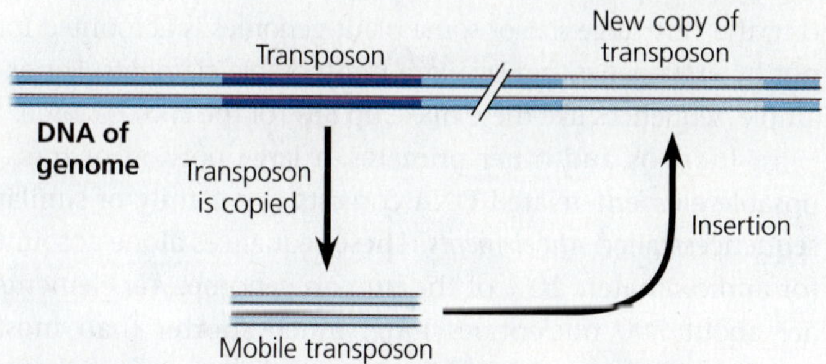

▲ **Figure 21.9 Transposon movement.** Movement of transposons by either the cut-and-paste mechanism or the copy-and-paste mechanism (shown here) involves a double-stranded DNA intermediate that is inserted into the genome.

? *How would this figure differ if it showed the cut-and-paste mechanism?*

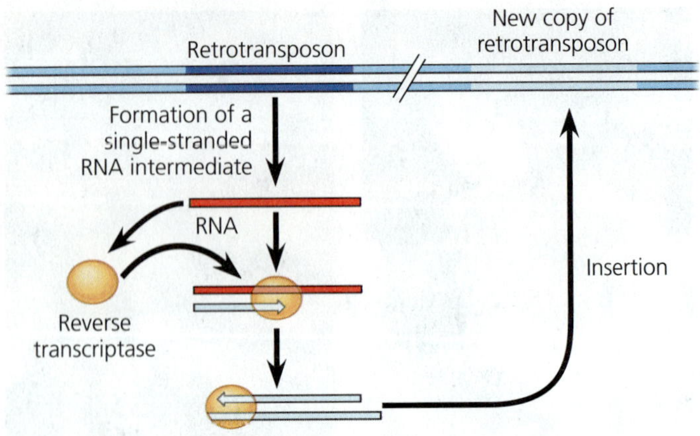

▲ **Figure 21.10 Retrotransposon movement.** Movement begins with formation of a single-stranded RNA intermediate. The remaining steps are essentially identical to part of the retrovirus replicative cycle (see Figure 19.8).

into an RNA intermediate **(Figure 21.10)**. To insert at another site, the RNA intermediate is first converted back to DNA by reverse transcriptase, an enzyme encoded by the retrotransposon. (Reverse transcriptase is also encoded by retroviruses, as you learned in Chapter 19. In fact, retroviruses may have evolved from retrotransposons.) Another cellular enzyme catalyzes insertion of the reverse-transcribed DNA at a new site.

Sequences Related to Transposable Elements

Multiple copies of transposable elements and sequences related to them are scattered throughout eukaryotic genomes. A single unit is usually hundreds to thousands of base pairs long, and the dispersed "copies" are similar but usually not identical to each other. Some of these are transposable elements that can move; the enzymes required for this movement may be encoded by any transposable element, including the one that is moving. Others are related sequences that have lost the ability to move altogether. Transposable elements and related sequences make up 25–50% of most mammalian genomes (see Figure 21.7) and even higher percentages in amphibians and many plants. In fact, the very large size of some plant genomes is accounted for not by extra genes, but by extra transposable elements. For example, sequences like these make up 85% of the corn genome!

In humans and other primates, a large portion of transposable element–related DNA consists of a family of similar sequences called *Alu elements*. These sequences alone account for approximately 10% of the human genome. *Alu* elements are about 300 nucleotides long, much shorter than most functional transposable elements, and they do not code for any protein. However, many *Alu* elements are transcribed into RNA; its cellular function, if any, is currently unknown.

An even larger percentage (17%) of the human genome is made up of a type of retrotransposon called *LINE-1*, or *L1*. These sequences are much longer than *Alu* elements—about 6,500 base pairs—and have a low rate of transposition. What might account for this low rate? Recent research has uncovered the presence of sequences within L1 that block the progress of RNA polymerase, which is necessary for transposition. An accompanying genomic analysis found L1 sequences within the introns of nearly 80% of the human genes that were analyzed, suggesting that L1 may help regulate gene expression. Other researchers have proposed that L1 retrotransposons may have differential effects on gene expression in developing neurons, contributing to the great diversity of neuronal cell types (see Chapter 48).

Although many transposable elements encode proteins, these proteins do not carry out normal cellular functions. Therefore, transposable elements are usually included in the "noncoding" DNA category, along with other repetitive sequences.

Other Repetitive DNA, Including Simple Sequence DNA

Repetitive DNA that is not related to transposable elements probably arises due to mistakes during DNA replication or recombination. Such DNA accounts for about 14% of the human genome (see Figure 21.7). About a third of this (5–6% of the human genome) consists of duplications of long stretches of DNA, with each unit ranging from 10,000 to 300,000 base pairs. The large segments seem to have been copied from one chromosomal location to another site on the same or a different chromosome and probably include some functional genes.

In contrast to scattered copies of long sequences, **simple sequence DNA** contains many copies of tandemly repeated short sequences, as in the following example (showing one DNA strand only):

. . . GTTACGTTACGTTACGTTACGTTACGTTAC . . .

In this case, the repeated unit (GTTAC) consists of 5 nucleotides. Repeated units may contain as many as 500 nucleotides, but often contain fewer than 15 nucleotides, as in this example. When the unit contains 2–5 nucleotides, the series of repeats is called a **short tandem repeat**, or **STR**; we discussed the use of STR analysis in preparing genetic profiles in Chapter 20. The number of copies of the repeated unit can vary from site to site within a given genome. There could be as many as several hundred thousand repetitions of the GTTAC unit at one site, but only half that number at another. STR analysis is performed on sites selected because they have relatively few repeats. The repeat number can vary from person to person, and since humans are diploid, each person has two alleles per site, which can differ. This diversity produces the variation represented in the genetic profiles that result from STR analysis. Altogether, simple sequence DNA makes up 3% of the human genome.

Much of a genome's simple sequence DNA is located at chromosomal telomeres and centromeres, suggesting that this DNA plays a structural role for chromosomes. The DNA at centromeres is essential for the separation of chromatids in cell division (see Chapter 12). Centromeric DNA, along with simple sequence DNA located elsewhere, may also help organize the chromatin within the interphase nucleus. The simple sequence DNA located at telomeres, at the tips of chromosomes, prevents genes from being lost as the DNA shortens with each round of replication (see Chapter 16). Telomeric DNA also binds proteins that protect the ends of a chromosome from degradation and from joining to other chromosomes.

Genes and Multigene Families

We finish our discussion of the various types of DNA sequences in eukaryotic genomes with a closer look at genes. Recall that DNA sequences that code for proteins or give rise to tRNA or rRNA compose a mere 1.5% of the human genome (see Figure 21.7). If we include introns and regulatory sequences associated with genes, the total amount of DNA that is gene-related—coding and noncoding—constitutes about 25% of the human genome. Put another way, only about 6% (1.5% out of 25%) of the length of the average gene is represented in the final gene product.

Like the genes of bacteria, many eukaryotic genes are present as unique sequences, with only one copy per haploid set of chromosomes. But in the human genome and the genomes of many other animals and plants, solitary genes make up less than half of the total gene-related DNA. The rest occur in **multigene families**, collections of two or more identical or very similar genes.

In multigene families that consist of *identical* DNA sequences, those sequences are usually clustered tandemly and, with the notable exception of the genes for histone proteins, have RNAs as their final products. An example is the family of identical DNA sequences that are the genes for the three largest rRNA molecules **(Figure 21.11a)**. These rRNA molecules are transcribed from a single transcription unit that is repeated tandemly hundreds to thousands of times in one or several clusters in the genome of a multicellular eukaryote. The many copies of this rRNA transcription unit help cells to quickly make the millions of ribosomes needed for active protein synthesis. The primary transcript is cleaved to yield the three rRNA molecules, which combine with proteins and one other kind of rRNA (5S rRNA) to form ribosomal subunits.

The classic examples of multigene families of *nonidentical* genes are two related families of genes that encode globins, a group of proteins that include the α and β polypeptide subunits of hemoglobin. One family, located on chromosome 16 in humans, encodes various forms of α-globin; the other, on chromosome 11, encodes forms of β-globin **(Figure 21.11b)**. The different forms of each globin subunit are expressed at different times in development, allowing hemoglobin to function effectively in the changing environment of the developing animal. In humans, for example, the embryonic and fetal forms of hemoglobin have a higher affinity for oxygen

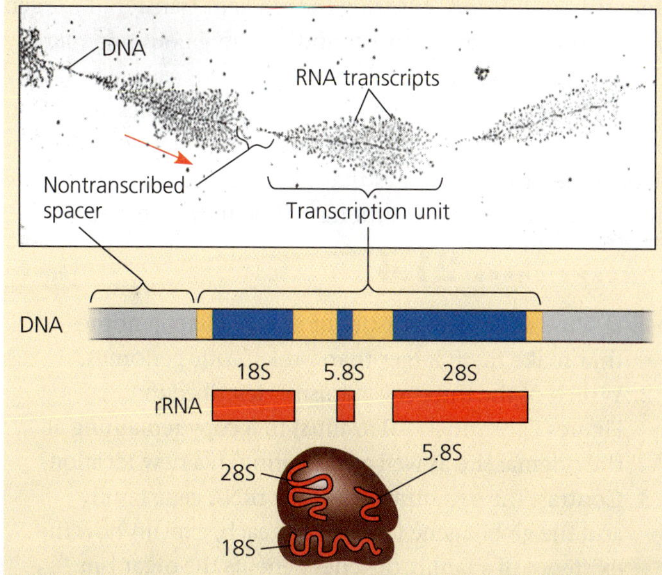

(a) Part of the ribosomal RNA gene family. The TEM at the top shows three of the hundreds of copies of rRNA transcription units in a salamander genome. Each "feather" corresponds to a single unit being transcribed by about 100 molecules of RNA polymerase (dark dots along the DNA), moving left to right (red arrow). The growing RNA transcripts extend from the DNA. In the diagram of a trascription unit below the TEM, the genes for three types of rRNA (blue) are adjacent to regions that are transcribed but later removed (yellow). A single trascript is processed to yield one of each of the three rRNAs (red), key components of the ribosome.

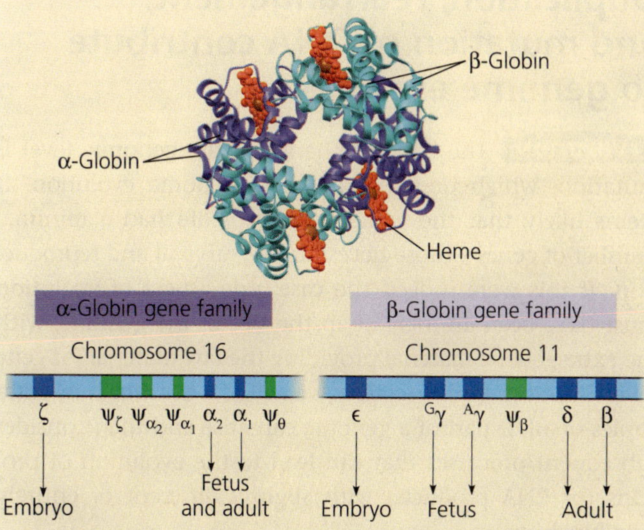

(b) The human α-globin and β-globin gene families. Adult hemoglobin is composed of two α-globin and two β-globin polypeptide subunits, as shown in the molecular model. The genes (dark blue) encoding α- and β-globins are found in two families, organized as shown here. The noncoding DNA separating the functional genes within each family includes pseudogenes (ψ; green), versions of the functional genes that no longer produce functional proteins. Genes and pseudogenes are named with Greek letters. Some genes are expressed only in the embryo or fetus.

▲ **Figure 21.11 Gene families.**

In (a), how could you determine the direction of transcription if it wasn't indicated by the red arrow?

than the adult forms, ensuring the efficient transfer of oxygen from mother to fetus. Also found in the globin gene family clusters are several pseudogenes.

The arrangement of the genes in gene families has given biologists insight into the evolution of genomes. We will consider some of the processes that have shaped the genomes of different species over evolutionary time in the next section.

CONCEPT CHECK 21.4

1. Discuss the characteristics of mammalian genomes that make them larger than prokaryotic genomes.
2. Which of the three mechanisms described in Figures 21.9 and 21.10 result(s) in a copy remaining at the original site as well as appearing in a new location?
3. Contrast the organizations of the rRNA gene family and the globin gene families. For each, explain how the existence of a family of genes benefits the organism.
4. **MAKE CONNECTIONS** Assign each DNA segment at the top of Figure 18.8 (p. 359) to a sector in the pie chart in Figure 21.7.

For suggested answers, see Appendix A.

CONCEPT **21.5**

Duplication, rearrangement, and mutation of DNA contribute to genome evolution

EVOLUTION The basis of change at the genomic level is mutation, which underlies much of genome evolution. It seems likely that the earliest forms of life had a minimal number of genes—those necessary for survival and reproduction. If this were indeed the case, one aspect of evolution must have been an increase in the size of the genome, with the extra genetic material providing the raw material for gene diversification. In this section, we will first describe how extra copies of all or part of a genome can arise and then consider subsequent processes that can lead to the evolution of proteins (or RNA products) with slightly different or entirely new functions.

Duplication of Entire Chromosome Sets

An accident in meiosis can result in one or more extra sets of chromosomes, a condition known as polyploidy. Although such accidents would most often be lethal, in rare cases they could facilitate the evolution of genes. In a polyploid organism, one set of genes can provide essential functions for the organism. The genes in the one or more extra sets can diverge by accumulating mutations; these variations may persist if the organism carrying them survives and reproduces. In this way, genes with novel functions can evolve. As long as one copy of an essential gene is expressed, the divergence of another copy can lead to its encoded protein acting in a novel way, thereby changing the organism's phenotype. The outcome of this accumulation of mutations may be the branching off of a new species, as happens often in flowering plants (see Chapter 24). Polyploid animals also exist, but they are much rarer; the tetraploid model organism *Xenopus laevis*, the African clawed frog, is an example.

Alterations of Chromosome Structure

Scientists have long known that sometime in the last 6 million years, when the ancestors of humans and chimpanzees diverged as species, the fusion of two ancestral chromosomes in the human line led to different haploid numbers for humans ($n = 23$) and chimpanzees ($n = 24$). The banding patterns in stained chromosomes suggested that the ancestral versions of current chimp chromosomes 12 and 13 fused end to end, forming chromosome 2 in an ancestor of the human lineage. With the recent explosion in genomic sequence information, we can now compare the chromosomal organizations of many different species on a much finer scale. This information allows us to make inferences about the evolutionary processes that shape chromosomes and may drive speciation. Sequencing and analysis of human chromosome 2 in 2005 provided very strong supporting evidence for the model we have just described **(Figure 21.12a)**.

In another study of broader scope, researchers compared the DNA sequence of each human chromosome with the whole-genome sequence of the mouse. **Figure 21.12b** shows the results of this comparison for human chromosome 16: Large blocks of genes on this chromosome are found on four mouse chromosomes, indicating that the genes in each block stayed together during the evolution of the mouse and human lineages.

Performing the same comparative analysis between chromosomes of humans and six other mammalian species allowed the researchers to reconstruct the evolutionary history of chromosomal rearrangements in these eight species. They found many duplications and inversions of large portions of chromosomes, the result of mistakes during meiotic recombination in which the DNA broke and was rejoined incorrectly. The rate of these events seems to have accelerated about 100 million years ago, around the time large dinosaurs became extinct and the number of mammalian species increased rapidly. The apparent coincidence is interesting because chromosomal rearrangements are thought to contribute to the generation of new species. Although two individuals with different arrangements could still mate and produce offspring, the offspring would have two nonequivalent sets of chromosomes, making meiosis inefficient or even impossible. Thus, chromosomal rearrangements would lead to two populations that could not successfully mate with each other, a step on the way to their becoming two separate species. (You'll learn more about this in Chapter 24.)

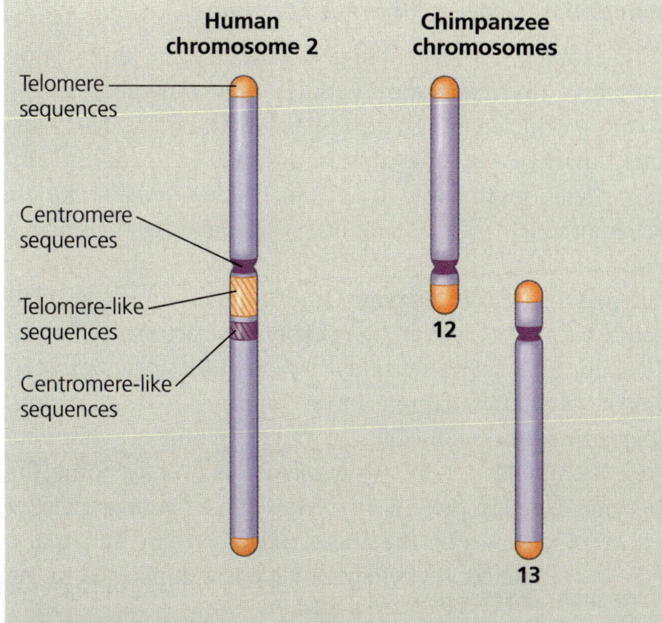

(a) Human and chimpanzee chromosomes. The positions of telomere-like and centromere-like sequences on human chromosome 2 (left) match those of telomeres on chimp chromosomes 12 and 13 and the centromere on chimp chromosome 13 (right). This suggests that chromosomes 12 and 13 in a human ancestor fused end to end to form human chromosome 2. The centromere from ancestral chromosome 12 remained functional on human chromosome 2, while the one from ancestral chromosome 13 did not. (Chimp chromosomes 12 and 13 have been renamed 2a and 2b, respectively.)

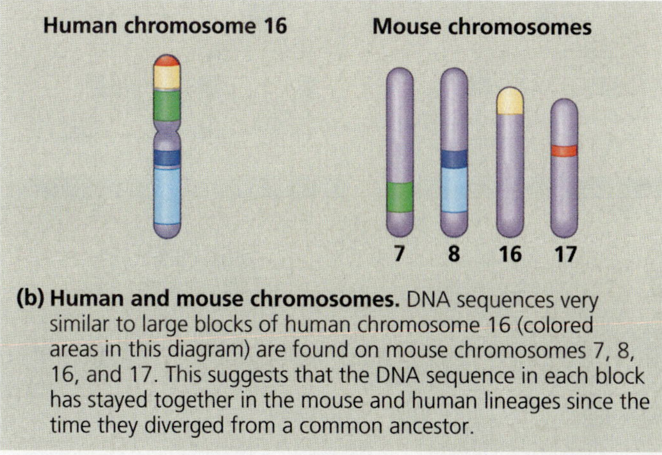

(b) Human and mouse chromosomes. DNA sequences very similar to large blocks of human chromosome 16 (colored areas in this diagram) are found on mouse chromosomes 7, 8, 16, and 17. This suggests that the DNA sequence in each block has stayed together in the mouse and human lineages since the time they diverged from a common ancestor.

▲ **Figure 21.12 Related chromosome sequences among mammals.**

Somewhat unexpectedly, the same study also unearthed a pattern with medical relevance. Analysis of the chromosomal breakage points associated with the rearrangements showed that they were not randomly distributed; specific sites were used over and over again. A number of these recombination "hot spots" correspond to locations of chromosomal rearrangements within the human genome that are associated with congenital diseases. Researchers are, of course, looking at the other sites as well for their possible association with as yet unidentified diseases.

Duplication and Divergence of Gene-Sized Regions of DNA

Errors during meiosis can also lead to the duplication of chromosomal regions that are smaller than the ones we've just discussed, including segments the length of individual genes. Unequal crossing over during prophase I of meiosis, for instance, can result in one chromosome with a deletion and another with a duplication of a particular gene. As illustrated in **Figure 21.13**, transposable elements can provide homologous sites where nonsister chromatids can cross over, even when other chromatid regions are not correctly aligned.

Also, slippage can occur during DNA replication, such that the template shifts with respect to the new complementary strand, and a part of the template strand is either skipped by the replication machinery or used twice as a template. As a result, a segment of DNA is deleted or duplicated. It is easy to imagine how such errors could occur in regions of repeats. The variable number of repeated units of simple sequence DNA at a given site, used for STR analysis, is probably due to errors like these. Evidence that unequal crossing over and template slippage during DNA replication lead to duplication of genes is found in the existence of multigene families, such as the globin family.

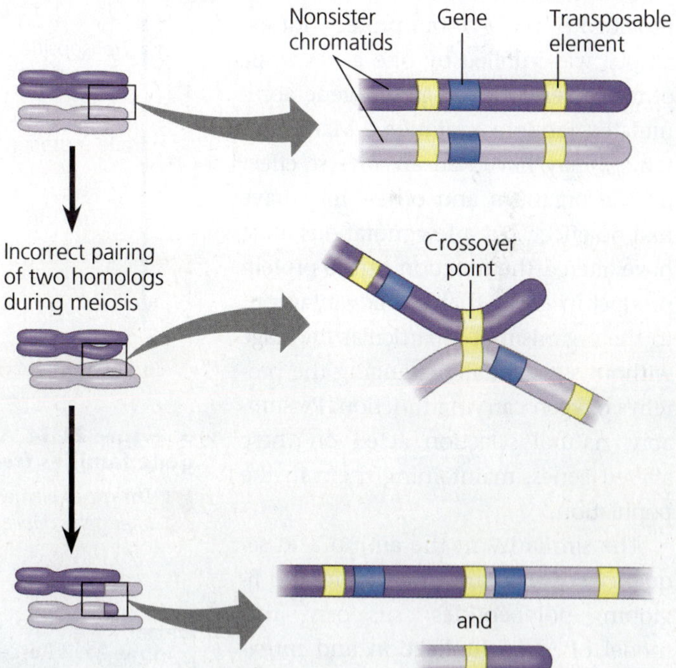

▲ **Figure 21.13 Gene duplication due to unequal crossing over.** One mechanism by which a gene (or other DNA segment) can be duplicated is recombination during meiosis between copies of a transposable element flanking the gene. Such recombination between misaligned nonsister chromatids of homologous chromosomes produces one chromatid with two copies of the gene and one chromatid with no copy.

MAKE CONNECTIONS *Examine how crossing over occurs in Figure 13.11 (p. 259). In the middle panel above, draw a line through the portions that result in the upper chromatid in the bottom panel. Use a different color to do the same for the other chromatid.*

Evolution of Genes with Related Functions: The Human Globin Genes

Duplication events can lead to the evolution of genes with related functions, such as those of the α-globin and β-globin gene families (see Figure 21.11b). A comparison of gene sequences within a multigene family can suggest the order in which the genes arose. This approach to re-creating the evolutionary history of the globin genes indicates that they all evolved from one common ancestral globin gene that underwent duplication and divergence into the α-globin and β-globin ancestral genes about 450–500 million years ago **(Figure 21.14)**. Each of these genes was later duplicated several times, and the copies then diverged from each other in sequence, yielding the current family members. In fact, the common ancestral globin gene also gave rise to the oxygen-binding muscle protein myoglobin and to the plant protein leghemoglobin. The latter two proteins function as monomers, and their genes are included in a "globin superfamily."

After the duplication events, the differences between the genes in the globin families undoubtedly arose from mutations that accumulated in the gene copies over many generations. The current model is that the necessary function provided by an α-globin protein, for example, was fulfilled by one gene, while other copies of the α-globin gene accumulated random mutations. Many mutations may have had an adverse effect on the organism and others may have had no effect, but a few mutations must have altered the function of the protein product in a way that was advantageous to the organism at a particular life stage without substantially changing the protein's oxygen-carrying function. Presumably, natural selection acted on these altered genes, maintaining them in the population.

The similarity in the amino acid sequences of the various α-globin and β-globin polypeptides supports this model of gene duplication and mutation **(Table 21.2)**. The amino acid sequences of the β-globins, for instance, are much more similar to each other than to the α-globin sequences. The existence of several pseudogenes among the functional globin genes provides additional evidence for this model (see Figure 21.11b): Random mutations in these "genes" over evolutionary time have destroyed their function.

Evolution of Genes with Novel Functions

In the evolution of the globin gene families, gene duplication and subsequent divergence produced family members whose protein products performed similar functions (oxygen transport). Alternatively, one copy of a duplicated gene can undergo alterations that lead to a completely new function for the protein product. The genes for lysozyme and α-lactalbumin are good examples.

Lysozyme is an enzyme that helps protect animals against bacterial infection by hydrolyzing bacterial cell walls; α-lactalbumin is a nonenzymatic protein that plays a role in milk production in mammals. The two proteins are quite similar in their amino acid sequences and three-dimensional structures. Both genes are found in mammals, whereas only the lysozyme gene is present in birds. These findings suggest that at some time after the lineages leading to mammals and birds had separated, the lysozyme gene was duplicated in the

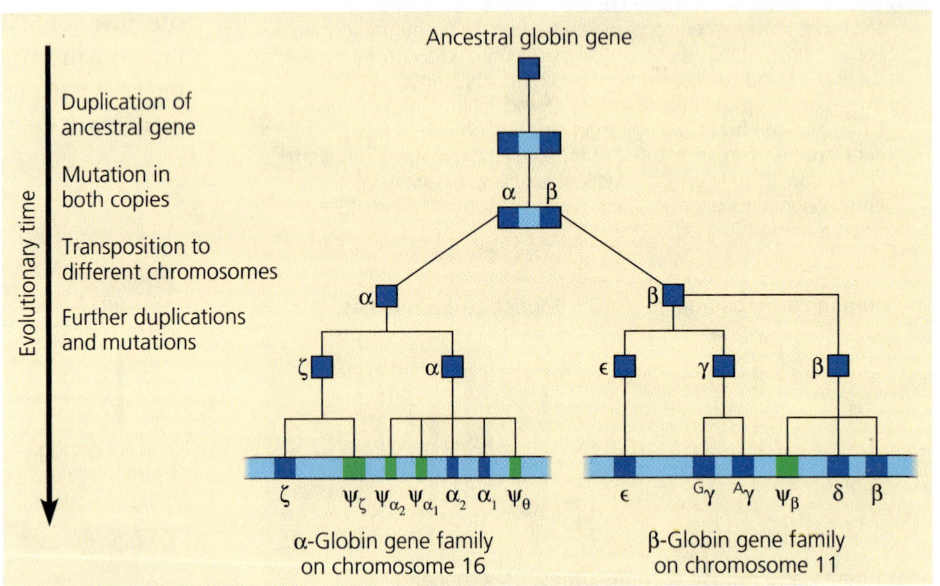

▲ **Figure 21.14 A model for the evolution of the human α-globin and β-globin gene families from a single ancestral globin gene.**

? *The green elements are pseudogenes. Explain how they could have arisen after gene duplication.*

Table 21.2 Percentage of Similarity in Amino Acid Sequence Between Human Globin Proteins

		α-Globins		β-Globins		
		α	ζ	β	γ	ε
α-Globins	α	—	58	42	39	37
	ζ	58	—	34	38	37
β-Globins	β	42	34	—	73	75
	γ	39	38	73	—	80
	ε	37	37	75	80	—

mammalian lineage but not in the avian lineage. Subsequently, one copy of the duplicated lysozyme gene evolved into a gene encoding α-lactalbumin, a protein with a completely different function.

Rearrangements of Parts of Genes: Exon Duplication and Exon Shuffling

Rearrangement of existing DNA sequences within genes has also contributed to genome evolution. The presence of introns in most genes of multicellular eukaryotes may have promoted the evolution of new and potentially useful proteins by facilitating the duplication or repositioning of exons in the genome. Recall from Chapter 17 that an exon often codes for a domain, a distinct structural or functional region of a protein.

We've already seen that unequal crossing over during meiosis can lead to duplication of a gene on one chromosome and its loss from the homologous chromosome (see Figure 21.13). By a similar process, a particular exon within a gene could be duplicated on one chromosome and deleted from the other. The gene with the duplicated exon would code for a protein containing a second copy of the encoded domain. This change in the protein's structure could augment its function by increasing its stability, enhancing its ability to bind a particular ligand, or altering some other property. Quite a few protein-coding genes have multiple copies of related exons, which presumably arose by duplication and then diverged. The gene encoding the extracellular matrix protein collagen is a good example. Collagen is a structural protein with a highly repetitive amino acid sequence, which is reflected in the repetitive pattern of exons in the collagen gene.

Alternatively, we can imagine the occasional mixing and matching of different exons either within a gene or between two different (nonallelic) genes owing to errors in meiotic recombination. This process, termed *exon shuffling,* could lead to new proteins with novel combinations of functions. As an example, let's consider the gene for tissue plasminogen activator (TPA). The TPA protein is an extracellular protein that helps control blood clotting. It has four domains of three types, each encoded by an exon; one exon is present in two copies. Because each type of exon is also found in other proteins, the gene for TPA is thought to have arisen by several instances of exon shuffling and duplication **(Figure 21.15)**.

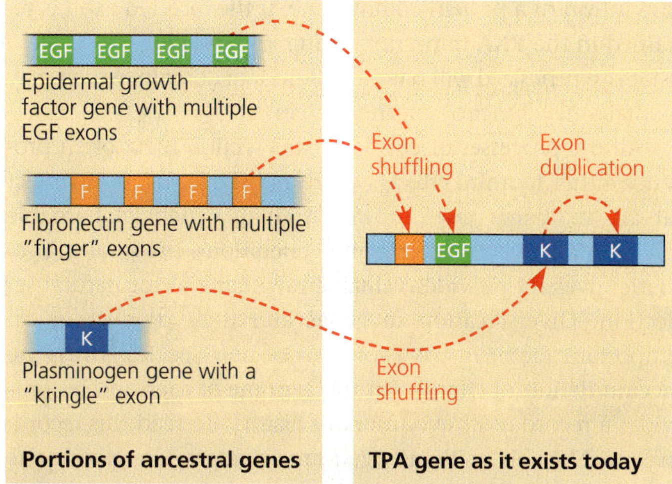

▲ **Figure 21.15 Evolution of a new gene by exon shuffling.** Exon shuffling could have moved exons, each encoding a particular domain, from ancestral forms of the genes for epidermal growth factor, fibronectin, and plasminogen (left) into the evolving gene for tissue plasminogen activator, TPA (right). Duplication of the "kringle" exon from the plasminogen gene after its movement could account for the two copies of this exon in the TPA gene.

? *How could the presence of transposable elements in introns have facilitated the exon shuffling shown here?*

How Transposable Elements Contribute to Genome Evolution

The persistence of transposable elements as a large fraction of some eukaryotic genomes is consistent with the idea that they play an important role in shaping a genome over evolutionary time. These elements can contribute to the evolution of the genome in several ways. They can promote recombination, disrupt cellular genes or control elements, and carry entire genes or individual exons to new locations.

Transposable elements of similar sequence scattered throughout the genome facilitate recombination between different chromosomes by providing homologous regions for crossing over. Most such recombination events are probably detrimental, causing chromosomal translocations and other changes in the genome that may be lethal to the organism. But over the course of evolutionary time, an occasional recombination event of this sort may be advantageous to the organism. (For the change to be heritable, of course, it must happen in a cell that will give rise to a gamete.)

The movement of a transposable element can have a variety of consequences. For instance, if a transposable element "jumps" into the middle of a protein-coding sequence, it will prevent the production of a normal transcript of the gene. If a transposable element inserts within a regulatory sequence, the transposition may lead to increased or decreased production of one or more proteins. Transposition caused both types of effects on the genes coding for pigment-synthesizing enzymes in McClintock's corn kernels. Again, while such changes are usually harmful, in the long run some may prove beneficial by providing a survival advantage.

During transposition, a transposable element may carry along a gene or group of genes to a new position in the genome. This mechanism probably accounts for the location of the α-globin and β-globin gene families on different human chromosomes, as well as the dispersion of the genes of certain other gene families. By a similar tag-along process, an exon from one gene may be inserted into another gene in a mechanism similar to that of exon shuffling during recombination. For example, an exon may be inserted by transposition into the intron of a protein-coding gene. If the inserted exon is retained in the RNA transcript during RNA splicing, the protein that is synthesized will have an additional domain, which may confer a new function on the protein.

All the processes discussed in this section most often produce either harmful effects, which may be lethal, or no effect at all. In a few cases, however, small beneficial heritable changes may occur. Over many generations, the resulting genetic diversity provides valuable raw material for natural selection. Diversification of genes and their products is an important factor in the evolution of new species. Thus, the accumulation of changes in the genome of each species provides a record of its evolutionary history. To read this record, we must be able to identify genomic changes. Comparing the genomes of different species allows us to do that and has increased our understanding of how genomes evolve. You will learn more about these topics in the final section.

CONCEPT CHECK 21.5

1. Describe three examples of errors in cellular processes that lead to DNA duplications.
2. Explain how multiple exons might have arisen in the ancestral EGF and fibronectin genes shown in Figure 21.15 (left).
3. What are three ways that transposable elements are thought to contribute to genome evolution?
4. **WHAT IF?** In 2005, Icelandic scientists reported finding a large chromosomal inversion present in 20% of northern Europeans, and they noted that Icelandic women with this inversion had significantly more children than women without it. What would you expect to happen to the frequency of this inversion in the Icelandic population in future generations?

For suggested answers, see Appendix A.

CONCEPT 21.6

Comparing genome sequences provides clues to evolution and development

EVOLUTION One researcher has likened the current state of biology to the Age of Exploration in the 15th century after major improvements in navigation and the building of faster ships. In the last 25 years, we have seen rapid advances in genome sequencing and data collection, new techniques for assessing gene activity across the whole genome, and refined approaches for understanding how genes and their products work together in complex systems. We are truly poised on the brink of a new world.

Comparisons of genome sequences from different species reveal much about the evolutionary history of life, from very ancient to more recent. Similarly, comparative studies of the genetic programs that direct embryonic development in different species are beginning to clarify the mechanisms that generated the great diversity of life-forms present today. In this final section of the chapter, we will discuss what has been learned from these two approaches.

Comparing Genomes

The more similar in sequence the genes and genomes of two species are, the more closely related those species are in their evolutionary history. Comparing genomes of closely related species sheds light on more recent evolutionary events, whereas comparing genomes of very distantly related species helps us understand ancient evolutionary history. In either case, learning about characteristics that are shared or divergent between groups enhances our picture of the evolution of life-forms and biological processes. As you learned in Chapter 1, the evolutionary relationships between species can be represented by a diagram in the form of a tree (often turned sideways), where each branch point marks the divergence of two lineages. **Figure 21.16** shows the evolutionary relationships of some groups and species we will be discussing. We will consider comparisons between distantly related species first.

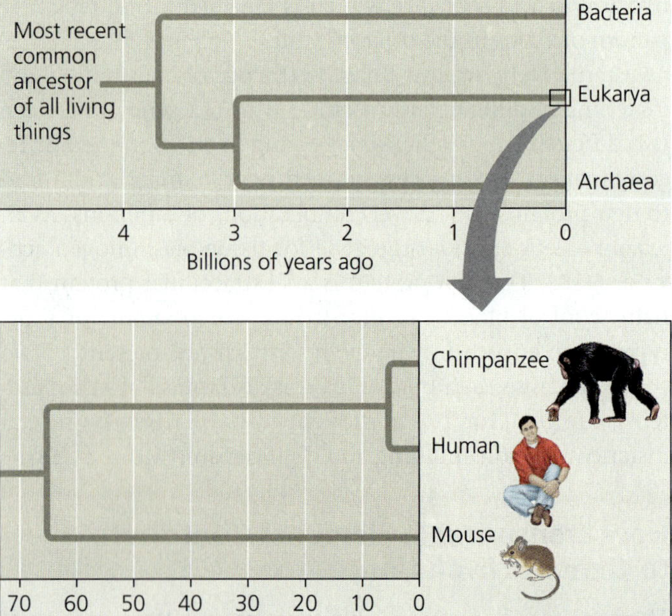

▲ **Figure 21.16 Evolutionary relationships of the three domains of life.** This tree diagram shows the ancient divergence of bacteria, archaea, and eukaryotes. A portion of the eukaryote lineage is expanded in the inset to show the more recent divergence of three mammalian species discussed in this chapter.

Comparing Distantly Related Species

Determining which genes have remained similar—that is, are *highly conserved*—in distantly related species can help clarify evolutionary relationships among species that diverged from each other long ago. Indeed, comparisons of the complete genome sequences of bacteria, archaea, and eukaryotes indicate that these three groups diverged between 2 and 4 billion years ago and strongly support the theory that they are the fundamental domains of life (see Figure 21.16).

In addition to their value in evolutionary biology, comparative genomic studies confirm the relevance of research on model organisms to our understanding of biology in general and human biology in particular. Genes that evolved a very long time ago can still be surprisingly similar in disparate species. As a case in point, several genes in yeast are so similar to certain human disease genes that researchers have deduced the functions of the disease genes by studying their yeast counterparts. This striking similarity underscores the common origin of these two distantly related species.

Comparing Closely Related Species

The genomes of two closely related species are likely to be organized similarly because of their relatively recent divergence. As we mentioned earlier, this allows the fully sequenced genome of one species to be used as a scaffold for assembling the genomic sequences of a closely related species, accelerating mapping of the second genome. For instance, using the human genome sequence as a guide, researchers were able to quickly sequence the chimpanzee genome.

The recent divergence of two closely related species also underlies the small number of gene differences that are found when their genomes are compared. The particular genetic differences can therefore be more easily correlated with phenotypic differences between the two species. An exciting application of this type of analysis is seen as researchers compare the human genome with the genomes of the chimpanzee, mouse, rat, and other mammals. Identifying the genes shared by all of these species but not by nonmammals should give clues about what it takes to make a mammal, while finding the genes shared by chimpanzees and humans but not by rodents should tell us something about primates. And, of course, comparing the human genome with that of the chimpanzee should help us answer the tantalizing question we asked at the beginning of the chapter: What genomic information makes a human or a chimpanzee?

An analysis of the overall composition of the human and chimpanzee genomes, which are thought to have diverged only about 6 million years ago (see Figure 21.16), reveals some general differences. Considering single nucleotide substitutions, the two genomes differ by only 1.2%. When researchers looked at longer stretches of DNA, however, they were surprised to find a further 2.7% difference due to insertions or deletions of larger regions in the genome of one or the other species; many of the insertions were duplications or other repetitive DNA. In fact, a third of the human duplications are not present in the chimpanzee genome, and some of these duplications contain regions associated with human diseases. There are more *Alu* elements in the human genome than in the chimpanzee genome, and the latter contains many copies of a retroviral provirus not present in humans. All of these observations provide clues to the forces that might have swept the two genomes along different paths, but we don't have a complete picture yet. We also don't know how these differences might account for the distinct characteristics of each species.

To discover the basis for the phenotypic differences between the two species, biologists are studying specific genes and types of genes that differ between humans and chimpanzees and comparing them with their counterparts in other mammals. This approach has revealed a number of genes that are apparently changing (evolving) faster in the human than in either the chimpanzee or the mouse. Among them are genes involved in defense against malaria and tuberculosis and at least one gene that regulates brain size. When genes are classified by function, the genes that seem to be evolving the fastest are those that code for transcription factors. This discovery makes sense because transcription factors regulate gene expression and thus play a key role in orchestrating the overall genetic program.

One transcription factor whose gene shows evidence of rapid change in the human lineage is called FOXP2. Several lines of evidence suggest that the *FOXP2* gene functions in vocalization in vertebrates. For one thing, mutations in this gene can produce severe speech and language impairment in humans. Moreover, the *FOXP2* gene is expressed in the brains of zebra finches and canaries at the time when these songbirds are learning their songs. But perhaps the strongest evidence comes from a "knock-out" experiment in which researchers disrupted the *FOXP2* gene in mice and analyzed the resulting phenotype (**Figure 21.17**, on the next page). The homozygous mutant mice had malformed brains and failed to emit normal ultrasonic vocalizations, and mice with one faulty copy of the gene also showed significant problems with vocalization. These results support the idea that the *FOXP2* gene product turns on genes involved in vocalization.

Expanding on this analysis, another research group more recently replaced the *FOXP2* gene in mice with a "humanized" copy coding for the human versions of two amino acids that differ between human and chimp; these are the changes potentially responsible for a human's ability to speak. Although the mice were generally healthy, they had subtly different vocalizations and showed changes in brain cells in circuits associated with speech in human brains.

The *FOXP2* story is an excellent example of how different approaches can complement each other in uncovering biological phenomena of widespread importance. The

▼ **Figure 21.17** **INQUIRY**

What is the function of a gene (*FOXP2*) that is rapidly evolving in the human lineage?

EXPERIMENT Several lines of evidence support a role for the *FOXP2* gene in the development of speech and language in humans and of vocalization in other vertebrates. In 2005, Joseph Buxbaum and collaborators at the Mount Sinai School of Medicine and several other institutions tested the function of *FOXP2*. They used the mouse, a model organism in which genes can be easily knocked out, as a representative vertebrate that vocalizes: Mice produce ultrasonic squeaks (whistles) to communicate stress. The researchers used genetic engineering to produce mice in which one or both copies of *FOXP2* were disrupted.

Wild type: two normal copies of *FOXP2*

Heterozygote: one copy of *FOXP2* disrupted

Homozygote: both copies of *FOXP2* disrupted

They then compared the phenotypes of these mice. Two of the characters they examined are included here: brain anatomy and vocalization.

Experiment 1: Researchers cut thin sections of brain and stained them with reagents that allow visualization of brain anatomy in a UV fluorescence microscope.

Experiment 2: Researchers separated each newborn pup from its mother and recorded the number of ultrasonic whistles produced by the pup.

RESULTS

Experiment 1: Disruption of both copies of *FOXP2* led to brain abnormalities in which the cells were disorganized. Phenotypic effects on the brain of heterozygotes, with one disrupted copy, were less severe. (Each color reveals a different cell or tissue type.)

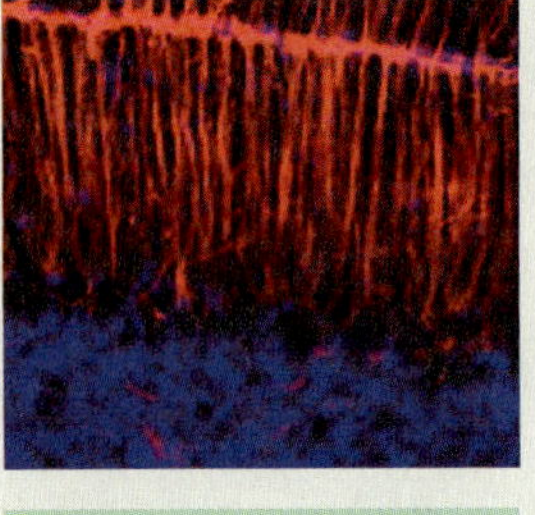

Wild type

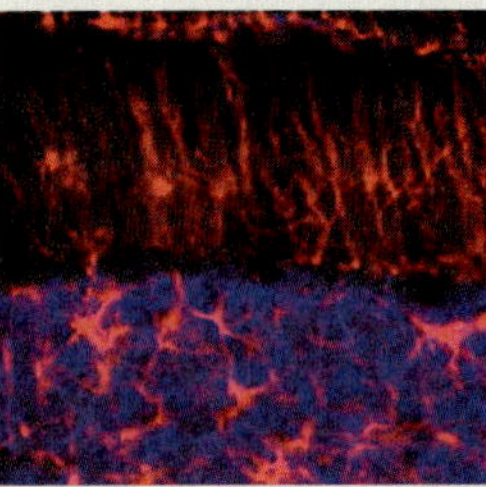

Heterozygote

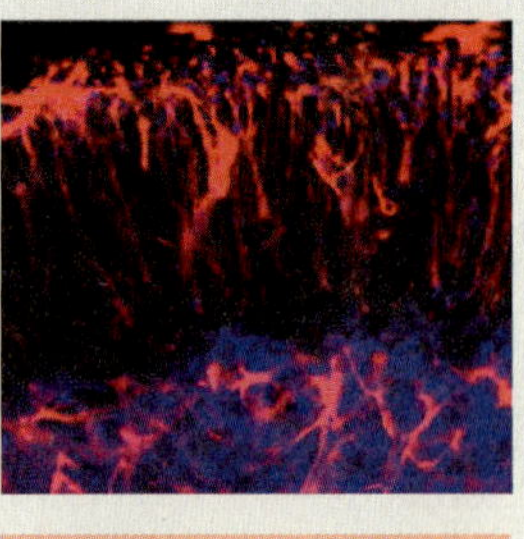

Homozygote

Experiment 2: Disruption of both copies of *FOXP2* led to an absence of ultrasonic vocalization in response to stress. The effect on vocalization in the heterozygote was also extreme.

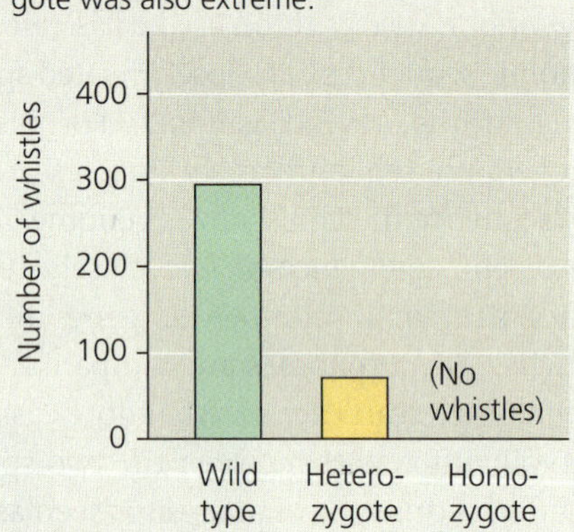

CONCLUSION *FOXP2* plays a significant role in the development of functional communication systems in mice. The results augment evidence from studies of birds and humans, supporting the hypothesis that *FOXP2* may act similarly in diverse organisms.

SOURCE W. Shu et al., Altered ultrasonic vocalization in mice with a disruption in the *Foxp2* gene, *Proceedings of the National Academy of Sciences* 102:9643–9648 (2005).

WHAT IF? Since the results support a role for mouse *FOXP2* in vocalization, you might wonder whether the human FOXP2 protein is a key regulator of speech. If you were given the amino acid sequences of wild-type and mutant human FOXP2 proteins and the wild-type chimpanzee FOXP2 protein, how would you investigate this question? What further clues could you obtain by comparing these sequences to that of the mouse FOXP2 protein?

FOXP2 experiments used mice as a model for humans because it would be unethical (as well as impractical) to carry out such experiments in humans. Mice and humans diverged about 65.5 million years ago (see Figure 21.16) and share about 85% of their genes. This genetic similarity can be exploited in studying human genetic disorders. If researchers know the organ or tissue that is affected by a particular genetic disorder, they can look for genes that are expressed in these locations in mice.

Further research efforts are under way to extend genomic studies to many more microbial species, additional primates, and neglected species from diverse branches of the tree of life.

These studies will advance our understanding of all aspects of biology, including health and ecology as well as evolution.

Comparing Genomes Within a Species

Another exciting consequence of our ability to analyze genomes is our growing understanding of the spectrum of genetic variation in humans. Because the history of the human species is so short—probably about 200,000 years—the amount of DNA variation among humans is small compared to that of many other species. Much of our diversity seems to be in the form of single nucleotide polymorphisms (SNPs, described in Chapter 20), usually detected by DNA sequencing. In the human genome, SNPs occur on average about once in 100–300 base pairs. Scientists have already identified the location of several million SNP sites in the human genome and continue to find more.

In the course of this search, they have also found other variations—including inversions, deletions, and duplications. The most surprising discovery has been the widespread occurrence of *copy-number variants* (*CNVs*), loci where some individuals have one or multiple copies of a particular gene or genetic region, rather than the standard two copies (one on each homolog). CNVs result from regions of the genome being duplicated or deleted inconsistently within the population. A 2010 study of 40 people found more than 8,000 CNVs involving 13% of the genes in the genome, and these CNVs probably represent just a small subset of the total. Since these variants encompass much longer stretches of DNA than the single nucleotides of SNPs, CNVs are more likely to have phenotypic consequences and to play a role in complex diseases and disorders. At the very least, the high incidence of copy-number variation casts doubt on the meaning of the phrase "a normal human genome."

Copy-number variants, SNPs, and variations in repetitive DNA such as short tandem repeats (STRs) will be useful genetic markers for studying human evolution. In 2010, the genomes of two Africans from different communities were sequenced: Archbishop Desmond Tutu, the South African civil rights advocate and a member of the Bantu tribe, the majority population in southern Africa; and !Gubi, a hunter-gatherer from the Khoisan community in Namibia, a minority African population that is probably the human group with the oldest known lineage. The comparison revealed many differences, as you might expect. The analysis was then broadened to compare the protein-coding regions of !Gubi's genome with those of three other Khoisan community members (self-identified Bushmen) living nearby. Remarkably, these four genomes differed more from each other than a European would from an Asian. These data highlight the extensive diversity among African genomes. Extending this approach will help us answer important questions about the differences between human populations and the migratory routes of human populations throughout history.

Comparing Developmental Processes

Biologists in the field of evolutionary developmental biology, or **evo-devo** as it is often called, compare developmental processes of different multicellular organisms. Their aim is to understand how these processes have evolved and how changes in them can modify existing organismal features or lead to new ones. With the advent of molecular techniques and the recent flood of genomic information, we are beginning to realize that the genomes of related species with strikingly different forms may have only minor differences in gene sequence or regulation. Discovering the molecular basis of these differences in turn helps us understand the origins of the myriad diverse forms that cohabit this planet, thus informing our study of evolution.

Widespread Conservation of Developmental Genes Among Animals

In Chapter 18, you learned about the homeotic genes in *Drosophila*, which specify the identity of body segments in the fruit fly (see Figure 18.20). Molecular analysis of the homeotic genes in *Drosophila* has shown that they all include a 180-nucleotide sequence called a **homeobox**, which specifies a 60-amino-acid *homeodomain* in the encoded proteins. An identical or very similar nucleotide sequence has been discovered in the homeotic genes of many invertebrates and vertebrates. The sequences are so similar between humans and fruit flies, in fact, that one researcher has whimsically referred to flies as "little people with wings." The resemblance even extends to the organization of these genes: The vertebrate genes homologous to the homeotic genes of fruit flies have kept the same chromosomal arrangement (**Figure 21.18**, on the next page). Homeobox-containing sequences have also been found in regulatory genes of much more distantly related eukaryotes, including plants and yeasts. From these similarities, we can deduce that the homeobox DNA sequence evolved very early in the history of life and was sufficiently valuable to organisms to have been conserved in animals and plants virtually unchanged for hundreds of millions of years.

Homeotic genes in animals were named *Hox* genes, short for *h*omeob*ox*-containing genes, because homeotic genes were the first genes found to have this sequence. Other homeobox-containing genes were later found that do not act as homeotic genes; that is, they do not directly control the identity of body parts. However, most of these genes, in animals at least, are associated with development, suggesting their ancient and fundamental importance in that process. In *Drosophila*, for example, homeoboxes are present not only in the homeotic genes but also in the egg-polarity gene *bicoid* (see Figures 18.21 and 18.22), in several of the segmentation genes, and in a master regulatory gene for eye development.

Researchers have discovered that the homeobox-encoded homeodomain is the part of a protein that binds to DNA

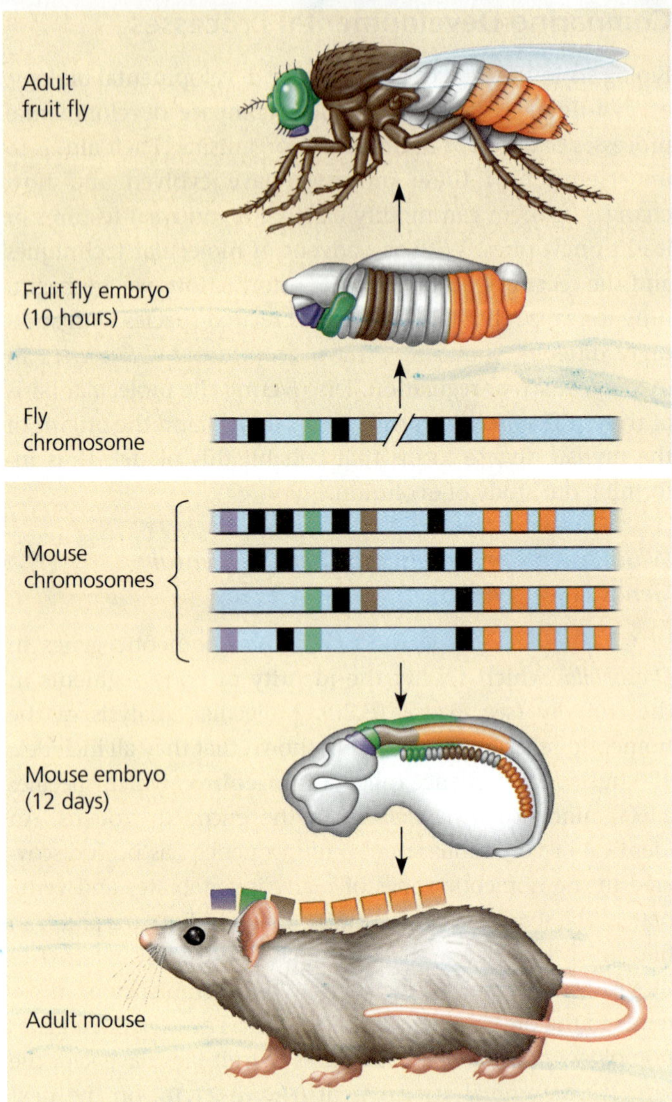

▲ **Figure 21.18 Conservation of homeotic genes in a fruit fly and a mouse.** Homeotic genes that control the form of anterior and posterior structures of the body occur in the same linear sequence on chromosomes in *Drosophila* and mice. Each colored band on the chromosomes shown here represents a homeotic gene. In fruit flies, all homeotic genes are found on one chromosome. The mouse and other mammals have the same or similar sets of genes on four chromosomes. The color code indicates the parts of the embryos in which these genes are expressed and the adult body regions that result. All of these genes are essentially identical in flies and mice, except for those represented by black bands, which are less similar in the two animals.

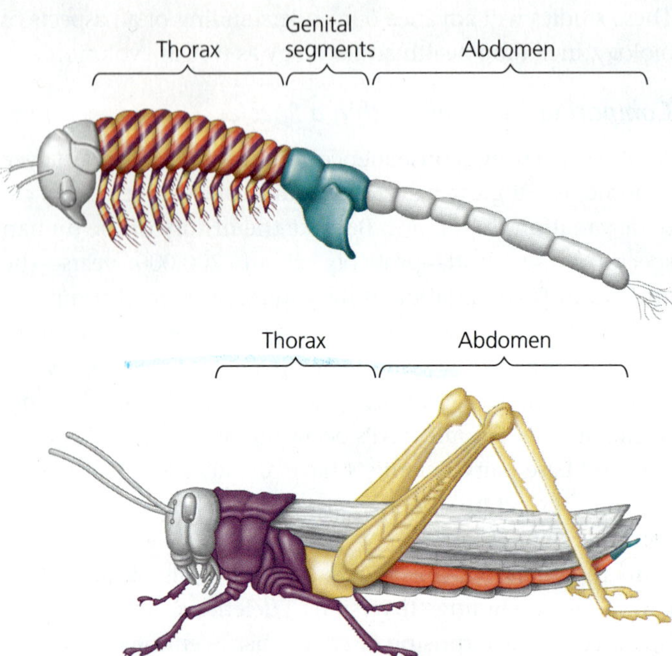

▲ **Figure 21.19 Effect of differences in *Hox* gene expression in crustaceans and insects.** Changes in the expression patterns of *Hox* genes have occurred over evolutionary time. These changes account in part for the different body plans of the brine shrimp *Artemia*, a crustacean (top), and the grasshopper, an insect. Shown here are regions of the adult body color-coded for expression of four *Hox* genes that determine formation of particular body parts during embryonic development. Each color represents a specific *Hox* gene. Colored stripes on the thorax of *Artemia* indicate co-expression of three *Hox* genes.

when the protein functions as a transcriptional regulator. However, the shape of the homeodomain allows it to bind to any DNA segment; its own structure is not specific for a particular sequence. Instead, other, more variable domains in a homeodomain-containing protein determine which genes the protein regulates. Interaction of these variable domains with still other transcription factors helps a homeodomain-containing protein recognize specific enhancers in the DNA. Proteins with homeodomains probably regulate development by coordinating the transcription of batteries of developmental genes, switching them on or off. In embryos of *Drosophila* and other animal species, different combinations of homeobox genes are active in different parts of the embryo. This selective expression of regulatory genes, varying over time and space, is central to pattern formation.

Developmental biologists have found that in addition to homeotic genes, many other genes involved in development are highly conserved from species to species. These include numerous genes encoding components of signaling pathways. The extraordinary similarity among particular developmental genes in different animal species raises a question: How can the same genes be involved in the development of animals whose forms are so very different from each other?

Ongoing studies are suggesting answers to this question. In some cases, small changes in regulatory sequences of particular genes cause changes in gene expression patterns that can lead to major changes in body form. For example, the differing patterns of expression of the *Hox* genes along the body axis in insects and crustaceans can explain the variation in number of leg-bearing segments among these segmented animals **(Figure 21.19)**. Also, recent research suggests that the same *Hox* gene product may have subtly dissimilar effects in different species, turning on new genes or turning on the same genes at higher or lower levels. In other cases, similar genes direct differing developmental processes in different organisms, resulting in diverse body shapes. Several *Hox* genes, for instance, are expressed in the embryonic and larval stages of the sea urchin, a nonsegmented animal that has a body plan quite different from those of insects and mice. Sea urchin adults make the pincushion-shaped shells you may have seen on the beach (see Figure 8.4). They are among the organisms long used in classical embryological studies (see Chapter 47).

Comparison of Animal and Plant Development

The last common ancestor of animals and plants was probably a single-celled eukaryote that lived hundreds of millions of years ago, so the processes of development must have evolved independently in the two multicellular lineages of organisms. Plants evolved with rigid cell walls, which rule out the morphogenetic movements of cells and tissues that are so important in animals. Instead, morphogenesis in plants relies primarily on differing planes of cell division and on selective cell enlargement. (You will learn about these processes in Chapter 35.) But despite the differences between animals and plants, there are similarities in the molecular mechanisms of development, which are legacies of their shared unicellular origin.

In both animals and plants, development relies on a cascade of transcriptional regulators turning on or turning off genes in a finely tuned series. For example, work on the small flowering plant *Arabidopsis thaliana* has shown that establishing the radial pattern of flower parts, like setting up the head-to-tail axis in *Drosophila*, involves a cascade of transcription factors (see Chapter 35). The genes that direct these processes, however, differ considerably in animals and plants. While quite a few of the master regulatory switches in *Drosophila* are homeobox-containing *Hox* genes, those in *Arabidopsis* belong to a completely different family of genes, called the *Mads-box* genes. And although homeobox-containing genes can be found in plants and *Mads-box* genes in animals, in neither case do they perform the same major roles in development that they do in the other group. Thus, molecular evidence supports the supposition that developmental programs evolved separately in animals and plants.

In this final chapter of the genetics unit, you have learned how studying genomic composition and comparing the genomes of different species can disclose much about how genomes evolve. Further, comparing developmental programs, we can see that the unity of life is reflected in the similarity of molecular and cellular mechanisms used to establish body pattern, although the genes directing development may differ among organisms. The similarities between genomes reflect the common ancestry of life on Earth. But the differences are also crucial, for they have created the huge diversity of organisms that have evolved. In the remainder of the book, we expand our perspective beyond the level of molecules, cells, and genes to explore this diversity on the organismal level.

CONCEPT CHECK 21.6

1. Would you expect the genome of the macaque (a monkey) to be more similar to the mouse genome or the human genome? Why?
2. The DNA sequences called homeoboxes, which help homeotic genes in animals direct development, are common to flies and mice. Given this similarity, explain why these animals are so different.
3. **WHAT IF?** There are three times as many *Alu* elements in the human genome as in the chimpanzee genome. How do you think these extra *Alu* elements arose in the human genome? Propose a role they might have played in the divergence of these two species.

For suggested answers, see Appendix A.

21 CHAPTER REVIEW

SUMMARY OF KEY CONCEPTS

CONCEPT 21.1

New approaches have accelerated the pace of genome sequencing (pp. 427–429)

- The **Human Genome Project** began in 1990, using a three-stage approach. In **linkage mapping**, the order of genes and other inherited markers in the genome and the relative distances between them can be determined from recombination frequencies. Next, **physical mapping** uses overlaps between DNA fragments to order the fragments and determine the distance in base pairs between markers. Finally, the ordered fragments are sequenced, providing the finished genome sequence.
- In the whole-genome shotgun approach, the whole genome is cut into many small, overlapping fragments that are sequenced; computer software then assembles the complete sequence. Correct assembly is made easier when mapping information is also available.

? *Why has the whole-genome shotgun approach been widely adopted for genome-sequencing projects?*

CONCEPT 21.2

Scientists use bioinformatics to analyze genomes and their functions (pp. 429–432)

- Websites on the Internet provide centralized access to genome sequence databases, analytical tools, and genome-related information.
- Computer analysis of genome sequences aids **gene annotation**, the identification of protein-coding sequences and determination of their function. Methods for determining gene function include comparing the sequences of newly discovered genes with those of known genes in other species and observing the phenotypic effects of experimentally inactivating genes of unknown function.
- In systems biology, scientists use the computer-based tools of **bioinformatics** to compare genomes and study sets of genes and proteins as whole systems (**genomics** and **proteomics**). Studies

include large-scale analyses of protein interactions, functional DNA elements, and genes contributing to medical conditions.

? *What was the most significant finding of the ENCODE pilot project? Why has the project been expanded to include other species?*

CONCEPT 21.3

Genomes vary in size, number of genes, and gene density (pp. 432–434)

	Bacteria	Archaea	Eukarya
Genome size	Most are 1–6 Mb		Most are 10–4,000 Mb, but a few are much larger
Number of genes	1,500–7,500		5,000–40,000
Gene density	Higher than in eukaryotes		Lower than in prokaryotes (Within eukaryotes, lower density is correlated with larger genomes.)
Introns	None in protein-coding genes	Present in some genes	Unicellular eukaryotes: present, but prevalent only in some species Multicellular eukaryotes: present in most genes
Other noncoding DNA	Very little		Can be large amounts; generally more repetitive noncoding DNA in multicellular eukaryotes

? *Compare genome size, gene number, and gene density (a) in the three domains and (b) among eukaryotes.*

CONCEPT 21.4

Multicellular eukaryotes have much noncoding DNA and many multigene families (pp. 434–438)

- Only 1.5% of the human genome codes for proteins or gives rise to rRNAs or tRNAs; the rest is noncoding DNA, including **pseudogenes** and **repetitive DNA** of unknown function.

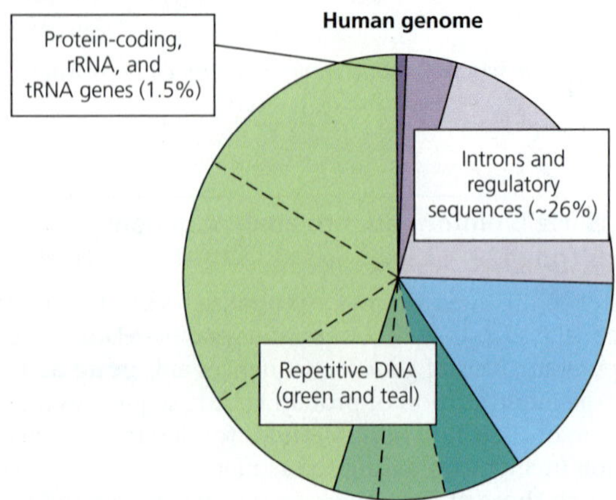

- The most abundant type of repetitive DNA in multicellular eukaryotes consists of **transposable elements** and related sequences. In eukaryotes, there are two types of transposable elements: **transposons**, which move via a DNA intermediate, and **retrotransposons**, which are more prevalent and move via an RNA intermediate.
- Other repetitive DNA includes short noncoding sequences that are tandemly repeated thousands of times (**simple sequence DNA**, which includes **STRs**); these sequences are especially prominent in centromeres and telomeres, where they probably play structural roles in the chromosome.
- Though many eukaryotic genes are present in one copy per haploid chromosome set, others (most, in some species) are members of a family of related genes, such as the human globin gene families:

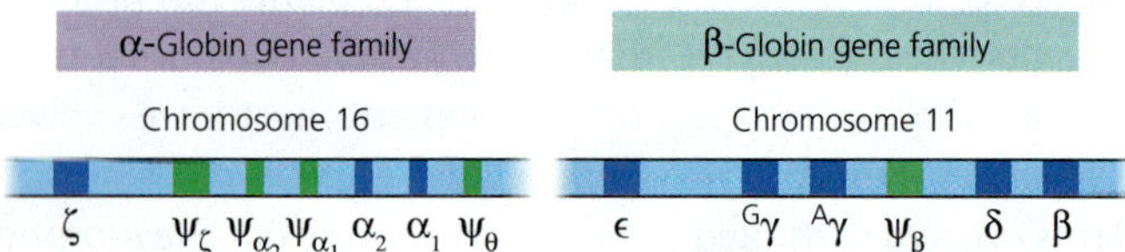

? *Explain how the function of transposable elements might account for their prevalence in human noncoding DNA.*

CONCEPT 21.5

Duplication, rearrangement, and mutation of DNA contribute to genome evolution (pp. 438–442)

- Accidents in cell division can lead to extra copies of all or part of entire chromosome sets, which may then diverge if one set accumulates sequence changes.
- The chromosomal organization of genomes can be compared among species, providing information about evolutionary relationships. Within a given species, rearrangements of chromosomes are thought to contribute to the emergence of new species.
- The genes encoding the various globin proteins evolved from one common ancestral globin gene, which duplicated and diverged into α-globin and β-globin ancestral genes. Subsequent duplication and random mutation gave rise to the present globin genes, all of which code for oxygen-binding proteins. The copies of some duplicated genes have diverged so much that the functions of their encoded proteins (such as lysozyme and α-lactalbumin) are now substantially different.
- Rearrangement of exons within and between genes during evolution has led to genes containing multiple copies of similar exons and/or several different exons derived from other genes.
- Movement of transposable elements or recombination between copies of the same element occasionally generates new sequence combinations that are beneficial to the organism. Such mechanisms can alter the functions of genes or their patterns of expression and regulation.

? *How could chromosomal rearrangements lead to the emergence of new species?*

CONCEPT 21.6

Comparing genome sequences provides clues to evolution and development (pp. 442–447)

- Comparative studies of genomes from widely divergent and closely related species provide valuable information about ancient and more recent evolutionary history, respectively. Human and chimpanzee sequences show about 4% difference, mostly due to insertions, deletions, and duplications in one lineage. Along with nucleotide variations in specific genes (such as *FOXP2*, a gene affecting speech), these differences may account for the distinct characteristics of the two species. Analysis of single nucleotide polymorphisms (SNPs) and copy-number variants (CNVs) among individuals in a species can also yield information about the evolution of that species.

- Evolutionary developmental (**evo-devo**) biologists have shown that homeotic genes and some other genes associated with animal development contain a **homeobox** region whose sequence is highly conserved among diverse species. Related sequences are present in the genes of plants and yeasts. During embryonic development in both plants and animals, a cascade of transcription regulators turns genes on or off in a carefully regulated sequence. However, the genes that direct analogous developmental processes differ in plants and animals as a result of their remote ancestry.

? *What type of information can be obtained by comparing the genomes of closely related species? Of very distantly related species?*

TEST YOUR UNDERSTANDING

LEVEL 1: KNOWLEDGE/COMPREHENSION

1. Bioinformatics includes all of the following *except*
   a. using computer programs to align DNA sequences.
   b. analyzing protein interactions in a species.
   c. using molecular biology to combine DNA from two different sources in a test tube.
   d. developing computer-based tools for genome analysis.
   e. using mathematical tools to make sense of biological systems.

2. One of the characteristics of retrotransposons is that
   a. they code for an enzyme that synthesizes DNA using an RNA template.
   b. they are found only in animal cells.
   c. they generally move by a cut-and-paste mechanism.
   d. they contribute a significant portion of the genetic variability seen within a population of gametes.
   e. their amplification is dependent on a retrovirus.

3. Homeotic genes
   a. encode transcription factors that control the expression of genes responsible for specific anatomical structures.
   b. are found only in *Drosophila* and other arthropods.
   c. are the only genes that contain the homeobox domain.
   d. encode proteins that form anatomical structures in the fly.
   e. are responsible for patterning during plant development.

LEVEL 2: APPLICATION/ANALYSIS

4. Two eukaryotic proteins have one domain in common but are otherwise very different. Which of the following processes is most likely to have contributed to this similarity?
   a. gene duplication
   b. RNA splicing
   c. exon shuffling
   d. histone modification
   e. random point mutations

5. **DRAW IT** Below are the amino acid sequences (using the single-letter code; see Figure 5.16) of four short segments of the FOXP2 protein from six species: chimpanzee, orangutan, gorilla, rhesus macaque, mouse, and human. These segments contain all of the amino acid differences between the FOXP2 proteins of these species.

   1. ATETI... PKSSD... TSSTT... NARRD
   2. ATETI... PKSSE... TSSTT... NARRD
   3. ATETI... PKSSD... TSSTT... NARRD
   4. ATETI... PKSSD... TSSNT... S ARRD
   5. ATETI... PKSSD... TSSTT... NARRD
   6. VTETI... PKSSD... TSSTT... NARRD

   Use a highlighter to color any amino acid that varies among the species. (Color that amino acid in all sequences.) Then answer the questions at the top of the next column.

   (a) The chimpanzee, gorilla, and rhesus macaque (C, G, R) sequences are identical. Which lines correspond to those sequences?
   (b) The human sequence differs from that of the C, G, R species at two amino acids. Which line corresponds to the human sequence? Underline the two differences.
   (c) The orangutan sequence differs from the C, G, R sequence at one amino acid (having valine instead of alanine) and from the human sequence at three amino acids. Which line corresponds to the orangutan sequence?
   (d) How many amino acid differences are there between the mouse and the C, G, R species? Circle the amino acid(s) that differ(s) in the mouse. How many amino acid differences are there between the mouse and the human? Draw a square around the amino acid(s) that differ(s) in the mouse.
   (e) Primates and rodents diverged between 60 and 100 million years ago, and chimpanzees and humans diverged about 6 million years ago. Knowing that, what can you conclude by comparing the amino acid differences between the mouse and the C, G, R species with the differences between the human and the C, G, R species?

LEVEL 3: SYNTHESIS/EVALUATION

6. **EVOLUTION CONNECTION**
   Genes important in the embryonic development of animals, such as homeobox-containing genes, have been relatively well conserved during evolution; that is, they are more similar among different species than are many other genes. Why is this?

7. **SCIENTIFIC INQUIRY**
   The scientists mapping the SNPs in the human genome noticed that groups of SNPs tended to be inherited together, in blocks known as haplotypes, ranging in length from 5,000 to 200,000 base pairs. There are as few as four or five commonly occurring combinations of SNPs per haplotype. Propose an explanation for this observation, integrating what you've learned throughout this chapter and this unit.

8. **WRITE ABOUT A THEME**
   The Genetic Basis of Life The continuity of life is based on heritable information in the form of DNA. In a short essay (100–150 words), explain how mutations in protein-coding genes and regulatory DNA contribute to evolution.

For selected answers, see Appendix A.

Animal Form and Function

An Interview with

Baldomero M. Olivera

Growing up in the Philippines, Baldomero (Toto) Olivera collected the shells of venomous cone snails as a hobby. Today, he is a leading expert on the use of cone snail toxins for exploring and modifying the activity of the mammalian nervous system. After graduating summa cum laude from the University of the Philippines, he earned a Ph.D. in Chemistry from the California Institute of Technology (Caltech). In 1970, Dr. Olivera joined the faculty at the University of Utah, where he is now a Distinguished Professor of Biology. In 2009, he became the first Filipino to be elected to the U.S. National Academy of Sciences.

How did you get started in science?

In high school I had a great teacher in both biology and chemistry. After I finished college, she encouraged me to go on to graduate school. She put an article in front of me about Caltech and told me to apply there. Fortunately, they accepted me.

When I got to Caltech, I assumed that since I'd taken five courses at a time as an undergraduate, I'd be expected to take even more in graduate school, so in my first semester I was taking seven courses. One was the only statistical mechanics course at Caltech, so not only chemists like me, but also physicists took it. The first day, the professor, Norman Davidson, reviewed all of classical mechanics, which I had never had. The second day he reviewed all of quantum mechanics. After the fourth lecture I went to him and said, "You know, Professor, I haven't understood a word you've said." He said, "Don't worry about it. Just hang in there." I did, and I ended up doing my Ph.D. research in his lab, studying DNA.

After graduate school I went to Stanford and continued to work on DNA as a postdoctoral fellow. I then took a faculty job in the Philippines, but soon found out that I probably would have no large lab equipment for at least a year. I decided to look for a project that we could start immediately with no equipment and, if at all possible, that would have some local advantages. Since I'd collected shells as a hobby, I knew that some of the local snails, called cone snails, had a venomous sting that could kill people. So we set out to purify the substances in the venom that were lethal to mammals.

Why do cone snails produce venom?

Not much was known about this until 1956, when a Yale graduate student, Alan Kohn, discovered that a cone snail he had in an aquarium hunted fish. (Kohn went on to become a leading expert on the ecology and taxonomy of cone snails.) We now know that about 100 of the 500–700 species of cone snails hunt fish. A similar number hunt other molluscs, and the rest hunt marine worms.

When a cone snail hunts, it first sniffs the fish by means of chemoreceptors, specialized cells that function in sensing smell or taste. Chemoreception by a cone snail in the marine environment is amazing. For example, fish-hunting cone snails in a large aquarium react the moment you drop in a fish. They move their siphon, which is sensory, back and forth so they can locate the fish. When the siphon touches fish skin, a disposable, harpoon-like tooth shoots out of the snail's proboscis and pierces the scales of the fish. The venom flows through the tooth into the fish. So the tooth, which is barbed, is both a harpoon and a hypodermic needle. The venom very rapidly immobilizes the harpooned fish, which the cone snail can then consume.

Some other animals hunt cone snails, which explains their cone-like shape. Among the major predators are crabs. If you're a crab, you grab the snail's shell and try to break it. But if the snail is cone-shaped, the crab claw slips, making it hard to break the shell. So the cone shape is a defensive adaptation. However, a cone-shaped snail shell that narrows to a point has an opening too small to allow a large prey animal to be pulled into the shell. So fish-hunting cone snails tend to be more cylindrical.

What did you hope to learn by studying cone snail venom?

We knew that venoms could be useful because the toxins of the pufferfish—*fugu* in Japanese—and the krait, a cobra-like snake, had been used to study ion channels in the nervous system. Ion channels are found on the surface of neurons, and they let ions in and out of the cell. This movement of ions mediates the electrical signaling within neurons and controls the transfer of a signal from one neuron to another. Venoms that interfere with that ion movement wipe out the prey animal's nervous system—that's why they're deadly.

The toxins in venoms are quite specific. The toxin of the pufferfish targets a sodium ion channel. The toxin of the krait targets a different type of ion channel, called a nicotinic receptor. By using these toxins in the lab to interfere with the function of particular kinds of ion channels, you can learn a great deal about how the nervous system normally functions.

How did you study the cone snail toxins?

In the Philippines, people eat marine snails, and the shells are by-products. Since I'd collected shells, I knew the people who sold them commercially and asked them to get me some live snails.

When you work with cone snails, you of course don't want to be stung. We put the snails on ice and they became very lethargic. We then dissected out the venom duct and pressed out the venom. The next step was to measure the activity of the venom. The medical literature said people stung by cone snails died because their diaphragms became paralyzed. So we assayed (tested) for paralysis. We would inject mice with various fractions of the venom and then immediately put them on a wire screen, upside down. Mice can actually stay attached this way indefinitely, but when paralysis sets in, they fall off. All we did was measure the time until they fell. It turns out that the falling time changes with the amount of toxin in a very reproducible way.

Using this falling-time assay, we purified an active peptide from two different fractions of the venom of the species *Conus geographus*. One of these peptides has only 13 amino acids; the other has 22. We found out that the peptide with 13 amino acids bound to exactly the same site on an ion channel as the krait toxin, which has about 80 amino acids. It was remarkable that a toxin had evolved in snails that was so much smaller but had the same activity.

How did the big breakthrough in your cone snail studies come about?

When I started as a full-time faculty member here at Utah, all our funding was still for experiments with DNA, and my graduate students all studied DNA. So we started recruiting undergraduates to work on the cone snail project—and that led to a very lucky break. One of the undergraduates, Craig Clark, said, "You guys are injecting into the body cavity of the mice. I think we should inject directly into the central nervous system." That was purely his idea. I thought it would probably just kill the mice.

At the time, we were beginning to use new techniques to separate the venom into many fractions, each containing different peptides. Using our falling-time assay, most fractions were inactive. But when Craig started injecting fractions into the central nervous system, the results were truly astonishing. Almost every fraction changed the mice's behavior in some way, and many of the effects were really bizarre. Some fractions made mice run around in circles. Others made mice scratch themselves. Still others made mice get up on their hind legs and move their front legs as if they were boxing. There was even a fraction that put juvenile, but not adult, mice into a sleeplike state. Craig purified the peptide with that activity, which he called the "sleeper peptide." I think the reason why university research is typically the most creative is that students sometimes do what they want, not what their professors tell them to do!

From then on, we let undergraduates pick which active fraction they wanted to study. Their job was to purify the peptide, which we would then synthesize and work on further.

Why and how did so much toxin diversity evolve in cone snails?

My suspicion is that if your survival depends on influencing the behavior of another animal and your strategy is purely chemical, it takes a very complex strategy for that to be successful in evolution. The adaptations that evolved in cone snails are in essence what we're just discovering in modern medical practice. To treat cancer or AIDS effectively, you can't use just one chemical; you use combination drug therapy. I think the evolution of venoms in cone snails followed the same principle.

For the cone snail toxins there are a few gene superfamilies, and they are subject to very accelerated evolution. There are parallels to the vertebrate adaptive immune system in that there are conserved and highly variable regions in the genes. The conserved regions are mostly all cut out. So, if you like, a cone snail peptide is equivalent to the highly variable part that makes an antibody very specific.

"The adaptations that evolved in cone snails are in essence what we're just discovering in modern medical practice."

How did your studies with snails lead to a discovery that was medically important?

After Craig discovered that you could study the effects of toxins by injecting them into the central nervous system of mice, he recruited a high school student, Michael McIntosh, to the lab. Mike decided to work with a different snail, called *Conus magus*, or the magician's cone. One of the most striking discoveries he made was a peptide that gave mice a characteristic tremor, or shake.

When Mike purified the "shaker peptide," we learned that it blocks a certain type of calcium channel. In fish these calcium channels control the release of the neurotransmitter that carries signals from nerve to muscle. When shaker peptide blocks these channels, the neurotransmitter is not released, and the fish becomes paralyzed. In mammals, this type of calcium channel is found only in the pain circuitry. When you experience severe pain, pain fibers send signals to a nerve cell in your spinal cord that takes the pain signal up to your brain. The pain fibers in mammals release signals to the nerves using the same type of calcium channel that in fish carries signals to muscles to direct movement. This difference in receptor location and function is why the shaker peptide paralyzes and kills fish, but not mammals.

The shaker peptide is now an approved drug. If a patient has very severe pain and the shaker peptide is injected where the pain fiber connects to the nerve, the perception of pain stops because the pain fiber no longer releases the chemical signal to the nerve. Doctors are using the shaker peptide in this way for people who are suffering from severe pain and no longer get pain relief from morphine due to tolerance built up over time. A surgically implanted pump introduces the shaker peptide to the spinal cord at the site where the pain fiber makes its connection.

What other discoveries do you hope to see come out of studying cone snails?

There are 500–700 species of cone snails, but the total biodiversity of molluscs with venom is probably 12,000 species. The biggest group is the turrids. You don't find many of them in shallow water, but at 400 meters, where there are no cone snails, you find hundreds of species of turrids. Many of them are just a few millimeters in length, and nobody knows anything about their biology. Fortunately, hobbyists are extremely interested in their tiny shells, so fishermen in the Philippines gather them. They tie all their broken fine-mesh fishing nets into bundles, which they sink in water for 3 to 6 months. When they lift these bundles, each will contain several thousand marine animals, almost all of them alive. For gastropods alone, a net will hold something like 250 different species, of which 40 are venomous. Each venom has 200 compounds. So this is a vast new world.

What advice would you give students interested in biology or medicine?

One piece of advice is to get as good a basic science background as you possibly can. I started out with very basic chemistry and very basic biology, and then of course the science grew as my research career grew. Having strong basic science training allows you to feel confident that even if you don't understand a problem at first (like quantum mechanics), you can still tackle it. A second thing is that you ought to know what it is in science that really turns you on. Third, keep in mind that it is always fun to learn something new.

Baldomero Olivera (center) with Jane Reece and Steve Wasserman (right)

40

Basic Principles of Animal Form and Function

▲ **Figure 40.1 How does a jackrabbit keep from overheating?**

KEY CONCEPTS

40.1 Animal form and function are correlated at all levels of organization

40.2 Feedback control maintains the internal environment in many animals

40.3 Homeostatic processes for thermoregulation involve form, function, and behavior

40.4 Energy requirements are related to animal size, activity, and environment

OVERVIEW

Diverse Forms, Common Challenges

The ears of the jackrabbit (*Lepus alleni*) in **Figure 40.1** are thin and remarkably large. They provide this hare with an acute sense of hearing, a primary defense against predators. The ears also help the jackrabbit shed excess heat. Blood flowing through each ear's network of vessels transfers heat to the surrounding air. However, when the air is warmer than the jackrabbit, blood passing through the ears could absorb heat, raising body temperature to a dangerous level. How, then, does a big-eared jackrabbit survive in the midday desert heat? To answer this question, we need to look more closely at the biological form, or **anatomy**, of the animal.

Over the course of its life, a jackrabbit faces the same fundamental challenges as any other animal, whether hydra, hawk, or human. All animals must obtain oxygen and nutrients, fight off infection, and produce offspring. Given that they share these and other basic requirements, why do species vary so enormously in makeup, complexity, organization, and appearance? The answer is adaptation: Natural selection favors those variations in a population that increase relative fitness (see Chapter 23). The solutions to the challenges of survival vary among environments and species, but they frequently result in a close match of form to function.

Because form and function are correlated, examining anatomy often provides clues to **physiology**—biological function. In the case of the jackrabbit, researchers noted that its large, pink-tinged ears turn pale when the air temperature exceeds 40°C (104°F), the normal temperature of the jackrabbit's body. The color change reflects a temporary narrowing of blood vessels in response to a hot environment. With their blood supply reduced, the ears can absorb heat without overheating the rest of the body. When the air cools, blood flow increases, and the large ears again help release excess heat.

In this chapter, we will begin our study of animal form and function by examining the levels of organization in the animal body and the systems for coordinating the activities of distinct body parts. Next, we will use the example of body temperature regulation to illustrate how animals control their internal environment. Finally, we will explore how anatomy and physiology relate to an animal's interactions with the environment and its management of energy use.

CONCEPT 40.1

Animal form and function are correlated at all levels of organization

An animal's size and shape are fundamental aspects of form that significantly affect the way the animal interacts with its environment. Although we may refer to size and shape as elements of a "body plan" or "design," this does not imply a process of conscious invention. The body plan of an animal is the result of a pattern of development programmed by the genome, itself the product of millions of years of evolution.

Evolution of Animal Size and Shape

EVOLUTION Many different body plans have arisen during the course of evolution, but these variations fall within certain bounds. Physical laws that govern strength, diffusion, movement, and heat exchange limit the range of animal forms.

As an example of how physical laws constrain evolution, let's consider how some properties of water limit the possible shapes for animals that are fast swimmers. Water is about a thousand times denser than air and also far more viscous. Therefore, any bump on an animal's body surface that causes drag impedes a swimmer more than it would a runner or flyer. Tuna and other fast ray-finned fishes can swim at speeds up to 80 km/hr (50 miles/hour). Sharks, penguins, dolphins, and seals are also fast swimmers. As is apparent in the examples in **Figure 40.2**, such animals share a streamlined body contour: a shape that is fusiform, meaning tapered on both ends. The similar shape found in these speedy vertebrates is an example of convergent evolution (see Chapter 22). Natural selection often results in similar adaptations when diverse organisms face the same environmental challenge, such as overcoming drag during swimming.

Physical laws also influence animal body plans with regard to maximum size. As body dimensions increase, thicker skeletons are required to maintain adequate support. This limitation affects internal skeletons, such as those of vertebrates, as well as external skeletons, such as those of insects and other arthropods. In addition, as bodies increase in size, the muscles required for locomotion must represent an ever-larger fraction of the total body mass. At some point, mobility becomes limited. By considering the fraction of body mass in leg muscles and the effective force such muscles generate, scientists can estimate maximum running speed for a wide range of body plans. Such calculations indicate that the dinosaur *Tyrannosaurus rex*, which stood more than 6 m tall, probably could reach speeds of 30 km/hr (19 miles/hour), about as fast as the fastest humans can run.

▲ **Figure 40.2 Convergent evolution in fast swimmers.**

Exchange with the Environment

Animals must exchange materials with their environment, and this requirement imposes limitations on their body plans (as it does for all multicellular organisms). Exchange occurs as substances dissolved in an aqueous solution move across the plasma membrane of each cell. The rates of exchange for nutrients, waste products, and gases are proportional to membrane surface area, whereas the amount of material that must be exchanged to sustain life is proportional to cell volume.

The opportunity for exchange depends on the number of cells in an organism's body. A single-celled organism, such as the amoeba in **Figure 40.3a**, has a sufficient membrane surface area in contact with its environment to carry out all necessary exchange. In contrast, an animal is composed of many cells, each with its own plasma membrane across which exchange must occur. A multicellular organization therefore works only if every cell has access to a suitable aqueous environment, either inside or outside the animal's body.

Many animals with a simple internal organization have body plans that enable direct exchange between almost all their cells and the external environment. For example, a pond-dwelling hydra, which has a saclike body plan, has a body wall only two cell layers thick **(Figure 40.3b)**. Because its

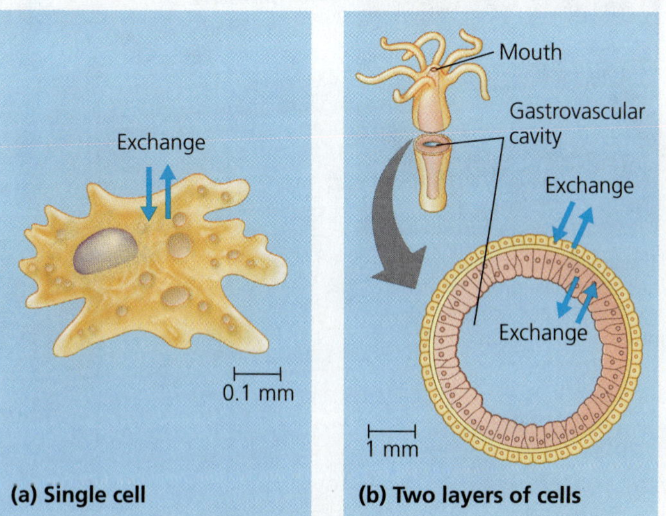

▲ **Figure 40.3 Contact with the environment. (a)** In a single-celled organism, such as an amoeba, the entire surface area contacts the environment. **(b)** Although all animals are multicellular, some have a simple organization in which all or nearly all cells contact the environment. For example, a hydra's body consists of two layers of cells. As fluid moves in and out of the hydra's mouth, every body cell can exchange material directly with the aqueous environment.

gastrovascular cavity opens to the external environment, both the outer and inner layers of cells are constantly bathed by pond water. Another common body plan that maximizes exposure to the surrounding medium is a flat shape. Consider, for instance, a parasitic tapeworm, which can reach several meters in length (see Figure 33.12). A thin, flat shape places most cells of the worm in direct contact with its particular environment—the nutrient-rich intestinal fluid of a vertebrate host.

The bodies of most animals are composed of compact masses of cells, with an internal organization much more complex than that of a hydra or a tapeworm. For such a body plan, increasing the number of cells decreases the ratio of outer surface area to total volume. As an extreme comparison, the ratio of outer surface to volume for a whale is hundreds of thousands of times smaller than that for a water flea (*Daphnia*). Nevertheless, every cell in the whale must be bathed in fluid and have access to oxygen, nutrients, and other resources. How is this accomplished?

In whales and most other animals, the evolutionary adaptations that enable sufficient exchange with the environment are specialized surfaces that are extensively branched or folded (**Figure 40.4**). In almost all cases, these exchange surfaces lie within the body, an arrangement that protects their delicate tissues from abrasion or dehydration and allows for streamlined body contours. In humans, the internal exchange surfaces of the digestive, respiratory, and circulatory systems each have an area more than 25 times that of the skin.

Internal body fluids link exchange surfaces to body cells. The spaces between cells are filled with fluid, in many animals called **interstitial fluid** (from the Latin for "stand between"). Complex body plans also include a circulatory fluid, such as blood. Exchange between the interstitial fluid and the circulatory fluid enables cells throughout the body to obtain nutrients and get rid of wastes (see Figure 40.4).

Despite the greater challenges of exchange with the environment, complex body plans have distinct benefits over simple

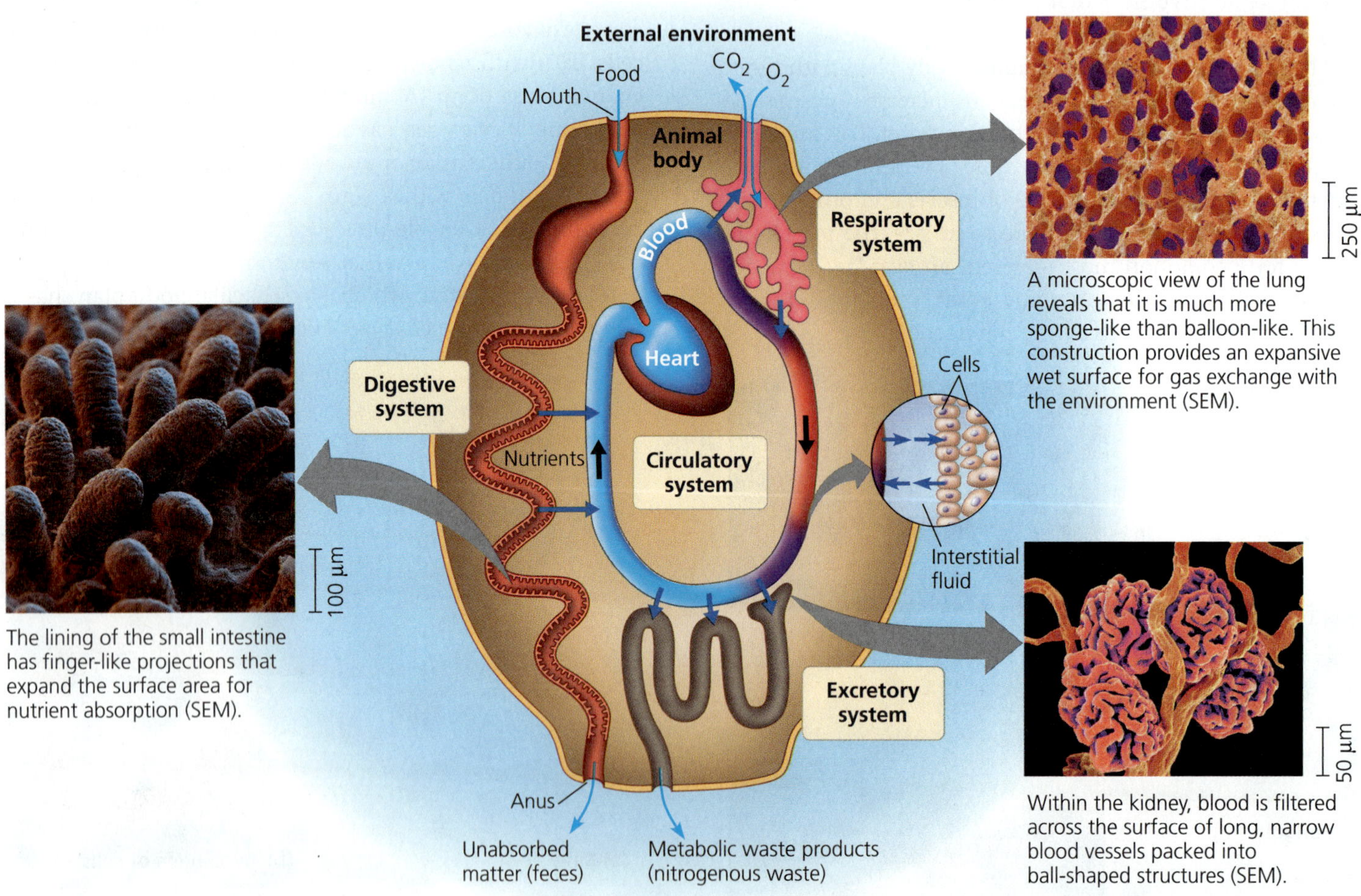

A microscopic view of the lung reveals that it is much more sponge-like than balloon-like. This construction provides an expansive wet surface for gas exchange with the environment (SEM).

The lining of the small intestine has finger-like projections that expand the surface area for nutrient absorption (SEM).

Within the kidney, blood is filtered across the surface of long, narrow blood vessels packed into ball-shaped structures (SEM).

▲ **Figure 40.4 Internal exchange surfaces of complex animals.** This diagram provides an overview of chemical exchange between an animal body and the environment. Most animals have surfaces that are specialized for exchanging chemicals with the surroundings. These exchange surfaces are usually internal but are connected to the environment via openings on the body surface (the mouth, for example). The exchange surfaces are finely branched or folded, giving them a very large area. The digestive, respiratory, and excretory systems all have such exchange surfaces. The circulatory system carries chemicals transported across these surfaces throughout the body.

? *In what sense are exchange surfaces such as the lining of the digestive system both internal and external?*

ones. For example, an external skeleton can protect against predators, and sensory organs can provide detailed information on the animal's surroundings. Internal digestive organs can break down food gradually, controlling the release of stored energy. In addition, specialized filtration systems can adjust the composition of the internal fluid that bathes the animal's body cells. In this way, an animal can maintain a relatively stable internal environment while living in a changeable external environment. A complex body plan is especially advantageous for animals living on land, where the external environment may be highly variable.

Hierarchical Organization of Body Plans

Cells form a functional animal body through their emergent properties. Recall from Chapter 1 that emergent properties arise by way of successive levels of structural and functional organization. Cells are organized into **tissues**, groups of cells with a similar appearance and a common function. Different types of tissues are further organized into functional units called **organs**. (The simplest animals, such as sponges, lack organs or even true tissues.) Groups of organs that work together provide an additional level of organization and coordination and make up an **organ system** (**Table 40.1**). Thus, for example, the skin is an organ of the integumentary system, which protects against infection and helps regulate body temperature.

Many organs contain tissues with distinct physiological roles. In some cases, the roles are different enough that we consider the organ to belong to more than one organ system. The pancreas, for instance, produces enzymes critical to the function of the digestive system and also regulates the level of sugar in the blood as a vital part of the endocrine system.

Just as viewing the body's organization from the "bottom up" (from cells to organ systems) reveals emergent properties, a "top-down" view of the hierarchy reveals the multilayered basis of specialization. Consider the human digestive system: the mouth, pharynx, esophagus, stomach, small and large intestines, accessory organs, and anus. Each organ has specific roles in digestion. One function of the stomach, for example, is to initiate the breakdown of proteins. This process requires a churning motion powered by stomach muscles, as well as digestive juices secreted by the stomach lining. Producing digestive juices, in turn, requires highly specialized cell types: One cell type secretes a protein-digesting enzyme, a second generates concentrated hydrochloric acid, and a third produces mucus, which protects the stomach lining.

The specialized and complex organ systems of animals are built from a limited set of cell and tissue types. For example, lungs and blood vessels have distinct functions but are lined by tissues that are of the same basic type and that therefore share many properties.

There are four main types of animal tissues: epithelial, connective, muscle, and nervous. **Figure 40.5**, on the next three pages, explores the structure and function of each type. In later chapters, we'll discuss how the tissues described here contribute to the functions of each organ system.

Table 40.1 Organ Systems in Mammals

Organ System	Main Components	Main Functions
Digestive	Mouth, pharynx, esophagus, stomach, intestines, liver, pancreas, anus	Food processing (ingestion, digestion, absorption, elimination)
Circulatory	Heart, blood vessels, blood	Internal distribution of materials
Respiratory	Lungs, trachea, other breathing tubes	Gas exchange (uptake of oxygen; disposal of carbon dioxide)
Immune and lymphatic	Bone marrow, lymph nodes, thymus, spleen, lymph vessels, white blood cells	Body defense (fighting infections and cancer)
Excretory	Kidneys, ureters, urinary bladder, urethra	Disposal of metabolic wastes; regulation of osmotic balance of blood
Endocrine	Pituitary, thyroid, pancreas, adrenal, and other hormone-secreting glands	Coordination of body activities (such as digestion and metabolism)
Reproductive	Ovaries or testes and associated organs	Reproduction
Nervous	Brain, spinal cord, nerves, sensory organs	Coordination of body activities; detection of stimuli and formulation of responses to them
Integumentary	Skin and its derivatives (such as hair, claws, skin glands)	Protection against mechanical injury, infection, dehydration; thermoregulation
Skeletal	Skeleton (bones, tendons, ligaments, cartilage)	Body support, protection of internal organs, movement
Muscular	Skeletal muscles	Locomotion and other movement

▼ **Figure 40.5**

Exploring Structure and Function in Animal Tissues

Epithelial Tissue

Occurring as sheets of cells, **epithelial tissues**, or **epithelia** (singular, *epithelium*), cover the outside of the body and line organs and cavities within the body. Because epithelial cells are closely packed, often with tight junctions (see Figure 6.32), they function as a barrier against mechanical injury, pathogens, and fluid loss. Epithelia also form active interfaces with the environment. For example, the epithelium that lines the nasal passages is crucial for olfaction, the sense of smell. Note how different cell shapes and arrangements correlate with distinct functions.

Stratified squamous epithelium

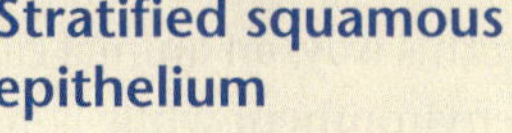

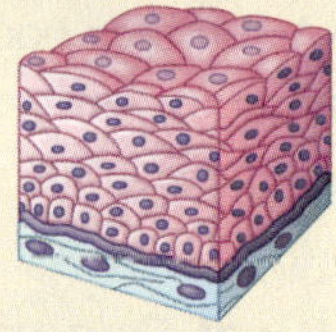

A stratified squamous epithelium is multilayered and regenerates rapidly. New cells formed by division near the basal lamina (see micrograph below) push outward, replacing cells that are sloughed off. This epithelium is commonly found on surfaces subject to abrasion, such as the outer skin and the linings of the mouth, anus, and vagina.

Cuboidal epithelium

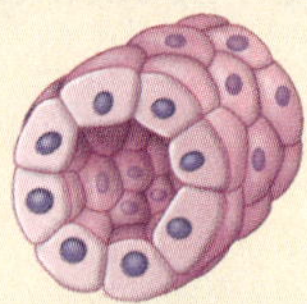

Cuboidal epithelium, with dice-shaped cells specialized for secretion, makes up the epithelium of kidney tubules and many glands, including the thyroid gland and salivary glands.

Simple columnar epithelium

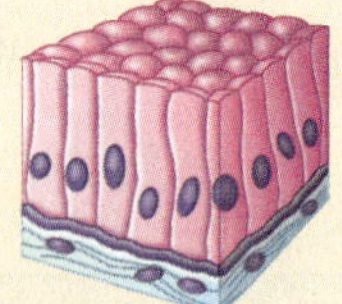

The large, brick-shaped cells of simple columnar epithelia are often found where secretion or active absorption is important. For example, a simple columnar epithelium lines the intestines, secreting digestive juices and absorbing nutrients.

Simple squamous epithelium

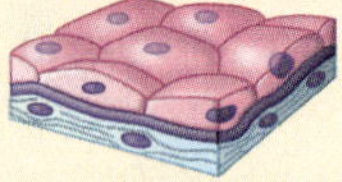

The single layer of platelike cells that form a simple squamous epithelium functions in the exchange of material by diffusion. This type of epithelium, which is thin and leaky, lines blood vessels and the air sacs of the lungs, where diffusion of nutrients and gases is critical.

Pseudostratified columnar epithelium

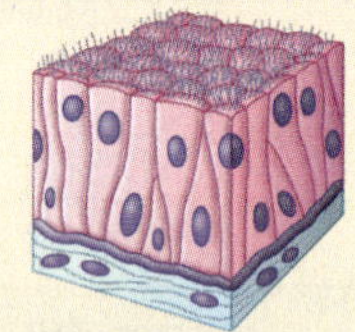

A pseudostratified epithelium consists of a single layer of cells varying in height. In many vertebrates, a pseudostratified epithelium of ciliated cells forms a mucous membrane that lines portions of the respiratory tract. The beating cilia sweep the film of mucus along the surface.

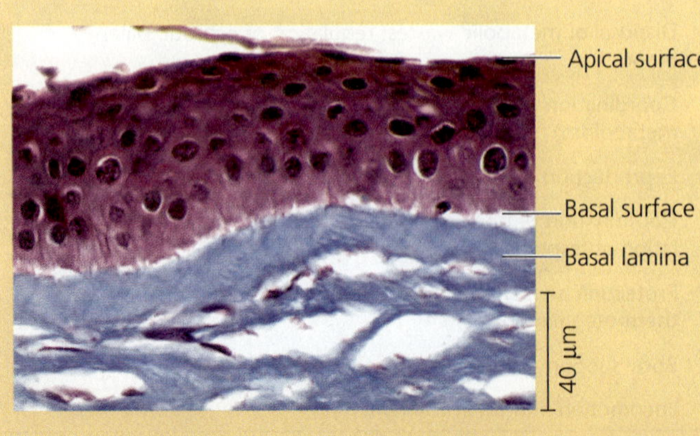

Polarity of epithelia

All epithelia are polarized, meaning that they have two different sides. The *apical* surface faces the lumen (cavity) or outside of the organ and is therefore exposed to fluid or air. Specialized projections often cover this surface. For example, the apical surface of the epithelium lining the small intestine is covered with microvilli, projections that increase the surface area available for absorbing nutrients. The opposite side of each epithelium is the *basal* surface. The basal surface is attached to a *basal lamina*, a dense mat of extracellular matrix that separates the epithelium from the underlying tissue.

(All photos in figure are LMs.)

Connective Tissue

Connective tissue, consisting of a sparse population of cells scattered through an extracellular matrix, holds many tissues and organs together and in place. The matrix generally consists of a web of fibers embedded in a liquid, jellylike, or solid foundation. Within the matrix are numerous cells called **fibroblasts**, which secrete fiber proteins, and **macrophages**, which engulf foreign particles and any cell debris by phagocytosis (see Chapter 6).

Connective tissue fibers are of three kinds: *Collagenous fibers* provide strength and flexibility, *reticular fibers* join connective tissue to adjacent tissues, and *elastic fibers* make tissues elastic. If you pinch a fold of tissue on the back of your hand, the collagenous and reticular fibers prevent the skin from being pulled far from the bone, whereas the elastic fibers restore the skin to its original shape when you release your grip. Different mixtures of fibers and foundation form the major types of connective tissue shown below.

Loose connective tissue

The most widespread connective tissue in the vertebrate body is *loose connective tissue,* which binds epithelia to underlying tissues and holds organs in place. Loose connective tissue gets its name from the loose weave of its fibers, which include all three types. It is found in the skin and throughout the body.

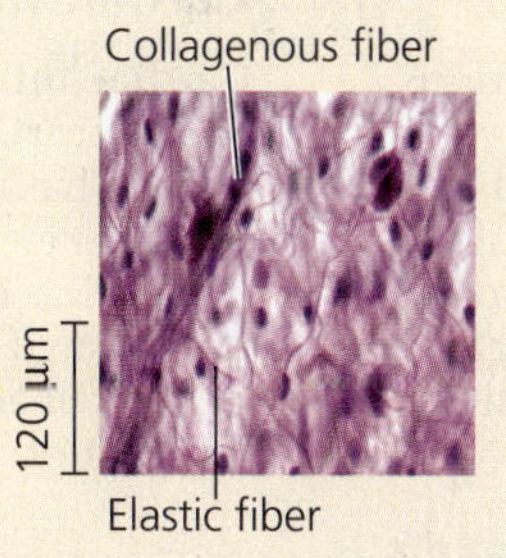

Fibrous connective tissue

Fibrous connective tissue is dense with collagenous fibers. It is found in **tendons**, which attach muscles to bones, and in **ligaments**, which connect bones at joints.

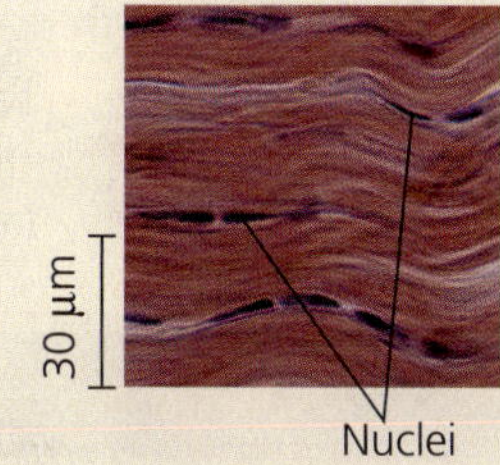

Bone

The skeleton of most vertebrates is made of **bone**, a mineralized connective tissue. Bone-forming cells called *osteoblasts* deposit a matrix of collagen. Calcium, magnesium, and phosphate ions combine into a hard mineral within the matrix. The microscopic structure of hard mammalian bone consists of repeating units called *osteons*. Each osteon has concentric layers of the mineralized matrix, which are deposited around a central canal containing blood vessels and nerves.

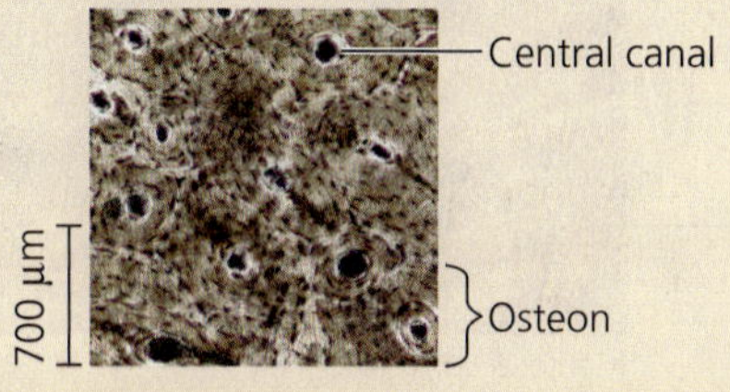

Adipose tissue

Adipose tissue is a specialized loose connective tissue that stores fat in adipose cells distributed throughout its matrix. Adipose tissue pads and insulates the body and stores fuel as fat molecules (see Figure 4.6). Each adipose cell contains a large fat droplet that swells when fat is stored and shrinks when the body uses that fat as fuel.

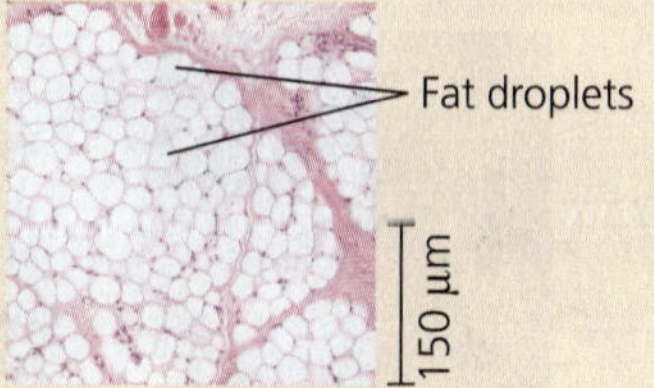

Blood

Blood has a liquid extracellular matrix called plasma, which consists of water, salts, and dissolved proteins. Suspended in plasma are erythrocytes (red blood cells), leukocytes (white blood cells), and cell fragments called platelets. Red cells carry oxygen, white cells function in defense, and platelets aid in blood clotting.

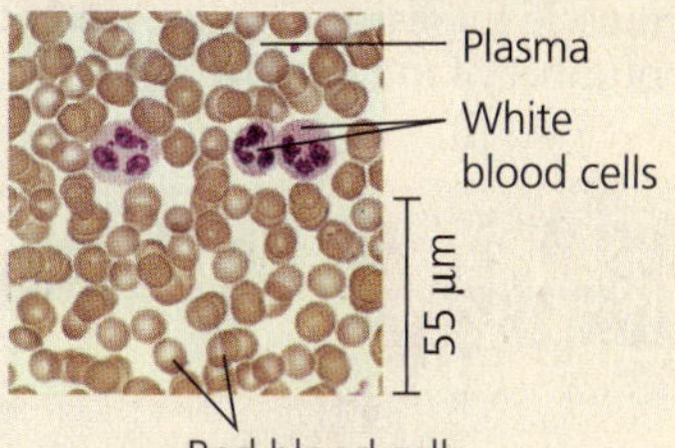

Cartilage

Cartilage contains collagenous fibers embedded in a rubbery protein-carbohydrate complex called chondroitin sulfate. Cells called *chondrocytes* secrete the collagen and chondroitin sulfate, which together make cartilage a strong yet flexible support material. The skeletons of many vertebrate embryos contain cartilage that is replaced by bone as the embryo matures. Cartilage remains in some locations, such as the disks that act as cushions between vertebrae.

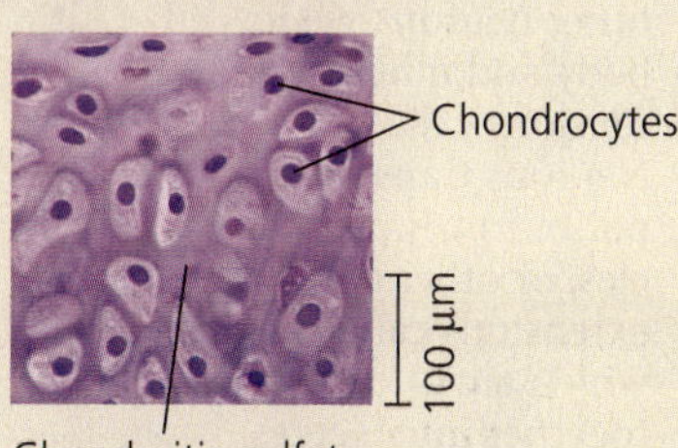

Exploring Structure and Function in Animal Tissues

Muscle Tissue

The tissue responsible for nearly all types of body movement is **muscle tissue**. All muscle cells consist of filaments containing the proteins actin and myosin, which together enable muscles to contract. There are three types of muscle tissue in the vertebrate body: skeletal, smooth, and cardiac.

Skeletal muscle

Attached to bones by tendons, **skeletal muscle**, or *striated muscle*, is responsible for voluntary movements. Skeletal muscle consists of bundles of long cells called muscle fibers. During development, skeletal muscle fibers form by the fusion of many cells, resulting in multiple nuclei in each muscle cell or fiber. The arrangement of contractile units, or sarcomeres, along the fibers gives the cells a striped (striated) appearance. In adult mammals, building muscle increases the size but not the number of muscle fibers.

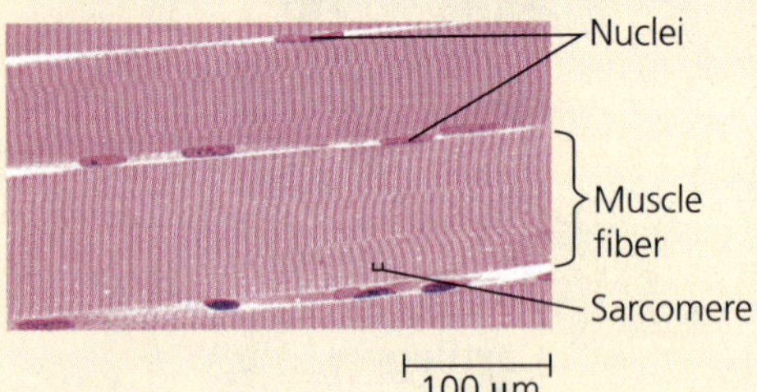

Smooth muscle

Smooth muscle, which lacks striations, is found in the walls of the digestive tract, urinary bladder, arteries, and other internal organs. The cells are spindle-shaped. Smooth muscles are responsible for involuntary body activities, such as churning of the stomach and constriction of arteries.

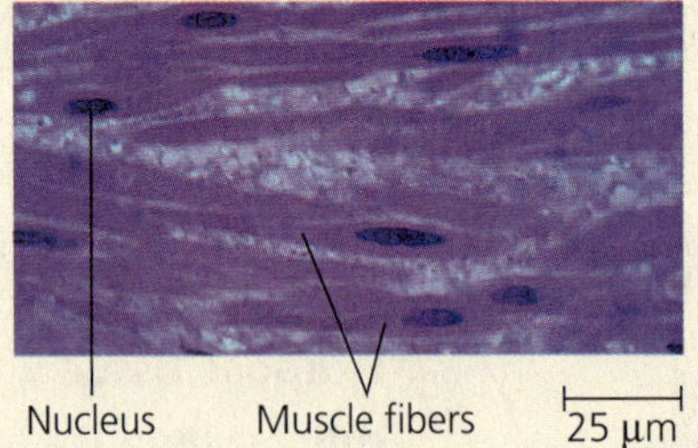

Cardiac muscle

Cardiac muscle forms the contractile wall of the heart. It is striated like skeletal muscle and has similar contractile properties. Unlike skeletal muscle, however, cardiac muscle has fibers that interconnect via intercalated disks, which relay signals from cell to cell and help synchronize heart contraction.

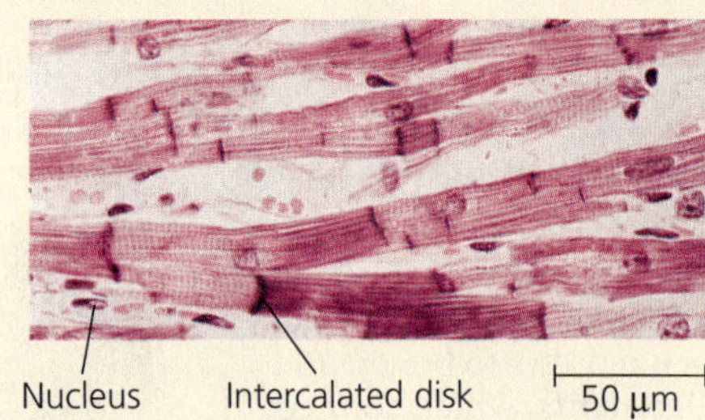

Nervous Tissue

Nervous tissue functions in the receipt, processing, and transmission of information. Nervous tissue contains **neurons**, or nerve cells, which transmit nerve impulses, as well as support cells called **glial cells**, or simply **glia**. In many animals, a concentration of nervous tissue forms a brain, an information-processing center.

Neurons

Neurons are the basic units of the nervous system. A neuron receives nerve impulses from other neurons via its cell body and multiple extensions called dendrites. Neurons transmit impulses to neurons, muscles, or other cells via extensions called axons, which are often bundled together into nerves.

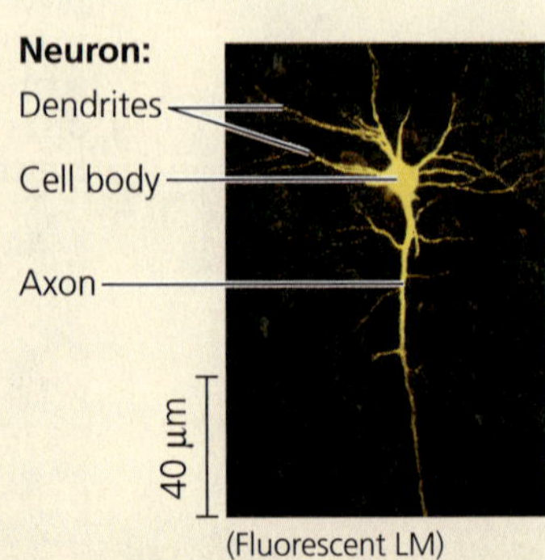

(Fluorescent LM)

Glia

The various types of glia help nourish, insulate, and replenish neurons, and in some cases, modulate neuron function.

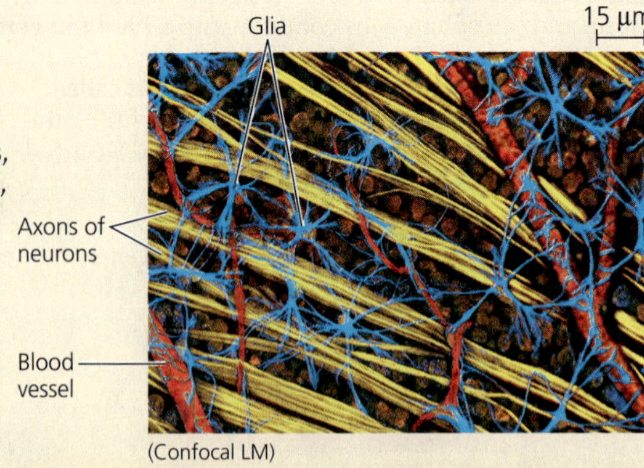

(Confocal LM)

Coordination and Control

An animal's tissues, organs, and organ systems must act in concert with one another. For example, during long dives, the harbor seal in Figure 40.2 slows its heart rate, collapses its lungs, and lowers its body temperature while propelling itself forward with its hind flippers. Coordinating activity across an animal's body in this way requires communication between different locations in the body. What signals are used? How do the signals move within the body? There are two sets of answers to these questions, reflecting the two major systems for controlling and coordinating responses to stimuli **(Figure 40.6)**.

In the endocrine system, signaling molecules released into the bloodstream by endocrine cells reach all locations in the body. In the nervous system, neurons transmit signals, called nerve impulses, between specific locations in the body. In each system, the type of pathway used is the same regardless of whether the signal reaches across the length of the body or ends up just a few cell diameters away.

▼ Figure 40.6 Signaling in the endocrine and nervous systems

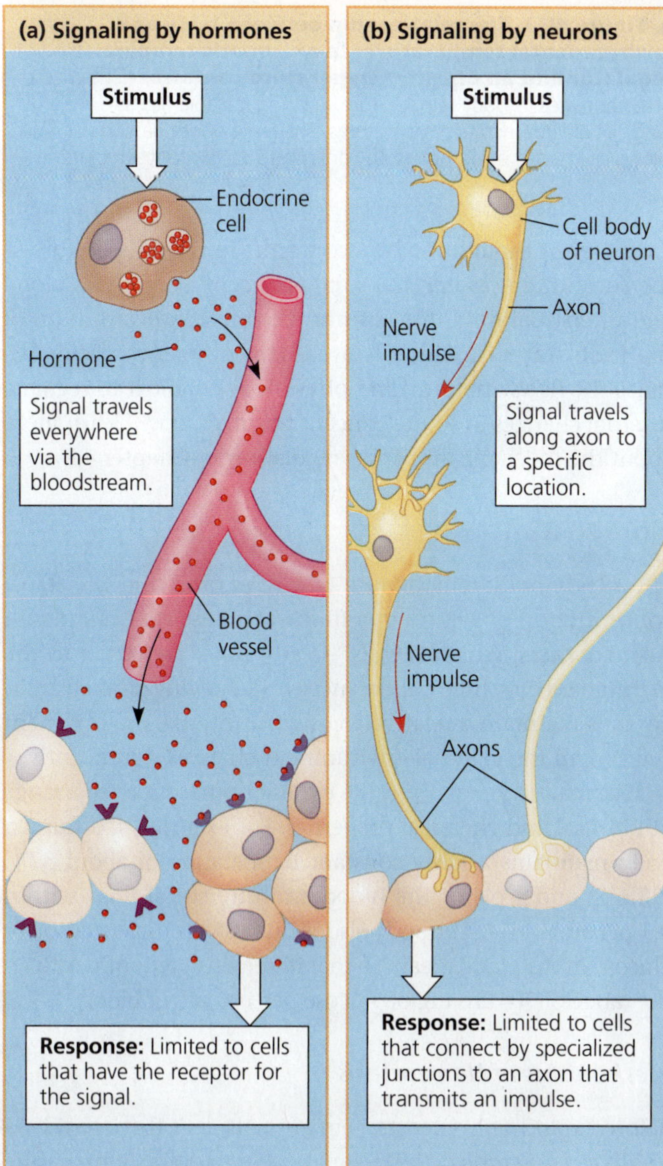

The signaling molecules broadcast throughout the body by the endocrine system are called **hormones**. Different hormones cause distinct effects, and only cells that have receptors for a particular hormone respond **(Figure 40.6a)**. Depending on which cells have receptors for that hormone, the hormone may have an effect in just a single location or in sites throughout the body. For example, only cells of the thyroid gland have the receptor for thyroid-stimulating hormone (TSH). Upon binding TSH, thyroid cells release thyroid hormone, which acts directly on cells in nearly every tissue to increase oxygen consumption and heat production.

Hormones are relatively slow acting. It takes many seconds for TSH and other hormones to be released into the bloodstream and carried throughout the body. The effects of hormones are often long-lasting, however, because hormones remain in the bloodstream for seconds, minutes, or even hours.

In the nervous system, signals are not broadcast throughout the entire body. Instead, each nerve impulse travels to specific target cells along dedicated communication lines consisting mainly of axons **(Figure 40.6b)**. Four types of cells can receive nerve impulses: other neurons, muscle cells, endocrine cells, and exocrine cells. Unlike the endocrine system, the nervous system conveys information by the *pathway* the signal takes. For example, a person can distinguish different musical notes because each note's frequency activates different neurons connecting the ear to the brain.

Communication in the nervous system usually involves more than one type of signal. Nerve impulses travel along axons, sometimes over long distances, as changes in voltage. But in many cases, passing information from one neuron to another involves very short-range chemical signals. Overall, transmission is extremely fast; nerve impulses take only a fraction of a second to reach the target and last only a fraction of a second.

Because the two major communication systems of the body differ in signal type, transmission, speed, and duration, they are adapted to different functions. The endocrine system is well suited for coordinating gradual changes that affect the entire body, such as growth and development, reproduction, metabolic processes, and digestion. The nervous system is well suited for directing immediate and rapid responses to the environment, especially in controlling fast locomotion and behavior.

Although the functions of the endocrine and nervous systems are distinct, the two systems often work in close coordination. Both contribute to maintaining a stable internal environment, our next topic of discussion.

CONCEPT CHECK 40.1

1. What properties are shared by all types of epithelia?
2. In cool weather, jackrabbits sometimes flatten their ears against their body. What advantage and disadvantage do you think this body posture offers for survival?
3. **WHAT IF?** Suppose you are standing at the edge of a cliff and suddenly slip—you barely manage to keep your balance and avoid falling. As your heart races, you feel a burst of energy, due in part to a surge of blood into dilated (widened) vessels in your muscles and an upward spike in the level of glucose in your blood. Why might you expect that this "fight-or-flight" response requires both the nervous and endocrine systems?

For suggested answers, see Appendix A.

CONCEPT 40.2

Feedback control maintains the internal environment in many animals

Imagine that your body temperature soared every time you took a hot shower or drank a freshly brewed cup of coffee. Managing the state of the internal environment is a major challenge for the animal body. Faced with environmental fluctuations, animals manage their internal environment by either regulating or conforming.

Regulating and Conforming

An animal is said to be a **regulator** for a particular environmental variable if it uses internal mechanisms to control internal change in the face of external fluctuation. The river otter in **Figure 40.7** is a regulator for temperature, keeping its body at a temperature that is largely independent of that of the water in which it swims.

An animal is said to be a **conformer** for a particular environmental variable if it allows its internal condition to change in accordance with external changes in the variable. The largemouth bass in Figure 40.7 conforms to the temperature of the lake it inhabits. As the water warms or cools, so does the body of the bass. Some animals conform to more constant environments. For example, many marine invertebrates, such as spider crabs of the genus *Libinia*, let their internal solute concentration conform to the relatively stable solute concentration (salinity) of their ocean environment.

Regulating and conforming represent extremes on a continuum. An animal may regulate some internal conditions while allowing others to conform to the environment. For example, even though the bass conforms to the temperature of the surrounding water, the solute concentration in its blood and interstitial fluid differs from the solute concentration of the fresh water in which it lives. This difference occurs because the fish's anatomy and physiology enable it to regulate internal changes in solute concentration. (You will learn more about the mechanisms of this regulation in Chapter 44.)

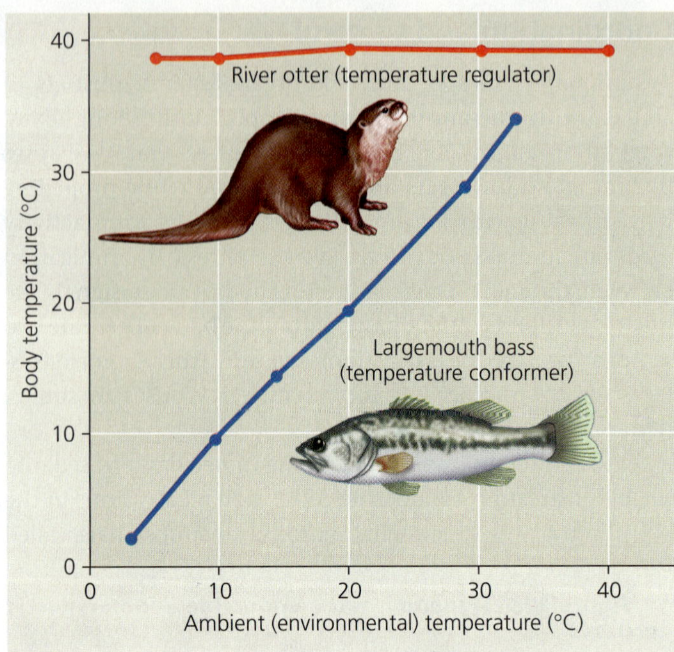

▲ **Figure 40.7 The relationship between body and environmental temperatures in an aquatic temperature regulator and an aquatic temperature conformer.** The river otter regulates its body temperature, keeping it stable across a wide range of environmental temperatures. The largemouth bass, meanwhile, allows its internal environment to conform to the water temperature.

Homeostasis

The steady body temperature of a river otter and the stable concentration of solutes in a freshwater bass are examples of **homeostasis**, which means "steady state," referring to the maintenance of internal balance. In achieving homeostasis, animals maintain a relatively constant internal environment even when the external environment changes significantly.

Like many animals, humans exhibit homeostasis for a range of physical and chemical properties. For example, the human body maintains a fairly constant temperature of about 37°C (98.6°F) and a pH of the blood and interstitial fluid within 0.1 pH unit of 7.4. The body also regulates the concentration of glucose in the bloodstream so that it remains predominantly in the range of 70–110 mg of glucose per 100 mL of blood.

Mechanisms of Homeostasis

Before exploring homeostasis in animals, let's first consider a nonliving example: the regulation of room temperature

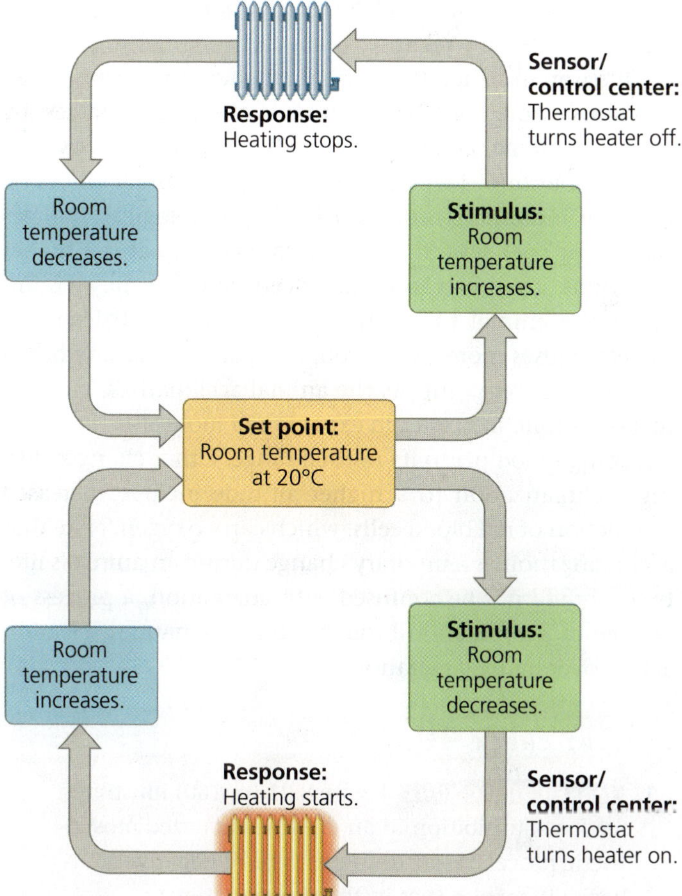

▲ **Figure 40.8 A nonliving example of temperature regulation: control of room temperature.** Regulating room temperature depends on a control center (a thermostat) that detects temperature change and activates mechanisms that reverse that change.

WHAT IF? *How would adding an air conditioner to the system contribute to homeostasis?*

(Figure 40.8). Let's assume you want to keep a room at 20°C (68°F), a comfortable temperature for normal activity. You adjust a control device—the thermostat—to 20°C and allow a thermometer in the thermostat to monitor temperature. If the room temperature falls below 20°C, the thermostat responds by turning on a radiator, furnace, or other heater. Heat is produced until the room reaches 20°C, at which point the thermostat switches off the heater. Whenever the temperature in the room again drifts below 20°C, the thermostat activates another heating cycle.

Like a home heating system, an animal achieves homeostasis by maintaining a variable, such as body temperature or solute concentration, at or near a particular value, or **set point**. Fluctuations in the variable above or below the set point serve as the **stimulus** detected by a receptor, or **sensor**. Upon receiving a signal from the sensor, a *control center* generates output that triggers a **response**, a physiological activity that helps return the variable to the set point. In the home heating example, a drop in temperature below the set point acts as a stimulus, the thermostat serves as the sensor and control center, and the heater produces the response.

Feedback Control in Homeostasis

Like the regulatory circuit shown in Figure 40.8, homeostasis in animals relies largely on **negative feedback**, a control mechanism that reduces, or "damps," the stimulus. For example, when you exercise vigorously, you produce heat, which increases body temperature. Your nervous system detects this increase and triggers sweating. As you sweat, the evaporation of moisture from your skin cools your body, helping return your body temperature to its set point.

Homeostasis is a dynamic equilibrium, the interplay between external factors that tend to change the internal environment and internal control mechanisms that oppose such changes. Note that physiological responses to stimuli are not instantaneous, just as switching on a furnace does not immediately warm a room. As a result, homeostasis moderates but doesn't eliminate changes in the internal environment. Additional fluctuation occurs if a variable has a *normal range*—an upper and lower limit—rather than a single set point. This is equivalent to a heating system that begins producing heat when the room temperature drops to 19°C (66°F) and stops heating when the temperature reaches 21°C (70°F). Regardless of whether there is a set point or a normal range, homeostasis is enhanced by adaptations that reduce fluctuations, such as insulation in the case of temperature and physiological buffers in the case of pH.

Unlike negative feedback, **positive feedback** is a control mechanism that amplifies rather than reduces the stimulus (see Figure 1.13). Positive-feedback loops in animals do not play a major role in homeostasis, but instead help drive processes to completion. During childbirth, for instance, the pressure of the baby's head against receptors near the opening of the mother's uterus stimulates the uterus to contract. These contractions result in greater pressure against the opening of the uterus, heightening the contractions and thereby causing even greater pressure, until the baby is born.

Alterations in Homeostasis

The set points and normal ranges for homeostasis can change under various circumstances. In fact, *regulated changes* in the internal environment are essential to normal body functions. Some regulated changes are associated with a particular stage in life, such as the radical shift in hormone balance that occurs during puberty. Other regulated changes are cyclic, such as the variation in hormone levels responsible for a woman's menstrual cycle (see Figure 46.14).

In all animals (and plants, too), certain cyclic alterations in metabolism reflect a **circadian rhythm**, a set of physiological changes that occur roughly every 24 hours. For example, your body temperature typically undergoes a cyclic rise and fall of more than 0.6°C (1°F) in every 24-hour period. Remarkably, a biological clock maintains this rhythm even when variations in human activity, room temperature, and light levels are minimized **(Figure 40.9a)**. A circadian rhythm is thus intrinsic to the body, although the biological clock is normally coordinated with the cycle of light and darkness in the environment **(Figure 40.9b)**. For example, the hormone melatonin is secreted at night, and more is released during the longer nights of winter. External stimuli can reset the biological clock, but the effect is not immediate. That is why flying across several time zones results in jet lag, a mismatch between the circadian rhythm and local environment that persists until the clock fully resets.

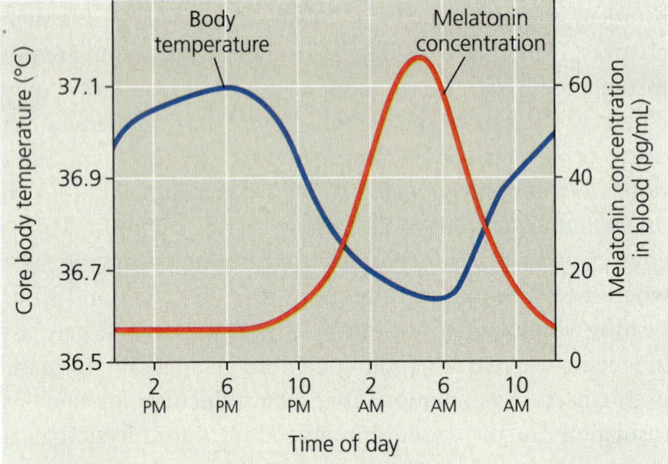

(a) Variation in core body temperature and melatonin concentration in blood. Researchers measured these two variables in resting but awake volunteers in an isolation chamber with constant temperature and low light. (Melatonin is a hormone that appears to be involved in sleep/wake cycles; see Chapter 45.)

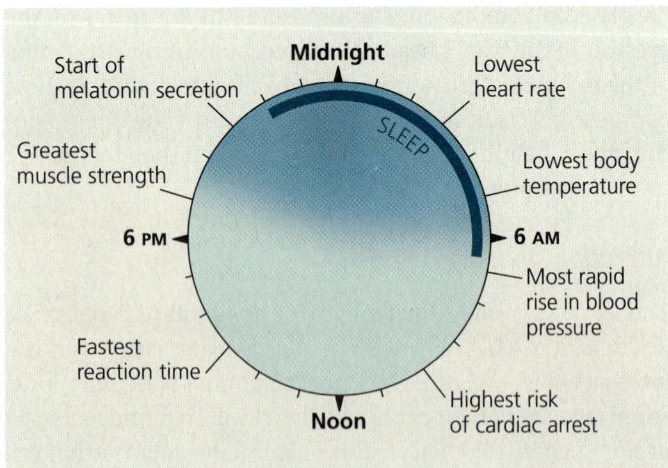

(b) The human circadian clock. Metabolic activities undergo daily cycles in response to the circadian clock. As illustrated for a typical individual who rises early in the morning, eats lunch around noon, and sleeps at night, this variation in metabolism is apparent both day and night.

▲ **Figure 40.9 Human circadian rhythm.**

One way in which the normal range of homeostasis may change is through **acclimatization**, the gradual process by which an animal adjusts to changes in its external environment. For example, when an elk or other mammal moves up into the mountains from sea level, physiological changes that occur over several days facilitate activity at the higher elevations. The lower oxygen concentration in the air stimulates the animal to breathe more rapidly and deeply. It therefore loses more CO_2 through exhalation, raising blood pH above its set point. As the animal acclimatizes, changes in kidney function result in excretion of more alkaline urine, returning blood pH to its normal range. Other changes during acclimatization to a higher altitude include increased production of red blood cells, which carry oxygen. Note that acclimatization, a temporary change during an animal's lifetime, should not be confused with adaptation, a process of change in a population brought about by natural selection acting over many generations.

CONCEPT CHECK 40.2

1. **MAKE CONNECTIONS** Figure 8.21 (p. 160) illustrates feedback inhibition in an enzyme-catalyzed biosynthetic process. How does this type of negative feedback differ from that in thermoregulation?
2. If you were deciding where to put the thermostat in a house, what factors would govern your decision? How do these factors relate to the fact that many homeostatic control sensors in humans are located in the brain?
3. **MAKE CONNECTIONS** Like animals, cyanobacteria have a circadian rhythm. By analyzing the genes that maintain biological clocks, scientists were able to conclude that the 24-hour rhythms of humans and cyanobacteria reflect convergent evolution (see Concept 26.2, pp. 540–541). What evidence would have supported this conclusion? Explain.

For suggested answers, see Appendix A.

CONCEPT 40.3

Homeostatic processes for thermoregulation involve form, function, and behavior

In this section, we will examine the regulation of body temperature as an example of how form and function work together in regulating an animal's internal environment. Later chapters in this unit will discuss other physiological systems involved in maintaining homeostasis.

Thermoregulation is the process by which animals maintain an internal temperature within a tolerable range. Thermoregulation is critical to survival because most biochemical and physiological processes are very sensitive to changes in body temperature. For every 10°C (18°F) decrease in temperature, the rates of most enzyme-mediated reactions decrease two- to threefold. Increases in temperature speed up reactions but cause some proteins to become less active. For instance, the oxygen carrier molecule hemoglobin becomes less effective at binding oxygen as temperature increases. Membranes can also change fluidity, becoming increasingly fluid or rigid as temperatures rise or fall, respectively.

Each animal species has an optimal temperature range. Thermoregulation helps maintain body temperature within that optimal range, enabling cells to function effectively even as the external temperature fluctuates.

Endothermy and Ectothermy

Internal metabolism and the external environment are the sources of heat for thermoregulation. Birds and mammals are mainly **endothermic**, meaning that they are warmed mostly by heat generated by metabolism. A few nonavian reptiles, some fishes, and many insect species are also mainly endothermic. In contrast, amphibians, lizards, snakes, turtles, many fishes, and most invertebrates are mainly **ectothermic**, meaning that they gain most of their heat from external sources.

Animals that are mainly endothermic are referred to as endotherms; those that are mainly ectothermic are known as ectotherms. Keep in mind, though, that endothermy and ectothermy are not mutually exclusive modes of thermoregulation. For example, a bird is mainly endothermic, but it may warm itself in the sun on a cold morning, much as an ectothermic lizard does.

Endotherms can maintain a stable body temperature even in the face of large fluctuations in the environmental temperature. For example, few ectotherms are active in the below-freezing weather that prevails during winter over much of Earth's surface, but many endotherms function very well in these conditions **(Figure 40.10a)**. In a cold environment, an endotherm generates enough heat to keep its body substantially warmer than its surroundings. In a hot environment, endothermic vertebrates have mechanisms for cooling their bodies, enabling them to withstand heat loads that are intolerable for most ectotherms.

Because their heat source is largely environmental, ectotherms generally need to consume much less food than endotherms of equivalent size—an advantage if food supplies are limited. Ectotherms also usually tolerate larger fluctuations in their internal temperature. Although ectotherms do not generate enough heat for thermoregulation, many adjust body temperature by behavioral means, such as seeking out shade or basking in the sun **(Figure 40.10b)**. Overall, ectothermy is an effective and successful strategy in most environments, as shown by the abundance and diversity of ectothermic animals.

(a) A walrus, an endotherm

(b) A lizard, an ectotherm

▲ **Figure 40.10 Endothermy and ectothermy.**

Variation in Body Temperature

Animals can have either a variable or a constant body temperature. An animal whose body temperature varies with its environment is called a *poikilotherm* (from the Greek *poikilos*, varied). In contrast, a *homeotherm* has a relatively constant body temperature. For example, the largemouth bass is a poikilotherm, and the river otter is a homeotherm (see Figure 40.7).

From the descriptions of ectotherms and endotherms, it might seem that all ectotherms are poikilothermic and all endotherms are homeothermic. In fact, there is no fixed relationship between the source of heat and the stability of body temperature. For example, many ectothermic marine fishes and invertebrates inhabit waters with such stable temperatures that their body temperature varies less than that of endotherms such as humans and other mammals. Conversely, the body temperature of a few endotherms varies considerably. For example, bats and hummingbirds may periodically enter an inactive state in which they maintain a lower body temperature.

It is a common misconception that ectotherms are "cold-blooded" and endotherms are "warm-blooded." Ectotherms do not necessarily have low body temperatures. In fact, when sitting in the sun, many ectothermic lizards have higher body temperatures than mammals. Thus, the terms *cold-blooded* and *warm-blooded* are misleading and are avoided in scientific communication.

Balancing Heat Loss and Gain

Thermoregulation depends on an animal's ability to control the exchange of heat with its environment. Any organism, like any object, exchanges heat by four physical processes: radiation, evaporation, convection, and conduction. **Figure 40.11** distinguishes these processes, which account for the flow of heat both within an organism and between an organism and its external environment. Note that heat is always transferred from an object of higher temperature to one of lower temperature.

The essence of thermoregulation is maintaining rates of heat gain that equal rates of heat loss. Animals do this through mechanisms that either reduce heat exchange overall or favor heat exchange in a particular direction. In mammals, several of these mechanisms involve the **integumentary system**, the outer covering of the body, consisting of the skin, hair, and nails (claws or hooves in some species).

Radiation is the emission of electromagnetic waves by all objects warmer than absolute zero. Here, a lizard absorbs heat radiating from the distant sun and radiates a smaller amount of energy to the surrounding air.

Evaporation is the removal of heat from the surface of a liquid that is losing some of its molecules as gas. Evaporation of water from a lizard's moist surfaces that are exposed to the environment has a strong cooling effect.

Convection is the transfer of heat by the movement of air or liquid past a surface, as when a breeze contributes to heat loss from a lizard's dry skin or when blood moves heat from the body core to the extremities.

Conduction is the direct transfer of thermal motion (heat) between molecules of objects in contact with each other, as when a lizard sits on a hot rock.

▲ **Figure 40.11 Heat exchange between an organism and its environment.**

Insulation

A major thermoregulatory adaptation in mammals and birds is insulation, which reduces the flow of heat between an animal and its environment. Sources of insulation include hair, feathers, and layers of fat formed by adipose tissue.

Many animals that rely on insulation to reduce overall heat exchange also adjust their insulating layers to help thermoregulate. Most land mammals and birds, for example, react to cold by raising their fur or feathers. This action traps a thicker layer of air, thereby increasing the insulating power of the fur or feather layer. To repel water that would reduce the insulating capacity of feathers or fur, some animals secrete oily substances, such as the oils that birds apply to their feathers during preening. Lacking feathers or fur, humans must rely primarily on fat for insulation. "Goose bumps" are a vestige of hair raising inherited from our furry ancestors.

Insulation is particularly important for marine mammals, such as whales and walruses. These animals swim in water colder than their body core, and many species spend at least part of the year in nearly freezing polar seas. The problem of thermoregulation is made worse by the fact that the transfer of heat to water occurs 50 to 100 times more rapidly than heat transfer to air. Just under their skin, marine mammals have a very thick layer of insulating fat called blubber. The insulation that blubber provides is so effective that marine mammals can maintain body core temperatures of about 36–38°C (97–100°F) without requiring much more energy from food than land mammals of similar size.

Circulatory Adaptations

Circulatory systems provide a major route for heat flow between the interior and exterior of the body. Adaptations that regulate the extent of blood flow near the body surface or that trap heat within the body core play a significant role in thermoregulation.

In response to changes in the temperature of their surroundings, many animals alter the amount of blood (and hence heat) flowing between their body core and their skin. Nerve signals that relax the muscles of the vessel walls result in *vasodilation*, a widening of superficial blood vessels (those near the body surface). As a consequence of the increase in vessel diameter, blood flow in the skin increases. In endotherms, vasodilation usually warms the skin and increases the transfer of body heat to the environment by radiation, conduction, and convection (see Figure 40.11). The reverse process, *vasoconstriction*, reduces blood flow and heat transfer by decreasing the diameter of superficial vessels. It is vasoconstriction in blood vessels of the ear that allows the jackrabbit shown in Figure 40.1 to avoid overheating on hot desert days.

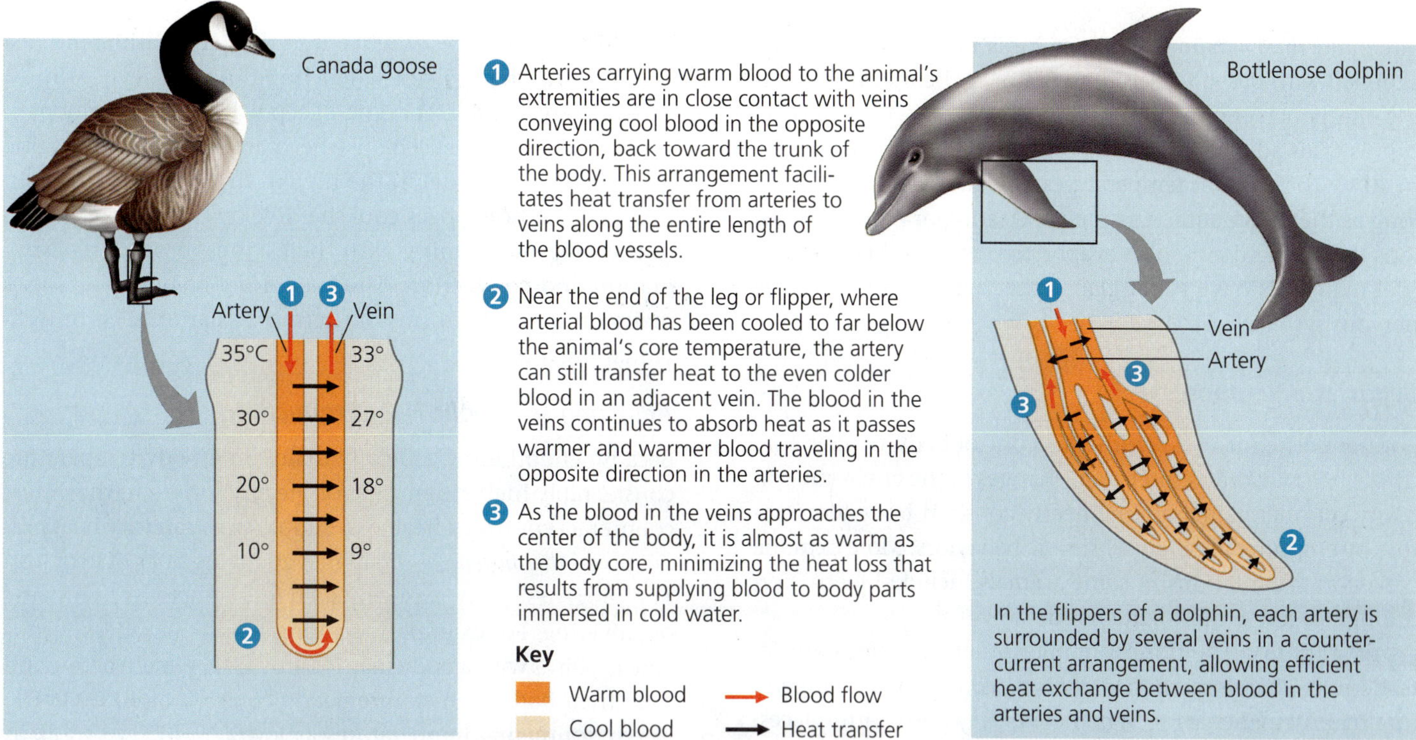

▲ **Figure 40.12 Countercurrent heat exchangers.** A countercurrent exchange system traps heat in the body core, thus reducing heat loss from the extremities, particularly when they are immersed in cold water or in contact with ice or snow. In essence, heat in the arterial blood emerging from the body core is transferred directly to the returning venous blood instead of being lost to the environment.

Like endotherms, some ectotherms control heat exchange by regulating blood flow. For example, when the marine iguana of the Galápagos Islands swims in the cold ocean, its superficial blood vessels undergo vasoconstriction. This process routes more blood to the central core of the iguana's body, conserving body heat.

In many birds and mammals, reducing heat loss from the body relies on **countercurrent exchange**, the transfer of heat (or solutes) between fluids that are flowing in opposite directions. In a countercurrent heat exchanger, arteries and veins are located adjacent to each other (**Figure 40.12**). As warm blood moves from the body core in the arteries, it transfers heat to the colder blood returning from the extremities in the veins. Because blood flows through the arteries and veins in opposite directions, heat is transferred along the entire length of the exchanger, maximizing the rate of heat exchange.

Certain sharks, fishes, and insects also use countercurrent heat exchange. Although most sharks and fishes are temperature conformers, countercurrent heat exchangers are found in some large, powerful swimmers, including great white sharks, bluefin tuna, and swordfish. By keeping the main swimming muscles several degrees warmer than tissues near the animal's surface, this adaptation enables the vigorous, sustained activity that is characteristic of these animals. Similarly, many endothermic insects (bumblebees, honeybees, and some moths) have a countercurrent exchanger that helps maintain a high temperature in their thorax, where flight muscles are located.

In controlling heat gain and loss, some species regulate the extent of blood flow to the countercurrent exchanger. By allowing blood to pass through the heat exchanger or diverting it to other blood vessels, these animals alter the rate of heat loss as their physiological state or environment changes. For example, insects flying in hot weather run the risk of overheating because of the large amount of heat produced by working flight muscles. In some species, the countercurrent mechanism can be "shut down," allowing muscle-produced heat to be lost from the thorax to the abdomen and then to the environment.

Cooling by Evaporative Heat Loss

Many mammals and birds live in places where thermoregulation requires cooling as well as warming. If the environmental temperature is above their body temperature, animals gain heat from the environment as well as from metabolism, and evaporation is the only way to keep body temperature from rising. Terrestrial animals lose water by evaporation from their skin and respiratory surfaces. Water absorbs considerable heat when it evaporates (see Chapter 3); this heat is carried away from the body surface with the water vapor.

Some animals have adaptations that can greatly augment the cooling effect of evaporation. Panting is important in birds and many mammals. Some birds have a pouch richly supplied with blood vessels in the floor of the mouth; fluttering the pouch increases evaporation. Pigeons, for example, can use this adaptation to keep their body temperature close to 40°C (104°F) in air temperatures as high as 60°C (140°F), as long as they have sufficient water. Sweating or bathing moistens the skin and enhances evaporative cooling. Many terrestrial mammals have sweat glands that are controlled by the nervous system.

Behavioral Responses

Both endotherms and ectotherms control body temperature through behavioral responses to changes in the environment. Many ectotherms maintain a nearly constant body temperature by engaging in relatively simple behaviors. More extreme behavioral adaptations in some animals include hibernation or migration to a more suitable climate.

All amphibians and most reptiles other than birds are ectothermic. Therefore, these organisms control body temperature mainly by behavior. When cold, they seek warm places, orienting themselves toward heat sources and expanding the portion of their body surface exposed to the heat source (see Figure 40.10b). When hot, they move to cool areas or turn in another direction.

Many terrestrial invertebrates can adjust internal temperature by the same behavioral mechanisms used by vertebrate ectotherms. The desert locust *(Schistocerca gregaria)*, for example, must reach a certain temperature to become active, and on cold days it orients itself in a direction that maximizes the absorption of sunlight. Other terrestrial invertebrates have certain postures that enable them to maximize or minimize their absorption of heat from the sun **(Figure 40.13)**.

Honeybees use a thermoregulatory mechanism that depends on social behavior. In cold weather, they increase heat production and huddle together, thereby retaining heat. Individuals move between the cooler outer edges of the cluster and the warmer center, thus circulating and distributing the heat. Even when huddling, honeybees must expend considerable energy to keep warm during long periods of cold weather. (This is the main function of storing large quantities of fuel in the hive in the form of honey.) In hot weather, honeybees cool the hive by transporting water to the hive and fanning with their wings, promoting evaporation and convection. Thus, a colony of honeybees uses many of the mechanisms of thermoregulation seen in individual organisms.

▶ **Figure 40.13** **Thermoregulatory behavior in a dragonfly.** This dragonfly's "obelisk" posture is an adaptation that minimizes the amount of body surface exposed to the sun. This posture helps reduce heat gain by radiation.

Adjusting Metabolic Heat Production

Because endotherms generally maintain a body temperature considerably higher than that of the environment, they must counteract continual heat loss. Endotherms can vary heat production—*thermogenesis*—to match changing rates of heat loss. Thermogenesis is increased by such muscle activity as moving or shivering. For example, shivering helps chickadees (genus *Poecile*), birds with a body mass of only 20 g, remain active and hold their body temperature nearly constant at 40°C (104°F) in environmental temperatures as low as −40°C (−40°F), as long as they have adequate food.

In some mammals, certain hormones can cause mitochondria to increase their metabolic activity and produce heat instead of ATP. This process, called *nonshivering thermogenesis,* takes place throughout the body; some mammals also have a tissue called *brown fat* in their neck and between their shoulders that is specialized for rapid heat production. (In human infants, brown fat represents about 5% of total body weight. In 2009, brown fat was found for the first time in human adults, with greater amounts being detected when outdoor temperatures were lower.) Through shivering and nonshivering thermogenesis, mammals and birds in cold environments can increase their metabolic heat production by as much as five to ten times the levels that occur in warm conditions.

A few large reptiles become endothermic in particular circumstances. In the early 1960s, Herndon Dowling documented this phenomenon for a female Burmese python (*Python molurus bivittatus*). Placing temperature-recording devices along the snake's coils, Dowling found that the snake maintained a body temperature roughly 6°C (11°F) above that of the surrounding air during the month when she was incubating eggs. Where did the heat come from? Further studies by Dowling and colleagues showed that pythons, like mammals and birds, can raise their body temperature through shivering **(Figure 40.14)**. These and other findings have led to new insights into thermoregulation in reptiles and have contributed to the idea, still under debate, that certain groups of Mesozoic dinosaurs were endothermic (see Chapter 34).

As mentioned earlier, many species of flying insects, such as bees and moths, are endothermic—the smallest of all endotherms. The capacity of such endothermic insects to elevate

▼ Figure 40.14 INQUIRY

How does a Burmese python generate heat while incubating eggs?

EXPERIMENT Herndon Dowling, graduate student Allen Vinegar, and the student's research supervisor, Victor Hutchison, at the Bronx Zoo in New York, observed that when a female Burmese python incubated eggs by wrapping her body around them, she raised her body temperature and frequently contracted the muscles in her coils. To learn if the contractions were elevating her body temperature, they placed the python and her eggs in a chamber. As they varied the chamber's temperature, they monitored the python's muscle contractions as well as her oxygen uptake, a measure of her rate of cellular respiration.

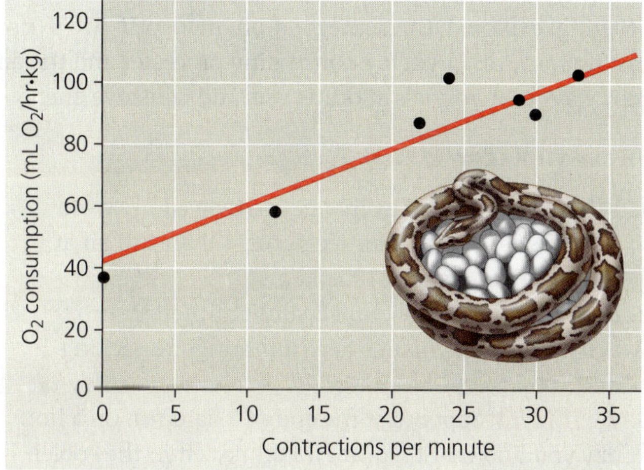

RESULTS The python's oxygen consumption increased when the temperature in the chamber decreased. Her oxygen consumption also increased with the rate of muscle contraction.

CONCLUSION Because oxygen consumption generates heat through cellular respiration and increases linearly with the rate of muscle contraction, the researchers concluded that the muscle contractions, a form of shivering, were the source of the Burmese python's elevated body temperature.

SOURCE V. H. Hutchison, H. G. Dowling, and A. Vinegar, Thermoregulation in a brooding female Indian python, *Python molurus bivittatus*, *Science* 151:694–696 (1966).

WHAT IF? Suppose you varied air temperature and measured oxygen consumption for a female Burmese python without a clutch of eggs. Since she would not show shivering behavior, how would you expect the snake's oxygen consumption to vary with environmental temperature?

body temperature depends on powerful flight muscles, which generate large amounts of heat when contracting. Many endothermic insects warm up by shivering before taking off. As they contract their flight muscles in synchrony, only slight wing movements occur, but considerable heat is produced. Chemical reactions, and hence cellular respiration, speed up in the warmed-up flight "motors," enabling these insects to fly even when the air is cold **(Figure 40.15)**.

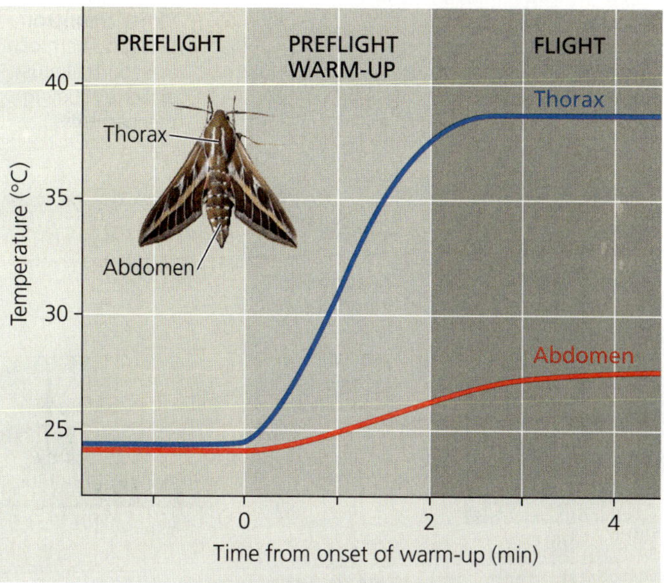

▲ Figure 40.15 Preflight warm-up in the hawkmoth. The hawkmoth (*Manduca sexta*) is one of many insect species that use a shivering-like mechanism for preflight warm-up of thoracic flight muscles. Warming up helps these muscles produce enough power to let the animal take off. Once the moth is airborne, flight muscle activity maintains a high thoracic temperature.

Acclimatization in Thermoregulation

Acclimatization contributes to thermoregulation in many animal species. In birds and mammals, acclimatization to seasonal temperature changes often includes adjusting insulation—growing a thicker coat of fur in the winter and shedding it in the summer, for example. These changes help endotherms keep a constant body temperature year-round.

Acclimatization in ectotherms often includes adjustments at the cellular level. Cells may produce variants of enzymes that have the same function but different optimal temperatures. Also, the proportions of saturated and unsaturated lipids in membranes may change; unsaturated lipids help keep membranes fluid at lower temperatures (see Figure 7.5). Some ectotherms that experience subzero body temperatures protect themselves by producing "antifreeze" compounds that prevent ice formation in their cells. In the Arctic Ocean and Southern (Antarctic) Ocean, these compounds enable certain fishes to survive in water as cold as −2°C (28°F), below the freezing point of unprotected body fluids (about −1°C, or 30°F).

Physiological Thermostats and Fever

The regulation of body temperature in humans and other mammals is brought about by a complex system based on feedback mechanisms. The sensors for thermoregulation are concentrated in a brain region called the **hypothalamus**. A group of nerve cells in the hypothalamus functions as a thermostat, responding to body temperatures outside a normal range by activating mechanisms that promote heat loss

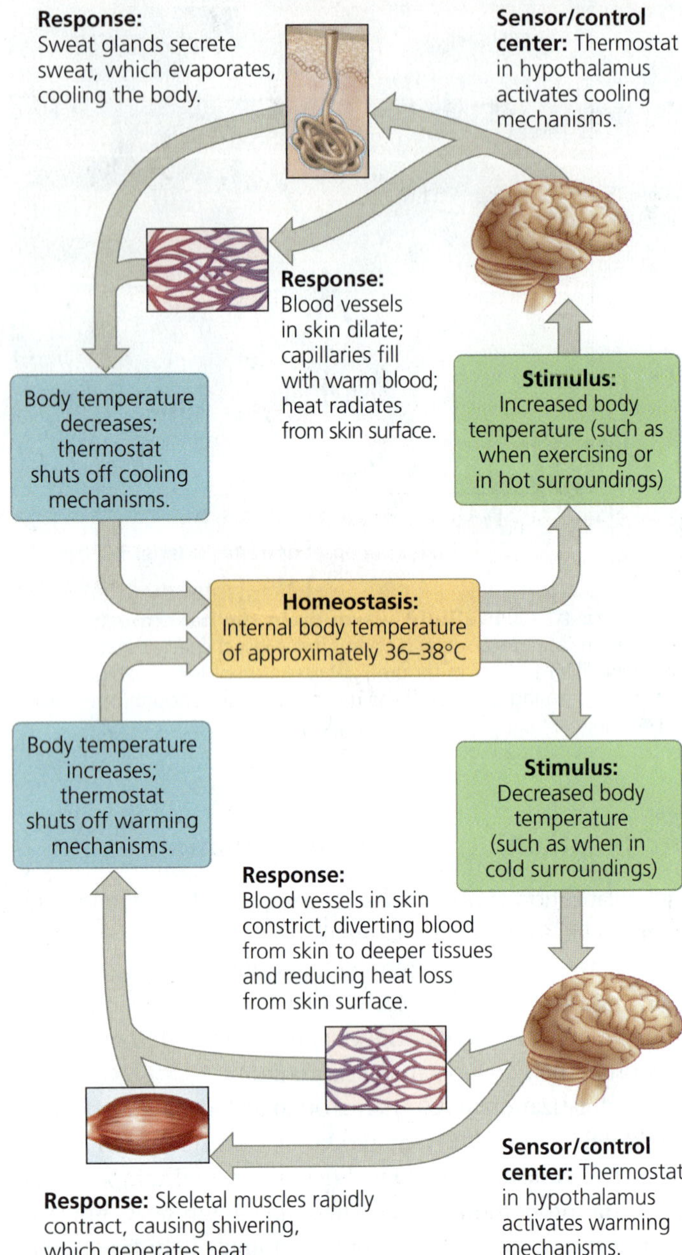

▲ **Figure 40.16 The thermostatic function of the hypothalamus in human thermoregulation.**

or gain **(Figure 40.16)**. Warm receptors signal the hypothalamic thermostat when temperatures increase; cold receptors signal when temperatures decrease. (Because the same blood vessel supplies the hypothalamus and ears, an ear thermometer records the temperature detected by the hypothalamic thermostat.) At body temperatures below the normal range, the thermostat inhibits heat loss mechanisms and activates heat-saving ones, such as vasoconstriction and the raising of fur, while stimulating heat-generating mechanisms (shivering and nonshivering thermogenesis). In response to elevated body temperature, the thermostat shuts down heat retention mechanisms and promotes cooling the body by vasodilation, sweating, or panting.

In the course of certain bacterial and viral infections, mammals and birds develop fever, an elevated body temperature. A variety of experiments have shown that fever reflects an increase in the set point for the biological thermostat. For example, artificially raising the temperature of the hypothalamus in an infected animal reduces fever in the rest of the body!

Although only endotherms develop fever, lizards exhibit a related response. When infected with certain bacteria, the desert iguana (*Dipsosaurus dorsalis*) seeks a warmer environment and then maintains a body temperature that is elevated by 2–4°C (4–7°F). Similar observations in fishes, amphibians, and even cockroaches indicate that this response to certain infections is a common feature of many animal species.

Having explored thermoregulation in depth, we'll now consider some other energy-consuming processes and the different ways that animals allocate, use, and conserve energy.

CONCEPT CHECK 40.3

1. What mode of heat exchange is involved in "wind chill," when moving air feels colder than still air at the same temperature? Explain.
2. Flowers differ in how much sunlight they absorb. Why might this matter to a hummingbird seeking nectar on a cool morning?
3. **WHAT IF?** Suppose at the end of a hard run on a hot day you find that there are no drinks left in the cooler. If, out of desperation, you dunk your head into the cooler, how might the ice-cold water affect the rate at which your body temperature returns to normal?

For suggested answers, see Appendix A.

CONCEPT **40.4**

Energy requirements are related to animal size, activity, and environment

One of the unifying themes of biology introduced in Chapter 1 is that life requires energy transfer and transformation. Like other organisms, animals use chemical energy for growth, repair, activity, and reproduction. The overall flow and transformation of energy in an animal—its **bioenergetics**—determines nutritional needs and is related to the animal's size, activity, and environment.

Energy Allocation and Use

As we have discussed in other chapters, organisms can be classified by how they obtain chemical energy. Most autotrophs, such as plants, use light energy to build energy-rich organic molecules and then use those organic molecules for fuel. Most heterotrophs, such as animals, must obtain their chemical

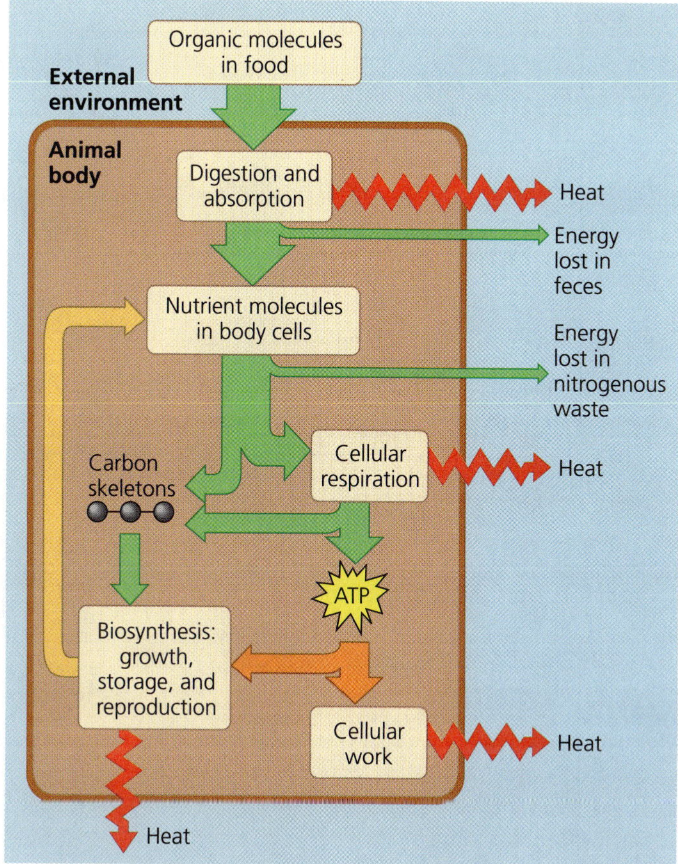

▲ **Figure 40.17 Bioenergetics of an animal: an overview.**

MAKE CONNECTIONS *Review the idea of energy coupling in Concept 8.3 (pp. 149–151). Then use that idea to explain why heat is produced in the absorption of nutrients, in cellular respiration, and in the synthesis of biopolymers.*

energy from food, which contains organic molecules synthesized by other organisms.

Animals use chemical energy harvested from the food they eat to fuel metabolism and activity **(Figure 40.17)**. Food is digested by enzymatic hydrolysis (see Figure 5.2b), and nutrients are absorbed by body cells. Most nutrient molecules are used to generate ATP. ATP produced by cellular respiration and fermentation powers cellular work, enabling cells, organs, and organ systems to perform the functions that keep an animal alive. Energy in the form of ATP is also used in biosynthesis, which is needed for body growth and repair, synthesis of storage material such as fat, and production of gametes. The production and use of ATP generates heat, which the animal eventually gives off to its surroundings.

Quantifying Energy Use

How much of the total energy an animal obtains from food does it need just to stay alive? How much energy must be expended to walk, run, swim, or fly from one place to another? What fraction of the energy intake is used for reproduction? Physiologists answer such questions by measuring the rate at which an animal uses chemical energy and how this rate changes in different circumstances.

▲ **Figure 40.18 Measuring the rate of oxygen consumption by a swimming shark.** A researcher monitors the drop in oxygen level in the recirculating water of a juvenile hammerhead's tank.

The amount of energy an animal uses in a unit of time is called its **metabolic rate**—the sum of all the energy used in biochemical reactions over a given time interval. Energy is measured in joules (J) or in calories (cal) and kilocalories (kcal). (A kilocalorie equals 1,000 calories, or 4,184 joules. The unit Calorie, with a capital C, as used by many nutritionists, is actually a kilocalorie.)

Metabolic rate can be determined in several ways. Because nearly all of the chemical energy used in cellular respiration eventually appears as heat, metabolic rate can be measured by monitoring an animal's rate of heat loss. For this approach, researchers use a calorimeter, which is a closed, insulated chamber equipped with a device that records an animal's heat loss. Metabolic rate can also be determined from the amount of oxygen consumed or carbon dioxide produced by an animal's cellular respiration **(Figure 40.18)**. To calculate metabolic rate over longer periods, researchers record the rate of food consumption, the energy content of the food (about 4.5–5 kcal per gram of protein or carbohydrate and about 9 kcal per gram of fat), and the chemical energy lost in waste products (feces and nitrogenous waste).

Minimum Metabolic Rate and Thermoregulation

Animals must maintain a minimum metabolic rate for basic functions such as cell maintenance, breathing, and heartbeat. Researchers measure this minimum metabolic rate differently for endotherms and ectotherms. The minimum metabolic rate of a nongrowing endotherm that is at rest, has an empty stomach, and is not experiencing stress is called the **basal metabolic rate (BMR)**. BMR is measured under a "comfortable" temperature range—a range that requires no generation or shedding of heat above the minimum. The minimum metabolic rate of ectotherms is determined at a specific temperature because changes in the environmental temperature alter body temperature and therefore metabolic rate. The metabolic rate of a fasting, nonstressed ectotherm at rest at a particular temperature is called its **standard metabolic rate (SMR)**.

Comparisons of minimum metabolic rates reveal that endothermy and ectothermy have different energy costs. The BMR for humans averages 1,600–1,800 kcal per day for adult males and 1,300–1,500 kcal per day for adult females. These BMRs are about equivalent to the rate of energy use by a 75-watt light bulb. In contrast, the SMR of an American alligator is only about 60 kcal per day at 20°C (68°F). Since this represents less than 1⁄20 the energy used by a comparably sized adult human, the lower energetic requirement of ectothermy is readily apparent.

Influences on Metabolic Rate

Metabolic rate is affected by many factors besides whether the animal is an endotherm or an ectotherm. Some key factors are age, sex, size, activity, temperature, and nutrition. Here we'll examine the effects of size and activity.

Size and Metabolic Rate

Larger animals have more body mass and therefore require more chemical energy. Remarkably, the relationship between overall metabolic rate and body mass is constant across a wide range of sizes and forms, as illustrated for various mammals in **Figure 40.19a**. In fact, for even more varied organisms ranging in size from bacteria to blue whales, metabolic rate remains roughly proportional to body mass to the three-quarter power ($m^{3/4}$). Scientists are still researching the basis of this relationship, which applies to ectotherms as well as endotherms.

The relationship of metabolic rate to size profoundly affects energy consumption by body cells and tissues. As shown in **Figure 40.19b**, the energy it takes to maintain each gram of body mass is inversely related to body size. Each gram of a mouse, for instance, requires about 20 times as many calories as a gram of an elephant, even though the whole elephant uses far more calories than the whole mouse. The smaller animal's higher metabolic rate per gram demands a higher rate of oxygen delivery. Correlated with its higher metabolic rate per gram, the smaller animal has a higher breathing rate, blood volume (relative to its size), and heart rate. Also, it must eat much more food per unit of body mass.

Bioenergetic considerations associated with body size provide a clear example of how trade-offs shape the evolution of body plans. As body size becomes smaller, each gram of tissue increases in energy cost. As body size increases, energy costs per gram of tissue decrease, but an ever-larger fraction of body tissue is required for exchange, support, and locomotion.

Activity and Metabolic Rate

For both ectotherms and endotherms, activity greatly affects metabolic rate. Even a person reading quietly at a desk or an insect twitching its wings consumes energy beyond the BMR or SMR. Maximum metabolic rates (the highest rates of ATP use) occur during peak activity, such as lifting heavy weights, sprinting, or high-speed swimming. In general, the maximum metabolic rate an animal can sustain is inversely related to the duration of activity.

For most terrestrial animals, the average daily rate of energy consumption is 2 to 4 times BMR (for endotherms) or SMR (for ectotherms). Humans in most developed countries have an unusually low average daily metabolic rate of about 1.5 times BMR—an indication of their relatively sedentary lifestyles.

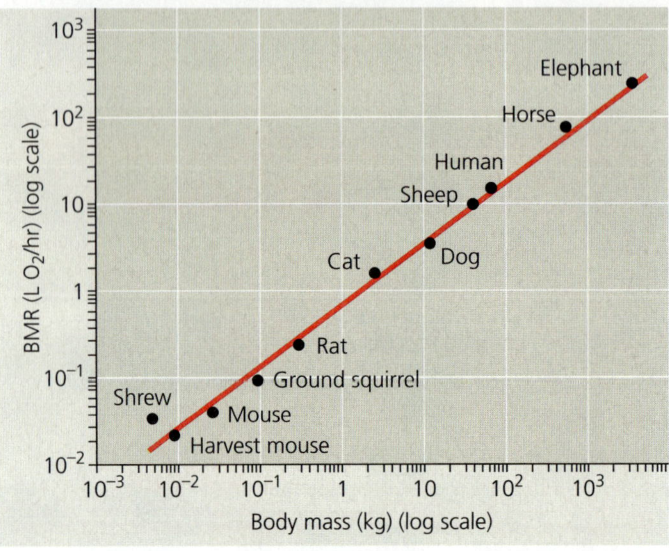

(a) Relationship of basal metabolic rate (BMR) to body size for various mammals. From shrew to elephant, size increases 1 millionfold.

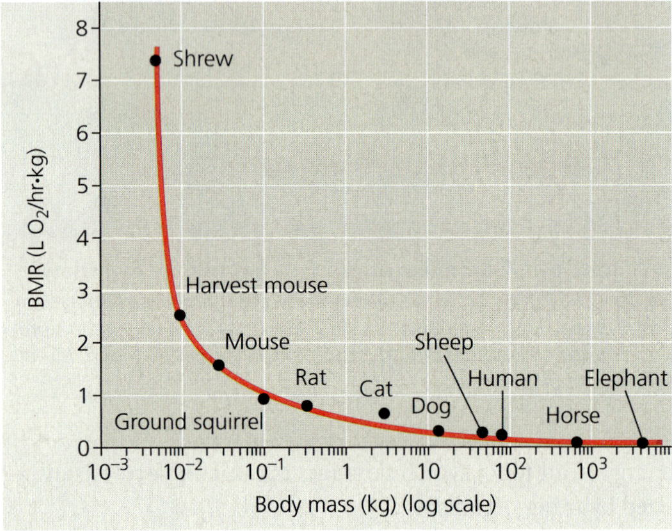

(b) Relationship of BMR per kilogram of body mass to body size for the same mammals as in (a).

▲ **Figure 40.19 The relationship of metabolic rate to body size.**

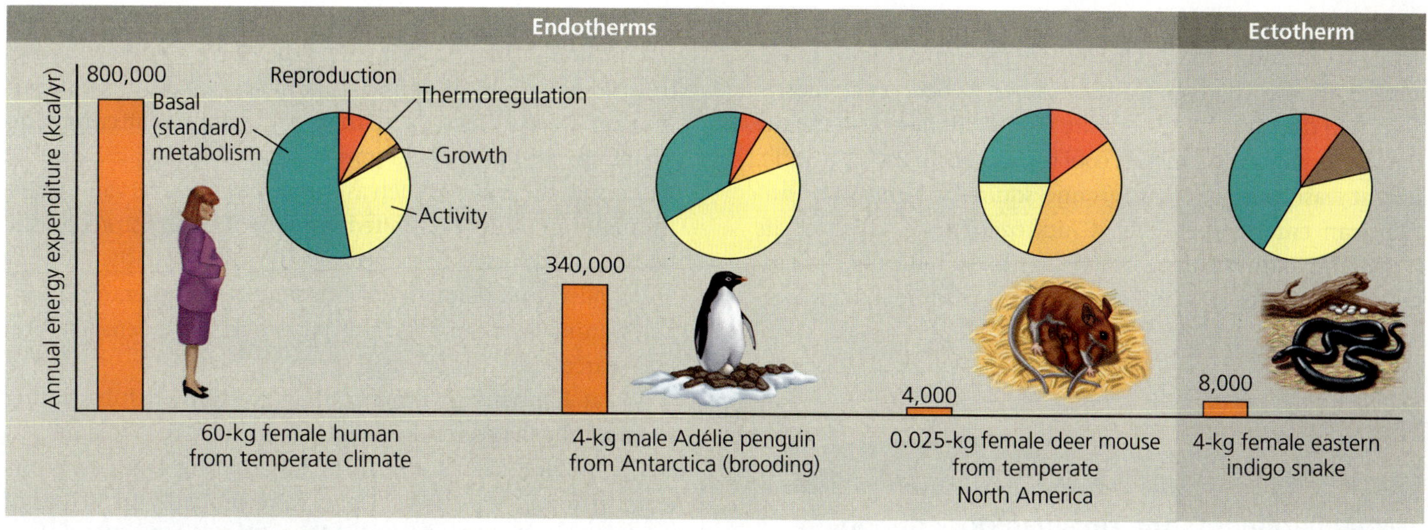

▲ **Figure 40.20 Energy budgets for four animals.** The slices of the pie charts indicate annual energy expenditures for various functions.

Energy Budgets

As we have seen, the ways in which animals use the chemical energy of food depend on environment, behavior, size, and thermoregulation. To understand how these influences affect bioenergetics in animal bodies, let's consider typical annual energy "budgets" of four terrestrial vertebrates varying in size and thermoregulatory strategy: a 60-kg female human, a 4-kg male Adélie penguin, a 25-g (0.025-kg) female deer mouse, and a 4-kg female eastern indigo snake **(Figure 40.20)**. Reproduction is included in these energy budgets because it can greatly influence energy allocation and is critical to species survival.

The female human, an endothermic mammal, spends the largest fraction of her annual energy budget for BMR and comparatively less for activity and thermoregulation. The small amount of growth, about 1%, is equivalent to adding about 1 kg of body fat or 5–6 kg of other tissues. (Growth is not shown in the budgets for the penguin and deer mouse because these animals don't typically gain weight year to year after they are adults.) The cost of nine months of pregnancy and several months of breast-feeding is only 5–8% of the mother's energy requirements for a year.

A male penguin spends the largest fraction of his energy for activity because he must swim to catch food. Being well insulated and fairly large, he has relatively low costs of thermoregulation in spite of living in the cold Antarctic. His reproductive costs, about 6% of annual energy expenditures, come mainly from incubating eggs (brooding) and bringing food to his chicks.

Despite living in a temperate climate, the female deer mouse spends a large fraction of her energy budget for temperature regulation. Because of the high surface-to-volume ratio that goes with small size, deer mice lose body heat rapidly and must constantly generate metabolic heat to maintain body temperature.

In contrast with these endothermic animals, the ectothermic snake has no thermoregulation costs. Like most snakes, she grows continuously throughout her life. In the example in Figure 40.20, the snake adds about 750 g of new body tissue in a year. She also produces about 650 g of eggs. The snake's economical ectothermic strategy is revealed by her very low energy expenditure, only 1⁄40 the energy expended by the similarly sized endothermic penguin.

For all the animals in Figure 40.20, locomotion and other activities are a major part of the energy budget. Some animals can conserve energy by temporarily decreasing their activity to a very low level, a process we will consider next.

Torpor and Energy Conservation

Despite their many adaptations for homeostasis, animals may encounter conditions that severely challenge their abilities to balance their heat, energy, and materials budgets. For example, at certain times of the day or year, their surroundings may be extremely hot or cold, or food may be unavailable. **Torpor**, a physiological state of decreased activity and metabolism, is an adaptation that enables animals to save energy while avoiding difficult and dangerous conditions.

Many small mammals and birds exhibit a daily torpor that seems to be adapted to feeding patterns. For instance, some bats feed at night and go into torpor in daylight. Chickadees and hummingbirds feed during the day and often go into torpor on cold nights; the body temperature of chickadees drops as much as 10°C (18°F) at night, and the temperature of hummingbirds can fall 25°C (45°F) or more. All endotherms that exhibit daily torpor are relatively small; when active, they have high metabolic rates and thus very high rates of energy consumption.

Hibernation is long-term torpor that is an adaptation to winter cold and food scarcity. When a mammal enters hibernation, its body temperature declines as its body's thermostat is turned down. The temperature reduction may be dramatic: Some hibernating mammals cool to as low as 1–2°C (34–36°F), and at least one, the Arctic ground squirrel (*Spermophilus parryii*), can enter a supercooled (unfrozen) state in which its body temperature dips below 0°C (32°F). Periodically, perhaps every two weeks or so, hibernating animals undergo arousal, raising their body temperature and becoming active briefly before resuming hibernation. Nevertheless, the energy savings from hibernation are huge: Metabolic rates during hibernation can be 20 times lower than if the animal attempted to maintain normal body temperatures of 36–38°C (97–100°F). As a result, hibernators such as the ground squirrel can survive through the winter on limited supplies of energy stored in the body tissues or as food cached in a burrow. Similarly, the slow metabolism and inactivity of *estivation*, or summer torpor, enables animals to survive long periods of high temperatures and scarce water supplies.

What happens to the circadian rhythm in hibernating animals? In the past, some researchers have reported detecting daily biological rhythms in hibernating animals. However, in some cases the animals were probably in a state of torpor, from which they could readily arouse, rather than "deep" hibernation. Recently, a group of researchers in France addressed this question in a different way, examining the machinery of the biological clock rather than the rhythms it controls (**Figure 40.21**). Working with the European hamster, they found that molecular components of the clock stopped oscillating during hibernation. These findings support the hypothesis that the circadian clock ceases operation during hibernation, at least in this species.

From discussing body shape to considering energy conservation, this chapter has focused on the whole animal. We surveyed common tissue types that make up organs and organ systems. We also investigated how body plans provide for exchange of materials with the environment, how some animals maintain a constant internal environment, and how size and activity affect metabolic rate. For much of the rest of this unit, we'll explore how specialized organs and organ systems enable animals to meet the basic challenges of life.

▼ Figure 40.21 INQUIRY

What happens to the circadian clock during hibernation?

EXPERIMENT To determine whether the 24-hour biological clock continues to run during hibernation, Paul Pévet and colleagues at the University of Louis Pasteur in Strasbourg, France, studied molecular components of the circadian clock in the European hamster (*Cricetus cricetus*). The researchers measured RNA levels for two clock genes—*Per2* and *Bmal1*—during normal activity (euthermia) and during hibernation in constant darkness. The RNA samples were obtained from the suprachiasmatic nuclei (SCN), a pair of structures in the mammalian brain that control circadian rhythms.

RESULTS

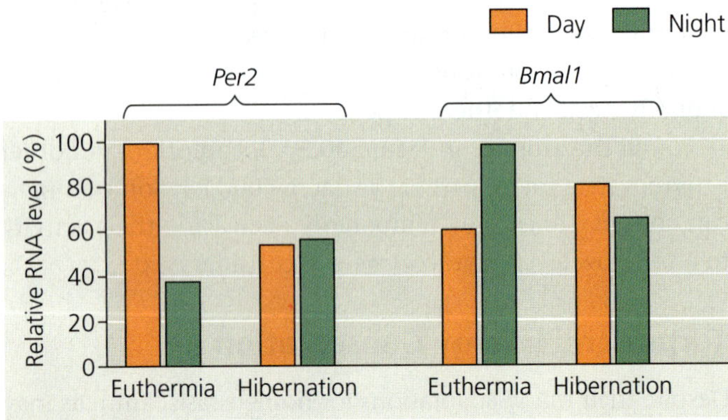

CONCLUSION Hibernation disrupted circadian variation in the hamster's clock gene RNA levels. Further experiments demonstrated that this disruption was not simply due to the dark environment during hibernation, since for nonhibernating animals RNA levels during a darkened daytime were the same as in daylight. The researchers concluded that the biological clock stops running in hibernating European hamsters and, perhaps, in other hibernators as well.

SOURCE F. G. Revel et al., The circadian clock stops ticking during deep hibernation in the European hamster, *Proceedings of the National Academy of Sciences USA* 104:13816–13820 (2007).

WHAT IF? Suppose you discovered a new hamster gene and found that the levels of RNA for this gene were constant during hibernation. What could you conclude about the day and night RNA levels for this gene during euthermia?

CONCEPT CHECK 40.4

1. If a mouse and a small lizard of the same mass (both at rest) were placed in experimental chambers under identical environmental conditions, which animal would consume oxygen at a higher rate? Explain.
2. Which animal must eat a larger proportion of its weight in food each day: a house cat or an African lion caged in a zoo? Explain.
3. **WHAT IF?** If you monitored energy allocation in the penguin in Figure 40.20 for just a few months instead of an entire year, you might find the "growth" category to be a significant part of the pie chart. Given that adult penguins don't grow from year to year, how would you explain this finding?

For suggested answers, see Appendix A.

40 CHAPTER REVIEW

SUMMARY OF KEY CONCEPTS

CONCEPT 40.1

Animal form and function are correlated at all levels of organization (pp. 852–860)

- Physical laws constrain the evolution of an animal's size and shape. These constraints contribute to convergent evolution, the similar but independent adaptations of different species to a common environmental challenge.
- Each animal cell must have access to an aqueous environment. Simple two-layered sacs and flat shapes maximize exposure to the surrounding medium. More complex body plans have highly folded internal surfaces specialized for exchanging materials.
- In the hierarchical organization of animal bodies, groups of cells with a common structure and function make up **tissues**. Different tissues make up **organs**, which together make up **organ systems**. Animal tissues fall into four main groups, each with distinct functions. **Epithelial tissue** forms active interfaces with the environment on external and internal surfaces of the body. **Connective tissue** binds and supports other tissues. **Muscle tissue** contracts, moving the parts of the body. **Nervous tissue** transmits nerve impulses throughout the body.
- The endocrine and nervous systems are the two means of communication between different locations in the body. The endocrine system broadcasts signaling molecules called **hormones** everywhere via the bloodstream, but only certain cells are responsive to each hormone. The nervous system uses dedicated cellular circuits involving electrical and chemical signals to send information to specific locations.

? *For a large animal, what challenges would a spherical shape pose for carrying out exchange with the environment?*

CONCEPT 40.2

Feedback control maintains the internal environment in many animals (pp. 860–862)

- Faced with environmental fluctuations, animals *regulate* (control) certain internal variables while allowing other internal variables to *conform* to (correspond to) external changes. **Homeostasis** is the maintenance of a steady state despite internal and external changes.
- Homeostatic mechanisms are usually based on **negative feedback**, in which the **response** reduces the **stimulus**.

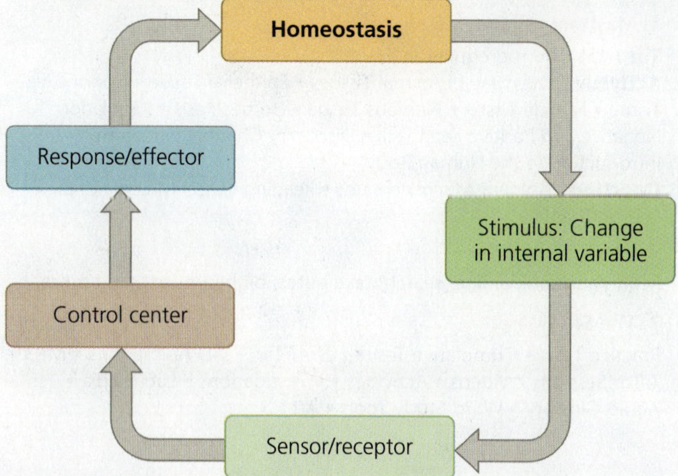

In contrast, **positive feedback** involves amplification of a stimulus by the response and often brings about a change in state, such as the transition from pregnancy to childbirth.

- Regulated change in the internal environment is essential to normal function. **Circadian rhythms** are daily fluctuations in metabolism and behavior tuned to the cycles of light and dark in the environment. Other environmental changes may trigger **acclimatization**, a temporary shift in the steady state.

? *Is it accurate to define homeostasis as a constant internal environment? Explain.*

CONCEPT 40.3

Homeostatic processes for thermoregulation involve form, function, and behavior (pp. 862–868)

- An animal maintains its internal temperature within a tolerable range by **thermoregulation**. **Endotherms** are warmed mostly by heat generated by metabolism. **Ectotherms** get most of their heat from external sources. Endothermy requires a greater expenditure of energy. Body temperature may vary with environmental temperature, as in *poikilotherms*, or be relatively constant, as in *homeotherms*.
- In thermoregulation, physiological and behavioral adjustments balance heat gain and loss, which occur through **radiation**, **evaporation**, **convection**, and **conduction**. Insulation and **countercurrent exchange** reduce heat loss, whereas panting, sweating, and bathing increase evaporation, cooling the body. Both ectotherms and endotherms adjust their rate of heat exchange with their surroundings by vasodilation or vasoconstriction and by behavioral responses.
- Many mammals and birds adjust their amount of body insulation in response to changes in environmental temperature. Ectotherms undergo a variety of changes at the cellular level to acclimatize to shifts in temperature.
- The **hypothalamus** acts as the thermostat in mammalian regulation of body temperature. Fever reflects a resetting of this thermostat to a higher set point in response to infection.

? *Given that humans thermoregulate, explain why your skin is cooler than your body core.*

CONCEPT 40.4

Energy requirements are related to animal size, activity, and environment (pp. 868–872)

- Animals obtain chemical energy from food, storing it for short-term use in ATP. The total amount of energy used in a unit of time defines an animal's **metabolic rate**. Metabolic rates are generally higher for endotherms than for ectotherms.
- Under similar conditions and for animals of the same size, the **basal metabolic rate** of endotherms is substantially higher than the **standard metabolic rate** of ectotherms. Minimum metabolic rate per gram is inversely related to body size among similar animals. Animals allocate energy for basal (or standard) metabolism, activity, homeostasis, growth, and reproduction.
- **Torpor**, a state of decreased activity and metabolism, conserves energy during environmental extremes. Animals may enter torpor during sleep periods (daily torpor), in winter (**hibernation**), or in summer (estivation).

? *Most hibernators are small. After reviewing Figure 40.19, suggest an explanation for this observation.*

TEST YOUR UNDERSTANDING

LEVEL 1: KNOWLEDGE/COMPREHENSION

1. The body tissue that consists largely of material located outside of cells is
   a. epithelial tissue.
   b. connective tissue.
   c. skeletal muscle.
   d. smooth muscle.
   e. nervous tissue.

2. Which of the following would increase the rate of heat exchange between an animal and its environment?
   a. feathers or fur
   b. vasoconstriction
   c. wind blowing across the body surface
   d. countercurrent heat exchanger
   e. blubber or fat layer

3. Consider the energy budgets for a human, an elephant, a penguin, a mouse, and a snake. The __________ would have the highest total annual energy expenditure, and the __________ would have the highest energy expenditure per unit mass.
   a. elephant; mouse
   b. elephant; human
   c. human; penguin
   d. mouse; snake
   e. penguin; mouse

LEVEL 2: APPLICATION/ANALYSIS

4. Compared with a smaller cell, a larger cell of the same shape has
   a. less surface area.
   b. less surface area per unit of volume.
   c. the same surface-to-volume ratio.
   d. a smaller average distance between its mitochondria and the external source of oxygen.
   e. a smaller cytoplasm-to-nucleus ratio.

5. An animal's inputs of energy and materials would exceed its outputs
   a. if the animal is an endotherm, which must always take in more energy because of its high metabolic rate.
   b. if it is actively foraging for food.
   c. if it is hibernating.
   d. if it is growing and increasing its mass.
   e. never; homeostasis makes these energy and material budgets always balance.

6. You are studying a large tropical reptile that has a high and relatively stable body temperature. How would you determine whether this animal is an endotherm or an ectotherm?
   a. You know from its high and stable body temperature that it must be an endotherm.
   b. You know that it is an ectotherm because it is not a bird or mammal.
   c. You subject this reptile to various temperatures in the lab and find that its body temperature and metabolic rate change with the ambient temperature. You conclude that it is an ectotherm.
   d. You note that its environment has a high and stable temperature. Because its body temperature matches the environmental temperature, you conclude that it is an ectotherm.
   e. You measure the metabolic rate of the reptile, and because it is higher than that of a related species that lives in temperate forests, you conclude that this reptile is an endotherm and its relative is an ectotherm.

7. Which of the following animals uses the largest percentage of its energy budget for homeostatic regulation?
   a. a hydra
   b. a marine jelly (an invertebrate)
   c. a snake in a temperate forest
   d. a desert insect
   e. a desert bird

8. **DRAW IT** Draw a model of the control circuit(s) required for driving an automobile at a fairly constant speed over a hilly road. Indicate each feature that represents a sensor, stimulus, or response.

LEVEL 3: SYNTHESIS/EVALUATION

9. **EVOLUTION CONNECTION**
   In 1847, the German biologist Christian Bergmann noted that mammals and birds living at higher latitudes (farther from the equator) are on average larger and bulkier than related species found at lower latitudes. Suggest an evolutionary hypothesis to explain this observation.

10. **SCIENTIFIC INQUIRY**
   Eastern tent caterpillars (*Malacosoma americanum*) live in large groups in silk nests, or tents, which they build in trees. They are among the first insects to be active in early spring, when daily temperature fluctuates from freezing to very hot. Over the course of a day, they display striking differences in behavior: Early in the morning, they rest in a tightly packed group on the tent's east-facing surface. In midafternoon, they are on its undersurface, each caterpillar hanging by a few of its legs. Propose a hypothesis to explain this behavior. How could you test it?

11. **SCIENCE, TECHNOLOGY, AND SOCIETY**
   Medical researchers are investigating artificial substitutes for various human tissues. Why might artificial blood or skin be useful? What characteristics would these substitutes need in order to function well in the body? Why do real tissues work better? Why not use the real tissues if they work better? What other artificial tissues might be useful? What problems do you anticipate in developing and applying them?

12. **WRITE ABOUT A THEME**
   Feedback Regulation In a short essay (about 100–150 words) focusing on feedback control in thermoregulation, explain why shivering is likely during the onset of a fever.

For selected answers, see Appendix A.

41

Animal Nutrition

▲ **Figure 41.1 How does a lean fish help a bear make fat?**

KEY CONCEPTS

41.1 An animal's diet must supply chemical energy, organic molecules, and essential nutrients

41.2 The main stages of food processing are ingestion, digestion, absorption, and elimination

41.3 Organs specialized for sequential stages of food processing form the mammalian digestive system

41.4 Evolutionary adaptations of vertebrate digestive systems correlate with diet

41.5 Feedback circuits regulate digestion, energy storage, and appetite

OVERVIEW

The Need to Feed

Dinnertime has arrived for the Kodiak bear in **Figure 41.1** (and for the salmon, though in quite a different sense). The skin, muscles, and other parts of the fish will be chewed into pieces, broken down by acid and enzymes in the bear's digestive system, and finally absorbed as small molecules into the body of the bear. Such a process is what is meant by animal **nutrition**: food being taken in, taken apart, and taken up.

Although a diet of fish plucked from a waterfall is not common, all animals eat other organisms—dead or alive, piecemeal or whole. Unlike plants, animals must consume food for both energy and the organic molecules used to assemble new molecules, cells, and tissues. Despite this shared need, animals have diverse diets. **Herbivores**, such as cattle, sea slugs, and termites, dine mainly on plants or algae. **Carnivores**, such as sharks, hawks, and spiders, mostly eat other animals. Bears and other **omnivores** (from the Latin *omni*, all) don't in fact eat everything, but they do regularly consume animals as well as plants or algae. We humans are typically omnivores, as are cockroaches and crows.

The terms *herbivore*, *carnivore*, and *omnivore* represent the kinds of food an animal usually eats. Keep in mind, however, that most animals are opportunistic feeders, eating foods outside their standard diet when their usual foods aren't available. For example, deer are herbivores, but in addition to feeding on grass and other plants, they occasionally eat insects, worms, or bird eggs. Note as well that microorganisms are an unavoidable "supplement" in every animal's diet.

Animals must eat. But to survive and reproduce, they must also balance their consumption, storage, and use of food. Bears, for example, store energy, largely in the form of body fat, in preparation for winter sleep. Eating too little food, too much food, or the wrong mixture of foods can endanger an animal's health. In this chapter, we will survey the nutritional requirements of animals, explore some of the diverse evolutionary adaptations for obtaining and processing food, and investigate the regulation of energy intake and expenditure.

CONCEPT **41.1**

An animal's diet must supply chemical energy, organic molecules, and essential nutrients

Overall, an adequate diet must satisfy three nutritional needs: chemical energy for cellular processes, organic building blocks for macromolecules, and essential nutrients.

The activities of cells, tissues, organs, and whole animals depend on sources of chemical energy in the diet. This energy is used to produce ATP, which powers processes ranging from DNA replication and cell division to vision and flight. To meet the continuous requirement for ATP, animals ingest and digest nutrients, including carbohydrates, proteins, and lipids, for use in cellular respiration and energy storage.

In addition to providing fuel for ATP production, an animal's diet must supply the raw materials needed for biosynthesis. To build the complex molecules it needs to grow, maintain itself, and reproduce, an animal must obtain two types of organic precursors from its food. Animals need a source of organic carbon (such as sugar) and a source of organic nitrogen (such as protein). Starting with these materials, animals can construct a great variety of organic molecules.

The materials that an animal's cells require but cannot synthesize are called **essential nutrients**. Obtained from dietary sources, these nutrients include certain minerals and preassembled organic molecules. Some nutrients are essential for all animals, whereas others are needed only by certain species. For instance, ascorbic acid (vitamin C) is an essential nutrient for humans and other primates, guinea pigs, and some birds and snakes, but not for most other animals.

▲ **Figure 41.2 Storing protein for growth.** Penguins, such as this Adélie from Antarctica, must make an abundance of new proteins when they molt and grow new feathers. Because of the temporary loss of their insulating coat of feathers, penguins cannot swim—or feed—when molting. How, then, do they obtain amino acids for production of feather protein? Before molting, a penguin greatly increases its muscle mass. The penguin then breaks down the extra muscle protein, which supplies the amino acids for growing new feathers.

MAKE CONNECTIONS *Taking into account the examples in Figure 5.15, what generalization can you make about the circumstances under which animals commonly use proteins for amino acid storage?*

Essential Nutrients

There are four classes of essential nutrients: essential amino acids, essential fatty acids, vitamins, and minerals.

Essential Amino Acids

Animals require 20 amino acids to make proteins (see Figure 5.16). The majority of animal species have the enzymes to synthesize about half of these amino acids, as long as their diet includes sulfur and organic nitrogen. The remaining amino acids must be obtained from food in prefabricated form and are therefore called **essential amino acids**. Most animals, including adult humans, require eight amino acids in their diet (infants also need a ninth, histidine).

The proteins in animal products such as meat, eggs, and cheese are "complete," which means that they provide all the essential amino acids in their proper proportions. In contrast, most plant proteins are "incomplete," being deficient in one or more essential amino acids. Corn (maize), for example, is deficient in tryptophan and lysine, whereas beans are lacking in methionine. However, vegetarians can easily obtain all of the essential amino acids by eating a varied diet of plant proteins.

Some animals have adaptations that help them through periods when their bodies demand extraordinary amounts of protein. In penguins, for example, muscle protein provides a source of amino acids for making new proteins when feathers are replaced after molting **(Figure 41.2)**.

Essential Fatty Acids

Animals produce the enzymes to synthesize most, but not all, of the fatty acids they need. The **essential fatty acids**, the ones they cannot make, are certain fatty acids that contain one or more double bonds and are thus unsaturated (see Figure 5.11). For example, humans require linoleic acid to make some membrane phospholipids. Because seeds, grains, and vegetables in the diets of humans and other animals generally furnish ample quantities of essential fatty acids, deficiencies in this class of nutrients are rare.

Vitamins

As Nobel Prize winner Albert Szent-Gyorgyi pointed out, "A vitamin is a substance that makes you ill if you *don't* eat it." **Vitamins** are organic molecules that have diverse functions and are required in the diet in very small amounts. Vitamin B_2, for example, is converted in the body to FAD, a coenzyme used in many metabolic processes, including cellular respiration (see Figure 9.12). For humans, 13 vitamins have been identified. Depending on the vitamin, the required amount ranges from about 0.01 to 100 mg per day.

Vitamins are classified as water-soluble or fat-soluble **(Table 41.1)**. The water-soluble vitamins include the B vitamins, which are compounds that generally function as coenzymes, and vitamin C, which is required for the production of connective tissue. Among the fat-soluble vitamins are vitamin A, which is incorporated into visual pigments of the eye, and vitamin K, which functions in blood clotting. Another is vitamin D, which aids in calcium absorption and bone formation.

Table 41.1 Vitamin Requirements of Humans

Vitamin	Major Dietary Sources	Major Functions in the Body	Symptoms of Deficiency
Water-Soluble Vitamins			
B_1 (thiamine)	Pork, legumes, peanuts, whole grains	Coenzyme used in removing CO_2 from organic compounds	Beriberi (tingling, poor coordination, reduced heart function)
B_2 (riboflavin)	Dairy products, meats, enriched grains, vegetables	Component of coenzymes FAD and FMN	Skin lesions, such as cracks at corners of mouth
B_3 (niacin)	Nuts, meats, grains	Component of coenzymes NAD^+ and $NADP^+$	Skin and gastrointestinal lesions, delusions, confusion
B_5 (pantothenic acid)	Meats, dairy products, whole grains, fruits, vegetables	Component of coenzyme A	Fatigue, numbness, tingling of hands and feet
B_6 (pyridoxine)	Meats, vegetables, whole grains	Coenzyme used in amino acid metabolism	Irritability, convulsions, muscular twitching, anemia
B_7 (biotin)	Legumes, other vegetables, meats	Coenzyme in synthesis of fat, glycogen, and amino acids	Scaly skin inflammation, neuromuscular disorders
B_9 (folic acid)	Green vegetables, oranges, nuts, legumes, whole grains	Coenzyme in nucleic acid and amino acid metabolism	Anemia, birth defects
B_{12} (cobalamin)	Meats, eggs, dairy products	Production of nucleic acids and red blood cells	Anemia, numbness, loss of balance
C (ascorbic acid)	Citrus fruits, broccoli, tomatoes	Used in collagen synthesis; antioxidant	Scurvy (degeneration of skin and teeth), delayed wound healing
Fat-Soluble Vitamins			
A (retinol)	Dark green and orange vegetables and fruits, dairy products	Component of visual pigments; maintenance of epithelial tissues	Blindness, skin disorders, impaired immunity
D	Dairy products, egg yolk	Aids in absorption and use of calcium and phosphorus	Rickets (bone deformities) in children, bone softening in adults
E (tocopherol)	Vegetable oils, nuts, seeds	Antioxidant; helps prevent damage to cell membranes	Nervous system degeneration
K (phylloquinone)	Green vegetables, tea; also made by colon bacteria	Important in blood clotting	Defective blood clotting

Our dietary requirement for vitamin D is variable because we synthesize this vitamin from other molecules when the skin is exposed to sunlight.

For people with poorly balanced diets, taking vitamin supplements that provide recommended daily levels is certainly reasonable. It is much less clear whether massive doses of vitamins confer any health benefits or are, in fact, safe. Moderate overdoses of water-soluble vitamins are probably harmless because excesses of these vitamins are excreted in urine. However, excesses of fat-soluble vitamins are deposited in body fat, so overconsumption may result in accumulating toxic levels of these compounds.

Minerals

Dietary **minerals** are inorganic nutrients, such as iron and sulfur, that are usually required in small amounts—from less than 1 mg to about 2,500 mg per day. As shown in **Table 41.2** on the next page, minerals have diverse functions in animal physiology. Some are cofactors built into the structure of enzymes; magnesium, for example, is present in enzymes that split ATP. In contrast, sodium, potassium, and chloride are important in the functioning of nerves and in maintaining osmotic balance between cells and the surrounding body fluid. Vertebrates use one mineral—iodine—specifically to make thyroid hormones, which regulate metabolic rate. Vertebrates also require relatively large quantities of calcium and phosphorus for building and maintaining bone.

Ingesting large amounts of some minerals can upset homeostatic balance and impair health. For example, excess salt (sodium chloride) intake can contribute to high blood pressure. This is a particular problem in the United States, where the typical person consumes enough salt to provide about 20 times the required amount of sodium. Packaged (prepared) foods often contain large amounts of sodium chloride, even if they do not taste very salty. Excessive consumption of iron can also endanger health: Liver damage due to iron overload affects as much as 10% of the human population in some regions of Africa where the water supply is especially iron-rich.

Table 41.2 Mineral Requirements of Humans*

	Mineral	Major Dietary Sources	Major Functions in the Body	Symptoms of Deficiency
Greater than 200 mg per day required	Calcium (Ca)	Dairy products, dark green vegetables, legumes	Bone and tooth formation, blood clotting, nerve and muscle function	Impaired growth, loss of bone mass
	Phosphorus (P)	Dairy products, meats, grains	Bone and tooth formation, acid-base balance, nucleotide synthesis	Weakness, loss of minerals from bone, calcium loss
	Sulfur (S)	Proteins from many sources	Component of certain amino acids	Impaired growth, fatigue, swelling
	Potassium (K)	Meats, dairy products, many fruits and vegetables, grains	Acid-base balance, water balance, nerve function	Muscular weakness, paralysis, nausea, heart failure
	Chlorine (Cl)	Table salt	Acid-base balance, formation of gastric juice, nerve function, osmotic balance	Muscle cramps, reduced appetite
	Sodium (Na)	Table salt	Acid-base balance, water balance, nerve function	Muscle cramps, reduced appetite
	Magnesium (Mg)	Whole grains, green leafy vegetables	Enzyme cofactor; ATP bioenergetics	Nervous system disturbances
	Iron (Fe)	Meats, eggs, legumes, whole grains, green leafy vegetables	Component of hemoglobin and of electron carriers; enzyme cofactor	Iron-deficiency anemia, weakness, impaired immunity
	Fluorine (F)	Drinking water, tea, seafood	Maintenance of tooth structure	Higher frequency of tooth decay
	Iodine (I)	Seafood, iodized salt	Component of thyroid hormones	Goiter (enlarged thyroid gland)

*Additional minerals required in trace amounts are chromium (Cr), cobalt (Co), copper (Cu), manganese (Mn), molybdenum (Mo), selenium (Se), and zinc (Zn). All of these minerals, as well as those in the table, are harmful when consumed in excess.

Dietary Deficiencies

A diet that lacks one or more essential nutrients or consistently supplies less chemical energy than the body requires results in *malnutrition,* a failure to obtain adequate nutrition. Malnutrition resulting from either type of dietary deficiency can have negative impacts on health and survival.

Deficiencies in Essential Nutrients

Insufficient intake of essential nutrients can cause deformities, disease, and even death. For example, cattle, deer, and other herbivores may develop dangerously fragile bones if they graze on plants growing in soil that lacks phosphorus. Some grazing animals obtain otherwise missing nutrients by consuming concentrated sources of salt or other minerals **(Figure 41.3)**. Among carnivores, spiders have been found to adjust for dietary deficiencies by switching to prey that restores nutritional balance.

▲ **Figure 41.3 Obtaining essential nutrients.** A juvenile chamois (*Rupicapra rupicapra*), an herbivore, licks exposed salts and minerals in its rocky alpine habitat. This behavior is common among herbivores living where soils and plants provide insufficient amounts of essential nutrients, such as sodium, calcium, phosphorus, and iron.

Like other animals, humans sometimes suffer from diets lacking in essential nutrients. A diet that provides insufficient amounts of one or more essential amino acids causes protein deficiency, the most common type of malnutrition among humans. For example, protein deficiency may arise if a child's diet shifts from consisting of breast milk to consisting solely of foods that provide almost all of their calories in the form of starch and other carbohydrates. Such children, if they survive infancy, often have impaired physical and mental development.

Among populations subsisting on simple rice diets, individuals are often afflicted with vitamin A deficiency, which can cause blindness or death. To overcome this problem, scientists have engineered a strain of rice to synthesize beta-carotene, the orange-colored pigment that is abundant in carrots. Once absorbed into the body, beta-carotene is converted to vitamin A.

The potential benefit of this "Golden Rice" (see Chapter 38) is enormous because 1–2 million young children worldwide die every year from vitamin A deficiency.

Undernutrition

A diet that fails to provide adequate sources of chemical energy results in *undernutrition*. When an animal is undernourished, a series of events unfold: The body uses up stored carbohydrates and fat and then begins breaking down its own proteins for fuel; muscles begin to decrease in size; and the brain may become protein-deficient. If energy intake remains less than energy expenditures, the animal will eventually die. Even if a seriously undernourished animal survives, some of the damage may be irreversible.

Human undernutrition is most common when drought, war, or another crisis severely disrupts the food supply. In sub-Saharan Africa, where the AIDS epidemic has crippled both rural and urban communities, approximately 200 million children and adults cannot obtain enough food.

Sometimes undernutrition occurs within well-fed human populations as a result of eating disorders. For example, anorexia nervosa leads individuals, usually female, to starve themselves compulsively.

Assessing Nutritional Needs

Determining the ideal diet for the human population is an important but difficult problem for scientists. As objects of study, people present many challenges. Unlike laboratory animals, humans are genetically diverse. They also live in settings far more varied than the stable and uniform environment that scientists use to facilitate comparisons in laboratory experiments. Ethical concerns present an additional barrier. For example, it is not acceptable to investigate the nutritional needs of children in a way that might harm a child's growth or development.

The methods used to study human nutrition have changed dramatically over time. To avoid harming others, several of the researchers who discovered vitamins a century ago used themselves as subject animals. Today, researchers typically rely on the study of genetic defects that disrupt food uptake, storage, or use. For example, a genetic disorder called hemochromatosis causes iron buildup in the absence of any abnormal iron consumption or exposure. Fortunately, this common disorder is remarkably easy to treat: Drawing blood regularly removes enough iron from the body to restore homeostasis. By studying the defective genes that can cause the disease, scientists have learned a great deal about the regulation of iron absorption.

Many insights into human nutrition have come from *epidemiology*, the study of human health and disease at the population level. In the 1970s, for instance, researchers discovered that children born to women of low socioeconomic status were more likely to have neural tube defects, which occur when tissue fails to enclose the developing brain and spinal cord (see Chapter 47). The English scientist Richard Smithells thought that malnutrition among these women might be responsible. As described in **Figure 41.4**, he found that vitamin supplementation greatly reduced the risk of neural tube defects. In other studies, he obtained evidence that folic acid (vitamin B_9) was the specific vitamin responsible, a finding confirmed by other researchers. Based on this evidence, the United States in 1998 began to require that folic acid be added to enriched grain products used to make bread, cereals, and other foods. Follow-up studies have documented the effectiveness of this program in reducing the frequency of neural tube defects. Thus, at a time when microsurgery and sophisticated diagnostic imaging dominate the headlines, a simple dietary change such as folic acid supplementation or consumption of Golden Rice may be among the greatest contributors to human health.

▼ **Figure 41.4** INQUIRY

Can diet influence the frequency of birth defects?

EXPERIMENT Richard Smithells, of the University of Leeds, in England, examined the effect of vitamin supplementation on the risk of neural tube defects. Women who had had one or more babies with such a defect were put into two study groups. The experimental group consisted of those who were planning a pregnancy and began taking a multivitamin at least four weeks before attempting conception. The control group, who were not given vitamins, included women who declined them and women who were already pregnant. The numbers of neural tube defects resulting from the pregnancies were recorded for each group.

RESULTS

Group	Number of infants/fetuses studied	Infants/fetuses with a neural tube defect
Vitamin supplements (experimental group)	141	1 (0.7%)
No vitamin supplements (control group)	204	12 (5.9%)

CONCLUSION This study provided evidence that vitamin supplementation protects against neural tube defects, at least after the first pregnancy. Follow-up trials demonstrated that folic acid alone provided an equivalent protective effect.

SOURCE R. W. Smithells et al., Possible prevention of neural-tube defects by periconceptional vitamin supplementation, *Lancet* 315: 339–340 (1980).

INQUIRY IN ACTION Read and analyze the original paper in *Inquiry in Action: Interpreting Scientific Papers*.

WHAT IF? Subsequent studies were designed to learn if folic acid supplements prevent neural tube defects during first-time pregnancies. To determine the required number of subjects, what type of additional information did the researchers need?

CONCEPT CHECK 41.1

1. All 20 amino acids are needed to make animal proteins. Why aren't they all essential to animal diets?
2. **MAKE CONNECTIONS** Review the discussion of enzymes in metabolic reactions in Concept 8.4 (pp. 152–156). Then explain why vitamins are required in very small amounts in the diet.
3. **WHAT IF?** If a zoo animal eating ample food shows signs of malnutrition, how might a researcher determine which nutrient is lacking in its diet?

For suggested answers, see Appendix A.

CONCEPT 41.2

The main stages of food processing are ingestion, digestion, absorption, and elimination

In this section, we turn from nutritional requirements to the mechanisms by which animals process food. Food processing can be divided into four distinct stages: ingestion, digestion, absorption, and elimination **(Figure 41.5)**. The first stage, **ingestion**, is the act of eating or feeding. **Figure 41.6** surveys and classifies the principal feeding mechanisms that have evolved in animals. Given the variation in food sources among animal species, it is not surprising that strategies for extracting resources from food also differ widely. We will focus, however, on the shared processes, pausing periodically to consider some adaptations to particular diets or environments.

In **digestion**, the second stage of food processing, food is broken down into molecules small enough for the body to absorb. Mechanical digestion, such as chewing, typically precedes chemical digestion. Mechanical digestion breaks food into smaller pieces, increasing the surface area available for chemical processes. Chemical digestion is necessary because animals cannot directly use the proteins, carbohydrates, nucleic acids, fats, and phospholipids in food. One problem is that these molecules are too large to pass through membranes and enter the cells of the animal. In addition, the large molecules in food are not all identical to those the animal needs for its particular tissues and functions. When large molecules in food are broken down into their components, however, the animal can use these smaller molecules to assemble the large molecules it needs. For example, although fruit flies and humans have very different diets, both convert proteins in their food to the same 20 amino acids from which they assemble all of the proteins specific for their species.

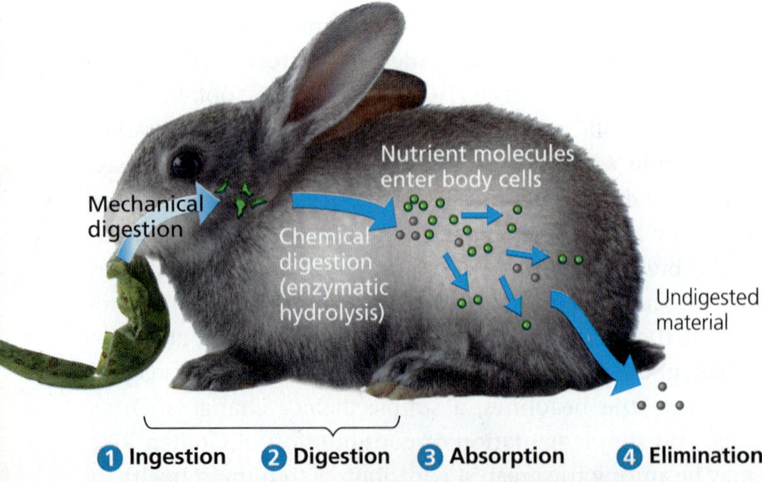

▲ **Figure 41.5 The four stages of food processing.**

Recall from Chapter 5 that a cell makes a macromolecule or fat by linking together smaller components; it does so by removing a molecule of water for each new covalent bond formed. Chemical digestion by enzymes reverses this process by breaking bonds with the addition of water (see Figure 5.2). This splitting process is called *enzymatic hydrolysis*. A variety of enzymes catalyze the digestion of large molecules in food. Polysaccharides and disaccharides are split into simple sugars; proteins are broken down into amino acids; and nucleic acids are cleaved into nucleotides and their components. Enzymatic hydrolysis also releases fatty acids and other components from fats and phospholipids.

The last two stages of food processing occur after the food is digested. In the third stage, **absorption**, the animal's cells take up (absorb) small molecules such as amino acids and simple sugars. **Elimination** completes the process as undigested material passes out of the digestive system.

Digestive Compartments

In our overview of food processing, we have seen that digestive enzymes hydrolyze the same biological materials (such as proteins, fats, and carbohydrates) that make up the bodies of the animals themselves. How, then, are animals able to digest food without digesting their own cells and tissues? The evolutionary adaptation found across a wide range of animal species is the processing of food within specialized compartments. Such compartments can be intracellular, in the form of food vacuoles, or extracellular, as in digestive organs and systems.

Intracellular Digestion

Food vacuoles—cellular organelles in which hydrolytic enzymes break down food—are the simplest digestive compartments. The hydrolysis of food inside vacuoles, called *intracellular digestion*, begins after a cell engulfs solid food by phagocytosis or liquid food by pinocytosis (see Figure 7.22). Newly formed food vacuoles fuse with lysosomes, organelles containing hydrolytic enzymes. This fusion of organelles brings food in contact with the enzymes, allowing digestion to occur safely within a compartment

▼ Figure 41.6

Exploring Four Main Feeding Mechanisms of Animals

Suspension Feeders and Filter Feeders

Many aquatic animals are **suspension feeders**, which eat small organisms or food particles suspended in the water. For example, clams and oysters feed on tiny morsels of food in the water that passes over their gills; cilia sweep the food particles to the animal's mouth in a film of mucus. **Filter feeders** such as the humpback whale shown above move water through a filtering structure to obtain food. Attached to the whale's upper jaw are comblike plates called baleen, which strain small invertebrates and fish from enormous volumes of water.

Substrate Feeders

Substrate feeders are animals that live in or on their food source. This leaf miner caterpillar, the larva of a moth, is eating through the soft tissue of an oak leaf, leaving a dark trail of feces in its wake. Some other substrate feeders include maggots (fly larvae), which burrow into animal carcasses.

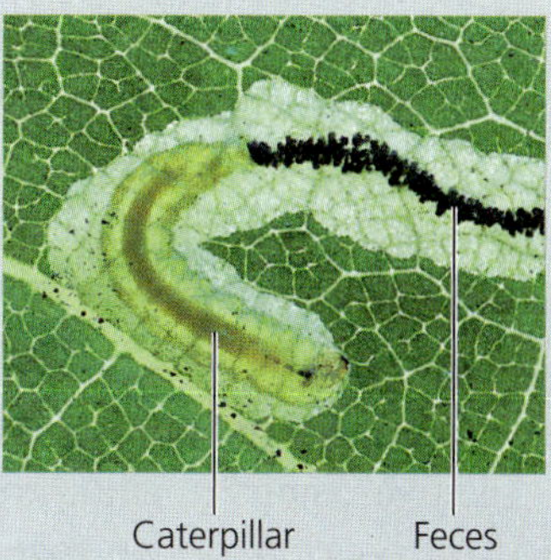

Fluid Feeders

Fluid feeders suck nutrient-rich fluid from a living host. This mosquito has pierced the skin of its human host with hollow, needlelike mouthparts and is consuming a blood meal (colorized SEM). Similarly, aphids are fluid feeders that tap the phloem sap of plants. In contrast to such parasites, some fluid feeders actually benefit their hosts. For example, hummingbirds and bees move pollen between flowers as they fluid-feed on nectar.

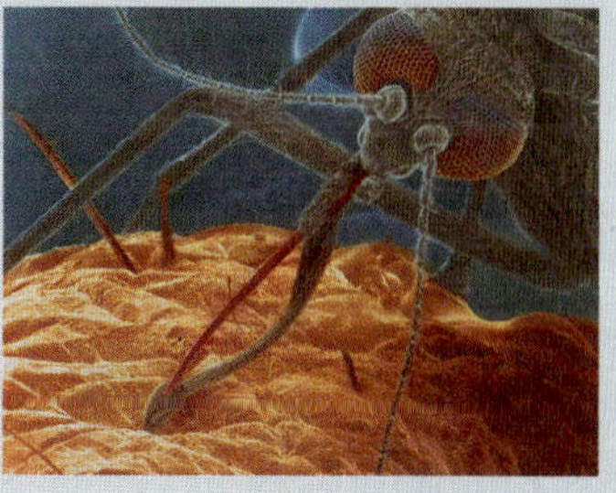

Bulk Feeders

Most animals, including humans, are **bulk feeders**, which eat relatively large pieces of food. Their adaptations include tentacles, pincers, claws, poisonous fangs, jaws, and teeth that kill their prey or tear off pieces of meat or vegetation. In this amazing scene, a rock python is beginning to ingest a gazelle it has captured and killed. Snakes cannot chew their food into pieces and must swallow it whole—even if the prey is much bigger than the diameter of the snake. They can do so because the lower jaw is loosely hinged to the skull by an elastic ligament that permits the mouth and throat to open very wide. After swallowing its prey, which may take more than an hour, the python will spend two weeks or more digesting its meal.

enclosed by a protective membrane. A few animals, such as sponges, digest their food entirely by this intracellular mechanism (see Figure 33.4).

Extracellular Digestion

In most animal species, at least some hydrolysis occurs by *extracellular digestion*, the breakdown of food in compartments that are continuous with the outside of the animal's body. Having one or more extracellular compartments for digestion enables an animal to devour much larger pieces of food than can be ingested by phagocytosis.

Many animals with relatively simple body plans have a digestive compartment with a single opening **(Figure 41.7)**. This pouch, called a **gastrovascular cavity**, functions in digestion as well as in the distribution of nutrients throughout the body (hence the *vascular* part of the term). The carnivorous cnidarians called hydras provide a good example of how a gastrovascular cavity works. A hydra uses its tentacles to stuff captured prey through its mouth into its gastrovascular cavity. Specialized gland cells of the hydra's gastrodermis, the tissue layer that lines the cavity, then secrete digestive enzymes that break the soft tissues of the prey into tiny pieces. Other cells of the gastrodermis engulf these food particles, and most of the hydrolysis of macromolecules occurs intracellularly, as in sponges. After a hydra has digested its meal, undigested materials that remain in the gastrovascular cavity, such as exoskeletons of small crustaceans, are eliminated through the same opening by which food entered. Many flatworms also have a gastrovascular cavity with a single opening (see Figure 33.10).

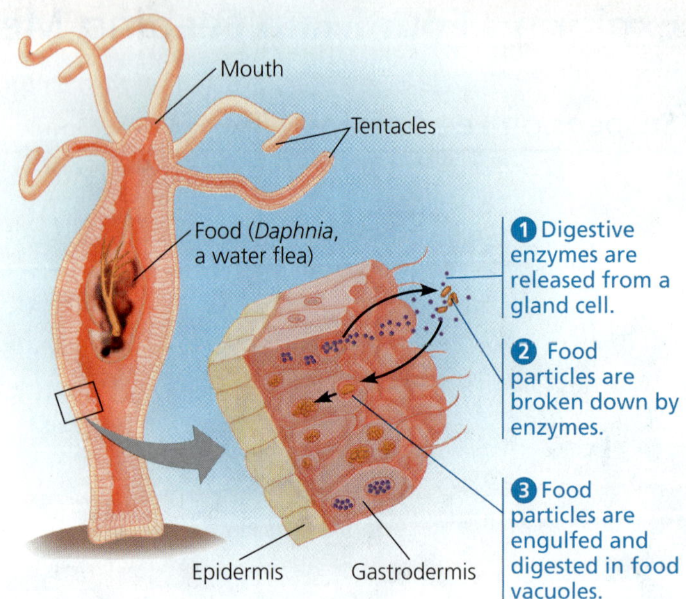

▲ **Figure 41.7 Digestion in a hydra.** Digestion begins in the gastrovascular cavity and is completed intracellularly after small food particles are engulfed by specialized cells of the gastrodermis.

In contrast with cnidarians and flatworms, most animals have a digestive tube extending between two openings, a mouth and an anus **(Figure 41.8)**. Such a tube is called a *complete digestive tract* or, more commonly, an **alimentary canal**. Because food moves along the alimentary canal in a single direction, the tube can be organized into specialized compartments that carry out digestion and nutrient absorption in a stepwise fashion. An animal with an alimentary canal can ingest food while earlier meals are still being digested, a feat that

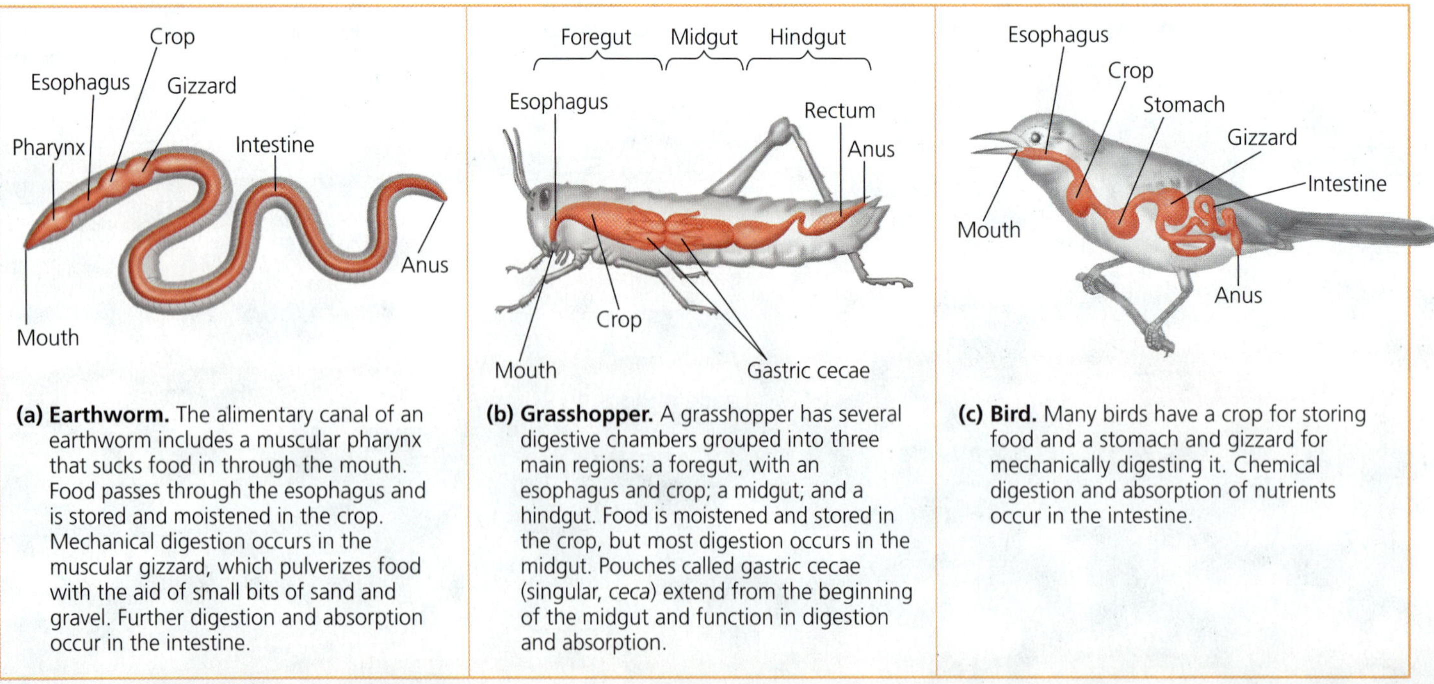

(a) Earthworm. The alimentary canal of an earthworm includes a muscular pharynx that sucks food in through the mouth. Food passes through the esophagus and is stored and moistened in the crop. Mechanical digestion occurs in the muscular gizzard, which pulverizes food with the aid of small bits of sand and gravel. Further digestion and absorption occur in the intestine.

(b) Grasshopper. A grasshopper has several digestive chambers grouped into three main regions: a foregut, with an esophagus and crop; a midgut; and a hindgut. Food is moistened and stored in the crop, but most digestion occurs in the midgut. Pouches called gastric cecae (singular, *ceca*) extend from the beginning of the midgut and function in digestion and absorption.

(c) Bird. Many birds have a crop for storing food and a stomach and gizzard for mechanically digesting it. Chemical digestion and absorption of nutrients occur in the intestine.

▲ **Figure 41.8 Variation in alimentary canals.**

is likely to be difficult or inefficient for animals with gastrovascular cavities. In the next section, we'll explore the spatial and functional organization of an alimentary canal.

CONCEPT CHECK 41.2

1. Distinguish the overall structure of a gastrovascular cavity from that of an alimentary canal.
2. In what sense are nutrients from a recently ingested meal not really "inside" your body prior to the absorption stage of food processing?
3. **WHAT IF?** Thinking in broad terms, what similarities can you identify between digestion in an animal body and the breakdown of gasoline in an automobile? (You don't have to know about auto mechanics.)

For suggested answers, see Appendix A.

CONCEPT 41.3

Organs specialized for sequential stages of food processing form the mammalian digestive system

Because most animals, including mammals, have an alimentary canal, we can use the mammalian digestive system as a representative example of the general principles of food processing. In mammals, the digestive system consists of the alimentary canal and various accessory glands that secrete digestive juices through ducts into the canal **(Figure 41.9)**. The accessory glands of the mammalian digestive system are three pairs of salivary glands, the pancreas, the liver, and the gallbladder.

Food is pushed along the alimentary canal by **peristalsis**, alternating waves of contraction and relaxation in the smooth muscles lining the canal. At some of the junctions between specialized compartments, the muscular layer forms ringlike valves called **sphincters**. Acting like drawstrings to close off the alimentary canal, sphincters regulate the passage of material between compartments.

Using the human digestive system as a model, let's now follow a meal through the alimentary canal. As we do so, we'll examine in more detail what happens to the food in each digestive compartment along the way.

The Oral Cavity, Pharynx, and Esophagus

Ingestion and the initial steps of digestion occur in the mouth, or **oral cavity**. Mechanical digestion begins as teeth of various shapes cut, mash, and grind food, making the food easier to swallow and increasing its surface area. Meanwhile, the presence of food stimulates a nervous reflex that causes the **salivary glands** to deliver saliva through ducts to the oral cavity. Saliva may also be released before food enters the mouth, triggered by a learned association between eating and the time of day, a cooking odor, or another stimulus.

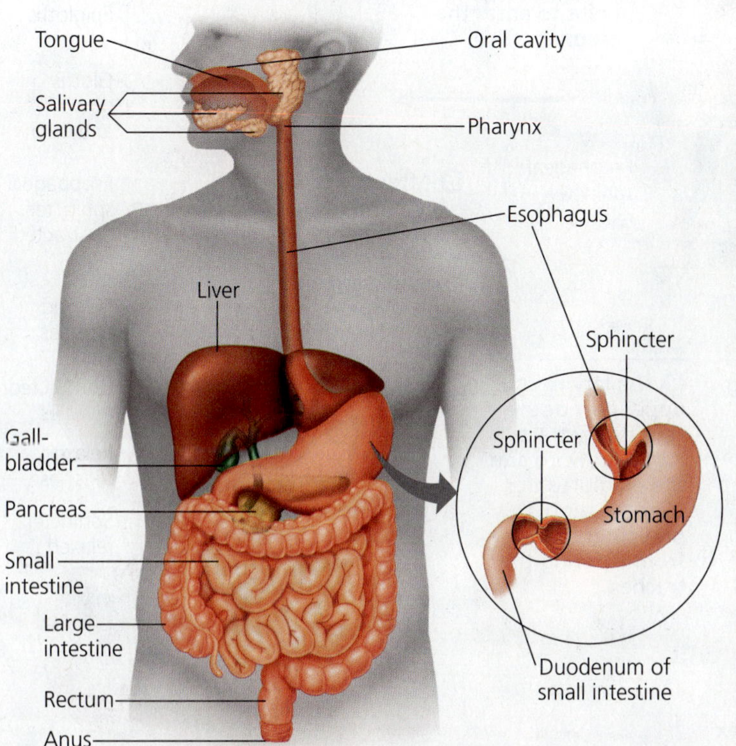

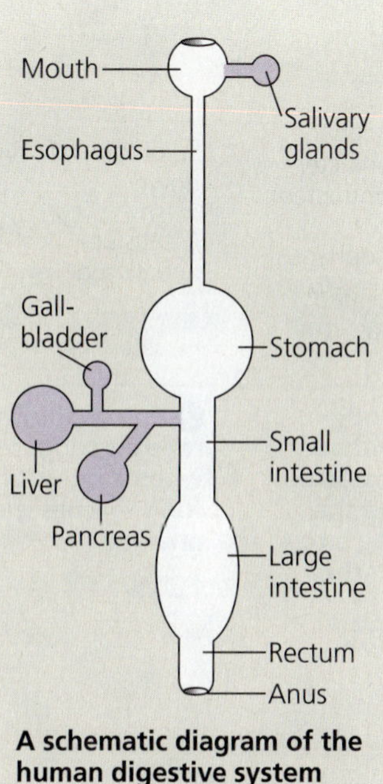

◄ **Figure 41.9 The human digestive system.** After food is chewed and swallowed, it takes 5–10 seconds for it to pass down the esophagus and into the stomach, where it spends 2–6 hours being partially digested. Final digestion and nutrient absorption occur in the small intestine over a period of 5–6 hours. In 12–24 hours, any undigested material passes through the large intestine, and feces are expelled through the anus.

Saliva initiates chemical digestion while also protecting the oral cavity. The enzyme **amylase**, found in saliva, hydrolyzes starch (a glucose polymer from plants) and glycogen (a glucose polymer from animals) into smaller polysaccharides and the disaccharide maltose. Much of the protective effect of saliva is provided by **mucus**, which is a viscous mixture of water, salts, cells, and slippery glycoproteins (carbohydrate-protein complexes) called mucins. Mucus in saliva protects the lining of the mouth from abrasion and lubricates food for easier swallowing. Additional components of saliva include buffers, which help prevent tooth decay by neutralizing acid, and antimicrobial agents (such as lysozyme; see Figure 5.18), which protect against bacteria that enter the mouth with food.

Much as a doorman screens and assists people entering a building, the tongue aids digestive processes by evaluating ingested material and then enabling its further passage. When food arrives at the oral cavity, the tongue plays a critical role in distinguishing which foods should be processed further. (See Chapter 50 for a discussion of the sense of taste.) After food is deemed acceptable and chewing commences, tongue movements manipulate the food, helping shape it into a ball called a **bolus**. During swallowing, the tongue provides further help, pushing the bolus to the back of the oral cavity and into the pharynx.

The **pharynx**, or throat region, opens to two passageways: the esophagus and the trachea (windpipe). The **esophagus** connects to the stomach, whereas the trachea leads to the lungs. Swallowing must therefore be carefully choreographed to keep food from entering and blocking the airway. When you swallow, a flap of cartilage called the *epiglottis* covers the *glottis*—the vocal cords and the opening between them. Guided by the movements of the *larynx*, the upper part of the respiratory tract, this swallowing reflex directs each bolus into the entrance of the esophagus (**Figure 41.10**, **1–4**). If the swallowing reflex fails, food or liquids can reach the trachea and cause choking, a blockage of the trachea. The resulting lack of airflow into the lungs can be fatal if the material is not dislodged by vigorous coughing, a series of back slaps, or a forced upward thrust of the diaphragm (the Heimlich maneuver).

The esophagus contains both striated and smooth muscle (see Figure 40.5). The striated muscle is situated at the top of the esophagus and is active during swallowing. Throughout the rest of the esophagus, smooth muscle functions in peristalsis. The rhythmic cycles of contraction move each bolus to the stomach (see Figure 41.10, 6). As with other parts of the digestive system, the form of the esophagus fits its function and varies among species. For example, fishes have no lungs to bypass and therefore have a very short esophagus. And it will come as no surprise that giraffes have a very long esophagus.

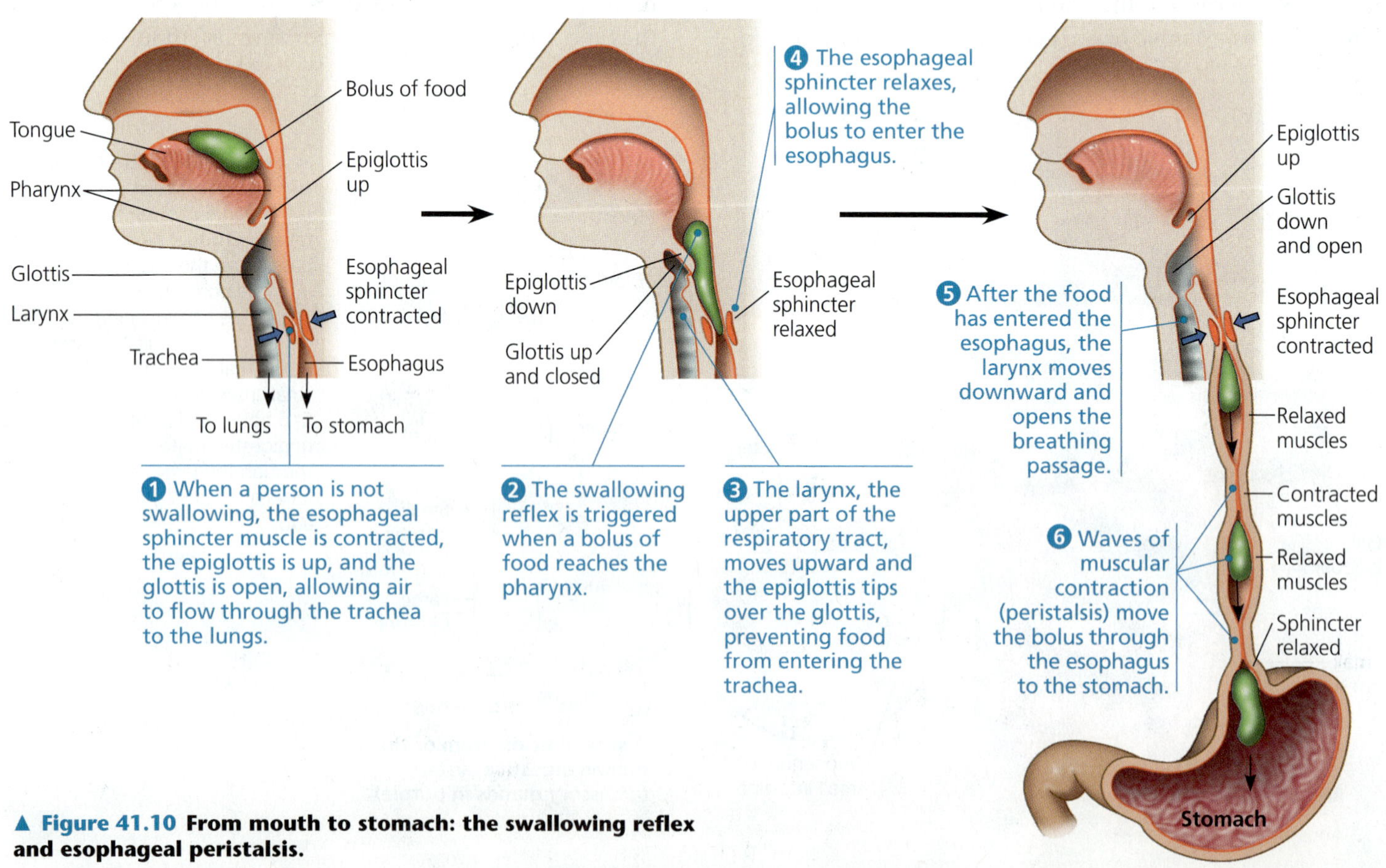

▲ **Figure 41.10 From mouth to stomach: the swallowing reflex and esophageal peristalsis.**

Digestion in the Stomach

The **stomach**, which is located just below the diaphragm, stores food and begins digestion of proteins. With accordion-like folds and a very elastic wall, this organ can stretch to accommodate about 2 L of food and fluid. The stomach secretes a digestive fluid called **gastric juice** and mixes this secretion with the food through a churning action. This mixture of ingested food and digestive juice is called **chyme**.

Chemical Digestion in the Stomach

Two components of gastric juice carry out chemical digestion. One is hydrochloric acid (HCl), which disrupts the extracellular matrix that binds cells together in meat and plant material. The concentration of HCl is so high that the pH of gastric juice is about 2, acidic enough to dissolve iron nails (and to kill most bacteria). This low pH denatures (unfolds) proteins in food, increasing exposure of their peptide bonds. The exposed bonds are attacked by the second component of gastric juice—a **protease**, or protein-digesting enzyme, called **pepsin**. Unlike most enzymes, pepsin works best in a strongly acidic environment. By breaking peptide bonds, it cleaves proteins into smaller polypeptides. Further digestion to individual amino acids occurs in the small intestine.

Why doesn't gastric juice destroy the stomach cells that make it? The answer is that the ingredients of gastric juice are kept inactive until they are released into the lumen (cavity) of the stomach. The components of gastric juice are produced by cells in the gastric glands of the stomach **(Figure 41.11)**. *Parietal cells* secrete hydrogen and chloride ions, which form HCl. Using an ATP-driven pump, the parietal cells expel hydrogen ions into the lumen. There, the hydrogen ions combine with chloride ions that diffuse into the lumen through specific membrane channels of the parietal cells. Meanwhile, *chief cells* release pepsin into the lumen in an inactive form called **pepsinogen**. HCl converts pepsinogen to active pepsin by clipping off a small portion of the molecule and exposing its active site. Through these processes, both HCl and pepsin form in the lumen of the stomach, not within the cells of the gastric glands.

After hydrochloric acid converts a small amount of pepsinogen to pepsin, pepsin itself helps activate the remaining pepsinogen. Pepsin, like HCl, can clip pepsinogen to expose the enzyme's active site. This generates more pepsin,

▶ **Figure 41.11 The stomach and its secretions.** The micrograph (colorized SEM) shows a gastric pit on the interior surface of the stomach, through which digestive juices are secreted.

10 µm

Esophagus

Sphincter

Stomach

Sphincter

Small intestine

Folds of epithelial tissue

Epithelium

Interior surface of stomach. The interior surface of the stomach wall is highly folded and dotted with pits leading into tubular gastric glands.

Gastric gland. The gastric glands have three types of cells that secrete different components of the gastric juice: mucous cells, chief cells, and parietal cells.

Mucous cells secrete mucus, which lubricates and protects the cells lining the stomach.

Chief cells secrete pepsinogen, an inactive form of the digestive enzyme pepsin.

Parietal cells secrete hydrochloric acid (HCl).

The production of gastric juice

1. Pepsinogen and HCl are secreted into the lumen of the stomach.
2. HCl converts pepsinogen to pepsin.
3. Pepsin then activates more pepsinogen, starting a chain reaction. Pepsin begins the chemical digestion of proteins.

Pepsinogen → Pepsin (active enzyme)

HCl

H^+

Cl^-

Chief cell

Parietal cell

which activates more pepsinogen, forming more active enzyme. This series of events is an example of positive feedback, which amplifies the effect of an initially small input.

When HCl and pepsin form within the stomach lumen, why aren't the cells that line the stomach damaged? Actually, these cells are vulnerable to gastric juice as well as to acid-tolerant pathogens in food or water. However, the stomach lining protects against self-digestion by secreting mucus. In addition, cell division adds a new epithelial layer every three days, replacing cells eroded by digestive juices. Despite these defenses, damaged areas of the stomach lining called gastric ulcers may appear. For decades, scientists thought they were caused by psychological stress and resulting excess acid secretion. In 1982, however, Australian researchers Barry Marshall and Robin Warren reported that infection by the acid-tolerant bacterium *Helicobacter pylori* causes ulcers. They also demonstrated that an antibiotic treatment could cure most gastric ulcers. For these findings, they were awarded the Nobel Prize in 2005.

Stomach Dynamics

Chemical digestion by gastric juice is facilitated by the churning action of the stomach. This coordinated series of muscle contractions and relaxations mixes the stomach contents about every 20 seconds. As a result of mixing and enzyme action, what begins as a recently swallowed meal becomes the acidic, nutrient-rich broth known as chyme. Most of the time, the stomach is closed off at both ends (see Figure 41.9). The sphincter between the esophagus and the stomach normally opens only when a bolus arrives. Occasionally, however, a person experiences acid reflux, a backflow of chyme from the stomach into the lower end of the esophagus. The resulting irritation of the esophagus is commonly called "heartburn."

The contents of the stomach typically pass into the small intestine within 2–6 hours after a meal. The sphincter located where the stomach opens to the small intestine helps regulate passage into the small intestine, allowing only one squirt of chyme at a time.

▼ Figure 41.12 Chemical digestion in the human digestive system.

? *Pepsin is resistant to the denaturing effect of the low pH environment of the stomach. Thinking about the different digestive processes that occur in the small intestine, what adaptation do you think the digestive enzymes in that compartment share?*

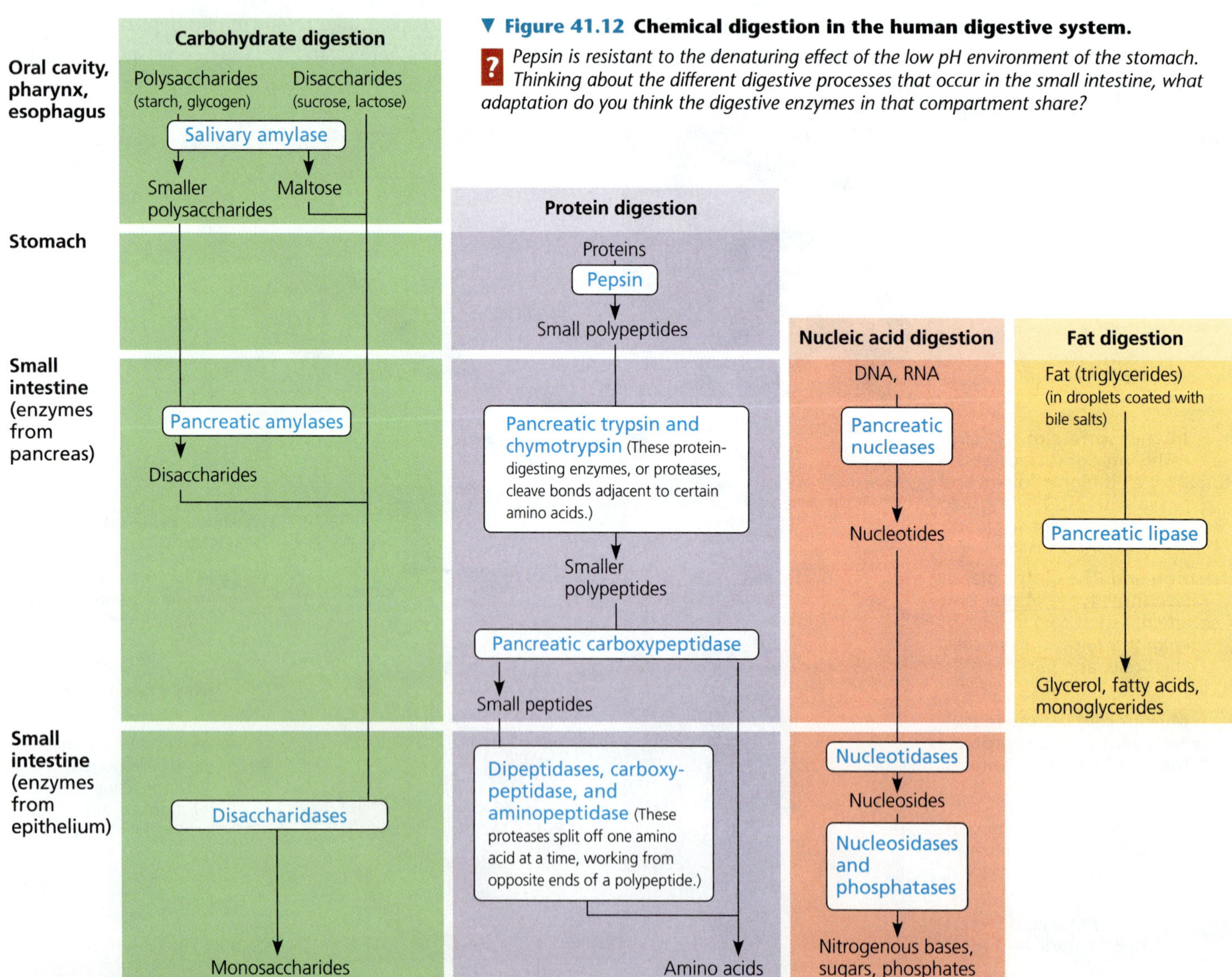

Digestion in the Small Intestine

Although chemical digestion of some nutrients begins in the oral cavity or stomach, most enzymatic hydrolysis of the macromolecules from food occurs in the **small intestine** (**Figure 41.12**, on p. 886). Over 6 m (20 feet) long in humans, the small intestine is the alimentary canal's longest compartment. Its name refers to its small diameter, compared with that of the large intestine. The first 25 cm (10 inches) or so of the small intestine forms the **duodenum**. It is here that chyme from the stomach mixes with digestive juices from the pancreas, liver, and gallbladder, as well as from gland cells of the intestinal wall itself. As you will see in Concept 41.5, hormones released by the stomach and duodenum control the digestive secretions into the alimentary canal.

Pancreatic Secretions

The **pancreas** aids chemical digestion by producing an alkaline solution rich in bicarbonate as well as several enzymes. The bicarbonate neutralizes the acidity of chyme and acts as a buffer. Among the pancreatic enzymes are trypsin and chymotrypsin, proteases secreted into the duodenum in inactive forms (see Figure 41.12). In a chain reaction similar to activation of pepsin, they are activated when safely located in the lumen within the duodenum.

Bile Production by the Liver

Digestion of fats and other lipids begins in the small intestine and relies on the production of **bile**, a mixture of substances that is made in the **liver**. Bile contains bile salts, which act as emulsifiers (detergents) that aid in digestion and absorption of lipids. Bile is stored and concentrated in the **gallbladder**.

Bile production is integral to one of the other vital functions of the liver: the destruction of red blood cells that are no longer fully functional. In producing bile, the liver incorporates some pigments that are by-products of red blood cell disassembly. These bile pigments are then eliminated from the body with the feces. In some liver or blood disorders, bile pigments accumulate in the skin, resulting in a characteristic yellowing called jaundice.

Secretions of the Small Intestine

The epithelial lining of the duodenum is the source of several digestive enzymes (see Figure 41.12). Some are secreted into the lumen of the duodenum, whereas others are bound to the surface of epithelial cells.

While enzymatic hydrolysis proceeds, peristalsis moves the mixture of chyme and digestive juices along the small intestine. Most digestion is completed in the duodenum. The remaining regions of the small intestine, called the *jejunum* and *ileum*, function mainly in the absorption of nutrients and water.

Absorption in the Small Intestine

To reach body tissues, nutrients in the lumen must first cross the lining of the alimentary canal. Most of this absorption occurs across the highly folded surface of the small intestine, as illustrated in **Figure 41.13**. Large folds in the lining encircle

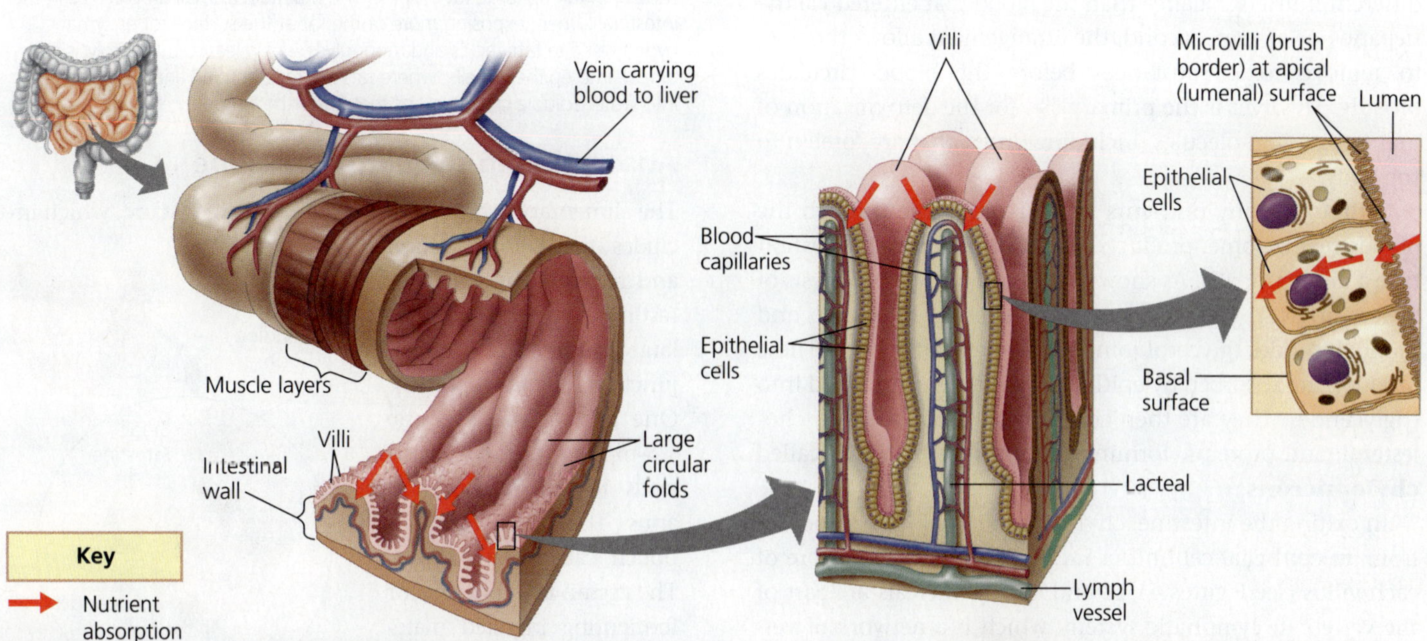

▲ **Figure 41.13 Nutrient absorption in the small intestine.**

? *Tapeworms sometimes infect humans, anchoring themselves to the wall of the small intestine. Based on how digestion is compartmentalized along the mammalian alimentary canal, what digestive functions would you expect these parasites to have?*

BioFlix Visit the Study Area at **www.masteringbiology.com** for the BioFlix® 3-D Animation on Membrane Transport.

the intestine and are studded with finger-like projections called **villi**. In turn, each epithelial cell of a villus has on its apical surface many microscopic projections, or **microvilli**, that are exposed to the intestinal lumen. The many side-by-side microvilli give cells of the intestinal epithelium a brush-like appearance—reflected in the name *brush border*. Together, the folds, villi, and microvilli of the small intestine have a surface area of 300 m^2, roughly the size of a tennis court. This enormous surface area is an evolutionary adaptation that greatly increases the rate of nutrient absorption.

Depending on the nutrient, transport across the epithelial cells can be passive or active (see Chapter 7). The sugar fructose, for example, moves by facilitated diffusion down its concentration gradient from the lumen of the small intestine into the epithelial cells. From there, fructose exits the basal surface and is absorbed into microscopic blood vessels, or capillaries, at the core of each villus. Other nutrients, including amino acids, small peptides, vitamins, and most glucose molecules, are pumped against concentration gradients by the epithelial cells of the villus. This active transport allows much more absorption of nutrients than would be possible with passive diffusion alone.

The capillaries and veins that carry nutrient-rich blood away from the villi all converge into the **hepatic portal vein**, a blood vessel that leads directly to the liver. From the liver, blood travels to the heart and then to other tissues and organs. This arrangement serves two major functions. First, it allows the liver to regulate the distribution of nutrients to the rest of the body. Because the liver can interconvert many organic molecules, blood that leaves the liver may have a very different nutrient balance than the blood that entered via the hepatic portal vein. Second, the arrangement allows the liver to remove toxic substances before the blood circulates broadly. The liver is the primary site for the detoxification of many organic molecules, including drugs, that are foreign to the body.

Although many nutrients leave the intestine through the bloodstream, some products of fat (triglyceride) digestion take a different path. As shown in **Figure 41.14**, hydrolysis of fats by lipase in the small intestine generates fatty acids and monoglycerides (glycerol joined to a single fatty acid). These products are absorbed by epithelial cells and recombined into triglycerides. They are then coated with phospholipids, cholesterol, and proteins, forming water-soluble globules called **chylomicrons**.

In exiting the intestine, chylomicrons are first transported from an epithelial cell into a **lacteal**, a vessel at the core of each villus (see Figures 41.13 and 41.14). Lacteals are part of the vertebrate lymphatic system, which is a network of vessels that are filled with a clear fluid called lymph. Starting at the lacteals, lymph containing the chylomicrons passes into the larger vessels of the lymphatic system and eventually into large veins that return the blood to the heart.

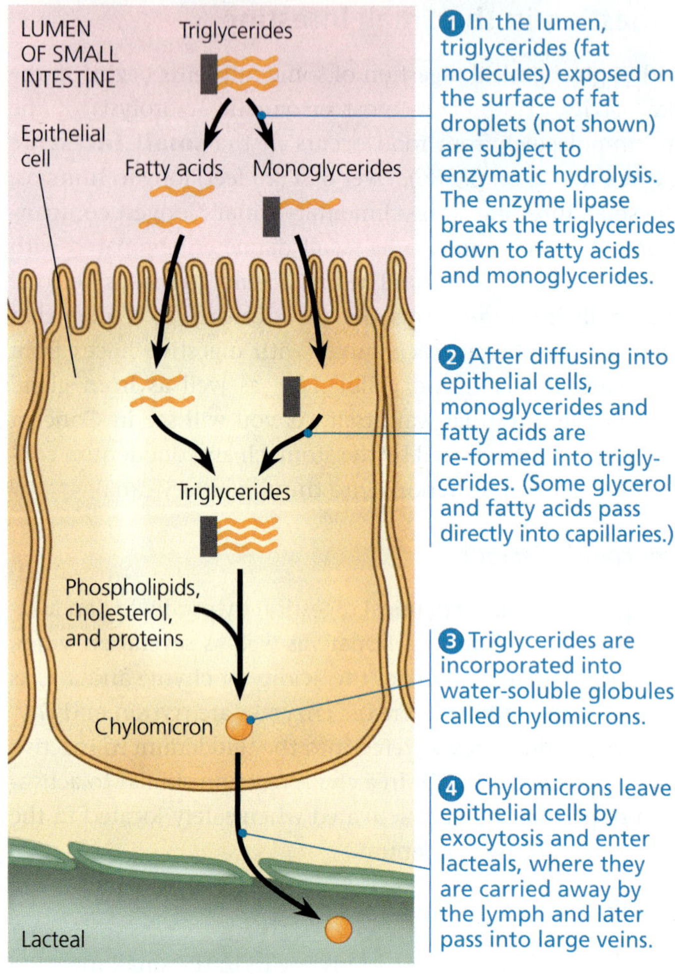

▲ **Figure 41.14 Absorption of fats.** Because fats are insoluble in water, adaptations are needed to digest and absorb them. Bile salts (not shown) break up large fat droplets and maintain a small droplet size in the intestinal lumen, exposing more of the fat at the surface for enzymatic hydrolysis. The fatty acids and monoglycerides released by hydrolysis can diffuse into epithelial cells, where fats are reassembled and incorporated into water-soluble chylomicrons that enter the lymphatic system.

Absorption in the Large Intestine

The alimentary canal ends with the **large intestine**, which includes the colon, cecum, and rectum. The small intestine connects to the large intestine at a T-shaped junction **(Figure 41.15)**. One arm of the T is the 1.5-m-long **colon**, which leads to the rectum and anus. The other arm is a pouch called the **cecum**. The cecum is important for fermenting ingested material, especially in animals that eat large amounts of plant material. Compared with many other mammals,

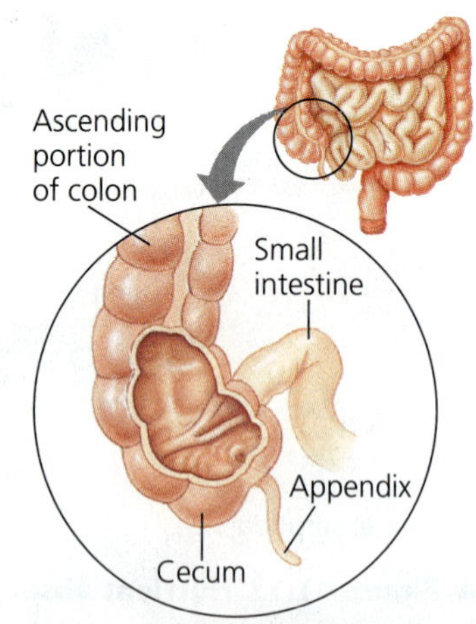

▲ **Figure 41.15 Junction of the small and large intestines.**

humans have a small cecum. The **appendix**, a finger-like extension of the human cecum, has a minor and dispensable role in immunity.

A major function of the colon is to recover water that has entered the alimentary canal as the solvent of digestive juices. About 7 L of fluid is secreted into the lumen of the alimentary canal each day, and about 90% of that is reabsorbed in the small intestine and colon. There is no mechanism for active transport of water. Instead, water is reabsorbed by osmosis when Na^+ and other ions are pumped out of the lumen of the colon.

The **feces**, the wastes of the digestive system, become increasingly solid as they are moved along the colon by peristalsis. It takes approximately 12–24 hours for material to travel the length of the colon. If the lining of the colon is irritated—by a viral or bacterial infection, for instance—less water than normal may be reabsorbed, resulting in diarrhea. The opposite problem, constipation, occurs when the feces move along the colon too slowly. An excess of water is reabsorbed, and therefore the feces become compacted.

A rich community of mostly harmless bacteria lives on unabsorbed organic material in the human colon, contributing approximately one-third of the dry weight of feces. One inhabitant, *Escherichia coli*, is so common in the human digestive system that its presence in lakes and streams is a useful indicator of contamination by untreated sewage. As by-products of their metabolism, many colon bacteria generate gases, including methane and hydrogen sulfide, which has an offensive odor. These gases and ingested air are expelled through the anus. Some bacteria produce vitamins, such as vitamin K, biotin, and folic acid, that supplement our dietary intake when absorbed into the blood.

Besides bacteria, feces contain undigested material, including cellulose fiber. Although it has no caloric value to humans, fiber helps move food along the alimentary canal.

The terminal portion of the large intestine is the **rectum**, where feces are stored until they can be eliminated. Between the rectum and the anus are two sphincters, the inner one being involuntary and the outer one being voluntary. Periodically, strong contractions of the colon create an urge to defecate. Because filling of the stomach triggers a reflex that increases the rate of contractions in the colon, the urge to defecate often follows a meal.

We have followed a meal from one opening (the mouth) of the alimentary canal to the other (the anus). Next we'll see how some digestive adaptations may have evolved.

CONCEPT CHECK 41.3

1. How does swallowed food reach the stomach of a weightless astronaut in orbit?
2. Explain why a proton pump inhibitor, such as the drug Prilosec, relieves the symptoms of acid reflux.
3. **WHAT IF?** If you mixed gastric juice with crushed food in a test tube, what would happen?

For suggested answers, see Appendix A.

CONCEPT **41.4**

Evolutionary adaptations of vertebrate digestive systems correlate with diet

EVOLUTION The digestive systems of mammals and other vertebrates are variations on a common plan, but there are many intriguing adaptations, often associated with the animal's diet. To highlight how form fits function, we'll examine a few of them.

Dental Adaptations

Dentition, an animal's assortment of teeth, is one example of structural variation reflecting diet (**Figure 41.16**). The evolutionary adaptation of teeth for processing different kinds of

▼ Figure 41.16 Dentition and diet.

Carnivore

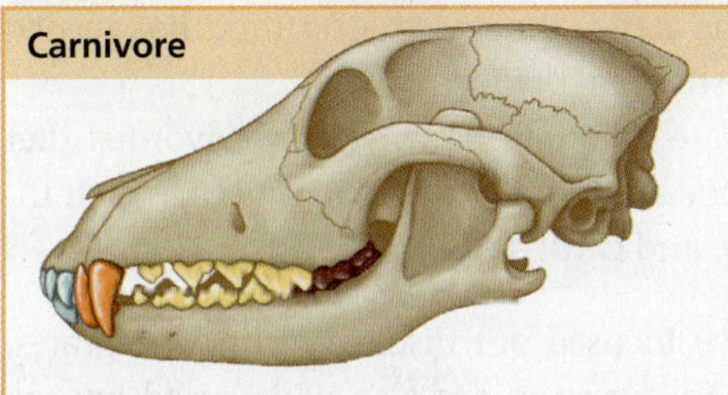

Carnivores, such as members of the dog and cat families, generally have large, pointed incisors and canines that can be used to kill prey and rip or cut away pieces of flesh. The jagged premolars and molars crush and shred food.

Herbivore

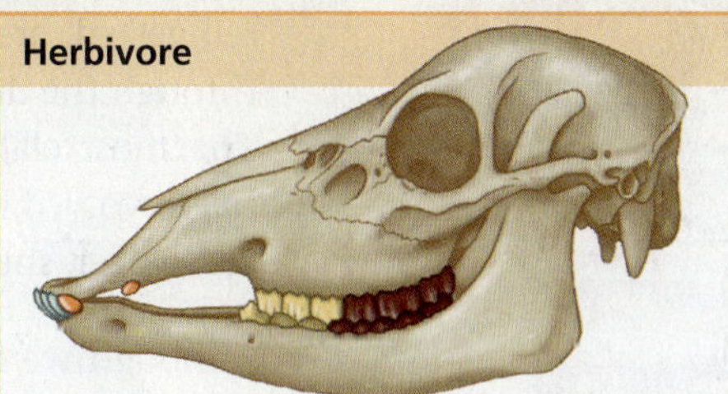

Herbivores, such as horses and deer, usually have premolars and molars with broad, ridged surfaces that grind tough plant material. The incisors and canines are generally modified for biting off pieces of vegetation. In some herbivores, canines are absent.

Omnivore

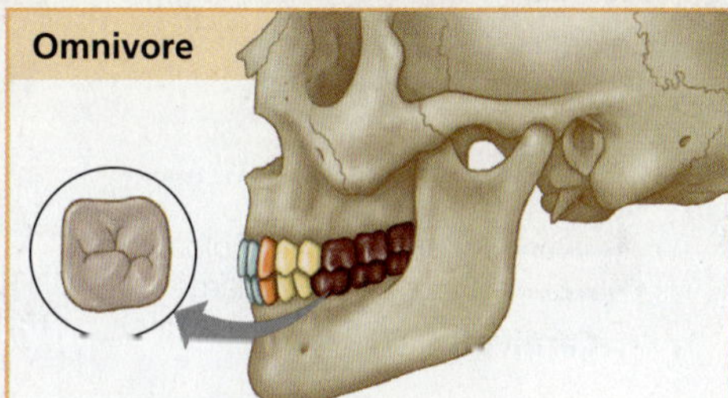

As omnivores, humans are adapted to eating both plants and meat. Adults have 32 teeth. From front to back along either side of the mouth are four bladelike incisors for biting, a pair of pointed canines for tearing, four premolars for grinding, and six molars for crushing (see inset, top view).

Key Incisors · Canines · Premolars · Molars

food is one of the major reasons mammals have been so successful. Nonmammalian vertebrates generally have less specialized dentition, but there are interesting exceptions. For example, poisonous snakes, such as rattlesnakes, have fangs, modified teeth that inject venom into prey. Some fangs are hollow, like syringes, whereas others drip the poison along grooves on the surfaces of the teeth.

Stomach and Intestinal Adaptations

Large, expandable stomachs are common in carnivorous vertebrates, which may go for a long time between meals and must eat as much as they can when they do catch prey. A 200-kg African lion can consume 40 kg of meat in one meal!

The length of the vertebrate digestive system is also correlated with diet. In general, herbivores and omnivores have longer alimentary canals relative to their body size than do carnivores. Vegetation is more difficult to digest than meat because it contains cell walls. A longer digestive tract furnishes more time for digestion and more surface area for the absorption of nutrients. As an example, consider the koala and coyote in **Figure 41.17**. Although these two mammals are about the same size, the koala's intestines are much longer, enhancing the processing of fibrous, protein-poor eucalyptus leaves from which the koala obtains virtually all its food and water.

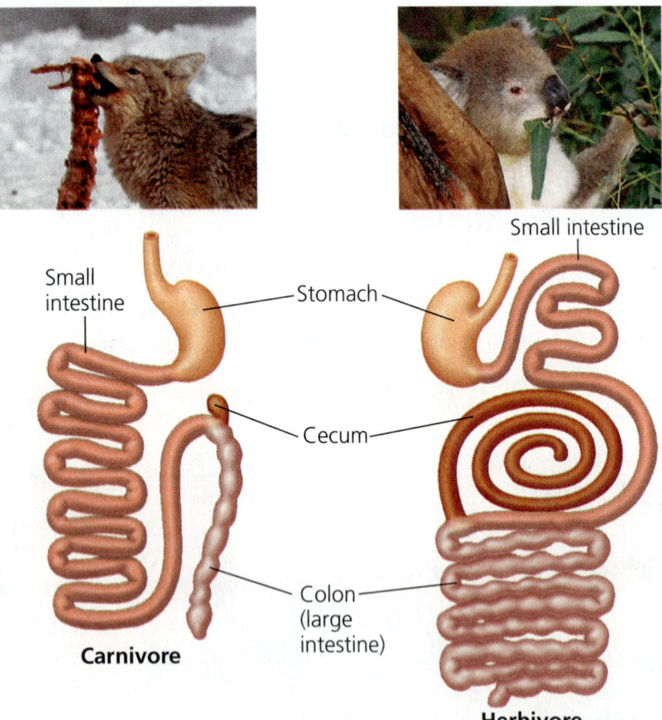

▲ **Figure 41.17 The alimentary canals of a carnivore (coyote) and herbivore (koala).** The koala's alimentary canal is specialized for digesting eucalyptus leaves. Extensive chewing chops the leaves into tiny pieces, increasing exposure to digestive juices. In the long cecum and the upper portion of the colon, symbiotic bacteria convert the shredded leaves to a more nutritious diet.

Mutualistic Adaptations

Some digestive adaptations involve mutualistic symbiosis, a mutually beneficial interaction between two species (see Chapter 54). For example, microorganisms help herbivores digest plants. Much of the chemical energy in herbivore diets comes from the cellulose of plant cell walls, but animals do not produce enzymes that hydrolyze cellulose. Instead, many vertebrates (as well as termites, whose wood diets consist largely of cellulose) house large populations of mutualistic bacteria and protists in fermentation chambers in their alimentary canals. These microorganisms have enzymes that can digest cellulose to simple sugars and other compounds that the animal can absorb. In many cases, the microorganisms also use the sugars from digested cellulose in the production of a variety of nutrients essential to the animal, such as vitamins and amino acids.

The location of mutualistic microbes in alimentary canals varies, depending on the type of herbivore. For example:

- The hoatzin, an herbivorous bird that lives in the South American rain forests, has a large, muscular crop (an esophageal pouch; see Figure 41.8) that houses mutualistic microorganisms. Hard ridges in the wall of the crop grind plant leaves into small fragments, and the microorganisms break down cellulose.
- Horses and many other herbivorous mammals house mutualistic microorganisms in a large cecum. The koala also has an enlarged cecum, where mutualistic bacteria ferment finely shredded eucalyptus leaves.
- In rabbits and some rodents, mutualistic bacteria live in the large intestine as well as in the cecum. Since most nutrients are absorbed in the small intestine, nourishing byproducts of fermentation by bacteria in the large intestine are initially lost with the feces. Rabbits and rodents recover these nutrients by *coprophagy* (from the Greek, meaning "dung eating"), feeding on some of their feces and then passing the food through the alimentary canal a second time. The familiar rabbit "pellets," which are not reingested, are the feces eliminated after food has passed through the digestive tract twice.
- The most elaborate adaptations for an herbivorous diet have evolved in the animals called **ruminants**, which include deer, sheep, and cattle **(Figure 41.18)**.

Although we have focused our discussion on vertebrates, adaptations related to digestion are also widespread among other animals. Some of the most remarkable examples are the giant tubeworms (over 3 m long) that live at pressures as high as 260 atmospheres around deep-sea hydrothermal vents (see Figure 52.16). These worms have no mouth or digestive system. Instead, they rely entirely on mutualistic bacteria to generate energy and nutrients from the carbon dioxide, oxygen,

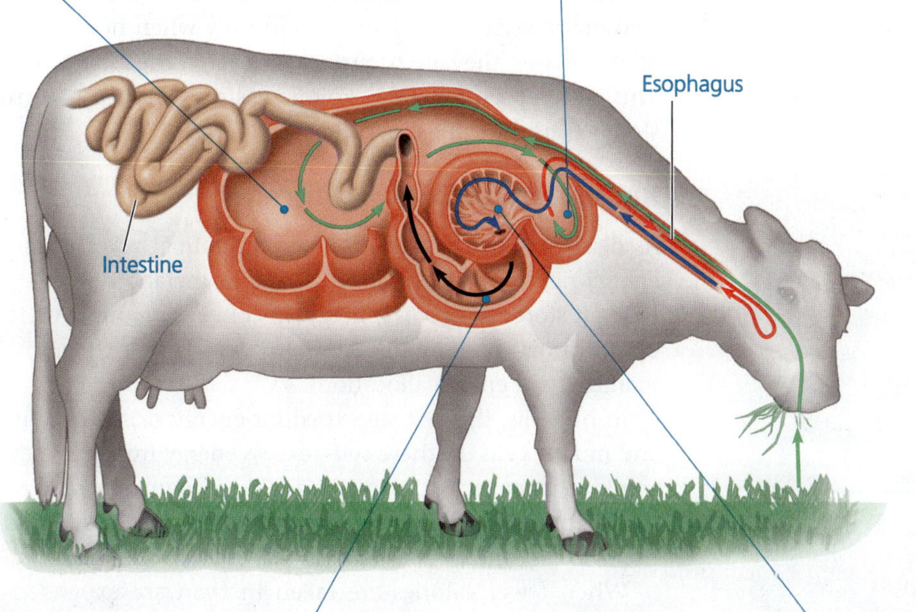

◄ **Figure 41.18 Ruminant digestion.** The stomach of a ruminant has four chambers. Because of the microbial action in the chambers, the diet from which a ruminant actually absorbs its nutrients is much richer than the grass the animal originally eats. In fact, a ruminant eating grass or hay obtains many of its nutrients by digesting the mutualistic microorganisms, which reproduce rapidly enough in the rumen to maintain a stable population.

hydrogen sulfide, and nitrate available at the vents. Thus, for invertebrates and vertebrates alike, mutualistic symbiosis has evolved as a general strategy for expanding the sources of nutrition available to animals.

Having examined how animals optimize their extraction of nutrients from food, we will next turn to the challenge of balancing the use of these nutrients.

CONCEPT CHECK 41.4

1. What are two advantages of a longer alimentary canal for processing plant material that is difficult to digest?
2. What features of a mammal's digestive system make it an attractive habitat for mutualistic microorganisms?
3. **WHAT IF?** "Lactose-intolerant" people have a shortage of lactase, the enzyme that breaks down lactose in milk. As a result, they sometimes develop cramps, bloating, or diarrhea after consuming dairy products. Suppose such a person ate yogurt containing bacteria that produce lactase. Why would eating yogurt likely provide at best only temporary relief of the symptoms?

For suggested answers, see Appendix A.

CONCEPT 41.5

Feedback circuits regulate digestion, energy storage, and appetite

Having examined the processes that enable an animal to obtain nutrients, we will finish our discussion of nutrition by considering how these processes are matched to circumstance and need.

Regulation of Digestion

Many animals go for long intervals between meals and do not need their digestive systems to be active continuously. Instead, each step in processing is activated as food reaches a new compartment in the alimentary canal. The arrival of food triggers the secretion of substances that promote the next stage of chemical digestion, as well as muscular contractions that propel food farther along the canal. For example, you learned earlier that nervous reflexes stimulate the release of saliva when food enters the oral cavity and orchestrate swallowing when a bolus of food reaches the pharynx. Similarly, the arrival of food in the stomach triggers churning and the release of gastric juices. A branch of the nervous system

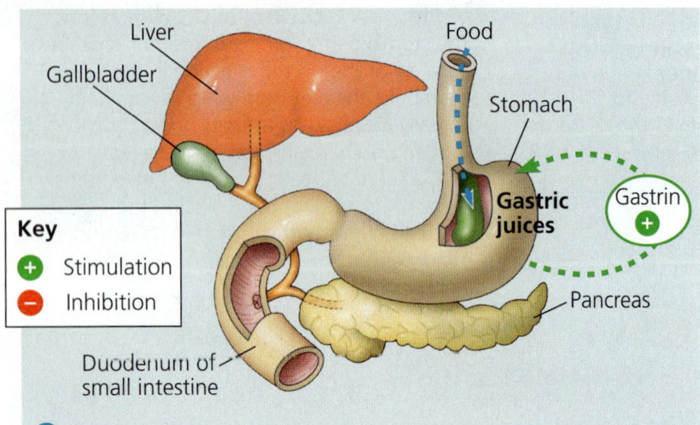

1 As food arrives at the stomach, it stretches the stomach walls, triggering release of the hormone **gastrin**. Gastrin circulates via the bloodstream back to the stomach, where it stimulates production of gastric juices.

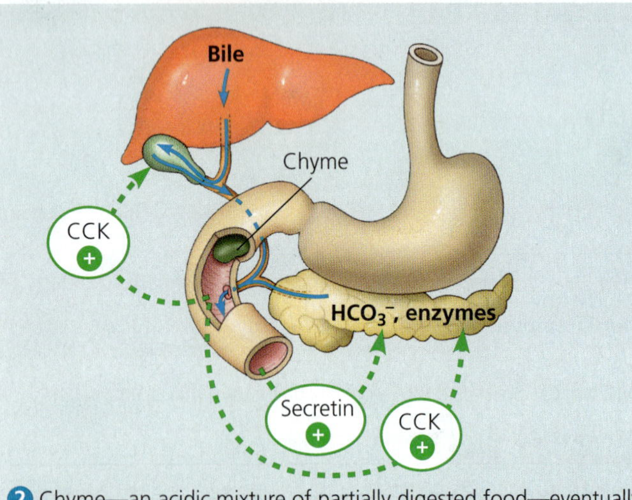

2 Chyme—an acidic mixture of partially digested food—eventually passes from the stomach to the duodenum. The duodenum responds to amino acids or fatty acids in the chyme by releasing the digestive hormones cholecystokinin and secretin. **Cholecystokinin (CCK)** stimulates the release of digestive enzymes from the pancreas and of bile from the gallbladder. **Secretin** stimulates the pancreas to release bicarbonate (HCO_3^-), which neutralizes chyme.

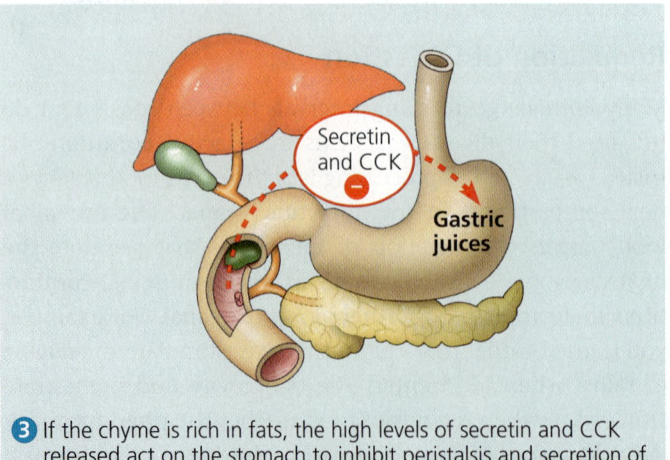

3 If the chyme is rich in fats, the high levels of secretin and CCK released act on the stomach to inhibit peristalsis and secretion of gastric juices, thereby slowing digestion.

▲ **Figure 41.19 Hormonal control of digestion.**

called the *enteric division*, which is dedicated to the digestive organs, regulates these events as well as peristalsis in the small and large intestines.

The endocrine system also plays a critical role in controlling digestion. As described in **Figure 41.19**, a series of hormones released by the stomach and duodenum help ensure that digestive secretions are present only when needed. Like all hormones, they are transported through the bloodstream. This is true even for the hormone gastrin, whose target (the stomach) is the same organ that secretes it.

Regulation of Energy Storage

As discussed in Chapter 40, when an animal takes in more energy-rich molecules than it needs for metabolism and activity, it stores the excess energy. In concluding our overview of nutrition, we'll examine some ways in which animals manage their energy allocation.

In humans, the first sites used for energy storage are liver and muscle cells. In these cells, excess energy from the diet is stored in glycogen, a polymer made up of many glucose units (see Figure 5.6b). Once glycogen depots are full, any additional excess energy is usually stored in fat in adipose cells.

When fewer calories are taken in than are expended—perhaps because of sustained heavy exercise or lack of food—the human body generally expends liver glycogen first and then draws on muscle glycogen and fat. Fats are especially rich in energy; oxidizing a gram of fat liberates about twice the energy liberated from a gram of carbohydrate or protein. For this reason, adipose tissue provides the most space-efficient way for the body to store large amounts of energy. Most healthy people have enough stored fat to sustain them through several weeks without food.

Glucose Homeostasis

The synthesis and breakdown of glycogen is central not only to energy storage, but also to maintaining metabolic balance through glucose homeostasis. Tissues throughout the body rely on the generation of ATP by oxidation of glucose to fuel cellular processes (see Chapter 9). The pancreatic hormones insulin and glucagon maintain glucose homeostasis by tightly regulating the synthesis and breakdown of glycogen.

The liver is a key site for glucose homeostasis **(Figure 41.20)**. When insulin levels rise after a carbohydrate-rich meal, glucose entering the liver in the hepatic portal vein is used to synthesize glycogen. Between meals, when blood in the hepatic portal vein has a much lower glucose concentration, glucagon stimulates the liver to break down glycogen, releasing glucose into the blood. Through the combined action of insulin and glucagon, blood exiting the liver has a glucose concentration of 70–110 mg per 100 mL at nearly all times.

We will return to the mechanism of glucose homeostasis (and explore the consequences when it fails) in our discussion of the endocrine system in Chapter 45.

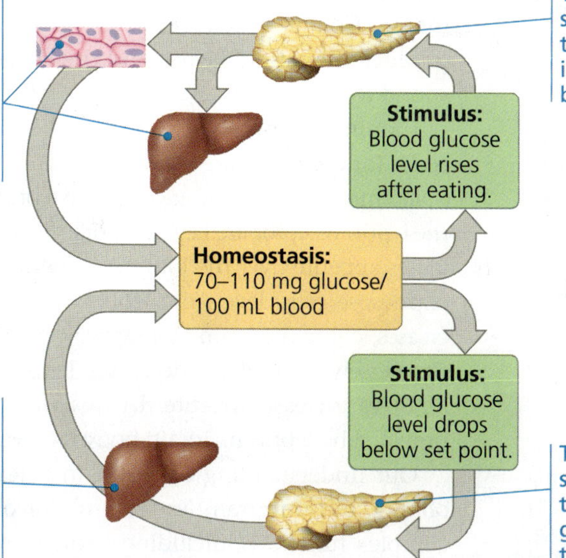

◀ **Figure 41.20 Homeostatic regulation of cellular fuel.** After a meal is digested, glucose and other monomers are absorbed into the blood from the digestive tract. The human body regulates the use and storage of glucose, a major cellular fuel.

MAKE CONNECTIONS *What form of feedback control does each of these regulatory circuits reflect (see Concept 40.2, p. 861)?*

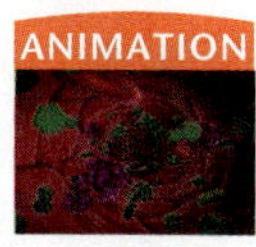

BioFlix Visit the Study Area at **www.masteringbiology.com** for the BioFlix® 3-D Animation on Homeostasis: Regulating Blood Sugar.

Regulation of Appetite and Consumption

Overnourishment, the consumption of more calories than the body needs for normal metabolism, causes obesity, the excessive accumulation of fat. Obesity, in turn, contributes to a number of health problems, including the most common type of diabetes (type 2), cancer of the colon and breast, and cardiovascular disease that can lead to heart attacks and strokes. It is estimated that obesity is a factor in about 300,000 deaths per year in the United States alone.

Researchers have discovered several homeostatic mechanisms that help regulate body weight. Operating as feedback circuits, these mechanisms control the storage and metabolism of fat. Several hormones regulate long-term and short-term appetite by affecting a "satiety center" in the brain **(Figure 41.21)**. In addition, a network of neurons relays and integrates information from the digestive system to regulate hormone release. To a large extent, this neuronal network functions independent of inputs from the central nervous system.

Mutations that cause mice to be chronically obese have played a key role in advancing our understanding of the satiety pathway. Mice with mutations in the *ob* or *db* gene eat voraciously and become much heavier than normal. Doug Coleman investigated how *ob* and *db* mutations disrupt normal

Secreted by the stomach wall, **ghrelin** is one of the signals that triggers feelings of hunger as mealtimes approach. In dieters who lose weight, ghrelin levels increase, which may be one reason it's so hard to stay on a diet.

A rise in blood sugar level after a meal stimulates the pancreas to secrete **insulin** (see Figure 41.20). In addition to its other functions, insulin suppresses appetite by acting on the brain.

Produced by adipose (fat) tissue, **leptin** suppresses appetite. When the amount of body fat decreases, leptin levels fall, and appetite increases.

The hormone **PYY**, secreted by the small intestine after meals, acts as an appetite suppressant that counters the appetite stimulant ghrelin.

Satiety center

Ghrelin +

Insulin −

Leptin −

PYY −

▲ **Figure 41.21 A few of the appetite-regulating hormones.** Secreted by various organs and tissues, the hormones reach the brain via the bloodstream. These signals act on a region of the brain that in turn controls the "satiety center," which generates the nervous impulses that make us feel either hungry or satiated ("full"). The hormone ghrelin is an appetite stimulant; the other three hormones shown here are appetite suppressants.

▼ **Figure 41.22** INQUIRY

What are the roles of the *ob* and *db* genes in appetite regulation?

EXPERIMENT Margaret Dickie, Katherine Hummel, and Doug Coleman, of the Jackson Laboratory in Bar Harbor, Maine, discovered that mice mutant for the *ob* gene (*ob ob*) or for the *db* gene (*db db*) eat voraciously and grow much more massive than mice with the wild-type (nonmutant) forms of both genes (designated ob^+ and db^+).

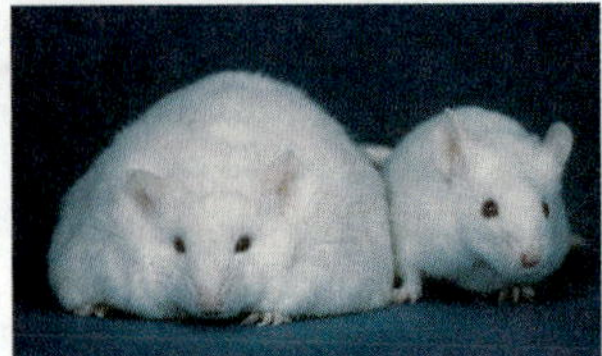

Obese mouse with mutant *ob* gene (left) next to wild-type mouse.

To explore further the roles of the two genes, Coleman measured the body masses of young mice with various genotypes and then surgically linked the circulatory system of each subject to that of another mouse. This procedure ensured that any factor circulating in the bloodstream of either mouse would be transferred to the other. After eight weeks, he again measured the mass of each subject.

RESULTS

Genotype pairing (red type indicates mutant genes)		Average change in body mass (g) of subject
Subject	Paired with	
ob^+ob^+, db^+db^+	ob^+ob^+, db^+db^+	8.3
ob ob, db^+db^+	*ob ob*, db^+db^+	38.7
ob ob, db^+db^+	ob^+ob^+, db^+db^+	8.2
ob ob, db^+db^+	ob^+ob^+, *db db*	−14.9*

*Due to pronounced weight loss and weakening, subjects in this pairing were reweighed after less than eight weeks.

CONCLUSION Because an *ob* mouse gains less weight when surgically joined with an ob^+ mouse than when joined with an *ob* mouse, Coleman concluded that the *ob* mouse fails to make a satiety factor but can respond to the factor when it is present. To explain the weight loss in an *ob* mouse that receives circulating factors from a *db* mouse, he reasoned that the *db* mutation blocks the response to the satiety factor but not its production, leading to an overproduction of the factor by the *db* mouse.

Subsequent molecular studies demonstrated the validity of both parts of Coleman's conclusion. The *ob* gene product is leptin, the satiety factor, whereas the *db* gene product is the leptin receptor. Thus, mice with the *ob* mutation cannot produce leptin, and mice with the *db* mutation produce leptin but cannot respond to it.

SOURCE D. L. Coleman, Effects of parabiosis of obese mice with diabetes and normal mice. *Diabetologia* 9:294–298 (1973).

MB See the related Experimental Inquiry Tutorial in MasteringBiology.

WHAT IF? Suppose you collected blood from a wild-type mouse and a *db* mouse over the course of a day. What changes would you expect in the concentration of leptin, the satiety factor, in each mouse? Explain your reasoning.

control of appetite (**Figure 41.22**). Based on his experiments, Coleman deduced that the *ob* gene is required to produce the satiety factor, and the *db* gene is required to respond to the factor.

Cloning of the *ob* gene led to the demonstration that it codes for the hormone now known as **leptin** (from the Greek *lepto*, thin). The *db* gene encodes the leptin receptor. Leptin and the leptin receptor are key components of the circuitry that regulates appetite over the long term. Because leptin is a product of adipose cells, levels rise when the amount of body fat increases, cuing the brain to suppress appetite (see Figure 41.20). Conversely, loss of fat decreases leptin levels, signaling the brain to increase appetite. In this way, the feedback signals provided by leptin maintain body fat levels within a set range.

Our understanding of leptin may lead to treatments for obesity, but uncertainties remain. For one thing, leptin has complex functions, including a role in how the nervous system develops. Also, most obese people have an abnormally high leptin level, which somehow fails to elicit a response from the brain's satiety center. Clearly, there is much to learn in this important area of human physiology.

Obesity and Evolution

EVOLUTION The relationship between fat storage and evolutionary adaptation in animals is sometimes complex. Consider the plump offspring of the seabirds called petrels (**Figure 41.23**). Their parents must fly long distances to find food. Most of the food that they bring to their chicks is very rich in lipids. The fact that fat has about twice as many calories per gram as other fuels minimizes the number of foraging trips. However, growing petrels need lots of protein for building new tissues, and there is relatively little in their oily

▲ **Figure 41.23 A plump petrel.** Too heavy to fly, the petrel chick (right) will have to lose weight before it takes wing. In the meantime, its stored fat provides energy during times when its parents fail to bring enough food.

diet. To get all the protein they need, young petrels must consume many more calories than they burn in metabolism, and consequently they become obese. Their fat depots nevertheless help them survive periods when their parents cannot find enough food. When food is plentiful, chicks at the end of the growth period weigh much more than their parents. The youngsters must then fast for several days to lose enough weight to be capable of flight.

Though fat hoarding in humans can be a health liability, it may have been an advantage in our evolutionary past. Our ancestors on the African savanna were hunter-gatherers who probably survived mainly on seeds and other plant products, a diet only occasionally supplemented by hunting game or scavenging meat from animals killed by other predators. In such a feast-or-famine existence, natural selection may have favored those individuals with a physiology that induced them to gorge on rich, fatty foods on those rare occasions when such treats were abundantly available. Individuals with genes promoting the storage of high-energy molecules during feasts may have been more likely to survive famines. Thus, our present-day taste for fats may be partly an evolutionary vestige of less nutritious times.

In the next chapter, we'll see that obtaining food, digesting it, and absorbing nutrients are parts of a larger story. Provisioning the body also involves distributing nutrients (circulation), and using nutrients for metabolism requires exchanging respiratory gases with the environment.

CONCEPT CHECK 41.5

1. Explain how people can become obese even if their intake of dietary fat is relatively low compared with carbohydrate intake.
2. After reviewing Figure 41.21, explain how PYY and leptin complement each other in regulating body weight.
3. **WHAT IF?** Suppose you were studying two groups of obese people with genetic abnormalities in the leptin pathway. In one group, the leptin levels are abnormally high; in the other group, they are abnormally low. How would each group's leptin levels change if both groups were placed on a low-calorie diet for an extended period? Explain.

For suggested answers, see Appendix A.

41 CHAPTER REVIEW

SUMMARY OF KEY CONCEPTS

- Animals have diverse diets. **Herbivores** mainly eat plants; **carnivores** mainly eat other animals; and **omnivores** eat both. Animals must balance consumption, storage, and use of food.

CONCEPT 41.1

An animal's diet must supply chemical energy, organic molecules, and essential nutrients (pp. 875–880)

- Food provides animals with energy for ATP production, carbon skeletons for biosynthesis, and **essential nutrients**—nutrients that must be supplied in preassembled form. Essential nutrients include certain amino acids and fatty acids that animals cannot synthesize; **vitamins**, which are organic molecules; and **minerals**, which are inorganic substances.
- Animals can suffer from two types of malnutrition: an inadequate intake of essential nutrients and a deficiency in sources of chemical energy. Studies of genetic defects and of disease at the population level help researchers determine human dietary requirements.

? *Propose a reason why the diet of many mammals doesn't need to include vitamin C, a substance that is important for collagen synthesis.*

CONCEPT 41.2

The main stages of food processing are ingestion, digestion, absorption, and elimination (pp. 880–883)

- Food processing in animals involves **ingestion** (eating), **digestion** (enzymatic breakdown of large molecules), **absorption** (uptake of nutrients by cells), and **elimination** (passage of undigested materials out of the body in feces).
- Animals differ in the ways they obtain and ingest food. Most animals are **bulk feeders**, eating large pieces of food.
- Compartmentalization is necessary to avoid self-digestion. In intracellular digestion, food particles are engulfed by endocytosis and digested within food vacuoles that have fused with lysosomes. In extracellular digestion, which is used by most animals, enzymatic hydrolysis occurs outside cells in a **gastrovascular cavity** or **alimentary canal**.

? *Propose an artificial diet that would eliminate the need for one of the first three steps in food processing.*

CONCEPT 41.3

Organs specialized for sequential stages of food processing form the mammalian digestive system (pp. 883–889)

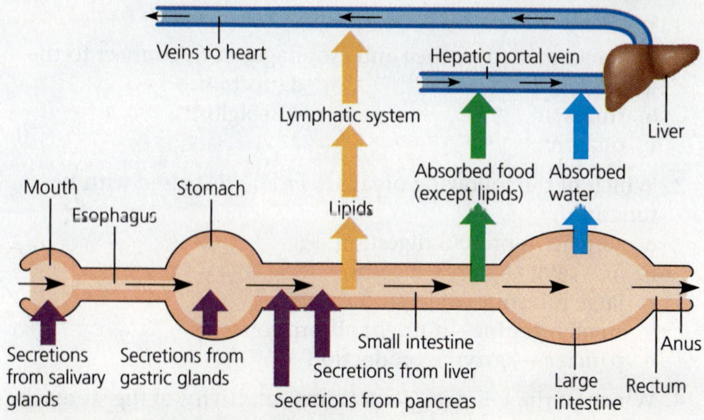

? *What structural feature of the small intestine makes it better suited for absorption of nutrients than the stomach?*

CONCEPT 41.4

Evolutionary adaptations of vertebrate digestive systems correlate with diet (pp. 889–891)

- Vertebrate digestive systems display many evolutionary adaptations associated with diet. For example, dentition, which is the assortment of teeth, generally correlates with diet. In addition, herbivores usually have longer alimentary canals than carnivores, reflecting the longer time needed to digest vegetation. Many herbivores, including cows, also have fermentation chambers where microorganisms digest cellulose, a form of mutualism.

? *How does our anatomy indicate that our ancestors were not vegetarians?*

CONCEPT 41.5

Feedback circuits regulate digestion, energy storage, and appetite (pp. 891–895)

- Nutrition is regulated at multiple levels. Food in the alimentary canal triggers nervous and hormonal responses that control the secretion of digestive juices and that promote the movement of ingested material through the canal. The availability of glucose for energy production is regulated by the hormones insulin and glucagon, which control the synthesis and breakdown of glycogen.
- Vertebrates store excess calories in glycogen (in liver and muscle cells) and in fat (in adipose cells). These energy stores can be tapped when an animal expends more calories than it consumes. If, however, an animal consumes more calories than it needs for normal metabolism, the resulting overnourishment can lead to the serious health problem of obesity.
- Several hormones, including leptin and insulin, regulate appetite by affecting the brain's satiety center. The problem of maintaining a healthy weight may stem partly from our evolutionary past, when fat hoarding may have been important for survival.

? *Explain why your stomach might make growling noises when you skip a meal.*

TEST YOUR UNDERSTANDING

LEVEL 1: KNOWLEDGE/COMPREHENSION

1. Which of the following animals is *incorrectly* paired with its feeding mechanism?
   a. lion—substrate feeder
   b. baleen whale—suspension feeder
   c. aphid—fluid feeder
   d. clam—suspension feeder
   e. snake—bulk feeder

2. The mammalian trachea and esophagus both connect to the
   a. large intestine.
   b. stomach.
   c. pharynx.
   d. rectum.
   e. epiglottis.

3. Which of the following organs is *incorrectly* paired with its function?
   a. stomach—protein digestion
   b. oral cavity—starch digestion
   c. large intestine—bile production
   d. small intestine—nutrient absorption
   e. pancreas—enzyme production

4. Which of the following is *not* a major activity of the stomach?
   a. mechanical digestion
   b. HCl secretion
   c. mucus secretion
   d. nutrient absorption
   e. enzyme secretion

LEVEL 2: APPLICATION/ANALYSIS

5. After surgical removal of an infected gallbladder, a person must be especially careful to restrict dietary intake of
   a. starch. b. protein. c. sugar. d. fat. e. water.

6. If you were to jog 1 km a few hours after lunch, which stored fuel would you probably tap?
   a. muscle proteins
   b. muscle and liver glycogen
   c. fat stored in the liver
   d. fat stored in adipose tissue
   e. blood proteins

LEVEL 3: SYNTHESIS/EVALUATION

7. **DRAW IT** Make a flowchart of the events that occur after partially digested food leaves the stomach. Use the following terms: bicarbonate secretion, circulation, decrease in acidity, secretin secretion, increase in acidity, signal detection. Next to each term, indicate the compartment(s) involved. You may use a term more than once.

8. **EVOLUTION CONNECTION**
   The human esophagus and trachea share a passage leading from the mouth and nasal passages, which can cause problems. After reviewing vertebrate evolution in Chapter 34, explain the evolutionary basis for this "imperfect" anatomy.

9. **SCIENTIFIC INQUIRY**
   In human populations of northern European origin, the disorder called hemochromatosis causes excess iron uptake from food and affects one in 200 adults. Men are ten times as likely as women to suffer from iron overload. Devise a hypothesis for the difference in the disease between the two sexes.

10. **WRITE ABOUT A THEME**
    Emergent Properties Hair is largely made up of the protein keratin. In a short essay (100–150 words), explain why a shampoo containing protein is not effective in replacing the protein in damaged hair.

For selected answers, see Appendix A.

www.masteringbiology.com

1. MasteringBiology® Assignments

Make Connections Tutorial Fat Absorption (Chapter 41) and Fat Structure (Chapter 5).

Experimental Inquiry Tutorial What Role Do Genes Play in Appetite Regulation?

BioFlix **Tutorials** Membrane Transport • Homeostasis: Regulating Blood Sugar

Tutorial Vitamins

Activities Digestive System Function • Hormonal Control of Digestion • The Digestion and Absorption of Food • Discovery Channel Video: Nutrition

Questions Student Misconceptions • Reading Quiz • Multiple Choice • End-of-Chapter

2. eText

Read your book online, search, take notes, highlight text, and more.

3. The Study Area

Practice Tests • Cumulative Test • ***BioFlix*** 3-D Animations • MP3 Tutor Sessions • Videos • Activities • Investigations • Lab Media • Audio Glossary • Word Study Tools • Art

42

Circulation and Gas Exchange

▲ **Figure 42.1 How does a feathery fringe help this animal survive?**

KEY CONCEPTS

42.1 Circulatory systems link exchange surfaces with cells throughout the body
42.2 Coordinated cycles of heart contraction drive double circulation in mammals
42.3 Patterns of blood pressure and flow reflect the structure and arrangement of blood vessels
42.4 Blood components function in exchange, transport, and defense
42.5 Gas exchange occurs across specialized respiratory surfaces
42.6 Breathing ventilates the lungs
42.7 Adaptations for gas exchange include pigments that bind and transport gases

OVERVIEW

Trading Places

The animal in **Figure 42.1** may look like a creature from a science fiction film, but it's actually an axolotl, a salamander native to shallow ponds in central Mexico. The feathery red appendages jutting out from the head of this albino adult are gills. Although external gills are uncommon in adult animals, they help satisfy the need shared by all animals to exchange substances with their environment.

Exchange between an axolotl or any other animal and its surroundings ultimately occurs at the cellular level. The resources that animal cells require, such as nutrients and oxygen (O_2), enter the cytoplasm by crossing the plasma membrane. Metabolic by-products, such as carbon dioxide (CO_2), exit the cell by crossing the same membrane. In unicellular organisms, exchange occurs directly with the external environment. For most multicellular organisms, however, direct transfer of materials between every cell and the environment is not possible. Instead, these organisms rely on specialized systems that carry out exchange with the environment and that transport materials between sites of exchange and the rest of the body.

The reddish color and branching structure of the axolotl's gills reflect the intimate association between exchange and transport. Tiny blood vessels lie close to the surface of each filament in the gills. Across this surface, there is a net diffusion of O_2 from the surrounding water into the blood and of CO_2 from the blood into the water. The short distances involved allow diffusion to be rapid. Pumping of the axolotl's heart propels the oxygen-rich blood from the gill filaments to all other tissues of the body. There, more short-range exchange occurs, involving nutrients and O_2 as well as CO_2 and other wastes.

Because internal transport and gas exchange are functionally related in most animals, not just axolotls, we will examine both circulatory and respiratory systems in this chapter. We will explore the remarkable variation in form and organization of these systems by considering examples from a number of species. We will also highlight the roles of circulatory and respiratory systems in maintaining homeostasis under a range of physiological and environmental conditions.

CONCEPT **42.1**

Circulatory systems link exchange surfaces with cells throughout the body

The molecular trade that an animal carries out with its environment—gaining O_2 and nutrients while shedding CO_2 and other waste products—must ultimately involve every cell in the body. As you learned in Chapter 7, small, nonpolar molecules such as O_2 and CO_2 can move between cells and their immediate surroundings by diffusion. But diffusion is very slow for distances of more than a few millimeters. That's because the time it takes for a substance to diffuse from one place to another is proportional to the *square* of the distance. For example, if it takes 1 second for a given quantity of glucose to diffuse 100 μm, it will take 100 seconds for the same quantity to diffuse 1 mm and almost 3 hours to diffuse 1 cm! This relationship between diffusion time and distance places a substantial constraint on the body plan of any animal.

Given that diffusion is rapid only over very small distances, how does each cell of an animal participate in exchange? Natural selection has resulted in two general solutions to this problem. The first solution is a body size and shape that keep many or all cells in direct contact with the environment. Each cell can thus exchange materials directly with the surrounding medium. This type of body plan is found only in certain invertebrates, including cnidarians and flatworms. The second solution, found in all other animals, is a circulatory system that moves fluid between each cell's immediate surroundings and the tissues where exchange with the environment occurs.

Gastrovascular Cavities

Let's begin by looking at animals that lack a distinct circulatory system. In hydras, jellies, and other cnidarians, a central gastrovascular cavity functions in the distribution of substances throughout the body and in digestion (see Figure 41.7). An opening at one end connects the cavity to the surrounding water. In a hydra, thin branches of the gastrovascular cavity extend into the animal's tentacles. In jellies and some other cnidarians, the gastrovascular cavity has a much more elaborate branching pattern (**Figure 42.2a**).

In animals with a gastrovascular cavity, fluid bathes both the inner and outer tissue layers, facilitating exchange of gases and cellular waste. Only the cells lining the cavity have direct access to nutrients released by digestion. However, because the body wall is a mere two cells thick, nutrients need diffuse only a short distance to reach the cells of the outer tissue layer.

Planarians and most other flatworms also survive without a circulatory system. Their combination of a gastrovascular cavity and a flat body is well suited for exchange with the environment (**Figure 42.2b**). A flat body optimizes diffusional exchange by increasing surface area and minimizing diffusion distances.

Evolutionary Variation in Circulatory Systems

EVOLUTION For animals with many cell layers, diffusion distances are too great for adequate exchange of nutrients and wastes by a gastrovascular cavity. In these organisms, a circulatory system minimizes the distances that substances must diffuse to enter or leave a cell.

General Properties of Circulatory Systems

A circulatory system has three basic components: a circulatory fluid, a set of interconnecting vessels, and a muscular pump, the **heart**. The heart powers circulation by using metabolic energy to elevate the hydrostatic pressure of the circulatory fluid, which then flows through the vessels and back to the heart.

By transporting fluid throughout the body, the circulatory system functionally connects the aqueous environment of the body cells to the organs that exchange gases, absorb nutrients, and dispose of wastes. In mammals, for example, O_2 from inhaled air diffuses across only two layers of cells in the lungs before reaching the blood. The circulatory system, powered by the heart, then carries the oxygen-rich blood to all parts of the body. As the blood streams throughout the body tissues in tiny blood vessels, O_2 in the blood diffuses only a short distance before entering the fluid that directly bathes the cells.

Several basic types of circulatory systems have arisen during evolution, each representing adaptations to constraints imposed by anatomy and environment. Circulatory systems are either open or closed, vary with regard to the number of circuits in the body, and rely on pumps that differ in structure and organization. We'll examine each of these variations and their physiological consequences in turn.

▶ **Figure 42.2 Internal transport in gastrovascular cavities.**

WHAT IF? *Suppose a gastrovascular cavity were open at two ends, with fluid entering one end and leaving the other. How would this affect the gastrovascular cavity's functions in gas exchange and digestion?*

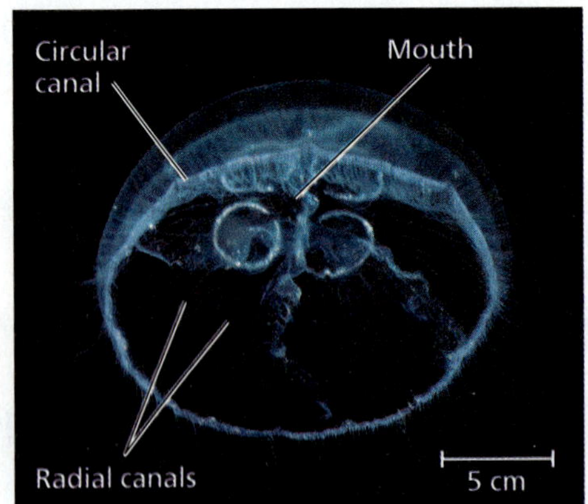

(a) The moon jelly *Aurelia*, a cnidarian. The jelly is viewed here from its underside (oral surface). The mouth leads to an elaborate gastrovascular cavity that consists of radial canals leading to and from a circular canal. Ciliated cells lining the canals circulate fluid within the cavity.

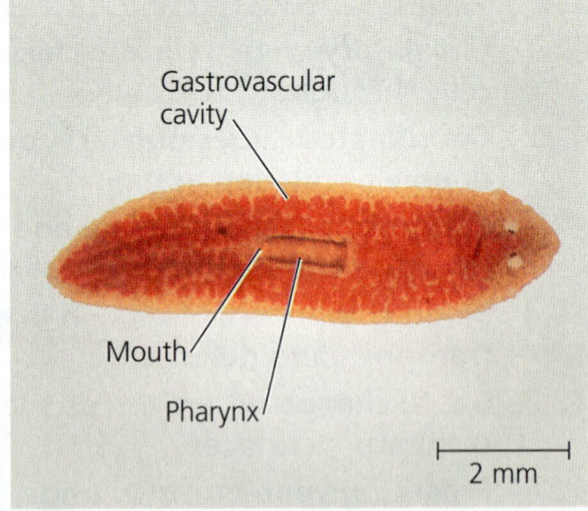

(b) The planarian *Dugesia*, a flatworm. The mouth and pharynx on the ventral side lead to the highly branched gastrovascular cavity, stained dark red in this specimen (LM).

Open and Closed Circulatory Systems

Arthropods and most molluscs have an **open circulatory system**, in which the circulatory fluid bathes the organs directly **(Figure 42.3a)**. In these animals, the circulatory fluid, called **hemolymph**, is also the *interstitial fluid* that bathes body cells. Contraction of one or more hearts pumps the hemolymph through the circulatory vessels into interconnected sinuses, spaces surrounding the organs. Within the sinuses, chemical exchange occurs between the hemolymph and body cells. Relaxation of the heart draws hemolymph back in through pores, which are equipped with valves that close when the heart contracts. Body movements help circulate the hemolymph by periodically squeezing the sinuses. The open circulatory system of larger crustaceans, such as lobsters and crabs, includes a more extensive system of vessels as well as an accessory pump.

▼ **Figure 42.3 Open and closed circulatory systems.**

(a) An open circulatory system

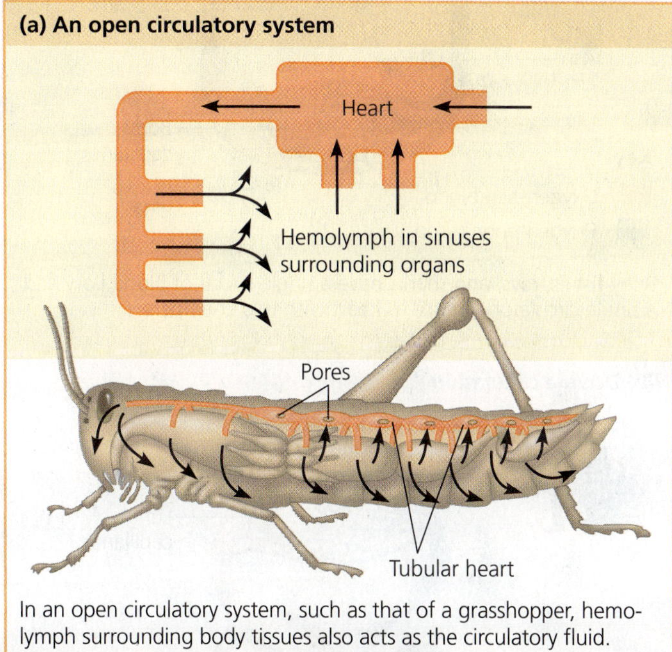

In an open circulatory system, such as that of a grasshopper, hemolymph surrounding body tissues also acts as the circulatory fluid.

(b) A closed circulatory system

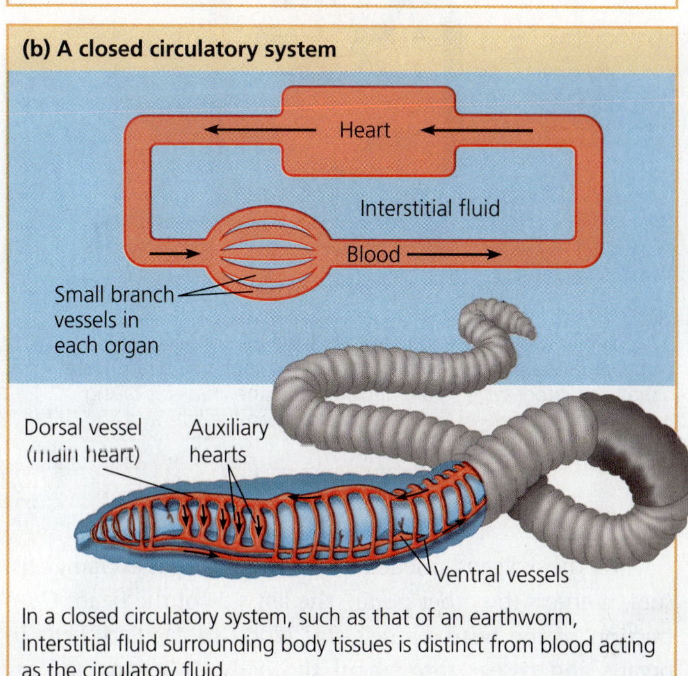

In a closed circulatory system, such as that of an earthworm, interstitial fluid surrounding body tissues is distinct from blood acting as the circulatory fluid.

In a **closed circulatory system**, a circulatory fluid called **blood** is confined to vessels and is distinct from the interstitial fluid **(Figure 42.3b)**. One or more hearts pump blood into large vessels that branch into smaller ones that infiltrate the organs. Chemical exchange occurs between the blood and the interstitial fluid, as well as between the interstitial fluid and body cells. Annelids (including earthworms), cephalopods (including squids and octopuses), and all vertebrates have closed circulatory systems.

The fact that both open and closed circulatory systems are widespread among animals suggests that there are advantages to each system. The lower hydrostatic pressures associated with open circulatory systems make them less costly than closed systems in terms of energy expenditure. In some invertebrates, open circulatory systems serve additional functions. For example, spiders use the hydrostatic pressure generated by their open circulatory system to extend their legs.

The benefits of closed circulatory systems include relatively high blood pressures, which enable the effective delivery of O_2 and nutrients to the cells of larger and more active animals. Among the molluscs, for instance, closed circulatory systems are found in the largest and most active species, the squids and octopuses. Closed systems are also particularly well suited to regulating the distribution of blood to different organs, as you'll learn later in this chapter. In examining closed circulatory systems in more detail, we will focus on the vertebrates.

Organization of Vertebrate Circulatory Systems

The closed circulatory system of humans and other vertebrates is often called the **cardiovascular system**. Blood circulates to and from the heart through an amazingly extensive network of vessels: The total length of blood vessels in an average human adult is twice Earth's circumference at the equator!

Arteries, veins, and capillaries are the three main types of blood vessels. Within each type, blood flows in only one direction. **Arteries** carry blood away from the heart to organs throughout the body. Within organs, arteries branch into **arterioles**, small vessels that convey blood to the capillaries. **Capillaries** are microscopic vessels with very thin, porous walls. Networks of these vessels, called **capillary beds**, infiltrate every tissue, passing within a few cell diameters of every cell in the body. Across the thin walls of capillaries, chemicals, including dissolved gases, are exchanged by diffusion between the blood and the interstitial fluid around the tissue cells. At their "downstream" end, capillaries converge into **venules**, and venules converge into **veins**, the vessels that carry blood back to the heart.

Arteries and veins are distinguished by the *direction* in which they carry blood, not by the O_2 content or other characteristics of the blood they contain. Arteries carry blood from the heart *toward* capillaries, and veins return blood to the heart *from* capillaries. The only exceptions are the portal veins, which carry blood between pairs of capillary beds. The hepatic portal vein, for example, carries blood from capillary beds in the digestive system to capillary beds in the liver (see Chapter 41). From the liver, blood passes into the hepatic veins, which conduct blood toward the heart.

The hearts of all vertebrates contain two or more muscular chambers. The chambers that receive blood entering the heart are called **atria** (singular, *atrium*). The chambers responsible for pumping blood out of the heart are called **ventricles**. The number of chambers and the extent to which they are separated from one another differ substantially among groups of vertebrates, as we will discuss next. These important differences reflect the close fit of form to function that arises from natural selection.

Single Circulation

In bony fishes, rays, and sharks, the heart consists of two chambers: an atrium and a ventricle. The blood passes through the heart once in each complete circuit, an arrangement called **single circulation** (Figure 42.4a). Blood entering the heart collects in the atrium before transfer to the ventricle. Contraction of the ventricle pumps blood to the gills, where there is a net diffusion of O_2 into the blood and of CO_2 out of the blood. As blood leaves the gills, the capillaries converge into a vessel that carries oxygen-rich blood to capillary beds throughout the body. Blood then returns to the heart.

In single circulation, blood that leaves the heart passes through two capillary beds before returning to the heart. When blood flows through a capillary bed, blood pressure drops substantially, for reasons we will explain shortly. The drop in blood pressure in the gills limits the rate of blood flow in the rest of the animal's body. As the animal swims, however, the contraction and relaxation of its muscles help accelerate the relatively sluggish pace of circulation.

Double Circulation

The circulatory systems of amphibians, reptiles, and mammals have two circuits, an arrangement called **double circulation** (Figure 42.4b). The pumps for the two circuits are combined into a single organ, the heart. Having both pumps within a single heart simplifies coordination of the pumping cycles.

One pump, the right side of the heart, delivers oxygen-poor blood to the capillary beds of the gas exchange tissues, where there is a net movement of O_2 into the blood and of CO_2 out of the blood. This part of the circulation is called a **pulmonary circuit** if the capillary beds involved are all in the lungs, as in reptiles and mammals. It is called a **pulmocutaneous circuit** if it includes capillaries in both the lungs and the skin, as in many amphibians.

▼ Figure 42.4 Single and double circulation in vertebrates.

(a) Single circulation

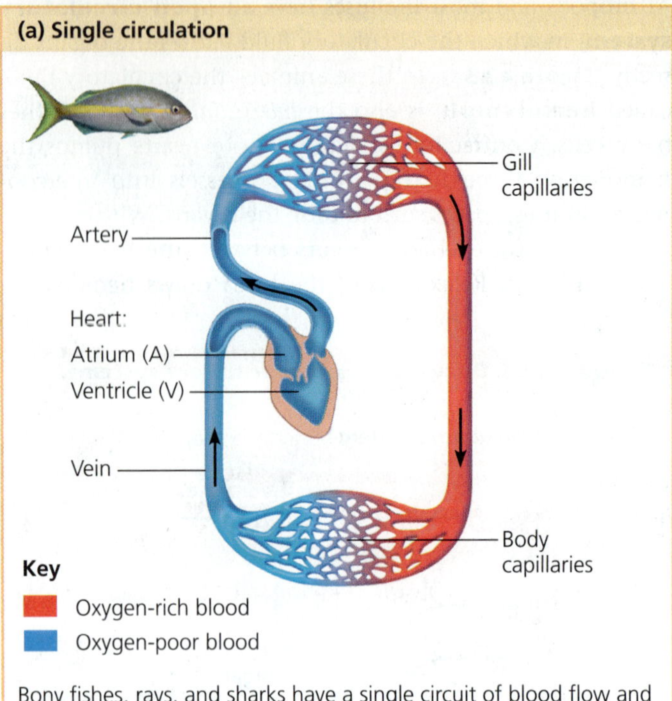

Bony fishes, rays, and sharks have a single circuit of blood flow and a single circulatory pump—a heart with two chambers.

(b) Double circulation

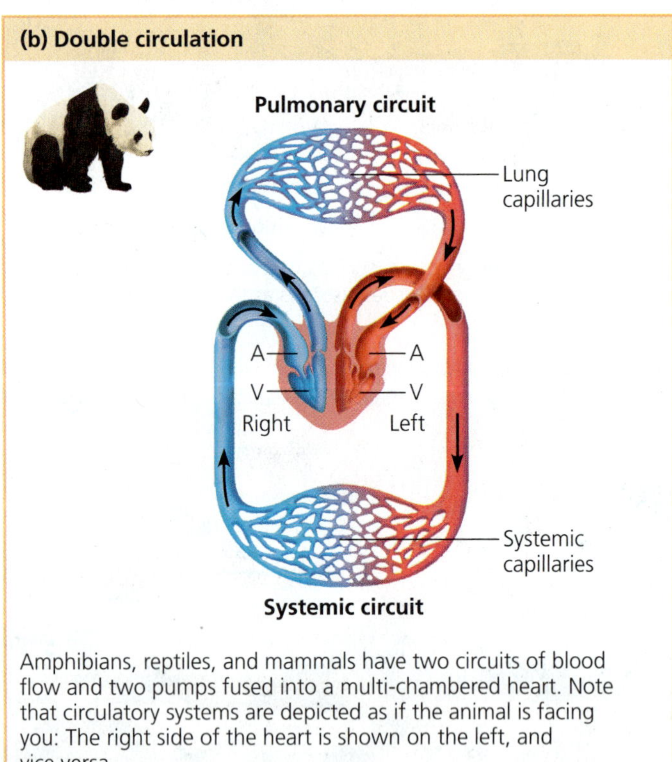

Amphibians, reptiles, and mammals have two circuits of blood flow and two pumps fused into a multi-chambered heart. Note that circulatory systems are depicted as if the animal is facing you: The right side of the heart is shown on the left, and vice versa.

After the oxygen-enriched blood leaves the gas exchange tissues, it enters the other pump, the left side of the heart. Contraction of the heart propels this blood to capillary beds in organs and tissues throughout the body. Following the exchange of O_2 and CO_2, as well as nutrients and waste products, the now oxygen-poor blood returns to the heart, completing the **systemic circuit**.

Double circulation provides a vigorous flow of blood to the brain, muscles, and other organs because the heart repressurizes the blood destined for these tissues after it passes through the capillary beds of the lungs or skin. Indeed, blood pressure is often much higher in the systemic circuit than in the gas exchange circuit. This contrasts sharply with single circulation, in which blood flows under reduced pressure directly from the gas exchange organs to other organs.

To explore the adaptations of double circulation that meet the particular needs of different vertebrates, we conclude our overview of circulatory systems with **Figure 42.5**. In the next section, we will restrict our focus to circulation in mammals and to the anatomy and physiology of the key circulatory organ—the heart.

▼ **Figure 42.5**

Exploring Double Circulation in Vertebrates

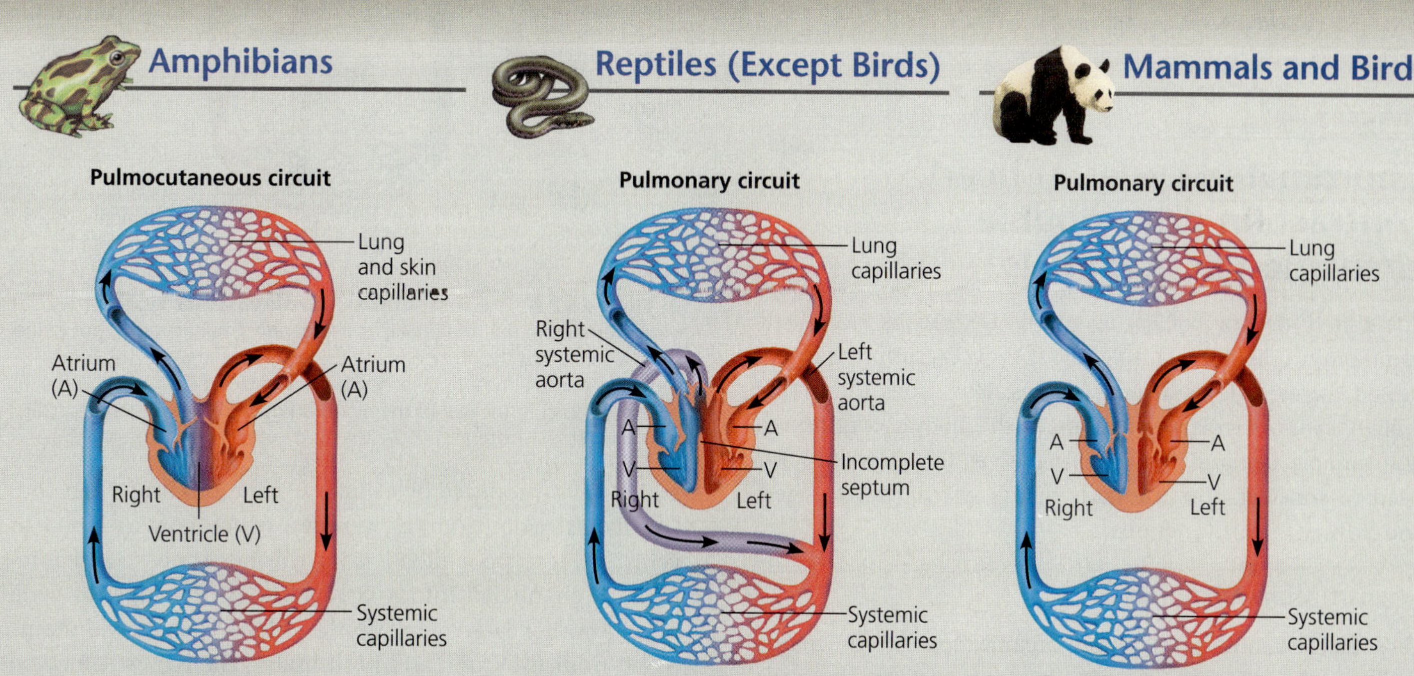

Amphibians

Frogs and other amphibians have a heart with three chambers: two atria and one ventricle. A ridge within the ventricle diverts most (about 90%) of the oxygen-poor blood from the right atrium into the pulmocutaneous circuit and most of the oxygen-rich blood from the left atrium into the systemic circuit. When underwater, a frog adjusts its circulation, for the most part shutting off blood flow to its temporarily ineffective lungs. Blood flow continues to the skin, which acts as the sole site of gas exchange while the frog is submerged.

Reptiles (Except Birds)

In the three-chambered heart of turtles, snakes, and lizards, an incomplete septum partially divides the single ventricle into separate right and left chambers. Two major arteries, called aortas, lead to the systemic circulation. The detailed anatomy of the heart varies among these three groups of reptiles, with some adaptations allowing control of the relative amount of blood flowing to the lungs and the body.

In alligators, caimans, and other crocodilians, the ventricles are divided by a complete septum (not shown), but the pulmonary and systemic circuits connect where the arteries exit the heart. This connection enables arterial valves to shunt blood flow away from the lungs temporarily, such as when the animal is underwater.

Mammals and Birds

In mammals and birds, there are two atria and two completely divided ventricles. The left side of the heart receives and pumps only oxygen-rich blood, while the right side receives and pumps only oxygen-poor blood. (In birds, the major vessels near the heart are slightly different than shown.) As endotherms, mammals and birds use about ten times as much energy as equal-sized ectotherms. Their circulatory systems therefore need to deliver about ten times as much fuel and O_2 to their tissues (and remove ten times as much CO_2 and other wastes). This large traffic of substances is made possible by separate and independently powered systemic and pulmonary circuits and by large hearts that pump the necessary volume of blood. A powerful four-chambered heart arose independently in the distinct ancestors of mammals and birds and thus reflects convergent evolution (see Chapter 34).

Key

Oxygen-rich blood

Oxygen-poor blood

CONCEPT CHECK 42.1

1. How is the flow of hemolymph through an open circulatory system similar to the flow of water through an outdoor fountain?
2. Three-chambered hearts with incomplete septa were once viewed as being less adapted to circulatory function than mammalian hearts. What advantage of such hearts did this viewpoint overlook?
3. **WHAT IF?** The heart of a normally developing human fetus has a hole between the left and right atria. In some cases, this hole does not close completely before birth. If the hole weren't surgically corrected, how would it affect the O_2 content of the blood entering the systemic circuit?

For suggested answers, see Appendix A.

CONCEPT 42.2

Coordinated cycles of heart contraction drive double circulation in mammals

The timely delivery of O_2 to the body's organs is critical: Some brain cells, for example, die if their O_2 supply is interrupted for as little as a few minutes. How does the mammalian cardiovascular system meet the body's continuous but variable demand for O_2? To answer this question, we need to consider how the parts of the system are arranged and how each part functions.

Mammalian Circulation

Let's first examine the overall organization of the mammalian cardiovascular system, beginning with the pulmonary circuit. (The circled numbers refer to corresponding locations in **Figure 42.6**.) ① Contraction of the right ventricle pumps blood to the lungs via ② the pulmonary arteries. As the blood flows through ③ capillary beds in the left and right lungs, it loads O_2 and unloads CO_2. Oxygen-rich blood returns from the lungs via the pulmonary veins to ④ the left atrium of the heart. Next, the oxygen-rich blood flows into ⑤ the heart's left ventricle, which pumps the oxygen-rich blood out to body tissues through the systemic circuit. Blood leaves the left ventricle via ⑥ the aorta, which conveys blood to arteries leading throughout the body. The first branches leading from the aorta are the coronary arteries (not shown), which supply blood to the heart muscle itself. Then branches lead to ⑦ capillary beds in the head and arms (forelimbs). The aorta then descends into the abdomen, supplying oxygen-rich blood to arteries leading to ⑧ capillary beds in the abdominal organs and legs (hind limbs). Within the capillaries, there is a net diffusion of O_2 from the blood to the tissues

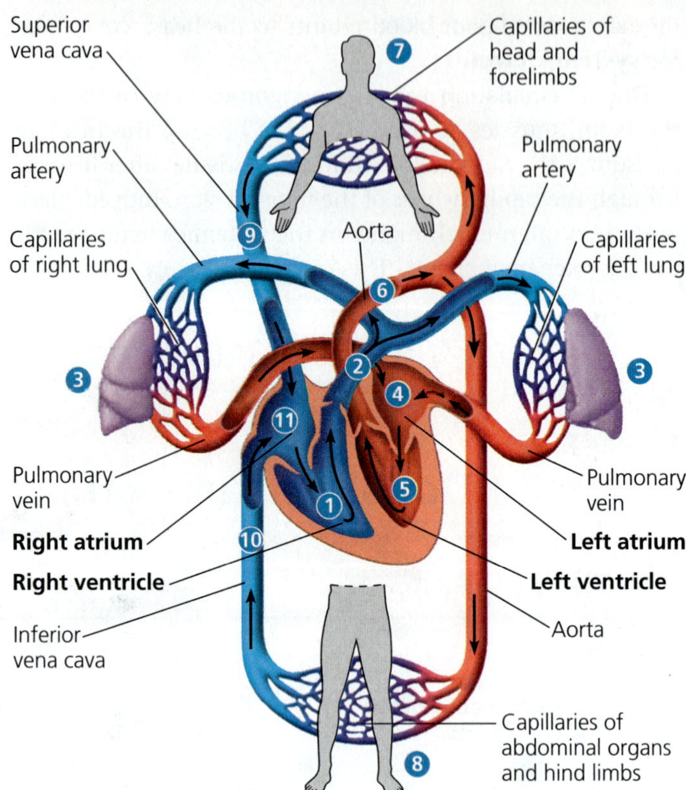

▲ **Figure 42.6 The mammalian cardiovascular system: an overview.** Note that the dual circuits operate simultaneously, not in the serial fashion that the numbering in the diagram suggests. The two ventricles pump almost in unison; while some blood is traveling in the pulmonary circuit, the rest of the blood is flowing in the systemic circuit.

and of CO_2 (produced by cellular respiration) into the blood. Capillaries rejoin, forming venules, which convey blood to veins. Oxygen-poor blood from the head, neck, and forelimbs is channeled into a large vein, ⑨ the superior vena cava. Another large vein, ⑩ the inferior vena cava, drains blood from the trunk and hind limbs. The two venae cavae empty their blood into ⑪ the right atrium, from which the oxygen-poor blood flows into the right ventricle.

The Mammalian Heart: *A Closer Look*

Using the human heart as an example, let's now take a closer look at how the mammalian heart works **(Figure 42.7)**. Located behind the sternum (breastbone), the human heart is about the size of a clenched fist and consists mostly of cardiac muscle (see Figure 40.5). The two atria have relatively thin walls and serve as collection chambers for blood returning to the heart from the lungs or other body tissues. Much of the blood that enters the atria flows into the ventricles while all heart chambers are relaxed. The remainder is transferred by contraction of the atria before the ventricles begin to contract. The ventricles have thicker walls and contract much more forcefully than the atria—especially the left ventricle, which pumps blood to all body organs through the systemic circuit. Although the left ventricle contracts with

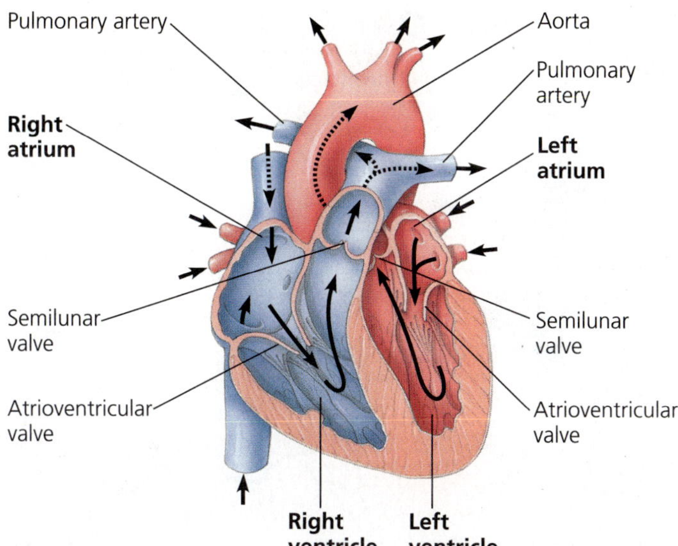

▲ **Figure 42.7 The mammalian heart: a closer look.** Notice the locations of the valves, which prevent backflow of blood within the heart. Also notice how the atria and left and right ventricles differ in the thickness of their muscular walls.

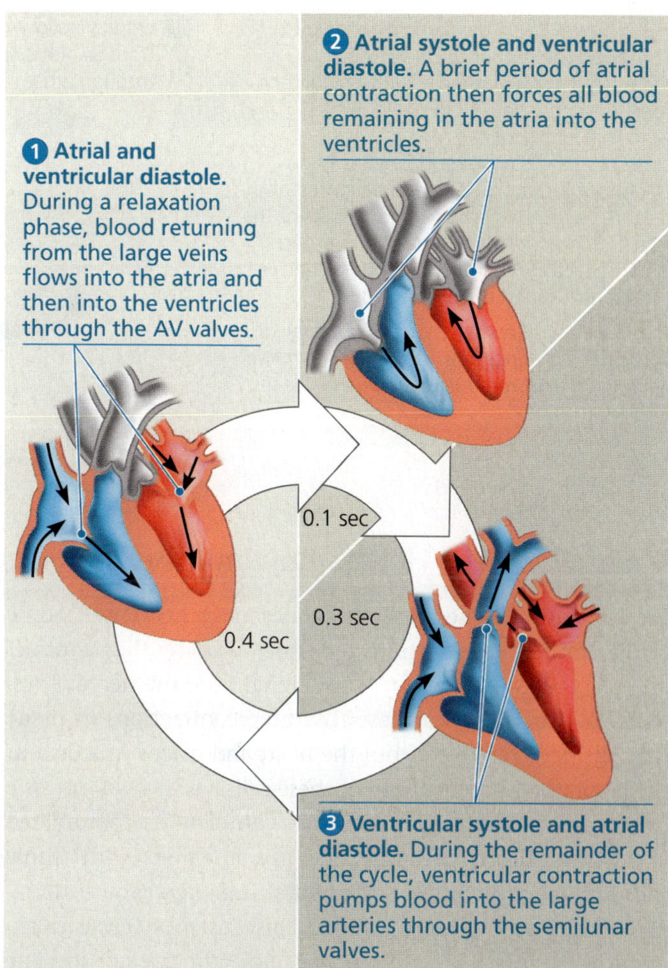

▲ **Figure 42.8 The cardiac cycle.** For an adult human at rest with a heart rate of about 72 beats per minute, one complete cardiac cycle takes about 0.8 second. Note that during all but 0.1 second of the cardiac cycle, the atria are relaxed and are filling with blood returning via the veins.

greater force than the right ventricle, it pumps the same volume of blood as the right ventricle during each contraction.

The heart contracts and relaxes in a rhythmic cycle. When it contracts, it pumps blood; when it relaxes, its chambers fill with blood. One complete sequence of pumping and filling is referred to as the **cardiac cycle**. The contraction phase of the cycle is called **systole**, and the relaxation phase is called **diastole** (**Figure 42.8**).

The volume of blood each ventricle pumps per minute is the **cardiac output**. Two factors determine cardiac output: the rate of contraction, or **heart rate** (number of beats per minute), and the **stroke volume**, the amount of blood pumped by a ventricle in a single contraction. The average stroke volume in humans is about 70 mL. Multiplying this stroke volume by a resting heart rate of 72 beats per minute yields a cardiac output of 5 L/min—about equal to the total volume of blood in the human body. During heavy exercise, cardiac output increases as much as fivefold.

Four valves in the heart prevent backflow and keep blood moving in the correct direction (see Figures 42.7 and 42.8). Made of flaps of connective tissue, the valves open when pushed from one side and close when pushed from the other. An **atrioventricular (AV) valve** lies between each atrium and ventricle. The AV valves are anchored by strong fibers that prevent them from turning inside out. Pressure generated by the powerful contraction of the ventricles closes the AV valves, keeping blood from flowing back into the atria. **Semilunar valves** are located at the two exits of the heart: where the aorta leaves the left ventricle and where the pulmonary artery leaves the right ventricle. These valves are pushed open by the pressure generated during contraction of the ventricles. When the ventricles relax, blood pressure built up in the aorta closes the semilunar valves and prevents significant backflow.

You can follow the closing of the two sets of heart valves either with a stethoscope or by pressing your ear tightly against the chest of a friend (or a friendly dog). The sound pattern is "lub-dup, lub-dup, lub-dup." The first heart sound ("lub") is created by the recoil of blood against the closed AV valves. The second sound ("dup") is produced by the recoil of blood against the closed semilunar valves.

If blood squirts backward through a defective valve, it may produce an abnormal sound called a **heart murmur**. Some people are born with heart murmurs; in others, the valves may be damaged by infection (from rheumatic fever, for instance). When a valve defect is severe enough to endanger health, surgeons may implant a mechanical replacement valve. However, not all heart murmurs are caused by a defect, and most valve defects do not reduce the efficiency of blood flow enough to warrant surgery.

▶ Figure 42.9 The control of heart rhythm. The sequence of electrical events in the heart is shown at the top; red highlights specialized muscle cells involved in the electrical control of the rhythm. The corresponding components of an electrocardiogram (ECG) are highlighted at the bottom in yellow. In step 4, the portion of the ECG to the right of the "spike" represents electrical activity that reprimes the ventricles for the next round of contraction.

WHAT IF? *If a doctor gave you a copy of your ECG recording, how could you determine what your heart rate had been during the test?*

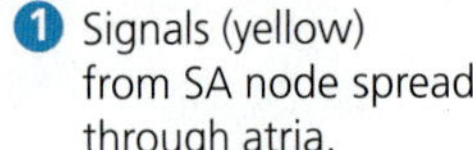

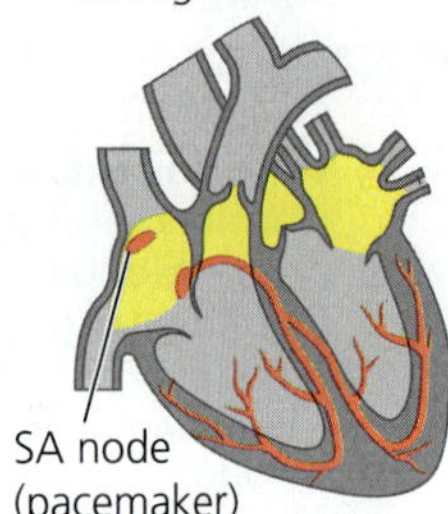

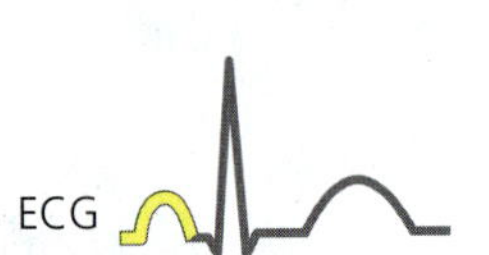

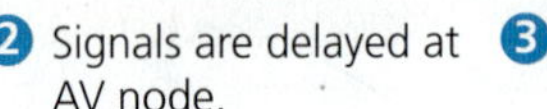

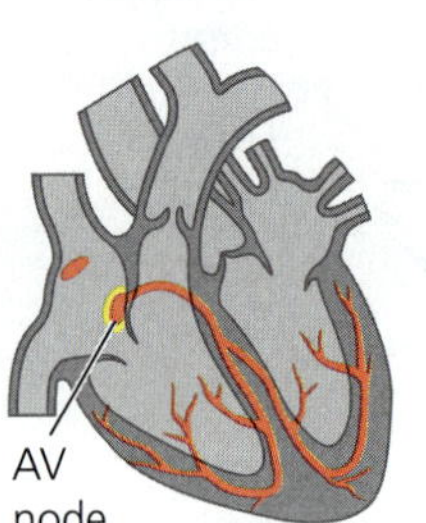

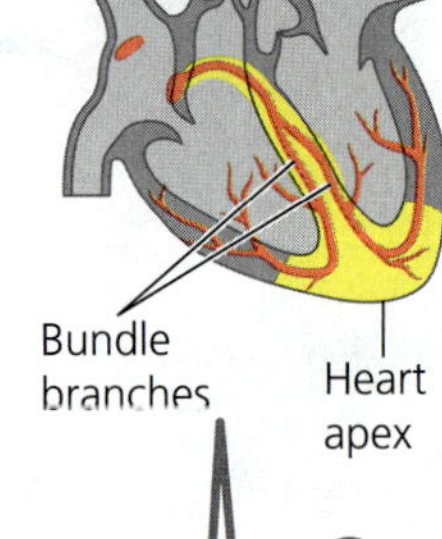

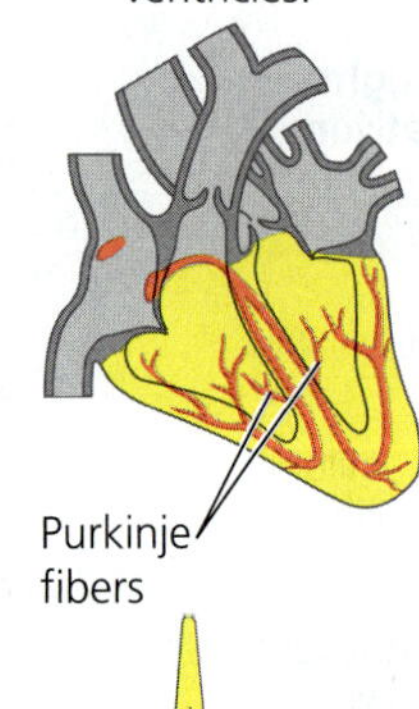

Maintaining the Heart's Rhythmic Beat

In vertebrates, the heartbeat originates in the heart itself. Some cardiac muscle cells are autorhythmic, meaning they contract and relax repeatedly without any signal from the nervous system. You can even see these rhythmic contractions in tissue that has been removed from the heart and placed in a dish in the laboratory! Because each of these cells has its own intrinsic contraction rhythm, how are their contractions coordinated in the intact heart? The answer lies in a group of autorhythmic cells located in the wall of the right atrium, near where the superior vena cava enters the heart. This cluster of cells is called the **sinoatrial (SA) node**, or *pacemaker*, and it sets the rate and timing at which all cardiac muscle cells contract. (In contrast to vertebrates, some arthropods have pacemakers located in the nervous system, outside the heart.)

The SA node generates electrical impulses much like those produced by nerve cells. Because cardiac muscle cells are electrically coupled through gap junctions (see Figure 6.32), impulses from the SA node spread rapidly within heart tissue. In addition, these impulses generate currents that are conducted to the skin via body fluids. In an **electrocardiogram (ECG** or, often, **EKG**, from the German spelling), these currents are recorded by electrodes placed on the skin. The resulting graph of current against time has a characteristic shape that represents the stages in the cardiac cycle (**Figure 42.9**).

Impulses from the SA node first spread rapidly through the walls of the atria, causing both atria to contract in unison. During atrial contraction, the impulses originating at the SA node reach other autorhythmic cells located in the wall between the left and right atria. These cells form a relay point called the **atrioventricular (AV) node**. Here the impulses are delayed for about 0.1 second before spreading to the heart apex. This delay allows the atria to empty completely before the ventricles contract. Then the signals from the AV node are conducted to the heart apex and throughout the ventricular walls by specialized muscle fibers called bundle branches and Purkinje fibers.

Physiological cues alter heart tempo by regulating the SA node. Two portions of the nervous system, the sympathetic and parasympathetic divisions, are largely responsible for this regulation. They function like the spurs and reins used in riding a horse: The sympathetic division speeds up the pacemaker, and the parasympathetic division slows it down. For example, when you stand up and start walking, the sympathetic division increases your heart rate, an adaptation that enables your circulatory system to provide the additional O_2 needed by the muscles that are powering your activity. If you then sit down and relax, the parasympathetic division decreases your heart rate, an adaptation that conserves energy. Hormones secreted into the blood also influence the pacemaker. For instance, epinephrine, the "fight-or-flight" hormone secreted by the adrenal glands, causes the heart rate to increase. A third type of input that affects the pacemaker is body temperature. An increase of only 1°C raises the heart rate by about 10 beats per minute. This is the reason your heart beats faster when you have a fever.

Having examined the operation of the circulatory pump, we turn in the next section to the forces and structures that influence blood flow in the vessels of each circuit.

CONCEPT CHECK 42.2

1. Explain why blood in the pulmonary veins has a higher O_2 concentration than blood in the venae cavae, which are also veins.
2. Why is it important that the AV node delay the electrical impulse moving from the SA node and the atria to the ventricles?
3. **WHAT IF?** After exercising regularly for several months, you find that your resting heart rate has decreased. What other change in the function of your heart at rest would you expect to find? Explain.

For suggested answers, see Appendix A.

CONCEPT 42.3

Patterns of blood pressure and flow reflect the structure and arrangement of blood vessels

The vertebrate circulatory system enables blood to deliver oxygen and nutrients and remove wastes throughout the body. In doing so, the circulatory system relies on a branching network of vessels much like the plumbing system that delivers fresh water to a city and removes its wastes. In fact, the same physical principles that govern the operation of plumbing systems apply to the functioning of blood vessels.

Blood Vessel Structure and Function

Blood vessels contain a central lumen (cavity) lined with an **endothelium**, a single layer of flattened epithelial cells. The smooth surface of the endothelium minimizes resistance to the flow of blood. Surrounding the endothelium are layers of tissue that differ in capillaries, arteries, and veins, reflecting the specialized functions of these vessels.

Capillaries are the smallest blood vessels, having a diameter only slightly greater than that of a red blood cell (**Figure 42.10**). Capillaries also have very thin walls, which consist of just the endothelium and its basal lamina. This structural organization facilitates the exchange of substances between the blood in capillaries and the interstitial fluid.

The walls of arteries and veins have a more complex organization than those of capillaries. Both arteries and veins have two layers of tissue surrounding the endothelium: an outer layer of connective tissue containing elastic fibers, which allow the vessel to stretch and recoil, and a middle layer containing smooth muscle and more elastic fibers. However, the walls of arteries and veins also differ, reflecting distinct adaptations of these vessels to their particular functions in circulation.

The walls of arteries are thick and strong, accommodating blood pumped at high pressure by the heart. Arterial walls also have an elastic recoil that helps maintain blood pressure and flow to capillaries when the heart relaxes between contractions. Signals from the nervous system and hormones circulating in the blood act on the smooth muscle in arteries and arterioles, dilating or constricting these vessels and thus controlling blood flow to different parts of the body.

Because veins convey blood back to the heart at a lower pressure, they do not require thick walls. For a given blood vessel diameter, a vein has a wall only about a third as thick as that of an artery. Valves inside the veins maintain a unidirectional flow of blood despite the low blood pressure.

We consider next how blood vessel diameter, vessel number, and pressure influence the speed at which blood flows in different locations within the body.

Blood Flow Velocity

To understand how blood vessel diameter influences blood flow, consider how water flows through a thick hose connected to a faucet. When the faucet is turned on, water flows at the same velocity at each point along the hose. However, if a narrow nozzle is attached to the end of the hose, the water

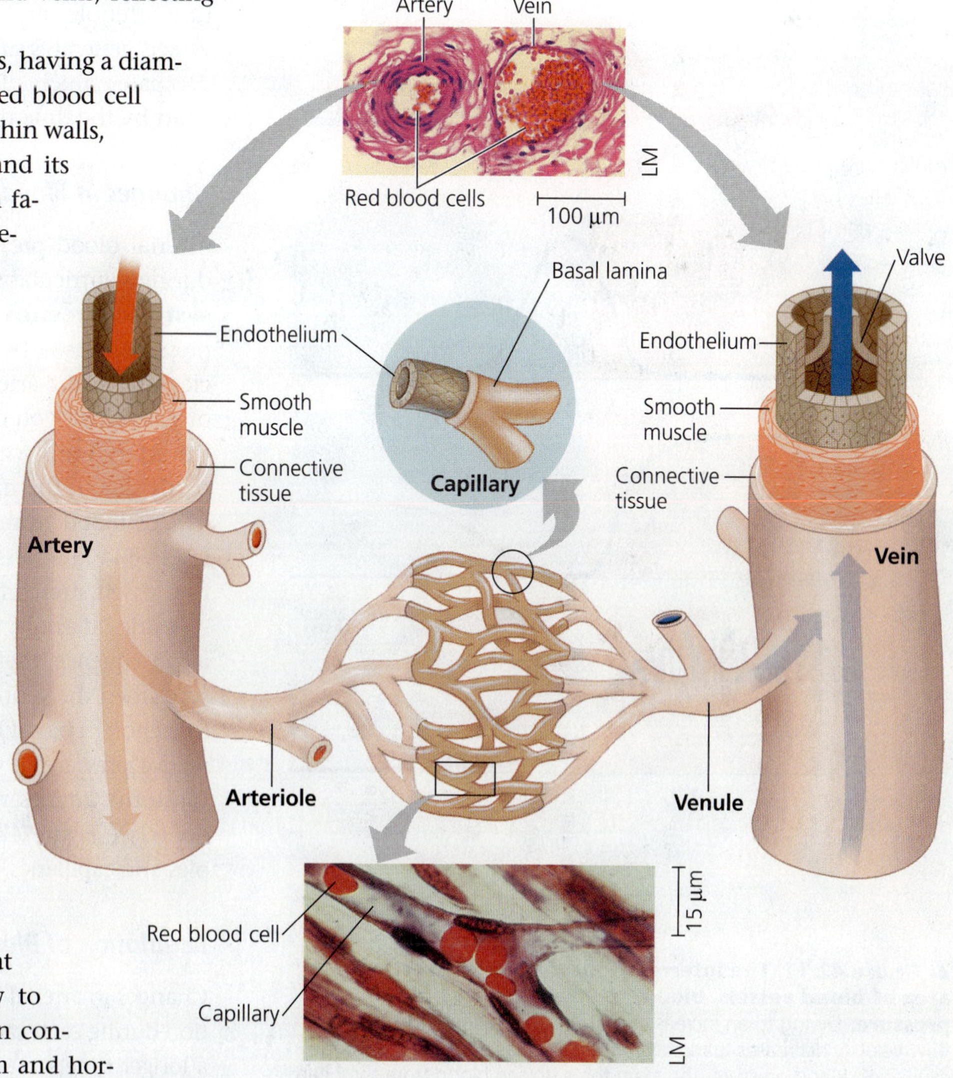

▲ **Figure 42.10 The structure of blood vessels.**

will exit the nozzle at a much greater velocity. Because water doesn't compress under pressure, the volume of water moving through the nozzle in a given time must be the same as the volume moving through the rest of the hose. The cross-sectional area of the nozzle is smaller than that of the hose, so the water speeds up in the nozzle.

An analogous situation exists in the circulatory system, but blood *slows* as it moves from arteries to arterioles to capillaries. Why? The reason is that the number of capillaries is enormous. Each artery conveys blood to so many capillaries that the *total* cross-sectional area is much greater in capillary beds than in the arteries or any other part of the circulatory system **(Figure 42.11)**. The result is a dramatic decrease in velocity from the arteries to the capillaries: Blood travels 500 times slower in the capillaries (about 0.1 cm/sec) than in the aorta (about 48 cm/sec).

The reduced velocity of blood flow in capillaries is essential to the function of the circulatory system. The exchange of substances between the blood and interstitial fluid occurs only in capillaries because only capillaries have walls thin enough to permit this transfer. Diffusion, however, is not instantaneous. The slower flow of blood through capillaries is thus necessary to provide time for exchange to occur. After passing through the capillaries, the blood speeds up as it enters the venules and veins, which have smaller *total* cross-sectional areas than the capillaries.

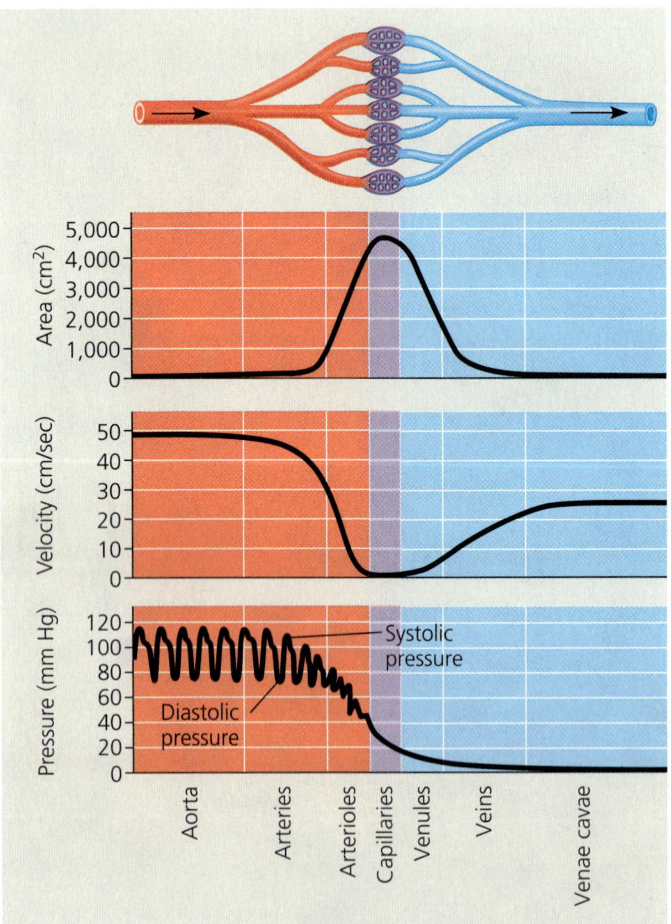

▲ **Figure 42.11 The interrelationship of cross-sectional area of blood vessels, blood flow velocity, and blood pressure.** Owing to an increase in total cross-sectional area, blood flow velocity decreases markedly in the arterioles and is lowest in the capillaries. Blood pressure, the main force driving blood from the heart to the capillaries, is highest in the aorta and other arteries.

Blood Pressure

Blood, like all fluids, flows from areas of higher pressure to areas of lower pressure. Contraction of a heart ventricle generates blood pressure, which exerts a force in all directions. The force directed lengthwise in an artery causes the blood to flow away from the heart, the site of highest pressure. The force exerted against the elastic wall of an artery stretches the wall, and the recoil of arterial walls plays a critical role in maintaining blood pressure, and hence blood flow, throughout the cardiac cycle. Once the blood enters the millions of tiny arterioles and capillaries, the narrow diameter of these vessels generates substantial resistance to flow. This resistance dissipates much of the pressure generated by the pumping heart by the time the blood enters the veins.

Changes in Blood Pressure During the Cardiac Cycle

Arterial blood pressure is highest when the heart contracts during ventricular systole. The pressure at this time is called **systolic pressure** (see Figure 42.11). The spikes in blood pressure caused by the powerful contractions of the ventricles stretch the arteries. By placing your fingers on the inside of your wrist, you can feel a **pulse**—the rhythmic bulging of the artery walls with each heartbeat. The surge of pressure is partly due to the narrow openings of arterioles impeding the exit of blood from the arteries. Thus, when the heart contracts, blood enters the arteries faster than it can leave, and the vessels stretch from the rise in pressure.

During diastole, the elastic walls of the arteries snap back. As a consequence, there is a lower but still substantial blood pressure when the ventricles are relaxed (**diastolic pressure**). Before enough blood has flowed into the arterioles to completely relieve pressure in the arteries, the heart contracts again. Because the arteries remain pressurized throughout the cardiac cycle (see Figure 42.11), blood continuously flows into arterioles and capillaries.

Regulation of Blood Pressure

Changes in arterial blood pressure are not limited to the oscillation during each cardiac cycle. Blood pressure also fluctuates on a longer time scale in response to signals that change the state of smooth muscles in arteriole walls. For example, physical or emotional stress can trigger nervous and hormonal responses that cause smooth muscles in arteriole walls to contract. When that happens, the arterioles narrow, a process called **vasoconstriction**. Narrowing of the arterioles increases blood pressure upstream in the arteries. When the smooth muscles relax, the arterioles undergo **vasodilation**, an increase in diameter that causes blood pressure in the arteries to fall.

Researchers have identified a gas, nitric oxide (NO), as a major inducer of vasodilation and a peptide, endothelin, as the most potent inducer of vasoconstriction. Both NO and endothelin are signaling molecules produced in blood vessels in response to cues from the nervous and endocrine systems. Each kind of molecule binds to a specific receptor, activating a signal transduction pathway that alters smooth muscle contraction and thus changes blood vessel diameter.

Vasoconstriction and vasodilation are often coupled to changes in cardiac output that also affect blood pressure. This coordination of regulatory mechanisms maintains adequate blood flow as the body's demands on the circulatory system change. During heavy exercise, for example, the arterioles in working muscles dilate, causing a greater flow of oxygen-rich blood to the muscles. By itself, this increased flow to the muscles would cause a drop in blood pressure (and therefore blood flow) in the body as a whole. However, cardiac output increases at the same time, maintaining blood pressure and supporting the necessary increase in blood flow.

Blood Pressure and Gravity

Blood pressure is generally measured for an artery in the arm at the same height as the heart **(Figure 42.12)**. For a healthy 20-year-old human at rest, arterial blood pressure in the systemic circuit is typically about 120 millimeters of mercury (mm Hg) at systole and 70 mm Hg at diastole, expressed as 120/70. (Arterial blood pressure in the pulmonary circuit is six to ten times lower.)

Gravity has a significant effect on blood pressure. When you are standing, for example, your head is roughly 0.35 m higher than your chest, and the arterial blood pressure in your brain is about 27 mm Hg less than that near your heart. If the blood pressure in your brain is too low to provide adequate blood flow, you will likely faint. By causing your body to collapse to the ground, fainting effectively places your head at the level of your heart, quickly increasing blood flow to your brain.

The challenge of pumping blood against gravity is particularly great for animals with very long necks. A giraffe, for example, requires a systolic pressure of more than 250 mm Hg near the heart to get blood to its head. When a giraffe lowers its head to drink, one-way valves and sinuses, along with feedback mechanisms that reduce cardiac output, prevent this high pressure from damaging its brain. We can calculate that a dinosaur with a neck nearly 10 m long would have required even greater systolic pressure—nearly 760 mm Hg—to pump blood to its brain when its head was fully raised. However, calculations based on anatomy and inferred metabolic rate suggest that dinosaurs did not have a heart powerful enough to generate such high pressure. Based on this evidence as well as studies of neck bone structure, some biologists have concluded that the long-necked dinosaurs fed close to the ground rather than on high foliage.

Gravity is also a consideration for blood flow in veins, especially those in the legs. Although blood pressure in veins is relatively low, several mechanisms assist the return of venous blood to the heart. First, rhythmic contractions of smooth muscles in the walls of venules and veins aid in the movement of the blood. Second, and more important, the contraction of skeletal muscles during exercise squeezes blood through the

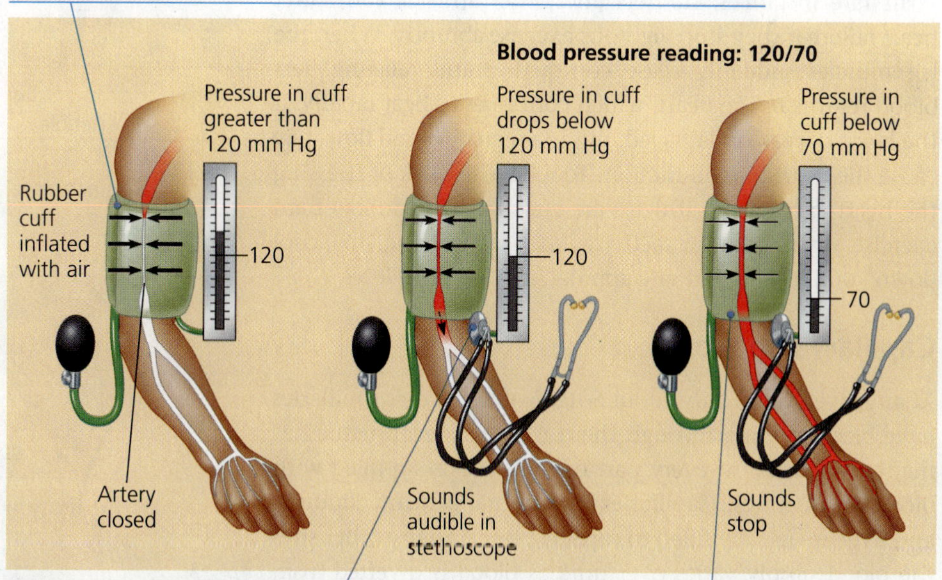

▲ **Figure 42.12 Measurement of blood pressure.** Blood pressure is recorded as two numbers separated by a slash. The first number is the systolic pressure; the second is the diastolic pressure.

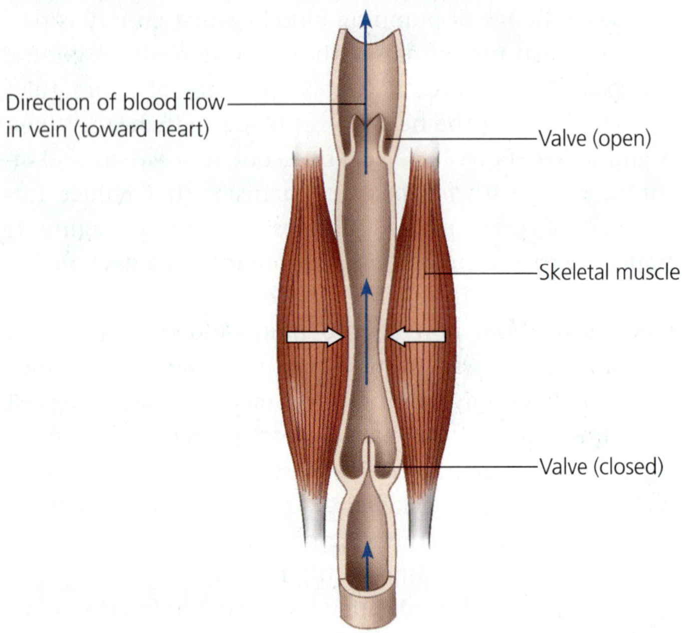

▲ **Figure 42.13 Blood flow in veins.** Skeletal muscle contraction squeezes and constricts veins. Flaps of tissue within the veins act as one-way valves that keep blood moving only toward the heart. If you sit or stand too long, the lack of muscular activity may cause your feet to swell as blood pools in your veins.

veins toward the heart **(Figure 42.13)**. Third, the change in pressure within the thoracic (chest) cavity during inhalation causes the venae cavae and other large veins near the heart to expand and fill with blood.

In rare instances, runners and other athletes can suffer heart failure if they stop vigorous exercise abruptly. When the leg muscles suddenly cease contracting and relaxing, less blood returns to the heart, which continues to beat rapidly. If the heart is weak or damaged, this inadequate blood flow may cause the heart to malfunction. To reduce the risk of stressing the heart excessively, athletes are encouraged to follow hard exercise with moderate activity, such as walking, to "cool down" until their heart rate approaches its resting level.

Capillary Function

At any given time, only about 5–10% of the body's capillaries have blood flowing through them. However, each tissue has many capillaries, so every part of the body is supplied with blood at all times. Capillaries in the brain, heart, kidneys, and liver are usually filled to capacity, but at many other sites the blood supply varies over time as blood is diverted from one destination to another. For example, blood flow to the skin is regulated to help control body temperature, and blood supply to the digestive tract increases after a meal. During strenuous exercise, blood is diverted from the digestive tract and supplied more generously to skeletal muscles and skin. This is one reason why exercising heavily immediately after eating a big meal may cause indigestion.

Given that capillaries lack smooth muscle, how is blood flow in capillary beds altered? There are two mechanisms, both of which rely on signals that regulate the flow into capillaries. One mechanism involves contraction of the smooth muscle in the wall of an arteriole, which reduces the vessel's diameter and decreases blood flow to the adjoining capillary beds. When the smooth muscle relaxes, the arterioles dilate, allowing blood to enter the capillaries. The other mechanism for altering flow, shown in **Figure 42.14**, involves the action of *precapillary sphincters*, rings of smooth muscle located at the entrance to capillary beds. The signals that regulate blood flow include nerve impulses, hormones traveling throughout the bloodstream, and chemicals produced locally. For example, the chemical histamine released by cells at a wound site causes smooth muscle relaxation, dilating blood vessels and increasing blood flow. The dilated vessels also give disease-fighting white blood cells greater access to invading microorganisms.

As you have read, the critical exchange of substances between the blood and interstitial fluid takes place across the

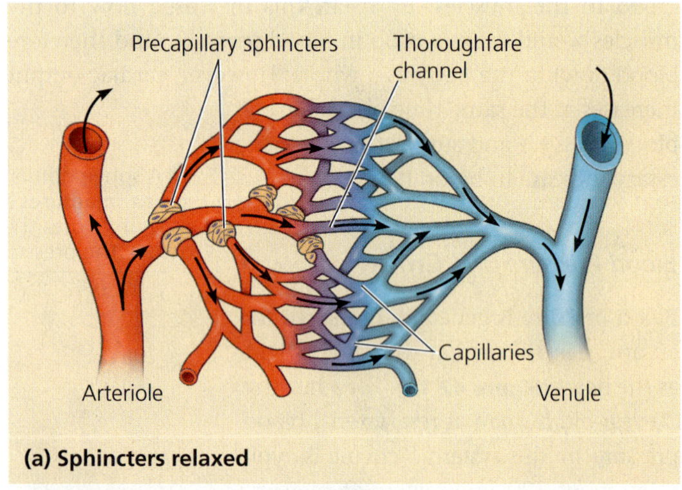

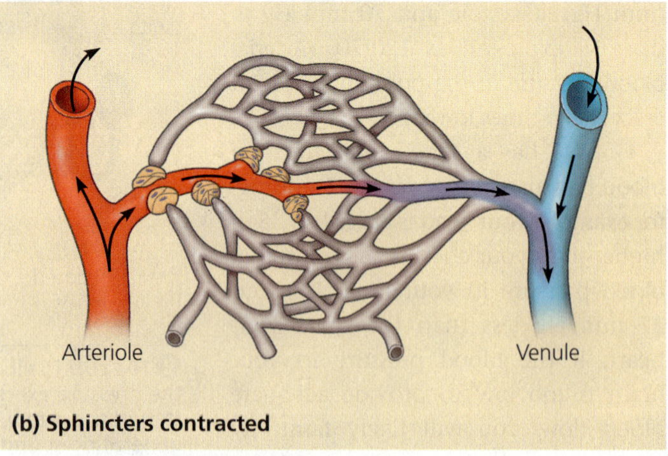

▲ **Figure 42.14 Blood flow in capillary beds.** Precapillary sphincters regulate the passage of blood into capillary beds. Some blood flows directly from arterioles to venules through capillaries called thoroughfare channels, which are always open.

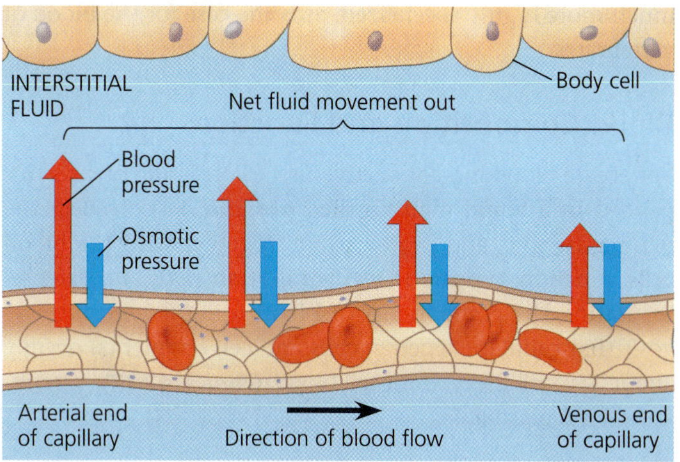

▲ **Figure 42.15 Fluid exchange between capillaries and the interstitial fluid.** This diagram shows a hypothetical capillary in which blood pressure exceeds osmotic pressure throughout the entire length of the capillary. In other capillaries, blood pressure may be lower than osmotic pressure along all or part of the capillary.

thin endothelial walls of the capillaries. Some substances are carried across the endothelium in vesicles that form on one side by endocytosis and release their contents on the opposite side by exocytosis. Small molecules, such as O_2 and CO_2, simply diffuse across the endothelial cells or, in some tissues, through microscopic pores in the capillary wall. These openings also provide the route for transport of small solutes such as sugars, salts, and urea, as well as for bulk flow of fluid into tissues driven by blood pressure within the capillary.

Two opposing forces control the movement of fluid between the capillaries and the surrounding tissues: Blood pressure tends to drive fluid out of the capillaries, and the presence of blood proteins tends to pull fluid back **(Figure 42.15)**. Many blood proteins (and all blood cells) are too large to pass readily through the endothelium, and they remain in the capillaries. These dissolved proteins are responsible for much of the blood's *osmotic pressure* (the pressure produced by the difference in solute concentration across a membrane). The difference in osmotic pressure between the blood and the interstitial fluid opposes fluid movement out of the capillaries. On average, blood pressure is greater than the opposing forces, leading to a net loss of fluid from capillaries. The net loss is generally greatest at the arterial end of these vessels, where blood pressure is highest.

Fluid Return by the Lymphatic System

Each day, the adult human body loses approximately 4–8 L of fluid from capillaries to the surrounding tissues. There is also some leakage of blood proteins, even though the capillary wall is not very permeable to large molecules. The lost fluid and proteins return to the blood via the **lymphatic system**, which includes a network of tiny vessels intermingled among capillaries of the cardiovascular system.

After entering the lymphatic system by diffusion, the fluid lost by capillaries is called **lymph**; its composition is about the same as that of interstitial fluid. The lymphatic system drains into large veins of the circulatory system at the base of the neck (see Figure 43.7). As you read in Chapter 41, this joining of the lymphatic and circulatory systems functions in the transfer of lipids from the small intestine to the blood.

The movement of lymph from peripheral tissues to the heart relies on much the same mechanisms that assist blood flow in veins. Lymph vessels, like veins, have valves that prevent the backflow of fluid. Rhythmic contractions of the vessel walls help draw fluid into the small lymphatic vessels. In addition, skeletal muscle contractions play a role in moving lymph.

Disorders that interfere with the lymphatic system highlight its role in maintaining proper fluid distribution in the body. Disruptions in the movement of lymph often cause edema, swelling resulting from the excessive accumulation of fluid in tissues. Severe blockage of lymph flow, as occurs when certain parasitic worms lodge in lymph vessels, results in extremely swollen limbs or other body parts, a condition known as elephantiasis.

Along a lymph vessel are organs called **lymph nodes** **(Figure 42.16)**. By filtering the lymph and by housing cells that attack viruses and bacteria, lymph nodes play an important role in the body's defense. Inside each lymph node is a honeycomb of connective tissue with spaces filled by white blood cells. When the body is fighting an infection, these cells multiply rapidly, and the lymph nodes become swollen and tender (which is why your doctor may check for swollen lymph nodes in your neck, armpits, or groin when you feel sick). Because lymph nodes have filtering and surveillance functions, doctors may examine the lymph nodes of cancer patients to detect the spread of diseased cells.

In recent years, evidence has surfaced demonstrating that the lymphatic system also plays a role in harmful immune responses, such as those responsible for asthma. Because of these and other findings, the lymphatic system, largely ignored until the 1990s, has become a very active and promising area of biomedical research.

▶ **Figure 42.16 Human lymph nodes and vessels.** In this colorized X-ray image of the groin, lymph nodes and vessels (yellow) are visible next to the upper thigh bone (femur).

CONCEPT CHECK 42.3

1. What is the primary cause of the low velocity of blood flow through capillaries?
2. What short-term changes in cardiovascular function might best enable skeletal muscles to help an animal escape from a dangerous situation?
3. **WHAT IF?** If you had additional hearts distributed throughout your body, what would be one likely advantage and one likely disadvantage?

For suggested answers, see Appendix A.

CONCEPT 42.4

Blood components function in exchange, transport, and defense

As we discussed earlier, the fluid transported by an open circulatory system is continuous with the fluid that surrounds all of the body cells and therefore has the same composition. In contrast, the fluid in a closed circulatory system can be much more highly specialized, as is the case for the blood of vertebrates.

Blood Composition and Function

Vertebrate blood is a connective tissue consisting of cells suspended in a liquid matrix called **plasma**. Dissolved in the plasma are ions and proteins that, together with the blood cells, function in osmotic regulation, transport, and defense. Separating the components of blood using a centrifuge reveals that cellular elements (cells and cell fragments) occupy about 45% of the volume of blood **(Figure 42.17)**. The remainder is plasma.

Plasma

Among the many solutes in plasma are inorganic salts in the form of dissolved ions, sometimes referred to as blood electrolytes (see Figure 42.17). Although plasma is about 90% water, the dissolved salts are an essential component of the blood. Some of these ions buffer the blood, which in humans normally has a pH of 7.4. Salts are also important in maintaining the osmotic balance of the blood. In addition, the concentration of ions in plasma directly affects the composition of

Plasma 55%	
Constituent	**Major functions**
Water	Solvent for carrying other substances
Ions (blood electrolytes) Sodium Potassium Calcium Magnesium Chloride Bicarbonate	Osmotic balance, pH buffering, and regulation of membrane permeability
Plasma proteins	
Albumin	Osmotic balance, pH buffering
Fibrinogen	Clotting
Immunoglobulins (antibodies)	Defense
Substances transported by blood Nutrients (such as glucose, fatty acids, vitamins) Waste products of metabolism Respiratory gases (O_2 and CO_2) Hormones	

Separated blood elements

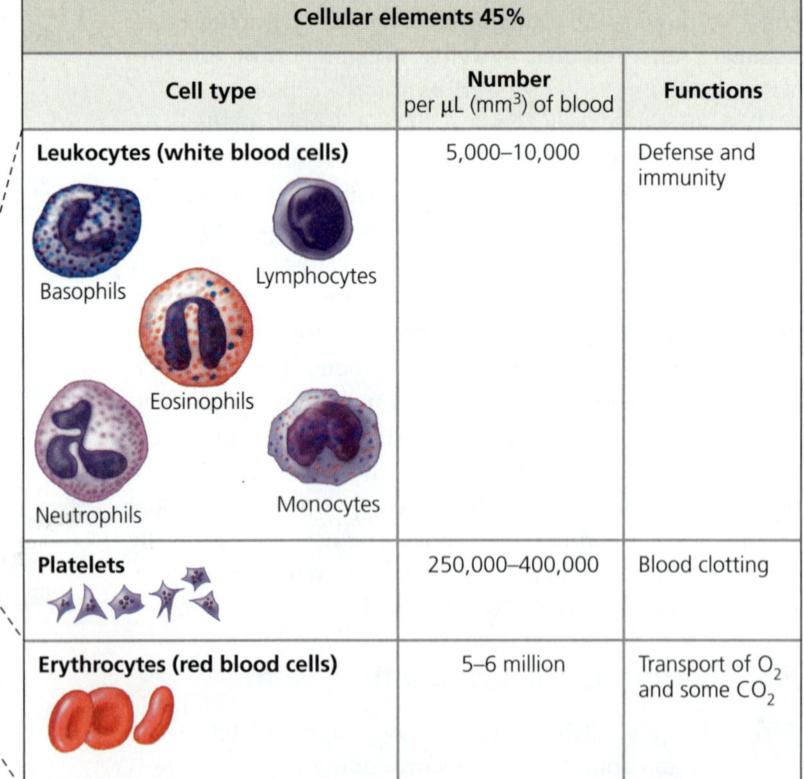

Cellular elements 45%		
Cell type	**Number** per µL (mm^3) of blood	**Functions**
Leukocytes (white blood cells) Basophils Lymphocytes Eosinophils Neutrophils Monocytes	5,000–10,000	Defense and immunity
Platelets	250,000–400,000	Blood clotting
Erythrocytes (red blood cells)	5–6 million	Transport of O_2 and some CO_2

▲ **Figure 42.17 The composition of mammalian blood.**

the interstitial fluid, where many of these ions have a vital role in muscle and nerve activity. To serve all of these functions, plasma electrolytes must be kept within narrow concentration ranges, a homeostatic function we will explore in Chapter 44.

Plasma proteins act as buffers against pH changes, help maintain the osmotic balance between blood and interstitial fluid, and contribute to the blood's viscosity (thickness). Particular plasma proteins have additional functions. The immunoglobulins, or antibodies, help combat viruses and other foreign agents that invade the body (see Chapter 43). Others are escorts for lipids, which are insoluble in water and can travel in blood only when bound to proteins. A third group of plasma proteins are clotting factors that help plug leaks when blood vessels are injured. (The term *serum* refers to blood plasma from which these clotting factors have been removed.)

Plasma also contains a wide variety of other substances in transit from one part of the body to another, including nutrients, metabolic wastes, respiratory gases, and hormones. Plasma has a much higher protein concentration than interstitial fluid, although the two fluids are otherwise similar. (Capillary walls, remember, are not very permeable to proteins.)

Cellular Elements

Blood contains two classes of cells: red blood cells, which transport O_2, and white blood cells, which function in defense (see Figure 42.17). Also suspended in blood plasma are **platelets**, fragments of cells that are involved in the clotting process.

Erythrocytes Red blood cells, or **erythrocytes**, are by far the most numerous blood cells. Each microliter (μL, or mm^3) of human blood contains 5–6 million red cells, and there are about 25 trillion of these cells in the body's 5 L of blood. Their main function is O_2 transport, and their structure is closely related to this function. Human erythrocytes are small disks (7–8 μm in diameter) that are biconcave—thinner in the center than at the edges. This shape increases surface area, enhancing the rate of diffusion of O_2 across their plasma membranes. Mature mammalian erythrocytes lack nuclei. This unusual characteristic leaves more space in these tiny cells for **hemoglobin**, the iron-containing protein that transports O_2 (see Figure 5.20). Erythrocytes also lack mitochondria and generate their ATP exclusively by anaerobic metabolism. Oxygen transport would be less efficient if erythrocytes were aerobic and consumed some of the O_2 they carry.

Despite its small size, an erythrocyte contains about 250 million molecules of hemoglobin. Because each molecule of hemoglobin binds up to four molecules of O_2, one erythrocyte can transport about a billion O_2 molecules. As erythrocytes pass through the capillary beds of lungs, gills, or other respiratory organs, O_2 diffuses into the erythrocytes and binds to hemoglobin. In the systemic capillaries, O_2 dissociates from hemoglobin and diffuses into body cells.

In **sickle-cell disease**, an abnormal form of hemoglobin (Hb^S) polymerizes into aggregates. Because the concentration of hemoglobin in erythrocytes is so high, these aggregates are large enough to distort the erythrocyte into an elongated, curved shape that resembles a sickle. As you learned in Chapter 5, this abnormality results from an alteration in the amino acid sequence of hemoglobin at a single position (see Figure 5.21).

Sickle-cell disease significantly impairs the function of the circulatory system. Sickled cells often lodge in arterioles and capillaries, preventing delivery of O_2 and nutrients and removal of CO_2 and wastes. Blood vessel blockage and resulting organ swelling often result in severe pain. In addition, sickled cells frequently rupture, reducing the number of red blood cells available for transporting O_2. The average life span of a sickled erythrocyte is only 20 days—one-sixth that of a normal erythrocyte. The rate of erythrocyte loss outstrips the replacement capacity of the bone marrow. Short-term therapy includes replacement of erythrocytes by blood transfusion; long-term treatments are generally aimed at inhibiting aggregation of Hb^S.

Leukocytes The blood contains five major types of white blood cells, or **leukocytes**. Their function is to fight infections. Some are phagocytic, engulfing and digesting microorganisms as well as debris from the body's own dead cells. As we will see in Chapter 43, other leukocytes, called lymphocytes, develop into specialized B cells and T cells that mount immune responses against foreign substances. Normally, 1 μL of human blood contains about 5,000–10,000 leukocytes; their numbers increase temporarily whenever the body is fighting an infection. Unlike erythrocytes, leukocytes are also found outside the circulatory system, patrolling both interstitial fluid and the lymphatic system.

Platelets Platelets are pinched-off cytoplasmic fragments of specialized bone marrow cells. They are about 2–3 μm in diameter and have no nuclei. Platelets serve both structural and molecular functions in blood clotting.

Blood Clotting

The occasional cut or scrape is not life-threatening because blood components seal the broken blood vessels. A break in a blood vessel wall exposes proteins that attract platelets and initiate coagulation, the conversion of liquid components of blood to a solid clot. The coagulant, or sealant, circulates in an inactive form called *fibrinogen*. In response to a broken blood vessel, platelets release clotting factors that trigger reactions leading to the formation of *thrombin*, an enzyme that converts fibrinogen to *fibrin*. Newly formed fibrin aggregates into threads that form the framework of the clot. Thrombin also activates a factor that catalyzes the formation of more thrombin, driving clotting to completion through positive feedback (see Chapter 40). The steps in the production of a

1. The clotting process begins when the endothelium of a vessel is damaged, exposing connective tissue in the vessel wall to blood. Platelets adhere to collagen fibers in the connective tissue and release a substance that makes nearby platelets sticky.

2. The platelets form a plug that provides emergency protection against blood loss.

3. This plug is reinforced by a fibrin clot when vessel damage is severe.

Collagen fibers
Platelet
Platelet plug
Fibrin clot
Red blood cell
5 µm

Clotting factors from:
Platelets
Damaged cells
Plasma (factors include calcium, vitamin K)

Enzymatic cascade
Prothrombin → Thrombin
Fibrinogen → Fibrin

Fibrin clot formation
Clotting factors released from the clumped platelets or damaged cells mix with clotting factors in the plasma, forming an enzymatic cascade that converts a plasma protein called prothrombin to its active form, thrombin. Thrombin itself is an enzyme that catalyzes the final step of the clotting process, the conversion of fibrinogen to fibrin. The threads of fibrin become interwoven into a clot (see colorized SEM above).

▲ **Figure 42.18 Blood clotting.**

blood clot are diagrammed in **Figure 42.18**. Any genetic mutation that blocks a step in the clotting process can cause hemophilia, a disease characterized by excessive bleeding and bruising from even minor cuts and bumps (see Chapter 15).

Anticlotting factors in the blood normally prevent spontaneous clotting in the absence of injury. Sometimes, however, clots form within a blood vessel, blocking the flow of blood. Such a clot is called a **thrombus**. We will explore how a thrombus forms and the danger that it poses later in this chapter.

Stem Cells and the Replacement of Cellular Elements

Erythrocytes, leukocytes, and platelets all develop from a common source: multipotent **stem cells** that are dedicated to replenishing the body's blood cell populations **(Figure 42.19)**. The stem cells that produce blood cells are located in the red marrow of bones, particularly the ribs, vertebrae, sternum, and pelvis. Multipotent stem cells are so named because they have the ability to form multiple types of cells—in this case, the myeloid and lymphoid cell lineages. When a stem cell divides, one daughter cell remains a stem cell while the other takes on a specialized function.

Throughout a person's life, erythrocytes, leukocytes, and platelets arising from stem cell divisions replace the worn-out

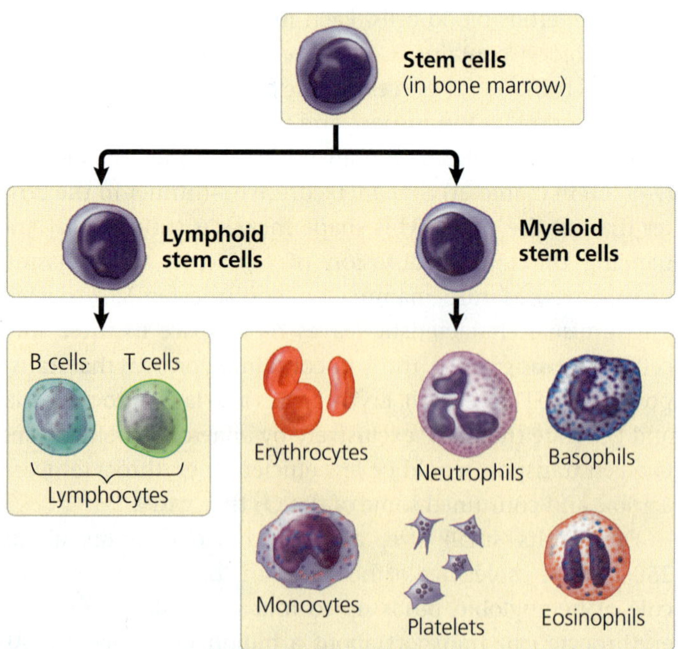

▲ **Figure 42.19 Differentiation of blood cells.** Some of the multipotent stem cells differentiate into lymphoid stem cells, which then develop into B cells and T cells, two types of lymphocytes that function in immunity (see Chapter 43). All other blood cells and platelets arise from myeloid stem cells.

cellular elements of blood. Erythrocytes, for example, circulate for only 120 days on average before being replaced; the old cells are consumed by phagocytic cells in the liver and spleen. The production of new erythrocytes involves recycling of materials, such as the use of iron scavenged from old erythrocytes in new hemoglobin molecules.

A negative-feedback mechanism, sensitive to the amount of O_2 reaching the body's tissues via the blood, controls erythrocyte production. If the tissues do not receive enough O_2, the kidneys synthesize and secrete a hormone called **erythropoietin (EPO)** that stimulates erythrocyte production. If the blood is delivering more O_2 than the tissues can use, the level of EPO falls and erythrocyte production slows. Physicians use synthetic EPO to treat people with health problems such as *anemia*, a condition of lower-than-normal erythrocyte or hemoglobin levels that lowers the oxygen-carrying capacity of the blood. Some athletes inject themselves with EPO to increase their erythrocyte levels, although this practice, a form of blood doping, has been banned by the International Olympic Committee and other sports organizations. In recent years, a number of well-known runners and cyclists have tested positive for EPO-related drugs and have forfeited both their records and their right to participate in future competitions.

Cardiovascular Disease

More than half of all human deaths in the United States are caused by cardiovascular diseases—disorders of the heart and blood vessels. Cardiovascular diseases range from a minor disturbance of vein or heart valve function to a life-threatening disruption of blood flow to the heart or brain.

Cholesterol metabolism plays a central role in cardiovascular disease. As you learned in Chapter 7, the presence of this steroid in animal cell membranes helps maintain normal membrane fluidity. Cholesterol travels in blood plasma mainly in particles that consist of thousands of cholesterol molecules and other lipids bound to a protein. One type of particle—**low-density lipoprotein (LDL)**—delivers cholesterol to cells for membrane production. Another type—**high-density lipoprotein (HDL)**—scavenges excess cholesterol for return to the liver. Individuals with a high ratio of LDL to HDL are at substantially increased risk for heart disease.

Another factor in cardiovascular disease is *inflammation*, the body's reaction to injury. As you will learn in the next chapter, tissue damage leads to recruitment of two types of circulating immune cells, macrophages and leukocytes. Signals released by these cells trigger a flow of fluid out of blood vessels at the site of injury, resulting in the tissue swelling characteristic of inflammation (see Figure 43.8). Although inflammation is often a normal and healthy response to injury, it can significantly disrupt circulatory function, as explained in the next section.

Atherosclerosis, Heart Attacks, and Stroke

Circulating cholesterol and inflammation can act together to produce a cardiovascular disease called **atherosclerosis**, the hardening of the arteries by accumulation of fatty deposits **(Figure 42.20)**. Healthy arteries have a smooth inner lining

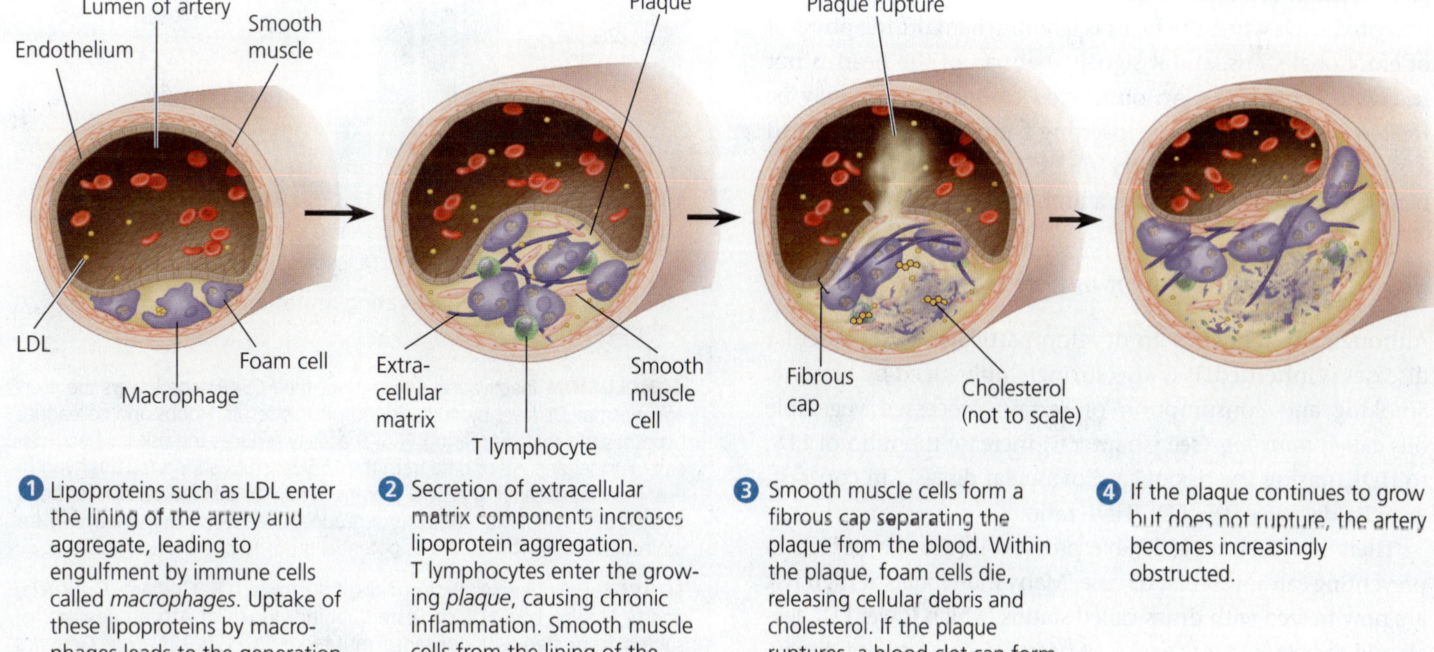

▲ **Figure 42.20 Atherosclerosis.** In atherosclerosis, thickening of an arterial wall by plaque formation can restrict blood flow through the artery. Fragments of a ruptured plaque can travel via the bloodstream and become lodged in other arteries. If those arteries supply the heart or brain, the resulting obstruction could cause a heart attack or stroke, respectively.

that reduces resistance to blood flow. Damage or infection can roughen the lining and lead to inflammation. Leukocytes are attracted to the damaged lining and begin to take up lipids, including cholesterol. A fatty deposit, called a plaque, grows steadily, incorporating fibrous connective tissue and additional cholesterol. As the plaque grows, the walls of the artery become thick and stiff, and the obstruction of the artery increases.

The result of untreated atherosclerosis is often a heart attack or a stroke. A **heart attack**, also called a *myocardial infarction*, is the damage or death of cardiac muscle tissue resulting from blockage of one or more coronary arteries, which supply oxygen-rich blood to the heart muscle. Because the coronary arteries are small in diameter, they are especially vulnerable to obstruction. Such blockage can destroy cardiac muscle quickly because the constantly beating heart muscle cannot survive long without O_2. If the heart stops beating, the victim may nevertheless survive if a heartbeat is restored by cardiopulmonary resuscitation (CPR) or some other emergency procedure within a few minutes of the attack. A **stroke** is the death of nervous tissue in the brain due to a lack of O_2. Strokes usually result from rupture or blockage of arteries in the head. The effects of a stroke and the individual's chance of survival depend on the extent and location of the damaged brain tissue. Rapid administration of a clot-dissolving drug may reduce the effects of a stroke or heart attack.

Although atherosclerosis often isn't detected until critical blood flow is disrupted, there can be warning signs. Partial blockage of the coronary arteries may cause occasional chest pain, a condition known as angina pectoris. The pain is most likely to be felt when the heart is laboring hard during physical or emotional stress, and it signals that part of the heart is not receiving enough O_2. An obstructed coronary artery may be treated surgically, either by inserting a metal mesh tube called a stent to expand the artery or by transplanting a healthy blood vessel from the chest or a limb to bypass the blockage.

Risk Factors and Treatment of Cardiovascular Disease

Although the tendency to develop particular cardiovascular diseases is inherited, it is also strongly influenced by lifestyle. Smoking and consumption of certain processed vegetable oils called *trans fats* (see Chapter 5) increase the ratio of LDL to HDL, raising the risk of cardiovascular disease. In contrast, exercise decreases the LDL/HDL ratio.

There has been considerable progress in the last decade in preventing cardiovascular disease. Many individuals at high risk are now treated with drugs called statins, which lower LDL levels and thereby reduce the risk of heart attacks. A recent discovery highlighted in **Figure 42.21** may lead to the development of additional drugs effective at lowering LDL levels in the blood.

The recognition that inflammation plays a central role in atherosclerosis and thrombus formation is also changing the

▼ Figure 42.21 INQUIRY

Can inactivating a liver enzyme lower plasma LDL levels?

EXPERIMENT In 2003, French researchers found that plasma LDL levels are higher in people who have mutations that increase the activity of a human liver enzyme called PCSK9. Helen Hobbs and co-workers in Dallas, Texas, then asked whether mutations that *inactivate* the *PCSK9* gene could *lower* LDL levels. By screening 15,000 participants in a 15-year study of cardiovascular disease, they discovered that 2% of individuals of African descent have mutations that inactivate one copy of the *PCSK9* gene. They then measured plasma LDL levels in individuals with one of these mutations and in control individuals.

RESULTS

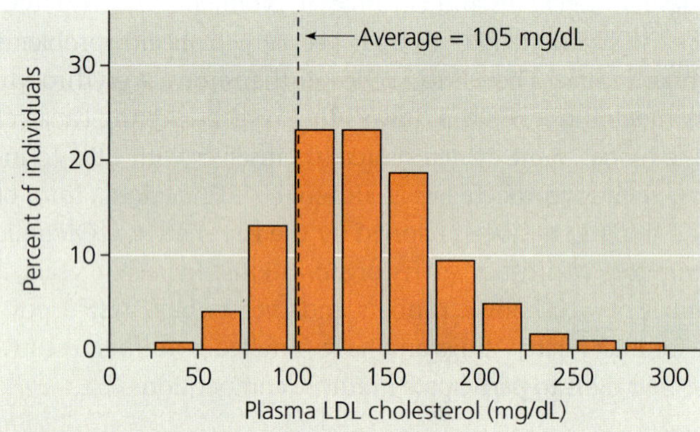

Individuals with two functional copies of *PCSK9* gene (control group)

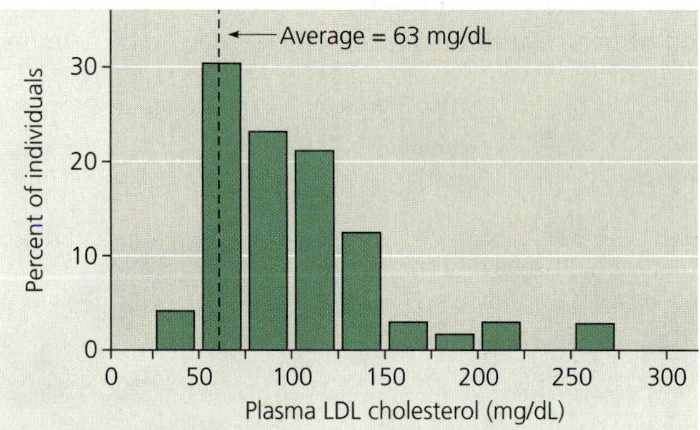

Individuals with an inactivating mutation in one copy of *PCSK9* gene

CONCLUSION Inactivating one copy of the *PCSK9* gene lowers the average plasma LDL level by 40%. Based on this result, Hobbs and colleagues hypothesized that decreasing PCSK9 activity reduces the risk for heart disease. Further analysis of data from the 15-year study supported this hypothesis: Individuals carrying *PCSK9* mutations had an 88% lower risk for heart disease compared with the control group. A search is now under way for molecules that inhibit PCSK9 as potential drugs to prevent heart disease.

SOURCE J. Cohen, A. Pertsemlidis, I. Kotowski, R. Graham, C. Garcia, and H. Hobbs, Low LDL cholesterol in individuals of African descent resulting from frequent nonsense mutations in *PCSK9*, *Nature Genetics* 37:161–165 (2005).

WHAT IF? Suppose you could measure the activity of PCSK9 in blood samples. How would you expect the activity to compare for the individuals studied by the French researchers and by Dr. Hobbs's team?

treatment of cardiovascular disease. For example, aspirin, which inhibits the inflammatory response, has been found to help prevent the recurrence of heart attacks and stroke. Researchers have also focused on C-reactive protein (CRP), which is produced by the liver and found in the blood during episodes of acute inflammation. Like a high level of LDL cholesterol, the presence of significant amounts of CRP in blood is a useful risk indicator for cardiovascular disease.

Hypertension (high blood pressure) is yet another contributor to heart attack and stroke as well as other health problems. According to one hypothesis, chronic high blood pressure damages the endothelium that lines the arteries, promoting plaque formation. The usual definition of hypertension in adults is a systolic pressure above 140 mm Hg or a diastolic pressure above 90 mm Hg. Fortunately, hypertension is simple to diagnose and can usually be controlled by dietary changes, exercise, medication, or a combination of these approaches.

CONCEPT CHECK 42.4

1. Explain why a physician might order a white cell count for a patient with symptoms of an infection.
2. Clots in arteries can cause heart attacks and strokes. Why, then, does it make sense to treat hemophiliacs by introducing clotting factors into their blood?
3. WHAT IF? Nitroglycerin (the key ingredient in dynamite) is sometimes prescribed for heart disease patients. Within the body, the nitroglycerin is converted to nitric oxide. Why would you expect nitroglycerin to relieve chest pain in these patients?
4. MAKE CONNECTIONS The allele that encodes Hb^S is codominant with the allele encoding normal hemoglobin (Hb) (see Concept 14.4, pp. 277–278). What can you deduce about the properties of Hb and Hb^S with regard to aggregate formation and sickling?
5. MAKE CONNECTIONS How do stem cells from the bone marrow of an adult differ from embryonic stem cells (see Concept 20.3, p. 415–416)?

For suggested answers, see Appendix A.

CONCEPT 42.5

Gas exchange occurs across specialized respiratory surfaces

In the remainder of this chapter, we will focus on the process of **gas exchange**. Although this process is often called respiratory exchange or respiration, it should not be confused with the energy transformations of cellular respiration. Gas exchange is the uptake of molecular O_2 from the environment and the discharge of CO_2 to the environment.

Partial Pressure Gradients in Gas Exchange

To understand the driving forces for gas exchange, we must calculate **partial pressure**, which is simply the pressure exerted by a particular gas in a mixture of gases. To do so, we need to know the pressure that the mixture exerts and the fraction of the mixture represented by a particular gas. Let's consider O_2 as an example. At sea level, the atmosphere exerts a downward force equal to that of a column of mercury (Hg) 760 mm high. Therefore, atmospheric pressure at sea level is 760 mm Hg. Since the atmosphere is 21% O_2 by volume, the partial pressure of O_2 is 0.21×760, or about 160 mm Hg. This value is called the *partial pressure* of O_2 (abbreviated P_{O_2}) because it is the part of atmospheric pressure contributed by O_2. The partial pressure of CO_2 (abbreviated P_{CO_2}) is much less, only 0.29 mm Hg at sea level.

Partial pressures also apply to gases dissolved in a liquid, such as water. When water is exposed to air, an equilibrium is reached in which the partial pressure of each gas in the water equals the partial pressure of that gas in the air. Thus, water exposed to air at sea level has a P_{O_2} of 160 mm Hg, the same as in the atmosphere. However, the *concentrations* of O_2 in the air and water differ substantially because O_2 is much less soluble in water than in air.

Once we have calculated partial pressures, we can readily predict the net result of diffusion at gas exchange surfaces: A gas always diffuses from a region of higher partial pressure to a region of lower partial pressure.

Respiratory Media

The conditions for gas exchange vary considerably, depending on whether the respiratory medium—the source of O_2—is air or water. As already noted, O_2 is plentiful in air, making up about 21% of Earth's atmosphere by volume. Compared to water, air is much less dense and less viscous, so it is easier to move and to force through small passageways. As a result, breathing air is relatively easy and need not be particularly efficient. Humans, for example, extract only about 25% of the O_2 in inhaled air.

Gas exchange with water as the respiratory medium is much more demanding. The amount of O_2 dissolved in a given volume of water varies but is always less than in an equivalent volume of air: Water in many marine and freshwater habitats contains only 4–8 mL of dissolved O_2 per liter, a concentration roughly 40 times less than in air. The warmer and saltier the water is, the less dissolved O_2 it can hold. Water's lower O_2 content, greater density, and greater viscosity mean that aquatic animals such as fishes and lobsters must expend considerable energy to carry out gas exchange. In the context of these challenges, adaptations have evolved that enable most aquatic animals to be very efficient in gas exchange. Many of these adaptations involve the organization of the surfaces dedicated to exchange.

Respiratory Surfaces

Specialization for gas exchange is apparent in the structure of the respiratory surface, the part of an animal's body where gas exchange occurs. Like all living cells, the cells that carry out gas exchange have a plasma membrane that must be in contact with an aqueous solution. Respiratory surfaces are therefore always moist.

The movement of O_2 and CO_2 across moist respiratory surfaces takes place entirely by diffusion. The rate of diffusion is proportional to the surface area across which it occurs and inversely proportional to the square of the distance through which molecules must move. In other words, gas exchange is fast when the area for diffusion is large and the path for diffusion is short. As a result, respiratory surfaces tend to be large and thin.

In some relatively simple animals, such as sponges, cnidarians, and flatworms, every cell in the body is close enough to the external environment that gases can diffuse quickly between all cells and the environment. In many animals, however, the bulk of the body's cells lack immediate access to the environment. The respiratory surface in these animals is a thin, moist epithelium that constitutes a respiratory organ.

The skin serves as a respiratory organ in some animals, including earthworms and some amphibians. Just below the skin, a dense network of capillaries facilitates the exchange of gases between the circulatory system and the environment. Because the respiratory surface must remain moist, earthworms and many other skin-breathers can survive for extended periods only in damp places.

The general body surface of most animals lacks sufficient area to exchange gases for the whole organism. The evolutionary solution to this limitation is a respiratory organ that is extensively folded or branched, thereby enlarging the available surface area for gas exchange. Gills, tracheae, and lungs are three such organs.

Gills in Aquatic Animals

Gills are outfoldings of the body surface that are suspended in the water. As illustrated in **Figure 42.22**, the distribution of gills over the body can vary considerably. Regardless of their distribution, gills often have a total surface area much greater than that of the rest of the body's exterior.

Movement of the respiratory medium over the respiratory surface, a process called **ventilation**, maintains the partial pressure gradients of O_2 and CO_2 across the gill that

(a) Marine worm. Many polychaetes (marine worms of the phylum Annelida) have a pair of flattened appendages called parapodia on each body segment. The parapodia serve as gills and also function in crawling and swimming.

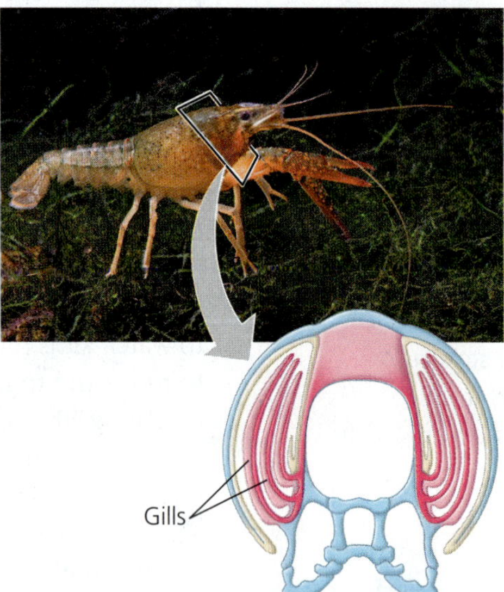

(b) Crayfish. Crayfish and other crustaceans have long, feathery gills covered by the exoskeleton. Specialized body appendages drive water over the gill surfaces.

(c) Sea star. The gills of a sea star are simple tubular projections of the skin. The hollow core of each gill is an extension of the coelom (body cavity). Gas exchange occurs by diffusion across the gill surfaces, and fluid in the coelom circulates in and out of the gills, aiding gas transport. The surfaces of a sea star's tube feet also function in gas exchange.

▲ **Figure 42.22 Diversity in the structure of gills, external body surfaces that function in gas exchange.**

MAKE CONNECTIONS *As shown in Figure 32.11, animals with bilateral symmetry are divided into three main lineages. What are those lineages? How many are represented by the gilled animals shown above?*

are necessary for gas exchange. To promote ventilation, most gill-bearing animals either move their gills through the water or move water over their gills. For example, crayfish and lobsters have paddle-like appendages that drive a current of water over the gills, whereas mussels and clams move water with cilia. Octopuses and squids ventilate their gills by taking in and ejecting water, with the side benefit of locomotion by jet propulsion. Fishes use the motion of swimming or coordinated movements of the mouth and gill covers to ventilate their gills. In both cases, a current of water enters the mouth, passes through slits in the pharynx, flows over the gills, and then exits the body **(Figure 42.23)**.

The arrangement of capillaries in a fish gill allows for **countercurrent exchange**, the exchange of a substance or heat between two fluids flowing in opposite directions. In a fish gill, this process maximizes gas exchange efficiency. Because blood flows in the direction opposite to that of water passing over the gills, at each point in its travel blood is less saturated with O_2 than the water it meets (see Figure 42.23). As blood enters a gill capillary, it encounters water that is completing its passage through the gill. Depleted of much of its dissolved O_2, this water nevertheless has a higher P_{O_2} than the incoming blood, and O_2 transfer takes place. As the blood continues its passage, its P_{O_2} steadily increases, but so does that of the water it encounters, since each successive position in the blood's travel corresponds to an earlier position in the water's passage over the gills. Thus, a partial pressure gradient favoring the diffusion of O_2 from water to blood exists along the entire length of the capillary.

Countercurrent exchange mechanisms are remarkably efficient. In the fish gill, more than 80% of the O_2 dissolved in the water is removed as it passes over the respiratory surface. In other settings, countercurrent exchange contributes to temperature regulation (see Chapter 40) and to the functioning of the mammalian kidney, as we will see in Chapter 44.

Gills are generally unsuitable for an animal living on land. An expansive surface of wet membrane exposed directly to air currents in the environment would lose too much water by evaporation. Furthermore, the gills would collapse as their fine filaments, no longer supported by water, stuck together. In most terrestrial animals, respiratory surfaces are enclosed within the body, exposed to the atmosphere only through narrow tubes.

Tracheal Systems in Insects

Although the most familiar respiratory structure among terrestrial animals is the lung, the most common is actually the **tracheal system** of insects. Made up of air tubes that branch throughout the body, this system is one variation on the theme of an internal respiratory surface. The largest

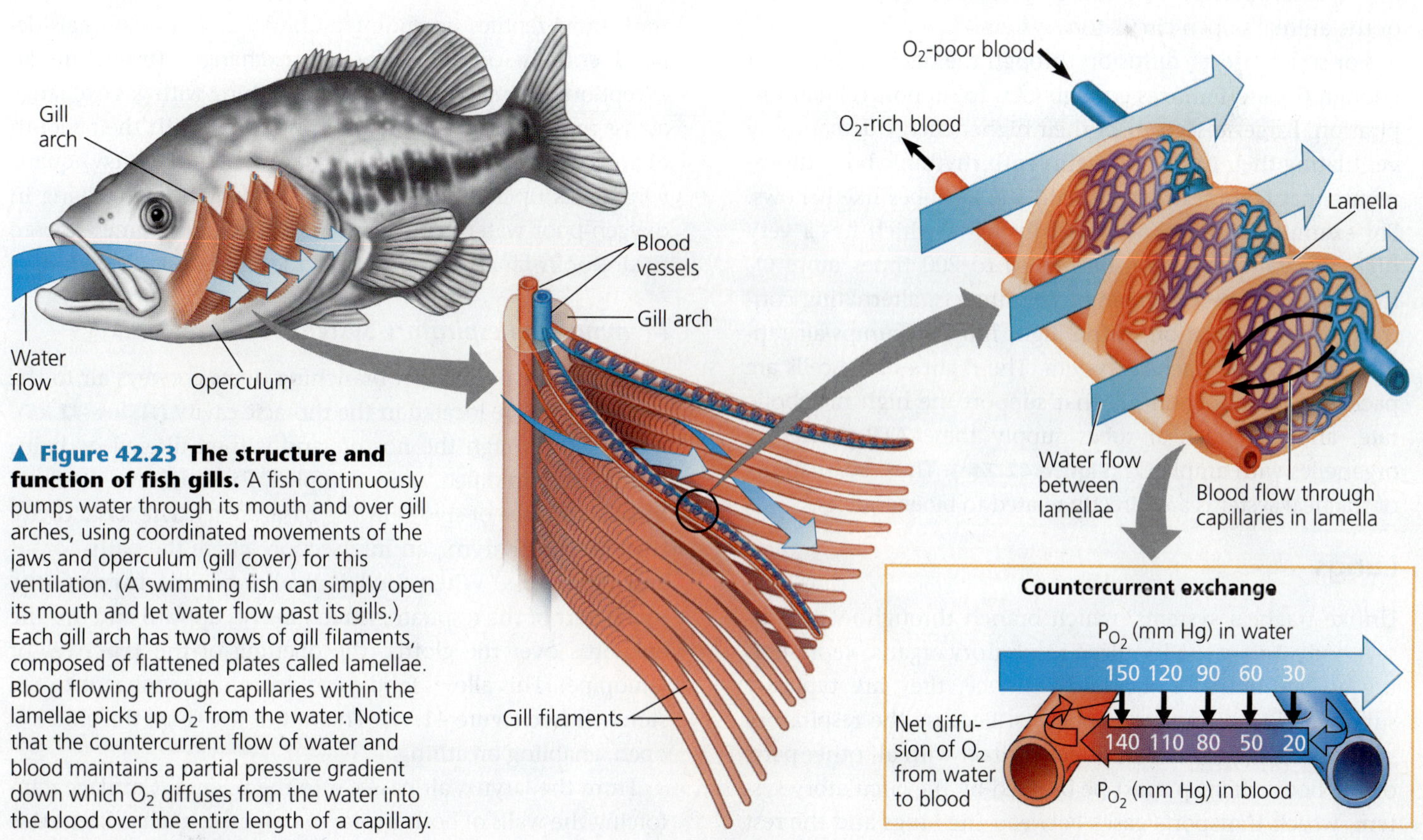

▲ **Figure 42.23 The structure and function of fish gills.** A fish continuously pumps water through its mouth and over gill arches, using coordinated movements of the jaws and operculum (gill cover) for this ventilation. (A swimming fish can simply open its mouth and let water flow past its gills.) Each gill arch has two rows of gill filaments, composed of flattened plates called lamellae. Blood flowing through capillaries within the lamellae picks up O_2 from the water. Notice that the countercurrent flow of water and blood maintains a partial pressure gradient down which O_2 diffuses from the water into the blood over the entire length of a capillary.

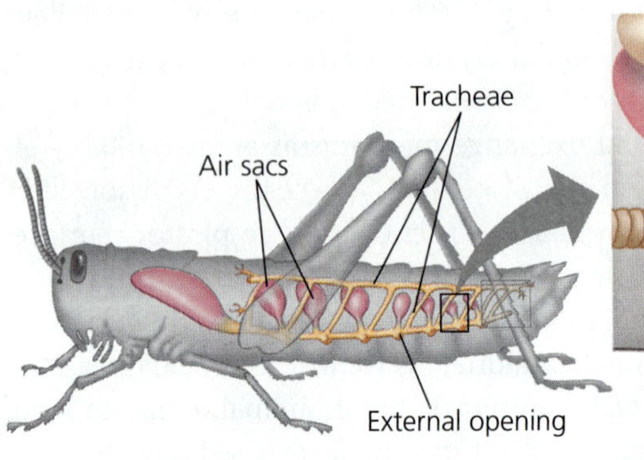

(a) The respiratory system of an insect consists of branched internal tubes. The largest tubes, called tracheae, connect to external openings spaced along the insect's body surface. Air sacs formed from enlarged portions of the tracheae are found near organs that require a large supply of oxygen.

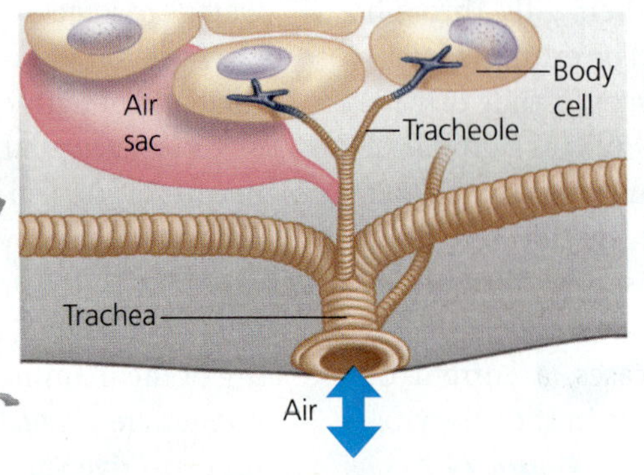

(b) Rings of chitin keep the tracheae open, allowing air to enter and pass into smaller tubes called tracheoles. The branched tracheoles deliver air directly to cells throughout the body. Tracheoles have closed ends filled with fluid (blue-gray). When the animal is active and using more O_2, most of the fluid is withdrawn into the body. This increases the surface area of air-filled tracheoles in contact with cells.

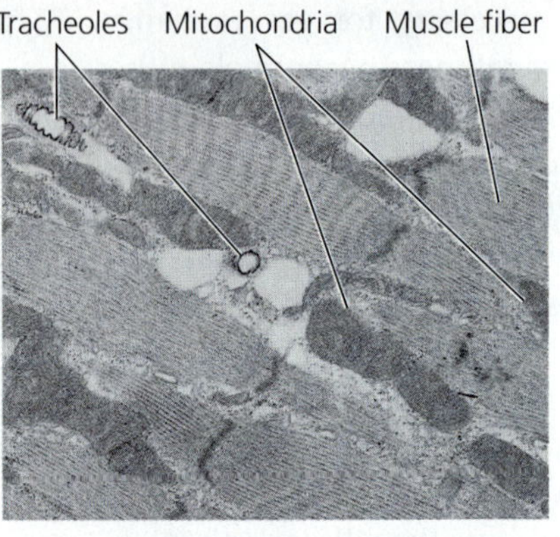

(c) The TEM above shows cross sections of tracheoles in a tiny piece of insect flight muscle. Each of the numerous mitochondria in the muscle cells lies within about 5 µm of a tracheole.

▲ **Figure 42.24 Tracheal systems.**

tubes, called tracheae, open to the outside **(Figure 42.24a)**. The finest branches extend close to the surface of nearly every cell, where gas is exchanged by diffusion across the moist epithelium that lines the tips of the tracheal branches **(Figure 42.24b)**. Because the tracheal system brings air within a very short distance of virtually every body cell in an insect, it can transport O_2 and CO_2 without the participation of the animal's open circulatory system.

For small insects, diffusion through the tracheae brings in enough O_2 and removes enough CO_2 to support cellular respiration. Larger insects meet their higher energy demands by ventilating their tracheal systems with rhythmic body movements that compress and expand the air tubes like bellows. For example, consider an insect in flight, which has a very high metabolic rate, consuming 10 to 200 times more O_2 than it does at rest. In many flying insects, alternating contraction and relaxation of the flight muscles pumps air rapidly through the tracheal system. The flight muscle cells are packed with mitochondria that support the high metabolic rate, and the tracheal tubes supply these ATP-generating organelles with ample O_2 **(Figure 42.24c)**. Thus, adaptations of tracheal systems are directly related to bioenergetics.

Lungs

Unlike tracheal systems, which branch throughout the insect body, **lungs** are localized respiratory organs. Representing an infolding of the body surface, they are typically subdivided into numerous pockets. Because the respiratory surface of a lung is not in direct contact with all other parts of the body, the gap must be bridged by the circulatory system, which transports gases between the lungs and the rest of the body. Lungs have evolved in organisms with open circulatory systems, such as spiders and land snails, as well as in vertebrates.

Among vertebrates that lack gills, the use of lungs for gas exchange varies. Amphibian lungs, when present, are relatively small and lack an extensive surface for exchange. Amphibians instead rely heavily on diffusion across other body surfaces, such as the skin, to carry out gas exchange. In contrast, most reptiles (including all birds) and all mammals depend entirely on lungs for gas exchange. Turtles are an exception; they supplement lung breathing with gas exchange across moist epithelial surfaces continuous with their mouth or anus. Lungs and air breathing have evolved in a few aquatic vertebrates (including lungfishes) as adaptations to living in oxygen-poor water or to spending part of their time exposed to air (for instance, when the water level of a pond recedes).

Mammalian Respiratory Systems: A Closer Look

In mammals, a system of branching ducts conveys air to the lungs, which are located in the thoracic cavity **(Figure 42.25)**. Air enters through the nostrils and is then filtered by hairs, warmed, humidified, and sampled for odors as it flows through a maze of spaces in the nasal cavity. The nasal cavity leads to the pharynx, an intersection where the paths for air and food cross. When food is swallowed, the **larynx** (the upper part of the respiratory tract) moves upward and tips the epiglottis over the glottis (the opening of the **trachea**, or windpipe). This allows food to go down the esophagus to the stomach (see Figure 41.11). The rest of the time, the glottis is open, enabling breathing.

From the larynx, air passes into the trachea. Cartilage reinforcing the walls of both the larynx and the trachea keeps this part of the airway open. Within the larynx of most mammals,

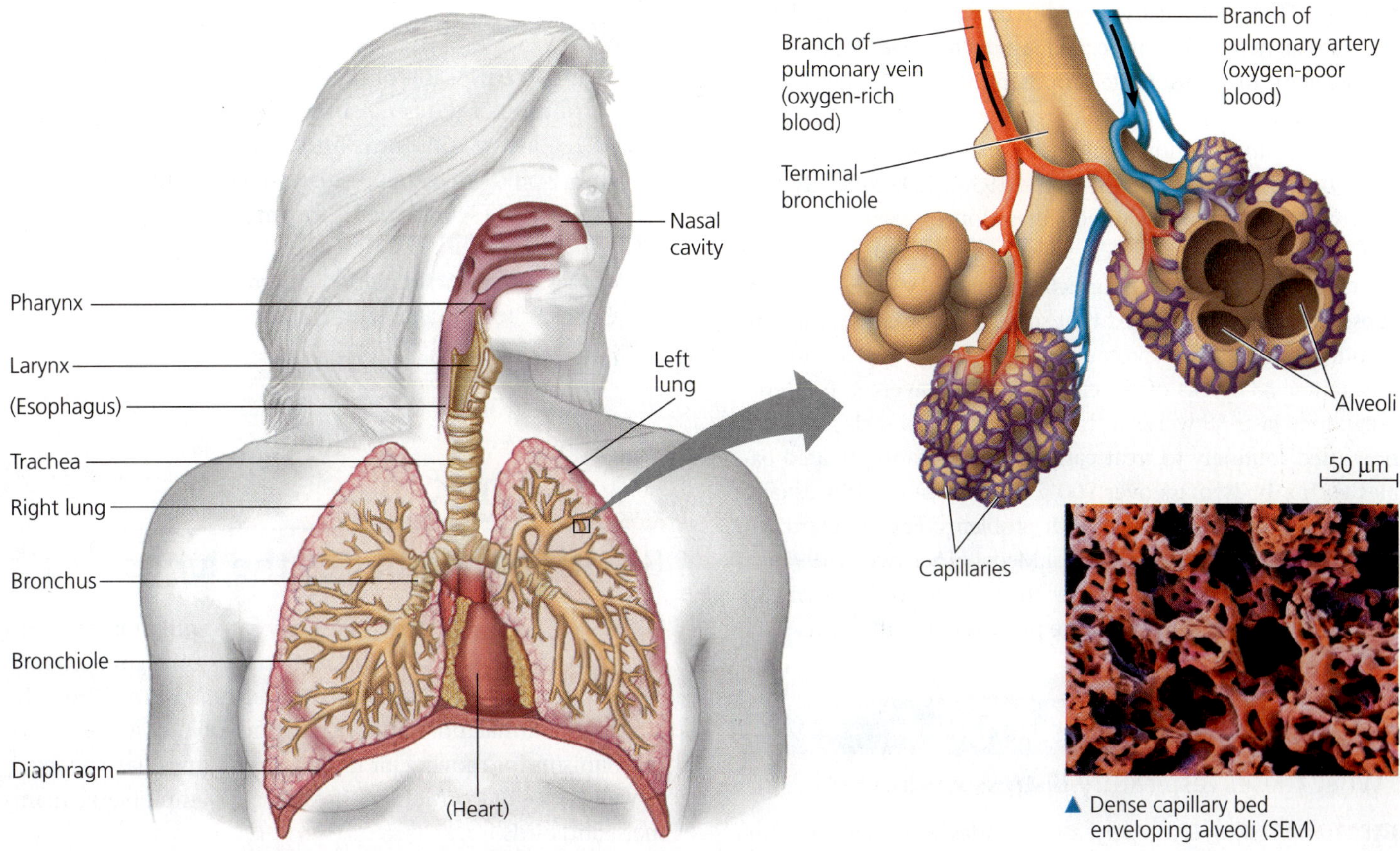

▲ Dense capillary bed enveloping alveoli (SEM)

▲ **Figure 42.25 The mammalian respiratory system.** From the nasal cavity and pharynx, inhaled air passes through the larynx, trachea, and bronchi to the bronchioles, which end in microscopic alveoli lined by a thin, moist epithelium. Branches of the pulmonary arteries convey oxygen-poor blood to the alveoli; branches of the pulmonary veins transport oxygen-rich blood from the alveoli back to the heart.

exhaled air rushes by a pair of elastic bands of muscle called *vocal folds*, or, in humans, *vocal cords*. Sounds are produced when muscles in the larynx are tensed, stretching the cords so they vibrate. High-pitched sounds result from tightly stretched cords vibrating rapidly; low-pitched sounds come from less tense cords vibrating slowly.

The trachea branches into two **bronchi** (singular, *bronchus*), one leading to each lung. Within the lung, the bronchi branch repeatedly into finer and finer tubes called **bronchioles**. The entire system of air ducts has the appearance of an inverted tree, the trunk being the trachea. The epithelium lining the major branches of this respiratory tree is covered by cilia and a thin film of mucus. The mucus traps dust, pollen, and other particulate contaminants, and the beating cilia move the mucus upward to the pharynx, where it can be swallowed into the esophagus. This process, sometimes referred to as the "mucus escalator," plays a crucial role in cleansing the respiratory system.

Gas exchange in mammals occurs in **alveoli** (singular, *alveolus*; see Figure 42.25), air sacs clustered at the tips of the tiniest bronchioles. Human lungs contain millions of alveoli, which together have a surface area of about 100 m^2, 50 times that of the skin. Oxygen in the air entering the alveoli dissolves in the moist film lining their inner surfaces and rapidly diffuses across the epithelium into a web of capillaries that surrounds each alveolus. Net diffusion of carbon dioxide occurs in the opposite direction, from the capillaries across the epithelium of the alveolus and into the air space.

Lacking cilia or significant air currents to remove particles from their surface, alveoli are highly susceptible to contamination. White blood cells patrol alveoli, engulfing foreign particles. However, if too much particulate matter reaches the alveoli, the defenses can be overwhelmed, leading to inflammation and irreversible damage. For example, particulates from cigarette smoke that enter alveoli can cause a permanent reduction in lung capacity. For coal miners, inhalation of large amounts of coal dust can lead to silicosis, a disabling, irreversible, and sometimes fatal lung disease.

The film of liquid that lines alveoli is subject to surface tension, an attractive force that acts to minimize the surface area of a liquid (see Chapter 3). Given their tiny diameter (about 0.25 mm), why don't alveoli collapse under high surface tension? Researchers reasoned that alveoli must be coated with a material that reduces surface tension. In 1955, English biophysicist Richard Pattle obtained experimental evidence for such a material, now called a **surfactant**, for *surf*ace-*act*ive agent. In addition, he proposed that the absence of surfactant might cause *respiratory distress syndrome (RDS)*, a disease common among preterm infants born 6 weeks or more before their due dates. In the 1950s, RDS killed 10,000 infants annually in the United States alone.

In the late 1950s, Mary Ellen Avery carried out the first experiment linking RDS to a surfactant deficiency **(Figure 42.26)**. Subsequent studies revealed that surfactant contains a mixture of phospholipids and proteins and typically appears in the lungs after 33 weeks of development. (The average full-term pregnancy lasts 38 weeks in humans.) Artificial surfactants are now used routinely to treat early preterm infants. Treated babies with a body mass over 900 g (2 pounds) at birth usually survive without long-term health problems. For her contributions, Avery received the National Medal of Science in 1991.

Having surveyed the route that air follows when we breathe, we will turn next to the process of breathing itself.

▼ Figure 42.26 **INQUIRY**

What causes respiratory distress syndrome?

EXPERIMENT Mary Ellen Avery, a research fellow working with Jere Mead at Harvard University Medical School, wondered whether a lack of surfactant caused respiratory distress syndrome (RDS) in preterm infants. She obtained autopsy samples of lungs from infants that had died of RDS and from infants that had died of other causes. She extracted material from the samples and allowed it to form a film on a water surface. Then Dr. Avery measured the tension (in dynes per centimeter) across the surface of the water and recorded the lowest surface tension observed for each sample.

RESULTS

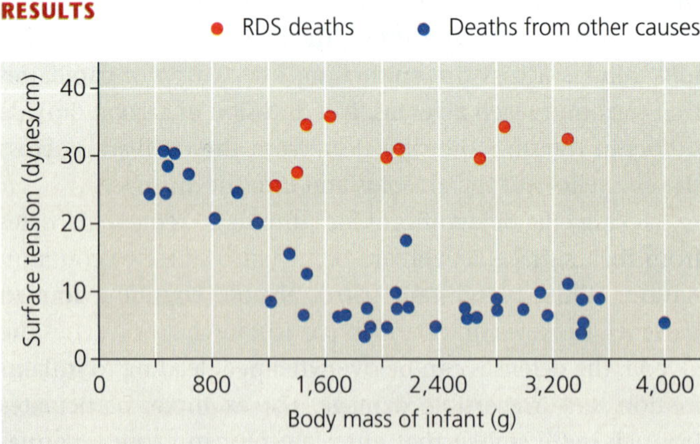

CONCLUSION The lungs of infants with a body mass over 1,200 g (2.7 pounds) contain a substance that reduces surface tension. That substance is absent in the lungs of infants with RDS.

SOURCE M. E. Avery and J. Mead, Surface properties in relation to atelectasis and hyaline membrane disease, *American Journal of Diseases of Children* 97:517–523 (1959).

WHAT IF? Suppose you repeated this experiment but instead measured the amount of surfactant in lung samples. Describe the graph you would expect if you plotted the amount of surfactant versus infant weight.

CONCEPT CHECK 42.5

1. Why is the position of lung tissues *within* the body an advantage for terrestrial animals?
2. After a heavy rain, earthworms come to the surface. How would you explain this behavior in terms of an earthworm's requirements for gas exchange?
3. **MAKE CONNECTIONS** Describe how countercurrent exchange can facilitate both thermoregulation (see Concept 40.3, p. 865) and respiration.

For suggested answers, see Appendix A.

CONCEPT **42.6**

Breathing ventilates the lungs

Like fishes, terrestrial vertebrates rely on ventilation to maintain high O_2 and low CO_2 concentrations at the gas exchange surface. The process that ventilates lungs is **breathing**, the alternating inhalation and exhalation of air. A variety of mechanisms for moving air in and out of lungs have evolved, as we will see by considering breathing in amphibians, mammals, and birds.

How an Amphibian Breathes

An amphibian such as a frog ventilates its lungs by **positive pressure breathing**, inflating the lungs with forced airflow. During the first stage of inhalation, muscles lower the floor of an amphibian's oral cavity, drawing in air through its nostrils. Next, with the nostrils and mouth closed, the floor of the oral cavity rises, forcing air down the trachea. During exhalation, air is forced back out by the elastic recoil of the lungs and by compression of the muscular body wall. When male frogs puff themselves up in aggressive or courtship displays, they disrupt this breathing cycle, taking in air several times without allowing any release.

How a Bird Breathes

Two features of ventilation in birds make it highly efficient. First, when birds breathe, they pass air over the gas exchange surface in only one direction. Second, incoming fresh air does not mix with air that has already carried out gas exchange.

To bring fresh air to their lungs, birds use eight or nine air sacs situated on either side of the lungs **(Figure 42.27)**. The air sacs do not function directly in gas exchange but act as bellows that keep air flowing through the lungs. Instead of alveoli, which are dead ends, the sites of gas exchange in bird lungs are tiny channels called *parabronchi*. Passage of air through the entire system—lungs and air sacs—requires two cycles of inhalation and exhalation. In some passageways, the direction in

Anterior air sacs

Posterior air sacs

Lungs

Airflow

Air tubes (parabronchi) in lung

1 mm

Posterior air sacs

Lungs

Anterior air sacs

Two cycles of inhalation and exhalation are required to pass one breath through the system:

1. First inhalation: Air fills the posterior air sacs.
2. First exhalation: Posterior air sacs contract, pushing air into lungs.
3. Second inhalation: Air passes through lungs and fills anterior air sacs.
4. Second exhalation: As anterior air sacs contract, air that entered body at first inhalation is pushed out of body.

▲ **Figure 42.27 The avian respiratory system.** This diagram traces a breath of air through the respiratory system of a bird. As shown, two cycles of inhalation and exhalation are required for the air to pass all the way through the system and out of the bird.

which air moves alternates (see Figure 42.27). Within the parabronchi, however, air always flows in the same direction.

How a Mammal Breathes

Unlike amphibians and birds, mammals employ **negative pressure breathing**—pulling, rather than pushing, air into their lungs **(Figure 42.28)**. Using muscle contraction to actively expand the thoracic cavity, mammals lower air pressure in their lungs below that of the air outside their body. Because gas flows from a region of higher pressure to a region of lower pressure, air rushes through the nostrils and mouth and down the breathing tubes to the alveoli. During exhalation, the muscles controlling the thoracic cavity relax, and the volume of the cavity is reduced. The increased air pressure in the alveoli forces air up the breathing tubes and out of the body. Thus, inhalation is always active and requires work, whereas exhalation is usually passive.

Expanding the thoracic cavity during inhalation involves the animal's rib muscles and the **diaphragm**, a sheet of skeletal muscle that forms the bottom wall of the cavity. Contracting the rib muscles expands the rib cage, the front wall of the thoracic cavity, by pulling the ribs upward and the sternum outward. At the same time, the diaphragm contracts, expanding the thoracic cavity downward. The effect of the descending diaphragm is similar to that of a plunger being drawn out of a syringe.

Within the thoracic cavity, a double membrane surrounds the lungs. The inner layer of this membrane adheres to the outside of the lungs, and the outer layer adheres to the wall of the thoracic cavity. A thin space filled with fluid separates the two layers. Surface tension in the fluid causes the two layers to stick together like two plates of glass separated by a film of water: The layers can slide smoothly past each other, but they cannot be pulled apart easily. Consequently, the volume of the thoracic cavity and the volume of the lungs change in unison.

Depending on activity level, additional muscles may be recruited to aid breathing. The rib muscles and diaphragm are sufficient to change lung volume when a mammal is at rest. During exercise, other muscles of the neck, back, and chest increase the volume of the thoracic cavity by raising the rib cage. In kangaroos and some other species, locomotion causes a rhythmic movement of organs in the abdomen, including the stomach and liver. The result is a piston-like pumping motion that pushes and pulls on the diaphragm, further increasing the volume of air moved in and out of the lungs.

Rib cage expands as rib muscles contract.

Air inhaled.

Lung

Diaphragm

1 Inhalation: Diaphragm contracts (moves down).

Rib cage gets smaller as rib muscles relax.

Air exhaled.

2 Exhalation: Diaphragm relaxes (moves up).

▲ **Figure 42.28 Negative pressure breathing.** A mammal breathes by changing the air pressure within its lungs relative to the pressure of the outside atmosphere.

WHAT IF? *The walls of alveoli contain elastic fibers that allow the alveoli to expand and contract with each breath. If alveoli lost their elasticity, how would that affect gas exchange in the lungs?*

The volume of air inhaled and exhaled with each breath is called **tidal volume**. It averages about 500 mL in resting humans. The tidal volume during maximal inhalation and exhalation is the **vital capacity**, which is about 3.4 L and 4.8 L for college-age women and men, respectively. The air that remains after a forced exhalation is called the **residual volume**. As we age, our lungs lose their resilience, and residual volume increases at the expense of vital capacity.

Because the lungs in mammals do not completely empty with each breath, and because inhalation occurs through the same airways as exhalation, each inhalation mixes fresh air with oxygen-depleted residual air. As a result, the maximum P_{O_2} in alveoli is always considerably less than in the atmosphere. The maximum P_{O_2} in lungs is also less for mammals than for birds, which renew the air in their lungs with every exhalation. This is one reason mammals function less well than birds at high altitude. For example, humans have great difficulty obtaining enough O_2 when climbing Earth's highest peaks, such as Mount Everest (8,850 m), in the Himalayas. However, bar-headed geese and several other bird species easily fly over the Himalayas during their migrations.

Control of Breathing in Humans

Although you can voluntarily hold your breath or breathe faster and deeper, most of the time your breathing is regulated by involuntary mechanisms. These control mechanisms ensure that gas exchange is coordinated with blood circulation and with metabolic demand.

The neurons mainly responsible for regulating breathing are in the medulla oblongata, near the base of the brain **(Figure 42.29)**. Neural circuits in the medulla form a *breathing control center* that establishes the breathing rhythm. When you breathe deeply, a negative-feedback mechanism prevents the lungs from overexpanding: During inhalation, sensors that detect stretching of the lung tissue send nerve impulses to the control circuits in the medulla, inhibiting further inhalation.

In regulating breathing, the medulla uses the pH of the surrounding tissue fluid as an indicator of blood CO_2 concentration. The reason pH can be used in this way is that blood CO_2 is the main determinant of the pH of *cerebrospinal fluid*, the fluid surrounding the brain and spinal cord. Carbon dioxide diffuses from the blood to the cerebrospinal fluid, where it reacts with water and forms carbonic acid (H_2CO_3). The H_2CO_3 can then dissociate into a bicarbonate ion (HCO_3^-) and a hydrogen ion (H^+):

$$CO_2 + H_2O \rightleftharpoons H_2CO_3 \rightleftharpoons HCO_3^- + H^+$$

Increased metabolic activity, such as occurs during exercise, lowers pH by increasing the concentration of CO_2 in the blood. Sensors in blood vessels and the medulla detect this pH change. In response, the medulla's control circuits increase the depth and rate of breathing. Both remain high until the excess CO_2 is eliminated in exhaled air and pH returns to a normal value.

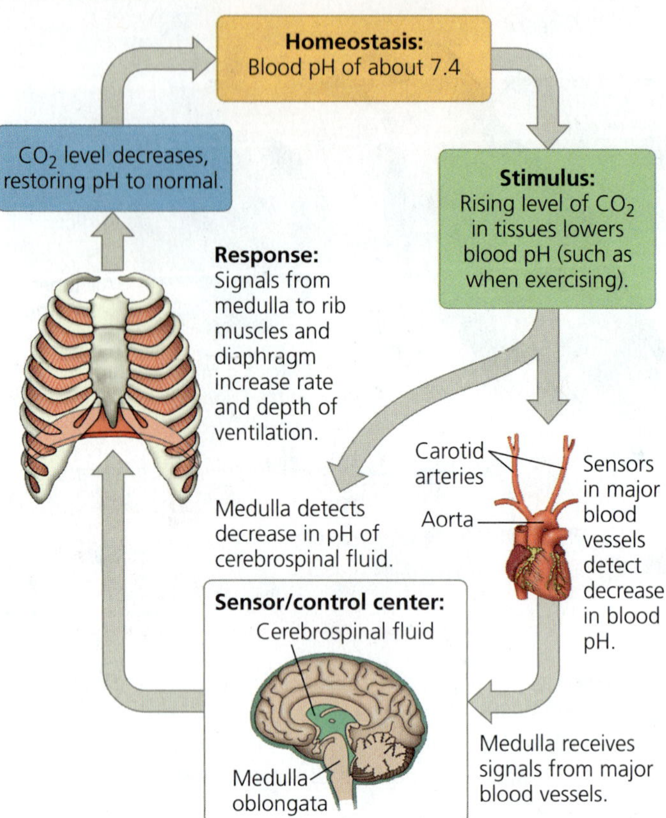

▲ **Figure 42.29 Homeostatic control of breathing.**

WHAT IF? *Suppose a person began breathing very rapidly while resting. Describe the effect on blood CO_2 levels and the steps by which the negative feedback circuit in this figure would restore homeostasis.*

The blood O_2 level usually has little effect on the breathing control centers. However, when the O_2 level drops very low (at high altitudes, for instance), O_2 sensors in the aorta and the carotid arteries in the neck send signals to the breathing control centers, which respond by increasing the breathing rate.

The pons, a part of the brain next to the medulla, also regulates breathing, although its exact role remains an open question. The pons may act in the regulatory circuit with the medulla or modulate the output of that circuit.

Breathing control is effective only if ventilation is matched to blood flow through alveolar capillaries. During exercise, for instance, such coordination couples an increased breathing rate, which enhances O_2 uptake and CO_2 removal, with an increase in cardiac output.

CONCEPT CHECK 42.6

1. How does an increase in the CO_2 concentration in the blood affect the pH of cerebrospinal fluid?
2. A drop in blood pH causes an increase in heart rate. What is the function of this control mechanism?
3. **WHAT IF?** If an injury tore a small hole in the membranes surrounding your lungs, what effect on lung function would you expect?

For suggested answers, see Appendix A.

CONCEPT 42.7

Adaptations for gas exchange include pigments that bind and transport gases

The high metabolic demands of many animals necessitate the exchange of large quantities of O_2 and CO_2. Here we'll examine how blood molecules called respiratory pigments facilitate this exchange through their interaction with O_2 and CO_2. We will also investigate physiological adaptations that enable animals to be active under conditions of high metabolic load or very limiting P_{O_2}. As a basis for exploring these topics, let's summarize the basic gas exchange circuit in humans.

Coordination of Circulation and Gas Exchange

The partial pressures of O_2 and CO_2 in the blood vary at different points in the circulatory system, as shown in **Figure 42.30**. Blood flowing through the alveolar capillaries has a lower P_{O_2} and a higher P_{CO_2} than the air in the alveoli. As a result, CO_2 diffuses down its partial pressure gradient from the blood to the air in the alveoli. Meanwhile, O_2 in the air dissolves in the fluid that coats the alveolar epithelium and diffuses into the blood. By the time the blood leaves the lungs in the pulmonary veins, its P_{O_2} has been raised and its P_{CO_2} has been lowered. After returning to the heart, this blood is pumped through the systemic circuit.

In the tissue capillaries, gradients of partial pressure favor the diffusion of O_2 out of the blood and CO_2 into the blood. These gradients exist because cellular respiration in the mitochondria of cells near each capillary removes O_2 from and adds CO_2 to the surrounding interstitial fluid. After the blood unloads O_2 and loads CO_2, it is returned to the heart and pumped to the lungs again.

Although this description faithfully characterizes the driving forces for gas exchange in different tissues, it omits the critical role of the specialized carrier proteins we will discuss next.

Respiratory Pigments

The low solubility of O_2 in water (and thus in blood) poses a problem for animals that rely on the circulatory system to deliver O_2. For example, a person requires almost 2 L of O_2 per minute during intense exercise, and all of it must be carried in the blood from the lungs to the active tissues. At normal body temperature and air pressure, however, only 4.5 mL of O_2 can dissolve into a liter of blood in the lungs. Even if 80% of the dissolved O_2 were delivered to the tissues (an unrealistically high percentage), the heart would still need to pump 555 L of blood per minute!

In fact, animals transport most of their O_2 bound to proteins called **respiratory pigments**. Respiratory pigments circulate with the blood or hemolymph and are often contained within specialized cells. The pigments greatly increase the amount of O_2 that can be carried in the circulatory fluid (to about 200 mL of O_2 per liter in mammalian blood).

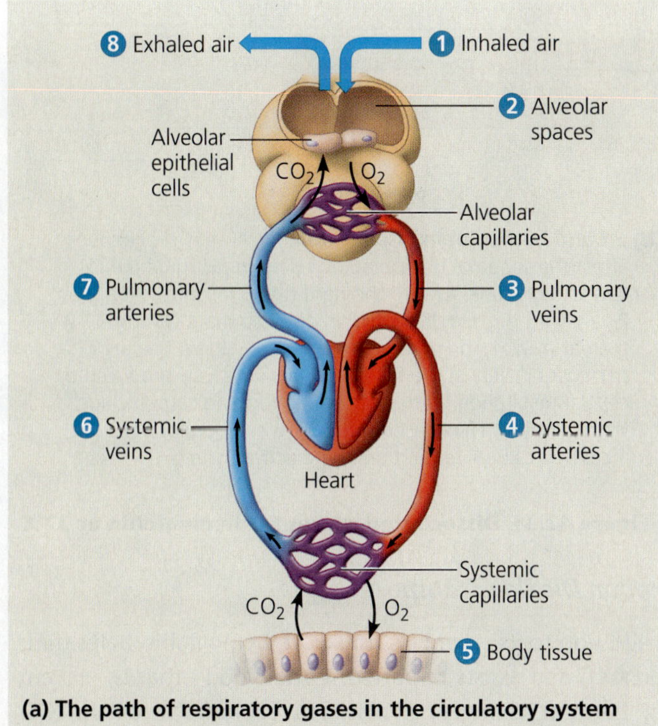

(a) The path of respiratory gases in the circulatory system

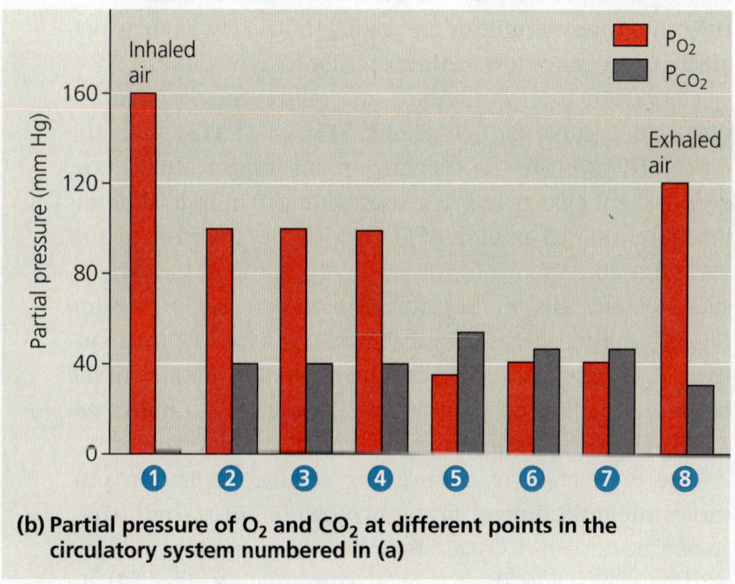

(b) Partial pressure of O_2 and CO_2 at different points in the circulatory system numbered in (a)

▲ **Figure 42.30 Loading and unloading of respiratory gases.**

WHAT IF? *If you consciously forced more air out of your lungs each time you exhaled, how would that affect the values shown in (b)?*

In our example of an exercising human with an O_2 delivery rate of 80%, the presence of a respiratory pigment reduces the cardiac output necessary for O_2 transport to a manageable 12.5 L of blood per minute.

A variety of respiratory pigments have evolved among the animal taxa. With a few exceptions, these molecules have a distinctive color (hence the term *pigment*) and consist of a protein bound to a metal. One example is the blue pigment *hemocyanin*, which has copper as its oxygen-binding component and is found in arthropods and many molluscs. The respiratory pigment of almost all vertebrates and many invertebrates is hemoglobin. In vertebrates, it is contained in the erythrocytes.

Hemoglobin

Vertebrate hemoglobin consists of four subunits (polypeptide chains), each with a cofactor called a heme group that has an iron atom at its center. Each iron atom binds one molecule of O_2; hence, a single hemoglobin molecule can carry four molecules of O_2. Like all respiratory pigments, hemoglobin binds O_2 reversibly, loading O_2 in the lungs or gills and unloading it in other parts of the body. This process depends on cooperativity between the hemoglobin subunits (see pp. 158–159). When O_2 binds to one subunit, the others change shape slightly, increasing their affinity for O_2. When four O_2 molecules are bound and one subunit unloads its O_2, the other three subunits more readily unload O_2, as an associated shape change lowers their affinity for O_2.

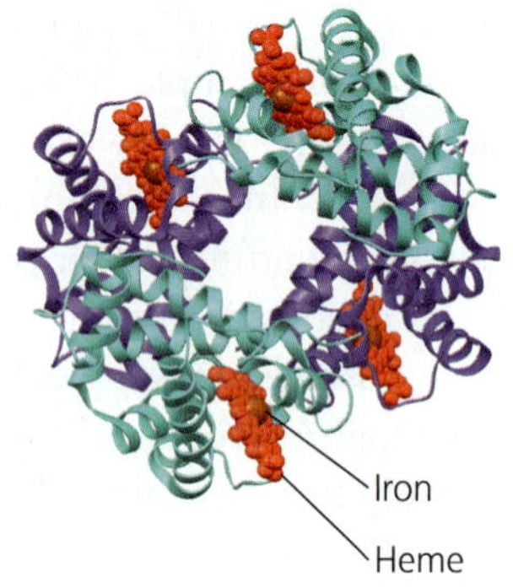

Hemoglobin

Cooperativity in O_2 binding and release is evident in the dissociation curve for hemoglobin **(Figure 42.31a)**. Over the range of P_{O_2} where the dissociation curve has a steep slope, even a slight change in P_{O_2} causes hemoglobin to load or unload a substantial amount of O_2. Notice that the steep part of the curve corresponds to the range of P_{O_2} found in body tissues. When cells in a particular location begin working harder—during exercise, for instance—P_{O_2} dips in their vicinity as the O_2 is consumed in cellular respiration. Because of the effect of subunit cooperativity, a slight drop in P_{O_2} causes a relatively large increase in the amount of O_2 the blood unloads.

The production of CO_2 during cellular respiration promotes the unloading of O_2 by hemoglobin in active tissues. As we have seen, CO_2 reacts with water, forming carbonic acid, which lowers the pH of its surroundings. Low pH, in turn, decreases the affinity of hemoglobin for O_2, an effect called the **Bohr shift (Figure 42.31b)**. Thus, where CO_2 production is greater, hemoglobin releases more O_2, which can then be used to support more cellular respiration.

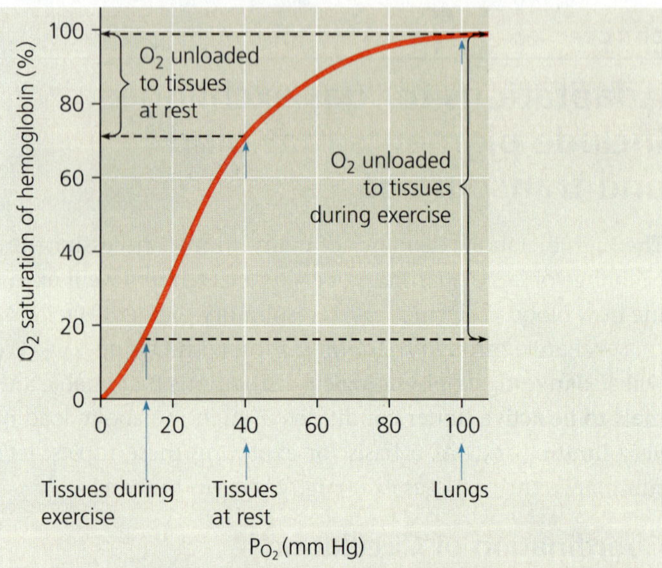

(a) P_{O_2} and hemoglobin dissociation at pH 7.4. The curve shows the relative amounts of O_2 bound to hemoglobin exposed to solutions with different P_{O_2}. At a P_{O_2} of 100 mm Hg, typical in the lungs, hemoglobin is about 98% saturated with O_2. At a P_{O_2} of 40 mm Hg, common in the vicinity of tissues at rest, hemoglobin is about 70% saturated. Hemoglobin can release additional O_2 to metabolically very active tissues, such as muscle tissue during exercise.

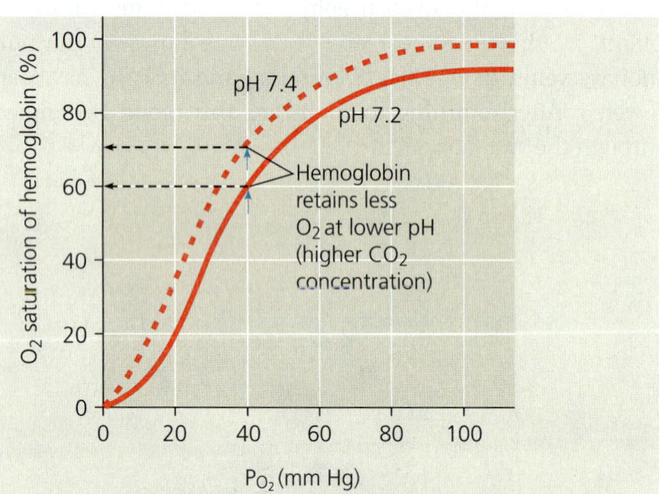

(b) pH and hemoglobin dissociation. Because hydrogen ions affect the shape of hemoglobin, a drop in pH shifts the O_2 dissociation curve toward the right (the Bohr shift). At a given P_{O_2}, say 40 mm Hg, hemoglobin gives up more O_2 at pH 7.2 than at pH 7.4, the normal pH of human blood. The pH decreases in very active tissues because the CO_2 produced by cellular respiration reacts with water, forming carbonic acid. Hemoglobin then releases more O_2, which supports the increased cellular respiration in the active tissues.

▲ **Figure 42.31 Dissociation curves for hemoglobin at 37°C.**

Carbon Dioxide Transport

In addition to its role in O_2 transport, hemoglobin helps transport CO_2 and assists in buffering the blood—that is, preventing harmful changes in pH. Only about 7% of the CO_2

released by respiring cells is transported in solution in blood plasma. Another 23% binds to the amino ends of the hemoglobin polypeptide chains, and about 70% is transported in the blood in the form of bicarbonate ions (HCO_3^-).

As shown in **Figure 42.32**, carbon dioxide from respiring cells diffuses into the blood plasma and then into erythrocytes. There the CO_2 reacts with water (assisted by the enzyme carbonic anhydrase) and forms H_2CO_3, which dissociates into H^+ and HCO_3^-. Most of the H^+ binds to hemoglobin and other proteins, minimizing the change in blood pH. The HCO_3^- diffuses into the plasma.

When blood flows through the lungs, the relative partial pressures of CO_2 favor the diffusion of CO_2 out of the blood. As CO_2 diffuses into alveoli, the amount of CO_2 in the blood decreases. This decrease shifts the chemical equilibrium in favor of the conversion of HCO_3^- to CO_2, enabling further net diffusion of CO_2 into alveoli. Overall, the P_{CO_2} gradient is sufficient to reduce P_{CO_2} by roughly 15% during passage of blood through the lungs.

Respiratory Adaptations of Diving Mammals

EVOLUTION Animals vary greatly in their ability to temporarily inhabit environments in which there is no access to their normal respiratory medium—for example, when an air-breathing mammal swims underwater. Whereas most humans, even well-trained divers, cannot hold their breath longer than 2 or 3 minutes or swim deeper than 20 m, the Weddell seal of Antarctica routinely plunges to 200–500 m and remains there for about 20 minutes (and sometimes for more than an hour). (Humans can remain submerged for comparable periods, but only with the aid of specialized gear and compressed air tanks.) Some whales and other species of seals make even more impressive dives. Elephant seals can reach depths of 1,500 m—almost a mile—and stay submerged for as long as 2 hours! One elephant seal carrying a recording device spent 40 days at sea,

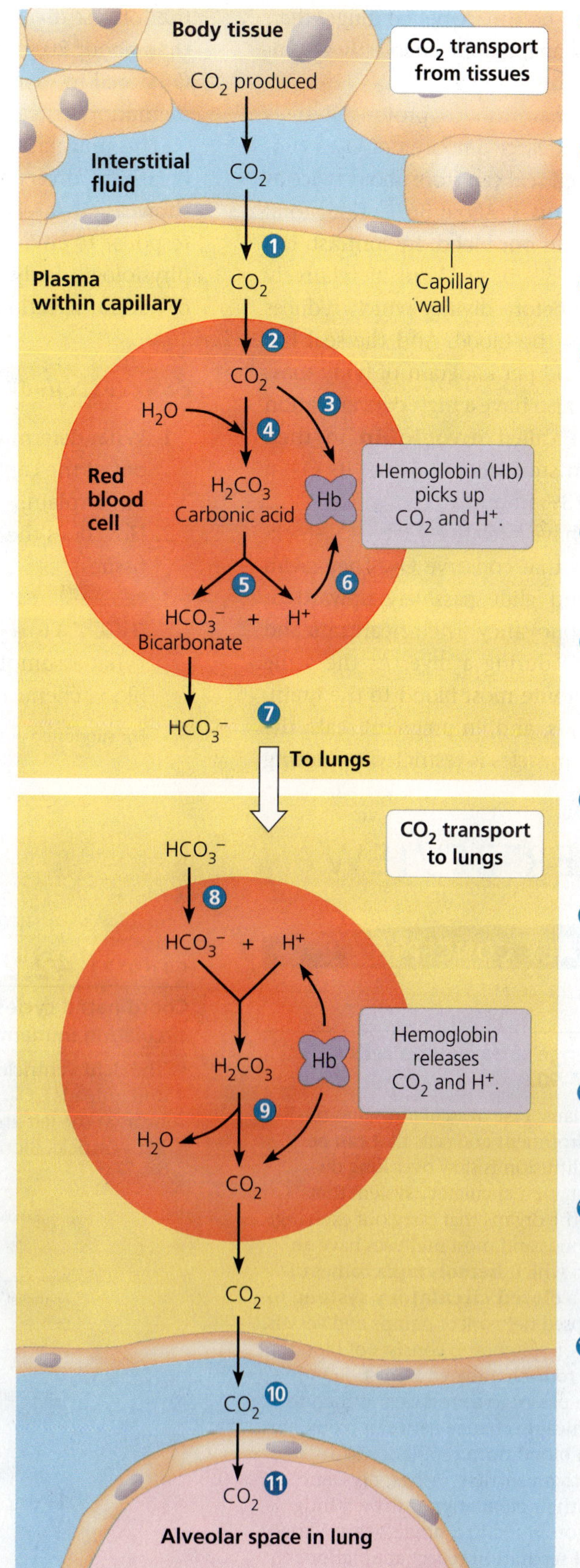

1 Carbon dioxide produced by body tissues diffuses into the interstitial fluid and the plasma.

2 Over 90% of the CO_2 diffuses into red blood cells, leaving only 7% in the plasma as dissolved CO_2.

3 Some CO_2 is picked up and transported by hemoglobin.

4 However, most CO_2 reacts with water in red blood cells, forming carbonic acid (H_2CO_3), a reaction catalyzed by carbonic anhydrase contained within red blood cells.

5 Carbonic acid dissociates into a bicarbonate ion (HCO_3^-) and a hydrogen ion (H^+).

6 Hemoglobin binds most of the H^+ from H_2CO_3, preventing the H^+ from acidifying the blood and thus preventing the Bohr shift.

7 Most of the HCO_3^- diffuses into the plasma, where it is carried in the bloodstream to the lungs.

8 In the lungs, HCO_3^- diffuses from the plasma into red blood cells, combining with H^+ released from hemoglobin and forming H_2CO_3.

9 Carbonic acid is converted back to CO_2 and water. CO_2 is also unloaded from hemoglobin.

10 CO_2 diffuses into the plasma and the interstitial fluid.

11 CO_2 diffuses into the alveolar space, from which it is expelled during exhalation. The reduction of CO_2 concentration in the plasma drives the breakdown of H_2CO_3 into CO_2 and water in the red blood cells (see 9), a reversal of the reaction that occurs near body tissues (see 4).

▲ **Figure 42.32 Carbon dioxide transport in the blood.**

? *In what three forms is CO_2 transported in the bloodstream?*

BioFlix Visit the Study Area at **www.masteringbiology.com** for the BioFlix® 3-D Animation on Gas Exchange.

diving almost continuously with no surface period longer than 6 minutes. What evolutionary adaptations enable these animals to perform such amazing feats?

One adaptation of diving mammals to prolonged stays underwater is an ability to store large amounts of O_2. Compared with humans, the Weddell seal can store about twice as much O_2 per kilogram of body mass. About 36% of our total O_2 is in our lungs, and 51% is in our blood. In contrast, the Weddell seal holds only about 5% of its O_2 in its relatively small lungs (and may exhale before diving, which reduces buoyancy), stockpiling 70% in the blood. And the seal has about twice the volume of blood per kilogram of body mass as a human. Diving mammals also have a high concentration of an oxygen-storing protein called **myoglobin** in their muscles. The Weddell seal can store about 25% of its O_2 in muscle, compared with only 13% in humans.

Diving mammals not only have a relatively large O_2 stockpile but also have adaptations that conserve O_2. They swim with little muscular effort and glide passively upward or downward by changing their buoyancy. Their heart rate and O_2 consumption rate decrease during a dive. At the same time, regulatory mechanisms route most blood to the brain, spinal cord, eyes, adrenal glands, and, in pregnant seals, the placenta. Blood supply to the muscles is restricted or, during the longest dives, shut off altogether. During dives of more than about 20 minutes, a Weddell seal's muscles deplete the O_2 stored in myoglobin and then derive their ATP from fermentation instead of respiration (see Chapter 9).

The unusual abilities of the Weddell seal and other air-breathing divers to power their bodies during long dives showcase two related themes in our study of organisms—the response to environmental challenges over the short term by physiological adjustments and over the long term as a result of natural selection.

CONCEPT CHECK 42.7

1. What determines whether O_2 and CO_2 diffuse into or out of the capillaries in the tissues and near the alveoli? Explain.
2. How does the Bohr shift help deliver O_2 to very active tissues?
3. **WHAT IF?** A doctor might give bicarbonate (HCO_3^-) to a patient who is breathing very rapidly. What assumption is the doctor making about the blood chemistry of the patient?

For suggested answers, see Appendix A.

42 CHAPTER REVIEW

SUMMARY OF KEY CONCEPTS

CONCEPT 42.1

Circulatory systems link exchange surfaces with cells throughout the body (pp. 897–902)

- In animals with simple body plans, gastrovascular cavities mediate exchange between the environment and cells that can be reached by diffusion. Because diffusion is slow over long distances, most complex animals have a circulatory system that moves fluid between cells and the organs that carry out exchange with the environment. Arthropods and most molluscs have an **open circulatory system**, in which **hemolymph** bathes organs directly. Vertebrates have a **closed circulatory system**, in which **blood** circulates in a closed network of pumps and vessels.
- The closed circulatory system of vertebrates consists of blood, **blood vessels**, and a two- to four-chambered **heart**. Blood pumped by a heart **ventricle** passes to **arteries** and then to **capillaries**, the sites of chemical exchange between blood and interstitial fluid. **Veins** return blood from capillaries to an **atrium**, which passes blood to a ventricle. Fishes, rays, and sharks have a single pump in their circulation. Air-breathing vertebrates have two pumps combined in a single heart. Variations in ventricle number and separation reflect adaptations to different environments and metabolic needs.

? *How does the flow of a fluid in a closed circulatory system differ from the movement of molecules between cells and their environment with regard to distance traveled, direction traveled, and driving force?*

CONCEPT 42.2

Coordinated cycles of heart contraction drive double circulation in mammals (pp. 902–904)

- The right ventricle pumps blood to the lungs, where it loads O_2 and unloads CO_2. Oxygen-rich blood from the lungs enters the heart at the left atrium and is pumped to the body tissues by the left ventricle. Blood returns to the heart through the right atrium.

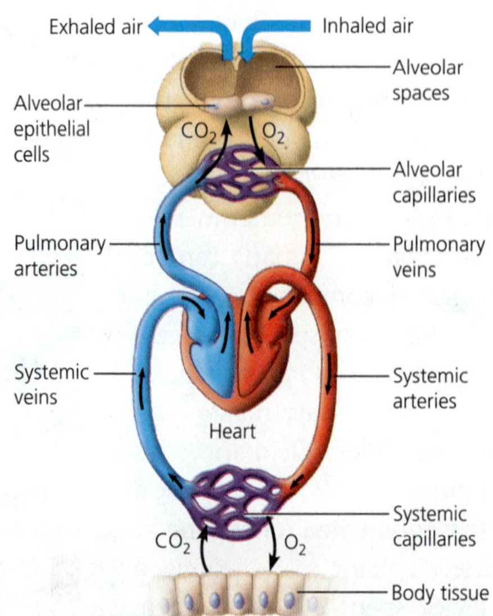

- The **cardiac cycle**, one complete sequence of the heart's pumping and filling, consists of a period of contraction, called **systole**, and a period of relaxation, called **diastole**. Heart function can be assessed by measuring the **pulse** (number of times the heart beats each minute) and **cardiac output** (volume of blood pumped by each ventricle per minute).
- The heartbeat originates with impulses at the **sinoatrial (SA) node** (pacemaker) of the right atrium. The impulses trigger contraction of both atria before passing to the **atrioventricular (AV) node**, where the impulses are temporarily delayed. They are then conducted along the bundle branches and Purkinje fibers, triggering contraction of the ventricles. The nervous system, hormones, and body temperature influence pacemaker activity.

? *What changes in cardiac function might you expect after surgical replacement of a defective heart valve?*

CONCEPT 42.3

Patterns of blood pressure and flow reflect the structure and arrangement of blood vessels (pp. 905–910)

- Blood vessels have structures well adapted to function. Capillaries have narrow diameters and thin walls that facilitate exchange. Arteries contain thick elastic walls that maintain blood pressure. Veins contain one-way valves that contribute to the return of blood to the heart.
- Physical laws governing the movement of fluids through pipes influence blood flow and blood pressure. The velocity of blood flow varies in the circulatory system, being lowest in the capillary beds as a result of their large total cross-sectional area. Blood pressure is altered by changes in cardiac output and by variable constriction of arterioles.
- Fluid leaks out of capillaries and is returned to blood by the **lymphatic system**. This system parallels the circulatory system in its extent and its mechanisms for fluid flow under low hydrostatic pressure. It also plays a vital role in defense against infection.

? *If you placed your forearm on your head, how, if at all, would the blood pressure in that arm change? Explain.*

CONCEPT 42.4

Blood components function in exchange, transport, and defense (pp. 910–915)

- Whole blood consists of cells and cell fragments (**platelets**) suspended in a liquid matrix called **plasma**. Plasma proteins influence blood pH, osmotic pressure, and viscosity, and they function in lipid transport, immunity (antibodies), and blood clotting (fibrinogen). Red blood cells, or **erythrocytes**, transport O_2. Five types of white blood cells, or **leukocytes**, function in defense against microbes and foreign substances in the blood. Platelets function in blood clotting, a cascade of reactions that converts plasma fibrinogen to fibrin.
- A variety of diseases impair function of the circulatory system. In **sickle-cell disease**, an aberrant form of **hemoglobin** disrupts erythrocyte shape and function, leading to blockage of small blood vessels and a decrease in the oxygen-carrying capacity of the blood. In cardiovascular disease, inflammation caused by damage to the lining of arteries enhances deposition of lipids and cells, resulting in the potential for life-threatening damage to the heart or brain.

? *In the absence of infection, what percentage of cells in human blood are leukocytes?*

CONCEPT 42.5

Gas exchange occurs across specialized respiratory surfaces (pp. 915–920)

- At all sites of **gas exchange**, a gas diffuses from where its **partial pressure** is higher to where it is lower. Air is more conducive to gas exchange than water because air has a higher O_2 content, lower density, and lower viscosity. Regardless of whether the respiratory medium is air or water, adequate diffusion of O_2 and CO_2 between the medium and an animal's cells requires large, moist respiratory surfaces.
- The structure and organization of respiratory surfaces differ among animal species. Gills are outfoldings of the body surface specialized for gas exchange in water. The effectiveness of gas exchange in some gills, including those of fishes, is increased by **ventilation** and **countercurrent exchange** between blood and water. Gas exchange in insects relies on a **tracheal system** consisting of tiny, branching tubes that penetrate the body, bringing O_2 directly to cells. Spiders, land snails, and most terrestrial vertebrates have internal **lungs**. In mammals, air inhaled through the nostrils passes through the pharynx into the **trachea, bronchi, bronchioles**, and dead-end **alveoli**, where gas exchange occurs.

? *Why does altitude have almost no effect on an animal's ability to rid itself of CO_2 through gas exchange?*

CONCEPT 42.6

Breathing ventilates the lungs (pp. 920–922)

- Breathing mechanisms vary substantially among vertebrates. An amphibian ventilates its lungs by **positive pressure breathing**, which forces air down the trachea. Birds use a system of air sacs as bellows to keep air flowing through the lungs in one direction only. Every exhalation completely renews the air in the lungs. Mammals ventilate their lungs by **negative pressure breathing**, which pulls air into the lungs. Lung volume increases as the rib muscles and **diaphragm** contract. Incoming and outgoing air mix, decreasing the efficiency of ventilation.
- Control centers in the medulla oblongata and pons of the human brain regulate the rate and depth of breathing. Sensors detect the pH of cerebrospinal fluid (reflecting CO_2 concentration in the blood), and the medulla adjusts breathing rate and depth to match metabolic demands. Secondary control over breathing is exerted by sensors in the aorta and carotid arteries that monitor blood levels of O_2 as well as CO_2 (via blood pH).

? *How does tidal volume differ from the volume of fresh air that enters the body during inspiration?*

CONCEPT 42.7

Adaptations for gas exchange include pigments that bind and transport gases (pp. 923–926)

- In the lungs, gradients of partial pressure favor the diffusion of O_2 into the blood and CO_2 out of the blood. The opposite situation exists in the rest of the body. **Respiratory pigments** transport O_2, greatly increasing the amount of O_2 that blood or hemolymph can carry. Many arthropods and molluscs have copper-containing hemocyanin; vertebrates and a wide variety of invertebrates have hemoglobin. Hemoglobin also helps transport CO_2 and assists in buffering the blood.

- Evolutionary adaptations enable some animals to satisfy extraordinary O_2 demands. Deep-diving air-breathers stockpile O_2 in blood and other tissues and deplete it slowly.

? *In what way is the role of a respiratory pigment like that of an enzyme?*

TEST YOUR UNDERSTANDING

LEVEL 1: KNOWLEDGE/COMPREHENSION

1. Which of the following respiratory systems is not closely associated with a blood supply?
   a. the lungs of a vertebrate
   b. the gills of a fish
   c. the tracheal system of an insect
   d. the skin of an earthworm
   e. the parapodia of a polychaete worm

2. Blood returning to the mammalian heart in a pulmonary vein drains first into the
   a. vena cava.
   b. left atrium.
   c. right atrium.
   d. left ventricle.
   e. right ventricle.

3. Pulse is a direct measure of
   a. blood pressure.
   b. stroke volume.
   c. cardiac output.
   d. heart rate.
   e. breathing rate.

4. When you hold your breath, which of the following blood gas changes first leads to the urge to breathe?
   a. rising O_2
   b. falling O_2
   c. rising CO_2
   d. falling CO_2
   e. rising CO_2 and falling O_2

5. One feature that amphibians and humans have in common is
   a. the number of heart chambers.
   b. the type of gas exchange tissues.
   c. a complete separation of circuits for circulation.
   d. the number of circuits for circulation.
   e. a low blood pressure in the systemic circuit.

LEVEL 2: APPLICATION/ANALYSIS

6. If a molecule of CO_2 released into the blood in your left toe is exhaled from your nose, it must pass through all of the following except
   a. the pulmonary vein.
   b. an alveolus.
   c. the trachea.
   d. the right atrium.
   e. the right ventricle.

7. Compared with the interstitial fluid that bathes active muscle cells, blood reaching these cells in arteries has a
   a. higher P_{O_2}.
   b. higher P_{CO_2}.
   c. greater bicarbonate concentration.
   d. lower pH.
   e. lower osmotic pressure.

8. Which of the following reactions prevails in red blood cells traveling through alveolar capillaries? (Hb = hemoglobin)
   a. $Hb + 4\ O_2 \rightarrow Hb(O_2)_4$
   b. $Hb(O_2)_4 \rightarrow Hb + 4\ O_2$
   c. $CO_2 + H_2O \rightarrow H_2CO_3$
   d. $H_2CO_3 \rightarrow H^+ + HCO_3^-$
   e. $Hb + 4\ CO_2 \rightarrow Hb(CO_2)_4$

LEVEL 3: SYNTHESIS/EVALUATION

9. **DRAW IT** Plot blood pressure against time for one cardiac cycle in humans, drawing separate lines for the pressure in the aorta, the left ventricle, and the right ventricle. Below the time axis, add a vertical arrow pointing to the time when you expect a peak in atrial blood pressure.

10. **EVOLUTION CONNECTION**
    One of the many mutant opponents that the movie monster Godzilla contends with is Mothra, a giant mothlike creature with a wingspan of several dozen meters. Science fiction creatures like these can be critiqued on the grounds of biomechanical and physiological principles. What problems of respiration and gas exchange would Mothra face? The largest insects that have ever lived are Paleozoic dragonflies with half-meter wingspans. Why do you think truly giant insects are improbable?

11. **SCIENTIFIC INQUIRY**
    The hemoglobin of a human fetus differs from adult hemoglobin. Compare the dissociation curves of the two hemoglobins in the graph at right. Propose a hypothesis to explain the benefit of this difference between these two hemoglobins.

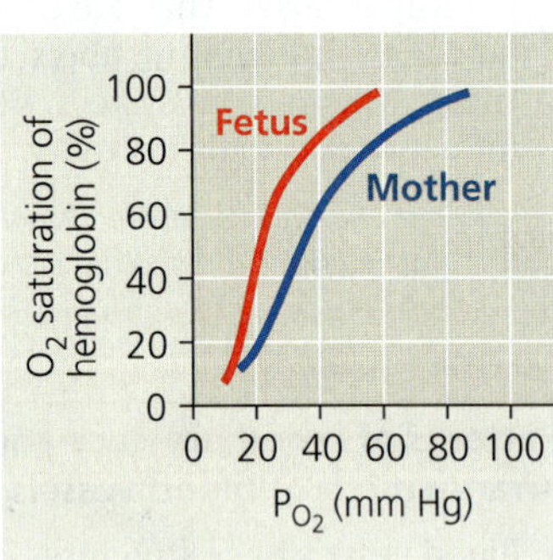

12. **SCIENCE, TECHNOLOGY, AND SOCIETY**
    Hundreds of studies have linked smoking with cardiovascular and lung disease. According to most health authorities, smoking is the leading cause of preventable, premature death in the United States. Antismoking groups have proposed that cigarette advertising in all media be banned entirely. What are some arguments in favor of a total ban on cigarette advertising? What are arguments in opposition? Do you favor or oppose such a ban? Defend your position.

13. **WRITE ABOUT A THEME**
    Environmental Interactions Some athletes prepare for competition at sea level by sleeping in a tent in which P_{O_2} is kept artificially low. When climbing very high peaks, some mountaineers breathe from bottles of pure O_2. In a short essay (100–150 words), relate these behaviors to the mechanism of O_2 transport in the human body and to our physiological interactions with our gaseous environment.

For selected answers, see Appendix A.

www.masteringbiology.com

1. MasteringBiology® Assignments

BioFlix **Tutorial** Gas Exchange
Tutorial Gas Transport in Blood
Activities Mammalian Cardiovascular System Structure • The Human Heart • Path of Blood Flow in Mammals • Mammalian Cardiovascular System Function • Discovery Channel Video: Blood • The Human Respiratory System • Gas Exchange in the Lungs and Tissues • Transport of Respiratory Gases
Questions Student Misconceptions • Reading Quiz • Multiple Choice • End-of-Chapter

2. eText

Read your book online, search, take notes, highlight text, and more.

3. The Study Area

Practice Tests • Cumulative Test • ***BioFlix*** 3-D Animations • MP3 Tutor Sessions • Videos • Activities • Investigations • Lab Media • Audio Glossary • Word Study Tools • Art

43

The Immune System

▲ **Figure 43.1 How do an animal's immune cells recognize foreign cells?**

KEY CONCEPTS

43.1 In innate immunity, recognition and response rely on traits common to groups of pathogens

43.2 In adaptive immunity, receptors provide pathogen-specific recognition

43.3 Adaptive immunity defends against infection of body fluids and body cells

43.4 Disruptions in immune system function can elicit or exacerbate disease

OVERVIEW

Recognition and Response

Pathogens, agents that cause disease, infect a wide range of animals. For a virus, bacterium, fungus, or other pathogen, the internal environment of an animal is a nearly ideal habitat. The animal body offers a ready source of nutrients, a protected setting for growth and reproduction, and a means of transport to new environments. From the perspective of a cold or flu virus, we are wonderful hosts. From our vantage point, things are not so ideal. Fortunately, adaptations have arisen over the course of evolution that protect animals against many invaders.

Dedicated immune cells in the body fluids and tissues of most animals specifically interact with and destroy pathogens. As shown in **Figure 43.1** (a colorized scanning electron micrograph), an immune cell called a macrophage (blue) can engulf a yeast cell (green). Additional responses to infection take many forms, including proteins that punch holes in bacterial membranes or block viruses from entering body cells. These and other defenses make up the **immune system**, which enables an animal to avoid or limit many infections. A foreign molecule or cell doesn't have to be pathogenic to elicit an immune response, but we'll focus here on the immune system's role in defending against pathogens.

All animals have **innate immunity**, a defense that is active immediately upon infection and is the same whether or not the pathogen has been encountered previously. Innate immunity includes an outer covering, such as a skin or shell, that provides a significant barrier to entry by microbes. Sealing off the entire body surface is impossible, however, because gas exchange, nutrition, and reproduction require openings to the environment. Chemical secretions that trap or kill microbes guard the body's entrances and exits, while the linings of the digestive tract, airway, and other exchange surfaces provide additional barriers to infection.

If a pathogen breaches barrier defenses and enters the body, the problem of how to fend off attack changes substantially. Housed within the body fluids and tissues, the invader is no longer an outsider. To fight infections, an animal's immune system must detect foreign particles and cells within the body. In other words, a properly functioning immune system distinguishes nonself from self. Detection of nonself is accomplished by *molecular recognition*, in which receptor molecules bind specifically to molecules from foreign cells or viruses.

In innate immunity, a small preset group of receptor proteins bind to molecules or structures that are absent from animal bodies but common to a group of viruses, bacteria, or other microbes. Binding of an innate immune receptor to a foreign molecule activates internal defenses, enabling responses to a very broad range of pathogens.

A different type of molecular recognition provides the basis for **adaptive immunity**, a defense found only in vertebrates. Animals with adaptive immunity produce a vast arsenal of receptors, each of which recognizes a feature typically found only on a particular part of a particular molecule in a particular pathogen. As a result, recognition and response in adaptive immunity occur with tremendous specificity.

The adaptive immune response, also known as the acquired immune response, is activated after the innate immune response and develops more slowly. The names *adaptive* and

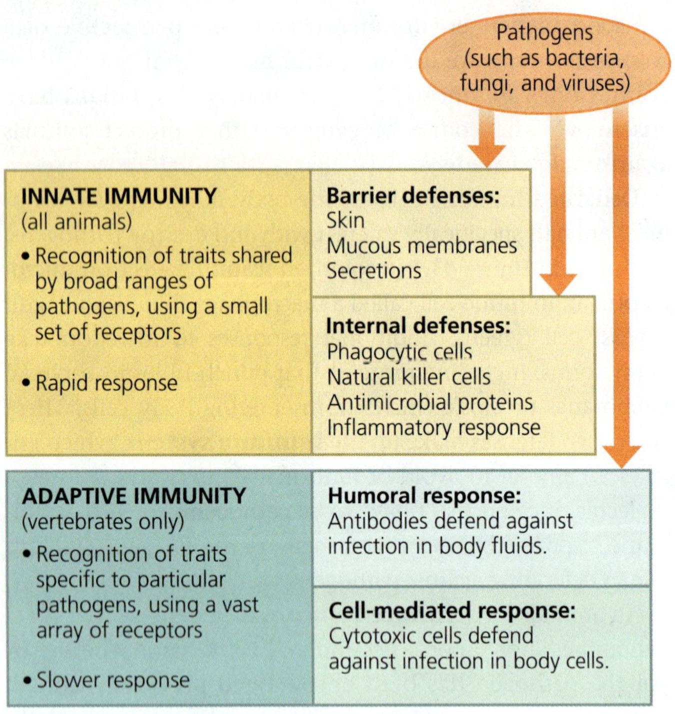

▲ **Figure 43.2 Overview of animal immunity.** Immune responses in animals can be divided into innate and adaptive immunity. Some components of innate immunity contribute to activation of adaptive immune defenses.

acquired reflect the fact that this immune response is enhanced by previous exposure to the infecting pathogen. Examples of adaptive responses include the synthesis of proteins that inactivate a bacterial toxin and the targeted killing of a virus-infected body cell.

Figure 43.2 provides an overview of the basic components of innate and adaptive immunity. In this chapter, you will learn how each type of immunity protects animals from disease. You will also examine how pathogens can avoid or overwhelm the immune system and how defects in the immune system can imperil an animal's health.

CONCEPT 43.1

In innate immunity, recognition and response rely on traits common to groups of pathogens

Innate immunity is found in all animals (as well as in plants). In exploring innate immunity, we'll begin with invertebrates, which repel and fight infection with only this type of immunity. We'll then turn to vertebrates, in which innate immunity serves both as an immediate defense against infection and as the foundation for adaptive immune defenses.

Innate Immunity of Invertebrates

The great success of insects in terrestrial and freshwater habitats teeming with diverse microbes highlights the effectiveness of invertebrate innate immunity. In each of these environments, insects rely on their exoskeleton as a first line of defense against infection. Composed largely of the polysaccharide chitin, the exoskeleton provides an effective barrier defense against most pathogens. A chitin-based barrier is also present in the insect intestine, where it blocks infection by many pathogens ingested with food. **Lysozyme**, an enzyme that breaks down bacterial cell walls, further protects the insect digestive system.

Any pathogen that breaches an insect's barrier defenses encounters a number of internal immune defenses. Immune cells called *hemocytes* travel throughout the body in the hemolymph, the insect circulatory fluid. Some hemocytes carry out a defense called **phagocytosis**, the cellular ingestion and digestion of bacteria and other foreign substances **(Figure 43.3)**. Other hemocytes trigger the production of chemicals that kill pathogens and help entrap large parasites, such as *Plasmodium*, the parasite of mosquitoes that causes malaria. In addition, encounters with pathogens in the hemolymph cause hemocytes and certain other cells to secrete *antimicrobial peptides*, which are short chains of amino acids. The antimicrobial peptides circulate throughout the body of the insect **(Figure 43.4)** and inactivate or kill fungi and bacteria by disrupting their plasma membranes.

Immune cells of insects bind to molecules found only in the outer layers of fungi or bacteria. Fungal cell walls contain certain unique polysaccharides, whereas bacterial cell walls have polymers containing combinations of sugars and amino

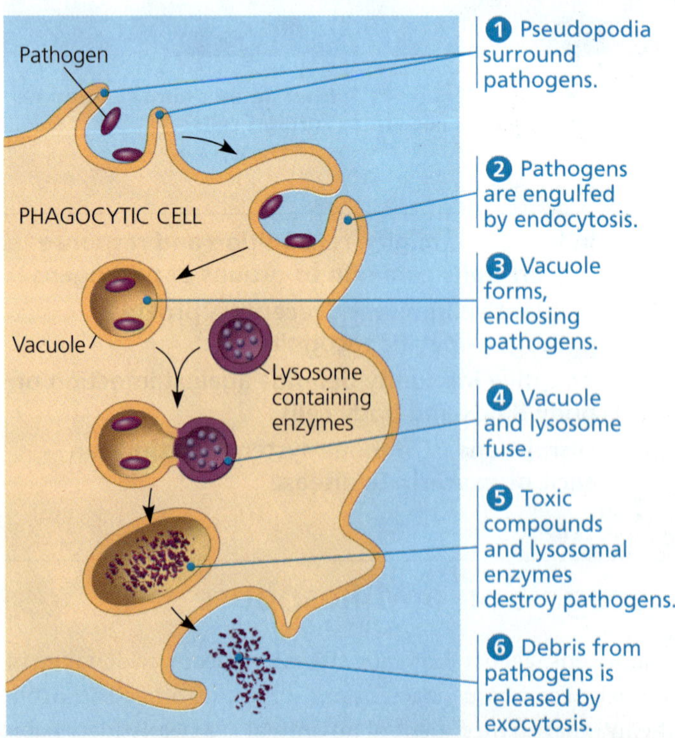

▲ **Figure 43.3 Phagocytosis.** This schematic depicts events in the ingestion and destruction of a microbe by a typical phagocytic cell.

▲ **Figure 43.4 An inducible innate immune response.** These fruit flies were engineered to express the green fluorescent protein (GFP) gene upon activation of the innate immune response. The fly on the top was injected with bacteria; the fly on the bottom was not. Only the infected fly activates antimicrobial peptide genes, produces GFP, and glows a bright green under fluorescent light.

acids not found in animal cells. Such macromolecules serve as identity tags in the process of pathogen recognition. Insect immune cells secrete specialized recognition proteins, each of which binds to a macromolecule characteristic of fungi or a broad class of bacteria.

Innate immune responses are distinct for different classes of pathogens. For example, when the fungus *Neurospora crassa* infects a fruit fly, pieces of the fungal cell wall bind a recognition protein. Together, the complex activates the protein Toll, a receptor on the surface of hemocytes. Signal transduction from the Toll receptor to the cell nucleus leads to synthesis of a set of antimicrobial peptides active against fungi. If the fly is instead infected by the bacterium *Micrococcus luteus*, a different recognition protein is activated, and the fly produces a different set of antimicrobial peptides effective against *M. luteus* and many related bacteria.

Because fruit flies secrete many distinct antimicrobial peptides in response to a single infection, it is difficult to study the activity of any one peptide. To get around this problem, Bruno Lemaitre and fellow researchers used modern genetic techniques to reprogram the fly immune system (**Figure 43.5**). They found that the synthesis of a single type of antimicrobial peptide in the fly's body could provide an effective immune defense. They also showed that particular antimicrobial peptides act against different kinds of pathogens.

▼ **Figure 43.5** INQUIRY

Can a single antimicrobial peptide protect fruit flies against infection?

EXPERIMENT In 2002, Bruno Lemaitre and colleagues in France devised a novel strategy to test the function of a single antimicrobial peptide. They began with a mutant fruit fly strain in which pathogens are recognized but the signaling that would normally trigger innate immune responses is blocked. As a result, the mutant flies do not make any antimicrobial peptides. The researchers then genetically engineered some of the mutant fruit flies to express significant amounts of a single antimicrobial peptide, either drosomycin or defensin. The scientists infected the various flies with the fungus *Neurospora crassa* and monitored survival over a five-day period. They repeated the procedure for infection by the bacterium *Micrococcus luteus*.

RESULTS

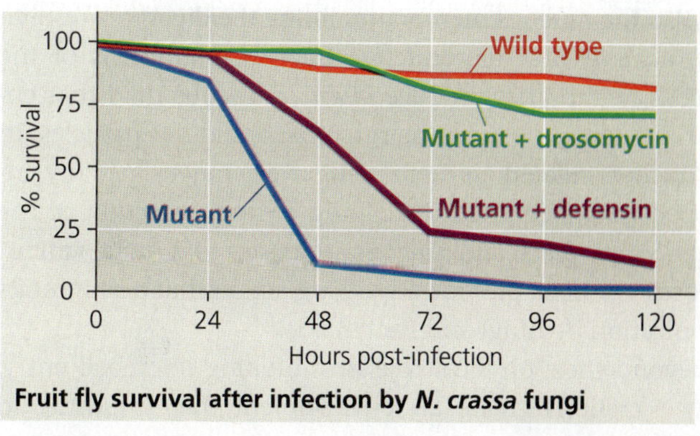

Fruit fly survival after infection by *N. crassa* fungi

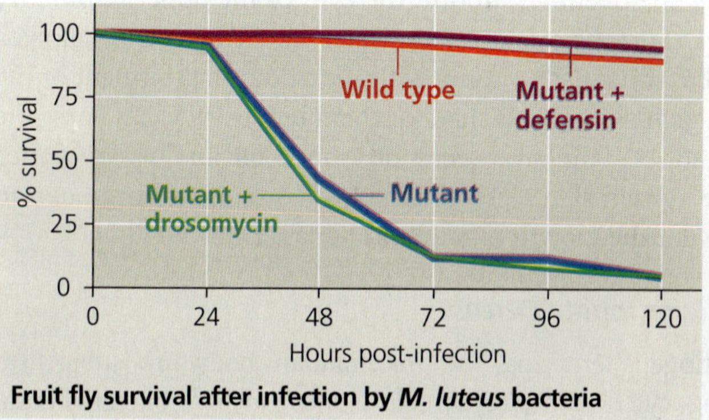

Fruit fly survival after infection by *M. luteus* bacteria

CONCLUSION Each of the two antimicrobial peptides provided a protective immune response. Furthermore, the different peptides defended against different pathogens. Drosomycin was effective against *N. crassa*, and defensin was effective against *M. luteus*.

SOURCE P. Tzou, J. Reichhart, and B. Lemaitre, Constitutive expression of a single antimicrobial peptide can restore wild-type resistance to infection in immunodeficient *Drosophila* mutants, *Proceedings of the National Academy of Sciences USA* 99:2152–2157 (2002).

WHAT IF? Even if a particular antimicrobial peptide showed no beneficial effect in such an experiment, why might it still be beneficial to flies?

Innate Immunity of Vertebrates

Among vertebrates, innate immune defenses coexist with the more recently evolved system of adaptive immunity. Because most of the recent discoveries regarding vertebrate innate immunity have come from studies of mice and humans, we'll focus here on mammals. We'll consider the innate defenses that are similar to those found among invertebrates: barrier defenses, phagocytosis, and antimicrobial peptides. We'll also examine some unique aspects of vertebrate innate immunity, such as natural killer cells, interferons, and the inflammatory response.

Barrier Defenses

In mammals, epithelial tissues block the entry of many pathogens. These barrier defenses include not only the skin but also the mucous membranes lining the digestive, respiratory, urinary, and reproductive tracts. Certain cells of the mucous membranes produce *mucus*, a viscous fluid that enhances defenses by trapping microbes and other particles. In the trachea, ciliated epithelial cells sweep mucus and any entrapped microbes upward, helping prevent infection of the lungs. Saliva, tears, and mucous secretions that bathe various exposed epithelia provide a washing action that also inhibits colonization by fungi and bacteria.

Beyond their physical role in inhibiting microbial entry, body secretions create an environment that is hostile to many microbes. Lysozyme in tears, saliva, and mucous secretions destroys the cell walls of susceptible bacteria as they enter the openings around the eyes or the upper respiratory tract. Microbes in food or water and those in swallowed mucus must also contend with the acidic environment of the stomach, which kills most of them before they can enter the intestines. Similarly, secretions from oil and sweat glands give human skin a pH ranging from 3 to 5, acidic enough to prevent the growth of many bacteria.

Cellular Innate Defenses

Pathogens entering the mammalian body are subject to phagocytosis. Phagocytic cells detect fungal or bacterial components using several types of receptors, some of which are very similar to the Toll receptor of insects. Each mammalian **Toll-like receptor (TLR)** binds to fragments of molecules characteristic of a set of pathogens **(Figure 43.6)**. For example, TLR3, on the inner surface of vesicles formed by endocytosis, is the sensor for double-stranded RNA, a form of nucleic acid characteristic of certain viruses. Similarly, TLR4, located on immune cell plasma membranes, recognizes lipopolysaccharide, a type of molecule found on the surface of many bacteria; and TLR5 recognizes flagellin, the main protein of bacterial flagella. In each case, the recognized macromolecule is normally absent from the vertebrate body and is an essential component of certain groups of pathogens.

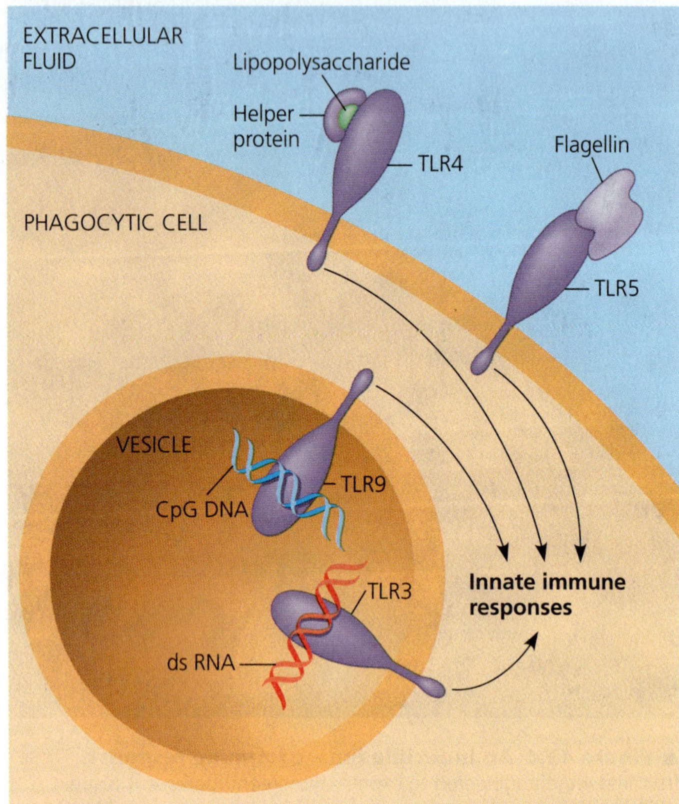

▲ **Figure 43.6 TLR signaling.** Each mammalian Toll-like receptor (TLR) recognizes a molecular pattern characteristic of a group of pathogens. Lipopolysaccharide, flagellin, CpG DNA (DNA containing unmethylated CG sequences), and double-stranded (ds) RNA are all found in bacteria, fungi, or viruses, but not in animal cells. Together with other recognition and response factors, TLR proteins trigger internal innate immune defenses.

? *Some TLR proteins are on the cell surface, whereas others are inside vesicles. Suggest a possible benefit of this distribution.*

After detecting invading pathogens, a phagocytic cell engulfs them, trapping them in a vacuole. The vacuole then fuses with a lysosome (see Figure 43.3), leading to destruction of the invaders in two ways. First, gases produced in the lysosome poison the engulfed pathogens. Second, lysozyme and other enzymes in the lysosome degrade the components of the pathogens.

The two main types of phagocytic cells in the mammalian body are neutrophils and macrophages. **Neutrophils**, which circulate in the blood, are attracted by signals from infected tissues and then engulf and destroy the infecting pathogens. **Macrophages** ("big eaters"), like the one shown in Figure 43.1, are larger phagocytic cells. Some migrate throughout the body, whereas others reside permanently in organs and tissues where they are likely to encounter pathogens. For example, some macrophages are located in the spleen, where pathogens in the blood become trapped.

Two other types of phagocytic cells—dendritic cells and eosinophils—provide additional functions in innate defense. **Dendritic cells** mainly populate tissues, such as skin, that contact the environment. They stimulate adaptive immunity against pathogens they encounter and engulf, as we'll explore shortly. *Eosinophils*, often found beneath mucosal surfaces, have low phagocytic activity but are important in defending against multicellular invaders, such as parasitic worms. Upon encountering such parasites, eosinophils discharge destructive enzymes.

Cellular innate defenses in vertebrates also involve **natural killer cells**. These cells circulate through the body and detect the abnormal array of surface proteins characteristic of some virus-infected and cancerous cells. Natural killer cells do not engulf stricken cells. Instead, they release chemicals that lead to cell death, inhibiting further spread of the virus or cancer.

Many cellular innate defenses of vertebrates involve the lymphatic system, a network that distributes the fluid called lymph throughout the body **(Figure 43.7)**. Some macrophages reside in the structures called lymph nodes, where they engulf pathogens that have flowed from the interstitial fluid into the lymph. Dendritic cells reside outside the lymphatic system but migrate to lymph nodes after interaction with pathogens. Within the lymph nodes, dendritic cells interact with other immune cells, stimulating adaptive immunity.

Antimicrobial Peptides and Proteins

In mammals, pathogen recognition triggers the production and release of a variety of peptides and proteins that attack pathogens or impede their reproduction. Some of these defense molecules function like the antimicrobial peptides of insects, damaging broad groups of pathogens by disrupting membrane integrity. Others, including the interferons and complement proteins, are unique to vertebrate immune systems.

Interferons are proteins that provide innate defense by interfering with viral infections. Virus-infected body cells secrete interferons, which induce nearby uninfected cells to produce substances that inhibit viral reproduction. In this way, interferons limit the cell-to-cell spread of viruses in the body, helping control viral infections such as colds and influenza. Some white blood cells secrete a different type of interferon that helps activate macrophages, enhancing their phagocytic ability. Pharmaceutical companies now use recombinant DNA technology to mass-produce interferons to help treat certain viral infections, such as hepatitis C.

The infection-fighting **complement system** consists of roughly 30 proteins in blood plasma. These proteins circulate in an inactive state and are activated by substances on the surface of many microbes. Activation results in a cascade of biochemical reactions that can lead to lysis (bursting) of invading cells. The complement system also functions in the inflammatory response, our next topic, as well as in the adaptive defenses discussed later in the chapter.

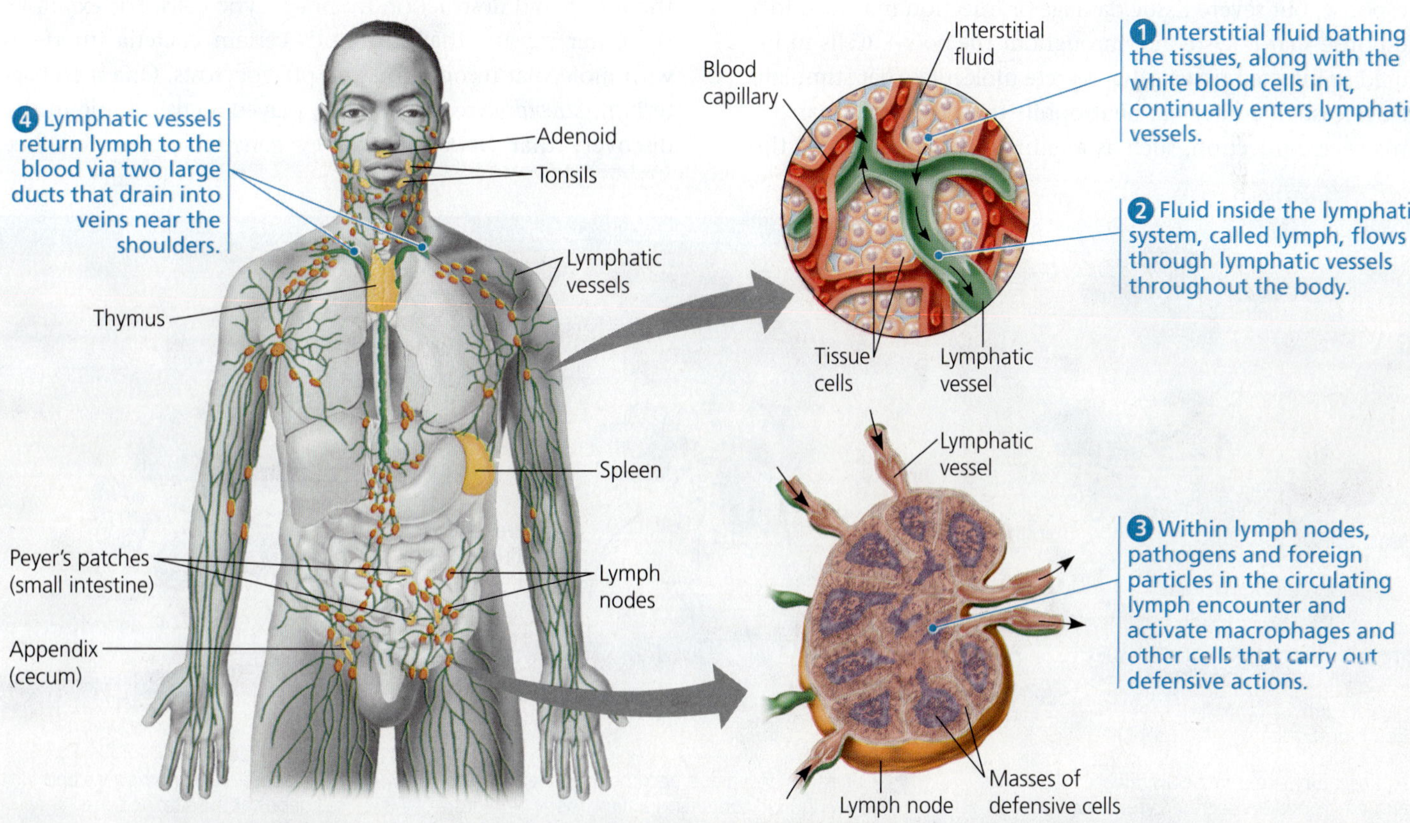

▲ **Figure 43.7 The human lymphatic system.** The lymphatic system consists of lymphatic vessels (shown in green), through which lymph travels, and structures that trap foreign substances. These structures include lymph nodes (orange) and lymphoid organs (yellow): the adenoids, tonsils, spleen, Peyer's patches, and appendix. Steps 1–4 trace the flow of lymph and illustrate the critical role of lymph nodes in activating adaptive immunity. (See also p. 909 for a description of the relationship between the lymphatic and circulatory systems.)

Inflammatory Response

The pain and swelling that alert you to a splinter under your skin are the result of a local **inflammatory response**, the changes brought about by signaling molecules released upon injury or infection **(Figure 43.8)**. One important inflammatory signaling molecule is **histamine**, which is stored in the granules (vesicles) of **mast cells**, found in connective tissue. Histamine released at sites of damage triggers nearby blood vessels to dilate and become more permeable. Activated macrophages and neutrophils discharge **cytokines**, signaling molecules that enhance an immune response. These cytokines promote blood flow to the site of injury or infection. The increase in local blood supply causes the redness and increased skin temperature typical of the inflammatory response (from the Latin *inflammare*, to set on fire). Blood-engorged capillaries leak fluid into neighboring tissues, causing swelling.

During inflammation, cycles of signaling and response transform the site. Activated complement proteins promote further release of histamine, attracting more phagocytic cells that enter injured tissues (see Figure 43.8) and carry out additional phagocytosis. At the same time, enhanced blood flow to the site helps deliver antimicrobial peptides. The result is an accumulation of *pus*, a fluid rich in white blood cells, dead pathogens, and cell debris from damaged tissue.

A minor injury or infection causes a local inflammatory response, but severe tissue damage or infection may lead to a response that is systemic (throughout the body). Cells in injured or infected tissue often secrete molecules that stimulate the release of additional neutrophils from the bone marrow. In a severe infection, such as meningitis or appendicitis, the number of white blood cells in the blood may increase severalfold within a few hours.

Another systemic inflammatory response is fever. In response to certain pathogens, substances released by activated macrophages cause the body's thermostat to reset to a higher temperature (see Chapter 40). The benefits of the resulting fever are still a subject of debate. One of several competing hypotheses is that an elevated body temperature may enhance phagocytosis and, by speeding up chemical reactions, accelerate tissue repair.

Certain bacterial infections can induce an overwhelming systemic inflammatory response, leading to a life-threatening condition called *septic shock*. Characterized by very high fever, low blood pressure, and poor blood flow through capillaries, septic shock occurs most often in the very old and the very young. It is fatal in more than one-third of cases and kills more than 90,000 people each year in the United States alone.

Chronic (ongoing) inflammation can also threaten human health. For example, millions of individuals worldwide suffer from Crohn's disease and ulcerative colitis, often debilitating disorders in which an unregulated inflammatory response disrupts intestinal function.

Evasion of Innate Immunity by Pathogens

Adaptations have evolved in some pathogens that enable them to avoid destruction by phagocytic cells. For example, the outer capsule that surrounds certain bacteria interferes with molecular recognition and phagocytosis. One such bacterium, *Streptococcus pneumoniae*, played a critical role in the discovery that DNA can convey genetic information (see

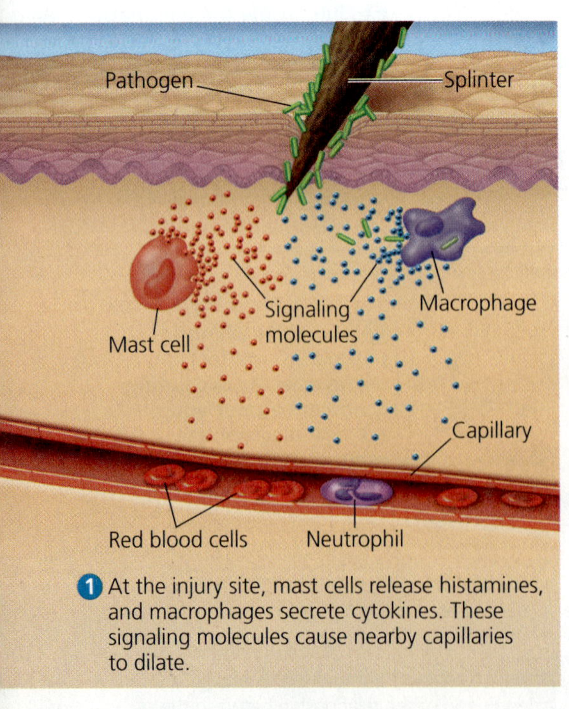

1 At the injury site, mast cells release histamines, and macrophages secrete cytokines. These signaling molecules cause nearby capillaries to dilate.

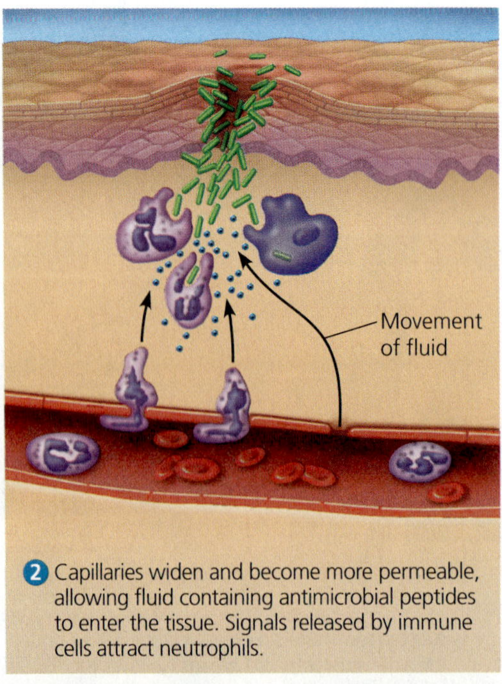

2 Capillaries widen and become more permeable, allowing fluid containing antimicrobial peptides to enter the tissue. Signals released by immune cells attract neutrophils.

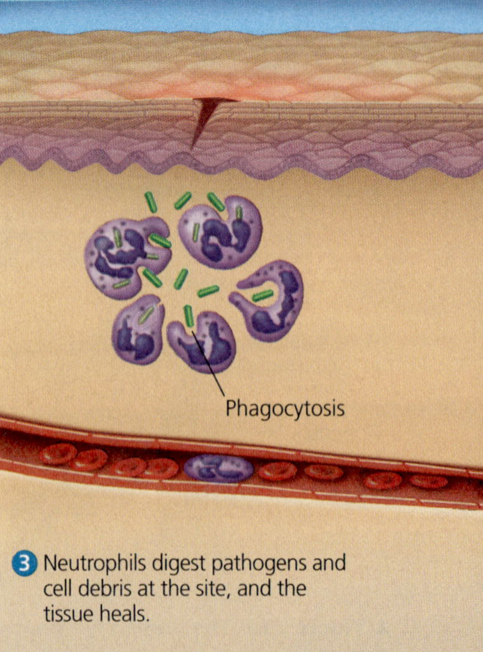

3 Neutrophils digest pathogens and cell debris at the site, and the tissue heals.

▲ **Figure 43.8 Major events in a local inflammatory response.**

Figure 16.2). Some bacteria, after being engulfed by a host cell, resist breakdown within lysosomes. An example is the bacterium that causes tuberculosis (TB). Rather than being destroyed within host cells, this bacterium grows and reproduces, effectively hidden from the body's innate immune defenses. These and other mechanisms that prevent destruction by the innate immune system make certain fungi and bacteria substantial pathogenic threats. Indeed, TB kills more than a million people a year worldwide.

CONCEPT CHECK 43.1

1. Although pus is often seen simply as a sign of infection, it is also an indicator of immune defenses in action. Explain.
2. MAKE CONNECTIONS How do the molecules that activate the vertebrate TLR signal transduction pathway differ from the ligands in most other pathways, such as those shown in Concept 11.2 (pp. 210–214)?
3. WHAT IF? Suppose humans were the major host for a bacterial species. What temperature would you predict would be optimal for growth of this species? Explain.

For suggested answers, see Appendix A.

CONCEPT 43.2

In adaptive immunity, receptors provide pathogen-specific recognition

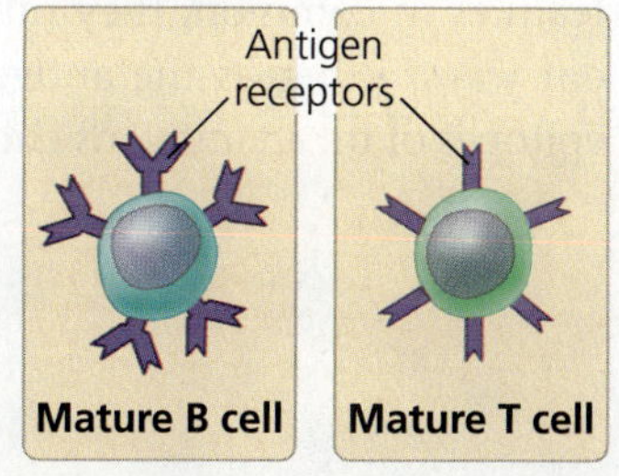

Vertebrates are unique in having adaptive immunity in addition to innate immunity. The adaptive response relies on T cells and B cells, which are types of white blood cells called **lymphocytes**. Like all blood cells, lymphocytes originate from stem cells in the bone marrow. Some lymphocytes migrate from the bone marrow to the **thymus**, an organ in the thoracic cavity above the heart (see Figure 43.7). These lymphocytes mature into **T cells**. Lymphocytes that remain and mature in the bone marrow develop as **B cells**. (Lymphocytes of a third type remain in the blood and become the natural killer cells active in innate immunity.)

Any substance that elicits a response from a B cell or T cell is called an **antigen**. In adaptive immunity, recognition occurs when a B cell or T cell binds to an antigen, such as a bacterial or viral protein, via a protein called an **antigen receptor**. An antigen receptor is specific enough to bind to just one part of one molecule from a particular pathogen, such as a species of bacteria or strain of virus. Although the cells of the immune system produce millions of different antigen receptors, all of the antigen receptors made by a single B or T cell are identical. Infection by a virus, bacterium, or other pathogen triggers activation of B and T cells with antigen receptors specific for parts of that pathogen. B and T cells are shown here with only a few antigen receptors, but there are actually about 100,000 antigen receptors on the surface of a single B or T cell.

Antigens are usually foreign and are typically large molecules, either proteins or polysaccharides. Many antigens protrude from the surface of foreign cells or viruses. Other antigens, such as toxins secreted by bacteria, are released into the extracellular fluid.

The small, accessible portion of an antigen that binds to an antigen receptor is called an **epitope**, or *antigenic determinant*. An example is a group of amino acids in a particular protein. A single antigen usually has several different epitopes, each binding a receptor with a different specificity. Because all antigen receptors produced by a single B cell or T cell are identical, they bind to the same epitope. Each B cell or T cell thus displays *specificity* for a particular epitope, enabling it to respond to any pathogen that produces molecules containing that same epitope.

The antigen receptors of B cells and T cells have similar components, but they encounter antigens in different ways. We'll consider the two processes in turn.

Antigen Recognition by B Cells and Antibodies

Each B cell antigen receptor is a Y-shaped molecule consisting of four polypeptide chains: two identical **heavy chains** and two identical **light chains**, with disulfide bridges linking the chains together (**Figure 43.9**). A transmembrane region near one end of each heavy chain anchors the receptor in the cell's plasma membrane. A short tail region at the end of the heavy chain extends into the cytoplasm.

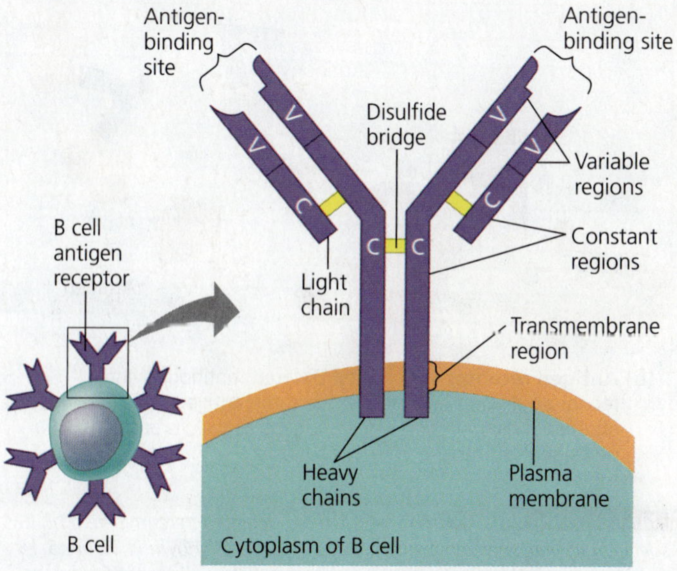

▲ **Figure 43.9 The structure of a B cell antigen receptor.**

The light and heavy chains each have a *constant (C) region*, where amino acid sequences vary little among the receptors on different B cells. The C region includes the cytoplasmic tail and transmembrane region of the heavy chain and all of the disulfide bridges. Within the two tips of the Y shape, the light and heavy chains each have a *variable (V) region*, so named because its amino acid sequence varies extensively from one B cell to another. Together, parts of a heavy-chain V region and a light-chain V region form an asymmetrical binding site for an antigen. As shown in Figure 43.9, each B cell antigen receptor has two identical antigen-binding sites.

The binding of a B cell antigen receptor to an antigen is an early step in B cell activation, leading eventually to formation of cells that secrete a soluble form of the receptor **(Figure 43.10a)**. This secreted protein is called an **antibody**, or **immunoglobulin (Ig)**. Antibodies have the same Y-shaped organization as B cell antigen receptors, but they are secreted rather than membrane bound. It is the antibodies, rather than the B cells themselves, that actually help defend against pathogens. Antibodies have distinct functions, as we'll see later.

The antigen-binding site of a membrane-bound receptor or antibody has a unique shape that provides a lock-and-key fit for a particular epitope. Many noncovalent bonds between an epitope and the binding surface provide a stable and specific interaction. Differences in the amino acid sequences of variable regions provide the variation in binding surfaces that enables this highly specific binding.

B cell antigen receptors and antibodies bind to intact antigens in the blood and lymph. As illustrated in **Figure 43.10b** for antibodies, they can bind to antigens on the surface of pathogens or free in body fluids. The antigen receptors of T cells function quite differently, as we'll see next.

▼ **Figure 43.10 Antigen recognition by B cells and antibodies.**

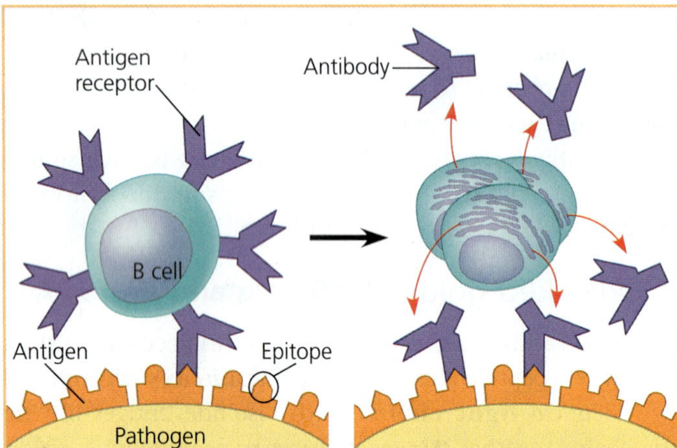

(a) B cell antigen receptors and antibodies. An antigen receptor of a B cell binds to an epitope, a particular part of an antigen. Following binding, the B cell gives rise to cells that secrete a soluble form of the antigen receptor. This soluble receptor, called an antibody, is specific for the same epitope as the original B cell.

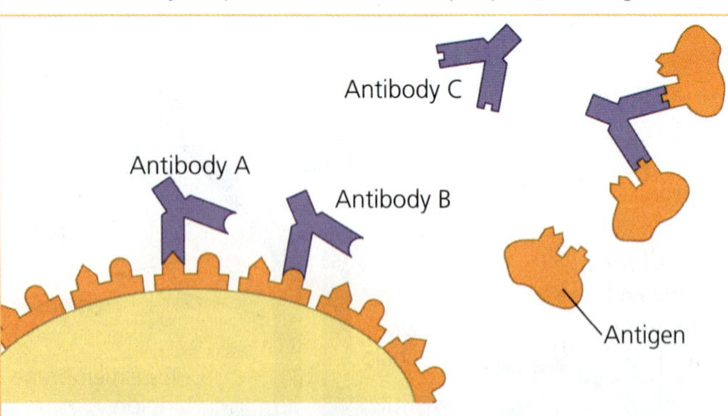

(b) Antigen receptor specificity. Different antibodies can recognize distinct epitopes on the same antigen. Furthermore, antibodies can recognize free antigens as well as antigens on a pathogen's surface.

MAKE CONNECTIONS *The interactions depicted here involve a highly specific binding between antigen and receptor, as shown in Figure 5.19 (p. 81). How is this similar to the enzyme-substrate interaction shown in Figure 8.14 (p. 154)?*

Antigen Recognition by T Cells

For a T cell, the antigen receptor consists of two different polypeptide chains, an *α chain* and a *β chain*, linked by a disulfide bridge **(Figure 43.11)**. Near the base of the T cell antigen receptor (often called simply a T cell receptor) is a transmembrane region that anchors the molecule in the cell's plasma membrane. At the outer tip of the molecule, the variable (V) regions of α and β chains together form a single antigen-binding site. The remainder of the molecule is made up of the constant (C) regions.

Although T cell and B cell antigen receptors have many features in common, they function in fundamentally different ways. Whereas the antigen receptors of B cells bind to epitopes of intact antigens circulating in body fluids, those of

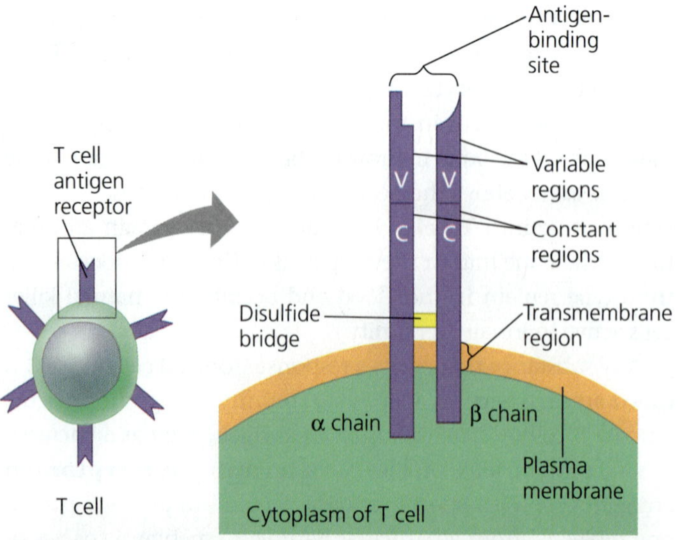

▲ **Figure 43.11 The structure of a T cell antigen receptor.**

T cells bind only to fragments of antigens that are displayed, or *presented*, on the surface of host cells. The host protein that displays the antigen fragment on the cell surface is called an **MHC (major histocompatibility complex) molecule**.

Recognition of protein antigens by T cells begins when a pathogen or part of a pathogen either infects or is taken in by a host cell **(Figure 43.12a)**. Inside the host cell, enzymes in the cell cleave the antigen into smaller peptides. Each peptide, called an *antigen fragment*, then binds to an MHC molecule inside the cell. Movement of the MHC molecule and bound antigen fragment to the cell surface results in **antigen presentation**, the display of the antigen fragment in an exposed groove of the MHC protein. **Figure 43.12b** shows a close-up view of antigen presentation, which advertises the fact that a host cell contains a foreign substance. If the cell displaying an antigen fragment encounters a T cell with the right specificity, the antigen receptor on the T cell can bind to both the antigen fragment and the MHC molecule. This interaction of an MHC molecule, an antigen fragment, and an antigen receptor is necessary for a T cell to participate in an adaptive immune response, as we'll see later.

▼ Figure 43.12 Antigen recognition by T cells.

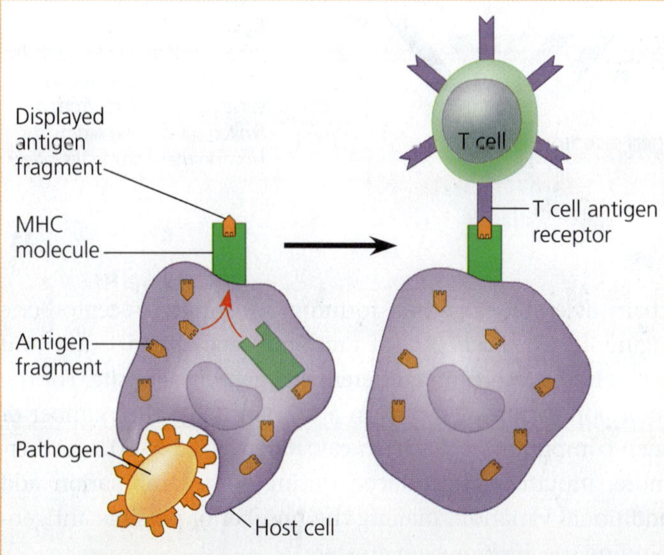

(a) Antigen recognition by a T cell. Inside the host cell, an antigen fragment from a pathogen binds to an MHC molecule and is brought up to the cell surface, where it is displayed. The combination of MHC molecule and antigen fragment is recognized by a T cell.

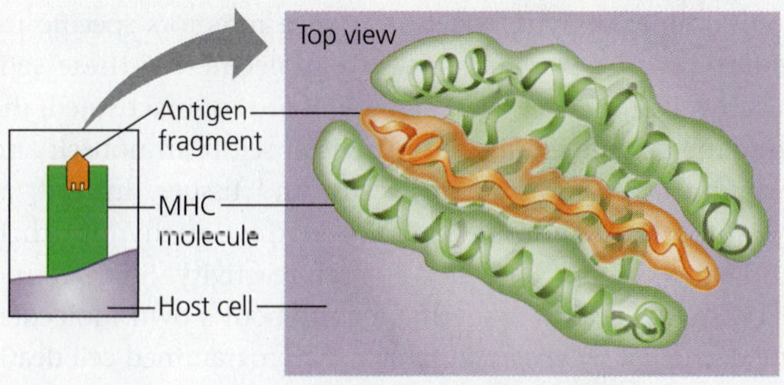

(b) A closer look at antigen presentation. As shown in this ribbon model, the top of the MHC molecule cradles an antigen fragment, like a bun holding a hot dog. An MHC molecule can display many different antigen fragments, but the antigen receptor of a T cell is specific for a single antigen fragment.

B Cell and T Cell Development

Now that you know how B cells and T cells recognize antigens, let's consider four major characteristics of adaptive immunity. First, there is an immense diversity of lymphocytes and receptors, enabling the immune system to detect pathogens never before encountered. Second, adaptive immunity normally has self-tolerance, the lack of reactivity against an animal's own molecules and cells. Third, cell proliferation triggered by activation greatly increases the number of B and T cells specific for an antigen. Fourth, there is a stronger and more rapid response to an antigen encountered previously, due to a feature known as *immunological memory*.

Receptor diversity and self-tolerance arise as a lymphocyte matures. Proliferation of cells and the formation of immunological memory occur later, after a mature lymphocyte encounters and binds to a specific antigen. We'll consider these four characteristics in the order in which they develop.

Generation of B and T Cell Diversity

Each person makes more than 1 million different B cell antigen receptors and 10 million different T cell antigen receptors. Yet there are only about 20,000 protein-coding genes in the human genome. How, then, do we generate such remarkable diversity in antigen receptors? The answer lies in combinations. Think of selecting a car with a choice of three interior colors and six exterior colors. There are 18 (3×6) color combinations to consider. Similarly, by combining variable elements, the immune system assembles many different receptors from a much smaller collection of parts.

To understand the origin of receptor diversity, let's consider an immunoglobulin (Ig) gene that encodes the light chain of both secreted antibodies (immunoglobulins) and membrane-bound B cell antigen receptors. Although we'll analyze only a single Ig light-chain gene, all B and T cell antigen receptor genes undergo very similar transformations.

The capacity to generate diversity is built into the structure of Ig genes. A receptor light chain is encoded by three gene segments: a variable (*V*) segment, a joining (*J*) segment, and a constant (*C*) segment. The *V* and *J* segments together encode the variable region of the receptor chain, while the *C* segment encodes the constant region. The light-chain gene contains a single *C* segment, 40 different *V* segments, and 5 different *J* segments. These alternative copies of the *V* and *J* segments are arranged within the gene in a

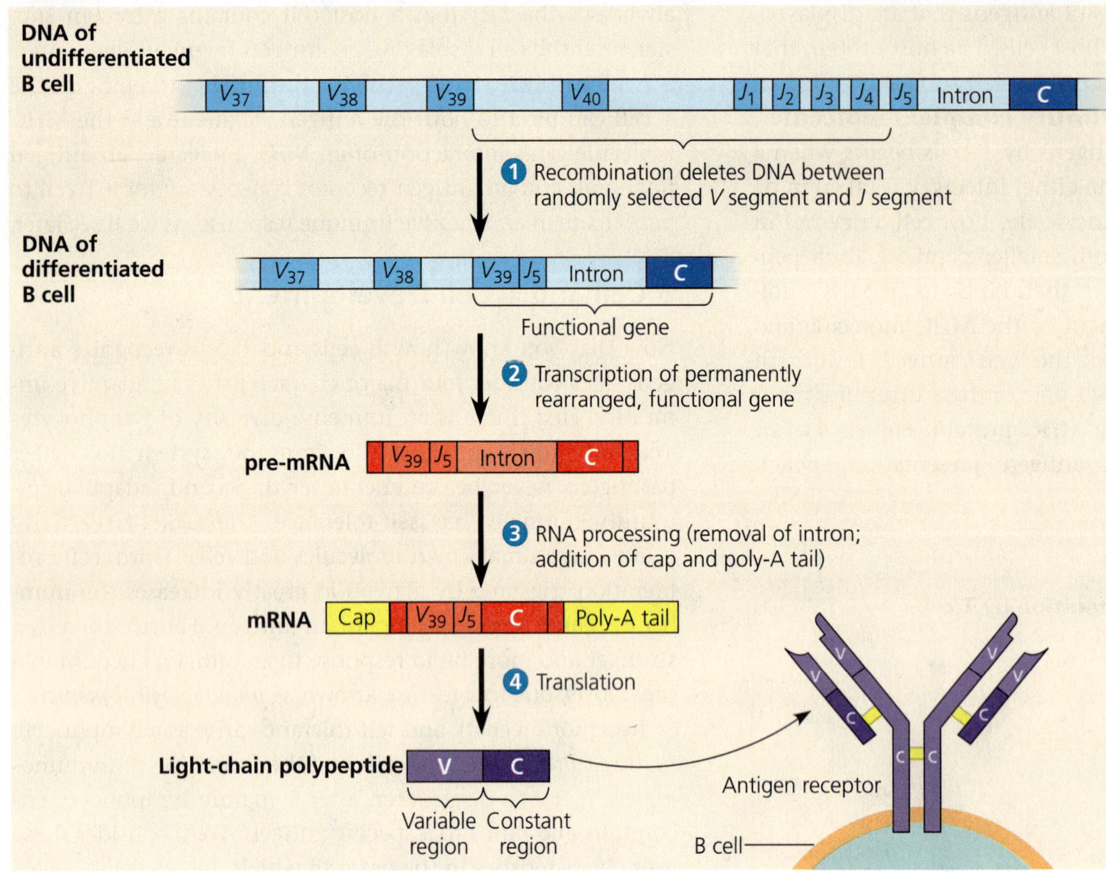

▼ Figure 43.13 Immunoglobulin (antibody) gene rearrangement. The joining of randomly selected V and J gene segments (V_{39} and J_5 in this example) results in a functional gene that encodes the light-chain polypeptide of a B cell antigen receptor. Transcription, splicing, and translation result in a light chain that combines with a polypeptide produced from an independently rearranged heavy-chain gene to form a functional receptor. Mature B cells (and T cells) are exceptions to the generalization that all nucleated cells in the body have exactly the same DNA.

MAKE CONNECTIONS *Both alternative splicing (see Figure 18.13 on p. 363) and joining of V and J segments by recombination generate diverse gene products from a limited set of gene segments. How do these processes differ?*

series **(Figure 43.13)**. Because a functional gene is built from one copy of each type of segment, the pieces can be combined in 200 different ways (40 $V \times 5\ J \times 1\ C$). The number of different heavy-chain combinations is even greater, resulting in even more diversity.

Assembling a functional Ig gene requires rearranging the DNA. Early in B cell development, an enzyme complex called *recombinase* links one light-chain V gene segment to one J gene segment. This recombination event eliminates the long stretch of DNA between the segments, forming a single exon that is part V and part J. Because there is only an intron between the J and C DNA segments, no further DNA rearrangement is required. Instead, the J and C segments of the RNA transcript will be joined when splicing removes the intervening RNA (see Figure 17.11 to review RNA splicing).

Recombinase acts randomly, linking any one of the 40 V gene segments to any one of the 5 J gene segments. Heavy-chain genes undergo a similar rearrangement. In any given cell, however, only one allele of a light-chain gene and one allele of a heavy-chain gene are rearranged. Furthermore, the rearrangements are permanent and are passed on to the daughter cells when the lymphocyte divides.

After both the light- and heavy-chain genes have rearranged, antigen receptors can be synthesized. The rearranged genes are transcribed, and the transcripts are processed for translation. Following translation, the light chain and heavy chain assemble together, forming an antigen receptor (see Figure 43.13). Each pair of randomly rearranged heavy and light chains results in a different antigen-binding site. For the total population of B cells in a human body, the number of such combinations has been calculated as 3.5×10^6. Furthermore, mutations introduced during VJ recombination add additional variation, making the number of possible antigen-binding specificities even greater.

Origin of Self-Tolerance

How does adaptive immunity distinguish self from nonself? Because antigen receptor genes are randomly rearranged, some immature lymphocytes produce receptors specific for epitopes on the organism's own molecules. If these self-reactive lymphocytes were not eliminated or inactivated, the immune system could not distinguish self from nonself and would attack body proteins, cells, and tissues. Instead, as lymphocytes mature in the bone marrow or thymus, their antigen receptors are tested for self-reactivity. Some B and T cells with receptors specific for the body's own molecules are destroyed by *apoptosis*, which is a programmed cell death (see Chapter 11). The remaining self-reactive lymphocytes are typically rendered nonfunctional, leaving only those that react to foreign molecules. Since the body normally lacks mature lymphocytes that can react against its own components, the immune system is said to exhibit *self-tolerance*.

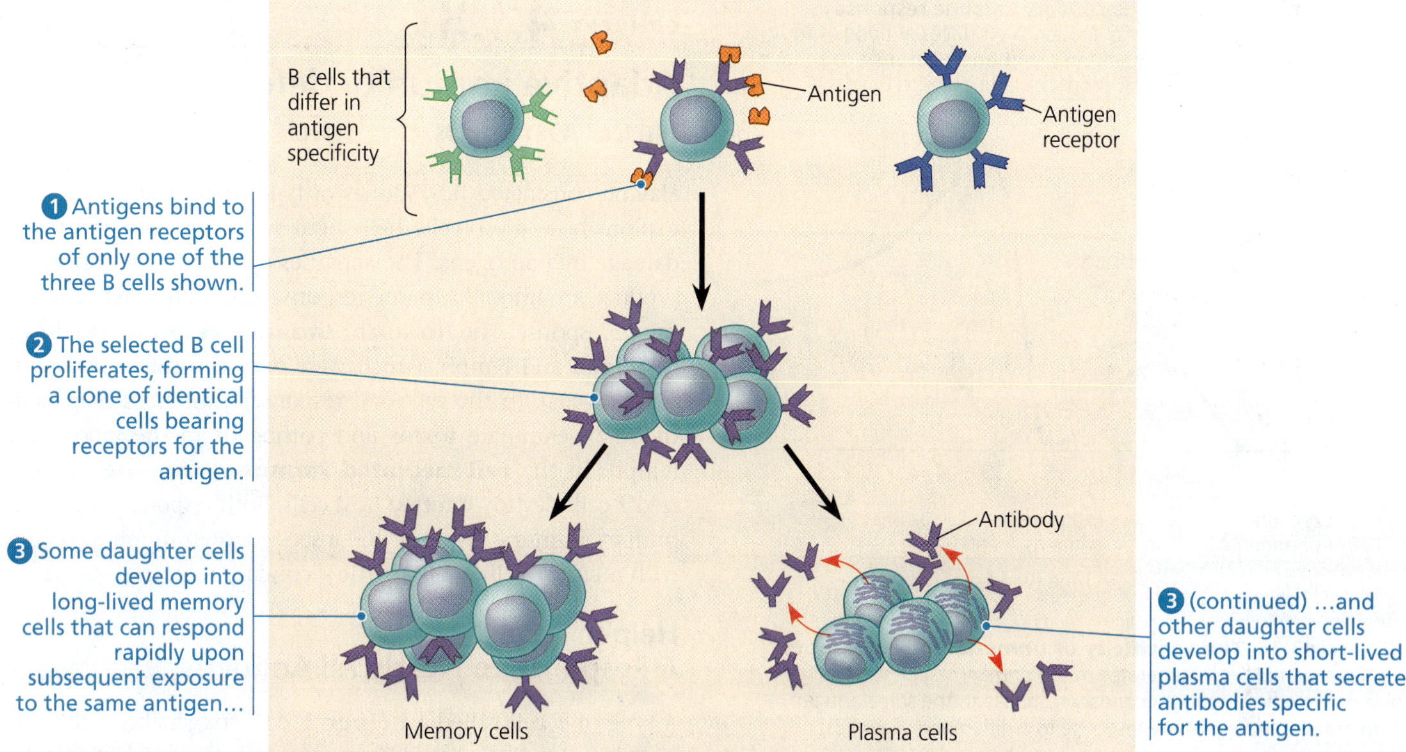

▲ **Figure 43.14 Clonal selection.** This figure illustrates clonal selection, using B cells as an example. In response to a specific antigen and to immune cell signals (not shown), one B cell divides and forms a clone of cells. The remaining B cells, which have antigen receptors specific for other antigens, do not respond. The clone of cells formed by the selected B cell gives rise to memory B cells and antibody-secreting plasma cells. T cells also undergo clonal selection, generating memory T cells and effector T cells (cytotoxic T cells and helper T cells).

Proliferation of B Cells and T Cells

Despite the enormous variety of antigen receptors, only a tiny fraction are specific for a given epitope. So how is adaptive immunity so effective? To begin with, an antigen is presented to a steady stream of lymphocytes in the lymph nodes (see Figure 43.7) until a match is made. A successful match then triggers changes in cell number and activity for the lymphocyte to which an antigen has bound.

The binding of an antigen receptor to an epitope initiates events that activate the lymphocyte. Once activated, a B cell or T cell undergoes multiple cell divisions. For each activated cell, the result of this proliferation is a clone, a population of cells that are identical to the original cell. Some cells from this clone become **effector cells**, short-lived cells that take effect immediately against the antigen and any pathogens producing that antigen. The effector forms of B cells are plasma cells, which secrete antibodies. The effector forms of T cells are helper T cells and cytotoxic T cells, whose roles we'll explore in Concept 43.3. The remaining cells in the clone become **memory cells**, long-lived cells that can give rise to effector cells if the same antigen is encountered later in the animal's life.

Figure 43.14 summarizes the proliferation of a lymphocyte into a clone of cells in response to binding to an antigen, using B cells as an example. This process is called **clonal selection** because an encounter with an antigen *selects* which lymphocyte will divide to produce a *clonal* population of thousands of cells specific for a particular epitope.

Immunological Memory

Immunological memory is responsible for the long-term protection that a prior infection or vaccination provides against many diseases, such as chickenpox. This type of protection was noted almost 2,400 years ago by the Greek historian Thucydides. He observed that individuals who had recovered from the plague could safely care for those who were sick or dying, "for the same man was never attacked twice—never at least fatally."

Prior exposure to an antigen alters the speed, strength, and duration of the immune response. The production of effector cells from a clone of lymphocytes during the first exposure to an antigen is the basis for the **primary immune response**. The primary response peaks about 10–17 days after the initial exposure. During this time, selected B cells and T cells give rise to their effector forms. If an individual is exposed again to the same antigen, the response is faster (typically peaking only 2–7 days after exposure), of greater magnitude, and more prolonged. This is the **secondary immune response**, a hallmark of adaptive, or acquired, immunity. Because selected B cells give

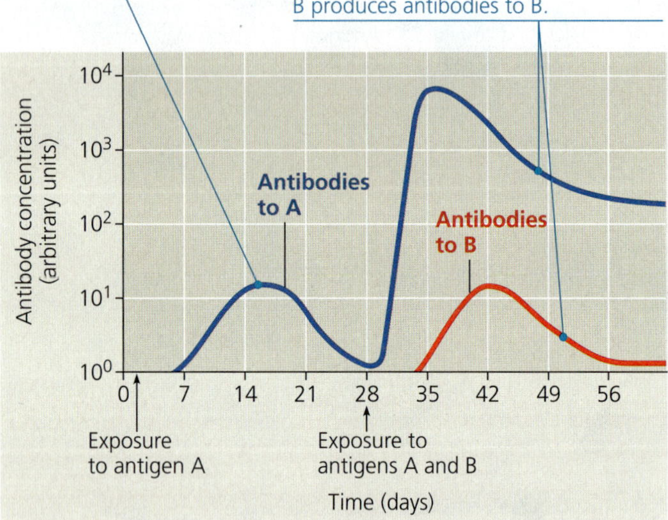

▲ **Figure 43.15 The specificity of immunological memory.** Long-lived memory cells generated in the primary response to antigen A give rise to a heightened secondary response to the same antigen, but do not affect the primary response to a different antigen (B).

rise to antibody-secreting effector cells, measuring the concentrations of specific antibodies in blood over time distinguishes the primary and secondary immune responses **(Figure 43.15)**.

The secondary immune response relies on the reservoir of T and B memory cells generated following initial exposure to an antigen. Because these cells are long-lived, they provide the basis for immunological memory, which can span many decades. (Effector cells have much shorter life spans, which is why the immune response diminishes after an infection is overcome.) If an antigen is encountered again, memory cells specific for that antigen enable the rapid formation of clones of thousands of effector cells also specific for that antigen, thus generating a greatly enhanced immune defense.

Although the processes for antigen recognition, clonal selection, and immunological memory are similar for B cells and T cells, these two classes of lymphocytes fight infection in different ways and in different settings, as we'll explore next.

CONCEPT CHECK 43.2

1. **DRAW IT** Sketch a B cell antigen receptor. Label the V and C regions of the light and heavy chains. Label the antigen-binding sites, disulfide bridges, and transmembrane region. Where are these features located relative to the V and C regions?
2. Explain two advantages of having memory cells when a pathogen is encountered for a second time.
3. **WHAT IF?** If both copies of a light-chain gene and a heavy-chain gene recombined in each (diploid) B cell, how would this affect B cell development?

For suggested answers, see Appendix A.

CONCEPT **43.3**

Adaptive immunity defends against infection of body fluids and body cells

Having considered how clones of lymphocytes arise, we now explore how these cells help fight infections and minimize damage by pathogens. The activities of B and T lymphocytes produce a humoral immune response and a cell-mediated immune response. The **humoral immune response** occurs in the blood and lymph, which were long ago called body humors (fluids). In the humoral response, antibodies help neutralize or eliminate toxins and pathogens in the blood and lymph. In the **cell-mediated immune response**, specialized T cells destroy infected host cells. Both responses include a primary immune response and a secondary immune response, with memory cells enabling the secondary response.

Helper T Cells: A Response to Nearly All Antigens

A type of T cell called a **helper T cell** triggers both the humoral and cell-mediated immune responses. Helper T cells themselves do not carry out those responses. Instead, signals from helper T cells initiate production of antibodies that neutralize pathogens and activate T cells that kill infected cells.

Two requirements must be met for a helper T cell to activate adaptive immune responses. First, a foreign molecule must be present that can bind specifically to the antigen receptor of the T cell. Second, this antigen must be displayed on the surface of an **antigen-presenting cell**. The antigen-presenting cell can be a dendritic cell, macrophage, or B cell.

When host cells are infected, they too display antigens on their surface. What then distinguishes an antigen-presenting cell? The answer lies in the existence of two classes of MHC molecules. Most body cells have only class I MHC molecules, but antigen-presenting cells have both class I and class II MHC molecules. The class II molecules provide a molecular signature by which an antigen-presenting cell is recognized.

A helper T cell and the antigen-presenting cell displaying its specific epitope have a complex interaction **(Figure 43.16)**. The antigen receptors on the surface of the helper T cell bind to the antigen fragment and to the class II MHC molecule displaying that fragment on the antigen-presenting cell. At the same time, an accessory protein on the helper T cell surface binds to the class II MHC molecule, helping keep the cells joined. As the two cells interact, signals in the form of cytokines are exchanged in both directions. For example, the cytokines secreted from a dendritic cell act in combination with the antigen to stimulate the helper T cell, causing it to produce its own set of cytokines. Also, extensive contact between the cell surfaces enables further information exchange.

The different types of antigen-presenting cells interact with helper T cells in distinct contexts. Antigen presentation

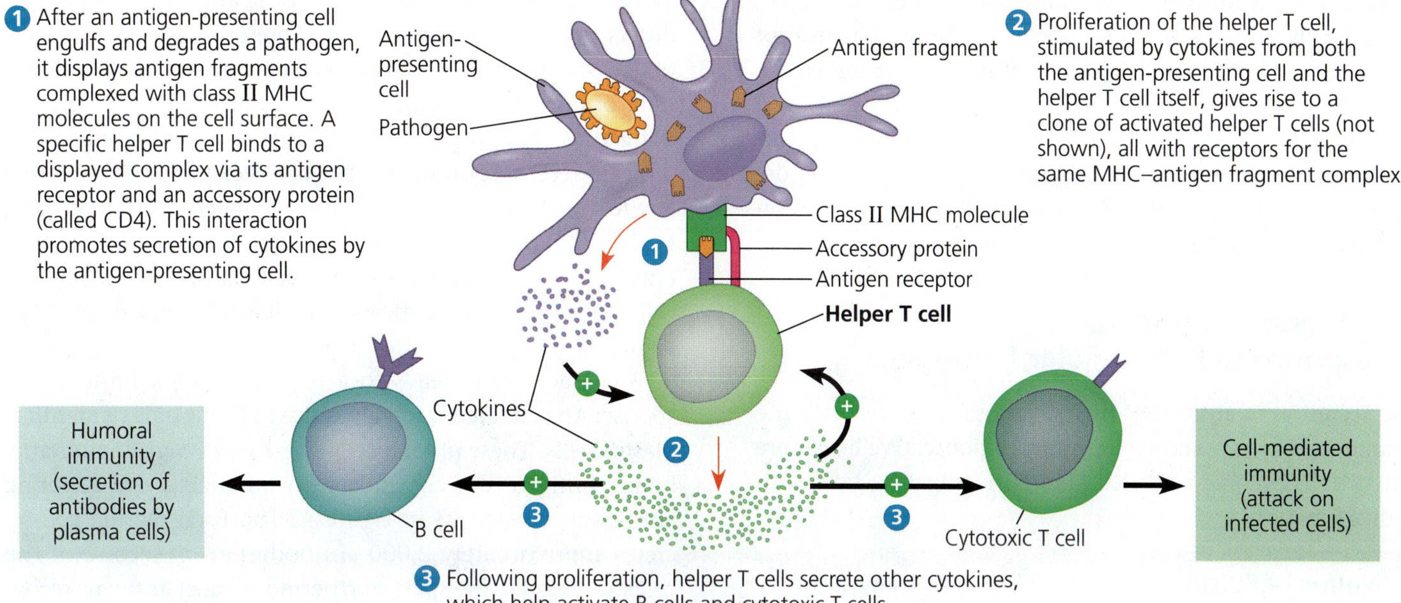

▲ **Figure 43.16 The central role of helper T cells in humoral and cell-mediated immune responses.** In this example, a helper T cell responds to a dendritic cell displaying a microbial antigen.

by a dendritic cell or macrophage activates a helper T cell. The helper T cell then proliferates, forming a clone of activated helper T cells. The B cells present antigens to *already* activated helper T cells, which in turn activate the B cells themselves. Activated helper T cells also help stimulate cytotoxic T cells, as we'll discuss next.

Cytotoxic T Cells: A Response to Infected Cells

In the cell-mediated immune response, **cytotoxic T cells** are the effector cells. The term *cytotoxic* refers to their use of toxic gene products to kill infected cells. To become active, they require signaling molecules from helper T cells as well as interaction with a cell that presents an antigen. Once activated, cytotoxic T cells can eliminate cells that are infected by viruses or other intracellular pathogens.

Fragments of foreign proteins produced in infected host cells associate with class I MHC molecules and are displayed on the cell surface, where they can be recognized by cytotoxic T cells **(Figure 43.17)**. As with helper T cells, cytotoxic T cells have an accessory protein that binds to the MHC molecule, helping keep the two cells in contact while the T cell is activated.

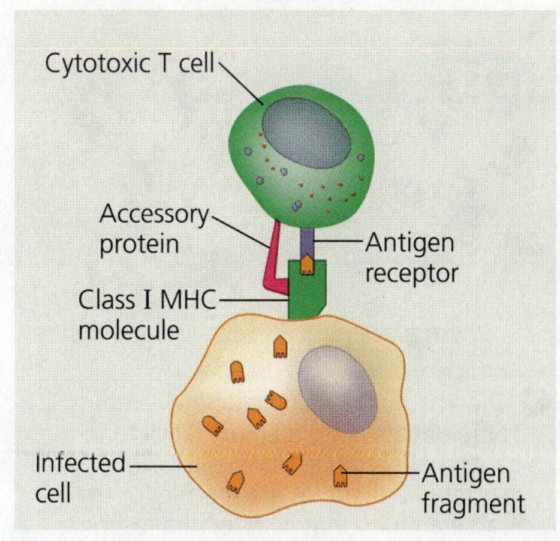

1 An activated cytotoxic T cell binds to a class I MHC–antigen fragment complex on an infected cell via its antigen receptor and an accessory protein (called CD8).

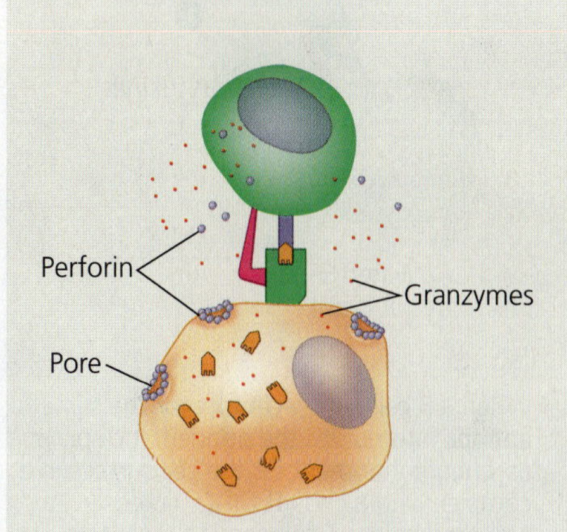

2 The T cell releases perforin molecules, which form pores in the infected cell membrane, and granzymes, enzymes that break down proteins. Granzymes enter the infected cell by endocytosis.

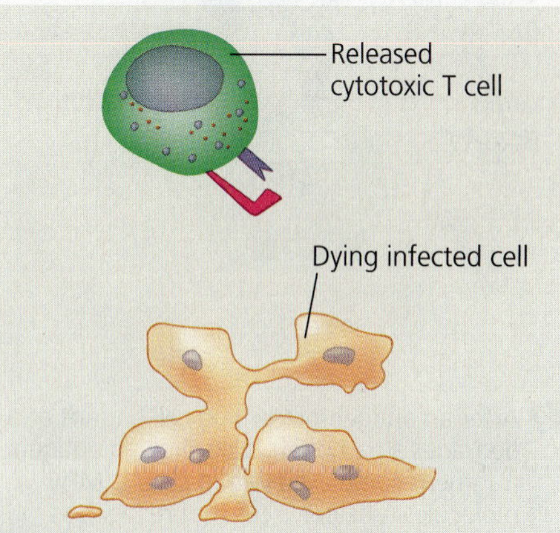

3 The granzymes initiate apoptosis within the infected cell, leading to fragmentation of the nucleus and cytoplasm and eventual cell death. The released cytotoxic T cell can attack other infected cells.

▲ **Figure 43.17 The killing action of cytotoxic T cells on an infected host cell.** An activated cytotoxic T cell releases molecules that make pores in an infected cell's membrane and enzymes that break down proteins, promoting the cell's death.

The targeted destruction of an infected host cell by a cytotoxic T cell involves the secretion of proteins that disrupt membrane integrity and trigger apoptosis (see Figure 43.17). The death of the infected cell not only deprives the pathogen of a place to reproduce, but also exposes cell contents to circulating antibodies, which mark them for disposal. After destroying an infected cell, the cytotoxic T cell can move on and kill other cells infected with the same pathogen.

B Cells and Antibodies: A Response to Extracellular Pathogens

The secretion of antibodies by clonally selected B cells is the hallmark of the humoral immune response. We'll explore how B cells become activated before investigating how antibodies function.

Activation of B Cells

Activation of the humoral immune response typically involves B cells and helper T cells as well as proteins on the surface of pathogens. As depicted in **Figure 43.18**, B cell activation by an antigen is aided by cytokines secreted from helper T cells that have encountered the same antigen. Stimulated by both an antigen and cytokines, the B cell proliferates and differentiates into memory B cells and antibody-secreting effector cells called **plasma cells**.

The pathway for antigen processing and display in B cells differs from that in other antigen-presenting cells. A macrophage or dendritic cell can present fragments from a wide variety of protein antigens, whereas a B cell presents only the antigen to which it specifically binds. When an antigen first binds to receptors on the surface of a B cell, the cell takes in a few foreign molecules by receptor-mediated endocytosis (see Figure 7.22). The class II MHC protein of the B cell then presents an antigen fragment to a helper T cell. This direct cell-to-cell contact is usually critical to B cell activation (see step 2 in Figure 43.18).

B cell activation leads to a robust humoral immune response: An activated B cell gives rise to thousands of identical plasma cells. These plasma cells stop expressing a membrane-bound antigen receptor and begin producing and secreting antibodies (see step 3 in Figure 43.18). Each plasma cell secretes approximately 2,000 antibodies every second of the cell's 4- to 5-day life span. Furthermore, most antigens recognized by B cells contain multiple epitopes. An exposure to a single antigen therefore normally activates a variety of B cells, with different plasma cells producing antibodies directed against different epitopes on the common antigen.

Antibody Function

Antibodies do not kill pathogens, but by binding to antigens, they mark pathogens in various ways for inactivation or

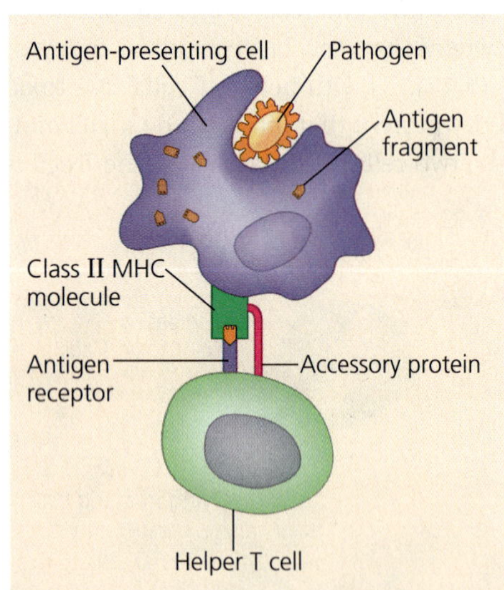

1 After an antigen-presenting cell engulfs and degrades a pathogen, it displays an antigen fragment complexed with a class II MHC molecule. A helper T cell that recognizes the complex is activated with the aid of cytokines secreted from the antigen-presenting cell.

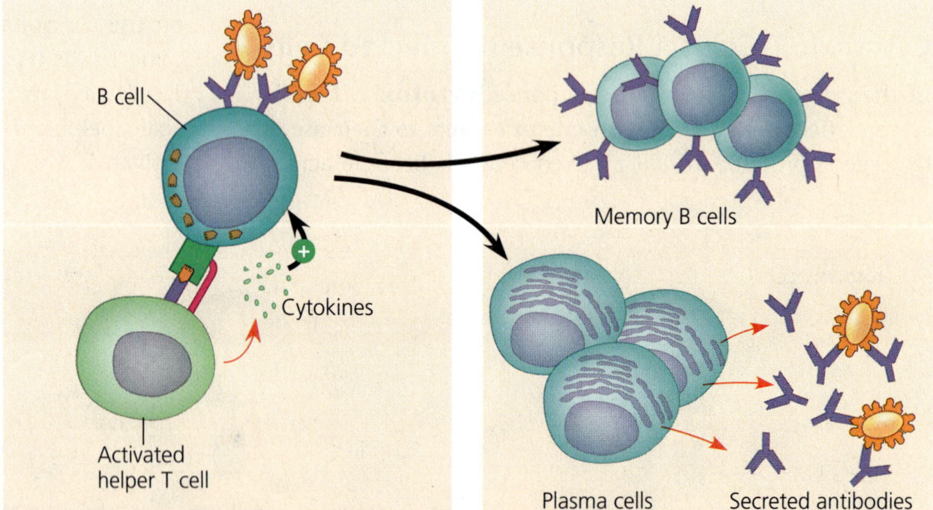

2 When a B cell with receptors for the same epitope internalizes the antigen, it displays an antigen fragment on the cell surface in a complex with a class II MHC molecule. An activated helper T cell bearing receptors specific for the displayed fragment binds to the B cell. This interaction, with the aid of cytokines from the T cell, activates the B cell.

3 The activated B cell proliferates and differentiates into memory B cells and antibody-secreting plasma cells. The secreted antibodies are specific for the same antigen that initiated the response.

▲ **Figure 43.18 Activation of a B cell in the humoral immune response.** Most protein antigens require activated helper T cells to trigger a humoral response. A macrophage (shown here) or a dendritic cell can activate a helper T cell, which in turn can activate a B cell to give rise to antibody-secreting plasma cells.

? *What function do cell-surface antigen receptors play for memory B cells?*

▼ Figure 43.19 Antibody-mediated mechanisms of antigen disposal.

Neutralization

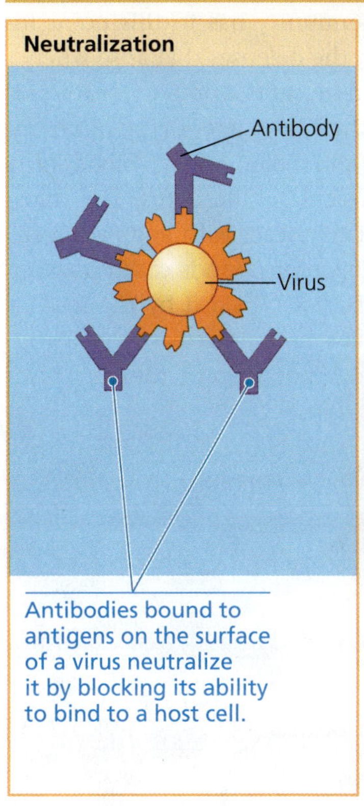

Opsonization

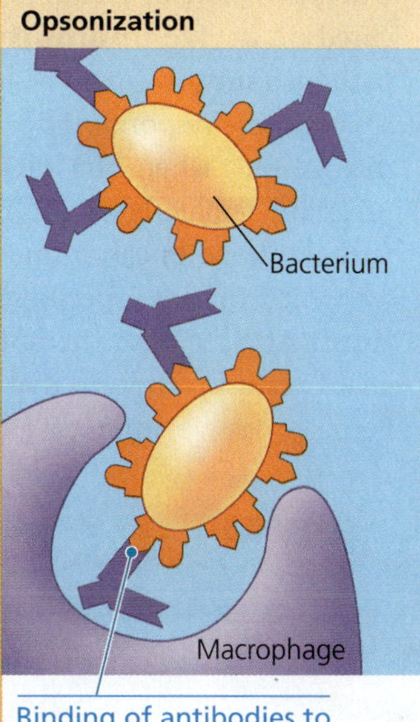

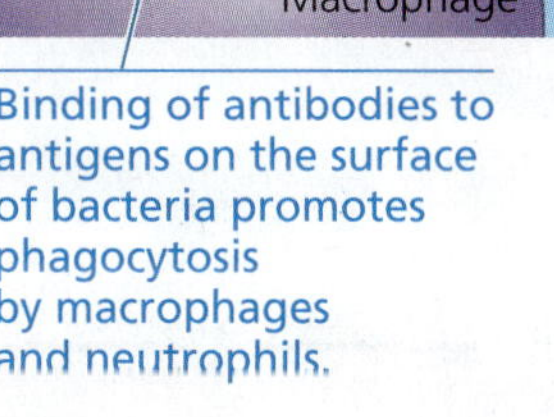

Activation of complement system and pore formation

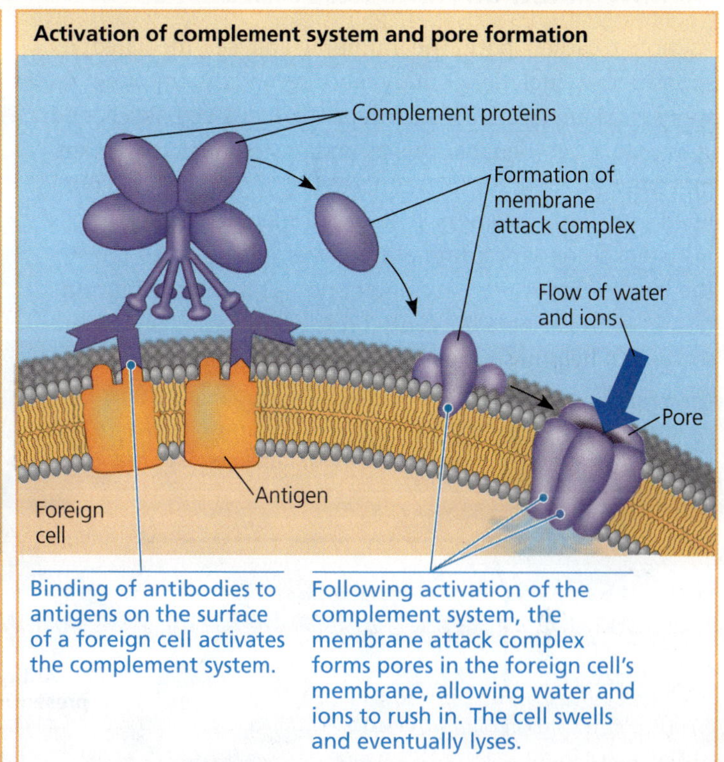

destruction. In the simplest of these activities, *neutralization*, antibodies bind to viral surface proteins (**Figure 43.19**, left). The bound antibodies prevent infection of a host cell, thus neutralizing the virus. Similarly, antibodies sometimes bind to toxins released in body fluids, preventing the toxins from entering body cells. In another process, called *opsonization*, antibodies bound to antigens on bacteria present a readily recognized structure for macrophages or neutrophils and therefore increase phagocytosis (**Figure 43.19**, middle). Because each antibody has two antigen-binding sites, antibodies sometimes also facilitate phagocytosis by linking bacterial cells, virus particles, or other foreign substances into aggregates.

Antibodies sometimes work together with the proteins of the complement system to dispose of pathogens. (The name *complement* reflects the fact that these proteins increase the effectiveness of antibody-directed attacks on bacteria.) Binding of a complement protein to an antigen-antibody complex on a foreign cell (or an enveloped virus) triggers a cascade in which each protein of the complement system activates the next protein. Ultimately, activated complement proteins generate a *membrane attack complex* that forms a pore in the membrane of the foreign cell. Ions and water rush into the cell, causing it to swell and lyse (**Figure 43.19**, right).Whether activated as part of innate defenses or as part of adaptive defenses, this cascade of complement protein activity results in the lysis of foreign cells and produces factors that promote inflammation or stimulate phagocytosis.

When antibodies facilitate phagocytosis (see Figure 43.19, middle), they also help fine-tune the humoral immune response. Recall that phagocytosis enables macrophages and dendritic cells to present antigens to and stimulate helper T cells, which in turn stimulate the very B cells whose antibodies contribute to phagocytosis. This positive feedback between innate and adaptive immunity contributes to a coordinated, effective response to infection.

Although antibodies are the cornerstone of the response in body fluids, there is also a mechanism by which they can bring about the death of infected body cells. When a virus uses a cell's biosynthetic machinery to produce viral proteins, these viral products can appear on the cell surface. If antibodies specific for epitopes on these viral proteins bind to the exposed proteins, the presence of bound antibody at the cell surface can recruit a natural killer cell. The natural killer cell then releases proteins that cause the infected cell to undergo apoptosis.

B cells can express five different forms of immunoglobulin (Ig). For a given B cell, each form or *class* has an identical antigen-binding specificity, but a distinct heavy-chain C region. The B cell antigen receptor, known as IgD, is membrane bound. The other four classes consist of soluble antibodies. IgM is the first class of soluble antibody produced. IgG, which follows next, is the most abundant antibody in blood. We will learn more about the function of IgG, as well as the two remaining antibody classes (IgA and IgE), as we further explore the role of antibodies in immunity and disease.

Summary of the Humoral and Cell-Mediated Immune Responses

As noted earlier, both the humoral and cell-mediated responses can include primary and secondary immune responses. Memory cells of each type—helper T cell, B cell, and cytotoxic T cell—enable the secondary response. For example, when body fluids are reinfected by a pathogen encountered previously, memory B cells and memory helper T cells initiate a secondary humoral response. **Figure 43.20** reviews the events that initiate humoral and cell-mediated immune responses, highlights the central role of the helper T cell, and serves as a helpful summary of adaptive immunity.

Active and Passive Immunization

Our discussion of adaptive immunity has to this point focused on **active immunity**, the defenses that arise when a pathogen infects the body and prompts a primary or secondary immune response. In contrast, a different type of immunity results when the IgG antibodies in the blood of a pregnant female cross the placenta to her fetus. The transferred antibodies can immediately react with any pathogens for which they are specific. This protection is called **passive immunity** because the antibodies provided by the mother guard against pathogens that have never infected the newborn. Because passive immunity does not involve the recipient's

Humoral (antibody-mediated) immune response
Cell-mediated immune response
Key
Stimulates
Gives rise to
Antigen (1st exposure)
Engulfed by
Antigen-presenting cell
B cell
Helper T cell
Cytotoxic T cell
Memory helper T cells
Antigen (2nd exposure)
Plasma cells
Memory B cells
Memory cytotoxic T cells
Active cytotoxic T cells
Secreted antibodies
Defend against extracellular pathogens in blood and lymph by binding to antigens, thereby neutralizing pathogens or making them better targets for phagocytic cells and complement proteins.
Defend against intracellular pathogens and cancer by binding to and lysing the infected cells or cancer cells.

▲ **Figure 43.20 An overview of the adaptive immune response.**

? *Identify each black or brown arrow as representing part of the primary or secondary response.*

BioFlix Visit the Study Area at **www.masteringbiology.com** for the BioFlix® 3-D Animation on Immunology.

B and T cells, it persists only as long as the transferred antibodies last (a few weeks to a few months).

After giving birth, a nursing mother continues to transfer protection against disease to her infant. IgA antibodies present in breast milk provide additional passive immunity to the infant's digestive tract while the infant's immune system develops. Later in life, IgA functions in active immunity: IgA antibodies secreted in tears, saliva, and mucus protect the mucous membranes of both males and females.

Both active immunity and passive immunity can be induced artificially. Active immunity can develop from the introduction of antigens into the body through **immunization**. In 1796, Edward Jenner noted that milkmaids who had cowpox, a mild disease usually seen only in cows, did not contract smallpox, a far more dangerous disease. In the first documented immunization (or **vaccination**, from the Latin *vacca*, cow), Jenner used the cowpox virus to induce adaptive immunity against the closely related smallpox virus. Today, many sources of antigen are used to make vaccines, including inactivated bacterial toxins, killed pathogens, parts of pathogens, weakened pathogens that generally do not cause illness, and even genes encoding microbial proteins. Because all of these agents induce a primary immune response and immunological memory, an encounter with the pathogen from which the vaccine was derived triggers a rapid and strong secondary immune response (see Figure 43.15).

Vaccination programs have been successful against many infectious diseases that once killed or incapacitated large numbers of people. A worldwide vaccination campaign led to eradication of smallpox in the late 1970s. In industrialized nations, routine active immunization of infants and children has dramatically reduced the incidence of sometimes devastating diseases, such as polio, measles, and whooping cough. Unfortunately, not all pathogens are easily managed by vaccination. Furthermore, some vaccines are not readily available in impoverished areas of the globe.

Misinformation about vaccine safety and disease risk has led some parents to refuse to immunize their children with available, effective vaccines. The consequence has been a substantial and growing public health problem. Consider measles as just one example. Side effects of immunization are remarkably rare: Fewer than one in a million children suffer a significant allergic reaction to the measles vaccine. The disease, however, is quite dangerous: Roughly one out of every 1,000 patients develop *encephalitis*, an inflammation of the brain. Worldwide, measles kills more than 200,000 people each year. Sadly, declines in measles vaccination rates in parts of the United Kingdom, Russia, and the United States have recently resulted in a number of measles outbreaks and significant numbers of preventable deaths.

In artificial passive immunization, antibodies from an immune animal are injected into a nonimmune animal. For example, humans bitten by venomous snakes are sometimes treated with antivenin, serum from sheep or horses that have been immunized against the venom of one or more species of venomous snakes. When injected immediately after a snakebite, the antibodies in antivenin can neutralize toxins in the venom before the toxins do massive damage.

Antibodies as Tools

The power of antibody specificity and antigen-antibody binding has been harnessed in research, diagnosis, and therapy. Some antibody tools are *polyclonal*: They are the products of many different clones of plasma cells, each specific for a different epitope **(Figure 43.21)**. Antibodies that an animal produces after exposure to a microbial antigen are polyclonal. In contrast, other antibody tools are *monoclonal*: They are prepared from a single clone of B cells grown in culture. The **monoclonal antibodies** produced by such a culture are identical and specific for the same epitope on an antigen.

Monoclonal antibodies have provided the basis for many recent advances in medical diagnosis and treatment. For example, home pregnancy kits use monoclonal antibodies to detect human chorionic gonadotropin (HCG). Because HCG is produced as soon as an embryo implants in the uterus (see Chapter 46), the presence of this hormone in a woman's urine is a reliable indicator for a very early stage of pregnancy. In the clinic, monoclonal antibodies are being used to treat many human diseases. For this type of therapy, researchers use mouse B cell clones to identify antibodies specific for an epitope on diseased cells. Next, the mouse antibody genes are altered to code for antibodies that appear less foreign to the human adaptive immune defenses. Scientists then use the "humanized" genes to produce large amounts of antibody for injecting into patients.

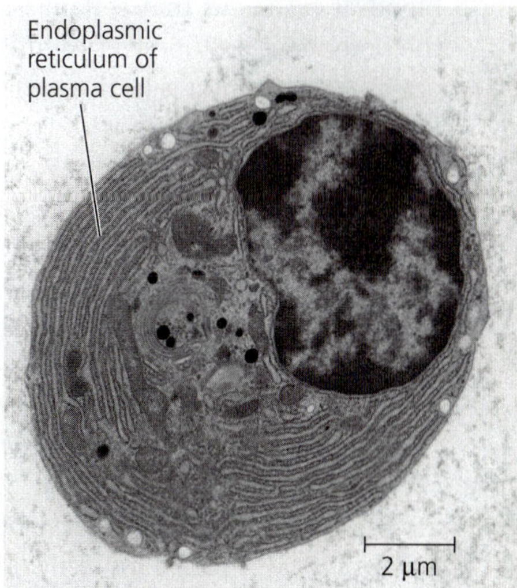

▲ **Figure 43.21 A plasma cell.** A plasma cell contains abundant endoplasmic reticulum, a common feature of cells dedicated to making proteins for secretion (TEM).

Immune Rejection

Like pathogens, cells from another person can be recognized as foreign and attacked by immune defenses. For example, skin transplanted from one person to a genetically nonidentical person will look healthy for a week or so but will then be destroyed (rejected) by the recipient's immune response. Keep in mind that the body's rejection of transplanted tissues or organs or of an incompatible blood transfusion is the expected reaction of a healthy immune system exposed to foreign antigens. (It remains a largely unanswered question why a pregnant woman does not reject her fetus as nonself tissue.)

Blood Groups

To avoid a blood transfusion being recognized as foreign by the recipient's immune system, the ABO blood groups of the donor and recipient must be taken into account. As discussed in Chapter 14, red blood cells are designated as type A if they have the type A carbohydrate on their surface. Similarly, the type B carbohydrate is found on type B red blood cells; both A and B carbohydrates are found on type AB red blood cells; and neither carbohydrate is found on type O red blood cells (see Figure 14.11).

To understand how ABO blood groups affect transfusions, let's consider the immune response of someone with type A blood. It turns out that certain bacteria normally present in the body have epitopes very similar to the A and B carbohydrates. By responding to the bacterial epitope similar to the B carbohydrate, a person with type A blood makes antibodies that will react with the type B carbohydrate. No antibodies are made against the bacterial epitope similar to the type A carbohydrate because lymphocytes reactive with the body's own molecules are inactivated or eliminated during development. If the person with type A blood receives a transfusion of type B blood, that person's anti-B antibodies cause an immediate and devastating transfusion reaction. The transfused red blood cells undergo lysis, which can lead to chills, fever, shock, and kidney malfunction. By the same token, anti-A antibodies in the donated type B blood will act against the recipient's type A red blood cells. Although such interactions prevent type O individuals from receiving transfusions of any other blood type, the recent discovery of enzymes that can cleave the A and B carbohydrates from red blood cells may eliminate this problem.

Tissue and Organ Transplants

In the case of tissue and organ transplants, or grafts, MHC molecules stimulate the immune response that leads to rejection. Each vertebrate species has many alleles for each MHC gene, enabling presentation of antigen fragments that vary in shape and net electrical charge. This diversity of MHC molecules almost guarantees that no two people, except identical twins, will have exactly the same set. Thus, in the vast majority of graft and transplant recipients, some MHC molecules on the donated tissue are foreign to the recipient. To minimize rejection, physicians use donor tissue bearing MHC molecules that match those of the recipient as closely as possible. In addition, the recipient takes medicines that suppress immune responses (but also leave the recipient more susceptible to infections).

Transplants of bone marrow from one person to another can also cause an immune reaction, but for a different reason. Bone marrow transplants are used to treat leukemia and other cancers as well as various hematological (blood cell) diseases. Prior to receiving transplanted bone marrow, the recipient is typically treated with radiation to eliminate his or her own bone marrow cells, thus destroying the source of abnormal cells. This treatment effectively obliterates the recipient's immune system, leaving little chance of graft rejection. However, lymphocytes in the donated marrow may react against the recipient. This *graft versus host reaction* is limited if the MHC molecules of the donor and recipient are well matched. Bone marrow donor programs continually seek volunteers because the great variability of MHC molecules makes a diverse pool of donors essential.

CONCEPT CHECK 43.3

1. If a child were born without a thymus, what cells and functions would be deficient? Explain.
2. Treatment of antibodies with a particular protease clips the heavy chains in half, releasing the two arms of the Y-shaped molecule. How might the antibodies continue to function?
3. **WHAT IF?** Suppose that a snake handler bitten by a particular venomous snake species was treated with antivenin. Why might the same treatment for a second such bite have different results?

For suggested answers, see Appendix A.

CONCEPT **43.4**

Disruptions in immune system function can elicit or exacerbate disease

Although adaptive immunity offers significant protection against a wide range of pathogens, it is not fail-safe. In this last section of the chapter, we'll first examine the problems that arise when adaptive immunity is blocked or misregulated. We'll then turn to some of the evolutionary adaptations of pathogens that diminish the effectiveness of host immune responses.

Exaggerated, Self-Directed, and Diminished Immune Responses

The highly regulated interplay among lymphocytes, other body cells, and foreign substances generates an immune response that provides extraordinary protection against many pathogens. When allergic, autoimmune, or immunodeficiency disorders disrupt this delicate balance, the effects are frequently severe and sometimes life-threatening.

Allergies

Allergies are exaggerated (hypersensitive) responses to certain antigens called **allergens**. The most common allergies involve antibodies of the IgE class. Hay fever, for instance, occurs when plasma cells secrete IgE antibodies specific for antigens on the surface of pollen grains **(Figure 43.22)**. Some IgE antibodies attach by their base to mast cells in connective tissues. Pollen grains that enter the body later attach to the antigen-binding sites of these IgE antibodies. This attachment links adjacent IgE molecules, inducing the mast cell to release histamine and other inflammatory chemicals from granules (vesicles). Acting on a variety of cell types, these signals bring about the typical allergy symptoms: sneezing, runny nose, teary eyes, and smooth muscle contractions that can result in breathing difficulty. Drugs called antihistamines diminish allergy symptoms (and inflammation) by blocking receptors for histamine.

An acute allergic response sometimes leads to *anaphylactic shock*, a whole-body, life-threatening reaction that can occur within seconds of exposure to an allergen. Anaphylactic shock develops when widespread release of mast cell contents triggers abrupt dilation of peripheral blood vessels, causing a precipitous drop in blood pressure, as well as constriction of bronchioles. Death may occur within minutes due to lack of blood flow and the inability to breathe. Allergic responses to bee venom or penicillin can lead to anaphylactic shock in people who are extremely allergic to these substances. Likewise, people very allergic to peanuts, fish, or shellfish can die from ingesting only tiny amounts of these allergens, which trigger reactions through interactions with mast cells on the surface of the digestive tract. People with severe hypersensitivities often carry syringes containing the hormone epinephrine, which counteracts this allergic response (see Figure 45.8).

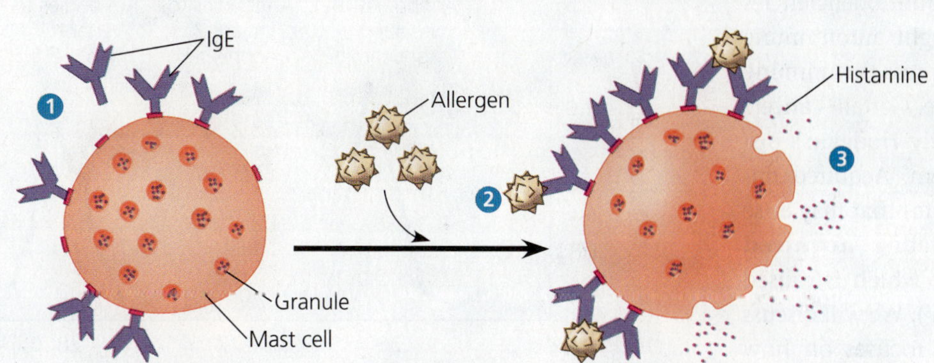

▲ **Figure 43.22 Mast cells, IgE, and the allergic response.** In this example, pollen grains act as the allergen.

Autoimmune Diseases

In some people, the immune system is active against particular molecules of the body, causing an **autoimmune disease**. Such a loss of self-tolerance has many forms. In *systemic lupus erythematosus*, commonly called *lupus*, the immune system generates antibodies against histones and DNA released by the normal breakdown of body cells. These self-reactive antibodies cause skin rashes, fever, arthritis, and kidney dysfunction. Another autoimmune disease, *rheumatoid arthritis*, leads to damage and painful inflammation of the cartilage and bone of joints **(Figure 43.23)**. In *type 1 diabetes mellitus*, the insulin-producing beta cells of the pancreas are the targets of autoimmune cytotoxic T cells. The most common chronic neurological disorder in developed countries is the autoimmune disease *multiple sclerosis*. In this disease, T cells infiltrate the central nervous system. The result is destruction of the myelin sheath that surrounds parts of many neurons (see Figure 48.12), leading to muscle paralysis through a disruption in neuron function.

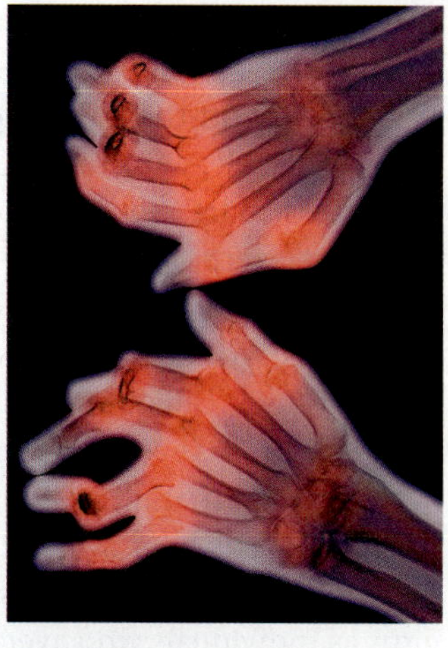

▲ **Figure 43.23 X-ray of hands deformed by rheumatoid arthritis.**

Gender, genetics, and environment all influence susceptibility to autoimmune disorders. For example, members of certain families show an increased susceptibility to particular autoimmune disorders. In addition, many autoimmune diseases afflict females more often than males. Women are two to three times more likely than men to suffer from multiple sclerosis and rheumatoid arthritis and nine times more likely to develop lupus. The cause of this sex bias, as well as the rise in autoimmune disease frequency in industrialized countries, is an area of active research and debate. Clearly, much remains to be learned about these often devastating disorders.

Exertion, Stress, and the Immune System

Many forms of exertion and stress influence immune system function. Consider, for example, susceptibility to the common cold and other infections of the upper respiratory tract. Moderate exercise improves immune system function and significantly reduces the risk of these infections. In contrast, exercise to the point of exhaustion leads to more frequent infections and to more severe symptoms. Studies of marathon runners support the conclusion that exercise intensity is the critical variable. On average, such runners get sick less often than their more sedentary peers during training, a time of moderate exertion, but have a marked increase in illness in the period immediately following the grueling race itself. Similarly, psychological stress has been shown to disrupt immune system regulation by altering the interplay of the hormonal, nervous, and immune systems (see Figure 45.21). Recent research also confirms that rest is important for immunity: Adults who averaged fewer than 7 hours of sleep a night got sick three times as often when exposed to a cold virus as individuals who averaged at least 8 hours of sleep.

Immunodeficiency Diseases

A disorder in which an immune system response to antigens is defective or absent is called an **immunodeficiency**. An *inborn immunodeficiency* results from a genetic or developmental defect in the immune system. An *acquired immunodeficiency* develops later in life following exposure to chemical or biological agents. Whatever its cause and nature, an immunodeficiency can lead to frequent and recurrent infections and increased susceptibility to certain cancers.

Inborn immunodeficiencies result from defects in the development of various immune system cells or defects in the production of specific proteins, such as antibodies or the proteins of the complement system. Depending on the specific genetic defect, either innate or adaptive defenses—or both—may be impaired. In severe combined immunodeficiency (SCID), functional lymphocytes are rare or absent. Lacking an adaptive immune response, SCID patients are susceptible to infections, such as pneumonia and meningitis, that can cause death in infancy. Treatments include bone marrow and stem cell transplantation.

Exposure to certain agents can cause immunodeficiencies that develop later in life. Drugs used to fight autoimmune diseases or prevent transplant rejection suppress the immune system, leading to an immunodeficient state. Certain cancers also suppress the immune system, especially Hodgkin's disease, which damages the lymphatic system. Acquired immunodeficiencies range from temporary states that may arise from physiological stress to the devastating **acquired immunodeficiency syndrome (AIDS)**, which is caused by the human immunodeficiency virus (HIV). We will discuss AIDS further in the next section, which focuses on how pathogens escape the adaptive immune response.

Evolutionary Adaptations of Pathogens That Underlie Immune System Avoidance

EVOLUTION Just as immune systems that ward off pathogens have evolved in animals, mechanisms that thwart immune responses have evolved in pathogens. Using human pathogens as examples, we'll examine some common mechanisms: antigenic variation, latency, and direct attack on the immune system.

Antigenic Variation

One mechanism for escaping the body's defenses is for a pathogen to alter how it appears to the immune system. Immunological memory is a record of the foreign epitopes an animal has encountered. If the pathogen that expressed those epitopes no longer does so, it can reinfect or remain in a host without triggering the rapid and robust response that memory cells provide. Such changes in epitope expression, which are called *antigenic variation*, are regular events for some viruses and parasites. The parasite that causes sleeping sickness (trypanosomiasis) provides one example. By periodically switching at random among 1,000 different versions of the protein found over its entire surface, this pathogen can persist in the body without facing an effective adaptive immune response **(Figure 43.24)**.

Antigenic variation is the major reason the influenza, or "flu," virus remains a major public health problem. As it replicates in one human host after another, the human influenza virus mutates. Because any change that lessens recognition by the immune system provides a selective advantage, the virus steadily accumulates such alterations. These changes in the surface proteins of the influenza virus are the reason that a new flu vaccine must be manufactured and distributed each year. Of much greater danger, however, is the fact that the human virus occasionally exchanges genes with influenza viruses that infect domesticated animals, such as pigs or chickens. When this occurs, influenza can take on such a radically different appearance that none of the memory cells in

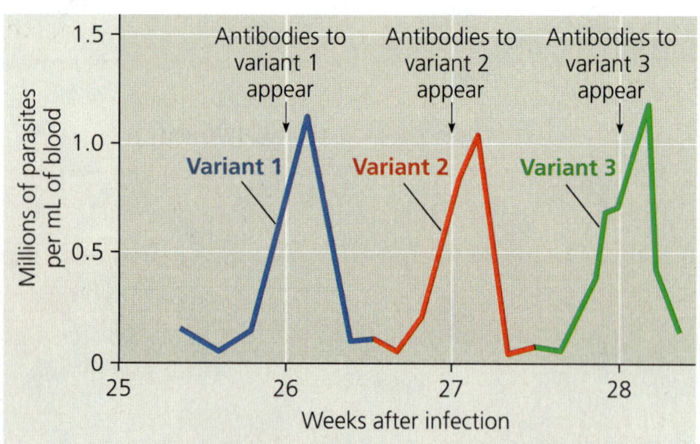

▲ **Figure 43.24 Antigenic variation in the parasite that causes sleeping sickness.** Blood samples taken from a patient during a chronic infection of sleeping sickness reveal cyclic variation in the surface coat protein of the parasite. The infection has become chronic because this weekly variation allows the parasite to evade the adaptive immune response.

the human population recognize the new strain. Such an event led to the influenza outbreak of 1918–1919, which killed more than half a million people in the United States (see Figure 19.9). Worldwide more than 20 million people died, a greater number than had died in World War I.

In 2009, an influenza virus called H1N1 appeared that contained a novel combination of genes from flu viruses that normally circulate in pigs, birds, and humans. The rapid spread of this flu across the human population caused a *pandemic*, an outbreak of worldwide proportions. Fortunately, a rapidly developed H1N1 vaccine soon provided public health officials with an excellent means of slowing the spread of this virus and reducing the impact of the outbreak.

Latency

After infecting a host, some viruses enter a largely inactive state called *latency*. Because such dormant viruses cease making most viral proteins and typically produce no free virus particles, they do not trigger an adaptive immune response. Nevertheless, the viral genome persists in the nuclei of infected cells, either as a separate small DNA molecule or as a copy integrated into the host genome. Latency typically persists until conditions arise that are favorable for viral transmission or unfavorable for host survival, such as when the host is infected by another pathogen. Such circumstances trigger the synthesis and release of virus particles that can infect new hosts.

Herpes simplex viruses, which establish themselves in human sensory neurons, provide a good example of latency. The type 1 virus causes most oral herpes infections, whereas the type 2 virus is responsible for most cases of genital herpes. Because sensory neurons express relatively few MHC I molecules, the infected cells are inefficient at presenting viral antigens to circulating lymphocytes. Stimuli such as fever, emotional stress, or menstruation reactivate the virus to reproduce and infect surrounding epithelial tissues. Activation of the type 1 virus can result in blisters around the mouth that are inaccurately called "cold" sores. The type 2 virus can cause genital sores, but people infected with either type 1 or type 2 virus often lack any apparent symptoms. Infections of the type 2 virus, which is sexually transmitted, pose a serious threat to the babies of infected mothers and can increase transmission of HIV, the virus that causes AIDS.

Attack on the Immune System: HIV

The human immunodeficiency virus (HIV), the pathogen that causes AIDS, both escapes and attacks the adaptive immune response. Once introduced into the body, HIV infects helper T cells with high efficiency. To infect these cells, the virus binds specifically to the CD4 accessory protein (see Figure 43.16). However, HIV also infects some cell types that have low levels of CD4, such as macrophages and brain cells. In the cell, the HIV RNA genome is reverse-transcribed, and the product DNA is integrated into the host cell's genome (see Figure 19.8). In this form, the viral genome can direct production of new virus particles.

Although the body responds to HIV with an immune response sufficient to eliminate most viral infections, some HIV invariably escapes. One reason HIV persists is antigenic variation. The virus mutates at a very high rate during replication. Altered proteins on the surface of some mutated viruses reduce interaction with antibodies and cytotoxic T cells. Such viruses survive, proliferate, and mutate further. The virus thus evolves within the body. The continued presence of HIV is also helped by latency. When the viral DNA integrates into the chromosome of a host cell but does not produce new virus proteins or particles, it is shielded from the immune system by the host cell. This inactive, or latent, viral DNA is also protected from antiviral agents currently used against HIV because they attack only actively replicating viruses.

Over time, an untreated HIV infection not only avoids the adaptive immune response but also abolishes it **(Figure 43.25)**. Viral reproduction and cell death triggered by the virus lead to loss of helper T cells, impairing both humoral and cell-mediated immune responses. The result is a progression to AIDS, characterized by a susceptibility to infections and cancers that a healthy immune system would usually defeat. For example, *Pneumocystis carinii*, a common fungus that does not cause disease in healthy individuals, can result in severe pneumonia in people with AIDS. Likewise, the Kaposi's sarcoma herpesvirus causes a cancer among AIDS patients that is extremely rare in individuals not infected with HIV. Such opportunistic diseases, as well as nerve damage and body wasting, are the primary causes of death in AIDS patients, not the HIV virus itself.

At present, HIV infection cannot be cured, although certain drugs can slow HIV reproduction and the progression to

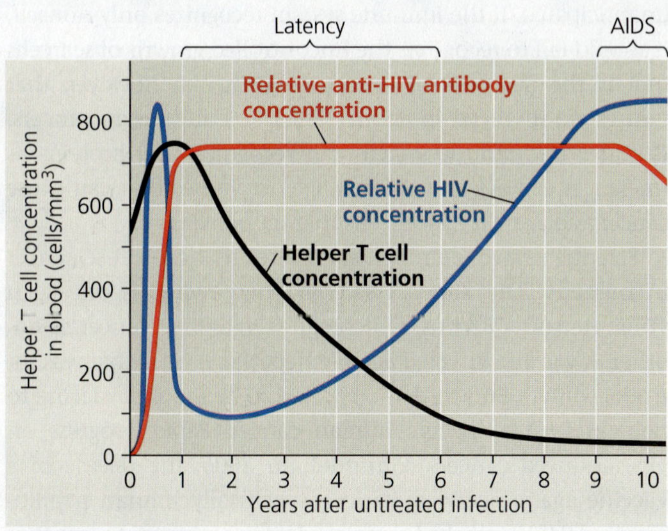

▲ **Figure 43.25 The progress of an untreated HIV infection.**

AIDS. Unfortunately, mutations that occur in each round of viral reproduction can generate strains of HIV that are drug resistant. The impact of such viral drug resistance can be reduced by the use of a combination of drugs; viruses newly resistant to one drug can be defeated by another. However, the appearance of strains resistant to multiple drugs reduces the effectiveness of such multidrug "cocktails" in some patients. Frequent mutations in genes for HIV surface antigens also have hampered efforts to develop an effective vaccine. Worldwide, the AIDS epidemic continues to grow. In 2008, approximately 2 million people died of AIDS, and the disease is now the leading cause of death in Africa.

Transmission of HIV requires the transfer of virus particles or infected cells from person to person via body fluids such as semen, blood, or breast milk. Unprotected sex (that is, without a condom) and transmission via HIV-contaminated needles (typically among intravenous drug users) account for the vast majority of HIV infections. The virus can enter the body through the mucosal linings of the vagina, vulva, penis, or rectum during intercourse or via the mouth during oral sex. The likelihood of transmission is increased by factors that may damage these linings, especially other sexually transmitted infections that cause ulcers or inflammation.

People infected with HIV can transmit the disease in the first few weeks of infection, *before* they express HIV-specific antibodies that can be detected in a blood test (see Figure 43.25). Currently, 10–50% of all new HIV infections appear to be caused by recently infected individuals.

Cancer and Immunity

When adaptive immunity is inactivated, the frequency of certain cancers increases dramatically. For example, the risk of developing Kaposi's sarcoma is 20,000 times greater for untreated AIDS patients than for healthy people. This observation was unanticipated. If the immune system recognizes only nonself, it should fail to recognize the uncontrolled growth of self cells that is the hallmark of cancer. It turns out, however, that viruses are involved in about 15–20% of all human cancers. Because the immune system can recognize viral proteins as foreign, it can act as a defense against viruses that can cause cancer and against cancer cells that harbor viruses.

Scientists have identified six viruses that can cause cancer in humans. The Kaposi's sarcoma herpesvirus is one such virus. Hepatitis B virus, which can trigger liver cancer, is another. A vaccine directed against hepatitis B virus that was introduced in 1986 was demonstrated to be the first vaccine to help prevent a specific human cancer. Rapid progress on virus-induced cancers continues. In 2006, the release of a vaccine against cervical cancer, specifically human papillomavirus (HPV), marked a major victory against a disease that afflicts more than half a million women worldwide every year **(Figure 43.26)**.

▼ **Figure 43.26**

IMPACT

Vaccinating Against Cervical Cancer

In the 1970s, Harald zur Hausen, working in Heidelberg, Germany, proposed that human papillomavirus (HPV) causes cervical cancer. Many scientists were skeptical that cancer could result from infection by HPV, the most common sexually transmitted pathogen. However, after more than a decade of work, zur Hausen isolated two particular types of HPV from patients with cervical cancer. He quickly made samples available to other scientists, leading to development of highly effective vaccines against cervical cancer. In 2008, zur Hausen shared the Nobel Prize in Physiology or Medicine for his discovery. This computer graphic image of an HPV particle illustrates the abundant copies of the capsid protein (yellow) that is used as the antigen in vaccination.

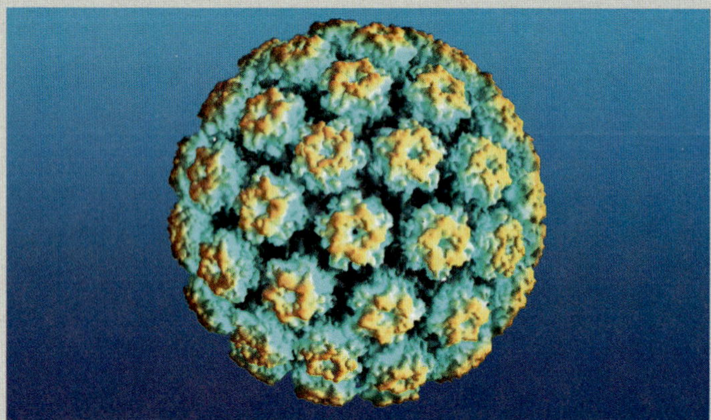

WHY IT MATTERS Cervical cancer kills more than 4,000 women annually in the United States and is the fifth-most common cause of cancer deaths among women worldwide. Administering an HPV vaccine, either Gardasil or Cervarix, to preteen girls and young women greatly reduces their chance of being infected with the HPV viruses that cause most cervical cancers.

FURTHER READING L. R. Badenet et al., Human papillomavirus vaccine: Opportunity and challenge, *New England Journal of Medicine* 356:1990–1991 (2007).

WHAT IF? Suppose you tracked the health of women infected with the types of HPV that cause cancer. Why might only a fraction of such women develop cervical cancer? (*Hint*: Refer to Figure 18.25 on p. 376 and the accompanying text.)

CONCEPT CHECK 43.4

1. In myasthenia gravis, antibodies bind to and block certain receptors on muscle cells, preventing muscle contraction. Is this disease best classified as an immunodeficiency disease, an autoimmune disease, or an allergic reaction? Explain.
2. People with herpes simplex type 1 viruses often get mouth sores when they have a cold or similar infection. How might this location benefit the virus?
3. **WHAT IF?** How would a macrophage deficiency likely affect a person's innate and adaptive defenses?

For suggested answers, see Appendix A.

43 CHAPTER REVIEW

SUMMARY OF KEY CONCEPTS

CONCEPT 43.1

In innate immunity, recognition and response rely on traits common to groups of pathogens (pp. 930–935)

- In both invertebrates and vertebrates, **innate immunity** is mediated by physical and chemical barriers as well as cell-based defenses. Activation of innate immune responses relies on recognition proteins specific for broad classes of **pathogens**. In insects, pathogens that penetrate barrier defenses are ingested by cells in the hemolymph that also release antimicrobial peptides.
- In vertebrates, intact skin and mucous membranes form barriers to pathogens. Mucus produced by membrane cells, the low pH of the skin and stomach, and degradation by **lysozyme** also deter pathogens. Microbes that penetrate barrier defenses are ingested by phagocytic cells, including **macrophages** and **dendritic cells**. Additional cellular defenses include **natural killer cells**, which can induce the death of virus-infected cells. **Complement system** proteins, **interferons**, and other antimicrobial peptides also act against microbes. In the **inflammatory response**, **histamine** and other chemicals released from cells at the injury site promote changes in blood vessels that allow fluid, more phagocytic cells, and antimicrobial peptides to enter tissues.
- Pathogens sometimes evade innate immune defenses. For example, some bacteria have an outer capsule that prevents recognition, while others are resistant to breakdown within lysosomes.

? *In what ways does innate immunity protect the mammalian digestive tract?*

CONCEPT 43.2

In adaptive immunity, receptors provide pathogen-specific recognition (pp. 935–940)

- **Adaptive immunity** relies on **lymphocytes** that arise from stem cells in the bone marrow and complete their maturation in the bone marrow (**B cells**) or in the **thymus** (**T cells**). Lymphocytes have cell-surface **antigen receptors** for foreign molecules. All receptor proteins on a single B or T cell are the same, but there are millions of B and T cells in the body that differ in the foreign molecules that their receptors recognize. Upon infection, B and T cells specific for the pathogen are activated. Some T cells help other lymphocytes; others kill infected host cells. B cells called **plasma cells** produce soluble receptor proteins called **antibodies**, which bind to foreign molecules and cells. The activated lymphocytes called **memory cells** defend against future infections by the same pathogen.
- Recognition of foreign molecules involves the binding of variable regions of receptors to an **epitope**, a small region of an antigen. B cells and antibodies recognize epitopes on the surface of antigens circulating in the blood or lymph. T cells recognize protein epitopes in small antigen fragments (peptides) that are presented on the surface of host cells, complexed with cell-surface proteins called **MHC (major histocompatibility complex) molecules**.
- The four major characteristics of B and T cell development are the generation of cell diversity, self-tolerance, proliferation, and immunological memory.

The following figure uses B cells to illustrate clonal selection:

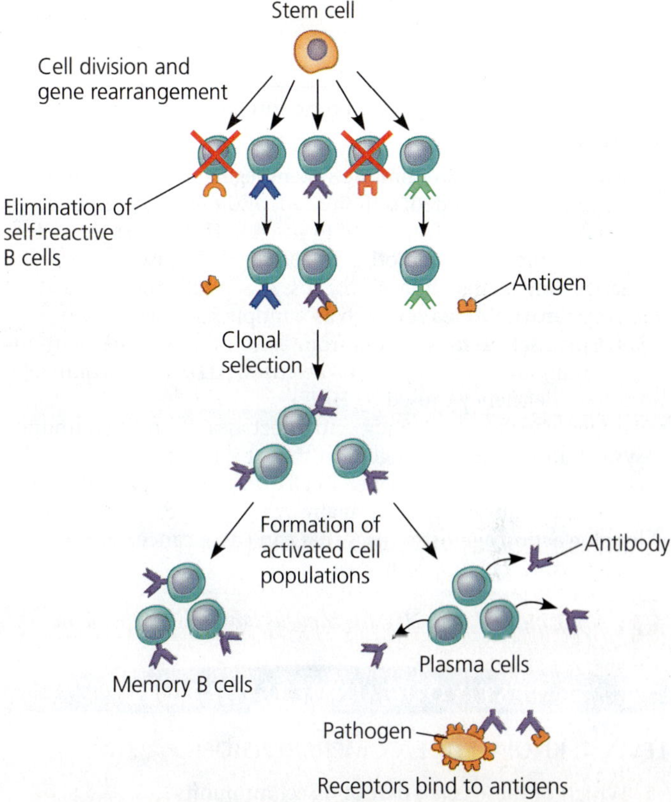

? *Why is the adaptive immune response to an initial infection slower than the innate response?*

CONCEPT 43.3

Adaptive immunity defends against infection of body fluids and body cells (pp. 940–946)

- **Helper T cells** interact with antigen fragments displayed by class II MHC molecules on the surface of dendritic cells, macrophages, and B cells (**antigen-presenting cells**). Activated helper T cells secrete **cytokines** that stimulate other lymphocytes as part of the response to nearly all antigens. **Cytotoxic T cells** bind to a complex of an antigen fragment and a class I MHC molecule on infected host cells. In the **cell-mediated immune response**, activated cytotoxic T cells secrete proteins that initiate destruction of infected cells. All T cells have an accessory protein that enhances binding to MHC–antigen fragment complexes.

  In the **humoral immune response**, B cell antigen receptors and antibodies bind to extracellular foreign substances in blood and lymph.The binding of antibodies helps eliminate antigens by phagocytosis and complement-mediated lysis. The five major antibody classes differ in distribution and function.
- **Active immunity** develops in response to infection or to immunization with a nonpathogenic form or part of a pathogen. Active immunity includes a response to and immunological memory for that pathogen. **Passive immunity**, which provides immediate, short-term protection, is conferred naturally when IgG crosses the placenta from mother to fetus or when IgA passes from mother to infant in breast milk. It also can be conferred artificially by injecting antibodies into a nonimmune person.

- Tissues or cells transferred from one person to another are subject to immune rejection. In tissue grafts and organ transplants, MHC molecules stimulate rejection. Lymphocytes in bone marrow transplants may cause a graft versus host reaction.

? *Is immunological memory after a natural infection fundamentally different from immunological memory after vaccination? Explain.*

CONCEPT 43.4

Disruptions in immune system function can elicit or exacerbate disease (pp. 946–950)

- Disruption of normal immune system regulation or function can result in an exaggerated, self-directed, or diminished response. In localized allergies, IgE attached to **mast cells** induces the cells to release histamine and other mediators that cause vascular changes and allergic symptoms. Loss of self-tolerance can lead to **autoimmune diseases**, such as multiple sclerosis. Inborn **immunodeficiencies** result from defects that interfere with innate, humoral, or cell-mediated defenses. **AIDS** is an acquired immunodeficiency caused by HIV.
- Antigenic variation, latency, and direct assault on the immune system allow some pathogens to thwart immune responses. HIV infection destroys helper T cells, leaving the patient prone to disease. Immune defense against cancer appears to primarily involve action against viruses that can cause cancer, as well as against cancer cells that harbor viruses.

? *Is being infected with HIV the same as having AIDS? Explain.*

TEST YOUR UNDERSTANDING

LEVEL 1: KNOWLEDGE/COMPREHENSION

1. Which of these is *not* part of insect immunity?
   a. enzyme activation of microbe-killing chemicals
   b. activation of natural killer cells
   c. phagocytosis by hemocytes
   d. production of antimicrobial peptides
   e. a protective exoskeleton
2. An epitope associates with which part of an antigen receptor or antibody?
   a. the disulfide bridge
   b. the heavy-chain constant regions only
   c. variable regions of a heavy chain and light chain combined
   d. the light-chain constant regions only
   e. the tail
3. Which statement best describes the difference in responses of effector B cells (plasma cells) and cytotoxic T cells?
   a. B cells confer active immunity; cytotoxic T cells confer passive immunity.
   b. B cells kill pathogens directly; cytotoxic T cells kill host cells.
   c. B cells secrete antibodies against a pathogen; cytotoxic T cells kill pathogen-infected host cells.
   d. B cells carry out the cell-mediated response; cytotoxic T cells carry out the humoral response.
   e. B cells respond the first time a pathogen is present; cytotoxic T cells respond subsequent times.

LEVEL 2: APPLICATION/ANALYSIS

4. Which of the following statements is *not* true?
   a. An antibody has more than one antigen-binding site.
   b. An antigen can have different epitopes.
   c. A pathogen makes more than one antigen.
   d. A lymphocyte has receptors for multiple different antigens.
   e. A liver cell makes one class of MHC molecule.
5. Which of the following should be the same in identical twins?
   a. the set of antibodies produced
   b. the set of MHC molecules produced
   c. the set of T cell antigen receptors produced
   d. the susceptibility to a particular virus
   e. the set of immune cells eliminated as self-reactive

LEVEL 3: SYNTHESIS/EVALUATION

6. Vaccination increases the number of
   a. different receptors that recognize a pathogen.
   b. lymphocytes with receptors that can bind to the pathogen.
   c. epitopes that the immune system can recognize.
   d. macrophages specific for a pathogen.
   e. MHC molecules that can present an antigen.
7. Which of the following would *not* help a virus avoid triggering an adaptive immune response?
   a. having frequent mutations in genes for surface proteins
   b. infecting cells that produce very few MHC molecules
   c. producing proteins very similar to those of other viruses
   d. infecting and killing helper T cells
   e. building the viral shell from host proteins
8. **DRAW IT** Consider a pencil-shaped protein with two epitopes, Y (the "eraser" end) and Z (the "point" end). They are recognized by antibodies A1 and A2, respectively. Draw and label a picture showing the antibodies linking proteins into a complex that could trigger endocytosis by a macrophage.
9. **MAKE CONNECTIONS** Contrast Lamarck's idea for the inheritance of acquired characteristics, discussed on pp. 454–455 of Concept 22.1, with the clonal selection of lymphocytes.
10. **EVOLUTION CONNECTION**
    Describe one invertebrate defense mechanism and discuss how it is an evolutionary adaptation retained in vertebrates.
11. **SCIENTIFIC INQUIRY**
    A diagnostic test for tuberculosis (TB) involves injecting antigen (from the bacterium that causes TB) under the skin and then waiting a few days for a reaction to appear. This test is *not* useful for diagnosing TB in AIDS patients. Why?
12. **WRITE ABOUT A THEME**
    The Genetic Basis of Life Among all nucleated body cells, only B and T cells lose DNA during their development and maturation. In a short essay (100–150 words), discuss the relationship between this loss and the theme of DNA as heritable biological information, focusing on similarities between cellular and organismal generations.

For selected answers, see Appendix A.

44

Osmoregulation and Excretion

▲ **Figure 44.1 How does an albatross drink salt water without ill effect?**

KEY CONCEPTS

44.1 Osmoregulation balances the uptake and loss of water and solutes
44.2 An animal's nitrogenous wastes reflect its phylogeny and habitat
44.3 Diverse excretory systems are variations on a tubular theme
44.4 The nephron is organized for stepwise processing of blood filtrate
44.5 Hormonal circuits link kidney function, water balance, and blood pressure

OVERVIEW

A Balancing Act

With a wingspan that can reach 3.5 m, the largest of any living bird, a wandering albatross (*Diomedea exulans*) soaring over the ocean is hard to miss **(Figure 44.1)**. Yet the albatross commands attention for more than just its size. This massive bird remains at sea day and night throughout the year, returning to land only to reproduce. A human with only seawater to drink would die of dehydration, but faced with the same conditions, the albatross thrives.

For both albatross and human, maintaining the fluid environment of their cells, tissues, and organs requires keeping relative concentrations of water and solutes within fairly narrow limits. In addition, ions such as sodium and calcium must be maintained at concentrations that permit normal activity of muscles, neurons, and other body cells. Homeostasis thus requires **osmoregulation**, the general term for the processes by which animals control solute concentrations and balance water gain and loss.

A number of strategies for water and solute control have arisen during evolution, reflecting the varied and often severe osmoregulatory challenges presented by an animal's surroundings. The arid environment of a desert, for instance, can quickly deplete an animal of body water. Despite a quite different environment, albatrosses and other marine animals also face potential dehydration. The success of animals in an ocean environment depends critically on conserving water and, for marine birds and fishes, eliminating excess salts. In contrast, freshwater animals live in an environment that threatens to flood and dilute their body fluids. These organisms survive by conserving solutes and absorbing salts from their surroundings.

In safeguarding their internal fluid environment, animals must also deal with a hazardous metabolite produced by the dismantling of proteins and nucleic acids. Breakdown of *nitrogenous* (nitrogen-containing) molecules releases ammonia, a very toxic compound. Several different mechanisms have evolved for **excretion**, the process that rids the body of nitrogenous metabolites and other metabolic waste products. Because systems for excretion and osmoregulation are structurally and functionally linked in many animals, we will consider both of these processes in this chapter.

CONCEPT 44.1

Osmoregulation balances the uptake and loss of water and solutes

Just as thermoregulation depends on balancing heat loss and gain (see Chapter 40), regulating the chemical composition of body fluids depends on balancing the uptake and loss of water and solutes. This process of osmoregulation is based largely on the controlled movement of solutes between internal fluids and the external environment. Because solute movement results in the movement of water by osmosis, the net effect is to regulate both solutes and water.

Osmosis and Osmolarity

All animals—regardless of habitat or type of waste produced—face the same need to balance water uptake and loss. If water

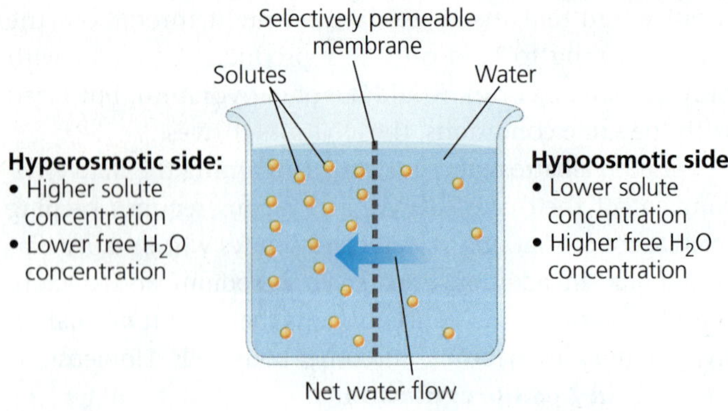

▲ **Figure 44.2 Solute concentration and osmosis.**

MAKE CONNECTIONS *Review types of membrane proteins and their functions in Concepts 7.1 and 7.2 (pp. 129–132). Which membrane proteins allow water, but not solutes, to diffuse across a lipid bilayer?*

uptake is excessive, animal cells swell and burst; if water loss is substantial, they shrivel and die (see Figure 7.15).

Water enters and leaves cells by osmosis. Recall from Chapter 7 that osmosis, a special case of diffusion, is the movement of water across a selectively permeable membrane. It occurs whenever two solutions separated by the membrane differ in osmotic pressure, or **osmolarity** (total solute concentration expressed as molarity, that is, moles of solute per liter of solution). The unit of measurement for osmolarity used in this chapter is milliOsmoles per liter (mOsm/L). Seawater has an osmolarity of about 1,000 mOsm/L (equivalent to a total solute concentration of 1 *M*), while the osmolarity of human blood is about 300 mOsm/L.

If two solutions separated by a selectively permeable membrane have the same osmolarity, they are said to be *isoosmotic*. Water molecules continually cross the membrane, but under these conditions they do so at equal rates in both directions. Thus, there is no *net* movement of water by osmosis between isoosmotic solutions. When two solutions differ in osmolarity, the one with the greater concentration of solutes is said to be *hyperosmotic*, and the more dilute solution is said to be *hypoosmotic* (**Figure 44.2**). Water flows by osmosis from a hypoosmotic solution to a hyperosmotic one.*

Osmotic Challenges

Given the chemical principles that govern osmotic flow, an animal can maintain water balance in two ways. One is to be an **osmoconformer**: to be isoosmotic with its surroundings. The second is to be an **osmoregulator**: to control internal osmolarity independent of that of its environment.

All osmoconformers are marine animals. Because an osmoconformer's internal osmolarity is the same as that of its environment, there is no tendency to gain or lose water. Many osmoconformers live in water that has a stable composition and hence have a constant internal osmolarity.

Osmoregulation enables animals to live in environments that are uninhabitable for osmoconformers, such as freshwater and terrestrial habitats. To survive in a hypoosmotic environment, an osmoregulator must discharge excess water. In a hyperosmotic environment, an osmoregulator must instead take in water to offset osmotic loss. Osmoregulation also allows many marine animals to maintain an internal osmolarity different from that of seawater.

Most animals, whether osmoconformers or osmoregulators, cannot tolerate substantial changes in external osmolarity and are said to be *stenohaline* (from the Greek *stenos*, narrow, and *halos*, salt). In contrast, *euryhaline* animals (from the Greek *eurys*, broad) can survive large fluctuations in external osmolarity. Euryhaline osmoconformers include many barnacles and mussels, which are continually covered and uncovered by ocean tides; examples of euryhaline osmoregulators are the striped bass and the various species of salmon.

Next we'll examine some adaptations for osmoregulation that have evolved in marine, freshwater, and terrestrial animals.

Marine Animals

Most marine invertebrates are osmoconformers. Their osmolarity is the same as that of seawater. They therefore face no substantial challenges in water balance. However, because these animals differ considerably from seawater in the concentrations of *specific* solutes, they must actively transport these solutes to maintain homeostasis. For example, although the concentration of magnesium ions (Mg^{2+}) in seawater is 50 m*M* (millimolar, or 10^{-3} mol/L), homeostatic mechanisms in the Atlantic lobster (*Homarus americanus*) result in a Mg^{2+} concentration of less than 9 m*M* in this animal's hemolymph (circulatory fluid).

Many marine vertebrates and some marine invertebrates are osmoregulators. For most of these animals, the ocean is a strongly dehydrating environment. For example, marine fishes, such as the cod in **Figure 44.3a**, constantly lose water by osmosis. Such fishes balance the water loss by drinking large amounts of seawater. In ridding themselves of salts, they make use of both their gills and kidneys. In the gills, specialized *chloride cells* actively transport chloride ions (Cl^-) out and allow sodium ions (Na^+) to follow passively. In the kidneys, excess calcium, magnesium, and sulfate ions are excreted with the loss of only small amounts of water.

A distinct osmoregulatory strategy evolved in marine sharks and most other chondrichthyans (cartilaginous animals; see Chapter 34). Like "bony fishes" (as we'll refer collectively to ray-finned and lobe-finned fishes in this chapter), sharks have an internal salt concentration much lower than that of seawater. Thus, salt tends to diffuse into their bodies from the water, especially across their gills. Unlike bony fishes, however, marine sharks are not hypoosmotic to seawater. The explanation

*In this chapter, we use the terms *isoosmotic, hypoosmotic,* and *hyperosmotic,* which refer specifically to osmolarity, instead of *isotonic, hypotonic,* and *hypertonic.* The latter set of terms applies to the response of animal cells—whether they swell or shrink—in solutions of known solute concentrations.

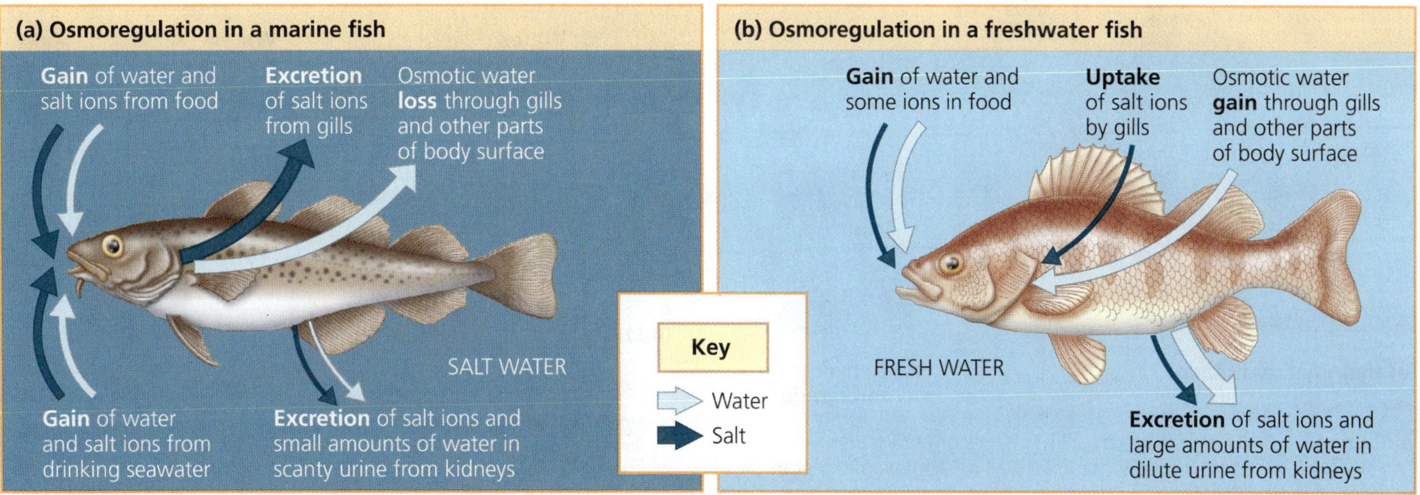

▲ **Figure 44.3 Osmoregulation in marine and freshwater bony fishes: a comparison.**

is that shark tissue contains high concentrations of urea, a nitrogenous waste product of protein and nucleic acid metabolism (see Figure 44.8). A shark's body fluids also contain trimethylamine oxide (TMAO), an organic molecule that protects proteins from damage by urea. Together, the salts, urea, TMAO, and other compounds maintained in the body fluids of sharks result in an osmolarity very close to that of seawater. For this reason, sharks are often considered osmoconformers. However, because the solute concentration in their body fluids is actually somewhat higher than 1,000 mOsm/L, water slowly *enters* the shark's body by osmosis and in food (sharks do not drink). This small influx of water is disposed of in urine produced by the shark's kidneys. The urine also removes some of the salt that diffuses into the shark's body; the rest is lost in feces or is secreted from a specialized gland.

▲ **Figure 44.4 Sockeye salmon (*Oncorhynchus nerka*), euryhaline osmoregulators.**

Freshwater Animals

The osmoregulatory problems of freshwater animals are the opposite of those of marine animals. The body fluids of freshwater animals must be hyperosmotic because animal cells cannot tolerate salt concentrations as low as that of lake or river water. Having internal fluids with an osmolarity higher than that of their surroundings, freshwater animals face the problem of gaining water by osmosis and losing salts by diffusion. Many freshwater animals, including bony fishes, solve the problem of water balance by drinking almost no water and excreting large amounts of very dilute urine. At the same time, salts lost by diffusion and in the urine are replenished by eating. Freshwater fishes, such as the perch in **Figure 44.3b**, also replenish salts by uptake across the gills. Chloride cells in the gills of the fish actively transport Cl^- into the body, and Na^+ follows.

Salmon and other euryhaline fishes that migrate between fresh water and seawater undergo dramatic changes in osmoregulatory status **(Figure 44.4)**. When living in rivers and streams, salmon osmoregulate like other freshwater fishes, producing large amounts of dilute urine and taking up salt from the dilute environment through their gills. When they migrate to the ocean, salmon acclimatize. They produce more of the steroid hormone cortisol, which increases the number and size of salt-secreting chloride cells. As a result of these and other physiological changes, salmon in salt water excrete excess salt from their gills and produce only small amounts of urine—just like bony fishes that spend their entire lives in salt water.

Animals That Live in Temporary Waters

Extreme dehydration, or *desiccation*, is fatal for most animals. However, a few aquatic invertebrates that live in temporary ponds and in films of water around soil particles can lose almost all their body water and survive. These animals enter a dormant state when their habitats dry up, an adaptation called **anhydrobiosis** ("life without water"). Among the

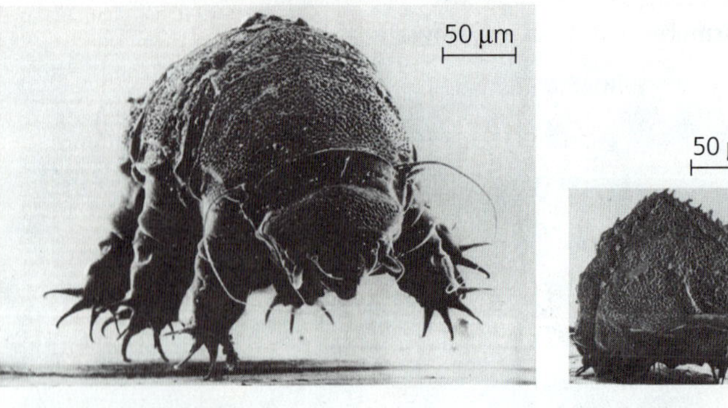

(a) Hydrated tardigrade **(b) Dehydrated tardigrade**

▲ **Figure 44.5 Anhydrobiosis.** SEM images of tardigrades (water bears), which inhabit temporary ponds as well as droplets of water in soil and on moist plants.

most striking examples are the tardigrades, or water bears **(Figure 44.5)**. Less than 1 mm long, these tiny invertebrates are found in marine, freshwater, and moist terrestrial environments. In their active, hydrated state, they contain about 85% water by weight, but they can dehydrate to less than 2% water and survive in an inactive state, dry as dust, for a decade or more. Just add water, and within hours the rehydrated tardigrades are moving about and feeding.

Anhydrobiosis requires adaptations that keep cell membranes intact. Researchers are just beginning to learn how tardigrades survive drying out, but studies of anhydrobiotic roundworms (phylum Nematoda; see Chapter 33) show that desiccated individuals contain large amounts of sugars. In particular, a disaccharide called trehalose seems to protect the cells by replacing the water that is normally associated with proteins and membrane lipids. Many insects that survive freezing in the winter also use trehalose as a membrane protectant, as do some plants resistant to desiccation.

Land Animals

The threat of dehydration is a major regulatory problem for terrestrial plants and animals. Humans, for example, die if they lose as little as 12% of their body water (desert camels can withstand approximately twice that level of dehydration). Adaptations that reduce water loss are key to survival on land. Much as a waxy cuticle contributes to the success of land plants, the body coverings of most terrestrial animals help prevent dehydration. Examples are the waxy layers of insect exoskeletons, the shells of land snails, and the layers of dead, keratinized skin cells covering most terrestrial vertebrates, including humans. Many terrestrial animals, especially desert-dwellers, are nocturnal, which reduces evaporative water loss because of the lower temperature and higher humidity of night air.

Despite these and other adaptations, most terrestrial animals lose water through many routes: in urine and feces, across their skin, and from the surfaces of gas exchange organs. Land animals maintain water balance by drinking and

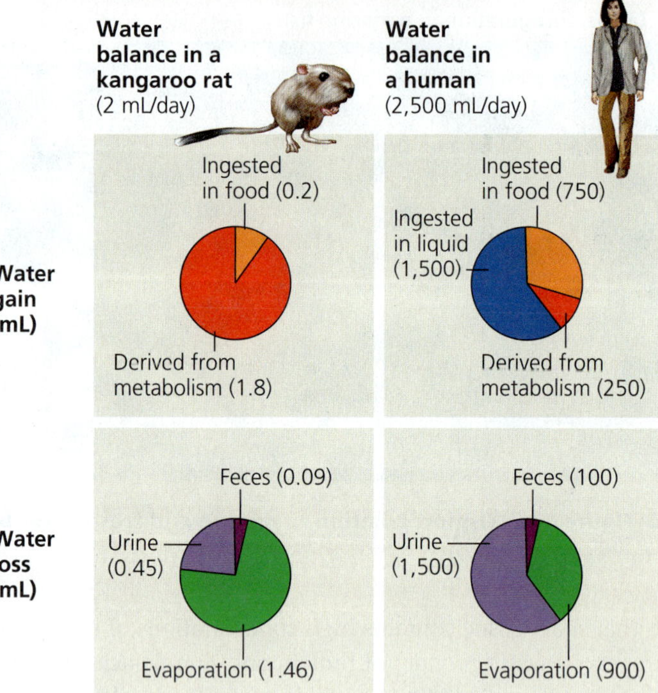

▲ **Figure 44.6 Water balance in two terrestrial mammals.** Kangaroo rats, which live in the American Southwest, eat mostly dry seeds and do not drink water. A kangaroo rat gains water mainly from cellular metabolism and loses water mainly by evaporation during gas exchange. In contrast, a human gains water in food and drink and loses the largest fraction of it in urine.

eating moist foods and by producing water metabolically through cellular respiration. A number of desert animals, including many insect-eating birds and other reptiles, are well enough adapted for minimizing water loss that they can survive for long periods of time without drinking. A noteworthy example is the kangaroo rat: It typically loses so little water that 90% is replaced by water it generates metabolically **(Figure 44.6)**; the remaining 10% comes from the small amount of water in its diet of seeds. During particularly hot periods, kangaroo rats supplement their diet with juicy insects, thereby maintaining their water balance.

Energetics of Osmoregulation

Maintaining an osmolarity difference between an animal's body and its external environment carries an energy cost. Because diffusion tends to equalize concentrations in a system, osmoregulators must expend energy to maintain the osmotic gradients that cause water to move in or out. They do so by using active transport to manipulate solute concentrations in their body fluids.

The energy cost of osmoregulation depends on how different an animal's osmolarity is from its surroundings, how easily water and solutes can move across the animal's surface, and how much work is required to pump solutes across the membrane. Osmoregulation accounts for 5% or more of the resting metabolic rate of many freshwater and marine bony fishes. For brine shrimp, small crustaceans that live in extremely salty lakes, the gradient between internal and external osmolarity is very large, and the cost of osmoregulation is correspondingly high—as much as 30% of the resting metabolic rate.

The energy cost to an animal of maintaining water and salt balance is minimized by having body fluids that are adapted to the salinity of the animal's habitat. Thus, the body fluids of most animals that live in fresh water (which has an osmolarity of 0.5–15 mOsm/L) have lower solute concentrations than the body fluids of their closest relatives that live in seawater (1,000 mOsm/L). For instance, whereas marine molluscs have body fluids with solute concentrations of approximately 1,000 mOsm/L, some freshwater molluscs maintain the osmolarity of their body fluids at just 40 mOsm/L. In each case, minimizing the osmotic difference between body fluids and the surrounding environment decreases the energy the animal expends for osmoregulation.

Transport Epithelia in Osmoregulation

The ultimate function of osmoregulation is to control solute concentrations in cells, but most animals do this indirectly by managing the solute content of an internal body fluid that bathes the cells. In insects and other animals with an open circulatory system, the fluid surrounding cells is hemolymph. In vertebrates and other animals with a closed circulatory system, the cells are bathed in an interstitial fluid that contains a mixture of solutes controlled indirectly by the blood. Maintaining the composition of such fluids depends on structures ranging from individual cells that regulate solute movement to complex organs such as the vertebrate kidney.

In most animals, osmoregulation and metabolic waste disposal rely on **transport epithelia**—one or more layers of epithelial cells specialized for moving particular solutes in controlled amounts in specific directions. Transport epithelia are typically arranged into complex tubular networks with extensive surface areas. Some transport epithelia face the outside environment directly, while others line channels connected to the outside by an opening on the body surface.

The transport epithelium that enables the albatross to survive on seawater remained undiscovered for many years. Some scientists suggested that marine birds do not actually drink water, asserting that although the birds take water into their mouths, they do not swallow. Questioning this idea, Knut Schmidt-Nielsen and colleagues at the Mount Desert Island Laboratory, in Maine, gave captive marine birds only seawater to drink. The researchers found that while very little salt appeared in the birds' urine, fluid dripping from the tip of their beaks was a concentrated solution of salt (NaCl). Where did this fluid come from? As Schmidt-Nielsen demonstrated, the salt solution was produced by a pair of structures called the nasal glands. Similar structures, called salt glands, eliminate excess salt from the bodies of sea turtles and marine iguanas.

As shown in **Figure 44.7**, the nasal gland removes excess NaCl (in the form of Na^+ and Cl^-) from the blood

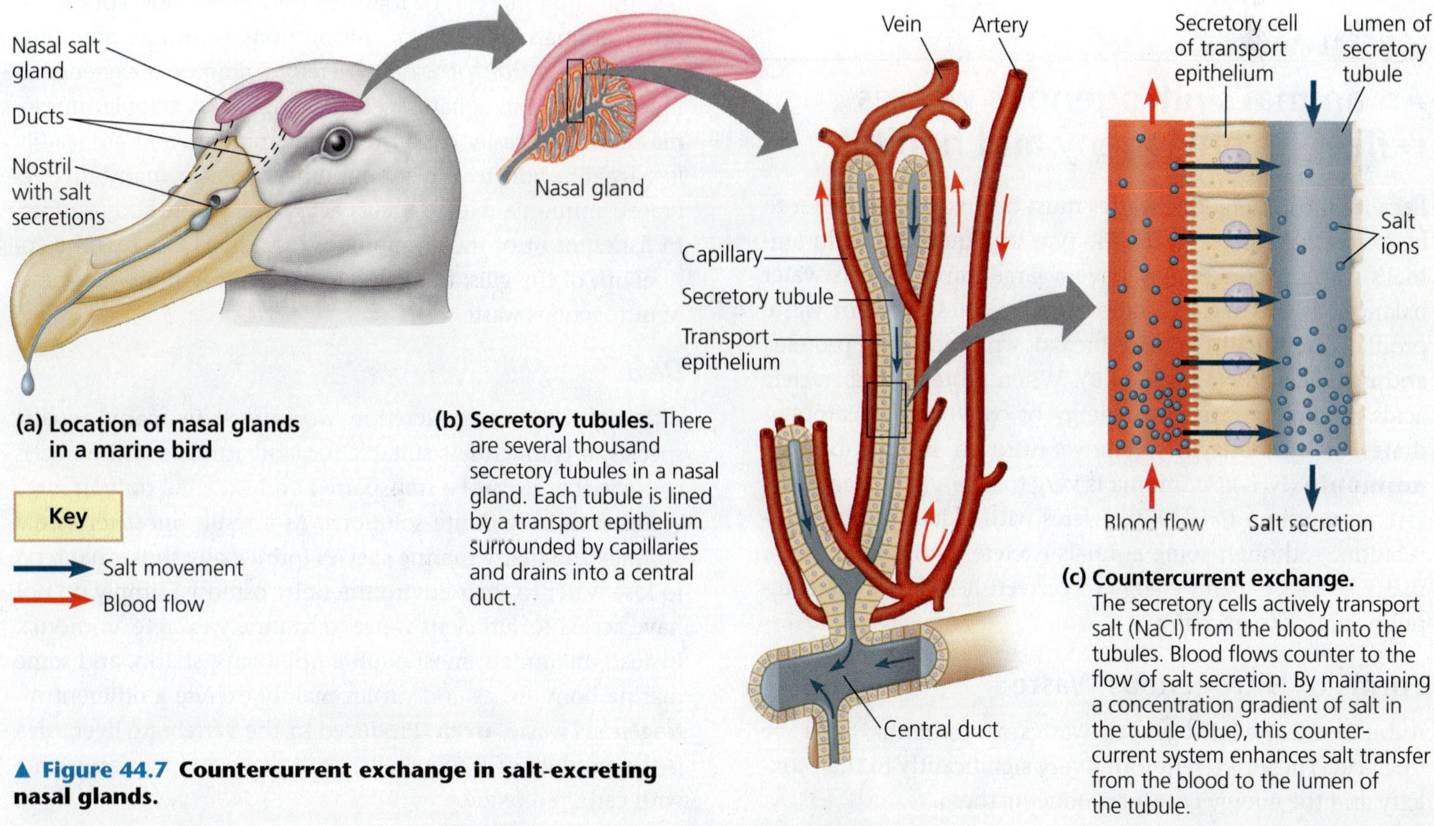

(a) Location of nasal glands in a marine bird

(b) Secretory tubules. There are several thousand secretory tubules in a nasal gland. Each tubule is lined by a transport epithelium surrounded by capillaries and drains into a central duct.

(c) Countercurrent exchange. The secretory cells actively transport salt (NaCl) from the blood into the tubules. Blood flows counter to the flow of salt secretion. By maintaining a concentration gradient of salt in the tubule (blue), this countercurrent system enhances salt transfer from the blood to the lumen of the tubule.

▲ **Figure 44.7 Countercurrent exchange in salt-excreting nasal glands.**

by countercurrent exchange. Recall from Chapter 40 that countercurrent exchange occurs between two fluids separated by one or more membranes and flowing in opposite directions. In the albatross's nasal gland, the net result is the secretion of fluid much saltier than the ocean. Thus, even though drinking seawater brings in a lot of salt, the bird achieves a net gain of water. By contrast, humans who drink a given volume of seawater must use a *greater* volume of water to excrete the salt load, with the result that they become dehydrated.

Transport epithelia that function in maintaining water balance also often function in disposal of metabolic wastes. We will see examples of this coordinated function in our upcoming consideration of earthworm and insect excretory systems as well as the vertebrate kidney.

CONCEPT CHECK 44.1

1. The movement of salt from the surrounding water to the blood of a freshwater fish requires the expenditure of energy in the form of ATP. Why?
2. Why aren't any freshwater animals osmoconformers?
3. **WHAT IF?** Researchers found that a camel standing in the sun required much more water when its fur was shaved off, although its body temperature remained the same. What can you conclude about the relationship between osmoregulation and the insulation provided by fur?

For suggested answers, see Appendix A.

CONCEPT 44.2

An animal's nitrogenous wastes reflect its phylogeny and habitat

Because most metabolic wastes must be dissolved in water to be excreted from the body, the type and quantity of an animal's waste products may have a large impact on its water balance. In this regard, some of the most significant waste products are the nitrogenous breakdown products of proteins and nucleic acids **(Figure 44.8)**. When proteins and nucleic acids are broken apart for energy or converted to carbohydrates or fats, enzymes remove nitrogen in the form of **ammonia** (NH_3). Ammonia is very toxic, in part because its ion, ammonium (NH_4^+), interferes with oxidative phosphorylation. Although some animals excrete ammonia directly, many species expend energy to convert it to less toxic compounds prior to excretion.

Forms of Nitrogenous Waste

Animals excrete nitrogenous wastes as ammonia, urea, or uric acid. These different forms vary significantly in their toxicity and the energy costs of producing them.

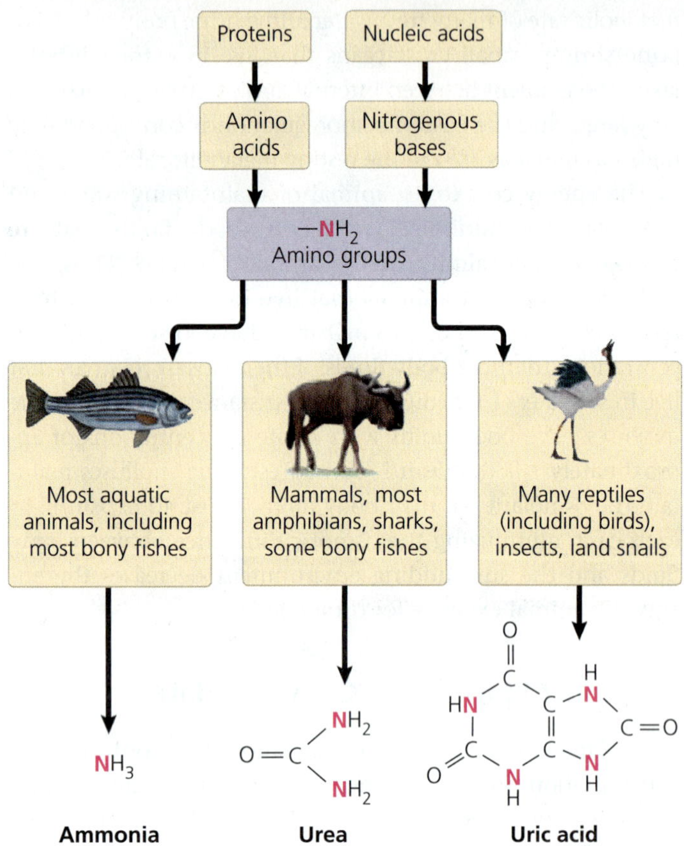

▲ **Figure 44.8 Forms of nitrogenous waste.**

Ammonia

Because ammonia can be tolerated only at very low concentrations, animals that excrete nitrogenous wastes as ammonia need access to lots of water. Therefore, ammonia excretion is most common in aquatic species. Being highly soluble, ammonia molecules easily pass through membranes and are readily lost by diffusion to the surrounding water. In many invertebrates, ammonia release occurs across the whole body surface. In fishes, most of the ammonia is lost as NH_4^+ across the epithelium of the gills; the kidneys excrete only minor amounts of nitrogenous waste.

Urea

Although ammonia excretion works well in many aquatic species, it is much less suitable for land animals. Ammonia is so toxic that it can be transported and excreted only in large volumes of very dilute solutions. As a result, most terrestrial animals and many marine species (principally those that tend to lose water to their environment by osmosis) simply do not have access to sufficient water to routinely excrete ammonia. Instead, mammals, most adult amphibians, sharks, and some marine bony fishes and turtles mainly excrete a different nitrogenous waste, **urea**. Produced in the vertebrate liver, urea is the product of a metabolic cycle that combines ammonia with carbon dioxide.

The main advantage of urea is its very low toxicity. Animals can transport urea in the circulatory system and store it safely at high concentrations. Furthermore, much less water is lost when a given quantity of nitrogen is excreted in a concentrated solution of urea rather than a dilute solution of ammonia.

The main disadvantage of urea is its energy cost: Animals must expend energy to produce urea from ammonia. From a bioenergetic standpoint, we would predict that animals that spend part of their lives in water and part on land would switch between excreting ammonia (thereby saving energy) and excreting urea (reducing excretory water loss). Indeed, many amphibians excrete mainly ammonia when they are aquatic tadpoles and switch largely to urea excretion when they become land-dwelling adults.

Uric Acid

Insects, land snails, and many reptiles, including birds, excrete **uric acid** as their primary nitrogenous waste. (Bird droppings, or *guano*, are a mixture of white uric acid and brown feces.) Uric acid is relatively nontoxic and does not readily dissolve in water. It therefore can be excreted as a semisolid paste with very little water loss. This is a great advantage for animals with little access to water, but there is a cost: Uric acid is even more energetically expensive to produce than urea, requiring considerable ATP for synthesis from ammonia.

Because uric acid releases nitrates to soil, bird guano can be used as fertilizer in agriculture. Before synthetic fertilizers were developed, this "waste" was so valued that nations fought wars over South American islands covered with piles of seabird guano as tall as 12-story buildings! Recently, interest in organic fertilizers has revived the commercial trade in guano **(Figure 44.9)**.

While not primarily uric acid producers, humans and some other animals generate a small amount of uric acid as a product of purine breakdown. Diseases that alter this process reflect the problems that can arise when a metabolic product is insoluble. For example, a genetic defect in purine metabolism predisposes Dalmatian dogs to form uric acid stones in their bladder. In humans, adult males are particularly susceptible to *gout*, a painful joint inflammation caused by deposits of uric acid crystals. Meals containing purine-rich animal tissues can increase the inflammation. Some dinosaurs appear to have been similarly affected: Fossilized bones of *Tyrannosaurus rex* exhibit joint damage characteristic of gout.

▼ **Figure 44.9 Recycling nitrogenous waste.** Seabirds nesting on islands off Peru annually produce 12,000 tons of guano, which workers collect for sale as organic fertilizer.

The Influence of Evolution and Environment on Nitrogenous Wastes

EVOLUTION In general, the kind of nitrogenous wastes an animal excretes depends on both the species' evolutionary history (phylogeny) and its habitat, especially the availability of water. For example, terrestrial turtles (which often live in dry areas) excrete mainly uric acid, whereas aquatic turtles excrete both urea and ammonia. Another factor affecting the primary type of nitrogenous waste produced by a particular group of animals is the immediate environment of the animal egg. For example, soluble wastes can diffuse out of a shell-less amphibian egg or be carried away from a mammalian embryo by the mother's blood. However, the shelled eggs produced by birds and other reptiles (see Figure 34.25) are permeable to gases but not to liquids, which means that soluble nitrogenous wastes released by an embryo would be trapped within the egg and could accumulate to dangerous levels. (Although urea is much less harmful than ammonia, it is toxic at very high concentrations.) Using uric acid as a waste product conveys a selective advantage because it precipitates out of solution and can be stored within the egg as a harmless solid left behind when the animal hatches.

Regardless of the type of nitrogenous waste, the amount produced is coupled to the animal's energy budget. Endotherms, which use energy at high rates, eat more food and produce more nitrogenous waste than ectotherms. The amount of nitrogenous waste is also linked to diet. Predators, which derive much of their energy from protein, excrete more nitrogen than animals that rely mainly on lipids or carbohydrates as energy sources.

Having surveyed the forms of nitrogenous waste and their interrelationship with evolutionary lineage, habitat, and energy consumption, we will turn next to the processes and systems animals use to excrete these and other wastes.

CONCEPT CHECK 44.2

1. What advantage does uric acid offer as a nitrogenous waste in arid environments?
2. **WHAT IF?** Suppose a bird and a human both have gout. Why might reducing purine in their diets help the human much more than the bird?

For suggested answers, see Appendix A.

CONCEPT **44.3**

Diverse excretory systems are variations on a tubular theme

Whether an animal lives on land, in salt water, or in fresh water, water balance depends on the regulation of solute movement between internal fluids and the external environment. Much of this movement is handled by excretory systems. These systems are central to homeostasis because they dispose of metabolic wastes and control body fluid composition. Before we describe particular excretory systems, let's consider a generalized version of the process of excretion.

Excretory Processes

Animals across a wide range of species produce a fluid waste called urine through the basic steps shown in **Figure 44.10**. In the first step, body fluid (blood, coelomic fluid, or hemolymph) is brought in contact with the selectively permeable membrane of a transport epithelium. In most cases, hydrostatic pressure (blood pressure in many animals) drives a process of **filtration**. Cells, as well as proteins and other large molecules, cannot cross the epithelial membrane and remain in the body fluid. In contrast, water and small solutes, such as salts, sugars, amino acids, and nitrogenous wastes, cross the membrane, forming a solution called the **filtrate**.

The filtrate is converted to a waste fluid by the specific transport of materials into or out of the filtrate. The process of selective **reabsorption** recovers useful molecules and water from the filtrate and returns them to the body fluids. Valuable solutes—including glucose, certain salts, vitamins, hormones, and amino acids—are reabsorbed by active transport. Nonessential solutes and wastes are left in the filtrate or are added to it by selective **secretion**, which also occurs by active transport. The pumping of various solutes adjusts the osmotic movement of water into or out of the filtrate. In the last step—excretion—the processed filtrate containing nitrogenous wastes is released from the body as urine.

Capillary

Filtrate

Excretory tubule

Urine

1 Filtration. The excretory tubule collects a filtrate from the blood. Water and solutes are forced by blood pressure across the selectively permeable membranes of a cluster of capillaries and into the excretory tubule.

2 Reabsorption. The transport epithelium reclaims valuable substances from the filtrate and returns them to the body fluids.

3 Secretion. Other substances, such as toxins and excess ions, are extracted from body fluids and added to the contents of the excretory tubule.

4 Excretion. The altered filtrate (urine) leaves the system and the body.

▲ **Figure 44.10 Key steps of excretory system function: an overview.** Most excretory systems produce a filtrate by pressure-filtering body fluids and then modify the filtrate's contents. This diagram is modeled after the vertebrate excretory system.

Survey of Excretory Systems

The systems that perform the basic excretory functions vary widely among animal groups. However, they are generally built on a complex network of tubules that provide a large surface area for the exchange of water and solutes, including nitrogenous wastes. We'll examine the excretory systems of flatworms, earthworms, insects, and vertebrates as examples of evolutionary variations on tubule networks.

Protonephridia

Flatworms (phylum Platyhelminthes) have excretory systems called **protonephridia** (singular, *protonephridium*), which form a network of dead-end tubules **(Figure 44.11)**. The

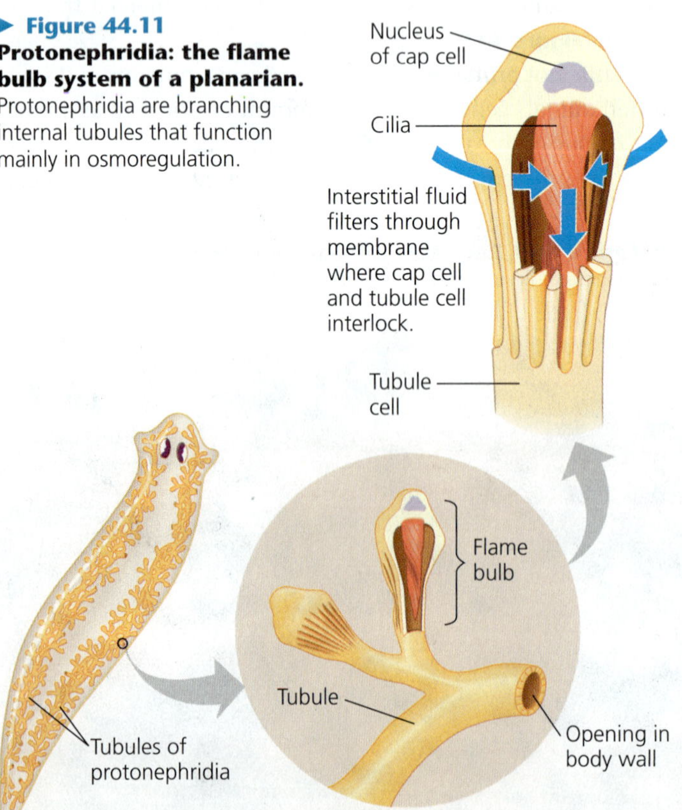

▶ **Figure 44.11 Protonephridia: the flame bulb system of a planarian.** Protonephridia are branching internal tubules that function mainly in osmoregulation.

tubules, which are connected to external openings, branch throughout the flatworm body, which lacks a coelom or body cavity. Cellular units called flame bulbs cap the branches of each protonephridium. Consisting of a tubule cell and a cap cell, each flame bulb has a tuft of cilia projecting into the tubule. During filtration, the beating of the cilia draws water and solutes from the interstitial fluid through the flame bulb, releasing filtrate into the tubule network. (The moving cilia resemble a flickering flame, hence the name *flame bulb*.) The processed filtrate then moves outward through the tubules and empties as urine into the external environment. The urine excreted by freshwater flatworms has a low solute concentration, helping to balance the osmotic uptake of water from the environment.

Protonephridia are also found in rotifers, some annelids, mollusc larvae, and lancelets (see Figure 34.4). Among these animals, the function of the protonephridia varies. In the freshwater flatworms, protonephridia serve chiefly in osmoregulation. Most metabolic wastes diffuse out of the animal across the body surface or are excreted into the gastrovascular cavity and eliminated through the mouth (see Figure 33.10). However, in some parasitic flatworms, which are isoosmotic to the surrounding fluids of their host organisms, the main function of protonephridia is the disposal of nitrogenous wastes. Natural selection has thus adapted protonephridia to different tasks in different environments.

Metanephridia

Most annelids, such as earthworms, have **metanephridia** (singular, *metanephridium*), excretory organs that collect fluid directly from the coelom **(Figure 44.12)**. Each segment of a worm has a pair of metanephridia, which are immersed in coelomic fluid and enveloped by a capillary network. A ciliated funnel surrounds the internal opening. As the cilia beat, fluid is drawn into a collecting tubule, which includes a storage bladder that opens to the outside.

The metanephridia of an earthworm have both excretory and osmoregulatory functions. As urine moves along the tubule, the transport epithelium bordering the lumen reabsorbs most solutes and returns them to the blood in the capillaries. Nitrogenous wastes remain in the tubule and are excreted to the outside. Earthworms inhabit damp soil and usually experience a net uptake of water by osmosis through their skin. Their metanephridia balance the water influx by producing urine that is dilute (hypoosmotic to body fluids).

Malpighian Tubules

Insects and other terrestrial arthropods have organs called **Malpighian tubules** that remove nitrogenous wastes and that also function in osmoregulation **(Figure 44.13)**. The Malpighian tubules extend from dead-end tips immersed in hemolymph (circulatory fluid) to openings into the digestive tract. The filtration step common to other excretory systems is absent. Instead, the transport epithelium that lines the tubules secretes certain solutes, including nitrogenous wastes, from the hemolymph into the lumen of the tubule. Water follows the solutes into the tubule by osmosis, and the fluid then passes into the rectum. There, most solutes are pumped back into the hemolymph, and water reabsorption by osmosis follows. The nitrogenous wastes—mainly insoluble

Coelom
Capillary network
Components of a metanephridium:
Collecting tubule
Internal opening
Bladder
External opening

▲ **Figure 44.12 Metanephridia of an earthworm.** Each segment of the worm contains a pair of metanephridia, which collect coelomic fluid from the adjacent anterior segment. The region highlighted in yellow illustrates the organization of one metanephridium of a pair; the other would be behind it.

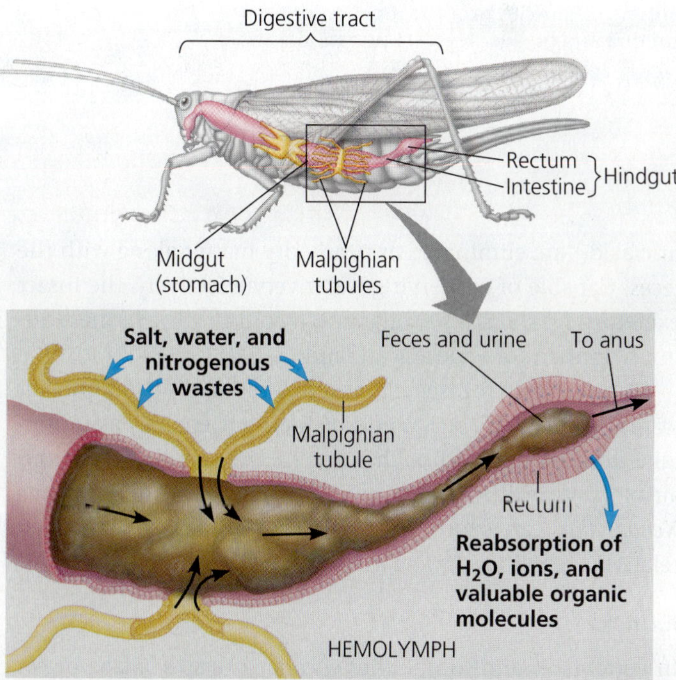

▲ **Figure 44.13 Malpighian tubules of insects.** Malpighian tubules are outpocketings of the digestive tract that remove nitrogenous wastes and function in osmoregulation.

▼ Figure 44.14

Exploring the Mammalian Excretory System

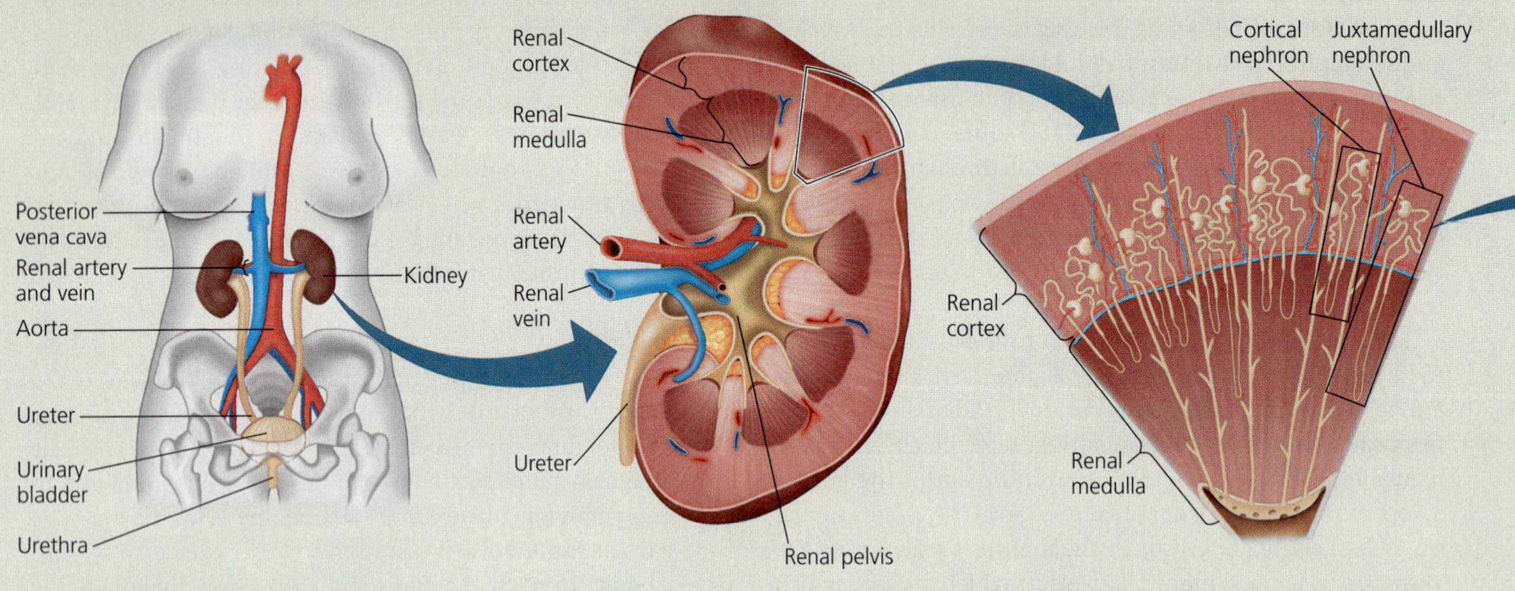

In humans, the excretory system consists of a pair of **kidneys**, bean-shaped organs about 10 cm in length, as well as organs for transporting and storing urine. Urine produced by each kidney exits through a duct called the **ureter**; the two ureters drain into a common sac called the **urinary bladder**. During urination, urine is expelled from the bladder through a tube called the **urethra**, which empties to the outside near the vagina in females and through the penis in males. Sphincter muscles near the junction of the urethra and bladder regulate urination.

Each kidney has an outer **renal cortex** and an inner **renal medulla**. Both regions are supplied with blood by a renal artery and drained by a renal vein. Within the cortex and medulla lie tightly packed excretory tubules and associated blood vessels. The inner **renal pelvis** collects urine from the excretory tubules and passes it to the urinary bladder.

Weaving back and forth across the renal cortex and medulla are the **nephrons**, the functional units of the vertebrate kidney. Of the roughly 1 million nephrons in a human kidney, 85% are **cortical nephrons**, which reach only a short distance into the medulla. The remainder, the **juxtamedullary nephrons**, extend deep into the medulla. Juxtamedullary nephrons are essential for production of urine that is hyperosmotic to body fluids, a key adaptation for water conservation in mammals.

uric acid—are eliminated as nearly dry matter along with the feces. Capable of conserving water very effectively, the insect excretory system is a key adaptation contributing to these animals' tremendous success on land.

Some terrestrial insects have an additional adaptation for water balance: The rectal end of their gut enables water uptake from the air. Although some species absorb water from air only when it is very humid, others, such as fleas (genus *Xenopsylla*), can capture water from the atmosphere when relative humidity is as low as 50%.

Kidneys

In vertebrates and some other chordates, a specialized organ called the kidney functions in both osmoregulation and excretion. Like the excretory organs of most animal phyla, kidneys consist of tubules. The numerous tubules of these compact organs are arranged in a highly organized manner and are closely associated with a network of capillaries. The vertebrate excretory system also includes ducts and other structures that carry urine from the tubules out of the kidney and, eventually, the body.

Vertebrate kidneys are typically nonsegmented. However, hagfishes, which are invertebrate chordates, have kidneys with segmentally arranged excretory tubules. This suggests that the excretory structures of vertebrate ancestors also may have been segmented.

Because kidney organization is integral to kidney function, we begin with **Figure 44.14**, an exploration of the anatomy of the mammalian kidney and associated structures. Familiarizing yourself with the terms and diagrams in this figure will provide you with a solid foundation for learning about filtrate processing in the kidney, our focus in the next concept.

Nephron Organization

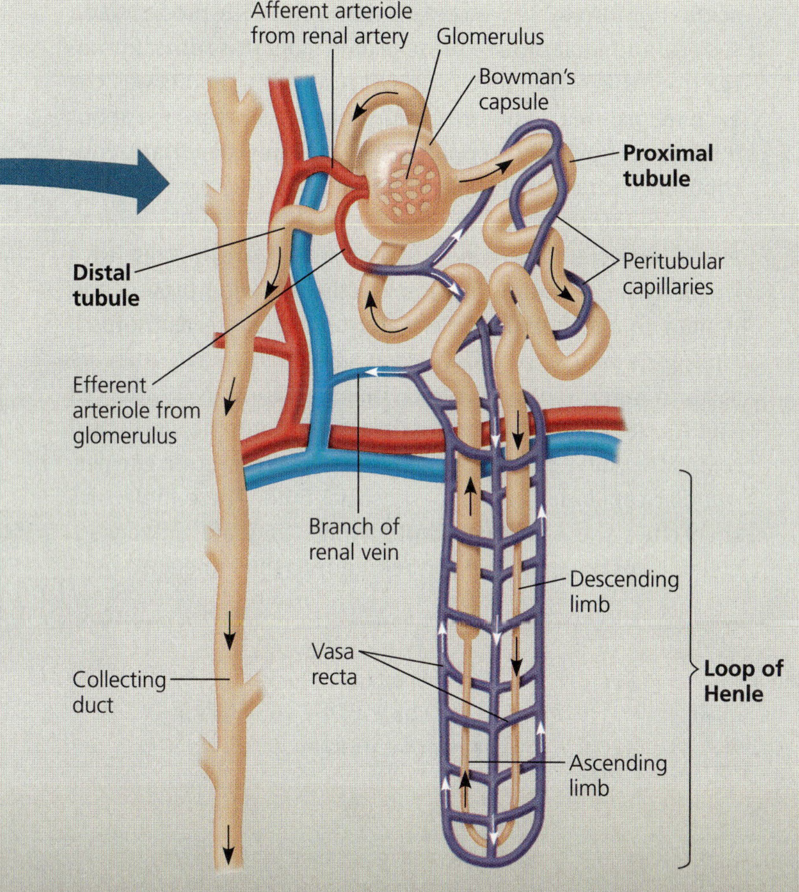

Each nephron consists of a single long tubule as well as a ball of capillaries called the **glomerulus**. The blind end of the tubule forms a cup-shaped swelling, called **Bowman's capsule**, which surrounds the glomerulus. Filtrate is formed when blood pressure forces fluid from the blood in the glomerulus into the lumen of Bowman's capsule. Processing occurs as the filtrate passes through three major regions of the nephron: the **proximal tubule**, the **loop of Henle** (a hairpin turn with a descending limb and an ascending limb), and the **distal tubule**. A **collecting duct** receives processed filtrate from many nephrons and transports it to the renal pelvis.

Each nephron is supplied with blood by an *afferent arteriole*, an offshoot of the renal artery that branches and forms the capillaries of the glomerulus. The capillaries converge as they leave the glomerulus, forming an *efferent arteriole*. Branches of this vessel form the **peritubular capillaries**, which surround the proximal and distal tubules. Other branches extend downward and form the **vasa recta**, hairpin-shaped capillaries that serve the renal medulla, including the long loop of Henle of juxtamedullary nephrons.

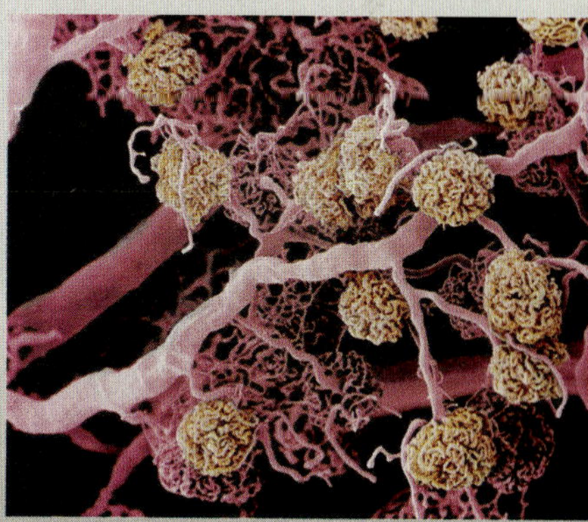

▶ In this SEM of densely packed blood vessels from a human kidney, arterioles and peritubular capillaries appear pink; the glomeruli appear yellow.

CONCEPT CHECK 44.3

1. Compare and contrast the different ways that metabolic waste products enter the excretory systems of flatworms, earthworms, and insects.
2. What is the function of the filtration step in excretory systems?
3. **WHAT IF?** Kidney failure is often treated by hemodialysis, in which blood diverted out of the body is filtered and then allowed to flow on one side of a semipermeable membrane. Fluid called dialysate flows in the opposite direction on the other side of the membrane. In replacing the reabsorption and secretion of solutes in a functional kidney, the makeup of the starting dialysate is critical. What initial solute composition would work well?

For suggested answers, see Appendix A.

CONCEPT **44.4**

The nephron is organized for stepwise processing of blood filtrate

We'll continue our exploration of the nephron with a discussion of filtrate processing. We will then focus on how tubules, capillaries, and surrounding tissue function together.

The porous capillaries and specialized cells of Bowman's capsule are permeable to water and small solutes, but not blood cells or large molecules, such as plasma proteins. Thus, the filtrate produced in the capsule contains salts, glucose, amino acids, vitamins, nitrogenous wastes, and other small molecules. Because such molecules pass freely between glomerular capillaries and Bowman's capsule, the concentrations of these substances in the initial filtrate are the same as those in blood plasma.

From Blood Filtrate to Urine: *A Closer Look*

In this section, we will follow filtrate along its path in the nephron and collecting duct, examining how each region contributes to the stepwise processing of filtrate into urine. The circled numbers correspond to the numbers in **Figure 44.15**.

1 Proximal tubule. Reabsorption in the proximal tubule is critical for the recapture of ions, water, and valuable nutrients from the huge volume of initial filtrate. NaCl (salt) in the filtrate diffuses into the cells of the transport epithelium, where Na^+ is actively transported into the interstitial fluid. This transfer of positive charge out of the tubule drives the passive transport of Cl^-, as well as the movement of more Na^+ from the lumen into the cells of the tubule wall by facilitated diffusion and cotransport mechanisms (see Figures 7.17 and 7.21).

As salt moves from the filtrate to the interstitial fluid, water follows by osmosis. The salt and water then diffuse from the interstitial fluid into the peritubular capillaries. Glucose, amino acids, potassium ions (K^+), and other essential substances are also actively or passively transported from the filtrate to the interstitial fluid and then into the peritubular capillaries.

Processing of filtrate in the proximal tubule helps maintain a relatively constant pH in body fluids. Cells of the transport epithelium secrete H^+ into the lumen of the tubule but also synthesize and secrete ammonia, which acts as a buffer to trap H^+ in the form of ammonium ions (NH_4^+). The more acidic the filtrate, the more ammonia the cells produce and secrete, and a mammal's urine usually contains some ammonia from this source (even though most nitrogenous waste is excreted as urea). The proximal tubules also reabsorb about 90% of the buffer bicarbonate (HCO_3^-) from the filtrate, contributing further to pH balance in body fluids.

As the filtrate passes through the proximal tubule, materials to be excreted become concentrated. Many wastes leave the body fluids during the nonselective filtration process and remain in the filtrate while water and salts are reabsorbed. Urea, for example, is reabsorbed at a much lower rate than are salt and water. Some other toxic materials are actively secreted into filtrate from surrounding tissues. For example, drugs and toxins that have been processed in the liver pass from the peritubular capillaries into the interstitial fluid. These molecules then enter the proximal tubule, where they are actively secreted from the transport epithelium into the lumen.

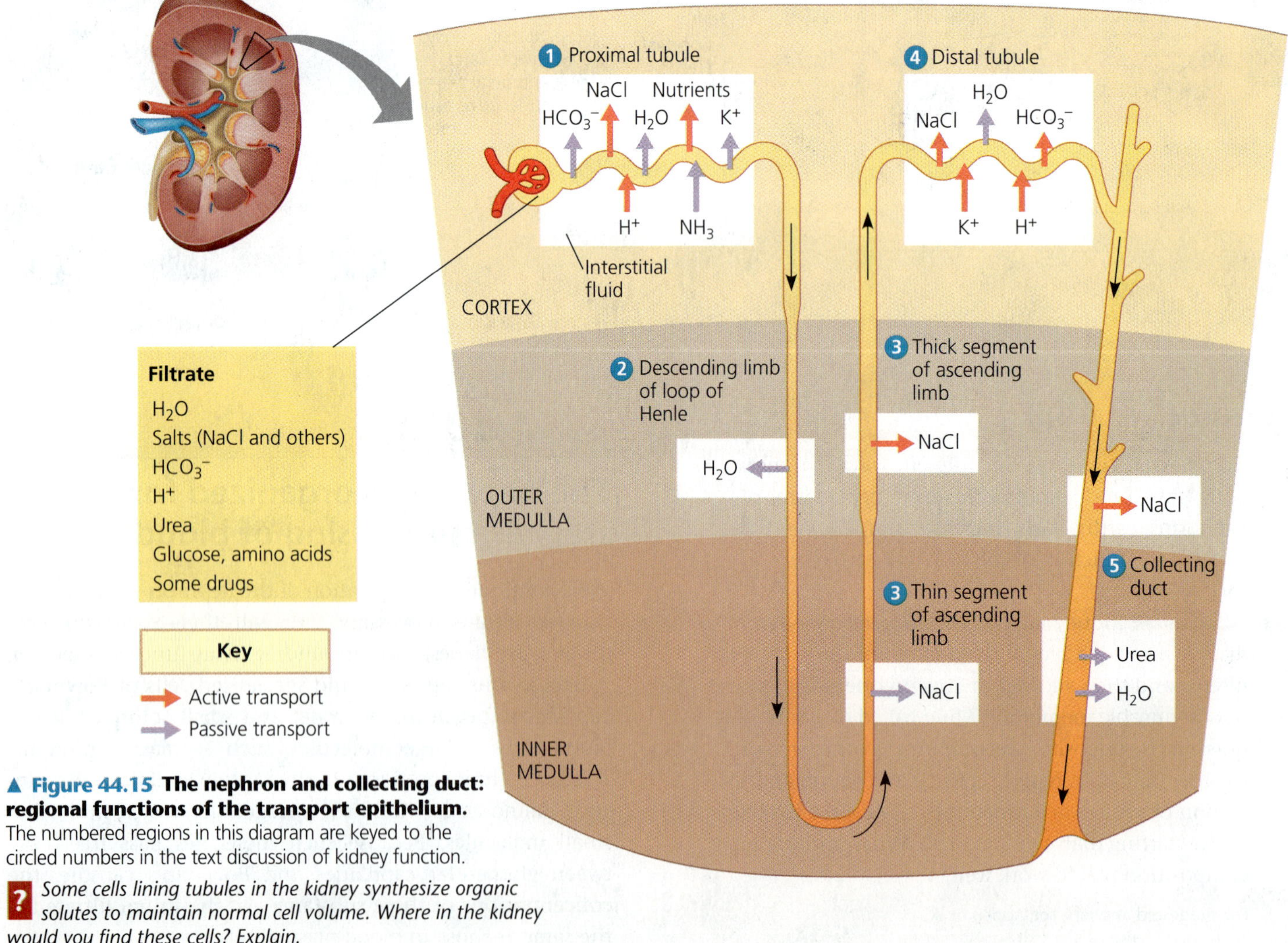

▲ **Figure 44.15 The nephron and collecting duct: regional functions of the transport epithelium.** The numbered regions in this diagram are keyed to the circled numbers in the text discussion of kidney function.

? *Some cells lining tubules in the kidney synthesize organic solutes to maintain normal cell volume. Where in the kidney would you find these cells? Explain.*

2 Descending limb of the loop of Henle. Reabsorption of water continues as the filtrate moves into the descending limb of the loop of Henle. Here numerous water channels formed by **aquaporin** proteins make the transport epithelium freely permeable to water. In contrast, there are almost no channels for salt and other small solutes, resulting in very low permeability for these substances.

For water to move out of the tubule by osmosis, the interstitial fluid bathing the tubule must be hyperosmotic to the filtrate. This condition is met along the entire length of the descending limb, because the osmolarity of the interstitial fluid increases progressively from the outer cortex to the inner medulla of the kidney. As a result, the filtrate loses water—and therefore its solute concentration increases—all along its journey down the descending limb.

3 Ascending limb of the loop of Henle. The filtrate reaches the tip of the loop and then travels within the ascending limb as it returns to the cortex. Unlike the descending limb, the ascending limb has a transport epithelium studded with ion channels, but not water channels. Indeed, this membrane is impermeable to water. Impermeability to water is very rare among biological membranes and is critical to the function of the ascending limb.

The ascending limb has two specialized regions: a thin segment near the loop tip and a thick segment adjacent to the distal tubule. As filtrate ascends in the thin segment, NaCl, which became concentrated in the descending limb, diffuses out of the permeable tubule into the interstitial fluid. This movement of NaCl out of the tubule helps maintain the osmolarity of the interstitial fluid in the medulla. In the thick segment of the ascending limb, the movement of NaCl out of the filtrate continues. Here, however, the epithelium actively transports NaCl into the interstitial fluid. As a result of losing salt but not water, the filtrate becomes progressively more dilute as it moves up to the cortex in the ascending limb of the loop.

4 Distal tubule. The distal tubule plays a key role in regulating the K^+ and NaCl concentration of body fluids. This regulation involves variation in the amount of K^+ secreted into the filtrate as well as the amount of NaCl reabsorbed from the filtrate. Like the proximal tubule, the distal tubule contributes to pH regulation by the controlled secretion of H^+ and reabsorption of HCO_3^-.

5 Collecting duct. The collecting duct carries the filtrate through the medulla to the renal pelvis. The transport epithelium of the nephron and collecting duct processes the filtrate, forming the urine. One of this epithelium's most important tasks is reabsorption of solutes and water. Under normal conditions, approximately 1,600 L of blood flows through a pair of human kidneys each day, about 300 times the total volume of blood in the body. From this enormous traffic of blood, the nephrons and collecting ducts process about 180 L of initial filtrate. Of this, about 99% of the water and nearly all of the sugars, amino acids, vitamins, and other organic nutrients are reabsorbed into the blood, leaving only about 1.5 L of urine to be transported to the bladder.

As filtrate passes along the transport epithelium of the collecting duct, hormonal control of permeability and transport determines the extent to which the urine becomes concentrated.

When the kidneys are conserving water, aquaporin channels in the collecting duct allow water molecules to cross the epithelium. At the same time, the epithelium remains impermeable to salt and, in the renal cortex, to urea. As the collecting duct traverses the gradient of osmolarity in the kidney, the filtrate becomes increasingly concentrated, losing more and more water by osmosis to the hyperosmotic interstitial fluid. In the inner medulla, the duct becomes permeable to urea. Because of the high urea concentration in the filtrate at this point, some urea diffuses out of the duct and into the interstitial fluid. Along with NaCl, this urea contributes to the high osmolarity of the interstitial fluid in the medulla. The net result is urine that is hyperosmotic to the general body fluids.

In producing dilute rather than concentrated urine, the kidney actively reabsorbs salts without allowing water to follow by osmosis. At these times, the epithelium lacks water channels, and NaCl is actively transported out of filtrate. As we will see shortly, the state of the collecting duct epithelium is controlled by hormones that together maintain homeostasis for osmolarity, blood pressure, and blood volume.

Solute Gradients and Water Conservation

The mammalian kidney's ability to conserve water is a key terrestrial adaptation. In humans, the osmolarity of blood is about 300 mOsm/L, but the kidney can excrete urine up to four times as concentrated—about 1,200 mOsm/L. Some mammals can do even better: Australian hopping mice, small marsupials that live in dry desert regions, can produce urine with an osmolarity of 9,300 mOsm/L, 25 times as concentrated as the animal's blood.

In a mammalian kidney, the production of hyperosmotic urine is possible only because considerable energy is expended for the active transport of solutes against concentration gradients. The nephrons—particularly the loops of Henle—can be thought of as energy-consuming machines that produce an osmolarity gradient suitable for extracting water from the filtrate in the collecting duct. The two primary solutes affecting osmolarity are NaCl, which is deposited in the renal medulla by the loop of Henle, and urea, which passes across the epithelium of the collecting duct in the inner medulla.

The Two-Solute Model

To better understand the physiology of the mammalian kidney as a water-conserving organ, let's retrace the flow of filtrate through the excretory tubule. This time, let's focus on how the juxtamedullary nephrons maintain an osmolarity gradient in the tissues that surround the loop of Henle and how they use that gradient to excrete a hyperosmotic urine **(Figure 44.16)**. Filtrate passing from Bowman's capsule to the proximal tubule has an osmolarity of about 300 mOsm/L, the same as blood. A large amount of water *and* salt is reabsorbed from the filtrate as it flows through the proximal tubule in the renal cortex. As a result, the filtrate's volume decreases substantially, but its osmolarity remains about the same.

As the filtrate flows from cortex to medulla in the descending limb of the loop of Henle, water leaves the tubule by osmosis. Solutes, including NaCl, become more concentrated, increasing the osmolarity of the filtrate. The highest osmolarity (about 1,200 mOsm/L) occurs at the elbow of the loop of Henle. This maximizes the diffusion of salt out of the tubule as the filtrate rounds the curve and enters the ascending limb, which is permeable to salt but not to water. NaCl diffusing from the ascending limb helps maintain a high osmolarity in the interstitial fluid of the renal medulla.

Notice that the loop of Henle has several qualities of a countercurrent system, such as the mechanisms that maximize oxygen absorption by fish gills (see Figure 42.22) or reduce heat loss in endotherms (see Figure 40.12). In those cases, the countercurrent mechanisms involve passive movement along either an oxygen concentration gradient or a heat gradient. In contrast, the countercurrent system involving the loop of Henle expends energy to actively transport NaCl from the filtrate in the upper part of the ascending limb of the loop. Such countercurrent systems, which expend energy to create concentration gradients, are called **countercurrent multiplier systems**. The countercurrent multiplier system involving the loop of Henle maintains a high salt concentration in the interior of the kidney, enabling the kidney to form concentrated urine.

What prevents the capillaries of the vasa recta from dissipating the gradient by carrying away the high concentration of NaCl in the medulla's interstitial fluid? As shown in Figure 44.14, the descending and ascending vessels of the vasa recta carry blood in opposite directions through the kidney's osmolarity gradient. As the descending vessel conveys blood toward the inner medulla, water is lost from the blood and NaCl is gained by diffusion. These fluxes are reversed as blood flows back toward the cortex in the ascending vessel, with water reentering the blood and salt diffusing out. Thus,

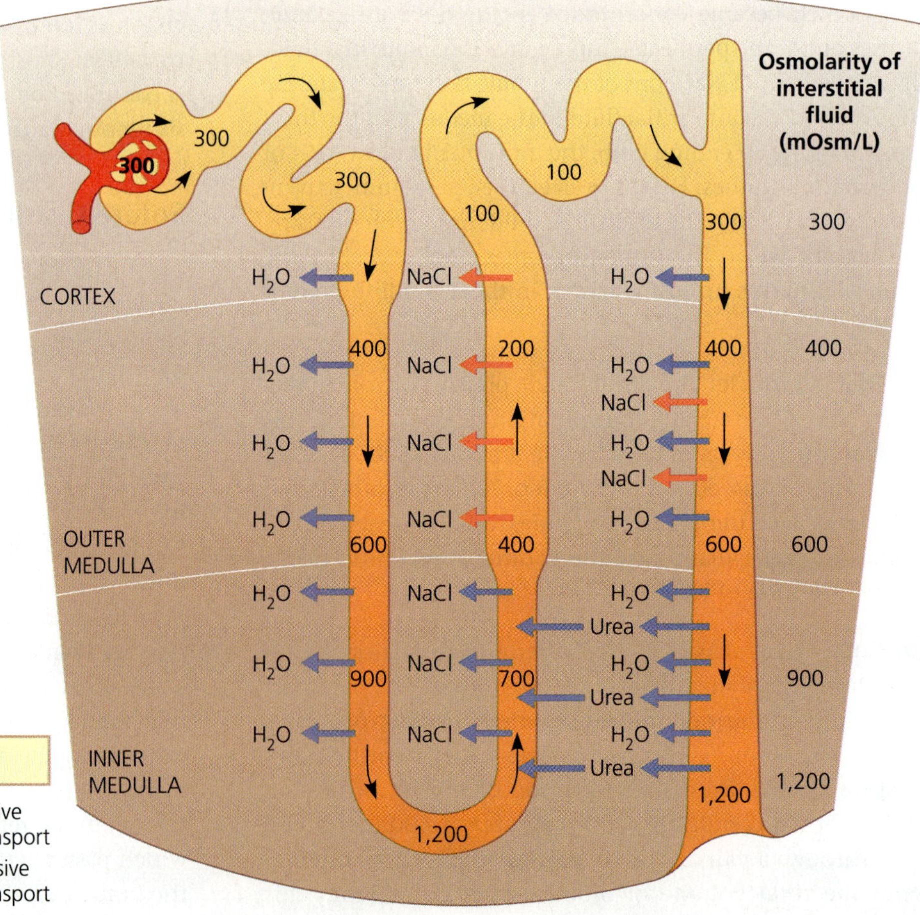

▶ **Figure 44.16 How the human kidney concentrates urine: the two-solute model.** Two solutes contribute to the osmolarity of the interstitial fluid: NaCl (used as shorthand here to refer collectively to Na^+ and Cl^-) and urea. The loop of Henle maintains the interstitial gradient of NaCl, which increases continuously in concentration from the cortex to the inner medulla. Urea diffuses into the interstitial fluid of the medulla from the collecting duct (most of the urea in the filtrate remains in the collecting duct and is excreted). The filtrate makes three trips between the cortex and medulla: first down, then up, and then down again in the collecting duct. As the filtrate flows in the collecting duct past interstitial fluid of increasing osmolarity, more water moves out of the duct by osmosis. The loss of water concentrates the solutes, including urea, that will be excreted in the urine.

WHAT IF? *The drug furosemide blocks the cotransporters for Na^+ and Cl^- in the ascending limb of the loop of Henle. What effect would you expect this drug to have on urine volume?*

the vasa recta can supply the kidney with nutrients and other important substances carried by the blood without interfering with the osmolarity gradient in the inner and outer medulla.

The countercurrent-like characteristics of the loop of Henle and the vasa recta help to generate the steep osmotic gradient between the medulla and cortex. However, diffusion will eventually eliminate any osmotic gradient within animal tissue unless gradient formation is supported by an expenditure of energy. In the kidney, this expenditure largely occurs in the thick segment of the ascending limb of the loop of Henle, where NaCl is actively transported out of the tubule. Even with the benefits of countercurrent exchange, this process—along with other renal active transport systems—consumes considerable ATP. Thus, for its size, the kidney has one of the highest metabolic rates of any organ.

As a result of active transport of NaCl out of the thick segment of the ascending limb, the filtrate is actually hypoosmotic to body fluids by the time it reaches the distal tubule. Next the filtrate descends again toward the medulla, this time in the collecting duct, which is permeable to water but not to salt. Therefore, osmosis extracts water from the filtrate as it passes from cortex to medulla and encounters interstitial fluid of increasing osmolarity. This process concentrates salt, urea, and other solutes in the filtrate. Some urea passes out of the lower portion of the collecting duct and contributes to the high interstitial osmolarity of the inner medulla. (This urea is recycled by diffusion into the loop of Henle, but continual leakage from the collecting duct maintains a high interstitial urea concentration.) When the kidney concentrates urine maximally, the urine reaches 1,200 mOsm/L, the osmolarity of the interstitial fluid in the inner medulla. Although *isoosmotic* to the inner medulla's interstitial fluid, the urine is *hyperosmotic* to blood and interstitial fluid elsewhere in the body. This high osmolarity allows the solutes remaining in the urine to be excreted from the body with minimal water loss.

Adaptations of the Vertebrate Kidney to Diverse Environments

EVOLUTION Vertebrate animals occupy habitats ranging from rain forests to deserts and from some of the saltiest bodies of water to the nearly pure waters of high mountain lakes. Variations in nephron structure and function equip the kidneys of different vertebrates for osmoregulation in their various habitats. The adaptations of the vertebrate kidney are made apparent by comparing species that inhabit a wide range of environments or by comparing the responses of different vertebrate groups to similar environmental conditions.

Mammals

The juxtamedullary nephron, with its urine-concentrating features, is a key adaptation to terrestrial life, enabling mammals to get rid of salts and nitrogenous wastes without squandering water. As we have seen, the remarkable ability of the mammalian kidney to produce hyperosmotic urine depends on the precise arrangement of the tubules and collecting ducts in the renal cortex and medulla. In this respect, the kidney is one of the clearest examples of how natural selection links the function of an organ to its structure.

Mammals that excrete the most hyperosmotic urine, such as Australian hopping mice, North American kangaroo rats, and other desert mammals, have loops of Henle that extend deep into the medulla. Long loops maintain steep osmotic gradients in the kidney, resulting in urine becoming very concentrated as it passes from cortex to medulla in the collecting ducts.

In contrast, beavers, muskrats, and other aquatic mammals that spend much of their time in fresh water and rarely face problems of dehydration have nephrons with relatively short loops, resulting in a much lower ability to concentrate urine. Terrestrial mammals living in moist conditions have loops of Henle of intermediate length and the capacity to produce urine intermediate in concentration to that produced by freshwater and desert mammals.

Birds and Other Reptiles

Most birds, including the albatross (see Figure 44.1) and the roadrunner **(Figure 44.17)**, live in environments that are dehydrating. Like mammals, birds have kidneys with juxtamedullary nephrons that specialize in conserving water. However, the nephrons of birds have loops of Henle that extend less far into the medulla than those of mammals. Thus, bird kidneys cannot concentrate urine to the high osmolarities achieved by mammalian kidneys. Although birds can produce hyperosmotic urine, their main water conservation adaptation is having uric acid as the nitrogen waste molecule. Since uric acid can be excreted as a paste, it reduces urine volume.

The kidneys of other reptiles, which have only cortical nephrons, produce urine that is isoosmotic or hypoosmotic to body fluids. However, the epithelium of the chamber from

▲ **Figure 44.17 The roadrunner (*Geococcyx californianus*), an animal well adapted to its dry environment.**

which urine and feces leave the body (the cloaca) helps conserve fluid by reabsorbing water from these wastes. Also like birds, most other reptiles excrete their nitrogenous wastes as uric acid.

Freshwater Fishes and Amphibians

Freshwater fishes are hyperosmotic to their surroundings, so they must excrete excess water continuously. In contrast to mammals and birds, freshwater fishes produce large volumes of very dilute urine. Their kidneys, which contain many nephrons, produce filtrate at a high rate. Freshwater fishes conserve salts by reabsorbing ions from the filtrate in their distal tubules, leaving water behind.

Amphibian kidneys function much like those of freshwater fishes. When in fresh water, the kidneys of frogs excrete dilute urine while the skin accumulates certain salts from the water by active transport. On land, where dehydration is the most pressing problem of osmoregulation, frogs conserve body fluid by reabsorbing water across the epithelium of the urinary bladder.

Marine Bony Fishes

The tissues of marine bony fishes gain excess salts from their surroundings and lose water. These environmental challenges are opposite to those faced by their freshwater relatives. Compared with freshwater fishes, marine fishes have fewer and smaller nephrons, and their nephrons lack a distal tubule. In addition, their kidneys have small glomeruli or lack glomeruli entirely. In keeping with these features, filtration rates are low and very little urine is excreted.

The main function of kidneys in marine bony fishes is to get rid of divalent ions (those with a charge of 2+ or 2−) such as calcium (Ca^{2+}), magnesium (Mg^{2+}), and sulfate (SO_4^{2-}). Marine fishes take in divalent ions by incessantly drinking seawater. They rid themselves of these ions by secreting them into the proximal tubules of the nephrons and excreting them in urine. Secretion by the gills maintains proper levels of monovalent ions (charge of 1+ or 1−) such as Na^+ and Cl^-.

CONCEPT CHECK 44.4

1. What do the number and length of nephrons in a fish's kidney indicate about the fish's habitat? How do they correlate with urine production?
2. Many medications make the epithelium of the collecting duct less permeable to water. How would taking such a drug affect kidney output?
3. **WHAT IF?** If blood pressure in the afferent arteriole leading to a glomerulus decreased, how would the rate of blood filtration within Bowman's capsule be affected? Explain.

For suggested answers, see Appendix A.

CONCEPT 44.5

Hormonal circuits link kidney function, water balance, and blood pressure

In mammals, both the volume and osmolarity of urine are adjusted according to an animal's water and salt balance and its rate of urea production. In situations of high salt intake and low water availability, a mammal can excrete urea and salt in small volumes of hyperosmotic urine with minimal water loss. If salt is scarce and fluid intake is high, the kidney can instead get rid of the excess water with little salt loss by producing large volumes of hypoosmotic urine. At such times, the urine can be as dilute as 70 mOsm/L, compared with an osmolarity of 300 mOsm/L for human blood.

The South American vampire bat shown in **Figure 44.18** illustrates the versatility of the mammalian kidney. Bats of this species feed at night on the blood of large birds and mammals. The bats use their sharp teeth to make a small incision in the prey's skin and then lap up blood from the wound (the prey animal is typically not seriously harmed). Anticoagulants in the bat's saliva prevent the blood from clotting. Because vampire bats often search for hours and fly long distances to locate a suitable victim, they benefit from consuming as much blood as possible when they do find prey—so much that after feeding, a bat could be too heavy to fly. However, the bat's kidneys offload much of the water absorbed from a blood meal by excreting large volumes of dilute urine as it feeds, up to 24% of body mass per hour. Having lost enough weight to take off, the bat can fly back to its roost in a cave or hollow tree, where it spends the day.

In the roost, the bat faces a different regulatory problem. Most of the nutrition it derives from blood comes in the form of protein. Digesting proteins generates large quantities of urea, but roosting bats lack access to the drinking water necessary to

▲ **Figure 44.18 A vampire bat (*Desmodus rotundas*), a mammal with a unique excretory situation.**

dilute it. Instead, their kidneys shift to producing small quantities of highly concentrated urine (up to 4,600 mOsm/L), an adjustment that disposes of the urea load while conserving as much water as possible. The vampire bat's ability to alternate rapidly between producing large amounts of dilute urine and small amounts of very hyperosmotic urine is an essential part of its adaptation to an unusual food source.

Antidiuretic Hormone

A combination of nervous and hormonal controls manages the osmoregulatory function of the mammalian kidney. One key hormone in this regulatory circuitry is **antidiuretic hormone (ADH)**, also called *vasopressin*. ADH is produced in the hypothalamus of the brain and stored in the posterior pituitary gland, located just below the hypothalamus. Osmoreceptor cells in the hypothalamus monitor the osmolarity of blood and regulate release of ADH from the posterior pituitary.

To understand the role of ADH, let's consider what occurs when blood osmolarity rises, such as after eating salty food or losing water through sweating (**Figure 44.19**). In response to an increase in osmolarity above the set point of 300 mOsm/L, more ADH is released into the bloodstream. When ADH reaches the kidney, its main targets are the collecting ducts. There, ADH brings about changes that make the epithelium more permeable to water. The resulting increase in water reabsorption concentrates urine, reduces urine volume, and lowers blood osmolarity back toward the set point. (Only the gain of additional water in food and drink can fully restore osmolarity to 300 mOsm/L.) As the osmolarity of the blood subsides, a negative-feedback mechanism reduces the activity of osmoreceptor cells in the hypothalamus, and ADH secretion is reduced (not shown in figure).

A reduction in blood osmolarity below the set point has the opposite set of effects. For example, intake of a large volume of water leads to a decrease in ADH secretion to a very low level. The resulting decrease in permeability of the collecting ducts reduces water reabsorption, resulting in discharge of large volumes of dilute urine. (Diuresis refers to increased urination, and ADH is called *anti*diuretic hormone because it opposes this state.)

ADH influences water uptake in the kidney's collecting ducts by regulating the water-selective channels formed by aquaporins. Binding of ADH to receptor molecules leads to a temporary increase in the number of aquaporin proteins in the membranes of collecting duct cells (**Figure 44.20**). Additional channels recapture more water, reducing urine volume.

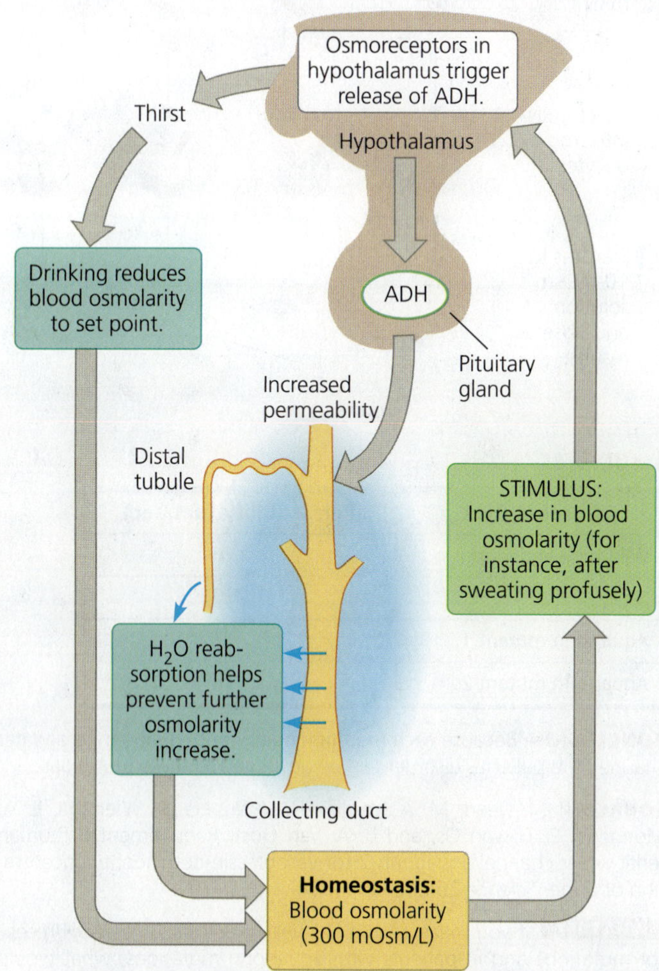

▲ **Figure 44.19 Regulation of fluid retention in the kidney by antidiuretic hormone (ADH).**

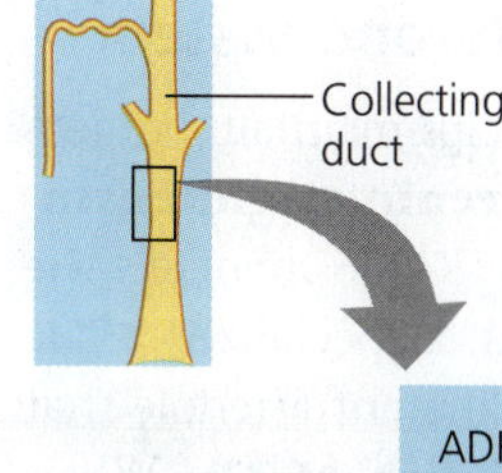

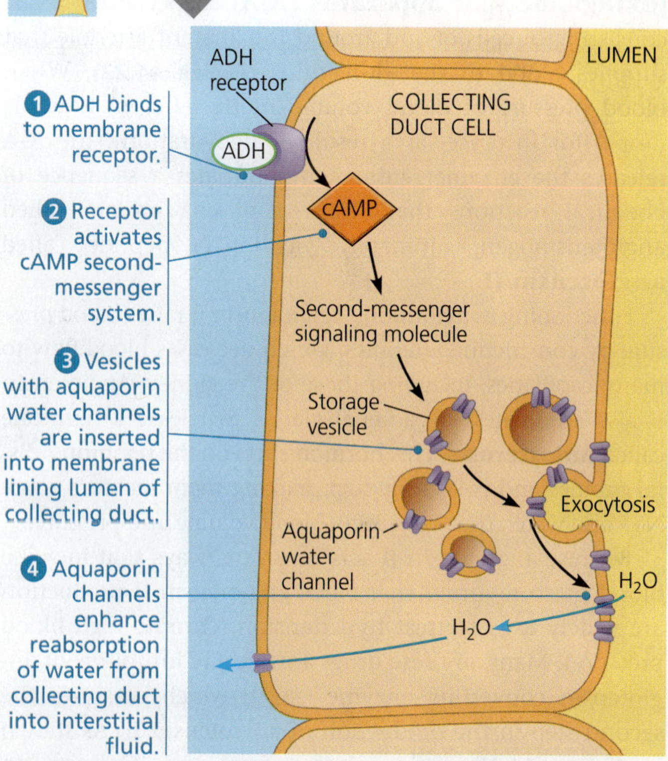

▲ **Figure 44.20 ADH response pathway in the collecting duct.**

Mutations that prevent ADH production or that inactivate the ADH receptor gene block the increase in channel number and thus the ADH response. The resulting disorder can cause severe dehydration and solute imbalance due to production of urine that is abnormally large in volume and very dilute. These symptoms give the condition its name: *diabetes insipidus* (from the Greek for "to pass through" and "having no flavor").

Researchers in the Netherlands wondered whether mutations in an aquaporin gene itself might also cause diabetes insipidus. Having found aquaporin gene mutations in a patient, they set out to determine whether the alterations led to nonfunctional water channels (**Figure 44.21**).

Taken together with previous studies, the experiment described in Figure 44.21 demonstrated that a wide variety of genetic defects can disrupt ADH regulation of water balance in the body. Even in the absence of such genetic changes, certain substances can alter the regulation of osmolarity. For example, alcohol can disturb water balance by inhibiting ADH release, leading to excessive urinary water loss and dehydration (which may cause some of the symptoms of a hangover). Normally, blood osmolarity, ADH release, and water reabsorption in the kidney are all linked in a feedback loop that contributes to homeostasis.

The Renin-Angiotensin-Aldosterone System

A second regulatory mechanism that helps maintain homeostasis by acting on the kidney is the **renin-angiotensin-aldosterone system (RAAS)**. The RAAS involves the **juxtaglomerular apparatus (JGA)**, a specialized tissue consisting of cells of and around the afferent arteriole that supplies blood to the glomerulus (**Figure 44.22**). When blood pressure or blood volume in the afferent arteriole drops (for instance, as a result of dehydration), the JGA releases the enzyme renin. Renin initiates a sequence of chemical reactions that cleave a plasma protein called angiotensinogen, ultimately yielding a peptide called **angiotensin II**.

Functioning as a hormone, angiotensin II raises blood pressure by constricting arterioles, which decreases blood flow to many capillaries, including those of the kidney. Angiotensin II also stimulates the adrenal glands to release a hormone called **aldosterone**. This hormone acts on the nephrons' distal tubules and collecting duct, making them reabsorb more Na^+ and water, thus increasing blood volume and pressure.

Because angiotensin II acts in several ways that increase blood pressure, drugs that block angiotensin II production are widely used to treat hypertension (chronic high blood pressure). Many of these drugs are specific inhibitors of angiotensin converting enzyme (ACE), which catalyzes the second step in the production of angiotensin II. As shown in Figure 44.21, renin released from the JGA acts on

▼ **Figure 44.21** INQUIRY

Can aquaporin mutations cause diabetes insipidus?

EXPERIMENT Bernard van Oost and colleagues at the University of Nijmegen were studying a patient with diabetes insipidus but a normal ADH receptor gene. Sequencing of the patient's DNA revealed two different mutations, one in each copy of an aquaporin gene. To determine whether each mutation blocked channel formation, they studied the mutant proteins in a cell that could be manipulated and studied outside the body. The cell they chose was the frog oocyte, which can be collected in large numbers from an adult female and will express foreign genes. The researchers synthesized messenger RNA from clones of the wild-type and mutant aquaporin genes and injected the synthetic mRNA into oocytes. Within the oocytes, the cellular machinery translated the mRNA into aquaporin proteins. To determine if the mutant aquaporin proteins made functional water channels in the plasma membrane, the investigators transferred the oocytes from a 200-mOsm to a 10-mOsm solution. They then measured swelling by light microscopy and calculated the permeability of the oocytes to water.

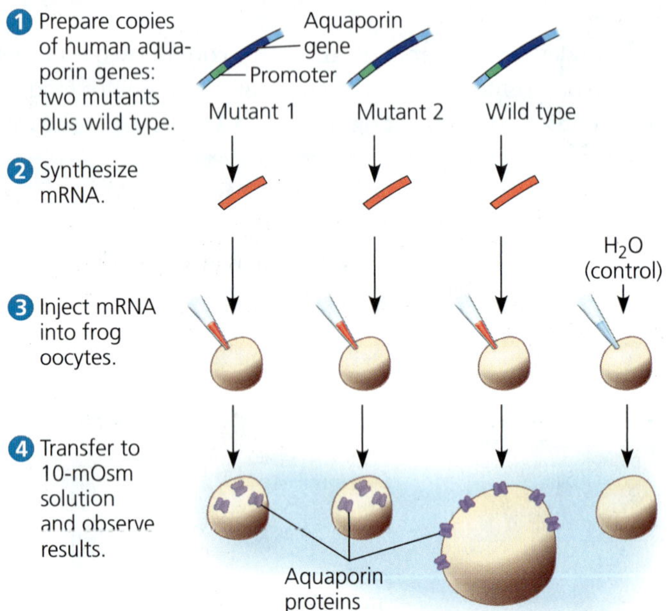

RESULTS

Injected RNA	Permeability (µm/sec)
Wild-type aquaporin	196
None	20
Aquaporin mutant 1	17
Aquaporin mutant 2	18

CONCLUSION Because each mutation inactivates aquaporin as a water channel, the patient's disorder can be attributed to these mutations.

SOURCE P. M. Deen, M. A. Verdijk, N. V. Knoers, B. Wieringa, L. A. Monnens, C. H. van Os, and B. A. van Oost, Requirement of human renal water channel aquaporin-2 for vasopressin-dependent concentration of urine, *Science* 264:92–95 (1994).

WHAT IF? If you measured ADH levels in patients with ADH receptor mutations and in patients with aquaporin mutations, what would you expect to find, compared with wild-type subjects?

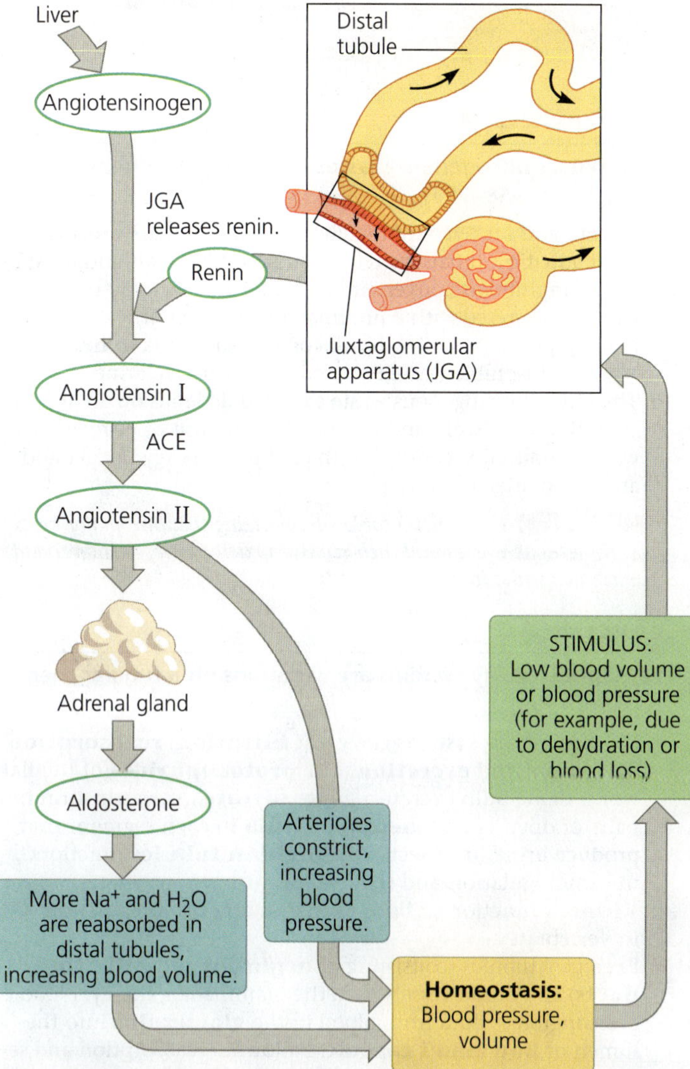

▲ **Figure 44.22 Regulation of blood volume and blood pressure by the renin-angiotensin-aldosterone system (RAAS).**

angiotensinogen (in the blood), forming angiotensin I. ACE in vascular endothelium, particularly in the lungs, then splits off two amino acids from angiotensin I, forming active angiotensin II. Blocking ACE activity with drugs prevents angiotensin II production and often lowers blood pressure into the normal range.

Homeostatic Regulation of the Kidney

The renin-angiotensin-aldosterone system operates as part of a complex feedback circuit that results in homeostasis. A drop in blood pressure and blood volume triggers renin release from the JGA. In turn, the rise in blood pressure and volume resulting from the various actions of angiotensin II and aldosterone reduces the release of renin.

The functions of ADH and the RAAS may seem to be redundant, but this is not the case. Both increase water reabsorption in the kidney, but they counter different osmoregulatory problems. The release of ADH is a response to an increase in blood osmolarity, as when the body is dehydrated from excessive water loss or inadequate water intake. However, a situation that causes an excessive loss of both salt and body fluids—a major wound, for example, or severe diarrhea—will reduce blood volume *without* increasing osmolarity. This will not affect ADH release, but the RAAS will respond to the drop in blood volume and pressure by increasing water and Na^+ reabsorption. Thus, ADH and the RAAS are partners in homeostasis. ADH alone would lower blood Na^+ concentration by stimulating water reabsorption in the kidney, but the RAAS helps maintain the osmolarity of body fluids at the set point by stimulating Na^+ reabsorption.

Another hormone, **atrial natriuretic peptide (ANP)**, opposes the RAAS. The walls of the atria of the heart release ANP in response to an increase in blood volume and pressure. ANP inhibits the release of renin from the JGA, inhibits NaCl reabsorption by the collecting ducts, and reduces aldosterone release from the adrenal glands. These actions lower blood volume and pressure. Thus, ADH, the RAAS, and ANP provide an elaborate system of checks and balances that regulate the kidney's ability to control the osmolarity, salt concentration, volume, and pressure of blood. The precise regulatory role of ANP is an area of active research.

In all animals, certain of the intricate physiological machines we call organs work continuously in maintaining solute and water balance and excreting nitrogenous wastes. The details that we have reviewed in this chapter only hint at the great complexity of the neural and hormonal mechanisms involved in regulating these homeostatic processes.

CONCEPT CHECK 44.5

1. How does alcohol affect regulation of water balance in the body?
2. Why could it be dangerous to drink a very large amount of water in a short period of time?
3. **WHAT IF?** Conn's syndrome is a condition caused by tumors of the adrenal cortex that secrete high amounts of aldosterone in an unregulated manner. What would you expect to be the major symptom of this disorder?
4. **MAKE CONNECTIONS** Compare the activity of renin and ACE in the renin-angiotensin-aldosterone system with that of the protein kinases in a phosphorylation cascade, such as the one shown in Figure 11.10 (p. 215). How are the roles of these enzymes similar and different in the two regulated response pathways?

For suggested answers, see Appendix A.

44 CHAPTER REVIEW

SUMMARY OF KEY CONCEPTS

CONCEPT 44.1

Osmoregulation balances the uptake and loss of water and solutes (pp. 953–958)

- Cells balance water gain and loss through **osmoregulation**, a process based on the controlled movement of solutes between internal fluids and the external environment and on the movement of water, which follows by osmosis. **Osmoconformers** are isoosmotic with their marine environment and do not regulate their **osmolarity**. In contrast, **osmoregulators** control water uptake and loss in a hypoosmotic or hyperosmotic environment, respectively. Water-conserving excretory organs help terrestrial animals to avoid desiccation. Animals that live in temporary waters may be **anhydrobiotic** for one stage of life.

Animal	Inflow/Outflow	Urine
Freshwater fish. Lives in water less concentrated than body fluids; fish tends to gain water, lose salt	Does not drink water Salt in (active transport by gills) H_2O in Salt out	▶ Large volume of urine ▶ Urine is less concentrated than body fluids
Marine bony fish. Lives in water more concentrated than body fluids; fish tends to lose water, gain salt	Drinks water Salt in H_2O out Salt out (active transport by gills)	▶ Small volume of urine ▶ Urine is slightly less concentrated than body fluids
Terrestrial vertebrate. Terrestrial environment; tends to lose body water to air	Drinks water Salt in (by mouth) H_2O and salt out	▶ Moderate volume of urine ▶ Urine is more concentrated than body fluids

- **Transport epithelia** contain specialized epithelial cells that regulate the solute movements required for waste disposal and for tempering changes in body fluids.

? *Under what environmental conditions does water move into a cell by osmosis?*

CONCEPT 44.2

An animal's nitrogenous wastes reflect its phylogeny and habitat (pp. 958–959)

- Protein and nucleic acid metabolism generates **ammonia**. Most aquatic animals excrete ammonia. Mammals and most adult amphibians convert ammonia to the less toxic **urea**, which is excreted with a minimal loss of water. Insects and many reptiles, including birds, convert ammonia to **uric acid**, a mostly insoluble waste excreted in a paste-like urine.
- The kind of nitrogenous waste excreted depends on an animal's evolutionary history and habitat. The amount of nitrogenous waste produced is coupled to the animal's energy budget and amount of dietary protein.

DRAW IT *Construct a table summarizing the three major types of nitrogenous wastes and their relative toxicity, energy content, and associated water loss during excretion.*

CONCEPT 44.3

Diverse excretory systems are variations on a tubular theme (pp. 960–963)

- Most excretory systems carry out **filtration**, **reabsorption**, **secretion**, and **excretion**. The **protonephridia** of the flatworm flame bulb excrete a dilute **filtrate**. An earthworm has pairs of open-ended **metanephridia** in each segment that produce urine. In insects, **Malpighian tubules** function in osmoregulation and removal of nitrogenous wastes. **Kidneys** function in both excretion and osmoregulation in vertebrates.
- Excretory tubules (consisting of **nephrons** and **collecting ducts**) and blood vessels pack the mammalian kidney. Blood pressure forces fluid from blood in the **glomerulus** into the lumen of **Bowman's capsule**. Following reabsorption and secretion, filtrate flows into a collecting duct. The **ureter** conveys urine from the **renal pelvis** to the **urinary bladder**.

? *Given that a typical excretory system selectively absorbs and secretes materials, what function does filtration serve?*

CONCEPT 44.4

The nephron is organized for stepwise processing of blood filtrate (pp. 963–968)

- Within the nephron, selective secretion and reabsorption in the **proximal tubule** alter filtrate volume and composition. The *descending limb* of the **loop of Henle** is permeable to water but not salt; water moves by osmosis into the interstitial fluid. The *ascending limb* is permeable to salt but not water; as the filtrate ascends, salt leaves by diffusion and by active transport. The **distal tubule** and collecting duct regulate K^+ and NaCl levels in body fluids. The collecting duct can respond to hormonal signals to reabsorb more water.
- In a mammalian kidney, a **countercurrent multiplier system** involving the loop of Henle maintains the gradient of salt concentration in the kidney interior. In response to hormonal signals, urine can be concentrated in the collecting duct. Urea, which leaves the collecting duct within the inner medulla, contributes to the osmotic gradient of the kidney.
- Natural selection has shaped the form and function of nephrons in various vertebrates to the osmoregulatory challenges of the animals' habitats. For example, desert mammals, which excrete

the most hyperosmotic urine, have loops of Henle that extend deep into the **renal medulla**, whereas mammals in moist habitats have shorter loops and excrete more dilute urine.

? *How do cortical and juxtamedullary nephrons differ with respect to reabsorbing nutrients and concentrating urine?*

CONCEPT 44.5

Hormonal circuits link kidney function, water balance, and blood pressure (pp. 968–971)

- The posterior pituitary gland releases **antidiuretic hormone (ADH)** when blood osmolarity rises above a set point, such as when water intake is inadequate. ADH increases permeability to water in collecting ducts through an increase in the number of epithelial water channels. When blood pressure or blood volume in the afferent arteriole drops, the **juxtaglomerular apparatus (JGA)** releases renin. **Angiotensin II** formed in response to **renin** constricts arterioles and triggers release of the hormone **aldosterone**, raising blood pressure and reducing the release of renin. This **renin-angiotensin-aldosterone system (RAAS)** has functions that overlap with those of ADH and are opposed by **atrial natriuretic peptide (ANP)**.

? *Why can only some patients with diabetes insipidus be treated effectively with ADH?*

TEST YOUR UNDERSTANDING

LEVEL 1: KNOWLEDGE/COMPREHENSION

1. *Unlike* an earthworm's metanephridia, a mammalian nephron
   a. is intimately associated with a capillary network.
   b. forms urine by changing fluid composition inside a tubule.
   c. functions in both osmoregulation and excretion.
   d. receives filtrate from blood instead of coelomic fluid.
   e. has a transport epithelium.

2. Which process in the nephron is *least* selective?
   a. filtration
   b. reabsorption
   c. active transport
   d. secretion
   e. salt pumping by the loop of Henle

3. Which of the following animals generally has the lowest volume of urine production?
   a. a vampire bat
   b. a salmon in fresh water
   c. a marine bony fish
   d. a freshwater bony fish
   e. a shark inhabiting freshwater Lake Nicaragua

LEVEL 2: APPLICATION/ANALYSIS

4. The high osmolarity of the renal medulla is maintained by all of the following *except*
   a. diffusion of salt from the thin segment of the ascending limb of the loop of Henle.
   b. active transport of salt from the upper region of the ascending limb.
   c. the spatial arrangement of juxtamedullary nephrons.
   d. diffusion of urea from the collecting duct.
   e. diffusion of salt from the descending limb of the loop of Henle.

5. Natural selection should favor the highest proportion of juxtamedullary nephrons in which of the following species?
   a. a river otter
   b. a mouse species living in a tropical rain forest
   c. a mouse species living in a temperate broadleaf forest
   d. a mouse species living in a desert
   e. a beaver

6. African lungfish, which are often found in small stagnant pools of fresh water, produce urea as a nitrogenous waste. What is the advantage of this adaptation?
   a. Urea takes less energy to synthesize than ammonia.
   b. Small stagnant pools do not provide enough water to dilute the toxic ammonia.
   c. The highly toxic urea makes the pool uninhabitable to potential competitors.
   d. Urea forms an insoluble precipitate.
   e. Urea makes lungfish tissue hypoosmotic to the pool.

LEVEL 3: SYNTHESIS/EVALUATION

7. **DRAW IT** Using Figure 44.3 as an example, sketch the exchange of salt (NaCl) and water between a shark and its marine environment.

8. **EVOLUTION CONNECTION**
   Merriam's kangaroo rats (*Dipodomys merriami*) live in North American habitats ranging from moist, cool woodlands to hot deserts. Assuming that natural selection has resulted in differences in water conservation between *D. merriami* populations, propose a hypothesis concerning the relative rates of evaporative water loss by populations that live in moist versus dry environments. Using a humidity sensor to detect evaporative water loss by kangaroo rats, how could you test your hypothesis?

9. **SCIENTIFIC INQUIRY**
   You are exploring kidney function in kangaroo rats. You measure urine volume and osmolarity, as well as the amount of chloride (Cl^-) and urea in the urine. If the water source provided to the animals were switched from tap water to a 2% NaCl solution, what change in urine osmolarity would you expect? How would you determine if this change was more likely due to a change in the excretion of Cl^- or urea?

10. **WRITE ABOUT A THEME**
    Structure and Function In a short essay (100–150 words), compare how membrane structures in the loop of Henle and collecting duct of the mammalian kidney enable water to be recovered from filtrate in the process of osmoregulation.

For selected answers, see Appendix A.

45

Hormones and the Endocrine System

▲ **Figure 45.1 What signals caused this butterfly to grow within the body of a caterpillar?**

KEY CONCEPTS

45.1 Hormones and other signaling molecules bind to target receptors, triggering specific response pathways

45.2 Feedback regulation and antagonistic hormone pairs are common in endocrine systems

45.3 The hypothalamus and pituitary are central to endocrine regulation

45.4 Endocrine glands respond to diverse stimuli in regulating homeostasis, development, and behavior

OVERVIEW

The Body's Long-Distance Regulators

To say that a butterfly, such as the anise swallowtail (*Papilio zelicaon*) in **Figure 45.1**, was once a caterpillar is only partly true. The adult cells that form the butterfly begin growing in the embryo. Within the larval caterpillar, they are nourished as islands of tissues that will eventually become the eyes, wings, brain, and other structures of the butterfly. Once the plump, crawling caterpillar becomes a stationary pupa, the adult cells take over. They complete their program of development, while many larval tissues undergo programmed cell death. The end result is a butterfly, a delicate, free-flying adult that bears little resemblance to the larval and pupal forms from which it arose.

What brings about such a complete change of body form, or *metamorphosis*? The answer for this and many other biological processes is a type of molecule called a **hormone** (from the Greek *horman*, to excite). In animals, hormones are secreted into the extracellular fluid, circulate in the hemolymph or blood, and communicate regulatory messages throughout the body. In the case of the caterpillar, a hormone called **ecdysteroid** stimulates the growth of adult cells, the programmed death of larval cells, and even the behaviors that bring about the motionless pupal stage. Communication within the body by ecdysteroid and other hormones also regulates the timing of metamorphosis and ensures that different parts of the swallowtail's adult body develop in unison.

Each hormone has specific receptors in the body. Although a given hormone can reach all cells of the body, only some cells have receptors for that hormone. A hormone elicits a response—such as a change in metabolism—only from specific *target cells*, those that have the matching receptor. Cells lacking a receptor for that particular hormone are unaffected.

Chemical signaling by hormones is the function of the **endocrine system**, one of the two basic systems of communication and regulation throughout the body. Hormones secreted by endocrine cells regulate reproduction, development, energy metabolism, growth, and behavior. The other major communication and control system is the **nervous system**, a network of specialized cells—neurons—that transmit signals along dedicated pathways. These signals in turn regulate neurons, muscle cells, and endocrine cells. Because signaling by neurons can regulate the release of hormones, the nervous and endocrine systems often overlap in function.

In this chapter, we'll begin with an overview of the different types of chemical signaling in animals and the ways in which the activities of the endocrine and nervous systems are coordinated. We will then explore how hormones regulate target cells, how hormone secretion is regulated, and how hormones help maintain homeostasis. We'll conclude by examining the role of hormones in regulating growth, development, and reproduction, topics we'll return to in Chapters 46 and 47.

CONCEPT 45.1

Hormones and other signaling molecules bind to target receptors, triggering specific response pathways

Endocrine signaling is just one of several ways information is transmitted between animal cells. Let's consider the similarities and differences in these various signaling processes.

Intercellular Communication

The ways in which signals are transmitted between animal cells are often classified by two criteria: the type of secreting cell and the route taken by the signal in reaching its target.

Endocrine Signaling

As illustrated in **Figure 45.2a**, hormones secreted into extracellular fluids by endocrine cells reach target cells via the bloodstream (or hemolymph). Endocrine signaling maintains homeostasis, mediates responses to environmental stimuli, and regulates growth and development. For example, hormones coordinate the body's responses to stress, dehydration, and low blood glucose levels. They also trigger behavioral and physical changes underlying sexual maturity and reproduction.

Paracrine and Autocrine Signaling

Many types of cells produce and secrete **local regulators**, molecules that act over short distances and reach their target cells solely by diffusion. Cytokines, for example, are local regulators that enable communication between immune cells (see Figures 43.16 and 43.18). Depending on the target cell, signaling by local regulators can be either paracrine or autocrine. In **paracrine** signaling (from the Greek *para*, to one side of), target cells lie near the secreting cell **(Figure 45.2b)**. In **autocrine** signaling (from the Greek *auto*, self), the target cell is the secreting cell itself **(Figure 45.2c)**. As we will discuss later in this chapter, paracrine and autocrine signaling play roles in many physiological processes, including blood pressure regulation, nervous system function, and reproduction.

Synaptic and Neuroendocrine Signaling

Secreted molecules are crucial for two types of signaling by neurons. In *synaptic signaling*, neurons form specialized junctions called synapses with target cells, such as other neurons and muscle cells. At synapses, neurons secrete molecules called **neurotransmitters** that diffuse a very short distance to bind to receptors on the target cells **(Figure 45.2d)**. Neurotransmitters are central to sensation, memory, cognition, and movement, as we will explore in Chapters 48–50.

In *neuroendocrine signaling*, specialized neurons called neurosecretory cells secrete molecules that diffuse from nerve cell endings into the bloodstream **(Figure 45.2e)**. These molecules, which travel through the bloodstream to target cells, are a class of hormone called **neurohormones**. One example is antidiuretic hormone, also known as vasopressin, a hormone essential to kidney function and water balance (see Chapter 44).

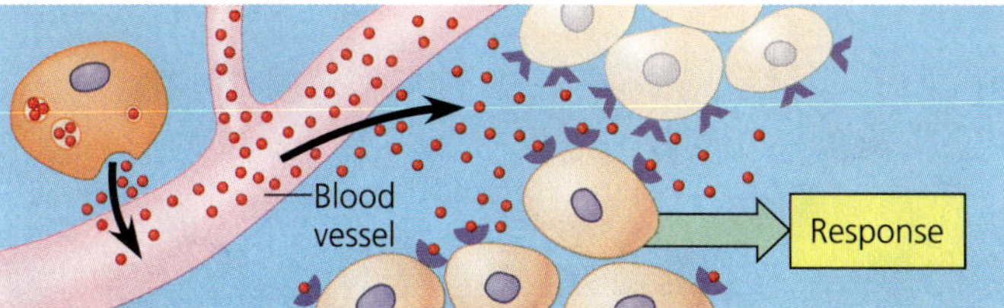

(a) In **endocrine signaling**, secreted molecules diffuse into the bloodstream and trigger responses in target cells anywhere in the body.

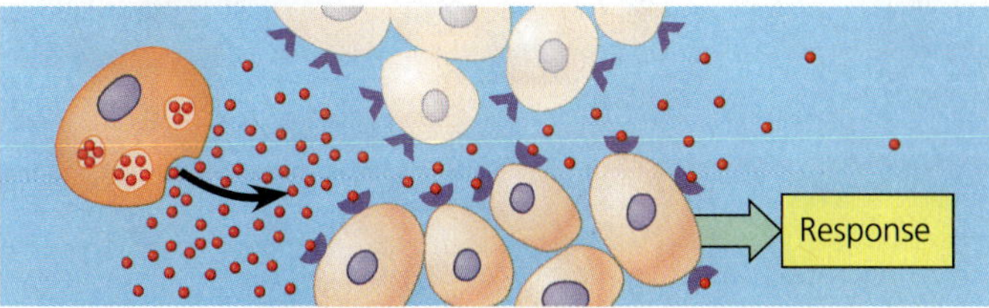

(b) In **paracrine signaling**, secreted molecules diffuse locally and trigger a response in neighboring cells.

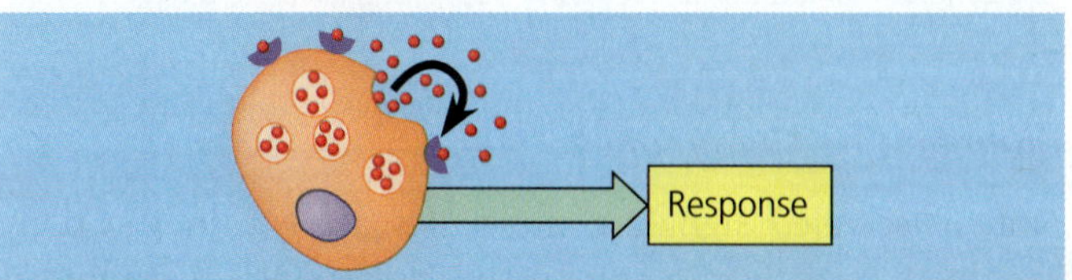

(c) In **autocrine signaling**, secreted molecules diffuse locally and trigger a response in the cells that secrete them.

(d) In **synaptic signaling**, neurotransmitters diffuse across synapses and trigger responses in cells of target tissues (neurons, muscles, or glands).

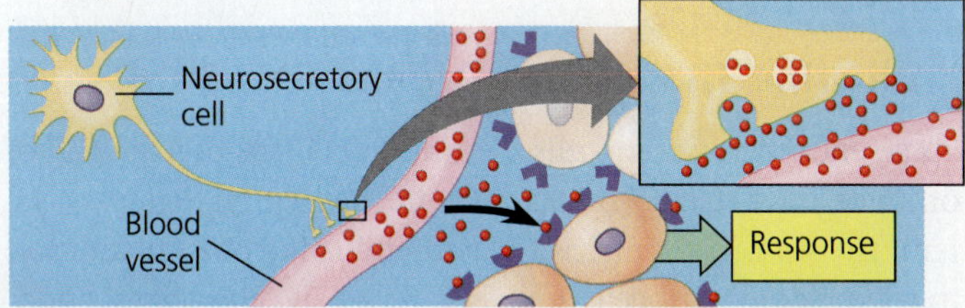

(e) In **neuroendocrine signaling**, neurohormones diffuse into the bloodstream and trigger responses in target cells anywhere in the body.

▲ **Figure 45.2 Intercellular communication by secreted molecules.** In each type of signaling, secreted molecules (•) bind to a specific receptor protein (◡) expressed by target cells. Some receptors are located inside cells, but for simplicity here, all are drawn on the cell surface.

▲ **Figure 45.3 Signaling by pheromones.** Using their lowered antennae, these Asian army ants (*Leptogenys distinguenda*) follow a pheromone-marked trail as they carry pupae and larvae to a new nest site.

Signaling by Pheromones

Not all secreted signaling molecules act within the body. Members of the same animal species sometimes communicate via **pheromones**, chemicals that are released into the external environment. For example, when a foraging ant discovers a new food source, it marks its path back to the nest with a pheromone. Ants also use pheromones for guidance when a colony migrates to a new location **(Figure 45.3)**.

Pheromones serve a wide range of functions that include defining territories, warning of predators, and attracting potential mates. The giant silk moth (*Antheraea polyphemus*) provides a noteworthy example: The sex pheromone released into the air by a female enables her to attract a male of the species from up to 4.5 km away.

Endocrine Tissues and Organs

Some endocrine system cells are found in organs that are part of other organ systems. For example, in the human digestive system, the stomach contains isolated endocrine cells in addition to the predominant cell and tissue types. In other cases, endocrine cells are grouped in ductless organs called **endocrine glands**, such as the thyroid and parathyroid glands of the neck. The various human glands and organs with endocrine function are illustrated in **Figure 45.4**, which will serve as a useful point of reference as you move through the chapter.

Note that endocrine glands secrete hormones directly into the surrounding fluid. Endocrine glands thus contrast with *exocrine glands*, such as salivary glands, which have ducts that carry secreted substances onto body surfaces or into body cavities. This distinction is reflected in their names: The Greek *endo* ("within") and *exo* ("out of") reflect secretion into or out of body fluids, while *crine* (from the Greek for "separate") reflects movement away from the secreting cell.

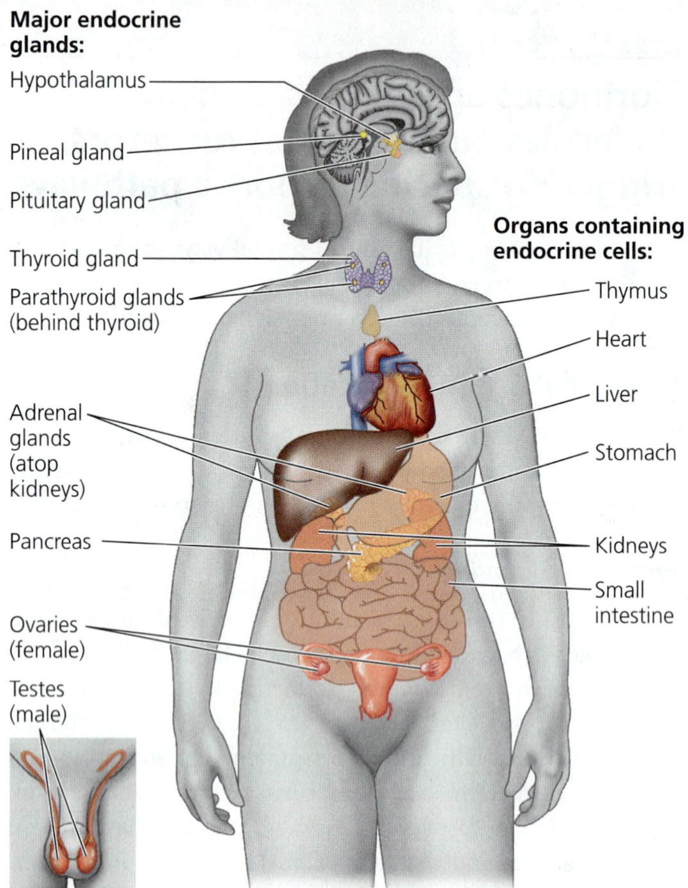

▲ **Figure 45.4 Major human endocrine glands.**

Chemical Classes of Hormones

Hormone molecules vary substantially in size and chemical properties. Some of these differences are apparent in examples drawn from the three major chemical classes of hormones: polypeptides (proteins and peptides), steroids, and amines **(Figure 45.5)**. The polypeptide hormone insulin is made up of two polypeptide chains. Like most hormones in this group, insulin is formed by cleavage of one long polypeptide chain. Steroid hormones, such as cortisol and ecdysteroid, are lipids that contain four fused carbon rings. All are derived from the steroid cholesterol (see Figure 5.14). Epinephrine and thyroxine are amine hormones, each synthesized from a single amino acid, either tyrosine or tryptophan.

As Figure 45.5 indicates, hormones vary in their solubility in aqueous and lipid-rich environments. Polypeptides and most amine hormones are water-soluble. Being insoluble in lipids, these hormones cannot pass through the plasma membranes of cells. Instead, they bind to cell-surface receptors that relay information to the nucleus through intracellular pathways. In contrast, steroid hormones, as well as other largely nonpolar (hydrophobic) hormones, such as thyroxine, are lipid-soluble and can pass through cell membranes readily. Receptors for lipid-soluble hormones typically reside in the cytoplasm or nucleus.

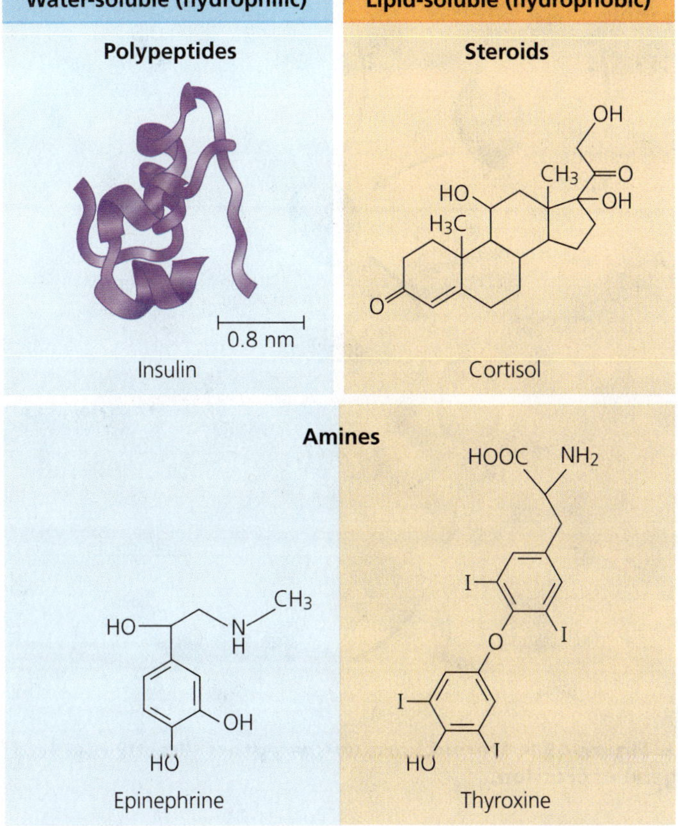

▲ **Figure 45.5 Hormones differ in structure and solubility.**

MAKE CONNECTIONS *The biosynthesis of epinephrine involves breaking just one carbon-carbon bond in the amino acid tyrosine (see Figure 5.16, p. 79). Which bond is it?*

Cellular Response Pathways

There are several differences between the response pathways for water-soluble and lipid-soluble hormones. One difference is the location of the target cells' signal receptors **(Figure 45.6)**. Water-soluble hormones are secreted by exocytosis, travel freely in the bloodstream, and bind to cell-surface signal receptors. Binding of such hormones to receptors induces changes in cytoplasmic molecules and sometimes alters gene transcription (synthesis of messenger RNA molecules). In contrast, lipid-soluble hormones diffuse out across the membranes of endocrine cells. Outside the cell, they bind to transport proteins that keep them soluble in the aqueous environment of the bloodstream. Upon leaving the bloodstream, they diffuse into target cells, bind to intracellular signal receptors, and trigger changes in gene transcription.

To follow the distinct cellular responses to water-soluble and lipid-soluble hormones, we'll examine the two response pathways in turn.

Pathway for Water-Soluble Hormones

The binding of a water-soluble hormone to a signal receptor protein triggers events at the plasma membrane that result in a cellular response. The response may be the activation of an enzyme, a change in the uptake or secretion of specific molecules,

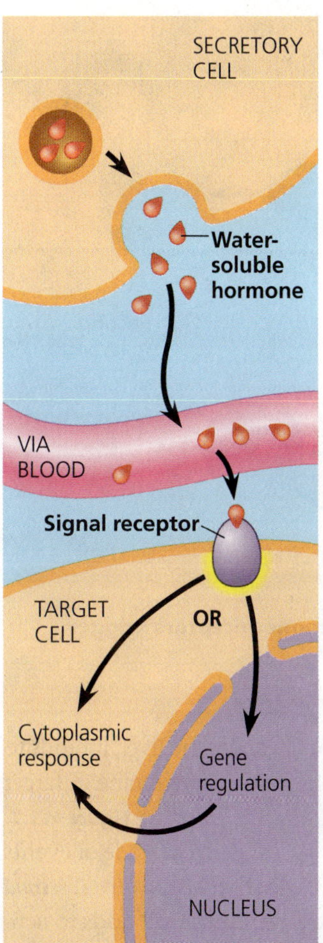

(a) Receptor in plasma membrane

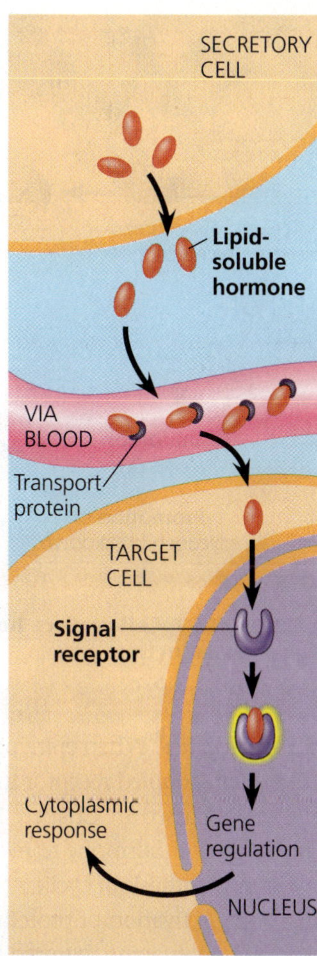

(b) Receptor in cell nucleus

▲ **Figure 45.6 Receptor location varies with hormone type.** **(a)** A water-soluble hormone binds to a signal receptor protein on the surface of a target cell. This interaction triggers events that lead to either a change in cytoplasmic function or a change in gene transcription in the nucleus. **(b)** A lipid-soluble hormone penetrates the target cell's plasma membrane and binds to an intracellular signal receptor, either in the cytoplasm or in the nucleus (shown here). The hormone-receptor complex acts as a transcription factor, typically activating gene expression.

? *Suppose you were studying a cell's response to a particular hormone, and you observed that the cell continued to respond to the hormone even when treated with a chemical that blocks transcription. What could you surmise about the hormone and its receptor?*

or a rearrangement of the cytoskeleton. In addition, some cell-surface receptors cause proteins in the cytoplasm to move into the nucleus and alter transcription of specific genes.

The series of changes in cellular proteins that converts the extracellular chemical signal to a specific intracellular response is called **signal transduction**. As described in Chapter 11, a signal transduction pathway typically involves multiple steps, each involving specific molecular interactions.

To explore the role of signal transduction in hormone signaling, consider one response to short-term stress. When you find yourself in a stressful situation, perhaps running to catch a bus, your adrenal glands secrete **epinephrine**, a hormone also

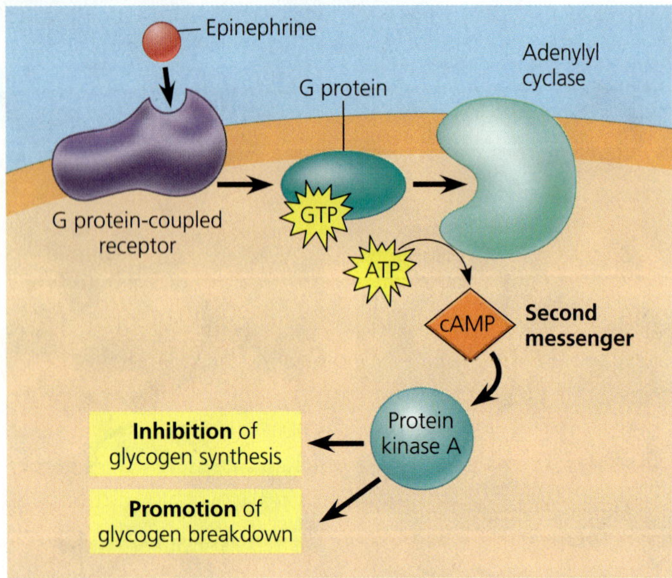

▲ **Figure 45.7 Cell-surface hormone receptors trigger signal transduction.**

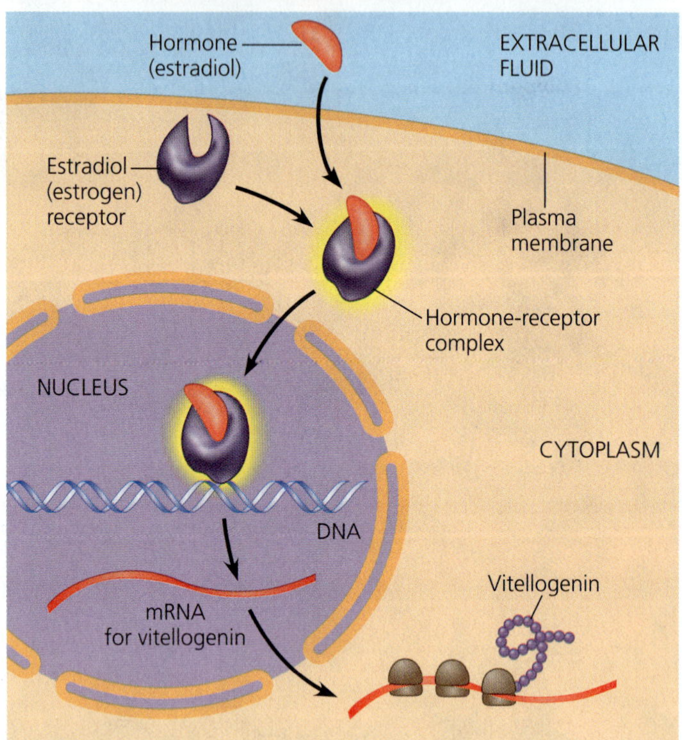

▲ **Figure 45.8 Steroid hormone receptors directly regulate gene expression.**

called *adrenaline*. When epinephrine reaches the liver, it binds to a G protein-coupled receptor in the plasma membrane of target cells, as discussed in Chapter 11 and reviewed in **Figure 45.7**. The binding of hormone to receptor triggers a cascade of events involving synthesis of cyclic AMP (cAMP) as a short-lived *second messenger*. Activation of protein kinase A by cAMP leads to activation of an enzyme required for glycogen breakdown and inactivation of an enzyme necessary for glycogen synthesis. The net result is that the liver releases glucose into the bloodstream, providing the fuel you need to chase the departing bus.

Pathway for Lipid-Soluble Hormones

Intracellular receptors for lipid-soluble hormones perform the entire task of transducing a signal within a target cell. The hormone activates the receptor, which then directly triggers the cell's response. In most cases, the response to a lipid-soluble hormone is a change in gene expression.

Steroid hormone receptors are located in the cytosol prior to binding to a hormone. When a steroid hormone binds to its cytosolic receptor, a hormone-receptor complex forms, which moves into the nucleus. There, the receptor portion of the complex alters transcription of particular genes by interacting with a specific DNA-binding protein or response element in the DNA (see Figure 18.9). Consider, for example, estrogens, steroid hormones necessary for female reproductive function in vertebrates. In female birds and frogs, estradiol, a form of estrogen, has a specific receptor in liver cells. Binding of estradiol to this receptor activates transcription of the gene for the protein vitellogenin **(Figure 45.8)**. Following translation of the messenger RNA, vitellogenin is secreted and transported in the blood to the reproductive system, where it is used to produce egg yolk.

Thyroxine, vitamin D, and other lipid-soluble hormones that are not steroid hormones have receptors that are typically located in the nucleus. These receptors bind hormone molecules that diffuse from the bloodstream across both the plasma membrane and nuclear envelope. Once bound by a hormone, the receptor binds to specific sites in the cell's DNA and stimulates the transcription of specific genes.

There is now substantial evidence that estrogens and some other lipid-soluble hormones sometimes trigger responses at the cell surface without entering the nucleus. How and when these responses arise are currently the subjects of active study.

Multiple Effects of Hormones

Many hormones elicit more than one type of response in the body. The effects brought about by a particular hormone can vary if target cells differ in the molecules that receive or produce the response to that hormone. Consider the effects of epinephrine in mediating the body's response to short-term stress **(Figure 45.9)**. Epinephrine simultaneously triggers glycogen breakdown in the liver, increased blood flow to major skeletal muscles, and decreased blood flow to the digestive tract. These varied effects enhance the rapid reactions of the body in emergencies.

Tissues vary in their response to epinephrine because they vary in their receptors or in their signal transduction pathways. Target cell recognition of epinephrine involves G protein-coupled receptors. Liver cells have a β-type epinephrine receptor that activates the enzyme protein kinase A, which in turn regulates enzymes in glycogen metabolism **(Figure 45.9a)**. In blood vessels supplying skeletal muscle, the same kinase activated by the same epinephrine receptor inactivates a muscle-specific enzyme. The result is smooth muscle relaxation and hence increased blood flow **(Figure 45.9b)**. In contrast, intestinal blood vessels have an α-type epinephrine receptor **(Figure 45.9c)**. Rather than activate protein kinase A, the α receptor triggers a distinct signaling pathway involving a different G protein and different enzymes. The result is smooth muscle contraction and restricted blood flow to the intestines.

Lipid-soluble hormones often exert different effects on different target cells as well. For example, the estrogen that stimulates a bird's liver to synthesize the yolk protein vitellogenin also stimulates its reproductive system to synthesize proteins that form the egg white.

Signaling by Local Regulators

Recall that local regulators are secreted molecules that link neighboring cells (paracrine signaling) or directly regulate the secreting cell (autocrine signaling). Once secreted, local regulators act on their target cells within seconds or even milliseconds, eliciting responses more quickly than do hormones. Nevertheless, the pathways by which local regulators trigger responses are the same as those activated by hormones. (Although the definition of hormones is sometimes broadened to include local regulators, in this chapter we use *hormone* to refer to chemicals that reach target cells through the bloodstream or hemolymph.)

Several types of chemical compounds function as local regulators. Polypeptide local regulators include cytokines, as mentioned, and also most **growth factors**, which stimulate cell proliferation and differentiation. Many types of cells grow, divide, and develop normally only when growth factors are present in their extracellular environment.

The gas **nitric oxide (NO)** functions in the body as both a neurotransmitter and a local regulator. When the level of oxygen (O_2) in the blood falls, endothelial cells in blood vessel walls synthesize and release NO. Nitric oxide activates an enzyme that relaxes the surrounding smooth muscle cells, resulting in vasodilation, which improves blood flow to tissues. Highly reactive and potentially toxic, NO usually triggers changes in a target cell within a few seconds of contact and then breaks down. In human males, NO's ability to promote vasodilation enables sexual function by increasing blood flow into the penis, producing an erection. The drug Viagra (sildenafil citrate), a treatment for male erectile dysfunction, sustains an erection by prolonging activity of the NO response pathway.

A group of local regulators called **prostaglandins** are modified fatty acids. They are so named because they were first discovered in prostate gland secretions that contribute to semen. Prostaglandins are produced by many cell types and have varied activities. In semen that reaches the reproductive tract of a female, prostaglandins stimulate the smooth muscles of the female's uterine wall to contract, helping sperm reach an egg. At the onset of childbirth, prostaglandin-secreting cells of the placenta cause the nearby muscles of the uterus to become more excitable, helping to induce labor (see Figure 46.18).

In the immune system, prostaglandins promote fever and inflammation and also intensify the sensation of pain. The anti-inflammatory and pain-relieving effects of aspirin and ibuprofen are due to the inhibition of prostaglandin synthesis by these drugs. Prostaglandins also help regulate the aggregation of platelets, one step in the formation of blood clots. Because blood clots can cause a heart attack by blocking blood flow in vessels that supply the heart (see Chapter 42), some physicians recommend that people at risk for a heart attack take aspirin on a regular basis. However, because prostaglandins also help maintain a protective lining in the stomach, long-term aspirin therapy can cause debilitating stomach irritation.

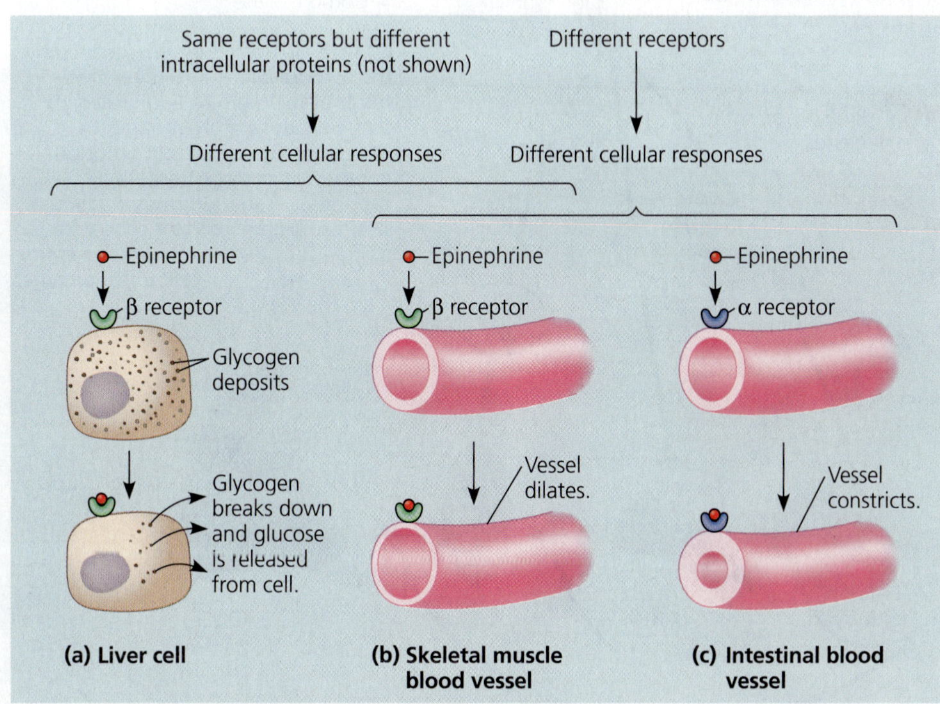

▲ **Figure 45.9 One hormone, different effects.** Epinephrine, the primary "fight-or-flight" hormone, produces different responses in different target cells. Target cells with the same receptor exhibit different responses if they have different signal transduction pathways and/or effector proteins; compare (a) with (b). Responses of target cells may also differ if they have different receptors for the hormone; compare (b) with (c).

Coordination of Neuroendocrine and Endocrine Signaling

In all animals but the simplest invertebrates, the endocrine and nervous systems act coordinately to control reproduction and development. As an example, we'll explore the life cycle of the butterfly, a process highlighted earlier in the chapter.

A butterfly larva grows in stages. Because its exoskeleton cannot stretch, the larva must periodically molt, shedding the old exoskeleton and secreting a new one. The signals that direct molting originate in the brain (**Figure 45.10**). There, neurosecretory cells produce *prothoracicotropic hormone (PTTH)*, a polypeptide neurohormone. In response to PTTH, a pair of endocrine glands behind the brain release ecdysteroid. Ecdysteroid triggers each successive molt, as well as the metamorphosis of larva into butterfly during the final molt.

Given that ecdysteroid triggers both molting and metamorphosis, what determines when metamorphosis takes place? The answer is a third molecule, *juvenile hormone*, secreted by another pair of endocrine glands behind the brain. As its name suggests, one of the many functions of juvenile hormone is to maintain larval (juvenile) characteristics. Juvenile hormone modulates the activity of ecdysteroid. As long as the level of juvenile hormone is high, ecdysteroid stimulates larval molting. When the juvenile hormone level drops, ecdysteroid-induced molting instead produces the pupal form, within which metamorphosis occurs.

Knowledge of endocrine signaling in insects has important applications for agricultural pest control. For example, synthetic chemicals that can bind to the ecdysteroid receptor cause insect larvae to molt prematurely and die.

CONCEPT CHECK 45.1

1. How do response mechanisms in target cells differ for water-soluble and lipid-soluble hormones?
2. In what way does one activity described for prostaglandins resemble that of a pheromone?
3. **MAKE CONNECTIONS** What parallels in properties and effects can you identify between epinephrine and the plant hormone auxin (see Concept 39.2, pp. 827–829)?

For suggested answers, see Appendix A.

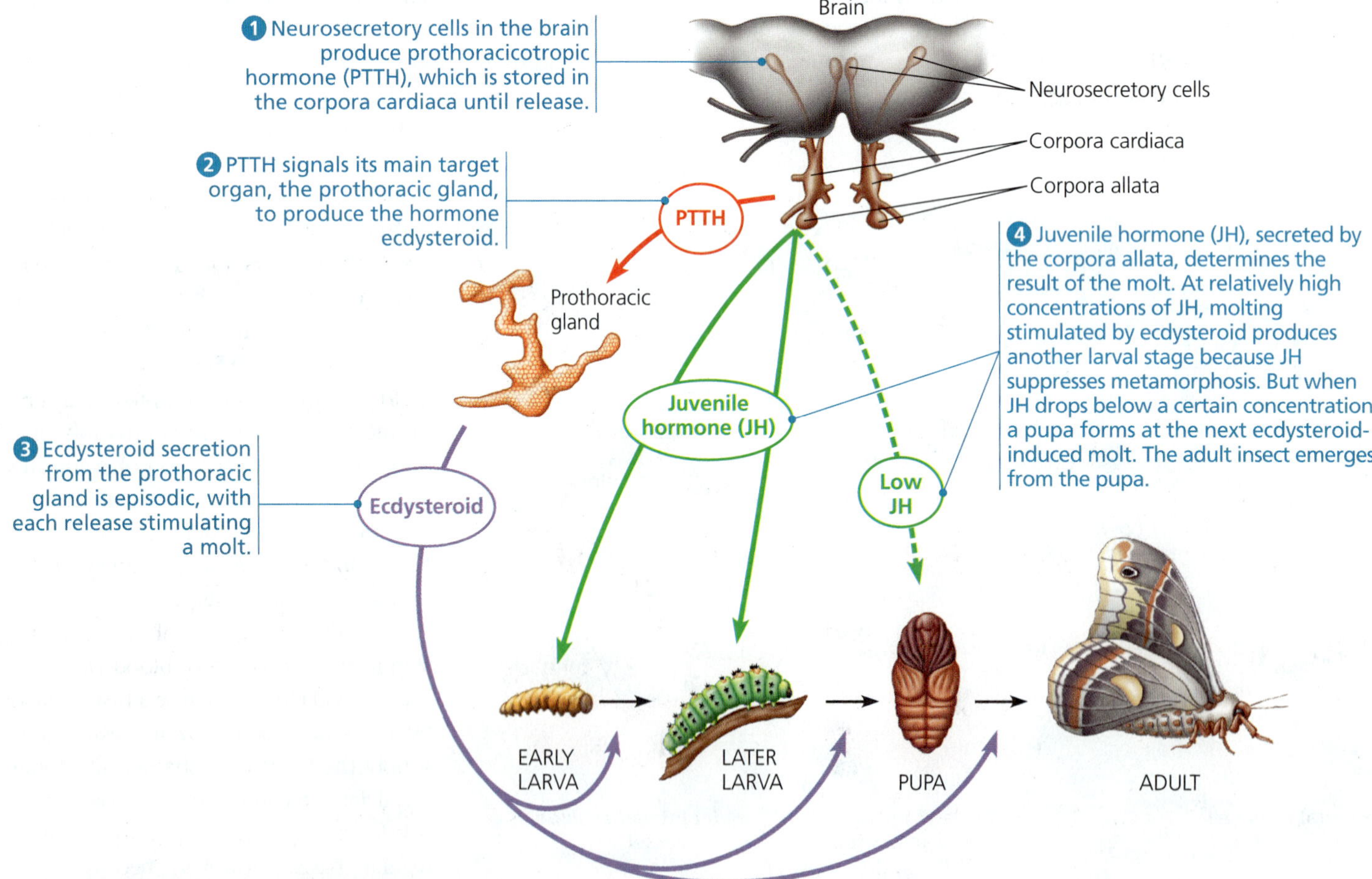

▲ **Figure 45.10 Regulation of insect development and metamorphosis.** As shown here for a moth, most insects go through a series of larval stages, with each molt (shedding of the old exoskeleton) leading to a larger larva. Molting of the final larval stage gives rise to a pupa, in which metamorphosis produces the adult form of the insect. Neurohormones and hormones control the progression of stages.

CONCEPT 45.2

Feedback regulation and antagonistic hormone pairs are common in endocrine systems

So far, we have explored forms of intercellular signaling as well as hormone structure, recognition, and response. We turn now to considering how regulatory pathways that control hormone secretion are organized.

Simple Hormone Pathways

In examining the regulation of hormone secretion, we begin with two basic types of organization—simple endocrine and simple neuroendocrine pathways. In a *simple endocrine pathway*, endocrine cells respond directly to an internal or environmental stimulus by secreting a particular hormone **(Figure 45.11)**. The hormone travels in the bloodstream to target cells, where it interacts with its specific receptors. Signal transduction within target cells brings about a physiological response.

In the example of a simple endocrine pathway shown in Figure 45.11, the stimulus is the release of the acidic contents of the stomach into the duodenum (the first part of the small intestine). Low pH in the duodenum stimulates certain endocrine cells there, called S cells, to secrete the hormone *secretin*. Secretin enters the bloodstream and travels to the **pancreas**, a gland located behind the stomach (see Figure 45.4). Target cells in the pancreas then release bicarbonate into ducts leading to the duodenum, where it raises the pH.

In a *simple neuroendocrine pathway*, the stimulus is received by a sensory neuron, which stimulates a neurosecretory cell **(Figure 45.12)**. The neurosecretory cell then secretes a neurohormone, which diffuses into the bloodstream and travels to target cells. Such a pathway regulates milk release during nursing in mammals. Suckling by an infant stimulates sensory neurons in the nipples, generating signals in the nervous system that reach the hypothalamus. Nerve impulses from the hypothalamus then trigger the release of the neurohormone **oxytocin** from the posterior pituitary gland. In response to circulating oxytocin, the mammary glands secrete milk.

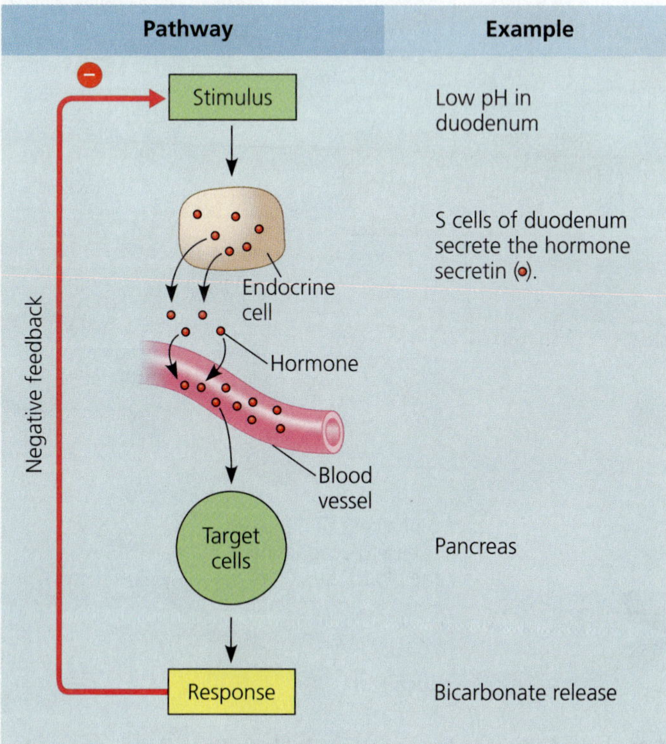

▲ **Figure 45.11 A simple endocrine pathway.** Endocrine cells respond to a change in some internal or external variable—the stimulus—by secreting hormone molecules that trigger a specific response by target cells. In the case of secretin signaling, the simple endocrine pathway is self-limiting because the response to secretin (bicarbonate release) reduces the stimulus (low pH) through negative feedback.

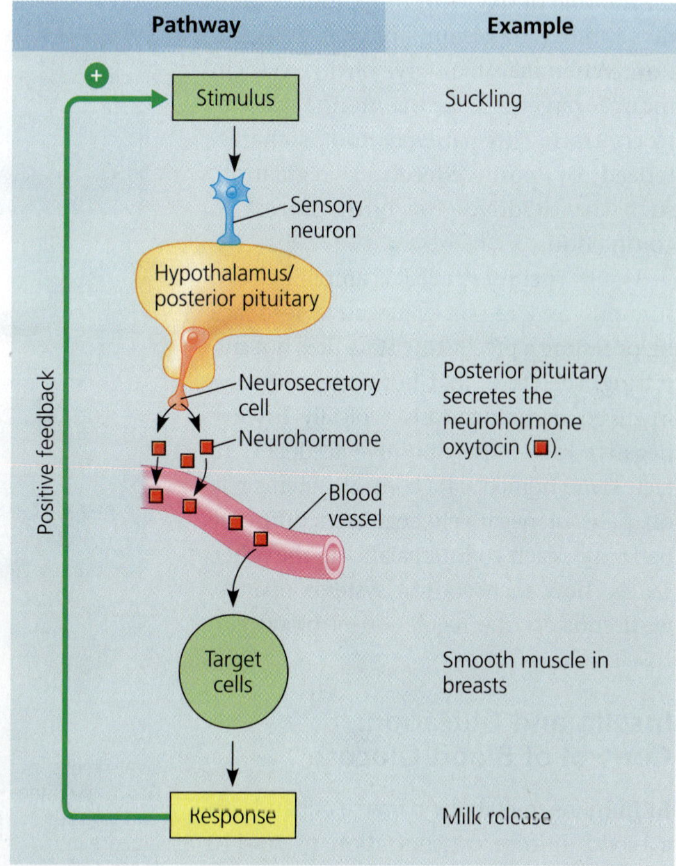

▲ **Figure 45.12 A simple neuroendocrine pathway.** Sensory neurons respond to a stimulus by sending nerve impulses to a neurosecretory cell, triggering secretion of a neurohormone. Upon reaching its target cells via the bloodstream, the neurohormone binds to its receptor, triggering signal transduction that results in a specific response. In the neuroendocrine pathway for oxytocin signaling, the response increases the stimulus, forming a positive-feedback loop that amplifies signaling in the pathway.

Feedback Regulation

A feedback loop linking the response back to the initial stimulus is characteristic of control pathways. For many hormones, the response pathway involves **negative feedback**, a loop in which the response reduces the initial stimulus. In the case of secretin signaling (see Figure 45.11), the release of bicarbonate by the pancreas increases pH in the intestine, eliminating the stimulus and thereby shutting off the pathway. By decreasing or abolishing hormone signaling, negative-feedback regulation prevents excessive pathway activity.

Whereas negative feedback dampens a stimulus, **positive feedback** reinforces a stimulus, leading to an even greater response. Consider, for instance, the oxytocin pathway outlined in Figure 45.12. In response to the circulating oxytocin, the mammary glands secrete milk. Milk released in response to the oxytocin leads to more suckling and therefore more stimulation. Activation of the pathway is sustained until the baby stops suckling.

The role of oxytocin in reproduction is not limited to mammary gland regulation. When mammals give birth, oxytocin induces target cells in the uterine muscles to contract. This pathway, too, is characterized by positive-feedback regulation, such that it drives the birth process to completion.

While positive feedback amplifies both stimulus and response, negative feedback helps restore a preexisting state. It is not surprising, therefore, that hormone pathways involved in homeostasis typically involve negative rather than positive feedback. In fact, some homeostatic control systems rely on pairs of negatively regulated hormone pathways, each counterbalancing the other. To see how such control systems operate, we'll consider the regulation of blood glucose levels.

Insulin and Glucagon: Control of Blood Glucose

In humans, metabolic balance depends on a blood glucose concentration of 70–110 mg/100 mL. Because glucose is a major fuel for cellular respiration and a key source of carbon skeletons for biosynthesis, maintaining blood glucose concentrations near this normal range is critical.

Two antagonistic (opposing) hormones, insulin and glucagon, regulate the concentration of glucose in the blood **(Figure 45.13)**. Each of these hormones operates in a simple endocrine pathway regulated by negative feedback. When blood glucose rises above the normal range, release of **insulin** triggers uptake of glucose from the blood into body cells, decreasing the blood glucose concentration. When blood glucose drops below the normal range, the release of **glucagon** promotes the release of glucose into the blood from energy stores, such as liver glycogen, increasing the blood glucose concentration. Because insulin and glucagon have opposing effects, the combined activity of these two hormones tightly controls the concentration of glucose in the blood.

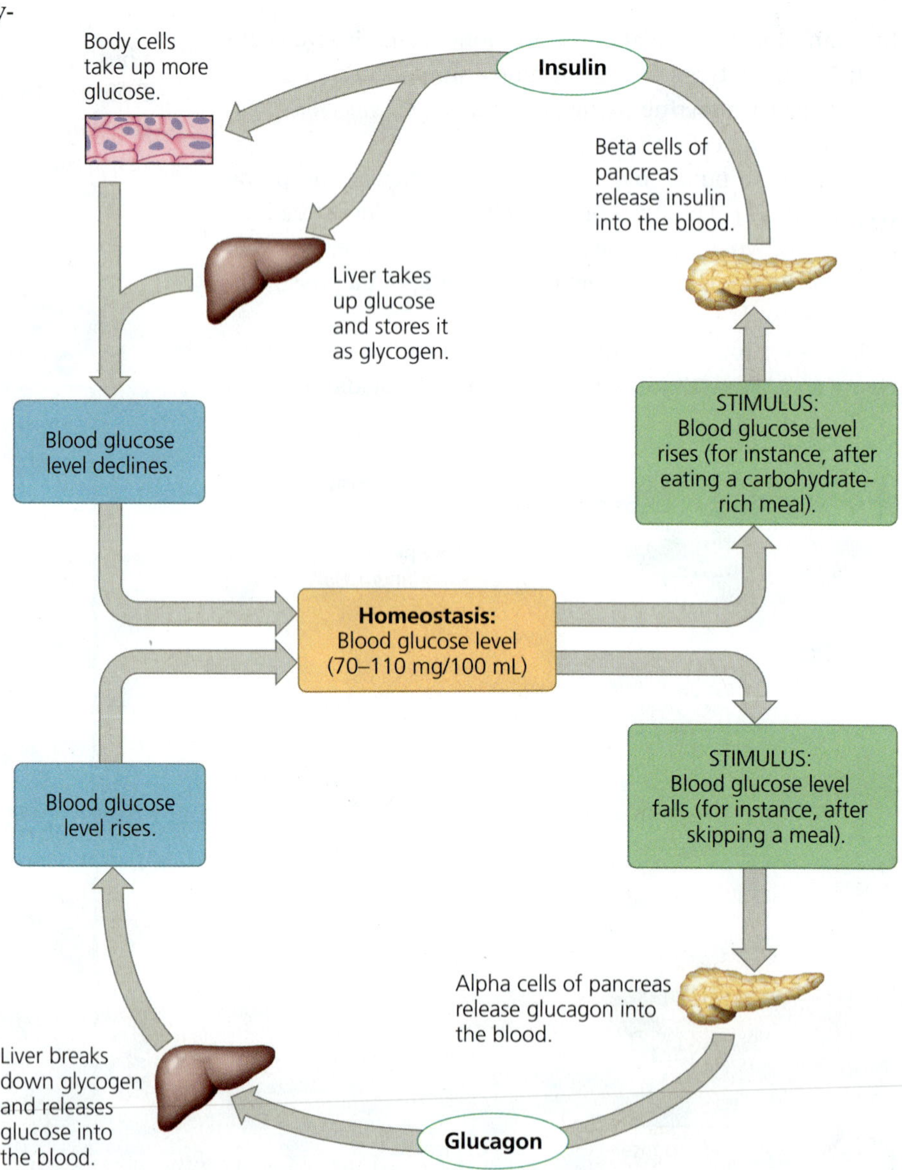

▲ **Figure 45.13 Maintenance of glucose homeostasis by insulin and glucagon.** The antagonistic effects of insulin and glucagon help keep blood glucose levels in the normal range.

BioFlix Visit the Study Area at **www.masteringbiology.com** for the BioFlix® 3-D Animation on Homeostasis: Regulating Blood Sugar.

Glucagon and insulin are produced in the pancreas. Scattered throughout this organ are clusters of endocrine cells called pancreatic islets. Each pancreatic islet has *alpha cells*, which make glucagon, and *beta cells*, which make insulin. Like all hormones, insulin and glucagon are secreted into the interstitial fluid and enter the circulatory system.

Overall, hormone-secreting cells make up only 1–2% of the mass of the pancreas. Other cells in the pancreas produce and secrete bicarbonate ions and digestive enzymes. These exocrine secretions are released into small ducts that empty into the pancreatic duct, which leads to the small intestine. Thus, the pancreas is both an endocrine gland and an exocrine gland and has functions in both the endocrine and digestive systems.

Target Tissues for Insulin and Glucagon

Insulin lowers blood glucose levels by stimulating nearly all body cells outside the brain to take up glucose from the blood. (Brain cells can take up glucose without insulin, so the brain almost always has access to circulating fuel.) Insulin also decreases blood glucose by slowing glycogen breakdown in the liver and inhibiting the conversion of glycerol (from fats) and amino acids to glucose.

Glucagon influences blood glucose levels mainly through its effects on target cells in the liver. The liver, skeletal muscles, and adipose tissues store large amounts of fuel. The liver and muscles store sugar as glycogen, whereas cells in adipose tissue convert sugars to fats. When the blood glucose level decreases to a level at or below the normal range (70–110 mg/100 mL), a primary effect of glucagon is to signal liver cells to increase glycogen hydrolysis, convert amino acids and glycerol to glucose, and release glucose into the bloodstream. The net result is a return of the blood glucose level to the normal range.

The antagonistic effects of glucagon and insulin are vital to managing fuel storage and consumption by body cells. For both hormones, as we've mentioned, the liver is a critical target. Recall from Chapter 41 that nutrients absorbed by blood vessels of the small intestine are transported directly to the liver by the hepatic portal vein. Within the liver, glucagon and insulin regulate nutrient processing in ways that support glucose homeostasis. However, glucose homeostasis also relies on responses to glucagon and insulin elsewhere in the body as well as responses to other hormones—growth hormone and glucocorticoids—discussed later in this chapter.

In discussing the role of insulin and glucagon in glucose homeostasis, we have focused exclusively on a healthy metabolic state. However, a number of disorders can disrupt glucose homeostasis with potentially serious consequences, especially for the heart, blood vessels, eyes, and kidneys. We'll discuss the best known and most prevalent of these disorders—diabetes mellitus—next.

Diabetes Mellitus

The disease **diabetes mellitus** is caused by a deficiency of insulin or a decreased response to insulin in target tissues. Blood glucose levels rise, but cells are unable to take up enough glucose to meet metabolic needs. Instead, fat becomes the main substrate for cellular respiration. In severe cases, acidic metabolites formed during fat breakdown accumulate in the blood, threatening life by lowering blood pH and depleting sodium and potassium ions from the body.

In people with diabetes mellitus, the level of glucose in blood may exceed the capacity of the kidneys to reabsorb this nutrient. Glucose that remains in the kidney filtrate is excreted. For this reason, the presence of sugar in urine is one test for this disorder. As glucose is concentrated in the urine, more water is excreted along with it, resulting in excessive volumes of urine. *Diabetes* (from the Greek *diabainein*, to pass through) refers to this copious urination; and *mellitus* (from the Greek *meli*, honey) refers to the presence of sugar in urine. (*Diabetes insipidus*, discussed in Chapter 44, is a rare disorder of kidney function that results in large volumes of dilute urine but no major disruption in glucose metabolism.)

There are two main types of diabetes mellitus. Each is marked by high blood glucose, but with very different causes. *Type 1 diabetes*, or insulin-dependent diabetes, is an autoimmune disorder in which the immune system destroys the beta cells of the pancreas. Type 1 diabetes, which usually appears during childhood, destroys the person's ability to produce insulin. Treatment consists of insulin, typically injected several times daily. In the past, insulin was extracted from animal pancreases, but now human insulin can be obtained from genetically engineered bacteria, a relatively inexpensive source (see Figure 20.2). Stem cell research may someday provide a cure for type 1 diabetes by generating replacement beta cells that restore insulin production by the pancreas.

Type 2 diabetes, or non-insulin-dependent diabetes, is characterized by a failure of target cells to respond normally to insulin. Insulin is produced, but target cells fail to take up glucose from the blood, and blood glucose levels remain elevated. Although heredity can play a role in type 2 diabetes, excess body weight and lack of exercise significantly increase the risk. This form of diabetes generally appears after age 40, but even children who are overweight and sedentary can develop the disease. More than 90% of people with diabetes have type 2. Many can control their blood glucose levels with regular exercise and a healthy diet; some require medications. Nevertheless, type 2 diabetes is the seventh most common cause of death in the United States and a growing public health problem worldwide.

The resistance to insulin signaling in type 2 diabetes is sometimes due to a genetic defect in the insulin receptor or the insulin response pathway. In many cases, however, events in target cells suppress activity of an otherwise functional response pathway. One source of this suppression appears to be inflammatory signals generated by the innate immune system (see Chapter 43). How obesity and inactivity relate to this suppression is being studied in both humans and laboratory animals.

CONCEPT CHECK 45.2

1. In a glucose tolerance test, periodic measurements of blood glucose level are taken after a person drinks a glucose-rich solution. In a healthy individual, blood glucose rises moderately at first but falls to near normal within 2–3 hours. Predict the results of this test in a person with diabetes mellitus. Explain your answer.
2. If a hormone pathway provides a transient response to a stimulus, how would shortening the stimulus duration affect the need for negative feedback?
3. **WHAT IF?** Consider a diabetes patient who has a family history of type 2 diabetes but is active and not obese. To identify genes that might be defective in the patient, which genes would you examine first?

For suggested answers, see Appendix A.

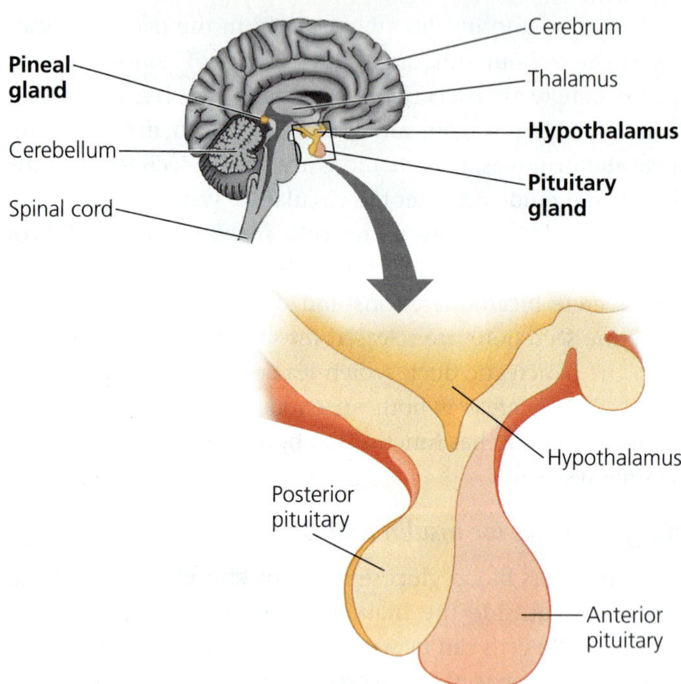

▲ **Figure 45.14 Endocrine glands in the human brain.** This side view of the brain indicates the position of the hypothalamus, the pituitary gland, and the pineal gland. (The pineal gland plays a role in regulating biorhythm.)

CONCEPT **45.3**

The hypothalamus and pituitary are central to endocrine regulation

Having reviewed the organization of hormone pathways, we return to the role of the nervous system in regulating endocrine pathways. In particular, we now turn our focus to the vertebrate brain and endocrine system.

Coordination of Endocrine and Nervous Systems in Vertebrates

In vertebrates, the **hypothalamus** plays a central role in integrating the endocrine and nervous systems. One of several endocrine glands located in the brain **(Figure 45.14)**, the hypothalamus receives information from nerves throughout the body, including the brain. In response, the hypothalamus initiates endocrine signaling appropriate to environmental conditions. In many vertebrates, for example, nerve signals from the brain pass sensory information to the hypothalamus about seasonal changes. The hypothalamus, in turn, regulates the release of reproductive hormones required during the breeding season.

Signals from the hypothalamus travel to the **pituitary gland**, a gland located at its base (see Figure 45.14). Roughly the size and shape of a lima bean, the pituitary has discrete posterior and anterior parts, or lobes, that secrete different sets of hormones.

The **posterior pituitary** is an extension of the hypothalamus. Hypothalamic axons that reach into the posterior pituitary secrete neurohormones synthesized in the hypothalamus. In contrast, the **anterior pituitary** is an endocrine gland that synthesizes and secretes hormones in response to signals from the hypothalamus. Many anterior pituitary hormones act as **tropic hormones**, meaning that they regulate the function of other endocrine cells or glands.

Posterior Pituitary Hormones

Neurosecretory cells of the hypothalamus synthesize the two posterior pituitary hormones: oxytocin and antidiuretic hormone. After traveling to the posterior pituitary within the long axons of the neurosecretory cells, the hormones are stored in pituitary cells, to be released in response to nerve impulses transmitted by the hypothalamus **(Figure 45.15)**.

As discussed in Concept 45.2 (see Figure 45.12), oxytocin regulates milk secretion by the mammary glands and also contractions of the uterus during birthing. In addition, oxytocin has targets in the brain, where it influences behaviors related to maternal care, pair bonding, and sexual activity.

Like oxytocin, **antidiuretic hormone (ADH)**, or *vasopressin*, regulates both physiology and behavior. As you read in Chapter 44, ADH is one of several hormones that regulate kidney function. In particular, ADH increases water retention in the kidneys, thus decreasing urine volume. The net result is to help maintain blood osmolarity within a normal range. ADH also plays an important role in social behavior, as detailed in Chapter 51.

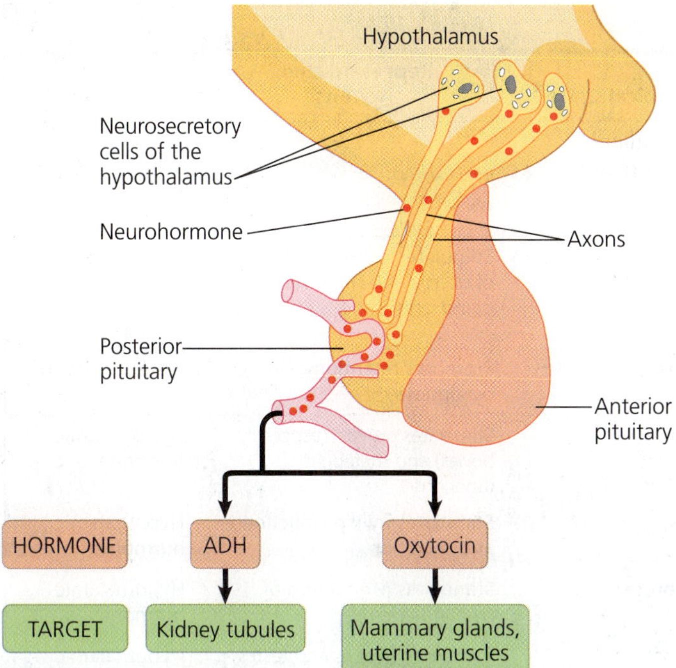

▲ **Figure 45.15 Production and release of posterior pituitary hormones.** The posterior pituitary gland is an extension of the hypothalamus. Certain neurosecretory cells in the hypothalamus make antidiuretic hormone (ADH) and oxytocin, which are transported to the posterior pituitary, where they are stored. Nerve signals from the brain trigger release of these neurohormones.

Anterior Pituitary Hormones

Endocrine signals generated by the hypothalamus regulate hormone secretion by the anterior pituitary **(Figure 45.16)**. Each hypothalamic hormone is either a *releasing hormone* or an *inhibiting hormone*, reflecting its role in promoting or inhibiting release of one or more specific hormones by the anterior pituitary. *Prolactin-releasing hormone*, for example, is a hypothalamic hormone that stimulates the anterior pituitary to secrete **prolactin**, which has activities that include stimulating milk production. Every anterior pituitary hormone is controlled by at least one releasing hormone. Some, such as prolactin, have both a releasing hormone and an inhibiting hormone.

The hypothalamic releasing and inhibiting hormones are secreted near capillaries at the base of the hypothalamus. The capillaries drain into short blood vessels, called portal vessels, which subdivide into a second capillary bed within the anterior pituitary. In this way, the releasing and inhibiting hormones have direct access to the gland they control.

Hormones secreted by the anterior pituitary regulate a diverse set of processes in the human body, including metabolism, osmoregulation, and reproductive activity. We turn next to an exploration of these hormones and the processes they govern, beginning with the hormones of the thyroid gland. **Table 45.1** (on the next page), which provides an overview of the major hormones of the endocrine system and their physiological functions, will serve as a useful point of reference for this discussion.

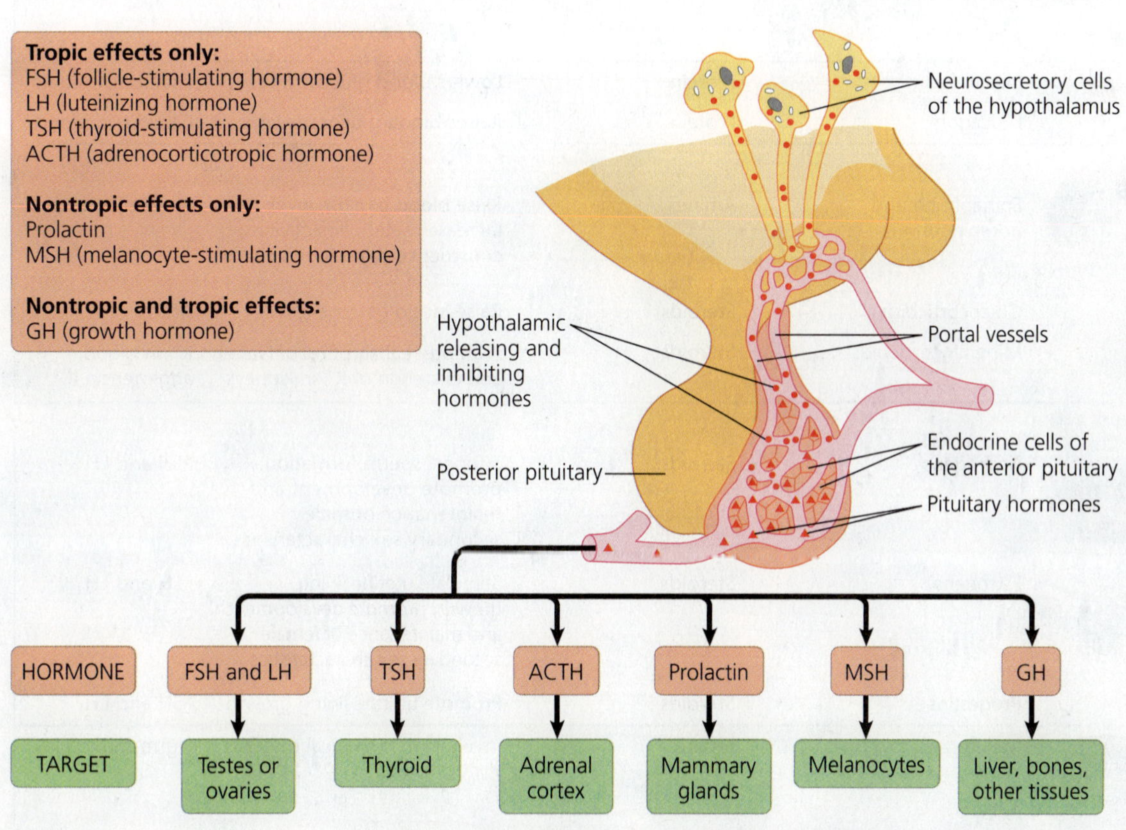

◀ **Figure 45.16 Production and release of anterior pituitary hormones.** The release of hormones synthesized in the anterior pituitary gland is controlled by hypothalamic releasing and inhibiting hormones. The hypothalamic hormones are secreted by neurosecretory cells and enter a capillary network within the hypothalamus. These capillaries drain into portal vessels that connect with a second capillary network in the anterior pituitary.

Table 45.1 Major Human Endocrine Glands and Some of Their Hormones

Gland	Hormone	Chemical Class	Representative Actions	Regulated By
Hypothalamus	Hormones released from the posterior pituitary and hormones that regulate the anterior pituitary (see below)			
Posterior pituitary gland (releases neurohormones made in hypothalamus)	Oxytocin	Peptide	Stimulates contraction of uterus and mammary gland cells	Nervous system
	Antidiuretic hormone (ADH)	Peptide	Promotes retention of water by kidneys	Water/salt balance
Anterior pituitary gland	Growth hormone (GH)	Protein	Stimulates growth (especially bones) and metabolic functions	Hypothalamic hormones
	Prolactin	Protein	Stimulates milk production and secretion	Hypothalamic hormones
	Follicle-stimulating hormone (FSH)	Glycoprotein	Stimulates production of ova and sperm	Hypothalamic hormones
	Luteinizing hormone (LH)	Glycoprotein	Stimulates ovaries and testes	Hypothalamic hormones
	Thyroid-stimulating hormone (TSH)	Glycoprotein	Stimulates thyroid gland	Hypothalamic hormones
	Adrenocorticotropic hormone (ACTH)	Peptide	Stimulates adrenal cortex to secrete glucocorticoids	Hypothalamic hormones
Thyroid gland	Triiodothyronine (T_3) and thyroxine (T_4)	Amines	Stimulate and maintain metabolic processes	TSH
	Calcitonin	Peptide	Lowers blood calcium level	Calcium in blood
Parathyroid glands	Parathyroid hormone (PTH)	Peptide	Raises blood calcium level	Calcium in blood
Pancreas	Insulin	Protein	Lowers blood glucose level	Glucose in blood
	Glucagon	Protein	Raises blood glucose level	Glucose in blood
Adrenal glands				
Adrenal medulla	Epinephrine and norepinephrine	Amines	Raise blood glucose level; increase metabolic activities; constrict certain blood vessels	Nervous system
Adrenal cortex	Glucocorticoids	Steroids	Raise blood glucose level	ACTH
	Mineralocorticoids	Steroids	Promote reabsorption of Na^+ and excretion of K^+ in kidneys	K^+ in blood; angiotensin II
Gonads				
Testes	Androgens	Steroids	Support sperm formation; promote development and maintenance of male secondary sex characteristics	FSH and LH
Ovaries	Estrogens	Steroids	Stimulate uterine lining growth; promote development and maintenance of female secondary sex characteristics	FSH and LH
	Progestins	Steroids	Promote uterine lining growth	FSH and LH
Pineal gland	Melatonin	Amine	Involved in biological rhythms	Light/dark cycles

Thyroid Regulation: A Hormone Cascade Pathway

Sets of hormones from the hypothalamus, the anterior pituitary, and a target endocrine gland are often organized into a *hormone cascade pathway* **(Figure 45.17)**. Signals to the brain stimulate the hypothalamus to secrete a hormone that stimulates or inhibits release of a tropic anterior pituitary hormone. The anterior pituitary hormone in turn acts on a target endocrine tissue, stimulating secretion of yet another hormone that exerts systemic metabolic or developmental effects.

To learn more about how a hormone cascade pathway works, let's consider activation of the thyroid gland when an infant is exposed to cold (see Figure 45.17). When a young child's body temperature drops, the hypothalamus secretes thyrotropin-releasing hormone (TRH). The anterior pituitary responds to TRH by secreting thyroid-stimulating hormone (TSH), also known as thyrotropin. TSH stimulates release of thyroid hormone by the **thyroid gland**, an organ consisting of two lobes on the ventral surface of the trachea (see Figure 42.24). As thyroid hormone accumulates, it increases metabolic rate, resulting in the release of thermal energy, which raises body temperature.

Like simple hormone pathways, hormone cascade pathways typically involve negative feedback. In the case of the thyroid hormone pathway, thyroid hormone itself carries out negative feedback. Because thyroid hormone blocks TSH release from the anterior pituitary and TRH release from the hypothalamus, the negative-feedback loop prevents overproduction of thyroid hormone. Overall, the hormone cascade pathway brings about a self-limiting response to the original stimulus in the target cells.

In humans and other mammals, thyroid hormone regulates bioenergetics; helps maintain normal blood pressure, heart rate, and muscle tone; and regulates digestive and reproductive functions. Too much or too little thyroid hormone in the blood can result in serious metabolic disorders.

Disorders of Thyroid Function and Regulation

In humans, hypothyroidism, a condition of too little thyroid function, can produce symptoms such as weight gain, lethargy, and intolerance to cold in adults. Excessive secretion of thyroid hormone, known as hyperthyroidism, can lead to high body temperature, profuse sweating, weight loss, irritability, and high blood pressure.

The most common form of hyperthyroidism is Graves' disease. Protruding eyes, caused by fluid accumulation behind the eyes, are a typical symptom. In this autoimmune disorder, the body produces antibodies that bind to and activate the receptor for TSH. The result is sustained thyroid hormone production.

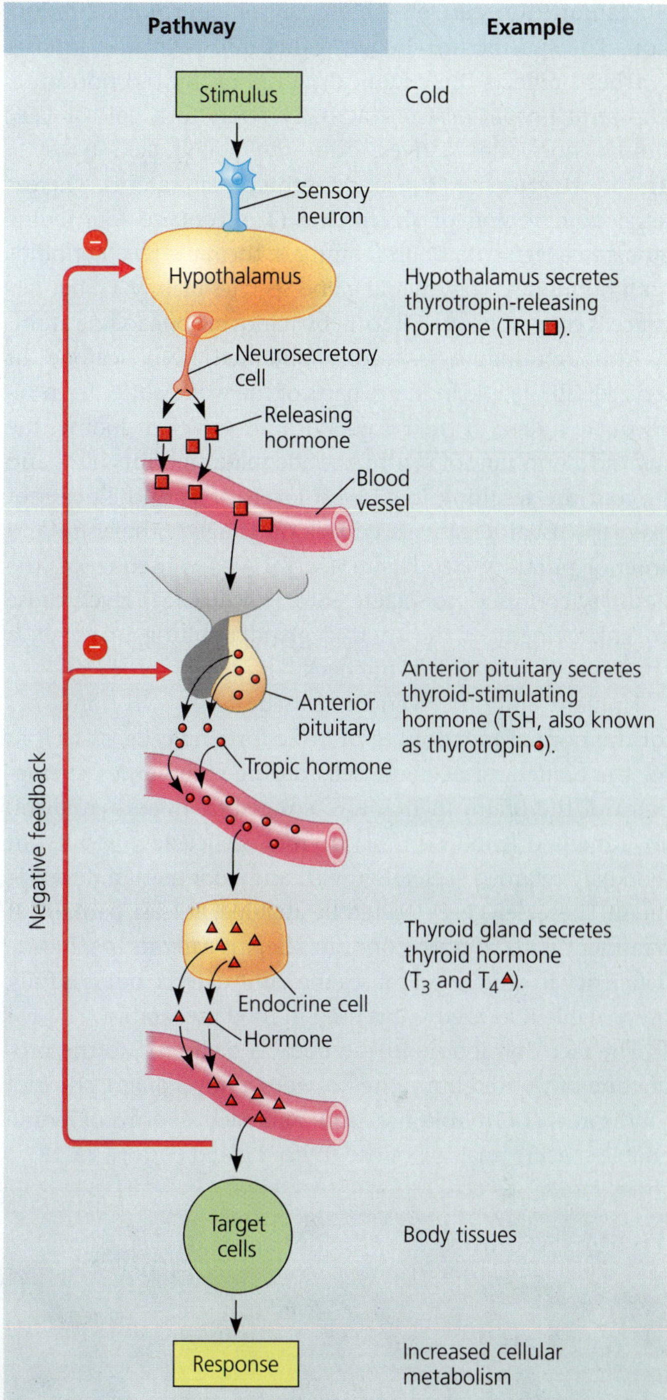

▲ **Figure 45.17 A hormone cascade pathway.** In response to the stimulus, the hypothalamus secretes a releasing hormone that targets the anterior pituitary. The anterior pituitary responds by secreting a second tropic hormone, which travels through the bloodstream to an endocrine gland. In response to this tropic hormone, the endocrine gland secretes a hormone that travels to target cells, where it induces a response. In the example of thyroid hormone regulation, thyroid hormone exerts negative feedback on the hypothalamus and anterior pituitary. This feedback inhibits release of TRH and TSH, preventing overreaction to the stimulus (such as low temperature in the case of a human infant).

? *Suppose a lab test of two patients, each diagnosed with excessive thyroid hormone production, revealed elevated levels of TSH in one but not the other. Was the diagnosis of one patient necessarily incorrect? Explain.*

Malnutrition can also alter thyroid hormone production. The specific link between diet and thyroid hormone synthesis reflects the chemical nature of thyroid hormone. The term *thyroid hormone* actually refers to a pair of very similar hormones derived from the amino acid tyrosine. **Triiodothyronine (T_3)** contains three iodine atoms, whereas tetraiodothyronine, or **thyroxine (T_4)**, contains four iodine atoms (see Figure 45.5). In mammals, the same receptor binds both hormones. The thyroid gland secretes mainly T_4, but target cells convert most of it to T_3 by removing one iodine atom.

Although iodine is readily obtained from seafood or iodized salt, people in many parts of the world suffer from inadequate iodine in their diet. Without sufficient iodine, the thyroid gland cannot synthesize adequate amounts of T_3 and T_4, and the resulting low blood levels of T_3 and T_4 cannot exert the usual negative feedback on the hypothalamus and anterior pituitary (see Figure 45.17). As a consequence, the pituitary continues to secrete TSH. Elevated TSH levels cause an enlargement of the thyroid gland resulting in goiter, a characteristic swelling of the neck.

Humans and other vertebrates require thyroid hormones for the normal functioning of bone-forming cells, as well as for the branching of nerve cells during embryonic development of the brain. In humans, congenital hypothyroidism, an inherited condition of thyroid deficiency, results in markedly retarded skeletal growth and poor mental development. These defects can often be avoided, at least partially, if treatment with thyroid hormones begins early in life. Iodine deficiency in childhood causes the same defects, but it is fully preventable if iodized salt is used in food preparation.

The fact that iodine in the body is dedicated to the production of thyroid hormone provides a novel diagnostic tool for disorders of thyroid function: Radioactive forms of iodine enable specific imaging of the thyroid gland **(Figure 45.18)**.

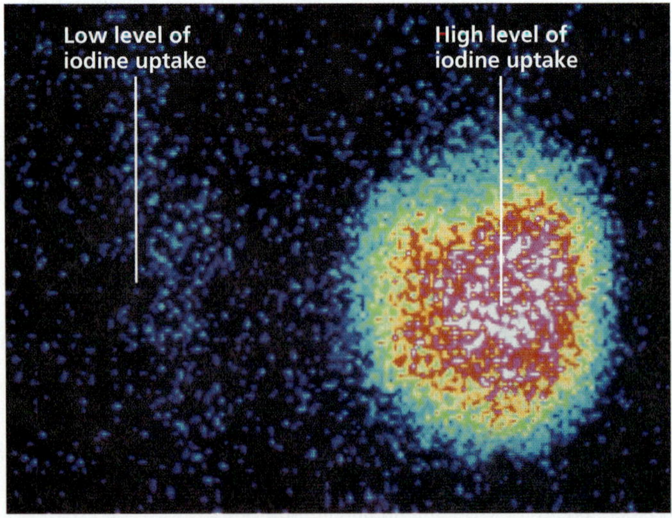

▲ **Figure 45.18 Thyroid scan.** Radioactive iodine enables doctors to identify abnormal patterns of iodine uptake that could indicate a thyroid disorder.

Evolution of Hormone Function

EVOLUTION Over the course of evolution, the functions of a given hormone often diverge between species. An example is thyroid hormone, which plays a role in regulating metabolism across many evolutionary lineages. In frogs, however, thyroid hormone (thyroxine) has taken on an apparently unique function: stimulating resorption of the tadpole's tail during metamorphosis **(Figure 45.19)**.

Diverse functions have also evolved for many other vertebrate hormones. Prolactin, a product of the anterior pituitary, has an especially broad range of activities. Prolactin stimulates mammary gland growth and milk synthesis in mammals, regulates fat metabolism and reproduction in birds, delays metamorphosis in amphibians, and regulates salt and water balance in freshwater fishes. These varied roles suggest that prolactin is an ancient hormone with functions that have diversified during the evolution of vertebrate groups.

Melanocyte-stimulating hormone (MSH) is another example of an anterior pituitary hormone with distinct functions in different evolutionary lineages. In amphibians, fishes, and reptiles, MSH regulates skin color by controlling pigment distribution in skin cells called melanocytes. In mammals, MSH functions in hunger and metabolism in addition to coloration.

The specialized action of MSH that has evolved in the mammalian brain may prove to be of particular medical importance. Many patients with late-stage cancer, AIDS, tuberculosis, and certain aging disorders suffer from a devastating wasting

▲ **Figure 45.19 Specialized role of a hormone in frog metamorphosis.** The hormone thyroxine is responsible for the resorption of the tadpole's tail as the frog develops into its adult form.

condition called cachexia. Characterized by weight loss, muscle atrophy, and loss of appetite, cachexia is only poorly responsive to existing therapies. However, it turns out that activation of one brain receptor for MSH stimulates metabolism of fat and severely decreases appetite, changes also seen in cachexia. This fact led scientists to hypothesize that activation of this MSH receptor causes cachexia. To test this idea, they studied mice with mutations that cause cancerous tumors to develop, triggering cachexia. When the mice were treated with drugs that inhibit the brain MSH receptor, tumors occurred, but not cachexia! Whether such drugs can be used to treat cachexia in humans is an area of active study.

Tropic and Nontropic Hormones

As we have seen, thyroid-stimulating hormone (TSH) regulates the thyroid gland. This activity makes TSH an example of a tropic hormone. Although MSH and prolactin don't regulate endocrine cells or glands and are thus nontropic, three other anterior pituitary hormones act primarily or exclusively as tropic hormones: **follicle-stimulating hormone (FSH)**, **luteinizing hormone (LH)**, and **adrenocorticotropic hormone (ACTH)**.

FSH and LH stimulate the activities of both the male and female gonads, the testes and ovaries. For this reason, FSH and LH are also known as *gonadotropins*, and they are both regulated by hypothalamic *gonadotropin-releasing hormone* (*GnRH*). In Chapter 46, we will discuss how gonadotropins regulate reproductive functions.

ACTH stimulates the production and secretion of steroid hormones by the adrenal cortex. We will take a closer look at the hormone pathway involving ACTH later in this chapter.

Growth hormone (GH), which is secreted by the anterior pituitary, stimulates growth through both tropic and nontropic effects. A major target, the liver, responds to GH by releasing *insulin-like growth factors* (*IGFs*), which circulate in the blood and directly stimulate bone and cartilage growth. (IGFs also appear to play a key role in aging in many animal species.) In the absence of GH, the skeleton of an immature animal stops growing. GH also exerts diverse metabolic effects that tend to raise blood glucose levels, thus opposing the effects of insulin.

Abnormal production of GH in humans can result in several disorders, depending on when the problem occurs and whether it involves hypersecretion (too much) or hyposecretion (too little). Hypersecretion of GH during childhood can lead to gigantism, in which the person grows unusually tall—as tall as 2.4 m (8 feet)—though body proportions remain relatively normal. Excessive GH production in adulthood stimulates bony growth in the few tissues that are still responsive to the hormone. Because remaining target cells are predominantly in the face, hands, and feet, the result is an overgrowth of the extremities called acromegaly (from the Greek *acros*, extreme, and *mega*, large).

Hyposecretion of GH in childhood retards long-bone growth and can lead to pituitary dwarfism. Individuals with this disorder are for the most part properly proportioned but generally reach a height of only about 1.2 m (4 feet). If diagnosed before puberty, pituitary dwarfism can be treated successfully with human GH (also called HGH). Since the mid-1980s, scientists have used recombinant DNA technology to produce HGH in bacteria (see Chapter 20). Treatment with this genetically engineered HGH is now fairly routine for affected children.

CONCEPT CHECK 45.3

1. How do the two fused glands of the pituitary gland differ in function?
2. Why does hypothalamic control of oxytocin not require a releasing factor?
3. **WHAT IF?** Propose an explanation for why people with defects in specific endocrine pathways typically have defects in the final gland in the pathway rather than in the hypothalamus or pituitary.

For suggested answers, see Appendix A.

CONCEPT 45.4

Endocrine glands respond to diverse stimuli in regulating homeostasis, development, and behavior

Now that we've seen how endocrine glands in the brain initiate hormone cascade pathways, we return to the broader question of how endocrine signaling regulates animal physiology. We'll focus on homeostasis, development, and behavior, leaving the topic of reproduction largely for later chapters. This section presents more examples of hormone regulation by metabolic stimuli, by nervous system input, and by hormones of the anterior pituitary. First we'll examine another simple hormone pathway, the regulation of calcium ion concentration in the circulatory system.

Parathyroid Hormone and Vitamin D: Control of Blood Calcium

Because calcium ions (Ca^{2+}) are essential to the normal functioning of all cells, homeostatic control of blood calcium level is critical. If the blood Ca^{2+} level falls substantially, skeletal muscles begin to contract convulsively, a potentially fatal condition called tetany. If the blood Ca^{2+} level rises substantially, precipitates of calcium phosphate can form in body tissues, leading to widespread organ damage.

In mammals, the **parathyroid glands**, a set of four small structures embedded in the posterior surface of the thyroid (see Figure 45.4), play a major role in blood Ca^{2+} regulation. When

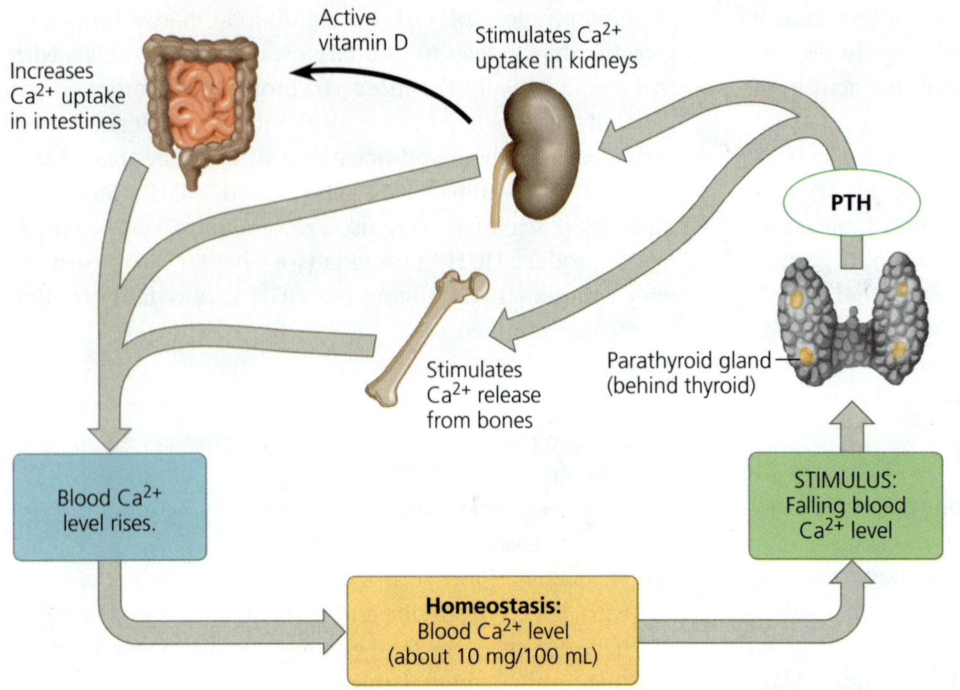

▲ **Figure 45.20 The roles of parathyroid hormone (PTH) in regulating blood calcium levels in mammals.**

blood Ca^{2+} falls below a set point of about 10 mg/100 mL, these glands release **parathyroid hormone (PTH)**.

PTH raises the level of blood Ca^{2+} by direct and indirect effects **(Figure 45.20)**. In bone, PTH causes the mineralized matrix to decompose and release Ca^{2+} into the blood. In the kidneys, PTH directly stimulates reabsorption of Ca^{2+} through the renal tubules. PTH also has an indirect effect on the kidneys, promoting the conversion of vitamin D to an active hormone. An inactive form of vitamin D, a steroid-derived molecule, is obtained from food or synthesized in the skin when exposed to sunlight. Vitamin D activation begins in the liver and is completed in the kidneys, the process stimulated by PTH. The active form of vitamin D acts directly on the intestines, stimulating the uptake of Ca^{2+} from food and thus augmenting the effect of PTH. As blood Ca^{2+} rises, a negative-feedback loop inhibits further release of PTH from the parathyroid glands (not shown in figure).

The thyroid gland can also contribute to calcium homeostasis. If blood Ca^{2+} rises above the set point, the thyroid gland releases **calcitonin**, a hormone that inhibits bone resorption and enhances Ca^{2+} release by the kidney. In fishes, rodents, and some other animals, calcitonin is required for Ca^{2+} homeostasis. In humans, however, it is apparently needed only during the extensive bone growth of childhood.

Adrenal Hormones: Response to Stress

The **adrenal glands** of vertebrates are associated with the kidneys (the *renal* organs). In mammals, each adrenal gland is actually made up of two glands with different cell types, functions, and embryonic origins: the *adrenal cortex*, the outer portion, and the *adrenal medulla*, the central portion. The adrenal cortex consists of true endocrine cells, whereas the secretory cells of the adrenal medulla derive from neural tissue during embryonic development. Thus, like the pituitary gland, each adrenal gland is a fused endocrine and neuroendocrine gland.

Catecholamines from the Adrenal Medulla

Imagine that while walking in the woods at night you hear a growling noise nearby. "A bear?" you wonder. Your heart beats faster, your breath quickens, your muscles tense, and your thoughts speed up. These and other rapid responses to perceived danger comprise the "fight-or-flight," or acute stress, response. This coordinated set of physiological changes is triggered by two hormones of the adrenal medulla, **norepinephrine** (also known as noradrenaline) and epinephrine (adrenaline). Both are **catecholamines**, a class of amine hormones synthesized from the amino acid tyrosine.

The adrenal medulla secretes epinephrine and norepinephrine in response to stress—whether extreme pleasure or life-threatening danger. A major activity of these hormones is to increase the amount of chemical energy available for immediate use. Both epinephrine and norepinephrine increase the rate of glycogen breakdown in the liver and skeletal muscles, promote glucose release by liver cells, and stimulate the release of fatty acids from fat cells. The released glucose and fatty acids circulate in the blood and can be used by body cells as fuel.

In addition to increasing the availability of energy sources, norepinephrine and epinephrine exert profound effects on the cardiovascular and respiratory systems. For example, they increase both the heart rate and stroke volume and dilate the bronchioles in the lungs, actions that raise the rate of oxygen delivery to body cells. For this reason, doctors may prescribe epinephrine as a heart stimulant or to open the airways during an asthma attack. The catecholamines also alter blood flow, causing constriction of some blood vessels and dilation of others (see Figure 45.9). The overall effect is to shunt blood away from the skin, digestive organs, and kidneys, while increasing the blood supply to the heart, brain, and skeletal muscles. Epinephrine generally has a stronger effect on heart and metabolic rates, while the primary role of norepinephrine is in modulating blood pressure.

Nerve signals carried from the brain via involuntary (autonomic) neurons regulate secretion by the adrenal medulla. In response to a stressful stimulus, nerve impulses travel to the adrenal medulla, where they trigger the

▼ **Figure 45.21 Stress and the adrenal gland.**

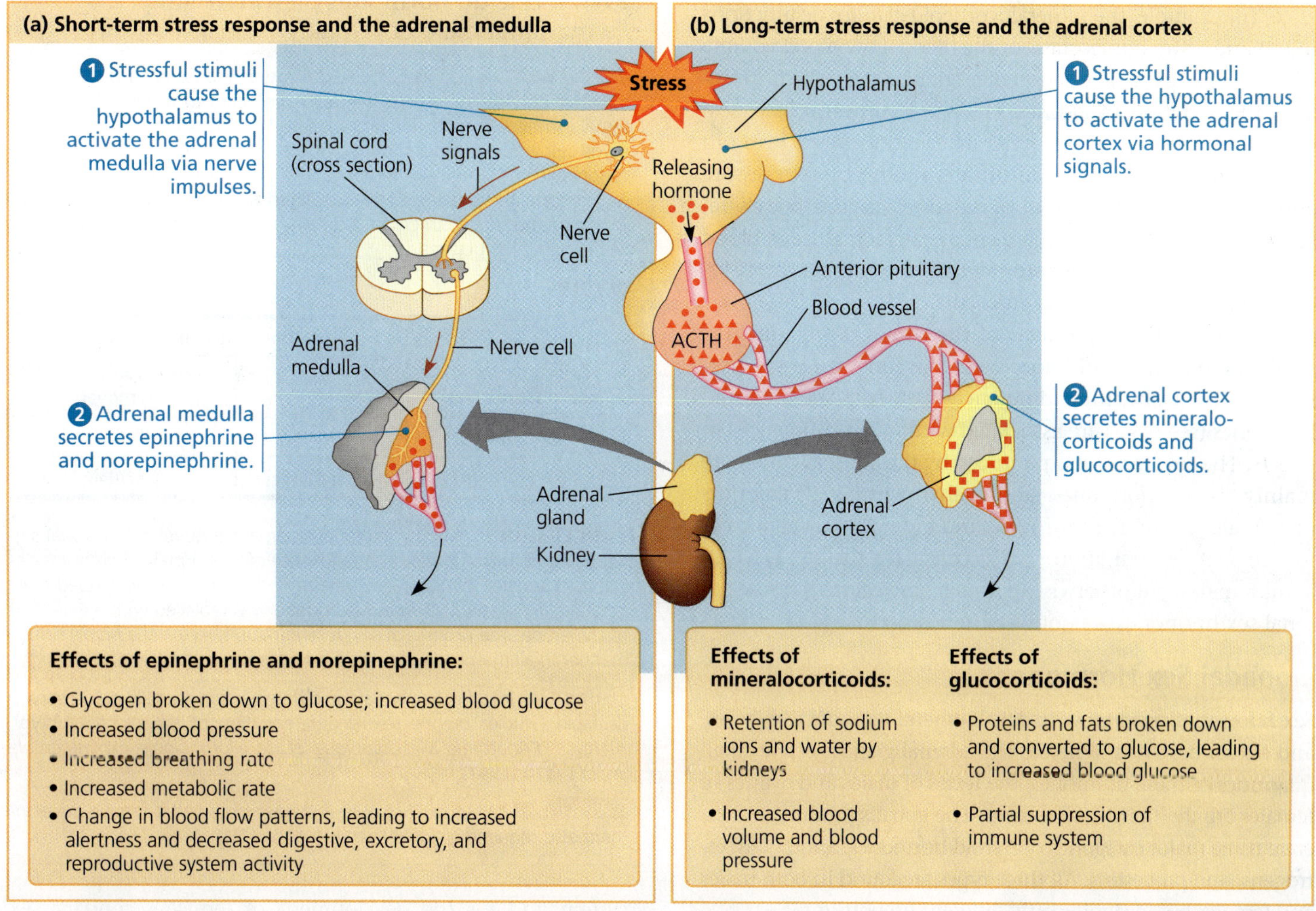

release of catecholamines from neurosecretory cells **(Figure 45.21a)**. Acting on target tissues, epinephrine and norepinephrine each function in a simple neurohormone pathway. As you will read in Chapter 48, epinephrine and norepinephrine also function as neurotransmitters.

Steroid Hormones from the Adrenal Cortex

Hormones from the adrenal cortex also function in the body's response to stress. But in contrast to the adrenal medulla, which reacts to nervous input, the adrenal cortex responds to endocrine signals. Stressful stimuli cause the hypothalamus to secrete a releasing hormone that stimulates the anterior pituitary to release the tropic hormone ACTH. When ACTH reaches the adrenal cortex via the bloodstream, it stimulates the endocrine cells to synthesize and secrete a family of steroids called **corticosteroids** **(Figure 45.21b)**. The two main types of corticosteroids in humans are glucocorticoids and mineralocorticoids.

As reflected in their name, **glucocorticoids** have a primary effect on glucose metabolism. Augmenting the fuel-mobilizing effects of glucagon from the pancreas, glucocorticoids promote glucose synthesis from noncarbohydrate sources, such as proteins, making more glucose available as fuel. Glucocorticoids, such as cortisol (see Figure 45.5), act on skeletal muscle, causing the breakdown of muscle proteins. The resulting amino acids are transported to the liver and kidneys, where they are converted to glucose and released into the blood. The synthesis of glucose from muscle proteins provides circulating fuel when the body requires more glucose than the liver can mobilize from its glycogen stores.

When glucocorticoids are introduced into the body at levels above those normally present, they suppress certain components of the body's immune system. Because of this anti-inflammatory effect, glucocorticoids are sometimes used to treat inflammatory diseases such as arthritis. However, long-term use can have serious side effects, reflecting the potent activity of glucocorticoids on metabolism. For these reasons, nonsteroidal anti-inflammatory drugs (NSAIDs), such as aspirin or ibuprofen, generally are preferred for treating chronic inflammatory conditions.

Mineralocorticoids, named for their effects on mineral metabolism, act principally in maintaining salt and water balance. For example, the mineralocorticoid *aldosterone* functions in ion and water homeostasis of the blood. Low blood volume or pressure leads to production of angiotensin II, which stimulates the secretion of aldosterone (see Figure 44.22). Aldosterone, in turn, stimulates cells in the kidneys to reabsorb sodium ions and water from filtrate, raising blood pressure and volume. Aldosterone also functions in the body's response to severe stress. In these circumstances, a rise in blood ACTH levels increases the rate at which the adrenal cortex secretes aldosterone as well as glucocorticoids.

The corticosteroid products of the adrenal cortex include small amounts of steroid hormones that function as sex hormones. Small structural differences between these steroid hormones (see p. 63) are associated with major differences in effects. The sex hormones produced by the adrenal cortex are mainly "male" hormones (androgens), with small amounts of "female" hormones (estrogens and progestins). There is evidence that adrenal androgens account for the sex drive in adult females, but otherwise the physiological roles of the adrenal sex hormones are not well understood.

Gonadal Sex Hormones

Sex hormones affect growth, development, reproductive cycles, and sexual behavior. Whereas the adrenal glands secrete small quantities of these hormones, the testes of males and ovaries of females are their principal sources. The gonads produce and secrete three major categories of steroid hormones: androgens, estrogens, and progestins. All three types are found in both males and females but in significantly different proportions.

The testes primarily synthesize **androgens**, the main one being **testosterone**. Testosterone first functions before birth, as shown in the 1940s by French researcher Alfred Jost. He was interested in how hormones determine whether an individual develops as a male or female. Working with rabbits, Jost carried out a surgical study that provided a simple and unexpected answer **(Figure 45.22)**. His studies established that for mammals (but not all animals), female development is the default process in embryos.

Androgens have a major role again at human puberty, when they are responsible for the development of human male secondary sex characteristics. High concentrations of androgen lead to a low voice and male patterns of hair growth, as well as increases in muscle and bone mass. The muscle-building, or anabolic, action of testosterone and related steroids has enticed some athletes to take them as supplements, despite prohibitions against their use in nearly all sports. Use of anabolic steroids, while effective in increasing muscle mass, can cause severe acne outbreaks and liver damage, as well as significant decreases in sperm count and testicular size.

Estrogens, of which the most important is **estradiol**, are responsible for the maintenance of the female reproductive system and for the development of female secondary sex characteristics. In mammals, **progestins**, which include **progesterone**, are primarily involved in preparing and maintaining tissues of the uterus required to support the growth and development of an embryo.

Estrogens and other gonadal sex hormones are components of hormone cascade pathways. Synthesis of these hormones is controlled by gonadotropins (FSH and LH) from the anterior pituitary gland (see Figure 45.16). FSH and LH secretion is in turn controlled by GnRH (gonadotropin-releasing hormone), a releasing hormone from the hypothalamus. We will examine the feedback relationships that regulate gonadal steroid secretion in detail in Chapter 46.

Endocrine Disruptors

Between 1938 and 1971, some pregnant women at risk for complications were prescribed a synthetic estrogen called diethylstilbestrol (DES). What was not known until 1971 was that exposure to DES can alter reproductive system development in the fetus. Collectively, daughters of women who took DES are more frequently afflicted with certain reproductive abnormalities, including a form of vaginal and cervical cancer, structural changes in the reproductive organs, and

▼ **Figure 45.22** INQUIRY

What role do hormones play in making a mammal male or female?

EXPERIMENT Alfred Jost, at the College de France in Paris, wondered whether gonadal hormones instruct an embryo to develop as male or female in accord with its chromosome set. Working with rabbit embryos still in the mother's uterus, at a stage before sex differences are observable, he surgically removed the portion of each embryo that would form the ovaries or testes. When the baby rabbits were born, Jost made note of both chromosomal sex and the sexual differentiation of the genital structures.

RESULTS

	Appearance of Genitalia	
Chromosome Set	**No surgery**	**Embryonic gonad removed**
XY (male)	Male	Female
XX (female)	Female	Female

CONCLUSION In rabbits, male development requires a hormonal signal from the male gonad. In the absence of this signal, all embryos develop as female. Jost later demonstrated that embryos developed male genitalia if the surgically removed gonad was replaced with a crystal of testosterone. The process of sex determination occurs in a highly similar manner in all mammals, including humans.

SOURCE A. Jost, Recherches sur la differenciation sexuelle de l'embryon de lapin (Studies on the sexual differentiation of the rabbit embryo), *Archives d'Anatomie Microscopique et de Morphologie Expérimentale* 36:271–316 (1947).

WHAT IF? What result would Jost have obtained if female development also required a signal from the gonad?

increased risk of miscarriage (spontaneous abortion). DES is now recognized as an *endocrine disruptor*, a foreign molecule that interrupts the normal function of a hormone pathway.

In recent years, it has been hypothesized that molecules in the environment also act as endocrine disruptors. Some estrogen-like molecules, such as those present in soybeans and other edible plant products, have been suggested to lower breast cancer risk. Others, such as bisphenol A, a chemical used in making some plastics, have been studied for potential interference with normal reproduction and development. Sorting out such effects has proved quite difficult, however, in part because enzymes in the liver change the properties of any such molecules entering the body through the digestive system.

Melatonin and Biorhythms

We conclude our discussion of the vertebrate endocrine system with the **pineal gland**, a small mass of tissue near the center of the mammalian brain (see Figure 45.14). The pineal gland is a primary source of the hormone **melatonin**, a modified amino acid.

Melatonin regulates functions related to light and to seasons marked by changes in day length. Although melatonin affects skin pigmentation in many vertebrates, its primary functions relate to biological rhythms associated with reproduction and with daily activity levels. Melatonin is secreted at night, and the amount released depends on the length of the night. In winter, for example, when days are short and nights are long, more melatonin is secreted. There is also good evidence that nightly increases in the levels of melatonin play a significant role in promoting sleep.

The release of melatonin by the pineal gland is controlled by a group of neurons in the hypothalamus called the suprachiasmatic nucleus (SCN). The SCN functions as a biological clock and receives input from specialized light-sensitive neurons in the retina of the eye. Although the SCN regulates melatonin production during the 24-hour light/dark cycle, melatonin also influences SCN activity. We will consider biological rhythms further in Chapter 49, where we analyze experiments on SCN function.

In the next chapter, we will look at reproduction in both vertebrates and invertebrates. There we will see that the endocrine system is central not only to the survival of the individual, but also to the propagation of the species.

CONCEPT CHECK 45.4

1. How does the fact that two adrenal hormones act as neurotransmitters relate to the developmental origin of the adrenal gland?
2. How would a decrease in the number of corticosteroid receptors in the hypothalamus affect levels of corticosteroids in the blood?
3. **WHAT IF?** Suppose you receive an injection of cortisone, a glucocorticoid, in an inflamed joint. What aspects of glucocorticoid activity would you be exploiting? If a glucocorticoid pill were also effective at treating the inflammation, why would it still be preferable to introduce the drug locally?

For suggested answers, see Appendix A.

45 CHAPTER REVIEW

SUMMARY OF KEY CONCEPTS

CONCEPT 45.1

Hormones and other signaling molecules bind to target receptors, triggering specific response pathways (pp. 975–980)

- The forms of communication between animal cells differ in the type of secreting cell and the route taken by the signal to its target. **Endocrine** signals, or **hormones**, are secreted into extracellular fluids by endocrine cells or ductless glands and reach target cells via circulatory fluids. **Paracrine** signals act on neighboring cells, whereas **autocrine** signals act on the secreting cell itself. **Neurotransmitters** also act locally, but **neurohormones** can act throughout the body. **Pheromones** are released into the environment for communication between animals of the same species.
- In insects, molting and development are controlled by PTTH; **ecdysteroid**, whose release is triggered by PTTH; and juvenile hormone. Coordination of signals from the nervous and endocrine systems and modulation of one hormone activity by another bring about the precise series of developmental stages that lead to an adult form.
- Distinct cellular responses are associated with water-soluble and lipid-soluble hormones. Polypeptide hormones and most amine hormones are water-soluble and bind to receptors embedded in the plasma membrane. Binding of water-soluble hormones to cell-surface receptors triggers intracellular **signal transduction**, leading to specific responses in the cytoplasm or changes in gene expression. In contrast, steroid and thyroid hormones are lipid-soluble and readily enter target cells. There they bind to specific protein receptors in the cytosol or nucleus. These complexes of a lipid-soluble hormone and its receptor act in the nucleus to regulate transcription of specific genes. The same hormone may have different effects on target cells that have different receptors for the hormone or different signal transduction pathways.
- **Local regulators**, which carry out paracrine and autocrine signaling, include cytokines and **growth factors** (proteins/peptides), **nitric oxide** (a gas), and **prostaglandins** (modified fatty acids).

? *Predict what would happen if you injected a water-soluble hormone directly into the cytosol of a target cell.*

CONCEPT 45.2

Feedback regulation and antagonistic hormone pairs are common in endocrine systems (pp. 981–984)

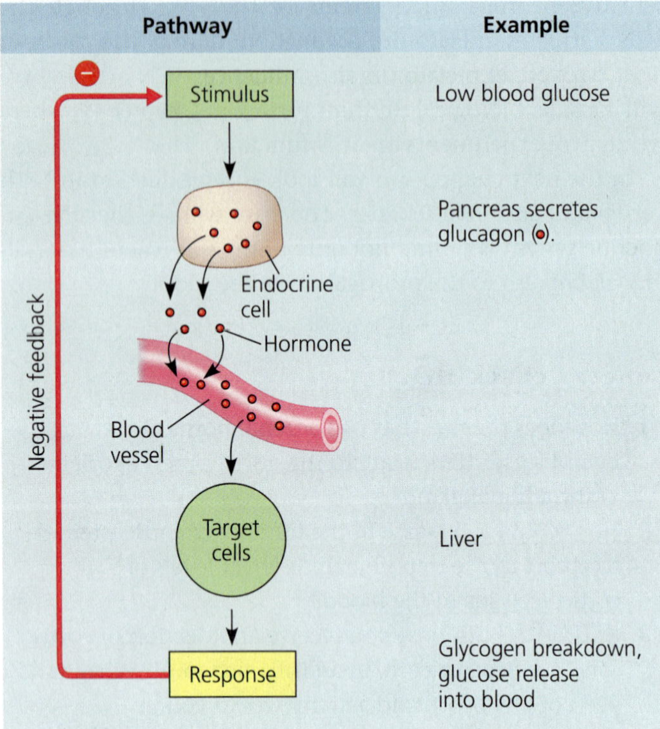

- Hormone pathways may be regulated by **negative feedback**, which dampens the stimulus, or **positive feedback**, which amplifies the stimulus and drives the response to completion. Negative-feedback pathways sometimes occur in antagonistic pairs, such as the maintainance of glucose homeostasis by **glucagon** (from alpha cells of the **pancreas**) and **insulin** (from beta cells of the pancreas). Insulin reduces blood glucose levels by promoting cellular uptake of glucose, glycogen formation in the liver, protein synthesis, and fat storage. The disorder **diabetes mellitus**, which is marked by elevated blood glucose levels, results from inadequate production of insulin (type 1) or loss of responsiveness of target cells to insulin (type 2).

? *Would taking a drug that blocks the action of glucagon lessen the symptoms of diabetes or make them worse? Explain.*

CONCEPT 45.3

The hypothalamus and pituitary are central to endocrine regulation (pp. 984–989)

- Some neurosecretory cells in the **hypothalamus** produce hormones secreted by the **posterior pituitary**. Other hypothalamic cells produce hormones that are transported by portal vessels to the **anterior pituitary**, where they stimulate or inhibit the release of particular hormones.
- The two hormones released from the **posterior pituitary** act directly on nonendocrine tissues. **Oxytocin** induces uterine contractions and release of milk from mammary glands, and **antidiuretic hormone** (ADH) enhances water reabsorption in the kidneys.
- Often, anterior pituitary hormones act in a cascade. In the case of thyrotropin, or thyroid-stimulating hormone (TSH), TSH secretion is regulated by thyrotropin-releasing hormone (TRH). TSH in turn induces the **thyroid gland** to secrete **thyroid hormone**, a combination of the iodine-containing hormones $\mathbf{T_3}$ and $\mathbf{T_4}$. Thyroid hormone stimulates metabolism and influences development and maturation.
- Hormones sometimes acquire distinct roles in different species over the course of evolution. **Prolactin** stimulates milk production in mammals but has diverse effects in different vertebrates. **Melanocyte-stimulating hormone** (MSH) influences skin pigmentation in some vertebrates and fat metabolism in mammals.
- Although prolactin and MSH act on nonendocrine targets, most anterior pituitary hormones are tropic, acting on endocrine tissues or glands to regulate hormone secretion. **Tropic hormones** of the anterior pituitary include TSH, **follicle-stimulating hormone** (FSH), **luteinizing hormone** (LH), and **adrenocorticotropic hormone** (ACTH). **Growth hormone** (GH) has both tropic and nontropic effects. It promotes growth directly, has diverse metabolic effects, and stimulates the production of growth factors by other tissues.

? *Which major endocrine organs are regulated independently of the hypothalamus and pituitary? Explain.*

CONCEPT 45.4

Endocrine glands respond to diverse stimuli in regulating homeostasis, development, and behavior (pp. 989–993)

- **Parathyroid hormone** (PTH), secreted by the **parathyroid glands**, causes bone to release Ca^{2+} into the blood and stimulates reabsorption of Ca^{2+} in the kidneys. PTH also stimulates the kidneys to activate vitamin D, which promotes intestinal uptake of Ca^{2+} from food. **Calcitonin**, secreted by the thyroid, has the opposite effects in bones and kidneys as PTH. Calcitonin is important for calcium homeostasis in adults of some vertebrates, but not humans.
- In response to stress, neurosecretory cells in the adrenal medulla release **epinephrine** and **norepinephrine**, which mediate various fight-or-flight responses. The adrenal cortex releases **glucocorticoids**, such as cortisol, which influence glucose metabolism and the immune system, as well as **mineralocorticoids**, primarily aldosterone, which help regulate salt and water balance.
- Although the adrenal cortex produces small amounts of sex hormones, the gonads—testes and ovaries—produce most of the body's sex hormones. All three types—**androgens**, **estrogens**, and **progestins**—are produced in males and females, but in different proportions.
- The **pineal gland**, located within the brain, secretes **melatonin**, which functions in biological rhythms related to reproduction and sleep. Release of melatonin is controlled by the SCN, the region of the brain that functions as a biological clock.

? *ADH and epinephrine act as hormones when released into the bloodstream and as neurotransmitters when released in synapses between neurons. What is similar about the endocrine glands that produce these two molecules?*

TEST YOUR UNDERSTANDING

LEVEL 1: KNOWLEDGE/COMPREHENSION

1. Which of the following is *not* an accurate statement?
   a. Hormones are chemical messengers that travel to target cells through the circulatory system.
   b. Hormones often regulate homeostasis through antagonistic functions.
   c. Hormones of the same chemical class usually have the same function.

d. Hormones are secreted by specialized cells usually located in endocrine glands.
e. Hormones are often regulated through feedback loops.

2. An example of antagonistic hormones controlling homeostasis is
a. thyroxine and parathyroid hormone in calcium balance.
b. insulin and glucagon in glucose metabolism.
c. progestins and estrogens in sexual differentiation.
d. epinephrine and norepinephrine in fight-or-flight responses.
e. oxytocin and prolactin in milk production.

3. Growth factors are local regulators that
a. are produced by the anterior pituitary.
b. are modified fatty acids that stimulate bone and cartilage growth.
c. are found on the surface of cancer cells and stimulate abnormal cell division.
d. bind to cell-surface receptors and stimulate growth and development of target cells.
e. convey messages between nerve cells.

4. Which hormone is *incorrectly* paired with its action?
a. oxytocin—stimulates uterine contractions during childbirth
b. thyroxine—stimulates metabolic processes
c. insulin—stimulates glycogen breakdown in the liver
d. ACTH—stimulates the release of glucocorticoids by the adrenal cortex
e. melatonin—affects biological rhythms, seasonal reproduction

LEVEL 2: APPLICATION/ANALYSIS

5. Steroid and peptide hormones typically have in common
a. the building blocks from which they are synthesized.
b. their solubility in cell membranes.
c. their requirement for travel through the bloodstream.
d. the location of their receptors.
e. their reliance on signal transduction in the cell.

6. Which of the following is the most likely explanation for hypothyroidism in a patient whose iodine level is normal?
a. greater production of T_3 than of T_4
b. hyposecretion of TSH
c. hypersecretion of TSH
d. hypersecretion of MSH
e. a decrease in the thyroid secretion of calcitonin

7. Shortly after ingesting a big plate of carbohydrate-rich pasta, you measure your blood's hormone levels. What results would you expect, compared to before the meal?
a. high insulin, low glucagon
b. low insulin, low glucagon
c. high insulin, high glucagon
d. low insulin, high glucagon
e. low insulin, no change in glucagon

8. The relationship between the insect hormones ecdysteroid and PTTH is an example of
a. an interaction of the endocrine and nervous systems.
b. homeostasis achieved by positive feedback.
c. how peptide-derived hormones have more widespread effects than steroid hormones.
d. homeostasis maintained by antagonistic hormones.
e. competitive inhibition of a hormone receptor.

9. **DRAW IT** In mammals, milk production by mammary glands is controlled by prolactin and prolactin-releasing hormone. Draw a simple sketch of this pathway, including glands and tissues, hormones, routes for hormone movement, and effects.

LEVEL 3: SYNTHESIS/EVALUATION

10. **EVOLUTION CONNECTION**
The intracellular receptors used by all the steroid and thyroid hormones are similar enough in structure that they are all considered members of one "superfamily" of proteins. Propose a hypothesis for how the genes encoding these receptors may have evolved. (*Hint*: See Figure 21.13.) How could you test your hypothesis using DNA sequence data?

11. **SCIENTIFIC INQUIRY**
Chronically high levels of glucocorticoids can result in obesity, muscle weakness, and depression, a combination of symptoms called Cushing's syndrome. Excessive activity of either the pituitary or the adrenal gland can be the cause. To determine which gland has abnormal activity in a particular patient, doctors use the drug dexamethasone, a synthetic glucocorticoid that blocks ACTH release. Based on the graph, which gland is affected in patient X?

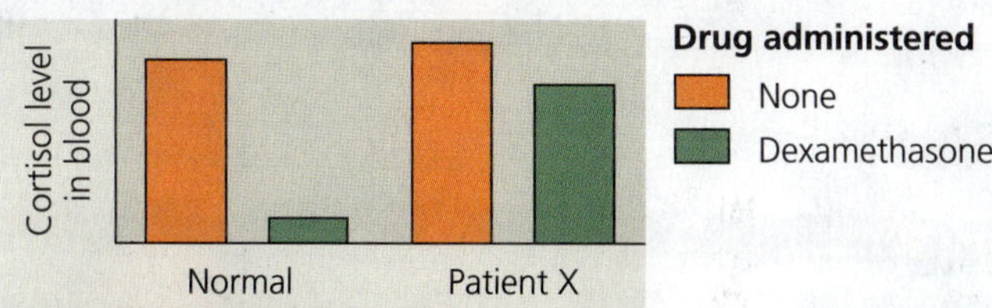

12. **WRITE ABOUT A THEME**
Environmental Interactions In a short essay (100–150 words), use specific examples to discuss the role of hormones in an animal's responses to changes in its environment.

For selected answers, see Appendix A.

MasteringBIOLOGY www.masteringbiology.com

1. MasteringBiology® Assignments:
BioFlix **Tutorial** Homeostasis: Regulating Blood Sugar
Tutorial Coordination of the Endocrine and Nervous Systems
Activities Steroid Hormone Action • Peptide Hormone Action • Endocrine System Anatomy • Human Endocrine Glands and Hormones • Hormone Actions on Target Cells • Discovery Channel Video: The Endocrine System
Questions Student Misconceptions • Reading Quiz • Multiple Choice • End-of-Chapter

2. eText
Read your book online, search, take notes, highlight text, and more.

3. The Study Area
Practice Tests • Cumulative Test • ***BioFlix*** 3-D Animations • MP3 Tutor Sessions • Videos • Activities • Investigations • Lab Media • Audio Glossary • Word Study Tools • Art

46

Animal Reproduction

▲ **Figure 46.1 How can each of these sea slugs be both male and female?**

KEY CONCEPTS

OVERVIEW

Pairing Up for Sexual Reproduction

The sea slugs, or nudibranchs (*Nembrotha rutilans*), in **Figure 46.1** are mating. If not disturbed, these marine molluscs may remain joined for hours. Sperm will be transferred and will fertilize eggs. A few weeks later, sexual reproduction will be complete. New individuals will hatch, but which parent is the mother? The answer is simple yet probably unexpected: both. In fact, each sea slug produces eggs *and* sperm.

As humans, we tend to think of reproduction in terms of the mating of males and females and the fusion of sperm and eggs. Animal reproduction, however, takes many forms. In some species, individuals change their sex during their lifetime; in other species, such as sea slugs, an individual is both male and female. There are animals that can fertilize their own eggs, as well as others that can reproduce without any form of sex. For certain species, such as honeybees, only a few individuals within a large population reproduce.

A population outlives its members only by reproduction, the generation of new individuals from existing ones. In this chapter, we will compare the diverse reproductive mechanisms that have evolved in the animal kingdom. We will then examine details of mammalian reproduction, particularly that of humans. We will focus on the physiology of reproduction mostly from the parents' perspective, deferring the details of embryonic development until the next chapter.

CONCEPT 46.1

Both asexual and sexual reproduction occur in the animal kingdom

There are two modes of animal reproduction—sexual and asexual. In **sexual reproduction**, the fusion of haploid gametes forms a diploid cell, the **zygote**. The animal that develops from a zygote can in turn give rise to gametes by meiosis (see Figure 13.8). The female gamete, the **egg**, is a large, nonmotile cell. The male gamete, the **sperm**, is generally a much smaller, motile cell. **Asexual reproduction** is the generation of new individuals without the fusion of egg and sperm. In most asexual animals, reproduction relies entirely on mitotic cell division.

For the vast majority of animals, reproduction is primarily or exclusively sexual. However, there are species that have a primarily asexual mode of reproduction, including a few all-female species for which reproduction is exclusively asexual. These include the microscopic bdelloid rotifer (see p. 677), as well as certain species of whiptail lizard (*Aspidoscelis*), which we will discuss shortly.

Mechanisms of Asexual Reproduction

Several forms of asexual reproduction are found only among invertebrates. One of these is **fission**, the separation of a parent organism into two individuals of approximately equal size **(Figure 46.2)**. Also common among invertebrates is **budding**, in which new individuals arise from outgrowths of existing ones (see Figure 13.2). In stony corals, for example, buds form and remain attached to the parent. The eventual result is a

▲ **Figure 46.2 Asexual reproduction of a sea anemone (*Anthopleura elegantissima*).** The large individual in the center of this photograph is undergoing fission, a type of asexual reproduction. Two smaller individuals will form as the parent divides approximately in half. Each offspring will be a genetic copy of the parent.

colony more than 1 m across, consisting of thousands of connected individuals. In another form of asexual reproduction, some invertebrates, including certain sponges, release specialized groups of cells that can grow into new individuals.

Another process of asexual reproduction involves two steps: *fragmentation*, the breaking of the body into several pieces, followed by *regeneration*, the regrowth of lost body parts. If more than one piece grows and develops into a complete animal, the net effect is reproduction. For example, certain annelid worms can split their body into several fragments, each regenerating a complete worm in less than a week. Numerous sponges, cnidarians, bristle worms, and sea squirts also reproduce by fragmentation and regeneration.

Parthenogenesis is asexual reproduction in which an egg develops without being fertilized. Among invertebrates, parthenogenesis occurs in certain species of bees, wasps, and ants. The progeny can be either haploid or diploid. If haploid, the offspring develop into adults that produce eggs or sperm without meiosis. In the case of honeybees, males (drones) are fertile haploid adults that arise by parthenogenesis. (In contrast, female honeybees, including both the sterile workers and the fertile queens, are diploid adults that develop from fertilized eggs.) Among vertebrates, parthenogenesis has been observed in about one in every thousand species. Recently, zookeepers discovered parthenogenesis in the Komodo dragon and in a species of hammerhead shark. In both cases, females had been kept completely isolated from males of their species but nevertheless produced offspring.

Sexual Reproduction: An Evolutionary Enigma

EVOLUTION Sex must enhance reproductive success or survival because it would otherwise rapidly disappear. To see why, consider an animal population in which half the females reproduce sexually and half reproduce asexually (**Figure 46.3**). We'll assume that the number of offspring per female is a constant, two in this case. The two offspring of an asexual female will both be daughters that will each give birth to two more reproductive daughters. In contrast, half of a sexual female's offspring will be male. The number of sexual offspring will remain the same at each generation, because both a male and a female are required to reproduce. Thus, the asexual condition will increase in frequency at each generation. Yet despite this "twofold cost," sex is maintained even in animal species that can also reproduce asexually.

What advantage does sex provide? The answer remains elusive. Most hypotheses focus on the unique combinations of parental genes formed during meiotic recombination and fertilization. By producing offspring of varied genotypes, sexual reproduction may enhance the reproductive success of parents when environmental factors, such as pathogens, change relatively rapidly. In contrast, asexual reproduction is expected to be most advantageous in stable, favorable environments because it perpetuates successful genotypes faithfully and precisely.

There are a number of reasons why the unique gene combinations formed during sexual reproduction might be advantageous. One is that beneficial gene combinations arising through recombination might speed up adaptation. Although this idea appears straightforward, the theoretical advantage is significant only when the rate of beneficial mutations is high and population size is small. Another idea is that the shuffling of genes during sexual reproduction might allow a population to rid itself of sets of harmful genes

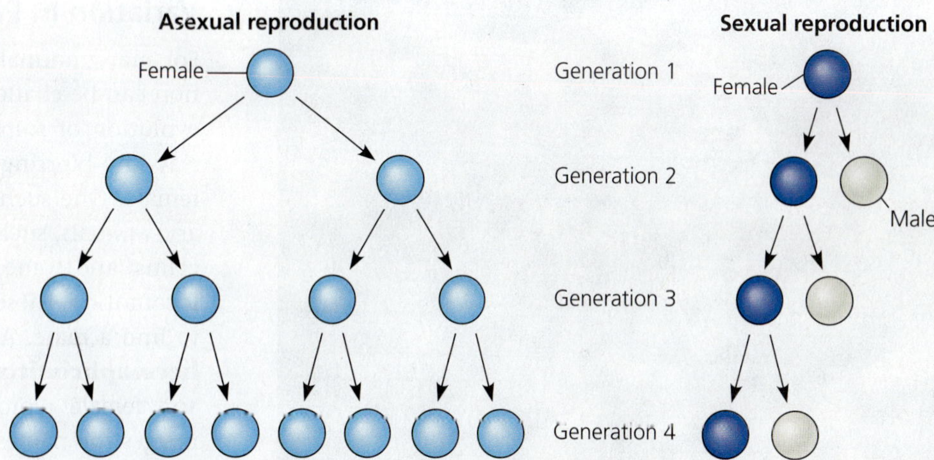

▲ **Figure 46.3 The "reproductive handicap" of sex.** These diagrams contrast the reproductive output of females (blue spheres) over four generations for asexual versus sexual reproduction, assuming two surviving offspring per female. The asexual population rapidly outgrows the sexual one.

more readily. Experiments to test these and other hypotheses are ongoing in many laboratories.

Reproductive Cycles

Most animals exhibit cycles in reproductive activity, often related to changing seasons. In this way, animals conserve resources, reproducing only when sufficient energy sources or stores are available and when environmental conditions favor the survival of offspring. For example, ewes (female sheep) have a reproductive cycle lasting 15–17 days. **Ovulation**, the release of mature eggs, occurs at the midpoint of each cycle. A ewe's cycle generally occurs only during fall and early winter, and the length of any resulting pregnancy is 5 months. Thus, most lambs are born in the early spring, when their chances of survival are optimal. Reproductive cycles are controlled by hormones, which in turn are regulated by environmental cues. Common environmental cues are changes in day length, seasonal temperature, rainfall, and lunar cycles.

Because seasonal temperature is often an important cue for reproduction, climate change can decrease reproductive success. Researchers in Denmark have demonstrated just such an effect on caribou (wild reindeer). In spring, caribou migrate to calving grounds to eat sprouting green plants, give birth, and care for their new calves (**Figure 46.4**). Changes in the length of daylight trigger the migration, while the seasonal rise in temperature that thaws the tundra causes plants to sprout. Prior to 1993, the arrival of the caribou at the calving grounds coincided with the brief period during which the plants were nutritious and digestible. Between 1993 and 2006, average spring temperatures in the calving grounds increased by more than 4°C, and the plants now sprout two weeks earlier. Since the length of daylight is unaffected by climate change, the timing of the caribou migration has not changed. The result is a timing mismatch between new plant growth and caribou birthing. Without adequate nutrition for the nursing females, production of caribou offspring has declined by 75%.

▲ **Figure 46.4 Caribou (*Rangifer tarandus*) mother and calf.** As a result of warming due to global climate change, the number of caribou offspring in a West Greenland study site has fallen fourfold.

Reproductive cycles are also found among animals that can reproduce both sexually and asexually. Consider, for instance, the water flea (genus *Daphnia*). A *Daphnia* female can produce eggs of two types. One type of egg requires fertilization to develop, but the other type does not and develops instead by parthenogenesis. Asexual reproduction occurs when environmental conditions are favorable, whereas sexual reproduction occurs during times of environmental stress. As a result, the switch between sexual and asexual reproduction is roughly linked to season.

A very different type of reproductive cycle is found among animals that only reproduce asexually. Several genera of fishes, amphibians, and reptiles engage in a complex form of parthenogenesis that involves the doubling of chromosomes after meiosis, producing diploid offspring. Among these are about 15 species of whiptail lizards in the genus *Aspidoscelis*. There are no males, but courtship and mating behaviors are typical of sexual species of the same genus. During the breeding season, one female of each mating pair mimics a male (**Figure 46.5a**). Each member of the pair alternates roles two or three times during the season. An individual adopts female behavior prior to ovulation, when the level of the female sex hormone estradiol is high, then switches to male-like behavior after ovulation, when the level of progesterone is highest (**Figure 46.5b**). Ovulation is more likely to occur if the individual is mounted during the critical time of the hormone cycle; isolated lizards lay fewer eggs than those that go through the motions of sex. These observations support the hypothesis that these parthenogenetic lizards evolved from species having two sexes and still require certain sexual stimuli for maximum reproductive success.

Variation in Patterns of Sexual Reproduction

For many animals, finding a partner for sexual reproduction can be challenging. Adaptations that arose during the evolution of some species meet this challenge in a novel way—by blurring the strict distinction between male and female. One such adaptation arose among sessile (stationary) animals, such as barnacles; burrowing animals, such as clams; and some parasites, including tapeworms. Lacking locomotion, these animals have a very limited opportunity to find a mate. An evolutionary solution to this problem is **hermaphroditism**, in which each individual has both male and female reproductive systems (the term *hermaphrodite* merges the names Hermes and Aphrodite, a Greek god and goddess). Because each hermaphrodite reproduces as both a male and a female, *any* two individuals can mate. Each animal donates and receives sperm during mating, as the sea slugs in Figure 46.1 are doing. In some species, hermaphrodites are

(a) Both lizards in this photograph are *A. uniparens* females. The one on top is playing the role of a male. Every two or three weeks during the breeding season, individuals switch sex roles.

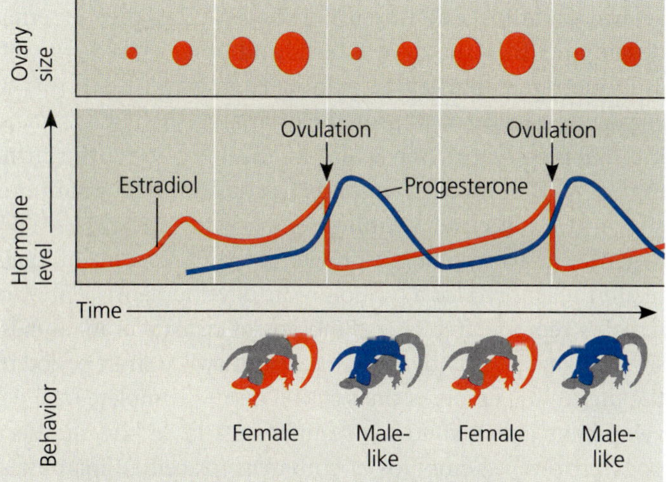

(b) The sexual behavior of *A. uniparens* is correlated with the cycle of ovulation mediated by sex hormones. As the blood level of estradiol rises, the ovaries grow, and the lizard behaves as a female. After ovulation, the estradiol level drops abruptly, and the progesterone level rises; these hormone levels correlate with male-like behavior.

▲ **Figure 46.5 Sexual behavior in parthenogenetic lizards.** The desert-grassland whiptail lizard (*Aspidoscelis uniparens*) is an all-female species. These reptiles reproduce by parthenogenesis, the development of an unfertilized egg. Nevertheless, ovulation is stimulated by mating behavior.

also capable of self-fertilization, allowing a form of sexual reproduction that doesn't require any partner.

The bluehead wrasse (*Thalassoma bifasciatum*), a coral reef fish, provides a well-studied example of a quite different variation in sexual reproduction. These wrasses live in harems, each consisting of a single male and several females. When the lone male dies, the opportunity for sexual reproduction would appear lost. Instead, a female wrasse undergoes sex reversal, a change in sex. Within a week, the transformed individual is producing sperm instead of eggs. Scientists have observed that it is the largest (and usually oldest) female in the harem that undergoes sex reversal. What advantage did this offer in the evolution of this wrasse? Because it is the male that defends a harem against intruders, a larger size may be more important for males than females in ensuring successful reproduction.

Certain oyster species also undergo sex reversal. In this case, individuals reproduce as males and then later as females, when their size is greatest. Since the number of gametes produced generally increases with size much more for females than for males, sex reversal in this direction maximizes gamete production. The result is enhanced reproductive success: Because oysters are sedentary animals and release their gametes into the surrounding water rather than mating directly, releasing more gametes tends to result in more offspring.

CONCEPT CHECK 46.1

1. Compare and contrast the outcomes of asexual and sexual reproduction.
2. Parthenogenesis is the most common form of asexual reproduction in animals that at other times reproduce sexually. What characteristic of parthenogenesis might explain this observation?
3. **WHAT IF?** If a hermaphrodite self-fertilizes, will the offspring be identical to the parent? Explain.
4. **MAKE CONNECTIONS** What examples of plant reproduction are most similar to asexual reproduction in animals? (See Concept 38.2, p. 812.)

For suggested answers, see Appendix A.

CONCEPT 46.2

Fertilization depends on mechanisms that bring together sperm and eggs of the same species

The union of sperm and egg—**fertilization**—can be either external or internal. In species with **external fertilization**, the female releases eggs into the environment, where the male then fertilizes them. Other species have **internal fertilization**: Sperm are deposited in or near the female reproductive tract, and fertilization occurs within the tract. (We'll discuss the cellular and molecular details of fertilization in Chapter 47.)

A moist habitat is almost always required for external fertilization, both to prevent the gametes from drying out and to allow the sperm to swim to the eggs. Many aquatic invertebrates simply shed their eggs and sperm into the surroundings, and fertilization occurs without the parents making physical contact. However, timing is crucial to ensure that mature sperm and eggs encounter one another.

Among some species with external fertilization, individuals clustered in the same area release their gametes into the water at the same time, a process known as *spawning*. In some cases, chemical signals that one individual generates in releasing gametes trigger others to release gametes. In other cases, environmental cues, such as temperature or day length, cause a whole population to release gametes at one time. For example, the palolo worm, native to coral reefs of the South Pacific, times its spawning to both the season and the lunar cycle. In spring, when the moon is in its last quarter, palolo worms break in half, releasing tail segments engorged with sperm or eggs. These packets rise to the ocean surface and burst in such vast numbers that the sea appears milky with gametes. The sperm quickly fertilize the floating eggs, and within hours, the palolo's once-a-year reproductive frenzy is complete.

When external fertilization is not synchronous across a population, individuals may exhibit specific mating behaviors leading to the fertilization of the eggs of one female by one male (**Figure 46.6**). Such "courtship" behavior has two important benefits: It allows mate choice (see Chapter 23) and, by triggering the release of both sperm and eggs, increases the probability of successful fertilization.

Internal fertilization is an adaptation that enables sperm to reach an egg efficiently, even when the environment is dry. It typically requires cooperative behavior that leads to copulation, as well as sophisticated and compatible reproductive systems. The male copulatory organ delivers sperm, and the female reproductive tract often has receptacles for storage and delivery of sperm to mature eggs.

No matter how fertilization occurs, the mating animals may make use of *pheromones*, chemicals released by one organism that can influence the physiology and behavior of other individuals of the same species. Pheromones are small, volatile or water-soluble molecules that disperse into the environment and, like hormones, are active in tiny amounts (see Chapter 45). Many pheromones function as mate attractants, enabling some female insects to be detected by males from as far away as a mile. (We will discuss mating behavior and pheromones further in Chapter 51.)

▲ **Figure 46.6 External fertilization.** Many species of amphibians reproduce by external fertilization. In most of these species, behavioral adaptations ensure that a male is present when the female releases eggs. Here, a female frog (on bottom) has released a mass of eggs in response to being clasped by a male. The male released sperm (not visible) at the same time, and external fertilization has already occurred in the water.

Ensuring the Survival of Offspring

Comparing internal and external fertilization across many species reveals that internal fertilization is typically associated with the production of fewer gametes but the survival of a higher fraction of zygotes. Better zygote survival is due in part to the fact that eggs fertilized internally are sheltered from potential predators. However, internal fertilization is also more often associated with mechanisms that provide greater protection of the embryos and parental care of the young. For example, the internally fertilized eggs of many species of terrestrial animals exhibit adaptations that protect against water loss and physical damage during their external development. In the case of birds and other reptiles, as well as monotremes (egg-laying mammals), the zygotes consist of eggs with calcium- and protein-containing shells and several internal membranes (see Figure 34.25). In contrast, the fertilized eggs of fishes and amphibians have only a gelatinous coat and lack internal membranes.

Rather than secreting a protective eggshell, some animals retain the embryo for a portion of its development within the female's reproductive tract. Embryos of marsupial mammals, such as kangaroos and opossums, spend only a short period in the uterus; the embryos then crawl out and complete fetal development attached to a mammary gland in the mother's pouch. However, embryos of eutherian (placental) mammals, such as humans, remain in the uterus throughout fetal development. There they are nourished by the mother's blood supply through a temporary organ, the placenta. The embryos of some fishes and sharks also complete development internally, although typically the embryo and mother in such species lack a connection dedicated to nutrient exchange.

When a baby eagle hatches out of an egg or when a human is born, the newborn is not yet capable of independent existence. Instead, adult birds feed their young and mammals nurse their offspring. Parental care is in fact much more widespread than you might suspect. For example, there are many invertebrates that provide parental care (**Figure 46.7**). Among vertebrates, the gastric brooding frogs (genus *Rheobatrachus*) of Australia provided a particularly unusual example prior to their extinction in the 1980s. During reproduction, the female frog would carry the tadpoles in her stomach until they underwent metamorphosis and hopped out of her mouth as young frogs.

Gamete Production and Delivery

Sexual reproduction in animals relies on sets of cells that are precursors for eggs and sperm. A group of cells dedicated to this purpose is often established very early in the formation of the embryo and remains in an inactive state while the body plan

▲ **Figure 46.7 Parental care in an invertebrate.** Compared with many other insects, giant water bugs of the genus *Belostoma* produce relatively few offspring, but offer much greater parental protection. Following internal fertilization, the female glues her fertilized eggs to the back of the male (shown here). The male carries them for days, frequently fanning water over them to keep the eggs moist, aerated, and free of parasites.

develops. Cycles of growth and mitosis then increase, or *amplify*, the number of cells available for making eggs or sperm.

In producing gametes from the amplified precursor cells and making them available for fertilization, animals employ a variety of reproductive systems. The simplest systems do not even include discrete **gonads**, the organs that produce gametes in most animals. The palolo and most other polychaete worms (phylum Annelida) have separate sexes but do not have distinct gonads; rather, the eggs and sperm develop from undifferentiated cells lining the coelom (body cavity). As the gametes mature, they are released from the body wall and fill the coelom. Depending on the species, mature gametes may be shed through the excretory opening, or the swelling mass of eggs may split a portion of the body open, spilling the eggs into the environment.

More elaborate reproductive systems include sets of accessory tubes and glands that carry, nourish, and protect the gametes and sometimes the developing embryos. Most insect species, for example, have separate sexes with complex reproductive systems (**Figure 46.8**). In the males, sperm develop in a pair of testes and are passed along a coiled duct to two seminal vesicles for storage. During mating, sperm are ejaculated into the female reproductive system. There, eggs develop in a pair of ovaries and are conveyed through ducts to the uterus. Eggs are fertilized in the uterus and then expelled for development outside the body. In many insect species, the female reproductive system includes one or more **spermathecae** (singular, spermatheca), sacs in which sperm may be stored for extended periods, a year or more in some species. Because the female releases male gametes from the spermatheca only in response to the appropriate stimuli, fertilization occurs

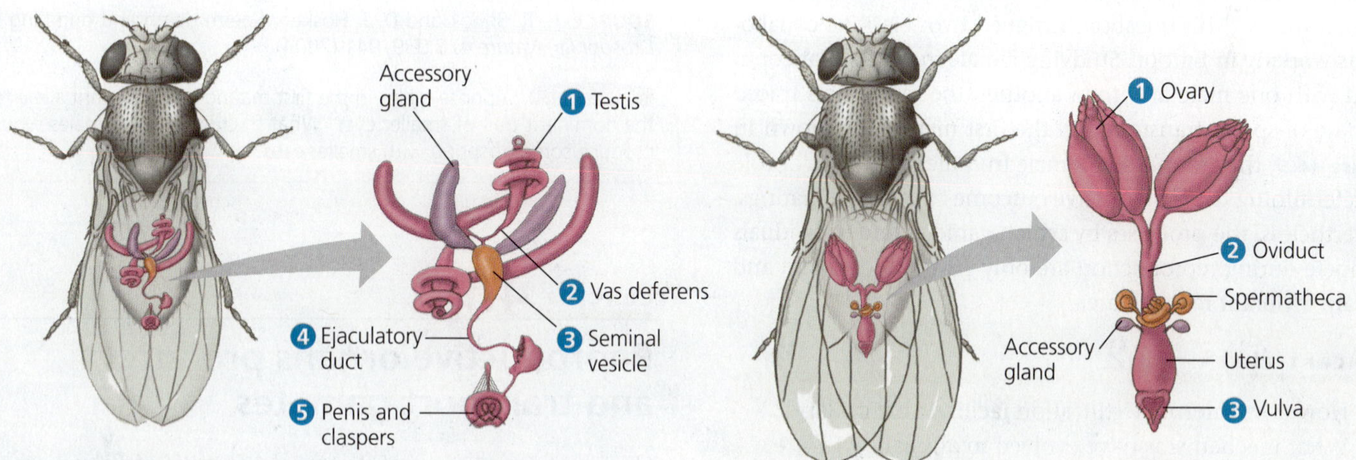

(a) Male fruit fly. Sperm form in the testes, pass through the sperm ducts (vas deferens), and are stored in the seminal vesicles. The male ejaculates sperm along with fluid from the accessory glands. (Males of some species of insects and other arthropods have appendages called claspers that grasp the female during copulation.)

(b) Female fruit fly. Eggs develop in the ovaries and then travel through the oviducts to the uterus. After mating, sperm are stored in the spermathecae, which are connected to the uterus by short ducts. The female uses a stored sperm to fertilize each egg as it enters the uterus before she passes the egg out through the vulva.

▲ **Figure 46.8 Insect reproductive anatomy.** Circled numbers indicate sequences of sperm and egg movement.

under conditions likely to be well suited to embryonic development. Even more complex reproductive systems can be found in some animals whose body plans are otherwise fairly simple, such as parasitic flatworms.

The basic plans of all vertebrate reproductive systems are quite similar, but there are some important variations. In many nonmammalian vertebrates, the digestive, excretory, and reproductive systems have a common opening to the outside, the **cloaca**, a structure that was probably also present in the ancestors of all vertebrates. In contrast, mammals generally lack a cloaca and have a separate opening for the digestive tract. In addition, most female mammals have separate openings for the excretory and reproductive systems. Among most vertebrates, the uterus is partly or completely divided into two chambers. However, in humans and other mammals that produce only one or a few young at a time, as well as in birds and many snakes, the uterus is a single structure. Male reproductive systems differ mainly in the copulatory organs. Many nonmammalian vertebrates, including all reptiles and amphibians, lack a well-developed penis and instead ejaculate sperm by turning the cloaca inside out.

Although fertilization involves the union of a single egg and sperm, animals often mate with more than one member of the other sex. Indeed, monogamy, the sustained sexual partnership of two individuals, is relatively rare among animals, including most mammals. Mechanisms have evolved, however, that enhance the reproductive success of a male with a particular female and diminish the chance of that female mating successfully with another partner. For example, some male insects transfer secretions that make a female less receptive to courtship, thereby reducing the likelihood of her mating again.

Can females also influence the relative reproductive success of their mates? This question intrigued two scientific collaborators working in Europe. Studying female fruit flies that copulated with one male and then another, the researchers traced the fate of sperm transferred in the first mating. As shown in **Figure 46.9**, they found that female fruit flies play a major role in determining the reproductive outcome of multiple matings. Nevertheless, the processes by which gametes and individuals compete during reproduction are only partly understood and remain a vibrant research area.

▼ **Figure 46.9** INQUIRY

Why is sperm usage biased when female fruit flies mate twice?

EXPERIMENT When a female fruit fly mates twice, 80% of the offspring result from the second mating. Scientists had postulated that ejaculate from the second mating displaces stored sperm. To test this hypothesis, Rhonda Snook, at the University of Sheffield, and David Hosken, at the University of Zurich, used mutant males with altered reproductive systems. "No-ejaculate" males mate, but do not transfer sperm or fluid to females. "No-sperm" males mate and ejaculate, but make no sperm. The researchers allowed females to mate with wild-type males and then mate with wild-type males, no-sperm males, or no-ejaculate males. As a control, some females were mated only once. The scientists then dissected each female under a microscope and recorded whether sperm were absent from the spermatheca, the major sperm storage organ.

RESULTS

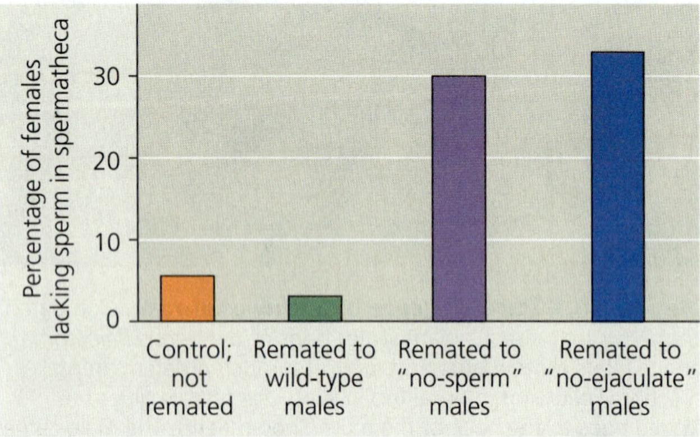

CONCLUSION Because remating reduces sperm storage when no sperm or fluids are transferred, the hypothesis that ejaculate from a second mating displaces stored sperm is incorrect. Instead, it appears that females sometimes get rid of stored sperm in response to remating. This might represent a way for females to replace stored sperm, possibly of diminished fitness, with fresh sperm.

SOURCE R. R. Snook and D. J. Hosken, Sperm death and dumping in *Drosophila*, *Nature* 428:939–941 (2004).

WHAT IF? Suppose males in the first mating had a mutant allele for the dominant trait of smaller eyes. What fraction of the females would produce some offspring with smaller eyes?

CONCEPT CHECK 46.2

1. How does internal fertilization facilitate life on land?
2. What mechanisms have evolved in animals with (a) external fertilization and (b) internal fertilization that help ensure that offspring survive to adulthood?
3. **MAKE CONNECTIONS** What are the shared and distinct functions of the uterus of an insect and the ovary of a flowering plant? (See Figure 38.6, p. 806.)

For suggested answers, see Appendix A.

CONCEPT 46.3

Reproductive organs produce and transport gametes

Having surveyed some of the general features of animal reproduction, we will focus the rest of the chapter on humans, beginning with the anatomy of the reproductive system in each sex.

Female Reproductive Anatomy

The female's external reproductive structures are the clitoris and two sets of labia, which surround the clitoris and vaginal opening. The internal organs are the gonads, which produce both eggs and reproductive hormones, and a system of ducts and chambers, which receive and carry gametes and house the embryo and fetus **(Figure 46.10)**.

Ovaries

The female gonads are a pair of ovaries that flank the uterus and are held in place in the abdominal cavity by ligaments. The outer layer of each ovary is packed with **follicles**, each consisting of an **oocyte**, a partially developed egg, surrounded by a group of support cells. The surrounding cells nourish and protect the oocyte during much of the formation and development of an egg. Although at birth the ovaries together contain about 1–2 million follicles, only about 500 follicles fully mature between puberty and menopause. During a typical 4-week menstrual cycle, one follicle matures and expels its egg, a process called ovulation. Prior to ovulation, cells of the follicle produce the primary female sex hormone, estradiol (a type of estrogen). After ovulation, the residual follicular tissue grows within the ovary, forming a mass called the **corpus luteum** ("yellow body"). The corpus luteum secretes additional estradiol, as well as progesterone, a hormone that helps maintain the uterine lining during pregnancy. If the egg cell is not fertilized, the corpus luteum degenerates, and a new follicle matures during the next cycle.

Oviducts and Uterus

An **oviduct**, or fallopian tube, extends from the uterus toward each ovary. The dimensions of this tube vary along its length, with the inside diameter near the uterus being as narrow as a human hair. At ovulation, the egg is released into the abdominal cavity near the funnel-like opening of the oviduct. Cilia on the epithelial lining of the duct help collect the egg by drawing fluid from the body cavity into the oviduct. Together with wavelike contractions of the oviduct, the cilia convey the egg down the duct to the **uterus**, also known as the womb. The uterus is a thick, muscular organ that can expand during pregnancy to accommodate a 4-kg fetus. The inner lining of the uterus, the **endometrium**, is richly supplied with blood vessels. The neck of the uterus, called the **cervix**, opens into the vagina.

Vagina and Vulva

The **vagina** is a muscular but elastic chamber that is the site for insertion of the penis and deposition of sperm during copulation. The vagina, which also serves as the birth canal through which a baby is born, opens to the outside at the **vulva**, the collective term for the external female genitalia.

A pair of thick, fatty ridges, the **labia majora**, encloses and protects the rest of the vulva. The vaginal opening and the separate opening of the urethra are located within a cavity bordered by a pair of slender skin folds, the **labia minora**. A thin piece of

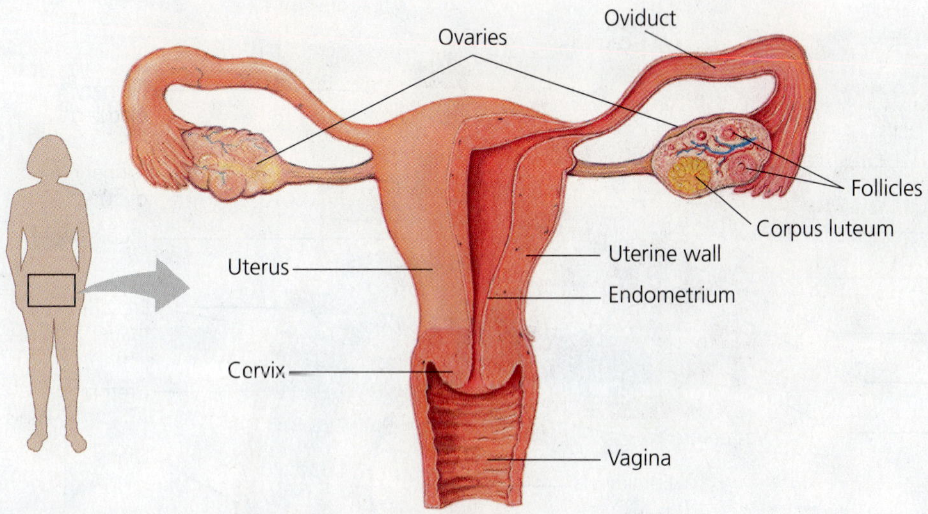

▲ **Figure 46.10 Reproductive anatomy of the human female.** Some nonreproductive structures are labeled in parentheses for orientation purposes.

tissue called the **hymen** partly covers the vaginal opening in humans at birth and usually until sexual intercourse or vigorous physical activity ruptures it. Located at the top of the labia minora, the **clitoris** consists of erectile tissue supporting a rounded **glans**, or head, covered by a small hood of skin, the **prepuce**. During sexual arousal, the clitoris, vagina, and labia minora all engorge with blood and enlarge. Richly supplied with nerve endings, the clitoris is one of the most sensitive points of sexual stimulation. Sexual arousal also induces the vestibular glands near the vaginal opening to secrete lubricating mucus, thereby facilitating intercourse.

Mammary Glands

The **mammary glands** are present in both sexes, but they normally produce milk only in females. Though not part of the reproductive system, the female mammary glands are important to reproduction. Within the glands, small sacs of epithelial tissue secrete milk, which drains into a series of ducts that open at the nipple. The breasts contain connective and fatty (adipose) tissue in addition to the mammary glands. Because the low level of estradiol in males limits the development of the fat deposits, male breasts usually remain small.

Male Reproductive Anatomy

The human male's external reproductive organs are the scrotum and penis. The internal reproductive organs consist of gonads that produce both sperm and reproductive hormones, accessory glands that secrete products essential to sperm movement, and ducts that carry the sperm and glandular secretions **(Figure 46.11)**.

Testes

The male gonads, or **testes** (singular, *testis*), produce sperm in highly coiled tubes called **seminiferous tubules**. The **Leydig cells**, scattered in connective tissue between the tubules, produce testosterone and other androgens (see Chapter 45).

Most mammals produce sperm properly only when the testes are cooler than normal body temperature. In humans and many other mammals, the **scrotum**, a fold of the body wall, maintains testis temperature about 2°C below that of the rest of the body. The testes develop in the abdominal cavity and descend into the scrotum just before birth (a testis within a scrotum is a *testicle*). In many rodents, the testes are drawn back into the cavity between breeding seasons, interrupting sperm maturation. Some mammals whose body temperature is low enough to allow sperm maturation—such as whales and elephants—retain the testes in the abdominal cavity.

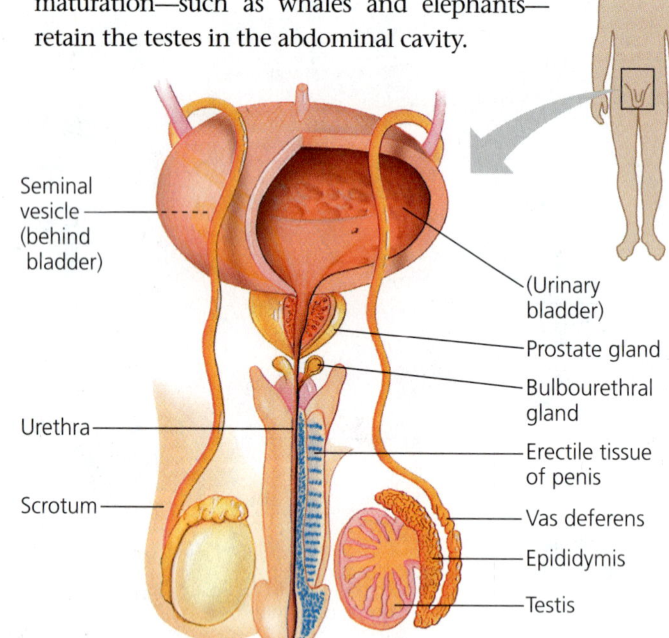

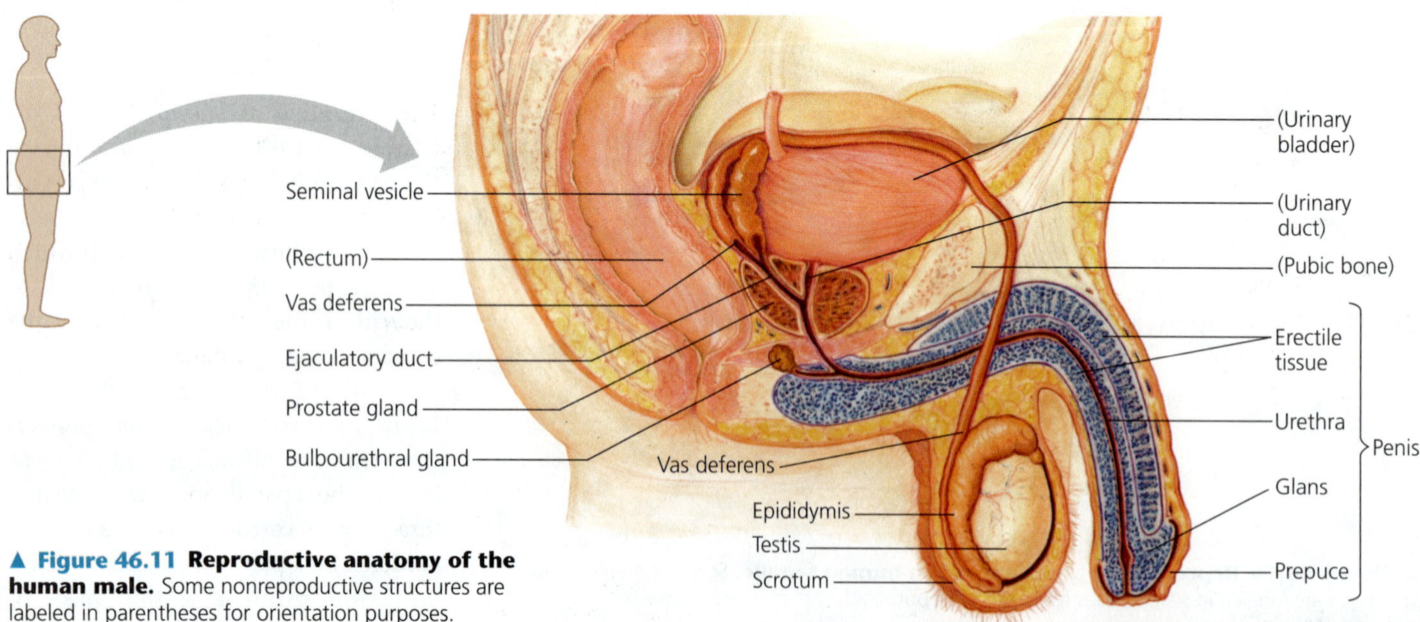

▲ **Figure 46.11 Reproductive anatomy of the human male.** Some nonreproductive structures are labeled in parentheses for orientation purposes.

Ducts

From the seminiferous tubules of a testis, the sperm pass into the coiled duct of an **epididymis**. In humans, it takes 3 weeks for sperm to pass through this 6-m-long duct. During this passage through the epididymis, the sperm complete maturation and become motile, although they acquire the ability to fertilize an egg only upon exposure to the chemical environment of the female reproductive system. During **ejaculation**, the sperm are propelled from each epididymis through a muscular duct, the **vas deferens**. Each vas deferens (one from each epididymis) extends around and behind the urinary bladder, where it joins a duct from the seminal vesicle, forming a short **ejaculatory duct**. The ejaculatory ducts open into the **urethra**, the outlet tube for both the excretory system and the reproductive system. The urethra runs through the penis and opens to the outside at the tip of the penis.

Accessory Glands

Three sets of accessory glands—the seminal vesicles, the prostate gland, and the bulbourethral glands—produce secretions that combine with sperm to form **semen**, the fluid that is ejaculated. Two **seminal vesicles** contribute about 60% of the volume of semen. The fluid from the seminal vesicles is thick, yellowish, and alkaline. It contains mucus, the sugar fructose (which provides most of the sperm's energy), a coagulating enzyme, ascorbic acid, and local regulators called prostaglandins (see Chapter 45).

The **prostate gland** secretes its products directly into the urethra through several small ducts. This fluid is thin and milky; it contains anticoagulant enzymes and citrate (a sperm nutrient). The prostate gland is the source of some of the most common medical problems of men over age 40. Benign (noncancerous) enlargement of the prostate occurs in more than half of all men in this age-group and in almost all men over 70. In addition, prostate cancer, which most often afflicts men 65 and older, is one of the most common human cancers.

The *bulbourethral glands* are a pair of small glands along the urethra below the prostate. Before ejaculation, they secrete clear mucus that neutralizes any acidic urine remaining in the urethra. Bulbourethral fluid also carries some sperm released before ejaculation, which is one reason for the high failure rate of the withdrawal method of birth control (coitus interruptus).

Penis

The human **penis** contains the urethra, as well as three cylinders of spongy erectile tissue. During sexual arousal, the erectile tissue, which is derived from modified veins and capillaries, fills with blood from the arteries. As this tissue fills, the increasing pressure seals off the veins that drain the penis, causing it to engorge with blood. The resulting erection enables the penis to be inserted into the vagina. Alcohol consumption, certain drugs, emotional issues, and aging all can cause a temporary inability to achieve an erection (erectile dysfunction). For individuals with long-term erectile dysfunction, drugs such as Viagra promote the vasodilating action of the local regulator nitric oxide (NO; see Chapter 45); the resulting relaxation of smooth muscles in the blood vessels of the penis enhances blood flow into the erectile tissues. Although all mammals rely on penile erection for mating, the penis of rodents, raccoons, walruses, whales, and several other mammals also contains a bone, the baculum, which probably further stiffens the penis for mating.

The main shaft of the penis is covered by relatively thick skin. The head, or glans, of the penis has a much thinner covering and is consequently more sensitive to stimulation. The human glans is covered by a fold of skin called the prepuce, or foreskin, which is removed if a male is circumcised.

Gametogenesis

Many of the differences in reproductive anatomy between males and females reflect the distinct structures and functions of the two types of gametes. Sperm are small and motile and must pass from the male to the female. In contrast, eggs, which provide the initial food stores for the embryo, are typically much larger and carry out their function within the female reproductive system. There they must mature in synchrony with the tissues that will support the embryo. Reflecting these differences, egg development and sperm development involve different patterns of meiotic division. We will highlight these differences as we explore **gametogenesis**, the production of gametes.

Spermatogenesis, the formation and development of sperm, is continuous and prolific in adult males. To produce hundreds of millions of sperm each day, cell division and maturation occur throughout the seminiferous tubules coiled within the two testes. For a single sperm, the process takes about 7 weeks from start to finish.

Oogenesis, the development of mature oocytes (eggs), is a prolonged process in the human female. Immature eggs form in the ovary of the female embryo but do not complete their development until years, and often decades, later.

Spermatogenesis differs from oogenesis in three significant ways. First, only in spermatogenesis do all four products of meiosis develop into mature gametes. In oogenesis, cytokinesis during meiosis is unequal, with almost all the cytoplasm segregated to a single daughter cell. This large cell is destined to become the egg; the other products of meiosis, smaller cells called polar bodies, degenerate. Second, spermatogenesis occurs throughout adolescence and adulthood. During oogenesis in human females, mitotic divisions are thought to be complete before birth, and the production of mature gametes ceases at about age 50. Third, spermatogenesis produces mature sperm from precursor cells in a continuous sequence, whereas oogenesis has long interruptions. **Figure 46.12**, on the next two pages, compares and contrasts the steps and organization of spermatogenesis and oogenesis in humans.

▼ **Figure 46.12**

Exploring Human Gametogenesis

Spermatogenesis

These drawings correlate the mitotic and meiotic divisions in sperm development with the microscopic structure of seminiferous tubules.

The initial or *primordial* germ cells of the embryonic testes divide and differentiate into stem cells that divide mitotically to form **spermatogonia**, which in turn generate spermatocytes, also by mitosis. Each spermatocyte gives rise to four spermatids through meiotic cell divisions that reduce the chromosome number from diploid ($2n = 46$ in humans) to haploid ($n = 23$). Spermatids undergo extensive changes in cell shape and organization in differentiating into sperm.

Within the seminiferous tubules, there is a concentric organization of the steps of spermatogenesis. Stem cells are situated near the outer edge of the tubules. As spermatogenesis proceeds, cells move steadily inward as they pass through the spermatocyte stage and the spermatid stage. In the last step, mature sperm are released into the lumen (fluid-filled cavity) of the tubule. The sperm travel along the tubule into the epididymis, where they become motile.

The structure of a sperm cell fits its function. In humans, as in most species, a head containing the haploid nucleus is tipped with a special vesicle, the **acrosome**, which contains enzymes that help the sperm penetrate an egg. Behind the head, the sperm cell contains large numbers of mitochondria (or one large mitochondrion in some species) that provide ATP for movement of the tail, which is a flagellum.

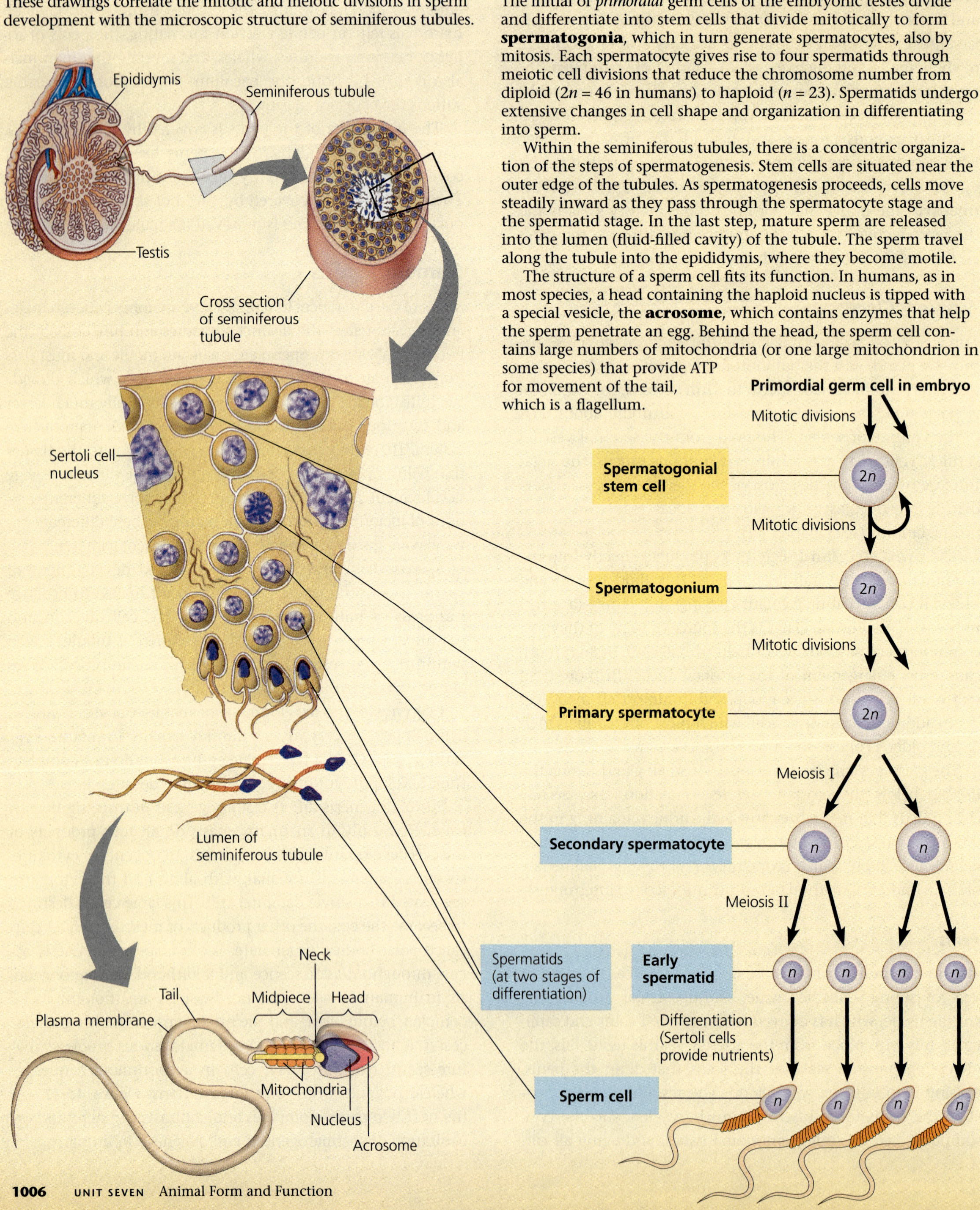

Oogenesis

Oogenesis begins in the female embryo with the production of **oogonia** from primordial germ cells. The oogonia divide by mitosis to form cells that begin meiosis, but stop the process at prophase I before birth. These developmentally arrested cells, called **primary oocytes**, each reside within a small follicle, a cavity lined with protective cells. Beginning at puberty, follicle-stimulating hormone (FSH) periodically stimulates a small group offollicles to resume growth and development. Typically, only one follicle fully matures each month, with its primary oocyte completing meiosis I. The second meiotic division begins, but stops at metaphase. Thus arrested in meiosis II, the **secondary oocyte** is released at ovulation, when its follicle breaks open. Only if a sperm penetrates the oocyte does meiosis II resume. (In other animal species, the sperm may enter the oocyte at the same stage, earlier, or later.) Each of the two meiotic divisions involves unequal cytokinesis, with the smaller cells becoming polar bodies that eventually degenerate (the first polar body may or may not divide again). Thus, the functional product of complete oogenesis is a single mature egg already containing a sperm head; fertilization is defined strictly as the fusion of the haploid nuclei of the sperm and secondary oocyte, although we often use it loosely to mean the entry of the sperm head into the egg.

The ruptured follicle left behind after ovulation develops into the corpus luteum. If the released oocyte is not fertilized and does not complete oogenesis, the corpus luteum degenerates.

To the best of our current knowledge, women are born with all the primary oocytes they will ever have. It is worth noting, however, that a similar conclusion regarding most other mammals was overturned by the discovery in 2004 of multiplying oogonia in the ovaries of adult mice that develop into oocytes. If the same turned out to be true of humans, it might be that the marked decline in fertility that occurs as women age results from both a depletion of oogonia and the degeneration of aging oocytes.

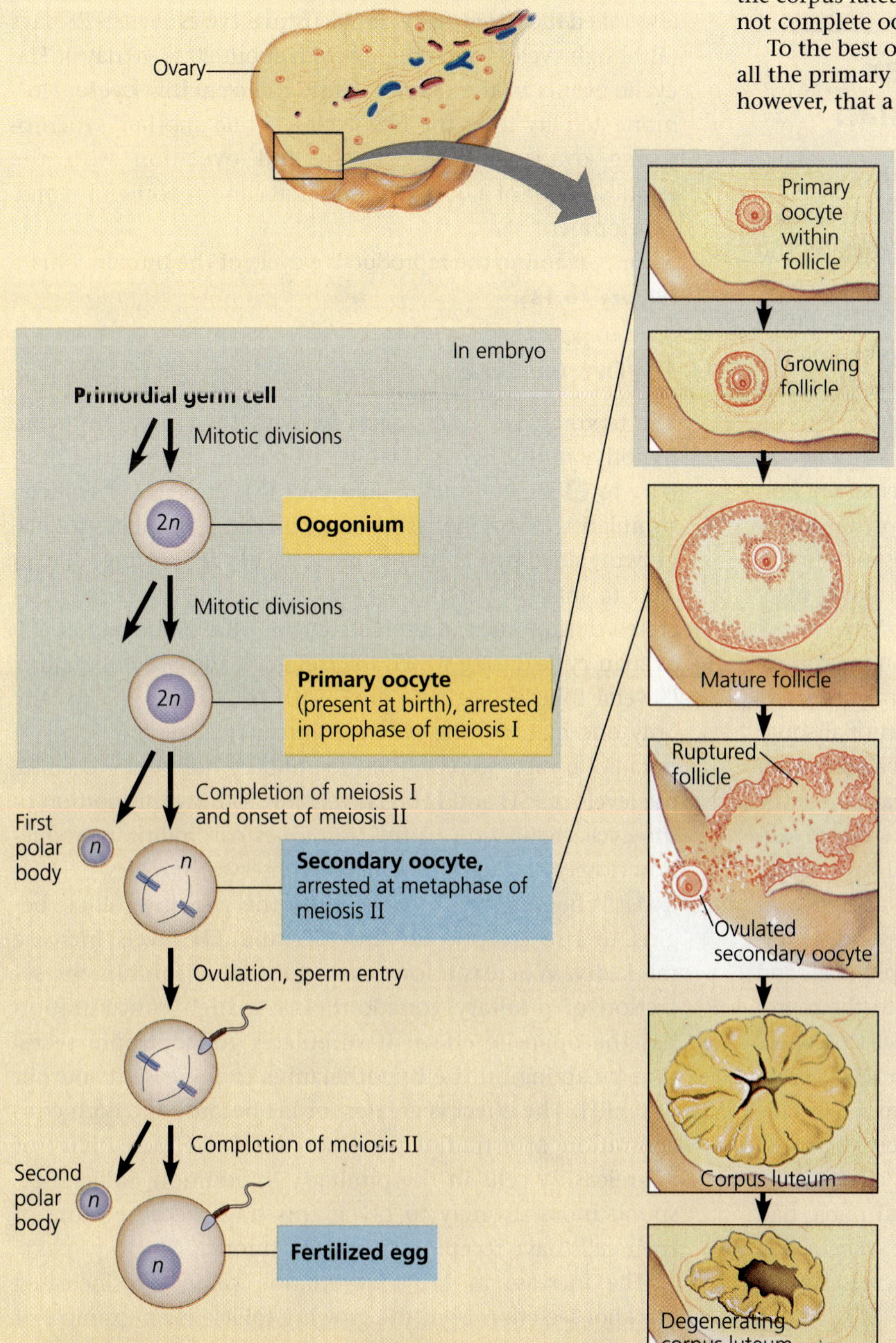

WHAT IF? Suppose you are analyzing the DNA from the polar bodies formed during human oogenesis. If the mother has a mutation in a known disease gene, would analyzing the polar body DNA allow you to infer whether the mutation is present in the mature oocyte? Explain.

CONCEPT CHECK 46.3

1. Why might using a hot tub frequently make it harder for a couple to conceive a child?
2. Oogenesis is often described as the production of a haploid egg by meiosis; but in some animals, including humans, this is not an entirely accurate description. Explain.
3. WHAT IF? If each vas deferens in a male was surgically sealed off, what changes would you expect in sexual response and ejaculate composition?

For suggested answers, see Appendix A.

CONCEPT 46.4

The interplay of tropic and sex hormones regulates mammalian reproduction

In both male and female humans, the coordinated actions of hormones from the hypothalamus, anterior pituitary, and gonads govern reproduction. The hypothalamus secretes gonadotropin-releasing hormone (GnRH), which then directs the anterior pituitary to secrete the gonadotropins, follicle-stimulating hormone (FSH) and luteinizing hormone (LH) (see Figure 45.16). These two hormones regulate gametogenesis directly, by targeting tissues in the gonads, as well as indirectly, by regulating sex hormone production. The principal sex hormones are steroid hormones: in males, androgens, especially testosterone; in females, estrogens, especially estradiol, and progesterone. Like the gonadotropins, the sex hormones regulate gametogenesis both directly and indirectly.

Sex hormones serve many functions in addition to promoting gamete production. In many vertebrates, androgens are responsible for male vocalizations, such as the territorial songs of birds and the mating calls of frogs. During development of the human embryo, androgens promote the appearance of the primary sex characteristics of males, the structures directly involved in reproduction. These include the seminal vesicles and associated ducts, as well as external reproductive anatomy. At puberty, sex hormones in both males and females induce formation of secondary sex characteristics, the physical and behavioral features that are not directly related to the reproductive system. In males, androgens cause the voice to deepen, facial and pubic hair to develop, and muscles to grow (by stimulating protein synthesis). Androgens also promote specific sexual behaviors and sex drive, as well as an increase in general aggressiveness. Estrogens similarly have multiple effects in females. At puberty, estradiol stimulates breast and pubic hair development. Estradiol also influences female sexual behavior, induces fat deposition in the breasts and hips, increases water retention, and alters calcium metabolism.

Hormonal Control of Female Reproductive Cycles

Upon reaching sexual maturity, human males carry out gametogenesis continuously, whereas human females produce gametes in cycles. Ovulation occurs only after the endometrium (lining of the uterus) has started to thicken and develop a rich blood supply, preparing the uterus for the possible implantation of an embryo. If pregnancy does not occur, the uterine lining is sloughed off, and another cycle begins. The cyclic shedding of the blood-rich endometrium from the uterus, a process that occurs in a flow through the cervix and vagina, is called **menstruation**.

There are two closely linked reproductive cycles in human females. Changes in the uterus define the **menstrual cycle**, also called the **uterine cycle**. Menstrual cycles average 28 days (although cycles vary, ranging from about 20 to 40 days). The cyclic events in the ovaries define the **ovarian cycle**. Hormone activity links the two cycles to one another, synchronizing ovarian follicle growth and ovulation with the establishment of a uterine lining that can support embryonic development.

Let's examine the reproductive cycle of the human female (**Figure 46.13**).

The Ovarian Cycle

The reproductive cycle begins ❶ with the release from the hypothalamus of GnRH, which stimulates the anterior pituitary to ❷ secrete small amounts of FSH and LH. ❸ Follicle-stimulating hormone (as its name implies) stimulates follicle growth, aided by LH, and ❹ the cells of the growing follicles start to make estradiol. There is a slow rise in estradiol secreted during most of the **follicular phase**, the part of the ovarian cycle during which follicles grow and oocytes mature. (Several follicles begin to grow with each cycle, but usually only one matures; the others disintegrate.) The low levels of estradiol inhibit secretion of the pituitary hormones, keeping the levels of FSH and LH relatively low. During this portion of the cycle, regulation of the hormones controlling reproduction closely parallels the regulation observed in males.

❺ When estradiol secretion by the growing follicle begins to rise steeply, ❻ the FSH and LH levels increase markedly. Whereas a low level of estradiol inhibits the secretion of pituitary gonadotropins, a high concentration has the opposite effect: It stimulates gonadotropin secretion by acting on the hypothalamus to increase its output of GnRH. The effect is greater for LH because the high concentration of estradiol increases the GnRH sensitivity of LH-releasing cells in the pituitary. In addition, follicles respond more strongly to LH at this stage because more of their cells have receptors for this hormone.

The increase in LH concentration caused by increased estradiol secretion from the growing follicle is an example of positive feedback. The result is final maturation of the follicle. 7 The maturing follicle, containing a fluid-filled cavity, enlarges, forming a bulge near the surface of the ovary. The follicular phase ends at ovulation, about a day after the LH surge. In response to the peak in LH levels, the follicle and adjacent wall of the ovary rupture, releasing the secondary oocyte. There is sometimes a distinctive pain in the lower abdomen at or near the time of ovulation; this pain is felt on the left or right side, corresponding to whichever ovary has matured a follicle during that cycle.

The **luteal phase** of the ovarian cycle follows ovulation. 8 LH stimulates the follicular tissue left behind in the ovary to transform into a corpus luteum, a glandular structure. Under continued stimulation by LH, the corpus luteum secretes progesterone and estradiol. As progesterone and estradiol levels rise, the combination of these steroid hormones exerts negative feedback on the hypothalamus and pituitary, reducing the secretion of LH and FSH to very low levels. This negative feedback prevents another egg from maturing when a pregnancy may already be under way.

Near the end of the luteal phase, low gonadotropin levels cause the corpus luteum to disintegrate, triggering a sharp decline in estradiol and progesterone concentrations. The decreasing levels of ovarian steroid hormones liberate the hypothalamus and pituitary from the negative-feedback effect of these hormones. The pituitary can then begin to secrete enough FSH to stimulate the growth of new follicles in the ovary, initiating the next ovarian cycle.

The Uterine (Menstrual) Cycle

Prior to ovulation, ovarian steroid hormones stimulate the uterus to prepare for support of an embryo. Estradiol secreted in increasing amounts by growing follicles signals the endometrium to thicken. In this way, the follicular phase of the ovarian cycle is coordinated with the **proliferative phase** of the uterine cycle. After ovulation, 9 estradiol and

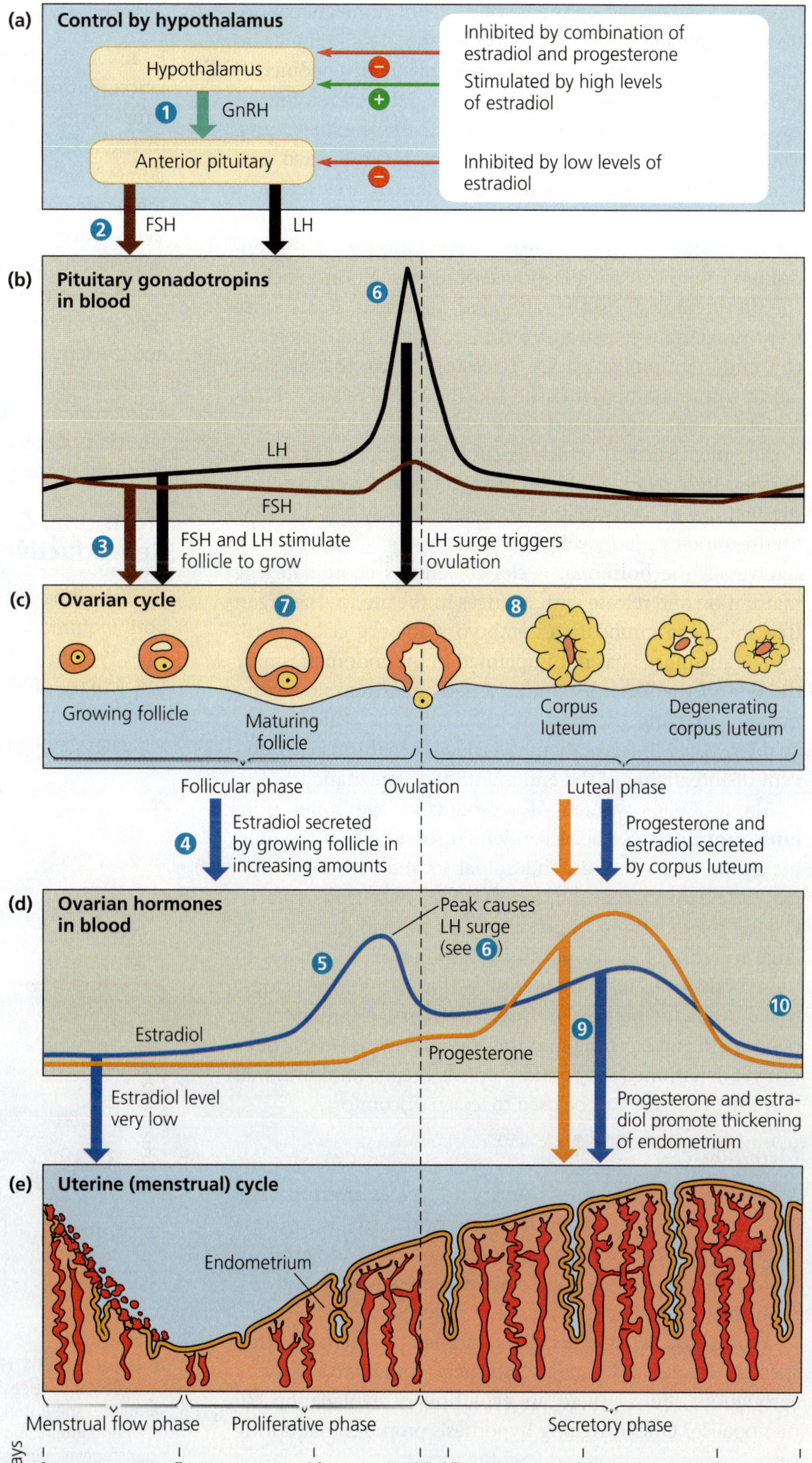

▲ **Figure 46.13 The reproductive cycle of the human female.** This figure shows how **(c)** the ovarian cycle and **(e)** the uterine (menstrual) cycle are regulated by changing hormone levels in the blood, depicted in parts **(a)**, **(b)**, and **(d)**. The time scale at the bottom of the figure applies to parts **(b)**–**(e)**.

progesterone secreted by the corpus luteum stimulate continued development and maintenance of the uterine lining, including enlargement of arteries and growth of endometrial glands. These glands secrete a nutrient fluid that can sustain an early embryo even before it implants in the uterine lining. Thus, the luteal phase of the ovarian cycle is coordinated with what is called the **secretory phase** of the uterine cycle.

Upon disintegration of the corpus luteum, the rapid drop ⑩ in ovarian hormone levels causes arteries in the endometrium to constrict. Deprived of its circulation, much of the uterine lining disintegrates, and the uterus, in response to prostaglandin secretion, contracts. Small blood vessels in the endometrium constrict, releasing blood that is shed along with endometrial tissue and fluid. The result is menstruation—the **menstrual flow phase** of the uterine cycle. During menstruation, which usually persists for a few days, a new group of ovarian follicles begin to grow. By convention, the first day of menstruation is designated day 1 of the new uterine (and ovarian) cycle.

Overall, the hormonal cycles in females coordinate egg maturation and release with changes in the uterus, the organ that must accommodate an embryo if the egg cell is fertilized. If an embryo has not implanted in the endometrium by the end of the secretory phase, a new menstrual flow commences, marking the start of the next cycle. Later in the chapter, you will learn about override mechanisms that prevent disintegration of the endometrium in pregnancy.

About 7% of women of reproductive age suffer from **endometriosis**, a disorder in which some cells of the uterine lining migrate to an abdominal location that is abnormal, or **ectopic** (from the Greek *ektopos*, away from a place). Having migrated to a location such as an oviduct, ovary, or large intestine, the ectopic tissue responds to hormones in the bloodstream. Like the uterine endometrium, the ectopic tissue swells and breaks down each ovarian cycle, resulting in pelvic pain and bleeding into the abdomen. Researchers have not yet determined why endometriosis occurs, but hormonal therapy or surgery can be used to lessen discomfort.

Menopause

After about 500 cycles, a woman undergoes **menopause**, the cessation of ovulation and menstruation. Menopause usually occurs between the ages of 46 and 54. During this interval, the ovaries lose their responsiveness to FSH and LH, resulting in a decline in estradiol production.

Menopause is an unusual phenomenon. In most other species, females and males retain their reproductive capacity throughout life. Is there an evolutionary explanation for menopause? One intriguing hypothesis proposes that during early human evolution, undergoing menopause after bearing several children allowed a mother to provide better care for her children and grandchildren, thereby increasing the survival of individuals who share much of her genetic makeup.

Menstrual Versus Estrous Cycles

In all female mammals, the endometrium thickens before ovulation, but only humans and some other primates have menstrual cycles. Other mammals have **estrous cycles**, in which in the absence of a pregnancy, the uterus reabsorbs the endometrium and no extensive fluid flow occurs. Whereas human females may engage in sexual activity throughout the menstrual cycle, mammals with estrous cycles usually copulate only during the period surrounding ovulation. This period, called estrus (from the Latin *oestrus*, frenzy, passion), is the only time the female is receptive to mating. It is often called "heat," and the female's temperature does increase slightly.

The length and frequency of estrous cycles vary widely among mammals. Bears and wolves have one estrous cycle per year; elephants have several. Rats have estrous cycles throughout the year, each lasting only 5 days.

Hormonal Control of the Male Reproductive System

In males, the FSH and LH secreted in response to GnRH are both required for normal spermatogenesis. Each acts on a distinct type of cell in the testis **(Figure 46.14)**. FSH promotes the activity of Sertoli cells. Within the seminiferous tubules, these cells nourish developing sperm (see Figure 46.12). LH regulates Leydig cells, located in the interstitial space between

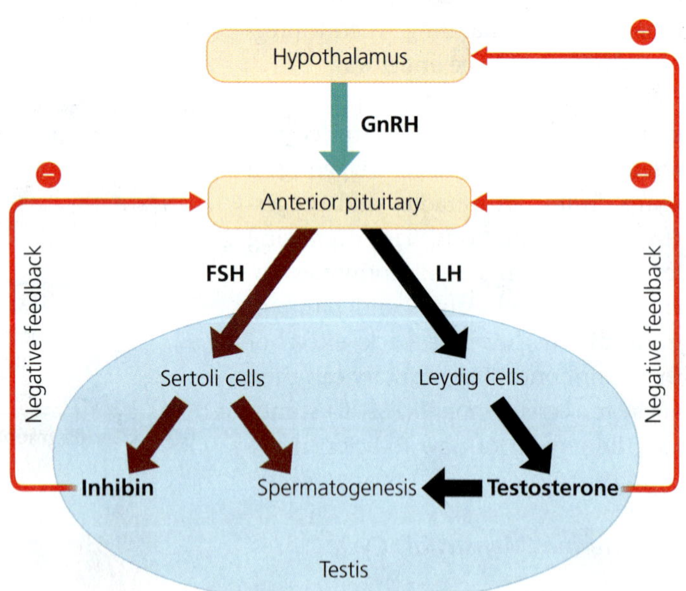

▲ **Figure 46.14 Hormonal control of the testes.** Gonadotropin-releasing hormone (GnRH) from the hypothalamus stimulates the anterior pituitary to secrete follicle-stimulating hormone (FSH) and luteinizing hormone (LH). FSH acts on Sertoli cells, which nourish developing sperm. LH acts on Leydig cells, which produce androgens, chiefly testosterone. Negative feedback by testosterone on the hypothalamus and anterior pituitary regulates blood levels of GnRH, LH, and FSH. FSH secretion is also subject to negative feedback by a hormone called inhibin, secreted by Sertoli cells.

the seminiferous tubules. In response to LH, Leydig cells secrete testosterone and other androgens, which promote spermatogenesis in the tubules. Both androgen secretion and spermatogenesis occur continuously from puberty onward.

Two negative-feedback mechanisms control sex hormone production in males (see Figure 46.14). Testosterone regulates blood levels of GnRH, FSH, and LH through inhibitory effects on the hypothalamus and anterior pituitary. In addition, **inhibin**, a hormone that in males is produced by Sertoli cells, acts on the anterior pituitary gland to reduce FSH secretion. Together, these negative-feedback circuits maintain androgen production at optimal levels.

Human Sexual Response

Whereas there is a wealth of information regarding the hormonal regulation of human oogenesis and spermatogenesis, comparable data regarding sexual desire and responses are scanty. Testosterone, prolactin, and oxytocin each appear to influence sexual function in males and females, but their precise roles have yet to be defined. Instead, the study of human sexual response has largely focused on the physiological changes associated with sexual activity.

As mentioned earlier, many animals exhibit elaborate mating behavior. The arousal of sexual interest in humans is particularly complex, involving a variety of psychological as well as physical factors. Reproductive structures in the male and female that are quite different in appearance often serve similar functions, reflecting their shared developmental origin. For example, the same embryonic tissues give rise to the glans of the penis and the clitoris, the scrotum and the labia majora, and the skin on the penis and the labia minora.

The general pattern of human sexual response is similar in males and females. Two types of physiological reactions predominate in both sexes: **vasocongestion**, the filling of a tissue with blood, and **myotonia**, increased muscle tension. Both skeletal and smooth muscle may show sustained or rhythmic contractions, including those associated with orgasm.

The sexual response cycle can be divided into four phases: excitement, plateau, orgasm, and resolution. An important function of the excitement phase is to prepare the vagina and penis for **coitus** (sexual intercourse). During this phase, vasocongestion is particularly evident in erection of the penis and clitoris; enlargement of the testicles, labia, and breasts; and vaginal lubrication. Myotonia may occur, resulting in nipple erection or tension of the arms and legs.

In the plateau phase, these responses continue as a result of direct stimulation of the genitalia. In females, the outer third of the vagina becomes vasocongested, while the inner two-thirds slightly expands. This change, coupled with the elevation of the uterus, forms a depression for receiving sperm at the back of the vagina. Breathing increases and heart rate rises, sometimes to 150 beats per minute—not only in response to the physical effort of sexual activity, but also as an involuntary response to stimulation of the autonomic nervous system (see Figure 49.8).

Orgasm is characterized by rhythmic, involuntary contractions of the reproductive structures in both sexes. Male orgasm has two stages. The first, emission, occurs when the glands and ducts of the reproductive tract contract, forcing semen into the urethra. Expulsion, or ejaculation, occurs when the urethra contracts and the semen is expelled. During female orgasm, the uterus and outer vagina contract, but the inner two-thirds of the vagina does not. Orgasm is the shortest phase of the sexual response cycle, usually lasting only a few seconds. In both sexes, contractions occur at about 0.8-second intervals and may also involve the anal sphincter and several abdominal muscles.

The resolution phase completes the cycle and reverses the responses of the earlier stages. Vasocongested organs return to their normal size and color, and muscles relax. Most of the changes of resolution are completed within 5 minutes, but some may take as long as an hour. Following orgasm, the male typically enters a refractory period, lasting anywhere from a few minutes to hours, during which erection and orgasm cannot be achieved. Females do not have a refractory period, making possible multiple orgasms within a short period of time.

CONCEPT CHECK 46.4

1. FSH and LH get their names from events of the female reproductive cycle, but they also function in males. How are their functions in females and males similar?
2. How does an estrous cycle differ from a menstrual cycle, and in what animals are the two types of cycles found?
3. **WHAT IF?** If a human female begins taking estradiol and progesterone immediately after the start of a new menstrual cycle, how will ovulation be affected? Explain.
4. **MAKE CONNECTIONS** A coordination of developmental events is characteristic of the reproductive cycles of a human female and an enveloped RNA virus (see Figure 19.7, p. 388). What is the nature of the coordination in each of these cycles?

For suggested answers, see Appendix A.

CONCEPT **46.5**

In placental mammals, an embryo develops fully within the mother's uterus

Having surveyed the ovarian and uterine cycles of human females, we turn now to reproduction itself, beginning with the events that transform an egg into a developing embryo.

Conception, Embryonic Development, and Birth

During human copulation, 2–5 mL of semen is transferred, with 70–130 million sperm in each milliliter. The alkalinity of the semen helps neutralize the acidic environment of the vagina, protecting the sperm and increasing their motility. When first ejaculated, the semen coagulates, which may serve to keep the ejaculate in place until sperm reach the cervix. Soon after, anticoagulants liquefy the semen, and the sperm begin swimming through the uterus and oviducts.

Fertilization—also called **conception** in humans—occurs when a sperm fuses with an egg (mature oocyte) in the oviduct **(Figure 46.15a)**. About 24 hours later, the resulting zygote begins dividing, a process called **cleavage**. After another 2–3 days, the embryo typically arrives at the uterus as a ball of 16 cells. By about 5 days after fertilization, cleavage has produced an embryonic stage called the **blastocyst**, a sphere of cells surrounding a central cavity.

Several days after blastocyst formation, the embryo implants into the endometrium **(Figure 46.15b)**. Only after implantation can an embryo develop into a fetus. The implanted embryo secretes hormones that signal its presence and regulate the mother's reproductive system. One embryonic hormone, **human chorionic gonadotropin (hCG)**, acts like pituitary LH in maintaining secretion of progesterone and estrogens by the corpus luteum through the first few months of pregnancy. In the absence of this hormonal override during pregnancy, the corpus luteum would deteriorate and progesterone levels would drop, resulting in menstruation and loss of the embryo. Levels of hCG in the maternal blood are so high that some is excreted in the urine, where its presence is the basis of many early pregnancy tests.

The condition of carrying one or more embryos in the uterus is called **pregnancy**, or **gestation**. Human pregnancy averages 266 days (38 weeks) from fertilization of the egg, or 40 weeks from the start of the last menstrual cycle. Duration of pregnancy in other placental mammals correlates with body size and the maturity of the young at birth. Many rodents have gestation periods of about 21 days, whereas those of dogs are closer to 60 days. In cows, gestation averages 270 days (almost the same as in humans), while in elephants it lasts more than 600 days.

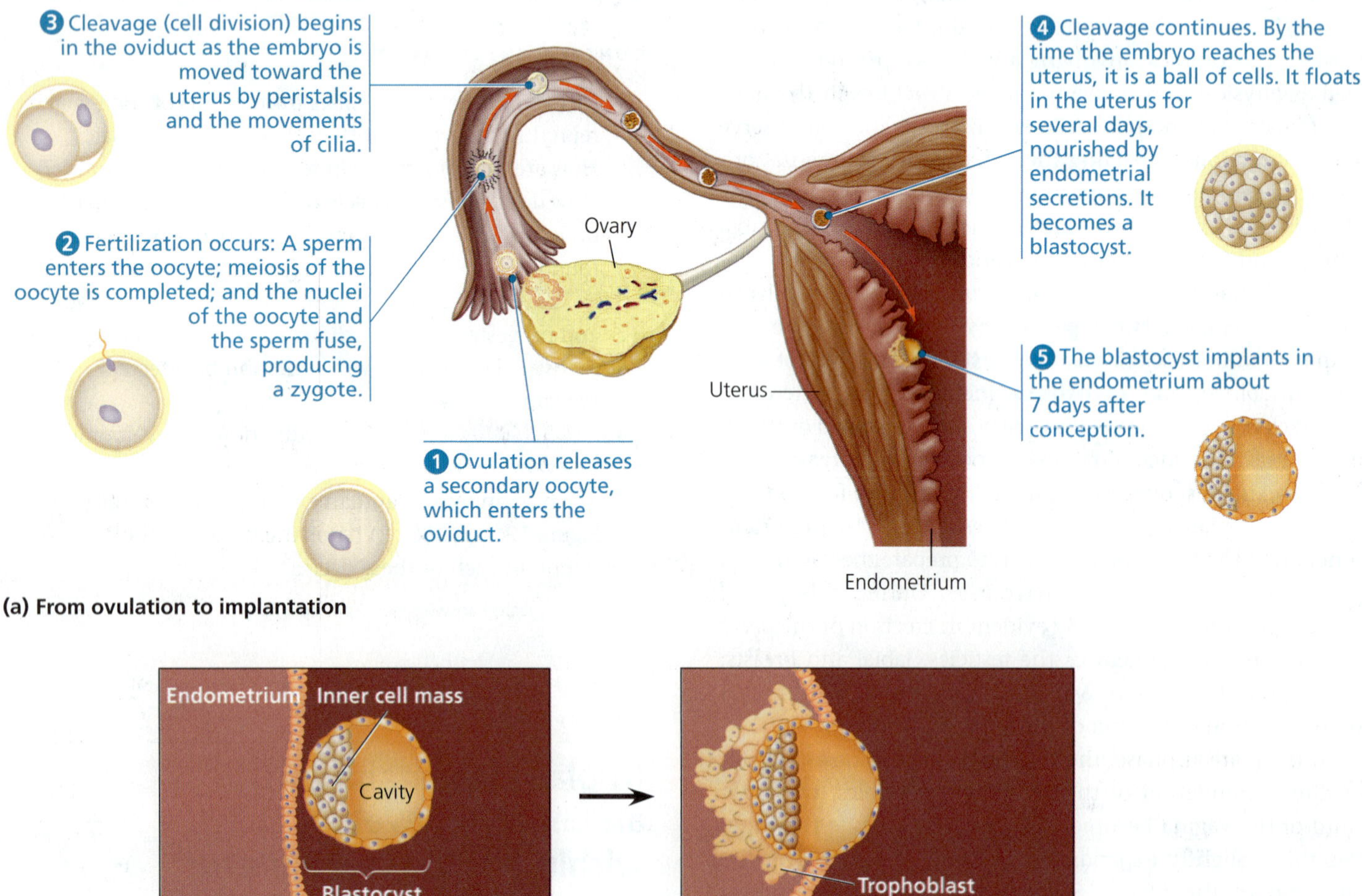

▲ **Figure 46.15 Formation of the zygote and early postfertilization events.**

Not all fertilized eggs are capable of completing development. Many pregnancies terminate spontaneously as a result of chromosomal or developmental abnormalities. Much less often, a fertilized egg lodges in the oviduct (fallopian tube), resulting in a tubal, or ectopic, pregnancy. Such pregnancies cannot be sustained and may rupture the oviduct, resulting in serious internal bleeding. A number of conditions, including endometriosis, increase the likelihood of tubal pregnancy. Bacterial infections arising during childbirth, from medical procedures, or as a *sexually transmitted disease* (*STD*) can scar the oviduct, making ectopic pregnancy more likely.

STDs are the most significant preventable causes of infertility. For women who are between 15 and 24 years of age, approximately 700,000 cases of chlamydia and gonorrhea are reported annually in the United States. The number of women infected is actually significantly higher because most women with these STDs have no symptoms and are therefore unaware of their infection. Among women who remain untreated for either chlamydia or gonorrhea, up to 40% develop an inflammatory disorder that can lead to infertility or to potentially fatal ectopic pregnancies.

First Trimester

Human gestation can be divided for convenience into three **trimesters** of about three months each. The first trimester is the time of most radical change for both the mother and the embryo. Upon implantation, the endometrium grows over the blastocyst. Cells and tissues of the embryo begin to differentiate into specialized body structures. (You will learn much more about embryonic development in Chapter 47.)

During its first 2–4 weeks of development, the embryo obtains nutrients directly from the endometrium. Meanwhile, the outer layer of the blastocyst, called the **trophoblast**, grows outward and mingles with the endometrium, eventually helping form the **placenta**. This disk-shaped organ, containing both embryonic and maternal blood vessels, can weigh close to 1 kg. Material diffusing between the maternal and embryonic circulatory systems supplies nutrients, provides immune protection, exchanges respiratory gases, and disposes of metabolic wastes for the embryo. Blood from the embryo travels to the placenta through the arteries of the umbilical cord and returns via the umbilical vein **(Figure 46.16)**.

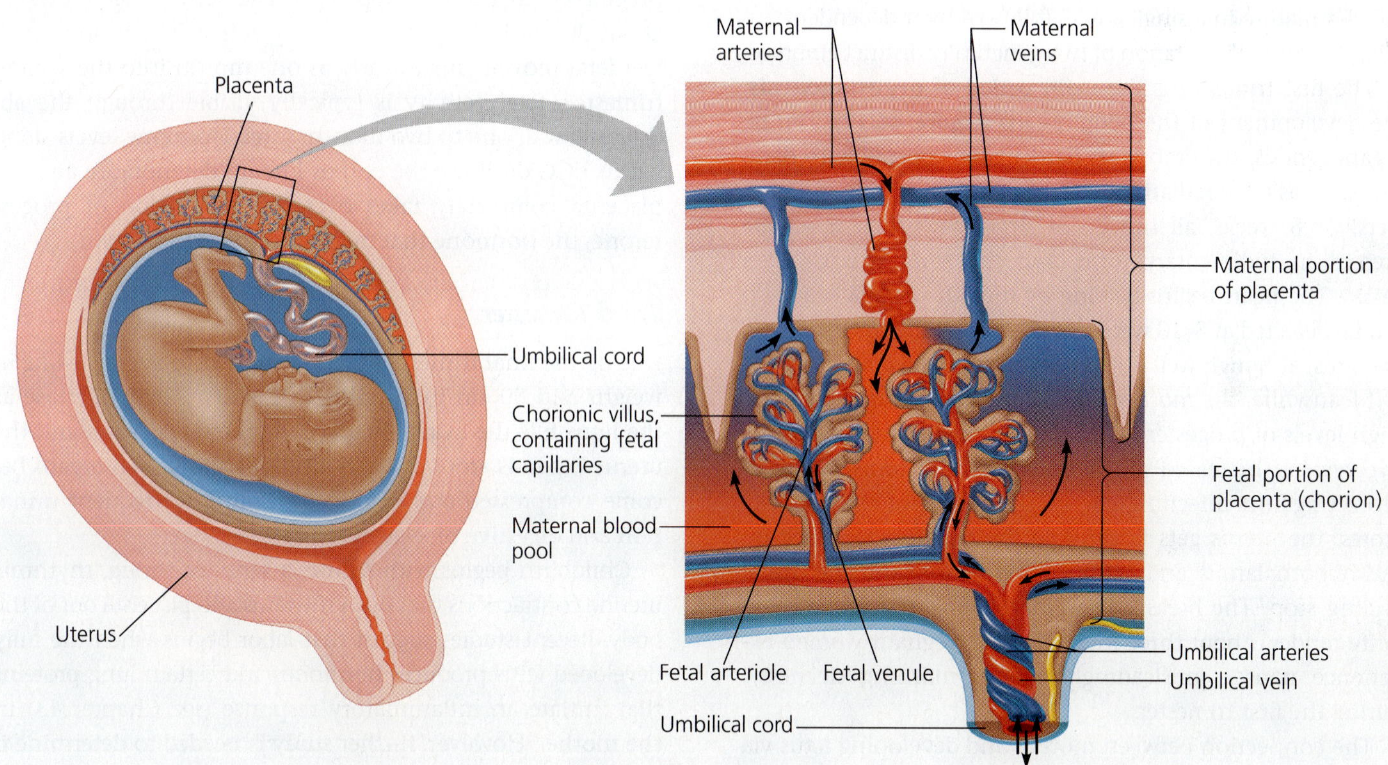

▲ **Figure 46.16 Placental circulation.** From the 4th week of development until birth, the placenta, a combination of maternal and embryonic tissues, transports nutrients, respiratory gases, and wastes between the embryo or fetus and the mother. Maternal blood enters the placenta in arteries, flows through blood pools in the endometrium, and leaves via veins. Embryonic or fetal blood, which remains in vessels, enters the placenta through arteries and passes through capillaries in finger-like chorionic villi, where oxygen and nutrients are acquired. As indicated in the drawing, the fetal (or embryonic) capillaries and villi project into the maternal portion of the placenta. Fetal blood leaves the placenta through veins leading back to the fetus. Materials are exchanged by diffusion, active transport, and selective absorption between the fetal capillary bed and the maternal blood pools.

? *In a rare genetic disorder, the absence of a particular enzyme leads to increased testosterone production. When the fetus has this disorder, the mother develops a male-like pattern of body hair during the pregnancy. Explain.*

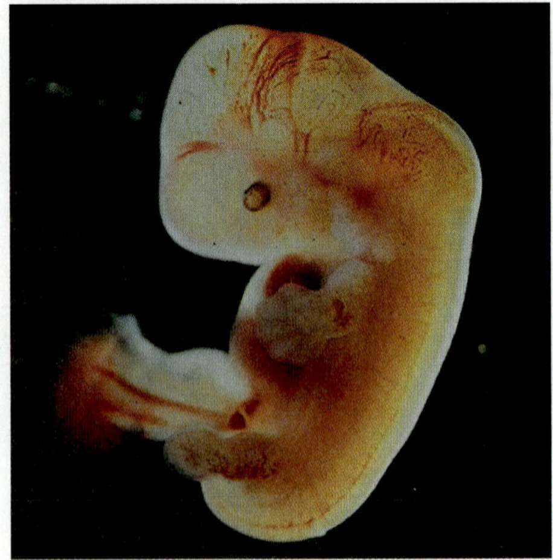

(a) 5 weeks. Limb buds, eyes, the heart, the liver, and rudiments of all other organs have started to develop in the embryo, which is only about 1 cm long.

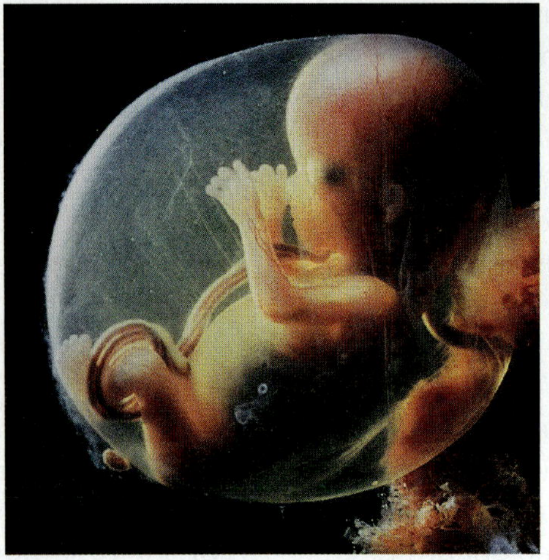

(b) 14 weeks. Growth and development of the offspring, now called a fetus, continue during the second trimester. This fetus is about 6 cm long.

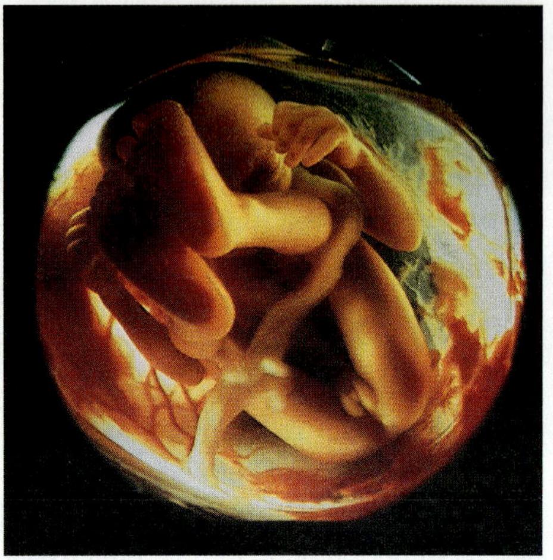

(c) 20 weeks. Growth to nearly 20 cm in length requires adoption of the fetal position (head at knees) due to the limited space available.

▲ **Figure 46.17 Human fetal development.**

Splitting of the embryo during the first month of development can result in identical, or *monozygotic* (one-egg), twins. Fraternal, or *dizygotic*, twins arise in a very different way: Two follicles mature in a single cycle, followed by independent fertilization and implantation of two genetically distinct embryos.

The first trimester is the main period of **organogenesis**, the development of the body organs **(Figure 46.17)**. During organogenesis, the embryo is particularly susceptible to damage, such as from radiation or drugs, that can lead to birth defects. At 8 weeks, all the major structures of the adult are present in rudimentary form, and the embryo is called a **fetus**. The heart begins beating by the 4th week; a heartbeat can be detected at 8–10 weeks. At the end of the first trimester, the fetus, although well differentiated, is only 5 cm long.

Meanwhile, the mother is also undergoing rapid changes. High levels of progesterone initiate changes in her reproductive system: Increased mucus in the cervix forms a plug to protect against infection, the maternal part of the placenta grows, the uterus gets larger, and (by negative feedback on the hypothalamus and pituitary) ovulation and menstrual cycling stop. The breasts also enlarge rapidly and are often quite tender. About three-fourths of all pregnant women experience nausea, misleadingly called "morning sickness," during the first trimester.

The connection between mother and developing fetus via the placenta allows harmful as well as beneficial substances to pass between them. For this reason, consuming alcohol during pregnancy poses a major risk. Alcohol that reaches the developing central nervous system of the fetus can cause fetal alcohol syndrome, a disorder that can result in mental retardation and other serious birth defects. Similarly, smoking during pregnancy is associated with high risk of low birth weight and other health problems.

Second Trimester

During the second trimester, the uterus grows enough for the pregnancy to become apparent. The fetus itself grows to about 30 cm in length and is very active. The mother may feel fetal movements as early as one month into the second trimester; fetal activity is typically visible through the abdominal wall one to two months later. Hormone levels stabilize as hCG declines; the corpus luteum deteriorates; and the placenta completely takes over the production of progesterone, the hormone that maintains the pregnancy.

Third Trimester

During the final trimester, the fetus grows to about 3–4 kg in weight and 50 cm in length. Fetal activity may decrease as the fetus fills the available space. As the fetus grows and the uterus expands around it, the mother's abdominal organs become compressed and displaced, leading to frequent urination and digestive blockages.

Childbirth begins with **labor**, a series of strong, rhythmic uterine contractions that push the fetus and placenta out of the body. Recent studies suggest that labor begins when the fully developed fetus produces hormones and certain lung proteins that initiate an inflammatory response (see Chapter 43) in the mother. However, further study is needed to determine if inflammation does in fact trigger labor.

Once labor begins, a complex interplay of local regulators (prostaglandins) and hormones (chiefly estradiol and oxytocin) induces and regulates further contractions of the uterus **(Figure 46.18)**. The action of oxytocin forms a positive-feedback loop (see Chapter 45), with uterine contractions stimulating secretion of oxytocin, which in turn stimulates further contractions.

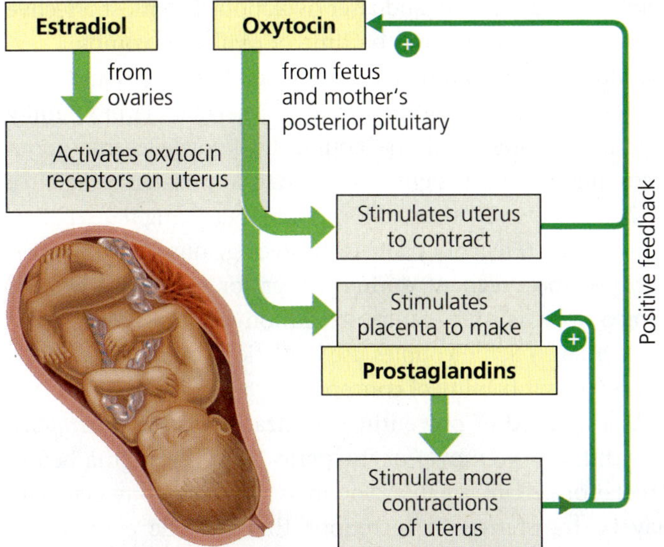

▲ **Figure 46.18 Positive feedback in labor.**

? *Predict the effect of a single dose of oxytocin on a pregnant woman at the end of 39 weeks gestation.*

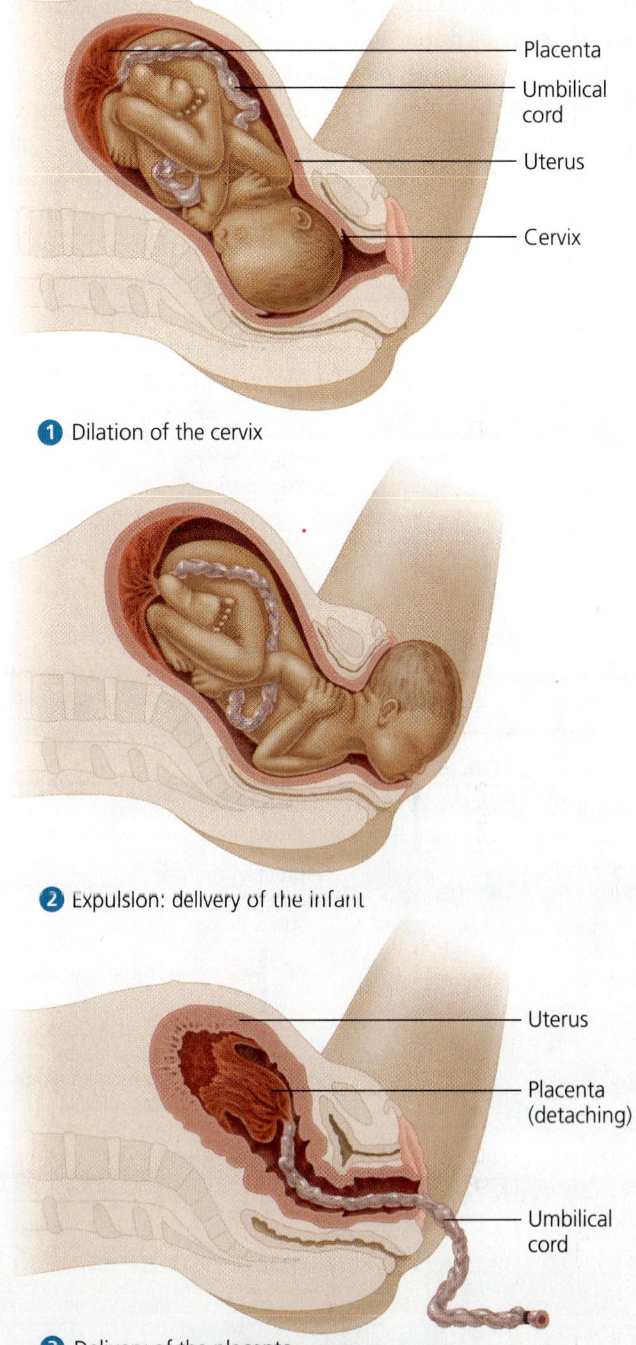

▲ **Figure 46.19 The three stages of labor.**

Labor is typically described as having three stages **(Figure 46.19)**. The first stage is the thinning and opening up (dilation) of the cervix. The second stage is the expulsion, or delivery, of the baby. Continuous strong contractions force the fetus out of the uterus and through the vagina. The final stage of labor is delivery of the placenta.

One aspect of postnatal care unique to mammals is **lactation**, the production of mother's milk. In response to suckling by the newborn, as well as changes in estradiol levels after birth, the hypothalamus signals the anterior pituitary to secrete prolactin, which stimulates the mammary glands to produce milk. Suckling also stimulates the secretion of oxytocin from the posterior pituitary, which triggers release of milk from the mammary glands (see Figure 45.15).

Maternal Immune Tolerance of the Embryo and Fetus

Pregnancy is an immunological puzzle. Half of the embryo's genes are inherited from the father; thus, many of the chemical markers present on the surface of the embryo are foreign to the mother. Why, then, does the mother not reject the embryo as a foreign body, as she would a tissue or organ graft from another person? One intriguing clue comes from the relationship between certain autoimmune disorders and pregnancy. For example, the symptoms of rheumatoid arthritis, an autoimmune disease of the joints, become less severe during pregnancy. Thus, the overall regulation of the immune system appears to be altered by the reproductive process. Sorting out these changes and how they might protect the developing fetus is an active area of research for immunologists.

Contraception and Abortion

Contraception, the deliberate prevention of pregnancy, can be achieved in a number of ways. Some contraceptive methods prevent gamete development or release from female or male gonads; others prevent fertilization by keeping sperm and egg apart; and still others prevent implantation of an embryo. For complete information on contraceptive methods, you should consult a health-care provider. The following brief introduction to the biology of the most common methods and the corresponding diagram in **Figure 46.20** make no pretense of being a contraception manual.

Fertilization can be prevented by abstinence from sexual intercourse or by any of several barriers that keep live sperm from contacting the egg. Temporary abstinence, often called the **rhythm method** of birth control or **natural family planning**, depends on refraining from intercourse when conception is most likely. Because the egg can survive in the oviduct for 24–48 hours and sperm for up to 5 days, a couple practicing temporary abstinence should not engage in intercourse for a number of days before and after ovulation. The most effective methods for determining the time of ovulation combine several indicators, including changes in cervical mucus and body temperature during the menstrual cycle. Thus, natural family planning requires that the couple be knowledgeable about these physiological signs. Note that a pregnancy rate of 10–20% is typically reported for couples practicing natural family planning. (Pregnancy rate is the average number of women who become pregnant during a year for every 100 women using a particular pregnancy prevention method, expressed as a percentage.) Some couples use ovulation-timing methods to *increase* the probability of conception.

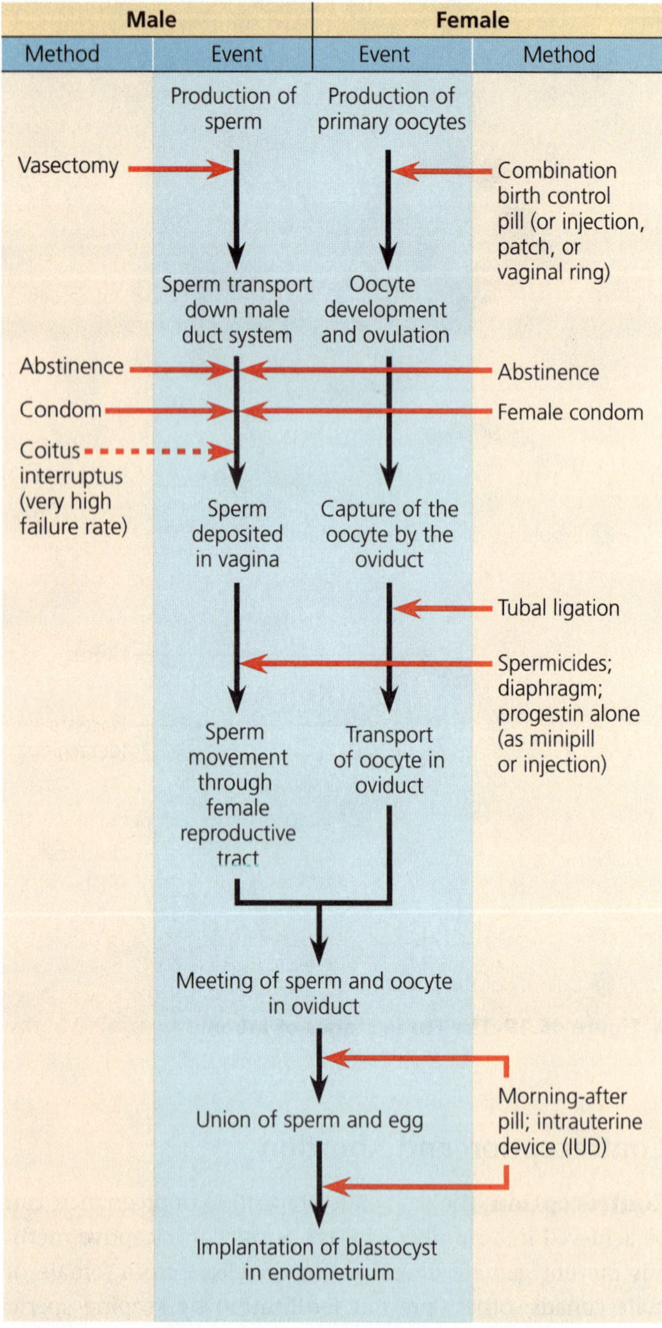

▲ **Figure 46.20 Mechanisms of several contraceptive methods.** Red arrows indicate where these methods, devices, or products interfere with events from the production of sperm and primary oocytes to an implanted, developing embryo.

As a method of preventing fertilization, *coitus interruptus*, or withdrawal (removal of the penis from the vagina before ejaculation), is unreliable. Sperm from a previous ejaculate may be transferred in secretions that precede ejaculation. Furthermore, a split-second lapse in timing or willpower can result in tens of millions of sperm being transferred before withdrawal.

The several barrier methods of contraception that block the sperm from meeting the egg have pregnancy rates of less than 10%. The **condom** is a thin, latex rubber or natural membrane sheath that fits over the penis to collect the semen. For sexually active individuals, latex condoms are the only contraceptives that are highly effective in preventing the spread of sexually transmitted diseases, including AIDS. (This protection is, however, not absolute.) Another common barrier device is the **diaphragm**, a dome-shaped rubber cap inserted into the upper portion of the vagina before intercourse. Both of these devices have lower pregnancy rates when used in conjunction with a spermicidal (sperm-killing) foam or jelly. Other barrier devices include the vaginal pouch, or "female condom."

Except for complete abstinence from sexual intercourse, the most effective means of birth control are sterilization, intrauterine devices (IUDs), and hormonal contraceptives. Sterilization (discussed later) is almost 100% effective. The IUD has a pregnancy rate of 1% or less and is the most commonly used reversible method of birth control outside the United States. Placed in the uterus by a doctor, the IUD interferes with fertilization and implantation. Hormonal contraceptives, most often in the form of **birth control pills**, also have pregnancy rates of 1% or less.

The most commonly prescribed birth control pills are a combination of a synthetic estrogen and a synthetic progestin (progesterone-like hormone). This combination mimics negative feedback in the ovarian cycle, stopping the release of GnRH by the hypothalamus and thus of FSH and LH by the pituitary. The prevention of LH release blocks ovulation. In addition, the inhibition of FSH secretion by the low dose of estrogens in the pills prevents follicles from developing. A similar combination of hormones is also available as an injection, as a ring inserted into the vagina, and as a skin patch. Combination birth control pills can also be used in high doses as "morning-after" pills. Taken within 3 days after unprotected intercourse, they prevent fertilization or implantation with an effectiveness of about 75%.

A different type of hormone-based contraceptive contains only progestin. Progestin causes thickening of a woman's cervical mucus so that it blocks sperm from entering the uterus. Progestin also decreases the frequency of ovulation and causes changes in the endometrium that may interfere with implantation if fertilization occurs. Progestin can be administered as injections that last for three months or as a tablet ("minipill") taken daily. Pregnancy rates for progestin treatment are very low.

Hormone-based contraceptives have both beneficial and harmful side effects. For women taking a combination pill, cardiovascular problems are the most serious concern. Women who regularly smoke cigarettes face a three to ten times greater risk of dying from cardiovascular disease if they also use oral contraceptives. Among nonsmokers, birth control pills slightly raise a woman's risk of abnormal blood clotting, high blood pressure, heart attack, and stroke. Although oral contraceptives increase the risk for these cardiovascular disorders, they eliminate the dangers of pregnancy; women on birth control pills have mortality rates about one-half those of pregnant women. Also, the pill decreases the risk of ovarian and endometrial cancers.

One elusive research goal has been a reversible chemical contraceptive for men. Recent strategies have focused on hormone combinations that suppress gonadotropin release and thereby block spermatogenesis. Testosterone included in such combinations has two desirable effects: inhibiting reproductive functions of the hypothalamus and pituitary and maintaining secondary sex characteristics. Although there have been some promising results, hormonal male contraceptives are still in the testing stage.

Sterilization is the permanent prevention of gamete production or release. **Tubal ligation** in women usually involves sealing shut or tying off (ligating) a section of each oviduct to prevent eggs from traveling into the uterus. Similarly, **vasectomy** in men is the cutting and tying off of each vas deferens to prevent sperm from entering the urethra. Both male and female sterilization procedures are relatively safe and free from harmful effects. Sex hormone secretion and sexual function are unaffected by both procedures, with no change in menstrual cycles in females or ejaculate volume in males. Although tubal ligation or vasectomy are considered permanent, both procedures can in many cases be reversed by microsurgery.

The termination of a pregnancy in progress is called **abortion**. Spontaneous abortion, or *miscarriage*, is very common; it occurs in as many as one-third of all pregnancies, often before the woman is even aware she is pregnant. In addition, each year about 850,000 women in the United States choose to have an abortion performed by a physician.

A drug called mifepristone, or RU486, can terminate a pregnancy nonsurgically within the first 7 weeks. RU486 blocks progesterone receptors in the uterus, thus preventing progesterone from maintaining the pregnancy. It is taken with a small amount of prostaglandin to induce uterine contractions.

Modern Reproductive Technologies

Recent scientific and technological advances have made it possible to address many reproductive problems, including genetic diseases and infertility.

Detecting Disorders During Pregnancy

Many genetic diseases and developmental problems can now be diagnosed while the fetus is in the uterus. Ultrasound imaging, which generates images using sound frequencies above the normal hearing range, is commonly used to analyze the fetus's size and condition. Amniocentesis and chorionic villus sampling are techniques in which a needle is used to obtain fetal cells from fluid or tissue surrounding the embryo; these cells then provide the basis for genetic analysis (see Figure 14.19). An alternative technique for obtaining fetal tissue relies on the fact that a few fetal blood cells leak across the placenta into the mother's bloodstream. A blood sample from the mother yields fetal cells that can be identified with specific antibodies (which bind to proteins on the surface of fetal cells) and then tested for genetic disorders.

Diagnosing genetic diseases in a fetus poses ethical questions. To date, almost all detectable disorders remain untreatable in the uterus, and many cannot be corrected even after birth. Parents may be faced with difficult decisions about whether to terminate a pregnancy or to raise a child who may have profound defects and a short life expectancy. These are complex issues that demand careful, informed thought and competent genetic counseling.

Treating Infertility

Infertility—an inability to conceive offspring—is quite common, affecting about one in ten couples both in the United States and worldwide. The causes of infertility are varied, and the likelihood of a reproductive defect is nearly the same for men and women. For women, however, the risk of reproductive difficulties, as well as genetic abnormalities of the fetus, increases steadily past age 35. Evidence suggests that the prolonged period of time oocytes spend in meiosis is largely responsible for this increased risk.

Reproductive technology can help with a number of fertility problems. Hormone therapy can sometimes increase sperm or egg production, and surgery can often correct ducts that have failed to form properly or have become blocked. Many infertile couples turn to **assisted reproductive technologies**, procedures that generally involve surgically removing eggs (secondary oocytes) from a woman's ovaries after hormonal stimulation, fertilizing the eggs, and returning early embryos to the woman's body. Unused eggs, sperm, and embryos are sometimes frozen for later pregnancy attempts.

The technique of ***in vitro* fertilization (IVF)** involves mixing oocytes and sperm in culture dishes. Fertilized eggs are incubated until they have formed at least eight cells and are then typically transferred to the woman's uterus for implantation. If mature sperm are defective, low in number (less than 20 million per milliliter of ejaculate), or even absent, fertility is often restored by a technique called **intracytoplasmic sperm injection (ICSI)**. In this form of IVF, the head of a spermatid or sperm is drawn up into a needle and injected directly into an oocyte to achieve fertilization.

Though costly, IVF procedures have enabled more than a million couples to conceive children. In some cases, these procedures are carried out with sperm or eggs from donors. To date, evidence indicates that abnormalities arising as a consequence of IVF procedures are rare.

By whatever means fertilization occurs, a developmental program follows that transforms the single-celled zygote into a multicellular organism. The mechanisms of this remarkable program of development in humans and other animals are the subject of Chapter 47.

CONCEPT CHECK 46.5

1. Why does testing for hCG (human chorionic gonadotropin) work as a pregnancy test early in pregnancy but not late in pregnancy? What is the function of hCG in pregnancy?
2. In what ways are tubal ligation and vasectomy similar?
3. **WHAT IF?** If a spermatid nucleus is used for ICSI, what steps of gametogenesis and conception are bypassed?

For suggested answers, see Appendix A.

46 CHAPTER REVIEW

SUMMARY OF KEY CONCEPTS

CONCEPT 46.1

Both asexual and sexual reproduction occur in the animal kingdom (pp. 996–999)

- Animals reproduce either asexually or sexually. **Sexual reproduction** requires the fusion of male and female gametes, forming a diploid **zygote**. **Asexual reproduction** is the production of offspring without gamete fusion. Fission, budding, fragmentation with regeneration, and **parthenogenesis** are mechanisms of asexual reproduction in various invertebrates. Facilitating selection for or against sets of genes may explain why sexual reproduction is widespread among animal species.
- Although most animals reproduce exclusively sexually or asexually, some alternate between the two. Variations on these two modes are made possible through parthenogenesis, **hermaphroditism**, and sex reversal. Hormones and environmental cues control reproductive cycles.

? *Would a pair of haploid offspring produced by parthenogenesis be genetically identical?*

CONCEPT 46.2

Fertilization depends on mechanisms that bring together sperm and eggs of the same species (pp. 999–1002)

- Fertilization can occur externally or internally with regard to the mother's body. In either case, fertilization requires coordinated timing, which may be mediated by environmental cues, pheromones, or courtship behavior. Internal fertilization is typically often associated both with relatively fewer offspring and with greater protection of offspring by the parents. Systems for gamete production and delivery range from undifferentiated cells in the body cavity to complex **gonads** with accessory tubes and glands that carry and protect gametes and embryos. Although sexual reproduction involves a partnership, it also provides an opportunity for competition between individuals and between gametes.

? *Identify which of the following are unique to mammals: a female uterus and a male vas deferens, extended internal development, parental care of newborns.*

CONCEPT 46.3

Reproductive organs produce and transport gametes (pp. 1002–1008)

- The reproductive system of the human female consists principally of the **labia** and the **glans** of the **clitoris** externally and the **vagina**, **uterus**, **oviducts**, and **ovaries** internally. Eggs are produced in the ovaries and upon fertilization develop in the uterus. In males, sperm are produced in **testes**, which are suspended outside the body in the **scrotum**. Ducts extending from the scrotum connect the testes to internal accessory glands and to the opening of the **penis**. Both males and females have **mammary glands**, but milk production occurs only in females. During intercourse, males and females each experience the erection of certain body tissues due to **vasocongestion** and **myotonia**, culminating in **orgasm**.
- Gametogenesis, or gamete production, consists of **oogenesis** in females and **spermatogenesis** in males. Meiosis generates one large egg in oogenesis, but four sperm in spermatogenesis. In humans, sperm develop continuously, whereas oocyte maturation is discontinuous and cyclic.

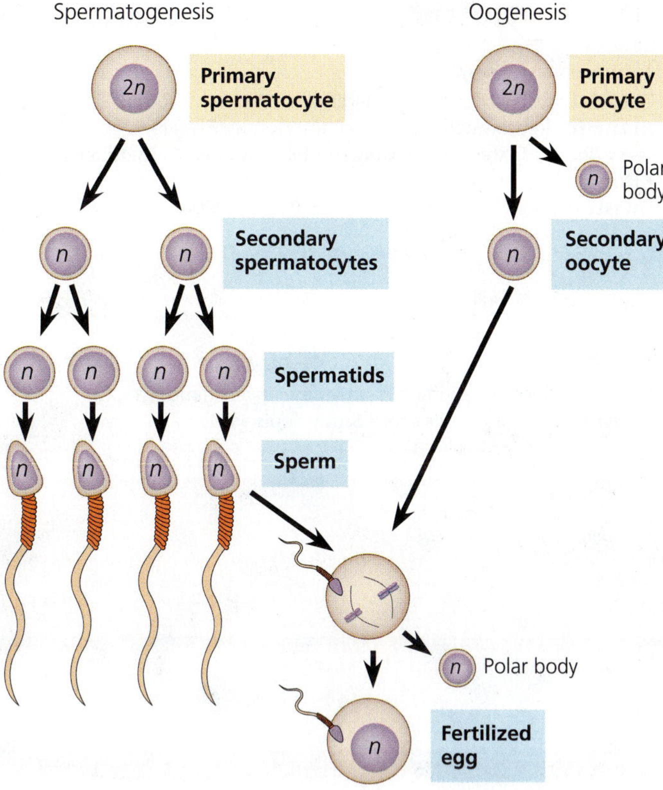

? *How does the difference in size and cellular contents between sperm and eggs relate to their specific functions in reproduction?*

CONCEPT 46.4

The interplay of tropic and sex hormones regulates mammalian reproduction (pp. 1008–1011)

- In human males, androgens (chiefly testosterone) from the testes cause the development of primary and secondary sex characteristics. Androgen secretion and sperm production are both controlled by hypothalamic and pituitary hormones.
- In human females, cyclic secretion of GnRH from the hypothalamus and FSH and LH from the anterior pituitary orchestrates the reproductive cycle. FSH and LH bring about changes in the ovary and uterus via estrogens, primarily estradiol, and progesterone. The developing follicle and the corpus luteum also secrete hormones, with positive and negative feedback coordinating the uterine and ovarian cycles.
- **Estrous cycles** differ from **menstrual cycles** in that the endometrial lining is reabsorbed rather than shed and sexual receptivity is limited to a heat period.

? *Why do anabolic steroids lead to reduced sperm count?*

CONCEPT 46.5

In placental mammals, an embryo develops fully within the mother's uterus (pp. 1011–1018)

- After fertilization and the completion of meiosis in the oviduct, the zygote undergoes cleavage and develops into a blastocyst before implantation in the endometrium. All major organs start developing by 8 weeks. A pregnant woman's acceptance of her "foreign" offspring likely reflects partial suppression of the maternal immune response.
- Contraceptive methods may prevent release of mature gametes from the gonads, fertilization, or implantation of the embryo. Reproductive technologies can assist infertile couples by hormonal methods or *in vitro* fertilization and can also help detect problems before birth.

? *What route would oxygen in the mother's blood follow to arrive at a body cell of the fetus?*

TEST YOUR UNDERSTANDING

LEVEL 1: KNOWLEDGE/COMPREHENSION

1. Which of the following characterizes parthenogenesis?
   a. An individual may change its sex during its lifetime.
   b. Specialized groups of cells grow into new individuals.
   c. An organism is first a male and then a female.
   d. An egg develops without being fertilized.
   e. Both mates have male and female reproductive organs.
2. In male mammals, excretory and reproductive systems share
   a. the testes.
   b. the urethra.
   c. the seminal vesicle.
   d. the vas deferens.
   e. the prostate.
3. Which of the following is *not* properly paired?
   a. seminiferous tubule—cervix
   b. Sertoli cells—follicle cells
   c. testosterone—estradiol
   d. scrotum—labia majora
   e. vas deferens—oviduct
4. Peaks of LH and FSH production occur during
   a. the menstrual flow phase of the uterine cycle.
   b. the beginning of the follicular phase of the ovarian cycle.
   c. the period just before ovulation.
   d. the end of the luteal phase of the ovarian cycle.
   e. the secretory phase of the menstrual cycle.
5. During human gestation, rudiments of all organs develop
   a. in the first trimester.
   b. in the second trimester.
   c. in the third trimester.
   d. while the embryo is in the oviduct.
   e. during the blastocyst stage.

LEVEL 2: APPLICATION/ANALYSIS

6. Which of the following is a true statement?
   a. All mammals have menstrual cycles.
   b. The endometrial lining is shed in menstrual cycles but reabsorbed in estrous cycles.
   c. Estrous cycles are more frequent than menstrual cycles.
   d. Estrous cycles are not controlled by hormones.
   e. Ovulation occurs before the endometrium thickens in estrous cycles.
7. For which of the following is the number the same in spermatogenesis and oogenesis?
   a. interruptions in meiotic divisions
   b. functional gametes produced by meiosis
   c. meiotic divisions required to produce each gamete
   d. gametes produced in a given time period
   e. different cell types produced by meiosis
8. Which statement about human reproduction is false?
   a. Fertilization occurs in the oviduct.
   b. Effective hormonal contraceptives are currently available only for females.
   c. An oocyte completes meiosis after a sperm penetrates it.
   d. The earliest stages of spermatogenesis occur closest to the lumen of the seminiferous tubules.
   e. Spermatogenesis and oogenesis require different temperatures.

LEVEL 3: SYNTHESIS/EVALUATION

9. **DRAW IT** In human spermatogenesis, mitosis of a stem cell gives rise to one cell that remains a stem cell and one cell that becomes a spermatogonium. (a) Draw four rounds of mitosis for a stem cell, and label the daughter cells. (b) For one spermatogonium, draw the cells it would produce from one round of mitosis followed by meiosis. Label the cells, and label mitosis and meiosis. (c) What would happen if stem cells divided like spermatogonia?

10. **EVOLUTION CONNECTION**
Hermaphroditism is often found in animals that are fixed to a surface. Motile species are less often hermaphroditic. Why?

11. **SCIENTIFIC INQUIRY**
You discover a new egg-laying worm species. You dissect four adults and find both oocytes and sperm in each. Cells outside the gonad contain five chromosome pairs. Lacking genetic variants, how would you determine whether the worms can self-fertilize?

12. **WRITE ABOUT A THEME**
Energy Transfer In reproducing, animals transfer energy to their offspring. In a short essay (100–150 words), discuss how distinct investments of energy by females contribute to the reproductive success of a frog, a chicken, and a human.

For selected answers, see Appendix A.

www.masteringbiology.com

1. MasteringBiology® Assignments

Tutorial Sex Hormones and Mammalian Reproduction
Activities Reproductive System of the Human Female • Reproductive System of the Human Male • Human Gametogenesis • Human Reproduction
Questions Student Misconceptions • Reading Quiz • Multiple Choice • End-of-Chapter

2. eText

Read your book online, search, take notes, highlight text, and more.

3. The Study Area

Practice Tests • Cumulative Test • *BioFlix* 3-D Animations • MP3 Tutor Sessions • Videos • Activities • Investigations • Lab Media • Audio Glossary • Word Study Tools • Art

47

Animal Development

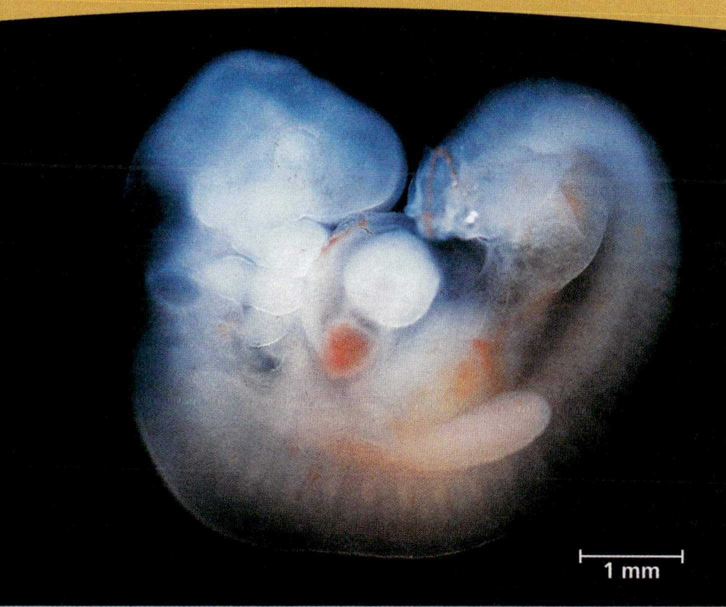

▲ **Figure 47.1 How did a single cell develop into this intricately detailed embryo?**

KEY CONCEPTS

47.1 Fertilization and cleavage initiate embryonic development

47.2 Morphogenesis in animals involves specific changes in cell shape, position, and survival

47.3 Cytoplasmic determinants and inductive signals contribute to cell fate specification

OVERVIEW

A Body-Building Plan

The 7-week-old human embryo in **Figure 47.1** has already achieved a remarkable number of milestones in its development. Its heart—the red spot in the center—is beating, and a digestive tract traverses the length of its body. Its brain is forming (at the upper left in the photo), while the blocks of tissue that will give rise to the vertebrae are lined up along its back.

Development occurs at many points in the life cycle of an animal **(Figure 47.2)**. In a frog, for example, a major developmental period is metamorphosis, when the larva (tadpole) is transformed into an adult. Other developmental events in the adult gonads produce sperm and eggs (gametes). In this chapter, our focus is on embryonic development.

Across a range of animal species, embryonic development involves common stages that occur in a set order. As shown in Figure 47.2, the first is fertilization, the fusion of sperm and egg, which forms a zygote. Development proceeds with the cleavage stage, during which a series of cell divisions divide, or cleave, the zygote into a many-celled embryo. These cleavage divisions, which typically are rapid and lack accompanying cell growth, convert the embryo to a hollow ball of cells called a blastula. Next, the blastula folds in on itself, rearranging into a three-layered embryo, the gastrula, in a process called gastrulation. During organogenesis, the last major stage of embryonic development, local changes in cell shape and large-scale changes in cell location generate the rudimentary organs from which adult structures grow.

By combining molecular genetics with classical embryology, developmental biologists have learned a great deal about the transformation of a fertilized egg into an adult. As an embryo develops, specific patterns of gene expression direct cells to adopt distinct fates. Although animals display

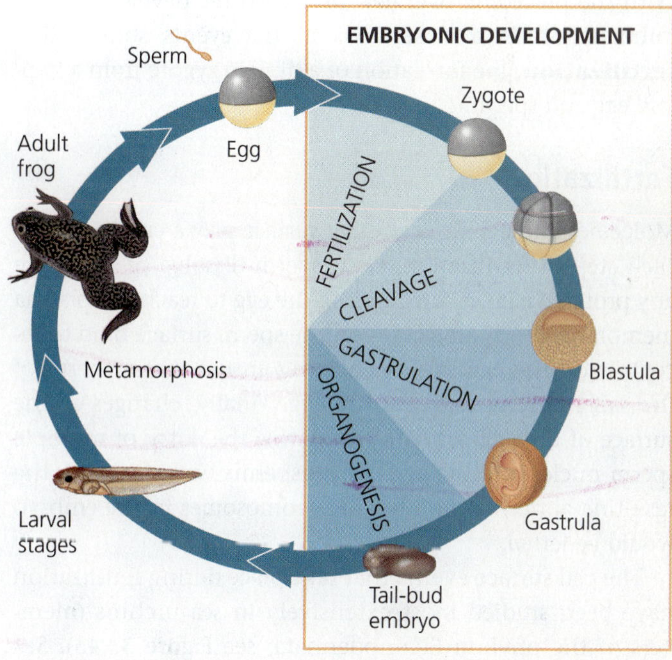

▲ **Figure 47.2 Developmental events in the life cycle of a frog.**

widely differing body plans, they share many basic mechanisms of development and use a common set of regulatory genes. For example, the gene that specifies heart location in a human embryo (such as the one in Figure 47.1) has a close counterpart with a nearly identical function in the fruit fly, *Drosophila*. (Noting that the fly heart fails to develop when this gene is defective, researchers named the *Drosophila* gene *tinman*, after the similarly affected character from *The Wizard of Oz*.)

In studying development, biologists frequently make use of **model organisms**, species chosen for the ease with which they can be studied in the laboratory. *Drosophila* is a useful model organism: Its life cycle is short, and mutants can be readily identified and studied (see Chapters 15 and 18). In this chapter, we will concentrate on four other model organisms: the sea urchin, the frog, the chick, and the nematode (roundworm). We will also explore some aspects of human embryonic development. Even though humans are not model organisms, we are, of course, intensely interested in our own species.

Our exploration of embryonic development will begin with a description of the basic stages common to most animals. We will then look at some of the cellular mechanisms that generate body form. Finally, we will consider how a cell becomes committed to a particular specialized role.

CONCEPT **47.1**

Fertilization and cleavage initiate embryonic development

With the preceding overview of embryonic development in mind, let's take a closer look at the events surrounding **fertilization**, the formation of a diploid zygote from a haploid egg and sperm.

Fertilization

Molecules and events at the egg surface play a crucial role in each step of fertilization. First, sperm dissolve or penetrate any protective layer surrounding the egg to reach the plasma membrane. Next, molecules on the sperm surface bind to receptors on the egg surface, helping ensure that a sperm of the same species fertilizes the egg. Finally, changes at the surface of the egg prevent *polyspermy*, the entry of multiple sperm nuclei into the egg. If polyspermy were to occur, the resulting abnormal number of chromosomes in the embryo would be lethal.

The cell surface events that take place during fertilization have been studied most extensively in sea urchins (members of the phylum Echinodermata; see Figure 33.43). Sea urchin gametes are easy to collect, and fertilization is external. As a result, researchers can observe fertilization and subsequent events simply by combining eggs and sperm in seawater in the laboratory. Furthermore, fertilization in sea urchins provides a good general model for the same process in vertebrates.

The Acrosomal Reaction

When sea urchins release their gametes into the water, the jelly coat that surrounds the egg exudes soluble molecules that attract the sperm, which swim toward the egg. As soon as the head of a sea urchin sperm contacts the jelly coat of a sea urchin egg, molecules in the jelly coat trigger the **acrosomal reaction** in the sperm **(Figure 47.3)**. This reaction begins with the discharge of hydrolytic enzymes from the **acrosome**, a specialized vesicle at the tip of the sperm. These enzymes partially digest the jelly coat, enabling a sperm structure called the *acrosomal process* to elongate and penetrate the coat. Protein molecules on the tip of the extended acrosomal process bind to specific receptor proteins that jut out from the egg plasma membrane. This "lock-and-key" recognition is especially important for sea urchins and other species with external fertilization because the surrounding water may contain gametes of other species.

Contact between the tip of the acrosomal process and the receptors on the egg leads to the fusion of the sperm and egg plasma membranes. The sperm nucleus then enters the egg cytoplasm as ion channels open in the egg's plasma membrane. Sodium ions diffuse into the egg and cause *depolarization*, a decrease in the membrane potential (see Chapter 7). The depolarization occurs within about 1–3 seconds after a sperm binds to an egg. By preventing additional sperm from fusing with the egg's plasma membrane, this depolarization acts as a **fast block to polyspermy**.

The Cortical Reaction

Membrane depolarization lasts for only a minute or so. A longer-lasting block to polyspermy is established by vesicles that lie just beneath the egg plasma membrane, in the rim of cytoplasm known as the *cortex*. Within seconds after a sperm binds to the egg, these vesicles, called cortical granules, fuse with the egg plasma membrane (see Figure 47.3, ❹). The contents of the cortical granules are released into the space between the plasma membrane and the surrounding *vitelline layer*, a structure formed by the extracellular matrix of the egg. Enzymes and other macromolecules from the granules trigger a cortical reaction, which lifts the vitelline layer away from the egg and hardens the layer into a protective fertilization envelope. Additional enzymes clip off and release the external portions of the remaining receptor proteins, along with any attached sperm. Together, the fertilization envelope

1 Contact. The sperm contacts the egg's jelly coat, triggering exocytosis of the sperm's acrosome.

2 Acrosomal reaction. Hydrolytic enzymes released from the acrosome make a hole in the jelly coat. Growing actin filaments form the acrosomal process, which protrudes from the sperm head and penetrates the jelly coat. Proteins on the surface of the acrosomal process bind to receptors in the egg plasma membrane.

3 Contact and fusion of sperm and egg membranes. Fusion triggers depolarization of the membrane, which acts as a fast block to polyspermy.

4 Cortical reaction. Cortical granules in the egg fuse with the plasma membrane. The secreted contents clip off sperm-binding receptors and cause the fertilization envelope to form. This acts as a slow block to polyspermy.

5 Entry of sperm nucleus.

Basal body (centriole)
Sperm head
Acrosome
Jelly coat
Sperm-binding receptors
Sperm nucleus
Sperm plasma membrane
Acrosomal process
Actin filament
Hydrolytic enzymes
Vitelline layer
Egg plasma membrane
Fused plasma membranes
Cortical granule
Perivitelline space
Fertilization envelope
EGG CYTOPLASM

▲ **Figure 47.3 The acrosomal and cortical reactions during sea urchin fertilization.** The events following contact of a single sperm and egg ensure that the nucleus of only one sperm enters the egg cytoplasm.

The icon above is a simplified drawing of an adult sea urchin. Throughout the chapter, this and other icons of an adult frog, chicken, nematode, and human indicate the animals whose embryos are featured in certain figures.

and other changes in the egg's surface impede the entry of additional sperm nuclei and thus act as a longer-term **slow block to polyspermy**.

Formation of the fertilization envelope requires a high concentration of calcium ions (Ca^{2+}) in the egg. Does a change in the Ca^{2+} concentration trigger the cortical reaction? To answer this question, researchers at the University of California, Berkeley, used a calcium-sensitive dye to assess the amount and distribution of Ca^{2+} in the egg during fertilization. As described in **Figure 47.4**, on the next page, they found that Ca^{2+} spread across the egg in a wave that correlated with the appearance of the fertilization envelope. Further studies demonstrated that release of Ca^{2+} into the cytosol from the endoplasmic reticulum is controlled by a signal transduction pathway activated by sperm binding. The resulting increase in Ca^{2+} levels causes cortical granules to fuse with the plasma membrane. Although understood in greatest detail in sea urchins, the cortical reaction triggered by Ca^{2+} also occurs in vertebrates such as fishes and mammals.

Egg Activation

A major function of fertilization is the combining of haploid sets of chromosomes from two individuals into a single diploid cell, the zygote. However, the events of fertilization also initiate metabolic reactions that trigger the onset of embryonic development, thus "activating" the egg. There is, for example, a marked increase in the rates of cellular respiration and protein synthesis in the egg following fertilization.

What triggers egg activation? Studies show that injecting Ca^{2+} into an unfertilized egg activates egg metabolism in many species, despite the absence of sperm. Researchers therefore conclude that the rise in Ca^{2+} concentration that causes the cortical reaction also causes egg activation. Further experiments have revealed that artificial activation is possible even if the nucleus has been removed from the egg. This finding indicates that egg activation requires only the proteins and mRNAs already present in the egg cytoplasm.

▼ **Figure 47.4** INQUIRY

Does the distribution of Ca^{2+} in an egg correlate with formation of the fertilization envelope?

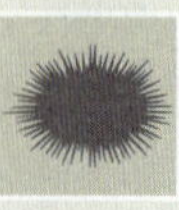

EXPERIMENT During fertilization, fusion of cortical granules with the egg plasma membrane causes the fertilization envelope to rise and spread around the egg from the point of sperm binding.

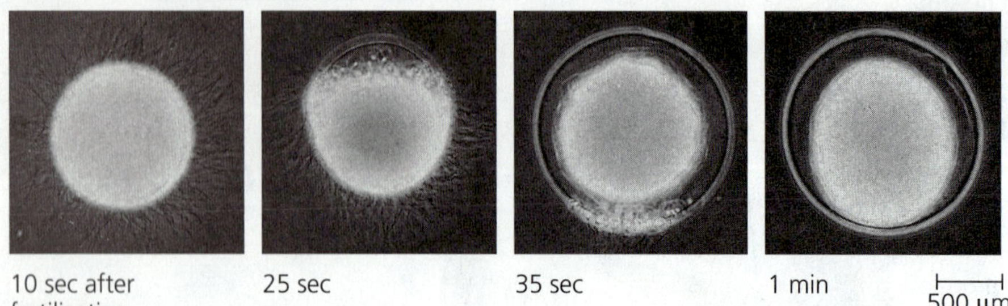

Calcium ion (Ca^{2+}) signaling is involved in fusion of vesicles with the plasma membrane during neurotransmitter release, insulin secretion, and plant pollen tube formation. Rick Steinhardt, Gerald Schatten, and colleagues, then at the University of California at Berkeley, hypothesized that an increase in Ca^{2+} levels similarly triggers cortical granule fusion. To test this hypothesis, they tracked the release of free Ca^{2+} in sea urchin eggs after sperm binding to see if it correlated with formation of the fertilization envelope. A fluorescent dye that glows when it binds free Ca^{2+} was injected into unfertilized eggs. The researchers then added sea urchin sperm and observed the eggs with a fluorescence microscope. Schatten and colleagues later repeated the experiment using a more sensitive dye, producing the results shown here.

RESULTS A rise in cytosolic Ca^{2+} concentration began at the point of sperm entry and spread in a wave to the other side of the egg. Soon after the wave passed, the fertilization envelope rose.

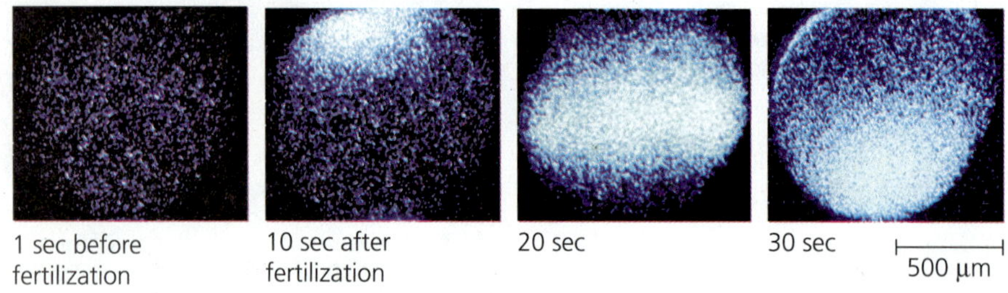

CONCLUSION The researchers concluded that Ca^{2+} release is correlated with the cortical reaction and formation of the fertilization envelope, supporting their hypothesis that an increase in Ca^{2+} levels triggers cortical granule fusion.

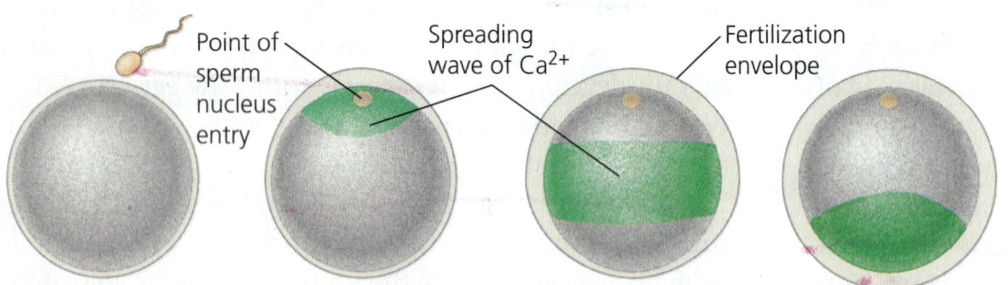

SOURCE R. Steinhardt et al., Intracellular calcium release at fertilization in the sea urchin egg, *Developmental Biology* 58:185–197 (1977). M. Hafner et al., Wave of free calcium at fertilization in the sea urchin egg visualized with Fura-2, *Cell Motility and the Cytoskeleton* 9:271–277 (1988).

MB See the related Experimental Inquiry Tutorial in MasteringBiology.

WHAT IF? Suppose you were given a chemical compound that could enter the egg and bind to Ca^{2+}, blocking its function. How would you use this compound to further test the hypothesis that a rise in Ca^{2+} levels triggers cortical granule fusion?

About 20 minutes after the sperm nucleus enters the sea urchin egg, the sperm and egg nuclei fuse. DNA synthesis begins, and the first cell division occurs after about 90 minutes, marking the end of the fertilization stage.

Fertilization in other species shares many features with the process in sea urchins. However, the timing of events differs, as does the stage of meiosis the egg has reached by the time it is fertilized. Sea urchin eggs have already completed meiosis when they are released from the female. In other species, eggs are arrested at a specific stage of meiosis and do not complete the meiotic divisions until fertilization occurs. Human eggs, for example, are arrested at metaphase of meiosis II prior to fertilization (see Figure 46.12).

Fertilization in Mammals

Unlike sea urchins and most other marine invertebrates, terrestrial animals, including mammals, fertilize eggs internally. Secretions in the mammalian female reproductive tract not only provide a moist environment for the sperm, but also bring about changes in sperm motility and structure. Only after these changes occur do sperm have the capacity to fertilize an egg. In humans, this process of *capacitation* occurs during the first 6 hours after the sperm enter the female reproductive tract.

Support cells of the developing follicle surround the mammalian egg and remain with it during and after ovulation (see Figure 46.12). A sperm must travel through this layer of follicle cells before it reaches the **zona pellucida**, the extracellular matrix of the egg. Within the zona pellucida is a component that functions as a receptor for sperm. Binding of a sperm to this receptor induces an acrosomal reaction, facilitating sperm passage through the zona pellucida to the egg. This binding also exposes a protein on the sperm that binds with the egg plasma membrane. At this point, the two cells fuse **(Figure 47.5)**.

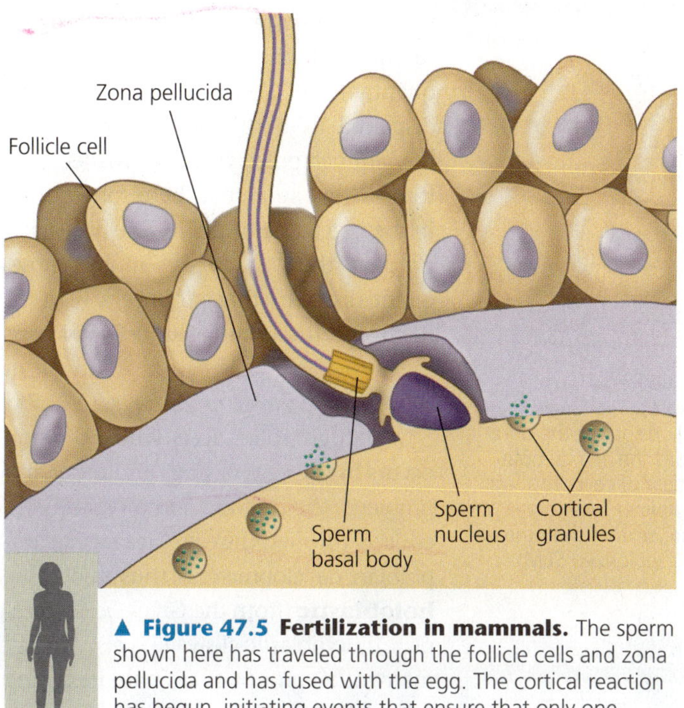

▲ **Figure 47.5 Fertilization in mammals.** The sperm shown here has traveled through the follicle cells and zona pellucida and has fused with the egg. The cortical reaction has begun, initiating events that ensure that only one sperm nucleus enters the egg.

As in sea urchin fertilization, sperm binding triggers changes within the mammalian egg that lead to a cortical reaction, the release of enzymes from cortical granules to the outside of the cell. These enzymes catalyze changes in the zona pellucida, which then functions as the slow block to polyspermy. (No fast block to polyspermy has been identified in mammals.)

After the egg and sperm membranes fuse, the whole sperm is taken into the egg. Once the envelopes of both haploid nuclei have dispersed, the sperm and egg chromosomes are organized onto a single mitotic spindle. Only after the first division is there a true diploid nucleus with a nuclear membrane.

Overall, fertilization is much slower in mammals than in sea urchins: The first cell division occurs 12–36 hours after sperm binding in mammals, compared with about 90 minutes in sea urchins. This cell division marks the end of fertilization and the beginning of the next stage, cleavage.

Cleavage

Once fertilization is complete, many animal species undergo a succession of rapid cell divisions that characterize the **cleavage** stage of early development. During cleavage, the cell cycle consists primarily of the S (DNA synthesis) and M (mitosis) phases. Cells essentially skip the G_1 and G_2 (gap) phases, and little or no protein synthesis occurs (see Figure 12.6 for a review of the cell cycle). As a result, cleavage partitions the cytoplasm of the large fertilized egg into many smaller cells called **blastomeres**, as shown in **Figure 47.6**.

The first five to seven cleavage divisions produce a hollow ball of cells, the **blastula**, surrounding a fluid-filled cavity called the **blastocoel** (see Figure 47.6).

Cleavage Patterns

In frogs and many other animals, the distribution of **yolk** (stored nutrients) is a key factor influencing the pattern of cleavage. Yolk is often concentrated toward one pole of the egg, called the **vegetal pole**. The yolk concentration decreases significantly toward the opposite pole, the **animal pole**. This difference in yolk distribution results

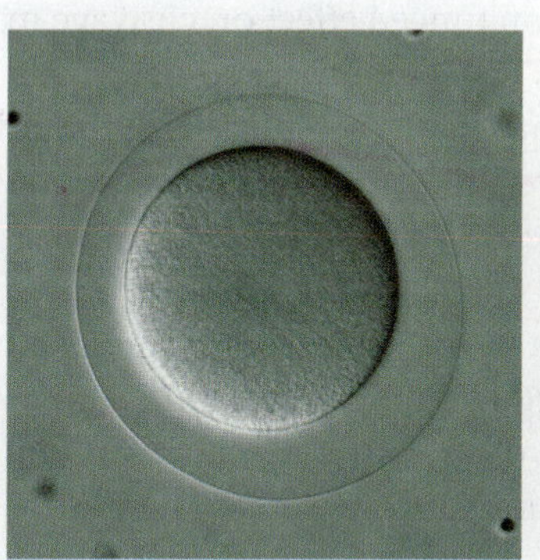

(a) Fertilized egg. Shown here is the zygote shortly before the first cleavage division, surrounded by the fertilization envelope.

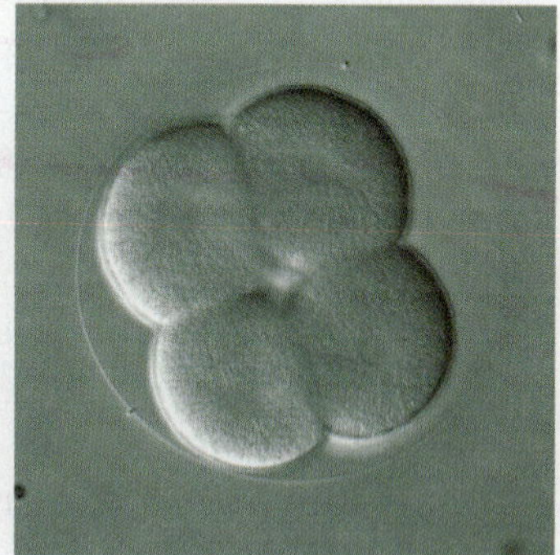

(b) Four-cell stage. Remnants of the mitotic spindle can be seen between the two pairs of cells that have just completed the second cleavage division.

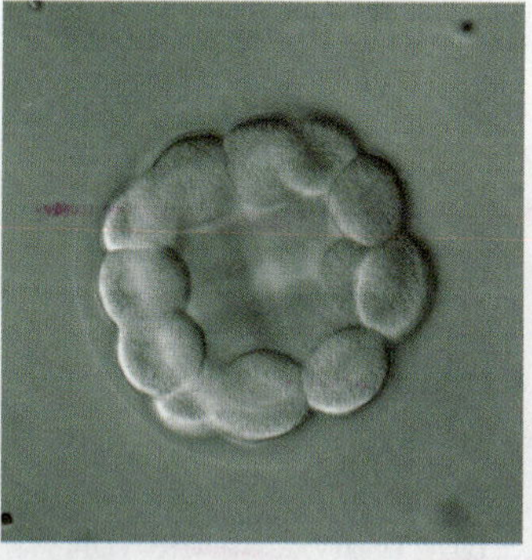

(c) Early blastula. After further cleavage divisions, the embryo is a multicellular ball that is still surrounded by the fertilization envelope. The blastocoel has begun to form in the center.

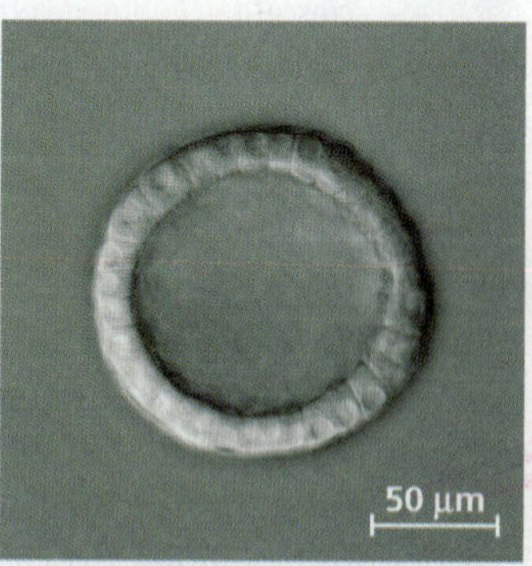

(d) Later blastula. A single layer of cells surrounds a large blastocoel. Although not visible here, the fertilization envelope is still present; the embryo will soon hatch from it and begin swimming.

▲ **Figure 47.6 Cleavage in an echinoderm embryo.** Cleavage is a series of mitotic cell divisions that transform the fertilized egg into a blastula, a hollow ball composed of cells called blastomeres. These light micrographs show the cleavage stages of a sand dollar embryo, which are virtually identical to those of a sea urchin.

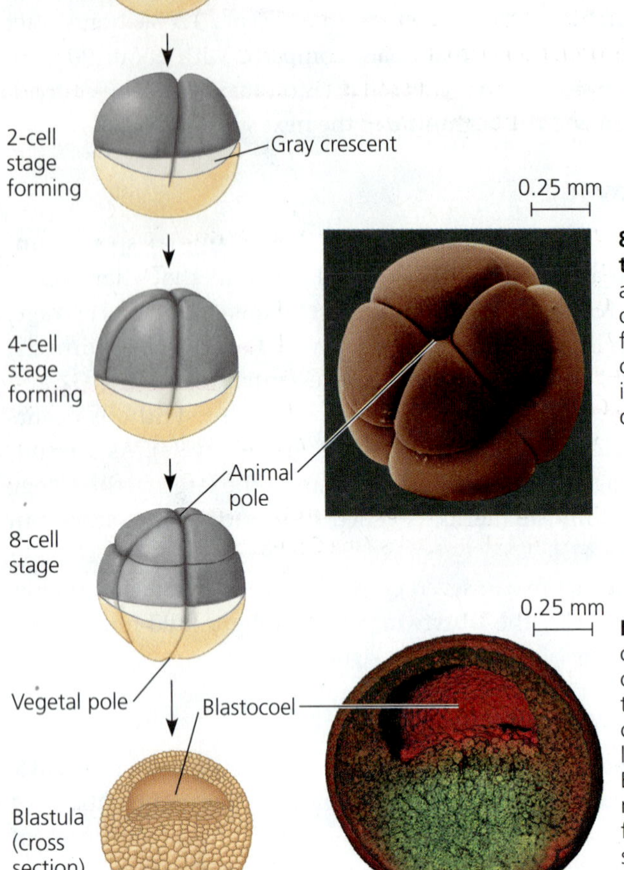

8-cell stage (viewed from the animal pole). The large amount of yolk displaces the third cleavage toward the animal pole, forming two tiers of cells. The four cells near the animal pole (closer, in this view) are smaller than the other four cells (colorized SEM).

Blastula (at least 128 cells). As cleavage continues, a fluid-filled cavity, the blastocoel, forms within the embryo. Because of unequal cell division, the blastocoel is located in the animal hemisphere. Both the drawing and the micrograph (assembled from fluorescence images) show cross sections of a blastula with about 4,000 cells.

▲ **Figure 47.7 Cleavage in a frog embryo.** The cleavage planes in the first and second divisions extend from the animal pole to the vegetal pole, but the third cleavage is perpendicular to the polar axis. In some species, the first division bisects the gray crescent, a lighter-colored region that appears opposite the site of sperm entry.

in animal and vegetal hemispheres that differ in appearance **(Figure 47.7)**.

During cell division, an indentation called a *cleavage furrow* forms in the cell surface as cytokinesis divides the cell in half. As shown in Figure 47.7, the first two cleavage furrows in the frog lie parallel to the line (or meridian) connecting the two poles. The second cell division begins before the first is complete, so the second cleavage furrow further divides the animal hemisphere while the first furrow is dividing the yolky cytoplasm of the vegetal hemisphere. Nevertheless, the first two divisions eventually produce four blastomeres of equal size, each extending from the animal pole to the vegetal pole.

During the third division of the frog egg, the asymmetric distribution of yolk in the embryo affects the relative size of cells produced in the two hemispheres. This division is equatorial (perpendicular to the line connecting the poles) and produces an eight-celled embryo. However, as each of the four blastomeres begins this division, the high concentration of yolk around the vegetal pole displaces the mitotic apparatus toward the animal pole. Consequently, the cleavage furrow is also displaced from the egg equator toward the animal pole, yielding smaller blastomeres in the animal hemisphere than in the vegetal hemisphere. The displacing effect of the yolk persists in the subsequent divisions that produce a blastula. In frogs, these unequal cell divisions cause the blastocoel to form entirely in the animal hemisphere (see Figure 47.7).

Although yolk affects where division occurs in the eggs of frogs and other amphibians, the cleavage furrow still passes entirely through the egg. Cleavage in amphibian development is thus said to be **holoblastic** (from the Greek *holos*, complete). Holoblastic cleavage is also seen in many other groups of animals, including echinoderms, mammals, and annelids. The orientation of the cleavage furrows varies within these groups, resulting in blastulas that vary considerably in appearance. In those animals whose eggs contain relatively little yolk, the blastocoel forms centrally and the blastomeres are often of similar size, particularly during the first few divisions (see Figure 47.6). This is the case for humans, whose embryos complete three divisions in the first 3 days after fertilization.

Yolk is most plentiful and has its most pronounced effect on cleavage in the eggs of birds, other reptiles, many fishes, and insects. In these animals, the volume of yolk is so great that cleavage furrows cannot pass through it, and only the region of the egg lacking yolk undergoes cleavage. This incomplete cleavage of a yolk-rich egg is said to be **meroblastic** (from the Greek *meros*, partial).

In birds, the part of the egg commonly called the yolk is actually the entire egg cell, swollen with yolk nutrients. Cell divisions are limited to a small whitish area at the animal pole. These divisions produce a cap of cells that sort into upper and lower layers. The cavity between these two layers is the avian version of the blastocoel.

In the eggs of *Drosophila* and most other insects, the sperm and egg nuclei fuse *within* a mass of yolk. Multiple rounds of mitosis occur without cytokinesis. In other words, no cell membranes form around the early nuclei. The first several hundred nuclei spread throughout the yolk and later migrate to the outer edge of the embryo. After several more rounds of mitosis, a plasma membrane forms around each nucleus, and the embryo, now the equivalent of a blastula, consists of a single layer of about 6,000 cells surrounding a mass of yolk (see Figure 18.22).

Regulation of Cleavage

The number of cleavage divisions varies among animals but appears to be controlled by a shared mechanism. Experimental results support the hypothesis that an animal embryo finishes the cleavage stage when the ratio of material in each nucleus to that in the cytoplasm is sufficiently large. One line of evidence comes from experiments in which researchers changed the starting amount of cytoplasm and then counted the cleavage divisions that occurred. For example, when half the normal amount of cytoplasm surrounds the newly formed zygotic nucleus, one fewer cleavage division occurs, consistent with the nuclear-cytoplasmic ratio reaching the threshold after one fewer cell cycle.

What is the adaptive advantage of linking the duration of the cleavage stage to the ratio of material in the nucleus and cytoplasm? The single nucleus in a newly fertilized egg has too little DNA to produce the amount of messenger RNA required to meet the cell's need for new proteins. Instead, the initial stages of development are carried out by RNA and proteins deposited in the egg during oogenesis. After cleavage, the egg cytoplasm has been divided among the many blastomeres, each with its own nucleus. Because each blastomere is much smaller than the entire egg or embryo, its nucleus can make enough RNA to program the cell's metabolism and its further development. The increase in the number of cells also sets the stage for morphogenesis, the transformation of embryo organization and shape.

CONCEPT CHECK 47.1

1. How does the fertilization envelope form in sea urchins? What is its function?
2. **WHAT IF?** Predict what would happen if you injected Ca^{2+} into an unfertilized sea urchin egg.
3. **MAKE CONNECTIONS** Review Figure 12.17 on page 240. Would you expect MPF activity to fluctuate or remain steady during cleavage? Explain your logic.

For suggested answers, see Appendix A.

CONCEPT 47.2

Morphogenesis in animals involves specific changes in cell shape, position, and survival

After cleavage, the rate of cell division slows considerably as the normal cell cycle is restored. The last two stages of embryonic development are responsible for **morphogenesis**, the cellular and tissue-based processes by which the animal body takes shape. During **gastrulation**, a set of cells at or near the surface of the blastula moves to an interior location, cell layers are established, and a primitive digestive tube is formed. Further transformation occurs during **organogenesis**, the formation of organs. We will discuss these two stages in turn, focusing in each case on the development of a few model organisms.

Gastrulation

Gastrulation is a dramatic reorganization of the hollow blastula into a two-layered or three-layered embryo called a **gastrula**. The cell layers produced by gastrulation are collectively called the embryonic **germ layers** (from the Latin *germen*, to sprout or germinate). In the late gastrula, **ectoderm** forms the outer layer and **endoderm** lines the embryonic digestive compartment or tract. In cnidarians and a few other radially symmetrical animals, only these two germ layers form during gastrulation. Such animals are called diploblasts (see Chapter 32). In contrast, animals with bilateral symmetry are triploblasts, having a third germ layer, the **mesoderm,** between the ectoderm and the endoderm.

Each germ layer contributes to a distinct set of structures in the adult animal **(Figure 47.8)**. Note that some organs and many organ systems of the adult derive from more than one germ layer. For example, the adrenal gland has both ectodermal and mesoderm tissue, and many other endocrine glands contain endodermal tissue.

ECTODERM (outer layer of embryo)

- Epidermis of skin and its derivatives (including sweat glands, hair follicles)
- Nervous and sensory systems
- Pituitary gland, adrenal medulla
- Jaws and teeth
- Germ cells

MESODERM (middle layer of embryo)

- Skeletal and muscular systems
- Circulatory and lymphatic systems
- Excretory and reproductive systems (except germ cells)
- Dermis of skin
- Adrenal cortex

ENDODERM (inner layer of embryo)

- Epithelial lining of digestive tract and associated organs (liver, pancreas)
- Epithelial lining of respiratory, excretory, and reproductive tracts and ducts
- Thymus, thyroid, and parathyroid glands

▲ **Figure 47.8 Major derivatives of the three embryonic germ layers in vertebrates.**

▶ **Figure 47.9 Gastrulation in a sea urchin embryo.** The movement of cells during gastrulation forms an embryo with a primitive digestive tube and three germ layers. Some of the mesodermal mesenchyme cells that migrate inward (step 1) will eventually secrete calcium carbonate and form a simple internal skeleton. Embryos in steps 1–3 are viewed from the front, those in 4 and 5 from the side.

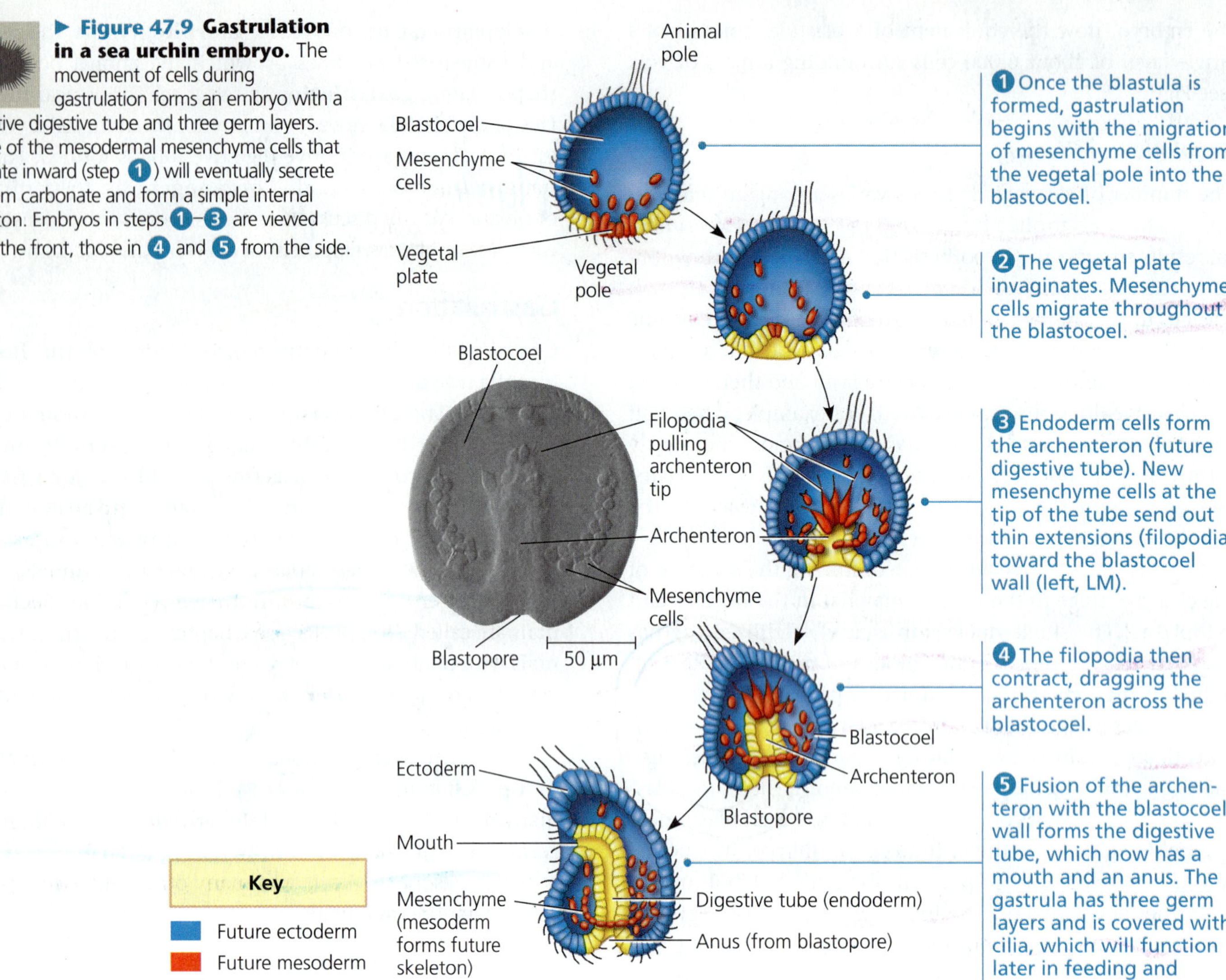

Gastrulation in Sea Urchins

Gastrulation in the sea urchin begins at the vegetal pole of the blastula **(Figure 47.9)**. There, cells called *mesenchyme cells* individually detach from the blastocoel wall and enter the blastocoel. The remaining cells near the vegetal pole flatten slightly and cause that end of the embryo to buckle inward as a result of cell shape changes we will discuss later. This process—the infolding of a sheet of cells into the embryo—is called *invagination*. Extensive rearrangement of cells transforms the shallow depression into a deeper, narrower, blind-ended tube called the **archenteron**. The open end of the archenteron, which will become the anus, is called the **blastopore**. A second opening, which will become the mouth, forms when the opposite end of the archenteron touches the inside of the ectoderm and the two layers fuse, producing a rudimentary digestive tube.

As you learned in Chapter 32, animals can be categorized by whether the mouth develops from the first opening that forms in the embryo (protostomes) or the second (deuterostomes). Sea urchins and other echinoderms are deuterostomes, as are chordates like ourselves and other vertebrates.

Upon completing gastrulation, sea urchin embryos develop into ciliated larvae that drift in ocean surface waters as zooplankton, feeding on bacteria and unicellular algae. Eventually, each larva metamorphoses into the adult form of the sea urchin, which takes up residence on the ocean floor.

Gastrulation in Frogs

The frog blastula contains large, yolk-laden cells in the vegetal hemisphere and a blastocoel wall that in most species is more than one cell thick. Recall from Chapter 32 that frogs and other bilaterally symmetrical animals have a dorsal (top) side and a ventral (bottom) side, a left side and a right side, and an anterior (front) end and a posterior (back) end. As shown in **Figure 47.10**, frog gastrulation begins when a group of cells on the dorsal side of the blastula begins to invaginate. This process forms a crease along the region where the gray crescent formed (see Figure 47.7). It may help to

1 Gastrulation begins when a small indented crease, the blastopore, appears on the dorsal side of the late blastula. The crease is formed by cells changing shape and invaginating. Sheets of outer cells then roll inward over the dorsal lip (involution) and move into the interior (shown by the dashed arrow), where they will form endoderm and mesoderm. Meanwhile, cells at the animal pole change shape and begin spreading over the outer surface.

2 The blastopore extends around both sides of the embryo (red arrows) as more cells invaginate. When the ends meet, the blastopore forms a circle that becomes smaller as ectoderm spreads downward over the surface. Internally, continued involution expands the endoderm and mesoderm, and the archenteron begins to form; as a result, the blastocoel becomes smaller.

3 Late in gastrulation, the endoderm-lined archenteron has completely replaced the blastocoel and the three germ layers are in place. The circular blastopore surrounds a plug of yolk-filled cells.

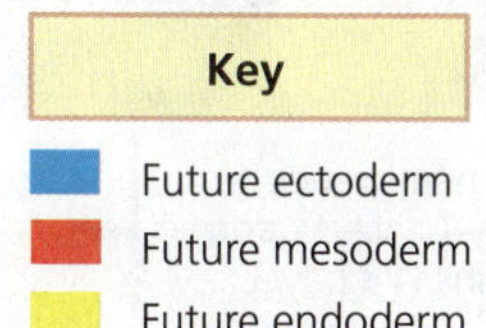

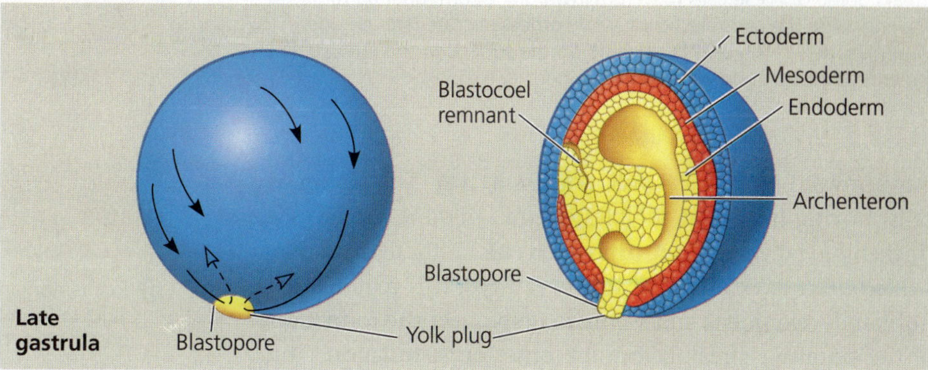

▲ **Figure 47.10 Gastrulation in a frog embryo.** In the frog blastula, the blastocoel is displaced toward the animal pole and is surrounded by a wall several cells thick. The cell movements that begin gastrulation occur on the dorsal side of the blastula, opposite where the sperm entered the egg.

think of this crease as the site where two thin lips are pressed together. The part above the crease becomes the dorsal side of the blastopore, called the **dorsal lip**.

As the blastopore is forming, a sheet of cells begins to spread out of the animal hemisphere. Some of these cells roll over the edge of the lip into the interior of the embryo, a process called *involution*. Once inside the embryo, these cells move away from the blastopore toward the animal pole and become organized into layers of endoderm and mesoderm, with the endoderm on the inside. Cells continue to spread over the gastrula surface, shifting and shrinking the blastopore. In the interior of the embryo, an archenteron forms and grows as the blastocoel shrinks and eventually disappears.

At the end of gastrulation, the cells remaining on the surface make up the ectoderm, the tube of endoderm is the innermost layer, and the mesoderm lies between them. As in the sea urchin, the frog's anus develops from the blastopore, and the mouth eventually breaks through at the opposite end of the archenteron.

Gastrulation in Chicks

The starting point for gastrulation in chicks is an embryo consisting of upper and lower layers—the *epiblast* and *hypoblast*—lying atop a yolk mass. All the cells that will form the embryo come from the epiblast. During gastrulation, some epiblast cells move toward the midline of the blastoderm, detach, and

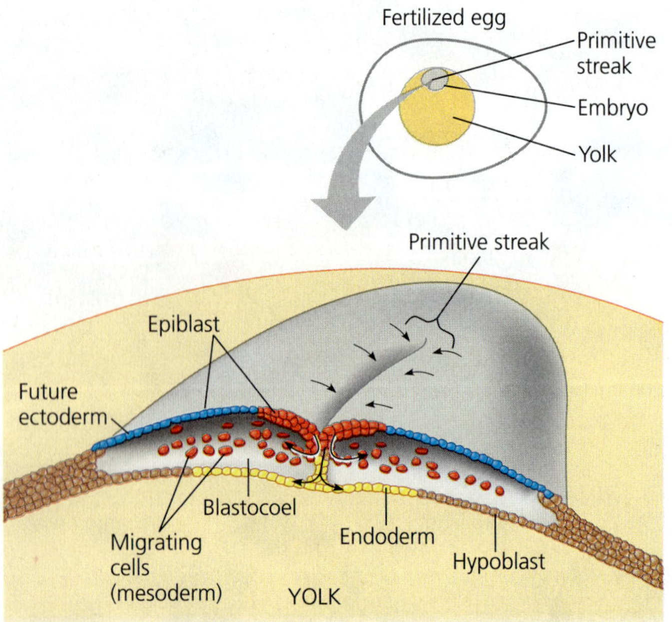

▲ **Figure 47.11 Gastrulation in a chick embryo.** The chick blastula consists of an upper layer of cells, the epiblast, and a lower layer, the hypoblast, with a space (the blastocoel) between them. This is a cross section at a right angle to the primitive streak, looking toward the anterior end of a gastrulating embryo. During gastrulation, some cells of the epiblast migrate (arrows) into the interior of the embryo through the primitive streak. Some of these cells move downward and form endoderm, pushing aside the hypoblast cells, while others migrate laterally and form mesoderm. The cells left behind on the surface of the embryo at the end of gastrulation will become ectoderm.

move inward toward the yolk (**Figure 47.11**). The pileup of cells moving inward at the blastoderm's midline produces a thickening called the **primitive streak**. Although the hypoblast contributes no cells to the embryo, it is required for normal development and seems to help direct the formation of the primitive streak before the onset of gastrulation. The hypoblast cells later segregate from the endoderm and eventually form part of the sac that surrounds the yolk and also part of the stalk that connects the yolk mass to the embryo.

Gastrulation in Humans

Unlike the large, yolky eggs of many vertebrates, human eggs are quite small, storing little in the way of food reserves. Fertilization takes place in the oviduct, and the earliest stages of development occur while the embryo completes its journey down the oviduct to the uterus (see Figure 46.15). Knowledge about gastrulation in humans is therefore largely based on what we can extrapolate from other mammals, such as the mouse, and on observation of very early human development following *in vitro* fertilization.

Figure 47.12 depicts development of the human embryo starting about 6 days after fertilization. The description on page 1031 follows the numbered stages in the figure.

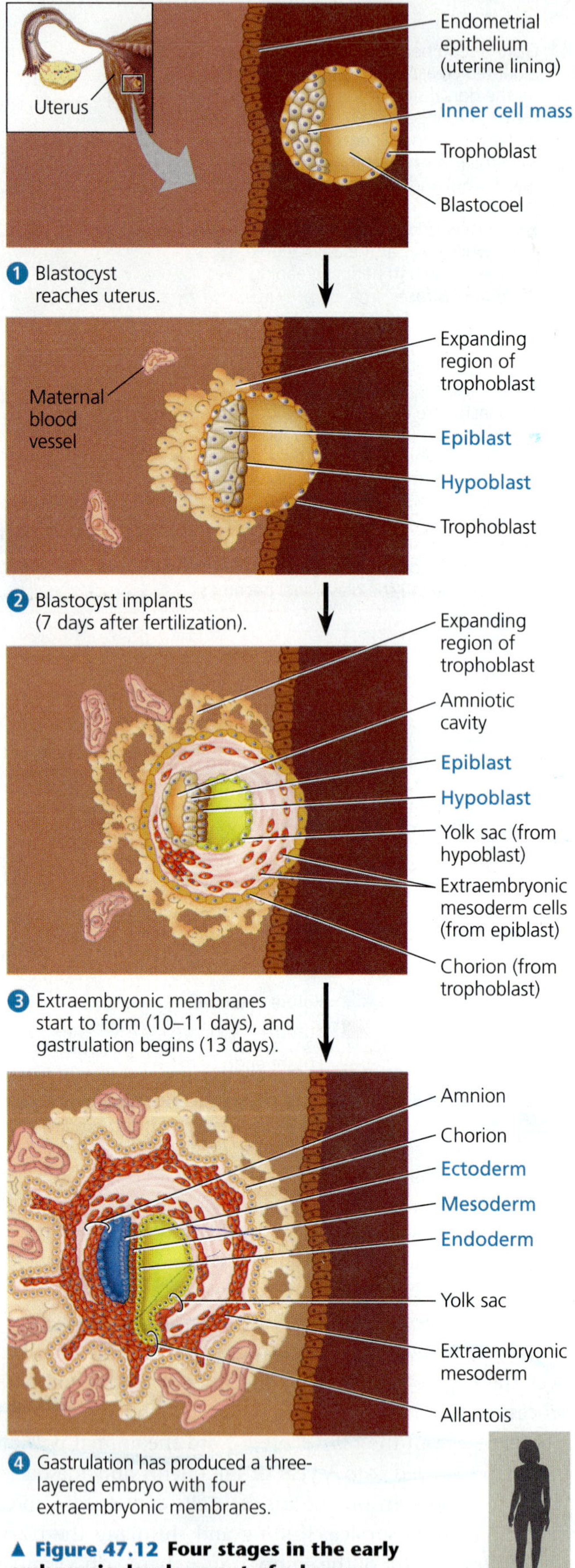

▲ **Figure 47.12 Four stages in the early embryonic development of a human.**

1 At the end of cleavage, the embryo has more than 100 cells arranged around a central cavity and has traveled down the oviduct to the uterus. At this stage of development, the embryo is called a **blastocyst**, the mammalian version of a blastula. Clustered at one end of the blastocyst cavity is a group of cells called the **inner cell mass**, which will develop into the embryo proper. It is the cells of the very early blastocyst stage that are the source of embryonic stem cell lines.

2 The **trophoblast**, the outer epithelium of the blastocyst, does not contribute to the embryo itself but instead supports embryo growth in a number of ways. It initiates implantation by secreting enzymes that break down molecules of the endometrium, the lining of the uterus. This allows the blastocyst to invade the endometrium. As the trophoblast thickens through cell division, it extends finger-like projections into the surrounding maternal tissue. Invasion by the trophoblast leads to erosion of capillaries in the endometrium, causing blood to spill out and bathe trophoblast tissues. Around the time of implantation, the inner cell mass of the blastocyst forms a flat disk with an upper layer of cells, the *epiblast*, and a lower layer, the *hypoblast*. As in birds, the human embryo develops almost entirely from epiblast cells.

3 Following implantation, the trophoblast continues to expand into the endometrium, and four new membranes appear. Although these **extraembryonic membranes** are formed by the embryo, they enclose specialized structures located outside the embryo. As implantation is completed, gastrulation begins. Cells move inward from the epiblast through a primitive streak and form mesoderm and endoderm, just as in the chick (see Figure 47.11).

4 By the end of gastrulation, the embryonic germ layers have formed. Extraembryonic mesoderm and the four extraembryonic membranes now surround the embryo. As development proceeds, the invading trophoblast, cells from the epiblast, and adjacent endometrial tissue will all contribute to formation of the placenta. This vital organ mediates exchange of nutrients, gases, and nitrogenous wastes between the embryo and the mother (see Figure 46.16).

Developmental Adaptations of Amniotes

EVOLUTION As you read in Chapter 34, birds and other reptiles, like mammals, form four extraembryonic membranes. In all these groups, such membranes provide a "life-support system" for further embryonic development. Why, then, did this adaptation appear in the evolutionary history of reptiles and mammals but not other vertebrates, such as fishes and amphibians? We can formulate a reasonable hypothesis by considering a few basic facts about embryonic development. All vertebrate embryos require an aqueous environment for their development. The embryos of fishes and amphibians usually develop in the surrounding sea or pond and need no specialized water-filled enclosure. However, the extensive colonization of land by vertebrates was possible only after the evolution of structures that would allow reproduction in dry environments. Two such structures exist today: (1) the shelled egg of birds and other reptiles as well as a few mammals (the monotremes) and (2) the uterus of marsupial and eutherian mammals. Inside the shell or uterus, the embryos of these animals are surrounded by fluid within a sac formed by one of the extraembryonic membranes, the amnion. Mammals and reptiles, including birds, are therefore called **amniotes** (see Chapter 34).

We can explore the evolution of extraembryonic membranes by comparing their functions in different groups of amniotes. For the purposes of this discussion, you may find it useful to refer to Figure 34.26, which describes the functions of the extraembryonic membranes in the egg of a reptile.

For the most part, the extraembryonic membranes have similar functions in mammals and reptiles, consistent with a common evolutionary origin. The chorion is the site of gas exchange, and the fluid within the amnion physically protects the developing embryo. (This amniotic fluid is released from the vagina when a pregnant woman's "water breaks" just before childbirth.) The allantois, which disposes of wastes in the reptilian egg, is incorporated into the umbilical cord in mammals. There it forms blood vessels that transport oxygen and nutrients from the placenta to the embryo and rid the embryo of carbon dioxide and nitrogenous wastes. The fourth extraembryonic membrane, the yolk sac, encloses yolk in the eggs of reptiles. In mammals it is a site of early formation of blood cells, which later migrate into the embryo proper. Thus, although the extraembryonic membranes of reptiles were conserved in mammals in the course of evolution, modifications appeared that were adapted to development within the uterus of the mother.

After gastrulation is complete and any extraembryonic membranes are formed, the next stage of embryonic development begins: organ formation.

Organogenesis

During organogenesis, regions of the three embryonic germ layers develop into the rudiments of organs. Whereas gastrulation involves mass movements of cells, organogenesis involves

(a) Neural plate formation. By this stage, the notochord has developed from dorsal mesoderm, and the dorsal ectoderm has thickened, forming the neural plate, in response to signals from other embryonic tissues. The neural folds are the two ridges that form the lateral edges of the neural plate. These folds are visible in the LM of a whole embryo.

(b) Neural tube formation. Infolding and pinching off of the neural plate generates the neural tube. Note the neural crest cells, which will migrate and give rise to numerous structures. (See also Figure 34.7.)

(c) Somites. The SEM is a side view of the whole embryo at the tail-bud stage. Part of the ectoderm has been removed to reveal the somites, blocks of tissue that will give rise to segmental structures such as vertebrae. The drawing shows a similar-stage embryo after formation of the neural tube, as if the embryo in the SEM were cut and viewed in cross section. By this time, the lateral mesoderm has begun to separate into two tissue layers that line the coelom, or body cavity. The somites, formed from mesoderm, flank the notochord.

▲ **Figure 47.13 Neurulation in a frog embryo.**

more localized changes. To illustrate the basic principles of this process, we'll focus on *neurulation*, the first steps in the formation of the brain and spinal cord in vertebrates.

Neurulation begins as cells from the dorsal mesoderm come together to form the **notochord**, the rod that extends along the dorsal side of the chordate embryo, seen in **Figure 47.13a** for the frog. Signaling molecules secreted by these mesodermal cells and other tissues induce the ectoderm above the notochord to become the *neural plate*. Next, the cells of the neural plate change shape, curving the neural plate inward. In this way, the neural plate rolls itself into the **neural tube**, which runs along the anterior-posterior axis of the embryo **(Figure 47.13b)**. The neural tube will become the brain in the head and the spinal cord along the rest of the body.

In vertebrate embryos, two sets of cells develop near the neural tube and then migrate elsewhere in the body. The first set is a band of cells called the **neural crest**, which develops along the borders where the neural tube pinches off from the ectoderm. Neural crest cells subsequently migrate to many parts of the embryo, forming a variety of tissues that include peripheral nerves as well as parts of the teeth and skull bones. The second set of migratory cells is formed when groups of cells located in strips of mesoderm lateral to the notochord separate into blocks called **somites** **(Figure 47.13c)**. The somites are arranged serially on both sides along the length of the notochord. Parts of the somites dissociate into mesenchyme cells, which migrate individually to new locations.

Somites play a major role in organizing the segmented structure of the vertebrate body. One of the major functions of the mesenchyme cells that leave the somites is formation of the vertebrae. Although the notochord disappears before birth, parts of the notochord persist as the inner portions of

▶ **Figure 47.14**
Organogenesis in a chick embryo.

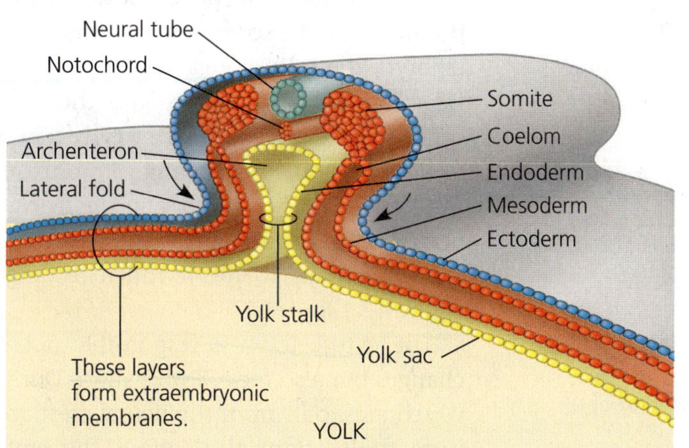

(a) Early organogenesis. The archenteron forms when lateral folds pinch the embryo away from the yolk. The embryo remains open to the yolk, attached by the yolk stalk, about midway along its length, as shown in this cross section. The notochord, neural tube, and somites subsequently develop much as they do in the frog. The germ layers lateral to the embryo itself form extraembryonic membranes.

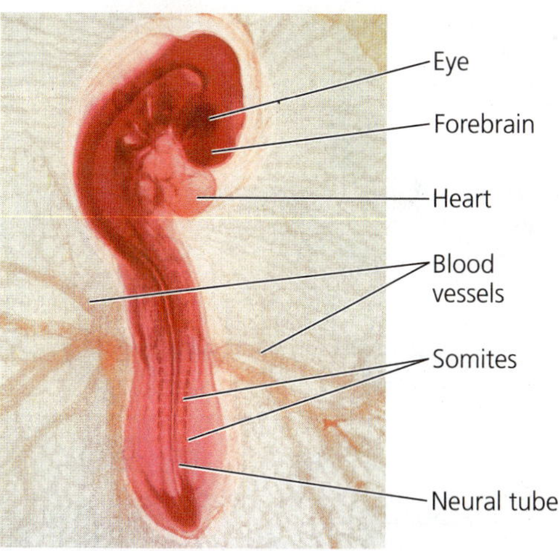

(b) Late organogenesis. Rudiments of most major organs have already formed in this chick embryo, which is 3 days old and about 2–3 mm long. The extraembryonic membranes eventually are supplied by blood vessels extending from the embryo; several major blood vessels are seen here (LM).

the vertebral disks in adults. (These are the disks that can herniate or rupture, causing back pain.) Somite cells that become mesenchymal later form the muscles associated with the vertebral column and the ribs. Through these processes, serially repeating structures of the embryo (somites) form repeated structures in the adult. Chordates can thus be described as segmented animals, although the segmentation becomes less obvious later in development. Lateral to the somites, the mesoderm splits into two layers that form the lining of the body cavity, or coelom (see Figure 32.8).

Early organogenesis in other vertebrates is quite similar to that in the frog. In the chick, for example, the borders of the blastoderm fold downward and come together, pinching the embryo into a three-layered tube joined under the middle of the body to the yolk **(Figure 47.14a)**. By the time the chick embryo is 3 days old, rudiments of the major organs, including the brain, eyes, and heart, are readily apparent **(Figure 47.14b)**.

In humans, an error in neural tube formation results in *spina bifida*, the most common disabling birth defect in the United States. In spina bifida, a portion of the neural tube fails to develop or close properly, leaving an opening in the spinal column and causing nerve damage. Although the opening can be surgically repaired shortly after birth, the nerve damage is permanent, resulting in varying degrees of leg paralysis.

Organogenesis is somewhat different in invertebrates, which is not surprising, given that their body plans diverge significantly from those of vertebrates. The underlying mechanisms, however, involve many of the same cellular activities: cell migration, cell signaling between different tissues, and cell shape changes generating new organs. In insects, for example, tissues of the nervous system form when ectoderm along the anterior-posterior axis rolls into a tube inside the embryo, similar to the vertebrate neural tube. Interestingly, the tube is on the ventral side of the insect embryo rather than the dorsal side, where it is in vertebrates. In spite of the different locations, the molecular signaling pathways that bring about the events in the two groups are very similar, underscoring their ancient shared evolutionary history.

As we have seen in our consideration of gastrulation and organogenesis, changes in cell shape and location are essential to early development. We will turn now to an exploration of how these changes take place.

Mechanisms of Morphogenesis

Morphogenesis is a major stage of development in both animals and plants, but only in animals does it involve the *movement* of cells. The rigid cell wall that surrounds plant cells prevents complex movements like those that occur during gastrulation and organogenesis. In animals, movement of parts of a cell can bring about changes in cell shape or enable a cell to migrate from one place to another within the embryo. Here we will consider some of the cellular components that contribute to these events. We'll begin with the roles of the microtubules and microfilaments that make up the cytoskeleton (see Table 6.1).

The Cytoskeleton in Morphogenesis

Reorganization of the cytoskeleton is a major force in changing cell shape during development. As an example, let's return to the topic of neurulation. At the onset of neural tube formation, microtubules oriented from dorsal to ventral in a sheet of ectodermal cells help lengthen the cells along that

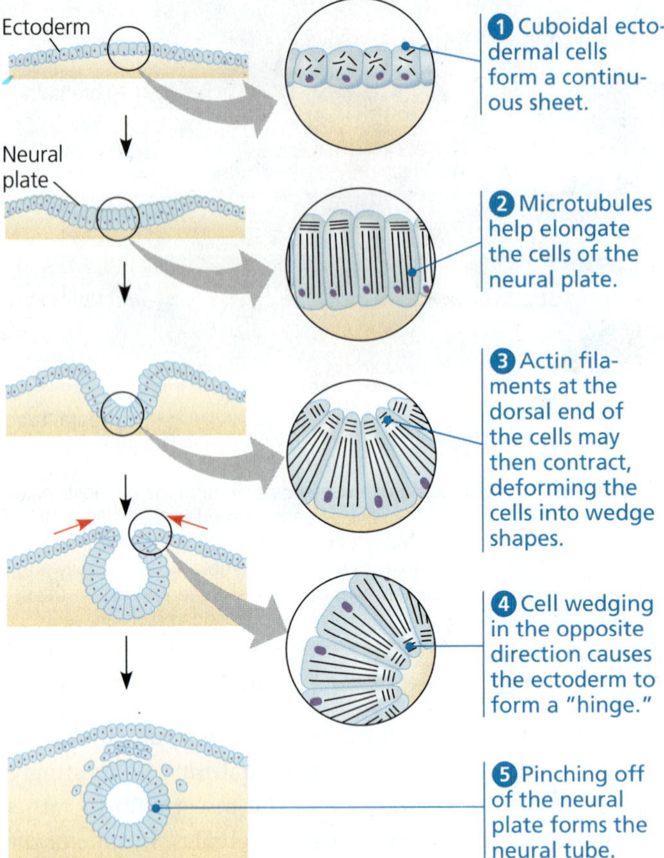

▲ **Figure 47.15 Change in cell shape during morphogenesis.** Reorganization of the cytoskeleton is associated with morphogenetic changes in embryonic tissues, as shown here for the formation of the neural tube in vertebrates.

axis **(Figure 47.15)**. At the dorsal end of each cell is a bundle of actin filaments (microfilaments) oriented crosswise. These actin filaments contract, giving the cells a wedge shape that bends the ectoderm layer inward. Similar changes in cell shape occur at the hinge regions where the neural tube is pinching off from the ectoderm. However, the generation of wedge-shaped cells is not limited to neurulation or even to vertebrates. In *Drosophila* gastrulation, for instance, the formation of wedge-shaped cells along the ventral surface is responsible for invagination of a tube of cells that form the mesoderm.

The cytoskeleton directs a different type of morphogenetic movement in promoting elongation of the archenteron in the sea urchin embryo (see Figure 47.9). In this case, cytoskeletal changes direct **convergent extension**, a rearrangement of the cells of a tissue layer that causes the sheet to become narrower (converge) while it becomes longer (extends). It's as if a crowd of people waiting to

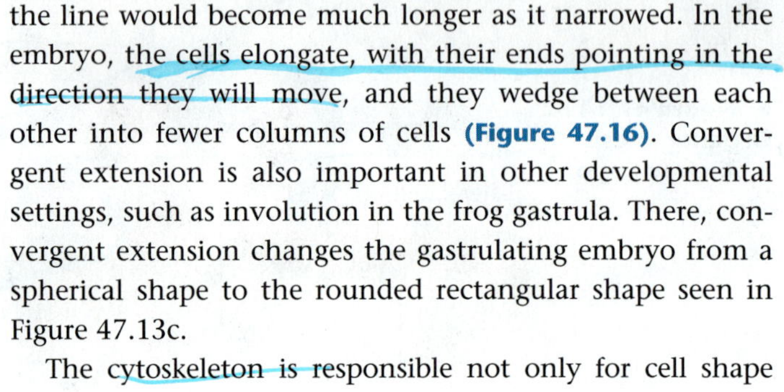

enter a theater for a concert began to form a single-file line; the line would become much longer as it narrowed. In the embryo, the cells elongate, with their ends pointing in the direction they will move, and they wedge between each other into fewer columns of cells **(Figure 47.16)**. Convergent extension is also important in other developmental settings, such as involution in the frog gastrula. There, convergent extension changes the gastrulating embryo from a spherical shape to the rounded rectangular shape seen in Figure 47.13c.

The cytoskeleton is responsible not only for cell shape changes but also for cell migration. During organogenesis in vertebrates, cells from the neural crest and from somites migrate to locations throughout the embryo. Cells "crawl" within the embryo by using cytoskeletal fibers to extend and retract cellular protrusions. This type of motility is akin to the amoeboid movement described in Figure 6.27b. Transmembrane glycoproteins called *cell adhesion molecules* play a key role in cell migration by promoting interaction between pairs of cells. Cell migration also involves the *extracellular matrix* (*ECM*), the meshwork of secreted glycoproteins and other macromolecules lying outside the plasma membranes of cells (see Figure 6.30). The ECM helps to guide cells in many types of movements, such as migration of individual cells and shape changes of cell sheets. Cells that line migration pathways regulate movement of migrating cells by secreting specific molecules into the ECM.

Programmed Cell Death

Just as certain cells of the embryo are programmed to change shape or location, others are programmed to die. A type of *programmed cell death* called **apoptosis** is in fact a common feature of animal development. At various times in development, individual cells, sets of cells, or whole tissues cease to develop and are engulfed by neighboring cells. In some cases, a structure functions in a larval or other immature form of the organism and then is eliminated during later development. One familiar example is provided by the cells in the tail of a tadpole, which undergo apoptosis during frog metamorphosis (see Figure 45.19). Apoptosis can also occur when cells compete with one another for survival. For instance, many more neurons are produced during development of the vertebrate nervous system than exist in the adult. In general, neurons survive if they make functional connections with other neurons and die if they do not.

Some cells that undergo apoptosis don't seem to have any function in the developing embryo. Why do such cells form? The answer can be found by considering the evolution of

▶ **Figure 47.16 Convergent extension of a sheet of cells.** In this simplified diagram, the cells elongate in a particular direction and crawl between each other (convergence) as the sheet becomes longer and narrower (extension).

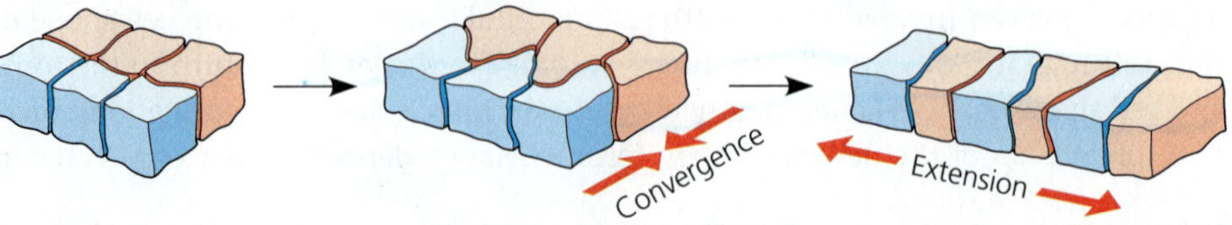

amphibians, birds, and mammals. When these groups began to diverge during evolution, the developmental program for making a vertebrate body was already in place. The differences in present-day body forms arose through modification of that common developmental program (which is why the early embryos of all vertebrates look so similar). As these groups evolved, many structures produced by the ancestral program that no longer offered a selective advantage were targeted for cell death. For example, the shared developmental program generates webbing between the embryonic digits, but in many birds and mammals the webbing is eliminated by apoptosis (see Figure 11.22).

As you have seen, cell behavior and the molecular mechanisms underlying it are crucial to the morphogenesis of the embryo. In the next section, you'll learn that a shared set of cellular and genetic processes ensure that the various types of cells end up in the right places in each embryo.

CONCEPT CHECK 47.2

1. In the frog embryo, convergent extension elongates the notochord. Explain how the words *convergent* and *extension* apply to this process.
2. **WHAT IF?** Predict what would happen if, just before neural tube formation, you treated embryos with a drug that blocks the function of microfilaments.
3. **MAKE CONNECTIONS** Unlike some other types of birth defects, neural tube defects are largely preventable. Explain (see Figure 41.4, p. 879).

For suggested answers, see Appendix A.

CONCEPT 47.3

Cytoplasmic determinants and inductive signals contribute to cell fate specification

During embryonic development, cells arise by division, take up particular locations in the body, and become specialized in structure and function. Where a cell resides, how it appears, and what it does define its development fate. Developmental biologists use the terms **determination** to refer to the process by which a cell or group of cells becomes committed to a particular fate and **differentiation** to refer to the resulting specialization in structure and function.

Every diploid cell formed during an animal's development has the same genome. With the exception of certain mature immune cells, the collection of genes present is the same throughout the cell's life. How, then, do cells acquire different fates? As discussed in Concept 18.4, particular tissues, and often cells within a tissue, differ from one another by expressing distinct sets of genes from their shared genome.

A major focus of developmental biology is to uncover the mechanisms that direct the differences in gene expression underlying developmental fates. As one step toward this goal, scientists often seek to trace tissues and cell types back to their origins in the early embryo.

Fate Mapping

One way to trace the ancestry of embryonic cells is direct observation through the microscope. Such studies produced the first **fate maps**, diagrams showing the structures arising from each region of an embryo. In the 1920s, German embryologist Walther Vogt used this approach to determine where groups of cells from the blastula end up in the gastrula **(Figure 47.17a)**. Later researchers developed techniques that allowed them to mark an individual blastomere during cleavage and then follow the marker as it was distributed to all the mitotic descendants of that cell **(Figure 47.17b)**.

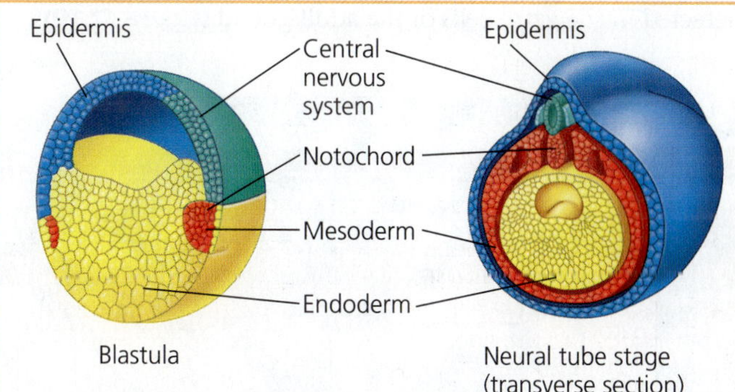

(a) Fate map of a frog embryo. The fates of groups of cells in a frog blastula (left) were determined in part by marking different regions of the blastula surface with nontoxic dyes of various colors. The embryos were sectioned at later stages of development, such as the neural tube stage shown on the right, and the locations of the dyed cells determined. The two embryonic stages shown here represent the result of numerous such experiments.

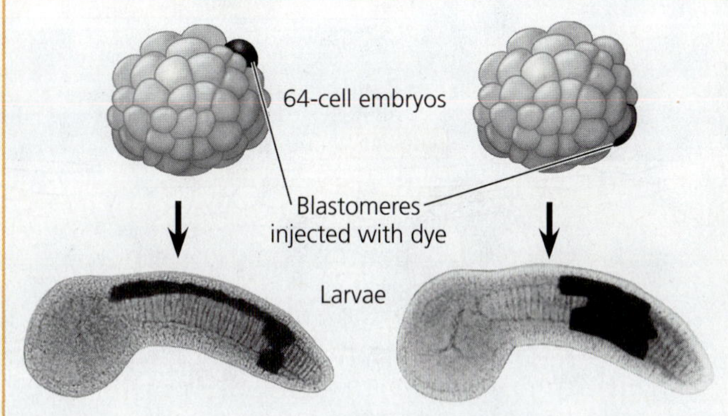

(b) Cell lineage analysis in a tunicate. In lineage analysis, an individual blastomere is injected with a dye during cleavage, as indicated in the drawings of 64-cell embryos of a tunicate, an invertebrate chordate (top). The dark regions in the light micrographs of larvae (bottom) correspond to the cells that developed from the two different blastomeres indicated in the drawings.

▲ **Figure 47.17 Fate mapping for two chordates.**

A much more comprehensive approach to fate mapping has been carried out on the soil-dwelling nematode *Caenorhabditis elegans*. This roundworm is about 1 mm long, has a simple, transparent body with only a few types of cells, and develops into a mature adult hermaphrodite in only 3½ days in the laboratory. These attributes allowed Sydney Brenner, Robert Horvitz, and John Sulston to determine the complete cell lineage of *C. elegans*. They found that every adult hermaphrodite has exactly 959 somatic cells, which arise from the fertilized egg in virtually the same way for every individual. Careful microscopic observations of worms at all stages of development, coupled with experiments in which particular cells or groups of cells were destroyed by a laser beam or through mutations, resulted in the cell lineage diagram shown in **Figure 47.18**.

As an example of a particular cell fate, we'll consider *germ cells*, the specialized cells that give rise to eggs or sperm. In all animals studied, complexes of RNA and protein are involved in the specification of germ cell fate. In *C. elegans*, such complexes, called *P granules*, persist throughout development and can be detected in the germ cells of the adult gonad **(Figure 47.19)**.

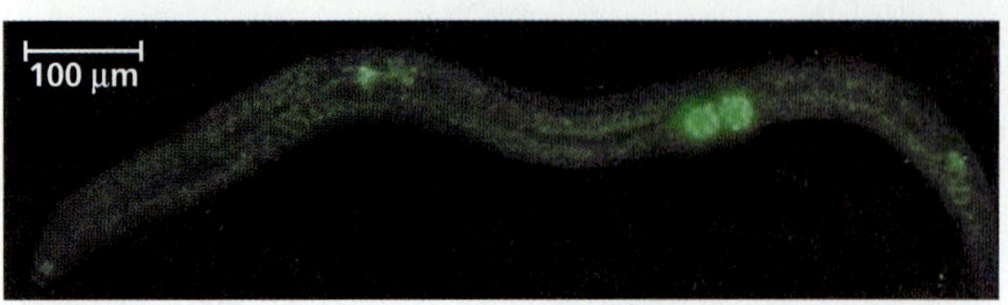

▲ **Figure 47.19 Determination of germ cell fate in *C. elegans*.** Labeling with an antibody specific for a *C. elegans* P granule protein (green) reveals the specific incorporation of P granules into the cells of the adult worm that will produce sperm or eggs.

Tracing the position of the P granules provides a dramatic illustration of cell fate specification during development. The P granules are distributed throughout the newly fertilized egg but move to the posterior end of the zygote before the first cleavage division **(Figure 47.20 1** and **2)**. As a result, only the posterior of the two cells formed by the first division contains P granules **(Figure 47.20 3)**. The P granules continue to be asymmetrically partitioned during subsequent divisions **(Figure 47.20 4)**. Thus, the P granules act as cytoplasmic determinants (see Concept 18.4), fixing germ cell fate at the earliest stage of *C. elegans* development.

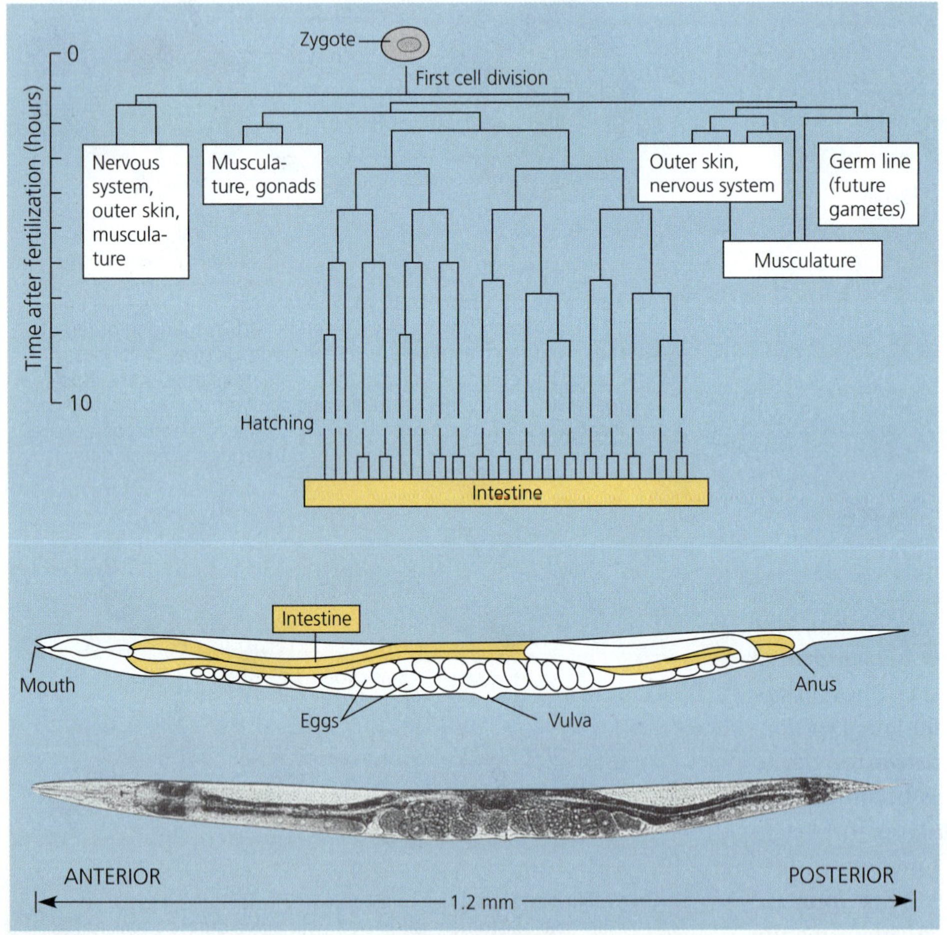

▲ **Figure 47.18 Cell lineage in *Caenorhabditis elegans*.** The *C. elegans* embryo is transparent, making it possible for researchers to trace the lineage of every cell, from the zygote to the adult worm (LM). The diagram shows a detailed lineage only for the intestine, which is derived exclusively from one of the first four cells formed from the zygote. The eggs will be fertilized internally and released through the vulva.

Fate mapping in *C. elegans* paved the way for major discoveries about programmed cell death. Lineage analysis demonstrated that exactly 131 cells die during normal *C. elegans* development. In the 1980s, researchers found that a mutation inactivating a single gene allows all 131 cells to live. Further research revealed that this gene is part of a pathway that controls and carries out apoptosis in a wide range of animals, including humans. In 2002, Brenner, Horvitz, and Sulston shared a Nobel Prize for their use of the *C. elegans* fate map in studies of programmed cell death and organogenesis.

Having established fate maps for early development, scientists were positioned to answer questions about underlying mechanisms, such as how the basic axes of the embryo are established, a process known as axis formation.

Axis Formation

A body plan with bilateral symmetry is found across a range of animals, including nematodes, echinoderms, and vertebrates (see Chapter 32). As shown for a frog tadpole in **Figure 47.21a**, this body plan exhibits asymmetry along the dorsal-ventral and anterior-posterior axes. The right-left axis is largely symmetrical, as

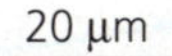

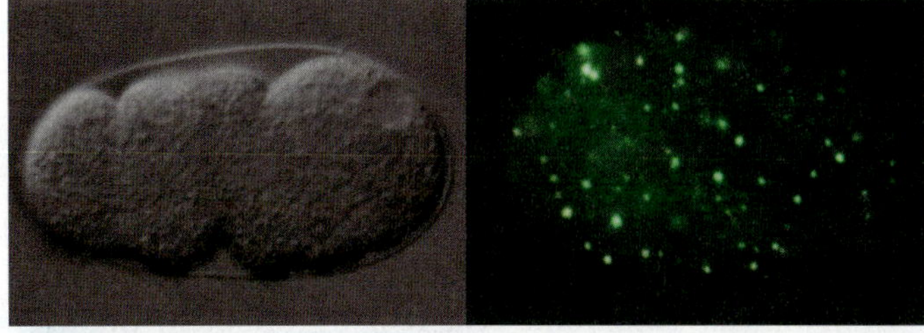

1 Newly fertilized egg

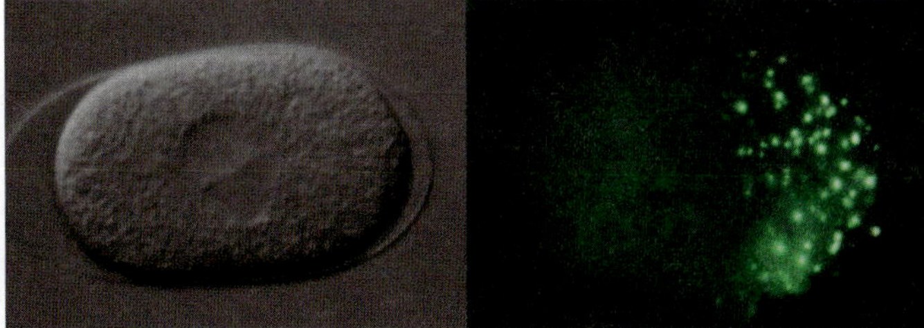

2 Zygote prior to first division

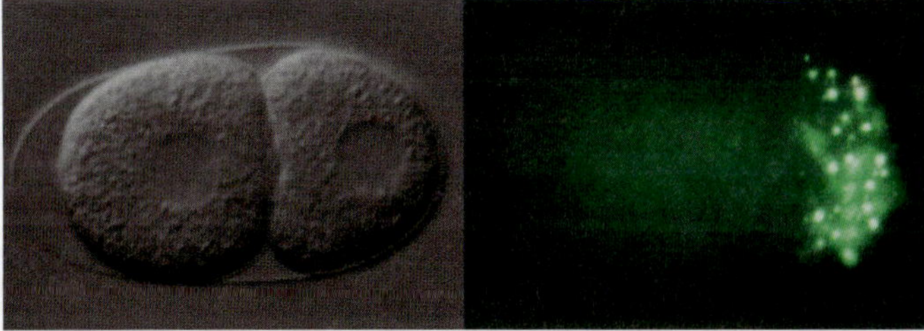

3 Two-cell embryo

4 Four-cell embryo

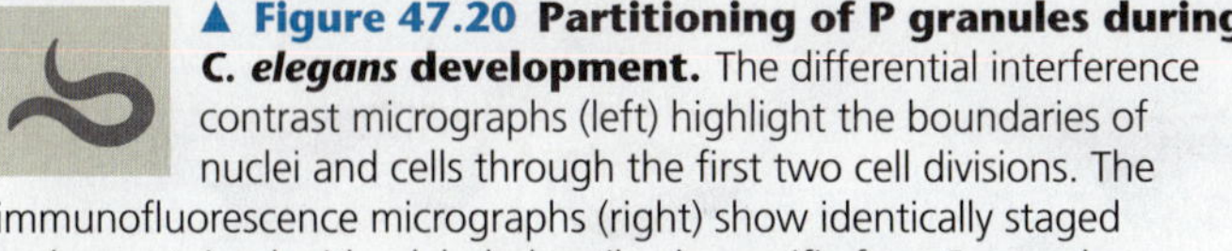

▲ **Figure 47.20 Partitioning of P granules during *C. elegans* development.** The differential interference contrast micrographs (left) highlight the boundaries of nuclei and cells through the first two cell divisions. The immunofluorescence micrographs (right) show identically staged embryos stained with a labeled antibody specific for a P granule protein.

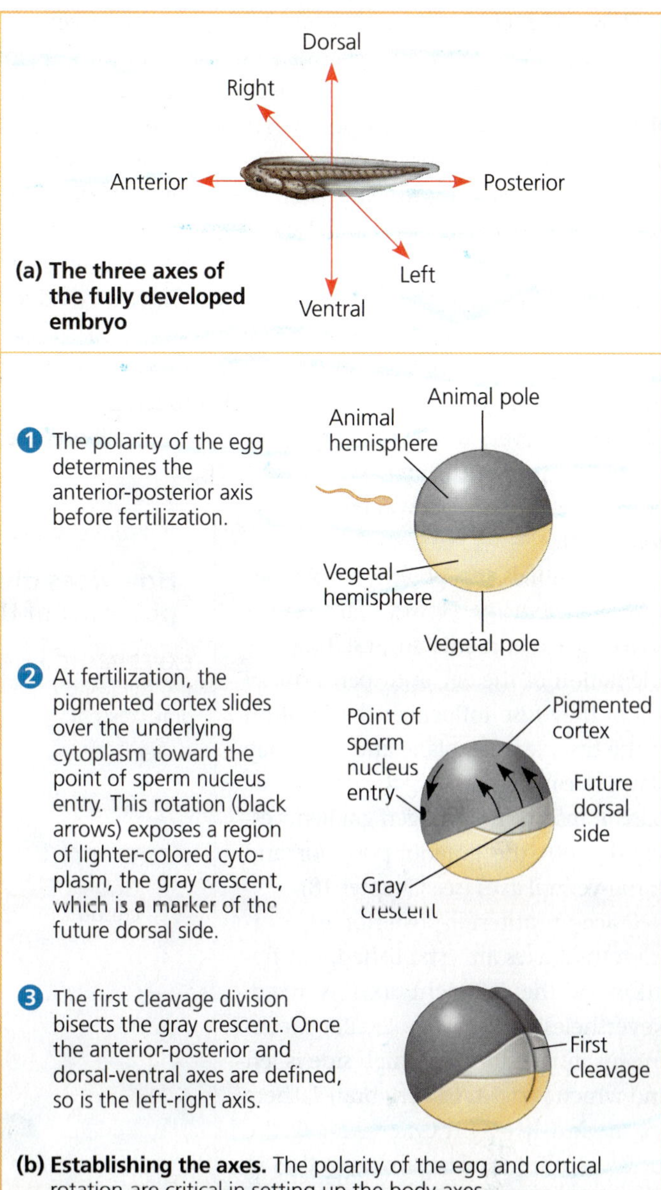

(b) Establishing the axes. The polarity of the egg and cortical rotation are critical in setting up the body axes.

▲ **Figure 47.21 The body axes and their establishment in an amphibian.** All three axes are established before the zygote begins to undergo cleavage.

WHAT IF? *To study axis establishment, researchers can block cortical rotation or force it to occur in a specific direction. One such study resulted in a two-headed embryo because the "back" developed on both sides. What do you think the researchers did to obtain such an embryo?*

the two sides are roughly mirror images of each other. These three body axes are established early in development.

The anterior-posterior axis of the frog embryo is determined during oogenesis. Asymmetry is apparent in the formation of two distinct hemispheres: Dark melanin granules are embedded in the cortex of the animal hemisphere, whereas a yellow yolk fills the vegetal hemisphere. This animal-vegetal asymmetry dictates where the anterior-posterior axis forms in the embryo. Note, however, that the anterior-posterior and animal-vegetal axes are not the same; that is, the head of the embryo does not form at the animal pole.

The dorsal-ventral axis of the frog embryo is not determined until fertilization. Upon fusion of the egg and the sperm, the egg surface—the plasma membrane and associated cortex—rotates with respect to the inner cytoplasm, a movement called *cortical rotation*. From the perspective of the animal pole, this rotation is always toward the point of sperm entry **(Figure 47.21b)**.

How does cortical rotation establish the dorsal-ventral axis? Cortical rotation allows molecules in one portion of the vegetal cortex to interact with molecules in the inner cytoplasm of the animal hemisphere. These inductive interactions activate regulatory factors in specific portions of the vegetal cortex, leading to expression of different sets of genes in dorsal and ventral regions of the embryo.

In chicks, gravity is apparently involved in establishing the anterior-posterior axis as the egg travels down the hen's oviduct before being laid. Later, pH differences between the two sides of the blastoderm cells establish the dorsal-ventral axis. If the pH is artificially reversed above and below the blastoderm, the cells' fates will be reversed: The side facing the egg white will become the ventral part of the embryo, whereas the side facing the yolk will become the dorsal part.

In mammals, no polarity is obvious until after cleavage. However, the results of recent experiments suggest that the orientation of the egg and sperm nuclei before they fuse influences the location of the first cleavage plane and thus may play a role in establishing the embryonic axes. In insects, morphogen gradients establish both the anterior-posterior and dorsal-ventral axes (see Chapter 18).

Once the anterior-posterior and dorsal-ventral axes are established, the position of the left-right axis is fixed. Nevertheless, specific molecular mechanisms must establish which side is left and which is right. In vertebrates, there are marked left-right differences in the location of internal organs as well as in the organization and structure of the heart and brain. Recent research has revealed that cilia are involved in setting up this left-right asymmetry. We will discuss this and other developmental roles of cilia at the end of this chapter.

Restricting Developmental Potential

Earlier we described determination in terms of commitment to a particular cell fate. Is cell fate commitment immediately irreversible, or is there a period of time during which cell fate can be modified? The German zoologist Hans Spemann addressed this question in 1938. By manipulating embryos to perturb normal development and then examining cell fate after the manipulation, he was able to assay a cell's *developmental potential*, the range of structures to which it can give rise (**Figure 47.22**). Spemann found that the fates of embryonic cells are affected by both the distribution of determinants and the pattern of cleavage relative to this distribution. Furthermore, the work of Spemann and others demonstrated that the first two blastomeres of the frog embryo are **totipotent**, meaning that they can each develop into all the different cell types of that species.

In mammals, embryonic cells remain totipotent through the eight-cell stage, much longer than in many other animals. Recent work, however, indicates that the very early cells (even the first two) are not actually equivalent in a normal embryo. Rather, their totipotency when isolated likely

▼ Figure 47.22 INQUIRY

How does distribution of the gray crescent affect the developmental potential of the first two daughter cells?

EXPERIMENT Hans Spemann, at the University of Freiburg-im-Breisgau, in Germany, carried out the following experiment in 1938 to test whether substances were located asymmetrically in the gray crescent.

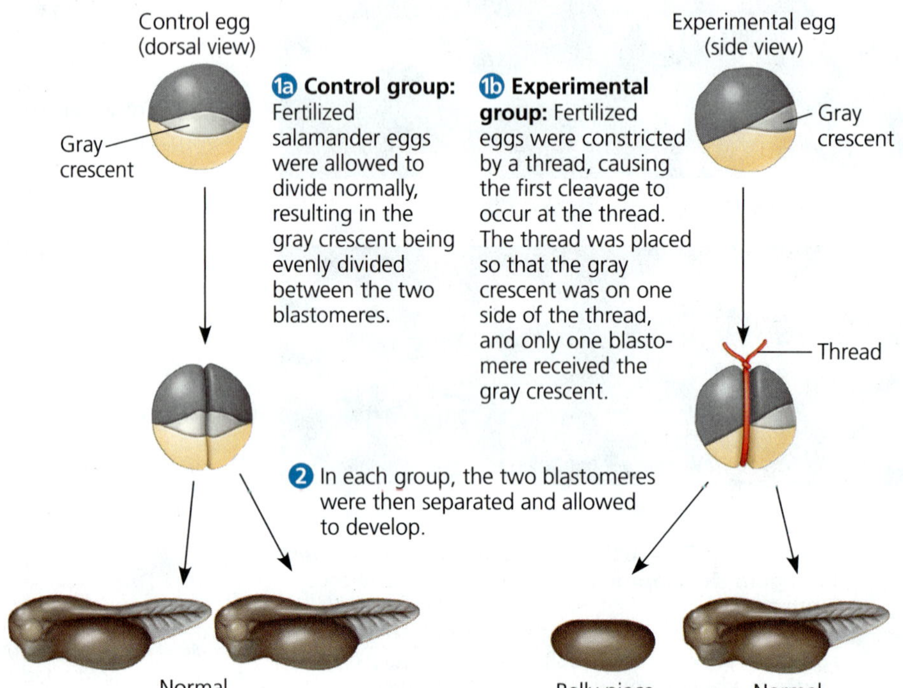

RESULTS Blastomeres that received half or all of the material in the gray crescent developed into normal embryos, but a blastomere that received none of the gray crescent gave rise to an abnormal embryo without dorsal structures. Spemann called it a "belly piece."

CONCLUSION The developmental potential of the two blastomeres normally formed during the first cleavage division depends on their acquisition of cytoplasmic determinants localized in the gray crescent.

SOURCE H. Spemann, *Embryonic Development and Induction*, Yale University Press, New Haven, CT (1938).

WHAT IF? In a similar experiment 40 years earlier, embryologist Hans Roux allowed the first cleavage to occur and then used a needle to kill just one blastomere. The embryo that developed from the remaining blastomere (plus remnants of the dead cell) was abnormal, resembling a half-embryo. Propose a hypothesis to explain why Roux's result differed from the control result in Spemann's experiment.

means that the cells can regulate their fate in response to their embryonic environment. Once the 16-cell stage is reached, mammalian cells are determined to form the trophoblast or the inner cell mass. Although the cells have a limited developmental potential from this point onward, their nuclei remain totipotent, as demonstrated in cloning experiments like that described in Figure 20.19.

As you learned in Chapter 46, identical (monozygotic) twins can develop when embryonic cells become separated. If the separation occurs before the trophoblast and inner cell mass become differentiated, two embryos grow, each with its own chorion and amnion. This is the case for about a third of identical twins. For the rest, the two embryos that develop share a chorion and, in very rare cases where separation is particularly late, an amnion as well.

Regardless of how uniform or varied early embryonic cells are in a particular species, the progressive restriction of developmental potential is a general feature of development in all animals. In general, the tissue-specific fates of cells are fixed in a late gastrula, but not always so in an early gastrula. For example, if the dorsal ectoderm of an early amphibian gastrula is experimentally replaced with ectoderm from some other location in the same gastrula, the transplanted tissue forms a neural plate. But if the same experiment is performed on a late-stage gastrula, the transplanted ectoderm does not respond to its new environment and does not form a neural plate.

Cell Fate Determination and Pattern Formation by Inductive Signals

As embryonic cells acquire distinct fates, the cells begin to influence each other's fates by induction. At the molecular level, the response to an inductive signal is usually to switch on a set of genes that make the receiving cells differentiate into a specific tissue. Here we will examine two examples of induction, an essential process in the development of many tissues in most animals.

The "Organizer" of Spemann and Mangold

Before his studies of totipotency in the fertilized frog egg, Spemann had investigated cell fate determination during gastrulation. In these experiments, he and his student Hilde Mangold transplanted tissues between early gastrulas. In their most famous such experiment, summarized in **Figure 47.23**, they made a remarkable discovery. Not only did a transplanted dorsal lip of the blastopore continue to be a blastopore lip, but it also triggered gastrulation of the surrounding tissue. They concluded that the dorsal lip of the blastopore in the early gastrula functions as an "organizer" of the embryo's body plan, inducing changes in surrounding tissue that direct formation of the notochord, the neural tube, and other organs.

Nearly a century later, developmental biologists are still actively studying the basis of induction by *Spemann's organizer*. An important clue has come from studies of a growth factor called bone morphogenetic protein 4 (BMP-4). (Bone morphogenetic proteins, a family of related proteins with a variety of developmental roles, derive their name from members of the family that are important in bone formation.) One major function of the cells of the organizer seems to be to *inactivate* BMP-4 on the dorsal side of the embryo. Inactivation of BMP-4 allows cells on

▼ Figure 47.23 INQUIRY

Can the dorsal lip of the blastopore induce cells in another part of the amphibian embryo to change their developmental fate?

EXPERIMENT In 1924, Hans Spemann and Hilde Mangold, at the University of Freiburg-im-Breisgau, in Germany, transplanted a piece of the dorsal lip from a pigmented newt gastrula to the ventral side of a nonpigmented newt gastrula to investigate the inductive ability of the dorsal lip. Cross sections of the gastrulas are shown here.

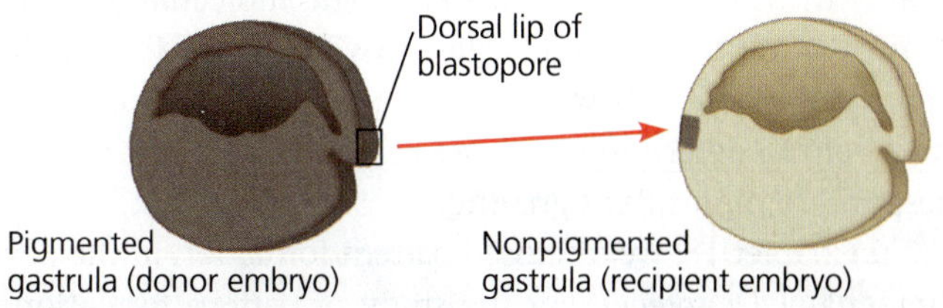

RESULTS The recipient embryo formed a second notochord and neural tube in the region of the transplant, and eventually most of a second embryo developed. Examination of the interior of the double embryo revealed that the secondary structures were formed partly, but not wholly, from recipient tissue.

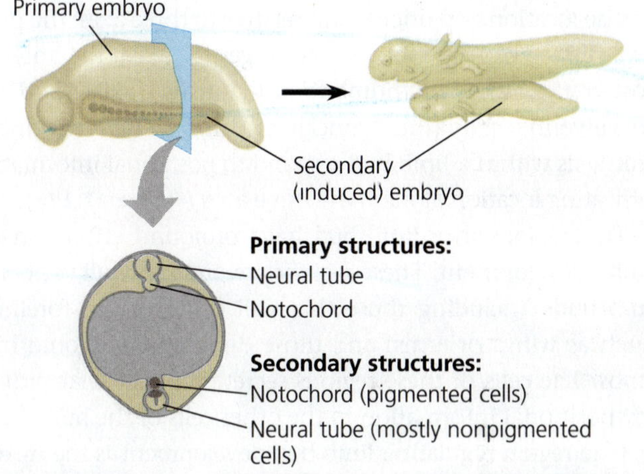

CONCLUSION The transplanted dorsal lip was able to induce cells in a different region of the recipient to form structures different from their normal fate. In effect, the transplanted dorsal lip "organized" the later development of an entire extra embryo.

SOURCE H. Spemann and H. Mangold, Induction of embryonic primordia by implantation of organizers from a different species, Trans. V. Hamburger (1924). Reprinted in *International Journal of Developmental Biology* 45:13–38 (2001).

WHAT IF? Because the transplanted dorsal lip caused the recipient tissue to become something it would not otherwise have become, a signal of some sort must have passed from the dorsal lip. If you identified a protein candidate for the signaling molecule, how could you test whether it actually functions in signaling?

the dorsal side to make dorsal structures, such as the notochord and neural tube. Proteins related to BMP-4 and its inhibitors are also found in other animals, including invertebrates such as the fruit fly, where they also regulate the dorsal-ventral axis.

Formation of the Vertebrate Limb

Inductive signals play a major role in **pattern formation**, the development of an animal's spatial organization, the arrangement of organs and tissues in their characteristic places in three-dimensional space. The molecular cues that control pattern formation, called **positional information**, tell a cell where it is with respect to the animal's body axes and help to determine how the cell and its descendants will respond to molecular signaling.

In Chapter 18, we discussed pattern formation in the development of *Drosophila*. For the study of pattern formation in vertebrates, a classic model system has been limb development in the chick. The wings and legs of chicks, like all vertebrate limbs, begin as limb buds, bumps of mesodermal tissue covered by a layer of ectoderm **(Figure 47.24a)**. Each component of a chick limb, such as a specific bone or muscle, develops with a precise location and orientation relative to three axes: the proximal-distal axis (the "shoulder-to-fingertip" axis), the anterior-posterior axis (the "thumb-to-little finger" axis), and the dorsal-ventral axis (the "knuckle-to-palm" axis). The embryonic cells within a limb bud respond to positional information indicating location along these three axes **(Figure 47.24b)**.

Two regions in a limb bud have profound effects on the limb's development. These regions are present in all vertebrate limb buds, including those that will develop into forelimbs (such as wings or arms) and those destined to become hind limbs. The cells of these regions secrete proteins that provide key positional information to the other cells of the bud.

One region regulating limb-bud development is the **apical ectodermal ridge (AER)**, a thickened area of ectoderm at the tip of the bud (see Figure 47.24a). Removing the AER blocks outgrowth of the limb along the proximal-distal axis. The cells of the AER secrete several protein signals in the fibroblast growth factor (FGF) family that promote limb-bud outgrowth. If the AER is surgically removed and beads soaked with FGF are put in its place, a nearly normal limb will develop. In 2006, researchers identified an FGF-secreting AER that appears to be responsible for building a shark's unpaired (median) fins. This finding suggests that the specific function of the AER predated the appearance of paired limbs in the vertebrate lineage.

The second major limb-bud regulatory region is the **zone of polarizing activity (ZPA)**, a block of mesodermal tissue located underneath the ectoderm where the posterior side of the bud is attached to the body (see Figure 47.24a). The ZPA is necessary for proper pattern formation along the anterior-posterior axis of the limb. Cells nearest the ZPA give rise to the posterior structures, such as the most posterior of the chick's three digits (positioned like our little finger); cells

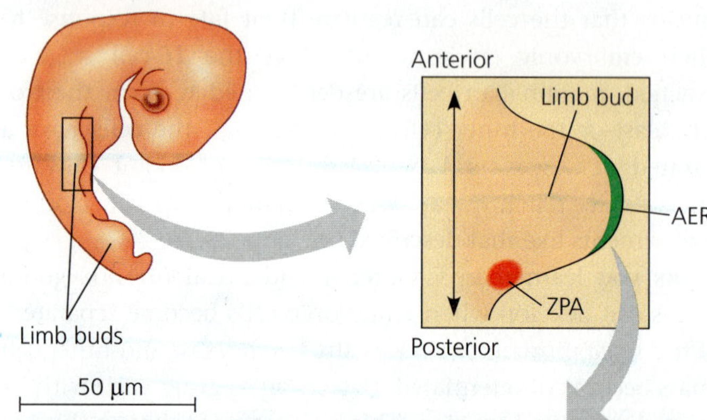

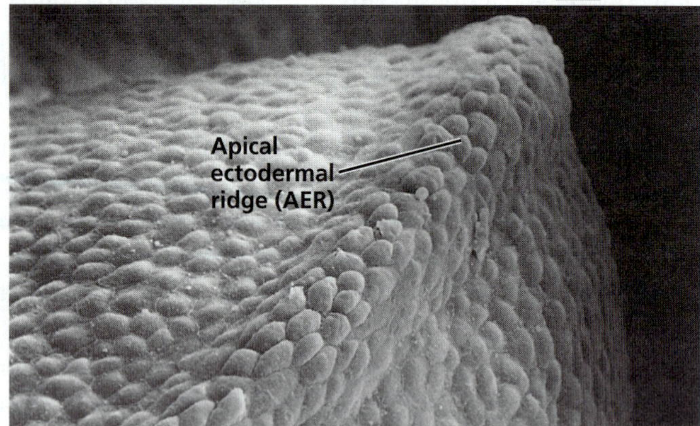

(a) Organizer regions. Vertebrate limbs develop from protrusions called limb buds, each consisting of mesoderm cells covered by a layer of ectoderm. Two regions in each limb bud, the apical ectodermal ridge (AER, shown in this SEM) and the zone of polarizing activity (ZPA), play key roles as organizers in limb pattern formation.

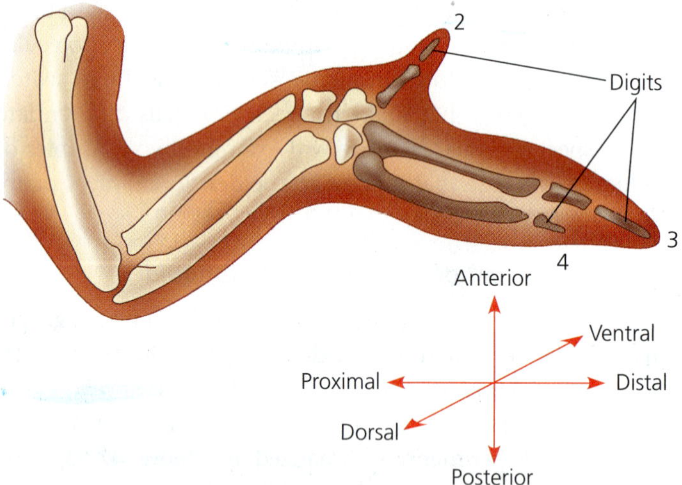

(b) Wing of chick embryo. As the bud develops into a limb, a specific pattern of tissues emerges. In the chick wing, for example, the digits are always present in the arrangement shown here. Pattern formation requires each embryonic cell to receive some kind of positional information indicating location along the three axes of the limb. The AER and ZPA secrete molecules that help provide this information. (Numbers are assigned to the digits based on a convention established for vertebrate limbs. The chicken wing has only four digits; the first digit points backward and is not shown in the diagram.)

▲ **Figure 47.24 Vertebrate limb development.**

farthest from the ZPA form anterior structures, including the most anterior digit (like our thumb).

The tissue transplantation experiment outlined in **Figure 47.25** supports the hypothesis that the inductive signal produced by the ZPA conveys positional information indicating "posterior." Indeed, researchers have discovered that the cells of the ZPA secrete a growth factor called Sonic hedgehog. (Sonic hedgehog gets its name from two sources: its similarity to a *Drosophila* protein called Hedgehog, which is involved in segmentation of the fly embryo, and a video game character.) If cells genetically engineered to produce large amounts of Sonic hedgehog are implanted in the anterior region of a normal limb bud, a mirror-image limb results—just as if a ZPA had been grafted there. Studies of the mouse version of Sonic hedgehog suggest that extra toes in mice—and perhaps also in humans—can result when this protein is produced in part of the limb bud where it is normally absent. Sonic hedgehog and other similar Hedgehog proteins function in many developmental settings and organisms, including pattern formation in *Drosophila* and regulation of cell fate and number in the vertebrate nervous system.

Signaling by Sonic hedgehog plays a vital role in limb-bud development, but what determines whether a limb bud develops into a forelimb or a hind limb? It turns out that the cells receiving the Hedgehog signals from the AER and ZPA respond according to their developmental histories. Before the AER or ZPA issues its signals, earlier developmental signaling sets up specific spatial patterns of *Hox* gene expression (see Figure 21.18). Differences in *Hox* gene expression cause cells of the forelimb and hind limb buds—and cells in different parts of each limb bud—to react differently to the same positional cues.

Hedgehog, FGF, and BMP-4 are examples of a much larger set of signaling molecules that govern cell fates in animals. Having mapped out many of the basic functions of these molecules in embryonic development, researchers are now addressing their role in organogenesis, focusing in particular on the development of the brain.

▼ **Figure 47.25** **INQUIRY**

What role does the zone of polarizing activity (ZPA) play in limb pattern formation in vertebrates?

EXPERIMENT In 1985, Dennis Summerbell and Lawrence Honig, then at the National Institute for Medical Research in Mill Hill, near London, were eager to investigate the nature of the zone of polarizing activity. They transplanted ZPA tissue from a donor chick embryo under the ectoderm in the anterior margin of a limb bud in another chick (the host).

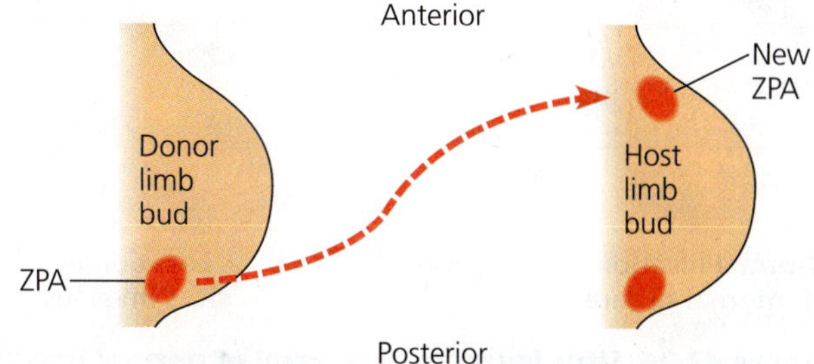

RESULTS The host limb bud developed extra digits from host tissue in a mirror-image arrangement to the normal digits, which also formed (compare with Figure 47.24b, which shows a normal chick wing).

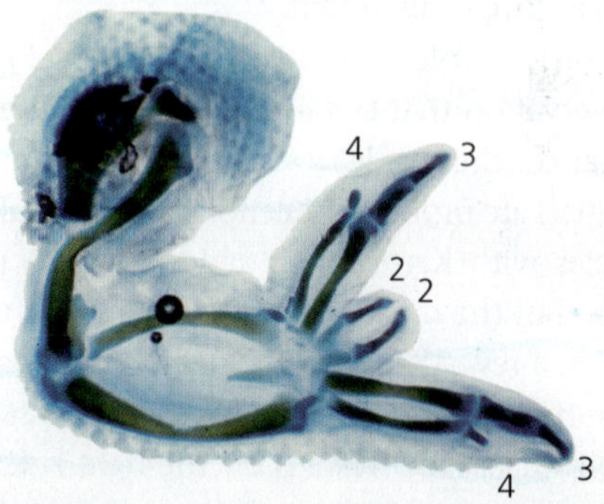

CONCLUSION The mirror-image duplication observed in this experiment suggests that ZPA cells secrete a signal that diffuses from its source and conveys positional information indicating "posterior." As the distance from the ZPA increases, the signal concentration decreases, and hence more anterior digits develop.

SOURCE L. S. Honig and D. Summerbell, Maps of strength of positional signaling activity in the developing chick wing bud, *Journal of Embryology and Experimental Morphology* 87:163–174 (1985).

WHAT IF? Suppose you learned that the ZPA forms after the AER, leading you to develop the hypothesis that the AER is necessary for formation of the ZPA. Given what you know about molecules expressed in the AER and ZPA (see the text), how could you test your hypothesis?

Cilia and Cell Fate

For many years, developmental biologists largely ignored the cellular organelles known as cilia. That is no longer the case. There is now good experimental evidence that ciliary function is essential for proper specification of cell fate in the human embryo.

Like other mammals, humans have stationary and motile cilia (see Figure 6.24). Stationary primary cilia, or *monocilia*, exist as a single projection on the surface of nearly all cells. Motile cilia are found on cells that propel fluid over their surface, such as the epithelial cells of airways, and on sperm (in the form of flagella that propel sperm movement). Both stationary and motile cilia play vital roles in development.

In 2003, geneticists discovered that certain mutations disrupting development of the mouse nervous system affect genes that function in the assembly of monocilia. Other researchers found that mutations responsible for a severe kidney disease in mice alter a gene important for the transport of materials up and down monocilia. Mutations that block the function of monocilia have also been linked to cystic kidney disease in humans.

Given that monocilia are stationary, how do they function in development? The answer is that each acts as an antenna on the cell surface, receiving signals from multiple signaling

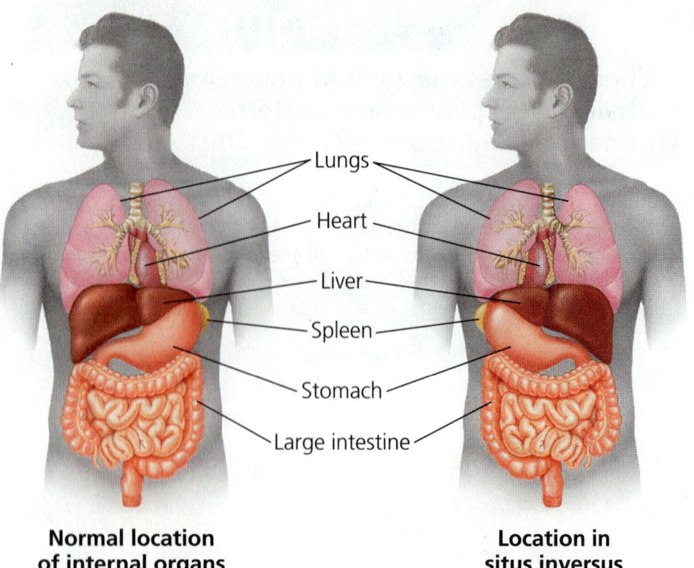

▲ **Figure 47.26 Situs inversus, a reversal of normal left-right asymmetry in the chest and abdomen.**

proteins, including Sonic hedgehog. When the monocilia are defective, signaling is disrupted.

Research on the role of motile cilia in development grew from the observation that certain individuals share a particular set of medical conditions, later named Kartagener's syndrome. Such individuals are prone to infections of the nasal sinuses and bronchi. Males with Kartagener's syndrome also produce immotile sperm. But the most intriguing feature of this syndrome is *situs inversus*, a reversal of the normal left-right asymmetry of the organs in the chest and abdomen **(Figure 47.26)**. For example, in situs inversus, the heart is on the right side rather than the left. (About one in 10,000 individuals have situs inversus, which causes no significant medical problems by itself.)

The conditions associated with Kartagener's syndrome all result from a defect that makes cilia immotile. Without motility, sperm tails cannot beat and airway cells cannot sweep mucus and microbes out of the airway. But what causes situs inversus in these individuals? The current model proposes that ciliary motion in a particular part of the embryo is essential for normal development. Evidence indicates that movement of the cilia generates a leftward fluid flow, breaking the symmetry between left and right sides. Without that flow, asymmetry along the left-right axis arises randomly, and half of the affected embryos develop situs inversus.

If we step back from the specification of particular cell fates to consider development as a whole, we see a sequence of events marked by cycles of signaling and differentiation. Initial cell asymmetries allow different types of cells to influence each other, resulting in the expression of specific sets of genes. The products of these genes then direct cells to differentiate into specific types. Through pattern formation and morphogenesis, differentiated cells ultimately produce a complex arrangement of tissues and organs, each functioning in its appropriate location and in coordination with other cells, tissues, and organs throughout the organism.

CONCEPT CHECK 47.3

1. How do axis formation and pattern formation differ?
2. MAKE CONNECTIONS How does a morphogen gradient differ from cytoplasmic determinants and inductive interactions with regard to the set of cells it affects (see Concept 18.4, p. 367)?
3. WHAT IF? If the ventral cells of an early frog gastrula are experimentally induced to express large amounts of a protein that inhibits BMP-4, could a second embryo develop? Explain.
4. WHAT IF? If you removed the ZPA from a limb bud and then placed a bead soaked in Sonic hedgehog in the middle of the limb bud, what would be the most likely result?

For suggested answers, see Appendix A.

47 CHAPTER REVIEW

SUMMARY OF KEY CONCEPTS

CONCEPT 47.1

Fertilization and cleavage initiate embryonic development (pp. 1022–1027)

- **Fertilization** brings together the nuclei of sperm and egg, forming a diploid zygote, and activates the egg, initiating embryonic development. The **acrosomal reaction**, which is triggered when the sperm meets the egg, releases hydrolytic enzymes that digest material surrounding the egg. Gamete contact and/or fusion depolarizes the egg cell membrane and sets up a **fast block to polyspermy** in many animals. Sperm-egg fusion also initiates the cortical reaction.

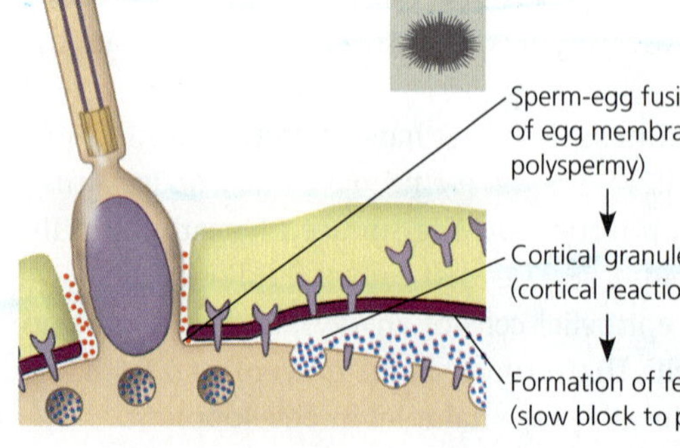

In mammalian fertilization, the cortical reaction modifies the zona pellucida as a **slow block to polyspermy**.

- Fertilization is followed by **cleavage**, a period of rapid cell division without growth, which results in the production of a large number of cells called **blastomeres**. In many species, cleavage creates a multicellular ball called the **blastula**, which contains a fluid-filled cavity, the **blastocoel**. **Holoblastic** cleavage (division of the entire egg) occurs in species whose eggs have little or moderate amounts of **yolk** (as in sea urchins, frogs, and mammals). **Meroblastic** cleavage (incomplete division of the egg) occurs in species with yolk-rich eggs (as in birds and other reptiles).

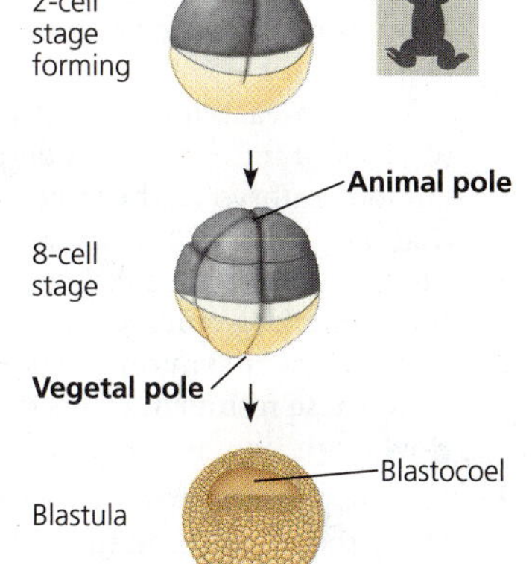

? *What cell-surface barrier prevents fertilization of an egg by a sperm of a different species?*

CONCEPT 47.2

Morphogenesis in animals involves specific changes in cell shape, position, and survival (pp. 1027–1035)

- **Gastrulation** converts the blastula to a **gastrula**, which has a primitive digestive cavity and three **germ layers**: **ectoderm** (blue), **mesoderm** (red), and **endoderm** (yellow).

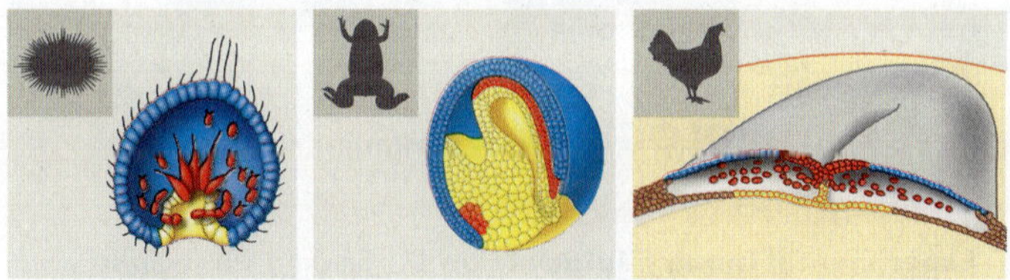

- Mammalian eggs are small, store few nutrients, exhibit holoblastic cleavage, and show no obvious polarity. However, gastrulation and organogenesis in mammals resemble the processes in birds and other reptiles. After fertilization and early cleavage in the oviduct, the **blastocyst** implants in the uterus. The **trophoblast** initiates formation of the fetal portion of the placenta, and the embryo proper develops from a single layer of cells, the epiblast, within the blastocyst.
- The embryos of birds, other reptiles, and mammals develop within a fluid-filled sac that is contained within a shell or the uterus. In these organisms, the three germ layers give rise not only to embryonic tissue but also to the four **extraembryonic membranes**: the amnion, chorion, yolk sac, and allantois.
- The organs of the animal body develop from specific portions of the three embryonic germ layers. Early events in **organogenesis** in vertebrates include neurulation: formation of the **notochord** by cells of the dorsal mesoderm and development of the **neural tube** from infolding of the ectodermal neural plate.

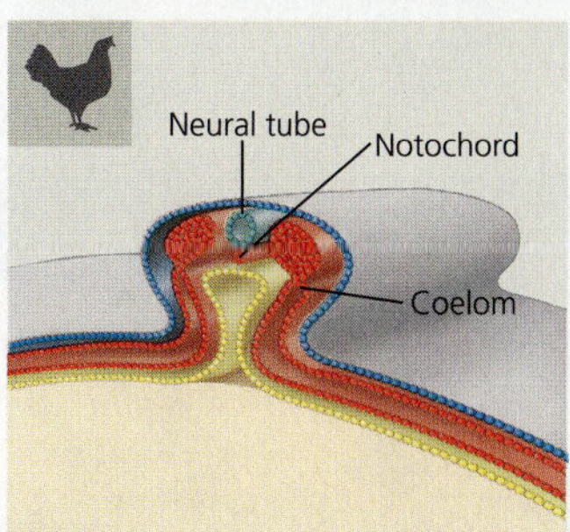

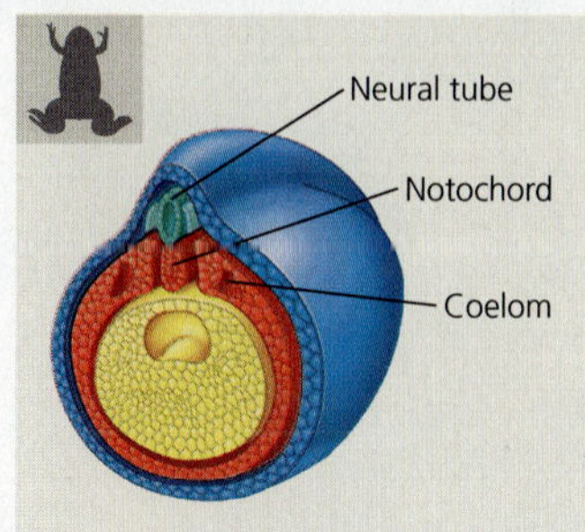

- Cytoskeletal rearrangements are responsible for changes in the shape of cells that underlie cell movements in gastrulation and organogenesis, including invaginations and **convergent extension**. The cytoskeleton is also involved in cell migration, which relies on cell adhesion molecules and the extracellular matrix to help cells reach specific destinations.

? *How does the neural tube form? How do neural crest cells arise?*

CONCEPT 47.3

Cytoplasmic determinants and inductive signals contribute to cell fate specification (pp. 1035–1042)

- Experimentally derived **fate maps** of embryos show that specific regions of the zygote or blastula develop into specific parts of older embryos. The complete cell lineage has been worked out for *C. elegans*. Mechanisms for establishing cellular asymmetries include morphogen gradients, localized determinants, and inductive interactions. As embryonic development proceeds, the developmental potential of cells becomes progressively more limited in all species.
- Cells in a developing embryo receive and respond to **positional information** that varies with location. This information is often in the form of signaling molecules secreted by cells in specific regions of the embryo, such as the dorsal lip of the blastopore in the amphibian gastrula and the **apical ectodermal ridge** and **zone of polarizing activity** of the vertebrate limb bud. The signaling molecules influence gene expression in the cells that receive them, leading to **differentiation** and the development of particular structures.

? *Suppose you found two classes of mouse mutations, one that affected limb development only and one that affected both limb and kidney development. Which class would be more likely to alter the function of monocilia? Explain.*

TEST YOUR UNDERSTANDING

LEVEL 1: KNOWLEDGE/COMPREHENSION

1. The cortical reaction of sea urchin eggs functions directly in
   a. the formation of a fertilization envelope.
   b. the production of a fast block to polyspermy.
   c. the release of hydrolytic enzymes from the sperm.
   d. the generation of an electrical impulse by the egg.
   e. the fusion of egg and sperm nuclei.
2. Which of the following is common to the development of both birds and mammals?
   a. holoblastic cleavage
   b. epiblast and hypoblast
   c. trophoblast
   d. yolk plug
   e. gray crescent
3. The archenteron develops into
   a. the mesoderm.
   b. the blastocoel.
   c. the endoderm.
   d. the placenta.
   e. the lumen of the digestive tract.
4. What structural adaptation in chickens allows them to lay their eggs in arid environments rather than in water?
   a. extraembryonic membranes
   b. yolk
   c. cleavage
   d. gastrulation
   e. development of the brain from ectoderm

LEVEL 2: APPLICATION/ANALYSIS

5. In an egg cell treated with EDTA, a chemical that binds calcium and magnesium ions,
   a. the acrosomal reaction would be blocked.
   b. the fusion of sperm and egg nuclei would be blocked.
   c. the fast block to polyspermy would not occur.
   d. the fertilization envelope would not form.
   e. the zygote would not contain maternal and paternal chromosomes.

6. In humans, identical twins are possible because
   a. cytoplasmic determinants are distributed unevenly in unfertilized eggs.
   b. extraembryonic cells interact with the zygote nucleus.
   c. convergent extension occurs.
   d. early blastomeres can form a complete embryo if isolated.
   e. the gray crescent divides the dorsal-ventral axis into new cells.

7. Cells transplanted from the neural tube of a frog embryo to the ventral part of another embryo develop into nervous system tissues. This result indicates that the transplanted cells were
   a. totipotent.
   b. determined.
   c. differentiated.
   d. mesenchymal.
   e. apoptotic.

8. **DRAW IT** Fill in the blanks in the figure below, and draw arrows showing the movement of ectoderm, mesoderm, and endoderm.

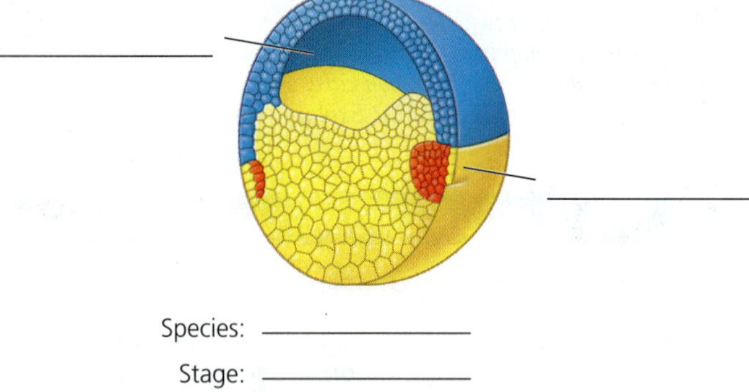

LEVEL 3: SYNTHESIS/EVALUATION

9. **EVOLUTION CONNECTION**
   Evolution in insects and vertebrates has involved the repeated duplication of body segments, followed by fusion of some segments and specialization of their structure and function. What parts of vertebrate anatomy reflect the vertebrate segmentation pattern?

10. **SCIENTIFIC INQUIRY**
    The "snout" of a frog tadpole bears a sucker. A salamander tadpole has a mustache-shaped structure called a balancer in the same area. Suppose that you perform an experiment in which you transplant ectoderm from the side of a young salamander embryo to the snout of a frog embryo. The tadpole that develops has a balancer. When you transplant ectoderm from the side of a slightly older salamander embryo to the snout of a frog embryo, the frog tadpole ends up with a patch of salamander skin on its snout. Suggest a hypothesis to explain these results in terms of developmental mechanisms. How might you test your hypothesis?

11. **SCIENCE, TECHNOLOGY, AND SOCIETY**
    Many scientists think that fetal tissue transplants offer great potential for treating Parkinson's disease, epilepsy, diabetes, Alzheimer's disease, and spinal cord injuries. Why might tissues from a fetus be particularly useful for replacing diseased or damaged cells in patients with such conditions? Some people would allow only tissues from miscarriages to be used in fetal transplant research. However, most researchers prefer to use tissues from surgically aborted fetuses. Why? Explain your position on this controversial issue.

12. **WRITE ABOUT A THEME**
    Emergent Properties In a short essay (100–150 words), describe how the emergent properties of the cells of the gastrula direct embryonic development.

For selected answers, see Appendix A.

www.masteringbiology.com

1. MasteringBiology® Assignments

Experimental Inquiry Tutorial How Do Calcium Ions Help to Prevent Polyspermy During Egg Fertilization?
Tutorial Embryonic Development
Activities Early Stages of Animal Development • Sea Urchin Development • Frog Development
Questions Student Misconceptions • Reading Quiz • Multiple Choice • End-of-Chapter

2. eText

Read your book online, search, take notes, highlight text, and more.

3. The Study Area

Practice Tests • Cumulative Test • *BioFlix* 3-D Animations • MP3 Tutor Sessions • Videos • Activities • Investigations • Lab Media • Audio Glossary • Word Study Tools • Art

48

Neurons, Synapses, and Signaling

▲ **Figure 48.1 What makes this snail such a deadly predator?**

KEY CONCEPTS

48.1 Neuron organization and structure reflect function in information transfer
48.2 Ion pumps and ion channels establish the resting potential of a neuron
48.3 Action potentials are the signals conducted by axons
48.4 Neurons communicate with other cells at synapses

OVERVIEW

Lines of Communication

The tropical cone snail (*Conus geographus*) in **Figure 48.1** is both beautiful and dangerous. A carnivore, this marine snail hunts, kills, and dines on fish. Injecting venom with a hollow, harpoon-like part of its mouth, the cone snail paralyzes its free-swimming prey in seconds. The venom is so deadly that unlucky scuba divers have died from just a single injection. What makes cone snail venom so fast acting and lethal? As Baldomero Olivera discusses in the interview opening the unit (pp. 850–851), the answer is a mixture of molecules that disable **neurons**, the nerve cells that transfer information within the body. Because the venom almost instantaneously disrupts neuronal control of locomotion and respiration, an animal attacked by the cone snail can neither defend itself nor escape.

Communication by neurons largely consists of long-distance electrical signals and short-distance chemical signals. The specialized structure of neurons allows them to use pulses of electrical current to receive, transmit, and regulate the flow of information over long distances within the body. In transferring information from one cell to another, neurons often rely on chemical signals that act over very short distances. The cone snail's venom is particularly potent because it interferes with both electrical and chemical signaling by neurons.

Neurons transmit sensory information, control heart rate, coordinate hand and eye movement, record memories, generate dreams, and much more. All of this information is transmitted within neurons as an electrical signal. The identity of the type of information being transmitted is encoded by the connections made by the active neuron. Interpreting signals in the nervous system therefore involves sorting a complex set of neuronal paths and connections. In more complex animals, this higher-order processing is carried out largely in groups of neurons organized into a **brain** or into simpler clusters called **ganglia**.

In this chapter, we examine the structure of a neuron and explore the molecules and physical principles that govern signaling by neurons. In Chapter 49, we will look at the organization of nervous systems and at higher-order information processing in vertebrates. In Chapter 50, we will investigate systems that detect environmental stimuli and systems that carry out the body's responses to those stimuli. Finally, in Chapter 51, we will consider how these nervous system functions are integrated into the activities and interactions that make up animal behavior.

CONCEPT 48.1

Neuron organization and structure reflect function in information transfer

Before delving into the activity of an individual neuron, let's take an overall look at how neurons function in the flow of information through the animal body. We'll use as our example the squid, an organism that has some extraordinarily large nerve cells that played a crucial role in the discovery of how neurons transmit signals.

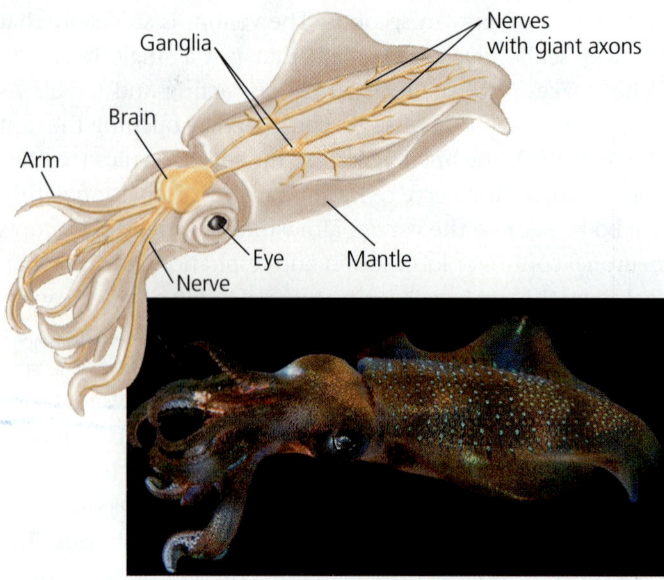

▲ **Figure 48.2 Overview of the squid nervous system.** Signals travel from the brain to the muscular mantle along *giant axons*, nerve cell extensions of unusually large diameter.

Introduction to Information Processing

Like the cone snail in Figure 48.1, the squid in **Figure 48.2** is an active predator. Using its brain to process information captured by its image-forming eyes, the squid surveys its environment. When the squid spots prey, signals travel from its brain to neurons in its mantle, causing muscle contractions that propel the squid forward.

Information processing by a nervous system occurs in three stages: sensory input, integration, and motor output **(Figure 48.3)**. In many animals, the neurons that carry out integration are organized in a **central nervous system (CNS)**, which includes the brain and a longitudinal nerve cord. The neurons that carry information into and out of the CNS constitute the **peripheral nervous system (PNS)**. When bundled together, such neurons form **nerves**.

In all but the simplest animals, specialized populations of neurons handle each stage of information processing. **Sensory neurons** transmit information from eyes and other sensors that detect external stimuli (light, sound, touch, heat, smell, and taste) or internal conditions (such as blood pressure, blood carbon dioxide level, and muscle tension). This information is sent to processing centers in the brain or ganglia. Neurons in the brain or ganglia integrate (analyze and interpret) the sensory input, taking into account the immediate context and the animal's experience. The vast majority of neurons in the brain are **interneurons**, which form the local circuits connecting neurons in the brain. Motor output relies on neurons that extend out of the processing centers and trigger muscle or gland activity. For example, **motor neurons** transmit signals to muscle cells, causing them to contract. In exploring how this transmission of information flows within the nervous system, we'll begin with the unique structure of neurons.

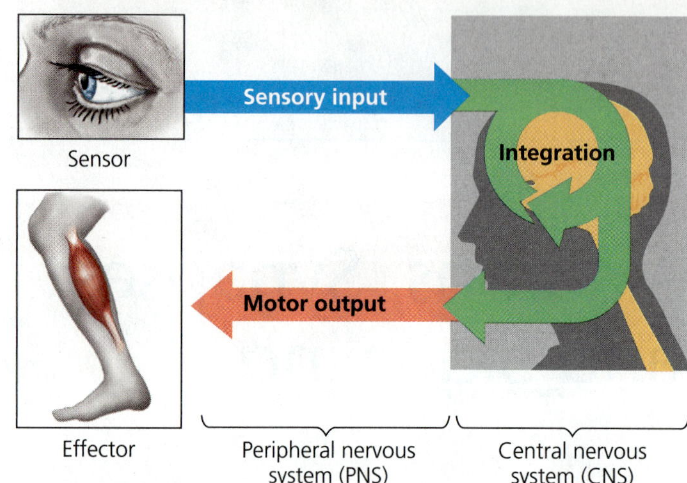

▲ **Figure 48.3 Summary of information processing.**

Neuron Structure and Function

The ability of a neuron to receive and transmit information is based on a highly specialized cellular organization **(Figure 48.4)**. Most of a neuron's organelles, including its nucleus, are located in the **cell body**. A typical neuron has numerous highly branched extensions called **dendrites** (from the Greek *dendron*, tree). Together with the cell body, the dendrites *receive* signals from other neurons. A neuron also has a single **axon**, an extension that *transmits* signals to other cells. Axons are often much longer than dendrites, and some, such as those that reach from the spinal cord of a giraffe to the muscle cells in its feet, are over a meter long. The cone-shaped base of an axon, the *axon hillock*, is typically where signals that travel down the axon are generated. Near its other end, an axon usually divides into many branches.

Each branched end of an axon transmits information to another cell at a junction called a **synapse** (see Figure 48.4). The part of each axon branch that forms this specialized junction is a *synaptic terminal*. At most synapses, chemical messengers called **neurotransmitters** pass information from the transmitting neuron to the receiving cell. In describing a synapse, we refer to the transmitting neuron as the *presynaptic cell* and the neuron, muscle, or gland cell that receives the signal as the *postsynaptic cell*.

Depending on the number of synapses a neuron has with other cells, its shape can vary from simple to quite complex **(Figure 48.5)**. Highly branched axons can transmit information to many target cells. Similarly, neurons with highly branched dendrites can receive input through large numbers of synapses, as many as 100,000 in the case of some interneurons.

The neurons of vertebrates and most invertebrates require supporting cells called **glial cells**, or **glia** (from a Greek word meaning "glue") **(Figure 48.6)**. Glia nourish neurons, insulate the axons of neurons, and regulate the extracellular fluid surrounding neurons. Overall, glia outnumber neurons in the mammalian brain 10- to 50-fold. We will examine the functions of specific glia later in this chapter and in Chapter 49.

CONCEPT CHECK 48.1

1. Describe the basic pathway of information flow through neurons that causes you to turn your head when someone calls your name.
2. **WHAT IF?** How might increased branching of an axon help coordinate responses to signals communicated by the nervous system?
3. **MAKE CONNECTIONS** Consider how communication occurs in a colony of bacteria (see Figure 11.3, p. 207). In what general ways is that communication similar to and different from transmission of a nerve impulse by a neuron?

For suggested answers, see Appendix A.

▲ **Figure 48.4 Neuron structure and organization.**

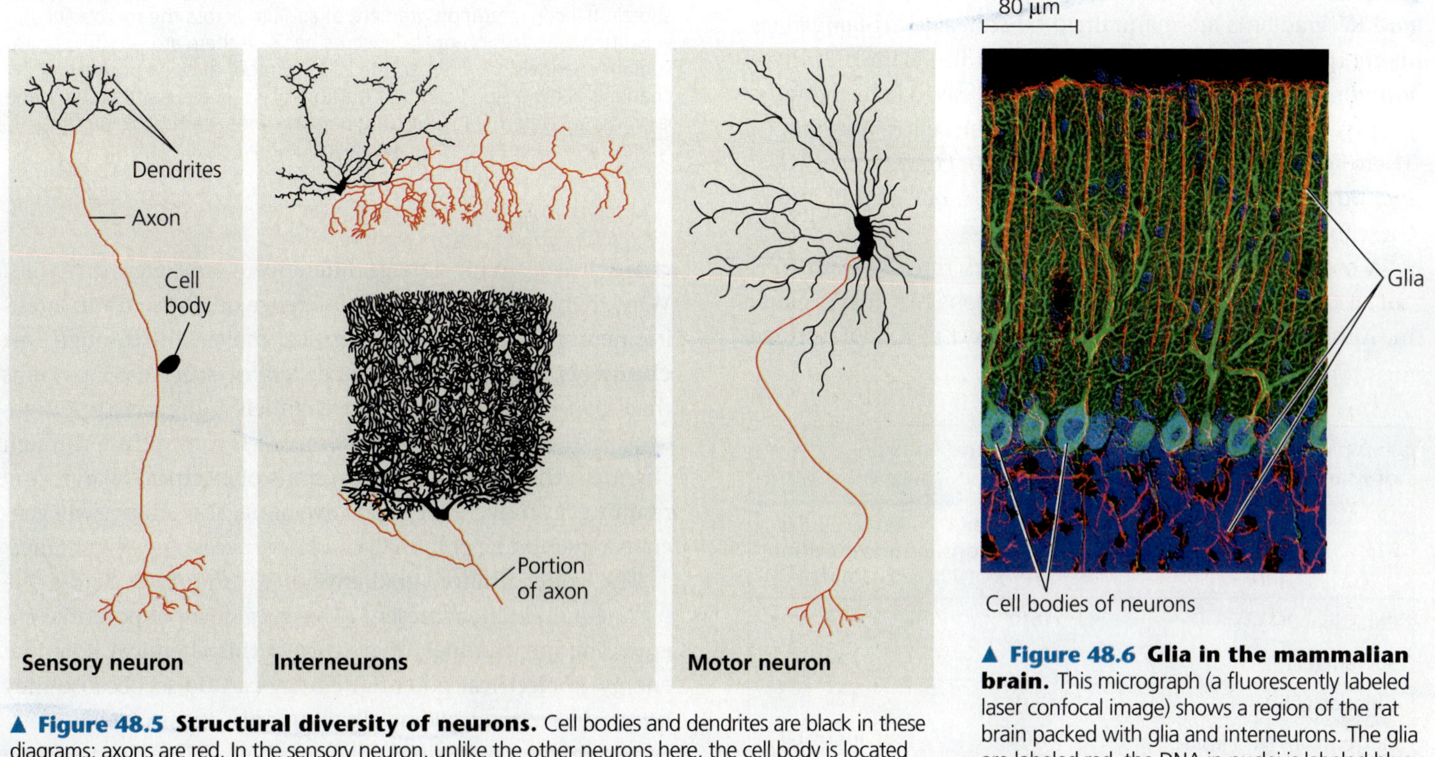

▲ **Figure 48.5 Structural diversity of neurons.** Cell bodies and dendrites are black in these diagrams; axons are red. In the sensory neuron, unlike the other neurons here, the cell body is located partway along the axon that conveys signals from the dendrites to the axon's terminal branches.

▲ **Figure 48.6 Glia in the mammalian brain.** This micrograph (a fluorescently labeled laser confocal image) shows a region of the rat brain packed with glia and interneurons. The glia are labeled red, the DNA in nuclei is labeled blue, and the dendrites of neurons are labeled green.

CONCEPT 48.2

Ion pumps and ion channels establish the resting potential of a neuron

As you read in Chapter 7, ions are unequally distributed between the interior of cells and the fluid that surrounds them. As a result, the inside of a cell is negatively charged relative to the outside. Because the attraction of opposite charges across the plasma membrane is a source of potential energy, this charge difference, or voltage, is called the **membrane potential**. The membrane potential of a resting neuron—one that is not sending a signal—is its **resting potential** and is typically between −60 and −80 mV (millivolts).

Inputs from other neurons or specific stimuli cause changes in the neuron's membrane potential that act as signals, transmitting and processing information. Rapid changes in membrane potential are what enable us to see a flower, read a book, or climb a tree. Thus, to understand how neurons function, we first need to examine how chemical and electrical forces form, maintain, and alter membrane potentials.

Formation of the Resting Potential

Potassium ions (K^+) and sodium ions (Na^+) play an essential role in the formation of the resting potential. Each type of ion has a concentration gradient across the plasma membrane of a neuron (**Table 48.1**). In the case of mammalian neurons, the concentration of K^+ is highest inside the cell, while the concentration of Na^+ is highest outside. These Na^+ and K^+ gradients are maintained by *sodium-potassium pumps* in the plasma membrane. As discussed in Chapter 7, these ion pumps use the energy of ATP hydrolysis to actively transport Na^+ out of the cell and K^+ into the cell (**Figure 48.7**). There are also concentration gradients for chloride ions (Cl^-) and other anions, as shown in Table 48.1, but we will ignore these for the moment.

A sodium-potassium pump transports three sodium ions out of the cell for every two potassium ions that it transports in. Although this pumping generates a net export of positive charge, the resulting voltage difference is only a few millivolts. Why, then, is there a voltage difference of 60–80 mV in a resting neuron? The answer lies in ion movement through **ion channels**, pores formed by clusters of specialized proteins that span the membrane. Ion channels allow ions to diffuse back and forth across the membrane. As ions diffuse through channels, they carry with them units of electrical charge. Any resulting *net* movement of positive or negative charge will generate a membrane potential, or voltage across the membrane.

The concentration gradients of K^+ and Na^+ across the plasma membrane represent a chemical form of potential energy. The ion channels that convert this chemical potential energy to electrical potential energy can do so because they have *selective permeability*, allowing only certain ions to pass. For example, a potassium channel allows K^+ to diffuse freely across the membrane, but not other ions, such as Na^+.

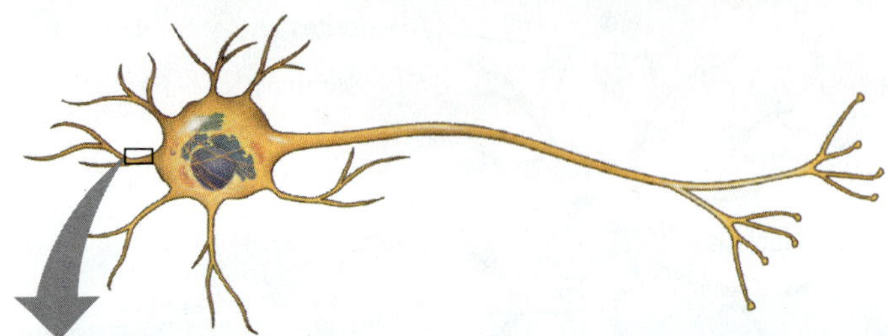

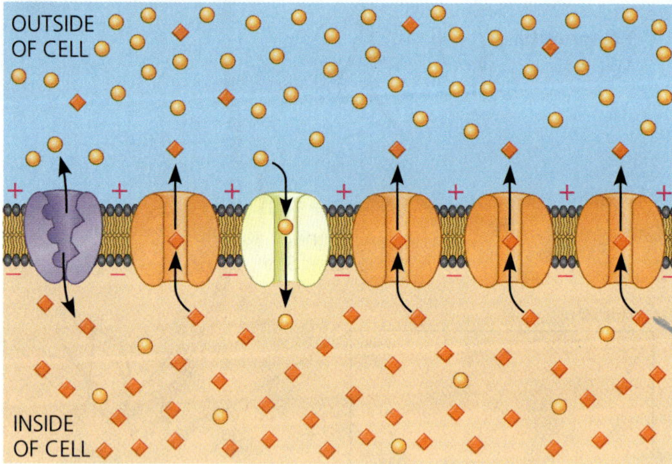

Key

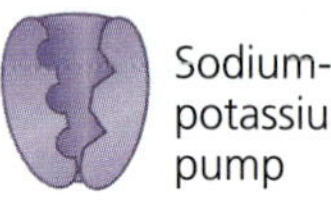

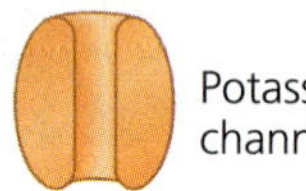

▲ **Figure 48.7 The basis of the membrane potential.** The sodium-potassium pump generates and maintains the ionic gradients of Na^+ and K^+ shown in Table 48.1. The pump uses ATP to actively transport Na^+ out of the cell and K^+ into the cell. Although there is a substantial concentration gradient of sodium across the membrane, very little net diffusion of Na^+ occurs because there are very few open sodium channels. In contrast, the large number of open potassium channels allow a significant net outflow of K^+. Because the membrane is only weakly permeable to chloride and other anions, this outflow of K^+ results in a net negative charge inside the cell.

Table 48.1 Ion Concentrations Inside and Outside of Mammalian Neurons

Ion	Intracellular Concentration (m*M*)	Extracellular Concentration (m*M*)
Potassium (K^+)	140	5
Sodium (Na^+)	15	150
Chloride (Cl^-)	10	120
Large anions (A^-) inside cell, such as proteins	100	(not applicable)

Diffusion of K^+ through open potassium channels is critical for formation of the resting potential. The K^+ concentration is 140 m*M* inside the cell, but only 5 m*M* outside. The chemical concentration gradient thus favors a net outflow of K^+. Furthermore, a resting neuron has many open potassium channels, but very few open sodium channels (see Figure 48.7). Because Na^+ and other ions can't readily cross the membrane, K^+ outflow leads to a net negative charge inside the cell. This buildup of negative charge within the neuron is the major source of the membrane potential.

What stops the buildup of negative charge? The excess negative charges inside the cell exert an attractive force that opposes the flow of additional positively charged potassium ions out of the cell. The separation of charge (voltage) thus results in an electrical gradient that counterbalances the chemical concentration gradient of K^+.

Modeling the Resting Potential

The net flow of K^+ out of a neuron proceeds until the chemical and electrical forces are in balance. How well do these two forces account for the resting potential in a mammalian neuron? To answer this question, let's consider a simple model consisting of two chambers separated by an artificial membrane **(Figure 48.8a)**. To begin, imagine that the membrane contains many open ion channels, all of which allow only K^+ to diffuse across. To produce a K^+ concentration gradient like that of a mammalian neuron, we place a solution of 140 m*M* potassium chloride (KCl) in the inner chamber and 5 m*M* KCl in the outer chamber. The K^+ will diffuse down its concentration gradient into the outer chamber. But because the chloride ions (Cl^-) lack a means of crossing the membrane, there will be an excess of negative charge in the inner chamber.

When our model neuron reaches equilibrium, the electrical gradient will exactly balance the chemical gradient, so that no further net diffusion of K^+ occurs across the membrane. The magnitude of the membrane voltage at equilibrium for a particular ion is called that ion's **equilibrium potential (E_{ion})**. For a membrane permeable to a single type of ion, E_{ion} can be calculated using a formula called the Nernst equation. At human body temperature (37°C) and for an ion with a net charge of 1+, such as K^+ or Na^+, the Nernst equation is

$$E_{ion} = 62\text{ mV}\left(\log \frac{[\text{ion}]_{\text{outside}}}{[\text{ion}]_{\text{inside}}}\right)$$

Plugging in the K^+ concentrations reveals that the equilibrium potential for K^+ (E_K) is −90 mV (see Figure 48.8a). The minus sign indicates that K^+ is at equilibrium when the inside of the membrane is 90 mV more negative than the outside.

Although the equilibrium potential for K^+ is −90 mV, the resting potential of a mammalian neuron is somewhat less negative. This difference reflects the small but steady movement of Na^+ across the few open sodium channels in a resting neuron. The concentration gradient of Na^+ has a direction opposite to that of K^+ (see Table 48.1). Na^+ therefore diffuses into the cell, making the inside of the cell less negative. If we model a membrane in which the only open channels are selectively permeable to Na^+, we find that a tenfold higher concentration of Na^+ in the outer chamber results in an equilibrium potential (E_{Na}) of +62 mV **(Figure 48.8b)**. In an actual neuron, the resting potential (−60 to −80 mV) is much closer to E_K than to E_{Na} because there are many open potassium channels but only a small number of open sodium channels.

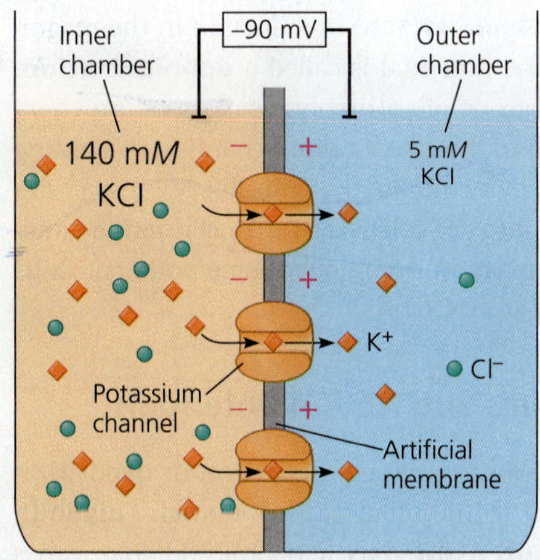

(a) Membrane selectively permeable to K^+

Nernst equation for K^+ equilibrium potential at 37°C:

$$E_K = 62\text{ mV}\left(\log\frac{5\text{ m}M}{140\text{ m}M}\right) = -90\text{ mV}$$

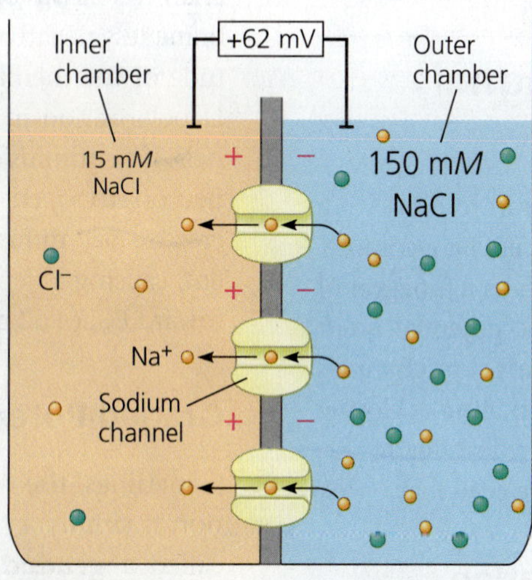

(b) Membrane selectively permeable to Na^+

Nernst equation for Na^+ equilibrium potential at 37°C:

$$E_{Na} = 62\text{ mV}\left(\log\frac{150\text{ m}M}{15\text{ m}M}\right) = +62\text{ mV}$$

◄ **Figure 48.8 Modeling a mammalian neuron.** Each container is divided into two chambers by an artificial membrane. Ion channels allow free diffusion for particular ions, resulting in the net ion flow represented by arrows. **(a)** The presence of open potassium channels makes the membrane selectively permeable to K^+, and the inner chamber contains a 28-fold higher concentration of K^+ than the outer chamber; at equilibrium, the inside of the membrane is −90 mV relative to the outside. **(b)** The membrane is selectively permeable to Na^+, and the inner chamber contains a tenfold lower concentration of Na^+ than the outer chamber; at equilibrium, the inside of the membrane is +62 mV relative to the outside.

WHAT IF? *Adding channels specific for one type of ion to the membrane in (b) would alter the membrane potential. Which ion would pass through these channels, and in what direction would the membrane potential change?*

Because neither K^+ nor Na^+ is at equilibrium in a resting neuron, each ion has a net flow (a current) across the membrane. The resting potential remains steady, which means that the K^+ and Na^+ currents are equal and opposite. Ion concentrations on either side of the membrane also remain steady. Keep in mind that the extent of ion movement required to generate the resting potential is extremely small (about 10^{-12} mole/cm^2 of membrane), far less than would be required to alter the chemical concentration gradient.

Under conditions that allow Na^+ to cross the membrane more readily, the membrane potential will move toward E_{Na} and away from E_K. As we will see in the next section, this is precisely what happens during the generation of a nerve impulse.

CONCEPT CHECK 48.2

1. Under what circumstances could ions flow through ion channels from regions of low ion concentration to regions of high ion concentration?
2. **WHAT IF?** Suppose a cell's membrane potential shifts from −70 mV to −50 mV. What changes in the cell's permeability to K^+ or Na^+ could cause such a shift?
3. **WHAT IF?** Ouabain, a plant substance used in some cultures to poison hunting arrows, disables the sodium-potassium pump. What change in the resting potential would you expect to see if you treated a neuron with ouabain? Explain.
4. **MAKE CONNECTIONS** Figure 7.13, on page 132, illustrates diffusion by dye molecules. Could diffusion eliminate the concentration gradient of a dye that has a net charge? Explain.

For suggested answers, see Appendix A.

CONCEPT 48.3

Action potentials are the signals conducted by axons

The membrane potential of a neuron changes in response to a variety of stimuli. Using the technique of intracellular recording, researchers can record and graph these changes as a function of time **(Figure 48.9)**. Changes in the membrane potential occur because neurons contain **gated ion channels**, ion channels that open or close in response to stimuli. The opening or closing of gated ion channels alters the membrane's permeability to particular ions, which in turn alters the membrane potential.

Hyperpolarization and Depolarization

To explore how the membrane potential changes, let's consider what happens when gated potassium channels that are closed in a resting neuron are stimulated to open. Opening

▼ **Figure 48.9** **RESEARCH METHOD**

Intracellular Recording

APPLICATION Electrophysiologists use intracellular recording to measure the membrane potential of neurons and other cells.

TECHNIQUE A microelectrode is made from a glass capillary tube filled with an electrically conductive salt solution. One end of the tube tapers to an extremely fine tip (diameter < 1 μm). While looking through a microscope, the experimenter uses a micropositioner to insert the tip of the microelectrode into a cell. A voltage recorder (usually an oscilloscope or a computer-based system) measures the voltage between the microelectrode tip inside the cell and a reference electrode placed in the solution outside the cell.

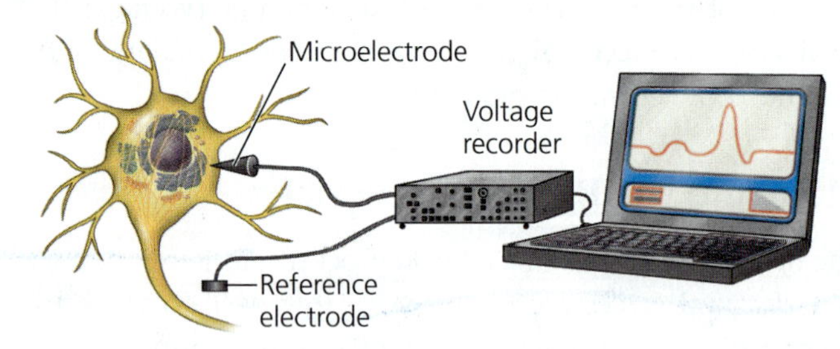

these potassium channels increases the membrane's permeability to K^+. Net diffusion of K^+ out of the neuron increases, shifting the membrane potential toward E_K (−90 mV at 37°C). This increase in the magnitude of the membrane potential, called a **hyperpolarization**, makes the inside of the membrane more negative **(Figure 48.10a)**. In a resting neuron, hyperpolarization results from any stimulus that increases the outflow of positive ions or the inflow of negative ions.

Although opening potassium channels in a resting neuron causes hyperpolarization, opening some other types of ion channels has an opposite effect, making the inside of the membrane less negative **(Figure 48.10b)**. A reduction in the magnitude of the membrane potential is called a **depolarization**. Depolarization in neurons often involves gated sodium channels. If a stimulus causes the gated sodium channels in a resting neuron to open, the membrane's permeability to Na^+ increases. Na^+ diffuses into the cell along its concentration gradient, causing a depolarization as the membrane potential shifts toward E_{Na} (+62 mV at 37°C).

Graded Potentials and Action Potentials

Sometimes, the response to hyperpolarization or depolarization is simply a shift in the membrane potential. This shift, called a **graded potential**, has a magnitude that varies with the strength of the stimulus, with a larger stimulus causing a greater change in the membrane potential. Graded potentials induce a small electrical current that leaks out of the neuron as it flows along the membrane. Graded potentials

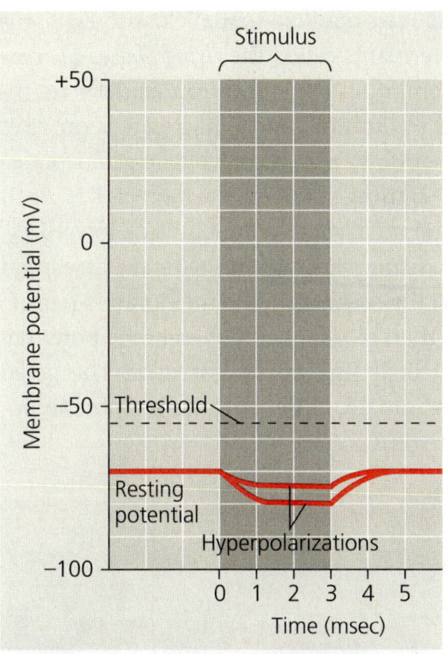

(a) Graded hyperpolarizations produced by two stimuli that increase membrane permeability to K^+. The larger stimulus produces a larger hyperpolarization.

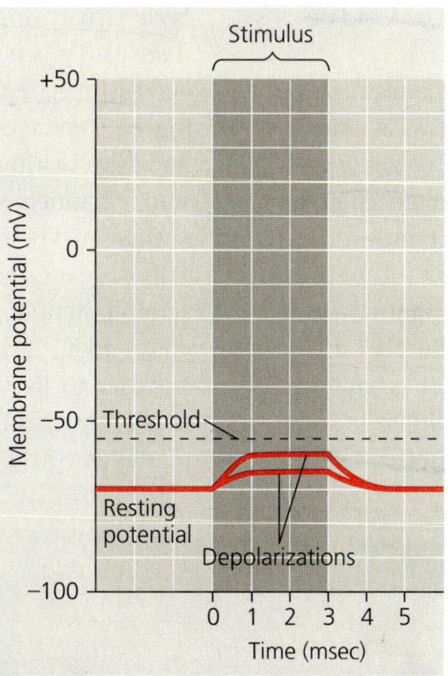

(b) Graded depolarizations produced by two stimuli that increase membrane permeability to Na^+. The larger stimulus produces a larger depolarization.

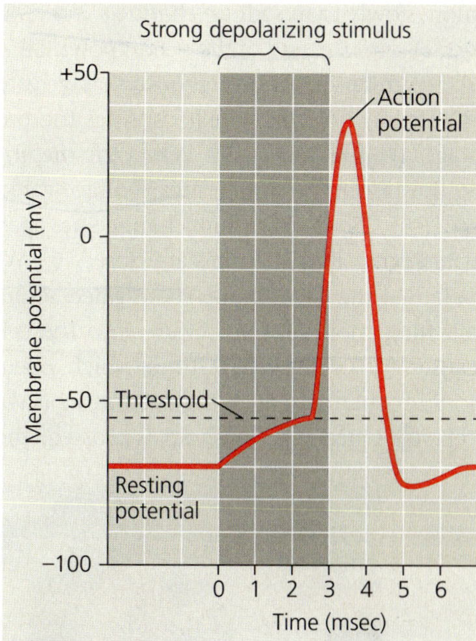

(c) Action potential triggered by a depolarization that reaches the threshold.

▲ **Figure 48.10 Graded potentials and an action potential in a neuron.**

DRAW IT *Redraw the graph in part (c), extending the y-axis. Then label the positions of E_K and E_{Na}.*

thus decay with distance from their source. Although graded potentials are not the nerve signals that travel along axons, they have a major effect on the generation of nerve signals.

If a depolarization shifts the membrane potential sufficiently, the result is a massive change in membrane voltage called an **action potential**. Unlike graded potentials, action potentials have a constant magnitude and can regenerate in adjacent regions of the membrane. Action potentials can therefore spread along axons, making them well suited for transmitting a signal over long distances.

Action potentials arise because some of the ion channels in neurons are **voltage-gated ion channels**, opening or closing when the membrane potential passes a particular level. If a depolarization opens voltage-gated sodium channels, the resulting flow of Na^+ into the neuron results in further depolarization. Because the sodium channels are voltage gated, an increased depolarization causes more sodium channels to open, leading to an even greater flow of current. The result is a process of *positive feedback* (see Figure 1.13) that triggers a very rapid opening of all voltage-gated sodium channels and the marked change in membrane potential that defines an action potential **(Figure 48.10c)**.

Action potentials occur whenever a depolarization increases the membrane voltage to a particular value, called the **threshold**. For mammalian neurons, the threshold is a membrane potential of about −55 mV. Once initiated, the action potential has a magnitude that is independent of the strength of the triggering stimulus. Because action potentials occur fully or not at all, they represent an *all-or-none* response to stimuli. This all-or-none property reflects the fact that depolarization opens voltage-gated sodium channels, and the opening of sodium channels causes further depolarization. The positive-feedback loop of depolarization and channel opening triggers an action potential whenever the membrane potential reaches the threshold.

The discovery of how action potentials are generated dates to the 1940s and 1950s, with the work of British scientists Andrew Huxley and Alan Hodgkin. Because no techniques were available for studying electrical events in small cells, they took electrical recordings from the giant neurons of the squid (see Figure 48.2). Their experiments led to a model, presented in the next section, that earned them a Nobel Prize in 1963.

Generation of Action Potentials: *A Closer Look*

The characteristic shape of the graph of an action potential (see Figure 48.10c) reflects the large change in membrane potential resulting from ion movement through voltage-gated sodium and potassium channels. Membrane depolarization opens both types of channels, but they respond independently and sequentially. Sodium channels open first, initiating the action potential. As the action potential proceeds, the sodium channels become inactivated: A loop of the channel protein moves, blocking ion flow through the opening. Sodium channels remain inactivated until after the membrane returns to the resting potential and the channels close. Potassium channels open more slowly than sodium channels, but remain open and functional until the end of the action potential.

To understand further how voltage-gated channels shape the action potential, we'll consider the process as a series of stages (**Figure 48.11**). ❶ When the membrane of the axon is at the resting potential, most voltage-gated sodium channels are closed. Some potassium channels are open, but most voltage-gated potassium channels are closed. ❷ When a stimulus depolarizes the membrane, some gated sodium channels open, allowing more Na^+ to diffuse into the cell. The Na^+ inflow causes further depolarization, which opens still more gated sodium channels, allowing even more Na^+ to diffuse into the cell. ❸ Once the threshold is crossed, the positive-feedback cycle rapidly brings the membrane potential close to E_{Na}. This stage of the action potential is called the *rising phase.* ❹ Two events prevent the membrane potential from actually reaching E_{Na}: Voltage-gated sodium channels inactivate soon after opening, halting Na^+ inflow; and most voltage-gated potassium channels open, causing a rapid outflow of K^+. Both events quickly bring the membrane potential back toward E_K. This stage is called the *falling phase.* ❺ In the final phase of an action potential, called the *undershoot*, the membrane's permeability to K^+ is higher than at rest, so the membrane potential is closer to E_K than it is at the resting potential. The gated potassium channels eventually close, and the membrane potential returns to the resting potential.

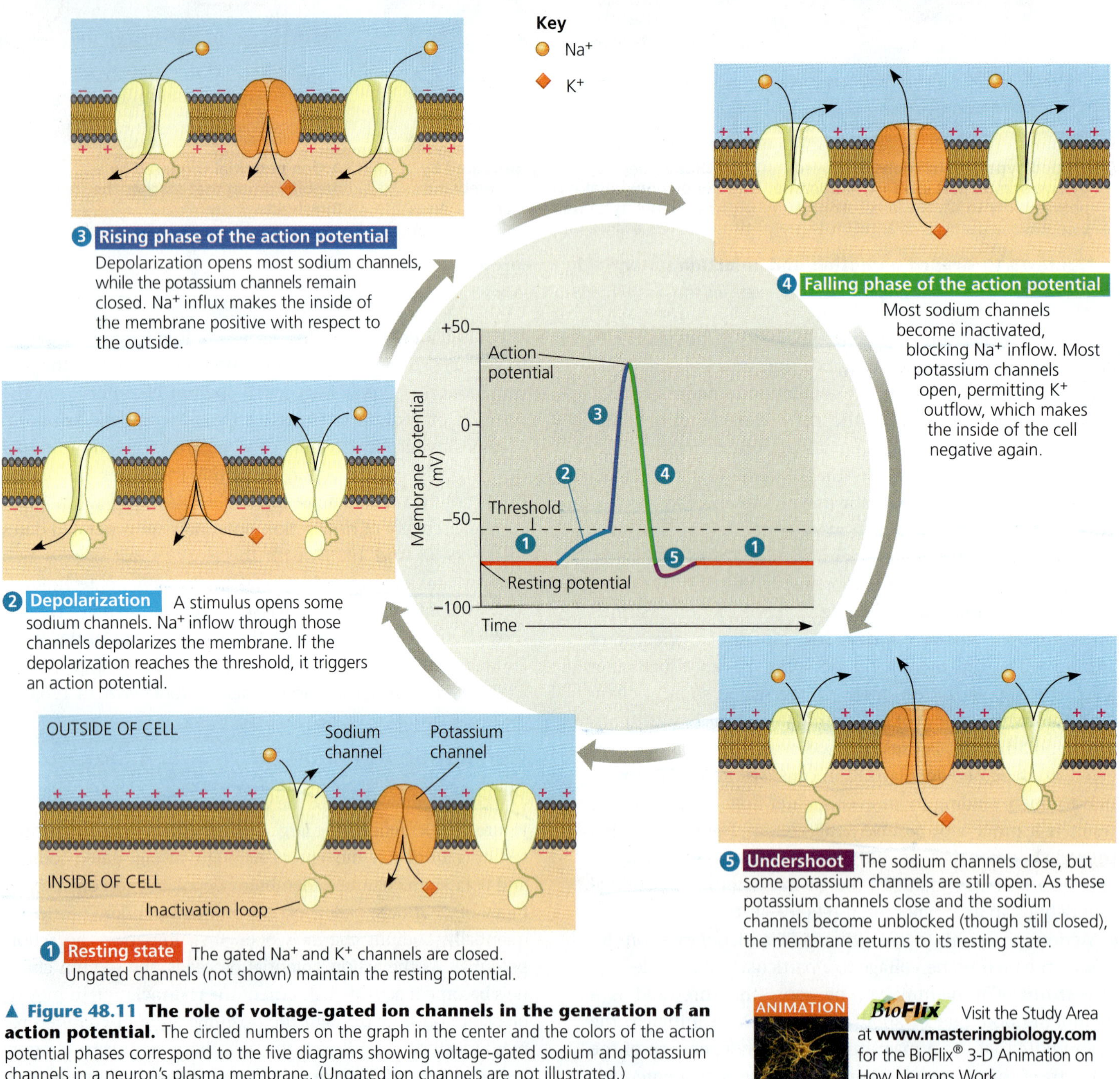

▲ **Figure 48.11 The role of voltage-gated ion channels in the generation of an action potential.** The circled numbers on the graph in the center and the colors of the action potential phases correspond to the five diagrams showing voltage-gated sodium and potassium channels in a neuron's plasma membrane. (Ungated ion channels are not illustrated.)

ANIMATION *BioFlix* Visit the Study Area at **www.masteringbiology.com** for the BioFlix® 3-D Animation on How Neurons Work.

The sodium channels remain inactivated during the falling phase and the early part of the undershoot. As a result, if a second depolarizing stimulus occurs during this period, it will be unable to trigger an action potential. The "downtime" when a second action potential cannot be initiated is called the **refractory period**. This interval sets a limit on the maximum frequency at which action potentials can be generated. As we will discuss shortly, the refractory period also ensures that all signals in an axon travel in one direction, from the cell body to the axon terminals.

Note that the refractory period is due to the inactivation of sodium channels, not to a change in the ion gradients across the plasma membrane. The flow of charged particles during an action potential involves far too few ions to change the concentration on either side of the membrane significantly.

For most neurons, the interval between the onset of an action potential and the end of the refractory period is only 1–2 milliseconds (msec). Because action potentials are so brief, a neuron can produce hundreds per second. Furthermore, the frequency with which a neuron generates action potentials varies in response to input. Such differences in action potential frequency convey information about signal strength. In hearing, for example, louder sounds result in more frequent action potentials in neurons connecting the ear to the brain. Differences in the time interval between action potentials are in fact the only variable in transmission of information by an axon.

Gated ion channels and action potentials have a central role in all nervous system function. As a consequence, mutations in genes that encode ion channel proteins can cause disorders affecting the nerves, muscles, brain, or heart. The type of disorder depends largely on where in the body the gene for the ion channel protein is expressed. For example, mutations affecting voltage-gated sodium channels in skeletal muscle cells can cause myotonia, a periodic spasming of those muscles; and mutations affecting sodium channels in the brain can cause epilepsy, in which excessive synchronized firing of groups of nerve cells causes seizures.

Conduction of Action Potentials

At the site where an action potential is initiated (usually the axon hillock), Na^+ inflow during the rising phase creates an electrical current that depolarizes the neighboring region of the axon membrane **(Figure 48.12)**. The depolarization in the neighboring region is large enough to reach the threshold, causing the action potential to be reinitiated there. This process is repeated many times along the length of the axon. Because an action potential is an all-or-none event, the magnitude and duration of the action potential remain constant at each position along the axon. The result is the movement of a nerve impulse from the cell body to the synaptic terminals, much like the cascade of events triggered by knocking over the first domino in a line.

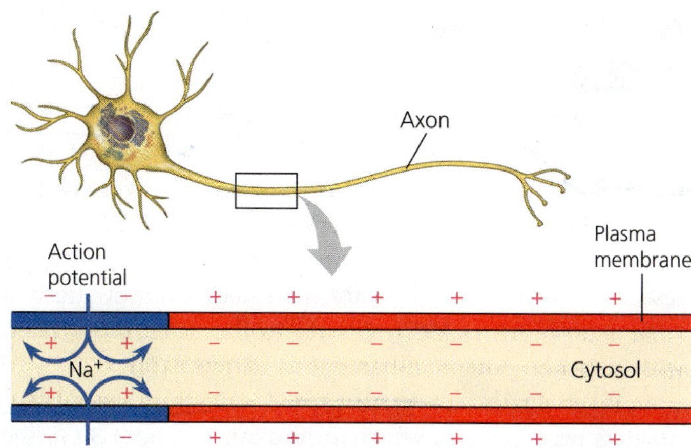

1 An action potential is generated as Na^+ flows inward across the membrane at one location.

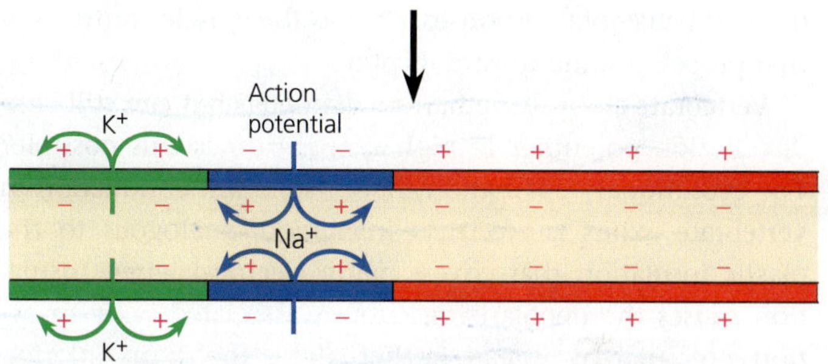

2 The depolarization of the action potential spreads to the neighboring region of the membrane, reinitiating the action potential there. To the left of this region, the membrane is repolarizing as K^+ flows outward.

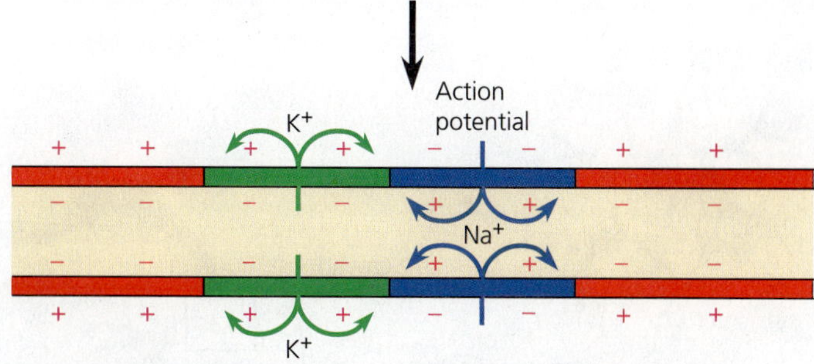

3 The depolarization-repolarization process is repeated in the next region of the membrane. In this way, local currents of ions *across* the plasma membrane cause the action potential to be propagated *along* the length of the axon.

▲ **Figure 48.12 Conduction of an action potential.** This figure shows events at three successive times as an action potential passes from left to right. At each point along the axon, voltage-gated ion channels go through the sequence of changes in Figure 48.10. Membrane colors correspond to the action potential phases in Figure 48.10.

An action potential that starts at the axon hillock moves along the axon only toward the synaptic terminals. Why? Immediately behind the traveling zone of depolarization caused by Na^+ inflow is a zone of repolarization caused by K^+ outflow. In the repolarized zone, the sodium channels remain inactivated. Consequently, the inward current that depolarizes the axon membrane *ahead* of the action potential cannot produce another action potential *behind* it. This prevents action potentials from traveling back toward the cell body.

Evolutionary Adaptations of Axon Structure

EVOLUTION Axon diameter is a major factor affecting the speed at which action potentials are conducted. One adaptation that increases conductance speed is an increased axon width. Resistance to electrical current flow is inversely proportional to the cross-sectional area of a conductor (such as a wire or an axon). In the same way that a wide hose offers less resistance to the flow of water than does a narrow hose, a wide axon provides less resistance to the current associated with an action potential than does a narrow axon.

In invertebrates, conduction speed varies from several centimeters per second in very narrow axons to about 30 m/sec in the giant axons of some arthropods and molluscs (see Figure 48.2). These giant axons (up to 1 mm wide) function in rapid behavioral responses, such as the muscle contraction that propels a squid toward its prey.

Vertebrate axons have narrow diameters but can still conduct action potentials at high speed. How is this possible? The evolutionary adaptation that enables fast conduction in vertebrate axons is electrical insulation, analogous to the plastic insulation that covers many electrical wires. Insulation causes the depolarizing current associated with an action potential to spread farther along the axon interior, bringing more distant regions to the threshold sooner.

The electrical insulation that surrounds vertebrate axons is called a **myelin sheath** **(Figure 48.13)**. Myelin sheaths are produced by two types of glia—**oligodendrocytes** in the CNS and **Schwann cells** in the PNS. During development, these specialized glia wrap axons in many layers of membrane. The membranes forming these layers are mostly lipid, which is a poor conductor of electrical current.

In myelinated axons, voltage-gated sodium channels are restricted to gaps in the myelin sheath called **nodes of Ranvier** (see Figure 48.13). The extracellular fluid is in contact with the axon membrane only at the nodes. As a result, action potentials are not generated in the regions between the nodes. Rather, the inward current produced during the rising phase of the action potential at a node travels all the way to the next node, where it depolarizes the membrane and regenerates the action potential **(Figure 48.14)**. Thus, the time-consuming process of opening and closing of ion channels occurs at only a limited number of positions along the axon. This mechanism for action potential propagation is called **saltatory conduction** (from the Latin *saltare*, to leap) because the action potential appears to jump along the axon from node to node.

The major selective advantage of myelination is its space efficiency. A myelinated axon 20 μm in diameter has a conduction speed faster than that of a squid giant axon with a diameter 40 times greater. Furthermore, more than 2,000 of those myelinated axons can be packed into the space occupied by just one giant axon.

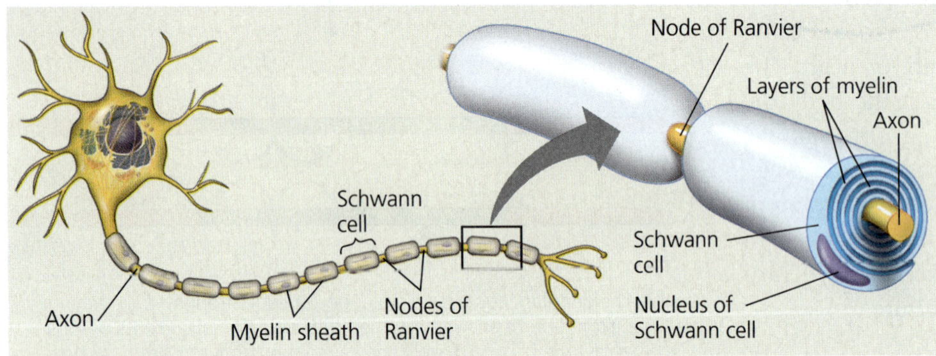

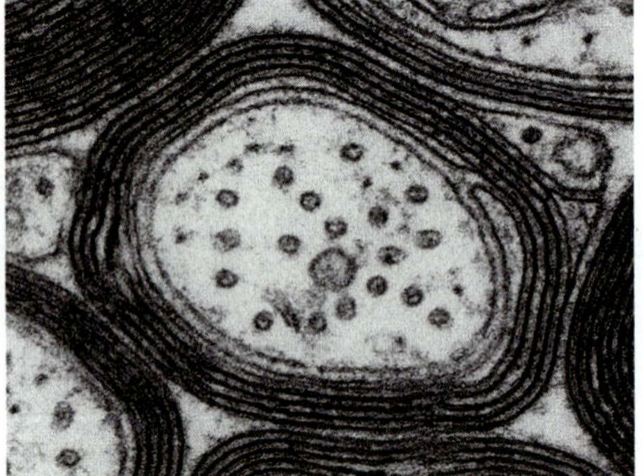

▲ **Figure 48.13 Schwann cells and the myelin sheath.** In the PNS, glia called Schwann cells wrap themselves around axons, forming layers of myelin. Gaps between adjacent Schwann cells are called nodes of Ranvier. The TEM shows a cross section through a myelinated axon.

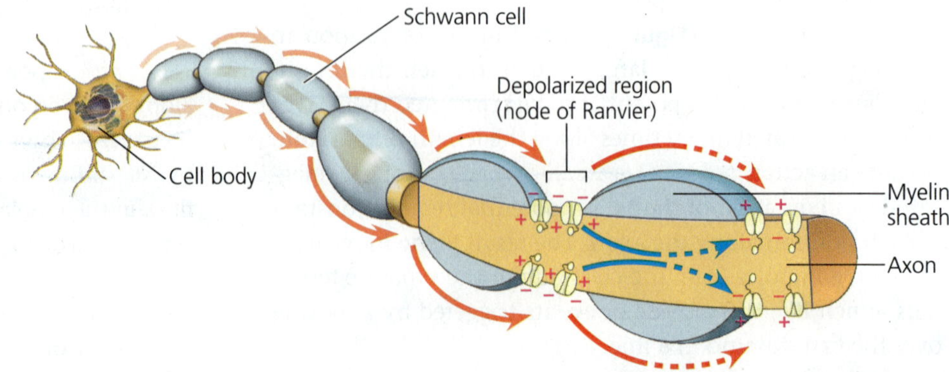

► **Figure 48.14 Saltatory conduction.** In a myelinated axon, the depolarizing current during an action potential at one node of Ranvier spreads along the interior of the axon to the next node (blue arrows), where voltage-gated sodium channels enable reinitiation. Thus, the action potential jumps from node to node as it travels along the axon (red arrows).

CONCEPT CHECK 48.3

1. How do action potentials and graded potentials differ?
2. In multiple sclerosis (from the Greek *skleros*, hard), myelin sheaths harden and deteriorate. How would this affect nervous system function?
3. **WHAT IF?** Suppose a mutation caused gated sodium channels to remain inactivated longer after an action potential. How would this affect the frequency at which action potentials could be generated? Explain.

For suggested answers, see Appendix A.

CONCEPT 48.4

Neurons communicate with other cells at synapses

In most cases, action potentials are not transmitted from neurons to other cells. However, information is transmitted, and this transmission occurs at the synapses. Some synapses, called *electrical synapses*, contain gap junctions (see Figure 6.32), which *do* allow electrical current to flow directly from one neuron to another. In both vertebrates and invertebrates, electrical synapses synchronize the activity of neurons responsible for certain rapid, unvarying behaviors. For example, electrical synapses associated with the giant axons of squids and lobsters facilitate the swift execution of escape responses. There are also many electrical synapses in the vertebrate brain.

The majority of synapses are *chemical synapses*, which involve the release of a chemical neurotransmitter by the presynaptic neuron. At each terminal, the presynaptic neuron synthesizes the neurotransmitter and packages it in multiple membrane-bounded compartments called *synaptic vesicles*. The arrival of an action potential at a synaptic terminal depolarizes the plasma membrane, opening voltage-gated channels that allow Ca^{2+} to diffuse into the terminal **(Figure 48.15)**. The resulting rise in Ca^{2+} concentration in the terminal causes some of the synaptic vesicles to fuse with the terminal membrane, releasing the neurotransmitter.

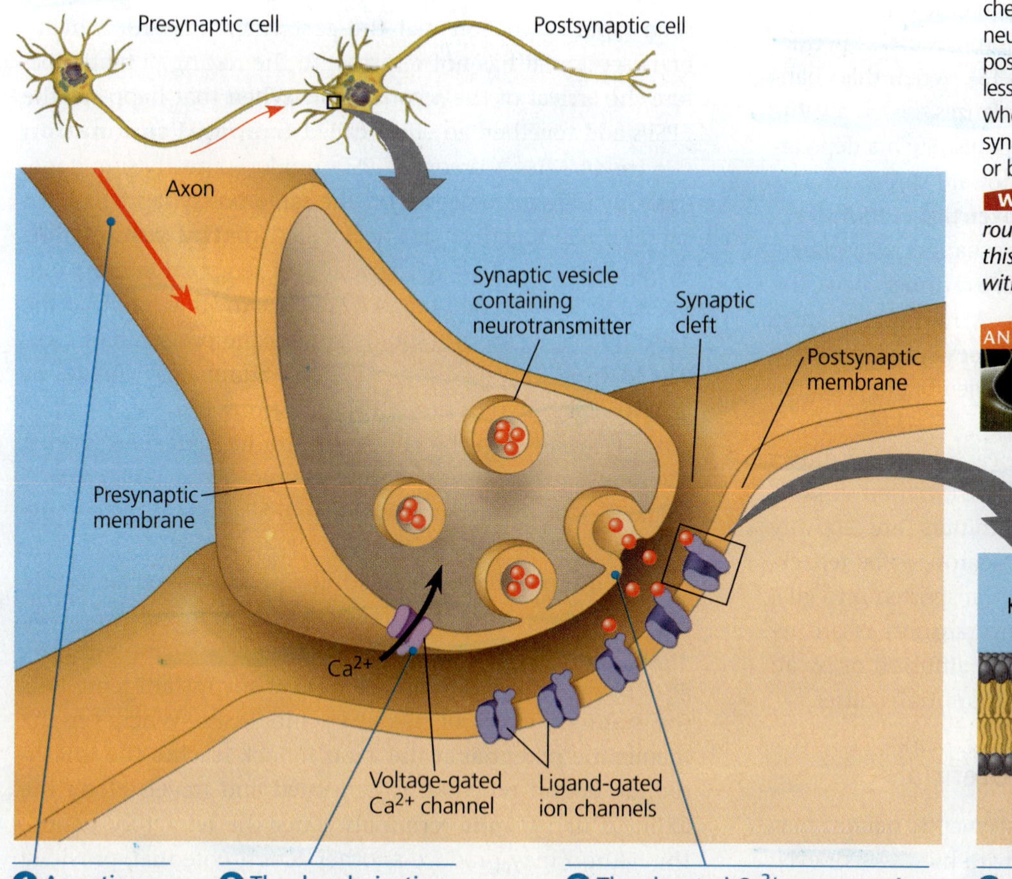

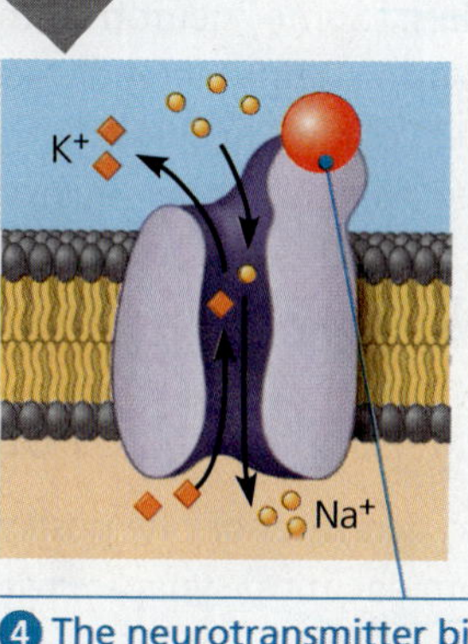

1 An action potential arrives, depolarizing the presynaptic membrane.

2 The depolarization opens voltage-gated channels, triggering an influx of Ca^{2+}.

3 The elevated Ca^{2+} concentration causes synaptic vesicles to fuse with the presynaptic membrane, releasing neurotransmitter into the synaptic cleft.

4 The neurotransmitter binds to ligand-gated ion channels in the postsynaptic membrane. In this example, binding triggers opening, allowing Na^+ and K^+ to diffuse through.

◄ **Figure 48.15 A chemical synapse.** This figure illustrates the sequence of events that transmits a nerve impulse across a chemical synapse. In response to binding of neurotransmitter, ligand-gated ion channels in the postsynaptic membrane open (as shown here) or, less commonly, close. Synaptic transmission ends when the neurotransmitter diffuses out of the synaptic cleft, is taken up by the synaptic terminal or by another cell, or is degraded by an enzyme.

WHAT IF? *If all the Ca^{2+} in the fluid surrounding a neuron were removed, how would this affect the transmission of information within and between neurons?*

BioFlix Visit the Study Area at **www.masteringbiology.com** for the BioFlix® 3-D Animation on How Synapses Work.

Once released, the neurotransmitter diffuses across the *synaptic cleft*, the gap that separates the presynaptic neuron from the postsynaptic cell. Diffusion time is very short because the gap is less than 50 nm across. Upon reaching the postsynaptic membrane, the neurotransmitter binds to and activates a specific receptor in the membrane.

Information transfer is much more readily modified at chemical synapses than at electrical synapses. A variety of factors can affect the amount of neurotransmitter that is released or the responsiveness of the postsynaptic cell. Such modifications underlie an animal's ability to alter its behavior in response to change and form the basis for learning and memory, as you will learn in Chapter 49.

Postsynaptic neuron

Synaptic terminals of presynaptic neurons

5 µm

▲ **Figure 48.16 Synaptic terminals on the cell body of a postsynaptic neuron** (colorized SEM).

Generation of Postsynaptic Potentials

At many chemical synapses, the receptor protein that binds and responds to neurotransmitters is a **ligand-gated ion channel**, often called an *ionotropic receptor*. These receptors are clustered in the membrane of the postsynaptic cell, directly opposite the synaptic terminal. Binding of the neurotransmitter (the receptor's ligand) to a particular part of the receptor opens the channel and allows specific ions to diffuse across the postsynaptic membrane. The result is a *postsynaptic potential*, a graded potential in the postsynaptic cell.

At some synapses, the ligand-gated ion channel is permeable to both K^+ and Na^+ (see Figure 48.15). When this channel opens, the membrane potential depolarizes toward a value roughly midway between E_K and E_{Na}. Because such a depolarization brings the membrane potential toward threshold, it is called an **excitatory postsynaptic potential (EPSP)**.

At other synapses, the ligand-gated ion channel is selectively permeable for only K^+ or Cl^-. When such a channel opens, the postsynaptic membrane hyperpolarizes. A hyperpolarization produced in this manner is an **inhibitory postsynaptic potential (IPSP)** because it moves the membrane potential further from threshold.

Various mechanisms that rapidly clear neurotransmitter molecules from the synaptic cleft limit the duration of postsynaptic potentials. Some neurotransmitters are actively transported back into the presynaptic neuron, to be repackaged into synaptic vesicles, or they are transported into glia, to be metabolized as fuel. Other neurotransmitters are removed from the synaptic cleft by simple diffusion or by an enzyme that catalyzes hydrolysis of the neurotransmitter.

Summation of Postsynaptic Potentials

The cell body and dendrites of one postsynaptic neuron may receive inputs from chemical synapses with hundreds or even thousands of synaptic terminals **(Figure 48.16)**. The magnitude of the postsynaptic potential at any one synapse varies with a number of factors, including the amount of neurotransmitter released by the presynaptic neuron. As a graded potential, a postsynaptic potential becomes smaller with distance from the synapse. Therefore, by the time a single EPSP reaches the axon hillock, it is usually too small to trigger an action potential in a postsynaptic neuron **(Figure 48.17a)**.

On some occasions, two EPSPs occur at a single synapse in such rapid succession that the postsynaptic neuron's membrane potential has not returned to the resting potential before the arrival of the second EPSP. When that happens, the EPSPs add together, an effect called **temporal summation** **(Figure 48.17b)**. Moreover, EPSPs produced nearly simultaneously by *different* synapses on the same postsynaptic neuron can also add together, an effect called **spatial summation** **(Figure 48.17c)**. Through spatial and temporal summation, several EPSPs can combine to depolarize the membrane at the axon hillock to the threshold, causing the postsynaptic neuron to produce an action potential. Summation applies as well to IPSPs: Two or more IPSPs occurring nearly simultaneously at synapses in the same region or in rapid succession at the same synapse have a larger hyperpolarizing effect than a single IPSP. Through summation, an IPSP can also counter the effect of an EPSP **(Figure 48.17d)**.

The interplay between multiple excitatory and inhibitory inputs is the essence of integration in the nervous system. The axon hillock is the neuron's integrating center, the region where the membrane potential at any instant represents the summed effect of all EPSPs and IPSPs. Whenever the membrane potential at the axon hillock reaches the threshold, an action potential is generated and travels along the axon to its synaptic terminals. After the refractory period, the neuron may produce another action potential, provided the membrane potential at the axon hillock once again reaches the threshold.

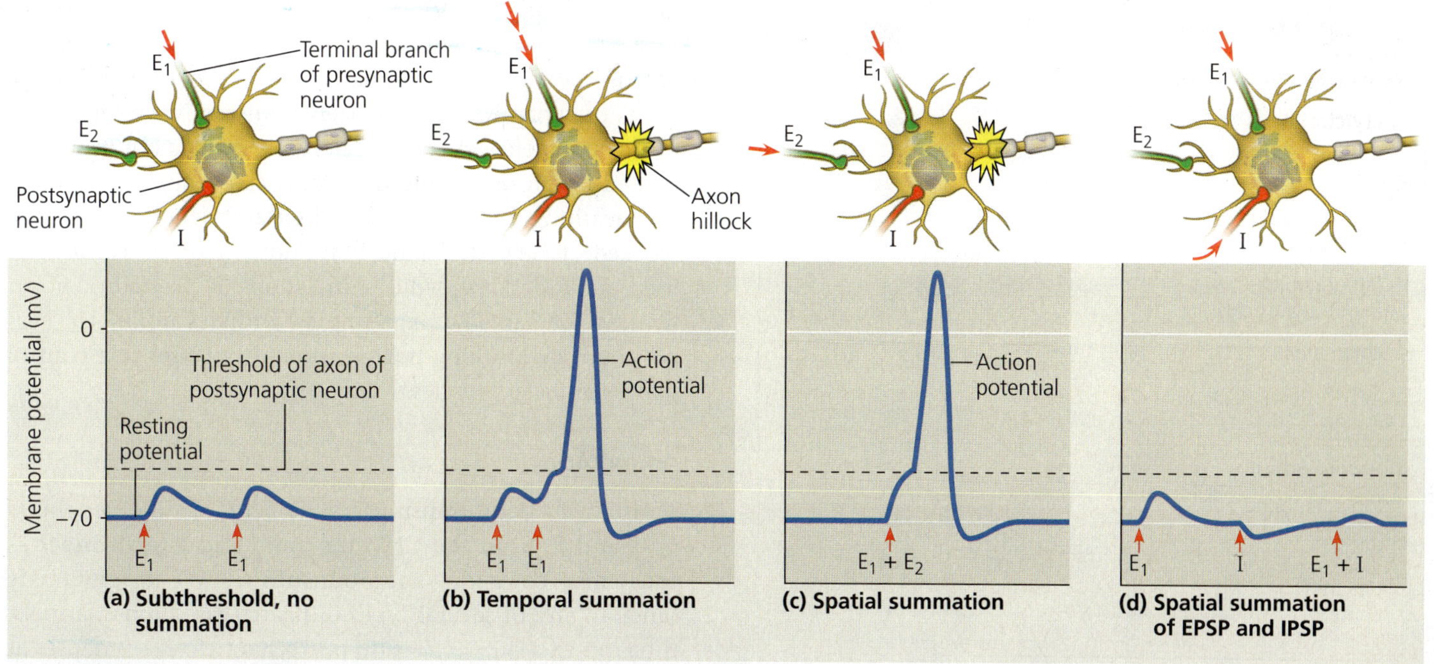

▲ **Figure 48.17 Summation of postsynaptic potentials.** These graphs trace changes in the membrane potential at a postsynaptic neuron's axon hillock. The arrows indicate times when postsynaptic potentials occur at two excitatory synapses (E_1 and E_2, green in the diagrams above the graphs) and at one inhibitory synapse (I, red). Like most EPSPs, those produced at E_1 or E_2 do not reach the threshold at the axon hillock without summation.

Modulated Signaling at Synapses

So far, we have focused on synapses where a neurotransmitter binds directly to an ion channel, causing the channel to open. However, there are also synapses in which the receptor for the neurotransmitter is *not* part of an ion channel. At these synapses, the neurotransmitter binds to a *metabotropic receptor*, so called because the resulting opening or closing of ion channels depends on one or more metabolic steps. Binding of a neurotransmitter to a metabotropic receptor activates a signal transduction pathway in the postsynaptic cell involving a second messenger (see Chapter 11). Compared with the postsynaptic potentials produced by ligand-gated channels, the effects of these second-messenger systems have a slower onset but last longer (minutes or even hours). Second messengers modulate the responsiveness of postsynaptic neurons to inputs in diverse ways, such as by altering the number of open potassium channels.

A variety of signal transduction pathways play a role in modulating synaptic transmission. One of the best-studied pathways involves cyclic AMP (cAMP) as a second messenger. For example, when the neurotransmitter norepinephrine binds to its metabotropic receptor, the neurotransmitter-receptor complex activates a G protein, which in turn activates adenylyl cyclase, the enzyme that converts ATP to cAMP (see Figure 11.11). Cyclic AMP activates protein kinase A, which phosphorylates specific ion channel proteins in the postsynaptic membrane, causing them to open or close. Because of the amplifying effect of the signal transduction pathway, the binding of a neurotransmitter molecule to a metabotropic receptor can open or close many channels.

Neurotransmitters

Researchers have identified more than 100 neurotransmitters belonging to five groups: acetylcholine, amino acids, biogenic amines, neuropeptides, and gases (**Table 48.2**, on the next page). The response triggered depends on the particular kind of receptor expressed by the postsynaptic cell. A single neurotransmitter may bind specifically to more than a dozen different receptors, including ionotropic and metabotropic types. Indeed, a particular neurotransmitter can excite postsynaptic cells expressing one receptor and inhibit postsynaptic cells expressing a different receptor. As an example, let's examine **acetylcholine**, a common neurotransmitter in both invertebrates and vertebrates.

Acetylcholine

Acetylcholine is vital for nervous system functions that include muscle stimulation, memory formation, and learning. In vertebrates, there are two major classes of acetylcholine receptor. One type is a ligand-gated ion channel. We know the most about its function at the *neuromuscular junction*, the site where motor neurons synapse with skeletal muscle cells. When acetylcholine released by motor neurons binds this receptor,

Table 48.2 Major Neurotransmitters

Neurotransmitter	Structure
Acetylcholine	$H_3C-C(=O)-O-CH_2-CH_2-N^+(CH_3)_3$
Amino Acids	
GABA (gamma-aminobutyric acid)	$H_2N-CH_2-CH_2-CH_2-COOH$
Glutamate	$H_2N-CH(COOH)-CH_2-CH_2-COOH$
Glycine	H_2N-CH_2-COOH
Biogenic Amines	
Norepinephrine	(HO)₂-benzene ring$-CH(OH)-CH_2-NH_2$
Dopamine	(HO)₂-benzene ring$-CH_2-CH_2-NH_2$
Serotonin	HO-indole ring$-C-CH_2-CH_2-NH_2$
Neuropeptides (a very diverse group, only two of which are shown)	
Substance P	Arg—Pro—Lys—Pro—Gln—Gln—Phe—Phe—Gly—Leu—Met
Met-enkephalin (an endorphin)	Tyr—Gly—Gly—Phe—Met
Gases	
Nitric oxide	N═O

the ion channel opens, producing an EPSP. This excitatory activity is soon terminated by acetylcholinesterase, an enzyme in the synaptic cleft that hydrolyzes the neurotransmitter.

The acetylcholine receptor active at the neuromuscular junction is also found elsewhere in the PNS, as well as in the CNS. There this ionotropic receptor can bind nicotine, a chemical found in tobacco and tobacco smoke. Nicotine's effects as a physiological and psychological stimulant result from its binding to this receptor.

A metabotropic acetylcholine receptor is found at locations that include the vertebrate CNS and heart. In heart muscle, acetylcholine released by neurons activates a signal transduction pathway. The G proteins in the pathway inhibit adenylyl cyclase and open potassium channels in the muscle cell membrane. Both effects reduce the rate at which the heart pumps. Thus, the effect of acetylcholine in heart muscle is inhibitory rather than excitatory.

A number of natural and synthetic toxins disrupt neurotransmission by acetylcholine. For example, the nerve gas sarin inhibits acetylcholinesterase, causing a buildup of acetylcholine to levels that trigger paralysis and typically death. In contrast, certain bacteria produce a toxin that inhibits presynaptic release of acetylcholine. This toxin causes a rare but severe form of food poisoning called botulism. Untreated botulism is typically fatal because muscles required for breathing fail to contract when acetylcholine release is blocked. Today, the same botulinum toxin is commonly used in cosmetic procedures. Injections of the toxin, known by the trade name Botox, minimize wrinkles around the eyes or mouth by blocking transmission at synapses that control particular facial muscles.

Amino Acids

Amino acid neurotransmitters are active in the vertebrate CNS and PNS. In the CNS, the amino acid **glutamate** is the most common neurotransmitter. When glutamate binds to any of several types of ligand-gated ion channels, it has an excitatory effect on postsynaptic cells. Synapses at which glutamate is the neurotransmitter have a key role in the formation of long-term memory, as we will discuss in Chapter 49.

The amino acid **gamma-aminobutyric acid (GABA)** is the neurotransmitter at most inhibitory synapses in the brain. Binding of GABA to receptors in postsynaptic cells increases membrane permeability to Cl^-, resulting in an IPSP. The widely prescribed drug diazepam (Valium) reduces anxiety through binding to a site on a GABA receptor.

A third amino acid, glycine, acts at inhibitory synapses in parts of the CNS that lie outside of the brain. There, glycine binds to an ionotropic receptor that is inhibited by strychnine, a chemical often used as a rat poison.

Biogenic Amines

The neurotransmitters grouped as **biogenic amines** are synthesized from amino acids and include **norepinephrine**, which is made from tyrosine. Norepinephrine is an excitatory neurotransmitter in the autonomic nervous system, a branch of the PNS discussed in Chapter 49. Outside the nervous system, norepinephrine has distinct but related functions as a hormone, as does the related biogenic amine *epinephrine* (see Chapter 45).

The biogenic amines **dopamine**, made from tyrosine, and **serotonin**, made from tryptophan, are released at many sites in the brain and affect sleep, mood, attention, and learning. Some psychoactive drugs, including LSD and mescaline, apparently produce their hallucinatory effects by binding to brain receptors for these neurotransmitters.

Biogenic amines have a central role in a number of nervous system disorders and treatments (see Chapter 49). The degenerative illness Parkinson's disease is associated with a lack of dopamine in the brain. In addition, depression is often treated with drugs that increase the brain concentrations of biogenic amines. Prozac, for instance, enhances the effect of serotonin by inhibiting its reuptake after release.

Neuropeptides

Several **neuropeptides**, relatively short chains of amino acids, serve as neurotransmitters that operate via metabotropic receptors. Such peptides are typically produced by cleavage of much larger protein precursors. The neuropeptide *substance P* is a key excitatory neurotransmitter that mediates our perception of pain, while other neuropeptides, called **endorphins**, function as natural analgesics, decreasing pain perception.

In the 1970s, Candace Pert, then a graduate student at Johns Hopkins University, and her research supervisor, Solomon Snyder, discovered endorphins as an outcome of their research on the biochemistry of behavior. Previous studies had suggested that the brain contains receptors for opiates, painkilling drugs such as morphine and heroin. To find these receptors, Pert and Snyder had the insight to apply existing knowledge about the activity of different drugs in the brain **(Figure 48.18)**. In a single, straightforward experiment, they provided the first demonstration that specific opiate receptors exist. Setting out to identify molecules normally present in the brain that could also activate these receptors, they discovered endorphins.

Endorphins are produced in the brain during times of physical or emotional stress, such as childbirth. In addition to relieving pain, they decrease urine output, depress respiration, and produce euphoria, as well as other emotional effects. Because opiates bind to the same receptor proteins as endorphins, opiates mimic endorphins and produce many of the same physiological effects (see Figure 2.18).

Gases

In common with many other types of cells, some neurons in vertebrates release dissolved gases, notably nitric oxide (NO), that act as local regulators. For example, during sexual arousal, certain neurons in human males release NO into the erectile tissue of the penis. In response, smooth muscle cells in the blood vessel walls of the erectile tissue relax, which causes the blood vessels to dilate and fill the spongy erectile tissue with blood, producing an erection. As you read in Chapter 45, the erectile dysfunction drug Viagra increases the ability to achieve and maintain an erection by inhibiting an enzyme that terminates the action of NO.

Unlike most neurotransmitters, NO is not stored in cytoplasmic vesicles but is instead synthesized on demand. NO

▼ Figure 48.18 INQUIRY

Does the brain have a specific protein receptor for opiates?

EXPERIMENT In 1973, Candace Pert and Solomon Snyder, of Johns Hopkins University, were searching for an opiate receptor in the mammalian brain. It was known that the drug naloxone antagonizes (opposes) the narcotic effect of opiates. Pert and Snyder reasoned that naloxone acts as an opiate antagonist by binding tightly to the opiate receptor without activating the receptor. They prepared radioactive naloxone and then incubated it with a protein mixture prepared from rodent brains. If proteins that could bind naloxone were present, the radioactivity would become stably associated with the protein mixture. Furthermore, the researchers could determine whether a specific receptor was present by comparing the ability of opiates and non-opiates to interfere with the binding activity.

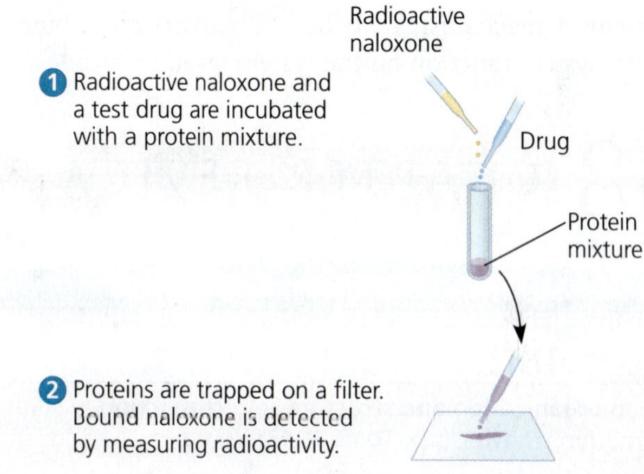

RESULTS

Drug	Opiate	Concentration That Blocked Naloxone Binding
Morphine	Yes	6×10^{-9} *M*
Methadone	Yes	2×10^{-8} *M*
Levorphanol	Yes	2×10^{-9} *M*
Phenobarbital	No	No effect at 10^{-4} *M*
Atropine	No	No effect at 10^{-4} *M*
Serotonin	No	No effect at 10^{-4} *M*

CONCLUSION Because opiates interfered with naloxone binding, but unrelated drugs did not, Pert and Snyder concluded that the binding activity had the specificity expected of the opiate receptor. They also found that the binding activity was present in tissue from regions of the brain involved in the sensation of pain, but not in tissue from the cerebellum, a brain region that coordinates motor activity.

SOURCE C. B. Pert and S. H. Snyder, Opiate receptor: demonstration in nervous tissue, *Science* 179:1011–1014 (1973).

WHAT IF? Suppose you found a drug that blocks naloxone binding at a concentration of 10^{-8} *M* but has no narcotic effect on animals. What are some possible explanations for this finding?

diffuses into neighboring target cells, produces a change, and is broken down—all within a few seconds. In many of its targets, including smooth muscle cells, NO works like many hormones, stimulating an enzyme to synthesize a second messenger that directly affects cellular metabolism.

Although inhaling air containing the gas carbon monoxide (CO) can be deadly, the vertebrate body produces small amounts of CO, some of which acts as a neurotransmitter. Carbon monoxide is generated by the enzyme heme oxygenase, one form of which is found in certain populations of neurons in the brain and PNS. In the brain, CO regulates the release of hypothalamic hormones. In the PNS, it acts as an inhibitory neurotransmitter that hyperpolarizes the plasma membrane of intestinal smooth muscle cells.

In the next chapter, we will consider how the cellular and biochemical mechanisms we have discussed contribute to nervous system function on the system level.

CONCEPT CHECK 48.4

1. How is it possible for a particular neurotransmitter to produce opposite effects in different tissues?
2. Organophosphate pesticides work by inhibiting acetylcholinesterase, the enzyme that breaks down the neurotransmitter acetylcholine. Explain how these toxins would affect EPSPs produced by acetylcholine.
3. **WHAT IF?** If a drug mimicked the activity of GABA in the CNS, what general effect on behavior might you expect? Explain.
4. **MAKE CONNECTIONS** A change in the concentration of calcium ions is important for fertilization in sea urchins and other animals (see Figure 47.3, on p. 1023). What membrane activity is common to fertilization and neurotransmitter release?

For suggested answers, see Appendix A.

48 CHAPTER REVIEW

SUMMARY OF KEY CONCEPTS

CONCEPT 48.1

Neuron organization and structure reflect function in information transfer (pp. 1045–1047)

- A **central nervous system (CNS)** and a **peripheral nervous system (PNS)** process information in three stages: sensory input, integration, and motor output to effector cells.
- Most neurons have branched **dendrites** that receive signals from other neurons and an **axon** that transmits signals to other cells at **synapses**. Neurons rely on **glia** for functions that include nourishment, insulation, and regulation.

? *How would severing an axon affect the flow of information in a neuron?*

CONCEPT 48.2

Ion pumps and ion channels establish the resting potential of a neuron (pp. 1048–1050)

- Ionic gradients generate a voltage difference, or **membrane potential**, across the plasma membrane of cells. The concentration of Na^+ is higher outside than inside; the reverse is true for K^+. In resting neurons, the plasma membrane has many open potassium channels but few open sodium channels. Diffusion of ions, principally K^+, through channels generates a **resting potential**, with the inside more negative than the outside.

? *Suppose you placed an isolated neuron in a solution similar to extracellular fluid and later transferred the neuron to a solution lacking any sodium ions. What change would you expect in the resting potential?*

CONCEPT 48.3

Action potentials are the signals conducted by axons (pp. 1050–1055)

- Neurons have gated ion channels that open or close in response to stimuli, leading to changes in the membrane potential. An increase in the magnitude of the membrane potential is a **hyperpolarization**; a decrease is a **depolarization**. Changes in membrane potential that vary continuously with the strength of a stimulus are known as **graded potentials**.
- An **action potential** is a brief, all-or-none depolarization of a neuron's plasma membrane. When a graded depolarization brings the membrane potential to the threshold, many **voltage-gated ion channels** open, triggering an inflow of Na^+ that rapidly brings the membrane potential to a positive value. A negative membrane potential is restored by the inactivation of sodium channels and by the opening of many voltage-gated potassium channels, which increases K^+ outflow. A **refractory period** follows, corresponding to the interval when the sodium channels are inactivated.

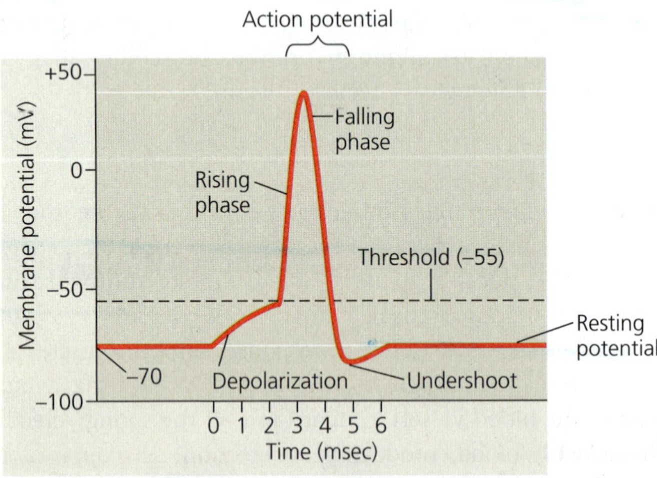

- A nerve impulse travels from the axon hillock to the synaptic terminals by propagation of a series of action potentials along the axon. The speed of conduction increases with the diameter of the axon and, in many vertebrate axons, with **myelination**. Action potentials in myelinated axons jump between the **nodes of Ranvier**, a process called **saltatory conduction**.

? *In what ways do both positive and negative feedback contribute to the shape of an action potential?*

CONCEPT 48.4

Neurons communicate with other cells at synapses (pp. 1055–1060)

- In an electrical **synapse**, electrical current flows directly from one cell to another. In a chemical synapse, depolarization causes synaptic vesicles to fuse with the terminal membrane and release **neurotransmitter** into the synaptic cleft.
- At many synapses, the neurotransmitter binds to **ligand-gated ion channels** in the postsynaptic membrane, producing an **excitatory** or **inhibitory postsynaptic potential** (**EPSP** or **IPSP**). The neurotransmitter then diffuses out of the cleft, is taken up by surrounding cells, or is degraded by enzymes. **Temporal** and **spatial summation** at the axon hillock determines whether a neuron generates an action potential.
- Different receptors for the same neurotransmitter produce different effects. Some neurotransmitter receptors activate signal transduction pathways, which can produce long-lasting changes in postsynaptic cells. Major neurotransmitters include acetylcholine; the amino acids GABA, glutamate, and glycine; biogenic amines; neuropeptides; and gases such as NO.

? *Why are many drugs used to treat nervous system diseases or affect brain function targeted to specific receptors rather than particular neurotransmitters?*

TEST YOUR UNDERSTANDING

LEVEL 1: KNOWLEDGE/COMPREHENSION

1. What happens when a resting neuron's membrane depolarizes?
   a. There is a net diffusion of Na^+ out of the cell.
   b. The equilibrium potential for K^+ (E_K) becomes more positive.
   c. The neuron's membrane voltage becomes more positive.
   d. The neuron is less likely to generate an action potential.
   e. The cell's inside is more negative than the outside.
2. A common feature of action potentials is that they
   a. cause the membrane to hyperpolarize and then depolarize.
   b. can undergo temporal and spatial summation.
   c. are triggered by a depolarization that reaches the threshold.
   d. move at the same speed along all axons.
   e. require the diffusion of Na^+ and K^+ through ligand-gated channels to propagate.
3. Where are neurotransmitter receptors located?
   a. the nuclear membrane
   b. the nodes of Ranvier
   c. the postsynaptic membrane
   d. synaptic vesicle membranes
   e. the myelin sheath
4. Temporal summation always involves
   a. both inhibitory and excitatory inputs.
   b. synapses at more than one site.
   c. inputs that are not simultaneous.
   d. electrical synapses.
   e. multiple inputs at a single synapse.

LEVEL 2: APPLICATION/ANALYSIS

5. Why are action potentials usually conducted in one direction?
   a. The nodes of Ranvier conduct potentials in one direction.
   b. The brief refractory period prevents reopening of voltage-gated Na^+ channels.
   c. The axon hillock has a higher membrane potential than the terminals of the axon.
   d. Ions can flow along the axon in only one direction.
   e. Voltage-gated channels for both Na^+ and K^+ open in only one direction.
6. Which of the following is a *direct* result of depolarizing the presynaptic membrane of an axon terminal?
   a. Voltage-gated calcium channels in the membrane open.
   b. Synaptic vesicles fuse with the membrane.
   c. The postsynaptic cell produces an action potential.
   d. Ligand-gated channels open, allowing neurotransmitters to enter the synaptic cleft.
   e. An EPSP or IPSP is generated in the postsynaptic cell.

LEVEL 3: SYNTHESIS/EVALUATION

7. **DRAW IT** Suppose a researcher inserts a pair of electrodes at two different positions along the middle of an axon dissected out of a squid. By applying a depolarizing stimulus, the researcher brings the plasma membrane at both positions to threshold. Using the drawing below as a model, create one or more drawings that illustrate where each action potential would terminate.

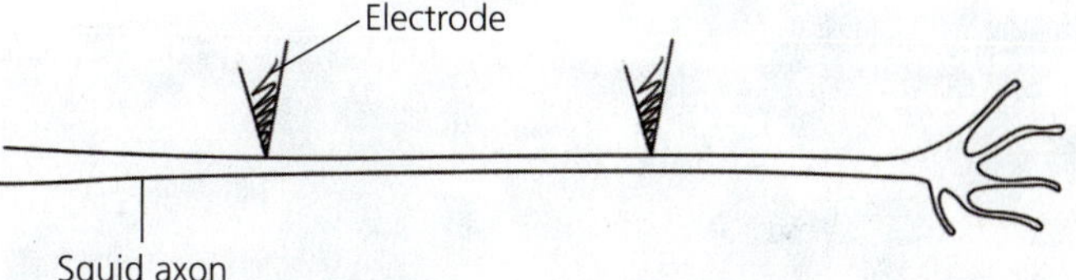

8. **EVOLUTION CONNECTION**
   An action potential is an all-or-none event. This on/off signaling is an evolutionary adaptation of animals that must sense and act in a complex environment. It is possible to imagine a nervous system in which the action potentials are graded, with the amplitude depending on the size of the stimulus. What evolutionary advantage might on/off signaling have over a graded (continuously variable) kind of signaling?
9. **SCIENTIFIC INQUIRY**
   From what you know about action potentials and synapses, propose two or three hypotheses for how various anesthetics might block pain.
10. **WRITE ABOUT A THEME**
    The Cellular Basis of Life In a short essay (100–150 words), describe how the structure and electrical properties of vertebrate neurons reflect similarities and differences with other animal cells.

For selected answers, see Appendix A.

MasteringBIOLOGY www.masteringbiology.com

1. MasteringBiology® Assignments

BioFlix **Tutorials** How Neurons Work: Neuron Structure and Resting Potential • The Action Potential • Conduction of an Action Potential; How Synapses Work: Chemical Synapses • Postsynaptic Potentials

Activities Neuron Structure • Membrane Potentials • Action Potentials • Nerve Signals: Action Potentials • Discovery Channel Video: Novelty Gene

Questions Student Misconceptions • Reading Quiz • Multiple Choice • End-of-Chapter

2. eText

Read your book online, search, take notes, highlight text, and more.

3. The Study Area

Practice Tests • Cumulative Test • ***BioFlix*** 3-D Animations • MP3 Tutor Sessions • Videos • Activities • Investigations • Lab Media • Audio Glossary • Word Study Tools • Art

49

Nervous Systems

▲ **Figure 49.1 How do scientists identify individual neurons in the brain?**

KEY CONCEPTS

49.1 Nervous systems consist of circuits of neurons and supporting cells
49.2 The vertebrate brain is regionally specialized
49.3 The cerebral cortex controls voluntary movement and cognitive functions
49.4 Changes in synaptic connections underlie memory and learning
49.5 Many nervous system disorders can be explained in molecular terms

OVERVIEW

Command and Control Center

What happens in your brain when you solve a math problem or listen to music? Until quite recently, scientists had little hope of answering that question. The human brain contains an estimated 10^{11} (100 billion) neurons. Interconnecting these brain cells are circuits more complex than those of even the most powerful supercomputers. Yet the circuitry of the brain has been largely hidden from view. That's no longer the case, thanks in part to several exciting new technologies.

One recent advance in exploring the brain relies on a method for expressing random combinations of colored proteins in brain cells—such that each cell shows up in a different color. The result is a "brainbow" like the one in **Figure 49.1**, which highlights neurons in the brain of a mouse. In this image, each neuron expresses one of more than 90 different color combinations of four fluorescent proteins. Using the brainbow technology, neuroscientists hope to develop detailed maps of the connections that transfer information between particular regions of the brain.

Another breakthrough came with the development of powerful imaging techniques that reveal activity in the working brain. Researchers can monitor multiple areas of the human brain while a subject is performing various tasks, such as speaking, looking at pictures, or forming a mental image of a person's face. They can use these techniques to look for a correlation between a particular task and activity in specific brain areas.

In this chapter, we will discuss the organization and evolution of animal nervous systems, exploring how groups of neurons function in specialized circuits dedicated to specific tasks. First we'll focus on specialization in regions of the vertebrate brain. We will then turn to the ways in which brain activity makes information storage and organization possible. Finally, we'll consider several disorders of the nervous system that are the subject of intense research today.

CONCEPT **49.1**

Nervous systems consist of circuits of neurons and supporting cells

The ability to sense and react originated billions of years ago with prokaryotes that could detect changes in their environment and respond in ways that enhanced their survival and reproductive success. For example, bacteria keep moving in a particular direction as long as they encounter increasing concentrations of a food source. Later in evolution, modification of simple recognition and response processes provided multicellular organisms with a mechanism for communication between cells of the body. By the time of the Cambrian explosion more than 500 million years ago (see Chapter 32), systems of neurons allowing animals to sense and move rapidly were present in essentially their current forms.

Hydras, jellies, and other cnidarians are the simplest animals with nervous systems. As you read in Chapters 33 and 41, these animals have radially symmetrical bodies organized around a central digestive compartment, the gastrovascular cavity. In most cnidarians, interconnected nerve cells form a diffuse **nerve net** (**Figure 49.2a**), which controls the contraction and expansion of the gastrovascular cavity. Unlike the nervous systems of other animals, the nerve net of cnidarians lacks clusters of neurons that perform specialized functions.

In more complex animals, the axons of multiple nerve cells are often bundled together, forming **nerves**. These fibrous structures channel and organize information flow along specific routes through the nervous system. For example, sea stars have a set of radial nerves connecting to a central nerve ring (**Figure 49.2b**). Within each arm of a sea star, the radial nerve is linked to a nerve net from which it receives input and to which it sends signals controlling muscle contraction.

Animals that have elongated, bilaterally symmetrical bodies have even more specialized nervous systems. Such animals exhibit cephalization, an evolutionary trend toward a clustering of sensory neurons and interneurons at the anterior (front) end of the body. These anterior neurons communicate with cells elsewhere in the body, including neurons located in one or more nerve cords extending toward the posterior (rear) end. In nonsegmented worms, such as the planarian shown in **Figure 49.2c**, a small brain and longitudinal nerve cords constitute the simplest clearly defined *central nervous system* (*CNS*). In some such animals, the entire nervous system is constructed from only a small number of cells, as shown by studies of another nonsegmented worm, the nematode *Caenorhabditis elegans*. In this species, an adult worm (hermaphrodite) has exactly 302 neurons, no more and no fewer. More complex invertebrates, such as segmented worms (annelids; **Figure 49.2d**) and arthropods (**Figure 49.2e**), have many more neurons. The behavior of such invertebrates is regulated by more complicated brains and by ventral nerve cords containing ganglia, segmentally arranged clusters of neurons.

Within an animal group, nervous system organization often correlates with lifestyle. Among the molluscs, for example, sessile and slow-moving species, such as clams and chitons, have relatively simple sense organs and little or no cephalization (**Figure 49.2f**). In contrast, active predatory molluscs, such as octopuses and squids (**Figure 49.2g**), have the most sophisticated nervous systems of any invertebrates, rivaling those of some vertebrates. With their large, image-forming eyes and a brain containing millions of neurons, octopuses can learn to discriminate between visual patterns and to perform complex tasks.

In vertebrates (**Figure 49.2h**), the brain and the spinal cord form the CNS; nerves and ganglia form the *peripheral nervous system* (*PNS*). Regional specialization is a hallmark of both systems, as we will see throughout the remainder of this chapter.

Organization of the Vertebrate Nervous System

In the vertebrate CNS, the functions of the brain and spinal cord are tightly coordinated. The brain provides the integrative power that underlies the complex behavior of vertebrates. The spinal cord, which runs lengthwise inside the vertebral column (spine), conveys information to and from the brain

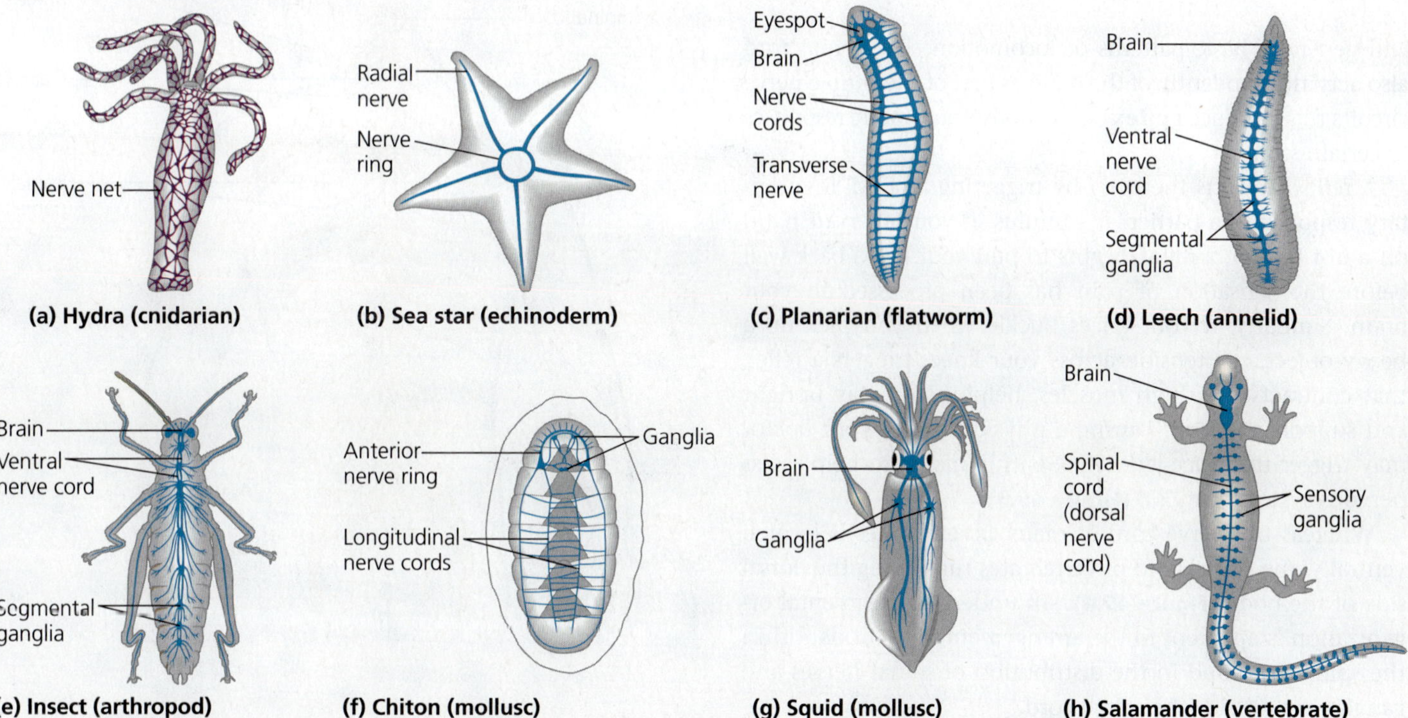

▲ **Figure 49.2 Nervous system organization.** **(a)** A hydra contains individual neurons (purple) organized in a diffuse nerve net. **(b–h)** Animals with more sophisticated nervous systems contain groups of neurons (blue) organized into nerves and often ganglia and a brain.

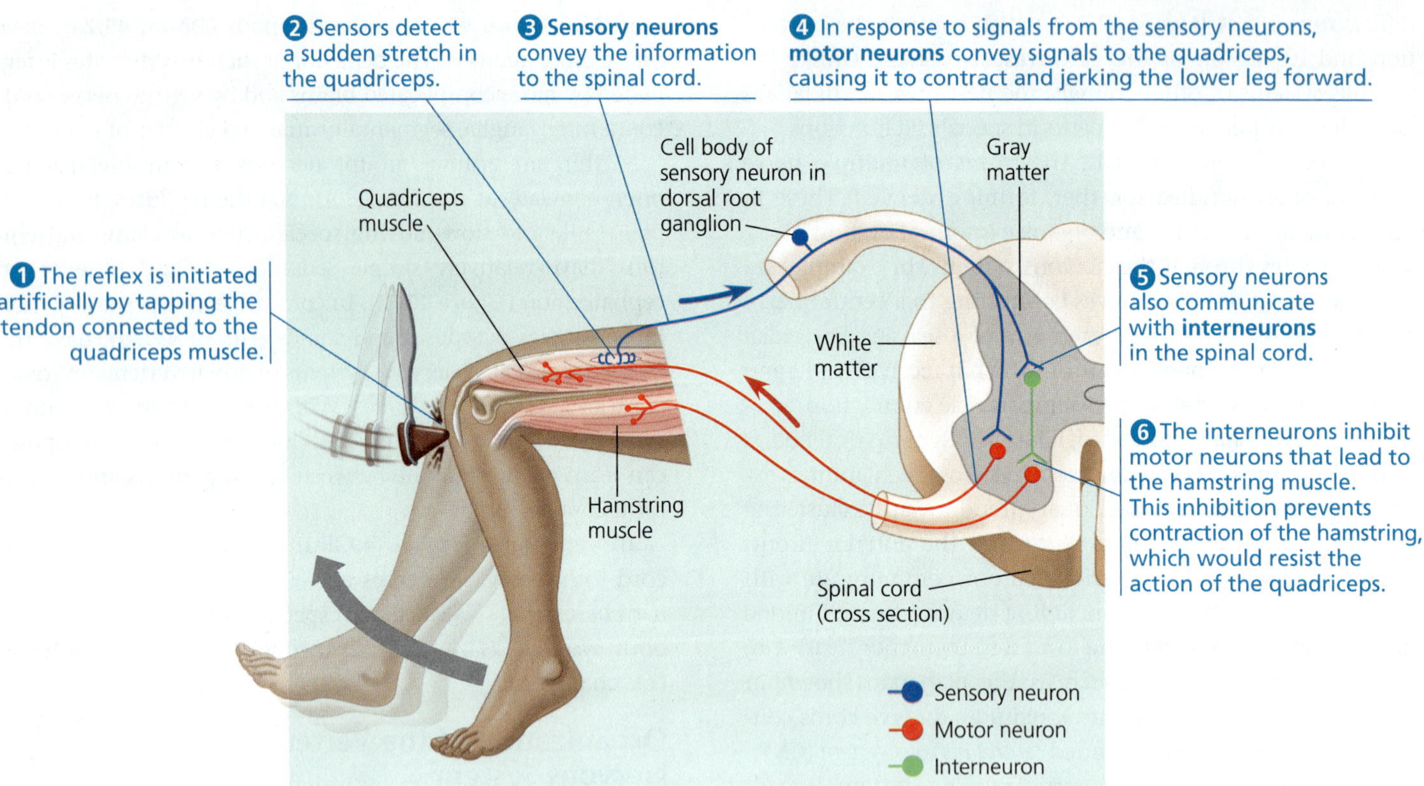

▲ **Figure 49.3 The knee-jerk reflex.** Many neurons are involved in the reflex, but for simplicity, only a few neurons are shown.

MAKE CONNECTIONS *Using the nerve signals to the hamstring and quadriceps in this reflex as an example, propose a model for regulation of smooth muscle activity in the esophagus during the swallowing reflex (see Figure 41.10, p. 884).*

and generates basic patterns of locomotion. The spinal cord also acts independently of the brain as part of the simple nerve circuits that produce **reflexes**, the body's automatic responses to certain stimuli.

A reflex protects the body by triggering a rapid, involuntary response to a particular stimulus. If you put your hand on a hot burner, a reflex begins to pull your hand back well before the sensation of pain has been processed in your brain. Similarly, if your knees buckle when you pick up a heavy object, the tension across your knees triggers a reflex that contracts the thigh muscles, helping you stay upright and support the load. During a physical exam, your doctor may trigger this knee-jerk reflex with a mallet to help assess nervous system function **(Figure 49.3)**.

Whereas the nerve cord of many invertebrates is located ventrally, the spinal cord of vertebrates runs along the dorsal side of the body **(Figure 49.4)**. An underlying segmental organization is apparent in the arrangement of neurons within the spinal cord and in the distribution of spinal nerves and ganglia just outside the spinal cord.

During embryonic development in vertebrates, the central nervous system develops from the hollow dorsal nerve cord—a hallmark of chordates (see Chapter 34). The cavity

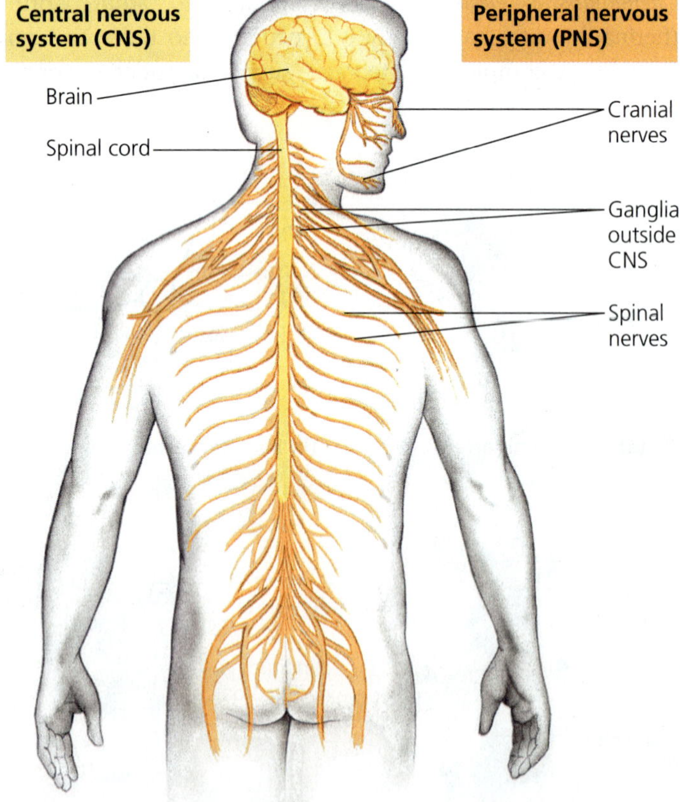

▲ **Figure 49.4 The vertebrate nervous system.** The central nervous system consists of the brain and spinal cord (yellow). Left-right pairs of cranial nerves, spinal nerves, and ganglia make up most of the peripheral nervous system (dark gold).

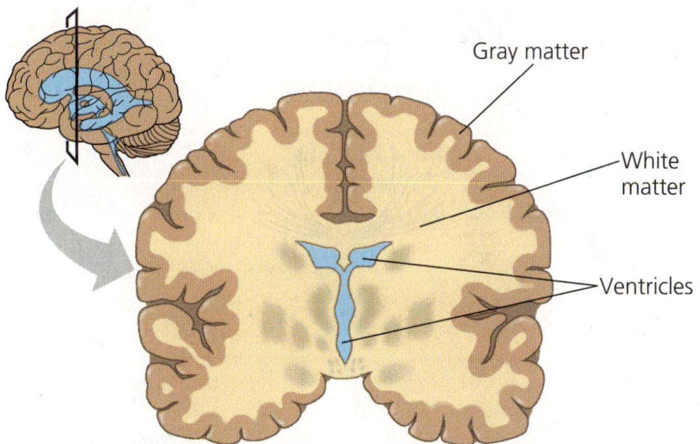

▲ **Figure 49.5 Ventricles, gray matter, and white matter.** Ventricles deep in the brain's interior contain cerebrospinal fluid. Most of the gray matter is on the surface of the brain, surrounding the white matter.

of the nerve cord gives rise to the narrow **central canal** of the spinal cord as well as the **ventricles** of the brain **(Figure 49.5)**. Both the canal and ventricles fill with **cerebrospinal fluid**, which is formed in the brain by filtration of arterial blood. The cerebrospinal fluid circulates slowly through the central canal and ventricles and then drains into the veins. This circulation supplies the brain with nutrients and hormones and carries away wastes. In mammals, the cerebrospinal fluid also cushions the brain and spinal cord by circulating between layers of connective tissue that surround the CNS.

In addition to these fluid-filled spaces, the brain and spinal cord contain gray matter and white matter (see Figure 49.5). **Gray matter** consists mainly of neuron cell bodies, dendrites, and unmyelinated axons. In contrast, **white matter** consists of bundled axons that have myelin sheaths, which give the axons a whitish appearance. White matter in the spinal cord lies on the outside, consistent with its function in linking the CNS to sensory and motor neurons of the PNS. White matter in the brain is predominantly on the inside, reflecting the role of signaling between neurons of the brain in learning, feeling emotions, processing sensory information, and generating commands.

Glia

The glia present throughout the vertebrate brain and spinal cord carry out functions crucial for the activity of the nervous system. **Figure 49.6** illustrates the major types of glia in the adult nervous system and provides an overview of the ways in which they nourish, support, and regulate the functioning of neurons.

CNS
PNS
VENTRICLE
Cilia
Neuron
Capillary

Ependymal cells line the ventricles and have cilia that promote circulation of the cerebrospinal fluid.

Oligodendrocytes myelinate axons in the CNS. Myelination greatly increases the conduction speed of action potentials.

Schwann cells myelinate axons in the PNS.

Microglia are immune cells that protect against pathogens.

Astrocytes (from the Greek *astron*, star) facilitate information transfer at synapses and in some instances release neurotransmitters. Astrocytes next to active neurons cause nearby blood vessels to dilate, increasing blood flow and enabling the neurons to obtain oxygen and glucose more quickly. Astrocytes also regulate extracellular concentrations of ions and neurotransmitters.

The green cells in this mammalian brain tissue are astrocytes labeled with a fluorescent antibody.

A blue dye that binds DNA in the nuclei of all cells reveals the intermingling of astrocytes with other cells, predominantly neurons.

50 µm

LM

▲ **Figure 49.6 Glia in the vertebrate nervous system.**

Glia also have an essential role in development of the nervous system. In embryos, cells called *radial glia* form tracks along which newly formed neurons migrate from the neural tube, the structure that gives rise to the CNS (see Figure 47.13). Later, astrocytes induce cells that line the capillaries in the CNS to form tight junctions (see Figure 6.32). The result is the *blood-brain barrier*, which controls the extracellular environment of the CNS by restricting the entry of most substances from the blood.

Both radial glia and astrocytes can also act as stem cells, generating new neurons and glia. Researchers view these multipotent precursors as a potential means for replacing neurons and glia that are lost to injury or disease, a topic we'll explore further in Concept 49.4.

The Peripheral Nervous System

The PNS transmits information to and from the CNS and plays a large role in regulating an animal's movement and internal environment **(Figure 49.7)**. Sensory information reaches the CNS along PNS neurons designated as *afferent* (from the Latin, meaning "to carry toward"). Following information processing within the CNS, instructions then travel to muscles, glands, and endocrine cells along PNS neurons designated as *efferent* (from the Latin, meaning "to carry away"). Most nerves contain both afferent and efferent neurons. One exception is the olfactory nerve, which conveys only sensory information from the nose to the brain.

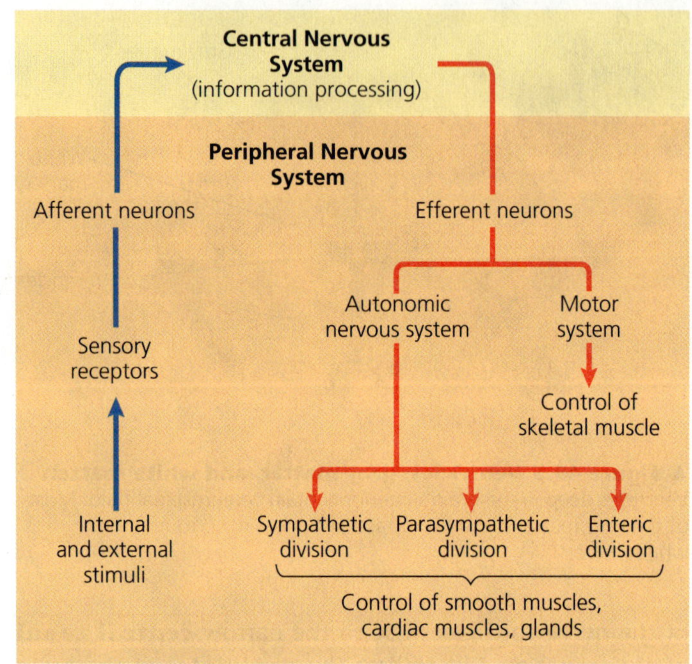

▲ **Figure 49.7 Functional hierarchy of the vertebrate peripheral nervous system.**

The PNS has two efferent components: the motor system and the autonomic nervous system (see Figure 49.7). The **motor system** consists of neurons that carry signals to skeletal muscles. This control of skeletal muscles can be voluntary, as when you raise your hand to ask a question, or involuntary, as in the knee-jerk reflex controlled by the spinal cord. In contrast, regulation of smooth and cardiac muscles by the **autonomic nervous system** is generally involuntary. The three divisions of the autonomic nervous system—sympathetic, parasympathetic, and enteric—together control the organs of the digestive, cardiovascular, excretory, and endocrine systems.

The sympathetic and parasympathetic divisions of the autonomic nervous system have largely antagonistic (opposite) functions in regulating organ function **(Figure 49.8)**. Activation of

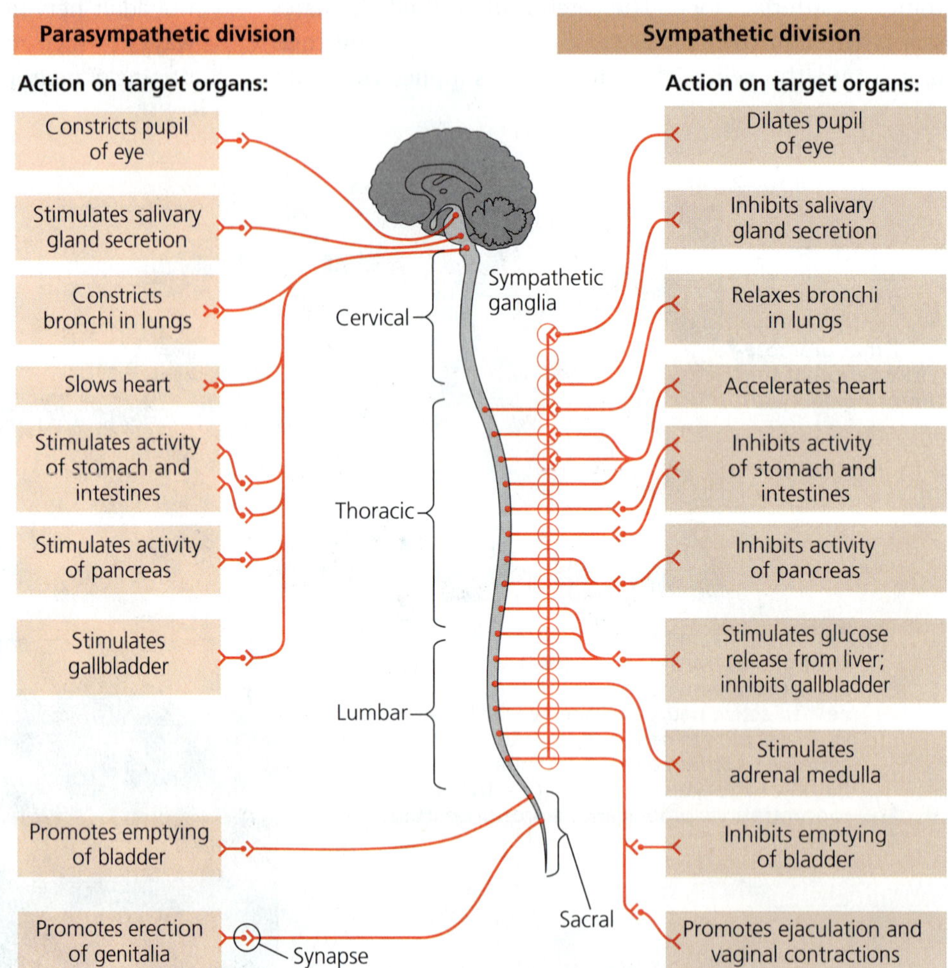

▲ **Figure 49.8 The parasympathetic and sympathetic divisions of the autonomic nervous system.** Most pathways in each division consist of preganglionic neurons (having cell bodies in the CNS) and postganglionic neurons (having cell bodies in ganglia in the PNS).

the **sympathetic division** corresponds to arousal and energy generation (the "fight-or-flight" response). For example, the heart beats faster, digestion is inhibited, the liver converts glycogen to glucose, and the adrenal medulla increases secretion of epinephrine (adrenaline) and norepinephrine. Activation of the **parasympathetic division** generally causes opposite responses that promote calming and a return to self-maintenance functions ("rest and digest"). Thus, heart rate decreases, digestion is enhanced, and glycogen production increases. In regulating reproductive activity, however, the parasympathetic division complements rather than antagonizes the sympathetic division (see Figure 49.8).

Networks of neurons that form the **enteric division** of the PNS are active in the digestive tract, pancreas, and gallbladder. Within these organs, the enteric division regulates secretion and peristalsis (see Chapter 41). The sympathetic and parasympathetic divisions normally regulate the enteric division, although it is capable of independent activity.

Homeostasis often relies on cooperation between the motor and autonomic nervous systems. In response to a drop in body temperature, for example, the hypothalamus signals the motor system to cause shivering, which increases heat production. At the same time, the hypothalamus signals the autonomic nervous system to constrict surface blood vessels, reducing heat loss.

CONCEPT CHECK 49.1

1. Which division of the autonomic nervous system would likely be activated if a student learned that an exam she had forgotten about would start in 5 minutes? Explain your answer.
2. The parasympathetic and sympathetic divisions of the PNS (see Figure 49.8) use the same neurotransmitters at the axon terminals of preganglionic neurons, but different neurotransmitters at the axon terminals of postganglionic neurons. How does this difference correlate with the function of the axons bringing signals into and out of the ganglia in the two divisions?
3. **WHAT IF?** Suppose a person had an accident that severed a small nerve required to move some of the fingers of the right hand. Would you also expect an effect on sensation from those fingers?
4. **MAKE CONNECTIONS** Most tissues regulated by the autonomic nervous system receive both sympathetic and parasympathetic input from postganglionic neurons. Responses are typically local. In contrast, the adrenal medulla receives input only from the sympathetic division and only from preganglionic neurons, yet responses are observed throughout the body. Explain why (see Figure 45.21, p. 991).

For suggested answers, see Appendix A.

CONCEPT 49.2

The vertebrate brain is regionally specialized

Having considered some of the basic functions of the PNS, we turn now to the brain. Images of the human brain in popular culture almost always focus on the cerebrum, the part of the brain whose surface lies just beneath the skull. The cerebrum is responsible for many activities we commonly associate with the brain, such as calculation, contemplation, and memory. Underneath the cerebrum, however, are additional brain structures with important and diverse activities, including homeostasis, coordination, and information transfer.

Figure 49.9, on pages 1068–1069, explores the origin, form, and function of major regions of the human brain. It outlines how brain structures arise during embryonic development, illustrates their size, shape, and location in the adult, and summarizes their best-understood functions. Figure 49.9 will serve as an introduction to the regional specialization in the brain and provide a useful point of reference for later discussions of specific brain functions.

To learn more about how brain organization relates to brain function, we'll first consider activity cycles of the brain and the physiological basis of emotion. Then, in Concept 49.3, we'll shift our attention to regional specialization within the cerebrum.

Arousal and Sleep

If you've ever drifted off to sleep while listening to a lecture (or reading a book), you know that your attentiveness and mental alertness can change rapidly. Such transitions are regulated by the brainstem and cerebrum, which control arousal and sleep. Arousal is a state of awareness of the external world. Sleep is a state in which external stimuli are received but not consciously perceived.

Contrary to appearances, sleep is an active state, at least for the brain. By placing electrodes at multiple sites on the scalp, we can record patterns of electrical activity called brain waves in an electroencephalogram (EEG). These recordings reveal that brain wave frequencies change as the brain progresses through distinct stages of sleep.

Although sleep is essential for survival, we still know very little about its function. One hypothesis is that sleep and dreams are involved in consolidating learning and memory. This hypothesis is supported by the finding that test subjects who are kept awake for 36 hours have a reduced ability to remember when particular events occurred, even if they first "perk up" with caffeine. Other experiments show that regions of the brain that are activated during a learning task can become active again during sleep.

▼ **Figure 49.9**

Exploring The Organization of the Human Brain

The brain is the most complex organ in the human body. Surrounded by the thick bones of the skull, the brain is divided into a set of distinctive structures, some of which are visible in the magnetic resonance image (MRI) of an adult's head shown at right. The diagram below traces the development of these structures in the embryo. Their major functions are explained on the facing page.

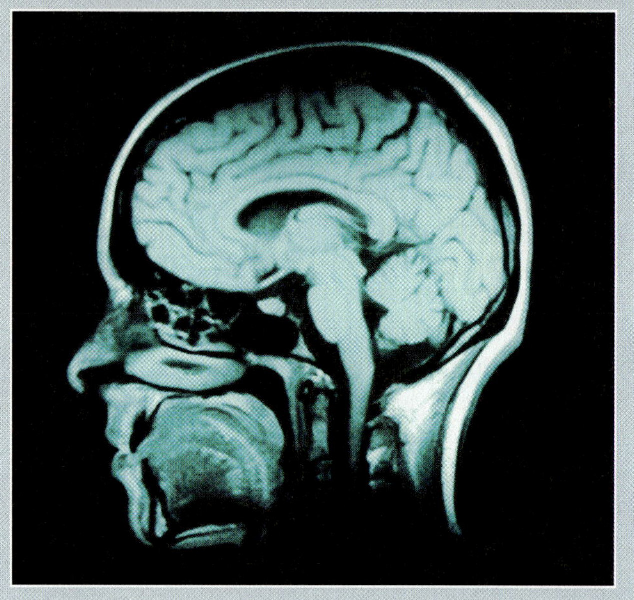

Human Brain Development

As a human embryo develops, the neural tube forms three anterior bulges—the **forebrain**, **midbrain**, and **hindbrain**—that together produce the adult brain. The midbrain and a part of the hindbrain give rise to the **brainstem**, a stalk that joins with the spinal cord at the base of the brain. The rest of the hindbrain gives rise to the **cerebellum**, which lies behind the brainstem. The third anterior bulge, the forebrain, develops into the diencephalon, including the neuroendocrine tissues of the brain, and the telencephalon, which becomes the **cerebrum**. Rapid, expansive growth of the telencephalon during the second and third months causes the outer portion, or cortex, of the cerebrum to extend over and around much of the rest of the brain.

Embryonic brain regions		Brain structures in child and adult
Forebrain	Telencephalon	Cerebrum (includes cerebral cortex, white matter, basal nuclei)
	Diencephalon	Diencephalon (thalamus, hypothalamus, epithalamus)
Midbrain	Mesencephalon	Midbrain (part of brainstem)
Hindbrain	Metencephalon	Pons (part of brainstem), cerebellum
	Myelencephalon	Medulla oblongata (part of brainstem)

Midbrain
Hindbrain
Forebrain

Embryo at 1 month

Mesencephalon
Metencephalon
Diencephalon
Myelencephalon
Spinal cord
Telencephalon

Embryo at 5 weeks

Cerebrum
Diencephalon
Midbrain
Pons
Medulla oblongata
Cerebellum
Spinal cord

Child

The Cerebrum

The cerebrum controls skeletal muscle contraction and is the center for learning, emotion, memory, and perception. It is divided into right and left **cerebral hemispheres**. The **cerebral cortex** is vital for perception, voluntary movement, and learning. Like the rest of the cerebrum, the cerebral cortex is divided into right and left sides. The left side receives information from, and controls the movement of, the right side of the body, and vice versa. A thick band of axons known as the **corpus callosum** enables the right and left cerebral cortices to communicate. Deep within the white matter, clusters of neurons called *basal nuclei* serve as centers for planning and learning movement sequences. Damage to these sites during fetal development can result in cerebral palsy, a disorder resulting from a disruption in the transmission of motor commands to the muscles.

The Cerebellum

The cerebellum coordinates movement and balance and helps in learning and remembering motor skills. The cerebellum receives sensory information about the positions of the joints and the lengths of the muscles, as well as input from the auditory (hearing) and visual systems. It also monitors motor commands issued by the cerebrum. The cerebellum integrates this information as it carries out coordination and error checking during motor and perceptual functions. Hand-eye coordination is an example of cerebellar control; if the cerebellum is damaged, the eyes can follow a moving object, but they will not stop at the same place as the object. Hand movement toward the object will also be erratic.

Adult brain viewed from the rear

The Diencephalon

The diencephalon gives rise to the thalamus, hypothalamus, and epithalamus. The **thalamus** is the main input center for sensory information going to the cerebrum. Incoming information from all the senses is sorted in the thalamus and sent to the appropriate cerebral centers for further processing. The thalamus is formed by two masses, each roughly the size and shape of a walnut. A much smaller structure, the **hypothalamus**, contains the body's thermostat as well as the central biological clock. Through its control of the pituitary gland, the hypothalamus regulates hunger and thirst, plays a role in sexual and mating behaviors, and controls the fight-or-flight response. The hypothalamus is also the source of posterior pituitary hormones and of releasing hormones that act on the anterior pituitary (see Figures 45.15 and 45.17). The *epithalamus* includes the pineal gland, the source of melatonin. It also contains one of several clusters of capillaries that generate cerebrospinal fluid from blood.

The Brainstem

The brainstem consists of the midbrain, the **pons**, and the **medulla oblongata** (commonly called the *medulla*). The midbrain receives and integrates several types of sensory information and sends it to specific regions of the forebrain. All sensory axons involved in hearing either terminate in the midbrain or pass through it on their way to the cerebrum. In addition, the midbrain coordinates visual reflexes, such as the peripheral vision reflex: The head turns toward an object approaching from the side without the brain having formed an image of the object.

A major function of the pons and medulla is to transfer information between the PNS and the midbrain and forebrain. The pons and medulla also help coordinate large-scale body movements, such as running and climbing. Most axons that carry instructions about these movements cross from one side of the CNS to the other in the medulla. As a result, the right side of the brain controls much of the movement of the left side of the body, and vice versa. An additional function of the medulla is the control of several automatic, homeostatic functions, including breathing, heart and blood vessel activity, swallowing, vomiting, and digestion. The pons also participates in some of these activities; for example, it regulates the breathing centers in the medulla.

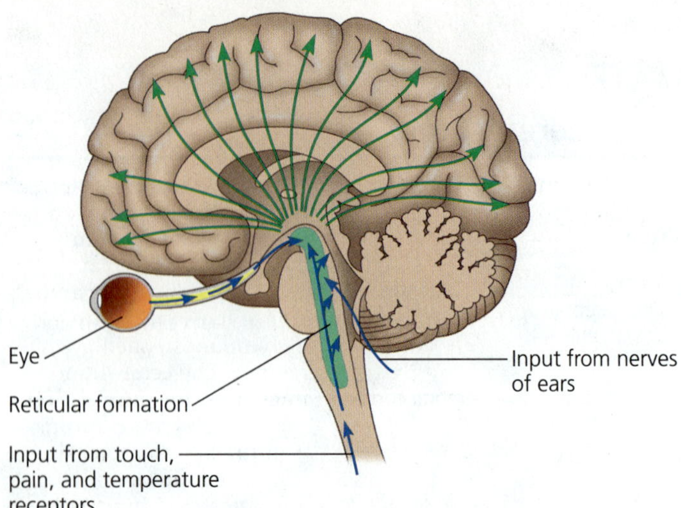

▲ **Figure 49.10 The reticular formation.** This system of neurons distributed throughout the core of the brainstem filters sensory input (blue arrows), blocking familiar and repetitive information that constantly enters the nervous system. It sends the filtered input to the cerebral cortex (green arrows).

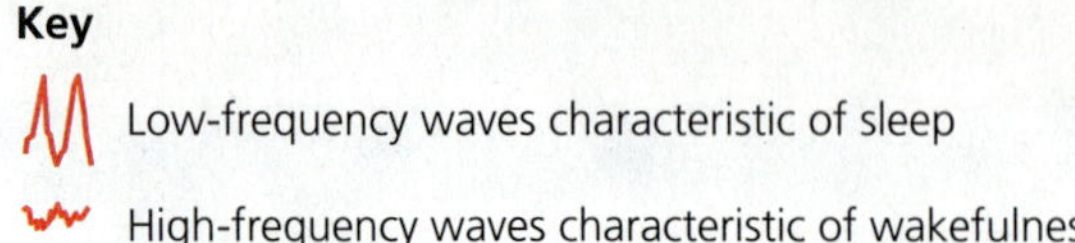

Location	Time: 0 hours	Time: 1 hour
Left hemisphere		
Right hemisphere		

▲ **Figure 49.11 Dolphins can be asleep and awake at the same time.** EEG recordings were made separately for the two sides of a dolphin's brain. Low-frequency activity was recorded in one hemisphere while higher-frequency activity typical of being awake was recorded in the other hemisphere.

Arousal and sleep are controlled in part by the **reticular formation**, a diffuse network of neurons in the core of the brainstem **(Figure 49.10)**. Acting as a sensory filter, the reticular formation determines which incoming information reaches the cerebrum. The more information the cerebrum receives, the more alert and aware a person is, although the brain often ignores certain stimuli while actively processing other inputs. Besides the diffuse reticular formation, there are also specific parts of the brainstem that regulate sleep and wakefulness: The pons and medulla contain centers that cause sleep when stimulated, and the midbrain has a center that causes arousal.

All birds and mammals show characteristic sleep/wake cycles. Melatonin, a hormone produced by the pineal gland, appears to play an important role in these cycles. As you read in Chapter 45, peak melatonin secretion occurs at night.

Some animals display evolutionary adaptations that allow for substantial activity during sleep. Bottlenose dolphins, for example, swim while sleeping, rising to the surface to breathe air on a regular basis. How do they manage this feat? A critical clue came from American physiologist John Lilly, who in 1964 observed that dolphins sleep with one eye open and one closed. As in humans and other mammals, the forebrain of dolphins is physically and functionally divided into two halves, the right and left hemispheres. Lilly suggested that a dolphin sleeping with one eye closed could mean that just one side of the brain was asleep. In 1977, Russian scientist Lev Mukhametov set out to test Lilly's hypothesis by collecting EEG recordings from each hemisphere of sleeping dolphins **(Figure 49.11)**. Mukhametov's findings demonstrate that dolphins do in fact sleep with one brain hemisphere at a time.

Biological Clock Regulation

Cycles of sleep and wakefulness are just one example of a circadian rhythm, a daily cycle of biological activity. Such cycles occur in organisms ranging from bacteria to fungi, plants, insects, birds, and humans. As in other organisms, circadian rhythms in mammals rely on a **biological clock**, a molecular mechanism that directs periodic gene expression and cellular activity. Although biological clocks are typically synchronized to the cycles of light and dark in the environment, they can maintain a roughly 24-hour cycle even in the absence of environmental cues (see Figure 40.9). For example, humans kept in a constant environment exhibit a cycle length of 24.2 hours, with very little variation among individuals.

What normally links an animal's biological clock to environmental cycles of light and dark? In mammals, circadian rhythms are coordinated by a group of neurons in the hypothalamus called the **suprachiasmatic nucleus**, or **SCN**. (Certain clusters of neurons in the CNS are referred to as "nuclei.") In response to transmission of sensory information by the eyes, the SCN acts as a pacemaker, synchronizing the biological clock in cells throughout the body to the natural cycles of day length. By surgically removing the SCN from laboratory animals and then observing their behavior, scientists demonstrated that the SCN is required for circadian rhythms: Animals without an SCN lack rhythmicity in behaviors and in electrical activity of the brain. These experiments did not, however, reveal whether rhythms originate in the SCN or elsewhere. In 1990, researchers answered this question with the aid of a mutation that changes the circadian rhythm of hamsters **(Figure 49.12)**. By transplanting brain tissue between normal and mutant hamsters, these scientists demonstrated that the SCN determines the circadian rhythm of the whole animal.

▼ **Figure 49.12** INQUIRY

Which cells control the circadian rhythm in mammals?

EXPERIMENT The τ (tau) mutation alters the period of the circadian rhythm in Syrian (golden) hamsters (*Mesocricetus auratus*). Whereas wild-type hamsters have a circadian cycle lasting 24 hours in the absence of external cues, hamsters homozygous for the τ mutation have a cycle lasting only about 20 hours. To determine if the SCN controls the circadian rhythm, Michael Menaker and colleagues at the University of Virginia surgically removed the SCN from wild-type and τ hamsters. Several weeks later, each of these hamsters received a transplant of an SCN from a hamster of the opposite genotype. The researchers then measured the circadian cycle period of the transplant recipients.

RESULTS In 80% of the hamsters in which the SCN had been removed, transplanting an SCN from another hamster restored rhythmic activity. For hamsters in which an SCN transplant restored a circadian rhythm, the net effect of the two procedures (SCN destruction and replacement) on circadian rhythm is graphed below. Each of the eight lines represents the change in the observed circadian cycle period for an individual hamster.

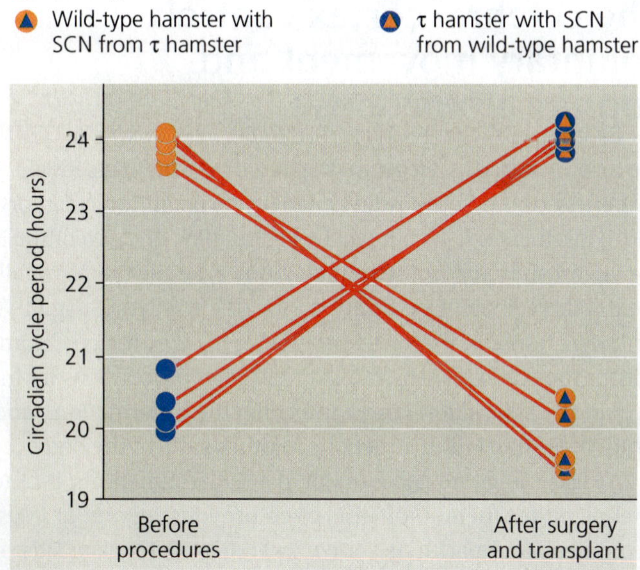

CONCLUSION Because the circadian rhythm of the animal that received the transplant was that of the donor animal, regardless of whether the recipient was wild-type or τ mutant, cells associated with the suprachiasmatic nucleus must determine the period of the circadian rhythm.

SOURCE M. R. Ralph, R. G. Foster, F. C. Davis, and M. Menaker, Transplanted suprachiasmatic nucleus determines circadian period, *Science* 247:975–978 (1990).

WHAT IF? Suppose in the course of your research you identified a hamster mutant that lacked rhythmic activity. How might you use this mutant in transplant experiments with wild-type or τ mutant hamsters to demonstrate that the mutation affected the pacemaker function of the SCN?

Emotions

Whereas a single structure in the brain controls the biological clock, the generation and experience of emotions depend on many brain structures, including the amygdala, hippocampus, and parts of the thalamus **(Figure 49.13)**. These structures border the brainstem in mammals and are therefore grouped as the *limbic system* (from the Latin *limbus*, border). The limbic system, however, is not dedicated solely to emotion. It also functions in motivation, olfaction (the sense of smell), behavior, and memory.

Generating emotion and experiencing emotion require parts of the brain in addition to the limbic system. For example, emotions that manifest themselves in behaviors such as laughing and crying involve an interaction of parts of the limbic system with sensory areas of the cerebrum. Structures in the forebrain also attach emotional "feelings" to basic, survival-related functions controlled by the brainstem, including aggression, feeding, and sexuality.

Emotional experiences are often stored as memories that can be recalled by similar circumstances. In the case of fear, emotional memory is stored separately from the memory system that supports explicit recall of events. The brain structure with the most important role in storage of emotional memory is the **amygdala**, an almond-shaped mass of nuclei (clusters of neurons) located near the base of the cerebrum.

To study the function of the human amygdala, researchers sometimes present adult subjects with an image followed by an unpleasant experience, such as a mild electrical shock. After several trials, study participants experience *autonomic arousal*—as measured by increased heart rate or sweating—if they see the image again. Subjects with brain damage confined to the amygdala can recall the image because their explicit memory is intact. However, they do not exhibit autonomic arousal, indicating that damage to the amygdala has resulted in a reduced capacity for emotional memory.

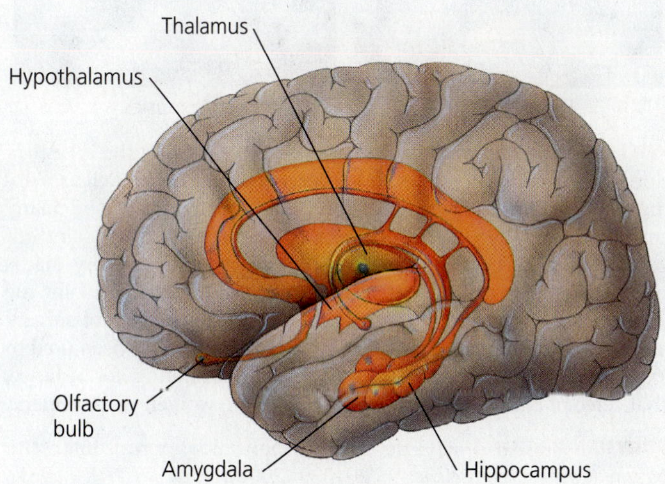

▲ **Figure 49.13 The limbic system.**

Today, the amygdala and other brain structures are being studied with functional imaging methods that are transforming our understanding of the normal and diseased brain **(Figure 49.14)**.

▼ Figure 49.14

IMPACT

Using Functional Brain Imaging to Map Activity in the Working Brain

Techniques for mapping brain activity have transformed the study of human brain function. The first widely used technique was positron-emission tomography (PET; see Figure 2.7). After injecting radioactive glucose into the blood of a subject, researchers can use PET scans to monitor metabolic activity across the brain. Further progress has come with functional magnetic resonance imaging (fMRI). In fMRI, the subject lies with his or her head in the center of a large, doughnut-shaped magnet. When the brain is scanned with electromagnetic waves, changes in blood oxygen concentration in active parts of the brain generate a signal that can be recorded.

Functional brain imaging has been applied to the study of human cognition, consciousness, and emotion. For example, functional imaging suggests that consciousness may be an emergent property of the brain based on activity in many areas of the cortex. In the experiment shown here, researchers explored differences in brain activity associated with music that listeners described as happy or sad. Listening to happy music activated the *nucleus accumbens*, a brain structure important for the perception of pleasure. In contrast, subjects who heard sad music had increased activity in the amygdala, a brain structure that serves as a center for emotional memory.

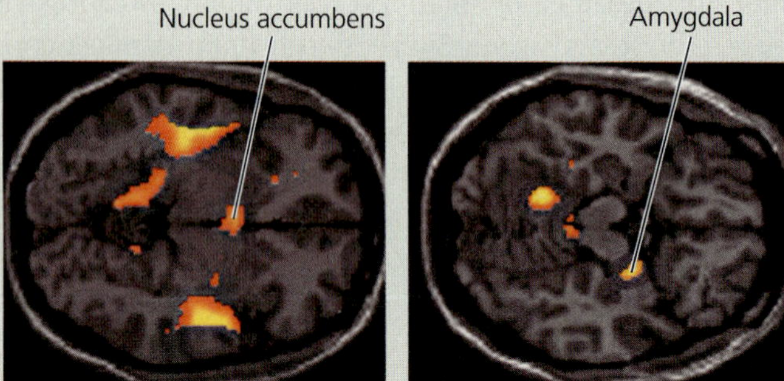

WHY IT MATTERS Functional brain imaging is aiding the investigation of recovery from stroke and other brain traumas, as well as helping map abnormalities in migraine headaches, dyslexia, and many psychiatric disorders. Functional imaging is also having a major impact on brain surgery. For example, for patients with epilepsy that is not responsive to drug therapy, functional imaging can pinpoint the region of abnormal function, increasing the effectiveness of surgery and enhancing recovery. Finally, functional imaging has been used to explore sex-based differences in the CNS, demonstrating, for instance, that cerebral blood flow is higher on average in women than in men.

FURTHER READING R. C. deCharms, Applications of real-time fMRI, *Nature Reviews Neuroscience* 9:720–729 (2008).

WHAT IF? In the experiment illustrated above, some regions of the brain showed activity under all conditions. What function might such regions carry out?

CONCEPT CHECK 49.2

1. When you wave your right hand, what part of your brain initiates the action?
2. When a police officer stops a driver for driving erratically and suspects that the person is intoxicated, the officer may ask the driver to close his or her eyes and touch his or her nose. What can you deduce from this test about one of the brain regions affected by alcohol?
3. **WHAT IF?** Suppose you examine individuals with damage to the CNS that has resulted in either coma (a prolonged state of unconsciousness) or general paralysis (a loss of skeletal muscle function throughout the body). Relative to the position of the reticular formation, where would you predict the site of injury to lie in each group of patients? Explain.

For suggested answers, see Appendix A.

CONCEPT **49.3**

The cerebral cortex controls voluntary movement and cognitive functions

We turn now to the cerebrum, the part of the brain essential for awareness of our surroundings, language, cognition, memory, and consciousness. As shown in Figure 49.9, the cerebrum is the largest structure in the human brain. Like the brain overall, it exhibits regional specialization. For the most part, cognitive functions reside in the cortex, the outer layer of the cerebrum. Within the cortex, sensory areas receive and process sensory information, association areas integrate the information, and motor areas transmit instructions to other parts of the body.

In discussing the location of particular functions in the cerebral cortex, neurobiologists often use four regions, or *lobes*, as physical landmarks. As shown in **Figure 49.15**, each side of the cerebral cortex has a frontal, temporal, occipital, and parietal lobe (each is named for a nearby bone of the skull).

Language and Speech

The mapping of higher cognitive functions to specific brain areas began in the 1800s when physicians learned that damage to particular regions of the cortex by injuries, strokes, or tumors can produce distinctive changes in a person's behavior. The French physician Pierre Broca conducted postmortem (after death) examinations of patients who had been able to understand language but unable to speak. He discovered that many of these patients had defects in a small region of the left frontal lobe, now known as *Broca's area*, that controls muscles in the face. The German physician Karl Wernicke found that damage to a posterior portion of the left temporal lobe, now

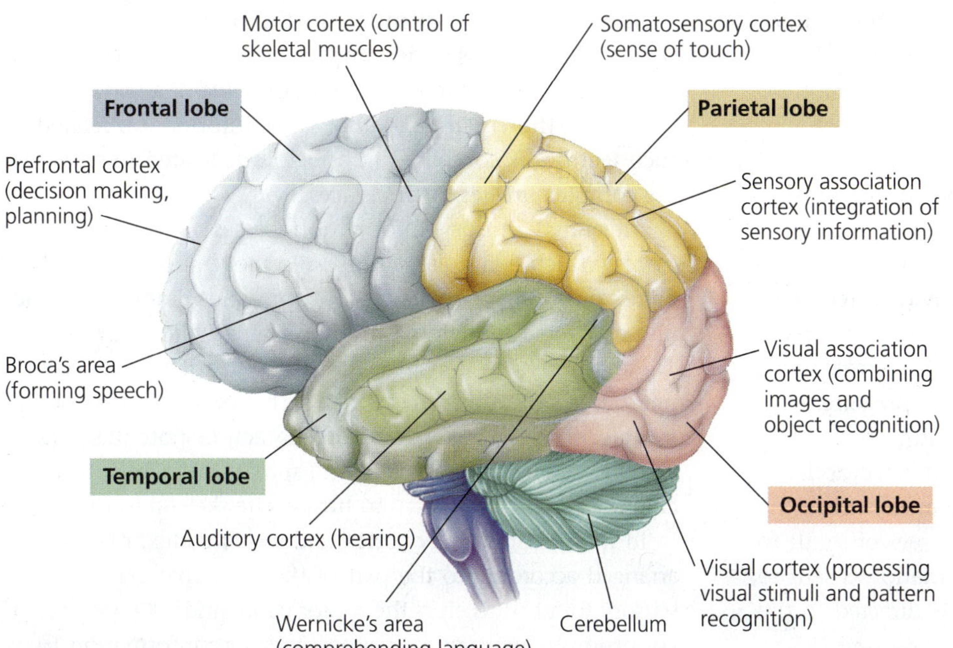

◀ **Figure 49.15 The human cerebral cortex.** Each side of the cerebral cortex is divided into four lobes, and each lobe has specialized functions, some of which are listed here. Some areas on the left side of the brain (shown here) have different functions from those on the right side (not shown).

called *Wernicke's area*, abolished the ability to comprehend speech but not the ability to speak.

More than a century after the discoveries of Broca and Wernicke, functional imaging studies confirmed that Broca's area is active during speech generation (**Figure 49.16**, lower left image) and Wernicke's area is active when speech is heard (Figure 49.16, upper left image). In addition, researchers have found that these areas belong to a much larger network of brain regions involved in language. Reading a printed word without speaking activates the visual cortex (Figure 49.16, upper right image), whereas reading a printed word out loud activates both the visual cortex and Broca's area. Frontal and temporal areas become active when meaning must be attached to words, such as when a person generates verbs to go with nouns or groups related words or concepts (Figure 49.16, lower right image).

Max
Min
Hearing words
Seeing words
Speaking words
Generating words

▲ **Figure 49.16 Mapping language areas in the cerebral cortex.** These PET images show regions with different activity levels in one person's brain during four activities, all related to speech.

Lateralization of Cortical Function

Both Broca's area and Wernicke's area reside in the left cortical hemisphere, reflecting a significantly greater role with regard to language for the left side of the cerebrum than for the right side. The two hemispheres also make distinct contributions to some other brain functions, although to a lesser degree than for language. For example, the left hemisphere is more adept at math and logical operations. In contrast, the right hemisphere appears to be dominant in the recognition of faces and patterns, spatial relations, and nonverbal thinking. The establishment of these differences in hemisphere function in humans is called **lateralization**.

At least some lateralization relates to handedness, the preference for using one hand for certain motor activities. Across human populations, roughly 90% of individuals are more skilled with their right hand than with their left hand. Studies using fMRI have revealed how language processing differs in relation to handedness. When subjects thought of words without speaking out loud, brain activity was localized to the left hemisphere in 96% of right-handed subjects but in only 76% of left-handed subjects.

The two hemispheres normally work together harmoniously, trading information back and forth through the fibers of the corpus callosum. The importance of this exchange is revealed in patients whose corpus callosum has been surgically severed (a treatment of last resort for the most extreme forms of epilepsy, a seizure disorder). Individuals with a severed corpus callosum exhibit a "split-brain" effect. When a familiar word appears in their left field of vision, they cannot read the word: The sensory information that travels from the left field of vision to the right hemisphere cannot reach the language centers in the left hemisphere. Each hemisphere in such patients functions independently of the other.

Information Processing

As you will learn further in Chapter 50, some of the sensory input to the cerebral cortex comes from groups of receptors clustered in dedicated sensory organs, such as the eyes and nose. Other sensory input originates in individual receptors in the hands, scalp, and elsewhere in the body. These somatic sensory, or *somatosensory*, receptors (from the Greek *soma*, body) provide information about touch, pain, pressure, temperature, and the position of muscles and limbs.

Most sensory information coming into the cortex is directed via the thalamus to primary sensory areas within the brain lobes. The thalamus directs different types of input to distinct locations. For example, visual information is sent to the occipital lobe, whereas auditory input is directed to the temporal lobe (see Figure 49.15).

Information received at the primary sensory areas is passed along to nearby association areas, which process particular features in the sensory input. In the occipital lobe, for instance, some groups of neurons in the primary visual area are specifically sensitive to rays of light oriented in a particular direction. In the visual association area, information related to such features is combined in a region dedicated to recognizing complex images, such as faces.

Integrated sensory information passes to the prefrontal cortex, which helps plan actions and movement. The cerebral cortex may then generate motor commands that cause particular behaviors—moving a limb or saying hello, for example. These commands consist of action potentials produced by neurons in the motor cortex, which lies at the rear of the frontal lobe (see Figure 49.15). The action potentials travel along axons to the brainstem and spinal cord, where they excite motor neurons, which in turn excite skeletal muscle cells.

In the somatosensory cortex and motor cortex, neurons are arranged according to the part of the body that generates the sensory input or receives the motor commands **(Figure 49.17)**. For example, neurons that process sensory information from the legs and feet lie in the region of the somatosensory cortex closest to the midline. Neurons that control muscles in the

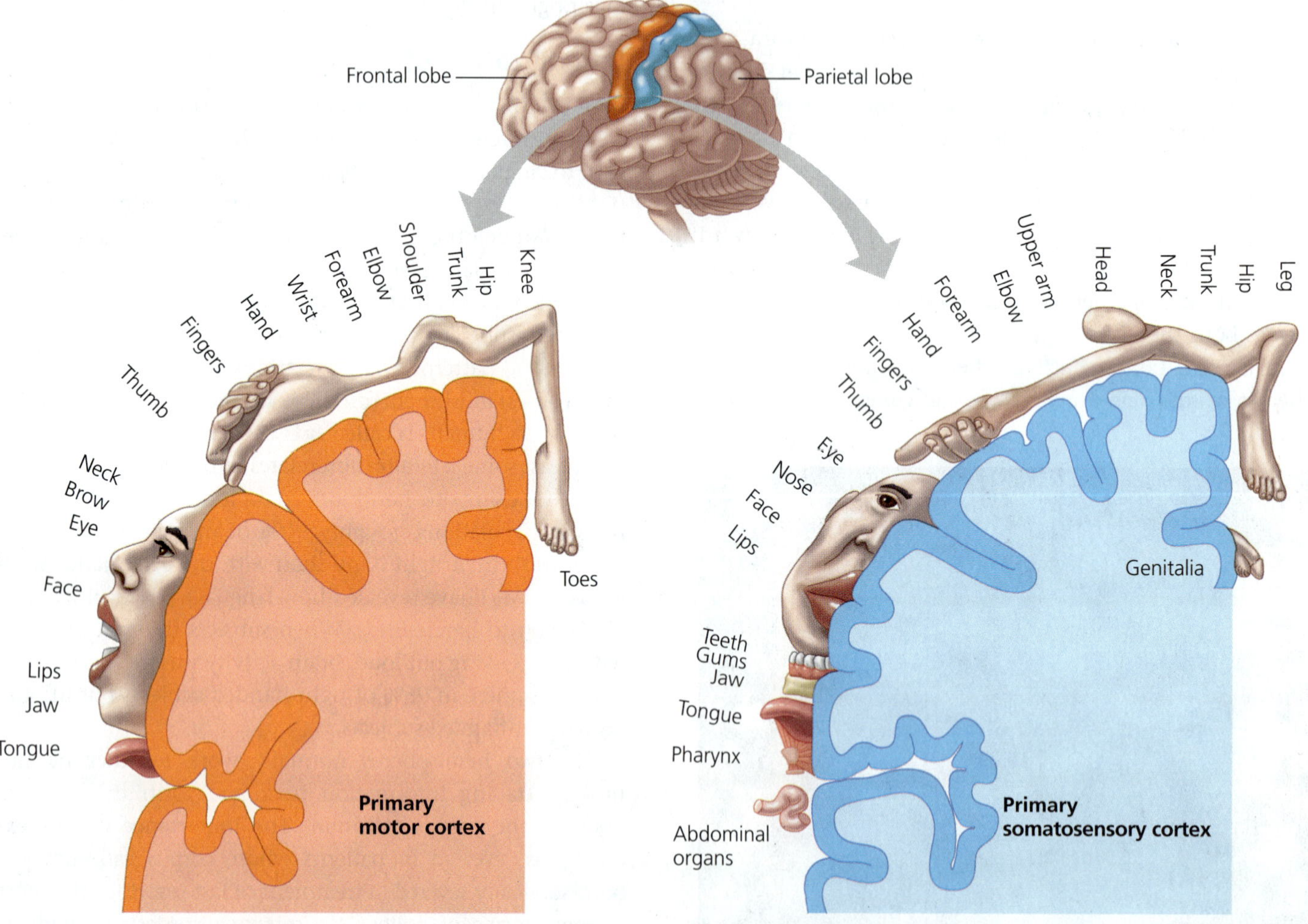

▲ **Figure 49.17 Body part representation in the primary motor and primary somatosensory cortices.** In these cross-sectional maps of the cortices, the cortical surface area devoted to each body part is represented by the relative size of that part in the cartoons.

legs and feet are located in the corresponding region of the motor cortex. Notice in Figure 49.17 that the cortical surface area devoted to each body part is not proportional to the size of the part. Instead, surface area correlates with the extent of neuronal control needed (for the motor cortex) or with the number of sensory neurons that extend axons to that part (for the somatosensory cortex). Thus, the surface area of the motor cortex devoted to the face is much larger than that devoted to the trunk, reflecting the extensive involvement of facial muscles in communication.

Frontal Lobe Function

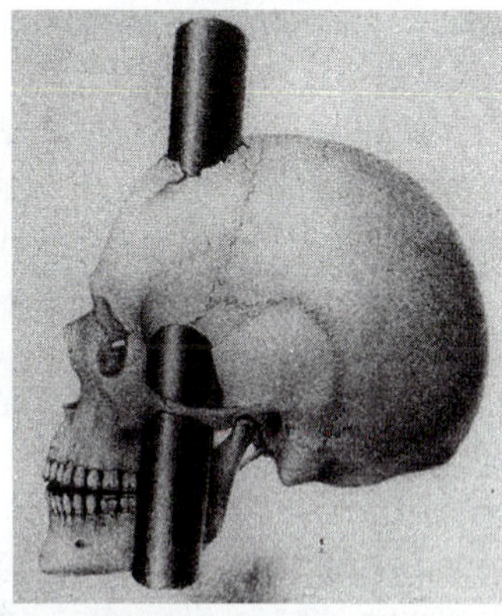

In 1848, a horrific workplace accident pointed to the role of the prefrontal cortex in temperament and decision making. Phineas Gage was working as the foreman of a railroad construction crew when an explosion drove a meter-long iron rod through his head. The rod, which was more than 3 cm in diameter at one end, entered his skull just below his left eye and exited through the top of his head, damaging large portions of his frontal lobe. Astonishingly, Gage recovered. His personality, however, changed dramatically. He became emotionally detached, impatient, and erratic in his behavior.

Although the connection between Gage's brain injury and his personality change is a subject of debate, some tumors that develop in the frontal lobe cause symptoms that are similar to those of Gage. Intellect and memory seem intact, but decision making is flawed and emotional responses are diminished. In the 20th century, the same problems were observed as a result of frontal lobotomy, a surgical procedure that severs the connection between the prefrontal cortex and the limbic system. Together, these observations provide evidence that the frontal lobes have a substantial influence on what are often called "executive functions."

Frontal lobotomy was once a common treatment for severe behavioral disorders but later was abandoned as a medical practice. Behavioral disorders are now typically treated with medications, as discussed later in this chapter.

Evolution of Cognition in Vertebrates

EVOLUTION In humans, the cerebral cortex accounts for about 80% of total brain mass and is highly convoluted (see Figure 49.9). The convolutions allow the cerebral cortex to have a large surface area and still fit inside the skull: Less than 5 mm thick, it has a surface area of approximately 1,000 cm^2. The outermost part of the human cerebral cortex forms the *neocortex*, six parallel layers of neurons arranged tangential to the brain surface.

It was long thought that a highly convoluted neocortex was required for advanced *cognition*, the perception and reasoning that constitute knowledge. Primates and cetaceans (whales, dolphins, and porpoises) possess an extensively convoluted neocortex. However, birds lack such a structure and were thought to have much lower intellectual capacity. Experiments in recent years have refuted this idea. Western scrub jays (*Aphelocoma californica*) can remember the relative period of time that has passed after they stored and hid specific food items. New Caledonian crows (*Corvus moneduloides*) are highly skilled at making and using tools, an ability otherwise well documented only for humans and some other apes. African gray parrots (*Psittacus erithacus*) understand numerical and abstract concepts, distinguishing between "same" and "different" and grasping the idea of "none."

The anatomical basis for sophisticated information processing in birds appears to be the grouping of nuclei within the *pallium*, the top or outer portion of the brain **(Figure 49.18)**. This arrangement is different from that in the human pallium—the cerebral cortex—which contains flat sheets of cells in six layers. Thus, there are two types of pallium, each of which supports complex and flexible brain function.

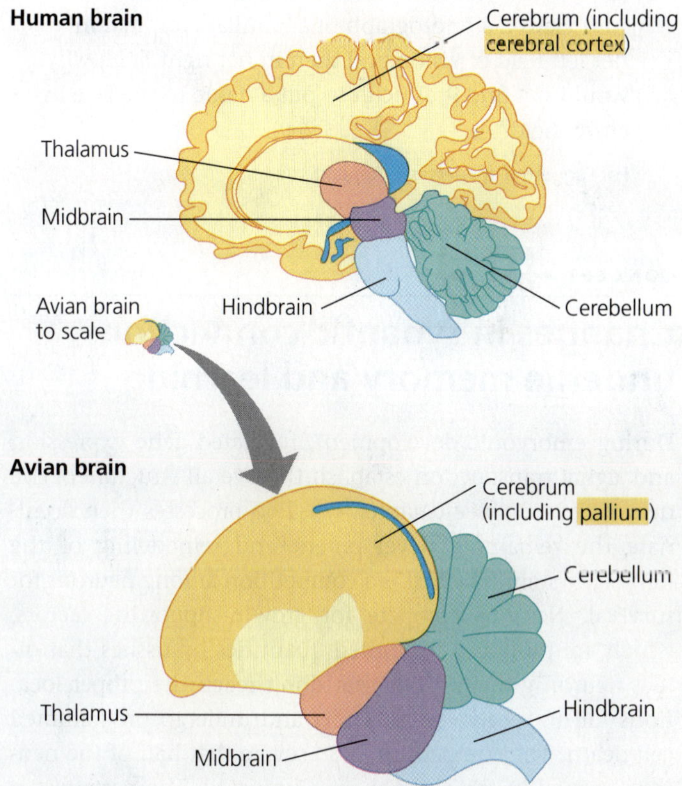

▲ **Figure 49.18 Comparison of regions for higher cognition in avian and human brains.** Although structurally different, the cerebral cortex of the human brain (top cross section) and the pallium of a songbird brain (bottom cross section) have similar roles in higher cognitive activities and make many similar connections with other brain structures.

How did the differences between the bird pallium and human pallium arise during evolution? The current consensus is that the common ancestor of birds and mammals had a pallium in which neurons were organized into nuclei, as is still found in birds. Early in mammalian evolution, this nuclear (clustered) organization of neurons was transformed into a layered one. Connectivity was maintained during this transformation such that, for example, the thalamus relays sensory input relating to sights, sounds, and touch to the pallium in both birds and mammals.

Sophisticated information processing depends not only on the overall organization of a brain but also on the very small-scale changes that enable learning and encode memory. We'll turn to these changes in the context of humans in the next section.

CONCEPT CHECK 49.3

1. How can studying individuals with damage to a particular brain region provide insight into the normal function of that region?
2. Two brain areas important in the generation or perception of speech are Broca's area and Wernicke's area. How is the function of each area related to the activity of the surrounding portion of the cerebral cortex?
3. **WHAT IF?** If a woman with a severed corpus callosum viewed a photograph of a familiar face, first in her left field of vision and then in her right field, why would she find it difficult to put a name to the face in either field?

For suggested answers, see Appendix A.

CONCEPT **49.4**

Changes in synaptic connections underlie memory and learning

During embryonic development, regulated gene expression and signal transduction establish the overall structure of the nervous system (see Chapter 47). Two processes then dominate the remaining development and remodeling of the nervous system. The first is a competition among neurons for survival. Neurons compete for growth-supporting factors, which are produced in limited quantities by tissues that direct neuron growth. Cells that don't reach the proper locations fail to receive such factors and undergo programmed cell death. The competition is so severe that half of the neurons formed in the embryo are eliminated. The net effect is the preferential survival of neurons that are located properly within the nervous system.

Synapse elimination is the second major process that shapes the nervous system. A developing neuron forms numerous synapses, more than are required for its proper function. The activity of that neuron then stabilizes some synapses and destabilizes others. By the end of embryonic development, neurons on average have lost more than half of their initial synapses, leaving behind the connections that survive into adulthood.

Together, neuron death and synapse elimination set up the basic network of cells and connections within the nervous system required throughout life.

Neural Plasticity

Although the overall organization of the CNS is established during embryonic development, it can change after birth. This capacity for the nervous system to be remodeled, especially in response to its own activity, is called **neural plasticity**.

Much of the reshaping of the nervous system occurs at synapses. When the activity of a synapse correlates with that of other synapses, changes may occur that reinforce that synaptic connection. Conversely, when the activity of a synapse fails to correlate with that of other synapses, the synaptic connection sometimes becomes weaker. In this way, synapses belonging to circuits that link information in useful ways are maintained, whereas those that convey bits of information lacking any context may be lost.

Figure 49.19a illustrates how these processes can result in either the addition or loss of a synapse. If you think of signals

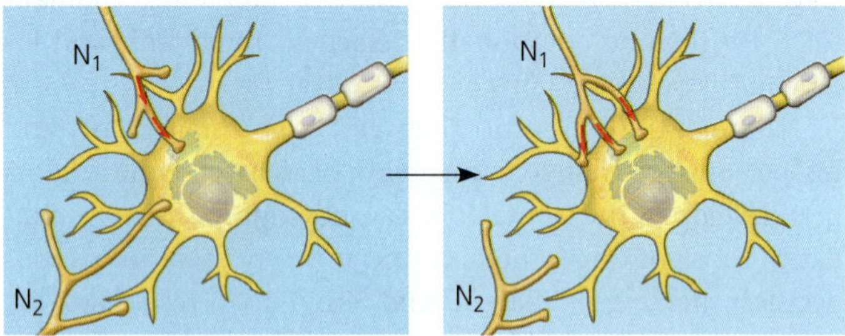

(a) Connections between neurons are strengthened or weakened in response to activity. High-level activity at the synapse of the post-synaptic neuron with presynaptic neuron N_1 leads to recruitment of additional axon terminals from that neuron. Lack of activity at the synapse with presynaptic neuron N_2 leads to loss of functional connections with that neuron.

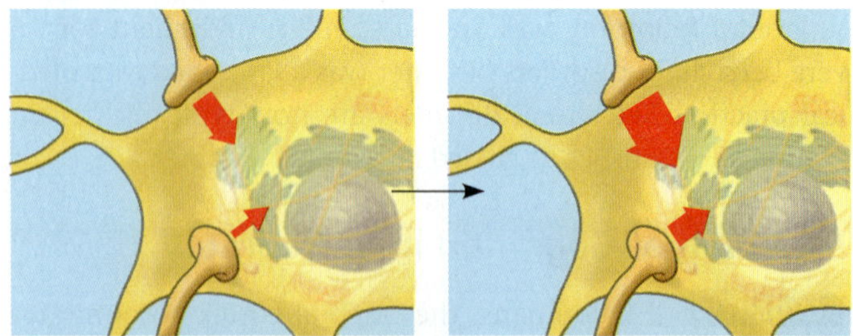

(b) If two synapses on the same postsynaptic cell are often active at the same time, the strength of the postsynaptic response may increase at both synapses.

▲ **Figure 49.19 Neural plasticity.** Synaptic connections can change over time, depending on the activity level at the synapse.

in the nervous system as traffic on a highway, such changes are comparable to adding or removing an entrance ramp. The net effect is to increase signaling between particular pairs of neurons and decrease signaling between other pairs. As shown in **Figure 49.19b**, changes can also strengthen or weaken signaling at a synapse. In our traffic analogy, this would be equivalent to widening or narrowing an entrance ramp.

Research indicates that *autism*, a developmental disorder that first appears early in childhood, involves a disruption of activity-dependent remodeling at synapses. Children affected with autism display impaired communication and social interaction, as well as stereotyped and repetitive behaviors.

Although the underlying causes of autism are unknown, there is a strong genetic contribution to this and related disorders. Extensive research has ruled out a link to vaccine preservatives, once proposed as a potential risk factor. Further understanding of the autism-associated disruption in synaptic plasticity may help efforts to better understand and treat this disorder.

Remodeling and refining of the nervous system occur in many contexts. For instance, soon after birth, the visual cortex of the mammalian brain undergoes reorganization in response to input from the optic nerve triggered by visual stimuli. Experiments have shown that this remodeling is a necessary step in the development of normal visual ability.

Remodeling of functional brain circuitry also occurs in diseases and injuries to the nervous system from which significant recovery is possible. One example is the treatment for a condition called phantom limb syndrome, in which a person feels pain or discomfort that seems to originate from an arm or leg that has been amputated. Having the patient view a reflection of the remaining limb in a mirrored box can reorganize the brain's neural connections in a way that eliminates the unpleasant feelings from the lost limb.

Memory and Learning

The formation of memories is another example of neural plasticity. Though we may not be aware of it, we are constantly checking what is happening against what just happened a few moments ago. We hold information for a time in **short-term memory** locations and then release it if it becomes irrelevant. If we wish to retain knowledge of a name, phone number, or other fact, the mechanisms of **long-term memory** are activated. If we later need to recall the name or number, we fetch it from long-term memory and return it to short-term memory.

Scientists have long wondered where in the brain short-term and long-term memories are located. We now know that both types of memory involve the storage of information in the cerebral cortex. In short-term memory, this information is accessed via temporary links formed in the hippocampus. When memories are made long-term, the links in the hippocampus are replaced by more permanent connections within the cerebral cortex itself. Some of this consolidation of memory is thought to occur during sleep. Furthermore, the reactivation of the hippocampus that is required for memory consolidation likely forms the basis for at least some of our dreams.

According to our current understanding of memory, the hippocampus is essential for acquiring new long-term memories but not for maintaining them. This hypothesis readily explains the symptoms of some individuals who suffer damage to the hippocampus: They cannot form any new lasting memories but can freely recall events from before their injury. In effect, their lack of normal hippocampal function traps them in their past.

What evolutionary advantage might be offered by organizing short-term and long-term memories differently? Current thinking is that the delay in forming connections in the cerebral cortex allows long-term memories to be integrated gradually into the existing store of knowledge and experience, providing a basis for more meaningful associations. Consistent with this idea, the transfer of information from short-term to long-term memory is enhanced by the association of new data with data previously learned and stored in long-term memory. For example, it's easier to learn a new card game if you already have "card sense" from playing other card games.

Motor skills, such as walking, tying your shoes, or writing, are usually learned by repetition. You can perform these skills without consciously recalling the individual steps required to do these tasks correctly. Learning skills and procedures, such as those required to ride a bicycle, appears to involve cellular mechanisms very similar to those responsible for brain growth and development. In such cases, neurons actually make new connections. In contrast, memorizing phone numbers, facts, and places—which can be very rapid and may require only one exposure to the relevant item—may rely mainly on changes in the strength of existing neuronal connections. Next we will consider one way that such changes in strength can take place.

Long-Term Potentiation

In searching for the physiological basis of memory, researchers have concentrated their attention on processes that can alter a synaptic connection, making the flow of communication either more efficient or less efficient. We will focus here on **long-term potentiation (LTP)**, a lasting increase in the strength of synaptic transmission.

First characterized in tissue slices from the hippocampus, LTP involves a presynaptic neuron that releases the excitatory neurotransmitter glutamate. For LTP to occur, there must be a high-frequency series of action potentials in this presynaptic neuron. In addition, these action potentials must arrive at the synaptic terminal at the same time that the postsynaptic cell receives a depolarizing stimulus at another synapse.

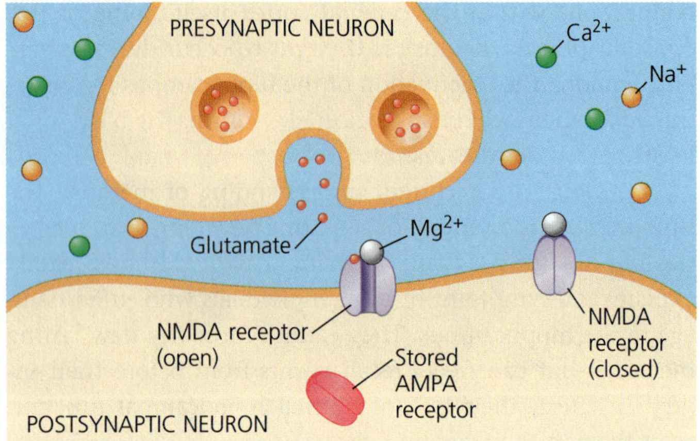

(a) Synapse prior to long-term potentiation (LTP). The NMDA glutamate receptors open in response to glutamate but are blocked by Mg^{2+}.

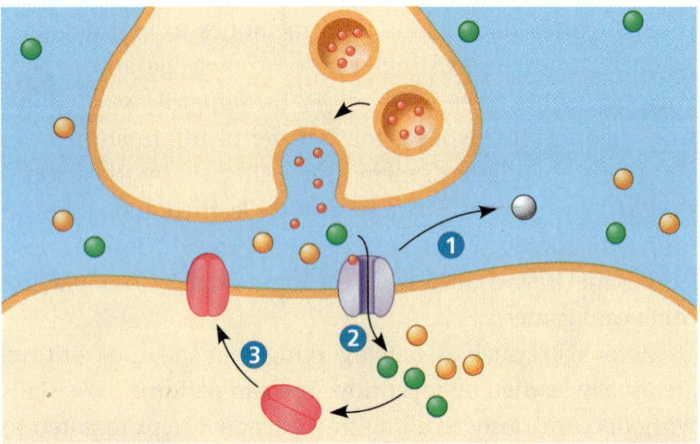

(b) Establishing LTP. Activity at nearby synapses (not shown) depolarizes the postsynaptic membrane, causing 1 Mg^{2+} release from NMDA receptors. The unblocked receptors respond to glutamate by allowing 2 an influx of Na^+ and Ca^{2+}. The Ca^{2+} influx triggers 3 insertion of stored AMPA glutamate receptors into the postsynaptic membrane.

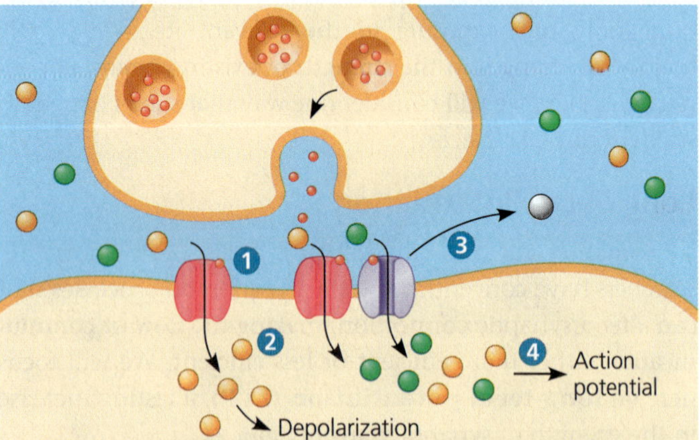

(c) Synapse exhibiting LTP. Glutamate release activates 1 AMPA receptors that trigger 2 depolarization. The depolarization unblocks 3 NMDA receptors. Together, the AMPA and NMDA receptors trigger postsynaptic potentials strong enough to initiate 4 action potentials without input from other synapses. Additional mechanisms (not shown) contribute to LTP, including receptor modification by protein kinases.

▲ **Figure 49.20 Long-term potentiation in the brain.**

LTP involves two types of glutamate receptors, each named for a molecule—NMDA or AMPA—that artificially activates that particular receptor. As shown in **Figure 49.20**, the set of receptors present on the postsynaptic membrane changes in response to an active synapse and a depolarizing stimulus. The result is LTP—a stable increase in the size of the postsynaptic potentials at the synapse. Because LTP can last for days or weeks in dissected tissue, it is thought to represent one of the fundamental processes by which memories are stored and learning takes place.

Stem Cells in the Brain

In 1998, Fred Gage, at the Salk Institute in California, and Peter Ericksson, at the Sahlgrenska University Hospital in Sweden, discovered that the adult human brain contains neural stem cells. Recall from Chapters 20 and 46 that stem cells retain the ability to divide indefinitely. While some of their progeny remain undifferentiated, others differentiate into specialized cells. Studies with mice reveal that stem cells in the brain give rise to neurons that mature, migrate to particular locations in the hippocampus, and become incorporated into the circuitry of the adult nervous system **(Figure 49.21)**. Evidence from other studies indicates that such neurons play an essential role in learning and memory. In this manner, adult neural stem cells contribute to the plasticity that enables remodeling of brain circuitry in response to experience.

Researchers are now tackling the challenge of finding a way to use neural stem cells as a means of replacing brain tissue

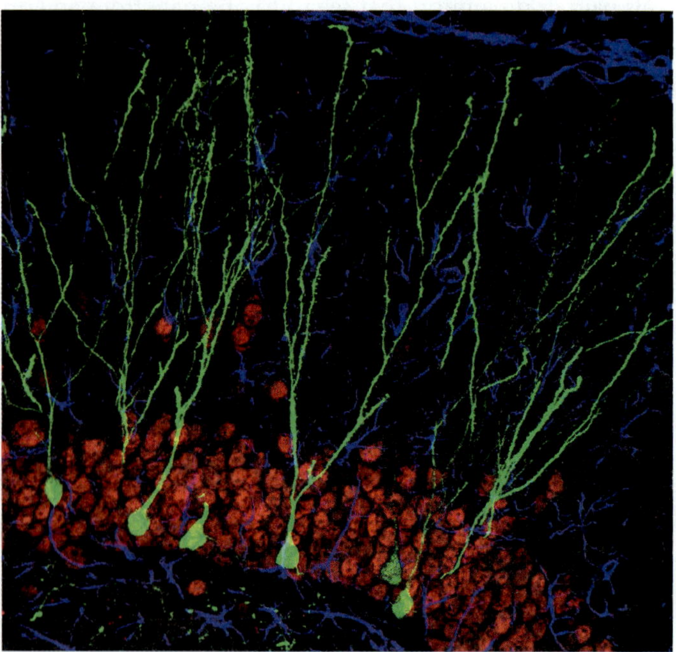

▲ **Figure 49.21 Newly born neurons in the hippocampus of an adult mouse.** In this light micrograph, new neurons derived from adult stem cells are labeled with green fluorescent protein (GFP), and all neurons are labeled with a red dye that binds DNA.

that has ceased to function properly. Unlike the PNS, the mammalian CNS cannot fully repair itself when damaged or diseased. Surviving neurons in the brain can make new connections and sometimes compensate for damage, as occurs in the remarkable recoveries of some stroke victims. Generally, however, brain and spinal cord injuries, strokes, and disorders that destroy CNS neurons, such as Alzheimer's disease and Parkinson's disease, have devastating and irreversible effects.

Although stem cell therapy for the brain is likely to be a long way off, the recent discovery that expression of just four particular genes converts differentiated adult cells to stem cells (see Chapter 20) represents significant progress in this endeavor.

CONCEPT CHECK 49.4

1. Outline two mechanisms by which the flow of information between two neurons in adults can increase.
2. Individuals with localized brain damage have been very useful in the study of many brain functions. Why is this unlikely to be true for consciousness?
3. **WHAT IF?** Suppose that a person with damage to the hippocampus is unable to acquire new long-term memories. Why might the acquisition of short-term memories also be impaired?

For suggested answers, see Appendix A.

CONCEPT 49.5

Many nervous system disorders can be explained in molecular terms

Disorders of the nervous system, including schizophrenia, depression, drug addiction, Alzheimer's disease, and Parkinson's disease, are a major public health problem. Together, they result in more hospitalizations in the United States than do heart disease or cancer. Until recently, hospitalization was typically the only available treatment, and many affected individuals were institutionalized for the rest of their lives. Today, many disorders that alter mood or behavior can be treated with medication, reducing average hospital stays for these disorders to only a few weeks. At the same time, societal attitudes are changing as awareness grows that nervous system disorders often result from chemical or anatomical changes in the brain. Many challenges remain, however, especially for Alzheimer's and other diseases that lead to nervous system degeneration.

Major research efforts are under way to identify genes that cause or contribute to disorders of the nervous system. Identifying such genes offers hope for identifying causes, predicting outcomes, and developing effective treatments. For most nervous system disorders, however, genetic contributions only partially account for which individuals are affected. The other significant contribution to disease comes from environmental factors. Unfortunately, environmental contributions are typically very difficult to identify.

To distinguish between genetic and environmental variables, scientists often carry out family studies. In such studies, researchers track how family members are related genetically, which individuals are affected, and which family members grew up in the same household. These studies are especially informative when one of the affected individuals has either an identical twin or an adopted sibling who is genetically unrelated. The results of family studies indicate that certain nervous system disorders, such as schizophrenia, have a very strong genetic component (**Figure 49.22**).

Schizophrenia

About 1% of the world's population suffer from **schizophrenia**, a severe mental disturbance characterized by psychotic episodes in which patients have a distorted perception of reality. People with schizophrenia typically experience hallucinations (such as "voices" that only they can hear) and delusions (for example, the idea that others are plotting to harm them). Despite the commonly held notion,

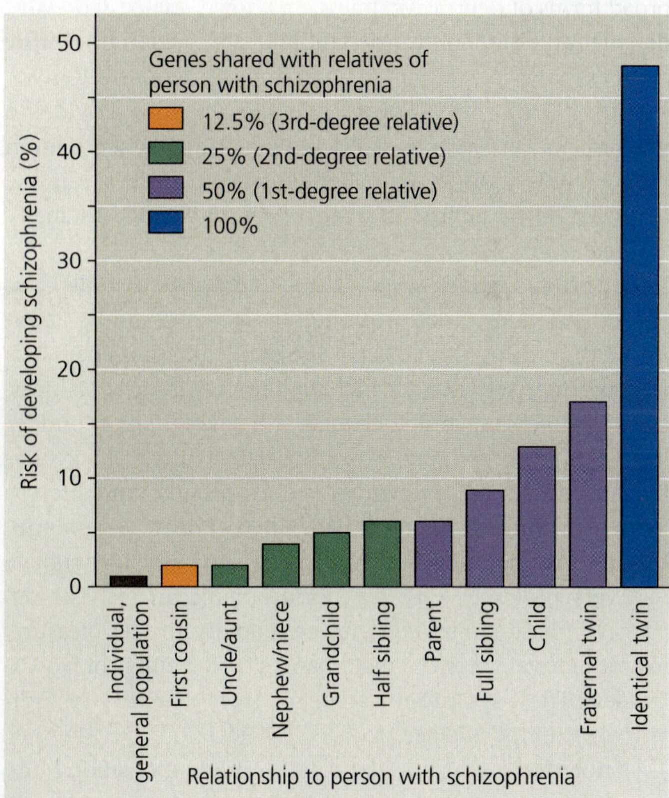

▲ **Figure 49.22 Genetic contribution to schizophrenia.** First cousins, uncles, and aunts of a person with schizophrenia have twice the risk of unrelated members of the population of developing the disease. The risks for closer relatives are many times greater.

schizophrenia does not necessarily result in multiple personalities. Rather, the name *schizophrenia* (from the Greek *schizo*, split, and *phren*, mind) refers to the fragmentation of what are normally integrated brain functions.

Two lines of evidence suggest that schizophrenia affects neuronal pathways that use dopamine as a neurotransmitter. First, the drug amphetamine ("speed"), which stimulates dopamine release, can produce the same set of symptoms as schizophrenia. Second, many of the drugs that alleviate the symptoms of schizophrenia block dopamine receptors. Schizophrenia may also alter glutamate signaling, since the street drug "angel dust," or PCP, blocks glutamate receptors and induces strong schizophrenia-like symptoms.

Fortunately, medications frequently can alleviate the major symptoms of schizophrenia. Although the first treatments developed often had substantial negative side effects, newer medications are equally effective and much safer to use. Ongoing research aimed at identifying the genetic mutations that contribute to schizophrenia may yield new insights about the causes of the disease and lead to even more effective therapies.

Depression

Depression is a disorder characterized by depressed mood, as well as abnormalities in sleep, appetite, and energy level. Two broad forms of depressive illness are known: major depressive disorder and bipolar disorder. Individuals affected by **major depressive disorder** undergo periods—often lasting many months—during which once enjoyable activities provide no pleasure and provoke no interest. One of the most common nervous system disorders, major depression affects about one in every seven adults at some point, and twice as many women as men.

Bipolar disorder, or manic-depressive disorder, involves swings of mood from high to low and affects about 1% of the world's population. The manic phase is characterized by high self-esteem, increased energy, a flow of ideas, overtalkativeness, and increased risk taking. In its milder forms, this phase is sometimes associated with great creativity, and some well-known artists, musicians, and literary figures (including Vincent Van Gogh, Robert Schumann, Virginia Woolf, and Ernest Hemingway, to name a few) have had very productive periods during manic phases. The depressive phase comes with lowered ability to feel pleasure, loss of motivation, sleep disturbances, and feelings of worthlessness. These symptoms can be so severe that affected individuals attempt suicide.

Major depressive and bipolar disorders are among the nervous system disorders for which available therapies are most effective. Many drugs used to treat depressive illness, including fluoxetine (Prozac), increase the activity of biogenic amines in the brain.

Drug Addiction and the Brain's Reward System

Drug addiction is a disorder characterized by compulsive consumption of a drug and loss of control in limiting intake. Addictive drugs include stimulants, such as cocaine and amphetamine, and sedatives, such as heroin. However, all of these drugs, as well as alcohol and nicotine, are addictive for the same reason: Each increases activity of the brain's reward system, neural circuitry that normally functions in pleasure, motivation, and learning.

In the absence of drug addiction, the reward system of the brain provides motivation for activities that enhance survival and reproduction, such as eating in response to hunger, drinking when thirsty, and engaging in sexual activity when aroused. In addicted individuals, "wanting" is instead directed toward further drug consumption.

As shown in **Figure 49.23**, inputs to the reward system are received by neurons in a region near the base of the brain called the *ventral tegmental area (VTA)*. When activated, these neurons release dopamine from their synaptic terminals in

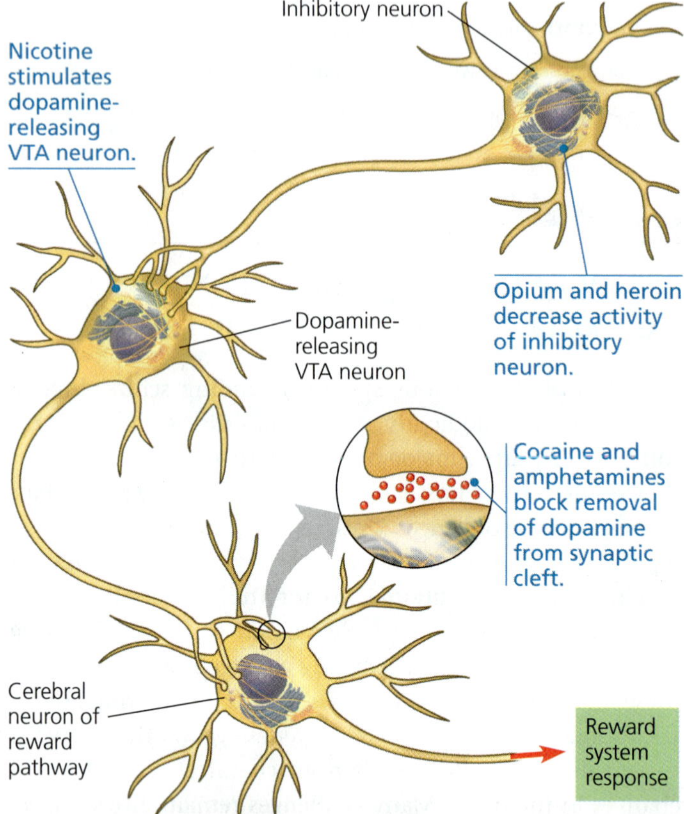

▲ **Figure 49.23 Effects of addictive drugs on the reward system of the mammalian brain.** Addictive drugs alter the transmission of signals in the pathway formed by neurons of the ventral tegmental area (VTA).

MAKE CONNECTIONS *Based on what you learned in Concept 48.3 (pp. 1050–1051), what effect would you expect if you depolarized the neurons in the VTA? Explain.*

specific regions of the cerebrum, including the *nucleus accumbens* (see Figure 49.14).

Addictive drugs affect the reward system in several ways. First, each drug has an immediate and direct effect that enhances the activity of the dopamine pathway (see Figure 49.23). As addiction develops, there are also long-lasting changes in the reward circuitry. The result is a craving for the drug independent of any pleasure associated with consumption.

Laboratory animals have proved especially useful in teaching us how the brain's reward system works and how particular drugs affect its function. Rats, for example, will provide themselves with cocaine, heroin, or amphetamine when given a dispensing system linked to a lever in their cage. Furthermore, they exhibit addictive behavior in such circumstances, continuing to self-administer the drug rather than seek food, even to the point of starvation.

As scientists expand their knowledge about the brain's reward system and the various forms of addiction, there is hope that the insights will lead to more effective prevention and treatment.

Alzheimer's Disease

Alzheimer's disease is a mental deterioration, or dementia, characterized by confusion and memory loss. Its incidence is age related, rising from about 10% at age 65 to about 35% at age 85. The disease is progressive, with patients gradually becoming less able to function and eventually needing to be dressed, bathed, and fed by others. Moreover, patients with Alzheimer's disease often lose their ability to recognize people, including their immediate family, and may treat them with suspicion and hostility.

Alzheimer's disease leads to the death of neurons in many areas of the brain, including the hippocampus and cerebral cortex. As a result, there is often massive shrinkage of brain tissue. Postmortem examination of the remaining brain tissue reveals two characteristic features—amyloid plaques and neurofibrillary tangles (**Figure 49.24**).

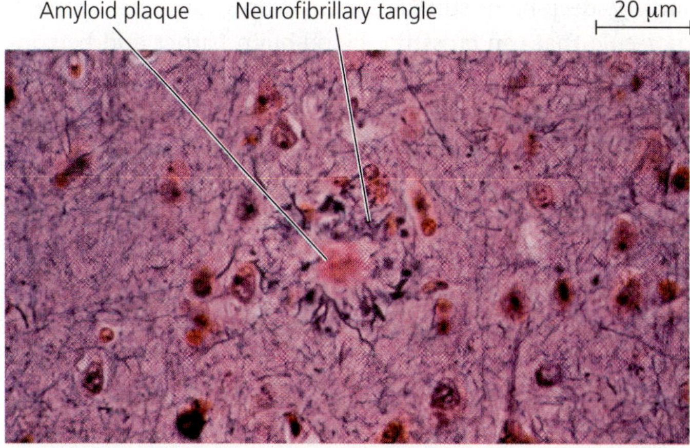

▲ **Figure 49.24 Microscopic signs of Alzheimer's disease.** A hallmark of Alzheimer's disease is the presence in brain tissue of neurofibrillary tangles surrounding plaques made of β-amyloid (LM).

The plaques are aggregates of β-amyloid, an insoluble peptide that is cleaved from the extracellular portion of a membrane protein found in neurons. Membrane enzymes, called secretases, catalyze the cleavage, causing β-amyloid to accumulate in plaques outside the neurons. It is these plaques that appear to trigger the death of surrounding neurons.

The neurofibrillary tangles observed in Alzheimer's disease are primarily made up of the tau protein. (This protein is unrelated to the tau mutation that affects circadian rhythm in hamsters.) The tau protein normally helps assemble and maintain microtubules that transport nutrients along axons. In Alzheimer's disease, tau undergoes changes that cause it to bind to itself, resulting in neurofibrillary tangles. There is evidence that changes in tau are associated with the appearance of early-onset Alzheimer's disease, a much less common disorder that affects relatively young individuals.

There is currently no cure for Alzheimer's disease, but an enormous effort has led to the recent development of drugs that are partially effective in relieving some of the symptoms. Doctors are also beginning to use functional brain imaging to diagnose Alzheimer's disease in patients exhibiting early signs of dementia.

Parkinson's Disease

Symptoms of **Parkinson's disease**, a motor disorder, include muscle tremors, poor balance, a flexed posture, and a shuffling gait. Facial muscles become rigid, limiting the ability of patients to vary their expressions. Like Alzheimer's disease, Parkinson's disease is a progressive brain illness and is more common with advancing age. The incidence of Parkinson's disease is about 1% at age 65 and about 5% at age 85. In the U.S. population, approximately 1 million people are afflicted.

The symptoms of Parkinson's disease result from the death of neurons in the midbrain that normally release dopamine at synapses in the basal nuclei. As with Alzheimer's disease, protein aggregates accumulate. Most cases of Parkinson's disease lack an identifiable cause; however, a rare form of the disease that appears in relatively young adults has a clear genetic basis. Molecular studies of mutations linked to this early-onset Parkinson's disease reveal disruption of genes required for certain mitochondrial functions. Researchers are investigating whether mitochondrial defects also contribute to the more common and later-onset form of the disease.

At present there is no cure for Parkinson's disease. Approaches used to manage the symptoms include brain surgery, deep-brain stimulation, and drugs such as L-dopa, a molecule that can cross the blood-brain barrier and be converted to dopamine in the CNS. One potential cure is to implant dopamine-secreting neurons, either in the midbrain or in the basal nuclei. Laboratory studies of this strategy show promise: In rats with an experimentally induced condition that mimics Parkinson's disease, implanting dopamine-secreting neurons can lead to a recovery of motor control. Whether this regenerative approach can also work in humans is one of many important questions in modern brain research.

CONCEPT CHECK 49.5

1. Compare Alzheimer's disease and Parkinson's disease.
2. How is dopamine activity related to schizophrenia, drug addiction, and Parkinson's disease?
3. **WHAT IF?** If you could detect early-stage Alzheimer's disease, would you expect to see brain changes that were similar to, although less extensive than, those seen in patients who have died of this disease? Explain.

For suggested answers, see Appendix A.

49 CHAPTER REVIEW

SUMMARY OF KEY CONCEPTS

CONCEPT 49.1

Nervous systems consist of circuits of neurons and supporting cells (pp. 1062–1067)

- Invertebrate nervous systems range in complexity from simple **nerve nets** to highly centralized nervous systems having complicated brains and ventral nerve cords.

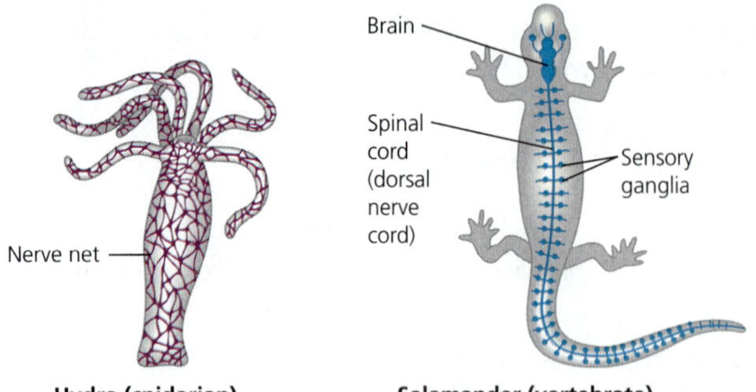

In vertebrates, the central nervous system (CNS), consisting of the brain and the spinal cord, integrates information, while the **nerves** of the peripheral nervous system (PNS) transmit sensory and motor signals between the CNS and the rest of the body. The simplest circuits in the vertebrate nervous system control **reflex** responses, in which sensory input is linked to motor output without involvement of the brain. Vertebrate neurons are supported by several types of glia, including **astrocytes**, oligodendrocytes, Schwann cells, and ependymal cells.

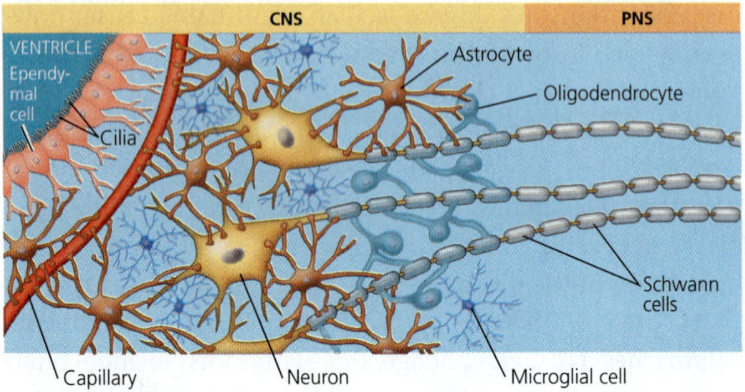

- Afferent neurons carry sensory signals to the CNS. Efferent neurons function in either the **motor system**, which carries signals to skeletal muscles, or the **autonomic nervous system**, which regulates smooth and cardiac muscles. The **sympathetic** and **parasympathetic divisions** of the autonomic nervous system have antagonistic effects on a diverse set of target organs, while the **enteric division** controls the activity of many digestive organs.

? *How does the circuitry of a reflex facilitate a rapid response?*

CONCEPT 49.2

The vertebrate brain is regionally specialized (pp. 1067–1072)

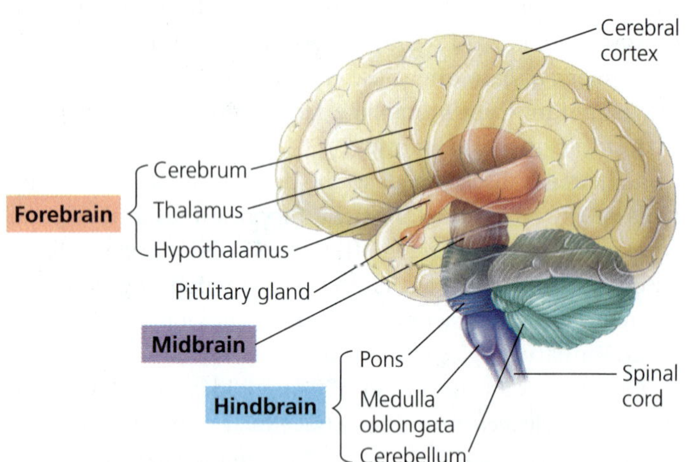

- The cerebrum has two hemispheres, each of which consists of cortical **gray matter** overlying **white matter** and basal nuclei, which are important in planning and learning movements. A thick band of axons, the **corpus callosum**, provides communication between the right and left cerebral cortices.
- Within each region of the brain, particular structures have specialized functions. The **pons** and **medulla oblongata** are relay stations for information traveling between the PNS and the cerebrum. The **reticular formation**, a network of neurons within the **brainstem**, regulates sleep and arousal. The **cerebellum** helps coordinate motor, perceptual, and cognitive functions. It is also involved in learning and remembering

motor skills. The **thalamus** is the main center through which sensory and motor information passes to the **cerebrum**. The **hypothalamus** regulates homeostasis and basic survival behaviors. Within the hypothalamus, the **suprachiasmatic nucleus (SCN)** acts as the pacemaker for circadian rhythms.
- The generation and experience of emotions involve many regions of the brain. However, the **amygdala** plays a key role in recognizing and recalling a number of emotions.

? *What role do particular regions of the brain play in vision and responses to visual input?*

CONCEPT 49.3

The cerebral cortex controls voluntary movement and cognitive functions (pp. 1072–1076)

- Each side of the **cerebral cortex** has four lobes—frontal, temporal, occipital, and parietal—that contain primary sensory areas and association areas. Specific types of sensory input enter the primary sensory areas. Association areas integrate information from different sensory areas.
- Portions of the frontal and temporal lobes, including Broca's area and Wernicke's area, are essential for generating and understanding language. These functions are concentrated in the left **cerebral hemisphere**, as are math and logic operations. The right hemisphere appears to be stronger at pattern recognition and nonverbal thinking. At least some of this **lateralization** of functions relates to handedness.
- In the somatosensory cortex and the motor cortex, neurons are distributed according to the part of the body that generates sensory input or receives motor commands.
- Primates and cetaceans, which are capable of higher cognition, have an extensively convoluted neocortex, the outermost part of the cerebral cortex. In birds, a brain region called the pallium contains clustered nuclei that carry out functions similar to those performed by the cerebral cortex of mammals. Some birds can solve problems and understand abstractions in a manner indicative of higher cognition.

? *After an accident, a patient has trouble with language and has paralysis on one side of the body. Which side would you expect to be paralyzed? Why?*

CONCEPT 49.4

Changes in synaptic connections underlie memory and learning (pp. 1076–1079)

- During development, more neurons and synapses form than will exist in the adult. The programmed death of neurons and elimination of synapses in embryos establish the basic structure of the nervous system. In the adult, reshaping of the nervous system can involve the loss or addition of synapses or the strengthening or weakening of signaling at synapses. This capacity for remodeling is termed **neural plasticity**. Defective remodeling of synapses is partly responsible for the developmental abnormalities of autism.
- **Short-term memory** relies on temporary links in the hippocampus. In **long-term memory**, these temporary links are replaced by connections within the cerebral cortex. This transfer of information from short-term to long-term memory is enhanced by the association of new data with that already in long-term memory. **Long-term potentiation (LTP)** is a lasting increase in the strength of synaptic transmission and appears to be an important process in memory storage and learning.
- The adult human brain contains stem cells that can differentiate into mature neurons. Therapy based on stem cells offers a potential method for replacing neurons lost to injury or disease.

? *Learning multiple languages is typically easier earlier in childhood than later in life. How does this fit with our understanding of neural development?*

CONCEPT 49.5

Many nervous system disorders can be explained in molecular terms (pp. 1079–1082)

- Research has identified the biochemical basis of a number of nervous system disorders. **Schizophrenia**, which is characterized by hallucinations, delusions, and other symptoms, affects neuronal pathways that use dopamine as a neurotransmitter. Drugs that increase the activity of biogenic amines in the brain can be used to treat **bipolar disorder** and **major depressive disorder**. The compulsive drug use that characterizes addiction reflects altered activity of the brain's reward system, which normally provides motivation for actions that enhance survival or reproduction.
- **Alzheimer's disease** and **Parkinson's disease** are neurodegenerative and typically age related. Alzheimer's disease is a dementia in which neurofibrillary tangles and amyloid plaques form in the brain. Parkinson's disease is a motor disorder caused by the death of dopamine-secreting neurons and associated with the presence of protein aggregates.

? *The fact that both amphetamines and PCP have effects similar to the symptoms of schizophrenia suggests a potentially complex basis for this disease. Explain.*

TEST YOUR UNDERSTANDING

LEVEL 1: KNOWLEDGE/COMPREHENSION

1. Wakefulness is regulated by the reticular formation, which is present in the
   a. basal nuclei.
   b. cerebral cortex.
   c. brainstem.
   d. limbic system.
   e. spinal cord.
2. Which of the following structures or regions is *incorrectly* paired with its function?
   a. limbic system—motor control of speech
   b. medulla oblongata—homeostatic control
   c. cerebellum—coordination of movement and balance
   d. corpus callosum—communication between the left and right cerebral cortices
   e. amygdala—emotional memory
3. Patients with damage to Wernicke's area have difficulty
   a. coordinating limb movement.
   b. generating speech.
   c. recognizing faces.
   d. understanding language.
   e. experiencing emotion.
4. The cerebral cortex plays a major role in all of the following *except*
   a. short-term memory.
   b. long-term memory.
   c. circadian rhythm.
   d. foot-tapping rhythm.
   e. breath holding.

LEVEL 2: APPLICATION/ANALYSIS

5. After suffering a stroke, a patient can see objects anywhere in front of him but pays attention only to objects in his right field of vision. When asked to describe these objects, he has difficulty judging their size and distance. What part of the brain was likely damaged by the stroke?
   a. the left frontal lobe
   b. the right frontal lobe
   c. the left parietal lobe
   d. the right parietal lobe
   e. the corpus callosum

6. Injury localized to the hypothalamus would most likely disrupt
   a. short-term memory.
   b. coordination during locomotion.
   c. executive functions, such as decision making.
   d. sorting of sensory information.
   e. regulation of body temperature.

7. **DRAW IT** The reflex that pulls your hand away when you prick your finger on a sharp object relies on a simple neuronal circuit with two synapses in the spinal cord. (a) Using a circle to represent a cross section of the spinal cord, draw the circuit, labeling the types of neurons, the direction of information flow in each, and the locations of synapses. (b) Draw a simple diagram of the brain indicating where pain would eventually be perceived.

LEVEL 3: SYNTHESIS/EVALUATION

8. **EVOLUTION CONNECTION**
   Scientists often use measures of "higher-order thinking" to assess intelligence in other animals. For example, birds are judged to have sophisticated thought processes because they can use tools and make use of abstract concepts. What problems do you see in defining intelligence in these ways?

9. **SCIENTIFIC INQUIRY**
   Consider an individual who had been fluent in American Sign Language before suffering damage to the left cerebral hemisphere. After the injury, this person could still understand signs, but could not readily generate signs that represented his thoughts. What two hypotheses could explain this finding, and how might you distinguish between them?

10. **SCIENCE, TECHNOLOGY, AND SOCIETY**
    With increasingly sophisticated methods for scanning brain activity, scientists are rapidly developing the ability to detect an individual's particular emotions and thought processes from outside the body. What benefits and problems do you envision when such technology becomes readily available?

11. **WRITE ABOUT A THEME**
    The Genetic Basis of Life In a short essay (100–150 words), explain how specification of the adult nervous system by the genome is incomplete.

For selected answers, see Appendix A.

www.masteringbiology.com

1. MasteringBiology® Assignments

Tutorial The Vertebrate Nervous System
Activities Discovery Channel Video: Teen Brains • Signal Transmission at a Chemical Synapse
Questions Student Misconceptions • Reading Quiz • Multiple Choice • End-of-Chapter

2. eText

Read your book online, search, take notes, highlight text, and more.

3. The Study Area

Practice Tests • Cumulative Test • *BioFlix* 3-D Animations • MP3 Tutor Sessions • Videos • Activities • Investigations • Lab Media • Audio Glossary • Word Study Tools • Art

50

Sensory and Motor Mechanisms

▲ **Figure 50.1 Is a star-shaped nose merely decorative?**

KEY CONCEPTS

50.1 Sensory receptors transduce stimulus energy and transmit signals to the central nervous system

50.2 The mechanoreceptors responsible for hearing and equilibrium detect moving fluid or settling particles

50.3 Visual receptors in diverse animals depend on light-absorbing pigments

50.4 The senses of taste and smell rely on similar sets of sensory receptors

50.5 The physical interaction of protein filaments is required for muscle function

50.6 Skeletal systems transform muscle contraction into locomotion

OVERVIEW

Sensing and Acting

The face of the star-nosed mole (*Condulura cristata*) is, in a word, astounding **(Figure 50.1)**. Eleven pairs of appendages protrude from its nose, forming a prominent pink star. Although they look a bit like fingers, these appendages are not used in grasping. Nor is the nose used to detect odors. Is the star, then, simply ornamental? No—it has a highly specialized function. Just below its surface lie 25,000 touch-sensitive receptors, more than are found in your whole hand.

Tunneling beneath the wetlands of eastern North America, the virtually blind mole lives in almost total darkness. But as 100,000 neurons relay tactile information from its nose to its brain, the mole finds and captures food with remarkable rapidity: A star-nosed mole can detect and eat prey in as little as 120 milliseconds (msec).

Detecting and processing sensory information and generating motor responses provide the physiological basis for all animal behavior. In this chapter, we will explore the processes of sensing and acting in both vertebrates and invertebrates. We will start with sensory processes that convey information about an animal's external and internal environment to its brain. We will then consider the structure and function of muscles and skeletons that carry out movements as instructed by the brain. Finally, we will investigate various mechanisms of animal movement. These topics will lead us naturally to our discussion of animal behavior in Chapter 51.

CONCEPT 50.1

Sensory receptors transduce stimulus energy and transmit signals to the central nervous system

All sensory processes begin with stimuli, and all stimuli represent forms of energy. A sensory receptor converts stimulus energy to a change in membrane potential and thereby regulates the output of action potentials to the central nervous system (CNS). Activating a sensory receptor does not necessarily require a large amount of stimulus energy. Indeed, some sensory receptors can detect the smallest possible unit of stimulus; most light receptors, for example, can detect a single quantum (photon) of light.

When a stimulus is received and processed by the nervous system, a motor response may be generated. One of the simplest stimulus-response circuits is a reflex, such as the knee-jerk reflex shown in Figure 49.3. Many other behaviors rely on more elaborate processing that involves integration of sensory input. As an example, consider how the star-nosed mole

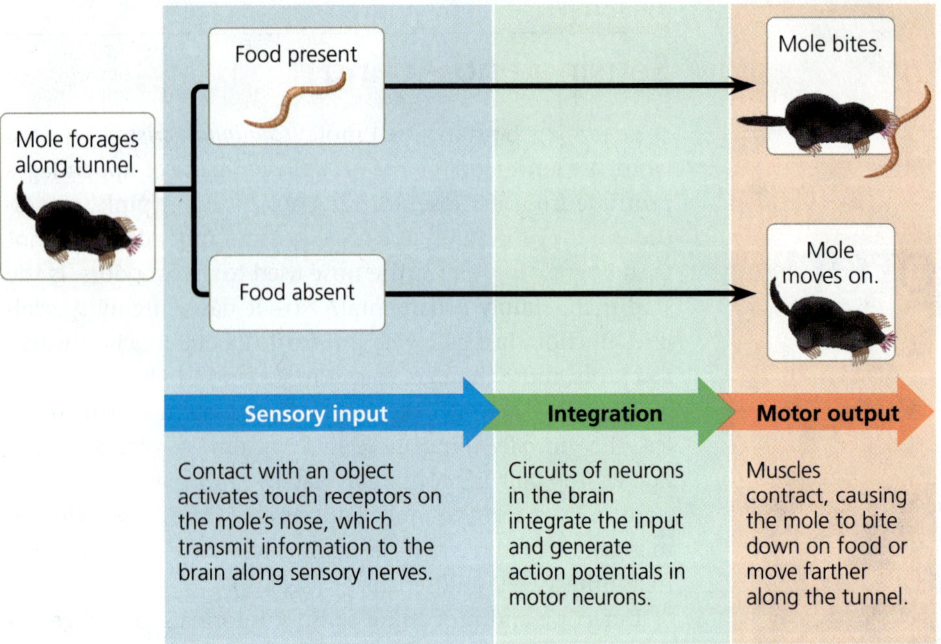

▲ **Figure 50.2 A simple response pathway: foraging by a star-nosed mole.**

forages for food in its tunnel environment **(Figure 50.2)**. When the mole's nose contacts an object in its tunnel, touch receptors in the nose are activated. These receptors transmit sensory information about the object to the mole's brain. Circuits in the brain integrate the input and initiate one of two response pathways, depending on whether food was detected. Motor output commands from the brain sent to skeletal muscles in the body cause the mole either to bite down with its teeth or to continue moving along the tunnel.

With this overview in mind, let's examine the general organization and activity of animal sensory systems.

Sensory Pathways

Sensory pathways have in common four basic functions: sensory reception, transduction, transmission, and perception.

Sensory Reception and Transduction

A sensory pathway begins with **sensory reception**, the detection of a stimulus by sensory cells. Most sensory cells are specialized neurons or epithelial cells. Some exist singly; others are collected in sensory organs, such as eyes and ears. The term **sensory receptor** is used to describe a sensory cell or organ, as well as the subcellular structure that interacts directly with stimuli. Many sensory receptors detect stimuli from outside the body, such as heat, light, pressure, and chemicals, but there are also receptors for stimuli from within the body, such as blood pressure and body position.

Although animals use a range of sensory receptors to detect widely varying stimuli, the effect in all cases is to open or close ion channels. Thus, for example, ion channels open or close when a substance outside the cell binds to a chemical receptor in the plasma membrane. The resulting flow of ions across the membrane changes the membrane potential.

The conversion of a physical or chemical stimulus to a change in the membrane potential of a sensory receptor is called **sensory transduction**, and the change in membrane potential itself is known as a **receptor potential**. Receptor potentials are graded potentials; their magnitude varies with the strength of the stimulus.

Transmission

Sensory information travels through the nervous system as nerve impulses, or action potentials. For many sensory receptors, transducing the energy in a stimulus into a receptor potential initiates **transmission** of action potentials to the CNS.

Some sensory receptors are themselves specialized neurons, whereas others are specialized cells that regulate neurons **(Figure 50.3)**. Neurons that act directly as sensory receptors produce action potentials and have an axon that extends into the CNS. Non-neuronal sensory receptor cells form chemical synapses with sensory (afferent) neurons and typically respond to stimuli by increasing the rate at

▼ **Figure 50.3 Classes of sensory receptors.**

(a) Receptor *is* afferent neuron.

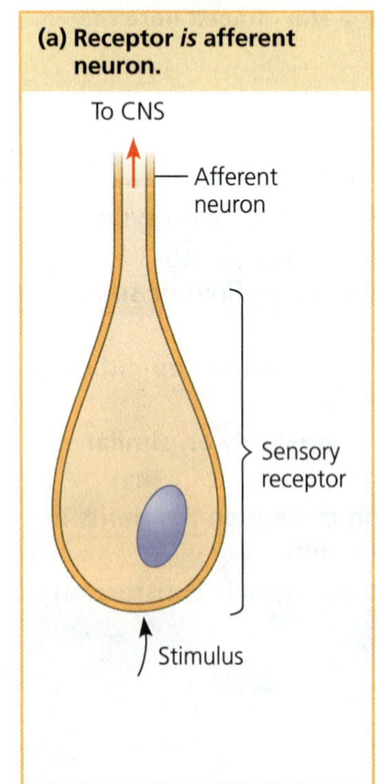

(b) Receptor *regulates* afferent neuron.

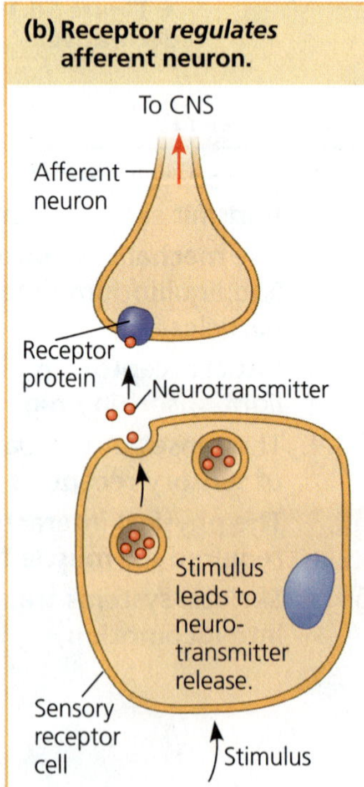

▼ Figure 50.4 Coding of stimulus intensity.

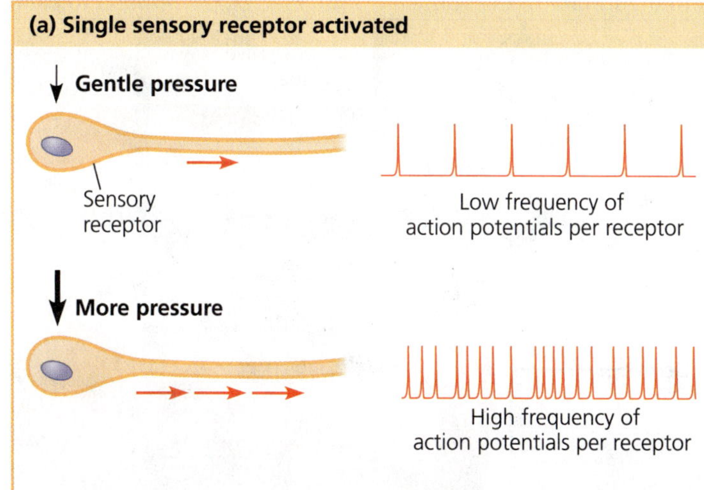

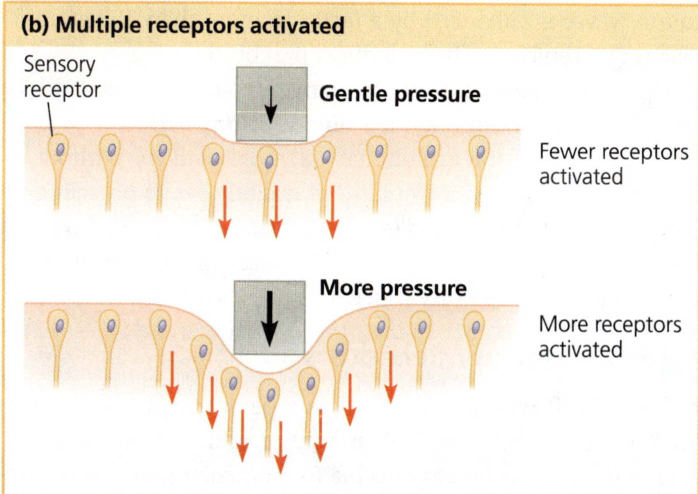

which the afferent neurons produce action potentials. (One exception is in the vertebrate visual system, discussed in Concept 50.3.)

The response of a sensory receptor varies with stimuli of different intensities. The primary difference is the magnitude of the receptor potential, which controls the rate at which action potentials are produced. If the receptor is a sensory neuron, a larger receptor potential results in more frequent action potentials **(Figure 50.4a)**. If the receptor is not a sensory neuron, a larger receptor potential causes more neurotransmitter to be released, which usually increases the production of action potentials by the postsynaptic neuron.

Many sensory neurons spontaneously generate action potentials at a low rate. In these neurons, a stimulus does not switch the production of action potentials on or off, but it does change *how often* an action potential is produced. In this manner, such neurons are also able to alert the nervous system to changes in stimulus intensity.

A difference in stimulus strength may not only alter the activity of individual receptors, but also affect the number of receptors that are activated **(Figure 50.4b)**. If a stronger stimulus triggers a response by more receptors, more axons transmit action potentials. This increase in the number of axons transmitting action potentials is then decoded by the nervous system as a stronger stimulus.

Processing of sensory information can occur before, during, and after transmission of action potentials to the CNS. In many cases, the *integration* of sensory information begins as soon as the information is received. Receptor potentials produced by stimuli delivered to different parts of a sensory receptor cell are integrated through summation, as are postsynaptic potentials in sensory neurons that form synapses with multiple receptors (see Figure 48.16). As we will discuss shortly, sensory structures such as eyes also provide higher levels of integration, and the brain further processes all incoming signals.

Perception

When action potentials reach the brain via sensory neurons, circuits of neurons process this input, generating the **perception** of the stimuli. Perceptions—such as colors, smells, sounds, and tastes—are constructions formed in the brain and do not exist outside it. So, if a tree falls and no animal is present to hear it, is there a sound? The falling tree certainly produces pressure waves in the air, but if sound is defined as a perception, then there is none unless an animal senses the waves and its brain perceives them.

Action potentials are all-or-none events (see Figure 48.10c). An action potential triggered by light striking the eye has the same properties as an action potential triggered by air vibrating in the ear. How, then, do we distinguish sights, sounds, and other stimuli? The answer lies in the connections that link sensory receptors to the brain. Action potentials from sensory receptors travel along neurons that are dedicated to a particular stimulus; these dedicated neurons synapse with particular neurons in the brain or spinal cord. As a result, the brain distinguishes sensory stimuli such as sight or sound solely by the path to the brain along which the action potentials have traveled.

Amplification and Adaptation

The transduction of stimuli by sensory receptors is subject to two types of modification—amplification and adaptation. **Amplification** refers to the strengthening of a sensory signal during transduction. The effect can be considerable. For example, an action potential conducted from the eye to the human brain has about 100,000 times as much energy as the few photons of light that triggered it.

Amplification that occurs in sensory receptor cells often requires signal transduction pathways involving second messengers. Because these pathways include enzyme-catalyzed reactions, they amplify signal strength through the formation of many product molecules by a single enzyme molecule.

Amplification may also take place in accessory structures of a complex sense organ, as when the pressure associated with sound waves is enhanced by a factor of more than 20 before reaching receptors in the innermost part of the ear.

Upon continued stimulation, many receptors undergo a decrease in responsiveness termed **sensory adaptation** (not to be confused with the evolutionary term *adaptation*). Without sensory adaptation, you would be constantly aware of feeling every beat of your heart and every bit of clothing on your body. Adaptation also enables you to see, hear, and smell changes in the environment that vary widely in stimulus intensity.

Types of Sensory Receptors

A sensory cell typically has a single type of receptor specific for a particular stimulus, such as light or cold. Often, distinct cells and receptors are responsible for particular qualities of a sensation, such as distinguishing red from blue. Before exploring these specializations, let's consider sensory receptor function at a more basic level. We can classify sensory receptors into five categories based on the nature of the stimuli they transduce: mechanoreceptors, chemoreceptors, electromagnetic receptors, thermoreceptors, and pain receptors.

Mechanoreceptors

Mechanoreceptors sense physical deformation caused by forms of mechanical energy such as pressure, touch, stretch, motion, and sound. Mechanoreceptors typically consist of ion channels that are linked to structures that extend outside the cell, such as "hairs" (cilia), as well as internal cell structures, such as the cytoskeleton. Bending or stretching of the external structure generates tension that alters the permeability of the ion channels. This change in ion permeability alters the membrane potential, resulting in a depolarization or hyperpolarization (see Chapter 48).

The familiar knee-jerk reflex (see Figure 49.3) is triggered by the vertebrate stretch receptor, a mechanoreceptor that detects muscle movement. Vertebrate stretch receptors are dendrites of sensory neurons that spiral around the middle of certain small skeletal muscle fibers. Groups of about 2 to 12 of these fibers, formed into a spindle shape and surrounded by connective tissue, are distributed throughout the muscle, parallel to other muscle fibers. When the muscle is stretched, the spindle fibers are stretched, depolarizing sensory neurons and triggering action potentials that are transmitted to the spinal cord.

The mammalian sense of touch also relies on mechanoreceptors that are the dendrites of sensory neurons. Touch receptors, such as those illustrated in Figure 50.4, are often embedded in layers of connective tissue. The structure of the connective tissue and the location of the receptors dramatically affect the type of mechanical energy (light touch, vibration, or strong pressure) that best stimulates them **(Figure 50.5)**. Receptors that detect a light touch or vibration are close to the surface of the skin; they transduce very slight inputs of mechanical energy into receptor

▲ **Figure 50.5 Sensory receptors in human skin.** Most receptors in the dermis are encapsulated by connective tissue. Receptors in the epidermis are naked dendrites, as are hair movement receptors that wind around the base of hairs in the dermis.

potentials. Receptors that respond to stronger pressure and vibrations are in deep skin layers.

Some animals use mechanoreceptors to literally get a feel for their environment. For example, cats as well as many rodents have extremely sensitive mechanoreceptors at the base of their whiskers. Because deflection of different whiskers triggers action potentials that reach different cells in the brain, an animal's whiskers provide detailed information about nearby objects.

Chemoreceptors

Chemoreceptors include both general receptors—those that transmit information about total solute concentration—and specific receptors—those that respond to individual kinds of molecules. Osmoreceptors in the mammalian brain, for example, detect changes in the total solute concentration of the blood and stimulate thirst when osmolarity increases (see Figure 44.19). Most animals also have receptors for specific molecules, including glucose, oxygen, carbon dioxide, and amino acids.

Two of the most sensitive and specific chemoreceptors known are found in the antennae of the male silkworm moth **(Figure 50.6)**; they detect the two chemical components of the female moth sex pheromone. For pheromones and other molecules detected by chemoreceptors, the stimulus molecule binds to the specific receptor on the membrane of the sensory cell and initiates changes in ion permeability.

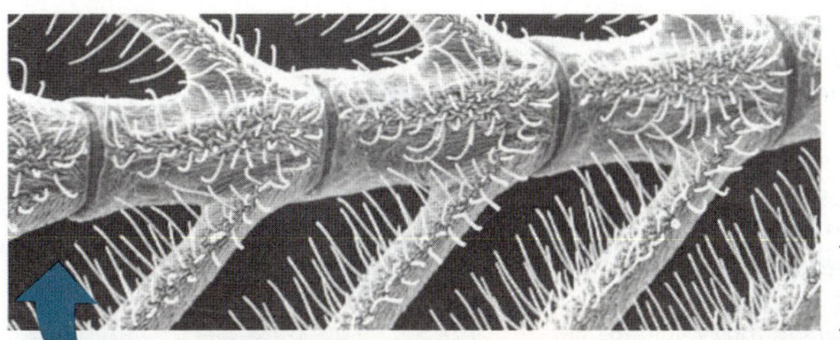

▲ **Figure 50.6 Chemoreceptors in an insect.** The antennae of the male silkworm moth *Bombyx mori* are covered with sensory hairs, visible in the SEM enlargement. The hairs have chemoreceptors that are highly sensitive to the sex pheromone released by the female.

(a) This rattlesnake and other pit vipers have a pair of infrared receptors, one anterior to and just below each eye. These organs are sensitive enough to detect the infrared radiation emitted by a warm mouse a meter away. The snake moves its head from side to side until the radiation is detected equally by the two receptors, indicating that the mouse is straight ahead.

(b) Some migrating animals, such as these beluga whales, apparently sense Earth's magnetic field and use the information, along with other cues, for orientation.

▲ **Figure 50.7 Specialized electromagnetic receptors.**

Electromagnetic Receptors

Electromagnetic receptors detect various forms of electromagnetic energy, such as visible light, electricity, and magnetism. For example, snakes have very sensitive infrared receptors that detect the body heat of prey (**Figure 50.7a**). Similarly, the platypus, a monotreme mammal (see Chapter 34), has electroreceptors on its bill that are thought to detect the electric field generated by the muscles of crustaceans, frogs, small fish, and other prey. In a few cases, the animal detecting an electromagnetic stimulus is also its source: Some fishes generate electric currents and then use their electroreceptors to locate prey or other objects that disturb those currents.

Many animals appear to use Earth's magnetic field lines to orient themselves as they migrate (**Figure 50.7b**). The iron-containing mineral magnetite is found in many vertebrates (including salmon, pigeons, sea turtles, and humans), in bees, in some molluscs, and in certain protists and prokaryotes that orient to Earth's magnetic field. Once collected by sailors to make compasses for navigation, magnetite may be part of an orienting mechanism in many animals (see Chapter 51).

Thermoreceptors

Thermoreceptors detect heat and cold. Located in the skin and in the anterior hypothalamus, thermoreceptor cells send information to the body's thermostat in the posterior hypothalamus. Our understanding of thermoreception has increased substantially recently, thanks to scientists with an appreciation for fiery foods. Jalapeno and cayenne peppers taste "hot" because they contain a natural product called capsaicin. It turns out that exposing sensory neurons to capsaicin triggers an influx of calcium ions. When scientists identified the receptor protein that binds capsaicin, they made a fascinating discovery: The receptor opens a calcium channel in response not only to capsaicin, but also to high temperatures (42°C or higher). In essence, spicy foods taste "hot" because they activate the same receptors as hot soup and coffee.

Mammals have a number of kinds of thermoreceptors, each specific for a particular temperature range. The capsaicin receptor and at least five other types of thermoreceptors belong to the TRP (transient receptor potential) family of ion channel proteins. Just as the TRP-type receptor specific for high temperature is sensitive to capsaicin, the receptor for temperatures below 28°C can be activated by menthol, a plant product that we perceive to have a "cool" flavor.

Pain Receptors

Extreme pressure or temperature, as well as certain chemicals, can damage animal tissues. To detect stimuli that reflect such noxious (or harmful) conditions, animals rely on **nociceptors** (from the Latin *nocere*, to hurt), also called **pain receptors**. By triggering defensive reactions, such as withdrawal from danger, the perception of pain serves an important function.

In humans, certain naked dendrites act as nociceptors by detecting noxious thermal, mechanical, or chemical stimuli. The capsaicin receptor is thus a thermoreceptor and also a nociceptor. Although nociceptor density is highest in skin, some pain receptors are associated with other organs.

Chemicals produced in an animal's body sometimes enhance the perception of pain. For example, damaged tissues produce prostaglandins, which act as local regulators of inflammation (see Chapter 45). Prostaglandins worsen pain by increasing nociceptor sensitivity to noxious stimuli. Aspirin and ibuprofen reduce pain by inhibiting the synthesis of prostaglandins.

Next we'll turn our focus to sensory systems, beginning with systems for maintaining balance and detecting sound.

CONCEPT CHECK 50.1

1. Which one of the five categories of sensory receptors is primarily dedicated to external stimuli?
2. Why can eating "hot" peppers cause a person to sweat?
3. **WHAT IF?** If you stimulated a sensory neuron electrically, how would that stimulation be perceived?

For suggested answers, see Appendix A.

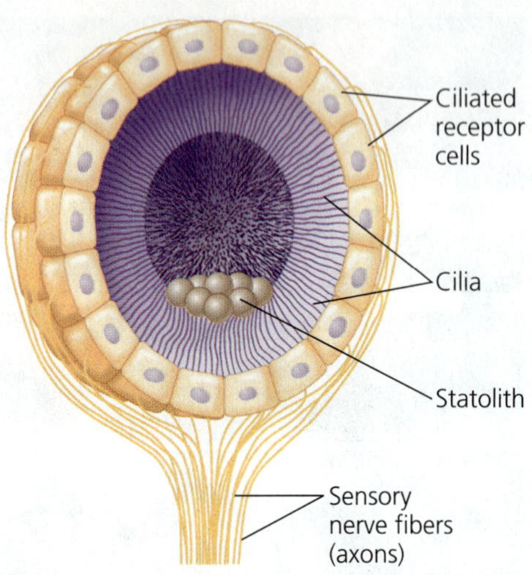

▶ **Figure 50.8 The statocyst of an invertebrate.** The settling of statoliths to the low point in the chamber bends cilia on receptor cells in that location, providing the brain with information about the orientation of the body with respect to gravity.

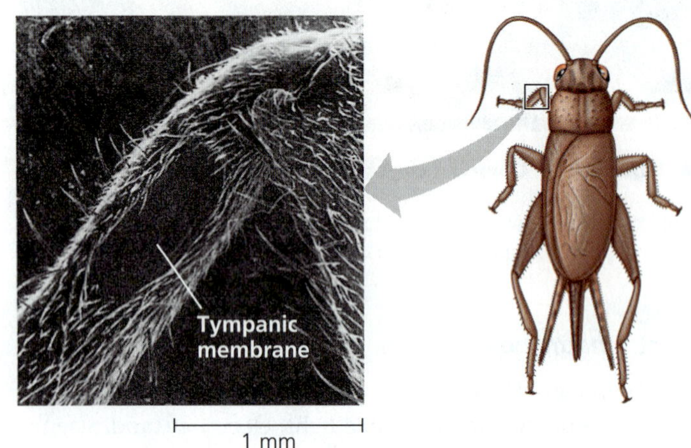

▲ **Figure 50.9 An insect's "ear"—on its leg.** The tympanic membrane, visible in this SEM of a cricket's front leg, vibrates in response to sound waves. The vibrations stimulate mechanoreceptors attached to the inside of the tympanic membrane.

CONCEPT 50.2

The mechanoreceptors responsible for hearing and equilibrium detect moving fluid or settling particles

Hearing and the perception of body equilibrium, or balance, are related in most animals. For both senses, mechanoreceptor cells produce receptor potentials when settling particles or moving fluid causes deflection of cell-surface structures.

Sensing of Gravity and Sound in Invertebrates

To sense gravity and maintain equilibrium, most invertebrates rely on mechanoreceptors located in organs called **statocysts** (Figure 50.8). In a typical statocyst, a layer of ciliated receptor cells surrounds a chamber that contains one or more **statoliths**, which are grains of sand or other dense granules. When statoliths settle to the low point in the chamber, they stimulate mechanoreceptors in that location. In experiments in which statoliths were replaced with metal shavings, researchers "tricked" crayfish into swimming upside down by using magnets to pull the shavings to the upper end of the statocysts located at the base of their antennae.

Many (perhaps most) insects have body hairs that vibrate in response to sound waves. Hairs of different stiffnesses and lengths vibrate at different frequencies. For example, fine hairs on the antennae of a male mosquito vibrate in a specific way in response to the hum produced by the beating wings of flying females. The importance of this sensory system in the attraction of males to a potential mate can be demonstrated very simply: A tuning fork vibrating at the same frequency as that of a female's wings will itself attract males.

Many insects also detect sound by means of "ears" consisting of a tympanic membrane (eardrum) stretched over an internal air chamber (Figure 50.9). Sound waves vibrate the tympanic membrane, stimulating receptor cells attached to the inside of the membrane and resulting in nerve impulses that are transmitted to the brain. Cockroaches lack such a tympanic membrane, but instead have vibration-sensitive organs located in each leg. These organs can provide enough warning for the insect to avoid being crushed by a descending human foot.

Hearing and Equilibrium in Mammals

In mammals, as in most other terrestrial vertebrates, the sensory organs for hearing and equilibrium are closely associated. Figure 50.10 explores the structure and function of these organs in the human ear.

▼ Figure 50.10

Exploring The Structure of the Human Ear

1 Overview of Ear Structure

The **outer ear** consists of the external pinna and the auditory canal, which collect sound waves and channel them to the **tympanic membrane** (eardrum), which separates the outer ear from the middle ear. In the **middle ear**, three small bones—the malleus (hammer), incus (anvil), and stapes (stirrup)—transmit vibrations to the **oval window**, which is a membrane beneath the stapes. The middle ear also opens into the **Eustachian tube**, which connects to the pharynx and equalizes pressure between the middle ear and the atmosphere. The **inner ear** consists of fluid-filled chambers, including the **semicircular canals**, which function in equilibrium, and the coiled **cochlea** (from the Latin meaning "snail"), a bony chamber that is involved in hearing.

2 The Cochlea

The cochlea has two large canals—an upper vestibular canal and a lower tympanic canal—separated by a smaller cochlear duct. Both canals are filled with fluid.

3 The Organ of Corti

The floor of the cochlear duct, the basilar membrane, bears the **organ of Corti**, which contains the mechanoreceptors of the ear, hair cells with hairs projecting into the cochlear duct. Many of the hairs are attached to the tectorial membrane, which hangs over the organ of Corti like an awning. Sound waves make the basilar membrane vibrate, which results in bending of the hairs and depolarization of the hair cells.

▲ Bundled hairs projecting from a single mammalian hair cell (SEM). Two shorter rows of hairs lie behind the tall hairs in the foreground.

4 Hair Cell

Projecting from each hair cell is a bundle of rod-shaped "hairs," each containing a core of actin filaments. Vibration of the basilar membrane in response to sound raises and lowers the hair cells, bending the hairs against the surrounding fluid and the tectorial membrane. When the hairs within the bundle are displaced, mechanoreceptors are activated, changing the membrane potential of the hair cell.

Hearing

Vibrating objects, such as a plucked guitar string or the vocal cords of your instructor, create pressure waves in the surrounding air. In *hearing*, the ear transduces this mechanical stimulus (pressure waves) into nerve impulses that the brain perceives as sound. To hear music, speech, or other sounds in our environment, we rely on **hair cells**, sensory receptors with hair-like projections on the cell surface that detect motion. Before the vibration waves reach the hair cells, however, they are amplified and transformed by several accessory structures.

The first steps in hearing involve structures in the ear that convert the vibrations of moving air to pressure waves in fluid. Upon reaching the outer ear, moving air causes the tympanic membrane to vibrate. The three bones of the middle ear transmit the vibrations to the oval window, a membrane on the cochlea's surface. When one of those bones, the stapes, vibrates against the oval window, it creates pressure waves in the fluid (called perilymph) inside the cochlea.

Upon entering the vestibular canal, the pressure waves push down on the cochlear duct and basilar membrane. In response, the basilar membrane and attached hair cells vibrate up and down. The hairs projecting from the moving hair cells are deflected by the tectorial membrane, which lies immediately above in a fixed position (see Figure 50.10). With each vibration, the hairs bend first in one direction and then the other. Mechanoreceptors in the hair cells respond by opening or closing ion channels. As shown in **Figure 50.11**, bending in one direction depolarizes hair cells, increasing neurotransmitter release and the frequency of action potentials directed to the brain along the auditory nerve. Bending the hairs in the other direction hyperpolarizes hair cells, reducing neurotransmitter release and the frequency of auditory nerve sensations.

What prevents pressure waves from reverberating within the ear and causing prolonged sensation? Once pressure waves travel through the vestibular canal, they pass around the apex (tip) of the cochlea. The waves then continue through the tympanic canal, dissipating as they strike the **round window** **(Figure 50.12a)**. This damping of sound waves resets the apparatus for the next vibrations that arrive.

The ear conveys information to the brain about two important sound variables: volume and pitch. *Volume* (loudness) is determined by the amplitude, or height, of the sound wave. A large-amplitude sound wave causes more vigorous vibration of the basilar membrane, greater bending of the hairs on hair cells, and more action potentials in the sensory neurons. *Pitch* is a function of a sound wave's frequency, the number of vibrations per unit time. High-frequency waves produce high-pitched sounds, whereas low-frequency waves produce low-pitched sounds. Pitch is commonly expressed in cycles per second, or hertz (Hz). Healthy young humans can hear in the range of 20–20,000 Hz; dogs can hear sounds as high as 40,000 Hz; and bats can emit and hear clicking sounds at frequencies above 100,000 Hz, using this ability to locate objects.

The cochlea can distinguish pitch because the basilar membrane is not uniform along its length: It is relatively narrow and stiff at the base of the cochlea near the oval window and wider and more flexible at the apex. Each region of the basilar membrane is tuned to a particular vibration frequency **(Figure 50.12b)**. At any instant, the region of the membrane

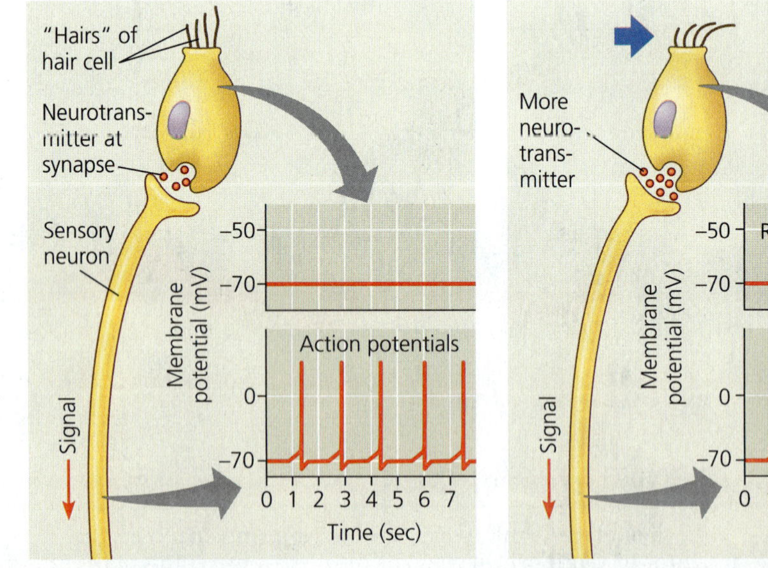
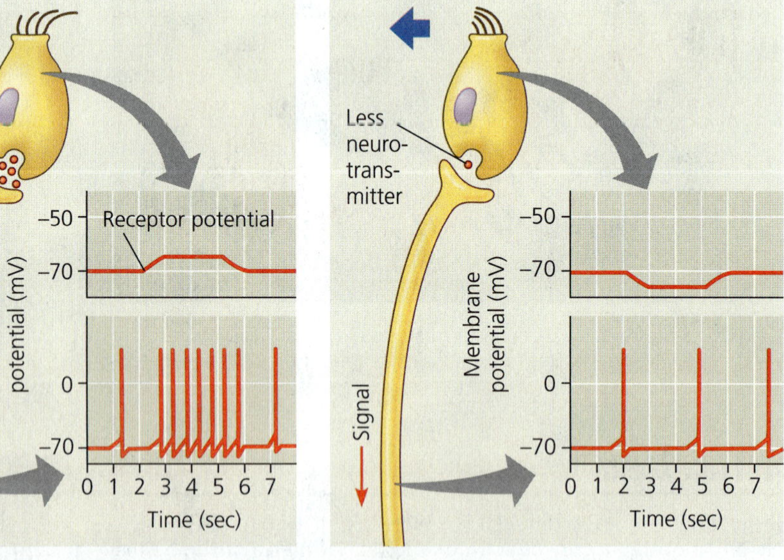

(a) No bending of hairs **(b) Bending of hairs in one direction** **(c) Bending of hairs in other direction**

▲ **Figure 50.11 Sensory reception by hair cells.** Vertebrate hair cells required for hearing and balance have "hairs" formed into a bundle that bends when surrounding fluid moves. Each hair cell releases an excitatory neurotransmitter at a synapse with a sensory neuron, which conducts action potentials to the CNS. Bending of the bundle in one direction depolarizes the hair cell, causing it to release more neurotransmitter and increasing the frequency of action potentials in the sensory neuron. Bending in the other direction has the opposite effect.

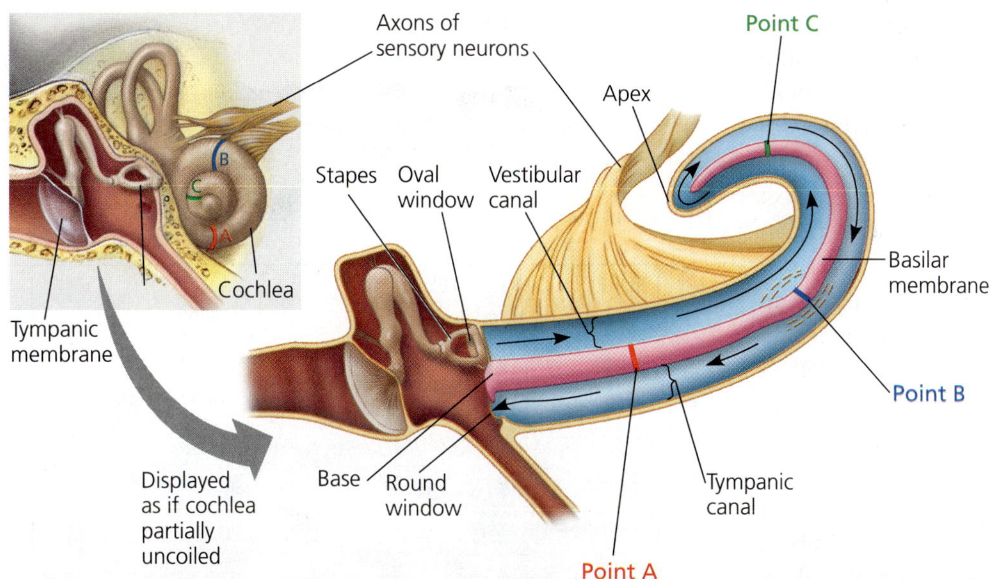

(a) Vibrations of the stapes against the oval window produce pressure waves (black arrows) in the fluid (perilymph; blue) of the cochlea. (For purposes of illustration, the cochlea on the right is drawn partially uncoiled.) The waves travel to the apex via the vestibular canal and back towards the base via the tympanic canal. The energy in the waves causes the basilar membrane (pink) to vibrate, stimulating hair cells (not shown). Because the basilar membrane varies in stiffness along its length, each point along the membrane vibrates maximally in response to waves of a particular frequency.

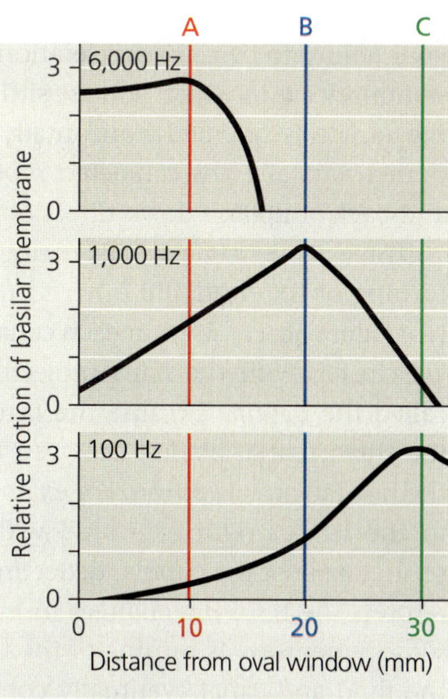

(b) These graphs show the patterns of vibration along the basilar membrane for three different frequencies, high (top), medium (middle), and low (bottom). The higher the frequency, the closer the vibration to the oval window.

▲ **Figure 50.12 Transduction in the cochlea.**

? *A musical chord consists of several notes, each formed by a sound wave of different frequency. When you hear a chord, where in your body are these notes combined?*

vibrating most vigorously triggers the highest frequency of action potentials in the neuronal pathway leading to the brain. There, within the cerebral cortex, the actual perception of pitch occurs. Axons in the auditory nerve project into auditory areas of the cerebral cortex according to the region of the basilar membrane in which the signal originated. When a particular site in our cortex is stimulated, we perceive the sound of a particular pitch.

Equilibrium

Several organs in the inner ear of humans and most other mammals detect body movement, position, and balance. Situated in a vestibule behind the oval window, the chambers called the **utricle** and **saccule** allow us to perceive position with respect to gravity or linear movement **(Figure 50.13)**. Each of these chambers contains a sheet of hair cells that project into a gelatinous material. Embedded in this gel are many small calcium carbonate particles called otoliths ("ear stones"). When you tilt your head, the otoliths press on the hairs protruding into the gel. Through the hair cell receptors, this deflection of the hairs is transformed into a change in the output of sensory neurons, signaling the brain that your head is at an angle. The otoliths are also responsible for

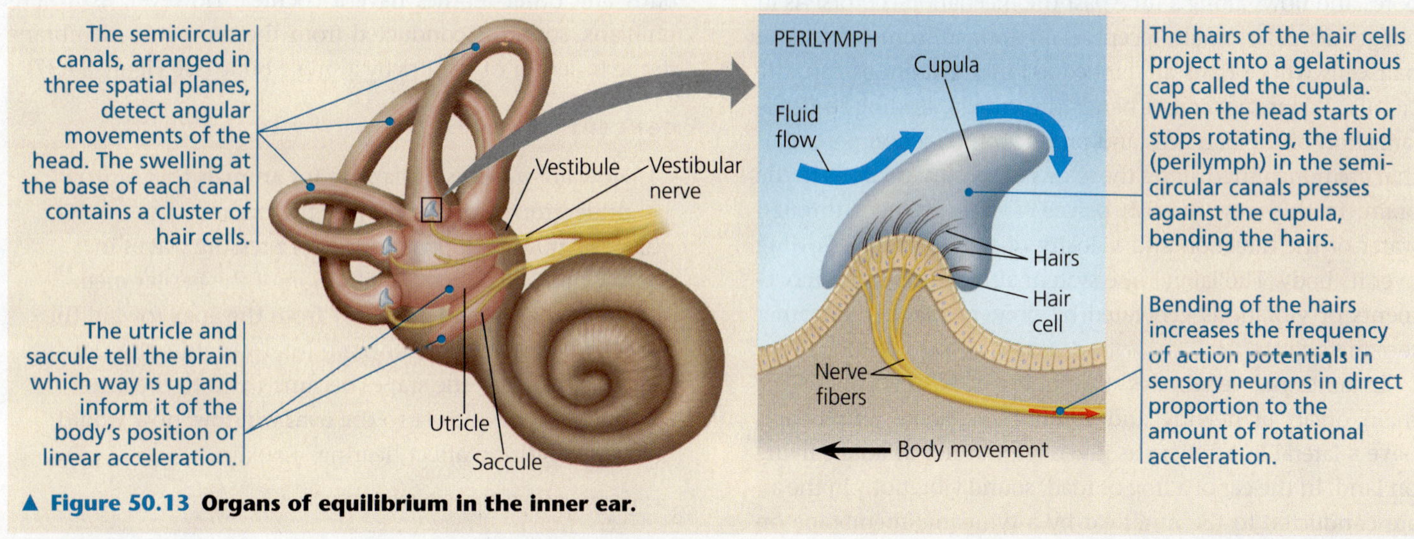

▲ **Figure 50.13 Organs of equilibrium in the inner ear.**

your ability to perceive acceleration, as, for example, when a stationary car in which you are sitting pulls forward. Because the utricle is oriented horizontally and the saccule is positioned vertically, you can detect motion in either the forward-and-back or up-and-down direction.

Three semicircular canals connected to the utricle detect turning of the head and other forms of angular acceleration (see Figure 50.11). Within each canal the hair cells form a single cluster, with the hairs projecting into a gelatinous cap called the cupula. Because the three canals are arranged in the three spatial planes, they can detect angular motion of the head in any direction. For example, if you turn your head to the left or right, the fluid within the horizontal canal pushes against the cupula, deflecting the hairs. The brain interprets the resulting changes in impulse production by the sensory neurons as turning of the head. If you spin in place, the fluid and canal eventually come to equilibrium and remain in that state until you stop. At that point, the moving fluid encounters a stationary cupula, triggering the false sensation of angular motion that we call dizziness.

Hearing and Equilibrium in Other Vertebrates

Unlike the mammalian hearing apparatus, the ear of a fish does not open to the outside of the body and has no eardrum or cochlea. The vibrations of the water caused by sound waves are conducted through the skeleton of the head to a pair of inner ears, setting otoliths in motion and stimulating hair cells. The fish's air-filled swim bladder (see Figure 34.16) also vibrates in response to sound. Some fishes, including catfishes and minnows, have a series of bones that conduct vibrations from the swim bladder to the inner ear.

As discussed in Chapter 34, most fishes and aquatic amphibians have a **lateral line system** along both sides of their body **(Figure 50.14)**. The system contains mechanoreceptors that detect low-frequency waves by a mechanism similar to that of the mammalian inner ear. Water from the animal's surroundings enters the lateral line system through numerous pores and flows along a tube past the mechanoreceptors. As in our semicircular canals, receptors are formed from a cluster of hair cells whose hairs are embedded in a gelatinous cap, the cupula. Water movement bends the cupula, leading to depolarization of the hair cells and production of action potentials that are transmitted along the axons of sensory neurons to the brain. In this way, the fish perceives its movement through water or the direction and velocity of water currents flowing over its body. The lateral line system also detects water movements or vibrations generated by prey, predators, and other moving objects.

In terrestrial vertebrates, the inner ear has evolved as the main organ of hearing and equilibrium. Some amphibians have a lateral line system as juveniles, but not as adults living on land. In the ear of a frog or toad, sound vibrations in the air are conducted to the inner ear by a tympanic membrane on the body surface and a single middle ear bone. Like mammals, birds and other reptiles have a cochlea. However, as in amphibians, sound is conducted from the tympanic membrane to the inner ear of reptiles by a single bone (see Figure 34.37).

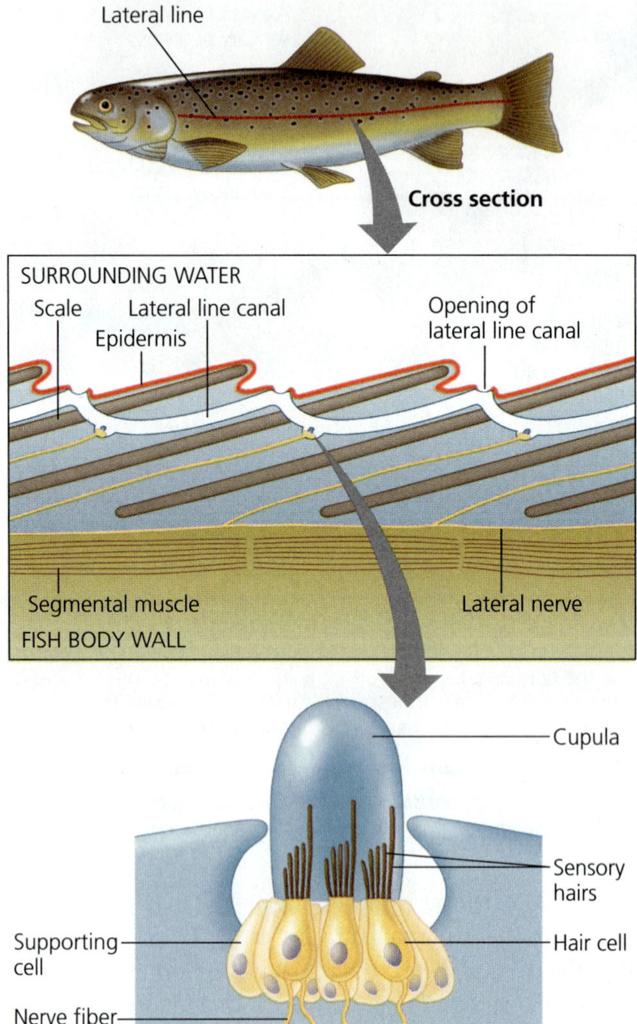

▲ **Figure 50.14 The lateral line system in a fish.** Water flowing through the system bends hair cells. The hair cells transduce the energy into receptor potentials, triggering action potentials that are conveyed to the brain. The lateral line system enables a fish to monitor water currents, pressure waves produced by moving objects, and low-frequency sounds conducted through the water.

CONCEPT CHECK 50.2

1. How are statocysts adaptive for animals that burrow underground or live deep in the ocean?
2. **WHAT IF?** Suppose a series of pressure waves in your cochlea caused a vibration of the basilar membrane that moves gradually from the apex toward the base. How would your brain interpret this stimulus?
3. **WHAT IF?** If the stapes became fused to the other middle ear bones or to the oval window, how would this condition affect hearing? Explain.

For suggested answers, see Appendix A.

CONCEPT 50.3

Visual receptors in diverse animals depend on light-absorbing pigments

The ability to detect light has a central role in the interaction of nearly all animals with their environment. Although animals use a diverse set of organs for vision, the underlying mechanism for capturing light is the same, suggesting a common evolutionary origin.

Evolution of Visual Perception

EVOLUTION Light detectors in the animal kingdom range from simple clusters of cells that detect only the direction and intensity of light to complex organs that form images. These diverse light detectors all contain **photoreceptors**, cells that contain light-absorbing pigment molecules. Furthermore, the genes that specify where and when photoreceptors arise during embryonic development are shared among animals as diverse as flatworms, annelids, arthropods, and vertebrates. It is thus very probable that the genetic underpinnings of all photoreceptors were already present in the earliest bilaterian animals.

Light-Detecting Organs

Most invertebrates have some kind of light-detecting organ. One of the simplest is that of planarians (Figure 50.15). A pair of ocelli (singular, *ocellus*), which are sometimes called eyespots, are located in the head region. A layer of darkly pigmented cells surrounds the ocelli on three sides, blocking light. Photoreceptors in each ocellus receive light only through the opening where there are no pigmented cells. Because the opening of one ocellus faces left and slightly forward and that of the other ocellus faces right and forward, light shining from one side of the planarian stimulates only the ocellus on that side. The planarian brain compares the rate of action potentials coming from the two ocelli and directs turning movements that minimize the stimulation of both ocelli. The result is that the planarian moves away from the light source until it reaches a shaded location, where a rock or other object is likely to hide the animal from predators.

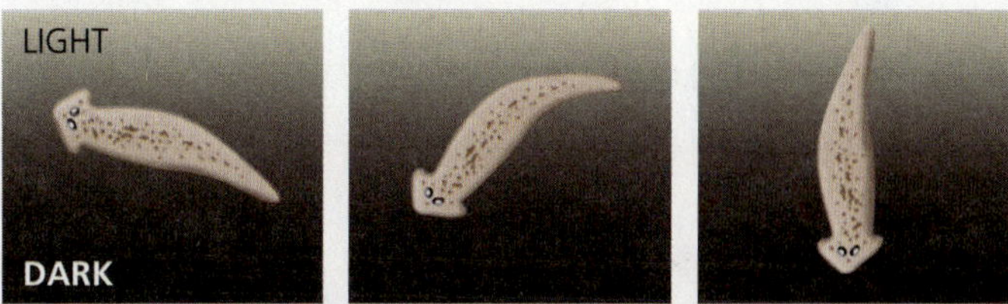

(a) The planarian's brain directs the body to turn until the sensations from the two ocelli are equal and minimal, causing the animal to move away from light.

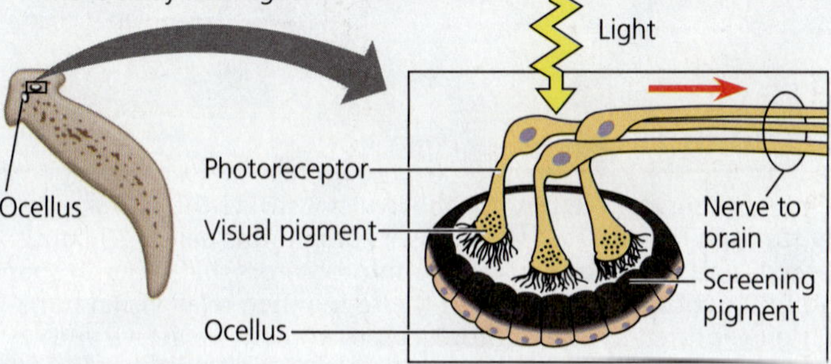

(b) Whereas light striking the front of an ocellus excites the photoreceptors, light striking the back is blocked by the screening pigment. In this way, the ocelli indicate the direction of a light source, triggering the light avoidance behavior.

▲ **Figure 50.15 Ocelli and orientation behavior of a planarian.**

Compound Eyes

Insects and crustaceans (phylum Arthropoda) have compound eyes, as do some polychaete worms (phylum Annelida). A **compound eye** consists of up to several thousand light detectors called **ommatidia** (the "facets" of the eye), each with its own light-focusing lens (Figure 50.16). Each ommatidium detects light from a tiny portion of the visual field. A compound eye is very effective at detecting movement, an important adaptation for flying insects and small animals constantly threatened with predation. Whereas the

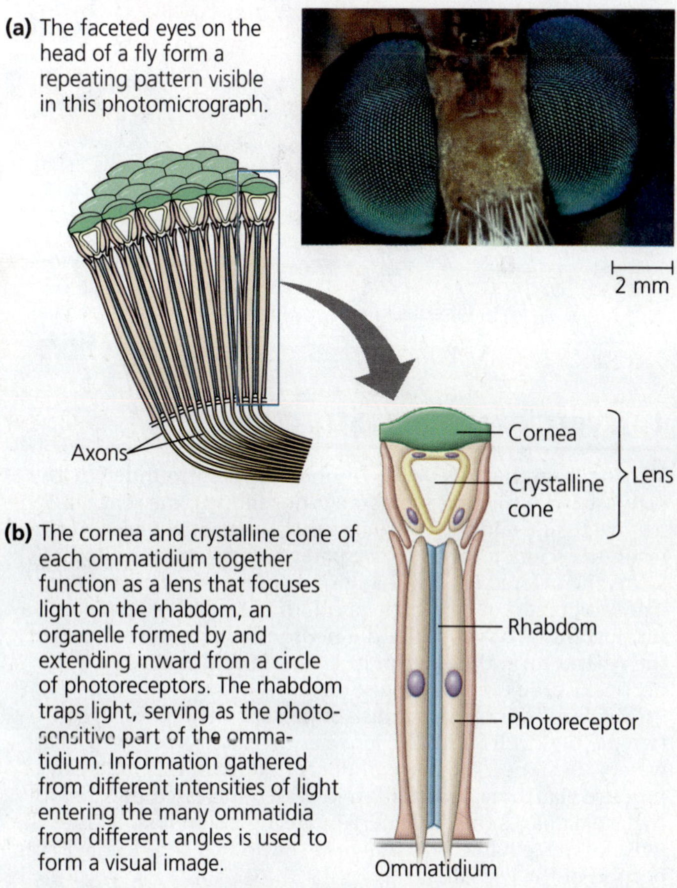

(a) The faceted eyes on the head of a fly form a repeating pattern visible in this photomicrograph.

(b) The cornea and crystalline cone of each ommatidium together function as a lens that focuses light on the rhabdom, an organelle formed by and extending inward from a circle of photoreceptors. The rhabdom traps light, serving as the photosensitive part of the ommatidium. Information gathered from different intensities of light entering the many ommatidia from different angles is used to form a visual image.

▲ **Figure 50.16 Compound eyes.**

human eye can distinguish only about 50 flashes of light per second, the compound eyes of some insects can detect flickering at six times that rate. (If they slipped into a movie theater, these insects could easily resolve each frame of the film being projected as a separate still image.) Insects also have excellent color vision, and some (including bees) can see into the ultraviolet (UV) range of the electromagnetic spectrum. Because UV light is invisible to humans, we miss seeing differences in the environment that bees and other insects detect. In studying animal behavior, we cannot simply extrapolate our sensory world to other species; different animals have different sensitivities and different brain organizations.

Single-Lens Eyes

Among invertebrates, **single-lens eyes** are found in some jellies and polychaete worms, as well as in spiders and many molluscs. A single-lens eye works somewhat like a camera. The eye of an octopus or squid, for example, has a small opening, the **pupil**, through which light enters. Like a camera's adjustable aperture, the **iris** contracts or expands, changing the

▼ **Figure 50.17**

Exploring The Structure of the Human Eye

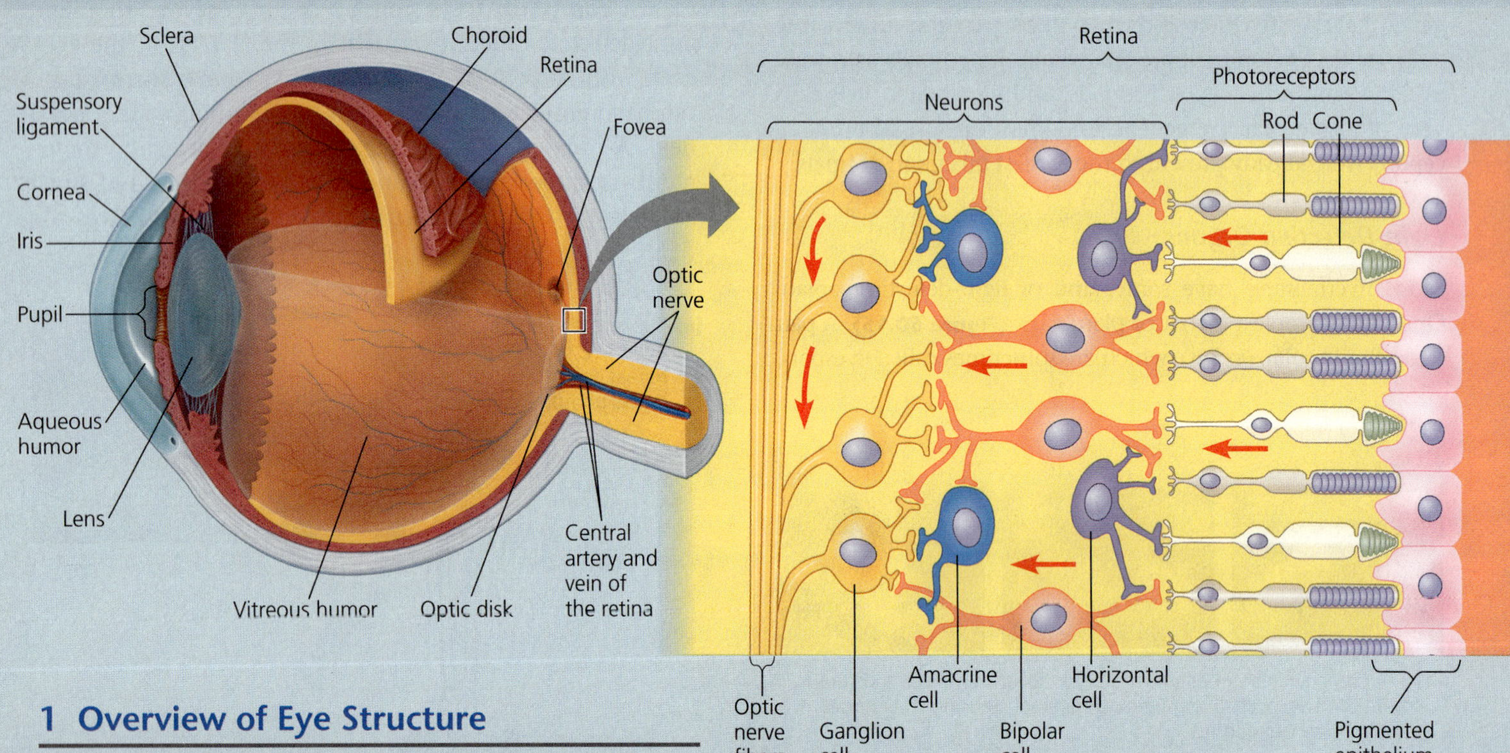

1 Overview of Eye Structure

Starting from the outside, the human eye is surrounded by the conjunctiva, a mucous membrane (not shown); the sclera, a connective tissue; and the choroid, a thin, pigmented layer. At the front, the sclera forms the transparent *cornea* and the choroid forms the colored *iris*. By changing size, the iris regulates the amount of light entering the pupil, the hole in the center of the iris. Just inside the choroid, the neurons and photoreceptors of the **retina** form the innermost layer of the eyeball. The optic nerve exits the eye at the optic disk.

The **lens**, a transparent disk of protein, divides the eye into two cavities. In front of the lens lies the *aqueous humor,* a clear watery substance. Blockage of ducts that drain this fluid can produce glaucoma, a condition in which increased pressure in the eye damages the optic nerve, causing vision loss. Behind the lens lies the jellylike *vitreous humor* (illustrated here in the lower portion of the eyeball).

2 The Retina

Light (coming from left in the above view) strikes the retina, passing through largely transparent layers of neurons before reaching the rods and cones, two types of photoreceptors that differ in shape and in function. The neurons of the retina then relay visual information captured by the photoreceptors to the optic nerve and brain along the pathways shown with red arrows. Each *bipolar cell* receives information from several rods or cones, and each *ganglion cell* gathers input from several bipolar cells. *Horizontal* and *amacrine cells* integrate information across the retina.

One region of the retina, the optic disk, lacks photoreceptors. As a result, this region forms a "blind spot" where light is not detected.

diameter of the pupil to let in more or less light. Behind the pupil, a single lens focuses light on a layer of photoreceptors. Similar to a camera's focusing action, muscles in an invertebrate's single-lens eye move the lens forward or backward, focusing on objects at different distances.

The eyes of all vertebrates have a single lens. In fishes, focusing is as in invertebrates, with the lens moving forward or backward. In other species, including mammals, focusing is achieved by changing the shape of the lens. We will learn about this mechanism, as well as explore visual perception in much more detail, as we shift our attention to the vertebrate visual system.

The Vertebrate Visual System

The human eye will serve as our model of vision in vertebrates. As detailed in **Figure 50.17,** vision begins when photons of light enter the eye and strike the rods and cones. There the energy of each photon is captured by a shift in configuration of a single chemical bond in retinal.

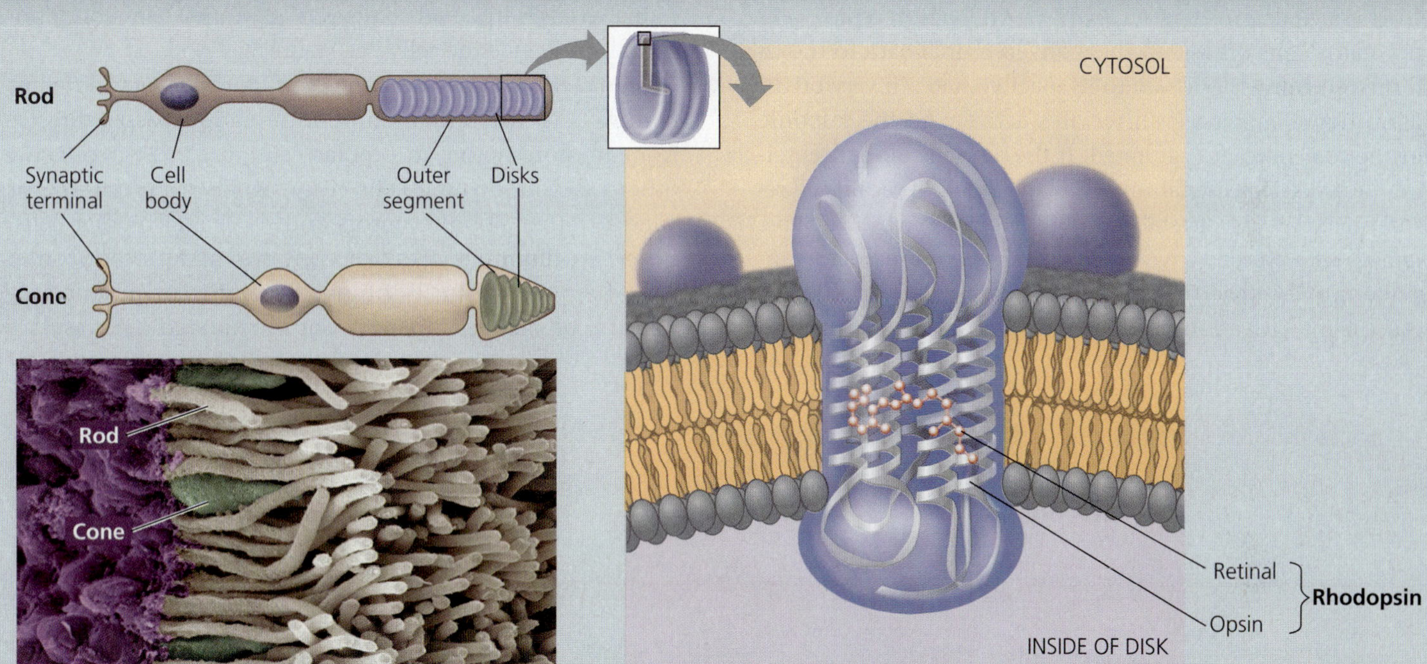

3 Photoreceptor Cells

Humans have two main types of photoreceptor cells: rods and cones. Within the outer segment of a rod or cone is a stack of membranous disks in which *visual pigments* are embedded. **Rods** are more sensitive to light but do not distinguish colors; they enable us to see at night, but only in black and white. **Cones** provide color vision, but, being less sensitive, contribute very little to night vision. There are three types of cones. Each has a different sensitivity across the visible spectrum, providing an optimal response to red, green, or blue light.

In the colorized SEM shown above, cones (green), rods (light tan), and adjacent neurons (purple) are visible. The pigmented epithelium, which was removed in this preparation, would be to the right.

4 Visual Pigments

Vertebrate visual pigments consist of a light-absorbing molecule called **retinal** (a derivative of vitamin A) bound to a membrane protein called an **opsin**. Seven α helices of each opsin molecule span the disk membrane. The visual pigment of rods, shown here, is called **rhodopsin**.

Retinal exists as two isomers. Absorption of light shifts one bond in retinal from a *cis* to a *trans* arrangement, converting the molecule from an angled shape to a straight shape. This change in configuration destabilizes and activates the opsin protein to which retinal is bound.

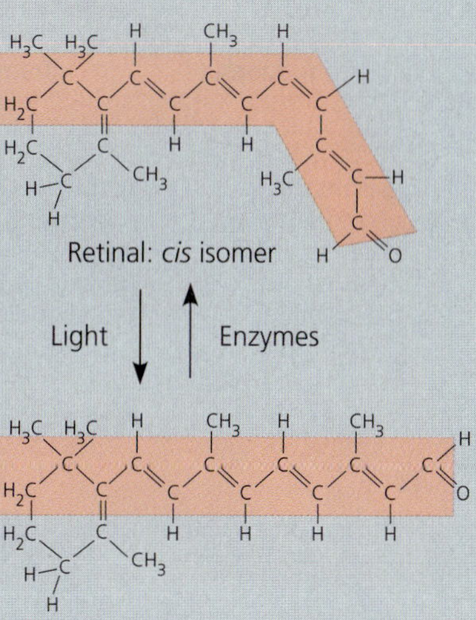

Although light detection in the eye is the first stage in vision, remember that it is actually the brain that "sees." Thus, to understand vision, we must examine how the capture of light by retinal changes the production of action potentials and then follow these signals to the visual centers of the brain, where images are perceived.

Sensory Transduction in the Eye

The transduction of visual information to the nervous system begins with the light-induced conversion of *cis*-retinal to *trans*-retinal. As shown in **Figure 50.18**, this conversion activates rhodopsin, which activates a G protein, which in turn activates an enzyme that can hydrolyze cyclic GMP. In the dark, cyclic GMP in photoreceptor cells binds to sodium ion (Na^+) channels and keeps them open. When the G protein-dependent pathway is activated, cyclic GMP is broken down, Na^+ channels close, and the cell becomes hyperpolarized.

The signal transduction pathway in photoreceptor cells normally shuts off as enzymes convert retinal back to the *cis* form, returning rhodopsin to its inactive state. In very bright light, however, rhodopsin remains active, and the response in the rods becomes saturated. If the amount of light entering the eyes decreases abruptly, the rods do not regain full responsiveness for several minutes. This is why you are temporarily blinded if you pass quickly from the bright sunshine into a movie theater or other dark environment. (Because light activation changes the color of rhodopsin from purple to yellow, rods in which the light response is saturated are often described as "bleached.")

Processing of Visual Information in the Retina

The processing of visual information begins in the retina itself, where both rods and cones form synapses with bipolar cells **(Figure 50.19)**. In the dark, rods and cones are depolarized and continually release the neurotransmitter glutamate at these synapses (see Table 48.2). Some bipolar cells depolarize in response to glutamate, whereas others hyperpolarize. Which of the two responses a bipolar cell exhibits depends on the type of glutamate receptor on its surface at the synapse. When light strikes the rods and cones, they hyperpolarize, shutting off their release of glutamate. In response, the bipolar cells that are depolarized by glutamate hyperpolarize, and those that are hyperpolarized by glutamate depolarize.

In addition to bipolar cells, information processing in the retina requires three other types of neurons—ganglion, horizontal, and amacrine cells (see Figure 50.17).

Signals from rods and cones can follow several different pathways in the retina. Some information passes directly from photoreceptors to bipolar cells to ganglion cells. In other cases, horizontal cells carry signals from one rod or cone to other photoreceptors and to several bipolar cells. When an illuminated rod or cone stimulates a horizontal cell, the horizontal cell inhibits more distant photoreceptors and bipolar cells that are not illuminated. The result is

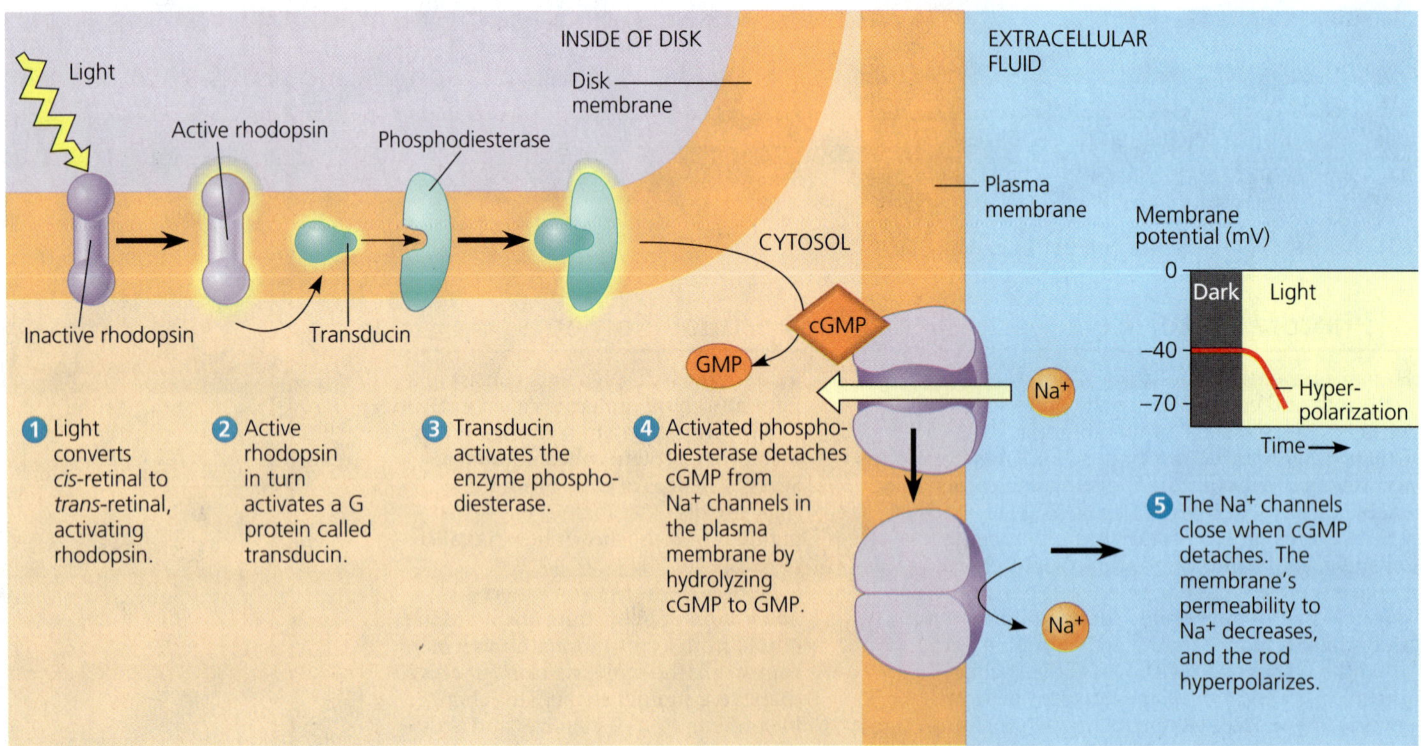

▲ **Figure 50.18 Production of the receptor potential in a rod cell.** In rods (and cones), the receptor potential triggered by light is a hyperpolarization, not a depolarization.

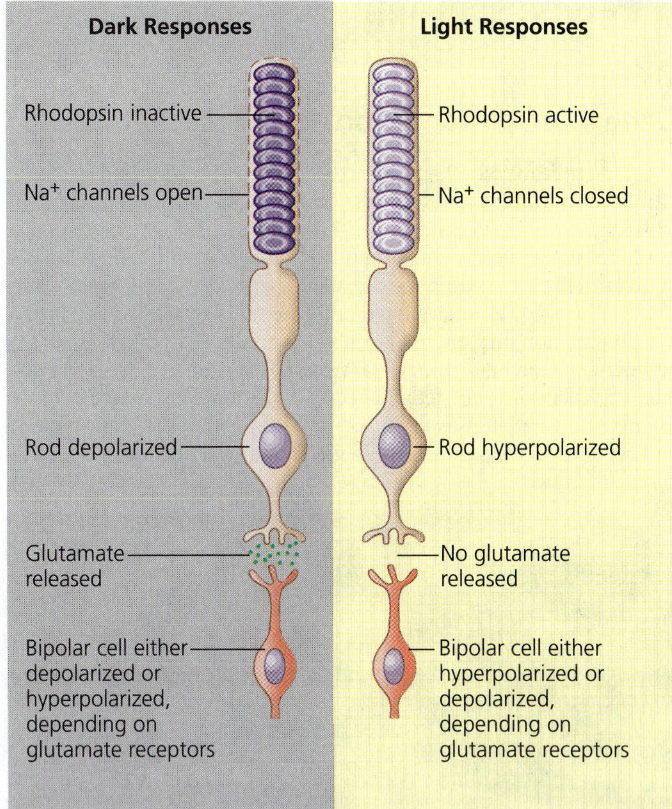

▲ **Figure 50.19 Synaptic activity of rod cells in light and dark.**

? *Like rods, cone cells are depolarized when rhodopsin is inactive. In the case of a cone, why might it be misleading to call this a dark response?*

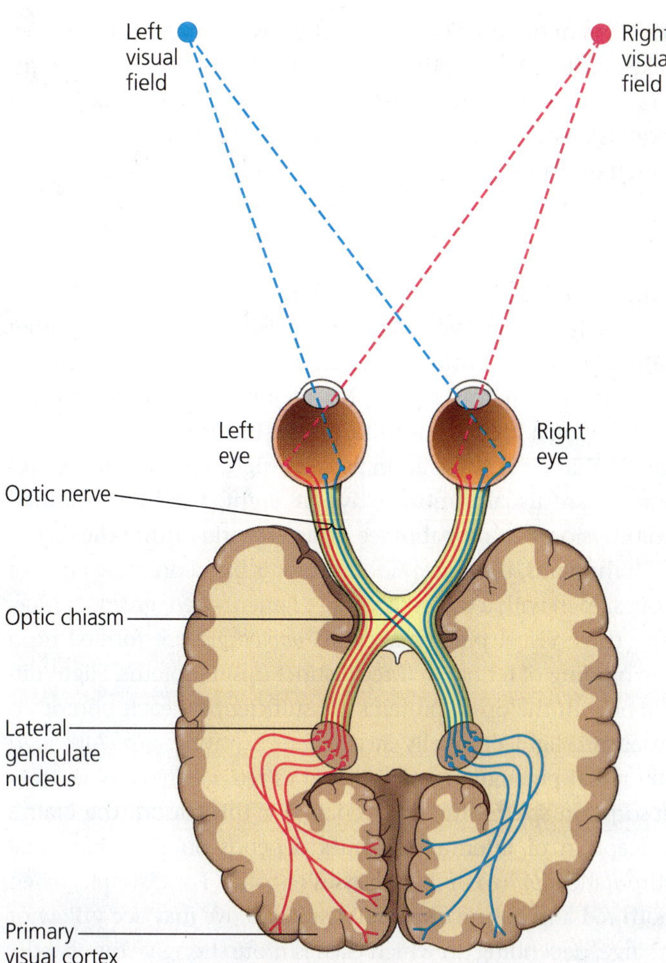

▲ **Figure 50.20 Neural pathways for vision.** Each optic nerve contains about a million axons that synapse with interneurons in the lateral geniculate nuclei. The nuclei relay sensations to the primary visual cortex, one of many brain centers that cooperate in constructing our visual perceptions.

that the region receiving light appears lighter and the dark surroundings even darker. This form of integration, called **lateral inhibition**, sharpens edges and enhances contrast in the image. Amacrine cells distribute some information from one bipolar cell to several ganglion cells. Lateral inhibition is repeated by the interactions of the amacrine cells with the ganglion cells and occurs at all levels of visual processing in the brain.

A single ganglion cell receives information from an array of rods and cones, each of which responds to light coming from a particular location. Together, the rods or cones that feed information to one ganglion cell define a *receptive field*—the part of the visual field to which the ganglion can respond. The fewer rods or cones that supply a single ganglion cell, the smaller the receptive field. A smaller receptive field results in a sharper image, because the information as to where light has struck the retina is more precise. The ganglion cells of the fovea have very small receptive fields, so visual acuity (sharpness) in the fovea is high.

Processing of Visual Information in the Brain

Axons of ganglion cells form the optic nerves that transmit sensations from the eyes to the brain **(Figure 50.20)**. The two optic nerves meet at the **optic chiasm** near the center of the base of the cerebral cortex. Axons in the optic nerves are routed at the optic chiasm such that sensations from the left visual field of both eyes are transmitted to the right side of the brain, and sensations from the right visual field are transmitted to the left side of the brain. (Note that each visual field, whether right or left, involves input from both eyes.)

Within the brain, most ganglion cell axons lead to the **lateral geniculate nuclei**, which have axons that reach the **primary visual cortex** in the cerebrum. Additional neurons carry the information to higher-order visual processing and integrating centers elsewhere in the cortex.

Point-by-point information in the visual field is projected along neurons onto the visual cortex. How does the cortex convert a complex set of action potentials representing two-dimensional images focused on the retina to three-dimensional perceptions of our surroundings? Researchers estimate that at least 30% of the cerebral cortex, comprising hundreds of millions of neurons in perhaps dozens of integrating centers, takes part in formulating what we actually "see." Determining how these centers integrate such components of our vision as color, motion, depth, shape, and detail is the focus of much exciting research.

Color Vision

Among vertebrates, most fishes, amphibians, and reptiles, including birds, have very good color vision. Humans and other primates also see color well, but are among the minority of mammals with this ability. Many mammals are nocturnal, and having a high proportion of rods in the retina is an adaptation that gives these animals keen night vision. Cats, for instance, are usually most active at night; they have limited color vision and probably see a pastel world during the day.

In humans, the perception of color is based on three types of cones, each with a different visual pigment—red, green, or blue. The three visual pigments, called *photopsins*, are formed from the binding of retinal to three distinct opsin proteins. Slight differences in the opsin proteins are sufficient for each photopsin to absorb light optimally at a different wavelength. Although the visual pigments are designated as red, green, or blue, their absorption spectra in fact overlap. For this reason, the brain's perception of intermediate hues depends on the differential stimulation of two or more classes of cones. For example, when both red and green cones are stimulated, we may see yellow or orange, depending on which class is more strongly stimulated.

Abnormal color vision typically results from alterations in the genes for one or more photopsin proteins. Because the human genes for the red and green pigments are located on the X chromosome, a single defective copy of either gene can disrupt color vision in males (see Figure 15.7 to review the genetics of sex-linked traits). For this reason, color blindness is more common in males than in females (5–8% of males, fewer than 1% of females) and nearly always disrupts perception of red or green (the gene for blue pigment is on human chromosome 7).

Color blindness is also more common among males than females in squirrel monkeys (*Saimiri sciureus*), providing a good experimental model for studying this disorder. In 2009, researchers studying color blindness in squirrel monkeys made a significant breakthrough in the field of gene therapy **(Figure 50.21)**.

▼ **Figure 50.21**

IMPACT

Gene Therapy for Vision

Seeking to learn whether a defect in color vision could be remedied in an adult animal, researchers chose to study squirrel monkeys, which have only two opsin genes. The opsin encoded by one gene is sensitive to blue light, while the other opsin is sensitive to either red or green light, depending on the allele. Because the red/green opsin gene is X-linked, all males have only the red-sensitive or the green-sensitive version and are red-green color-blind. In a gene therapy experiment, researchers injected a virus containing the gene for the missing version into the retina of adult male monkeys. After 20 weeks, the new opsin allele was being expressed in cones and the monkeys had begun to distinguish red from green in a field of colored dots.

WHY IT MATTERS These experiments demonstrate that the neural circuits required to process visual information can be generated or activated even in adults, opening up the possibility for treating a range of vision disorders by gene therapy. Indeed, gene therapy has already been used to treat Leber's congenital amaurosis (LCA), an inherited retinal degenerative disease that causes severe loss of vision at birth. After using gene therapy to restore vision in dogs and mice with LCA, researchers successfully treated the disease in humans by injecting the functional LCA gene in a viral vector.

FURTHER READING F. P. M. Cremers and R. W. J. Collin, Promises and challenges of genetic therapy for blindness, *The Lancet* 374:1569–1570 (2009).

MAKE CONNECTIONS Red-green color blindness is X-linked in squirrel monkeys and humans (see Figure 15.7, p. 291). Why is the inheritance pattern in humans not apparent in squirrel monkeys?

The Visual Field

The brain not only processes visual information, but also controls what information is captured. One important type of control is focusing, which, as noted earlier and illustrated in **Figure 50.22**, occurs by changing the shape of the lens. When you focus on a close object, your lens becomes almost spherical. When you view a distant object, your lens is flattened. By turning your head and pointing your eyes in a particular direction, your brain also determines what lies in your field of vision.

Although our peripheral vision allows us to see objects over a nearly 180° range, the distribution of photoreceptors across the eye limits both what we see and how well we see it. Overall, the human retina contains about 125 million rods and about 6 million cones. At the **fovea**, the center of the visual field, there are no rods but a very high density of cones—about 150,000 cones per square millimeter. The ratio of rods to cones increases with distance from the fovea, with the peripheral regions having only rods. In daylight, you achieve your sharpest vision by looking directly at an object, such that light shines on the tightly packed cones in your fovea.

▼ Figure 50.22 Focusing in the mammalian eye. Ciliary muscles control the shape of the lens, which bends light and focuses it on the retina. The thicker the lens, the more sharply the light is bent.

(a) Near vision (accommodation)

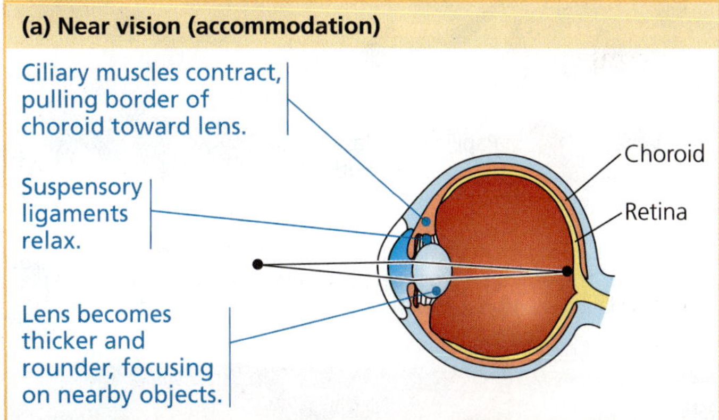

(b) Distance vision

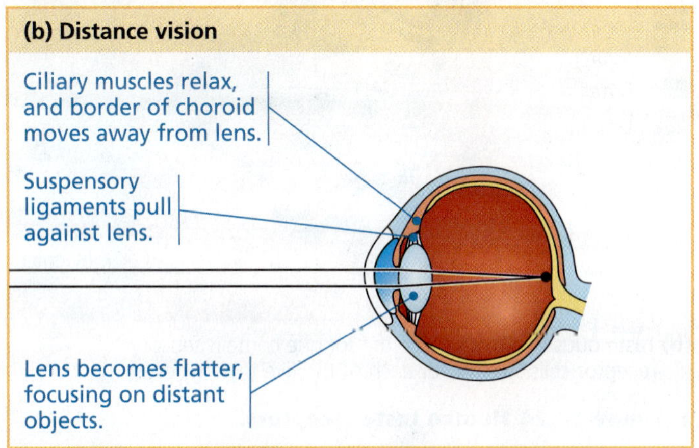

At night, looking directly at a dimly lit object is ineffective, since the rods—the more sensitive light receptors—are found outside the fovea. Thus, for example, you see a dim star best by focusing on a point just to one side of it.

CONCEPT CHECK 50.3

1. Contrast the light-detecting organs of planarians and flies. How is each organ adaptive for the lifestyle of the animal?
2. In a condition called presbyopia, the eyes' lenses lose much of their elasticity and maintain a flat shape. Explain how this condition affects a person's vision.
3. **WHAT IF?** If you perceive an object floating across your field of view, how might you determine whether the image represents a real object rather than a disturbance in your eye or in a neural circuit of your brain?
4. **MAKE CONNECTIONS** Compare the function of retinal in the eye with that of the pigment chlorophyll in a plant photosystem (see Concept 10.2, pp. 190–194).

For suggested answers, see Appendix A.

CONCEPT **50.4**

The senses of taste and smell rely on similar sets of sensory receptors

Many animals use their chemical senses to find mates (as when male silk moths respond to pheromones emitted by females), to recognize territory that has been marked by some chemical substance (as when dogs and cats sniff boundaries that have been staked out by their spraying neighbors), and to help navigate during migration (as when salmon use the unique scent of their streams of origin to return there for breeding). Animals such as ants and bees that live in large social groups rely extensively on chemical "conversation." In all animals, chemical senses are important in feeding behavior. For example, a hydra retracts its tentacles toward its mouth when it detects the compound glutathione, which is released from prey captured by the tentacles.

The perceptions of **gustation** (taste) and **olfaction** (smell) both depend on chemoreceptors that detect specific chemicals in the environment. In the case of terrestrial animals, taste is the detection of chemicals called **tastants** that are present in a solution, and smell is the detection of **odorants** that are carried through the air. There is no distinction between taste and smell in aquatic animals.

The taste receptors of insects are located within sensory hairs located on the feet and in mouthparts. These animals use their sense of taste to select food. A tasting hair contains several chemoreceptors, each especially responsive to a particular class of tastant, such as sugar or salt. Insects are also capable of smelling airborne odorants using olfactory hairs, usually located on their antennae (see Figure 50.6). The chemical DEET (*N,N*-diethyl-meta-toluamide), sold as an insect "repellant," actually protects against bites by blocking the olfactory receptor in mosquitoes that detects human scent.

Taste in Mammals

Humans and other mammals recognize five types of tastants. Four represent the familiar taste perceptions—sweet, sour, salty, and bitter. The fifth, called umami (Japanese for "delicious"), is elicited by the amino acid glutamate. Often used as a flavor enhancer, monosodium glutamate (MSG) occurs naturally in foods such as meat and aged cheese, imparting a quality sometimes described as savory. Researchers have identified the receptor proteins for all of the tastes except salty.

For decades, many researchers assumed that a taste cell could have more than one type of receptor. An alternative idea is that each taste cell has a single receptor type, programming the cell to recognize only one of the five tastes. Which hypothesis is correct? In 2005, scientists at the University of California, San Diego, used a cloned bitter taste receptor to genetically

▼ **Figure 50.23** INQUIRY

How do mammals detect different tastes?

EXPERIMENT To investigate the basis of mammalian taste perception, Ken Mueller, Nick Ryba, and Charles Zuker used a chemical called phenyl-β-D-glucopyranoside (PBDG). Humans find the taste of PBDG extremely bitter. Mice, however, appear to lack a receptor for PBDG. Whereas mice avoid drinking water containing other bitter tastants, they show no aversion to water that contains PBDG.

Using a molecular cloning strategy, Mueller generated mice that made the human PBDG receptor in cells that normally make either a sweet receptor or a bitter receptor. The mice were given a choice of two bottles, one filled with pure water and one filled with water containing PBDG at varying concentrations. The researchers then observed whether the mice had an attraction or an aversion to PBDG.

RESULTS

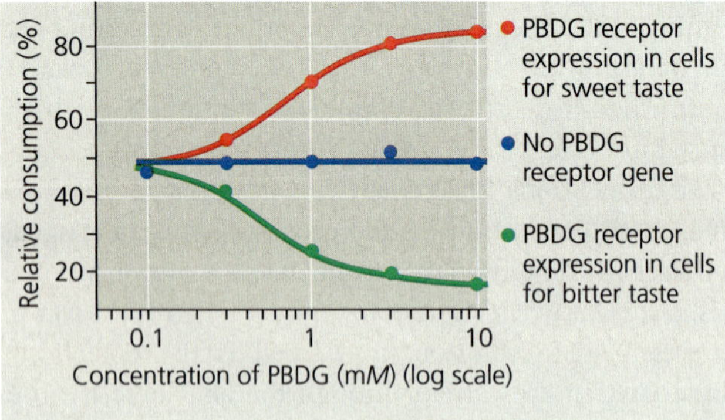

Relative consumption = (Fluid intake from bottle containing PBDG ÷ Total fluid intake) × 100%

CONCLUSION The researchers found that the presence of a bitter receptor in sweet taste cells is sufficient to cause mice to be attracted to a bitter chemical. They concluded that the mammalian brain must therefore perceive sweet or bitter taste solely on the basis of which sensory neurons are activated.

SOURCE K. L. Mueller et al., The receptors and coding logic for bitter taste, *Nature* 434:225–229 (2005).

WHAT IF? Suppose instead of the PBDG receptor the researchers had used a receptor specific for a sweetener that humans crave but mice ignore. How would the results of the experiment have differed?

reprogram gustation in a mouse **(Figure 50.23)**. Based on these and other studies, the researchers concluded that an individual taste cell expresses a single receptor type and detects tastants representing only one of the five tastes.

The receptor cells for taste in mammals are modified epithelial cells organized into **taste buds**, which are scattered in several areas of the tongue and mouth **(Figure 50.24)**. Most taste buds on the tongue are associated with nipple-shaped projections called papillae. Any region of the tongue with taste buds can detect any of the five types of taste. (The frequently reproduced "taste maps" of the tongue are thus not accurate.)

Taste receptors fall into two categories, each evolutionarily related to receptors for other senses. The sensation of sweet,

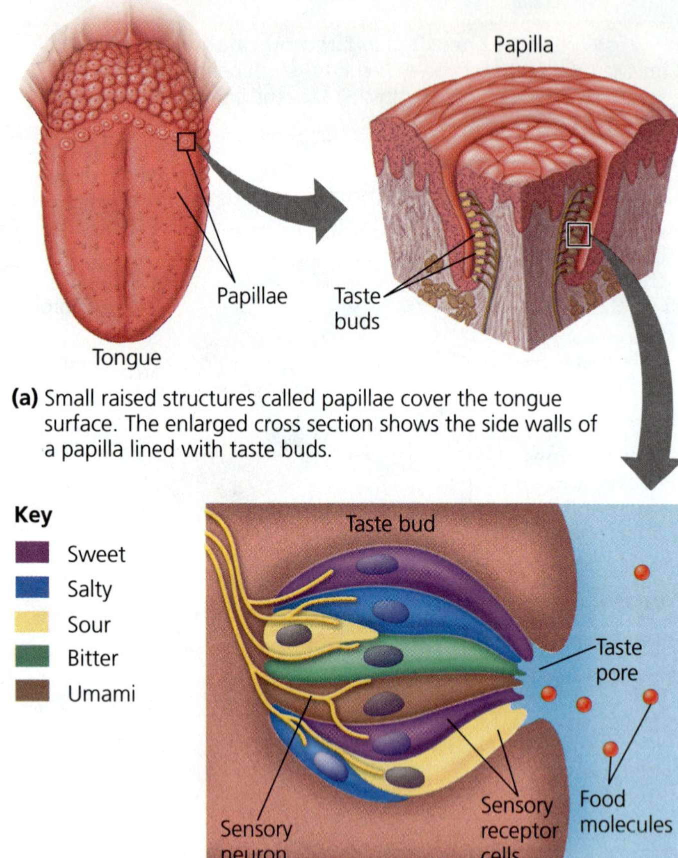

(a) Small raised structures called papillae cover the tongue surface. The enlarged cross section shows the side walls of a papilla lined with taste buds.

(b) Taste buds in all regions of the tongue contain sensory receptor cells specific for each of the five taste types.

▲ **Figure 50.24 Human taste receptors.**

umami, and bitter tastes requires a G protein-coupled receptor, or GPCR (see Figure 11.7). In humans, there are more than 30 different receptors for bitter taste, each able to recognize multiple bitter tastants. In contrast, humans have one type of sweet receptor and one type of umami receptor, each assembled from a different pair of GPCR proteins. Other GPCR proteins are critical for the sense of smell, as we will discuss shortly.

Unlike the other identified taste receptors, the receptor for sour tastants belongs to the TRP family (see p. 1089). Formed from a pair of TRP proteins, the sour receptor is similar to the capsaicin receptor and other thermoreceptor proteins. In taste buds, the TRP proteins of the sour receptor assemble into a channel in the plasma membrane of the taste cell. Binding of an acid or other sour-tasting substance to the receptor triggers a change in the ion channel. Depolarization occurs, activating a sensory neuron.

Smell in Humans

In olfaction, unlike gustation, the sensory cells are neurons. Olfactory receptor cells line the upper portion of the nasal cavity and send impulses along their axons directly to the olfactory bulb of the brain **(Figure 50.25)**. The receptive ends

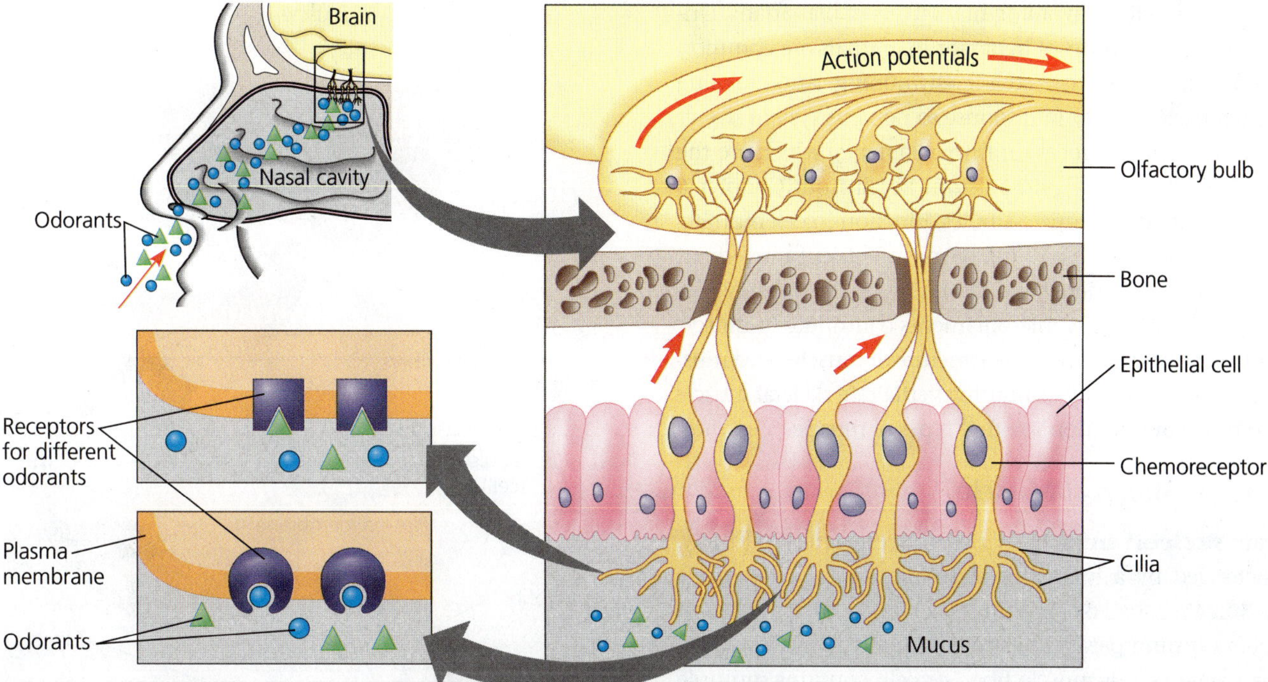

▲ **Figure 50.25 Smell in humans.** Odorant molecules bind to specific receptor proteins in the plasma membrane of olfactory receptor cells, triggering action potentials.

WHAT IF? *If you spray an "air freshener" in a musty room, would you be affecting detection, transmission, or perception of the odorants responsible for the musty smell?*

of the cells contain cilia that extend into the layer of mucus coating the nasal cavity. When an odorant diffuses into this region, it binds to a specific GPCR protein called an odorant receptor (OR) on the plasma membrane of the olfactory cilia. These events trigger signal transduction leading to the production of cyclic AMP. In olfactory cells, cyclic AMP opens channels in the plasma membrane that are permeable to both Na^+ and Ca^{2+}. The flow of these ions into the receptor cell leads to depolarization of the membrane, generating action potentials.

Humans can distinguish thousands of different odors, each caused by a structurally distinct odorant. This level of sensory discrimination requires many different ORs. In 1991, Richard Axel and Linda Buck, working at Columbia University, discovered a family of more than 1,000 OR genes—about 3% of all human genes. Each olfactory receptor cell appears to express one OR gene. Cells selective for different odorants are interspersed in the nasal cavity. Those cells that express the same OR gene transmit action potentials to the same small region of the olfactory bulb. In 2004, Axel and Buck shared a Nobel Prize for their studies of the gene family and receptors that function in olfaction.

Although the receptors and brain pathways for taste and smell are independent, the two senses do interact. Indeed, much of the complex flavor humans experience when eating is due to our sense of smell. If the olfactory system is blocked, as occurs when you have a head cold, the perception of taste is sharply reduced.

CONCEPT CHECK 50.4

1. Explain why some taste receptor cells and all olfactory receptor cells use G protein-coupled receptors, yet only olfactory receptor cells produce action potentials.
2. Pathways involving G proteins provide an opportunity for an increase in signal strength in the course of signal transduction, a change referred to as amplification. How might this be beneficial in olfaction?
3. **WHAT IF?** If you discovered a mutation in mice that disrupted the ability to taste sweet, bitter, and umami, but not sour or salty, what might you predict about where this mutation acts in the signaling pathways used by these receptors?

For suggested answers, see Appendix A.

CONCEPT 50.5

The physical interaction of protein filaments is required for muscle function

Throughout our discussion of sensory mechanisms, we have seen how sensory inputs to the nervous system result in specific behaviors: the touch-guided foraging of a star-nosed mole, the upside-down swimming of a crayfish with manipulated statocysts, and the light-avoiding maneuvers of planarians. Underlying the diverse forms of behavior in animals are common fundamental mechanisms: Feeding, swimming, and crawling all require muscle activity in response to nervous system input.

Muscle cell function relies on microfilaments, which are the actin components of the cytoskeleton. Recall from Chapter 6 that microfilaments, like microtubules, function in cell motility. Muscle contraction is the product of microfilament movement powered by chemical energy; muscle extension occurs only passively. To understand how microfilaments contribute to muscle contraction, we must analyze the structure of muscles and muscle fibers. We will begin by examining vertebrate skeletal muscle and then turn our attention to other types of muscle.

Vertebrate Skeletal Muscle

Vertebrate **skeletal muscle**, which moves bones and body, is characterized by a hierarchy of smaller and smaller units **(Figure 50.26)**. Most skeletal muscles consist of a bundle of long fibers running parallel to the length of the muscle. Each fiber is a single cell. A muscle fiber, or cell, contains multiple nuclei, reflecting its formation by the fusion of many embryonic cells. Inside the fiber lies a bundle of smaller **myofibrils** arranged longitudinally. The myofibrils, in turn, are composed of thin filaments and thick filaments. **Thin filaments** consist of two strands of actin and two strands of a regulatory protein (not shown here) coiled around one another. **Thick filaments** are staggered arrays of myosin molecules.

Skeletal muscle is also called **striated muscle** because the regular arrangement of the filaments creates a pattern of light and dark bands. Each repeating unit is a **sarcomere**, the basic contractile unit of the muscle. The borders of the sarcomere are lined up in adjacent myofibrils and contribute to the striations visible with a light microscope. Thin filaments are attached at the Z lines and project toward the center of the sarcomere, while thick filaments are attached at the M lines centered in the sarcomere. In a muscle fiber at rest, thick and thin filaments only partially overlap. Near the edge of the sarcomere there are only thin filaments, whereas the zone in the center contains only thick filaments. This arrangement is the key to how the sarcomere, and hence the whole muscle, contracts.

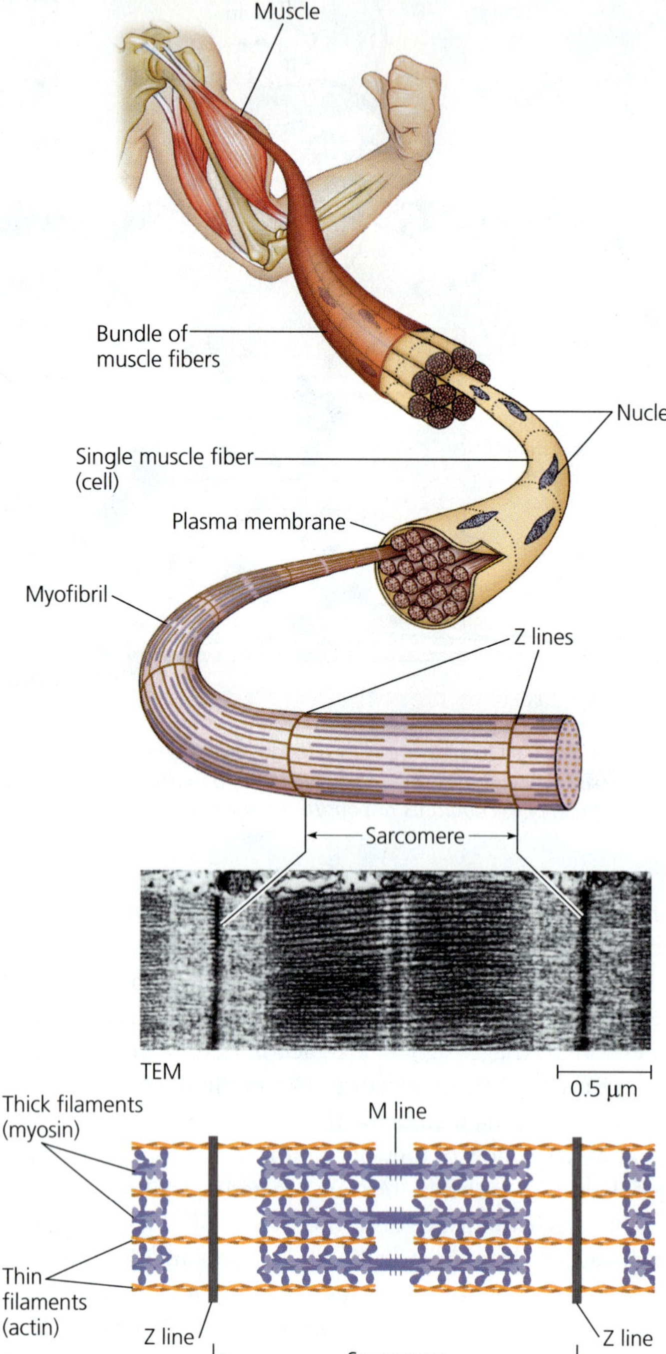

▲ **Figure 50.26 The structure of skeletal muscle.**

The Sliding-Filament Model of Muscle Contraction

We can explain much of what happens during the contraction of a whole muscle by focusing on the contraction of a single sarcomere **(Figure 50.27)**. According to the **sliding-filament model**, the filaments do not change in length when the sarcomere shortens. Instead, the thin and thick filaments slide past each other, increasing their overlap.

The longitudinal sliding of the filaments relies on the interaction of actin and myosin. Each myosin molecule has a long "tail" region and a globular "head" region. The tail adheres to the tails of other myosin molecules that form the thick filament. The head, which extends to the side, can bind ATP and hydrolyze it to ADP and inorganic phosphate. As shown in **Figure 50.28**, hydrolysis of ATP converts myosin to a high-energy form. This form of myosin binds to actin, forms a cross-bridge, and pulls the thin filament toward the center of the sarcomere. The cross-bridge is broken when a new molecule of ATP binds to the myosin head.

Muscle contraction requires repeated cycles of binding and release. In each cycle, the myosin head freed from a cross-bridge cleaves the newly bound ATP and binds again to actin. Because the thin filament moved toward the center of

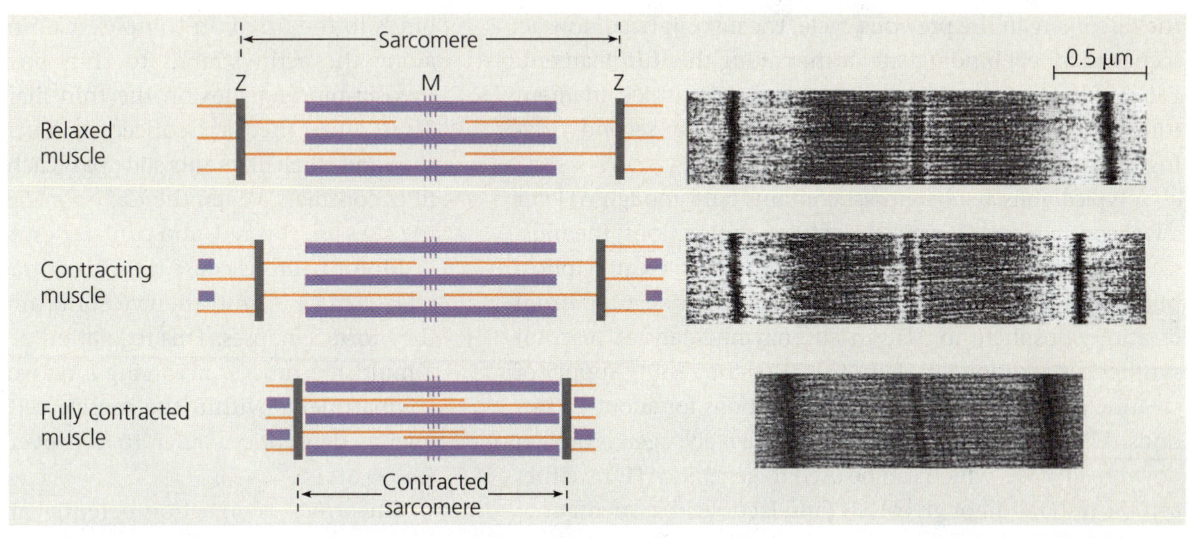

▶ **Figure 50.27 The sliding-filament model of muscle contraction.** The drawings on the left show that the lengths of the thick (myosin) filaments (purple) and thin (actin) filaments (orange) remain the same as a muscle fiber contracts.

Thick filament

Thin filaments

1 Starting here, the myosin head is bound to ATP and is in its low-energy configuration.

Thin filament

ATP

Myosin head (low-energy configuration)

Thick filament

2 The myosin head hydrolyzes ATP to ADP and inorganic phosphate (Ⓟ$_i$) and is in its high-energy configuration.

Actin

Myosin-binding sites

ADP

Ⓟ$_i$

Myosin head (high-energy configuration)

3 The myosin head binds to actin, forming a cross-bridge.

ADP

Ⓟ$_i$

Cross-bridge

ADP + Ⓟ$_i$

4 Releasing ADP and Ⓟ$_i$, myosin returns to its low-energy configuration, sliding the thin filament.

Thin filament moves toward center of sarcomere.

Myosin head (low-energy configuration)

5 Binding of a new molecule of ATP releases the myosin head from actin, and a new cycle begins.

ATP

▲ **Figure 50.28 Myosin-actin interactions underlying muscle fiber contraction.**

? *When ATP binds, what prevents the filaments from sliding back into their original positions?*

ANIMATION **BioFlix** Visit the Study Area at **www.masteringbiology.com** for the BioFlix® 3-D Animation on Muscle Contraction.

the sarcomere in the previous cycle, the myosin head now attaches to a new binding site farther along the thin filament. Each of the approximately 350 heads of a thick filament forms and re-forms about five cross-bridges per second, driving filaments past each other.

A typical muscle fiber at rest contains only enough ATP for a few contractions. To power repetitive contractions, the muscle cell relies on two other storage compounds: creatine phosphate and glycogen. Transfer of a phosphate group from creatine phosphate to ADP in an enzyme-catalyzed reaction synthesizes additional ATP. In this way, the resting supply of creatine phosphate can sustain contractions for about 15 seconds. ATP stores are also replenished when glycogen is broken down to glucose, which can be used to generate ATP, by either aerobic respiration or glycolysis (and lactic acid fermentation; see Chapter 9). Using a typical muscle fiber's glycogen store, glycolysis can support about 1 minute of sustained contraction, whereas aerobic respiration can power contractions for nearly an hour.

The Role of Calcium and Regulatory Proteins

Calcium ions (Ca^{2+}) and proteins bound to actin play crucial roles in both muscle cell contraction and relaxation. **Tropomyosin**, a regulatory protein, and the **troponin complex**, a set of additional regulatory proteins, are bound to the actin strands of thin filaments. In a muscle fiber at rest, tropomyosin covers the myosin-binding sites along the thin filament, preventing actin and myosin from interacting **(Figure 50.29a)**. When Ca^{2+} accumulates in the cytosol, it

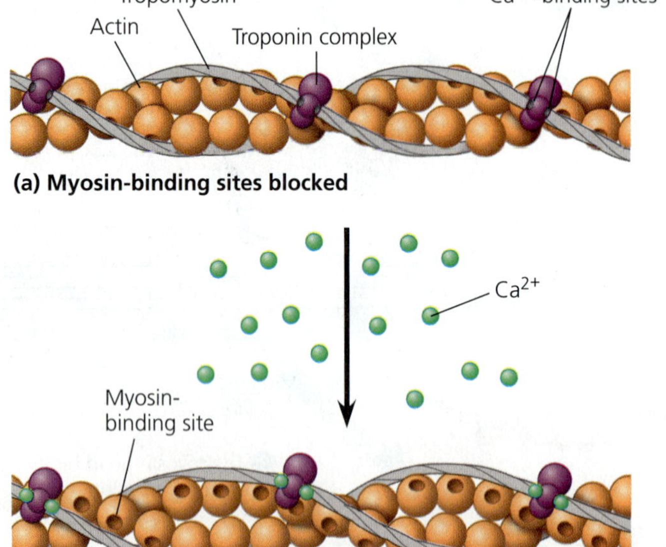

▲ **Figure 50.29 The role of regulatory proteins and calcium in muscle fiber contraction.** Each thin filament consists of two strands of actin, tropomyosin, and the troponin complex.

binds to the troponin complex, causing tropomyosin bound along the actin strands to shift position and expose the myosin-binding sites on the thin filament **(Figure 50.29b)**. Thus, when the Ca^{2+} concentration rises in the cytosol, the thin and thick filaments slide past each other, and the muscle fiber contracts. When the Ca^{2+} concentration falls, the binding sites are covered, and contraction stops.

Motor neurons cause muscle contraction by triggering the release of Ca^{2+} into the cytosol of muscle cells with which they form synapses. This regulation of Ca^{2+} concentration is a multistep process involving a network of membranes and compartments within the muscle cell. As you read the following description, refer to the overview and diagram in **Figure 50.30**.

The arrival of an action potential at the synaptic terminal of a motor neuron causes release of the neurotransmitter acetylcholine. Binding of acetylcholine to receptors on the muscle fiber leads to a depolarization, triggering an action potential. Within the muscle fiber, the action potential spreads deep into the interior, following infoldings of the plasma membrane called **transverse (T) tubules**. The T tubules make close contact with the **sarcoplasmic reticulum (SR)**, a specialized endoplasmic reticulum. As the action potential spreads along the T tubules, it triggers changes in the SR, opening Ca^{2+} channels. Calcium ions stored in the interior of the SR flow through these open channels into the cytosol and bind to the troponin complex, initiating contraction of the muscle fiber.

When motor neuron input stops, the muscle cell relaxes. As it relaxes, the filaments slide back to their starting position. During this phase, proteins in the cell reset the muscle for the next cycle of contraction. Relaxation begins as transport proteins in the SR pump Ca^{2+} in from the cytosol. When the Ca^{2+} concentration in the cytosol drops to a low level, the regulatory proteins bound to the thin filament shift back to their starting position, once again blocking the myosin-binding sites. At the same time, the Ca^{2+} pumped from the cytosol accumulates in the SR, providing the stores needed to respond to the next action potential.

Several diseases cause paralysis by interfering with the excitation of skeletal muscle fibers by motor neurons. In amyotrophic lateral sclerosis (ALS), also called Lou Gehrig's disease, motor neurons in the spinal cord and brainstem degenerate, and the muscle fibers with which they synapse atrophy. ALS is progressive and usually fatal within five years after symptoms appear; currently there is no cure or treatment. Myasthenia gravis is an autoimmune disease in which a person produces antibodies to the acetylcholine receptors on skeletal muscle fibers. As the number of these receptors decreases, synaptic transmission between motor neurons and muscle fibers declines. Fortunately, effective treatments are available for mysathenia gravis.

▼ **Figure 50.30**

Exploring The Regulation of Skeletal Muscle Contraction

The electrical, chemical, and molecular events regulating skeletal muscle contraction are shown in a cutaway view of a muscle cell and in the enlarged diagram below. Action potentials (red arrows) triggered by the motor neuron sweep across the muscle fiber and into it along the transverse (T) tubules, initiating the movements of calcium (green dots) that regulate muscle activity.

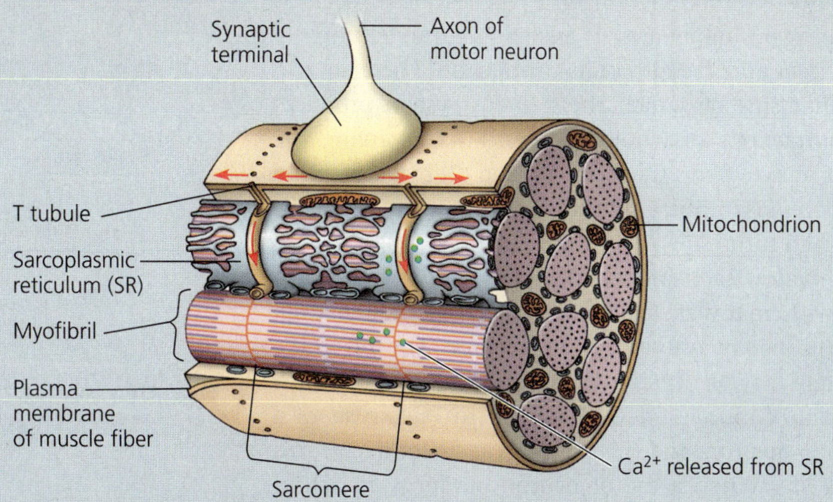

Synaptic terminal of motor neuron

Synaptic cleft

T tubule

Plasma membrane

ACh

1 Acetylcholine (ACh) released at synaptic terminal diffuses across synaptic cleft and binds to receptor proteins on muscle fiber's plasma membrane, triggering an action potential in muscle fiber.

2 Action potential is propagated along plasma membrane and down T tubules.

Sarcoplasmic reticulum (SR)

3 Action potential triggers Ca^{2+} release from SR.

Ca^{2+}

Ca^{2+} pump

ATP

CYTOSOL

Ca^{2+}

4 Calcium ions bind to troponin in thin filament; myosin-binding sites exposed.

5 Cycles of myosin cross-bridge formation and breakdown, coupled with ATP hydrolysis, slide thin filament toward center of sarcomere.

6 Cytosolic Ca^{2+} is removed by active transport into SR after action potential ends.

7 Tropomyosin blockage of myosin-binding sites is restored; contraction ends, and muscle fiber relaxes.

Nervous Control of Muscle Tension

Whereas contraction of a single skeletal muscle fiber is a brief all-or-none twitch, contraction of a whole muscle, such as the biceps in your upper arm, is graded; you can voluntarily alter the extent and strength of its contraction. There are two basic mechanisms by which the nervous system produces graded contractions of whole muscles: (1) by varying the number of muscle fibers that contract and (2) by varying the rate at which muscle fibers are stimulated. Let's consider each mechanism in turn.

In vertebrates, each branched motor neuron may form synapses with many skeletal muscle fibers, although each fiber is controlled by only one motor neuron. For the whole muscle, there may be hundreds of motor neurons, each with its own pool of muscle fibers. A **motor unit** consists of a single motor neuron and all the muscle fibers it controls **(Figure 50.31)**. When a motor neuron produces an action potential, all the muscle fibers in its motor unit contract as a group. The strength of the resulting contraction depends on how many muscle fibers the motor neuron controls.

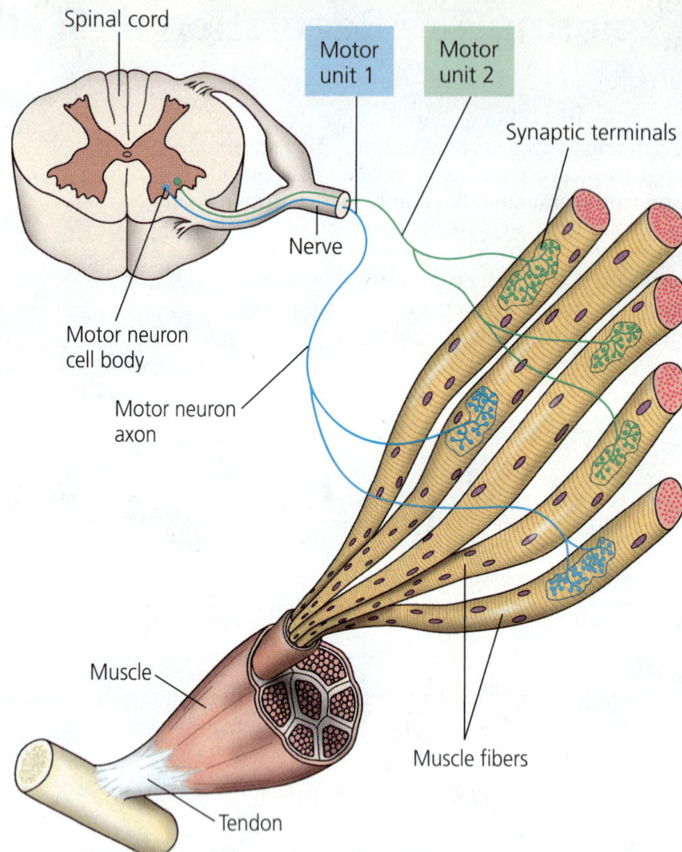

▲ **Figure 50.31 Motor units in a vertebrate skeletal muscle.** Each muscle fiber (cell) forms synapses with only one motor neuron, but each motor neuron typically synapses with many muscle fibers. A motor neuron and all the muscle fibers it controls constitute a motor unit.

In most muscles, the number of muscle fibers in different motor units ranges from a few to hundreds. The nervous system can thus regulate the strength of contraction in a muscle by determining how many motor units are activated at a given instant and by selecting large or small motor units to activate. The force (tension) developed by a muscle progressively increases as more and more of the motor neurons controlling the muscle are activated, a process called *recruitment* of motor neurons. Depending on the number of motor neurons your brain recruits and the size of their motor units, you can lift a fork or something much heavier, like your biology textbook.

Some muscles, especially those that hold up the body and maintain posture, are almost always partially contracted. In such muscles, the nervous system may alternate activation among the motor units, reducing the length of time any one set of fibers is contracted. Prolonged contraction can result in muscle fatigue due to the depletion of ATP and dissipation of ion gradients required for normal electrical signaling. Although accumulation of lactate (see Figure 9.17) may also contribute to muscle fatigue, recent research actually points to a beneficial effect of lactate on muscle function.

The nervous system regulates muscle contraction not only by controlling which motor units are activated, but also by varying the rate of muscle fiber stimulation. A single action potential produces a twitch lasting about 100 msec or less. If a second action potential arrives before the muscle fiber has completely relaxed, the two twitches add together, resulting in greater tension **(Figure 50.32)**. Further summation occurs as the rate of stimulation increases. When the rate is so high that the muscle fiber cannot relax at all between stimuli, the twitches fuse into one smooth, sustained contraction called **tetanus**. Motor neurons usually deliver their action potentials in rapid-fire volleys, and the resulting summation of tension results in the smooth contraction typical of tetanus rather than the jerky actions of individual twitches. (Although a smooth, sustained contraction is part of normal muscle function, tetanus is also the name of a disease of uncontrolled muscle contraction caused by a bacterial toxin.)

The increase in tension during summation and tetanus occurs because skeletal muscle fibers are connected to bones via tendons and connective tissues. When a muscle fiber contracts, it stretches these elastic structures, which then transmit tension to the bones. In a single twitch, the muscle fiber begins to relax before the elastic structures are fully stretched. During summation, however, the high-frequency action potentials maintain an elevated concentration of Ca^{2+} in the muscle fiber's cytosol, prolonging cross-bridge cycling and causing greater stretching of the elastic structures. During tetanus, the elastic structures are fully stretched, and all of the tension generated by the muscle fiber is transmitted to the bones.

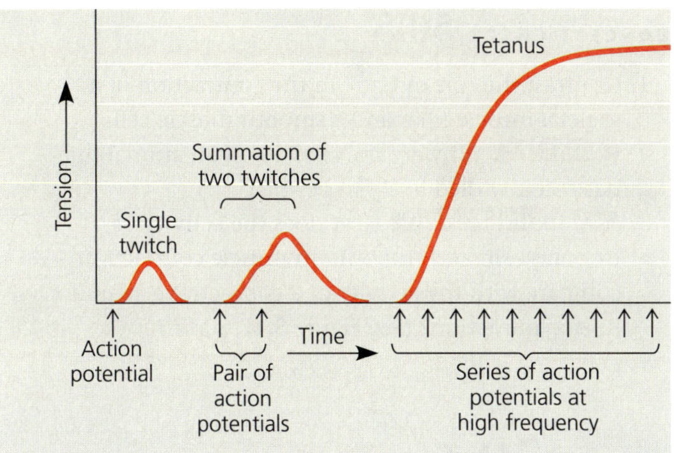

▲ **Figure 50.32 Summation of twitches.** This graph illustrates how the number of action potentials in a short period of time influences the tension developed in a muscle fiber.

? *How could the nervous system cause a skeletal muscle to produce the most forceful contraction it is capable of?*

Types of Skeletal Muscle Fibers

Our discussion to this point has focused on the general properties of vertebrate skeletal muscles. There are, however, several distinct types of skeletal muscle fibers, each of which is adapted to a particular set of functions. Scientists typically classify these varied fiber types either by the source of ATP used to power muscle activity or by the speed of muscle contraction. We'll consider each of the two classification schemes.

Oxidative and Glycolytic Fibers Fibers that rely mostly on aerobic respiration are called oxidative fibers. Such fibers are specialized in ways that enable them to make use of a steady energy supply: They have many mitochondria, a rich blood supply, and a large amount of an oxygen-storing protein called **myoglobin**. A brownish red pigment, myoglobin binds oxygen more tightly than does hemoglobin, enabling oxidative fiber to extract oxygen from the blood efficiently. In contrast to oxidative fibers, glycolytic fibers use glycolysis as their primary source of ATP. They have a larger diameter and less myoglobin than oxidative fibers and thus fatigue much more readily. These different fiber types are readily apparent in the muscle of poultry and fish: The dark meat is made up of oxidative fibers rich in myoglobin, and the light meat is composed of glycolytic fibers.

Fast-Twitch and Slow-Twitch Fibers Muscle fibers vary in the speed with which they contract: **Fast-twitch fibers** develop tension two to three times faster than **slow-twitch fibers**. Fast fibers enable brief, rapid, powerful contractions. Slow fibers, often found in muscles that maintain posture, can sustain long contractions. A slow fiber has less sarcoplasmic reticulum and pumps Ca^{2+} more slowly than a fast fiber. Because Ca^{2+} remains in the cytosol longer, a muscle twitch in a slow fiber lasts about five times as long as one in a fast fiber.

The difference in contraction speed between slow-twitch and fast-twitch fibers mainly reflects the rate at which their myosin heads hydrolyze ATP. However, there isn't a one-to-one relationship between contraction speed and ATP source. Whereas all slow-twitch fibers are oxidative, fast-twitch fibers can be either glycolytic or oxidative.

Most human skeletal muscles contain both fast- and slow-twitch fibers, although the muscles of the eye and hand are exclusively fast-twitch. In a muscle that has a mixture of fast and slow fibers, the relative proportions of each are genetically determined. However, if such a muscle is used repeatedly for activities requiring high endurance, some fast glycolytic fibers can develop into fast oxidative fibers. Because fast oxidative fibers fatigue more slowly than fast glycolytic fibers, the result will be a muscle that is more resistant to fatigue.

Some vertebrates have skeletal muscle fibers that twitch at rates far faster than any human muscle. For example, superfast muscles produce a rattlesnake's rattle and a dove's coo. The fastest such muscles, however, surround the gas-filled swim bladder inside the male toadfish **(Figure 50.33)**. In producing its characteristic "boat whistle" mating call, the toadfish can contract and relax these muscles more than 200 times per second!

Other Types of Muscle

Although all muscles share the same fundamental mechanism of contraction—actin and myosin filaments sliding past each other—there are many different types of muscle. Vertebrates, for example, have cardiac muscle and smooth muscle in addition to skeletal muscle (see Figure 40.5).

▲ **Figure 50.33 Specialization of skeletal muscles.** The male toadfish (*Opsanus tau*) uses superfast muscles to produce its mating call.

Vertebrate **cardiac muscle** is found in only one part of the body: the heart. Like skeletal muscle, cardiac muscle is striated. However, structural differences between skeletal and cardiac muscle fibers result in differences in their electrical and membrane properties. Whereas skeletal muscle fibers do not produce action potentials unless stimulated by a motor neuron, cardiac muscle cells have ion channels in their plasma membrane that cause rhythmic depolarizations, triggering action potentials without input from the nervous system. Action potentials of cardiac muscle cells last up to 20 times longer than those of the skeletal muscle fibers. Plasma membranes of adjacent cardiac muscle cells interlock at specialized regions called **intercalated disks**, where gap junctions (see Figure 6.32) provide direct electrical coupling between the cells. Thus, the action potential generated by specialized cells in one part of the heart spreads to all other cardiac muscle cells, causing the whole heart to contract. A long refractory period prevents summation and tetanus.

Smooth muscle in vertebrates is found mainly in the walls of hollow organs, such as blood vessels and organs of the digestive tract. Smooth muscle cells lack striations because their actin and myosin filaments are not regularly arrayed along the length of the cell. Instead, the thick filaments are scattered throughout the cytoplasm, and the thin filaments are attached to structures called dense bodies, some of which are tethered to the plasma membrane. There is less myosin than in striated muscle fibers, and the myosin is not associated with specific actin strands. Some smooth muscle cells contract only when stimulated by neurons of the autonomic nervous system. Others can generate action potentials without input from neurons—they are electrically coupled to one another. Smooth muscles contract and relax more slowly than striated muscles.

Although Ca^{2+} regulates smooth muscle contraction, the mechanism for regulation is different from that in skeletal and cardiac muscle. Smooth muscle cells have no troponin complex or T tubules, and their sarcoplasmic reticulum is not well developed. During an action potential, Ca^{2+} enters the cytosol mainly through the plasma membrane. Calcium ions cause contraction by binding to the protein calmodulin, which activates an enzyme that phosphorylates the myosin head, enabling cross-bridge activity.

Invertebrates have muscle cells similar to vertebrate skeletal and smooth muscle cells, and arthropod skeletal muscles are nearly identical to those of vertebrates. However, because the flight muscles of insects are capable of independent, rhythmic contraction, the wings of some insects can actually beat faster than action potentials can arrive from the central nervous system. Another interesting evolutionary adaptation has been discovered in the muscles that hold a clam's shell closed. The thick filaments in these muscles contain a protein called paramyosin that enables the muscles to remain contracted for as long as a month with only a low rate of energy consumption.

CONCEPT CHECK 50.5

1. Contrast the role of Ca^{2+} in the contraction of a skeletal muscle fiber and a smooth muscle cell.
2. **WHAT IF?** Why are the muscles of an animal that has recently died likely to be stiff?
3. **MAKE CONNECTIONS** How does the activity of tropomyosin and troponin in muscle contraction compare with the activity of a competitive inhibitor in enzyme action? (See Figure 8.17, p. 156.)

For suggested answers, see Appendix A.

CONCEPT 50.6

Skeletal systems transform muscle contraction into locomotion

Converting muscle contraction to movement requires a skeleton—a rigid structure to which muscles can attach. An animal changes its rigidity, shape, or location by contracting muscles connecting two parts of its skeleton.

Because muscles exert force only during contraction, moving a body part back and forth typically requires two muscles attached to the same section of the skeleton. We can see such an arrangement of muscles in the upper portion of a human arm or grasshopper leg **(Figure 50.34)**. Although we call such

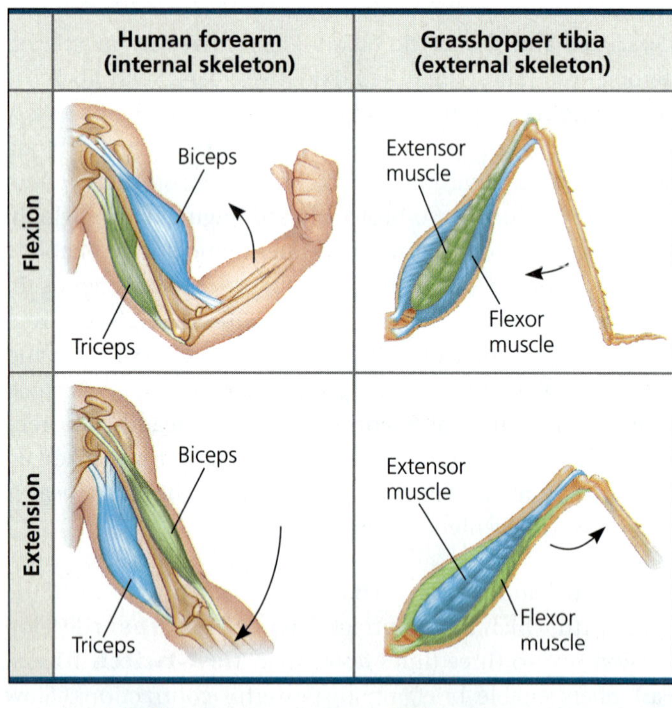

▲ **Figure 50.34 The interaction of muscles and skeletons in movement.** Back-and-forth movement of a body part is generally accomplished by antagonistic muscles. This arrangement works with either an internal skeleton, as in mammals, or an external skeleton, as in insects.

muscles an antagonistic pair, their function is actually cooperative, coordinated by the nervous system. For example, when you extend your arm, motor neurons trigger your triceps muscle to contract while the absence of neuronal input allows your biceps to relax.

Vital for movement, the skeletons of animals also function in support and protection. Most land animals would collapse if they had no skeleton to support their mass. Even an animal living in water would be formless without a framework to maintain its shape. In many animals, a hard skeleton also protects soft tissues. For example, the vertebrate skull protects the brain, and the ribs of terrestrial vertebrates form a cage around the heart, lungs, and other internal organs.

Types of Skeletal Systems

Although we tend to think of skeletons only as interconnected sets of bones, skeletons come in many different forms. Hardened support structures can be external (as in exoskeletons), internal (as in endoskeletons), or even absent (as in fluid-based, or hydrostatic, skeletons).

Hydrostatic Skeletons

A **hydrostatic skeleton** consists of fluid held under pressure in a closed body compartment. This is the main type of skeleton in most cnidarians, flatworms, nematodes, and annelids (see Chapter 33). These animals control their form and movement by using muscles to change the shape of fluid-filled compartments. Among the cnidarians, for example, a hydra elongates by closing its mouth and using contractile cells in its body wall to constrict its central gastrovascular cavity. Because water cannot be compressed very much, decreasing the diameter of the cavity forces the cavity to become longer.

Worms use hydrostatic skeletons in diverse ways to move through their environment. In planarians and other flatworms, movement results mainly from muscles in the body wall exerting localized forces against the interstitial fluid. In nematodes (roundworms), longitudinal muscles contracting around the fluid-filled body cavity move the animal forward by undulations, or wavelike motions. In earthworms and many other annelids, circular and longitudinal muscles act together to change the shape of individual fluid-filled segments, which are divided by septa. These shape changes bring about **peristalsis**, a movement produced by rhythmic waves of muscle contractions passing from front to back **(Figure 50.35)**.

Hydrostatic skeletons are well suited for life in aquatic environments. On land, they provide support for crawling and burrowing and may cushion internal organs from shocks. However, a hydrostatic skeleton cannot support walking or running, in which an animal's body is held off the ground.

Exoskeletons

The clam shell you find on a beach once served as an **exoskeleton**, a hard encasement deposited on an animal's surface. The shells of clams and most other molluscs are made of calcium carbonate secreted by the mantle, a sheet-like extension of the body wall (see Figure 33.15). Clams and other bivalves close their hinged shell using muscles attached to the inside of this exoskeleton. As the animal grows, it enlarges its shell by adding to the outer edge.

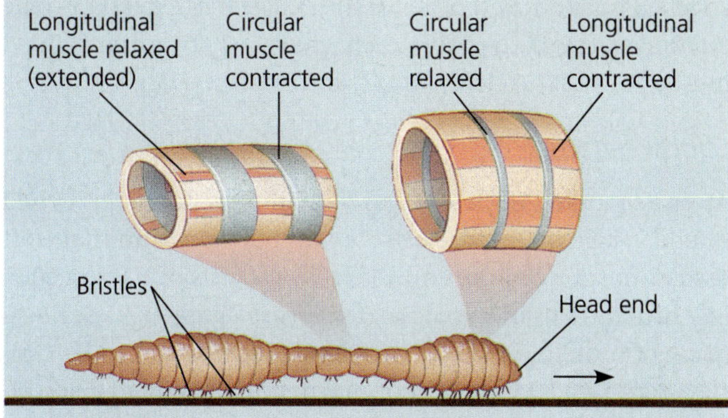

1 At the moment depicted, body segments at the earthworm's head end and just in front of the rear end are short and thick (longitudinal muscles contracted; circular muscles relaxed) and are anchored to the ground by bristles. The other segments are thin and elongated (circular muscles contracted; longitudinal muscles relaxed).

2 The head has moved forward because circular muscles in the head segments have contracted. Segments behind the head and at the rear are now thick and anchored, thus preventing the worm from slipping backward.

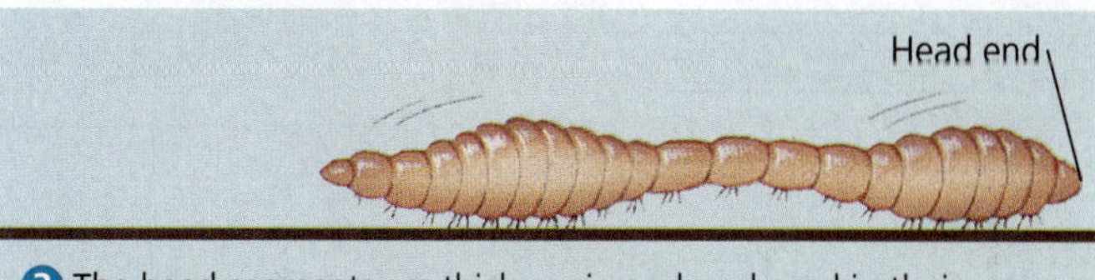

3 The head segments are thick again and anchored in their new positions. The rear segments have released their hold on the ground and have been pulled forward.

▲ **Figure 50.35 Crawling by peristalsis.** Contraction of the longitudinal muscles thickens and shortens the earthworm; contraction of the circular muscles constricts and elongates it.

Insects and other arthropods have a jointed exoskeleton called a cuticle, a nonliving coat secreted by the epidermis. About 30–50% of the arthropod cuticle consists of **chitin**, a polysaccharide similar to cellulose (see Figure 5.9). Fibrils of chitin are embedded in a protein matrix, forming a composite material that combines strength and flexibility. The cuticle may be hardened with organic compounds that cross-link the proteins of the matrix, and in crustaceans such as lobsters, calcium salts may also be added. In body parts that must be flexible, such as leg joints, the cuticle remains unhardened. Muscles are attached to knobs and plates of the cuticle that extend into the interior of the body. With each growth spurt, an arthropod must shed its exoskeleton (molt) and produce a larger one.

Endoskeletons

Animals ranging from sponges to mammals have a hardened internal skeleton, or **endoskeleton**, buried within their soft tissues. In sponges, the endoskeleton consists of hard needle-like structures of inorganic material (see Figure 33.4) or fibers made of protein. Echinoderms' bodies are reinforced by ossicles, hard plates composed of magnesium carbonate and calcium carbonate crystals. Whereas the ossicles of sea urchins are tightly bound, the ossicles of sea stars are more loosely linked, allowing a sea star to change the shape of its arms.

Chordates have an endoskeleton consisting of cartilage, bone, or some combination of these materials (see Figure 40.5). The mammalian skeleton is built from more than 200 bones, some fused together and others connected at joints by ligaments that allow freedom of movement **(Figures 50.36 and 50.37)**.

Size and Scale of Skeletons

An exoskeleton needs to cover and protect an animal's body, but how thick does an endoskeleton need to be? We can begin to answer this question by applying ideas from civil

▼ **Figure 50.36 Bones and joints of the human skeleton.**

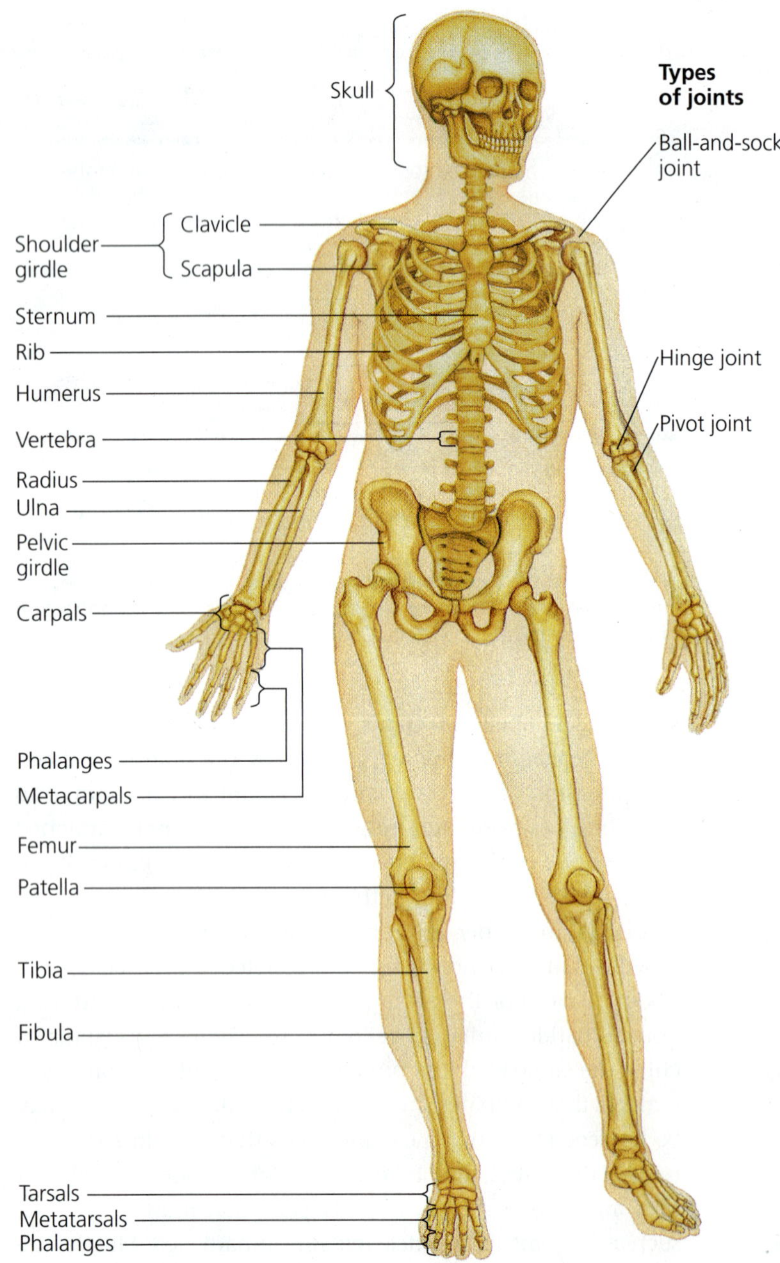

▼ **Figure 50.37 Types of joints.**

Ball-and-socket joint

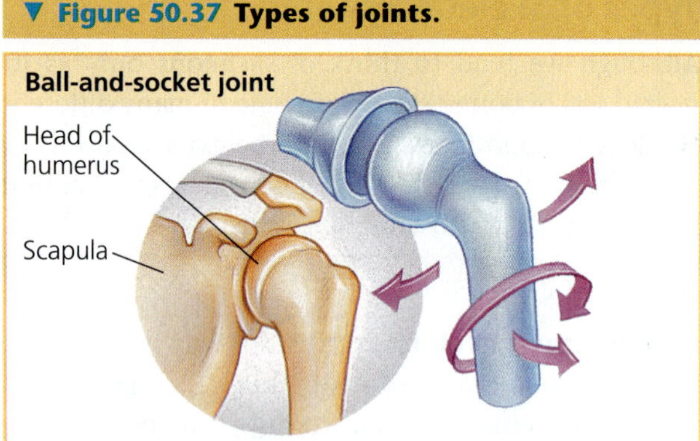

Ball-and-socket joints are found where the humerus contacts the shoulder girdle and where the femur contacts the pelvic girdle. These joints enable the arms and legs to rotate and move in several planes.

Hinge joint

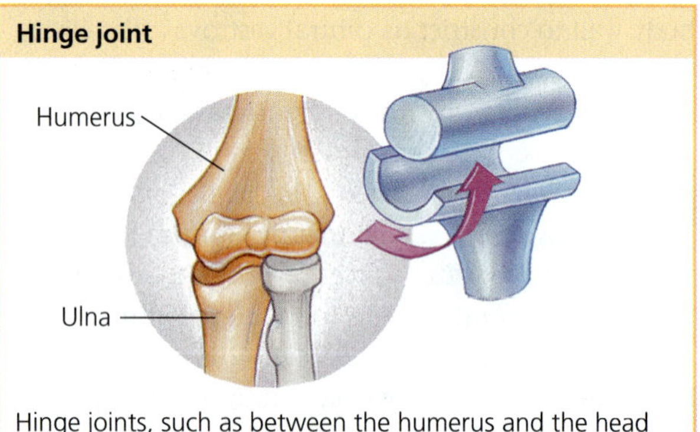

Hinge joints, such as between the humerus and the head of the ulna, restrict movement to a single plane.

Pivot joint

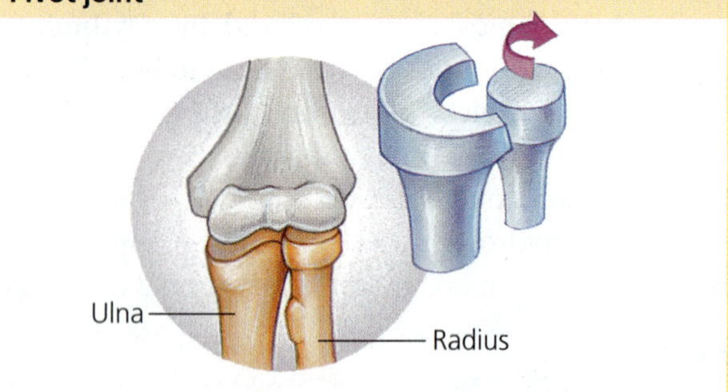

Pivot joints enable rotating the forearm at the elbow and turning the head from side to side.

engineering. For example, the weight of a building increases with the cube of its dimensions. However, the strength of a building support depends on its cross-sectional area, which only increases with the square of its diameter. We can thus predict that if we scaled up a mouse to the size of an elephant, the legs of the giant mouse would be too thin to support its weight. Indeed, large animals do have very different body proportions from those of small animals.

In applying the building analogy, we might also predict that the size of leg bones should be directly proportional to the strain imposed by body weight. However, our prediction would be inaccurate, in part because animal bodies are complex and nonrigid. In supporting body weight, it turns out that body posture—the position of the legs relative to the main body—is more important than leg size, at least in mammals and birds. Muscles and tendons (connective tissue that joins muscle to bone) hold the legs of large mammals relatively straight and positioned under the body and actually bear most of the load.

▲ **Figure 50.38 Energy-efficient locomotion on land.** Members of the kangaroo family travel from place to place mainly by leaping on their large hind legs. Kinetic energy momentarily stored in tendons after each leap provides a boost for the next leap. In fact, a large kangaroo hopping at 30 km/hr uses no more energy per minute than it does at 6 km/hr. The large tail helps balance the kangaroo when it leaps as well as when it sits.

Types of Locomotion

Movement is a hallmark of animals. Even animals fixed to a surface move their body parts: Sponges use beating flagella to generate water currents that draw and trap small food particles, and sessile cnidarians wave tentacles that capture prey (see Chapter 33). Most animals, however, are mobile and spend a considerable portion of their time and energy actively searching for food, escaping from danger, and seeking mates. These activities involve **locomotion**, or active travel from place to place.

Friction and gravity tend to keep an animal stationary and thus oppose locomotion. To move, an animal must expend energy to overcome these two forces. As we will see next, the amount of energy required to oppose friction or gravity is often reduced by an animal body plan adapted for movement in a particular environment.

Locomotion on Land

On land, a walking, running, hopping, or crawling animal must be able to support itself and move against gravity, but air poses relatively little resistance, at least at moderate speeds. When a land animal walks, runs, or hops, its leg muscles expend energy both to propel it and to keep it from falling down. With each step, the animal's leg muscles must overcome inertia by accelerating a leg from a standing start. For moving on land, powerful muscles and strong skeletal support are more important than a streamlined shape.

Diverse adaptations for traveling on land have evolved in various vertebrates. For example, kangaroos have large, powerful muscles in their hind legs, suitable for locomotion by hopping **(Figure 50.38)**. As a kangaroo lands after each leap, tendons in its hind legs momentarily store energy. The farther the animal hops, the more energy its tendons store. Analogous to the energy in a compressed spring, the energy stored in the tendons is available for the next jump and reduces the total amount of energy the animal must expend to travel. The legs of an insect, dog, or human also retain some energy during walking or running, although a considerably smaller share than those of a kangaroo.

Maintaining balance is another prerequisite for walking, running, or hopping. A kangaroo's large tail helps balance its body during leaps and also forms a stable tripod with its hind legs when the animal sits or moves slowly. Illustrating the same principle, a walking cat, dog, or horse keeps three feet on the ground. Bipedal animals, such as humans and birds, keep part of at least one foot on the ground when walking. When an animal runs, all four feet (or both feet for bipeds) may be off the ground briefly, but at running speeds it is momentum more than foot contact that keeps the body upright.

Crawling poses a very different situation. Because much of its body is in contact with the ground, a crawling animal must exert considerable effort to overcome friction. You have read how earthworms crawl by peristalsis. Many snakes crawl by undulating their entire body from side to side. Assisted by large, movable scales on its underside, a snake's body pushes against the ground, propelling the animal forward. Boa constrictors and pythons creep straight forward, driven by muscles that lift belly scales off the ground, tilt the scales forward, and then push them backward against the ground.

Swimming

Because most animals are reasonably buoyant in water, overcoming gravity is less of a problem for swimming animals than for species that move on land or through the air. On the other hand, water is a much denser and more viscous medium than air, and thus drag (friction) is a major problem for aquatic animals. A sleek, fusiform (torpedo-like) shape is a common adaptation of fast swimmers (see Figure 40.2).

Although most animal phyla include species that swim, swimming occurs in diverse ways. For instance, many insects and four-legged vertebrates use their legs as oars to push against the water. Squids, scallops, and some cnidarians are jet-propelled, taking in water and squirting it out in bursts. Sharks and bony fishes swim by moving their body and tail from side to side, while whales and dolphins move by undulating their body and tail up and down.

Flying

Active flight (in contrast to gliding downward from a tree) has evolved in only a few animal groups: insects, reptiles (including birds), and, among the mammals, bats. One group of flying reptiles, the pterosaurs, died out millions of years ago, leaving birds and bats as the only flying vertebrates.

Gravity poses a major problem for a flying animal because its wings must develop enough lift to overcome gravity's downward force. The key to flight is wing shape. All wings are airfoils—structures whose shape alters air currents in a way that helps animals or airplanes stay aloft. As for the body to which the wings attach, a fusiform shape helps reduce drag in air as it does in water.

Flying animals are relatively light, with body masses ranging from less than a gram for some insects to about 20 kg for the largest flying birds. Many flying animals have structural adaptations that contribute to low body mass. Birds, for example, have no urinary bladder or teeth and have relatively large bones with air-filled regions that help lessen the bird's weight (see Chapter 34).

Energy Costs of Locomotion

During the 1960s, three scientists at Duke University became interested in the bioenergetics of locomotion. Physiologists typically determine an animal's rate of energy use during locomotion by measuring oxygen consumption or carbon dioxide production (see Chapter 40). To apply such a strategy to flight, Vance Tucker trained parakeets to fly in a wind tunnel while wearing a face mask **(Figure 50.39)**. By connecting the mask to a tube that collected the air the bird exhaled as it flew, he could measure rates of gas exchange and calculate energy expenditure. In the meantime, Dick Taylor and Knut Schmidt-Nielsen measured energy consumption at rest and during locomotion for animals of widely varying body sizes. Schmidt-Nielsen then calculated an energy cost for locomotion: the amount of fuel it takes to transport a given amount of body weight over a set distance.

Schmidt-Nielsen's analysis demonstrated that the energy cost of locomotion depends on the mode of locomotion and the environment **(Figure 50.40)**. Swimming is the most energy-efficient mode of locomotion (assuming that an animal has adaptations that facilitate swimming). Running animals generally expend more energy per meter traveled than equivalently sized swimming animals, partly because running and walking require energy to overcome gravity. If we compare

▲ **Figure 50.39 Measuring energy usage during flight.** The tube connected to the plastic face mask collects the gases this parakeet exhales during flight in a wind tunnel.

▼ **Figure 50.40** INQUIRY

What are the energy costs of locomotion?

EXPERIMENT Knut Schmidt-Nielsen wondered whether there were general principles governing the energy costs of different types of locomotion among diverse animal species. To answer this question, he drew on his own studies as well as the scientific literature for measurements made when animals swam in water flumes, ran on treadmills, or flew in wind tunnels. He converted all of these data to a common set of units and graphed the results.

RESULTS

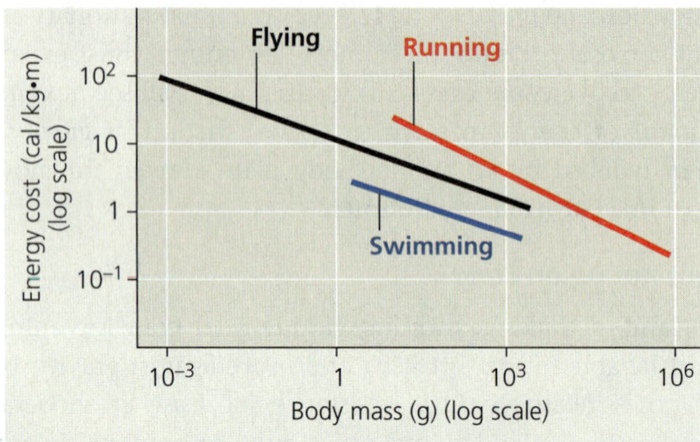

This graph plots the energy cost, in calories per kilogram of body mass per meter traveled, against body mass for animals specialized for running, flying, and swimming. Note that both axes are plotted on logarithmic scales.

CONCLUSION For most animals of a given body mass, swimming is the most energy-efficient and running the least energy-efficient mode of locomotion. In addition, a small animal typically expends more energy per kilogram of body mass than a large animal, regardless of the type of locomotion used.

SOURCE K. Schmidt-Nielsen, Locomotion: Energy cost of swimming, flying, and running, *Science* 177:222–228 (1972).

WHAT IF? If you plotted the efficiency of a duck as a swimmer on this graph, where might you expect it to fall, and why?

the energy consumption per minute rather than per meter, we find that flying animals use more energy than swimming or running animals with the same body mass.

Figure 50.40 also reveals that a larger animal travels more efficiently than a smaller animal adapted to the same mode of transport. This relationship of size to energy expenditure during locomotion is apparent in the downward slope of each line on the graph. For example, a 450-kg horse expends less energy *per kilogram of body mass* than a 4-kg cat running the same distance. Of course, the total amount of energy expended in locomotion is greater for the larger animal.

Energy from food that is used for locomotion is unavailable for other activities, such as growth and reproduction. Thus, structural and behavioral adaptations that maximize the efficiency of locomotion increase an organism's evolutionary fitness.

Although we have discussed sensory receptors and muscles separately in this chapter, they are part of a single integrated system linking brain, body, and the external world. An animal's behavior is the product of this system. In Chapter 51, we'll discuss behavior in the context of animal form and function and also link it to ecology, the study of how organisms interact with their environment.

CONCEPT CHECK 50.6

1. In what way are septa an important feature of the earthworm skeleton?
2. Contrast swimming and flying in terms of the main problems they pose and the adaptations that allow animals to overcome those problems.
3. **WHAT IF?** When using your arms to lower yourself into a chair, you bend your arms without using your biceps. Explain how this is possible. (*Hint*: Think about gravity as an antagonistic force.)

For suggested answers, see Appendix A.

50 CHAPTER REVIEW

SUMMARY OF KEY CONCEPTS

CONCEPT 50.1

Sensory receptors transduce stimulus energy and transmit signals to the central nervous system (pp. 1085–1090)

- **Sensory receptors** are usually specialized neurons or epithelial cells that detect external or internal stimuli. The detection of a stimulus by sensory cells precedes **sensory transduction**, the change in the membrane potential of a sensory receptor in response to a stimulus. The resulting receptor potential controls **transmission** of action potentials to the CNS, where sensory information is integrated to generate perceptions. The frequency of action potentials in an axon and the number of axons activated determine stimulus strength. The identity of the axon carrying the signal encodes the nature or quality of the stimulus. Signal transduction pathways in receptor cells often amplify the signal, which causes the receptor cell either to produce action potentials or to release neurotransmitter at a synapse with a sensory neuron.
- There are five basic types of sensory receptors. **Mechanoreceptors** respond to stimuli such as pressure, touch, stretch, motion, and sound. **Chemoreceptors** detect either total solute concentrations or specific molecules. **Electromagnetic receptors** detect different forms of electromagnetic radiation. Various types of **thermoreceptors** signal surface and core temperatures of the body. Pain is detected by a group of **nociceptors** that respond to excess heat, pressure, or specific classes of chemicals.

? *To simplify the classification of sensory receptors, why might it make sense to eliminate nociceptors as a distinct class?*

CONCEPT 50.2

The mechanoreceptors responsible for hearing and equilibrium detect moving fluid or settling particles (pp. 1090–1094)

- Most invertebrates sense their orientation with respect to gravity by means of **statocysts**. Specialized **hair cells** form the basis for hearing and balance in mammals and for detection of water movement in fishes and aquatic amphibians. In mammals, the **tympanic membrane** (eardrum) transmits sound waves to three small bones of the middle ear, which transmit the waves through the oval window to the fluid in the coiled **cochlea** of the inner ear. Pressure waves in the fluid vibrate the **basilar membrane**, depolarizing hair cells and triggering action potentials that travel via the auditory nerve to the brain. Each region of the basilar membrane vibrates most vigorously at a particular frequency and leads to excitation of a specific auditory area of the cerebral cortex. Receptors in the inner ear function in balance and equilibrium.

? *What quality of sound determines the direction of displacement of a particular hair cell in the ear, and how is that quality encoded in signals sent to the brain?*

CONCEPT 50.3

Visual receptors in diverse animals depend on light-absorbing pigments (pp. 1095–1101)

- Invertebrates have varied light detectors, including simple light-sensitive eyespots, image-forming compound eyes, and single-lens eyes. In the vertebrate eye, a single lens is used to focus light on **photoreceptors** in the **retina**. Both **rods** and **cones** contain a pigment, **retinal**, bonded to a protein (opsin).

Absorption of light by retinal triggers a signal transduction pathway that hyperpolarizes the photoreceptors, causing them to release less neurotransmitter. Synapses transmit information from photoreceptors to cells that integrate information and convey it to the brain along axons that form the optic nerve.

? *How does processing of sensory information sent to the vertebrate brain in vision differ from that in hearing or olfaction?*

CONCEPT 50.4

The senses of taste and smell rely on similar sets of sensory receptors (pp. 1101–1103)

- Both taste (**gustation**) and smell (**olfaction**) depend on the stimulation of chemoreceptors by small dissolved molecules that bind to proteins on the plasma membrane. In humans, sensory cells within taste buds express a single receptor type specific for one of the five taste perceptions—sweet, sour, salty, bitter, and umami (elicited by glutamate). Olfactory receptor cells line the upper part of the nasal cavity and extend axons to the olfactory bulb of the brain. More than 1,000 genes code for membrane proteins that bind to specific classes of odorants, and each receptor cell appears to express only one of those genes.

? *Why do foods taste bland when you have a head cold?*

CONCEPT 50.5

The physical interaction of protein filaments is required for muscle function (pp. 1103–1110)

- The muscle cells (fibers) of vertebrate skeletal muscle contain myofibrils composed of **thin filaments** of (mostly) actin and **thick filaments** of myosin. Together with accessory proteins, these filaments are organized into repeating units called **sarcomeres**. Myosin heads, energized by the hydrolysis of ATP, bind to the thin filaments, forming cross-bridges, then release upon binding ATP anew. As this cycle repeats, the thick and thin filaments slide past each other, shortening the sarcomere and contracting the muscle fiber.

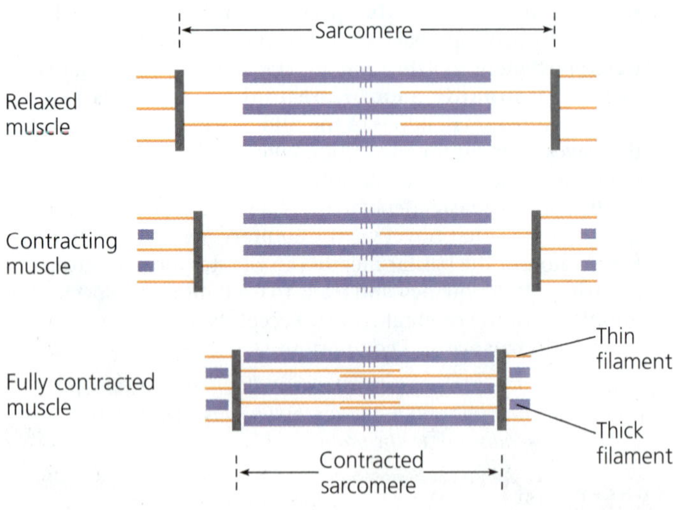

- Motor neurons release acetylcholine, triggering action potentials that penetrate the muscle fiber along the T tubules and stimulate the release of Ca^{2+} from the **sarcoplasmic reticulum**. When the Ca^{2+} binds the **troponin complex, tropomyosin** repositions on the thin filaments, exposing the myosin-binding sites on actin and thus initiating cross-bridge formation. A **motor unit** consists of a motor neuron and the muscle fibers it controls. Recruiting multiple motor units results in stronger contractions. A twitch results from a single action potential in a motor neuron. Skeletal muscle fibers can be slow-twitch or fast-twitch and oxidative or glycolytic.
- Cardiac muscle, found only in the heart, consists of striated cells that are electrically connected by intercalated disks and that can generate action potentials without input from neurons. In smooth muscles, contractions are slow and may be initiated by the muscles themselves or by stimulation from neurons in the autonomic nervous system.

? *What are two major functions of ATP hydrolysis in skeletal muscle activity?*

CONCEPT 50.6

Skeletal systems transform muscle contraction into locomotion (pp. 1110–1115)

- Skeletal muscles, often in antagonistic pairs, bring about movement by contracting and pulling against the skeleton. Skeletons may be **hydrostatic** and maintained by fluid pressure, as in worms; hardened into **exoskeletons**, as in insects; or in the form of **endoskeletons**, as in vertebrates.
- Each form of **locomotion**—swimming, movement on land, or flying—presents a particular challenge. For example, swimmers need to overcome friction, but face less of a challenge from gravity than do animals that move on land or fly. Animals specialized for swimming expend less energy per distance traveled than similarly sized animals specialized for flying or running. For any of the three major modes of locomotion, larger animals are more efficient than smaller ones.

? *Explain how microscopic and macroscopic anchoring of muscle filaments enables you to bend your elbow.*

TEST YOUR UNDERSTANDING

LEVEL 1: KNOWLEDGE/COMPREHENSION

1. Which of the following sensory receptors is *incorrectly* paired with its category?
   a. hair cell—mechanoreceptor
   b. muscle spindle—mechanoreceptor
   c. taste receptor—chemoreceptor
   d. rod—electromagnetic receptor
   e. olfactory receptor—electromagnetic receptor

2. The middle ear converts
   a. air pressure waves to fluid pressure waves.
   b. fluid pressure waves to air pressure waves.
   c. air pressure waves to nerve impulses.
   d. fluid pressure waves to nerve impulses.
   e. pressure waves to hair cell movements.

3. During the contraction of a vertebrate skeletal muscle fiber, calcium ions
   a. break cross-bridges by acting as a cofactor in the hydrolysis of ATP.
   b. bind with troponin, changing its shape so that the myosin-binding sites on actin are exposed.
   c. transmit action potentials from the motor neuron to the muscle fiber.
   d. spread action potentials through the T tubules.
   e. re-establish the polarization of the plasma membrane following an action potential.

LEVEL 2: APPLICATION/ANALYSIS

4. Which sensory distinction is *not* encoded by a difference in neuron identity?
   a. white and red
   b. red and green
   c. loud and faint
   d. salty and sweet
   e. spicy and cool

5. The transduction of sound waves into action potentials takes place
   a. within the tectorial membrane as it is stimulated by the hair cells.
   b. when hair cells are bent against the tectorial membrane, causing them to depolarize and release neurotransmitter that stimulates sensory neurons.
   c. as the basilar membrane becomes more permeable to sodium ions and depolarizes, initiating an action potential in a sensory neuron.
   d. as the basilar membrane vibrates at different frequencies in response to the varying volume of sounds.
   e. within the middle ear as the vibrations are amplified by the malleus, incus, and stapes.

LEVEL 3: SYNTHESIS/EVALUATION

6. Although some sharks close their eyes just before they bite, their bites are on target. Researchers have noted that sharks often misdirect their bites at metal objects and that they can find batteries buried under sand. This evidence suggests that sharks keep track of their prey during the split second before they bite in the same way that
   a. a rattlesnake finds a mouse in its burrow.
   b. a male silkworm moth locates a mate.
   c. a bat finds moths in the dark.
   d. a platypus locates its prey in a muddy river.
   e. a flatworm avoids light places.

7. **DRAW IT** Based on the information in the text, fill in the following graph. Use one line for rods and another line for cones.

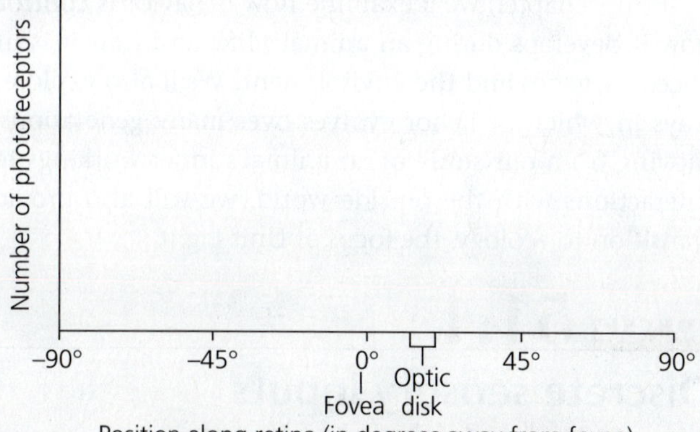

8. **EVOLUTION CONNECTION**
   In general, locomotion on land requires more energy than locomotion in water. By integrating what you have learned about animal form and function in Unit 7, discuss some of the evolutionary adaptations of mammals that support the high energy requirements for moving on land.

9. **SCIENTIFIC INQUIRY**
   Although skeletal muscles generally fatigue fairly rapidly, clam shell muscles have a protein called paramyosin that allows them to sustain contraction for up to a month. From your knowledge of the cellular mechanism of contraction, propose a hypothesis to explain how paramyosin might work. How would you test your hypothesis experimentally?

10. **WRITE ABOUT A THEME**
    Structure and Function In a short essay (100–150 words), describe at least three ways in which the structure of the lens of the human eye is well adapted to its function in vision.

For selected answers, see Appendix A.

51

Animal Behavior

▲ **Figure 51.1 What prompts a fiddler crab to wave its giant claw?**

KEY CONCEPTS

51.1 Discrete sensory inputs can stimulate both simple and complex behaviors

51.2 Learning establishes specific links between experience and behavior

51.3 Selection for individual survival and reproductive success can explain most behaviors

51.4 Inclusive fitness can account for the evolution of behavior, including altruism

OVERVIEW

The How and Why of Animal Activity

Unlike most animals, male fiddler crabs (genus *Uca*) are highly asymmetrical: One claw grows to giant proportions, half the mass of the entire body (**Figure 51.1**). The name *fiddler* comes from the crab's appearance as it feeds on algae from the mudflats where it lives: The smaller front claw moves to and from the mouth in front of the enlarged claw. Sometimes, however, the male waves his large claw in the air. What triggers this behavior? What purpose does it serve?

Claw waving by a male fiddler crab has two functions. Waving the claw, which can be used as a weapon, helps the crab *repel* other males wandering too close to his burrow. Vigorous claw waving also helps him *attract* females, who wander through the crab colony in search of a mate. After the male fiddler crab lures a female to his burrow, he seals her in with mud or sand in preparation for mating.

Animal behavior, be it solitary or social, fixed or variable, is based on physiological systems and processes. An individual **behavior** is an action carried out by muscles under control of the nervous system in response to a stimulus. Examples include an animal using its chest and throat muscles to produce a song, releasing a scent to mark its territory, or simply waving a claw. Behavior is an essential part of acquiring nutrients for digestion and finding a partner for sexual reproduction. Behavior also contributes to homeostasis, as in honeybees huddling to conserve heat (see Chapter 40). In short, all of animal physiology contributes to behavior, and animal behavior influences all of physiology.

Being essential for survival and reproduction, behavior is subject to substantial natural selection over time. This evolutionary process of selection also affects anatomy because the recognition and communication that underlie many behaviors depend on body form and appearance. Thus, the enlarged claw of the male fiddler crab is an adaptation that enables the display essential for recognition by other members of the species. Similarly, the positioning of the eyes on stalks held well above the crab's head enables him to see intruders from far off.

In this chapter, we'll examine how behavior is controlled, how it develops during an animal's life, and how it is influenced by genes and the environment. We'll also explore the ways in which behavior evolves over many generations. In moving from our study of an animal's inner workings to its interactions with the outside world, we will also provide a transition to ecology, the focus of Unit Eight.

CONCEPT **51.1**

Discrete sensory inputs can stimulate both simple and complex behaviors

What approach do biologists use to determine how behaviors arise and what functions they serve? The Dutch scientist Niko Tinbergen, a pioneer in the study of animal behavior, suggested that understanding any behavior requires answering four questions, which can be summarized as follows:

1. What stimulus elicits the behavior, and what physiological mechanisms mediate the response?
2. How does the animal's experience during growth and development influence the response?

3. How does the behavior aid survival and reproduction?
4. What is the behavior's evolutionary history?

Tinbergen's first two questions ask about *proximate causation*: "how" a behavior occurs or is modified. The last two questions ask about *ultimate causation*: "why" a behavior occurs in the context of natural selection.

Today, Tinbergen's questions and the associated ideas of causation underlie **behavioral ecology**, the study of the ecological and evolutionary basis for animal behavior. As we explore this vibrant area of modern biological research, we will also review studies by Tinbergen and two other early researchers—Karl von Frisch and Konrad Lorenz—that earned the three scientists a Nobel Prize in 1973.

In addressing Tinbergen's first question, the nature of the stimuli that trigger behavior, we'll begin with behavioral responses to well-defined stimuli, starting with an example from Tinbergen's own experiments.

Fixed Action Patterns

As part of his research, Tinbergen kept fish tanks containing three-spined sticklebacks (*Gasterosteus aculeatus*). Male sticklebacks, which have red bellies, attack other males that invade their nesting territories. Tinbergen noticed that his male sticklebacks also behaved aggressively when a red truck passed in front of their tank. Inspired by this chance observation, he carried out experiments showing that the red color of an intruder's underside is what provokes the attack behavior. A male stickleback will not attack a fish lacking red coloration (note that female sticklebacks never have red bellies), but will attack even unrealistic models if they contain areas of red color **(Figure 51.2)**.

The territorial response of male sticklebacks is an example of a **fixed action pattern**, a sequence of unlearned acts directly linked to a simple stimulus. Fixed action patterns are essentially unchangeable and, once initiated, usually carried to completion. The trigger for the behavior is an external cue called a **sign stimulus**, such as a red object prompting the male stickleback's aggressive behavior.

Migration

Environmental stimuli not only trigger behaviors but also provide cues that animals use to carry out those behaviors. For example, a wide variety of birds, fishes, and other animals use environmental cues to guide **migration**—a regular, long-distance change in location **(Figure 51.3)**.

Many migrating animals pass through environments they have not previously encountered. How, then, do they find their way in these foreign settings?

Some migrating animals track their position relative to the sun, even though the sun's position relative to Earth changes throughout the day. Animals can adjust for these changes by means of a *circadian clock*, an internal mechanism that maintains a 24-hour activity rhythm or cycle (see Chapter 49). For

(a) A male stickleback fish attacks other male sticklebacks that invade its nesting territory. The red belly of the intruding male (left) acts as the sign stimulus that releases the aggressive behavior.

(b) The realistic model at the top, without a red underside, produces no aggressive response in a male three-spined stickleback. The other models, with red undersides, produce strong responses.

▲ **Figure 51.2 Sign stimuli in a classic fixed action pattern.**

? *Suggest an explanation for why this behavior evolved (its ultimate causation).*

▲ **Figure 51.3 Migration.** Each spring, snow geese (*Chen caerulescens*) migrate from their wintering grounds, which may be as far south as Mexico, to their breeding grounds in Greenland, Canada, and Alaska. In the autumn, they return to their wintering grounds.

example, experiments have shown that migrating birds orient differently relative to the sun at distinct times of the day. Nocturnal animals can instead use the North Star, which has a constant position in the night sky.

Although the sun and stars can provide useful clues for navigation, these landmarks can be obscured by clouds. How do migrating animals overcome this problem? A simple experiment with homing pigeons provides one answer. On an overcast day, placing a small magnet on the head of a homing pigeon prevents it from returning efficiently to its roost. Researchers concluded that pigeons can sense their position relative to Earth's magnetic field and thereby navigate without solar or celestial cues.

The way in which animals detect Earth's magnetic field remains a matter of debate. It is known that the heads of migrating fishes and birds contain bits of magnetite, a magnetic iron mineral. This fact leads some scientists to hypothesize that Earth's pull on magnetite-containing structures triggers transmission of nerve impulses to the brain. Others propose that migrating animals are guided by the effects of Earth's magnetic field on photoreceptors in the eye. The idea that animals "see" the magnetic field is supported by experiments showing that light of particular wavelengths must be present for birds to orient in a magnetic field during the day or night.

Behavioral Rhythms

Although the circadian clock plays a small but significant role in navigation by some migrating species, it has a major role in the daily activity of all animals. As discussed in Chapters 40 and 49, the output of the clock is a circadian rhythm, a daily cycle of rest and activity with far-reaching effects on behavioral physiology. The clock is normally synchronized with the light and dark cycles of the environment but can maintain rhythmic activity under constant environmental conditions, such as during hibernation.

Some behaviors, such as migration and reproduction, reflect biological rhythms with a longer cycle, or period, than the circadian rhythm. Behavioral rhythms linked to the yearly cycle of seasons are called *circannual rhythms*. Although migration and reproduction typically correlate with food availability, these behaviors are not a direct response to changes in food intake. Instead, circannual rhythms, like circadian rhythms, are influenced by the periods of daylight and darkness in the environment. For example, studies with several bird species have shown that an artificial environment with extended daylight can induce out-of-season migratory behavior.

Not all biological rhythms are linked to the light and dark cycles in the environment. Consider, for instance, the fiddler crab shown in Figure 51.1. The male's claw-waving courtship behavior is linked not to day length but to the timing of the new and full moon. Why? Fiddler crabs begin their lives as plankton, settling in the mudflats after several larval stages. By courting at the time of the new or full moon, crabs link their reproduction to the times of greatest tidal movement. The tides disperse larvae to deeper waters, where they complete early development in relative safety before returning to the tidal flats.

Animal Signals and Communication

Claw waving by fiddler crabs during courtship is an example of one animal (the male crab) generating the stimulus that guides the behavior of another animal (the female crab). A stimulus transmitted from one animal to another is called a **signal**. The transmission and reception of signals constitute animal **communication**, an essential element of interactions between individuals.

Forms of Animal Communication

Let's consider the courtship behavior of *Drosophila melanogaster*, the fruit fly **(Figure 51.4)**, as an introduction to the four

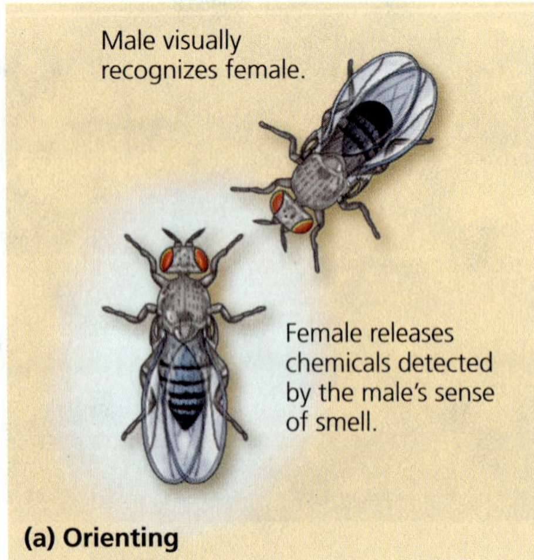

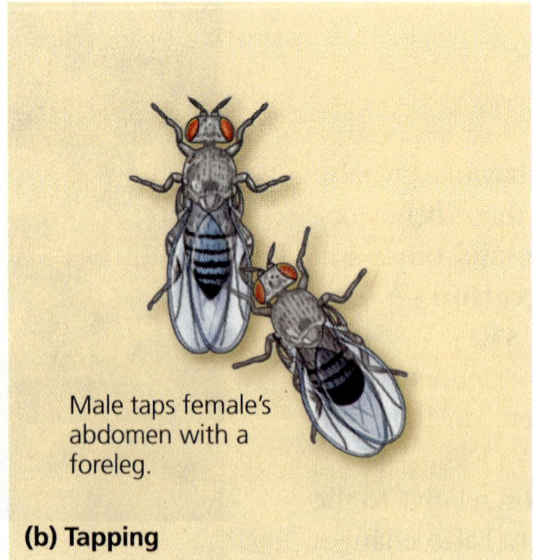

▲ **Figure 51.4 Courtship behavior of the fruit fly.** Fruit fly courtship involves a fixed set of behaviors that follow one another in a fixed order.

common modes of animal communication: visual, chemical, tactile, and auditory.

Fruit fly courtship constitutes a *stimulus-response chain*, in which the response to each stimulus is itself the stimulus for the next behavior. In the first step, a male identifies a female of the same species and then orients his body toward hers. When the male sees and turns toward the female, he relies on *visual communication*. In addition, the male's sense of smell, or olfactory system, detects chemicals released into the air by the female. This is an example of *chemical communication*, the transmission and reception of signals in the form of specific molecules. Having recognized the female, the male approaches and taps the female with a foreleg. This touching, or *tactile communication*, alerts the female to the male's presence. In the process, chemicals on her abdomen are transferred to the male, providing further chemical confirmation of her species identity. In the third stage of courtship, the male extends and vibrates his wing, producing a specific courtship song. This singing, an example of *auditory communication*, informs the female that the male is of the same species. Only if all of these forms of communication are successful will a female allow the male to attempt copulation.

The information content of animal communication varies considerably. One of the most remarkable examples is the symbolic language of the European honeybee (*Apis mellifera*), discovered in the early 1900s by Austrian researcher Karl von Frisch. Using glass-walled observation hives, he and his students spent several decades observing these bees. Methodical recordings of bee movements enabled von Frisch to decipher a "dance language" that returning foragers use to inform other bees about the distance and direction of travel to food sources.

As illustrated in **Figure 51.5**, a returning bee quickly becomes the center of attention for other bees, called followers. If the food source is close to the hive (less than 50 m away), the returning bee moves in tight circles while waggling its abdomen from side to side. This behavior, called the "round dance," motivates the follower bees to leave the hive and search for nearby food.

When the food source is farther from the nest, the returning bee instead performs a "waggle dance." This dance, consisting of a half-circle swing in one direction, a straight run during which the bee waggles its abdomen, and a half-circle swing in the other direction, communicates to the follower bees both the direction and distance of the food source in relation to the hive. The angle of the straight run relative to the hive's vertical surface is the same as the horizontal angle of the food in relation to the sun. For example, if the returning bee runs at a 30° angle to the right of vertical, the follower bees leaving the hive fly 30° to the right of the horizontal direction of the sun. A dance with a longer straight run, and

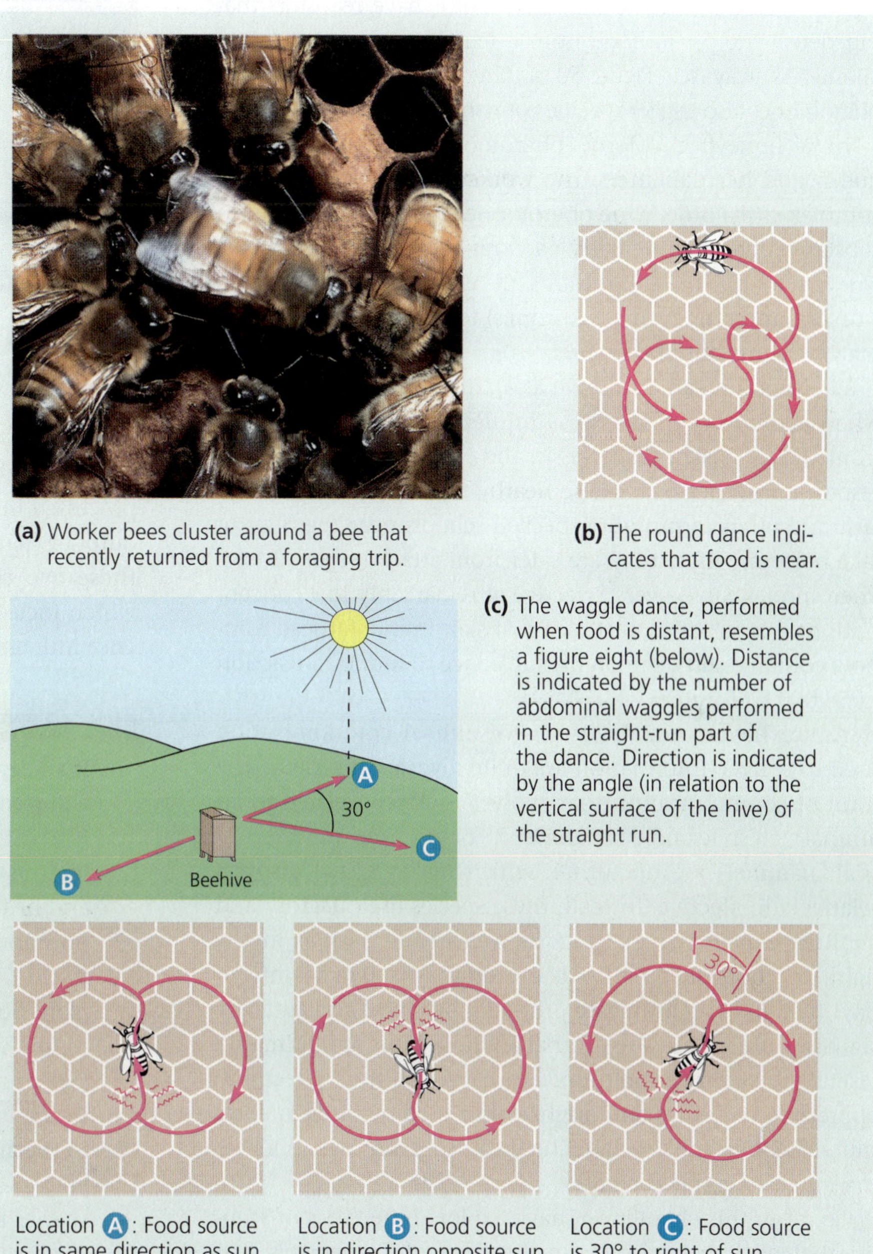

▲ **Figure 51.5 Honeybee dance language.** Honeybees returning to the hive communicate the location of food sources through the symbolic language of a dance.

therefore more abdominal waggles per run, indicates a greater distance to the food source. As follower bees exit the hive, they fly almost directly to the area indicated by the waggle dance. By using flower odor and other clues, they locate the food source within this area.

Pheromones

Animals that communicate through odors or tastes emit chemical substances called **pheromones**. Pheromones are especially common among mammals and insects and often relate to reproductive behavior. For example, pheromones are the basis for the chemical communication in fruit fly courtship (see Figure 51.4). Pheromones are not limited to short-distance signaling, however. Male silkworm moths have receptors that can detect the pheromone from a female moth from several kilometers away (see Figure 50.6). After the moths are together, pheromones also trigger specific courtship behaviors.

In a honeybee colony, pheromones produced by the queen and her daughters, the workers, maintain the hive's complex social order. One pheromone (once called the queen substance) has a particularly wide range of effects. It attracts workers to the queen, inhibits development of ovaries in workers, and attracts males (drones) to the queen during her mating flights out of the hive.

Pheromones can also serve as alarm signals. For example, when a minnow or catfish is injured, a substance released from the fish's skin disperses in the water, inducing a fright response in other fish. These nearby fish become more vigilant and often form tightly packed schools near the river or lake bottom, where they are safer from attack **(Figure 51.6)**. Pheromones can be very effective at remarkably low concentrations. For instance, just 1 cm^2 of skin from a fathead minnow contains sufficient alarm substance to induce a reaction in 58,000 L of water.

As we have seen, the forms of animal communication used to convcy information are quite diverse. In general, the form of communication that evolves is closely related to an animal's lifestyle and environment. For example, most terrestrial mammals are nocturnal, which makes visual displays relatively ineffective. Instead, these species use olfactory and auditory signals, which work as well in the dark as in the light. In contrast, most birds are diurnal (active mainly in daytime) and communicate primarily by visual and auditory signals. Humans are also diurnal and, like birds, use primarily visual and auditory communication. We can thus detect and appreciate the songs and bright colors used by birds to communicate but miss many chemical cues on which other mammals base their behavior.

So far in this chapter, we have explored the types of stimuli that elicit behaviors—the first part of Tinbergen's first question. The second part of that question—the physiological mechanisms that mediate responses—was the focus of

(a) Minnows are widely dispersed in an aquarium before an alarm substance is introduced.

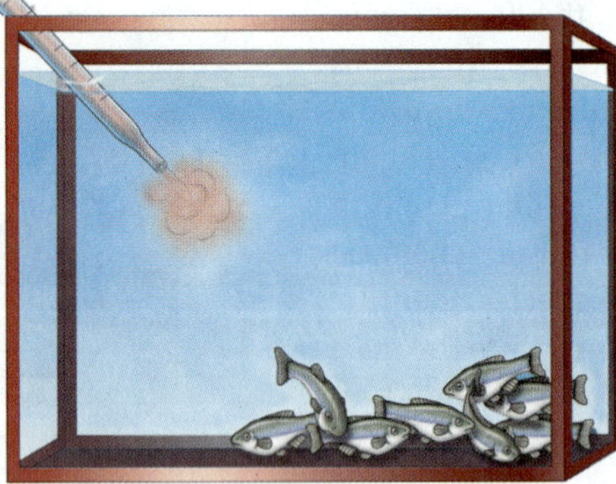

(b) Within seconds of the alarm substance being introduced, minnows aggregate near the bottom of the aquarium and reduce their movement.

▲ **Figure 51.6 Minnows responding to the presence of an alarm substance.**

Chapters 49 and 50. Stimuli activate sensory systems, are processed in the central nervous system, and result in motor outputs that constitute behavior. You may want to review those two chapters before proceeding to the next concept, which focuses on Tinbergen's second question—how experience influences behavior.

CONCEPT CHECK 51.1

1. If an egg rolls out of the nest, a mother graylag goose will retrieve it by nudging it with her beak and head. If researchers remove the egg or substitute a ball during this process, the goose continues to bob her beak and head while she moves back to the nest. Explain how and why this behavior occurs.
2. **MAKE CONNECTIONS** How is the seasonal timing of plant flowering similar in mechanism and function to the lunar-linked rhythm of fiddler crab courtship? (See pp. 839–841 of Concept 39.3.)
3. **WHAT IF?** Suppose you exposed various fish species to the alarm substance from minnows. Based on what you know about natural selection, suggest why some species might respond like minnows, some might increase activity, and some might show no change.

For suggested answers, see Appendix A.

CONCEPT 51.2

Learning establishes specific links between experience and behavior

For many behaviors—such as a fixed action pattern, a courtship stimulus-response chain, and pheromone signaling—nearly all individuals in a population exhibit virtually the same behavior, despite internal and environmental differences during development and throughout life. Behavior that is developmentally fixed in this way is known as **innate behavior**. In other cases, behavior is variable, depending on experience.

Experience and Behavior

Tinbergen's second question asks how an animal's experiences during growth and development influence the response to stimuli. How do researchers answer this question? One informative approach is a **cross-fostering study**, in which the young of one species are placed in the care of adults from another species. The extent to which the offspring's behavior changes in such a situation provides one measure of how the social and physical environment influences behavior.

The males of certain mouse species have behavioral differences that are well suited for cross-fostering experiments. Males of the species *Peromyscus californicus*, commonly called California mice, are highly aggressive toward other mice and provide extensive parental care. In contrast, male white-footed mice (*Peromyscus leucopus*) are less aggressive and engage in little parental care. When the pups of each species were placed in the nests of the other species, the cross-fostering altered some behaviors of both species **(Table 51.1)**. For instance, male California mice raised by white-footed mice were less aggressive toward intruders. Thus, experience during development can strongly influence aggressive behavior in these rodents.

Table 51.1 Influence of Cross-Fostering on Male Mice*

Species	Aggression Toward an Intruder	Aggression in Neutral Situation	Paternal Behavior
California mice fostered by white-footed mice	Reduced	No difference	Reduced
White-footed mice fostered by California mice	No difference	Increased	No difference

*Comparisons are with mice raised by parents of their own species.

One of the most important findings of the cross-fostering experiments with mice was that the influence of experience on behavior can be passed on to progeny: When the cross-fostered California mice became parents, they spent less time retrieving offspring who wandered off than did California mice raised by their own species. Thus, experience during development can modify physiology in a way that alters parental behavior, extending the influence of environment to a subsequent generation.

For humans, the influence of genetics and environment on behavior can be explored by a **twin study**, in which researchers compare the behavior of identical twins raised apart with the behavior of those raised in the same household. Twin studies have been instrumental in studying human behavioral disorders, such as schizophrenia, anxiety disorders, and alcoholism. As discussed in Chapter 49, these investigations have revealed that both genetics and environment (nature *and* nurture) contribute significantly to the behaviors that characterize these disorders in humans.

Learning

One of the most powerful ways that environmental conditions can influence behavior is through **learning**, the modification of behavior based on specific experiences. We will now consider a number of different types of learning, beginning with a form of learning first explored by Austrian biologist Konrad Lorenz.

Imprinting

A type of behavior that includes both learned and innate components is **imprinting**, the formation at a specific stage in life of a long-lasting behavioral response to a particular individual or object. Imprinting is distinguished from other types of learning by having a **sensitive period**, also called a critical period, a limited developmental phase when this type of learning can occur. During the sensitive period, the young imprint on their parent and learn the basic behaviors of their species, while the parent learns to recognize its offspring. Among gulls, for instance, the sensitive period for a parent to bond with its young lasts one to two days. If bonding does not occur, the parent will not care for the infant, leading to death for the offspring and a decrease in reproductive success for the parent.

But how do the young know on whom—or what—to imprint? For example, how do young birds know that they should follow their mother? The tendency to respond is innate in the birds; the outside world provides the *imprinting stimulus*, something to which the response will be directed. Experiments with many species of waterfowl indicate that they have no innate recognition of "mother." Rather, they identify with the first object they encounter that has certain key characteristics. In classic experiments done in the 1930s, Lorenz showed that the principal imprinting stimulus in graylag geese (*Anser anser*) is a nearby object that is moving away from the young. When incubator-hatched goslings spent their first few hours with Lorenz rather than with a goose, they imprinted on him and steadfastly followed him from then on **(Figure 51.7a)**. Furthermore, they showed no recognition of their biological mother or other adults of their own species.

(a) These young greylag geese imprinted on ethologist Konrad Lorenz.

(b) A pilot wearing a crane suit and flying an ultralight plane acts as a surrogate parent to direct the migration of whooping cranes.

▲ **Figure 51.7 Imprinting.** Imprinting can be altered to **(a)** investigate animal behavior or **(b)** direct animal behavior.

WHAT IF? *Suppose the geese following Lorenz were bred to each other. How might their imprinting on Lorenz affect their offspring? Explain.*

Imprinting has become an important component of efforts to save endangered species, such as the whooping crane (*Grus americana*). Scientists tried raising whooping cranes in captivity by using sandhill cranes (*Grus canadensis*) as foster parents. However, because the whooping cranes imprinted on their foster parents, none formed a *pair-bond* (strong attachment) with a whooping crane mate. To avoid such problems, captive breeding programs now isolate young cranes, exposing them to the sights and sounds of members of their own species.

Scientists have made further use of imprinting to teach cranes born in captivity to migrate along safe routes. Young whooping cranes are imprinted on humans in "crane suits" and then allowed to follow these "parents" as they fly ultralight aircraft along selected migration routes **(Figure 51.7b)**. Importantly, these cranes still form mating pair-bonds with other whooping cranes, indicating that the crane costumes have the features required to direct "normal" imprinting.

Spatial Learning and Cognitive Maps

Every natural environment has spatial variation, as in locations of nest sites, hazards, food, and prospective mates. Therefore, an organism's fitness may be enhanced by the capacity for **spatial learning**, the establishment of a memory that reflects the environment's spatial structure.

The idea of spatial learning intrigued Tinbergen while he was a graduate student in the Netherlands. At that time, he was studying the female digger wasp (*Philanthus triangulum*), which nests in small burrows dug into sand dunes. Tinbergen noticed that when a wasp left her nest to go hunting, she hid the entrance from potential intruders by covering it with sand. Upon her return, she flew directly to her hidden nest, despite the presence of hundreds of other burrows in the area. Tinbergen hypothesized that a wasp locates her nest by learning its position relative to visible landmarks, or location indicators. To test this hypothesis, he carried out an experiment in the wasps' natural habitat **(Figure 51.8)**. By manipulating objects around nest entrances, he demonstrated that digger wasps engage in spatial learning. This experiment was so simple and informative that it could be summarized very concisely. In fact, at 32 pages, Tinbergen's Ph.D. thesis from 1932 is still the shortest ever approved at Leiden University.

In many animal species, spatial learning can be quite sophisticated. Some animals guide their activity by a **cognitive map**, a representation in the nervous system of the spatial relationships between objects in an animal's surroundings. Rather than relying solely on moving from landmark to landmark, animals using cognitive maps can navigate more flexibly and efficiently by relating landmark positions to one another.

One striking example of cognitive mapping is found in the Clark's nutcracker (*Nucifraga columbiana*). Nutcrackers are corvids, the bird family that also includes ravens, crows, and jays. In the fall, a single nutcracker stores as many as 30,000 pine seeds in thousands of hiding places called caches, distributed over an area as large as 35 km^2. During the winter, the birds relocate many of their caches. By experimentally varying the distance between landmarks, researchers demonstrated that birds could identify the halfway point between landmarks. This behavior suggests that nutcrackers employ an abstract geometric rule, which we can approximate as "Seed caches are found halfway between particular landmarks." Such rules, a fundamental property of cognitive maps, reduce the amount of detail required to remember an object's location. As we discussed in Chapter 49, corvids also display other forms of higher nervous system function.

▼ Figure 51.8 INQUIRY

Does a digger wasp use landmarks to find her nest?

EXPERIMENT A female digger wasp covers the entrance to her nest while foraging for food, but finds the correct wasp nest reliably upon her return 30 minutes or more later. Niko Tinbergen wanted to test the hypothesis that a wasp learns visual landmarks that mark her nest before she leaves on hunting trips. First, he marked one nest with a ring of pinecones while the wasp was in the burrow. After leaving the nest to forage, the wasp returned to the nest successfully.

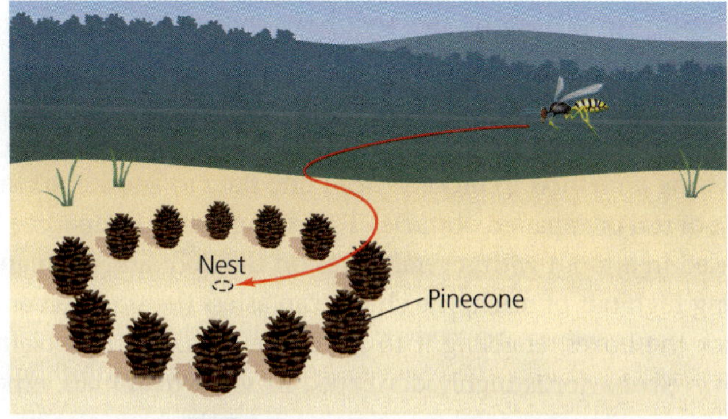

Two days later, after the wasp had again left, Tinbergen shifted the ring of pinecones away from the nest. Then he waited to observe the wasp's behavior.

RESULTS When the wasp returned, she flew to the center of the pinecone circle instead of to the nearby nest. Repeating the experiment with many wasps, Tinbergen obtained the same results.

CONCLUSION The experiment supported the hypothesis that digger wasps use visual landmarks to keep track of their nests.

SOURCE N. Tinbergen, *The Study of Instinct*, Clarendon Press, Oxford (1951).

WHAT IF? Suppose the digger wasp had returned to her original nest site, despite the pinecones having been moved. What alternative hypotheses might you propose regarding how the wasp finds her nest and why the pinecones didn't misdirect the wasp?

Associative Learning

Learning often involves making associations between experiences. Consider, for example, a blue jay (*Cyanocitta cristata*) that ingests a brightly colored monarch butterfly (*Danaus plexippus*). Substances that the monarch accumulates from milkweed plants cause the blue jay to vomit almost immediately **(Figure 51.9)**. Following such experiences, blue jays avoid attacking monarchs and similar-looking butterflies. The ability to associate one environmental feature (such as a color) with another (such as a foul taste) is called **associative learning**.

Among animal behaviors, associative learning is particularly suited to laboratory studies because they typically involve either classical conditioning or operant conditioning. In *classical conditioning*, an arbitrary stimulus becomes associated with a particular outcome. Russian physiologist Ivan Pavlov carried out early experiments in classical conditioning, demonstrating that if he always rang a bell just before feeding a dog, the dog would eventually salivate when the bell sounded, anticipating food. In *operant conditioning*, also called trial-and-error learning, an animal first learns to associate one of its behaviors with a reward or punishment and then tends to repeat or avoid that behavior (see Figure 51.9). B. F. Skinner, an American pioneer in the study of operant conditioning, explored this process in the laboratory by, for example, having a rat learn through trial-and-error to obtain food by pressing a lever.

▲ Figure 51.9 Associative learning. Having ingested and vomited a monarch butterfly, a blue jay has probably learned to avoid this species.

Studies reveal that animals can learn to link many pairs of features of their environment, but not all. For example, pigeons can learn to associate danger with a sound but not with a color. However, they can learn to associate a color with food. What does this mean? The development and organization of the pigeon's nervous system apparently restrict the associations that can be formed. Moreover, such restrictions are not limited to birds. Rats, for example, can learn to avoid illness-inducing foods on the basis of smells, but not on the basis of sights or sounds.

If we take into account the evolution of behavior, the fact that some animals can't learn to make particular associations appears logical. The associations an animal can readily form typically reflect relationships likely to occur in nature. Conversely, associations that can't be formed are those unlikely to be of selective advantage in a native environment. In the case of a rat's diet in the wild, for example, a harmful food is far more likely to have a certain odor than to be associated with a particular sound.

Cognition and Problem Solving

The most complex forms of learning involve **cognition**—the process of knowing that involves awareness, reasoning, recollection, and judgment. Although it was once argued that only primates and certain marine mammals have high-level thought processes, many other groups of animals, including insects, appear to exhibit cognition in controlled laboratory studies. For example, an experiment using Y-shaped mazes tested whether honeybees can distinguish between "same" and "different." One maze had different colors, and one had different black-and-white striped patterns, either vertical or horizontal bars. Two groups of honeybees were trained in the color maze. Upon entering, a bee would see a sample color and could then choose between an arm of the maze with the same color or an arm with a different color. Only one arm contained a food reward. The first group of bees were rewarded for flying into the arm with the *same* color as the sample **(Figure 51.10a)**; the second group were rewarded for choosing the arm with the *different* color. Next, they were tested in the bar maze, which had no food reward. After encountering a sample black-and-white pattern of bars, a bee could choose between an arm with the same pattern or an arm with a different pattern. The bees in the first group most often chose the arm with the same pattern **(Figure 51.10b)**, whereas those in the second group typically chose the arm with the different pattern.

The maze experiments provide strong experimental support for the hypothesis that honeybees can distinguish on the basis of "same" and "different." Remarkably, research published in 2010 indicates that honeybees can also learn to distinguish between human faces.

The information-processing ability of a nervous system can also be revealed in **problem solving**, the cognitive activity of devising a method to proceed from one state to another in the face of real or apparent obstacles. For example, if a chimpanzee is placed in a room with several boxes on the floor and a banana hung high out of reach, the chimp can assess the situation and stack the boxes, enabling it to reach the food. Such problem-solving behavior is highly developed in some mammals, especially primates and dolphins. Notable examples have also been observed in some bird species, especially corvids. In one study, ravens were confronted with food hanging from a branch by a string. After failing to grab the food in flight, one raven flew to the branch and alternately pulled up and stepped on the string until the food was within reach. A number of other ravens eventually arrived at similar solutions. Nevertheless, some ravens failed to solve the problem, indicating that problem-solving success in this species, as in others, varies with individual experience and abilities.

(a) Bees were trained in a color maze. As shown here, one group were rewarded for choosing the same color as the stimulus.

(b) Bees were tested in a pattern maze. If previously rewarded for choosing the same color, bees most often chose lines oriented the same way as the stimulus.

▲ **Figure 51.10 A maze test of abstract thinking by honeybees.** These mazes are designed to test whether honeybees can distinguish "same" from "different."

Development of Learned Behaviors

Most of the learned behaviors we have discussed develop over a relatively short time. Some behaviors develop more gradually. For example, some bird species learn songs in stages.

In the case of the white-crowned sparrow (*Zonotrichia leucophrys*), the first stage of song learning takes place early in life, when the fledgling sparrow first hears the song. If a fledgling is prevented from hearing real sparrows or recordings of sparrow songs during the first 50 days of its life, it fails to develop the adult song of its species. Although the young bird does not sing during the sensitive period, it memorizes the song of its species by listening to other white-crowned sparrows sing. During the sensitive period, fledglings chirp more in response to songs of their own species than to songs of other species. Thus, although young white-crowned sparrows learn the songs they will sing as adults, learning appears to be bounded by genetically controlled preferences.

The sensitive period when a white-crowned sparrow memorizes its species' song is followed by a second learning phase when the juvenile bird sings tentative notes called a subsong. The juvenile bird hears its own singing and compares it with the song memorized during the sensitive period. Once a sparrow's own song matches the one it memorized, the song "crystallizes" as the final song, and the bird sings only this adult song for the rest of its life.

The song-learning process is quite different for canaries than for white-crowned sparrows. Canaries, for example, do not have a single sensitive period for song learning. A young canary begins with a subsong, but the full song does not crystallize in the same way as in white-crowned sparrows. Between breeding seasons, the song becomes flexible again, and an adult male may learn new song "syllables" each year, adding to the song it already sings.

Song learning is one of many examples of how animals learn from other members of their species. In finishing our exploration of learning, we'll look at several more examples that reflect the more general phenomenon of social learning.

Social Learning

Many animals learn to solve problems by observing the behavior of other individuals. Young wild chimpanzees, for example, learn how to crack open oil palm nuts with two stones by copying experienced chimpanzees **(Figure 51.11)**. This type of learning through observing others is called **social learning**.

▲ **Figure 51.11 A young chimpanzee learning to crack oil palm nuts by observing an experienced elder.**

Another example of how social learning can modify behavior comes from studies of the vervet monkeys (*Cercopithecus aethiops*) in Amboseli National Park, Kenya. Vervet monkeys, which are about the size of a domestic cat, produce a complex set of alarm calls. Amboseli vervets give distinct alarm calls for leopards, eagles, or snakes, all of which prey on vervets. When a vervet sees a leopard, it gives a loud barking sound; when it sees an eagle, it gives a short double-syllable cough; and the snake alarm call is a "chutter." Upon hearing a particular alarm call, other vervets in the group behave in an appropriate way: They run up a tree on hearing the alarm for a leopard (vervets are nimbler than leopards in the trees); look up on hearing the alarm for an eagle; and look down on hearing the alarm for a snake **(Figure 51.12)**.

Infant vervet monkeys give alarm calls, but in a relatively undiscriminating way. For example, they give the "eagle" alarm on seeing any bird, including harmless birds such as bee-eaters. With age, the monkeys improve their accuracy. In fact, adult vervet monkeys give the eagle alarm only on seeing an eagle belonging to either of the two species that eat vervets. Infants probably learn how to give the right call by observing other members of the group and receiving social confirmation. For instance, if the infant gives the call on the right occasion—say, an eagle alarm when there is an eagle overhead—another member of the group will also give the eagle call. But if the infant gives the call when a bee-eater flies by, the adults in the group are silent. Thus, vervet monkeys have an initial, unlearned tendency to give calls upon seeing potentially threatening objects in the environment. Learning fine-tunes the call so that adult vervets give calls only in response to genuine danger and can fine-tune the alarm calls of the next generation.

▲ **Figure 51.12 Vervet monkeys learning correct use of alarm calls.** On seeing a python (foreground), vervet monkeys give a distinct "snake" alarm call (inset), and the members of the group stand upright and look down.

Social learning forms the roots of **culture**, which can be defined as a system of information transfer through social learning or teaching that influences the behavior of individuals in a population. Cultural transfer of information can alter behavioral phenotypes and thereby influence the fitness of individuals.

Changes in behavior that result from natural selection occur on a much longer time scale than does learning. In Concept 51.3, we'll examine the relationship between particular behaviors and the processes of selection related to survival and reproduction.

CONCEPT CHECK 51.2

1. How might associative learning explain why different species of distasteful or stinging insects have similar colors?
2. **WHAT IF?** How might you position and manipulate a few objects in a lab to test whether an animal can use a cognitive map to remember the location of a food source?
3. **MAKE CONNECTIONS** How can a learned behavior contribute to speciation? (See Concept 24.1, pp. 488–492.)

For suggested answers, see Appendix A.

CONCEPT 51.3

Selection for individual survival and reproductive success can explain most behaviors

We turn now from the physiology of behavior (how animals behave) to the benefits to a species from a particular behavior (why animals behave the way they do). In particular, we will address Tinbergen's third question—how behavior enhances survival and reproductive success in a population. We'll begin with an activity essential for both types of success: gathering food.

Foraging Behavior

Because adequate nutrition is essential to an animal's survival and reproductive success, we should expect natural selection to refine behaviors that enhance the efficiency of feeding. Food-obtaining behavior, or **foraging**, includes not only eating but also any activities an animal uses to search for, recognize, and capture food items.

Evolution of Foraging Behavior

The fruit fly (*Drosophila melanogaster*) has provided an opportunity to examine how foraging behavior might have evolved. Variation in a gene called *forager* (*for*) dictates the food-search behavior of fruit fly larvae. On average, larvae carrying the for^R ("Rover") allele travel nearly twice as far while feeding as do larvae with the for^s ("sitter") allele. Experiments have shown that the enzyme encoded by the *forager* locus is more active in for^R larvae than in for^s larvae and has properties typical of an enzyme in a signal transduction pathway (see Chapter 45).

Both the for^R and for^s alleles are present in natural populations. What circumstances might favor one or the other allele? The answer became apparent in experiments when flies were kept for many generations at either low or high population densities. The larvae from the two samples clearly diverged in behavior, as measured by differences in average length of their foraging paths **(Figure 51.13)**. Larvae maintained for many generations at a low density foraged over shorter distances than those kept at high density. Furthermore, genetic tests indicated that the for^s allele had increased in frequency in the low-density populations, whereas the for^R allele had increased in frequency in the high-density group. These changes make sense. At low population density, short-distance foraging yields sufficient food, while long-distance foraging would result in unnecessary energy expenditure. Under crowded conditions, however, long-distance foraging could enable larvae to move beyond areas of food depletion. In summary, there was an observable and interpretable evolutionary change in behavior in the laboratory populations.

Optimal Foraging Model

To study the proximate and ultimate causation of diverse foraging strategies, biologists sometimes apply a type of cost-benefit analysis used in economics. This idea proposes that foraging behavior is a compromise between the benefits of nutrition and the costs of obtaining food. These costs might include the energy expenditure of foraging as well as the risk of being eaten while foraging. According to this **optimal foraging model**, natural selection should favor a foraging behavior that minimizes the costs of foraging and maximizes the benefits.

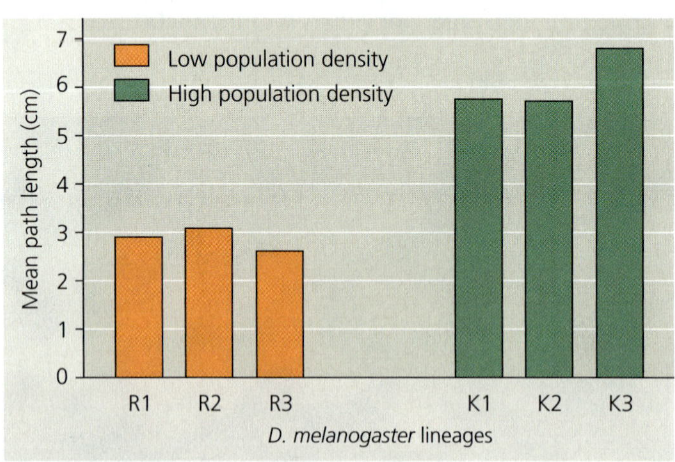

▲ **Figure 51.13 Evolution of foraging behavior by laboratory populations of *Drosophila melanogaster*.** After 74 generations of living at low population density, *D. melanogaster* larvae (populations R1–R3) followed foraging paths significantly shorter than those of *D. melanogaster* larvae that had lived at high density (populations K1–K3).

As an example of how the optimal foraging model can be applied, let's consider the feeding behavior of the Northwestern crow (*Corvus caurinus*). On islands off British Columbia, these crows search rocky tide pools for sea snails called whelks. After spotting a whelk, the crow picks the whelk up in its beak, flies upward, and drops the whelk onto the rocks. If the drop is successful, the shell breaks and the crow can dine on the whelk's soft parts. If not, the crow flies up and drops the whelk again and again until the shell breaks. What determines how high the crow flies? The higher the crow flies, the greater the force with which the whelk strikes the rocks, increasing the chance the shell will break. Flying higher, however, means consuming more energy.

If energetic considerations dominated selection for the crow's foraging behavior, the average drop height might reflect a trade-off between the cost of flying higher and the benefit of more frequent success. To test this idea, researchers dropped whelks from different heights and noted the number of drops required to break a shell. For each height, they calculated the average number of drops and the average *total flight height*, the drop height times the average number of drops **(Figure 51.14)**. A drop height of about 5 m turned out to be optimal, breaking the shells with the lowest total flight height—in other words, with the least work. The actual average flight height for crows in their whelk-eating behavior is 5.23 m, very close to the prediction based on an optimal trade-off between energy gained (food) and energy expended.

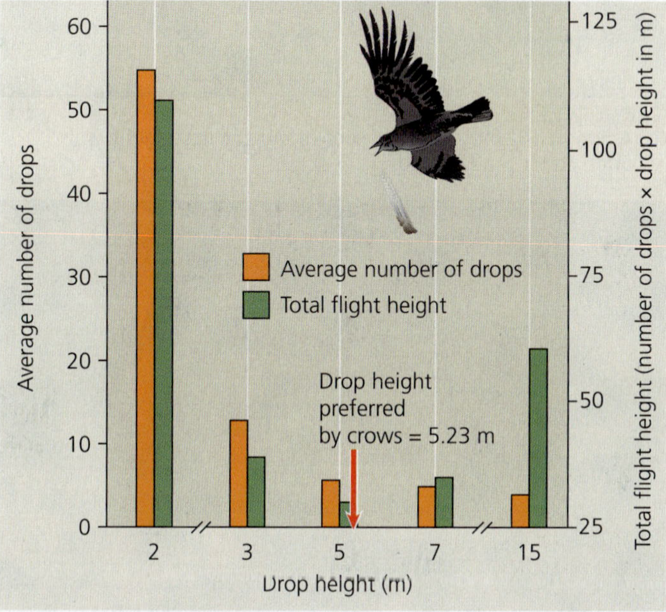

▲ **Figure 51.14 Energy costs and benefits in foraging behavior.** Experimental results indicate that dropping shells from a height of 5 m results in breakage with the least amount of work. The actual drop height preferred by crows corresponds almost exactly to the height that minimizes total flight height.

The close agreement between the predicted and actual flight heights suggests that the optimal foraging model reflects the selective forces shaping the evolution of this behavior. However, other models could explain the findings equally well. For example, the average flight height could minimize the average *time* necessary to break open a whelk. Further experiments are needed to evaluate these possibilities.

Balancing Risk and Reward

One of the most significant potential costs to a forager is risk of predation. Maximizing energy gain and minimizing energy costs are of little benefit if the behavior makes the forager a likely meal for a predator. It seems logical, therefore, that predation risk would influence foraging behavior. Such appears to be the case for the mule deer (*Odocoileus hemionus*), which lives in the mountains of western North America. Researchers found that the food available for mule deer was fairly uniform across the potential foraging areas, although somewhat lower in open, nonforested areas. In contrast, the risk of predation differed greatly; mountain lions (*Puma concolor*), the major predator, killed large numbers of mule deer at forest edges and only a small number in open areas and forest interiors.

How does mule deer foraging behavior reflect the differences in predation risk in particular areas? Mule deer feed predominantly in open areas. Thus, it appears that mule deer foraging behavior reflects the large variation in predation risk and not the smaller variation in food availability. This result underscores the point that behavior typically reflects a compromise between competing selective pressures.

Mating Behavior and Mate Choice

Just as foraging is crucial for individual survival, mating behavior and mate choice play a major role in determining reproductive success. These behaviors include seeking or attracting mates, choosing among potential mates, competing for mates, and caring for offspring. Although we tend to think of mating simply as the union of a male and female, the mating relationship between males and females varies greatly from species to species, defining a number of distinct mating systems.

Mating Systems and Sexual Dimorphism

Mating systems vary with regard to both the length and number of relationships. In many animal species, mating is **promiscuous**, with no strong pair-bonds. In species in which the mates remain together for a longer period, the relationship may be **monogamous** (one male mating with one female) or **polygamous** (an individual of one sex mating with several of the other). Polygamous relationships most often involve a single male and many females, a system called *polygyny*, though some species exhibit *polyandry*, in which a single female mates with several males.

The extent to which males and females differ in appearance, a characteristic known as *sexual dimorphism*, typically varies with the type of mating system **(Figure 51.15)**. Among monogamous species, males and females are often so much alike morphologically that they may be difficult or impossible to distinguish based on external characteristics. In contrast, polygynous species are generally dimorphic, with males being showier and often larger than females. Polyandrous species are also dimorphic, but the females are generally more ornamented and larger than the males.

Mating Systems and Parental Care

The needs of the young are an important factor constraining the evolution of mating systems. Most newly hatched birds, for instance, cannot care for themselves. Rather, they require a large, continuous food supply, a need that is difficult for a single parent to meet. In such cases, a male that stays with and helps a single mate may ultimately have more viable offspring than it would by going off to seek additional mates. This may explain why most birds are monogamous. In contrast, for birds with young that can feed and care for themselves almost immediately after hatching, the males derive less benefit from staying with their partner. Males of these species, such as pheasants and quail, can maximize their reproductive success by seeking other mates, and polygyny is relatively common in such birds. In the case of mammals, the lactating female is often the only food source for the young; males usually play no role in raising the young. In mammalian species where males protect the females and young, such as lions, a male or small group of males typically takes care of a harem of many females at the same time.

Another factor influencing mating behavior and parental care is *certainty of paternity*. Young born to or eggs laid by a female definitely contain that female's genes. However, even within a normally monogamous relationship, a male other than the female's usual mate may have fathered that female's offspring. The certainty of paternity is relatively low in most species with internal fertilization because the acts of mating and birth (or mating and egg laying) are separated over time. This could explain why exclusively male parental care is rare in bird and mammal species. However, the males of many species with internal fertilization engage in behaviors that appear to increase their certainty of paternity. These behaviors include guarding females, removing any sperm from the female reproductive tract before copulation, and introducing large quantities of sperm that displace the sperm of other males.

Certainty of paternity is high when egg laying and mating occur together, as in external fertilization. This may explain why parental care in aquatic invertebrates, fishes, and amphibians, when it occurs at all, is at least as likely to be by males as

▼ **Figure 51.15 Relationship between mating system and male and female forms.**

(a) In monogamous species, such as these western gulls, males and females are difficult to distinguish using external characteristics only.

(b) Among polygynous species, such as elk, the male (right) is often highly ornamented.

(c) In polyandrous species, such as these Wilson's phalaropes, females (top) are generally more ornamented than males.

▲ **Figure 51.16 Paternal care by a male jawfish.** The male jawfish, which lives in tropical marine environments, holds the eggs it has fertilized in its mouth, keeping them aerated and protecting them from egg predators until the young hatch.

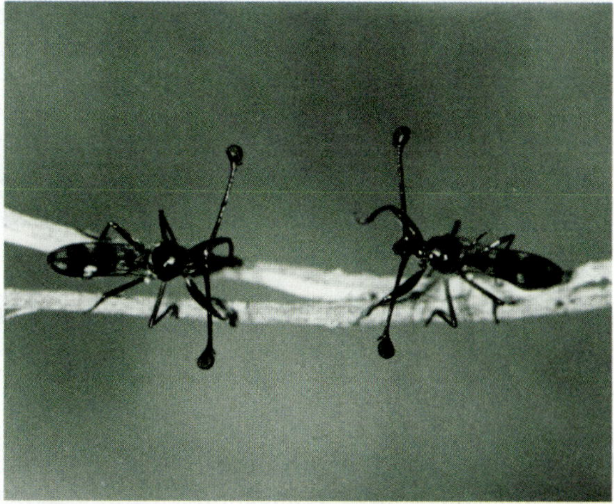

▲ **Figure 51.17 Male stalk-eyed flies.** Male eye span plays a role in mate selection by females and, as shown here, in ritualized contests between males. In such contests, two males face off, with the male whose eye span is smaller very often retreating without any combat taking place.

by females (**Figure 51.16**; see also Figure 46.7). Among fishes and amphibians, parental care occurs in only 7% of species with internal fertilization but in 69% of species with external fertilization.

It is important to point out that certainty of paternity does not mean that animals are aware of those factors when they behave a certain way. Parental behavior correlated with certainty of paternity exists because it has been reinforced over generations by natural selection. Nevertheless, the relationship between certainty of paternity and male parental care remains an area of active research, enlivened by controversy.

Sexual Selection and Mate Choice

As you read in Chapter 23, the degree of sexual dimorphism within a species results from sexual selection, a form of natural selection in which differences in reproductive success among individuals are a consequence of differences in mating success. Recall from that chapter that sexual selection can take the form of *intersexual selection*, in which members of one sex choose mates on the basis of characteristics of the other sex, such as courtship songs, or *intrasexual selection*, which involves competition between members of one sex for mates. Let's look at some experimental evidence for sexual selection.

Mate Choice by Females Mate preferences of females may play a central role in the evolution of male behavior and anatomy through intersexual selection. As an example, let's consider the courtship behavior of stalk-eyed flies. The eyes of these insects are at the tips of stalks, which are longer in males than in females **(Figure 51.17)**. During courtship, a male approaches the female headfirst. Researchers have shown that females are more likely to mate with males that have relatively long eyestalks. Why would females favor this seemingly arbitrary trait? As discussed in Chapter 23, ornaments such as long eyestalks in these flies and bright coloration in male birds correlate in general with the male's health and vitality. A female whose mate choice is a healthy male is likely to produce more offspring that survive to reproduce.

Mate choice can also be influenced by imprinting, as revealed by experiments carried out with zebra finches. Both male and female zebra finches normally lack any feather crest on their head **(Figure 51.18)**. To explore whether parental appearance affects mate preference in offspring independent of any genetic influence, researchers provided zebra finches with artificial ornamentation. A 2.5-cm-long red feather was taped to the forehead feathers of either or both zebra finch parents when their chicks were 8 days old, approximately 2 days before they opened their eyes. A control group of zebra finches were raised by unadorned parents. When the chicks matured,

▲ **Figure 51.18 Appearance of zebra finches in nature.** The male zebra finch (left) is more patterned and colorful than the female zebra finch.

they were presented with prospective mates that were either artificially ornamented with a red feather or non-ornamented (**Figure 51.19**). Males showed no preference. Females also showed no preference if they were raised by a male parent that was not ornamented. However, females raised by an ornamented male parent preferred ornamented males as their own mates. Thus, female finches apparently take cues from their fathers in choosing mates.

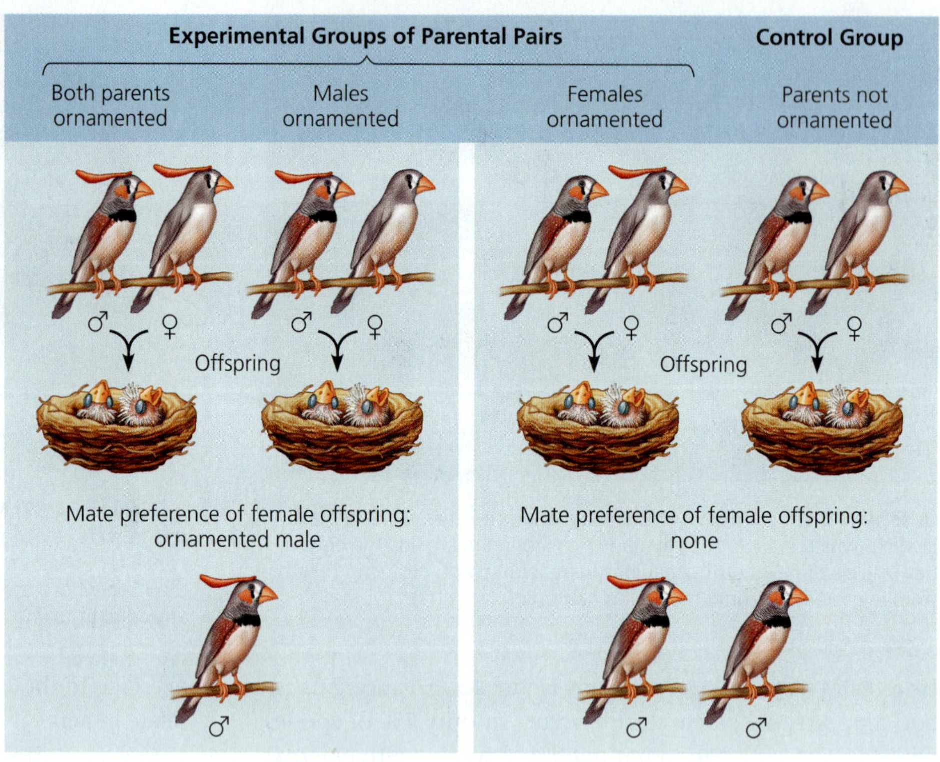

▲ **Figure 51.19 Sexual selection influenced by imprinting.** Experiments demonstrated that female zebra finch chicks that had imprinted on artificially ornamented fathers preferred ornamented males as adult mates. For all experimental groups, male offspring showed no preference for either ornamented or non-ornamented female mates.

Mate-choice copying, a behavior in which individuals in a population copy the mate choice of others, has been studied in the guppy *Poecilia reticulata*. When a female guppy chooses between males with no other females present, the female almost always chooses the male with more orange coloration. To explore if the behavior of other females could influence this preference, an experiment was set up using both living females and artificial model females (**Figure 51.20**). If a female guppy observed the model "courting" a male with less extensive orange markings, she often copied the preference of the model female. That is, the female chose the male that had been presented in association with a model female rather than a more orange alternative. The exceptions were also informative. Mate-choice behavior typically did not change when the difference in coloration was particularly large. Mate-choice copying can thus mask genetically controlled female preference below a certain threshold of difference, in this case for male color.

Mate-choice copying, a form of social learning, has also been observed in several other fish and bird species. What is the selective pressure for such a mechanism? One possibility is that a female that mates with males that are attractive to other females increases the probability that her male offspring will also be attractive and have high reproductive success.

Male Competition for Mates The previous examples show how female choice can select for one best type of male in a given situation, resulting in low variation among males. Male competition for mates also can reduce variation among males.

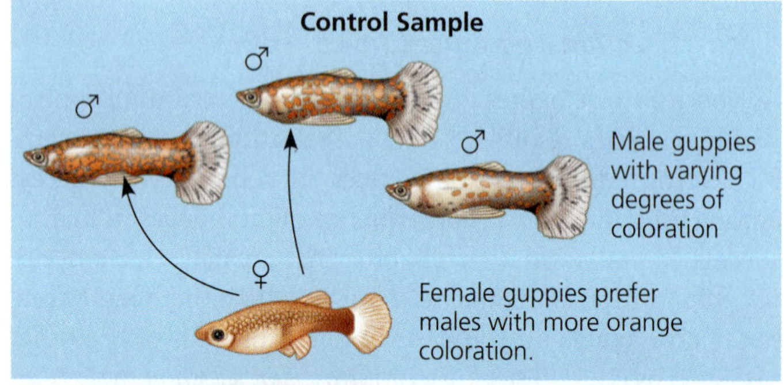

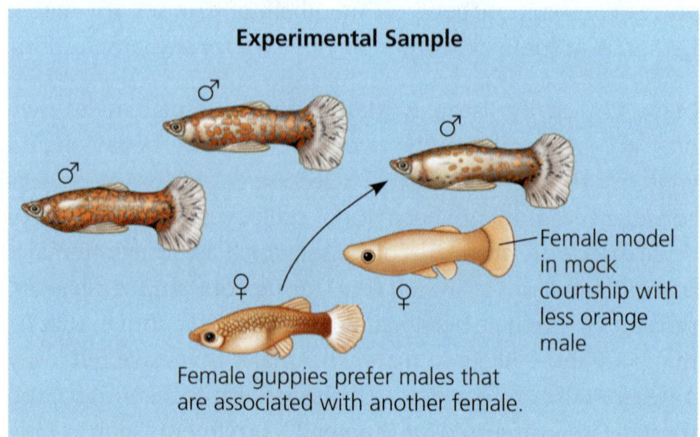

▶ **Figure 51.20 Mate choice copying by female guppies (*Poecilia reticulata*).** Female guppies generally choose the males with more orange coloration. But when males were matched for orange or differed in the amount of orange by 12% or 24%, the females in the experimental group chose the *less* orange male that was presented with a model female. Females ignored the apparent choice of the model female only where the alternative male had 40% more orange coloration.

Such competition may involve *agonistic behavior*, an often-ritualized contest that determines which competitor gains access to a resource, such as food or mates **(Figure 51.21)**. The outcomes of such contests are often determined by strength or size, but the consequences may nevertheless be psychological rather than physical (see Figure 51.17).

Despite the potential for male competition to select for reduced variation, behavioral and morphological variation in males is extremely high in some vertebrate species, including species of fish and deer, as well as in a wide variety of invertebrates. In some species, sexual selection has led to the evolution of alternative male mating behavior and morphology. How do scientists analyze situations where more than one mating behavior can result in successful reproduction? One approach relies on the rules that govern games.

Applying Game Theory

Often, the fitness of a particular behavioral phenotype is influenced by other behavioral phenotypes in the population. In studying such situations, behavioral ecologists use a range of tools, including game theory. Developed by American mathematician John Nash and others to model human economic behavior, **game theory** evaluates alternative strategies in situations where the outcome depends on the strategies of all the individuals involved.

▲ **Figure 51.21 Agonistic interaction.** Male eastern grey kangaroos (*Macropus giganteus*) often "box" in contests that determine which male is most likely to mate with an available female. Typically, one male snorts loudly before striking the other around the head and throat with his forelimbs. Further snorting and cuffing, as well as grappling, often follow. If the male under attack does not retreat, the fight may escalate, with each male balancing on his tail while attempting to kick his rival with the sharp toenails of a hind leg.

▲ **Figure 51.22 Male polymorphism in the side-blotched lizard (*Uta stansburiana*).** An orange-throat male, left; a blue-throat male, center; a yellow-throat male, right.

As an example of applying game theory to mating behavior, let's consider the side-blotched lizard (*Uta stansburiana*) of California. Males can have orange, blue, or yellow throats **(Figure 51.22)**. Each throat color is associated with a different pattern of behavior. Orange-throat males are the most aggressive and defend large territories that contain many females. Blue-throat males are also territorial but defend smaller territories and fewer females. Yellow-throats are nonterritorial males that mimic females and use "sneaky" tactics to gain the chance to mate.

Evidence indicates that the mating success of each male lizard type is influenced by the relative abundance of the other types, an example of frequency-dependent selection. In one study population, the most frequent throat coloration changed over a period of several years from blue to orange to yellow and back to blue.

By comparing the competition between side-blotched lizard males to the children's game of rock-paper-scissors, scientists devised an explanation for the cycles of variation in the lizard population. In the game, paper defeats rock, rock defeats scissors, and scissors defeats paper. Each hand symbol thus wins one matchup but loses the other. Similarly, each type of male lizard has an advantage over one of the other two types. When blue-throats are abundant, they can defend the few females in their territories from the advances of the sneaky yellow-throat males. However, blue-throats cannot defend their territories against the hyperaggressive orange-throats. Once the orange-throats become the most abundant, the larger number of females in each territory provides the opportunity for the yellow-throats to have greater mating success. The yellow-throats become more frequent, but then give way to the blue-throats, whose tactic of guarding small territories once again allows them the most success.

Game theory provides a way to think about complex evolutionary problems in which relative performance (reproductive success relative to other phenotypes), not absolute performance, is the key to understanding the evolution of behavior. This makes game theory an important tool because the relative performance of one phenotype compared with others is a measure of Darwinian fitness.

CONCEPT CHECK 51.3

1. Why does the mode of fertilization correlate with the presence or absence of male parental care?
2. **MAKE CONNECTIONS** Balancing selection can maintain variation at a locus (see Concept 23.4, pp. 483–484). Based on the foraging experiments described in this chapter, devise a simple hypothesis to explain the presence of both for^R and for^s alleles in natural fly populations.
3. **WHAT IF?** Suppose an infection in a side-blotched lizard population killed many more males than females. What would be the immediate effect on male competition for reproductive success?

For suggested answers, see Appendix A.

CONCEPT 51.4

Inclusive fitness can account for the evolution of behavior, including altruism

EVOLUTION We'll now explore issues related to the focus of Tinbergen's fourth question—the evolutionary history of behaviors. We will first look at examples that reveal the genetic underpinnings of behavior. Next, we will examine the genetic variation underlying the evolution of particular behaviors. Finally, we will see how expanding the definition of fitness beyond individual survival can help explain "selfless" behavior.

Genetic Basis of Behavior

In exploring the behavioral basis of behavior, we'll begin with the courtship behavior of the male fruit fly, diagrammed in Figure 51.4. During courtship, the male fly carries out a complex series of actions in response to multiple sensory stimuli. Genetic studies have revealed that a single gene called *fru* controls this entire courtship ritual. If the *fru* gene is mutated to an inactive form, males do not court or mate with females. (The name *fru* is short for *fruitless*, reflecting the absence of offspring from the mutant males.) Normal male and female flies express distinct forms of the *fru* gene. When females are genetically manipulated to express the male form of *fru*, they court other females, performing the role normally played by the male. How can a single gene control so many behaviors and actions? Experiments carried out cooperatively in several laboratories demonstrated that *fru* is a master regulatory gene that directs the expression and activity of many genes with narrower functions. Together, genes that are controlled by the *fru* gene bring about sex-specific development of the fly nervous system. In effect, *fru* programs the fly for male courtship behavior by overseeing a male-specific wiring of the central nervous system.

Researchers have also used insect courtship to explore genetic variation underlying differences in behavior. One well-studied example involves the courtship song of the green lacewing **(Figure 51.23)**. Found throughout central to northern Eurasia and North America, these insects include at least

▼ Figure 51.23 INQUIRY

Are the songs of green lacewing species under the control of multiple genes?

EXPERIMENT Charles Henry, Lucía Martínez, and Kent Holsinger crossed males and females of *Chrysoperla plorabunda* and *Chrysoperla johnsoni*, two morphologically identical species of lacewings that sing different courtship songs.

SOUND RECORDINGS

Chrysoperla plorabunda parent:

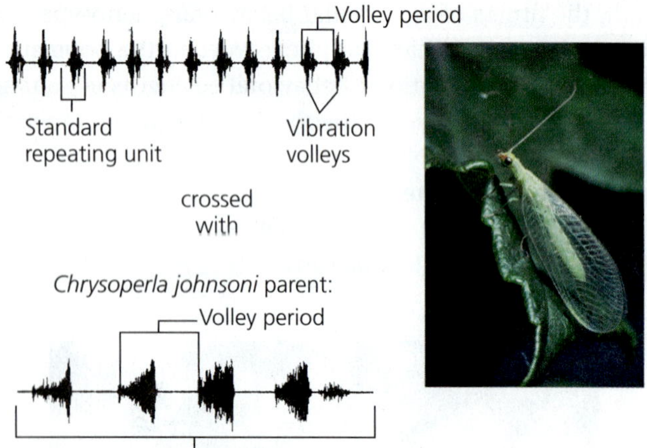

The researchers compared the songs of the male and female parents with those of the hybrid offspring that had been raised in isolation from other lacewings.

RESULTS The F_1 hybrid offspring sang a song in which the length of the standard repeating unit was similar to that sung by the *Chrysoperla plorabunda* parent, but the volley period—the interval between vibration volleys—was more similar to that of the *Chrysoperla johnsoni* parent.

F_1 hybrids, typical phenotype:

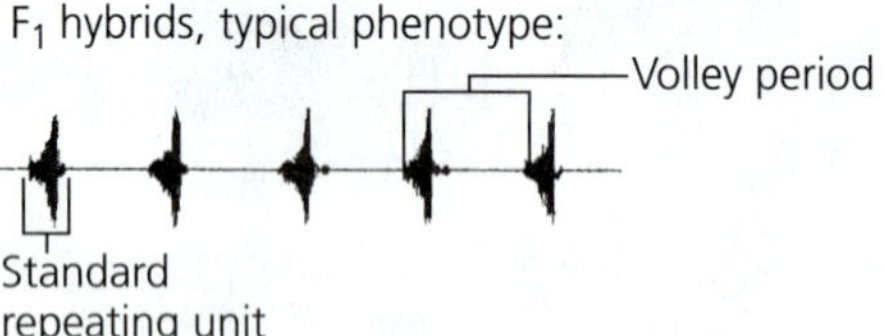

CONCLUSION Since the song of the hybrid offspring has features of the songs of both parents, the results indicate that the songs sung by *Chrysoperla plorabunda* and *Chrysoperla johnsoni* are under the control of more than one gene.

SOURCE C. S. Henry et al., The inheritance of mating songs in two cryptic, sibling lacewing species, *Genetica* 116:269–289 (2002).

WHAT IF? Suppose the hybrids generated in this experiment were fertile. Would the appearance of the hybrid song shown in the figure be likely to lead to the formation of a new species? Explain your answer.

15 species, identical in appearance but having different courtship songs. When researchers at the University of Connecticut reared lacewings in isolation in the laboratory, they found that the lacewings performed the song specific to their species. Thus, they concluded, the courtship song must be genetically controlled. They then crossed different green lacewing species in the laboratory and analyzed the songs produced by the hybrid offspring. These experiments demonstrated that a different gene governs each component or property of the courtship song. Furthermore, the distinct courtship song of each green lacewing species reflects genetic differences at multiple, independent loci.

Whereas variation in multiple genes can bring about distinct behaviors, as is true for the lacewing courtship song, variation in a single locus is sometimes sufficient to bring about dramatic differences in behavior. One striking example is the behavior of two closely related species of voles, which are small, mouse-like rodents. Male meadow voles (*Microtus pennsylvanicus*) are solitary and do not form lasting relationships with mates. Following mating, they pay little attention to their pups. In contrast, male prairie voles (*Microtus ochrogaster*) form a pair-bond with a single female after they mate **(Figure 51.24)**. Male prairie voles hover over their young pups, licking them and carrying them, while acting aggressively toward intruders.

Research suggested that a neurotransmitter released during mating is critical for the partnering and parental behavior of male voles. Known as ADH or vasopressin (see Chapter 44), this peptide binds to a specific receptor in the central nervous system. When male prairie voles are treated with a drug that inhibits the brain receptor for vasopressin, they fail to form pair-bonds after mating. Scientists have also observed that the vasopressin receptor gene of prairie voles is highly expressed in the brain, whereas that of meadow voles is not.

◀ **Figure 51.24 A pair of prairie voles (*Microtus ochrogaster*) huddling.** Male North American prairie voles associate closely with their mates, as shown here, and contribute substantially to the care of young.

To test whether the amount of the vasopressin receptor present in the brain regulates the postmating behavior of voles, researchers inserted the vasopressin receptor gene from prairie voles into male meadow voles. The meadow voles carrying this gene not only developed brains with higher levels of the vasopressin receptor but also showed many of the same mating behaviors as male prairie voles, such as pair-bonding. Thus, although many genes influence pair-bond formation and parenting among voles, the level of the vasopressin receptor alone determines which behavioral pattern develops.

Genetic Variation and the Evolution of Behavior

Behavioral differences between closely related species, such as meadow and prairie voles, are common. Significant differences in behavior can also be found *within* a species but are often less obvious. When behavioral variation between populations of a species corresponds to variation in environmental conditions, it may be evidence of past evolution.

Case Study: Variation in Prey Selection

An example of genetically based behavioral variation within a species involves prey selection by the western garter snake (*Thamnophis elegans*). The natural diet of this species differs widely across its range in California. Coastal populations feed predominantly on banana slugs (*Ariolimus californicus*) **(Figure 51.25)**. Inland populations feed on frogs, leeches, and fish, but not on banana slugs. In fact, banana slugs are rare or absent in the inland habitats.

▲ **Figure 51.25 Western garter snake from a coastal habitat eating a banana slug.** Experiments indicate that the preference of these snakes for banana slugs may be influenced mainly by genetics rather than by environment.

When researchers offered banana slugs to snakes from each wild population, most coastal snakes readily ate them, whereas inland snakes tended to refuse. To what extent does genetic variation contribute to a snake's fondness for banana slugs? To answer this question, researchers collected pregnant snakes from each wild population and housed them in separate cages in the laboratory. While still very young, the offspring were offered a small piece of banana slug on each of ten days. More than 60% of the young snakes from coastal mothers ate banana slugs on eight or more of the ten days. In contrast, fewer than 20% of the young snakes from inland mothers ate a piece of banana slug even once. Perhaps not surprisingly, banana slugs thus appear to be a genetically acquired taste.

How did a genetically determined difference in feeding preference come to match the snakes' habitats so well? It turns out that the coastal and inland populations also vary with respect to their ability to recognize and respond to odor molecules produced by banana slugs. Researchers hypothesize that when inland snakes colonized coastal habitats more than 10,000 years ago, some of them could recognize banana slugs by scent. Because these snakes took advantage of this food source, they had higher fitness than snakes in the population that ignored the slugs. Over hundreds or thousands of generations, the capacity to recognize the slugs as prey increased in frequency in the coastal population. The marked variation in behavior observed today between the coastal and inland populations may be evidence of this past evolutionary change.

Case Study: Variation in Migratory Patterns

Another species suited to the study of behavioral variation is the blackcap (*Sylvia atricapilla*), a small migratory warbler. Blackcaps that breed in Germany generally migrate southwest to Spain and then south to Africa for the winter. In the 1950s, a few blackcaps began to spend their winters in Britain, and over time the population of blackcaps wintering in Britain grew to many thousands. Leg bands showed that some of these birds had migrated westward from central Germany. Why were there now two patterns of migration from Germany? To answer this question, researchers at the Max Planck Research Center in Radolfzell, Germany, devised a strategy to study migratory orientation in the laboratory **(Figure 51.26)**. The results demonstrated that the two patterns of migration reflect genetic differences between the two populations.

The study of western European blackcaps indicated that the change in their migratory behavior occurred both recently and rapidly. Before the year 1950, there were no known westward-migrating blackcaps in Germany. By the 1990s, westward migrants made up 7–11% of the blackcap populations of Germany. Once westward migration began, it persisted and increased in frequency, perhaps due to the widespread use of winter bird feeders in Britain, as well as shorter migration distances.

▼ Figure 51.26 **INQUIRY**

Are differences in migratory orientation within a species genetically determined?

EXPERIMENT Peter Berthold and colleagues in southern Germany raised two sets of young birds for their study. One group consisted of the offspring of blackcaps captured while wintering in Britain and then bred in Germany in an outdoor cage. The other group consisted of young birds collected from nests near the laboratory and then raised in cages. In the autumn, Berthold's team placed the blackcaps captured in Britain and the young birds raised in cages in large, glass-covered funnel cages lined with carbon-coated paper for 1.5–2 hours. When the funnels were placed outside at night, the birds moved around, making marks on the paper that indicated the direction in which they were trying to "migrate."

RESULTS The wintering adult birds captured in Britain and their laboratory-raised offspring both attempted to migrate to the west. In contrast, the young birds collected from nests in southern Germany attempted to migrate to the southwest.

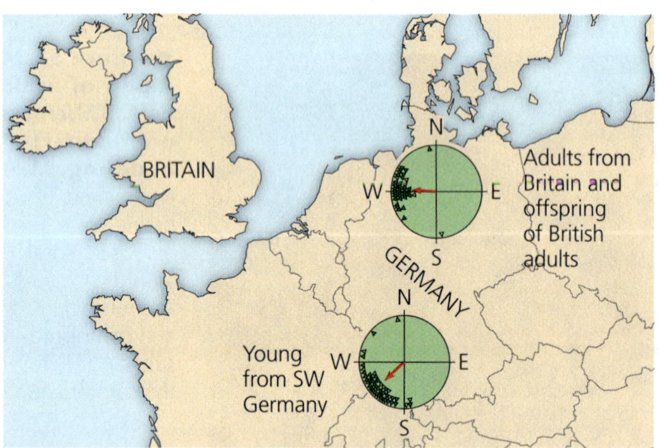

CONCLUSION The young of the British blackcaps and the young birds from Germany (the control group) were raised under similar conditions but showed very different migratory orientations, indicating that migratory orientation has a genetic basis.

SOURCE P. Berthold et al., Rapid microevolution of migratory behavior in a wild bird species, *Nature* 360:668–690 (1992).

WHAT IF? Suppose the birds had not shown a difference in orientation in these experiments. Could you conclude that the behavior was not genetically based? Explain.

Altruism

Reconstructing the evolutionary basis of a behavior requires an understanding of the behavior's genetic basis and of the selective advantage conferred by the behavior. Typically, we assume that behaviors are selfish; that is, they benefit the individual at the expense of others, especially competitors. For example, superior foraging ability by one individual may leave less food for others. The problem comes, however, with "unselfish" behaviors. How can such behaviors arise through natural selection? To answer this question, let's look more closely at some examples of unselfish behavior and then consider how such behaviors might arise.

In discussing selflessness, we will use the term **altruism** to describe a behavior that reduces an animal's individual fitness but increases the fitness of other individuals in the population. Consider the Belding's ground squirrel, which lives in some mountainous regions of the western United States and is vulnerable to predators such as coyotes and hawks. A squirrel that sees a predator approach often gives a high-pitched alarm call that alerts unaware individuals to retreat to their burrows. Note that for the squirrel that warns others, the conspicuous alarm behavior increases the risk of being killed because it brings attention to the caller's location.

Another example of altruistic behavior occurs in honeybee societies, in which the workers are sterile. The workers themselves never reproduce, but they labor on behalf of a single fertile queen. Furthermore, the workers sting intruders, a behavior that helps defend the hive but results in the death of those workers.

Altruism is also observed in naked mole rats (*Heterocephalus glaber*), highly social rodents that live in underground chambers and tunnels in southern and northeastern Africa. The naked mole rat, which is almost hairless and nearly blind, lives in colonies of 75 to 250 or more individuals (**Figure 51.27**). Each colony has only one reproducing female, the queen, who mates with one to three males, called kings. The rest of the colony consists of nonreproductive females and males who forage for underground roots and tubers and care for the queen, the kings, and new offspring. The nonreproductive individuals may sacrifice their own lives in trying to protect the queen or kings from snakes or other predators that invade the colony.

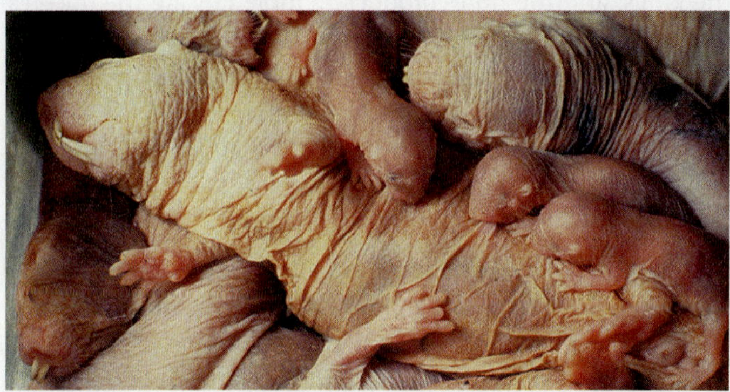

▲ **Figure 51.27 Naked mole rats, a species of colonial mammal that exhibits altruistic behavior.** Pictured here is a queen nursing offspring while surrounded by other members of the colony.

Inclusive Fitness

How can a Belding's ground squirrel, a worker honeybee, or a naked mole rat enhance its fitness by aiding members of the population that may be its closest competitors? How can altruistic behavior be maintained by evolution if it does not enhance the survival and reproductive success of the self-sacrificing individuals?

The selection for altruistic behavior is most readily apparent in the case of parents sacrificing for their offspring. When parents sacrifice their own well-being to produce and aid offspring, this actually increases the fitness of the parents because it maximizes their genetic representation in the population. However, individuals sometimes help others who are not their offspring.

Biologist William Hamilton proposed that an animal could increase its genetic representation in the next generation by "altruistically" helping close relatives other than its own offspring. Like parents and offspring, full siblings have half their genes in common. Therefore, selection might also favor helping siblings or helping one's parents produce more siblings. This idea led to Hamilton's idea of **inclusive fitness**, the total effect an individual has on proliferating its genes by producing its own offspring *and* by providing aid that enables other close relatives, who share many of those genes, to produce offspring.

Hamilton's Rule and Kin Selection

According to Hamilton, the three key variables in an act of altruism are the benefit to the recipient, the cost to the altruist, and the coefficient of relatedness. The benefit, B, is the average number of *extra* offspring that the beneficiary of an altruistic act produces. The cost, C, is how many *fewer* offspring the altruist produces. The **coefficient of relatedness**, r, equals the fraction of genes that, on average, are shared. Natural selection favors altruism when the benefit to the recipient multiplied by the coefficient of relatedness exceeds the cost to the altruist—in other words, when $rB > C$. This statement is called **Hamilton's rule**.

To better understand Hamilton's rule, let's apply it to a human population in which the average individual has two children. We'll imagine that a young man is close to drowning in heavy surf, and his sister risks her life to swim out and pull her sibling to safety. If the young man had drowned, his reproductive output would have been zero; but now, if we use the average, he can father two children. The benefit to the brother is thus two offspring ($B = 2$). What about the risk taken by his sister? Let's say that the sister has a 25% chance of drowning in attempting to rescue her brother. We can then calculate the cost of the altruistic act to the sister as 0.25 times 2, the number of offspring she would be expected to have if she had stayed on shore ($C = 0.25 \times 2 = 0.5$). Finally, we note that a brother and sister share half their genes on average ($r = 0.5$). One way to see this is in terms of the segregation of homologous chromosomes that occurs during meiosis of gametes (**Figure 51.28**; see also Chapter 13).

We can now use our values of B, C, and r to evaluate whether natural selection would favor the altruistic act in our imaginary scenario. For the surf rescue, $rB = 0.5 \times 2 = 1$, whereas $C = 0.5$. Because rB is greater than C, Hamilton's rule is satisfied; thus, natural selection would favor this altruistic act of the sister attempting to save her brother. Averaging over many individuals and generations, any particular gene in a sister faced with this situation will be passed on to more offspring if she risks the rescue than if she does not. Furthermore, among genes propagated in this way may be some that contribute to altruistic behavior. The natural selection that favors altruistic behavior by enhancing reproductive success of relatives is called **kin selection**.

Kin selection weakens with hereditary distance. Siblings have an r of 0.5, but between an aunt and her niece, $r = 0.25$ (¼), and between first cousins, $r = 0.125$ (⅛). Notice that as the degree of relatedness decreases, the rB term in the Hamilton inequality also decreases. Would natural selection favor rescuing a cousin? Not unless the surf were less treacherous. For the original conditions, $rB = 0.125 \times 2 = 0.25$, which is only half the value of C (0.5). British geneticist J. B. S. Haldane appears to have anticipated these ideas when he jokingly stated that he would not lay down his life for one brother, but would do so for two brothers or eight cousins.

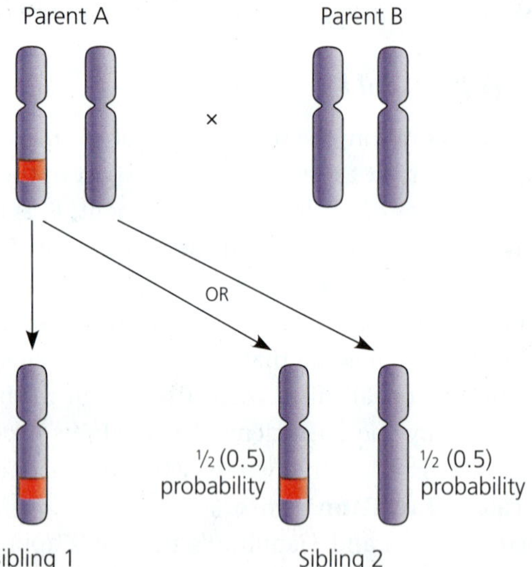

▲ **Figure 51.28 The coefficient of relatedness between siblings.** The red band indicates a particular allele (version of a gene) present on one chromosome, but not its homolog, in parent A. Sibling 1 has inherited the allele from parent A. There is a probability of ½ that sibling 2 will also inherit this allele from parent A. Any allele present on one chromosome of either parent will behave similarly. The coefficient of relatedness between the two siblings is thus ½, or 0.5.

WHAT IF? *The coefficient of relatedness of an individual to a full (nontwin) sibling or to either parent is the same, 0.5. Does this value also hold true in the cases of polyandry and polygyny?*

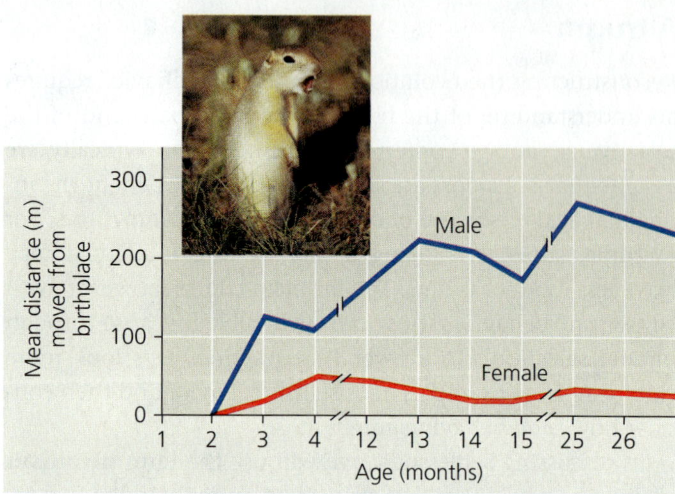

▲ **Figure 51.29 Kin selection and altruism in Belding's ground squirrels.** This graph helps explain the male-female difference in altruistic behavior of ground squirrels. Once weaned (pups are nursed for about one month), females are more likely than males to live near close relatives. Alarm calls that warn these relatives increase the inclusive fitness of the female altruist.

If kin selection explains altruism, then the examples of unselfish behavior we observe among diverse animal species should involve close relatives. This is apparently the case, but often in complex ways. Like most mammals, female Belding's ground squirrels settle close to their site of birth, whereas males settle at distant sites **(Figure 51.29)**. Since nearly all alarm calls are given by females, they are most likely aiding close relatives. In the case of worker bees, who are all sterile, anything they do to help the entire hive benefits the only permanent member who is reproductively active—the queen, who is their mother.

In the case of naked mole rats, DNA analyses have shown that all the individuals in a colony are closely related. Genetically, the queen appears to be a sibling, daughter, or mother of the kings, and the nonreproductive mole rats are the queen's direct descendants or her siblings. Therefore, when a nonreproductive individual enhances a queen's or king's chances of reproducing, the altruist increases the chance that some genes identical to its own will be passed to the next generation.

Reciprocal Altruism

Some animals occasionally behave altruistically toward others who are not relatives. A baboon may help an unrelated companion in a fight, or a wolf may offer food to another wolf even though they share no kinship. Such behavior can be adaptive if the aided individual returns the favor in the future. This sort of exchange of aid, called **reciprocal altruism**, is commonly invoked to explain altruism that occurs between unrelated humans. Reciprocal altruism is rare in other animals; it is limited largely to species (such as chimpanzees) with social groups stable enough that individuals have many chances to exchange aid. It is generally thought to occur when individuals are likely to meet again and when there would be negative consequences associated with not returning favors to individuals who had been helpful in the past, a pattern of behavior that behavioral ecologists refer to as "cheating."

Since cheating may benefit the cheater substantially, how could reciprocal altruism evolve? Game theory provides a possible answer in the form of a behavioral strategy called *tit for tat*. In the tit-for-tat strategy, an individual treats another in the same way it was treated the last time they met. Individuals adopting this behavior are always altruistic, or cooperative, on the first encounter with another individual and will remain so as long as their altruism is reciprocated. When their cooperation is not reciprocated, however, individuals employing tit for tat will retaliate immediately but return to cooperative behavior as soon as the other individual becomes cooperative. The tit-for-tat strategy has been used to explain the few apparently reciprocal altruistic interactions observed in animals—ranging from blood sharing between nonrelated vampire bats to social grooming in primates.

Evolution and Human Culture

Human culture is related to evolutionary theory in the discipline of **sociobiology**. The main premise of sociobiology is that certain behavioral characteristics exist because they are expressions of genes that have been perpetuated by natural selection. In his seminal 1975 book *Sociobiology: The New Synthesis*, E. O. Wilson speculated about the evolutionary basis of certain kinds of social behavior. By including a few examples from human culture, he sparked a debate that remains heated today.

The spectrum of human social behaviors may be influenced by our genetic makeup, but this is very different from saying that genes are rigid determinants of behavior. This distinction is at the core of the debate about evolutionary perspectives on human behavior. Skeptics fear that evolutionary interpretations of human behavior could be used to justify the status quo in human society, thus rationalizing current social injustices. Evolutionary biologists argue that this is a gross oversimplification and misunderstanding of what the data tell us about human biology. Evolutionary explanations of human behavior do not reduce us to robots stamped out of rigid genetic molds. Just as individuals vary extensively in anatomical features, we should expect inherent variations in behavior as well. Environment intervenes in the pathway from genotype to phenotype for physical traits and even more so for behavioral traits. And because of our capacity for learning and our versatility, human behavior is probably more plastic than that of any other animal. Over our recent evolutionary history, we have built up a diversity of structured societies with governments, laws, cultural values, and religions that define what is acceptable behavior and what is not, even when unacceptable behavior might enhance an individual's Darwinian fitness. Perhaps it is our social and cultural institutions that make us distinct and that provide those qualities in which there is the least continuum between humans and other animals.

CONCEPT CHECK 51.4

1. Explain why geographic variation in garter snake prey choice might indicate that the behavior evolved by natural selection.
2. **WHAT IF?** If an animal were unable to distinguish close from distant relatives, would the concept of inclusive fitness still be applicable? Explain.
3. **WHAT IF?** Suppose you applied Hamilton's logic to a situation in which one individual is past reproductive age. Could there still be a selection for an altruistic act?

For suggested answers, see Appendix A.

51 CHAPTER REVIEW

SUMMARY OF KEY CONCEPTS

CONCEPT 51.1

Discrete sensory inputs can stimulate both simple and complex behaviors (pp. 1118–1122)

- **Behavior** is the sum of responses to external and internal stimuli and includes muscular as well as nonmuscular activity. Tinbergen developed a set of questions that highlight the complementary nature of two perspectives. Proximate, or "how," questions focus on the environmental stimuli, if any, that trigger a behavior, as well as the genetic, physiological, and anatomical mechanisms underlying a behavioral act. Ultimate, or "why," questions address the evolutionary significance of a behavior.
- A **fixed action pattern** is a largely invariant behavior triggered by a simple cue known as a **sign stimulus**. Migratory movements involve navigation, which can be based on orientation relative to the sun, the stars, or Earth's magnetic field. Animal behavior is sometimes synchronized to the daily, or circadian, cycle of light and dark in the environment or to environmental cues that cycle over the seasons.

- The transmission and reception of signals constitute animal **communication**. Animals use visual, auditory, chemical (usually olfactory), and tactile signals, sometimes as part of a stimulus-response chain that governs a complex behavior. Chemical substances called pheromones transmit species-specific information through the environment in behaviors ranging from foraging to courtship.

? *How is migration based on circannual rhythms poorly suited for adaptation to global climate change?*

CONCEPT 51.2

Learning establishes specific links between experience and behavior (pp. 1123–1128)

- Cross-fostering studies can be used to measure the influence of social environment and experience on behavior.
- **Learning**, the modification of behavior based on experience, can take many forms:

? *How do imprinting in geese and song development in sparrows differ with regard to the resulting behavior?*

CONCEPT 51.3

Selection for individual survival and reproductive success can explain most behaviors (pp. 1128–1134)

- An **optimal foraging model** is based on the idea that natural selection should favor foraging behavior that minimizes the costs of foraging and maximizes the benefits.
- Sexual dimorphism correlates with the type of mating relationship between males and females. These include **monogamous** and **polygamous** mating systems. Variation in mating system and variation in the mode of fertilization affect certainty of paternity, which in turn has a significant influence on mating behavior and parental care.
- Game theory provides a way of thinking about evolution in situations where the fitness of a particular behavioral phenotype is influenced by other behavioral phenotypes in the population.

? *In some spider species, the female eats the male immediately after copulation. How might you explain this behavior from an evolutionary perspective?*

CONCEPT 51.4

Inclusive fitness can account for the evolution of behavior, including altruism (pp. 1134–1139)

- Genetic studies in insects have revealed the existence of master regulatory genes that control complex behaviors. Within the underlying hierarchy, multiple genes influence specific behaviors, such as a courtship song. Research with two species of voles has revealed that variation in a single gene can determine differences in complex behaviors involved in both mating and parenting.
- When behavioral variation within a species corresponds to variation in environmental conditions, it may be evidence of past evolution. Field and laboratory studies have documented the genetic basis for a change in migratory behavior of certain birds and revealed behavioral differences in snakes that correlate with geographic variation in prey availability.
- On occasion, animals exhibit **altruism**. This behavior can be explained by the concept of **inclusive fitness**, the total effect an individual has on proliferating its genes by producing its own offspring *and* by providing aid that enables close relatives to produce offspring. **The coefficient of relatedness** and **Hamilton's rule** provide a way of measuring the strength of the selective forces favoring altruism against the potential cost of the "selfless" behavior. Kin selection favors altruistic behavior by enhancing the reproductive success of relatives. Altruistic behavior toward unrelated individuals can be adaptive if the aided individual returns the favor in the future, an exchange of aid called reciprocal altruism.

? *Suppose you studied the genetics of the lacewing courtship song, but not the effects of courtship mutations in flies or of variation in the vasopressin receptor gene of voles. What insight about the genetic basis of behavior would you likely have missed?*

TEST YOUR UNDERSTANDING

LEVEL 1: KNOWLEDGE/COMPREHENSION

1. Which of the following is true of innate behaviors?
   a. Their expression is only weakly influenced by genes.
   b. They occur with or without environmental stimuli.
   c. They are limited to invertebrate animals.
   d. They are expressed in most individuals in a population.
   e. They occur in invertebrates and some vertebrates but not mammals.

2. According to Hamilton's rule,
   a. natural selection does not favor altruistic behavior that causes the death of the altruist.
   b. natural selection favors altruistic acts when the resulting benefit to the beneficiary, corrected for relatedness, exceeds the cost to the altruist.
   c. natural selection is more likely to favor altruistic behavior that benefits an offspring than altruistic behavior that benefits a sibling.
   d. the effects of kin selection are larger than the effects of direct natural selection on individuals.
   e. altruism is always reciprocal.

3. Female spotted sandpipers aggressively court males and, after mating, leave the clutch of young for the male to incubate. This sequence may be repeated several times with different males until no available males remain, forcing the female to incubate her last clutch. Which of the following terms best describes this behavior?
   a. monogamy
   b. polygyny
   c. polyandry
   d. promiscuity
   e. certainty of paternity

LEVEL 2: APPLICATION/ANALYSIS

4. A region of the canary forebrain shrinks during the nonbreeding season and enlarges when breeding season begins. This change is probably associated with the annual
   a. addition of new syllables to a canary's song repertoire.
   b. crystallization of subsong into adult songs.
   c. sensitive period in which canary parents imprint on new offspring.
   d. renewal of mating and nest-building behaviors.
   e. elimination of the memorized template for songs sung the previous year.

5. Although many chimpanzees live in environments containing oil palm nuts, members of only a few populations use stones to crack open the nuts. The likely explanation is that
   a. the behavioral difference is caused by genetic differences between populations.
   b. members of different populations have different nutritional requirements.
   c. the cultural tradition of using stones to crack nuts has arisen in only some populations.
   d. members of different populations differ in learning ability.
   e. members of different populations differ in manual dexterity.

6. Which of the following is *not* required for a behavioral trait to evolve by natural selection?
   a. In each individual, the form of the behavior is determined entirely by genes.
   b. The behavior varies among individuals.
   c. An individual's reproductive success depends in part on how the behavior is performed.
   d. Some component of the behavior is genetically inherited.
   e. An individual's genotype influences its behavioral phenotype.

LEVEL 3: SYNTHESIS/EVALUATION

7. **DRAW IT** You are considering two optimal foraging models for the behavior of a mussel-feeding shorebird, the oystercatcher. In model A, the energetic reward increases solely with mussel size. In model B, you take into consideration that larger mussels are more difficult to open. Draw a graph of reward (energy benefit on a scale of 0–10) versus mussel length (scale of 0–70 mm) for each model. Assume that mussels under 10 mm provide no benefit and are ignored by the birds. Also assume that mussels start becoming difficult to open when they reach 40 mm in length and impossible to open when 70 mm long. Considering the graphs you have drawn, how could you distinguish between the models by observation and measurement in the oystercatcher's habitat?

8. **EVOLUTION CONNECTION**
   We often explain our behavior in terms of subjective feelings, motives, or reasons, but evolutionary explanations are based on reproductive fitness. What is the relationship between the two kinds of explanation? For instance, is a human explanation for behavior, such as "falling in love," incompatible with an evolutionary explanation?

9. **SCIENTIFIC INQUIRY**
   Scientists studying scrub jays found that "helpers" often assist mated pairs of birds in raising their young. The helpers lack territories and mates of their own. Instead, they help the territory owners gather food for their offspring. Propose a hypothesis to explain what advantage there might be for the helpers to engage in this behavior instead of seeking their own territories and mates. How would you test your hypothesis? If it is correct, what results would you expect your tests to yield?

10. **SCIENCE, TECHNOLOGY, AND SOCIETY**
    Researchers are very interested in studying identical twins separated at birth and raised apart. So far, the data reveal that such twins frequently have similar personalities, mannerisms, habits, and interests. What general question do you think researchers hope to answer by studying such twins? Why do identical twins make good subjects for this research? What are the potential pitfalls of this research? What abuses might occur if the studies are not evaluated critically?

11. **WRITE ABOUT A THEME**
    The Genetic Basis of Life Learning is defined as a change in behavior based on experience. In a short essay (100–150 words), describe the role of heritable information in the acquisition of learning, using some examples from imprinting and associative learning.

For selected answers, see Appendix A.

MasteringBIOLOGY www.masteringbiology.com

1. MasteringBiology® Assignments

Tutorial Animal Behavior and Learning

Activities Honeybee Waggle Dance Video • Homing Behavior in Digger Wasps

Questions Student Misconceptions • Reading Quiz • Multiple Choice • End-of-Chapter

2. eText

Read your book online, search, take notes, highlight text, and more.

3. The Study Area

Practice Tests • Cumulative Test •  3-D Animations • MP3 Tutor Sessions • Videos • Activities • Investigations • Lab Media • Audio Glossary • Word Study Tools • Art

APPENDIX A Answers

Chapter 1

Figure Questions

Figure 1.7 The arrangement of fingers and opposable thumb in the human hand, combined with fingernails and a complex system of nerves and muscles, allows the hand to grasp and manipulate objects with great dexterity. **Figure 1.13** Substance B would be made continuously and would accumulate in large amounts. Neither C nor D would be made, so D would not be able to inhibit Enzyme 1 and regulate the pathway. **Figure 1.27** The percentage of brown artificial snakes attacked would probably be higher than the percentage of artificial kingsnakes attacked in all areas (whether or not inhabited by coral snakes).

Concept Check 1.1

1. Examples: A molecule consists of *atoms* bonded together. Each organelle has an orderly arrangement of *molecules*. Photosynthetic plant cells contain *organelles* called chloroplasts. A tissue consists of a group of similar *cells*. Organs such as the heart are constructed from several *tissues*. A complex multicellular organism, such as a plant, has several types of *organs*, such as leaves and roots. A population is a set of *organisms* of the same species. A community consists of *populations* of the various species inhabiting a specific area. An ecosystem consists of a biological *community* along with the nonliving factors important to life, such as air, soil, and water. The biosphere is made up of all of Earth's *ecosystems*. **2.** (a) Structure and function are correlated. (b) Cells are an organism's basic units, *and* the continuity of life is based on heritable information in the form of DNA. (c) Organisms interact with other organisms and with the physical environment, *and* life requires energy transfer and transformation. **3.** Some possible answers: *Emergent properties*: The ability of a human heart to pump blood requires an intact heart; it is not a capability of any of the heart's tissues or cells working alone. *Environmental interactions*: A mouse eats food, such as nuts or grasses, and deposits some of the food material as feces and urine. Construction of a nest rearranges the physical environment and may hasten degradation of some of its components. The mouse may also act as food for a predator. *Energy transfer*: A plant, such as a grass, absorbs energy from the sun and transforms it into molecules that act as stored fuel. Animals can eat parts of the plant and use the food for energy to carry out their activities. *Structure and function*: The strong, sharp teeth of a wolf are well suited to grasping and dismembering its prey. *The cellular basis of life*: The digestion of food is made possible by chemicals (chiefly enzymes) made by cells of the digestive tract. *The genetic basis of life*: Human eye color is determined by the combination of genes inherited from the two parents. *Feedback regulation*: When your stomach is full, it signals your brain to decrease your appetite. *Evolution*: All plants have chloroplasts, indicating their descent from a common ancestor.

Concept Check 1.2

1. An address pinpoints a location by tracking from broader to narrower categories—a state, city, zip, street, and building number. This is analogous to the groups-subordinate-to-groups structure of biological taxonomy. **2.** The naturally occurring heritable variation in a population is "edited" by natural selection because individuals with heritable traits better suited to the environment survive and reproduce more successfully than others. Over time, better-suited individuals persist and their percentage in the population increases, while less suited individuals become less prevalent—a type of population editing.
3.

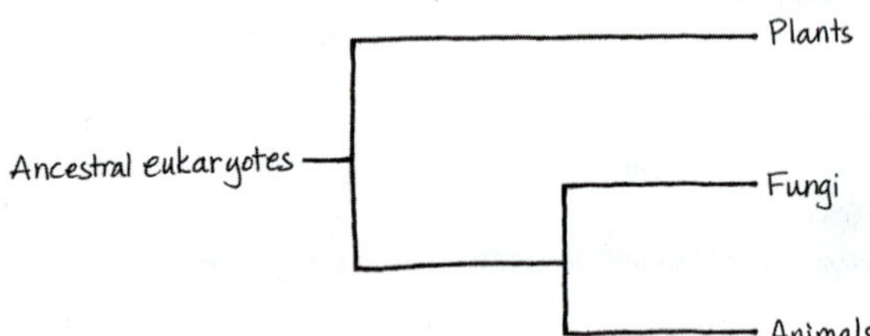

Concept Check 1.3

1. Inductive reasoning derives generalizations from specific cases; deductive reasoning predicts specific outcomes from general premises. **2.** The coloration pattern on the snakes **3.** Compared to a hypothesis, a scientific theory is usually more general and substantiated by a much greater amount of evidence. Natural selection is an explanatory idea that applies to all kinds of organisms and is supported by vast amounts of evidence of various kinds. **4.** Based on the results shown in Figure 1.27, you might predict that the colorful artificial snakes would be attacked more often than the brown ones, simply because they are easier to see. This prediction assumes that the area in Virginia where you are working has predators that attack snakes but no poisonous snakes that resemble the colorful artificial snakes.

Concept Check 1.4

1. Science aims to understand natural phenomena and how they work, while technology involves application of scientific discoveries for a particular purpose or to solve a specific problem. **2.** Natural selection could be operating. Malaria is present in sub-Saharan Africa, so there might be an advantage to people with the sickle-cell disease form of the gene that makes them more able to survive and pass on their genes to offspring. Among those of African descent living in the United States, where malaria is absent, there would be no advantage, so they would be selected against more strongly, resulting in fewer individuals with the sickle-cell disease form of the gene.

Summary of Key Concepts Questions

1.1 Evolution explains the most fundamental aspects of all life on earth. It accounts for the common features shared by all forms of life due to descent from a common ancestor, while also providing an explanation for how the great diversity of living organisms on the planet has arisen. **1.2** Ancestors of this plant may have exhibited variation in how well their leaves conserved water. Because not much soil is present in the crevices where these plants are found, the variant plants that could conserve water may have survived better and been able to produce more offspring. Over time, a higher and higher proportion of individuals in the population would have had the adaptation of thick, water-conserving leaves. **1.3** Inductive reasoning is used in forming hypotheses, while deductive reasoning leads to predictions that are used to test hypotheses. **1.4** Different approaches taken by scientists studying natural phenomena at different levels complement each other, so more is learned about each problem being studied. A diversity of backgrounds among scientists may lead to fruitful ideas in the same way that important innovations have often arisen where a mix of cultures coexist.

Test Your Understanding

1. b **2.** d **3.** a **4.** c **5.** c **6.** c **7.** b **8.** c **9.** c **10.** d
11. Your figure should show: (1) For the biosphere, the Earth with an arrow coming out of a tropical ocean; (2) for the ecosystem, a distant view of a coral reef; (3) for the community, a collection of reef animals and algae, with corals, fishes, some seaweed, and any other organisms you can think of; (4) for the population, a group of fish of the same species; (5) for the organism, one fish from your population; (6) for the organ, the fish's stomach, and for the organ system, the whole digestive tract (see Chapter 41 for help); (7) for a tissue, a group of similar cells from the stomach; (8) for a cell, one cell from the tissue, showing its nucleus and a few other organelles; (9) for an organelle, the nucleus, where most of the cell's DNA is located; and (10) for a molecule, a DNA double helix. Your sketches can be very rough!

Chapter 2

Figure Questions

Figure 2.2 The most significant difference in the results would be that the two *Cedrela* saplings inside each garden would show similar amounts of dying leaf tissue because a poisonous chemical released from the *Duroia* trees would presumably reach the saplings via the air or soil and would not be blocked by the insect barrier. The *Cedrela* saplings planted outside the gardens would not show damage unless *Duroia* trees were nearby. Also, any ants present on the unprotected *Cedrela* saplings inside the gardens would probably not be observed making injections into the leaves. However, formic acid would likely still be found in the ants' glands, as it is for most species of ants. **Figure 2.9** Atomic number = 12; 12 protons, 12 electrons; 3 electron shells; 2 valence electrons **Figure 2.16** One possible answer:

Figure 2.19 The plant is submerged in water (H_2O), in which the CO_2 is dissolved. The sun's energy is used to make sugar, which is found in the plant and can act as food for the plant itself, as well as for animals that eat the plant. The oxygen (O_2) is present in the bubbles.

Concept Check 2.1

1. Table salt (sodium chloride) is made up of sodium and chlorine. We are able to eat the compound, showing that it has different properties from those of a metal (sodium) and a poisonous gas (chlorine). **2.** Yes, because an organism requires trace elements, even though only in small amounts **3.** A person with an iron deficiency will probably show fatigue and other effects of a low oxygen level in the blood. (The condition is called anemia and can also result from too few red blood cells or abnormal hemoglobin.) **4.** Variant ancestral plants that could tolerate the toxic elements could grow and reproduce in serpentine soils. (Plants that were well adapted to nonserpentine soils would not be expected to survive in serpentine areas.) The offspring of the variants would also vary, with those most capable of thriving under serpentine conditions growing best and reproducing most. Over many generations, this probably led to the serpentine-adapted species we see today.

Concept Check 2.2

1. 7 **2.** $^{15}_{7}N$ **3.** 9 electrons; two electron shells; 1*s*, 2*s*, 2*p* (three orbitals); 1 electron is needed to fill the valence shell. **4.** The elements in a row all have the same number of electron shells. In a column, all the elements have the same number of electrons in their valence shells.

Concept Check 2.3

1. Each carbon atom has only three covalent bonds instead of the required four. **2.** The attraction between oppositely charged ions, forming ionic bonds **3.** If you could synthesize molecules that mimic these shapes, you might be able to treat diseases or conditions caused by the inability of affected individuals to synthesize such molecules.

Concept Check 2.4

1.

H : H + O :: O → :O: H (H)
H : H   :O: H (H)

$2H_2$ O_2 $2H_2O$

2. At equilibrium, the forward and reverse reactions occur at the same rate. **3.** $C_6H_{12}O_6 + 6\ O_2 \rightarrow 6\ CO_2 + 6\ H_2O$ + Energy. Glucose and oxygen react to form carbon dioxide and water, releasing energy. We breathe in oxygen because we need it for this reaction to occur, and we breathe out carbon dioxide because it is a by-product of this reaction. (This reaction is called cellular respiration, and you will learn more about it in Chapter 9.)

Summary of Key Concepts Questions

2.1 Iodine (part of a thyroid hormone) and iron (part of hemoglobin in blood) are both trace elements, required in minute quantities. Calcium and phosphorus (components of bones and teeth) are needed by the body in much greater quantities.
2.2

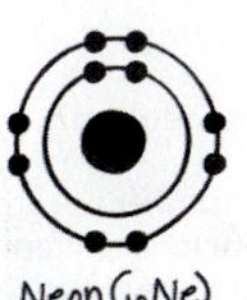
Neon ($_{10}Ne$)

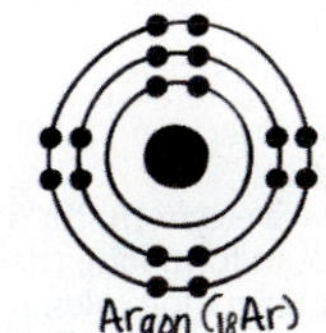
Argon ($_{18}Ar$)

Both neon and argon have completed valence shells, containing 8 electrons. They do not have unpaired electrons that could participate in chemical bonds. **2.3** Electrons are shared equally between the two atoms in a nonpolar covalent bond. In a polar covalent bond, the electrons are drawn closer to the more electronegative atom. In the formation of ions, an electron is completely transferred from one atom to a much more electronegative atom. **2.4** The concentration of products would increase as the added reactants were converted to products. Eventually, an equilibrium would again be reached in which the forward and reverse reactions were proceeding at the same rate and the relative concentrations of reactants and products returned to where they were before the addition of more reactants.

Test Your Understanding

1. a **2.** e **3.** b **4.** a **5.** d **6.** b **7.** c **8.** e

9. a. O :: C : H   This structure doesn't make sense because the valence shell of carbon is incomplete; carbon can form 4 bonds.

b. H : O : C : C :: O (with H, H, H)   This structure makes sense because all valence shells are complete, and all bonds have the correct number of electrons.

c. H : C : H . C :: O (with H, H, H)   This structure doesn't make sense because H has only 1 electron to share, so it cannot form bonds with 2 atoms.

d. This structure doesn't make sense for several reasons:
:O: The valence shell of oxygen is incomplete; oxygen can form 2 bonds.
H : N .. H   H has only 1 electron to share, so it cannot form a double bond.
Nitrogen usually makes only 3 bonds. It does not have enough electrons to make 2 single bonds, make a double bond, and complete its valence shell.

Chapter 3

Figure Questions

Figure 3.2 One possible answer:

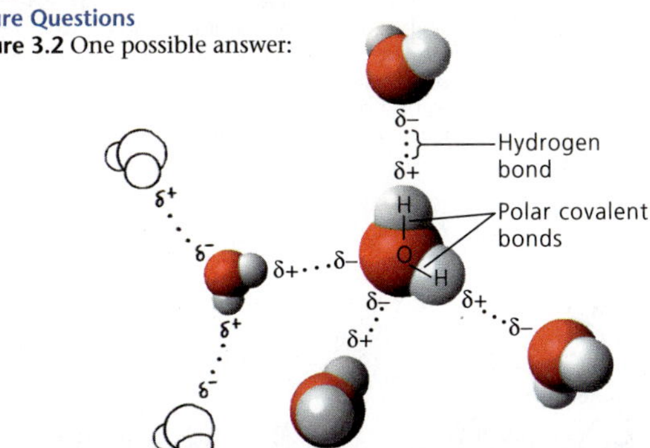

Figure 3.6 Without hydrogen bonds, water would behave like other small molecules, and the solid phase (ice) would be denser than liquid water. The ice would sink to the bottom and would no longer insulate the whole body of water, which would eventually freeze because the average annual temperature at the South Pole is –50°C. The krill could not survive. **Figure 3.7** Heating the solution would cause the water to evaporate faster than it is evaporating at room temperature. At a certain point, there wouldn't be enough water molecules to dissolve the salt ions. The salt would start coming out of solution and re-forming crystals. Eventually, all the water would evaporate, leaving behind a pile of salt like the original pile. **Figure 3.12** By causing the loss of coral reefs, a decrease in the ocean's carbonate concentration would have a ripple effect on noncalcifying organisms. Some of these organisms depend on the reef structure for protection, while others feed on species associated with reefs.

Concept Check 3.1

1. Electronegativity is the attraction of an atom for the electrons of a covalent bond. Because oxygen is more electronegative than hydrogen, the oxygen atom in H_2O pulls electrons toward itself, resulting in a partial negative charge on the oxygen atom and partial positive charges on the hydrogen atoms. Atoms in neighboring water molecules with opposite partial charges are attracted to each other, forming a hydrogen bond. **2.** The hydrogen atoms of one molecule, with their partial positive charges, would repel the hydrogen atoms of the adjacent molecule. **3.** The covalent bonds of water molecules would not be polar, and water molecules would not form hydrogen bonds with each other.

Concept Check 3.2

1. Hydrogen bonds hold neighboring water molecules together. This cohesion helps the chain of water molecules move upward against gravity in water-conducting cells as water evaporates from the leaves. Adhesion between water molecules and the walls of the water-conducting cells also helps counter gravity. **2.** High humidity hampers cooling by suppressing the evaporation of sweat. **3.** As water freezes, it expands because water molecules move farther apart in forming ice crystals. When there is water in a crevice of a boulder, expansion due to freezing may crack the boulder. **4.** A liter of blood would contain 7.8×10^{13} molecules of ghrelin (1.3×10^{-10} moles per liter $\times\ 6.02 \times 10^{23}$ molecules per mole). **5.** The hydrophobic substance repels water, perhaps helping to keep the ends of the legs from becoming coated with water and breaking through the surface. If the legs were coated with a hydrophilic substance, water would be drawn up them, possibly making it more difficult for the water strider to walk on water.

Concept Check 3.3

1. 10^5, or 100,000 **2.** $[H^+] = 0.01\ M = 10^{-2} M$, so pH = 2. **3.** $CH_3COOH \rightarrow CH_3COO^- + H^+$. CH_3COOH is the acid (the H^+ donor), and CH_3COO^- is the base (the H^+ acceptor). **4.** The pH of the water should decrease from 7 to about 2; the pH of the acetic acid solution will decrease only a small amount, because the reaction shown for question 3 will shift to the left, with CH_3COO^- accepting the influx of H^+ and becoming CH_3COOH molecules.

Summary of Key Concepts Questions

3.1

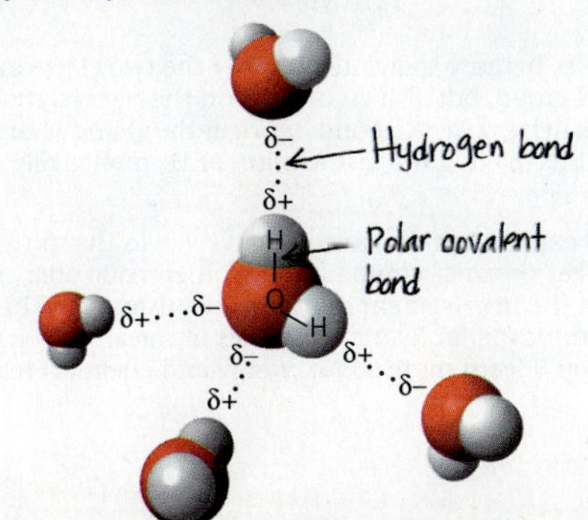

Each water molecule can make four hydrogen bonds with neighboring molecules. **3.2** Ions dissolve in water when polar water molecules form a hydration shell around them. Polar molecules dissolve as water molecules form hydrogen bonds with them and surround them. Solutions are homogeneous mixtures of solute and solvent. Colloids form when particles that are too large to dissolve remain suspended in a liquid. **3.3** CO_2 reacts with H_2O to form carbonic acid (H_2CO_3), which dissociates into H^+ and bicarbonate (HCO_3^-). Although the carbonic acid–bicarbonate reaction is a buffering system, adding CO_2 drives the reaction to the right, releasing more H^+ and lowering pH. The excess protons combine with CO_3^{2-} to form bicarbonate, lowering the concentration of carbonate available for the formation of calcium carbonate (calcification) by corals.

Test Your Understanding

1. d **2.** b **3.** c **4.** e **5.** c **6.** a **7.** e **8.** d

9.

10. Both global warming and ocean acidification are caused by increasing levels of carbon dioxide in the atmosphere, the result of burning fossil fuels. **11.** Due to intermolecular hydrogen bonds, water has a high specific heat (the amount of heat required to increase the temperature of water by 1°C). When water is heated, much of the heat is absorbed in breaking hydrogen bonds before the water molecules increase their motion and the temperature increases. Conversely, when water is cooled, many H bonds are formed, which releases a significant amount of heat. This release of heat can provide some protection against freezing of the plants' leaves, thus protecting the cells from damage.

Chapter 4

Figure Questions

Figure 4.2 Because the concentration of the reactants influences the equilibrium (as discussed in Chapter 2), there might have been more HCN relative to CH_2O, since there would have been a higher concentration of the reactant gas containing nitrogen.

Figure 4.4

Figure 4.6 The tails of fats contain only carbon-hydrogen bonds, which are relatively nonpolar. Because the tails occupy the bulk of a fat molecule, they make the molecule as a whole nonpolar and therefore incapable of forming hydrogen bonds with water.

Figure 4.7

Figure 4.9 Molecule b, because there are not only the two electronegative oxygens of the carboxyl group, but also an oxygen on the next (carbonyl) carbon. All of these oxygens help make the bond between the O and H of the —OH group more polar, thus making the dissociation of H^+ more likely.

Concept Check 4.1

1. Prior to Wöhler's experiment, the prevailing view was that only living organisms could synthesize "organic" compounds. Wöhler made urea, an organic compound, without the involvement of living organisms. **2.** The spark provided energy needed for the inorganic molecules in the atmosphere to react with each other. (You'll learn more about energy and chemical reactions in Chapter 8.)

Concept Check 4.2

1.

2. The forms of C_4H_{10} in (b) are structural isomers, as are the butenes in (c). **3.** Both consist largely of hydrocarbon chains. **4.** No. There is not enough diversity in the atoms. It can't form structural isomers because there is only one way for three carbons to attach to each other (in a line). There are no double bonds, so *cis-trans* isomers are not possible. Each carbon has at least two hydrogens attached to it, so the molecule is symmetrical and cannot have enantiomers.

Concept Check 4.3

1. It has both an amino group (—NH_2), which makes it an amine, and a carboxyl group (—COOH), which makes it a carboxylic acid. **2.** The ATP molecule loses a phosphate, becoming ADP. **3.** A chemical group that can act as a base has been replaced with a group that can act as an acid, increasing the acidic properties of the molecule. The shape of the molecule would also change, likely changing the molecules with which it can interact. The original cysteine molecule has an asymmetric carbon in the center. After replacement of the amino group with a carboxyl group, this carbon is no longer asymmetric.

Summary of Key Concepts Questions

4.1 Miller showed that organic molecules could form under the physical and chemical conditions believed to have been present on early Earth. This abiotic synthesis of organic molecules would have been a first step in the origin of life. **4.2** Acetone and propanal are structural isomers. Acetic acid and glycine have no asymmetric carbons, whereas glycerol phosphate has one. Therefore, glycerol phosphate can exist as forms that are enantiomers, but acetic acid and glycine cannot. **4.3** The methyl group is nonpolar and not reactive. The other six groups are called functional groups. They are each hydrophilic, increasing the solubility of organic compounds in water, and can participate in chemical reactions.

Test Your Understanding

1. b **2.** b **3.** d **4.** d **5.** a **6.** b **7.** a **8.** The molecule on the right; the middle carbon is asymmetric.

9. Si has 4 valence electrons, the same number as carbon. Therefore, silicon would be able to form long chains, including branches, that could act as skeletons for large molecules. It would clearly do this much better than neon (with no valence electrons) or aluminum (with 3 valence electrons).

Chapter 5

Figure Questions

Figure 5.3 Glucose and fructose are structural isomers.

Figure 5.4

Note that the oxygen on carbon 5 lost its proton and that the oxygen on carbon 2, which used to be the carbonyl oxygen, gained a proton. Four carbons are in the fructose ring, and two are not. (The latter two carbons are attached to carbons 2 and 5, which are in the ring.) The fructose ring differs from the glucose ring, which has five carbons in the ring and one that is not. (Note that the orientation of this fructose molecule is flipped relative to that of the one in Figure 5.5b.)

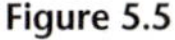

Figure 5.5

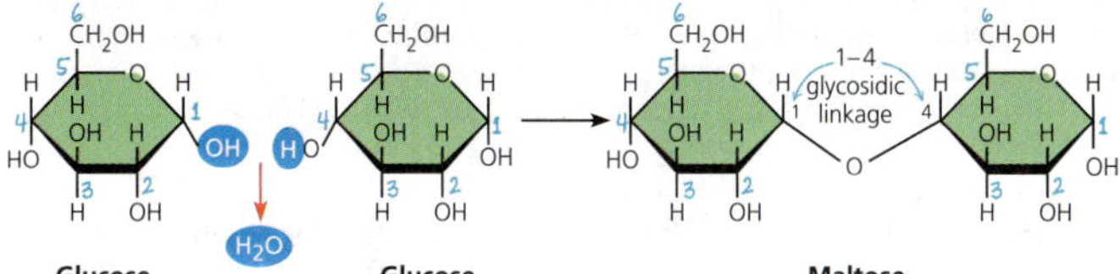

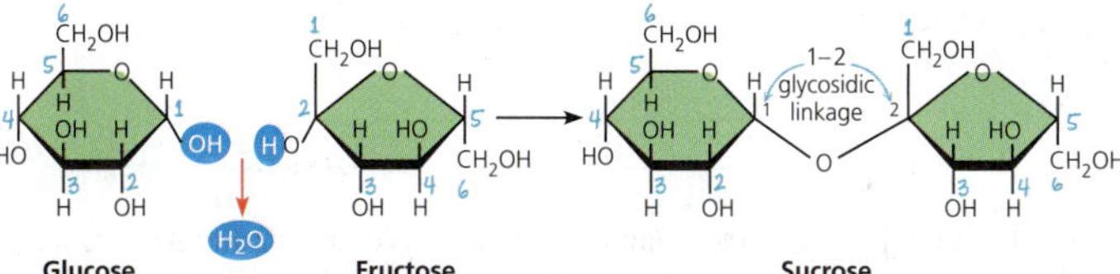

Figure 5.12

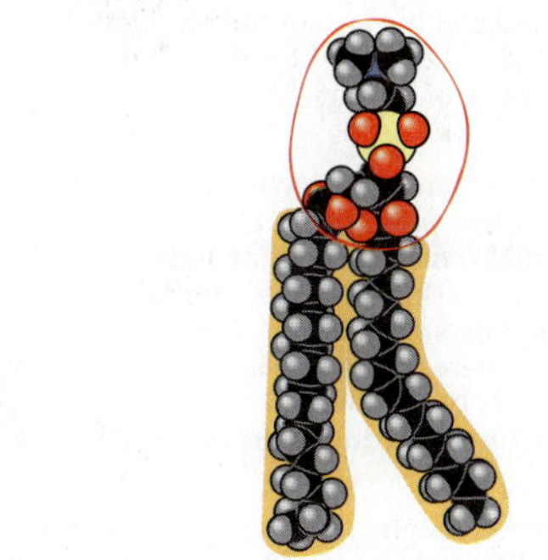

Figure 5.14

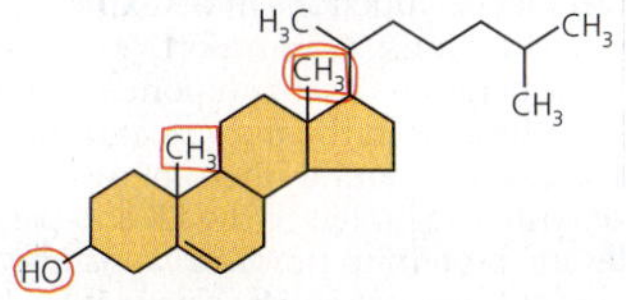

Figure 5.17

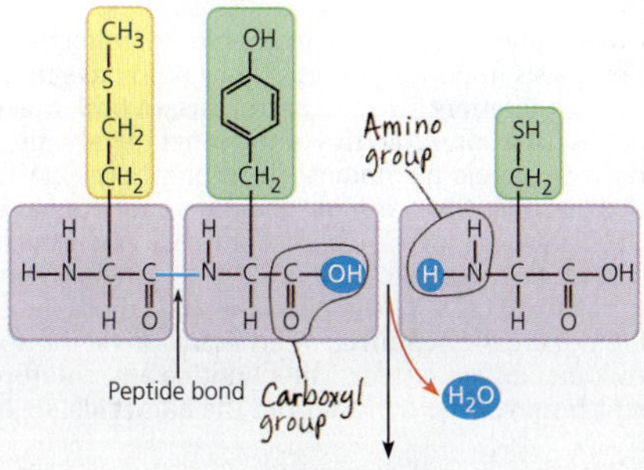

Figure 5.21 The R group on glutamic acid is acidic and hydrophilic, whereas that on valine is nonpolar and hydrophobic. Therefore, it is unlikely that valine can participate in the same intramolecular interactions that glutamic acid can. A change in these interactions causes a disruption of molecular structure.
Figure 5.24 The spirals are α helices.

Concept Check 5.1

1. The four main classes are proteins, carbohydrates, lipids, and nucleic acids. Lipids are not polymers. **2.** Nine, with one water molecule required to hydrolyze each connected pair of monomers **3.** The amino acids in the fish protein must be released in hydrolysis reactions and incorporated into other proteins in dehydration reactions.

Concept Check 5.2

1. $C_3H_6O_3$ **2.** $C_{12}H_{22}O_{11}$ **3.** The antibiotic treatment is likely to have killed the cellulose-digesting prokaryotes in the cow's stomach. The absence of these prokaryotes would hamper the cow's ability to obtain energy from food and could lead to weight loss and possibly death. Thus, prokaryotic species are reintroduced, in appropriate combinations, in the gut culture given to treated cows.

Concept Check 5.3

1. Both have a glycerol molecule attached to fatty acids. The glycerol of a fat has three fatty acids attached, whereas the glycerol of a phospholipid is attached to two fatty acids and one phosphate group. **2.** Human sex hormones are steroids, a type of hydrophobic compound. **3.** The oil droplet membrane could consist of a single layer of phospholipids rather than a bilayer, because an arrangement in which the hydrophobic tails of the membrane phospholipids were in contact with the hydrocarbon regions of the oil molecules would be more stable.

Concept Check 5.4

1. The function of a protein is a consequence of its specific shape, which is lost when a protein becomes denatured. **2.** Secondary structure involves hydrogen bonds between atoms of the polypeptide backbone. Tertiary structure involves interactions between atoms of the side chains of the amino acid subunits. **3.** These are all nonpolar amino acids, so you would expect this region to be located in the interior of the folded polypeptide, where it would not contact the aqueous environment inside the cell.

Concept Check 5.5

1.

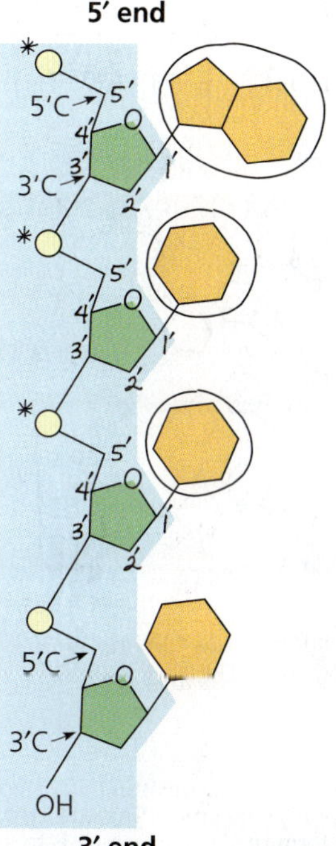

2. 5′–T A G G C C T–3′
3′–A T C C G G A–5′

3. a. Mismatch
5′–T A A G C C T–3′
3′–A T C C G G A–5′

b. 3′–A T T C G G A–5′

Summary of Key Concepts Questions

Concept 5.1 The polymers of carbohydrates, proteins, and nucleic acids are built from three different types of monomers: monosaccharides, amino acids, and nucleotides, respectively. **Concept 5.2** Both starch and cellulose are polymers of glucose, but the glucose monomers are in the α configuration in starch and the β configuration in cellulose. The glycosidic linkages thus have different geometries, giving the polymers different shapes and thus different properties. Starch is an energy-storage compound in plants; cellulose is a structural component of plant cell walls. Humans can hydrolyze starch to provide energy but cannot hydrolyze cellulose. Cellulose aids in the passage of food through the digestive tract. **Concept 5.3** Lipids are not polymers because they do not exist as a chain of linked monomers. They are not considered macromolecules because they do not reach the giant size of many polysaccharides, proteins, and nucleic acids. **Concept 5.4** A polypeptide, which may consist of hundreds of amino acids in a specific sequence (primary structure), has regions of coils and pleats (secondary structure), which are then folded into irregular contortions (tertiary structure) and may be noncovalently associated with other polypeptides (quaternary structure). The linear order of amino acids, with the varying properties of their side chains (R groups), determines what secondary and tertiary structures will form to produce a protein. The resulting unique three-dimensional shapes of proteins are key to their specific and diverse functions. **Concept 5.5** The complementary base pairing of the two strands of DNA makes possible the precise replication of DNA every time a cell divides, ensuring that genetic information is faithfully transmitted. In some types of RNA, complementary base pairing enables

RNA molecules to assume specific three-dimensional shapes that facilitate diverse functions.

Test Your Understanding

1. d **2.** a **3.** b **4.** a **5.** b **6.** c **7.** d

8.

	Monomers or Components	Polymer or larger molecule	Type of linkage
Carbohydrates	Monosaccharides	Polysaccharides	Glycosidic linkages
Lipids	Fatty acids	Triacylglycerols	Ester linkages
Proteins	Amino acids	Polypeptides	Peptide bonds
Nucleic acids	Nucleotides	Polynucleotides	Phosphodiester linkages

9.

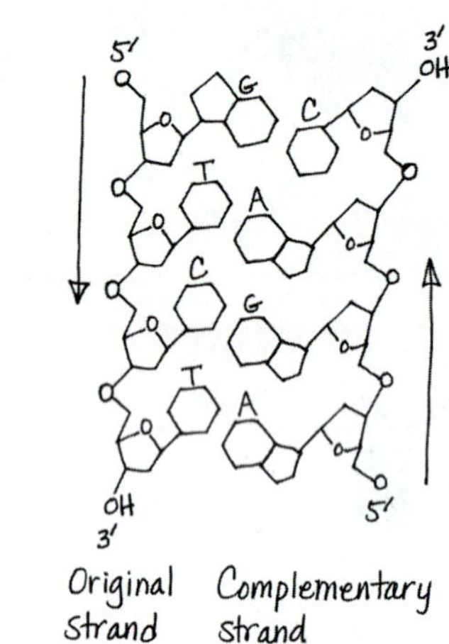

Chapter 6

Figure Questions

Figure 6.6 A phospholipid is a lipid, consisting of a glycerol molecule joined to two fatty acids and one phosphate group. Together, the glycerol and phosphate end of the phospholipid form the "head," which is hydrophilic, while the hydrocarbon chains on the fatty acids form hydrophobic "tails." The presence in a single molecule of both a hydrophilic and a hydrophobic region makes the molecule ideal as the main building block of a membrane. **Figure 6.9** The DNA in a chromosome dictates synthesis of a messenger RNA (mRNA) molecule, which then moves out to the cytoplasm. There, the information is used for the production, on ribosomes, of proteins that carry out cellular functions. **Figure 6.10** Any of the bound ribosomes (attached to the endoplasmic reticulum) could be circled, because any could be making a protein that will be secreted. **Figure 6.22** Each centriole has 9 sets of 3 microtubules, so the entire centrosome (two centrioles) has 54 microtubules. Each microtubule consists of a helical array of tubulin dimers (as shown in Table 6.1).

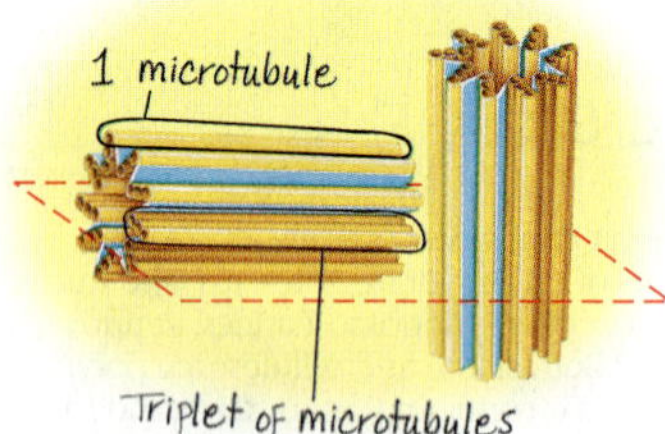

Figure 6.24

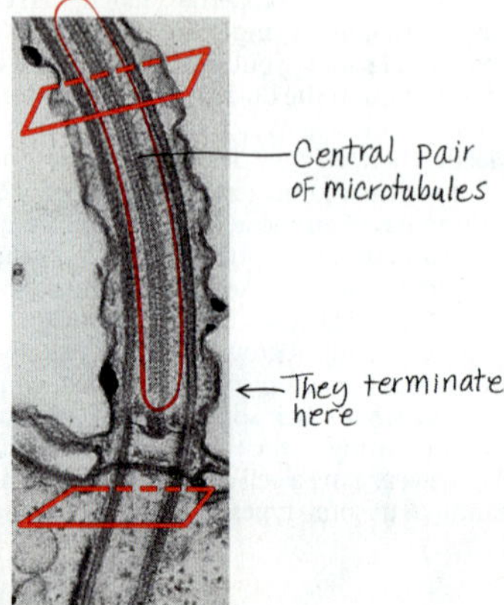

The two central microtubules terminate above the basal body, so they aren't present at the level of the cross section through the basal body, indicated by the lower red rectangle in (a). **Figure 6.29** The microtubules would reorient, and based on the earlier results, the cellulose synthase proteins would also change their path, orienting along the repositioned microtubules. (This is, in fact, what was observed.)

Concept Check 6.1

1. Stains used for light microscopy are colored molecules that bind to cell components, affecting the light passing through, while stains used for electron microscopy involve heavy metals that affect the beams of electrons passing through. **2.** (a) Light microscope, (b) scanning electron microscope

Concept Check 6.2

1. See Figure 6.8.

2. 125 × 1 × 1

This cell would have the same volume as the cells in columns 2 and 3 but proportionally more surface area than that in column 2 and less than that in column 3. Thus, the surface-to-volume ratio should be greater than 1.2 but less than 6. To obtain the surface area, you would add the area of the six sides (the top, bottom, sides, and ends): $125 + 125 + 125 + 125 + 1 + 1 = 502$. The surface-to-volume ratio equals 502 divided by a volume of 125, or 4.0.

Concept Check 6.3

1. Ribosomes in the cytoplasm translate the genetic message, carried from the DNA in the nucleus by mRNA, into a polypeptide chain. **2.** Nucleoli consist of DNA and the ribosomal RNA (rRNA) made according to its instructions, as well as proteins imported from the cytoplasm. Together, the rRNA and proteins are assembled into large and small ribosomal subunits. (These are exported through nuclear pores to the cytoplasm, where they will participate in polypeptide synthesis.) **3.** No. Each chromosome is present whether its chromatin is relatively diffuse (when the cell is not dividing) or condensed (when the cell is dividing).

Concept Check 6.4

1. The primary distinction between rough and smooth ER is the presence of bound ribosomes on the rough ER. Both types of ER make phospholipids, but membrane proteins and secretory proteins are all produced on the ribosomes of the rough ER. The smooth ER also functions in detoxification, carbohydrate metabolism, and storage of calcium ions. **2.** Transport vesicles move membranes and substances they enclose between other components of the endomembrane system. **3.** The mRNA is synthesized in the nucleus and then passes through a nuclear pore to be translated on a bound ribosome, attached to the rough ER. The protein is synthesized into the lumen of the ER and perhaps modified there. A transport vesicle carries the protein to the Golgi apparatus. After further modification in the Golgi, another transport vesicle carries it back to the ER, where it will perform its cellular function.

Concept Check 6.5

1. Both organelles are involved in energy transformation, mitochondria in cellular respiration and chloroplasts in photosynthesis. They both have multiple membranes that separate their interiors into compartments. In both organelles, the innermost membranes—cristae, or infoldings of the inner membrane, in mitochondria, and the thylakoid membranes in chloroplasts—have large surface areas with embedded enzymes that carry out their main functions. **2.** Yes. Plant cells are able to make their own sugar by photosynthesis, but mitochondria in these eukaryotic cells are the organelles that are able to generate energy from sugars, a function required in all cells. **3.** Mitochondria and chloroplasts are not derived from the ER, nor are they connected physically or via transport vesicles to organelles of the endomembrane system. Mitochondria and chloroplasts are structurally quite different from vesicles derived from the ER, which are bounded by a single membrane.

Concept Check 6.6

1. Both systems of movement involve long filaments that are moved in relation to each other by motor proteins that grip, release, and grip again adjacent polymers. **2.** Dynein arms, powered by ATP, move neighboring doublets of microtubules relative to each other. Because they are anchored within the organelle and with respect to one another, the doublets bend instead of sliding past each other. Synchronized bending of the nine microtubule doublets brings about bending of both structures. **3.** Such individuals have defects in the microtubule-based movement of cilia and flagella. Thus, the sperm can't move because of malfunctioning or nonexistent flagella, and the airways are compromised because cilia that line the trachea malfunction or don't exist, and so mucus cannot be cleared from the lungs.

Concept Check 6.7

1. The most obvious difference is the presence of direct cytoplasmic connections between cells of plants (plasmodesmata) and animals (gap junctions). These connections result in the cytoplasm being continuous between adjacent cells. **2.** The cell would not be able to function properly and would probably soon die, as the cell wall or ECM must be permeable to allow the exchange of matter between the cell and its external environment. Molecules involved in energy production and use must be allowed entry, as well as those that provide information about the cell's environment. Other molecules, such as products synthesized by the cell for export and the by-products of cellular respiration, must be allowed to exit. **3.** The parts of the protein that face aqueous regions would be expected to have polar or charged (hydrophilic) amino acids, while the parts that go through the membrane would be expected to have nonpolar (hydrophobic) amino acids. You would predict polar or charged amino acids at each end (tail), in the region of the cytoplasmic loop, and in the regions of the two extracellular loops. You would predict nonpolar amino acids in the four regions that go through the membrane between the tails and loops.

Summary of Key Concepts Questions

6.1 Both light and electron microscopy allow cells to be studied visually, thus helping us understand internal cellular structure and the arrangement of cell components. Cell fractionation techniques separate out different groups of cell components, which can then be analyzed biochemically to determine their function. Performing microscopy on the same cell fraction helps to correlate the biochemical function of the cell with the cell component responsible. **6.2** The separation of different functions in different organelles has several advantages. Reactants and enzymes can be concentrated in one area instead of spread throughout the cell. Reactions that require specific conditions, such as a lower pH, can be compartmentalized. And enzymes for specific reactions are often embedded in the membranes that enclose or partition an organelle. **6.3** The nucleus contains the genetic material of the cell in the form of DNA, which codes for messenger RNA, which in turn provides instructions for the synthesis of proteins (including the proteins that make up part of the ribosomes). DNA also codes for ribosomal RNA, which is combined with proteins in the nucleolus into the subunits of ribosomes. Within the cytoplasm, ribosomes join with mRNA to build polypeptides, using the genetic information in the mRNA. **6.4** Transport vesicles move proteins and membranes synthesized by the rough ER to the Golgi for further processing and then to the plasma membrane, lysosomes, or other locations in the cell, including back to the ER. **6.5** According to the endosymbiont theory, mitochondria originated from an oxygen-using prokaryotic cell that was engulfed by an ancestral eukaryotic cell. Over time, the host and endosymbiont evolved into a single organism. Chloroplasts originated when at least one of these eukaryotic cells containing mitochondria engulfed and then retained a photosynthetic prokaryote. **6.6** Inside the cell, motor proteins interact with components of the cytoskeleton to move cellular parts. Motor proteins may "walk" vesicles along microtubules. The movement of cytoplasm within a cell involves interactions of the motor protein myosin and microfilaments (actin filaments). Whole cells can be moved by the rapid bending of flagella or cilia, which is caused by the motor-protein-powered sliding of microtubules within these structures. Cell movement can also occur when pseudopodia form at one end of a cell (caused by actin polymerization into a filamentous network), followed by contraction of the cell toward that end; this is powered by interactions of microfilaments with myosin. Interactions of motor proteins and microfilaments in muscle cells can propel whole organisms. **6.7** A plant cell wall is primarily composed of microfibrils of cellulose embedded in other polysaccharides and proteins. The ECM of animal cells is primarily composed of collagen and other glycoprotein fibers, such as fibronectins. These fibers are embedded in a network of carbohydrate-rich proteoglycans. A plant cell wall provides structural support for the cell and, collectively, for the plant body. In addition to giving support, the ECM of an animal cell allows for communication of environmental changes into the cell.

Test Your Understanding

1. b **2.** d **3.** b **4.** e **5.** a **6.** d **7.** c **8.** See Figure 6.8.

Chapter 7

Figure Questions

Figure 7.2

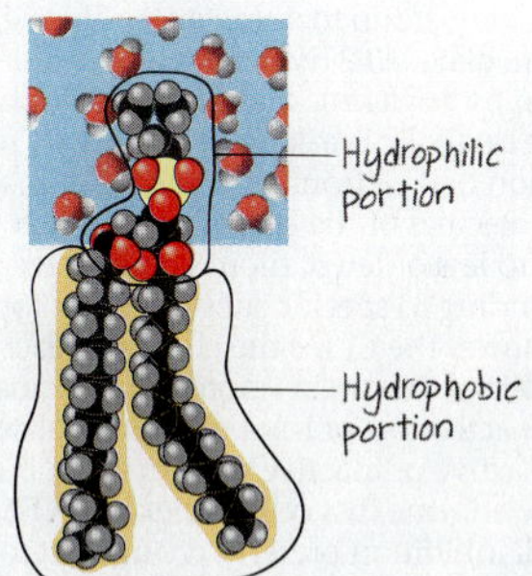

The hydrophilic portion is in contact with an aqueous environment (cytosol or extracellular fluid), and the hydrophobic portion is in contact with the hydrophobic portions of other phospholipids in the interior of the bilayer. **Figure 7.7** You couldn't rule out movement of proteins within membranes of the same species. You might propose that the membrane lipids and proteins from one species weren't able to mingle with those from the other species because of some incompatibility. **Figure 7.10** A transmembrane protein like the dimer in (f) might change its shape upon binding to a particular ECM molecule. The new shape might enable the interior portion of the protein to bind to a second, cytoplasmic protein that would relay the message to the inside of the cell, as shown in (c). **Figure 7.11** The shape of a protein on the HIV surface is likely to be complementary to the shape of the receptor (CD4) and also to that of the co-receptor (CCR5). A molecule that was a similar shape to the HIV surface protein could bind CCR5, blocking HIV binding. (Another alternative would be a molecule that bound to CCR5 and changed the shape of CCR5 so it could no longer bind HIV.)

Figure 7.12

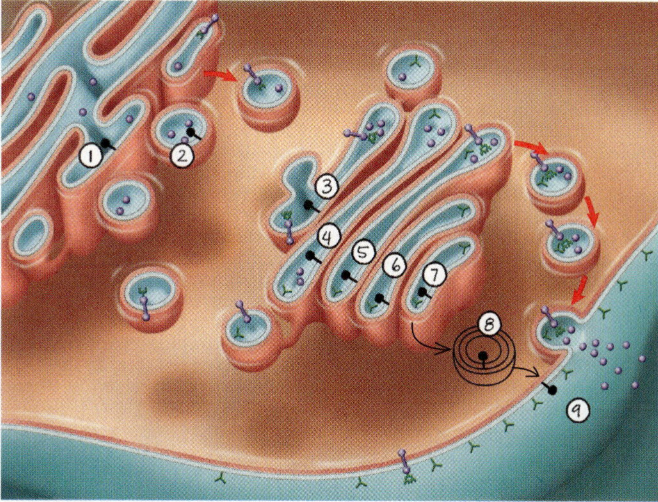

The protein would contact the extracellular fluid. **Figure 7.14** The orange dye would be evenly distributed throughout the solution on both sides of the membrane. The solution levels would not be affected because the orange dye can diffuse through the membrane and equalize its concentration. Thus, no additional osmosis would take place in either direction. **Figure 7.19** The diamond solutes are moving into the cell (down), and the round solutes are moving out of the cell (up); both are moving against their concentration gradient.

Concept Check 7.1

1. They are on the inner side of the transport vesicle membrane. **2.** The grasses living in the cooler region would be expected to have more unsaturated fatty acids in their membranes because those fatty acids remain fluid at lower temperatures. The grasses living immediately adjacent to the hot springs would be expected to have more saturated fatty acids, which would allow the fatty acids to "stack" more closely, making the membranes less fluid and therefore helping them to stay intact at higher temperatures. (Cholesterol could not be used to moderate the effects of temperature on membrane fluidity because it is not found within plant cell membranes.)

Concept Check 7.2

1. O_2 and CO_2 are both nonpolar molecules that can easily pass through the hydrophobic interior of a membrane. **2.** Water is a polar molecule, so it cannot pass very rapidly through the hydrophobic region in the middle of a phospholipid bilayer. **3.** The hydronium ion is charged, while glycerol is not. Charge is probably more significant than size as a basis for exclusion by the aquaporin channel.

Concept Check 7.3

1. CO_2 is a nonpolar molecule that can diffuse through the plasma membrane. As long as it diffuses away so that the concentration remains low outside the cell, it will continue to exit the cell in this way. (This is the opposite of the case for O_2, described in this section.) **2.** The water is hypotonic to the plant cells, so the plant cells take up water. Thus, the cells of the vegetable remain turgid rather than plasmolyzing, and the vegetable (for example, lettuce or spinach) remains crisp and not wilted. **3.** The activity of *Paramecium caudatum*'s contractile vacuole will decrease. The vacuole pumps out excess water that accumulates in the cell; this accumulation occurs only in a hypotonic environment.

Concept Check 7.4

1. The pump uses ATP. To establish a voltage, ions have to be pumped against their gradients, which requires energy. **2.** Each ion is being transported against its electrochemical gradient. If either ion were transported down its electrochemical gradient, this *would* be considered cotransport. **3.** The internal environment of a lysosome is acidic, so it has a higher concentration of H^+ than does the cytoplasm. Therefore, you might expect the membrane of the lysosome to have a proton pump such as that shown in Figure 7.20 to pump H^+ into the lysosome.

Concept Check 7.5

1. Exocytosis. When a transport vesicle fuses with the plasma membrane, the vesicle membrane becomes part of the plasma membrane.
2.

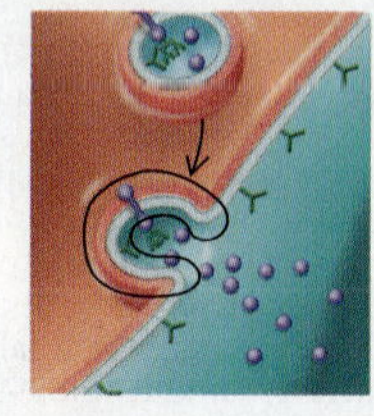

3. The glycoprotein would be synthesized in the ER lumen, move through the Golgi apparatus, and then travel in a vesicle to the plasma membrane, where it would undergo exocytosis and become part of the ECM.

Summary of Key Concepts Questions

7.1 Plasma membranes define the cell by separating the cellular components from the external environment. This allows conditions inside cells to be controlled by membrane proteins, which regulate entry and exit of molecules and even cell function (see Figure 7.10). The processes of life can be carried out inside the controlled environment of the cell, so membranes are crucial. In eukaryotes, membranes also function to subdivide the cytoplasm into different compartments where distinct processes can occur, even under differing conditions such as pH. **7.2** Aquaporins are channel proteins that greatly increase the permeability of a membrane to water molecules, which are polar and therefore do not readily diffuse through the hydrophobic interior of the membrane. **7.3** There will be a net diffusion of water out of a cell into a hypertonic solution. The free water concentration is higher inside the cell than in the solution (where water molecules are not free, but are clustered around the higher concentration of solute particles). **7.4** One of the solutes moved by the cotransporter is actively transported against its concentration gradient. The energy for this transport comes from the concentration gradient of the other solute, which was established by an electrogenic pump that used energy to transport the other solute across the membrane. **7.5** In receptor-mediated endocytosis, specific molecules act as ligands when they bind to receptors on the plasma membrane. The cell can acquire bulk quantities of those molecules when a coated pit forms a vesicle and carries the bound molecules into the cell.

Test Your Understanding

1. b **2.** c **3.** a **4.** d **5.** b

6. (a)

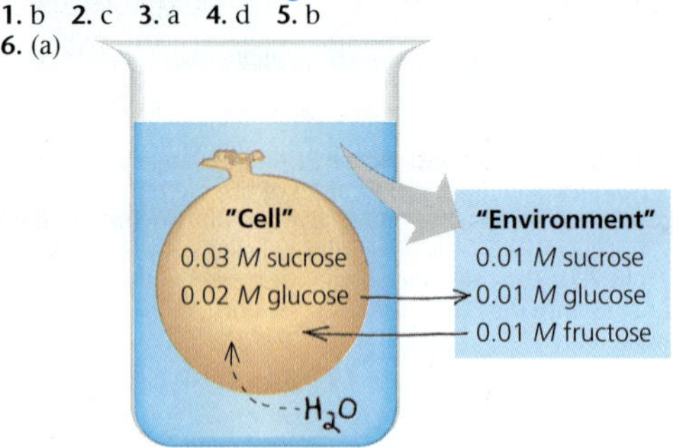

(b) The solution outside is hypotonic. It has less sucrose, which is a nonpenetrating solute. (c) See answer for (a). (d) The artificial cell will become more turgid. (e) Eventually, the two solutions will have the same solute concentrations. Even though sucrose can't move through the membrane, water flow (osmosis) will lead to isotonic conditions.

Chapter 8

Figure Questions

Figure 8.12 **Figure 8.16**

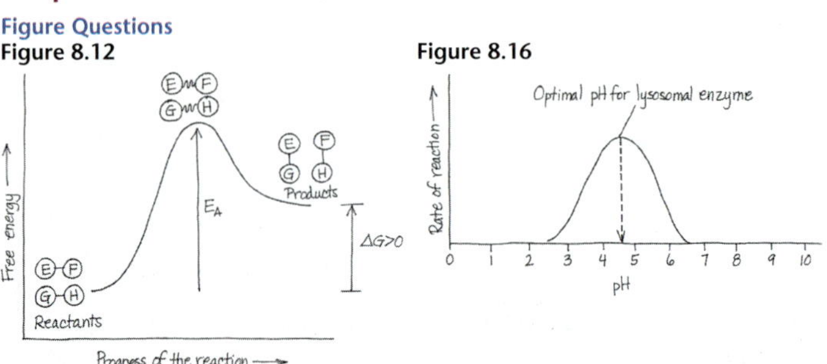

Figure 8.20 Because the affinity of the caspase for the inhibitor is very low (as is expected of an allosterically inhibited enzyme), the inhibitor is likely to diffuse away. Because no additional source of the inhibitory compound is present and the concentration of inhibitor is very low, the inhibitor is unlikely to bind again to the allosteric binding site once the covalent linkage is broken. Therefore, the activity of the enzyme would most likely be normal. (In fact, this is what the researchers observed when they broke the disulfide linkage.)

Concept Check 8.1

1. The second law is the trend toward randomization, or increasing entropy. When the concentrations of a substance on both sides of a membrane are equal, the distribution is more random than when they are unequal. Diffusion of a substance to a region where it is initially less concentrated increases entropy, making it an energetically favorable (spontaneous) process as described by the second law. This explains the process seen in Figure 7.13. **2.** The apple has potential energy in its position hanging on the tree, and the sugars and other nutrients it contains have chemical energy. The apple has kinetic energy as it falls from the tree to the ground. Finally, when the apple is digested and its molecules broken down, some of the chemical energy is used to do work, and the rest is lost as thermal energy. **3.** The sugar crystals become less ordered (entropy increases) as they dissolve and become randomly spread out in the water. Over time, the water evaporates, and the crystals form again because the water volume is insufficient to keep them in solution. While the reappearance of sugar crystals may represent a "spontaneous" increase in order (decrease in entropy), it is balanced by the decrease in order (increase in entropy) of the water molecules, which changed from a relatively compact arrangement as liquid water to a much more dispersed and disordered form as water vapor.

Concept Check 8.2

1. Cellular respiration is a spontaneous and exergonic process. The energy released from glucose is used to do work in the cell or is lost as heat. **2.** When the H^+ concentrations are the same, the system is at equilibrium and can do no work. Hydrogen ions can perform work only if their concentrations on each side of a membrane differ—in other words, when a gradient is present. This is consistent with Figure 7.20, which shows that an energy input (provided by ATP hydrolysis) is required to establish the concentration gradient (the H^+ gradient) that can in turn perform work. **3.** The reaction is exergonic because it releases energy—in this case, in the form of light. (This is a nonbiological version of the bioluminescence seen in Figure 8.1.)

Concept Check 8.3

1. ATP usually transfers energy to endergonic processes by phosphorylating (adding phosphate groups to) other molecules. (Exergonic processes phosphorylate ADP to regenerate ATP.) **2.** A set of coupled reactions can transform the first combination into the second. Since this is an exergonic process overall, ΔG is negative and the first combination must have more free energy (see Figure 8.9). **3.** Active transport: The solute is being transported against its concentration gradient, which requires energy, provided by ATP hydrolysis.

Concept Check 8.4

1. A spontaneous reaction is a reaction that is exergonic. However, if it has a high activation energy that is rarely attained, the rate of the reaction may be low. **2.** Only the specific substrate(s) will fit properly into the active site of an enzyme, the part of the enzyme that carries out catalysis. **3.** In the presence of malonate, increase the concentration of the normal substrate (succinate) and see whether the rate of reaction increases. If it does, malonate is a competitive inhibitor. **4.** If lactose wasn't present in the environment as a source of food and the fucose-containing disaccharide was available, bacteria that could digest the latter would be better able to grow and multiply than those that could not.

Concept Check 8.5

1. The activator binds in such a way that it stabilizes the active form of an enzyme, whereas the inhibitor stabilizes the inactive form. **2.** An inhibitor that binds to the active site of the enzyme you want to inhibit could also bind to and block the enzymes with similar structures, causing significant side effects. For this reason, you would be better off choosing to screen chemical compounds that bind allosterically to the enzyme in question, because allosteric regulatory sites are less likely to share similarity with other enzymes.

Summary of Key Concepts Questions

8.1 The process of "ordering" a cell's structure is accompanied by an increase in the entropy or disorder of the universe. For example, an animal cell takes in highly ordered organic molecules as the source of matter and energy used to build and maintain its structures. In the same process, however, the cell releases heat and the simple molecules of carbon dioxide and water to the surroundings. The increase in entropy of the latter process offsets the entropy decrease in the former. **8.2** A spontaneous reaction has a negative ΔG and is exergonic. For a chemical reaction to proceed with a net release of free energy ($-\Delta G$), the enthalpy or total energy of the system must decrease ($-\Delta H$), and/or the entropy or disorder must increase (yielding a more negative term, $-T\Delta S$). Spontaneous reactions supply the energy to perform cellular work. **8.3** The free energy released from the hydrolysis of ATP may drive endergonic reactions through the transfer of a phosphate group to a reactant molecule, forming a more reactive phosphorylated intermediate. ATP hydrolysis also powers the mechanical and transport work of a cell, often by powering shape changes in the relevant motor proteins. Cellular respiration, the catabolic breakdown of glucose, provides the energy for the endergonic regeneration of ATP from ADP and Ⓟ$_i$. **8.4** Activation energy barriers prevent the complex molecules of the cell, which are rich in free energy, from spontaneously breaking down to less ordered, more stable molecules. Enzymes permit a regulated metabolism by binding to specific substrates and forming enzyme-substrate complexes that selectively lower the E_A for the chemical reactions in a cell. **8.5** A cell tightly regulates its metabolic pathways in response to fluctuating needs for energy and materials. The binding of activators or inhibitors to regulatory sites on allosteric enzymes stabilizes either the active or inactive form of the subunits. For example, the binding of ATP to a catabolic enzyme in a cell with excess ATP would inhibit that pathway. Such types of feedback inhibition preserve chemical resources within a cell. If ATP supplies are depleted, binding of ADP to the regulatory site of catabolic enzymes would activate that pathway.

Test Your Understanding

1. b **2.** c **3.** b **4.** a **5.** c **6.** e **7.** c

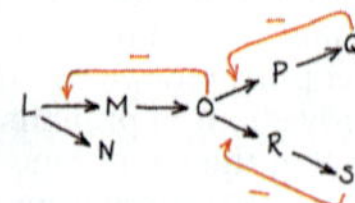

9.

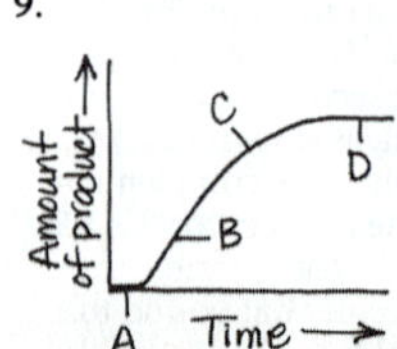

A. The substrate molecules are entering the cells, so no product is made yet.
B. There is sufficient substrate, so the reaction is proceeding at a maximum rate.
C. As the substrate is used up, the rate decreases (the slope is less steep).
D. The line is flat because no new substrate remains and thus no new product appears.

Chapter 9

Figure Questions

Figure 9.7 Because there is no external source of energy for the reaction, it must be exergonic, and the reactants must be at a higher energy level than the products. **Figure 9.9** The removal would probably stop glycolysis, or at least slow it down, since it would push the equilibrium for step 5 toward the left. If less (or no) glyceraldehyde 3-phosphate were available, step 6 would slow down (or be unable to occur). **Figure 9.15** At first, some ATP could be made, since electron transport could proceed as far as complex III, and a small H^+ gradient could be built up. Soon, however, no more electrons could be passed to complex III because it could not be reoxidized by passing its electrons to complex IV. **Figure 9.16** First, there are 2 NADH from the oxidation of pyruvate plus 6 NADH from the citric acid cycle (CAC); 8 NADH $\times$ 2.5 ATP/NADH = 20 ATP. Second, there are 2 $FADH_2$ from the CAC; 2 $FADH_2 \times$ 1.5 ATP/$FADH_2$ = 3 ATP. Third, the 2 NADH from glycolysis enter the mitochondrion through one of two types of shuttle. They pass their electrons either to 2 FAD, which become $FADH_2$ and result in 3 ATP, or to 2 NAD^+, which become NADH and result in 5 ATP. Thus, 20 + 3 + 3 = 26 ATP, or 20 + 3 + 5 = 28 ATP from all NADH and $FADH_2$.

Concept Check 9.1

1. Both processes include glycolysis, the citric acid cycle, and oxidative phosphorylation. In aerobic respiration, the final electron acceptor is molecular oxygen (O_2); in anaerobic respiration, the final electron acceptor is a different substance. **2.** $C_4H_6O_5$ would be oxidized and NAD^+ would be reduced.

Concept Check 9.2

1. NAD^+ acts as the oxidizing agent in step 6, accepting electrons from glyceraldehyde 3-phosphate, which thus acts as the reducing agent. **2.** Since the overall process of glycolysis results in net production of ATP, it would make sense for the process to slow down when ATP levels have increased substantially. Thus, we would expect ATP to allosterically inhibit phosphofructokinase.

Concept Check 9.3

1. NADH and $FADH_2$; they will donate electrons to the electron transport chain. **2.** CO_2 is released from the pyruvate that is the end product of glycolysis, and CO_2 is also released during the citric acid cycle. **3.** In both cases, the precursor molecule loses a CO_2 molecule and then donates electrons to an electron carrier in an oxidation step. Also, the product has been activated due to the attachment of a CoA group.

Concept Check 9.4

1. Oxidative phosphorylation would eventually stop entirely, resulting in no ATP production by this process. Without oxygen to "pull" electrons down the electron transport chain, H^+ would not be pumped into the mitochondrion's intermembrane space and chemiosmosis would not occur. **2.** Decreasing the pH means addition of H^+. This would establish a proton gradient even without the function of the electron transport chain, and we would expect ATP synthase to function and synthesize ATP. (In fact, it was experiments like this that provided support for chemiosmosis as an energy-coupling mechanism.) **3.** One of the components of the electron transport chain, ubiquinone (Q), must be able to diffuse within the membrane. It could not do so if the membrane were locked rigidly into place.

Concept Check 9.5

1. A derivative of pyruvate, such as acetaldehyde during alcohol fermentation, or pyruvate itself during lactic acid fermentation; oxygen **2.** The cell would need to consume glucose at a rate about 16 times the consumption rate in the aerobic environment (2 ATP are generated by fermentation versus up to 32 ATP by cellular respiration).

Concept Check 9.6

1. The fat is much more reduced; it has many $—CH_2—$ units, and in all these bonds the electrons are equally shared. The electrons present in a carbohydrate molecule are already somewhat oxidized (shared unequally in bonds), as quite a few of them are bound to oxygen. **2.** When we consume more food than necessary for metabolic processes, our body synthesizes fat as a way of storing energy for later use. **3.** Glycogen is a storage polysaccharide in liver and muscle cells. When energy is needed, glucose units are hydrolyzed from glycogen. Glycolysis in the cytosol breaks down glucose to two pyruvate molecules, which are transported into the mitochondrion. Here they are further oxidized, ultimately producing the needed ATP. **4.** AMP will accumulate, stimulating phosphofructokinase, and thus increasing the rate of glycolysis. Since oxygen is not present, the cell will convert pyruvate to lactate in lactic acid fermentation, providing a supply of ATP. **5.** When oxygen is present, the fatty acid chains containing most of the energy of a fat are oxidized and fed into the citric acid cycle and the electron transport chain. During intense exercise, however, oxygen is scarce in muscle cells, so ATP must be generated by glycolysis alone. A very small part of the fat molecule, the glycerol backbone, can be oxidized via glycolysis, but the amount of energy released by this portion is insignificant compared to that released by the fatty acid chains. (This is why moderate exercise, staying below 70% maximum heart rate, is better for burning fat—because enough oxygen remains available to the muscles.)

Summary of Key Concepts Questions

9.1 Most of the ATP produced in cellular respiration comes from oxidative phosphorylation, in which the energy released from redox reactions in an electron transport chain is used to produce ATP. In substrate-level phosphorylation, an enzyme directly transfers a phosphate group to ADP from an intermediate substrate. All ATP production in glycolysis occurs by substrate-level phosphorylation; this form of ATP production also occurs at one step in the citric acid cycle. **9.2** The oxidation of the three-carbon sugar, glyceraldehyde 3-phosphate, yields energy. In this oxidation, electrons and H^+ are transferred to NAD^+, forming NADH, and a phosphate group is attached to the oxidized substrate. ATP is then formed by substrate-level phosphorylation when this phosphate group is transferred to ADP. **9.3** The release of six molecules of CO_2 represents the complete oxidation of glucose. During the processing of two pyruvates to acetyl CoA, the fully oxidized carboxyl group ($—COO^-$) is given off as CO_2. The remaining four carbons are released as CO_2 in the citric acid cycle as citrate is oxidized back to oxaloacetate. **9.4** The flow of H^+ through the ATP synthase complex causes the rotor and attached rod to rotate, exposing catalytic sites in the knob portion that produce ATP from ADP and Ⓟ$_i$. ATP synthases are found in the inner mitochondrial membrane, the plasma membrane of prokaryotes, and membranes within chloroplasts. **9.5** Anaerobic respiration yields more ATP. The 2 ATP produced by substrate-level phosphorylation in glycolysis represent the total energy yield of fermentation. NADH passes its "high-energy" electrons to pyruvate or a derivative of pyruvate, recycling NAD^+ and allowing glycolysis to continue. Anaerobic respiration uses an electron transport chain to capture the energy of the electrons in NADH via a series of redox reactions; ultimately, the electrons are transferred to an electronegative molecule other than oxygen. And additional molecules of NADH are produced in anaerobic respiration as pyruvate is oxidized. **9.6** The ATP produced by catabolic pathways is used to drive anabolic pathways. Also, many of the intermediates of glycolysis and the citric acid cycle are used in the biosynthesis of a cell's molecules.

Test Your Understanding

1. d **2.** c **3.** c **4.** a **5.** e **6.** a **7.** b

8.

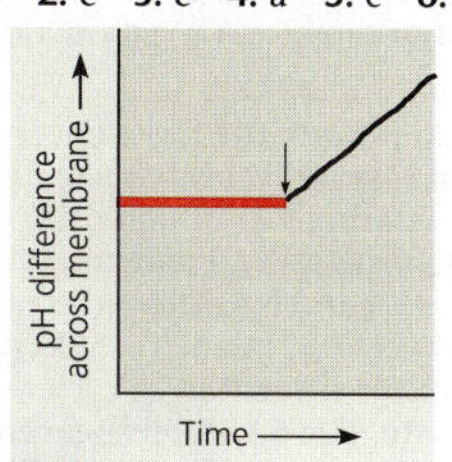

Chapter 10

Figure Questions

Figure 10.3 Situating containers of algae near sources of CO_2 emissions makes sense because algae need CO_2 to carry out photosynthesis. The higher their rate of photosynthesis, the more plant oil they will produce. At the same time, algae would be absorbing the CO_2 emitted from industrial plants or from car engines, reducing the amount of CO_2 entering the atmosphere. **Figure 10.10** Red, but not violet-blue, wavelengths would pass through the filter, so the bacteria would not congregate where the violet-blue light normally comes through. Therefore, the left "peak" of bacteria would not be present, but the right peak would be observed because the red wavelengths passing through the filter would be used for photosynthesis. **Figure 10.12** In the leaf, most of the chlorophyll electrons excited by photon absorption are used to power the reactions of photosynthesis. **Figure 10.16** The person at the top of the photosystem I tower would not turn and throw his electron into the bucket. Instead, he would throw it onto the top of the ramp right next to the photosystem II tower. The electron would then roll down the ramp, get energized by a photon, and return to him. This cycle would continue as long as light was available. (This is why it's called cyclic electron flow.) **Figure 10.19**

3 × 1 (=3)
3 × 6 (=18)
6 × 3 (=18)
6 × 3 (=18)
6 × 3 (=18)
5 × 3 (=15)
3 × 5 (=15)
1 × 3 (=3)

Three carbon atoms enter the cycle, one by one, as individual CO_2 molecules, and leave the cycle in one three-carbon molecule (G3P) per three turns of the cycle. **Figure 10.22**

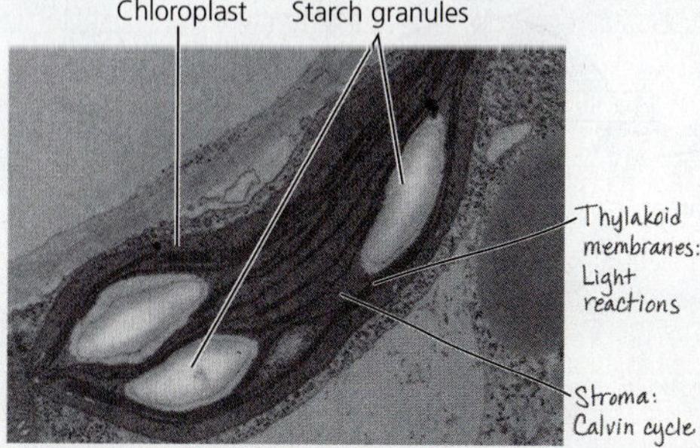

The photosystems that carry out the light reactions are embedded in the thylakoid membranes, and the ATP and NADPH that are formed are released into the stroma. There, they are used for the reactions of the Calvin cycle, which produces G3P. Excess sugar molecules that are not used by the plant can be converted to glucose, then stored in the form of starch.

Concept Check 10.1

1. CO_2 enters leaves via stomata, and water enters via roots and is carried to leaves through veins. **2.** Using ^{18}O, a heavy isotope of oxygen, as a label, researchers were able to confirm van Niel's hypothesis that the oxygen produced during photosynthesis originates in water, not in carbon dioxide. **3.** The light reactions could *not* keep producing NADPH and ATP without the $NADP^+$, ADP, and Ⓟ$_i$ that the Calvin cycle generates. The two cycles are interdependent.

Concept Check 10.2

1. Green, because green light is mostly transmitted and reflected—not absorbed—by photosynthetic pigments **2.** In chloroplasts, light-excited electrons are trapped by a primary electron acceptor, which prevents them from dropping back to the ground state. In isolated chlorophyll, there is no electron acceptor, so the photoexcited electrons immediately drop back down to the ground state, with the emission of light and heat. **3.** Water (H_2O) is the initial electron donor; $NADP^+$ accepts electrons at the end of the electron transport chain, becoming reduced to NADPH. **4.** In this experiment, the rate of ATP synthesis would slow and eventually stop. Because the added compound would not allow a proton gradient to build up across the membrane, ATP synthase could not catalyze ATP production.

Concept Check 10.3

1. 6, 18, 12 **2.** The more potential energy a molecule stores, the more energy and reducing power is required for the formation of that molecule. Glucose is a valuable energy source because it is highly reduced, storing lots of potential energy in its electrons. To reduce CO_2 to glucose, much energy and reducing power are required in the form of large numbers of ATP and NADPH molecules, respectively. **3.** The light reactions require ADP and $NADP^+$, which would not be formed in sufficient quantities from ATP and NADPH if the Calvin cycle stopped. **4.** In glycolysis, G3P acts as an intermediate. The 6-carbon sugar fructose 1,6-bisphosphate is cleaved into two 3-carbon sugars, one of which is G3P. The other is an isomer called dihydroxyacetone phosphate, which can be converted to G3P by an isomerase. Because G3P is the substrate for the next enzyme, it is constantly removed, and the reaction equilibrium is pulled in the direction of conversion of dihydroxyacetone phosphate to more G3P. In the Calvin cycle, G3P acts as both an intermediate and a product. For every three CO_2 molecules that enter the cycle, six G3P molecules are formed, five of which must remain in the cycle and become rearranged to regenerate three 5-carbon RuBP molecules. The one remaining G3P is a product, which can be thought of as the result of "reducing" the three CO_2 molecules that entered the cycle into a 3-carbon sugar that can later be used to generate energy.

Concept Check 10.4

1. Photorespiration decreases photosynthetic output by adding oxygen, instead of carbon dioxide, to the Calvin cycle. As a result, no sugar is generated (no carbon is fixed), and O_2 is used rather than generated. **2.** Without PS II, no O_2 is generated in bundle-sheath cells. This avoids the problem of O_2 competing with CO_2 for binding to rubisco in these cells. **3.** Both problems are caused by a drastic change in Earth's atmosphere due to burning of fossil fuels. The increase in CO_2 concentration affects ocean chemistry by decreasing pH, thus affecting calcification by marine organisms. On land, CO_2 concentration and air temperature are conditions that plants have become adapted to, and changes in these characteristics have a strong effect on photosynthesis by plants. Thus, alteration of these two fundamental factors could have critical effects on organisms all around the planet, in all different habitats. **4.** C_4 and CAM species would replace many of the C_3 species.

Summary of Key Concepts Questions

10.1 CO_2 and H_2O are the products of respiration; they are the reactants in photosynthesis. In respiration, glucose is oxidized to CO_2 as electrons are passed through an electron transfer chain from glucose to O_2, producing H_2O. In photosynthesis, H_2O is the source of electrons, which are energized by light, temporarily stored in NADPH, and used to reduce CO_2 to carbohydrate. **10.2** The action spectrum of photosynthesis shows that some wavelengths of light that are not absorbed by chlorophyll *a* are still effective at promoting photosynthesis. The light-harvesting complexes of photosystems contain accessory pigments such as chlorophyll *b* and carotenoids, which absorb different wavelengths and pass the energy to chlorophyll *a*, broadening the spectrum of light useful for photosynthesis.
10.3

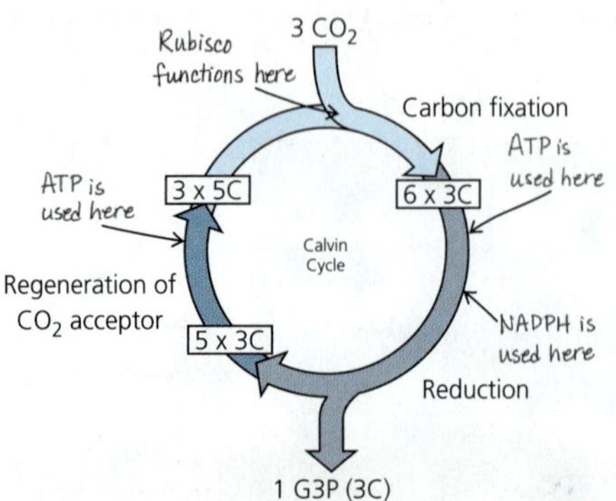

In the reduction phase of the Calvin cycle, ATP phosphorylates a 3-carbon compound, and NADPH then reduces this compound to G3P. ATP is also used in the regeneration phase, when five molecules of G3P are converted to three molecules of the 5-carbon compound RuBP. Rubisco catalyzes the first step of carbon fixation—the addition of CO_2 to RuBP. **10.4** Both C_4 and CAM photosynthesis involve initial fixation of CO_2 to produce a 4-carbon compound (in mesophyll cells in C_4 plants and at night in CAM plants). These compounds are then broken down to release CO_2 (in the bundle-sheath cells in C_4 plants and during the day in CAM plants). ATP is required for recycling the molecule that is used initially to combine with CO_2. These pathways avoid the photorespiration that consumes ATP and reduces the photosynthetic output of C_3 plants when they close stomata on hot, dry, bright days. Thus, hot, arid climates would favor C_4 and CAM plants.

Test Your Understanding

1. d **2.** b **3.** c **4.** d **5.** c **6.** b **7.** d
9.

High H^+ concentration (pH=4)
H^+
Low H^+ concentration (pH=8)
ADP + Ⓟ$_i$
H^+
ATP

The ATP would end up outside the thylakoid. The thylakoids were able to make ATP in the dark because the researchers set up an artificial proton concentration gradient across the thylakoid membrane; thus, the light reactions were not necessary to establish the H^+ gradient required for ATP synthesis by ATP synthase.

Chapter 11

Figure Questions

Figure 11.6 Epinephrine is a signaling molecule; presumably, it binds to a cell-surface receptor protein. **Figure 11.7** Figure 7.1 shows a potassium channel, which, according to the description on p. 135, opens in response to an electrical stimulus, allowing potassium ions to rush out of the cell. Thus, it is a voltage-gated ion channel. **Figure 11.8** When the receptor is actively transmitting a signal to the inside of the cell, it is bound to a G protein. To determine a structure corresponding to that state, it might work to crystallize the receptor in the presence of many copies of the G protein. (In fact, the researchers planned to try this approach next. Another research group also used this approach successfully with a related G protein-coupled receptor the following year.) **Figure 11.9** The testosterone molecule is hydrophobic and can therefore pass directly through the lipid bilayer of the plasma membrane into the cell. (Hydrophilic molecules cannot do this.) **Figure 11.10** The active form of protein kinase 2 **Figure 11.11** The signaling molecule (cAMP) would remain in its active form and would continue to signal. **Figure 11.17** In the model, the directionality of growth is determined by the association of Fus3 with the membrane near the site of receptor activation. Thus, the development of shmoos would be severely compromised, and the affected cell would likely resemble the ΔFus3 and Δformin cells. **Figure 11.18** The signaling pathway shown in Figure 11.14 leads to the splitting of PIP_2 into the second messengers DAG and IP_3, which produce different responses. (The response elicited by DAG is mentioned but not shown.) The pathway shown for cell B is similar in that it branches and leads to two responses.

Concept Check 11.1

1. The two cells of opposite mating type (**a** and **α**) each secrete a certain signaling molecule, which can only be bound by receptors carried on cells of the opposite mating type. Thus, the **a** mating factor cannot bind to another **a** cell and cause it to grow toward the first **a** cell. Only an **α** cell can "receive" the signaling molecule and respond by directed growth (see Figure 11.17 for more information). **2.** The secretion of neurotransmitter molecules at a synapse is an example of local signaling. The electrical signal that travels along a very long nerve cell and is passed to the next nerve cell can be considered an example of long-distance signaling. (Note, however, that local signaling at the synapse between two cells is necessary for the signal to pass from one cell to the next.) **3.** Glucose 1-phosphate is not generated, because the activation of the enzyme requires an intact cell, with an intact receptor in the membrane and an intact signal transduction pathway. The enzyme cannot be activated directly by interaction with the signaling molecule in the test tube. **4.** Glycogen phosphorylase acts in the third stage, the response to epinephrine signaling.

Concept Check 11.2

1. NGF is water-soluble (hydrophilic), so it cannot pass through the lipid membrane to reach intracellular receptors, as steroid hormones can. Therefore, you'd expect the NGF receptor to be in the plasma membrane—which is, in fact, the case. **2.** The cell with the faulty receptor would not be able to respond appropriately to the signaling molecule when it was present. This would most likely have dire consequences for the cell, since regulation of the cell's activities by this receptor would not occur appropriately. **3.** Binding of a ligand to a receptor changes the shape of the receptor, altering the ability of the receptor to transmit a signal. Binding of an allosteric regulator to an enzyme changes the shape of the enzyme, either promoting or inhibiting enzyme activity.

Concept Check 11.3

1. A protein kinase is an enzyme that transfers a phosphate group from ATP to a protein, usually activating that protein (often a second type of protein kinase). Many signal transduction pathways include a series of such interactions, in which each phosphorylated protein kinase in turn phosphorylates the next protein kinase in the series. Such phosphorylation cascades carry a signal from outside the cell to the cellular protein(s) that will carry out the response. **2.** Protein phosphatases reverse the effects of the kinases. **3.** The signal that is being transduced is the information that a signaling molecule is bound to the cell-surface receptor. Information is transduced by way of sequential protein-protein interactions that change protein shapes, causing them to function in a way that passes the signal along. **4.** The IP_3-gated channel opens, allowing calcium ions to flow out of the ER, which raises the cytosolic Ca^{2+} concentration.

Concept Check 11.4

1. At each step in a cascade of sequential activations, one molecule or ion may activate numerous molecules functioning in the next step. **2.** Scaffolding proteins hold molecular components of signaling pathways in a complex with each other. Different scaffolding proteins would assemble different collections of proteins, leading to different cellular responses in the two cells. **3.** A malfunctioning protein phosphatase would not be able to dephosphorylate a particular receptor or relay protein. As a result, the signaling pathway, once activated, would not be able to be terminated. (In fact, one study found altered protein phosphatases in cells from 25% of colorectal tumors.)

Concept Check 11.5

1. In formation of the hand or paw in mammals, cells in the regions between the digits are programmed to undergo apoptosis. This serves to shape the digits of the hand or paw so that they are not webbed. **2.** If a receptor protein for a death-signaling molecule were defective such that it was activated even in the absence of the death signal, this would lead to apoptosis when it wouldn't normally occur. Similar defects in any of the proteins in the signaling pathway, which would activate these relay or response proteins in the absence of interaction with the previous protein or second messenger in the pathway, would have the same effect. Conversely, if any protein in the pathway were defective in its ability to respond to an interaction with an early protein or other molecule or ion, apoptosis would not occur when it normally should. For example, a receptor protein for a death-signaling ligand might not be able to be activated, even when ligand was bound. This would stop the signal from being transduced into the cell.

Summary of Key Concepts Questions

11.1 A cell is able to respond to a hormone only if it has a receptor protein on the cell surface or inside the cell that can bind to the hormone. The response to a hormone depends on the specific cellular activity that a signal transduction pathway triggers within the cell. The response can vary for different types of cells. **11.2** Both GPCRs and RTKs have an extracellular binding site for a signaling molecule (ligand) and an α helix region of the polypeptide that spans the membrane. GPCRs usually trigger a single transduction pathway, whereas the multiple activated tyrosines on an RTK dimer may trigger several different transduction pathways at the same time. **11.3** A protein kinase is an enzyme that adds a phosphate group to another protein. Protein kinases are often part of a phosphorylation cascade that transduces a signal. A second messenger is a small, nonprotein molecule or ion that rapidly diffuses and relays a signal throughout a cell. Both protein kinases and second messengers can operate in the same pathway. For example, the second messenger cAMP often activates protein kinase A, which then phosphorylates other proteins. **11.4** In G protein-coupled pathways, the GTPase portion of a G protein converts GTP to GDP and inactivates the G protein. Protein phosphatases remove phosphate groups from activated proteins, thus stopping a phosphorylation cascade of protein kinases. Phosphodiesterase converts cAMP to AMP, thus reducing the effect of cAMP in a signal transduction pathway. **11.5** The basic mechanism of controlled cell suicide evolved early in eukaryotic evolution, and the genetic basis for these pathways has been conserved during animal evolution. Such a mechanism is essential to the development and maintenance of all animals.

Test Your Understanding

1. c **2.** d **3.** a **4.** b **5.** a **6.** d **7.** c **8.** c **9.** This is one possible drawing of the pathway. (Similar drawings would also be correct.)

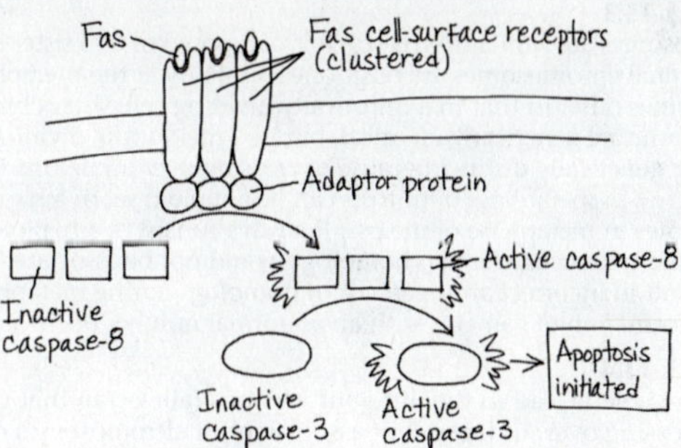

Chapter 12

Figure Questions

Figure 12.4

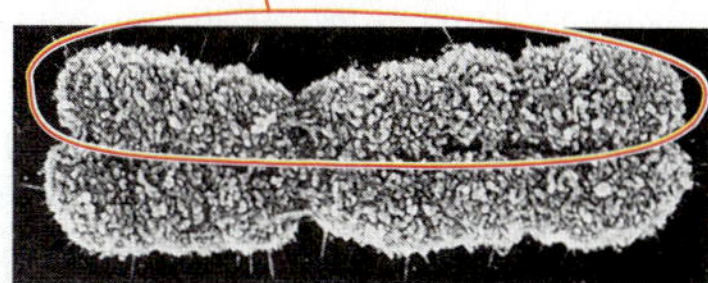

Circling the other chromatid instead would also be correct. **Figure 12.5** The chromosome has four arms. **Figure 12.7** 12; 2; 2; 1
Figure 12.8

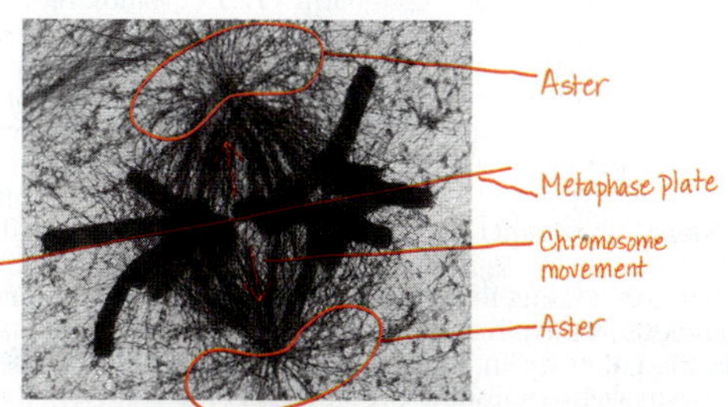

Figure 12.9 The mark would have moved toward the nearer pole. The lengths of fluorescent microtubules between that pole and the mark would have decreased, while the lengths between the chromosomes and the mark would have remained the same. **Figure 12.14** In both cases, the G_1 nucleus would have remained in G_1 until the time it normally would have entered the S phase. Chromosome condensation and spindle formation would not have occurred until the S and G_2 phases had been completed. **Figure 12.16** The cell would divide under conditions where it was inappropriate to do so. If the daughter cells and their descendants also ignored the checkpoint and divided, there would soon be an abnormal mass of cells. (This type of inappropriate cell division can contribute to the development of cancer.) **Figure 12.17** Passing the G_2 checkpoint in the diagram corresponds to the beginning of the "Time" axis of the graph, and entry into the mitotic phase (yellow background on the diagram) corresponds to the peaks of MPF activity and cyclin concentration on the graph (see the yellow M banner over the peaks). During G_1 and S phase in the diagram, Cdk is present without cyclin, so on the graph both cyclin concentration and MPF activity are low. The curved purple arrow in the diagram shows increasing cyclin concentration, seen on the graph during the end of S phase and throughout G_2 phase. Then the cell cycle begins again. **Figure 12.18** The cells in the vessel with PDGF would not be able to respond to the growth factor signal and thus would not divide. The culture would resemble that without the added PDGF. **Figure 12.21** The intracellular estrogen receptor, once activated, would be able to act as a transcription factor in the nucleus, turning on genes that may cause the cell to pass a checkpoint and divide. The HER2 receptor, when activated by a ligand, would form a dimer, and each subunit of the dimer would phosphorylate the other. This would lead to a series of signal transduction steps, ultimately turning on genes in the nucleus. As in the case of the estrogen receptor, the genes would code for proteins necessary to commit the cell to divide.

Concept Check 12.1

1. 2 **2.** 39; 39; 78

Concept Check 12.2

1. 6 chromosomes, duplicated; 12 chromatids **2.** Following mitosis, cytokinesis results in two genetically identical daughter cells in both plant cells and animal cells. However, the mechanism of dividing the cytoplasm is different in animals and plants. In an animal cell, cytokinesis occurs by cleavage, which divides the parent cell in two with a contractile ring of actin filaments. In a plant cell, a cell plate forms in the middle of the cell and grows until its membrane fuses with the plasma membrane of the parent cell. A new cell wall grows inside the cell plate. **3.** They elongate the cell during anaphase. **4.** During eukaryotic cell division, tubulin is involved in spindle formation and chromosome movement, while actin functions during cytokinesis. In bacterial binary fission, it's the opposite: Tubulin-like molecules are thought to act in daughter cell separation, and actin-like molecules are thought to move the daughter bacterial chromosomes to opposite ends of the cell. **5.** Microtubules made up of tubulin in the cell provide "rails" along which vesicles and other organelles can travel, based on interactions of motor proteins with tubulin in the microtubules. In muscle cells, actin in microfilaments interacts with myosin filaments to cause muscle contraction. **6.** From the end of S phase in interphase through the end of metaphase in mitosis

Concept Check 12.3

1. The nucleus on the right was originally in the G_1 phase; therefore, it had not yet duplicated its chromosome. The nucleus on the left was in the M phase, so it had already duplicated its chromosome. **2.** A sufficient amount of MPF has to exist for a cell to pass the G_2 checkpoint; this occurs through the accumulation of cyclin proteins, which combine with Cdk to form MPF. **3.** Most body cells are in a nondividing state called G_0. **4.** Both types of tumors consist of abnormal cells, but their characteristics are different. A benign tumor stays at the original site and can usually be surgically removed; the cells have some genetic and cellular changes from normal, non-tumor cells. Cancer cells from a malignant tumor have more significant genetic and cellular changes, can spread from the original site by metastasis, and may impair the functions of one or more organs. **5.** The cells might divide even in the absence of PDGF. In addition, they would not stop when the surface of the culture vessel was covered; they would continue to divide, piling on top of one another.

Summary of Key Concepts Questions

12.1 The DNA of a eukaryotic cell is packaged into structures called *chromosomes*. Each chromosome is a long molecule of DNA, which carries hundreds to thousands of genes, with associated proteins that maintain chromosome structure and help control gene activity. This DNA-protein complex is called *chromatin*. The chromatin of each chromosome is long and thin when the cell is not dividing. Prior to cell division, each chromosome is duplicated, and the resulting sister *chromatids* are attached to each other by proteins at the centromeres and, for many species, all along their lengths (sister chromatid cohesion). **12.2** Chromosomes exist as single DNA molecules in G_1 of interphase and in anaphase and telophase of mitosis. During S phase, DNA replication produces sister chromatids, which persist during G_2 of interphase and through prophase, prometaphase, and metaphase of mitosis. **12.3** Checkpoints allow cellular surveillance mechanisms to determine whether the cell is prepared to go to the next stage. Internal and external signals move a cell past these checkpoints. The G_1 checkpoint, called the "restriction point" in mammalian cells, determines whether a cell will complete the cell cycle and divide or switch into the G_0 phase. The signals to pass this checkpoint often are external—such as growth factors. Passing the G_2 checkpoint requires sufficient numbers of active MPF complexes, which in turn orchestrate several mitotic events. MPF also initiates degradation of its cyclin component, terminating the M phase. The M phase will not begin again until sufficient cyclin is produced during the next S and G_2 phases. The signal to pass the M phase checkpoint is not activated until all chromosomes are attached to kinetochore fibers and are aligned at the metaphase plate. Only then will sister chromatid separation occur.

Test Your Understanding

1. b **2.** a **3.** c **4.** c **5.** a **6.** b
7. See Figure 12.7 for a description of major events.

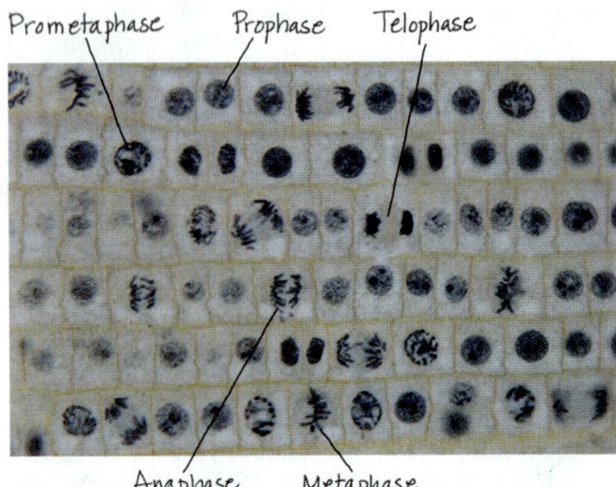

Only one cell is indicated for each stage, but other correct answers are also present in this micrograph.

8. a **9.** e
10.

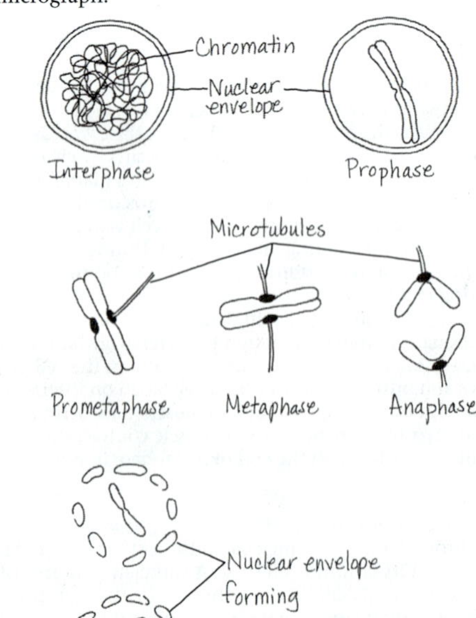

Chapter 13

Figure Questions

Figure 13.4 The haploid number, *n*, is 3. A set is always haploid.
Figure 13.7

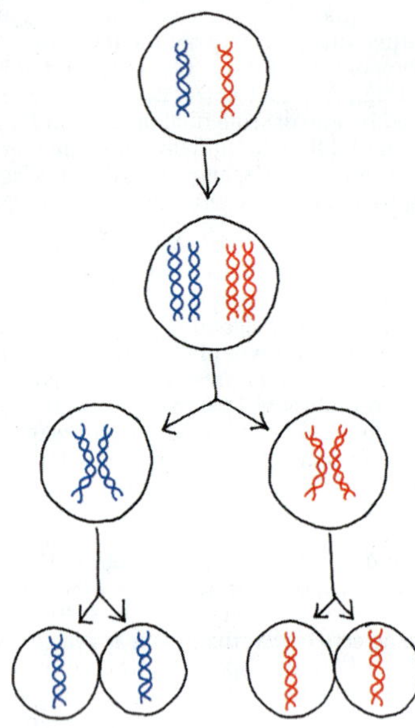

(A short strand of DNA is shown here for simplicity, but each chromosome or chromatid contains a very long coiled and folded DNA molecule.)
Figure 13.8 If the two cells in Figure 12.7 underwent another round of mitosis, each of the four resulting cells would have six chromosomes, while the four cells resulting from meiosis in Figure 13.8 each have three chromosomes. In mitosis, DNA replication (and thus chromosome duplication) precedes each prophase, ensuring that daughter cells have the same number of chromosomes as the parent cell. In meiosis, in contrast, DNA replication occurs only before prophase I (not prophase II). Thus, in two rounds of mitosis, the chromosomes duplicate twice and divide twice, while in meiosis, the chromosomes duplicate once and divide twice. **Figure 13.9** Yes. Each of the six chromosomes (three per cell) shown in telophase I has one nonrecombinant chromatid and one recombinant chromatid. Therefore, eight possible sets of chromosomes can be generated for the cell on the left and eight for the cell on the right.

Concept Check 13.1

1. Parents pass genes to their offspring; the genes program cells to make specific enzymes and other proteins, whose cumulative action produces an individual's inherited traits. **2.** Such organisms reproduce by mitosis, which generates offspring whose genomes are exact copies of the parent's genome (in the absence of mutation). **3.** She should clone it. Cross-breeding it with another plant would generate offspring that have additional variation, which she no longer desires now that she has obtained her ideal orchid.

Concept Check 13.2

1. Each of the six chromosomes is duplicated, so each contains two DNA double helices. Therefore, there are 12 DNA molecules in the cell. **2.** In meiosis, the chromosome count is reduced from diploid to haploid; the union of two haploid gametes in fertilization restores the diploid chromosome count. **3.** The haploid number (*n*) is 7; the diploid number (2*n*) is 14. **4.** This organism has the life cycle shown in Figure 13.6c. Therefore, it must be a fungus or a protist, perhaps an alga.

Concept Check 13.3

1. The chromosomes are similar in that each is composed of two sister chromatids, and the individual chromosomes are positioned similarly at the metaphase plate. The chromosomes differ in that in a mitotically dividing cell, sister chromatids of each chromosome are genetically identical, but in a meiotically dividing cell, sister chromatids are genetically distinct because of crossing over in meiosis I. Moreover, the chromosomes in metaphase of mitosis can be a diploid set or a haploid set, but the chromosomes in metaphase of meiosis II always consist of a haploid set. **2.** If crossing over did not occur, the two homologs would not be associated in any way. This might result in incorrect arrangement of homologs during metaphase I and ultimately in formation of gametes with an abnormal number of chromosomes.

Concept Check 13.4

1. Mutations in a gene lead to the different versions (alleles) of that gene.
2. Without crossing over, independent assortment of chromosomes during meiosis I theoretically can generate 2^n possible haploid gametes, and random fertilization can produce $2^n \times 2^n$ possible diploid zygotes. Because the haploid number (*n*) of grasshoppers is 23 and that of fruit flies is 4, two grasshoppers would be expected to produce a greater variety of zygotes than would two fruit flies. **3.** If the segments of the maternal and paternal chromatids that undergo

crossing over are genetically identical and thus have the same two alleles for every gene, then the recombinant chromosomes will be genetically equivalent to the parental chromosomes. Crossing over contributes to genetic variation only when it involves the rearrangement of different alleles.

Summary of Key Concepts Questions

13.1 Genes program specific traits, and offspring inherit their genes from each parent, accounting for similarities in their appearance to one or the other parent. Humans reproduce sexually, which ensures new combinations of genes (and thus traits) in the offspring. Consequently, the offspring are not clones of their parents (which would be the case if humans reproduced asexually). **13.2** Animals and plants both reproduce sexually, alternating meiosis with fertilization. Both have haploid gametes that unite to form a diploid zygote, which then goes on to divide mitotically, forming a diploid multicellular organism. In animals, haploid cells become gametes and don't undergo mitosis, while in plants, the haploid cells resulting from meiosis undergo mitosis to form a haploid multicellular organism, the gametophyte. This organism then goes on to generate haploid gametes. (In plants such as trees, the gametophyte is quite reduced in size and not obvious to the casual observer.) **13.3** At the end of meiosis I, the two members of a homologous pair end up in different cells, so they cannot pair up and undergo crossing over. **13.4** First, during independent assortment in metaphase I, each pair of homologous chromosomes lines up independent of each other pair at the metaphase plate, so a daughter cell of meiosis I randomly inherits either a maternal or paternal chromosome. Second, due to crossing over, each chromosome is not exclusively maternal or paternal, but includes regions at the ends of the chromatid from a nonsister chromatid (a chromatid of the other homolog). (The nonsister segment can also be in an internal region of the chromatid if a second crossover occurs beyond the first one before the end of the chromatid.) This provides much additional diversity in the form of new combinations of alleles. Third, random fertilization ensures even more variation, since any sperm of a large number containing many possible genetic combinations can fertilize any egg of a similarly large number of possible combinations.

Test Your Understanding

1. a **2.** d **3.** b **4.** a **5.** d **6.** c **7.** d
8. (a)

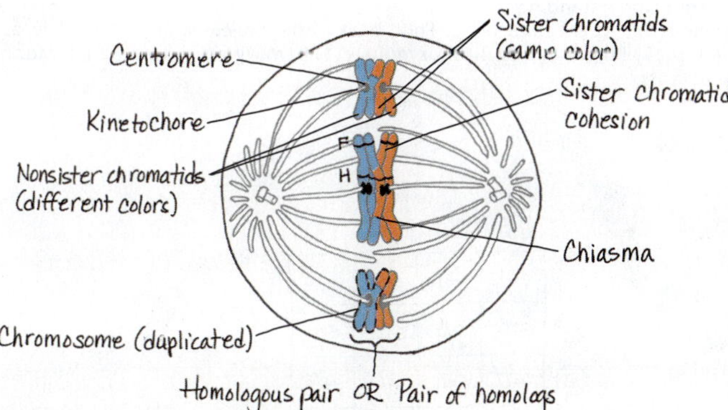

The chromosomes of one color make up a haploid set.
All red and blue chromosomes together make up a diploid set.

(b) The chromosomes of one color make up a haploid set. (In cases where crossovers have occurred, a haploid set of one color may include segments of chromatids of the other color.) All red and blue chromosomes together make up a diploid set.
(c) Metaphase I **9.** This cell must be undergoing meiosis because homologous chromosomes are associated with each other at the metaphase plate; this does not occur in mitosis.

Chapter 14

Figure Questions

Figure 14.3 All offspring would have purple flowers. (The ratio would be one purple to zero white.) The P generation plants are true-breeding, so mating two purple-flowered plants produces the same result as self-pollination: All the offspring have the same trait.

Figure 14.8

If dependent assortment:
Parents
YyRr × yyrr
Sperm from YyRr plant
1/2 (YR) 1/2 (yr)
Eggs from yyrr plant (yr): YyRr | yyrr
1/2 yellow-round : 1/2 green-wrinkled
1 yellow-round : 1 green-wrinkled
Phenotypic ratio

If independent assortment:
Parents
YyRr × yyrr
Sperm from YyRr plant
1/4 (YR) 1/4 (Yr) 1/4 (yR) 1/4 (yr)
Eggs from yyrr plant (yr): YyRr | Yyrr | yyRr | yyrr
1/4 yellow-round : 1/4 yellow-wrinkled:
1/4 green-round : 1/4 green-wrinkled
1 yellow-round : 1 yellow-wrinkled : 1 green-round : 1 green-wrinkled
Phenotypic ratio

Yes, this cross would also have allowed Mendel to make different predictions for the two hypotheses, thereby allowing him to distinguish the correct one.
Figure 14.10 Your classmate would probably point out that the F_1 generation hybrids show an intermediate phenotype between those of the homozygous parents, which supports the blending hypothesis. You could respond that crossing the F_1 hybrids results in the reappearance of the white phenotype, rather than identical pink offspring, which fails to support the idea of traits blending during inheritance.
Figure 14.11 Both the I^A and I^B alleles are dominant to the i allele, which results in no attached carbohydrate. The I^A and I^B alleles are codominant; both are expressed in the phenotype of I^AI^B heterozygotes, who have type AB blood.
Figure 14.13

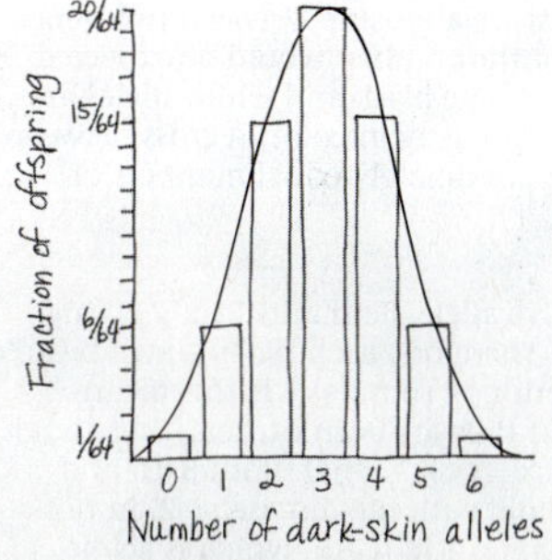

The majority of individuals have intermediate phenotypes (skin color in the middle range), while fewer individuals have phenotypes at either end (very dark or very light skin). (As you may know, this is called a "bell curve" and represents a "normal distribution.") **Figure 14.16** In the Punnett square, two of the three individuals with normal coloration are carriers, so the probability is ⅔. (Note that you must take into account everything you know when you calculate probability: You know she is not *aa*, so there are only three possible genotypes to consider.) **Figure 14.18** If one parent tests negative for the recessive allele, then the probability is zero that the offspring will have the disease and ½ that the offspring will be a carrier. If the first child is a carrier, the probability of the next child being a carrier is still ½ because the two births are independent events.

Concept Check 14.1

1. According to the law of independent assortment, 25 plants (1/16 of the offspring) are predicted to be *aatt*, or recessive for both characters. The actual result is likely to differ slightly from this value.

Parents
AaTt × AaTt

Sperm from AaTt plant; Eggs from AaTt plant

	AT	At	aT	at
AT	AATT	AATt	AaTT	AaTt
At	AATt	AAtt	AaTt	Aatt
aT	AaTT	AaTt	aaTT	aaTt
at	AaTt	Aatt	aaTt	aatt

2. The plant could make eight different gametes (*YRI, YRi, YrI, Yri, yRI, yRi, yrI,* and *yri*). To fit all the possible gametes in a self-pollination, a Punnett square would need 8 rows and 8 columns. It would have spaces for the 64 possible unions of gametes in the offspring. **3.** Self-pollination is sexual reproduction because meiosis is involved in forming gametes, which unite during fertilization. As a result, the offspring in self-pollination are genetically different from the parent. (As mentioned in the footnote on p. 263, we have simplified the explanation in referring to the single pea plant as a parent. Technically, the gametophytes in the flower are the two "parents.")

Concept Check 14.2

1. ½ homozygous dominant (*AA*), 0 homozygous recessive (*aa*), and ½ heterozygous (*Aa*) **2.** ¼ *BBDD*; ¼ *BbDD*; ¼ *BBDd*; ¼ *BbDd* **3.** The genotypes that fulfill this condition are *ppyyIi, ppYyii, Ppyyii, ppYYii,* and *ppyyii*. Use the multiplication rule to find the probability of getting each genotype, and then use the addition rule to find the overall probability of meeting the conditions of this problem:

ppyyIi	½ (probability of pp) × ¼ (yy) × ½ (ii)	= 1/16
ppYyii	½ (pp) × ½ (Yy) × ½ (ii)	= 2/16
Ppyyii	½ (Pp) × ¼ (yy) × ½ (ii)	= 1/16
ppYYii	½ (pp) × ¼ (YY) × ½ (ii)	= 1/16
ppyyii	½ (pp) × ¼ (yy) × ½ (ii)	= 1/16
Fraction predicted to have at least two recessive traits		= 6/16 or 3/8

Concept Check 14.3

1. Incomplete dominance describes the relationship between two alleles of a single gene, whereas epistasis relates to the genetic relationship between two genes (and the respective alleles of each). **2.** Half of the children would be expected to have type A blood and half type B blood. **3.** The black and white alleles are incompletely dominant, with heterozygotes being gray in color. A cross between a gray rooster and a black hen should yield approximately equal numbers of gray and black offspring.

Concept Check 14.4

1. 1/9 (Since cystic fibrosis is caused by a recessive allele, Beth and Tom's siblings who have CF must be homozygous recessive. Therefore, each parent must be a carrier of the recessive allele. Since neither Beth nor Tom has CF, this means they each have a ⅔ chance of being a carrier. If they are both carriers, there is a ¼ chance that they will have a child with CF. ⅔ × ⅔ × ¼ = 1/9); 0 (Both Beth and Tom would have to be carriers to produce a child with the disease.) **2.** In normal hemoglobin, the sixth amino acid is glutamic acid (Glu), which is acidic (has a negative charge on its side chain). In sickle-cell hemoglobin, Glu is replaced by valine (Val), which is a nonpolar amino acid, very different from Glu. The primary structure of a protein (its amino acid sequence) ultimately determines the shape of the protein and thus its function. The substitution of Val for Glu enables the hemoglobin molecules to interact with each other and form long fibers, leading to the protein's deficient function and the deformation of the red blood cell. **3.** Joan's genotype is *Dd*. Because the allele for polydactyly (*D*) is dominant to the allele for five digits per appendage (*d*), the trait is expressed in people with either the *DD* or *Dd* genotype. But because Joan's father does not have polydactyly, his genotype must be *dd*, which means that Joan inherited a *d* allele from him. Therefore Joan, who does have the trait, must be heterozygous. **4.** In the monohybrid cross involving flower color, the ratio is 3.15 purple : 1 white, while in the human family in the pedigree, the ratio in the third generation is 1 free : 1 attached earlobe. The difference is due to the small sample size (two offspring) in the human family. If the second-generation couple in this pedigree were able to have 929 offspring as in the pea plant cross, the ratio would likely be closer to 3:1. (Note that none of the pea plant crosses in Table 14.1 yielded *exactly* a 3:1 ratio.)

Summary of Key Concepts Questions

14.1 Alternative versions of genes, called alleles, are passed from parent to offspring during sexual reproduction. In a cross between purple- and white-flowered homozygous parents, the F_1 offspring are all heterozygous, each inheriting a purple allele from one parent and a white allele from the other. Because the purple allele is dominant, it determines the phenotype of the F_1 offspring to be purple, and the expression of the white allele is masked. Only in the F_2 generation is it possible for a white allele to exist in a homozygous state, which causes the white trait to be expressed.

14.2

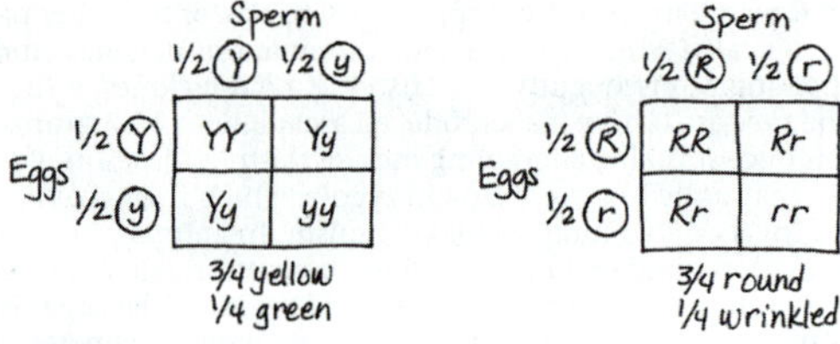

3/4 yellow × 3/4 round = 9/16 yellow-round
3/4 yellow × 1/4 wrinkled = 3/16 yellow-wrinkled
1/4 green × 3/4 round = 3/16 green-round
1/4 green × 1/4 wrinkled = 1/16 green-wrinkled
= 9 yellow-round : 3 yellow-wrinkled : 3 green-round : 1 green-wrinkled

14.3 The ABO blood group is an example of multiple alleles because this single gene has more than two alleles (I^A, I^B, and *i*). Two of the alleles, I^A and I^B, exhibit codominance, since both carbohydrates (A and B) are present when these two alleles exist together in a genotype. I^A and I^B each exhibit complete dominance over the *i* allele. This situation is not an example of incomplete dominance because each allele affects the phenotype in a distinguishable way, so the result is not intermediate between the two phenotypes. Because this situation involves a single gene, it is not an example of epistasis or polygenic inheritance. **14.4** The chance of the fourth child having cystic fibrosis is ¼, as it was for each of the other children, because each birth is an independent event. We already know both parents are carriers, so whether their first three children are carriers or not has no bearing on the probability that their next child will have the disease. The parents' genotypes provide the only relevant information.

Test Your Understanding

1. Gene, l. Allele, e. Character, g. Trait, b. Dominant allele, j. Recessive allele, a. Genotype, k. Phenotype, h. Homozygous, c. Heterozygous, f. Testcross, i. Monohybrid cross, d.

2.

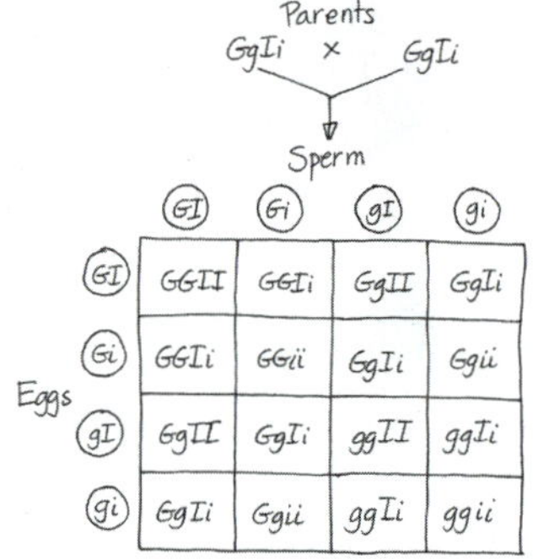

3. Parental cross is $AAC^RC^R \times aaC^WC^W$. F_1 genotype is AaC^RC^W, phenotype is all axial-pink. F_2 genotypes are 1 AAC^RC^R : 2 AAC^RC^W : 1 AAC^WC^W : 2 AaC^RC^R : 4 AaC^RC^W : 2 AaC^WC^W : 1 aaC^RC^R : 2 aaC^RC^W : 1 aaC^WC^W. F_2 phenotypes are 3 axial-red : 6 axial-pink : 3 axial-white : 1 terminal-red : 2 terminal-pink : 1 terminal-white. **4.** Man I^Ai; woman I^Bi; child *ii*. Genotypes for future children are predicted to be ¼ I^AI^B, ¼ I^Ai, ¼ I^Bi, ¼ *ii*. **5.** ½ **6.** A cross of *Ii* × *ii* would yield offspring with a genotypic ratio of 1 *Ii* : 1 *ii* (2:2 is an equivalent answer) and a phenotypic ratio of 1 inflated : 1 constricted (2:2 is equivalent).

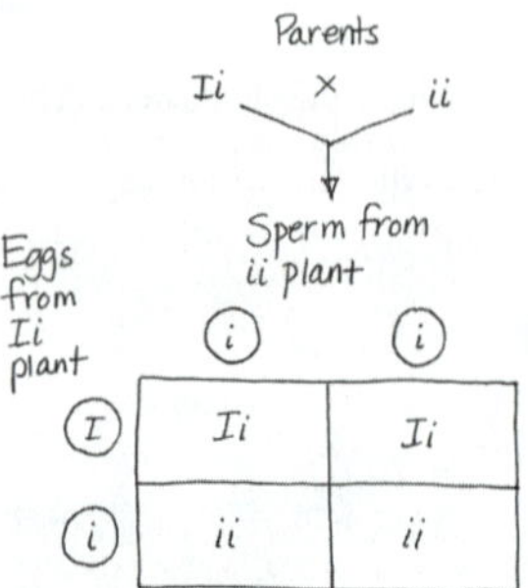

7. (a) $\frac{1}{64}$; (b) $\frac{1}{64}$; (c) $\frac{1}{8}$; (d) $\frac{1}{32}$ **8.** Albino (*b*) is a recessive trait; black (*B*) is dominant. First cross: parents *BB* × *bb*; gametes *B* and *b*; offspring all *Bb* (black coat). The black guinea pig in the second cross is a heterozygote. Second cross: parents *Bb* × *bb*; gametes $\frac{1}{2}$ *B* and $\frac{1}{2}$ *b* (heterozygous parent) and *b*; offspring $\frac{1}{2}$ *Bb* and $\frac{1}{2}$ *bb*. **9.** (a) *PPLl* × *PPLl*, *PPLl* × *PpLl*, or *PPLl* × *ppLl*; (b) *ppLl* × *ppLl*; (c) *PPLL* × any of the 9 possible genotypes or *PPll* × *ppLL*; (d) *PpLl* × *Ppll*; (e) *PpLl* × *PpLl* **10.** (a) $\frac{3}{4} \times \frac{3}{4} \times \frac{3}{4} = \frac{27}{64}$; (b) $1 - \frac{27}{64} = \frac{37}{64}$; (c) $\frac{1}{4} \times \frac{1}{4} \times \frac{1}{4} = \frac{1}{64}$; (d) $1 - \frac{1}{64} = \frac{63}{64}$ **11.** (a) $\frac{1}{256}$; (b) $\frac{1}{16}$; (c) $\frac{1}{256}$; (d) $\frac{1}{64}$; (e) $\frac{1}{128}$ **12.** (a) 1; (b) $\frac{1}{32}$; (c) $\frac{1}{8}$; (d) $\frac{1}{2}$ **13.** $\frac{1}{9}$ **14.** Matings of the original mutant cat with true-breeding noncurl cats will produce both curl and noncurl F_1 offspring if the curl allele is dominant, but only noncurl offspring if the curl allele is recessive. You would obtain some true-breeding offspring homozygous for the curl allele from matings between the F_1 cats resulting from the original curl × noncurl crosses whether the curl trait is dominant or recessive. You know that cats are true-breeding when curl × curl matings produce only curl offspring. As it turns out, the allele that causes curled ears is dominant. **15.** $\frac{1}{16}$ **16.** 25%, or $\frac{1}{4}$, will be cross-eyed; all (100%) of the cross-eyed offspring will also be white. **17.** The dominant allele *I* is epistatic to the *P*/*p* locus, and thus the genotypic ratio for the F_1 generation will be 9 *I_P_* (colorless) : 3 *I_pp* (colorless) : 3 *iiP_* (purple) : 1 *iipp* (red). Overall, the phenotypic ratio is 12 colorless : 3 purple : 1 red. **18.** Recessive. All affected individuals (Arlene, Tom, Wilma, and Carla) are homozygous recessive *aa*. George is *Aa*, since some of his children with Arlene are affected. Sam, Ann, Daniel, and Alan are each *Aa*, since they are all unaffected children with one affected parent. Michael also is *Aa*, since he has an affected child (Carla) with his heterozygous wife Ann. Sandra, Tina, and Christopher can each have either the *AA* or *Aa* genotype. **19.** $\frac{1}{6}$ **20.** 9 *B_A_* (agouti) : 3 *B_aa* (black) : 3 *bbA_* (white) : 1 *bbaa* (white). Overall, 9 agouti : 3 black : 4 white.

Chapter 15

Figure Questions

Figure 15.2 The ratio would be 1 yellow-round : 1 green-round : 1 yellow-wrinkled : 1 green-wrinkled. **Figure 15.4** About $\frac{3}{4}$ of the F_2 offspring would have red eyes and about $\frac{1}{4}$ would have white eyes. About half of the white-eyed flies would be female and half would be male; similarly, about half of the red-eyed flies would be female and half would be male. **Figure 15.7** All the males would be color-blind, and all the females would be carriers. **Figure 15.9** The two largest classes would still be the parental-type offspring (offspring with the phenotypes of the true-breeding P generation flies), but now they would be gray-vestigial and black-normal because those were the specific allele combinations in the P generation. **Figure 15.10** The two chromosomes below, left, are like the two chromosomes inherited by the F_1 female, one from each P generation fly. They are passed by the F_1 female intact to the offspring and thus could be called "parental" chromosomes. The other two chromosomes result from crossing over during meiosis in the F_1 female. Because they have combinations of alleles not seen in either of the F_1 female's chromosomes, they can be called "recombinant" chromosomes. (Note that in this example, the alleles on the recombinant chromosomes, $b^+ \, vg^+$ and *b vg*, are the allele combinations that were on the parental chromosomes in the cross shown in Figures 15.9 and 15.10. The basis for calling them parental chromosomes is the combination of alleles that was present on the P generation chromosomes.)

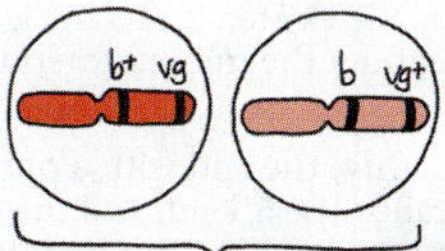

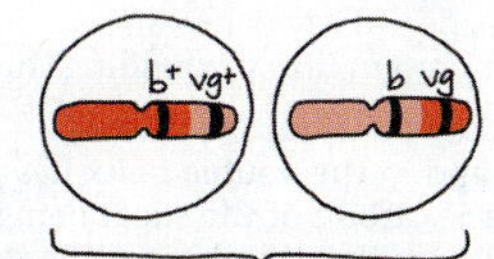

Concept Check 15.1

1. The law of segregation relates to the inheritance of alleles for a single character. The law of independent assortment of alleles relates to the inheritance of alleles for two characters. **2.** The physical basis for the law of segregation is the separation of homologs in anaphase I. The physical basis for the law of independent assortment is the alternative arrangements of homologous chromosome pairs in metaphase I. **3.** To show the mutant phenotype, a male needs to possess only one mutant allele. If this gene had been on a pair of autosomes, *two* mutant alleles would have had to be present for an individual to show the recessive mutant phenotype, a much less probable situation.

Concept Check 15.2

1. Because the gene for this eye-color character is located on the X chromosome, all female offspring will be red-eyed and heterozygous ($X^{w^+} X^w$); all male offspring will inherit a Y chromosome from the father and be white-eyed ($X^w Y$). **2.** $\frac{1}{4}$ ($\frac{1}{2}$ chance that the child will inherit a Y chromosome from the father and be male × $\frac{1}{2}$ chance that he will inherit the X carrying the disease allele from his mother) If the child is a boy, there is a $\frac{1}{2}$ chance he will have the disease; a female would have zero chance (but $\frac{1}{2}$ chance of being a carrier). **3.** With a disorder caused by a dominant allele, there is no such thing as a "carrier," since those with the allele have the disorder. Because the allele is dominant, the females lose any "advantage" in having two X chromosomes, since one disorder-associated allele is sufficient to result in the disorder. All fathers who have the dominant allele will pass it along to *all* their daughters, who will also have the disorder. A mother who has the allele (and thus the disorder) will pass it to half of her sons and half of her daughters.

Concept Check 15.3

1. Crossing over during meiosis I in the heterozygous parent produces some gametes with recombinant genotypes for the two genes. Offspring with a recombinant phenotype arise from fertilization of the recombinant gametes by homozygous recessive gametes from the double-mutant parent. **2.** In each case, the alleles contributed by the female parent determine the phenotype of the offspring because the male in this cross contributes only recessive alleles. **3.** No. The order could be *A-C-B* or *C-A-B*. To determine which possibility is correct, you need to know the recombination frequency between *B* and *C*.

Concept Check 15.4

1. In meiosis, a combined 14-21 chromosome will behave as one chromosome. If a gamete receives the combined 14-21 chromosome and a normal copy of chromosome 21, trisomy 21 will result when this gamete combines with a normal gamete during fertilization. **2.** No. The child can be either $I^A I^A i$ or $I^A ii$. A sperm of genotype $I^A I^A$ could result from nondisjunction in the father during meiosis II, while an egg with the genotype *ii* could result from nondisjunction in the mother during either meiosis I or meiosis II. **3.** Activation of this gene could lead to the production of too much of this kinase. If the kinase is involved in a signaling pathway that triggers cell division, too much of it could trigger unrestricted cell division, which in turn could contribute to the development of a cancer (in this case, a cancer of one type of white blood cell).

Concept Check 15.5

1. Inactivation of an X chromosome in females and genomic imprinting. Because of X inactivation, the effective dose of genes on the X chromosome is the same in males and females. As a result of genomic imprinting, only one allele of certain genes is phenotypically expressed. **2.** The genes for leaf coloration are located in plastids within the cytoplasm. Normally, only the maternal parent transmits plastid genes to offspring. Since variegated offspring are produced only when the female parent is of the B variety, we can conclude that variety B contains both the wild-type and mutant alleles of pigment genes, producing variegated leaves. (Variety A contains only the wild-type allele of pigment genes.) **3.** The situation is similar to that for chloroplasts. Each cell contains numerous mitochondria, and in affected individuals, most cells contain a variable mixture of normal and mutant mitochondria. The normal mitochondria carry out enough cellular respiration for survival.

Summary of Key Concepts Questions

15.1 Because the sex chromosomes are different from each other and because they determine the sex of the offspring, Morgan could use the sex of the offspring as a phenotypic characteristic to follow the parental chromosomes. (He could also have followed them under a microscope, as the X and Y chromosomes look different.) At the same time, he could record eye color to follow the eye-color alleles. **15.2** Males have only one X chromosome, along with a Y chromosome, while females have two X chromosomes. The Y chromosome has very few genes on it, while the X has about 1,000. When a recessive X-linked allele that causes a disorder is inherited by a male on the X from his mother, there isn't a second allele present on the Y (males are hemizygous), so the male has the disorder. Because females have two X chromosomes, they must inherit two recessive alleles in order to have the disorder, a rarer occurrence. **15.3** Crossing over results in new combinations of alleles. Crossing over is a random occurrence, and the more distance there is between two genes, the more chances there are for crossing over to occur, leading to a new allele combination. **15.4** In inversions and reciprocal translocations, the same genetic material is present in the same relative amount but just organized differently. In aneuploidy, duplications, deletions, and nonreciprocal translocations, the balance of genetic material is upset, as large segments are either missing or present in more than one copy. Apparently, this type of imbalance is very damaging to the organism. (Although it isn't lethal in the developing embryo, the reciprocal translocation that produces the Philadelphia chromosome can lead to a serious condition, cancer, by altering the expression of important genes.) **15.5** In these cases, the sex of the parent contributing an allele affects the inheritance pattern. For imprinted genes, either the paternal or the maternal allele is expressed, depending on the imprint. For mitochondrial and chloroplast genes, only the maternal contribution will affect offspring phenotype because the offspring inherit these organelles from the mother, via the egg cytoplasm.

Test Your Understanding

1. 0; $\frac{1}{2}$; $\frac{1}{16}$ **2.** Recessive; if the disorder were dominant, it would affect at least one parent of a child born with the disorder. The disorder's inheritance is sex-linked because it is seen only in boys. For a girl to have the disorder, she would have to inherit recessive alleles from *both* parents. This would be very rare, since males with the recessive allele on their X chromosome die in their early teens. **3.** 17% **4.** The disorder would always be inherited from the mother. **5.** Between *T* and *A*, 12%; between *A* and *S*, 5% **6.** Between *T* and *S*, 18%; sequence of genes is *T–A–S* **7.** $\frac{1}{4}$ for each daughter ($\frac{1}{2}$ chance that child will be female × $\frac{1}{2}$ chance of a homozygous recessive genotype); $\frac{1}{2}$ for first son. **8.** 6%; wild-type heterozygous for normal wings and red eyes) × recessive homozygous for vestigial wings and purple eyes
9.

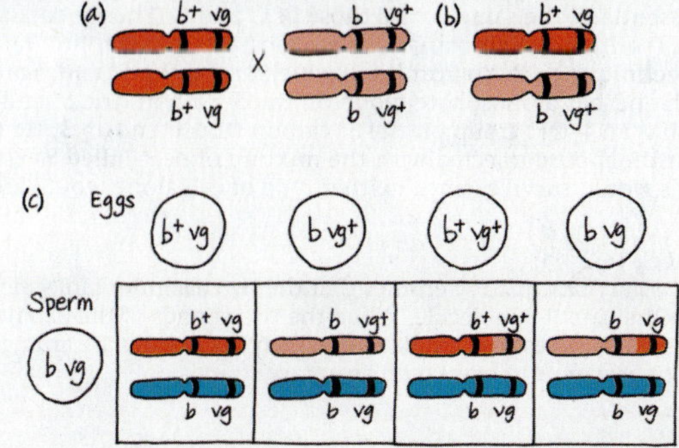

(d) 41.5% gray body, vestigial wings
41.5% black body, normal wings
8.5% gray body, normal wings
8.5% black body, vestigial wings

10. The inactivation of two X chromosomes in XXX women would leave them with one genetically active X, as in women with the normal number of chromosomes. Microscopy should reveal two Barr bodies in XXX women. **11.** *D–A–B–C* **12.** Fifty percent of the offspring will show phenotypes resulting from crossovers. These results would be the same as those from a cross where *A* and *B* were *not* on the same chromosome. Further crosses involving other genes on the same chromosome would reveal the genetic linkage and map distances. **13.** 450 each of blue-oval and white-round (parentals) and 50 each of blue-round and white-oval (recombinants) **14.** About one-third of the distance from the vestigial-wing locus to the brown-eye locus **15.** Because bananas are triploid, homologous pairs cannot line up during meiosis. Therefore, it is not possible to generate gametes that can fuse to produce a zygote with the triploid number of chromosomes.

Chapter 16

Figure Questions

Figure 16.2 The living S cells found in the blood sample were able to reproduce to yield more S cells, indicating that the S trait is a permanent, heritable change, rather than just a one-time use of the dead S cells' capsules. **Figure 16.4** The radioactivity would have been found in the pellet when proteins were labeled (batch 1) because proteins would have had to enter the bacterial cells to program them with genetic instructions. It's hard for us to imagine now, but the DNA might have played a structural role that allowed some of the proteins to be injected while it remained outside the bacterial cell (thus no radioactivity in the pellet in batch 2). **Figure 16.11** The tube from the first replication would look the same, with a middle band of hybrid ^{15}N-^{14}N DNA, but the second tube would not have the upper band of two light blue strands. Instead it would have a bottom band of two dark blue strands, like the bottom band in the result predicted after one replication in the conservative model. **Figure 16.12** In the bubble at the top in (b), arrows should be drawn pointing left and right to indicate the two replication forks. **Figure 16.14** Looking at any of the DNA strands, we see that one end is called the 5′ end and the other the 3′ end. If we proceed from the 5′ end to the 3′ end on the left-most strand, for example, we list the components in this order: phosphate group → 5′ C of the sugar → 3′ C → phosphate → 5′ C → 3′ C. Going in the opposite direction on the same strand, the components proceed in the reverse order: 3′ C → 5′ C → phosphate. Thus, the two directions are distinguishable, which is what we mean when we say that the strands have directionality. (Review Figure 16.5 if necessary.)
Figure 16.17

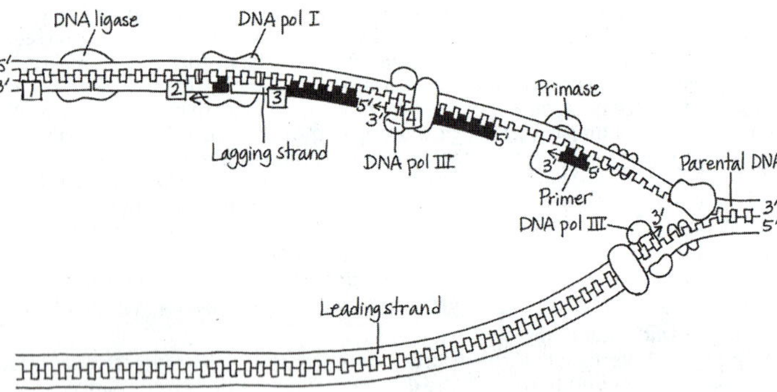

Figure 16.23 The two members of a homologous pair (which would be the same color) would be associated tightly together at the metaphase plate. In metaphase of mitosis, however, each chromosome would be lined up individually, so the two chromosomes of the same color would be in different places at the metaphase plate.

Concept Check 16.1

1. Chargaff's rule about base ratios states that in DNA, the percentages of A and T are essentially the same, as are those of G and C. The fly data are consistent with that rule. (Slight variations are most likely due to limitations of analytical technique.) **2.** You can't tell which end is the 5′ end. You need to know which end has a phosphate group on the 5′ carbon (the 5′ end) or which end has an —OH group on the 3′ carbon (the 3′ end). **3.** He was expecting that the mouse injected with the mixture of heat-killed S cells and living R cells would survive, since neither type of cell alone would kill the mouse.

Concept Check 16.2

1. Complementary base pairing ensures that the two daughter molecules are exact copies of the parental molecule. When the two strands of the parental molecule separate, each serves as a template on which nucleotides are arranged, by the base-pairing rules, into new complementary strands.

2.

Protein	Function
Helicase	Unwinds parental double helix at replication forks
Single-strand binding protein	Binds to and stabilizes single-stranded DNA until it can be used as a template
Topoisomerase	Relieves "overwinding" strain ahead of replication forks by breaking, swiveling, and rejoining DNA strands
Primase	Synthesizes an RNA primer at 5′ end of leading strand and at 5′ end of each Okazaki fragment of lagging strand
DNA pol III	Using parental DNA as a template, synthesizes new DNA strand by covalently adding nucleotides to 3′ end of a pre-existing DNA strand or RNA primer
DNA pol I	Removes RNA nucleotides of primer from 5′ end and replaces them with DNA nucleotides
DNA ligase	Joins 3′ end of DNA that replaces primer to rest of leading strand and joins Okazaki fragments of lagging strand

3. In the cell cycle, DNA synthesis occurs during the S phase, between the G_1 and G_2 phases of interphase. DNA replication is therefore complete before the mitotic phase begins. **4.** Synthesis of the leading strand is initiated by an RNA primer, which must be removed and replaced with DNA, a task that could not be performed if the cell's DNA pol I were nonfunctional. In the overview box in Figure 16.17, just to the left of the top origin of replication, a functional DNA pol I would replace the RNA primer of the leading strand (shown in red) with DNA nucleotides (blue).

Concept Check 16.3

1. A nucleosome is made up of eight histone proteins, two each of four different types, around which DNA is wound. Linker DNA runs from one nucleosome to the next. **2.** Euchromatin is chromatin that becomes less compacted during interphase and is accessible to the cellular machinery responsible for gene activity. Heterochromatin, on the other hand, remains quite condensed during interphase and contains genes that are largely inaccessible to this machinery. **3.** The nuclear lamina is a netlike array of protein filaments that provides mechanical support just inside the nuclear envelope and thus maintains the shape of the nucleus. Considerable evidence also supports the existence of a nuclear matrix, a framework of protein fibers extending throughout the nuclear interior.

Summary of Key Concepts Questions

16.1 Each strand in the double helix has polarity, the end with a phosphate group on the 5′ carbon of the sugar being called the 5′ end, and the end with an —OH group on the 3′ carbon of the sugar being called the 3′ end. The two strands run in opposite directions, so each end of the molecule has both a 5′ and a 3′ end. This arrangement is called "antiparallel." If the strands were parallel, they would both run 5′ → 3′ in the same direction, so an end of the molecule would have either two 5′ ends or two 3′ ends. **16.2** On both the leading and lagging strands, DNA polymerase adds onto the 3′ end of an RNA primer synthesized by primase, synthesizing DNA in the 5′ → 3′ direction. Because the parental strands are antiparallel, however, only on the leading strand does synthesis proceed continuously into the replication fork. The lagging strand is synthesized bit by bit in the direction away from the fork as a series of shorter Okazaki fragments, which are later joined together by DNA ligase. Each fragment is initiated by synthesis of an RNA primer by primase as soon as a given stretch of single-stranded template strand is opened up. Although both strands are synthesized at the same rate, synthesis of the lagging strand is delayed because initiation of each fragment begins only when sufficient template strand is available. **16.3** Most of the chromatin in an interphase nucleus is uncondensed. Much is present as the 30-nm fiber, with some in the form of the 10-nm fiber and some as looped domains of the 30-nm fiber. (These different levels of chromatin packing may reflect differences in gene expression occurring in these regions.) Also, a small percentage of the chromatin, such as that at the centromeres and telomeres, is highly condensed heterochromatin.

Test Your Understanding

1. c **2.** c **3.** b **4.** d **5.** c **6.** d **7.** b **8.** a

9. Like histones, the *E. coli* proteins would be expected to contain many basic (positively charged) amino acids, such as lysine and arginine, which can form weak bonds with the negatively charged phosphate groups on the sugar-phosphate backbone of the DNA molecule. **10.** Each species' DNA has a slightly different percentage of a given base. For example, the percentage of A ranges from 24.7% for *E. coli* to 32.8% for sea urchin. This illustrates Chargaff's rule that the DNA of different species varies in its base composition. Chargaff's other rule states that in any given species, the percentage of A is roughly equal to that of T, and the percentage of C is roughly equal to that of G. For example, sea urchins have about 32–33% each of A and T and about 17% of G and C. (Your answer may use any similar examples from the table.)

12.

Chapter 17

Figure Questions

Figure 17.2 The previously presumed pathway would have been wrong. The new results would support this pathway: precursor → citrulline → ornithine → arginine. They would also indicate that class I mutants have a defect in the second step and class II mutants have a defect in the first step. **Figure 17.4** The mRNA sequence (5′-UGGUUUGGCUCA-3′) is the same as the nontemplate DNA strand sequence (5′-TGGTTTGGCTCA-3′), except there is U in the mRNA and T in the DNA. **Figure 17.7** The processes are similar in that polymerases form polynucleotides complementary to an antiparallel DNA template strand. In replication, however, both strands act as templates, whereas in transcription, only one DNA strand acts as a template. **Figure 17.8** The RNA polymerase would bind directly to the promoter, rather than depending on the previous binding of other factors. **Figure 17.25** The mRNA on the right (the longest one) started transcription first. The ribosome at the top, closest to the DNA, started translating first and thus has the longest polypeptide.

Concept Check 17.1

1. Recessive **2.** A polypeptide made up of 10 Gly (glycine) amino acids

3.

Template sequence (from problem):	3′-TTCAGTCGT-5′
Nontemplate sequence:	5′-AAGTCAGCA-3′
mRNA sequence:	5′-AAGUCAGCA-3′

The nontemplate and mRNA nucleotide sequences are the same except that there is T in the nontemplate strand of DNA wherever there is U in the mRNA.

4.

"Template sequence" (from nontemplate sequence in problem, written 3′ → 5′):	3′-ACGACTGAA-5′
mRNA sequence:	5′-UGCUGACUU-3′
Translated:	Cys-STOP-Leu

(Remember that the mRNA is antiparallel to the DNA strand.) A protein translated from the nontemplate sequence would have a completely different amino acid sequence and would most likely be nonfunctional. (It would also be shorter because of the stop signal shown in the mRNA sequence above—and possibly others earlier in the mRNA sequence.)

Concept Check 17.2

1. Both assemble nucleic acid chains from monomer nucleotides whose order is determined by complementary base pairing to a template strand. Both synthesize in the 5′ → 3′ direction, antiparallel to the template. DNA polymerase requires a primer, but RNA polymerase can start a nucleotide chain from scratch. DNA polymerase uses nucleotides with the sugar deoxyribose and the base T, whereas RNA polymerase uses nucleotides with the sugar ribose and the base U. **2.** The promoter is the region of DNA to which RNA polymerase binds to begin transcription, and it is at the upstream end of the gene (transcription unit). **3.** In a bacterial cell, RNA polymerase recognizes the gene's promoter and binds to it. In a eukaryotic cell, transcription factors mediate the binding of RNA polymerase to the promoter. In both cases, sequences in the promoter bind precisely to the RNA polymerase, so the enzyme is in the right location and orientation. **4.** The transcription factor that recognizes the TATA sequence would be unable to bind, so RNA polymerase could not bind and transcription of that gene probably would not occur.

Concept Check 17.3

1. Due to alternative splicing of exons, each gene can result in multiple different mRNAs and can thus direct synthesis of multiple different proteins. **2.** In editing a video, segments are cut out and discarded (like introns), and the remaining segments are joined together (like exons) so that the regions of joining ("splicing") are not noticeable. **3.** Once the mRNA has exited the nucleus, the cap prevents it from being degraded by hydrolytic enzymes and facilitates its attachment to ribosomes. If the cap were removed from all mRNAs, the cell would no longer be able to synthesize any proteins and would probably die.

Concept Check 17.4

1. First, each aminoacyl-tRNA synthetase specifically recognizes a single amino acid and attaches it only to an appropriate tRNA. Second, a tRNA charged with its specific amino acid binds only to an mRNA codon for that amino acid. **2.** The structure and function of the ribosome seem to depend more on the rRNAs than on the ribosomal proteins. Because it is single-stranded, an RNA molecule can hydrogen-bond with itself and with other RNA molecules. RNA molecules make up the interface between the two ribosomal subunits, so presumably RNA-RNA binding helps hold the ribosome together. The binding site for mRNA in the ribosome includes rRNA that can bind the mRNA. Also, complementary bonding within an RNA molecule allows it to assume a particular three-dimensional shape and, along with the RNA's functional groups, presumably enables rRNA to catalyze peptide bond formation during translation. **3.** A signal peptide on the leading end of the polypeptide being synthesized is recognized by a signal-recognition particle that brings the ribosome to the ER membrane. There the ribosome attaches and continues to synthesize the polypeptide, depositing it in the ER lumen. **4.** Because of wobble, the tRNA could bind to either 5′-GCA-3′ or 5′-GCG-3′, both of which code for alanine (Ala). Alanine would be attached to the tRNA.

Concept Check 17.5

1. In the mRNA, the reading frame downstream from the deletion is shifted, leading to a long string of incorrect amino acids in the polypeptide, and in most cases, a stop codon will arise, leading to premature termination. The polypeptide will most likely be nonfunctional. **2.** Heterozygous individuals, said to have sickle-cell trait, have a copy each of the wild-type allele and the sickle-cell allele. Both alleles will be expressed, so these individuals will have both normal and sickle-cell hemoglobin molecules. Apparently, having a mix of the two forms of β-globin has no effect under most conditions, but during prolonged periods of low blood oxygen (such as at higher altitudes), these individuals can show some signs of sickle-cell disease.

3.

Normal DNA sequence (template strand is on top):	3′-TACTTGTCCGATATC-5′ 5′-ATGAACAGGCTATAG-3′
mRNA sequence:	5′-AUGAACAGGCUAUAG-3′
Amino acid sequence:	Met-Asn-Arg-Leu-STOP
Mutated DNA sequence (template strand is on top):	3′-TACTTGTCCAATATC-5′ 5′-ATGAACAGGTTATAG-3′
mRNA sequence:	5′-AUGAACAGGUUAUAG-3′
Amino acid sequence:	Met-Asn-Arg-Leu-STOP

No effect: The amino acid sequence is Met-Asn-Arg-Leu both before and after the mutation because the mRNA codons 5′-CUA-3′ and 5′-UUA-3′ both code for Leu. (The fifth codon is a stop codon.)

Concept Check 17.6

1. No, transcription and translation are separated in space and time in a eukaryotic cell, a result of the eukaryotic cell's nuclear compartment. **2.** When one ribosome terminates translation and dissociates, the two subunits would be very close to the cap. This could facilitate their rebinding and initiating synthesis of a new polypeptide, thus increasing the efficiency of translation.

Summary of Key Concepts Questions

17.1 A gene contains genetic information in the form of a nucleotide sequence. The gene is first transcribed into an RNA molecule, and a messenger RNA molecule is ultimately translated into a polypeptide. The polypeptide makes up part or all of a protein, which performs a function in the cell and contributes to the phenotype of the organism. **17.2** Both bacterial and eukaryotic genes have promoters, regions where RNA polymerase ultimately binds and begins transcription. In bacteria, RNA polymerase binds directly to the promoter; in eukaryotes, transcription factors bind first to the promoter, and then RNA polymerase binds to the transcription factors and promoter together. **17.3** Both the 5′ cap and the poly-A tail help the mRNA exit from the nucleus and then, in the cytoplasm, help ensure mRNA stability and allow it to bind to ribosomes. **17.4** tRNAs function as translators between the nucleotide-based language of mRNA and the amino-acid-based language of polypeptides. A tRNA carries a specific amino acid, and the anticodon on the tRNA is complementary to the codon on the mRNA that codes for that amino acid. In the ribosome, the tRNA binds to the A site, where the polypeptide being synthesized is joined to the new amino acid, which becomes the new (C-terminal) end of the polypeptide. Next, the tRNA moves to the P site. When the next amino acid is added via transfer of the polypeptide to the new tRNA, the now empty tRNA moves to the E site, where it exits the ribosome. **17.5** When a nucleotide base is altered chemically, its base-pairing characteristics may be changed. When that happens, an incorrect nucleotide is likely to be incorporated into the complementary strand during the next replication of the DNA, and successive rounds of replication will perpetuate the mutation. Once the gene is transcribed, the mutated codon may code for a different amino acid that inhibits or changes the function of a protein. If the chemical change in the base is detected and repaired by the DNA repair system before the next replication, no mutation will result. **17.6** The presence of a nuclear envelope in eukaryotes means that transcription and translation are separated in space and therefore in time. This separation allows other processes (specifically, RNA processing) to occur and provides other steps at which gene expression can be regulated.

Test Your Understanding

1. b **2.** d **3.** a **4.** a **5.** b **6.** d **7.** e

8.

Type of RNA	Functions
Messenger RNA (mRNA)	Carries information specifying amino acid sequences of proteins from DNA to ribosomes
Transfer RNA (tRNA)	Serves as translator molecule in protein synthesis; translates mRNA codons into amino acids
Ribosomal RNA (rRNA)	Plays catalytic (ribozyme) roles and structural roles in ribosomes
Primary transcript	Is a precursor to mRNA, rRNA, or tRNA, before being processed; some intron RNA acts as a ribozyme, catalyzing its own splicing
Small nuclear RNA (snRNA)	Plays structural and catalytic roles in spliceosomes, the complexes of protein and RNA that splice pre-mRNA

Chapter 18

Figure Questions

Figure 18.3 As the concentration of tryptophan in the cell falls, eventually there will be none bound to repressor molecules. These will then take on their inactive shapes and dissociate from the operator, allowing transcription of the operon to resume. The enzymes for tryptophan synthesis will be made, and they will begin to synthesize tryptophan again in the cell. **Figure 18.11** The albumin gene enhancer has the three control elements colored yellow, gray, and red. The sequences in the liver and lens cells would be identical, since the cells are in the same organism. **Figure 18.18** Even if the mutant MyoD protein couldn't activate the *myoD* gene, it could still turn on genes for the other proteins in the pathway (other transcription factors, which would turn on the genes for muscle-specific proteins, for example). Therefore, some differentiation would occur. But unless there were other activators that could compensate for the loss of the MyoD protein's activation of the *myoD* gene, the cell would not be able to maintain its differentiated state. **Figure 18.22** Normal Bicoid protein would be made in the anterior end and compensate for the presence of mutant *bicoid* mRNA put into the egg by the mother. Development should be normal, with a head present. **Figure 18.24** The mutation is likely to be recessive because it is more likely to have an effect if both copies of the gene are mutated and code for nonfunctional proteins. If one normal copy of the gene is present, its product could inhibit the cell cycle. (However, there are also known cases of dominant *p53* mutations.)

Concept Check 18.1

1. Binding by the *trp* corepressor (tryptophan) activates the *trp* repressor, shutting off transcription of the *trp* operon; binding by the *lac* inducer (allolactose) inactivates the *lac* repressor, leading to transcription of the *lac* operon. **2.** When glucose is scarce, cAMP is bound to CAP and CAP is bound to the promoter, favoring the binding of RNA polymerase. However, in the absence of lactose, the repressor is bound to the operator, blocking RNA polymerase binding to the promoter. Therefore, the operon genes are not transcribed. **3.** The cell would continuously produce β-galactosidase and the two other enzymes for lactose utilization, even in the absence of lactose, thus wasting cell resources.

Concept Check 18.2

1. Histone acetylation is generally associated with gene expression, while DNA methylation is generally associated with lack of expression. **2.** General transcription factors function in assembling the transcription initiation complex at the promoters for all genes. Specific transcription factors bind to control elements associated with a particular gene and, once bound, either increase (activators) or decrease (repressors) transcription of that gene. **3.** The three genes should have some similar or identical sequences in the control elements of their enhancers. Because of this similarity, the same specific transcription factors in muscle cells could bind to the enhancers of all three genes and stimulate their expression coordinately. **4.** Degradation of the mRNA, regulation of translation, activation of the protein (by chemical modification, for example), and protein degradation **5.** Expression of the gene encoding the yellow activator (YA) must be regulated at one of the steps shown in Figure 18.6. The YA gene might be transcribed only in liver cells because the necessary activators for the enhancer of the YA gene are found only in liver cells.

Concept Check 18.3

1. Both miRNAs and siRNAs are small, single-stranded RNAs that associate with a complex of proteins and then can base-pair with mRNAs that have a complementary sequence. This base pairing leads to either degradation of the mRNA or blockage of its translation. Some siRNAs, in association with other proteins, can bind back to the chromatin in a certain region, causing chromatin changes that affect transcription. Both miRNAs and siRNAs are processed from double-stranded RNA precursors by the enzyme Dicer. All miRNAs are specified by genes in the cell's genome, and the single transcript folds back on itself to form one or more double-stranded hairpins, each of which is processed into an miRNA. In contrast, siRNAs arise from a longer stretch of linear double-stranded RNA, which may be introduced into the cell by a virus or an experimenter. Alternatively, in some cases, a cellular gene specifies one RNA strand of the precursor molecule, and an enzyme then synthesizes the complementary strand. **2.** The mRNA would persist and be translated into the cell division–promoting protein, and the cell would probably divide. If the intact miRNA is necessary for inhibition of cell division, then division of this cell might be inappropriate. Uncontrolled cell division could lead to formation of a mass of cells (tumor) that prevents proper functioning of the organism and could contribute to the development of cancer. **3.** The *XIST* RNA is transcribed from the *XIST* gene on the X chromosome that will be inactivated. It then binds to that chromosome and induces heterochromatin formation. A likely model is that the *XIST* RNA somehow recruits chromatin modification enzymes that lead to formation of heterochromatin.

Concept Check 18.4

1. Cells undergo differentiation during embryonic development, becoming different from each other; in the adult organism, there are many highly specialized cell types. **2.** By binding to a receptor on the receiving cell's surface and triggering a signal transduction pathway, involving intracellular molecules such as second messengers and transcription factors that affect gene expression **3.** Because their products, made and deposited into the egg by the mother, determine the head and tail ends, as well as the back and belly, of the embryo (and eventually the adult fly) **4.** The lower cell is synthesizing signaling molecules because the gene encoding them is activated, meaning that the appropriate specific transcription factors are binding to the gene's enhancer. The genes encoding these specific transcription factors are also being expressed in this cell because the transcriptional activators that can turn them on were expressed in the precursor to this cell. A similar explanation also applies to the cells expressing the receptor proteins. This scenario began with specific cytoplasmic determinants localized in specific regions of the egg. These cytoplasmic determinants were distributed unevenly to daughter cells, resulting in cells going down different developmental pathways.

Concept Check 18.5

1. Apoptosis is signaled by p53 protein when a cell has extensive DNA damage, so apoptosis plays a protective role in eliminating a cell that might contribute to cancer. If mutations in the genes in the apoptotic pathway blocked apoptosis, a cell with such damage could continue to divide and might lead to tumor formation. **2.** When an individual has inherited an oncogene or a mutant allele of a tumor-suppressor gene **3.** A cancer-causing mutation in a proto-oncogene usually makes the gene product overactive, whereas a cancer-causing mutation in a tumor-suppressor gene usually makes the gene product nonfunctional.

Summary of Key Concepts Questions

18.1 A corepressor and an inducer are both small molecules that bind to the repressor protein in an operon, causing the repressor to change shape. In the case of a corepressor (like tryptophan), this shape change allows the repressor to bind to the operator, blocking transcription. In contrast, an inducer causes the repressor to dissociate from the operator, allowing transcription to begin. **18.2** The chromatin must not be tightly condensed because it must be accessible to transcription factors. The appropriate specific transcription factors (activators) must bind to the control elements in the enhancer of the gene, while repressors must not be bound. The DNA must be bent by a bending protein so the activators can contact the mediator proteins and form a complex with general transcription factors at the promoter. Then RNA polymerase must bind and begin transcription. **18.3** miRNAs do not "code" for the amino acids of a protein—they are never translated. Each miRNA is cleaved from its hairpin RNA structure and then trimmed by Dicer. Next, one strand is degraded while the other associates with a group of proteins to form a complex. Binding of the complex to an mRNA with a complementary sequence causes that mRNA to be degraded or blocks its translation. This is considered gene regulation because it controls the amount of a particular mRNA that can be translated into a functional protein. **18.4** The first process involves cytoplasmic determinants, including mRNAs and proteins, placed into specific locations in the egg by the mother. The cells that are formed from different regions in the egg during early cell divisions will have different proteins in them, which will direct different programs of gene expression. The second process involves the cell in question responding to signaling molecules secreted by neighboring cells. The signaling pathway in the responding cell also leads to a different pattern of gene expression. The coordination of these two processes results in each cell following a unique pathway in the developing embryo. **18.5** The protein product of a proto-oncogene is usually involved in a pathway that stimulates cell division. The protein product of a tumor-suppressor gene is usually involved in a pathway that inhibits cell division.

Test Your Understanding

1. d **2.** a **3.** a **4.** a **5.** c **6.** d **7.** c **8.** e **9.** b **10.** b
11. a.

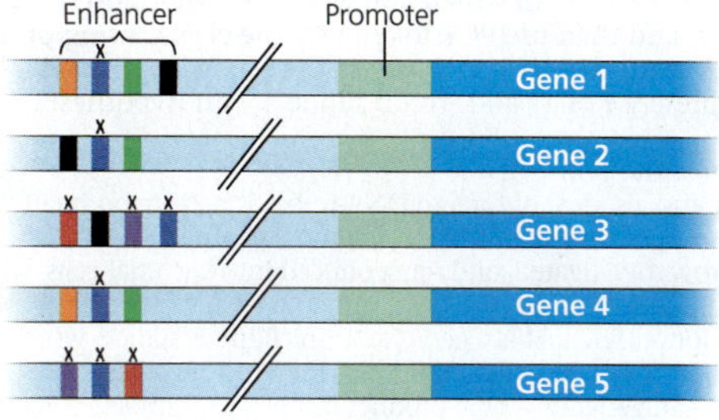

The purple, blue, and red activator proteins would be present.
b.

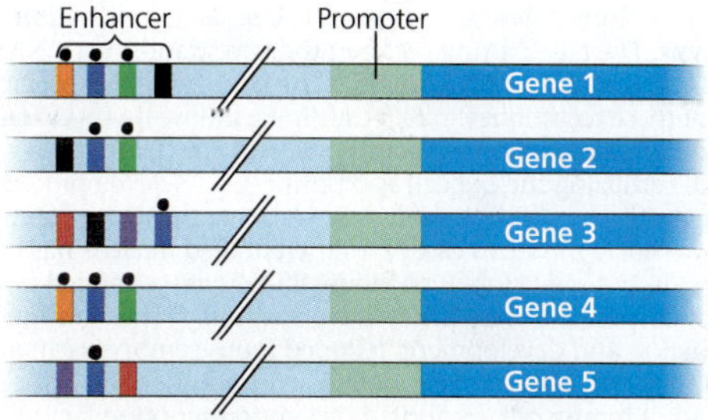

Only gene 4 would be transcribed.
c. In nerve cells, the orange, blue, green, and black activators would have to be present, thus activating transcription of genes 1, 2, and 4. In skin cells, the red, black, purple, and blue activators would have to be present, thus activating genes 3 and 5.

Chapter 19

Figure Questions

Figure 19.2 Beijerinck might have concluded that the agent was a toxin produced by the plant that was able to pass through a filter but that became more and more dilute. In this case, he would have concluded that the infectious agent could not replicate. **Figure 19.4** Top vertical arrow: Infection. Left upper arrow: Replication. Right upper arrow: Transcription. Right middle arrow: Translation. Lower left and right arrows: Self-assembly. Bottom middle arrow: Exit. **Figure 19.7** Any class V virus, including the viruses that cause influenza (flu), measles, and mumps **Figure 19.8** The main protein on the cell surface that HIV binds to is called CD4. However, HIV also requires a "co-receptor," which in many cases is a protein called CCR5. HIV binds to both of these proteins together and then is taken into the cell. Researchers discovered this requirement by studying individuals who seemed to be resistant to HIV infection, despite multiple exposures. These individuals turned out to have mutations in the gene that encodes CCR5 such that the protein apparently cannot act as a co-receptor, and so HIV can't enter and infect cells.

Concept Check 19.1

1. TMV consists of one molecule of RNA surrounded by a helical array of proteins. The influenza virus has eight molecules of RNA, each surrounded by a helical array of proteins, similar to the arrangement of the single RNA molecule in TMV. Another difference between the viruses is that the influenza virus has an outer envelope and TMV does not. **2.** The T2 phages were an excellent choice for use in the Hershey-Chase experiment because they consist of only DNA surrounded by a protein coat, and DNA and protein were the two candidates for macromolecules that carried genetic information. Hershey and Chase were able to radioactively label each type of molecule alone and follow it during separate infections of *E. coli* cells with T2. Only the DNA entered the bacterial cell during infection, and only labeled DNA showed up in some of the progeny phage. Hershey and Chase concluded that the DNA must carry the genetic information necessary for the phage to reprogram the cell and produce progeny phages.

Concept Check 19.2

1. Lytic phages can only carry out lysis of the host cell, whereas lysogenic phages may either lyse the host cell or integrate into the host chromosome. In the latter case, the viral DNA (prophage) is simply replicated along with the host chromosome. Under certain conditions, a prophage may exit the host chromosome and initiate a lytic cycle. **2.** Both the viral RNA polymerase and the RNA polymerase in Figure 17.9 synthesize an RNA molecule complementary to a template strand. However, the RNA polymerase in Figure 17.9 uses one of the strands of the DNA double helix as a template, whereas the viral RNA polymerase uses the RNA of the viral genome as a template. **3.** Because it synthesizes DNA from its RNA genome. This is the reverse ("retro") of the usual DNA → RNA information flow. **4.** There are many steps that could be interfered with: binding of the virus to the cell, reverse transcriptase function, integration into the host cell chromosome, genome synthesis (in this case, transcription of RNA from the integrated provirus), assembly of the virus inside the cell, and budding of the virus. (Many of these, if not all, are targets of actual medical strategies to block progress of the infection in HIV-infected people.)

Concept Check 19.3

1. Mutations can lead to a new strain of a virus that can no longer be effectively fought by the immune system, even if an animal had been exposed to the original strain; a virus can jump from one species to a new host; and a rare virus can spread if a host population becomes less isolated. **2.** In horizontal transmission, a plant is infected from an external source of virus, which could enter through a break in the plant's epidermis due to damage by herbivores. In vertical transmission, a plant inherits viruses from its parent either via infected seeds (sexual reproduction) or via an infected cutting (asexual reproduction). **3.** Humans are not within the host range of TMV, so they can't be infected by the virus.

Summary of Key Concepts Questions

19.1 Viruses are generally considered nonliving, because they are not capable of replicating outside of a host cell. To replicate, they depend completely on host enzymes and resources. **19.2** Single-stranded RNA viruses require an RNA polymerase that can make RNA using an RNA template. (Cellular RNA polymerases make RNA using a DNA template.) Retroviruses require reverse transcriptases to make DNA using an RNA template. (Once the first DNA strand has been made, the same enzyme can promote synthesis of the second DNA strand.) **19.3** The mutation rate of RNA viruses is higher than that of DNA viruses because RNA polymerase has no proofreading function, so errors in replication are not corrected. Their higher mutation rate means that RNA viruses change faster than DNA viruses, leading to their being able to have an altered host range and to evade immune defenses in possible hosts.

Test Your Understanding

1. c **2.** d **3.** c **4.** d **5.** b
6. As shown below, the viral genome would be translated into capsid proteins and envelope glycoproteins directly, rather than after a complementary RNA copy was made. A complementary RNA strand would still be made, however, that could be used as a template for many new copies of the viral genome.

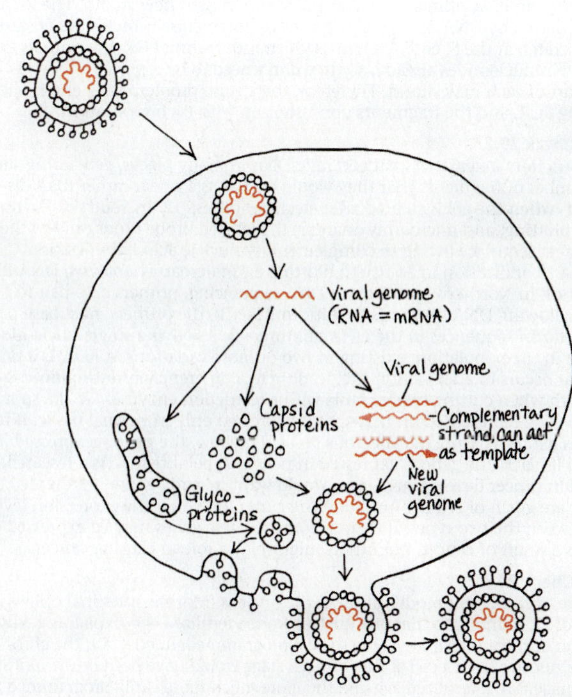

Appendix A Answers

Chapter 20

Figure Questions

Figure 20.3

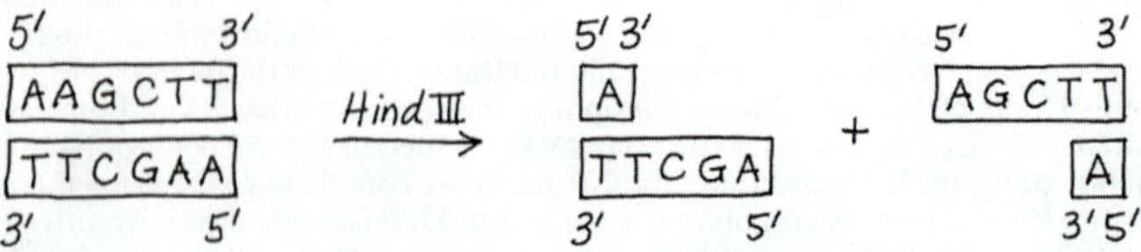

Figure 20.4 Cells containing no plasmid at all would be able to grow; these colonies would be white because they would lack functional *lacZ* genes. **Figure 20.10** Grow each clone of cells in culture. Isolate the plasmids from each and cut them with the restriction enzyme originally used to make the clone (see Figure 20.4). Run each sample on an electrophoretic gel, and recover the DNA of the insert from the gel band. **Figure 20.16** Crossing over, which causes recombination, is a random event. The chance of crossing over occurring between two loci increases as the distance between them increases. The SNP is located very close to an unknown disease-causing allele, and therefore crossing over rarely occurs between the SNP and the allele, so the SNP is a genetic marker indicating the presence of the particular allele. **Figure 20.18** None of the eggs with the transplanted nuclei from the four-cell embryo at the upper left would have developed into a tadpole. Also, the result might include only some of the tissues of a tadpole, which might differ, depending on which nucleus was transplanted. (This assumes that there was some way to tell the four cells apart, as one can in some frog species.) **Figure 20.22** Using converted iPS cells would not carry the same risk, which is its major advantage. Because the donor cells would come from the patient, they would be perfectly matched. The patient's immune system would recognize them as "self" cells and would not mount an attack (which is what leads to rejection).

Concept Check 20.1

1. The covalent sugar-phosphate bonds of the DNA strands **2.** Yes, *Pvu*I will cut the molecule.

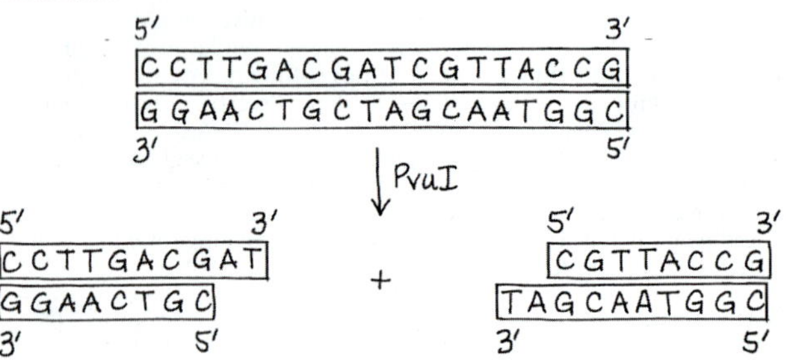

3. Some human genes are too large to be incorporated into bacterial plasmids. Bacterial cells lack the means to process RNA transcripts into mRNA, and even if the need for RNA processing is avoided by using cDNA, bacteria lack enzymes to catalyze the post-translational processing that many human proteins require to function properly. **4.** During the replication of the ends of linear DNA molecules (see Figure 16.20), an RNA primer is used at the 5' end of each new strand. The RNA must be replaced by DNA nucleotides, but DNA polymerase is incapable of starting from scratch at the 5' end of a new DNA strand. During PCR, the primers are made of DNA nucleotides already, so they don't need to be replaced—they just remain as part of each new strand. Therefore, there is no problem with end replication during PCR, and the fragments don't shorten with each replication.

Concept Check 20.2

1. Any restriction enzyme will cut genomic DNA in many places, generating such a large number of fragments that they would appear as a smear rather than distinct bands when the gel is stained after electrophoresis. **2.** In Southern blotting, Northern blotting, and microarray analysis, the labeled probe binds only to the specific target sequence owing to complementary nucleic acid hybridization (DNA-DNA hybridization in Southern blotting and microarray analysis, DNA-RNA hybridization in Northern blotting). In DNA sequencing, primers base-pair to the template, allowing DNA synthesis to start. In RT-PCR, the primers must base-pair with their target sequences in the DNA mixture. **3.** A SNP is a single nucleotide that varies in the population, existing in two or more variations. A RFLP is a type of SNP that occurs in a restriction site, leading to a difference in restriction fragment length when cutting two variants with a restriction enzyme. **4.** If a spot is green, the gene represented on that spot is expressed only in normal tissue. If red, the gene is expressed only in cancerous tissue. If yellow, the gene is expressed in both. And if black, the gene is expressed in neither type of tissue. As a researcher interested in cancer development, you would want to study genes represented by spots that are green or red because these are genes for which the expression level differs between the two types of tissues. Some of these genes may be expressed differently as a result of cancer, but others might play a role in causing cancer.

Concept Check 20.3

1. The state of chromatin modification in the nucleus from the intestinal cell was undoubtedly less similar to that of a nucleus from a fertilized egg, explaining why many fewer of these nuclei were able to be reprogrammed. In contrast, the chromatin in a nucleus from a cell at the four-cell stage would have been much more like that of a nucleus in a fertilized egg and therefore much more easily programmed to direct development. **2.** No, primarily because of subtle (and perhaps not so subtle) differences in their environments **3.** A technique would have to be worked out for turning a human iPS cell into a pancreatic cell (probably by inducing expression of pancreas-specific regulatory genes in the cell). **4.** The carrot cell has much more potential. The cloning experiment shows that an individual carrot cell can generate all the tissues of an adult plant. The muscle cell, on the other hand, will always remain a muscle cell because of its genetic program (it expresses the *myoD* gene, which ensures continued differentiation). The muscle cell is like other fully differentiated animal cells: It will remain fully differentiated on its own unless it is reprogrammed into an iPS cell using the new techniques described here. (This would be quite difficult to accomplish because a muscle cell has multiple nuclei.)

Concept Check 20.4

1. Stem cells continue to reproduce themselves. **2.** Herbicide resistance, pest resistance, disease resistance, salinity resistance, and delayed ripening **3.** Because hepatitis A is an RNA virus, you could isolate RNA from the blood and try to detect copies of hepatitis A RNA by one of three methods. First, you could run the RNA on a gel and then do a Northern blot using probes complementary to hepatitis A genome sequences. A second approach would be to use reverse transcriptase to make cDNA from the RNA in the blood, run the cDNA on a gel, and do a Southern blot using the same probe. However, neither of these methods would be as sensitive as RT-PCR, in which you would reverse-transcribe the blood RNA into cDNA and then use PCR to amplify the cDNA, using primers specific to hepatitis A sequences. If you then ran the products on an electrophoretic gel, the presence of a band would support your hypothesis.

Summary of Key Concepts Questions

20.1 A plasmid vector and a source of foreign DNA to be cloned are both cut with the same restriction enzyme, generating restriction fragments with sticky ends. These fragments are mixed together, ligated, and reintroduced into bacterial cells, which are then grown on the antibiotic ampicillin. The plasmid has two genes that allow selection of recombinant clones. The first is a gene for ampicillin resistance, which only allows cells to grow if they have taken up a plasmid. The second is a gene for β-galactosidase, which can generate a blue product if the gene is intact. The cloning site is within this gene, so only nonrecombinant colonies will be blue, and recombinant plasmids will be found in cells that are in white colonies. **20.2** Many techniques used to analyze genes and their expression involve nucleic acid hybridization: Southern and Northern blotting, DNA sequencing, PCR, *in situ* hybridization, and DNA microarray analysis. The base pairing between the two strands of a DNA molecule or between a DNA strand and an RNA strand is the key to finding specific nucleic acid sequences in all of these techniques. **20.3** Cloning a mouse involves transplanting a nucleus from a differentiated mouse cell into a mouse egg cell that has had its own nucleus removed. Fertilizing the egg cell and promoting its development into an embryo in a surrogate mother results in a mouse that is genetically identical to the mouse that donated the nucleus. In this case, the differentiated nucleus has been reprogrammed by factors in the egg cytoplasm. Mouse ES cells are generated from inner cells in mouse blastocysts, so in this case the cells are "naturally" reprogrammed by the process of reproduction and development. (Cloned mouse embryos can also be used as a source of ES cells.) iPS cells can be generated without the use of embryos from a differentiated adult mouse cell, by adding certain transcription factors into the cell. In this case, the transcription factors are reprogramming the cells to become pluripotent. **20.4** First, the disease must be caused by a single gene, and the molecular basis of the problem must be understood. Second, the cells that are going to be introduced into the patient must be cells that will integrate into body tissues and continue to multiply (and provide the needed gene product). Third, the gene must be able to be introduced into the cells in question in a safe way, as there have been instances of cancer resulting from some gene therapy trials. (Note that this will require testing the procedure in mice; moreover, the factors that determine a safe vector are not yet well understood. Maybe one of you will go on to solve this problem!)

Test Your Understanding

1. b **2.** c **3.** b **4.** c **5.** d **6.** b **7.** a **8.** c
9.

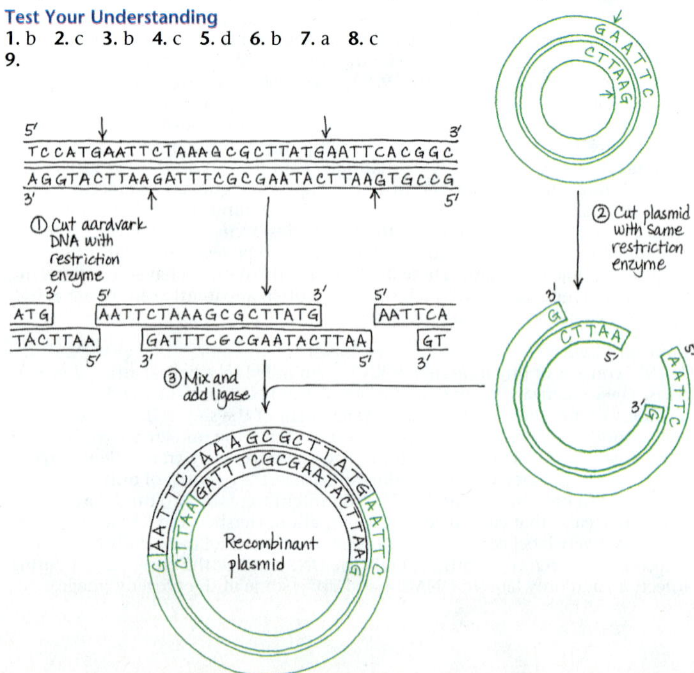

10. A cDNA library, made using mRNA from human lens cells, which would be expected to contain many copies of mRNA for the crystallin of interest

Chapter 21

Figure Questions

Figure 21.3 The fragments in stage 2 of this figure are like those in stage 2 of Figure 21.2, but in this figure their order relative to each other is not known and will be determined later by computer. The order of the fragments in Figure 21.2 is completely known before sequencing begins. (Determining the order takes longer but makes the eventual sequence assembly much easier.) **Figure 21.9** The transposon would be cut out of the DNA at the original site rather than copied, so the figure would show the original stretch of DNA without the transposon after the mobile transposon had been cut out. **Figure 21.11** The RNA transcripts extending from the DNA in each transcription unit are shorter on the left and longer on the right. This means that RNA polymerase must be starting on the left end of the unit and moving toward the right.
Figure 21.13

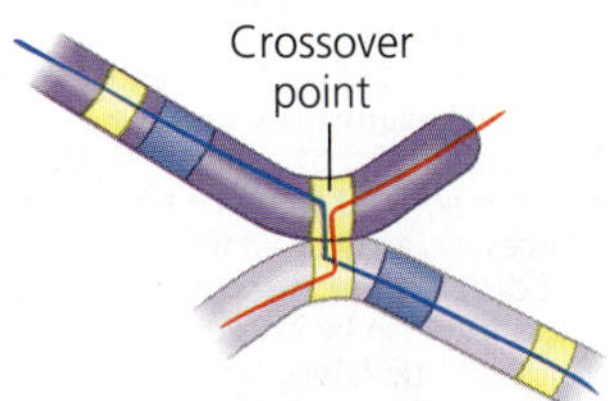

Figure 21.14 Pseudogenes are nonfunctional. They could have arisen by any mutations in the second copy that made the gene product unable to function. Examples would be base changes that introduce stop codons in the sequence, alter amino acids, or change a region of the gene promoter so that the gene can no longer be expressed. **Figure 21.15** Let's say a transposable element (TE) existed in the intron to the left of the indicated EGF exon in the EGF gene, and the same TE was present in the intron to the right of the indicated F exon in the fibronectin gene. During meiotic recombination, these TEs could cause nonsister chromatids on homologous chromosomes to pair up incorrectly, as seen in Figure 21.13. One gene might end up with an F exon next to an EGF exon. Further mistakes in pairing over many generations might result in these two exons being separated from the rest of the gene and placed next to a single or duplicated K exon. In general, the presence of repeated sequences in introns and between genes facilitates these processes because it allows incorrect pairing of nonsister chromatids, leading to novel exon combinations. **Figure 21.17** Since you know that chimpanzees do not speak but humans do, you'd probably want to know how many amino acid differences there are between the human wild-type FOXP2 protein and that of the chimpanzee and whether these changes affect the function of the protein. (As we explain later in the text, there are two amino acid differences.) You know that humans with mutations in this gene have severe language impairment. You would want to learn more about the human mutations by checking whether they affect the same amino acids in the gene product that the chimpanzee sequence differences affect. If so, those amino acids might play an important role in the function of the protein in language. Going further, you could analyze the differences between the chimpanzee and mouse FOXP2 proteins. You might ask: Are they more similar than the chimpanzee and human proteins? (It turns out that the chimpanzee and mouse proteins have only one amino acid difference and thus are more similar than the chimpanzee and human proteins, which have two differences, and more similar than the human and mouse proteins, which have three differences.)

Concept Check 21.1

1. In a linkage map, genes and other markers are ordered with respect to each other, but only the relative distances between them are known. In a physical map, the actual distances between markers, expressed in base pairs, are known. **2.** The three-stage approach employed in the Human Genome Project involves linkage mapping, physical mapping, and then sequencing of short, overlapping fragments that previously have been ordered relative to each other (see Figure 21.2). The whole-genome shotgun approach eliminates the linkage mapping and physical mapping stages; instead, short fragments generated by multiple restriction enzymes are sequenced and then ordered by computer programs that identify overlapping regions (see Figure 21.3).

Concept Check 21.2

1. The Internet allows centralization of databases such as GenBank and software resources such as BLAST, making them freely accessible. Having all the data in a central database, easily accessible on the Internet, minimizes the possibility of errors and of researchers working with different data. It streamlines the process of science, since all researchers are able to use the same software programs, rather than each having to obtain their own, possibly different, software. It speeds up dissemination of data and ensures as much as possible that errors are corrected in a timely fashion. These are just a few answers; you can probably think of more. **2.** Cancer is a disease caused by multiple factors. To focus on a single gene or a single defect would ignore other factors that may influence the cancer and even the behavior of the single gene being studied. The systems approach, because it takes into account many factors at the same time, is more likely to lead to an understanding of the causes and most useful treatments for cancer. **3.** Some of the transcribed region is accounted for by introns. The rest is transcribed into noncoding RNAs, including small RNAs, such as microRNAs (miRNAs). These RNAs help regulate gene expression by blocking translation, causing degradation of mRNA, binding to the promoter and repressing transcription, or causing remodeling of chromatin structure. The functions of the remainder are not yet known. **4.** Genome-wide association studies use the systems biology approach in that they consider the correlation of many single nucleotide polymorphisms (SNPs) with particular diseases, such as heart disease and diabetes, in an attempt to find patterns of SNPs that correlate with each disease.

Concept Check 21.3

1. Alternative splicing of RNA transcripts from a gene and post-translational processing of polypeptides **2.** The total number of completed genomes is found by clicking on "Published Complete Genomes." Add the figures for bacterial, archaeal, and eukaryotic "ongoing genomes" to get the number "in progress." Finally, look at the top of the Published Complete Genomes page to get numbers of completed genomes for each domain. (*Note*: You can click on the "Size" column and the table will be re-sorted by genome size. Scroll down to get an idea of relative sizes of genomes in the three domains. Remember, though, that most of the sequenced genomes are bacterial.) **3.** Prokaryotes are generally smaller cells than eukaryotic cells, and they reproduce by binary fission. The evolutionary process involved is natural selection for more quickly reproducing cells: The faster they can replicate their DNA and divide, the more likely they will be able to dominate a population of prokaryotes. The less DNA they have to replicate, then, the faster they will reproduce.

Concept Check 21.4

1. The number of genes is higher in mammals, and the amount of noncoding DNA is greater. Also, the presence of introns in mammalian genes makes them larger, on average, than prokaryotic genes. **2.** In the copy-and-paste transposon mechanism and in retrotransposition **3.** In the rRNA gene family, identical transcription units for the three different RNA products are present in long, tandemly repeated arrays. The large number of copies of the rRNA genes enable organisms to produce the rRNA for enough ribosomes to carry out active protein synthesis, and the single transcription unit ensures that the relative amounts of the different rRNA molecules produced are correct. Each globin gene family consists of a relatively small number of nonidentical genes. The differences in the globin proteins encoded by these genes result in production of hemoglobin molecules adapted to particular developmental stages of the organism. **4.** The exons would be classified as exons (1.5%); the enhancer region containing the distal control elements, the region closer to the promoter containing the proximal control elements, and the promoter itself would be classified as regulatory sequences (5%); and the introns would be classified as introns (20%).

Concept Check 21.5

1. If meiosis is faulty, two copies of the entire genome can end up in a single cell. Errors in crossing over during meiosis can lead to one segment being duplicated while another is deleted. During DNA replication, slippage backward along the template strand can result in segment duplication. **2.** For either gene, a mistake in crossing over during meiosis could have occurred between the two copies of that gene, such that one ended up with a duplicated exon. This could have happened several times, resulting in the multiple copies of a particular exon in each gene. **3.** Homologous transposable elements scattered throughout the genome provide sites where recombination can occur between different chromosomes. Movement of these elements into coding or regulatory sequences may change expression of genes. Transposable elements also can carry genes with them, leading to dispersion of genes and in some cases different patterns of expression. Transport of an exon during transposition and its insertion into a gene may add a new functional domain to the originally encoded protein, a type of exon shuffling. (For any of these changes to be heritable, they must happen in germ cells, cells that will give rise to gametes.) **4.** Because more offspring are born to women who have this inversion, it must provide some advantage. It would be expected to persist and spread in the population. (In fact, evidence in the study allowed the researchers to conclude that it has been increasing in proportion in the population. You'll learn more about population genetics in the next unit.)

Concept Check 21.6

1. Because both humans and macaques are primates, their genomes are expected to be more similar than the macaque and mouse genomes are. The mouse lineage diverged from the primate lineage before the human and macaque lineages diverged. **2.** Homeotic genes differ in their *non*homeobox sequences, which determine the interactions of homeotic gene products with other transcription factors and hence which genes are regulated by the homeotic genes. These nonhomeobox sequences differ in the two organisms, as do the expression patterns of the homeobox genes. **3.** *Alu* elements must have undergone transposition more actively in the human genome for some reason. Their increased numbers may have then allowed more recombination errors in the human genome, resulting in more or different duplications. The divergence of the organization and content of the two genomes presumably made the chromosomes of each genome less homologous to those of the other, thus accelerating divergence of the two species by making matings less and less likely to result in fertile offspring.

Summary of Key Concepts Questions

21.1 Considering the sequencing of the human genome as an example, less time was required to sequence the first human genome using the whole-genome shotgun approach. Although this approach relied in part on data resulting from the three-stage approach used by the public consortium, the whole-genome shotgun approach was (and still is) faster and more efficient than the more labor-intensive three-stage process. The whole-genome shotgun approach was facilitated in large part by significant advances in computing power. **21.2** The most significant finding was that more than 90% of the human genomic region studied was transcribed, which suggested that the transcribed RNA (and thus the DNA from which it was produced) was performing some unknown functions. The project has been expanded to include other species because to determine the functions of these transcribed DNA elements, it is necessary to carry out this type of analysis on the genomes of species that can be used in laboratory experiments. **21.3** (a) In general, bacteria and archaea have smaller genomes, lower numbers of genes, and higher gene density than eukaryotes. (b) Among eukaryotes, there is no apparent systematic relationship between genome size and phenotype. The number of genes is often lower than would be expected from the size of the genome—in other words, the gene density is often lower in larger genomes. (Humans are an example.) **21.4** Transposable element-related sequences can move from place to place in the genome, and a subset of these sequences make a new copy of themselves when they do so. Thus, it is not surprising that they make up a significant percentage of the genome, and this percentage might be expected to increase over evolutionary time. **21.5** Chromosomal rearrangements within a species lead to some individuals having different chromosomal arrangements. Each of these individuals could still undergo meiosis and produce gametes, and fertilization involving gametes with different chromosomal arrangements could result in viable offspring. However, during meiosis in the offspring, the maternal and paternal chromosomes might not be able to pair up, causing gametes with incomplete sets of chromosomes to form. Most often, when zygotes are produced from such gametes, they do not survive. Ultimately, a new species could form if two different chromosomal arrangements became prevalent within a population and individuals could mate successfully only with other individuals having the same arrangement. **21.6** Comparing the genomes of two closely related species can reveal information about more recent evolutionary events, perhaps events that resulted in the distinguishing characteristics of the two species. Comparing the genomes of very distantly related species can tell us about evolutionary events that occurred a very long time ago. For example, genes that are shared between two distantly related species must have arisen before the two species diverged.

Test Your Understanding

1. c **2.** a **3.** a **4.** c

5.

1. ATETI...PKSSD...TSSTT...NARRD
2. ATETI...PKSSE...TSSTT...NARRD
3. ATETI...PKSSD...TSSTT...NARRD
4. ATETI...PKSSD...TSSNT...SARRD
5. ATETI...PKSSD...TSSTT...NARRD
6. VTETI...PKSSD...TSSTT...NARRD

(a) Lines 1, 3, and 5 are the C, G, R species. (b) Line 4 is the human sequence. (c) Line 6 is the orangutan sequence. (d) There is one amino acid difference between the mouse (the E on line 2) and the C, G, R species (which have a D in that position). There are three amino acid differences between the mouse and the human. (The E, T, and N in the mouse sequence are instead D, N, and S, respectively, in the human sequence.) (e) Because only one amino acid difference arose during the 60–100 million years since the mouse and C, G, R species diverged, it is somewhat surprising that two additional amino acid differences resulted during the 6 million years since chimpanzees and humans diverged. This indicates that the *FOXP2* gene has been evolving faster in the human lineage than in the lineages of other primates.

Chapter 22

Figure Questions

Figure 22.6 The cactus-eater is more closely related to the seed-eater; Figure 1.22 shows that they share a more recent common ancestor (a seed-eater) than the cactus-eater shares with the insect-eater. **Figure 22.8** More than 5.5 million years ago. **Figure 22.12** The colors and body forms of these mantids allow them to blend into their surroundings, providing an example of how organisms are well matched to life in their environments. The mantids also share features with one another (and with all other mantids), such as six legs, grasping forelimbs, and large eyes. These shared features illustrate another key observation about life: the unity of life that results from descent from a common ancestor. Over time, as these mantids diverged from a common ancestor, they accumulated different adaptations that made them well suited for life in their different environments. Eventually, these differences became large enough that new species were formed, thus contributing to the great diversity of life. **Figure 22.13** These results show that being reared from the egg stage on one plant species or the other did not result in the adult having a beak length appropriate for that host; instead, adult beak lengths were determined primarily by the population from which the eggs were obtained. Because an egg from a balloon vine population likely had long-beaked parents, while an egg from a goldenrain tree population likely had short-beaked parents, these results indicate that beak length is an inherited trait. **Figure 22.14** Both strategies should increase the time it takes *S. aureus* to become resistant to a new drug. If a drug that harms *S. aureus* does not harm other bacteria, natural selection will not favor resistance to that drug in the other species. This would decrease the chance that *S. aureus* would acquire resistance genes from other bacteria—thus slowing the evolution of resistance. Similarly, selection for resistance to a drug that slows the growth but does not kill *S. aureus* is much weaker than selection for resistance to a drug the kills *S. aureus*—again slowing the evolution of resistance. **Figure 22.17** Based on this evolutionary tree, crocodiles are more closely related to birds than to lizards because they share a more recent common ancestor with birds (ancestor 5) than with lizards (ancestor 4). **Figure 22.20** Hind limb structure changed first. *Rodhocetus* lacked flukes, but its pelvic bones and hind limbs had changed substantially from how those bones were shaped and arranged in *Pakicetus*. For example, in *Rodhocetus*, the pelvis and hind limbs appear to be oriented for paddling, whereas they were oriented for walking in *Pakicetus*.

Concept Check 22.1

1. Hutton and Lyell proposed that events in the past were caused by the same processes operating today. This principle suggested that Earth must be much older than a few thousand years, the age that was widely accepted at that time. Hutton and Lyell also thought that geologic change occurs gradually, stimulating Darwin to reason that the slow accumulation of small changes could ultimately produce the profound changes documented in the fossil record. In this context, the age of Earth was important to Darwin, because unless Earth was very old, he could not envision how there would have been enough time for evolution to occur. **2.** By these criteria, Cuvier's explanation of the fossil record and Lamarck's hypothesis of evolution are both scientific. Cuvier thought that species did not evolve over time. He also suggested that catastrophes and the resulting extinctions were usually confined to local regions and that such regions were later repopulated by a different set of species that immigrated from other areas. These assertions can be tested against the fossil record, and his assertion that species do not evolve has been found to be false. With respect to Lamarck, his principle of use and disuse can be used to make testable predictions for fossils of groups such as whale ancestors as they adapted to a new habitat. Lamarck's principle of use and disuse and his associated principle of the inheritance of acquired characteristics can also be tested directly in living organisms (these principles have been found to be false).

Concept Check 22.2

1. Organisms share characteristics (the unity of life) because they share common ancestors; the great diversity of life occurs because new species have repeatedly formed when descendant organisms gradually adapted to different environments, becoming different from their ancestors. **2.** The fossil mammal species (or its ancestors) would most likely have colonized the Andes from within South America, whereas ancestors of mammals currently found in African mountains would most likely have colonized those mountains from other parts of Africa. As a result, the Andes fossil species would share a more recent common ancestor with South American mammals than with mammals in Africa. Thus, for many of its traits, the fossil mammal species would probably more closely resemble mammals that live in South American jungles than mammals that live on African mountains. It is also possible, however, that the fossil mammal species could resemble the African mountain mammals by convergent evolution (even though they were only distantly related to one another). **3.** As long as the white phenotype (encoded by the genotype *pp*) continues to be favored by natural selection, the frequency of the *p* allele will likely increase over time in the population. The explanation is that if the proportion of white individuals increases relative to purple individuals, the frequency of the recessive *p* allele will also increase relative to that of the *P* allele, which only appears in purple individuals (some of whom also carry a *p* allele).

Concept Check 22.3

1. An environmental factor such as a drug does not create new traits, such as drug resistance, but rather selects for traits among those that are already present in the population. **2.** (a) Despite their different functions, the forelimbs of different mammals are structurally similar because they all represent modifications of a structure found in the common ancestor. (b) Convergent evolution: The similarities between the sugar glider and flying squirrel indicate that similar environments selected for similar adaptations despite different ancestry. **3.** At the time that dinosaurs originated, Earth's landmasses formed a single large continent, Pangaea. Because many dinosaurs were large and mobile, it is likely that early members of these groups lived on many different parts of Pangaea. When Pangaea broke apart, fossils of these organisms would have moved with the rocks in which they were deposited. As a result, we would predict that fossils of early dinosaurs would have a broad geographic distribution (this prediction has been upheld).

Summary of Key Concepts Questions

Concept 22.1 Darwin thought that descent with modification occurred as a gradual, steplike process. The age of Earth was important to him because if Earth were only a few thousand years old (as conventional wisdom suggested), there wouldn't have been sufficient time for major evolutionary change. **Concept 22.2** All species have the potential to produce more offspring (overreproduce) than can be supported by the environment. This ensures that there will be what Darwin called a "struggle for existence" in which many of the offspring are eaten, starved, diseased, or unable to reproduce for a variety of other reasons. Members of a population exhibit a range of heritable variations, some of which make it likely that their bearers will leave more offspring than other individuals (for example, the bearer may escape predators more effectively or be more tolerant of the physical conditions of the environment). Over time, natural selection resulting from factors such as predators, lack of food, or the physical conditions of the environment can increase the proportion of individuals with favorable traits in a population (evolutionary adaptation). **Concept 22.3** The hypothesis that cetaceans originated from a terrestrial mammal and are closely related to even-toed ungulates is supported by several lines of evidence. For example, fossils document that early cetaceans had hind limbs, as expected for organisms that descended from a land mammal; these fossils also show that cetacean hind limbs became reduced over time. Other fossils show that early cetaceans had a type of ankle bone that is otherwise found only in even-toed ungulates, providing strong evidence that even-toed ungulates are the land mammals to which cetaceans are most closely related. DNA sequence data also indicate that even-toed ungulates are the land mammals to which cetaceans are most closely related.

Test Your Understanding
1. b **2.** d **3.** d **4.** c **5.** a
7. (a)

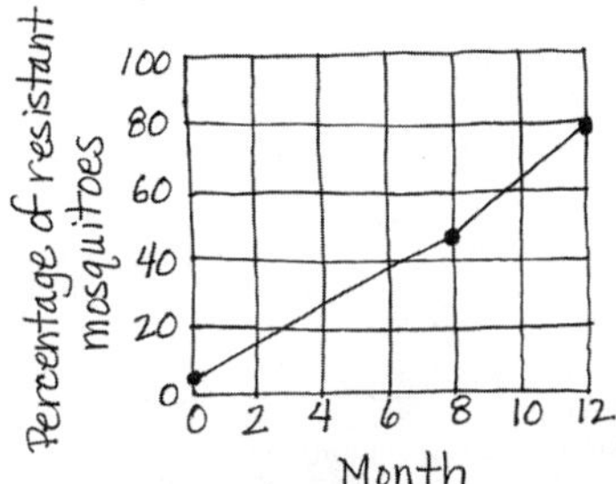

(b) The rapid rise in the percentage of mosquitoes resistant to DDT was most likely caused by natural selection in which mosquitoes resistant to DDT could survive and reproduce while other mosquitoes could not. (c) In India—where DDT resistance first appeared—natural selection would have caused the frequency of resistant mosquitoes to increase over time. If resistant mosquitoes then migrated from India (for example, transported by wind or in planes, trains, or ships) to other parts of the world, the frequency of DDT resistance would increase there as well.

Chapter 23

Figure Questions
Figure 23.8 The predicted frequencies are 36% C^RC^R, 48% C^RC^W, and 16% C^WC^W. **Figure 23.12** Local survival should increase in both populations. The increase would probably occur more rapidly among birds born into the central population because gene flow from the mainland is higher in that population. **Figure 23.13** Directional selection. Goldenrain tree has smaller fruit than does the native host, balloon vine. Thus, in soapberry bug populations feeding on goldenrain tree, bugs with shorter beaks had an advantage, resulting in directional selection for shorter beak length. **Figure 23.16** Crossing a single female's eggs with both an SC and an LC male's sperm allowed the researchers to directly compare the effects of the males' contribution to the next generation, since both batches of offspring had the same maternal contribution. This isolation of the male's impact enabled researchers to draw conclusions about differences in genetic "quality" between the SC and LC males. **Figure 23.18** The researchers measured the percentages of successfully reproducing adults in the breeding adult population that had each phenotype. This approach of determining which phenotype was favored by selection assumes that reproduction was a sufficient indicator of relative fitness (as opposed to counting the number of eggs laid or offspring hatched, for example) and that mouth phenotype was the driving factor determining the fish's ability to reproduce.

Concept Check 23.1
1. (a) Within a population, genetic differences among individuals provide the raw material on which natural selection and other mechanisms can act. Without such differences, allele frequencies could not change over time—and hence the population could not evolve. (b) Genetic differences between separate populations can result from natural selection if different alleles are favored in different populations; this might occur, for example, if the different populations experienced different environmental conditions (as in Figure 23.4). Genetic differences between populations can also result from chance events (genetic drift) if the genetic changes have few or no phenotypic effects (as in Figure 23.3). **2.** Many mutations occur in somatic cells, which do not produce gametes and so are lost when the organism dies. Of mutations that do occur in cell lines that produce gametes, many do not have a phenotypic effect on which natural selection can act. Others have a harmful effect and are thus unlikely to increase in frequency because they decrease the reproductive success of their bearers. **3.** Its genetic variation (whether measured at the level of the gene or at the level of nucleotide sequences) would probably drop over time. During meiosis, crossing over and the independent assortment of chromosomes produce many new combinations of alleles. In addition, a population contains a vast number of possible mating combinations, and fertilization brings together the gametes of individuals with different genetic backgrounds. Thus, via crossing over, independent assortment of chromosomes, and fertilization, sexual reproduction reshuffles alleles into fresh combinations each generation. Without sexual reproduction, the rate of forming new combinations of alleles would be vastly reduced, causing the overall amount of genetic variation to drop.

Concept Check 23.2
1. 30,000. Half the loci (10,000) are fixed, meaning only one type of allele exists for each locus: $10{,}000 \times 1 = 10{,}000$. There are two types of alleles each for the other loci: $10{,}000 \times 2 = 20{,}000$. $10{,}000 + 20{,}000 = 30{,}000$. **2.** $p^2 + 2pq$; p^2 represents homozygotes with two *A* alleles, and $2pq$ represents heterozygotes with one *A* allele. **3.** There are 120 individuals in the population, so there are 240 alleles. Of these, there are 124 *A* alleles—32 from the 16 *AA* individuals and 92 from the 92 *Aa* individuals. Thus, the frequency of the *A* allele is $p = 124/240 = 0.52$; hence, the frequency of the *a* allele is $q = 0.48$. Based on the Hardy-Weinberg equation, if the population were not evolving, the frequency of genotype *AA* should be $p^2 = 0.52 \times 0.52 = 0.27$; the frequency of genotype *Aa* should be $2pq = 2 \times 0.52 \times 0.48 = 0.5$; and the frequency of genotype *aa* should be $q^2 = 0.48 \times 0.48 = 0.23$. In a population of 120 individuals, these expected genotype frequencies lead us to predict that there would be 32 *AA* individuals (0.27×120), 60 *Aa* individuals (0.5×120), and 28 *aa* individuals (0.23×120). The actual numbers for the population (16 *AA*, 92 *Aa*, 12 *aa*) deviate from these expectations (fewer homozygotes and more heterozygotes than expected). This indicates that the population is not in Hardy-Weinberg equilibrium and hence may be evolving at this locus.

Concept Check 23.3
1. Natural selection is more "predictable" in that it alters allele frequencies in a nonrandom way: It tends to increase the frequency of alleles that increase the organism's reproductive success in its environment and decrease the frequency of alleles that decrease the organism's reproductive success. Alleles subject to genetic drift increase or decrease in frequency by chance alone, whether or not they are advantageous. **2.** Genetic drift results from chance events that cause allele frequencies to fluctuate at random from generation to generation; within a population, this process tends to decrease genetic variation over time. Gene flow is the exchange of alleles between populations, a process that can introduce new alleles to a population and hence may increase its genetic variation (albeit slightly, since rates of gene flow are often low). **3.** Selection is not important at this locus; furthermore, the populations are not small, and hence the effects of genetic drift should not be pronounced. Gene flow is occurring via the movement of pollen and seeds. Thus, allele and genotype frequencies in these populations should become more similar over time as a result of gene flow.

Concept Check 23.4
1. Zero, because fitness includes reproductive contribution to the next generation, and a sterile mule cannot produce offspring. **2.** Although both gene flow and genetic drift can increase the frequency of advantageous alleles in a population, they can also decrease the frequency of advantageous alleles or increase the frequency of harmful alleles. Only natural selection *consistently* results in an increase in the frequency of alleles that enhance survival or reproduction. Thus, natural selection is the only mechanism that consistently causes adaptive evolution. **3.** The three modes of natural selection (directional, stabilizing, and disruptive) are defined in terms of the selective advantage of different *phenotypes*, not different genotypes. Thus, the type of selection represented by heterozygote advantage depends on the phenotype of the heterozygotes. In this question, because heterozygous individuals have a more extreme phenotype than either homozygote, heterozygote advantage represents directional selection. **4.** Under prolonged low-oxygen conditions, some of the red blood cells of a heterozygote may sickle, leading to harmful effects (see Chapter 14). This does not occur in individuals with two normal hemoglobin alleles, suggesting that there may be selection against heterozygotes in malaria-free regions (where heterozygote advantage does not occur). However, since heterozygotes are healthy under most conditions, selection against them is unlikely to be strong.

Summary of Key Concepts Questions
23.1 One reason biologists estimate gene variability and nucleotide variability is to assess whether populations have enough genetic variation for evolution to occur. Gene variability indicates the extent to which individuals differ genetically at the whole-gene level. Nucleotide variability provides a measure of genetic variation at the DNA sequence level. **23.2** No, this is not an example of circular reasoning. Calculating *p* and *q* from observed genotype frequencies does not imply that those genotype frequencies must be in Hardy-Weinberg equilibrium. Consider a population that has 195 individuals of genotype *AA*, 10 of genotype *Aa*, and 195 of genotype *aa*. Calculating *p* and *q* from these values yields $p = q = 0.5$. Using the Hardy-Weinberg equation, the predicted equilibrium frequencies are $p^2 = 0.25$ for genotype *AA*, $2pq = 0.5$ for genotype *Aa*, and $q^2 = 0.25$ for genotype *aa*. Since there are 400 individuals in the population, these predicted genotype frequencies indicate there should be 100 *AA* individuals, 200 *Aa* individuals, and 100 *aa* individuals—numbers that differ greatly from the values that we used to calculate *p* and *q*. **23.3** It is unlikely that two such populations would evolve in similar ways. Since their environments are very different, the alleles favored by natural selection would probably differ between the two populations; although genetic drift may have important effects in each of these small populations, drift causes unpredictable changes in allele frequencies, so it is unlikely that drift would cause the populations to evolve in similar ways; both populations are geographically isolated, suggesting that little gene flow will occur between them (again making it less likely that they will evolve in similar ways). **23.4** Compared to males, it is likely that the females of such species would be larger, more colorful, endowed with more elaborate ornamentation (for example, a large morphological feature such as the peacock's tail), and more apt to engage in behaviors intended to attract mates or prevent other members of their sex from obtaining mates.

Test Your Understanding
1. e **2.** c **3.** e **4.** b **5.** a **6.** d

7. Although natural selection can improve the match between organisms and their environments, the evolutionary process can also lead to imperfections in organisms. A central reason for this is that evolution does not design organisms from scratch to match their environments and ways of life but works instead by a process of descent with modification: Organisms inherit a basic form from their ancestors, and that form is modified by natural selection over time. As a result, a flying mammal such as a bat has wings that are not perfectly designed, but rather represent modifications of forelimbs that bat ancestors used for walking. Imperfections in organisms result from a variety of other constraints, such as a lack of genetic variation for the trait in question, and the fact that adaptations often represent compromises (since organisms must do many different things, and a "perfect" design for one activity might impair the performance of another activity). **8.** The frequency of the *lap*^{94} allele forms a cline, decreasing as one moves from southwest to northeast across Long Island Sound.

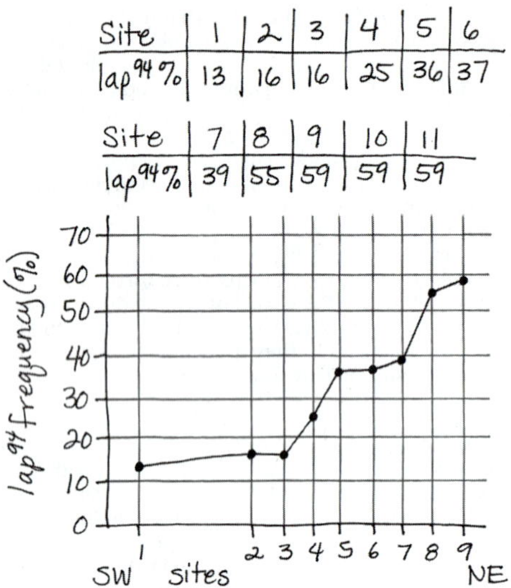

A hypothesis that explains the cline and accounts for the observations stated in the question is that the cline is maintained by an interaction between selection and gene flow. Under this hypothesis, in the southwest portion of the Sound, salinity is relatively low, and selection against the *lap*^{94} allele is strong. Moving toward the northeast and into the open ocean, where salinity is relatively high, selection favors a high frequency of the *lap*^{94} allele. However, because mussel larvae disperse long distances, gene flow prevents the *lap*^{94} allele from becoming fixed in the open ocean or from declining to zero in the southwestern portion of Long Island Sound.

Chapter 24

Figure Questions

Figure 24.10 This was done to remove the possibility that the flies could differentiate among potential mates by detecting what those potential mates had eaten as larvae. If this had not been done, the strong preference of "starch flies" and "maltose flies" to mate with like-adapted flies could have occurred simply because the flies could detect (for example, by sense of smell) what their potential mates had eaten as larvae—and they preferred to mate with flies that had a similar smell to their own. **Figure 24.12** Such results would suggest that mate choice based on coloration does not provide a reproductive barrier between these two cichlid species. **Figure 24.13** The graph suggests there has been gene flow of some fire-bellied toad alleles into the range of the yellow-bellied toad. Otherwise, all individuals located to the left of the hybrid zone portion of the graph would have allele frequencies close to 1.0. **Figure 24.14** Because the populations had only just begun to diverge from one another at this point in the process, it is likely that any existing barriers to reproduction would weaken over time. **Figure 24.19** No. Over time, the chromosomes of the experimental hybrids came to resemble those of *H. anomalus*. This occurred even though conditions in the laboratory differed greatly from conditions in the field, where *H. anomalus* is found, suggesting that selection for laboratory conditions was not strong. Thus, it is unlikely that the observed rise in the fertility of the experimental hybrids was due to selection for life under laboratory conditions. **Figure 24.20** The presence of *M. cardinalis* plants that carry the *M. lewisii yup* allele would make it more likely that bumblebees would transfer pollen between the two monkey flower species. As a result, we would expect the number of hybrid offspring to increase.

Concept Check 24.1

1. (a) All except the biological species concept can be applied to both asexual and sexual species because they define species on the basis of characteristics other than ability to reproduce. In contrast, the biological species concept can be applied only to sexual species. (b) The easiest species concept to apply in the field would be the morphological species concept because it is based only on the appearance of the organism. Additional information about its ecological habits, evolutionary history, and reproduction are not required. **2.** Because these birds live in fairly similar environments and can breed successfully in captivity, the reproductive barrier in nature is probably prezygotic; given the species' differences in habitat preference, this barrier could result from habitat isolation.

Concept Check 24.2

1. In allopatric speciation, a new species forms while in geographic isolation from its parent species; in sympatric speciation, a new species forms in the absence of geographic isolation. Geographic isolation greatly reduces gene flow between populations, whereas ongoing gene flow is more likely in sympatric populations. As a result, sympatric speciation is less common than allopatric speciation. **2.** Gene flow between subsets of a population that live in the same area can be reduced in a variety of ways. In some species—especially plants—changes in chromosome number can block gene flow and establish reproductive isolation in a single generation. Gene flow can also be reduced in sympatric populations by habitat differentiation (as seen in the apple maggot fly, *Rhagoletis*) and sexual selection (as seen in Lake Victoria cichlids). **3.** Allopatric speciation would be less likely to occur on a nearby island than on an isolated island of the same size. The reason we expect this result is that continued gene flow between mainland populations and those on a nearby island reduces the chance that enough genetic divergence will take place for allopatric speciation to occur. **4.** If all of the homologs failed to separate during anaphase I of meiosis, some gametes would end up with an extra set of chromosomes (and others would end up with no chromosomes). If a gamete with an extra set of chromosomes fused with a normal gamete, a triploid would result; if two gametes with an extra set of chromosomes fused with each other, a tetraploid would result.

Concept Check 24.3

1. Hybrid zones are regions in which members of different species meet and mate, producing some offspring of mixed ancestry. Such regions can be viewed as "natural laboratories" in which to study speciation because scientists can directly observe factors that cause (or fail to cause) reproductive isolation. **2.** (a) If hybrids consistently survive and reproduce poorly compared to the offspring of intraspecific matings, reinforcement could occur. If it did, natural selection would cause prezygotic barriers to reproduction between the parent species to strengthen over time, decreasing the production of unfit hybrids and leading to a completion of the speciation process. (b) If hybrid offspring survived and reproduced as well as the offspring of intraspecific matings, indiscriminate mating between the parent species would lead to the production of large numbers of hybrid offspring. As these hybrids mated with each other and with members of both parent species, the gene pools of the parent species could fuse over time, reversing the speciation process.

Concept Check 24.4

1. The time between speciation events includes (1) the length of time that it takes for populations of a newly formed species to begin diverging reproductively from one another and (2) the time it takes for speciation to be complete once this divergence begins. Although speciation can occur rapidly once populations have begun to diverge from one another, it may take millions of years for that divergence to begin. **2.** Investigators transferred alleles at the *yup* locus (which influences flower color) from each parent species to the other. *M. lewisii* plants with an *M. cardinalis yup* allele received many more visits from hummingbirds than usual; hummingbirds usually pollinate *M. cardinalis* but avoid *M. lewisii*. Similarly, *M. cardinalis* plants with an *M. lewisii yup* allele received many more visits from bumblebees than usual; bumblebees usually pollinate *M. lewisii* and avoid *M. cardinalis*. Thus, alleles at the *yup* locus can influence pollinator choice, which in these species provides the primary barrier to interspecific mating. Nevertheless, the experiment does not prove that the *yup* locus alone controls barriers to reproduction between *M. lewisii* and *M. cardinalis*; other genes might enhance the effect of the *yup* locus (by modifying flower color) or cause entirely different barriers to reproduction (for example, gametic isolation or a postzygotic barrier). **3.** Crossing over. If crossing over did not occur, each chromosome in an experimental hybrid would remain as in the F_1 generation: composed entirely of DNA from one parent species or the other.

Summary of Key Concepts Questions

24.1 According to the biological species concept, a species is a group of populations whose members interbreed and produce viable, fertile offspring; thus, gene flow occurs between populations of a species. In contrast, members of different species do not interbreed and hence no gene flow occurs between their populations. Overall, then, in the biological species concept, species can be viewed as designated by the *absence* of gene flow—making gene flow of central importance to the biological species concept. **24.2** Sympatric speciation can be promoted by factors such as polyploidy, habitat shifts, and sexual selection, all of which can reduce gene flow between the subpopulations of a larger population. But such factors can also occur in allopatric populations and hence can also promote allopatric speciation. **24.3** If the hybrids are selected against, the hybrid zone could persist if individuals from the parent species regularly travel into the zone, where they mate to produce hybrid offspring. If hybrids are not selected against, there is no cost to the continued production of hybrids, and large numbers of hybrid offspring may be produced. However, natural selection for life in different environments may keep the gene pools of the two parent species distinct—thus preventing the loss (by fusion) of the parent species and once again causing the hybrid zone to be stable over time. **24.4** As the goatsbeard plant, Bahamas mosquitofish, and apple maggot fly examples illustrate, speciation continues to happen today. A new species can begin to form whenever gene flow is reduced between populations of the parent species. Such reductions in gene flow can occur in many ways: A new, geographically isolated population may be founded by a few colonists; some members of the parent species may begin to utilize new habitat; and sexual selection may isolate formerly connected populations or subpopulations. These and many other such events are happening today.

Test Your Understanding

1. b **2.** c **3.** c **4.** a **5.** e **6.** d **7.** e
10. Here is one possibility:

(2n=14) AA × BB (2n=14)

↓

AB (sterile)

↓ meiotic error

(2n=28) AABB × DD (2n=14)

↓

ABD (sterile)

↓ meiotic error

AA BB DD (2n=42)

Chapter 25

Figure Questions

Figure 25.2 Proteins are almost always composed of the 20 amino acids shown in Figure 5.16. However, many other amino acids could potentially form in this or any other experiment. For example, any molecule that had a different R group than those listed in Figure 5.16 (yet still contained an α carbon, an amino group, and a carboxyl group) would be an amino acid—yet it would not be one of the 20 amino acids commonly found in nature. **Figure 25.5** Because uranium-238 has a half-life of 4.5 billion years, the *x*-axis would be relabeled (in billions of years) as 4.5, 9, 13.5, and 18. **Figure 25.10** You should have circled the node, shown in the tree diagram at approximately 580 million years ago (mya), that leads to the echinoderm/chordate lineage and to the lineage that gave rise to brachiopods, annelids, molluscs, and arthropods. Although the 580 mya date is estimated, this common ancestor must be at least as old as any of its descendants. Since fossil molluscs date to about 555 mya, the common ancestor represented by the circled branch point must be at least 555 million years old. **Figure 25.15** The blue curve is for marine animal families. Families often contain many species, so we would expect the percentage of families that became extinct to be lower than the percentage of species that became extinct. **Figure 25.25** The coding sequence of the *Pitx1* gene would differ between the marine and lake populations, but patterns of gene expression would not.

Concept Check 25.1

1. The hypothesis that conditions on early Earth could have permitted the synthesis of organic molecules from inorganic ingredients **2.** In contrast to random mingling of molecules in an open solution, segregation of molecular systems by membranes could concentrate organic molecules, assisting biochemical reactions. **3.** Today, genetic information usually flows from DNA to RNA, as when the DNA sequence of a gene is used as a template to synthesize the mRNA encoding a particular protein. However, the life cycle of retroviruses such as HIV shows that genetic information can flow in the reverse direction (from RNA to DNA). In these viruses, the enzyme reverse transcriptase uses RNA as a template for DNA synthesis, suggesting that a similar enzyme could have played a key role in the transition from an RNA world to a DNA world.

Concept Check 25.2

1. 22,920 years (four half-lives: 5,730 × 4) **2.** The fossil record shows that different groups of organisms dominated life on Earth at different points in time and that many organisms once alive are now extinct; specific examples of these points can be found in Figure 25.4. The fossil record also indicates that new groups of organisms can arise via the gradual modification of previously existing organisms, as illustrated by fossils that document the origin of mammals from their cynodont ancestors. **3.** The discovery of such a (hypothetical) fossil organism would indicate that aspects of our current understanding of the origin of mammals are not correct because mammals are thought to have originated much more recently (see Figure 25.6). For example, such a discovery could suggest that the dates of previous fossil discoveries are not correct or that the lineages shown in Figure 25.6 shared features with mammals but were not their direct ancestors. Such a discovery would also suggest that radical changes in multiple aspects of the skeletal structure of organisms could arise suddenly—an idea that is not supported by the known fossil record.

Concept Check 25.3

1. Free oxygen attacks chemical bonds and can inhibit enzymes and damage cells. As a result, prokaryotes that had thrived in anaerobic environments would have survived and reproduced poorly in oxygen-rich environments, driving many species to extinction. **2.** All eukaryotes have mitochondria or remnants of these organelles, but not all eukaryotes have plastids. **3.** A fossil record of life today would include many organisms with hard body parts (such as vertebrates and many marine invertebrates), but might not include some species we are very familiar with, such as those that have small geographic ranges and/or small population sizes (for example, endangered species such as the giant panda, tiger, and several rhinoceros species).

Concept Check 25.4

1. Continental drift alters the physical geography and climate of Earth, as well as the extent to which organisms are geographically isolated. Because these factors affect extinction and speciation rates, continental drift has a major impact on life on Earth. **2.** Mass extinctions; major evolutionary innovations; the diversification of another group of organisms (which can provide new sources of food); migration to new locations where few competitor species exist **3.** In principle, fossils of both common and rare species would be present right up to the time of the catastrophic event, then disappear. Reality is more complicated because the fossil record is not perfect. So the most recent fossil for a species might be a million years before the mass extinction—even though the species did not become extinct *until* the mass extinction. This complication is especially likely for rare species because few of their fossils will form and be discovered. Hence, for many rare species, the fossil record would not document that the species was alive immediately before the extinction (even if it was).

Concept Check 25.5

1. Heterochrony can cause a variety of morphological changes. For example, if the onset of sexual maturity changes, a retention of juvenile characteristics (paedomorphosis) may result. Paedomorphosis can be caused by small genetic changes that result in large changes in morphology, as seen in the axolotl salamander. **2.** In animal embryos, *Hox* genes influence the development of structures such as limbs and feeding appendages. As a result, changes in these genes—or in the regulation of these genes—are likely to have major effects on morphology. **3.** From genetics, we know that gene regulation is altered by how well transcription factors bind to noncoding DNA sequences called control elements. Thus, if changes in morphology are often caused by changes in gene regulation, portions of noncoding DNA that contain control elements are likely to be strongly affected by natural selection.

Concept Check 25.6

1. Complex structures do not evolve all at once, but in increments, with natural selection selecting for adaptive variants of the earlier versions. **2.** Although the myxoma virus is highly lethal, initially some of the rabbits are resistant (0.2% of infected rabbits are not killed). Thus, assuming resistance is an inherited trait, we would expect the rabbit population to show a trend for increased resistance to the virus. We would also expect the virus to show an evolutionary trend toward reduced lethality. We would expect this trend because a rabbit infected with a less lethal virus would be more likely to live long enough for a mosquito to bite it and hence potentially transmit the virus to another rabbit. (A virus that kills its rabbit host before a mosquito transmits the virus to another rabbit dies with its host.)

Summary of Key Concepts Questions

Concept 25.1 Particles of montmorillonite clay may have provided surfaces on which organic molecules became concentrated and hence were more likely to react with one another. Montmorillonite clay particles may also have facilitated the transport of key molecules, such as short strands of RNA, into vesicles. These vesicles can form spontaneously from simple precursor molecules, "reproduce" and "grow" on their own, and maintain internal concentrations of molecules that differ from those in the surrounding environment. These features of vesicles represent key steps in the emergence of protocells and (ultimately) the first living cells. **Concept 25.2** One challenge is that organisms do not use radioisotopes that have long half-lives to build their bones or shells. As a result, fossils older than 75,000 years cannot be dated directly. Fossils are often found in sedimentary rock, but those rocks typically contain sediments of different ages, again posing a challenge when trying to date old fossils. To circumvent these challenges, geologists date layers of volcanic rock that surround old fossils and that use radioisotopes with long half-lives. This approach provides minimum and maximum estimates for the ages of fossils sandwiched between two layers of volcanic rock. **Concept 25.3** The "Cambrian explosion" refers to a relatively short interval of time (535–525 million years ago) during which large forms of many present-day animal phyla first appear in the fossil record. The evolutionary changes that occurred during this time, such as the appearance of large predators and well-defended prey, were important because they set the stage for many of the key events in the history of life over the last 500 million years. **Concept 25.4** The broad evolutionary changes documented by the fossil record reflect the rise and fall of major groups of organisms. In turn, the rise or fall of any particular group results from a balance between speciation and extinction rates: A group increases in size when the rate at which its members produce new species is greater than the rate at which its member species are lost to extinction, while a group shrinks in size if extinction rates are greater than speciation rates. **Concept 25.5** Yes. A change to the sequence or regulation of a developmental gene can produce major morphological changes. In some cases, such morphological changes may enable organisms to perform new functions or live in new environments—thus potentially leading to an adaptive radiation and the formation of a new group of organisms. **Concept 25.6** Evolutionary change results from interactions between organisms and their current environments. No goal is involved in this process. As environments change over time, the features of organisms favored by natural selection may also change. When this happens, what once may have seemed like a "goal" of evolution (for example, improvements in the function of a feature previously favored by natural selection) may cease to be beneficial or may even be harmful.

Test Your Understanding

1. c **2.** a **3.** e **4.** b **5.** d

6.

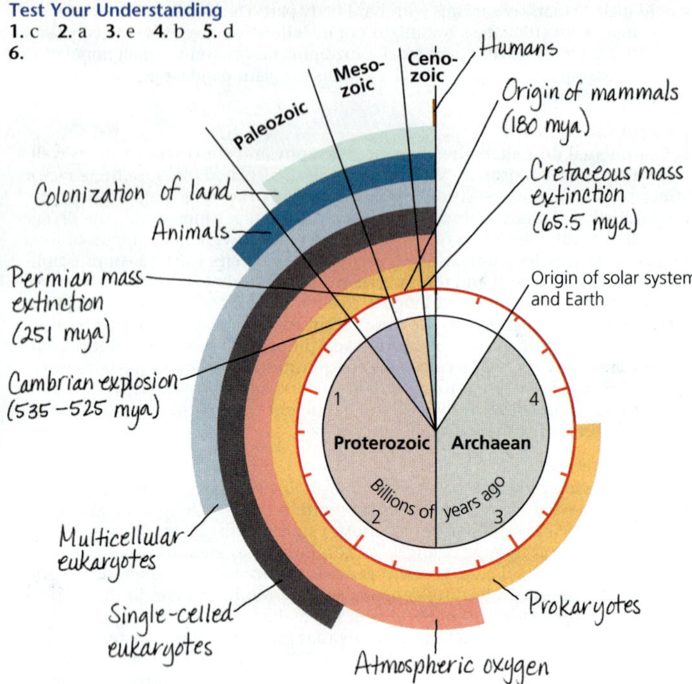

7. c **8.** b

Chapter 26

Figure Questions

Figure 26.5 This new version does not alter any of the evolutionary relationships shown in Figure 26.5. For example, B and C remain sister taxa, taxon A is still as closely related to taxon B as it is to taxon C, and so on.

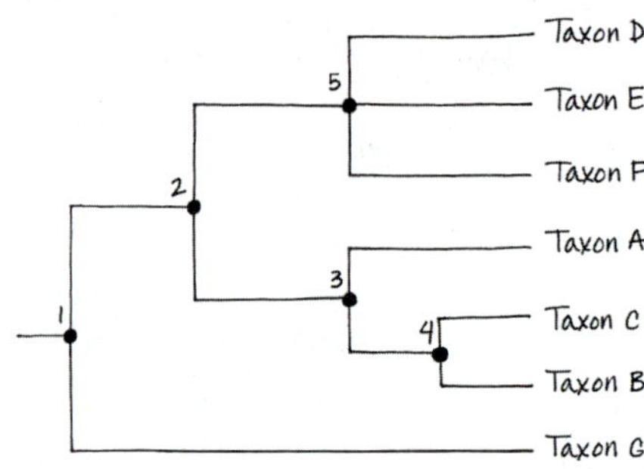

Figure 26.6 Unknown 1b (a portion of sample 1) and Unknowns 9–13 all would have to be located on the branch of the tree that currently leads to Minke (Southern Hemisphere) and Unknowns 1a and 2–8. **Figure 26.9** There are four possible bases (A, C, G, T) at each nucleotide position. If the base at each position depends on chance, not common descent, we would expect roughly one out of four (25%) of them to be the same. **Figure 26.11** You should have drawn a box around the frog, turtle, and leopard lineages, along with their most recent common ancestor. **Figure 26.12** The zebrafish lineage; of the five vertebrate lineages shown, its branch is the longest. **Figure 26.16** The lizard and snake lineage is the most basal taxon shown (closest to the root of the tree). Among the descendants of the common ancestor indicated by the blue dot, the crocodilian lineage is the most basal. **Figure 26.19** The molecular clock indicates that the divergence time is roughly 45–50 million years. **Figure 26.21** As shown in this diagram, Bacteria was the first to emerge and Archaea is the sister domain to Eukarya.

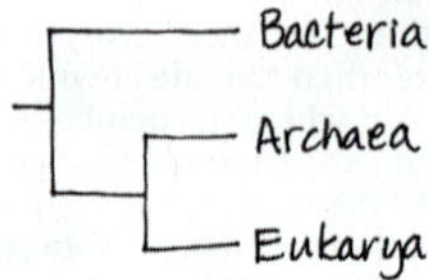

Concept Check 26.1

1. We are classified the same from the domain level to the class level; both the leopard and human are mammals. Leopards belong to order Carnivora, whereas humans do not. **2.** The branching pattern of the tree indicates that the badger and the wolf share a common ancestor that is more recent than the ancestor that these two animals share with the leopard. **3.** The tree in (c) shows a different pattern of evolutionary relationships. In (c), C and B are sister taxa, whereas C and D are sister taxa in (a) and (b).

4.

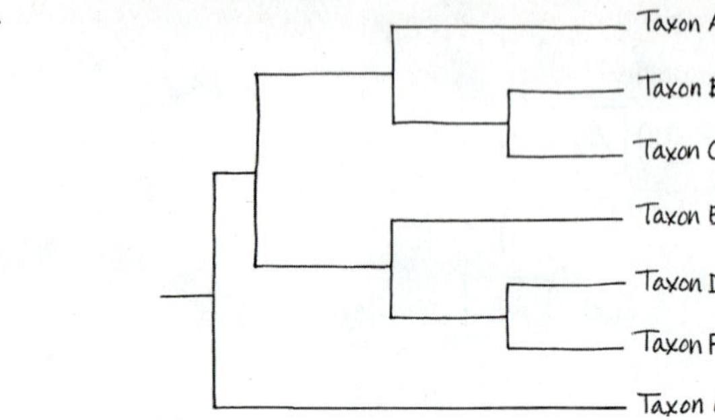

Concept Check 26.2

1. (a) Analogy, since porcupines and cacti are not closely related and since most other animals and plants do not have similar structures; (b) homology, since cats and humans are both mammals and have homologous forelimbs, of which the hand and paw are the lower part; (c) analogy, since owls and hornets are not closely related and since the structure of their wings is very different **2.** Species 2 and 3 are more likely to be closely related. Small genetic changes (as between species 2 and 3) can produce divergent physical appearances, but if many genes have diverged greatly (as in species 1 and 2), then the lineages have probably been separate for a long time.

Concept Check 26.3

1. No; when comparing groups of mammals, hair is a shared ancestral character common to all mammals and thus is not helpful in distinguishing different mammalian subgroups. **2.** The principle of maximum parsimony states that the hypothesis about nature we investigate first should be the simplest explanation found to be consistent with the facts. Actual evolutionary relationships may differ from those inferred by parsimony owing to complicating factors such as convergent evolution. **3.** The traditional classification provides a poor match to evolutionary history, thus violating the basic principle of cladistics—that classification should be based on common descent. Both birds and mammals originated from groups traditionally designated as reptiles, making reptiles (as traditionally delineated) a paraphyletic group. These problems can be addressed by removing *Dimetrodon* and cynodonts from the reptiles and by considering birds as a group of reptiles (specifically, as a group of dinosaurs).

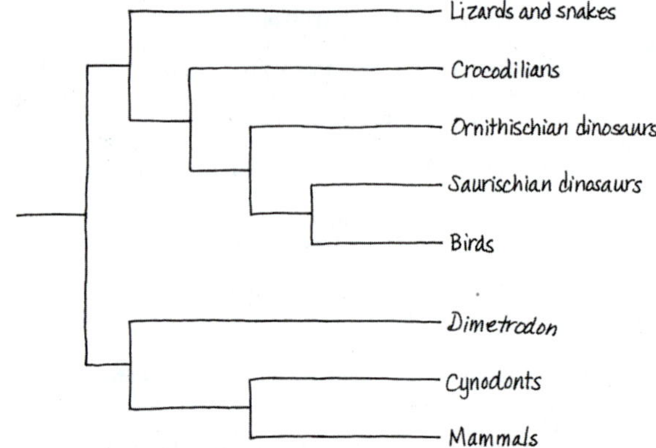

Concept Check 26.4

1. Proteins are gene products. Their amino acid sequences are determined by the nucleotide sequences of the DNA that codes for them. Thus, differences between comparable proteins in two species reflect underlying genetic differences that have accumulated as the species diverged from one another. As a result, differences between the proteins can reflect the evolutionary history of the species. **2.** In RNA processing, the exons or coding regions of a gene can be spliced together in different ways, yielding different mRNAs and hence different protein products. As a result, different proteins could potentially be produced from the same gene in different tissues, thereby enabling the gene to perform different functions in these different tissues. **3.** These observations suggest that the evolutionary lineages leading to species 1 and species 2 diverged from one another before a gene duplication event in species 1 produced gene B from gene A.

Concept Check 26.5

1. A molecular clock is a method of estimating the actual time of evolutionary events based on numbers of base changes in orthologous genes. It is based on the assumption that the regions of genomes being compared evolve at constant rates. **2.** There are many portions of the genome that do not code for genes; many base changes in these regions could accumulate through drift without affecting an organism's fitness. Even in coding regions of the genome, some mutations may not have a critical effect on genes or proteins. **3.** The gene (or genes) used for the molecular clock may have evolved more slowly in these two taxa than in the species used to calibrate the clock; as a result, the clock would underestimate the time at which the taxa diverged from each other.

Concept Check 26.6

1. The kingdom Monera included bacteria and archaea, but we now know that these organisms are in separate domains. Kingdoms are subsets of domains, so a single kingdom (like Monera) that includes taxa from different domains is not valid. **2.** Because of horizontal gene transfer, some genes in eukaryotes are more closely related to bacteria, while others are more closely related to archaea; thus, depending on which genes are used, phylogenetic trees constructed from DNA data can yield conflicting results.

3.

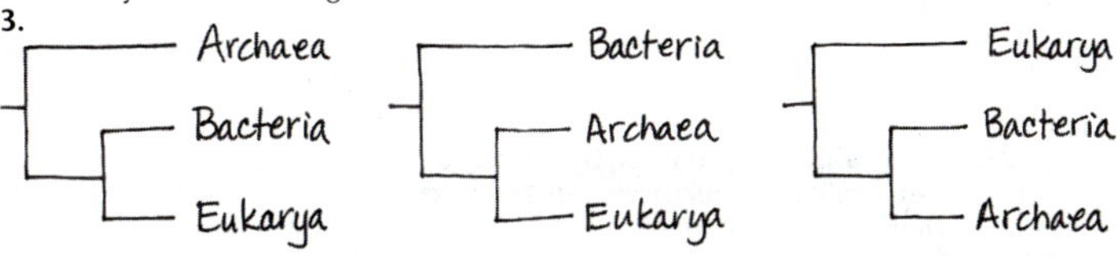

The fossil record indicates that prokaryotes originated long before eukaryotes. This suggests that the third tree, in which the eukaryotic lineage diverged first, is not accurate and hence is not likely to receive support from genetic data.

Summary of Key Concepts Questions

26.1 The fact that humans and chimpanzees are sister species indicates that we share a more recent common ancestor with chimpanzees than we do with any other living primate species. But that does not mean that humans evolved from chimpanzees, or vice versa; instead, it indicates that both humans and chimpanzees are descendants of that common ancestor. **26.2** Homologous characters result from shared ancestry. As organisms diverge over time, some of their homologous characters will also diverge. The homologous characters of organisms that diverged long ago typically differ more than do the homologous characters of organisms that diverged more recently. As a result, differences in homologous characters can be used to infer phylogeny. In contrast, analogous characters result from convergent evolution, not shared ancestry, and hence can give misleading estimates of phylogeny. **26.3** All features of organisms arose at some point in the history of life. In the group in which a new feature first arises, that feature is a shared derived character that is unique to that clade. The group in which each shared derived character first appears can be determined, and the resulting nested pattern can be used to infer evolutionary history. **26.4** Orthologous genes should be used; for such genes, the homology results from speciation and hence reflects evolutionary history. **26.5** A key assumption of molecular clocks is that nucleotide substitutions occur at fixed rates and hence the number of nucleotide differences between two DNA sequences is proportional to the time since the sequences diverged from each other. Some limitations of molecular clocks: No gene marks time with complete precision; natural selection can favor certain DNA changes over others; nucleotide substitution rates can change over long periods of time (causing molecular clocks estimates of when events in the distant past occurred to be highly uncertain); and the same gene can evolve at different rates in different organisms **26.6** Genetic data indicated that many prokaryotes differed as much from each other as they did from eukaryotes. This indicated that organisms should be grouped into three "super-kingdoms," or domains (Archaea, Bacteria, Eukarya). These data also indicated that the previous kingdom Monera (which had contained all the prokaryotes) did not make biological sense and should be abandoned. Later genetic and morphological data also indicated that the former kingdom Protista (which had primarily contained single-celled organisms) should be abandoned because it was polyphyletic.

Test Your Understanding

1. b **2.** d **3.** a **4.** d **5.** d **6.** c **7.** d

9. (a)

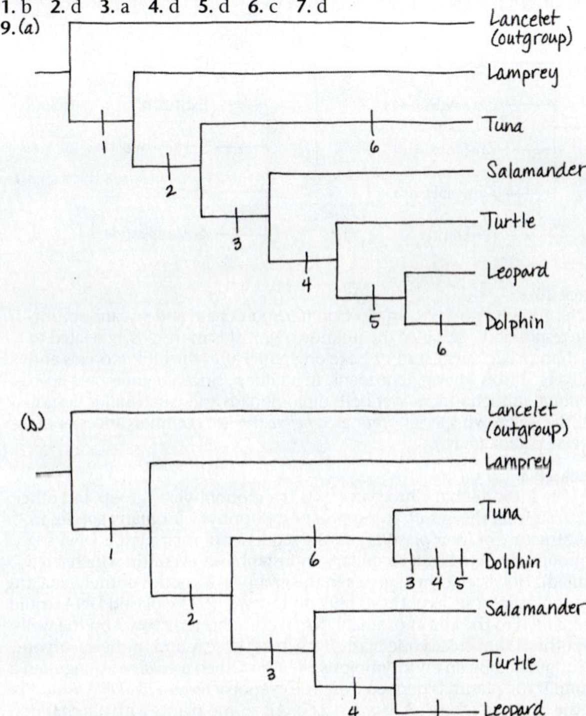

(c) The tree in (a) requires seven evolutionary changes, while the tree in (b) requires nine evolutionary changes. Thus, the tree in (a) is more parsimonious, since it requires fewer evolutionary changes.

Chapter 27

Figure Questions

Figure 27.10 It is likely that the expression or sequence of genes that affect glucose metabolism may have changed; genes for metabolic processes no longer needed by the cell also may have changed. **Figure 27.11** Transduction results in horizontal gene transfer when the host and recipient cells are members of different species. **Figure 27.16** Thermophiles live in very hot environments, so it is likely that their enzymes can continue to function normally at much higher temperatures than do the enzymes of other organisms. At low temperatures, however, the enzymes of thermophiles may not function as well as the enzymes of other organisms. **Figure 27.18** From the graph, plant uptake can be estimated as 0.7, 0.6, and 0.95 (mg K) for strains 1, 2, and 3, respectively. These values average to 0.75 mg K. If bacteria had no effect, the average plant uptake of potassium for strains 1, 2, and 3 should be close to 0.5 mg K, the value observed for plants grown in bacteria-free soil.

Concept Check 27.1

1. Adaptations include the capsule (shields prokaryotes from host's immune system) and endospores (enable cells to survive harsh conditions and to revive when the environment becomes favorable). **2.** Prokaryotic cells generally lack the internal compartmentalization of eukaryotic cells. Prokaryotic genomes have much less DNA than eukaryotic genomes, and most of this DNA is contained in a single ring-shaped chromosome located in the nucleoid rather than within a true membrane-bounded nucleus. In addition, many prokaryotes also have plasmids, small ring-shaped DNA molecules containing a few genes. **3.** Plastids such as chloroplasts are thought to have evolved from an endosymbiotic photosynthetic prokaryote. More specifically, the phylogenetic tree shown in Figure 26.21 indicates that plastids are closely related to cyanobacteria. Hence, we can hypothesize that the thylakoid membranes of chloroplasts resemble those of cyanobacteria because chloroplasts evolved from a cyanobacterium endosymbiont.

Concept Check 27.2

1. Prokaryotes have extremely large population sizes, in part because they have short generation times. The large number of individuals in prokaryotic populations makes it likely that in each generation there will be thousands of individuals that have new mutations at any particular gene, thereby adding considerable genetic diversity to the population. **2.** In transformation, naked, foreign DNA from the environment is taken up by a bacterial cell. In transduction, phages carry bacterial genes from one bacterial cell to another. In conjugation, a bacterial cell directly transfers plasmid or chromosomal DNA to another cell via a mating bridge that temporarily connects the two cells. **3.** The population that includes individuals capable of conjugation would probably be more successful, since some of its members could form recombinant cells whose new gene combinations might be advantageous in a novel environment. **4.** Yes. Genes for antibiotic resistance could be transferred (by transformation, transduction, or conjugation) from the nonpathogenic bacterium to a pathogenic bacterium; this could make the pathogen an even greater threat to human health. In general, transformation, transduction, and conjugation tend to increase the spread of resistance genes.

Concept Check 27.3

1. A phototroph derives its energy from light, while a chemotroph gets its energy from chemical sources. An autotroph derives its carbon from a form of CO_2, while a heterotroph gets its carbon from organic nutrients such as glucose. Thus, there are four nutritional modes: photoautotrophic, photoheterotrophic (unique to prokaryotes), chemoautotrophic (unique to prokaryotes), and chemoheterotrophic. **2.** Chemoheterotrophy; the bacterium must rely on chemical sources of energy, since it is not exposed to light, and it must be a heterotroph if it requires a source of carbon other than CO_2 (or a related compound, such as bicarbonate). **3.** If humans could fix nitrogen, we could build proteins using atmospheric N_2 and hence would not need to eat high-protein foods such as meat, fish, or soy. Our diet would, however, need to include a source of carbon, along with minerals and water. Thus, a typical meal might consist of carbohydrates as a carbon source, along with fruits and vegetables to provide essential minerals (and additional carbon).

Concept Check 27.4

1. Before molecular systematics, taxonomists classified prokaryotes according to phenotypic characters that did not clarify evolutionary relationships. Molecular comparisons—of DNA in particular—indicate key divergences in prokaryotic lineages. **2.** By not requiring that organisms be cultured in the laboratory, genetic prospecting has revealed an immense diversity of previously unknown prokaryotic species. Over time, the ongoing discovery of new species by genetic prospecting is likely to alter our understanding of prokaryotic phylogeny greatly. **3.** At present, all known methanogens are archaea in the clade Euryarchaeota; this suggests that this unique metabolic pathway probably arose in ancestral species within Euryarchaeota. Since Bacteria and Archaea have been separate evolutionary lineages for billions of years, the discovery of a methanogen from the domain Bacteria would suggest that adaptations that enabled the use of CO_2 to oxidize H_2 may have evolved twice—once in Archaea (within Euryarchaeota) and once in Bacteria. (It is also possible that a newly discovered bacterial methanogen could have acquired the genes for this metabolic pathway by horizontal gene transfer from a methanogen in domain Archaea. However, horizontal gene transfer is not a likely explanation because of the large number of genes involved and because gene transfers between species in different domains are rare.)

Concept Check 27.5

1. Although prokaryotes are small, their large numbers and metabolic abilities enable them to play key roles in ecosystems by decomposing wastes, recycling chemicals, and affecting the concentrations of nutrients available to other organisms **2.** Cyanobacteria produce oxygen when water is split in the light reactions of photosynthesis. The Calvin cycle incorporates CO_2 from the air into organic molecules, which are then converted to sugars.

Concept Check 27.6

1. Sample answers: eating fermented foods such as yogurt, sourdough bread, or cheese; receiving clean water from sewage treatment; taking medicines produced by bacteria **2.** No. If the poison is secreted as an exotoxin, live bacteria could be transmitted to another person. But the same is true if the poison is an endotoxin—only in this case, the live bacteria that are transmitted may be descendants of the (now-dead) bacteria that produced the poison. **3.** Some of the many different species of prokaryotes that live in the human gut compete with one another for resources (in the food you eat). Because different prokaryotic species have different adaptations, a change in diet may alter which species can grow most rapidly, thus altering species abundance.

Summary of Key Concepts Questions

27.1. Prokaryotes are small, they have short generation times, and their populations can consist of trillions of individuals. As a result, populations of prokaryotes can evolve substantially in short periods of time, enabling them to adapt to a wide range of environments. Specific structural features that enable prokaryotes to thrive in diverse environments include their cell walls (which provide shape and protection), flagella (which function in directed movement), and ability to form endospores (which enable tolerance of harsh conditions). Prokaryotes also possess biochemical adaptations for growth in varied conditions, such as those that enable them to tolerate extremely hot or salty environments. **27.2** Prokaryotes reproduce extremely rapidly, and their populations can number in the trillions. As a result, even though mutations are rare, every day thousands of offspring are produced that have new mutations at particular gene loci. In addition, even though prokaryotes reproduce asexually and hence the vast majority of offspring are genetically identical to their parent, the genetic variation of their populations can be increased by transduction, transformation, and conjugation. Each of these (nonreproductive) processes can increase genetic variation by transferring DNA from one cell to another—even among cells that are of different species. **27.3** Prokaryotes have an exceptionally broad range of metabolic adaptations. As a group, prokaryotes perform all four modes of nutrition (photoautotrophy, chemoautotrophy, photoheterotrophy, and chemoheterotrophy), whereas eukaryotes perform only two of these (photoautotrophy and chemoheterotrophy). Prokaryotes are also able to metabolize nitrogen in a wide variety of forms (again unlike eukaryotes), and they frequently cooperate with other prokaryotic cells of the same or different species. **27.4** Phenotypic criteria such as shape, motility, and nutritional mode do not provide a clear picture of the evolutionary history of the prokaryotes. In contrast, molecular data have elucidated relationships among major groups of prokaryotes. Molecular data have also allowed researchers to sample genes directly from the environment; using such genes to construct phylogenies has led to the discovery of major new groups of prokaryotes. **27.5** Prokaryotes play key roles in the chemical cycles on which life depends. For example, prokaryotes are important decomposers, breaking down corpses and waste materials, thereby releasing nutrients to the environment where they can be used by other organisms. Prokaryotes also convert inorganic compounds to forms that other organisms can use. With respect to their ecological interactions, many prokaryotes form life-sustaining mutualisms with other species. In some cases, such as hydrothermal vent communities, the metabolic activities of prokaryotes provide an energy source on which hundreds of other species depend; in the absence of the prokaryotes, the community collapses. **27.6** Human well-being depends on our associations with mutualistic prokaryotes, such as the many species that live in our intestines and digest food that we cannot. Humans also can harness the remarkable metabolic capabilities of prokaryotes to produce a wide range of useful products. Negative effects of prokaryotes result primarily from bacterial pathogens that cause disease.

Test Your Understanding

1. e **2.** a **3.** d **4.** d **5.** b **6.** a

8. (a)

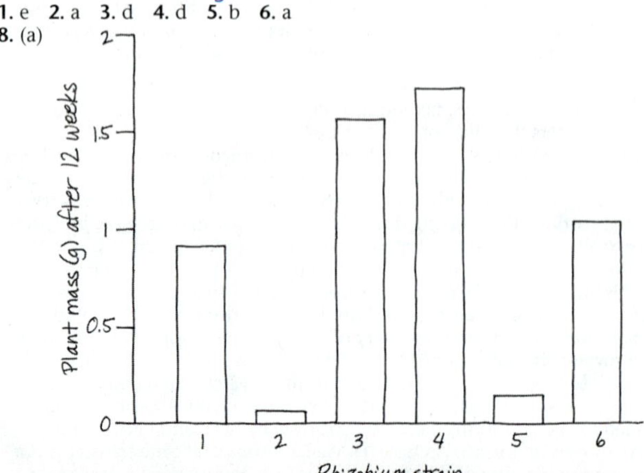

(b) Some *Rhizobium* strains are much more effective at promoting plant growth than other *Rhizobium* strains; the most ineffective strains have little positive effect (plant growth with these strains differs little from plant growth in the absence of *Rhizobium*). The ineffective strains may transfer relatively little nitrogen to their plant host, limiting plant growth.

Chapter 28

Figure Questions

Figure 28.10 Merozoites are produced by the asexual (mitotic) cell division of haploid sporozoites; similarly, gametocytes are produced by the asexual cell division of merozoites. Hence, it is likely that individuals in these three stages have the same complement of genes and that morphological differences between them result from changes in gene expression. **Figure 28.16** The sperm cells in the diagram are produced by the asexual (mitotic) division of cells in a single male gametophyte, which was itself produced by the asexual (mitotic) division of a single zoospore. Thus, the sperm cells are all derived from a single zoospore and so are genetically identical to one another. However, the male gametophyte that produced the sperm developed from one zoospore, while the female gametophyte that produced the egg developed from a different zoospore. Zoospores are produced by meiosis, so each zoospore differs genetically from the others. Hence, the sperm cells (which can be traced back to one zoospore) differ genetically from the egg cell (which can be traced back to another zoospore). **Figure 28.22** The following stage should be circled: step 6, where a mature cell undergoes mitosis and forms four or more daughter cells. In step 7, the zoospores eventually grow into mature haploid cells, but they do not produce new daughter cells. Likewise, in step 2, a mature cell develops into a gamete, but it does not produce new daughter cells. **Figure 28.23** If the assumption is correct, then their results indicate that the DHFR-TS gene fusion may be a derived trait shared by members of four supergroups of eukaryotes (Excavata, Chromalveolata, Rhizaria, and Archaeplastida). However, if the assumption is not correct, the presence or absence of the gene fusion may tell little about phylogenetic history. For example, if the genes fused multiple times, groups could share the trait because of convergent evolution rather than common descent. If the genes were secondarily split, a group with such a split could be placed (incorrectly) in Unikonta rather than its correct placement in one of the other four supergroups. **Figure 28.28** If diatom populations decrease in size as global temperatures warm, less carbon dioxide would be "pumped" from surface waters to the deep ocean floor. Atmospheric carbon dioxide levels might increase as a result, thereby potentially causing further warming. If this process continues, a positive-feedback loop would result in which warming causes diatom populations to drop, thereby causing additional warming, further drops in diatom populations, and so on.

Concept Check 28.1

1. Sample response: Protists include unicellular, colonial, and multicellular organisms; photoautotrophs, heterotrophs, and mixotrophs; species that reproduce asexually, sexually, or both ways; and organisms with diverse physical forms and adaptations. **2.** Strong evidence shows that eukaryotes acquired mitochondria after an early eukaryote first engulfed and then formed an endosymbiotic association with an alpha proteobacterium. Similarly, chloroplasts in red and green algae appear to have descended from a photosynthetic cyanobacterium that was engulfed by an ancient heterotrophic eukaryote. Secondary endosymbiosis also played an important role: Various protist lineages acquired plastids by engulfing unicellular red or green algae. **3.** The modified tree would look as follows:

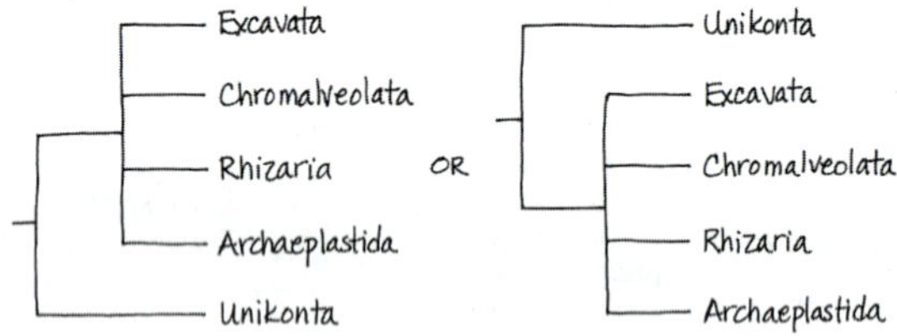

Concept Check 28.2

1. Their mitochondria do not have an electron transport chain and so cannot function in aerobic respiration. **2.** Since the unknown protist is more closely related to diplomonads than to euglenids, it must have originated after the diplomonads and parabasalids diverged from the euglenozoans. In addition, since the unknown species has fully functional mitochondria—yet both diplomonads and parabasalids do not—it is likely that the unknown species originated *before* the last common ancestor of the diplomonads and parabasalids.

Concept Check 28.3

1. Some DNA data indicate that Chromalveolata is a monophyletic group, but other DNA data fail to support this result. In support of monophyly, for many species in the group, the structure of their plastids and the sequence of their plastid DNA suggest that the group originated by a secondary endosymbiosis event (in which a red alga was engulfed). However, other species in the group lack plastids entirely, making the secondary endosymbiosis hypothesis difficult to test. **2.** The plastid DNA would likely be more similar to the chromosomal DNA of cyanobacteria based on the well-supported hypothesis that eukaryotic plastids (such as those found in the eukaryotic groups listed) originated by an endosymbiosis event in which a eukaryote engulfed a cyanobacterium. If the plastid is derived from the cyanobacterium, its DNA would be derived from the bacterial DNA. **3.** Figure 13.6b. Algae and plants with alternation of generations have a multicellular haploid stage *and* a multicellular diploid stage. In the other two life cycles, either the haploid stage or the diploid stage is unicellular.

Concept Check 28.4

1. Because foram tests are hardened with calcium carbonate, they form long-lasting fossils in marine sediments and sedimentary rocks. **2.** Convergent evolution. The different organisms have come to display similar morphological adaptations over time owing to their similar lifestyles. **3.** During photosynthesis, aerobic algae produce O_2 and use CO_2. O_2 is produced as a by-product of the light reactions, while CO_2 is used as an input to the Calvin cycle (the end products of which are sugars). Aerobic algae also perform cellular respiration, which uses O_2 as an input and produces CO_2 as a waste product.

Concept Check 28.5

1. Many red algae contain an accessory pigment called phycoerythrin, which gives them a reddish color and allows them to carry out photosynthesis in relatively deep coastal water. Also unlike brown algae, red algae have no flagellated stages in their life cycle and must depend on water currents to bring gametes together for fertilization. **2.** *Ulva*'s thallus contains many cells and is differentiated into leaflike blades and a rootlike holdfast. *Caulerpa*'s thallus is composed of multinucleate filaments without cross-walls, so it is essentially one large cell. **3.** Red algae have no flagellated stages in their life cycle and hence must depend on water currents to bring their gametes together. This feature of their biology might increase the difficulty of reproducing on land. In contrast, the gametes of green algae are flagellated, making it possible for them to swim in thin films of water. In addition, a variety of green algae contain compounds in their cytoplasm, cell wall, or zygote coat that protect against intense sunlight and other terrestrial conditions. Such compounds may have increased the chance that descendants of green algae could survive on land.

Concept Check 28.6

1. Amoebozoans have lobe-shaped pseudopodia, whereas forams have threadlike pseudopodia. **2.** Slime molds are fungus-like in that they produce fruiting bodies that aid in the dispersal of spores, and they are animal-like in that they are motile and ingest food. However, slime molds are more closely related to gymnamoebas and entamoebas than to fungi or animals. **3.** Support. Unikonts lack the unique cytoskeletal features shared by many excavates (see Concept 28.2). Thus, if the unikonts were the first group of eukaryotes to diverge from other eukaryotes (as shown in Figure 28.23), it would be unlikely that the eukaryote common ancestor had the cytoskeletal features found today in many excavates. Such a result would strengthen the case that many excavates share cytoskeletal features because they are members of a monophyletic group, the Excavata.

Concept Check 28.7

1. Because photosynthetic protists constitute the base of aquatic food webs, many aquatic organisms depend on them for food, either directly or indirectly. (In addition, a substantial percentage of the oxygen produced in photosynthesis on Earth is made by photosynthetic protists.) **2.** Protists form mutualistic and parasitic associations with other organisms. Examples include photosynthetic dinoflagellates that form a mutualistic symbiosis with coral polyps, parabasalids that form a mutualistic symbiosis with termites, and the oomycete *Phytophthora ramorum*, a parasite of oak trees. **3.** Corals depend on their dinoflagellate symbionts for nourishment, so coral bleaching would be expected to cause the corals to die. As the corals die, less food will be available for fishes and other species that eat coral. As a result, populations of these species may decline, and that, in turn, might cause populations of their predators to decline.

Summary of Key Concepts Questions

28.1 Sample response: Protists, plants, animals, and fungi are similar in that their cells have a nucleus and other membrane-bounded organelles, unlike the cells of prokaryotes. These membrane-bounded organelles make the cells of eukaryotes more complex than the cells of prokaryotes. With respect to differences between protists and other eukaryotes, most protists are unicellular, unlike animals, plants, and most fungi. Protists also have greater nutritional diversity than other eukaryotes. **28.2** Unique cytoskeletal features are shared by many excavates. In addition, some members of Excavata have an "excavated" feeding groove for which the group was named. DNA evidence does not strongly support or refute Excavata as a group. Overall, evidence for the group is relatively weak. **28.3** We can infer that the common ancestor of the group had a plastid, in this case of red algal origin. Thus, we would expect that members of Chromalveolata either would have plastids or would have lost their plastids over the course of evolution. **28.4** The main subgroups are the radiolarians, the forams, and the cercozoans. They are united as a clade by DNA similarities. **28.5** Red algae, green algae, and land plants are placed in the same supergroup because considerable evidence indicates that these organisms all descended from the same ancestor, an ancient heterotrophic protist that acquired a cyanobacterial endosymbiont. **28.6** The unikonts are a diverse group of eukaryotes that includes many protists, along with animals and fungi. Most of the protists in Unikonta are amoebozoans, a clade of amoebas that have lobe- or tube-shaped pseudopodia (as opposed to the threadlike pseudopodia of rhizarians). Other protists in Unikonta include several groups that are closely related to fungi and several other groups that are closely related to animals. **28.7** Sample response: Ecologically important protists include photosynthetic dinoflagellates that provide essential sources of energy to their symbiotic partners, the corals that build coral reefs. Other important protistan symbionts include those that enable termites to digest wood and *Plasmodium*, the pathogen that causes malaria. Photosynthetic protists such as diatoms are among the most important producers in aquatic communities; as such, many other species in aquatic environments depend on them for food.

Test Your Understanding

1. d **2.** b **3.** c **4.** e **5.** d **6.** d

7.

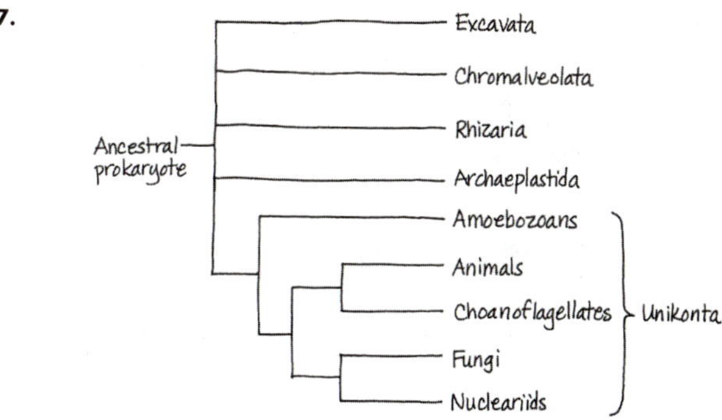

Pathogens that share a relatively recent common ancestor with humans should also share metabolic and structural characteristics with humans. Because drugs target the pathogen's metabolism or structure, developing drugs that harm the pathogen but not the patient should be most difficult for pathogens with whom we share the most recent evolutionary history. Working backward in time, we can use the phylogenetic tree to determine the order in which humans shared a common ancestor with pathogens in different taxa. This process leads to the prediction that it should be hardest to develop drugs to combat animal pathogens, followed by choanoflagellate pathogens, fungal and nucleariid pathogens, amoebozoans, other protists, and finally prokaryotes.

Chapter 29

Figure Questions

Figure 29.5 The life cycle in Figure 13.6b has alternation of generations; the others do not. Unlike the animal life cycle (Figure 13.6a), in alternation of generations, meiosis produces spores, not gametes. These spores then divide repeatedly by mitosis, ultimately forming a multicellular haploid individual that produces gametes. There is no multicellular haploid stage in the animal life cycle. An alternation of generations life cycle also has a multicellular diploid stage, whereas the life cycle shown in Figure 13.6c does not. **Figure 29.8** Yes. As shown in the diagram, the sperm cell and the egg cell that fuse each resulted from the mitotic division of spores produced by the same sporophyte. However, these spores would differ genetically from one another because they were produced by meiosis, a cell division process that generates genetic variation among the offspring cells. **Figure 29.10** Because the moss reduces nitrogen loss from the ecosystem, species that typically colonize the soils after the moss probably experience higher soil nitrogen levels than they otherwise would. The resulting increased availability of nitrogen may benefit these species because nitrogen is an essential nutrient that often is in short supply. **Figure 29.13** A fern that had wind-dispersed sperm would not require water for fertilization, thus removing a difficulty that ferns face when they live in arid environments. The fern would also be under strong selection to produce sperm above ground (as opposed to the current situation, where some fern gametophytes are located below ground). **Figure 29.16** When not reproducing, the lycophyte trees that dominate this forest would resemble poles covered by short leaves (microphylls). Since the trees would not have a branched canopy at their tops, the forest would be very open, and a considerable amount of light would reach the ground level.

Concept Check 29.1

1. Land plants share some key traits only with charophytes: rings of cellulose-synthesizing complexes, presence of peroxisome enzymes, similarity in sperm structure, and the formation of a phragmoplast in cell division. Comparisons of nuclear and chloroplast genes also point to a common ancestry. **2.** Spore walls toughened by sporopollenin (protects against harsh environmental conditions); multicellular, dependent embryos (provides nutrients and protection to the developing embryo); cuticle (reduces water loss) **3.** The multicellular diploid stage of the life cycle would not produce gametes. Instead, both males and females would produce haploid spores by meiosis. These spores would give rise to multicellular male and female haploid stages—a major change from the single-celled haploid stages (sperm and eggs) that we actually have. The multicellular haploid stages would produce gametes and reproduce sexually. An individual at the multicellular haploid stage of the human life cycle might look like us, or it might look completely different. **4.** Land plants, vascular plants, and seed plants are monophyletic because each of these groups includes the common ancestor of the group and all of the descendants of that common ancestor. The other two categories of plants, the nonvascular plants and the seedless vascular plants, are paraphyletic: These groups do not include all of the descendants of the group's most recent common ancestor.

Concept Check 29.2

1. Bryophytes do not have a vascular transport system, and their life cycle is dominated by gametophytes rather than sporophytes. **2.** Answers may include the following: Large surface area of protonema enhances absorption of water and minerals; the vase-shaped archegonia protect eggs during fertilization and transport nutrients to the embryos via placental transfer cells; the stalk-like seta conducts nutrients from the gametophyte to the capsule, where spores are produced; the peristome enables gradual spore discharge; stomata enable CO_2/O_2 exchange while minimizing water loss; lightweight spores are readily dispersed by wind.

3.

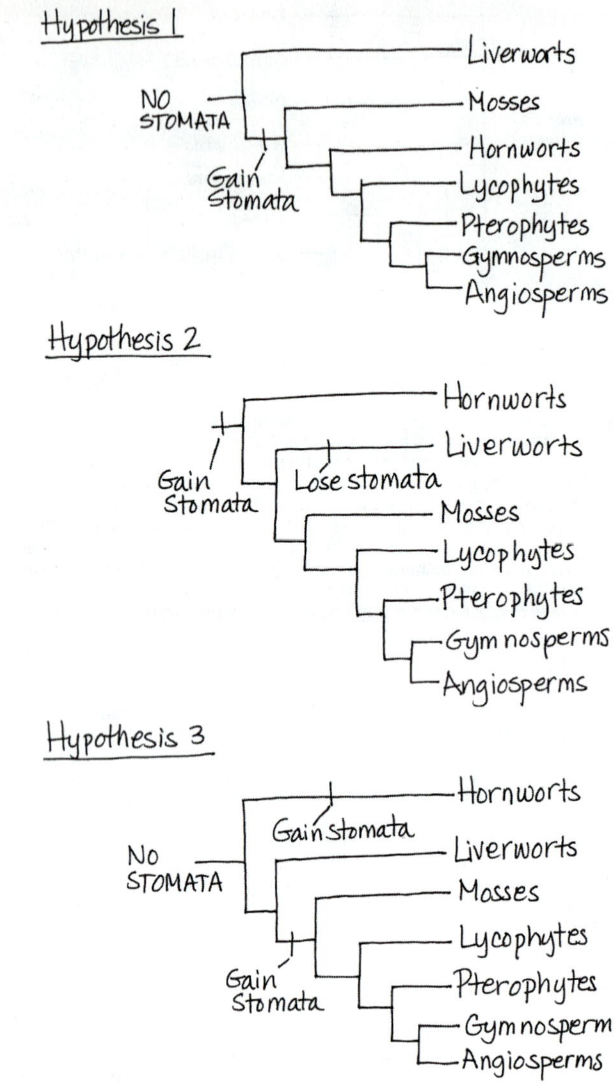

Concept Check 29.3

1. Lycophytes have microphylls, whereas seed plants and pterophytes (ferns and their relatives) have megaphylls. Pterophytes and seed plants also share other traits not found in lycophytes, such as overtopping growth and the initiation of new root branches at various points along the length of an existing root. **2.** Both seedless vascular plants and bryophytes have flagellated sperm that require moisture for fertilization; this shared similarity poses challenges for these species in arid regions. With respect to key differences, seedless vascular plants have lignified, well-developed vascular tissue, a trait that enables the sporophyte to grow tall and that has transformed life on Earth (via the formation of forests). Seedless vascular plants also have true leaves and roots, which, when compared with bryophytes, provide increased surface area for photosynthesis and improve their ability to extract nutrients from soil. **3.** If lycophytes and pterophytes formed a clade, the traits shared by pterophytes and seed plants might have been present in the common ancestor of all vascular plants, but lost in the lycophytes. Alternatively, the common ancestor of all vascular plants may have lacked the traits shared by pterophytes and seed plants; in this case, pterophytes and seed plants would share these traits as a result of convergent evolution. **4.** Three mechanisms contribute to the production of genetic variation in sexual reproduction: independent assortment of chromosomes, crossing over, and random fertilization. If fertilization were to occur between gametes from the same gametophyte, all of the offspring would be genetically identical. This would be the case because all of the cells produced by a gametophyte—including its sperm and egg cells—are the descendants of a single spore and hence are genetically identical. Genetic variation would continue to be generated by the first two mechanisms mentioned, but overall, the amount of genetic variation produced by sexual reproduction would drop.

Summary of Key Concepts Questions

29.1

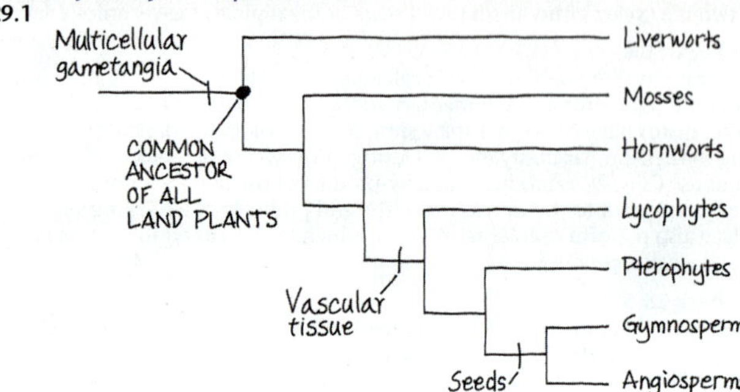

29.2 Some mosses colonize bare, sandy soils, leading to the increased retention of nitrogen in these otherwise low-nitrogen environments. Other mosses harbor nitrogen-fixing cyanobacteria that increase the availability of nitrogen in the ecosystem. The moss *Sphagnum* is often a major component of deposits of peat (partially decayed organic material). Boggy regions with thick layers of peat, known as peatlands, cover broad geographic regions and contain large reservoirs of carbon. By storing large amounts of carbon—in effect, removing CO_2 from the atmosphere—peatlands affect the global climate, making them of considerable ecological importance.
29.3 Lignified vascular tissue provided the strength needed to support a tall plant against gravity, as well as a means to transport water and nutrients to plant parts located high above ground. Roots were another key trait, anchoring the plant to the ground and providing additional structural support for plants that grew tall. Tall plants could shade shorter plants, thereby outcompeting them for light. Because the spores of a tall plant disperse farther than the spores of a short plant, it is also likely that tall plants could colonize new habitats more rapidly than short plants.

Test Your Understanding

1. b **2.** e **3.** d **4.** c **5.** a **6.** a. diploid; b. haploid; c. haploid; d. diploid; e. haploid **7.** c
8. Based on our current understanding of the evolution of major plant groups, the phylogeny has the four branch points shown here:

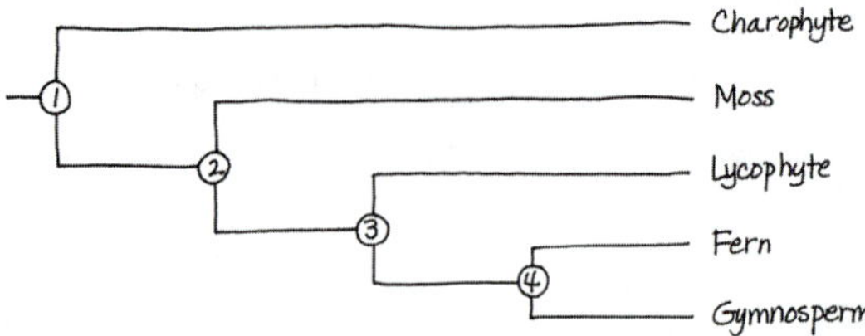

Derived characters unique to the charophyte and land plant clade (indicated by branch point 1) include rings of cellulose-synthesizing complexes, peroxisome enzymes, flagellated sperm structure, and a phragmoplast. Derived characters unique to the land plant clade (branch point 2) include apical meristems, alternation of generations, walled spores produced in sporangia, and multicellular gametangia. Derived characters unique to the vascular plant clade (branch point 3) include life cycles with dominant sporophytes, complex vascular systems (xylem and phloem), and well-developed roots and leaves. Derived characters unique to the pterophyte and seed plant clade (branch point 4) include megaphylls and overtopping growth.
9. (a)

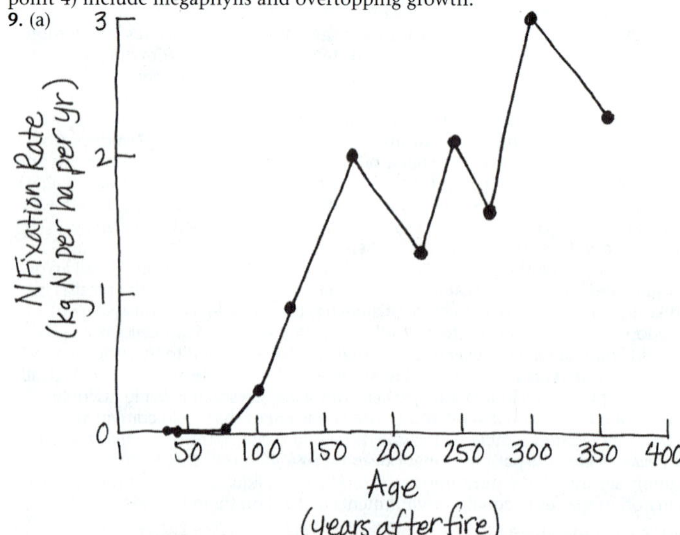

(b) In the first 40 years after a fire, nitrogen fixation rates were below 0.01 kg per ha per yr, which was less than 1% of the amount of nitrogen deposited from the atmosphere. Thus, in the initial decades after a fire, the moss *Pleurozium* and the nitrogen-fixing bacteria it harbors had relatively little effect on the amount of nitrogen added to the forest. With time, however, *Pleurozium* and its symbiotic, nitrogen-fixing bacteria became increasingly important. By 170 years after a fire, the percentage of the ground surface covered by the moss had increased to about 70%, leading to a corresponding increase in populations of the symbiotic bacteria. As would be predicted from this result, in older forests considerably more nitrogen (130–300%) was added by nitrogen fixation than was deposited from the atmosphere.

Chapter 30

Figure Questions

Figure 30.2 Retaining the gametophyte within the sporophyte shields the egg-containing gametophyte from UV radiation. UV radiation is a mutagen. Hence, we would expect fewer mutations to occur in the egg cells produced by a gametophyte retained within the body of a sporophyte. Most mutations are harmful. Thus, the fitness of embryos should increase because fewer embryos would carry harmful mutations. **Figure 30.3** Three generations: (1) the current sporophyte (cells of ploidy $2n$, found in the seed coat and in the megasporangium remnant that surrounds the spore wall); (2) the female gametophyte (cells of ploidy n, found in the food supply); and (3) the sporophyte of the next generation (cells of ploidy $2n$, found in the embryo) **Figure 30.6** Mitosis. A single haploid megaspore divides by mitosis to produce a multicellular, haploid female gametophyte. (Likewise, a single haploid microspore divides by mitosis to produce a multicellular male gametophyte.) **Figure 30.12** No. The branching order shown could still be correct if *Amborella* and other early angiosperms had originated prior to 150 million years ago, but angiosperm fossils of that age had not yet been discovered. In such a situation, the 140-million-year-old date for the origin of the angiosperms shown on the phylogeny would be incorrect. **Figure 30.16** Temperatures and the amount of sunlight striking the forest floor would both be likely to increase, whereas rainfall would likely decrease. Each of these changes could have dramatic effects on forest species that live near the edges of remaining forest fragments.

Concept Check 30.1

1. To reach the eggs, the flagellated sperm of seedless plants must swim through a film of water, usually over a distance of no more than a few centimeters. In contrast, the sperm of seed plants do not require water because they are produced within pollen grains that can be transported long distances by wind or by animal pollinators. Although flagellated in some species, the sperm of seed plants do not require mobility because pollen tubes convey them from the point at which the pollen grain is deposited (near the ovules) directly to the eggs. **2.** The reduced gametophytes of seed plants are nurtured by sporophytes and protected from stress, such as drought conditions and UV radiation. Pollen grains, with walls containing sporopollenin, provide protection during transport by wind or animals. Seeds have one or two layers of protective tissue, the seed coat, that improve survival by providing more protection from environmental stresses than do the walls of spores. Seeds also contain a stored supply of food, which provides nourishment for growth after dormancy is broken and the embryo emerges as a seedling. **3.** If a seed could not enter dormancy, the embryo would continue to grow after it was fertilized. As a result, the embryo might rapidly become too large to be dispersed, thus limiting its transport. The embryo's chance of survival might also be reduced because it could not delay growth until conditions become favorable.

Concept Check 30.2

1. Although gymnosperms are similar in not having their seeds enclosed in ovaries and fruits, their seed-bearing structures vary greatly. For instance, cycads have large cones, whereas some gymnosperms, such as *Ginkgo* and *Gnetum*, have small cones that look somewhat like berries, even though they are not fruits. Leaf shape also varies greatly, from the needles of many conifers to the palmlike leaves of cycads to *Gnetum* leaves that look like those of flowering plants. **2.** The pine life cycle illustrates heterospory, as ovulate cones produce megaspores and pollen cones produce microspores. The reduced gametophytes are evident in the form of the microscopic pollen grains that develop from microspores and the microscopic female gametophyte that develops from the megaspore. The egg is shown developing within an ovule, and a pollen tube is shown conveying the sperm. The figure also shows the protective and nutritive features of a seed. **3.** No. Fossil evidence indicates that gymnosperms originated at least 305 million years ago, but this does not mean that angiosperms are that old—only that the most recent common ancestor of gymnosperms and angiosperms must be that old.

Concept Check 30.3

1. In the oak's life cycle, the tree (the sporophyte) produces flowers, which contain gametophytes in pollen grains and ovules; the eggs in ovules are fertilized; the mature ovaries develop into dry fruits called acorns. We can view the oak's life cycle as starting when the acorn seeds germinate, resulting in embryos giving rise to seedlings and finally to mature trees, which produce flowers—and then more acorns. **2.** Pine cones and flowers both have sporophylls, modified leaves that produce spores. Pine trees have separate pollen cones (with pollen grains) and ovulate cones (with ovules inside cone scales). In flowers, pollen grains are produced by the anthers of stamens, and ovules are within the ovaries of carpels. Unlike pine cones, many flowers produce both pollen and ovules. **3.** The fact that the clade with bilaterally symmetrical flowers had more species establishes a correlation between flower shape and the rate of plant speciation. Flower shape is not necessarily responsible for the result because the shape (that is, bilateral or radial symmetry) may have been correlated with another factor that was the actual cause of the observed result. Note, however, that flower shape was associated with increased speciation rates when averaged across 19 different pairs of plant lineages. Since these 19 lineage pairs were independent of one another, this association suggests—but does not establish—that differences in flower shape cause differences in speciation rates. In general, strong evidence for causation can come from controlled, manipulative experiments, but such experiments are usually not possible for studies of past evolutionary events.

Concept Check 30.4

1. Plant diversity can be considered a resource because plants provide many important benefits to humans; as a resource, plant diversity is nonrenewable because if a species is lost to extinction, that loss is permanent. **2.** A detailed phylogeny of the seed plants would identify many different monophyletic groups of seed plants. Using this phylogeny, researchers could look for clades that contained species in which medicinally useful compounds had already been discovered. Identification of such clades would allow researchers to concentrate their search for new medicinal compounds among clade members—as opposed to searching for new compounds in species that were selected at random from the more than 250,000 existing species of seed plants.

Summary of Key Concepts Questions

30.1 The integument of an ovule develops into the protective seed coat of a seed. The ovule's megaspore develops into a haploid female gametophyte, and two parts of the seed are related to that gametophyte: The food supply of the seed is derived from haploid gametophyte cells, and the embryo of the seed develops after the female gametophyte's egg cell is fertilized by a sperm cell. A remnant of the ovule's megasporangium surrounds the spore wall that encloses the seed's food supply and embryo. **30.2** Gymnosperms arose about 305 million years ago, making them a successful group in terms of their evolutionary longevity. Gymnosperms have the five derived traits common to all seed plants (reduced gametophytes, heterospory, ovules, pollen, and seeds), making them well adapted for life on land. Finally, because gymnosperms dominate immense geographic regions today, the group is also highly successful in geographic distribution. **30.3** The origin of flowering plants is puzzling because their distinctive features, flowers and fruits, bear little resemblance to structures in living gymnosperms. This makes it difficult to infer how flowers and fruits arose, leading Darwin to refer to the origin of flowering plants as an "abominable mystery." Progress has been made toward solving this mystery, particularly in our understanding of angiosperm phylogeny, but the mystery has yet to be solved. We still do not know, for example, which extinct group of seed plants is most closely related to flowering plants. **30.4** The loss of tropical forests could contribute to global warming (which would have negative effects on many human societies) and reduce agricultural production in some of the world's poorest regions. People also depend on Earth's biodiversity for many products and services and hence would be harmed by the loss of species that would occur if the world's remaining tropical forests were cut down. With respect to a possible mass extinction, tropical forests harbor at least 50% of the species on Earth. If the remaining tropical forests were destroyed, large numbers of these species could be driven to extinction, thus rivaling the losses that occurred in the five mass extinction events documented in the fossil record.

Test Your Understanding

1. d **2.** a **3.** b **4.** a **5.** d

6.

Charophyte green algae
Mosses
Ferns
Gymnosperms
Angiosperms

8. (a)

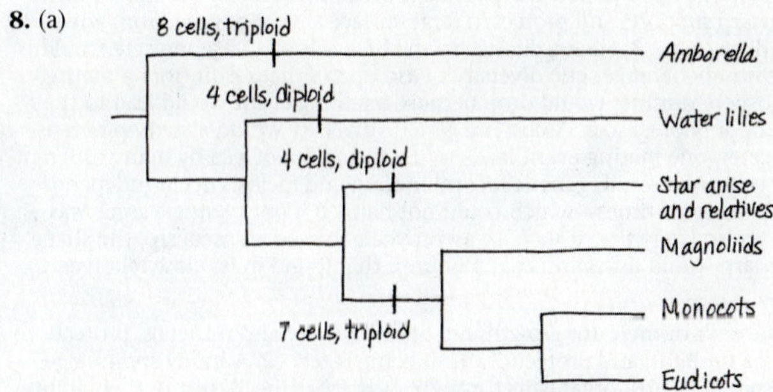

(b) The phylogeny indicates that basal angiosperms differed from other angiosperms in terms of the number of cells in female gametophytes and the ploidy of the endosperm. The ancestral state of the angiosperms cannot be determined from these data alone. It is possible that the common ancestor of angiosperms had seven-celled female gametophytes and triploid endosperm and hence that the eight-celled and four-celled conditions found in basal angiosperms represent derived traits for those lineages. Alternatively, either the eight-celled or four-celled condition may represent the ancestral state.

Chapter 31

Figure Questions

Figure 31.2 DNA from each of these mushrooms would be identical if each mushroom is part of a single hyphal network, as is likely. **Figure 31.16** One or both of the following would apply to each species: DNA analyses would reveal that it is a member of the ascomycete clade, or aspects of its sexual life cycle would indicate that it is an ascomycete (for example, it would produce asci and ascospores). **Figure 31.21** Two possible controls would be E−P− and E+P−. Results from an E−P− control could be compared with results from the E−P+ experiment, and results from an E+P− control could be compared with results from the E+P+ experiment. Together, these two comparisons would indicate whether the addition of the pathogen causes an increase in leaf mortality. Results from an E−P− experiment could also be compared with results from the second control (E+P−) to determine whether adding the endophytes has a negative effect on the plant. **Figure 31.26** The correspondence between when the chytrid fungi arrived to the region and when frog populations decreased greatly suggests that the chytrid was the cause of the decline, not just correlated to it. The fact that by 2009, the only surviving frogs were in the two lakes treated with fungicides provides additional support for causation.

Concept Check 31.1

1. Both a fungus and a human are heterotrophs. Many fungi digest their food externally by secreting enzymes into the food and then absorbing the small molecules that result from digestion. Other fungi absorb such small molecules directly from their environment. In contrast, humans (and most other animals) ingest relatively large pieces of food and digest the food within their bodies. **2.** The ancestors of such a mutualist most likely secreted powerful enzymes to digest the body of their insect host. Since such enzymes would harm a living host, it is likely that the mutualist would not produce such enzymes or would restrict their secretion and use. **3.** Carbon that enters the plant through stomata is fixed into sugar through photosynthesis. Some of these sugars are absorbed by the fungus that partners with the plant to form mycorrhizae; others are transported within the plant body and used in the plant. Thus, the carbon may be deposited in either the body of the plant or the body of the fungus.

Concept Check 31.2

1. The majority of the fungal life cycle is spent in the haploid stage, whereas the majority of the human life cycle is spent in the diploid stage. **2.** The two mushrooms might be reproductive structures of the same mycelium (the same organism). Or they might be parts of two separate organisms that have arisen from a single parent organism through asexual reproduction and thus carry the same genetic information.

Concept Check 31.3

1. DNA evidence indicates that fungi, animals, and their protistan relatives form a clade, the opisthokonts. Furthermore, an early-diverging fungal lineage, the chytrids, have posterior flagella, as do most other opisthokonts. This suggests that other fungal lineages lost their flagella after diverging from chytrids. **2.** This indicates that fungi had already established mutualistic relationships with plants by the date the fossils of the earliest vascular plants had formed. **3.** Fungi are heterotrophs. Prior to the colonization of land by plants, terrestrial fungi would have lived where other organisms (or their remains) were present and provided a source of food. Thus, if fungi had colonized land before plants, they could have fed on any prokaryotes or protists that lived on land or by the water's edge—but not on the plants or animals on which many fungi feed today.

Concept Check 31.4

1. Flagellated spores; molecular evidence also suggests that chytrids are an early-diverging fungal lineage. **2.** Possible answers include the following: In zygomycetes, the sturdy, thick-walled zygosporangium can withstand harsh conditions and then undergo karyogamy and meiosis when the environment is favorable for reproduction. In glomeromycetes, the hyphae have a specialized morphology that enables the fungi to form arbuscular mycorrhizae with plant roots. In ascomycetes, the asexual spores (conidia) are often produced in chains or clusters at the tips of conidiophores, where they are easily dispersed by wind. The often cup-shaped ascocarps house the sexual spore-forming asci. In basidiomycetes, the basidiocarp supports and protects a large surface area of basidia, from which spores are dispersed. **3.** Such a change to the life cycle of an ascomycete would reduce the number and genetic diversity of ascospores that result from a mating event. Ascospore number would drop because a mating event would lead to the formation of only one ascus. Ascospore genetic diversity would also drop because in ascomycetes, one mating event leads to the formation of asci by many different dikaryotic cells. As a result, genetic recombination and meiosis occur independently many different times—which could not happen if only a single ascus was formed. It is also likely that if such an ascomycete formed an ascocarp, the shape of the ascocarp would differ considerably from that found in its close relatives.

Concept Check 31.5

1. A suitable environment for growth, retention of water and minerals, protection from intense sunlight, and protection from being eaten **2.** A hardy spore stage enables dispersal to host organisms through a variety of mechanisms; their ability to grow rapidly in a favorable new environment enables them to capitalize on the host's resources. **3.** Many different outcomes might have occurred. Organisms that currently form mutualisms with fungi might have gained the ability to perform the tasks currently done by their fungal partners, or they might have formed similar mutualisms with other organisms (such as bacteria). Alternatively, organisms that currently form mutualisms with fungi might be less effective at living in their present environments. For example, the colonization of land by plants might have been more difficult. And if plants did eventually colonize land without fungal mutualists, natural selection might have favored plants that formed more highly divided and extensive root systems (in part replacing mycorrhizae).

Summary of Key Concepts Questions

31.1 The body of a multicellular fungus typically consists of thin filaments called hyphae. These filaments form an interwoven mass (mycelium) that penetrates the substrate on which the fungus grows and feeds. Because the individual filaments are thin, the surface-to-volume ratio of the mycelium is maximized, making nutrient absorption highly efficient.

31.2

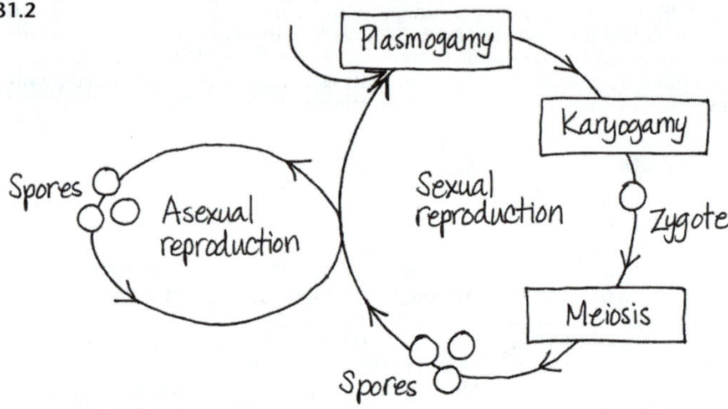

31.3 Phylogenetic analyses show that fungi and animals are more closely related to each other than either is to other multicellular eukaryotes (such as plants or multicellular algae). These analyses also show that fungi are more closely related to single-celled protists called nucleariids than they are to animals, whereas animals are more closely related to a different group of single-celled protists, the choanoflagellates, than they are to fungi. In combination, these results indicate that multicellularity evolved in fungi and animals independently, from different single-celled ancestors.

31.4

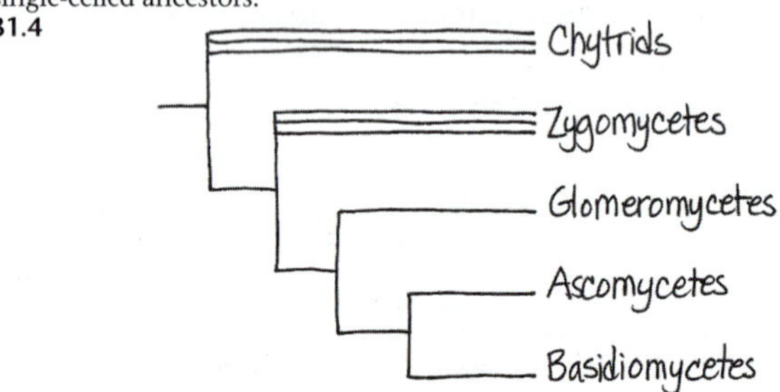

31.5 As decomposers, fungi break down the bodies of dead organisms, thereby recycling elements between the living and nonliving environments. Without the activities of fungi and bacterial decomposers, essential nutrients would remain tied up in organic matter, and life would cease. As an example of their key role as mutualists, fungi form mycorrhizal associations with plants. These associations improve the growth and survival of plants, thereby indirectly affecting the many other species (humans included) that depend on plants. As pathogens, fungi harm other species. In some cases, fungal pathogens have caused their host populations to decline across broad geographic regions, as seen for the American chestnut.

Test Your Understanding

1. b **2.** c **3.** d **4.** b **5.** a **6.** e

7.

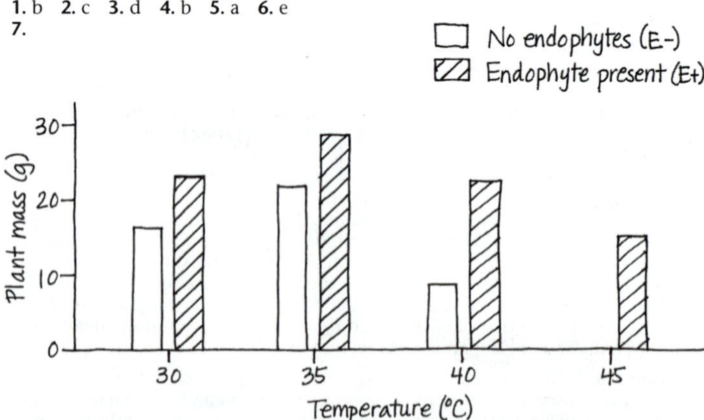

As indicated by the raw data and bar graph, grass plants with endophytes (E+) produced more new shoots and had greater biomass than did grass plants that lacked endophytes (E−). These differences were especially pronounced at the highest soil temperature, where E− grass plants produced no new shoots and had a biomass of zero (indicating they were dead).

Chapter 32

Figure Questions

Figure 32.3 As described in **1** and **2**, choanoflagellates and a broad range of animals have collar cells. Since collar cells have never been observed in plants, fungi, or non-choanoflagellate protists, this suggests that choanoflagellates may be more closely related to animals than to other eukaryotes. If choanoflagellates are more closely related to animals than to any other group of eukaryotes, choanoflagellates and animals should share other traits that are not found in other eukaryotes. The data described in ❸ are consistent with this prediction. **Figure 32.6** The sea anemone embryos could be infused with a protein that can bind to β-catenin's DNA-binding site, thereby limiting the extent to which β-catenin activates the transcription of genes necessary for gastrulation. Such an experiment would provide an independent check of the results shown in step 4. **Figure 32.9** The cells of an early embryo with deuterostome development typically are not committed to a particular developmental fate, whereas the cells of an early embryo with protostome development typically are committed to a particular developmental fate. As a result, an embryo with deuterostome development would be more likely to contain stem cells that could give rise to cells of any type. **Figure 32.10** Ctenophora is the sister phylum in this figure, while Cnidaria is the sister phylum in Figure 32.11.

Concept Check 32.1

1. In most animals, the zygote undergoes cleavage, which leads to the formation of a blastula. Next, in gastrulation, one end of the embryo folds inward, producing layers of embryonic tissue. As the cells of these layers differentiate, a wide variety of animal forms result. Despite the diversity of animal forms, animal development is controlled by a similar set of *Hox* genes across a broad range of taxa. **2.** The imaginary plant would require tissues composed of cells that were analogous to the muscle and nerve cells found in animals: "Muscle" tissue would be necessary for the plant to chase prey, and "nerve" tissue would be required for the plant to coordinate its movements when chasing prey. To digest captured prey, the plant would need to either secrete enzymes into one or more digestive cavities (which could be modified leaves, as in a Venus flytrap) or secrete enzymes outside of its body and feed by absorption. To extract nutrients from the soil—yet be able to chase prey—the plant would need something other than fixed roots, perhaps retractable "roots" or a way to ingest soil. To conduct photosynthesis, the plant would require chloroplasts. Overall, such an imaginary plant would be very similar to an animal that had chloroplasts and retractable roots. **3.** As described in Chapter 18, miRNAs help regulate gene expression by binding to and hence blocking the translation of particular mRNA molecules, their "target mRNAs." Because each miRNA can bind to many target mRNAs, an increase in the number of miRNA molecules could potentially add many layers of control to the regulation of gene expression. As a result of a more varied and complex control of gene expression, the construction of a complex body form would be more likely in an organism with many miRNAs than in an organism with few miRNAs, even though both organisms had roughly the same total number of genes.

Concept Check 32.2

1. c, b, a, d **2.** We cannot infer whether animals originated before or after fungi. If correct, the date provided for the most recent common ancestor of fungi and animals would indicate that animals originated some time within the last billion years. The fossil record indicates that animals originated at least 565 million years ago. Thus, we could conclude only that animals originated sometime between 1 billion years ago and 565 million years ago.

Concept Check 32.3

1. Grade-level characteristics are those that multiple lineages share regardless of evolutionary history. Some grade-level characteristics may have evolved multiple times independently. Features that unite clades are derived characteristics that originated in a common ancestor and were passed on to the various descendants. **2.** A snail has a spiral and determinate cleavage pattern; a human has radial, indeterminate cleavage. In a snail, the coelomic cavity is formed by splitting of mesoderm masses; in a human, the coelom forms from folds of archenteron. In a snail, the mouth forms from the blastopore; in a human, the anus develops from the blastopore. **3.** Most coelomate triploblasts have two openings to their digestive tract, a mouth and an anus. As such, their bodies have a structure that is analogous to that of a doughnut: The digestive tract (the hole of the doughnut) runs from the mouth to the anus and is surrounded by various tissues (the solid part of the doughnut). The doughnut analogy is most obvious at early stages of development (see Figure 32.9c).

Concept Check 32.4

1. Cnidarians possess true tissues, while sponges do not. Also unlike sponges, cnidarians exhibit body symmetry, though it is radial and not bilateral as in other animal phyla. **2.** The morphology-based tree divides Bilateria into two major clades: Deuterostomia and Protostomia. The molecular-based tree recognizes three major clades: Deuterostomia, Ecdysozoa, and Lophotrochozoa. **3.** The phylogeny in Figure 32.11 indicates that molluscs are members of Lophotrochozoa, one the three main groups of bilaterians (the others being Deuterostomia and Ecdysozoa). As seen in Figure 25.10, the fossil record shows that molluscs were present tens of millions of years before the Cambrian explosion. Thus, long before the Cambrian explosion, the lophotrochoan clade had formed and was evolving independently of the evolutionary lineages leading to Deuterostomia and Ecdysozoa. Based on the phylogeny in Figure 32.11, we can also conclude that the lineages leading to Deuterostomia and Ecdysozoa were independent of one another before the Cambrian explosion. Since the lineages leading to the three main clades of bilaterians were evolving independently of one another prior to the Cambrian explosion, that explosion could be viewed as consisting of three "explosions," not one.

Summary of Key Concepts Questions

Concept 32.1 Unlike animals, which are heterotrophs that ingest their food, plants are autotrophs, and fungi are heterotrophs that grow on their food and feed by absorption. Animals lack cell walls, which are found in both plants and fungi. Animals also have muscle tissue and nerve tissue, which are not found in either plants or fungi. In addition, the sperm and egg cells of animals are produced by meiotic division, unlike what occurs in plants and fungi (where reproductive cells such as sperm and eggs are produced by mitotic division). Finally, animals regulate the development of body form with *Hox* genes, a unique group of genes that is not found in either plants or fungi. **Concept 32.2** Current hypotheses about the cause of the Cambrian explosion include new predator-prey relationships, an increase in atmospheric oxygen, and an increase in developmental flexibility provided by the origin of *Hox* genes and other genetic changes. **Concept 32.3** Body plans provide a helpful way to compare and contrast key features of organisms. However, phylogenetic analyses show that similar body plans have arisen independently in different groups of organisms. As such, similar body plans may have arisen by convergent evolution and hence may not be informative about evolutionary relationships. **Concept 32.4** To reconstruct the evolutionary history of animal life, researchers collect morphological and molecular data and use cladistic methods to analyze that data. In a cladistic approach, shared derived (morphological and molecular) characters are used to place organisms into a nested hierarchy of monophyletic clades.

Test Your Understanding

1. a **2.** d **3.** e **4.** c **5.** b

7.

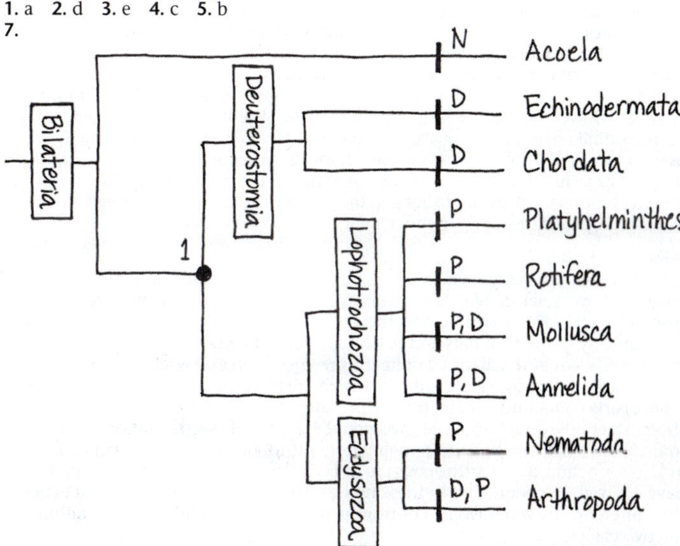

Based on the phylogeny, the ancestral condition in bilaterians may have been similar to that in Acoela, where the blastopore closes and the mouth forms elsewhere (N); however, it is also possible that blastopore fate in Acoela is a derived trait and hence is not informative about the ancestral condition. Although the phylogeny indicates that blastopore fate has changed multiple times over the course of evolution, a precise estimate cannot be made. For example, if we assume that the common ancestor of all non-Acoela bilaterians (marked with a 1 on the tree) exhibited protostomy, then blastopore fate has changed at least five times: once in common ancestor 1, once in Deuterostomia, at least once in Mollusca, at least once in Annelida, and at least once in Arthropoda. Other assumptions would lead to different estimates.

Chapter 33

Figure Questions

Figure 33.8 Within a reproductive polyp, a cell that gives rise to a medusa would have to divide by meiosis. A resulting haploid cell would then divide repeatedly (by mitosis), forming a haploid medusa. Later, cells in the medusa's gonads would divide by mitosis, forming the haploid eggs and sperm. **Figure 33.11** Adding fertilizer to the water supply would probably increase the abundance of algae, and that, in turn, would likely increase the abundance of snails (which eat algae). If the water was also contaminated with infected human feces, an increase in the number of snails would likely lead to an increase in the abundance of blood flukes (which require snails as an intermediate host). As a result, the occurrence of schistosomiasis might increase. **Figure 33.22** The extinctions of freshwater bivalves might lead to an increase in the abundance of photosynthetic protists and bacteria. Because these organisms are at the base of aquatic food webs, increases in their abundance could have major effects on aquatic communities (including both increases and decreases in the abundance of other species). **Figure 33.29** Such a result would be consistent with the *Ubx* and *abd-A Hox* genes having played a major role in the evolution of increased body segment diversity in arthropods. However, by itself, such a result would simply show that the presence of the *Ubx* and *abd-A Hox* genes was *correlated with* an increase in body segment diversity in arthropods; it would not provide direct experimental evidence that the acquisition of the *Ubx* and *adb-A* genes *caused* an increase in arthropod body segment diversity.

Concept Check 33.1

1. The flagella of choanocytes draw water through their collars, which trap food particles. The particles are engulfed by phagocytosis and digested, either by choanocytes or by amoebocytes. **2.** The collar cells of sponges (and some other animals—see Chapter 32) bear a striking resemblance to a choanoflagellate cell. This suggests that the last common ancestor of animals and their protist sister group may have resembled a choanoflagellate. Nevertheless, mesomycetozoans could still be the sister group of animals. If this is the case, the lack of collar cells in mesomycetozoans would indicate that over time their structure evolved in ways that caused it to no longer resemble a choanoflagellate cell. It is also possible that choanoflagellates and sponges share similar-looking collar cells as a result of convergent evolution.

Concept Check 33.2

1. Both the polyp and the medusa are composed of an outer epidermis and an inner gastrodermis separated by a gelatinous layer, the mesoglea. The polyp is a cylindrical form that adheres to the substrate by its aboral end; the medusa is a flattened, mouth-down form that moves freely in the water. **2.** Cnidarian stinging cells (cnidocytes) function in defense and prey capture. They contain capsule-like organelles (cnidae), which in turn contain coiled threads. The threads either inject poison or stick to and entangle small prey. **3.** Evolution is not goal oriented; hence, it would not be correct to argue that cnidarians were not "highly evolved" simply because their form had changed relatively little over the past 560 million years. Instead, the fact that cnidarians have persisted for hundreds of millions of years indicates that the cnidarian body plan is a highly successful one.

Concept Check 33.3

1. Tapeworms can absorb food from their environment and release ammonia into their environment through their body surface because their body is very flat, due in part to the lack of a coelom. **2.** The inner tube is the alimentary canal, which runs the length of the body. The outer tube is the body wall. The two tubes are separated by the coelom. **3.** All molluscs have inherited a foot from their common ancestor. However, in different groups of molluscs, the structure of the foot has been modified over time (by natural selection) in ways that reflect how the foot is used in locomotion by members of each clade. In gastropods, the foot is used as a holdfast or to move slowly on the substrate. In cephalopods, the foot has been modified into part of the tentacles and into an excurrent siphon, through which water is propelled (resulting in movement in the opposite direction).

Concept Check 33.4

1. Nematodes lack body segments and a true coelom; annelids have both. **2.** The arthropod exoskeleton, which had already evolved in the ocean, allow terrestrial species to retain water and support their bodies on land. Wings allow them to disperse quickly to new habitats and to find food and mates. The tracheal system allows for efficient gas exchange despite the presence of an exoskeleton. **3.** Arthropod mouthparts are modified appendages, which are bilaterally paired. As a result, the mouthparts come into contact by moving laterally, not up and down. **4.** Yes. Under the traditional hypothesis, we would expect body segmentation to be controlled by similar *Hox* genes in annelids and arthropods. However, if annelids are in Lophotrochozoa and arthropods are in Ecdysozoa, body segmentation may have evolved independently in these two groups. In such a case, we might expect that different *Hox* genes would control the development of body segmentation in the two clades.

Concept Check 33.5

1. Each tube foot consists of an ampulla and a podium. When the ampulla squeezes, it forces water into the podium, which causes the podium to expand and contact the substrate. Adhesive chemicals are then secreted from the base of the podium, thereby attaching the podium to the substrate. **2.** Both insects and nematodes are members of Ecdysozoa, one of the three major clades of bilaterians. Therefore, a characteristic shared by *Drosophila* and *Caenorhabditis* may be informative for other members of their clade—but not necessarily for members of Deuterostomia. Instead, Figure 33.2 suggests that a species within Echinodermata or Chordata might be a more appropriate invertebrate model organism from which to draw inferences about humans and other vertebrates. **3.** Echinoderms include species with a wide range of body forms, some of which are shown in Figure 33.40. However, even echinoderms that look very different from one another, such as sea stars and sea cucumbers, share characteristics unique to their phylum, including a water vascular system and tube feet. The differences between echinoderm species illustrate the diversity of life, while the characteristics they share illustrate the unity of life. The match between organisms and their environments can be seen in such echinoderm features as the eversible stomachs of sea stars (enabling them to digest prey that are larger than their mouth) and the complex, jaw-like structure that sea urchins use to eat seaweed.

Summary of Key Concepts Questions

Concept 33.1 The sponge body consists of two layers of cells, both of which are in contact with water. As a result, gas exchange and waste removal occur as substances diffuse into and out of the cells of the body. Choanocytes and amoebocytes ingest food particles from the surrounding water. Choanocytes also release food particles to amoebocytes, which then digest the food particles and deliver nutrients to other cells. **Concept 33.2** The cnidarian body plan consists of a sac with a central digestive compartment, the gastrovascular cavity. The single opening to this compartment serves as both a mouth and an anus. The two main variations on this body plan are sessile polyps (which adhere to the substrate at the end of the body opposite to the mouth/anus) and motile medusae (which move freely through the water and resemble flattened, mouth-down versions of polyps). **Concept 33.3** No. Some lophotrochozoans have a crown of ciliated tentacles that function in feeding (called a lophophore), while others go through a distinctive developmental stage known as trochophore larvae. Many other lophotrochozoans do not have either of these features. As a result, the clade is defined primarily by DNA similarities, not morphological similarities. **Concept 33.4** Many nematode species live in soil and in sediments on the bottom of bodies of water. These free-living species play important roles in decomposition and nutrient cycling. Other nematodes are parasites, including many species that attack the roots of plants and some that attack animals (including humans). Arthropods have profound effects on all aspects of ecology. In aquatic environments, crustaceans play key roles as grazers (of algae), scavengers, and predators, and some species, such as krill, are important sources of food for whales and other vertebrates. On land, it is difficult to think of features of the natural world that are not affected in some way by insects and other arthropods, such as spiders and ticks. There are more than 1 million species of insects, many of which have enormous ecological effects as herbivores, predators, parasites, decomposers, and vectors of disease. Insects are also key sources of food for many organisms, including humans in some regions of the world. **Concept 33.5** Echinoderms and chordates are both members of Deuterostomia, one of the three main clades of bilaterian animals. As such, chordates (including humans) are more closely related to echinoderms than we are to animals in any of the other phyla covered in this chapter. Nevertheless, echinoderms and chordates have evolved independently for over 500 million years. This statement does not contradict the close relationship of echinoderms and chordates, but it does make clear that "close" is a relative term indicating that these two phyla are more closely related to each other than either is to animal phyla not in Deuterostomia.

Test Your Understanding

1. a **2.** d **3.** b **4.** e **5.** c **6.** d
7.

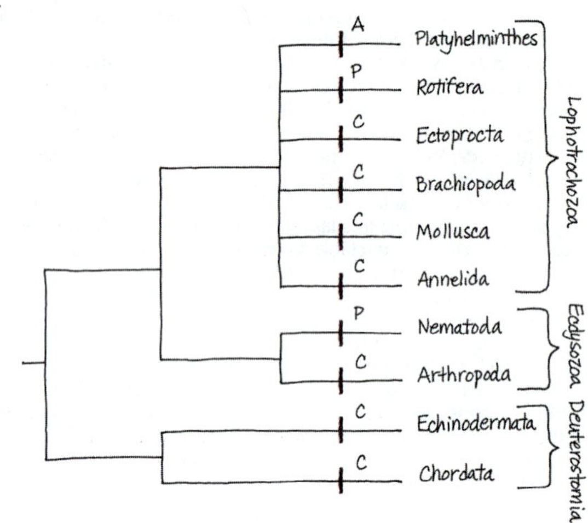

(a) Both phyla in Deuterostomia are coelomates, suggesting that their most recent common ancestor had a true coelom. Lophotrochozoa contains one phylum of acoelomates (Platyhelminthes), one phylum of pseudocoelomates (Rotifera), and four phyla of coelomates (Ectoprocta, Brachiopoda, Mollusca, Annelida); thus, we cannot, from this information alone, infer the condition of the most recent common ancestor shared by these phyla. Similarly, since Ecdysozoa contains one phylum of pseudocoelomates (Nematoda) and one phylum of coelomates (Arthropoda), we cannot infer whether their most recent common ancestor had a true coelom or not. (b) Depending on whether or not the last common ancestor of Bilateria had a true coelom, the presence of a true coelom has either been lost or gained multiple times during the evolutionary history of bilaterians. Thus, the presence of a true coelom appears to have changed over the course of evolution.

Chapter 34

Figure Questions

Figure 34.6 Results in these figures suggest that specific *Hox* genes, as well as the order in which they are expressed, have been highly conserved over the course of evolution. **Figure 34.20** *Tiktaalik* was a lobe-fin fish that had both fish and tetrapod characters. Like a fish, *Tiktaalik* had fins, scales, and gills. As described by Darwin's concept of descent with modification, such shared characters can be attributed to descent from ancestral species—in this case, *Tiktaalik's* descent from fish ancestors. *Tiktaalik* also had traits that were unlike a fish, but like a tetrapod, including a flat skull, a neck, a full set of ribs, and the skeletal structure of its fin. These characters illustrate the second part of descent with modification, showing how ancestral features had become modified over time. **Figure 34.21** Sometime between 380 mya and 340 mya. We can infer this because amphibians must have originated after the most recent common ancestor of *Tulerpeton* and living tetrapods (and that ancestor originated 380 mya), but no later than the date of the earliest known fossils of amphibians (shown in the figure as 340 mya). **Figure 34.25** Crocodilians. Among extant amniotes, crocodilians are the sister group of birds. Hence, it is likely that DNA sequences in birds are more similar to those in crocodilians than they are to those of more distantly related amniotes. **Figure 34.37** In general, the process of exaptation occurs as a structure that had one function acquires a different function via a series of intermediate stages. Each of these intermediate stages typically has some function in the organism in which it is found. The incorporation of articular and quadrate bones into the mammalian ear illustrates exaptation because these bones originally evolved as part of the jaw, where they functioned as the jaw hinge, but over time they became co-opted for another function, namely the transmission of sound. **Figure 34.43** The phylogeny shows humans as the sister group to the lineage that contains chimpanzees and

bonobos. This relationship is not consistent with humans as having descended from either chimpanzees or bonobos. If humans had descended from chimpanzees, for example, the human lineage would be nested within the chimpanzee lineage, much as birds are nested within the reptile clade (see Figure 34.25).
Figure 34.50 No. The phylogeny shown in RESULTS does not include any information on when the branch points occurred, nor is it scaled by time. Thus, although the phylogeny shows the relative order in which lineages diverged, we cannot infer when those events took place.

Concept Check 34.1

1. The four characters are a notochord; a dorsal, hollow nerve chord; pharyngeal slits or clefts; and a muscular, post-anal tail. **2.** In humans, these characters are present only in the embryo. The notochord becomes disks between the vertebrae, the tail is almost completely lost, and the pharyngeal clefts develop into various adult structures. **3.** Not necessarily. It would be possible that the chordate common ancestor had this gene, which was then lost in the lancelet lineage and retained in other chordates. However, it would also be possible that the chordate common ancestor lacked this gene; this could occur if the gene originated after lancelets diverged from other chordates but before tunicates diverged from other chordates.

Concept Check 34.2

1. Hagfishes have a head and skull made of cartilage, plus a small brain, sensory organs, and tooth-like structures. They have a neural crest, gill slits, and more extensive organ systems. In addition, hagfishes have slime glands that ward off predators and may repel competing scavengers. **2.** *Myllokunmingia*. Fossils of this organism provide evidence of ear capsules and eye capsules; these structures are part of the skull. Thus, *Myllokunmingia* is considered a craniate, as are humans. *Haikouella* did not have a skull. **3.** Such a finding suggests that early organisms with a head were favored by natural selection in several different evolutionary lineages. However, while a logical argument can be made that having a head was advantageous, fossils alone do not constitute proof.

Concept Check 34.3

1. Lampreys have a round, rasping mouth, which they use to attach to fish. Conodonts had two sets of mineralized dental elements, which may have been used to impale prey and cut it into smaller pieces. **2.** In armored jawless vertebrates, bone served as external armor that may have provided protection from predators. Some species also had mineralized mouthparts, which could be used for either predation or scavenging. Still others had mineralized fin rays, which may have enabled them to swim more rapidly and with greater steering control.

Concept Check 34.4

1. Both are gnathostomes and have jaws, four clusters of *Hox* genes, enlarged forebrains, and lateral line systems. Shark skeletons consist mainly of cartilage, whereas tuna have bony skeletons. Sharks also have a spiral valve. Tuna have an operculum and a swim bladder, as well as flexible rays supporting their fins.
2. Aquatic gnathostomes have jaws (an adaptation for feeding) and paired fins and a tail (adaptations for swimming). Aquatic gnathostomes also typically have streamlined bodies for efficient swimming and swim bladders or other mechanisms (such as oil storage in sharks) for buoyancy. **3.** Yes, that could have happened. The paired appendages of aquatic gnathostomes other than the lobe-fins could have served as a starting point for the evolution of limbs. The colonization of land by aquatic gnathostomes other than the lobe-fins might have been facilitated in lineages that possessed lungs, as that would have enabled those organisms to breathe air.

Concept Check 34.5

1. Tetrapods are thought to have originated about 365 million years ago when the fins of some lobe-fins evolved into the limbs of tetrapods. In addition to their four limbs with digits—a key derived trait for which the group is named—other derived traits of tetrapods include a neck (consisting of vertebrae that separate the head from the rest of the body), a pelvic girdle that is fused to the backbone, and a lack of gill slits. **2.** Some fully aquatic species are paedomorphic, retaining larval features for life in water as adults. Species that live in dry environments may avoid dehydration by burrowing or living under moist leaves, and they protect their eggs with foam nests, viviparity, and other adaptations. **3.** Many amphibians spend part of their life cycle in aquatic environments and part on land. Thus, they may be exposed to a wide range of environmental problems, including water and air pollution and the loss or degradation of aquatic and/or terrestrial habitats. In addition, amphibians have highly permeable skin, providing relatively little protection from external conditions, and their eggs do not have a protective shell.

Concept Check 34.6

1. The amniotic egg provides protection to the embryo and allows the embryo to develop on land, eliminating the necessity of a watery environment for reproduction. Another key adaptation is rib cage ventilation, which improves the efficiency of air intake and may have allowed early amniotes to dispense with breathing through their skin. Finally, not breathing through their skin allowed amniotes to develop relatively impermeable skin, thereby conserving water. **2.** Yes. Although snakes lack limbs, they descended from lizards with legs. Some snakes retain vestigial pelvic and leg bones, providing evidence of their descent from an ancestor with legs. **3.** Birds have weight-saving modifications, including the absence of teeth, a urinary bladder, and a second ovary in females. The wings and feathers are adaptations that facilitate flight, and so are efficient respiratory and circulatory systems that support a high metabolic rate.

4.

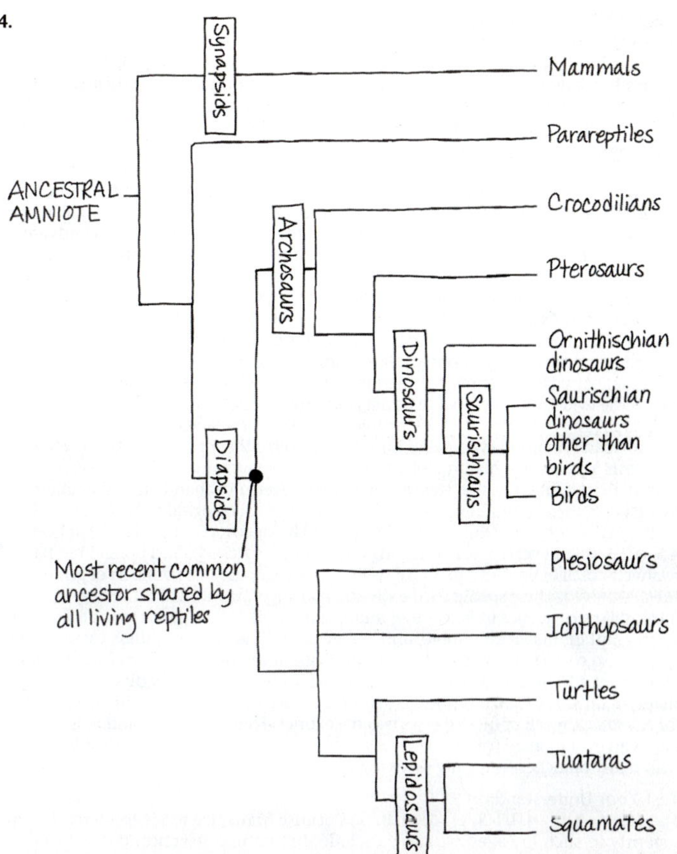

Under this convention, the reptiles would consist of all groups in Figure 34.25 except parareptiles and mammals.

Concept Check 34.7

1. Monotremes lay eggs. Marsupials give birth to very small live young that attach to a nipple in the mother's pouch, where they complete development. Eutherians give birth to more developed live young. **2.** Hands and feet adapted for grasping, flat nails, large brain, forward-looking eyes on a flat face, parental care, and movable big toe and thumb **3.** Mammals are endothermic, enabling them to live in a wide range of habitats. Milk provides young with a balanced set of nutrients, and hair and a layer of fat under the skin help mammals retain heat. Mammals have differentiated teeth, enabling them to eat many different kinds of food. Mammals also have relatively large brains, and many species are capable learners. Following the mass extinction at the end of the Cretaceous period, the absence of large terrestrial dinosaurs may have opened many new ecological niches to mammals, promoting their adaptive radiation. Continental drift also isolated many groups of mammals from one another, promoting the formation of many new species.

Concept Check 34.8

1. Hominins are a clade within the ape clade that includes humans and all species more closely related to humans than other apes. The derived characters of hominins include bipedal locomotion and relatively larger brains. **2.** In hominins, bipedal locomotion evolved long before large brain size. *Homo ergaster*, for example, was fully upright, bipedal, and as tall as modern humans, but its brain was significantly smaller than that of modern humans. **3.** Yes, both can be correct. *Homo sapiens* may have established populations outside of Africa as early as 115,000 years ago, as indicated by the fossil record. However, those populations may have left few or no descendants today. Instead, all living humans may have descended from Africans that spread from Africa roughly 50,000 years ago, as indicated by genetic data.

Summary of Key Concepts Questions

Concept 34.1 Lancelets are the most basal group of living chordates, and as adults they have key derived characters of chordates. This suggests that the chordate common ancestor may have resembled a lancelet in having an anterior end with a mouth along with the following four derived characters: a notochord; a dorsal, hollow nerve cord; pharyngeal slits or clefts; and a muscular, post-anal tail. **Concept 34.2** Craniates have a head and a more extensive muscular system than lancelets or tunicates. These features enable craniates to coordinate and perform more complex behaviors than found in lancelets and tunicates. Basal craniates (the hagfishes), for example, are scavengers that locate and feed on worms and dead or sick fishes. In contrast, lancelets and tunicates feed by filtering food items from the water. **Concept 34.3** Conodonts, among the earliest vertebrates in the fossil record, were very abundant for over 300 million years. While jawless, their well-developed teeth provide early signs of bone formation. Other species of jawless vertebrates developed armor on the outside of their bodies, which probably helped protect them from predators. Like lampreys, these species had paired fins for locomotion and an inner ear with semicircular canals that provided a sense of balance. There were many species of these armored jawless vertebrates, but they all became extinct by the close of the Devonian period, 359 million years ago. **Concept 34.4** The origin of jaws altered how fossil gnathostomes obtained food, which in turn had large effects on ecological interactions. Predators could use their jaws to grab prey or remove chunks of flesh, stimulating the evolution of increasingly sophisticated means of defense in prey species. Evidence for these changes can be found in the fossil record, which includes fossils of 10-m-long predators with remarkably powerful jaws, as well as lineages of well-defended prey species whose bodies were covered by armored plates. **Concept 34.5** Amphibians require water for reproduction; their bodies can lose water rapidly through their moist, highly permeable skin; and amphibian eggs do not have a shell and hence are vulnerable to desiccation. **Concept 34.6** Birds are descended from theropod dinosaurs, and dinosaurs are nested within the archosaur lineage, one of the two main reptile lineages. Thus, the other living archosaur reptiles, the crocodilians, are more closely related to birds than they are to non-archosaur reptiles such as lizards. As a result, birds are considered reptiles. (Note that if reptiles were defined as excluding birds, the reptiles would not form a clade; instead, the reptiles would be a paraphyletic group.) **Concept 34.7** Mammals are members of a group of amniotes called synapsids. Early (nonmammalian) synapsids laid eggs and had a sprawling gait. Fossil evidence shows that mammalian features arose gradually over a period of more than 100 million years. For example, the jaw was modified over time in nonmammalian synapsids, eventually coming to resemble that of a mammal. By 180 million years ago, the first mammals had appeared. There were many species of early mammals, but most of them were small, and they were not abundant or dominant members of their community. Mammals did not rise to ecological dominance until after the extinction of the dinosaurs. **Concept 34.8** The fossil record shows that from 4.5 to 2.5 million years ago, a wide range of hominin species walked upright but had relatively small brain sizes. About 2.5 million years ago, the first members of genus *Homo* emerged. These species used tools and had larger brains than those of earlier hominins. Fossil evidence indicates that multiple members of our genus were alive at any given point in time. Furthermore, until about 1.3 million years ago, these various *Homo* species also coexisted with members of earlier hominin lineages, such as *Paranthropus*. The different hominins alive at the same periods of time varied in body size, body shape, brain size, dental morphology, and the capacity for tool use. Ultimately, except for *Homo sapiens*, all of these species became extinct. Thus, human evolution is viewed not as an evolutionary path leading to *H. sapiens*, but rather as an evolutionary tree with many branches—the only surviving lineage of which is our own.

Test Your Understanding

1. e **2.** c **3.** a **4.** b **5.** c **6.** c **8.** (a) Because brain size tends to increase consistently in such lineages, we can conclude that natural selection favored the evolution of larger brains and hence that the benefits outweighed the costs. (b) As long as the benefits of brains that are large relative to body size are greater than the costs, large brains can evolve. Natural selection might favor the evolution of brains that are large relative to body size because such brains confer an advantage in obtaining mates and/or an advantage in survival.
(c)

Mortality rate (0 to 1.0) vs. Deviation of brain size from expected (−3 to 4)

Adult mortality tends to be lower in birds with larger brains.

Chapter 35

Figure Questions

Figure 35.9 The finding might suggest that the tan-colored trichomes deter the beetles by some means other than physically obstructing the beetles. Perhaps they contain a chemical that is harmful or distasteful to the beetles. **Figure 35.17** Pith and cortex are defined, respectively, as ground tissue that is internal and ground tissue that is external to vascular tissue. Since vascular bundles of monocot stems are scattered throughout the ground tissue, there is no clear distinction between internal and external relative to the vascular tissue. **Figure 35.19** The vascular cambium produces growth that increases the diameter of a stem or root. The tissues that are exterior to the vascular cambium cannot keep pace with the growth because their cells no longer divide. As a result, these tissues rupture. **Figure 35.31** Every root epidermal cell would develop a root hair. **Figure 35.33** Another example of homeotic gene mutation is the *Drosophila* mutation depicted in Figure 18.20, in which a mutation in a *Hox* gene causes legs to form in place of antennae. **Figure 35.34** The flower would consist of nothing but carpels.

Concept Check 35.1

1. The vascular tissue system connects leaves and roots, allowing sugars to move from leaves to roots in the phloem and allowing water and minerals to move to the leaves in the xylem. **2.** (a) large axillary buds; (b) petioles; (c) a bulb, an underground shoot with a small stem and large storage leaves; (d) storage roots **3.** To get sufficient energy from photosynthesis, we would need lots of surface area exposed to the sun. This large surface-to-volume ratio, however, would create a new problem—evaporative water loss. We would have to be permanently connected to a water source—the soil, also our source of minerals. In short, we would probably look and behave very much like plants. **4.** As plant cells enlarge, they typically form a huge central vacuole that contains a dilute watery sap. Central vacuoles enable plant cells to become large with only a minimal investment of new cytoplasm. The orientation of the cellulose microfibrils in plant cell walls affects the growth pattern of cells.

Concept Check 35.2

1. Primary growth arises from apical meristems and involves production and elongation of organs. Secondary growth arises from lateral meristems and adds to the girth of roots and stems. **2.** Your dividing cells are normally limited in the types of cells they can form. In contrast, the products of cell division in a plant meristem can differentiate into all the types of plant cells. **3.** The largest, oldest leaves would be lowest on the shoot. Since they would probably be heavily shaded, they would not photosynthesize much regardless of their size. **4.** No, the carrot roots will probably be smaller at the end of the second year because the food stored in the root will be used to produce flowers, fruits, and seeds.

Concept Check 35.3

1. In roots, primary growth occurs in three successive stages, moving away from the tip of the root: the zones of cell division, elongation, and differentiation. In shoots, it occurs at the tip of apical buds, with leaf primordia arising along the sides of an apical meristem. Most growth in length occurs in older internodes below the shoot tip. **2.** No. Because vertically oriented leaves, such as maize, can capture light equally on both sides of the leaf, you would expect them to have mesophyll cells that are not differentiated into palisade and spongy layers. This is typically the case. Also, vertical leaves usually have stomata on both leaf surfaces. **3.** Root hairs are cellular extensions that increase the surface area of the root epidermis, thereby enhancing the absorption of minerals and water. Microvilli are extensions that increase the absorption of nutrients by increasing the surface area of the gut.

Concept Check 35.4

1. The sign will still be 2 m above the ground because this part of the tree is no longer growing in length (primary growth); it is now growing only in thickness (secondary growth). **2.** Stomata must be able to close because evaporation is much more intensive from leaves than from the trunks of woody trees as a result of the higher surface-to-volume ratio in leaves. **3.** Since there is little temperature variation in the tropics, the growth rings of a tree from the tropics would be difficult to discern unless the tree came from an area that had pronounced wet and dry seasons. **4.** Girdling removes an entire ring of secondary phloem (part of the bark), completely preventing transport of sugars and starches from the shoots to the roots.

Concept Check 35.5

1. Although all the living vegetative cells of a plant have the same genome, they differentiate to have different forms and functions because of differential gene expression. **2.** Plants show indeterminate growth; juvenile and mature phases are found on the same individual plant; cell differentiation in plants is more dependent on final position than on lineage. **3.** In theory, tepals could arise if *B* gene activity was present in all three of the outer whorls of the flower.

Summary of Key Concepts Questions

35.1 Here are a few examples: The cuticle of leaves and stems protects these structures from desiccation. Collenchyma and sclerenchyma cells have thick walls that provide support for plants. Strong, branching root systems help anchor the plant in the soil. **35.2** All plant organs and tissues are ultimately derived by meristematic activity. **35.3** Lateral roots emerge from the pericycle and destroy plant cells as they emerge. In stems, branches arise from axillary buds and do not destroy any cells. **35.4** With the evolution of secondary growth, plants were able to grow taller and shade competitors. **35.5** The orientation of cellulose microfibrils in the innermost layers of the cell wall causes this growth along one axis. Microtubules play a key role in regulating the plane of cell expansion. It is the orientation of microtubules in the cell's outermost cytoplasm that determines the orientation of cellulose microfibrils.

Test Your Understanding

1. d **2.** c **3.** c **4.** a **5.** b **6.** e **7.** d
8.

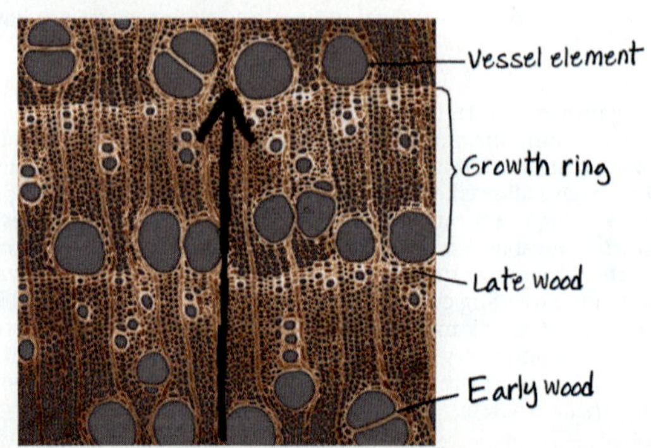

Chapter 36

Figure Questions

Figure 36.3 The leaves are being produced in a counterclockwise spiral. **Figure 36.4** A higher leaf area index will not necessarily increase photosynthesis because of upper leaves shading lower leaves. **Figure 36.7** A proton pump inhibitor would depolarize the membrane potential because fewer H^+ ions would be pumped out across the plasma membrane. The immediate effect of an inhibitor of the H^+/sucrose transporter would be to hyperpolarize the membrane potential because fewer H^+ ions would be leaking back into the cell through these cotransporters. An inhibitor of the H^+/NO_3^- cotransporter would have no effect on the membrane potential because the simultaneous cotransport of a positively charged ion and a negatively charged ion has no net effect on charge difference across the membrane. An inhibitor of the K^+ ion channels would decrease the membrane potential because additional positively charged ions would not be accumulating outside the cell. **Figure 36.10** The Casparian strip blocks water and minerals from moving between endodermal cells or moving around an endodermal cell via the cell's wall. Therefore, water and minerals must pass through an endodermal cell's plasma membrane. **Figure 36.19** Because the xylem is under negative pressure (tension), excising a stylet that had been inserted into a tracheid or vessel element would probably introduce air into the cell. No xylem sap would exude unless positive root pressure was predominant.

Concept Check 36.1

1. Vascular plants must transport minerals and water absorbed by the roots to all the other parts of the plant. They must also transport sugars from sites of production to sites of use. **2.** Many features of plant architecture affect self-shading, including leaf arrangement, leaf and stem orientation, and leaf area index. **3.** Increased stem elongation would raise the plant's upper leaves. Erect leaves and reduced lateral branching would make the plant less subject to shading by the encroaching neighbors. **4.** As discussed in Chapter 35 (p. 741), pruning shoot tips removes apical dominance, resulting in axillary buds growing into lateral shoots (branches). This branching produces a bushier plant with a higher leaf area index. **5.** Fungal hyphae are long, thin filaments that form a large interwoven network in the soil. Their high surface-to-volume ratio is an adaptation that enhances the absorption of materials from the soil.

Concept Check 36.2

1. The cell's ψ_P is 0.7 MPa. In a solution with a ψ of –0.4 MPa, the cell's ψ_P at equilibrium would be 0.3 MPa. **2.** The cells would still adjust to changes in their osmotic environment, but their responses would be slower. Although aquaporins do not affect the water potential gradient across membranes, they allow for more rapid osmotic adjustments. **3.** If tracheids and vessel elements were alive at maturity, their cytoplasm would impede water movement, preventing rapid long-distance transport. **4.** The protoplasts would burst. Because the cytoplasm has many dissolved solutes, water would enter the protoplast continuously without reaching equilibrium. (When present, the cell wall prevents rupturing by excessive expansion of the protoplast.)

Concept Check 36.3

1. Because water-conducting xylem cells are dead at maturity and form essentially hollow tubes, they offer little resistance to water flow, and their thick walls prevent the cells from collapsing from the negative pressure inside. **2.** At dawn, a drop is exuded because the xylem is under positive pressure due to root pressure. At noon, the xylem is under negative pressure tension when it is cut and the xylem sap is pulled away from the cut surface up into the stem. Root pressure cannot keep pace with the increased rate of transpiration at noon. **3.** The endodermis regulates the passage of water-soluble solutes by requiring all such molecules to cross a selectively permeable membrane. Presumably, the inhibitor never reaches the plant's photosynthetic cells. **4.** Perhaps greater root mass helps compensate for the lower water permeability of the plasma membranes. **5.** The Casparian strip and tight junctions both prevent movement of fluid between cells.

Concept Check 36.4

1. Stomatal opening at dawn is controlled mainly by light, CO_2 concentrations, and a circadian rhythm. Environmental stresses such as drought, high temperature, and wind can stimulate stomata to close during the day. Water deficiency can trigger release of the plant hormone abscisic acid, which signals guard cells to close stomata. **2.** The activation of the proton pump of stomatal cells would cause the guard cells to take up K^+. The increased turgor of the guard cells would lock the stomata open and lead to extreme evaporation from the leaf. **3.** After the flowers are cut, transpiration from any leaves and from the petals (which are modified leaves) will continue to draw water up the xylem. If cut flowers are transferred directly to a vase, air pockets in xylem vessels prevent delivery of water from the vase to the flowers. Cutting stems again underwater, a few centimeters from the original cut, will sever the xylem above the air pocket. The water droplets prevent another air pocket from forming while placing the flowers in a vase. **4.** Water molecules are in constant motion, traveling at different rates. The average speed of these particles depends on the water's temperature. If water molecules gain enough energy, the most energetic molecules near the liquid's surface will impart sufficient speed, and therefore sufficient kinetic energy, to cause water molecules to propel away from the liquid in the form of gaseous molecules or, more simply, as water vapor. As the particles with the highest kinetic energy levels evaporate, the average kinetic energy of the remaining liquid decreases. Because a liquid's temperature is directly related to the average kinetic energy of its molecules, the liquid cools as it evaporates.

Concept Check 36.5

1. In both cases, the long-distance transport is a bulk flow driven by a pressure difference at opposite ends of tubes. Pressure is generated at the source end of a sieve tube by the loading of sugar and resulting osmotic flow of water into the phloem, and this pressure *pushes* sap from the source end to the sink end of the tube. In contrast, transpiration generates a negative pressure potential (tension) as a force that *pulls* the ascent of xylem sap. **2.** The main sources are fully grown leaves (by photosynthesis) and fully developed storage organs (by breakdown of starch). Roots, buds, stems, expanding leaves, and fruits are powerful sinks because they are actively growing. A storage organ may be a sink in the summer when accumulating carbohydrates, but a source in the spring when breaking down starch into sugar for growing shoot tips. **3.** Positive pressure, whether it be in the xylem when root pressure predominates or in the sieve-tube elements of the phloem, requires active transport. Most long-distance transport in the xylem depends on bulk flow driven by negative pressure potential generated ultimately by the evaporation of water from the leaf and does not require living cells. **4.** The spiral slash prevents optimal bulk flow of the phloem sap to the root sinks. Therefore, more phloem sap can move from the source leaves to the fruit sinks, making them sweeter.

Concept Check 36.6

1. Plasmodesmata, unlike gap junctions, have the ability to pass RNA, proteins, and viruses from cell to cell. **2.** Long-distance signaling is critical for the integrated functioning of all large organisms, but the speed of such integration is much less critical to plants because their responses to the environment, unlike those of animals, do not typically involve rapid movements. **3.** Although this strategy would eliminate the systemic spread of viral infections, it would also severely impact the development of the plants.

Summary of Key Concepts Questions

36.1 Plants with tall shoots and elevated leaf canopies generally had an advantage over shorter competitors. A consequence of the selective pressure for tall shoots was the further separation of leaves from roots. This separation created problems for the transport of materials between root and shoot systems. Plants with xylem cells were more successful at supplying their shoot systems with soil resources (water and minerals). Similarly, those with phloem cells were more successful at supplying sugar sinks with carbohydrates. **36.2** Xylem sap is usually pulled up the plant by transpiration, much more often than it is pushed up the plant by root pressure. **36.3** Hydrogen bonds are necessary for the cohesion of water molecules to each other and for the adhesion of water to other materials, such as cell walls. Both adhesion and cohesion of water molecules are involved in the ascent of xylem sap under conditions of negative pressure. **36.4** Although stomata account for most of the water lost from plants, they are necessary for exchange of gases—for example, for the uptake of carbon dioxide needed for photosynthesis. **36.5** Although the movement of phloem sap depends on bulk flow, the pressure gradient that drives phloem transport depends on the osmotic uptake of water in response to the loading of sugars into sieve-tube elements at sugar sources. Phloem loading depends on H^+ cotransport processes that ultimately depend on H^+ gradients established by active H^+ pumping. **36.6** Voltage between cells, cytoplasmic pH, cytoplasmic calcium concentration, and viral movement proteins all affect symplastic communication, as do developmental changes in the number of plasmodesmata.

Test Your Understanding

1. c **2.** a **3.** b **4.** b **5.** c **6.** e **7.** c **8.** a **9.** d
10.

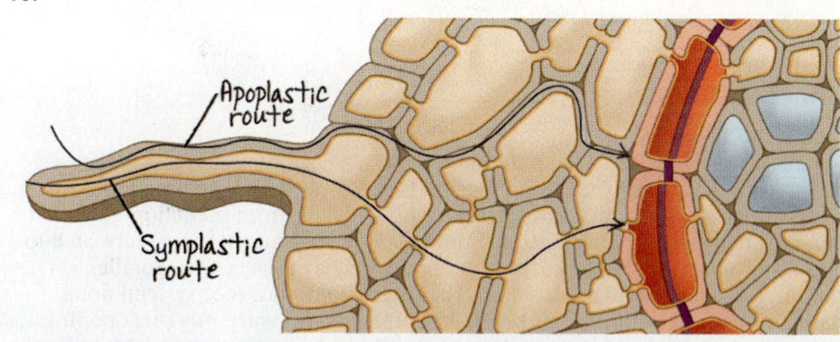

Chapter 37

Figure Questions

Figure 37.3 Anions. Because cations are bound to soil particles, they are less likely to be lost from the soil following heavy rains. **Table 37.1** Fluorine, selenium, and chromium. Plants may contain more than 50 elements, but only a few are essential for the plant to complete its life cycle. The others, including fluorine, selenium and chromium, are present but not essential for survival. **Figure 37.11** The legume plants benefit because the bacteria fix nitrogen that is absorbed by their roots. The bacteria benefit because they acquire photosynthetic products from the plants. **Figure 37.12** All three plant tissue systems are affected. Root hairs (dermal tissue) are modified to allow *Rhizobium* penetration. The cortex (ground tissue) and pericycle (vascular tissue) proliferate during nodule formation. The vascular tissue of the nodule connects to the vascular cylinder of the root to allow for efficient nutrient exchange. **Figure 37.14** If phosphate were the only limiting mineral, then native tree growth would be less severely impacted by the reduction in mycorrhizal associations caused by garlic mustard. Consequently, the competitive advantage of garlic mustard would be reduced by the addition of phosphate to the soil.

Concept Check 37.1

1. Overwatering deprives roots of oxygen. Overfertilizing is wasteful and can lead to soil salinization and water pollution. **2.** As lawn clippings decompose, they restore mineral nutrients to the soil. If they are removed, the minerals lost from the soil must be replaced by fertilization. **3.** Because of their small size and negative charge, clay particles would increase the number of binding sites for cations and water molecules and would therefore increase cation exchange and water retention in the soil. **4.** Due to hydrogen bonding between water molecules, water expands when it freezes, and this causes mechanical fracturing of rocks. Water also coheres to many objects, and this cohesion combined with other forces, such as gravity, can help tug particles from rock. Finally, water, because it is polar, is an excellent solvent that allows many substances, including ions, to become dissolved in solution.

Concept Check 37.2

1. Table 37.1 shows that CO_2 is the source of 90% of a plant's dry weight, supporting Hales's view that plants are nourished mostly by air. **2.** No, because even though macronutrients are required in greater amounts, all essential elements are necessary for the plant to complete its life cycle. **3.** No. The fact that the addition of an element results in an increase in the growth rate of a crop does not mean that the element is strictly required for the plant to complete its life cycle. **4.** Waterlogging displaces air from the soil, leading to low O_2 conditions. Such conditions promote the anaerobic process of alcoholic fermentation in plants, the end product of which is ethanol.

Concept Check 37.3

1. The rhizosphere is a narrow zone in the soil immediately adjacent to living roots. This zone is especially rich in both organic and inorganic nutrients and has a microbial population that is many times greater than the bulk of the soil. **2.** Soil bacteria and mycorrhizae enhance plant nutrition by making certain minerals more available to plants. For example, many types of soil bacteria are involved in the nitrogen cycle, and the hyphae of mycorrhizae provide a large surface area for the absorption of nutrients, particularly phosphate ions. **3.** Saturating rainfall may deplete the soil of oxygen. A lack of soil oxygen would inhibit nitrogen fixation by the peanut root nodules and decrease the nitrogen available to the plant. Alternatively, heavy rain may leach nitrate from the soil. A symptom of nitrogen deficiency is yellowing of older leaves.

Summary of Key Concepts Questions

37.1 The term *ecosystem* refes to the communities of organisms within a given area and their interactions with the physical environment around them. Soil is teeming with many communities of organisms, including bacteria, fungi, animals, and the root systems of plants. The vigor of these individual communities depends on nonliving factors in the soil environment, such as minerals, oxygen, and water, as well as on interactions, both positive and negative, between different communities of organisms. **37.2** No, plants can complete their life cycle when grown hydroponically, that is, in aerated salt solutions containing the proper ratios of all the minerals needed by plants. **37.3** No, some parasitic plants obtain their energy by siphoning off carbon nutrients from other organisms.

Test Your Understanding

1. b **2.** b **3.** a **4.** e **5.** b **6.** b **7.** d **8.** c **9.** d
10.

H^+ K^+ Soil particle Root hair

Chapter 38

Figure Questions

Figure 38.4 Having a specific pollinator is more efficient because less pollen gets delivered to flowers of the wrong species. However, it is also a risky strategy: If the pollinator population suffers to an unusual degree from predation, disease, or climate change, then the plant may not be able to produce seeds. **Figure 38.8** In addition to having a single cotyledon, monocots have leaves with parallel leaf venation, scattered vascular bundles in their stems, a fibrous root system, floral parts in threes or multiples of threes, and pollen grains with only one opening. In contrast, dicots have two cotyledons, netlike leaf venation, vascular bundles in a ring, taproots, floral parts in fours or fives or multiples thereof, and pollen grains with three openings. **Figure 38.9** Beans use a hypocotyl hook to push through the soil. The delicate leaves and shoot apical meristem are also protected by being sandwiched between two large cotyledons. The coleoptile of maize seedlings helps protect the emerging leaves. **Figure 38.17** The crown gall bacterium (*Agrobacterium tumefaciens*) normally causes cancer-like growths in susceptible plants. *Agrobacterium* inserts its own genes into plant cells by means of plasmids. These plasmids have been genetically engineered to retain their ability to insert genes into plant cells without causing cancerous growth.

Concept Check 38.1

1. In angiosperms, pollination is the transfer of pollen from an anther to a stigma. Fertilization is the fusion of the egg and sperm to form the zygote; it cannot occur until after the growth of the pollen tube from the pollen grain. **2.** Seed dormancy prevents the premature germination of seeds. A seed will germinate only when the environmental conditions are optimal for the survival of its embryo as a young seedling. **3.** Long styles help to weed out pollen grains that are genetically inferior and not capable of successfully growing long pollen tubes. **4.** No. The haploid (gametophyte) generation of plants is multicellular and arises from spores. The haploid phase of the animal life cycles is a single-celled gamete (egg or sperm) that arises directly from meiosis: There are no spores.

Concept Check 38.2

1. Asexually propagated crops lack genetic diversity. Genetically diverse populations are less likely to become extinct in the face of an epidemic because there is a greater likelihood that a few individuals in the population are resistant. **2.** In the short term, selfing may be advantageous in a population that is so dispersed and sparse that pollen delivery is unreliable. In the long term, however, selfing is an evolutionary dead end because it leads to a loss of genetic diversity that may preclude adaptive evolution. **3.** This might be possible, but satisfactory results would be very unlikely. Both tubers and fruits are tremendous energy sinks. Each plant has only a finite amount of energy to divide between sexual and asexual reproduction. Although a tomato-potato hybrid could, in theory, produce an offspring that makes fruits and tubers equally, these fruits and tubers would be of inferior quality or low yielding.

Concept Check 38.3

1. Traditional breeding and genetic engineering both involve artificial selection for desired traits. However, genetic engineering techniques facilitate faster gene transfer and are not limited to transferring genes between closely related varieties or species. **2.** GM crops may be more nutritious and less susceptible to insect damage or pathogens that invade insect-damaged plants. They also may not require as much chemical spraying. However, unknown risks may include adverse effects on human health and nontarget organisms and the possibility of transgene escape. **3.** *Bt* maize suffers less insect damage; therefore, *Bt* maize plants are less likely to be infected by fumonisin-producing fungi that infect plants through wounds. **4.** In such species, engineering the transgene into the chloroplast DNA would not prevent its escape in pollen; such a method requires that the chloroplast DNA be found only in the egg. An entirely different method of preventing transgene escape would therefore be needed, such as male sterility, apomixis, or self-pollinating closed flowers.

Summary of Key Concepts Questions

38.1 After pollination, a flower typically changes into a fruit. The petals, sepals, and stamens typically fall off the flower. The stigma of the pistil withers and the ovary begins to swell. The ovules (embryonic seeds) inside the ovary begin to mature. **38.2** Asexual reproduction can be advantageous in a stable environment because individual plants that are well suited to that environment pass on all their genes to offspring. Also, asexual reproduction generally results in offspring that are less fragile than the seedlings produced by sexual reproduction. However, sexual reproduction offers the advantage of dispersal of tough seeds. Moreover, sexual reproduction produces genetic variety, which may be advantageous in an unstable environment. The likelihood is better that at least one offspring of sexual reproduction will survive in a changed environment. **38.3** "Golden Rice" has been engineered to produce more vitamin A, thereby raising the nutritional value of rice. A protoxin gene from a soil bacterium has been engineered into *Bt* maize. This protoxin is lethal to invertebrates but harmless to vertebrates. *Bt* crops require less pesticide spraying and have lower levels of fungal infection. The nutritional value of cassava is being increased in many ways by genetic engineering. Enriched levels of protein, iron, and beta-carotene (a vitamin A precursor) have been achieved, and cyanide-producing chemicals have been almost eliminated from the roots.

Test Your Understanding

1. c **2.** a **3.** c **4.** e **5.** c **6.** c **7.** d **8.** c **9.** d
10.

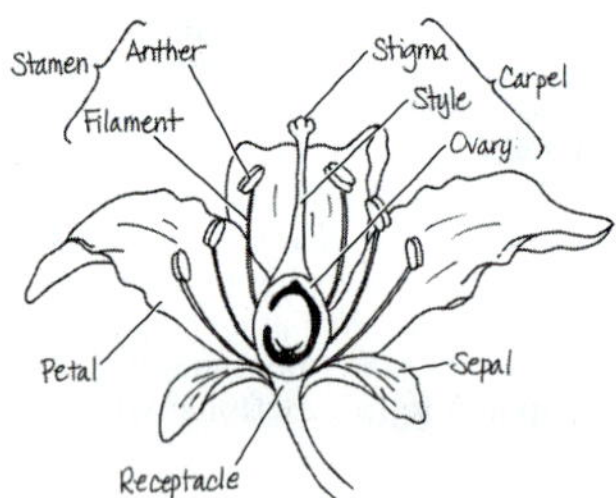

Chapter 39

Figure Questions

Figure 39.4 Panel B in Figure 11.18 shows a branching signal transduction pathway that resembles the branching phytochrome-dependent pathway involved in de-etiolation. **Figure 39.5** To determine which wavelengths of light are most effective in phototropism, you could use a glass prism to split white light into its component colors and see which colors cause the quickest bending (the answer is blue; see Figure 39.16). **Figure 39.6** More auxin would move down the side without the TIBA-containing agar bead, causing greater elongation on this side and, consequently, bending of the coleoptile toward the side with the bead. **Figure 39.7** No. Polar auxin transport depends on the distribution of auxin transport proteins at the basal ends of cells. **Figure 39.17** Yes. The white light, which contains red light, would stimulate seed germination in all treatments. **Figure 39.22** The short-day plant would not flower. The long-day plant would flower. **Figure 39.23** If this were true, florigen would be an inhibitor of flowering, not an inducer.

Concept Check 39.1

1. Dark-grown seedlings are etiolated: They have long stems, underdeveloped root systems, and unexpanded leaves, and their shoots lack chlorophyll. Etiolated growth is beneficial to seeds sprouting under the dark conditions they would encounter underground. By devoting more energy to stem elongation and less to leaf expansion and root growth, a plant increases the likelihood that the shoot will reach the sunlight before its stored foods run out. **2.** Cycloheximide should inhibit de-etiolation by preventing the synthesis of new proteins necessary for de-etiolation. **3.** No. Applying Viagra, like injecting cyclic GMP as described in the text, should cause only a partial de-etiolation response. Full de-etiolation would require activation of the calcium branch of the signal transduction pathway.

Concept Check 39.2

1. Because cytokinins delay leaf senescence and floral parts are modified leaves, cytokinins also delay the senescence of cut flowers. **2.** Fusicoccin's ability to cause an increase in plasma H^+ pump activity has an auxin-like effect and promotes stem cell elongation. **3.** The plant will exhibit a constitutive triple response. Because the kinase that normally prevents the triple response is dysfunctional, the plant will undergo the triple response regardless of whether ethylene is present or the ethylene receptor is functional. **4.** Since ethylene often stimulates its own synthesis, it is under positive-feedback regulation.

Concept Check 39.3

1. Not necessarily. Many environmental factors, such as temperature and light, change over a 24-hour period in the field. To determine whether the enzyme is under circadian control, a scientist would have to demonstrate that its activity oscillates even when environmental conditions are held constant. **2.** Flowering of the species may have been day-neutral or required multiple exposures to short nights. **3.** You might determine which wavelengths of light are most effective and plot an action spectrum. If the action spectrum indicates phytochrome, you could do further experiments to test for red/far-red photosensitivity. **4.** It is impossible to say. To establish that this species is a short-day plant, it would be necessary to establish the critical night length for flowering and that this species only flowers when the night is longer than the critical night length. **5.** According to the action spectrum of photosynthesis, red and blue light are the most effective in photosynthesis. Thus, it is not surprising that plants assess their light environment using blue- and red-light-absorbing photoreceptors.

Concept Check 39.4

1. A plant that overproduces ABA would undergo less evaporative cooling because its stomata would not open as widely. **2.** Plants close to the aisles may be more subject to mechanical stresses caused by passing workers and air currents. The plants nearer to the center of the bench may also be taller as a result of shading and less evaporative stress. **3.** No. Because root caps are involved in sensing gravity, roots that have their root caps removed are almost completely insensitive to gravity.

Concept Check 39.5

1. Some insects increase plants' productivity by eating harmful insects or aiding in pollination. **2.** Mechanical damage breaches a plant's first line of defense against infection, its protective dermal tissue. **3.** No. Pathogens that kill their hosts would soon run out of victims and might themselves go extinct. **4.** Perhaps the breeze dilutes the local concentration of a volatile defense compound that the plants produce.

Summary of Key Concepts Questions

39.1 Signal transduction pathways often activate protein kinases, enzymes that phosphorylate other proteins. Protein kinases can directly activate certain preexisting enzymes by phosphorylating them, or they can regulate gene transcription (and enzyme production) by phosphorylating specific transcription factors. **39.2** Yes, there is truth to the old adage that one bad apple spoils the whole bunch. Ethylene, a gaseous hormone that stimulates ripening, is produced by damaged, infected, or overripe fruits. Ethylene can diffuse to healthy fruit in the "bunch" and stimulate their rapid ripening. **39.3** Plant physiologists proposed the existence of a floral-promoting factor (florigen) based on the fact that a plant induced to flower could induce flowering in a second plant to which it was grafted, even though the second plant was not in an environment that would normally induce flowering in that species. **39.4** Plants subjected to drought stress are often more resistant to freezing stress because the two types of stress are quite similar. Freezing of water in the extracellular spaces causes free water concentrations outside the cell to decrease. This, in turn, causes free water to leave the cell by osmosis, leading to the dehydration of cytoplasm, much like what is seen in drought stress. **39.5** Chewing insects make plants more susceptible to pathogen invasion by disrupting the waxy cuticle of shoots, thereby creating an opening for infection. Moreover, substances released from damaged cells can serve as nutrients for the invading pathogens.

Test Your Understanding

1. e **2.** c **3.** d **4.** e **5.** b **6.** b **7.** c **8.** e

9.

	Control	Ethylene added	Ethylene synthesis inhibitor
Wild-type			
Ethylene insensitive (ein)			
Ethylene overproducing (eto)			
Constitutive triple response (ctr)			

Chapter 40

Figure Questions

Figure 40.4 Such exchange surfaces are internal in the sense that they are inside the body. However, they are also continuous with openings on the external body surface that contact the environment. **Figure 40.8** The air conditioner would form a second control circuit, cooling the house when air temperature exceeded the set point. Such opposing, or antagonistic, pairs of control circuits increase the effectiveness of a homeostatic mechanism. **Figure 40.14** If a female Burmese python were not incubating eggs, her oxygen consumption would decrease with decreasing temperature, as for any other ectotherm. **Figure 40.17** The transport of nutrients across membranes and the synthesis of RNA and protein are coupled to ATP hydrolysis. These processes proceed spontaneously because there is an overall drop in free energy, with the excess energy given off as heat. Similarly, less than half of the free energy in glucose is captured in the coupled reactions of cellular respiration. The remainder of the energy is released as heat. **Figure 40.21** Nothing. Although genes that show a circadian variation in expression during euthermia exhibit constant RNA levels during hibernation, a gene that shows constant expression during hibernation might also show constant expression during euthermia.

Concept Check 40.1

1. All types of epithelia consist of cells that line a surface, are tightly packed, are situated on top of a basal lamina, and form an active and protective interface with the external environment. **2.** By flattening its ears along its body, the jackrabbit can reduce the exposed surface area of its body and hence the amount of heat lost. However, by placing its ears flat, the jackrabbit reduces its ability to detect potential predators. **3.** You need the nervous system to perceive the danger and provoke a split-second muscular response to keep from falling. The nervous system, however, does not make a direct connection with blood vessels or glucose-storing cells in the liver. Instead, the nervous system triggers the release of a hormone (called epinephrine, or adrenaline) by the endocrine system, bringing about a change in these tissues in just a few seconds.

Concept Check 40.2

1. In the enzyme-catalyzed biosynthetic process, the product of a pathway (in this case, isoleucine) inhibits the pathway that generated it. In thermoregulation, the product of the pathway (a change in temperature) decreases pathway activity by reducing the stimulus. **2.** You would want to put the thermostat close to where you would be spending time, where it would be protected from environmental perturbations, such as direct sunshine, and not right in the path of the output of the heating system. Similarly, the sensors for homeostasis located in the human brain are separated from environmental influences and can monitor conditions in a vital and sensitive tissue. **3.** In convergent evolution, the same biological trait arises independently in two or more species. Gene analysis can provide evidence for an independent origin. In particular, if the genes responsible for the trait in one species lack significant sequence similarity to the corresponding genes in another species, scientists conclude that there is a separate genetic basis for the trait in the two species and thus an independent origin. In the case of circadian rhythms, the clock genes in cyanobacteria appear unrelated to those in humans.

Concept Check 40.3

1. "Wind chill" involves heat loss through convection, as the moving air contributes to heat loss from the skin surface. **2.** The hummingbird, being a very small endotherm, has a very high metabolic rate. If by absorbing sunlight certain flowers warm their nectar, a hummingbird feeding on these flowers is saved the metabolic expense of warming the nectar to its body temperature. **3.** The ice water would cool tissues in your head, including blood that would then circulate throughout your body. This effect would accelerate the return to a normal body temperature. If, however, the ice water reached the eardrum and cooled the blood vessel that supplies the hypothalamus, the hypothalamic thermostat would respond by inhibiting sweating and constricting blood vessels in the skin, slowing cooling elsewhere in the body.

Concept Check 40.4

1. The mouse would consume oxygen at a higher rate because it is an endotherm, so its basal metabolic rate is higher than the ectothermic lizard's standard metabolic rate. **2.** The house cat; smaller animals have a higher metabolic rate per unit body mass and a greater demand for food per unit body mass. **3.** Although penguins do not grow as adults, they increase and decrease in size as they repeatedly form and use energy stores. A significant amount of energy might be stored in fat during part of the year but be used later in the year. Monitoring energy allocation only during the period when energy is stored in fat would lead to the erroneous conclusion that the penguin is growing.

Summary of Key Concepts Questions

40.1 Animals exchange materials with their environment across their body surface, and a spherical shape has the minimum surface area per unit volume. As body size increases, the ratio of surface area to body volume decreases. **40.2** No; even though an animal regulates some aspects of its internal environment, the internal environment fluctuates slightly around set points. Homeostasis is a dynamic state. Furthermore, there are sometimes programmed changes in set points, such as those resulting in radical increases in hormone levels at particular times in development. **40.3** Heat exchange across the skin is a primary mechanism for the regulation of body core temperature, with the result that the skin is cooler than the body core. **40.4** Because small animals have the highest ratio of BMR per unit mass, they have the greatest energetic savings per unit mass during hibernation. Consequently, the selective pressure favoring hibernation during evolution is especially strong for small animals.

Test Your Understanding

1. b **2.** c **3.** a **4.** b **5.** d **6.** c **7.** e

8.

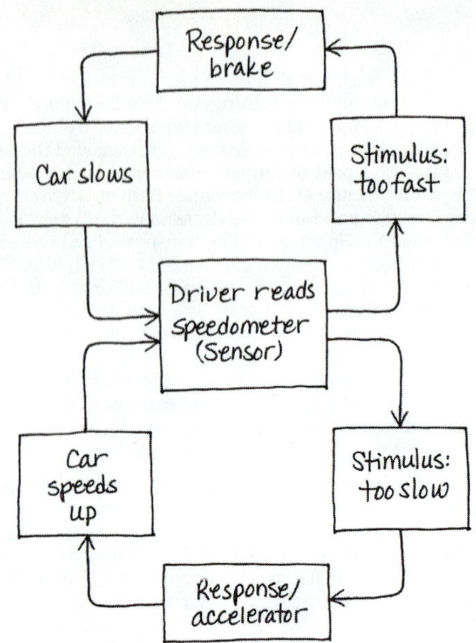

Chapter 41

Figure Questions

Figure 41.2 Just as penguin muscle protein provides amino acids for regrowth of large numbers of feathers, ovalbumin and casein provide amino acids for growth of a fertilized egg or nursing mammal. Thus, the common circumstance is a period of rapid developmental growth. **Figure 41.4** As in the described study, they needed a sample size large enough that they could expect a significant number of neural tube defects in the control group. The information needed to determine the appropriate sample size was the frequency of neural tube defects in first-time pregnancies in the general population. **Figure 41.12** Since enzymes are proteins, and proteins are hydrolyzed in the small intestine, the digestive enzymes in that compartment need to be resistant to enzymatic cleavage other than the cleavage required to activate them. **Figure 41.13** None. Since digestion is completed in the small intestine, tapeworms simply absorb predigested nutrients through their large body surface. **Figure 41.20** Both insulin and glucagon are involved in negative feedback circuits. **Figure 41.22** The wild-type mouse produces leptin after a meal. As the mouse depletes its fat stores, leptin production drops. The mouse eventually regains its appetite, eats another meal, and makes another burst of leptin. Thus, leptin levels would oscillate during the day. Because the *db* mouse cannot respond to leptin, its fat stores are constantly built up through excessive consumption. As a result, leptin is produced continuously and builds up to a high, steady concentration in the blood.

Concept Check 41.1

1. The only essential amino acids are those that an animal cannot synthesize from other molecules. **2.** Many vitamins serve as enzyme cofactors, which, like enzymes themselves, are unchanged by the chemical reactions in which they participate. Therefore, only very small amounts of vitamins are needed. **3.** To identify the essential nutrient missing from an animal's diet, a researcher could supplement the diet with individual nutrients and determine which nutrient eliminates the signs of malnutrition.

Concept Check 41.2

1. A gastrovascular cavity is a digestive pouch with a single opening that functions in both ingestion and elimination; an alimentary canal is a digestive tube with a separate mouth and anus at opposite ends. **2.** As long as nutrients are within the cavity of the alimentary canal, they are in a compartment that is continuous with the outside environment via the mouth and anus and have not yet crossed a membrane to enter the body. **3.** Just as food remains outside the body in a digestive tract, gasoline moves from the fuel tank to the engine, and waste products exit through the exhaust without ever entering the passenger compartment of the automobile. In addition, gasoline, like food, is broken down in a specialized compartment, so that the rest of the automobile (or body) is protected from disassembly. In both cases, high-energy fuels are consumed, complex molecules are broken down into simpler ones, and waste products are eliminated.

Concept Check 41.3

1. By peristalsis, which can squeeze food through the esophagus even without the help of gravity **2.** Because parietal cells in the stomach pump hydrogen ions to produce HCl, a proton pump inhibitor reduces the acidity of chyme and thus the irritation that occurs when chyme enters the esophagus. **3.** Proteins would be denatured and digested into peptides. Further digestion, to individual amino acids, would require enzymatic secretions found in the small intestine. No digestion of carbohydrates or lipids would occur.

Concept Check 41.4

1. The increased time for transit through the alimentary canal allows for more extensive processing, and the increased surface of the canal area provides greater opportunity for absorption. **2.** A mammal's digestive system provides mutualistic microbes with an environment that is protected against other microbes by saliva and gastric juice, that is held at a constant temperature conducive to enzyme action, and that provides a steady source of nutrients. **3.** For the yogurt treatment to be effective, the bacteria from yogurt would have to establish a mutualistic relationship with the small intestine, where disaccharides are broken down and sugars are absorbed. Conditions in the small intestine are likely to be very different from those in a yogurt culture. The bacteria might be killed before they reach the small intestine, or they might not be able to grow there in sufficient numbers to aid in digestion.

Concept Check 41.5

1. Over the long term, the body stores excess calories in fat, whether those calories come from fat, carbohydrate, or protein in food. **2.** Both hormones have appetite-suppressing effects on the brain's satiety center. During the course of a day, PYY, secreted by the intestine, suppresses appetite after meals. Over the longer term, leptin, produced by adipose tissue, normally reduces appetite as fat storage increases. **3.** In normal individuals, leptin levels decline during fasting. Individuals in the group with low levels of leptin are likely to be defective in leptin production, so leptin levels would remain low regardless of food intake. Individuals in the group with high leptin levels are likely to be defective in responding to leptin, but they still should shut off leptin production as fat stores are used up.

Summary of Key Concepts Questions

41.1 Since collagen is found in all mammals, a likely explanation is that mammals other than primates and guinea pigs can synthesize vitamin C from other organic molecules. **41.2** A liquid diet containing glucose, amino acids, and other building blocks could be ingested and absorbed without the need for mechanical or chemical digestion. **41.3** The small intestine has a much larger surface area than the stomach. **41.4** The assortment of teeth in our mouth and the short length of our cecum suggest that our ancestors' digestive systems were not specialized for digesting plant material. **41.5** When mealtime arrives, nervous inputs from the brain signal the stomach to prepare to digest food through secretions and churning.

Test Your Understanding

1. a **2.** c **3.** c **4.** d **5.** d **6.** b

7.

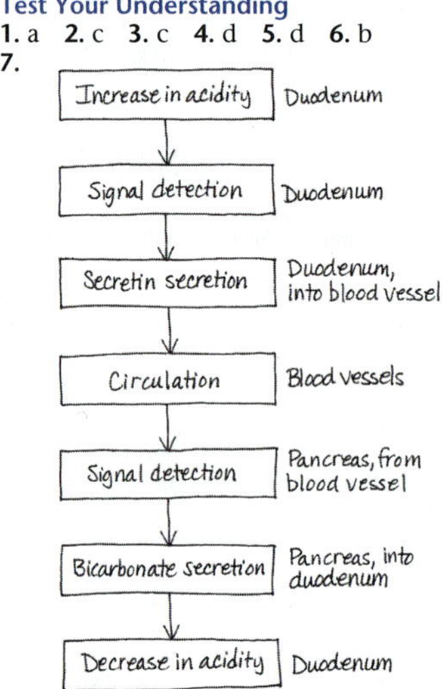

Chapter 42

Figure Questions

Figure 42.2 Although gas exchange might be improved by a steady, one-way flow of fluid, there would likely be inadequate time for food to be digested and nutrients absorbed if fluid flowed through the cavity in this manner. **Figure 42.9** Each feature of the ECG recording, such as the sharp upward spike, occurs once per cardiac cycle. Using the *x*-axis to measure the time in seconds between successive spikes and dividing that number into 60 would yield the heart rate as the number of cycles per minute. **Figure 42.21** The mutations studied by Dr. Hobbs's team inactivate the enzyme. Carriers for these mutations should have roughly half the normal level of PCSK9 activity. The mutations studied by the French researchers have the opposite effect on LDL levels as the inactivating mutations. It is therefore likely that individuals carrying these mutations would have increased PCSK9 activity. **Figure 42.22** The three main lineages are Deuterostomia, Lophotrochozoa, and Ecdysozoa. All three are represented by the animals shown in Figure 42.22: Polychaetes (phylum Annelida) are lophotrochozoans, crayfish (phylum Arthropoda) are ecdysozoans, and sea stars (phylum Echinodermata) are deuterostomes. **Figure 42.26** The reduction in surface tension results from the presence of surfactant. Therefore, for all the infants that had died of RDS, you would expect the amount of surfactant to be near zero. For infants that had died of other causes, you would expect the amount of surfactant to be near zero for weights less than 1,200 g but much greater than zero for weights above 1,200 g. **Figure 42.28** Since exhalation is largely passive, the recoil of the elastic fibers in alveoli helps force air out of the lungs. When alveoli lose their elasticity, as occurs in the disease emphysema, less air is exhaled. Because more air is left in the lungs, less fresh air can be inhaled. With a smaller volume of air exchanged, there is a decrease in the partial pressure gradient that drives gas exchange. **Figure 42.29** Breathing at a rate greater than that needed to meet metabolic demand (hyperventilation) would lower blood CO_2 levels. Sensors in major blood vessels and the medulla would signal the breathing control centers to decrease the rate of contraction of the diaphragm and rib muscles, decreasing the breathing rate and restoring normal CO_2 levels in the blood and other tissues. **Figure 42.30** The resulting increase in tidal volume would enhance ventilation within the lungs, increasing P_{O_2} and decreasing P_{CO_2} in the alveoli. **Figure 42.32** Some CO_2 is dissolved in plasma, some is bound to hemoglobin, and some is converted to bicarbonate ion (HCO_3^-), which is dissolved in plasma.

Concept Check 42.1

1. In both an open circulatory system and a fountain, fluid is pumped through a tube and then returns to the pump after collecting in a pool. **2.** The ability to shut off blood supply to the lungs when the animal is submerged **3.** The O_2 content would be abnormally low because some oxygen-depleted blood returned to the right atrium from the systemic circuit would mix with the oxygen-rich blood in the left atrium.

Concept Check 42.2

1. The pulmonary veins carry blood that has just passed through capillary beds in the lungs, where it accumulated O_2. The venae cavae carry blood that has just passed through capillary beds in the rest of the body, where it lost O_2 to the tissues. **2.** The delay allows the atria to empty completely, filling ventricles fully before they contract. **3.** The heart, like any other muscle, becomes stronger through regular exercise. You would expect a stronger heart to have a greater stroke volume, which would allow for the decrease in heart rate.

Concept Check 42.3

1. The large total cross-sectional area of the capillaries **2.** An increase in blood pressure and cardiac output combined with the diversion of more blood to the skeletal muscles would increase the capacity for action by increasing the rate of blood circulation and delivering more O_2 and nutrients to the skeletal muscles. **3.** Additional hearts could be used to improve blood return from the legs. However, it might be difficult to coordinate the activity of multiple hearts and to maintain adequate blood flow to hearts far from the gas exchange organs.

Concept Check 42.4

1. An increase in the number of white blood cells (leukocytes) may indicate that the person is combating an infection. **2.** Clotting factors do not initiate clotting but are essential steps in the clotting process. Also, the clots that form a thrombus typically result from an inflammatory response to an atherosclerotic plaque, not from clotting at a wound site. **3.** The chest pain results from inadequate blood flow in coronary arteries. Vasodilation promoted by nitric oxide from nitroglycerin increases blood flow, providing the heart muscle with additional oxygen and thus relieving the pain. **4.** When a mutant allele is codominant with the wild-type allele, the phenotype of heterozygotes is intermediate between that of wild-type and mutant homozygotes. Therefore, in the presence of wild-type Hb, the aggregation of Hb^S that causes sickling must be significantly reduced. Based on this fact, some therapies for sickle-cell disease are aimed at boosting adult expression of another hemoglobin gene in the body, such as that normally expressed only in the fetus. **5.** Embryonic stem cells are pluripotent rather than multipotent, meaning that they can give rise to many rather than a few different cell types.

Concept Check 42.5

1. Their interior position helps them stay moist. If the respiratory surfaces of lungs extended out into the terrestrial environment, they would quickly dry out, and diffusion of O_2 and CO_2 across these surfaces would stop. **2.** Earthworms need to keep their skin moist for gas exchange, but they need air outside this moist layer. If they stay in their waterlogged tunnels after a heavy rain, they will suffocate because they cannot get as much O_2 from water as from air. **3.** In the extremities of some vertebrates, blood flows in opposite directions in neighboring veins and arteries; this countercurrent arrangement maximizes the recapture of heat from blood leaving the body core in arteries, which is important for thermoregulation in cold environments. Similarly, in the gills of fish, water passes over the gills in the direction opposite to that of blood flowing through the gill capillaries, maximizing the extraction of oxygen from the water along the length of the exchange surface.

Concept Check 42.6

1. An increase in blood CO_2 concentration causes an increase in the rate of CO_2 diffusion into the cerebrospinal fluid, where the CO_2 combines with water to form carbonic acid. Dissociation of carbonic acid releases hydrogen ions, decreasing the pH of the cerebrospinal fluid. **2.** Increased heart rate increases the rate at which CO_2-rich blood is delivered to the lungs, where CO_2 is removed. **3.** A hole would allow air to enter the space between the inner and outer layers of the double membrane, resulting in a condition called a pneumothorax. The two layers would no longer stick together, and the lung on the side with the hole would collapse and cease functioning.

Concept Check 42.7

1. Differences in partial pressure; gases diffuse from a region of higher partial pressure to a region of lower partial pressure. **2.** The Bohr shift causes hemoglobin to release more O_2 at a lower pH, such as found in the vicinity of tissues with high rates of cellular respiration and CO_2 release. **3.** The doctor is assuming that the rapid breathing is the body's response to low blood pH. Metabolic acidosis, the lowering of blood pH, can have many causes, including complications of certain types of diabetes, shock (extremely low blood pressure), and poisoning.

Summary of Key Concepts Questions

42.1 In a closed circulatory system, an ATP-driven muscular pump generally moves fluids in one direction on a scale of millimeters to meters. Exchange between cells and their environment relies on diffusion, which involves random movements of molecules. Concentration gradients of molecules across exchange surfaces can drive rapid net diffusion on a scale of 1 mm or less. **42.2** Replacement of a defective valve should increase stroke volume. A lower heart rate would therefore be sufficient to maintain the same cardiac output. **42.3** Blood pressure in the arm would fall by 25–30 mm Hg, the same difference as is normally seen between your heart and your brain. **42.4.** One microliter of blood contains about 5 million erythrocytes and 5,000 leukocytes, so leukocytes make up only about 0.1% of the cells in the absence of infection. **42.5.** Because CO_2 is such a small fraction of atmospheric gas (0.29 mm Hg/760 mm Hg, or less than 0.04%), the partial pressure gradient of CO_2 between the respiratory surface and the environment always strongly favors the release of CO_2 to the atmosphere. **42.6.** Because the lungs do not empty completely with each breath, incoming and outgoing air mix, so not all of the tidal volume represents fresh air. **42.7.** An enzyme speeds up a reaction without changing the equilibrium and without being consumed. Similarly, a respiratory pigment speeds up the movement of gases in the body without changing the equilibrium and without being consumed.

Test Your Understanding

1. c **2.** b **3.** d **4.** c **5.** d **6.** a **7.** a **8.** a
9.

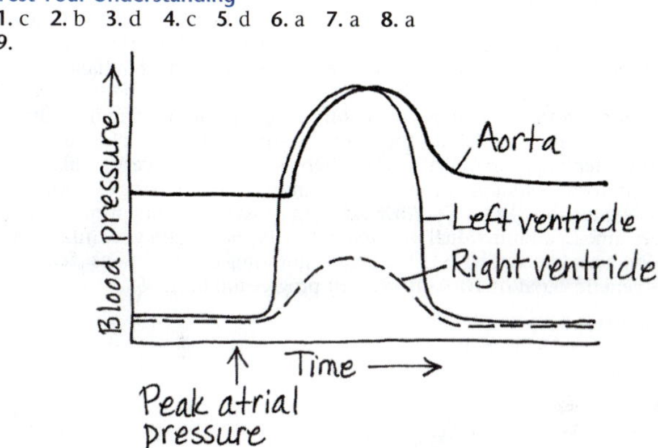

Chapter 43

Figure Questions

Figure 43.5 The seemingly inactive peptides might offer protection against pathogens other than those studied. Also, some antimicrobial peptides might work best in combination. **Figure 43.6** Cell-surface TLRs recognize pathogens identifiable by surface molecules, whereas TLRs in vesicles recognize pathogens identifiable by internal molecules after the pathogens are broken down. **Figure 43.10** Part of the enzyme or antigen receptor provides a structural "backbone" that maintains overall shape, while interaction occurs at a surface with a close fit to the substrate or antigen. The combined effect of multiple noncovalent interactions at the active site or binding site is a high-affinity interaction of tremendous specificity. **Figure 43.13** After gene rearrangement, a lymphocyte and its daughter cells make a single version of the antigen receptor. In contrast, alternative splicing is not heritable and can give rise to diverse gene products in a single cell. **Figure 43.18** These receptors enable memory cells to present antigen on their cell surface to a helper T cell. This presentation of antigen is required to activate memory cells in a secondary immune response. **Figure 43.20** Primary response: arrows extending from Antigen (1st exposure), Antigen-presenting cell, Helper T cell, B cell, Plasma cells, Cytotoxic T cell, and Active cytotoxic T cells; secondary response: arrows extending from Antigen (2nd exposure), Memory helper T cells, Memory B cells, and Memory cytotoxic T cells. **Figure 43.26** The loss of growth control that characterizes cancer involves many changes in gene regulation. HPV and other viruses can bring about some of these changes, but other mutations must occur in an infected cell to transform the cell to a cancerous state.

Concept Check 43.1

1. Because pus contains white blood cells, fluid, and cell debris, it indicates an active and at least partially successful inflammatory response against invading microbes. **2.** Whereas the ligand for the TLR receptor is a foreign molecule, the ligand for many signal transduction pathways is a molecule produced by the animal itself. **3.** Bacteria with a human host would be likely to grow optimally at normal body temperature or, if fever were often induced, at a temperature a few degrees higher.

Concept Check 43.2

1. See Figure 43.9. The transmembrane regions lie within the C regions, which also form the disulfide bridges. In contrast, the antigen-binding sites are in the V regions. **2.** Generating memory cells ensures both that a receptor specific for a particular epitope will be present and that there will be more lymphocytes with this specificity than in a host that had never encountered the antigen. **3.** If each B cell produced two different light and heavy chains for its antigen receptor, different combinations would make four different receptors. If any one was self-reactive, the lymphocyte would be eliminated in the generation of self-tolerance. For this reason, many more B cells would be eliminated, and those that could respond to a foreign antigen would be less effective at doing so due to the variety of receptors (and antibodies) they express.

Concept Check 43.3

1. A child lacking a thymus would have no functional T cells. Without helper T cells to help activate B cells, the child would be unable to produce antibodies against extracellular bacteria. Furthermore, without cytotoxic T cells or helper T cells, the child's immune system would be unable to kill virus-infected cells. **2.** Since the antigen-binding site is intact, the antibody fragments could neutralize viruses and opsonize bacteria. **3.** If the handler developed immunity to proteins in the antivenin, another injection could provoke a severe immune response. The handler's immune system might also now produce antibodies that could neutralize the venom.

Concept Check 43.4

1. Myasthenia gravis is considered an autoimmune disease because the immune system produces antibodies against self molecules (certain receptors on muscle cells). **2.** A person with a cold is likely to produce oral and nasal secretions that facilitate viral transfer. In addition, since sickness can cause incapacitation or death, a virus that is programmed to exit the host when there is a physiological stress has the opportunity to find a new host at a time when the current host may cease to function. **3.** A person with a macrophage deficiency would have frequent infections. The causes would be poor innate responses, due to diminished phagocytosis and inflammation, and poor adaptive responses, due to the lack of macrophages to present antigens to helper T cells.

Summary of Key Concepts Questions

43.1 Lysozyme in saliva destroys bacterial cell walls; the viscosity of mucus helps trap bacteria; acidic pH in the stomach kills many bacteria; and the tight packing of cells lining the gut provides a physical barrier to infection. **43.2** Sufficient numbers of cells to mediate an innate immune response are always present, whereas an adaptive response requires selection and proliferation of an initially very small cell population specific for the infecting pathogen. **43.3** No. Immunological memory after a natural infection and after vaccination are very similar. There may be minor differences in the particular antigens that can be recognized in a subsequent infection. **43.4** No. AIDS refers to a loss of immune function that can occur over time in an individual infected with HIV, not to the viral infection itself. For individuals infected with HIV, certain multidrug combinations ("cocktails") or rare genetic variations usually prevent progression to AIDS.

Test Your Understanding

1. b **2.** c **3.** c **4.** d **5.** b **6.** b **7.** c
8. One possible answer:

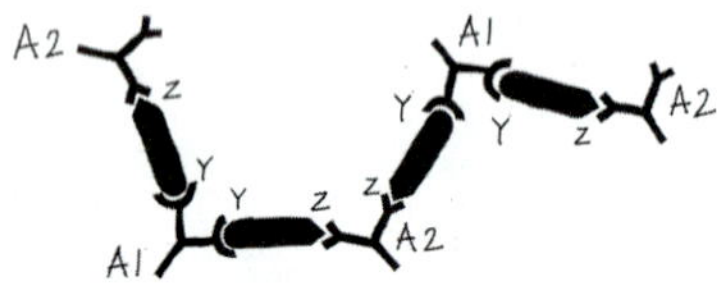

Chapter 44

Figure Questions

Figure 44.2 Aquaporins, which act as water channels **Figure 44.15** You would expect to find these cells lining tubules where they pass through the renal medulla. Because the extracellular fluid of the renal medulla has a very high osmolarity, production of solutes by tubule cells in this region keeps intracellular osmolarity high, with the result that these cells maintain normal volume. **Figure 44.16** Furosemide increases urine volume. The absence of ion transport in the ascending limb leaves the filtrate too concentrated for substantial volume reduction in the distal tubule and collecting duct. **Figure 44.21** The ADH levels would likely be elevated in both sets of patients with mutations because either defect prevents the recapture of water that restores blood osmolarity to normal levels.

Concept Check 44.1

1. Because the salt is moved against its concentration gradient, from low concentration (fresh water) to high concentration (blood) **2.** A freshwater osmoconformer would have body fluids too dilute to carry out life's processes. **3.** Without a layer of insulating fur, the camel must use the cooling effect of evaporative water loss to maintain body temperature, thus linking thermoregulation and osmoregulation.

Concept Check 44.2

1. Because uric acid is largely insoluble in water, it can be excreted as a semisolid paste, thereby reducing an animal's water loss. **2.** Humans produce uric acid from purine breakdown, and reducing purines in the diet often lessens the severity of gout. Birds, however, produce uric acid as a waste product of general nitrogen metabolism. They would therefore need a diet low in all nitrogen-containing compounds, not just purines.

Concept Check 44.3

1. In flatworms, ciliated cells draw interstitial fluids containing waste products into protonephridia. In earthworms, waste products pass from interstitial fluids into the coelom. From there the cilia move the wastes into metanephridia via a funnel surrounding an internal opening to the metanephridia. In insects, the Malpighian tubules pump fluids from the hemolymph, which receives waste products during exchange with interstitial fluids in the course of circulation. **2.** Filtration produces a fluid for exchange processes that is free of cells and large molecules, which are of benefit to the animal and could not readily be reabsorbed. **3.** The presence of Na^+ and other ions (electrolytes) in the dialysate would limit the extent to which they would be removed from the filtrate during dialysis. Adjusting the electrolytes in the starting dialysate can thus lead to the restoration of proper electrolyte concentrations in the plasma. Similarly, the absence of urea and other waste products in the starting dialysate results in their efficient removal from the filtrate.

Concept Check 44.4

1. The numerous nephrons and well-developed glomeruli of freshwater fishes produce urine at a high rate, while the small numbers of nephrons and smaller glomeruli of marine fishes produce urine at a low rate. **2.** The kidney medulla would absorb less water; thus, the drug would increase the amount of water lost in the urine. **3.** A decline in blood pressure in the afferent arteriole would reduce the rate of filtration by moving less material through the vessels.

Concept Check 44.5

1. Alcohol inhibits the release of ADH, causing an increase in urinary water loss and increasing the chance of dehydration. **2.** The consumption of a large amount of water in a very short period of time, coupled with an absence of solute intake, can reduce sodium levels in the blood below tolerable levels. This condition, called hyponatremia, leads to disorientation and, sometimes, respiratory distress. It has occurred in some marathon runners who drink water rather than sports drinks. (It has also caused the death of a fraternity pledge as a consequence of a water hazing ritual and the death of a contestant in a water-drinking competition.) **3.** High blood pressure **4.** Each molecule of renin or ACE activates multiple molecules of the next protein in the pathway. The same is true for the protein kinases. The proteases differ from the protein kinases in at least two ways. First, their action is irreversible. Second, they do not require activation by another enzyme molecule.

Summary of Key Concepts Questions

44.1 Water moves into a cell by osmosis when the fluid outside the cells is hypoosmotic (has a lower solute concentration than the cytosol).
44.2

Waste Attribute	Ammonia	Urea	Uric Acid
Toxicity	High	Very low	Low
Energy content	Low	Moderate	High
Water loss in excretion	High	Moderate	Low

44.3 Filtration retains large molecules that would be difficult to transport across membranes. **44.4** Both types of nephrons have proximal tubules that can reabsorb nutrients, but only juxtamedullary nephrons have loops of Henle that extend deep into the renal medulla. Thus, only kidneys containing juxtamedullary nephrons can produce urine that is more concentrated than the blood. **44.5** Patients who don't produce ADH have symptoms relieved by treatment with the hormone, but many patients with diabetes insipidus lack functional receptors for ADH.

Test Your Understanding

1. d **2.** a **3.** c **4.** e **5.** d **6.** b
7.

Water and salt in food
Osmotic water gain through gills and other surfaces
Uptake of salt ions by gills
Ions and water lost in urine
Salt excreted by rectal gland

Chapter 45

Figure Questions

Figure 45.5 Synthesizing epinephrine requires breaking the bond between the carboxyl group (−COOH) and the α-carbon in tyrosine. **Figure 45.6** The hormone is water-soluble and has a cell-surface receptor. Such receptors, unlike those for lipid-soluble hormones, can cause observable changes in cells without hormone-dependent gene transcription. **Figure 45.17** Both diagnoses could be correct. In one case, the thyroid gland may produce excess thyroid hormone despite normal hormonal input from the hypothalamus and anterior pituitary. In the other, abnormally elevated hormonal input may be the cause of the overactive thyroid gland. **Figure 45.22** The result of the surgery would have been the same for both sexes—an absence of sexual differentiation in the genitalia.

Concept Check 45.1

1. Water-soluble hormones, which cannot penetrate the plasma membrane, bind to cell-surface receptors. This interaction triggers an intracellular signal transduction pathway that ultimately alters the activity of a preexisting protein in the cytoplasm and/or changes transcription of specific genes in the nucleus. Steroid hormones are lipid-soluble and can cross the plasma membrane into the cell interior, where they bind to receptors located in the cytosol or nucleus. The hormone-receptor complex then functions directly as a transcription factor that changes transcription of specific genes. **2.** Prostaglandins in semen that induce contractions in the uterus are acting as signaling molecules that are transferred from one individual to another of the same species (like pheromones), thus aiding in reproduction. **3.** Epinephrine in animals and auxin in plants act as hormones that trigger specific cellular responses that vary among different tissues of the organism.

Concept Check 45.2

1. In a healthy person, insulin released in response to the initial rise in blood glucose stimulates uptake of glucose by body cells. In a person with diabetes, however, inadequate production of insulin or nonresponsiveness of target cells decreases the body's ability to clear excess glucose from the blood. The initial increase in blood glucose is therefore greater in a person with diabetes, and it remains high for a prolonged period. **2.** If the function of the pathway is to provide a transient response, a short-lived stimulus would be less dependent on negative feedback. **3.** Since patients with type 2 diabetes produce insulin but fail to maintain normal glucose levels, you might predict that there could be mutations in the genes for the insulin receptor or the signal transduction pathway it activates. Such mutations have in fact been found in type 2 patients.

Concept Check 45.3

1. The posterior pituitary, an extension of the hypothalamus that contains the axons of neurosecretory cells, is the storage and release site for two neurohormones, oxytocin and antidiuretic hormone (ADH). The anterior pituitary contains endocrine cells that make at least six different hormones. Secretion of anterior pituitary hormones is controlled by hypothalamic hormones that travel via portal vessels to the anterior pituitary. **2.** Because oxytocin responses involve positive feedback (via nerve cells) from suckling, the pathway does not require a sustained hormonal input stimulus. **3.** The hypothalamus and pituitary glands function in many different endocrine pathways. Many defects in these glands, such as those affecting growth or organization, would therefore disrupt many hormone pathways. Only a very specific defect, such as a mutation affecting a particular hormone receptor, would alter just one endocrine pathway. The situation is quite different for the final gland in a pathway, such as the thyroid gland. In this case, a wide range of defects that disrupt gland function would disrupt only the one pathway or small set of pathways in which that gland functions.

Concept Check 45.4

1. The adrenal medulla is derived from neural tissue during development. Reflecting this origin, it is an endocrine organ that produces two molecules—epinephrine and norepinephrine—that act both as hormones and as neurotransmitters. **2.** The levels of these hormones in the blood would become very high. This would be due to the diminished negative feedback on the hypothalamic neurons that secrete the releasing hormone that stimulates the secretion of ACTH by the anterior pituitary. **3.** By applying glucocorticoids to tissue by local injection, you exploit their anti-inflammatory activity. Local injection avoids the effects on glucose metabolism that would occur if glucocorticoids were taken orally and transported throughout the body in the bloodstream.

Summary of Key Concepts Questions

45.1 Because receptors for water-soluble hormones are located on the cell surface, facing the extracellular space, injecting the hormone into the cytoplasm would not trigger a response. **45.2** It would lessen the symptoms. Glucagon acts antagonistically with insulin, so lowering the effects of glucagon would be similar to increasing the levels or activity of insulin. **45.3** The pancreas, parathyroid glands, and pineal gland **45.4** Both the pituitary and the adrenal glands are formed by fusion of neural and nonneural tissue. ADH is secreted by the neurosecretory portion of the pituitary gland, and epinephrine is secreted by the neurosecretory portion of the adrenal gland.

Test Your Understanding

1. c **2.** b **3.** d **4.** c **5.** c **6.** b **7.** a **8.** a

9.

Stimulus
Hypothalamus
Neurosecretory cell
Hypothalamus secretes prolactin-releasing hormone (□)
Blood vessel
Anterior pituitary secretes prolactin (○)
Blood vessel
Mammary glands
Milk production

Chapter 46

Figure Questions

Figure 46.9 According to the graph, about one-third of the females rid themselves of all sperm from the first mating. Thus, two-thirds retain some sperm from the first mating. We would therefore predict that two-thirds of the females would have some offspring exhibiting the small-eye phenotype of the dominant mutation carried by the males with which the females mated first. **Figure 46.12** The analysis would be informative because the polar bodies contain all of the maternal chromosomes that don't end up in the mature egg. For example, finding two copies of the disease gene in the polar bodies would indicate its absence in the egg. This method of genetic testing is sometimes carried out when oocytes collected from a female are fertilized with sperm in a laboratory dish. **Figure 46.16** Testosterone can pass from fetal blood to maternal blood via the placental circulation, temporarily upsetting the hormonal balance in the mother. **Figure 46.18** Oxytocin would most likely induce labor, starting a positive-feedback loop that would direct labor to completion. Synthetic oxytocin is in fact frequently used to induce labor when prolonged pregnancy might endanger the mother or fetus.

Concept Check 46.1

1. The offspring of sexual reproduction are more genetically diverse. However, asexual reproduction can produce more offspring over multiple generations. **2.** Unlike other forms of asexual reproduction, parthenogenesis involves gamete production. By controlling whether or not haploid eggs are fertilized, species such as honeybees can readily switch between asexual and sexual reproduction. **3.** No. Owing to random assortment of chromosomes during meiosis, the offspring may receive the same copy or different copies of a particular parental chromosome from the sperm and the egg. Furthermore, genetic recombination during meiosis will result in reassortment of genes between pairs of parental chromosomes. **4.** Both fragmentation and budding in animals have direct counterparts in the asexual reproduction of plants.

Concept Check 46.2

1. Internal fertilization allows the sperm to reach the egg without either gamete drying out. **2.** (a) Animals with external fertilization tend to release many gametes at once, resulting in the production of enormous numbers of zygotes. This increases the chances that some will survive to adulthood. (b) Animals with internal fertilization produce fewer offspring but generally exhibit greater care of the embryos and the young. **3.** Like the uterus of an insect, the ovary of a plant is the site of fertilization. Unlike the plant ovary, the uterus is not the site of egg production, which occurs in the insect ovary. In addition, the fertilized insect egg is expelled from the uterus, whereas the plant embryo develops within a seed in the ovary.

Concept Check 46.3

1. Spermatogenesis occurs normally only when the testicles are cooler than normal body temperature. Extensive use of a hot tub (or of very tight-fitting underwear) can cause a decrease in sperm quality and number. **2.** In humans, the secondary oocyte combines with a sperm before it finishes the second meiotic division. Thus, oogenesis is completed after, not before, fertilization. **3.** The only effect of sealing off each vas deferens is an absence of sperm in the ejaculate. Sexual response and ejaculate volume are unchanged. The cutting and sealing off of these ducts, a *vasectomy*, is a common surgical procedure for men who do not wish to produce any (more) offspring.

Concept Check 46.4

1. In the testis, FSH stimulates the Sertoli cells, which nourish developing sperm. LH stimulates the production of androgens (mainly testosterone), which in turn stimulate sperm production. In both females and males, FSH encourages the growth of cells that support and nourish developing gametes (follicle cells in females and Sertoli cells in males), and LH stimulates the production of sex hormones that promote gametogenesis (estrogens, primarily estradiol, in females and androgens, especially testosterone, in males). **2.** In estrous cycles, which occur in most female mammals, the endometrium is reabsorbed (rather than shed) if fertilization does not occur. Estrous cycles often occur just one or a few times a year, and the female is usually receptive to copulation only during the period around ovulation. Menstrual cycles are found only in humans and some other primates. **3.** The combination of estradiol and progesterone would have a negative-feedback effect on the hypothalamus, blocking release of GnRH. This would interfere with LH secretion by the pituitary, thus preventing ovulation. This is in fact one basis of action of the most common hormonal contraceptives. **4.** In the viral reproductive cycle, the production of new viral genomes is coordinated with capsid protein expression and with the production of phospholipids for viral coats. In the case of the human female, there is hormonally based coordination of egg maturation with the development of support tissues of the uterus.

Concept Check 46.5

1. hCG secreted by the early embryo stimulates the corpus luteum to make progesterone, which helps maintain the pregnancy. During the second trimester, however, hCG production drops, the corpus luteum disintegrates, and the placenta completely takes over progesterone production. **2.** Both tubal ligation and vasectomy block the movement of gametes from the gonads to a site where fertilization could take place. **3.** By introducing a spermatid nucleus directly into an oocyte, ICSI bypasses the sperm's acquisition of motility in the epididymis, its swimming to meet the egg in the oviduct, and its fusion with the egg.

Summary of Key Concepts Questions

46.1. No. Because parthenogenesis involves meiosis, the mother would pass on to each offspring a random and therefore typically distinct combination of the chromosomes she inherited from her mother and father. **46.2.** None **46.3.** The small size and lack of cytoplasm characteristic of a sperm are adaptations well suited to its function as a delivery vehicle for DNA. The large size and rich cytoplasmic contents of eggs support the growth and development of the embryo. **46.4.** Circulating anabolic steroids mimic the feedback regulation of testosterone, turning off pituitary signaling to the testes and thereby blocking release of signals required for spermatogenesis. **46.5.** Oxygen-rich blood in maternal arteries flows into pools in the endometrium, passes into fetal capillaries in the chorionic villi of the placenta, and from there travels throughout the circulatory system of the fetus.

Test Your Understanding

1. d **2.** b **3.** a **4.** c **5.** a **6.** b **7.** c **8.** d

9.

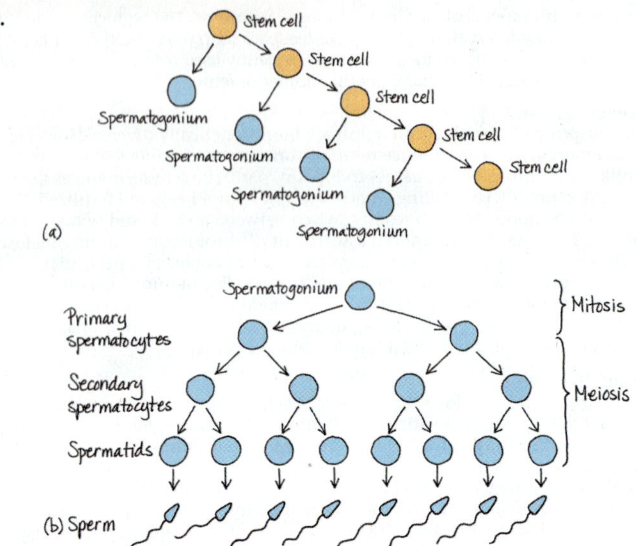

(c) The supply of stem cells would be used up and spermatogenesis would not be able to continue.

Chapter 47

Figure Questions

Figure 47.4 You could inject the compound into an unfertilized egg, expose the egg to sperm, and see whether the fertilization envelope forms. **Figure 47.21** The researchers allowed normal cortical rotation to occur, resulting in activation of the "back-forming" determinants. Then they forced the opposite rotation to occur, which established the back on the opposite side as well. Because the molecules on the normal side were already activated, forcing the opposite rotation apparently did not "cancel out" the establishment of the back side by the first rotation. **Figure 47.22** In Spemann's control, the two blastomeres were physically separated, and each grew into a whole embryo. In Roux's experiment, remnants of the dead blastomere were still contacting the live blastomere, which developed into a half-embryo. Therefore, molecules present in the dead cell's remnants may have been signaling to the live cell, inhibiting it from making all the embryonic structures. **Figure 47.23** You could inject the isolated protein or an mRNA encoding it into ventral cells of an earlier gastrula. If dorsal structures form on the ventral side, that would support the idea that the protein is the signaling molecule secreted or presented by the dorsal lip. You should also do a control experiment to make sure the injection process alone did not cause dorsal structures to form. **Figure 47.25** You could remove the AER and look for Sonic hedgehog mRNA or protein as a marker of the ZPA. If either was absent, that would support your hypothesis. You could also block FGF function and see whether the ZPA formed (by looking for Sonic hedgehog).

Concept Check 47.1

1. The fertilization envelope forms after cortical granules release their contents outside the egg, causing the vitelline membrane to rise and harden. The fertilization envelope serves as a barrier to fertilization by more than one sperm.
2. The increased Ca^{2+} concentration in the egg would cause the cortical granules to fuse with the plasma membrane, releasing their contents and causing a fertilization envelope to form, even though no sperm had entered. This would prevent fertilization. **3.** You would expect it to fluctuate. The fluctuation of MPF drives the transition between DNA replication (S phase) and mitosis (M phase), which is still required in the abbreviated cleavage cell cycle.

Concept Check 47.2

1. The cells of the notochord migrate toward the midline of the embryo (converge), rearranging themselves so there are fewer cells across the notochord, which thus becomes longer overall (extends; see Figure 47.16). **2.** Because microfilaments would not be able to contract and decrease the size of one end of the cell, both the inward bending in the middle of the neural tube and the outward bending of the hinge regions at the edges would be blocked. Therefore, the neural tube probably would not form. **3.** Dietary intake of the vitamin folic acid dramatically reduces the frequency of neural tube defects.

Concept Check 47.3

1. Axis formation establishes the location and polarity of the three axes that provide the coordinates for development. Pattern formation positions particular tissues and organs in the three-dimensional space defined by those coordinates.
2. Morphogen gradients act by specifying cell fates across a field of cells through variation in the level of a determinant. Morphogen gradients thus act more globally than cytoplasmic determinants or inductive interactions between pairs of cells. **3.** Yes, a second embryo could develop because inhibiting BMP-4 activity would have the same effect as transplanting an organizer. **4.** The limb that developed probably would have a mirror-image duplication, with the most posterior digits in the middle and the most anterior digits at either end.

Summary of Key Concepts Questions

47.1 The binding of a sperm to a receptor on the egg surface is very specific and likely would not occur if the two gametes were from different species. Without sperm binding, the sperm and egg membranes would not fuse. **47.2** The neural tube forms when the neural plate, a band of ectodermal tissue oriented along the anterior-posterior axis on the dorsal side of the embryo, rolls into a tube and pinches off from the rest of the ectoderm. Neural crest cells arise as groups of cells in the regions between the edges of the neural tube and the surrounding ectoderm migrate away from the neural tube. **47.3** Mutations that affected both limb and kidney development would be more likely to alter the function of monocilia because these organelles are important in several signaling pathways. Mutations that affected limb development but not kidney development would more likely alter a single pathway, such as Hedgehog signaling.

Test Your Understanding

1. a **2.** b **3.** e **4.** a **5.** d **6.** d **7.** b
8.

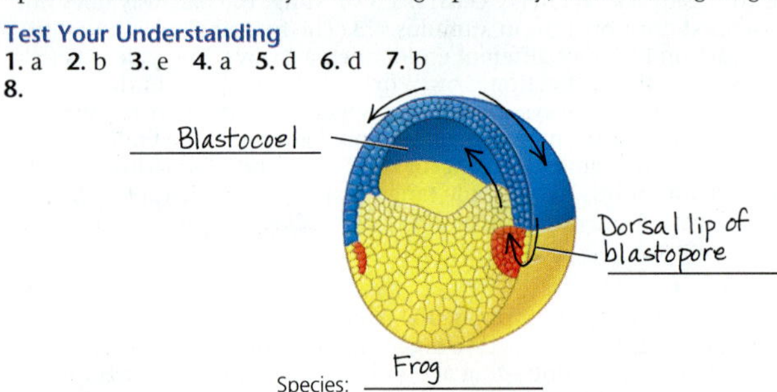

Chapter 48

Figure Questions

Figure 48.8 Adding chloride channels makes the membrane potential less positive. Adding sodium or potassium channels would have no effect, because sodium movement is already at equilibrium and there are no potassium ions present.
Figure 48.10

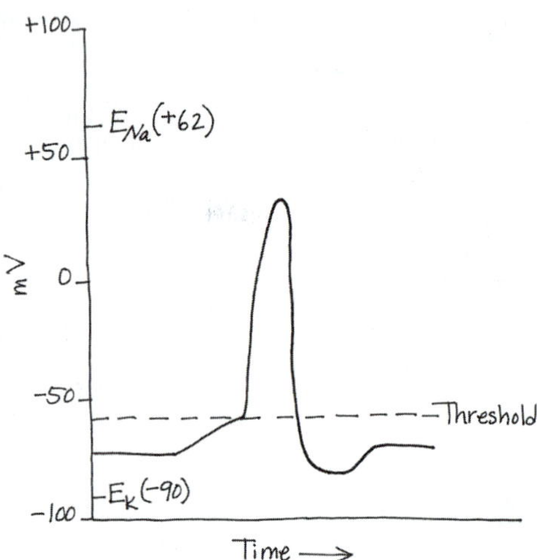

Figure 48.15 The production and transmission of action potentials would be unaffected. However, action potentials arriving at chemical synapses would be unable to trigger release of neurotransmitter. Signaling at such synapses would thus be blocked. **Figure 48.18** The drug might be destroyed very quickly in the body, not reach the central nervous system, or bind to but not activate the receptor.

Concept Check 48.1

1. Sensors in your ear transmit information to your brain. There the activity of interneurons in processing centers enables you to recognize your name. In response, signals transmitted via motor neurons cause contraction of muscles that turn your neck.
2. Increased branching would allow control of a greater number of postsynaptic cells, enhancing coordination of responses to nervous system signals. **3.** Communication by bacteria involves all the cells in a colony, whereas communication by neurons involves just a few cells in the animal body. In addition, neurons direct signals from one location to another, whereas bacterial cells communicate in all directions.

Concept Check 48.2

1. Ions can flow against a chemical concentration gradient if there is an opposing electrical gradient of greater magnitude. **2.** A decrease in permeability to K^+, an increase in permeability to Na^+, or both **3.** The activity of the sodium-potassium pump is essential to maintain the resting potential. With the pump inactivated, the sodium and potassium concentration gradients would gradually disappear, resulting in a greatly reduced resting potential. **4.** Charged dye molecules could equilibrate only if other charged molecules could also cross the membrane. If not, a membrane potential would develop that would counterbalance the chemical gradient.

Concept Check 48.3

1. A graded potential has a magnitude that varies with stimulus strength, whereas an action potential has an all-or-none magnitude that is independent of stimulus strength. **2.** Loss of the insulation provided by myelin sheaths leads to a disruption of action potential propagation along axons. Voltage-gated sodium channels are restricted to the nodes of Ranvier, and without the insulating effect of myelin, the inward current produced at one node during an action potential cannot depolarize the membrane to the threshold at the next node. **3.** The maximum frequency would decrease because the refractory period would be extended.

Concept Check 48.4

1. It can bind to different types of receptors, each triggering a specific response in postsynaptic cells. **2.** These toxins would prolong the EPSPs that acetylcholine produces because the neurotransmitter would remain longer in the synaptic cleft. **3.** Since GABA is an inhibitory neurotransmitter in the CNS, this drug would be expected to decrease brain activity. A decrease in brain activity might be expected to slow down or reduce behavioral activity. Many sedative drugs act in this fashion. **4.** Membrane fusion

Summary of Key Concepts Questions

48.1 It would prevent information from being transmitted away from the cell body along the axon. **48.2** There are very few open sodium channels in a resting neuron, so the resting potential either would not change or would become slightly more negative (hyperpolarization). **48.3** Positive feedback is responsible for the rapid opening of many voltage-gated sodium channels, causing the rapid outflow of sodium ions responsible for the rising phase of the action potential. As the membrane potential becomes positive, voltage-gated potassium channels open in a form of negative feedback that helps bring about the falling phase of the action potential. **48.4** A given neurotransmitter can have many receptors that differ in their location and activity. Drugs that target receptor activity rather than neurotransmitter release or stability are therefore likely to exhibit greater specificity and potentially have fewer undesirable side effects.

Test Your Understanding

1. c **2.** c **3.** c **4.** e **5.** b **6.** a **7.** As shown in this pair of drawings, a pair of action potentials would move outward in both directions from each electrode. (Action potentials are unidirectional only if they begin at one end of an axon.) However, because of the refractory period, the two action potentials between the electrodes both stop where they meet. Thus, only one action potential reaches the synaptic terminals.

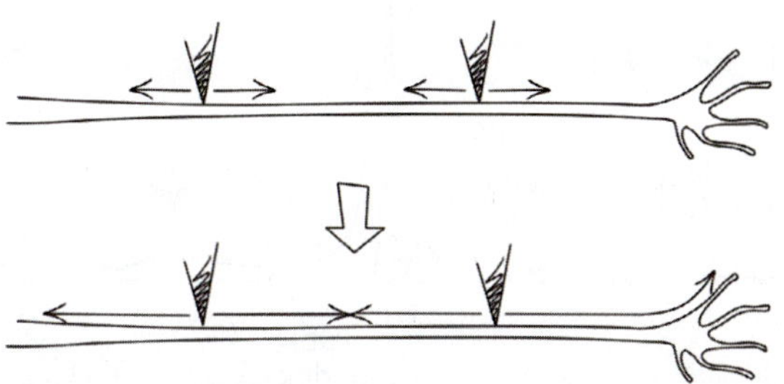

Chapter 49

Figure Questions

Figure 49.3 During swallowing, muscles along the esophagus alternately contract and relax, resulting in peristalsis. One model to explain this alternation is that each section of muscle receives nerve impulses that alternate between excitation and inhibition, just as the quadriceps and hamstring receive opposing signals in the knee-jerk reflex. **Figure 49.12** If the new mutation disrupted only pacemaker function, you should be able to restore rhythmic activity by removing the SCN and replacing it with an SCN transplant from either a wild-type or τ mutant hamster. Using the new mutant as the donor would not be as informative, since both failed transplants and successful ones would result in a lack of rhythmic activity. **Figure 49.14** Regions you would expect to be active regardless of the type of music played would include ones that are important for processing and interpreting sounds. **Figure 49.23** If the depolarization brings the membrane potential to or past threshold, it should initiate action potentials that cause dopamine release from the VTA neurons. This should mimic natural stimulation of the brain reward system, resulting in positive and perhaps pleasurable sensations.

Concept Check 49.1

1. The sympathetic division would likely be activated. It mediates the "fight-or-flight" response in stressful situations. **2.** The preganglionic neurons use the same neurotransmitter and function similarly in each division (to stimulate postganglionic neurons). The postganglionic neurons use different neurotransmitters that generally bring about opposing functions in the same target tissues. **3.** Nerves contain bundles of axons, some that belong to motor neurons, which send signals outward from the CNS, and some that belong to sensory neurons, which bring signals into the CNS. Therefore, you would expect effects on both motor control and sensation. **4.** Neurosecretory cells of the adrenal medulla secrete the hormones epinephrine and norepinephrine in response to preganglionic input from sympathetic neurons. These hormones travel in the circulation throughout the body, triggering responses in many tissues.

Concept Check 49.2

1. The cerebral cortex on the left side of the brain initiates voluntary movement of the right side of the body. **2.** Alcohol diminishes function of the cerebellum. **3.** Paralysis reflects an inability to carry out motor commands transmitted from the cerebrum to the spinal cord. You would expect these patients to have injuries in the portion of the CNS extending from the spinal cord up to but not including the reticular formation. A coma reflects a disruption in the cycles of sleep and arousal regulated by communication between the reticular formation and the cerebrum. You would expect these patients to have injuries in the reticular formation or in the portion of the brain on the opposite side of the reticular formation from the spinal cord.

Concept Check 49.3

1. Brain damage that disrupts behavior, cognition, memory, or other functions provides evidence that the portion of the brain affected by the damage is important for the normal activity that is blocked or altered. **2.** Broca's area, which is active during the generation of speech, is located near the part of the primary motor cortex that controls muscles in the face. Wernicke's area, which is active when speech is heard, is located near the part of the temporal lobe that is involved in hearing. **3.** Each cerebral hemisphere is specialized for different parts of this task—the right for face recognition and the left for language. Without an intact corpus callosum, neither hemisphere can take advantage of the other's processing abilities.

Concept Check 49.4

1. There can be an increase in the number of synapses between the neurons or an increase in the strength of existing synaptic connections. **2.** If consciousness is an emergent property resulting from the interaction of many different regions of the brain, then it is unlikely that localized brain damage will have a discrete effect on consciousness. **3.** The hippocampus is responsible for organizing newly acquired information. Without hippocampal function, the links necessary to retrieve information from the neocortex will be lacking, and no functional memory, short- or long-term, will be formed.

Concept Check 49.5

1. Both are progressive brain diseases whose risk increases with advancing age. Both result from the death of brain neurons and are associated with the accumulation of peptide or protein aggregates. **2.** The symptoms of schizophrenia can be mimicked by a drug that stimulates dopamine-releasing neurons. The brain's reward system, which is involved in drug addiction, is composed of dopamine-releasing neurons that connect the ventral tegmental area to regions in the cerebrum. Parkinson's disease results from the death of dopamine-releasing neurons. **3.** Not necessarily. It might be that the plaques, tangles, and missing regions of the brain seen at death reflect secondary effects, the consequence of other unseen changes that are actually responsible for the alterations in brain function.

Summary of Key Concepts Questions

49.1 Because reflex circuits involve only a few neurons—the simplest consist of a sensory neuron and a motor neuron—the path for information transfer is short and simple, increasing the speed of the response. **49.2** The pons and medulla (the midbrain) coordinate visual reflexes; the cerebellum controls coordination of movement that depends on visual input; the thalamus serves as a routing center for visual information; and the cerebrum is essential for converting visual input to a visual image. **49.3** You would expect the right side of the body to be paralyzed because it is controlled by the left cerebral hemisphere, where language generation and interpretation are localized. **49.4** Learning a new language likely requires the maintenance of synapses that are formed during early development but are otherwise lost prior to adulthood. **49.5** Whereas amphetamines stimulate dopamine release, PCP blocks glutamate receptors, suggesting that schizophrenia does not reflect a defect in the function of just one neurotransmitter.

Test Your Understanding

1. c **2.** a **3.** d **4.** c **5.** d **6.** e
7.

(a)

Pain receptor in finger
Afferent sensory neuron from pain receptor in finger
Pin pricks finger
Spinal cord
Interneurons
Synapses in arm muscles
Hand pulls away
Efferent motor neurons to arm muscles

(b)

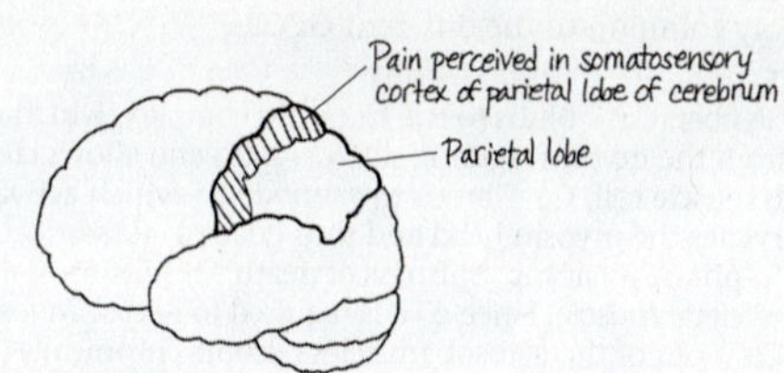

Chapter 50

Figure Questions

Figure 50.12 In the brain. Each note is detected separately in the ear, with each causing vibration of the basilar membrane and deflection of hair cells in a distinct location. Sensory neurons in each location provide output in the form of action potentials that travel along distinct axons in the auditory nerve. It is not until the information reaches the brain that the individual notes are detected and the perception of the chord is generated. **Figure 50.19** Each of the three types of cones is most sensitive to a different wavelength of light. A cone might be fully depolarized when there is light present if the light is of a wavelength far from its optimum. **Figure 50.21** In humans, an X chromosome with a defect in the red or green opsin gene is much less common than a wild-type X chromosome. Color blindness therefore typically skips a generation as the defective allele passes from an affected male to a carrier daughter and back to an affected grandson. In squirrel monkeys, no X chromosome can confer full color vision. As a result, all males are color-blind and no unusual inheritance pattern is observed. **Figure 50.23** The results of the experiment would have been identical. What matters is the activation of particular sets of neurons, not the manner in which they are activated. Any signal from a bitter cell will be interpreted by the brain as a bitter taste, regardless of the nature of the compound and the receptor involved. **Figure 50.25** Only perception. Binding of an odorant to its receptor will cause action potentials to be sent to the brain. Although an excess of that odorant might cause a diminished response through adaptation, another odorant can mask the first only at the level of perception in the brain. **Figure 50.28** Hundreds of myosin heads participate in sliding each pair of thick and thin filaments past each other. Because cross-bridge formation and breakdown are not synchronized, many myosin heads are exerting force on the thin filaments at all times during muscle contraction. **Figure 50.32** By causing all of the motor neurons that control the muscle to generate action potentials at a rate high enough to produce tetanus in all of the muscle fibers **Figure 50.40** Since a duck is more specialized for flying than for swimming, you might expect that it would consume more energy per unit body mass and distance in swimming than would, for example, a fish. (In fact, if the value for a 10^3-g swimming duck were plotted on this graph, it would appear well above the line for swimmers and just above the line for runners.)

Concept Check 50.1

1. Electromagnetic receptors in general detect only external stimuli. Nonelectromagnetic receptors, such as chemoreceptors or mechanoreceptors, can act as either internal or external sensors. **2.** The capsaicin present in the peppers activates the thermoreceptor for high temperatures. In response to the perceived high temperature, the nervous system triggers sweating to achieve evaporative cooling. **3.** You would perceive the electrical stimulus as if the sensory receptors that regulate that neuron had been activated. For example, electrical stimulation of the sensory neuron controlled by the thermoreceptor activated by menthol would likely be perceived as a local cooling.

Concept Check 50.2

1. Statocysts detect the animal's orientation with respect to gravity, providing information that is essential in environments such as these, where light cues are absent. **2.** As a sound that changes gradually from a very low to a very high pitch **3.** The stapes and the other middle ear bones transmit vibrations from the tympanic membrane to the oval window. Fusion of these bones (as occurs in a disease called otosclerosis) would block this transmission and result in hearing loss.

Concept Check 50.3

1. Planarians have ocelli that cannot form images but can sense the intensity and direction of light, providing enough information to enable the animals to find protection in shaded places. Flies have compound eyes that form images and excel at detecting movement. **2.** The person can focus on distant objects but not close objects (without glasses) because close focusing requires the lens to become almost spherical. This problem is common after age 50. **3.** Close each eye in turn. An object floating on the surface of an eyeball will appear only when that eye is open. **4.** Absorption of light by retinal converts a structure isomer in the *cis* configuration to the isomer in the *trans* configuration, initiating the process of light detection. In contrast, a photon absorbed by chlorophyll does not bring about isomerization, but instead boosts an electron to a higher energy orbital, initiating the electron flow that generates ATP and NADPH.

Concept Check 50.4

1. Both taste cells and olfactory cells have receptor proteins in their plasma membrane that bind certain substances, leading to membrane depolarization through a signal transduction pathway involving a G protein. However, olfactory cells are sensory neurons, whereas taste cells are not. **2.** Since animals rely on chemical signals for behaviors that include finding mates, marking territories, and avoiding dangerous substances, it is adaptive for the olfactory system to have a robust response to a very small number of molecules of a particular odorant. **3.** Because the sweet, bitter, and umami tastes involve GPCR proteins but the sour taste does not, you might predict that the mutation is in a molecule that acts in the signal transduction pathway common to the different GPCRs.

Concept Check 50.5

1. In a skeletal muscle fiber, Ca^{2+} binds to the troponin complex, which moves tropomyosin away from the myosin-binding sites on actin and allows cross-bridges to form. In a smooth muscle cell, Ca^{2+} binds to calmodulin, which activates an enzyme that phosphorylates the myosin head and thus enables cross-bridge formation. **2.** *Rigor mortis*, a Latin phrase meaning "stiffness of death," results from the complete depletion of ATP in skeletal muscle. Since ATP is required to release myosin from actin and to pump Ca^{2+} out of the cytosol, muscles become chronically contracted beginning about 3–4 hours after death. **3.** A competitive inhibitor binds to the same site as the substrate for the enzyme. In contrast, the troponin and tropomyosin complex masks, but does not bind to, the myosin-binding sites on actin.

Concept Check 50.6

1. Septa provide the divisions of the coelom that allow for peristalsis, a form of locomotion requiring independent control of different body segments. **2.** The main problem in swimming is drag; a fusiform body minimizes drag. The main problem in flying is overcoming gravity; wings shaped like airfoils provide lift, and adaptations such as air-filled bones reduce body mass. **3.** When you grasp the sides of the chair, you are using a contraction of the triceps to keep your arms extended against the pull of gravity on your body. As you lower yourself slowly into the chair, you gradually decrease the number of motor units in the triceps that are contracted. Contracting your biceps would jerk you down, since you would no longer be opposing gravity.

Summary of Key Concepts Questions

50.1 Nociceptors overlap with other classes of receptors in the type of stimulus they detect. They differ from other receptors only in how a particular stimulus is perceived. **50.2** The direction of displacement of a hair cell is determined by stimulus intensity, which is encoded by the frequency of action potentials transmitted to the brain. **50.3** Our olfactory sense is responsible for most of what we describe as distinct tastes. A head cold or other source of congestion blocks odorant access to receptors lining portions of the nasal cavity. **50.4** The major difference is that neurons in the retina integrate information from multiple sensory receptors (photoreceptors) before transmitting information to the central nervous system. **50.5** Hydrolysis of ATP is required to convert myosin to a high-energy configuration for binding to actin and to power the Ca^{2+} pump that removes cytosolic Ca^{2+} during muscle relaxation. **50.6** Human body movements rely on the contraction of muscles anchored to a rigid endoskeleton. Tendons attach muscles to bones, which in turn are composed of fibers built up from a basic organizational unit, the sarcomere. The thin and thick filaments have separate points of attachment within the sarcomere. In response to nervous system motor output, the formation and breakdown of cross-bridges between myosin heads and actin ratchet the thin and thick filaments past each other. Because the filaments are anchored, this sliding movement shortens the muscle fibers. Furthermore, because the fibers themselves are part of the muscles attached at each end to bones, muscle contraction moves bones of the body relative to each other. In this way, the structural anchoring of muscles and filaments enables muscle function, such as the bending of an elbow by contraction of the biceps.

Self Quiz

1. e **2.** a **3.** b **4.** c **5.** b **6.** d
7.

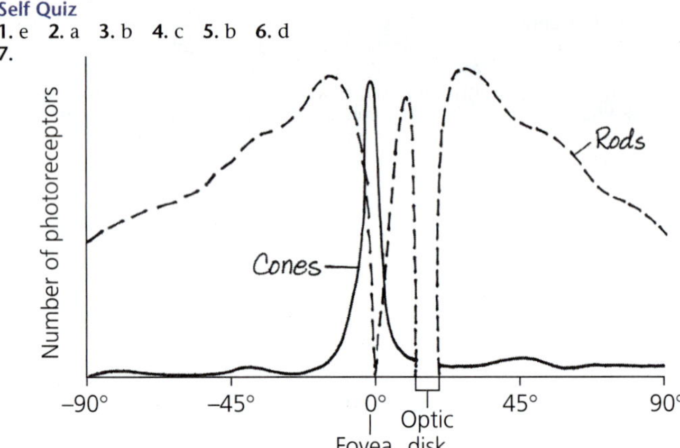

The answer shows the actual distribution of rods and cones in the human eye. Your graph may differ but should have the following properties: Only cones at the fovea; fewer cones and more rods at both ends of the *x*-axis; no photoreceptors in the optic disk.

Chapter 51

Figure Questions

Figure 51.2 The fixed action pattern based on the sign stimulus of a red belly ensures that the male will chase away any invading males of his species. By chasing away such males, the defender decreases the chance that another male will fertilize eggs laid in his nesting territory. **Figure 51.7** There should be no effect. Imprinting is an innate behavior that is carried out anew in each generation. Assuming the nest was not disturbed, the offspring of the Lorenz followers would imprint on the mother goose. **Figure 51.8** Perhaps the wasp doesn't use visual cues. It might also be that wasps recognize objects native to their environment, but not foreign objects, such as the pinecones. Tinbergen addressed these ideas before carrying out the pinecone study. When he swept away the pebbles and sticks around the nest, the wasps could no longer find their nests. If he shifted the natural objects in their natural arrangement, the shift in the landmarks caused a shift in the site to which the wasps returned. Finally, if natural objects around the nest site were replaced with pinecones while the wasp was in the burrow, the wasp nevertheless found her way back to the nest site. **Figure 51.23** Courtship song generation must be coupled to courtship song recognition. Unless the genes that control generation of particular song elements also control recognition, the hybrids might be unlikely to find mating partners, depending on what aspects of the songs are important for mate recognition and acceptance. **Figure 51.26** It might be that the birds require stimuli during flight to exhibit their migratory preference. If this were true, the birds would show the same orientation in the funnel experiment despite their distinct genetic programming. **Figure 51.28** It holds true for some, but not all individuals. If a parent has more than one reproductive partner, the offspring of different partners will have a coefficient of relatedness less than 0.5.

Concept Check 51.1

1. The proximate explanation for this fixed action pattern might be that nudging and rolling are released by the sign stimulus of an object outside the nest, and the behavior is carried to completion once initiated. The ultimate explanation might be that ensuring that eggs remain in the nest increases the chance of producing healthy offspring. **2.** In both cases, the detection of periodic variation in the environment results in a reproductive cycle timed to environmental conditions that optimize the opportunity for success. **3.** There might be selective pressure for other prey fish to detect an injured fish because the source of the injury might threaten them as well. Among predators, there might be selection for those that are attracted to the alarm substance because they would be more likely to encounter crippled prey. Fish with adequate defenses might show no change because they have a selective advantage if they do not waste energy responding to the alarm substance.

Concept Check 51.2

1. Natural selection would tend to favor convergence in color pattern because a predator learning to associate a pattern with a sting or bad taste would avoid all other individuals with that same color pattern, regardless of species. **2.** You might move objects around to establish an abstract rule, such as "past landmark A, the same distance as A is from the starting point," while maintaining a minimum of fixed metric relationships, that is, avoiding having the food directly adjacent to or a set distance from a landmark. As you might surmise, designing an informative experiment of this kind is not easy. **3.** Learned behavior, just like innate behavior, can contribute to reproductive isolation and thus to speciation. For example, learned bird songs contribute to species recognition during courtship, thereby helping ensure that only members of the same species mate.

Concept Check 51.3

1. Certainty of paternity is higher with external fertilization. **2.** Balancing selection could maintain the two alleles at the *forager* locus if population density fluctuated from one generation to another. At times of low population density, the energy-conserving sitter larvae (carrying the for^s allele) would be favored, while at higher population density, the more mobile Rover larvae (for^R allele) would have a selective advantage. **3.** Because females would now be present in much larger numbers than males, all three types of males should have some reproductive success. Nevertheless, since the advantage that the blue-throats rely on—a limited number of females in their territory—will be absent, the yellow-throats are likely to increase in frequency in the short term.

Concept Check 51.4

1. Because this geographic variation corresponds to differences in prey availability between two garter snake habitats, it seems likely that snakes with characteristics enabling them to feed on the abundant prey in their locale would have had increased survival and reproductive success. In this way, natural selection would have resulted in the divergent foraging behaviors. **2.** Yes. Kin selection does not require any recognition or awareness of relatedness. **3.** The older individual cannot be the beneficiary because he or she cannot have extra offspring. However, the cost is low for an older individual performing the altruistic act because that individual has already reproduced (but perhaps is still caring for a child or grandchild). There can therefore be selection for an altruistic act by a postreproductive individual that benefits a young relative.

Summary of Key Concepts Questions

51.1 Circannual rhythms are typically based on the cycles of light and dark in the environment. As the global climate changes, animals that migrate in response to these rhythms may shift to a location before or after local environmental conditions are optimal for reproduction and survival. **51.2** For the goose, all that is acquired is an object at which the behavior is directed. In the case of the sparrow, learning takes place that will give shape to the behavior itself. **51.3** Because feeding the female is likely to improve her reproductive success, the genes from the sacrificed male are likely to appear in a greater number of progeny. **51.4** You would likely have missed the idea that changes in a single gene can have large-scale effects on even complex behaviors.

Test Your Understanding

1. d **2.** b **3.** c **4.** a **5.** c **6.** a
7.

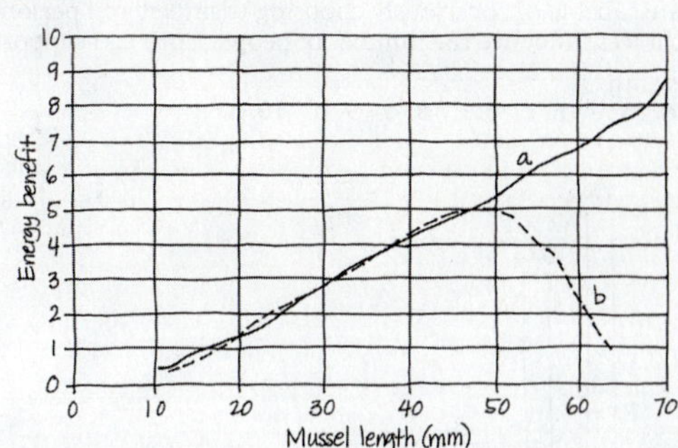

You could measure the size of mussels that oystercatchers successfully open and compare that with the size distribution in the habitat.

Chapter 52

Figure Questions

Figure 52.7 Dispersal limitations, the activities of people (such as a broad-scale conversion of forests to agriculture or selective harvesting), or many other factors, including those discussed later in the chapter (see Figure 52.18) **Figure 52.18** Some factors, such as fire, are relevant only for terrestrial systems. At first glance, water availability is primarily a terrestrial factor, too. However, species living along the intertidal zone of oceans or along the edge of lakes also suffer desiccation. Salinity stress is important for species in some aquatic and terrestrial systems. Oxygen availability is an important factor primarily for species in some aquatic systems and in soils and sediments. **Figure 52.20** When only urchins were removed, limpets may have increased in abundance and reduced seaweed cover somewhat (the difference between the purple and blue lines on the graph).

Concept Check 52.1

1. In the tropics, high temperatures evaporate water and cause warm, moist air to rise. The rising air cools and releases much of its water as rain over the tropics. The remaining dry air descends at approximately 30° north and south, causing deserts to occur in those regions. **2.** The microclimate around the stream will be cooler, moister, and shadier than that around the unplanted agricultural field. **3.** Trees that require a long time to reach reproductive age are likely to evolve more slowly than annual plants in response to climate change, constraining the potential ability of such trees to respond to rapid climate change. **4.** Plants with C_4 photosynthesis are likely to expand their range globally as Earth's climate warms. As described in Concept 10.4, C_4 photosynthesis minimizes photorespiration and enhances sugar production, an advantage that is especially useful in warmer regions where C_4 plants are found today.

Concept Check 52.2

1. Temperate broadleaf forests have higher mean annual precipitation. **2.** Answers will vary by location but should be based on the information and maps in Figure 52.12. How much your local area has been altered from its natural state will influence how much it reflects the expected characteristics of your biome, particularly the expected plants and animals. **3.** Northern coniferous forest is likely to replace tundra along the boundary between these biomes. To see why, note that northern coniferous forest is adjacent to tundra throughout North America, northern Europe, and Asia (see Figure 52.9) and that the temperature range for northern coniferous forest is just above that for tundra (see Figure 52.10).

Concept Check 52.3

1. In the oceanic pelagic zone, the ocean bottom lies below the photic zone, so there is too little light to support benthic algae or rooted plants. **2.** As explained in Concept 44.1, aquatic organisms either gain or lose water by osmosis if the osmolarity of their environment differs from their internal osmolarity. Water gain can cause cells to swell, and water loss can cause them to shrink. To avoid excessive changes in cell volume, organisms that live in estuaries must be able to compensate for both water gain (under freshwater conditions) and water loss (under saltwater conditions). **3.** In a river below a dam, the fish are more likely to be species that prefer colder water. In summer, the deep layers of a reservoir are colder than the surface layers, so a river below a dam will be colder than an undammed river.

Concept Check 52.4

1. (a) Humans might transplant a species to a new area that it could not previously reach because of a geographic barrier. (b) Humans might eliminate a predator or herbivore species, such as sea urchins, from an area. **2.** One test would be to build a fence around a plot of land in an area that has trees of that species, excluding all deer from the plot. You could then compare the abundance of tree seedlings inside and outside the fenced plot over time. **3.** Because the ancestor of the silverswords reached isolated Hawaii early in the islands' existence, it likely faced little competition and was able to occupy many unfilled niches. The cattle egret, in contrast, arrived in the Americas only recently and has to compete with a well-established group of species. Thus, its opportunities for adaptive radiation have probably been much more limited.

Summary of Key Concepts Questions

52.1 Because dry air would descend at the equator instead of at 30° north and south latitude (where deserts exist today), deserts would be more likely to exist along the equator (see Figure 52.3). **52.2** Because tundra is much cooler than deserts (see Figure 52.10), less water evaporates during the growing season and the tundra stays more moist. **52.3** An aphotic zone is most likely to be found in the deep waters of a lake, the oceanic pelagic zone, or the marine benthic zone. **52.4** You could arrange a flowchart that begins with abiotic limitations—first determining the physical and chemical conditions under which a species could survive—and then moves through the other factors listed in the flowchart.

Test Your Understanding

1. c **2.** b **3.** d **4.** e **5.** d **6.** d **7.** c **8.** a **9.** a **10.** c
11.

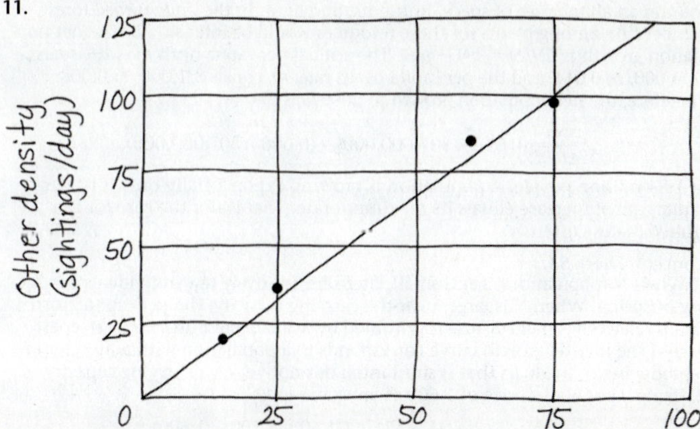

Based on what you learned from Figure 52.20 and on the positive relationship you observed in the field between kelp abundance and otter density, you could hypothesize that otters lower sea urchin density, reducing feeding of the urchins on kelp.

Chapter 53

Figure Questions

Figure 53.4 The dispersion of the penguins would likely appear clumped as you flew over densely populated islands and sparsely populated ocean. **Figure 53.13** If male European kestrels provided no parental care, brood size should not affect their survival. Therefore, the three bars representing male survival in Figure 53.13 should have similar heights. In contrast, female survival should still decline with increasing brood size, as shown in the current figure.

Figure 53.15

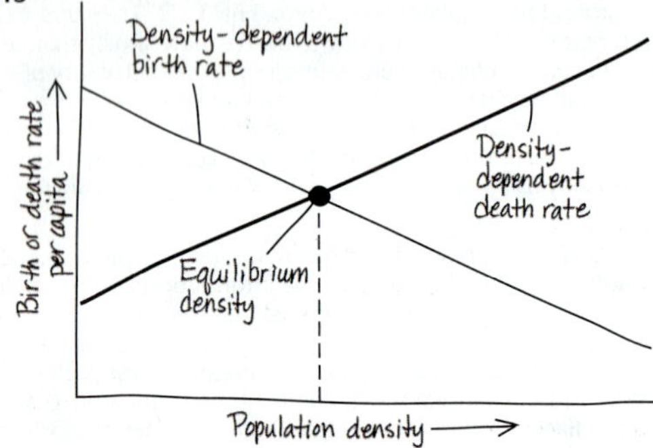

Figure 53.19 Hare numbers typically peaked slightly before lynx numbers did. The lynx depend on the hares for food, but there is a delay between increased food availability and increased reproduction by the lynx. **Figure 53.20** *Dictyostelium* slugs are more vulnerable to predation by animals than are amoebas.

Concept Check 53.1

1.

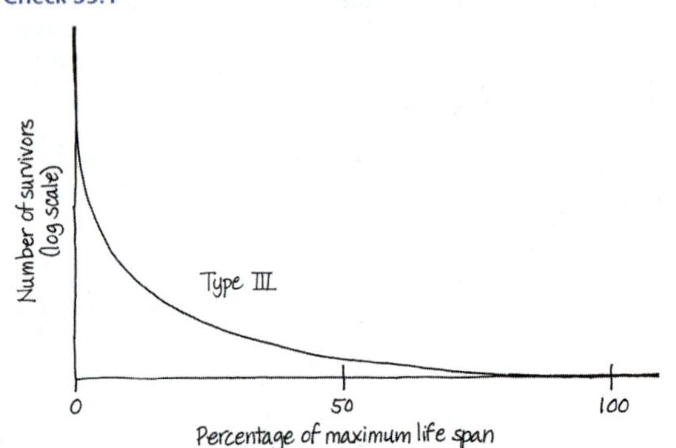

A type III survivorship curve is most likely because very few of the young probably survive. **2.** If an animal is captured by attracting it with food, it may be more likely to be recaptured if it seeks the same food. The number of marked animals captured (x) would be an overestimate, and because the population size (N) is equal to sn/x, N would be an underestimate. Alternatively, if an animal has a negative experience during capture and learns from that experience, it may be less likely to be recaptured. In this case, x would be an underestimate and N would be an overestimate. **3.** Male sticklebacks would likely have a uniform pattern of dispersion, with antagonistic interactions maintaining a relatively constant spacing between them.

Concept Check 53.2

1. Though r_{max} is constant, N, the population size, is increasing. As r_{max} is applied to an increasingly large N, population growth ($r_{max}N$) accelerates, producing the J-shaped curve. **2.** Exponential growth is more likely in the area where a forest was destroyed by fire. The first plants that found suitable habitat there would encounter an abundance of space, nutrients, and light. In the undisturbed forest, competition among plants for these resources would be intense. **3.** The net population growth is $\Delta N/\Delta t = bN - mN$. The annual per capita birth rate, b, equals 14/1,000, or 0.014, and the per capita death rate, m, equals 8/1,000, or 0.008. Therefore, the net population growth in 2009 was

$$\frac{\Delta N}{\Delta t} = (0.014 \times 307{,}000{,}000) - (0.008 \times 307{,}000{,}000)$$

or 1.84 million people. A population is growing exponentially only if its per capita rate of increase equals its maximum rate. That is not the case for the United States currently.

Concept Check 53.3

1. When N (population size) is small, there are relatively few individuals producing offspring. When N is large, near the carrying capacity, the per capita growth rate is relatively small because it is limited by available resources. The steepest part of the logistic growth curve corresponds to a population with a number of reproducing individuals that is substantial but not yet near carrying capacity. **2.** Using a population size of 1,600 as an example,

$$\frac{dN}{dt} = r_{max}N\frac{(K-N)}{K} = \frac{1(1{,}600)(1{,}500 - 1{,}600)}{1{,}500}$$

and the population "growth" rate is –107 individuals per year. The population shrinks even faster when N is farther from the carrying capacity; when N equals 1,750 and 2,000 individuals, the population shrinks by 292 and 667 individuals per year, respectively. These negative growth rates correspond most closely to the time when the *Daphnia* population has overshot its carrying capacity and is shrinking, about days 65–100 in Figure 53.10b. **3.** If a population becomes too crowded, the likelihood of disease and mortality may increase because of the effects of pathogens. Thus, pathogens can reduce the long-term carrying capacity of a population.

Concept Check 53.4

1. The constant, spring-fed stream. In more constant physical conditions, populations are more stable and competition for resources is more likely. In such conditions, larger, well-provisioned young typical of iteroparous species have a better chance of surviving. **2.** By preferentially investing in the eggs it lays in the nest, the peacock wrasse increases their probability of survival. The eggs it disperses widely and does not provide care for are less likely to survive, at least some of the time, but require a lower investment by the adults. (In this sense, the adults avoid the risk of placing all their eggs in one basket.) **3.** If a parent's survival is compromised greatly by bearing young during times of stress, the animal's fitness may increase if it abandons its current young and survives to produce healthier young at a later time.

Concept Check 53.5

1. Three attributes are the size, quality, and isolation of patches. A patch that is larger or of higher quality is more likely to attract individuals and to be a source of individuals for other patches. A patch that is relatively isolated will undergo less exchange of individuals with other patches. **2.** You would need to study the population for more than one cycle (longer than 10 years and probably at least 20) before having sufficient data to examine changes through time. Otherwise, it would be impossible to know whether an observed decrease in the population size reflected a long-term trend or was part of the normal cycle. **3.** In negative feedback, the output, or product, of a process slows that process. In populations that have a density-dependent birth rate, such as dune fescue grass, an accumulation of product (more individuals, resulting in a higher population density) slows the process (population growth) by decreasing the birth rate.

Concept Check 53.6

1. A bottom-heavy age structure, with a disproportionate number of young people, portends continuing growth of the population as these young people begin reproducing. In contrast, a more evenly distributed age structure predicts a more stable population size, and a top-heavy age structure predicts a decrease in population size because relatively fewer young people are reproducing. **2.** The growth rate of Earth's human population has dropped by half since the 1960s, from 2.2% in 1962 to 1.2% today. Nonetheless, growth has not slowed much because the smaller growth rate is counterbalanced by increased population size; the number of extra people on Earth each year remains enormous—approximately 79 million. **3.** Each of us influences our ecological footprint by how we live—what we eat, how much energy we use, and the amount of waste we generate—as well as by how many children we have. Making choices that reduce our demand for resources makes our ecological footprint smaller.

Summary of Key Concepts Questions

53.1 Ecologists can potentially estimate birth rates by counting the number of young born each year, and they can estimate death rates by seeing how the number of adults changes each year. **53.2** Under the exponential model, both populations will continue to grow to infinite size, regardless of the specific value of r_{max} (see Figure 53.7). **53.3** There are many things you can do to increase the carrying capacity of the species, including increasing its food supply, protecting it from predators, and providing more sites for nesting or reproduction. **53.4** Two key factors appear to be the survival rate of the offspring and the chance that adults will live long enough to reproduce again. **53.5** An example of a biotic factor would be disease caused by a pathogen; natural disasters, such as floods and storms, are examples of abiotic factors. **53.6** Humans are unique in our potential ability to reduce global population through contraception and family planning. Humans also are capable of consciously choosing their diet and personal lifestyle, and these choices influence the number of people Earth can support.

Test Your Understanding

1. b **2.** a **3.** c **4.** d **5.** e **6.** c **7.** c **8.** d **9.** b **10.** a

11.

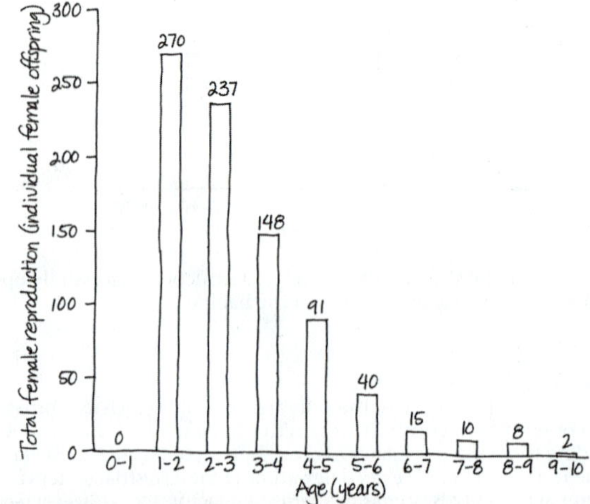

The total number of female offspring produced is greatest in females 1–2 years of age. Sample calculation for females of this age group: 252 indiv. × 1.07 female offspring/indiv. = 270 female offspring.

Chapter 54

Figure Questions

Figure 54.3 Its realized and fundamental niches would be similar, unlike those of *Chthamalus*. **Figure 54.16** The low-productivity treatment had the shortest food chain, so that food chain should be the most stable. **Figure 54.17** The death of individuals of *Mytilus*, a dominant species, should open up space for other species and increase species richness even in the absence of *Pisaster*. **Figure 54.23** At the earliest stages of primary succession, free-living prokaryotes in the soil would reduce atmospheric N_2 to NH_3. Symbiotic nitrogen fixation could not occur until plants were present at the site. **Figure 54.28** Other factors not included in the model must contribute to the number of species. **Figure 54.29** Shrew populations in different locations and habitats might show substantial genetic variation in their susceptibility to the Lyme pathogen. Further studies would be needed to test the generality of the results shown in Figure 54.29.

Concept Check 54.1

1. Interspecific competition has negative effects on both species (−/−). In predation, the predator population benefits at the expense of the prey population (+/−). Mutualism is a symbiosis in which both species benefit (+/+). **2.** One of the competing species will become locally extinct because of the greater reproductive success of the more efficient competitor. **3.** By specializing in eating seeds of a single plant species, individuals of the two finch species may be less likely to come into contact in the separate habitats, reinforcing a reproductive barrier to hybridization.

Concept Check 54.2

1. Species richness, the number of species in the community, and relative abundance, the proportions of the community represented by the various species, both contribute to species diversity. Compared to a community with a very high proportion of one species, one with a more even proportion of species is considered more diverse. **2.** The energetic hypothesis suggests that the length of a food chain is limited by the inefficiency of energy transfer along the chain, while the dynamic stability hypothesis proposes that long food chains are less stable than short chains. The energetic hypothesis predicts that food chains will be longer in habitats with higher primary productivity. The dynamic stability hypothesis predicts that food chains will be longer in more predictable environments. **3.** According to the bottom-up model, adding extra predators would have little effect on lower trophic levels, particularly vegetation. If the top-down model applied, increased bobcat numbers would decrease raccoon numbers, increase snake numbers, decrease grasshopper numbers, and increase plant biomass.

Concept Check 54.3

1. High levels of disturbance are generally so disruptive that they eliminate many species from communities, leaving the community dominated by a few tolerant species. Low levels of disturbance permit competitively dominant species to exclude other species from the community. But moderate levels of disturbance can facilitate coexistence of a greater number of species in a community by preventing competitively dominant species from becoming abundant enough to eliminate other species from the community. **2.** Early successional species can facilitate the arrival of other species in many ways, including increasing the fertility or water-holding capacity of soils or providing shelter to seedlings from wind and intense sunlight. **3.** The absence of fire for 100 years would represent a change to a low level of disturbance. According to the intermediate disturbance hypothesis, this change should cause diversity to decline as competitively dominant species gain sufficient time to exclude less competitive species.

Concept Check 54.4

1. Ecologists propose that the greater species richness of tropical regions is the result of their longer evolutionary history and the greater solar energy input and water availability in tropical regions. **2.** Immigration of species to islands declines with distance from the mainland and increases with island area. Extinction of species is lower on larger islands and on less isolated islands. Since the number of species on islands is largely determined by the difference between rates of immigration and extinction, the number of species will be highest on large islands near the mainland and lowest on small islands far from the mainland. **3.** Because of their greater mobility, birds disperse to islands more often than snakes and lizards, so birds should have greater richness.

Concept Check 54.5

1. Pathogens are microorganisms, viruses, viroids, or prions that cause disease. **2.** To keep the rabies virus out, you could ban imports of all mammals, including pets. Potentially, you could also attempt to vaccinate all dogs in the British Isles against the virus. A more practical approach might be to quarantine all pets brought into the country that are potential carriers of the disease, the approach the British government actually takes.

Summary of Key Concepts Questions

54.1 Competition: a fox and a bobcat competing for prey. Predation: an orca eating a sea otter. Herbivory: a bison grazing in a prairie. Parasitism: a parasitoid wasp that lays its eggs on a caterpillar. Mutualism: a fungus and an alga that make up a lichen. Commensalism: a remora attached to a whale. Facilitation: a flowering plant and its pollinator. **54.2** Not necessarily if the more species-rich community is dominated by only one or a few species **54.3** Because of the presence of species initially, the disturbance would initiate secondary succession in spite of its severe appearance. **54.4** Glaciations have severely reduced diversity in northern temperate, boreal, and Arctic ecosystems, compared to tropical ecosystems. **54.5** A host is required to complete the pathogen's life cycle, but a vector is not. Vectors are intermediate species that merely transport a pathogen to its host.

Test Your Understanding

1. e **2.** d **3.** c **4.** a **5.** b **6.** c **7.** d **8.** b **9.** Community 1: $H = -(0.05 \ln 0.05 + 0.05 \ln 0.05 + 0.85 \ln 0.85 + 0.05 \ln 0.05) = 0.59$. Community 2: $H = -(0.30 \ln 0.30 + 0.40 \ln 0.40 + 0.30 \ln 0.30) = 1.1$. Community 2 is more diverse.

10. Crab numbers should increase, reducing the abundance of eelgrass.

Kemp's Ridley sea turtle
Human
Blue crab
(Cannibalism)
Eelgrass
Clam

Chapter 55

Figure Questions

Figure 55.6 Wetlands, coral reefs, and coastal zones cover areas too small to show up clearly on global maps. **Figure 55.7** The availability of nutrients, particularly nitrogen, phosphorus, and iron, as well as temperature, is likely to limit primary production in the oceans. **Figure 55.8** If the new duck farms made nitrogen available in rich supply, as phosphorus already is, then adding extra nitrogen in the experiment would not increase phytoplankton density. **Figure 55.15** Water availability is probably another factor that varied across the sites. Such factors not included in the experimental design could make the results more difficult to interpret. Multiple factors can also covary in nature, so ecologists must be careful that the factor they are studying is actually causing the observed response and is not just correlated with it.

Concept Check 55.1

1. Energy passes through an ecosystem, entering as sunlight and leaving as heat. It is not recycled within the ecosystem. **2.** You would need to know how much biomass the wildebeests ate from your plot and how much nitrogen was contained in that biomass. You would also need to know how much nitrogen they deposited in urine or feces. **3.** The second law states that in any energy transfer or transformation, some of the energy is dissipated to the surroundings as heat. This "escape" of energy from an ecosystem is offset by the continuous influx of solar radiation.

Concept Check 55.2

1. Only a fraction of solar radiation strikes plants or algae, only a portion of that fraction is of wavelengths suitable for photosynthesis, and much energy is lost as a result of reflection or heating of plant tissue. **2.** By manipulating the level of the factors of interest, such as phosphorus availability or soil moisture, and measuring responses by primary producers **3.** The enzyme rubisco, which catalyzes the first step in the Calvin cycle, is the most abundant protein on Earth. Photosynthetic organisms require considerable nitrogen to make rubisco. Phosphorus is also needed as a component of several metabolites in the Calvin cycle and as a component of both ATP and NADPH (see Figure 10.19).

Concept Check 55.3

1. 20 J; 40% **2.** Nicotine protects the plant from herbivores. **3.** Unlike for the woman pictured in Figure 40.20, the caterpillar would have no energy devoted to thermoregulation or to reproduction. Its relative contribution to growth, however, would be much higher than for the woman.

Concept Check 55.4

1. For example, for the carbon cycle:

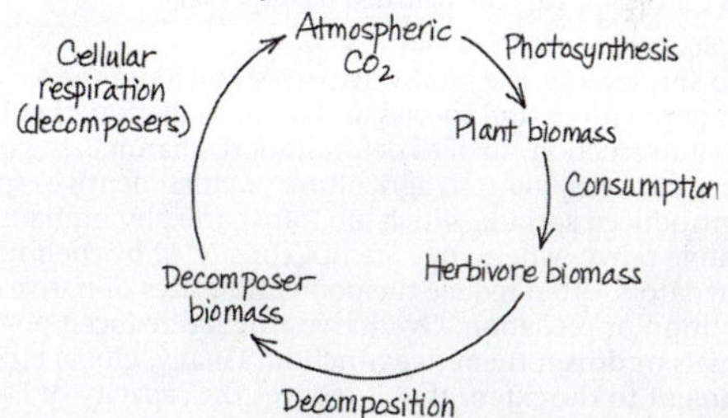

Cycling of a carbon atom

2. Removal of the trees stops nitrogen uptake from the soil, allowing nitrate to accumulate there. The nitrate is washed away by precipitation and enters the streams. **3.** Most of the nutrients in a tropical rain forest are contained in the trees, so removing the trees by logging rapidly depletes nutrients from the ecosystem. The nutrients that remain in the soil are quickly carried away into streams and groundwater by the abundant precipitation.

Concept Check 55.5

1. The main goal is to restore degraded ecosystems to a more natural state. **2.** Bioremediation uses organisms, generally prokaryotes, fungi, or plants, to detoxify or remove pollutants from ecosystems. Biological augmentation uses organisms, such as nitrogen-fixing plants, to add essential materials to degraded ecosystems. **3.** The Kissimmee River project returns the flow of water to the original channel and restores natural flow, a self-sustaining outcome. Ecologists at the Maungatautari reserve will need to maintain the integrity of the fence indefinitely, an outcome that is not self-sustaining in the long term.

Summary of Key Concepts Questions

55.1 Because energy conversions are inefficient, with some energy inevitably lost as heat, you would expect that a given mass of primary producers would support a smaller biomass of secondary producers. **55.2** For estimates of NEP, you need to measure the respiration of all organisms in an ecosystem, not just the respiration of primary producers. In a sample of ocean water, primary producers and other organisms are usually mixed together, making their respective respirations hard to separate. **55.3** The runner would typically burn many more calories through respiration, reducing his or her production efficiency. **55.4** Factors other than temperature, including a shortage of water and nutrients, slow decomposition in hot deserts. **55.5** If the topsoil and deeper soil are kept separate, you could return the deeper soil to the site first and then apply the more fertile topsoil to improve the success of revegetation and other restoration efforts.

Test Your Understanding

1. c **2.** b **3.** d **4.** d **5.** c **6.** e **7.** a **8.** e

9.

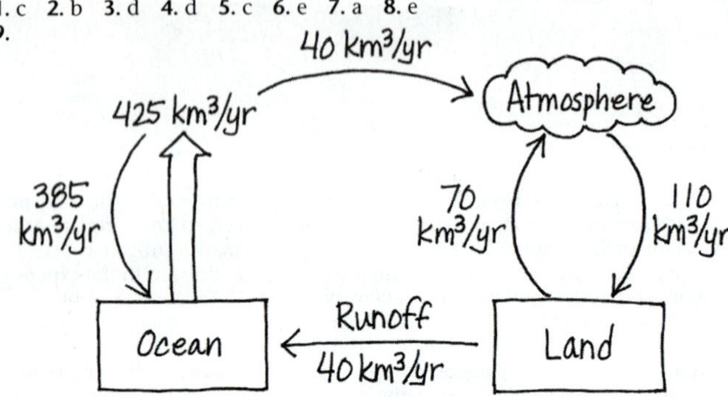

Based on these global numbers, approximately 110 km^3 of precipitation falls over land each year.

Chapter 56

Figure Questions

Figure 56.4 You would need to know the complete range of the species and that it is missing across all of that range. You would also need to be certain that the species isn't hidden, as might be the case for an animal that is hibernating underground or a plant that is present in the form of seeds or spores. **Figure 56.9** The two examples are similar in that segments of DNA from the harvested samples were analyzed and compared with segments from specimens of known origin. One difference is that the whale researchers investigated relatedness at species and population levels to determine whether illegal activity had occurred, whereas the elephant investigators determined relatedness at the population level to determine the precise location of the poaching. Another difference is that mtDNA was used for the whale study, whereas STRs were used for the elephant study. The primary limitations of such approaches are the need to have (or generate) a reference database and the requirement that the organisms have sufficient variation in their DNA to reveal the relatedness of samples. **Figure 56.13** Because the population of Illinois birds has a different genetic makeup than birds in other regions, you would want to maintain to the greatest extent possible the frequency of beneficial genes or alleles found only in that population. In restoration, preserving genetic diversity in a species is as important as increasing organism numbers. **Figure 56.15** The natural disturbance regime in this habitat includes frequent fires that clear undergrowth but do not kill mature pine trees. Without these fires, the undergrowth quickly fills in and the habitat becomes unsuitable for red-cockaded woodpeckers.

Concept Check 56.1

1. In addition to species loss, the biodiversity crisis includes the loss of genetic diversity within populations and species and the degradation of entire ecosystems. **2.** Habitat destruction, such as deforestation, channelizing of rivers, or conversion of natural ecosystems to agriculture or cities, deprives species of places to live. Introduced species, which are transported by humans to regions outside their native range, where they are not controlled by their natural pathogens or predators, often reduce the population sizes of native species through competition or predation. Overharvesting has reduced populations of plants and animals or driven them to extinction. Finally, global change is altering the environment to the extent that it reduces the capacity of Earth to sustain life. **3.** If both populations breed separately, then gene flow between the populations would not occur and genetic differences between them would be greater. As a result, the loss of genetic diversity would be greater than if the populations interbreed.

Concept Check 56.2

1. Reduced genetic variation decreases the capacity of a population to evolve in the face of change. **2.** The effective population size, N_e, was $4(15 \times 5)/(15 + 5) = 15$ birds. **3.** Because millions of people use the greater Yellowstone ecosystem each year, it would be impossible to eliminate all contact between people and bears. Instead, you might try to reduce the kinds of encounters where bears are killed. You might recommend lower speed limits on roads in the park, adjust the timing or location of hunting seasons (where hunting is allowed outside the park) to minimize contact with mother bears and cubs, and provide financial incentives for livestock owners to try alternative means of protecting livestock, such as using guard dogs.

Concept Check 56.3

1. A small area supporting numerous endemic species as well as a large number of endangered and threatened species **2.** Zoned reserves may provide sustained supplies of forest products, water, hydroelectric power, educational opportunities, and income from tourism. **3.** Habitat corridors can increase the rate of movement or dispersal of organisms between habitat patches and thus the rate of gene flow between subpopulations. They thus help prevent a decrease in fitness attributable to inbreeding. They can also minimize interactions between organisms and humans as the organisms disperse; in cases involving potential predators, such as bears or large cats, minimizing such interactions is desirable.

Concept Check 56.4

1. Adding nutrients causes population explosions of algae and the organisms that feed on them. Increased respiration by algae and consumers, including detritivores, depletes the lake's oxygen, which the fish require. **2.** Because higher temperatures lead to faster decomposition, organic matter in these soils could be quickly decomposed to CO_2, speeding up global warming. **3.** Reduced concentrations of ozone in the atmosphere increase the amount of UV radiation that reaches Earth's surface and the organisms living there. UV radiation can cause mutations by producing disruptive thymine dimers in DNA.

Concept Check 56.5

1. Sustainable development is an approach to development that works toward the long-term prosperity of human societies and the ecosystems that support them, which requires linking the biological sciences with the social sciences, economics, and humanities. **2.** Biophilia, our sense of connection to nature and all forms of life, may act as a significant motivation for the development of an environmental ethic that resolves not to allow species to become extinct or ecosystems to be destroyed. Such an ethic is necessary if we are to become more attentive and effective custodians of the environment. **3.** At a minimum, you would want to know the size of the population and the average reproductive rate of individuals in it. To develop the fishery sustainably, you would seek a harvest rate that maintains the population near its original size and maximizes its harvest in the long term rather than the short term.

Summary of Key Concepts Questions

56.1. Nature provides us with many beneficial services, including a supply of reliable, clean water, the production of food and fiber, and the dilution and detoxification of our pollutants. **56.2.** A more genetically diverse population is better able to withstand pressures from disease or environmental change, making it less likely to become extinct over a given period of time. **56.3.** Habitat fragmentation can isolate populations, leading to inbreeding and genetic drift, and it can make populations more susceptible to local extinctions resulting from the effects of pathogens, parasites, or predators. **56.4.** It's healthier to feed at a lower trophic level because biological magnification increases the concentration of toxins at higher levels. **56.5.** One goal of conservation biology is to preserve as many species as possible. Sustainable approaches that maintain the quality of habitats are required for the long-term survival of organisms.

Test Your Understanding

1. d **2.** d **3.** e **4.** a **5.** c **6.** a

7.

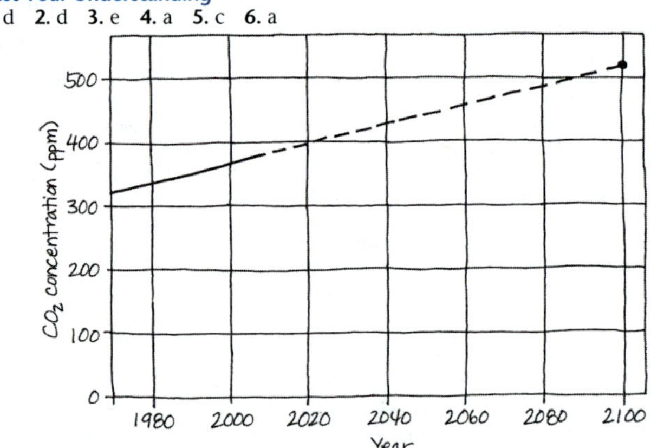

Between 1974 and 2009, Earth's atmospheric CO_2 concentration increased from approximately 330 ppm to 390 ppm. If this rate of increase of 1.7 ppm/yr continues, the concentration in 2100 will be about 540 ppm. The actual rise in CO_2 concentration could be larger or smaller, depending on Earth's human population, per capita energy use, and the extent to which societies take steps to reduce CO_2 emissions, including replacing fossil fuels with renewable or nuclear fuels. Additional scientific data will be important for many reasons, including determining how quickly greenhouse gases such as CO_2 are removed from the atmosphere by the biosphere.

9.

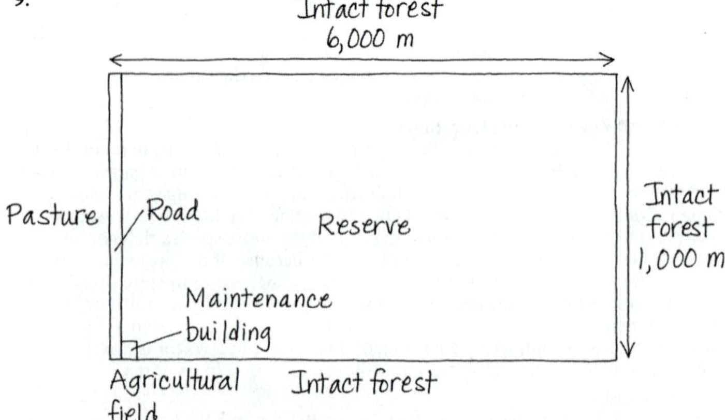

To minimize the area of forest into which the cowbirds penetrate, you should locate the road along one edge of the reserve. Any other location would increase the area of affected habitat. Similarly, the maintenance building should be in a corner of the reserve to minimize the area susceptible to cowbirds.

APPENDIX B Periodic Table of the Elements

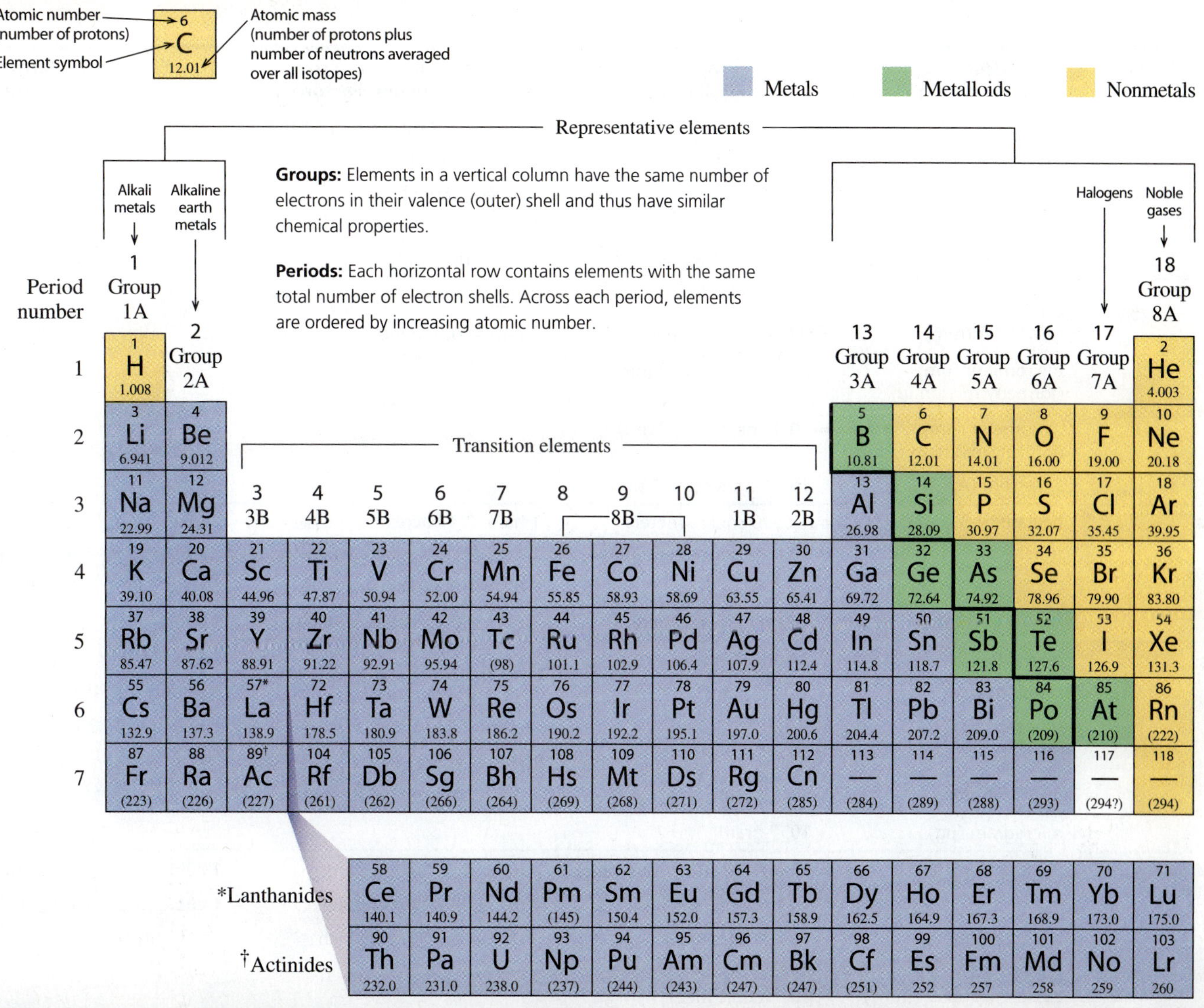

Name (Symbol)	Atomic Number	Name (Symbol)	Atomic Number	Name (Symbol)	Atomic Number	Name (Symbol)	Atomic Number	Name (Symbol)	Atomic Number
Actinium (Ac)	89	Copernicium (Cn)	112	Iridium (Ir)	77	Palladium (Pd)	46	Sodium (Na)	11
Aluminum (Al)	13	Copper (Cu)	29	Iron (Fe)	26	Phosphorus (P)	15	Strontium (Sr)	38
Americium (Am)	95	Curium (Cm)	96	Krypton (Kr)	36	Platinum (Pt)	78	Sulfur (S)	16
Antimony (Sb)	51	Darmstadtium (Ds)	110	Lanthanum (La)	57	Plutonium (Pu)	94	Tantalum (Ta)	73
Argon (Ar)	18	Dubnium (Db)	105	Lawrencium (Lr)	103	Polonium (Po)	84	Technetium (Tc)	43
Arsenic (As)	33	Dysprosium (Dy)	66	Lead (Pb)	82	Potassium (K)	19	Tellurium (Te)	52
Astatine (At)	85	Einsteinium (Es)	99	Lithium (Li)	3	Praseodymium (Pr)	59	Terbium (Tb)	65
Barium (Ba)	56	Erbium (Er)	68	Lutetium (Lu)	71	Promethium (Pm)	61	Thallium (Tl)	81
Berkelium (Bk)	97	Europium (Eu)	63	Magnesium (Mg)	12	Protactinium (Pa)	91	Thorium (Th)	90
Beryllium (Be)	4	Fermium (Fm)	100	Manganese (Mn)	25	Radium (Ra)	88	Thulium (Tm)	69
Bismuth (Bi)	83	Fluorine (F)	9	Meitnerium (Mt)	109	Radon (Rn)	86	Tin (Sn)	50
Bohrium (Bh)	107	Francium (Fr)	87	Mendelevium (Md)	101	Rhenium (Re)	75	Titanium (Ti)	22
Boron (B)	5	Gadolinium (Gd)	64	Mercury (Hg)	80	Rhodium (Rh)	45	Tungsten (W)	74
Bromine (Br)	35	Gallium (Ga)	31	Molybdenum (Mo)	42	Roentgenium (Rg)	111	Uranium (U)	92
Cadmium (Cd)	48	Germanium (Ge)	32	Neodymium (Nd)	60	Rubidium (Rb)	37	Vanadium (V)	23
Calcium (Ca)	20	Gold (Au)	79	Neon (Ne)	10	Ruthenium (Ru)	44	Xenon (Xe)	54
Californium (Cf)	98	Hafnium (Hf)	72	Neptunium (Np)	93	Rutherfordium (Rf)	104	Ytterbium (Yb)	70
Carbon (C)	6	Hassium (Hs)	108	Nickel (Ni)	28	Samarium (Sm)	62	Yttrium (Y)	39
Cerium (Ce)	58	Helium (He)	2	Niobium (Nb)	41	Scandium (Sc)	21	Zinc (Zn)	30
Cesium (Cs)	55	Holmium (Ho)	67	Nitrogen (N)	7	Seaborgium (Sg)	106	Zirconium (Zr)	40
Chlorine (Cl)	17	Hydrogen (H)	1	Nobelium (No)	102	Selenium (Se)	34		
Chromium (Cr)	24	Indium (In)	49	Osmium (Os)	76	Silicon (Si)	14		
Cobalt (Co)	27	Iodine (I)	53	Oxygen (O)	8	Silver (Ag)	47		

APPENDIX C The Metric System

Measurement	Unit and Abbreviation	Metric Equivalent	Metric-to-English Conversion Factor	English-to-Metric Conversion Factor
Length	1 kilometer (km)	= 1,000 (10^3) meters	1 km = 0.62 mile	1 mile = 1.61 km
	1 meter (m)	= 100 (10^2) centimeters	1 m = 1.09 yards	1 yard = 0.914 m
		= 1,000 millimeters	1 m = 3.28 feet	1 foot = 0.305 m
			1 m = 39.37 inches	
	1 centimeter (cm)	= 0.01 (10^{-2}) meter	1 cm = 0.394 inch	1 foot = 30.5 cm
				1 inch = 2.54 cm
	1 millimeter (mm)	= 0.001 (10^{-3}) meter	1 mm = 0.039 inch	
	1 micrometer (μm) (formerly micron, μ)	= 10^{-6} meter (10^{-3} mm)		
	1 nanometer (nm) (formerly millimicron, mμ)	= 10^{-9} meter (10^{-3} μm)		
	1 angstrom (Å)	= 10^{-10} meter (10^{-4} μm)		
Area	1 hectare (ha)	= 10,000 square meters	1 ha = 2.47 acres	1 acre = 0.405 ha
	1 square meter (m^2)	= 10,000 square centimeters	1 m^2 = 1.196 square yards	1 square yard = 0.8361 m^2
			1 m^2 = 10.764 square feet	1 square foot = 0.0929 m^2
	1 square centimeter (cm^2)	= 100 square millimeters	1 cm^2 = 0.155 square inch	1 square inch = 6.4516 cm^2
Mass	1 metric ton (t)	= 1,000 kilograms	1 t = 1.103 tons	1 ton = 0.907 t
	1 kilogram (kg)	= 1,000 grams	1 kg = 2.205 pounds	1 pound = 0.4536 kg
	1 gram (g)	= 1,000 milligrams	1 g = 0.0353 ounce	1 ounce = 28.35 g
			1 g = 15.432 grains	
	1 milligram (mg)	= 10^{-3} gram	1 mg = approx. 0.015 grain	
	1 microgram (μg)	= 10^{-6} gram		
Volume (solids)	1 cubic meter (m^3)	= 1,000,000 cubic centimeters	1 m^3 = 1.308 cubic yards	1 cubic yard = 0.7646 m^3
			1 m^3 = 35.315 cubic feet	1 cubic foot = 0.0283 m^3
	1 cubic centimeter (cm^3 or cc)	= 10^{-6} cubic meter	1 cm^3 = 0.061 cubic inch	1 cubic inch = 16.387 cm^3
	1 cubic millimeter (mm^3)	= 10^{-9} cubic meter		
		= 10^{-3} cubic centimeter		
Volume (liquids and gases)	1 kiloliter (kL or kl)	= 1,000 liters	1 kL = 264.17 gallons	
	1 liter (L or l)	= 1,000 milliliters	1 L = 0.264 gallons	1 gallon = 3.785 L
			1 L = 1.057 quarts	1 quart = 0.946 L
	1 milliliter (mL or ml)	= 10^{-3} liter	1 mL = 0.034 fluid ounce	1 quart = 946 mL
		= 1 cubic centimeter	1 mL = approx. ¼ teaspoon	1 pint = 473 mL
			1 mL = approx. 15–16 drops (gtt.)	1 fluid ounce = 29.57 mL
				1 teaspoon = approx. 5 mL
	1 microliter (μL or μl)	= 10^{-6} liter (10^{-3} milliliters)		
Time	1 second (s or sec)	= 1/60 minute		
	1 millisecond (ms or msec)	= 10^{-3} second		
Temperature	Degrees Celsius (°C) (Absolute zero, when all molecular motion ceases, is −273.15°C. The Kelvin [K] scale, which has the same size degrees as Celsius, has its zero point at absolute zero. Thus, 0 K = −273.15°C.)		°F = 9/5°C + 32	°C = 5/9 (°F − 32)

APPENDIX **D**

A Comparison of the Light Microscope and the Electron Microscope

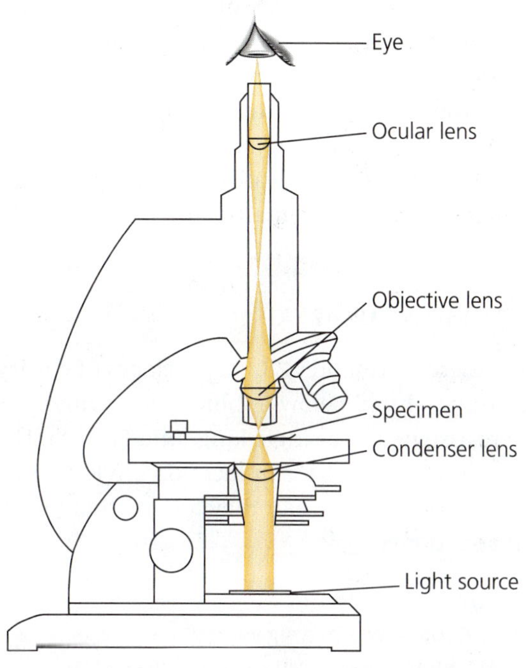

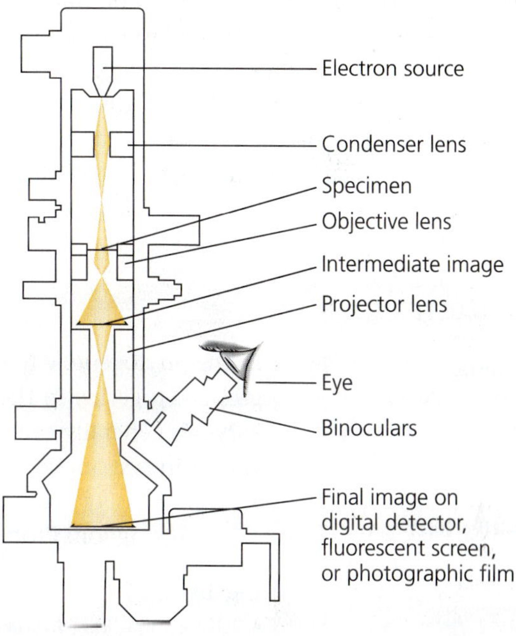

Light Microscope

In light microscopy, light is focused on a specimen by a glass condenser lens; the image is then magnified by an objective lens and an ocular lens, for projection on the eye, digital camera, digital video camera, or photographic film.

Electron Microscope

In electron microscopy, a beam of electrons (top of the microscope) is used instead of light, and electromagnets are used instead of glass lenses. The electron beam is focused on the specimen by a condenser lens; the image is magnified by an objective lens and a projector lens for projection on a digital detector, fluorescent screen, or photographic film.

APPENDIX E Classification of Life

This appendix presents a taxonomic classification for the major extant groups of organisms discussed in this text; not all phyla are included. The classification presented here is based on the three-domain system, which assigns the two major groups of prokaryotes, bacteria and archaea, to separate domains (with eukaryotes making up the third domain).

Various alternative classification schemes are discussed in Unit Five of the text. The taxonomic turmoil includes debates about the number and boundaries of kingdoms and about the alignment of the Linnaean classification hierarchy with the findings of modern cladistic analysis. In this review, asterisks (*) indicate currently recognized phyla thought by some systematists to be paraphyletic.

DOMAIN BACTERIA

- **Proteobacteria**
- **Chlamydia**
- **Spirochetes**
- **Gram-positive Bacteria**
- **Cyanobacteria**

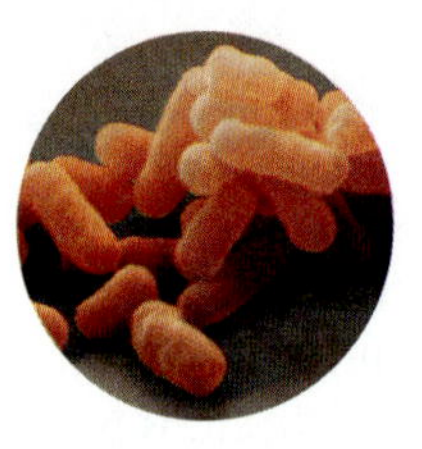

DOMAIN ARCHAEA

- **Korarchaeota**
- **Euryarchaeota**
- **Crenarchaeota**
- **Nanoarchaeota**

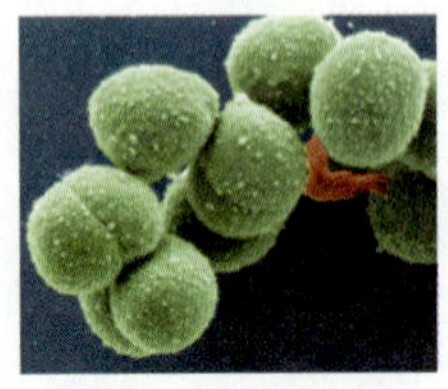

DOMAIN EUKARYA

In the phylogenetic hypothesis we present in Chapter 28, major clades of eukaryotes are grouped together in the five "supergroups" listed below in blue type. Formerly, all the eukaryotes generally called protists were assigned to a single kingdom, Protista. However, advances in systematics have made it clear that Protista is in fact polyphyletic: Some protists are more closely related to plants, fungi, or animals than they are to other protists. As a result, the kingdom Protista has been abandoned.

Excavata

- Diplomonadida (diplomonads)
- Parabasala (parabasalids)
- Euglenozoa (euglenozoans)
  - Kinetoplastida (kinetoplastids)
  - Euglenophyta (euglenids)

Chromalveolata

- Alveolata (alveolates)
  - Dinoflagellata (dinoflagellates)
  - Apicomplexa (apicomplexans)
  - Ciliophora (ciliates)
- Stramenopila (stramenopiles)
  - Bacillariophyta (diatoms)
  - Chrysophyta (golden algae)
  - Phaeophyta (brown algae)
  - Oomycota (water molds)

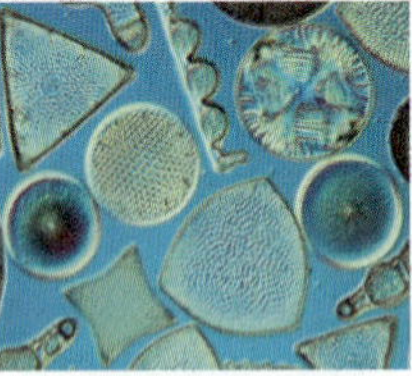

Archaeplastida

- Rhodophyta (red algae)
- Chlorophyta (green algae: chlorophytes)
- Charophyta (green algae: charophytes)
- Plantae
  - Nonvascular plants (bryophytes):
    - Phylum Hepatophyta (liverworts)
    - Phylum Bryophyta (mosses)
    - Phylum Anthocerophyta (hornworts)
  - Seedless vascular plants:
    - Phylum Lycophyta (lycophytes)
    - Phylum Pterophyta (ferns, horsetails, whisk ferns)
  - Seed plants:
    - Gymnosperms:
      - Phylum Ginkgophyta (ginkgo)
      - Phylum Cycadophyta (cycads)
      - Phylum Gnetophyta (gnetophytes)
      - Phylum Coniferophyta (conifers)
    - Angiosperms:
      - Phylum Anthophyta (flowering plants)

Rhizaria

- Radiolaria (radiolarians)
- Foraminifera (forams)
- Cercozoa (cercozoans)

DOMAIN EUKARYA, continued

Unikonta

- Amoebozoa (amoebozoans)
  - Myxogastrida (plasmodial slime molds)
  - Dictyostelida (cellular slime molds)
  - Gymnamoeba (gymnamoebas)
  - Entamoeba (entamoebas)
- Nucleariida (nucleariids)
- Fungi
  - *Phylum Chytridiomycota (chytrids)
  - *Phylum Zygomycota (zygomycetes)
  - Phylum Glomeromycota (glomeromycetes)
  - Phylum Basidiomycota (club fungi)
  - Phylum Ascomycota (sac fungi)

- Choanoflagellata (choanoflagellates)
- Animalia
  - Phylum Porifera (sponges)
  - Phylum Ctenophora (comb jellies)
  - Phylum Cnidaria (cnidarians)
    - Hydrozoa (hydrozoans)
    - Scyphozoa (jellies)
    - Cubozoa (box jellies and sea wasps)
    - Anthozoa (sea anemones and most corals)
  - Phylum Acoela (acoel flatworms)
  - Phylum Placozoa (placozoans)

  <u>Lophotrochozoa (lophotrochozoans)</u>
  - Phylum Kinorhyncha (kinorhynchs)
  - Phylum Platyhelminthes (flatworms)
    - Catenulida (chain worms)
    - Rhabditophora (planarians, flukes, tapeworms)
  - Phylum Nemertea (proboscis worms)
  - Phylum Ectoprocta (ectoprocts)
  - Phylum Brachiopoda (brachiopods)
  - Phylum Phoronida (phoronids)
  - Phylum Rotifera (rotifers)
  - Phylum Cycliophora (cycliophorans)
  - Phylum Mollusca (molluscs)
    - Polyplacophora (chitons)
    - Gastropoda (gastropods)
    - Bivalvia (bivalves)
    - Cephalopoda (cephalopods)
  - Phylum Annelida (segmented worms)
    - Polychaeta (polychaetes)
    - Oligochaeta (oligochaetes)
  - Phylum Acanthocephala (spiny-headed worms)

  <u>Ecdysozoa (ecdysozoans)</u>
  - Phylum Loricifera (loriciferans)
  - Phylum Priapula (priapulans)
  - Phylum Nematoda (roundworms)
  - Phylum Arthropoda (This survey groups arthropods into a single phylum, but some zoologists now split the arthropods into multiple phyla.)
    - Subphylum Chelicerata (horseshoe crabs, arachnids)
    - Subphylum Myriapoda (millipedes, centipedes)
    - Subphylum Hexapoda (insects, springtails)
    - Subphylum Crustacea (crustaceans)
  - Phylum Tardigrada (tardigrades)
  - Phylum Onychophora (velvet worms)

  <u>Deuterostomia (deuterostomes)</u>
  - Phylum Hemichordata (hemichordates)
  - Phylum Echinodermata (echinoderms)
    - Asteroidea (sea stars, sea daisies)
    - Ophiuroidea (brittle stars)
    - Echinoidea (sea urchins and sand dollars)
    - Crinoidea (sea lilies)
    - Holothuroidea (sea cucumbers)
  - Phylum Chordata (chordates)
    - Subphylum Cephalochordata (cephalochordates: lancelets)
    - Subphylum Urochrodata (urochordates: tunicates)
    - Subphylum Craniata (craniates)
      - Myxini (hagfishes)
      - Vertebrates:
        - Petromyzontida (lampreys)
        - Chondrichthyes (sharks, rays, chimaeras)
        - Actinopterygii (ray-finned fishes)
        - Actinistia (coelacanths)
        - Dipnoi (lungfishes)
        - Amphibia (amphibians)
        - Reptilia (tuataras, lizards, snakes, turtles, crocodilians, birds)
        - Mammalia (mammals)

Credits

Photo Credits

Cover Image "Succulent I" © 2005 Amy Lamb, www.amylamb.com

Unit Opening Interviews **UNIT 1** Justina Thorsen, Showcase Reflections; **UNIT 2** Brian Wilson, Princeton University; **UNIT 3** William K. Sacco, Yale ITS Photo + Design Services; **UNIT 4** Steve Bonnel, Bonnel Photography; **UNIT 5** Jacob Mailman; **UNIT 6** Michael Starghill, Michael Starghill Photography; **UNIT 7** River Healey, Rusty Healey Photography; **UNIT 8** Marsha Miller, Director of Photography, University of Texas at Austin.

About the Author Photos River Healey, Rusty Healey Photography.

Detailed Contents **UNIT I** Photo by T. Naeser, from Patrick Cramer Laboratory, Gene Center Munich, Ludwig-Maximilians-Universität München, Munich, Germany; **UNIT II** Biophoto Associates/Photo Researchers; **UNIT III** From: Multicolor Spectral Karyotyping of Human Chromosomes. E. Schröck, S. du Manoir, T. Veldman, B. Schoell, J. Wienberg, M. A. Ferguson-Smith, Y. Ning, D. H. Ledbetter, I. Bar-Am, D. Soenksen, Y. Garini, T. Ried. *Science.* 1996 Jul 26;273(5274):494-7.; **UNIT IV** © Ted Daeschler/ Academy of Natural Sciences/VIREO; **UNIT V** Yufeng Zhou/iStockphoto; **UNIT VI** PD photo: John Walker (www.fourmilab.ch); **UNIT VII** David Wall/Alamy; David Doubilet/Getty Images; Kenneth Catania; **UNIT VIII** Daniel Mosquin.

Chapter 1 **1.1** Amy Lamb Studio; **1.2** Walter Teague; **1.3.2** R.Dirscherl/FLPA; **1.3.3** Kim Taylor and Jane Burton/Dorling Kindersley; **1.3.4** Malcolm Schuyl/FLPA; **1.3.5** Frans Lanting/Corbis; **1.3.6** Michael & Patricia Fogden/Corbis; **1.3.7** Joe McDonald/ Corbis; **1.3.8** ImageState/International Stock Photography Ltd.; **1.4.2** WorldSat International/Photo Researchers, Inc.; **1.4.3** Bill Brooks/Alamy; **1.4.4** Linda Freshwaters Arndt/Alamy; **1.4.5** Michael Orton/Photographer's Choice/Getty Images; **1.4.6** Ross M. Horowitz/Getty Images; **1.4.7** Photodisc/Getty Images; **1.4.9** Jeremy Burgess/ SPL/Photo Researchers; **1.4.11** John Durham/Photo Researchers; **1.4.13** Electron micrograph by Wm. P. Wergin, courtesy of E. H. Newcomb, University of Wisconsin; **1.5.2** James Balog/Aurora Creative/Getty Images; **1.6.2** Anup Shah/Nature Picture Library; **1.6.3** Anup Shah/Nature Picture Library; **1.7.2** Photodisc/Getty Images; **1.7.3** Janice Sheldon; **1.8.2** S. C. Holt/Biological Photo Service; **1.8.3** Steve Gschmeissner/ Photo Researchers; **1.9.2** Conly L. Rieder; **1.9.3** Conly L. Rieder; **1.10.2** Camille Tokerud/ Stone/Getty Images; **1.11.2** Photodisc/Getty Images; **1.12** Roy Kaltschmidt, Lawrence Berkeley National Laboratory; **1.15.2** Oliver Meckes/Nicole Ottawa/Photo Researchers; **1.15.3** Eye of Science/Photo Researchers; **1.15.4** Kunst & Scheidulin/AGE Fotostock; **1.15.5** Peter Lilja/Taxi/Getty Images; **1.15.6** Anup Shah/Nature Picture Library; **1.15.7** D. P. Wilson/Photo Researchers; **1.16.2** VVG/SPL/Photo Researchers; **1.16.3** W. L. Dentler/Biological Photo Service; **1.16.4** OMIKRON/Photo Researchers; **1.17** Photo by Dede Randrianarisata, from Kristi Curry Rogers, Macalester College, St. Paul, MN; **1.18** ARCHIV/Photo Researchers; **1.19.2** Michael P. Fogden/Bruce Coleman/ Alamy; **1.19.3** Matt T. Lee; **1.19.4** Hal Horwitz/Corbis; **1.21** Frank Greenaway/Dorling Kindersley; **1.23.2** Karl Ammann/Corbis; **1.23.3** Tim Ridley/Dorling Kindersley, Courtesy of the Jane Goodall Institute, Clarendon Park, Hampshire; **1.25.2** Breck P. Kent; **1.25.3** Photo Researchers; **1.26.2** David Pfennig; **1.26.3** David Pfennig; **1.28** Gary L. Firestone's Lab researchers meeting in the Dept. of Molecular and Cell Biology, University of California at Berkeley. Photo: Seelevel.com, Pearson Science; **1.29** Tim Sharp/AP Images; **p. 25, top right** James Balog/Aurora Creative/Getty Images; **p. 25, bottom left** Anup Shah/Nature Picture Library; **p. 25, bottom left** Anup Shah/Nature Picture Library; **p.26, top** PhotoDisc/Getty Images; **p. 26, 2nd from top** S. C. Holt/ Biological Photo Service; **p. 26, 3rd from top** Steve Gschmeissner/Photo Researchers.

Chapter 2 **2.1** Martin Dohrn/BBC Natural History Unit; **2.2.3** Martin Dohrn/BBC Natural History Unit; **2.3.2** Chip Clark; **2.4.2** C. Michael Hogan; **2.4.3** Rick York and California Native Plant Society (www.cnps.org); **2.4.4** Andrew Alden; **2.6.2** Clayton T. Hamilton, Stanford University; **2.7.2** National Library of Medicine; **p. 41, left** Jerry Young/Dorling Kindersley; **2.19** Nigel Cattlin/Photo Researchers; **p. 45** Rolf Nussbaumer/Nature Picture Library.

Chapter 3 **3.1** Alexander/Fotoloia; **3.3.2** N.C. Brown Center for Ultrastructure Studies, SUNY-Environmental Science & Forestry, Syracuse, NY; **3.4** iStockphoto; **3.6.2** Jan van Franeker, IMARES, Alfred Wegener Institute for Polar and Marine Research; **3.9** NASA/JPL-Caltech/University of Arizona/Texas A&M University; **3.10.2** Jakub Semeniuk/ iStockphoto; **3.10.3** Feng Yu/iStockphoto; **3.10.4** Monika Wisniewska/iStockphoto; **3.10.5** Beth Van Trees/Shutterstock; **3.12.2** From "Coral Reefs Under Rapid Climate Change and Ocean Acidification." O. Hoegh-Guldberg, et al. *Science* 14 December 2007: 318 (5857):1737–1742. Photos by Ove Hoegh-Guldberg, Centre for Marine Studies, The University of Queensland; **3.12.3** From "Coral Reefs Under Rapid Climate Change and Ocean Acidification." O. Hoegh-Guldberg, et al. *Science* 14 December 2007: 318 (5857):1737–1742. Photos by Ove Hoegh-Guldberg, Centre for Marine Studies, The University of Queensland; **3.12.4** From "Coral Reefs Under Rapid Climate Change and Ocean Acidification." O. Hoegh-Guldberg, et al. *Science* 14 December 2007: 318 (5857):1737–1742. Photos by Ove Hoegh-Guldberg, Centre for Marine Studies, The University of Queensland.

Chapter 4 **4.1** Shutterstock; **4.6.2** David M. Phillips/Photo Researchers.

Chapter 5 **5.1** Photo by T. Naeser, from Patrick Cramer Laboratory, Gene Center Munich, Ludwig-Maximilians-Universität München, Munich, Germany; **5.6.2** John N. A. Lott/Biological Photo Service; **5.6.3** H. Shio and P. B. Lazarow; **5.8.2** Alexey Repka/iStockphoto; **5.8.3** John Durham/Photo Researchers; **5.8.4** Biophoto Associates/Photo Researchers; **5.9.2** F. Collet/Photo Researchers; **5.9.3** Corbis; **5.11.2** Dorling Kindersley; **5.11.3** Dorling Kindersley; **5.15.2** Andrey Stratilatov/ Shutterstock; **5.15.3** Nina Zanetti; **5.15.4** Nina Zanetti; **5.19.2** Reproduced by permission from Tulip WR, Varghese JN, Laver WG, Webster RG, Colman PM. Refined crystal structure of the influenza virus N9 neuraminidase-NC41 Fab complex. *J Mol Biol.* September 5; 227(1):122–48. Copyright © 1992 by Elsevier Science Ltd. **5.20.3** Dieter Hopf/AGE Fotostock; **5.20.8** Monika Wisniewska/iStockphoto; **5.21.2** Eye of Science/Photo Researchers; **5.21.3** Eye of Science/Photo Researchers; **5.23.2** Reprinted by permission from *Nature*. P. B. Sigler from Z. Xu, A. L. Horwich, and P. B. Sigler. 388:741–750 Copyright © 1997 Macmillan Magazines Limited; **5.24.2** Dave Bushnell; **5.24.4** Dave Bushnell; **p. 90** Dorling Kindersley.

Chapter 6 **6.1** Eye of Science/Photo Researchers; **6.3.2** Elisabeth Pierson, FNWI-Radboud University Nijmegen, Pearson Science; **6.3.3** Elisabeth Pierson, FNWI-Radboud University Nijmegen, Pearson Science; **6.3.4** Elisabeth Pierson, FNWI-Radboud University Nijmegen, Pearson Science; **6.3.5** Elisabeth Pierson, FNWI-Radboud University Nijmegen, Pearson Science; **6.3.6** Michael W. Davidson/The Florida State University Research Foundation; **6.3.7** Karl Garsha, Beckman Institute for Advanced Science and Technology, University of Illinois; **6.3.9** Macrophage fluorescently stained for tubulin (yellow), actin (red) and the nucleus (DAPI, blue). Top part of the image: data recorded with a widefield microscope and visualized with the Simulated Fluorescence Process (SFP) volume rendering algorithm. Bottom part: the same dataset, deconvolved using Huygens Professional (Scientific Volume Imaging, Hilversum, The Netherlands) and again rendered with the SFP algorithm. Data courtesy Dr. James G. Evans, Whitehead Institute, MIT Boston MA, USA; **6.3.11** From "Motile Cilia of Human Airway Epithelia Are Chemosensory." Alok S. Shah, Yehuda Ben-Shahar, T.s O. Moninger, J. N. Kline, M. J. Welsh. Science Express on 23 July 2009. *Science* 28 August 2009:325 (5944): 1131–1134 (cover). Pseudocolored scanning electron micrograph by Tom Moninger (epithelia generated by Phil Karp); **6.3.12** William Dentler/Biological Photo Service; **6.3.13** From "STED microscopy reveals that synaptotagmin remains clustered after synaptic vesicle exocytosis." Katrin I. Willig, Silvio O. Rizzoli, Volker Westphal, Reinhard Jahn & . . . Stefan W. Hell. *Nature,* 440 (13) April 2006. doi:10.1038/nature04592, Letters; **6.5.2** S. C. Holt/Biological Photo Service; **6.6.2** Daniel Friend; **6.8.4** S. Cinti/Photo Researchers; **6.8.6** SPL/Photo Researchers; **6.8.8** A. Barry Dowsett/Photo Researchers; **6.8.10** Biophoto Associates/ Photo Researchers; **6.8.12** SPL/Photo Researchers; **6.8.14** From "Flagellar Microtubule Dynamics in *Chlamydomonas:* Cytochalasin D Induces Periods of Microtubule Shortening and Elongation; and Colchicine Induces Disassembly of the Distal, but Not Proximal, Half of the Flagellum." W. L. Dentler and C. Adams. *The Journal of Cell Biology,* 117(6): 1289–1298, Copyright © 1992 by The Rockefeller University Press; **6.9.2** Reproduced by permission from L.Orci and A.Perelet, Freeze-Etch Histology. (Heidelberg: Springer-Verlag, 1975). Copyright ©1975 by Springer-Verlag GmbH & Co KG; **6.9.3** Reproduced by permission from A. C. Faberge, *Cell Tiss. Res.* 151 Copyright © 1974 by Springer-Verlag GmbH & Co KG; **6.9.4** Reprinted by permission from *Nature* 323. U. Aebi et al. Copyright © 1996: 560–564, figure 1a. Used with permission. Macmillan Magazines Limited; **6.10.2** D. W. Fawcett/Photo Researchers; **6.11.2** R. Bolender, D. Fawcett/Photo Researchers; **6.12.2** Don W. Fawcett/ Photo Researchers; **6.13.2** Daniel S. Friend; **6.13.3** Daniel S. Friend; **6.14.2** E. H. Newcomb; **6.17.2** Daniel S. Friend; **6.17.3** From "The shape of mitochondria and the number of mitochondrial nucleoids during the cell cycle of *Euglena gracilis.*" Y. Hayashi and K. Ueda. *Journal of Cell Science,* 93:565–570, Copyright © 1989 by Company of Biologists; **6.18.2** Courtesy of W.P. Wergin and E.H. Newcomb, University of Wisconsin/Biological Photo Service; **6.18.3** Franz Grolig, Philipps-University Marburg, Germany. Image acquired with the confocal microscope Leica TCS SP2; **6.19.2** From S. E. Fredrick and E. H. Newcomb, *The Journal of Cell Biology* 43 (1969):343. Provided by E. H. Newcomb; **6.20** Albert Tousson, High Resolution Imaging Facility, University of Alabama at Birmingham; **6.21.2** Dr. Bruce J. Schnapp; **p. 113, left** Mary Osborn; **p. 113, middle** Frank Solomon; **p. 113, right** Mark S. Ladinsky and J. Richard McIntosh, University of Colorado; **6.22.2** Kent L. McDonald; **6.23.2** Biophoto Associates/Photo Researchers; **6.23.3** Oliver Meckes & Nicole Ottawa/Eye of Science/Photo Researchers; **6.24.2** OMIKRON/Science Source/Photo Researchers; **6.24.3** W. L. Dentler/Biological Photo Service; **6.24.4** Linck RW, Stephens RE. Functional protofilament numbering of ciliary, flagellar, and centriolar microtubules. *Cell Motil Cytoskeleton.* 2007 Jul;64(7):489–95; cover. Micrograph by D. Woodrum Hensley; **6.26.2** From Hirokawa Nobutaka, *The Journal of Cell Biology* 94 (1982):425 by copyright permission of The Rockefeller University Press; **6.27.2** Clara Franzini-Armstrong, University of Pennsylvania; **6.27.3** M. I. Walker/Photo Researchers; **6.27.4** Michael Clayton, University of Wisconsin-Madison; **6.28.2** G. F. Leedale/Photo Researchers; **6.29.2** David Ehrhardt; **6.29.3** From "Visualization of cellulose synthase demonstrates functional association with microtubules." A. R. Paredez, C. R. Somerville, D. W. Ehrhardt. *Science.* 2006 Jun 9;312(5779):1491–5. Epub 2006 Apr 20; **6.31.2** Micrograph by W. P. Wergin, provided by E. H. Newcomb; **6.32.2** From Douglas J. Kelly, *The Journal of Cell Biology* 28 (1966): 51. Fig.17. Reproduced by copyright permission of The Rockefeller University Press; **6.32.3** From Douglas J. Kelly, *The Journal of Cell Biology* 28 (1966):51 by copyright permission of The Rockefeller University Press; **6.32.5** Reproduced by permission from L. Orci and A. Perrelet, Freeze-Etch Histology. (Heidelberg: Springer-Verlag). Copyright © 1975 by Springer-Verlag GmbH & Co KG; **6.32.7** From C. Peracchia and A. F. Dulhunty, *The Journal of Cell Biology* 70 (1976):419 by copyright permission of The Rockefeller University Press; **6.33.2** Lennart Nilsson/ Scanpix; **p. 123, top** Courtesy E. H. Newcomb; **p. 123, bottom** From S. E. Fredrick and E. H. Newcomb, *The Journal of Cell Biology* 43 (1969): 343.

Chapter 7 **7.1** Roderick Mackinnon; **7.4.3** D. W. Fawcett/Photo Researchers; **7.4.4** D. W. Fawcett/Photo Researchers; **7.16.2.** Michael Abbey/Photo Researchers; **7.22.3** H. S. Pankratz, T.C. Beaman & P. Gerhardt/Biological Photo Service; **7.22.6** D. W. Fawcett/Photo Researchers; **7.22.9** M. M. Perry and A. B. Gilbert, *J. Cell Science* 39 (1979) 257. Copyright 1979 by The Company of Biologists Ltd. **7.22.10** M. M. Perry and A. B. Gilbert, *J. Cell Science* 39 (1979) 257. Copyright 1979 by The Company of Biologists Ltd.

Chapter 8 **8.1** Photoshot/NHPA; **8.2.2** Jupiter Images; **8.3.2** Robert N. Johnson/RnJ Photography; **8.3.3** Robert N. Johnson/RnJ Photography; **8.4.1** Brandon Blinkenberg/iStockphoto; **8.4.2** Bridget Lazenby/iStockphoto; **8.14.2** Thomas A. Steitz, Yale University; **8.14.3** Thomas A. Steitz, Yale University; **8.20.3** Scheer JM, Romanowski MJ, Wells JA. A common allosteric site and mechanism in caspases. *Proc Natl Acad Sci U S A.* 2006 May 16;103(20):7595–600; Fig. 4a; **8.22.2** Nicolae Simionescu.

Chapter 9 **9.1** Anup Shah/Nature Picture Library.

Chapter 10 **10.1** Bob Rowan, Progressive Image/Corbis; **10.2.2** Jean-Paul Nacivet/AGE Fotostock; **10.2.3** Lawrence Naylor/Photo Researchers; **10.2.4** M. I. Walker/Photo Researchers; **10.2.5** Susan M. Barns; **10.2.6** National Library of Medicine; **10.3** Robert Clark Photography, robertclark.com; **10.4.2** Image courtesy Andreas Holzenburg and Stanislav Vitha, Dept. of Biology and Microscopy & Imaging Center, Texas A&M University; **10.4.3** E.H Newcomb & WP Wergin/Biological Photo Service; **10.12.2** Christine L. Case, Skyline College; **10.13.2** From "Architecture of the photosynthetic oxygen-evolving center."; . . . K. N. Ferreira,T. M. Iverson, K. Maghlaoui, J. Barber, S. Iwata. *Science.* 2004 Mar 19;303(5665):1831–8. Epub 2004 Feb 5; **10.21.2** David Muench/Corbis; **10.21.3** Dave Bartruff/Corbis.

Chapter 11 **11.1** Winfried Wisniewski/Corbis; **11.3.2** A. Dale Kaiser, Stanford University; **11.8.2** From "High-resolution crystal structure of an engineered human beta2-adrenergic G protein-coupled receptor." V. Cherezov, D. M. Rosenbaum, M. A. Hanson, S. G. F. Rasmussen, F. S. Thian, T. S. Kobilka, H-J. Choi, P. Kuhn, W. I. Weis, B. K. Kobilka, R. C. Stevens. *Science.* 2007 Nov 23;318(5854):1258–65. Epub 2007 Oct 25; **11.17.3** Matheos D, Metodiev M, Muller E, Stone D, Rose MD. Pheromone-induced polarization is dependent on the Fus3p MAPK acting through the formin Bni1p. *J Cell Biol.* 2004 April; 165(1);99–109; Fig.9. Reproduced by copyright permission of The Rockefeller University Press; **11.20.2** Gopal Murti/Photo Researchers; **11.22.2** *Development* 127, 5245–5252 (2000). Mesenchymal cells engulf and clear apoptotic footplate cells in macrophageless PU.1 null mouse embryos. William Wood, Mark Turmaine, Roberta Weber, Victoria Camp, Richard A. Maki, Scott R. McKercher and Paul Martin; **11.22.3** *Development* 127, 5245–5252 (2000). Mesenchymal cells engulf and clear apoptotic footplate cells in macrophageless PU.1 null mouse embryos. William Wood, Mark Turmaine, Roberta Weber, Victoria Camp, Richard A. Maki, Scott R. McKercher and Paul Martin; **11.22.4** *Development* 127, 5245–5252 (2000). Mesenchymal cells engulf and clear apoptotic footplate cells in macrophageless PU.1 null mouse embryos. William Wood, Mark Turmaine, Roberta Weber, Victoria Camp, Richard A. Maki, Scott R. McKercher and Paul Martin.

Chapter 12 **12.1** Jan-Michael Peters/Silke Hauf; **12.2.2** Biophoto Associates/Photo Researchers; **12.2.3** C.R. Wyttenbach/Biological Photo Service; **12.2.4** Biophoto/Science Source/Photo Researchers; **12.3.2** John Murray; **12.4.2** Biophoto/Photo Researchers; **12.5.2** Biophoto/Photo Researchers; **12.7.2** Conly L. Rieder; **12.7.3** Conly L. Rieder; **12.7.4** Conly L. Rieder; **12.7.6** Conly L. Rieder; **12.7.7** Conly L. Rieder; **12.7.8** Conly L. Rieder; **12.8.2** J. Richard McIntosh, University of Colorado at Boulder; **12.8.3** Reproduced by permission from Matthew Schibler, from *Protoplasma* 137. Copyright © 1987: 29–44 by Springer-Verlag GmbH & Co KG; **12.10.2** Don W. Fawcett/Photo Researchers; **12.10.3** B. A. Palevitz, Courtesy of E. H. Newcomb, University of Wisconsin; **12.11.2** Elisabeth Pierson, FNWI-Radboud University Nijmegen, Pearson Science; **12.11.3** Elisabeth Pierson, FNWI-Radboud University Nijmegen, Pearson Science; **12.11.4** Elisabeth Pierson, FNWI-Radboud University Nijmegen, Pearson Science; **12.11.5** Elisabeth Pierson, FNWI-Radboud University Nijmegen, Pearson Science; **12.11.6** Elisabeth Pierson, FNWI-Radboud University Nijmegen, Pearson Science; **12.18.2** Guenter Albrecht-Buehler; **12.19.2** Lan Bo Chen; **12.19.3** Lan Bo Chen; **12.21** Anne Weston, LRI, CRUK, Wellcome Images; **p. 245** USDA/ARS/Agricultural Research Service.

Chapter 13 **13.1** Steve Granitz/WireImage/Getty Images; **13.2.2** Roland Birke/OKAPIA/Photo Researchers; **13.2.3** SuperStock; **13.3.2** Veronique Burger/Phanie Agency/Photo Researchers; **13.3.3** CNRI/Photo Researchers; **13.11.2** Mark Petronczki and Maria Siomos; **13.12.2** John Walsh, Micrographia.com.

Chapter 14 **14.1** Mendel Museum, Augustinian Abbey, Brno; **14.14.2** Altrendo nature/Getty Images; **14.14.3** PictureNet Corporation/Corbis; **14.15.2** Photodisc/Getty Images; **14.15.3** Photodisc/Getty Images; **14.15.4** Anthony Loveday; **14.15.5** Anthony Loveday; **14.16.2** Rick Guidotti and Diane McLean/Positive Exposure; **14.17.2** Michael Ciesielski Photography; **14.18** Douglas C. Pizac/AP Images; **14.19.2** CNRI/Photo Researchers; **p. 285** Norma Jubinville.

Chapter 15 **15.1** David C. Ward; **15.3.2** From: "Learning to Fly: Phenotypic Markers in *Drosophila*." A poster of common phenotypic markers used in *Drosophila* genetics. Jennifer Childress, Richard Behringer, and Georg Halder. 2005. *Genesis* 43(1). Cover illustration; **15.3.3** From: "Learning to Fly: Phenotypic Markers in *Drosophila*." A poster of common phenotypic markers used in *Drosophila* genetics. Jennifer Childress, Richard Behringer, and Georg Halder. 2005. *Genesis* 43(1). Cover illustration; **15.5** Andrew Syred/Photo Researchers; **15.8.2** Dave King/Dorling Kindersley; **15.15.2** Lauren Shear/SPL/Photo Researchers; **15.15.3** CNRI/SPL/Photo Researchers; **15.18** Geoff Kidd/Photo Researchers; **p. 304** James K. Adams, Biology, Dalton State College, Dalton, Georgia.

Chapter 16 **16.1** National Institutes of Health; **16.3.2** Oliver Meckes/Photo Researchers; **16.6.2** Courtesy of the Library of Congress; **16.6.3** From "The Double Helix" by James D. Watson, Atheneum Press, N.Y., 1968, p. 215. © 1968. Courtesy CSHL Archive; **16.12.2** Jerome Vinograd; **16.12.3** From D. J. Burks and P. J. Stambrook, *The Journal of Cell Biology 77* (1978). 762, fig. 6 by copyright permission of The Rockefeller University Press. Photo provided by P. J. Stambrook, **16.21** Peter Lansdorp; **16.22.2** S. C. Holt/Biological Photo Service; **16.22.3** Victoria E. Foe; **16.22.4** Barbara Hamkalo; **16.22.5** From J. R. Paulsen and U. K. Laemmli, *Cell* 12 (1977):817–828; **16.22.6** Biophoto/Photo Researchers; **16.23.2** From: Multicolor Spectral Karyotyping of Human Chromosomes. E. Schröck, S. du Manoir, T. Veldman, B. Schoell, J. Wienberg, M. A. Ferguson-Smith, Y. Ning, D. H. Ledbetter, I. Bar-Am, D. Soenksen, Y. Garini, T. Ried. *Science.* 1996 Jul 26;273(5274):494–7; **16.23.3** From: The new cytogenetics: blurring the boundaries with molecular biology; . . . M. R. Speicher, N. P. Carter. *Nat Rev Genet.* 2005 Oct;6(10):782–92; **p. 324** Thomas A. Steitz, Yale University.

Chapter 17 **17.1** Deutscher Fotodienst GmbH; **17.6.2** Keith V. Wood, University of California, San Diego; **17.6.3** AP Images; **17.16.2** Thomas Steitz; **17.17.2** Joachim Frank; **17.21.2** B. Hamkalo and O. Miller, Jr; **17.25.2** Reprinted with permission from O. L. Miller , Jr., B. A. Hamkalo, and C. A Thomas, Jr., *Science* 169 (1970):392. Copyright © 1970 American Association for the Advancement of Science.

Chapter 18 **18.1** Reproduced by permission from Cook O, Biehs B, Bier E. Brinker and optomotor-blind act coordinately to initiate development of the L5 wing vein primordium in *Drosophila. Development.* 2004 May: 131(9):2113–24; **18.12.2** From: The new cytogenetics: blurring the boundaries with molecular biology; M. R. Speicher, N. P. Carter. *Nat Rev Genet.* 2005 Oct;6(10):782–92; **18.16.2** Mike Wu; **18.16.3** Hans Pfletschinger/Peter Arnold/PhotoLibrary; **18.20.2** F. R. Turner, Indiana University; **18.20.3** F. R. Turner, Indiana University; **18.21.2** Wolfgang Driever, University of Freiburg, Freiburg, Germany; **18.21.3** Wolfgang Driever, University of Freiburg, Freiburg, Germany; **18.22.2** Ruth Lehmann, The Whitehead Institution; **18.22.3** Ruth Lehmann, The Whitehead Institution; **18.26** Roy Kaltschmidt, Lawrence Berkeley National Laboratory.

Chapter 19 **19.1** Science Photo Library/Photo Researchers; **19.2.2** Peter von Sengbusch/Botanik; **19.2.3** Peter von Sengbusch/Botanik; **19.2.4** Peter von Sengbusch/Botanik; **19.3.2** Robley C. Williams/Biological Photo Service; **19.3.3** R.C. Valentine and H.G. Pereira, "Antigens and Structure of the Adenovirus," *Journal of Molecular Biology* 13: 13–20 (1965); **19.3.4** Hazel Appleton, Health Protection Agency Centre for Infections/Photo Researchers; **19.3.5** Robley C. Williams/ Biological Photo Service; **19.8.2** C. Dauguet/Institute Pasteur/Photo Researchers; **19.8.3** C. Dauguet/Institute Pasteur/Photo Researchers; **19.8.4** C. Dauguet/Institute Pasteur/Photo Researchers; **19.8.5** C. Dauguet/Institute Pasteur/Photo Researchers; **19.8.6** C. Dauguet/Institute Pasteur/Photo Researchers; **19.9.2** NIBSC/Photo Researchers; **19.9.3** Seo Myung-gon/AP Images; **19.9.4** National Museum of Health and Medicine/Armed Forces Institute of Pathology; **19.10.2** Dennis E. Mayhew; **19.10.3** Thomas A. Zitter; **19.10.4** A. Vogler/Shutterstock.

Chapter 20 **20.1** Reproduced with permission from R.F. Service, *Science* (1998) 282:396–3999. Copyright 1998 American Association for the Advancement of Science. Incyte Pharmaceuticals, Inc., Palo Alto, CA; **20.5.2** L. Brent Selinger, Pearson Science; **20.9.2** Repligen Corporation; **20.14** Ethan Bier; **20.15.2** Reproduced with permission from R.F. Service, *Science* (1998) 282:396–3999. Copyright 1998 American Association for the Advancement of Science. Incyte Pharmaceuticals, Inc., Palo Alto, CA; **20.20** Pat Sullivan/AP Images; **20.24.2** Brad DeCecco Photography; **20.24.3** Brad DeCecco Photography; **20.25.2** Steve Helber/AP Images.

Chapter 21 **21.1** Karen Huntt/Corbis; **21.5.2** From "The genetic landscape of a cell." M. Costanzo, et al. *Science.* 2010 Jan 22;327(5964):425–31; **21.6** GeneChip Human Genome U133 Plus 2.0 Array, image courtesy of Affymetrix; **21.8.2** AP Images; **21.8.3** Courtesy of Virginia Walbot, Stanford University; **21.11.2** Courtesy of O. L. Miller Jr., Dept. of Biology, University of Virginia; **21.17.2** Shu W, Cho JY, Jiang Y, Zhang M, Weisz D, Elder GA, Schmeidler J, De Gasperi R, Sosa MA, Rabidou D, Santucci AC, Perl D, Morrisey E, Buxbaum JD. Altered ultrasonic vocalization in mice with a disruption in the *Foxp2* gene. : *Proc Natl Acad Sci U S A.* 2005 Jul 5; 102(27):9643–8; Fig. 3. Image supplied by Joseph Buxbaum; **21.17.3** Shu W, Cho JY, Jiang Y, Zhang M, Weisz D, Elder GA, Schmeidler J, De Gasperi R, Sosa MA, Rabidou D, Santucci AC, Perl D, Morrisey E, Buxbaum JD. Altered ultrasonic vocalization in mice with a disruption in the *Foxp2* gene. : *Proc Natl Acad Sci U S A.* 2005 Jul 5;102(27):9643–8; Fig. 3. Image supplied by Joseph Buxbaum; **21.17.4** Shu W, Cho JY, Jiang Y, Zhang M, Weisz D, Elder GA, Schmeidler J, De Gasperi R, Sosa MA, Rabidou D, Santucci AC, Perl D, Morrisey E, Buxbaum JD. Altered ultrasonic vocalization in mice with a disruption in the *Foxp2* gene. : *Proc Natl Acad Sci U S A.* 2005 Jul 5; 102(27):9643–8; Fig. 3. Image supplied by Joseph Buxbaum; **21.17.5** Joe McDonald/Corbis.

Chapter 22 **22.1** Olivier Grunewald; **22.2.2** PD image: Skeleton of the "Rhinoceros unicorne de Java" in the Paris Museum of Natural History. From G. Cuvier, Recherches sur les ossemens fossiles, *Atlas*, pl. 17. 1836; **22.2.3** Eileen Tweedy/Picture Desk/Kobal Collection; **22.2.4** Wayne Lynch/AGE Fotostock; **22.2.5** Neg./Transparency no. 330300. Courtesy Dept. of Library Services, American Museum of Natural History; **22.4** Michael S. Yamashita/Corbis; **22.5.2** George Richmond/ARCHIV/Photo Researchers; **22.5.3** National Maritime Museum; **22.6.2** Michel Gunther/PhotoLibrary; **22.6.3** David Hosking/FLPA; **22.6.4** David Hosking/Alamy; **22.7** Darwin's Tree of life sketch, MS.DAR.121:p.36. Reproduced with the permission of the Cambridge University Library; **22.9.2** Gerard Schulz/Naturphoto; **22.9.3** Robert Sarno/iStockphoto; **22.9.4** Paul Rapson/Alamy; **22.9.5** Izaokas Sapiro/Shutterstock; **22.9.6** YinYang/iStockphoto; **22.9.7** floricica buzlea/iStockphoto; **22.10** Laura Jesse, Extension Entomologist, Iowa State University; **22.11.2** Richard Packwood/Oxford Scientific/Jupiter Images; **22.12.2** E. S. Ross, California Academy of Sciences; **22.12.3** Mark Taylor/Nature Picture Library; **22.13.1** Scott P. Carroll; **22.16.2** Dr. Keith Wheeler/Photo Researchers; **22.16.3** Lennart Nilsson/Scanpix; **22.18.2** Visible Earth (http://visibleearth.nasa.gov/), NASA; **22.19.2** Chris Linz, Thewissen lab, Northeastern Ohio Universities College of Medicine; **22.19.3** Chris Linz, Thewissen lab, Northeastern Ohio Universities College of Medicine; **22.19.4** Chris Linz, Thewissen lab, Northeastern Ohio Universities College of Medicine; **22.19.5** Chris Linz, Thewissen lab, Northeastern Ohio Universities College of Medicine.

Chapter 23 **23.1** Rosemary B. Grant; **23.3.2** Erick Greene, University of Montana; **23.3.3** Erick Greene, University of Montana; **23.4.2** Janice Britton-Davidian, ISEM, UMR 5554 CNRS, Universite Montpellier II. Reprinted by permission from *Nature*, Vol. 403, 13 January 2000, p. 158. © 2000 Macmillan Magazines Ltd.; **23.4.3** Janice Britton-Davidian, ISEM, UMR 5554 CNRS, Universite Montpellier II. Reprinted by permission from *Nature*, Vol. 403, 13 January 2000, p. 158. © 2000 Macmillan Magazines Ltd.; **23.4.4** Steve Gorton/Dorling Kindersley; **23.5.2** New York State Department of Environmental Conservation; **23.6.2** Gary Schultz/Photoshot; **23.6.3** James L. Davis/ProWildlife; **23.11.2** William Ervin/SPL/Photo Researchers; **23.12.2** Jan Visser; **23.14.2** John Visser/Photoshot; **23.15** Dave Blackey/PhotoLibrary; **23.16.2** Allison M. Welch; **23.19** Merlin D. Tuttle, Bat Conservation International, www.batcon.org.

Chapter 24 **24.1** Mark Jones/AGE Fotostock; **24.2.2** Malcolm Schuyl/Alamy; **24.2.3** Wave Rf/PhotoLibrary; **24.2.4** Robert Kneschke/iStockphoto; **24.2.5** Justin Horrocks/iStockphoto; **24.2.6** Photodisc/Getty Images; **24.2.7** Photodisc/Getty Images; **24.2.8** Photodisc/Getty Images; **24.2.9** Masterfile; **24.3.a2** Joe McDonald/Photoshot; **24.3.b2** Joe McDonald/Corbis; **24.3.c2** USDA/APHIS/Animal and Plant Health Inspection Service; **24.3.d2** Stephen Krasemann/Photo Researchers; **24.3.e1** Michael Dietrich/

imagebroker/Alamy; **24.3.f2** Takahito Asami; **24.3.g2** William E. Ferguson; **24.3.h2** Charles W. Brown; **24.3.i2** Photodisc/Getty Images; **24.3.j2** Corbis; **24.3.k2** DawnYL/Fotolia; **24.3.l2** Kazutoshi Okuno; **24.4.2** CLFProductions/Shutterstock; **24.4.3** Boris Karpinski/Alamy; **24.4.4** Troy Maben/AP Images; **24.6.2** John Shaw/ Bruce Coleman/Photoshot; **24.6.3** Michael Fogden/Bruce Coleman/Photoshot; **24.6.4** Corbis; **24.7.2** From: Morphology, performance, fitness: functional insight into a post-Pleistocene radiation of mosquitofish. R. B. Langerhans. *Biology Letters* 2009;5(4):488–491; **24.7.3** From: Morphology, performance, fitness: functional insight into a post-Pleistocene radiation of mosquitofish. R. B. Langerhans. *Biology Letters* 2009;5(4):488–491; **24.8.2** Visible Earth (http://visibleearth.nasa.gov/), NASA; **24.8.3** Arthur Anker; **24.12** Ole Seehausen; **24.13.2** Jeroen Speybroeck, Research Institute for Nature and Forest, Belgium; **24.13.3** Jeroen Speybroeck, Research Institute for Nature and Forest, Belgium; **24.16.2** Ole Seehausen; **24.16.3** Ole Seehausen; **24.16.4** Ole Seehausen; **24.18** Jason Rick; **24.20.2** Reprinted by permission from *Nature* Bradshaw HD, Schemske DW. Allele substitution at a flower colour locus produces a pollinator shift in monkeyflowers. *Nature.* 2003 November 12; 426(6963):176–8 Copyright © 2003. Macmillan Magazines Limited; **24.20.3** Reprinted by permission from *Nature* Bradshaw HD, Schemske DW. Allele substitution at a flower colour locus produces a pollinator shift in monkeyflowers. *Nature.* 2003 November 12; 426(6963):176–8 Copyright © 2003. Macmillan Magazines Limited; **24.20.4** Reprinted by permission from *Nature* Bradshaw HD, Schemske DW. Allele substitution at a flower colour locus produces a pollinator shift in monkeyflowers. *Nature.* 2003 November 12; 426(6963):176–8 Copyright © 2003. Macmillan Magazines Limited; **24.20.5** Reprinted by permission from *Nature* Bradshaw HD, Schemske DW. Allele substitution at a flower colour locus produces a pollinator shift in monkeyflowers. *Nature.* 2003 November 12; 426(6963):176–8 Copyright © 2003. Macmillan Magazines Limited.

Chapter 25 **25.1** Gerhard Boeggemann; **p. 507.UN1** Rebecca Hunt; **25.2.2** UPI Photo/Landov; **25.3.2** Courtesy of F. M. Menger and Kurt Gabrielson, Emory University; **25.3.3** M. Hanczyc; **25.4.10** Specimen No 12478, Markus Moser, Staatliches Museum fur Naturkunde Stuttgart; **25.4.2** S. M. Awramik/Biological Photo Service; **25.4.3** Sinclair Stammers/Photo Researchers; **25.4.4** Andrew H. Knoll, Harvard University; **25.4.5** Lisa-Ann Gershwin, University of California-Berkeley, Museum of Paleontology; **25.4.6** Chip Clark; **25.4.7** Ted Daeschler/Academy of Natural Sciences/ VIREO; **25.4.8** Roger Jones; **25.4.9** Seelevel.com; **25.11.2** Shuhai Xiao, Tulane University; **25.11.3** Shuhai Xiao, Tulane University; **25.20.2** Gerald D. Carr; **25.20.3** Gerald D. Carr; **25.20.4** Gerald D. Carr; **25.20.5** Gerald D. Carr; **25.20.6** Gerald D. Carr; **25.20.7** Bruce G. Baldwin; **25.21.2** Jean Kern; **25.21.3** Jean Kern; **25.22** Juniors Bildarchiv/Alamy; **25.23** Reprinted from "The Origin of Form" by Sean B. Carroll, *Natural History*, November 2005. Burke, A.C. 2000, "*Hox*" Genes and the Global Patterning of the Somitic Mesoderm. In Somitogenesis. C. Ordahl (ed.) "Current Topics in Developmental Biology", Vo. 47. Academic Press; **25.25.1** Oxford Scientific/ PhotoLibrary; **25.25.3** Shapiro MD, Marks ME, Peichel CL, Blackman BK, Nereng KS, Jonsson B, Schluter D, Kingsley DM. Genetic and developmental basis of evolutionary pelvic reduction in threespine sticklebacks. *Nature.* Erratum. 2006 Feb 23;439(7079):1014; Fig. 1; **25.25.4** Shapiro MD, Marks ME, Peichel CL, Blackman BK, Nereng KS, Jonsson B, Schluter D, Kingsley DM. Genetic and developmental basis of evolutionary pelvic reduction in threespine sticklebacks. *Nature.* Erratum. 2006 Feb 23;439(7079):1014; Fig. 1.

Chapter 26 **26.1** Ken Griffiths/NHPA/Photoshot; **26.2.2** Ryan McVay/Photodisc/ Getty Images; **26.2.3** Neil Fletcher/Dorling Kindersley; **26.2.4** Dorling Kindersley; **26.17.2** Ed Heck; **26.17.3** Courtesy Dept. of Library Services, American Museum of Natural History.

Chapter 27 **27.1** Bonnie K. Baxter, Great Salt Lake Institute, Westminster College, Utah; **27.2.2** CDC; **27.2.3** Dr. Kari Lounatmaa/Photo Researchers; **27.2.4** Stem Jems/ Photo Researchers; **27.3.2** L. Brent Selinger, Pearson Science; **27.4.2** Dr. Immo Rantala/ Photo Researchers; **27.5.2** Kwangshin Kim/Photo Researchers; **27.6.2** Julius Adler; **27.7.2** S. W. Watson; **27.7.3** Norma J. Lang/Biological Photo Service; **27.8.2** Huntington Potter, Byrd Alzheimer's Institute and University of South Florida, David Dressler, Oxford University and Balliol College; **27.9.2** H.S. Pankratz, T.C. Beaman/ Biological Photo Service; **27.12.2** Charles C. Brinton, Jr., University of Pittsburgh; **27.14.2** Susan M. Barns; **27.16** Jack Dykinga/Stone/Getty Images; **27.17.3** L. Evans Roth/Biological Photo Service; **27.17.5** Yuichi Suwa; **27.17.7** National Library of Medicine; **27.17.9** From "Scanning electron microscopy of fruiting body formation by myxobacteria." P. L. Grilione and J. Pangborn. *J. Bacteriol.* 1975 December; 124(3): 1558–1565; **27.17.13** Photo Researchers; **27.17.15** Moredon Animal Health/SPL/Photo Researchers; **27.17.17** CNRI/SPL/Photo Researchers; **27.17.19** Culture Collection CCALA, Institute of Botany, Academy of Sciences Dukelska, Czech Republic; **27.17.21** Paul Hoskisson, Strathclyde Institute of Pharmacy and Biomedical Sciences, Glasgow, Scotland; **27.17.22** David M. Phillips/Photo Researchers; **27.18.2** Pascale Frey-Klett; **27.19** Ken Lucas/Biological Photo Service; **27.20.2** Scott Camazine/Photo Researchers; **27.20.3** David M. Phillips/Photo Researchers; **27.20.4** James Marshall/The Image Works; **27.21** Metabolix; **27.21.3** Courtesy of Exxon Mobil Corporation; **27.21.4** Seelevel.com.

Chapter 28 **28.1** Brian S. Leander; **28.3.3** Joel Mancuso, University of California, Berkeley; **28.3.5** M. I. Walker/NHPA/Photoshot; **28.3.7** Howard Spero, University of California-Davis; **28.3.8** NOAA; **28.3.10** Kim Taylor/Nature Picture Library; **28.3.11** David J. Patterson/micro*scope; **28.3.13** Tom Stack/PhotoLibrary; **28.4.2** David M. Phillips/The Population Council/Photo Researchers; **28.5.2** David J. Patterson; **28.6.2** Meckes/Ottawa/Photo Researchers; **28.7.2** D. J. Patterson, L. Amaral-Zettler, M. Peglar and T. Nerad, http://micro*scope.mbl.edu; **28.8.2** Guy Brugerolle, Universitad Clearmont, Ferrand; **28.9.2** Virginia Institute of Marine Science; **28.10.2** Masamichi Aikawa, Tokai University School of Medicine, Japan; **28.11.2** M. I. Walker/ Photo Researchers; **28.12.2** Centers for Disease Control & Prevention; **28.13.2** Steve Gschmeissner/Photo Researchers; **28.14.2** Stephen Durr; **28.15.2** Colin Bates; **28.16.2** J. Robert Waaland/Biological Photo Service; **28.17.2** Fred Rhoades; **28.18.2** Robert Brons/Biological Photo Service; **28.19.2** Eva Nowack; **28.20.2** D. P. Wilson, Eric & David Hosking/Photo Researchers; **28.20.3** Michael D. Guiry; **28.20.4** Biophoto Associates/Photo Researchers; **28.20.6** Michael Yamashita/IPN/Aurora Photos; **28.20.7** David Murray/Dorling Kindersley; **28.21.2** Laurie Campbell/NHPA; **28.21.3** Marine Sciences, University of Puerto Rico; **28.22.2** William L. Dentler; **28.24.2** George Barron; **28.24.3** Mikel Tapia (www.argazkik.com); **28.25.2** Robert Kay; **28.25.3** Robert Kay; **28.26.2** Kevin Carpenter and Patrick Keeling.

Chapter 29 **29.1** Martin Rugner/AGE Fotostock; **29.2.2** S. C. Mueller and R. M. Brown, Jr; **29.3.2** Natural Visions; **29.3.3** Linda Graham, University of Wisconsin-Madison; **29.5.11** Ed Reschke; **29.5.12** CDC; **29.5.3** Linda Graham, University of Wisconsin-Madison; **29.5.4** Karen S. Renzaglia; **29.5.6** Alan S. Heilman; **29.5.7** Michael Clayton, University of Wisconsin-Madison; **29.5.9** David John Jones (mybitoftheplanet.com); **29.6.2** Charles H. Wellman; **29.6.3** Charles H. Wellman; **29.8.2** Laurie Knight (www.laurieknight.net); **29.9.2** Linda Graham, University of Wisconsin-Madison; **29.9.3** Alvin E. Staffan/National Audubon Society/Photo Researchers; **29.9.4** Hidden Forest; **29.9.6** Hidden Forest; **29.9.8** Tony Wharton, Frank Lane Picture Agency/Corbis; **p. 609** From "Mosses and Other Bryophytes, an Illustrated Glossary" (2006) by Bill and Nancy Malcolm; **29.11.1** Brian Lightfoot/AGE Fotostock; **29.11.2** Chris Lisle/Corbis; **29.15.2** Jody Banks, Purdue University; **29.15.3** Murray Fagg, Australian National Botanic Gardens; **29.15.4** Helga & Kurt Rasbach; **29.15.6** Jon Meier/iStockphoto; **29.15.8** Milton Rand/Tom Stack & Associates; **29.15.9** Francisco Javier Yeste Garcia (www.flickr.com/photos/fryega/); **29.15.10** Francisco Javier Yeste Garcia (www.flickr.com/photos/fryega/); **29.16.2** The Open University; **p. 616, top** Ed Reschke; **p. 616, bottom** Michael Clayton.

Chapter 30 **30.1** National Museum of Natural History, Smithsonian Institution; **30.5.1** Johannes Greyling/iStockphoto; **30.5.3** CC-BY-SA photo: Kurt Stueber (www.biolib.de); **30.5.4** Jeroen Peys/iStockphoto; **30.5.5** Michael Clayton; **30.5.6** Thomas Schoepke; **30.5.7** Bob Gibbons/FLPA; **30.5.8** Raymond Gehman/Corbis; **30.5.9** Adam Jones/Getty Images; **30.5.10** David Muench/Corbis; **30.5.11** Gunter Marx Photography/Corbis; **30.5.12** Royal Botanic Gardens Sydney; **30.5.13** Jaime Plaza/ AP Images; **30.5.14** Mario Verin/Photolibrary; **30.8.2** Dave King/ Dorling Kindersley; **30.8.3** Andy Crawford/Dorling Kindersley; **30.8.4** Dave King/ Dorling Kindersley; **30.8.5** Maria Dryfhout/iStockphoto; **30.8.6** Peter Rees/Getty Images; **30.9.2** PIXTAL/AGE Fotostock; **30.9.3** Hans Dieter Brandl, Frank Lane Picture Agency/Corbis; **30.9.4** Scott Camazine/Photo Researchers; **30.9.5** Derek Hall/ Dorling Kindersley; **30.11.2** David L. Dilcher; **30.13.1** Howard Rice/Dorling Kindersley; **30.13.2** Jack Scheper, Floridata.com; **30.13.3** Stephen McCabe; **30.13.4** Andrew Butler/Dorling Kindersley; **30.13.6** Eric Crichton/Dorling Kindersley; **30.13.7** John Dransfield; **30.13.8** Dorling Kindersley; **30.13.10** Terry W. Eggers/Corbis; **30.13.11** CC-BY-SA photo: Artslave; **30.13.12** Matthew Ward/Dorling Kindersley; **30.13.13** Tony Wharton, Frank Lane Picture Agency/Corbis; **30.13.14** Howard Rice/ Dorling Kindersley; **30.13.15** Gerald D. Carr; **30.14** kkaplin/Shutterstock; **30.15** D. Wilder; **30.16.2** NASA's Earth Observatory; **30.16.3** NASA's Earth Observatory.

Chapter 31 **31.1** Georg Müller; **31.2.2** Fred Rhoades/Mycena Consulting; **31.2.3** Hans Reinhard/Taxi/Getty Images; **31.2.4** George Barron, University of Guelph, Canada; **31.4.2** © N. Allin & G.L. Barron, University of Guelph/Biological Photo Service; **31.6.2** Biophoto Associates/Photo Researchers; **31.6.3** Popovaphoto/ dreamstime; **31.7.2** Stephen J. Kron; **31.9.2** Dirk Redecker, Robin Kodner, and Linda E. Graham. Glomalean Fungi from the Ordovician. *Science* 15 September 2000; 289: 1920–1921; **31.10.2** CDC; **31.11.2** John Taylor; **31.11.3** Ray Watson; **31.11.5** Reproduced by permission from Kiers ET, van der Heijden MG. Mutualistic stability in the arbuscular mycorrhizal symbiosis: exploring hypotheses of evolutionary cooperation. *Ecology.* 2006 July: 87(7):1627–36: Fig. 1a. Image by Marcel van der Heijden, Swiss Federal Research Station for Agroecology and Agriculture. Copyright © 2006, Ecological Society of America; **31.11.6** Frank Young/Papilio/Corbis; **31.11.7** Phil Dotson/Photo Researchers; **31.12.2** William E. Barstow; **31.13.2** Antonio D'Albore/ iStockphoto; **31.13.3** Alena Kubátová (http://botany.natur.cuni.cz/cs/sbirka-kulturhub-ccf); **31.13.4** Ed Reschke/Peter Arnold/PhotoLibrary; **31.13.5** George Barron; **31.14.2** G.L. Barron/Biological Photo Service; **31.15.2** M. F. Brown/Biological Photo Service; **31.16.2** Viard/Jacana/Photo Researchers; **31.16.3** Douglas Adams/iStockphoto; **31.17.2** Fred Spiegel; **31.18.2** Fletcher and Baylis/Photo Researchers; **31.18.3** Michael Fogden/Photolibrary; **31.18.4** Frank Paul/Alamy; **31.19.2** Biophoto Associates/Photo Researchers; **31.20** University of Tennessee Entomology and Plant Pathology; **31.22** Mark Bowler/Photo Researchers; **31.23.2** Ralph Lee Hopkins/Getty Images; **31.23.3** Geoff Simpson/naturepl.com; **31.23.4** Wild-Worlders of Europe/ Benvie/naturepl.com; **31.24.2** Eye of Science/Photo Researchers; **31.25.2** Alamy; **31.25.3** Peter Chadwick/Dorling Kindersley; **31.25.4** Hecker-Sauer/AGE Fotostock; **31.26.2** Vance T. Vredenburg, San Francisco State University; **31.27.2** Christine Case.

Chapter 32 **32.1** Jeff Hunter/Image Bank/Getty Images; **p. 656** Biological Photo Service; **32.4.2** © The Museum Board of South Australia 2004 Photographer: Dr. J. Gehling; **32.4.3** © The Museum Board of South Australia 2004 Photographer: Dr. J. Gehling; **32.5** J. Sibbick/The Natural History Museum, London; **32.6.2** Wikramanayake AH, Hong M, Lee PN, Pang K, Byrum CA, Bince JM, Xu R, Martindale MQ. An ancient role for nuclear beta-catenin in the evolution of axial polarity and germ layer segregation. *Nature.* 2003 Nov 27;426(6965):446–50; Fig. 2, 3 and 4; **32.6.3** Wikramanayake AH, Hong M, Lee PN, Pang K, Byrum CA, Bince JM, Xu R, Martindale MQ. An ancient role for nuclear beta-catenin in the evolution of axial polarity and germ layer segregation. *Nature.* 2003 Nov 27;426(6965):446–50; Fig. 2, 3 and 4; **32.6.4** Wikramanayake AH, Hong M, Lee PN, Pang K, Byrum CA, Bince JM, Xu R, Martindale MQ. An ancient role for nuclear beta-catenin in the evolution of axial polarity and germ layer segregation. *Nature.* 2003 Nov 27;426(6965):446–50; Fig. 2, 3 and 4; **32.6.5** Wikramanayake AH, Hong M, Lee PN, Pang K, Byrum CA, Bince JM, Xu R, Martindale MQ. An ancient role for nuclear beta-catenin in the evolution of axial polarity and germ layer segregation. *Nature.* 2003 Nov 27; 426(6965):446–50; Fig. 2, 3 and 4; **32.12** Kent Wood/Photo Researchers; **32.13.2** Hecker/Sauer/AGE Fotostock.

Chapter 33 **33.1** C. Wolcott Henry III/National Geographic/Getty Images; **33.3.1** Andrew J. Martinez/Photo Researchers; **33.3.2** Robert Brons/Biological Photo Service; **33.3.3** Teresa (Zubi) Zuberbühler; **33.3.4** Stephen Dellaporta; **33.3.5** Gregory G. Dimijian/Photo Researchers; **33.3.6** Ed Robinson/Pacific Stock/Photolibrary; **33.3.7** Hecker/Sauer/AGE Fotostock; **33.3.8** W. I. Walker/Photo Researchers; **33.3.9** Kåre Telnes/Image Quest Marine; **33.3.10** PD image: Anilocra/Taken by Neil Campbell, University of Aberdeen, Scotland, UK; **33.3.11** Erling Svensen/UWPhoto ANS; **33.3.12** Peter Funch; **33.3.13** photonimo/iStockphoto; **33.3.14** Peter Batson/Image Quest Marine; **33.3.15** Reinhart Mobjerg Kristensen; **33.3.16** Erling Svensen/

UWPhoto ANS; **33.3.17** Andrew Syred/Photo Researchers; **33.3.18** Reproduced with permission from A. Eizinger and R. Sommer, Max Planck Institut fur entwicklungsbiologie, Tubingen. Copyright 2000 American Association for the Advancement of Science. Cover 278(5337) 17 Oct 97; **33.3.19** Thomas Stromberg; **33.3.20** Tim Flach/Stone/Getty Images; **33.3.21** Heather Angel/Natural Visions; **33.3.22** Robert Harding World Imagery/Alamy; **33.3.23** Robert Brons/Biological Photo Service; **33.4.2** Andrew J. Martinez/Photo Researchers; **33.7.2** Andrew J. Martinez/Photo Researchers; **33.7.3** Robert Brons/Biological Photo Service; **33.7.4** Commonwealth of Australia (GBRMPA); **33.7.5** Neil G. McDaniel/Photo Researchers; **33.8.2** Robert Brons/Biological Photo Services; **33.9.2** Ed Robinson/Pacific Stock/PhotoLibrary; **33.11.2** CDC; **33.12.2** Eye of Science/Photo Researchers; **33.13.2** W. I. Walker/Photo Researchers; **33.14.2** Hecker/Sauer/AGE Fotostock; **33.14.3** Kåre Telnes/Image Quest Marine; **33.16** Jeff Foott/Tom Stack and Associates; **33.17.2** Amruta Bhelke/Dreamstime.com; **33.17.3** Corbis; **33.19** H. W. Pratt/Biological Photo Service; **33.21.2** photonimo/iStockphoto; **33.21.3** Mark Conlin/Image Quest Marine; **33.21.4** Jonathan Blair/Corbis; **33.22.2** Photograph courtesy of the U.S. Bureau of Fisheries (1919), and Illinois State Museum; **33.22.3** © Zoological Society of London (ZSL); **33.23.2** Peter Batson/Image Quest Marine; **33.24.2** A.N.T./NHPA/Photoshot; **33.25** Astrid & Hanns-Frieder Michler/Photo Researchers; **33.26.2** Reproduced with permission from A. Eizinger and R. Sommer, Max Planck Institut fur entwicklungsbiologie, Tubingen. Copyright 2000 American Association for the Advancement of Science. Cover 278(5337) 17 Oct 97; **33.27.2** SPL/Photo Researchers; **33.28** Collection of Dan Cooper; **33.29.2** Grenier JK, Garber TL, Warren R, Whitington PM, Carroll S. Evolution of the entire arthropod *Hox* gene set predated the origin and radiation of the onychophoran/arthropod clade. *Curr Biol.* 1997 Aug 1;7(8):547–53; Fig. 3c; **33.31** Mark Newman/FLPA; **33.32.2** Tim Flach/Stone/Getty Images; **33.32.3** Andrew Syred/Photo Researchers; **33.32.4** Eric Lawton/iStockphoto; **33.34.2** Premaphotos/Nature Picture Library; **33.34.3** Tom McHugh/Photo Researchers; **33.36** Meul/ARCO/Nature Picture Library; **33.37.2** John Shaw/Tom Stack and Associates; **33.37.3** John Shaw/Tom Stack and Associates; **33.37.4** John Shaw/Tom Stack and Associates; **33.37.5** John Shaw/Tom Stack and Associates; **33.37.6** John Shaw/Tom Stack and Associates; **33.38.2** Dr. John Brackenbury/Photo Researchers; **33.38.3** Perry Babin; **33.38.4** PREMAPHOTOS/Nature Picture Library; **33.38.5** Dante Fenolio/Photo Researchers; **33.38.6** John Cancalosi/Nature Picture Library; **33.38.7** Hans Christoph Kappel/Nature Picture Library; **33.38.8** Michael & Patricia Fogden/CORBIS; **33.38.9** CC-BY-SA photo: Bruce Marlin (www.cirrusimage.com/fly_whale-tail.htm); **33.39.2** Maximilian Weinzierl/Alamy; **33.39.3** Peter Herring/Image Quest Marine; **33.39.4** Peter Parks/Image Quest Marine; **33.40.2** Andrey Nekrasov/Image Quest Marine; **33.41** Daniel Janies; **33.42** Jeff Rotman/Photo Researchers; **33.43** Robert Harding World Imagery/Alamy; **33.44** Jurgen Freund/Nature Picture Library; **33.45** Hal Beral/Corbis.

Chapter 34 **34.1** Xian-guang H, Aldridge RJ, Siveter DJ, Siveter DJ, Xiang-hong F. New evidence on the anatomy and phylogeny of the earliest vertebrates. *Proceedings of the Royal Society of London-B-Biological Sciences,* Sept. 22, 2002; 269 (1503) 1865–1869; Fig. 1c; **34.4.2** Oxford Scientific/PhotoLibrary; **34.5.2** Robert Brons/Biological Photo Service; **34.8.2** Nanjing Institute of Geology and Palaeontology; **34.9.2** Tom McHugh/Photo Researchers; **34.10.2** A Hartl/AGE Fotostock; **34.10.3** Marevision/AGE Fotostock; **34.14.2** The Field Museum, #GEO82014; **34.15.2** Carlos Villoch/Image Quest Marine; **34.15.3** Masa Ushioda/Image Quest Marine; **34.15.4** Andy Murch/V&W/Image Quest Marine; **34.17.2** James D. Watt/Image Quest Marine; **34.17.3** Jez Tryner/Image Quest Marine; **34.17.4** George Grall/Getty Images; **34.17.5** Fred McConnaughey/Photo Researchers; **34.18.2** From "The oldest - articulated osteichthyan reveals mosaic gnathostome characters." M. Zhu et al. *Nature.* 2009 Mar 26;458(7237):469–74; **34.19** Arnaz Mehta; **34.20.2** © Ted Daeschler/Academy of Natural Sciences/VIREO; **34.20.3** © Ted Daeschler/Academy of Natural Sciences/VIREO; **34.20.4** © Kalliopi Monoyios; **34.22.2** Alberto Fernandez/AGE Fotostock America, Inc. **34.22.3** Michael Fogden/Bruce Coleman/Photoshot; **34.22.4** Michael Fogden/Bruce Coleman/Photoshot; **34.23.2** John Cancalosi/Peter Arnold/PhotoLibrary; **34.23.3** Stephen Dalton/Photo Researchers; **34.23.4** Hans Pfletschinger/Peter Arnold/PhotoLibrary; **34.24** Michael Fogden/OSF/PhotoLibrary; **34.27** Nobumichi Tamura; **34.28** Michael Fogden/OSF/PhotoLibrary; **34.29.2** Natural Visions/Alamy; **34.29.3** Matt T. Lee; **34.29.5** Medford Taylor/National Geographic Image Collection; **34.29.6** Carl & Ann Purcell/Corbis; **34.30.2** Visceralimage/dreamstime; **34.30.3** Janice Sheldon; **34.32** Russell Mountford/Alamy; **34.33** DLILLC/Corbis; **34.34** Yufeng Zhou/iStockphoto; **34.35.1** McPHOTO/AGE Fotostock; **34.35.2** paolo barbanera/AGE Fotostock; **34.36** Gianpiero Ferrari/FLPA; **34.36** Gianpiero Ferrari/FLPA; **34.38.2** D. Parer and E. Parer Cook/Auscape International Proprietary Ltd. **34.38.3** Mervyn Griffiths/Commonwealth Scientific and Industrial Research Organization; **34.39.2** John Cancalosi/Alamy; **34.39.3** Wells Bert & Babs/OSF/PhotoLibrary; **34.42** Frans Lanting/Corbis; **34.44.2** Kevin Schafer/AGE Fotostock America, Inc. **34.44.3** J & C Sohns/Photolibrary; **34.45.2** Morales/AGE Fotostock America, Inc. **34.45.3** Anup Shah/ImageState/Alamy Images; **34.45.4** T. J. Rick/Nature Picture Library; **34.45.5** E. A. Janes/AGE Fotostock America, Inc. **34.45.6** Frans Lanting/Corbis; **34.47** Fossilized bone, partial skeleton, of Ardipithecus ramidus, articulated, with bones laid in their approximate positions. Housed in National Museum of Ethiopia, Addis Ababa. Photo © T. White 2009, From "Ardipithecus ramidus and the Paleobiology of Early Hominids." T. White et al. *Science.* 2009 Oct 2;326(5949):75–86; **34.48.2** John Reader/SPL/Photo Researchers; **34.48.3** John Gurche Studios; **34.49.2** Alan Walker @ National Museums of Kenya. Printed with permission; **34.51** David L Brill/Brill Atlanta; **34.52** C. Henshilwood & F. d'Errico/Professor Chris Henshilwood.

Chapter 35 **35.1** PD photo: John Walker (www.fourmilab.ch); **35.3** Robert & Linda Mitchell; **35.4.2** CC-BY-SA photo: Forest & Kim Starr; **35.4.3** Rob Walls/ Alamy; **35.4.4** YinYang/iStockphoto; **35.4.5** Robert Holmes/Corbis; **35.4.6** Geoff Tompkinson/Science Photo Library/Photo Researchers; **35.5.2** Donald Gregory Clever; **35.5.3** Gusto Productions/SPL/Photo Researchers; **35.5.4** Dorling Kindersley; **35.5.5** Aflo Foto Agency/Alamy; **35.7.2** Neil Cooper/Alamy; **35.7.3** Martin Ruegner/ jupiterimages; **35.7.4** Mike Zens/Corbis; **35.7.5** Jerome Wexler/Photo Researchers; **35.7.6** Kathy Piper/iStockphoto; **35.9.2** Purdue Extension Entomology; **35.10.2** Brian Capon; **35.10.4** © Clouds Hill Imaging/www.lastrefuge.co.uk; **35.10.6** Graham Kent, Pearson Science; **35.10.7.** Graham Kent, Pearson Science; **35.10.9** N.C. Brown Center for Ultrastructure Studies, SUNY-Environmental Science & Forestry, Syracuse, NY; **35.10.11** Graham Kent; **35.10.12** Reproduced with permission from "Plant Cell Biology on CD," by B E S Gunning, www.plantcellbiologyonCD.com; **35.10.13** Professor Ray F. Evert; **35.13.2** From "Arabidopsis TCP20 links regulation of growth and cell division control pathways." C. Li et al. *Proc Natl Acad Sci U S A.* 2005 Sep 6;102(36):12978–83. Epub 2005 Aug 25. Photo: Peter Doerner; **35.14.2** Natalie B. Bronstein; **35.14.3** Ed Reschke; **35.14.4** Ed Reschke; **35.15.2** Michael Clayton; **35.16.2** Michael Clayton; **35.17.2** Ed Reschke; **35.17.3** Ed Reschke; **35.18.2** Ed Reschke; **35.18.3** Ed Reschke; **35.19.2** Michael Clayton; **35.19.3** Alison W. Roberts; **35.21.2** Dr. Edward R. Cook; **35.23** California Historical Society Collection (CHS-1177), University of Southern California on behalf of the USC Specialized Libraries and Archival Collections; **p. 756** Reproduced by permission from Janet Braam, *Cell* 60 (9 February 1990): Cover. Copyright ©1990 Cell Press. Image courtesy of Elsevier Sciences Ltd. **35.25.2** From "Microtubule plus-ends reveal essential links between intracellular polarization and localized modulation of endocytosis during division-plane establishment in plant cells." P. Dhonukshe et al. *BMC Biol.* 2005 Apr 14;3:11; **35.26.2** From "The tangled-1 mutation alters cell division orientations throughout maize leaf development without altering leaf shape." L. G. Smith et al. *Development.* 1996 Feb;122(2):481–9; **35.26.3** From "The tangled-1 mutation alters cell division orientations throughout maize leaf development without altering leaf shape." L. G. Smith et al. *Development.* 1996 Feb;122(2):481–9; **35.28** From figure 1a in U. Mayer et al., *Development* 117 (1): 149–162. © 1993 The Company of Biologists Ltd. **35.29.2** From "Microtubule plus-ends reveal essential links between intracellular polarization and localized modulation of endocytosis during division-plane establishment in plant cells." P. Dhonukshe et al. *BMC Biol.* 2005 Apr 14;3:11; **35.29.3** B. Wells and K. Roberts; **35.30.2** Reproduced by permission from figure 1 in D. Hareven et al, *Cell* 84 (5): 735–744. Copyright © 1996. by Elsevier Science Ltd. **35.30.3** Reproduced by permission from figure 1 in D. Hareven et al, *Cell* 84 (5): 735–744. Copyright © 1996. by Elsevier Science Ltd. **35.31.2** Reproduced by permission from Figure 2g in Hung et al, *Plant Physiology* 117:73–84. Copyright © 1998 by the American Society of Plant Biologists. Image courtesy of John Schiefelbein/Univesity of Michigan; **35.32.2** Dr. Gerald D. Carr, PhD; **35.33.2** Dr. E. M. Meyerowitz and John Bowman, *Development* 112 1991:1–231.2. Division of Biology, California Institute of Technolgy; **35.33.3** Dr. E. M. Meyerowitz and John Bowman, *Development* 112 1991:1–231.2. Division of Biology, California Institute of Technolgy.

Chapter 36 **36.1** Peggy Heard/FLPA/Alamy; **36.3.2** Rolf Rutishauser; **36.3.3** Rolf Rutishauser; **36.5.2** Dr. Jeremy Burgess/SPL/Photo Researchers; **p. 770, top** Nigel Cattlin/Holt Studios International/Photo Researchers; **p. 770, bottom** Nigel Cattlin/Holt Studios International/Photo Researchers; **36.11** Scott Camazine/Photo Researchers; **36.14.2** Graham Kent; **36.14.3** Graham Kent; **36.16.2** Mlane/Dreamstime.com; **36.16.3** Kate Shane, Southwest School of Botanical Medicine; **36.16.4** Frans Lanting/Corbis; **36.16.5** Natalie Bronstein; **36.16.6** Andrew de Lory/Dorling Kindersley; **36.16.7** Danita Delimont/Alamy; **36.19.2** M. H. Zimmermann, courtesy of Professor P. B. Tomlinson, Harvard University; **36.19.3** M. H. Zimmermann, courtesy of Professor P. B. Tomlinson, Harvard University; **36.19.4** M. H. Zimmermann, courtesy of Professor P. B. Tomlinson, Harvard University; **36.20.2** From "A coiled-coil interaction mediates cauliflower mosaic virus cell-to-cell movement." L. Stavolone et al. *Proc Natl Acad Sci U S A.* 2005 Apr 26;102(17):6219–24. Epub 2005 Apr 18.

Chapter 37 **37.1** Chris Mattison/Alamy; **37.2.2** USDA/ARS/Agricultural Research Service; **37.4** National Oceanic and Atmospheric Administration NOAA; **37.5** U.S. Geological Survey, Denver; **37.6** Kevin Horan/Stone/Getty Images; **37.9.2** White et al., Plant Physiology, June 2003; **37.11.2** Scimat/Photo Researchers; **37.11.3** E. H. Newcomb and S. R. Tandon/Biological Photo Service; **37.13.2** Hugues B. Massicotte, University of Northern British Columbia Ecosystem Science and Management Program, Prince George, B.C., Canada; **37.13.3** Mark Brundrett (http://mycorrhizas.info); **37.13.4** Mark Brundrett (http://mycorrhizas.info); **37.14.2** Elizabeth J. Czarapata/The Park People; **37.15.1** Wolfgang Kaehler/Corbis; **37.15.2** Ruud de Man/iStockphoto; **37.15.3** Kevin Schafer/Corbis; **37.15.4** Gary W. Carter/Corbis; **37.15.5** Kim Taylor and Jane Burton/Dorling Kindersley; **37.15.6** Biophoto Associates/Photo Researchers; **37.15.7** Philip Blenkinsop/Dorling Kindersley; **37.15.8** Paul A. Zahl/Photo Researchers; **37.15.9** Fritz Polking, Frank Lane Picture Agency/Corbis.

Chapter 38 **38.1** Pierre-Michel Blais; **38.3.2** Ed Reschke/Peter Arnold/PhotoLibrary; **38.3.3** Ed Reschke/Peter Arnold/PhotoLibrary; **38.3.4** David Scharf/Peter Arnold/PhotoLibrary; **38.4.2** Marianne Wiora; **38.4.3** Stephen Dalton/NHPA Limited/Photoshot; **38.4.4** Bjorn Rorslett Photographe; **38.4.5** Bjorn Rorslett Photographe; **38.4.7** Doug Backlund; **38.4.8** Martin Pieter Heigan; **38.4.9** Rolf Nussbaumer/Nature Picture Library; **38.4.10** Merlin D. Tuttle, Bat Conservation International, www.batcon.org; **38.5** Photo: © W. Barthlott/W. Rauh; **38.11.2** Kevin Schafer/Alamy; **38.11.3** Nature Production; **38.11.4** Brian Gordon Green/National Geographic Image Collection; **38.11.5** Steve Bloom Images/Alamy; **38.11.6** Aaron McCoy/Botanica/Photolibrary; **38.11.7** California Department of Food and Agriculture's Plant Health and Pest Prevention Services; **38.11.8** Kim A. Cabrera Photographer; **38.11.9** Steve Shattuck/CSIRO Entomology; **38.11.10** Alan Williams/ Alamy; **38.12.11** Dennis Frates/Alamy; **38.13.2** Marcel E. Dorken; **38.13.3** Marcel E. Dorken; **38.13.4** Nobumitsu Kawakubo, Gifu University, Japan; **38.14.2** Bruce Iverson, Photomicrography; **38.14.3** Bruce Iverson, Photomicrography; **38.14.4** Meriel G. Jones, University of Liverpool, UK; **38.15.2** Sinclair Stammers/Photo Researchers; **38.16.2** Andrew McRobb/Dorling Kindersley; **38.16.3** Andrew McRobb/Dorling Kindersley; **38.17** John Van Hasselt/Corbis.

Chapter 39 **39.1** *Plant Physiol.* 1999 Jul;120(3): Cover. By permission of the American Society of Plant Physiologists. Illustration by Niemeyer MI and Fernandez MC; **39.2** Natalie B. Bronstein; **39.7.2** "Regulation of Polar Auxin transport ATPIN1 in Arabidopsis Vascular Tissue," by Leo Galweiler, et al. *Science* 18 DEC 1998, vol. 282; **39.7.3** "Regulation of Polar Auxin transport ATPIN1 in Arabidopsis Vascular Tissue," by Leo Galweiler, et al. *Science* 18 DEC 1998, vol. 282; **39.9.2** Malcolm B. Wilkins, University of Glasgow, Glasgow, Scotland, U.K. **39.9.3** Malcolm B. Wilkins, University of Glasgow, Glasgow, Scotland, U.K. **39.9.4** Malcolm B. Wilkins, University of Glasgow, Glasgow, Scotland, U.K. **39.10.2** Dr. Richard Amasino; **39.10.3** Fred Jensen; **39.12.2** Mia Molvray; **39.12.3** Karen E. Koch; **39.14.2** Kurt Stepnitz, DOE

Plant Research Laboratory, Michigan State University; **39.14.3** Joe Kieber, University of North Carolina; **39.15.2** Ed Reschke; **39.16.3** Malcolm B. Wilkins, University of Glasgow, Glasgow, Scotland, U.K. **39.16.4** Malcolm B. Wilkins, University of Glasgow, Glasgow, Scotland, U.K. **39.17.2** Malcolm B. Wilkins, University of Glasgow, Glasgow, Scotland, U.K. **39.17.3** Malcolm B. Wilkins, University of Glasgow, Glasgow, Scotland, U.K. **39.17.4** Malcolm B. Wilkins, University of Glasgow, Glasgow, Scotland, U.K. **39.17.5** Malcolm B. Wilkins, University of Glasgow, Glasgow, Scotland, U.K. **39.17.6** Malcolm B. Wilkins, University of Glasgow, Glasgow, Scotland, U.K. **39.20.2** Malcolm B. Wilkins, University of Glasgow, Glasgow, Scotland, U.K. **39.20.3** Malcolm B. Wilkins, University of Glasgow, Glasgow, Scotland, U.K. **39.24.2** Michael Evans, Ohio State University; **39.24.3** Michael Evans, Ohio State University; **39.24.4** Michael Evans, Ohio State University; **39.24.5** Michael Evans, Ohio State University; **39.25** Reproduced by permission from Janet Braam, *Cell* 60 (9 February 1990): Cover. Copyright © 1990 Cell Press. Image courtesy of Elsevier Sciences Ltd. **39.26.2** Martin Shields/Photo Researchers; **39.26.3** Martin Shields/Photo Researchers; **39.26.4** From K. Esau. "Anatomy of Seed Plants," 2nd ed. (New York: John Wiley and Sons, 1977), fig. 19.4, p.358; **39.26.5** From K. Esau. "Anatomy of Seed Plants," 2nd ed. (New York: John Wiley and sons, 1977), fig. 19.4, p.358; **39.27.2** J. L. Basq and M. C. Drew; **39.27.3** J. L. Basq and M. C. Drew; **39.29.2** New York State Agricultural Experiment Station (NYSAES)/Cornell.

Chapter 40 **40.1** JOEL SARTORE/National Geographic Stock; **40.2.2** Brandon Cole/ PhotoLibrary; **40.2.3** Duncan Usher/Alamy; **40.2.4** Frank Greenaway/Dorling Kindersley/Getty Images; **40.4.2** Susumu Nishinaga/Photo Researchers; **40.4.3** Eye of Science/Photo Researchers; **40.4.4** Susumu Nishinaga/Photo Researchers; **40.5.7** CNRI/SPL/Photo Researchers; **40.5.11** Dr. Gopal Murti/SPL/Photo Researchers; **40.5.13** Chuck Brown/Photo Researchers; **40.5.15** Nina Zanetti; **40.5.17** Nina Zanetti; **40.5.19** Nina Zanetti; **40.5.21** Alamy; **40.5.24** Nina Zanetti; **40.5.26** Ed Reschke/Peter Arnold Images/PhotoLibrary; **40.5.28** Manfred Kage/Peter Arnold/Photolibrary; **40.5.31** Ulrich Gartner; **40.5.33** Thomas Deerinck/National Center for Microscopy and Imaging Research, University of California, San Diego; **40.10.2** Patricio Robles Gil/ naturepl.com; **40.10.3** Matt T. Lee; **40.13** Robert Ganz; **40.18** Jeff Rotman/Alamy.

Chapter 41 **41.1** Michael deYoung/Corbis; **41.2** Roland Seitre/Peter Arnold/ PhotoLibrary; **41.3** Stefan Huwiler/Rolf Nussbaumer Photography/Alamy; **41.5.2** cameilia/Shutterstock; **41.6.2** Hervey Bay Whale Watch (www.herveybaywhalewatch .com.au); **41.6.4** Thomas Eisner; **41.6.5** Photo Lennart Nilsson/Scanpix; **41.6.6** Gunter Ziesler/Peter Arnold/PhotoLibrary; **41.11.2** Visuals Unlimited/Corbis; **41.17.2** Fritz Polking/Peter Arnold, Inc./Alamy; **41.17.3** EyeWire Collection/ Photodisc/Getty Images; **41.22** Photo courtesy of The Jackson Laboratory, Bar Harbor, Maine; **41.23** Wolfgang Kaehler/Corbis.

Chapter 42 **42.1** Stephen Dalton/Photo Researchers; **42.2.2** Reinhard dirscheri/ PhotoLibrary; **42.2.3** Eric Grave/Photo Researchers; **42.10.2** From "Human Histology Photo CD." Image courtesy Indigo Instruments (www.indigo.com); **42.10.3** Photo Lennart Nilsson/Scanpix; **42.16** Biophoto Associates/Photo Researchers; **42.18.2** Eye of Science/Photo Researchers; **42.22.2** Peter Batson/Image Quest Marine; **42.22.3** Olgysha/Shutterstock; **42.22.4** Jez Tryner/Image Quest Marine; **42.24.2** Prepared by Dr. Hong Y. Yan, University of Kentcky and Dr. Peng Chai, University of Texas; **42.25.2** Motta & Macchiarelli/Anatomy Dept., Univ. La Sapienza, Rome/Photo Researchers; **42.27.2** Hans-Rainer Duncker, University of Giessen, Germany.

Chapter 43 **43.1** Biology Media/Science Source/Photo Researchers; **43.4** Dominique Ferrandon; **43.21.2** Steve Gschmeissner/Photo Researchers; **43.23** CNRI/Photo Researchers; **43.26** The Laboratory of Structural Cell Biology, headed by Stephen C. Harrison, Harvard Medical School/HHMI.

Chapter 44 **44.1** David Wall/Alamy; **44.4** Mark Conlin/Image Quest Marine; **44.5.2** Dr. John Crowe, University of California, Davis; **44.5.3** Dr. John Crowe, University of California, Davis; **44.9** AFP/Getty Images; **44.14.5** Steve Gschmeissner/Photo Researchers; **44.17** John Cancalosi/Peter Arnold Images/PhotoLibrary; **44.18** Michael Fodgen/OSF/PhotoLibrary.

Chapter 45 **45.1** Ralph A. Clevenger/Corbis; **p. 974** Stuart Wilson/Photo Researchers; **45.3** Volker Witte; **45.18.2** Astier/BSIP/Photo Researchers; **45.19.2** Photoshot Holdings Ltd/Alamy; **45.19.3** Jurgen & Christine Sohns/FLPA.

Chapter 46 **46.1** David Doubilet/Getty Images; **46.2** David Wrobel; **46.4** Chris Wallace Photography, photographersdirect.com; **46.5.2** P. de Vries, courtesy of David Crews; **46.6** Andy Sands/naturepl.com; **46.7** John Cancalosi/Peter Arnold/PhotoLibrary; **46.17.2** Photo Lennart Nilsson/Scanpix; **46.17.3** Photo Lennart Nilsson/Scanpix; **46.17.4** Photo Lennart Nilsson/Scanpix.

Chapter 47 **47.1** Photo Lennart Nilsson/Scanpix; **47.4.2** Vacquier VD, Payne JE. Methods for quantitating sea urchin sperm-egg binding. *Exp Cell Res.* 1973 Nov;82(1):227–35; **47.4.3** Vacquier VD, Payne JE. Methods for quantitating sea urchin sperm-egg binding. *Exp Cell Res.* 1973 Nov;82(1):227–35; **47.4.4** Vacquier VD, Payne JE. Methods for quantitating sea urchin sperm-egg binding. *Exp Cell Res.* 1973 Nov;82(1):227–35; **47.4.5** Vacquier VD, Payne JE. Methods for quantitating sea urchin sperm-egg binding. *Exp Cell Res.* 1973 Nov;82(1):227–35; **47.4.7** Hafner, M., Petzelt, C., Nobiling, R., Pawley, J., Kramp, D. and G. Schatten. Wave of Free Calcium at Fertilization in the Sea Urchin Egg Visualized with Fura-2. *Cell Motil. Cytoskel.*, 9:271–277 (1988); **47.4.8** Hafner, M., Petzelt, C., Nobiling, R., Pawley, J., Kramp, D. and G. Schatten. Wave of Free Calcium at Fertilization in the Sea Urchin Egg Visualized with Fura-2. *Cell Motil. Cytoskel.*, 9:271–277 (1988); **47.4.9** Hafner, M., Petzelt, C., Nobiling, R., Pawley, J., Kramp, D. and G. Schatten. Wave of Free Calcium at Fertilization in the Sea Urchin Egg Visualized with Fura-2. *Cell Motil. Cytoskel.*, 9:271–277 (1988); **47.4.10** Hafner, M., Petzelt, C., Nobiling, R., Pawley, J., Kramp, D. and G. Schatten. Wave of Free Calcium at Fertilization in the Sea Urchin Egg Visualized with Fura-2. *Cell Motil. Cytoskel.*, 9:271–277 (1988); **47.6.2** George von Dassow; **47.6.3** George von Dassow; **47.6.4** George von Dassow; **47.6.5** George von Dassow; **47.7.2** Jürgen Berger/Max Planck Institute for Developmental Biology, Tübingen, Germany; **47.7.3** Andrew J. Ewald, Johns Hopkins Medical School; **47.9.2** Charles A. Ettensohn; **47.13.2** Huw Williams; **47.13.3** Thomas Poole, SUNY Health Science Center; **47.14.2** Dr. Keith Wheeler/Photo Researchers; **47.17.2** Hiroki Nishida, *Developmental Biology* 121 (1987): 526. Reprinted by permission of Academic Press; **47.17.3** Hiroki Nishida, *Developmental Biology* 121 (1987): 526. Reprinted by permission of Academic Press; **47.18.2** J. E. Sulston and H. R. Horvitz, *Dev. Biol.* 56 (1977):110–156; **47.19.2** Adapted from Strome (International Review of Cytology 114: 81–123, 1989); **47.20.2** Adapted from Strome (International Review of Cytology 114: 81–123, 1989); **47.20.3** Adapted from Strome (International Review of Cytology 114: 81–123, 1989); **47.20.4** Adapted from Strome (International Review of Cytology 114: 81–123, 1989); **47.20.5** Adapted from Strome (International Review of Cytology 114: 81–123, 1989); **47.24.2** Kathryn W. Tosney, University of Michigan; **47.25.2** Dennis Summerbell.

Chapter 48 **48.1** Marinethemes.com; **48.2.2** David Fleetham/Alamy; **48.6.2** Thomas Deerinck; **48.13.2** Bear, Connors, and Paradiso, "Neuroscience: Exploring the Brain" © 1996, p. 43; **48.16.2** Edwin R. Lewis, University of California at Berkeley.

Chapter 49 **49.1** Brainbow mouse cerebellum. Image by Tamily Weissman, Harvard University. The Brainbow mouse was produced by Livet J, Weissman TA, Kang H, Draft RW, Lu J, Bennis RA, Sanes JR, Lichtman JW. *Nature* (2007) 450:56–62; **49.6.2** N. Kedersha/Photo Researchers; **49.9** Larry Mulvehill/Corbis; **49.14.2** From "A functional MRI study of happy and sad affective states induced by classical music."; . . . M. T. Mitterschiffthaler et al. *Hum Brain Mapp.* 2007 Nov;28(11):1150–62; **49.16.2** Marcus E. Raichle, M.D., Washington University Medical Center; **p. 1075** From "Dr. Harlow's Case of Recovery from the passage of an Iron Bar through the Head." H. Bigelow. *Am. Journal of the Med. Sci.* July 1850;XXXIX. Images from the History of Medicine (NLM); **49.21** Image by Sebastian Jessberger. Fred H. Gage, Laboratory of Genetics LOG-G, The Salk Institute for Biological Studies; **49.24.2** Martin M. Rotker/ Photo Researchers.

Chapter 50 **50.1** Kenneth Catania; **50.6.1** CSIRO; **50.6.2** R. A. Steinbrecht; **50.7.2** James Gerholdt/Photolibrary; **50.7.3** Splashdown Direct/ OSF/PhotoLibrary; **50.9.2** From Richard Elzinga, "Fundamentals of Entomology" 3rd. © 1987, p. 185. Reprinted by permission of Prentice-Hall, Upper Saddle River, NJ; **50.10.2** SPL/Photo Researchers; **50.16.2** USDA/APHIS Animal and Plant Health Inspection Service; **50.17.3** STEVE GSCHMEISSNER/SPL/Photo Researchers; **50.21** From "Gene therapy for red-green colour blindness in adult primates."; . . . K. Mancuso et al. *Nature.* 2009 Oct 8; 461(7265):784–7. Photo: Neitz Laboratory; **50.26.2** Clara Franzini-Armstrong, University of Pennsylvania; **50.27.2** Courtesy of Dr. H. E. Huxley; **50.27.3** Courtesy of Dr. H. E. Huxley; **50.27.4** Courtesy of Dr. H. E. Huxley; **50.33** George Cathcart Photography, photographersdirect.com; **50.38** Dave Watts/NHPA/Photo Researchers; **50.39** Vance A. Tucker.

Chapter 51 **51.1** Michael Nichols/National Geographic/Getty Images; **51.3** Susan Lee Powell; **51.5.2** Kenneth Lorenzen, UC Davis; **51.7.2** Thomas McAvoy/Life Magazine/Getty Images; **51.7.3** Operation Migration Inc.; **51.9.2** Lincoln Brower, Sweet Briar College; **51.9.3** Lincoln Brower, Sweet Briar College; **51.9.4** Lincoln Brower, Sweet Briar College; **51.11** Clive Bromhall/OSF/PhotoLibrary; **51.12.2** Alissa Crandall/Corbis; **51.12.3** Richard Wrangham; **51.15.2** Matt T. Lee; **51.15.3** David Osborn/Alamy; **51.15.4** Bill Schmoker; **51.16.2** James D Watt/Image Quest Marine; **51.17** Courtesy of Gerald S. Wilkinson; from G.S. Wilkinson and G. N. Dodson, in J. Choe and B. Crespi (eds)., "The Evolution of Mating Systems in Insects and Arachnids," Cambridge University Press, Cambridge (1997), pp. 310–328; **51.18** Cyril Laubscher/Dorling Kindersley; **51.21** Martin Harvey/Peter Arnold Images/ PhotoLibrary; **51.22** Erik Svensson, Lund University, Sweden; **51.23.2** Robert Pickett/ Corbis; **51.24** Lowell L. Getz and Lisa Davis; **51.25** Rory Doolin; **51.27** Jennifer Jarvis; **51.29.2.** Stephen J. Krasemann/Peter Arnold/PhotoLibrary.

Chapter 52 **52.1** Dr. Paul A. Zahl/Photo Researchers; **52.2.2** James D. Watt/ Stephen Frink Collection/Alamy; **52.2.3** Gianni Tortoli/Photo Researchers; **52.2.4** Tom Bean/Corbis; **52.2.5** B. Tharp/Photo Researchers; **52.2.6** Yann Arthus-Bertrand/ Corbis; **52.2.7** NASA/Goddard Space Flight Center; **52.11.1** JTB Photo Communications, Inc./Alamy; **52.11.2** imagebroker/Alamy; **52.12.2** Frans Lanting/Corbis; **52.12.4** Gordon Whitten/Corbis; **52.12.6** Wolfgang Kaehler/Corbis; **52.12.8** The California Chaparral Institute (http://californiachaparral.net). Photo supplied by Richard Halsey; **52.12.10** Tom Bean/Image Bank/ Getty Images; **52.12.12** Shutterstock; **52.12.14** Kennan Ward/Corbis; **52.12.16** Darrell Gulin/Corbis; **52.16.1** Allen Russell/PhotoLibrary; **52.16.2** AfriPics.com/Alamy; **52.16.3** David Tipling/ Nature Picture Library; **52.16.4** Ron Watts/Corbis; **52.16.5** Photononstop/SuperStock; **52.16.6** James Randklev/Image Bank/Getty Images; **52.16.7** Stuart Westmorland/ Corbis; **52.16.8** Stuart Westmorland/Corbis; **52.16.9** Digital Vision/Getty Images; **52.16.10** William Lange/Woods Hole Oceanographic Institution; **52.17.2** Geoff Dann/ Dorling Kindersley; **52.19.2** Peter Llewellyn/Alamy; **52.21** Daniel Mosquin.

Chapter 53 **53.1** Arpat Ozgul; **53.2** Todd Pusser/Naturepl.com; **53.4.2** Bernard Castelein/ Nature Picture Library/Alamy; **53.4.3** Frans Lanting/Corbis; **53.4.4** Niall Benvie/Corbis; **53.8.2** Hansjoerg Richter/iStockphoto; **53.11** Photodisc/White/PhotoLibrary; **53.12** Tom Bean/Corbis; **53.13.2** H. Willcox/Wildlife Picture/Peter Arnold/Photolibrary; **53.14.2** Jean Louis Batt/Taxi/Getty Images; **53.14.3** Christine Osborne/ Corbis; **53.14.4** Edward Parker/Alamy; **53.17.2** fotoVoyager/iStockphoto; **53.17.3** Adrian Bailey/Aurora Photos; **53.17.4** Joe Raedle/Getty Images; **53.17.5** Patrick Clayton, www.fisheyeguyphotography.com; **53.17.6** JOZSEF SZENTPETERI/NGS Image Collection; **53.17.7** Andrew Syred/Photo Researchers; **53.17.8** Nicholas Bergkessel, Jr./ Photo Researchers; **53.19.2** Joe McDonald/Corbis; **53.20.2** Robert Kay; **53.21.2** Niclas Fritzén.

Chapter 54 **54.1** Hal Beral VWPics/SuperStock; **54.2.2** Joseph T. Collins/Photo Researchers; **54.2.3** National Museum of Natural History/Smithsonian Institution; **p. 1196** Frank W Lane/FLPA; **54.5.2** Barry Mansell/Nature Picture Library; **54.5.3** Fogden/Corbis; **54.5.4** Stephen J. Krasemann/Photo Researchers; **54.5.5** Robert Pickett/ Papilio/Alamy; **54.5.6** Edward S. Ross, California Academy of Sciences; **54.5.7** © James K. Lindsey; **54.6** Douglas Faulkner/Photo Researchers; **54.7.2** Fogden/ Corbis; **54.7.3** Dan Janzen, Department of Biology, University of Pennsylvania; **54.8** Peter Johnson/Corbis; **54.9.2** Sally D. Hacker; **54.12** Cedar Creek Ecosystem Science Reserve, University of Minnesota; **54.17.2** Genny Anderson; **54.19** SuperStock; **54.21.2** Ron Landis Photography, www.ronlandisphotography.co; **54.21.3** Scott T. Smith/Corbis; **54.22.2** Charles Mauzy/Corbis; **54.22.3** Keith Boggs; **54.22.4** Glacier Bay National Park Photo/Glacier Bay National Park and Preserve; **54.22.5** Terry Donnelly, Mary Liz Austin; **54.24.2** R. Grant Gilmore, Dynamac Corporation; **54.24.3** Lance Horn, National Undersea Research Center, University of North Carolina-Wilmington; **54.29** Nelish Pradhan, Bates College, Lewiston, ME; **54.30** Josh Spice.

Chapter 55 **55.1** Hassan Basagic; **55.2** Stone Nature Photography/Alamy; **55.3** Justus de Cuveland/AGE Fotostock; **55.16.2** Hubbard Brook Research Foundation/ USDA Forest Service; **55.16.3** USDA Forest Service; **55.17.2** Mark Gallagher;

55.17.3 Mark Gallagher; **55.18.2** U.S. Department of Energy; **55.19.2** Stewart Rood, University of Lethbridge; **55.19.3** Photo provided by Kissimmee Division staff, South Florida Water Management District (WPB); **55.19.4** Tim Day, Xcluder Pest Proof Fencing Company; **55.19.5** Daniel H. Janzen, University of Pennsylvania; **55.19.6** Bert Boekhoven; **55.19.7** Jean Hall/Holt Studios/Photo Researchers; **55.19.8** Kenji Morita/Environment Division, Tokyo Kyuei Co., Ltd.

Chapter 56 **56.1** Stephen J Richards; **56.2** Wayne Lawler/Ecoscene/Corbis; **56.4.2** Neil Lucas/Nature Picture Library; **56.4.3** Mark Carwardine/Still Pictures/Peter Arnold/PhotoLibrary; **56.4.4** Nazir Foead; **56.5** Merlin D. Tuttle, Bat Conservation International, www.batcon.org; **56.6** Scott Camazine/Photo Researchers; **56.7** Michael Edwards/Getty Images; **56.8.2** Bruce Cowell, www.brucecowellphotographer.com; **56.8.3** Robert Ginn/PhotoEdit Inc. **56.9** Benezeth Mutayoba, photo provided by the University of Washington; **56.10** Richard Vogel/Liaison/Getty Images; **56.13.2** William Ervin/SPL/Photo Researchers; **56.14** Craighead Environmental Research Institute; **56.15.2** Tim Thompson/Corbis; **56.15.3** Chuck Bargeron, University of Georgia; **56.15.4** William D. Boyer, USDA Forest Service; **56.16.2** Yann Arthus-Bertrand/Corbis; **56.16.3** James P. Blair/National Geographic Image Collection; **56.17** R. O. Bierregaard, Jr., Biology Dept., University of North Carolina, Charlotte; **56.18** SPL/Photo Researchers; **56.21.2** Edwin Giesbers/naturepl.com; **56.22.2** Mark Chiappone and Steven Miller, Center for Marine Science, University of North Carolina-Wilmington, Key Largo, FL; **56.23** Nigel Cattlin/Photo Researchers; **56.24.2** NASA; **56.24.3** NASA; **56.26** Erich Hartmann/Magnum Photos; **56.28** Prof. William H. Schlesinger; **56.31.2** NASA; **56.31.3** NASA; **56.33.2** Serge de Sazo/Photo Researchers; **56.33.3** Hilde Jensen, University of Tubingen/Nature Magazine/AP Photo; **56.33.4** Gabriel Rojo/Nature Picture Library; **56.33.5** Titus Lacoste/Getty Images. **Appendix A** **p. A-5** OMIKRON/Science Source/Photo Researchers; **p. A-8** John N. A. Lott/Biological Photo Service; **p. A-10, top** Biophoto/Photo Researchers; **p. A-10, 2nd from top** Conly L. Rieder; **p. A-11** USDA/ARS/Agricultural Research Service; **p. A-35** Peter Kitin; **Appendix E** **p. E-1, top left** Dr. Kari Lounatmaa/Photo Researchers; **p. E-1, bottom left** Eye of Science/Photo Researchers; **p. E-1, middle** M. I. Walker/NHPA/Photoshot; **p. E-1 right** Kathy Piper/iStockphoto; **p. E-2, left** Douglas Adams/iStockphoto; **p. E-2, right** McPHOTO/AGE Fotostock.

Illustration and Text Credits

4.6b, 9.9, 17.17b and **c** are adapted from C. K. Matthews and K. E. van Holde, *Biochemistry*, 2nd ed. Copyright © 1996 Pearson Education, Inc., publishing as Pearson Benjamin Cummings. **4.7, 6.6b, 11.7, 11.12, 17.11, 18.25, 20.8, 21.9, and 21.10** are adapted from W. M. Becker, J. B. Reece, and M. F. Poenie, *The World of the Cell*, 3rd ed. Copyright © 1996 Pearson Education, Inc., publishing as Pearson Benjamin Cummings. **Table 6.1a** Adapted from W. M. Becker, L. J. Kleinsmith, and J. Hardin, *The World of the Cell*, 4th ed. p. 753. Copyright © 2000 Pearson Education, Inc., publishing as Pearson Benjamin Cummings. **6.8** and **6.23a** and cell organelle drawings in **6.11** and **6.12** are adapted from illustrations by Tomo Narashima in E. N. Marieb, *Human Anatomy and Physiology*, 5th ed. Copyright © 2001 Pearson Education, Inc., publishing as Pearson Benjamin Cummings. **6.9a, 50.12,** and **50.13** are also from *Human Anatomy and Physiology*, 5th ed. Copyright © 2001 Pearson Education, Inc., publishing as Pearson Benjamin Cummings. **30.4, 30.13i,** and **39.13** are adapted from M. W. Nabors, *Introduction to Botany*, Copyright © 2004 Pearson Education, Inc., publishing as Pearson Benjamin Cummings. **42.30a, 46.16, 49.8, 49.10, 50.26, and 50.30** are adapted from E. N. Marieb, *Human Anatomy and Physiology*, 4th ed. Copyright © 1998 Pearson Education, Inc., publishing as Pearson Benjamin Cummings. **42.30a** From Campbell et al., *Biology: Concepts and Connections*, 6th ed., fig. 22.10, p. 462. Copyright © 2009 Pearson Education, Inc., publishing as Pearson Benjamin Cummings. **43.8** Adapted from Gerard J. Tortora, Berdell R. Funke, and Christine L. Case. 1998. *Microbiology: An Introduction*, 6th ed. Copyright © 1998 Pearson Education, Inc., publishing as Pearson Benjamin Cummings. **44.8** and **51.8** are adapted from L. G. Mitchell, J. A. Mutchmor, and W. D. Dolphin. *Zoology*. Copyright © 1988 Pearson Education, Inc., publishing as Pearson Benjamin Cummings.

Chapter 1 **1.25** Map provided courtesy of David W. Pfennig, University of North Carolina at Chapel Hill; **1.27** Data in bar graph based on D. W. Pfennig et al. 2001. Frequency-dependent Batesian mimicry. *Nature 410*:323.

Chapter 2 **2.2 (bottom)** Reprinted by permission of Macmillan Publishers Ltd. *Nature*. M.E. Frederickson et al. 'Devil's gardens' bedeviled by ants, *437*:495, 9/22/05. Copyright © 2005.

Chapter 3 **3.8a** Adapted from *Scientific American*, Nov. 1998, p.102.

Chapter 5 **5.12** Adapted from *Biology: The Science of Life*, 3rd ed. by Robert Wallace et al. Copyright © 1991. Reprinted by permission of Pearson Education, Inc; **5.15 and 5.20F** PDB ID 1CGD: J. Bella, B. Brodsky, and H. M. Berman. 1995. Hydration structure of a collagen peptide. *Structure 3*: 893–906; **5.18** Adapted from D. W. Heinz et al. 1993. How amino-acid insertions are allowed in an alpha-helix of T4 lysozyme. *Nature 361*:561; **5.20D** PDB ID 3GS0: Palaninathan, S.K., Mohamedmohaideen, N.N., Orlandini, E., Ortore, G., Nencetti, S., Lapucci, A., Rossello, A., Freundlich, J.S., Sacchettini, J.C. 2009. Novel transthyretin amyloid fibril formation inhibitors: synthesis, biological evaluation, and X-ray structural analysis. *Public Library of Science One 4*: e6290–e6290; **5.20G, 21.10b, Un42.1** PDB ID 2HHB: G. Fermi, M. F. Perutz, B. Shaanan, R. Fourme. 1984. The crystal structure of human deoxyhaemoglobin at 1.74 A resolution. *J.Mol.Biol. 175*: 159–174.

Chapter 7 **7.9** PDB ID 3HAO: N. H. Joh, A. Oberai, D. Yang, J. P. Whitelegge, J. U. Bowie. 2009. Similar energetic contributions of packing in the core of membrane and water-soluble proteins. *J. Am.Chem.Soc. 131*: 10846–10847.

Chapter 8 **8.18** PDB ID 3e1f: D H. Juers, B. Rob, M. L. Dugdale, N. Rahimzadeh, C. Giang, M. Lee, B. W. Matthews, R.E. Huber. 2009. Direct and indirect roles of His-418 in metal binding and in the activity of beta-galactosidase (E. coli). *Protein Sci. 18:* 1281–1292; **8.19** PDB ID 1MDY: P. C. Ma, M.A. Rould, H. Weintraub, C.O. Pabo. 1994. Crystal structure of MyoD bHLH domain-DNA complex: perspectives on DNA recognition and implications for transcriptional activation. *Cell* (Cambridge, Mass.) *77*: 451–459; **8.20** Figures 4a and 4e from "A common allosteric site and mechanism in caspases" by J.M. Scheer et al., in *Proceedings of the National Academy of Sciences, 103*, no. 20: 7595–7600, May 16, 2006. Copyright © 2006 National Academy of Sciences, U.S.A. Used by permission.

Chapter 9 **9.5** From *Molecular Biology of the Cell,* 4th edition, by Bruce Alberts et al., fig. 2.69, p. 92. Copyright © 2002 by Bruce Alberts, Alexander Johnson, Julian Lewis, Martin Raff, Keith Roberts, and Peter Walter. Used by permission

Chapter 10 **10.13b** Figure 1a from "Architecture of the photosynthetic Oxygen-evolving center" by K.N. Ferreira et al., in *Science, 303,* No. 5665: 1831–1838, March 19, 2004. Copyright © 2004, The American Association for the Advancement of Science. Reprinted with permission from AAAS; **10.15** Adapted from Richard and David Walker. Energy, Plants, and Man, fig. 4.1, p. 69. Sheffield: University of Sheffield. Oxygraphics http://www.oxygraphics.co.uk. © Richard Walker. Used with permission.

Chapter 12 **12.13** From *Molecular Biology of the Cell,* 4th edition, by Bruce Alberts et al., fig. 18.41, p. 1059. Copyright © 2002 by Bruce Alberts, Alexander Johnson, Julian Lewis, Martin Raff, Keith Roberts, and Peter Walter. Used by permission.

Chapter 17 **17.13** Figure 10-45 from *Principles of Cell and Molecular Biology,* 2nd edition, by Valerie M. Kish and Lewis J. Kleinsmith. Copyright © 1995 HarperCollins College Publishers. Reprinted by permission of Pearson Education.

Chapter 18 **18.15a** Figure 1d from "An Abundant Class of Tiny RNAs with Probable Regulatory Roles in *Caenorhabditis elegans*" by N.C. Lau et al., in *Science, 294,* No. 5543: 858–862, Oct. 1, 2001. Copyright © 2001, The American Association for the Advancement of Science. Reprinted with permission from AAAS.

Chapter 20 **20.10** Figure 15.24, p. 481, from *Genetics*, 5th ed., by Peter Russell. Copyright © 1998 Pearson Education, Inc., publishing as Pearson Benjamin Cummings. Used by permission of the publisher.

Chapter 21 **21.2** Adapted from a figure by Chris A. Kaiser and Erica Beade; **21.5b** Figure 2B from "The Genetic Landscape of a Cell" by M. Costanzo et al., in *Science, 327,* No. 5964: 425–431, Jan. 22, 2010. Copyright © 2010, The American Association for the Advancement of Science. Reprinted with permission from AAAS; **21.17** Adapted from an illustration by William McGinnis in Peter Radetsky, "The homeobox: Something very precious that we share with flies, from egg to adult." Bethesda, MD: Howard Hughes Medical Institute, 1992, p. 92. Reprinted by permission of William McGinnis; **21.18** Adapted from M. Akam, *"Hox* genes and the evolution of diverse body plans," *Philosophical Transactions of the Royal Society B*, 1995, *349:* 313–319, fig. 3. © Royal Society of London. Reprinted by permission.

Chapter 22 **22.8** © Utako Kikutani 2007. Used with permission; **22.13** Figure 4a from "Host Race Radiation in the Soapberry Bug: Natural History with the History" by S.P. Carroll and C. Boyd, in Evolution, Vol. 46 No. 4, p. 1060, Aug. 1992. Reproduced with permission of Blackwell Publishing Ltd.; **22.14** Reprinted from *Trends in Microbiology, 16,* Issue 8: 361–369. B.A. Diep and M. Otto, "The role of virulence determinants in community-associated MRSA pathogenesis," Copyright © 2008, with permission from Elsevier.

Chapter 23 **23.5** Figure 3 from "Genetic mechanisms for adapting to a changing environment" by D.A. Powers et al., from *Annual Review of Genetics, 25,* Dec. 1991. Copyright © 1991 by Annual Reviews. Reprinted by permission; **23.11** Figure 20.6 (maps only) from *Discover Biology,* Second Edition by Michael L. Cain, Hans Damman, Robert A. Lue & Carol Kaesuk Loon, Editors. Copyright © 2002 by Sinauer Associates, Inc. Used by permission of W.W. Norton & Company, Inc.; **23.12** Reprinted by permission from Macmillan Publishers Ltd.: *Nature*. E. Postma and A. J. van Noordijk, Gene flow maintains a large genetic difference in clutch size at a small spatial scale, *433,* 1/6/05. Copyright © 2005; **23.14** Adapted from many sources including D. J. Futuyma. 2005. *Evolution*, fig. 11.3. Sunderland, MA: Sinauer Associates and from R. L. Carroll, 1988. *Vertebrate Paleontology and Evolution.* W.H. Freeman & Co.; **23.16** Adapted from A. M. Welch et al. 1998. Call duration as an indicator of genetic quality in male gray tree frogs. *Science* 280:1928–1930; **23.17** Adapted from A. C. Allison. 1961. Abnormal hemoglobin and erythrocyte enzyme-deficiency traits. In *Genetic Variation in Human Populations,* ed. G.A. Harrison. Oxford: Elsevier Science and from S. I. Hay et al., A world malaria map: *Plasmodium falciparum* endemicity in 2007. *PLoS Medicine 6:* fig. 3, p. 291; **23.18** Figure 2a from "Frequency-Dependent Natural Selection in the Handedness of Scale-Eating Cichlid Fish" by Michio Hori in *Science, 260,* No. 5105: 216–219, April 9, 1993. Copyright © 1993, The American Association for the Advancement of Science. Reprinted with permission from AAAS; **Un 23.2** Data from R. K. Koehn and T. J. Hilbish. 1987. The adaptive importance of genetic variation. *American Scientist* 75: 134–141.

Chapter 24 **24.7** Figure 3 from "Ecological Speciation in *Gambusia* Fishes" by R.B. Langerhans et al., from *Evolution, 61,* No. 9, July 2007, published by The Society for the Study of Evolution. Copyright © 2007 R.B. Langerhans, M.E. Gifford, E.O. Joseph. Reprinted by permission; **24.9** From figure 2 in "Correspondence between sexual isolation and allozyme differentiation" *Proceedings of the National Academy of Science, 87:* 2715–2719, 1990, p. 2718. Copyright © 1990 Stephen G. Tilley, Paul A. Verrell, Steven J. Arnold. Used with permission; **24.10a** Adapted from D. M. B. Dodd, 1989. Reproductive isolation as a consequence of adaptive divergence in *Drosophila pseudoobscura. Evolution* 43: 1308–1311; **24.13** *Hybrid Zone and the Evolutionary Process* edited by R.G. Harrison (1993): Map of Bombina hybrid zone (p. 263) and figure 10.1 (p. 278) from chapter "Analysis of hybrid zones with bombina" by J. M. Szymura. By permission of Oxford University Press; **24.15** Reprinted by permission of Macmillan Publishers Ltd.: *Nature*. G. P. Saetre et al. A sexually selected character displacement in flycatchers reinforces premating isolation, *387:*589–591, fig. 2, 6/5/97. Copyright © 1997; **24.19b** From fig. 2 in L. H. Rieseberg et al. 1996. Role of Gene Interactions in Hybrid Speciation: Evidence from Ancient and Experimental Hybrids. *Science* 272: 741–745. Copyright © 1996. Reprinted with permission from AAAS.

Chapter 25 **25.3a** Graph from "Experimental Models of Primitive Cellular Compartments Encapsulation Growth and Division" by M.M. Hanczyc et al., in *Science, 302,* No. 5645: 618–622. Oct. 24, 2003. Copyright © 2003, The American Association for the Advancement of Science. Reprinted with permission from AAAS; **25.5** From Don L. Eicher, *Geologic Time,* 1st edition, © 1968. Printed and Electronically reproduced by permission of Pearson Education, Inc., Upper Saddle River, New Jersey; **25.6a–d** Adapted from many sources including D. J. Futuyma. 2005. *Evolution*, fig. 4.10. Sunderland, MA: Sinauer Associates and from R. L. Carroll, 1988. *Vertebrate*

Paleontology and Evolution. W.H. Freeman & Co.; **25.6e** Adapted from Luo et al. 2001. A new mammalia form from the Early Jurassic and evolution of mammalian characteristics. *Science* 292: 1535; **25.7** Adapted from D. J. Des Marais. September 8, 2000. When did photosynthesis emerge on Earth? *Science* 289: 1703–1705; **25.8** Reprinted by permission from Macmillan Publishers Ltd.: *Nature.* L. R. Kump. The rise of atmospheric oxygen, *451:*277–278, 1/17/08. Copyright © 2008; **25.13** Map adapted from http://gelology.er.usgs.gov/eastern/plates.html; **25.15** Graph created from D. M. Raup and J. J. Sepkoski, Jr. 1982. Mass extinctions in the marine fossil record. *Science.* 215: 1501–1503 and J. J. Sepkoski, Jr. 1984. A kinetic model of Phanerozoic taxonomic diversity. III. Post-Paleozoic families and mass extinctions. *Paleobiology.* Vol 10, No. 2, pp. 246–267 in D. J. Futuyma, fig. 7.3a, p. 143 and fig. 7.6, p. 145, Sunderland, MA: Sinauer Associates; **25.17** From Mayhew, P. J. et al. 2008. A long-term association between global temperature and biodiversity, origination and extinction in the fossil record. *Proceedings of the Royal Society B, 275:* 47–53, fig. 3b. Reprinted by permission; **25.18** Figure 3 from "Anatomical and ecological constraints on Phanerozoic animal diversity in the marine realm" by R.K. Bambach et al., in *Proceedings of the National Academy of Sciences, 99,* no. 10: 6854–6859, May 14, 2002. Copyright © 2002 National Academy of Sciences, U.S.A. Used by permission; **25.19** Adapted from Hickman, Roberts, and Larson.1997, *Zoology,* 10th ed, Wm. C. Brown, fig. 31.1; **25.24** Reprinted by permission from Macmillan Publishers Ltd: *Nature.* M. Ronshaugen et al. Hox protein mutation and macroevolution of the insect body plan, *415:*914–917, fig. 1a. Copyright © 2002; **25.26** Adapted from M. Strickberger, 1990. *Evolution,* Boston: Jones & Bartlett.

Chapter 26 **26.6** Figure 1 from "Which Whales Are Hunted? A Molecular Genetic Approach to Monitoring Whaling" by C.S. Baker and S.R. Palumbi, in *Science, 265,* No. 5178:1538–1539, Sep. 9, 1994. Copyright © 1994, The American Association for the Advancement of Science. Reprinted with permission from AAAS; **26.12** With kind permission from Springer Science+Business Media: *Development Genes and Evolution,* "The evolution of the hedgehog gene family in chordates: insights from amphioxus hedgehog," vol. 209, 1999, pp. 40–47, Jan. 1999, S.M. Shimeld, fig. 3; **26.19** Figure 4.3c, p. 124 from *Molecular Markers, Natural History, and Evolution,* 2nd edition, by John Avise. Copyright © 2004 Sinauer Associates. Used by permission; **26.20** Figure from "Timing the Ancestor of the HIV-1 Pandemic Strains" by B. Korber et al., in *Science, 288,* No. 5472: 1789–1796, Jun 9, 2000. Copyright © 2000, The American Association for the Advancement of Science. Reprinted with permission from AAAS; **26.21** Figure 4.1, p. 45, "The three domains of life" by S.L. Baldauf et al., from *Assembling the Tree of Life,* edited by Joel Cracraft and Michael Donoghue. By permission of Oxford University Press, Inc. **26.22** Adapted from S. Blair Hedges. The origin and evolution of model organisms. *Nature Reviews Genetics 3:*838–848, fig. 1, p. 840.

Chapter 27 **27.10** Reprinted by permission from Macmillan Publishers Ltd.: *Nature.* V.S. Cooper and R.E. Lenski. The population genetics of ecological specialization in evolving *E. coli* populations, *407:*736–739, fig. 1. Copyright © 2000; **27.18** Graph created from data in C. Calvaruso et al. 2006. Root-associated bacteria contribute to mineral weathering and to mineral nutrition in trees: A budgeting analysis. *Applied and Environmental Microbiology* 72: 1258–1266.

Chapter 28 **28.2** Reprinted from *Trends in Genetics,18,* No. 11, J.M. Archibald and P.J. Keeling, "Recycled plastids: a 'green movement' in eukaryotic evolution," Copyright © 2002, with permission from Elsevier; **28.11** Figure 12.7, p. 350, from *Microbiology,* by R. W. Bauman. Copyright © 2004 Pearson Education, Inc., publishing as Benjamin Cummings; **28.23** Data from A. Stechman and T. Cavalier-Smith. 2002. Rooting the eukaryote tree by using a derived gene fusion, *Science* 297: 89–91; **28.28** Reprinted by permission from Macmillan Publishers Ltd.: *Nature.* M.J. Behrenfeld et al. Climate-driven trends in contemporary ocean productivity, *44:* 752–755, fig. 3. Copyright © 2006.

Chapter 29 **29.10** Source: R. D. Bowden. 1991. Inputs, outputs and accumulation of nitrogen in an early successional moss *(Polytrichum)* ecosystem. *Ecological Monographs* 61: 207–223; **29.14** Adapted from Raven et al. *Biology of Plants,* 6th ed., fig. 19.7, W. H. Freeman & Co.

Chapter 30 **30.12a** From P. R. Crane. 1985. Phylogenetic analysis of seed plants and the origin of angiosperms. *Annals of the Missouri Botanical Garden, 72:*716–793, fig. 11a, p. 738. Used by permission of Missouri Botanical Garden Press; **30.12b** Figure 2.3, p. 28, from *Phylogeny and Evolution of Angiosperms* by Douglas E. Soltis et al. Copyright © 2005 Sinauer Associates. Used by permission. **Table 30.1** Adapted from Randy Moore et al., *Botany,* 2nd ed. Dubuque, IA: Brown, 1998, Table 2.2, p. 37.

Chapter 31 **31.21** Figures 4 and 5 from "Fungal endophytes limit pathogen damage in a tropical tree" by A.E. Arnold et al., in *Proceedings of the National Academy of Sciences, 100,* no. 26: 15652–15653, December 23, 2003. Copyright © 2003 National Academy of Sciences, U.S.A. Used by permission; **31.26** Figure 1 from "Reversing introduced species effects: Experimental removal of introduced fish leads to rapid recovery of a declining frog" by Vance T. Vredenburg from *Proceedings of the National Academy of Sciences 101:* 7646–7650. Copyright © 2004 National Academy of Sciences, U.S.A. Used by permission.

Chapter 33 **33.22** C. Lydeard et al., The Global Decline of Nonmarine Mollusks, *BioScience,* Vol. 54, No. 4: 321–330. © 2004, American Institute of Biological Sciences. Used by permission. All rights reserved. (Updated data are from International Union for Conservation of Nature, 2008.); **33.29a** Reprinted from *Current Biology, 7,* Issue 8, J.K. Grenier, S. Carroll et al., "Evolution of the entire arthropod *Hox* gene set predated the origin and radiation of the onychophoran/arthropod clade," p. 551, fig. 2a, Copyright © 1987, with permission from Elsevier.

Chapter 34 **34.8b** From "Fossil sister group of craniates: Predicted and found" by J. Mallatt and J. Chen, from *Journal of Morphology, 258,* Issue 1, May 15, 2003. Copyright © 2003 Wiley-Liss, Inc. Reprinted with permission; **34.12** From *Vertebrates: Comparative Anatomy, Function, Evolution,* 3/e by Kenneth Kardong. © 2002 McGraw-Hill Science/Engineering/Mathematics. Reprinted by permission of The McGraw-Hill Companies, Inc. **34.18 (bottom)** Reprinted by permission of Macmillan Publishers Ltd.: *Nature.* M. Zhu et al. The oldest articulated osteichthyan reveals mosaic gnathostome characters, *458:*469–474. Copyright © 2009; **34.21 (left)** Reproduced by permission of the Royal Society of Edinburgh from Transactions of the Roya Society of Edinburgh: Earth Sciences, volume 87 (1996), pp. 363–421; **34.21 (right)** Reprinted by permission of Macmillan Publishers Ltd.: *Nature.* N.H. Shubin et al. The pectoral fin of *Tiktaalik roseae* and the origin of the tetrapod limb, *440:*768, fig. 4. Copyright © 2006; **34.37a** Adapted from many sources including D. J. Futuyma. 2005. *Evolution,* 1/e, fig. 4.10. Sunderland, MA: Sinauer Associates and from R. L. Carroll, 1988. *Vertebrate Paleontology and Evolution.* W.H. Freeman & Co. **34.47** Drawn from many photos of fossils. Some sources are *O. tugenensis* photo in Michael Balter, Early hominid sows division, *ScienceNow,* Feb. 22, 2001, © 2001 American Association for the Advancement of Science. *A. garhi,* and *H. neanderthalensis* adapted from *The Human Evolution Coloring Book. K platyops* drawn from photo in Meave Leakey et al., New hominid genus from eastern Africa shows diverse middle Pliocene lineages, *Nature,* March 22, 2001, 410: 433. *P. boisei* drawn from a photo by David Bill. *H. ergaster* drawn from a photo at www.inhandmuseum.com. *S. tchadensis* drawn from a photo in Michel Brunet et al., A new hominid from the Upper Miocene of Chad, Central Africa, *Nature,* July 11, 2002, 418: 147, fig. 1b; **34.50 (a/b)** Reprinted by permission of Macmillan Publishers Ltd.: *Nature.* I.V. Ovchinnikov et al. Molecular analysis of Neanderthal DNA from the northern Caucasus, *404:* 492, fig.3a and b. Copyright © 2000.

Chapter 35 **35.21** Figure 2b from "Mongolian Tree Rings and 20th-Century Warming" by G.C. Jacoby et al., in *Science, 273,* No. 5276: 771–773, Aug. 9, 1996. Copyright © 1996, The American Association for the Advancement of Science. Reprinted with permission from AAAS.

Chapter 39 **39.16 (top)** Adapted from M. Wilkins. 1988. *Plant Watching,* Facts of File Publ. **39.28** Figure "No Free Lunch" from "Plant Biology: New fatty acid-based signals: A lesson from the plant world" by Edward Farmer in *Science, 276,* No. 5314: 912–913, May 9, 1997. Copyright © 1997, The American Association for the Advancement of Science. Reprinted with permission from AAAS.

Chapter 40 **40.14** Figure 2 from "Thermoregulation in a brooding female Indian python, Python molurus bivittatus" by V.H. Hutchison et al., in *Science, 151,* No. 3711: 694–695, Feb. 11, 1966. Copyright © 1966, The American Association for the Advancement of Science. Reprinted with permission from AAAS; **40.15** Figure 7 from "Thermoregulation in Endothermic Insects" by Bernd Heinrich, in *Science, 185,* No. 4153: 747–756, August 30, 1974. Copyright © 1974, The American Association for the Advancement of Science. Reprinted with permission from AAAS; **40.21** Adapted from figures 2b and 2c from "The circadian clock stops ticking during deep hibernation in the European hamster" by F.G. Revel et al., in *Proceedings of the National Academy of Sciences, 104,* no. 34: 13816–13820, Aug. 21, 2007. Copyright © 2007 National Academy of Sciences, U.S.A. Used by permission.

Chapter 41 **41.9a** Figure 23.1 from *Human Anatomy and Physiology,* 8e, by Elaine Marieb and Katja Hoehn. Copyright © 2010 Pearson Education, Inc., publishing as Pearson Benjamin Cummings. Used by permission of the publisher; **41.9b** Figure 22-1 from Rhoades, *Human Physiology,* 3e. © 1996 Brooks/Cole, a part of Cengage Learning, Inc. Reproduced by permission. www.cengage.com/permissions; **41.21** Figure "Appetite Controllers" from "Cellular Warriors at the Battle of the Bulge" by Jean Marx, from *Science, 299:* p. 86, Feb. 7, 2003. Illustration by Kathleen Sutliff. Copyright © 2003, The American Association for the Advancement of Science. Reprinted with permission from AAAS.

Chapter 42 **42.20** Reprinted by permission from Macmillan Publishers Ltd.: *Nature.* D.J. Rader and A. Daugherty. Translating molecular discoveries into new therapies for atherosclerosis, *451:*904–913, fig. 1, 2/21/08. Copyright © 2008; **42.21** From J.C. Cohen et al., "Sequence variations in PCSK9, low LDL, and protection against coronary heart disease," *New England Journal of Medicine.* 2006 Mar 23; *354:*1264–72, fig. 1A. Copyright © 2006 Massachusetts Medical Society. Used by permission. All rights reserved; **42.26** Adapted from "Surface properties in relation to atelectasis and hyaline membrane disease" by M.E. Avery and J. Mead, from A.M.A. *American Journal of Diseases of Children 97:*517–523 (June 1959). Copyright © 1959 American Medical Association. Used by permission. All rights reserved.

Chapter 43 **43.5** From figures 2a and 4a in Phoebe Tzou et al., "Constitutive expression of a single antimicrobial peptide can restore wild-type resistance to infection in immuno-deficient *Drosophila* mutants," *PNAS, 99:* 2152–2157. Copyright © 2002 National Academy of Sciences, U.S.A. Used with permission; **43.7** Figures 20.4 and 20.5 from *Human Anatomy and Physiology,* 8e, by Elaine Marieb and Katja Hoehn. Copyright © 2010 Pearson Education, Inc., publishing as Pearson Benjamin Cummings. Used by permission of the publisher.

Chapter 44 **44.6** Kangaroo rat data adapted from K. B. Schmidt-Nielson. 1990. *Animal Physiology: Adaptation and Environment,* 4th ed., p. 339. Cambridge: Cambridge University Press; **44.7a** Adapted from K. B. Schmidt-Nielsen et al. 1958. Extrarenal salt excretion in birds. *American Journal of Physiology* 193: 101–107; **44.14B and 44.15** Figure 25.3b from *Human Anatomy and Physiology,* 8e, by Elaine Marieb and Katja Hoehn. Copyright © 2010 Pearson Education, Inc., publishing as Pearson Benjamin Cummings. Used by permission of the publisher; **44.21** Table 1 from "Requirement of human renal water channel aquaporin-2 for vasopressin-dependent concentration in urine" by P.M. Deen et al., in *Science, 264,* No. 5155: 92–95, April 1, 1994. Copyright © 1994, The American Association for the Advancement of Science. Reprinted with permission from AAAS; **44EOC** From W.S. Beck et al., *Life: An Introduction to Biology,* 3rd edition, copyright © 1991. Reprinted and electronically reproduced by permission of Pearson Education, Inc., Upper Saddle River, New Jersey.

Chapter 46 **46.9** Reprinted by permission of Macmillan Publishers Ltd.: *Nature.* R.R. Snook and D.J. Hosken, "Sperm death and dumping in *Drosophila,*" *428:*939–941, fig. 2. Copyright © 2004.

Chapter 47 **47.16** Figures 1.10 and 8.25 from *Principles of Development* by Lewis Wolpert (1998). By permission of Oxford University Press, Inc. **47.17a** Figure 21-70 from *Molecular Biology of the Cell,* 4th ed, by Bruce Alberts, used with permission from Garland Science - Books, permission conveyed through Copyright Clearance Center, Inc. Adapted for use from "Cell commitment and gene expression in the axolotl embryo" by T.J. Mohun from *Cell 22:* 9–15 (1980), by permission of the author; **47.17b** Reprinted from *Developmental Biology, 121,* Issue 2, Hiroki Nishida, "Cell lineage analysis in ascidian embryos by intracellular injection of a tracer enzyme: III. Up to the tissue restricted stage," p. 526, Copyright © 1987, with permission from Elsevier; **47.18** From *Molecular Biology of the Cell,* 4th edition, by Bruce Alberts et al., fig. 21.17, p. 1172. Copyright © 2002 by Bruce Alberts, Alexander Johnson, Julian Lewis, Martin Raff, Keith Roberts, and Peter Walter. Used by permission; **47.23 (experiment, results left)** Figures 1.10 and 8.25 from *Principles of Development* by Lewis Wolpert (1998). By permission of Oxford University Press, Inc. **47.23 (results, right)** Figure 15.12, p. 604, from *Developmental Biology,* 5th edition, by Gilbert et al. Copyright © 1997 Sinauer Associates. Used by permission; **47.26** Figure 23.1 from *Human Anatomy and Physiology,* 8e, by Elaine Marieb and Katja Hoehn. Copyright © 2010 Pearson Education, Inc., publishing as Pearson Benjamin Cummings. Used by permission of the publisher.

Chapter 48 **48.11** Adapted from G. Matthews, 2003. *Cellular Physiology of Nerve and Muscle,* 4th edition, fig. 6-2d, p. 61. Cambridge, MA: Blackwell Scientific Publications. Reprinted by permission of Wiley Blackwell; **48.18** Table 1 from "Opiate Receptor: Demonstration in nervous tissue" by C.B. Pert and S.H. Snyder in *Science, 179,* No. 4077: 1011–1014, March 9, 1973. Copyright © 1973, The American Association for the Advancement of Science. Reprinted with permission from AAAS.

Chapter 49 **49.11** Adapted from L. M. Mukhametov. 1984. Sleep in marine mammals. In *Sleep Mechanisms,* by A. A. Borbély and J. L. Valatx (eds.). Munich: Springer-Verlag, pp 227–238; **49.12** Figure 2a from "Transplanted suprachiasmatic nucleus determines circadian period" by M.R. Ralph et al., in *Science, 247,* No. 4945: 975–978, Feb. 23, 1990. Copyright © 1990, The American Association for the Advancement of Science. Reprinted with permission from AAAS; **49.18** Adapted from E. D. Jarvis et al. 2005. Avian brains and a new understanding of vertebrate brain evolution. *Nature Reviews Neuroscience* 6: 151–159, fig. 1c; **49.22** Adapted from Gottesman, I. I. and Wolfgram, D. (1991). *Schizophrenia genesis: The origins of madness,* fig. 10, p. 96. New York: Freeman.

Chapter 50 **50.17A** Figure 15.4(a) from *Human Anatomy and Physiology,* 8e, by Elaine Marieb and Katja Hoehn. Copyright © 2010 Pearson Education, Inc., publishing as Pearson Benjamin Cummings. Used by permission of the publisher; **50.17B** Figure 15.15 from *Human Anatomy and Physiology,* 8e, by Elaine Marieb and Katja Hoehn. Copyright © 2010 Pearson Education, Inc., publishing as Pearson Benjamin Cummings. Used by permission of the publisher; **50.23** Reprinted by permission from Macmillan Publishers Ltd.: *Nature.* K.L. Mueller et al. The receptors and coding logic for bitter taste, *434:*225–229, fig. 4b. Copyright © 2005; **50.24a** Figure 15.23(a) and (b) from *Human Anatomy and Physiology,* 8e, by Elaine Marieb and Katja Hoehn. Copyright © 2010 Pearson Education, Inc., publishing as Pearson Benjamin Cummings. Used by permission of the publisher; **50.34** Grasshopper adapted from Hickman et al. 1993. *Integrated Principles of Zoology,* 9th ed., Fig. 22.6, p. 518. New York: McGraw-Hill Higher Education. © 1993 The McGraw-Hill Companies; **50.40** Figure 4 from "Locomotion: Energy Cost of Swimming, Flying, and Running" by K. Schmidt-Nielsen, in *Science, 177,* No. 4045: 222–228, July 21, 1972. Copyright © 1972, The American Association for the Advancement of Science. Reprinted with permission from AAAS.

Chapter 51 **51.2b** Figure 20, p. 28, from *Study of Instinct* by N. Tinbergen (1989). By permission of Oxford University Press, Inc.; **51.4** Reprinted by permission from Macmillan Publishing Ltd.: *Nature Reviews: Genetics:* M.B. Sokolowski. Drosophila: Genetics meets behavior, *2:*881, fig. 1. Copyright © 2001; **51.10** Figure 3a from "Prospective and retrospective learning in honeybees" by M. Biurfa and J. Benard, in *International Journal of Comparative Psychology, 19,* Issue 3: 358–367, 2006; Reprinted with permission; **51.13** Figure 2a from "Evolution of foraging behavior in Drosophila by density-dependent selection" by M.B. Sokolowski et al., in *Proceedings of the National Academy of Sciences, 94,* no. 14: 7373–7377, July 8, 1997. Copyright © 2007 National Academy of Sciences, U.S.A. Used by permission; **51.19** K. Witte and N. Sawka. 2003. Sexual imprinting on a novel trait in the dimorphic zebra finch: sexes differ. *Animal Behaviour* 65: 195–203. Art adapted from http://www.uni-bielefeld.de/biologue/vhf/KW/Forschungsprojekte2.html; **51.23 (left, bottom)** With kind permission from Springer Science+Business Media: *Genetica,* "The inheritance of mating songs in two cryptic, sibling lacewing species," vol. 116, 2002, pp. 269–289, C. S. Henry, fig. 2; **51.26 (top)** Adapted from a photograph by Jonathan Blair in Alcock, 2002. *Animal Behavior,* 7th ed. Sinauer Associates, Inc., Publiishers; **51.26 (bottom)** Reprinted by permission from Macmillan Publishers Ltd.: *Nature.* P. Berthold et al. Rapid microevolution of migratory behaviour in a wild bird species, *360:*668, fig. 1, 12/17/92. Copyright © 1992.

Chapter 52 **52.7** Adapted from L. Roberts. 1989. How fast can trees migrate? *Science* 243: 736, fig. 2. © 1989 by the American Association for the Advancement of Science; **52.8** Reprinted by permission from Macmillan Publishers Ltd.: *Nature.* C. Parmesan et al. Poleward shift of butterfly species' ranges associated with regional warming, *399:*579–583, fig. 3. Copyright © 1999; **52.9** Adapted from Heinrich Walter and Siegmar-Walter Breckle. 2003. *Walter's Vegetation of the Earth,* fig. 16, p. 36. Springer-Verlag, © 2003; **52.17** Figure 1.7, p. 9 from *Kangaroos, Their Ecology and Management in the Sheep Rangelands of Australia,* edited by Graeme Caughley, Neil Shepherd, Jeff Short. © Cambridge University Press 1987. Reprinted with the permission of Cambridge University Press; **52.19** Map adapted from R. L. Smith. 1974. *Ecology and Field Biology,* fig. 11.19, p. 353. Harper and Row Publishers. Map updated from D. A. Sibley. 2000. National Audubon Society *The Sibley Guide to Birds,* Alfred A. Knopf: New York; **52.20** Data from W. J. Fletcher. 1987. Interactions among subtidal Australian sea urchins, gastropods and algae: effects of experimental removals. *Ecological Monographs* 57: 89–109; **Un. 52.2** Data from J. Clausen, D. D. Keck, and W. M. Hiesey. 1948. Experimental studies on the nature of species. III. Environmental responses of climatic races of *Achillea.* Carnegie Institution of Washington Publication 581

Chapter 53 **53.5** Adapted from P. W. Sherman and M. L. Morton. 1984. "Demography of Belding's ground squirrels," *Ecology,* Vol. 65, No. 5, p. 1622, fig. 1a, 1984. Copyright © 1984 Ecological Society of America. Used by permission; **53.15** Adapted from J. T. Enright. 1976. Climate and population regulation: The biogeographer's dilemma. *Oecologia* 24: 295–310; **53.16** Figure 3.5b, page 59, from *Soay Sheep: Dynamics and Selection in an Island Population* edited by T.H. Clutton-Brock and J.M. Pemberton. Copyright © Cambridge University Press 2004. Reprinted with the permission of Cambridge University Press; **53.18** Data courtesy of Rolf O. Peterson, Michigan Technological University; **53.23** Data from U. S. Census Bureau International Data Base; **53.24** Data from U. S. Census Bureau International Data Base; **53.25** Data from U. S. Census Bureau International Data Base 2008; **53.26** Source: Used courtesy of UNEP/GRID-Arendel at http://maps.grida.no/go/graphic/energy_consumption_per_capita_2004; **Tables 53.1** and **53.2** Data from P. W. Sherman and M. L. Morton, "Demography of Belding's ground squirrels," *Ecology,* Vol. 65, No. 5, p. 1622, fig. 1a, 1984. Copyright © 1984 Ecological Society of America.

Chapter 54 **54.2** From "The anoles of La Palma: aspects of their ecological relationships" by A.S. Rand and E. E. Williams in *Breviora 327:* 1–19, 1969. Copyright © 1969 by the President and Fellows of Harvard College. Used with permission from the Museum of Comparative Zoology, Harvard University; **54.9** Data for graph from S. D. Hacker and M. D. Bertness. 1999. Experimental evidence for factors maintaining plant species diversity in a New England salt marsh. *Ecology* 80: 2064–2073; **54.11** Adapted from N. Fierer and R. B. Jackson. 2006. The diversity and biogeography of soil bacterial communities. *Proceedings of the National Academy of Sciences USA 103:* 626–631 fig. 1a; **54.14** Adapted from E. A. Knox. 1970. Antarctic marine ecosystems. In *Antarctic Ecology,* ed. M. W. Holdgate, 69–96. London: Academic Press; **54.15** Adapted from D. L. Breitburg et al. 1997. Varying effects of low dissolved oxygen on trophic interactions in an estuarine food web. *Ecological Monographs* 67: 490. Copyright © 1997 Ecological Society of America; **54.16** Adapted from B. Jenkins. 1992. Productivity, disturbance and food web structure at a local spatial scale in experimental container habitats. *Oikos* 65: 252. Copyright © 1992 Oikos, Sweden; **54.17** Adapted from R. T. Paine. 1966. Food web complexity and species diversity. *American Naturalist* 100: 65–75; **54.18** From J. A. Estes et al. 1998. Killer whale predation on sea otters linking oceanic and nearshore ecosystems. *Science 282:* 474, fig. 1. Copyright © 1998 by the American Association for the Advancement of Science; **54.20** Graph adapted from A. R. Townsend et al. 1997. The intermediate disturbance hypothesis, refugia, and diversity in streams. *Limnology and Oceanography* 42:938–949. **54.22** Adapted from R. L. Crocker and J. Major. 1955. Soil Development in relation to vegetation and surface age at Glacier Bay, Alaska. *Journal of Ecology* 43: 427–448; **54.23** Data from F. S. Chapin, III, et al. 1994. Mechanisms of primary succession following deglaciation at Glacier Bay, Alaska. *Ecological Monographs* 64: 149–175; **54.25** Adapted from D. J. Currie. 1991. Energy and large-scale patterns of animal- and plant-species richness. *American Naturalist* 137: 27–49; **54.26** Adapted from F. W. Preston. 1960. Time and space and the variation of species. *Ecology 41:* 611–627; **54.28** Adapted from F. W. Preston. 1962. The canonical distribution of commonness and rarity. *Ecology 43:*185–215, 410–432.

Chapter 55 **55.4** and **Un 55.1** Adapted from D. L. DeAngelis. 1992. *Dynamics of Nutrient Cycling and Food Webs.* New York: Chapman & Hall; **55.7** Adapted from National Oceanic and Atmospheric Administration's National Data Buoy Center Voluntary Observing Ship Project, www.vos.noaa.gov/MWL/dec_06/Images/OCP_fig4.jpg; **55.8** Figure 2 from "Nitrogen, phosphorus, and eutrophication in the coastal marine environment" by J.H. Ryther, in *Science, 171,* No. 3975: 1008–1013, Mar. 12, 1971. Copyright © 1971, The American Association for the Advancement of Science. Reprinted with permission from AAAS; **55.9** Fig. 4.1 p. 82 from *Communities and Ecosystems,* 1e by Robert H. Whittaker. Copyright © 1970 Robert H. Whittaker. Reprinted by permission of Pearson Education; **55.14** From *The Economy of Nature,* 4th edition. © 1997 by W. H. Freeman and Company. Used with permission; **55.15a** Adapted from J.A. Trofymow et al., *The Canadian Intersite Decomposition Experiment: Project and Site Establishment Report,* Information Report BC-X-378, page 2, Natural Resources Canada, Canadian Forest Service (1998). Reproduced with permission from the Minister of Public Works and Government Services, Canada, 2010; **55.15b** Adapted from: T.R. Moore et al., Litter decomposition rates in Canadian forests, *Global Change Biology* 5: 75–82 (1999), copyright © 2001, 1998 Blackwell Science Ltd. Reproduced with permission from the Minister of Public Works and Government Services, Canada, 2010 and Wiley Blackwell; **55.18b** Data adapted from Wu, W-M et al. 2006. Pilot-scale in situ bioremediation of uranium in a highly contaminated aquifer. 2. Reduction of U(VI) and geochemical control of a U(VI) bioavailability. *Environ Sci. Technol. 40:* 3986–3995, fig. 1D; **Table 55.1** Data from Menzel and Ryther. 1961. *Deep Sea Ranch* 7: 276–281.

Chapter 56 **56.12** Figure 19.1 from *Ecology,* 5e by C.J. Krebs. Copyright © 2001 Pearson Education, Inc. Reprinted by permission; **56.13** Figure 2 from "Tracking the Long-Term Decline and Recovery of an Isolated Population" by R.L. Westemeier et al., in *Science, 282,* No. 5394: 1695–1698, Nov. 27, 1998. Copyright © 1998, The American Association for the Advancement of Science. Reprinted with permission from AAAS; **56.19** Reprinted by permission from Macmillan Publishers, Ltd.: *Nature.* N. Myers et al. Biodiversity hotspots for conservation priorities, *403:* 853–858, fig. 1, 2/24/00. Copyright © 2000; **56.20** Reprinted from W. D. Newmark, 1985. Legal and biotic boundaries of western North American national parks: a problem of congruence. *Biological Conservation* 33: 197–208, fig. 1, p. 199. © 1985, with permission from Elsevier; **56.21a** Map adapted from W. Purves and G. Orians, *Life, The Science of Biology,* 5th ed., fig. 55.23, p. 1239. © 1998 by Sinauer Associates, Inc. Reprinted with permission; **56.27** CO_2 data from www.esrl.noaa.gov/gmd/ccgg/trends. Temperature data from www.giss.nasa.gov/gistemps/graphs/Fig.A.lrg.gif; **56.29** Data from ozonewatch.gsfc.nasa.gov/facts/history/html; **56.32** Data from Instituto Nacional de Estadistica y Censos de Costa Rica and Centro Centroamericano de Poblacion, Universidad de Costa Rica.

Glossary

Pronunciation Key

Pronounce

ā	as in	ace
a/ah		ash
ch		chose
ē		meet
e/eh		bet
g		game
ī		ice
i		hit
ks		box
kw		quick
ng		song
ō		robe
o		ox
oy		boy
s		say
sh		shell
th		thin
ū		boot
u/uh		up
z		zoo

′ = primary accent

′ = secondary accent

5′ cap A modified form of guanine nucleotide added onto the 5′ end of a pre-mRNA molecule.

A site One of a ribosome's three binding sites for tRNA during translation. The A site holds the tRNA carrying the next amino acid to be added to the polypeptide chain. (A stands for aminoacyl tRNA.)

ABC hypothesis A model of flower formation identifying three classes of organ identity genes that direct formation of the four types of floral organs.

abiotic (ā′-bī-ot′-ik) Nonliving; referring to the physical and chemical properties of an environment.

abortion The termination of a pregnancy in progress.

abscisic acid (ABA) (ab-sis′-ik) A plant hormone that slows growth, often antagonizing the actions of growth hormones. Two of its many effects are to promote seed dormancy and facilitate drought tolerance.

absorption The third stage of food processing in animals: the uptake of small nutrient molecules by an organism's body.

absorption spectrum The range of a pigment's ability to absorb various wavelengths of light; also a graph of such a range.

abyssal zone (uh-bis′-ul) The part of the ocean's benthic zone between 2,000 and 6,000 m deep.

acanthodian (ak′-an-thō′-dē-un) Any of a group of ancient jawed aquatic vertebrates from the Silurian and Devonian periods.

accessory fruit A fruit, or assemblage of fruits, in which the fleshy parts are derived largely or entirely from tissues other than the ovary.

acclimatization (uh-klī′-muh-tī-zā′-shun) Physiological adjustment to a change in an environmental factor.

acetyl CoA Acetyl coenzyme A; the entry compound for the citric acid cycle in cellular respiration, formed from a fragment of pyruvate attached to a coenzyme.

acetylcholine (as′-uh-til-kō′-lēn) One of the most common neurotransmitters; functions by binding to receptors and altering the permeability of the postsynaptic membrane to specific ions, either depolarizing or hyperpolarizing the membrane.

acid A substance that increases the hydrogen ion concentration of a solution.

acid precipitation Rain, snow, or fog that is more acidic than pH 5.2.

acoelomate (uh-sē′-lō-māt) A solid-bodied animal lacking a cavity between the gut and outer body wall.

acrosomal reaction (ak′-ruh-sōm′-ul) The discharge of hydrolytic enzymes from the acrosome, a vesicle in the tip of a sperm, when the sperm approaches or contacts an egg.

acrosome (ak′-ruh-sōm) A vesicle in the tip of a sperm containing hydrolytic enzymes and other proteins that help the sperm reach the egg.

actin (ak′-tin) A globular protein that links into chains, two of which twist helically about each other, forming microfilaments (actin filaments) in muscle and other kinds of cells.

action potential An electrical signal that propagates (travels) along the membrane of a neuron or other excitable cell as a nongraded (all-or-none) depolarization.

action spectrum A graph that profiles the relative effectiveness of different wavelengths of radiation in driving a particular process.

activation energy The amount of energy that reactants must absorb before a chemical reaction will start; also called free energy of activation.

activator A protein that binds to DNA and stimulates gene transcription. In prokaryotes, activators bind in or near the promoter; in eukaryotes, activators generally bind to control elements in enhancers.

active immunity Long-lasting immunity conferred by the action of B cells and T cells and the resulting B and T memory cells specific for a pathogen. Active immunity can develop as a result of natural infection or immunization.

active site The specific region of an enzyme that binds the substrate and that forms the pocket in which catalysis occurs.

active transport The movement of a substance across a cell membrane against its concentration or electrochemical gradient, mediated by specific transport proteins and requiring an expenditure of energy.

adaptation Inherited characteristic of an organism that enhances its survival and reproduction in a specific environment.

adaptive immunity A vertebrate-specific defense that is mediated by B lymphocytes (B cells) and T lymphocytes (T cells). It exhibits specificity, memory, and self-nonself recognition. Also called acquired immunity.

adaptive radiation Period of evolutionary change in which groups of organisms form many new species whose adaptations allow them to fill different ecological roles in their communities.

addition rule A rule of probability stating that the probability of any one of two or more mutually exclusive events occurring can be determined by adding their individual probabilities.

adenosine triphosphate *See* ATP (adenosine triphosphate).

adenylyl cyclase (uh-den′-uh-lil) An enzyme that converts ATP to cyclic AMP in response to an extracellular signal.

adhesion The clinging of one substance to another, such as water to plant cell walls by means of hydrogen bonds.

adipose tissue A connective tissue that insulates the body and serves as a fuel reserve; contains fat-storing cells called adipose cells.

adrenal gland (uh-drē′-nul) One of two endocrine glands located adjacent to the kidneys in mammals. Endocrine cells in the outer portion (cortex) respond to adrenocorticotropic hormone (ACTH) by secreting steroid hormones that help maintain homeostasis during long-term stress. Neurosecretory cells in the central portion (medulla) secrete epinephrine and norepinephrine in response to nerve signals triggered by short-term stress.

adrenocorticotropic hormone (ACTH) A tropic hormone that is produced and secreted by the anterior pituitary and that stimulates the production and secretion of steroid hormones by the adrenal cortex.

aerobic respiration A catabolic pathway for organic molecules, using oxygen (O_2) as the final electron acceptor in an electron transport chain and ultimately producing ATP. This is the most efficient catabolic pathway and is carried out in most eukaryotic cells and many prokaryotic organisms.

age structure The relative number of individuals of each age in a population.

aggregate fruit A fruit derived from a single flower that has more than one carpel.

AIDS (acquired immunodeficiency syndrome) The symptoms and signs present during the late stages of HIV infection, defined by a specified reduction in the number of T cells and the appearance of characteristic secondary infections.

alcohol fermentation Glycolysis followed by the reduction of pyruvate to ethyl alcohol, regenerating NAD^+ and releasing carbon dioxide.

aldosterone (al-dos′-tuh-rōn) A steroid hormone that acts on tubules of the kidney to regulate the transport of sodium ions (Na^+) and potassium ions (K^+).

algae A diverse grade of photosynthetic protists, including unicellular and multicellular forms. Algal species are included in three of the five eukaryote supergroups (Chromalveolata, Rhizaria, and Archaeplastida).

alimentary canal (al′-uh-men′-tuh-rē) A complete digestive tract, consisting of a tube running between a mouth and an anus.

allele (uh-lē′-ul) Any of the alternative versions of a gene that may produce distinguishable phenotypic effects.

allergen An antigen that triggers an exaggerated immune response.

allopatric speciation (al′-uh-pat′-rik) The formation of new species in populations that are geographically isolated from one another.

allopolyploid (al′-ō-pol′-ē-ployd) A fertile individual that has more than two chromosome sets as a result of two different species interbreeding and combining their chromosomes.

allosteric regulation The binding of a regulatory molecule to a protein at one site that affects the function of the protein at a different site.

alpha (α) helix (al′-fuh hē′-liks) A coiled region constituting one form of the secondary structure of proteins, arising from a specific pattern of hydrogen bonding between atoms of the polypeptide backbone (not the side chains).

alternation of generations A life cycle in which there is both a multicellular diploid form, the sporophyte, and a multicellular haploid form, the gametophyte; characteristic of plants and some algae.

alternative RNA splicing A type of eukaryotic gene regulation at the RNA-processing level in which different mRNA molecules are produced from the same primary transcript, depending on which RNA segments are treated as exons and which as introns.

altruism (al′-trū-iz-um) Selflessness; behavior that reduces an individual's fitness while increasing the fitness of another individual.

alveolate (al-vē′-uh-let) A protist with membrane-bounded sacs (alveoli) located just under the plasma membrane.

alveolus (al-vē′-uh-lus) (plural, **alveoli**) One of the dead-end air sacs where gas exchange occurs in a mammalian lung.

Alzheimer's disease (alts′-hī-merz) An age-related dementia (mental deterioration) characterized by confusion and memory loss.

amacrine cell (am′-uh-krin) A neuron of the retina that helps integrate information before it is sent to the brain.

amino acid (uh-mēn′-ō) An organic molecule possessing both a carboxyl and an amino group. Amino acids serve as the monomers of polypeptides.

amino group A chemical group consisting of a nitrogen atom bonded to two hydrogen atoms; can act as a base in solution, accepting a hydrogen ion and acquiring a charge of 1+.

aminoacyl-tRNA synthetase An enzyme that joins each amino acid to the appropriate tRNA.

ammonia A small, toxic molecule (NH_3) produced by nitrogen fixation or as a metabolic waste product of protein and nucleic acid metabolism.

ammonite A member of a group of shelled cephalopods that were important marine predators for hundreds of millions of years until their extinction at the end of the Cretaceous period (65.5 million years ago).

amniocentesis (am′-nē-ō-sen-tē′-sis) A technique associated with prenatal diagnosis in which amniotic fluid is obtained by aspiration from a needle inserted into the uterus. The fluid and the fetal cells it contains are analyzed to detect certain genetic and congenital defects in the fetus.

amniote (am′-nē-ōt) Member of a clade of tetrapods named for a key derived character, the amniotic egg, which contains specialized membranes, including the fluid-filled amnion, that protect the embryo. Amniotes include mammals as well as birds and other reptiles.

amniotic egg An egg that contains specialized membranes that function in protection, nourishment, and gas exchange. The amniotic egg was a major evolutionary innovation, allowing embryos to develop on land in a fluid-filled sac, thus reducing the dependence of tetrapods on water for reproduction.

amoeba (uh-mē′-buh) A protist grade characterized by the presence of pseudopodia.

amoebocyte (uh-mē′-buh-sīt′) An amoeba-like cell that moves by pseudopodia and is found in most animals. Depending on the species, it may digest and distribute food, dispose of wastes, form skeletal fibers, fight infections, or change into other cell types.

amoebozoan (uh-mē′-buh-zō′-an) A protist in a clade that includes many species with lobe- or tube-shaped pseudopodia.

amphibian Member of the tetrapod class Amphibia, including salamanders, frogs, and caecilians.

amphipathic (am′-fē-path′-ik) Having both a hydrophilic region and a hydrophobic region.

amplification The strengthening of stimulus energy during transduction.

amygdala (uh-mig′-duh-luh) A structure in the temporal lobe of the vertebrate brain that has a major role in the processing of emotions.

amylase (am′-uh-lās′) An enzyme that hydrolyzes starch (a glucose polymer from plants) and glycogen (a glucose polymer from animals) into smaller polysaccharides and the disaccharide maltose.

anabolic pathway (an′-uh-bol′-ik) A metabolic pathway that consumes energy to synthesize a complex molecule from simpler molecules.

anaerobic respiration (an-er-ō′-bik) A catabolic pathway in which inorganic molecules other than oxygen accept electrons at the "downhill" end of electron transport chains.

analogous Having characteristics that are similar because of convergent evolution, not homology.

analogy (an-al′-uh-jē) Similarity between two species that is due to convergent evolution rather than to descent from a common ancestor with the same trait.

anaphase The fourth stage of mitosis, in which the chromatids of each chromosome have separated and the daughter chromosomes are moving to the poles of the cell.

anatomy The structure of an organism.

anchorage dependence The requirement that a cell must be attached to a substratum in order to initiate cell division.

androgen (an′-drō-jen) Any steroid hormone, such as testosterone, that stimulates the development and maintenance of the male reproductive system and secondary sex characteristics.

aneuploidy (an′-yū-ploy′-dē) A chromosomal aberration in which one or more chromosomes are present in extra copies or are deficient in number.

angiosperm (an′-jē-ō-sperm) A flowering plant, which forms seeds inside a protective chamber called an ovary.

angiotensin II A peptide hormone that stimulates constriction of precapillary arterioles and increases reabsorption of NaCl and water by the proximal tubules of the kidney, increasing blood pressure and volume.

anhydrobiosis (an-hī′-drō-bī-ō′-sis) A dormant state involving loss of almost all body water.

animal pole The point at the end of an egg in the hemisphere where the least yolk is concentrated; opposite of vegetal pole.

anion (an′-ī-on) A negatively charged ion.

anterior Pertaining to the front, or head, of a bilaterally symmetrical animal.

anterior pituitary A portion of the pituitary that develops from nonneural tissue; consists of endocrine cells that synthesize and secrete several tropic and nontropic hormones.

anther In an angiosperm, the terminal pollen sac of a stamen, where pollen grains containing sperm-producing male gametophytes form.

antheridium (an-thuh-rid′-ē-um) (plural, **antheridia**) In plants, the male gametangium, a moist chamber in which gametes develop.

anthropoid (an′-thruh-poyd) Member of a primate group made up of the monkeys and the apes (gibbons, orangutans, gorillas, chimpanzees, bonobos, and humans).

antibody A protein secreted by plasma cells (differentiated B cells) that binds to a particular antigen; also called immunoglobulin. All antibodies have the same Y-shaped structure and in their monomer form consist of two identical heavy chains and two identical light chains.

anticodon (an′-tī-kō′-don) A nucleotide triplet at one end of a tRNA molecule that base-pairs with a particular complementary codon on an mRNA molecule.

antidiuretic hormone (ADH) (an′-tī-dī-yū-ret′-ik) A peptide hormone, also known as vasopressin, that promotes water retention by the kidneys. Produced in the hypothalamus and released from the posterior pituitary, ADH also functions in the brain.

antigen (an′-ti-jen) A substance that elicits an immune response by binding to receptors of B cells, antibodies, or of T cells.

antigen presentation The process by which an MHC molecule binds to a fragment of an intracellular protein antigen and carries it to the cell surface, where it is displayed and can be recognized by a T cell.

antigen receptor The general term for a surface protein, located on B cells and T cells, that binds to antigens, initiating adaptive immune responses. The antigen receptors on B cells are called B cell receptors, and the antigen receptors on T cells are called T cell receptors.

antigen-presenting cell A cell that upon ingesting pathogens or internalizing pathogen proteins generates peptide fragments that are bound by class II MHC molecules and subsequently displayed on the cell surface to T cells. Macrophages, dendritic cells, and B cells are the primary antigen-presenting cells.

antiparallel Referring to the arrangement of the sugar-phosphate backbones in a DNA double helix (they run in opposite $5' \rightarrow 3'$ directions).

aphotic zone (ā′-fō′-tik) The part of an ocean or lake beneath the photic zone, where light does not penetrate sufficiently for photosynthesis to occur.

apical bud (ā′-pik-ul) A bud at the tip of a plant stem; also called a terminal bud.

apical dominance (ā′-pik-ul) Tendency for growth to be concentrated at the tip of a plant shoot, because the apical bud partially inhibits axillary bud growth.

apical ectodermal ridge (AER) A thickened area of ectoderm at the tip of a limb bud that promotes outgrowth of the limb bud.

apical meristem (ā′-pik-ul mār′-uh-stem) Embryonic plant tissue in the tips of roots and buds of shoots. The dividing cells of an apical meristem enable the plant to grow in length.

apicomplexan (ap′-ē-kom-pleks′-un) A protist in a clade that includes many species that parasitize animals. Some apicomplexans cause human disease.

apomixis (ap′-uh-mik′-sis) The ability of some plant species to reproduce asexually through seeds without fertilization by a male gamete.

apoplast (ap′-ō-plast) Everything external to the plasma membrane of a plant cell, including cell walls, intercellular spaces, and the space within dead structures such as xylem vessels and tracheids.

apoptosis (ā-puh-tō′-sus) A type of programmed cell death, which is brought about by activation of enzymes that break down many chemical components in the cell.

aposematic coloration (ap′-ō-si-mat′-ik) The bright warning coloration of many animals with effective physical or chemical defenses.

appendix A small, finger-like extension of the vertebrate cecum; contains a mass of white blood cells that contribute to immunity.

aquaporin A channel protein in the plasma membrane of a plant, animal, or microorganism cell that specifically facilitates osmosis, the diffusion of free water across the membrane.

aqueous solution (ā′-kwē-us) A solution in which water is the solvent.

arachnid A member of a major arthropod group, the chelicerates. Arachnids include spiders, scorpions, ticks, and mites.

arbuscular mycorrhiza (ar-bus′-kyū-lur mī′-kō-rī′-zuh) Association of a fungus with a plant root system in which the fungus causes the invagination of the host (plant) cells' plasma membranes.

arbuscular mycorrhizal fungus A symbiotic fungus whose hyphae grow through the cell wall of plant roots and extend into the root cell (enclosed in tubes formed by invagination of the root cell plasma membrane).

Archaea (ar′-kē′-uh) One of two prokaryotic domains, the other being Bacteria.

Archaeplastida (ar′-kē-plas′-tid-uh) One of five supergroups of eukaryotes proposed in a current hypothesis of the evolutionary history of eukaryotes. This monophyletic group, which includes red algae, green algae, and land plants, descended from an ancient protist ancestor that engulfed a cyanobacterium. *See also* Excavata, Chromalveolata, Rhizaria, and Unikonta.

archegonium (ar-ki-gō′-nē-um) (plural, **archegonia**) In plants, the female gametangium, a moist chamber in which gametes develop.

archenteron (ar-ken′-tuh-ron) The endoderm-lined cavity, formed during gastrulation, that develops into the digestive tract of an animal.

archosaur (ar′-kō-sōr) Member of the reptilian group that includes crocodiles, alligators and dinosaurs, including birds.

arteriole (ar-ter′-ē-ōl) A vessel that conveys blood between an artery and a capillary bed.

artery A vessel that carries blood away from the heart to organs throughout the body.

arthropod A segmented ecdysozoan with a hard exoskeleton and jointed appendages. Familiar examples include insects, spiders, millipedes, and crabs.

artificial selection The selective breeding of domesticated plants and animals to encourage the occurrence of desirable traits.

ascocarp The fruiting body of a sac fungus (ascomycete).

ascomycete (as′-kuh-mī′-sēt) Member of the fungal phylum Ascomycota, commonly called sac fungus. The name comes from the saclike structure in which the spores develop.

ascus (plural, **asci**) A saclike spore capsule located at the tip of a dikaryotic hypha of a sac fungus.

asexual reproduction The generation of offspring from a single parent that occurs without the fusion of gametes (by budding, division of a single cell, or division of the entire organism into two or more parts). In most cases, the offspring are genetically identical to the parent.

assisted migration The translocation of a species to a favorable habitat beyond its native range for the purpose of protecting the species from human-caused threats.

assisted reproductive technology A fertilization procedure that generally involves surgically removing eggs (secondary oocytes) from a woman's ovaries after hormonal stimulation, fertilizing the eggs, and returning them to the woman's body.

associative learning The acquired ability to associate one environmental feature (such as a color) with another (such as danger).

aster A radial array of short microtubules that extends from each centrosome toward the plasma membrane in an animal cell undergoing mitosis.

astrocyte A glial cell with diverse functions, including providing structural support for neurons, regulating the interstitial environment, facilitating synaptic transmission, and assisting in regulating the blood supply to the brain.

atherosclerosis A cardiovascular disease in which fatty deposits called plaques develop in the inner walls of the arteries, obstructing the arteries and causing them to harden.

atom The smallest unit of matter that retains the properties of an element.

atomic mass The total mass of an atom, which is the mass in grams of 1 mole of the atom.

atomic nucleus An atom's dense central core, containing protons and neutrons.

atomic number The number of protons in the nucleus of an atom, unique for each element and designated by a subscript to the left of the elemental symbol.

ATP (adenosine triphosphate) (a-den′-ō-sēn trī-fos′-fāt) An adenine-containing nucleoside triphosphate that releases free energy when its phosphate bonds are hydrolyzed. This energy is used to drive endergonic reactions in cells.

ATP synthase A complex of several membrane proteins that functions in chemiosmosis with adjacent electron transport chains, using the energy of a hydrogen ion (proton) concentration gradient to make ATP. ATP synthases are found in the inner mitochondrial membranes of eukaryotic cells and in the plasma membranes of prokaryotes.

atrial natriuretic peptide (ANP) (ā′-trē-ul na′-trē-yū-ret′-ik) A peptide hormone secreted by cells of the atria of the heart in response to high blood pressure. ANP's effects on the kidney alter ion and water movement and reduce blood pressure.

atrioventricular (AV) node A region of specialized heart muscle tissue between the left and right atria where electrical impulses are delayed for about 0.1 second before spreading to both ventricles and causing them to contract.

atrioventricular (AV) valve A heart valve located between each atrium and ventricle that prevents a backflow of blood when the ventricle contracts.

atrium (ā′-trē-um) (plural, **atria**) A chamber of the vertebrate heart that receives blood from the veins and transfers blood to a ventricle.

autocrine Referring to a secreted molecule that acts on the cell that secreted it.

autoimmune disease An immunological disorder in which the immune system turns against self.

autonomic nervous system (ot′-ō-nom′-ik) An efferent branch of the vertebrate peripheral nervous system that regulates the internal environment; consists of the sympathetic, parasympathetic, and enteric divisions.

autopolyploid (ot′-ō-pol′-ē-ployd) An individual that has more than two chromosome sets that are all derived from a single species.

autosome (ot′-ō-sōm) A chromosome that is not directly involved in determining sex; not a sex chromosome.

autotroph (ot′-ō-trōf) An organism that obtains organic food molecules without eating other organisms or substances derived from other organisms. Autotrophs use energy from the sun or from oxidation of inorganic substances to make organic molecules from inorganic ones.

auxin (ôk′-sin) A term that primarily refers to indoleacetic acid (IAA), a natural plant hormone that has a variety of effects, including cell elongation, root formation, secondary growth, and fruit growth.

average heterozygosity (het′-er-ō-zī-gō′-si-tē) The percentage, on average, of a population's loci that are heterozygous in members of the population.

avirulent Describing a pathogen that can mildly harm, but not kill, the host.

axillary bud (ak′-sil-ār-ē) A structure that has the potential to form a lateral shoot, or branch. The bud appears in the angle formed between a leaf and a stem.

axon (ak′-son) A typically long extension, or process, of a neuron that carries nerve impulses away from the cell body toward target cells.

B cells The lymphocytes that complete their development in the bone marrow and become effector cells for the humoral immune response.

Bacteria One of two prokaryotic domains, the other being Archaea.

bacterial artificial chromosome (BAC) A large plasmid that acts as a bacterial chromosome and can carry inserts of 100,000 to 300,000 base pairs (100–300 kb).

bacteriophage (bak-tēr′-ē-ō-fāj) A virus that infects bacteria; also called a phage.

bacteroid A form of the bacterium *Rhizobium* contained within the vesicles formed by the root cells of a root nodule.

balancing selection Natural selection that maintains two or more phenotypic forms in a population.

bark All tissues external to the vascular cambium, consisting mainly of the secondary phloem and layers of periderm.

Barr body A dense object lying along the inside of the nuclear envelope in cells of female mammals, representing a highly condensed, inactivated X chromosome.

basal angiosperm A member of one of three clades of early-diverging lineages of flowering plants. Examples are *Amborella*, water lilies, and star anise and its relatives.

basal body (bā′-sul) A eukaryotic cell structure consisting of a "9 + 0" arrangement of microtubule triplets. The basal body may organize the microtubule assembly of a cilium or flagellum and is structurally very similar to a centriole.

basal metabolic rate (BMR) The metabolic rate of a resting, fasting, and nonstressed endotherm at a comfortable temperature.

basal taxon In a specified group of organisms, a taxon whose evolutionary lineage diverged early in the history of the group.

base A substance that reduces the hydrogen ion concentration of a solution.

basidiocarp Elaborate fruiting body of a dikaryotic mycelium of a club fungus.

basidiomycete (buh-sid′-ē-ō-mī′-sēt) Member of the fungal phylum Basidiomycota, commonly called club fungus. The name comes from the club-like shape of the basidium.

basidium (plural, **basidia**) (buh-sid′-ē-um, buh-sid′-ē-ah) A reproductive appendage that produces sexual spores on the gills of mushrooms (club fungi).

Batesian mimicry (bāt′-zē-un mim′-uh-krē) A type of mimicry in which a harmless species looks like a species that is poisonous or otherwise harmful to predators.

behavior Individually, an action carried out by muscles or glands under control of the nervous system in response to a stimulus; collectively, the sum of an animal's responses to external and internal stimuli.

behavioral ecology The study of the evolution of and ecological basis for animal behavior.

benign tumor A mass of abnormal cells with specific genetic and cellular changes such that the cells are not capable of surviving at a new site and generally remain at the site of the tumor's origin.

benthic zone The bottom surface of an aquatic environment.

benthos (ben′-thōz) The communities of organisms living in the benthic zone of an aquatic biome.

beta (β) pleated sheet One form of the secondary structure of proteins in which the polypeptide chain folds back and forth. Two regions of the chain lie parallel to each other and are held together by hydrogen bonds between atoms of the polypeptide backbone (not the side chains).

beta oxidation A metabolic sequence that breaks fatty acids down to two-carbon fragments that enter the citric acid cycle as acetyl CoA.

bicoid A maternal effect gene that codes for a protein responsible for specifying the anterior end in *Drosophila melanogaster*.

bilateral symmetry Body symmetry in which a central longitudinal plane divides the body into two equal but opposite halves.

bilaterian (bī′-luh-ter′-ē-uhn) Member of a clade of animals with bilateral symmetry and three germ layers.

bile A mixture of substances that is produced in the liver and stored in the gallbladder; enables formation of fat droplets in water as an aid in the digestion and absorption of fats.

binary fission A method of asexual reproduction by "division in half." In prokaryotes, binary fission does not involve mitosis, but in single-celled eukaryotes that undergo binary fission, mitosis is part of the process.

binomial The two-part, latinized format for naming a species, consisting of the genus and specific epithet; a binomen.

biodiversity hot spot A relatively small area with numerous endemic species and a large number of endangered and threatened species.

bioenergetics (1) The overall flow and transformation of energy in an organism. (2) The study of how energy flows through organisms.

biofilm A surface-coating colony of one or more species of prokaryotes that engage in metabolic cooperation.

biofuel A fuel produced from dry organic matter or combustible oils produced by plants.

biogenic amine A neurotransmitter derived from an amino acid.

biogeochemical cycle Any of the various chemical cycles, which involve both biotic and abiotic components of ecosystems.

biogeography The study of the past and present geographic distribution of species.

bioinformatics The use of computers, software, and mathematical models to process and integrate biological information from large data sets.

biological augmentation An approach to restoration ecology that uses organisms to add essential materials to a degraded ecosystem.

biological clock An internal timekeeper that controls an organism's biological rhythms. The biological clock marks time with or without environmental cues but often requires signals from the environment to remain tuned to an appropriate period. *See also* circadian rhythm.

biological magnification A process in which retained substances become more concentrated at each higher trophic level in a food chain.

biological species concept Definition of a species as a group of populations whose members have the potential to interbreed in nature and produce viable, fertile offspring, but do not produce viable, fertile offspring with members of other such groups.

biology The scientific study of life.

biomanipulation An approach that applies the top-down model of community organization to alter ecosystem characteristics. For example, ecologists can prevent algal blooms and eutrophication by altering the density of higher-level consumers in lakes instead of by using chemical treatments.

biomass The total mass of organic matter comprising a group of organisms in a particular habitat.

biome (bī′-ōm) Any of the world's major ecosystem types, often classified according to the predominant vegetation for terrestrial biomes and the physical environment for aquatic biomes and characterized by adaptations of organisms to that particular environment.

bioremediation The use of organisms to detoxify and restore polluted and degraded ecosystems.

biosphere The entire portion of Earth inhabited by life; the sum of all the planet's ecosystems.

biotechnology The manipulation of organisms or their components to produce useful products.

biotic (bī-ot′-ik) Pertaining to the living factors—the organisms—in an environment.

bipolar cell A neuron that relays information between photoreceptors and ganglion cells in the retina.

bipolar disorder A depressive mental illness characterized by swings of mood from high to low; also called manic-depressive disorder.

birth control pill A chemical contraceptive that inhibits ovulation, retards follicular development, or alters a woman's cervical mucus to prevent sperm from entering the uterus.

Glossary

blade (1) A leaflike structure of a seaweed that provides most of the surface area for photosynthesis. (2) The flattened portion of a typical leaf.

blastocoel (blas'-tuh-sēl) The fluid-filled cavity that forms in the center of a blastula.

blastocyst (blas'-tuh-sist) The blastula stage of mammalian embryonic development, consisting of an inner cell mass, a cavity, and an outer layer, the trophoblast. In humans, the blastocyst forms 1 week after fertilization.

blastomere An early embryonic cell arising during the cleavage stage of an early embryo.

blastopore (blas'-tō-pōr) In a gastrula, the opening of the archenteron that typically develops into the anus in deuterostomes and the mouth in protostomes.

blastula (blas'-tyū-luh) A hollow ball of cells that marks the end of the cleavage stage during early embryonic development in animals.

blood A connective tissue with a fluid matrix called plasma in which red blood cells, white blood cells, and cell fragments called platelets are suspended.

blue-light photoreceptor A type of light receptor in plants that initiates a variety of responses, such as phototropism and slowing of hypocotyl elongation.

body cavity A fluid- or air-filled space between the digestive tract and the body wall.

body plan In animals, a set of morphological and developmental traits that are integrated into a functional whole—the living animal.

Bohr shift A lowering of the affinity of hemoglobin for oxygen, caused by a drop in pH. It facilitates the release of oxygen from hemoglobin in the vicinity of active tissues.

bolus A lubricated ball of chewed food.

bone A connective tissue consisting of living cells held in a rigid matrix of collagen fibers embedded in calcium salts.

book lung An organ of gas exchange in spiders, consisting of stacked plates contained in an internal chamber.

bottleneck effect Genetic drift that occurs when the size of a population is reduced, as by a natural disaster or human actions. Typically, the surviving population is no longer genetically representative of the original population.

bottom-up model A model of community organization in which mineral nutrients influence community organization by controlling plant or phytoplankton numbers, which in turn control herbivore numbers, which in turn control predator numbers.

Bowman's capsule (bō'-munz) A cup-shaped receptacle in the vertebrate kidney that is the initial, expanded segment of the nephron where filtrate enters from the blood.

brachiopod (bra'-kē- uh-pod') A marine lophophorate with a shell divided into dorsal and ventral halves. Brachiopods are also called lamp shells.

brain Organ of the central nervous system where information is processed and integrated.

brainstem A collection of structures in the vertebrate brain, including the midbrain, the pons, and the medulla oblongata; functions in homeostasis, coordination of movement, and conduction of information to higher brain centers.

branch point The representation on a phylogenetic tree of the divergence of two or more taxa from a common ancestor. A branch point is usually shown as a dichotomy in which a branch representing the ancestral lineage splits (at the branch point) into two branches, one for each of the two descendant lineages.

brassinosteroid A steroid hormone in plants that has a variety of effects, including inducing cell elongation, retarding leaf abscission, and promoting xylem differentiation.

breathing Ventilation of the lungs through alternating inhalation and exhalation.

bronchiole (brong'-kē-ōl') A fine branch of the bronchi that transports air to alveoli.

bronchus (brong'-kus) (plural, **bronchi**) One of a pair of breathing tubes that branch from the trachea into the lungs.

brown alga A multicellular, photosynthetic protist with a characteristic brown or olive color that results from carotenoids in its plastids. Most brown algae are marine, and some have a plantlike body (thallus).

bryophyte (brī'-uh-fīt) An informal name for a moss, liverwort, or hornwort; a nonvascular plant that lives on land but lacks some of the terrestrial adaptations of vascular plants.

budding Asexual reproduction in which outgrowths from the parent form and pinch off to live independently or else remain attached to eventually form extensive colonies.

buffer A solution that contains a weak acid and its corresponding base. A buffer minimizes changes in pH when acids or bases are added to the solution.

bulk feeder An animal that eats relatively large pieces of food.

bulk flow The movement of a fluid due to a difference in pressure between two locations.

bundle-sheath cell In C_4 plants, a type of photosynthetic cell arranged into tightly packed sheaths around the veins of a leaf.

C_3 plant A plant that uses the Calvin cycle for the initial steps that incorporate CO_2 into organic material, forming a three-carbon compound as the first stable intermediate.

C_4 plant A plant in which the Calvin cycle is preceded by reactions that incorporate CO_2 into a four-carbon compound, the end product of which supplies CO_2 for the Calvin cycle.

calcitonin (kal'-si-tō'-nin) A hormone secreted by the thyroid gland that lowers blood calcium levels by promoting calcium deposition in bone and calcium excretion from the kidneys; nonessential in adult humans.

callus A mass of dividing, undifferentiated cells growing in culture.

calorie (cal) The amount of heat energy required to raise the temperature of 1 g of water by 1°C; also the amount of heat energy that 1 g of water releases when it cools by 1°C. The Calorie (with a capital C), usually used to indicate the energy content of food, is a kilocalorie.

Calvin cycle The second of two major stages in photosynthesis (following the light reactions), involving fixation of atmospheric CO_2 and reduction of the fixed carbon into carbohydrate.

CAM plant A plant that uses crassulacean acid metabolism, an adaptation for photosynthesis in arid conditions. In this process, carbon dioxide entering open stomata during the night is converted to organic acids, which release CO_2 for the Calvin cycle during the day, when stomata are closed.

Cambrian explosion A relatively brief time in geologic history when many present-day phyla of animals first appeared in the fossil record. This burst of evolutionary change occurred about 535–525 million years ago and saw the emergence of the first large, hard-bodied animals.

cAMP *See* cyclic AMP (cAMP).

canopy The uppermost layer of vegetation in a terrestrial biome.

capillary (kap'-il-ār'-ē) A microscopic blood vessel that penetrates the tissues and consists of a single layer of endothelial cells that allows exchange between the blood and interstitial fluid.

capillary bed A network of capillaries in a tissue or organ.

capsid The protein shell that encloses a viral genome. It may be rod-shaped, polyhedral, or more complex in shape.

capsule (1) In many prokaryotes, a dense and well-defined layer of polysaccharide or protein that surrounds the cell wall and is sticky, protecting the cell and enabling it to adhere to substrates or other cells. (2) The sporangium of a bryophyte (moss, liverwort, or hornwort).

carbohydrate (kar'-bō-hī'-drāt) A sugar (monosaccharide) or one of its dimers (disaccharides) or polymers (polysaccharides).

carbon fixation The initial incorporation of carbon from CO_2 into an organic compound by an autotrophic organism (a plant, another photosynthetic organism, or a chemoautotrophic prokaryote).

carbonyl group (kar-buh-nēl') A chemical group present in aldehydes and ketones and consisting of a carbon atom double-bonded to an oxygen atom.

carboxyl group (kar-bok'-sil) A chemical group present in organic acids and consisting of a single carbon atom double-bonded to an oxygen atom and also bonded to a hydroxyl group.

cardiac cycle (kar'-dē-ak) The alternating contractions and relaxations of the heart.

cardiac muscle A type of striated muscle that forms the contractile wall of the heart. Its cells are joined by intercalated disks that relay the electrical signals underlying each heartbeat.

cardiac output The volume of blood pumped per minute by each ventricle of the heart.

cardiovascular system A closed circulatory system with a heart and branching network of arteries, capillaries, and veins. The system is characteristic of vertebrates.

carnivore An animal that mainly eats other animals.

carotenoid (kuh-rot'-uh-noyd' An accessory pigment, either yellow or orange, in the chloroplasts of plants and in some prokaryotes. By absorbing wavelengths of light that chlorophyll cannot, carotenoids broaden the spectrum of colors that can drive photosynthesis.

carpel (kar′-pul) The ovule-producing reproductive organ of a flower, consisting of the stigma, style, and ovary.

carrier In genetics, an individual who is heterozygous at a given genetic locus for a recessively inherited disorder. The heterozygote is generally phenotypically normal for the disorder but can pass on the recessive allele to offspring.

carrying capacity The maximum population size that can be supported by the available resources, symbolized as *K*.

cartilage (kar′-til-ij) A flexible connective tissue with an abundance of collagenous fibers embedded in chondroitin sulfate.

Casparian strip (ka-spār′-ē-un) A water-impermeable ring of wax in the endodermal cells of plants that blocks the passive flow of water and solutes into the stele by way of cell walls.

catabolic pathway (kat′-uh-bol′-ik) A metabolic pathway that releases energy by breaking down complex molecules to simpler molecules.

catalyst (kat′-uh-list) A chemical agent that selectively increases the rate of a reaction without being consumed by the reaction.

catastrophism (kuh-tas′-truh-fiz′-um) The principle that events in the past occurred suddenly and were caused by different mechanisms than those operating today. *See* uniformitarianism.

catecholamine (kat′-uh-kōl′-uh-mēn) Any of a class of neurotransmitters and hormones, including the hormones epinephrine and norepinephrine, that are synthesized from the amino acid tyrosine.

cation (cat′-ī-on) A positively charged ion.

cation exchange A process in which positively charged minerals are made available to a plant when hydrogen ions in the soil displace mineral ions from the clay particles.

cDNA library A gene library containing clones that carry complementary DNA (cDNA) inserts. The library includes only the genes that were transcribed in the cells whose mRNA was isolated to make the cDNA.

cecum (sē′-kum) (plural, **ceca**) The blind pouch forming one branch of the large intestine.

cell body The part of a neuron that houses the nucleus and most other organelles.

cell cycle An ordered sequence of events in the life of a cell, from its origin in the division of a parent cell until its own division into two. The eukaryotic cell cycle is composed of interphase (including G_1, S, and G_2 subphases) and M phase (including mitosis and cytokinesis).

cell cycle control system A cyclically operating set of molecules in the eukaryotic cell that both triggers and coordinates key events in the cell cycle.

cell division The reproduction of cells.

cell fractionation The disruption of a cell and separation of its parts by centrifugation at successively higher speeds.

cell plate A membrane-bounded, flattened sac located at the midline of a dividing plant cell, inside which the new cell wall forms during cytokinesis.

cell wall A protective layer external to the plasma membrane in the cells of plants, prokaryotes, fungi, and some protists. Polysaccharides such as cellulose (in plants and some protists), chitin (in fungi), and peptidoglycan (in bacteria) are important structural components of cell walls.

cell-mediated immune response The branch of adaptive immunity that involves the activation of cytotoxic T cells, which defend against infected cells.

cellular respiration The catabolic pathways of aerobic and anaerobic respiration, which break down organic molecules and use an electron transport chain for the production of ATP.

cellular slime mold A type of protist characterized by unicellular amoeboid cells and aggregated reproductive bodies in its life cycle.

cellulose (sel′-yū-lōs) A structural polysaccharide of plant cell walls, consisting of glucose monomers joined by β glycosidic linkages.

Celsius scale (sel′-sē-us) A temperature scale (°C) equal to $^5/_9$(°F – 32) that measures the freezing point of water at 0°C and the boiling point of water at 100°C.

central canal The narrow cavity in the center of the spinal cord that is continuous with the fluid-filled ventricles of the brain.

central nervous system (CNS) The portion of the nervous system where signal integration occurs; in vertebrate animals, the brain and spinal cord.

central vacuole In a mature plant cell, a large membranous sac with diverse roles in growth, storage, and sequestration of toxic substances.

centriole (sen′-trē-ōl) A structure in the centrosome of an animal cell composed of a cylinder of microtubule triplets arranged in a 9 + 0 pattern. A centrosome has a pair of centrioles.

centromere (sen′-trō-mēr) In a duplicated chromosome, the region on each sister chromatid where they are most closely attached to each other by proteins that bind to specific DNA sequences; this close attachment causes a constriction in the condensed chromosome. (An uncondensed, unduplicated chromosome has a single centromere, identified by its DNA sequence.)

centrosome (sen′-trō-sōm) A structure present in the cytoplasm of animal cells that functions as a microtubule-organizing center and is important during cell division. A centrosome has two centrioles.

cephalization (sef′-uh-luh-zā′-shun) An evolutionary trend toward the concentration of sensory equipment at the anterior end of the body.

cercozoan An amoeboid or flagellated protist that feeds with threadlike pseudopodia.

cerebellum (sār′-ruh-bel′-um) Part of the vertebrate hindbrain located dorsally; functions in unconscious coordination of movement and balance.

cerebral cortex (suh-rē′-brul) The surface of the cerebrum; the largest and most complex part of the mammalian brain, containing nerve cell bodies of the cerebrum; the part of the vertebrate brain most changed through evolution.

cerebral hemisphere The right or left side of the cerebrum.

cerebrospinal fluid (suh-rē′-brō-spī′-nul) Blood-derived fluid that surrounds, protects against infection, nourishes, and cushions the brain and spinal cord.

cerebrum (suh-rē′-brum) The dorsal portion of the vertebrate forebrain, composed of right and left hemispheres; the integrating center for memory, learning, emotions, and other highly complex functions of the central nervous system.

cervix (ser′-viks) The neck of the uterus, which opens into the vagina.

chaparral A scrubland biome of dense, spiny evergreen shrubs found at midlatitudes along coasts where cold ocean currents circulate offshore; characterized by mild, rainy winters and long, hot, dry summers.

chaperonin (shap′-er-ō′-nin) A protein complex that assists in the proper folding of other proteins.

character An observable heritable feature that may vary among individuals.

character displacement The tendency for characteristics to be more divergent in sympatric populations of two species than in allopatric populations of the same two species.

checkpoint A control point in the cell cycle where stop and go-ahead signals can regulate the cycle.

chelicera (kē-lih′-suh-ruh) (plural, **chelicerae**) One of a pair of clawlike feeding appendages characteristic of chelicerates.

chelicerate (kē-lih-suh′-rāte) An arthropod that has chelicerae and a body divided into a cephalothorax and an abdomen. Living chelicerates include sea spiders, horseshoe crabs, scorpions, ticks, and spiders.

chemical bond An attraction between two atoms, resulting from a sharing of outer-shell electrons or the presence of opposite charges on the atoms. The bonded atoms gain complete outer electron shells.

chemical energy Energy available in molecules for release in a chemical reaction; a form of potential energy.

chemical equilibrium In a chemical reaction, the state in which the rate of the forward reaction equals the rate of the reverse reaction, so that the relative concentrations of the reactants and products do not change with time.

chemical reaction The making and breaking of chemical bonds, leading to changes in the composition of matter.

chemiosmosis (kem′-ē-oz-mō′-sis) An energy-coupling mechanism that uses energy stored in the form of a hydrogen ion gradient across a membrane to drive cellular work, such as the synthesis of ATP. Under aerobic conditions, most ATP synthesis in cells occurs by chemiosmosis.

chemoautotroph (kē′-mō-ot′-ō-trōf) An organism that obtains energy by oxidizing inorganic substances and needs only carbon dioxide as a carbon source.

chemoheterotroph (kē′-mō-het′-er-ō-trōf) An organism that requires organic molecules for both energy and carbon.

chemoreceptor A sensory receptor that responds to a chemical stimulus, such as a solute or an odorant.

chiasma (plural, **chiasmata**) (kī-az′-muh, kī-az′-muh-tuh) The X-shaped, microscopically visible region where crossing over has occurred earlier in prophase I between homologous nonsister chromatids. Chiasmata become visible after synapsis ends, with the two homologs remaining associated due to sister chromatid cohesion.

chitin (kī′-tin) A structural polysaccharide, consisting of amino sugar monomers, found in many fungal cell walls and in the exoskeletons of all arthropods.

chlorophyll (klōr′-ō-fil) A green pigment located in membranes within the chloroplasts of plants and algae and in the membranes of certain prokaryotes. Chlorophyll *a* participates directly in the light reactions, which convert solar energy to chemical energy.

chlorophyll *a* A photosynthetic pigment that participates directly in the light reactions, which convert solar energy to chemical energy.

chlorophyll *b* An accessory photosynthetic pigment that transfers energy to chlorophyll *a*.

chloroplast (klōr′-ō-plast) An organelle found in plants and photosynthetic protists that absorbs sunlight and uses it to drive the synthesis of organic compounds from carbon dioxide and water.

choanocyte (kō-an′-uh-sīt) A flagellated feeding cell found in sponges. Also called a collar cell, it has a collar-like ring that traps food particles around the base of its flagellum.

cholesterol (kō-les′-tuh-rol) A steroid that forms an essential component of animal cell membranes and acts as a precursor molecule for the synthesis of other biologically important steroids, such as many hormones.

chondrichthyan (kon-drik′-thē-an) Member of the class Chondrichthyes, vertebrates with skeletons made mostly of cartilage, such as sharks and rays.

chordate Member of the phylum Chordata, animals that at some point during their development have a notochord; a dorsal, hollow nerve cord; pharyngeal slits or clefts; and a muscular, post-anal tail.

chorionic villus sampling (CVS) (kōr′-ē-on′-ik vil′-us) A technique associated with prenatal diagnosis in which a small sample of the fetal portion of the placenta is removed for analysis to detect certain genetic and congenital defects in the fetus.

Chromalveolata One of five supergroups of eukaryotes proposed in a current hypothesis of the evolutionary history of eukaryotes. Chromalveolates may have originated by secondary endosymbiosis and include two large protist clades, the alveolates and the stramenopiles. *See also* Excavata, Rhizaria, Archaeplastida, and Unikonta.

chromatin (krō′-muh-tin) The complex of DNA and proteins that makes up eukaryotic chromosomes. When the cell is not dividing, chromatin exists in its dispersed form, as a mass of very long, thin fibers that are not visible with a light microscope.

chromosome (krō′-muh-sōm) A cellular structure carrying genetic material, found in the nucleus of eukaryotic cells. Each chromosome consists of one very long DNA molecule and associated proteins. (A bacterial chromosome usually consists of a single circular DNA molecule and associated proteins. It is found in the nucleoid region, which is not membrane bounded.) *See also* chromatin.

chromosome theory of inheritance A basic principle in biology stating that genes are located at specific positions (loci) on chromosomes and that the behavior of chromosomes during meiosis accounts for inheritance patterns.

chylomicron (kī′-lō-mī′-kron) A lipid transport globule composed of fats mixed with cholesterol and coated with proteins.

chyme (kīm) The mixture of partially digested food and digestive juices formed in the stomach.

chytrid (kī′-trid) Member of the fungal phylum Chytridiomycota, mostly aquatic fungi with flagellated zoospores that represent an early-diverging fungal lineage.

ciliate (sil′-ē-it) A type of protist that moves by means of cilia.

cilium (sil′-ē-um) (plural, **cilia**) A short appendage containing microtubules in eukaryotic cells. A motile cilium is specialized for locomotion or moving fluid past the cell; it is formed from a core of nine outer doublet microtubules and two inner single microtubules (the "9 + 2" arrangement) ensheathed in an extension of the plasma membrane. A primary cilium is usually nonmotile and plays a sensory and signaling role; it lacks the two inner microtubules (the "9 + 0" arrangement).

circadian rhythm (ser-kā′-dē-un) A physiological cycle of about 24 hours that persists even in the absence of external cues.

***cis-trans* isomer** One of several compounds that have the same molecular formula and covalent bonds between atoms but differ in the spatial arrangements of their atoms owing to the inflexibility of double bonds; formerly called a geometric isomer.

citric acid cycle A chemical cycle involving eight steps that completes the metabolic breakdown of glucose molecules begun in glycolysis by oxidizing acetyl CoA (derived from pyruvate) to carbon dioxide; occurs within the mitochondrion in eukaryotic cells and in the cytosol of prokaryotes; together with pyruvate oxidation, the second major stage in cellular respiration.

clade (klayd) A group of species that includes an ancestral species and all of its descendants.

cladistics (kluh-dis′-tiks) An approach to systematics in which organisms are placed into groups called clades based primarily on common descent.

class In Linnaean classification, the taxonomic category above the level of order.

cleavage (1) The process of cytokinesis in animal cells, characterized by pinching of the plasma membrane. (2) The succession of rapid cell divisions without significant growth during early embryonic development that converts the zygote to a ball of cells.

cleavage furrow The first sign of cleavage in an animal cell; a shallow groove around the cell in the cell surface near the old metaphase plate.

climate The long-term prevailing weather conditions at a given place.

climograph A plot of the temperature and precipitation in a particular region.

cline A graded change in a character along a geographic axis.

clitoris (klit′-uh-ris) An organ at the upper intersection of the labia minora that engorges with blood and becomes erect during sexual arousal.

cloaca (klō-ā′-kuh) A common opening for the digestive, urinary, and reproductive tracts found in many nonmammalian vertebrates but in few mammals.

clonal selection The process by which an antigen selectively binds to and activates only those lymphocytes bearing receptors specific for the antigen. The selected lymphocytes proliferate and differentiate into a clone of effector cells and a clone of memory cells specific for the stimulating antigen.

clone (1) A lineage of genetically identical individuals or cells. (2) In popular usage, an individual that is genetically identical to another individual. (3) As a verb, to make one or more genetic replicas of an individual or cell. *See also* gene cloning.

cloning vector In genetic engineering, a DNA molecule that can carry foreign DNA into a host cell and replicate there. Cloning vectors include plasmids and bacterial artificial chromosomes (BACs), which move recombinant DNA from a test tube back into a cell, and viruses that transfer recombinant DNA by infection.

closed circulatory system A circulatory system in which blood is confined to vessels and is kept separate from the interstitial fluid.

cnidocyte (nī′-duh-sīt) A specialized cell unique to the phylum Cnidaria; contains a capsule-like organelle housing a coiled thread that, when discharged, explodes outward and functions in prey capture or defense.

cochlea (kok′-lē-uh) The complex, coiled organ of hearing that contains the organ of Corti.

codominance The situation in which the phenotypes of both alleles are exhibited in the heterozygote because both alleles affect the phenotype in separate, distinguishable ways.

codon (kō′-don) A three-nucleotide sequence of DNA or mRNA that specifies a particular amino acid or termination signal; the basic unit of the genetic code.

coefficient of relatedness The fraction of genes that, on average, are shared by two individuals.

coelom (sē′-lōm) A body cavity lined by tissue derived only from mesoderm.

coelomate (sē′-lō-māt) An animal that possesses a true coelom (a body cavity lined by tissue completely derived from mesoderm).

coenocytic fungus (sē′-no-si′-tic) A fungus that lacks septa and hence whose body is made up of a continuous cytoplasmic mass that may contain hundreds or thousands of nuclei.

coenzyme (kō-en′-zīm) An organic molecule serving as a cofactor. Most vitamins function as coenzymes in metabolic reactions.

coevolution The joint evolution of two interacting species, each in response to selection imposed by the other.

cofactor Any nonprotein molecule or ion that is required for the proper functioning of an enzyme. Cofactors can be permanently bound to the active site or may bind loosely and reversibly, along with the substrate, during catalysis.

cognition The process of knowing that may include awareness, reasoning, recollection, and judgment.

cognitive map A neural representation of the abstract spatial relationships between objects in an animal's surroundings.

cohesion The linking together of like molecules, often by hydrogen bonds.

cohesion-tension hypothesis The leading explanation of the ascent of xylem sap. It states that transpiration exerts pull on xylem sap, putting the sap under negative pressure or tension, and that the cohesion of water molecules transmits this pull along the entire length of the xylem from shoots to roots.

cohort A group of individuals of the same age in a population.

coitus (kō′-uh-tus) The insertion of a penis into a vagina; also called sexual intercourse.

coleoptile (kō′-lē-op′-tul) The covering of the young shoot of the embryo of a grass seed.

coleorhiza (kō′-lē-uh-rī′-zuh) The covering of the young root of the embryo of a grass seed.

collagen A glycoprotein in the extracellular matrix of animal cells that forms strong fibers, found extensively in connective tissue and bone; the most abundant protein in the animal kingdom.

collecting duct The location in the kidney where processed filtrate, called urine, is collected from the renal tubules.

collenchyma cell (kō-len′-kim-uh) A flexible plant cell type that occurs in strands or cylinders that support young parts of the plant without restraining growth.

colloid A mixture made up of a liquid and particles that (because of their large size) remain suspended rather than dissolved in that liquid.

colon (kō′-len) The largest section of the vertebrate large intestine; functions in water absorption and formation of feces.

commensalism (kuh-men′-suh-lizm) A symbiotic relationship in which one organism benefits but the other is neither helped nor harmed.

communication In animal behavior, a process involving transmission of, reception of, and response to signals. The term is also used in connection with other organisms, as well as individual cells of multicellular organisms.

community All the organisms that inhabit a particular area; an assemblage of populations of different species living close enough together for potential interaction.

community ecology The study of how interactions between species affect community structure and organization.

companion cell A type of plant cell that is connected to a sieve-tube element by many plasmodesmata and whose nucleus and ribosomes may serve one or more adjacent sieve-tube elements.

competitive exclusion The concept that when populations of two similar species compete for the same limited resources, one population will use the resources more efficiently and have a reproductive advantage that will eventually lead to the elimination of the other population.

competitive inhibitor A substance that reduces the activity of an enzyme by entering the active site in place of the substrate, whose structure it mimics.

complement system A group of about 30 blood proteins that may amplify the inflammatory response, enhance phagocytosis, or directly lyse extracellular pathogens.

complementary DNA (cDNA) A double-stranded DNA molecule made *in vitro* using mRNA as a template and the enzymes reverse transcriptase and DNA polymerase. A cDNA molecule corresponds to the exons of a gene.

complete digestive tract A digestive tube that runs between a mouth and an anus; also called an alimentary canal.

complete dominance The situation in which the phenotypes of the heterozygote and dominant homozygote are indistinguishable.

complete flower A flower that has all four basic floral organs: sepals, petals, stamens, and carpels.

complete metamorphosis The transformation of a larva into an adult that looks very different, and often functions very differently in its environment, than the larva.

compound A substance consisting of two or more different elements combined in a fixed ratio.

compound eye A type of multifaceted eye in insects and crustaceans consisting of up to several thousand light-detecting, focusing ommatidia.

concentration gradient A region along which the density of a chemical substance increases or decreases.

conception The fertilization of an egg by a sperm in humans.

condom A thin, latex rubber or natural membrane sheath that fits over the penis to collect semen.

conduction The direct transfer of thermal motion (heat) between molecules of objects in direct contact with each other.

cone A cone-shaped cell in the retina of the vertebrate eye, sensitive to color.

conformer An animal for which an internal condition conforms to (changes in accordance with) changes in an environmental variable.

conidium (plural, **conidia**) A haploid spore produced at the tip of a specialized hypha in ascomycetes during asexual reproduction.

conifer Member of the largest gymnosperm phylum. Most conifers are cone-bearing trees, such as pines and firs.

conjugation (kon′-jū-gā′-shun) (1) In prokaryotes, the direct transfer of DNA between two cells that are temporarily joined. When the two cells are members of different species, conjugation results in horizontal gene transfer. (2) In ciliates, a sexual process in which two cells exchange haploid micronuclei but do not reproduce.

connective tissue Animal tissue that functions mainly to bind and support other tissues, having a sparse population of cells scattered through an extracellular matrix.

conodont An early, soft-bodied vertebrate with prominent eyes and dental elements.

conservation biology The integrated study of ecology, evolutionary biology, physiology, molecular biology, and genetics to sustain biological diversity at all levels.

contraception The deliberate prevention of pregnancy.

contractile vacuole A membranous sac that helps move excess water out of certain freshwater protists.

control element A segment of noncoding DNA that helps regulate transcription of a gene by serving as a binding site for a transcription factor. Multiple control elements are present in a eukaryotic gene's enhancer.

controlled experiment An experiment in which an experimental group is compared with a control group that varies only in the factor being tested.

convection The mass movement of warmed air or liquid to or from the surface of a body or object.

convergent evolution The evolution of similar features in independent evolutionary lineages.

convergent extension A process in which the cells of a tissue layer rearrange themselves in such a way that the sheet of cells becomes narrower (converges) and longer (extends).

cooperativity A kind of allosteric regulation whereby a shape change in one subunit of a protein caused by substrate binding is transmitted to all the other subunits, facilitating binding of additional substrate molecules to those subunits.

copepod (cō′-puh-pod) Any of a group of small crustaceans that are important members of marine and freshwater plankton communities.

coral reef Typically a warm-water, tropical ecosystem dominated by the hard skeletal structures secreted primarily by corals. Some coral reefs also exist in cold, deep waters.

corepressor A small molecule that binds to a bacterial repressor protein and changes the protein's shape, allowing it to bind to the operator and switch an operon off.

cork cambium (kam′-bē-um) A cylinder of meristematic tissue in woody plants that replaces the epidermis with thicker, tougher cork cells.

corpus callosum (kor′-pus kuh-lō′-sum) The thick band of nerve fibers that connects the right and left cerebral hemispheres in mammals, enabling the hemispheres to process information together.

corpus luteum (kor′-pus lū′-tē-um) A secreting tissue in the ovary that forms from the collapsed follicle after ovulation and produces progesterone.

cortex (1) The outer region of cytoplasm in a eukaryotic cell, lying just under the plasma membrane, that has a more gel-like consistency than the inner regions due to the presence of multiple microfilaments. (2) In plants, ground tissue that is between the vascular tissue and dermal tissue in a root or eudicot stem.

cortical nephron In mammals and birds, a nephron with a loop of Henle located almost entirely in the renal cortex.

corticosteroid Any steroid hormone produced and secreted by the adrenal cortex.

cotransport The coupling of the "downhill" diffusion of one substance to the "uphill" transport of another against its own concentration gradient.

cotyledon (kot′-uh-lē′-dun) A seed leaf of an angiosperm embryo. Some species have one cotyledon, others two.

countercurrent exchange The exchange of a substance or heat between two fluids flowing in opposite directions. For example, blood in a fish gill flows in the opposite direction of water passing over the gill, maximizing diffusion of oxygen into and carbon dioxide out of the blood.

countercurrent multiplier system A countercurrent system in which energy is expended in active transport to facilitate exchange of materials and generate concentration gradients.

covalent bond (kō-vā′-lent) A type of strong chemical bond in which two atoms share one or more pairs of valence electrons.

craniate A chordate with a head.

crassulacean acid metabolism (CAM) An adaptation for photosynthesis in arid conditions, first discovered in the family Crassulaceae. In this process, a plant takes up CO_2 and incorporates it into a variety of organic acids at night; during the day, CO_2 is released from organic acids for use in the Calvin cycle.

crista (plural, **cristae**) (kris′-tuh, kris′-tē) An infolding of the inner membrane of a mitochondrion. The inner membrane houses electron transport chains and molecules of the enzyme catalyzing the synthesis of ATP (ATP synthase).

critical load The amount of added nutrient, usually nitrogen or phosphorus, that can be absorbed by plants without damaging ecosystem integrity.

crop rotation The practice of planting nonlegumes one year and legumes in alternating years to restore concentrations of fixed nitrogen in the soil.

cross-fostering study A behavioral study in which the young of one species are placed in the care of adults from another species.

crossing over The reciprocal exchange of genetic material between nonsister chromatids during prophase I of meiosis.

cross-pollination In angiosperms, the transfer of pollen from an anther of a flower on one plant to the stigma of a flower on another plant of the same species.

crustacean (kruh-stā′-shun) A member of a subphylum of mostly aquatic arthropods that includes lobsters, crayfishes, crabs, shrimps, and barnacles.

cryptic coloration Camouflage that makes a potential prey difficult to spot against its background.

culture A system of information transfer through social learning or teaching that influences the behavior of individuals in a population.

cuticle (kyū′-tuh-kul) (1) A waxy covering on the surface of stems and leaves that prevents desiccation in terrestrial plants. (2) The exoskeleton of an arthropod, consisting of layers of protein and chitin that are variously modified for different functions. (3) A tough coat that covers the body of a nematode.

cyclic AMP (cAMP) Cyclic adenosine monophosphate, a ring-shaped molecule made from ATP that is a common intracellular signaling molecule (second messenger) in eukaryotic cells. It is also a regulator of some bacterial operons.

cyclic electron flow A route of electron flow during the light reactions of photosynthesis that involves only photosystem I and that produces ATP but not NADPH or O_2.

cyclin (sī′-klin) A cellular protein that occurs in a cyclically fluctuating concentration and that plays an important role in regulating the cell cycle.

cyclin-dependent kinase (Cdk) A protein kinase that is active only when attached to a particular cyclin.

cystic fibrosis (sis′-tik fī-brō′-sis) A human genetic disorder caused by a recessive allele for a chloride channel protein; characterized by an excessive secretion of mucus and consequent vulnerability to infection; fatal if untreated.

cytochrome (sī′-tō-krōm) An iron-containing protein that is a component of electron transport chains in the mitochondria and chloroplasts of eukaryotic cells and the plasma membranes of prokaryotic cells.

cytogenetic map A map of a chromosome that locates genes with respect to chromosomal features distinguishable in a microscope.

cytokine (sī′-tō-kīn′) Any of a group of small proteins secreted by a number of cell types, including macrophages and helper T cells, that regulate the function of other cells.

cytokinesis (sī′-tō-kuh-nē′-sis) The division of the cytoplasm to form two separate daughter cells immediately after mitosis, meiosis I, or meiosis II.

cytokinin (sī′-tō-kī′-nin) Any of a class of related plant hormones that retard aging and act in concert with auxin to stimulate cell division, influence the pathway of differentiation, and control apical dominance.

cytoplasm (sī′-tō-plaz′-um) The contents of the cell bounded by the plasma membrane; in eukaryotes, the portion exclusive of the nucleus.

cytoplasmic determinant A maternal substance, such as a protein or RNA, that when placed into an egg influences the course of early development by regulating the expression of genes that affect the developmental fate of cells.

cytoplasmic streaming A circular flow of cytoplasm, involving interactions of myosin and actin filaments, that speeds the distribution of materials within cells.

cytoskeleton A network of microtubules, microfilaments, and intermediate filaments that extend throughout the cytoplasm and serve a variety of mechanical, transport, and signaling functions.

cytosol (sī′-tō-sol) The semifluid portion of the cytoplasm.

cytotoxic T cell A type of lymphocyte that, when activated, kills infected cells as well as certain cancer cells and transplanted cells.

dalton A measure of mass for atoms and subatomic particles; the same as the atomic mass unit, or amu.

data Recorded observations.

day-neutral plant A plant in which flower formation is not controlled by photoperiod or day length.

decapod A member of the group of crustaceans that includes lobsters, crayfishes, crabs, and shrimps.

decomposer An organism that absorbs nutrients from nonliving organic material such as corpses, fallen plant material, and the wastes of living organisms and converts them to inorganic forms; a detritivore.

deductive reasoning A type of logic in which specific results are predicted from a general premise.

deep-sea hydrothermal vent A dark, hot, oxygen-deficient environment associated with volcanic activity on or near the seafloor. The producers in a vent community are chemoautotrophic prokaryotes.

de-etiolation The changes a plant shoot undergoes in response to sunlight; also known informally as greening.

dehydration reaction A chemical reaction in which two molecules become covalently bonded to each other with the removal of a water molecule.

deletion (1) A deficiency in a chromosome resulting from the loss of a fragment through breakage. (2) A mutational loss of one or more nucleotide pairs from a gene.

demographic transition In a stable population, a shift from high birth and death rates to low birth and death rates.

demography The study of changes over time in the vital statistics of populations, especially birth rates and death rates.

denaturation (dē-nā′-chur-ā′-shun) In proteins, a process in which a protein loses its native shape due to the disruption of weak chemical bonds and interactions, thereby becoming biologically inactive; in DNA, the separation of the two strands of the double helix. Denaturation occurs under extreme (noncellular) conditions of pH, salt concentration, or temperature.

dendrite (den′-drīt) One of usually numerous, short, highly branched extensions of a neuron that receive signals from other neurons.

dendritic cell An antigen-presenting cell, located mainly in lymphatic tissues and skin, that is particularly efficient in presenting antigens to helper T cells, thereby initiating a primary immune response.

density The number of individuals per unit area or volume.

density dependent Referring to any characteristic that varies with population density.

density independent Referring to any characteristic that is not affected by population density.

density-dependent inhibition The phenomenon observed in normal animal cells that causes them to stop dividing when they come into contact with one another.

deoxyribonucleic acid (DNA) (dē-ok′-sē-rī′-bō-nū-klā′-ik) A nucleic acid molecule, usually a double-stranded helix, in which each polynucleotide strand consists of nucleotide monomers with a deoxyribose sugar and the nitrogenous bases adenine (A), cytosine (C), guanine (G), and thymine (T); capable of being replicated and determining the inherited structure of a cell's proteins.

deoxyribose (dē-ok′-si-rī′-bōs) The sugar component of DNA nucleotides, having one fewer hydroxyl group than ribose, the sugar component of RNA nucleotides.

depolarization A change in a cell's membrane potential such that the inside of the membrane is made less negative relative to the outside. For example, a neuron membrane is depolarized if a stimulus decreases its voltage from the resting potential of -70 mV in the direction of zero voltage.

dermal tissue system The outer protective covering of plants.

desert A terrestrial biome characterized by very low precipitation.

desmosome A type of intercellular junction in animal cells that functions as a rivet, fastening cells together.

determinate cleavage A type of embryonic development in protostomes that rigidly casts the developmental fate of each embryonic cell very early.

determinate growth A type of growth characteristic of most animals and some plant organs, in which growth stops after a certain size is reached.

determination The progressive restriction of developmental potential in which the possible fate of each cell becomes more limited as an embryo develops. At the end of determination, a cell is committed to its fate.

detritivore (deh-trī′-tuh-vōr) A consumer that derives its energy and nutrients from nonliving organic material such as corpses, fallen plant material, and the wastes of living organisms; a decomposer.

detritus (di-trī′-tus) Dead organic matter.

deuteromycete (dū′-tuh-rō-mī′-sēt) Traditional classification for a fungus with no known sexual stage.

deuterostome development (dū′-tuh-rō-stōm′) In animals, a developmental mode distinguished by the development of the anus from the blastopore; often also characterized by radial cleavage and by the body cavity forming as outpockets of mesodermal tissue.

development The events involved in an organism's changing gradually from a simple to a more complex or specialized form.

diabetes mellitus (dī′-uh-bē′-tis mel′-uh-tus) An endocrine disorder marked by an inability to maintain glucose homeostasis. The type 1 form results from autoimmune destruction of insulin-secreting cells; treatment usually requires daily insulin injections. The type 2 form most commonly results from reduced responsiveness of target cells to insulin; obesity and lack of exercise are risk factors.

diacylglycerol (DAG) (dī-a′-sil-glis′-er-ol) A second messenger produced by the cleavage of the phospholipid PIP_2 in the plasma membrane.

diaphragm (dī′-uh-fram′) (1) A sheet of muscle that forms the bottom wall of the thoracic cavity in mammals. Contraction of the diaphragm pulls air into the lungs. (2) A dome-shaped rubber cup fitted into the upper portion of the vagina before sexual intercourse. It serves as a physical barrier to the passage of sperm into the uterus.

diapsid (dī-ap′-sid) Member of an amniote clade distinguished by a pair of holes on each side of the skull. Diapsids include the lepidosaurs and archosaurs.

diastole (dī-as′-tō-lē) The stage of the cardiac cycle in which a heart chamber is relaxed and fills with blood.

diastolic pressure Blood pressure in the arteries when the ventricles are relaxed.

dicot A term traditionally used to refer to flowering plants that have two embryonic seed leaves, or cotyledons. Recent molecular evidence indicates that dicots do not form a clade; species once classified as dicots are now grouped into eudicots, magnoliids, and several lineages of basal angiosperms.

differential gene expression The expression of different sets of genes by cells with the same genome.

differentiation The process by which a cell or group of cells become specialized in structure and function.

diffusion The spontaneous movement of a substance down its concentration or electrochemical gradient, from a region where it is more concentrated to a region where it is less concentrated.

digestion The second stage of food processing in animals: the breaking down of food into molecules small enough for the body to absorb.

dihybrid (dī′-hī′-brid) An organism that is heterozygous with respect to two genes of interest. All the offspring from a cross between parents doubly homozygous for different alleles are dihybrids. For example, parents of genotypes *AABB* and *aabb* produce a dihybrid of genotype *AaBb*.

dihybrid cross A cross between two organisms that are each heterozygous for both of the characters being followed (or the self-pollination of a plant that is heterozygous for both characters).

dikaryotic (dī′-kār-ē-ot′-ik) Referring to a fungal mycelium with two haploid nuclei per cell, one from each parent.

dinoflagellate (dī′-nō-flaj′-uh-let) Member of a group of mostly unicellular photosynthetic algae with two flagella situated in perpendicular grooves in cellulose plates covering the cell.

dinosaur Member of an extremely diverse clade of reptiles varying in body shape, size, and habitat. Birds are the only extant dinosaurs.

dioecious (dī-ē′-shus) In plant biology, having the male and female reproductive parts on different individuals of the same species.

diploblastic Having two germ layers.

diploid cell (dip′-loyd) A cell containing two sets of chromosomes ($2n$), one set inherited from each parent.

diplomonad A protist that has modified mitochondria, two equal-sized nuclei, and multiple flagella.

directional selection Natural selection in which individuals at one end of the phenotypic range survive or reproduce more successfully than do other individuals.

disaccharide (dī-sak′-uh-rīd) A double sugar, consisting of two monosaccharides joined by a glycosidic linkage formed by a dehydration reaction.

dispersal The movement of individuals or gametes away from their parent location. This movement sometimes expands the geographic range of a population or species.

dispersion The pattern of spacing among individuals within the boundaries of a population.

disruptive selection Natural selection in which individuals on both extremes of a phenotypic range survive or reproduce more successfully than do individuals with intermediate phenotypes.

distal tubule In the vertebrate kidney, the portion of a nephron that helps refine filtrate and empties it into a collecting duct.

disturbance A natural or human-caused event that changes a biological community and usually removes organisms from it. Disturbances, such as fires and storms, play a pivotal role in structuring many communities.

disulfide bridge A strong covalent bond formed when the sulfur of one cysteine monomer bonds to the sulfur of another cysteine monomer.

DNA (deoxyribonucleic acid) (dē-ok′-sē-rī′-bō-nū-klā′-ik) A double-stranded, helical nucleic acid molecule, consisting of nucleotide monomers with a deoxyribose sugar and the nitrogenous bases adenine (A), cytosine (C), guanine (G), and thymine (T); capable of being replicated and determining the inherited structure of a cell's proteins.

DNA ligase (lī′-gās) A linking enzyme essential for DNA replication; catalyzes the covalent bonding of the 3′ end of one DNA fragment (such as an Okazaki fragment) to the 5′ end of another DNA fragment (such as a growing DNA chain).

DNA methylation The presence of methyl groups on the DNA bases (usually cytosine) of plants, animals, and fungi. (The term also refers to the process of adding methyl groups to DNA bases.)

DNA microarray assay A method to detect and measure the expression of thousands of genes at one time. Tiny amounts of a large number of single-stranded DNA fragments representing different genes are fixed to a glass slide and tested for hybridization with samples of labeled cDNA.

DNA polymerase (puh-lim′-er-ās) An enzyme that catalyzes the elongation of new DNA (for example, at a replication fork) by the addition of nucleotides to the 3′ end of an existing chain. There are several different DNA polymerases; DNA polymerase III and DNA polymerase I play major roles in DNA replication in *E. coli*.

DNA replication The process by which a DNA molecule is copied; also called DNA synthesis.

domain (1) A taxonomic category above the kingdom level. The three domains are Archaea, Bacteria, and Eukarya. (2) A discrete structural and functional region of a protein.

dominant allele An allele that is fully expressed in the phenotype of a heterozygote.

dominant species A species with substantially higher abundance or biomass than other species in a community. Dominant species exert a powerful control over the occurrence and distribution of other species.

dopamine A neurotransmitter that is a catecholamine, like epinephrine and norepinephrine.

dormancy A condition typified by extremely low metabolic rate and a suspension of growth and development.

dorsal Pertaining to the top of an animal with radial or bilateral symmetry.

dorsal lip The region above the blastopore on the dorsal side of the amphibian embryo.

double bond A double covalent bond; the sharing of two pairs of valence electrons by two atoms.

double circulation A circulatory system consisting of separate pulmonary and systemic circuits, in which blood passes through the heart after completing each circuit.

double fertilization A mechanism of fertilization in angiosperms in which two sperm cells unite with two cells in the female gametophyte (embryo sac) to form the zygote and endosperm.

double helix The form of native DNA, referring to its two adjacent antiparallel polynucleotide strands wound around an imaginary axis into a spiral shape.

Down syndrome A human genetic disease usually caused by the presence of an extra chromosome 21; characterized by developmental delays and heart and other defects that are generally treatable or non-life-threatening.

Duchenne muscular dystrophy (duh-shen′) A human genetic disease caused by a sex-linked recessive allele; characterized by progressive weakening and a loss of muscle tissue.

duodenum (dū′-uh-dēn′-um) The first section of the small intestine, where chyme from the stomach mixes with digestive juices from the pancreas, liver, and gallbladder as well as from gland cells of the intestinal wall.

duplication An aberration in chromosome structure due to fusion with a fragment from a homologous chromosome, such that a portion of a chromosome is duplicated.

dynamic stability hypothesis The concept that long food chains are less stable than short chains.

dynein (dī′-nē-un) In cilia and flagella, a large motor protein extending from one microtubule doublet to the adjacent doublet. ATP hydrolysis drives changes in dynein shape that lead to bending of cilia and flagella.

E site One of a ribosome's three binding sites for tRNA during translation. The E site is the place where discharged tRNAs leave the ribosome. (E stands for exit.)

ecdysozoan Member of a group of animal phyla identified as a clade by molecular evidence. Many ecdysozoans are molting animals.

ecdysteroid A steroid hormone, secreted by the prothoracic glands, that triggers molting in arthropods.

echinoderm (i-kī′-nō-derm) A slow-moving or sessile marine deuterostome with a water vascular system and, in larvae, bilateral symmetry. Echinoderms include sea stars, brittle stars, sea urchins, feather stars, and sea cucumbers.

ecological footprint The aggregate land and water area required by a person, city, or nation to produce all of the resources it consumes and to absorb all of the wastes it generates.

ecological niche (nich) The sum of a species' use of the biotic and abiotic resources in its environment.

ecological species concept A definition of species in terms of ecological niche, the sum of how members of the species interact with the nonliving and living parts of their environment.

ecological succession Transition in the species composition of a community following a disturbance; establishment of a community in an area virtually barren of life.

ecology The study of how organisms interact with each other and their environment.

ecosystem All the organisms in a given area as well as the abiotic factors with which they interact; one or more communities and the physical environment around them.

ecosystem ecology The study of energy flow and the cycling of chemicals among the various biotic and abiotic components in an ecosystem.

ecosystem engineer An organism that influences community structure by causing physical changes in the environment.

ecosystem service A function performed by an ecosystem that directly or indirectly benefits humans.

ecotone The transition from one type of habitat or ecosystem to another, such as the transition from a forest to a grassland.

ectoderm (ek′-tō-durm) The outermost of the three primary germ layers in animal embryos; gives rise to the outer covering and, in some phyla, the nervous system, inner ear, and lens of the eye.

ectomycorrhiza (ek′-tō-mī′-kō-rī′-zuh) Association of a fungus with a plant root system in which the fungus surrounds the roots but does not cause invagination of the host (plant) cells' plasma membranes.

ectomycorrhizal fungus A symbiotic fungus that forms sheaths of hyphae over the surface of plant roots and also grows into extracellular spaces of the root cortex.

ectoparasite A parasite that feeds on the external surface of a host.

ectopic Occurring in an abnormal location.

ectoproct A sessile, colonial lophophorate; also called a bryozoan.

ectothermic Referring to organisms for which external sources provide most of the heat for temperature regulation.

Ediacaran biota (ē′-dē-uh-keh′-run bī-ō′-tuh) An early group of soft-bodied, multicellular eukaryotes known from fossils that range in age from 565 million to 550 million years old.

effective population size An estimate of the size of a population based on the numbers of females and males that successfully breed; generally smaller than the total population.

effector cell (1) A muscle cell or gland cell that performs the body's response to stimuli as directed by signals from the brain or other processing center of the nervous system. (2) A lymphocyte that has undergone clonal selection and is capable of mediating an adaptive immune response.

egg The female gamete.

egg-polarity gene A gene that helps control the orientation (polarity) of the egg; also called a maternal effect gene.

ejaculation The propulsion of sperm from the epididymis through the muscular vas deferens, ejaculatory duct, and urethra.

ejaculatory duct In mammals, the short section of the ejaculatory route formed by the convergence of the vas deferens and a duct from the seminal vesicle. The ejaculatory duct transports sperm from the vas deferens to the urethra.

electrocardiogram (ECG or **EKG)** A record of the electrical impulses that travel through heart muscle during the cardiac cycle.

electrochemical gradient The diffusion gradient of an ion, which is affected by both the concentration difference of an ion across a membrane (a chemical force) and the ion's tendency to move relative to the membrane potential (an electrical force).

electrogenic pump An active transport protein that generates voltage across a membrane while pumping ions.

electromagnetic receptor A receptor of electromagnetic energy, such as visible light, electricity, or magnetism.

electromagnetic spectrum The entire spectrum of electromagnetic radiation, ranging in wavelength from less than a nanometer to more than a kilometer.

electron A subatomic particle with a single negative electrical charge and a mass about 1/2,000 that of a neutron or proton. One or more electrons move around the nucleus of an atom.

electron microscope (EM) A microscope that uses magnets to focus an electron beam on or through a specimen, resulting in a practical resolution of a hundredfold greater than that of a light microscope using standard techniques. A transmission electron microscope (TEM) is used to study the internal structure of thin sections of cells. A scanning electron microscope (SEM) is used to study the fine details of cell surfaces.

electron shell An energy level of electrons at a characteristic average distance from the nucleus of an atom.

electron transport chain A sequence of electron carrier molecules (membrane proteins) that shuttle electrons down a series of redox reactions that release energy used to make ATP.

electronegativity The attraction of a given atom for the electrons of a covalent bond.

electroporation A technique to introduce recombinant DNA into cells by applying a brief electrical pulse to a solution containing the cells. The pulse creates temporary holes in the cells' plasma membranes, through which DNA can enter.

element Any substance that cannot be broken down to any other substance by chemical reactions.

elimination The fourth and final stage of food processing in animals: the passing of undigested material out of the body.

embryo sac (em′-brē-ō) The female gametophyte of angiosperms, formed from the growth and division of the megaspore into a multicellular structure that typically has eight haploid nuclei.

embryonic lethal A mutation with a phenotype leading to death of an embryo or larva.

embryophyte Alternate name for land plants that refers to their shared derived trait of multicellular, dependent embryos.

emergent properties New properties that arise with each step upward in the hierarchy of life, owing to the arrangement and interactions of parts as complexity increases.

emigration The movement of individuals out of a population.

enantiomer (en-an′-tē-ō-mer) One of two compounds that are mirror images of each other and that differ in shape due to the presence of an asymmetric carbon.

endangered species A species that is in danger of extinction throughout all or a significant portion of its range.

endemic (en-dem′-ik) Referring to a species that is confined to a specific geographic area.

endergonic reaction (en′-der-gon′-ik) A nonspontaneous chemical reaction, in which free energy is absorbed from the surroundings.

endocrine gland (en′-dō-krin) A ductless gland that secretes hormones directly into the interstitial fluid, from which they diffuse into the bloodstream.

endocrine system The internal system of communication involving hormones, the ductless glands that secrete hormones, and the molecular receptors on or in target cells that respond to hormones; functions in concert with the nervous system to effect internal regulation and maintain homeostasis.

endocytosis (en′-dō-sī-tō′-sis) Cellular uptake of biological molecules and particulate matter via formation of vesicles from the plasma membrane.

endoderm (en′-dō-durm) The innermost of the three primary germ layers in animal embryos; lines the archenteron and gives rise to the liver, pancreas, lungs, and the lining of the digestive tract in species that have these structures.

endodermis In plant roots, the innermost layer of the cortex that surrounds the vascular cylinder.

endomembrane system The collection of membranes inside and surrounding a eukaryotic cell, related either through direct physical contact or by the transfer of membranous vesicles; includes the plasma membrane, the nuclear envelope, the smooth and rough endoplasmic reticulum, the Golgi apparatus, lysosomes, vesicles, and vacuoles.

endometriosis (en′-dō-mē-trē-ō′-sis) The condition resulting from the presence of endometrial tissue outside of the uterus.

endometrium (en′-dō-mē′-trē-um) The inner lining of the uterus, which is richly supplied with blood vessels.

endoparasite A parasite that lives within a host.

endophyte A fungus that lives inside a leaf or other plant part without causing harm to the plant.

endoplasmic reticulum (ER) (en′-dō-plaz′-mik ruh-tik′-yū-lum) An extensive membranous network in eukaryotic cells, continuous with the outer nuclear membrane and composed of ribosome-studded (rough) and ribosome-free (smooth) regions.

endorphin (en-dōr′-fin) Any of several hormones produced in the brain and anterior pituitary that inhibit pain perception.

endoskeleton A hard skeleton buried within the soft tissues of an animal.

endosperm In angiosperms, a nutrient-rich tissue formed by the union of a sperm with two polar nuclei during double fertilization. The endosperm provides nourishment to the developing embryo in angiosperm seeds.

endospore A thick-coated, resistant cell produced by some bacterial cells when they are exposed to harsh conditions.

endosymbiont theory The theory that mitochondria and plastids, including chloroplasts, originated as prokaryotic cells engulfed by an ancestral eukaryotic cell. The engulfed cell and its host cell then evolved into a single organism.

endosymbiosis A process in which a unicellular organism (the "host") engulfs another cell, which lives within the host cell and ultimately becomes an organelle in the host cell. *See also* endosymbiont theory.

endothelium (en′-dō-thē′-lē-um) The simple squamous layer of cells lining the lumen of blood vessels.

endothermic Referring to organisms that are warmed by heat generated by their own metabolism. This heat usually maintains a relatively stable body temperature higher than that of the external environment.

endotoxin A toxic component of the outer membrane of certain gram-negative bacteria that is released only when the bacteria die.

energetic hypothesis The concept that the length of a food chain is limited by the inefficiency of energy transfer along the chain.

energy The capacity to cause change, especially to do work (to move matter against an opposing force).

energy coupling In cellular metabolism, the use of energy released from an exergonic reaction to drive an endergonic reaction.

enhancer A segment of eukaryotic DNA containing multiple control elements, usually located far from the gene whose transcription it regulates.

enteric division One of three divisions of the autonomic nervous system; consists of networks of neurons in the digestive tract, pancreas, and gallbladder; normally regulated by the sympathetic and parasympathetic divisions of the autonomic nervous system.

entropy A measure of disorder, or randomness.

enzymatic hydrolysis The process in digestion that splits macromolecules from food by the enzymatic addition of water.

enzyme (en′-zīm) A macromolecule serving as a catalyst, a chemical agent that increases the rate of a reaction without being consumed by the reaction. Most enzymes are proteins.

enzyme-substrate complex A temporary complex formed when an enzyme binds to its substrate molecule(s).

epicotyl (ep′-uh-kot′-ul) In an angiosperm embryo, the embryonic axis above the point of attachment of the cotyledon(s) and below the first pair of miniature leaves.

epidemic A general outbreak of a disease.

epidermis (1) The dermal tissue system of nonwoody plants, usually consisting of a single layer of tightly packed cells. (2) The outermost layer of cells in an animal.

epididymis (ep′-uh-did′-uh-mus) A coiled tubule located adjacent to the mammalian testis where sperm are stored.

epigenetic inheritance Inheritance of traits transmitted by mechanisms not directly involving the nucleotide sequence of a genome.

epinephrine (ep′-i-nef′-rin) A catecholamine that, when secreted as a hormone by the adrenal medulla, mediates "fight-or-flight" responses to short-term stresses; also released by some neurons as a neurotransmitter; also known as adrenaline.

epiphyte (ep′-uh-fīt) A plant that nourishes itself but grows on the surface of another plant for support, usually on the branches or trunks of trees.

epistasis (ep′-i-stā′-sis) A type of gene interaction in which the phenotypic expression of one gene alters that of another independently inherited gene.

epithelial tissue (ep′-uh-thē′-lē-ul) Sheets of tightly packed cells that line organs and body cavities as well as external surfaces.

epithelium An epithelial tissue.

epitope A small, accessible region of an antigen to which an antigen receptor or antibody binds; also called an antigenic determinant.

EPSP *See* excitatory postsynaptic potential.

equilibrium potential (E_{ion}) The magnitude of a cell's membrane voltage at equilibrium; calculated using the Nernst equation.

erythrocyte (eh-rith′-ruh-sīt) A blood cell that contains hemoglobin, which transports oxygen; also called a red blood cell.

erythropoietin (EPO) (eh-rith′-rō-poy′-uh-tin) A hormone that stimulates the production of erythrocytes. It is secreted by the kidney when body tissues do not receive enough oxygen.

esophagus (eh-sof′-uh-gus) A muscular tube that conducts food, by peristalsis, from the pharynx to the stomach.

essential amino acid An amino acid that an animal cannot synthesize itself and must be obtained from food in prefabricated form.

essential element A chemical element required for an organism to survive, grow, and reproduce.

essential fatty acid An unsaturated fatty acid that an animal needs but cannot make.

essential nutrient A substance that an organism cannot synthesize from any other material and therefore must absorb in preassembled form.

estradiol (es′-truh-dī′-ol) A steroid hormone that stimulates the development and maintenance of the female reproductive system and secondary sex characteristics; the major estrogen in mammals.

estrogen (es′-trō-jen) Any steroid hormone, such as estradiol, that stimulates the development and maintenance of the female reproductive system and secondary sex characteristics.

estrous cycle (es′-trus) A reproductive cycle characteristic of female mammals except humans and certain other primates, in which the nonpregnant endometrium is reabsorbed rather than shed, and sexual response occurs only during mid-cycle at estrus.

estuary The area where a freshwater stream or river merges with the ocean.

ethylene (eth′-uh-lēn) A gaseous plant hormone involved in responses to mechanical stress, programmed cell death, leaf abscission, and fruit ripening.

etiolation Plant morphological adaptations for growing in darkness.

euchromatin (yū-krō′-muh-tin) The less condensed form of eukaryotic chromatin that is available for transcription.

eudicot (yū-dī′-kot) Member of a clade that contains the vast majority of flowering plants that have two embryonic seed leaves, or cotyledons.

euglenid (yū′-glen-id) A protist, such as *Euglena* or its relatives, characterized by an anterior pocket from which one or two flagella emerge.

euglenozoan Member of a diverse clade of flagellated protists that includes predatory heterotrophs, photosynthetic autotrophs, and pathogenic parasites.

Eukarya (yū-kar′-ē-uh) The domain that includes all eukaryotic organisms.

eukaryotic cell (yū′-ker-ē-ot′-ik) A type of cell with a membrane-enclosed nucleus and membrane-enclosed organelles. Organisms with eukaryotic cells (protists, plants, fungi, and animals) are called eukaryotes.

eumetazoan (yū′-met-uh-zō′-un) Member of a clade of animals with true tissues. All animals except sponges and a few other groups are eumetazoans.

eurypterid (yur-ip′-tuh-rid) An extinct carnivorous chelicerate; also called a water scorpion.

Eustachian tube (yū-stā′-shun) The tube that connects the middle ear to the pharynx.

eutherian (yū-thēr′-ē-un) Placental mammal; mammal whose young complete their embryonic development within the uterus, joined to the mother by the placenta.

eutrophic lake (yū-trōf′-ik) A lake that has a high rate of biological productivity supported by a high rate of nutrient cycling.

eutrophication A process by which nutrients, particularly phosphorus and nitrogen, become highly concentrated in a body of water, leading to increased growth of organisms such as algae or cyanobacteria.

evaporation The process by which a liquid changes to a gas.

evaporative cooling The process in which the surface of an object becomes cooler during evaporation, a result of the molecules with the greatest kinetic energy changing from the liquid to the gaseous state.

evapotranspiration The total evaporation of water from an ecosystem, including water transpired by plants and evaporated from a landscape, usually measured in millimeters and estimated for a year.

evo-devo Evolutionary developmental biology; a field of biology that compares developmental processes of different multicellular organisms to understand how these processes have evolved and how changes can modify existing organismal features or lead to new ones.

evolution Descent with modification; the idea that living species are descendants of ancestral species that were different from the present-day ones; also defined more narrowly as the change in the genetic composition of a population from generation to generation.

evolutionary tree A branching diagram that reflects a hypothesis about evolutionary relationships among groups of organisms.

Excavata One of five supergroups of eukaryotes proposed in a current hypothesis of the evolutionary history of eukaryotes. Excavates have unique cytoskeletal features, and some species have an "excavated" feeding groove on one side of the cell body. *See also* Chromalveolata, Rhizaria, Archaeplastida, and Unikonta.

excitatory postsynaptic potential (EPSP) An electrical change (depolarization) in the membrane of a postsynaptic cell caused by the binding of an excitatory neurotransmitter from a presynaptic cell to a postsynaptic receptor; makes it more likely for a postsynaptic cell to generate an action potential.

excretion The disposal of nitrogen-containing metabolites and other waste products.

exergonic reaction (ek′-ser-gon′-ik) A spontaneous chemical reaction, in which there is a net release of free energy.

exocytosis (ek′-sō-sī-tō′-sis) The cellular secretion of biological molecules by the fusion of vesicles containing them with the plasma membrane.

exon A sequence within a primary transcript that remains in the RNA after RNA processing; also refers to the region of DNA from which this sequence was transcribed.

exoskeleton A hard encasement on the surface of an animal, such as the shell of a mollusc or the cuticle of an arthropod, that provides protection and points of attachment for muscles.

exotoxin (ek′-sō-tok′-sin) A toxic protein that is secreted by a prokaryote or other pathogen and that produces specific symptoms, even if the pathogen is no longer present.

expansin Plant enzyme that breaks the cross-links (hydrogen bonds) between cellulose microfibrils and other cell wall constituents, loosening the wall's fabric.

exponential population growth Growth of a population in an ideal, unlimited environment, represented by a J-shaped curve when population size is plotted over time.

expression vector A cloning vector that contains a highly active bacterial promoter just upstream of a restriction site where a eukaryotic gene can be inserted, allowing the gene to be expressed in a bacterial cell. Expression vectors are also available that have been genetically engineered for use in specific types of eukaryotic cells.

external fertilization The fusion of gametes that parents have discharged into the environment.

extinction vortex A downward population spiral in which inbreeding and genetic drift combine to cause a small population to shrink and, unless the spiral is reversed, become extinct.

extracellular digestion The breakdown of food in compartments that are continuous with the outside of an animal's body.

extracellular matrix (ECM) The meshwork surrounding animal cells, consisting of glycoproteins, polysaccharides, and proteoglycans synthesized and secreted by the cells.

extraembryonic membrane One of four membranes (yolk sac, amnion, chorion, and allantois) located outside the embryo that support the developing embryo in reptiles and mammals.

extreme halophile An organism that lives in a highly saline environment, such as the Great Salt Lake or the Dead Sea.

extreme thermophile An organism that thrives in hot environments (often 60–80°C or hotter).

extremophile An organism that lives in environmental conditions so extreme that few other species can survive there. Extremophiles include extreme halophiles ("salt lovers") and extreme thermophiles ("heat lovers").

F factor In bacteria, the DNA segment that confers the ability to form pili for conjugation and associated functions required for the transfer of DNA from donor to recipient. The F factor may exist as a plasmid or be integrated into the bacterial chromosome.

F plasmid The plasmid form of the F factor.

F_1 generation The first filial, hybrid (heterozygous) offspring arising from a parental (P generation) cross.

F_2 generation The offspring resulting from interbreeding (or self-pollination) of the hybrid F_1 generation.

facilitated diffusion The passage of molecules or ions down their electrochemical gradient across a biological membrane with the assistance of specific transmembrane transport proteins, requiring no energy expenditure.

facilitation An interaction in which one species has a positive effect on the survival and reproduction of another species without the intimate association of a symbiosis.

facultative anaerobe (fak′-ul-tā′-tiv an′-uh-rōb) An organism that makes ATP by aerobic respiration if oxygen is present but that switches to anaerobic respiration or fermentation if oxygen is not present.

family In Linnaean classification, the taxonomic category above genus.

fast block to polyspermy The depolarization of the egg plasma membrane that begins within 1–3 seconds after a sperm binds to an egg membrane protein. The depolarization lasts about 1 minute and prevents additional sperm from fusing with the egg during that time.

fast-twitch fiber A muscle fiber used for rapid, powerful contractions.

fat A lipid consisting of three fatty acids linked to one glycerol molecule; also called a triacylglycerol or triglyceride.

fate map A territorial diagram of embryonic development that displays the future derivatives of individual cells and tissues.

fatty acid A carboxylic acid with a long carbon chain. Fatty acids vary in length and in the number and location of double bonds; three fatty acids linked to a glycerol molecule form a fat molecule, also known as a triacylglycerol or triglyceride.

feces (fē′-sēz) The wastes of the digestive tract.

feedback inhibition A method of metabolic control in which the end product of a metabolic pathway acts as an inhibitor of an enzyme within that pathway.

fermentation A catabolic process that makes a limited amount of ATP from glucose (or other organic molecules) without an electron transport chain and that produces a characteristic end product, such as ethyl alcohol or lactic acid.

fertilization (1) The union of haploid gametes to produce a diploid zygote. (2) The addition of mineral nutrients to the soil.

fetus (fē′-tus) A developing mammal that has all the major structures of an adult. In humans, the fetal stage lasts from the 9th week of gestation until birth.

fiber A lignified cell type that reinforces the xylem of angiosperms and functions in mechanical support; a slender, tapered sclerenchyma cell that usually occurs in bundles.

fibroblast (fī′-brō-blast) A type of cell in loose connective tissue that secretes the protein ingredients of the extracellular fibers.

fibronectin An extracellular glycoprotein secreted by animal cells that helps them attach to the extracellular matrix.

filament In an angiosperm, the stalk portion of the stamen, the pollen-producing reproductive organ of a flower.

filtrate Cell-free fluid extracted from the body fluid by the excretory system.

filtration In excretory systems, the extraction of water and small solutes, including metabolic wastes, from the body fluid.

fimbria (plural, **fimbriae**) A short, hairlike appendage of a prokaryotic cell that helps it adhere to the substrate or to other cells.

first law of thermodynamics The principle of conservation of energy: Energy can be transferred and transformed, but it cannot be created or destroyed.

fission The separation of an organism into two or more individuals of approximately equal size.

fixed action pattern In animal behavior, a sequence of unlearned acts that is essentially unchangeable and, once initiated, usually carried to completion.

flaccid (flas′-id) Limp. Lacking turgor (stiffness or firmness), as in a plant cell in surroundings where there is a tendency for water to leave the cell. (A walled cell becomes flaccid if it has a higher water potential than its surroundings, resulting in the loss of water.)

flagellum (fluh-jel′-um) (plural, **flagella**) A long cellular appendage specialized for locomotion. Like motile cilia, eukaryotic flagella have a core with nine outer doublet microtubules and two inner single microtubules (the "9 + 2" arrangement) ensheathed in an extension of the plasma membrane. Prokaryotic flagella have a different structure.

florigen A flowering signal, probably a protein, that is made in leaves under certain conditions and that travels to the shoot apical meristems, inducing them to switch from vegetative to reproductive growth.

flower In an angiosperm, a specialized shoot with up to four sets of modified leaves, bearing structures that function in sexual reproduction.

fluid feeder An animal that lives by sucking nutrient-rich fluids from another living organism.

fluid mosaic model The currently accepted model of cell membrane structure, which envisions the membrane as a mosaic of protein molecules drifting laterally in a fluid bilayer of phospholipids.

follicle (fol′-uh-kul) A microscopic structure in the ovary that contains the developing oocyte and secretes estrogens.

follicle-stimulating hormone (FSH) A tropic hormone that is produced and secreted by the anterior pituitary and that stimulates the production of eggs by the ovaries and sperm by the testes.

follicular phase That part of the ovarian cycle during which follicles are growing and oocytes maturing.

food chain The pathway along which food energy is transferred from trophic level to trophic level, beginning with producers.

food vacuole A membranous sac formed by phagocytosis of microorganisms or particles to be used as food by the cell.

food web The interconnected feeding relationships in an ecosystem.

foot (1) The portion of a bryophyte sporophyte that gathers sugars, amino acids, water, and minerals from the parent gametophyte via transfer cells. (2) One of the three main parts of a mollusc; a muscular structure usually used for movement. *See also* mantle, visceral mass.

foraging The seeking and obtaining of food.

foram (foraminiferan) An aquatic protist that secretes a hardened shell containing calcium carbonate and extends pseudopodia through pores in the shell.

forebrain One of three ancestral and embryonic regions of the vertebrate brain; develops into the thalamus, hypothalamus, and cerebrum.

fossil A preserved remnant or impression of an organism that lived in the past.

founder effect Genetic drift that occurs when a few individuals become isolated from a larger population and form a new population whose gene pool composition is not reflective of that of the original population.

fovea (fō′-vē-uh) The place on the retina at the eye's center of focus, where cones are highly concentrated.

fragmentation A means of asexual reproduction whereby a single parent breaks into parts that regenerate into whole new individuals.

frameshift mutation A mutation occurring when nucleotides are inserted in or deleted from a gene and the number inserted or deleted is not a multiple of three, resulting in the improper grouping of the subsequent nucleotides into codons.

free energy The portion of a biological system's energy that can perform work when temperature and pressure are uniform throughout the system. The change in free energy of a system (ΔG) is calculated by the equation $\Delta G = \Delta H - T\Delta S$, where ΔH is the change in enthalpy (in biological systems, equivalent to total energy), T is the absolute temperature, and ΔS is the change in entropy.

frequency-dependent selection Selection in which the fitness of a phenotype depends on how common the phenotype is in a population.

fruit A mature ovary of a flower. The fruit protects dormant seeds and often aids in their dispersal.

functional group A specific configuration of atoms commonly attached to the carbon skeletons of organic molecules and involved in chemical reactions.

G protein A GTP-binding protein that relays signals from a plasma membrane signal receptor, known as a G protein-coupled receptor, to other signal transduction proteins inside the cell.

G protein-coupled receptor (GPCR) A signal receptor protein in the plasma membrane that responds to the binding of a signaling molecule by activating a G protein. Also called a G protein-linked receptor.

G_0 phase A nondividing state occupied by cells that have left the cell cycle, sometimes reversibly.

G_1 phase The first gap, or growth phase, of the cell cycle, consisting of the portion of interphase before DNA synthesis begins.

G_2 phase The second gap, or growth phase, of the cell cycle, consisting of the portion of interphase after DNA synthesis occurs.

gallbladder An organ that stores bile and releases it as needed into the small intestine.

game theory An approach to evaluating alternative strategies in situations where the outcome of a particular strategy depends on the strategies used by other individuals.

gametangium (gam′-uh-tan′-jē-um) (plural, **gametangia**) Multicellular plant structure in which gametes are formed. Female gametangia are called archegonia, and male gametangia are called antheridia.

gamete (gam′-ēt) A haploid reproductive cell, such as an egg or sperm. Gametes unite during sexual reproduction to produce a diploid zygote.

gametogenesis The process by which gametes are produced.

gametophore (guh-mē′-tō-fōr) The mature gamete-producing structure of a moss gametophyte.

gametophyte (guh-mē′-tō-fīt) In organisms (plants and some algae) that have alternation of generations, the multicellular haploid form that produces haploid gametes by mitosis. The haploid gametes unite and develop into sporophytes.

gamma-aminobutyric acid (GABA) An amino acid that functions as a CNS neurotransmitter in the central nervous system of vertebrates.

ganglia (gang′-glē-uh) (singular, **ganglion**) Clusters (functional groups) of nerve cell bodies in a centralized nervous system.

ganglion cell A type of neuron in the retina that synapses with bipolar cells and transmits action potentials to the brain via axons in the optic nerve.

gap junction A type of intercellular junction in animal cells, consisting of proteins surrounding a pore that allows the passage of materials between cells.

gas exchange The uptake of molecular oxygen from the environment and the discharge of carbon dioxide to the environment.

gastric juice A digestive fluid secreted by the stomach.

gastrovascular cavity A central cavity with a single opening in the body of certain animals, including cnidarians and flatworms, that functions in both the digestion and distribution of nutrients.

gastrula (gas′-trū-luh) An embryonic stage in animal development encompassing the formation of three layers: ectoderm, mesoderm, and endoderm.

gastrulation (gas′-trū-lā′-shun) In animal development, a series of cell and tissue movements in which the blastula-stage embryo folds inward, producing a three-layered embryo, the gastrula.

gated channel A transmembrane protein channel that opens or closes in response to a particular stimulus.

gated ion channel A gated channel for a specific ion. The opening or closing of such channels may alter a cell's membrane potential.

gel electrophoresis (ē-lek′-trō-fōr-ē′-sis) A technique for separating nucleic acids or proteins on the basis of their size and electrical charge, both of which affect their rate of movement through an electric field in a gel made of agarose or another polymer.

gene A discrete unit of hereditary information consisting of a specific nucleotide sequence in DNA (or RNA, in some viruses).

gene annotation Analysis of genomic sequences to identify protein-coding genes and determine the function of their products.

gene cloning The production of multiple copies of a gene.

gene expression The process by which information encoded in DNA directs the synthesis of proteins or, in some cases, RNAs that are not translated into proteins and instead function as RNAs.

gene flow The transfer of alleles from one population to another, resulting from the movement of fertile individuals or their gametes.

gene pool The aggregate of all copies of every type of allele at all loci in every individual in a population. The term is also used in a more restricted sense as the aggregate of alleles for just one or a few loci in a population.

gene therapy The introduction of genes into an afflicted individual for therapeutic purposes.

gene-for-gene recognition A widespread form of plant disease resistance involving recognition of pathogen-derived molecules by the protein products of specific plant disease resistance genes.

genetic drift A process in which chance events cause unpredictable fluctuations in allele frequencies from one generation to the next. Effects of genetic drift are most pronounced in small populations.

genetic engineering The direct manipulation of genes for practical purposes.

genetic map An ordered list of genetic loci (genes or other genetic markers) along a chromosome.

genetic profile An individual's unique set of genetic markers, detected most often today by PCR or, previously, by electrophoresis and nucleic acid probes.

genetic recombination General term for the production of offspring with combinations of traits that differ from those found in either parent.

genetic variation Differences among individuals in the composition of their genes or other DNA segments.

genetically modified (GM) organism An organism that has acquired one or more genes by artificial means; also known as a transgenic organism.

genetics The scientific study of heredity and hereditary variation.

genome (jē′-nōm) The genetic material of an organism or virus; the complete complement of an organism's or virus's genes along with its noncoding nucleic acid sequences.

genome-wide association study A large-scale analysis of the genomes of many people having a certain phenotype or disease, with the aim of finding genetic markers that correlate with that phenotype or disease.

genomic imprinting A phenomenon in which expression of an allele in offspring depends on whether the allele is inherited from the male or female parent.

genomic library A set of cell clones containing all the DNA segments from a genome, each within a plasmid, BAC, or other cloning vector.

genomics (juh-nō′-miks) The study of whole sets of genes and their interactions within a species, as well as genome comparisons between species.

genotype (jē′-nō-tīp) The genetic makeup, or set of alleles, of an organism.

genus (jē′-nus) (plural, **genera**) A taxonomic category above the species level, designated by the first word of a species' two-part scientific name.

geographic variation Differences between the gene pools of geographically separate populations or population subgroups.

geologic record The division of Earth's history into time periods, grouped into three eons—Archaean, Proterozoic, and Phanerozoic—and further subdivided into eras, periods, and epochs.

germ layer One of the three main layers in a gastrula that will form the various tissues and organs of an animal body.

gestation (jes-tā′-shun) Pregnancy; the state of carrying developing young within the female reproductive tract.

gibberellin (jib′-uh-rel′-in) Any of a class of related plant hormones that stimulate growth in the stem and leaves, trigger the germination of seeds and breaking of bud dormancy, and (with auxin) stimulate fruit development.

glans The rounded structure at the tip of the clitoris or penis that is involved in sexual arousal.

glia (glial cells) Cells of the nervous system that support, regulate, and augment the functions of neurons.

global climate change Increase in temperature and change in weather patterns all around the planet, due mostly to increasing atmospheric CO_2 levels from the burning of fossil fuels. The increase in temperature, called global warming, is a major aspect of global climate change.

global ecology The study of the functioning and distribution of organisms across the biosphere and how the regional exchange of energy and materials affects them.

glomeromycete (glō′-mer-ō-mī′-sēt) Member of the fungal phylum Glomeromycota, characterized by a distinct branching form of mycorrhizae called arbuscular mycorrhizae.

glomerulus (glō-mār′-yū-lus) A ball of capillaries surrounded by Bowman's capsule in the nephron and serving as the site of filtration in the vertebrate kidney.

glucagon (glū′-kuh-gon) A hormone secreted by pancreatic alpha cells that raises blood glucose levels. It promotes glycogen breakdown and release of glucose by the liver.

glucocorticoid A steroid hormone that is secreted by the adrenal cortex and that influences glucose metabolism and immune function.

glutamate An amino acid that functions as a neurotransmitter in the central nervous system.

glyceraldehyde 3-phosphate (G3P) (glis′-er-al′-de-hīd) A three-carbon carbohydrate that is the direct product of the Calvin cycle; it is also an intermediate in glycolysis.

glycogen (glī′-kō-jen) An extensively branched glucose storage polysaccharide found in the liver and muscle of animals; the animal equivalent of starch.

glycolipid A lipid with one or more covalently attached carbohydrates.

glycolysis (glī-kol′-uh-sis) A series of reactions that ultimately splits glucose into pyruvate. Glycolysis occurs in almost all living cells, serving as the starting point for fermentation or cellular respiration.

glycoprotein A protein with one or more covalently attached carbohydrates.

glycosidic linkage A covalent bond formed between two monosaccharides by a dehydration reaction.

gnathostome (na′-thu-stōm) Member of the vertebrate subgroup possessing jaws.

golden alga A biflagellated, photosynthetic protist named for its color, which results from its yellow and brown carotenoids.

Golgi apparatus (gol′-jē) An organelle in eukaryotic cells consisting of stacks of flat membranous sacs that modify, store, and route products of the endoplasmic reticulum and synthesize some products, notably noncellulose carbohydrates.

gonads (gō′-nadz) The male and female sex organs; the gamete-producing organs in most animals.

grade A group of organisms that share the same level of organizational complexity or share a key adaptation.

graded potential In a neuron, a shift in the membrane potential that has an amplitude proportional to signal strength and that decays as it spreads.

Gram stain A staining method that distinguishes between two different kinds of bacterial cell walls; may be used to help determine medical response to an infection.

gram-negative Describing the group of bacteria that have a cell wall that is structurally more complex and contains less peptidoglycan than the cell wall of gram-positive bacteria. Gram-negative bacteria are often more toxic than gram-positive bacteria.

gram-positive Describing the group of bacteria that have a cell wall that is structurally less complex and contains more peptidoglycan than the cell wall of gram-negative bacteria. Gram-positive bacteria are usually less toxic than gram-negative bacteria.

granum (gran′-um) (plural, **grana**) A stack of membrane-bounded thylakoids in the chloroplast. Grana function in the light reactions of photosynthesis.

gravitropism (grav′-uh-trō′-pizm) A response of a plant or animal to gravity.

gray matter Regions of dendrites and clustered neuron cell bodies within the CNS.

green alga A photosynthetic protist, named for green chloroplasts that are similar in structure and pigment composition to those of land plants. Green algae are a paraphyletic group, some of whose members are more closely related to land plants than they are to other green algae.

greenhouse effect The warming of Earth due to the atmospheric accumulation of carbon dioxide and certain other gases, which absorb reflected infrared radiation and reradiate some of it back toward Earth.

gross primary production (GPP) The total primary production of an ecosystem.

ground tissue system Plant tissues that are neither vascular nor dermal, fulfilling a variety of functions, such as storage, photosynthesis, and support.

growth An irreversible increase in size or biomass.

growth factor (1) A protein that must be present in the extracellular environment (culture medium or animal body) for the growth and normal development of certain types of cells. (2) A local regulator that acts on nearby cells to stimulate cell proliferation and differentiation.

growth hormone (GH) A hormone that is produced and secreted by the anterior pituitary and that has both direct (nontropic) and tropic effects on a wide variety of tissues.

guard cells The two cells that flank the stomatal pore and regulate the opening and closing of the pore.

gustation The sense of taste.

guttation The exudation of water droplets from leaves, caused by root pressure in certain plants.

gymnosperm (jim′-nō-sperm) A vascular plant that bears naked seeds—seeds not enclosed in protective chambers.

hair cell A mechanosensory cell that alters output to the nervous system when hairlike projections on the cell surface are displaced.

half-life The amount of time it takes for 50% of a sample of a radioactive isotope to decay.

Hamilton's rule The principle that for natural selection to favor an altruistic act, the benefit to the recipient, devalued by the coefficient of relatedness, must exceed the cost to the altruist.

haploid cell (hap′-loyd) A cell containing only one set of chromosomes (n).

Hardy-Weinberg principle The principle that frequencies of alleles and genotypes in a population remain constant from generation to generation, provided that only Mendelian segregation and recombination of alleles are at work.

haustorium (plural, **haustoria**) (ho-stōr′-ē-um, ho-stōr′-ē-uh) In certain symbiotic fungi, a specialized hypha that can penetrate the tissues of host organisms.

heart A muscular pump that uses metabolic energy to elevate the hydrostatic pressure of the circulatory fluid (blood or hemolymph). The fluid then flows down a pressure gradient through the body and eventually returns to the heart.

heart attack The damage or death of cardiac muscle tissue resulting from prolonged blockage of one or more coronary arteries.

heart murmur A hissing sound that most often results from blood squirting backward through a leaky valve in the heart.

heart rate The frequency of heart contraction (in beats per minute).

heat The total amount of kinetic energy due to the random motion of atoms or molecules in a body of matter; also called thermal energy. Heat is energy in its most random form.

heat of vaporization The quantity of heat a liquid must absorb for 1 g of it to be converted from the liquid to the gaseous state.

heat-shock protein A protein that helps protect other proteins during heat stress. Heat-shock proteins are found in plants, animals, and microorganisms.

heavy chain One of the two types of polypeptide chains that make up an antibody molecule and B cell receptor; consists of a variable region, which contributes to the antigen-binding site, and a constant region.

helicase An enzyme that untwists the double helix of DNA at replication forks, separating the two strands and making them available as template strands.

helper T cell A type of T cell that, when activated, secretes cytokines that promote the response of B cells (humoral response) and cytotoxic T cells (cell-mediated response) to antigens.

hemoglobin (hē′-mō-glō′-bin) An iron-containing protein in red blood cells that reversibly binds oxygen.

hemolymph (hē′-mō-limf′) In invertebrates with an open circulatory system, the body fluid that bathes tissues.

hemophilia (hē′-muh-fil′-ē-uh) A human genetic disease caused by a sex-linked recessive allele resulting in the absence of one or more blood-clotting proteins; characterized by excessive bleeding following injury.

hepatic portal vein A large vessel that conveys nutrient-laden blood from the small intestine to the liver, which regulates the blood's nutrient content.

herbivore (hur′-bi-vōr′) An animal that mainly eats plants or algae.

herbivory An interaction in which an organism eats parts of a plant or alga.

heredity The transmission of traits from one generation to the next.

hermaphrodite (hur-maf′-ruh-dīt′) An individual that functions as both male and female in sexual reproduction by producing both sperm and eggs.

hermaphroditism (hur-maf′-rō-dī-tizm) A condition in which an individual has both female and male gonads and functions as both a male and female in sexual reproduction by producing both sperm and eggs.

heterochromatin (het′-er-ō krō′ muh tin) Eukaryotic chromatin that remains highly compacted during interphase and is generally not transcribed.

heterochrony (het′-uh-rok′-ruh-nē) Evolutionary change in the timing or rate of an organism's development.

heterocyst (het′-er-ō-sist) A specialized cell that engages in nitrogen fixation in some filamentous cyanobacteria; also called a *heterocyte*.

heterokaryon (het′-er-ō-kār′-ē-un) A fungal mycelium that contains two or more haploid nuclei per cell.

heteromorphic (het′-er-ō-mōr′-fik) Referring to a condition in the life cycle of plants and certain algae in which the sporophyte and gametophyte generations differ in morphology.

heterosporous (het-er-os′-pōr-us) Referring to a plant species that has two kinds of spores: microspores, which develop into male gametophytes, and megaspores, which develop into female gametophytes.

heterotroph (het′-er-ō-trōf) An organism that obtains organic food molecules by eating other organisms or substances derived from them.

heterozygote advantage Greater reproductive success of heterozygous individuals compared with homozygotes; tends to preserve variation in a gene pool.

heterozygous (het′-er-ō-zī′-gus) Having two different alleles for a given gene.

hexapod An insect or closely related wingless, six-legged arthropod.

hibernation A long-term physiological state in which metabolism decreases, the heart and respiratory system slow down, and body temperature is maintained at a lower level than normal.

high-density lipoprotein (HDL) A particle in the blood made up of thousands of cholesterol molecules and other lipids bound to a protein. HDL scavenges excess cholesterol.

hindbrain One of three ancestral and embryonic regions of the vertebrate brain; develops into the medulla oblongata, pons, and cerebellum.

histamine (his′-tuh-mēn) A substance released by mast cells that causes blood vessels to dilate and become more permeable in inflammatory and allergic responses.

histone (his′-tōn) A small protein with a high proportion of positively charged amino acids that binds to the negatively charged DNA and plays a key role in chromatin structure.

histone acetylation The attachment of acetyl groups to certain amino acids of histone proteins.

HIV (human immunodeficiency virus) The infectious agent that causes AIDS. HIV is a retrovirus.

holdfast A rootlike structure that anchors a seaweed.

holoblastic (hō′-lō-blas′-tik) Referring to a type of cleavage in which there is complete division of the egg; occurs in eggs that have little yolk (such as those of the sea urchin) or a moderate amount of yolk (such as those of the frog).

homeobox (hō′-mē-ō-boks′) A 180-nucleotide sequence within homeotic genes and some other developmental genes that is widely conserved in animals. Related sequences occur in plants and yeasts.

homeostasis (hō′-mē-ō-stā′-sis) The steady-state physiological condition of the body.

homeotic gene (hō-mē-o′-tik) Any of the master regulatory genes that control placement and spatial organization of body parts in animals, plants, and fungi by controlling the developmental fate of groups of cells.

hominin (hō′-mi-nin) A member of the human branch of the evolutionary tree. Hominins include *Homo sapiens* and our ancestors, a group of extinct species that are more closely related to us than to chimpanzees.

homologous chromosomes (hō-mol′-uh-gus) A pair of chromosomes of the same length, centromere position, and staining pattern that possess genes for the same characters at corresponding loci. One homologous chromosome is inherited from the organism's father, the other from the mother. Also called homologs, or a homologous pair.

homologous structures Structures in different species that are similar because of common ancestry.

homology (hō-mol′-ō-jē) Similarity in characteristics resulting from a shared ancestry.

homoplasy (hō′-muh-play′-zē) A similar (analogous) structure or molecular sequence that has evolved independently in two species.

homosporous (hō-mos′-puh-rus) Referring to a plant species that has a single kind of spore, which typically develops into a bisexual gametophyte.

homozygous (hō′-mō-zī′-gus) Having two identical alleles for a given gene.

horizontal cell A neuron of the retina that helps integrate the information that is sent to the brain.

horizontal gene transfer The transfer of genes from one genome to another through mechanisms such as transposable elements, plasmid exchange, viral activity, and perhaps fusions of different organisms.

hormone In multicellular organisms, one of many types of secreted chemicals that are formed in specialized cells, travel in body fluids, and act on specific target cells in other parts of the body, changing the target cells' functioning. Hormones are thus important in long-distance signaling.

hornwort A small, herbaceous, nonvascular plant that is a member of the phylum Anthocerophyta.

host The larger participant in a symbiotic relationship, often providing a home and food source for the smaller symbiont.

host range The limited number of species whose cells can be infected by a particular virus.

human chorionic gonadotropin (hCG) (kōr′-ē-on′-ik gō-na′-dō-trō′-pin) A hormone secreted by the chorion that maintains the corpus luteum of the ovary during the first three months of pregnancy.

Human Genome Project An international collaborative effort to map and sequence the DNA of the entire human genome.

humoral immune response (hyū′-mer-ul) The branch of adaptive immunity that involves the activation of B cells and that leads to the production of antibodies, which defend against bacteria and viruses in body fluids.

humus (hyū′-mus) Decomposing organic material that is a component of topsoil.

Huntington's disease A human genetic disease caused by a dominant allele; characterized by uncontrollable body movements and degeneration of the nervous system; usually fatal 10 to 20 years after the onset of symptoms.

hybrid Offspring that results from the mating of individuals from two different species or from two true-breeding varieties of the same species.

hybrid zone A geographic region in which members of different species meet and mate, producing at least some offspring of mixed ancestry.

hybridization In genetics, the mating, or crossing, of two true-breeding varieties.

hydration shell The sphere of water molecules around a dissolved ion.

hydrocarbon An organic molecule consisting only of carbon and hydrogen.

hydrogen bond A type of weak chemical bond that is formed when the slightly positive hydrogen atom of a polar covalent bond in one molecule is attracted to the slightly negative atom of a polar covalent bond in another molecule or in another region of the same molecule.

hydrogen ion A single proton with a charge of 1+. The dissociation of a water molecule (H_2O) leads to the generation of a hydroxide ion (OH^-) and a hydrogen ion (H^+); in water, H^+ is not found alone but associates with a water molecule to form a hydronium ion.

hydrolysis (hī-drol′-uh-sis) A chemical reaction that breaks bonds between two molecules by the addition of water; functions in disassembly of polymers to monomers.

hydronium ion A water molecule that has an extra proton bound to it; H_3O^+, commonly represented as H^+.

hydrophilic (hī′-drō-fil′-ik) Having an affinity for water.

hydrophobic (hī′-drō-fō′-bik) Having no affinity for water; tending to coalesce and form droplets in water.

hydrophobic interaction A type of weak chemical interaction caused when molecules that do not mix with water coalesce to exclude water.

hydroponic culture A method in which plants are grown in mineral solutions rather than in soil.

hydrostatic skeleton A skeletal system composed of fluid held under pressure in a closed body compartment; the main skeleton of most cnidarians, flatworms, nematodes, and annelids.

hydroxide ion A water molecule that has lost a proton; OH^-.

hydroxyl group (hī-drok′-sil) A chemical group consisting of an oxygen atom joined to a hydrogen atom. Molecules possessing this group are soluble in water and are called alcohols.

hymen A thin membrane that partly covers the vaginal opening in the human female. The hymen is ruptured by sexual intercourse or other vigorous activity.

hyperpolarization A change in a cell's membrane potential such that the inside of the membrane becomes more negative relative to the outside. Hyperpolarization reduces the chance that a neuron will transmit a nerve impulse.

hypersensitive response A plant's localized defense response to a pathogen, involving the death of cells around the site of infection.

hypertension A disorder in which blood pressure remains abnormally high.

hypertonic Referring to a solution that, when surrounding a cell, will cause the cell to lose water.

hypha (plural, **hyphae**) (hī′-fuh, hī′-fē) One of many connected filaments that collectively make up the mycelium of a fungus.

hypocotyl (hī′-puh-cot′-ul) In an angiosperm embryo, the embryonic axis below the point of attachment of the cotyledon(s) and above the radicle.

hypothalamus (hī′-pō-thal′-uh-mus) The ventral part of the vertebrate forebrain; functions in maintaining homeostasis, especially in coordinating the endocrine and nervous systems; secretes hormones of the posterior pituitary and releasing factors that regulate the anterior pituitary.

hypothesis (hī-poth′-uh-sis) A testable explanation for a set of observations based on the available data and guided by inductive reasoning. A hypothesis is narrower in scope than a theory.

hypotonic Referring to a solution that, when surrounding a cell, will cause the cell to take up water.

imbibition The physical adsorption of water onto the internal surfaces of structures.

immigration The influx of new individuals into a population from other areas.

immune system An animal body's system of defenses against agents that cause disease.

immunization The process of generating a state of immunity by artificial means. In active immunization, also called vaccination, an inactive or weakened form of a pathogen is administered, inducing B and T cell responses and immunological memory. In passive immunization, antibodies specific for a particular microbe are administered, conferring immediate but temporary protection.

immunodeficiency A disorder in which the ability of an immune system to protect against pathogens is defective or absent.

immunoglobulin (Ig) (im′-yū-nō-glob′-yū-lin) Any of the class of proteins that function as antibodies. Immunoglobulins are divided into five major classes that differ in their distribution in the body and antigen disposal activities.

imprinting In animal behavior, the formation at a specific stage in life of a long-lasting behavioral response to a specific individual or object. *See also* genomic imprinting.

***in situ* hybridization** A technique using nucleic acid hybridization with a labeled probe to detect the location of a specific mRNA in an intact organism.

***in vitro* fertilization (IVF)** (vē′-trō) Fertilization of oocytes in laboratory containers followed by artificial implantation of the early embryo in the mother's uterus.

***in vitro* mutagenesis** A technique used to discover the function of a gene by cloning it, introducing specific changes into the cloned gene's sequence, reinserting the mutated gene into a cell, and studying the phenotype of the mutant.

inclusive fitness The total effect an individual has on proliferating its genes by producing its own offspring and by providing aid that enables other close relatives to increase production of their offspring.

incomplete dominance The situation in which the phenotype of heterozygotes is intermediate between the phenotypes of individuals homozygous for either allele.

incomplete flower A flower in which one or more of the four basic floral organs (sepals, petals, stamens, or carpels) are either absent or nonfunctional.

incomplete metamorphosis A type of development in certain insects, such as grasshoppers, in which the young (called nymphs) resemble adults but are smaller and have different body proportions. The nymph goes through a series of molts, each time looking more like an adult, until it reaches full size.

indeterminate cleavage A type of embryonic development in deuterostomes in which each cell produced by early cleavage divisions retains the capacity to develop into a complete embryo.

indeterminate growth A type of growth characteristic of plants, in which the organism continues to grow as long as it lives.

induced fit Caused by entry of the substrate, the change in shape of the active site of an enzyme so that it binds more snugly to the substrate.

inducer A specific small molecule that binds to a bacterial repressor protein and changes the repressor's shape so that it cannot bind to an operator, thus switching an operon on.

induction The process in which one group of embryonic cells influences the development of another, usually by causing changes in gene expression.

inductive reasoning A type of logic in which generalizations are based on a large number of specific observations.

inflammatory response An innate immune defense triggered by physical injury or infection of tissue involving the release of substances that promote swelling, enhance the infiltration of white blood cells, and aid in tissue repair and destruction of invading pathogens.

inflorescence A group of flowers tightly clustered together.

ingestion The first stage of food processing in animals: the act of eating.

ingroup A species or group of species whose evolutionary relationships we seek to determine.

inhibin A hormone produced in the male and female gonads that functions in part by regulating the anterior pituitary by negative feedback.

inhibitory postsynaptic potential (IPSP) An electrical change (usually hyperpolarization) in the membrane of a postsynaptic neuron caused by the binding of an inhibitory neurotransmitter from a presynaptic cell to a postsynaptic receptor; makes it more difficult for a postsynaptic neuron to generate an action potential.

innate behavior Animal behavior that is developmentally fixed and under strong genetic control. Innate behavior is exhibited in virtually the same form by all individuals in a population despite internal and external environmental differences during development and throughout their lifetimes.

innate immunity A form of defense common to all animals that is active immediately upon exposure to pathogens and that is the same whether or not the pathogen has been encountered previously.

inner cell mass An inner cluster of cells at one end of a mammalian blastocyst that subsequently develops into the embryo proper and some of the extraembryonic membranes.

inner ear One of three main regions of the vertebrate ear; includes the cochlea (which in turn contains the organ of Corti) and the semicircular canals.

inositol trisphosphate (IP_3) (in-ō′-suh-tol) A second messenger that functions as an intermediate between certain signaling molecules and a subsequent second messenger, Ca^{2+}, by causing a rise in cytoplasmic Ca^{2+} concentration.

inquiry The search for information and explanation, often focusing on specific questions.

insertion A mutation involving the addition of one or more nucleotide pairs to a gene.

insulin (in′-suh-lin) A hormone secreted by pancreatic beta cells that lowers blood glucose levels. It promotes the uptake of glucose by most body cells and the synthesis and storage of glycogen in the liver and also stimulates protein and fat synthesis.

integral protein A transmembrane protein with hydrophobic regions that extend into and often completely span the hydrophobic interior of the membrane and with hydrophilic regions in contact with the aqueous solution on one or both sides of the membrane (or lining the channel in the case of a channel protein).

integrin In animal cells, a transmembrane receptor protein with two subunits that interconnects the extracellular matrix and the cytoskeleton.

integument (in-teg′-yū-ment) Layer of sporophyte tissue that contributes to the structure of an ovule of a seed plant.

integumentary system The outer covering of a mammal's body, including skin, hair, and nails, claws, or hooves.

intercalated disk (in-ter′-kuh-lā′-ted) A specialized junction between cardiac muscle cells that provides direct electrical coupling between the cells.

interferon (in′-ter-fēr′-on) A protein that has antiviral or immune regulatory functions. Interferon-α and interferon-β, secreted by virus-infected cells, help nearby cells resist viral infection; interferon-γ, secreted by T cells, helps activate macrophages.

Glossary

intermediate disturbance hypothesis The concept that moderate levels of disturbance can foster greater species diversity than low or high levels of disturbance.

intermediate filament A component of the cytoskeleton that includes filaments intermediate in size between microtubules and microfilaments.

internal fertilization The fusion of eggs and sperm within the female reproductive tract. The sperm are typically deposited in or near the tract.

interneuron An association neuron; a nerve cell within the central nervous system that forms synapses with sensory and/or motor neurons and integrates sensory input and motor output.

internode A segment of a plant stem between the points where leaves are attached.

interphase The period in the cell cycle when the cell is not dividing. During interphase, cellular metabolic activity is high, chromosomes and organelles are duplicated, and cell size may increase. Interphase often accounts for about 90% of the cell cycle.

intersexual selection Selection whereby individuals of one sex (usually females) are choosy in selecting their mates from individuals of the other sex; also called mate choice.

interspecific competition Competition for resources between individuals of two or more species when resources are in short supply.

interspecific interaction A relationship between individuals of two or more species in a community.

interstitial fluid The fluid filling the spaces between cells in most animals.

intertidal zone The shallow zone of the ocean adjacent to land and between the high- and low-tide lines.

intracytoplasmic sperm injection (ICSI) The fertilization of an egg in the laboratory by the direct injection of a single sperm.

intrasexual selection Selection in which there is direct competition among individuals of one sex for mates of the opposite sex.

introduced species A species moved by humans, either intentionally or accidentally, from its native location to a new geographic region; also called non-native or exotic species.

intron (in′-tron) A noncoding, intervening sequence within a primary transcript that is removed from the transcript during RNA processing; also refers to the region of DNA from which this sequence was transcribed.

invasive species A species, often introduced by humans, that takes hold outside its native range.

inversion An aberration in chromosome structure resulting from reattachment of a chromosomal fragment in a reverse orientation to the chromosome from which it originated.

invertebrate An animal without a backbone. Invertebrates make up 95% of animal species.

ion (ī′-on) An atom or group of atoms that has gained or lost one or more electrons, thus acquiring a charge.

ion channel A transmembrane protein channel that allows a specific ion to diffuse across the membrane down its concentration or electrochemical gradient.

ionic bond (ī-on′-ik) A chemical bond resulting from the attraction between oppositely charged ions.

ionic compound A compound resulting from the formation of an ionic bond; also called a salt.

IPSP *See* inhibitory postsynaptic potential.

iris The colored part of the vertebrate eye, formed by the anterior portion of the choroid.

isomer (ī′-sō-mer) One of several compounds with the same molecular formula but different structures and therefore different properties. The three types of isomers are structural isomers, *cis-trans* isomers, and enantiomers.

isomorphic Referring to alternating generations in plants and certain algae in which the sporophytes and gametophytes look alike, although they differ in chromosome number.

isopod A member of one of the largest groups of crustaceans, which includes terrestrial, freshwater, and marine species. Among the terrestrial isopods are the pill bugs, or wood lice.

isotonic (ī′-sō-ton′-ik) Referring to a solution that, when surrounding a cell, causes no net movement of water into or out of the cell.

isotope (ī′-sō-tōp′) One of several atomic forms of an element, each with the same number of protons but a different number of neutrons, thus differing in atomic mass.

iteroparity Reproduction in which adults produce offspring over many years; also known as repeated reproduction.

joule (J) A unit of energy: 1 J = 0.239 cal; 1 cal = 4.184 J.

juxtaglomerular apparatus (JGA) (juks′-tuh-gluh-mār′-yū-ler) A specialized tissue in nephrons that releases the enzyme renin in response to a drop in blood pressure or volume.

juxtamedullary nephron In mammals and birds, a nephron with a loop of Henle that extends far into the renal medulla.

karyogamy (kār′-ē-og′-uh-mē) In fungi, the fusion of haploid nuclei contributed by the two parents; occurs as one stage of sexual reproduction, preceded by plasmogamy.

karyotype (kār′-ē-ō-tīp) A display of the chromosome pairs of a cell arranged by size and shape.

keystone species A species that is not necessarily abundant in a community yet exerts strong control on community structure by the nature of its ecological role or niche.

kidney In vertebrates, one of a pair of excretory organs where blood filtrate is formed and processed into urine.

kilocalorie (kcal) A thousand calories; the amount of heat energy required to raise the temperature of 1 kg of water by 1°C.

kin selection Natural selection that favors altruistic behavior by enhancing the reproductive success of relatives.

kinetic energy (kuh-net′-ik) The energy associated with the relative motion of objects. Moving matter can perform work by imparting motion to other matter.

kinetochore (kuh-net′-uh-kōr) A structure of proteins attached to the centromere that links each sister chromatid to the mitotic spindle.

kinetoplastid A protist, such as a trypanosome, that has a single large mitochondrion that houses an organized mass of DNA.

kingdom A taxonomic category, the second broadest after domain.

***K*-selection** Selection for life history traits that are sensitive to population density; also called density-dependent selection.

labia majora A pair of thick, fatty ridges that encloses and protects the rest of the vulva.

labia minora A pair of slender skin folds that surrounds the openings of the vagina and urethra.

labor A series of strong, rhythmic contractions of the uterus that expels a baby out of the uterus and vagina during childbirth.

lactation The continued production of milk from the mammary glands.

lacteal (lak′-tē-ul) A tiny lymph vessel extending into the core of an intestinal villus and serving as the destination for absorbed chylomicrons.

lactic acid fermentation Glycolysis followed by the reduction of pyruvate to lactate, regenerating NAD^+ with no release of carbon dioxide.

lagging strand A discontinuously synthesized DNA strand that elongates by means of Okazaki fragments, each synthesized in a $5' \rightarrow 3'$ direction away from the replication fork.

lancelet Member of the clade Cephalochordata, small blade-shaped marine chordates that lack a backbone.

landscape An area containing several different ecosystems linked by exchanges of energy, materials, and organisms.

landscape ecology The study of how the spatial arrangement of habitat types affects the distribution and abundance of organisms and ecosystem processes.

large intestine The portion of the vertebrate alimentary canal between the small intestine and the anus; functions mainly in water absorption and the formation of feces.

larva (lar′-vuh) (plural, **larvae**) A free-living, sexually immature form in some animal life cycles that may differ from the adult animal in morphology, nutrition, and habitat.

larynx (lār′-inks) The portion of the respiratory tract containing the vocal cords; also called the voice box.

lateral geniculate nucleus One of a pair of structures in the brain that are the destination for most of the ganglion cell axons that form the optic nerves.

lateral inhibition A process that sharpens the edges and enhances the contrast of a perceived image by inhibiting receptors lateral to those that have responded to light.

lateral line system A mechanoreceptor system consisting of a series of pores and receptor units along the sides of the body in fishes and aquatic amphibians; detects water movements made by the animal itself and by other moving objects.

lateral meristem (mār′-uh-stem) A meristem that thickens the roots and shoots of woody plants. The vascular cambium and cork cambium are lateral meristems.

lateral root A root that arises from the pericycle of an established root.

lateralization Segregation of functions in the cortex of the left and right cerebral hemispheres.

law of conservation of mass A physical law stating that matter can change form but cannot be created or destroyed. In a closed system, the mass of the system is constant.

law of independent assortment Mendel's second law, stating that each pair of alleles segregates, or assorts, independently of each other pair during gamete formation; applies when genes for two characters are located on different pairs of homologous chromosomes or when they are far enough apart on the same chromosome to behave as though they are on different chromosomes.

law of segregation Mendel's first law, stating that the two alleles in a pair segregate (separate from each other) into different gametes during gamete formation.

leading strand The new complementary DNA strand synthesized continuously along the template strand toward the replication fork in the mandatory $5' \rightarrow 3'$ direction.

leaf The main photosynthetic organ of vascular plants.

leaf primordium A finger-like projection along the flank of a shoot apical meristem, from which a leaf arises.

learning The modification of behavior based on specific experiences.

lens The structure in an eye that focuses light rays onto the photoreceptors.

lenticel (len′-ti-sel) A small raised area in the bark of stems and roots that enables gas exchange between living cells and the outside air.

lepidosaur (leh-pid′-uh-sōr) Member of the reptilian group that includes lizards, snakes, and two species of New Zealand animals called tuataras.

leptin A hormone produced by adipose (fat) cells that acts as a satiety factor in regulating appetite.

leukocyte (lū′-kō-sīt′) A blood cell that functions in fighting infections; also called a white blood cell.

Leydig cell (lī′-dig) A cell that produces testosterone and other androgens and is located between the seminiferous tubules of the testes.

lichen The mutualistic association between a fungus and a photosynthetic alga or cyanobacterium.

life cycle The generation-to-generation sequence of stages in the reproductive history of an organism.

life history The traits that affect an organism's schedule of reproduction and survival.

life table An age-specific summary of the survival pattern of a population.

ligament A fibrous connective tissue that joins bones together at joints.

ligand (lig′-und) A molecule that binds specifically to another molecule, usually a larger one.

ligand-gated ion channel A transmembrane protein containing a pore that opens or closes as it changes shape in response to a signaling molecule (ligand), allowing or blocking the flow of specific ions; also called an ionotropic receptor.

light chain One of the two types of polypeptide chains that make up an antibody molecule and B cell receptor; consists of a variable region, which contributes to the antigen-binding site, and a constant region.

light microscope (LM) An optical instrument with lenses that refract (bend) visible light to magnify images of specimens.

light reactions The first of two major stages in photosynthesis (preceding the Calvin cycle). These reactions, which occur on the thylakoid membranes of the chloroplast or on membranes of certain prokaryotes, convert solar energy to the chemical energy of ATP and NADPH, releasing oxygen in the process.

light-harvesting complex A complex of proteins associated with pigment molecules (including chlorophyll *a*, chlorophyll *b*, and carotenoids) that captures light energy and transfers it to reaction-center pigments in a photosystem.

lignin (lig′-nin) A hard material embedded in the cellulose matrix of vascular plant cell walls that provides structural support in terrestrial species.

limiting nutrient An element that must be added for production to increase in a particular area.

limnetic zone In a lake, the well-lit, open surface waters far from shore.

linear electron flow A route of electron flow during the light reactions of photosynthesis that involves both photosystems (I and II) and produces ATP, NADPH, and O_2. The net electron flow is from H_2O to $NADP^+$.

linkage map A genetic map based on the frequencies of recombination between markers during crossing over of homologous chromosomes.

linked genes Genes located close enough together on a chromosome that they tend to be inherited together.

lipid (lip′-id) Any of a group of large biological molecules, including fats, phospholipids, and steroids, that mix poorly, if at all, with water.

littoral zone In a lake, the shallow, well-lit waters close to shore.

liver A large internal organ in vertebrates that performs diverse functions, such as producing bile, maintaining blood glucose level, and detoxifying poisonous chemicals in the blood.

liverwort A small, herbaceous, nonvascular plant that is a member of the phylum Hepatophyta.

loam The most fertile soil type, made up of roughly equal amounts of sand, silt, and clay.

lobe-fin Member of the vertebrate clade Sarcopterygii, osteichthyans with rod-shaped muscular fins, including coelacanths, lungfishes, and tetrapods.

local regulator A secreted molecule that influences cells near where it is secreted.

locomotion Active motion from place to place.

locus (lō′-kus) (plural, **loci**) A specific place along the length of a chromosome where a given gene is located.

logistic population growth Population growth that levels off as population size approaches carrying capacity.

long-day plant A plant that flowers (usually in late spring or early summer) only when the light period is longer than a critical length.

long-term memory The ability to hold, associate, and recall information over one's lifetime.

long-term potentiation (LTP) An enhanced responsiveness to an action potential (nerve signal) by a receiving neuron.

loop of Henle The hairpin turn, with a descending and ascending limb, between the proximal and distal tubules of the vertebrate kidney; functions in water and salt reabsorption.

lophophore (lof′-uh-fōr) In some lophotrochozoan animals, including brachiopods, a crown of ciliated tentacles that surround the mouth and function in feeding.

lophotrochozoan Member of a group of animal phyla identified as a clade by molecular evidence. Lophotrochozoans include organisms that have lophophores or trochophore larvae.

low-density lipoprotein (LDL) A particle in the blood made up of thousands of cholesterol molecules and other lipids bound to a protein. LDL transports cholesterol from the liver for incorporation into cell membranes.

lung An infolded respiratory surface of a terrestrial vertebrate, land snail, or spider that connects to the atmosphere by narrow tubes.

luteal phase That portion of the ovarian cycle during which endocrine cells of the corpus luteum secrete female hormones.

luteinizing hormone (LH) (lū′-tē-uh-nī′-zing) A tropic hormone that is produced and secreted by the anterior pituitary and that stimulates ovulation in females and androgen production in males.

lycophyte (lī′-kuh-fīt) An informal name for a member of the phylum Lycophyta, which includes club mosses, spike mosses, and quillworts.

lymph The colorless fluid, derived from interstitial fluid, in the lymphatic system of vertebrates.

lymph node An organ located along a lymph vessel. Lymph nodes filter lymph and contain cells that attack viruses and bacteria.

lymphatic system A system of vessels and nodes, separate from the circulatory system, that returns fluid, proteins, and cells to the blood.

lymphocyte A type of white blood cell that mediates immune responses. The two main classes are B cells and T cells.

lysogenic cycle (lī′-sō-jen′-ik) A type of phage replicative cycle in which the viral genome becomes incorporated into the bacterial host chromosome as a prophage, is replicated along with the chromosome, and does not kill the host.

lysosome (lī′-suh-sōm) A membrane-enclosed sac of hydrolytic enzymes found in the cytoplasm of animal cells and some protists.

lysozyme (lī′-sō-zīm) An enzyme that destroys bacterial cell walls; in mammals, found in sweat, tears, and saliva.

Glossary

lytic cycle (lit′-ik) A type of phage replicative cycle resulting in the release of new phages by lysis (and death) of the host cell.

macroclimate Large-scale patterns in climate; the climate of an entire region.

macroevolution Evolutionary change above the species level. Examples of macroevolutionary change include the origin of a new group of organisms through a series of speciation events and the impact of mass extinctions on the diversity of life and its subsequent recovery.

macromolecule A giant molecule formed by the joining of smaller molecules, usually by a dehydration reaction. Polysaccharides, proteins, and nucleic acids are macromolecules.

macronutrient An essential element that an organism must obtain in relatively large amounts. *See also* micronutrient.

macrophage (mak′-rō-fāj) A phagocytic cell present in many tissues that functions in innate immunity by destroying microbes and in acquired immunity as an antigen-presenting cell.

magnoliid Member of the angiosperm clade that is most closely related to the combined eudicot and monocot clades. Extant examples are magnolias, laurels, and black pepper plants.

major depressive disorder A mood disorder characterized by feelings of sadness, lack of self-worth, emptiness, or loss of interest in nearly all things.

major histocompatibility complex (MHC) molecule A host protein that functions in antigen presentation. Foreign MHC molecules on transplanted tissue can trigger T cell responses that may lead to rejection of the transplant.

malignant tumor A cancerous tumor containing cells that have significant genetic and cellular changes and are capable of invading and surviving in new sites. Malignant tumors can impair the functions of one or more organs.

Malpighian tubule (mal-pig′-ē-un) A unique excretory organ of insects that empties into the digestive tract, removes nitrogenous wastes from the hemolymph, and functions in osmoregulation.

mammal Member of the class Mammalia, amniotes that have hair and mammary glands (glands that produce milk).

mammary gland An exocrine gland that secretes milk to nourish the young. Mammary glands are characteristic of mammals.

mandible One of a pair of jaw-like feeding appendages found in myriapods, hexapods, and crustaceans.

mantle One of the three main parts of a mollusc; a fold of tissue that drapes over the mollusc's visceral mass and may secrete a shell. *See also* foot, visceral mass.

mantle cavity A water-filled chamber that houses the gills, anus, and excretory pores of a mollusc.

map unit A unit of measurement of the distance between genes. One map unit is equivalent to a 1% recombination frequency.

marine benthic zone The ocean floor.

mark-recapture method A sampling technique used to estimate the size of animal populations.

marsupial (mar-sū′-pē-ul) A mammal, such as a koala, kangaroo, or opossum, whose young complete their embryonic development inside a maternal pouch called the marsupium.

mass extinction The elimination of a large number of species throughout Earth, the result of global environmental changes.

mass number The sum of the number of protons and neutrons in an atom's nucleus.

mast cell A vertebrate body cell that produces histamine and other molecules that trigger inflammation in response to infection and in allergic reactions.

mate-choice copying Behavior in which individuals in a population copy the mate choice of others, apparently as a result of social learning.

maternal effect gene A gene that, when mutant in the mother, results in a mutant phenotype in the offspring, regardless of the offspring's genotype. Maternal effect genes, also called egg-polarity genes, were first identified in *Drosophila melanogaster.*

matter Anything that takes up space and has mass.

maximum likelihood As applied to molecular systematics, a principle that states that when considering multiple phylogenetic hypotheses, one should take into account the hypothesis that reflects the most likely sequence of evolutionary events, given certain rules about how DNA changes over time.

maximum parsimony A principle that states that when considering multiple explanations for an observation, one should first investigate the simplest explanation that is consistent with the facts.

mechanoreceptor A sensory receptor that detects physical deformation in the body's environment associated with pressure, touch, stretch, motion, or sound.

medulla oblongata (meh-dul′-uh ōb′-long-go′-tuh) The lowest part of the vertebrate brain, commonly called the medulla; a swelling of the hindbrain anterior to the spinal cord that controls autonomic, homeostatic functions, including breathing, heart and blood vessel activity, swallowing, digestion, and vomiting.

Medusa (plural, **medusae**) (muh-dū′-suh) The floating, flattened, mouth-down version of the cnidarian body plan. The alternate form is the polyp.

megapascal (MPa) (meg′-uh-pas-kal′) A unit of pressure equivalent to about 10 atmospheres of pressure.

megaphyll (meh′-guh-fil) A leaf with a highly branched vascular system, characteristic of the vast majority of vascular plants. *See* microphyll.

megaspore A spore from a heterosporous plant species that develops into a female gametophyte.

meiosis (mī-ō′-sis) A modified type of cell division in sexually reproducing organisms consisting of two rounds of cell division but only one round of DNA replication. It results in cells with half the number of chromosome sets as the original cell.

meiosis I The first division of a two-stage process of cell division in sexually reproducing organisms that results in cells with half the number of chromosome sets as the original cell.

meiosis II The second division of a two-stage process of cell division in sexually reproducing organisms that results in cells with half the number of chromosome sets as the original cell.

melanocyte-stimulating hormone (MSH) A hormone produced and secreted by the anterior pituitary with multiple activities, including regulating the behavior of pigment-containing cells in the skin of some vertebrates.

melatonin A hormone that is secreted by the pineal gland and that is involved in the regulation of biological rhythms and sleep.

membrane potential The difference in electrical charge (voltage) across a cell's plasma membrane due to the differential distribution of ions. Membrane potential affects the activity of excitable cells and the transmembrane movement of all charged substances.

memory cell One of a clone of long-lived lymphocytes, formed during the primary immune response, that remains in a lymphoid organ until activated by exposure to the same antigen that triggered its formation. Activated memory cells mount the secondary immune response.

menopause The cessation of ovulation and menstruation marking the end of a human female's reproductive years.

menstrual cycle (men′-strū-ul) In humans and certain other primates, a type of reproductive cycle in which the nonpregnant endometrium is shed through the cervix into the vagina; also called the uterine cycle.

menstrual flow phase That portion of the uterine (menstrual) cycle when menstrual bleeding occurs.

menstruation The shedding of portions of the endometrium during a uterine (menstrual) cycle.

meristem (mār′-uh-stem) Plant tissue that remains embryonic as long as the plant lives, allowing for indeterminate growth.

meristem identity gene A plant gene that promotes the switch from vegetative growth to flowering.

meroblastic (mār′-ō-blas′-tik) Referring to a type of cleavage in which there is incomplete division of a yolk-rich egg, characteristic of avian development.

mesoderm (mez′-ō-derm) The middle primary germ layer in a triploblastic animal embryo; develops into the notochord, the lining of the coelom, muscles, skeleton, gonads, kidneys, and most of the circulatory system in species that have these structures.

mesohyl (mez′-ō-hīl) A gelatinous region between the two layers of cells of a sponge.

mesophyll (mez′-ō-fil) Leaf cells specialized for photosynthesis. In C_3 and CAM plants, mesophyll cells are located between the upper and lower epidermis; in C_4 plants, they are located between the bundle-sheath cells and the epidermis.

messenger RNA (mRNA) A type of RNA, synthesized using a DNA template, that attaches to ribosomes in the cytoplasm and specifies the primary structure of a protein. (In eukaryotes, the primary RNA transcript must undergo RNA processing to become mRNA.)

metabolic pathway A series of chemical reactions that either builds a complex molecule (anabolic pathway) or breaks down a complex molecule to simpler molecules (catabolic pathway).

metabolic rate The total amount of energy an animal uses in a unit of time.

metabolism (muh-tab′-uh-lizm) The totality of an organism's chemical reactions, consisting of catabolic and anabolic pathways, which manage the material and energy resources of the organism.

metagenomics The collection and sequencing of DNA from a group of species, usually an environmental sample of microorganisms. Computer software sorts partial sequences and assembles them into genome sequences of individual species making up the sample.

metamorphosis (met′-uh-mōr′-fuh-sis) A developmental transformation that turns an animal larva into either an adult or an adult-like stage that is not yet sexually mature.

metanephridium (met′-uh-nuh-frid′-ē-um) (plural, **metanephridia**) An excretory organ found in many invertebrates that typically consists of tubules connecting ciliated internal openings to external openings.

metaphase The third stage of mitosis, in which the spindle is complete and the chromosomes, attached to microtubules at their kinetochores, are all aligned at the metaphase plate.

metaphase plate An imaginary structure located at a plane midway between the two poles of a cell in metaphase on which the centromeres of all the duplicated chromosomes are located.

metapopulation A group of spatially separated populations of one species that interact through immigration and emigration.

metastasis (muh-tas′-tuh-sis) The spread of cancer cells to locations distant from their original site.

methanogen (meth-an′-ō-jen) An organism that produces methane as a waste product of the way it obtains energy. All known methanogens are in domain Archaea.

methyl group A chemical group consisting of a carbon bonded to three hydrogen atoms. The methyl group may be attached to a carbon or to a different atom.

microclimate Climate patterns on a very fine scale, such as the specific climatic conditions underneath a log.

microevolution Evolutionary change below the species level; change in the allele frequencies in a population over generations.

microfilament A cable composed of actin proteins in the cytoplasm of almost every eukaryotic cell, making up part of the cytoskeleton and acting alone or with myosin to cause cell contraction; also known as an actin filament.

micronutrient An essential element that an organism needs in very small amounts. *See also* macronutrient.

microphyll (mī′-krō-fil) In lycophytes, a small leaf with a single unbranched vein. *See* megaphyll.

micropyle A pore in the integuments of an ovule.

microRNA (miRNA) A small, single-stranded RNA molecule, generated from a hairpin structure on a precursor RNA transcribed from a particular gene. The miRNA associates with one or more proteins in a complex that can degrade or prevent translation of an mRNA with a complementary sequence.

microspore A spore from a heterosporous plant species that develops into a male gametophyte.

microtubule A hollow rod composed of tubulin proteins that makes up part of the cytoskeleton in all eukaryotic cells and is found in cilia and flagella.

microvillus (plural, **microvilli**) One of many fine, finger-like projections of the epithelial cells in the lumen of the small intestine that increase its surface area.

midbrain One of three ancestral and embryonic regions of the vertebrate brain; develops into sensory integrating and relay centers that send sensory information to the cerebrum.

middle ear One of three main regions of the vertebrate ear; in mammals, a chamber containing three small bones (the malleus, incus, and stapes) that convey vibrations from the eardrum to the oval window.

middle lamella (luh-mel′-uh) In plants, a thin layer of adhesive extracellular material, primarily pectins, found between the primary walls of adjacent young cells.

migration A regular, long-distance change in location.

mineral In nutrition, a simple nutrient that is inorganic and therefore cannot be synthesized in the body.

mineralocorticoid A steroid hormone secreted by the adrenal cortex that regulates salt and water homeostasis.

minimum viable population (MVP) The smallest population size at which a species is able to sustain its numbers and survive.

mismatch repair The cellular process that uses specific enzymes to remove and replace incorrectly paired nucleotides.

missense mutation A nucleotide-pair substitution that results in a codon that codes for a different amino acid.

mitochondrial matrix The compartment of the mitochondrion enclosed by the inner membrane and containing enzymes and substrates for the citric acid cycle, as well as ribosomes and DNA.

mitochondrion (mī′-tō-kon′-drē-un) (plural, **mitochondria**) An organelle in eukaryotic cells that serves as the site of cellular respiration; uses oxygen to break down organic molecules and synthesize ATP.

mitosis (mī-tō′-sis) A process of nuclear division in eukaryotic cells conventionally divided into five stages: prophase, prometaphase, metaphase, anaphase, and telophase. Mitosis conserves chromosome number by allocating replicated chromosomes equally to each of the daughter nuclei.

mitotic (M) phase The phase of the cell cycle that includes mitosis and cytokinesis.

mitotic spindle An assemblage of microtubules and associated proteins that is involved in the movement of chromosomes during mitosis.

mixotroph An organism that is capable of both photosynthesis and heterotrophy.

model organism A particular species chosen for research into broad biological principles because it is representative of a larger group and usually easy to grow in a lab.

molarity A common measure of solute concentration, referring to the number of moles of solute per liter of solution.

mold Informal term for a fungus that grows as a filamentous fungus, producing haploid spores by mitosis and forming a visible mycelium.

mole (mol) The number of grams of a substance that equals its molecular weight in daltons and contains Avogadro's number of molecules.

molecular clock A method for estimating the time required for a given amount of evolutionary change, based on the observation that some regions of genomes evolve at constant rates.

molecular mass The sum of the masses of all the atoms in a molecule; sometimes called molecular weight.

molecular systematics A scientific discipline that uses nucleic acids or other molecules to infer evolutionary relationships between different species.

molecule Two or more atoms held together by covalent bonds.

molting A process in ecdysozoans in which the exoskeleton is shed at intervals, allowing growth by the production of a larger exoskeleton.

monoclonal antibody (mon′-ō-klōn′-ul) Any of a preparation of antibodies that have been produced by a single clone of cultured cells and thus are all specific for the same epitope.

monocot Member of a clade consisting of flowering plants that have one embryonic seed leaf, or cotyledon.

monogamous (muh-nog′-uh-mus) Referring to a type of relationship in which one male mates with just one female.

monohybrid An organism that is heterozygous with respect to a single gene of interest. All the offspring from a cross between parents homozygous for different alleles are monohybrids. For example, parents of genotypes *AA* and *aa* produce a monohybrid of genotype *Aa*.

monohybrid cross A cross between two organisms that are heterozygous for the character being followed (or the self-pollination of a heterozygous plant).

monomer (mon′-uh-mer) The subunit that serves as the building block of a polymer.

monophyletic (mon′-ō-fī-let′-ik) Pertaining to a group of taxa that consists of a common ancestor and all of its descendants. A monophyletic taxon is equivalent to a clade.

monosaccharide (mon′-ō-sak′-uh-rīd) The simplest carbohydrate, active alone or serving as a monomer for disaccharides and polysaccharides. Also known as simple sugars, monosaccharides have molecular formulas that are generally some multiple of CH_2O.

monosomic Referring to a diploid cell that has only one copy of a particular chromosome instead of the normal two.

monotreme An egg-laying mammal, such as a platypus or echidna. Like all mammals, monotremes have hair and produce milk, but they lack nipples.

morphogen A substance, such as Bicoid protein in *Drosophila*, that provides positional information in the form of a concentration gradient along an embryonic axis.

morphogenesis (mōr′-fō-jen′-uh-sis) The cellular and tissue-based processes by which an animal body takes shape.

morphological species concept A definition of species in terms of measurable anatomical criteria.

moss A small, herbaceous, nonvascular plant that is a member of the phylum Bryophyta.

motor neuron A nerve cell that transmits signals from the brain or spinal cord to muscles or glands.

motor protein A protein that interacts with cytoskeletal elements and other cell components, producing movement of the whole cell or parts of the cell.

motor system An efferent branch of the vertebrate peripheral nervous system composed of motor neurons that carry signals to skeletal muscles in response to external stimuli.

motor unit A single motor neuron and all the muscle fibers it controls.

movement corridor A series of small clumps or a narrow strip of quality habitat (usable by organisms) that connects otherwise isolated patches of quality habitat.

MPF Maturation-promoting factor (or M-phase-promoting factor); a protein complex required for a cell to progress from late interphase to mitosis. The active form consists of cyclin and a protein kinase.

mucus A viscous and slippery mixture of glycoproteins, cells, salts, and water that moistens and protects the membranes lining body cavities that open to the exterior.

Müllerian mimicry (myū-lār′-ē-un) Reciprocal mimicry by two unpalatable species.

multifactorial Referring to a phenotypic character that is influenced by multiple genes and environmental factors.

multigene family A collection of genes with similar or identical sequences, presumably of common origin.

multiple fruit A fruit derived from an entire inflorescence.

multiplication rule A rule of probability stating that the probability of two or more independent events occurring together can be determined by multiplying their individual probabilities.

muscle tissue Tissue consisting of long muscle cells that can contract, either on its own or when stimulated by nerve impulses.

mutagen (myū′-tuh-jen) A chemical or physical agent that interacts with DNA and can cause a mutation.

mutation (myū-tā′-shun) A change in the nucleotide sequence of an organism's DNA or in the DNA or RNA of a virus.

mutualism (myū′-chū-ul-izm) A symbiotic relationship in which both participants benefit.

mycelium (mī-sē′-lē-um) The densely branched network of hyphae in a fungus.

mycorrhiza (mī′-kō-rī′-zuh) (plural, **mycorrhizae**) A mutualistic association of plant roots and fungus.

mycosis (mī-kō′-sis) General term for a fungal infection.

myelin sheath (mī′-uh-lin) Wrapped around the axon of a neuron, an insulating coat of cell membranes from Schwann cells or oligodendrocytes. It is interrupted by nodes of Ranvier, where action potentials are generated.

myofibril (mī′-ō-fī′-bril) A longitudinal bundle in a muscle cell (fiber) that contains thin filaments of actin and regulatory proteins and thick filaments of myosin.

myoglobin (mī′-uh-glō′-bin) An oxygen-storing, pigmented protein in muscle cells.

myosin (mī′-uh-sin) A type of motor protein that associates into filaments that interact with actin filaments to cause cell contraction.

myotonia (mī′-uh-tō′-nī-uh) Increased muscle tension, characteristic of sexual arousal in certain human tissues.

myriapod (mir′-ē-uh-pod′) A terrestrial arthropod with many body segments and one or two pairs of legs per segment. Millipedes and centipedes are the two major groups of living myriapods.

NAD^+ Nicotinamide adenine dinucleotide, a coenzyme that cycles easily between oxidized (NAD^+) and reduced (NADH) states, thus acting as an electron carrier.

$NADP^+$ Nicotinamide adenine dinucleotide phosphate, an electron acceptor that, as NADPH, temporarily stores energized electrons produced during the light reactions.

natural family planning A form of contraception that relies on refraining from sexual intercourse when conception is most likely to occur; also called the rhythm method.

natural killer cell A type of white blood cell that can kill tumor cells and virus-infected cells as part of innate immunity.

natural selection A process in which individuals that have certain inherited traits tend to survive and reproduce at higher rates than other individuals *because of* those traits.

negative feedback A form of regulation in which accumulation of an end product of a process slows the process; in physiology, a primary mechanism of homeostasis, whereby a change in a variable triggers a response that counteracts the initial change.

negative pressure breathing A breathing system in which air is pulled into the lungs.

nematocyst (nem′-uh-tuh-sist′) In a cnidocyte of a cnidarian, a capsule-like organelle containing a coiled thread that when discharged can penetrate the body wall of the prey.

nephron (nef′-ron) The tubular excretory unit of the vertebrate kidney.

neritic zone The shallow region of the ocean overlying the continental shelf.

nerve A fiber composed primarily of the bundled axons of PNS neurons.

nerve net A weblike system of neurons, characteristic of radially symmetrical animals, such as hydras.

nervous system The fast-acting internal system of communication involving sensory receptors, networks of nerve cells, and connections to muscles and glands that respond to nerve signals; functions in concert with the endocrine system to effect internal regulation and maintain homeostasis.

nervous tissue Tissue made up of neurons and supportive cells.

net ecosystem production (NEP) The gross primary production of an ecosystem minus the energy used by all autotrophs and heterotrophs for respiration.

net primary production (NPP) The gross primary production of an ecosystem minus the energy used by the producers for respiration.

neural crest In vertebrates, a region located along the sides of the neural tube where it pinches off from the ectoderm. Neural crest cells migrate to various parts of the embryo and form pigment cells in the skin and parts of the skull, teeth, adrenal glands, and peripheral nervous system.

neural plasticity The capacity of a nervous system to change with experience.

neural tube A tube of infolded ectodermal cells that runs along the anterior-posterior axis of a vertebrate, just dorsal to the notochord. It will give rise to the central nervous system.

neurohormone A molecule that is secreted by a neuron, travels in body fluids, and acts on specific target cells, changing their functioning.

neuron (nyūr′-on) A nerve cell; the fundamental unit of the nervous system, having structure and properties that allow it to conduct signals by taking advantage of the electrical charge across its plasma membrane.

neuropeptide A relatively short chain of amino acids that serves as a neurotransmitter.

neurotransmitter A molecule that is released from the synaptic terminal of a neuron at a chemical synapse, diffuses across the synaptic cleft, and binds to the postsynaptic cell, triggering a response.

neutral theory The hypothesis that much evolutionary change in genes and proteins has no effect on fitness and therefore is not influenced by natural selection.

neutral variation Genetic variation that does not provide a selective advantage or disadvantage.

neutron A subatomic particle having no electrical charge (electrically neutral), with a mass of about 1.7×10^{-24} g, found in the nucleus of an atom.

neutrophil The most abundant type of white blood cell. Neutrophils are phagocytic and tend to self-destruct as they destroy foreign invaders, limiting their life span to a few days.

nitric oxide (NO) A gas produced by many types of cells that functions as a local regulator and as a neurotransmitter.

nitrogen cycle The natural process by which nitrogen, either from the atmosphere or from decomposed organic material, is converted by soil bacteria to compounds assimilated by plants. This incorporated nitrogen is then taken in by other organisms and subsequently released, acted on by bacteria, and made available again to the nonliving environment.

nitrogen fixation The conversion of atmospheric nitrogen (N_2) to ammonia (NH_3). Biological nitrogen fixation is carried out by certain prokaryotes, some of which have mutualistic relationships with plants.

nociceptor (nō′-si-sep′-tur) A sensory receptor that responds to noxious or painful stimuli; also called a pain receptor.

node A point along the stem of a plant at which leaves are attached.

node of Ranvier (ron′-vē-ā′) Gap in the myelin sheath of certain axons where an action potential may be generated. In saltatory conduction, an action potential is regenerated at each node, appearing to "jump" along the axon from node to node.

nodule A swelling on the root of a legume. Nodules are composed of plant cells that contain nitrogen-fixing bacteria of the genus *Rhizobium*.

noncompetitive inhibitor A substance that reduces the activity of an enzyme by binding to a location remote from the active site, changing the enzyme's shape so that the active site no longer effectively catalyzes the conversion of substrate to product.

nondisjunction An error in meiosis or mitosis in which members of a pair of homologous chromosomes or a pair of sister chromatids fail to separate properly from each other.

nonequilibrium model A model that maintains that communities change constantly after being buffeted by disturbances.

nonpolar covalent bond A type of covalent bond in which electrons are shared equally between two atoms of similar electronegativity.

nonsense mutation A mutation that changes an amino acid codon to one of the three stop codons, resulting in a shorter and usually nonfunctional protein.

norepinephrine A catecholamine that is chemically and functionally similar to epinephrine and acts as a hormone or neurotransmitter; also known as noradrenaline.

norm of reaction The range of phenotypes produced by a single genotype, due to environmental influences.

Northern blotting A technique that enables specific nucleotide sequences to be detected in samples of mRNA. It involves gel electrophoresis of RNA molecules and their transfer to a membrane (blotting), followed by nucleic acid hybridization with a labeled probe.

northern coniferous forest A terrestrial biome characterized by long, cold winters and dominated by cone-bearing trees.

no-till agriculture A plowing technique that minimally disturbs the soil, thereby reducing soil loss.

notochord (nō′-tuh-kord′) A longitudinal, flexible rod made of tightly packed mesodermal cells that runs along the anterior-posterior axis of a chordate in the dorsal part of the body.

nuclear envelope In a eukaryotic cell, the double membrane that surrounds the nucleus, perforated with pores that regulate traffic with the cytoplasm. The outer membrane is continuous with the endoplasmic reticulum.

nuclear lamina A netlike array of protein filaments that lines the inner surface of the nuclear envelope and helps maintain the shape of the nucleus.

nucleariid Member of a group of unicellular, amoeboid protists that are more closely related to fungi than they are to other protists.

nuclease An enzyme that cuts DNA or RNA, either removing one or a few bases or hydrolyzing the DNA or RNA completely into its component nucleotides.

nucleic acid (nū-klā′-ik) A polymer (polynucleotide) consisting of many nucleotide monomers; serves as a blueprint for proteins and, through the actions of proteins, for all cellular activities. The two types are DNA and RNA.

nucleic acid hybridization The process of base pairing between a gene and a complementary sequence on another nucleic acid molecule.

nucleic acid probe In DNA technology, a labeled single-stranded nucleic acid molecule used to locate a specific nucleotide sequence in a nucleic acid sample. Molecules of the probe hydrogen-bond to the complementary sequence wherever it occurs; radioactive, fluorescent, or other labeling of the probe allows its location to be detected.

nucleoid (nū′-klē-oyd) A non-membrane-bounded region in a prokaryotic cell where the DNA is concentrated.

nucleolus (nū-klē′-ō-lus) (plural, **nucleoli**) A specialized structure in the nucleus, consisting of chromosomal regions containing ribosomal RNA (rRNA) genes along with ribosomal proteins imported from the cytoplasm; site of rRNA synthesis and ribosomal subunit assembly. *See also* ribosome.

nucleosome (nū′-klē-ō-sōm′) The basic, bead-like unit of DNA packing in eukaryotes, consisting of a segment of DNA wound around a protein core composed of two copies of each of four types of histone.

nucleotide (nū′-klē-ō-tīd′) The building block of a nucleic acid, consisting of a five-carbon sugar covalently bonded to a nitrogenous base and one or more phosphate groups.

nucleotide excision repair A repair system that removes and then correctly replaces a damaged segment of DNA using the undamaged strand as a guide.

nucleotide-pair substitution A type of point mutation in which one nucleotide in a DNA strand and its partner in the complementary strand are replaced by another pair of nucleotides.

nucleus (1) An atom's central core, containing protons and neutrons. (2) The organelle of a eukaryotic cell that contains the genetic material in the form of chromosomes, made up of chromatin. (3) A cluster of neurons.

nutrition The process by which an organism takes in and makes use of food substances.

obligate aerobe (ob′-lig-et ār′-ōb) An organism that requires oxygen for cellular respiration and cannot live without it.

obligate anaerobe (ob′-lig-et an′-uh-rōb) An organism that only carries out fermentation or anaerobic respiration. Such organisms cannot use oxygen and in fact may be poisoned by it.

ocean acidification Decreasing pH of ocean waters due to absorption of excess atmospheric CO_2 from the burning of fossil fuels.

oceanic pelagic zone Most of the ocean's waters far from shore, constantly mixed by ocean currents.

odorant A molecule that can be detected by sensory receptors of the olfactory system.

Okazaki fragment (ō′-kah-zah′-kē) A short segment of DNA synthesized away from the replication fork on a template strand during DNA replication. Many such segments are joined together to make up the lagging strand of newly synthesized DNA.

olfaction The sense of smell.

oligodendrocyte A type of glial cell that forms insulating myelin sheaths around the axons of neurons in the central nervous system.

oligotrophic lake A nutrient-poor, clear lake with few phytoplankton.

ommatidium (ōm′-uh-tid′-ē-um) (plural, **ommatidia**) One of the facets of the compound eye of arthropods and some polychaete worms.

omnivore An animal that regularly eats animals as well as plants or algae.

oncogene (on′-kō-jēn) A gene found in viral or cellular genomes that is involved in triggering molecular events that can lead to cancer.

oocyte A cell in the female reproductive system that differentiates to form an egg.

oogenesis (ō′-uh-jen′-uh-sis) The process in the ovary that results in the production of female gametes.

oogonium (ō′-uh- gō′-nē-em) (plural, **oogonia**) A cell that divides mitotically to form oocytes.

oomycete (ō′-uh-mī′-sēt) A protist with flagellated cells, such as a water mold, white rust, or downy mildew, that acquires nutrition mainly as a decomposer or plant parasite.

open circulatory system A circulatory system in which fluid called hemolymph bathes the tissues and organs directly and there is no distinction between the circulating fluid and the interstitial fluid.

operator In bacterial and phage DNA, a sequence of nucleotides near the start of an operon to which an active repressor can attach. The binding of the repressor prevents RNA polymerase from attaching to the promoter and transcribing the genes of the operon.

operculum (ō-per′-kyuh-lum) In aquatic osteichthyans, a protective bony flap that covers and protects the gills.

operon (op′-er-on) A unit of genetic function found in bacteria and phages, consisting of a promoter, an operator, and a coordinately regulated cluster of genes whose products function in a common pathway.

opisthokont (uh-pis′-thuh-kont′) Member of the diverse clade Opisthokonta, organisms that descended from an ancestor with a posterior flagellum, including fungi, animals, and certain protists.

opposable thumb A thumb that can touch the ventral surface of the fingertips of all four fingers.

opsin A membrane protein bound to a light-absorbing pigment molecule.

optic chiasm The place where the two optic nerves meet and axons representing distinct sides of the visual field are segregated from one another before reaching the brain.

optimal foraging model The basis for analyzing behavior as a compromise between feeding costs and feeding benefits.

oral cavity The mouth of an animal.

orbital The three-dimensional space where an electron is found 90% of the time.

order In Linnaean classification, the taxonomic category above the level of family.

organ A specialized center of body function composed of several different types of tissues.

organ identity gene A plant homeotic gene that uses positional information to determine which emerging leaves develop into which types of floral organs.

organ of Corti The actual hearing organ of the vertebrate ear, located in the floor of the cochlear duct in the inner ear; contains the receptor cells (hair cells) of the ear.

organ system A group of organs that work together in performing vital body functions.

organelle (ōr-guh-nel′) Any of several membrane-enclosed structures with specialized functions, suspended in the cytosol of eukaryotic cells.

organic chemistry The study of carbon compounds (organic compounds).

organismal ecology The branch of ecology concerned with the morphological, physiological, and behavioral ways in which individual organisms meet the challenges posed by their biotic and abiotic environments.

organogenesis (ōr-gan′-ō-jen′-uh-sis) The process in which organ rudiments develop from the three germ layers after gastrulation.

orgasm Rhythmic, involuntary contractions of certain reproductive structures in both sexes during the human sexual response cycle.

origin of replication Site where the replication of a DNA molecule begins, consisting of a specific sequence of nucleotides.

orthologous genes Homologous genes that are found in different species because of speciation.

osculum (os′-kyuh-lum) A large opening in a sponge that connects the spongocoel to the environment.

osmoconformer An animal that is isoosmotic with its environment.

osmolarity (oz′-mō-lār′-uh-tē) Solute concentration expressed as molarity.

osmoregulation Regulation of solute concentrations and water balance by a cell or organism.

osmoregulator An animal that controls its internal osmolarity independent of the external environment.

osmosis (oz-mō′-sis) The diffusion of free water across a selectively permeable membrane.

osteichthyan (os′-tē-ik′-thē-an) Member of a vertebrate clade with jaws and mostly bony skeletons.

outer ear One of three main regions of the ear in reptiles (including birds) and mammals; made up of the auditory canal and, in many birds and mammals, the pinna.

outgroup A species or group of species from an evolutionary lineage that is known to have diverged before the lineage that contains the group of species being studied. An outgroup is selected so that its members are closely related to the group of species being studied, but not as closely related as any study-group members are to each other.

oval window In the vertebrate ear, a membrane-covered gap in the skull bone, through which sound waves pass from the middle ear to the inner ear.

ovarian cycle (ō-vār′-ē-un) The cyclic recurrence of the follicular phase, ovulation, and the luteal phase in the mammalian ovary, regulated by hormones.

ovary (ō′-vuh-rē) (1) In flowers, the portion of a carpel in which the egg-containing ovules develop. (2) In animals, the structure that produces female gametes and reproductive hormones.

oviduct (ō′-vuh-duct) A tube passing from the ovary to the vagina in invertebrates or to the uterus in vertebrates, where it is also known as a fallopian tube.

oviparous (ō-vip′-uh-rus) Referring to a type of development in which young hatch from eggs laid outside the mother's body.

ovoviviparous (ō′-vō-vī-vip′-uh-rus) Referring to a type of development in which young hatch from eggs that are retained in the mother's uterus.

ovulation The release of an egg from an ovary. In humans, an ovarian follicle releases an egg during each uterine (menstrual) cycle.

ovule (o′-vyūl) A structure that develops within the ovary of a seed plant and contains the female gametophyte.

oxidation The complete or partial loss of electrons from a substance involved in a redox reaction.

oxidative phosphorylation (fos′-fōr-uh-lā′-shun) The production of ATP using energy derived from the redox reactions of an electron transport chain; the third major stage of cellular respiration.

oxidizing agent The electron acceptor in a redox reaction.

oxytocin (ok′-si-tō′-sen) A hormone produced by the hypothalamus and released from the posterior pituitary. It induces contractions of the uterine muscles during labor and causes the mammary glands to eject milk during nursing.

P generation The true-breeding (homozygous) parent individuals from which F_1 hybrid offspring are derived in studies of inheritance; P stands for "parental."

P site One of a ribosome's three binding sites for tRNA during translation. The P site holds the tRNA carrying the growing polypeptide chain. (P stands for peptidyl tRNA.)

***p53* gene** A tumor-suppressor gene that codes for a specific transcription factor that promotes the synthesis of proteins that inhibit the cell cycle.

paedomorphosis (pē′-duh-mōr′-fuh-sis) The retention in an adult organism of the juvenile features of its evolutionary ancestors.

pain receptor A sensory receptor that responds to noxious or painful stimuli; also called a nociceptor.

paleoanthropology The study of human origins and evolution.

paleontology (pā′-lē-un-tol′-ō-jē) The scientific study of fossils.

pancreas (pan′-krē-us) A gland with exocrine and endocrine tissues. The exocrine portion functions in digestion, secreting enzymes and an alkaline solution into the small intestine via a duct; the ductless endocrine portion functions in homeostasis, secreting the hormones insulin and glucagon into the blood.

pandemic A global epidemic.

Pangaea (pan-jē′-uh) The supercontinent that formed near the end of the Paleozoic era, when plate movements brought all the landmasses of Earth together.

parabasalid A protist, such as a trichomonad, with modified mitochondria.

paracrine Referring to a secreted molecule that acts on a neighboring cell.

paralogous genes Homologous genes that are found in the same genome as a result of gene duplication.

paraphyletic (pār′-uh-fī-let′-ik) Pertaining to a group of taxa that consists of a common ancestor and some, but not all, of its descendants.

parareptile A basal group of reptiles, consisting mostly of large, stocky quadrupedal herbivores. Parareptiles died out in the late Triassic period.

parasite (pār′-uh-sīt) An organism that feeds on the cell contents, tissues, or body fluids of another species (the host) while in or on the host organism. Parasites harm but usually do not kill their host.

parasitism (pār′-uh-sit-izm) A symbiotic relationship in which one organism, the parasite, benefits at the expense of another, the host, by living either within or on the host.

parasympathetic division One of three divisions of the autonomic nervous system; generally enhances body activities that gain and conserve energy, such as digestion and reduced heart rate.

parathyroid gland One of four small endocrine glands, embedded in the surface of the thyroid gland, that secrete parathyroid hormone.

parathyroid hormone (PTH) A hormone secreted by the parathyroid glands that raises blood calcium level by promoting calcium release from bone and calcium retention by the kidneys.

parenchyma cell (puh-ren′-ki-muh) A relatively unspecialized plant cell type that carries out most of the metabolism, synthesizes and stores organic products, and develops into a more differentiated cell type.

parental type An offspring with a phenotype that matches one of the true-breeding parental (P generation) phenotypes; also refers to the phenotype itself.

Parkinson's disease A progressive brain disease characterized by difficulty in initiating movements, slowness of movement, and rigidity.

parthenogenesis (par'-thuh-nō'-jen'-uh-sis) A form of asexual reproduction in which females produce offspring from unfertilized eggs.

partial pressure The pressure exerted by a particular gas in a mixture of gases (for instance, the pressure exerted by oxygen in air).

passive immunity Short-term immunity conferred by the transfer of antibodies, as occurs in the transfer of maternal antibodies to a fetus or nursing infant.

passive transport The diffusion of a substance across a biological membrane with no expenditure of energy.

pathogen An organism, virus, viroid, or prion that causes disease.

pattern formation The development of a multicellular organism's spatial organization, the arrangement of organs and tissues in their characteristic places in three-dimensional space.

peat Extensive deposits of partially decayed organic material often formed primarily from the wetland moss *Sphagnum*.

pedigree A diagram of a family tree with conventional symbols, showing the occurrence of heritable characters in parents and offspring over multiple generations.

pelagic zone The open-water component of aquatic biomes.

penis The copulatory structure of male mammals.

PEP carboxylase An enzyme that adds CO_2 to phosphoenolpyruvate (PEP) to form oxaloacetate in mesophyll cells of C_4 plants. It acts prior to photosynthesis.

pepsin An enzyme present in gastric juice that begins the hydrolysis of proteins.

pepsinogen The inactive form of pepsin secreted by chief cells located in gastric pits of the stomach.

peptide bond The covalent bond between the carboxyl group on one amino acid and the amino group on another, formed by a dehydration reaction.

peptidoglycan (pep'-tid-ō-glī'-kan) A type of polymer in bacterial cell walls consisting of modified sugars cross-linked by short polypeptides.

perception The interpretation of sensory system input by the brain.

pericycle The outermost layer in the vascular cylinder, from which lateral roots arise.

periderm (pār'-uh-derm') The protective coat that replaces the epidermis in woody plants during secondary growth, formed of the cork and cork cambium.

peripheral nervous system (PNS) The sensory and motor neurons that connect to the central nervous system.

peripheral protein A protein loosely bound to the surface of a membrane or to part of an integral protein and not embedded in the lipid bilayer.

peristalsis (pār'-uh-stal'-sis) (1) Alternating waves of contraction and relaxation in the smooth muscles lining the alimentary canal that push food along the canal. (2) A type of movement on land produced by rhythmic waves of muscle contractions passing from front to back, as in many annelids.

peristome A ring of interlocking, tooth-like structures on the upper part of a moss capsule (sporangium), often specialized for gradual spore discharge.

peritubular capillary One of the tiny blood vessels that form a network surrounding the proximal and distal tubules in the kidney.

permafrost A permanently frozen soil layer.

peroxisome (puh-rok'-suh-sōm') An organelle containing enzymes that transfer hydrogen atoms from various substrates to oxygen (O_2), producing and then degrading hydrogen peroxide (H_2O_2).

petal A modified leaf of a flowering plant. Petals are the often colorful parts of a flower that advertise it to insects and other pollinators.

petiole (pet'-ē-ōl) The stalk of a leaf, which joins the leaf to a node of the stem.

pH A measure of hydrogen ion concentration equal to $-\log [H^+]$ and ranging in value from 0 to 14.

phage (fāj) A virus that infects bacteria; also called a bacteriophage.

phagocytosis (fag'-ō-sī-tō'-sis) A type of endocytosis in which large particulate substances or small organisms are taken up by a cell. It is carried out by some protists and by certain immune cells of animals (in mammals, mainly macrophages, neutrophils, and dendritic cells).

pharyngeal cleft (fuh-rin'-jē-ul) In chordate embryos, one of the grooves that separate a series of pouches along the sides of the pharynx and may develop into a pharyngeal slit.

pharyngeal slit (fuh-rin'-jē-ul) In chordate embryos, one of the slits that form from the pharyngeal clefts and communicate to the outside, later developing into gill slits in many vertebrates.

pharynx (fār'-inks) (1) An area in the vertebrate throat where air and food passages cross. (2) In flatworms, the muscular tube that protrudes from the ventral side of the worm and ends in the mouth.

phase change A shift from one developmental phase to another.

phenotype (fē'-nō-tīp) The observable physical and physiological traits of an organism, which are determined by its genetic makeup.

pheromone (fār'-uh-mōn) In animals and fungi, a small molecule released into the environment that functions in communication between members of the same species. In animals, it acts much like a hormone in influencing physiology and behavior.

phloem (flō'-em) Vascular plant tissue consisting of living cells arranged into elongated tubes that transport sugar and other organic nutrients throughout the plant.

phloem sap The sugar-rich solution carried through a plant's sieve tubes.

phosphate group A chemical group consisting of a phosphorus atom bonded to four oxygen atoms; important in energy transfer.

phospholipid (fos'-fō-lip'-id) A lipid made up of glycerol joined to two fatty acids and a phosphate group. The hydrocarbon chains of the fatty acids act as nonpolar, hydrophobic tails, while the rest of the molecule acts as a polar, hydrophilic head. Phospholipids form bilayers that function as biological membranes.

phosphorylated intermediate A molecule (often a reactant) with a phosphate group covalently bound to it, making it more reactive (less stable) than the unphosphorylated molecule.

photic zone (fō'-tic) The narrow top layer of an ocean or lake, where light penetrates sufficiently for photosynthesis to occur.

photoautotroph (fō'-tō-ot'-ō-trōf) An organism that harnesses light energy to drive the synthesis of organic compounds from carbon dioxide.

photoheterotroph (fō'-tō-het'-er-ō-trōf) An organism that uses light to generate ATP but must obtain carbon in organic form.

photomorphogenesis Effects of light on plant morphology.

photon (fō'-ton) A quantum, or discrete quantity, of light energy that behaves as if it were a particle.

photoperiodism (fō'-tō-pēr'-ē-ō-dizm) A physiological response to photoperiod, the relative lengths of night and day. An example of photoperiodism is flowering.

photophosphorylation (fō'-tō-fos'-fōr-uh-lā'-shun) The process of generating ATP from ADP and phosphate by means of chemiosmosis, using a proton-motive force generated across the thylakoid membrane of the chloroplast or the membrane of certain prokaryotes during the light reactions of photosynthesis.

photoreceptor An electromagnetic receptor that detects the radiation known as visible light.

photorespiration A metabolic pathway that consumes oxygen and ATP, releases carbon dioxide, and decreases photosynthetic output. Photorespiration generally occurs on hot, dry, bright days, when stomata close and the O_2/CO_2 ratio in the leaf increases, favoring the binding of O_2 rather than CO_2 by rubisco.

photosynthesis (fō'-tō-sin'-thi-sis) The conversion of light energy to chemical energy that is stored in sugars or other organic compounds; occurs in plants, algae, and certain prokaryotes.

photosystem A light-capturing unit located in the thylakoid membrane of the chloroplast or in the membrane of some prokaryotes, consisting of a reaction-center complex surrounded by numerous light-harvesting complexes. There are two types of photosystems, I and II; they absorb light best at different wavelengths.

photosystem I (PS I) A light-capturing unit in a chloroplast's thylakoid membrane or in the membrane of some prokaryotes; it has two molecules of P700 chlorophyll *a* at its reaction center.

photosystem II (PS II) One of two light-capturing units in a chloroplast's thylakoid membrane or in the membrane of some prokaryotes; it has two molecules of P680 chlorophyll *a* at its reaction center.

phototropism (fō'-tō-trō'-pizm) Growth of a plant shoot toward or away from light.

phragmoplast (frag'-mō-plast') An alignment of cytoskeletal elements and Golgi-derived vesicles that forms across the midline of a dividing plant cell.

phyllotaxy (fil'-uh-tak'-sē) The pattern of leaf attachment to the stem of a plant.

PhyloCode Proposed system of classification of organisms based on evolutionary relationships: Only groups that include a common ancestor and all of its descendants are named.

phylogenetic species concept A definition of species as the smallest group of individuals that share a common ancestor, forming one branch on the tree of life.

phylogenetic tree A branching diagram that represents a hypothesis about the evolutionary history of a group of organisms.

phylogeny (fī-loj'-uh-nē) The evolutionary history of a species or group of related species.

phylum (fī'-lum) (plural, **phyla**) In Linnaean classification, the taxonomic category above class.

physical map A genetic map in which the actual physical distances between genes or other genetic markers are expressed, usually as the number of base pairs along the DNA.

physiology The processes and functions of an organism.

phytochrome (fī'-tuh-krōm) A type of light receptor in plants that mostly absorbs red light and regulates many plant responses, such as seed germination and shade avoidance.

phytoremediation An emerging technology that seeks to reclaim contaminated areas by taking advantage of some plant species' ability to extract heavy metals and other pollutants from the soil and to concentrate them in easily harvested portions of the plant.

pilus (plural, **pili**) (pī'-lus, pī'-lī) In bacteria, a structure that links one cell to another at the start of conjugation; also known as a sex pilus or conjugation pilus.

pineal gland (pī'-nē-ul) A small gland on the dorsal surface of the vertebrate forebrain that secretes the hormone melatonin.

pinocytosis (pī'-nō-sī-tō'-sis) A type of endocytosis in which the cell ingests extracellular fluid and its dissolved solutes.

pistil A single carpel or a group of fused carpels.

pith Ground tissue that is internal to the vascular tissue in a stem; in many monocot roots, parenchyma cells that form the central core of the vascular cylinder.

pituitary gland (puh-tū'-uh-tār'-ē) An endocrine gland at the base of the hypothalamus; consists of a posterior lobe, which stores and releases two hormones produced by the hypothalamus, and an anterior lobe, which produces and secretes many hormones that regulate diverse body functions.

placenta (pluh-sen'-tuh) A structure in the pregnant uterus for nourishing a viviparous fetus with the mother's blood supply; formed from the uterine lining and embryonic membranes.

placental transfer cell A plant cell that enhances the transfer of nutrients from parent to embryo.

placoderm A member of an extinct group of fishlike vertebrates that had jaws and were enclosed in a tough outer armor.

planarian A free-living flatworm found in ponds and streams.

plasma (plaz'-muh) The liquid matrix of blood in which the blood cells are suspended.

plasma cell The antibody-secreting effector cell of humoral immunity. Plasma cells arise from antigen-stimulated B cells.

plasma membrane The membrane at the boundary of every cell that acts as a selective barrier, regulating the cell's chemical composition.

plasmid (plaz'-mid) A small, circular, double-stranded DNA molecule that carries accessory genes separate from those of a bacterial chromosome; in DNA cloning, used as vectors carrying up to about 10,000 base pairs (10 kb) of DNA. Plasmids are also found in some eukaryotes, such as yeasts.

plasmodesma (plaz'-mō-dez'-muh) (plural, **plasmodesmata**) An open channel through the cell wall that connects the cytoplasm of adjacent plant cells, allowing water, small solutes, and some larger molecules to pass between the cells.

plasmodial slime mold (plaz-mō'-dē-ul) A type of protist that has amoeboid cells, flagellated cells, and a plasmodial feeding stage in its life cycle.

plasmodium A single mass of cytoplasm containing many diploid nuclei that forms during the life cycle of some slime molds.

plasmogamy (plaz-moh'-guh-mē) In fungi, the fusion of the cytoplasm of cells from two individuals; occurs as one stage of sexual reproduction, followed later by karyogamy.

plasmolysis (plaz-mol'-uh-sis) A phenomenon in walled cells in which the cytoplasm shrivels and the plasma membrane pulls away from the cell wall; occurs when the cell loses water to a hypertonic environment.

plastid One of a family of closely related organelles that includes chloroplasts, chromoplasts, and amyloplasts. Plastids are found in cells of photosynthetic eukaryotes.

plate tectonics The theory that the continents are part of great plates of Earth's crust that float on the hot, underlying portion of the mantle. Movements in the mantle cause the continents to move slowly over time.

platelet A pinched-off cytoplasmic fragment of a specialized bone marrow cell. Platelets circulate in the blood and are important in blood clotting.

pleiotropy (plī'-o-truh-pē) The ability of a single gene to have multiple effects.

pluripotent Describing a cell that can give rise to many, but not all, parts of an organism.

point mutation A change in a single nucleotide pair of a gene.

polar covalent bond A covalent bond between atoms that differ in electronegativity. The shared electrons are pulled closer to the more electronegative atom, making it slightly negative and the other atom slightly positive.

polar molecule A molecule (such as water) with an uneven distribution of charges in different regions of the molecule.

polarity A lack of symmetry; structural differences in opposite ends of an organism or structure, such as the root end and shoot end of a plant.

pollen grain In seed plants, a structure consisting of the male gametophyte enclosed within a pollen wall.

pollen tube A tube that forms after germination of the pollen grain and that functions in the delivery of sperm to the ovule.

pollination (pol'-uh-nā'-shun) The transfer of pollen to the part of a seed plant containing the ovules, a process required for fertilization.

poly-A tail A sequence of 50–250 adenine nucleotides added onto the 3′ end of a pre-mRNA molecule.

polygamous Referring to a type of relationship in which an individual of one sex mates with several of the other.

polygenic inheritance (pol'-ē-jen'-ik) An additive effect of two or more genes on a single phenotypic character.

polymer (pol'-uh-mer) A long molecule consisting of many similar or identical monomers linked together by covalent bonds.

polymerase chain reaction (PCR) (puh-lim'-uh-rās) A technique for amplifying DNA *in vitro* by incubating it with specific primers, a heat-resistant DNA polymerase, and nucleotides.

polynucleotide (pol'-ē-nū'-klē-ō-tīd) A polymer consisting of many nucleotide monomers in a chain. The nucleotides can be those of DNA or RNA.

polyp The sessile variant of the cnidarian body plan. The alternate form is the medusa.

polypeptide (pol'-ē-pep'-tīd) A polymer of many amino acids linked together by peptide bonds.

polyphyletic (pol'-ē-fī-let'-ik) Pertaining to a group of taxa derived from two or more different ancestors.

polyploidy (pol'-ē-ploy'-dē) A chromosomal alteration in which the organism possesses more than two complete chromosome sets. It is the result of an accident of cell division.

polyribosome (polysome) (pol'-ē-rī'-buh-sōm') A group of several ribosomes attached to, and translating, the same messenger RNA molecule.

polysaccharide (pol'-ē-sak'-uh-rīd) A polymer of many monosaccharides, formed by dehydration reactions.

polytomy (puh-lit'-uh-mē) In a phylogenetic tree, a branch point from which more than two descendant taxa emerge. A polytomy indicates that the evolutionary relationships between the descendant taxa are not yet clear.

pons A portion of the brain that participates in certain automatic, homeostatic functions, such as regulating the breathing centers in the medulla.

population A group of individuals of the same species that live in the same area and interbreed, producing fertile offspring.

population dynamics The study of how complex interactions between biotic and abiotic factors influence variations in population size.

population ecology The study of populations in relation to their environment, including environmental influences on population density and distribution, age structure, and variations in population size.

positional information Molecular cues that control pattern formation in an animal or plant embryonic structure by indicating a cell's location relative to the organism's body axes. These cues elicit a response by genes that regulate development.

positive feedback A form of regulation in which an end product of a process speeds up that process; in physiology, a control mechanism in which a change in a variable triggers a response that reinforces or amplifies the change.

positive pressure breathing A breathing system in which air is forced into the lungs.

posterior Pertaining to the rear, or tail end, of a bilaterally symmetrical animal.

posterior pituitary An extension of the hypothalamus composed of nervous tissue that secretes oxytocin and antidiuretic hormone made in the hypothalamus; a temporary storage site for these hormones.

postzygotic barrier (pōst′-zī-got′-ik) A reproductive barrier that prevents hybrid zygotes produced by two different species from developing into viable, fertile adults.

potential energy The energy that matter possesses as a result of its location or spatial arrangement (structure).

predation An interaction between species in which one species, the predator, eats the other, the prey.

pregnancy The condition of carrying one or more embryos in the uterus.

prepuce (prē′-pyūs) A fold of skin covering the head of the clitoris or penis.

pressure potential (Ψ_P) A component of water potential that consists of the physical pressure on a solution, which can be positive, zero, or negative.

prezygotic barrier (prē′-zī-got′-ik) A reproductive barrier that impedes mating between species or hinders fertilization if interspecific mating is attempted.

primary cell wall In plants, a relatively thin and flexible layer that surrounds the plasma membrane of a young cell.

primary consumer An herbivore; an organism that eats plants or other autotrophs.

primary electron acceptor In the thylakoid membrane of a chloroplast or in the membrane of some prokaryotes, a specialized molecule that shares the reaction-center complex with a pair of chlorophyll *a* molecules and that accepts an electron from them.

primary growth Growth produced by apical meristems, lengthening stems and roots.

primary immune response The initial adaptive immune response to an antigen, which appears after a lag of about 10 to 17 days.

primary oocyte (ō′-uh-sīt) An oocyte prior to completion of meiosis I.

primary producer An autotroph, usually a photosynthetic organism. Collectively, autotrophs make up the trophic level of an ecosystem that ultimately supports all other levels.

primary production The amount of light energy converted to chemical energy (organic compounds) by the autotrophs in an ecosystem during a given time period.

primary structure The level of protein structure referring to the specific linear sequence of amino acids.

primary succession A type of ecological succession that occurs in an area where there were originally no organisms present and where soil has not yet formed.

primary transcript An initial RNA transcript from any gene; also called pre-mRNA when transcribed from a protein-coding gene.

primary visual cortex The destination in the occipital lobe of the cerebrum for most of the axons from the lateral geniculate nuclei.

primase An enzyme that joins RNA nucleotides to make a primer during DNA replication, using the parental DNA strand as a template.

primer A short stretch of RNA with a free 3′ end, bound by complementary base pairing to the template strand and elongated with DNA nucleotides during DNA replication.

primitive streak A thickening along the future anterior-posterior axis on the surface of an early avian or mammalian embryo, caused by a piling up of cells as they congregate at the midline before moving into the embryo.

prion An infectious agent that is a misfolded version of a normal cellular protein. Prions appear to increase in number by converting correctly folded versions of the protein to more prions.

problem solving The cognitive activity of devising a method to proceed from one state to another in the face of real or apparent obstacles.

producer An organism that produces organic compounds from CO_2 by harnessing light energy (in photosynthesis) or by oxidizing inorganic chemicals (in chemosynthetic reactions carried out by some prokaryotes).

product A material resulting from a chemical reaction.

production efficiency The percentage of energy stored in assimilated food that is not used for respiration or eliminated as waste.

progesterone A steroid hormone that prepares the uterus for pregnancy; the major progestin in mammals.

progestin Any steroid hormone with progesterone-like activity.

progymnosperm (prō′-jim′-nō-sperm) An extinct seedless vascular plant that may be ancestral to seed plants.

prokaryotic cell (prō′-kār′-ē-ot′-ik) A type of cell lacking a membrane-enclosed nucleus and membrane-enclosed organelles. Organisms with prokaryotic cells (bacteria and archaea) are called prokaryotes.

prolactin A hormone produced and secreted by the anterior pituitary with a great diversity of effects in different vertebrate species. In mammals, it stimulates growth of and milk production by the mammary glands.

proliferative phase That portion of the uterine (menstrual) cycle when the endometrium regenerates and thickens.

prometaphase The second stage of mitosis, in which the nuclear envelope fragments and the spindle microtubules attach to the kinetochores of the chromosomes.

promiscuous Referring to a type of relationship in which mating occurs with no strong pair-bonds or lasting relationships.

promoter A specific nucleotide sequence in the DNA of a gene that binds RNA polymerase, positioning it to start transcribing RNA at the appropriate place.

prophage (prō′-fāj) A phage genome that has been inserted into a specific site on a bacterial chromosome.

prophase The first stage of mitosis, in which the chromatin condenses into discrete chromosomes visible with a light microscope, the mitotic spindle begins to form, and the nucleolus disappears but the nucleus remains intact.

prostaglandin (pros′-tuh-glan′-din) One of a group of modified fatty acids secreted by virtually all tissues and performing a wide variety of functions as local regulators.

prostate gland (pros′-tāt) A gland in human males that secretes an acid-neutralizing component of semen.

protease An enzyme that digests proteins by hydrolysis.

proteasome A giant protein complex that recognizes and destroys proteins tagged for elimination by the small protein ubiquitin.

protein (prō′-tēn) A biologically functional molecule consisting of one or more polypeptides folded and coiled into a specific three-dimensional structure.

protein kinase An enzyme that transfers phosphate groups from ATP to a protein, thus phosphorylating the protein.

protein phosphatase An enzyme that removes phosphate groups from (dephosphorylates) proteins, often functioning to reverse the effect of a protein kinase.

proteoglycan (prō′-tē-ō-glī′-kan) A large molecule consisting of a small core protein with many carbohydrate chains attached, found in the extracellular matrix of animal cells. A proteoglycan may consist of up to 95% carbohydrate.

proteomics (prō′-tē-ō′-miks) The systematic study of the full protein sets (proteomes) encoded by genomes.

protist An informal term applied to any eukaryote that is not a plant, animal, or fungus. Most protists are unicellular, though some are colonial or multicellular.

protocell An abiotic precursor of a living cell that had a membrane-like structure and that maintained an internal chemistry different from that of its surroundings.

proton (prō′-ton) A subatomic particle with a single positive electrical charge, with a mass of about 1.7×10^{-24} g, found in the nucleus of an atom.

proton pump An active transport protein in a cell membrane that uses ATP to transport hydrogen ions out of a cell against their concentration gradient, generating a membrane potential in the process.

protonema (plural, **protonemata**) A mass of green, branched, one-cell-thick filaments produced by germinating moss spores.

protonephridium (prō′-tō-nuh-frid′-ē-uhm) (plural, **protonephridia**) An excretory system, such as the flame bulb system of flatworms, consisting of a network of tubules lacking internal openings.

proton-motive force The potential energy stored in the form of a proton electrochemical gradient, generated by the pumping of hydrogen ions (H^+) across a biological membrane during chemiosmosis.

proto-oncogene (prō′-tō-on′-kō-jēn) A normal cellular gene that has the potential to become an oncogene.

protoplast The living part of a plant cell, which also includes the plasma membrane.

protoplast fusion The fusing of two protoplasts from different plant species that would otherwise be reproductively incompatible.

protostome development In animals, a developmental mode distinguished by the development of the mouth from the blastopore; often also characterized by spiral cleavage and by the body cavity forming when solid masses of mesoderm split.

provirus A viral genome that is permanently inserted into a host genome.

proximal tubule In the vertebrate kidney, the portion of a nephron immediately downstream from Bowman's capsule that conveys and helps refine filtrate.

pseudocoelomate (sū′-dō-sē′-lō-māt) An animal whose body cavity is lined by tissue derived from mesoderm and endoderm.

pseudogene (sū′-dō-jēn) A DNA segment very similar to a real gene but which does not yield a functional product; a DNA segment that formerly functioned as a gene but has become inactivated in a particular species because of mutation.

pseudopodium (sū′-dō-pō′-dē-um) (plural, **pseudopodia**) A cellular extension of amoeboid cells used in moving and feeding.

pterophyte (ter′-uh-fīt) An informal name for a member of the phylum Pterophyta, which includes ferns, horsetails, and whisk ferns and their relatives.

pterosaur Winged reptile that lived during the Mesozoic era.

pulmocutaneous circuit A branch of the circulatory system in many amphibians that supplies the lungs and skin.

pulmonary circuit The branch of the circulatory system that supplies the lungs.

pulse The rhythmic bulging of the artery walls with each heartbeat.

punctuated equilibria In the fossil record, long periods of apparent stasis, in which a species undergoes little or no morphological change, interrupted by relatively brief periods of sudden change.

Punnett square A diagram used in the study of inheritance to show the predicted genotypic results of random fertilization in genetic crosses between individuals of known genotype.

pupil The opening in the iris, which admits light into the interior of the vertebrate eye. Muscles in the iris regulate its size.

purine (pyū′-rēn) One of two types of nitrogenous bases found in nucleotides, characterized by a six-membered ring fused to a five-membered ring. Adenine (A) and guanine (G) are purines.

pyrimidine (puh-rim′-uh-dēn) One of two types of nitrogenous bases found in nucleotides, characterized by a six-membered ring. Cytosine (C), thymine (T), and uracil (U) are pyrimidines.

quantitative character A heritable feature that varies continuously over a range rather than in an either-or fashion.

quaternary structure (kwot-er-nār-ē) The particular shape of a complex, aggregate protein, defined by the characteristic three-dimensional arrangement of its constituent subunits, each a polypeptide.

R plasmid A bacterial plasmid carrying genes that confer resistance to certain antibiotics.

radial cleavage A type of embryonic development in deuterostomes in which the planes of cell division that transform the zygote into a ball of cells are either parallel or perpendicular to the vertical axis of the embryo, thereby aligning tiers of cells one above the other.

radial symmetry Symmetry in which the body is shaped like a pie or barrel (lacking a left side and a right side) and can be divided into mirror-imaged halves by any plane through its central axis.

radiation The emission of electromagnetic waves by all objects warmer than absolute zero.

radicle An embryonic root of a plant.

radioactive isotope An isotope (an atomic form of a chemical element) that is unstable; the nucleus decays spontaneously, giving off detectable particles and energy.

radiolarian A protist, usually marine, with a shell generally made of silica and pseudopodia that radiate from the central body.

radiometric dating A method for determining the absolute age of rocks and fossils, based on the half-life of radioactive isotopes.

radula A straplike scraping organ used by many molluscs during feeding.

***ras* gene** A gene that codes for Ras, a G protein that relays a growth signal from a growth factor receptor on the plasma membrane to a cascade of protein kinases, ultimately resulting in stimulation of the cell cycle.

ratite (rat′-īt) Member of the group of flightless birds.

ray-finned fish Member of the class Actinopterygii, aquatic osteichthyans with fins supported by long, flexible rays, including tuna, bass, and herring.

reabsorption In excretory systems, the recovery of solutes and water from filtrate.

reactant A starting material in a chemical reaction.

reaction-center complex A complex of proteins associated with a special pair of chlorophyll *a* molecules and a primary electron acceptor. Located centrally in a photosystem, this complex triggers the light reactions of photosynthesis. Excited by light energy, the pair of chlorophylls donates an electron to the primary electron acceptor, which passes an electron to an electron transport chain.

reading frame On an mRNA, the triplet grouping of ribonucleotides used by the translation machinery during polypeptide synthesis.

receptacle The base of a flower; the part of the stem that is the site of attachment of the floral organs.

receptor potential An initial response of a receptor cell to a stimulus, consisting of a change in voltage across the receptor membrane proportional to the stimulus strength.

receptor tyrosine kinase (RTK) A receptor protein spanning the plasma membrane, the cytoplasmic (intracellular) part of which can catalyze the transfer of a phosphate group from ATP to a tyrosine on another protein. Receptor tyrosine kinases often respond to the binding of a signaling molecule by dimerizing and then phosphorylating a tyrosine on the cytoplasmic portion of the other receptor in the dimer. The phosphorylated tyrosines on the receptors then activate other signal transduction proteins within the cell.

receptor-mediated endocytosis (en′-dō-sī-tō′-sis) The movement of specific molecules into a cell by the inward budding of vesicles containing proteins with receptor sites specific to the molecules being taken in; enables a cell to acquire bulk quantities of specific substances.

recessive allele An allele whose phenotypic effect is not observed in a heterozygote.

reciprocal altruism Altruistic behavior between unrelated individuals, whereby the altruistic individual benefits in the future when the beneficiary reciprocates.

recombinant chromosome A chromosome created when crossing over combines DNA from two parents into a single chromosome.

recombinant DNA A DNA molecule made *in vitro* with segments from different sources.

recombinant type (recombinant) An offspring whose phenotype differs from that of the true-breeding P generation parents; also refers to the phenotype itself.

rectum The terminal portion of the large intestine, where the feces are stored prior to elimination.

red alga A photosynthetic protist, named for its color, which results from a red pigment that masks the green of chlorophyll. Most red algae are multicellular and marine.

redox reaction (rē′-doks) A chemical reaction involving the complete or partial transfer of one or more electrons from one reactant to another; short for **red**uction-**ox**idation reaction.

reducing agent The electron donor in a redox reaction.

reduction The complete or partial addition of electrons to a substance involved in a redox reaction.

reflex An automatic reaction to a stimulus, mediated by the spinal cord or lower brain.

refractory period (rē-frakt′-ōr-ē) The short time immediately after an action potential in which the neuron cannot respond to another stimulus, owing to the inactivation of voltage-gated sodium channels.

regulator An animal for which mechanisms of homeostasis moderate internal changes in a particular variable in the face of external fluctuation of that variable.

regulatory gene A gene that codes for a protein, such as a repressor, that controls the transcription of another gene or group of genes.

reinforcement In evolutionary biology, a process in which a process in which natural selection strengthens prezygotic barriers to reproduction, thus reducing the chances of hybrid formation. Such a process is likely to occur only if hybrid offspring are less fit than members of the parent species.

relative abundance The proportional abundance of different species in a community.

relative fitness The contribution an individual makes to the gene pool of the next generation, relative to the contributions of other individuals in the population.

renal cortex The outer portion of the vertebrate kidney.

renal medulla The inner portion of the vertebrate kidney, beneath the renal cortex.

renal pelvis The funnel-shaped chamber that receives processed filtrate from the vertebrate kidney's collecting ducts and is drained by the ureter.

renin-angiotensin-aldosterone system (RAAS) A hormone cascade pathway that helps regulate blood pressure and blood volume.

repetitive DNA Nucleotide sequences, usually noncoding, that are present in many copies in a eukaryotic genome. The repeated units may be short and arranged tandemly (in series) or long and dispersed in the genome.

replication fork A Y-shaped region on a replicating DNA molecule where the parental strands are being unwound and new strands are being synthesized.

repressor A protein that inhibits gene transcription. In prokaryotes, repressors bind to the DNA in or near the promoter. In eukaryotes, repressors may bind to control elements within enhancers, to activators, or to other proteins in a way that blocks activators from binding to DNA.

reproductive isolation The existence of biological factors (barriers) that impede members of two species from producing viable, fertile offspring.

reproductive table An age-specific summary of the reproductive rates in a population.

reptile Member of the clade of amniotes that includes tuataras, lizards, snakes, turtles, crocodilians, and birds.

residual volume The amount of air that remains in the lungs after forceful exhalation.

resource partitioning The division of environmental resources by coexisting species such that the niche of each species differs by one or more significant factors from the niches of all coexisting species.

respiratory pigment A protein that transports oxygen in blood or hemolymph.

response (1) In cellular communication, the change in a specific cellular activity brought about by a transduced signal from outside the cell. (2) In feedback regulation, a physiological activity triggered by a change in a variable.

resting potential The membrane potential characteristic of a nonconducting excitable cell, with the inside of the cell more negative than the outside.

restriction enzyme An endonuclease (type of enzyme) that recognizes and cuts DNA molecules foreign to a bacterium (such as phage genomes). The enzyme cuts at specific nucleotide sequences (restriction sites).

restriction fragment A DNA segment that results from the cutting of DNA by a restriction enzyme.

restriction fragment length polymorphism (RFLP) A single nucleotide polymorphism (SNP) that exists in the restriction site for a particular enzyme, thus making the site unrecognizable by that enzyme and changing the lengths of the restriction fragments formed by digestion with that enzyme. A RFLP can be in coding or noncoding DNA.

restriction site A specific sequence on a DNA strand that is recognized and cut by a restriction enzyme.

reticular formation (re-tik′-yū-ler) A diffuse network of neurons in the core of the brainstem that filters information traveling to the cerebral cortex.

retina (ret′-i-nuh) The innermost layer of the vertebrate eye, containing photoreceptor cells (rods and cones) and neurons; transmits images formed by the lens to the brain via the optic nerve.

retinal The light-absorbing pigment in rods and cones of the vertebrate eye.

retrotransposon (re′-trō-trans-pō′-zon) A transposable element that moves within a genome by means of an RNA intermediate, a transcript of the retrotransposon DNA.

retrovirus (re′-trō-vī′-rus) An RNA virus that replicates by transcribing its RNA into DNA and then inserting the DNA into a cellular chromosome; an important class of cancer-causing viruses.

reverse transcriptase (tran-skrip′-tās) An enzyme encoded by certain viruses (retroviruses) that uses RNA as a template for DNA synthesis.

reverse transcriptase–polymerase chain reaction (RT-PCR) A technique for determining expression of a particular gene. It uses reverse transcriptase and DNA polymerase to synthesize cDNA from all the mRNA in a sample and then subjects the cDNA to PCR amplification using primers specific for the gene of interest.

Rhizaria (rī-za′-rē-uh) One of five supergroups of eukaryotes proposed in a current hypothesis of the evolutionary history of eukaryotes; a morphologically diverse protist clade that is defined by DNA similarities. *See also* Excavata, Chromalveolata, Archaeplastida, and Unikonta.

rhizobacterium A soil bacterium whose population size is much enhanced in the rhizosphere, the soil region close to a plant's roots.

rhizoid (rī′-zoyd) A long, tubular single cell or filament of cells that anchors bryophytes to the ground. Unlike roots, rhizoids are not composed of tissues, lack specialized conducting cells, and do not play a primary role in water and mineral absorption.

rhizosphere The soil region close to plant roots and characterized by a high level of microbiological activity.

rhodopsin (rō-dop′-sin) A visual pigment consisting of retinal and opsin. Upon absorbing light, the retinal changes shape and dissociates from the opsin.

rhythm method A form of contraception that relies on refraining from sexual intercourse when conception is most likely to occur; also called natural family planning.

ribonucleic acid (RNA) (rī′-bō-nū-klā′-ik) A type of nucleic acid consisting of a polynucleotide made up of nucleotide monomers with a ribose sugar and the nitrogenous bases adenine (A), cytosine (C), guanine (G), and uracil (U); usually single-stranded; functions in protein synthesis, gene regulation, and as the genome of some viruses.

ribose The sugar component of RNA nucleotides.

ribosomal RNA (rRNA) (rī′-buh-sō′-mul) RNA molecules that, together with proteins, make up ribosomes; the most abundant type of RNA.

ribosome (rī′-buh-sōm′) A complex of rRNA and protein molecules that functions as a site of protein synthesis in the cytoplasm; consists of a large and a small subunit. In eukaryotic cells, each subunit is assembled in the nucleolus. *See also* nucleolus.

ribozyme (rī′-buh-zīm) An RNA molecule that functions as an enzyme, such as an intron that catalyzes its own removal during RNA splicing.

RNA interference (RNAi) A technique used to silence the expression of selected genes. RNAi uses synthetic double-stranded RNA molecules that match the sequence of a particular gene to trigger the breakdown of the gene's messenger RNA.

RNA polymerase An enzyme that links ribonucleotides into a growing RNA chain during transcription, based on complementary binding to nucleotides on a DNA template strand.

RNA processing Modification of RNA primary transcripts, including splicing out of introns, joining together of exons, and alteration of the 5′ and 3′ ends.

RNA splicing After synthesis of a eukaryotic primary RNA transcript, the removal of portions of the transcript (introns) that will not be included in the mRNA and the joining together of the remaining portions (exons).

rod A rodlike cell in the retina of the vertebrate eye, sensitive to low light intensity.

root An organ in vascular plants that anchors the plant and enables it to absorb water and minerals from the soil.

root cap A cone of cells at the tip of a plant root that protects the apical meristem.

root hair A tiny extension of a root epidermal cell, growing just behind the root tip and increasing surface area for absorption of water and minerals.

root pressure Pressure exerted in the roots of plants as the result of osmosis, causing exudation from cut stems and guttation of water from leaves.

root system All of a plant's roots, which anchor it in the soil, absorb and transport minerals and water, and store food.

rooted Describing a phylogenetic tree that contains a branch point (often, the one farthest to the left) representing the most recent common ancestor of all taxa in the tree.

rough ER That portion of the endoplasmic reticulum with ribosomes attached.

round window In the mammalian ear, the point of contact where vibrations of the stapes create a traveling series of pressure waves in the fluid of the cochlea.

***r*-selection** Selection for life history traits that maximize reproductive success in uncrowded environments; also called density-independent selection.

rubisco (rū-bis'-kō) Ribulose bisphosphate (RuBP) carboxylase, the enzyme that catalyzes the first step of the Calvin cycle (the addition of CO_2 to RuBP).

ruminant (rū'-muh-nent) An animal, such as a cow or a sheep, with multiple stomach compartments specialized for an herbivorous diet.

S phase The synthesis phase of the cell cycle; the portion of interphase during which DNA is replicated.

saccule In the vertebrate ear, a chamber in the vestibule behind the oval window that participates in the sense of balance.

salicylic acid (sal'-i-sil'-ik) A signaling molecule in plants that may be partially responsible for activating systemic acquired resistance to pathogens.

salivary gland A gland associated with the oral cavity that secretes substances that lubricate food and begin the process of chemical digestion.

salt A compound resulting from the formation of an ionic bond; also called an ionic compound.

saltatory conduction (sol'-tuh-tōr'-ē) Rapid transmission of a nerve impulse along an axon, resulting from the action potential jumping from one node of Ranvier to another, skipping the myelin-sheathed regions of membrane.

sarcomere (sar'-kō-mēr) The fundamental, repeating unit of striated muscle, delimited by the Z lines.

sarcoplasmic reticulum (SR) (sar'-kō-plaz'-mik ruh-tik'-yū-lum) A specialized endoplasmic reticulum that regulates the calcium concentration in the cytosol of muscle cells.

saturated fatty acid A fatty acid in which all carbons in the hydrocarbon tail are connected by single bonds, thus maximizing the number of hydrogen atoms that are attached to the carbon skeleton.

savanna A tropical grassland biome with scattered individual trees and large herbivores and maintained by occasional fires and drought.

scaffolding protein A type of large relay protein to which several other relay proteins are simultaneously attached, increasing the efficiency of signal transduction.

scanning electron microscope (SEM) A microscope that uses an electron beam to scan the surface of a sample, coated with metal atoms, to study details of its topography.

schizophrenia (skit'-suh-frē'-nē-uh) A severe mental disturbance characterized by psychotic episodes in which patients have a distorted perception of reality.

Schwann cell A type of glial cell that forms insulating myelin sheaths around the axons of neurons in the peripheral nervous system.

science An approach to understanding the natural world.

scion (sī'-un) The twig grafted onto the stock when making a graft.

sclereid (sklār'-ē-id) A short, irregular sclerenchyma cell in nutshells and seed coats. Sclereids are scattered throughout the parenchyma of some plants.

sclerenchyma cell (skluh-ren'-kim-uh) A rigid, supportive plant cell type usually lacking a protoplast and possessing thick secondary walls strengthened by lignin at maturity.

scrotum A pouch of skin outside the abdomen that houses the testes; functions in maintaining the testes at the lower temperature required for spermatogenesis.

second law of thermodynamics The principle stating that every energy transfer or transformation increases the entropy of the universe. Usable forms of energy are at least partly converted to heat.

second messenger A small, nonprotein, water-soluble molecule or ion, such as a calcium ion (Ca^{2+}) or cyclic AMP, that relays a signal to a cell's interior in response to a signaling molecule bound by a signal receptor protein.

secondary cell wall In plant cells, a strong and durable matrix that is often deposited in several laminated layers around the plasma membrane and provides protection and support.

secondary consumer A carnivore that eats herbivores.

secondary endosymbiosis A process in eukaryotic evolution in which a heterotrophic eukaryotic cell engulfed a photosynthetic eukaryotic cell, which survived in a symbiotic relationship inside the heterotrophic cell.

secondary growth Growth produced by lateral meristems, thickening the roots and shoots of woody plants.

secondary immune response The adaptive immune response elicited on second or subsequent exposures to a particular antigen. The secondary immune response is more rapid, of greater magnitude, and of longer duration than the primary immune response.

secondary oocyte (ō'-uh-sīt) An oocyte that has completed the first of the two meiotic divisions.

secondary production The amount of chemical energy in consumers' food that is converted to their own new biomass during a given time period.

secondary structure Regions of repetitive coiling or folding of the polypeptide backbone of a protein due to hydrogen bonding between constituents of the backbone (not the side chains).

secondary succession A type of succession that occurs where an existing community has been cleared by some disturbance that leaves the soil or substrate intact.

secretion (1) The discharge of molecules synthesized by a cell. (2) The discharge of wastes from the body fluid into the filtrate.

secretory phase That portion of the uterine (menstrual) cycle when the endometrium continues to thicken, becomes more vascularized, and develops glands that secrete a fluid rich in glycogen.

seed An adaptation of some terrestrial plants consisting of an embryo packaged along with a store of food within a protective coat.

seed coat A tough outer covering of a seed, formed from the outer coat of an ovule. In a flowering plant, the seed coat encloses and protects the embryo and endosperm.

seedless vascular plant An informal name for a plant that has vascular tissue but lacks seeds. Seedless vascular plants form a paraphyletic group that includes the phyla Lycophyta (club mosses and their relatives) and Pterophyta (ferns and their relatives).

selective permeability A property of biological membranes that allows them to regulate the passage of substances across them.

self-incompatibility The ability of a seed plant to reject its own pollen and sometimes the pollen of closely related individuals.

semelparity Reproduction in which an organism produces all of its offspring in a single event; also known as big-bang reproduction.

semen (sē'-mun) The fluid that is ejaculated by the male during orgasm; contains sperm and secretions from several glands of the male reproductive tract.

semicircular canals A three-part chamber of the inner ear that functions in maintaining equilibrium.

semiconservative model Type of DNA replication in which the replicated double helix consists of one old strand, derived from the parental molecule, and one newly made strand.

semilunar valve A valve located at each exit of the heart, where the aorta leaves the left ventricle and the pulmonary artery leaves the right ventricle.

seminal vesicle (sem'-i-nul ves'-i-kul) A gland in males that secretes a fluid component of semen that lubricates and nourishes sperm.

seminiferous tubule (sem'-i-nif'-er-us) A highly coiled tube in the testis in which sperm are produced.

senescence (se-nes'-ens) The growth phase in a plant or plant part (as a leaf) from full maturity to death.

sensitive period A limited phase in an animal's development when learning of particular behaviors can take place; also called a critical period.

sensor In homeostasis, a receptor that detects a stimulus.

sensory adaptation The tendency of sensory neurons to become less sensitive when they are stimulated repeatedly.

sensory neuron A nerve cell that receives information from the internal or external environment and transmits signals to the central nervous system.

sensory reception The detection of a stimulus by sensory cells.

sensory receptor An organ, cell, or structure within a cell that responds to specific stimuli from an organism's external or internal environment.

sensory transduction The conversion of stimulus energy to a change in the membrane potential of a sensory receptor cell.

sepal (sē′-pul) A modified leaf in angiosperms that helps enclose and protect a flower bud before it opens.

septum (plural, **septa**) One of the cross-walls that divide a fungal hypha into cells. Septa generally have pores large enough to allow ribosomes, mitochondria, and even nuclei to flow from cell to cell.

serial endosymbiosis A hypothesis for the origin of eukaryotes consisting of a sequence of endosymbiotic events in which mitochondria, chloroplasts, and perhaps other cellular structures were derived from small prokaryotes that had been engulfed by larger cells.

serotonin (ser′-uh-tō′-nin) A neurotransmitter, synthesized from the amino acid tryptophan, that functions in the central nervous system.

set point In homeostasis in animals, a value maintained for a particular variable, such as body temperature or solute concentration.

seta (sē′-tuh) (plural, **setae**) The elongated stalk of a bryophyte sporophyte.

sex chromosome A chromosome responsible for determining the sex of an individual.

sex-linked gene A gene located on either sex chromosome. Most sex-linked genes are on the X chromosome and show distinctive patterns of inheritance; there are very few genes on the Y chromosome.

sexual dimorphism (dī-mōr′-fizm) Differences between the secondary sex characteristics of males and females.

sexual reproduction A type of reproduction in which two parents give rise to offspring that have unique combinations of genes inherited from both parents via the gametes.

sexual selection A form of selection in which individuals with certain inherited characteristics are more likely than other individuals to obtain mates.

Shannon diversity An index of community diversity symbolized by H and represented by the equation $H = -(p_A \ln p_A + p_B \ln p_B + p_C \ln p_C + \ldots)$, where A, B, C . . . are species, p is the relative abundance of each species, and ln is the natural logarithm.

shared ancestral character A character, shared by members of a particular clade, that originated in an ancestor that is not a member of that clade.

shared derived character An evolutionary novelty that is unique to a particular clade.

shoot system The aerial portion of a plant body, consisting of stems, leaves, and (in angiosperms) flowers.

short tandem repeat (STR) Simple sequence DNA containing multiple tandemly repeated units of two to five nucleotides. Variations in STRs act as genetic markers in STR analysis, used to prepare genetic profiles.

short-day plant A plant that flowers (usually in late summer, fall, or winter) only when the light period is shorter than a critical length.

short-term memory The ability to hold information, anticipations, or goals for a time and then release them if they become irrelevant.

sickle-cell disease A recessively inherited human blood disorder in which a single nucleotide change in the β-globin gene causes hemoglobin to aggregate, changing red blood cell shape and causing multiple symptoms in afflicted individuals.

sieve plate An end wall in a sieve-tube element, which facilitates the flow of phloem sap in angiosperm sieve tubes.

sieve-tube element A living cell that conducts sugars and other organic nutrients in the phloem of angiosperms; also called a sieve-tube member. Connected end to end, they form sieve tubes.

sign stimulus An external sensory cue that triggers a fixed action pattern by an animal.

signal In animal behavior, transmission of a stimulus from one animal to another. The term is also used in the context of communication in other kinds of organisms and in cell-to-cell communication in all multicellular organisms.

signal peptide A sequence of about 20 amino acids at or near the leading (amino) end of a polypeptide that targets it to the endoplasmic reticulum or other organelles in a eukaryotic cell.

signal transduction The linkage of a mechanical, chemical, or electromagnetic stimulus to a specific cellular response.

signal transduction pathway A series of steps linking a mechanical, chemical, or electrical stimulus to a specific cellular response.

signal-recognition particle (SRP) A protein-RNA complex that recognizes a signal peptide as it emerges from a ribosome and helps direct the ribosome to the endoplasmic reticulum (ER) by binding to a receptor protein on the ER.

silent mutation A nucleotide-pair substitution that has no observable effect on the phenotype; for example, within a gene, a mutation that results in a codon that codes for the same amino acid.

simple fruit A fruit derived from a single carpel or several fused carpels.

simple sequence DNA A DNA sequence that contains many copies of tandemly repeated short sequences.

single bond A single covalent bond; the sharing of a pair of valence electrons by two atoms.

single circulation A circulatory system consisting of a single pump and circuit, in which blood passes from the sites of gas exchange to the rest of the body before returning to the heart.

single nucleotide polymorphism (SNP) A single base-pair site in a genome where nucleotide variation is found in at least 1% of the population.

single-lens eye The camera-like eye found in some jellies, polychaete worms, spiders, and many molluscs.

single-strand binding protein A protein that binds to the unpaired DNA strands during DNA replication, stabilizing them and holding them apart while they serve as templates for the synthesis of complementary strands of DNA.

sinoatrial (SA) node A region in the right atrium of the heart that sets the rate and timing at which all cardiac muscle cells contract; the pacemaker.

sister chromatids Two copies of a duplicated chromosome attached to each other by proteins at the centromere and, sometimes, along the arms. While joined, two sister chromatids make up one chromosome. Chromatids are eventually separated during mitosis or meiosis II.

sister taxa Groups of organisms that share an immediate common ancestor and hence are each other's closest relatives.

skeletal muscle A type of striated muscle that is generally responsible for the voluntary movements of the body.

sliding-filament model The idea that muscle contraction is based on the movement of thin (actin) filaments along thick (myosin) filaments, shortening the sarcomere, the basic unit of muscle organization.

slow block to polyspermy The formation of the fertilization envelope and other changes in an egg's surface that prevent fusion of the egg with more than one sperm. The slow block begins about 1 minute after fertilization.

slow-twitch fiber A muscle fiber that can sustain long contractions.

small interfering RNA (siRNA) One of multiple small, single-stranded RNA molecules generated by cellular machinery from a long, linear, double-stranded RNA molecule. The siRNA associates with one or more proteins in a complex that can degrade or prevent translation of an mRNA with a complementary sequence. In some cases, siRNA can also block transcription by promoting chromatin modification.

small intestine The longest section of the alimentary canal, so named because of its small diameter compared with that of the large intestine; the principal site of the enzymatic hydrolysis of food macromolecules and the absorption of nutrients.

smooth ER That portion of the endoplasmic reticulum that is free of ribosomes.

smooth muscle A type of muscle lacking the striations of skeletal and cardiac muscle because of the uniform distribution of myosin filaments in the cells; responsible for involuntary body activities.

social learning Modification of behavior through the observation of other individuals.

sociobiology The study of social behavior based on evolutionary theory.

sodium-potassium pump A transport protein in the plasma membrane of animal cells that actively transports sodium out of the cell and potassium into the cell.

Glossary

soil horizon A soil layer with physical characteristics that differ from those of the layers above or beneath.

solute (sol′-yūt) A substance that is dissolved in a solution.

solute potential (Ψ_S) A component of water potential that is proportional to the molarity of a solution and that measures the effect of solutes on the direction of water movement; also called osmotic potential, it can be either zero or negative.

solution A liquid that is a homogeneous mixture of two or more substances.

solvent The dissolving agent of a solution. Water is the most versatile solvent known.

somatic cell (sō-mat′-ik) Any cell in a multicellular organism except a sperm or egg or their precursors.

somite One of a series of blocks of mesoderm that exist in pairs just lateral to the notochord in a vertebrate embryo.

soredium (plural, **soredia**) In lichens, a small cluster of fungal hyphae with embedded algae.

sorus (plural, **sori**) A cluster of sporangia on a fern sporophyll. Sori may be arranged in various patterns, such as parallel lines or dots, which are useful in fern identification.

Southern blotting A technique that enables specific nucleotide sequences to be detected in samples of DNA. It involves gel electrophoresis of DNA molecules and their transfer to a membrane (blotting), followed by nucleic acid hybridization with a labeled probe.

spatial learning The establishment of a memory that reflects the environment's spatial structure.

spatial summation A phenomenon of neural integration in which the membrane potential of the postsynaptic cell is determined by the combined effect of EPSPs or IPSPs produced nearly simultaneously by different synapses.

speciation (spē′-sē-ā′-shun) An evolutionary process in which one species splits into two or more species.

species (spē′-sēz) A population or group of populations whose members have the potential to interbreed in nature and produce viable, fertile offspring, but do not produce viable, fertile offspring with members of other such groups.

species diversity The number and relative abundance of species in a biological community.

species richness The number of species in a biological community.

species-area curve The biodiversity pattern that shows that the larger the geographic area of a community is, the more species it has.

specific heat The amount of heat that must be absorbed or lost for 1 g of a substance to change its temperature by 1°C.

spectrophotometer An instrument that measures the proportions of light of different wavelengths absorbed and transmitted by a pigment solution.

sperm The male gamete.

spermatheca (sper′-muh-thē′-kuh) In many insects, a sac in the female reproductive system where sperm are stored.

spermatogenesis The continuous and prolific production of mature sperm cells in the testis.

spermatogonium (plural, **spermatogonia**) A cell that divides mitotically to form spermatocytes.

sphincter (sfink′-ter) A ringlike band of muscle fibers that controls the size of an opening in the body, such as the passage between the esophagus and the stomach.

spiral cleavage A type of embryonic development in protostomes in which the planes of cell division that transform the zygote into a ball of cells are diagonal to the vertical axis of the embryo. As a result, the cells of each tier sit in the grooves between cells of adjacent tiers.

spliceosome (splī′-sō-sōm) A large complex made up of proteins and RNA molecules that splices RNA by interacting with the ends of an RNA intron, releasing the intron and joining the two adjacent exons.

spongocoel (spon′-jō-sēl) The central cavity of a sponge.

spontaneous process A process that occurs without an overall input of energy; a process that is energetically favorable.

sporangium (spōr-an′-jē-um) (plural, **sporangia**) A multicellular organ in fungi and plants in which meiosis occurs and haploid cells develop.

spore (1) In the life cycle of a plant or alga undergoing alternation of generations, a haploid cell produced in the sporophyte by meiosis. A spore can divide by mitosis to develop into a multicellular haploid individual, the gametophyte, without fusing with another cell. (2) In fungi, a haploid cell, produced either sexually or asexually, that produces a mycelium after germination.

sporocyte A diploid cell, also known as a spore mother cell, that undergoes meiosis and generates haploid spores.

sporophyll (spō′-ruh-fil) A modified leaf that bears sporangia and hence is specialized for reproduction.

sporophyte (spō-ruh-fīt′) In organisms (plants and some algae) that have alternation of generations, the multicellular diploid form that results from the union of gametes. The sporophyte produces haploid spores by meiosis that develop into gametophytes.

sporopollenin (spōr-uh-pol′-eh-nin) A durable polymer that covers exposed zygotes of charophyte algae and forms the walls of plant spores, preventing them from drying out.

stabilizing selection Natural selection in which intermediate phenotypes survive or reproduce more successfully than do extreme phenotypes.

stamen (stā′-men) The pollen-producing reproductive organ of a flower, consisting of an anther and a filament.

standard metabolic rate (SMR) Metabolic rate of a resting, fasting, and nonstressed ectotherm at a particular temperature.

starch A storage polysaccharide in plants, consisting entirely of glucose monomers joined by α glycosidic linkages.

start point In transcription, the nucleotide position on the promoter where RNA polymerase begins synthesis of RNA.

statocyst (stat′-uh-sist′) A type of mechanoreceptor that functions in equilibrium in invertebrates by use of statoliths, which stimulate hair cells in relation to gravity.

statolith (stat′-uh-lith′) (1) In plants, a specialized plastid that contains dense starch grains and may play a role in detecting gravity. (2) In invertebrates, a dense particle that settles in response to gravity and is found in sensory organs that function in equilibrium.

stele (stēl) The vascular tissue of a stem or root.

stem A vascular plant organ consisting of an alternating system of nodes and internodes that support the leaves and reproductive structures.

stem cell Any relatively unspecialized cell that can produce, during a single division, one identical daughter cell and one more specialized daughter cell that can undergo further differentiation.

steroid A type of lipid characterized by a carbon skeleton consisting of four fused rings with various chemical groups attached.

sticky end A single-stranded end of a double-stranded restriction fragment.

stigma (plural, **stigmata**) The sticky part of a flower's carpel, which receives pollen grains.

stimulus In feedback regulation, a fluctuation in a variable that triggers a response.

stipe A stemlike structure of a seaweed.

stock The plant that provides the root system when making a graft.

stoma (stō′-muh) (plural, **stomata**) A microscopic pore surrounded by guard cells in the epidermis of leaves and stems that allows gas exchange between the environment and the interior of the plant.

stomach An organ of the digestive system that stores food and performs preliminary steps of digestion.

stramenopile A protist in which a "hairy" flagellum (one covered with fine, hairlike projections) is paired with a shorter, smooth flagellum.

stratum (strah′-tum) (plural, **strata**) A rock layer formed when new layers of sediment cover older ones and compress them.

striated muscle Muscle in which the regular arrangement of filaments creates a pattern of light and dark bands.

strigolactones A class of plant hormone that inhibits shoot branching, triggers the germination of parasitic plant seeds, and stimulates the association of plant roots with mycorrhizal fungi.

strobilus (strō-bī′-lus) (plural, **strobili**) The technical term for a cluster of sporophylls known commonly as a cone, found in most gymnosperms and some seedless vascular plants.

stroke The death of nervous tissue in the brain, usually resulting from rupture or blockage of arteries in the head.

stroke volume The volume of blood pumped by a heart ventricle in a single contraction.

stroma (strō′-muh) The dense fluid within the chloroplast surrounding the thylakoid membrane and containing ribosomes and DNA; involved in the synthesis of organic molecules from carbon dioxide and water.

stromatolite Layered rock that results from the activities of prokaryotes that bind thin films of sediment together.

structural isomer One of several compounds that have the same molecular formula but differ in the covalent arrangements of their atoms.

style The stalk of a flower's carpel, with the ovary at the base and the stigma at the top.

substrate The reactant on which an enzyme works.

substrate feeder An animal that lives in or on its food source, eating its way through the food.

substrate-level phosphorylation The enzyme-catalyzed formation of ATP by direct transfer of a phosphate group to ADP from an intermediate substrate in catabolism.

sugar sink A plant organ that is a net consumer or storer of sugar. Growing roots, shoot tips, stems, and fruits are examples of sugar sinks supplied by phloem.

sugar source A plant organ in which sugar is being produced by either photosynthesis or the breakdown of starch. Mature leaves are the primary sugar sources of plants.

sulfhydryl group A chemical group consisting of a sulfur atom bonded to a hydrogen atom.

suprachiasmatic nucleus (SCN) A group of neurons in the hypothalamus of mammals that functions as a biological clock.

surface tension A measure of how difficult it is to stretch or break the surface of a liquid. Water has a high surface tension because of the hydrogen bonding of surface molecules.

surfactant A substance secreted by alveoli that decreases surface tension in the fluid that coats the alveoli.

survivorship curve A plot of the number of members of a cohort that are still alive at each age; one way to represent age-specific mortality.

suspension feeder An aquatic animal, such as a sponge, clam, or baleen whale, that feeds by sifting small organisms or food particles from the water.

sustainable agriculture Long-term productive farming methods that are environmentally safe.

sustainable development Development that meets the needs of people today without limiting the ability of future generations to meet their needs.

swim bladder In aquatic osteichthyans, an air sac that enables the animal to control its buoyancy in the water.

symbiont (sim′-bē-ont) The smaller participant in a symbiotic relationship, living in or on the host.

symbiosis An ecological relationship between organisms of two different species that live together in direct and intimate contact.

sympathetic division One of three divisions of the autonomic nervous system; generally increases energy expenditure and prepares the body for action.

sympatric speciation (sim-pat′-rik) The formation of new species in populations that live in the same geographic area.

symplast In plants, the continuum of cytoplasm connected by plasmodesmata between cells.

synapse (sin′-aps) The junction where a neuron communicates with another cell across a narrow gap via a neurotransmitter or an electrical coupling.

synapsid Member of an amniote clade distinguished by a single hole on each side of the skull. Synapsids include the mammals.

synapsis (si-nap′-sis) The pairing and physical connection of duplicated homologous chromosomes during prophase I of meiosis.

systematics A scientific discipline focused on classifying organisms and determining their evolutionary relationships.

systemic acquired resistance A defensive response in infected plants that helps protect healthy tissue from pathogenic invasion.

systemic circuit The branch of the circulatory system that supplies oxygenated blood to and carries deoxygenated blood away from organs and tissues throughout the body.

systems biology An approach to studying biology that aims to model the dynamic behavior of whole biological systems based on a study of the interactions among the system's parts.

systole (sis′-tō-lē) The stage of the cardiac cycle in which a heart chamber contracts and pumps blood.

systolic pressure Blood pressure in the arteries during contraction of the ventricles.

T cells The class of lymphocytes that mature in the thymus; they include both effector cells for the cell-mediated immune response and helper cells required for both branches of adaptive immunity.

taproot A main vertical root that develops from an embryonic root and gives rise to lateral (branch) roots.

tastant Any chemical that stimulates the sensory receptors in a taste bud.

taste bud A collection of modified epithelial cells on the tongue or in the mouth that are receptors for taste in mammals.

TATA box A DNA sequence in eukaryotic promoters crucial in forming the transcription initiation complex.

taxis (tak′-sis) An oriented movement toward or away from a stimulus.

taxon (plural, **taxa**) A named taxonomic unit at any given level of classification.

taxonomy (tak-son′-uh-mē) A scientific discipline concerned with naming and classifying the diverse forms of life.

Tay-Sachs disease A human genetic disease caused by a recessive allele for a dysfunctional enzyme, leading to accumulation of certain lipids in the brain. Seizures, blindness, and degeneration of motor and mental performance usually become manifest a few months after birth, followed by death within a few years.

technology The application of scientific knowledge for a specific purpose, often involving industry or commerce but also including uses in basic research.

telomerase An enzyme that catalyzes the lengthening of telomeres in eukaryotic germ cells.

telomere (tel′-uh-mēr) The tandemly repetitive DNA at the end of a eukaryotic chromosome's DNA molecule. Telomeres protect the organism's genes from being eroded during successive rounds of replication. *See also* repetitive DNA.

telophase The fifth and final stage of mitosis, in which daughter nuclei are forming and cytokinesis has typically begun.

temperate broadleaf forest A biome located throughout midlatitude regions where there is sufficient moisture to support the growth of large, broadleaf deciduous trees.

temperate grassland A terrestrial biome that exists at midlatitude regions and is dominated by grasses and forbs.

temperate phage A phage that is capable of replicating by either a lytic or lysogenic cycle.

temperature A measure of the intensity of heat in degrees, reflecting the average kinetic energy of the molecules.

template strand The DNA strand that provides the pattern, or template, for ordering, by complementary base pairing, the sequence of nucleotides in an RNA transcript.

temporal summation A phenomenon of neural integration in which the membrane potential of the postsynaptic cell in a chemical synapse is determined by the combined effect of EPSPs or IPSPs produced in rapid succession.

tendon A fibrous connective tissue that attaches muscle to bone.

terminator In bacteria, a sequence of nucleotides in DNA that marks the end of a gene and signals RNA polymerase to release the newly made RNA molecule and detach from the DNA.

territoriality A behavior in which an animal defends a bounded physical space against encroachment by other individuals, usually of its own species.

tertiary consumer (ter-shē-ār′-ē) A carnivore that eats other carnivores.

tertiary structure The overall shape of a protein molecule due to interactions of amino acid side chains, including hydrophobic interactions, ionic bonds, hydrogen bonds, and disulfide bridges.

testcross Breeding an organism of unknown genotype with a homozygous recessive individual to determine the unknown genotype. The ratio of phenotypes in the offspring reveals the unknown genotype.

testis (plural, **testes**) The male reproductive organ, or gonad, in which sperm and reproductive hormones are produced.

testosterone A steroid hormone required for development of the male reproductive system, spermatogenesis, and male secondary sex characteristics; the major androgen in mammals.

tetanus (tet′-uh-nus) The maximal, sustained contraction of a skeletal muscle, caused by a very high frequency of action potentials elicited by continual stimulation.

tetrapod A vertebrate clade whose members have limbs with digits. Tetrapods include mammals, amphibians, and birds and other reptiles.

thalamus (thal′-uh-mus) An integrating center of the vertebrate forebrain. Neurons with cell bodies in the thalamus relay neural input to specific areas in the cerebral cortex and regulate what information goes to the cerebral cortex.

thallus (plural, **thalli**) A seaweed body that is plantlike, consisting of a holdfast, stipe, and blades, yet lacks true roots, stems, and leaves.

theory An explanation that is broader in scope than a hypothesis, generates new hypotheses, and is supported by a large body of evidence.

thermal energy *See* heat.

thermocline A narrow stratum of abrupt temperature change in the ocean and in many temperate-zone lakes.

thermodynamics (ther′-mō-dī-nam′-iks) The study of energy transformations that occur in a collection of matter. *See* first law of thermodynamics; second law of thermodynamics.

thermoreceptor A receptor stimulated by either heat or cold.

thermoregulation The maintenance of internal body temperature within a tolerable range.

theropod Member of a group of dinosaurs that were bipedal carnivores.

thick filament A filament composed of staggered arrays of myosin molecules; a component of myofibrils in muscle fibers.

thigmomorphogenesis A response in plants to chronic mechanical stimulation, resulting from increased ethylene production. An example is thickening stems in response to strong winds.

thigmotropism (thig-mo′-truh-pizm) A directional growth of a plant in response to touch.

thin filament A filament consisting of two strands of actin and two strands of regulatory protein coiled around one another; a component of myofibrils in muscle fibers.

threatened species A species that is considered likely to become endangered in the foreseeable future.

threshold The potential that an excitable cell membrane must reach for an action potential to be initiated.

thrombus A fibrin-containing clot that forms in a blood vessel and blocks the flow of blood.

thylakoid (thī′-luh-koyd) A flattened, membranous sac inside a chloroplast. Thylakoids often exist in stacks called grana that are interconnected; their membranes contain molecular "machinery" used to convert light energy to chemical energy.

thymus (thī′-mus) A small organ in the thoracic cavity of vertebrates where maturation of T cells is completed.

thyroid gland An endocrine gland, located on the ventral surface of the trachea, that secretes two iodine-containing hormones, triiodothyronine (T_3) and thyroxine (T_4), as well as calcitonin.

thyroxine (T_4) One of two iodine-containing hormones that are secreted by the thyroid gland and that help regulate metabolism, development, and maturation in vertebrates.

Ti plasmid A plasmid of a tumor-inducing bacterium (the plant pathogen *Agrobacterium*) that integrates a segment of its DNA (T DNA) into a chromosome of a host plant. The Ti plasmid is frequently used as a vector for genetic engineering in plants.

tidal volume The volume of air a mammal inhales and exhales with each breath.

tight junction A type of intercellular junction between animal cells that prevents the leakage of material through the space between cells.

tissue An integrated group of cells with a common structure, function, or both.

tissue system One or more tissues organized into a functional unit connecting the organs of a plant.

Toll-like receptor (TLR) A membrane receptor on a phagocytic white blood cell that recognizes fragments of molecules common to a set of pathogens.

tonicity The ability of a solution surrounding a cell to cause that cell to gain or lose water.

top-down model A model of community organization in which predation influences community organization by controlling herbivore numbers, which in turn control plant or phytoplankton numbers, which in turn control nutrient levels; also called the trophic cascade model.

topoisomerase A protein that breaks, swivels, and rejoins DNA strands. During DNA replication, topoisomerase helps to relieve strain in the double helix ahead of the replication fork.

topsoil A mixture of particles derived from rock, living organisms, and decaying organic material (humus).

torpor A physiological state in which activity is low and metabolism decreases.

torsion In gastropods, a developmental process in which the visceral mass rotates up to 180°, causing the animal's anus and mantle cavity to be positioned above its head.

totipotent (tō′-tuh-pōt′-ent) Describing a cell that can give rise to all parts of the embryo and adult, as well as extraembryonic membranes in species that have them.

trace element An element indispensable for life but required in extremely minute amounts.

trachea (trā′-kē-uh) The portion of the respiratory tract that passes from the larynx to the bronchi; also called the windpipe.

tracheal system In insects, a system of branched, air-filled tubes that extends throughout the body and carries oxygen directly to cells.

tracheid (trā′-kē-id) A long, tapered water-conducting cell found in the xylem of nearly all vascular plants. Functioning tracheids are no longer living.

trait One of two or more detectable variants in a genetic character.

trans fat An unsaturated fat, formed artificially during hydrogenation of oils, containing one or more *trans* double bonds.

transcription The synthesis of RNA using a DNA template.

transcription factor A regulatory protein that binds to DNA and affects transcription of specific genes.

transcription initiation complex The completed assembly of transcription factors and RNA polymerase bound to a promoter.

transcription unit A region of DNA that is transcribed into an RNA molecule.

transduction (1) A process in which phages (viruses) carry bacterial DNA from one bacterial cell to another. When these two cells are members of different species, transduction results in horizontal gene transfer. (2) In cellular communication, the conversion of a signal from outside the cell to a form that can bring about a specific cellular response; also called *signal transduction*.

transfer RNA (tRNA) An RNA molecule that functions as a translator between nucleic acid and protein languages by carrying specific amino acids to the ribosome, where they recognize the appropriate codons in the mRNA.

transformation (1) The conversion of a normal animal cell to a cancerous cell. (2) A change in genotype and phenotype due to the assimilation of external DNA by a cell. When the external DNA is from a member of a different species, transformation results in horizontal gene transfer.

transgenic Pertaining to an organism whose genome contains a gene introduced from another organism of the same or a different species.

translation The synthesis of a polypeptide using the genetic information encoded in an mRNA molecule. There is a change of "language" from nucleotides to amino acids.

translocation (1) An aberration in chromosome structure resulting from attachment of a chromosomal fragment to a nonhomologous chromosome. (2) During protein synthesis, the third stage in the elongation cycle, when the RNA carrying the growing polypeptide moves from the A site to the P site on the ribosome. (3) The transport of organic nutrients in the phloem of vascular plants.

transmission The passage of a nerve impulse along axons.

transmission electron microscope (TEM) A microscope that passes an electron beam through very thin sections stained with metal atoms and is primarily used to study the internal ultrastructure of cells.

transpiration The evaporative loss of water from a plant.

transport epithelium One or more layers of specialized epithelial cells that carry out and regulate solute movement.

transport protein A transmembrane protein that helps a certain substance or class of closely related substances to cross the membrane.

transport vesicle A small membranous sac in a eukaryotic cell's cytoplasm carrying molecules produced by the cell.

transposable element A segment of DNA that can move within the genome of a cell by means of a DNA or RNA intermediate; also called a transposable genetic element.

transposon A transposable element that moves within a genome by means of a DNA intermediate.

transverse (T) tubule An infolding of the plasma membrane of skeletal muscle cells.

triacylglycerol (trī-as′-ul-glis′-uh-rol) A lipid consisting of three fatty acids linked to one glycerol molecule; also called a fat or triglyceride.

triiodothyronine (T_3) (trī′-ī-ō′-dō-thī′-rō-nēn) One of two iodine-containing hormones that are secreted by the thyroid gland and that help regulate metabolism, development, and maturation in vertebrates.

trimester In human development, one of three 3-month-long periods of pregnancy.

triple response A plant growth maneuver in response to mechanical stress, involving slowing of stem elongation, thickening of the stem, and a curvature that causes the stem to start growing horizontally.

triplet code A genetic information system in which a set of three-nucleotide-long words specify the amino acids for polypeptide chains.

triploblastic Possessing three germ layers: the endoderm, mesoderm, and ectoderm. Most eumetazoans are triploblastic.

trisomic Referring to a diploid cell that has three copies of a particular chromosome instead of the normal two.

trochophore larva (trō′-kuh-fōr) Distinctive larval stage observed in some lophotrochozoan animals, including some annelids and molluscs.

trophic efficiency The percentage of production transferred from one trophic level to the next.

trophic structure The different feeding relationships in an ecosystem, which determine the route of energy flow and the pattern of chemical cycling.

trophoblast The outer epithelium of a mammalian blastocyst. It forms the fetal part of the placenta, supporting embryonic development but not forming part of the embryo proper.

tropic hormone A hormone that has an endocrine gland or cells as a target.

tropical dry forest A terrestrial biome characterized by relatively high temperatures and precipitation overall but with a pronounced dry season.

tropical rain forest A terrestrial biome characterized by relatively high precipitation and temperatures year-round.

tropics Latitudes between 23.5° north and south.

tropism A growth response that results in the curvature of whole plant organs toward or away from stimuli due to differential rates of cell elongation.

tropomyosin The regulatory protein that blocks the myosin-binding sites on actin molecules.

troponin complex The regulatory proteins that control the position of tropomyosin on the thin filament.

true-breeding Referring to organisms that produce offspring of the same variety over many generations of self-pollination.

tubal ligation A means of sterilization in which a woman's two oviducts (fallopian tubes) are tied closed to prevent eggs from reaching the uterus. A segment of each oviduct is removed.

tube foot One of numerous extensions of an echinoderm's water vascular system. Tube feet function in locomotion and feeding.

tumor-suppressor gene A gene whose protein product inhibits cell division, thereby preventing the uncontrolled cell growth that contributes to cancer.

tundra A terrestrial biome at the extreme limits of plant growth. At the northernmost limits, it is called arctic tundra, and at high altitudes, where plant forms are limited to low shrubby or matlike vegetation, it is called alpine tundra.

tunicate Member of the clade Urochordata, sessile marine chordates that lack a backbone.

turgid (ter′-jid) Swollen or distended, as in plant cells. (A walled cell becomes turgid if it has a lower water potential than its surroundings, resulting in entry of water.)

turgor pressure The force directed against a plant cell wall after the influx of water and swelling of the cell due to osmosis.

turnover The mixing of waters as a result of changing water-temperature profiles in a lake.

turnover time The time required to replace the standing crop of a population or group of populations (for example, of phytoplankton), calculated as the ratio of standing crop to production.

twin study A behavioral study in which researchers compare the behavior of identical twins raised apart with that of identical twins raised in the same household.

tympanic membrane Another name for the eardrum, the membrane between the outer and middle ear.

uniformitarianism The principle that mechanisms of change are constant over time. *See* catastrophism.

Unikonta (yū′-ni-kon′-tuh) One of five supergroups of eukaryotes proposed in a current hypothesis of the evolutionary history of eukaryotes. This clade, which is supported by studies of myosin proteins and DNA, consists of amoebozoans and opisthokonts. *See also* Excavata, Chromalveolata, Rhizaria, and Archaeplastida.

unsaturated fatty acid A fatty acid that has one or more double bonds between carbons in the hydrocarbon tail. Such bonding reduces the number of hydrogen atoms attached to the carbon skeleton.

urea A soluble nitrogenous waste produced in the liver by a metabolic cycle that combines ammonia with carbon dioxide.

ureter (yū-rē′-ter) A duct leading from the kidney to the urinary bladder.

urethra (yū-rē′-thruh) A tube that releases urine from the mammalian body near the vagina in females and through the penis in males; also serves in males as the exit tube for the reproductive system.

uric acid A product of protein and purine metabolism and the major nitrogenous waste product of insects, land snails, and many reptiles. Uric acid is relatively nontoxic and largely insoluble.

urinary bladder The pouch where urine is stored prior to elimination.

uterine cycle The changes that occur in the uterus during the reproductive cycle of the human female; also called the menstrual cycle.

uterus A female organ where eggs are fertilized and/or development of the young occurs.

utricle In the vertebrate ear, a chamber in the vestibule behind the oval window that opens into the three semicircular canals.

vaccination *See* immunization.

vaccine A harmless variant or derivative of a pathogen that stimulates a host's immune system to mount defenses against the pathogen.

vacuole (vak′-yū-ōl′) A membrane-bounded vesicle whose specialized function varies in different kinds of cells.

vagina Part of the female reproductive system between the uterus and the outside opening; the birth canal in mammals. During copulation, the vagina accommodates the male's penis and receives sperm.

valence The bonding capacity of a given atom; usually equals the number of unpaired electrons required to complete the atom's outermost (valence) shell.

valence electron An electron in the outermost electron shell.

valence shell The outermost energy shell of an atom, containing the valence electrons involved in the chemical reactions of that atom.

van der Waals interactions Weak attractions between molecules or parts of molecules that result from transient local partial charges.

variation Differences between members of the same species.

vas deferens In mammals, the tube in the male reproductive system in which sperm travel from the epididymis to the urethra.

vasa recta The capillary system in the kidney that serves the loop of Henle.

vascular cambium A cylinder of meristematic tissue in woody plants that adds layers of secondary vascular tissue called secondary xylem (wood) and secondary phloem.

vascular plant A plant with vascular tissue. Vascular plants include all living plant species except liverworts, mosses, and hornworts.

vascular tissue Plant tissue consisting of cells joined into tubes that transport water and nutrients throughout the plant body.

vascular tissue system A transport system formed by xylem and phloem throughout a vascular plant. Xylem transports water and minerals; phloem transports sugars, the products of photosynthesis.

vasectomy The cutting and sealing of each vas deferens to prevent sperm from entering the urethra.

vasocongestion The filling of a tissue with blood, caused by increased blood flow through the arteries of that tissue.

vasoconstriction A decrease in the diameter of blood vessels caused by contraction of smooth muscles in the vessel walls.

vasodilation An increase in the diameter of blood vessels caused by relaxation of smooth muscles in the vessel walls.

vector An organism that transmits pathogens from one host to another.

vegetal pole The point at the end of an egg in the hemisphere where most yolk is concentrated; opposite of animal pole.

vegetative reproduction Cloning of plants by asexual means.

vein (1) In animals, a vessel that carries blood toward the heart. (2) In plants, a vascular bundle in a leaf.

ventilation The flow of air or water over a respiratory surface.

ventral Pertaining to the underside, or bottom, of an animal with radial or bilateral symmetry.

ventricle (ven′-tri-kul) (1) A heart chamber that pumps blood out of the heart. (2) A space in the vertebrate brain, filled with cerebrospinal fluid.

venule (ven′-yūl) A vessel that conveys blood between a capillary bed and a vein.

vernalization The use of cold treatment to induce a plant to flower.

vertebrate A chordate animal with a backbone, including sharks and rays, ray-finned fishes, coelacanths, lungfishes, amphibians, reptiles, and mammals.

vesicle (ves′-i-kul) A membranous sac in the cytoplasm of a eukaryotic cell.

vessel A continuous water-conducting micropipe found in most angiosperms and a few nonflowering vascular plants.

vessel element A short, wide water-conducting cell found in the xylem of most angiosperms and a few nonflowering vascular plants. Dead at maturity, vessel elements are aligned end to end to form micropipes called vessels.

vestigial structure A feature of an organism that is a historical remnant of a structure that served a function in the organism's ancestors.

villus (plural, **villi**) (1) A finger-like projection of the inner surface of the small intestine. (2) A finger-like projection of the chorion of the mammalian placenta. Large numbers of villi increase the surface areas of these organs.

viral envelope A membrane, derived from membranes of the host cell, that cloaks the capsid, which in turn encloses a viral genome.

viroid (vī′-royd) A plant pathogen consisting of a molecule of naked, circular RNA a few hundred nucleotides long.

virulent Describing a pathogen against which an organism has little specific defense.

virulent phage A phage that replicates only by a lytic cycle.

virus An infectious particle incapable of replicating outside of a cell, consisting of an RNA or DNA genome surrounded by a protein coat (capsid) and, for some viruses, a membranous envelope.

visceral mass One of the three main parts of a mollusc; the part containing most of the internal organs. *See also* foot, mantle.

visible light That portion of the electromagnetic spectrum that can be detected as various colors by the human eye, ranging in wavelength from about 380 nm to about 750 nm.

vital capacity The maximum volume of air that a mammal can inhale and exhale with each breath.

vitamin An organic molecule required in the diet in very small amounts. Many vitamins serve as coenzymes or parts of coenzymes.

viviparous (vī-vip′-uh-rus) Referring to a type of development in which the young are born alive after having been nourished in the uterus by blood from the placenta.

voltage-gated ion channel A specialized ion channel that opens or closes in response to changes in membrane potential.

vulva Collective term for the female external genitalia.

water potential (Ψ) The physical property predicting the direction in which water will flow, governed by solute concentration and applied pressure.

water vascular system A network of hydraulic canals unique to echinoderms that branches into extensions called tube feet, which function in locomotion and feeding.

wavelength The distance between crests of waves, such as those of the electromagnetic spectrum.

wetland A habitat that is inundated by water at least some of the time and that supports plants adapted to water-saturated soil.

white matter Tracts of axons within the CNS.

wild type The phenotype most commonly observed in natural populations; also refers to the individual with that phenotype.

wilting The drooping of leaves and stems as a result of plant cells becoming flaccid.

wobble Flexibility in the base-pairing rules in which the nucleotide at the 5′ end of a tRNA anticodon can form hydrogen bonds with more than one kind of base in the third position (3′ end) of a codon.

xerophyte A plant adapted to an arid climate.

X-linked gene A gene located on the X chromosome; such genes show a distinctive pattern of inheritance.

X-ray crystallography A technique used to study the three-dimensional structure of molecules. It depends on the diffraction of an X-ray beam by the individual atoms of a crystallized molecule.

xylem (zī′-lum) Vascular plant tissue consisting mainly of tubular dead cells that conduct most of the water and minerals upward from the roots to the rest of the plant.

xylem sap The dilute solution of water and dissolved minerals carried through vessels and tracheids.

yeast Single-celled fungus that reproduces asexually by binary fission or by the pinching of small buds off a parent cell. Some species exhibit cell fusion between different mating types.

yolk Nutrients stored in an egg.

zero population growth (ZPG) A period of stability in population size, when additions to the population through births and immigration are balanced by subtractions through deaths and emigration.

zona pellucida The extracellular matrix surrounding a mammalian egg.

zone of polarizing activity (ZPA) A block of mesoderm located just under the ectoderm where the posterior side of a limb bud is attached to the body; required for proper pattern formation along the anterior-posterior axis of the limb.

zoned reserve An extensive region that includes areas relatively undisturbed by humans surrounded by areas that have been changed by human activity and are used for economic gain.

zoonotic pathogen A disease-causing agent that is transmitted to humans from other animals.

zoospore Flagellated spore found in chytrid fungi and some protists.

zygomycete (zī′-guh-mī′-sēt) Member of the fungal phylum Zygomycota, characterized by the formation of a sturdy structure called a zygosporangium during sexual reproduction.

zygosporangium (zī′-guh-spōr-an′-jē-um) In zygomycete fungi, a sturdy multinucleate structure in which karyogamy and meiosis occur.

zygote (zī′-gōt) The diploid cell produced by the union of haploid gametes during fertilization; a fertilized egg.

Index

NOTE: A page number in regular type indicates where a topic is discussed in text (topic may also be in a figure on that page); a **bold** page number indicates where a term is bold and defined; an *f* following a page number indicates a figure (topic may also be discussed in text on that page); a *t* following a page number indicates a table (topic may also be discussed in text on that page).

C

D

E

J

K

L

U

V